T0182898

Evolution of Tertiary Mammals of North America

This second volume completes the unique survey of North American Tertiary mammals, and covers all the remaining taxa not contained in Volume 1. It provides a database of mammalian diversity over time and space, and evaluates the effect of biogeography and climatic change on evolutionary patterns and faunal transitions.

As with Volume 1, this book lays out, in a standardized format, the distribution in time and space of each taxon. It summarizes the current state of the systematics of the various mammal groups, and it discusses their paleobiology and evolutionary patterns. It contains six summary chapters that integrate systematic and biogeographic information for higher taxa, and provides a detailed account of the patterns of occurrence for different species at hundreds of different fossil localities, with the inclusion of many more localities than were contained in the first volume.

With over 30 chapters, each written by leading authorities, and an addendum that updates the occurrence and systematics of all of the groups covered in Volume 1, this will be a valuable reference for paleontologists and zoologists.

CHRISTINE M. JANIS is Professor of Biology in the Department of Ecology and Evolutionary Biology at Brown University. She is on the editorial board of *Journal of Mammalian Evolution* and *Acta Paleontologica Polonica*, a member of the Society for Vertebrate Paleontology, the President of the Society for the Study of Mammalian Evolution, and a past editor for the journal *Evolution*. Professor Janis was also editor-in-chief for *Evolution of Tertiary Mammals of North America* Vol. 1: *Terrestrial Carnivores, Ungulates, and Ungulatelike Mammals* (Cambridge University Press 1998).

GREGG F. GUNNELL is Associate Research Scientist in the Museum of Paleontology at the University of Michigan. He is Associate Editor for the journal *Palaois* and was Associate Editor for *Journal of Human Evolution*. He is a Member of the Society for Vertebrate Paleontology, the Paleontological Society, and the American Society of Mammalogists.

MARK D. UHEN is a Senior Postdoctoral Fellow in the Department of Paleobiology at the National Museum of Natural History, the chair of the Society of Vertebrate Paleontology Patterson Award Committee, and a member of the Advisory Board and the Vertebrate Working Group for the Paleobiology Database.

This volume is dedicated to out elders and betters
in the areas of small mammal and marine mammal paleontology:
Malcolm McKenna, Leigh Van Valen, and Charles Repenning

Evolution of Tertiary Mammals of North America

Volume 2: Small Mammals, Xenarthrans, and Marine Mammals

CHRISTINE M. JANIS
Brown University, Providence, RI, USA

GREGG F. GUNNELL
Museum of Paleontology, University of Michigan, Ann Arbor, MI, USA

MARK D. UHEN
Smithsonian Institution, Washington, DC, USA

CAMBRIDGE
UNIVERSITY PRESS

University Printing House, Cambridge CB2 8BS, United Kingdom

One Liberty Plaza, 20th Floor, New York, NY 10006, USA

477 Williamstown Road, Port Melbourne, VIC 3207, Australia

314-321, 3rd Floor, Plot 3, Splendor Forum, Jasola District Centre, New Delhi - 110025, India

79 Anson Road, #06-04/06, Singapore 079906

Cambridge University Press is part of the University of Cambridge.

It furthers the University's mission by disseminating knowledge in the pursuit of education, learning and research at the highest international levels of excellence.

www.cambridge.org
Information on this title: www.cambridge.org/9781108462082

First published 2008
First paperback edition 2018

A catalogue record for this publication is available from the British Library

ISBN 978-0-521-78117-6 Hardback
ISBN 978-1-108-46208-2 Paperback

Contents

Contributors

Deborah Anderson
Department of Biology, St. Norbert College, 100 Grant Street, De Pere, WI 54115, USA

Lawrence G. Barnes
Department of Vertebrate Paleontology, Natural History Museum of Los Angeles County, 900 Exposition Boulevard, Los Angeles, CA 90007, USA

Jonathan I. Bloch
Florida Museum of Natural History, University of Gainsville, FL 32611–2710, USA

Thomas M. Bown
Erathem-Vanir Geological, 2021 Ardella Drive, Pocatello, ID 83201, USA

Douglas J. Boyer
Department of Anatomical Sciences, Health Sciences Center, Stony Brook University, Stony Brook, NY 11794-8081, USA

Nicholas J. Czaplewski
Oklahoma Museum of Natural History, University of Oklahoma, 2401 Chautauqua Avenue, Norman, OK 73072-7029, USA

Mary R. Dawson
Section of Vertebrate Paleontology, Carnegie Museum of Natural History, 4400 Forbes Avenue, Pittsburgh, PA 15213-4080, USA

Daryl P. Domning
Department of Anatomy, Howard University, 520 W Street NW, Washington, DC 20059, USA

Lawrence J. Flynn
Peabody Museum, Harvard University, 11 Divinity Avenue, Cambridge, MA 02138, USA

R. Ewan Fordyce
Department of Geology, University of Otago, PO Box 56, Dunedin, New Zealand

H. Thomas Goodwin
Department of Biology, Andrews University, Berrien Springs, MI 49104, USA

Gregg F. Gunnell
Museum of Paleontology, University of Michigan, 1109 Geddes Avenue, Ann Arbor, MI 48109-1079, USA

Richard C. Hulbert Jr.
Florida Museum of Natural History, University of Florida, Dickinson Hall, Gainesville, FL 32611-7800, USA

J. Howard Hutchison
Museum of Paleontology, University of California at Berkeley, Berkeley, CA 94720, USA

Louis L. Jacobs
Department of Geological Sciences, Shuler Museum of Paleontology, Southern Methodist University, 310 Heroy Hall, Dallas, TX 75275, USA

Christine M. Janis
Department of Ecology and Evolutionary Biology, Brown University, PO Box G-B207, Providence, RI 02912, USA

Irina A. Koretsky
Department of Anatomy, Howard University, Washington, DC 20059, USA

William W. Korth
Rochester Institute of Vertebrate Paleontology, 265 Carling Road, Rochester, NY 14610, USA

David W. Krause
Department of Anatomical Sciences, Health Sciences Center, Stony Brook, State University of New York, NY 11794-8081, USA

Everett H. Lindsay
Department of Geosciences, University of Arizona, Tucson, AZ 85721, USA

H. Gregory McDonald
Park Museum Management Program, National Park Service, 1201 Oakridge Drive, Fort Collins, CO 80525, USA

Robert A. Martin
Department of Biological Sciences, Murray State University, Murray, KY 42071, USA

Samuel A. McLeod
Department of Vertebrate Paleontology, Los Angeles County Museum of Natural History, 900 Exposition Blvd, Los Angeles, CA 90007, USA

Matthew C. Mihlbachler
Department of Anatomy, New York College of Osteopathic Medicine, Old Westbury, NY 11568, USA

Gary S. Morgan
Department of Vertebrate Paleontology, New Mexico Museum of Natural History, 1801 Mountain Road NW, Albuquerque, NM 87104-1375, USA

Virginia L. Naples
Department of Biological Sciences, Northern Illinois University, DeKalb, IL 60115-2861, USA

Read D. Porter
Pratt Museum of Natural History, Amherst College, MA 01002, USA

D. Tab Rasmussen
Department of Anthropology, Washington University, Campus Box 1114, 1 Brookings Drive, St. Louis, MO 63130, USA

Kenneth D. Rose
Center for Functional Anatomy and Evolution, Johns Hopkins University School of Medicine, 1830 East Monument Street, Baltimore, MD 21205, USA

Mary T. Silcox
Department of Anthropology, University of Winnipeg, 515 Portage Avenue, Winnipeg, MB R3B 2E9, Canada

John E. Storer
6937 Porpoise Drive, Sechelt, BC V0N 3A4, Canada

Mark D. Uhen
Smithsonian Institution, PO Box 3701, Washington, DC 20013-7012, USA

Stephen L. Walsh (deceased)
Late of the Department of Paleontology, San Diego Natural History Museum, PO Box 121390, San Diego, CA 92112, USA

Anne H. Walton
Pratt Museum of Natural History, Amherst College, MA 01002, USA

Anne Weil
Department of Anatomy and Cell Biology, Oklahoma State University Center for Health Sciences, 1111 West 17th Street, Tulsa, OK 74107-1898, USA

0 Introduction

CHRISTINE M. JANIS,[1] GREGG F. GUNNELL[2] and MARK D. UHEN[3]

[1]*Brown University, Providence, RI, USA*
[2]*Museum of Paleontology, University of Michigan, Ann Arbor, MI, USA*
[3]*Smithsonian Institution, Washington, DC, USA*

AIMS OF VOLUME 2

This enterprise was originally conceived of as a single volume. However, after a span of 10 years from its original conception, the current senior editor (Christine Janis), and the then junior editors (Kathleen Scott and Louis Jacobs) realized that it would be more realistic to proceed with chapters then in hand, which could more or less be assembled into the conceptually useful, if taxonomically paraphyletic, rubric of "Terrestrial Carnivores and Ungulates" (Janis, Scott, and Jacobs, 1998). This in part reflected the chapters that had been assembled to date, although it should be noted that some of the chapters in this current volume, most notably those by Darryl Domning on sirenians and desmostylians, were among the first ones received almost 20 years ago.

The mammals covered in this volume are thus those remaining from Volume 1, and they can more or less be grouped into two conceptual (and again paraphyletic) groupings: small mammals (aka "vermin") and marine mammals. The only group of large terrestrial mammals considered in this volume are the xenarthrans, which do not appear until the latest Miocene. Also in this volume, new editors came on board (although Louis Jacobs continued as a co-author on two rodent chapters). Gregg Gunnell, who wrote the chapter on Hyaenodontidae in Volume 1, was a welcome addition as someone familiar with many of these small mammal groups, especially the primates. Mark D. Uhen, a mere teenager when this project was first conceived, was an essential addition as an expert in marine mammals.

As in Volume 1, the taxonomic level of interest in this volume is typically the genus, but locality information is (usually) provided at the level of the species. The faunal localities have been standardized throughout the chapters and are listed in Appendix I (see explanation below), and the locality references are available in Appendix II. For the purposes of standardization, and to provide equal quality of information across each chapter, the stratigraphic range charts in the chapter are presented according to a standardized format, and the institutional abbreviations have also been standardized and are listed in an appendix (Appendix III).

THE STANDARDIZED LAYOUT OF EACH CHAPTER

The chapters are laid out in a similar fashion to those in Volume 1. The contributors were requested to adhere to a common layout for each chapter, in order to provide uniform information throughout the book. The "Introduction" for each chapter introduces the group. The "Defining features" section lays out the basic cranial, dental, and postcranial features of the taxon. The term "defining features" was used, rather than the cladistically preferred term "diagnostic features," as this section was intended to be a general introduction to the characters of the group as a whole, plesiomorphic as well as apomorphic. Due to the constraints of production costs, contributors were generally requested to limit their illustrations to one taxon for pictures of the skull, dentition, and skeleton.

The section on "Systematics" includes a "Suprataxon" section that deals with the history of the ideas of the relationships of the taxon in question among mammals in general, and an "Infrataxon" section that deals with interrelationships within the group, including a cladogram. Rather than have a more general "suprataxon" cladogram in each chapter, a single concensus cladogram is presented in the summary chapter for each section: Chapters 1 (non-eutherian mammals), 4 (insectivorous mammals), 8 ("Edentata"), 11 (Archonta), 16 (Glires), and 30 (marine mammals).

The "Included genera" section includes a brief description of each genus, including the listing of the type species and type specimen, and a listing of the valid species, including the localities at which each species was found. We also requested contributors to provide an average dental length measurement for each genus, m2 if it was available; if not, some other tooth. This was to provide some size

Evolution of Tertiary Mammals of North America, Vol. 2. ed. C. M. Janis, G. F. Gunnell, and M. D. Uhen. Published by Cambridge University Press.

estimate for the taxon, as dental length measurements are a good proxy of body mass (see Damuth and MacFadden, 1990), and m2 length is the most reliable measurement, at least in ungulates (Janis, 1990). However, marine mammals have highly derived dentitions, and the link between molar size and body size is lost (as is often the ability even to identify a tooth as m2); so instead we have chosen to use occipital condyle breadth as an indicator for body size in pinnipeds and cetaceans, following Marino *et al.* (2000).

We have also retained the style, as in Volume 1, of putting "a" or "b" in the reference for taxonomic groups where appropriate (e.g., on p. 32, *Neoliotomus* Jepsen, 1930a). We acknowledge that, as the reference is actually part of the official taxonomic name, that the "a" does not strictly belong there. However, the problems that would ensue with other references, and the issue of the correct identification of the taxonomic reference in the bibliography, led us to decide to retain this style in this volume.

Finally, the "Biology and evolutionary patterns" section provides a synopsis of the paleobiology and evolutionary trends of the group. This section includes the standardized temporal range chart for each taxon. The biogeographic range charts (which may combine a number of taxa) are in the summary chapters (see below).

One difference from Volume 1 is that we no longer have reconstructions of extinct mammals in each chapter. This is partially because our previous artist, Brian Regal, has now changed careers. Additionally, it seemed that for many of the small mammals reconstructions were not known, and they would pretty much all look the same in any case! As we had to find funds to pay a new artist, Marguette Dongvillo, we decided to limit the art work to the summary chapters.

Another difference in the chapter layout is the way in which the synonyms have been handled. Stephen Walsh pointed out to us that the previous mode of noting taxonomic synonymies was phylogenetically suspect, and we have adopted a new standardized way of doing this, following his suggestions. In addition, we were not so anxious in this volume to note all of the known synonymies for each genus, as such information is now readily available in McKenna and Bell (1997).

THE UNIFIED LOCALITY LISTINGS

THE CREATION OF THE LISTING AND THE USE OF THE APPENDICES

The original unified listing in Volume 1 was created from the lists of localities supplied by the authors, supplemented with lists derived from Woodburne (1987). The localities in the individual chapters, (e.g., CP1, NP5), must be looked up in Appendix I. This saves space in the volume, as well as providing an overall unification. Despite extensive checking and cross checking by both editors and authors, it is impossible to have complete confidence that these listings are totally error free, but every attempt has been made to minimize errors.

A locality number (e.g., CP101) encompasses an entire formation. Subdivisions within that formation are then numbered A, B, C, etc., according to relative age. For the purposes of numbering, as well as for the creation of the biogeographic range charts, the localities are grouped into various biogeographic regions (see below). Within each biogeographic region, the localities are numbered according to stratigraphic position. The biogeographic regions are themselves ordered in a general west to east fashion, except for the Pacific Northwest and Northern Great Plains localities, which are listed after those of the Central Great Plains (see ordering in the figures in the summary chapters). A few localities appear to be slightly out of order; this is because information about the exact age was later revised after the creation of the list.

The unification of the localities necessitated a certain degree of grouping of sites. Sometimes this involved grouping of the quarries within a single time horizon in a formation (e.g., the quarries in the Miocene Valentine Formation, localities CP114A–CP114D). At other times, localities that were in a similar location at a similar time were grouped together (e.g., the North Coalinga Local Fauna and Domengine Creek, in the Temblor Formation of the Miocene of California, both contained within locality CC23). To list every single fossil-containing site as its own separate locality would have increased the number of individually listed localities by at least an order of magnitude. As references are provided for each locality, it should be possible in most cases for a concerned researcher to reconstruct finer detail.

Because the original numbering of the localities was accomplished before all final contributions were received, revisions had to be made to the listings that made the final more cumbersome than we would have preferred. In the case of new formations (primary locality numbers), additions were made by creating an intermediate locality between two existing ones, affixing the suffix II to the younger of the two localities (e.g., NP19, and NP19II). In the case of new subdivisions within formations, double letters were created (e.g., NP10B, NP10BB, NP10B2). This rather cumbersome mode of renumbering localities as "work in progress" proved to be more practical than renumbering localities throughout, which would have then necessitated renumbering the localities for the individual taxa that had already been processed (not only of the numbered locality itself, but of all younger localities within the region).

The locality list has grown dramatically over the past decade. Many new localities have been added, either ones that are completely new or ones that are new additions within existing localities. To maintain continuity with Volume 1, we have made the new localities fit in within the preexisting scheme. This, unfortunately, has only added to the cumbersome nature of the listing, but this could not be avoided: we considered that cumbersome was preferable to incompatible.

All new localities have been noted in boldface in Appendix I. Also noted in boldface are other changes that were made. In some instances, localities were moved and given different numbers (see discussion below about certain Mexican localities). In some instances, a locality became subdivided, and the original site contained within that locality was now given the suffix A or B, etc., depending on its age relative to the added sublocality. (For example, locality GC5 (Lower Fleming Formation) originally only contained

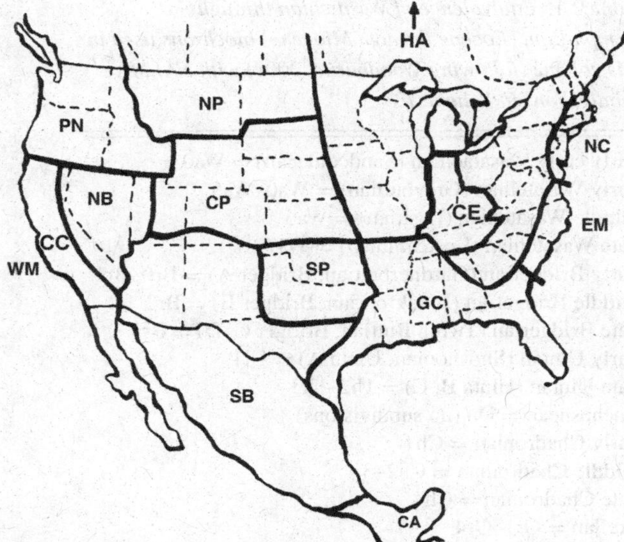

Figure 0.1. Map of North America showing the biogeographic regions employed in this volume. Key: CA, Central America; CC, California Central and Coastal Ranges; CE, Central East America; CP, Central Great Plains; EM, East Coast Marine; GC, Gulf Coast; HA, Canadian High Arctic; NB, Northern Great Basin; NC, Northern East Coast; NP, Northern Great Plains; PN, Pacific Northwest; SB, Southern Great Basin; SP, Southern Great Plains; WM, West Coast Marine.

the Carnahan Bayou Member; with the addition of the earlier Toledo Bend Local Fauna [now GC5A], the Carnahan Bayou Member now became GC5B.) Finally, some localities (fortunately only a few) have had their date changed since Volume 1: for example, the date of the Friars Formation (locality CC4) has been changed from early Uintan (Ui1) to middle Uintan (Ui2).

BIOGEOGRAPHIC REGIONS

ORIGINAL DIVISION OF THE REGIONS IN VOLUME 1

The localities were originally divided into biogeographic regions, so that biogeographical variation as well as stratigraphic ranges could be seen (see Figure 0.1). The biogeographic regions were originally based on those presented by Tedford *et al.* (1987). The division of the Mexican localities (as to inclusion with the Southern Great Basin, California Coast or the Central American region) followed distributional maps in chapters in MacFadden (1984) (but see later revisions for Volume 2).

The "West Coast Marine" localities (prefix WM) include coastal faunas of Washington, Oregon, and California. Terrestrial localities are placed into the "Pacific Northwest" (prefix PN; including Washington, Oregon, and Idaho) or "California Central and Coast" (prefix CC) sections according to latitude, the latter also including localities in Baja California, Mexico. The "Gulf Coast" (prefix GC) includes the Texas Gulf Coast, Florida, and the southern East Coast (Louisiana, Mississippi, Georgia, and North and South Carolina).

The "Northern Great Basin" (prefix NB) includes southeastern California, Nevada, and southwestern Utah. The "Southern Great Basin" (prefix SB) includes Arizona, New Mexico, Texas Big Bend area, southwestern Colorado (i.e., the Paleocene Animas Formation and the Eocene Huerfano Formation), and northern Mexico. The Texas Big Bend area includes all the Paleogene Texas localities (except for the Duchesnean Yegua Formation, grouped with the Gulf Coast), the Miocene Delaho, Rawls, and Banta Shut-In Formations, and the Pliocene Camp Rice and Love Formations.

The "Southern Great Plains" (prefix SP) includes the Texas/Oklahoma panhandles. The "Central Great Plains" (prefix CP) includes Nebraska, South Dakota, Wyoming, Kansas, most of Utah, and northern/northeastern Colorado (i.e., the Paleocene Denver and Wasatach Formations, the Eocene Wasatach, DeBeque, Uinta, and Washakie Formations, and all Oligocene and Neogene sites). The "Northern Great Plains" (prefix NP) includes Montana, North Dakota, western Canada (Alberta, Saskatchewan, and British Columbia). The "Northern East Coast" (prefix NC) includes the East Coast north of the Carolinas.

It is important to emphasize that, because all the individual information has been preserved in this volume (in the form of Appendices I and II), any scheme that we have used to lump together information, for the purposes of diagrams or discussion, has not been lost. The original information is retained for others to reconstruct their own biogeographic scheme.

CHANGES IN VOLUME 2

Some new biogeographic regions have been added, and some old regions have had some boundary changes (see Figure 0.1). Boundary changes include the following. The Gulf Coast region has now been extended to include northern Mississippi and western Tennessee, to include marine localities that form part of the Mississippi Embayment (see further discussion later about the nature of the marine localities). The East Coast Marine/Gulf Coast Region boundary has been more firmly delimited as the boundary between South Carolina and Georgia. A number of Mexican localities, previously included in the Central American region (prefix CA), have now been transferred to the Southern Great Basin region. Mexican localities remaining in the Central American region include the Mexican Gulf Coast – Oaxaca, Chiapas, and the Yucatan Peninsula.

The new biogeographic regions both contain relatively few faunas. They include the Canadian High Arctic (prefix HA) and Central East America (prefix CE). The Canadian High Arctic includes Ellesmere Island, Axel Heiberg Island, and Devon Island. In Volume 1 only a single site, Iceberg Bay Formation on Ellesmere Island, was included, and listed (for convenience more than anything else) with the Northern Great Plains localities. The Central East America region includes newly discovered late Tertiary sites from Tennessee (Gray Fossil Site) and Indiana (Pipe Creek Sinkhole).

Finally, we note that some sites in eastern Oregon that were included within the Pacific Northwest are in fact in close proximity to some of the northwestern Nevada sites included in the Northern Great Basin. Woodburne (2004) considered this to be a single biogeographic region, the Columbia Plateau. We have not made this a separate biogeographic region in this volume, but note in Appendix I which NB and PN sites this applies to.

THE SUMMARY BIOGEOGRAPHIC CHARTS

The biogeographic charts in the summary chapters represent the combined information from the authors' original contributions and information added (with consultation with the original authors) from the published literature and records from museum collections. These charts are presented in the summary chapters, rather than in the individual chapters, for the following reasons. First, space was saved by combining taxa. Second, overall diversity trends were more easily visible when closely related taxa were grouped together. Finally, the summary chapters proved to be a suitable venue in the book to discuss supratenon evolutionary trends, and the biogeographic charts provide the appropriate illustrations.

THE GEOCHRONOLOGICAL TIME SCALE AND NALMA DIVISIONS

In Volume 1, the time scale and the divisions of the North American Land-Mammal Ages (NALMAs) were adopted from the first edition of Woodburne (1987). The Introduction in Volume 1 discussed the various controversies that existed with dating and NALMA boundaries at that time. In adjusting the time scale for this volume, we followed the second edition of Woodburne (2004) throughout. We acknowledge that there are controversies surrounding some of these changes, and some updates to epoch boundary ages in the past three years, but we decided to make our volume concordant with this publication. One specific issue, that of the division between early and late Blancan, is discussed below.

One profound change that has occurred since Volume 1 is the new division of the NALMAs into biochrons, or numbered units. At the time Volume 1 was published, the Paleocene had already been divided into biochrons rather than descriptive time periods (e.g., Puercan 1, 2, and 3, rather than early, middle, and late), but the other epochs had not yet received this treatment. During the past decade, the Wasatchian through Hemphillian epochs have received formal biochron subdivisions, which have been incorporated here, and we have also updated the ages of NALMA and NALMA subdivision boundaries. Table 0.1 lists the correspondence of these new biochrons with the descriptive units in Volume 1.

Table 0.1. *Equivalence of Wasatchian through Hemphillian (Eocene through Miocene) biochrons used in this volume (following Woodburne, 2004) with NALMA subdivisions in Volume 1*

Early early Wasatachian (Sandcouleean) = Wa0
Early Wasatchian (Greybullian) = Wa0–Wa5
Middle Wasatchian (Lysitian) = Wa6
Late Wasatchian (Lostcabinian) = Wa7
Early Bridgerian (Gardnerbuttian, Bridger A) = Br0–Br1
Middle Bridgerian (Blackforkian, Bridger B) = Br2
Late Bridgerian (Twinn Buttian, Bridger C, D) = Br3
Early Uintan (Shoshonian, Uinta A) = Ui1
Late Uintan (Uinta B, C) = Ui2–Ui3
Duchesnean = Du (no subdivisions)
Early Chadronian = Ch1
Middle Chadronian = Ch2–3
Late Chadronian = Ch4
Orellan = Or1–Or4
Whitneyan = Wh1–Wh2
Early early Arikareean = Ar1
Late early Arikareean = Ar2
Early late Arikareean = Ar3
Late late Arikareean = Ar4
Early Hemingfordian = He1
Late Hemingfordian = He2
Early Barstovian = Ba1
Early late Barstovian = Ba2
Late late Barstovian = Ba2
Barstovian–Clarendonian boundary = Cl1
Early Clarendonian = Cl1–Cl2
Late Clarendonian = Cl2–Cl3.
Early early Hemphillian = Hh1
Late early Hemphillian = Hh2
Late Hemphillian = Hh3.
Latest Hemphillian = Hh4

Shifts have also occurred in the position of the Oligo-Miocene and Mio-Pliocene boundaries. The Oligo-Miocene boundary, previously considered to fall between early and late Arikareean (i.e., between Ar2 and Ar3) is now considered to fall within late Ar2 (i.e., at 23.8 Ma rather than 23 Ma). Similarly, the Mio-Pliocene boundary, previously considered to fall between late and latest Hemphillian (i.e., between Hh3 and Hh4), is now considered to fall within Hh4, at 5.3 Ma. With regards to the Plio-Pleistocene boundary, we have included a few early Irvingtonian faunas in this volume as these faunas are now considered to be included within the Tertiary (although controversy remains). Our biogeographic charts now include these faunas in a "latest Blancan/earliest Irvingtonian" unit.

In Volume 1, we followed the scheme in the earlier edition of Woodburne (Woodburne, 1987) of setting the early/late Blancan boundary at 2.5 Ma, between the Gauss and Matuyama chrons. This meant that the early Blancan included the microtine rodent units Bl I to Bl IV, and the late Blancan included Bl V. We have followed this division in this volume, although the date of some faunas has been adjusted. Our usage of "early Blancan" includes the "middle Blancan" of many authors. The late Blancan, as defined in this fashion, is based on the appearance of certain Great American Interchange mammals, such as the xenarthrans *Dasypus* and *Holmensina*. Note that Flynn *et al.* (2005) discussed the fact that certain Interchange mammals appear earlier in central Mexico than in the United States, and discussed how this might affect the designation of the early/late Blancan boundary (although it is not surprising to us that these immigrants should appear sooner in more southern regions).

Robert Martin (personal communication) would prefer a division into early, middle, and late Blancan, based on his work on Meade Basin rodents (e.g., Martin, Honey, and Peláez-Campomanes, 2000) (see also discussion in Chapter 28). Bearing all these issues in mind, we have retained the early/late Blancan boundary at 2.5 Ma and would alert the reader to the fact that our way of noting individual localities means, as previously, that the position of individual taxa is tied to individual localities, not to particular time units.

FURTHER NEW ISSUES IN VOLUME 2

ISSUES WITH MARINE LOCALITIES AND MARINE MAMMALS

Inclusion of marine mammals in this volume posed some particular problems in that descriptions of marine deposits usually refer to marine time scales for placement in geologic time. Occasionally, terrestrial mammals have been found in these marine localities, providing a direct link to the NALMA time scale used elsewhere in this volume. In other cases, we had to rely on less direct means of relating marine and terrestrial time scales, including interfingering of marine and terrestrial deposits, dating of ash beds, and paleomagnetic dating.

In addition, many older specimens, including many type specimens of important taxa, are poorly placed in time and space. We have placed them into our locality listing system as best we can, but in some cases it was not possible to do so. These instances are clearly noted in each chapter.

THE ADDENDUM

A final contribution to this volume is the Addendum, which provides (minor) corrections from Volume 1 and updates information about taxa from Volume 1, including locality, systematic, and paleobiological information. This was originally planned to be the sole work of the senior editor, but Richard Hulbert was coopted to help with the extensive revision of equids, and Matt Mihlbachler was coopted to add his thesis work on revision of brontothere taxonomy and locality information. This Addendum proved to be an enormous undertaking, with a significant amount of new information, even though the search for new information was by no means exhaustive. New range charts and biogeographic charts are provided for those taxa that had the most revisions: borhyaenid canids, brontotheres, equids, and rhinocerotids.

ACKNOWLEDGEMENTS

A number of publications were indispensable in the production of this book. In addition to the edited volume by Woodburne (2004) already mentioned, we would have been lost without the books by Korth (1994) and McKenna and Bell (1997). We are also extremely grateful to Phil Gingerich for use of his office space and library at the University of Michigan for various editorial meetings during the production of this volume. The late Steve Walsh contributed much to this volume; his knowledge and wisdom will be missed.

The following people helped with reviewing individual chapters: Andrea Bair, Annalisa Berta, Doug Boyer, Bill Clemens, Richard Fox, Jonathan Geisler, Bill Korth, Nancy Simmons, Dale and Alisa Winkler, and Rick Zakrzewski. The following people helped with providing locality information: Tom Goodwin, Scott Foss, Richard Hulbert, Bob Martin, Don Prothero, and Steve Walsh. Mary Dawson, Don Prothero, Margaret Stevens, and Samantha Hopkins provided copies of preprinted papers for help with chapters. Julia Keller, Caitlyn Thompson, and Megan Dawson helped with editing details, and Jamie Lemon helped with obtaining figure permissions. CMJ thanks the Bushnell Foundation for help with some travel and editing costs.

We also acknowledge the extreme patience of certain authors who prepared manuscripts for inclusion close to two decades ago, notably Tom Bown, Daryl Domning, Larry Flynn, Howard Hutchinson, David Krause, Sam McLeod, and Ken Rose, although we are also delighted to have had the opportunity to work with many junior people who have come of age since Volume 1 was published. We also note the sterling contributions of Larry Flynn to this volume in writing the bulk of the rodent chapters, a truly stupendous task! Special thanks are owed to our copy editor, Jane Ward, whose diligence, patience, and extraordinary attention to detail meant that many embarrassing errors were caught and fixed, and she also greatly improved the consistency of the text.

Finally, we thank our own North American mammals for their company during the production of this volume: felines Diego, Sherman, Mimi, and Critter; canines Ronnie and Boswell; and equines Duster and Mel.

REFERENCES

Damuth, J. and MacFadden, B. J. (ed.) (1990). *Body Size in Mammalian Paleobiology: Estimation and Biological Implications*. Cambridge: Cambridge University Press.

Flynn, J. J., Kowallis, B. J., Nuñez, C., *et al.* (2005). Geochronology of Hemphillian–Blancan aged strata, Guanajuato, Mexico, and implications for the timing of the Great American Biotic Interchange. *Journal of Geology*, 113, 287–307.

Janis, C. M. (1990). Correlation of cranial and dental variables with body size in ungulates and macropodoids. In *Body Size in Mammalian Paleobiology: Estimation and Biological Implications*, ed. J. Damuth and B. J. MacFadden, pp. 255–99. Cambridge: Cambridge University Press.

Janis, C. M., Scott, K. M., and Jacobs, L. L. (1998). *Evolution of Tertiary Mammals of North America*. Vol. 1: *Terrestrial Carnivores, Ungulates, and Ungulatelike Mammals*. Cambridge: Cambridge University Press.

Korth, W. W. (1994). *The Tertiary Record of Rodents in North America*. New York: Plenum Press.

MacFadden, B. J. (ed.) (1984). The origin and evolution of the Cenozoic vertebrate fauna of middle America. *Journal of Vertebrate Paleontology*, 4, 169–283.

Marino, L., Uhen, M. D., Frohlich B., *et al.* (2000). Endocranial volume of mid-late Eocene archaeocetes (Order: Cetacea) revealed by computed tomography: implications for cetacean brain evolution. *Journal of Mammalian Evolution*, 7, 81–94.

Martin, R. A., Honey, J. G., and Peláez-Campomanes (2000). The Meade Basin rodent project: a progress report. *Paludicola*, 3, 1–32.

McKenna, M. C. and Bell, S. K. (1997). *Classification of Mammals Above the Species Level*. New York: Columbia University Press.

Tedford, R. H., Skinner, M. F., Fields, R. W., *et al.* (1987). Faunal succession and biochronology of the Arikareean through Hemphillian Interval (late Oligocene through earliest Pliocene epochs) in North America. In *Cenozoic Mammals of North America: Geochronology and Biostratigraphy*, ed. M. O. Woodburne, pp. 153–210. Berkeley, CA: University of California Press.

Woodburne, M. O. (ed.) (1987). *Cenozoic Mammals of North America: Geochronology and Biostratigraphy*. Berkeley, CA: University of California Press.

(2004). *Late Cretaceous and Cenozoic Mammals of North America: Biostratigraphy and Geochronology*. New York: Columbia University Press.

Part I: Non-eutherian mammals

1 Non-eutherian mammals summary

CHRISTINE M. JANIS[1] and ANNE WEIL[2]
[1]*Brown University, Providence, RI, USA*
[2]*Oklahoma State University Center for Health Sciences, Tulsa, OK, USA*

INTRODUCTION

"Non-eutherian mammals" is obviously a paraphyletic grouping. Metatheria (Huxley, 1880: extant marsupials and their extinct relatives that fall outside of the extant crown group) and Eutheria (Gill, 1872: extant placentals and their extinct relatives that fall outside of the extant crown group) have long been considered to belong in the Theria (Parker and Haswell, 1897), exclusive of both the multituberculates and the monotremes (although see below for discussion of past and present notions of the "Marsupionta", uniting marsupials and monotremes in a clade).

MULTITUBERCULATA

As discussed in Chapter 2, multituberculates have long been recognized as a distinctive group of mammals. Simpson (1945) ranked them as the subclass Allotheria (Marsh, 1880), one of three mammalian subclasses (the other two being Prototheria, or monotremes, and Theria). McKenna and Bell (1997), by comparison, recognized Allotheria as an infraclass within subclass Theriiformes; subclass Prototheria is retained, and subclass Theriiformes includes infraclasses Allotheria, Triconodonta, and Holotheria, the last including Theria as a supercohort (note that this classification differs from the one shown in Figure 1.2, below). Multituberculates are commonly known as "the rodents of the Mesozoic," and it is probable that they filled a rodent-like niche as small omnivores and herbivores prior to the evolution of the rodents in the early Tertiary. The probable paleobiology of multituberculates, including possible reasons for their extinction, is discussed in Chapter 2, this volume.

Multituberculates are usually considered as the order Multituberculata (Cope, 1884, originally proposed as a suborder of Marsupialia, now an order within the Allotheria). As such, they represent the longest-lived order of mammals, ranging from the Late Triassic (if the haramiyids are included, around 205 Ma) (Butler and Hooker, 2005) or the Middle Jurassic (if the haramiyids are excluded, around 160 Ma), to the late Eocene (around 35 Ma). The haramiyids are a paraphyletic and problematical group of mammals, with rather multituberculate-like cheek teeth, known from the Late Triassic to Late Jurassic (see further discussion below). Kielan-Jaworowska, Cifelli, and Luo (2004) considered the haramiyids to be a separate order, Haramiyida, within the Allotheria, rather than included within the Multituberculata. Kielan-Jaworowska, Cifelli, and Luo (2004) also rejected the notion that the rather multituberculate-like Gondwanatheria (known from the Late Cretaceous to early Paleocene of South America, Africa, India, and Madagascar) belong with the multituberculates, classifying them as Mammalia incertae sedis: the distribution of multituberculates is thus confined to the Northern Hemisphere, with the exception of a few fragmentary teeth referred from Morroco (Sigogneau-Russell, 1991) and Madagascar (Krause, *et al.*, 1999).

Kielan-Jaworowska, Cifelli, and Luo (2004; Chapter 8) provided a general summary of multituberculates, focusing primarily on the Mesozoic radiation, and the following text summarizes some of their major points. The main radiation of multituberculates was in the Cretaceous, with the earliest Tertiary multituberculates representing a diversity considerably reduced by the end Cretaceous extinctions. There are two major groups within the multituberculates: the paraphyletic "Plagiaulacidae," and the Cimolodonta, which are both composed of around 10 families (depending on the classification scheme). The plagiaulacids were less derived in their dentition than the cimolodonts, and are all of fairly small size; their major radiation was in the Late Jurassic of Eurasia and North America, although some lineages persisted into the Early Cretaceous (where one lineage is also found in Morocco). The cimolodonts include taxa of larger size (such as the wombat-sized taeniolabidids): they are more derived than plagiaulacids in a number of features, including the loss of the first upper incisor and the transformation of the lower fourth premolar into an arcuate, bladelike tooth. Cimolodonts first appeared

Evolution of Tertiary Mammals of North America, Vol. 2. ed. C. M. Janis, G. F. Gunnell, and M. D. Uhen. Published by Cambridge University Press.

Figure 1.1. Restoration of the early Tertiary marsupial *Peradectes* (by Marguette Dongvillo).

in the Early Cretaceous, were probably at their most diverse during the Late Cretaceous, and were known to occur throughout the Northern Hemisphere. Five families extend into the Tertiary: the Eucosmodontidae (North America and Europe), the Microcosmodontidae (North America), the Taeniolabididae (North America and Asia), the Ptilodontidae (North America, Europe, and Asia), and the Neoplagiaulacidae (North America, Europe, and Asia). The genus *Cimexomys*, placed incertae sedis within Cimolodonta, survived into the Paleocene of North America.

Hurum, Luo, and Kielan-Jaworowska (2006) noted that a monotreme-like os calcaris is present in several multituberculates from the Late Cretaceous of Mongolia. In the platypus, this bone is associated with the spur and poison gland of males, and an os calcaris is also seen in some other Mesozoic mammals such as *Gobioconodon* and *Zhangheotherium*. They, therefore, concluded that this is a primitive mammalian feature, not a monotreme autapomorphy as previously thought.

THERIA

MARSUPIAL/PLACENTAL SIMILARITIES AND DIFFERENCES

Extant therians are united by many morphological features. Osteological characteristics include tribosphenic molars, presence of a scapular spine and supraspinous fossa (see Sánchez-Villagra and Maier, 2003), middle ear bones fully enclosed (now determined to have occurred independently of the condition in monotremes [Rich *et al.*, 2005]) with a cochlea of two and a half coils, and numerous features of the basicranium (Wible and Hopson, 1993). Soft anatomy features include the presence of a dually innervated (cranial nerves V and VII) digastric jaw-opening muscle (as opposed to the detrahans muscle of monotremes, innervated by cranial nerve V, or the depressor mandibulae of other tetrapods, innervated by cranial nerve VII), scapular sling muscles derived from the hypaxial layer (e.g., rhomboideus and serratus muscles), an external ear (pinna), nipples, and various features of the urogenital system. Urogenital features include the rerouting of the ureters into the bladder (from the cloaca, evidently done convergently between marsupials and placentals as the position of the ureters relative to the reproductive ducts differs), separate openings for alimentary and urogenital systems (i.e., loss of the cloaca), descent of the testes into a scrotum (clearly accomplished convergently, as the scrotum, if present, is postpenile in marsupials and prepenile in [most] placentals), and a penis that is now used for urination as well as sperm transmission (see Renfree, 1993, for a summary of mammalian reproductive differences). Marsupials and placentals also share the derived feature of viviparity, but it is not clear if this condition arose once, or convergently between the two groups, because the formation of a uterus more derived than the monotreme condition was clearly evolved convergently (Renfree, 1993). Cifelli (1993) and Kielan-Jaworowska, Cifelli, and Luo (2004) have discussed the osteological attributes of extinct clades also considered to belong within the Theria.

Both placentals and marsupials possess unique features, and one general consideration about the original divergence, based on the postcranial skeleton, is that marsupials were originally more arboreal while placentals were more terrestrial (see Szalay, 1994). Placental apomorphies include the following features: the loss of the epipubic bones (although note the presence of these bones in some Cretaceous eutherians [Novacek *et al.*, 1997]), a corpus callosum linking the two cerebral hemispheres, the retention of the young in the uterus past a single estrus cycle, the fusion of the Müllerian ducts into a midline uterus, vasa deferentia that loop over the ureters, and a scrotum (if present) that is (usually) placed behind the penis (lagomorphs are an exception). Eutherians also possess various detailed derived features of the cranium, dentition, and postcranial skeleton (see Kielan-Jaworowska, Cifelli, and Luo, 2004), including the reduction of the number of molars to three, and (except for the most primitive forms) a reduced number of incisors with three or fewer in each jaw half. Additionally, although not all placentals have large brains, it is only among placental mammals that large brains (encephalization quotient significantly greater than one) have arisen (convergently, in many different clades).

Marsupials have a number of derived features relating to their unique mode of reproduction, including the presence of a pseudovaginal canal, and others that are discussed below. Marsupials are also unique in the possession of end arteries on the surface of the brain (Lillegraven, 1984). In addition, there are a diversity of derived features of the cranium, dentition, and postcranial skeleton (see Kielan-Jaworowska, Cifelli, and Luo, 2004). Extant marsupials have an auditory bulla, if present, made from the alisphenoid bone (the placental auditory bulla may be derived from a variety of sources, but never from the alisphenoid). A distinctive feature of almost all metatherians is the shelflike inflected angle of the dentary. However, this feature is absent from the supposed first metatherian, *Sinodelphys* (Luo *et al.*, 2003); a slight inflection is seen in early eutherians, and the angle is secondarily reduced in some extant marsupials, in the koala (*Phascolarctos*), the numbat (*Myrmecobius*), and the honey possum (*Tarsipes*). Other "typical" marsupial cranial features, used to distinguish extant forms from placentals, such as diamond-shaped nasals, palatal vacuities, and the exclusion of the jugal from the jaw glenoid, are all primitive therian features that may be variously observed among extant placentals. The marsupial dentition is apomorphic in the reduction of the number of premolars to three, and the condition of virtual monophyodonty, where the only tooth to be replaced is the last premolar.

Finally, despite the popularity of textbook figures showing ecomorphological convergence between extant marsupials and placentals (e.g., marsupial "wolf" etc.), marsupials exhibit some ecomorphological types not seen among placentals: no placental has evolved in a large (> 10 kg) ricochetal (hopping) form like the diversity of kangaroos, and there is no non-volant nectivore among placentals. All nectivorous placentals are bats, whereas the marsupial nectivore is the honey possum, or noolbenger (*Tarsipes rostratus*).

We discuss below the contribution of molecular biology to higher-level mammalian systematics. We also note here that molecular biology has also provoked controversy in the discussion of when the splits occurred between major mammalian lineages. For example, while the fossil record shows the earliest eutherians and metatherians to be in the Early Cretaceous, around 120 Ma (e.g., Ji *et al.*, 2002; Luo *et al.*, 2003), some molecular studies propose a split as early as the Late Jurassic (e.g., Kumar and Hedges, 1998). In addition, numerous authors have proposed the diversification of the major placental orders deep within the Cretaceous, in contrast to fossil record evidence that would place this diversification in the latest Cretaceous at the earliest (e.g., Hedges *et al.*, 1996; Eizirik, Murphy, and O'Brien, 2001; Madsen *et al.*, 2001; Murphy *et al.*, 2001a,b; but for estimates much closer to the K/T boundary, see Kitozoe [2007; molecular data] and Wible *et al.* [2007; morphological data]). This issue is further discussed in Kielan-Jaworowska, Cifelli, and Luo (2004: Chapter 15); Hunter and Janis (2006) also discussed this issue and the paleobiogeography of early placentals.

MARSUPIALS

Although marsupials are thought of as quintessentially Australian mammals, with perhaps a second outpost in South America, the first definitive metatherians are known from the Late Cretaceous of North America (Cifelli and Muizon, 1997; Cifelli, 1999). However, a candidate for the earliest metatherian, *Sinodelphys szalayi*, is known from the Early Cretaceous of Asia (Luo *et al.*, 2003), in the same deposits (the Yixian Formation of China) as the earliest known eutherian, *Eomaia scansoria* (Ji *et al.*, 2002). During the Tertiary, marsupials are found not only in North America but also during the Eocene and Oligocene in Europe, Asia, and Africa (possibly extending into the middle Miocene in Asia; see McKenna and Bell, 1997). However, these Old World marsupials were not a diverse radiation and are known by only a few individual fossils.

Present-day marsupials make up only 6% of mammalian species; however, marsupials exhibit a great degree of morphological diversity, and their low taxonomic diversity is explained, at least in part, by the area of land that they occupy today (Kirsch, 1977). Note, however, that the marsupials found in the Northern Hemisphere during the Tertiary were uniformly small to medium-sized, fairly generalized mammals, resembling Recent didelphids (opossums) in their ecomorphology.

Marsupials are usually distinguished from placentals by their mode of parturition. While placentals carry their young in the uterus past a single estrus cycle, marsupials all eject their young at the end of the estrus cycle. The neonates are highly altricial and make their way up their mother's ventral side to attach onto a nipple, which in more derived forms is enclosed within a pouch (or marsupium), where they complete their development. The marsupial form of reproduction was once thought to be some primitive intermediate stage between the oviparity of monotremes and the form of viviparity seen in placentals. Marsupials were reported to lack the chorioallantoic placenta of placentals, instead relying on the yolk sac (choriovitalline) placenta, which is also seen in the early stages of gestation in placentals; they are also primitive in retaining vestigies of the egg shell membrane. Marsupials have been assumed to be constrained in their taxonomic and morphological diversity by their reproductive mode (e.g., Lillegraven, 1975). However, in more recent years, this issue of "marsupial inferiority" has been reexamined (see Sears [2004] for a review).

For a start, the apparent lack of a chorioallantoic placenta in marsupials is not a primitive condition (indeed, this could hardly be the case as a chorioallantoic membrane is present in all amniotes). Some marsupials (e.g., bandicoots, koalas, and wombats) do indeed show evidence of a transitory chorioallantoic placenta at the end of gestation, and developmental studies show that the outgrowth of the chorioallantoic membrane is actually *suppressed* in marsupials (see Smith, 2001) (i.e., this apparent lack of a chorioallantoic membrane is a derived feature, not a primitive one). Furthermore, marsupial neonates are not merely undeveloped versions of placental neonates but show many derived features. In ontogeny, the development of the forelimb and craniofacial structures have been accelerated, at the expense of the later development of the nervous system (Smith, 1997), and marsupial neonates have a unique cartilaginous "shoulder arch," made up in part from retained interclavicle and coracoid elements (otherwise lost in adult therians), that aids in the crawl to the nipple (Sánchez-Villagra and Maier, 2003; Sears, 2004).

However, while it appears that marsupials should not be considered inferior to placentals because of their different mode of reproduction, it is likely that constraints on neonate forelimb anatomy have led to constraints on adult locomotor mode (see Sears, 2004). No marsupial has greatly reduced the numbers of fingers, as do many placental ungulates. Constraints on forelimb anatomy might also prevent the evolution of flippers in aquatic marsupials (Lillegraven, 1975), but it seems that the evolution of a fully aquatic marsupial might be more constrained by the need to carry young in a pouch, and to give birth on land. (Note that there is one semi-aquatic marsupial: the South American yapok, or water-opossum, *Chironectes minimus*, which can seal the pouch during brief underwater forays.)

SYSTEMATICS

MULTITUBERCULATES

Multituberculates are without close living relatives, and their placement within Mammalia is contentious. Most parsimony analyses place them within crown-group Mammalia (Rowe, 1988; Wible and Hopson, 1993; Rougier, Wible, and Novacek, 1996; Luo, Cifelli, and Kielan-Jaworowska, 2001; Woodburne, 2003; but see Wible [1991] and Miao [1993] for characters that would seem to place them outside). Among these, one school of thought is that multituberculates are more closely related to monotremes than to Theria. This hypothesis of relationship is supported by some braincase morphology (Kielan-Jaworowska, 1971), and the shape, position, and orientation of the ear ossicles (Meng and Wyss, 1995). It was also supported by a single character in Wible and Hopson's (1993) phylogeny of basicrania, although the authors thought that the relationship to monotremes was unlikely. The other, perhaps more widely accepted hypothesis, that Multituberculata is more closely related to Theria than to Monotremata, is the better supported when cranial and postcranial characters are combined (Rowe, 1988; Rougier, Wible, and Novacek, 1996; Luo, Cifelli, and Kielan-Jaworowska, 2001; Luo, Kielan-Jaworowska, and Cifelli, 2002) (as shown in Figure 1.2, below).

Both of these hypotheses seem problematic when basal mammalian dentitions are considered, because there are no identifiable cusp homologies between multituberculate teeth and tribosphenic teeth. The most common objection this raises is that multituberculate molars cannot be derived from a tribosphenic pattern. However, Krause (1982) noted that some murid rodents converge on the multituberculate pattern of longitudinal rows of cusps. Meng and Wyss (1995) pointed out that a similar molar form has evolved in the bat *Harpyionycteris* (illustrated in Nowak and Paradiso, 1983, p. 186); rodents and bats are placental groups with primitively tribosphenic molars. Moreover, recent phylogenies (Luo, Cifelli, and Kielan-Jaworowska, 2001; Luo, Kielan-Jaworowska, and Cifelli, 2002) have raised the possibility that tribosphenic molars arose separately in the monotreme and therian lineages (see discussion below). Alternatively, Woodburne (2003) argued that monotremes do not in fact have tribosphenic teeth. Either of these last hypotheses would obviate the need for any derivation of the multituberculate molar pattern from a tribosphenic morphology.

The significant problem posed is instead that, without identifiable homologies, many phylogenetic characters of the dentition are not applicable to multituberculates. For example, Luo, Kielan-Jaworowska, and Cifelli (2002) listed 55 molar morphology and 12 molar wear characters; this adds up to about one quarter of the 271 informative characters in their analysis of mammalian phylogeny, most of which are not applicable to multituberculates. The problem thus introduced is not resolvable by the analysis software used; the computer cannot distinguish between data that are missing (owing, for instance, to non-preservation) and characters that are inapplicable. As a result, the true most parsimonious trees may be rejected (Maddison, 1993). Any placement of Multituberculata on the tree of basal mammals should, therefore, be considered provisional and somewhat unreliable.

One precladistic classification of early mammalian relationships placed Multituberculata in the "Allotheria" with the Haramiyida. Haramiyids are known from the Late Triassic and until recently (Jenkins *et al.*, 1997) were known only from isolated teeth, on which multiple cusps are arranged in rows, similar to those of multituberculate teeth. The discovery of a dentary, premaxilla, and maxillary fragment in Greenland, however, revealed that haramiyids retained relatively substantial postdentary bones and dental specializations that preclude them from being directly ancestral to, or closely related to, multituberculates (Jenkins *et al.*, 1997). An experiment in which Multituberculata and Haramiyida were constrained to be related in phylogenetic analysis resulted in both "Allotheria" and eutriconodonts being pulled outside crown-group Mammalia. However, in phylogenetic analyses of the same matrix in which Multituberculata and Haramiyida were considered independently, haramiyids branch off far below crown-group Mammalia (Luo, Kielan-Jaworowska, and Cifelli, 2002).

MARSUPIALS

Kielan-Jaworowska, Cifelli, and Luo (2004; their Chapter 15) provided an extensive review of previous and current hypotheses and controversies concerning the interrelationships of various basal groups of mammals. Here we largely summarize their discussion of the relationship of marsupials to other mammals.

While tribosphenic molars were long considered to be a synapomorphy of therian mammals, various Mesozoic mammals have been found in the Southern Hemisphere since the mid 1990s that apparently possess tribosphenic molars (including an early monotreme, *Steropodon*), and there is considerable controversy about whether or not tribosphenic molars evolved once or convergently between Mesozoic mammals in Northern and Southern Hemispheres. A single evolution of the tribosphenic condition (e.g., Rich *et al.*, 1997, 2002; Woodburne, Rich, and Springer, 2003) would imply that monotremes are much more closely related to therians than previously supposed. In contrast, Luo and colleagues (Luo, Cifelli, and Kielan-Jaworowska, 2001; Luo, Kielan-Jaworowska, and Cifelli, 2002) considered that tribosphenic molars arose independently in the southern, monotreme-related group (their Australosphenida) and in the northern, therian group (their Boreosphenida), with monotremes and therians belonging to very separate clades that diverged from each other deep within the early mammalian radiation (see Figure 1.2).

The derived features uniting therians were discussed previously, and therian monophyly is strongly supported by a number of recent morphological analyses (e.g., Zeller, 1999; Szalay and Sargis, 2001; Luo, Kielan-Jarowoska, and Cifelli, 2002). Most molecular studies, either using nuclear DNA sequences (e.g., Retief, Winkfein, and Dixon, 1993; Kullander, Carlson, and Hallbrook, 1997; Lee *et al.*, 1999; Gilbert and Labuda, 2000; Killian *et al.*, 2001) or protein sequences (Messer *et al.*, 1998; Belov, Hellman, and Cooper, 2002), also support therian monophyly. However, studies of mitochondrial genomes (e.g., Janke, Xu, and Arnason, 1997; Kirsch,

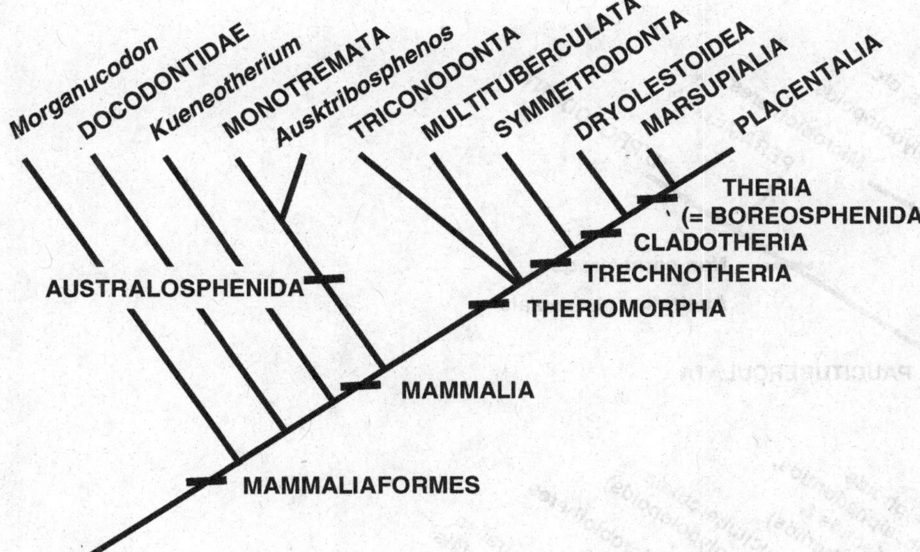

Figure 1.2. Interrelationships of the major lineages of mammals (modified from Rich *et al.* [2005] and Kielan-Jaworowska, Cifelli, and Luo [2004; their Figure 15.1]). Note that an alternative cladogram (displayed in Kielan-Jaworowska, Cifelli, and Luo [2004; their Figure 15.2]) would place *Kueneotherium* as the sister taxon of *Morganucodon*, below docodonts.

Lapointe, and Springer, 1997; Penny and Hasegawa, 1997; Kirsch and Mayer, 1998; Janke *et al.*, 2002) have instead supported the notion of a sister-group relationship between monotremes and marsupials, the old notion of the "Marsupionta" (Haeckel, 1898; Gregory, 1947). Springer *et al.* (2001) and Baker *et al.* (2004) both discussed this molecular controversy; they supported therian monophyly and advanced the notion that nuclear exon genes are more informative than mitochondrial genes for resolving deep branches within mammalian phylogeny. Baker *et al.* (2004) also provided a critique of some of the statistical methodologies applied by other workers.

Case, Goin, and Woodburne (2005) reviewed the higher-level classification of the Marsupialia, and the discussion below summarizes their work. Simpson (1945) originally classified marsupials (as order Marsupialia) into two major groups (superfamilies): the American Didelphoidea and the Australian Dasyuroidea. Ride (1964, 1970) raised marsupials to an equivalent taxonomic rank with placentals, including four orders: the Marsupicarnivora (including the North and South American, and Australian insectivorous taxa: didelphids, stagodontids, pediomyids, borhyenids, thylacines, and dasyurids), the Paucituberculata (the South American caenolestids, polydolopids, groeberiids, and argyrolagids), the Peramelina (the Australian bandicoots and bilbies), and the Diprotodonta (the Australian herbivores, including possums, koalas, wombats, kangaroos, etc.). The Didelphidae (including the North American Cretaceous and early Tertiary marsupials) was assumed to be the basal marsupicarnivoran family, from which all other marsupials were ultimately derived. However, within these early marsupials, Crochet (1979) distinguished between the South American Cenozoic Didelphidae and the predominately North American Peradectidae.

The work of Szalay (see Szalay, 1994) on tarsal morphology showed that the South American microbiotheres (including the extant genus *Dromiciops*) were actually united with the Australian marsupials in what he termed the cohort Australidelphia, leaving the Ameridelphia as a paraphyletic American grouping; this proposition was later backed up by molecular evidence (e.g., Kirsch *et al.*, 1991; Springer *et al.*, 1998). Marshall, Case, and Woodburne (1990) proposed a third cohort to define the major clades of marsupials, the Alphadelphia, which distinguished the North American Cretaceous and early Cenozoic taxa from the Cenozoic South American Ameridelphia.

McKenna and Bell (1997) recognized the magnorder Ameridelphia (including orders Didelphimorphia and Paucituberculata [including polydolopids, see below]) and magnorder Australidelphia (including superorders Microbiotheria and Eometatheria [all Australian marsupials]). The taxa in the proposed Ameridelphia are dealt with as follows. The North American families Stagodontidae and Pediomyidae (which are confined to the Cretaceous) are left outside of this magnorder grouping, united in the suborder Archimetatheria. In contrast, alphadontids (Late Cretaceous of North America), peradectids, and herpetotheriids (both from the Late Cretaceous and Tertiary of North America) are now included as subfamilies within the Didelphidae. Thus the notion of a distinct grouping of Alphadelphia (whether paraphyletic or not) is abandoned. This is the classification essentially followed by Korth (Figure 3.1; see Chapter 3).

It was previously supposed that the alphadelphian marsupials dispersed to South America, giving rise to the ameridelphians (Case and Woodburne, 1986). However, recent discoveries of Cretaceous marsupials in North America (Case, Goin, and Woodburne, 2005) have led to a revision of this hypothesis. It appears that some ameridelphians were actually present in the latest Cretaceous of North America: thus the cohort must have had its origins on this continent. The Ameridelphia remains as a paraphyletic grouping, and at least two South American clades (Didelphoidea and Polydolopimorphia) originated in the Late Cretaceous of North America

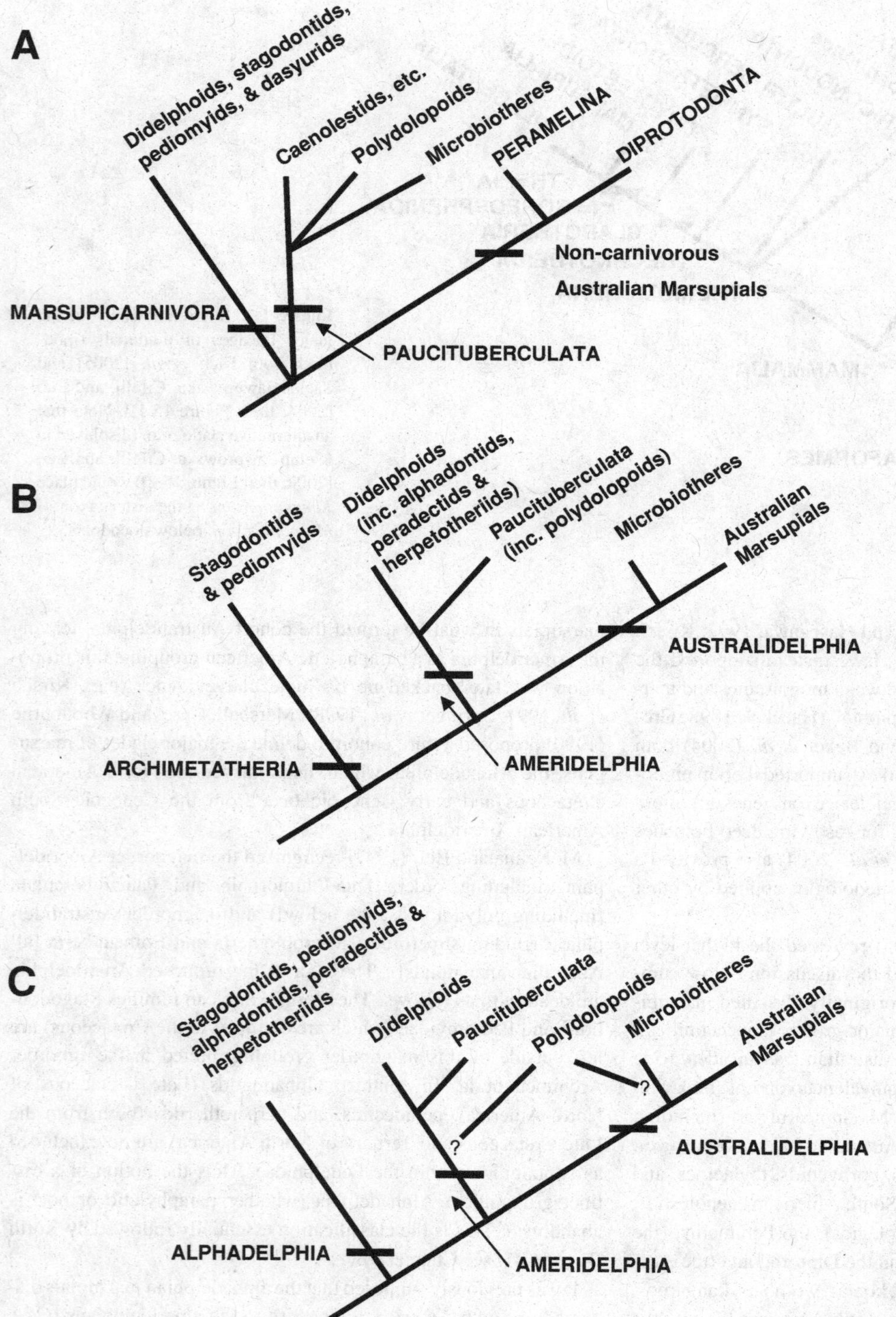

Figure 1.3. Interrelationships of major marsupial clades. A. Scheme of Ride (1964, 1970). B. Scheme of McKenna and Bell (1997). C. Scheme of Case, Goin, and Woodburne (2005).

(or were derived from taxa from that place and time) and subsequently migrated to South America. It has also been suggested that the polydolopimorphians (which are also known from Antarctica) might be related to the microbiotheres (and hence be members of the cohort Australidelphia) (Goin, 2003), which would further complicate hypotheses of marsupial dispersal and biogeography.

EVOLUTIONARY AND BIOGEOGRAPHIC PATTERNS

Non-eutherian mammals are essentially an early Tertiary radiation, with the exception of the reimmigration of the extant North American opossum, *Didelphis*, in the Pleistocene. However, there is a difference in the diversity patterns between multituberculates and

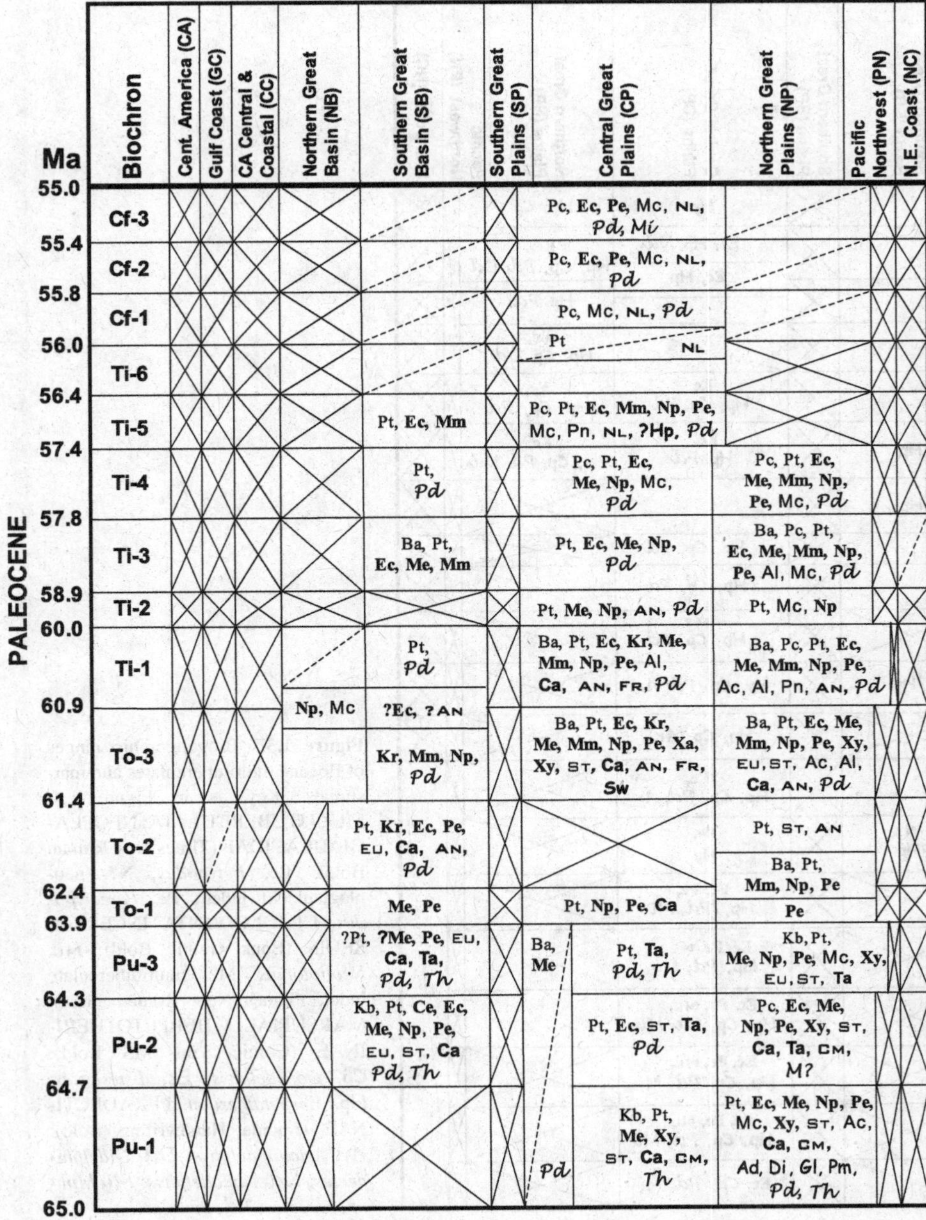

Figure 1.4. Biogeographic ranges of Paleocene multituberculates and marsupials. Key: A "box" (for a particular time period in a particular biogeographic region) that has a cross through it means no fossil mammal localities are known for that time period from that area; a single dashed line through the box means only scant fossil mammal information is available (usually only a single, small locality). MULTITUBERCULATA: PTILODONTIDAE (Times New Roman Plain): Ba, *Baiotomeus*; Kb, *Kimbetohia*; Pc, *Prochetodon*; Pt, *Ptilodus*. NEOPLAGIAULACIDAE (Times New Roman Bold): **Ce**, *Cernaysia*; **Ec**, *Ectypodus*; **Kr**, *Krauseia*; **Me**, *Mesodma*; **Mm**, *Mimetodon*; **Np**, *Neoplagiaulax*; **Pe**, *Parectypodus*; **Xa**, *Xanclomys*; **Xy**, *Xyronomys*. EUCOSMODONTIDAE (Bank Gothic Plain): EU, *Eucosmodon*; ST, *Stygimys*. MICROCOSMODONTIDAE (Arial Plain): Ac, *Acheronodon*; Al, *Allocosmodon*; Mc, *Microcosmodon*; Pn, *Pentacosmodon*. TAENIOLABIDIDAE (Arial Bold): **Ca**, *Catopsalis*; **Ta**, *Taeniolabis*. CIMOLODONTA INCERTAE SEDIS (Bank Gothic Bold): **AN**, *Anconodon*; **CM**, *Cimexomys*; **FR**, *Fractinus*; **NL**, *Neoliotomus*. MARSUPIALIA: MESOZOIC HOLDOVERS (Comic Sans MS Plain): Ad, *Alphadon*; Di, *Didelphodon*; Gl, *Glasbius*; Pm, *Pediomys*. HERPETOTHERIINAE (Comic Sans MS Bold): **Sw**, *Swaindelphys*. PERADECTINAE (Lucida Handwriting Bold): **Mi**, *Mimoperadectes*; **Pd**, *Peradectes*; **Th**, *Thylacodon*; M?, marsupial indet. (Lucida Handwriting Plain).

marsupials. Tertiary multituberculates reach their greatest diversity in the Paleocene; only three genera extend into the Eocene, and none extend past the end of the epoch. In contrast, marsupials are most diverse in the Eocene, three genera extend into the Oligocene, and there are a few occurrences in the Neogene.

PALEOCENE

The earliest Paleocene (Puercan 1) contains representatives of all of the surviving multituberculate families, and also some Mesozoic marsupial holdovers, as well as the marsupial peradectine taxa *Peradectes* and *Thylacodon*. *Thylacodon* does not persist past the Puercan, while *Peradectes* is the longest-lived Tertiary marsupial genus, persisting until the end of the Eocene. Among the multituberculates, *Kimbetohia, Cernaysia*, *Taeniolabis*, and *Cimexomys*, are confined to the Puercan (Figure 1.4).

In terms of Puercan biogeographic patterns, marsupials are initially known primarily from the Northern Great Plains region, but past Puercan 1 are absent from this area, and are found in the Central Great Plains and Southern Great Basin regions. The multituberculates are more evenly distributed through these three regions, but the cimolodont incertae sedis *Cimexomys* and the neoplagiaulacid *Xyronomys* are not described from the Southern Great Basin (and both are confined to the Northern Great Plains after Puercan 1), whereas the eucosmodontid *Eucosmodon* is absent from the

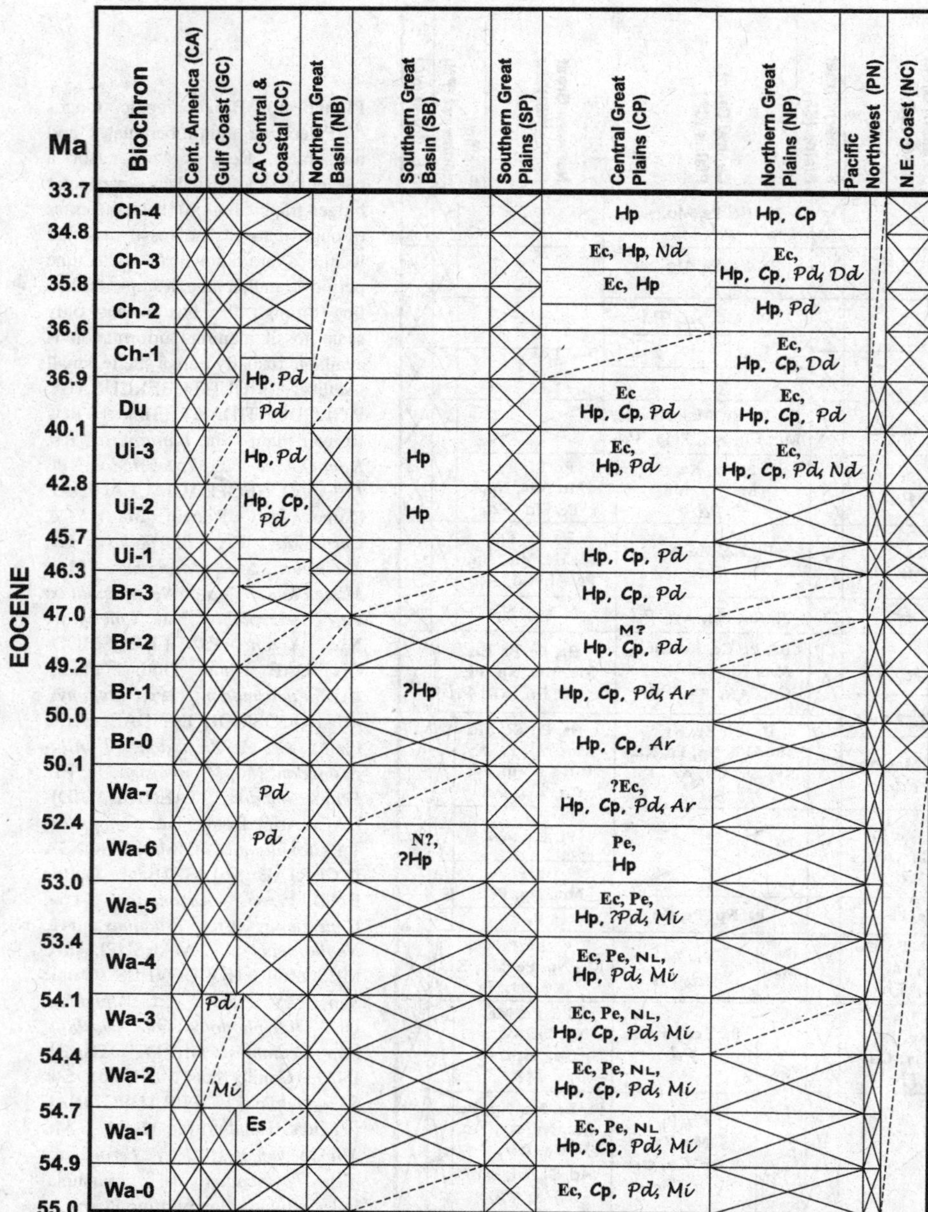

Figure 1.5. Biogeographic ranges of Eocene multituberculates and marsupials. Key as in Figure 1.4. MULTITUBERCULATA: NEOPLAGIAULACIDAE (Times New Roman Bold): **Ec**, *Ectypodus*; **N?**, neoplagiaulacid indet.; **Pe**, *Parectypodus*. CIMOLODONTA INCERTAE SEDIS (Bank Gothic Bold): **NL**, *Neoliotomus*. M?, multituberculate indet. (Times New Roman Plain). MARSUPIALIA: HERPETOTHERIINAE (Comic Sans MS Bold): **Cp**, *Copedelphys*; **Es**, *Esteslestes*; **Hp**, *Herpetotherium*. PERADECTINAE (Lucida Handwriting Bold): *Ar*, *Armintodelphys*; *Dd*, *Didelphidectes*; *Nd*, *Nanodelphys*; *Mi*, *Mimoperadectes*; *Pd*, *Peradectes*.

Central Great Plains. Several genera (*Kimbetohia, Stigymys, Catopsalis, Mesodma*) are known principally from the Central Great Plains in Puercan 1 but are absent from this region (although present in others) in the later Puercan. The ptilodontid *Baiotomeus*, primarily a genus of the later Paleocene, is known only from an undifferentiated Puercan site in the Central Great Plains.

The marsupial *Peradectes* is briefly joined in the Torrejonian by the herpetotherine *Swaindelphys* (known only from Torrejonian 3), but otherwise marsupial diversity remains low. *Peradectes* is found throughout the known biogeographic regions in the Torrejonian, but *Swaindelphys* is known only from the Central Great Plains. Several new multituberculate genera appear in the Torrejonian: *Krauseia, Mimetodon, Xanclomys*, *Allocosmodon*, and *Fractinus*. With the exception of *Xanclomys*, all of these taxa also range into the Tiffanian. However, several other genera go extinct at the end of the Torrejonian: *Xyronomys*, *Eucosmodon*, and *Stygimys*. *Baiotomeus, Xanoclomys*, and *Xyronomys* are absent from the Southern Great Basin, and *Krauseia* is absent from the Northern Great Plains region.

Peradectes is the only marsupial genus known in the Tiffanian (aside from a possible occurrence of *Herpetotherium* in Ti5), and it retains a fairly cosmopolitan biogeographic distribution, although it is more rarely found in the Southern Great Basin than elsewhere. A couple of new multituberculate genera appear in the Tiffanian: *Pentacosmodon* at the start of this land mammal age (and confined to it), and *Neoliotomus* towards its end. However, the

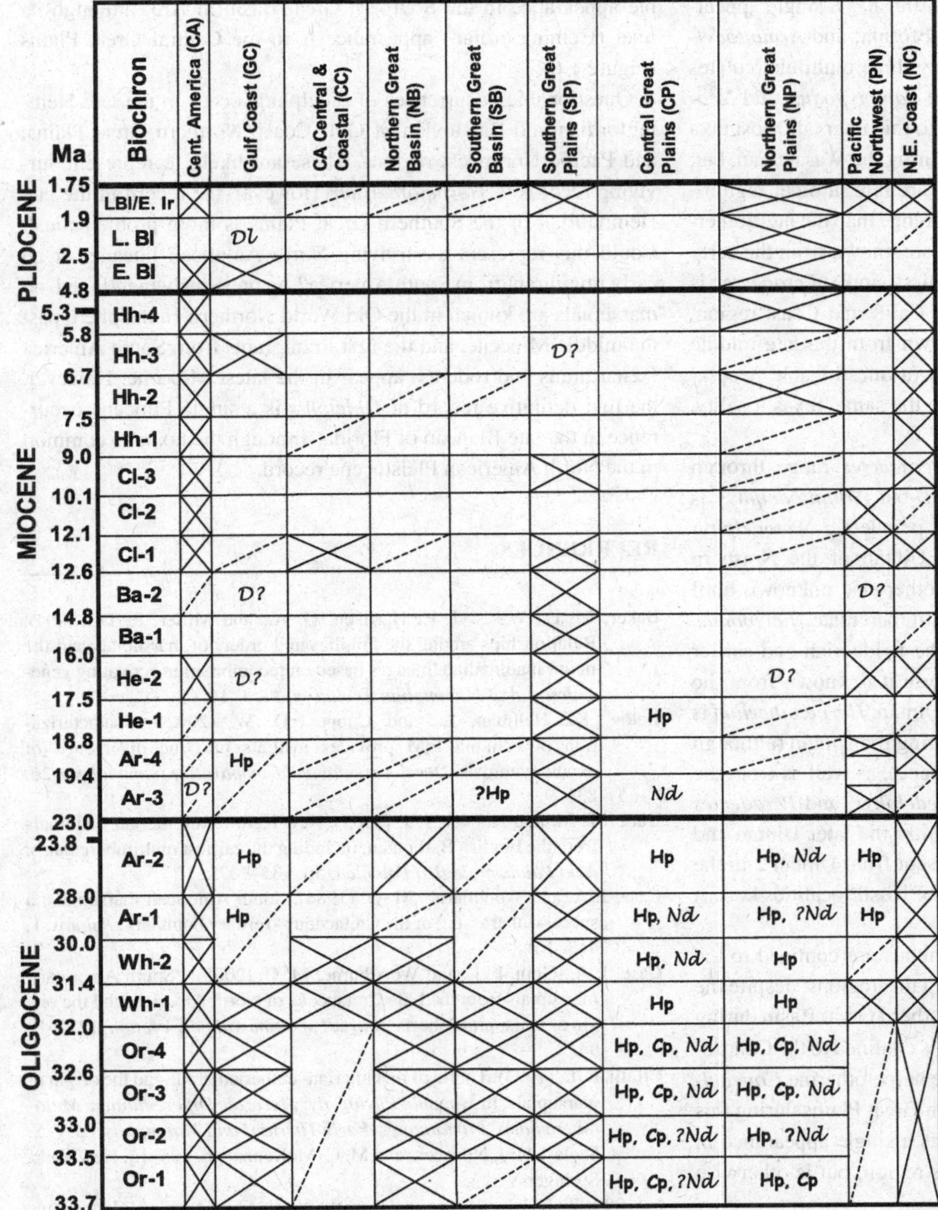

Figure 1.6. Oligocene–Pliocene marsupials: Key as in Figure 1.4. HERPETOTHERIINAE (Comic Sans MS Bold): **Cp**, *Copedelphys*; **Hp**, *Herpetotherium*. PERADECTINAE (Lucida Handwriting Bold): **Nd**, *Nanodelphys*. DIDELPHINAE (Lucida Handwriting Plain): Dl, *Didelphis*; D?, didelphid indet.

majority of earlier multituberculate genera are extinct by the end of the Tiffanian, including such long-ranging and widely distributed taxa as *Ptilodus* and *Neoplagiaulax*, and all members of the families Eucosmodontidae and Taeniolabididae. In terms of Tiffanian biogeography, the microcosmodontids are absent from the Southern Great Basin, and neoplagiaulacids rare in this region. There appears to be little difference between Central Great Plains and Northern Great Plains regions, although *Mimetodon* is confined to the Northern Great Plains during the middle portion of the Tiffanian, (Ti2 to Ti4), and *Neoliotomus* is known only from the Central Great Plains.

The peradectine marsupial *Peradectes* is joined by the peradectine *Mimoperadectes* in the latest Clarkforkian. Only five multituberculate genera are known from the Clarkforkian. With the exception of *Neoliotomus*, they are all taxa that first appeared in the Puercan (*Prochetodon, Ectypodus, Parectypodus*, and *Microcosmodon*). Two of these genera (*Prochetodon*, the last surviving ptilodontid, and *Microcosmodon*, the last surviving microcosmodontid) do not survive into the Eocene. All of these taxa are known only from the Central Great Plains during this time period, but fossil deposits from other areas are scarce.

EOCENE

Marsupial diversity increases in the early Eocene. In addition to the Paleocene peradectine survivors, *Peradectes* and *Mimoperadectes*, a number of herpetotherines are known from the early Wasatchian (Wa): *Copedelphys* first appears in Wa0; *Herpetotherium* makes its first definitive appearance in Wa1; *Esteslestes* has a single appearance in the early Wasatchian of Baja California; and *Armintodelphys* appears in the late Wasatchian. Only three multituberculates are known in the Wasatchian, *Ectypodus, Parectypodus*, and *Neoliotomus*, and the latter two are confined to this interval. Most taxa are confined to the Central Great Plains during the Wasatchian, but, as with the Clarkforkian, fossil localities are rare in other regions. The marsupials apparently have a broader range than the multituberculates. *Peradectes* and *Mimoperadectes* are known from the early Wasatchian of the Gulf Coast region (Mississippi), *Peradectes* is known from the later Wasatchian of the California Coast region, and *Herpetotherium* is questionably known from the late middle Wasatchian of the Southern Great Basin (a questionable neoplagiaulacid multituberculate is known from the same Texas locality, SB24) (Figure 1.5).

Herpetotherium, Copedelphys, and *Peradectes* range through the Bridgerian, Uintan, and Duchesnean, but *Armintodelphys* is unknown after the early Bridgerian. The peradectid *Nanodelphys* makes a single appearance in the late Uintan of the Northern Great Plains (in Saskatechewan) but is otherwise unknown until the Chadronian. The only surviving multituberculate, *Ectypodus*, is absent from the fossil record during the Bridgerian and earlier Uintan but must have been present because it is known from the late Wasatchian and reappears in the late Uintan. *Herpetotherium* is known from the Southern Great Basin during the Uintan (although is absent from this region in the later Eocene), as well as from the other regions, and *Herpetotherium, Copedelphys*, and *Peradectes* are all known from the Gulf Coast region in the later Uintan and Duchesnean. Marsupials are strangely absent from Uintan 2 in the Central Great Plains, despite the number of fossil localities known from this region.

In the Chadronian, non-eutherian mammals are confined to the Central Great Plains and Northern Great Plains regions, despite the fact that localities are known from the Southern Great Basin during this time. The peradectine *Didelphidectes* is confined to the Chadronian of the Northern Great Plains, and the herpetotherine *Copedelphys* is also only known from the Northern Great Plains during this time. The herpetotherine *Nanodelphys* has a single appearance in Chadronian 3 of the Central Great Plains region, but is otherwise primarily an Oligocene taxon.

OLIGOCENE–PLIOCENE

No multituberculate survives past the end of the Eocene (the latest survivor is known from Chadronian 3). Three marsupial genera persist past the Eocene, although both are rarely known as fossils past the early Oligocene: the herpetotherine *Copedelphys* is confined to the Orellan, the peradectine *Nanodelphys* survives into the earliest Miocene (Ar3) and the herpetotherine *Herpetotherium* survives into the late early Miocene (He1). During the Orellan, all of these taxa are confined to the Central Great Plains and Northern Great Plains, although there are few fossil localities known from outside these regions at this time. *Herpetotherium* is more widely ranging during the Arikareean, being found in the Gulf Coast and Pacific Northwest regions during this time interval, with a possible appearance in the Southern Great Basin in Ar3, although its final (Hemingfordian) appearance is in the Central Great Plains (Figure 1.6).

Questionable occurrences of didelphids occur in the late Hemingfordian and Barstovian of Gulf Coast, Northern Great Plains, and Pacific Northwest regions. These are likely to represent surviving records of *Herpetotherium*. However, the record in the late Hemphillian of the Southern Great Plains is more problematical. Could this represent a surviving North American lineage, or an early immigrant from South America? As previously mentioned, no marsupials are known in the Old World Northern Hemisphere past the middle Miocene, and the first immigrants from South America (xenarthrans and rodents) appear in the latest Miocene. However, the first definitive record of *Didelphis* is a single Pliocene occurrence in the late Blancan of Florida, although the taxon is common in the North American Pleistocene record.

REFERENCES

Baker, M. L., Wares, J. P., Harrison, G. A., and Miller, R. D. (2004). Relationships among the families and orders of marsupials and the major mammalian lineages based on recombination activating gene-1. *Journal of Mammalian Evolution*, 11, 1–16.

Belov, K., Hellman, L., and Cooper, D. W. (2002). Characterization of echidna IgM provides insights into the divergence of extant mammals. *Developmental and Comparative Immunology*, 26, 831–9.

Butler, P. M. and Hooker, J. J. (2005). New teeth of allotherian mammals from the English Bathonian, including the earliest multituberculates. *Acta Palaeontologica Polonica*, 50, 185–207.

Case, J. A. and Woodburne, M. O. (1986). South American marsupials: a successful crossing of the Cretaceous–Tertiary boundary. *Palaios*, 1, 413–16.

Case, J. A., Goin, F. J., and Woodburne, M. O. (2005). "South American" marsupials from the Late Cretaceous of North America, and the origin of marsupial cohorts. *Journal of Mammalian Evolution*, 12, 461–94.

Cifelli, R. L. (1993). Theria of metatherian–eutherian grade and the origin of marsupials. In *Mammal Phylogeny: Mesozoic Differentiation, Multituberculates, Monotremes, Early Therians and Marsupials*, ed. F. S. Szalay, M. J. Novacek, and M. C. McKenna, pp. 205–15. New York: Springer-Verlag.

(1999). Tribosphenic mammal from the North American early Cretaceous. *Nature*, 401, 363–6.

Cifelli, R. L. and Muizon, C., de (1997). Dentition and jaw of *Kokopellia juddi*, a primitive marsupial or near-marsupial from the medial Cretaceous of Utah. *Journal of Mammalian Evolution*, 4, 241–58.

Cope, E. D. (1884). Second addition to the knowledge of the Puerco fauna. *Proceedings of the American Philosophical Society*, 21, 309–24.

Crochet, J.-Y. (1979). *Les Marsupiaux du Tertiare d'Europe*. Paris: Éditions de la Foundation Singer-Polignac.

Eizirik, E., Murphy, W. J., and O'Brien, S. J. (2001). Molecular dating and biogeography of the early placental mammal radiation. *Journal of Heredity*, 92, 212–19.

Gilbert, N. and Labuda, D. (2000). Evolutionary inventions and continuity of CORE-SINES in mammals. *Journal of Molecular Biology*, 298, 365–77.

Gill, T. (1872). Arrangement of the families of mammals with analytical tables. *Smithsonian Miscellaneous Collections*, 11, 1–98.

Goin, F. J. (2003). Early marsupial radiations in South America. In *Predators with Pouches: The Biology of Carnivorous Marsupials*, ed. M. Jones,

C. Dickman, and M. Archer, pp. 30–42. Collingwood, Australia: CSIRO.

Gregory, W. K. (1947). The monotremes and the palimpsest theory. *Bulletin of the American Museum of Natural History*, 88, 1–52.

Haeckel, E. (1898). *Ueber unsere gegenwärtige Kenntniss vom Ursprung des Menschen*. Bonn: Vortrag gehalten auf dem 4en Internationalen Zoologen-Congress, Cambridge.

Hedges, S. B., Parker, P., Sibley, G., and Kumar, S. (1996). Continental breakup and ordinal diversification of birds and mammals. *Nature*, 381, 226–8.

Hunter, J. P. and Janis, C. M. (2006). Spiny Norman in the "Garden of Eden"? *Ausktribosphenos* and the geography of placental mammalian origins. *Journal of Mammalian Evolution*, 8, 107–24.

Hurum, J. H., Luo, Z.-X., and Kielan-Jaworowska, Z. (2006). Were mammals originally venomous? *Acta Palaeontologica Polonica*, 51, 1–11.

Huxley, T. H. (1880). On the application of the laws of evolution to the arrangement of Vertebrata, and more particularly of the Mammalia. *Proceedings of the Zoological Society of London*, 1880, 649–62.

Janke, A., Xu, X., and Arnason, U. (1997). The complete mitochondrial genome of the wallaroo (*Macropus robustus*) and the phylogenetic relationship among Monotremata, Marsupialia and Eutheria. *Proceedings of the National Academy of Sciences, USA*, 94, 1276–81.

Janke, A., Magnell, O., Wieczorek, G., Westerman, M., and Arnason, U. (2002). Phylogenetic analysis of 18S RNA and the mitochondrial genomes of the wombat, *Vombatus ursinus*, and the spiny anteater, *Tachyglossus aculeatus*: increased support for the Marsupionta hypothesis. *Journal of Molecular Evolution*, 54, 71–80.

Jenkins, F. A., Gatesy, S. M., Shubin, N. H., and Amaral, W. W. (1997). Haramiyids and Triassic mammalian evolution. *Nature*, 385, 715–18.

Ji, Q., Luo, Z.-X., Yuan, C.-X., *et al.* (2002). The earliest known eutherian mammal. *Nature*, 416, 816–22.

Kielan-Jaworowska, Z. (1971). Skull structure and affinities of the Multituberculata. *Acta Paleontologia Polonica*, 25, 5–41.

Kielan-Jaworowska, Z., Cifelli, R. L., and Luo, Z.-X. (2004). *Mammals from the Age of Dinosaurs: Origins, Evolution, and Structure*. New York: Columbia University Press.

Killian, J. K., Buckley, T. R., Stewart, N., Munday, B. L., and Jirtle, R. L. (2001). Marsupials and eutherians reunited: genetic evidence for the Theria hypothesis of mammalian evolution. *Mammalian Genome*, 12, 513–17.

Kirsch, J. A. W. (1977). The six percent solution: second thoughts about the adaptiveness of marsupials. *American Scientist*, 65, 276–88.

Kirsch, J. A. W. and Mayer, G. C. (1998). The platypus is not a rodent: DNA hybridization, amniote phylogeny and the palimpsest theory. *Philosophical Transactions of the Royal Society London, B*, 353, 1221–37.

Kirsch, J. A. W., Dickerman, A. A., Reig, O. A., and Springer, M. S. (1991). DNA hybridization evidence for the Australasian affinity of *Dromiciops australis*. *Proceedings of the National Academy of Sciences, USA*, 88, 10 465–9.

Kirsch, J. A. W., Lapointe, F.-J., and Springer, M. S. (1997). DNA-hybridisation studies of marsupials and their implications for metatherian classification. *Australian Journal of Zoology*, 45, 211–80.

Kitazoe, Y., Hirohisa, H., Waddell, P. J., *et al.* (2007). Robust time estimate reconciles views of the antiquity of placental mammals. *PLoS Online*, 2, e384. doi: 10.1371/journal.pone.0000384.

Krause, D. W. (1982). Jaw movement, dental function, and diet in the Paleocene multituberculate *Ptilodus*. *Paleobiology*, 8, 265–81.

Krause, D. W., Rogers, R. R., Forster, C. A., *et al.* (1999). The Late Cretaceous vertebrate fauna of Madagascar: implications for Gondwanan paleobiogeography. *GSA Today*, 9, 1–7.

Kullander, K., Carlson, B., and Hallbrook, F. (1997). Molecular phylogeny and evolution of the neurotropins from monotremes and marsupials. *Journal of Molecular Evolution*, 45, 311–21.

Kumar, S. and Hedges, S. B. (1998). A molecular timescale for vertebrate evolution. *Nature*, 392, 917–20.

Lee, M.-H., Schroff, R., Cooper, S. J. B., and Hope, R. (1999). Evolution and molecular characterization of a β-globin gene from the Australian echidna *Tachyglossus aculeatus* (Monotremata). *Molecular Phylogenetics and Evolution*, 12, 205–14.

Lillegraven, J. A. (1975). Biological considerations of the marsupial–placental dichotomy. *Evolution*, 29, 707–22.

(1984). Why was there a "marsupial-placental dichotomy." [*In Mammals: Notes for a Short Course*, ed. P. D. Gingerich and C. Badgley.] *University of Tennessee, Studies in Geology*, 8, 72–86.

Luo, Z.-X., Cifelli, R. L., and Kielan-Jaworowska, Z. (2001). Dual origin of tribosphenic mammals. *Nature*, 409, 53–7.

Luo, Z.-X., Kielan-Jaworowska, Z., and Cifelli, R. L. (2002). In quest for a phylogeny of Mesozoic mammals. *Acta Paleontologica Polonica*, 47, 1–78.

Luo, Z.-X., Ji, Q., Wible, J. R., and Yuan, C.-X. (2003). An Early Cretaceous tribosphenic mammal and metatherian evolution. *Science*, 302, 1934–40.

Maddison, W. P. (1993). Missing data versus missing characters in phylogenetic analysis. *Systematic Biology*, 42, 576–81.

Madsen, O., Scally, M., Douady, C. J., *et al.* (2001). Parallel adaptive radiations in two major clades of placental mammals. *Nature*, 409, 610–14.

Marsh, O. C. (1880). Notice of Jurassic mammals representing two new orders. *American Journal of Science, Series 3*, 20, 235–9.

Marshall, L. G., Case, J. A., and Woodburne, M. O. (1990). Phylogenetic relationships of the families of marsupials. In *Current Mammalogy*, Vol. 2, ed. H. Genoways, pp. 33–405. New York: Plenum Press.

McKenna, M. C. and Bell, S. K. (1997). *Classification of Mammals Above the Species Level*. New York: Columbia University Press.

Meng, J. and Wyss, A. R. (1995). Monotreme affinities and low-frequency hearing suggested by multituberculate ear. *Nature*, 377, 141–4.

Messer, M., Weiss, A. S., Shaw, D. C., and Westerman, M. (1998). Evolution of the monotremes: phylogenetic relationship to marsupials and eutherians, and estimation of divergence time based on α-lactalbumin amino acid sequences. *Journal of Mammalian Evolution*, 5, 95–105.

Miao, D. (1993). Cranial morphology and multituberculate relationships. In *Mammal Phylogeny: Mesozoic Differentiation, Multituberculates, Monotremes, Early Therians and Marsupials*, ed. F. S. Szalay, M. J. Novacek, and M. C. McKenna, pp. 63–74. New York: Springer-Verlag.

Murphy, W. J., Eizirik, E., Johnson, W. E., *et al.* (2001a). Molecular phylogenetics and the origins of placental mammals. *Nature*, 409, 614–18.

Murphy, W. J., Eizirik, E., O'Brien, S. J., *et al.* (2001b). Resolution of the early placental mammal radiation using Bayesian phylogenetics. *Science*, 294, 2348–51.

Novacek, M. J., Rougier, G. W., Wible, J. R., *et al.* (1997). Epipubic bones in eutherian mammals from the Late Cretaceous of Mongolia. *Nature*, 389, 483–6.

Nowak, R. M., and Paradiso, J. L. (1983). *Walker's Mammals of the World*. Baltimore, MD: Johns Hopkins University Press.

Parker, T. J. and Haswell, W. A. (1897). *A Text-book of Zoology*. London: Macmillan.

Penny, D. and Hasegawa, M. (1997). The platypus put in its place. *Nature*, 387, 549–50.

Renfree, M. R. (1993). Ontogeny, genetic control, and phylogeny of female reproduction in monotreme and marsupial mammals. In *Mammal Phylogeny: Mesozoic Differentiation, Multituberculates, Monotremes, Early Therians and Marsupials*, ed. F. S. Szalay, M. J. Novacek, and M. C. McKenna, pp. 4–20. New York: Springer-Verlag.

Retief, J. D., Winkfein, R. J., and Dixon, G. H. (1993). Evolution of the monotremes: the sequences of the protamine P1 genes in the platypus and echidna. *European Journal of Biochemistry*, 218, 457–61.

Rich, T. H., Vickers-Rich, P., Constantine, A., *et al.* (1997). A tribosphenic mammal from the Mesozoic of Australia. *Science*, 278, 1438–42.

Rich, T. H., Flannery, T. F., Trusler, P., *et al.* (2002). Evidence that monotremes and ausktribosphenids are not sister groups. *Journal of Vertebrate Paleontology*, 22, 466–9.

Rich, T. H., Hopson, J. A., Musser, A. M., Flannery, T. F., and Vickers-Rich, P. (2005). Independent origins of middle ear bones in monotremes and therians. *Science*, 307, 910–14.

Ride, W. D. L. (1964). A review of Australian fossil marsupials. *Journal of the Proceedings of the Royal Society of Western Australia*, 46, 97–131.

(1970). *A Guide to the Native Mammals of Australia*. Oxford: Oxford University Press.

Rougier, G. W., Wible, J. R., and Novacek, M. J. (1996). Multituberculate phylogeny. *Nature*, 379, 406–67.

Rowe, T. (1988). Definition, diagnosis, and origin of Mammalia. *Journal of Vertebrate Paleontology*, 8, 241–64.

Sánchez-Villagra, M. R., and Maier, W. (2003). Ontogenesis of the scapula in marsupial mammals, with special emphasis on perinatal stages of *Didelphis* and remarks on the origin of the therian scapula. *Journal of Morphology*, 258, 115–29.

Sears, K. E. (2004). Constraints on the morphological evolution of marsupial shoulder girdles. *Evolution*, 58, 2535–70.

Sigogneau-Russell, D. (1991). First evidence of Multituberculata (Mammalia) in the Mesozoic of Africa. *Neues Jahrbuch fur Palaontologie, Monatshefte*, 1991, 119–25.

Simpson, G. G. (1945). The principles of classification and a classification of mammals. *Bulletin of the American Museum of Natural History*, 85, 1–350.

Smith, K. K. (1997). Comparative patterns of craniofacial development in eutherian and metatherian mammals. *Evolution*, 51, 1663–78.

(2001). The evolution of mammalian development. *Bulletin of the Museum of Comparative Zoology*, 156, 119–35.

Springer, M. S., Westerman, M., Kavanagh, J. R., *et al.* (1998). The origin of the Australasian marsupial fauna and the phylogenetic affinities of the enigmatic Monito de Monte and the marsupial mole. *Proceedings of the Royal Society of London, B*, 265, 2381–6.

Springer, M. S., DeBry, R. W., Douady, G., *et al.* (2001). Mitochondrial versus nuclear gene sequences in deep-level mammalian phylogeny reconstruction. *Molecular Biology and Evolution*, 18, 132–43.

Szalay, F. S. (1994). *Evolutionary History of the Marsupials and an Analysis of Osteological Characters*. Cambridge, UK: Cambridge University Press.

Szalay, F. S. and Sargis, E. J. (2001). Model-based analysis of postcranial osteology of marsupials from the Palaeocene of Itaboraí (Brazil) and the phylogenetics and biogeography of the Metatheria. *Geodiversitas*, 23, 139–302.

Wible, J. R. (1991). Origin of Mammalia: the craniodental evidence reexamined. *Journal of Vertebrate Paleontology*, 11, 1–18.

Wible, J. R. and Hopson, J. A. (1993). Basicranial evidence for early mammal phylogeny. In *Mammal Phylogeny: Mesozoic Differentiation, Multituberculates, Monotremes, Early Therians, and Marsupials*, ed. F. S. Szalay, M. J. Novacek, and M. C. McKenna, pp. 45–62. New York: Springer-Verlag.

Wible, J. R., Rougier, G. W., Novacek, U. J., and Asher, R. J. (2007). Cretaceous eutherians and Laurasian origin for placental mammals near the K/T boundary. *Nature*, 447, 1003–6.

Woodburne, M. O. (2003). Monotremes as pretribosphenic mammals. *Journal of Mammalian Evolution*, 10, 195–248.

Woodburne, M. O., Rich, T. H., and Springer, M. S. (2003). The evolution of tribospheny and the antiquity of mammalian clades. *Molecular Phylogeny and Evolution*, 28, 360–85.

Zeller, U. (1999). Phylogeny and systematic relations of the Monotremata: why we need an integrative approach. *Courier Forschungsinstitut Senckenberg*, 215, 227–32.

2 Multituberculata

ANNE WEIL[1] and DAVID W. KRAUSE[2]
[1]*Oklahoma State University Center for Health Sciences, Tulsa, OK, USA*
[2]*Health Sciences Center, Stony Brook, State University of New York, USA*

INTRODUCTION

The Multituberculata are named for their unusual teeth, which have multiple molar cusps, or "tubercles," arranged in longitudinal rows. Although now extinct, multituberculates were among the most successful of mammals by any criterion. Multituberculata is the longest-lived order within Mammalia, with a range extending from at least the Kimmeridgian (Late Jurassic) to the Chadronian (late Eocene). Multituberculates were widely distributed and are known from throughout Laurasia. Isolated teeth and tooth fragments suggest that they were also present in the Cretaceous of Africa (Sigogneau-Russell, 1991; Hahn and Hahn, 2003) and Madagascar (Krause *et al.*, 2006) and possibly from Argentina (Kielan-Jaworowska *et al.*, 2004). Multituberculates were diverse and common – so much so that they are commonly employed as stratigraphic index fossils in the Late Cretaceous and Paleocene of North America (Krause, 1982a; Savage and Russell, 1983; Lillegraven and McKenna, 1986; Sloan, 1987).

The greatest known diversity of multituberculates for any single area and time period is that of the North American middle Paleocene, but in recent years burgeoning numbers of newly described taxa from the Cretaceous of North America (Eaton, 1995; Eaton and Cifelli, 2001), and the Cretaceous of Asia (Kielan-Jaworowska and Nessov, 1992; Rougier, Novacek, and Dashzeveg, 1997) have suggested that our documentation of the group remains woefully incomplete. Where they occur in North America, multituberculates are common. They make up, on average, a little less than 50% of the individual mammals in Late Cretaceous localities, and at least 20% of the individuals in every well-sampled locality of the middle Paleocene. Multituberculates have been found in widely divergent facies and depositional systems, indicating that they were widespread across environments.

While articulated specimens are rare in North America, there is evidence that at least two Paleocene taxa were arboreal (Jenkins and Krause, 1983), and abundant remains from the semi-arid areas of Asia have led to reconstructions of those animals as fossorial or desert adapted (Miao, 1988; Kielan-Jaworowska and Gambaryan, 1994). Most multituberculates were the size of a mouse or rat, but at least one genus, *Taeniolabis*, attained a size probably slightly larger than that of a modern-day beaver. Their dentition is specialized for herbivory, but they probably were not strictly herbivorous, and their tooth morphologies indicate a variety of dietary adaptations.

All Tertiary multituberculates belong to the derived clade Cimolodonta. Earlier, more primitive multituberculates are referred to the paraphyletic Plagiaulacida. The last known plagiaulacidans are about 100 million years old; they are not discussed here.

DEFINING FEATURES OF THE ORDER MULTITUBERCULATA (SUBORDER CIMOLODONTA)

CRANIAL

The only described, relatively complete skulls of Tertiary North American multituberculates are those of *Ptilodus* and *Taeniolabis*, studied in detail by Simpson (Simpson, 1937; see also Kielan-Jaworowska, 1971), and *Ectypodus* (briefly described by Sloan, 1979). Excellent cranial material of multituberculates is known from the Late Cretaceous and Paleocene of Asia (e.g., Kielan-Jaworowska, 1974a; Miao, 1988; Wible and Rougier, 2000) a single skull of a Cretaceous multituberculate is known from Europe (Râdulescu and Samson, 1996), and a single skull is known from the Late Cretaceous of North America (Weil and Tomida, 2001), but it is not yet fully described. Information from these other skulls, particularly the well-described Asian fossils, has been incorporated in the list of cranial features below.

The skull is dorsoventrally compressed and wide posteriorly. The snout is short (except in the European *Kogaionon*) and strongly

Evolution of Tertiary Mammals of North America, Vol. 2. ed. C. M. Janis, G. F. Gunnell, and M. D. Uhen. Published by Cambridge University Press.

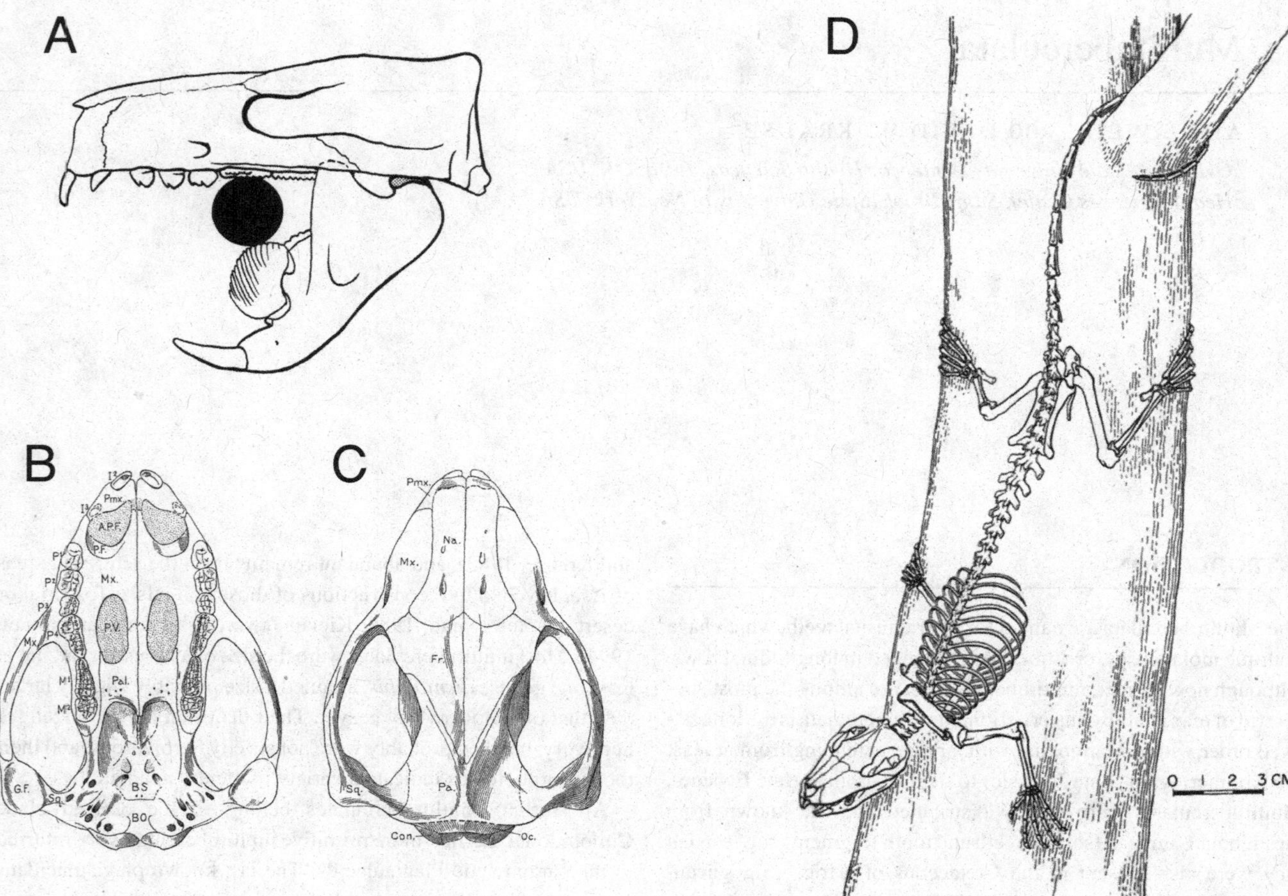

Figure 2.1. *Ptilodus*. A. Lateral view of the skull, illustrating use of the p4 (modified from Wall and Krause, 1992). B. Ventral view of the skull and upper dentition (modified from Simpson, 1937). C. Dorsal view of the skull (modified from Simpson, 1937). D. Reconstruction of the skeleton (modified from Krause and Jenkins, 1983). (A: Courtesy of the Society of Vertebrate Paleontology.)

tapered anteriorly, giving the skull a triangular shape (Figure 2.1). The parietals are expanded to form the entire posterior cranial roof. Most of the lateral wall of the braincase is formed by the anterior lamina of the petrosal and the orbitosphenoid. The orbitosphenoid also forms part of the intraorbital wall, with the maxilla and frontal. In contrast to that of therian mammals, the laterally directed orbit does not have a floor but does have a roof (formed by the frontals and sometimes parietals). A postorbital process may be present or absent; an orbital pocket for pars anterior of the medial masseter (Gambaryan and Kielan-Jaworowska, 1995) is usually present anterior to the orbit (Wible and Rougier, 2000). The nasal bones are large and expanded posteriorly. The premaxilla has an extensive palatal process. Some taxa have large palatal vacuities; all have an incisive foramen. The infraorbital foramen is large and may have one or two exits. The zygomatic arch is robust. Contra Simpson (1937), a jugal is present (Hopson, Kielan-Jaworowska, and Allin, 1989). The cochlea is short, straight, and has a tiny hook at the end. The middle ear cavity is not enclosed ventrally by an ossified bulla, and three auditory ossicles are present. The auditory vestibule is large. The brain has large, anteriorly tapering olfactory bulbs (Krause and Kielan-Jaworowska, 1993). The mandibular fossa is flat, shallow and ovoid, and it is situated lateral to the petrosal (rather than anterior to it, as in therians). The mandibular condyle is large, wide anteriorly and narrower posteriorly, and extends far inferiorly on the posterior surface of the dentary. The lower jaw consists entirely of the dentary and has a large and well-defined masseteric fossa. This masseteric fossa extends farther forward on the dentary than in any therian mammal (Gambaryan and Kielan-Jaworowska, 1995). The mandibular symphysis is unfused in smaller species but fused anteroventrally in *Taeniolabis* (Figure 2.2) and *Catopsalis*. The fossa for the medial pterygoid on medial and posteroinferior surfaces of the mandible is characteristically deep, and is bordered posteroventrally by a prominent flange that functions as an angular process (Miao, 1988).

DENTAL

The greatest number of teeth found in cimolodontan multituberculates are of the dental formula I2/1, C0/0, P4/2, M2/2; some Cenozoic taxa lose a number of premolars. The primitive

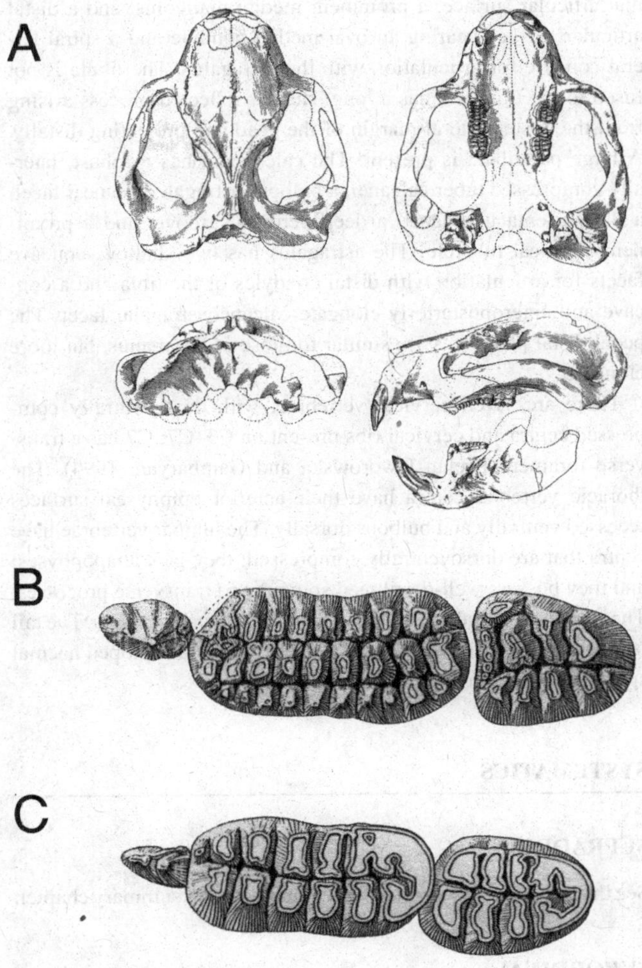

Figure 2.2. *Taeniolabis.* A. Dorsal, ventral, and posterior views of the skull, and lateral view of the skull and mandible (modified from Simpson, 1937). B. Occlusal view of upper cheek teeth (modified from Granger and Simpson, 1929). C. Occlusal view of lower cheek teeth (modified from Granger and Simpson, 1929). (Courtesy of the American Museum of Natural History.)

multituberculate dental formula of I3/1, C0/0, P4-5/3-4, M2/2 is present in Late Jurassic and Early Cretaceous taxa, but not in any Cenozoic multituberculate. The homologies of the anterior teeth of the more primitive Plagiaulacida to those of Cimolodonta are uncertain (see Kielan-Jaworowska and Hurum [2001] for discussion).

The upper central incisors (I2) may be unicuspid or bicuspid, and rarely even polycuspid. Position of the upper lateral incisor (I3) varies with the size of the incisive foramen; if the foramen is small, the I3 emerges from the anterior palate behind I2, but in taxa in which the incisive foramen is large, the I3 is situated on the palatal margin. The lower incisors are elongate, with roots extending below the roots of the second molar. Primitively the lower incisors are slender, tapered, and completely covered with enamel. Restriction of enamel to the ventrolabial surface of the lower incisor has evolved more than once within Multituberculata, however, and occurs in Tertiary North American Taeniolabididae, Eucosmodontidae, and Microcosmodontidae. There is a prominent diastema between the lower incisor and the cheek teeth. The p3 is sometimes absent, particularly in later members of many lineages. When present it is proportionately small, single-rooted, and peg shaped, nestling under and perhaps supporting the anterior margin of the p4 crown. The p4 itself varies widely in shape and in size relative to the rest of the dentition; it may be large and bladelike with a serrate apical margin, small and rounded with cusplike serrations, or reduced to a triangular peg. The upper anterior premolars (P1-3) are small relative to P4 and usually bear two to four cusps, although the P3 has more in a few taxa. The P4 is longer than the P3 and invariably has more cusps. Multituberculates have only two upper and two lower molars. The first molars of Cenozoic taxa are always longer than the second molars, although ratios vary greatly. Molar cusps are approximately equal in height and are arranged in parallel or subparallel rows; the cusps vary in shape from conical to pyramidal to crescentic, sometimes on a single tooth. When crescentic or subcrescentic, lower molar cusps are concave posteriorly and upper molar cusps are concave anteriorly. The derived Cimolodonta is distinguished in part by the presence of three cusp rows on the upper molars. The derived or "added" row is the most lingual on the M1, and it may run the length of the tooth or be present only on the posterior portion. On the M2, the third row occurs anterolabially (Figure 2.3).

Primitively, molar wear occurs along the sides of the cusp rows, deepening and widening the valleys between cusps. Some groups, however, notably the Taeniolabididae, have bulbous cusps on which only the tips wear, creating enamel cross-lophs similar to those of some rodents. As in rodents, the length of the upper dentition (measured from incisor tip to the posterior end of the second molar) is greater than that of the lower dentition. Diphyodonty has been demonstrated for lower incisors and some anterior upper premolars (e.g., Szalay, 1965), and Greenwald (1988) described a six-stage sequence of dental ontogeny. The permanent incisors erupt posterior

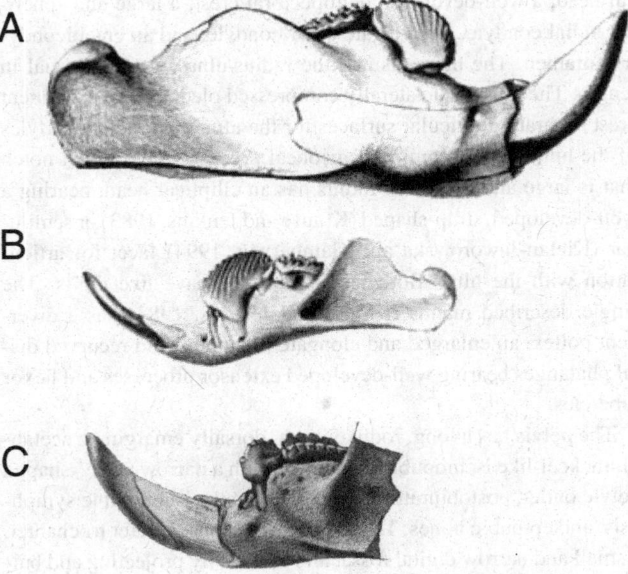

Figure 2.3. Mandibles of Paleocene multituberculates (modified from Jepsen, 1940). A. *Stygimys kuszmauli.* B. *Mesodma ambigua.* C. *Microcosmodon conus.* (Courtesy of the American Philosophical Society.)

to the deciduous incisors, and in the Paleocene Asian *Lambdopsalis* (the only cimolodontan for which an extensive growth series has been recovered), the distribution and color of enamel on the permanent lower incisor differ from those of the deciduous incisor (Miao, 1986). The p3 may be lost during ontogeny in some species (Weil, 1998), but it is not known whether it is mono- or diphyodont. The upper and lower fourth premolars are monophyodont (Clemens and Kielan-Jaworowska, 1979; Sloan, 1979; Miao, 1986) and are probably homologous to dP4/4 of diphyodont plagiaulacoids (Greenwald, 1988). The second molars erupt after the first molars, and in *Lambdopsalis* the lower molars move anteriorly with increasing age (Miao, 1986). Tooth enamel may have large, arcuate prisms, small circular prisms, or small prisms with a mixture of arcuate and circular prisms (Carlson and Krause, 1985; Krause and Carlson, 1986); large prisms appear to be primitive, and smaller prisms may have evolved more than once (Krause and Carlson, 1987).

POSTCRANIAL

The postcranium of Tertiary North American multituberculates is represented by only three incomplete specimens, two of *Ptilodus* and one questionably referred to *Eucosmodon* (or perhaps *Catopsalis*; see Williamson and Lucas, 1993). Additional isolated elements from Tertiary sediments have been attributed to *Ectypodus*, *Stygimys*, *Taeniolabis* (Krause and Jenkins, 1983), and *Catopsalis* (Middleton, 1982). Detailed study of Late Cretaceous multituberculates of Asia has provided a wider perspective of morphological diversity within Cimolodonta and is incorporated here.

The scapulocoracoid has a shallow glenoid and a reduced coracoid and lacks a supraspinous fossa; the scapular spine lies along the anterior margin. A triangular interclavicle is present between the clavicles and the manubrium sternum (Sereno and McKenna, 1995). The sternebrae are ossified. The humerus has a large, hemispherical head, a well-developed deltopectoral crest, a large and spherical radial condyle, an elongate ulnar condyle, and an entepicondylar foramen. The humerus and the radius/ulna are about equal in length. The ulna has a laterally compressed olecranon, a prominent crest separating articular surfaces for the ulnar and radial condyles of the humerus, a prominent anconeal process, and a radial notch that is large and oval. The radius has an elliptical head, bearing a well-developed, strap-shaped (Krause and Jenkins, 1983) or semilunar (Kielan-Jaworowska and Gambaryan, 1994) facet for articulation with the ulna. Both manus and pes have five digits. The single described manus (Krause and Jenkins, 1983) has a divergent pollex, an enlarged and elongate prepollex, and recurved distal phalanges bearing well-developed extensor processes and flexor tubercles.

The pelvis has a long, rodlike ilium, dorsally emarginate acetabulum, keel-like ischiopubic symphysis with a narrow and V-shaped pelvic outlet, postobturator foramen within the ischiopubic symphysis, and epipubic bones. The femur has a robust greater trochanter, a small and narrow digital fossa, and a ventrally projecting and bulbous lesser trochanter. A patella is present. The tibia has a large, flat facet on the posterolateral margin of the head for articulation with the fibula, a deep excavation posteriorly beneath the proximal articular surface, a prominent medial malleolus, and a distal articular surface bearing an oval medial condyle and a spiral lateral condyle for articulation with the astragalus. The fibula is not fused to the tibia and has a long, slender reflected process arising from the posterolateral margin of the head and projecting distally. A large parafibula is present. The calcaneum has a robust, laterally compressed tuber calcanei, a bulbous astragalocalcaneal facet, a flat sustentacular facet, a deep peroneal groove, and a prominent peroneal tubercle. The astragalus has two shallow, concave facets for articulation with distal condyles of the tibia and a concave and anteroposteriorly elongate calcaneoastragalar facet. The pedal distal phalanges are similar to those of the manus, but more elongate.

There are seven cervical vertebrae, with dorsoventrally compressed centra and cervical ribs present on C3–C7; C7 has a transverse foramen (Kielan-Jaworowska and Gambaryan, 1994). The thoracic vertebral centra have their anterior epiphyseal surfaces recessed ventrally and bulbous dorsally. The lumbar vertebrae have centra that are dorsoventrally compressed; they lack anapophyses; and they possess well-developed spinous and transverse processes. There are four sacral vertebrae, with high spinous processes. The tail is long and the caudal vertebrae stout, with well-developed haemal spines and transverse processes.

SYSTEMATICS

SUPRAORDINAL

See the discussion in Chapter 1, the non-eutherian summary chapter.

SUBORDINAL

Relationships among the multituberculates are no more certain than those of multituberculates to other mammals (Figure 2.4). Within the last few years, the group has been the focus of many phylogenetic studies (Simmons, 1993; Kielan-Jaworowska and Hurum, 1997; Rougier, Novacek, and Dashzeveg, 1997; Weil, 1999; Wible and Rougier, 2000; Kielan-Jaworowska and Hurum, 2001) and has undergone repeated systematic revision (McKenna and Bell, 1997; Fox, 1999; Kielan-Jaworowska and Hurum, 2001). Because so many taxa are known only from isolated teeth, phylogenetic efforts focusing on North American taxa have produced little resolution. Perhaps the only thing that these authors agree on is that all multituberculates of Tertiary North America belong to the derived clade Cimolodonta. Cimolodontans are distinguished by a reduced number of lower premolars to a maximum of two, and by development of a third cusp row on the upper molars (Figure 2.2).

Cimolodonta was long divided into two suborders, the Taeniolabidoidea and Ptilodontoidea, differentiated by lower incisor structure; taeniolabidoids had enamel restricted to the ventrolabial surface, which produced a chisel-like cutting tip, as in rodents, while ptilodontoids had a long, slender, tapered incisor fully coated with enamel. As a result of both intense interest in multituberculate phylogeny and many fossil discoveries, it is recognized that

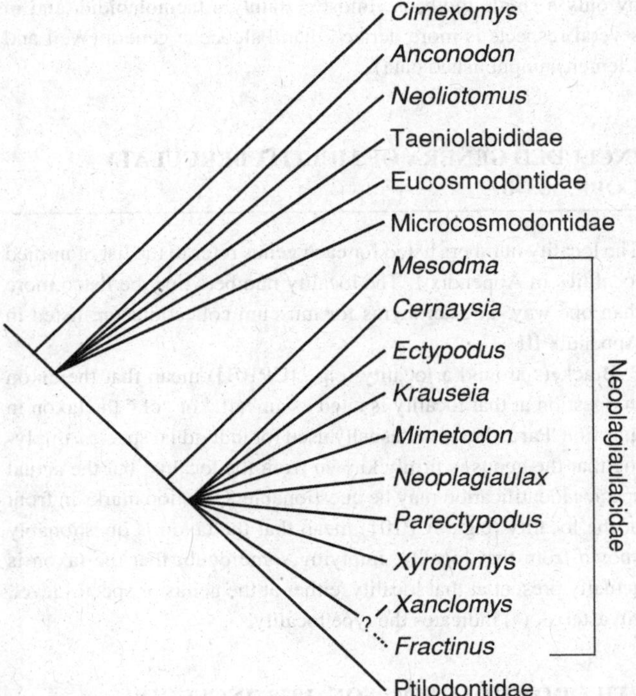

Figure 2.4. Highly hypothetical interrelationships of the Cenozoic multituberculate taxa. This tree represents a "best guess" rather than a strict phylogenetic analysis. Because many of the taxa are incomplete, limited mostly to isolated teeth or mandibles, the Cenozoic Multituberculata is not amenable to phylogenetic analyses, which can produce hundreds of thousands of shortest trees. Some genera may not be monophyletic.

this convenient division was an artifact of limited information that is disappearing as we learn more. In particular, we now recognize that restricted enamel evolved more than once within Multituberculata, including at least once in non-cimolodontan multituberculates (Engelmann and Callison, 1999) and in an endemic Asian superfamily, Djadochtatheroidea (Kielan-Jaworowska and Hurum, 1997; Rougier, Novacek, and Dashzeveg, 1997). Also, many of the trees resulting from phylogenetic studies show Ptilodontoidea (sensu Hahn and Hahn, 1983) to be paraphyletic, polyphyletic, or both. The volume and pace of recent and ongoing work on multituberculates is high, as demonstrated by three substantial systematic revisions in five years. Higher-order relationships seem likely to remain unstable for some time to come. We, therefore, present multituberculate families without suprafamilial groupings and refrain from making formal revisions here.

PTILODONTIDAE

Ptilodontid multituberculates were among the first North American Paleocene mammals to be described (Cope, 1881). The family was first diagnosed in 1928 by Simpson, although the name had been listed without definition in at least two earlier works. The number of genera included within the family has varied greatly since its diagnosis and, at one time or another, included genera now assigned to the families Eucosmodontidae, Cimolodontidae, Cimolomyidae, and Neoplagiaulacidae. Sloan and Van Valen (1965) considerably restricted the generic content of the family to include only *Ptilodus*, *Prochetodon*, *Kimbetohia*, *Essonodon*, *Cimolodon*, *Anconodon*, and *Liotomus*. The latter three genera were placed in the resurrected family Cimolodontidae by Sloan (in Van Valen and Sloan, 1966; their Fig. 2.4) and *Essonodon* was questionably allied with the Cimolomyidae by Archibald (1982). This left only *Ptilodus*, *Prochetodon*, and *Kimbetohia* until the addition of *Baiotomeus* (Krause, 1987b). McKenna and Bell (1997) considered the taxon to be of subfamilial rank, but Kielan-Jaworowska and Hurum (2001) restored it to familial rank.

NEOPLAGIAULACIDAE

Isolated teeth of neoplagiaulacids are among the most difficult to assign to lower taxa among Cenozoic multituberculates. The p4 is the easiest tooth to identify, but even in this case the purportedly diagnostic features are very subtle. There has been little attempt to define the genera on the basis of shared, derived characters, which renders phylogenetic analyses at the generic level (e.g., Simmons, 1993; Rougier, Novacek, and Dashzeveg, 1997) unreliable with respect to these taxa. Systematics of neoplagiaulacid species have been particularly dependent on typological definitions; the family remains in need of revision, with careful attention to both quantitative and qualitative morphological variability.

The generic content of the Neoplagiaulacidae was initially the result of work by R. E. Sloan. Sloan and Van Valen (1965) erected the family Ectypodidae, which was emended to Ectypodontidae by Van Valen and Sloan (1966) and was later replaced by the senior synonym Neoplagiaulacidae Ameghino, 1890 (see Clemens, 1973a, p. 77). Hahn and Hahn (1983) argued that the name Neoplagiaulacidae is a nomen oblitum, but it remains in use because of its general acceptance. The Neoplagiaulacidae initially included *Cimexomys*, *Ectypodus*, *Mesodma*, *Mimetodon*, *Neoplagiaulax*, and *Parectypodus*. To this list has been added *Xanclomys* (see Rigby, 1980) and *Xyronomys* (originally assigned to the Eucosmodontidae by Rigby [1980] but allied with the Neoplagiaulacidae by Johnston and Fox [1984] and Sloan [1987]). *Cimexomys* was recognized as more primitive and removed (Archibald, 1982). Vianey-Liaud (1986) made several reassignments within Neoplagiaulacidae, notably assignment of the species *Anconodon russelli*, *Parectypodus vanvaleni*, and probably *Parectypodus sinclairi* to the genus *Liotomus*; assignment of the species *Parectypodus clemensi* to the new genus *Krauseia*; and transfer of the P4 of *Parectypodus vanvaleni* to the new species *Cernaysia davidi*. While we regard these assignments as questionable, they have been adopted by some subsequent authors. McKenna and Bell (1997) revised the Neoplagiaulacidae to subfamily rank within Ptilodontidae and included the North American genera *Mesodma*, *Parectypodus*, *Ectypodus*, *Neoplagiaulax*, *Cernaysia*, *Krauseia*, *Xyronomys*, *Xanclomys*, *Ectypodus*, and *Mimetodon*, in addition to the Asian *Mesodmops*. Kielan-Jaworowska and Hurum (2001) did not examine relationships within Ptilodontoidea sensu McKenna and Bell (1997), and repeated this list.

EUCOSMODONTIDAE

The Eucosmodontinae was named by Jepsen (1940) as a subfamily of the Ptilodontidae but was elevated to familial rank within Taeniolabidoidea by Sloan and Van Valen (1965). The subfamilial name Eucosmodontinae was resurrected when Holtzman and Wolberg (1977) named a new subfamily, the Microcosmodontinae. Nonmicrocosmodontine eucosmodontids were placed in the Eucosmodontinae. More recently, Fox (1999) elevated Microcosmodontinae to familial rank, thus removing it from the Eucosmodontidae. He also removed the Eucosmodontidae from Taeniolabidoidea, a scheme that has been followed by Kielan-Jaworowska and Hurum (2001).

MICROCOSMODONTIDAE

Microcosmodon and *Pentacosmodon* were first recognized as differing substantially from other eucosmodontids by Holtzman and Wolberg (1977), who placed them in the subfamily Microcosmodontinae. *Acheronodon* was later described and recognized to be closely related by Archibald (1982). One result of Simmons' (1993) phylogenetic analysis was the assignment of *Pentacosmodon* and *Microcosmodon* to widely different clades, although they vary in only a single scored character. Following Simmons, McKenna and Bell (1997) placed *Pentacosmodon* in the otherwise Asian Chulsanbaatarinae, while retaining *Microcosmodon* and *Acheronodon* in the Microcosmodontinae. Weil (1998) reinstated *Pentacosmodon* in the monophyletic Microcosmodontinae, and Fox (1999, 2005) elevated the group to familial rank, removing it from the Eucosmodontidae. Fox (2005) made a differential diagnosis with the benefit of relatively complete specimens, and also named a fourth genus, *Allocosmodon*, to the family.

TAENIOLABIDIDAE

Taeniolabidids include the largest multituberculates known and are characterized by a highly derived dentition. The first taeniolabidids to be described were placed in the family Plagiaulacidae until Cope (1884a) named the Polymastodontidae. As reviewed by Hahn and Hahn (1983), the family name was changed to Taeniolabididae when it was recognized that the type genus *Polymastodon* Cope (1882b) was synonymous with *Taeniolabis* (Cope, 1882a). Again following Simmons (1993), McKenna and Bell (1997) included the genera *Buginbaatar*, *Meniscoessus*, *Prionessus*, *Lambdopsalis*, and *Sphenopsalis* in Taeniolabididae. More recent phylogenetic analyses (Kielan-Jaworowska and Hurum, 1997; Rougier, Novacek, and Dashzeveg, 1997) showed that *Buginbaatar* is not closely related to North American taeniolabidids, and Kielan-Jaworowska and Hurum (2001) tentatively removed it to Cimolomyidae. *Meniscoessus* is a cimolomyid, and McKenna and Bell's (1997) reassignment has not been accepted (Weil, 1999; Kielan-Jaworowska and Hurum, 2001). The major recent reclassifications of Multituberculata (McKenna and Bell, 1997; Kielan-Jaworowska and Hurum, 2001) have omitted the genus *Bubodens*, described by Wilson (1987) from the Late Cretaceous of South Dakota and placed by him in Multituberculata suborder and family incertae sedis. This taxon, represented thus far by only a single tooth, is almost certainly a taeniolabidid, and in several respects is more derived than Paleocene genera (Weil and Clemens, unpublished data).

INCLUDED GENERA OF MULTITUBERCULATA COPE, 1884b

The locality numbers listed for each genus refer to the list of unified localities in Appendix I. The locality numbers may be listed more than one way. The acronyms for museum collections are listed in Appendix III.

Brackets around a locality (e.g., [CP101]) mean that the taxon in question at that locality is cited as an "aff." or "cf." the taxon in question. Parentheses are usually used for individual species, implying that the genus is firmly known from the locality, but the actual species identification may be questionable. Question marks in front of the locality (e.g., ?CP101) mean that the taxon is questionably known from that locality, implying some doubt that the taxon is actually present at that locality, either at the genus or species level. An asterisk (*) indicates the type locality.

PTILODONTIDAE SIMPSON, 1928 (INCLUDING CHIROGIDAE COPE, 1887; ECTYPODIDAE SLOAN AND VAN VALEN, 1965; AND ECTYPODONTIDAE SLOAN AND VAN VALEN, 1965)

Characteristics (modified from Krause, 1982b): Larger than most Neoplagiaulacidae and Cimolodontidae. Further differ from neoplagiaulacids and cimolodontids in presence of anterolabial bulge on P4 (this bulge may bear cusps, but usually does not); in having essentially straight ventral margin of P4 in lateral profile (the middle row of cusps is not elevated posteriorly as in many similar taxa); in having relatively large p4 (ratio between lengths of p4 and m1 is usually approximately 2.0 but ranges from 1.7 to 2.4).

Baiotomeus Krause, 1987a

Type species: *Baiotomeus douglassi* (Simpson, 1935) (originally described as ?*Ptilodous douglassi*).

Type specimen: USNM 9795, right dentary with p4–m2 from the late Torrejonian of Montana.

Characteristics (from Krause, 1987a): Small genus of Ptilodontidae, equivalent in size to *Kimbetohia* and some species of *Ptilodus*, but much smaller than *Prochetodon*. Differs from *Kimbetohia* in having much narrower P4 with relatively small anterolabial bulge; p4 with higher first serration but lacking ventral bifurcations on labial ridges. Differs from *Ptilodus* in having relatively poorly developed anterolabial bulge but well-developed labial row of prominent cusps on P4. Differs from both *Ptilodus* and *Prochetodon* in having relatively low p4 that variably bears incipient serrations on anterior apical margin; in having prominent and angular anterolabial lobe on p4; in having m1 that is relatively longer compared with the length of either p4. In addition, m1 appears to be relatively longer than m2 in *Baiotomeus* in comparison with *Ptilodus* and relatively shorter in comparison with *Prochetodon*.

Average length of p4: 6.36 mm.

Average length of m1: 3.80 mm.

Included species: *B. douglassi* (known from localities SB39B, CP13B, CP14A, NP19C*); *B. lamberti* Krause, 1987a (localities CP13IIA, B, NP20A*); *B. rhothonion* Scott, 2003 (localities NP1B, NP3A0*); *B. russelli* Scott, Fox, and Youzwyshyn, 2002 (locality NP1C*).

Baiotomeus sp. is also known from localities CP13A, CP14B, NP2, NP3B, G.

Kimbetohia Simpson, 1936a

Type species: *Kimbetohia campi* Simpson, 1936a.

Type specimen: UCMP 31305, fragmentary maxilla with broken P1, P2–P4 from the Puercan of New Mexico.

Characteristics (from Krause, 1982b): Small genus of ptilodontids, equivalent in size to *Baiotomeus* and some species of *Ptilodus*, but much smaller than *Prochetodon*. Differs from *Baiotomeus* in having much wider P4 with a well-developed anterolabial bulge; p4 with lower first serration and ventral bifurcations on some of the labial ridges; plane of occlusal surfaces of upper premolars of *Kimbetohia* forming dorsally concave arc in side view; in *Ptilodus* and *Prochetodon*, occlusal surfaces forming nearly straight line. Differs from *Ptilodus* and *Prochetodon* in having only three cusps (one labially and two lingually) and one root on P2 (*Prochetodon* occasionally has only three cusps on P2, but they are arranged in a different pattern with two labially and one lingually); relatively wide P4; p4 with relatively low crown and straight posterior slope; bifurcations on one to three of the labial ridges descending from serrations six to nine; relatively low first serration.

Average length of p4: 6.08 mm.

Average length of m1: unknown.

Included species: *K. campi* (known from locality SB23A*); *K.? mziae* Middleton and Dewar, 2004 (locality CP61A*); *K.* n. sp. of Van Valen and Sloan (1966) (locality NP16C).

Comments: Krause (1992) removed specimens from the Late Cretaceous Lance Formation from *Kimbetohia campi* and erected a new genus and species, *Clemensodon megaloba*. It is questionable whether *Kimbetohia* occurs at all in the Late Cretaceous; although specimens have been mentioned (Clemens, 1973b), none have been described (Krause, 1992).

Prochetodon Jepsen, 1940

Type species: *Prochetodon cavus* Jepsen, 1940.

Type specimen: YPM-PU 13925, right dentary with base of i1, p3, and fragmentary p4 from the Tiffanian of Wyoming.

Characteristics (from Krause, 1987b): Large ptilodontids, much larger than *Kimbetohia* and *Baiotomeus* but equivalent in size to some species of *Ptilodus*. Differ from *Kimbetohia*, *Baiotomeus*, and *Ptilodus* in possession of lenticular, rather than conical, cusps labially on P1–3; arcuate arrangement of anterior and labial cusps on P2 (condition unknown in *Baiotomeus*); presence of eight to nine cusps on P3; relatively long and narrow P4 with persistent absence of cusps on anterolabial bulge; p4 with broad, flat anterior face, reduced anterolabial lobe, posteriorly canted shape in lateral view; presence of shallow vertical grooves labially on labial cusps of m1.

Average length of p4: 7.83 mm.

Average length of m1: 3.74 mm.

Included species: *P. cavus* (known from localities CP13F, G*); *P. foxi* Krause, 1987b (localities CP13F*, CP15B, [NP3C], NP7D, NP47B2); *P. speirsae* Scott, 2004 (localities NP1C*, NP3A, B, D); *P. taxus* Krause, 1987b (localities CP13H*, CP17A, B).

Prochetodon sp. is also known from localities? CP24B, NP3C, F, NP4, NP7C.

Ptilodus Cope, 1881 (including *Chirox* Cope, 1884b)

Type species: *Ptilodus mediaevus* Cope, 1881.

Type specimen: AMNH 3019, isolated p4 from the ?Torrejonian of the San Juan Basin of New Mexico.

Characteristics (from Krause, 1982b): Differs from *Kimbetohia* in having four cusps and two roots on P2; relatively narrow P4; and p4 with relatively high crown and arched posterior slope; no bifurcations on labial ridges relatively high first serration. Differs from *Baiotomeus* in having well-developed anterolabial bulge on p4; relatively high-crowned p4 with a less angular anterolabial lobe almost never bearing pseudoserrations; an m1 that is short relative to the lengths of p4 and m2. Differs from *Prochetodon* in having conical rather than lenticular cusps on the labial sides of P1–3; cusps of P2 arranged in two longitudinal pairs; fewer cusps on P3 (there are a few instances of eight cusps on P3 of *Ptilodus*, however); relatively short and broad P4 with cusps occasionally present on anterolabial bulge; p4 with transversely rounded anterior margin, large anterolabial lobe, an uncanted lateral profile; m1 without grooved labial side of cusps in labial view.

Average length of p4: 6.99 mm.

Average length of m1: 3.38 mm.

Included species: *P. mediaevus* (known from localities SB23E*, SB39A-C, CP1C, CP13IIA, CP14A); *P. fractus* Dorr, 1952 (localities CP22B, CP26A); *P. gnomus* Scott, Fox, and Youzwyshyn, 2002 (localities CP13IIA, CP16A, NP1C*, NP3A0); *P. kummae* Krause, 1977 (localities SB20A, CP13E, F, CP13IIA, CP14C, CP15B, CP24A, NP3F, NP7D*, NP47C, NP48B); *P. montanus* Douglass, 1908 (localities CP11IIA, [CP21A], [CP22A], NP3A0, NP19B, C*, NP20D, NP47B); *P. tsosiensis* Sloan, 1981 (localities SB23A*, [CP1A], [CP11IIA, G], [NP7B], NP16C); *P. wyomingensis* Jepsen, 1940 (localities CP13B, NP47B2, C); *Ptilodus* sp. "C" Krause 1982b (localities CP13E, EE, CP13IIA, CP16A, CP21B, CP22A, CP26A,

NP1C, NP3A, NP4, NP19IIC, NP20C, D, NP47A1, B, B2); *Ptilodus* sp. "T" Krause 1982b (localities CP13IIA, NP1C, NP3A, C, G, NP19IIA).

Ptilodus sp. is also known from localities SB39IIA, ?CPIC, CP11IIG, CP13C, D, CP15A, CP62A2, NP1C, NP2, NP3[A0, A], B–D, G, NP19IIA, C, NP20B, E, NP47C.

Indeterminate ptilodontids
Fossil material referred to Ptilodontidae has also been reported from localities SB23GG, CP22B, and CP26A.

NEOPLAGIAULACIDAE AMEGHINO, 1890

Characteristics (modified from Sloan and Van Valen [1965], Hahn and Hahn [1983], and Vianey-Liaud [1986]): Generally small; I1 pointed, gracile, and covered with enamel; tendency for labial cusp row of P4 to be reduced, but rarely altogether absent; lateral profile of P4 usually close to an isosceles triangle with the posterior edge shorter; p4 usually 1.4 to 2.0 times length of m1; lateral profile of p4 arcuate, anterior portion of profile curved, and highest point on profile occurring at midlength, above line of cusps of m1 and m2; first labial ridge of p4 begins at first serration; second through fifth ridges abut the first at an angle.

***Cernaysia* Vianey-Liaud, 1986 (including *Parectypodus*, in part)**

Type species: *Cernaysia manueli* Vianey-Liaud, 1986.

Type specimen: CRL 897, right P4 from the late Paleocene (Cernaysian) of France.

Characteristics (modified from Vianey-Liaud, 1986): P4 low and compact; ultimate cusp highest; cusps of labial row well developed; posterobasal concavity and cingulum present; wear surface horizontal.

Average length of p4: not known for *C. davidi*.

Average length of m1: not known for *C. davidi*.

Included species: *C. davidi* (including *Parectypodus vanvaleni* Sloan 1981 in part) only; known from locality SB23A* only.

Comments: The type species, *C. manueli*, is known only from the late Paleocene of Europe. The formal diagnosis of *Cernaysia* does not effectively distinguish it from other neoplagiaulacids and does not accurately encompass specimens referred to *C. davidi*. Nevertheless, validity of the taxon has not been formally challenged, and recent systematic revisions of Multituberculata (McKenna and Bell, 1997; Kielan-Jaworowska and Hurum, 2001) included it as a valid taxon.

***Ectypodus* Matthew and Granger, 1921 (including *Charlesmooria* Kühne, 1969)**

Type species: *Ectypodus musculus* Matthew and Granger, 1921.

Type specimen: AMNH 17373, maxilla from the Tiffanian of the San Juan Basin of Colorado.

Characteristics (modified from Sloan, 1981): Ultimate cusp in lingual row of P4 highest on crown; P4 very trenchant with both anterior and posterior slopes straight in profile; posterior basal cuspule usually weak or absent. Early species have p3; crown of p4 relatively low, fourth serration highest; relative height of first serration between one-third and one-half length of p4; height of enamel at anterolabial lobe of p4 less than crown length; posterior angle between plane of occlusion of molars and anterior face of p4 approximately a right angle.

Average length of p4: 3.22 mm.

Average length of m1: 1.90 mm.

Included species: *E. musculus* (known from localities SB20A*, SB39B, C, [CP16A]); *E. aphronorus* Sloan, 1987 (localities CP13B, CP15A, NP19C*); *E. childei* (Kuehne, 1969; originally described as *Charlesmooria childei*) (European, but may occur in localities CP27D, CP63); *E. elaphus* Scott, 2005 (localities NP3B, C*, D); *E. lovei* Sloan, 1966 (localities CP29C, D, CP39B, CP44, CP98B, NP8, NP9A, B, NP10B, Bi, NP23A, NP49II); *E. powelli* Jepsen, 1940 (localities CP13[E], G*, H, [CP14C, E], CP17B, CP21A, CP22B, [CP24A], CP26A, [NP3B], [NP7D], [NP19IIC], [NP20E]); *E. szalayi* Sloan, 1981 (localities SB23F, CP13IIA, CP14A, [NP1C], [NP3A0], NP19C*); *E. tardus* Jepsen, 1930a (localities CP15B, CP19AA–C, CP20A, BB, [CP27A], CP63).

Ectypodus sp. is also known from localities ?SB23H, CP11IIE, CP44, NP3C, G, NP7B, NP17, NP20D, NP47B.

Comments: *Ectypodus* is also known from the late Paleocene to early Eocene of Europe.

***Krauseia* Vianey-Liaud, 1986 (including *Parectypodus*, in part; *Ectypodus*, in part)**

Type species: *Krauseia clemensi* (Sloan, 1981) (originally described as *Parectypodus clemensi*).

Type specimen: KU 16001, crushed rostrum with P1–M2, right dentary with broken i1, p3–m1, and fragmentary left dentary from the Puercan of New Mexico.

Characteristics (modified from Vianey-Liaud, 1986): P4 large; lingual side very steep and labial side vertical to labially inclined; ultimate cusp highest; posterior edge abrupt but more inclined than that of *Ectypodus*, bearing slight posterobasal cingulum; mandible shorter and thicker than that of *Mimetodon*, with raised horizontal ramus; mental foramen below p3 alveolus; short diastema; robust i1 completely covered with enamel, although enamel thinner on dorsolingual side; anterior edge of p4 gently rounded; height at first serration 38–40% of crown length, crest of p4 highest between third and fifth serrations.

Average length of p4: 3.4 mm.

Average length of m1: 2.7 mm.

Included species: *K. clemensi* (also including *Ectypodus* sp. C of Rigby, 1980) only (known from localities SB23F*, GG, CP13IIA. CP14A).

Mesodma Jepsen, 1940

Type species: *Mesodma ambigua* Jepsen, 1940.

Type specimen: YMP-PU 14414, left dentary with i1–m1 from the Puercan of Wyoming.

Characteristics (modified from Clemens, 1963; Sloan, 1981; Hahn and Hahn, 1983): As in *Ectypodus* and *Mimetodon*, the ultimate cusp of P4 highest; p3 present; p4 crown relatively low (height of blade at first serration less than one-third crown length), rising anteriorly, then declining gradually until level with occlusal surface of m1; molar cusps semi-crescentic or crescentic. Includes the smallest species of Tertiary multituberculate, *M. pygmaea*.

Average length of p4: 3.46 mm.

Average length of m1: 2.32 mm.

Included species: *M. ambigua* (known from localities CP11IIA, IIB, CP12A*, [CP61A], [NP16C]); *M. formosa* Marsh, 1889 (localities SB23A, CP11IIB, NP7B, C, NP16C, NP17 and additional Late Cretaceous localities); *M. garfieldensis* Archibald, 1982 (localities SB39IIA, [CP11IIA], NP16A*); *M. hensleighi*: Lillegraven, 1969 (locality CP11IIB, and additional Late Cretaceous localities); *M. pygmaea* Sloan, 1987 (localities [SB39B], CP13B, ?C, CP13IIA, CP14C, CP15A, CP16A, NP1C, NP3A0, C, D, F, NP4, NP7D, NP19C*, NP19IIA, C, NP20A, NP47B); *M. thompsoni* Clemens, 1963 (localities SB23A, SB39IIA, NP6, NP7A-C, NP17 and additional Late Cretaceous localities).

Mesodma sp. is also known from localities CP11IIB, CP13A, CP15B, CP24A, NP7C, NP15A-C, NP16A, B, NP47A0, ?B.

Comments: In addition to Late Cretaceous occurrences of *M. hensleighi*, *M. formosa*, and *M. thompsoni*, there are two species (*M. primaeva* and *M. senecta*) that are entirely Cretaceous. *Mesodma* may be paraphyletic. Sloan (1981, 1987) hypothesized that *Neoplagiaulax*, *Parectypodus*, and *Mimetodon* arose from different species of *Mesodma*, and this hypothesis has not been ruled out by phylogenetic analyses. *Mesodma* molars retain several primitive characteristics, and where *Mesodma* species co-occur with *Cimexomys* of similar size, m1, m2, and M2 of the two genera cannot be reliably distinguished.

Mimetodon Jepsen, 1940

Type species: *Mimetodon churchilli* Jepsen, 1940.

Type specimen: YPM-PU 14525, left dentary with i1, p3-m1 from the Tiffanian of Wyoming.

Characteristics (modified from Jepsen, 1940; Hahn and Hahn, 1983): As in *Ectypodus* and *Mesodma*, ultimate cusp on P4 highest; P4 bearing three posteroventrally directed grooves on posterolabial surface; lower incisor disproportionately robust compared with those of other neoplagiaulacids; dentary deep below p4 to accommodate i1; p3 present.

Average length of p4: 4.16 mm.

Average length of m1: 2.71 mm.

Included species: *M. churchilli* (known from locality CP13G*); *M. krausei* Sloan, 1981 (locality SB23GG*); *M. silberlingi* Simpson, 1935 (localities SB39B, C, CP13[B], C, [CP15A], NP1C, NP3A0, A, C, D, NP7D, NP19C*, NP20E, NP48B).

Mimetodon sp. is also known from localities NP2, NP19IIA.

Comments: *Mimetodon* is also questionably known from the late Paleocene of Europe (McKenna and Bell, 1997).

Neoplagiaulax Lemoine, 1882 (including _Plagiaulax_, in part)

Type species: *Neoplagiaulax eocaenus* (Lemoine, 1880) (originally described as *Plagiaulax eocaenus*).

Type specimen: CRL 936, fragmentary left dentary with p4–m2 from the late Paleocene (Cernaysian) of France.

Characteristics (modified from Sloan, 1981): Includes the largest species of Neoplagiaulacidae. As in *Parectypodus*, penultimate or antepenultimate cusp highest on crown of P4; posterior slope of middle cusp row (between ultimate cusp and crown base) short and straight, rather than concave as in *Parectypodus*; p3 present in all species except *N. eocaenus*; anterior profile of p4 less convex than that of *Ectypodus*; serrations five to seven highest on p4; relative height of first serration 0.30–0.45 times crown length.

Average length of p4: 4.58 mm.

Average length of m1: 2.65 mm (not known from all spp.).

Included species: *N. cimolodontoides* Scott, 2005 (known from localities NP3B, C*); *N. grangeri* Simpson, 1935 (localities CP13C, ?CP21A, NP19C*); *N. hazeni* Jepsen, 1940 (localities CP13G*, CP24A, [NP3[A], B–D, F], [NP7D], NP47C, NP48B); *N. hunteri* Simpson, 1936b (localities CP13E, CP24A, NP1C, NP3A0, [C, D], F, NP4, NP7D, NP19IIA, C*, NP20D, NP47A1, B, B2, [C]); *N. jepi* Sloan, 1987 (localities CP13E*, CP13IIA); *N. kremnus* Johnston and Fox, 1984 (localities NP7B*, NP17, [NP17IIA]); *N. macintyrei* Sloan, 1981 (localities SB23A*, CP1C); *N. macrotomeus* Wilson, 1956 (locality SB23GG*); *N. mckennai* Sloan, 1987 (localities CP13F, CP26A*, NP20E, NP47C); *N. nanophus* Holtzman, 1978 (localities [NP2], NP47B2*); *N. nelsoni* Sloan, 1987 (localities CP13IIB, CP14A, CP15A, NP1C, NP2, NP3A0, NP16C*, NP19IIA); *N. paskapooensis* Scott, 2005 (localities NP3A–C*, NP4); *N. serrator* Scott, 2005 (localities NP3A- C*, D, NP4).

Neoplagiaulax sp. is also known from localities NB1A, CP13C, D, CP14C, NP3E, NP17, NP17IIA. NP19IIA, NP20B, NP47B, B2.

Comments: One species questionably referred to *Neoplagiaulax* (?*N. burgessi*) is known from the Late Cretaceous of North America. Five species (*N. annae*, *N. eocaenus*, *N. copei*, *N. nicolae*, and *N. sylvani*) are known from the late Paleocene of Europe.

Parectypodus Jepsen, 1930a

Type species: *Parectypodus simpsoni* Jepsen, 1930a.

Type specimen: YPM-PU 13242, fragmentary left dentary with i1–m1 from the Wasatchian of Wyoming.

Characteristics (modified from Sloan, 1981): Antepenultimate cusp of P4 usually highest; posterobasal cusp usually absent; posterior slope of middle cusp row between ultimate cusp row and crown base of P4 short, steep, and slightly concave; anterior slope convex; p3 absent in later species; relative height of first serration greater than 0.45 times crown length of p4; height of enamel at anterolabial lobe of p4 approximately equal to or greater than crown length; posterior angle between plane of occlusion of molars and anterior face of p4 a right or (usually) acute angle; third or fourth serration of p4 highest.

Average length of p4: 3.74 mm.

Average length of m1: 2.28 mm (not known from all species).

Included species: *P. simpsoni* (known from localities CP19B, C, CP20A*, BB); *P. armstrongi* Johnston and Fox, 1984 (locality NP7B*); *P. corystes* Scott, 2003 (localities NP1C, NP2*, NP3A0); *P. laytoni* Jepsen, 1940 (localities CP13G*, H, CP17B, [NP20D]); *P. lunatus* Krause, 1982a (localities CP19A, CP20A, CP25B, CP27A, CP28A, CP63*); *P. sinclairi* Simpson, 1935 (localities CP13IIA, CP15A, NP1C, NP4, NP19A, C*); *P. sloani* Schiebout, 1974 (localities SB39B, C); *P. sylviae* Rigby, 1980 (localities CP13C, CP13IIA, B, CP14A*, CP15A, [NP1C], [NP3A0]); *P. trovessartianus* Cope, 1882b (locality SB23F*); *P. vanvaleni* Sloan, 1981 (locality SB23A*).

Parectypodus sp. is also known from localities CP1C, CP25A, NP2, NP3A0, NP7C, NP16C, NP17, NP47B, B2.

Comments: One species, *P. foxi*, is known only from the Late Cretaceous. *P. pattersoni* was synonymized with *P. sylviae* by Secord (1998). This synonymy was ignored without any explanation by Higgins (2003a), and we have followed Secord here.

Xanclomys Rigby, 1980

Type species: *Xanclomys mcgrewi* Rigby, 1980.

Type specimen: AMNH 87859, fragmentary dentary with p3–p4 from the Torrejonian of Wyoming.

Characteristics (modified from Rigby, 1980): P4 with subquadrate to triangular occlusal outline; ultimate cusps of P4 fused and highest; anterolabial row form two sides of triangle and with single major elevation of one or two cusps at most laterally extended portion, with cusps descending anteriorly and posteriorly; p3 present. Differs from all other multituberculate genera in shape of p4; crown of p4 triangular or sickle shaped in lateral outline, with long, flat, gently inclined anterior slope bearing one to three serrations near apex, and short, nearly vertical posterior slope with serrations.

Average length of p4: 4.90 mm.

Average length of m1: unknown.

Included species: *X. mcgrewi* only; known from locality CP14A only.

Comments: It is unclear on what basis Rigby referred *Xanclomys* to Neoplagiaulacidae. The fourth premolars are unique. The molars, which are all isolated teeth, are questionably referred and seem not to differ from those of *Neoplagiaulax*.

Xyronomys Rigby, 1980

Type species: *Xyronomys swainae* Rigby, 1980.

Type specimen: AMNH 87897a, isolated right p4 from the Torrejonian of Wyoming.

Characteristics (modified from Rigby, 1980; Middleton, 1983; Johnston and Fox, 1984): p4 lower than those of comparably sized *Mesodma*, with little or no anterolabial lobe. Lower molars similar to those of *Mesodma* but proportionally narrower.

Average length of p4: 3.05 mm.

Average length of m1 (known only from *X. robinsoni*): 2.15 mm.

Included species: *X. swainae* (known from locality CP14A*); *X. robinsoni* Middleton and Dewar, 2004 (locality CP61A*).

Xyronomys sp. is also known from localities NP3A0, NP7B, C, NP16B.

Comments: Rigby (1980) had only isolated premolars and placed *Xyronomys* in the Eucosmodontidae on the basis of its low p4, which is similar in shape to that of *Stygimys*. Middleton's (1983) discovery of a fragmentary dentary with p4–m2 of a second species confirmed the suspicion of Johnston (mentioned in his 1980 M.Sc. dissertation, and published in Johnston and Fox [1984]) that *Xyronomys* is a neoplagiaulacid. Unlike the eucosmodontids *Eucosmodon* and *Stygimys*, *Xyronomys* has small, circular tooth enamel prisms (Carlson and Krause, 1985; Krause and Carlson, 1986).

Indeterminate neoplagiaulacids

Fossil material referred to Neoplagiaulacidae has also been recovered at localities SB24, CP13B, CP61B, NP3A0, NP7C, and NP19C.

EUCOSMODONTIDAE JEPSEN, 1940 (INCLUDING BOFFIDAE HAHN AND HAHN, 1983)

Characteristics: I2 with two cusps; I3 palatal; P1 (when present), P2, and P3 single-rooted; i1 strongly laterally compressed, with enamel restricted to ventrolabial side; p3 absent; p4 proportionally large and elongate, as long or longer than m1 and with at least seven serrations; crest of p4 continuous with occlusal surface of m1.

Eucosmodon Matthew and Granger, 1921 (including *Neoplagiaulax*, in part)

Type species: *Eucosmodon americanus* (Cope, 1885) (originally described as *Neoplagiaulax americanus*).

Type specimen: AMNH 3028, fragmentary left dentary with p4 and broken i1 from the Puercan of the San Juan Basin of New Mexico.

Characteristics: Very similar to *Stygimys* but larger; P1 absent; i1 less compressed than that of *Stygimys*; p4 often over 1 cm in length, about twice as long as m1; anterolabial lobe of p4 not pronounced.

Average length of p4: 11.53 mm.

Average length of m1: 5.6 mm.

Included species: *E. americanus* (known from locality SB23B*); *E. molestus* (Cope, 1886) (originally described as *Neoplagiaulax molestus*) (localities SB23A, NP1B); *E. primus* Sloan, 1981 (locality SB23A*).

Eucosmodon sp. is also known from localities SB23E and NP16B (Weil, 1999).

Stygimys Sloan and Van Valen, 1965

Type species: *Stygimys kuszmauli* Sloan and Van Valen, 1965.

Type specimen: UMVP 1478, fragmentary left dentary with i1 from mixed Cretaceous–Tertiary sediments of Montana.

Characteristics (modified from Sloan and Van Valen, 1965): I2 bicuspid; P1 present; i1 more laterally compressed than that of *Eucosmodon*; lateral profile of posterior portion of p4 nearly straight; p4 slightly longer than m1; p4 roots large, the anterior one being larger and curved forward.

Average length of p4: 4.65 mm.

Average length of m1: 4.24 mm (but not known from the larger species.).

Included species: *S. kuszmauli* (including *S. cupressus* Fox, 1989; see Lofgren *et al.*, 2005) (known from localities CP1A, CP12A, NP7A, NP15A*, B, C, NP16A); *S. camptorhiza* Johnston and Fox, 1984 (localities NP7B*, [NP17]); *S. jepseni* Simpson, 1935 (localities NP19B, C*); *S. teilhardi* Granger and Simpson, 1929 (locality SB23A); *S. vastus* Lofgren *et al.*, 2005 (locality CP39IIB).

Stygimys sp. is also known from localities CP13B, [NP3A0], NP7C, NP16A-C.

Comments: Lofgren (1995) recognized that the type of *Stygimys gratus* belonged to *Cimexomys "hausoi"* and assigned the remaining specimens of *Stygimys* from Harbicht Hill (locality NP15C), and Mantua Lentil (locality CP12A) to *S. kuszmauli*. Fox (1989) described a new species, *S. cupressus*, from the Ravenscrag Formation (locality NP4), which Lofgren *et al.* (2005) synonymized with *S. kuszmauli*.

Indeterminate eucosmodontids

Fossil material referred to Eucosmodontidae has also been recovered at localities SB23GG and NP15C.

MICROCOSMODONTIDAE HOLTZMAN AND WOLBERG, 1977

Characteristics: Tooth i1 proportionally large and strongly laterally compressed, with enamel restricted to ventrolabial surface for most of incisor length (enamel cap, quickly worn away, may be present on unworn incisors); root extending posterior to p4; crown of p4 with three to six serrations; anterior serration (lost in *Microcosmodon rosei*) strong; posterior serrations cusplike; m1 and m2 both with notch posterior to last cusp of internal cusp row. In some species, first molars with accessory roots. In the three species for which dentaries are known, pterygoid fossa very large and deep.

Allocosmodon Fox, 2005

Type species: *Allocosmodon woodi* (Holtzman and Wolberg, 1977) (originally described as *Microcosmodon woodi*).

Type specimen: MCZ 19963, right p4 from Ti1 of Wyoming.

Characteristics (based on Holtzman and Wolberg, 1977; Weil, 1998; Fox, 2005): I2 bicuspid with thick anterior enamel, exposed dentine medially, labial groove separating cusps; P4 large with five or six large cusps, ultimate cusp highest; M1 cusps conical to pyramidal, accessory roots present i1 laterally compressed, with enamel engulfing tip but not covering the entire dorsolingual surface; p3 present; p4 longer than m1, with five or six serrations, lingual and labial ridges that are short but strong, differs from other microcosmodontids in long, low blade.

Average length of p4: 3.1 mm.

Average length of m1: 2.4 mm.

Included species: *A. woodi* only (known from localities CP15A*, NP3A–C, NP19IIA, NP20D, NP47B).

Comments: Holtzman and Wolberg (1977) placed this species in *Microcosmodon*. Weil (1998) removed it from *Microcosmodon* and Microcosmodontinae (= Microcosmodontidae Fox 1999, 2005) but did not give it a new generic name or refer it to any named genus. Fox (2005), working with considerably better specimens, replaced it within Microcosmodontidae, named the new genus, and provided a revised, differential diagnosis.

Acheronodon Archibald, 1982

Type species: *Acheronodon garbani* Archibald, 1982 (emended by Fox, 2005).

Type specimen: UCMP 116953, fragmentary p4 from the Puercan of Montana.

Characteristics (modified from Archibald [1982] with reference to Fox [2005]): The p4 has a very distinct, cusplike, ultimate serration and a wide posterolabial shelf.

Average length of p4: 1.90 mm.

Average length of m1: 2.3 mm (not known for *A. garbani*).

Included species: *A. garbani* (known from locality NP16A*); *A. vossae* Fox, 2005 (localities NP1C*, [NP3A]).

Acheronodon sp. is also known from locality NP3A0.

Comments: The type of, and for a long time the only specimen referred to, this genus is a fragmentary p4, and it may well be that the name *Acheronodon* is a nomen nudum. The type species was named *A. garbani*, by Archibald (1982), but the Latin was emended to *garbanii* by Fox (2005). Scott

(2003) refer an m1 with length of 1.7 mm from locality NP3A0 to *Acheronodon*, on the basis of a personal communication with Fox, apparently concerning the specimens described in Fox (2005).

***Microcosmodon* Jepsen, 1930b**

Type species: *Microcosmodon conus* Jepsen, 1930b.

Type specimen: YPM-PU 13331, left dentary with i1 and p4 from the Tiffanian of Wyoming.

Characteristics (modified from Weil, 1998; Fox, 2005): The p3 may be present in all species but *M. rosei*; p4 posterior crown wider than anterior; p4 shorter than m1, with high, arcuate, serrated edge; molar cusps semi-crescentic to strongly recurved; molars with more cusps than those of *Pentacosmodon*; enamel prisms a mix of small, circular and large, arcade shaped.

Average length of p4: 2.05 mm.

Average length of m1: 2.11 mm.

Included species: *M. conus* (known from localities CP13G*, CP14E, [CP15B], NP3C, F, G, NP4, NP7D) "*M.*" *arcuatus* Johnston and Fox, 1984 (locality NP7B*); "*M.*" *harleyi* Weil, 1998 (locality NP16B); *M. rosei* Krause, 1980 (localities CP13H, CP17A, B*).

Microcosmodon sp. is also known from localities NB1A, NP47A1.

Comments: Fox (2005) removed "*M.*" *arcuatus* Johnston and Fox, 1984 and "*M.*" *harleyi* Weil, 1998 from *Microcosmodon* but did not refer them to another genus or create a new generic name. *Microcosmodon* is also known from the early Eocene of Europe (McKenna and Bell, 1997).

***Pentacosmodon* Jepsen, 1940**

Type species: *Pentacosmodon pronus* Jepsen, 1940.

Type specimen: YPM-PU 14085, left dentary with broken i1, p4, and m1–2 from the Tiffanian of Wyoming.

Characteristics (modified from Jepsen, 1940; Hahn and Hahn, 1983): The p3 absent; p4 less reduced than those of later species of *Microcosmodon*, and with strong lateral ridges, crown shape appears truncated posteriorly; molars with fewer cusps than those of *Microcosmodon*, molar cusps subcrescentic with anteroposteriorly elongated bases; large, arcade-shaped enamel prisms.

Average length of p4: 1.9 mm.

Average length of m1: 5.4 mm.

Included species: *P. pronus* (known from locality CP13G*); *P. bowensis* Fox, 2005 (locality NP1C*).

Indeterminate microcosmodontines

Fossil material referred to Microcosmodontinae has been recovered from localities NP3G and NP7A.

TAENIOLABIDIDAE GRANGER AND SIMPSON, 1929

Characteristics: Tooth enamel gigantoprismatic. Large in size relative to multituberculate genera with which they co-occur. P1-3 absent; P4 extremely reduced; i1 enamel ridged and restricted to ventrolabial side of tooth; i1 not laterally compressed as in Eucosmodontidae and Microcosmodontidae; p3 absent; p4 extremely reduced, altered from slicing tooth in more primitive members of family to single-pointed tooth in more derived members; molar cusps bulbous or transversely expanded, wearing first on apices rather than on sides as in other North American multituberculates; M1 tending to develop lingual cusp row running entire length of tooth.

***Catopsalis* Cope, 1882c**

Type species: *Catopsalis foliatus* Cope, 1882c.

Type specimen: AMNH 3035 from the Nacimiento Formation of New Mexico.

Characteristics: *Catopsalis* is paraphyletic (Simmons and Desui, 1986) and it is difficult to distinguish larger species of *Catopsalis* from smaller species of *Taeniolabis*. Simmons (1987) found m1 and M1 of *C. calgariensis* to overlap in size with those of *Taeniolabis*; however, she found i1, I2, m2, and M2 of *Catopsalis* spp. to be smaller in all cases than those of *Taeniolabis*. Buckley (1995) described an even larger species of *Catopsalis*, and distinguished it from *Taeniolabis* on the basis of low cusp count.

Average length of p4: 4.3 mm (not known from all species).

Average length of m1: 11.8 mm.

Included species: *C. foliatus* (known from localities SB23A, NP7A); *C. alexanderi* Middleton, 1982 (localites CP12A, CP61A*, NP15C, NP16A); *C. calgariensis* Russell, 1926 (localities CP13IIA, CP15A, NP1B*); *C. fissidens* Cope, 1884a (localities SB23E, CP1C, CP13IIA); *C. joyneri* Sloan and Van Valen, 1965 (localities CP11IIA, NP6, [NP7A], NP15A*, B, NP16A); *C. waddleae* Buckley, 1995 (locality NP17*).

Catopsalis sp. is also known from localities SB23B, CP13B, NP7B, C, NP16A.

***Taeniolabis* Cope, 1882b**

Type species: *Taeniolabis taoensis* Cope, 1882b.

Type specimen: AMNH 3036, fragmentary right maxilla with M1–2 and skull fragments from the Puercan of New Mexico.

Characteristics (modified from Simmons, 1987): Includes the largest known multituberculate species; i1, I2, m2, M2 larger than those of any other genus, although M1/1 dimensions may overlap those of some species of *Catopsalis*; M1 with nine or more cusps in labial and lingual cusp rows; M2 with four or more cusps in medial cusp row; m1 with seven or more cusps in labial row and six or more cusps in lingual row; m2 with four or more cusps in lingual row. Length ratio of p4/m1 less than 0.40.

Average length of p4: 6.2 mm.

Average length of m1: 17.9 mm.

Included species: *T. taoensis* (known from localities SB23B*, [CP1A], CP11IIF, NP7C); *T. lamberti* Simmons, 1987 (locality NP16B*).

Taeniolabis sp. is also known from localities CP1B, NP16C, NP18.

CIMOLODONTA INCERTAE SEDIS

Anconodon Jepsen, 1940 (including *Ptilodus*, in part)

Type species: *Anconodon gidleyi* (Simpson, 1935) (originally described as ?*Ptildous gidleyi*).

Type specimen: USNM 9763, left dentary with p4 and fragmentary m1 from the Torrejonian of Montana.

Characteristics (modified from Jepsen, 1940; Hahn and Hahn, 1983): Lingual cusp row of P4 more strongly developed into cutting edge than in *Cimolodon* and less so than in *Liotomus*, the two other genera that historically have been included in Cimolodontidae. The p4 highly arched; distance between anterior "beak" and first serration straight and greater than one-half length of tooth; p4 "squared" above basal concavity in anterior view; anterior ridges on lingual side tend to branch.

Average length of p4: 5.31 mm.

Average length of m1: 2.70 mm.

Included species: *A. gidleyi* (known from localities SB23F, CP13B, NP1B, C, NP19B, C*); *A. cochranensis* Russell, 1929 (localities CP13B, [C], D, CP13IIA, CP16A, NP1B*, C, NP3A0, NP19C, NP19IIA); *A. lewisi* Sloan, 1987 (localities CP13IIA, CP15A*, NP19IIA).

Anconodon sp. is also known from localities ?SB23H, CP13C, CP13IIB, NP20B.

Comments: *Anconodon* is generally considered to belong to the Cimolodontidae, a family that was never formally diagnosed and is hazily united in having a slender i1 that is completely covered with enamel, and a round, high arch of the cutting edge of p4. Hahn and Hahn (1983) included *Anconodon* and *Cimolodon* from the Late Cretaceous of North America, and *Liotomus* from the late Paleocene of Europe. Vianey-Liaud (1986) assigned the species *A. russelli* to *Liotomus*, but we have followed Sloan (1987) in synonymizing *A. russelli* with *A. cochranensis*. Simmons' (1993) parsimony analysis did not include *Anconodon* but found *Cimolodon* and *Liotomus* to be phylogenetically disparate. Accordingly, McKenna and Bell (1997) removed *Liotomus* to the Eucosmodontidae (they also followed Vianey-Liaud and listed an early Paleocene North American occurrence of *Liotomus*) and retained *Anconodon* in Cimolodontidae. Kielan-Jaworowska and Hurum (2001) included all three genera in Cimolodontidae. The difficulty with all these assignments is that the type genus, *Cimolodon*, is non-monophyletic (Weil, 1999). Characters used to diagnose *Cimolodon* (Clemens, 1963) are widely distributed among Cimolodonta, and many are primitive. Considering that the type genus is polyphyletic and that the only analysis to test monophyly of the Cimolodontidae indicated that it was also non-monophyletic, we have listed *Anconodon* as Cimolodonta incertae sedis.

Cimexomys Sloan and Van Valen, 1965

Type species: *Cimexomys minor* Sloan and Van Valen, 1965.

Type specimen: SPSM 62-2115, fragmentary left dentary with p3–4 from mixed Cretaceous–Tertiary sediments of Montana.

Characteristics (from Archibald, 1982; Montellano, 1992): Height of P4 lower than in comparably sized species of *Mesodma*. M1 cusp formula greater than that of *Paracimexomys*, with which it co-occurs in the Cretaceous, but less than that of *Mesodma*; M1 cusp rows not divergent anteriorly; lingual cusp row 50% or less of M1 length, and tending to be ridgelike with indistinctly divided cusps; p4 with fewer serrations (8–10) than in comparably sized species of *Mesodma*, arcuate in outline, last two or three serrations formed into distinct cusps without lingual or labial ridges; m1 cusp formula (5–7:4–5) greater than that of *Paracimexomys*, and less than or equal to that of *Mesodma*; molar cusps tending to be subcrescentic to crescentic; molars not waisted.

Average length of p4: 3.53 mm.

Average length of m1: 3.27 mm (but does not include *C. minor*).

Included species: *C. minor* (known from localities CP61A, NP7A, B, NP15A*, B, C, NP16A and additional Late Cretaceous localities); *C. arapahoensis* Middleton and Dewar, 2004 (locality CP61A*); *C. gratus* (Jepsen [1930b], originally described as *C. hausoi* see Lofgren, [1995]) (localities CP12A*, [NP7A, B, C], NP15A, C, NP16A).

Cimexomys sp. may also be present at locality NP17.

Comments: Three species (*C. antiquus*, *C. gregoryi*, and *C. judithae*) are exclusively Late Cretaceous. Sloan and Van Valen (1965) placed *Cimexomys* in the Neoplagiaulacidae, but Archibald (1982) removed it to Cimolodonta incertae sedis because it is more primitive than those multituberculates that have been placed in Taeniolabidoidea and Ptilodontoidea. Isolated teeth of *Cimexomys* can be difficult to distinguish from those of other genera that have retained some primitive characters. The M1 is the only molar of *Cimexomys* that is consistently differentiable from those of co-occurring *Mesodma* of similar size (Lofgren, 1995; Montellano, Weil, and Clemens, 2000). However, the M1 of *Cimexomys* is very similar to (and perhaps indistinguishable from) M1s attributed to *Microcosmodon* (Krause, 1980; Weil, 1998) and to the Cenomanian *Dakotamys* (Eaton, 1995).

Fractinus Higgins, 2003b

Type species: *Fractinus palmorum* Higgins, 2003b.

Type specimen: UW 27063, left p4 from the Tiffanian of Wyoming.

Characteristics: Very distinct from any other multituberculate. p4 with five serrations, serrations rounded in unworn state, first serration at highest point on arc of p4 in profile; anterior edge of crown only slightly convex. Differs from *Xanclomys* in having all serrations except the first in a straight line when viewed in profile. Differs from other taxa with similarly low numbers of serrations in having a prominent anterolabial lobe.

Average length of p4: 4.92 mm.

Average length of m1: unknown.

Included species: *F. palmorum* only, known from locality CP13IIA* only.

Comments: Known from only the type and the anterior half of a second p4.

***Neoliotomus* Jepsen, 1930a**

Type species: *Neoliotomus conventus* Jepsen, 1930a.

Type specimen: YPM-PU 13297, fragmentary left dentary with broken i1, root of p3, and p4 from the Clarkforkian of Wyoming.

Characteristics: Relatively large size, with p4 greater than 1 cm in length; I2 unicuspid; P1–4 present; all have two roots; i1 laterally compressed, with enamel restricted to ventrolabial surface; p3 present; p4 longer than m1, and in unworn state higher than m1; enamel prisms small and circular.

Average length of p4: 12.52 mm.

Average length of m1: 6.99 mm.

Included species: *N. conventus* (known from localities CP13G, H, CP14E, CP17A, B, CP22B, CP62A2, B); *N. ultimus* Granger and Simpson, 1928 (localities CP19A, B, CP20A, CP25B, CP63).

Comments: Because of its laterally compressed, gliriform incisor, in combination with its relatively large p4, *Neoliotomus* was long considered to belong in Eucosmodontidae. It differs, however, from Eucosmodontidae in the following respects: enamel prisms are small and circular; p3 is present; p4 crest is higher than the occlusal surface of m1–2; I2 has only one cusp; and P1–3 are all present and have two roots. As a result of her phylogenetic study, Simmons (1993) removed *Neoliotomus* from Eucosmodontidae. McKenna and Bell (1997) placed it in the Boffiidae sensu Hahn and Hahn, 1983, but changed the rank to that of tribe (and the taxon name to Boffiini) within the family Eucosmodontidae. Following the result of their parsimony analysis, Kielan-Jaworowska and Hurum (2001) placed *Neoliotomus* in Ptilodontoidea incertae sedis.

INDETERMINATE MULTITUBERCULATES

Fossil material referred to Multituberculata has also been recovered from localities SB39IIA, CP17B, CP31E.

BIOLOGY AND EVOLUTIONARY PATTERNS

Multituberculates have for a long time been referred to as the ecological vicars of rodents. Comparisons with rodents are inviting: multituberculates and rodents evolved numerous similar cranial and dental adaptations convergently, multituberculates were the most common small mammals of the early Tertiary as rodents are today, and the size range of multituberculates was about the same as that of living rodents. The smallest known North American Cenozoic multituberculate is *Mesodma pygmaea*, which was similar in size to small shrews living today; its lower first molar is only about 1.5 mm long. By contrast, *Taeniolabis taoensis*, the largest known multituberculate with a lower first molar about 20 mm long (skull length about 160 mm), was probably slightly larger than a modern beaver (notwithstanding that Cope [1884a] described it as being the same size as a large kangaroo) (Figure 2.5).

Despite characterizations of multituberculates as the first mammalian herbivores, it is likely that most were not strictly herbivorous. Most multituberculates were small, and living mammals of comparable size require more protein than is provided in a herbivorous diet (Krause, 1982c; Wing and Tiffney, 1987). Since there are no close living relatives of multituberculates, their diet is difficult to infer from their relationships. The closest living structural analogs are some small phalangeroid marsupials (e.g., *Aepyprymnus*, *Bettongia*, *Burramys*, *Hypsiprymnodon*), all of which are omnivorous (Clemens and Kielan-Jaworowska, 1979; Krause, 1982c), and it has been suggested that the pointed lower incisors of some taxa may have been used to stab insects, as are the incisors of living caenolestid marsupials.

One trend among Tertiary North American multituberculates is that of proportionate enlargement of the p4 to an exaggerated degree not seen in the Late Cretaceous. Those multituberculates that had enlarged lower fourth premolars may have ingested hard food items (Krause, 1982c; Wall and Krause, 1992), perhaps fruits and seeds. After a food item was ingested, it could have been held against the multicusped upper premolars (primarily P4) and sliced or wedged apart by the enlarged, laterally compressed and serrated p4 as the jaw closed. Following this slicing action, food passed posteriorly to the grinding molars. The molar chewing cycle of multituberculates is noteworthy because the power stroke was directed posteriorly, a unique or nearly unique condition among mammals. The two rows of lower molar cusps, which tend to be concave posteriorly, fit between the three rows of upper molar cusps, which tend to be concave anteriorly. This arrangement provided for a highly efficient mechanism of en-echelon shear as the lower molars were drawn posteriorly along the uppers (Krause, 1982c; Wall and Krause, 1992).

Taeniolabidids elaborated a different set of dental specializations: chisel-shaped lower incisors, reduction of the p4 and P4, and molars with large, often bulbous or transversely expanded cusps, on which wear did not occur in the valleys between cusps, but on the cusp apices. Cusp-tip-wearing multituberculates appear first in North America in the latest Cretaceous. Interestingly, both the Late Cretaceous and Paleocene taxa are the largest multituberculates in their faunas. Their dental morphology and large size suggest an increased

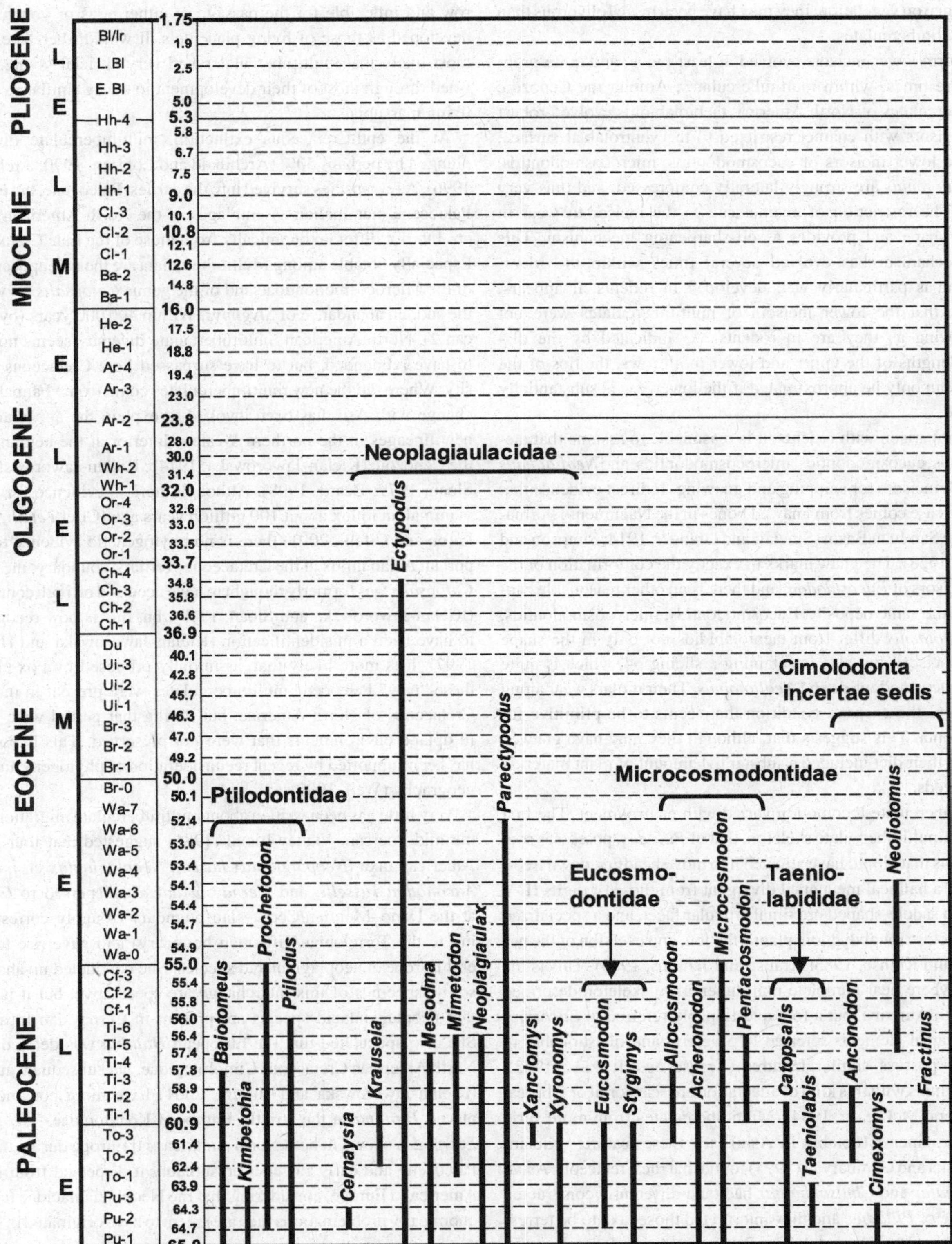

Figure 2.5. Temporal ranges of North American Tertiary multituberculate genera.

dependence on vegetation; they may have been more folivorous than other multituberculates.

Gliriform lower incisors evolved at least twice, and quite probably more times, within multituberculates. Among the Cenozoic multituberculates of North America, taeniolabidids evolved robust lower incisors with enamel restricted to the ventrolabial surface, while the lower incisors of eucosmodontids, microcosmodontids, and *Neoliotomus* are strongly laterally compressed, and thus very narrow. The restriction of enamel causes the incisor to wear in a chisel shape and provides a self-sharpening mechanism. This same mechanism has evolved several times within the Mammalia and is particularly well developed in rodents. It appears, however, that the lower incisors of multituberculates were not ever-growing as they are in rodents. As indicated by the discrepant lengths of the upper and lower tooth rows, the tips of the incisors can only be approximated if the lower jaw is substantially protruded.

By comparison with rodents, it is reasonable to assume that taeniolabidids, eucosmodontids, microcosmodontids, and *Neoliotomus* used their incisors for snipping and gnawing. Indirect evidence that this is the case comes from gnawed bones in the Nacimiento Formation of the San Juan Basin (Sinclair and Granger, 1914; Simpson and Elftman, 1928). The gnaw marks fit exactly the conformation of the lower incisors of *Eucosmodon*, and there is no other reasonable candidate in the same deposits. Eucosmodontids, microcosmodontids, and *Neoliotomus* differ from taeniolabidids not only in the shape of their incisors, but also in retaining a slicing p4, which is quite large in eucosmodontids and *Neoliotomus*. Their molars wear along the sides of the cusps, a condition that seems to be primitive for Cimolodonta. This suggests that, although they may have gnawed on bones, their diet included a substantial amount of plant material, such as seeds.

Multituberculate locomotion varied with environment. The first described multituberculate skeleton, that of *Ptilodus*, possesses specializations that would have allowed it to move headfirst down trees. These are a hallucal metatarsal divergent from those of digits II–V and with a saddle-shaped proximal articular facet, and a specialized tarsal structure for abduction, plantarflexion, and eversion of the pes (Jenkins and Krause, 1983; Krause and Jenkins, 1983). This is the only anywhere near complete multituberculate skeleton described from North America, but a foot questionably referred to *Eucosmodon* and tarsal elements referred to *Stygimys* and questionably to *Mesodma* possess similar features (Krause and Jenkins, 1983). Extant animals with this kind of adaptation are scansorial or arboreal (Jenkins and McClearn, 1984). Multituberculates from more xeric paleoenvironments (notably in Asia) lack these features. Kielan-Jaworowska and Gambaryan (1994) found that the Cretaceous Asian *Nemegtbaatar* and *Chulsanbaatar* had tarsi differently constructed from those of *Ptilodus*, and they interpreted those taxa to be terrestrial and possibly ricochetal. The Paleocene Asian *Lambdopsalis*, included within Taeniolabididae, is interpreted as fossorial (Miao, 1988).

Multituberculate reproduction is also a subject of speculation. Kielan-Jaworowska (1979) thought that the narrow, V-shaped pelvis of multituberculates, with its long, keeled symphysis, was too narrow and inflexible for the passage of either eggs or young as far developed as those of living placentals. It would, therefore, seem most likely that multituberculates had very altricial young, nourished through most of their development in a way similar to that of living marsupials.

At the end-Cretaceous extinction, multituberculate diversity plunged by perhaps 50% (Archibald and Lofgren, 1990; Archibald, 1996). A few species survived into the earliest Paleocene, but earliest Paleocene mammalian assemblages of the North American Western Interior differ taxonomically from those of the Late Cretaceous. Especially notable among Multituberculata are the first appearances of the Microcosmodontidae and of the genus *Catopsalis*, as well as the sudden abundance of *Stygimys*. Within 400 000 years (by Puercan 2), North American multituberculate diversity seems not only to have rebounded, but to have surpassed Late Cretaceous diversity. Where did the new multituberculates come from? Faunal interchange with Asia has been invoked to explain the appearance of new lineages in the northern Western Interior at the beginning of the Cenozoic (Kielan-Jaworowska, 1974b; Kielan-Jaworowska and Sloan, 1979; Beard, 1998). Although there is evidence for Asian mammalian influx about 100 million years ago (Cifelli *et al.*, 1997; Eaton and Cifelli, 2001) there is no phylogenetic evidence to support an Asian influx at the Cretaceous–Tertiary boundary; the genus *Catopsalis* was formerly thought to have occurred on both continents (Kielan-Jaworowska and Sloan, 1979), but this is now recognized to have been a misidentification (Kielan-Jaworowska and Hurum, 1997). It is more likely that, as initially proposed by Fox (1968), these "new" Paleocene multituberculates were present in the Late Cretaceous of North America, but during that period were living in upland environments that were not preserved. This hypothesis has been supported by recent reconsideration of phylogeny and biogeography (Weil, 1999).

Very little has been written about multituberculate migration after the mid-Puercan. Vianey-Liaud (1986) suggested that four North American taxa (*Neoplagiaulax hazeni*, *Neoplagiaulax* cf. *hunteri*, *Anconodon russelli*, and *Cernaysia davidi*) dispersed to Europe at the Dano–Montian/Cernaysian boundary (roughly corresponding to the Torrejonian/Tiffianian boundary) and gave rise to several European neoplagiaulacid species. She conducted no analysis, so the specifics of this interchange are speculative, but it is clear that neoplagiaulacid species are present in Cernaysian localities. She also speculated that the European *Hainina* was derived from North American Cretaceous Cimolomyidae, but subsequent authors (Kielan-Jaworowska and Hurum, 2001) have more convincingly placed *Hainina* in the strictly European Kogaionidae. The genus *Ectypodus* occurs in both North America and Europe during the late Paleocene and early Eocene; presumably it dispersed from North America to Europe, considering that the Neoplagiaulacidae (as taxonomically problematic as they are) seem to have originated in North America.

This is not to say that dispersal between continents played no role in multituberculate evolution; the immigration of rodents may have played a key role in multituberculate extinction. The similarities between multituberculates and rodents led several early workers (e.g., Matthew, 1897) to suggest that the decline and extinction of multituberculates was related to the adaptive radiation of rodents. More recent attempts to test this hypothesis of competitive exclusion and taxonomic displacement have, in fact, found more evidence in support of it (Van Valen and Sloan, 1966; Krause, 1986). High taxonomic diversity was reestablished by the middle Paleocene. However, North American multituberculates underwent a decline in diversity in the latest Paleocene and earliest Eocene, at the time during which rodents are thought to have dispersed from Asia into North America. Competitive exclusion of multituberculates by rodents can be inferred from inverse correlations of relative abundance and generic richness between the two groups. Although multituberculates lived on into the Chadronian Land-Mammal Age, they were a minor component of North American faunas during the Eocene.

REFERENCES

Ameghino, F. (1890). Los plagiaulácidos Argentinos y sus relaciones zoológicas, geológicas y geográficas. *Boletin del Instituto Geográphico Argentino*, 11, 143–208.

Archibald, J. D. (1982). A study of Mammalia and geology across the Cretaceous–Tertiary boundary in Garfield County, Montana. *University of California Publications in Geological Sciences*, 122, 1–286.

(1996). *Dinosaur Extinction and the End of an Era: What the Fossils Say*. New York: Columbia University Press.

Archibald, J. D. and Lofgren, D. L. (1990). Mammalian zonation near the Cretaceous–Tertiary boundary. [In *The Dawn of the Age of Mammals in the Northern Part of the Rocky Mountain Interior, North America*, ed. T. M. Bown and K. D. Rose.] *Geological Society of America, Special Paper*, 243, 31–50.

Beard, K. C. (1998). East of Eden: Asia as an important center of taxonomic origination in mammalian evolution. [In *Dawn of the Age of Mammals in Asia*, ed. K. C. Beard and M. R. Dawson.] *Bulletin of Carnegie Museum of Natural History*, 34, 5–39.

Buckley, G. A. (1995). The multituberculate *Catopsalis* from the early Paleocene of the Crazy Mountains Basin in Montana. *Acta Palaeontologica Polonica*, 40, 389–98.

Carlson, S. J., and Krause, D. W. (1985). Enamel ultrastructure of multituberculate mammals: an investigation of variability. *Contributions from the Museum of Paleontology, University of Michigan*, 27, 1–50.

Cifelli, R. L., Kirkland, J. I., Weil, A., Deino, A. L., and Kowallis, B. J. (1997). High-precision $^{40}Ar/^{39}Ar$ geochronology and the advent of North America's Late Cretaceous terrestrial fauna. *Proceedings of the National Academy of Sciences, USA*, 94, 11 163–7.

Clemens, W. A. (1963). Fossil mammals of the type Lance Formation Wyoming. Part I. Introduction and Multituberculata. *University of California Publications in Geological Sciences*, 48, 1–105.

(1973a). Fossil mammals of the type Lance Formation Wyoming. Part III. Eutheria and summary. *University of California Publications in Geological Sciences*, 94, 1–102.

(1973b). The roles of fossil vertebrates in interpretation of Late Cretaceous stratigraphy of the San Juan Basin, New Mexico, In *Cretaceous and Tertiary Rocks of the Southern Colorado Plateau*, ed. J. E. Fassett, pp. 154–67. Durango, CO: Four Corners Geological Society.

Clemens, W. A. and Kielan-Jaworowska, Z. (1979). Multituberculata, In *Mesozoic Mammals: The First Two-thirds of Mammalian History*, ed. J. A. Lillegraven, Z. Kielan-Jaworowska, and W. A. Clemens, W. A., pp. 99–149. Berkeley, CA: University of California Press.

Cope, E. D. (1881). Eocene Plagiaulacidae. *The American Naturalist*, 15, 921–2.

(1882a). A new genus of Taeniodonta. *The American Naturalist*, 16, 604–65.

(1882b). A second genus of Eocene Plagiaulacidae. *The American Naturalist*, 16, 416–17.

(1884a). The Tertiary Marsupialia. *The American Naturalist*, 18, 686–97.

(1884b). Second addition to the knowledge of the Puerco fauna. *Proceedings of the American Philosophical Society*, 21, 309–24.

(1885). Marsupials from the Lower Eocene of New Mexico. *The American Naturalist*, 19, 493–4.

(1886). Plagiaulacidae of the Puerco epoch. *The American Naturalist*, 20, 451.

(1887). The marsupial genus *Chirox*. *The American Naturalist*, 21, 566–7.

Dorr, J. A., Jr. (1952). Early Cenozoic stratigraphy and vertebrate paleontology of the Hoback Basin, Wyoming. *Bulletin of the Geological Society of America*, 63, 59–94.

Douglass, E. (1908). Vertebrate fossils from the Fort Union beds. *Annals of the Carnegie Museum*, 5, 11–26.

Eaton, J. G. (1995). Cenomanian and Turonian (early Late Cretaceous) multituberculate mammals from southwestern Utah. *Journal of Vertebrate Paleontology*, 15, 761–84.

Eaton, J. G. and Cifelli, R. L. (2001). Multituberculate mammals from near the Early–Late Cretaceous boundary, Cedar Mountain Formation, Utah. *Acta Paleontologica Polonica*, 46, 453–518.

Engelmann, G. F. and Callison, G. (1999). *Glirodon grandis*, a new multituberculate mammal from the Upper Jurassic Morrison Formation. In *Vertebrate Paleontology in Utah*, ed. D. D. Gillette, pp. 161–77. Salt Lake City, UT: Utah Geological Survey.

Fox, R. (1968). Studies of Late Cretaceous vertebrates. II. Generic diversity among multituberculates. *Systematic Zoology*, 17, 339–42.

(1989). The Wounded Knee local fauna and mammalian evolution near the Cretaceous–Tertiary boundary, Saskatchewan, Canada. *Palaeontographica Abteilung A*, 208, 11–59.

(1999). The monophyly of the Taeniolabidoidea (Mammalia: Multituberculata). *Proceedings of the VII International Symposium on Mesozoic Terrestrial Ecosystems*, p. 26.

(2005). Microcosmodontid multituberculates (Allotheria, Mammalia) from the Paleocene and Late Cretaceous of western Canada. *Palaeontographica Canadiana*, 23, 1–109.

Gambaryan, P. P. and Kielan-Jaworowska, Z. (1995). Masticatory musculature of Asian taeniolabidoid multituberculate mammals. *Acta Palaeontologica Polonica*, 40, 45–108.

Granger, W., and Simpson, G. G. (1928). Multituberculates in the Wasatch Formation. *American Museum Novitates*, 312, 1–4.

(1929). A revision of the Tertiary Multituberculata. *Bulletin of the American Museum of Natural History*, 56, 601–79.

Greenwald, N. S. (1988). Patterns of tooth eruption and replacement in multituberculate mammals. *Journal of Vertebrate Paleontology*, 8, 265–77.

Hahn, G. and Hahn, R. (1983). *Multituberculata, Fossilium Catalogus I: Animalia*. Amsterdam: Kugler.

(2003). New multituberculate teeth from the Early Cretaceous of Morocco. *Acta Palaeontologica Polonica*, 48, 349–56.

Higgins, P. (2003a). A Wyoming succession of Paleocene mammal-bearing localities bracketing the boundary between the Torrejonian and Tiffanian North American Land Mammal "Ages." *Rocky Mountain Geology*, 38, 247–80.

(2003b). A new species of Paleocene multituberculate (Mammalia: Allotheria) from the Hanna Basin, South-Central Wyoming. *Journal of Vertebrate Paleontology*, 23, 468–70.

Holtzman, R. C. (1978). Late Paleocene mammals of the Tongue River Formation, western North Dakota. *North Dakota Geological Survey Report of Investigations*, 65, 1–88.

Holtzman, R. C. and Wolberg, D. L. (1977). The Microcosmodontinae and *Microcosmodon woodi*, new multituberculate taxa (Mammalia) from the Late Paleocene of North America. *Scientific Publications of the Science Museum of Minnesota*, 4, 1–13.

Hopson, J. A., Kielan-Jaworowska, Z., and Allin, E. F. (1989). The cryptic jugal of multituberculates. *Journal of Vertebrate Paleontology*, 9, 201–9.

Jenkins, F. A. and Krause, D. W. (1983). Adaptations for climbing in North American multituberculates (Mammalia). *Science*, 220, 712–15.

Jenkins, F. A., Jr. and McClearn, D. (1984). Mechanisms of hind foot reversal in climbing mammals. *Journal of Morphology*, 182, 197–219.

Jepsen, G. (1930a). New vertebrate fossils from the lower Eocene of the Bighorn Basin, Wyoming. *Proceedings of the American Philosophical Society*, 69, 117–31.

(1930b). Stratigraphy and paleontology of the Paleocene of northeastern Park County, Wyoming. *Proceedings of the American Philosophical Society*, 83, 463–528.

(1940). Paleocene faunas of the Polecat Bench Formation, Park County, Wyoming. *Proceedings of the American Philosophical Society*, 83, 217–338.

Johnston, P. A. and Fox, R. C. (1984). Paleocene and Late Cretaceous mammals from Saskatchewan, Canada. *Palaeontographica Abteilung A*, 186, 163–222.

Kielan-Jaworowska, Z. (1971). Skull structure and affinities of the Multituberculata. *Acta Palaeontologia Polonica*, 25, 5–41.

(1974a). Multituberculate succession in the Late Cretaceous of the Gobi desert (Mongolia). [*Results of the Polish–Mongolian Palaeontological Expeditions, Part V*, ed. Kielan-Jaworowska.] *Palaeontologia Polonica*, 30, 23–44.

(1974b). Migrations of the Multituberculata and the Late Cretaceous connections between Asia and North America. *Annals of the South African Museum*, 64, 231–43.

(1979). Pelvic structure and nature of reproduction in Multituberculata. *Nature*, 277, 402–3.

Kielan-Jaworowska, Z. and Gambaryan, P. P. (1994). Postcranial anatomy and habits of Asian multituberculate mammals. *Fossils and Strata*, 36, 1–92.

Kielan-Jaworowska, Z. and Hurum, J. H. (1997). Djadochtatheria; a new suborder of multituberculate mammals. *Acta Palaeontologica Polonica*, 42, 201–42.

(2001). Phylogeny and systematics of multituberculate mammals. *Palaeontology*, 44, 389–429.

Kielan-Jaworowska, Z. and Nessov, L. A. (1992). Multituberculate mammals from the Cretaceous of Uzbekistan. *Acta Palaeontologica Polonica*, 37, 1–17.

Kielan-Jaworowska, Z. and Sloan, R. E. (1979). *Catopsalis* (Multituberculata) from Asia and North America and the problem of taeniolabidid dispersal in the Late Cretaceous. *Acta Palaeontologica Polonica*, 24, 187–97.

Kielan-Jaworowska, Z., Cifelli, R. L., and Luo, Z. (2004). *Mammals from the Age of Dinosaurs: Origins, Evolution, and Structure*. New York: Columbia University Press.

Krause, D. W. (1977). Paleocene multituberculates (Mammalia) of the Roche Percée Local Fauna, Ravenscrag Formation, Saskatchewan, Canada. *Palaeontographica Abteilung A*, 159, 1–36.

(1980). Multituberculates from the Clarkforkian Land-Mammal Age, Late Paleocene–Early Eocene, of western North America. *Journal of Paleontology*, 54, 1163–83.

(1982a). Multituberculates from the Wasatchian Land-Mammal Age, early Eocene, of Western North America. *Journal of Paleontology*, 56, 271–94.

(1982b). Evolutionary history and paleobiology of Early Cenozoic Multituberculata (Mammalia) with emphasis on the family Ptilodontidae. Ph.D. Thesis, University of Michigan, Ann Arbor.

(1982c). Jaw movement, dental function, and diet in the Paleocene multituberculate *Ptilodus*. *Paleobiology*, 8, 265–81.

(1986). Competitive exclusion and taxonomic displacement in the fossil record: the case of rodents and multituberculates in North America. [In *Vertebrates, Phylogeny, and Philosophy*, ed. K. M. Flanagan and J. A. Lillegraven.] *Contributions to Geology, University of Wyoming Special Paper*, 3, 95–118.

(1987a). *Baiotomeus*, a new ptilodontid multituberculate (Mammalia) from the middle Paleocene of western North America. *Journal of Paleontology*, 61, 595–603.

(1987b). Systematic revision of the genus *Prochetodon* (Ptilodontidae, Multituberculata) from the late Paleocene and early Eocene of Western North America. *Contributions from the Museum of Paleontology, University of Michigan*, 27, 221–36.

(1992). *Clemensodon megaloba*, a new genus and species of Multituberculata (Mammalia) from the Upper Cretaceous type Lance Formation, Powder River Basin, Wyoming. *PaleoBios*, 14, 1–8.

Krause, D. W. and Carlson, S. J. (1986). The enamel ultrastructure of multituberculate mammals: a review. *Scanning Electron Microscopy*, IV, 1591–607.

(1987). Prismatic enamel in multituberculate mammals: tests of homology and polarity. *Journal of Mammalogy*, 68, 755–65.

Krause, D. W. and Jenkins, F. A., Jr. (1983). The postcranial skeleton of North American multituberculates. *Bulletin of the Museum of Comparative Zoology*, 150, 199–246.

Krause, D. W. and Kielan-Jaworowska, Z. (1993). The endocranial cast and encephalization quotient of *Ptilodus* (Multituberculata, Mammalia). *Palaeovertebrata*, 22, 99–112.

Krause, D. W., O'Connor, P. M., Curry Rogers, K., *et al.* (2006). Late Cretaceous terrestrial vertebrates from Madagascar: implications for Latin American biogeography. *Annals of the Missouri Botanical Garden*, 93, 178–208.

Kühne, W. G. (1969). A multituberculate from the Eocene of the London basin. *Proceedings of the London Geological Society*, 1658, 199–202.

Lemoine, V. (1880). Sur les ossements fossiles des terrains tertiaires inférieurs des environs de Reims. *Comptes Rendus Association Française pour l'avancement des Sciences*, VIII, 585–94.

(1882). Sur deux *Plagiaulax* tertiaires, recueilleis aux environs de Reims. *Comptes Rendus Academie des Sciences*, XCV, 1009–11.

Lillegraven, J. A. (1969). Latest Cretaceous mammals of the upper part of Edmonton Formation of Alberta, Canada, and review of marsupial–placental dichotomy in mammalian evolution. *University of Kansas Paleontological Contributions*, 50, 1–122.

Lillegraven, J. A. and McKenna, M. C. (1986). Fossil mammals from the "Mesaverde" Formation (Late Cretaceous, Judithian) of the Bighorn and Wind River Basins, Wyoming, with Definitions of Late Cretaceous North American Land Mammal "Ages." *American Museum Novitates*, 2840, 1–68.

Lofgren, D. L. (1995). The Bug Creek problem and the Cretaceous-Tertiary transition at McGuire Creek, Montana. *University of California Publications in Geological Sciences*, 140, 1–185.

Lofgren, D. L., Scherer, B. E., Clark, C. K., and Standhardt, B. (2005). First record of *Stygimys* (Mammalia, Multituberculata, Eucosmodontidae) from the Paleocene (Puercan) part of the North Horn Formatin, Utah, and a review of the genus. *Journal of Mammalian Evolution*, 12, 77–97.

Marsh, O. C. (1889). Discovery of Cretaceous Mammalia, Part II. *American Journal of Science*, 3, 177–80.

Matthew, W. D. (1897). A revision of the Puerco fauna. *Bulletin of the American Museum of Natural History*, 9, 259–323.

Matthew, W. D. and Granger, W. (1921). New genera of Paleocene mammals. *American Museum Novitates*, 13, 1–7.

McKenna, M. C. and Bell, S. K. (1997). *Classification of Mammals Above the Species Level*. New York: Columbia University Press.

Miao, D. (1986). Dental anatomy and ontogeny of *Lambdopsalis bulla* (Mammalia, Multituberculata). *Contributions to Geology, University of Wyoming*, 24, 65–76.

(1988). Skull morphology of *Lambdopsalis bulla* (Mammalia, Multituberculata) and its implications to mammalian evolution. *Contributions to Geology, University of Wyoming, Special Paper* 4, 1–104.

Middleton, M. D. (1982). A new species and additional material of *Catopsalis* (Mammalia: Multituberculata) from the Western Interior of North America. *Journal of Paleontology*, 56, 1197–206.

(1983). Early Paleocene vertebrates of the Denver Basin, Colorado. Thesis, Ph.D. University of Colorado, Boulder.

Middleton, M. D. and Dewar, E. W. (2004). New mammals from the Early Paleocene Littleton Fauna (Denver Formation, Colorado). [In *Paleogene Mammals*, ed. S. G. Lucas and K. E. Zeigler.] *Bulletin of the New Mexico Museum of Natural History and Science*, 26, 59–80.

Montellano, M. (1992). Mammalian fauna of the Judith River Formation (Late Cretaceous, Judithian), north central Montana. *University of California Publications in Geological Sciences*, 136, 1–115.

Montellano, M., Weil, A., and Clemens, W. A. (2000). An exceptional specimen of *Cimexomys judithae* (Mammalia: Multituberculata) from the Campanian Two Medicine Formation of Montana, and the phylogenetic status of *Cimexomys*. *Journal of Vertebrate Paleontology*, 20, 333–40.

Râdulescu, C. and Samson, P.-M. (1996). The first multituberculate skull from the Late Cretaceous (Maastrichtian) of Europe (Hateg Basin, Romania). *Anuarul Institutului de Geologie al României, Supplement 1*, 69, 177–8.

Rigby, J. K., Jr. (1980). Swain Quarry of the Fort Union Formation, middle Paleocene (Torrejonian), Carbon County, Wyoming. Geologic setting and mammalian fauna. *Evolutionary Monographs*, 3, 1–162.

Rougier, G. W., Novacek, M. J., and Dashzeveg, D. (1997). A new multituberculate from the Late Cretaceous locality Ukhaa Tolgod, Mongolia. Considerations on multituberculate interrelationships. *American Museum Novitates*, 3191, 1–26.

Russell, L. S. (1926). A new genus of the species *Catopsalis* from the Paskapoo Formation of Alberta. *American Journal of Science*, 12, 230–4.

(1929). Paleocene vertebrates from Alberta. *American Journal of Science*, 17, 162–78.

Savage, D. E. and Russell, D. E. (1983). *Mammalian Paleofaunas of the World*. Reading, MA: Addison-Wesley.

Schiebout, J. A. (1974). Vertebrate paleontology and paleoecology of the Paleocene Black Peaks Formation, Big Bend National Park, Texas. *Bulletin of the Texas Memorial Museum*, 24, 1–88.

Scott, C. S. (2003). Late Torrejonian (Middle Paleocene) mammals from south central Alberta, Canada. *Journal of Paleontology*, 77, 745–68.

(2004). A new species of the ptilodontid multituberculate *Prochetodon* (Mammalia, Allotheria) from the Paleocene Paskapoo Formation of Alberta, Canada. *Canadian Journal of Earth Sciences*, 41, 237–46.

(2005). New neoplagiaulacid multituberculates (Mammalia: Allotheria) from the Paleocene of Alberta, Canada. *Journal of Paleontology*, 1189–213.

Scott, C. S., Fox, R. C., and Youzwyshyn, G. P. (2002). New earliest Tiffanian (late Paleocene) mammals from Cochrane 2, southwestern Alberta, Canada. *Acta Palaeontologica Polonica*, 47, 691–704.

Secord, R. (1998). Paleocene mammalian biostratigraphy of the Carbon Basin, southeastern Wyoming, and age constraints on local phases of tectonism. *Rocky Mountain Geology*, 33, 119–54.

Sereno, P. C. and McKenna, M. C. (1995). Cretaceous multituberculate skeleton and the early evolution of the mammalian shoulder girdle. *Nature*, 377, 144–7.

Sigogneau-Russell, D. (1991). First evidence of Multituberculata (Mammalia) in the Mesozoic of Africa. *Neues Jahrbuch fur Palaontologie, Monatshefte*, 1991, 119–25.

Simmons, N. B. (1987). A revision of *Taeniolabis* (Mammalia: Multituberculata), with a new species from the Puercan of eastern Montana. *Journal of Paleontology*, 61, 794–808.

(1993). Phylogeny of Multituberculata. In *Mammal Phylogeny: Mesozoic Differentiation, Multituberculates, Monotremes, Early Therians and Marsupials*, ed. F. S. Szalay, M. J. Novacek, and M. C. McKenna, pp. 146–64. New York: Springer-Verlag.

Simmons, N. B. and Desui, M. (1986). Paraphyly in *Catopsalis* (Mammalia: Multituberculata) and its biogeographic implications. [In *Vertebrates, Phylogeny, and Philosophy, Contributions to Geology*, ed, K. M. Flanagan and J. A. Lillegraven.] *Contributions to Geology, University of Wyoming, Special Paper* 3, 87–94.

Simpson, G. G. (1928). A new mammalian fauna from the Fort Union of southern Montana. *American Museum Novitates*, 297, 1–15.

(1935). New Paleocene mammals from the Fort Union of Montana. *Proceedings of the United States National Museum*, 83, 221–44.

(1936a). Additions to the Puerco Fauna, Lower Paleocene. *American Museum Novitates*, 848, 1–11.

(1936b). A new fauna from the Fort Union of Montana. *American Museum Novitates*, 873, 1–27.

(1937). Skull structure of the Multituberculata. *Bulletin of the American Museum of Natural History*, 73, 727–63.

Simpson, G. G. and Elftman, H. O. (1928). Hind limb musculature and habits of a Paleocene multituberculate. *American Museum Novitates*, 333, 1–19.

Sinclair, W. J. and Granger, W. (1914). Paleocene deposits of the San Juan Basin, New Mexico. *Bulletin of the American Museum of Natural History*, 33, 297–316.

Sloan, R. E. (1966). Paleontology and geology of the Badwater Creek area, central Wyoming, Part 2. The Badwater multituberculate. *Annals of the Carnegie Museum*, 38, 309–15.

(1979). Multituberculata. In *The Encyclopedia of Paleontology*, ed. R. W. Fairbridge and D. Jablonski, pp. 492–8. New York: Dowden, Hutchinson and Ross.

(1981). Systematics of Paleocene multituberculates from the San Juan Basin, New Mexico. In *Advances in San Juan Basin Paleontology*, ed. S. G. Lucas, K. J. Rigby, Jr., and B. S. Kues, pp. 127–60. Albuquerque, NM: University of New Mexico Press.

(1987). Paleocene and latest Cretaceous mammal ages, biozones, magnetozones, rates of sedimentation, and evolution. [In *The Cretaceous-Tertiary Boundary in the San Juan and Raton Basins, New Mexico and Colorado*, ed. J. E. Fassett and J. K. Rigby, Jr.] *Geological Society of America Special Paper* 209, 165–200.

Sloan, R. E. and Van Valen, L. (1965). Cretaceous mammals from Montana. *Science*, 148, 220–7.

Szalay, F. S. (1965). First evidence of tooth replacement in the subclass Allotheria (Multituberculata). *American Museum Novitates*, 2226, 1–12.

Van Valen, L. and Sloan, R. E. (1966). The extinction of the multituberculates. *Systematic Zoology*, 15, 261–78.

Vianey-Liaud, M. (1986). Les Multituberculés Thanetiens de France, et leur rapports avec le Multituberculés Nord-Americains. *Palaeontographica Abteilung A*, 191, 85–171.

Wall, C. E. and Krause, D. W. (1992). A biomechanical analysis of the masticatory apparatus of *Ptilodus* (Multituberculata). *Journal of Vertebrate Paleontology*, 12, 172–87.

Weil, A. (1998). A new species of *Microcosmodon* (Mammalia: Multituberculata) from the Paleocene Tullock Formation of Montana, and an argument for the Microcosmodontinae. *PaleoBios*, 18, 1–15.

(1999). Multituberculate phylogeny and mammalian biogeography in the Late Cretaceous and earliest Paleocene western interior of North America. Ph.D. Thesis, University of California, Berkeley.

Weil, A. and Tomida, Y. (2001). First description of the skull of *Meniscoessus robustus* expands known morphological diversity of Multituberculata and deepens phylogenetic mystery. *Journal of Vertebrate Paleontology*, 31(suppl. to no. 3), p. 112A.

Wible, J. R. and Rougier, G. W. (2000). Cranial anatomy of *Kryptobaatar dashzevegi* (Mammalia, Multituberculata), and its bearing on the evolution of mammalian characters. *Bulletin of the American Museum of Natural History*, 247, 1–124.

Williamson, T. E. and Lucas, S. G. (1993). Paleocene vertebrate paleontology of the San Juan Basin, New Mexico. [In *Vertebrate Paleontology in New Mexico*, ed. S. G. Lucas and J. Zidek.] *Bulletin of the New Mexico Museum of Natural History and Science*, 2, 105–35.

Wilson, R. W. (1956). A new multituberculate from the Paleocene Torrejon fauna of New Mexico. *Transactions of the Kansas Academy of Science*, 59, 76–84.

(1987). Late Cretaceous (Fox Hills) multituberculates from the Red Owl Local Fauna of western South Dakota. [In *Papers in Vertebrate Paleontology in Honor of Morton Green*, ed. J. E. Martin and G. E. Ostrander. *Dakoterra*, 3, 118–32.

Wing, S. L. and Tiffney, B. H. (1987). Interactions of angiosperms and herbivorous tetrapods through time. In *The Origins of Angiosperms and their Biological Consequences*, ed. E. M. Friis and W. L. Crepet, pp. 203–24. Cambridge, UK: Cambridge University Press.

3 Marsupialia

WILLIAM W. KORTH
Rochester Institute of Vertebrate Paleontology, Rochester, NY, USA

INTRODUCTION

Marsupials have been identified first in North America beginning in the Cretaceous (Clemens, 1979; Johanson, 1996; Cifelli and Muizon, 1997; Cifelli, 1999). The Tertiary record begins in the Paleocene and extends to the early middle Miocene (Barstovian – locality GC4C; Slaughter, 1978). There is no further record of marsupials until the late Pliocene (Kurtén and Anderson, 1980).

North American Tertiary marsupials were never very diverse either morphologically or taxonomically. Although marsupials are included in nearly all Paleocene through Oligocene faunal samples, they represent only a small percentage of the collected specimens from any fossil locality. Nearly all of the Tertiary record of marsupials consists of dental elements. By the early Miocene, marsupials are extremely rare. The early middle Miocene (Barstovian) record consists of only a single isolated molar (Slaughter, 1978; Korth, 1994). The latest reviews of North American marsupials were presented by Krishtalka and Stucky (1983a) for the Paleocene and Eocene species (Tiffanian–Duchesnean) and Korth (1994) for the Oligocene and Miocene (Chadronian–Hemingfordian).

DEFINING FEATURES OF THE COHORT MARSUPIALIA

There are a number of major differences in physiology and overall morphology of the skeleton and body between the marsupials and all placental mammals (e.g., Lillegraven, 1969; Nowak, 1991; Szalay, 1993). However, the fossil record of marsupials in North America consists almost entirely of dental elements. For this reason, the features discussed here will be limited to teeth and jaws (Figure 3.1). All of these taxa probably had a generalized type of terrestrial or scansorial postcranial morphology, like that of extant didelphid marsupials.

The dental formula for all North American marsupials, as far as it is known, is the primitive formula for all marsupials: I5(?)/4, C1/1, P3/3, M4/4. This is easily distinguishable from even the most primitive eutherian mammals (primitive dental formula I3/3, C1/1, P4/4, M3/3), which have fewer molars and incisors and more premolars. All North American marsupials retain pointed cuspate cheek teeth with little or no modification from the primitive condition. The crowns of the teeth never become lophate or hypsodont in any species. The premolars never attain the level of molarization of many eutherians, maintaining a conservative premolariform shape.

The lower molars of marsupials can be distinguished from those of eutherians by the close positioning of the entoconid and hypoconulid on the talonids, often referred to as "twinning." The upper molars of marsupials differ from those of eutherian mammals in having a much wider stylar shelf (generally with an ectoflexus) and well-developed stylar cusps. Marsupial upper molars also lack a hypocone. The mandible is generally very slender with multiple mental foramina and a characteristic medial flange of the angle of the dentary.

SYSTEMATICS

SUPRAFAMILIAL

North American Tertiary marsupials are all very small-sized species and show very conservative changes throughout their fossil record occurrence. They can be divided into three recognizable groups based on dental morphology: herpetotheriines, peradectines, and didelphines. These groups either have been classified as tribes or subfamilies of the family Didelphidae (Crochet, 1980; McKenna and Bell, 1997); or the peradectines have been included in their own family Peradectidae (Reig, Kirsch, and Marshall, 1985), or even as a superfamily (Marshall, Case, and Woodburne, 1990) (see also discussion in Case, Godin, and Woodburne, 2005).

Evolution of Tertiary Mammals of North America, Vol. 2. ed. C. M. Janis, G. F. Gunnell, and M. D. Uhen. Published by Cambridge University Press.

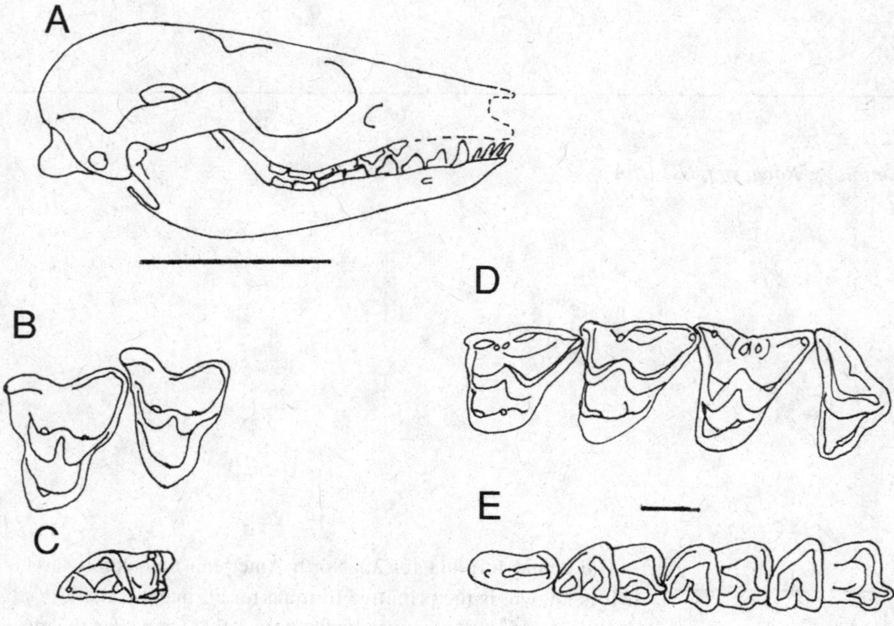

Figure 3.1. Skull and dentitions of Tertiary marsupials. A. Reconstruction of the skull of Orellan peradectine *Nanodelphys hunti* (based on USNM 11955, as per description by Gazin [1935]). B. Left M2–3 of the Orellan peradectine *Nanodelphys hunti* (from McGrew, 1937, his Fig. 3.4). C. Right m1 of the Duchesnean peradectine *Peradectes californicus* (from Rothecker and Storer, 1996, their Fig. 3.1Q). D. Left M1–4 of the Orellan herpetotheriine *Herpetotherium fugax* (redrawn from Green and Martin, 1976, their Fig. 3.2). E. Right p3–m3 of *H. fugax* (redrawn from Korth, 1994, his Fig. 3.3). All figures not to same scale. Scale bar below A = 1 cm. Others = 1 mm. B and C to same scale (below left), D and E to same scale. (A, E: courtesy of the Society for Sedimentary Geology. B: courtesy of the University of Chicago Press. C: courtesy of the Society of Vertebrate Paleontology.)

The systematics of North American marsupials above the family level recently has been in question. The variation in the systematics has been based on the allocation of the peradectines and Cretaceous genera rather than the distinction between the recognizable Tertiary subfamilies (or families). The herpetotheriines and didelphines have been considered as subfamilies of the Didelphidae by all authors.

When first proposed, the Peradectinae was intended only to include the Tertiary genus *Peradectes* (Crochet, 1979). Reig, Kirsch, and Marshall (1985, 1987) raised the rank of the peradectines to the family level and included the Cretaceous genera *Alphadon* and *Albertatherium*. Marshall, Case, and Woodburne (1990) erected a superordinal rank (cohort), Ameridelphia, which included most of the marsupials from North and South America and Eurasia, but excluded the peradectines *Alphadon* and *Albertatherium*. The Cretaceous genera were included with the Tertiary peradectines in the cohort Alphadelphia. Johanson (1996) demonstrated that there were shared, derived morphologies of the peradectines and didelphids that separated the former from the Cretaceous alphadontines, and included the peradectines along with the didelphids in the Ameridelphia, leaving the Cretaceous species as a separate group, the Alphadelphia. The most recent classification of marsupials (McKenna and Bell, 1997) included the alphadontines, peradectines, didelphines, and herpetotheriines as subfamilies of the Didelphidae.

INFRAFAMILIAL

Crochet (1979) demonstrated a number of dental morphologies that separated herpetotheriines from peradectines. On the upper molars: (1) herpetotheriines have a wider stylar shelf with more pronounced stylar cusps; (2) herpetotheriines have a V-shaped centrocrista (dilambdodonty) that is lacking in peradectines; (3) the conules are larger on herpetotheriines; (4) the metacone is taller than the paracone (cusps subequal in peradectines on lower molars); and (5) the hypoconulid is posteriorly projecting on herpetotheriines and nearly vertical on peradectines (Figure 3.2).

In addition to the different morphologies of the cheek teeth, there may also be some differences in the anterior dentitions of peradectines and herpetotheriines. The lower incisors of *Herpetotherium* and *Copedelphys* are specialized over those of the peradectines. In both of the former herpetotheriines, the first two lower incisors are greatly enlarged (compared with i3 and i4) and procumbent (Fox, 1983; Korth, 1994). In *Peradectes*, the first two incisors are the largest but are closer in size to the last two incisors and nearly vertically implanted (Fox, 1983). Also, in another peradectine, *Nanodelphys*, the first three lower incisors are subequal in size (not greatly enlarged) and also nearly vertically oriented (Gazin, 1935). However, the enlargement and horizontal orientation of the first two incisors of the North American herpetotheriines is not necessarily the same in European didelphids, where these incisors are enlarged but not procumbent (Crochet, 1980).

Herpetotheriines and peradectines appear to be separate throughout their occurrence in the Tertiary. Species of these subfamilies commonly co-occur, but peradectines, although they appear first in the fossil record (Cretaceous), become extinct by the latest Oligocene (Arikareean [Martin, 1973]), before the last occurrence of didelphids in the Barstovian (Slaughter, 1978).

INCLUDED GENERA OF MARSUPIALIA ILLIGER, 1811

The locality numbers listed for each genus refer to the list of unified localities in Appendix I. The locality numbers may be listed more

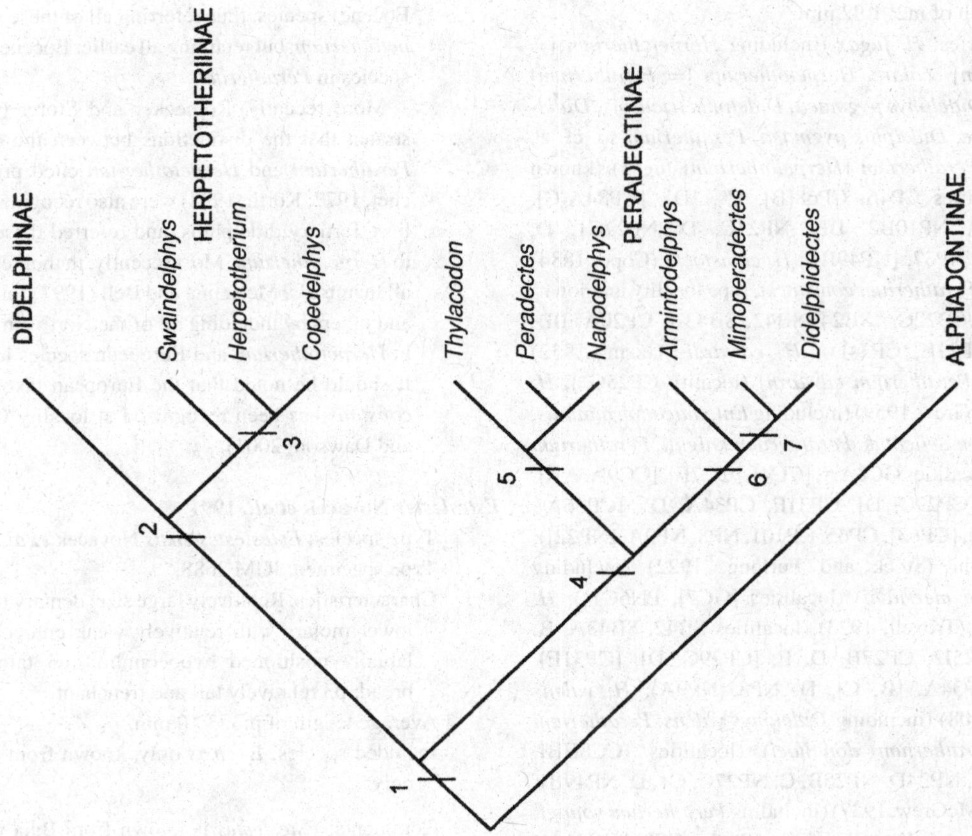

Figure 3.2. Suggested generic relationships of North American Tertiary marsupials. 1. Paracone lower than metacone, stylar cusp A smaller and more closely approximated to B at level of stylar shelf; preparacrista running from paracone to a point anterior to cusp B; conules and internal conular cristae reduced in size; absence of postmetaconular crista; cristid obliqua meeting posterior wall of trigonid labial to protocristid notch (Johanson, 1996, his Fig. 3.3). 2. V-shaped centrocrista on upper molars (dilambdodonty); metacone markedly larger than paracone; hypoconulid projecting posteriorly on lower molars. 3. First two lower incisors enlarged and procumbent. 4. Reduction of styles and conules on upper molars; entoconid on lower molars vertically oriented and closely appressed to hypoconulid. 5. Further progressive reduction of stylar cusps and conules on upper molars. 6. Entoconid on lower molars reduced. 7. Cheek teeth and mandible more massive, premolars crowded.

than one way. The acronyms for museum collections are listed in Appendix II.

Brackets around a locality (e.g., [CP101]) mean that the taxon in question at that locality is cited as an "aff." or "cf." the taxon in question. Parentheses are usually used for individual species, implying that the genus is firmly known from the locality, but the actual species identification may be questionable. Question marks in front of the locality (e.g., ?CP101) mean that the taxon is questionably known from that locality, implying some doubt that the taxon is actually present at that locality, either at the genus or species level. An asterisk (*) indicates the type locality.

HERPETOTHERIINAE

The changes in *Herpetotherium* and *Copedelphys* through their Tertiary record is mainly in overall size and relative size of the stylar cusps of the upper molars (Korth, 1994). In both, their overall size is reduced through time. In *Herpetotherium*, stylar cusp B (primitively the largest) is progressively reduced and the central stylar cusp becomes dominant. It is likely that *Copedelphys* was ultimately derived from an Eocene species of *Herpetotherium*.

Another genus of herpetotheriine, *Swaindelphys*, appears much earlier in the fossil record and is distinct from later herpetotheriines in the more primitive condition of the molars: shorter talonid, hypoconulid not directed posteriorly, no dominant stylar cusp on upper molars (Johanson, 1996). There is no morphology of *Swaindelphys* that would bar it from an ancestral position with either *Herpetotherium* or *Copedelphys*.

Herpetotherium Cope, 1873 (including *Peratherium*, in part; *Didelphis* [including *Didelphys*], in part; *Entomaconodon*; *Centraconodon*)

Type species: *Herpetotherium fugax* Cope, 1873.

Type specimen: AMNH 5254.

Characteristics: Central, dominant stylar cusp on upper molars (cusp D on M1–2, cusp C on M3–4); stylar cusps C and D progressively fuse on M1–2; stylar cusp B progressively reduced; i1–2 enlarged and procumbent.

Average length of m2: 1.92 mm.

Included species: *H. fugax* (including *Herpetotherium* [= *Peratherium*] *scalare*, *Herpetotherium* [= *Peratherium*] *tricuspis*, *Didelphys pygmaea*, *Didelphis tricuspis*, *Didelphis scalare*, *Didelphis pygmaea*, *Peratherium* sp. cf. *P. spindleri*, *Peratherium* [*Herpetotherium*] *fugax*) (known from localities CP46, CP68[B], C*, [D], [CP84A-C], CP98B, C, NP10B2, BB, NP24C, D, NP27C1, D, NP32B, C, NP37, [NP49II]); *H. comstocki* (Cope, 1884) (including *Peratherium comstocki*, type locality unknown) (localities ?SB22C, ?SB24, SB42, SB43A, CP20A, BB, CP25A, CP27E, CP34D); *H. edwardi* (Gazin, 1952) (including *Peratherium edwardi*) (locality CP25G*); *H. knighti* (McGrew, 1959) (including *Entomacodon minutus*, *Centracodon delicatus*, *Peratherium knighti*, *Peratherium morrisi*) (localities GC8AA, [CC4], [CC7E], [CC9A, AA], CP25I, J, [CP29C, D], CP31E, CP34A, D*, [CP36A], [CP38A, B], [CP63], CP65, CP101, NP8, NP9A, [NP22]); *H. merriami* (Stock and Furlong, 1922) (including *Peratherium merriami*) (localities [GC7], PN6C1*); *H. marsupium* (Troxell, 1923) (localities SB42, SB43A, B, CP5A, CP25I2, CP27B, D, E, [CP29C, D], [CP31E], [CP32], CP34A, [B, C], D, NP8, NP9A); *H. valens* (Lambe, 1908) (including *Didelphys valens*, *Peratherium valens*, *Peratherium donahoei*) (localities [CC8IIB], NP10B, B2, NP24D, NP25B, C, NP27C, C1, D, NP49II); *H. youngi* (McGrew, 1937) (including *Peratherium youngi*, *Peratherium spindleri*) (localities CP85C, CP86B, CP101, CP103B*, NP10CC).

Herpetotherium (including *Peratherium*) sp. is also known from a large number of localities including GC8A, AB, CC6A, CC7A, B, CC8, CC9IVC, ?SB39IIA, SB44A, ?SB46, ?CP13G, CP25A, GG, J2, CP31A, C, CP32B, CP34C, E, CP35, CP36B, ?CP38C, CP39?B, C, ?F, CP64A, ?CP83C, CP84B, C, CP88, CP99C, NP9B, NP10Bi, Bii, C, C2, CC, NP23A, NP25A, NP34A, NP36A, B, D, PN6D3.

Comments: Cope (1873) first named *Herpetotherium* but considered it an insectivore. Later, Cope (1884) recognized that it was a marsupial and referred all of the species he had proposed to the previously named European genus *Peratherium* Aymard (1846), an allocation followed by most later authors for nearly a century. Crochet (1977, 1980) referred all of the North American "*Peratherium*" to Cope's original genus, leaving only the European species in *Peratherium*. Fox (1983) described the anterior dentition of *Herpetotherium*, noting its difference from the European species and allocated only the type species to *Herpetotherium*. This practice was followed by a number of authors (Krishtalka and Stucky, 1983a; Reig, Kirsch, and Marshall, 1985). Korth (1994) described the anterior dentition of a number of North American species, demonstrating that the characters used by Fox (1983) were consistent with several post-Duchesnean (early late Eocene) species, thus referring all of these species to *Herpetotherium*, but retaining all earlier Eocene and Paleocene species in *Peratherium*.

More recently, Rothecker and Storer (1996) demonstrated that the distinctions between the cheek teeth of *Peratherium* and *Herpetotherium* cited previously (Crochet, 1977; Korth, 1994) were also recognizable in the earliest Tertiary didelphids, and referred these early species to *Herpetotherium*. Most recently, in their classification of all mammals, McKenna and Bell (1997) followed Crochet and others by including all of the North American species in *Herpetotherium* and European species in *Peratherium*. It should be noted that the European taxon *Peratherium constans* has been recognized at locality GC21II (Beard and Dawson, 2001).

Esteslestes Novacek _et al._, 1991

Type species: *Esteslestes ensis* Novacek *et al.*, 1991.

Type specimen: IGM 3688.

Characteristics: Relatively large size; dentary relatively deep; lower molars with relatively weak entoconid notch and labially positioned hypoconulid; m3 talonid relatively broad; p3 relatively tall and trenchant.

Average length of m3: 3.10 mm.

Included species: *E. ensis* only, known from locality CC50 only.

Comments: *Esteslestes* is known from Baja California and is represented by the type specimen only. Novacek *et al.* (1991) originally placed it in the tribe Dildelphini but their concept of that tribe is similar to what is recognized here as Herpetotheriinae (these authors also included North American *Herpetotherium* within their Didelphini). *Esteslestes* does differ from other known herpetotheriines in having a much deeper and more robust dentary and more labially positioned lower molar hypoconulids, both features that are in common with extant *Didelphis*. Novacek *et al.* (1991) also noted some character states that are similar to South American marsupials, especially *Mirandotherium*. It is possible that *Esteslestes* belongs in a separate subfamily from other North American herpetotheriines but this determination must await the discovery of more complete material.

Copedelphys Korth, 1994 (including _Peratherium_ [including _Herpetotherium_], in part; _Herpetotherium_, in part; _Nanodelphys_, in part)

Type species: *Copedelphys titanelix* (Matthew, 1903) (originally described as *Peratherium titanelix*).

Type specimen: AMNH 9603.

Characteristics: Small size; stylar cusps B, C, and D on upper molars equal in size; stylar cusp D elongated and obliquely oriented; M1 wider relative to length; conules on upper molars minute; p2 and p3 subequal in size; p3 markedly smaller than m1; trigonid of m1 more widely open lingually; i1 and 12 enlarged and procumbent as in *Herpetotherium*.

Average length of m2: 1.26 mm.

Included species: *C. titanelix* (Matthew, 1903) (including *Peratherium titanelix, Peratherium titanohelix*) (known from localities NP10B, NP24C*, D, [NP25B, C], NP27D); *C. innominata* (Simpson, 1928) (including *Peratherium innominatum, Peratherium mcgrewi*) (localities [CC4], CP5A, CP5II, CP20AA, CP25I1, CP27D, E, CP29C, D, [CP31E], CP33B, CP34A–B, C, D, CP63, CP65, NP8, NP9A); *C. stevensoni* (Cope, 1873) (including *Herpetotherium stevensonii, Peratherium huntii*, in part, *Didelphis huntii*, in part, *Herpetotherium stavensoni, Nanodelphys* n. sp., in part) (localities CP46, CP68C*, [NP32B, C]).

Copedelphys sp. is also possibly known from locality CP67.

Comments: The Cedar Ridge Fauna of Setoguchi (1978) (locality CP46), which contained referred specimens of *Copedelphys stevensoni*, is not Whitneyan in age as originally cited but is rather Orellan (Korth, 1989). Although *Peratherium innominatum* Simpson (1928) has been allocated exclusively to the European genus *Peratherium*, Rothecker and Storer (1996) demonstrated that it was a more primitive species of *Copedelphys*, thus extending the record of the genus into the Duchesnean.

***Swaindelphys* Johanson, 1996**

Type species: *Swaindelphys cifellii* Johanson, 1996.

Type specimen: AMNH 100417C, right m3.

Characteristics: Upper molars with greater transverse width to length ratio on M2 and M3; ectoflexus present on M2, small and symmetrical ectoflexus present on M3; anterior stylar shelf wider on M2 and M3; posterior stylar shelf more posterolabially directed than posteriorly directed on M2 and M3; stylar cusps A and B separated by a distinct notch; stylar cusps C and D approximately equal in size; cusp B larger than C and D but no stylar cusp clearly dominant on M1–3; talonid anteroposteriorly short on lower molars; hypoconulid lower than other herpetotheriines and directed posterolabially (talonid basin not open posteriorly).

Average length of m2: 2.05 mm.

Included species: *S. cifellii* only, known from locality CP14A only.

Comments: *Swaindelphys* is the earliest and most primitive species of herpetotheriine from North America. It occurs in the Torrejonian (Johanson, 1996) whereas the earliest occurrence of any other herpetotheriine in North America is Wasatchian (Krishtalka and Stucky, 1983a).

PERADECTINAE

Unlike the herpetotheriines, there is little difference in the size of peradectines throughout their Tertiary occurrence. One lineage or clade of peradectines includes *Thylacodon, Peradectes*, and *Nanodelphys*. Among these genera, there is a reduction in the development of stylar cusps and conules on the upper molars. *Thylacodon*, the early Paleocene genus, has much more distinct stylar cusps and conules (Archibald, 1982), whereas these features are progressively reduced in Eocene *Peradectes* (Krishtalka and Stucky, 1983a) and nearly completely lost in Oligocene *Nanodelphys* (Korth, 1994). There is also a change in the proportions of the upper molars. In *Nanodelphys*, the upper molars are much more transversely elongated than in earlier genera.

Three distinctive genera of peradectines, *Didelphidectes, Armintodelphys*, and *Mimoperadectes*, share the unique feature of a reduced entoconid on m4. Both *Didelphidectes* and *Mimoperadectes* are also characterized by more robust cheek teeth and dentaries, features that are lacking in *Armintodelphys* (Bown and Rose, 1979; Krishtalka and Stucky, 1983b; Korth, 1994). These three genera are clearly distinct and separable from the *Thylacodon–Peradectes–Nanodelphys* clade, but it is uncertain whether they form a distinct clade of their own.

***Peradectes* Matthew and Granger, 1921 (including *Peratherium* [including *Herpetotherium*], in part; *Nanodelphys*, in part)**

Type species: *Peradectes elegans* Matthew and Granger, 1921.

Type specimen: AMNH 17376.

Characteristics: Talonids on m1–3 short; labially positioned cristid obliqua; entoconid and hypoconulid subequal in height on lower molars; M1–3 paracone and metacone subequal in size; stylar cusps and conules on upper molars weak; posterolingual part of the base of protocone not expanded on upper molars.

Average length of m2: 1.82 mm.

Included species: *P. elegans* (known from localities SB20A*, CP13E, G, CP16A, CP24A, CP63, NP3A, C, D, NP4, NP15C, NP16A, NP47B); *P. californicus* (Stock, 1936) (including *P. californicum, Nanodelphys californicus*) (localities CC4, CC6B, CC7B, D, E, CC8, CC8IIA, CC9A*, [AA], B2, CC9IVA, CC10, CP29C, D, NP8, NP9A); *P. chesteri* Gazin, 1952 (localities CP5A, [CP18B], CP20A, ?D, CP25G*, ?I, CP27D, CP34A, B, D); *P. pauli* Gazin, 1956 (localities CP16A*, [NP19IIA]); *P. protinnominatus* McKenna, 1960 (including *Peradectes chesteri*, in part, *Peradectes* cf. *chesteri*) (localities CP14E, CP15B, CP17B, CP20AA, [BB], CP25B, CP27A, CP63*, NP4, [NP20E]).

Peradectes sp. is known from localities CC1, CC4, CC6A, CC7A, CC8IIB, CC9A, SB23A, B, F, H, CP13A, [CP14D], [CP15A], [CP17A], CP20BB, CP25GG, J2, CP31C, CP34C, E, CP36B, NP1C, [NP20B], NP47C, NP49II.

Comments: *Peradectes* is also known from the Late Cretaceous to early Paleocene of South America, from the early Eocene of Europe, and also possibly from the Late Cretaceous of North America (some of these instances may be of *Thylacodon*, see below) (McKenna and Bell, 1997).

Thylacodon Matthew and Granger, 1921 (including _Peradectes_, in part)

Type species: *Thylacodon pusillus* Matthew and Granger, 1921.

Type specimen: AMNH 16414.

Characteristics: Large size for a peradectine; conules and stylar cusps on upper molars large; stylar cusp B is the largest, cusps C and D small to absent; entoconid on lower molars more posterior than in other peradectines, closer to the hypoconulid.

Average length of m2: 2.47 mm.

Included species: *T. pusillus* only (including *Peradectes* cf. *P. pusillus*) (known from localities SB23A*, B, [CP11IIB, G], CP12A, [NP16A]).

Comments: Archibald (1982) considered *Thylacodon* a synonym of *Peradectes*. This suggested synonymy has been followed by some later authors (Marshall, Case, and Woodburne, 1990; McKenna and Bell, 1997) and not by others (Krishtalka and Stucky, 1983a; Johanson, 1996). Krishtalka and Stucky (1983a) felt that the difference between *Thylacodon* and *Peradectes* was so great that they suggested that the former was likely referable to a different subfamily.

Nanodelphys McGrew, 1937 (including _Herpetotherium_, in part; _Miothen_ [= _Domnina_] in part; _Domnina_, in part; _Peratherium_ [= _Herpetotherium_], in part; _Peradectes_, in part)

Type species: *Nanodelphys hunti* (Cope, 1873) (originally described as *Herpetotherium huntii*).

Type specimen: AMNH 5266.

Characteristics: Small size; upper molars relatively wider than long; stylar cusp B only major distinguishable stylar cusp on upper molars, remainder reduced to a low rim along buccal edge of stylar shelf; conules minute to absent on upper molars; trigonids of lower molars narrower and more elongated; entoconids of lower molars not reduced.

Average length of m2: 1.48 mm.,

Included species: *N. hunti* only (including *Herpetotherium huntii*, *Miothen gracile*, *Domnina gracilis*, *Peratherium huntii*, *Didelphis huntii*, *Nanodelphys minutus*, *Herpetotherium fugax*, in part, *Peradectes minutus*) (known from localities CP46, CP68C*, CP84C, CP98B, CP99A, B, [NP8], [NP10BB], NP32C, NP36?A, ?B, D).

Nanodelphys sp. is also known from localities CP41B, CP101 (Martin, 1973), NP10CC.

Mimoperadectes Bown and Rose, 1979

Type species: *Mimoperadectes labrus* Bown and Rose, 1979.

Type specimen: UM 66144.

Characteristics: Large size; cheek teeth more robust than other North American marsupials; paraconids on lower molars larger than metaconids; paraconid less removed anteriorly; metaconid posterolingual to protoconid; trigonid both longer and wider than talonid; talonid width decreases from m2 to m4; entoconids reduced on lower molars; stylar cusp B largest on upper molars, all others small; conules reduced.

Average length of m2: 2.97 mm.

Included species: *M. labrus* only (known from localities GC21II, ?CP18B, CP19A*, A, CP20AA, A, CP62C, CP64B).

Armintodelphys Krishtalka and Stucky, 1983b

Type species: *Armintodelphys blacki* Krishtalka and Stucky, 1983b.

Type specimen: CM 41159.

Characteristics: Entoconid lower and smaller than hypoconulid; talonid narrower than the trigonid on m1–2.

Average length of m2: 2.0 mm.

Included species: *A. blacki* (known from localities CP27D*, E); *A. dawsoni* Krishtalka and Stucky, 1983b (localities CP5A, CP27D, E).

Cf. *Armintodelphys* sp. is also known from locality CP34A.

Didelphidectes Hough, 1961

Type species: *Didelphidectes pumilis* Hough, 1961.

Type specimen: USNM 20084.

Characteristics: Cheek teeth more robust than alphadontines, less robust than *Mimoperadectes*; trigonids on lower molars subequal in width to talonids; entoconids on lower molars reduced; trigonid cusps subequal in size; lower premolars massive, p2 longer than p3; m4 longest of the lower molars.

Average length of m2: 1.65 mm.

Included species: *D. pumilis* only (known from localities NP10B, NP24C*).

Comments: McKenna and Bell (1997) recently considered *Didelphidectes* a junior synonym of *Nanodelphys*. However, the more robust nature of the cheek teeth and reduction of the entoconid on the lower molars of *Didelphidectes* is not present in species of *Nanodelphys*, making the former a distinct genus (Korth, 1994).

MISCELLANEOUS DIDELPHIDAE

Typically Mesozoic marsupials, including *Alphadon rhaister*, *Pediomys hatcheri*, *Pediomys elegans*, *Pediomys florencae*, *Pediomys krejcii*, *Glasbius twitchelli*, and *Didelphodon vorax*, have been recognized from earliest Paleocene localities (NP15C, NP16A) in Montana (Lofgren, 1995).

Unidentified didelphids have been reported from Tertiary localities GC4C, GC8DB, CC1, ?CC9BB, ?SP1D, NP6, NP17, NP34D, PN9B. Other than its numerous occurrences in the Pleistocene and widespread extant distribution, the only Tertiary record of *Didelphis virginiana* in North America is from late Blancan deposits in Florida (locality GC14A).

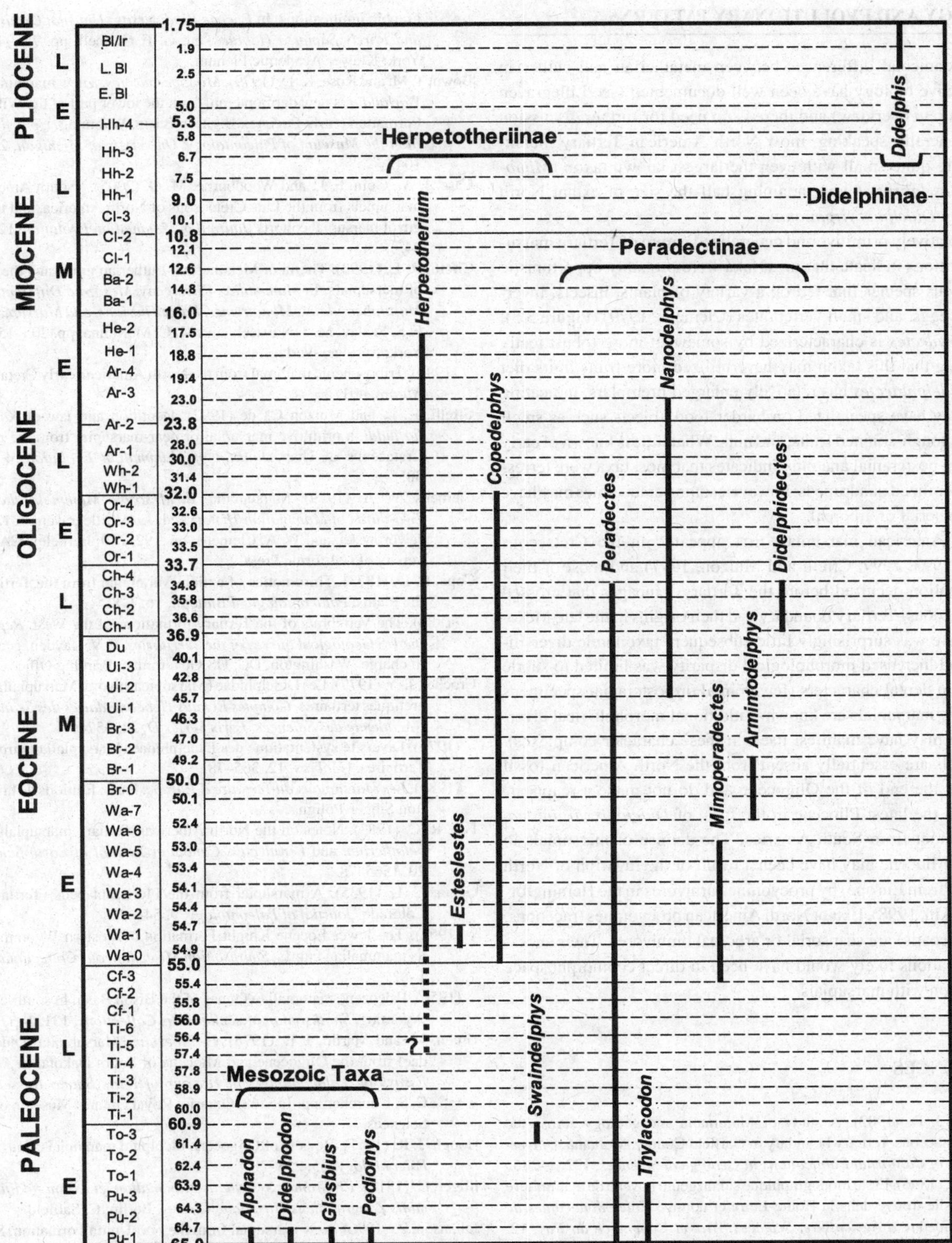

Figure 3.3. Temporal ranges of North American marsupial genera.

BIOLOGY AND EVOLUTIONARY PATTERNS

The fundamental differences between metatherian and eutherian reproductive biology have been well documented (see Lillegraven [1979] for an overview) and there is no need for further discussion here. Generally speaking, most North American Tertiary marsupials were quite small with even the largest known taxon (*Mimoperadectes labrus*) barely attaining half the size of extant North American *Didelphis virginiana*.

The relatively primitive and generalized nature of Tertiary marsupial dentitions indicate that most taxa were probably opportunistic omnivorous species that fed on a variety of plants, insects, invertebrates, eggs, and small vertebrates (Clemens, 1979) (Figure 3.3). *Mimoperadectes* is characterized by somewhat more robust teeth, suggesting that this taxon may have utilized more fruits in its diet while *Didelphidectes* has relatively enlarged premolars, indicating that it may have specialized on harder food objects such as small shellfish, tough-skinned fruits, and nuts. What little is known of early marsupial postcranial anatomy indicates that most taxa were terrestrial, generalized quadrupeds; however, some peradectines may have been scansorial or arboreal.

North American marsupials first appeared in the Cretaceous (Cifelli, 1993, 1999; Cifelli and Muizon, 1997) and most of their diversification occurred before the Tertiary. Lineages that crossed the Cretaceous–Tertiary boundary had their origins in the late Mesozoic. There was surprisingly little subsequent taxonomic diversification and increased morphological disparity was limited to subtle changes in dental characters (few cranial or postcranial specimens of Tertiary marsupials are known – additional morphological diversification may have manifest itself in these character complexes). Marsupials are essentially absent from the North American fossil record by the end of the Oligocene and do not make a reappearance until the latest Pliocene in the form of *Didelphis virginiana*, an immigrant from South America. The demise of marsupials in the early Miocene may have been a result of the invasion of North America (from Europe) by procyonine carnivores in the Hemingfordian (Baskin, 1998). Extant North American procyonines (raccoons, ringtails, coatis) are scansorial or arboreal omnivores. Their ancestral populations likely would have been in direct ecomorphospace competition with marsupials.

REFERENCES

Archibald, J. D. (1982). A study of Mammalia and geology across the Cretaceous-Tertiary boundary in Garfield County, Montana. *University of California Publications in Geological Sciences*, 122, 1–286.

Aymard, A. (1846). Essai monographique sur un genre noveau de mammifére fossile trouvé dans la Houte-Loire et nommé *Entelodon*. *Annes de la Societe d'Agriculture, Sciences, Aret et Commerce du Puy*, 12, 227–67.

Baskin, J. A. (1998). Procyonidae. In *Evolution of Tertiary Mammals of North America*, Vol. 1: *Terrestrial Carnivores, Ungulates, and Ungulatelike Mammals*, ed. C. M. Janis, K. M. Scott, and L. L. Jacobs, pp. 144–51. Cambridge, UK: Cambridge University Press.

Beard, K. C. and Dawson, M. R. (2001). Early Wasatchian mammals from the Gulf Coastal Plain of Mississippi: biostratigraphic and paleobiogeographic implications. In *Eocene Biodiversity: Unusual Occurrences and Rarely Sampled Habitats*, ed. G. F. Gunnell, pp. 75–94. New York: Kluwer Academic/Plenum.

Bown, T. M. and Rose, K. D. (1979). *Mimoperadectes*, a new marsupial, and *Worlandia*, a new dermopteran, from the lower part of the Willwood Formation (early Eocene), Bighorn Basin, Wyoming. *Contributions from the Museum of Paleontology, University of Michigan*, 25, 89–104.

Case, J. A., Goin, F. J., and Woodburne, M. O. (2005). "South American" marsupials from the Late Cretaceous of North America, and the origin of marsupial cohorts. *Journal of Mammalian Evolution*, 12, 461–94.

Cifelli, R. L. (1993). Theria of Metatherian–Eutherian grade and the origin of marsupials. In *Mammalian Phylogeny, Mesozoic Differentiation, Multituberculates, Monotremes, Early Therians, and Marsupials*, ed. F. S. Szalay, M. J. Novacek, and M. C. McKenna, pp. 205–15. New York: Springer-Verlag.

(1999). Tribosphenic mammal from the North American early Cretaceous. *Nature*, 401, 363–6.

Cifelli, R. L. and Muizon C., de (1997). Dentition and jaw of *Kokopellia juddi*, a primitive marsupial or near-marsupial from the medial Cretaceous of Utah. *Journal of Mammalian Evolution*, 4, 241–58.

Clemens, W. A. (1979). Marsupialia. In *Mesozoic Mammals, the First Two-thirds of Mammalian History*, ed. J. A. Lillegraven, Z. Kielan-Jaworowska, and W. A. Clemens, pp. 192–220. Berkeley, CA: University of California Press.

Cope, E. D. (1873). Third notice of extinct Vertebrata from the Tertiary of the Plains. *Palaeontological Bulletin*, 16, 1–8.

(1884). The Vertebrata of the Tertiary formations of the West. *Report of the US Geological Survey of the Territories*, F. V. Hayden, geologist in charge. Washington, DC: US Government Printing Office.

Crochet, J.-Y. (1977). Les Didelphidae (Marsupicarnivora, Marsupialia) holarctiques tertiaires. *Comptes Rendus Hebdomadaires des Seances de l'Academie des Sciences, Paris Series D*, 24, 357–60.

(1979). Diversité systématique des Didelphidae (Marsupialia) européens Tertiaires. *Géobios*, 12, 365–78.

(1980). *Les Marsupiaux du Tertiaire d'Europe*. Paris: Editions de la Fondtion Singer-Polignac.

Fox, R. C. (1983). Notes on the North American Tertiary marsupials *Herpetotherium* and *Peradectes*. *Canadian Journal of Earth Sciences*, 20, 1565–78.

Gazin, C. L. (1935). A marsupial from the Florissant beds (Tertiary) of Colorado. *Journal of Paleontology*, 9, 54–62.

(1952). The lower Eocene Knight Formation of western Wyoming and its mammalian faunas. *Smithsonian Miscellaneous Collections*, 117, 1–82.

(1956). Paleocene mammalian faunas of the Bison Basin in south-central Wyoming. *Smithsonian Miscellaneous Collections*, 131, 1–57.

Green, M. and Martin, J. E. (1976). *Peratherium* (Marsupialia: Didelphidae) from the Oligocene and Miocene of South Dakota. In *Athlon: Essays on Palaeontology in Honour of Loris Shano Russell*, ed. C. S. Churcher, pp. 155–68. Ontario: Royal Ontario Museum of Life Sciences.

Hough, J. R. (1961). Review of Oligocene didelphid marsupials. *Journal of Paleontology*, 35, 218–28.

Illiger, C. (1811). *Prodromus Systematis Mammalium et Avium Additis Terminis Zoographicus Utriudque Classis*. Berlin: C. Salfield.

Johanson, Z. (1996). New marsupial from the Fort Union Formation, Swain Quarry, Wyoming. *Journal of Paleontology*, 70, 1023–31.

Korth, W. W. (1989). Stratigraphic occurrence of rodents and lagomorphs in the Orella Member, Brule Formation (Oligocene), northwestern Nebraska. *Contributions to Geology, University of Wyoming*, 27, 15–20.

(1994). Middle Tertiary marsupials (Mammalia) from North America. *Journal of Paleontology*, 68, 376–97.

Krishtalka, L. and Stucky, R. K. (1983a). Paleocene and Eocene marsupials of North America. *Annals of the Carnegie Museum*, 52, 229–63.

(1983b). Revision of the Wind River Faunas, early Eocene of central Wyoming. Part 3. Marsupialia. *Annals of the Carnegie Museum*, 52, 205–27.

Kurtén, B. and Anderson, E. (1980). *Pleistocene Mammals of North America*. New York: Columbia University Press.

Lambe, L. M. (1908). The Vertebrata of the Oligocene of the Cypress Hills, Saskatchewan. *Contributions to Canadian Paleontology*, 3, 1–65.

Lillegraven, J. A. (1969). Latest Cretaceous mammals of upper part of Edmonton Formation of Alberta, Canada, and review of marsupial–placental dichotomy in mammalian evolution. *University of Kansas Paleontological Contributions, Vertebrata*, 12, 1–122.

(1979). Reproduction in Mesozoic mammals. In *Mesozoic Mammals: The First Two-Thirds of Mammalian History*, ed. J. A. Lillegraven, Z. Kielan-Jaworowska, and W. A. Clemens, pp. 259–76. Berkeley, CA: University of California Press.

Lofgren, D. L. (1995). The Bug Creek problem and the Cretaceous-Tertiary transition at McGuire Creek, Montana. *University of California Publications in Geological Sciences*, 140, 1–185.

Marshall, L. G., Case, J. A., and Woodburne, M. O. (1990). Phylogenetic relationships of the families of marsupials. In *Current Mammalogy*, Vol. 2, ed. H. H. Genoways, pp. 433–505. New York: Plenum Press.

Martin, L. D. (1973). The mammalian fauna of the Lower Miocene Gering Formation of Western Nebraska and the early evolution of the North American Cricetidae. Ph.D. Thesis, University of Kansas, Lawrence.

Matthew, W. D. (1903). The fauna of the *Titanotherium* beds at Pipestone Springs, Montana. *Bulletin of the American Museum of Natural History*, 19, 1–19.

Matthew, W. D. and Granger, W. (1921). New genera of Paleocene mammals. *American Museum Novitates*, 13, 1–17.

McGrew, P. O. (1937). New marsupials from the Tertiary of Nebraska. *Journal of Geology*, 45, 448–55.

(1959). The geology and paleontology of the Elk Mountain and Tabernacle Butte area, Wyoming. *Bulletin of the American Museum of Natural History*, 117, 121–76.

McKenna, M. C. (1960). Fossil Mammalia from the early Wasatchian Four Mile Fauna, Eocene of northwest Colorado. *University of California Publications in Geological Sciences*, 37, 1–130.

McKenna, M. C. and Bell, S. K. (1997). *Classification of Mammals Above the Species Level*. New York: Columbia University Press.

Novacek, M. J., Ferrusquía-Villafranca, I., Flynn, J. J., Wyss, A. R., and Norell, M. (1991). Wasatchian (early Eocene) mammals and other vertebrates from Baja California, Mexico: the Lomas Las Tetas de Cabra Fauna. *Bulletin of the American Museum of Natural History*, 208, 1–88.

Nowak, R. M. (1991). *Walker's Mammals of the World*, 5th edn. Baltimore, MD: Johns Hopkins University Press.

Reig, O. A., Kirsch, J. A. W., and Marshall, L. G. (1985). New conclusions on the relationships of the opossum-like marsupials, with an annotated classification of the Didelphimorphia. *Ameghiniana*, 21, 335–43.

(1987). Systematic relationships of the living and neo-Cenozoic American "opossum-like" marsupials. In *Possums and Opossums; Studies in Evolution*, ed. M. Archer, pp. 1–89. Sydney: Surrey Beatty and Sons and the Royal Zoological Society of New South Wales.

Rothecker, J. and Storer, J. E. (1996). The marsupials of the Lac Pelletier lower fauna, middle Eocene (Duchesnean) of Saskatchewan. *Journal of Vertebrate Paleontology*, 16, 770–4.

Setoguchi, T. (1978). Paleontology and geology of the Badwater Creek area, central Wyoming. Part 16. The Cedar Ridge local fauna (late Oligocene). *Bulletin of the Carnegie Museum of Natural History*, 9, 1–61.

Simpson, G. G. (1928). American Eocene didelphids. *American Museum Novitates*, 307, 1–7.

Slaughter, B. H. (1978). Occurrence of didelphine marsupials from the Eocene and Miocene of the Texas Gulf Coastal Plain. *Journal of Paleontology*, 52, 744–6.

Stock, C. (1936). Sespe Eocene didelphids. *Proceedings of the National Academy of Sciences, Philadelphia*, 22, 122–4.

Stock, C. and Furlong, E. L. (1922). A marsupial from the John Day Oligocene of Logan Butte, eastern Oregon. *University of California Publications in Geological Sciences*, 13, 311–17.

Szalay, F. S. (1993). Metatherian taxon phylogeny: evidence and interpretation from the cranioskeletal system. In *Mammalian Phylogeny, Mesozoic Differentiation, Multituberculates, Monotremes, Early Therians, and Marsupials*, ed. F. S. Szalay, M. J. Novacek, and M. C. McKenna, pp. 216–42. New York: Springer-Verlag.

Troxell, E. L. (1923). A new marsupial. *American Journal of Science*, 5, 507–10.

Part II: Insectivorous mammals

4 Insectivorous mammals summary

GREGG F. GUNNELL[1] and JONATHAN I. BLOCH[2]

[1]*Museum of Paleontology, University of Michigan, Ann Arbor, MI, USA*
[2]*Florida Museum of Natural History, University of Florida, Gainesville, FL, USA*

INTRODUCTION

GENERAL CONSIDERATIONS

The idea of "Insectivora" serving as a waste-basket taxon for a variety of extinct and extant small, insectivorous mammals has had a relatively long history (Symonds, 2005). For the purposes of this book, we have chosen to recognize three groups of North American insectivorous mammals: the potentially monophyletic Leptictida and Lipotyphla, and the clearly polyphyletic "Proteutheria." Leptictids and proteutherians are only represented by extinct taxa, while lipotyphlans are abundant and diverse from the Paleocene to the Recent. It is nearly certain that these three groups do not share common ancestry except perhaps at a very basal level within the mammalian tree where insectivore-grade stem lineages may have given rise to a number of different mammalian groups (Asher, 2005).

Modern lipotyphlans are represented by nearly 400 species included in four families, Erinaceidae, Solenodontidae, Soricidae, and Talpidae. We consider the families Tenrecidae, Potamogalidae, and Chrysochloridae to be members of Afrotheria and distinct from Lipotyphla (Asher, Novacek, and Geisler, 2003; Springer *et al.*, 2005). Among living mammals, lipotyphlan diversity is eclipsed only by bats (approximately 1000 species) and rodents (over 2000 species). Lipotyphlans are common on all of the northern continents, have modest representation in Africa and Madagascar, and have recently colonized South America (represented by the soricid *Cryptotis*). Lipotyphlans are lacking in Australia and Antarctica but are present on the islands of the Philippines and the Greater Antilles (Symonds, 2005).

FEATURES UNITING INSECTIVOROUS MAMMALS

Butler (1972, p. 254) noted that there is no way to define an order Insectivora "except by exclusion of the specializations which distinguish the other orders." In gross anatomical terms, living insectivorous mammals can be described as being relatively small (ranging from 2 g to about 1 kg), having elongate snouts (often extremely mobile), relatively small eyes and ears, relatively small brains with smooth cerebral hemispheres, pentadactyl and plantigrade feet, and rudimentary endothermy (Eisenberg, 1980; Symonds, 2005).

However, searching for shared and derived characters that unite insectivorous mammals is a difficult task. Asher (2005), in a comprehensive review of the morphological similarities that have been used to define an order Insectivora, found only two characters that were potentially consistent with insectivore monophyly (including lipotyphlans, tenrecs, chrysochlorids, and potamogalids): a reduced pubic symphysis and a simplified intestinal tract including absence of the cecum. Yet both of these characters can be found in other mammals (in a variety of groups for the former and in some bats, pangolins, and some dolphins for the latter), although the absence of a cecum may well represent a shared and derived character of Lipotyphla (and convergently Afrosoricoidea [also known as Tenrecoidea] as well). An additional potential synapomorphy for Lipotyphla is a posteriorly expanded maxilla found in conjunction with a ventrally restricted palatine (Asher, Novacek, and Geisler, 2003; Asher, 2005).

The addition of fossil taxa makes any definition of insectivorous mammals even more difficult, especially since so many early fossil forms are represented only by a handful of teeth and jaws. In some cases, fossil taxa have simply been referred to the insectivoran-grade taxa Soricomorpha or Erinaceomorpha based on phenetic similarity in tooth and dental patterns. In other cases, it is not possible to assign fossil taxa to any living higher-level grouping either because they lack characters linking them with any extant group or because they represent clades that left no living descendants.

Evolution of Tertiary Mammals of North America, Vol. 2. ed. C. M. Janis, G. F. Gunnell, and M. D. Uhen. Published by Cambridge University Press.

FEATURES DISTINGUISHING LIPOTYPHLA, LEPTICTIDA, AND "PROTEUTHERIA"

BODY SIZE

In general, all insectivorous mammals are quite small (500 g or less). However, some fossil groups attained relatively larger body sizes, especially within the proteutherian families Pantolestidae and Pentacodontidae, some members of which probably having reached body weights of 4–5 kg. The largest living insectivorans are *Erinaceus* and *Solenodon*, both of which may range up to 3 kg while the smallest living forms (*Suncus*) barely reach 2 g and are among the smallest known living mammals. Interestingly, at least one fossil insectivore seems to have achieved a relatively smaller body size within the lipotyphlan family Geolabididae, with a body mass of about 1.3 g (Bloch, Rose, and Gingerich, 1998).

LOCOMOTION AND POSTCRANIAL SPECIALIZATIONS

All living lipotyphlans are pentadactyl, plantigrade, and quadrupedal, and they usually have forelimbs shorter than hindlimbs. Most talpids and some shrews have extremely specialized forelimbs and pectoral musculature adapted for efficient digging. Most shrews have distally fused tibiae and fibulae. Proscalopids have a number of postcranial specializations reflecting extreme digging adaptations (including head digging). These specializations include a shortened neck with some co-ossification of cervical vertebrae, fusion of the first rib and manubrium, elevated deltoid crest of scapula, long clavicle, rotary burrowing modifications of the humerus and forelimb, a laterally compressed humeral head, a broad distal humerus, pectoral crest of humerus terminating in a spike or plate, teres tubercle extending distally beyond midshaft, transversely expanded ulnar olecranon process, and tibia and fibula unfused (Barnosky, 1982).

Lipotyphlans practice terrestrial, fossorial, or semi-aquatic locomotor patterns. Where known, fossil lipotyphlans are similar to their living relatives except perhaps in the case of the soricomorph family Nyctitheriidae. Hooker (2001) has suggested that at least some nyctitheriids were arboreal, based on the structure of isolated tarsal elements from the late Eocene of England.

Leptictids have shorter forelimbs than hindlimbs and have distally fused tibiae–fibulae. They were probably terrestrial mammals that were capable of rapid running and quadrupedal jumping (Rose, 1999). The structure of the forelimb indicates that leptictids may have been accomplished diggers as well (Cavigelli, 1997; Rose, 1999).

In general, most proteutherians are poorly represented by postcranial material. A single partial forelimb is known of a palaeoryctid and indicates that these animals may have been fossorial diggers although not nearly as specialized as talpids (Van Valen, 1966).

Pantolestids are small to moderate in size and in general, larger than other proteutherians. In overall limb proportions, *Pantolestes* resembles extant otters and was probably otter-like in habits (Matthew, 1909). Pfretzchner (1999) noted aquatic adaptations in the European pantolestid *Buxolestes piscator*. Following Koenigswald (1980), he noted osteological features indicative of strong neck and tail musculature including a long spinous process on C2, broad and enlarged vertebral mammillary processes, and spinous processes that expanded craniocaudally. Rose and Koenigswald (2005) described a nearly complete skeleton of *Palaeosinopa*, the oldest known pantolestid skeleton, from the Eocene Green River Formation of Wyoming. They noted strong similarities to *Buxolestes* and suggested that pantolestids as a group were semi-aquatic mammals with a propensity for digging, most similar among extant mammals to river otters (*Lutra* and *Lontra*) and beavers (*Castor*). They hypothesized that during swimming, *Palaeosinopa* propelled itself by hindlimb paddling and dorsoventral tail undulation.

Apatemyids such as *Apatemys* (Koenigswald *et al.*, 2005) and *Labidolemur* (Bloch and Boyer, 2001, Bloch, Boyer, and Houde, 2004) are known by nearly complete skeletons. They document the arboreal nature of apatemyids and also indicate that these taxa have specialized finger elongation like that seen in the European apatemyid *Heterohyus*. Apatemyids may have been similar to extant *Dactylopsila* (the striped possum) and *Daubentonia* (the aye-aye lemur), which utilize elongate hand digits for percussive feeding in search of tree-boring insects (Koenigswald and Schierning, 1987; Koenigswald, 1990; Fleagle, 1999; Bloch and Boyer, 2001; Bloch, Boyer, and Houde, 2004; Kalthoff, Koenigswald, and Kurz, 2004; Koenigswald *et al.*, 2005). These mammals have been perceived as taking a similar ecological niche to woodpeckers: it is interesting to note that apatemyids did not survive past the appearance of woodpeckers (birds) in the mid Cenozoic, but places where the striped possum and the aye-aye are found (Australia and Madagascar, respectively) are islands where there have never been woodpeckers.

DIET AND CRANIODENTAL SPECIALIZATIONS

Lipotyphlan skulls indicate animals that rely predominantly on olfaction for locating insect prey. The skulls are generally low with laterally facing small orbits, small braincases, and long snouts. Auditory bullae range from unossified to solidly ossified with minor inflation. Proscalopid skulls appear to be functionally convergent on Chrysochloridae (Barnosky, 1981), animals that use their heads in burrowing. Proscalopid skulls have fused cranial bones (in adults), a long rostrum, broad and deep cranium with ventrally placed condyles, premaxillae with prominent lateral shelves, and slightly inflated tympanic bullae. Unlike proscalopids, talpids do not use their heads in digging. Their skulls are relatively long and shallow with unfused cranial bones and thin and low zygomatic arches and they lack jugals. Geolabidid skulls have a very long, tubular rostrum and lack zygomatic arches. Apternodontids have well-developed sagittal crests and develop lambdoid plates on the posterolateral braincase (Asher *et al.*, 2002). Soricids tend to have inflated braincases, have either weak or absent zygomatic arches, lack jugals, and have relatively large ectotympanics that lie horizontally. Erinaceids have complete zygomatic arches and relatively large orbits.

Teeth vary from acutely cuspate to more flattened with rounded, low cusps. In some forms, posterior premolars become specialized for crushing and grinding while in others anterior incisors become

Figure 4.1. Restoration of the early Tertiary leptictid, *Leptictis* (by Marguette Dongvillo).

enlarged for specialized food acquisition. In general, insectivoran teeth indicate dietary specializations predominantly for invertebrate prey, although the teeth in some forms indicate a more omnivorous diet (many erinaceomorphs) or other specializations (fruit and fish feeding in pantolestids; Koenigswald, 1980; Richter, 1987; Rose and Koenigswald, 2005).

Erinaceomorphs have quadritubercular upper molars and lower molars with distinct trigonids and well-developed talonids. Many erinaceomorphs have molars that reduce in size from first to last. Proscalopids have enlarged anterior incisors and exhibit loss of anterior cheek teeth in more derived taxa. Proscalopid molars range from brachydont to hypsodont in derived forms. Talpids are dilambdodont, with primitive forms retaining all teeth but later forms losing anterior cheek teeth and often having enlarged incisors. Apternodontids, parapternodontids, and oligoryctids all have very specialized zalambdodont dentitions. Apternodontids have enlarged anterior incisors while oligoryctids have extremely reduced lower molar talonids. In plesiosoricids, the second lower incisor is enlarged and procumbent and upper molars lack dilambdodonty. Soricids have dilambdodont upper molars, a distinct, enlarged, hooklike upper incisor, simple unicuspid teeth between the first upper incisor and the upper fourth premolar, and an enlarged and procumbent anterior lower incisor.

In general cranial proportions, leptictids were similar to extant tupaiids and macroscelidians and probably had elongate, mobile noses (see Figure 4.1) (Novacek, 1986). Dentitions indicate that leptictids were primarily insectivorous and omnivorous.

Palaeoryctids have relatively low and elongate skulls with distinct but low sagittal crests, incomplete zygomatic arches, and ossified auditory bullae (Thewissen and Gingerich, 1989; Bloch, Secord, and Gingerich, 2004). Upper molars were protozalambdodont with closely appressed, connate paracones and metacones, while lower molars had high trigonids and narrow talonids. Shearing crests were well developed on cheek teeth indicating an insectivorous diet.

Cimolestids are poorly represented by cranial material. What little is known indicates that these animals are characterized by low skulls with broad nasals, small lacrimal exposure on the face, a single sagittal crest, flaring nuchal crests, and lacking an ossified bulla (Van Valen, 1966). Cimolestids have small and procumbent incisors, enlarged canines, premolariform premolars, upper molars lacking hypocones, and lower molars with high trigonids. Lower molars increase in size posteriorly.

Pantolestid skulls are generally characterized by having a short rostrum, an elongate and low neurocranium, a distinct postorbital constriction, a single sagittal crest, and flaring nuchal crests. The basicranium is broad, with pantolestines lacking, and pentacodontines having, ossified auditory bullae (Matthew, 1909; Gingerich, Houde, and Krause, 1983; Pfretzchner, 1999; Rose and Lucas 2000; Boyer and Bloch, 2003). Pantolestids have an enlarged and bladelike upper third incisor, large and massive canines, fourth premolars either premolariform (pantolestines) or semimolariform and enlarged (pentacodontines), and molars with relatively low and massive cusps. (Matthew, 1909; Simpson, 1937a,b; Gazin, 1959, 1969; Van Valen, 1967; Gingerich, Houde, and Krause, 1983; Pfretzchner, 1999; Rose and Lucas 2000; Boyer and Bloch, 2003).

Apatemyids have skulls that lack ossified auditory bullae, possess a groove on the promontorium marking the course of the internal carotid artery, and have a large, free ectotympanic ring, an infraorbital foramen placed above the first molar, and small foramina on the skull roof (Jepsen, 1934; McKenna, 1963; Koenigswald, 1990; Bloch, Boyer, and Houde, 2004). Apatemyid dentitions are specialized in possessing an enlarged and procumbent first lower incisor that has a root extending posteriorly to the last molar, in lacking canines, in having a lower third premolar bladelike and single rooted, lower molars low crowned with labial paraconids, simple upper molars lacking mesostyles, and a very large mental foramen placed below the lower first molar.

OTHER ANATOMICAL FEATURES

Living insectivorans exhibit some features that are uncommon in most other groups of mammals. Pinnae (external ears) are often very small or absent and eyes are always small, and sometimes very small and essentially non-functional.

Among erinaceids, hedgehogs have spiny guard hairs and specialized "drawstring muscles" that allow them to roll into a defensive ball, but gymnures have neither spines nor paniculus carnosus muscles (Feldhamer *et al.*, 2004). Some hedgehog species that live in relatively harsh, seasonal habitats may estivate or even practice true hibernation, unlike any other known insectivoran.

Several species of living insectivorans practice rudimentary forms of echolocation. Desmans actively use echolocation for hunting and maneuvering in water (Richard, 1973). Solenodons emit high-pitched clicking noises, which may be used for prey detection, while many species of shrews emit high-frequency sounds, which presumably are used for communication, orientation, and hunting (Feldhamer *et al.*, 2004).

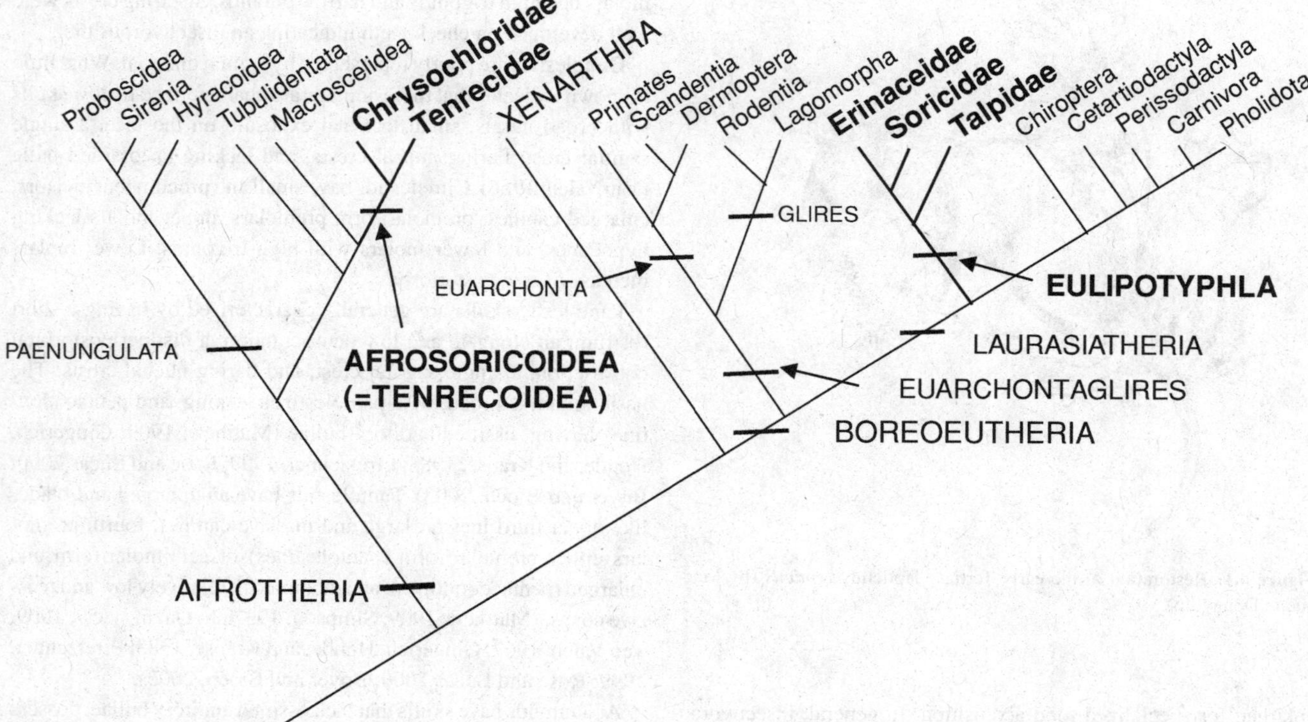

Figure 4.2. Cladogram showing the interrelationships of extant insectivorous mammals with other placental mammals. "Insectivorans" are indicated by bold face type.

Solenodons and some shrews are the only extant mammals (in addition to the platypus) that produce toxins to immobilize prey (Churchfield, 1990; Feldhamer *et al.*, 2004). In the case of *Solenodon*, there is a deep groove along the inner surface of the enlarged i2 that accommodates toxins produced by the submaxillary gland. Recently, Fox and Scott (2005) described a specimen of *Bisonalveus* (pentacodontine) from the Paleocene of Canada that potentially has a venom-delivery system via a groove in the upper canine.

SYSTEMATICS

ORIGIN AND AFFINITIES OF LIPOTYPHLA, LEPTICTIDA, AND "PROTEUTHERIA"

EXTANT FORMS

Extant animals included in this summary are represented only by the lipotyphlan families Erinaceidae, Soricidae, Talpidae, and Solenodontidae. Among these families, most authorities would agree that talpids and soricids share sister-group status while the relationships of the other two families are less clear (Figure 4.2). Based on morphological evidence alone, Butler (1956) placed erinaceids as the basal clade in his concept of Insectivora which also included tenrecs (Tenrecidae) and golden moles (Chrysochloridae). He grouped solenodontids with tenrecids as a sister clade to chrysochlorids, and that clade as sister group to a soricid–talpid clade. Butler (1988) changed his insectivore topology slightly by moving solenodontids to sister-group status with soricids and talpids but maintained the other relationships as before. Other variations on these arrangements based on interpretation of morphological evidence were offered by McDowell (1958), Van Valen (1967), Eisenberg (1980), and MacPhee and Novacek (1993).

Shortly after the landmark study of MacPhee and Novacek (1993) had questioned the status of "Insectivora" and the position of Chrysochloridae within that "order," Springer *et al.* (1997) and Stanhope *et al.* (1998) published papers based on mitochondrial gene sequence data indicating that golden moles and tenrecs were not closely related to soricids, talpids, erinaceids, and solenodontids, but instead were members of an African endemic mammalian radiation referred to as Afrotheria (also including aardvarks, elephant shrews, elephants, sirenians, hyraxes, and potentially some fossil groups such as ptolemaiids). In nearly every subsequent molecular phylogenetic analysis since 1998, Afrotheria has been supported, although it has proven difficult to find any morphological synapomorphies that unambiguously support this African clade (Seiffert and Simons, 2001; Asher, Novacek, and Geisler, 2003; Asher, 2005).

Recent molecular phylogenies normally support the sister-group status of soricids and talpids (Arnason *et al.*, 2002), but erinaceids and solenodontids appear in a variety of different places on phylogenetic trees (Emerson *et al.*, 1999; Douady *et al.*, 2002; Waddell and Shelley, 2003). Roca *et al.* (2004) presented molecular evidence that indicated a very early (Late Cretaceous) divergence for the extant Greater Antillean *Solenodon*. These studies suggest that there is at least a biphyletic "Insectivora" if not a triphyletic or multiphyletic "Order Insectivora" (Symonds, 2005).

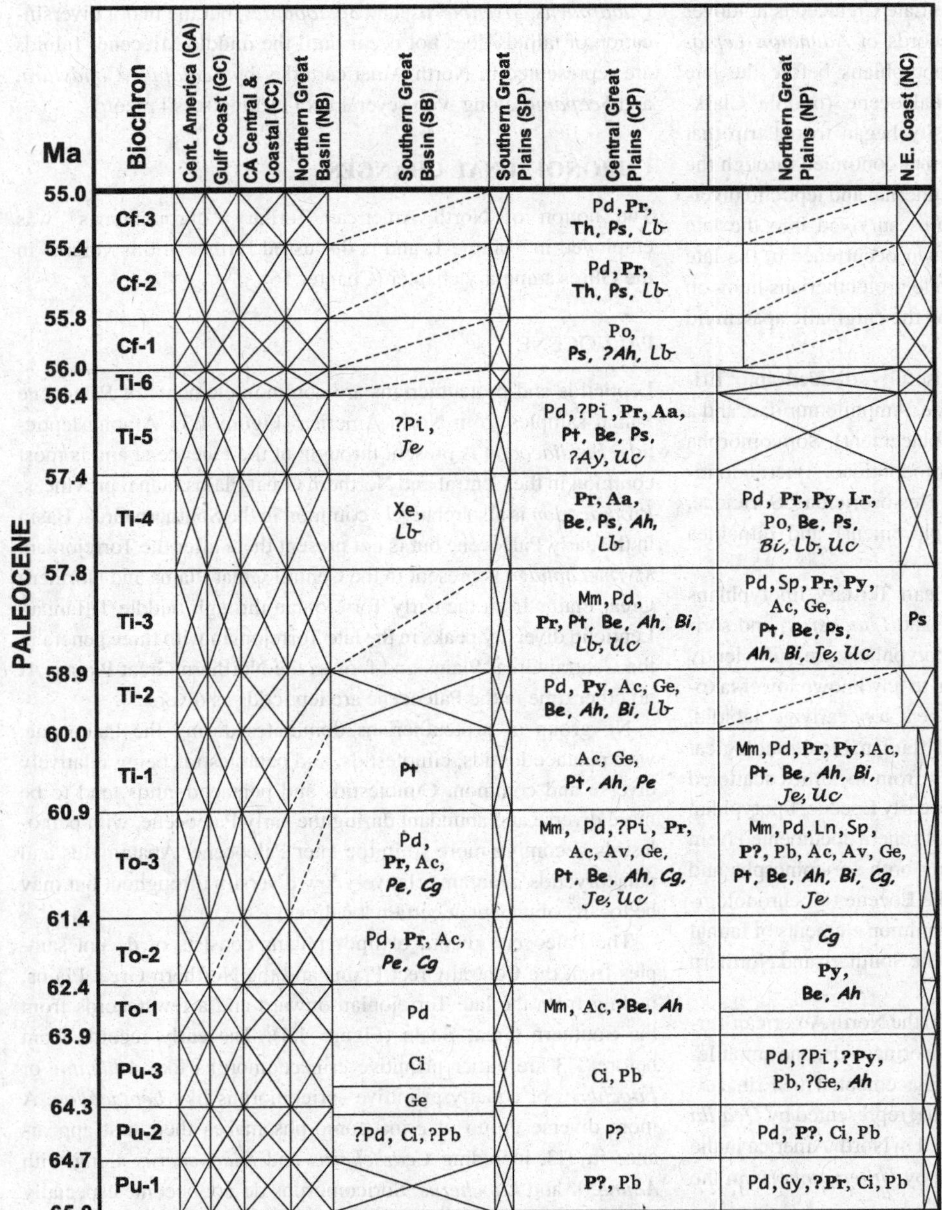

Figure 4.3. Biogeographic ranges of Paleocene proteutherians. Key: A "box" (for a particular time period in a particular biogeographic region) that has a cross through it means no fossil mammal localities are known for that time period from that area; a single dashed line through the box means only scant fossil mammal information is available (usually only a single, small locality). LEPTICTIDA: LEPTICTIDAE (Monaco Plain): Ln, *Leptonysson*; L?, Leptictid indet.; Mm, *Myrmecoboides*; Pd, *Prodiacodon*; Pi, *Palaeictops*; Xe, *Xenacodon*. GYPSONICTOPIDAE (Monaco Bold): **Gy**, *Gypsonictops*; **Sp**, *Stilpnodon*. PALAEORYCTIDAE (Georgia Bold): **Aa**, *Aaptoryctes*; **Lr**, *Lainoryctes*; **Pr**, *Palaeoryctes*; **Py**, *Pararyctes*; **P?**, Palaeoryctid indet. CIMOLESTIDAE and CIMOLESTA INDET. (Georgia Plain): Ac, *Acmeodon*; Av, *Avunculus*; Ci, *Cimolestes*; Ge, *Gelastops*; Pb, *Procerberus*; Po, *Protentomodon*. PANTOLESTA: PANTOLESTIDAE: PANTOLESTINAE (Comic Sans MS Bold): **Be**, *Bessoecetor*; **Ps**, *Palaeosinopa*; **Pt**, *Paleotomus*; **Th**, *Thelysia*. PENTACODONTINAE (Comic Sans MS Bold Italics): ***Ah***, *Aphronorus*; ***Bi***, *Bisonalveus*; ***Cg***, *Coriphagus*; ***Pe***, *Pentacodon*. PANTOLESTA INCERTAE SEDIS (Comic Sans MS Plain): P?, Pantolestid indet. APATEMYIDAE (Lucida Handwriting): *Ay*, *Apatemys*; *Je*, *Jepsenella*; *Lb*, *Labidolemur*; *Uc*, *Unuchinia*.

FOSSIL RELATIVES

The fossil record of insectivorous mammals is ancient, abundant, diverse, geographically widespread and fraught with incompleteness and complexities. Lipotyphlans (both erinaceomorphs and soricomorphs) are known beginning in the earliest Tertiary and almost certainly originated in the Cretaceous from uncertain basal placental lineage(s). The fossil record of solenodontids is not temporally deep (Pleistocene and Recent), but extant solenodons, both morphologically and genetically, appear to be relicts of an ancient (perhaps even Cretaceous) radiation.

Insectivorous groups only known by fossils, including proteutherians, leptictids, and a variety of early lipotyphlans, are difficult to place taxonomically. The origins of most can be traced in a vague way to the archaic roots of the mammalian tree but few (if any) can be traced to specific ancestral lineages. Nearly all probably had their origins in the Cretaceous of Laurasia, either in North America or Asia.

EVOLUTIONARY AND BIOGEOGRAPHIC PATTERNS

GENERAL TRENDS

Figures 4.3–4.9, below, summarize the biogeographic ranges of insectivorous mammals in the North American fossil record. Most of the earliest appearances in the Cenozoic are represented by proteutherians or leptictids with only the Late Cretaceous holdover taxon *Batodon* and relatively sparse records of *Adunator, Leptacodon*, and *Litocherus* representing lipotyphlans before the late Paleocene. It is not until the latest Paleocene (middle Clarkforkian onward) that lipotyphlan diversity began to outstrip that of proteutherians and leptictids, a trend that continued through the Eocene. By the end of the Eocene, proteutherian and leptictid diversity was greatly reduced but both groups survived into the late Oligocene. Leptictids had their last known occurrence in the late Whitneyan, represented by *Leptictis*, while proteutherians hung on until the early Arikareean, represented by the enigmatic apatemyid *Sinclairella*.

The lipotyphlan radiation can be usefully divided into Erinaceomorpha (Erinaceidae, Sespedectidae, Amphilemuridae, and a variety of erinaceomorphs of uncertain placement), Soricomorpha (Geolabididae, Nyctitheriidae, Micropternodontidae, Apternodontidae, Parapternodontidae, Oligoryctidae, Plesiosoricidae, Soricidae, and various soricomorphs of uncertain placement), and Talpoidea (Proscalopidae and Talpidae).

The earliest records of North American Tertiary lipotyphlans include both erinaceomorphs (*Adunator* and *Litocherus*) and soricomorphs (*Batodon*). These earliest lipotyphlans are sufficiently primitive that it is not easy to assign them to any known lower taxonomic grouping, although *Batodon* may be a very early geolabidid. Beginning in the late Paleocene (middle Tiffanian, biochronological zone Ti3) lipotyphlan diversity increases from localities scattered along the Rocky Mountain corridor. In the early Eocene, lipotyphlan generic diversity reached a maximum of 18 genera documented from the Central Great Plains and representing both soricomorphs and erinaceomorphs. By the end of the middle Eocene (biochronological zones Ui3 and Du), lipotyphlans are common elements of faunal assemblages from California across both the Southern and Northern Great Plains.

Erinaceomorphs are represented early in the North American Tertiary record by sespedectids, three possible erinaceids, one amphilemurid, and a number of taxa of uncertain placement. True Erinaceidae do not appear until late early Oligocene, represented by *Ocajila* and *Proterix*. Erinaceomorphs last occurred in North America in the early late Hemphillian (Hh3) represented by *Untermannerix* in the Central Great Plains.

The early record of soricomorphs in North America is dominated by primitive families such as Geolabididae, Nyctitheriidae, Micropternodontidae, Parapternodontidae, Apternodontidae, and Oligoryctidae. More advanced soricids make their first appearance in the late middle Eocene (Uintan biochronological zone Ui3) represented by *Domnina* and then later by *Pseudotrimylus*, first appearing at the beginning of the Oligocene (Orellan biochronological zone Or1). Soricids underwent diversification through the Miocene with four of those genera (*Sorex, Notiosorex, Cryptotis*, and *Blarina*) surviving to the present.

Talpoids have their earliest occurrence in North America in the latest Eocene (Chadronian biochronological zone Ch3), represented by the proscalopids *Oligoscalops* and *Proscalops*. Proscalopids disappear from the North American record by the end of the Barstovian. Talpids are first represented in the earliest Arikareean (Ar1) by *Quadrodens, Mystipterus*, and *Scalopoides*, but the major diversification of talpids does not occur until the middle Miocene. Talpids are represented in North America today by *Scalopus, Condylura*, and *Scapanus* along with several taxa lacking fossil records.

CHRONOFAUNAL CHANGES

The notion of North American Tertiary “chronofaunas” was employed in Volume 1, and is discussed further in this volume in the Glires summary chapter (Chapter 16).

PALEOCENE

Leptictids and proteutherians are common elements of Paleocene faunal samples from North America (Figure 4.3). Among leptictids, *Prodiacodon* is present throughout the Paleocene and is most common in the Central and Northern Great Plains faunal provinces. *Prodiacodon* is also relatively common in the Southern Great Basin in the early Paleocene but is not present there after the Torrejonian. *Myrmecophaga* is present in the Central Great Plains and Northern Great Plains from the early Torrejonian through middle Tiffanian. Leptictid diversity peaks in the late Torrejonian with three genera in the Central Great Plains and four in the Northern Great Plains. At no other time in the Paleocene are leptictids as diverse.

No group of proteutherians dominates during the Paleocene, with pentacodontids, cimolestids, and pantolestids being relatively diverse and common. Cimolestids and pentacodontids tend to be more diverse and abundant during the early Paleocene, with pantolestids becoming more so in the later Paleocene. Apatemyids and palaeoryctids maintain relatively low diversity throughout but may be locally abundant at certain localities.

The Paleocene record of lipotyphlans consists of decent samples from the Central Great Plains and the Northern Great Plains, mainly from the late Torrejonian onward and a few records from the Southern Great Basin (Figure 4.4). The early records from before Ti3 are either primitive erinaceomorphs like *Adunator* or *Litocherus* or equally primitive soricomorphs like *Leptacodon*. A more diverse group of erinaceomorphs makes their first appearance in Ti3, including *Cedrocherus* and *Diacocherus* along with *Adunator* and *Litocherus*. Soricomorphs do not become especially diverse until the middle Clarkforkian (Cf2) when *Leptacodon, Pontifactor, Plagioctenodon, Limaconyssus, Wyonycteris*, and *Ceutholestes* all co-occur. This soricomorph diversity, consisting of nyctitheriids only, persists to the end of the Paleocene, while erinaceomorphs maintain lower diversity, represented by only two genera in Cf3.

EARLY TO MIDDLE EOCENE

As in most of the Paleocene, leptictids maintain a relatively low diversity through the early and middle Eocene, never having more than two co-occurring genera in any biochron. From the earliest Wasatchian (Wa0) through Ui2, the only leptictids in the record are *Prodiacodon* and *Palaeictops*. *Leptictis* makes its first appearance near the end of the Uintan in Ui3 along with *Palaeictops*. *Prodiacodon* makes its last appearance in the latest Wasatchian (Wa7).

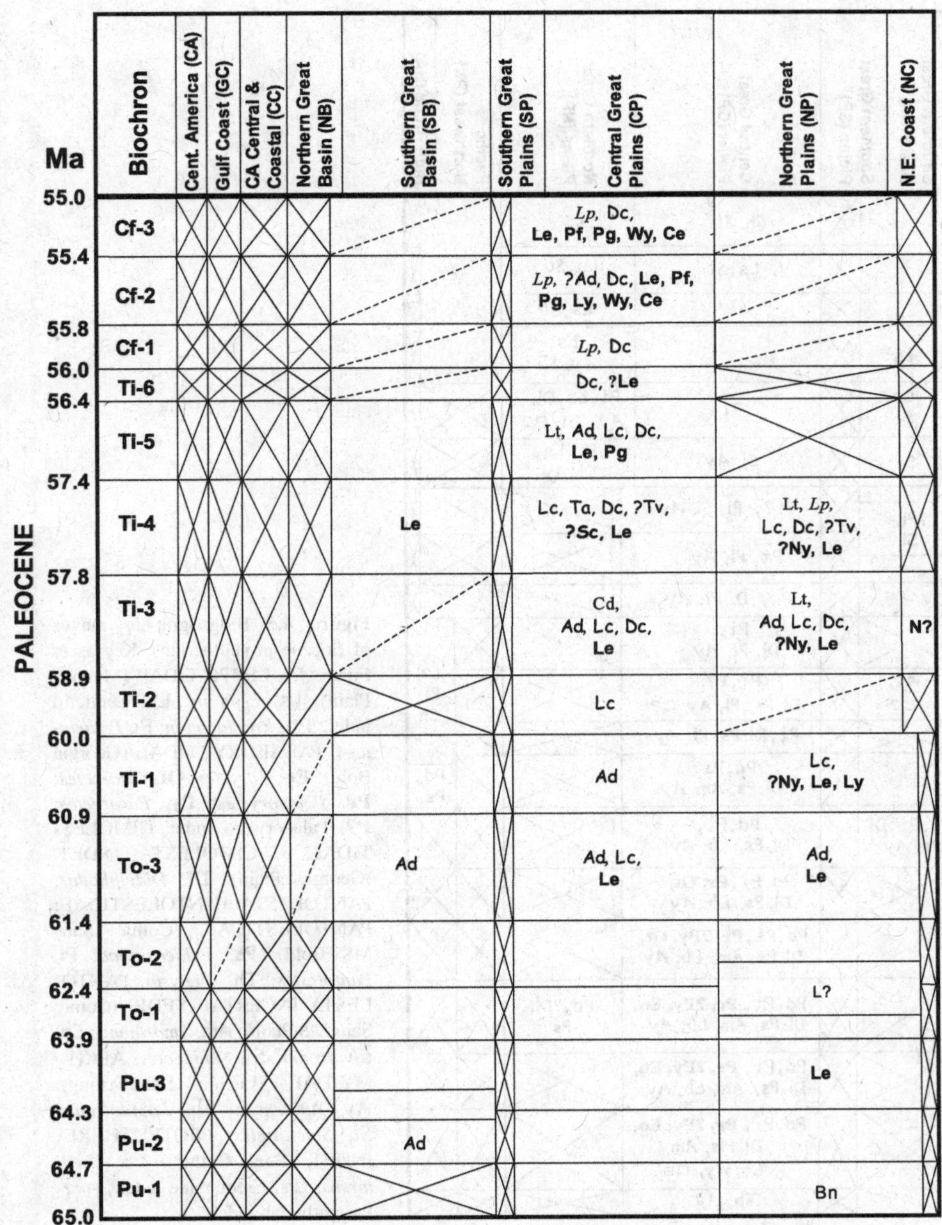

Figure 4.4. Biogeographic ranges of Paleocene lipotyphlans. Key as in Figure 4.3. ERINACIDAE INCERTAE SEDIS (Times New Roman Plain): Cd, *Cedrocherus*; Lt, *Litolestes*. ERINACEOIDEA INSERTAE SEDIS (Times New Roman Italics): *Lp*, *Leipsanolestes*. ERINACEOMORPHA INCERTAE SEDIS (Comic Sans MS Plain): Ad, *Adunator*; Dc, *Diacocherus*; Lc, *Litocherus*; Ta, *Talpavoides*; Tv, *Talpavus*. SORICOIDEA: GEOLABIDIDAE (Arial Plain): Bn, *Batodon*. NYCTITHERIIDAE (Arial Bold): **Ce**, *Ceutholestes*; **Le**, *Leptacodon*; **Ly**, *Limaconyssus*; **Ny**, *Nyctitherium*; **N?**, Nyctitheriid indet.; **Pf**, *Pontifactor*; **Pg**, *Plagioctenodon*; **Wy**, *Wyonycteris*.

Unlike the Paleocene where pentacodontids and cimolestids were well represented, the former is gone by the Eocene and the latter is only represented by a single, yet persistent genus, *Didelphodus*. The best samples of Eocene proteutherians come from the Central Great Plains and are dominated by palaeoryctids until the beginning of Wa6, when palaeoryctids essentially disappear from the record (Figure 4.5). Pantolestids and apatemyids maintain low but nearly constant diversity throughout the early and middle Eocene, fluctuating between one and two genera through the entire sequence.

Early and middle Eocene lipotyphlans are well represented in the Central Great Plains and, in the middle and late Uintan, in California Central and Coast and the Northern Great Plains as well (Figure 4.6). There are also a few records from the Southern Great Basin and one locality with lipotyphlans in the Gulf Coast Faunal Province that spans Biochrons Wa2/3. Nyctitheriids are diverse early in the Eocene and maintain moderate diversity through the Bridgerian but are only represented by *Nyctitherium* by the end of the Uintan. Other soricomorphs such as parapternodontids and geolabidids are less diverse than nyctitheriids through Wa4 but are equally as persistent. Parapternodontids disappear after Wa4 and nyctitheriid and geolabidid diversity evens out through the Bridgerian. In the Uintan, geolabidids remain modestly diverse in California Central and Coast but are only represented by *Centetodon* (a very speciose genus) in the North American interior while oligoryctids, apternodontids, and micropternodontids are all present

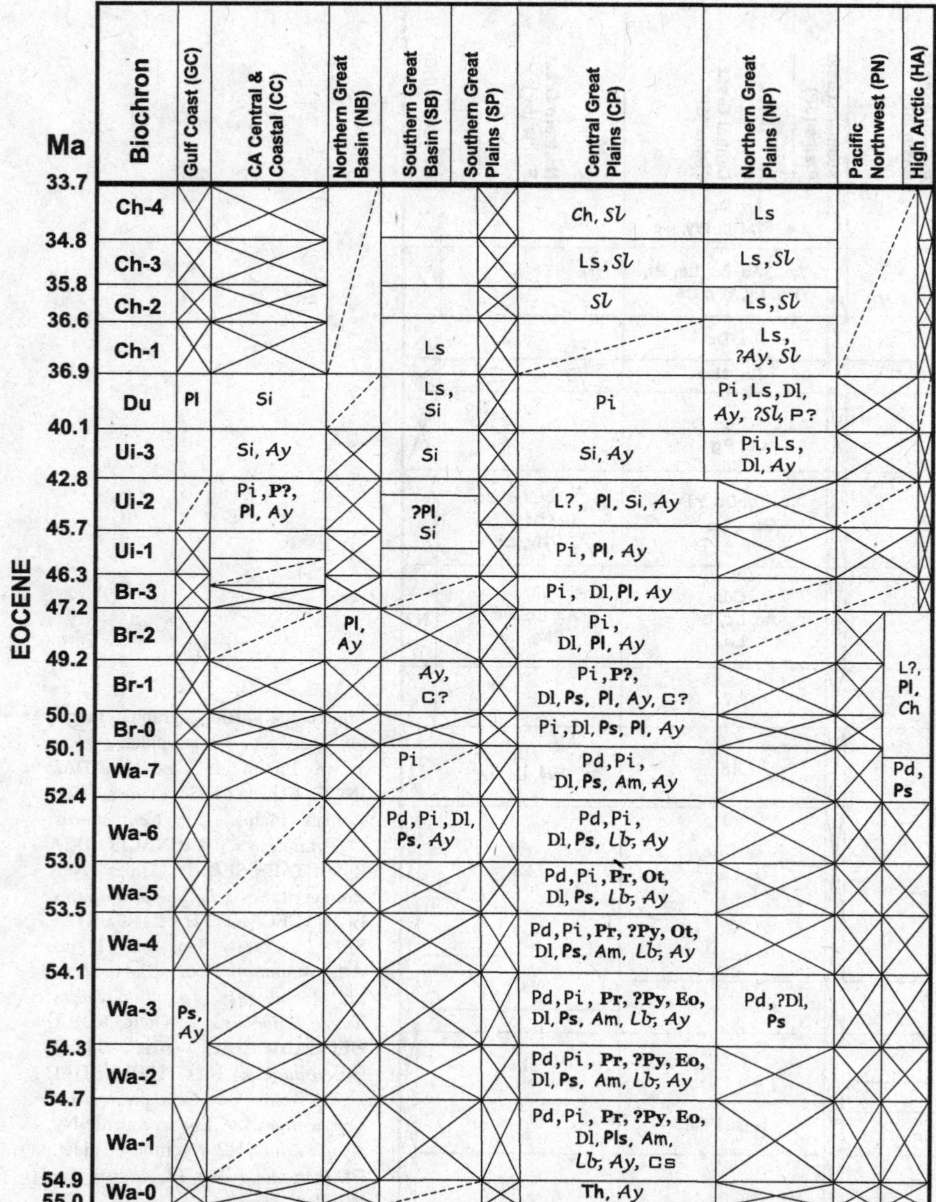

Figure 4.5. Biogeographic ranges of Eocene proteutherians. Key as in Figure 4.3. LEPTICTIDAE (Monaco Plain): Ls, *Leptictis*; L?, Leptictid indet.; Pd, *Prodiacodon*; Pi, *Palaeictops*. PALAEORYCTIDAE (Georgia Bold): **Eo**, *Eoryctes*; **Ot**, *Ottoryctes*; **Pr**, *Palaeoryctes*; **Py**, *Pararyctes*; **P?**, Palaeoryctid indet. CIMOLESTIDAE & CIMOLESTA INDET. (Georgia Plain): Dl, *Didelphodus*. PANTOLESTA: PANTOLESTIDAE: PANTOLESTINAE (Comic Sans MS Bold): **Ps**, *Palaeosinopa*; **Pl**, *Pantolestes*; **Th**, *Thelysia*. PANTOLESTA INCERTAE SEDIS (Comic Sans MS Plain): Am, *Amaramnis*; Ch, *Chadronia*; Si, *Simidectes*. APATEMYIDAE (Lucida Handwriting): *Ay*, *Apatemys*; *Lb*, *Labidolemur*; *Sl*, *Sinclairella*. PROTEUTHERIA INDET. (Bank Gothic): CS, *Creotarsus*; C?, Creotarsine indet.; PT?, Proteutherian indet.

at low diversities, especially in California Central and Coast, Central Great Plains, and Northern Great Plains localities. Erinaceomorphs are diverse through the Wasatchian, being represented by sespedectids like *Macrocranion* and *Scenopagus* and a variety of taxa of uncertain placement within Erinaceomorpha. By the late Uintan, the first records of heterosoricine shrews (represented by *Domnina*) appear in the Central Great Plains and Northern Great Plains.

WHITE RIVER CHRONOFAUNA

Leptictids and proteutherians are present in the White River Chronofauna but are gone from the North American record by the end of Ar1. In the Duchesnean, leptictids are represented by both *Palaeictops* and *Leptictis*, but *Palaeictops* disappears from the record after the Duchesnean while *Leptictis* survives into the early Arikareean in the Central Great Plains (Figure 4.7).

Proteutherian diversity is greatly diminished through this period as well. In the Duchesnean, pantolestids are represented by *Simidectes* in Central and Coastal California (California Central and Coast) and in the Southern Great Basin, while the cimolestid *Didelphodus*, the apatemyids *Apatemys*, and ?*Sinclairella*, and a possible palaeoryctid, still survive in the Northern Great Plains. There is a single occurrence of the possible pantolestid *Chadronia* in Ch4 in the Central Great Plains representing the last occurrence of North American pantolestids. *Sinclairella* remains a small part of mammalian faunal samples until the end of the early Arikareean in

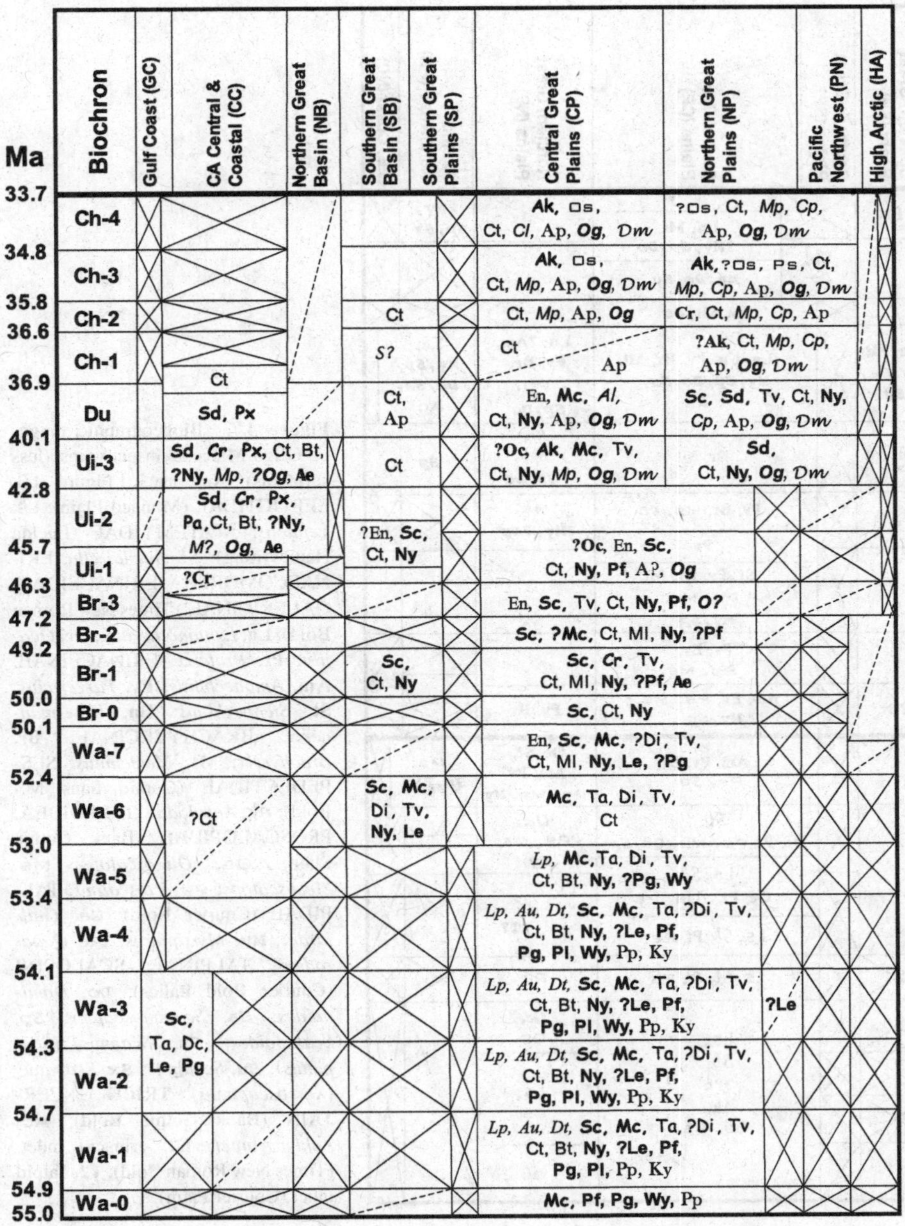

Ma	Biochron	Gulf Coast (GC)	CA Central & Coastal (CC)	Northern Great Basin (NB)	Southern Great Basin (SB)	Southern Great Plains (SP)	Central Great Plains (CP)	Northern Great Plains (NP)	Pacific Northwest (PN)	High Arctic (HA)
33.7	Ch-4						Ak, Os, Ct, Cl, Ap, Og, Dm	?Os, Ct, Mp, Cp, Ap, Og, Dm		
34.8	Ch-3						Ak, Os, Ct, Mp, Ap, Og, Dm	Ak, ?Os, Ps, Ct, Mp, Cp, Ap, Og, Dm		
35.8	Ch-2				Ct		Ct, Mp, Ap, Og	Cr, Ct, Mp, Cp, Ap		
36.6	Ch-1		Ct		S?		Ct Ap	?Ak, Ct, Mp, Cp, Ap, Og, Dm		
36.9	Du		Sd, Px		Ct, Ap		En, Mc, Al, Ct, Ny, Ap, Og, Dm	Sc, Sd, Tv, Ct, Ny, Cp, Ap, Og, Dm		
40.1	Ui-3		Sd, Cr, Px, Ct, Bt, ?Ny, Mp, ?Og, Ae		Ct		?Oc, Ak, Mc, Tv, Ct, Ny, Mp, Og, Dm	Sd, Ct, Ny, Og, Dm		
42.8	Ui-2		Sd, Cr, Px, Pa, Ct, Bt, ?Ny, M?, Og, Ae		?En, Sc, Ct, Ny					
45.7	Ui-1		?Cr				?Oc, En, Sc, Ct, Ny, Pf, A?, Og			
46.3	Br-3						En, Sc, Tv, Ct, Ny, Pf, O?			
47.2	Br-2						Sc, ?Mc, Ct, Ml, Ny, ?Pf			
49.2	Br-1				Sc, Ct, Ny		Sc, Cr, Tv, Ct, Ml, Ny, ?Pf, Ae			
50.0	Br-0						Sc, Ct, Ny			
50.1	Wa-7						En, Sc, Mc, ?Di, Tv, Ct, Ml, Ny, Le, ?Pg			
52.4	Wa-6		?Ct		Sc, Mc, Di, Tv, Ny, Le		Mc, Ta, Di, Tv, Ct			
53.0	Wa-5						Lp, Mc, Ta, Di, Tv, Ct, Bt, Ny, ?Pg, Wy			
53.4	Wa-4						Lp, Au, Dt, Sc, Mc, Ta, ?Di, Tv, Ct, Bt, Ny, ?Le, Pf, Pg, Pl, Wy, Pp, Ky			
54.1	Wa-3	Sc, Ta, Dc, Le, Pg					Lp, Au, Dt, Sc, Mc, Ta, ?Di, Tv, Ct, Bt, Ny, ?Le, Pf, Pg, Pl, Wy, Pp, Ky	?Le		
54.3	Wa-2						Lp, Au, Dt, Sc, Mc, Ta, ?Di, Tv, Ct, Bt, Ny, ?Le, Pf, Pg, Pl, Wy, Pp, Ky			
54.7	Wa-1						Lp, Au, Dt, Sc, Mc, Ta, ?Di, Tv, Ct, Bt, Ny, ?Le, Pf, Pg, Pl, Pp, Ky			
54.9	Wa-0						Mc, Pf, Pg, Wy, Pp			
55.0										

EOCENE

Figure 4.6. Biogeographic ranges of Eocene lipotyphlans. Key as in Figure 4.3. ERINACEOMORPHA: ERINACIDAE INCERTAE SEDIS (Times New Roman Plain): En, *Entomolestes*. ERINACEOIDEA INSERTAE SEDIS (Times New Roman Italics): *Au*, *Auroralestes*; *Dt*, *Dartonius*; *Lp*, *Leipsanolestes*. SESPEDECTIDAE (Comic Sans MS Bold): **Ak**, *Ankylodon*; **Cr**, *Crypholestes*; **Mc**, *Macrocranion*; **Pa**, *Patriolestes*; **Px**, *Proterixoides*; **Sc**, *Scenopagus*; **Sd**, *Sespedectes*. AMPHILEMURIDAE (Comic Sans MS Italics). *Al*, *Amphilemur*. ERINACEOMORPHA INCERTAE SEDIS (Comic Sans MS Plain): Dc, *Diacocherus*; Di, *Diacodon*; Ta, *Talpavoides*, Tv, *Talpavus*. TALPOIDEA: PROSCALOPIDAE (Bank Gothic). Os, *Oligoscalops*; Ps, *Proscalops*. SORICOIDEA: GEOLABIDIDAE (Arial Plain): Ct, *Centetodon*; Bt, *Batodonoides*; Ml, *Marsholestes*. NYCTITHERIIDAE (Arial Bold): **Le**, *Leptacodon*; **Ny**, *Nyctitherium*; **Pf**, *Pontifactor*; **Pg**, *Plagioctenodon*; **Pl**, *Plagioctenoides*; **Wy**, *Wyonycteris*. MICROPTERNODONTIDAE (Arial Italics): *Cl*, *Clinopternodus*; *Cp*, *Cryptoryctes*; *Mp*, *Micropternodus*; *M?*, Micropternodontid indet. APTERNODONTIDAE (Baskerville Old Face Bold): **Ap**, *Apternodus*; **A?**, apternodontine indet. PARAPTERNODONTIDAE (Baskerville Old Face Plain): Ky, *Koniaryctes*; Pp, *Parapternodus*. OLIGORYCTIDAE (Arial Bold Italics): ***Og***, *Oligoryctes*; ***O?***, oligoryctid indet. LIPOTYPHLAN INCERTAE SEDIS (Brittanic Bold): **Ae**, *Aethomylos*.

the Central Great Plains and represents the last surviving proteutherian in North America.

Lipotyphlans are dominated by soricomorphs from the Duchesnean through the end of the Chadronian mostly represented by several species of *Centetodon* and a variety of apternodontids, micropternodontids, and oligoryctids, all of which survive through the Orellan in the Central Great Plains and Northern Great Plains. Nyctitheriids have disappeared from the North American record by the beginning of the Chadronian while *Centetodon* survives into the early late Arikareean (Ar3) in the Central Great Plains and California Central and Coast. Oligoryctids are present until the end of the Orellan in the Northern Great Plains and through the Whitneyan in the Central Great Plains, while micropternodontids make a late appearance in the early Arikareean (Ar1/2) in the Pacific Northwest.

Erinaceomorphs are never very diverse after the Duchesnean but are present through the entire White River Chronofauna, mostly represented by *Ankylodon* early and then by *Ocajila* and *Proterix* later. Erinaceomorphs in general are more common from the Central Great Plains than the Northern Great Plains throughout the White River Chronofauna.

Soricids first appear in North America in Ui3, represented by *Domnina*, which is present throughout the White River Chronofauna in the Central Great Plains, Northern Great Plains, and Pacific Northwest. *Pseudotrimylus* and *Limnoecus* join *Domnina* in the Central

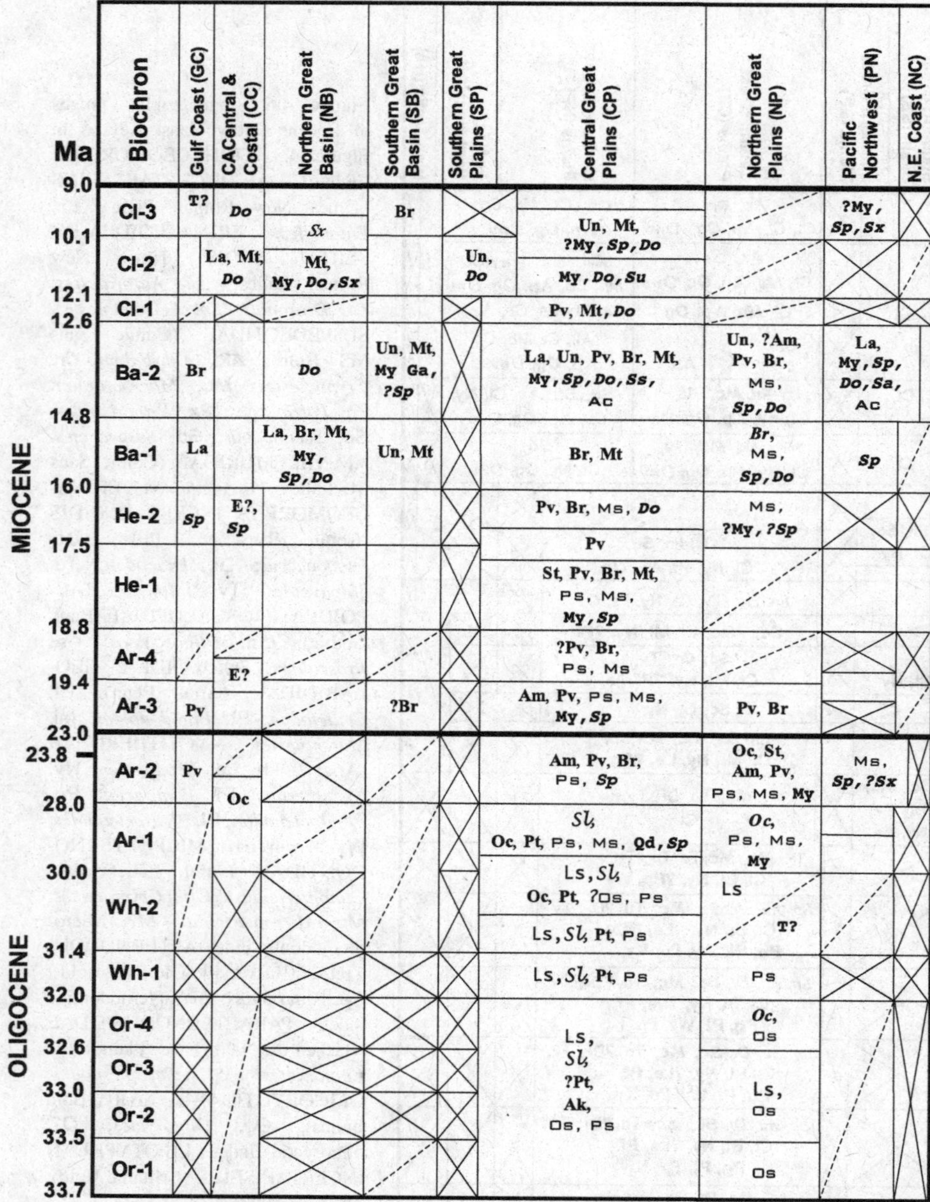

Figure 4.7. Biogeographic ranges of Oligo-Miocene insectivores (less soricoids). Key as in Figure 4.3. LEPTICTIDAE (Monaco Plain): Ls, *Leptictis*. APATEMYIDAE (Lucida Handwriting): Sl, *Sinclairella*. ERINAECOMORPHA: ERINACEIDAE: GALERICINAE: (Times New Roman Bold): **La**, *Lantanotherium*; **Oc**, *Ocajila*; **Pt**, *Proterix*. ERINACEINAE: **Am**, *Amphechinus*; **Pv**, *Parvericius*; **St**, *Stenoechinus*; **Un**, *Untermannerix*. BRACHYERCINAE: **Br**, *Brachyerix*; **Mt**, *Metechinus*. SESPEDECTIDAE (Comic Sans MS Bold): **Ak**, *Ankylodon*. TALPOIDEA: PROSCALOPIDAE (Bank Gothic Plain). Os, *Oligoscalops*; Ms, *Mesoscalops*; Ps, *Proscalops*. TALPIDAE (Courier Bold): **Ga**, *Gaillardia*; **My**, *Mystipterus*; **Qd**, *Quadrodens*. TALPINAE: SCALOPINI (Courier Bold Italics): ***Do***, *Dominoides*; ***Sa***, *Scapanoscapter*; ***Sp***, *Scalapoides*; ***Ss***, *Scapnus (Scapanus)*; ***Su***, *Scalopus*; ***Sx***, *Scapanus (Xereoscapanus)*. TRIBE UNCERTAIN (Bank Gothic Bold): **Ac**, *Achlyoscapter*. **E?**, erinacid indet. (Times New Roman Bold). **T?**, Talpid indet (Courier Plain).

Great Plains in the Orellan and early Arikareean, respectively, and *Pseudotrimylus* in the early Arikareean.

Talpoids first appear in Ch3 in the Central Great Plains and Northern Great Plains, represented by *Oligoscalops* (Central Great Plains, Northern Great Plains) and *Proscalops* (Northern Great Plains only). These taxa are present through most of the White River Chronofauna and are joined by *Mesoscalops*, *Quadrodens, Scalopoides*, and *Mystipterus* in the early Arikareean. In general, talpoid and soricid diversity is relatively low through the White River Chronofauna until the late Whitneyan or early Arikareean, when modest diversification happens in both groups.

RUNNINGWATER CHRONOFAUNA

The Runningwater Chronofauna is the first to be dominated by modern lipotyphlan groups as the last geolabidids and micropternodontids are gone from the North American record by the end of the early late Arikareean (Ar3). *Plesiosorex* makes its first appearance in the late Arikareean in the Central Great Plains and also appears in the Pacific Northwest and Northeast Coast in the early Hemingfordian. Among soricids, *Domnina* persists until Ar3 in the Central Great Plains and into the early Hemingfordian from California Central and Coast along with *Limnoecus*, which is also known from the Gulf Coast in He1. The highest diversity of soricids in the

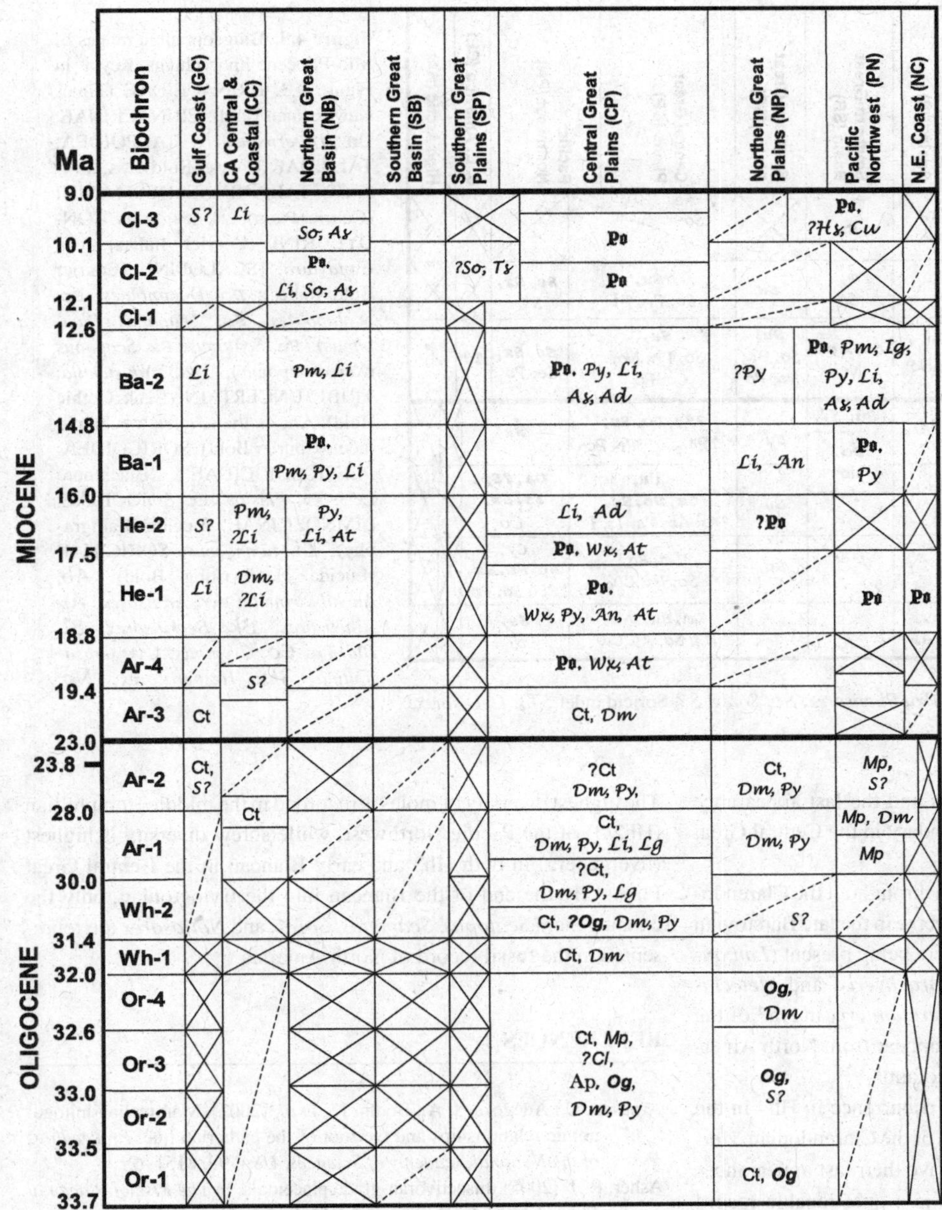

Figure 4.8. Biogeographic ranges of Oligo-Miocene soricoids. Key as in Figure 4.3. GEOLABIDIDAE (Arial Plain): Ct, *Centetodon*. MICROPTERNODONTIDAE (Arial Italics): *Cl*, *Clinopternodus*; *Mp*, *Micropternodus*. APTERNODONTIDAE (Baskerville Old Face Bold): **Ap**, *Apternodus*. OLIGORYCTIDAE (Arial Bold Italics): ***Og***, *Oligoryctes*. PLESIOSORICIDAE (Blackmore Let): **Po**, *Plesiosorex*. SORICIDAE: HETEROSORICINAE (Lucida Handwriting Plain): *Dm*, *Domnina*; *Ig*, *Ingentisorex*; *Pm*, *Paradomnina*; *Py*, *Pseudotrimylus*; *Wx*, *Wilsonosorex*. LIMNOECINAE (Lucida Calligraphy): *An*, *Angustidens*; *Lg*, limnoecine new genus; *Li*, *Limnoecus*. SORICINAE (Lucida Handwriting Bold): *Ad*, *Adeloblarina*; *As*, *Alluvisorex*; *At*, *Antesorex*; *Cu*, *Crusafontina*; *Hs*, *Hesperosorex*; *So*, *Sorex*; *S?*, Soricid indet.; *Ts*, *Tregosorex*.

Runningwater Chronofauna occurs in He1 in Central Great Plains, where four genera (*Wilsonosorex, Pseudotrimylus, Angustidens*, and *Antesorex*) are present.

Talpoids are well represented from the Runningwater Chronofauna in the Central Great Plains, with two to four genera being present throughout (*Proscalops, Mesoscalops, Mystipterus*, and *Scalopoides*). The only other talpoid records from the Runningwater Chronofauna are those from the Pacific Northwest and Northeast Coast where three mole genera are present in the late early Arikareean (Ar2).

Erinaceomorphs have a relatively widespread distribution during the Runningwater Chronofauna with records known throughout the Arikareean from Gulf Coast, California Central and Coast, Southern Great Basin, Central Great Plains, and Northern Great Plains. The only Hemingfordian records of erinaceomorphs are from the Central Great Plains, where *Stenoechinus, Parvericius, Brachyerix*, and *Metechinus* occur.

CLARENDONIAN CHRONOFAUNA

Proscalopidae disappear during the Clarendonian Chronofauna, with *Mesoscalops* last appearing in the late Barstovian (Ba2) in the Northern Great Plains. The only non-Hemphillian occurrence of *Gaillardia* is also recorded in this chronofauna from Ba2 in the Southern Great Basin. The late Barstovian also has the highest diversity of talpoids from this time period, with five genera being found in the Central Great Plains and Pacific Northwest. The first appearance of the modern genus *Scalopus* is recorded from the middle

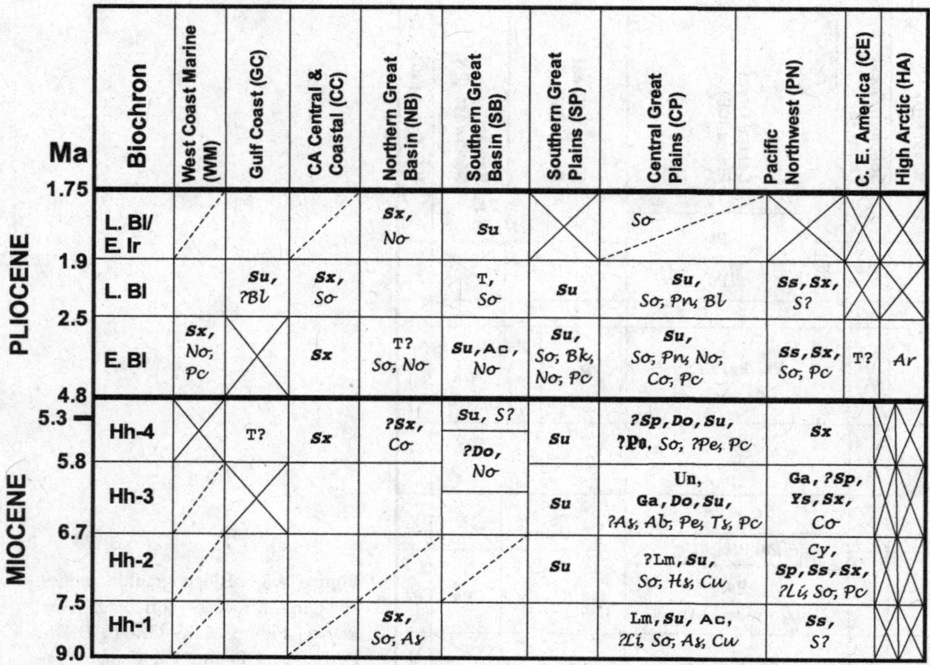

Ma	Biochron	West Coast Marine (WM)	Gulf Coast (GC)	CA Central & Coastal (CC)	Northern Great Basin (NB)	Southern Great Basin (SB)	Southern Great Plains (SP)	Central Great Plains (CP)	Pacific Northwest (PN)	C. E. America (CE)	High Arctic (HA)
PLIOCENE 1.75–1.9	L. Bl/ E. Ir				Sx, No	Su		So			
1.9–2.5	L. Bl		Su, ?Bl	Sx, So		T, So	Su	Su, So, Pn, Bl	Ss, Sx, S?		
2.5–4.8	E. Bl	Sx, No, Pc		Sx	T? So, No	Su, Ac, No	Su, So, Bk, No, Pc	Su, So, Pn, No, Co, Pc	Ss, Sx, So, Pc	T?	Ar
MIOCENE 5.3–5.8	Hh-4		T?	Sx	?Sx, Co	Su, S?; ?Do, No	Su	?Sp, Do, Su, ?Po, So, ?Pe, Pc	Sx		
5.8–6.7	Hh-3						Su	Un, Ga, Do, Su, ?Ax, Ab, Pe, Tx, Pc	Ga, ?Sp, Ys, Sx, Co		
6.7–7.5	Hh-2						Su	?Lm, Su, So, Hx, Cu	Cy, Sp, Ss, Sx, ?Li, So, Pc		
7.5–9.0	Hh-1				Sx, So, Ax			Lm, Su, Ac, ?Li, So, Ax, Cu	Ss, S?		

Figure 4.9. Biogeographic ranges of Mio-Pliocene lipotyphlans. Key as in Figure 4.3. ERINACEIDAE (Times New Roman Bold): ERINACEINAE: **Un**, *Untermannerix*. TALPOIDEA: TALPIDAE (Courier Bold): **Ga**, *Gaillardia*. TALPININAE: DESMANINI (Courier Plain). Lm, *Lemoynea*. CONDYLURINI (Courier Italics). *Cy*, *Condylura*. SCALOPINI (Courier Bold Italics): ***Do***, *Dominoides*; ***Sp***, *Scalapoides*; ***Ss***, *Scapanus (Scapanus)*; ***Su***, *Scalopus*; ***Sx***, *Scapanus (Xereoscapanus)*; ***Ys***, *Yanshuella*. TRIBE UNCERTAIN (Bank Gothic Bold): **AC**, *Achlyoscapter*. **T?**, Talpid indet (Courier Bold). SORICOIDEA: PLESIOSORICIDAE (Blackmore Let): **Po**, *Plesiosorex*. SORICIDAE: LIMNOECINAE (Lucida Calligraphy): *Li*, *Limnoecus*. SORICINAE (Lucida Handwriting Bold): *Ab*, *Anchiblarinella*; *Ar*, *Arcitsorex*; *Ax*, *Alluvisorex*; *Bk*, *Beckiasorex*; *Bl*, *Blarina*; *Co*, *Cryptotis*; *Cu*, *Crusafontina*; *Hx*, *Hesperosorex*; *No*, *Notiosorex*; *Pc*, *Paracryptotis*; *Pe*, *Petenyia*; *Pn*, *Planisorex*; *So*, *Sorex*; *S?*, Soricid indet.; *Tx*, *Tregosorex*.

Clarendonian in the Central Great Plains and the last appearance of *Mystipterus* occurs near the Cl2/3 boundary in the Central Great Plains.

Erinaceomorphs are relatively common throughout the Clarendonian Chronofauna. They are especially diverse in the late Barstovian of the Central Great Plains, with five genera being present (*Lantanotherium, Untermannerix, Parvericius, Brachyerix*, and *Metechinus*). Except for a late occurrence of *Untermannerix* in Hh3 of the Central Great Plains, erinaceomorphs disappear from North America by the end of the Clarendonian Chronofauna.

Plesiosorex, except for a questionable occurrence in Hh4 in the Central Great Plains, disappears at the end of the Clarendonian. Heterosoricine and limnoecine shrews also have their last appearances in the Clarendonian Chronofauna (there is a questionable record of *Limnoecus* from the earliest Hemphillian in the Central Great Plains). Soricid diversity is relatively stable through this time period with two or three genera normally being represented from any given biochron (Figure 4.8). However, as was the case with erinaceomorphs and talpoids, soricid diversity peaks in the late Barstovian with four genera being known from the Central Great Plains and five from the Pacific Northwest.

MIO-PLIOCENE CHRONOFAUNA

The Mio-Pliocene North American Tertiary lipotyphlans are represented by talpids and soricine shrews. The most complete temporal sample is from the Central Great Plains, where shrews are more diverse throughout except in the latest Hemphillian where both shrews and moles are represented by three genera (Figure 4.9). The highest diversity of moles is recorded in the middle Hemphillian (Hh2/3) of the Pacific Northwest, while shrew diversity is highest (five genera) in both Hh3 and early Blancan in the Central Great Plains. By the end of the Blancan into the Irvingtonian, only the extant genera *Scalopus, Scapanus, Sorex*, and *Notiosorex* are represented in the fossil record of North America.

REFERENCES

Arnason, U., Adegoke, J. A., Bodin, K., *et al.* (2002). Mammalian mitogenomic relationships and the root of the eutherian tree. *Proceedings of the National Academy of Sciences, USA*, 99, 8151–6.

Asher, R. J. (2005). Insectivoran-grade placentals. In *The Rise of Placental Mammals*, ed. K. D. Rose and J. D. Archibald, pp. 50–70. Baltimore, MD: Johns Hopkins University Press.

Asher, R. J., McKenna, M. C., Emry, R. J., Tabrum, A. R., and Kron, D. G. (2002). Morphology and relationships of *Apternodus* and other extinct, zalambdodont, placental mammals. *Bulletin of the American Museum of Natural History*, 273, 1–117.

Asher, R. J., Novacek, M. J., and Geisler, J. H. (2003). Relationships of endemic African mammals and their fossil relatives based on morphological and molecular evidence. *Journal of Mammalian Evolution*, 10, 131–63.

Barnosky, A. D. (1981). A skeleton of *Mesoscalops* (Mammalia, Insectivora) from the Miocene Deep River Formation, Montana, and a review of the proscalopid moles: evolutionary, functional, and stratigraphic relationships. *Journal of Vertebrate Paleontology*, 1 285–339.

(1982). A new species of *Proscalops* (Mammalia, Insectivora) from the Arikareean Deep River Formation, Meagher County, Montana. *Journal of Paleontology*, 56, 103–11.

Bloch, J. I. and Boyer, D. M. (2001). Taphonomy of small mammals in freshwater limestones from the Paleocene of the Clarks Fork Basin. [In

Paleocene–Eocene Stratigraphy and Biotic Change in the Bighorn and Clarks Fork Basins, Wyoming, ed. P. D. Gingerich.] *University of Michigan Papers on Paleontology*, 33, 185–98.

Bloch, J. I., Rose, K. D., and Gingerich, P. D. (1998). New species of *Batodonoides* (Lipotyphla, Geolabididae) from the early Eocene of Wyoming: smallest known mammal? *Journal of Mammalogy*, 79, 804–27.

Bloch, J. I., Boyer, D. M., and Houde, P. (2004). New skeletons of Paleocene–Eocene *Labidolemur kayi* (Mammalia, Apatemyidae): ecomorphology and relationship of apatemyids to primates and other mammals. *Journal of Vertebrate Paleontology*, 24(suppl. to no. 3), p. 40A.

Bloch, J. I., Secord, R., and Gingerich, P. D. (2004). Systematics and phylogeny of late Paleocene and early Eocene Palaeoryctinae (Mammalia, Insectivora) from the Clarks Fork and Bighorn Basins, Wyoming. *Contributions from the Museum of Paleontology, University of Michigan*, 31, 119–54.

Boyer, D. M. and Bloch, J. I. (2003). Comparative anatomy of the pentacodontid *Aphronorus orieli* (Mammalia, Pantolesta) from the Paleocene of the western Crazy Mountains Basin, Montana. *Journal of Vertebrate Paleontology*, 23(suppl. to no. 3), p. 36A.

Butler, P. M. (1956). The skull of *Ictops* and the classification of the Insectivora. *Proceedings of the Zoological Society of London*, 126, 453–81.

(1972). The problem of insectivore classification. In *Studies in Vertebrate Evolution*, ed. K. A. Joysey and T. S. Kemp, pp. 253–65. New York: Winchester.

(1988). Phylogeny of the insectivores. In *The Phylogeny and Classification of the Tetrapods,* Vol. 2: *Mammals*, ed. M. J. Benton, pp. 117–41. Oxford: Clarendon Press.

Cavigelli, J. P. (1997). A preliminary description of a *Leptictis* skeleton from the White River Formation of eastern Wyoming. *Tate Geological Museum Guidebook*, 2, 101–18.

Churchfield, S. (1990). *The Natural History of Shrews*. Ithaca, NY: Cornell University Press.

Douady, C. J., Chatelier, P. I., Madsen, O., *et al.* (2002). Molecular phylogenetic evidence confirming the Eulipotyphla concept and in support of hedgehogs as the sister group to shrews. *Molecular Phylogenetics and Evolution*, 25, 200–9.

Eisenberg, J. F. (1980). Biological strategies of living conservative mammals. In *Comparative Physiology: Primitive Mammals*, ed. K. Schmidt-Nielsen, L. Bolis, and C. R. Taylor, pp. 13–30. Cambridge, UK: Cambridge University Press.

Emerson, G. L., Kilpatrick, C. W., McNiff, B. E., Ottenwalder, J., and Allard, M. W. (1999). Phylogenetic relationships of the order Insectivora based on complete 12S rRNA sequences from mitochondria. *Cladistics*, 15, 221–30.

Feldhamer, G. A., Drickamer, L. C., Vessey, S. H., and Merritt, J. F. (2004). *Mammalogy: Adaptation, Diversity, Ecology*, 2nd edn. New York: McGraw-Hill.

Fleagle, J. G. (1999). *Primate Adaptations and Evolution*, 2nd edn. London: Academic Press.

Fox, R. C. and Scott, C. S. (2005). First evidence of a venom delivery apparatus in extinct mammals. *Nature*, 435, 1091–3.

Gazin, C. L. (1959). Early Tertiary *Apheliscus* and *Phenacodaptes* as pantolestid insectivores. *Smithsonian Miscellaneous Collections*, 139, 1–7.

(1969). A new occurrence of Paleocene mammals in the Evanston Formation, southwestern Wyoming. *Smithsonian Contributions to Paleobiology*, 2, 1–16.

Gingerich, P. D., Houde, P., and Krause, D. W. (1983). A new earliest Tiffanian (Late Paleocene) mammalian fauna from Bangtail Plateau, Western Crazy Mountain Basin, Montana. *Journal of Paleontology*, 57, 957–70.

Hooker, J. J. (2001). Tarsals of the extinct insectivoran family Nyctitheriidae (Mammalia): evidence for archontan relationships. *Zoological Journal of the Linnaean Society*, 132, 501–29.

Jepsen, G. L. (1934). A revision of the American Apatemyidae and the description of a new genus, *Sinclairella*, from the White River Oligocene of South Dakota. *Proceedings of the American Philosophical Society*, 74, 287–305.

Kalthoff, D. C., Koenigswald, W., von, and Kurz, C. (2004). A new specimen of *Heterohyus nanus* (Apatemyidae, Mammalia) from the Eocene of Messel (Germany) with unusual soft-part preservation. *Courier Forschungs-Institut Senckenberg*, 252, 1–12.

Koenigswald, W., von (1980). Das skelett eines Pantolestiden (Proteutheria, Mamm.) aus dem mittleren Eozan von Messel bei Darmstadt. *Palaontologische Zeitschrift*, 54, 267–87.

(1990). Die Palaobiologie der Apatemyiden (Insectivora s. l.) und die Ausdeutung der Skelettfunde von *Heterohyus nanus* aus dem Mitteleozan von Messel bei Darmstadt. *Palaeontographica Abteilung A Palaeozoologie-Stratigraphie*, 210, 41–77.

Koenigswald, W., von and Schierning, H.-P. (1987). The ecological niche of an extinct group of mammals, the Early Tertiary apatemyids. *Nature*, 326, 595–7.

Koenigswald, W., von, Rose, K. D., Grande, L., and Martin, R. (2005). First apatemyid skeleton from the Lower Eocene Fossil Butte Member, Wyoming (USA), compared to the European apatemyid from Messel, Germany. *Palaeontographica A*, 272, 149–69.

MacPhee, R. D. E. and Novacek, M. J. (1993). Definition and relationships of Lipotyphla. In *Mammal Phylogeny,* Vol. 2: *Placentals*, ed. F. Szalay, M. J. Novacek, and M. C. McKenna, pp. 13–31. New York: Springer.

Matthew, W. D. (1909). The Carnivora and Insectivora of the Bridger Basin, middle Eocene. *American Museum of Natural History Memoirs*, 9, 289–567.

McDowell, S. B. (1958). The greater Antillean insectivores. *Bulletin of the American Museum of Natural History*, 115, 113–214.

McKenna, M. C. (1963). Primitive Paleocene and Eocene Apatemyidae (Mammalia, Insectivora) and the primate–insectivore boundary. *American Museum Novitates*, 2160, 1–39.

Novacek, M. J. (1986). The skull of leptictid insectivorans and the higher-level classification of eutherian mammals. *Bulletin of the American Museum of Natural History*, 183, 1–112.

Pfretzschner, H. U. (1999). *Buxolestes minor* n. sp.: ein neuer Pantolestide (Mammalia, Proteutheria) aus der eozänen Messel-Formation. *Courier Forschungs-Institut Senckenberg*, 216, 19–29.

Richard, P. B. (1973). Capture, transport, and husbandry of the Pyrenian desman *Galemys pyrenaicus*. *International Zoo Yearbook*, 13, 175–7.

Richter, G. (1987). Untersuchungen zur Ernährung eozäner Säuger aus der Fossilfundstätte Messel bei Darmstadt. *Courier Forschungsinstitut Senckenberg*, 91, 1–33.

Roca, A. L., Bar-Gal, G. K., Eizirik, E., *et al.* (2004). Mesozoic origin for West Indian insectivores. *Nature*, 429, 649–51.

Rose, K. D. (1999). Postcranial skeleton of Eocene Leptictidae (Mammalia), and its implications for behavior and relationships. *Journal of Vertebrate Paleontology*, 19, 355–72.

Rose, K. D. and Lucas, S. G. (2000). An early Paleocene palaeanodont (Mammalia,?Pholidota) from New Mexico, and the origin of Palaeanodonta. *Journal of Vertebrate Paleontology*, 20, 139–56.

Rose, K. D. and Koenigswald, W., von (2005). An exceptionally complete skeleton of *Palaeosinopa* (Mammalia, Cimolesta, Pantolestidae) from the Green River Formation, and other postcranial elements of the Pantolestidae from the Eocene of Wyoming (USA). *Palaeontographica Abteilung A*, 273, 55–96.

Seiffert, E. R. and Simons, E. L. (2001). *Widanelfarasia*, a diminutive new placental from the late Eocene of Egypt. *Proceedings of the National Academy of Sciences, USA*, 97, 2646–51.

Simpson, G. G. (1937a). The Fort Union of the Crazy Mountain Field, Montana, and its mammalian faunas. *Bulletin of the United States National Museum*, 169, 1–278.

(1937b). Additions to the upper Paleocene fauna of the Crazy Mountain Field. *American Museum Novitates*, 940, 1–15.

Springer, M. S., Cleven, G. C., Madsen, O., *et al.* (1997). Endemic African mammals shake the phylogenetic tree. *Nature*, 388, 61–4.

Springer, M. S., Murphy, W. J., Eizirik, E., and O'Brien, S. J. (2005). Molecular evidence for major placental clades. In *The Rise of Placental Mammals*, ed. K. D. Rose and J. D. Archibald, pp. 37–49. Baltimore, MD: Johns Hopkins University Press.

Stanhope, M. J., Waddell, V. G., Madsen, O., *et al.* (1998). Molecular evidence for multiple origins of Insectivora and for a new order of endemic African insectivore mammals. *Proceedings of the National Academy of Sciences, USA*, 95, 9967–72.

Symonds, M. R. E. (2005). Phylogeny and life histories of the "Insectivora": controversies and consequences. *Biological Reviews*, 80, 93–128.

Thewissen, H. and Gingerich, P. D. (1989). Skull and endocranial cast of *Eoryctes melanus*, a new palaeoryctid (Mammalia: Insectivora) from the early Eocene of western North America. *Journal of Vertebrate Paleontology*, 9, 459–70.

Van Valen, L. (1966). Deltatheridia, a new order of mammals. *Bulletin of the American Museum of Natural History*, 132, 1–126.

(1967). New Paleocene insectivores and insectivore classification. *Bulletin of the American Museum of Natural History*, 135, 217–84.

Waddell, P. J. and Shelley, S. (2003). Evaluating placental interordinal phylogenies with novel sequences including RAG1, γ-fibrinogen, ND6, and mt-tRNA, plus MCMC-driven nucleotide, amino acid, and codon models. *Molecular Phylogenetics and Evolution*, 28, 197–224.

5 "Proteutheria"

GREGG F. GUNNELL,[1] THOMAS M. BOWN,[2] JONATHAN I. BLOCH[3] and DOUGLAS M. BOYER[4]

[1]*Museum of Paleontology, University of Michigan, Ann Arbor, MI, USA*
[2]*Erathem-Vanir Geological, Pocatello, ID, USA*
[3]*Florida Museum of Natural History, University of Florida, Gainesville, FL, USA*
[4]*Stony Brook University, NY, USA*

INTRODUCTION

Romer (1966) proposed a new suborder, "Proteutheria," for insectivorous mammals that had no clear relationship to living insectivorans. Romer's concept of "Proteutheria" included leptictids, zalambdalestids, anagalids, paroxyclaenids, pantolestids, ptolemaiids, tupaiids, pentacodontids, apatemyids, and macroscelidians. "Proteutheria" as constituted by Romer is an unnatural grouping and cannot be sustained phylogenetically. According to McKenna and Bell (1997), the lowest-level grouping that contains all of the mammals discussed in this chapter is the Magnorder Epitheria (cohort Placentalia). Within the Epitheria, Romer's "proteutherians" are distributed unevenly in the Superorders Leptictida (leptictids) and Preptotheria.

"Proteutheria" is perhaps still best thought of as a paraphyletic group of archaic insectivorous mammals traditionally not placed in Lipotyphla. Bloch, Rose, and Gingerich (1998) showed that taxa included in this group generally exhibit larger body size than those grouped in Lipotyphla and argued that the term was still useful in representing an ecologically coherent subset of Paleogene faunas. We include Palaeoryctidae, Cimolestidae, Pantolestidae, and Apatemyidae as members of this informal group.

Palaeoryctidae are known from the early Paleocene through early Eocene in North America. Some species from the Late Cretaceous of Asia and Europe, and from the late Paleocene to early Eocene of Africa, may also be palaeoryctids (McKenna and Bell, 1997).

Cimolestids first appear in the Late Cretaceous in North America, represented by *Cimolestes*, and survive through the Duchesnean, last represented by *Didelphodus*. Cimolestids are otherwise known from the late Paleocene of Europe and Africa and the early Eocene of Europe and Asia and may be represented in the early Paleocene in South America (McKenna and Bell, 1997).

Pantolestids are a geographically widespread group with representatives from North America, Europe, Asia (e.g., Dashzeveg and Russell, 1985), and Africa (Gheerbrant, 1991). Pantolestids are known from the Torrejonian through the end of the Eocene (Chadronian) in North America but survive to the end of the Oligocene in Asia (*Oboia*; Gabunia, 1989).

Apatemyids range from the late Paleocene into the early Arikareean in North America. They are also represented in the Paleocene and Eocene in Europe but remain unknown from elsewhere (McKenna and Bell, 1997).

DEFINING FEATURES OF "PROTEUTHERIANS'

FAMILY PALAEORYCTIDAE

CRANIAL

Skull relatively low and elongate; rostrum and neurocranium of approximately the same size; postorbital constriction moderate; single sagittal crest distinct but low; incomplete zygomatic arches; auditory bulla ossified; rostral tympanic processes of petrosal that contribute to bulla to varying degrees; intrabullar annular ectotympanic; no pyriform fenestra; large epitympanic recess (Thewissen and Gingerich, 1989; Bloch, Secord, and Gingerich, 2004).

DENTAL

Upper molars protozalambdodont (Figure 5.1) with closely appressed, connate paracones and metacones; lower molars with high trigonids and narrow talonids; shearing crests well developed on cheek teeth; three upper and lower premolars except in *Pararyctes* (Fox, 2004).

Evolution of Tertiary Mammals of North America, Vol. 2. ed. C. M. Janis, G. F. Gunnell, and M. D. Uhen. Published by Cambridge University Press.

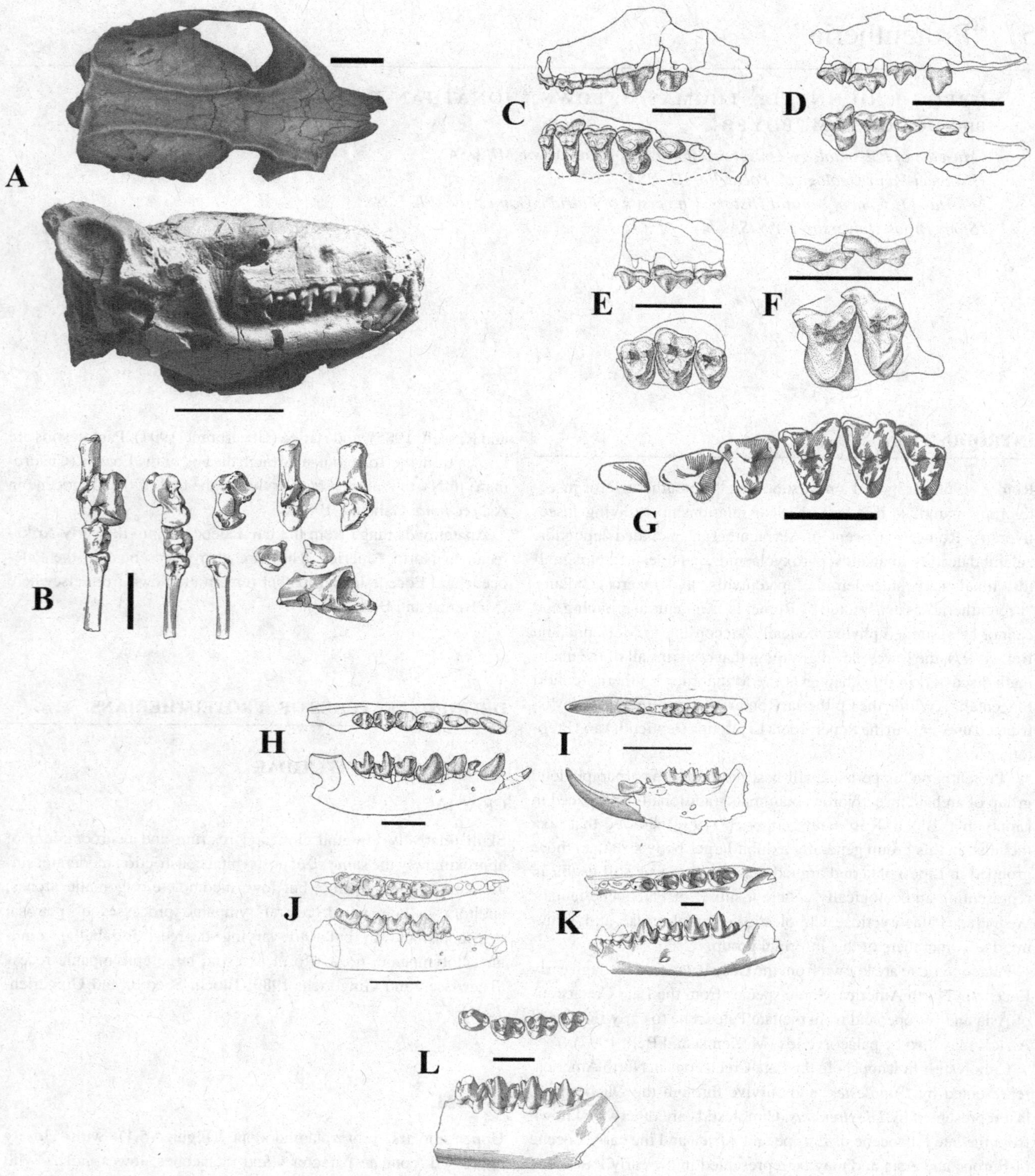

Figure 5.1. A. Skulls of apatemyid *Sinclairella dakotensis* (top, from Jepsen, 1934) and of palaeoryctid *Ottoryctes winkleri* (below, from Bloch, Secord, and Gingerich, 2004). B. Postcrania of creotarsine *Creotarsus lepidus* (from Matthew, 1918). Scale bars for A and B = 1 cm. C–G. Upper dentitions of selected "proteutherians;" C. Palaeoryctid *Aaptoryctes ivyi* (from Gingerich, 1982a). D. Apatemyid *Labidolemur kayi* (from Gingerich and Rose, 1982). E. Pentacodontine *Bisonalveus browni* (from Krause and Gingerich, 1983). F. Pantolestid *Bessoecetor septentrionalis* (from Krause and Gingerich, 1983). G. Cimolestid *Didelphodus absarokae* (from Matthew, 1918). Scale bars = 5 mm. H–L. Lower dentitions of selected "proteutherians;" H. Palaeoryctid *Aaptoryctes ivyi* (from Gingerich, 1982a). I. Apatemyid *Labidolemur kayi* (from Gingerich and Rose, 1982). J. Pentacodontine *Bisonalveus browni* (from Krause and Gingerich, 1983). K. Pantolestine *Palaeosinopa veterrima* (from Matthew, 1918). L. Cimolestid *Didelphodus absarokae* (from Matthew, 1918). Scale bars = 5 mm. (A (top): courtesy of the American Philosophical Society. A (bottom), C–F, H–J: courtesy of the University of Michigan. B, G, K, L: Courtesy of the American Museum of Natural History.)

POSTCRANIAL

The only known postcranial material for the group is that associated with *Palaeoryctes punctatus*, represented by a distal humerus and proximal ulna: humerus with prominent medial epicondyle that is proximally inflected with a hook; trochlea mediolaterally broad with a small arc of curvature; lacks fossa seen in talpids; ulna with shallow trochlea, truncated coronoid, and anteriorly inflected olecranon process of moderate length (Van Valen, 1966).

FAMILY CIMOLESTIDAE

CRANIAL

Broad nasals that may contact lacrimals; lacrimal with small exposure on face and prominent foramen; possible palatine–lacrimal contact; palatine extends to P4, lacks postpalatine spine, and has large tori; low, single sagittal crest; occipital condyles dorsally shielded by flaring nuchal crests; infraorbital foramen large, situated at level of P3–4; parietal foramina for tributaries of sagittal venous sinuses present; no evidence for ossified bulla; no development of caudal tympanic process of petrosal – foramen rotundum exposed; promontorium with three grooves for "auricular nerve" (running mediolateral), promontory branch of internal carotid artery (running anterolaterally) and entocarotid artery (running anteroposterior on medial aspect of promontorium); tympanohyal element fused to posterolateral wall of stapedius fossa as well as to small mastoid process; facial canal closed by crista parotica; pyriform fenestra present; glenoid fossa flat with prominent postglenoid and entoglenoid processes; postglenoid foramen present posterior to postglenoid process; no separate vidian foramen or groove preserved (based on *Puercolestes* [= *Cimolestes*] and *Didelphodus*; Van Valen 1966).

DENTAL

Dental formula I?/3, C?/1, P?/4, M3/3; incisors small, peglike and procumbent; canine large and slightly procumbent; premolars premolariform; upper molars very transverse with expanded stylar shelves and no hypocones; lower molars with high trigonids; m1 < m2 < m3.

POSTCRANIAL

Astragalus with posteriorly extensive, grooved tibial facet exhibiting sharp medial and lateral margins with lateral half wider than medial half, fibular facet with dorsally facing shelf, superior astragalar foramen absent, neck relatively long, distinctly separate navicular and sustentacular facets; calcaneum with reduced calcaneofibular facet, peroneal tuberosity and plantar tubercle distally positioned, no distinct groove between cuboid facet and plantar tuberosity marking calcaneocuboid ligament insertion; ungual phalanx straight and flat (based on *Didelphodus* [Van Valen, 1966] and *Procerberus* and *Cimolestes* [Szalay, 1977]).

FAMILY PANTOLESTIDAE

Only pantolestine members of the Pantolestidae are currently known to exhibit aquatic adaptations. Several nearly complete pantolestid skeletons have been recovered from lacustrine limestones and oil shales (Pfretzchner, 1993).

CRANIAL

Pantolestidae: Facial portion of skull short; neurocranium elongate and low; marked postorbital constriction; infraorbital foramen relatively large; no postpalatine spine on palate; basicranium broad; single sagittal crest; occipital condyles dorsally shielded by flaring nuchal crest; anterior nares very large and posteriorly extensive (based on *Pantolestes*, Matthew, 1909; *Buxolestes*, Pfretzchner, 1993; Rose and Lucas 2000; *Aphronorus*, Gingerich, Houde, and Krause, 1983; Boyer and Bloch, 2003).

Pantolestinae: Nares extends to level of M2; auditory bulla unossified; basioccipital mediolaterally broad (based on *Pantolestes*, Matthew, 1909; *Buxolestes*, Pfretzchner, 1993).

Pentacodontinae: Promontorium with bifurcating groove on lateral aspect; ossified bulla lacking contribution from basisphenoid (*Aphronorus*, Boyer and Bloch, 2003)

DENTAL

Pantolestidae: Dental formula I3/3, C1/1, P4/4, M3/3; I3 enlarged and bladelike; canines large and massive; p4 premolariform; premolars and molars with low, massive cusps; stylar shelves reduced or absent; external cingula on lower molars reduced or absent; mandibular angle robust and blunt (Matthew, 1909; Simpson, 1937a; *Buxolestes*: Pfretzchner, 1993; Rose and Lucas 2000; *Aphronorus*: Gingerich, Houde, and Krause, 1983; Boyer and Bloch, 2003).

Pantolestinae: Upper incisors peglike, separated, except for I3, which is enlarged and bladelike; distinct mental foramen below m1 (Matthew, 1909; Simpson, 1937a; *Buxolestes*, Pfretzchner, 1993).

Pentacodontinae: P4/p4 semimolariform and greatly enlarged in most taxa; p4 with very heavy, posteriorly angled protoconid, well-developed metaconid, and basined heel; P4 with massive, conical, paracone, smaller but sharply distinct metacone, styles small or lacking, and large, low protocone with widely expanded anterior and posterior cingula; trigonids relatively low; m3 trigonid smaller than m1–2 (Simpson, 1937b; Gazin, 1959, 1969; Van Valen, 1967; Gingerich, Houde, and Krause, 1983).

POSTCRANIAL

Pantolestidae: Humerus with strong deltoid, pectoral, and supinator crests and wide distal end; radius and ulna not fused; tibia and fibula united distally; astragalus with short, deeply grooved trochlea, and convex, pyriform head; superior astragalar foramen present, but small and located on tibial facet; pes habitually everted (based on *Pantolestes*: Matthew, 1909; *Buxolestes*: Pfretzchner, 1993; Rose and Lucas 2000; *Aphronorus*: Boyer and Bloch, 2003).

Pantolestinae: Craniocaudally extensive spinus process on C2; lumbar zygopophyses flat; tail vertebrae massive, tail long; humerus with deltoid crest that is sharp and faces laterally, and wide distal end; radius and ulna not fused; manus and pes with elongate, flattened claws; femur with flattened shaft, possessing third trochanter, short and broad patellar trochlea; tibia very curved with cnemial crest reduced to process and placed nearly at mid shaft; fibular–calcaneal articulation large; astragalus with short neck, confluent navicular and sustentacular facets; forelimbs and especially metacarpals much shorter than hindlimbs and metatarsals, respectively (based on *Pantolestes*: Matthew, 1909; *Buxolestes*: Pfretzchner, 1993; *Palaeosinopa*: Rose and Von Koenigswald, 2005).

Pentacodontinae: Humerus with prominent deltoid tuberosity expanded into shelflike projection that is concave and facing anteriorly; ulna with deep shaft strongly grooved on lateral aspect; radius with elliptical head, shaft flattened mediolaterally towards distal end, and arcuate ridge on posterolateral aspect marking origin of deep digital flexors; tibial shaft mediolaterally flattened; astragalus with distinctly separate navicular and sustentacular facets (Boyer and Bloch, 2003).

FAMILY APATEMYIDAE

CRANIAL

Lack ossified bulla; groove on promontorium marking course of internal carotid artery; large, free ectotympanic ring; infraorbital foramen above M1; small foramina on skull roof (Jepsen, 1934; McKenna, 1963; Koenigswald, 1990; Bloch, Boyer, and Houde, 2004).

DENTAL

First lower incisor enlarged and procumbent with root extending posteriorly to m3; I1 lacking lateral enamel; canines absent; p3 bladelike and single-rooted; lower molars low, crowned with labial paraconids; P3–4 double-rooted but unicuspate; upper molars simple, lacking mesostyles; depression on mandible under p4 or m1, posterior mental foramen generally very large below m1.

POSTCRANIAL

Distal femora anteroposteriorly deep; astragalus grooved; metatarsals elongate; unguals mediolaterally narrow; elongate second and third hand digits (Koenigswald and Schierning, 1987; Koenigswald, 1990; Bloch and Boyer, 2001; Bloch, Boyer, and Houde, 2004; Kalthoff, Koenigswald, and Kurz, 2004; Koenigswald *et al.*, 2004, 2005).

SYSTEMATICS

PALAEORYCTIDAE

The systematic position of Palaeoryctidae is uncertain. Van Valen (1966) included the subfamilies Deltatheridiinae, Palaeoryctinae, and Micropternodontinae in Palaeoryctidae and classified the family in the Deltatheridia. Later publications removed deltatheridiines (Butler and Kielan-Jaworowska, 1973) and micropternodontines (Butler, 1972) from Palaeoryctidae. McKenna (1975) classified Palaeoryctidae in Kennalestida, Szalay (1977) in Leptictimorpha, and Butler (1972), Novacek (1976), Kielan-Jaworowska (1981), and Bown and Schankler (1982) in Proteutheria. While Gingerich (1982a) and Thewissen and Gingerich (1989) placed Palaeoryctidae in Insectivora, and McKenna, Xue, and Zhou (1984) placed it in Soricomorpha, others have argued that Palaeoryctidae are not directly related to Lipotyphla (McDowell, 1958; Van Valen, 1966, 1967; Lillegraven, 1969; Szalay, 1977; Butler, 1988). Thewissen and Gingerich (1989) included palaeoryctids in Insectivora (sensu Novacek, 1986) in order to recognize a close relationship between taxa included in Palaeoryctidae, Leptictidae, and Lipotyphla. This broad classification is at least supported by the recent phylogenetic hypotheses published by Asher *et al.* (2002) in which "Lipotyphla" is paraphyletic with respect to a number of fossil taxa, including the palaeoryctids *Eoryctes* and *Pararyctes*.

Palaeoryctidae as recognized here includes four species of *Palaeoryctes* (Matthew, 1913; Van Valen, 1966; Gunnell, 1994; Bloch, Secord, and Gingerich, 2004), two species of *Pararyctes* (Van Valen, 1966), one species of *Aaptoryctes* (Gingerich, 1982a), one species of *Eoryctes* (Thewissen and Gingerich, 1989), one species of *Ottoryctes* (Bloch, Secord, and Gingerich, 2004), and one species of *Lainoryctes* (Fox, 2004). This is similar to the classification of McKenna and Bell (1997) except that it includes *Pararyctes* (also excluded by Kellner and McKenna, 1996) in order to recognize, given our current state of phylogenetic resolution, that excluding taxa like *Pararyctes* that appear closely related will only complicate later phylogenetic studies and classification (Bloch, Secord, and Gingerich, 2004) (Figure 5.2).

CIMOLESTIDAE

Cimolestids include some of the most primitive members of Tertiary "proteutherians" as recognized here. *Cimolestes* includes several Late Cretaceous species and cimolestids have been proposed as possible ancestral or sister taxa for a variety of other groups including Taeniodonta, Carnivora, Creodonta, Apatemyidae, Palaeoryctidae, and Condylarthra (Kielan-Jaworowska, Bown, and Lillegraven, 1979). We follow McKenna and Bell (1997) in recognizing Cimolestidae as a family of the suborder Didelphodonta. Except for *Paleotomus*, which we consider to be a pantolestine, the North American members of Cimolestidae recognized here are the same as those included by McKenna and Bell (1997).

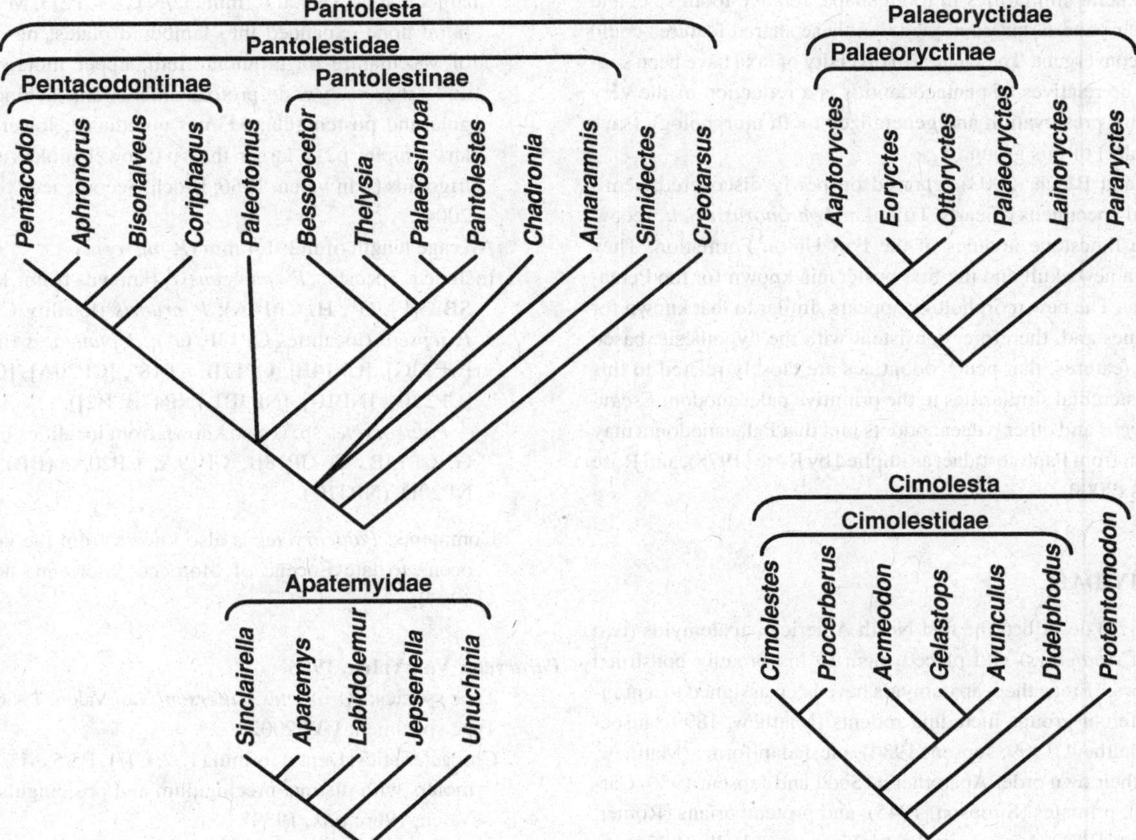

Figure 5.2. Proposed relationships of Pantolesta, Palaeoryctidae, Apatemyidae, and Cimolesta. See text for characters supporting labeled clades.

PANTOLESTIDAE

Matthew (1899) initially suggested a primate or creodont affinity for pantolestids. Following his description of a relatively complete specimen of *Pantolestes natans*, Matthew (1909) rejected all previous hypotheses of relationship and asserted that pantolestids were "insectivores of a peculiar type." He noted that *Pentacodon* probably belonged in the Pantolestidae, partly because of the position of the mental foramen (below m1). He did, however, also suggest a broader relationship of pantolestids to aquatic carnivores, stating: "the arrangement of the fronto-nasal maxillary sutures, the large infraorbital foramen, the arrangement of the upper incisors and canines, the form of the astragalus, etc. compare with the walrus . . . in such a way as to suggest some degree of actual affinity." However, Matthew also suggested that other features of *Pantolestes* were probably convergent on Pinnipedia. He noted "flattening of the distal part of the femur, reduction of the cnemial crest and internal malleolus of the tibia and the facet for the tibia on the neck of the astragalus, the elongate pisiform, elongation of the metapodials and especially of proximal phalanges, thickening of the metapodials and proximal phalangeal shafts, flattening of the distal phalangeal shafts and unguals and above all, the conversion of the ball-and-socket joints of the metapodials and phalanges into hinge-joints" as "probably adaptive." Why he felt that some features were "synapomorphies," while regarding other shared features as convergences, is unclear. He also noted that the odd wear of the teeth in combination with the body size probably indicated a diet consisting of fish, turtles, and clams.

The Pentacodontinae (Simpson, 1937b) is a fairly diverse group recognized here as a subfamily of Pantolestidae, which is as Simpson (1937b) originally conceived it (see also McKenna and Bell, 1997). Their fossils are known from Torrejonian (*Coriphagus*) through middle Tiffanian (*Bisonalveus*). Pentacodontines were originally differentiated from pantolestines on the basis of a hypertrophied submolariform fourth premolar, as well as by relatively lower molar trigonids. Doubt regarding the precise relationships of this group began to accrue, leading Van Valen (1966) to elevate it to the family level. Previously, Gazin (1959) suggested that the condition of the premolars was phylogenetically significant and indicated that some hyopsodontid condylarths (*Phenacodaptes* and *Apheliscus*) should be considered closer relatives of pentacodontids than to other condylarths. Other researchers (Saban, 1954, 1958; McKenna, 1960; Van Valen, 1966) agreed with this interpretation, although Van Valen believed the premolar similarity to be convergent. Rose (1978) noted the possibility of a close relationship to palaeanodonts on the basis of very general similarities in tooth shape, relative tooth size, and dental wear pattern, but cautioned that these shared features could easily be convergent. The fact that a diversity of taxa have been suggested to be relatives of pentacodontids is a reflection of the very fragmentary preservation and generalized tooth morphology (save the premolar) of this group.

Boyer and Bloch (2003) reported on newly discovered, semi-articulated specimens of early Tiffanian *Aphronorus orieli*, recovered from limestone nodules of the Fort Union Formation. They represent a new skull and the first postcrania known for the Pentacodontidae. The new morphology appears similar to that known for pantolestines and, therefore, consistent with the hypothesis, based on dental features, that pentacodontines are closely related to this group. Postcranial similarities to the primitive palaeanodont *Escavadodon zygus* and other palaeanodonts hint that Palaeanodonta may have arisen from Pantolestidae, as implied by Rose (1978), and Rose and Lucas (2000).

APATEMYIDAE

Marsh (1872) described the first North American apatemyids (two species of *Apatemys*) and placed them in his broadly construed "Insectivora." Since then, apatemyids have been assigned to a number of different groups including rodents (Matthew, 1899), insectivores (Matthew, 1909; Jepsen, 1934), plesiadapiforms (Matthew, 1918), in their own order Apatotheria (Scott and Jepsen, 1936; Carroll, 1988), primates (Simpson, 1945), and proteutherians (Romer, 1966; Rose, 1981). Most recently, McKenna and Bell (1997) recognized Apatotheria as a suborder of Cimolesta.

INCLUDED GENERA OF "PROTEUTHERIANS"

The locality numbers listed for each genus refer to the list of unified localities in Appendix I. The locality numbers may be listed more than one way. The acronyms for museum collections are listed in Appendix III.

Parentheses around the locality (e.g., [CP101]) mean that the taxon in question at that locality is cited as an "aff." or "cf." the taxon in question. Parentheses are usually used for individual species, implying that the genus is firmly known from the locality, but the actual species identification may be questionable. Question marks in front of the locality (e.g., ?CP101) mean that the taxon is questionably known from that locality, implying some doubt that the taxon is actually present at that locality, either at the genus or species level. An asterisk (*) indicates the type locality.

CIMOLESTA MCKENNA, 1975

PALAEORYCTIDAE WINGE, 1917

Palaeoryctes Matthew, 1913

Type species: *Palaeoryctes puercensis* Matthew, 1913.

Type specimen: AMNH 15923.

Characteristics: Dental formula I?/3, C1/1, P3/3, M3/3; temporal bone expanded into lambdoid plates; no evidence of vasculature on promontorium; upper molars narrow and transversely wide, protozalambdodont, lacking precingula and postcingula; M3/m3 unreduced; lower premolars simple; p2–3 larger than p4; lower molars with tall trigonids (Van Valen, 1966; Bloch, Secord, and Gingerich, 2004).

Average length of m2: 1.4 mm (*P. puercensis*).

Included species: *P. puercensis* (known from localities SB23[F], G*, H, CP14A); *P. cruoris* (locality CP24A*); *P. jepseni* (localities CP13F, G*); *P. punctatus* (localities [CP13G], [CP14E], CP17B, CP18*, [CP20A], [CP24A], [CP25B], [NP1C], [NP3B], [NP47B, B2]).

Palaeoryctes sp. is also known from localities CP13(E), G, CP14B, C, CP18B, CP19A, CP20A, (BB), NP7B, NP20D, (NP47C).

Comments: *Palaeoryctes* is also known from the late Paleocene to late Eocene of Morocco (McKenna and Bell, 1997).

Pararyctes Van Valen, 1966

Type species: *Pararyctes pattersoni* Van Valen, 1966.

Type specimen: UW 2002.

Characteristics: Dental formula I?/?, C1/1, P5/5, M3/3; upper molars with distinct precingulum and postcingulum (Van Valen, 1966; Fox, 1983).

Average length of m1: 1.6 mm.

Included species: *P. pattersoni* (known from localities CP16A*, NP1C, NP3B–D, NP4, NP19IIC, NP47B); *P. rutherfordi* (locality NP1C*).

Pararyctes sp. is also known from localities (CP20A), NP1C, NP2, NP3A–C, F, NP7D, (NP16C), (NP47C).

Aaptoryctes Gingerich, 1982a

Type species: *Aaptoryctes ivyi* Gingerich, 1982a.

Type specimen: UMMP 77291.

Characteristics: Dental formula I?/3, C1/1, P3/3, M3/3; upper molars simple, protozalambdodont with broad stylar shelf, connate paracone and metacone, no precingulum or postcingulum; lower molars with broad high trigonids and narrow talonids; premolars enlarged and inflated (especially P4/p4) (Gingerich, 1982a).

Average length of m2: 2.5 mm.

Included species: *A. ivyi* only (known from localities CP13F, G*).

Aaptoryctes sp. is also known from locality CP22B.

Eoryctes Thewissen and Gingerich, 1989

Type species: *Eoryctes melanus* Thewissen and Gingerich, 1989.

Type specimen: UM 68074.

Characteristics: Dental formula I?/?, C?/1, P?/3, M3/3; temporal bone expanded into lambdoid plates; middle ear arteries enclosed in bony tubes; bulla ossified; retroarticular foramen small; P4 uninflated, transversely elongate with relatively small parastyle; upper molars with strongly inflected labial cingulum and weak lingual cingula; p1 absent; p4 uninflated and lacking paraconid; lower molars with relatively high trigonid and labially placed hypoconid (Thewissen and Gingerich, 1989).

Average length of m1: 2.3 mm.

Included species: *E. melanus* only (known from localities CP19A, B).

***Lainoryctes* Fox, 2004**

Type species: *Lainoryctes youzwyshyni* Fox, 2004.

Type specimen: UALVP 43003.

Characteristics: Dental formula I?/?, C1/?, P3/?, M3/?; relatively large size; P3 lacking protocone; P4 lingually slender; M1 paracone and metacone moderately connate; upper molars with conules (Fox, 2004).

Average length of M2: 2.6 mm.

Included species: *L. youzwyshyni* only, known from locality NP3E only.

***Ottoryctes* Bloch, Secord, and Gingerich, 2004**

Type species: *Ottoryctes winkleri* Bloch, Secord, and Gingerich, 2004.

Type specimen: UM 72624.

Characteristics: Dental formula I3/3, C1/1, P3/3, M3/3; temporal bone expanded into lambdoid plates; C1 double-rooted with distinct anterior and posterior accessory cuspules; P2 double-rooted; P3 lacking protocone; P4 with metacone; M3 metacone reduced; c1 relatively small but larger than p2; p3 with anterior cuspule; p4 with narrow and pointed protoconid and bladelike unbasined talonid; m1–3 with relatively expanded talonids (Bloch, Secord, and Gingerich, 2004).

Average length of m2: 1.95 mm.

Included species: *O. winkleri* only, known from locality CP19C* only.

Indeterminate palaeoryctids

In addition to the taxa above, indeterminate or undescribed palaeoryctids are known from localities CC4, CP5A, CP11IIA, CP15B, NP1C, NP3A0, D, NP7B, NP16C, NP17, NP19IIA, NP48B.

CIMOLESTA INCERTAE SEDIS

***Protentomodon* Simpson, 1928**

Type species: *Protentomodon ursirivalis* Simpson, 1928.

Type specimen: AMNH 22164.

Characteristics: m2–3 trigonid elevated well above heel, cusps acute and angulate, metaconid anterointernal to protoconid, large, and not as high as protoconid, paraconid distinct, acute, low and median on tooth, united by sharp ridge to protoconid; m2 talonid short, low, about as wide as trigonid, basin surrounded by continuous, sharp, raised rim with entoconid, hypoconulid, and hypoconid as elevations of equal prominence, hypoconulid somewhat closer to hypoconid; m3 talonid narrower and longer, hypoconulid median and elevated above other more indistinct cusps; m2–3 with cingulum running sharply downward and externally from near paraconid around the external base of protoconid; dentary very slender and elongate, with posterior mental foramen below m1 (Simpson, 1928).

Average length of m2: 2.0 mm.

Included species: *P. ursirivalis* only (known from localities CP17A*, NP20E).

Comments: Simpson originally placed *Protentomodon* in a new family, Nyctitheriidae (which it does not at all resemble) while Van Valen (1967) suggested affinities with Pentacodontidae. McKenna (1960) suggested that the genus might have apatemyid affinities while McKenna and Bell (1997) placed it in Cimolestidae. The genus is too poorly known to ascertain its relationship positively, but the idea of an affinity with apatemyids may have merit given the presence of elevated trigonids that are canted somewhat posteriorly, the medial paraconid, talonid structure, mental foramen beneath m1, and the extension of the symphysis posteriorly to a point beneath the p3–4 boundary.

DIDELPHODONTA MCKENNA, 1975

***Cimolestes* Marsh, 1889 (including *Nyssodon*; *Puercolestes*)**

Type species: *Cimolestes incisus* Marsh, 1889.

Type specimen: YPM 11775.

Characteristics: Dental formula I?/2, C?/1, P?/4, M3/3; P3–4 with protocones; P4 lacks or has a small metacone; upper molars with relatively broad stylar shelves, small stylar cusps, closely approximated, high, connate paracone and metacone, postparaconule and premetaconule wings extending from conules to bases of paracone and metacone, and precingula and postcingula absent or small; p4 trenchant with small anterior accessory cusp, unicuspid talonid, and lacking metaconid; lower molar paraconids lingual positions; metaconids slightly lower and smaller than protoconids (Clemens, 1973).

Average length of m2: 3.4 mm (*C. incisus*).

Included species: *C. incisus* (known from localities NP15A, [NP17]); *C. cerberoides* (localities [NP6], [NP7A]); *C. magnus* (locality NP15A); *C. simpsoni* (localities SB23A*, B, [NP7B]); *C. stirtoni* (locality [NP7A]).

Cimolestes sp. is also known from localities SB23A, B, CP11IIG, NP7C, NP17.

Comments: The holotypes of *C. incisus, C. magnus*, and *C. stirtoni* are from the Late Cretaceous Lance Formation in Montana, that of *C. cerberoides* is from the Late Cretaceous Scollard Formation in Alberta. *Cimolestes* is also known from the Paleocene of Africa and questionably from the Paleocene of South America (McKenna and Bell, 1997).

Procerberus Sloan and Van Valen, 1965

Type species: *Procerberus formicarum* Sloan and Van Valen, 1965.

Type specimen: UMVP 1460.

Characteristics: Relatively small size; P3–4 submolariform with shearing metacristae, lingual cingula and conules absent; upper molars not transverse, lingual cingula small to absent, paracone and metacone connate, conules small; p3 simple with paraconid and single-cusped talonid; p4 submolariform, narrow with anterior paraconid, weak paralophid, metaconid slightly posterior to protoconid; lower molars with large, lingual paraconids, paraconid and metaconid of subequal height, and trigonid only moderately higher than talonid (Sloan and Van Valen, 1965).

Average length of m2: 2.8 mm.

Included species: *P. formicarum* (known from localities [NP7A], NP15A*-C, NP16A); *P. plutonis* (locality NP16C*).

Procerberus sp. is also known from localities ?SB23A, CP11IIG, CP61A, NP3, NP7B, C, NP16A, NP17.

Comments: When originally described by Sloan and Van Valen (1965), the type locality of *P. formicarum*, Bug Creek Anthills (locality NP15A), was thought to be in the Late Cretaceous, as were the other two localities (Bug Creek West and Harbicht Hill, localities NP15B and NP15C, respectively) where *P. formicarum* was recognized. All of these localities are now placed in the early Paleocene (Lofgren *et al.*, 2004).

Acmeodon Matthew and Granger, 1921

Type species: *Acmeodon secans* Matthew and Granger, 1921.

Type specimen: AMNH 16599.

Characteristics: p4 low with deeply basined postvallid and talonid; m1 hypoconulid close to hypoconid.

Average length of m2: 3.0 mm (*A. secans*).

Included species: *A. secans* (known from localities SB23F, G, H*, CP13C, D, CP13IIA, CP14B); *A. hyoni* (localities CP14A*, [B]).

Acmeodon sp. is also known from localities CP1C, NP19IIA, NP20B, D.

Avunculus Van Valen, 1966

Type species: *Avunculus didelphodontidi* Van Valen, 1966.

Type specimen: AMNH 35297.

Characteristics: c1 alveolus about half as long as that of p4; p1 single-rooted; p3 much shorter and smaller than p4 with relatively prominent paraconid; p4 protostylid and metaconid absent and talonid unbasined; m1 metacristid absent from at least m1 (Van Valen, 1966).

Average length of m1: 2.8 mm.

Included species: *A. didelphodontidi* only (known from localities [CP13C], NP19C*).

Gelastops Simpson, 1935 (including _Emperodon_)

Type species: *Gelastops parcus* Simpson, 1935.

Type specimens: USNM 6148.

Characteristics: Dental formula I?/?, C?/1, P?/4, M3/3; M2 metaconule vestigial, hypocone absent; c1 erect; lower premolars crowded; p4 with distinct and subequal paraconid and metaconid; m1 trigonid long relative to talonid; m2–3 trigonids shorter and more elevated; m2 and particularly m3 smaller relative to m1 (Simpson, 1935; Van Valen, 1966).

Average length of m2: 2.5 – 2.9 mm (*G. parcus*).

Included species: *G. parcus* (including *Emperodon acmeodontoides* Simpson, 1935) (known from localities CP13B, [C, D], [CP14B], NP19C*); *G. joni* (localities [CP13IIB], CP14A*).

Gelastops sp. is also known from localities SB39IIA, CP15A, NP3A0, C, (NP16C).

Didelphodus Cope, 1882 (including _Didelphyodus_; _Phenacops_; _Deltatherium_, in part)

Type species: *Didelphodus absarokae* (Cope, 1881) (originally described as *Deltatherium absarokae*).

Type specimen: AMNH 4228.

Characteristics: Dental formula I?/3, C1/1, P4/4, M3/3; lower premolars simple and uncrowded; p3 moderately reduced relative to p4; m1 trigonid short relative to talonid; m2–3 moderately reduced relative to m1 (Matthew, 1909; Van Valen, 1966).

Average length of m2: 3.8 mm (*D. absarokae*).

Included species: *D. absarokae* (including *Didelphodus ventanus* Matthew, 1918) (known from localities SB24, CP19A, CP20A-D, CP25B, CP27C, D, CP63, CP64A-C); *D. altidens* (including *Phenacops incerta* Matthew, 1909) (localities CP27C, [D, E], CP34B, D*, CP63, CP64C); *D. rheos* (locality NP9A*); *D. serus* (locality NP8*).

Didelphodus sp. is also known from localities CP25F, G, CP27B, D, E, CP31E, NP9B, (NP49).

Comments: The precise locality of Cope's holotype of *Didelphodus absarokae* is unknown. The specimen was collected by Jacob Wortman in 1881. According to Gingerich (1980), Wortman's work in the Bighorn Basin that summer was probably concentrated around the Dorsey Creek area south of the Greybull River. If the holotype of *D. absarokae* was collected from this area it was found somewhere in the CP20 sequence.

Didelphodus is also known from the Eocene of Europe (McKenna and Bell, 1997).

Indeterminate didelophodontids

An indeterminate didelphodontid is known from locality NP1C.

PANTOLESTA MCKENNA, 1975

PANTOLESTIDAE COPE, 1884

PANTOLESTINAE COPE, 1884

***Paleotomus* Van Valen, 1967 (including *Niphredil*; *Palaeosinopa*, in part)**

Type species: *Paleotomus senior* (Simpson, 1937a) (originally described as *Palaeosinopa senior*).

Type specimen: AMNH 33990.

Characteristics: Lower molar paraconid moderately high and sectorial but not shifted forward, trigonid high, labial border of protoconid forms circular arc, protoconid considerably higher than metaconid, paralophid and protolophid with deep carnassial notches, talonid narrower than trigonid, entoconid distinct and higher than hypoconid, prevallid and postvallid shear well developed; p4 ~ m1 < m2 ~ m3; m3 trigonid higher than that of m1–2 (Van Valen, 1967).

Average length of m3: 5.0 mm (*P. senior*).

Included species: *P. senior* (including *Palaeosinopa simpsoni* Van Valen, 1967) (known from localities SB39A, CP22B, NP1C, NP19IIA, C*, [NP47A, B]); *P. carbonensis* (localities CP13IIA, B*); *P. junior* (localities NP1C*, NP3A0); *P. milleri* (locality CP14A*); *P. radagasti* (Van Valen, 1978) (originally described as *Niphredil radagasti*) (localities CP13E*, [CP22B]).

Paleotomus sp. is also known from localities CP13IIA–B, CP24A, NP1C.

Comments: McKenna and Bell (1997) placed *Paleotomus* in Cimolestidae along with *Procerberus* and *Cimolestes* while Scott, Fox, and Youzwyshyn (2002) assigned *Paleotomus* to family incertae sedis. We choose to retain it as a pantolestid because of the similarities between it and *Palaeosinopa*.

***Bessoecetor* Simpson, 1936 (including *Propalaeosinopa*; *Thylacodon*, in part; *Palaeosinopa*, in part; *Palaeictops*, in part; *Diacodon*, in part)**

Type species: *Bessoecetor septentrionalis* (Russell, 1929) (originally described as *Diacodon septentrionalis*).

Type specimen: UALVP 126.

Characteristics: Upper molars relatively transverse, anterolingual cingulum weak, hypocone moderate to small; M3 only slightly reduced with distinct metacone; p4 elongate with distinct anterior basal cusp, talonid short and weakly basined; lower molars with low paraconid shelf, talonid cusps distinct, especially on m3; p4 ~ m1 < m2 ~ m3; m3 trigonid higher than that of m1–2 (Simpson, 1936).

Average length of m2: 2.3 mm.

Included species: *B. septentrionalis* (including *Propalaeosinopa diluculi* Simpson, 1935; *Propalaeosinopa thomsoni* Simpson, 1936; *Thylacodon* sp. nov. Russell [in Rutherford, 1927]; *Palaeictops septentrionalis*, Van Valen, 1967) only (known from localities CP13C, D, [E], CP14A, C, D, [CP24A], CP26A, NP1C*, NP3A0, NP3B, D, NP4, NP14D, NP19C, NP19IIA, C, NP20D, E, NP47A, B, BB).

Bessoecetor sp. is also known from localities?CP1B, NP2, NP3A0, F, NP7D, NP20B, NP47C.

Comments: Scott, Fox, and Youzwyshyn (2002) noted that Simpson's (1927) holotype of *Propalaeosinopa albertensis* is non-diagnostic, rendering the genotype species a nomen dubium and leaving the genus *Propalaeosinopa* unavailable. These authors also recognized that Russell's (1929) holotype of *Diacodon septentrionalis* represents the same species as Simpson's (1935) "*Propalaeosinopa diluculi*," giving it priority over the latter species. Since *Propalaeosinopa* has no taxonomic status, Scott, Fox, and Youzwyshyn (2002) referred Russell's taxon to *Bessoecetor* as the type species, placing both "*Propalaeosinopa diluculi*" and "*Propalaeosinopa thomsoni*" in synonymy with *Bessoecetor septentrionalis*.

Bessoecetor is also questionably known from the late Paleocene of Europe (McKenna and Bell, 1997).

***Thelysia* Gingerich, 1982a**

Type species: *Thelysia artemia* Gingerich, 1982a.

Type specimen: UMMP 68281.

Characteristics: Dentary shallow; lower molars with anteroposteriorly short, high trigonids, reduced paraconids, and narrow talonids (especially on m2–3) lacking entocristids and open lingually (Gingerich, 1982a).

Average length of m2: 3.9 mm.

Included species: *T. artemia* only, known from locality CP17B* only.

Thelysia sp. is known from locality CP19AA.

***Palaeosinopa* Matthew, 1901**

Type species: *Palaeosinopa veterrima* Matthew, 1901.

Type specimen: AMNH 95.

Characteristics: Dental formula I3/3, C1/1, P4/4, M3/3; P3–4 simple with small P3 protocone and strong P4 protocone; P4 anterobuccal and posterobuccal basal cuspules small to tiny; M1 metastyle rudimentary, moderately extended on M2; upper molars tritubercular with strong posterolingual cingular ledge, protocone high and crescentic, paracone and metacone conical, sharp, somewhat inset from buccal border, well-developed buccal cingular shelf, crest curving anterobuccally from paracone apex to posterobuccal angle, narrow anterior cingulum; M3 transverse with vestigial metacone; m1–3 trigonid cusps subequal, paraconid distinct, talonid larger than trigonid, hypoconid and entoconid strong, hypoconulid present but strong only on m3 (Matthew, 1901, 1918).

Average length of m2: 5.4 mm (*P. didelphoides*).

Included species: *P. veterrima* (known from localities CP20A*, B, BB, D, CP27D); *P. didelphoides* (localities SB24, [CP25H, I], CP26C, CP27D, E*, [NP20E]);

P. dorri (including *P. simpsoni*, in part, Dorr, 1977) (locality CP22B*); *P. incerta* (localities CP19A*, CP20A, B, BB, CP25A, [CP27A], CP63, [CP64B]); *P. lutreola* (localities GC21II, CP20A, B*, [C], [CP25B, I2], CP27D); *P. nunavutensis* (locality NP14A2*).

Palaeosinopa sp. is also known from localities (CP13H), CP14C–E, CP17A, B, CP18B, CP20A, C, CP25B, C, G, NP20D, NP49, (NC0).

Comments: *Palaeosinopa* is also known from the early Eocene of Europe (McKenna and Bell, 1997).

***Pantolestes* Cope, 1872 (including *Passalacodon*; *Anisacodon*)**

Type species: *Pantolestes natans* Cope, 1872.

Type specimen: AMNH 12153.

Characteristics: As for family except lower molar paraconid reduced or absent.

Average length of m2: 6.5 mm (*P. natans*).

Included species: *P. natans* (known from localities [GC23A], CP34[C], D*, [CP38B], CP65, [NP14A]); *P. elegans* (Marsh, 1872) (originally described as *Anisacodon elegans*) (localities CP34C*, [CP38B]); *P. intermedius* (locality CP34C*); *P. longicaudus* (including *Passalacodon litoralis* Marsh, 1872) (localities NB14, [CP5A], CP27E, CP34A, C*, [CP38B]); *P. phocipes* (locality CP34D*).

Pantolestes sp. is also known from localities (CC4), (SB43A), CP6AA, CP27E, CP34C, E.

Comments: *Pantolestes* is also known from the middle Eocene of Asia (McKenna and Bell, 1997).

Indeterminate pantolestines

In addition to the taxa discussed above, indeterminate or undescribed pantolestids are known from localities CP13E, CP14A, CP20BB, CP25GG, J2, CP34C, NP9A.

PENTACODONTINAE SIMPSON, 1937b

In addition to the genera discussed below, two other genera, Paleocene *Phenacodaptes* and early Eocene *Apheliscus*, are commonly placed in the Pentacodontidae, a family here treated as a subfamily of the Pantolestidae. In Volume 1, the latter two genera were included in the Condylarthra, and our inclination would also be to place the pentacodontids there. Gazin (1959) questioned Simpson's (1936) assignment of *Phenacodaptes* to the Hyposodontidae, but McKenna (1960) placed *Apheliscus* there. Though there is no doubt that *Phenacodaptes* and *Apheliscus* are very closely related, they seem to be no more comfortably situated within Hyopsodontidae than within Pentacodontinae. The similarities with the latter group are in the premolars, specifically P4/p4, which are pervasive. More dissimilarity is seen in the molars, which do resemble those in the hyopsodontid *Haplomylus*. Recently reported postcranial evidence indicates remarkable similarities between *Apheliscus, Haplomylus*, and extant macroscelidians (Penkrot *et al.*, 2003; Zack *et al.*, 2005).

***Aphronorus* Simpson, 1935**

Type species: *Aphronorus fraudator* Simpson, 1935.

Type specimen: USNM 6177.

Characteristics: Dental formula I3/3, C1/1, P3/3, M3/3; P4 with metacone well differentiated, protoconule distinct; M1, and to lesser degree M2, slender and relatively transverse; p4 with weak anterobasal distension, talonid distinctly basined with distinct entoconid; p4 > m1 > m2 > m3; m1–2 trigonids relatively short, entoconids relatively high; m3 with three distinct talonid cusps (Simpson, 1937b).

Average length of m2: 2.7 mm (*A. fraudator*).

Included species: *A. fraudator* (known from localities CP13C, D, CP15A, B, NP19C*); *A. orieli* (localities CP13B, CP15A, CP21A*, NP19IIB, NP20B); *A. ratatoski* (localities [CP13C, D], CP14B*); *A. simpsoni* (locality CP1C*).

Aphronorus sp. is also known from localities CP13B, ?CP17A, NP2, NP3C, (NP16C), (NP19C), NP19IIA, [NP47A].

***Pentacodon* Scott, 1892 (including *Chriacus*, in part)**

Type species: *Pentacodon inversus* (Cope, 1888) (originally described as *Chriacus inversus*).

Type specimen: AMNH 3129.

Characteristics: I?/? C?/? P?/3 M3/3; P4 metacone not well differentiated and protoconule not distinct; M1 relatively wide and weakly transverse; p4 with distinct anterobasal distension, talonid weakly basined with poorly differentiated entoconid; p4 > m1 > m2 > m3; m1–2 trigonids relatively long, entoconids relatively low; m2–3 reduced relative to m1; m3 talonid cusps weakly defined.

Average length of m2: 4.5 mm.

Included species: *P. inversus* (known from localities SB23E*, F, G); *P. occultus* (locality SB23H*).

Pentacodon sp. is also known from localities SB23F, CP15A.

***Bisonalveus* Gazin, 1956**

Type species: *Bisonalveus browni* Gazin, 1956.

Type specimen: USNM 20928.

Characteristics: Dental formula I?/?, C?/1, P?/3, M3/3; p4 small with small paraconid and developed, posterolingual entoconid; p4 < m1 < m2 ~ m3; m1–3 paraconids relatively high and lingual; m1–2 hypoconulid indistinct, entoconid relatively anterior; m3 hypoconulid weakly projecting with small cuspule present on crest anterolingual to hypoconid (Gazin, 1956).

Average length of m2: 2.6 mm (*B. browni*).

Included species: *B. browni* (known from localities CP16A*, NP1C, [NP3A], NP19IIA); *B. holtzmani* (localities CP13E*, NP47B).

Bisonalveus sp. is also known from localities NP3B, C, G, NP20(B), D, (NP48).

***Coriphagus* Douglass, 1908 (including *Pantomimus*; *Mixoclaenus*)**

Type species: *Coriphagus montanus* Douglass, 1908.
Type specimen: CM 1669.
Characteristics: Dental formula I?/?, C1/1, P3/3, M3/3; dentary moderately long and slender, masseteric fossa large and deep; teeth heavy with low cusps; p2–4 double-rooted, cusps simple, oblong, lens-shaped in horizontal section; p4 < m1 > m2 > m3; m1–3 with low trigonid and lower talonid, external cusps higher than internal cusps; medial buttress (mylohyoid line) of mandible pronounced (Douglass, 1908).
Average length of m2: 2.5 mm.
Included species: *C. montanus* (including *Pantomimus leari* Van Valen, 1967) (known from localities CP13C, NP19B, C*); *C. encinensis* (Matthew and Granger, 1921) (originally described as *Mixoclaenus encinensis*) (localities SB23E, F, G*, GG, [CP13B]).

Coriphagus sp. is also known from locality (SB23H).

Indeterminate pentacodontines

Indeterminate pentacodontines are known from localities NP1C and NP19IIA.

PANTOLESTIDAE INCERTAE SEDIS

***Chadronia* Cook, 1954 (including *Cymaprimodon*)**

Type species: *Chadronia margaretae* Cook, 1954.
Type specimen: AMNH HC 750.
Characteristics: p3 much smaller than p4; m1 much longer than m2, and m2 much longer than m3; paraconid nearly connate with metaconid on m2; trigonids relatively low with respect to talonids, talonids less well defined; m1 trigonid much larger with respect to talonid than m2 or m3; upper canine immense; M1 and M2 of approximately equal breadth.
Average length of m2: 6.3 mm.
Included species: *C. margaretae* (including *Cymaprimodon kenni* Clark, 1968) only, known from locality CP98C* only.

Chadronia sp. is also known from locality NP14A.

***Amaramnis* Gazin, 1962**

Type species: *Amaramnis gregoryi* Gazin, 1962.
Type specimen: YPM 14702.
Characteristics: Lower molars relatively slender, trigonids anteroposteriorly elongated; m2–3 trigonids with acute anteroexternal angle between paraconid and paracristid, paraconid closely joined to metaconid, talonid basin shallow, cusps low, cristid obliqua joins postvallids buccally (Gazin, 1962).
Average length of m2: 3.4 mm.
Included species: *A. gregoryi* only, known from locality CP25C* only.

Amaramnis sp. is also known from localities CP25A, CP27D.

PANTOLESTA INCERTAE SEDIS

***Simidectes* Stock, 1933 (including *Pleurocyon*, in part; *Sespecyon*; *Petersonella*)**

Type species: *Simidectes magnus* (Peterson, 1919) (originally described as *Pleurocyon magnus*).
Type specimen: CM 2928.
Characteristics: Dental formula I3/3, C1/1, P4/4, M3/3; canine large, compressed transversely; C1/c1–P1/p1 diastemata absent; upper and lower first two premolars reduced; p1 single-rooted; p2 double-rooted with roots coalesced; upper and lower third premolar premolariform; upper and lower fourth premolar premolariform and as large as molars; upper and lower molars decrease in size anterior to posterior with reduced cingula; M1–3 paracone larger than metacone, paracone and metacone appressed, hypocone absent; m1–3 trigonids low crowned; m1–2 talonids trenchant between hypoconid and entoconid; m3 lacking entoconid; mental foramina below posterior p2 and posterior p3; masseteric fossa deep; mandibular articular process semi-cylindrical; angular process strong and angled inward; humerus with entepicondylar foramen and large deltoid crest; femur with third trochanter; pes with five metatarsals and unfused tarsals (Coombs, 1971).
Average length of m2: 9.9 mm (*S. merriami*).
Included species: *S. magnus* (known from localities [SB42], SB43A–C, CP6B*); *S. medius* (localities [CC7C], [CC8], CP6A*); *S. merriami* (localities CC7C, CC9BB*).

Simidectes sp. is also known from localities (CC8), (CC9A).

Comments: *Simidectes* has a rather complex taxonomic history. The genus was originally named *Pleurocyon* by Peterson (1919), with the type species, *P. magnus*, from the late Uintan of Utah (CP6B). Stock (1933) recognized a new species of the same genus from California (CC9BB) that he chose to designate as a subgenus, *Simidectes*, of Peterson's *Pleurocyon* (Stock, in the caption to his Plate I, referred to the subgenus as *Sespecyon*, not *Simidectes*, but he used *Simidectes* throughout the paper otherwise – *Sespecyon* was presumably used by mistake and is, therefore, a nomen nudum). Kraglievich (1948), recognizing that *Pleurocyon* was an occupied name (*Pleurocyon* Mercerat, 1917 [which in turn is a junior synonym of *Theriodictis* Mercerat, 1891], a South American Pleistocene canid), proposed *Petersonella* to replace *Pleurocyon* Peterson, 1919 (not Mercerat, 1917). Van Valen (1965) raised *Simidectes* from subgenus to genus level and included

Pleurocyon Peterson, 1919 in it as a junior synonym (and by extension *Petersonella* as well).

APATOTHERIA SCOTT AND JEPSEN, 1936

APATEMYIDAE MATTHEW, 1909

APATEMYINAE MATTHEW, 1909

***Jepsenella* Simpson, 1940**

Type species: *Jepsenella praepropera* Simpson, 1940.

Type specimen: AMNH 35292.

Characteristics: One lower incisor only; lower molar trigonids elevated above talonids, protoconids and metaconids high and subequal, paraconids lower but strong and distinct; m3 paraconid small and median, talonid short; squaring of anteroexternal trigonid rim not pronounced (Simpson, 1940).

Average length of m2: 1.6 mm.

Included species: *J. praepropera* only (known from localities CP13B, [CPC], CP14A, [NP1C], NP3[A0], B, NP19C*).

Jepsenella sp. is also known from locality SB39C.

***Labidolemur* Matthew and Granger, 1921 (including *Jepsenella*, in part)**

Type species: *Labidolemur soricoides* Matthew and Granger, 1921.

Type specimen: AMNH 17400.

Characteristics: Lower dental formula i1, c0, p3, m3; one lower incisor only, root extending beneath m3; dentary short and of moderate depth; mental foramen below anterior edge of m2 (Matthew and Granger, 1921; West, 1973a; Gingerich, 1982b; Gingerich and Rose, 1982).

Average length of m2: 1.9 mm (*L. kayi*).

Included species: *L. soricoides* (known from localities SB20A*, CP13E, CP24A, NP47C); *L. kayi* (localities CP14E, CP15B, CP17A*, B, CP20A, C, CP63); *L. major* (West, 1972) (originally described as *Jepsenella major*) (locality CP16A*); *L. serus* (localities CP19B*, CP20B, CP63).

Labidolemur sp. is also known from localities CP17B, CP18B, NP20E.

***Apatemys* Marsh, 1872 (including *Stehlinius*; *Stehlinella*; *Teilhardella*; *Labidolemur*, in part)**

Type species: *Apatemys bellus* Marsh, 1872.

Type specimen: YPM 13512.

Characteristics: Lower molar trigonids low relative to talonids; dentary short, dentition crowded; only two lower premolars retained.

Average length of m2: 2.3 mm (*A. bellus*).

Included species: *A. bellus* (known from localities [CC4], CC9A, NB14, [CP5A], CP20[BB], D, CP25J, CP27B, D, [E], [CP29C], [CP31E], CP34A–D*, CP35, [CP38A, B]); *A. bellulus* (including *A. hurzeleri* Gazin, 1962; *A. whitakeri* Guthrie, 1967) (localities SB24, [CP5A], CP20A, BB, C, CP25G, CP27C, [D, E], [CP31E], CP34A-D*); *A. chardini* (Jepsen, 1930) (originally described as *Teilhardella chardini*) (localities CP20A*, BB, D, [CP25B]); *A. downsi* (localities [CC4], CC7B, CC9A*); *A. hendryi* (localities CP29C*, NP8, NP9A); *A. rodens* (localities [CP5A], CP20A], [BB], B, C, CP34C, D*); *A. uintensis* (Matthew, 1921) (originally described as *Stehlinius uintensis*, later assigned to new genus *Stehlinella* Matthew, 1929) (localities CC7C, CP6AA, A*).

Apatemys sp. is also known from localities GC21II, CC8, SB22C, CP13G, CP20AA, B, BB, CP25A, F, GG, J2, CP27D, CP31C, CP34C–E, CP36B, NP9A, ?NP10B.

Comments: *Apatemys* is also known from the early Eocene of Europe (McKenna and Bell, 1997).

***Sinclairella* Jepsen, 1934**

Type species: *Sinclairella dakotensis* Jepsen, 1934.

Type specimen: YPM-PU 13585.

Characteristics: Relatively large size; upper molars squared, ectoflexus absent, hypocone large; p4 tiny; lower molar paraconids greatly reduced; m3 elongate with expanded hypoconulid; lower molar row about four times length of premolar row; single mental foramen below posterior root of m1; small labial fossa on dentary below p4 (West, 1973b).

Average length of m2: 3.7 mm.

Included species: *S. dakotensis* only (known from localities CP39C, CP68C, CP83C*, [CP84B, C], [CP85A], CP98B, CP101).

Sinclairella sp. is also known from localities CP39B, C, CP84B, ?NP9A, NP10B, NP49II.

Comments: The holotype, and most complete, specimen of *Sinclairella* has been lost for over 30 years.

UNUCHINIINAE VAN VALEN, 1966

***Unuchinia* Simpson, 1937c (including *Apator* Simpson, 1936, not *Apator* Semenow, 1898 [coleopteran])**

Type species: *Unuchinia asaphes* (Simpson, 1936) (originally described as *Apator asaphes*).

Type specimen: AMNH 33894.

Characteristics: Two lower incisors present; p4 tall, simple, and single-rooted; lower molars primitive with tall trigonids, bladelike paraconids, straight paralophids, and talonids with distinct cusps (Van Valen and McKenna in Van Valen, 1967; Holtzman, 1978; Gunnell, 1988).

Average length of m2: 2.4 mm (*U. asaphes*).

Included species: *U. asaphes* (known from locality NP19IIC*); *U. diaphanes* (localities CP13B*, C); *U. dysmathes* (localities CP24A, NP47B*, BB).

Unuchinia sp. is also known from locality CP13G.

Indeterminate apatemyids

Indeterminate apatemyids are known from localities NP3B, C.

INDETERMINATE "PROTEUTHERIANS"

CREOTARSINAE HAY, 1930

Creotarsus Matthew, 1918

Type species: *Creotarsus lepidus* Matthew, 1918.

Type specimen: AMNH 16169.

Characteristics: Lower dental formula i?, c?, p4, m3; p1 single-rooted, p2-4 double-rooted, subequal, and spaced; p4 with strong, well-separated metaconid, small paraconid and broad, short heel; m1–2 with protoconid and metaconid of equal height and widely separated; m1 paraconid small, low, internal in position, doubtfully present on m2; m1–2 trigonids relatively high and wide, talonid basin equal in width to trigonid, hypoconid strong and marginal, entoconid small and marginal, hypoconulid rudimentary; m1–2 similar in size and construction; m3 longer with (?narrow) heel; astragalus with oblique, moderately grooved trochlea, distinct astragalar foramen; calcaneum with distinct fibular facet; cuboid astragalar facet moderately wide.

Average length of m2: 3.6 mm.

Included species: *C. lepidus* only, from locality CP20A* only.

Comments: *Creotarsus* remains an enigmatic taxon. When Matthew (1918) described the genus he was uncertain as to where it belonged, noting resemblances with Artiodactyla, Condylarthra, Leptictida, Creodonta, Mesonychia, and "Insectivora." Simpson (1945) placed *Creotarsus* in Creodonta incertae sedis following Camp, Taylor, and Welles (1942). Van Valen (1967) assigned *Creotarsus* to the Erinaceoidea, including it in the subfamily Creotarsinae along with a variety of other insectivorans from North America, Europe, and Asia. Following Van Valen, McKenna and Bell (1997) felt that *Creotarsus* was an erinaceomorph lipotyphlan. We have chosen to place it questionably within "Proteutheria" pending discovery of more complete specimens and completion of more thorough study.

Indeterminate creotarsines

An indeterminate creotarsine is known from locality CP34B.

In addition to the taxa discussed above, indeterminate or undescribed proteutherians are known from localities SB23 (sublocality unknown), CP1A, CP11IIB, CP13C, NP20E (originally questionably assigned to the European taxon *Pagonomus*).

McKenna and Bell (1997) placed *Ravenictis*, represented by a single upper molar from the Puercan of Saskatchewan (locality NP7B) in Cimolesta incertae sedis. Fox and Youzwyshyn (1994) originally described *Ravenictis* as a basal carnivore of uncertain affinities. *Ravenictis* was briefly mentioned by Flynn (1998, Volume 1) wherein he placed it in Carnivora incertae sedis. We agree with the latter authors that *Ravenictis* cannot be a member of Cimolesta and may well represent a basal carnivore.

BIOLOGY AND EVOLUTIONARY PATTERNS

There are no apomorphic characters uniting "Proteutheria." Proteutherians, as recognized in this chapter, exhibit broad ranges of cranial, dental, and postcranial adaptations indicating widely divergent ecological roles (Figure 5.3). With further fossil discoveries and more study, it is likely that the "Proteutheria" will be documented to include several orders.

Palaeoryctidae were relatively small, insectivorous forms that may have been fossorial (Van Valen, 1966; Thewissen and Gingerich, 1989). They are characterized by protozalambdodont dentitions. Based on the skull of *Eoryctes*, Thewissen and Gingerich (1989) suggested that palaeoryctids probably had poor eyesight (presence of a small optic foramen), acute hearing (large bulla and epitympanic recess), and a well-developed sense of smell (large infraorbital canal). All of these features would suggest a nocturnal, if not fossorial, lifestyle. Palaeoryctids make their first appearance in the early Puercan (locality NP7B) (earliest Paleocene), from a probable cimolestid ancestry. The group shows its greatest diversity in the late Tiffanian (late Paleocene), with five genera present in Tiffanian zone Ti4 and four in Ti5. Palaeoryctids decrease in diversity through the earliest Eocene and disappear from the fossil record after the early middle Wasatchian (Wasatchian zone Wa5).

Tertiary cimolestids are generally somewhat larger than palaeoryctids although they are morphologically similar in comparable elements. Most taxa are represented by dental elements only, and these were likely to have been insectivorous or omnivorous. Like palaeoryctids, they probably were nocturnal or crepuscular and, like their Late Cretaceous relatives, terrestrial or scansorial (Kielan-Jaworowska, Bown, and Lillegraven, 1979). Cimolestids first appear in the Late Cretaceous and are most diverse in the Torrejonian (late early Paleocene). Except for *Didelphodus*, which may have survived through the Duchesnean, all other cimolestids were extinct by the end of the middle Tiffanian (zone Ti3).

Pantolestids are small to moderate in size and, in general, larger than other proteutherians. Based on the morphology of *Pantolestes*, Matthew (1909) suggested that pantolestids were aquatic omnivores or carnivores (Figure 5.4). In overall skull and limb proportions, *Pantolestes* resembles extant otters and was probably otter-like in habits. Pantolestid teeth are bluntly cusped and they wear relatively rapidly, suggesting that these animals were feeding on relatively hard objects, perhaps shellfish and grit-covered aquatic vegetation.

Dorr (1977) developed an argument for malacophagy and a semi-aquatic lifestyle in *Pantolestes natans*. He noted that the skull is streamlined and flat, so that drag will be reduced during swimming. Furthermore, he claimed that the marked degree of postorbital constriction is consistent with increased area and improved mechanical advantage for the anterior portion of the temporalis muscles. The presence of a robust anterior dentition that exhibits heavy apical wear is also consistent with malacophagy (mollusk eating).

Pfretzchner (1993) noted that aquatic adaptations were abundant and unambiguous in the Messel pantolestid *Buxolestes piscator*. His principle argument was derived from the anatomy of the neck, lumbar vertebrae, and skull and corroborated conclusions based on previously identified and analyzed material (Koenigswald, 1980)

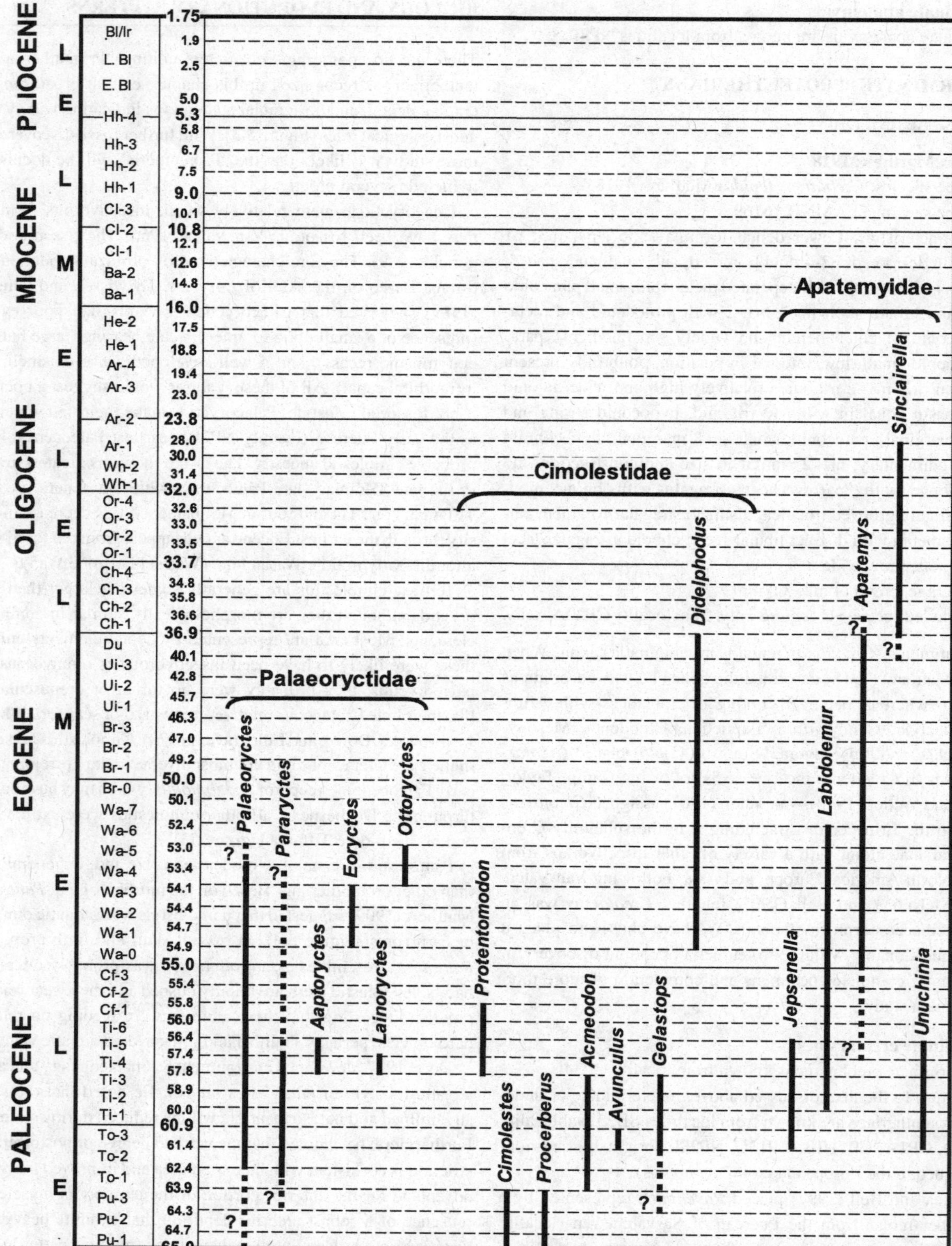

Figure 5.3. Temporal ranges of North American genera of "proteutherian" families Palaeoryctidae, Cimolestidae, and Apatemyidae.

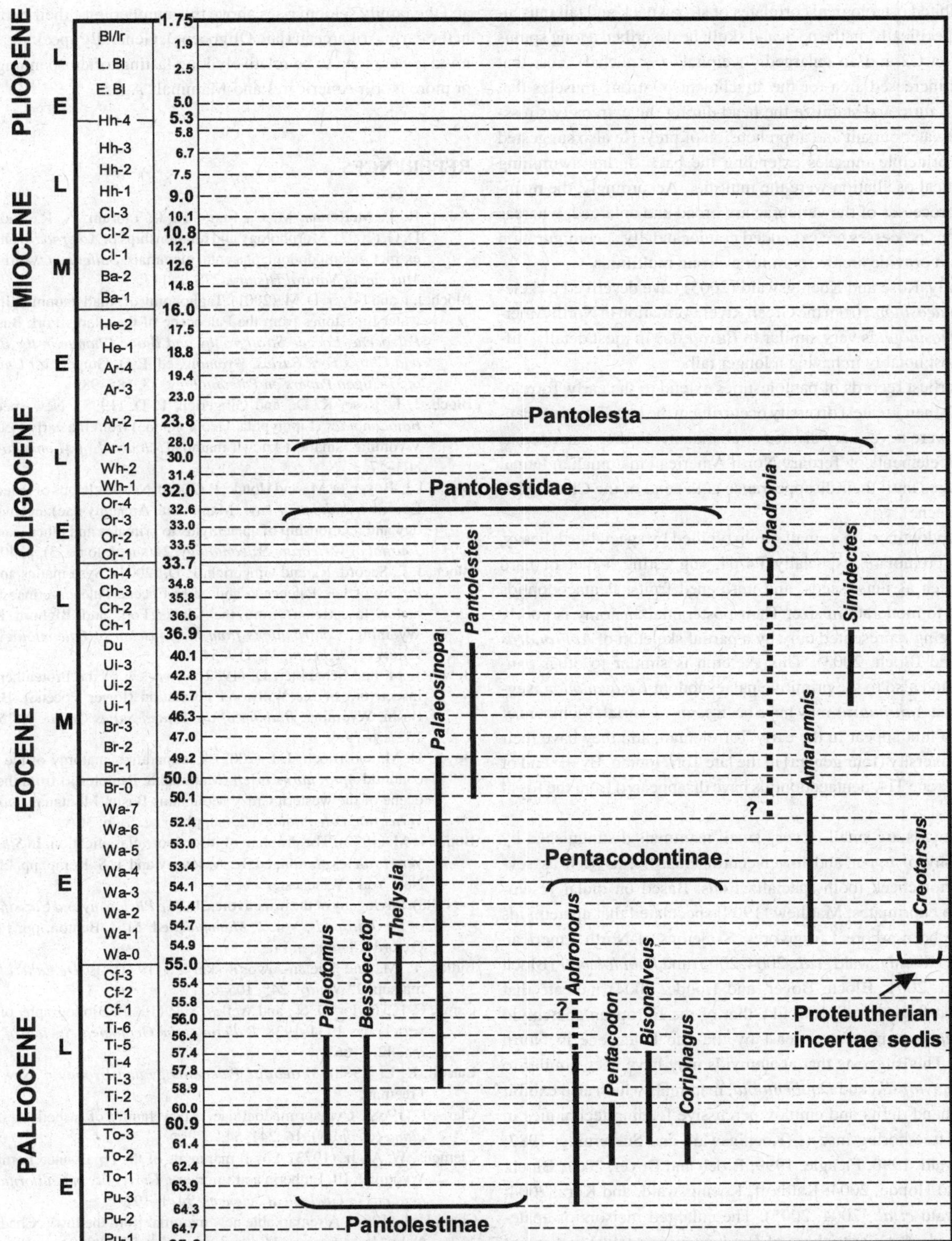

Figure 5.4. Temporal ranges of North American Pantolesta genera.

that exhibited osteological correlates of strong neck and tail musculature. Specifically, in the neck and skull, he described a long spinus process on C2 and an enlarged "occipital" (= nuchal) crest that provides increased area for the attachment of strong muscles that could lift, turn, and stabilize the head during the extremely stressful underwater pursuit and apprehension of prey. He also suggested that the principle muscles extending the back during swimming with vertical oscillation were the multifidi. Accordingly, the mammillary processes of the vertebrae are broad and enlarged, whereas the spinous processes are expanded craniocaudally, a configuration that would provide ample room for a strong multifidus.

Recently, Rose and Koenigswald (2005) have described a skeleton of *Palaeosinopa* from the Green River Formation in North America. *Palaeosinopa* is very similar to *Buxolestes* in most details, differing most notably in having a longer tail.

The earliest records of pantolestines extend to the early Torrejonian, with their greatest diversity occurring in the late Tiffanian. Pantolestids were never very diverse or abundant, but they remained as persistent elements of Tertiary North American mammalian faunal assemblages until their disappearance at the end of the Chadronian (latest Eocene).

Pentacodontines are a small group typified by specialized, heavy, crushing premolars, especially P4/p4, suggesting a diet of hard objects such as nuts, seeds, and unripened fruits. Pentacodontids are small to moderate in size. Their postcranial anatomy is poorly known, being represented only by a partial skeleton of *Aphronorus* (Boyer and Bloch, 2003). This skeleton is similar to other pantolestids and also to the primitive palaeanodont *Escavadodon*, suggesting that *Aphronorus* may have been semi-fossorial. Members of this group first appear in the early Torrejonian, and they have their greatest diversity (four genera) in the late Torrejonian. By the end of Tiffanian zone Ti4, pentacodontids have disappeared from the fossil record.

Apatemyids are small to very small mammals distinguished by hypertrophy of upper and lower central incisors, as well as several anterior cheek tooth specializations. Based on molar resemblances to euprimates, Matthew (1909) speculated that apatemyids may have been arboreal frugivores. Skeletons of North American *Apatemys* (Koenigswald *et al.*, 2004, 2005) and *Labidolemur* (Bloch and Boyer, 2001; Bloch, Boyer, and Houde, 2004) indicate that specialized finger elongation like that of the European apatemyid *Heterohyus* was present at least by the late Paleocene in North America. This suggests that apatemyids may have been similar to extant *Dactylopsila* and *Daubentonia*, mammals which also exhibit elongate hand digits and employ percussive feeding techniques in search of tree-boring insects (Koenigswald and Schierning, 1987; Koenigswald, 1990; Fleagle, 1999; Bloch and Boyer, 2001; Bloch, Boyer, and Houde, 2004; Kalthoff, Koenigswald, and Kurz, 2004; Koenigswald *et al.*, 2004, 2005). The enlarged incisors of apatemyids are also similar to those of *Daubentonia* and, like that extant animal, may have been used to gnaw tree bark in search of insect larvae.

Apatemyids first appear in the earliest Tiffanian, where the family is represented by three genera. By the beginning of the Clarkforkian (latest Paleocene), apatemyid diversity is reduced to a single genus, and the family seldom rises above this number until their extinction in the early Arikareean (late Oligocene). Generally speaking, apatemyid genera tend to be relatively long lasting, often spanning three or more North American Land-Mammal "Ages."

REFERENCES

Asher, R. J., McKenna, M. C., Emry, R. J., Tabrum, A. R., and Kron, D. G. (2002). Morphology and relationships of *Apternodus* and other extinct, zalambdodont, placental mammals. *Bulletin of the American Museum of Natural History*, 273, 1–117.

Bloch, J. I. and Boyer, D. M. (2001). Taphonomy of small mammals in freshwater limestones from the Paleocene of the Clarks Fork Basin. [In *Paleocene–Eocene Stratigraphy and Biotic Change in the Bighorn and Clarks Fork Basins, Wyoming*, ed. P. D. Gingerich.] *University of Michigan Papers on Paleontology*, 33, 185–98.

Bloch, J. I., Rose, K. D., and Gingerich, P. D. (1998). New species of *Batodonoides* (Lipotyphla, Geolabididae) from the early Eocene of Wyoming: smallest known mammal? *Journal of Mammalogy*, 79, 804–27.

Bloch, J. I., Boyer, D. M., and Houde, P. (2004). New skeletons of Paleocene–Eocene *Labidolemur kayi* (Mammalia, Apatemyidae): ecomorphology and relationship of apatemyids to primates and other mammals. *Journal of Vertebrate Paleontology*, 24(suppl. to no. 3), p. 40A.

Bloch, J. I., Secord, R., and Gingerich, P. D. (2004). Systematics and phylogeny of late Paleocene and early Eocene Palaeoryctinae (Mammalia, Insectivora) from the Clarks Fork and Bighorn Basins, Wyoming. *Contributions from the Museum of Paleontology, University of Michigan*, 31, 119–54.

Bown, T. M. and Schankler, D. (1982). A review of the Proteutheria and Insectivora of the Willwood Formation (lower Eocene), Bighorn Basin, Wyoming. *Bulletin of the United States Geological Survey*, 1523, 1–79.

Boyer, D. M. and Bloch, J. I. (2003). Comparative anatomy of the pentacodontid *Aphronorus orieli* (Mammalia, Pantolesta) from the Paleocene of the western Crazy Mountains Basin, Montana. *Journal of Vertebrate Paleontology*, 23(suppl. to no. 3), p. 36A.

Butler, P. M. (1972). The problem of insectivore classification. In *Studies in Vertebrate Evolution*, ed, K. A. Joysey and T. S. Kemp, pp. 253–65. New York: Winchester.

(1988). Phylogeny of the insectivores. In *The Phylogeny and Classification of the Tetrapods,* Vol. 2: *Mammals*, ed. M. J. Benton, pp. 117–41. Oxford: Clarendon Press.

Butler, P. M. and Kielan-Jaworowska, Z. (1973). Is *Deltatheridium* a marsupial? *Nature*, 245, 105–6.

Camp, C. L., Taylor, D. N., and Welles, S. P. (1942). Bibliography of fossil vertebrates 1934–1938. *Bulletin of the Geological Society of America*, 42, 1–663.

Carroll, R. L. (1988). *Vertebrate Paleontology and Evolution*. New York: Freeman.

Clark, J. (1968). Cymaprimadontidae, a new family of insectivores. *Fieldiana (Geology)*, 16, 241–54.

Clemens, W. A., Jr. (1973). Fossil mammals of the type Lance Formation, Wyoming, III. Eutheria and summary. *University of California Publications in Geological Sciences*, 94, 1–102.

Cook, H. J. (1954). A remarkable new mammal from the lower Chadron of Nebraska. *American Midland Naturalist*, 52, 388–91.

Coombs, M. C. (1971). Status of *Simidectes* (Insectivora, Pantolestoidea) of the late Eocene of North America. *American Museum Novitates*, 2455, 1–41.

Cope, E. D. (1872). Second account of new Vertebrata from the Bridger Eocene. *Paleontological Bulletin, American Philosophical Society*, 2, 466–8.

(1881). The temporary dentition of a new creodont. *The American Naturalist*, 15, 667–9.

(1882). Contributions to the history of the Vertebrata of the lower Eocene of Wyoming and New Mexico, made during 1881. *Paleontological Bulletin, American Philosophical Society*, 20, 139–97.

(1884). The Creodonta. *The American Naturalist*, 17, 255–67, 344–53, 478–85.

(1888). Synopsis of the vertebrate fauna of the Puerco series. *Transactions of the American Philosophical Society*, 16, 298–361.

Dashzeveg, D. and Russell, D. E. (1985) A new middle Eocene insectivore from the Mongolian People's Republic. *Geobios*, 18, 871–5.

Dorr, J. A., Jr. (1977). Partial skull of *Palaeosinopa simpsoni* (Mammalia, Insectivora), latest Paleocene Hoback Formation, central western Wyoming, with some general remarks on the family Pantolestidae. *Contributions from the Museum of Paleontology, University of Michigan*, 24, 281–307.

Douglass, E. (1908). Vertebrata fossils from the Fort Union beds. *Annals of the Carnegie Museum*, 5, 11–26.

Fleagle, J. G. (1999). *Primate Adaptations and Evolution*, 2nd edn. San Diego, CA: Academic Press.

Flynn, J. J. (1998). Early Cenozoic Carnivora ("Miacoidea"). In *Evolution of Tertiary Mammals of North America*, Vol. 1, ed. C. M. Janis, K. M. Scott, and L. L. Jacobs, pp. 110–23. Cambridge, UK: Cambridge University Press.

Fox, R. C. (1983). Evolutionary implications of tooth replacement in the Paleocene mammal *Pararyctes*. *Canadian Journal of Earth Sciences*, 20, 19–22.

(2004). A new palaeoryctid (Insectivora: Mammalia) from the late Paleocene of Alberta, Canada. *Journal of Paleontology*, 78, 612–16.

Fox, R. C. and Youzwyshyn, G. P. (1994). New primitive carnivorans (Mammalia) from the Paleocene of western Canada, and their bearing on relationships of the order. *Journal of Vertebrate Paleontology*, 14, 382–404.

Gabunia, L. K. (1989). O pervoy nakhodke pantolestid (Pantolestidae, Insectivora) v Paleogene SSSR. [Concerning the first find of a pantolestid (Pantolestidae, Insectivora) in the Paleogene of the USSR.] [In Russian with English and Georgian summaries.] *Bulletin of the Academy of Sciences, Georgia, SSR*, 136, 177–80.

Gazin, C. L. (1956). Paleocene mammalian faunas of the Bison Basin in south-central Wyoming. *Smithsonian Miscellaneous Collections*, 131, 1–47.

(1959). Early Tertiary *Apheliscus* and *Phenacodaptes* as pantolestid insectivores. *Smithsonian Miscellaneous Collections*, 139, 1–7.

(1962). A further study of the lower Eocene mammalian faunas of Southwestern Wyoming. *Smithsonian Miscellaneous Collections*, 144, 1–98.

(1969). A new occurrence of Paleocene mammals in the Evanston Formation, southwestern Wyoming. *Smithsonian Contributions to Paleobiology*, 2, 1–16.

Gheerbrant, E. (1991). *Todralestes variabilis* n.g., n.sp., new proteutherian (Eutheria, Todralestidae fam. nov.) from the Paleocene of Morocco. *Comptes Rendus Hebdomadaires des Seances de l'Academie des Sciences, Paris*, 312, 1249–55.

Gingerich, P. D. (1980). History of early Cenozoic vertebrate paleontology in the Bighorn Basin. [In *Early Cenozoic Paleontology and Stratigraphy of the Bighorn Basin, Wyoming 1880–1980*, ed. P. D. Gingerich.] *University of Michigan Papers on Paleontology*, 24, 7–24.

(1982a). *Aaptoryctes* (Palaeoryctidae) and *Thelysia* (Palaeoryctidae?): new insectivorous mammals from the late Paleocene and early Eocene of western North America. *Contributions from the Museum of Paleontology, University of Michigan*, 26, 37–47.

(1982b). Studies on Paleocene and early Eocene Apatemyidae (Mammalia, Insectivora), II. *Labidolemur* and *Apatemys* from the early Wasatchian of the Clark's Fork Basin, Wyoming. *Contributions from the Museum of Paleontology, University of Michigan*, 26, 56–69.

Gingerich, P. D. and Rose, K. D. (1982). Studies of Paleocene and early Eocene Apatemyidae (Mammalia, Insectivora), I. Dentition of Clarkforkian *Labidolemur kayi*. *Contributions from the Museum of Paleontology, University of Michigan*, 26, 49–55.

Gingerich, P. D., Houde, P., and Krause, D. W. (1983). A new earliest Tiffanian (Late Paleocene) mammalian fauna from Bangtail Plateau, Western Crazy Mountain Basin, Montana. *Journal of Paleontology*, 57, 957–70.

Gunnell, G. F. (1988). New species of *Unuchinia* (Mammalia: Insectivora) from the middle Paleocene of North America. *Journal of Paleontology*, 62, 139–41.

(1994). Paleocene mammals and faunal analysis of the Chappo Type Locality (Tiffanian), Green River Basin, Wyoming. *Journal of Vertebrate Paleontology*, 14, 81–104.

Guthrie, D. A. (1967). The mammalian fauna of the Lysite Member, Wind River Formation (early Eocene), Wyoming. *Memoir of the Southern California Academy of Science*, 5, 1–53.

Hay, O. P. (1930). *Second Bibliography and Catalogue of the Fossil Vertebrata of North America*, Vol. 2, Washington, DC: Carnegie Institution of Washington.

Holtzman, R. C. (1978). Late Paleocene mammals of the Tongue River Formation, western North Dakota. *North Dakota Geological Survey Report of Investigation*, 65, 1–88.

Jepsen, G. L. (1930). New vertebrate fossils from the lower Eocene of the Bighorn Basin, Wyoming. *Proceedings of the American Philosophical Society*, 69, 117–31.

(1934). A revision of the American Apatemyidae and the description of a new genus, *Sinclairella*, from the White River Oligocene of South Dakota. *Proceedings of the American Philosophical Society*, 74, 287–305.

Kalthoff, D. C., Koenigswald, W. von, and Kurz, C. (2004). A new specimen of *Heterohyus nanus* (Apatemyidae; Mammalia) from the Eocene of Messel (Germany) with unusual soft-part preservation. *Courier Forschungs-Institut Senckenberg*, 252, 1–12.

Kellner, A. W. A. and McKenna, M. C. (1996). A leptictid mammal from the Hsanda Gol Formation (Oligocene), central Mongolia, with comments on some Palaeoryctidae. *American Museum Novitates*, 3168, 1–13.

Kielan-Jaworowska, Z. (1981). Evolution of the therian mammals in the Late Cretaceous of Asia. Part IV. Skull structure in *Kennalestes* and *Asioryctes*. *Acta Palaeontologia Polonica*, 42, 25–78.

Kielan-Jaworowska, Z., Bown, T. M., and Lillegraven, J. A. (1979). Eutheria. In *Mesozoic Mammals: The First Two-Thirds of Mammalian History*, ed. J. A. Lillegraven, Z. Kielan-Jaworowska, and W. A. Clemens, pp. 221–58. Berkeley, CA: University of California Press.

Koenigswald, W. von (1980). Das skelett eines Pantolestiden (Proteutheria, Mamm.) aus dem mittleren Eozan von Messel bei Darmstadt. *Palaontologische Zeitschrift*, 54, 267–87.

(1990). Die Palaobiologie der Apatemyiden (Insectivora s. l.) und die Ausdeutung der Skelettfunde von *Heterohyus nanus* aus dem Mitteleozan von Messel bei Darmstadt. *Palaeontographica Abteilung A Palaeozoologie-Stratigraphie*, 210, 41–77.

Koenigswald, W. von and Schierning, H.-P. (1987). The ecological niche of an extinct group of mammals, the Early Tertiary apatemyids. *Nature*, 326, 595–7.

Koenigswald, W. von, Rose, K. D., Grande, L., and Martin, R. (2004). Apatemyid and pantolestid skeletons from the Eocene Fossil Butte Member (Wyoming) compared to those from Messel (Germany). *Journal of Vertebrate Paleontology*, 24(suppl. to no. 3), p. 25A.

Koenigswald, W. von, Rose, K. D., Grande, L., and Martin, R. (2005). First apatemyid skeleton from the Lower Eocene Fossil Butte Member, Wyoming (USA), compared to the European apatemyid from Messel, Germany. *Palaeontographica A*, 272, 149–69.

Kraglievich, L. J. (1948). Substitucion de un nombre generico. *Anales de la Sociedad Científica Argentina*, 146, 161–2.

Krause, D. W. and Gingerich, P. D. (1983). Mammalian fauna from Douglass Quarry, earliest Tiffanian (Late Paleocene) of the eastern Crazy Mountain Basin, Montana. *Contributions from the Museum of Paleontology, University of Michigan*, 26, 157–96.

Lillegraven, J. A. (1969). Latest Cretaceous mammals in the upper part of Edmonton Formation of Alberta, Canada, and review of marsupial–placental dichotomy in mammalian evolution. *University of Kansas Paleontological Contributions*, 50, 1–122.

Lofgren, D. L., Lillegraven, J. A., Clemens, W. A., Gingerich, P. D., and Williamson, T. E. (2004). Paleocene biochronology of North America: the Puercan through Clarkforkian land-mammal ages. In *Late Cretaceous and Cenozoic Mammals of North America*, ed. M. O. Woodburne, pp. 43–105. New York: Columbia University Press.

Marsh, O. C. (1872). Preliminary description of new Tertiary mammals. *American Journal of Science*, 38, 81–92.

(1889). Discovery of Cretaceous Mammalia. *American Journal of Science*, 38, 81–92.

Matthew, W. D. (1899). A provisional classification of fresh-water Tertiary of the West. *Bulletin of the American Museum of Natural History*, 12, 19–75.

(1901). Additional observations on the Creodonta. *Bulletin of the American Museum of Natural History*, 14, 1–37.

(1909). The Carnivora and Insectivora of the Bridger Basin, middle Eocene. *American Museum of Natural History Memoirs*, 9, 289–567.

(1913). A zalambdodont insectivore from the basal Eocene. *Bulletin of the American Museum of Natural History*, 32, 307–14.

(1918). Part V. Insectivora (continued), Glires, Edentata. *Bulletin of the American Museum of Natural History*, 38, 565–657.

(1921). *Stehlinius*, a new Eocene insectivore. *American Museum Novitates*, 14, 1–5.

(1929). Preoccupied names. *Journal of Mammalogy*, 10, 171.

Matthew, W. D. and Granger, W. (1921). New genera of Paleocene mammals. *American Museum Novitates*, 13, 1–7.

McDowell, S. B. (1958). The greater Antillean insectivores. *Bulletin of the American Museum of Natural History*, 115, 113–214.

McKenna, M. C. (1960). Fossil Mammalia from the early Wasatchian Four Mile Fauna, Eocene of northwest Colorado. *University of California Publications in Geological Sciences*, 37, 1–130.

(1963). Primitive Paleocene and Eocene Apatemyidae (Mammalia, Insectivora) and the primate–insectivore boundary. *American Museum Novitates*, 2160, 1–39.

(1975). Toward a phylogenetic classification of the Mammalia, In *Phylogeny of the Primates*, ed. W. P. Luckett and F. S. Szalay, pp. 21–46. New York: Plenum Press.

McKenna, M. C. and Bell, S. K. (1997). *Classification of Mammals Above the Species Level*. New York: Columbia University Press.

McKenna, M. C., Xue, X., and Zhou, M. (1984). *Prosarcodon lonanensis*, a new Paleocene micropternodontid palaeoryctoid insectivore from Asia. *American Museum Novitates*, 2780, 1–17.

Mercerat, A. (1891). Caracteres diagnósticos de algunas especies de Creodonta conservadas en el Museo de La Plata. *Revista del Museo de La Plata*, 2, 51–6.

(1917). *Notas sobre algunos carnívoros fósiles y actuels de la América del Sud*. Buenos Aires, 1–12.

Novacek, M. J. (1976). Insectivora and Proteutheria of the later Eocene (Uintan) of San Diego County, California. *Contributions in Science, Los Angeles County Museum of Natural History*, 283, 1–52.

(1986). The skull of leptictid insectivorans and the higher-level classification of eutherian mammals. *Bulletin of the American Museum of Natural History*, 183, 1–112.

Penkrot, T., Zack, S., Rose, K. D., and Bloch, J. I. (2003). Postcrania of early Eocene *Apheliscus* and *Haplomylus* (Mammalia: "Condylarthra"). *Journal of Vertebrate Paleontology*, 23(suppl. to no. 3), p. 86A.

Peterson, O. A. (1919). Report upon the material discovered in the upper Eocene of the Uinta Basin by Earl Douglass in the years 1908–1909, and by O. A. Peterson in 1912. *Annals of the Carnegie Museum*, 12, 40–168.

Pfretzchner, H.-U. (1993). Muscle reconstruction and aquatic locomotion in the Middle Eocene *Buxolestes piscator* from Messel near Darmstadt. *Kaupia, Darmstader Beitrage zur Naturgeschichte*, 3, 75–87.

Romer, A. S. (1966). *Vertebrate Paleontology*. Chicago, IL: University of Chicago Press.

Rose, K. D. (1978). A new Paleocene epoicotheriid (Mammalia), with comments on the Palaeanodonta. *Journal of Paleontology*, 52, 658–74.

(1981). The Clarkforkian Land-Mammal Age and mammalian faunal composition across the Paleocene–Eocene boundary. *University of Michigan Papers on Paleontology*, **26**, 1–197.

Rose, K. D. and Koenigswald, W., von (2005). An exceptionally complete skeleton of *Palaeosinopa* (Mammalia, Cimolesta, Pantolestidae) from the Green River Formation, and other postcranial elements of the Pantolestidae from the Eocene of Wyoming (USA). *Palaeontographica Abteilung A*, 273, 55–96.

Rose, K. D. and Lucas, S. G. (2000). An early Paleocene palaeanodont (Mammalia,?Pholidota) from New Mexico, and the origin of Palaeanodonta. *Journal of Vertebrate Paleontology*, 20, 139–56.

Russell, L. S. (1929). Paleocene vertebrates from Alberta. *American Journal of Science*, 17, 162–78.

Rutherford, R. L. (1927). Geology along the Bow Rivér between Cochrane and Kananaskis, Alberta. *Scientific and Industrial Research Council of Alberta Report*, 17, 1–29.

Saban, R. (1954). Phylogenie des insectivores. *Bulletin of the Natural History Museum of Paris*, 26, 419–32.

(1958). Insectivora. In *Traite de Paleontologie*, Vol. 6, ed. J. Piveteau, pp. 822–909. Paris: Masson et Cie.

Scott, C. S., Fox, R. C., and Youzwyshyn, G. P. (2002). New earliest Tiffanian (late Paleocene) mammals from Cochrane 2, southwestern Alberta, Canada. *Acta Palaeontologica Polonica*, 47, 691–704.

Scott, W. B. (1892). A revision of the North American Creodonts with notes on some genera which have been referred to that group. *Proceedings of the Academy of Natural Sciences, Philadelphia*, 1892, 291–323.

Scott, W. B. and Jepsen, G. L. (1936). The mammalian fauna of the White River Oligocene. Part I. Insectivora and Carnivora. *Transactions of the American Philosophical Society*, 28, 1–153.

Semenow, A. P. (1898). Nota 153. *Bulletin of the Moscow Society of Naturalists*, 1, 132.

Simpson, G. G. (1927). Mammalian fauna and correlation of the Paskapoo Formation of Alberta. *American Museum Novitates*, 268, 1–10.

(1928). A new mammalian fauna from the Fort Union of southern Montana. *American Museum Novitates*, 297, 1–15.

(1935). New Paleocene mammals from the Fort Union of Montana. *Proceedings of the United States National Museum*, 83, 221–44.

(1936). A new fauna from the Fort Union of Montana. *American Museum Novitates*, 873, 1–27.

(1937a). Additions to the upper Paleocene fauna of the Crazy Mountain Field. *American Museum Novitates*, 940, 1–15.

(1937b). The Fort Union of the Crazy Mountain Field, Montana, and its mammalian faunas. *Bulletin of the United States National Museum*, 169, 1–278.

(1937c). *Unuchinia*, a new name for *Apator* Simpson, not Semenow. *Journal of Paleontology*, 11, 78.

(1940). Studies on the earliest primates. *Bulletin of the American Museum of Natural History*, 77, 185–212.

(1945). The principles of classification and a classification of mammals. *Bulletin of the American Museum of Natural History*, 85, 1–350.

Sloan, R. E. and Van Valen, L. (1965). Cretaceous mammals from Montana. *Science*, 148, 220–7.

Stock, C. (1933). A miacid from the Sespe, upper Eocene, California. *Proceedings of the National Academy of Sciences, Philadelphia*, 19, 481–6.

Szalay, F. S. (1977). Phylogenetic relationships and a classification of the eutherian Mammalia. In *Major Patterns in Vertebrate Evolution*, ed. M. K. Hecht, P. C. Goody, and B. M. Hecht, pp. 315–74. New York: Plenum Press.

Thewissen, H. and Gingerich, P. D. (1989). Skull and endocranial cast of *Eoryctes melanus*, a new palaeoryctid (Mammalia: Insectivora) from the early Eocene of western North America. *Journal of Vertebrate Paleontology*, 9, 459–70.

Van Valen, L. (1965). Paroxyclaenidae, an extinct family of Eurasian mammals. *Journal of Mammalogy*, 46, 388–97.

(1966). Deltatheridia, a new order of mammals. *Bulletin of the American Museum of Natural History*, 132, 1–126.

(1967). New Paleocene insectivores and insectivore classification. *Bulletin of the American Museum of Natural History*, 135, 217–84.

(1978). The beginning of the age of mammals. *Evolutionary Theory*, 4, 45–80.

West, R. M. (1972). A new late Paleocene apatemyid (Mammalia, Insectivora) from Bison Basin, central Wyoming. *Journal of Paleontology*, 46, 714–18.

(1973a). Antemolar dentitions of the Paleocene apatemyid insectivorans *Jepsenella* and *Labidolemur*. *Journal of Mammalogy*, 54, 33–40.

(1973b). Review of the North American Eocene and Oligocene Apatemyidae (Mammalia, Insectivora). *Special Publications of the Museum of Technical Texas University*, 3, 1–42.

Winge, H. (1917). Udsigt over Insectaedernes inbrydes Slaegtskab. *Videnskabelige Meddelelser fra Dansk naturhistorisk Forening, Copenhagen*, 68, 83–203.

Zack, S. P., Penkrot, T. A., Bloch, J. I., and Rose, K. D. (2005). Affinities of "hyopsodontids" to elephant-shrews and a Holartic origin of Afrotheria. *Nature*, 434, 497–501.

6 Leptictida

GREGG F. GUNNELL,[1] THOMAS M. BOWN,[2] and JONATHAN I. BLOCH[3]

[1]*Museum of Paleontology, University of Michigan, Ann Arbor, MI, USA*
[2]*Erathem-Vanir Geological, Pocatello, ID, USA*
[3]*Florida Museum of Natural History, University of Florida, Gainesville, FL, USA*

INTRODUCTION

Leptictida is an enigmatic group of insectivorous mammals that has been viewed as ancestral or related to a variety of mammalian groups including Lipotyphla, Erinaceomorpha, Tupaiidae, Primates, Macroscelididae, Apatemyidae, Microsyopoidea, Pantolestidae, and Rodentia among others (Gregory, 1910; Matthew, 1918, 1937; Simpson, 1945; Butler, 1956, 1972; McDowell, 1958; Van Valen, 1965, 1967; Lillegraven, 1969; McKenna, 1969, 1975; Szalay, 1969, 1977; Novacek, 1986; McKenna and Bell, 1997). Leptictidans first appeared in the Late Cretaceous of North America, with their record continuing from the early Paleocene through the early Arikareean (late Oligocene). Leptictids are also known from the late Paleocene of Europe and the early Oligocene of Asia (McKenna and Bell, 1997).

DEFINING FEATURES OF THE ORDER LEPTICTIDA

CRANIAL

Skull with elongate snout and nasal bones, lacrimal foramen small, and infraorbital canal short; jugal present and postglenoid process well developed; subsquamosal foramen present; maxilla with large orbital wing; auditory bullae formed by entotympanic and not completely covering the tympanic cavities; flange of periotic at posterior wall of tympanic chamber absent and fenestra rotunda fully exposed; stylomastoid foramen not isolated from tympanic chamber by tympanohyal; small, triangular exposure of parietal on posterior occiput; deep antorbital fossa for snout muscles; inferior ramus of stapedial artery exits tympanic cavity near forward apex of facial canal; alisphenoid broadly exposed in orbital wall; short alisphenoid canal with anterior exit well behind sphenorbital fissure; subarcuate fossa very excavated (Novacek, 1977, 1986). Known leptictidan crania include those of Eocene leptictid *Palaeictops* (Matthew, 1918; Novacek, 1986) and Oligocene leptictid *Leptictis* (Butler, 1956; Novacek, 1986) (Figure 6.1).

DENTAL

Dental formula I2/3-2, C1/1, P5-4/5-4, M3/3 (tooth nomenclature for leptictids follows Novacek, 1986); canines and dP1/dp1 single-rooted; p4 double-rooted and trenchant without basined talonid; p5 semi-molariform lacking paraconid or molariform with paraconid; p5 talonid basin fully formed or nearly trenchant; lower molar trigonids anteroposteriorly compressed and tall, talonid basins relatively large and shallow with entoconulid often present; m3 with moderately expanded hypoconulid; P4 with paracone, metacone, protocone (secondarily lost in some taxa), anterior accessory cuspule, and posterior cingulum; molariform P5 with posterior and anterior cingula; upper molars with labially positioned paracone and metacone, narrow ectocingulum, paraconule and metaconule, pre- and postcingula, and weak-to-present hypocone (Clemens, 1973; Novacek, 1977, 1986).

POSTCRANIAL

Forelimb much shorter than hindlimb; robust humerus with prominent deltopectoral crest; radius and ulna relatively slender with relatively short olecranon process; stout metacarpals with extensor tuberosities on proximodorsal surface of metacarpals II and III; strongly keeled manubrium sterni; pelvis with prominent tuberosity for muscularis rectus femoris; slender femur, with narrow, elevated patellar trochlea and posteriorly directed lesser trochanter; variable degree of distal tibiofibular synostosis; deep astragalar trochlea; moderately elongate astragalar neck and metatarsals; pronounced peroneal process on metacarpal I; ungual phalanges with large extensor and flexor processes (Rose, 1999). Known leptictidan postcrania

Evolution of Tertiary Mammals of North America, Vol. 2. ed. C. M. Janis, G. F. Gunnell, and M. D. Uhen. Published by Cambridge University Press.

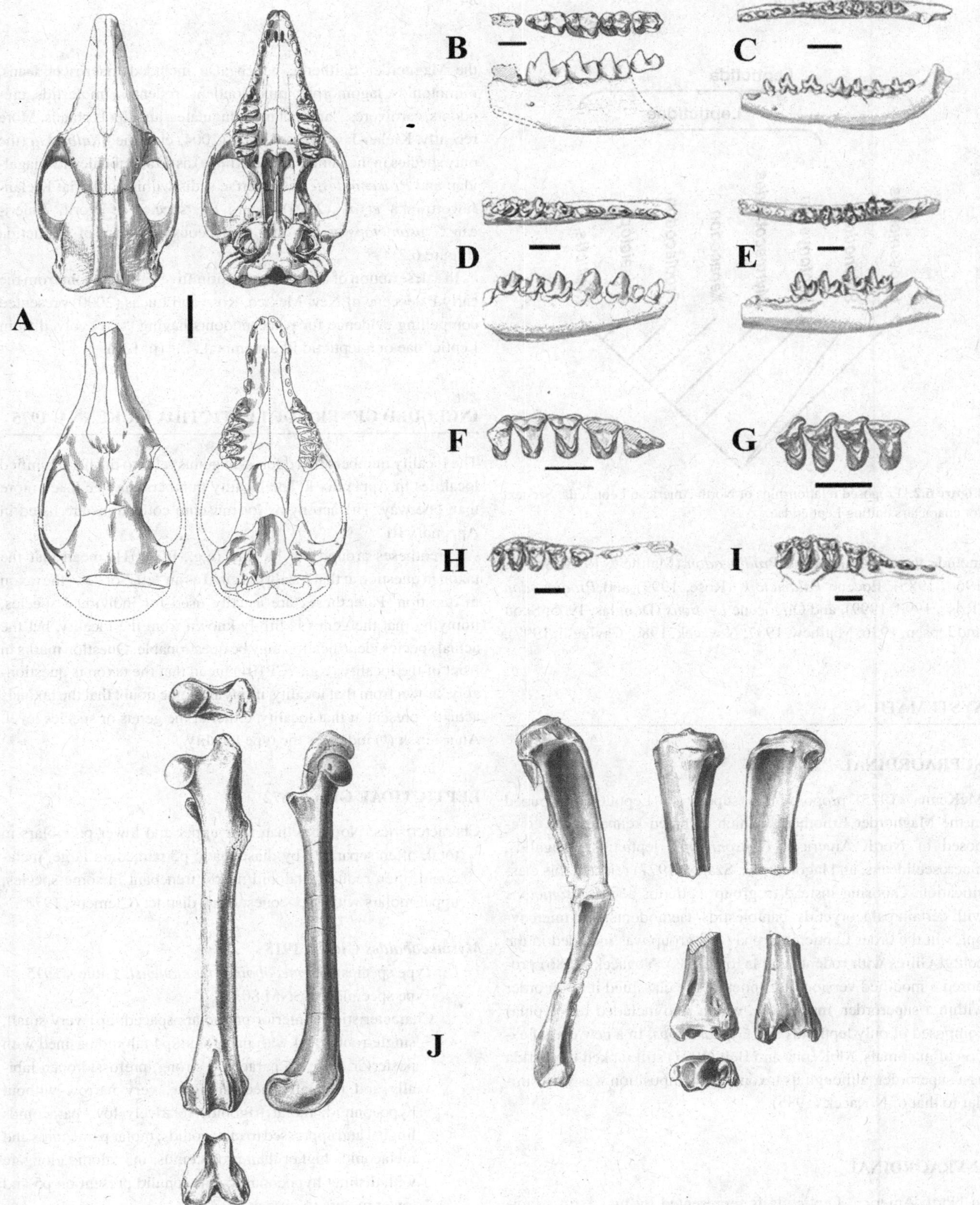

Figure 6.1. A. Skulls of *Leptictis dakotensis* (top, adapted from Novacek, 1986) and *Palaeictops bicuspis* (below, adapted from Matthew, 1918). Scale bar = 1 cm. B–I. Lower and upper dentitions. B. *Leptictis douglassi* (from Novacek, 1976). C. *Myrmecoboides montanensis* (from Novacek, 1977). D. *Palaeictops matthewi* (from Novacek, 1977). E. *Prodiacodon puercensis* (from Novacek, 1977). F. *Leptictis wilsoni* (from Novacek, 1976). G. *Myrmecoboides montanensis* (from Simpson, 1937). H. *Palaeictops matthewi* (from Novacek, 1977). I. *Prodiacodon puercensis* (from Novacek, 1977). Scale bars = 4 mm. J. Leptictid postcrania, including femur (left) and tibiofibula (right) (adapted from Rose, 1999). (A: courtesy of Samuel D. McDowell (top) and the American Museum of Natural History (bottom). B, F: courtesy of the University of Texas. D,E,G, H: courtesy of the University of California Museum of Paleontology. J: courtesy of the Society of Vertebrate Paleontology.)

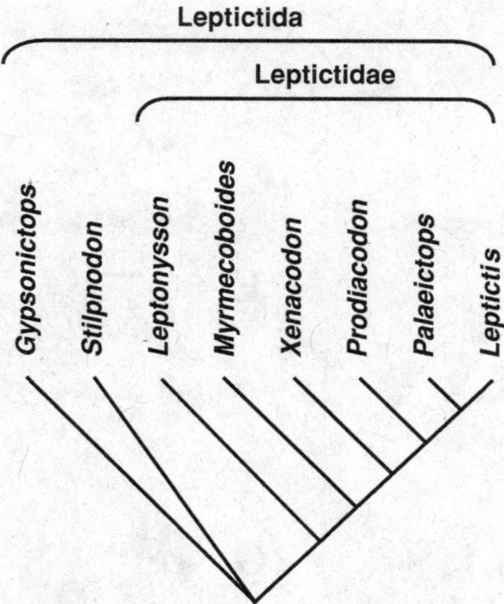

Figure 6.2. Proposed relationships of North American Leptictida. See text for characters uniting Leptictidae.

include those of Paleocene *Prodiacodon* (Matthew, 1918; Szalay, 1966, 1985), Eocene *Palaeictops* (Rose, 1999) and *Prodiacodon* (Rose, 1990, 1999), and Oligocene *Leptictis* (Douglas, 1905; Scott and Jepsen, 1936; Matthew, 1937; Novacek, 1980; Cavigelli, 1997).

SYSTEMATICS

SUPRAORDINAL

McKenna (1975) proposed the superorder Leptictida (grouped in the Magnorder Ernotheria, which included kennalestids) composed of North American *Gypsonictops*, leptictids, anagalids, macroscelideans, and lagomorphs. Szalay (1977) rejected this classification, choosing instead to group leptictids and *Gypsonictops* with certain palaeoryctids, pantolestids, taeniodonts, and microsyopids in the order Leptictomorpha (this group was included in the cohort Glires with rodents and lagomorphs). Novacek (1986) proposed a modified version of Leptictida (he classified it as an order within a superorder, Insectivora, which also included Lipotyphla) composed of only leptictids and *Gypsonictops*. In a new classification of mammals, McKenna and Bell (1997) still ranked Leptictida as a superorder, although its taxonomic composition was very similar to that of Novacek (1986).

INFRAORDINAL

In North America, Leptictida is represented by two extinct families, Gypsonictopidae and Leptictidae. Also included in Leptictida by McKenna and Bell (1997) are Asian Kulbeckiidae (Nessov, 1993) and Didymoconidae (Kretzoi, 1943), and European Pseudorhynchocyoninae (Sigé, 1974; included as a subfamily of Leptictidae). McKenna and Bell (1997) classified Leptictida within the Magnorder Epitheria, which also included macroscelideans, mimotonids, lagomorphs, mixodontians, rodents, cimolestids, creodonts, carnivores, "archontans," ungulates, and lipotyphlans. More recently, Kielan-Jaworowska *et al.* (2004) classified *Kulbeckia* (the only species in the family Kulbeckiidae) as a zalambdalestid anagalidan and *Prokennalestes* as incertae sedis within Eutheria. Kielan-Jaworowska *et al.* (2004) regard the species of North American *Gypsonictops* as the only Cretaceous members of Leptictida (Figure 6.2).

In a description of a skeleton of a primitive palaeanodont from the early Paleocene of New Mexico, Rose and Lucas (2000) presented compelling evidence for palaeanodonts having ". . . evolved from Leptictidae or a leptictid like mammal . . ." (p. 139).

INCLUDED GENERA OF LEPTICTIDA MCKENNA, 1975

The locality numbers listed for each genus refer to the list of unified localities in Appendix I. The locality numbers may be listed more than one way. The acronyms for museum collections are listed in Appendix III.

Parentheses around the locality (e.g., [CP101]) mean that the taxon in question at that locality is cited as an "aff." or "cf." the taxon in question. Parentheses are usually used for individual species, implying that the genus is firmly known from the locality, but the actual species identification may be questionable. Question marks in front of the locality (e.g., ?CP101) mean that the taxon is questionably known from that locality, implying some doubt that the taxon is actually present at that locality, either at the genus or species level. An asterisk (*) indicates the type locality.

LEPTICTIDAE GILL, 1872

Characteristics: No more than four upper and lower premolars in total, often separated by diastemata; p5 paraconid large, metaconid often reduced, talonid nearly trenchant in some species; upper molars with hypocones, often distinct (Clemens, 1973).

Myrmecoboides Gidley, 1915

Type species: *Myrmecoboides montanensis* Gidley, 1915.

Type specimen: USNM 8037.

Characteristics: Anterior premolars spaced; dp1 very small, single-rooted; p4 with paraconid; p4 talonid basined with posterior cusp; p5 paraconid strong, prefossid open labially and lingually, heel elongate, very narrow without hypoconulid; molar trigonids relatively low, paraconids lingual and appressed to metaconids; molar paraconids and metaconids higher than protoconids; m3 talonid elongate with distinct hypoconulid; entoconulid present on p5 and lower molars (Novacek, 1977).

Average length of m2: 2.4 mm.

Included species: *M. montanensis* (known from localities CP14A, NP1C, NP19C*, NP19IIA, B).

Myrmecoboides sp. is also known from localities CP1C, CP13B, E.

Xenacodon Matthew and Granger, 1921

Type species: *Xenacodon mutilatus* Matthew and Granger, 1921.

Type specimen: AMNH 17407.

Characteristics: Lower dental formula ?i3, c1, p4, m3; incisors and canine small; all premolars two-rooted; p5 with large metaconid, paraconid small and basal, talonid very short and not distinctly basined; m2–3 trigonids relatively low, protoconid slightly larger than metaconid, hypoconid large, united by crest with hypoconulid, entoconid very small, isolated, and conical; m3 talonid reduced, short, very narrow, hypoconulid not projecting (Simpson, 1935a).

Average length of m2: 3.5 mm.

Included species: *Xenacodon mutilatus* only, known from locality SB20A* only.

Xenacodon sp. is also known from locality SB20A.

Prodiacodon Matthew, 1929 (including _Diacodon_, in part)

Type species: *Prodiacodon tauricinerei* (Jepsen, 1930) (originally described as *Diacodon tauricinerei*).

Type specimen: YPM-PU 13104.

Characteristics: P5 and upper molars transverse, anteroposteriorly compressed with sharp cusps; P5 hypocone vestigial or absent; precingula on P5 and molars long, nearly reaching anterolingual corner of crown; upper molar hypocones very short relative to protocones; M1–2 paraconules lingual to metaconules, paraconules doubled; parastylar lobes on M1–2 prominent with deep extoflexi; lower molars with sharp cusps, trigonids relatively high and anteroposteriorly compressed, paraconids not closely appressed to metaconids and not higher than protoconids (Novacek, 1977).

Average length of m2: 2.5 mm (*P. tauricinerei*).

Included species: *P. tauricinerei* (known from localities SB24, [CP17B], [CP18], CP20A*, [BB], C, D, CP25B, [C], [CP63], CP64A, B, [NP49]); *P. concordiarcensis* (including *Diacodon pearcei* Gazin, 1956) (localities [CP13G], CP14A, CP16A, [CP24A], NP1C, NP19C*, NP19IIA); *P. crustulum* (locality NP16C*); *P. furor* (localities [CP24A], NP1C, [NP3A0], NP19C*, NP19IIA); *P. puercensis* (localities SB23C, F, H, [CP14A], [NP1C]).

Prodiacodon sp. is also known from localities (SB23A), CP13C, CP14A, CP20A, (CP25B), (CP26A), NP3C, NP7B, C, NP14A2, NP16B, NP48B.

Comments: The precise locality of the holotype of *P. puercensis* is unknown. It was described as coming from the Upper Paleocene Torrejon of New Mexico by Matthew (1918), which suggests that it may be from either locality SB23F or SB23H.

Palaeictops Matthew, 1899 (including _Stypolophus_, in part; _Ictops_, in part; _Diacodon_, in part; _Parictops_; _Hypictops_)

Type species: *Palaeictops bicuspis* (Cope, 1880) (originally described as *Stypolophus bicuspis*).

Type specimen: AMNH 4802.

Characteristics: P5 and upper molars not transverse nor anteroposteriorly compressed, with low bulbous cusps, narrow stylar shelves, small but distinct hypocones, short anterior cingula, conules at labial base of paracone and metacone; M2 not wider than M1; M2–3 parastylar spurs weak and ectoflexi shallow; p5 paraconid well developed; lower molar trigonids relatively low; skull (where known) with single median sagittal crest and narrow zygomatic arch (Novacek, 1977).

Average length of m2: 2.9 mm (*P. bicuspis*).

Included species: *P. bicuspis* (including *Diacodon bicuspis* Matthew, 1918; *Ictops bicuspis*, Cope, 1881) (known from localities SB22A, CP20A, [BB], D, CP25D*, I2, J, [CP64B, C]); *P. bridgeri* (Simpson, 1959; originally described as *Diacodon bridgeri*, including *Hypictops syntaphus* Gazin, 1949) (localities CP5A, [CP31E], [CP32B], CP34D*); *P. borealis* (localities NP8*, [NP9A]); *P. matthewi* (localities SB22B, CP64C*); *P. multicuspis* (Granger, 1910; originally described as *Parictops multicuspis*) (locality CP27E*); *P. pineyensis* (Gazin, 1952; originally described as *Diacodon pineyensis*) (localities CP25G*, [H], [CP27B]).

Palaeictops sp. is also known from localities (CC4), (SB22B]), (SB39C), (CP13B, G), CP25A, (F), CP27E, CP29D, CP31C, CP36B, NP16C.

Leptictis Leidy, 1868 (including _Ictops_; _Nanohyus_; _Isacus_; _Isacis_; _Mesodectes_; _Ictidops_)

Type species: *Leptictis haydeni* Leidy 1868.

Type specimen: None designated.

Characteristics: Lower molar paraconids medial, not appressed to metaconids; lower molar entoconulids present; skull with twinned sagittal crest; upper molar conules set at labial bases of paracone and metacone; upper molars transverse with narrow stylar shelves and large hypocones; p4–5 similar in size; P4 with complex lingual moiety.

Average length of m2: 3.0 mm (*L. acutidens*).

Included species: *L. haydeni* (known from locality CP84A*); *L. acutidens* (localities CP98B, C, [NP10B], NP24C*, D); *L. bullatus* (locality CP84A*); *L. dakotensis* (Cope, 1875; originally described as *Leptictis* [*Mesodectes*] *caniculus*) (localities CP84A*, [B, C]); *L. douglassi* (localities SB44B*, C); *L. major* (locality NP25C*); *L. montanus* (including *L. intermedius* Douglass, 1905; *L. tenuis* Douglass, 1905) (locality NP25C*); *L. thompsoni* (locality NP24C*); *L. wilsoni* (locality SB44C*).

Leptictis sp. is also known from localities CP41B, CP43, CP46, CP68C, CP84A, CP85A, CP99A, NP8, NP9A, B, NP10Bi, BB, C2, NP25B, NP49II.

Comments: See Novacek (1976) for discussion of taxonomy of *Leptictis*.

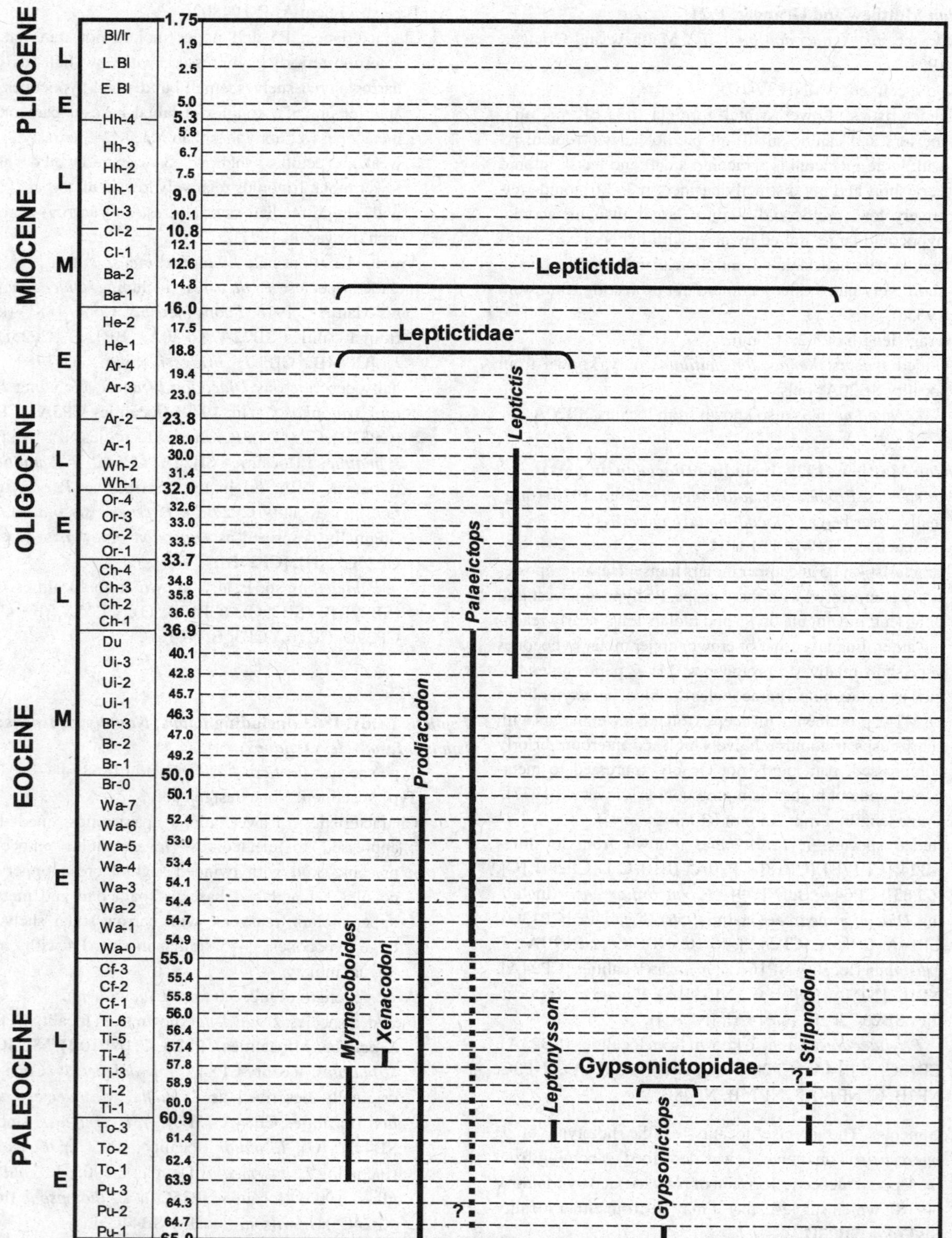

Figure 6.3. Temporal ranges of North American leptictid genera.

LEPTICTIDAE INCERTAE SEDIS

Leptonysson Van Valen, 1967

Type species: *Leptonysson basiliscus* Van Valen, 1967.

Type specimen: AMNH 35295.

Characteristics: Canine small, perhaps incisiform; molar trigonids relatively tall, with carnassial notches in paralophid and protolophid; paraconid relatively central on m1 and m2; lingual face of protoconid strongly concave; entocristid complete; mesoconid absent; entoconid taller than hypoconid; hypoconulid at least as large as entoconid on m1 and m2; m3 about as long as m2; m3 metaconid somewhat lower than protoconid. Other known characters similar to *Procerberus*, but differing in having paraconid less anterior on m1; paralophid stronger; more differentiation among molars from m1 to m3; cristid obliqua less oblique; and talonid cusps more distinct from each other (Van Valen, 1967).

Average length of m2: 3.4 mm.

Included species: *L. basiliscus* only, known from locality NP19C* only.

GYPSONICTOPIDAE VAN VALEN, 1967

Characteristics: Four or five upper and lower premolars present, crowded together with no diastemata; p4 paraconid small or absent, metaconid strong and subequal to protoconid, talonid basins fully formed; upper molars lacking distinct hypocones (Clemens, 1973).

Gypsonictops Simpson, 1927

Type species: *Gypsonictops hypoconus* Simpson, 1927.

Type specimen: YPM 13662.

Characteristics: Dental formula I?/2, C1/1, P?4/5, M3/3; lower premolars double-rooted; p3 metaconid absent with trenchant or shallowly basined talonid; P3 triple-rooted with distinct protocone and paracone, metacone normally present; P4 nearly molariform; upper molars with protocone crescentic, paracone and metacone transversely compressed, small conules, narrow but sharply defined buccal cingulum that is stronger posteriorly, and very weak anterior cingulum (Simpson, 1927; Clemens, 1973).

Average length of m2: 1.9 mm (*G. hypoconus*).

Included Tertiary species: *G. illuminatus* only (known from localities NP7A, NP15C, NP16A).

Comments: *Gypsonictops* is a relatively common faunal element in North American Late Cretaceous deposits but is extremely rare in the Tertiary.

LEPTICTIDAE OR GYPSONICTOPIDAE, INCERTAE SEDIS

Stilpnodon Simpson, 1935b

Type species: *Stilpnodon simplicidens* Simpson, 1935b.

Type specimen: USNM 9629.

Characters: p4 with high, slender main cusp, rudimentary anterior basal cuspule, metaconid absent, simple nonbasined talonid with single cuspule; m3 reduced and low, median paraconid, trigonid erect, moderately elevated protoconid large, trigonid nearly as long as short talonid (Simpson, 1935b).

Average length of m1: 1.4 mm.

Included species: *S. simplicidens* only, known from locality NP19C* only.

Stilpnodon sp. is also questionably known from locality NP3D.

INDETERMINATE LEPTICTIDS

In addition to the taxa discussed above, indeterminate or undescribed leptictids are also present at localities CC9BB, SB23 (precise sublocality unknown), SB23H, CP6AA, CP13B, CP20BB, CP25J2, CP38B, C, F, CP65, NP14A, NP29C.

BIOLOGY AND EVOLUTIONARY PATTERNS

Leptictida first appear in the fossil record in the Late Cretaceous in North America (Figure 6.3). In the North American Tertiary, leptictids are first represented by *Gypsonictops* in the early Puercan (Pu1). By the late Puercan, four leptictid genera are represented, with the long-surviving *Prodiacodon* (through Wa7) and *Palaeictops* (through the Duchesnean) making their first appearances. By the end Tiffanian zone Ti4, only *Prodiacodon* and *Palaeictops* are present until the appearance of *Leptictis* in the late Uintan (Ui3). *Leptictis* survives through the earliest Arikareean, at which time leptictids disappear from the North American fossil record.

Leptictids are relatively small mammals with estimated body masses ranging from about 400 to 700 g (Rose, 1999). In general cranial proportions (although differing dramatically in detail), leptictids are similar to extant tupaiids and macroscelidians, and probably had elongate, mobile noses (Novacek, 1986). Dentitions indicate that leptictids were primarily insectivorous and omnivorous. Postcranially, leptictids have shorter forelimbs than hindlimbs and have distally fused tibiae–fibulae, similar to those of *Erinaceus* and *Rhynchocyon* (Rose, 1999). Rose (1999) suggested that leptictids were terrestrial mammals that were capable of rapid running and quadrupedal jumping but were less likely to have exhibited bipedal hopping behaviors. The structure of the forelimb, similar to many extant lipotyphlans, indicates that leptictids often dug for food or shelter (Cavigelli, 1997; Rose, 1999). In many ways, primitive leptictids are quite similar to primitive palaeanodonts (Rose and Lucas, 2000; Rose, Chapter 9, this volume) and it is quite possible that palaeanodont ancestry can be traced to leptictids.

REFERENCES

Butler, P. M. (1956). The skull of *Ictops* and the classification of the Insectivora. *Proceedings of the Zoological Society of London*, 126, 453–81.

(1972). The problem of insectivore classification. In *Studies in Vertebrate Evolution*, ed. K. A. Joysey and T. S. Kemp, pp. 253–65. New York: Winchester.

Cavigelli, J. P. (1997). A preliminary description of a *Leptictis* skeleton from the White River Formation of eastern Wyoming. *Tate Geological Museum Guidebook*, 2, 101–18.

Clemens, W. A., Jr. (1973). Fossil mammals of the type Lance Formation, Wyoming, III. Eutheria and summary. *University of California Publications in Geological Sciences*, 94, 1–102.

Cope, E. D. (1875). *Systematic Catalogue of Vertebrata of the Eocene of New Mexico, Collected in 1874. Report to the Engineer, Department, United States Army*, in charge, Lt. George M. Wheeler, pp. 5–37. Washington, DC: US Government Printing Office.

(1880). The northern Wasatch fauna. *The American Naturalist*, 14, 908.

(1881). On the Vertebrata of the Eocene Wind River beds of Wyoming. *Bulletin of United States Geological Survey of the Territories*, 6, 183–202.

Douglass, E. (1905) The Tertiary of Montana. *Memoirs of the Carnegie Museum*, 2, 212–25.

Gazin, C. L. (1949). A leptictid insectivore from the middle Eocene Bridger Formation of Wyoming. *Journal of the Washington Academy of Sciences*, 39, 220–3.

(1952). The lower Eocene Knight Formation of western Wyoming and its mammalian faunas. *Smithsonian Miscellaneous Collections*, 117, 1–82.

(1956). Paleocene mammalian faunas of the Bison Basin in south-central Wyoming. *Smithsonian Miscellaneous Collections*, 131, 1–47.

Gidley, J. W. (1915). An extinct marsupial from the Fort Union with notes on the Myrmecobidae and other families of this group. *Proceedings of the United States National Museum*, 48, 395–402.

Gill, T. (1872). Arrangement of the families of mammals with analytical tables. *Smithsonian Miscellaneous Collections*, 11, 1–98.

Granger, W. (1910). Tertiary faunal horizons in the Wind River Basin, Wyoming, with descriptions of new Eocene mammals. *Bulletin of the American Museum of Natural History*, 28, 235–51.

Gregory, W. K. (1910). The orders of mammals. *Bulletin of the American Museum of Natural History*, 27, 1–524.

Jepsen, G. L. (1930). New vertebrate fossils from the lower Eocene of the Bighorn Basin, Wyoming. *Proceedings of the American Philosophical Society*, 69, 117–31.

Kielan-Jaworowska, Z., Cifelli, R. K., and Luo, Z. (2004). *Mammals from the Age of Dinosaurs*. New York: Columbia University Press.

Kretzoi, M. (1943). *Kochictis centenii* n.g., n.sp. az egeresi felsö Oligocénböl. *Földtani Közlöny*, 73, 10–17.

Leidy, J. (1868). Notice of some remains of extinct Insectivora from Dakota. *Proceedings of the Academy of Natural Sciences, Philadelphia*, 1868, 315–16.

Lillegraven, J. A. (1969). Latest Cretaceous mammals of the upper part of the Edmonton Formation of Alberta, Canada, and a review of the marsupial–placental dichotomy in mammalian evolution. *University of Kansas Paleontological Contributions*, 50, 1–122.

Matthew, W. D. (1899). A provisional classification of the fresh-water Tertiary of the west. Bulletin of the American Museum of Natural History, 12, 19–75.

(1918). Part V. Insectivora (continued), Glires, Edentata. *Bulletin of the American Museum of Natural History*, 38, 565–657.

(1937). Paleocene faunas of the San Juan Basin, New Mexico. *Transactions of the American Philosophical Society*, 30, 1–510.

Matthew, W. D. and Granger, W. (1921). New genera of Paleocene mammals. *American Museum Novitates*, 13, 1–7.

McDowell, S. B. (1958). The greater Antillean insectivores. *Bulletin of the American Museum of Natural History*, 115, 113–214.

McKenna, M. C. (1969). The origin and early differentiation of therian mammals. *Annals of the New York Academy of Sciences*, 167, 217–40.

(1975). Toward a phylogenetic classification of the Mammalia. In *Phylogeny of the Primates*, ed. W. P. Luckett and F. S. Szalay, pp. 21–46. New York: Plenum Press.

McKenna, M. C. and Bell, S. K. (1997). *Classification of Mammals Above the Species Level*. New York: Columbia University Press.

Nessov, L. A. (1993). New Mesozoic mammals of middle Asia and Kazakhstan and comments about evolution of theriofaunas of Cretaceous coastal plains of Asia. *Trudy Zoological Institute*, 249, 105–33.

Novacek, M. J. (1976). Early Tertiary vertebrate faunas, Vieja Group, Trans-Pecos Texas: Insectivora. *Pearce-Sellards Series, Texas Memorial Museum*, 23, 1–18.

(1977). A review of Paleocene and Eocene Leptictidae (Eutheria: Mammalia) from North America. *PaleoBios*, 24, 1–42.

(1980). Cranioskeletal features in tupaiids and selected eutherians as phylogenetic evidence. In *Comparative Biology and Evolutionary Relationships of Tree Shrews*, ed. W. P. Luckett, pp. 35–93. New York: Plenum Press.

(1986). The skull of leptictid insectivorans and the higher-level classification of eutherian mammals. *Bulletin of the American Museum of Natural History*, 183, 1–112.

Rose, K. D. (1990). Postcranial remains and adaptations in early Eocene mammals from the Willwood Formation, Bighorn Basin, Wyoming. *Geological Society of America Special Paper*, 243, 107–33.

(1999). Postcranial skeleton of Eocene Leptictidae (Mammalia), and its implications for behavior and relationships. *Journal of Vertebrate Paleontology*, 19, 355–72.

Rose, K. D. and Lucas, S. G. (2000). An early Paleocene palaeanodont (Mammalia, ?Pholidota) from New Mexico, and the origin of Palaeanodonta. *Journal of Vertebrate Paleontology*, 20, 139–56.

Scott, W. B. and Jepsen, G. L. (1936). The mammalian fauna of the White River Oligocene. Part I. Insectivora and Carnivora. *Transactions of the American Philosophical Society*, 28, 1–153.

Sigé, B. (1974). *Pseudorhyncocyon cayluxi* Filhol, 1892; Insectivore géant des Phosphorites du Quercy. *Palaeovertebrata*, 6, 33–46.

Simpson, G. G. (1927). Mammalian fauna of the Hell Creek Formation of Montana. *American Museum Novitates*, 267, 1–7.

(1935a). The Tiffany fauna, upper Paleocene, I: Multituberculata, Marsupialia, Insectivora, and ?Chiroptera. *American Museum of Novitates*, 795, 1–19.

(1935b). New Paleocene mammals from the Fort Union of Montana. *Proceedings of the United States National Museum*, 83, 221–44.

(1937). The Fort Union of the Crazy Mountain Field, Montana and its mammalian faunas. *Bulletin of the United States National Museum* 169, 1–287.

(1945). The principles of classification and a classification of mammals. *Bulletin of the American Museum of Natural History*, 85, 1–350.

(1959). Two new records from the Bridger middle Eocene of Tabernacle Butte, Wyoming. *American Museum Novitates*, 1966, 1–5.

Szalay, F. S. (1966). The tarsus of the Paleocene leptictid *Prodiacodon* (Insectivora, Mammalia). *American Museum Novitates*, 2267, 1–13.

(1969). Mixodectidae, Microsyopidae, and the insectivore–primate transition. *Bulletin of the American Museum of Natural History*, 140, 197–330.

(1977). Phylogenetic relationships and a classification of the eutherian Mammalia. In *Major Patterns in Vertebrate Evolution*, ed. M. K. Hecht, P. C. Goody, and B. M. Hecht, pp. 315–74. New York: Plenum.

(1985). Rodentia and lagomorph morphotype adaptations, origins, and relationships: some postcranial attributes analyzed. In *Evolutionary Relationships among Rodents: a Multidisciplinary Analysis*, ed. W. P. Luckett and J. L. Hartenberger, pp. 83–132. New York: Plenum.

Van Valen, L. (1965). Treeshrews, primates, and fossils. *Evolution*, 19, 137–51.

(1967). New Paleocene insectivores and insectivore classification. *Bulletin of the American Museum of Natural History*, 135, 217–84.

7 Lipotyphla

GREGG F. GUNNELL,[1] THOMAS M. BOWN,[2] J. HOWARD HUTCHISON,[3]
JONATHAN I. BLOCH[4]

[1]*Museum of Paleontology, University of Michigan, Ann Arbor, MI, USA*
[2]*Erathem-Vanir Geological, Pocatello, ID, USA*
[3]*Museum of Paleontology, University of California at Berkeley, CA, USA*
[4]*Florida Museum of Natural History, University of Florida, Gainesville, FL, USA*

INTRODUCTION

Used since its inception (Bowdich, 1821) as a "waste-basket" for morphologically disparate, putatively "primitive," and generally ancient groups of small- to moderate-sized mammals, the traditional "Order Insectivora" of Simpson (1945) simply does not exist as an internally consistent or unified taxonomic grouping. This is true in both a morphological and a phylogenetic sense, and it reflects the considerable, though commonly understated, adaptive breadth of "insectivores." Collectively, the "insectivorous" mammals are generally referred to by the appellation "primitive." However, the cranial and dental specializations of hedgehogs (Erinaceidae), and shrews, moles, and apternodontids (Soricomorpha), are anything but primitive. Similarly, though fossil postcrania for most of these groups is poorly known among North American forms, those of moles are uniquely specialized by any definition.

Most of the mammals discussed herein are relatively rare components of fossil vertebrate faunas. A few are known only from the type materials and many more only from incomplete upper or lower dentitions. Therefore, the dental anatomy of insectivorous mammals has both largely determined their diagnoses from related forms as well as played an instrumental role in their phylogenetic reconstruction. Notable exceptions to the constraints posed by limited knowledge of the anatomy of fossil forms are the importance of the anatomy of the humerus and the mandible to the taxonomy and phylogeny of, respectively, the moles and shrews.

Of the groups of insectivorous mammals discussed herein, only Solenodontidae (including Nesophontidae), Talpidae, and Soricidae survive in North America. The latter two families are also extant in the Old World, where they are joined by living representatives of Erinaceidae, Potamogalidae, Chrysochloridae, Tenrecidae, Macroscelididae, and Tupaiidae.

DEFINING FEATURES OF THE ORDER LIPOTYPHLA

CRANIAL

Brain small with smooth cerebral hemispheres; cerebrum not extended over cerebellum; postorbital bar absent; orbital wing of palatine very small, confined to floor of orbit; maxilla in orbit strongly expanded, broadly contacts frontal and completely excludes palatine from contact with lacrimal; jugal reduced or absent, does not contact lacrimals; fenestra rotunda partly concealed ventrally, opening into pit defined posteriorly by raised rim of petromastoid; no true postglenoid process (replaced in some lipotyphlans by expanded entoglenoid process of squamosal); optic foramen much smaller than sphenorbital fissure; suprameatal foramen absent (Haeckel, 1866; McDowell, 1958; Novacek, 1986) (Figure 7.1).

DENTAL

Teeth vary from acutely cuspate to more flattened with rounded, low cusps. In some forms, posterior premolars become specialized for crushing and grinding while in others anterior incisors become enlarged for specialized food acquisition. Erinaceomorphs have quadritubercular upper molars and lower molars with distinct trigonids and well-developed talonids. Many erinaceomorphs have molars that reduce in size from first to last. Proscalopids have enlarged anterior incisors and exhibit loss of anterior cheek teeth in more derived taxa. Proscalopid molars range from brachydont to hypsodont in derived forms. Talpids have dilambdodont dentitions – primitive forms retain all their teeth but more derived taxa lose anterior cheek teeth and often have enlarged incisors. Many nyctitheriids have acute, cuspate cheek teeth and specialized, multi-tined incisors.

Evolution of Tertiary Mammals of North America, Vol. 2. ed. C. M. Janis, G. F. Gunnell, and M. D. Uhen. Published by Cambridge University Press.

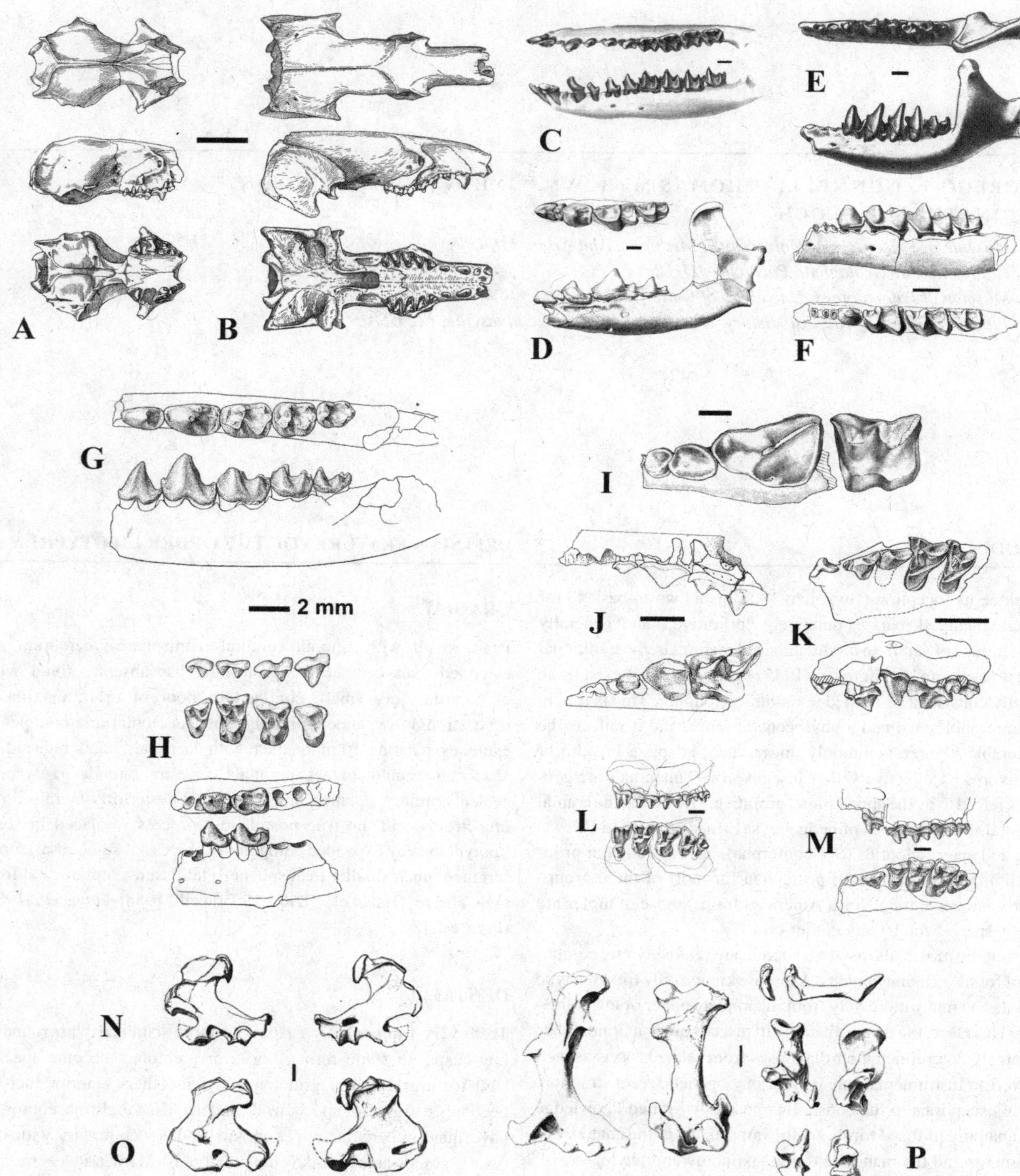

Figure 7.1. A, B. Lipotyphlan skulls. A. Erinaceid *Brachyerix macrotis* (adapted from Matthew and Mook, 1933). B. Apternodontid *Apternodus gregoryi* (adapted from Schlaikjer, 1933). Scale bars = 10 mm. C–F. Lipotyphlan lower dentitions. C. Nyctitheriid *Ceutholestes dolosus* (adapted from Rose and Gingerich, 1987). D. Soricid *Ingentisorex tumididens* (adapted from Hutchison, 1966). E. Oligoryctid *Oligoryctes cameronensis* (adapted from Asher *et al.*, 2002). F. Talpid *Mystipterus pacificus* (adapted from Hutchison, 1968). Scale bars = 1 mm. G, H. Erinaceomorph dentitions. G. *Litocherus zygeus* (adapted from Gingerich, 1983). H. *Macrocranion junnei* (adapted from Smith *et al.*, 2002). Scale bar = 2 mm. I–M. Lipotyphlan upper dentitions. I. Soricid *Ingentisorex tumididens* (reversed, adapted from Hutchison, 1966). J. Talpid *Mystipterus pacificus* (reversed, adapted from Hutchison, 1968). K. Geolabidid *Batodonoides vanhouteni* (adapted from Bloch, Rose, and Gingerich, 1998). L. Nyctitheriid *Leptacodon rosei* (adapted from Gingerich, 1987). M. Erinaceomorph *Diacocherus minutus* (adapted from Gingerich, 1983). Scale bars = 1 mm. N–P. Lipotyphlan postcrania. N. Micropternodontid *Cryptoryctes kayi* (adapted from C. R. Reed, 1956). O. Proscalopid *Proscalops evelynae* (adapted from C. R. Reed, 1956). P. Talpid *Gaillardia thomsoni* (adapted from Hutchison, 1968). Scale bars = 2 mm. (A, E: courtesy of the American Museum of Natural History. C: © 1987 American Society of Mammalogists, from *Journal of Mammalogy* by K. D. Rose and P. D. Gingerich, reprinted by permission of Alliance Communications Group, a division of Allen Press, Inc. D,F,I,J,P: courtesy of the University of Oregon. G,H,L,M: courtesy of the University of Michigan. K: ©1998 American Society of Mammalogists, from *Journal of Mammalogy* by J. I. Bloch, K. D. Rose, and P. D. Gingerich, reprinted by permission of Alliance Communications Group, a division of Allen Press, Inc. N,O: courtesy of the Field Museum of Natural History.)

Apternodontids, parapternodontids, and oligoryctids all have very specialized zalambdodont dentitions. Apternodontids have enlarged anterior incisors while oligoryctids have extremely reduced lower molar talonids. In plesiosoricids, the second lower incisor is enlarged and procumbent and upper molars lack dilambdodonty. Soricids have dilambdodont upper molars; a distinct, enlarged, hooklike upper incisor; simple, unicuspid teeth between the first upper incisor and the upper fourth premolar; and an enlarged and procumbent anterior lower incisor.

POSTCRANIAL

Feet plantigrade and pentadactyl; intestinal cecum absent; pubic symphysis reduced or absent.

SYSTEMATICS

SUPRAORDINAL

Recent molecular studies (Springer *et al.*, 2004) are in conflict with relationships supported by morphological data. Based on morphological analysis, "Insectivora" is often placed as the sister taxon to all other eutherians except Xenarthra (Springer *et al.*, 2004). Molecular analysis places Eulipotyphla (hedgehogs, moles, shrews) within Laurasiatheria as the sister group to all other laurasiatheres (carnivores, pangolins, perissodactyls, artiodactyls, whales, and bats).

INFRAORDINAL

Simpson (1945) classified insectivorous mammals in 10 families of "Insectivora": Leptictidae, Palaeoryctidae, Pantolestidae, Mixodectidae, Deltatheridiidae, Solenodontidae, Erinaceidae, Nyctitheriidae, Soricidae, and Talpidae. Use of a new suborder, "Proteutheria" (Romer, 1966), for certain of the more problematic forms with no clear relationship to living "insectivores" did nothing to clarify the origins or interrelationships of these animals; grouping, for example, disparate forms such as leptictids, zalambdalestids, ptolemaiids, apatemyids, and tupaiids. Nonetheless, historically, such terms had utility for workers at the moment less concerned with phylogenetic considerations than with concisely (if not correctly) grouping relatively rare mammals in their faunas (Figures 7.2 and 7.3).

Van Valen's classifications of the Deltatheridia (1966) and Insectivora (1967) recognized and dealt with some of the fundamental problems. Though his studies were quite detailed and comparative (and thereby substantial improvements over the synoptic overviews of Simpson and Romer), they continued to imply somewhat closer interrelationships of these groups than can be supported by either morphological or biostratigraphic evidence.

It is probable that most, if not all, of the several forms earlier grouped with "insectivores," and/or collectively with the "proteutherians," have little to do with one another phylogenetically and are not at all close in terms of recency of origin. Whereas some are now believed to be more closely related to other archaic mammals than to other members of the "Insectivora" sensu lato, still others remain taxonomically isolated with no known ancestors. In the most recent published classifications treating their higher-level relationships (McKenna, 1975; Novacek, 1986; Carroll, 1988; McKenna and Bell, 1997), many of Simpson's (1945) families of insectivores are accorded at least ordinal or subordinal status.

Erinaceomorpha are represented among living mammals by hedgehogs and moon rats (gymnures). Much of the early radiation of erinaceomorphs is represented by enigmatic taxa that are difficult to place taxonomically. Interestingly, this early radiation is predominantly a North American (and European) one, although extant erinaceomorphs are not represented in the New World (unless moles are included in this order) (Figure 7.4, below). Galericines (gymnures) do not appear in the fossil record until the middle Eocene (Asia) and much of their radiation takes place in the Old World, while erinaceines (hedgehogs) first appear slightly later (early Oligocene in Asia and Europe) and are predominantly known from the Miocene and Plio-Pleistocene of both the New and Old Worlds.

Talpids have usually been placed either within the Erinaceomorpha (McDowell, 1958; Van Valen, 1967; McKenna and Bell, 1997) or Soricomorpha (Butler, 1972) but are generally regarded as transitional between these large groupings. The taxonomic limits of most of the major subgroupings of the extant Talpidae were defined by Dobson as early as 1883 with only minor adjustments in rank in the classifications of Simpson (1945) and Van Valen (1967). The Proscalopidae were originally included as a subfamily by K. Reed (1961) but were elevated to family rank by Reed and Turnbull (1965) and Barnosky (1981). Hutchison (1968) removed *Gaillardia* from the Desmaninae and placed it in its own subfamily. The classification used here is modified from those of Van Valen (1967), Hutchison (1968, 1974), and McKenna and Bell (1997). Because of the highly specialized nature of the postcranial skeleton and preservational durability of many of the limb bones, non-cranial skeletal morphology plays a critical role in talpid classification.

Extant Soricomorpha are represented by the nearly ubiquitous shrews and the much less widespread tenrecs. Shrews are known from all of the northern continents and Africa and have a small presence in Central and South America as well. Tenrecs are restricted to Africa and Madagascar. The earliest records of soricomorphs may date from the Late Cretaceous and there are relatively large radiations in the early Tertiary in both North America and Asia with a slightly smaller diversity documented from Europe. The earliest records of Soricidae (shrews) are from the middle Eocene of North America and there are large radiations of this family in the Miocene of the northern continents. Tenrecs first appear in the early Miocene in Africa but most of the known diversity of tenrecs is represented by recent taxa. Tenrecs may belong to Afrotheria instead of Lipotyphla (Chapter 4).

The Soricidae form a well-defined group and their anatomy is extensively known from extant forms. They form the core family of the Soricomorpha but the expressed relationships to the other members of this group vary among authors. McDowell (1958) explicitly suggested that the Solenodontidae (including Nesophontidae) is the sister group of the Soricidae with the Tenrecidae plus Chrysochloridae and Talpidae as progressively more distant groups. A similar arrangement is supported by the character suite given by MacPhee (1981), but other arrangements derived from data of Novacek (1986). Stehlin (1940) and Gureev (1971) included the

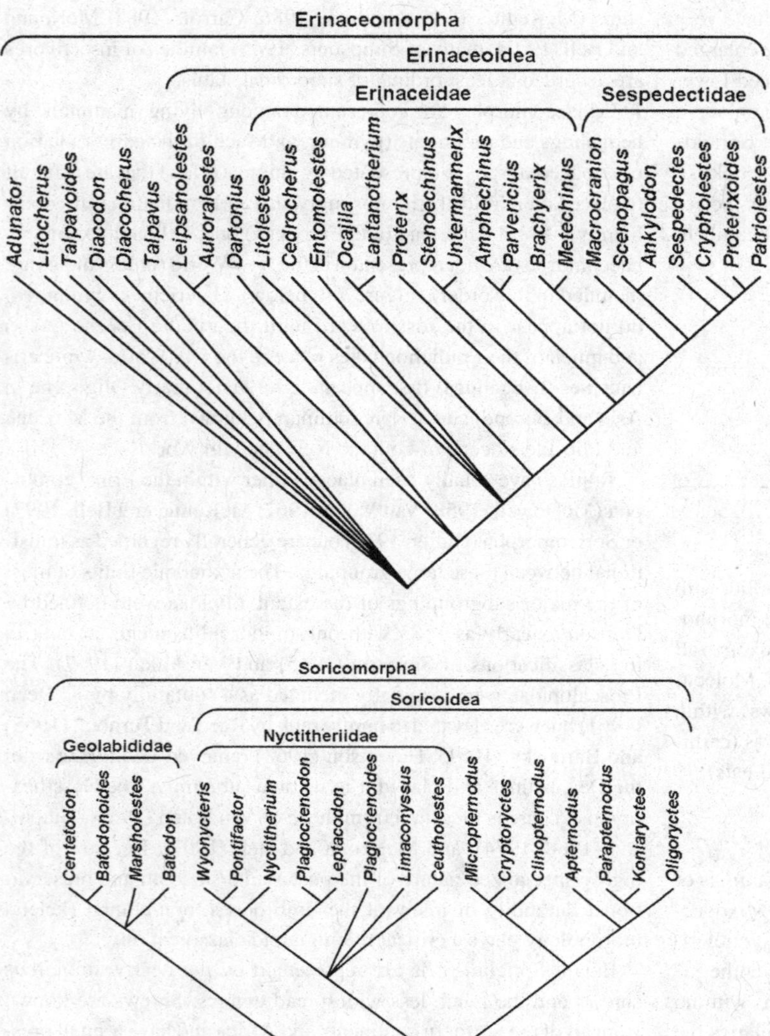

Figure 7.2. Proposed relationships of Erinaceomorpha and Soricomorpha. See text for discussion of characters supporting labeled clades.

nyctithere *Saturninia* Stehlin within the Soricidae and proposed it as the sister taxon of all other soricids, but this was rejected by Repenning (1967), Robinson (1968), and others. Although the inclusion of *Plesiosorex* by Winge (1917) within the Soricidae was not accepted by subsequent authors, Van Valen (1967) suggested that the Soricidae arose from a stock that would be included in the Plesiosoricidae (as might have Talpidae as well). He noted that this is not supported by direct evidence, and the relationships of the Plesiosoricidae to other members of the Soricidae remain obscure. Repenning (1967) and Gureev (1971) provided the most comprehensive reviews of the fossil shrews and their relationships. These two works differ significantly in taxonomic philosophy but that of Repenning is more clearly based on detailed morphology of the dentition and dentary.

McKenna and Bell (1997) recognized both the Solenodontidae and Anthony's (1916) Nesophontidae, in contrast to McDowell (1958), who largely ignored significant dental differences. The dental disparities alone warrant familial separation. The Solendontidae here include the doubtlessly valid nesophontids to incorporate some of McDowell's (1958) cranial information, which is useful in distinguishing these unusual Antillean insectivores (as a group) from other Soricomorpha (see also discussion in McKenna, 1960a).

INCLUDED GENERA OF LIPOTYPHLA HAECKEL, 1866

The locality numbers listed for each genus refer to the list of unified localities in Appendix I. The locality numbers may be listed more than one way. The acronyms for museum collections are listed in Appendix III.

Parentheses around the locality (e.g., [CP101]) mean that the taxon in question at that locality is cited as an "aff." or "cf." the taxon in question. Parentheses are usually used for individual species, implying that the genus is firmly known from the locality, but the actual species identification may be questionable. Question marks in front of the locality (e.g., ?CP101) mean that the taxon is

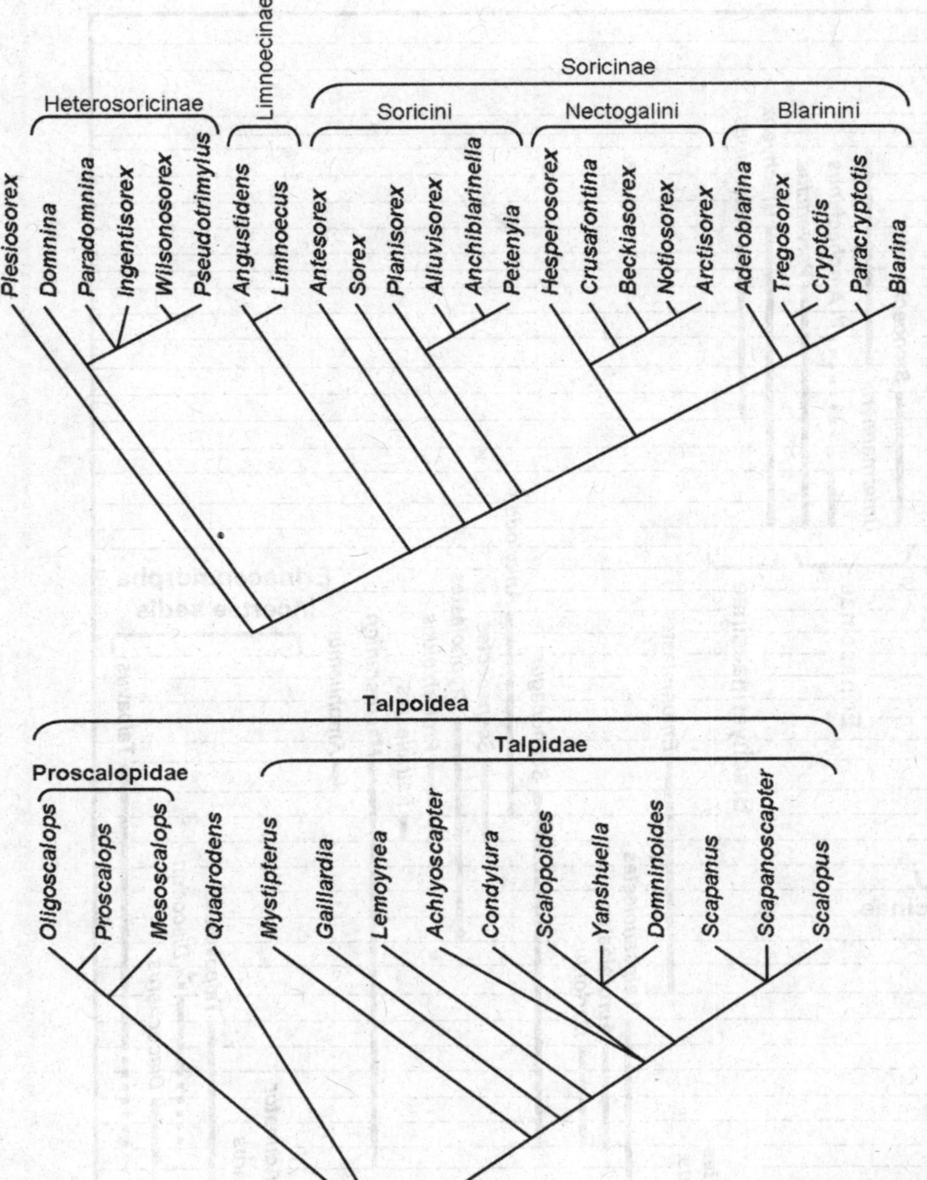

Figure 7.3. Proposed relationships of Soricidae and Talpoidea. See text for discussion of characters supporting labeled clades.

questionably known from that locality, implying some doubt that the taxon is actually present at that locality, either at the genus or species level. An asterisk (*) indicates the type locality.

ERINACEOMORPHA GREGORY, 1910

ERINACEIDAE FISCHER VON WALDHEIM, 1817

Characteristics: Zygomatic arches complete (based on extant forms). Dental formula I2–3/3, C1/1, P3–4/2–4, M3/3; incisors often enlarged; lower molars progressively reduced from m1 to m3; lower molars rectangular with small or absent hypoconulids and weak exodaenodonty; m1 with strong, anteriorly angled paraconid; upper molars quadrate, often bunodont; M1–2 with well-developed hypocones; M3 reduced in size and oval in outline. Feet plantigrade.

GALERICINAE POMEL, 1848

GALERICINI POMEL, 1848

Ocajila Macdonald, 1963

Type species: *Ocajila makpiyahe* Macdonald, 1963.

Type specimen: SDSM 56105.

Characteristics: Lower molar trigonids compressed anteroposteriorly; trigonid and talonid subequal in length; paraconid reduced to ridge (Macdonald, 1963).

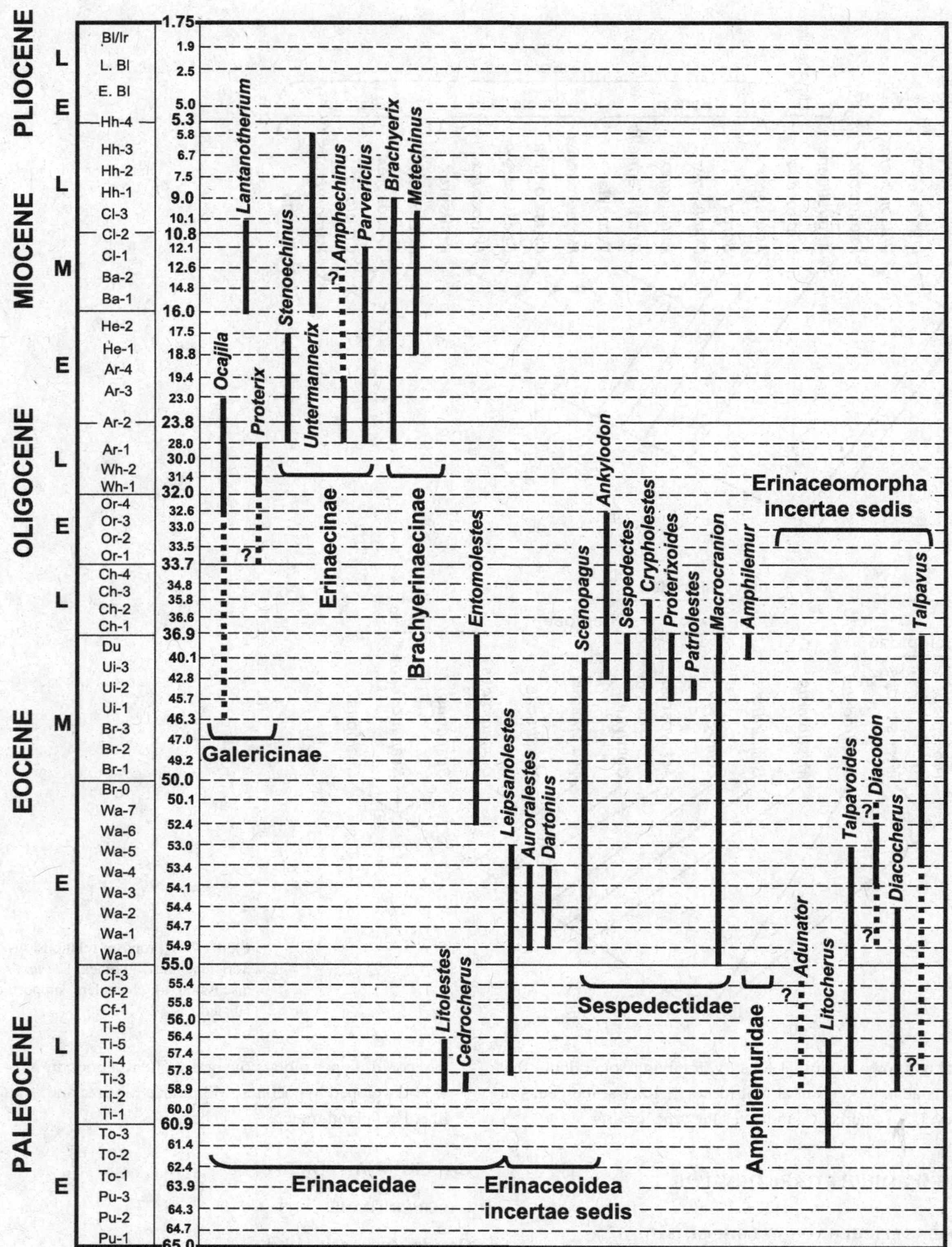

Figure 7.4. Temporal ranges of North American Erinaceomorpha genera.

Average length of m2: 1.78 mm.

Included species: *O. makpiyahe* (known from localities [CP84C], CP85C*, [NP32C], [NP36B]).

Ocajila sp. is also known from localities (CC12), (CP36B), (CP37B), NP32C, NP36A.

***Lantanotherium* Filhol, 1888**

Type species: *Lantanotherium sansaniense*, Filhol, 1888.

Type specimen: MN Paris.

Characteristics: m1 paralophid elongate; m2–3 trigonids open with hypoconid and entoconid each connected to trigonid by longitudinal ridge (Seeman, 1938; Viret, 1940; Butler, 1948; Webb, 1961).

Average length of m2: 2.4–2.9 mm (*L. sawini*).

Included species: *L. dehmi* (known from locality CC17G*); *L. sawini* (localities CC17G*, NB6D).

Lantanotherium sp. is also known from localities GC9C, (CP114B), PN8B.

Comments: *Lantanotherium* is also known from the Miocene of Europe, Asia, and Africa (McKenna and Bell, 1997).

PROTERICINI BUTLER, 1948

***Proterix* Matthew, 1903a**

Type species: *Proterix loomisi* Matthew, 1903a.

Type specimen: AMNH 9756.

Characteristics: Anterior part of skull short; palate completely ossified; well-developed supraorbital crests and postorbital processes; temporal crests form strong sagittal crest; premaxillae do not meet frontals; elongate nasals extending posterior to postorbital processes; deep depressions anterior to orbits for snout musculature; strong zygomatic arches; extremely large, completely ossified bullae; massive, flat, vertical lambdoid plates; no basisphenoid pit; vertebrae with sides covered by large, overlapping, vertical plates; number of upper and lower premolars reduced; P3–M3 three-rooted; small, triangular M3; M1 quadrate, large, with subequal paracone and metacone, subequal protocone and hypocone, small metaconule, small metastylar crest; P4 with large paracone, short, strong metastylar crest, protocone with small hypocone appressed to it; C1 double-rooted; I1/i1 enlarged; i2 (and i3 if present) procumbent; c1 long, peglike, and procumbent; p2, if present, small, simple cone; p3 small, with posterior cingulum; p4 large, with large protoconid, small paraconid crest, and narrow talonid; m1 five-cusped, with low, elongate paraconid; m2 and m3 four-cusped, with relatively broad cingulum (Rich, 1981).

Average length of m2: 1.8 mm (*P. loomisi*).

Included species: *P. loomisi* (known from locality CP84B*); *P. bicuspis* (locality CP84B*).

Proterix sp. is also known from localities (CP41B), CP84B, C, CP85A, CP101.

ERINACEINAE FISCHER VON WALDHEIM, 1817

ERINACEINI FISCHER VON WALDHEIM, 1817

***Stenoechinus* Rich and Rasmussen, 1973**

Type species: *Stenoechinus tantalus* Rich and Rasmussen, 1973.

Type specimen: UKMNH 18001.

Characteristics: M1 wide relative to length; i1 root short; lower molars with reduced paraconids lower than metaconid, much lower than protoconid; m1 trigonid anteroposteriorly compressed; m3 with prominent postcingulum (Rich, 1981).

Average length of m2: 1.6 mm.

Included species: *S. tantalus* only (known from localities CP74A, NP36[B], D*).

***Untermannerix* Rich, 1981**

Type species: *Untermannerix copiosus* Rich, 1981.

Type specimen: F:AMNH 76703.

Characteristics: P3 larger than P2; M3 lacking metacone; p2 single-rooted; p4 paraconid only slightly lower than protoconid; talonids of lower molars relatively short anteroposteriorly with hypoconid and entoconid large and well separated; m1-2 talonids about as wide as trigonids, m3 talonid narrower; large mental foramen below anterior root of p4; dentary deep and stout (Gawne, 1968; see also Matthew, 1903a).

Average length of M2: 2.9 mm.

Included species: *U. copiosus* only (known from localities SB32B, C, D*, SP1A, CP114A, B, CP116B, CP123E, NP11).

AMPHECHININI GUREEV, 1979

***Amphechinus* Aymard, 1850**

Type species: *Amphechinus arvernensis* (Blainville, 1840).

Type specimen: No number (BMNH 27825 was taken for the type by Lydekker [1885] but, as Schlosser [1887] observed, the type is not this specimen [Butler, 1948]).

Characteristics: M1 long relative to width; M3 lacks metacone; p2 single-rooted; p4 paraconid only slightly lower than protoconid; m1 trigonid anteroposteriorly elongate; m3 lacks talonid or postcingulum; mandibular angle elongate; anterior surface of ascending ramus posteriorly inclined (Rich, 1981).

Average length of m2: 3.1 mm (*A. horncloudi).*

Included species: *A. horncloudi* only (known from localities CP52, CP69II, CP86B*, NP36B, D).

Amphechinus sp. is also known from localities NP11, NP36B.

Comments: *Amphechinus* is also known from the Oligocene to mid Miocene of Asia, the Oligocene to Pliocene (possibly Pleistocene) of Europe, and the Miocene of Africa (McKenna and Bell, 1997).

***Parvericius* Koerner, 1940**

Type species: *Parvericius montanus* Koerner, 1940.

Type specimen: YPM 19356.

Characteristics: M1 long relative to width; M3 metacone absent; p2 single-rooted; p4 paraconid only slightly lower than protoconid; m1–2 trigonids anteroposteriorly compressed; m3 lacking postcingulum or talonid; palate solid (not fenestrated); palate long relative to width; anterior surface of ascending ramus vertical; articular condyle relatively high; mandibular angle relatively reduced (Rich, 1981).

Average length of m2: 2.8 mm.

Included species: *P. montanus* (known from localities GC8AA, CP54B, CP74A, [CP86B], [CP88], CP89, CP103A, CP105, CP114A, B, [NP10CC], [NP11], NP34C*, NP36B); *P. voorhiesi* (localities CP103A, CP103A, B*).

Parvericius sp. is also known from localities GC8AA, GC8B, [CP86D, F], CP114D, NP10CC.

Comments: *Parvericius* is also known from the late Oligocene of Asia (McKenna and Bell, 1997).

BRACHYERICINAE BUTLER, 1948

***Brachyerix* Matthew, in Matthew and Mook, 1933**

Type species: *Brachyerix macrotis* Matthew and Mook, 1933.

Type specimen: AMNH 21335.

Characteristics: Basicranial lateral margins nearly parallel; auditory bullae large and inflated, close to one another; flange present separating posterior and concave lateral surfaces of squamous region of zygoma; ventral tips of interparietal expanded anteriorly between parietal and mastoid; lower partition formed by mastoid between stylomastoid and jugular foramina; M1 postmetaconule crista joined to metacone; m1 trigonid expanded anteroposteriorly; m2 shorter relative to m1 (Rich and Rich, 1971; Rich, 1981).

Average length of m2: 2.0 mm (*B. incertis*).

Included species: *B. macrotis* (known from localities CP54B, CP74A, CP104A, CP105, CP107, CP108A, NP34C*); *B. incertis* (localities [GC4DD], NB6C, NB23B, SB33B, [CP89], CP110*, [NP11], NP41B); ?*B. hibbardi* (locality SB46*).

Brachyerix sp. is also known from localities CP71, CP75C, CP86B, CP110.

Comments: *Brachyerix macrotis* includes "*Brachyerix montanus*," the latter being a nomen nudum. In the original description of *B. macrotis*, the holotype skull (AMNH 21336) was referred to in the figure caption as *B. montanus*, presumably in error. This paper was published by C. C. Mook after Matthew died – it is possible that Matthew's original intention was to name the type species *B. montanus*, but the diagnosis and description refer to *B. macrotis*.

***Metechinus* Matthew, 1929 (including *Brachyerix*, in part)**

Type species: *Metechinus nevadensis* Matthew, 1929.

Type specimen: UCMP 29600.

Characteristics: Basicranial lateral margins converge posteriorly at large angle; auditory bullae relatively small and less inflated, separated from one another; absence of flange separating posterior and flat lateral surfaces of squamous region of zygoma; ventral tips of interparietal not intercalated between parietal and mastoid; tall partition formed from mastoid between stylomastoid and jugular foramina; M1 postmetaconule crista separated from metacone by a prominent notch; m1 trigonid anteroposteriorly compressed (Rich, 1981).

Average length of m2: 3.7 mm (*M. nevadensis*).

Included species: *M. nevadensis* (known from locality NB23C*); *M. amplior* (localities CC17G, SB32B, C, D*, CP74A0, CP76B, CP111, CP114A, B, CP114D, CP116B); *M. fergusoni* (locality NB23B*) (including *Brachyerix incertis*, Rich, 1981); *M. marslandensis* (locality [CP71*]).

Metechinus sp. is also known from CP114C, D, CP116B.

ERINACEIDAE INCERTAE SEDIS

***Litolestes* Jepsen, 1930**

Type species: *Litolestes ignotus* Jepsen, 1930.

Type specimen: YPM-PU 13362.

Characteristics: Mandible slender, deepest beneath m1; I2/i2 with digitate crowns; canines small; p4 as large as m1; p4 premolariform with small metaconid; m1 paraconid compressed but more distinct than on m2–3 (Jepsen, 1930; Krishtalka, 1976a).

Average length of m2: 1.48 mm.

Included species: *L. ignotus* only (known from localities CP13G*, NP3C-F, [NP7D]).

Litolestes sp. is also known from localities NP4 and CP13G. A possible *Litolestes* is represented at NP3C.

***Cedrocherus* Gingerich, 1983**

Type species: *Cedrocherus ryani* Gingerich, 1983.

Type specimen: UMMP 82028.

Characteristics: Very steep gradient of decreasing molar size from m1 to m3; m1 entoconid squared; m2 entoconid rounded (Gingerich, 1983).

Average length of m2: 1.5 mm (*C. ryani*).

Included species: *C. ryani* (known from locality CP13E*); *C. aceratus* (locality CP24A*).

***Entomolestes* Matthew, 1909 (including *Macrocranion*, in part)**

Type species: *Entomolestes grangeri* Matthew, 1909.

Type specimen: AMNH 11485.

Characteristics: p4/m1 ratio relatively large; lower molars with long, relatively straight prevallid, exodaenodont labial margins, and relatively large and linguolabially compressed entocristids.

Average length of m2: 1.4 mm.

Included species: *E. grangeri* only (known from localities CP34C*, CP34E).

Entomolestes sp. is also known from localities (SB43A), CP25F, CP29D, CP34C, E. A previously published record of *Entomolestes* from NP49 is incorrect, instead representing an indeterminate lipotyphlan.

ERINACEOIDEA INCERTAE SEDIS

Leipsanolestes Simpson, 1928 (including *Entomolestes*, in part)

Type species: *Leipsanolestes siegfriedti* Simpson, 1928.

Type specimen: AMNH 22157.

Characteristics: p4 semi-molariform; lower molars with only moderate size decrease posteriorly (Simpson, 1928; Krishtalka, 1976a).

Average length of m2: 1.6 mm.

Included species: *L. siegfriedti* (known from localities CP17A*, B, [CP20A], CP25B, CP26C).

Leipsanolestes sp. is also known from localities CP14E, CP17A, B, CP18B, CP19A, (CP20A), NP47C.

Auroralestes Holroyd, Bown and Schankler, 2004 (including *Leipsanolestes*, in part; *Eolestes* Bown and Schankler, 1982, not *Eolestes* Cockerell, 1940 [dragonfly])

Type species: *Auroralestes simpsoni* (Bown, 1979) (originally described as *Leipsanolestes simpsoni*).

Type specimen: UW 9616.

Characteristics: p4 talonid reduced, trigonid relatively long and metaconid relatively small; p4/m1 ratio relatively large; m1 with lingual paraconid; lower molar talonids relatively narrow, entocristids tall, paraconids cuspidate, trigonids anteroposteriorly compressed, cristid obliqua joins metaconid, and metaconid twinned (Bown and Schankler, 1982).

Average length of m2: 1.2 mm.

Included species: *A. simpsoni* only, known from locality CP20A* only.

Dartonius Novacek, Bown, and Schankler, 1985 (including *Leptacodon*, in part)

Type species: *Dartonius jepseni* (McKenna, 1960b) (originally described as *Leptacodon jepseni*)

Type specimen: UCMP 45949.

Characteristics: small size; p3/p4 and p4/m1 ratios small; lower molars with short, curved prevallid, weakly exodaenodont labial margins, small, weakly linguolabially compressed entocristids, cristid obliqua not connected to metaconid, and an acute paracristid notch; p4 talonid basin developed (Novacek, Bown, and Schankler, 1985).

Average length of m2: 1.1 mm.

Included species: *D. jepseni* only (known from localities CP20A, CP25B, CP63*).

ERINACEOMORPHA INCERTAE SEDIS

Adunator Russell, 1964 (including *Mckennatherium*; *Leptacodon*, in part)

Type species: *Adunator lehmani* Russell, 1964.

Type specimen: Halle Wa/368 (Geologisch–Paläontologisches Institut, Halle, Walbeck Locality).

Characteristics: m1-2 subequal in size; m2 paraconid appressed to metaconid; lower molar metaconids large with expanded anterior bases; p4 paraconid shorter than metaconid; p3 with strong paraconid.

Average length of m2: 1.7 mm.

Included species: *A. ladae* (including *Mckennatherium libitum* Van Valen, 1965) (known from localities [SB23H], [CP13B, C, E], CP14B, CP21A, [CP22B], [CP26A], NP19C*); *A. martinezi* (locality CP14A*); *A. fredricki* (locality CP14A*).

Adunator sp. is also known from localities SB23A, CP13B, (CP14E), NP3C.

Comments: *Adunator* is also known from the late Paleocene of Europe (McKenna and Bell, 1997).

Litocherus Gingerich, 1983 (including *Litolestes*, in part)

Type species: *Litocherus zygeus* Gingerich, 1983.

Type specimen: UMMP 64508.

Characteristics: Canines large and projecting; lower premolars relatively long and narrow; p4 premolariform with small, narrow talonid; lower molars with reduced, often crestlike paraconids, more rounded talonid basins with more rounded talonid cusps; lower molar length decreases moderately from m1 to m3 (Gingerich, 1983).

Average length of m2: 1.9-2.1 mm.

Included species: *L. zygeus* (known from localities CP13E*, [NP3A, [C, D], NP47B, BB); *L. lacunatus* (including *Litolestes lacunatus*) (localities [CP13C], CP14C, D, CP16[A], B, C*, CP22A, CP24A, [NP1C], NP3F, NP7D, NP20D); *L. notissimus* (localities [CP26A], [NP3B–D], NP4, NP19IIC*, NP20E].

Litocherus sp. is also known from localities NP1C and NP4.

Talpavoides Bown and Schankler, 1982

Type species: *Talpavoides dartoni* Bown and Schankler, 1982.

Type specimen: UW 9624.

Characteristics: p4 relatively small and semi-molariform; p4 talonid basin weak with poorly formed cristid obliqua; m1–2 entoconids tall, acute, and slightly displaced lingually, molar entoconids situated at posterolingual talonid margins (Bown and Schankler, 1982).

Average length of m2: 1.1–1.4 mm.

Included species: *T. dartoni* only (known from localities GC21II, CP14C, CP20A*, B, [CP27A]).

Talpavoides sp. is also known from locality CP20BB.

Diacodon Cope, 1875

Type species: *Diacodon alticuspis* Cope, 1875.

Type specimen: USNM 1098.

Characteristics: Relatively large size; p2–3 relatively large; p3 with small talonid; p4 metaconid distinct, paraconid low and cuspate, talonid relatively wide; lophidlike paraconid on m1–3, which wears to two distinct facets; m3 relatively large compared with m1–2; m1–2 talonid with very short postcingulum descending from hypoconulid to base of hypoconid; P3 with a projecting anterolabial spur; P4 with incipient metacone; upper molars transverse; lacks progressive reduction in size from m1 to m3 (Novacek, 1982).

Average length of m2: 3.0 mm.

Included species: *D. alticuspis* only (known from localities SB24*, [CP25E], [CP64B]).

Diacodon sp. is also known from localities (CP25B, G).

Diacocherus Gingerich, 1983 (including _Diacodon_, in part; _Scenopagus_, in part; _Leptacodon_, in part; _Adunator_, in part; _Elpidophorus_, in part)

Type species: *Diacocherus minutus* (Jepsen, 1930) (originally described as *Diacodon minutus*).

Type specimen: YPM-PU 13360.

Characteristics: P3–4 with metacones; p4 submolariform with relatively high paraconid; m2 equal to or larger than m1; m2–3 with reduced paraconid closely appressed to metaconid (Gingerich, 1983).

Average length of m2: 1.73 mm (*D. meizon*).

Included species: *D. minutus* (including *Leptacodon packi*, in part, Dorr, 1952; *Leptacodon* near *L. ladae*, Dorr, 1958; *Elpidophorus minutulus* Dorr, 1958; "*Diacodon*" *minutus*, in part, Krishtalka, 1976a; ?*Scenopagus proavus* Winterfeld, 1982; *Adunator minutus*, in part, Winterfeld, 1982) (known from localities CP13E, G*, CP14C–E, CP15B, [CP17A, B], [CP18B], CP22B, NP3C, D, NP48B); *D. meizon* (localities CP13E*, CP24A, NP3E, F, [NP7D]).

Diacocherus sp. is also known from localities GC21II, CP13G, CP17A, B, CP18A, B.

Talpavus Marsh, 1872

Type species: *Talpavus nitidus* Marsh, 1872.

Type specimen: YPM 13511.

Characteristics: m2 paraconid compressed anteroposteriorly into thin, high crest that flares anterolingually; lower molar entoconids much higher than hypoconids, talonids and trigonids of equal width, hypoconulids small and slightly lingual occurring at the posterolabial base of the entoconid and delineated from that cusp by a sharp notch (Robinson, 1968; Krishtalka, 1976a).

Average length of m2: 1.5 mm (*T. nitidus*).

Included species: *T. nitidus* (known from localities CP5A, [CP20BB], CP33C, CP34D*, [CP35]); *T. conjectus* (locality NP9A*); *T. duplus* (localities CP6B*, CP29C); *T. sullivani* (locality CP27B*).

Talpavus sp. is also known from localities SB22A, (CP15B), (CP19A), (CP20A), CP27B, D, CP34C, CP63, NP9B, (NP20E).

Comments: *Talpavus* is also questionably known from the early Eocene of Europe (McKenna and Bell, 1997).

SESPEDECTIDAE NOVACEK, 1985

Characteristics: p1-2 reduced, single-rooted, often procumbent; p4 premolariform with small paraconid and metaconid and weakly basined talonid; m1-3 with transverse, crestiform paraconids closely appressed to metaconids, trigonids anteriorly inclined and anteroposteriorly compressed; P4 with strong metastylar crest, weak to absent metacone, and weak posterolingual expansion of cingulum; M1-2 with well-developed hypocone (Novacek, 1985).

SCENOPAGINAE NOVACEK, 1985

Scenopagus McKenna and Simpson, 1959 (including _Nyctitherium_, in part; _Diacodon_, in part)

Type species: *Scenopagus mcgrewi* McKenna and Simpson, 1959.

Type specimen: AMNH 56035.

Characteristics: P3 longer than wide and triangular with small protocone; P4 lacking separate metacone, paracone with shearing metastylar crest, incipient, low hypocone on posterior crest from protocone; M1–2 transverse, not truly quadrate, trigon somewhat compressed anteroposteriorly, small crescentic protoconule with wings to paracone base and to parastyle, small metaconule, prominent shelf buccal to paracone and metacone (especially on M2), projecting parastylar and metastylar lobes, edges emarginated between them, no other styles, small hypocone as elevated lingual end of posterior cingulum forming small lobe more basal than protocone and lacking crests, lacking separate hypoconal root; M3 apparently not reduced and with moderately strong parastylar lobe; no groove in palatal surface and maxillary bone immediately lingual to molar roots (McKenna and Simpson, 1959; Robinson, 1968; Krishtalka, 1976a).

Average length of m2: 2.2 mm (*S. edenensis*).

Included species: *S. mcgrewi* (known from locality CP34D*); *S. curtidens* (localities CC4, SB24, CP20E, CP31E, CP32B, CP34A, C, D*); *S. edenensis* (localities SB22C, CP5A, CP27D, E, CP31E, CP34A, B, C, D*); *S. hewettensis* (locality CP20A*); *S. priscus* (localities [CC4], SB22C, SB43A, CP5A, CP25J, CP27D, E, [CP31E], CP34A, B, D*, E, CP35).

Scenopagus sp. is also known from localities GC21II, CC7B, SB24, SB43A, (CP15B), CP20A, CP25GG, J2, CP27(D), E, (CP31B), CP34C, E, NP23A.

Comments: *Scenopagus* is also questionably known from the early Eocene of Europe (McKenna and Bell, 1997).

***Ankylodon* Patterson and McGrew, 1937**

Type species: *Ankylodon annectens* Patterson and McGrew, 1937.

Type specimen: FMNH P15326.

Characteristics: P3 relatively long and wide with relatively large protocone; P4 only slightly larger than P3; upper molars with strong conules, lacking anterior cingula, and with prominent hypocone; lower first premolars relatively wide; p3 protoconid with doubly ridged posterior surface; p4 relatively short and broad and lacking talonid ridge; lower molars relatively broad; mental foramina relatively anterior; coronoid process relatively tall; teeth anterior to P3 not set more dorsally than M1–3 (Patterson and McGrew, 1937; Lillegraven, McKenna, and Krishtalka, 1981).

Average length of m2: 2.2 mm.

Included species: *A. annectens* (known from localities [CP46], CP68C*); *A. progressus* (locality CP68C*).

Ankylodon sp. is also known from localities CP29C, CP39C, F, CP98B, (NP10B), [NP49II].

Comments: An additional species of *Ankylodon* from locality CP29D has been proposed but exists only in manuscript form at this time.

SESPEDECTINAE NOVACEK, 1985

Characteristics: p3 with well-developed protoconid and short heel; P3 similar in structure to P4 with prominent metastylar crest, paracone, protocone, and posterolingual cingulum; upper molars bunodont with large hypocones and distinct conules; lower molars bunodont (Novacek, 1985).

***Sespedectes* Stock, 1935**

Type species: *Sespedectes singularis* Stock, 1935.

Type specimen: LACM-CIT 1785.

Characteristics: Relatively small size; p4 talonid basin only slightly excavated with a single small cusp; m1-3 with crestiform paraconids; P3 relatively transverse and resembling P4; P3-4 with squared lingual border; molar cusps bulbous; distinct metacrista on M2 and expanded metaconule on M1-2; upper molars anteroposteriorly elongate (especially M2); labial spurs on M1-2 not prominent; weak or absent precingulum on M1-3; inflated hypocone on M2 (Novacek, 1985).

Average length of m2: 1.69 mm (*S. singularis*).

Included species: *S. singularis* (known from localities CC4, CC6B, CC7B, E, CC8, CC8IIA, CC9A0-B, BB*, B2, CC9IVA); *S. stocki* (locality CC7E*).

Sespedectes sp. is also known from localities CC7B–E, [CC9A, BB], CC10, NP8, NP9A.

***Crypholestes* Novacek, 1980 (including *Cryptolestes* Novacek, 1976, not *Cryptolestes* Ganglbaur, 1899 [coleopteran], not *Cryptolestes* Tate, 1934 [marsupial])**

Type species: *Crypholestes vaughni* (Novacek, 1976).

Type specimen: UCMP 103912.

Characteristics: DP4 three-rooted with paracone, metacone, protocone, prominent metastylar lobe, and lingual cingula; P3 not greatly reduced relative to P4 and similar in morphology; upper molars, especially M2, transverse with sharp cusps, paracone and metacone with steep labial faces, labial lobes, particularly the parastylar lobe of M2, prominent, a distinct precingulum is consistently present; hypocone of M1–2 well developed, but lower than protocone, paraconule and metaconule are distinct and subequal in size, postmetaconule wing continues labially as a metacingulum; M3 with a paraconule; three mental foramina are present (Novacek, 1976).

Average length of m2: 1.58 mm.

Included species: *C. vaughni* (known from localities CC4, CC6A, CC7A, B, CC8*).

Crypholestes sp. is also known from localities?CC3, ?CC4, ?CC7B, CP5A.

***Proterixoides* Stock, 1935**

Type species: *Proterixoides davisi* Stock, 1935.

Type specimen: LACM-CIT 1673.

Characteristics: Relatively large size; p4 talonid basin excavated; m1–3 paraconids lophid, rather than crestiform; m1–3 quadrate in outline; P3 triangular, similar to but not matching P4 in outline and construction; P3–4 lingual borders with convex outline; conical molar cusps; M2 metacrista weak or absent and metaconule weakly inflated (Novacek, 1985).

Average length of m2: 3.3 mm.

Included species: *P. davisi* (known from localities CC6B, CC7B–D, [E], CC8, CC8IIA, CC9A–B, BB*).

Proterixoides sp. is also known from locality (CC8).

***Patriolestes* Walsh, 1998 (including *Proterixoides*, in part)**

Type species: *Patriolestes novaceki* Walsh, 1998.

Type specimen: SDSNH 49250.

Characteristics: No anterior diastemata in lower dentition; i2 larger than i1; i1 larger than i3; c1 single–rooted; dp1, p2–3 single-rooted and relatively small; p4 longer than m1 and bulbous, lacking paraconid and metaconid, talonid short with a single posterolingual cusp; lower molar trigonids erect and anteroposteriorly compressed, paraconids bladelike; I1-2 somewhat enlarged with shovel-shaped crowns, I3 small; C1 double-rooted with bulbous crown; dP1–2 single-rooted; P3 lacking metacone and hypocone, much smaller than P4; P4 triple-rooted, large and transverse, lacking metacone and hypocone; M1 with hypocone extending further lingually than protocone; M1–3 with strong pre- and postparaconule and pre- and postmetaconule wings (Walsh, 1998).

Average length of m2: 2.87 mm.

Included species: *P. novaceki* only (including *Proterixoides davisi*, in part, Novacek, 1985) (known from localities CC4*, CC7B).

SESPEDECTIDAE INCERTAE SEDIS

Macrocranion Weitzel, 1949 (including *Entomolestes*, in part)

Type species: *Macrocranion tupaiodon* Weitzel, 1949.

Type specimen: ME 4403a.

Characteristics: p4 metaconid separated from protoconid, cristid obliqua relatively weak, posterior talonid shelf elevated; m1–3 relatively low with flattened labial bases, talonids wider and longer relative to trigonids, cusps relatively low and bulbous, paraconids anterior and distinct from metaconids; m1 trigonid open lingually, hypoconid weakly anteroposteriorly compressed, cristid obliqua long and meeting trigonid lingually; P4 with convex labial margin and weak and low paracone–metastylar crest; upper molars rectangular with weakly emarginated labial margins, stylar areas unexpanded; hypocones tall relative to protocones (Krishtalka, 1976a).

Average length of m2: 1.96 mm (*M. nitens*).

Included species: *M. junnei* (known from localities CP19AA*, CP20AA); *M. nitens* (from localities SB24, CP20A, BB, C*, D, [CP25B], CP27[A], B, D, [CP63], CP64B); *M. robinsoni* (localities CP29C*, D).

Macrocranion sp. is also known from localities SB24, CP27B, D, (CP31E).

Comments: *Macrocranion* is also known from the early to middle Eocene of Europe (McKenna and Bell, 1997).

Indeterminate sespedectines

In addition to the taxa discussed above, indeterminate or undescribed sespedectids are also present at localities CP6AA, CP20A, CP31E, CP34E.

AMPHILEMURIDAE HILL, 1953

The genus *Amphilemur* Heller (1935) has been reported from locality CP29D. This family is otherwise only represented in the middle Eocene of Europe.

INDETERMINATE ERINACEOMORPHA

In addition to the taxa discussed above, indeterminate or undescribed erinaceomorphs are also present at localities GC21II, CC9AA, CC9IIIA, CC9IVC, CC19, NB7A, CP29C, D, CP31A, CP105, NP3C, D, NP8, PN9B.

TALPOIDEA FISCHER VON WALDHEIM, 1817

PROSCALOPIDAE K. M. REED, 1961

Cranial characteristics: Skull functionally convergent on Chrysochloridae (Barnosky, 1981) in which the head is used in burrowing; cranial bones fuse in adults; rostrum distinctly longer than cranium; cranium broad and deep with condyles placed ventrally and well below plane of dentaries; premaxillae marked with prominent lateral shelves that extend anterior to nares; tympanic bulla, where known, ossified but only slightly inflated.

Dental characteristics: Anterior-most incisors enlarged, piercing teeth in most genera; seven or less antemolars with at least some premolars single-rooted; additional teeth and roots lost in later taxa; lower molar cingulids incomplete labially; molars brachydont to hypsodont in latest forms.

Postcranial characteristics: Neck shortened, some cervical vertebrae co-ossify to form rigid base; first rib fuses with manubrium; deltoid crest of scapula greatly elevated and arches broadly over glenoid, contacts long clavicle; humerus and forelimb highly modified for rotary burrowing (Barnosky, 1982); humeral head laterally compressed and situated near lateral border, distal end greatly expanded with distinct fossa for insertion of ligament for flexor digitorum profundus, pectoral crest terminates in projecting spike or plate, teres tubercle situated halfway or more distally; olecranon process of ulna transversely expanded; tibia and fibula unfused.

Oligoscalops K. Reed, 1961 (including *Arctoryctes*, in part)

Type species: *Oligoscalops galbreathi* (C. A. Reed, 1956) (originally described as *Arctoryctes galbreathi*).

Type specimen: USNM 21310.

Characteristics: Antemolar region length relatively long with seven antemolar teeth; teeth relatively brachydont; P4 with weak lingual shelf; M1–2 with weak expression of metaconules; coronoid process of dentary high and spicular; depth of horizontal ramus under antemolars does not exceed that under molars; anterior lower incisor enlarged but not gliriform; lower molars with open talonid valleys; humerus with relatively straight shaft, relatively short lesser tuberosity, pronounced greater tuberosity, and large fossa near trochlea for ligament of flexor digitorum profundus; distal part of pectoral process spicular and directed anteromedialy; teres tubercle relatively distally situated and separated from medial epicondyle by deep but narrower notch.

Average length of m2: 1.95–2.1 mm.

Average length of humerus: 8.2–8.8 mm.

Included species: *O. galbreathi* (including *Oligoscalops whitmanensis* K. Reed, 1961) only (known from localities CP68B, C*, CP83B, NP50B).

Oligoscalops sp. is also known from localities ?CP46, CP68C, CP83C, CP84A, ?C, CP98B, ?NP24C–D, NP32C.

Comments: *Oligoscalops* is also questionably known from the early Oligocene of Asia (McKenna and Bell, 1997).

***Proscalops* Matthew, 1901 (including *Arctoryctes*, in part; *Domninoides*, in part)**

Type species: *Proscalops miocaenus* Matthew, 1901.

Type specimen: AMNH 8949a.

Characteristics: Antemolar region relatively short with six antemolar teeth; molars relatively high crowned; dentary deeper under antemolars than molars, coronoid process high and spicular; anterior lower incisor gliriform; lower molars brachyhypsodont to hypsodont with open to nearly closed talonid valleys; P4 parastyle absent and lingual shelf enlarged; upper molar metaconules and paracones prominent; humerus teres tubercle and medial epicondyle overlapped or fused together, the lesser tuberosity relatively long.

Average length of m2: 2.8 mm.

Average length of humerus: 10.4–11.4 mm.

Included species: *P. miocaenus* (known from localities CP46, CP68D*); *P. evelynae* (including *Arctoryctes terrenus* Matthew, 1907; *Domninoides evelynae* Macdonald, 1963) (localities CP85C*, CP86D); *P. intermedius* (locality NP34B*); *P. secundus* (localities [CP47], CP51B, [CP71], CP86A*, B, D, CP103A, B, NP10B2, CC); *P. tertius* (localities CP84B, C, ?CP99A, CP101).

Proscalops sp. is also known from localities?CP41B, CP49A, CP68C, ?CP71, CP84(A), B, CP85A, C, CP86B, NP10C, CC, NP34B, NP36B, D.

Comments: The locality of the holotype of *Proscalops tertius* (AMNH 19420) is uncertain. It was collected by G. L. Jepsen from the "White River Formation, Badlands, South Dakota." K. Reed (1961) noted that it might have come from the Brule Formation, but no precise information is available. The holotype of *Proscalops secundus* (AMNH 13798) has never been fully described. The species was named and figured by Matthew (1909) in his Plate 51 with a brief mention in the text (p. 538) but with no formal diagnosis being provided for the type and only known specimen. The type locality is also unclear. It was cited in the caption for Plate 51 by Matthew (1909) as being from the "Lower Rosebud, Pine Ridge Agency, South Dakota." K. Reed (1961) cited the type locality as Bear-in-the-lodge Creek, South Dakota and in a footnote indicated that a personal communication from J. R. Macdonald suggested that the type may have come from either the Monroe Creek or Harrison Formation.

***Mesoscalops* K. Reed, 1960**

Type species: *Mesoscalops scopelotemos* K. Reed, 1960.

Type specimen: ACM 10461.

Characteristics: Antemolar region short with five antemolar teeth; P4 with expanded and rectangular lingual shelf, lacks a parastyle; upper molar metaconules expanded posteromedially and impart a rectangular shaft to the lingual shelf; dentary is deepened anteriorly, with a short coronoid process; lower anterior incisor gliriform; cheek teeth hypsodont; lower molar talonids closed lingually by ento- and metacristids; humerus teres tubercle and entepicondyle are always fused.

Average length of m2: 2.8–3.3 mm.

Average length of humerus: 13.2 mm.

Included species: *M. scopelotemos* (known from localities CP49A, CP54B*, CP105); *M. montanensis* (locality NP34B*).

Mesoscalops sp. is also known from localities CP51A, CP71, CP88, CP103A, NP10E, NP40B, NP42, PN6G.

Indeterminate proscalopines

In addition to the taxa discussed above, an indeterminate proscalopid is also present at locality NP10C2.

TALPIDAE FISCHER VON WALDHEIM, 1817

Although the group is typically thought of and frequently distinguished by its burrowing adaptations, at least one group, the Uropsilinae, is generally shrew-like and exhibits no notable burrowing specializations. The diverse aquatic and burrowing specializations seen among family members are within-group phenomena.

Cranial characteristics: The head plays little role in digging; cranium relatively long, shallow, and unfused; neck flexible and relatively long; bullar ossification ranges from primitively unossified to fully ossified but not inflated in advanced burrowing forms; zygoma thin and low, lacking jugal; muscularis levator labii superioris proprius origin located on zygomatic process of maxilla.

Dental characteristics: A full complement of incisors, canines, premolars, and molars in primitive forms; first premolars not replaced, representing retained Dp1–dp1 (Ziegler, 1971); premolars and canines double-rooted; P4 semi-molariform with reduced lingual shelf supporting only low protocone; upper molars with W-shaped ectolophs, simple protocones, which may be complicated by development of conules, no hypocones; lower molars with basic W-shaped lophid pattern, well-defined cingula primitively, lack distinct hypoconulids, postcristids unite entoconids and hypoconids directly; common trends in various lineages include loss of antemolar teeth and reduction in number of antemolar roots with development of i2 and I1 as enlarged, piercing teeth.

Postcranial characteristics: Primitively the skeleton is shrew-like and unspecialized (Campbell, 1939; C. Reed, 1951; Yalden, 1966); desmanines and talpines with clavicle articulating directly with humerus, bypassing scapula; talpines with shortened clavicles and enlarged manubrium; great ligament of muscularis flexor ligamentum profundus functions as tensile element during rotation of forearm with salient insertion pit on humerus.

SUBFAMILY UNCERTAIN

***Quadrodens* Macdonald, 1970 (including *Palaeoscalopus*; *Domnina*, in part)**

Type species: *Quadrodens wilsoni* Macdonald, 1970.

Type specimen: LACM 9331.

Characteristics: Dentary shortened and tapered anteriorly; p4 inflated, lacking cingula; lower molars brachydont with labially situated cristid oblique on m2; m2 rectangular in occlusal view with trigonid cusps clustered medially with open trigonid valley; humerus with long, medially directed distal pectoral process.

Average length of m2: 2.06 mm.

Included species: *Quadrodens wilsoni* (including *Palaeoscalopus pineridgensis* Macdonald, 1970; *Domnina greeni* Macdonald, 1963) only, known from locality CP85C* only.

UROPSILINAE DOBSON, 1883

Characteristics: Small size; skeleton unspecialized and shrew-like, antebrachium long and slender, pelves broadly separated; teeth brachydont, I1–i1 enlarged, M1–2 with well-developed metaconule and concave posterior margin; lower antemolars reduced by one or more teeth; unlike other talpids, zygomatic bar arches dorsally with a distinct subzygomatic process on maxilla; lacrimal foramen is as large as infraorbital foramen; bulla unossified; humerus articulates normally with scapula.

***Mystipterus* Hall, 1930 (including *Mydecodon*)**

Type species: *Mystipterus vespertilio* Hall, 1930.

Type specimen: UCMP 29604.

Characteristics: Antemolars less reduced in number and in premolar root number than extant *Uropsilus*; p2 single- or double-rooted; lower molar cristid obliqua join metalophids lingually.

Average length of m2: 1.54–1.77 mm.

Included species: *M. vespertilio* (known from locality NB23C*); *M. martini* (R. W. Wilson, 1960) (originally described as *Mydecodon martini*) (localities CP71*, [CP88], CP105); *M. pacficius* (localities PN7II, [PN8B], PN9B*).

Mystipterus sp. is also known from localities NB23B, SB32D, CP103A, CP114A, B, CP116A, (B), ?NP10E, NP36A, B, (PN10).

GAILLARDIINAE HUTCHISON, 1968

Characteristics: Aquatic moles; lower molars with anteroposteriorly compressed trigonids and talonids and wide separation of metaconid from metastylid by hypoflexid; dentary with small coronoid and angular processes; clavicle simple, long, but robust; humerus with bicipital canal open throughout, broadly oval head, lesser tuberosity lower than head, teres tubercle protruding from shaft, greatest width distal, and no fossa for the ligament of muscularis flexor digitorum profundus; femur with third and greater trochanters continuous, medial epicondyle spikelike, and lateral epicondylar sesamoid fused to epicondyle; tibiofibula long and robust and deeply grooved distally; metatarsals long and twisted.

***Gaillardia* Matthew, 1932 (including *Hydroscapheus*)**

Type species: *Gaillardia thompsoni* Matthew, 1932.

Type specimen: AMNH 20508

Characteristics: Same as for subfamily.

Average length of m2: 2.59 mm

Average length of humerus: 18.9 mm.

Included species: *G. thompsoni* (including *Hydroscapheus americanus* Shotwell, 1956) only (known from localities CP115D*, PN14).

Gaillardia sp. is also known from locality SB32D.

TALPININAE (NEW)

Characteristics: Clavicle articulates directly with humerus; humerus with laterally compressed head, lesser trochanter elevated above the head and greater trochanter, distinct fossa for the ligament of the muscularis flexor digitorum profundus, and part of bicipital canal enclosed by flanges of lesser trochanter; lacrimal foramen smaller than infraorbital foramen; C1 and p1 primitively enlarged but frequently reduced or lost in several lineages, retained in Old World tribes Talpini and Scaptonychini.

DESMANINI MIVART, 1871 (NEW RANK)

Characteristics: Aquatic moles; dentition complete; I1 enlarged; mesostyles of upper molars distinctly twinned; dentary with tall and straight coronoid and short angular processes; manubrium not extended anteriorly; clavicle simple, long, and moderately slender; humerus with pectoral processes and lesser tuberosity fused above bicipital canal, transverse width of distal end distinctly greater than that across the teres and greater trochanters, teres tubercle protruding from shaft only slightly; pubes widely separated but joined by bony rod; femur with wide anteroposteriorly compressed shaft and wide distal end; tibiofibula robust, long, and deeply grooved distally; metatarsals long and twisted; pes webbed.

***Lemoynea* Bown, 1980**

Type species: *Lemoynea biradicularis* Bown, 1980.

Type specimen: UNSM 27518.

Characteristics: Dentition complete; p1-4 double-rooted and complex; M1-3 mesostyles separated by groove; m1-3 metaconid–metastylid cusp robust but not twinned.

Average length of m2: 2.23 mm.
Included species: *L. biradicularis* only (known from localities CP116C*, [D]).
Lemoynea sp. is also known from CP116C.

TALPINAE FISCHER VON WALDHEIM, 1817

Characteristics: Clavicles shortened, with ventral process; humerus with transverse width and across teres tubercle and greater tuberosity about equal or greater than that across distal end, enclosure of the bicipital canal by flanges of lesser tuberosity and pectoral crest (exception *Condylura*), and prominent protrusion of the teres tubercle from shaft; manubrium elongate anteriorly with deep ventral keel; scapula narrow, lacking metacromion process; pubes close together or touching; femur not compressed.

CONDYLURINI TROUESSART, 1879

Characteristics: Semi-aquatic and semi-fossoral; enlarged pes for swimming; fingerlike fleshy projections radiate from rhinarium in extant species; antemolar region long including diastemata between premolars; dentition complete; p1–4 semi-molariform; clavicle with separate process for origin of part of muscularis acromiodeltoideus; scapula without suprascapular foramen; humerus with flanges enclosing bicipital canal and fused, axis of head directed laterally from shaft; pubes hooked ventrally at posterior end; tibiofibula long.

Condylura Illiger, 1811 (including *Sorex*, in part)

Type species: *Condylura cristata* (Linnaeus, 1758) (originally described as *Sorex cristatus*), the extant star-nosed mole.
Type specimen: None designated.
Characteristics: As for tribe.
Average length of humerus: 12.8–13.5 mm.
Included species: *Condylura* sp. is known from locality PN12B.
Condylura is also known from Pleistocene records in Maryland and West Virginia. A previous record of *Condylura* from the Upper Pawnee Creek Formation (locality CP75C) is in error and instead represents *Domninoides*.

Comments: The genus *Condylura* is also known from the Pliocene of Europe and questionably from the late Miocene or early Pliocene of Asia (McKenna and Bell, 1997).

SCALOPINI TROUESSART, 1879

Characteristics: p1 reduced in size to adjacent teeth, i2 and I1 enlarged; scapula with reduced acromion process that is pierced by suprascapular nerve; humerus with axis of head parallel to shaft or directed medially; tibiofibula short.

Scalopoides R. W. Wilson, 1960

Type species: *Scalopoides isodens* R. W. Wilson, 1960.
Type specimen: KU 10067.
Characteristics: p1 absent; i2 moderately enlarged; p2 single- or double-rooted; p3-4 double-rooted; m2-3 with well-developed metaconids; humerus relatively slender, axis of head parallel to shaft, teres tubercle extends posteriorly less than half total length; acromion process of scapula relatively high with a deep infraspinatus trough.
Average length of m2: 1.96–2.30 mm.
Average length of humerus: 9.24–9.49 mm.
Included species: *S. isodens* (known from localities CP71*, [CP88], [CP105]); *S. ripafodiator* (localities CP116B, [PN7II], [PN8A, B], PN9B*).
Scalopoides sp. is also known from localities GC8DB, CC19, NB18A, NB23B, ?SB32D, CP45E, CP86B, CP88, (CP89), CP101, CP103B, CP114(A), C, CP116B, (F), ?NP10E, NP11, NP40A, B, NP42, PN6G, PN10, PN11, ?PN14.

Comments: *Scalopoides* is also questionably known from the middle Miocene of Europe (McKenna and Bell, 1997).

Yanshuella Storch and Qui, 1983 (including *Neurotrichus*, *Scaptochirus*)

Type species: *Yanshuella primaeva* (Schlosser, 1924) (originally described as *Scaptochirus primaevus*).
Lectotype specimen: Schlosser, 1924, P1.1, Fig. 8.
Characteristics: Dentary shortened and tapering anteriorly; p1 absent; P3 small and double-rooted; P4 double-rooted; lower molars with relatively open trigonids and lacking metastylids on m2–3.
Average length of m2: 2.05–2.07 mm.
Average length of humerus: 11.4–12.1 mm.
Included species: *Y. columbiana* (originally described as *Neurotrichus columbianus* Hutchison, 1968) only, known from locality PN14* only.

Comments: *Yanshuella* is also known from the middle Miocene to Pleistocene of Asia (McKenna and Bell, 1997).

Domninoides Green, 1956 (including *Talpa*, in part)

Type species: *Domninoides riparensis* Green, 1956.
Type specimen: SDSM 53107.
Characteristics: Dental formula reduced by 1 to 3 teeth; antemolar region shortened and crowded, i2 enlarged; P2 and p3 double-rooted but p2 single-rooted; upper molars with distinctly divided mesostyles and distinct conules; m1-3 cristid obliqua joined to metastylid, talonids open lingually; humerus robust, teres tubercle relatively distally placed and elongate; clavicle with hooklike ventral process; radius, ulna, and metacarpals shortened.
Average length of m2: 2.26–3.3 mm.
Average length of humerus: 11.7–23.3 mm.
Included species: *D. riparensis* (known from localities [NB23C], CP90B*); *D. hessei* (locality SP1A*); *?D. knoxjonesi* (locality SP1A*); *D. mimcus* (localities

[CP114D], CP116B, CP123E*); *D. platybrachys* (Douglass, 1903) (originally described as *Talpa platybrachys*) (locality NP42*); *D. valentinensis* (including *D. storeri* Russell, 1976) (localities [CP45E], CP114A, B*).

Domninoides sp. is also known from localities CC17G, H, NB6E, NB23B, ?SB11, CP76B, CP87II, CP90A, CP108B, CP113A, CP114A, C, CP116A, F, CP123E, NP11, NP41B, PN7II, PN9B.

Comments: *Domninoides* is also known from the late Miocene of Europe (McKenna and Bell, 1997).

***Scapanoscapter* Hutchison, 1968**

Type species: *Scapanoscapter simplicidens* Hutchison, 1968.

Type specimen: UO 24286.

Characteristics: Dental formula complete; i2 not notably enlarged; p1–4 double-rooted, antemolar region long and angled down anteriorly; m1–3 trigonids strongly compressed anteroposteriorly; forelimb elements similar to *Scapanus*.

Average length of m2: 2.33 mm.

Average length of humerus: 12.5 mm.

Included species: *S. simplicidens* only (known from localities PN7III, PN8B*).

***Scapanus* (*Scapanus*) Pomel, 1848 (including *Scalopus*, in part)**

Type species: *Scapanus townsendii* (Bachman, 1839) (originally described as *Scalops townsendii*).

Type specimen: ANSP 449.

Characteristics: Antemolar region relatively long, teeth less crowded than *Xeroscapheus*; p1-4 single-rooted; i2 moderately enlarged; clavicle with knoblike ventral process, not pierced by vena cava; radius, ulna, metacarpals, and non-ungual phalanges short.

Average length of humerus: 12.8–18.0 mm.

Included species: *S.* (*S.*) *townsendii* (known from locality [PN23C]); *S.* (*S*). *hagermanensis* (known from localities CP114B, PN23A*).

Scapanus (*S*). sp. is also known from locality PN11.

***Scapanus* (*Xeroscapheus*) Hutchison, 1968**

Type species: *Scapanus* (*Xeroscapheus*) *proceridens* Hutchison, 1968.

Type specimen: UO 22508.

Characteristics: Dentary shortened, antemolars crowded; p1–3 single-rooted but p4 double-rooted or indication of fused roots; molars high crowned to hypsodont; lower molar trigonids anteroposteriorly compressed; scapula with a narrow infraspinatus fossa that extends as a narrow groove to acromion process; radius, ulna and non-terminal phalanges of manus relatively shorter than in *S.* (*Scapanus*).

Average length of m2: 2.25–2.7 mm.

Average length of humerus: 15.4 mm.

Included species: *S. proceridens* (known from localities NB31, [PN11], PN12B, PN14*, PN15, [PN17]); *S. latimanus* (locality [PN15II]); *S. malatinus* (locality NB13C*); *S. schultzi* (localities NB7C*, D, E, [PN10]).

Scapanus (*X.*) sp. is also known from localities WM26, CC26C, CC37, CC46II, (CC46III), (NB9), (NB10), NB13C, NB31, PN3C, ?PN11, PN14, PN23A, C.

Scapanus sp. is also present in the Pleistocene of California.

***Scalopus* Geoffroy Saint-Hilaire, 1803 (including *Hesperoscalops*)**

Type species: *Scalopus aquaticus* (Linnaeus, 1758) (originally described as *Sorex aquaticus*), the extant Eastern mole.

Type specimen: None designated.

Characteristics: DP1–dp1 absent; one or two other antemolars may be lost; molars brachyhypsodont to hypsodont; postcranium resembles *Scapanus* but ulna with medial olecranon crest reduced or absent.

Average length of m2: 2.0–3.5 mm.

Average length of humerus: 14.2–17.8 mm.

Included species: *S. aquaticus* (known from localities GC15A, SB49, SP4B, CP118, CP131); *S. blancoensis* (Dalquest, 1975) (originally described as *Hesperoscalops blancoensis*) (localities [SB37D], SP5*); *S. rexroadi* (Hibbard, 1941) (originally described as *Hesperoscalops rexroadi*) (localities SP1F, CP117B, CP128[AA], A, B, C*); *S. mcgrewi* (localities SP1F, CP116F*); *S. ruficervus* (localities SP3B*, SP4A). *S. sewardensis* (K. Reed, 1962) (originally described as *Hesperoscalops sewardensis*) (locality CP128AA*).

Scalopus sp. is also known from localities SB34III, SB37D, SB38IIC, SP1H, CP90A, CP116D, (E), CP123C, D, CP126. *Scalopus* is known from Pleistocene localities in Arkansas, Florida, Georgia, Kansas, Nebraska, Pennsylvania, and Texas.

TALPINAE INCERTAE SEDIS

***Achlyoscapter* Hutchison, 1968**

Type species: *Achlyoscapter longirostris* Hutchison, 1968.

Type specimen: UO 22412.

Characteristics: Dentition complete and elongate; i1–c1 simple and unenlarged; p1–4 double-rooted; m1–3 brachydont with incomplete cingulids; P4 elongate and partly bladelike.

Average length of m2: 1.64 mm.

Included species: *A. longirostris* only (known from localities SB49, PN8B, PN9B*).

Achlyoscapter sp. is also known from localities CP114B, CP116C.

Indeterminate talpids

In addition to the taxa discussed above, indeterminate or undescribed talpids are also present at localities GC8DB, GC11A, GC13B, CC17G, NB13B, NB23B, SB14F, CP34C, CP56, CP86D, CP114A,

CP116C, E, NP10E, NP11, NP32C, NP34B, D, NP36A, PN6G, PN9B, CE2.

SORICOMORPHA GREGORY, 1910

GEOLABIDIDAE MCKENNA, 1960A

Cranial characteristics: Snout very long, tubular, abruptly narrowing above P3 in late forms; zygomatic arch apparently lost; lacrimal foramen very large, opening into orbit; lacrimal canal running anteromedially and dorsally, increasing in diameter as it enters the nasal cavity; infraorbital canal short; nasals elongate, widest above P3, not fused to other skull bones but may coalesce with each other posteriorly; palatine apparently excluded from orbit by expanded orbital process of maxilla; alveolar border of rostrum sinuous in late forms; scar for origin of muscularis levator labii superioris proprius apparently anterodorsal to orbit as in *Solenodon* and *Nesophontes*; palate terminates at transverse crest; no palatine vacuities; long, occasionally roofed groove for subdivision of major palatine artery present in palatal surface of maxillary bone just lingual to molar roots.

Dental characteristics: Dental formula I3/3, C1/1, P4/4, M3/3; I1 enlarged; P1 double-rooted; P4 lacking metacone, paracone progressively hypertrophied; M1-3 with very strong stylar shelf (stylar cusp posterior to parastyle in [*Centetodon marginalis*]), appressed paracone and metacone, conules very weak or absent; M3 with strong parastyle, weak metacone; M1-2 with separate hypoconal root; lower premolars double-rooted; p3 becoming progressively reduced; p4 with progressively more simplified and lingual heel, finally reduced to one weak lingual cusp; lower molars with appressed high trigonid cusps, hypoconulids suppressed or fusing with entoconid (McKenna, 1960a).

Centetodon Marsh, 1872 (including *Geolabis*; *Metacodon*; *Embassis*; *Hypacodon*; *Diacodon*, in part; *Herpetotherium*, in part; *Peratherium*, in part; *Didelphys*, in part)

Type species: *Centetodon pulcher* Marsh, 1872.

Type specimen: YPM 13507.

Characteristics: Small soricoid; snout abruptly narrowed at level of P3; palatal margin elevated anterior to midpoint of P4 (Lillegraven, McKenna, and Krishtalka, 1981).

Average length of m2: 1.6–1.9 mm.

Included species: *C. pulcher* (including *Hypacodon praecursor* McKenna, 1960a) (known from localities SB42, SB43A, CP5A, CP34[B], C, D*); *C. aztecus* (localities CC4*, [CC9A, AA], NP8, NP9A); *C. bacchanalis* (McGrew, 1959) (originally described as *Diacodon bacchanalis*) (locality CP34D*); *C. bembicophagus* (localities [CC4], [CC6A], CP5A, CP27D, E, CP34A, C, D*); *C. chadronensis* (localities CP39B, C*, F, G, CP39IIA, CP98B, C, NP49II); *C. divaricatus* (localities CP103B*, NP10B2, CC); *C. hendryi* (localities CP29C*, D, CP98B); *C. kuenzii* (localities NP23A, C, NP24C*, D, [NP25B], NP27C1, D); *C. magnus* (Clark, 1936) (originally described as *Metacodon magnus*) (localities GC8AA, [CC8IIB], CP29C-E, [CP42A], CP46, CP68C, CP83C*, [CP84B, C], ?CP86B, CP101, NP9A, B, [NP10B], NP24C, NP25C, NP49II); *C. marginalis* (including *Embassis alternans* Cope, 1873; *Herpetotherium marginale* Cope, 1873; *Embassis marginalis* Cope, 1874; *Didelphys marginalis* Cope, 1879; *Peratherium alternans* Cope, 1884; *Peratherium marginale* Cope, 1884; *Geolabis rhynchaeus* Cope, 1884; *Metacodon mellingeri* Patterson and McGrew, 1937) (localities CP29E, [CP34D], CP68C*, CP84A, CP99A, NP50B); *C. neashami* (localities CP20[A], B*, BB); *C. patratus* (localities CP20A, [BB], CP63*); *C. wolffi* (Macdonald, 1965) (originally described as *Geolabis wolffi*) (localities CP84A, B*).

Centetodon sp. is also known from localities GC8B, ?CC1, CC4, CC6A, B, CC7B, CC8, CC8IIA, CC12, SB22C, SB43A, SB44B, D, CP20A, C, CP27D, E, CP29C, D, CP34C–E, CP41B, CP63, CP65, CP84A, B, (C), CP85A, NP10Bii.

Comments: Butler (1972) raised McKenna's (1960a) Geolabidinae to family status and Lillegraven, McKenna, and Krishtalka (1981) thoroughly reviewed Bridgerian and younger *Centetodon*, making it one of the most completely studied early Tertiary insectivores. However, neither paper provided a diagnosis for the family, even though only three additional genera (*Marsholestes*, *Batodon*, and *Batodonoides*) are currently thought to belong there. McKenna's diagnosis for the Geolabidinae is used here although, as acknowledged by that author, it is based largely on the morphology of *Geolabis* (= *Centetodon*).

Marsholestes McKenna and Haase, 1992 (including "*Myolestes*" Matthew, 1909, not *Myolestes* Brèthes, 1904 [dipteran])

Type species: *Marsholestes dasypelix* (Matthew, 1909) (originally described as *Myolestes dasypelix*).

Type specimen: AMNH 11490.

Characteristics: Mental foramen beneath anterior root of m1; coronoid process with anterolingual "boss"; first lower premolar somewhat procumbent, double-rooted, bladelike with weak, slightly lingual paraconid and small but distinct posterior cusp; p4 with metaconid opposite to and lower than protoconid and not connected to that cusp; lower molars with compressed, crestlike paraconids, tricuspid talonids with cusps relatively low and indistinct (McKenna, 1960a; Robinson, 1968; McKenna and Haase, 1992).

Average length of m2: 1.5 mm.

Included species: *M. dasypelix* only (known from localities CP27D, CP34B, C*).

Marsholestes sp. is also known from localities (CP31C, E).

Comments: Despite being known for nearly 80 years, no good diagnosis is available for Matthew's taxon. *Marsholestes* most closely resembles *Centetodon* in comparable morphology. McKenna's (1960a) description of a specimen referred to "cf. *Myolestes dasypelix* Matthew, 1909D" is the best available characterization of this taxon and is used here along with some aspects of the holotype noted by Robinson (1968) and McKenna and Haase (1992).

***Batodonoides* Novacek, 1976**

Type species: *Batodonoides powayensis* Novacek, 1976.

Type specimen: UCMP 96459.

Characteristics: Among smallest known mammals; upper molars with high, piercing cusps, wide stylar shelf, strong, sharp metacrista, prominent lingual paraconule, no metaconule, narrow precingulum and postcingulum not connected across the lingual base of protocone, and closely appressed paracone and metacone; p1 elongate anteriorly; P4 lacking metacone and stylar cuspules; m3 smaller than m1-2; m3 with weak to absent entoconid (Novacek, 1976; Bloch, Rose, and Gingerich, 1998).

Average length of m2: 0.72 mm (*B. vanhouteni*).

Included species: *B. powayensis* (known from localities CC4*, CC6A, B, CC7A, B, CC8, CC9A); *B. vanhouteni* (localities CP19C*, CP20A, BB).

Batodonoides sp. is also known from locality (CC8).

Comments: Novacek (1976) believed Uintan *Batodonoides powayensis* to be a geolabidid sharing close similarity with Late Cretaceous *Batodon tenuis*. These are indeed strikingly close forms, leaving, however, a dearth of geolabidid-like forms in the Paleocene. As suggested by Van Valen (1967), *Stilpnodon* (Simpson, 1935) may fill this gap. Though Simpson (1935, 1945) placed the genus in his Nyctitheriidae, Romer (1966) transferred it to the Adapisoricidae and Carroll (1988) referred *Stilpnodon* to the Palaeoryctidae. McKenna (1960a) discussed the genus with respect to members of his then new subfamily Geolabidinae, stating that if *Stilpnodon* is a member of that group "... it would be necessary to regard it as more advanced than *Geolabis* [= *Centetodon*] itself." McKenna and Bell (1997) assigned *Stilpnodon* to Gypsonictopidae but noted that its placement there was doubtful, a view endorsed here.

***Batodon* Marsh, 1892**

Type species: *Batodon tenuis* Marsh, 1892

Type specimen: USNM 2139.

Characteristics: Lower canine relatively large; p1 single-rooted; crowns of p2-3 anteriorly inclined; p4 larger than p3 with a small metaconid and a trenchant talonid; lower molars with high protoconids, shorter metaconids, low paraconids, and centrally placed hypoconulids (Clemens, 1973).

Average length of m2: 1.25 mm.

Included Tertiary species: *Batodon* is a Late Cretaceous taxon represented by *B. tenuis* in the Tertiary only at locality NP16A.

Indeterminate geolabidids

In addition to the taxa discussed above, indeterminate or undescribed geolabidids are also present at locality CP7C (including *Protictops alticuspidens* Peterson, 1934, a nomen dubium, possibly representing *Centetodon* [Lillegraven, McKenna, and Krishtalka, 1981]) and CP86B.

SORICOIDEA FISCHER VON WALDHEIM, 1817

NYCTITHERIIDAE SIMPSON, 1928

Cranial characteristics: Infraorbital canal short, similar to later occurring soricomorphs (and in contrast to erinaceomorphs; see Butler, 1972); cranial morphology otherwise unknown for the Nyctitheriidae.

Dental characteristics: Dental formula I3/3, C1/1, P4/4, M3/3 (*Leptacodon tener* may have had five upper premolars [McKenna, 1968; but see Fox, 1983]); P4 semi-molariform with high paracone, lower metacone joined, but not incorporated into metastylar crest, distinct protocone with strong pre- and postprotocrista; M1–2 with high paracone, metacone and protocone, strong conules on pre- and postprotocristae, strong pre- and postpara metaconular and postmetaconular cristae, posterolingual hypoconal shelf with small hypocone, and small precingulum; M3 with large parastyle, metacone smaller than paracone, no metastyle, strong conules, low protocone, and reduced pre- and postcingula with no hypocone; i1–3 multi lobed; p4 semi- to completely molariform with low paraconid; m1–3 exodaenodont, lingually canted, with sharp cusps, distinct cuspate paraconid, deep hypoflexid, talonid equal or greater in width than trigonid, large hypoconulid closer to entoconid than hypoconid, and higher hypoconid than entoconid.

Nyctitheriids have a complex taxonomic history. Marsh (1872) described *Nyctitherium velox*, *Nyctitherium priscus*, and *Nyctilestes serotinus*, from the Eocene of Wyoming, as the first known fossil bats from North America. Matthew (1909) added *Nyctitherium curtidens* from the middle Eocene of Wyoming and synonymized *Talpavus* Marsh 1872 and *Nyctilestes* Marsh 1872 with *Nyctitherium*, recognizing five species: *N. velox*, *N. priscum*, *N. serotinum*, *N. nitidus*, and *N. curtidens*. Matthew (1909) disagreed with Marsh's (1872) claim that these taxa represented fossil bats, instead referring them to a separate insectivore group related to Talpidae. Matthew (1918) referred Cope's (1875) *Diacodon celatus*, from the Eocene of New Mexico, to *Nyctitherium*. In an interesting turn of events, Matthew (1918) referred *N. celatus* to "?Soricidea or Chiroptera" based on a new specimen from the Eocene of the Bighorn Basin, Wyoming. Matthew (1918) also suggested that *Anaptomorphus minimus* Loomis 1906 might be synonymous with *N. celatus*.

Simpson (1928) proposed the family Nyctitheriidae to include *Nyctitherium*, *Entomacodon*, *Centetodon*, *Myolestes*, and *Protentomodon*, with uncertain affinities to soricomorph insectivores. Simpson (1935) described *Stilpnodon simplicidens* as a nyctitheriid from the Paleocene of Montana and later (Simpson, 1937) documented a ?Nyctitheriidae from the Paleocene of Wyoming. McKenna (1960b) removed *N. nitidus*, *N. curtidens*, and *N. celatum* from the genus *Nyctitherium*, with the idea that they might be closely related to *Entomacodon minutus* (now = *Herpetotherium knighti*, see Chapter 3). McKenna (1960b) regarded Simpson's (1928) Nyctitheriidae as an unnatural grouping of marsupials, geolabidids, and Insectivora incertae sedis. For this reason, McKenna (1960b) preferred not to use the term Nyctitheriidae but rather referred to a group that included some species of *Nyctitherium*, *Entomolestes*, *Sespedectes*, *Amphilemur*, *Macrocranion*, and *Aculeodens*, which he called Amphilemuridae. Robinson (1966) referred several specimens from the Eocene of Colorado to *Nyctitherium* sp. cf. *N. velox*.

Matthew and Granger (1921) described *Leptacodon tener* from the Paleocene of Colorado as a leptictid insectivore. Jepsen (1930) later described *Leptacodon packi* from the Paleocene of Wyoming, and Simpson (1935) noted the occurrence of *L.* cf. *tener* from the Paleocene of Montana. McKenna (1960b) named *Leptacodon jepseni* from the early Eocene of Colorado and questionably included *Leptacodon* in the insectivore family Metacodontidae. However, McKenna (1968), in a revision of the genus, placed *Leptacodon* in Nyctitheriidae (suggesting affinities with early erinaceoid insectivores such as *Entomolestes* and *Nyctitherium*). Robinson (1968) regarded *Leptacodon*, *Nyctitherium*, and *Saturninia* Stehlin 1940 as the only genera in Nyctitheriidae. Butler (1972) allied nyctitheriids with geolabidids, regarding both as primitive soricomorph lipotyphlans. West (1974) described a new genus and species of nyctitheriid, *Pontifactor bestiola*, from the middle Eocene Bridger Formation, Wyoming. Unlike Butler (1972), he regarded Nyctitheriidae to be erinaceoid insectivores. West (1974) felt that Nyctitheriidae included *Nyctitherium*, *Remiculus*, *Saturninia*, *Pontifactor*, and possibly *Leptacodon*.

In a revision of Nyctitheriidae, Krishtalka (1976b) recognized three North American genera: *Leptacodon*, *Nyctitherium*, and *Pontifactor*. Krishtalka (1976b) included *L. tener*, *L. catulus* (new species), and provisionally *L. munusculum* and *L. packi*, in *Leptacodon*. He excluded "*Leptacodon*" *jepseni* and *Myolestes dasypelix* from the Nyctitheriidae. Krishtalka (1976b) concluded that there are two distinct groups of nyctitheriids in North America – one which includes *L. tener*, *L. catulus*, *L. packi*, and the species of *Nyctitherium*, and a second including *L. munusculum* and the species of *Pontifactor*. He further speculated that *Saturninia* and *Remiculus*, respectively, might represent these two North American groups in Europe.

Bown (1979) described *Plagioctenodon krausae* and *Plagioctenoides microlestes*, from the Wasatchian of Wyoming, as new adapisoricid insectivores. Bown and Schankler (1982) formally assigned *Plagioctenodon* and *Plagioctenoides* to Nyctitheriidae and added a new species, *Plagioctenodon savagei*, from the Wasatchian of Wyoming. Bown and Schankler (1982) suggested a division of Nyctitheriidae into two groups: the first composed of *Leptacodon tener*, *Plagioctenodon*, *Plagioctenoides*, and *Saturninia beata*, and the second composed of "*Leptacodon*" *munusculum*, "*Leptacodon*" *packi*, *Saturninia gracilis*, *Amphidozotherium*, *Pontifactor*, and *Nyctitherium*. However, they cautioned that a suite of "crossing characters" indicated that these may not be natural phylogenetic groupings.

Fox (1984) tentatively referred Cretaceous *Paranyctoides* to Nyctitheriidae. He noted that this genus is also similar to the Cretaceous condylarth *Protungulatum*, perhaps suggesting a common ancestor for Cenozoic lipotyphlans, ungulates, and primates among the Cretaceous Erinaceoidea. However, McKenna and Bell (1997) did not include *Paranyctoides* in Nyctitheriidae, instead classifying it as a soricomorph lipotyphlan with no inference of familial affinity.

Rose and Gingerich (1987) described a new genus and species, *Ceutholestes dolosus*, from the Clarkforkian (late Paleocene) of Wyoming. They placed *Ceutholestes* in a new family, Ceutholestidae (including only the type genus), but noted that this taxon is closely allied with Nyctitheriidae. McKenna and Bell (1997) included *C. dolosus* in the Nyctitheriidae.

Gingerich (1987) described *Leptacodon rosei* from the Clarkforkian of Wyoming based on exceptionally complete jaws, maxillae, and a possibly associated metatarsal. Gingerich (1987) used the long, straight metapodial as evidence that *Leptacodon* was digitigrade and possibly saltatorial. Gingerich (1987) also described a new nyctitheriid genus and species, *Limaconyssus habrus*, from the Clarkforkian of Wyoming and noted the presence of other nyctitheriid specimens that he referred to Cf. *Leptacodon* sp. and cf. *Plagioctenodon krausae*.

Gingerich (1987) described *Wyonycteris chalix* as the earliest known record (Clarkforkian) of Chiroptera. He referred specimens previously thought to be *Pontifactor* (*Pontifactor* sp. of Krishtalka, 1976b, Bown, 1979, and Rose, 1981) to Cf. *Wyonycteris* sp. and suggested that *Pontifactor* itself might be a primitive chiropteran. Hand *et al.* (1994) disagreed with Gingerich's (1987) referral of *Wyonycteris* to Chiroptera, pointing out that it lacks certain key bat synapomorphies, such as a well-developed buccal cingulum on lower molars and marked reduction of para- and metaconules on upper molars. Smith (1995), in his description of a new species of *Wyonycteris* from the late Paleocene and early Eocene of Belgium, suggested affinities with the lipotyphlan family Adapisoriculidae. McKenna and Bell (1997) included *Wyonycteris*, along with later occurring but morphologically similar *Pontifactor*, in the Nyctitheriidae.

Butler (1988) stated that Nyctitheriidae represents the oldest and most primitive soricomorph lipotyphlans. He noted that the family is probably paraphyletic and would exclude *Pontifactor* from this group and include only those genera with submolariform last premolars (such as *Leptacodon*, *Plagioctenodon*, *Nyctitherium*, *Saturninia*, *Scraeva*, and *Amphidozotherium*). McKenna and Bell (1997) put *Remiculus*, *Saturninia*, *Leptacodon*, *Pontifactor*, *Oedolius*, *Nyctitherium*, and *Scraeva* in the nyctitheriid subfamily Nyctitheriinae Simpson, 1928 (= Nyctitheriidae, Simpson, 1928; Nyctitheriinae, Van Valen, 1967; including Saturniniinae, Gureev, 1971). They further placed North American *Plagioctenoides* with the European genera *Amphidozotherium*, *Paradoxonycteris*, and *Darbonetus*, in the nyctitheriid subfamily Amphidozotheriinae Sigé, 1976. They also included *Voltaia*, *Plagioctenodon*, *Ceutholestes*, *Limaconyssus*, and *Wyonycteris* in Nyctitheriidae but not in either of the established subfamilies.

For descriptions of European Nyctitheriidae, see Crochet (1974), Sigé (1976, 1978, 1988, 1997), Smith (1995, 1996, 1997), Estravis (1996) and Hooker (1996, 2001). For descriptions of Asian Nyctitheriidae, see Russell and Dashzeveg (1986), Nessov (1987), Averianov (1995), Dashzeveg *et al.* (1998), Meng *et al.* (1998), Ting (1998), and Tong and Wang (1998).

Nyctitherium Marsh, 1872 (including *Nyctolestes*)

Type species: *Nyctitherium velox* Marsh, 1872.

Type specimen: YPM 13510.

Characteristics: Semi-molariform p4, with compressed talonid basin, hypoconid and hypoconulid present, and entoconid usually absent; lower molars with forward projecting, uncompressed paraconid, small hypoconulid positioned lingual of the midline but never behind the entoconid; P4 semi-molariform with a small metacone, and a large posteriorly projecting hypocone; M1–2 with large sectorial paracone and metacone, well-defined pre- and post-protocrtista, well-defined pre- and post-metacrista, small mesostyle, large protocone, large hypocone that extends anteriorly, and V-shaped paraconule and metaconule.

Average length of m2: 1.6 mm (*N. velox*).

Included species: *N. velox* (known from localities [SB22C], SB43A, CP34C, D*); *N. celatum* (locality CP23B*); *N. cristopheri* (localities CP29C, D*); *N. serotinum* (Marsh, 1872) (originally described as *Nyctilestes serotinum*) (localities [SB24], CP5A, CP25J, [CP27D, E], CP31E, CP33B, CP34A, C*, CP35).

Nyctitherium sp. is also known from localities (CC4), (CC7C), SB24, CP6B, CP25A, CP27D, E, CP29D, (CP31E), CP34B, C, E, CP36B, [CP38B, C], ?CP63, (NP1C), (NP4), NP8, NP9A, B, NP20E.

Comments: Krishtalka and Setoguchi (1977) cited *Nyctitherium* from the middle and late Eocene (Bridgerian through Duchesnean NALMAs). However, Fox (1988) and Wolberg (1979) cited possible occurrences of *Nyctitherium* in the late Paleocene (Tiffanian NALMA; localities NP4 and NP20E, respectively). Because of difficulty in the classification of insectivorous mammals, and because of the uncertainty with regards to the identification of these specimens, we regard *Nyctitherium* as an exclusively Eocene genus, although we recognize that its range may be extended into the Paleocene with further taxonomic study of Paleogene lipotyphlans.

Krishtalka and Setoguchi (1977) described a new species, *Macrocranion robinsoni*, from the Uintan Badwater Creek Locality 6 (locality CP29C). Maas (1985) listed *Nyctitherium robinsoni* from Duchesnean Badwater Creek Locality 20 (locality CP29D), a species never before described in the literature. It is possible that Maas (1985) may have been informally proposing a new combination for *M. robinsoni*. If so, this new combination has never been discussed (only implied in a faunal list). *Macrocranion robinsoni* is here considered to still be a valid combination, with *Nyctitherium robinsoni* a possible synonym or a nomen nudum.

Leptacodon Matthew and Granger, 1921 (including *Diacodon*, in part)

Type species: *Leptacodon tener* Matthew and Granger, 1921.

Type specimen: AMNH 17179.

Characteristics: Cheek teeth low crowned; upper molars relatively triangular, lacking mesostyles, profound talon, and large stylar shelf; p2–3 not particularly procumbent; p4 semi-molariform with relatively low metaconid, paraconid situated relatively low on anterior face of trigonid; lower molars with central hypoconulids; molars do not decrease markedly in size from m1 to m3; m1–2 hypoconids subequal in height to entoconids, entoconids situated relatively anteriorly (McKenna, 1960b, 1968; Krishtalka, 1976b; Bown and Schankler, 1982; Rose and Gingerich, 1987).

Average length of m2: 1.3 mm (*L. tener*).

Included species: *L. tener* (known from localities SB20A*, [CP13E], [CP14A, C, D], [NP1C], NP3A0, [A], F, NP4, (NP19IIC), NP20[D], E, NP47B, ?BB); *L. catulus* (locality SB24*); *L. munusculum* (localities [CP13B, E], [CP24A], NP1C, NP3A0, A, NP19C*, NP19IIA, B); *L. rosei* (locality CP17B*); *L. packi* (including *Diacodon packi*, Van Valen, 1967) (localities CP13G*, [CP17B], [CP18B], [CP22B], [NP1C], NP3[B], C, E, NP4, NP7D, [NP19IIB]); *L. proserpinae* (locality NP16C*).

Leptacodon sp. is also known from localities GC21II, SB20A, SB24, (CP13G), (CP17B), (CP18A), (CP22A), CP25(A, B), F, CP26A, (CP63), NP1C, NP47B, C, ?NP49.

Comments: Van Valen (1978) described *L. proserpinae* from the early Paleocene of Montana based on a single p4. If *Leptacodon*, this specimen is the only record from the Puercan NALMA and represents the oldest known occurrence of the genus. More complete material of *L. proserpinae* is needed to assess its generic affinities. McKenna (1960b) has suggested that *L. munusculum* might be generically distinct from *L. tener*. Krishtalka (1976b) shared that opinion and believed that *L. munusculum* might even share affinities with a specimen he referred to as *Pontifactor* sp. (AMNH 15103). Bown and Schankler (1982) believed that *L. munusculum* is generically distinct from *L. tener*, but they did not agree with Krishtalka (1976b) that it should be synonymized with *Pontifactor*. Whatever the case, *L. munusculum* has never been formally assigned to any other genus and thus is included in *Leptacodon*. McKenna (1960b) also believed that *L. packi* might be distinct from

Leptacodon, suggesting that it might be a leptictid, a view supported by Van Valen (1967), who placed this taxon in the leptictid genus *Diacodon*. This synonymy has not been followed by subsequent workers who still refer *L. packi* to the genus *Leptacodon* (e.g. Krishtalka, 1976b; Novacek, 1977; and Gingerich, Houde, and Krause, 1983). However, Krishtalka (1976b) suggested that *L. packi* might represent an intermediate form between *L. tener* and *Nyctitherium* and that it might be more properly assigned to the genus *Nyctitherium* once its upper dentition was known. Based on similarities in premolar morphology with early geolabidids, Bown and Schankler (1982) suggested that *L. packi* might be a basal soricoid and not a synonym of *Nyctitherium*. As in the case of *L. munusculum*, *L. packi* has not been formally assigned to any other genus (since its revalidation by Krishtalka, 1976b) and, therefore, is retained in *Leptacodon*, a group whose monophyly seems seriously in doubt.

Leptacodon is also known from the early Eocene of Europe (McKenna and Bell, 1997).

Pontifactor West, 1974

Type species: *Pontifactor bestiola* West, 1974.

Type specimen: AMNH 91784.

Characteristics: Small size; P4 triangular with large paracone, small metacone, low protocone, long curving postparacrista; upper molars with strongly developed W-shaped ectoloph characterized by a stylocone and a prominent mesostyle, conules present but small; hypocone relatively small on M1–2, absent on M3; M1 with well-developed postmetacrista and metastyle; M2 more rectangular with larger ectoflexi; M3 simple, triangular (West, 1974).

Average length of M2: 1.4 mm.

Included species: *P. bestiola* only (known from localities [CP17B], [CP20A], CP34D*).

Pontifactor sp. is also known from localities CP17B, CP19B, CP20AA, A, (CP31C, E), CP34E.

Comments: Bown (1979) and Bown and Schankler (1982) referred 12 specimens from the Wasatchian to *Pontifactor* sp. (locality CP20A). See comments on *Wyonycteris* for discussion of Paleocene and other early Eocene specimens referred to *Pontifactor* sp. and cf. *Pontifactor bestiola* by Krishtalka (1976b), and Rose (1981).

Pontifactor is also questionably known from the early Eocene of Europe (McKenna and Bell, 1997).

Plagioctenodon Bown, 1979

Type species: *Plagioctenodon krausae* Bown, 1979.

Type specimen: UW 9682.

Characteristics: Cheek tooth crowns not especially tall; p2–3 anteriorly inclined with p2 larger than p3; no p2–3 diastema; p4 semi-molariform, elongate anteroposteriorly, narrow transversely, with paraconid large, anteriorly projecting, talonid with two or three well-defined cusps and well-developed basin; m1-2 about equal in size, m3 slightly smaller than m2; m1-2 entoconids significantly taller than hypoconids, entoconids do not project lingually and are placed at posterolingual margins of molars; no vespiform construction between trigonid and talonid of molars (Bown and Schankler, 1982).

Average length of m2: 1.2–1.5 mm (*P. krausae*).

Included species: *P. krausae* (known from localities [CP17B], CP20A*); *P. savagei* (localities CP20AA, A*, [BB]).

Plagioctenodon sp. is also known from localities GC21II, CP13G, CP14E, CP17B, (CP20BB), (CP27D).

Plagioctenoides Bown, 1979

Type species: *Plagioctenoides microlestes* Bown, 1979.

Type specimen: UW 9694.

Characteristics: Extremely small size; p1-3 reduced, single-rooted; p4 relatively narrow with a paraconid very small and high on the anterior face of the protoconid and a weakly defined talonid with the hypoconid being the only distinct cusp; lower molars with angular (less basined) molar talonids and hypoconulids closely appressed to entoconids.

Average length of m2: 1.0 mm.

Included species: *P. microlestes* only, known from locality CP20A* only.

Limaconyssus Gingerich, 1987

Type species: *Limaconyssus habrus* Gingerich, 1987.

Type specimen: UM 86724.

Characteristics: Cheek teeth high crowned; p4 with large, narrow trigonid; prominent paraconid displaced anteriorly, high metaconid but lower than protoconid, and short and narrow talonid; lower molars with acutely angled trigonid crests, making molar trigonids triangular in occlusal view (Gingerich, 1987).

Average length of m2: 1.2 mm.

Included species: *L. habrus* only, known from locality CP17B* only.

Limaconyssus sp. is also known from localities CP14E, NP1C.

Wyonycteris Gingerich, 1987

Type species: *Wyonycteris chalix* Gingerich, 1987.

Type specimen: UM 76910.

Characteristics: Upper molars with strongly developed W-shaped ectoloph characterized by a small mesostyle, strong conules, and a distinct pericone; p4 with small metaconid and reduced talonid; lower molars (especially m2) set obliquely in dentary, protoconids curved, cristid obliqua extends to metaconid tip, hypoconulid connected to hypoconid by postcristid (nyctalodont) (Gingerich, 1987).

Average length of m2: 1.3 mm.

Included species: *W. chalix* only (known from localities CP17B*, CP18B).

Wyonycteris sp. is also known from localities CP14E, CP20AA, BB.

Comments: Krishtalka (1976b) referred a specimen from the early Eocene (Wasatchian NALMA) to *Pontifactor* sp. (locality CP19). Rose (1981) referred a specimen from the late Paleocene (Clarkforkian NALMA) to cf. *Pontifactor bestiola* (locality CP17B). Gingerich (1987) referred Rose's (1981) cf. *Pontifactor bestiola* and Krishtalka's (1976b) *Pontifactor* sp. to Cf. *Wyonycteris* sp., a new Clarkforkian genus that he regarded as the first Paleocene record of Chiroptera. The chiropteran affinities of *Wyonycteris* have since been disputed by Hand *et al.* (1994), who claimed that it lacks certain key bat synapomorphies, such as a well-developed buccal cingulum on lower molars and marked reduction of para- and metaconules on upper molars. Smith (1995), in his description of a new species of *Wyonycteris* from the late Paleocene and early Eocene of Belgium, suggested affinities with the lipotyphlan family Adapisoriculidae. We follow McKenna and Bell (1997) in including *Wyonycteris*, along with later occurring but morphologically similar *Pontifactor*, in the Nyctitheriidae. However, we believe that retention of primitive characteristics (and lack of derived ones) does not preclude this clade from being at, or close to, the ancestry of bats. Discovery of more-complete fossils (preferably postcrania) would go a long way towards addressing this question.

Wyonycteris is also known from the late Paleocene of Europe (McKenna and Bell, 1997).

Ceutholestes Rose and Gingerich, 1987

Type species: *Ceutholestes dolosus* Rose and Gingerich, 1987.

Type specimen: UMMP 82503.

Characteristics: Lower dental formula i1–3, c1, p1–4, m1–3; i1–3 anteriorly inclined, with digitate lateral margins, none markedly enlarged but i2 larger than i1 and i3; p3 premolariform and larger than p2, both double-rooted; p4 molariform with anteroposteriorly compressed but broad trigonid and broad, basined talonid with distinct entoconid, hypoconid, and hypoconulid; lower molars relatively broad with anteroposteriorly compressed trigonids as broad as talonids, protoconid and metaconid widely separated, with protocristid composed of short protoconid segment and much longer metaconid segment, metaconid large and sharply elevated, protoconid and hypoconid markedly constricted on buccal aspect, entoconid acute and higher than hypoconid; all molar cusps, but especially metaconid, strongly canted lingually, cristid obliqua very low, entocristid much higher, hypoflexid deep (Rose and Gingerich, 1987).

Average length of m2: approximately 1.75 mm.

Included species: *C. dolosus* only, known from locality CP17B* only.

Ceutholestes sp. is also known from locality CP14E.

Comments: Rose and Gingerich (1987) noted similarities between *C. dolosus* and nyctitheriid insectivores. However, owing to a number of derived characteristics, they established a new family, Ceutholestidae. We follow McKenna and Bell (1997) in including *C. dolosus* in the Nyctitheriidae, but we recognize that the exact affinities of this taxon remain as ambiguous as when it was originally described by Rose and Gingerich (1987).

Indeterminate nyctitheriids

In addition to the taxa discussed above, indeterminate or undescribed nyctitheriids are also present at localities CP20A, BB, CP25A, NP1C, NP3A0, NP3F, NC0.

MICROPTERNODONTIDAE STIRTON AND RENSBERGER, 1964

Characteristics: About the size of moles, probably fossorial. Rostrum long, moderately wide and deep (or short, wide, and deep); sutures tend to fuse early; infraorbital foramen short and large; no lacrimal tubercle; upper cheek teeth hypsibrachydont; ectolophs slant strongly lingually; I1 greatly enlarged; P1 absent; P3 with extremely reduced protocone; P4 submolariform with small hooklike cusp behind and above metastyle at posterior end of sectorial blade of ectoloph; M1–2 much larger than p2, somewhat submolariform; p4 and lower molars with very high and anteroposteriorly narrow trigonids with upper half of crowns curving posteriorly (Stirton and Rensberger, 1964).

Micropternodus Matthew, 1903b (including *Kentrogomphios*)

Type species: *Micropternodus borealis* Matthew, 1903b.

Type specimen: AMNH 9602.

Characteristics: Relatively small; c1 and crowns of p3–m1 not especially procumbent or lingually curved, tips of crowns more recumbent; p2 present although small; p3 relatively high crowned; m1 talonid almost equal to height of trigonid (Stirton and Rensberger, 1964).

Average length of m2: 2.4 mm (*M. morgani*).

Included species: *M. borealis* (known from localities CP39E, CP98B, NP23E, NP24C*, D, NP29C, NP30C); *M. morgani* (locality PN6C2*); *M. montrosensis* (locality CP98B*); *M. strophensis* (White, 1954) (originally described as *Kentrogomphios strophensis*) (locality NP29C*).

Micropternodus sp. is also known from localities CP6B, CP46, NP10B, NP25C, PN6D, D3.

Cryptoryctes C. Reed, 1954

Type species: *Cryptoryctes kayi* C. Reed, 1954.

Type specimen: FMNH PM1009.

Characteristics: Known from humerus only; distal part of pectoral process platelike and directed proximomedially; teres tubercle separated from medial epicondyle by deep notch; shaft skewed so that distal end more laterally positioned than proximal end; trochlea well separated from fossa of great ligament of flexor digitorum profundus muscle.

Average length of humerus (teeth unknown): 12.2 mm.

Included species: *C. kayi* only (known from localities NP24C*, D, E, NP25B, C, NP29C).

Cryptoryctes sp. is also known from locality NP10B.

Comments: *Cryptoryctes* is still only represented by humeri and it is distinctly possible that these humeri actually represent *Micropternodus borealis* (Russell, 1960; A. Tabrum, personal communication), in which case *Cryptoryctes* would be a junior synonym of *Micropternodus*.

Clinopternodus Clark, 1937 (including *Clinodon* Scott and Jepsen, 1936, not *Clinodon* Regan, 1920 [cichlid fish])

Type species: *Clinopternodus gracilis* Clark, 1937.

Type specimen: YPM-PU 13835.

Characteristics: Larger than *Micropternodus*; lower canine, crowns of p3–m1 relatively procumbent, lingually curved; tips of crowns not recumbent; p2 absent; p3 with relatively low crown; m1 talonid equal to two-thirds height of trigonid (Stirton and Rensberger, 1964).

Average length of m1: 2.5 mm.

Included species: *C. gracilis* only, known from locality CP83C* only.

Clinopternodus sp. is also known from locality ?CP41B.

Indeterminate micropternodontids

In addition to the taxa discussed above, an indeterminate micropternodontid is present at locality CC4.

APTERNODONTIDAE MATTHEW, 1910

The anatomy of the Apternodontidae is well known and several good, rather complex assessments of its relationships exist (Schlaikjer, 1933; McDowell, 1958; McKenna, 1960a; Reed and Turnbull, 1965; Asher *et al.*, 2002). They are relatively small soricomorph insectivores; almost certainly fossorial, based on scant postcranial material.

Cranial characteristics: Lacrimal foramen enlarged and laterally facing; anterior margin of infraorbital canal concave; sagittal crest present, continuous with strong nuchal crest; ethmoid foramen and sinus canal open in orbitotemporal region exiting in the superior margin of the sphenorbital fissure; posterolateral braincase expanded into lambdoid plates; entoglenoid process enlarged; postglenoid process absent; zygomatic arch incomplete; ossified auditory bulla absent (Asher *et al.*, 2002).

Dental characteristics: Dental formula I2/3, C1/1, P3/3, M3/3; anterior incisors enlarged and procumbent; P4/p4 molariform; upper molars zalambdodont with metacones absent; coronoid process extends anteriorly (Asher *et al.*, 2002).

Apternodus Matthew, 1903b

Type species: *Apternodus mediaevus* Matthew, 1903b.

Type specimen: AMNH 9601.

Characteristics: As for family.

Average length of m2: 2.1 mm. (*A. iliffensis*).

Included species: *A. mediaevus* (known from localities CP39G, CP83C, NP24C*, D, E, [NP25C], NP27C, NP29C, NP30C); *A. baladontus* (localities NP24D, NP25B, C*); *A. brevirostris* (localities [SB44B], ?CP39B, C, E, F, G*); *A. dasophylakas* (locality CP39G*); *A. gregoryi* (localities CP39B, [39C], E, F, CP42C*); *A. iliffensis* (localities [CP29D], CP39IIA, CP68B*); *A. major* (locality CP39B*).

Apternodus sp. is also known from localities CP41B, CP43, CP45A, CP98B, NP9A, NP10B.

Indeterminate apternodontids

In addition to the taxa discussed above, indeterminate or undescribed apternodontids are also present at localities CP29D, CP65.

PARAPTERNODONTIDAE ASHER *ET AL.*, 2002

Characteristics: Small tooth size; anterior incisor enlarged with root extending posterior to p3; molars zalambdodont; lower molar talonids reduced and unbasined and lacking buccal cingulids; m3 posterior cusp absent (Asher *et al.*, 2002).

Parapternodus Bown and Schankler, 1982

Type species: *Parapternodus antiquus* Bown and Schankler, 1982.

Type specimen: YPM 31169

Characteristics: p4 premolariform; m1–2 with reduced talonids and weak talonid cuspule; m2 postvallid oriented transversely; m3 postvallid oriented anterolabially–posterolingually; m2–3 lacking precingulids; m3 short without talonid cusp; medially pocketed coronoid process (Bown and Schankler, 1982; Asher *et al.*, 2002).

Average length of m2: 1.15 mm.

Included species: *P. antiquus* only (known from localities CP19A, CP20A*).

Parapternodus sp. is also known from locality CP20AA.

Koniaryctes Robinson and Kron, 1998

Type species: *Koniaryctes paulus* Robinson and Kron, 1998.

Type specimen: UCM 58291.

Characteristics: Molar talonids reduced to crests with no basins and lacking talonid cuspules; lower molars lacking buccal cingulids and notches below paraconids; m3 anteroposteriorly short (Asher *et al.*, 2002).

Average length of m2: 1.2mm.

Included species: *K. paulus* only, known from locality CP25B only.

OLIGORYCTIDAE ASHER *ET AL.*, 2002

Cranial characteristics: Posterior braincase unspecialized, lacking lateral lambdoid plates; prominent entoglenoid process present; foramen ovale enlarged and alisphenoid canal absent; ethmoid foramen and sinus canal exit anterior to sphenorbital fissure; lacrimal foramen large and laterally oriented (Asher *et al.*, 2002).

Dental characteristics: Molars zalambdodont; upper molars with distinct protocones and anterior cingula and lacking metacones; i2–3 tricuspid; lower molars with reduced talonid basins; m3 talonid cusp taller than paraconid; coronoid process pocketed medially (Asher *et al.*, 2002).

Oligoryctes Hough, 1956

Type species: *Oligoryctes cameronensis* Hough, 1956.

Type specimen: USNM 19909.

Characteristics: As for family.

Average length of m2: 1.0 mm.

Included species: *O. cameronensis* (known from localities CP39C, E*, CP98B, NP23E, NP24C, D, [NP25B]); *O. altitalonidus* (localities CP39B, C, [C], CP83C*, NP24C, D, [NP32C], NP50B).

Oligoryctes sp. is also known from localities (CC4), CC7B, (CC8), CP29C, D, CP36B, CP68C, (CP85A), NP8, NP9A, NP10B, (NP22), NP32C.

Indeterminate oligoryctids

In addition, an undescribed oligoryctid is present at localities CP34C–E.

PLESIOSORICIDAE WINGE, 1917

Characteristics: i2 hypertrophied into procumbent, piercing tooth; p4 with low, short talonid; M1–2 not dilambdodont, cuspidate stylar shelf, well-developed hypoconal shelf or hypocone, divided protoconal crests in conule regions; dentary with sharply upright or slightly anteriorly inclined coronoid process and transversely cylindrical condyle; symphysis procumbent and extends to anterior root of p4.

Plesiosorex Pomel, 1848 (including *Erinaceus*, in part; *Hibbarderix*; *Meterix*; *Theridosorex*)

Type species: *Plesiosorex soricinoides* (Blainville, 1840) (originally described as *Erinaceus soricinoides*).

Type specimen: Clermont–Ferrand Julien collection No. 3.

Characteristics: Jugal absent, zygoma incomplete; lacrimal foramen anterior to infraorbital canal; skull bones fused in adults; origin of musculars levator labii superioris anterior to level of infraorbital canal; well-defined postpalatal torus present but no suboptic foramen; dental formula complete or reduced by one antemolar; molars distinctly graded in size from large M1/m1; C1/c1 reduced or absent; i2 and I1 enlarged, piercing teeth; I1 ever-growing; P3–4 and p4 semi-molariform; P4 and m1 with carnassial-like shear; lower molar hypocones reduced and centrally located on hypolophids; incisors to P2–p2 and usually p3 single-rooted and unicuspid.

Average length of m2: 2.7–4.5 mm.

Included species: *P. coloradensis* (known from localities CP71*, CP88, [PN17II], [NC2B]); *P. donroosai* (localities CP89*, CP114A, B, [PN8B], [PN9B]); *P. latidens* (Hall, 1929) (originally described as *Meterix latidens*) (localities NB23B, C*, [PN8B], PN9B); *P. obfuscatus* (Martin and Green, 1984) (originally described as *Hibbarderix obfuscatus*) (localities CP86D*, F, CP88).

Plesiosorex sp. is also known from localities NB23B, ?CP71, CP88, CP89, CP90A, CP105, CP114C, CP116B, F, [NP10E], PN8A, PN10.

Comments: *Plesiosorex* is clearly the best known member and central to any definition of this small family. Butler (1972) argued for inclusion of the European Oligocene genus *Butselia* Quinet and Misonne, 1965. The dentition of this form is more zalambdodont and presumably more primitive than that of *Plesiosorex* and only the dentary and some teeth are known. Green (1977) formally synonymized *Meterix* Hall 1929 and *Plesiosorex*, thus reducing the content of the family to *Plesisorex* and *Butselia*. *Plesiosorex* has been shuttled between the erinaceomorphs and soricomorphs in previous studies, but more recent studies (Wilson, 1960; Butler, 1972) and additional material clearly confirm inclusion in the Soricomorpha.

Plesiosorex is also known from the late Oligocene to late Miocene of Europe, and the middle Miocene of Anatolia (McKenna and Bell, 1997).

Indeterminate plesiosoricids

In addition to *Plesiosorex*, an indeterminate plesiosoricid is also present at locality SB51.

SORICIDAE FISCHER VON WALDHEIM, 1817

Cranial characteristics: Skull with relatively short rostrum in which bones fuse early; cranium, although small, is inflated; origin of muscularis levator labii superoris proprius from the tip of top of frontal near midline; lacrimal duct curves upward and forward from a small ventrally exiting foramen on infraorbital bridge; strong postpaltine torus with lateral corners; alisphenoid canal and postglenoid foramen absent; zygomatic arches thin (Heterosoricinae) or absent, jugal absent; large pyriform fenestra excludes squamosal from tympanic roof; tympanic ring large, lying nearly horizontal and close to skull; bulla not ossified; basisphenoid lacking tympanic wing (process); paroccipital processes greatly reduced; sphenopalatine foramen lies well forward over molars; usually with well-developed pyriform fenestra over otic chamber;

Dental characteristics: Dentary tapers anteriorly with long procumbent symphysis; coronoid process oriented vertical to horizontal ramus; articular condylar facet wide and moderately to distinctly bipartite with posterodorsal and posteroventral articular surfaces; deciduous dentition unclarified and non-functional; I1 hook-shaped; teeth between I1 and P4 unicuspid and single-rooted; P4 semi-molariform with well-developed and basined lingual shelf and large paracone extended posteriorly into prominent blade; M1–2 dilambdodont, roughly rectangular, with well-developed cingular hypocone and crested protocone; upper dentition reduced by one or more antemolars and lower dentition by two or more antemolars; anteromost lower incisor hypotrophied and procumbent; lower antemolars primitively unicuspid and single-rooted; p4 not molariform; m1-2 with relatively low and subequally elevated talonid and trigonid, talonid basined with distinct entoconid and postcristid extending posterior to entoconid (but may fuse with entoconid in derived condition); m3 primitive with well-developed and basined talonid; teeth usually pigmented red at least on the tips of the crowns but this may be lost in some Old World groups.

Postcranial characteristics: Little is known of the postcranium of most fossil shrews. The skeleton of shrews in general is unspecialized (C. Reed, 1951) with the exception of fusion of the tibia and fibula distally – there is some evidence that this may be absent in Heterosoricinae.

HETEROSORICINAE VIRET AND ZAPFE, 1951

Characteristics: Zygomatic arch slender and complete, formed only by maxilla and squamosal (where known); dentary with masseteric fossa, shallow, unpocketed internal temporal fossa, and lingual postsymphysial foramen; four or more lower antemolars; dorsal part of dentary condyle labial to ventral part of condyle; P4 with about equally well-developed posterolabial and posterolingual crests; P4–M2 not emarginated posteriorly.

Domnina Cope, 1873 (including *Protosorex*, *Miothen*)

Type species: *Domnina gradata* Cope, 1873.

Type specimen: AMNH 5353.

Characteristics: Least specialized shrew; only weakly differentiated condyle; five or six lower antemolars; m1-3 entoconid crest high and joined to metaconid but distinctly separated from postcristid.

Average length of m2: 1.5–2.5 mm.

Included species: *D. gradata* (known from localities [CP29C, D], [CP46], CP68B, C*, CP84A-C, CP98B, C, CP99A, B, [NP9A], [NP10B2], [NP23E], [NP32C]); *D. dakotensis* (localities CP85C*, CP100III, CP101, [NP10B2, CC]); *D. greeni* (localities CP85C, D*); *D. thompsoni* (localities CP98B, NP24C*, D, [NP25B], NP29C, NP30C).

Domnina sp. is also known from localities CC17A, CP29D, CP41B, CP44, CP84A, B, CP85A, D, CP103A, NP8, NP10B, NP23A, NP32C, NP36A, B, D, NP49II, PN16.

Paradomnina Hutchison, 1966

Type species: *Paradomnina relictus* Hutchison, 1966.

Type specimen: UO 24279.

Characteristics: Resembles *Domina* in general conformation of dentary; six lower antemolars present; entoconid joined to metaconid by only a low crest, not separated from postcrisitid.

Average length of m2: 1.81–1.91 mm.

Included species: *P. relictus* only, known from localities [CC22A], [NB6C, E], PN8B*.

Ingentisorex Hutchison, 1966

Type species: *Ingentisorex tumididens* Hutchison, 1966.

Type specimen: UO 21960.

Characteristics: Largest heterosoricines; P4 comparable in volume to m2, with two anteroposteriorly and distinctly separate roots; m1 trigonid width greater than talonid width; dental formula reduced to six/four antemolars; only two upper and lower molars present; P4 elongate and carnassialiform; dentary condyle not sharply offset.

Average length of m2: 3.04 mm.

Included species: *I. tumididens* only, known from locality PN9B* only.

Wilsonosorex Martin, 1978

Type species: *Wilsonosorex conulatus* Martin, 1978.

Type specimen: KU 10036.

Characteristics: Upper molars with distinct paraconules and metaconules, divided mesostyle, lingual cusps anteroposteriorly compressed into a narrow V-shape; lower molars bulbous, entoconid joined to postcristid.

Average length of m2: 1.5–1.6 mm.

Included species: *W. conulatus* (known from locality CP71*); *W. bateslandensis* (locality CP88*).

Wilsonosorex sp. is also known from localities CP86D, F.

Pseudotrimylus Gureev, 1971 (including *Domnina*, in part; *Trimylus*, in part; *Heterosorex*)

Type species: *Pseudotrimylus roperi* (R. W. Wilson, 1960) (originally described as *Heterosorex roperi*).

Type specimen: KU 10008.

Characteristics: Teeth bulbous with lower molar cingulae inflated; lower antemolars reduced to five; I1 most enlarged within heterosoricines; p4 smaller than second antemolar; lower molar entoconid joined to postcristid with entoconid crest short or absent; dentary massive with shortened antemolar region, mental foramen placed in deep depression on lateral face, condylar facets widely separated and sharply offset.

Average length of m2: 1.6 –2.1 mm.

Included species: *P. roperi* (known from locality CP71*); *P. compressus* (Galbreath, 1953) (originally described as *Domnina compressa*) (known from locality CP68C); *P. dakotensis* (Repenning, 1967) (originally described as *Trimylus dakotensis*) (locality CP88); *P. mawbyi* (Repenning, 1967) (originally described as *Trimylus mawbyi*) (localities PN7II, PN8A).

Pseudotrimylus sp. is also known from localities NB17, NB18A, NB23B, CP84C, CP85A, C, CP86B, CP114A, ?NP11, NP36A, D, PN8B.

Indeterminate heterosoricines

In addition to the taxa discussed above, indeterminate heterosoricines are present at localities CC9IVC, CP84C and NP11.

LIMNOECINAE REPENNING, 1967

Characteristics: Condylar articular facets vertically aligned, joined medially and embayed labially; internal temporal fossa deep and pocketed; four to five lower antemolars (fifth miniscule when present); P4 with posterolingual crest reduced over that in heterosoricines; m1 with metaconid and protoconid very close together.

***Limnoecus* Stirton, 1930**

Type species: *Limnoecus tricuspis* Stirton, 1930.

Type specimen: UCMP 31047.

Characteristics: p4 with posterolabial crest lying along midline.

Average length of m2: 1.14 mm.

Included species: *L. tricuspis* (known from localities CC17H, NB6B-D, E*, NB23C, [PN11]); *L. niobrarensis* (localities GC4DD, CP89, CP114B*).

Limnoecus sp. is also known from localities GC8D, [CC17A], [CC19], NB7A, NB23B, CP49A, CP54B, ?CP114C, [CP116C], NP41B, PN8B.

Comments: *Limnoecus* is also known from the late Miocene of Europe (McKenna and Bell, 1997).

***Angustidens* Repenning, 1967 (including *Sorex*, in part)**

Type species: *Angustidens vireti* (R. W. Wilson, 1960) (originally described as *Sorex vireti*).

Type specimen: KU 10037.

Characteristics: Condyle massive; p4 with posterolabial crest labial to midline.

Average length of m2: 1.5–1.7 mm.

Included species: *A. vireti* only (known from localities CP71* and [NP41B]).

Comments: *Angustidens* is also known from the late Miocene of Europe (McKenna and Bell, 1997).

Indeterminate or undescribed limnoecines

A new genus of limnoecine is reported from locality CP101.

SORICINAE FISCHER VON WALDHEIM, 1817

Characteristics: Condylar facets vertically aligned with lingual emargination between condyles if any; internal temporal fossa deep and pocketed; P4–M2 usually emarginated posteriorly; four to five lower antemolars (fifth minute when present); p4 with posterolingual crest extremely reduced or absent, posterolabial crest prominent and hooked lingually in its posterior part.

SORICINI FISCHER VON WALDHEIM, 1817

Characteristics: Least specialized of Soricinae; articular facets of condyle continuous or only slightly separated; m1–2 retain an entoconid crest that joins the entoconid to the metaconid.

***Antesorex* Repenning, 1967 (including *Sorex*, in part)**

Type species: *Antesorex compressus* (R. W. Wilson, 1960) (originally described as *Sorex compressus*).

Type specimen: KU 10050.

Characteristics: Articular facets confluent but condylar surface emarginated lingually; mental foramen under p4; lower molar protoconid very close to metaconid.

Average length of m2: 1.1 mm.

Included species: *A. compressus* (known from localities NB17, CP71*, [CB86D, F]).

Antesorex sp. is also known from locality CP88.

***Sorex* Linnaeus, 1758**

Type species: *Sorex araneus* Linnaeus, 1758, extant common shrew.

Type specimen: None designated.

Characteristics: Generally primitive for subfamily; differing from *Antesorex* in definition of articular facets and placement of mental foramina below m1.

Average length of m2: 1.0–1.48 mm.

Included Tertiary species: *S. cinereus* (known from locality [CP131]); *S. edwardsi* (localities CP116C*, [D], PN12B); *S. hagermanensis* (locality PN23A*); *S. leahyi* (localities CC47D, CP131II*); *S. meltoni* (localities [PN4], PN23A*); *S. palustris* (PN23A*); *S. powersi* (localities PN3C, PN23A*); *S. rexroadensis* (localities CP128A*, [PN23A]); *S. sandersi* (localities CP118, CP130*, [CP131]); *S. taylori* (localities SB14E, SP1F, SP5, CP128C*, CP132B); *S. yatkolai* (locality CP116C*).

Sorex sp. is also known from localities NB7B–E, NB13C, NB35IV, SB15A, SB18B, SP1?A, H, CP116C, CP117B, CP131II, PN4.

Comments: North American Pleistocene records of *Sorex* are known from Alaska, Arkansas, California, Colorado, Florida, Georgia, Idaho, Illinois, Iowa, Kansas, Maryland, Nebraska, New Mexico, Oklahoma, Pennsylvania, South Dakota, Texas, Utah, and West Virginia. The genus *Sorex* is also known from the late Miocene to Recent of Asia and the early Pliocene to Recent of Europe (McKenna and Bell, 1997).

***Planisorex* Hibbard in Skinner and Hibbard, 1972 (including *Sorex*, in part)**

Type species: *Planisorex dixonensis* (Hibbard, 1956) (originally described as *Sorex dixonensis*).

Type specimen: UMMP V31986.

Characteristics: m3 basined; M1–2 with distinct hypocones.

Average length of m2: 1.54–1.68 mm.

Included species: *P. dixonensis* only (known from localities CP118, CP119, CP131II*). *Planisosorex* is also known from the Pleistocene.

***Alluvisorex* Hutchison, 1966 (including *Hesperosorex*, in part)**

Type species: *Alluvisorex arcadentes* Hutchison, 1966.

Type specimen: UO 22307.

Characteristics: Cheek teeth stout and dentary robust, mental foramen centered below m1; P4–M2 with intermediate degree of posterior emargination; lower molars with metaconid close to protoconid; m1–2 occlusal outline distinctly rectangular; m3 with longitudinal metalophid and low entoconid ridge.

Average length of m2: 1.22–1.40 mm.

Included species: *A. arcadentes* (known from localities [CP116C], PN8B, PN9B*); *A. chasseae* (Tedford, 1961) (originally described as *Hesperosorex chasseae*) (localities NB7B, C*, D, NB23C).

Alluvisorex sp. is also known from localities CP114A, B, [CP116E].

***Anchiblarinella* Hibbard and Jammot, 1971**

Type species: *Anchiblarinella wakeeneyensis* Hibbard and Jammot, 1971.

Type specimen: UMMP V60446.

Characteristics: Closely resembles *Petenyia*; internal temporal fossa more triangular; lower molars with lower entoconid crest and entoconid.

Average length of m2: 1.31 mm.

Included species: *A. wakeeneyensis* only, known from locality CP123E* only.

***Petenyia* Kormos, 1934 (including *Parydrosorex*)**

Type species: *Petenyia hungarica* Kormos, 1934.

Type specimen: Kungliga Ungarischen geologischen Reichsanstalt Number Ob/3684.

Characteristics: Mental foramen below talonid of m1; P4–M1 with only slight emargination of posterior outline; m3 greatly reduced with no trace of entoconid crest.

Average length of m2: 1.22 mm (topotype).

Included species: *P. concise* (R. L. Wilson, 1968) (originally described as *Parydrosorex concisa*) only, known from locality CP123E* only.

Petenyia sp. is also known from locality (CP116F).

Comments: *Petenyia* is also known from the late Miocene to Pleistocene of Europe, and the late Pliocene of Asia (McKenna and Bell, 1997).

NECTOGALINI ANDERSON, 1879

Characteristics: Similar to Blarinini in enlargement of condylar process and lateral offset of articular facets but lower facet is separated by a groove from body of dentary.

***Hesperosorex* Hibbard, 1957**

Type species: *Hesperosorex lovei* Hibbard, 1957.

Type specimen: USNM 21384.

Characteristics: General conformation of dentary, m1–3, and condylar facets with relatively broad contact *Sorex*-like but condyle enlarged and offset lingually as in Nectogalini.

Average length of m2: 1.4 mm.

Included species: *H. lovei* only, known from locality CP59A* only.

Hesperosorex sp. is also known from locality ?PN10.

***Crusafontina* Gibert, 1975 (including *Anouroneomys*)**

Type species: *Crusafontina endemica* Gibert, 1975.

Type specimen: NR 9009.

Characteristics: Dental formula I1/1, C0/0, P4/2, M3/3; M3/m3 reduced but present; I1 with buccal cingulum; P4 pentagonal with projecting parastyle and distinct posterior emargination; M1 parastyle and metastyle subequally projecting buccally, distinct posterior emargination; M2 with complete talon; i1 single- to triple-cusped; p4 with posterolingual depression; m1–2 with entocristid; upper condylar facet triangular or oval, lower oblong; internal temporal fossa subtriangular to round or oval (van Dam, 2004).

Average length of m2: 1.7 mm (*C. endemica*).

Included species: *C. magna* (known from localities CP116C*, D); *C. minima* (locality PN10*).

Comments: *Anouroneomys* Hutchison and Bown, in Bown, 1980, was placed in *Crusafontina* as a junior synonym by van Dam (2004). *Crusafontina* is also known from the late Miocene of Europe (McKenna and Bell, 1997).

***Beckiasorex* Dalquest, 1972**

Type species: *Beckiasorex hibbardi* Dalquest, 1972.

Type specimen: UMMP V60449.

Characteristics: Upper condylar facet triangular in posterior view; coronoid process inclined anteriorly; internal temporal fossa constricted and ovoid; m1–2 lacking entoconid crests; m3 talonid with single central cusp.

Average length of m2: 1.14 mm.

Included species: *B. hibbardi* only, known from locality SP1F* only.

***Notiosorex* Baird, in Coues, 1877 (including *Megasorex*)**

Type species: *Notiosorex crawfordi* (Baird, in Coues, 1877) (originally described as *Sorex* [*Notiosorex*] *crawfordi* Baird, in Coues, 1877), the extant Crawford shrew.

Type specimen: None designated.

Characteristics: Dentary with anterior inclination of coronoid process and deep ovoid internal temporal fossa; m2 talonid weakly reduced and basined; m1–2 with entoconid ridge.

Average length of m2: 1.3–1.6 mm.

Included species: *N. jacksoni* (including *Megasorex gigas*, Hibbard, 1950) (known from localities NB13[B], C,

SB14A, SP1F, CP128A*, B); *N. repenningi* (locality SB62*).

Notiosorex sp. is also known from localities WM26, NB33A, SB11, ?SB14D, SB15A. *Notiosorex* is also known from Pleistocene localities in Arizona, California, Kansas, and Texas.

Arctisorex Hutchison and Harington, 2002

Type species: *Arctisorex polaris* Hutchison and Harington, 2002.

Type specimen: CMN 51090.

Characteristics: Coronoid expanded distally and inclined anteriorly; m3 longer than m2; m3 talonid longer than trigonid; m2-3 talonids with postentoconid valleys.

Average length of m2: 2.01 mm.

Included species: *A. polaris* only, known from locality HA3* only.

BLARININI STIRTON, 1930

Characteristics: Articular facets widely separated but not separated from body of dentary by groove; m1-2 with entoconid crest very low or absent.

Adeloblarina Repenning, 1967

Type species: *Adeloblarina berklandi* Repenning, 1967.

Type specimen: USNM. 23098.

Characteristics: Articular facets widely separated; mental foramen below trigonid of m1; internal temporal fossa triangular but divided by strong limula.

Average length of m2: unknown.

Included species: *A. berklandi* (known from localities [CP89], PN7III*); *A.* n. sp. (locality CP108A).

Tregosorex Hibbard and Jammot, 1971

Type species: *Tregosorex holmani* Hibbard and Jammot, 1971.

Type specimen: UMMP V60444.

Characteristics: Internal temporal fossa weakly constricted; sharp groove extends along ventral half of lateral side of condyle; lower molars with low entoconid crests.

Average length of m2: 1.68 mm.

Included species: *T. holmani* only (known from localities SP1A, CP123E*).

Cryptotis Pomel, 1848 (including *Blarina*, in part; *Sorex*, in part)

Type species: *Cryptotis parva* (Say, 1823) (originally described as *Sorex parvus* Say, 1823), the extant least shrew.

Type specimen: None designated (see Miller, 1912).

Characteristics: Coronoid deflected laterally; internal temporal fossa triangular but somewhat filled in dorsally, lower facet mostly posterior to lower sigmoid notch; m3 talonid reduced with a single cusp.

Average length of m2: 1.4 mm.

Included species: *C. adamsi* (Hibbard, 1953) (originally described as *Blarina adamsi*) (known from localities CP128A*, B, PN15); *C. kansasensis* (locality CP130*); *C? meadensis* (locality CP128A*).

Cryptotis sp. is also known from locality NB10. *Cryptotis* is also known from Pleistocene localities in Florida, Kansas, Nebraska, and Texas, and from the Pleistocene to Recent of South America.

Paracryptotis Hibbard, 1950 (including *Blarina*, in part)

Type species: *Paracryptotis rex* Hibbard, 1950.

Type specimen: UMMP 25172.

Characteristics: Relatively large taxon; lower articular facet relatively large and anterior to lower sigmoid notch, upper facet turned more medially; internal temporal fossa reduced to an ovoid opening; m1–3 robust; m3 talonid with entoconid ridge and postcristid.

Average length of m2: 1.68–1.92 mm.

Included species: *P. rex* (known from localities SP1F, CP128[AA], A*, C, PN3C, PN11); *P. gidleyi* (Gazin, 1933) (originally described as *Blarina gidleyi*) (locality PN23A*).

Paracryptotis sp. is also known from localities WM26, CP116E, F.

Comments: *Paracryptotis* is also questionably known from the late Miocene of Europe (McKenna and Bell, 1997).

Blarina Gray, 1838

Type species: *Blarina talpoides* (Gapper, 1830) (originally described as *Sorex talpoides*), the extant short-tailed shrew.

Type specimen: None designated.

Included Tertiary species: *B. brevicauda* only (known from localities [CP131], [CP131II]).

Blarina sp. is also known from localities (GC15A), ?CP119. *Blarina* is also known from the Pleistocene of Alabama, Arkansas, Florida, Georgia, Illinois, Iowa, Kansas, Maryland, Nebraska, Oklahoma, Pennsylvania, South Dakota, Texas, and West Virginia.

INDETERMINATE SORICIDS

In addition to the taxa above, indeterminate or undescribed soricids are also present at localities GC8AB, DB, GC11A, NB13A, B, SB12, SB34III, SB52, SB60, CP29C, D, CP96, CP114A, CP116F, CP128B, NP8, NP9A, B, NP10C2, NP32B, PN6G, PN11, PN23C.

LIPOTYPHLA INCERTAE SEDIS

Aethomylos Novacek, 1976

Type species: *Aethomylos simplicidens* Novacek, 1976.

Type specimen: UCMP 96133.

Characteristics: Upper molars triangular in outline viewed from labial aspect, M1-3 with ectoflexus, very narrow to absent stylar shelf, conical paracone, metacone, protocone, and large, deeply excavated, protofossa, lacking conules, hypocones, and precingula; p4 (or Dp4) elongate and molariform with large talonid basin, three talonid cusps, and a "steplike" structure on back of trigonid; lower molars with low trigonids carrying shelflike paraconids, no anterior cingulum, trigonids with semi-rectangular outline when viewed from dorsal aspect, talonids with deeply excavated basin bordered by three cusps, hypoconids much higher than hypoconulids (Novacek, 1976).

Average length of m2: 2.25 mm (m1 or m2).

Included species: *A. simplicidens* (known from localities CC4*, CC6B, CC7B, CC8, CP5A).

Aethomylos sp. is also known from locality CC4.

INDETERMINATE LIPOTYPHLANS

In addition to the taxa discussed above, indeterminate lipotyphlans are also present at localities GC5A, CP34C, CP128C, NP2, NP8, NP9A, NP25B, NP49.

GREATER ANTILLEAN LIPOTYPHLANS

Any assessment of North American insectivorous mammals would be incomplete without mention of the Antillean lipotyphlans *Solenodon* and *Nesophontes* of the family Solenodontidae Gill, 1872. Though the record of these forms is confined to the late Quaternary (*Nesophontes*, six species) and Recent (*Solenodon*, commonly called alamiqui, survives today represented by two highly endangered species), these unusual mammals are likely to be survivors of a very ancient group of soricomorphs that differentiated from some unknown lineage as early as the early Tertiary.

Included characteristics of solenodontids: Lacrimal foramen very large, trumpet-bell-shaped, placed on orbital margin, leading to straight lacrimal duct; tympanohyal small, separated from tympanic and promontorium; alisphenoid and transverse canals present; postglenoid foramen present; muscularis levator labii superioris proprius arising from maxilla dorsal to lacrimal foramen; venous condyloid canal present; foramen magnum very large, extending well above dorsal extremity of occipital condyle; deciduous dentition calcified, functional; upper molars with strongly developed buccal stylar wall, forming cutting crest divided into anterior and posterior portions by narrow fissure, posterior portion curved lingually onto occlusal surface anteriorly; hypocone (*Solenodon*) or narrow cingulum (*Nesophontes*), not filling interdental embrasure; metacromion developed as triangular, flaplike projection of the scapular spine (McDowell, 1958).

BIOLOGY AND EVOLUTIONARY PATTERNS

Summarizing the biology and evolutionary patterns of such a long-lived and diverse group as Lipotyphla is not easily done in a few pages (Figures 7.4–7.7). Many (if not most) fossil lipotyphlans are only known from jaws and teeth, leaving much of their behavioral repertoire open to interpretation. In general, as Simpson's "Insectivora" implies, all members of Lipotyphla were relatively small, insect-eating forms, with dental specializations for consumption of both soft- and hard-bodied insects being prevalent throughout the group. Where known, fossil lipotyphlan skulls indicate animals that relied predominantly on olfaction for locating insect prey. The skulls are generally low with laterally facing orbits, small braincases, and long snouts. Teeth vary from acutely cuspate with high molar trigonids and very short talonids to more flattened molars with rounded cusps and well-developed talonid basins. In some forms, posterior premolars become specialized for crushing and grinding, whereas in others anterior incisors become enlarged for specialized food acquisition. Postcrania of fossil lipotyphlans are poorly known. Those taxa that are represented by postcrania are mostly generalized quadrupeds that were probably either scansorial or terrestrial. Most moles and some shrews have highly derived anterior limbs specialized for digging and were clearly fossorial, while others, such as desmanine moles, became semi-aquatic (Figure 7.5).

Erinaceomorphs have relatively low and bunodont teeth and often exhibit reduction in molar size from anterior to posterior. Lower molars are tribosphenic, but paraconids are often reduced to shelves, especially in the posterior molars, and hypoconulids are normally small or absent. Upper molars are quadritubercular with well-developed hypocones on M1–2. Premolars tend to be low but P4/p4 may be enlarged and specialized. Erinaceomorphs were probably more omnivorous than strictly insectivorous.

Erinaceomorphs first appear in the middle Puercan (Pu2) in North America, represented by the erinaceid *Adunator*, but most of their early diversity does not occur until the middle Tiffanian, where they are represented by up to six other erinaceids. Erinaceids are present throughout the Eocene and most of the Oligocene at rather low diversity and then undergo a subsequent diversification, beginning in the late early Arikareean (Ar2), and continuing through most of the Miocene. Sespedectids make their first appearance at the Paleocene–Eocene boundary and are present at low diversity through the early and early middle Eocene until the Uintan (Ui2), where they undergo a modest diversification. Sespedectids are most diverse in the late Uintan, where they are represented by seven taxa, and the family disappears from the North American record near the end of the Orellan. The latest occurring North American fossil erinaceomorph is the erinaceid *Untermannerix*, last appearing in the late Hemphillian (Hh4).

Unlike erinaceomorphs, which tend to be rather conservative dentally, soricomorphs are typified by more derived and specialized dentitions, which indicate a more strictly insectivorous diet. Many soricomorphs have molars with high, pointed cusps on tall trigonids, with either relatively low and short talonids or well-developed and broad talonids with distinct hypoconulids. Upper molars tend to be tritubercular with small or absent hypocones, and tall and sharply pointed cusps in more primitive forms. Advanced soricomorphs have either zalambdodont (apternodontids, parapternodontids, oligoryctids, solenodontids) or dilambdodont (soricids) dentitions. Skulls, where known, tend to be low with short infraorbital canals, enlarged lacrimal canals, relatively short rostrae, and

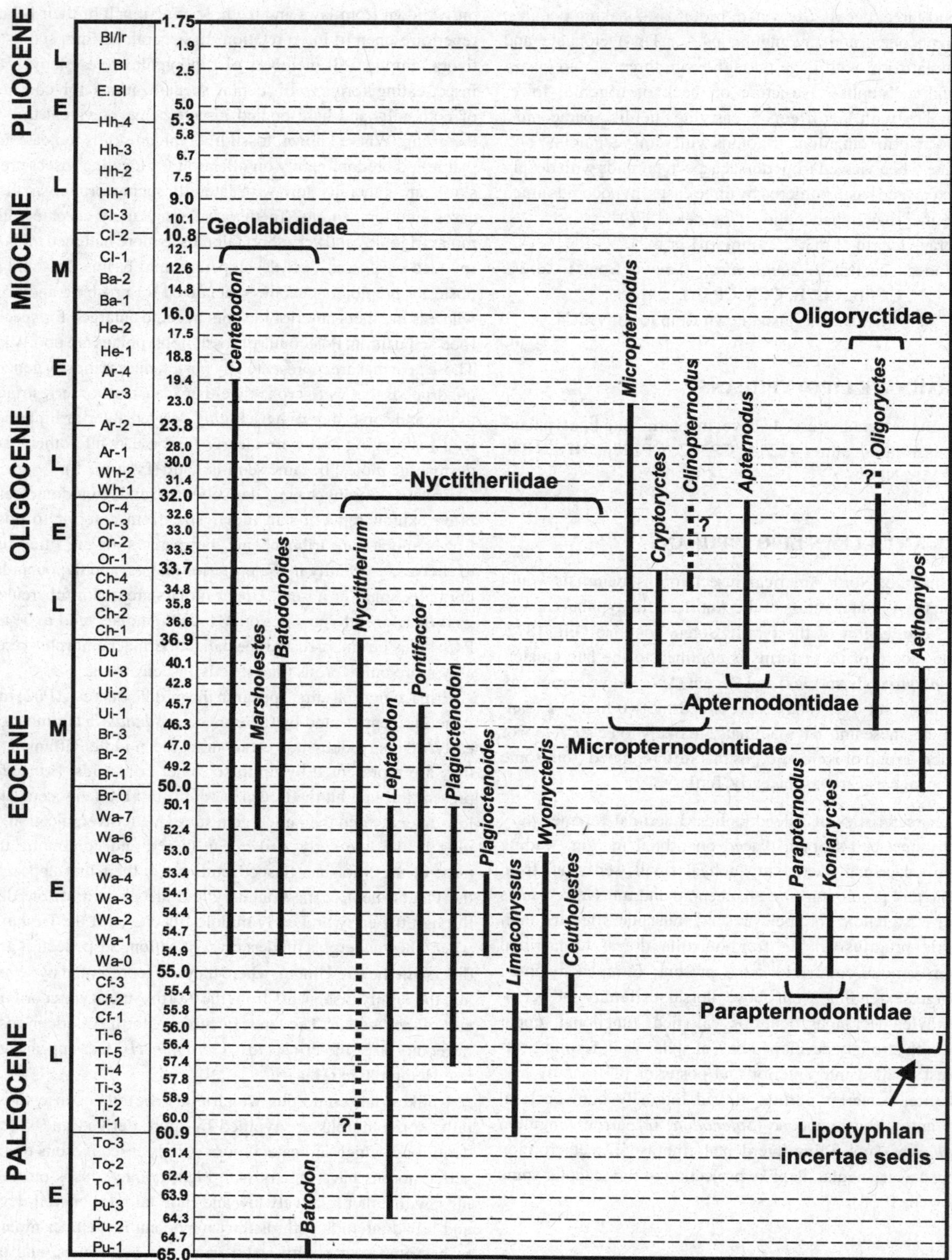

Figure 7.5. Temporal ranges of North American early Soricomorpha genera.

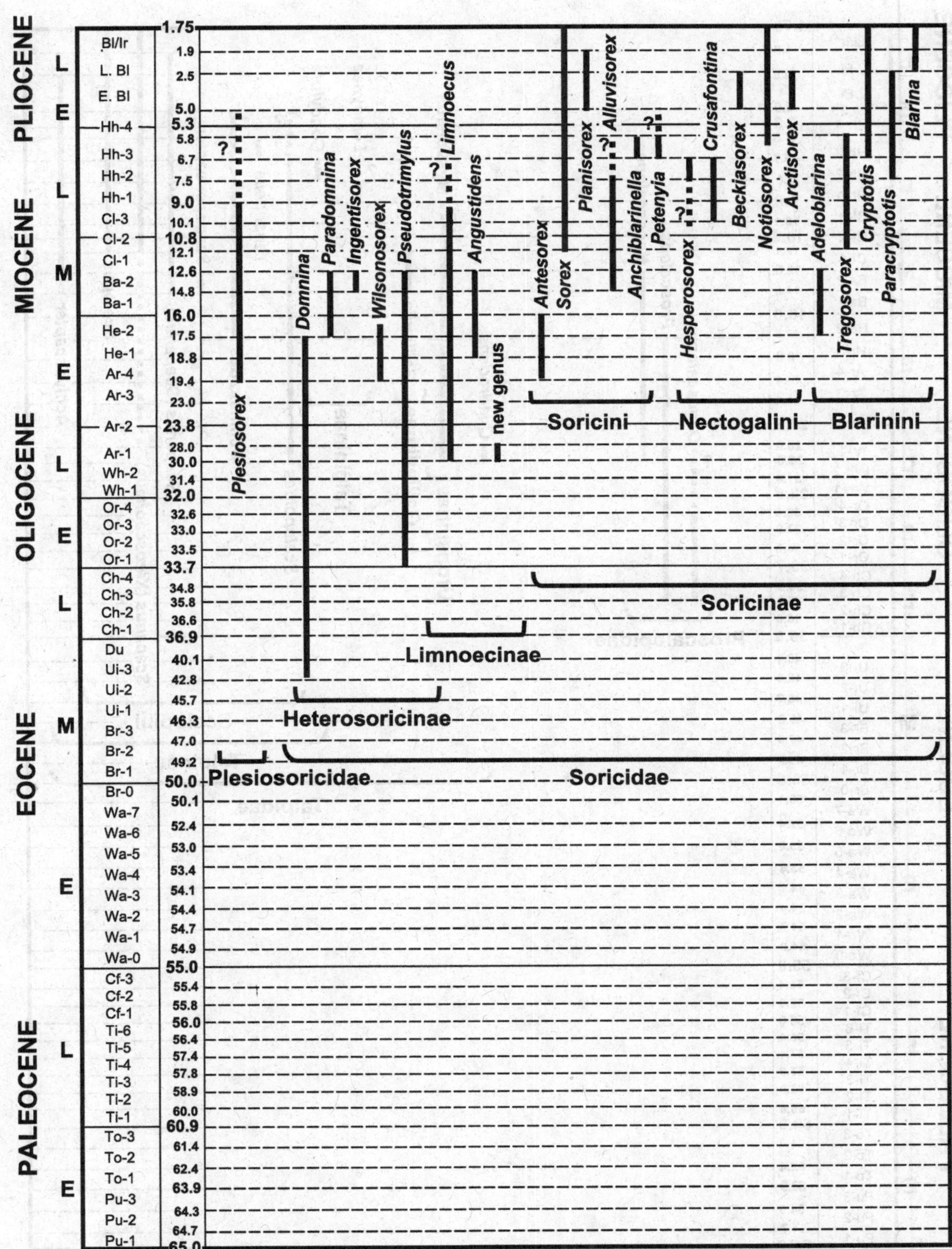

Figure 7.6. Temporal ranges of North American later Soricomorpha genera.

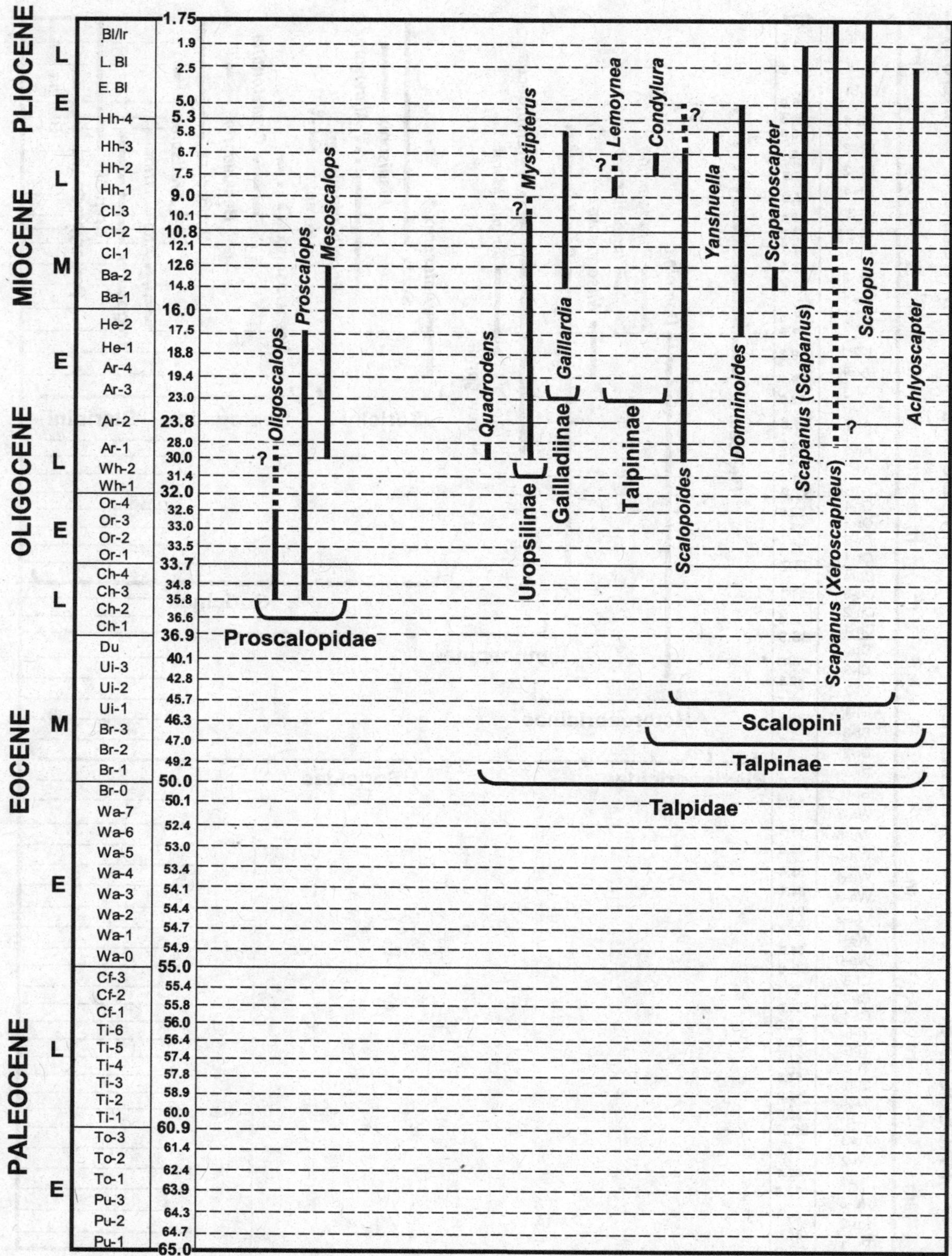

Figure 7.7. Temporal ranges of North American Talpoidea genera.

relatively globular braincases. Zygomatic arches are often absent or weakly formed, and there is no ossified auditory bulla.

There are few postcrania of fossil soricomorphs known except for some possible fore- and hindlimb elements of nyctitheriids. Matthew (1918, p. 603) referred *Nyctitherium celatus* to the "?Soricidea or Chiroptera" based on "three long slender bones, suitable in size and proportions for the shafts of chiropteran fore limb bones . . ." Gingerich (1987) questionably referred an elongate metatarsal to *Leptacodon rosei*. He concluded that *Leptacodon* was certainly digitigrade and possibly saltatorial. There are indications that at least some nyctitheriids may have been arboreal (Hooker, 2001).

The earliest known North American soricomorph is the geolabidid *Batodon*, known from the latest Cretaceous and earliest Paleocene (Pu1). All other Paleocene records of soricomorphs are represented by nyctitheriids, whereas, in the Eocene, both geolabidids and nyctitheriids are represented by modest radiations. Apternodontids are only known from the early Eocene. Nyctitheriids and geolabidids are essentially gone by the end of the Eocene (except for *Centetodon*, which survives almost to the end of the Arikareean). Small radiations of oligoryctids, apternodontids, and micropternodontids begin in the late Uintan, but none is known after the early middle Arikareean (Ar2). Soricids first appear in the late Uintan (*Domnina*), and there are a couple of other early occurrences in the late Orellan and earliest Arikareean. However, it is not until the late Arikareean that the large radiation of soricids begins (Figure 7.6). A broad diversity of soricids is evident through most of the Miocene and Pliocene, with two genera surviving to the Recent. Plesiosoricids first appear in the late Arikareean as well, and survive as a lone genus until the latest Hemphillian.

Proscalopidae are the earliest known Lipotyphla to exhibit highly specialized burrowing adaptations, and all known proscalopids were apparently burrowers to some degree. The group was initially placed in the Talpidae, but Barnosky (1981) followed earlier suggestions and elevated them to family level. Proscalopids and talpids exhibit several peculiar or unique specializations of the humerus such as the compression of the head, development of the distinct fossa for the hypertrophied ligament of the muscularis flexor digitorum profundus, and a large and distally situated teres tubercle. These features have generally been regarded by talpid specialists as independently derived and related to lateral stroke digging (Reed and Turnbull, 1965; Barnosky, 1981). Proscalopids have broad and deep skulls with ossified but uninflated auditory bullae. Talpids have long and shallow skulls with auditory bullae ranging from unossified to fully ossified, but uninflated. Talpids have dilambdodont dentitions.

The first talpoids appear in the Chadronian in North America (Figure 7.7), represented by the talpid *Mystipterus* (Ch1) and the proscalopid *Oligoscalops* (Ch2). Proscalopids are never very diverse, represented by at most two genera throughout their range. The last known occurrence of a proscalopid in North America is the late Barstovian (Ba2). Talpid diversity increases beginning in the late Barstovian and reaches its maximum in the early Hemphillian (Hh1). Relatively high talpid diversity is maintained through the Pliocene.

REFERENCES

Anderson, J. (1879) *Anatomical and Zoological Researches: Comprising an Account of the Zoological Results of the Two Expeditions to Western Yunnan in 1868 and 1875*. London: Bernard Quaritch.

Anthony, H. E. (1916). Preliminary diagnosis of an apparently new family of insectivores. *Bulletin of the American Museum of Natural History*, 35, 725–8.

Asher, R. J., McKenna, M. C., Emry, R. J., Tabrum, A. R., and Kron, D. G. (2002). Morphology and relationships of *Apternodus* and other extinct zalambdodont, placental mammals. *Bulletin of the American Museum of Natural History*, 273, 1–117.

Averianov, A. (1995). Nyctitheriid insectivores from the upper Paleocene of southern Kazakhstan (Mammalia, Lipotyphla). *Senckenbergiana Lethaea*, 75, 215–19.

Aymard, A. (1850). Concernant les restes the mammifères fossiles recueillis dans la calcaire Miocène de environs du Puy. *Annales de la Societe d'Agriculture du Puy*, 14, 104–14.

Bachman, J. (1839) Additional remarks on the genus *Lepus* with corrections of a former paper and descriptions of other species of quadrupeds found in North America. *Journal of the Academy of Natural Science, Philadelphia*, 8, 75–101.

Barnosky, A. D. (1981). A skeleton of *Mesoscalops* (Mammalia, Insectivora) from the Miocene Deep River Formation, Montana, and a review of the proscalopid moles: evolutionary, functional, and stratigraphic relationships. *Journal of Vertebrate Paleontology*, 1, 285–339.

(1982). A new species of *Proscalops* (Mammalia, Insectivora) from the Arikareean Deep River Formation, Meagher County, Montana. *Journal of Paleontology*, 56, 103–11.

Blainville, H., de (1840). *Osteographie des Mammifères, 1. Primates-Secundates. Libración 6-H. Osteographie des Mammifères Insectivores*. Paris: J. B. Baillière.

Bloch, J. I., Rose, K. D., and Gingerich, P. D. (1998). New species of *Batodonoides* (Lipotyphla, Geolabididae) from the early Eocene of Wyoming: smallest known mammal? *Journal of Mammalogy*, 79, 804–27.

Bowdich, T. E. (1821). *An Analysis of the Natural Classification of Mammalia for the Use of Students and Travelers*. Paris: J. Smith.

Bown, T. M. (1979). Geology and mammalian paleontology of the Sand Creek facies, lower Willwood Formation (lower Eocene), Washakie County, Wyoming. *Geological Survey of Wyoming Memoir*, 2, 1–151.

(1980). The fossil Insectivora of Lemoyne Quarry (Ash Hollow Formation, Hemphillian, Keith County, Nebraska). *Transactions of the Nebraska Academy of Sciences*, 8, 99–122.

Bown, T. M. and Schankler, D. (1982). A review of the Proteutheria and Insectivora of the Willwood Formation (lower Eocene), Bighorn Basin, Wyoming. *United States Geological Survey Bulletin*, 1523, 1–79.

Brèthes, J. (1904). Insectos de Tucumán. *Anales del Museo Nacional de Buenos Aires, Series* 3, 4, 329–47.

Butler, P. M. (1948). On the evolution of the skull and teeth in the Erinaceidae, with special reference to the material in the British Museum. *Proceedings of the Zoological Society of London*, 118, 446–500.

(1972). The problem of insectivore classification. In *Studies in Vertebrate Evolution*, ed. K. A. Joysey and T. S. Kemp, pp. 253–65. New York: Winchester.

(1988). Phylogeny of the insectivores. In *The Phylogeny of Tetrapods,* Vol. 2: *Mammals*, ed. M. J. Benton, pp. 117–41. Oxford: Clarendon Press.

Campbell, B. (1939). The shoulder anatomy of the moles. A study in phylogeny and adaptation. *American Journal of Anatomy*, 64, 1–39.

Carroll, R. L. (1988). *Vertebrate Paleontology and Evolution*. New York: W. H. Freeman.

Clark, J. (1936). Diagnosis of *Metacodon* and description of *M. magnus*. [In *The Mammalian Fauna of the White River Oligocene. Part 1: Insectivora and Carnivora*, ed. W. B. Scott and G. L. Jepsen] *Transactions of the American Philosophical Society*, 28, p. 22.

(1937). The stratigraphy and paleontology of the Chadron Formation in the Big Badlands of South Dakota. *Annals of the Carnegie Museum*, 25, 261–350.

Clemens, W. A., Jr. (1973). Fossil mammals of the type Lance Formation, Wyoming, III. Eutheria and summary. *University of California Publications in Geological Sciences*, 94, 1–102.

Cockerell, T. D. A. (1940). A dragon-fly from the Eocene of Colorado; Odonata, Agrionidae. *Entomological News*, 51, 103–5.

Cope, E. D. (1873). *Synopsis of New Vertebrata from the Tertiary of Colorado, Obtained during the Summer of 1873*. Washington, DC: Government Printing Office.

(1874). Report on the vertebrate paleontology of Colorado. *7th Annual Report of the United States Geological and Geographical Survey of the Territories*, pp. 427–533. Washington, DC: US Government Printing Office.

(1875). *Systematic Catalogue of Vertebrata of the Eocene of New Mexico, Collected in 1874. Report to the Engineer, Department, United States Army*, in charge, Lt. George M. Wheeler, pp. 5–37. Washington, DC: US Government Printing Office.

(1879). The relations of the horizons of extinct Vertebrata of Europe and North America. *Bulletin of the United States Geological and Geographical Survey of the Territories* (1880), 5, 33–54.

(1884). The Vertebrata of the Tertiary formations of the West. Book 1. *United States Geological Survey of the Territories Report*, 1–1002.

Coues, E. (1877). Precursory notes on American insectivorous mammals, with descriptions of new species. *Bulletin of the United States Geological and Geographical Survey of the Territories*, 3, 631–53.

Crochet, J.-Y. (1974). Les Insectivores des phosphorites du Quercy. *Palaeovertebrata*, 6, 109–59.

Dalquest, W. W. (1972). A new genus and species of shrew from the upper Pliocene of Texas. *Journal of Mammalogy*, 53, 570–3.

(1975). Vertebrate fossils from the Blanco local fauna of Texas. *Occasional Papers of the Museum of Texas Technical University*, 30, 1–52.

Dashzeveg, D., Hartenberger, J., Martin, T., and Legendre, S. (1998). A peculiar minute glires (Mammalia) from the early Eocene of Mongolia. [In *Dawn of the Age of Mammals in Asia*, ed. K. C. Beard and M. R. Dawson.] *Bulletin of Carnegie Museum of Natural History*, 34, 194–209.

Dobson, G. E. (1883). *A Monograph of the Insectivora, Systematic and Anatomical*, Pt. 2. London: John Van Voorst.

Dorr, J. A., Jr. (1952). Early Cenozoic stratigraphy and vertebrate paleontology of the Hoback Basin, Wyoming. *Bulletin of the Geological Society of America*, 63, 59–93.

(1958). Early Cenozoic vertebrate paleontology, sedimentation, and orogeny in central western Wyoming. *Bulletin of the Geological Society of America*, 69, 1217–43.

Douglass, E. (1903). New vertebrates from the Montana Territory. *Annals of the Carnegie Museum*, 2, 145–99.

Estravis, C. (1996). *Leptacodon nascimentoi* n. sp., un nouveau Nyctitheriidae (Mammalia, Lipotyphla) de l'Eocene inferieur de Silveirinha (Baixo Mondego, Portugal). *Palaeovertebrata*, 25, 279–86.

Filhol, H. (1888). Sur un nouveau genre d'insectivore. *Bulletin de la Société Philomathique de Paris*, 12, 24–5.

Fischer von Waldheim, G. (1817). Adversaria zoological. *Memoires, Societe imperiale des sciences Naturelles, Moscow*, 5, 368–428.

Fox, R. C. (1983). Evolutionary implications of tooth replacement in the Paleocene mammal *Pararyctes*. *Canadian Journal of Earth Sciences*, 20, 19–22.

(1984). *Paranyctoides maleficus* (new species), an early eutherian mammal from the Cretaceous of Alberta. *Carnegie Museum of Natural History Special Publication*, 9, 9–20.

(1988). Late Cretaceous and Paleocene mammal localities of southern Alberta. *Occasional Papers, Tyrrell Museum of Palaeontology*, 6, 1–38.

Galbreath, E. C. (1953). A contribution to the Tertiary geology and paleontology of northeastern Colorado. *University of Kansas Paleontological Contributions*, 4, 1–120.

Ganglbaur, L. (1899). *Die Käfer von Mitteleuropa*, Vol. 3, *Coleoptera: Cucujidae*. Wien.

Gapper, Dr. (1830). Observations on the quadrupeds found in the District of Upper Canada extending between York and Lake Semcoe, with the view of illustrating their geographical distribution, as well as of describing some species hitherto unnoticed. *Zoological Journal*, 5, 201–7.

Gawne, C. E. (1968). The genus *Proterix* (Insectivora, Erinaceidae) of the upper Oligocene of North America. *American Museum Novitates*, 2315, 1–26.

Gazin, C. L. (1933). A new shrew from the upper Pliocene of Idaho. *Journal of Mammalogy*, 14, 142–4.

Geoffroy Saint-Hilaire, E. F. (1803). *Catalogue des Mammifères du Museum d'Histoire Naturelle, Paris*.

Gibert, J. (1975). New insectivores from the Miocene of Spain; I and II. *Proceedings of the Koninklijke Nederlandse Akadamie van Wetenschappen, Series B*, 78, 108–33.

Gill, T. (1872). Arrangement of the families of mammals with analytical tables. *Smithsonian Miscellaneous Collections*, 11, 1–98.

Gingerich, P. D. (1983). New Adapisoricidae, Pentacodontidae, and Hyopsodontidae (Mammalia, Insectivora and Condylarthra) from the late Paleocene of Wyoming and Colorado. *Contributions from the Museum of Paleontology, University of Michigan*, 26, 227–55.

(1987). Early Eocene bats (Mammalia, Chiroptera) and other vertebrates in freshwater limestones of the Willwood Formation, Clark's Fork Basin, Wyoming. *Contributions from the Museum of Paleontology, University of Michigan*, 27, 275–320.

Gingerich, P. D., Houde, P., and Krause, D. W. (1983). A new earliest Tiffanian (Late Paleocene) mammalian fauna from Bangtail Plateau, western Crazy Mountain Basin, Montana. *Journal of Paleontology*, 75, 957–70.

Gray, J. E. (1837). Revison of the genus *Sorex* Linnaeus. *Proceedings of the Zoological Society of London*, 5, 123–6.

Green, M. (1956). The lower Pliocene Ogallala–Wolf Creek vertebrate fauna, South Dakota. *Journal of Paleontology*, 30, 146–59.

(1977). A new species of *Plesiosorex* (Mammalia, Insectivora) from the Miocene of South Dakota. *Neues Jahrbuch für Geologie und Paleontologie*, 154, 189–98.

Gregory, W. K. (1910). The orders of mammals. *Bulletin of the American Museum of Natural History*, 27, 1–524.

Gureev, A. A. (1971). *Zemleroiki (Soricidae) Fauny Mira*. [*Shrews (Soricidae) of the World Fauna*.] Leningrad: "Nauka" Press (Leningrad Zoological Institute).

(1979). Mammalia, Insectivora: hedgehogs, moles and shrews (Erinaceidae, Talpidae, Soricidae). *Fauna SSSR*, 120, 1–502.

Haeckel, E. (1866). *Generelle Morphologie der Organismen. Allgemeine Grundzüge der organischen Formen-Wissenschaft, mechanisch begründet durch die von Charles Darwin reformierte Deszendenz-Theorie. II. Allgemeine Entwicklungsgeschichte der Organismen*. Berlin: G. Reimer.

Hall, E. R. (1929). A second new genus of hedgehog from the Pliocene of Nevada. *University of California Publications, Bulletin of the Department of Geological Sciences*, 18, 227–31.

(1930). A new genus of bat from the later Tertiary of Nevada. *University of California Publications, Bulletin of the Department of Geological Sciences*, 19, 319–20.

Hand, S., Novacek, M., Godthelp, H., and Archer, M. (1994). First Eocene bat from Australia. *Journal of Vertebrate Paleontology*, 14, 375–81.

Heller, F. (1935). *Amphilemur eocaenicus* n. g. et n. sp., ein primitiver Primate aus dem Mitteleozän des Geiseltales bei Halle a. S. *Nova Acta Academiae Caesareae Leopoldino*, (new series), 2, 293–300.

Hibbard, C. W. (1941). Mammals of the Rexroad fauna from the upper Pliocene of southwestern Kansas. *Transactions of the Kansas Academy of Science*, 44, 265–313.

(1950). Mammals of the Rexroad formation from Fox Canyon, Meade County, Kansas. *Contributions from the Museum of Paleontology, University of Michigan*, 8, 113–92.

(1953). The insectivores of the Rexroad fauna, upper Pliocene of Kansas. *Journal of Paleontology*, 27, 21–32.

(1956). Vertebrate fossils from the Meade formation of southwestern Kansas. *Michigan Academy of Science Papers*, 41, 145–203.

(1957). Notes on late Cenozoic shrews. *Kansas Academy of Science Transactions*, 60, 327–36.

Hibbard, C. W. and Jammot, D. (1971). The shrews of the Wakeeney local fauna lower Pliocene of Trego County, Kansas. *Contributions from the Museum of Paleontology, University of Michigan*, 23, 377–80.

Hill, W. C. O. (1953). *Primates: Comparative Anatomy and Taxonomy. I: Strepsirhini*. Edinburgh: Edinburgh University Press.

Holroyd, P. A., Bown, T. M., and Schankler, D. M. (2004). *Auroralestes*, gen. nov., a replacement name for *Eolestes* Bown and Schankler, 1982, a preoccupied name. *Journal of Vertebrate Paleontology*, 24, 979.

Hooker, J. J. (1996). A primitive emballonurid bat (Chiroptera, Mammalia) from the earliest Eocene of England. *Palaeovertebrata*, 25, 287–300.

(2001). Tarsals of the extinct insectivoran family Nyctitheriidae (Mammalia): evidence for archontan relationships. *Zoological Journal of the Linnaean Society*, 132, 501–29.

Hough, J. (1956). A new insectivore from the Oligocene of the Wind River Basin, Wyoming, with notes on taxonomy of the Oligocene Tenrecoidea. *Journal of Paleontology*, 30, 531–41.

Hutchison, J. H. (1966). Notes on some upper Miocene shrews of Oregon. *Bulletin of the Museum of Natural History, University of Oregon*, 2, 1–23.

(1968). Fossil Talpidae (Insectivora, Mammalia) from the later Tertiary of Oregon. *Bulletin of the Museum of Natural History, University of Oregon*, 11, 1–117.

(1974). Notes on type specimens of European Miocene Talpidae and a tentative classification of Old World Tertiary Talpidae (Insectivora: Mammalia). *Geobios*, 7, 211–56.

Hutchison, J. H. and Harington, C. R. (2002). A peculiar new fossil shrew (Lipotyphla, Soricidae) from the High Arctic of Canada. *Canadian Journal of Earth Science*, 39, 439–43.

Illiger, C. (1811). *Prodromus Systematics Mammalian et Avium Additis Terminis Zoographicus Utriusque Classis, Eorumque Versione Germanica*. Berlin: C. Salfeld.

Jepsen, G. L. (1930). Stratigraphy and paleontology of the Paleocene of northeastern Park County, Wyoming. *Proceedings of the American Philosophical Society*, 69, 463–528.

Koerner, H. E. (1940). The geology and vertebrate paleontology of the Fort Logan and Deep River formations of Montana, I. New vertebrates. *American Journal of Science*, 238, 837–62.

Kormos, T. (1934). Neue Insektenfresser, Fledermause, und Nager aus dem Oberpliozan der Villanyer Gegend. *Földtani Közlöny*, 64, 296–321.

Krishtalka, L. (1976a). Early Tertiary Adapisoricidae and Erinaceidae (Mammalia, Insectivora) of North America. *Bulletin of the Carnegie Museum of Natural History*, 1, 1–40.

(1976b). North American Nyctitheriidae (Mammalia, Insectivora). *Annals of the Carnegie Museum*, 46, 7–28.

Krishtalka, L. and Setoguchi, T. (1977). Paleontology and geology of the Badwater Creek area, central Wyoming, 13. The late Eocene Insectivora and Dermoptera. *Annals of the Carnegie Museum*, 46, 71–99.

Lillegraven, J. A., McKenna, M. C., and Krishtalka, L. (1981). Evolutionary relationships of middle Eocene and younger species of *Centetodon* (Mammalia, Insectivora, Geolabididae) with a description of the dentition of *Ankylodon* (Adapisoricidae). *University of Wyoming Publications*, 45, 1–115.

Linnaeus, C. (1758). *Systema Naturae per Regna Tria Naturae, Secundum Classes, Ordines, Genera, Species, cum Characteribus, Differentiis, Synonymis, Locis.* Vol. 1: *Regnum Animale. Editio Decima, Reformata*. Stockholm: Laurentii Salvia.

Loomis, F. B. (1906). Wasatch and Wind River primates. *American Journal of Science*, 21, 277–85.

Lydekker, R. (1885). *Catalogue of the Fossil Mammalia in the British Museum*. London: British Museum Publications.

Maas, M. C. (1985). Taphonomy of a Late Eocene microvertebrate locality, Wind River Basin, Wyoming (USA). *Palaeogeography, Palaeoclimatology, and Palaeoecology*, 52, 123–42.

Macdonald, J. R. (1963). The Miocene faunas from the Wounded Knee area of western South Dakota. *Bulletin of the American Museum of Natural History*, 125, 139–238.

(1965). *Geolabis wolffi*, a new fossil insectivore from the late Oligocene of South Dakota. *Los Angeles County Museum, Contributions in Science*, 88, 1–6.

(1970). Review of the Miocene Wounded Knee faunas of southwestern South Dakota. *Los Angeles County Museum Bulletin*, 8, 1–82.

MacPhee, R. D. E. (1981). Auditory regions of primates and eutherian insectivores. Morphology, ontogeny, and character analysis. *Contributions to Primatology*, 18, 1–282.

Marsh, O. C. (1872). Preliminary description of new Tertiary mammals. *American Journal of Science*, 38, 81–92.

(1892). Discovery of Cretaceous Mammalia. Part III. *American Journal of Science*, 3, 249–62.

Martin, J. E. (1978). A new and unusual shrew (Soricidae) from the Miocene of Colorado and South Dakota. *Journal of Paleontology*, 52, 636–41.

Martin, J. E. and Green, M. F. (1984). Insectivora, Sciuridae, and Cricetidae from the early Miocene Rosebud Formation in South Dakota. *Carnegie Museum of Natural History, Special Publication* 9, 28–40.

Matthew, W. D. (1901). Fossil mammals of the Tertiary of northeastern Colorado. *American Museum of Natural History Memoirs*, 1, 355–447.

(1903a) A fossil hedgehog from the American Oligocene. *Bulletin of the American Museum of Natural History*, 19, 227–9.

(1903b). The fauna of the *Titanotherium* beds at Pipestone Springs, Montana. *Bulletin of the American Museum of Natural History*, 19, 197–226.

(1907). A lower Miocene fauna from South Dakota. *Bulletin of the American Museum of Natural History*, 23, 169–219.

(1909). The Carnivora and Insectivora of the Bridger Basin, middle Eocene. *American Museum of Natural History Memoirs*, 9, 289–567.

(1910). On the skull of *Apternodus* and the skeleton of a new artiodactyl. *Bulletin of the American Museum of Natural History*, 28, 33–42.

(1918). Part V. Insectivora (continued), Glires, Edentata. *Bulletin of the American Museum of Natural History*, 38, 565–657.

(1929). A new and remarkable hedgehog from the later Tertiary of Nevada. *University of California Publications in Geological Sciences*, 18, 93–102.

(1932). New fossil mammals from the Snake Creek quarries. *American Museum Novitates*, 540, 1–8.

Matthew, W. D. and Granger, W. (1921). New genera of Paleocene mammals. *American Museum Novitates*, 13, 1–7.

Matthew, W. D. and Mook, C. C. (1933). New fossil mammals from the Deep River beds of Montana. *American Museum Novitates*, 601, 1–7.

McDowell, S. B. (1958). The greater Antillean insectivores. *Bulletin of the American Museum of Natural History*, 115, 113–214.

McGrew, P. O. (1959). The geology and paleontology of the Elk Mountain and Tabernacle Butte Area, Wyoming. *Bulletin of the American Museum of Natural History*, 117, 117–76.

McKenna, M. C. (1960a). The Geolabidinae, a new subfamily of Early Cenozoic erinaceoid insectivores. *University of California Publications in Geological Sciences*, 37, 131–64.

(1960b). Fossil Mammalia from the early Wasatchian Four Mile Fauna, Eocene of northwest Colorado. *University of California Publications in Geological Sciences*, 37, 1–130.

(1968). *Leptacodon*, an American Paleocene nyctithere (Mammalia, Insectivora). *American Museum Novitates*, 2317, 1–12.

(1975). Toward a phylogenetic classification of the Mammalia. In *Phylogeny of the Primates*, ed. W. P. Luckett and F. S. Szalay, pp. 21–46. New York: Plenum Press.

McKenna, M. C. and Bell, S. K. (1997). *Classification of Mammals Above the Species Level*. New York: Columbia University Press.

McKenna, M. C. and Haase, F. (1992). *Marsholestes*, a new name for the Eocene insectivoran *Myolestes* Matthew, not *Myolestes* Brèthes, 1904. *Journal of Vertebrate Paleontology*, 12, 256.

McKenna, M. C. and Simpson, G. G. (1959). A new insectivore from the middle Eocene of Tabernacle Butte, Wyoming. *American Museum Novitates*, 1952, 1–12.

Meng, J., Zhai, R., and Wyss, A. R. (1998). The late Paleocene Bayun Alan Fauna of Inner Mongolia, China. [In *Dawn of the Age of Mammals in Asia*, ed. K. C. Beard and M. R. Dawson.] *Bulletin of Carnegie Museum of Natural History*, 34, 148–85.

Miller, G. S., Jr. (1912). List of North American land mammals in the United States National Museum. *Bulletin of the United States National Museum*, 79, 1–455.

Mivart, St. George (1871). On *Hemicentetes*, a new genus of Insectivora, with some additional remarks on the osteology of that order. *Proceedings of the Zoological Society of London*, 1871, 58–79.

Nessov, L. A. (1987). Results of search and investigation of Cretaceous and early Paleogene mammals on the territory of the USSR. *Ezhegodnik Vsesoyuznogo Paleontologicheskogo Obshchestva*, 30, 199–218.

Novacek, M. J. (1976). Insectivora and Proteutheria of the later Eocene (Uintan) of San Diego County, California. *Contributions in Science, Los Angeles County Museum of Natural History*, 283, 1–52.

(1977). A review of Paleocene and Eocene Leptictidae (Eutheria: Mammalia) from North America. *PaleoBios*, 24, 1–42.

(1980). *Crypholestes*, a new name for the late Eocene insectivore *Cryptolestes* (Novacek, 1976). *Journal of Paleontology*, 54,1135.

(1982). *Diacodon alticuspus*, an erinaceomorph insectivore from the early Eocene of northern New Mexico. *Contributions to Geology, University of Wyoming*, 20, 135–49.

(1985). The Sespedectinae, a new subfamily of hedgehog-like insectivores. *American Museum Novitates*, 2822, 1–24.

(1986). The skull of leptictid insectivorans and the higher-level classification of eutherian mammals. *Bulletin of the American Museum of Natural History*, 183, 1–112.

Novacek, M. J., Bown, T. M., and Schankler, D. (1985). On the classification of the early Tertiary Erinaceomorpha (Insectivora, Mammalia). *American Museum Novitates*, 2813, 1–22.

Patterson, B. and McGrew, P. O. (1937). A soricid and two erinaceids from the White River Oligocene. *Field Museum of Natural History Geological Series*, 6, 245–72.

Peterson, O. A. (1934). List of species and description of new material from the Duchesne River Oligocene, Uinta Basin, Utah. *Annals of the Carnegie Museum*, 23, 373–89.

Pomel, A. (1848). Etudes sur less carnassiers insectivores. Second Partie, classification des insectivores. *Bibliotheque Universelle de Geneve, Archives des Sciences Physiques at Naturelles*, 9, 244–51.

Quinet, G. E. and Misonne, X. (1965). Les insectivores zalambdodontes de l'Oligocene inferieur Belge. *Bulletin de l'Institut royal des Sciences naturelles de Belgique*, 41, 1–15.

Reed, C. A. (1951). Locomotion and appendicular anatomy in three soricoid insectivores. *The American Midland Naturalist*, 45, 513–671.

(1954). Some fossorial mammals from the Tertiary of western North America. *Journal of Paleontology*, 28, 102–11.

(1956). A new species of the fossil mammal *Arctoryctes* from the Oligocene of Colorado. *Fieldiana (Geology)*, 10, 305–11.

Reed, C. A. and Turnbull, W. D. (1965). The mammalian genera *Arctoryctes* and *Cryptoryctes* from the Oligocene and Miocene of North America. *Fieldiana (Geology)*, 15, 95–170.

Reed, K. M. (1960). Insectivora of the middle Miocene Split Rock local fauna, Wyoming. *Breviora, Museum of Comparative Zoology*, 116, 1–11.

(1961). The Proscalopinae, a new subfamily of talpid insectivores. *Bulletin of the Museum of Comparative Zoology*, 125, 471–94.

(1962). Two new species of fossil talpid insectivores. *Breviora, Museum of Comparative Zoology*, 168, 1–6.

Regan, C. T. (1920). The classification of the fishes of the family Cichlidae. I. The Tanganyika genera. *Annals and Magazine of Natural History*, 9, 33–53.

Repenning, C. A. (1967). Subfamilies and genera of the Soricidae. *United States Geological Survey Professional Paper*, 565, 1–74.

Rich, T. H. V. (1981). Origin and history of the Erinaceinae and Brachyericinae (Mammalia, Insectivora) in North America. *Bulletin of the American Museum of Natural History*, 171, 1–116.

Rich, T. H. V. and Rasmussen, D. L. (1973). New North American erinaceine hedgehogs (Mammalia: Insectivora). *Occasional Papers of the Museum of Natural History, the University of Kansas*, 21, 1–54.

Rich, T. H. V. and Rich, P. V. (1971). *Brachyerix*, a Miocene hedgehog from western North America, with a description of the tympanic regions of *Parechinus* and *Podogymnura*. *American Museum Novitates*, 2477, 1–58.

Robinson, P. (1966). Fossil Mammalia of the Huerfano Formation, Eocene of Colorado. *Bulletin of the Yale Peabody Museum*, 21, 1–85.

(1968). Nyctitheriidae (Mammalia, Insectivora) from the Bridger Formation of Wyoming. *Contributions to Geology, University of Wyoming*, 7, 129–38.

Robinson, P. and Kron, D. G. (1998). *Koniaryctes*, a new genus of apternodontid insectivore from Lower Eocene rocks of the Powder River Basin, Wyoming. *Contributions to Geology, University of Wyoming*, 32, 187–90.

Romer, A. S. (1966). *Vertebrate Paleontology*. Chicago, IL: University of Chicago Press.

Rose, K. D. (1981). The Clarkforkian Land-Mammal Age and mammalian faunal composition across the Paleocene–Eocene boundary. *University of Michigan Papers on Paleontology*, 26, 1–197.

Rose, K. D. and Gingerich, P. D. (1987). A new insectivore from the Clarkforkian (earliest Eocene) of Wyoming. *Journal of Mammalogy*, 68, 17–27.

Russell, D. A. (1960). A review of the Oligocene insectivore *Micropternodus borealis*. *Journal of Paleontology*, 34, 940–9.

(1976). A new species of talpid insectivore from the Miocene of Saskatchewan. *Canadian Journal of Earth Sciences*, 13, 1602–7.

Russell, D. E. (1964). Les mammifères Paleocenes d'Europe. *Memoir Museum National d'Histoire Naturelle*, 13, 1–324.

Russell, D. E. and Dashzeveg, D. (1986). Early Eocene insectivores (Mammalia) from the People's Republic of Mongolia. *Palaeontology*, 29, 269–91.

Say, T. (1823). *Account of an Expedition from Pittsburgh to the Rocky Mountains Performed in the Years 1819 and 1820*. Philadelphia, PA: H. C. Carey and I. Lea.

Schlaikjer, E. M. (1933). Contributions to the stratigraphy and paleontology of the Goshen Hole area, Wyoming, 1. A detailed study of the structure and relationships of a new zalambdodont insectivore from the middle Oligocene. *Bulletin of the Museum of Comparative Zoology*, 76, 1–27.

Schlosser, M. (1887–1890). Die Affen, Lemuren, Insectivoren, Marsupialier, Creodonten und Carnivoren des Europaischen Tertiars. *Beiträge Palaeontologie Oesterreich-Ungarns*, 6–8, 1–492.

(1924). Tertiary vertebrates from Mongolia. *Paleontologia Sinica Series C*, 1, 1–119.

Scott, W. B. and Jepsen, G. L. (1936). The mammalian fauna of the White River Oligocene. Part I. Insectivora and Carnivora. *Transactions of the American Philosophical Society*, 28, 1–153.

Seeman, I. (1938). Die Insektenfresser, Fledermäuse, und Nager aus der Obermiokänen Braunkohle von Viehhausen bei Regensburg. *Palaeontographica*, 89, 1–55.

Shotwell, J. A. (1956). Hemphillian mammalian assemblage from northeastern Oregon. *Bulletin of the Geological Society of America*, 67, 717–38.

Sigé, B. (1976). Insectivores primitifs de l'Eocène supérieur et Oligocène inférieur d'Europe occidentale. Nyctithériidés. *Mémoirs du Muséum National D'Histoire Naturelle*, 34, 1–140.

(1978). La poche à phosphate de Sainte-Néboule (Lot) et sa faune de Vertébrés du Ludien supérieur. 8. Insectivores et chiroptères. *Palaeovertebrata*, 8, 243–68.

(1988). Le Gisement du Bretou (Phosphorites du Quercy, Tarn-et-Garonne, France) et sa faune de vertebrates de l'Eocene superieur. IV. Insectivores et chiropteres. *Palaeontographica Abteilung A*, 205, 69–102.

(1997). Les mammifères insectivores des nouvelles collections de Sossis et sites associés (Eocène supérieur, Espagne). *Geobios*, 30, 91–113.

Simpson, G. G. (1928). A new mammalian fauna from the Fort Union of southern Montana. *American Museum Novitates*, 297, 1–15.

(1935). New Paleocene mammals from the Fort Union of Montana. *Proceedings of the United States National Museum*, 83, 221–44.

(1937). Notes on the Clark Fork, upper Paleocene, fauna. *American Museum Novitates*, 954, 1–24.

(1945). The principles of classification and a classification of mammals. *Bulletin of the American Museum of Natural History*, 85, 1–350.

Skinner, M. F. and Hibbard, C. W. (1972). Early Pleistocene preglacial and glacial rocks and faunas of north-central Nebraska. *Bulletin of the American Museum of Natural History*, 148, 1–148.

Smith, T. (1995). Présence du genre *Wyonycteris* (Mammalia, Lipotyphla) à la limite Paléocène-Eocène en Europe. *Comptes Rendus de l'Académie des Sciences, Paris*, 321, 923–30.

(1996). *Leptacodon dormaalensis* (Mammalia, Lipotyphla), un nyctithère primitif de la transition Paléocène-Eocène de Belgique. *Belgian Journal of Zoology*, 126, 153–67.

(1997). Les insectivores (Mammalia, Lipotyphla) de la transition Paléocène-Eocène de Dormaal (MP 7, Belgique): implications biochronologiques et Paleogeographiques. Actes du Congrès BiochroM'97. *Memoire Travail Ecole Prat Hautes Etudes, Institut Montpellier*, 21, 687–96.

Smith, T., Bloch, J. L., Strait, S. G., and Gingerich, P. D. (2002). New species of *Macrocranion* (Mammalia, Lipotyphla) from the earliest Eocene of North America and its biogeographic implications. *Contributions from the Museum of Paleontology, University of Michigan*, 30, 373–84.

Springer, M. S., Stanhope, M. J., Madsen, O., and de Jong, W. W. (2004). Molecules consolidate the placental mammal tree. *Trends in Ecology and Evolution*, 19, 430–8.

Stehlin, H. G. (1940). Zur Stammesgeschichte der Soriciden. *Eclogae Geologicae Helvetiae*, 33, 298–306.

Stirton, R. A. (1930). A new genus of Soricidae from the Barstow Miocene of California. *University of California Publications in Geological Sciences*, 19, 217–28.

Stirton, R. A. and Rensberger, J. M. (1964). Occurrence of the insectivore genus *Micropternodus* in the John Day Formation of central Oregon. *Bulletin of the Southern California Academy of Sciences*, 63, 57–80.

Stock, C. (1935). Insectivora from the Sespe uppermost Eocene, California. *Proceedings of the National Academy of Sciences*, 21, 214–19.

Storch, G. and Qui, Z. (1983). The Neogene mammalian faunas of Ertemte an Harr Obo in Inner Mongolia (Nei Mongol), China. 2. Moles-Insectivora: Talpidae. *Senckenbergiana Lethaea*, 64, 89–127.

Tate, G. H. H. (1934). New generic names for two South American marsupials. *Journal of Mammalogy*, 15, 154.

Tedford, R. H. (1961). Clarendonian Insectivora from the Ricardo Formation, Kern County, California. *Bulletin of the Southern California Academy of Sciences*, 60, 57–76.

Ting, S. (1998). Paleocene and early Eocene land-mammal ages of Asia. [In *Dawn of the Age of Mammals in Asia*, ed. K. C. Beard and M. R. Dawson.] *Bulletin of Carnegie Museum of Natural History*, 34, 124–47.

Tong, Y. and Wang, J. (1998). A preliminary report on the early Eocene mammals of the Wutu Fauna, Shandong Province, China. [*Dawn of the Age of Mammals.*] *Bulletin of Carnegie Museum of Natural History*, 34, 186–93.

Trouessart, E. L. (1879). Revisión des musaraignes (Soricidae) d'Europe et notes sur les Insectivores en general avec l'indication des especes qui se trouvent en France. *Bulletin de Societe d'etudes Scientifiques d'Angers*, 179–202.

van Dam, J. A. (2004). Anourosoricini (Mammalia: Soricidae) from the Mediterranean region: a pre-Quaternary example of recurrent climate-controlled north-south range shifting. *Journal of Paleontology*, 78, 741–64.

Van Valen, L. (1965). A middle Paleocene primate. *Nature*, 207, 435–6.

(1966). Deltatheridia, a new order of mammals. *Bulletin of the American Museum of Natural History*, 132, 1–126.

(1967). New Paleocene insectivores and insectivore classification. *Bulletin of the American Museum of Natural History*, 135, 217–84.

(1978). The beginning of the Age of Mammals. *Evolutionary Theory*, 4, 45–80.

Viret, J. (1940). Etude sur quelques Erinaceides fossils. *Travail de Laboratoire Geologique d'Universite de Lyon*, 39, 33–65.

Viret, J. and Zapfe, H. (1951). Sur quelques Soricides Miocenes. *Ecologae Geologicae Helvetiae*, 44, 411–26.

Walsh, S. L. (1998). Notes on the anterior dentition and skull of *Proterixoides* (Mammalia: Insectivora: Dormaaliidae), and a new dormaaliid genus from the early Uintan (Middle Eocene) of Southern California. *Proceedings of the San Diego Society of Natural History*, 34, 1–26.

Webb, S. D. (1961). The first American record of *Lantanotherium* Filhol. *Journal of Paleontology*, 35, 1085–7.

Weitzel, K. (1949). Neue Wirbeltiere (Rodentia, Insectivora, Testudinata) aus den Mitteleozan von Messel bei Darmstadt. *Abhandlung Senckenbergische Naturforschende Gesellschaft*, 480, 1–24.

West, R. M. (1974). New North American middle Eocene nyctithere (Mammalia, Insectivora). *Journal of Paleontology*, 48, 983–7.

White, T. E. (1954). Preliminary analysis of the fossil vertebrates of the Canyon Ferry Reservoir area. *Proceedings of the United States National Museum*, 103, 395–438.

Wilson, R. L. (1968). Systematics and faunal analysis of a lower Pliocene vertebrate assemblage from Trego County, Kansas. *Contributions from the Museum of Paleontology, University of Michigan*, 22, 75–126.

Wilson, R. W. (1960). Early Miocene rodents and insectivores from northeastern Colorado. *Kansas University Publications, Paleontology Contributions, Vertebrata*, 7, 1–92.

Winge, H. (1917). Udsigt over Insectaedernes inbrydes Slaegtskab. *Videnskabelige Meddelelser Naturhistorisk Forening, Copenhagen*, 68, 83–203.

Winterfeld, G. F. (1982). Mammalian paleontology of the Fort Union Formation (Paleocene), eastern Rock Springs Uplift, Sweetwater County, Wyoming. *Contributions to Geology, University of Wyoming*, 21, 73–112.

Wolberg, D. L. (1979). Late Paleocene (Tiffanian) mammalian fauna of two localities in eastern Montana. *Northwest Geology*, 8, 82–93.

Yalden, D. W. (1966). The anatomy of mole locomotion. *Journal of Zoology, London*, 149, 55–64.

Ziegler, A. C. (1971). Dental homologies and possible relationships of Recent Talpidae. *Journal of Mammalogy*, 52, 50–68.

Part III: "Edentata"

8 "Edentata" summary

GREGG F. GUNNELL[1] and KENNETH D. ROSE[2]

[1]*Museum of Paleontology, University of Michigan, Ann Arbor, MI, USA*
[2]*Johns Hopkins University School of Medicine, Baltimore, MD, USA*

INTRODUCTION

Edentata was originally employed (Cuvier, 1798) to include living members of what are now recognized as three distinct mammalian clades, Xenarthra (armadillos, anteaters, sloths), Pholidota (pangolins), and Tubulidentata (aardvarks). Since that time, various combinations of these taxa, their fossil relatives, the extinct groups Palaeanodonta and Ernanodonta, and some enigmatic fossil forms have been included within Edentata (Rose *et al.*, 2005). In general, "Edentata" is now favored as an informal term to include xenarthrans, pholidotans, and palaeanodonts, although Novacek (1986, 1992) and Novacek and Wyss (1986) have argued that a clade consisting of Pholidota and Xenarthra can be diagnosed and should be formally recognized as the cohort Edentata.

GENERAL CONSIDERATIONS

The North American Tertiary record of edentates consists solely of Paleocene and Eocene palaeanodonts and pangolins and late Miocene to Recent xenarthrans. A recent summary of edentates can be found in Rose *et al.* (2005).

FEATURES UNITING "EDENTATA"

As the name implies, many edentates either lack teeth completely (pholidotans, anteaters) or have reduced and simplified dentitions. When postcanine teeth are present, they are normally simple, peg-like, and lack enamel coverings (except in most palaeanodonts). Edentates in general have relatively robust skeletons with specialized forelimbs adapted for digging. Forelimb specializations include humeri with long and well-developed deltopectoral crests, large medial epicondyles, robust supinator crests, ulnae with elongate, medially inflected olecranon processes, and hands with enlarged middle digits and robust claws. Many, if not all, of these forelimb characters are indicative of a fossorial or semi-fossorial lifestyle and do not necessarily support any close phylogenetic ties between edentate groups.

FEATURES DISTINGUISHING XENARTHRA, PHOLIDOTA, AND PALAEANODONTA

BODY SIZE

Living xenarthrans range in body mass from around 50 g in fairy armadillos (*Chlamyphorus*) to 50 kg in giant armadillos (*Priodontes*), but some extinct forms, such as *Megatherium*, reached truly gigantic size (over 6000 kg) (Bargo *et al.*, 2000). Extant pholidotans (there are at least seven species of *Manis*) range between 2 and 35 kg, while fossil pholidotans (Emry, 1970) were no larger than moderate-sized living forms (5–10 kg). Palaeanodonts were all fairly small, ranging from small epoicotheriids (10 g to 2 kg) to modestly sized metacheiromyids (50 g to 5 kg) and escavadodontids (1.5 kg).

LOCOMOTION AND POSTCRANIAL SPECIALIZATIONS

Living xenarthrans exhibit a variety of locomotor styles ranging from terrestrial quadrupedalism, often in conjunction with fossorial adaptations (armadillos), arboreal quadrupedalism (most anteaters), modified terrestrial knuckle walking (giant anteaters; see Orr, 2005), and suspensory arborealism (sloths). Xenarthrans all share accessory or xenarthrous joints between posterior thoracic and lumbar vertebrae (glyptodonts secondarily lost these joints because of extensive vertebral fusion with body armor). Other skeletal features of xenarthrans include the presence of an ossified larynx, ossified sternal ribs, ribs that are expanded anteroposteriorly, reduced numbers of lumbar vertebrae, incorporation of caudal vertebrae into the

Evolution of Tertiary Mammals of North America, Vol. 2. ed. C. M. Janis, G. F. Gunnell, and M. D. Uhen. Published by Cambridge University Press.

Figure 8.1. Restoration of the giant ground sloth *Eremotherium*, by Marguette Dongvillo.

sacrum, an elevated scapular spine, elongate scapular acromion, and a secondary scapular spine associated with a well-developed fossa for the teres major muscle (Rose *et al.*, 2005).

Cingulates (armadillos and glyptodonts) share the features of fused cervical vertebrae posterior to the atlas, a femur with an enlarged greater trochanter that extends proximally beyond the femoral head, and fusion of the tibia and fibula (Engelmann, 1985). Vermilinguans (anteaters) have prehensile tails, except for the knuckle-walking *Myrmecophaga* (giant anteater), and all have manual third digits with enlarged unguals. Phyllophagans (sloths) have astragalar trochlea that are reduced to a single medial ridge and forelimbs longer than hindlimbs (except for extinct mylodontids). Large, fossil ground sloths have robust limbs with round proximal radial articulations, laterally flaring ilia, and elongate thoraxes (Rose *et al.*, 2005). Extant tree sloths have more gracile limbs and are two- or three-toed, with toes having claws modified into hooks to facilitate arboreal suspension and suspensory locomotion. Sloths often have variable numbers of cervical vertebrae, ranging from five to eight in two-toed sloths and eight or nine in three-toed forms (Feldhamer *et al.*, 2004).

Pholidotans are characterized skeletally by loss of the scapular coracoid process, presence of an unciform process on the hamate, a reduced obturator foramen, deeply fissured unguals, and scapholunar fusion (the latter two characters are apparently absent in some fossil pangolins).

Palaeanodonts have relatively robust skeletons adapted for a fossorial lifestyle. The scapula has a distinct and elevated spine and a bifurcate acromion. The humerus has a long, distinct, and often shelflike deltopectoral crest, an enlarged and medially directed entepicondyle, and a very prominent supinator crest. The ulna has a long and medially directed olecranon process. Palaeanodonts have robust manual digits with extensor tubercles developed on enlarged second and third metacarpals (Rose *et al.*, 2005).

DIET AND CRANIODENTAL SPECIALIZATIONS

Xenarthrans either have lost all of their teeth (anteaters) or have reduced numbers of teeth. An exception is the giant armadillo, which may have up to 100 small, vestigial cheek teeth (Feldhamer *et al.*, 2004). All but the most primitive and earliest known forms (Simpson, 1948) have teeth that have open roots and have lost all enamel. Most xenarthrans are either myrmecophagous (anteaters, some armadillos) or herbivorous (glyptodonts, some armadillos, sloths), although some primitive early members may have been omnivorous (Vizcaíno and Bargo, 1998; Rose *et al.*, 2005). Cingulates also are characterized cranially by a large postglenoid fossa that contains the postglenoid foramen.

Skulls of anteaters are elongate, tubular, and edentulous. The hard palate is extended posteriorly to the basicranium (Gaudin and Branham, 1998). In living forms, the bulla is ossified and is connected to an enlarged hypotympanic sinus in the pterygoid.

Sloths retain some functional teeth, normally having five uppers and four lowers with the anterior-most in each arcade being generally caniniform (but unlike true canines in other placental mammals, in sloths the uppers occlude anterior to the lowers). Teeth posterior to the caniniforms are cylindrical in living forms but vary in shape in extinct sloths. Distinctive skull characteristics include an enlarged descending lamina of the pterygoid, incomplete zygomatic arches, and large jugal processes for attachment of muscles for chewing (Feldhamer *et al.*, 2004; Rose *et al.*, 2005).

Pholidotan cranial characters include a tubular and edentulous rostrum, anteriorly extended bony prongs on either side of the mandibular symphysis, a deep median palatal groove, incomplete zygomatic arches, and presence of a large epitympanic sinus in the squamosal (Gaudin and Wible, 1999). The last feature is not present in the North American fossil pangolin *Patriomanis* (Gaudin, Emry, and Pogue, 2006).

Palaeanodont skulls have short rostra and broad occiputs. The mastoid region is inflated and the epitympanic sinus extends into the squamosal sinus as in pholidotans. Dentaries are thickened or buttressed medially and postcanine teeth tend to be simple and peglike and reduced in number. Canines are usually robust and triangular in cross section and develop distinct honing facets. Palaeanodont canines are true canines (upper occludes behind lower), unlike the caniniform teeth of sloths.

OTHER ANATOMICAL FEATURES

Most cingulates (armadillos and glyptodonts) have extensive body armor. It is composed of ossified dermal plates that form a carapace covering the head, dorsal part of the body, outer legs, and tail. This carapace is then covered by non-overlapping, keratinized epidermal scales.

Pholidotans are similarly armored but their protective covering is formed in a different manner. Pangolin scales are overlapping (imbricate) and are composed of keratinized epidermis. Pangolins retain the same number of scales throughout their lives and their scales grow in conjunction with body growth (Feldhamer *et al.*, 2004).

Anteaters and pangolins have extremely long tongues. Anteaters have tongues that are anchored on the sternum and when fully extended are longer than the elongate skull. Their tongues are covered by tiny, posteriorly directed barbs and a viscous secretion produced in the submandibular gland, both of which aid in the capture of ants and termites. Pangolins have an even longer tongue, which is greater in length than that of the head and body together. The tongue muscles are anchored on the xiphisternum and the tongue is covered by a sticky saliva produced by a large salivary gland in the chest cavity (Feldhamer *et al.*, 2004).

Most xenarthrans have relatively low metabolic rates and lower body temperatures than other placental mammals. Body temperatures of anteaters and armadillos average around 34 °C, while those of other mammals range between 36 and 38 °C (Feldhamer *et al.*, 2004). Sloths have even lower body temperatures (ranging between 24 and 33 °C) that track ambient environmental temperatures much more closely than most other mammals. All xenarthrans have limited thermoregulatory capacities compared with other mammals, a factor in their restricted geographic distribution.

SYSTEMATICS

ORIGIN AND AFFINITIES OF "EDENTATA"

EXTANT FORMS

Living edentates consist of the New World xenarthrans and Old World pholidotans. Xenarthrans are divided into two higher-level groupings, Cingulata (armadillos) and Pilosa, with the latter being further subdivided into Vermilingua (anteaters) and Phyllophaga (sloths). Cingulates are the most diverse group of living xenarthrans, being represented by eight genera. Vermilinguans are represented by three extant genera and phyllophagans by two genera. Living xenarthrans are mostly restricted to Central and South America although the nine-banded armadillo (*Dasypus*) does range into North America (Mexico and the southern United States).

Pholidotans are much less diverse than xenarthrans. Two subfamilies are recognized (McKenna and Bell, 1997): the Maninae (including the Asiatic pangolin *Manis*) and the Smutsiinae (African pangolins represented by three living genera).

The only other living animal that has been occasionally included within edentates is the aardvark (*Orycteropus*). Aardvarks now live only in Africa and are placed in their own order, Tubulidentata, now included in the Afrotheria.

FOSSIL RELATIVES

Living xenarthrans represent but a small relict of what was once a much more diverse and widespread radiation. The earliest possible xenarthrans known come from the Paleocene of South America and probably represent cingulates (there is evidence of dermal plate body armor: Simpson, 1948; Bergqvist and Oliveira, 1995; Oliveira and Bergqvist, 1998, 1999). The first possible pilosans come from the middle Eocene of Europe and the late Eocene of Antarctica and Chile (Storch, 1981, 2003; Engelmann, 1987; Wyss *et al.*, 1990, 1994; Vizcaíno and Scillato-Yané, 1995). The European record is based on *Eurotamandua* from the Messel Oil Shales in Germany, which was originally described as a myrmecophagous xenarthran (anteater) (Storch, 1981). *Eurotamandua* since has been variously interpreted as a possible sister taxon to palaeanodonts (Rose and Emry, 1993; Rose, 1999; Rose *et al.*, 2005), a sister taxon to all pilosans (Gaudin and Branham, 1998), a pholidotan (McKenna and Bell, 1997), or as a member of a distinct clade, Afredentata (Szalay and Schrenk, 1998). The Antarctic records of possible pilosans are fragmentary and await further corroboration. By the Miocene, xenarthrans were extremely diverse (over 130 genera) and widespread throughout much of the New World.

The fossil record of pholidotans is much less diverse than that of xenarthrans but manids were geographically widespread in the Eocene. The earliest record of a pangolin is *Eomanis* from the middle Eocene Messel Oil Shales (Storch, 1978). Late Eocene records include *Patriomanis* from North America (Emry, 1970), *Necromanis* from Europe, and *Cryptomanis* from East Asia (Gaudin, Emry and Pogue, 2006). As noted above, *Eurotamandua* from Messel has also been interpreted as a possible manid by some authorities (McKenna and Bell, 1997). *Necromanis* survives into the later Miocene in Europe and there are possible manid remains from Oligocene deposits in North Africa (Gebo and Rasmussen, 1985). *Manis* is known from the late Miocene to present day in Asia and there are also scattered Pleistocene remains known from Europe and the East Indies (McKenna and Bell, 1997). Among the living African pangolins, *Smutsia* has a limited fossil record in the Pliocene while *Uromanis* and *Phataginus* are unknown from fossils.

Fossil aardvarks are not especially common. There are two possible records of aardvarks from Oligocene deposits in Europe (McKenna and Bell, 1997). Three genera of aardvarks are known from the Miocene of Africa but only extant *Orycteropus* survived past the Miocene.

Extinct edentates may have included palaeanodonts, ernanodonts, and possibly enigmatic genera such as *Asiabradypus*, *Chungchienia*, and *Plesiorycteropus* (Rose *et al.*, 2005). Palaeanodonts are a Paleocene and Eocene group that is mainly North American (Rose *et al.*, 2005). Their North American radiation is reviewed in more detail below. Possible palaeanodonts or sister taxa to palaeanodonts from outside North America include *Eurotamandua*, *Ernanodon*, and pholidotans. As we have seen, *Eurotamandua* remains difficult to place (Rose *et al.*, 2005). *Ernanodon*, from the late Paleocene of China (Ding, 1987), was originally thought to be a xenarthran but more recently has been assigned to a separate suborder (McKenna and Bell, 1997). *Ernanodon* shares some

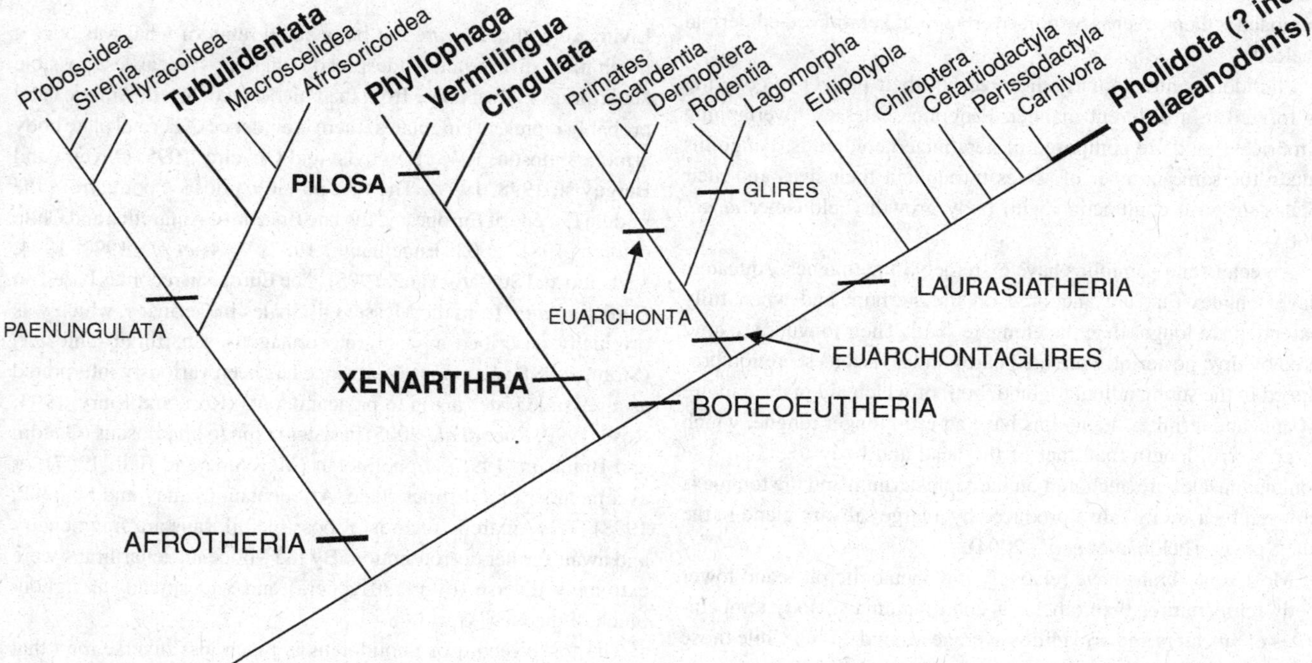

Figure 8.2. Cladogram showing the relationships of the various "edentates" to other placental mammals. "Edentates" are in bold face type.

characteristics with xenarthrans, pholidotans, and palaeanodonts, but overall it appears to be more closely related to palaeanodonts than to any other group (Rose *et al.*, 2005).

Among other possible edentates, *Asiabradypus*, from the late Paleocene of Kazakhstan, was originally described as a possible sloth (Nessov, 1987), but reexamination led Rose *et al.* (2005) to conclude that the single broken specimen is not a xenarthran. A possible *Asiabradypus* specimen from the earliest Eocene of Wyoming (Gingerich, 1989), therefore, is not an edentate whatever it might otherwise represent. *Chungchienia* (middle Eocene, China) was originally described as a sloth (Chow, 1963) but has more recently been shown to be a tillodont (Chow, Wang, and Meng, 1996). *Plesiorycteropus*, a Holocene subfossil form from Madagascar that was initially thought to be an aardvark (Patterson, 1975), was more recently (MacPhee, 1994) put in its own order (Bibymalagasia) to reflect its distinctiveness from other mammals. It does, however, show some striking osteological similarities to the armadillo *Dasypus*.

"EDENTATA" PHYLOGENETICS

EDENTATE POLYPHYLY AND INTERRELATIONSHIPS OF EDENTATE GROUPS

Nearly all molecular phylogenetic analyses have failed to support the monophyly of Edentata (Xenarthra and Pholidota together, and these groups together with Tubulidentata). Most molecular studies (Springer *et al.*, 1997, 2005; Eizirik, Murphy, and O'Brien, 2001; Madsen *et al.*, 2001; Murphy *et al.*, 2001; Delsuc *et al.*, 2002) support a monophyletic Xenarthra as a distinct clade from all other mammals (or as a sister taxon to Afrotheria). A recent molecular analysis specifically focused on Xenarthra (Delsuc *et al.*, 2002) placed the xenarthran clade as sister group to Boreoeutheria (Laurasiatheria and Euarchontoglires) with Afrotheria being the sister taxon to the combined Boreoeutheria and Xenarthra clade. Pangolins were nested within Laurasiatheria as the sister group to Carnivora, and aardvarks were nested within Afrotheria as the sister group to a clade consisting of elephant shrews, cape golden moles and tenrecs (Figure 8.2).

Anatomical evidence is less certain and has been interpreted in a number of different ways. Novacek (1986, 1992) and Novacek and Wyss (1986) found evidence to support a xenarthran–pholidotan clade (Edentata) separate from all other placental mammals (Epitheria), but most of the features used by these authors to link the two groups have been questioned by others (e.g. Rose and Emry, 1993). Engelmann (1985) suggested that Pilosa and Pholidota (specifically anteaters and pangolins) share a close relationship, but most of their shared morphology more likely arose through convergence (Patterson *et al.*, 1992; Gaudin and Wible, 1999). In general, anatomical evidence supports the distinctiveness of Xenarthra, Pholidota, and Tubulidentata but does not provide convincing evidence to determine the higher-level relationships of these groups.

Among the best-supported hypotheses based on anatomical evidence is a proposed relationship between palaeanodonts and pholidotans (Emry, 1970; Rose and Emry, 1993; Storch, 2003; Rose *et al.*, 2005). Characters shared between palaeanodonts and Eocene pholidotans include presence of a mandibular medial

Epoch	Ma	Biochron	Cent. America (CA)	Gulf Coast (GC)	CA Central & Coastal (CC)	Northern Great Basin (NB)	Southern Great Basin (SB)	Southern Great Plains (SP)	Central Great Plains (CP)	Northern Great Plains (NP)	Pacific Northwest (PN)	N.E. Coast (NC)
EOCENE	33.7	Ch-4							Xe			
	34.8	Ch-3							Ep, Xe, ***Pm***	***Pm***		
	35.8	Ch-2							Ep, ***Pm***	Ep		
	36.6	Ch-1								Ep		
	36.9	Du								Ep		
	40.1	Ui-3								Ep		
	42.8	Ui-2										
	45.7	Ui-1							?Tt			
	46.3	Br-3							**Mc**, Tt			
	47.0	Br-2				Tt			**Mc**, Tt			
	49.2	Br-1							**Mc**, **Br**, Tt			
	50.0	Br-0							Tu, Pt, Dp			
	50.1	Wa-7							?**Pl**, Tu, Pt, Dp			
	52.4	Wa-6			Tu		?**Pl**		**Pl**, Tu, Al, Pt, Dp			**Ar**
	53.0	Wa-5							**Pl**, Tu, Al			
	53.4	Wa-4							**Pl**, Tu			
	54.1	Wa-3							**Pl**	?**Pl**		
	54.4	Wa-2							**Pl**			
	54.7	Wa-1							**Pl**			
	54.9	Wa-0							**Pl**			
PALEOCENE	55.0	Cf-3							**Pl**			
	55.4	Cf-2							**Pl**			
	55.8	Cf-1										
	56.0	Ti-6										
	56.9	Ti-5							**Pp**, **My**			
	57.4	Ti-4							Am			
	57.8	Ti-3								**Me**		
	58.9	Ti-2										
	60.0	Ti-1										
	60.9	To-3					Es					
	61.9											

Figure 8.3. Biogeographic ranges of palaeanodonts. Key: A "box" (for a particular time period in a particular biogeographic region) that has a cross through it means no fossil mammal localities are known for that time period from that area; a single dashed line through the box means only scant fossil mammal information is available (usually only a single, small locality). PALAEANODONTA: ESCAVADODONTIDAE (Times New Roman Plain): Es, *Escavadodon*. METACHEIROMYIDAE (Times New Roman Bold): **Br**, *Brachianodon*; **Mc**, *Metacheiromys*; **My**, *Mylanodon*; **Pl**, *Palaeanodon*; **Pp**, *Propalaeanodon*. EPOICOTHERIIDAE (Arial Plain): Al, *Alocodontulum*; Am, *Amelotabes*; Dp, *Dipassalus*; Ep, *Epoicotherium*; Pt, *Pentapassalus*; Tt, *Tetrapassalus*; Tu, *Tubulodon*; Xe, *Xenocranium*. PALAEANODONTA INCERTAE SEDIS (Arial Bold). **Ar**, *Arctican-odon*; **Me**, *Melaniella*. PHOLIDOTA: MANIDAE (Arial Bold Italics): ***Pm***, *Patriomanis*.

buttress; a loosely attached, C-shaped premaxilla; similarly shaped scapula with elevated spine and a long acromion; humerus with robust and broad deltopectoral crest, prominent supinator crest and expanded entepicondyle; distally expanded radius; short and broad metapodials; and enlarged, curved, unfissured distal manual phalanges (Rose *et al.*, 2005).

The origin of palaeanodonts can be traced through *Escavadodon* to primitive leptictids (Rose and Lucas, 2000). Palaeanodonts appear to share a close relationship, perhaps even sister-group status, with pantolestids. If so, this palaeanodont–pantolestid clade may have been derived from leptictids. The origin of other edentate groups remains vague at this time although if molecular phylogenies correctly depict the mammalian branching sequence, then Xenarthra must be an ancient radiation with an origin in the Cretaceous, and Pholidota may be nearly as old.

EVOLUTIONARY AND BIOGEOGRAPHIC PATTERNS

GENERAL TRENDS

The North American Tertiary record of edentates is relatively modest and is summarized in Figures 8.3 and 8.4. Palaeanodonts are known through most of the Paleocene (though rare in this epoch) and Eocene but are only well represented in the Central Great Plains. There are isolated records from the Canadian High Arctic, Central

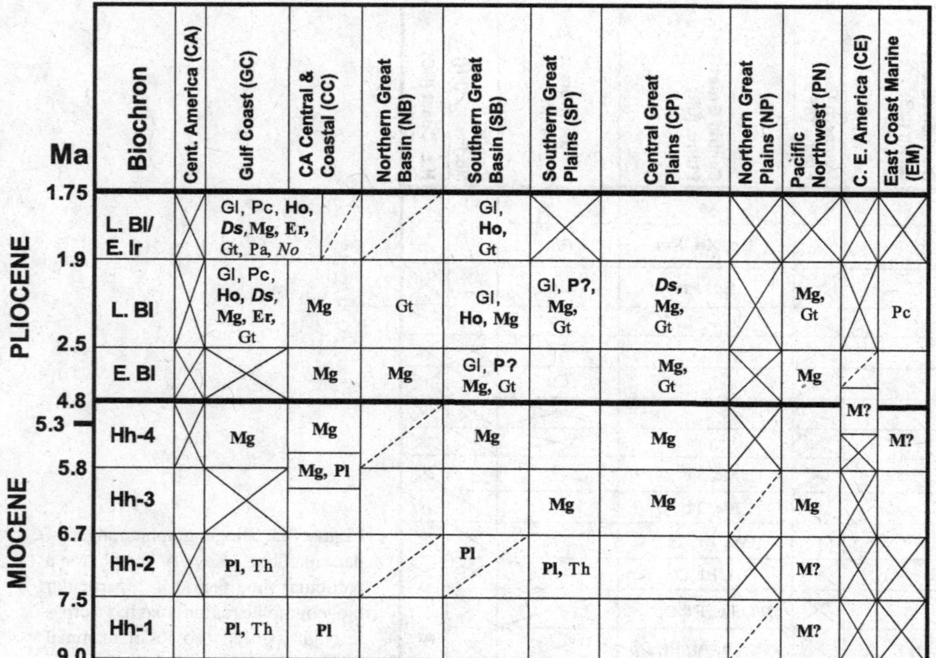

Epoch	Ma	Biochron	Cent. America (CA)	Gulf Coast (GC)	CA Central & Coastal (CC)	Northern Great Basin (NB)	Southern Great Basin (SB)	Southern Great Plains (SP)	Central Great Plains (CP)	Northern Great Plains (NP)	Pacific Northwest (PN)	C. E. America (CE)	East Coast Marine (EM)
PLIOCENE	1.75–1.9	L. Bl/ E. Ir		Gl, Pc, **Ho**, ***Ds***, **Mg**, **Er**, Gt, Pa, *No*			Gl, **Ho**, Gt						
PLIOCENE	1.9–2.5	L. Bl		Gl, Pc, **Ho**, ***Ds***, **Mg**, **Er**, Gt	**Mg**	Gt	Gl, **Ho**, **Mg**	Gl, **P?**, **Mg**, Gt	***Ds***, **Mg**, Gt		**Mg**, Gt		Pc
PLIOCENE	2.5–4.8	E. Bl			**Mg**	**Mg**	Gl, **P?** **Mg**, Gt		**Mg**, Gt		**Mg**		
MIOCENE	5.3–5.8	Hh-4		**Mg**	**Mg**		**Mg**		**Mg**			**M?**	**M?**
MIOCENE	5.8–6.7	Hh-3			**Mg, Pl**			**Mg**	**Mg**		**Mg**		
MIOCENE	6.7–7.5	Hh-2		**Pl**, Th			**Pl**	**Pl**, Th			**M?**		
MIOCENE	7.5–9.0	Hh-1		**Pl**, Th	**Pl**						**M?**		

Figure 8.4. Biogeographic ranges of xenarthrans. Key as in Figure 8.3. CINGULATA: GLYPTODONTIDAE (Arial Plain): Gl, *Glyptotherium*; Pc, *Pachyarmatherium*. PAMPATHERIIDAE (Arial Bold): **Ho**, *Holmesina*; **P?**, pampathere indet. DASYPODIDAE (Arial Bold Italics): ***Ds***, *Dasypus*. PILOSA: MEGALONYCHIDAE (Times New Roman Bold): **Mg**, *Megalonyx*; **M?**, Megalonychid indet.; **Pl**, *Pliometanastes*. MEGATHERIIDAE (Times New Roman Bold Italics): ***Er***, *Eremotherium*. MYLODONTIDAE (Times New Roman Plain): Gt, *Glossotherium*; Pa, *Paramyladon*; Th, *Thinobadistes*.

and Coastal California, and the Northern and Southern Great Basins, and a few records (especially in the late middle and late Eocene) from the Northern Great Plains. No palaeanodonts are known in North America (or anywhere else for that matter) after the late Eocene.

North American pholidotans are only known from isolated occurrences of *Patriomanis* in the late Eocene from the Central Great Plains and Northern Great Plains. The remainder of the North American edentate record begins in the late Miocene (early Hemphillian) and consists exclusively of xenarthrans. North American xenarthrans are best represented from the Gulf Coast and Southern Great Basin, with less abundant and diverse samples coming from the Central Great Plains, Southern Great Plains, California Central and Coast, Northern Great Basin, and Pacific Northwest.

CHRONOFAUNAL CHANGES

The notion of North American Tertiary "chronofaunas" was employed in Volume 1 and is discussed further in this volume in Chapter 16.

PALEOCENE

The earliest record of a palaeanodont is that of *Escavadodon* from the late Torrejonian in the Southern Great Basin (Figure 8.3). There is also a doubtful palaeanodont, *Melaniella*, from the middle Tiffanian (Ti3) in the Northern Great Plains. The only other Paleocene records are from the Central Great Plains and include the earliest known epoicotheriid *Amelotabes* and the early metacheiromyids *Mylanodon* and *Propalaeanodon* from Ti4 and Ti5, respectively. By the end of the Paleocene, only *Palaeanodon* is known in the Central Great Plains.

EARLY TO MIDDLE EOCENE

As with the Paleocene record, early and middle Eocene palaeanodonts are mostly known from the Central Great Plains. The metacheiromyid *Palaeanodon* is the only palaeanodont known from the Central Great Plains until Wa4, when it is joined by the epoicotheriid *Tubulodon*. From Wa5 through the earliest Bridgerian (Br0), epoicotheriid diversity is higher than that of metacheiromyids (as many as four genera to one) as *Palaeanodon* continues to be the lone representative of the latter family until the early Bridgerian (Br1). Epoicotheriid diversity reduces to a single genus (*Tetrapassalus*) beginning in Br1 while metacheiromyids are represented by two genera early in the Bridgerian and then a single genus (*Metacheiromys*) after that. Metacheiromyids disappear from the North American record by the beginning of the Uintan.

Isolated records of metacheiromyids are known from Wa6 in the Southern Great Basin (?*Palaeanodon*), from Wa3 in the Northern Great Plains (?*Palaeanodon*), and from the Wasatchian of the Canadian High Arctic (*Arcticanodon*). Epoicotheriids are also known from Wa5/6 in the California Central and Coast (*Tubulodon*), Br2 in the Northern Great Basin (*Tetrapassalus*), and Ui3 in Northern Great Plains (*Epoicotherium*).

WHITE RIVER CHRONOFAUNA

The White River Chronofauna sample of palaeanodonts consists solely of two very derived epoicotheriids, *Epoicotherium* and *Xenocranium*. *Epoicotherium* is known from the Duchesnean through early middle Chadronian (Ch2) in the Northern Great Plains and from the middle Chadronian (Ch2/3) in the Central Great Plains. *Xenocranium* is only represented from the middle late and late Chadronian (Ch3/4) in the Central Great Plains.

In addition to these epoicotheriids, the only North American pholidotan (*Patriomanis*) is known from the middle Chadronian in the Central Great Plains (Ch2/3) and from the middle late Chadronian (Ch3) in the Northern Great Plains. No other edentates are known from North America until the late Miocene.

MIO-PLIOCENE CHRONOFAUNA

Xenarthrans make their appearance in North America beginning in the early Hemphillian (Hh1) (Figure 8.4). North American Tertiary records of xenarthrans include only cingulates (armadillos, pampatheres, and glyptodonts) and phyllophagans (sloths). The only North American fossil anteaters known come from Pleistocene deposits in Mexico.

The Hemphillian record from all North American faunas consists entirely of sloths. *Megalonyx, Pliometanastes*, and *Thinobadistes* are the most common sloths through the Hemphillian with *Glossotherium* making an early appearance in the late Hemphillian (Hh4) in the Southern Great Basin (in Mexico). The early Blancan is dominated by *Megalonyx* and *Glossotherium*, but the first cingulate also makes its appearance at that time, represented by the glyptodont *Glyptotherium* from the Southern Great Basin (again from Mexico). Cingulates become much more common in the late Blancan into the Irvingtonian, especially in Gulf Coast localities. The late Blancan also records the first appearance of the armadillo *Dasypus*, the pampathere *Holmesina*, the glyptodont *Pachyarmatherium*, and the megatherid sloth *Eremotherium*. Other sloths (*Paramylodon* and *Nothrotheriops*) do not appear until the earliest Irvingtonian. Except in the Gulf Coast and Southern Great Basin regions, xenarthrans are gone from the North American fossil record by the Irvingtonian.

REFERENCES

Bargo, M. S., Vizcaino, S. F., Archuby, F. M., and Blanco, R. E. (2000). Limb bone proportions, strength and digging in some Lujanian (Late Pleistocene–Early Holocene) mylodontid ground sloths (Mammalia, Xenarthra). *Journal of Vertebrate Paleontology*, 20, 601–10.

Bergqvist, L. P. and Oliveira, E. V. (1995). Novo material póscraniano de Cingulata (Mammalia–Xenarthra) do Paleoceno médio (Itaboraiense) do Brasil. *14th Congresso Brasileiro de Paleontologia (Sociedade Brasiliera de Paleontologia, Uberaba)*, p. 16–17.

Chow, M. (1963). A xenarthran-like mammal from the Eocene of Honan. *Scientia Sinica*, 12, 1889–93.

Chow, M., Wang, J., and Meng, J. (1996). A new species of *Chungchienia* (Tillodontia, Mammalia) from the Eocene of Lushi, China. *American Museum Novitates*, 3171, 1–10.

Cuvier, G. L. (1798). *Tableau Élémentaire de l'Histoire Naturelle des Animaux*. Paris: J. B. Baillière.

Delsuc, F., Scally, M., Madsen, O., *et al.* (2002). Molecular phylogeny of living xenarthrans and the impact of character and taxon sampling on the placental tree rooting. *Molecular Biology and Evolution*, 19, 1656–71.

Ding, S.-Y. (1987). A Paleocene edentate from Nanxiong Basin, Guangdong. *Palaeontologia Sinica, New Series C*, 173, 1–118.

Eizirik, E., Murphy, W. J., and O'Brien, S. J. (2001). Molecular dating and biogeography of the early placental mammal radiation. *Journal of Heredity*, 92, 212–9.

Emry, R. J. (1970). A North American Oligocene pangolin and other additions to the Pholidota. *Bulletin of the American Museum of Natural History*, 142, 455–510.

Engelmann, G. F. (1985). The phylogeny of the Xenarthra. In *The Evolution and Ecology of Armadillos, Sloths, and Vermilinguas*, ed. G. G. Montgomery, pp. 51–64. Washington, DC: Smithsonian Institution Press.

(1987). A new Deseadan sloth (Mammalia: Xenarthra) from Salla, Bolivia, and its implications for the primitive condition of the dentition in edentates. *Journal of Vertebrate Paleontology*, 7, 217–23.

Feldhamer, G. A., Drickamer, L. C., Vessey, S. H., and Merritt, J. F. (2004). *Mammalogy: Adaptation, Diversity, Ecology*, 2nd edn. New York: McGraw-Hill.

Gaudin, T. J. and Branham, D. G. (1998). The phylogeny of the Myrmecophagidae (Mammalia, Xenarthra, Vermilingua) and the relationship of *Eurotamandua* to the Vermilingua. *Journal of Mammalian Evolution*, 5, 237–65.

Gaudin, T. J. and Wible, J. R. (1999). The entotympanic of pangolins and the phylogeny of the Pholidota (Mammalia). *Journal of Mammalian Evolution*, 6, 39–65.

Gaudin, T. J., Emry, R. J., and Pogue, B. (2006). A new genus and species of pangolin (Mammalia, Pholidota) from the late Eocene of Inner Mongolia, China. *Journal of Vertebrate Paleontology*, 26, 146–59.

Gebo, D. L. and Rasmussen, D. T. (1985). The earliest fossil pangolin (Pholidota: Manidae) from Africa. *Journal of Mammalogy*, 66, 538–40.

Gingerich, P. D. (1989). New earliest Wasatchian mammalian fauna from the Eocene of northwestern Wyoming: Composition and diversity in a rarely sampled high-floodplain assemblage. *University of Michigan Papers on Paleontology*, 28, 1–97.

MacPhee, R. D. E. (1994). Morphology, adaptations, and relationships of *Plesiorycteropus*, and a diagnosis of a new order of eutherian mammals. *Bulletin of the American Museum of Natural History*, 220, 1–214.

Madsen, O., Scally, M., Douady, C. J., *et al.* (2001). Parallel adaptive radiations in two major clades of placental mammals. *Nature*, 409, 610–14.

McKenna, M. C. and Bell, S. K. (1997). *Classification of Mammals Above the Species Level*. New York: Columbia University Press.

Murphy, W. J., Eizirik, E., Johnson, W. E., *et al.* (2001). Molecular phylogenetics and the origins of placental mammals. *Nature*, 409, 614–18.

Nessov, L. (1987). Rezultaty poiskov I issledovanija melovych I rannepaleogenovych mlekopitajushchi ich na territorii SSSR. *Akademiya Nauk SSSR*, 30, 199–218.

Novacek, M. J. (1986). The skull of leptictid insectivorans and the higher-level classification of eutherian mammals. *Bulletin of the American Museum of Natural History*, 183, 1–112.

(1992). Mammalian phylogeny: Shaking the tree. *Nature*, 356, 121–5.

Novacek, M. J. and Wyss, A. (1986). Higher-level relationships of the recent eutherian orders: morphological evidence. *Cladistics*, 2, 257–87.

Oliveira, E. V. and Bergqvist, L. P. (1998). A new Paleocene armadillo (Mammalia, Dasypodoidea) from the Itaboraí Basin, Brazil. *Asociación Paleontológica Argentina (Buenos Aires), Publicación Especial*, 5, 35–40.

(1999). A new Paleocene armadillo (Mammalia, Xenarthra, Astegotherini) from Itaboraí, Brazil, and phylogeny of early Tertiary astegotherines. *Anais da Academia Brasileira de Ciências*, 71, 814–15.

Orr, C. M. (2005). Knuckle-walking anteater: a convergence test of adaptation for purported knuckle-walking features of African Hominidae. *American Journal of Physical Anthropology*, 128, 639–58.

Patterson, B. (1975). The fossil aardvarks (Mammalia: Tubulidentata). *Bulletin of the Museum of Comparative Zoology*, 147, 185–237.

Patterson, B., Segall, W., Turnbull, W., and Gaudin, T. (1992). The ear region in xenarthrans (= Edentata: Mammalia). Part II. Pilosa (sloths, anteaters), palaeanodonts, and a miscellany. *Fieldiana Geology, New Series*, 24, 1–79.

Rose, K. D. (1999). *Eurotamandua* and Palaeanodonta: convergent or related? *Paläeontologische Zeitschrift*, 73, 395–401.

Rose, K. D. and Emry, R. J. (1993). Relationships of Xenarthra, Pholidota, and fossil "edentates": the morphological evidence. In *Mammal Phylogeny: Placentals*, ed. F. S. Szalay, M. J. Novacek, and M. C. McKenna, pp. 81–102. New York: Springer-Verlag.

Rose, K. D. and Lucas, S. G. (2000). An early Paleocene palaeanodont (Mammalia,?Pholidota) from New Mexico, and the origin of Palaeanodonta. *Journal of Vertebrate Paleontology*, 20, 139–56.

Rose, K. D., Emry, R. J., Gaudin, T. J., and Storch, G. (2005). Xenarthra and Pholidota. In *The Rise of Placental Mammals*, ed. K. D. Rose and J. D. Archibald, pp. 106–26. Baltimore, MD: Johns Hopkins University Press.

Simpson, G. G. (1948). The beginning of the Age of Mammals in South America. *Bulletin of the American Museum of Natural History*, 91, 1–232.

Springer, M. S., Cleven, G. C., Madsen, O., *et al.* (1997). Endemic African mammals shake the phylogenetic tree. *Nature*, 388, 61–4.

Springer, M. S., Murphy, W. J., Eizirik, E., and O'Brien, S. J. (2005). Molecular evidence for major placental clades. In *The Rise of Placental Mammals*, ed. K. D. Rose and J. D. Archibald, pp. 37–49. Baltimore, MD: Johns Hopkins University Press.

Storch, G. (1978). *Eomanis waldi*, ein Schuppentier aus dem Mittel-Eozän der "Grube Messel" bei Darmstadt (Mammalia: Pholidota). *Senckenbergiana Lethaea*, 59, 503–29.

(1981). *Eurotamandua joresi*, ein Myrmecophagide aus dem Eozän der "Grube Messel" bei Darmstadt (Mammalia: Xenarthra). *Senckenbergiana Lethaea*, 61, 247–89.

(2003). Fossil Old World "edentates." *Senckenbergiana Biologica* 83, 51–60.

Szalay, F. S. and Schrenk, F. (1998). The middle Eocene *Eurotamandua* and a Darwinian phylogenetic analysis of "edentates." *Kaupia*, 7, 97–186.

Vizcaíno, S. F. and Bargo, M. S. (1998). The masticatory apparatus of the armadillo *Eutatus* (Mammalia: Cingulata) and some allied genera: paleobiology and evolution. *Paleobiology*, 24, 371–83.

Vizcaíno, S. F. and Scillato-Yané, G. J. (1995). An Eocene tardigrade (Mammalia, Xenarthra) from Seymour Island, West Antarctica. *Antarctic Science*, 7, 407–8.

Wyss, A. R., Norell, M. A., Flynn, J. J., *et al.* (1990). A new early Tertiary mammal fauna from central Chile: Implications for Andean stratigraphy and tectonics. *Journal of Vertebrate Paleontology*, 10, 518–22.

Wyss, A. R., Flynn, J. J., Norell, M. A., *et al.* (1994). Paleogene mammals from the Andes of Central Chile: a preliminary taxonomic, biostratigraphic, and geochronological assessment. *American Museum Novitates*, 3098, 1–31.

9 Palaeanodonta and Pholidota

KENNETH D. ROSE
Johns Hopkins University School of Medicine, Baltimore, MD, USA

INTRODUCTION

During the Cenozoic, North America served as homeland to a diversity of what might be broadly called "edentate" mammals. True Xenarthra, largely confined to South America, did not arrive in North America until the late Miocene (Hemphillian). Earlier in the Tertiary, the resident "edentates" belonged to the Palaeanodonta and the Pholidota.

Palaeanodonts were a moderately successful radiation of small, robust, fossorial mammals known mainly from early Paleocene (Torrejonian) through latest Eocene (Chadronian) rocks of the Rocky Mountain region. In recent years, a few specimens have been found in lower Eocene sediments of China and lower Oligocene beds of Europe (Heissig, 1982; Tong and Wang, 1997; Storch and Rummel, 1999). Most representatives show dental reduction and modifications of the jaws and palate believed to be associated with myrmecophagy, or at least a diet of small invertebrates.

Although never abundant or diverse, palaeanodonts are well represented in some early and middle Eocene faunas, and their highly distinctive anatomy is at once recognizable. Jaw fragments are common in some deposits, but the teeth are reduced or vestigial in most species and are practically never found isolated. By comparison, skulls and partial skeletons – typically rare in the fossil record – are not unusual for palaeanodonts, perhaps because they died and were buried in burrows.

Despite relatively complete osteological knowledge of several species, the relationships of palaeanodonts remain controversial. They are usually considered to be related to either Xenarthra or Pholidota (or both), but conclusive evidence of their affinities is lacking. Consequently, Pholidota and Palaeanodonta are treated separately in this chapter.

The fossil record of unequivocal Pholidota (pangolins or scaly anteaters) is meager, although a few exceptional fossils are known. The best records are from the Cenozoic of Europe and Africa (e.g., Koenigswald, 1969; Patterson, 1978; Storch, 1978; Koenigswald and Martin, 1990; Storch and Richter, 1992). A single taxon is known from North America, the Chadronian *Patriomanis americanus*. It is variously assigned to the extant family Manidae (by authors who include all fossil and living pangolins in a single family) or to the paraphyletic stem family Patriomanidae (Szalay and Schrenk, 1998).

DEFINING FEATURES OF THE ORDER PALAEANODONTA

CRANIAL

The skull in most palaeanodonts is low and caudally very broad, with a prominent lambdoid crest, large and inflated bullae, and inflated squamosals (Figure 9.1A,B). The palate (e.g., in metacheiromyids and *Pentapassalus*) possesses a broad, shallow median groove, suggesting the presence of a protrusile tongue. As far as is known, all palaeanodonts have a shallow but robust dentary with a spout-like symphysis, and there is only one small incisor present; but the front of the jaw is unknown in several forms, including *Escavadodon*, the most primitive genus. Posteromedially, the dentary in most forms bears a prominent shelf known as the medial buttress, which Matthew (1918) suggested may have supported a horny plate.

DENTAL

The postcanine teeth are reduced in all except the most primitive forms, but to variable degrees, with progressive reduction in enamel, crown size and complexity, and number of teeth. Reduction in postcanine tooth size is accompanied by fusion or loss of roots. In the dentally most derived species, there are only one or two vestigial peglike postcanines in the anterior part of the dentary. The canines,

Evolution of Tertiary Mammals of North America, Vol. 2. ed. C. M. Janis, G. F. Gunnell, and M. D. Uhen. Published by Cambridge University Press.

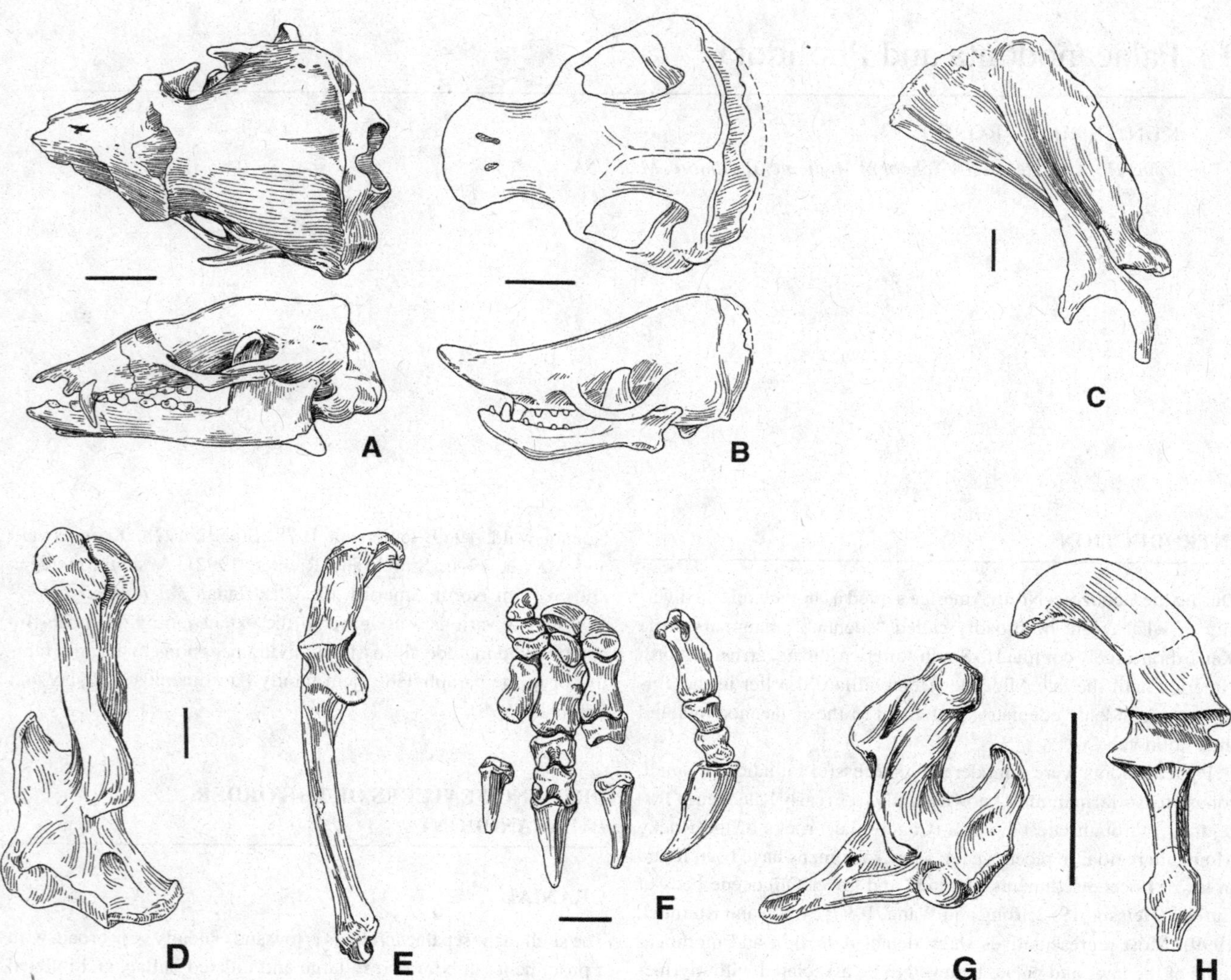

Figure 9.1. Skeletal elements of palaeanodonts. A. Skull and mandible of *Pentapassalus pearcei* (after Gazin, 1952). B. Skull and mandible of *Xenocranium pileorivale* (after Rose and Emry, 1983). C. Right scapula of *Metacheiromys dasypus* (after Simpson, 1931). D. Right humerus of *M. dasypus* (after Simpson, 1931). E. Right ulna of *M. dasypus* (after Simpson, 1931). F. Partial right manus of *M. dasypus*; with lateral view of third digit (after Simpson, 1931). G. Left humerus of *Epoicotherium unicum* (after Rose, Krishtalka, and Stucky, 1991). H. Left ulna of *E. unicum* (after Rose and Emry, 1983). Scale bars = 1 cm. (C,D,E,F: courtesy of the American Museum of Natural History. G: courtesy of the Carnegie Museum of Natural History.)

however, are always prominent and, in most forms, distinctly triangular in cross section at the base, with well-developed honing surfaces.

POSTCRANIAL

Palaeanodont limb elements are relatively short and stout, with very prominent crests and processes (Figure 9.1C–H). The following characteristics are typical: scapula with elevated spine, elongate, bifid acromion, and thickened caudal border or secondary spine; humerus (primitively) with broad, elevated, and long deltopectoral crest, exaggerated supinator crest, and salient entepicondyle; distal humerus very broad; ulna with very long, broad, medially inflected olecranon; radius thickened distally; femur stout with pronounced third trochanter; tibia bowed and, in the most specialized species, fused at both ends with the fibula; astragalus comparatively broad, with a shallow trochlea; metapodials and phalanges very short and stout; unguals generally large, curved, and somewhat laterally compressed, but smaller, broader, and flatter on the hind feet.

DEFINING FEATURES OF THE ORDER PHOLIDOTA

The most obvious distinctions of the pholidotes are their edentulous jaws and their body covering of epidermally derived scales. The skull is conical and relatively smooth, with crests and ridges weakly expressed or absent, and the orbital and temporal fossae are confluent. The squamosal overlaps the mastoid, a condition termed amastoidy (Novacek and Wyss, 1986). Associated with the absence of teeth, the dentary is substantially reduced, and the angular and coronoid processes are reduced or absent. Pholidotans have a robust postcranial skeleton adapted for digging and/or climbing; scapholunar fusion; and a functionally tridactyl manus with deeply fissured terminal phalanges bearing long, curved claws. The tongue is exceptionally long and protrusile, attaching to the xiphisternum rather than the hyoid. (Mostly after Barlow [1984], who lists additional traits.)

SYSTEMATICS

SUPRAORDINAL

Ever since their initial discovery, the unusual anatomy of palaeanodonts has generated confusion about their broader relationships. Wortman (1903) classified the first known palaeanodont, *Metacheiromys*, as a primate, but this assignment turned out to have been based on a mistakenly associated tibia. More complete skeletons led Osborn (1904) to proclaim that *Metacheiromys* was an armadillo, and since that time palaeanodonts have usually been included in either Edentata (variously considered to consist of Xenarthra plus Pholidota, or Xenarthra alone) or Pholidota. Consequently, considerations of the relationships of palaeanodonts and pholidotans have often been intertwined.

There have been at least two exceptions to this generality. *Epoicotherium* was first considered to be a monotreme (Douglass, 1905), and then a chrysochlorid insectivore (Matthew, 1906). Interesting, even striking, resemblances to both of these groups do exist, but they are now universally regarded as convergent. Later, Jepsen (1932) concluded that *Tubulodon* was a fossil tubulidentate, based largely on the presence of dentinal tubules in the teeth. This assignment was rejected by Colbert (1941), who questioned the homology of the tubules with those in tubulidentates. Gazin (1952) subsequently hinted at a relationship with the epoicothere *Pentapassalus*, but it took several more years before *Tubulodon*'s palaeanodont affinities were formally recognized (Simpson, 1959).

Edentate relationships for palaeanodonts were reinforced by Matthew (1918) and especially Simpson (1927, 1931) and became widely accepted for many decades thereafter. Palaeanodonta has generally been ranked as a suborder of Edentata (e.g., Matthew, 1918; Colbert, 1942; Simpson, 1945, 1959; Romer, 1966; Gingerich, 1989), but the meaning of Edentata has been inconsistent (Xenarthra plus Pholidota [e.g., Matthew, 1918; Novacek, Wyss, and McKenna, 1988] or Xenarthra only [e.g., McKenna, 1975]) or, in many cases, ambiguous. Simpson (1931) clearly used the term in the current sense of Xenarthra. Many of the resemblances between palaeanodonts and Xenarthra, however, now seem to be convergences resulting from fossorial habits and a tendency toward myrmecophagy (e.g., Emry, 1970). Strengthening this probability is Gaudin's (1999) confirmation that palaeanodonts lack the derived vertebral structure diagnostic of Xenarthra. Consequently, there has been a growing consensus that close relationship to Xenarthra is unlikely. One recent study, however, adduced evidence from the auditory region that again supports the notion of palaeanodont–xenarthran ties, to the exclusion of Pholidota (Patterson *et al.*, 1992).

An alternative view was proposed by Emry (1970), who specifically excluded relationship with Xenarthra and instead transferred the palaeanodont families to the Pholidota, without formally recognizing Palaeanodonta. The same arrangement was adopted by McKenna and Bell (1997). As Emry (1970) observed, palaeanodonts such as metacheiromyids share derived characters with pholidotans, suggesting a close relationship between the two (see also Rose and Emry, 1993). Nonetheless, it is possible that these resemblances also are convergent, and that palaeanodonts belong to an independent radiation of fossorial mammals not specially related to either Xenarthra or Pholidota (Rose, 1979; Rose and Emry, 1993).

The association of Xenarthra and Pholidota in a clade Edentata separate from other eutherians (= Epitheria) (e.g., Novacek and Wyss, 1986; Novacek, Wyss, and McKenna, 1988; McKenna, 1992) is rather weak, being based on a relatively small number of characters, many of which are now believed to be homoplasious, inaccurate, or ambiguous (Rose and Emry, 1993). Pholidota have also been linked with Carnivora, again weakly, in some morphological and molecular studies (e.g., Shoshani, 1986; de Jong, Leunissen, and Wistow, 1993; Sarich, 1993; Shoshani and McKenna, 1998). McKenna and Bell (1997) included Pholidota as a suborder of their order Cimolesta. These varied assessments underscore that the phylogenetic position of Pholidota remains very unstable. The problematic Palaeanodonta aside, living and fossil pholidotans are currently classified in three families: Eomanidae (for middle Eocene *Eomanis*), Patriomanidae (*Patriomanis* and Oligo-Miocene *Necromanis*), and the extant Manidae (Szalay and Schrenk, 1998; Storch, 2003).

Dental and postcranial anatomy of the oldest and most primitive palaeanodonts resembles that of leptictids and pantolestoid cimolestans, suggesting that palaeanodonts evolved from or shared a common ancestor with one of these groups (Szalay, 1977; Rose, 1978; Rose and Lucas, 2000). This will have obvious implications for the phylogenetic position of Pholidota, if further evidence corroborates a close relationship between palaeanodonts and pholidotans. The oldest known leptictids, pantolestoids, and palaeanodonts are all from North America. Nothing now known would preclude a North American origin of palaeanodonts, and there is no convincing evidence to suggest their origin elsewhere.

Existing evidence indicates dispersal of palaeanodonts from North America to Asia and Europe by the early Eocene, which could be related to the origin of Pholidota sensu stricto, presumably in the Old World. In view of the Paleocene age (Itaboraian South American Land-Mammal Age) of the oldest true xenarthrans (e.g., Cifelli, 1983; Oliveira and Bergqvist, 1998), any relationship between palaeanodonts and Xenarthra presumably would have required an early Paleocene (or older) dispersal to South America.

INFRAORDINAL

Three families of palaeanodonts have been described: the primitive and probably paraphyletic Escavadodontidae, and the more derived Metacheiromyidae and Epoicotheriidae. All share a similarly specialized fossorial skeleton, but Metacheiromyidae and

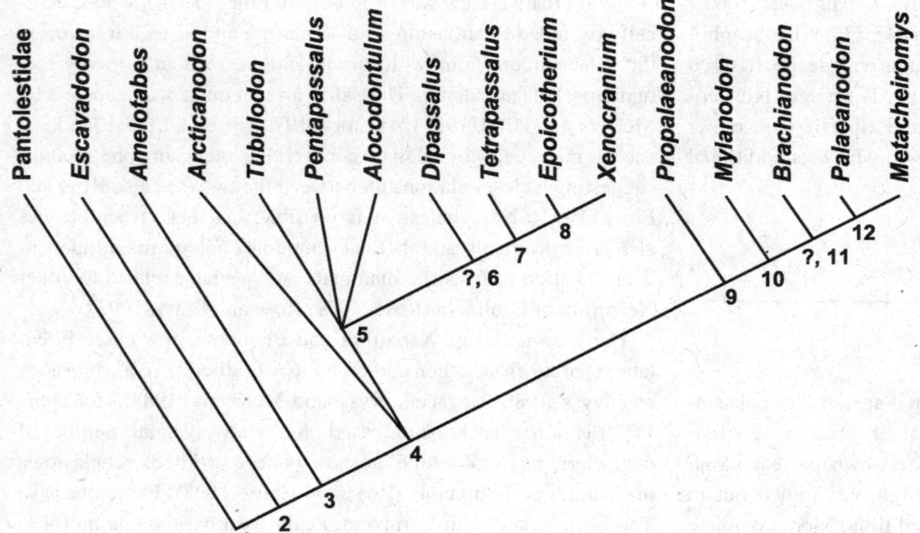

Figure 9.2. Cladogram of palaeanodonts, based mainly on Rose, Krishtalka, and Stucky (1991) and Rose and Lucas (2000), with modifications. *Pentapassalus* and *Alocodontulum* are excluded but are considered very close to or synonymous with *Tubulodon*. For reasons discussed in the text, positions of *Dipassalus* and *Brachianodon* are ambiguous. Characteristics at nodes 1–12 are as follows. 1. Humerus with long, elevated deltopectoral crest (more than half total humeral length), prominent and medially extended entepicondyle, very large supinator crest; ulna with elongate and medially inflected olecranon process. 2. Upper molars with stylar shelf and ectocingulum reduced or absent; upper canines with prominent honing facet and triangular in cross section; manubrium sterni thick and with weak to moderate keel, distal end of deltopectoral crest broad and rounded; humerus with prominent teres tubercle; distal radius expanded; tibial shaft bowed concave laterally. 3. Cheek teeth with degenerate crown morphology and thin enamel; lower molars with low trigonids and lacking any remnants of cingula; dentary robust with medial buttress posteriorly. 4. Cheek teeth reduced in size, simplified, and single-rooted or with coalesced roots (except p4 and m1, which may be two-rooted). 5. Six or more postcanines (at least one postcanine lost); supinator crest with hooklike proximal projection; olecranon process relatively longer and metacarpals shorter and more robust than in metacheiromyids. 6. Dentary foreshortened; five or fewer postcanines, reduced to single-rooted pegs separated by wide diastemata; head of humerus proximodistally elongate, lesser tuberosity distally displaced and medially projecting, deltopectoral crest less elevated; trochanteric fossa reduced, tibia and fibula synostosed at both ends. 7. Greater tubercle reduced and lower than humeral head, deltopectoral crest low, pectoral part curved (convex medially) and ending in very distal pectoral process; cervical vertebrae 2–5 synostosed. 8. Hooklike process of supinator crest very long, olecranon process wide and extremely long. 9. Postcanines further reduced in size, more simplified, and separated by short diastemata. 10. Six or fewer postcanines, with diastema behind last postcanine. 11. Five or fewer postcanines, relatively smaller, single-rooted, and quickly worn to simple pegs. 12. Longer edentulous region at back of dentary.

Epoicotheriidae were more derived both postcranially and dentally than Escavadodontidae. Moreover, Metacheiromyidae typically achieved greater dental reduction than contemporaneous epoicotheres, whereas epoicotheres attained more specialized postcranial skeletons.

Proposed interrelationships among palaeanodont genera are depicted in Figure 9.2. Torrejonian *Escavadodon* is the most primitive known palaeanodont and is plausibly the sister taxon or ancestor of all later palaeanodonts. Tiffanian *Amelotabes* is the next most primitive known palaeanodont. Although it retains seven relatively unreduced postcanines, it shows progressive dental features that ally it with later epoicotheriids. *Propalaeanodon* and later metacheiromyids may have evolved independently from *Escavadodon* or a similar form, but nothing now known about *Amelotabes* precludes it from the ancestry of metacheiromyids (which would render Epoicotheriidae paraphyletic as well).

Most epoicotheriid genera are monotypic, raising the question whether the family is taxonomically oversplit. There is mounting evidence that *Tubulodon*, *Pentapassalus*, and *Alocodontulum* are closely allied and may be congeneric; still, inconsistencies between them in dental formula, crown morphology, tooth spacing, and number of roots justify maintaining these genera until these matters are resolved. Among later epoicotheres, *Epoicotherium* and *Xenocranium* are probably sister taxa (though European *Molaetherium* may be even closer to *Epoicotherium*), with *Tetrapassalus* closely allied. *Dipassalus* seems to belong with epoicotheres, but its unique dental configuration and unusual postcranial skeleton make its phylogenetic position uncertain.

The metacheiromyid genera *Propalaeanodon–Palaeanodon–Metacheiromys* constitute a chronologic and structural sequence that plausibly reflects phylogeny. Despite the presence in *Propalaeanodon* of seven postcanines extending posteriorly in the dentary (a primitive feature shared with *Amelotabes*), these teeth were already reduced to the small pegs characteristic of Metacheiromyidae. *Brachianodon* seems to share the dental configuration found in *Palaeanodon*, but the apparent retention of multicusped postcanines, as well as certain postcranial traits, complicates its phylogenetic position.

Because of the reduction in size and number of postcanines, comparison of m2 lengths in the next section is irrelevant in all but the most dentally conservative genera.

INCLUDED GENERA OF PALAEANODONTA AND PHOLIDOTA

The locality numbers listed for each genus refer to the list of unified localities in Appendix I. The locality numbers may be listed more than one way. The acronyms for museum collections are listed in Appendix III.

Parentheses around the locality (e.g., [CP101]) mean that the taxon in question at that locality is cited as an "aff." or "cf." the taxon in question. Parentheses are usually used for individual species, implying that the genus is firmly known from the locality, but the actual species identification may be questionable. Question marks in front of the locality (e.g., ?CP101) mean that the taxon is questionably known from that locality, implying some doubt that the taxon is actually present at that locality, either at the genus or species level.

PALAEANODONTA MATTHEW, 1918

ESCAVADODONTIDAE ROSE AND LUCAS, 2000

Characteristics: Lower dental formula i?1, c1, p4, m3; differ from other palaeanodonts in having tribosphenic upper molars and posterior premolars that lack a hypocone, with the stylar shelf very reduced, and cingula weak or absent: lower molariform teeth leptictid-like, with distinct trigonid and talonid; last lower premolar molariform, with extended trigonid as in leptictids. Skeleton robust and fossorially specialized, but less so than in other palaeanodonts: deltopectoral crest broad and shelflike but relatively shorter, radius not as expanded distally, and metacarpals less robust: distal tibiofibular joint synostosed, in contrast to most other primitive palaeanodonts.

Escavadodontidae is based solely on *Escavadodon zygus*, but it is possible that *Amelotabes* belongs here as well. The family is distinguished from other palaeanodonts primarily by its primitive features, and it may be paraphyletic. Although *Escavadodon* shares a number of derived postcranial features with other palaeanodonts, its molariform posterior premolars and lower molar structure, and the distally fused tibiofibula, suggest ties with Leptictidae.

Escavadodon Rose and Lucas, 2000

Type species: *Escavadodon zygus* Rose and Lucas, 2000.

Type specimen: NMMNH P22051; dentaries, fragmentary maxillae, several teeth, and associated partial skeleton.

Characteristics: Same as for family.

Length of m2 in referred specimen: 2.6 mm.

Included species: *E. zygus* only, known from locality SB23H* only.

METACHEIROMYIDAE WORTMAN, 1903

Characteristics: Dentaries long, shallow, and robust; one small lower incisor present, apparently no upper incisors present; canines large and typically triangular in cross section at the base; postcanine teeth reduced to simple, rounded pegs separated by wide diastemata (in later forms reduced in number and absent from posterior part of the jaw, perhaps replaced by a horny plate): skull relatively longer and occiput narrower than in epoicotheriids; palate with shallow median longitudinal groove; postcrania robust and fossorially specialized, as for Palaeanodonta. (Based mostly on Matthew, 1918, and Simpson, 1931.)

Propalaeanodon Rose, 1979

Type species: *Propalaeanodon schaffi* Rose, 1979.

Type specimen: MCZ 20122; left dentary with p2-3 and alveoli for p1, p4, m1-3.

Characteristics: Primitive metacheiromyid retaining seven postcanines which extend to the back of the horizontal ramus; tooth crowns simplified, peglike, and all but p4 single-rooted; p4 two-rooted, m1-2 and possibly p3 with bilobed single root; humerus (only known postcranial element) similar to that in *Palaeanodon*.

Included species: *P. schaffi* only, known from locality CP13G* only.

Mylanodon Secord, Gingerich, and Bloch, 2002

Type species: *Mylanodon rosei* Secord, Gingerich, and Bloch, 2002.

Type specimen: UM 109174; left dentary with two cheek teeth.

Characteristics: Palaeanodont with lower postcanine teeth relatively larger and more complex than in any other metacheiromyid except *Propalaeanodon*, but relatively smaller and less complex than in the epoicotheres *Amelotabes*, *Tubulodon*, *Alocodontulum*, and *Pentapassalus*; apparently six postcanines retained, separated by short diastemata; p2, p3, and m2 single-rooted; m3 absent and short edentulous section present behind m2, in contrast to *Propalaeanodon*. Differs from other palaeanodonts in having a molariform lower first molar with only a protoconid on the trigonid and only a hypoconid on the talonid. Enamel lacking on all postcanine teeth. (Based mainly on Secord *et al.*, 2002.)

Included species: *M. rosei* only, known from locality CP13G* only.

Comments: *Mylanodon* is both morphologically and stratigraphically intermediate between *Amelotabes* and *Palaeanodon* and it lacks any specializations that would prevent it from being directly ancestral to the later-occurring *Palaeanodon*. *Mylanodon* has a double-rooted lower first molar like *Amelotabes* but unlike *Palaeanodon*, in which all postcanines are single-rooted. Comparison with *Propalaeanodon* is more difficult because no comparable teeth are present. *Propalaeanodon* has seven postcanines separated by short diastemata and extending almost to the ascending ramus. The p4 is the largest postcanine and is double-rooted, whereas m1 has a weakly bilobed root not unlike that in *Mylanodon*. Without preserved tooth crowns, it is impossible to know if p4–m1 of *Mylanodon* are more complex than their homologues in *Propalaeanodon*. However, existing evidence suggests that these two taxa are very closely related and that *Mylanodon* represents a plausible structural and phylogenetic intermediate between *Propalaeanodon* and *Palaeanodon*.

Palaeanodon Matthew, 1918

Type species: *Palaeanodon ignavus* Matthew, 1918.

Type specimen: AMNH 15086; incomplete skull and skeleton.

Characteristics: Postcanine teeth vestigial, peglike, with rounded crowns, and mostly single-rooted; five in dentary, at least four in maxilla (back of dentary and maxilla edentulous); canines robust and crown distinctly triangular in cross section; dentary with prominent medial buttress.

Included species: *P. ignavus* (known from localities CP20A, B*, C, D); *P. nievelti* (localities CP19AA*–C, CP20A); *P. parvulus* (localities [CP14E], CP17B, CP18B*).

Palaeanodon sp. is also known from localities?SB24, CP19A, CP20A -BB, [CP27D], ?NP49.

***Metacheiromys* Wortman, 1903**

Type species: *Metacheiromys marshi* Wortman, 1903.

Type specimen: YPM 12903; left dentary and associated fragments.

Characteristics: Teeth more reduced than in any other palaeanodont: canine and one postcanine above, incisor, canine, and two postcanines below, all single-rooted; canines remain large, more laterally compressed than in other palaeanodonts, other teeth vestigial; enamel present but very thin; most of the dentary and palate edentulous, medial buttress smaller than in *Palaeanodon*; ossified bullae complete; postcranial skeleton not markedly modified from that in *Palaeanodon.* (Mainly after Simpson, 1931.)

Included species: *M. marshi* (including *M. tatusia*) (known from locality CP34C*); *M. dasypus* (including *M. osborni*) (localities CP34C*, D).

Metacheiromys sp. is also known from locality CP33B.

Comments: Simpson (1931) recognized four species of *Metacheiromys*, but the known variation seems to fit better within two somewhat variable or sexually dimorphic species (Rose, 1978; Schoch, 1984).

***Brachianodon* Gunnell and Gingerich, 1993**

Type species: *Brachianodon westorum* Gunnell and Gingerich, 1993.

Type specimen: UM 98743; dentaries, cranial fragments, and most of skeleton.

Characteristics: At least five postcanine teeth in dentary; crowns with enamel, and with three apical cuspules present on unworn postcanines; medial buttress of dentary weaker than in *Palaeanodon*; tibia, astragalar neck, and distal calcaneus all relatively shorter than in other metacheiromyids; humeral head longer and narrower than in *Metacheiromys*. (Based on Gunnell and Gingerich, 1993.)

Included species: *B. westorum* only, known from locality CP34A* only.

Comments: The only tooth preserved in the holotype is heavily worn, revealing no evidence of apical cusps. The inference that the postcanines originally had three cuspules is, therefore, based on a referred specimen, which also preserves only one tooth. As noted by Gunnell and Gingerich (1993), certain features that separate *Brachianodon* from other metacheiromyids suggest that it may be more closely related to epoicotheriids (particularly the presence of multicusped postcanines that extend posteriorly almost to the ascending ramus).

Indeterminate metacheiromyds

Metacheiromys sp. is also known from locality CP25J2.

EPOICOTHERIIDAE SIMPSON, 1927

Characteristics: Skull usually shorter and caudally broader, and jaws shorter, than in metacheiromyids; postcanine teeth variably reduced in size and number and in crown morphology, but present all along horizontal ramus in nearly all forms, often relatively larger than in contemporary metacheiromyids but reduced to single-rooted pegs in the most derived forms; crowns often worn obliquely or gabled, as in dasypodids; postcranial skeleton more specialized for fossorial habits than in metacheiromyids.

***Amelotabes* Rose, 1978**

Type species: *Amelotabes simpsoni* Rose, 1978.

Type specimen: PU 14855; right dentary with p2–3, m1-2, and alveoli for incisor, canine, p1, p4, and m3.

Characteristics: Primitive palaeanodont having at least one small incisor, a large canine, and seven postcanines (four premolars and three molars, all double-rooted except p1) extending to the back of the horizontal ramus; all teeth with thin enamel and separated by very short diastemata; molars relatively low crowned, more so lingually than buccally, with relatively flat occlusal surfaces; despite wear on the occlusal surfaces, a recognizable, albeit poorly defined, cusp pattern retained in *Amelotabes*, typically obliterated during the earliest stages of wear in all other palaeanodonts except *Tubulodon*, *Alocodontulum*, and *Mylanodon*; dentary shallow and robust, with well-developed medial buttress.

Length of m2 in holotype: 2.6 mm.

Included species: *A. simpsoni* only, known from locality CP13F* only.

***Tubulodon* Jepsen, 1932**

Type species: *Tubulodon taylori* Jepsen, 1932.

Type specimen: PU 13418l right dentary with m1–2, left dentary with p4-m2 and associated fragments, reported by Jepsen as Lostcabinian in age but now known to be from earliest Bridgerian (Gardnerbuttean) of the Wind River Basin.

Characteristics: At least five lower postcanines (front of dentary unknown), without obvious diastemata; molars extending to back of horizontal ramus, occlusal surfaces oval, with little relief; p4–m3 with poorly defined, peripheral cusps; enamel thin on sides, absent from or worn off occlusal surface; roots of p4–m2 appear single or faintly bilobate at base of crown but may separate into two closely appressed roots within alveolus; m3 single-rooted.

Length of m2 in holotype: 2.4 mm.

Included species: *T. taylori* only (known from localities CP27B-C, CP27D*, E).

Tubulodon sp. is also known from localities (CC1) and CP20BB.

Comments: New specimens of *Tubulodon* from the Wind River Basin and *Alocodontulum* from the Bighorn Basin indicate that these two genera are anatomically very similar to each other and to *Pentapassalus*. The latter differs from the others primarily in having more heavily worn teeth. It is probable, therefore, that the three genera are congeneric, the latter two being junior synonyms of *Tubulodon* (Rose, Krishtalka, and Stucky, 1991; K. D. Rose and R. J. Emry, unpublished data).

Alocodontulum Rose, Bown, and Simons, 1978 (including _Alocodon_ Rose, Bown, and Simons, 1977, non _Alocodon_ Thulborn, 1973)

Type species: *Alocodontulum atopum* (Rose, Bown, and Simons, 1977).

Type specimen: YPM 30790; concretion containing fragmented right maxilla with canine, P3–4, M1–3, root of P2, and questionably P1 root and incisor alveolus; right dentary with canine and one postcanine (misidentified by Rose, Bown, and Simons [1977] as the left maxilla); and associated postcrania.

Characteristics: Upper dental formula reported to be I1, C1, P4, M3, but the presence of the incisor and P1 ambiguous, and no other anterior upper dentitions found to confirm their presence. Canines of typical palaeanodont form: triangular in cross section and with matching honing facets; P3 (and P4?) distinctly premolariform, apparently three-rooted, and crown enamel covered; M1-2 longer than wide and with distinct longitudinal furrow, cusps poorly defined and peripheral, and enamel thin and restricted to sides of crown; M3 vestigial and single-rooted; P3–M3 not separated by diastemata; dentary with large canine (incisor region unknown) and six postcanines, p4 and m1 with two closely appressed roots, other teeth single-rooted; limb elements show fossorial adaptations more or less intermediate between those of *Palaeanodon* and *Pentapassalus* (Rose, 1990; Rose, Emry, and Gingerich, 1992).

Included species: *A. atopum* only, known from locality CP20B* only.

Pentapassalus Gazin, 1952 (including _Palaeanodon_, in part)

Type species: *Pentapassalus pearcei* Gazin, 1952.

Type specimen: USNM 20028; skull and dentaries with nearly complete dentition, and substantial part of skeleton.

Characteristics: About the size of *Tubulodon*; one small lower incisor present, status of upper incisors unknown; canines typically large and triangular in cross section; five upper and six lower postcanines, separated by short diastemata anteriorly but essentially no diastemata present from last premolar through molars; skull triangular, very broad posteriorly; postcranial skeleton distinctively adapted for fossorial habits: limb elements more derived than those of metacheiromyids in being comparatively shorter and more robust; olecranon process relatively longer relative to distal ulna.

Included species: *P. pearcei* (known from locality CP25G*); *P. woodi* (localities CP27B*-E).

Comments: *Pentapassalus woodi* was initially placed in *Palaeanodon* by Guthrie (1967), but it is clearly an epoicotheriid, not a metacheiromyid (Rose, 1978). Additional material of *P. woodi* was reported by Rose, Krishtalka, and Stucky (1991), but *P. pearcei* is still known only from the holotype. As noted above, *Pentapassalus* is a probable synonym of *Tubulodon*.

Tetrapassalus Simpson, 1959

Type species: *Tetrapassalus mckennai* Simpson, 1959.

Type specimen: AMNH 56030; right dentary with two postcanines and alveoli for canine and two other postcanines.

Characteristics: Diminutive epoicotheriid with five single-rooted, peglike postcanines preceded by a somewhat larger canine, all separated by short diastemata; weak medial buttress.

Included species: *T. mckennai* (known from localities [NB14], CP34D*); *T. proius* (locality CP34C*).

Tetrapassalus sp. is also known from localities [CP5A], CP33B, CP34A, [D], [CP36B].

Comments: Simpson (1959) defined the type species as having four postcanines (hence the generic name), but the holotype is broken and may initially have held five postcanines; four upper and five lower postcanines are present in cf. *T. mckennai* from the Elderberry Canyon Local Fauna, Nevada (locality NB14; Emry, 1990).

Epoicotherium Simpson, 1927 (including _Xenotherium_ Douglass, 1905; _Pseudochrysochloris_ Reed and Turnbull, 1967)

Type species: *Epoicotherium unicum* (Douglass, 1905) (originally described as *Xenotherium unicum*, also equivalent to *Pseudochrysochloris yoderensis* Reed and Turnbull, 1967).

Type specimen: CM 1018; edentulous skull with alveoli for canine and five postcanines.

Characteristics: Very small palaeanodont; cheek teeth single-rooted and cylindrical, maxilla containing canine and five postcanines, dentary holding incisor, canine, and five postcanines; occipital region distinctly domed; known postcrania similar to those of *Xenocranium* (though smaller), but humerus with relatively even larger supinator crest and entepicondyle, and ulna with longer olecranon. (Based on Rose and Emry, 1983.)

Included species: *E. unicum* only (known from localities CP39B, C, F, CP42A, NP25C*).

Epoicotherium sp. is also known from localities NP10B, NP22, NP25B.

Comments: *Epoicotherium* is questionably known from the early Oligocene of Europe (McKenna and Bell, 1997).

Xenocranium Colbert, 1942

Type species: *Xenocranium pileorivale* Colbert, 1942.

Type specimen: ANSP 14984; skull (lacking snout) and dentaries.

Characteristics: Approximately twice the linear dimensions of *Epoicotherium* and generally similar; dentition as in *Epoicotherium* but only four upper postcanines; skull with upturned, spatulate snout, very reduced orbits, very broad and domed occipital region, wide lambdoid crests, and greatly inflated squamosals enclosing what appears to be the hypertrophied head of the malleus. Postcranial skeleton possessing some of the most extreme fossorial specializations known in any mammal, including cervical vertebrae 2–5 synostosed; scapula with very elevated spine, very large bifid acromion, secondary spine, and postscapular fossa; humerus with very long supinator crest and entepicondyle and very distal pectoral process; ulna with extraordinarily large, inflected olecranon; large carpal sesamoid; very short and robust metapodials and phalanges, except for manual unguals, which are large, curved, and laterally compressed; tibia and fibula fused at both ends. (Based on Rose and Emry, 1983.)

Included species: *X. pileorivale* only (known from localities CP39C, CP40A, CP98BB).

Comments: The precise locality of the holotype is uncertain. Colbert (1942) reported that it came from Brule (= Orellan) age sediments "about 15 miles east of Hat Creek Post Office, Wyoming," but all other known specimens are Chadronian.

Dipassalus Rose, Krishtalka, and Stucky, 1991

Type species: *Dipassalus oryctes* Rose, Krishtalka, and Stucky, 1991.

Type specimen: CM 43286; partial skeleton including snout occluded with dentaries.

Characteristics: Small epoicothere about the size of *Epoicotherium*; jaws short, with large canine and only two postcanines in both lower and upper jaws; back of dentary edentulous as in most metacheiromyids, and with well-developed medial buttress; postcranial elements showing fossorial specializations approximately intermediate between those of *Pentapassalus* and *Epoicotherium*.

Included species: *D. oryctes* only (known from localities CP27B, D, E*).

Indeterminate epoicotheres

Epoicotherium spp. are also known from localities CP25GG, J2, CP34C.

?PALAEANODONTA, INCERTAE SEDIS

Melaniella Fox, 1984

Type species: *Melaniella timosa* Fox, 1984.

Type specimen: UA 21509; right dentary with one premolariform tooth and alveolus for large canine.

Characteristics: "Robust but shallow dentary lacking incisors, and possessing a large, laterally compressed canine followed by a long diastema; anterior-most cheek tooth two rooted, having a primitive premolariform crown, and followed by a second diastema" (Fox, 1984).

Included species: *M. timosa* only, known from locality NP3C only.

Comments: As Fox (1984) observed, the unique specimen of this taxon differs from palaeanodonts in having a more primitive first postcanine and longer diastemata. Its assignment to palaeanodonts must be considered doubtful.

Arcticanodon Rose, Eberle, and McKenna, 2004

Type species: *Arcticanodon dawsonae* Rose, Eberle, and McKenna, 2004.

Type specimen: NUFV 10 (= CMN 51389); left dentary with canine, root of incisor, and alveoli for p2–4, m1–2.

Characteristics: "Small palaeanodont with six postcanine teeth: p2–3 one-rooted, p4 and m1–2 clearly double-rooted (m3 root number unknown); teeth closely spaced except for short diastema between canine and p2. Slightly larger than half the size of *Tubulodon* (= *Alocodontulum* and *Pentapassalus*). Differs from *Tubulodon* in having relatively larger p2–3, no diastema around p3, and distinctly double-rooted m2. Differs from *Escavadodon, Amelotabes*, and *Auroratherium* in lacking p1, and in root configuration of postcanines. Further differs from *Escavadodon* in having relatively more robust dentary, and from *Amelotabes* in having teeth behind p2 more closely spaced. Differs from metacheiromyids in having closely-spaced, relatively larger postcanines, presumably with more complex crowns" (Rose *et al.*, 2004).

Included species: *A. dawsonae* only, known from locality HA1A0 only.

Comments: This small palaeanodont is similar in mandibular morphology to *Tubulodon*, but the resemblances are predominantly plesiomorphous. The Ellesmere record of *Arcticanodon* is the most northern occurrence of a palaeanodont.

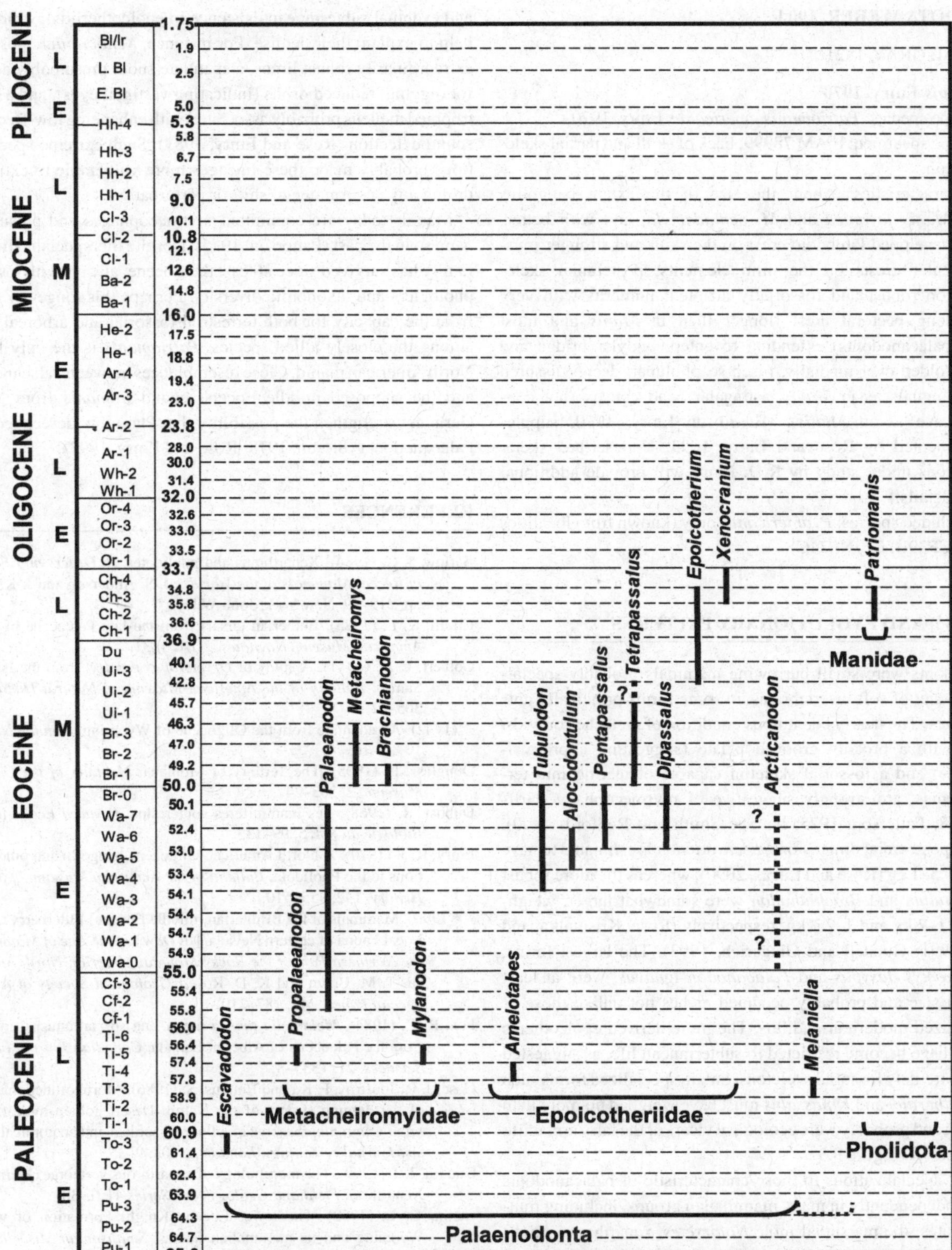

Figure 9.3. Temporal ranges of North American palaeanodont genera.

PHOLIDOTA WEBER, 1904

MANIDAE GRAY, 1821

***Patriomanis* Emry, 1970**

Type species: *Patriomanis americanus* Emry, 1970.

Type specimen: F:AM 78999; back of skull and partial skeleton.

Characteristics: About the size of the extant pangolins *Manis javanica* and *M. pentadactyla*, and with better-developed lambdoid crest on the skull and stronger muscular crests on the limb elements; postcranial skeleton robust and fossorially adapted; humerus with very long pectoral crest (longer than in *Manis* and most palaeanodonts) extending to entepicondylar bridge and folded over medially, fused scapholunar, deeply fissured unguals as in *Manis*, astragalar head convex (not concave as in *Manis*). (Based on Emry, 1970, supplemented by Rose and Emry, 1993. Undescribed skeletons under study by R. J. Emry will provide additional details.)

Included species: *P. americanus* only (known from localities CP39B, C*, NP24C).

BIOLOGY AND EVOLUTIONARY PATTERNS

Palaeanodonts were small burrowing mammals probably specialized for a diet of soft invertebrates, in some forms principally ants and termites. The nearly edentulous condition of metacheiromyids, coupled with a broadly grooved palate (suggesting a protrusible tongue) and a fossorial skeleton capable of tearing into termite colonies, are strongly suggestive of myrmecophagy (Griffiths, 1968; Patterson, 1975; but see Smith and Redford, 1990). Based on postcranial dimensions, *Escavadodon* is estimated to have weighed 0.5–1 kg (Rose and Lucas, 2000), whereas the more robust *Alocodontulum* and *Brachianodon* were somewhat larger, weighing about 1–2 kg and 1–2.7 kg, respectively (Rose, Krishtalka, and Stucky, 1991; Gunnell and Gingerich, 1993). The largest forms, *Metacheiromys dasypus* and *Palaeanodon ignavus*, were at least twice that size and probably occupied niches not unlike those of medium-sized modern armadillos. The more derived epoicotheres may well have become restricted to subterranean life, as suggested by their extremely fossorial skeletons. Small, specialized forms such as *Epoicotherium* and *Dipassalus* must have weighed no more than 100–200 g and probably were ecological vicars of the fairy armadillo *Chlamyphorus* (Figure 9.3).

Similar specializations to those characteristic of palaeanodonts evolved independently in many mammalian groups, including multituberculates, the marsupial mole *Notoryctes*, xenarthrans, pholidotans, and chrysochlorid insectivores (e.g., Dubost, 1968; Gasc, Jouffroy, and Renous, 1986; Kielan-Jaworowska, 1989). The most specialized of them tend to be rare and of restricted geographic and temporal range. Palaeanodonts were never very diverse, and known lineages became progressively more committed to specialized diets (myrmecophagy in metacheiromyids) or habitats (digging and eventual subterranean existence in epoicotheriids), which may help to explain their demise. For instance, *Xenocranium* possessed extraordinarily robust limbs, a spatulate snout presumably adapted for digging, reduced orbits (indicating vestigial eyes), and a hypertrophied malleus probably associated with enhancing low-frequency sound detection (Rose and Emry, 1983). Such extreme specializations probably made these lineages more vulnerable to extinction during any environmental shift or perturbation.

Manids today are committed myrmecophages and presumably were so in the past (Patterson, 1975). Despite this specialization, the family has survived since at least the Eocene, always with low morphological and taxonomic diversity. Perhaps this longevity results from the capacity for both terrestrial/fossorial and arboreal habits among the closely allied species. *Patriomanis* is the only known North American manid. Close resemblances between palaeanodonts and the supposed middle Eocene manid *Eomanis* from Messel Germany strengthen the possibility that Pholidota descended from Palaeanodonta (Storch, 1978; Rose and Emry, 1993).

REFERENCES

Barlow, J. C. (1984). Xenarthrans and pholidotes. In *Orders and Families of Recent Mammals of the World*, ed. S. Anderson and J. K. Jones, pp. 219–39. New York: John Wiley.

Cifelli, R. L. (1983). Eutherian tarsals from the late Paleocene of Brazil. *American Museum Novitates*, 2761, 1–31.

Colbert, E. H. (1941). A study of *Orycteropus gaudryi* from the Island of Samos. *Bulletin of the American Museum of Natural History*, 78, 305–51.

(1942). An edentate from the Oligocene of Wyoming. *Notulae Naturae*, 109, 1–16.

Douglass, E. (1905). The Tertiary of Montana. *Memoirs of the Carnegie Museum*, 2, 203–24.

Dubost, R. (1968). Les mammifères souterrains. *Revue d'Ecologie et de Biologie du Sol*, 5, 99–133.

Emry, R. J. (1970). A North American Oligocene pangolin and other additions to the Pholidota. *Bulletin of the American Museum of Natural History*, 142, 455–510.

(1990). Mammals of the Bridgerian (middle Eocene) Elderberry Canyon Local Fauna of eastern Nevada. [In *Dawn of the Age of Mammals in the Northern Part of the Rocky Mountain Interior, North America*, ed. T. M. Bown and K. D. Rose.] *Geological Society of America Special Paper*, 243, 187–210.

Fox, R. C. (1984). *Melaniella timosa* n.gen. and sp., an unusual mammal from the Paleocene of Alberta, Canada. *Canadian Journal of Earth Sciences*, 21, 1335–8.

Gasc, J. P., Jouffroy, F. K., and Renous, S. (1986). Morphofunctional study of the digging system of the Namib Desert golden mole (*Eremitalpa granti namibensis*): cinefluorographical and anatomical analysis. *Journal of Zoology, London A*, 208, 9–35.

Gaudin, T. J. (1999). The morphology of xenarthrous vertebrae (Mammalia: Xenarthra). *Fieldiana Geology New Series* 41, 1–38.

Gazin, C. L. (1952). The lower Eocene Knight Formation of western Wyoming and its mammalian faunas. *Smithsonian Miscellaneous Collections*, 117, 1–82.

Gingerich, P. D.(1989). New earliest Wasatchian mammalian fauna from the Eocene of northwestern Wyoming: composition and diversity in a rarely sampled high-floodplain assemblage. *University of Michigan Papers on Paleontology*, 28, 1–97.

Gray, J. E. (1821). On the natural arrangement of vertebrose animals. *London Medical Repository*, 15, 296–310.

Griffiths, M. (1968). *Echidnas*. Oxford: Pergamon Press.

Gunnell, G. F. and Gingerich, P. D. (1993). Skeleton of *Brachianodon westorum*, a new middle Eocene metacheiromyid (Mammalia, Palaeanodonta) from the early Bridgerian (Bridger A) of the southern Green River Basin, Wyoming. *Contributions from the Museum of Paleontology, University of Michigan*, 28, 365–92.

Guthrie, D. A. (1967). The mammalian fauna of the Lysite Member, Wind River Formation, (Early Eocene) of Wyoming. *Memoirs of the Southern California Academy of Sciences*, 5, 1–53.

Heissig, K. (1982). Ein Edentate aus dem Oligozan Suddeutschlands. *Mitteilungen der Bayerischen Staatssammlung für Paläontologie und historische Geologie*, 22, 91–6.

Jepsen, G. L. (1932). *Tubulodon taylori*, a Wind River Eocene tubulidentate from Wyoming. *Proceedings of the American Philosophical Society*, 71, 255–74.

Jong, W. W. de, Leunissen, J. A. M., and Wistow, G. J. (1993). Eye lens crystallins and the phylogeny of placental orders: evidence for a macroscelid–paenungulate clade? In *Mammal Phylogeny: Placentals*, ed. F. S. Szalay, M. J. Novacek and M. C. McKenna, pp. 5–12. New York: Springer-Verlag.

Kielan-Jaworowska, Z. (1989). Postcranial skeleton of a Cretaceous multituberculate mammal. *Acta Palaeontologica Polonica*, 34, 75–85.

Koenigswald, W. von (1969). Die Maniden (Pholidota, Mamm.) des europaischen Tertiars. *Mittelungen der Bayerischen Staatssammlung für Palaeontologie und Historische Geologie*, 9, 61–71.

Koenigswald, W. von and Martin, T. (1990). Ein Skelett von *Necromanis franconica*, einem Schuppentier (Pholidota, Mammalia) aus dem Aquitan von Saulcet im Allier-Becken (Frankreich). *Ecologae Geologicae Helvetiae*, 83, 845–64.

McKenna, M. C. (1975). Toward a phylogenetic classification of the Mammalia. In *Phylogeny of the Primates*, ed. W. P. Luckett and F. S. Szalay, pp. 21–46. New York: Plenum Press.

(1992). The alpha crystallin A chain of the eye lens and mammalian phylogeny. *Annales Zoologici Fennici*, 28, 349–60.

McKenna, M. C. and Bell, S. K. (1997). *Classification of Mammals Above the Species Level*. New York: Columbia University Press.

Matthew, W. D. (1906). Fossil Chrysochloridae in North America. *Science*, 24, 786–8.

(1918). A revision of the lower Eocene Wasatch and Wind River faunas. Part V. Insectivora (continued), Glires, Edentata. *Bulletin of the American Museum of Natural History*, 38, 565–657.

Novacek, M. J. and Wyss, A. (1986). Higher-level relationships of the recent eutherian orders: morphological evidence. *Cladistics*, 2, 257–87.

Novacek, M. J., Wyss, A., and McKenna, M. C. (1988). The major groups of eutherian mammals. In *The Phylogeny and Classification of Tetrapods,* Vol. 2: *Mammals*, ed. M. J. Benton, pp. 31–71. [*Systematics Association Special Volume*, 35B.] Oxford: Clarendon Press.

Oliveira, E. V. and Bergqvist, L. P. (1998). A new Paleocene armadillo (Mammalia, Dasypodoidea) from the Itaboraí Basin, Brazil. *Asociación Paleontológica Argentina, Publicación Especial* 5, 35–40.

Osborn, H. F. (1904). An armadillo from the Middle Eocene (Bridger) of North America. *Bulletin of the American Museum of Natural History*, 20, 163–5.

Patterson, B. (1975). The fossil aardvarks (Mammalia: Tubulidentata). *Bulletin of the Museum of Comparative Zoology*, 147, 185–237.

(1978). Pholidota and Tubulidentata. In *Evolution of African Mammals*, ed. V. J. Maglio and H. B. S. Cooke, pp. 268–78. Cambridge, MA: Harvard University Press.

Patterson, B., Segall, W., Turnbull, W. D., and Gaudin, T. J. (1992). The ear region in xenarthrans (= Edentata: Mammalia) Part II. Pilosa (sloths, anteaters), palaeanodonts, and a miscellany. *Fieldiana, Geology New Series*, 24, 1–79.

Romer, A. S. (1966). *Vertebrate Paleontology*. Chicago: University of Chicago Press.

Rose, K. D. (1978). A new Paleocene epoicotheriid (Mammalia), with comments on the Palaeanodonta. *Journal of Paleontology*, 52, 658–74.

(1979). A new Paleocene palaeanodont and the origin of the Metacheiromyidae (Mammalia). *Breviora*, 455, 1–14.

(1990). Postcranial skeletal remains and adaptations in early Eocene mammals from the Willwood Formation, Bighorn Basin, Wyoming. [In *Dawn of the Age of Mammals in the Northern Part of the Rocky Mountain Interior, North America*, ed. T. M. Bown and K. D. Rose.] *Geological Society of America Special Paper*, 243, 107–33.

Rose, K. D. and Emry, R. J. (1983). Extraordinary fossorial adaptations in the Oligocene palaeanodonts *Epoicotherium* and *Xenocranium* (Mammalia). *Journal of Morphology*, 175, 33–56.

Rose, K. D. and Lucas, S. G. (2000). An early Paleocene palaeanodont (Mammalia,?Pholidota) from New Mexico, and the origin of Palaeanodonta. *Journal of Vertebrate Paleontology*, 20, 133–50.

Rose, K. D., Bown, T. M., and Simons, E. L. (1977). An unusual new mammal from the early Eocene of Wyoming. *Postilla, Yale Peabody Museum*, 172, 1–10.

(1978). *Alocodontulum*, a new name for *Alocodon* Rose, Bown and Simons 1977, non Thulborn, 1973. *Journal of Paleontology*, 52, 1162.

Rose, K. D., Krishtalka, L., and Stucky, R. K. (1991). Revision of the Wind River faunas, early Eocene of central Wyoming. Part 11. Palaeanodonta (Mammalia). *Annals of the Carnegie Museum*, 60, 63–82.

Rose, K. D., Emry, R. J., and Gingerich, P. D. (1992). Skeleton of *Alocodontulum atopum*, an early Eocene epoicotheriid (Mammalia, Palaeanodonta) from the Bighorn Basin, Wyoming. *Contributions from the Museum of Paleontology, University of Michigan*, 28, 221–45.

(1993). Relationships of Xenarthra, Pholidota, and fossil "edentates": the morphological evidence. In *Mammal Phylogeny: Placentals*, ed. F. S. Szalay, M. J. Novacek, and M. C. McKenna, pp. 81–102. New York: Springer-Verlag.

Rose, K. D., Eberle, J. J., and McKenna, M. C. (2004). *Arcticanodon dawsonae*, a primitive new palaeanodont from the lower Eocene of Ellesmere Island, Canadian High Arctic. *Canadian Journal of Earth Sciences*, 41, 757–63.

Sarich, V. M. (1993). Mammalian systematics: twenty-five years among their albumins and transferrins. In *Mammal Phylogeny: Placentals*, ed. F. S. Szalay, M. J. Novacek and M. C. McKenna, pp. 103–14. New York: Springer-Verlag.

Schoch, R. M. (1984). Revision of *Metacheiromys* Wortman, 1903 and a review of the Palaeanodonta. *Postilla, Yale Peabody Museum*, 192, 1–28.

Secord, R., Gingerich, P. D., and Bloch, J. I. (2002). *Mylanodon rosei*, a new metacheiromyid (Mammalia, Palaeanodonta) from the late Tiffanian (late Paleocene) of northwestern Wyoming. *Contributions from the Museum of Paleontology, The University of Michigan*, 30, 385–99.

Shoshani, J. (1986). Mammalian phylogeny: comparison of morphological and molecular results. *Molecular Biology and Evolution*, 3, 222–42.

Shoshani, J. and McKenna, M. C. (1998). Higher taxonomic relationships among extant mammals based on morphology, with selected comparisons of results from molecular data. *Molecular Phylogenetics and Evolution*, 9, 572–84.

Simpson, G. G. (1927). A North American Oligocene edentate. *Annals of the Carnegie Museum*, 17, 283–98.

(1931). *Metacheiromys* and the Edentata. *Bulletin of the American Museum of Natural History*, 59, 295–381.

(1945). The principles of classification and a classification of the mammals. *Bulletin of the American Museum of Natural History*, 85, 1–350.

(1959). A new middle Eocene edentate from Wyoming. *American Museum Novitates*, 1950, 1–8.

Smith, K. and Redford, K. H. (1990). The anatomy and function of the feeding apparatus in two armadillos (Dasypoda): anatomy is not destiny. *Journal of Zoology, London*, 222, 27–47.

Storch, G. (1978). *Eomanis waldi*, ein Schuppentier aus dem Mittel-Eozan der "Grube Messel" bei Darmstadt (Mammalia: Pholidota). *Senckenbergiana Lethaea*, 59, 503–29.

(2003). Fossil Old World "edentates." *Senckenbergiana Biologica*, 83, 51–60.

Storch, G. and Richter, G. (1992). Pangolins: almost unchanged for 50 million years. In *Messel: An Insight into the History of Life and of the Earth*, ed. S. Schaal and W. Ziegler, pp. 201–7. Oxford: Clarendon Press.

Storch, G. and Rummel, M. (1999). *Molaetherium heissigi* n. gen., n.sp., an unusual mammal from the early Oligocene of Germany (Mammalia: Palaeanodonta). *Paläontologische Zeitschrift*, 73, 179–85.

Szalay, F S. (1977). Phylogenetic relationships and a classification of the eutherian Mammalia. In *Major Patterns in Vertebrate Evolution*, ed. M. K. Hecht, P. C. Goody, and B. M. Hecht, pp. 315–74. New York: Plenum Press.

Szalay, F. S. and Schrenk, F. (1998). The middle Eocene *Eurotamandua* and a Darwinian phylogenetic analysis of "edentates." *Kaupia-Darmstädter Beiträge zur Naturgeschichte*, 7, 97–186.

Thulborn, R. A. (1973). Teeth of ornithischian dinosaurs from the Upper Jurassic of Portugal. *Mémoires des Services Géologique du Portugal*, 22, 89–134.

Tong, Y. and Wang, J. (1997). A new palaeanodont (Mammalia) from the early Eocene of Wutu Basin, Shandong Province. *Vertebrata Palasiatica*, 35, 110–20.

Turnbull, W. D. and Reed, C. A. (1967). *Pseudochrysochloris*, a specialized burrowing mammal from the early Oligocene of Wyoming. *Journal of Paleontology*, 41, 623–31.

Weber, M. (1904). *Die Säugetiere. Einführing in die Anatomie und Systematik der recenten und fossilen Mammalian.* Jena: Gustav Fischer.

Wortman, J. L. (1903). Studies of Eocene Mammalia in the Marsh collection, Peabody Museum. *American Journal of Science*, 16, 345–68.

10 Xenarthra

H. GREGORY MCDONALD[1] and VIRGINIA L. NAPLES[2]

[1]*Park Museum Management Program, National Park Service, Fort Collins, CO, USA*
[2]*Northern Illinois University, DeKalb, IL, USA*

INTRODUCTION

The Xenarthra includes the living armadillos, tree sloths, and anteaters. It is divided into the Cingulata which includes the armadillos, with living and extinct species and the extinct pampatheres and glyptodonts, and the Pilosa, which includes sloths and anteaters. The Pilosa in North America were represented by four groups of sloths (mylodonts, megatheres, nothrotheres, and megalonychids) and the anteaters, while Cingulates in the North American Tertiary included glyptodonts, pampatheres, and armadillos.

While the xenarthrans originated in South America and most of their evolutionary history is recorded on that continent, beginning in the late Tertiary members of the order began dispersing into North America. Since Thomas Jefferson's description of the sloth *Megalonyx* from a cave in West Virginia in 1799, representatives of every major group within the order have been described as present in North America. Most xenarthrans in North America, both the pilosans (sloths) and cingulates (armadillos, glyptodonts and pampatheres) survived until the end of the Pleistocene. The only representative of the order present in temperate North America today is the armadillo, *Dasypus novemcinctus*, while tree sloths, anteaters, and additional genera of armadillos occur in the tropical portions from Mexico south to Panama.

The diets of the members of the order range from herbivory, including both browsers and grazers, among the sloths, glyptodonts, and pampatheres to omnivory in the armadillos to obligate specialized insectivory (myrmecophagy) in the anteaters. While the broad feeding categories among the Xenarthra would appear to overlap with many of the previously established mammalian species in North America, members of the order seem to have "insinuated" themselves into the North American mammalian fauna (Patterson and Pascual, 1963; Marshall, 1981). Their success indicates they were sufficiently different ecologically to minimize competition or ecological overlap with other North American mammals.

Unlike many lineages of North American mammals that radiated and diversified following their entry into South America, there was little or no diversification of the xenarthran lineages in North America. Many groups had multiple species, but these were chronospecies reflective of gradualistic morphological changes, and not radiation nor diversification. Only in a couple of instances was there sufficient morphological change from the invading ancestral species to the terminal species that the end members are placed in separate genera.

DEFINING FEATURES OF THE ORDER XENARTHRA

CRANIAL

The skull is simple and tubular and the zygomatic arch may be incomplete (sloths) (Figure 10.1) or greatly reduced retaining only the zygomatic process on the maxilla and temporal (anteaters). In both sloths and glyptodonts, the zygomatic arch has an enlarged ascending and descending process, suggesting a complex jaw musculature (Naples, 1987).

CINGULATA

The skull of glyptodonts is deep, robust, and foreshortened and the maxilla is deep, unlike the low and elongated skull of pampatheres and armadillos. The zygomatic arch is complete in all three groups but glyptodonts are distinguished from the others by the development of an elongated, mediolaterally broadened and anteroposteriorly compressed flange for the origin of the masseter muscles.

PILOSA

Primitively the zygomatic arch may be reduced or incomplete, in that the jugal does not contact the zygomatic process of the squamosal

Evolution of Tertiary Mammals of North America, Vol. 2. ed. C. M. Janis, G. F. Gunnell, and M. D. Uhen. Published by Cambridge University Press.

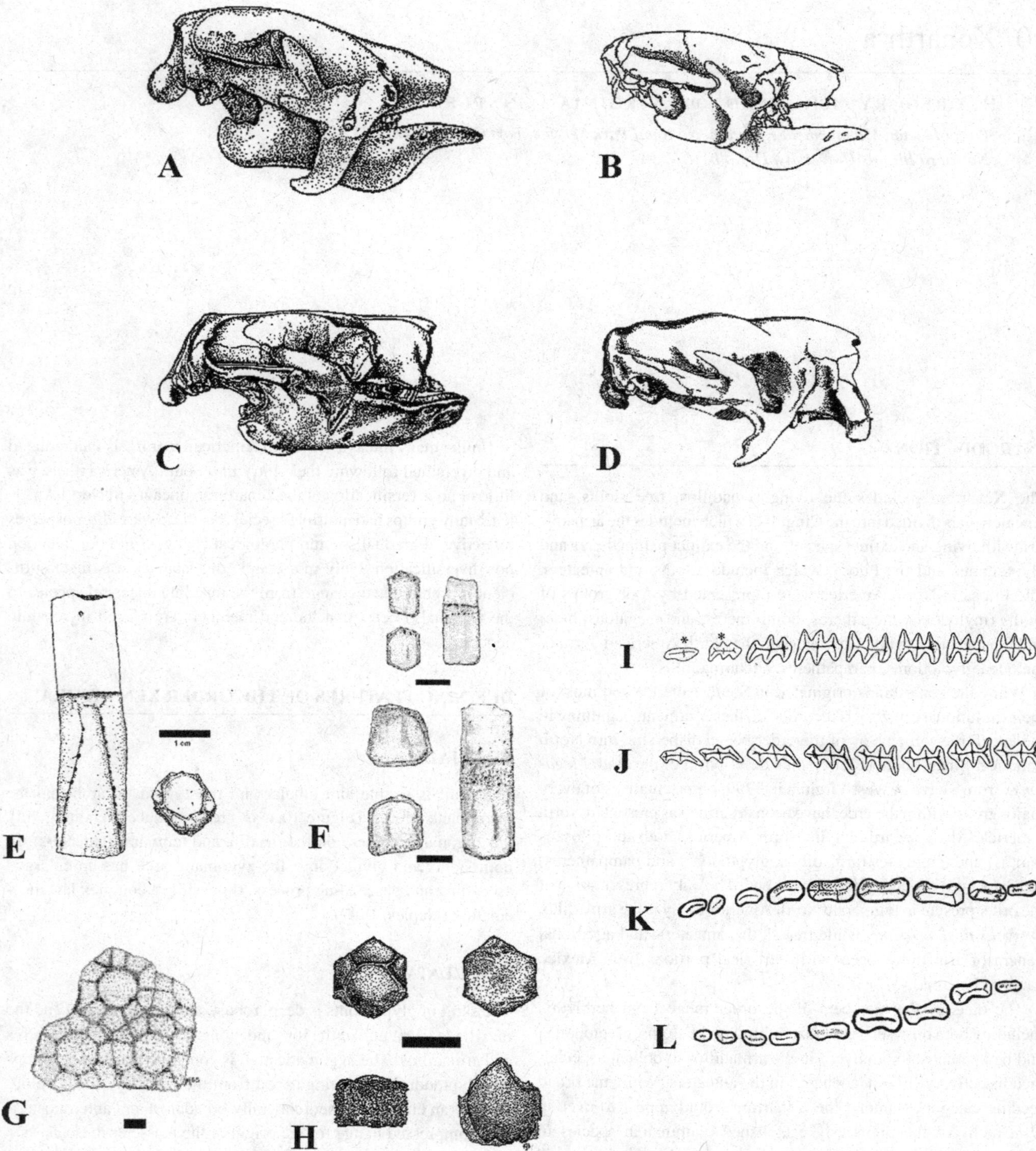

Figure 10.1. A–D. Skulls of North American ground sloths. A. *Eremotherium laurillardi* (after Gazin, 1957). B. *Nothrotheriops shastensi* (after Stock, 1925). C. *Paramylodon harlani* (after Stock, 1925). D. *Megalonyx jeffersonii* (after McDonald, 1977). E–H. Osteoderms of North American cingulates. E. Imbricating osteoderm and buckler osteoderm of *Dasypus bellus* (after Voorhies, 1987). F. Imbricating osteoderm and buckler osteoderms of *Holmesina floridanus* (upper) and buckler osteoderms of *Holmesina septentrionalis* (after Edmund, 1985a). G. Carapace osteoderms of *Glyptotherium texanum* (after Akersten, 1972). H. Carapace osteoderm of *Pachyarmatherium leiseyi* (after Downing and White, 1995). Scale bars = 1 cm. I–L. Occlusal view of dentitions of North American cingulates. Anterior is to the left. I. *Glyptotherium arizonae* upper left dentition (modified from Gillette and Ray, 1981). J. *Glyptotherium arizonae* lower left dentition (modified from Gillette and Ray, 1981). K. *Holmesina septentrionalis* upper left dentition (modified from Edmund, 1985b). L. *Holmesina septentrionalis* lower left dentition (modified from Edmund, 1985b). (B,C: courtesy of the Carnegie Institution of Washington. E: courtesy of the Southwestern Association of Naturalists. F, G: courtesy of the University of Texas.)

such as in *Nothrotheriops*. A complete zygomatic arch may be secondarily acquired as in *Megalonyx* and *Eremotherium*.

The skull in the mylodonts is relatively broader than in other sloths and expanded anteriorly. The ascending process of the jugal is oriented more posterodorsally than in other sloths and the descending process is robust and also posteriorly oriented. The anterior portion of the squamosal process fits between the ascending and descending processes but is not fused to them. The orbits are low on the lateral surface of the skull. The predental spout of the mandible is wider than long. The condyle is approximately at the level of the tooth row.

The pilosan anteaters are characterized by an extreme elongation of the skull, loss of dentition, and a dramatic reduction of the zygomatic arch.

DENTAL

An older term for the order, "Edentata," referred to the absence of teeth in the anteaters or Vermilinguas. Except for the anteaters, all members of the order have teeth. The teeth differ from most other mammals in the absence of enamel and are composed only of dentine and cementum. There is a harder outer dentine and the softer inner dentine core, which results in differential wear that shapes the occlusal surface of the tooth during mastication (Naples, 1995). Morphology of the teeth is simple and columnar and the teeth are ever-growing. Except for some armadillos, xenarthrans lack milk dentition; the absence of genetically determined cusps and their simple columnar structure prevents determination of homology with other mammals. The number of teeth is reduced (e.g., sloths) or lost altogether in others (e.g., anteaters); there is an increase in the number of teeth in some armadillos.

CINGULATA

The number of cheek teeth in glyptodonts is eight, and in pampatheres nine (Figure 10.1I–L), while the number varies in armadillos. The teeth tend to be homodont and only vary slightly in morphology. Glyptodont teeth are distinctively trilobate. The shape of the tooth in pampatheres changes from oval to reniform (kidney-shaped) to bilobate from front to back in the tooth series. The first tooth in *Holmesina* is in the premaxilla. The teeth of armadillos are simple and peglike.

PILOSA

In addition to the simplification of the dentition and loss of enamel seen also in the cingulates, the sloths have a reduced number of teeth with a maximum of five upper and four lower, although the front tooth may be lost so the number is reduced to four upper and three lower. The dentition is not as homodont as in the cingulates and in many forms the anterior-most tooth may be either chisel-shaped or caniniform, a strictly functional term to distinguish it from the molariform cheek teeth. The caniniform is triangular in cross section with the occlusal surface oblique to the long axis of the tooth in *Pliometanastes* but in *Megalonyx* it is oval with the occlusal surface forming at a right angle to the axis of the tooth. The molariforms in megalonychids are triangular with the mesial and distal edges of the occlusal surface forming lophs. The ever-growing teeth erupt as simple cones and acquire their characteristic adult form exclusively through wear (Naples, 1995).

The dentition of the Tertiary North American mylodont sloths is five upper and four lower teeth. The first tooth is a caniniform. The molariforms are lobate with a flat occlusal surface.

The dentition in *Eremotherium* is quadrangular with the occlusal surface of each tooth having a pair of sharp transverse lophs separated by a V-shaped trough.

POSTCRANIAL

There are numerous distinctive features of the skeleton that define members of the order (Figure 10.2). The order's name, Xenarthra, refers to the presence of accessory articulations on the lumbar vertebrae, although the morphology and number varies among the various members of the order (Gaudin, 1999). All members of the order have a synsacrum formed by the fusion of the transverse processes of the proximal caudals to the ischium. The sternum is segmented but each sternebrae supports two articular facets, which provide for a movable articulation with the ossified sternal ribs, another character distinctive of the order.

All xenarthrans retain the clavicle. It is generally stoutly constructed and provides support and a large surface area of major muscles that facilitate digging and climbing. The scapula has an enlarged secondary spine along its ventral border paralleling the true spine. The blade ventral to this secondary spine serves as the origin of the teres minor muscle but the degree of development varies among the different groups. The head of the humerus is round or oval, permitting significant freedom of movement of the forelimbs of sloths and anteaters.

In the xenarthrans, the proximal and middle phalangeal joints are capable of little motion while the ungual and middle phalanx of each digit can be flexed to greater than 90 degrees, particularly in sloths and anteaters. The unguals may be either claws or hooves. Claws are present in the sloths, anteaters, and armadillos on both manus and pes, and the claw on the third digit is usually the largest. In the pampatheres, the manus has claws, but the unguals on the pes are hooflike. In the glyptodonts, unguals on both manus and pes are hooflike.

As the names Pilosa and Cingulata suggest, one of the primary distinguishing features between the two groups is the presence of a hairy coat in the former while the latter group possess a bony armored carapace. The presence of cingulates in a fauna is often indicated by the presence of disarticulated and scattered individual segments of the armor. Fortunately the ornamentation of the armor is distinctive in different groups making identification even to species possible (Edmund, 1985a). Small dermal ossicles are present in the North American mylodont *Glossotherium chapadmalense* but have not been recorded in *Thinobadistes*. They are not present in any of the other North American sloths.

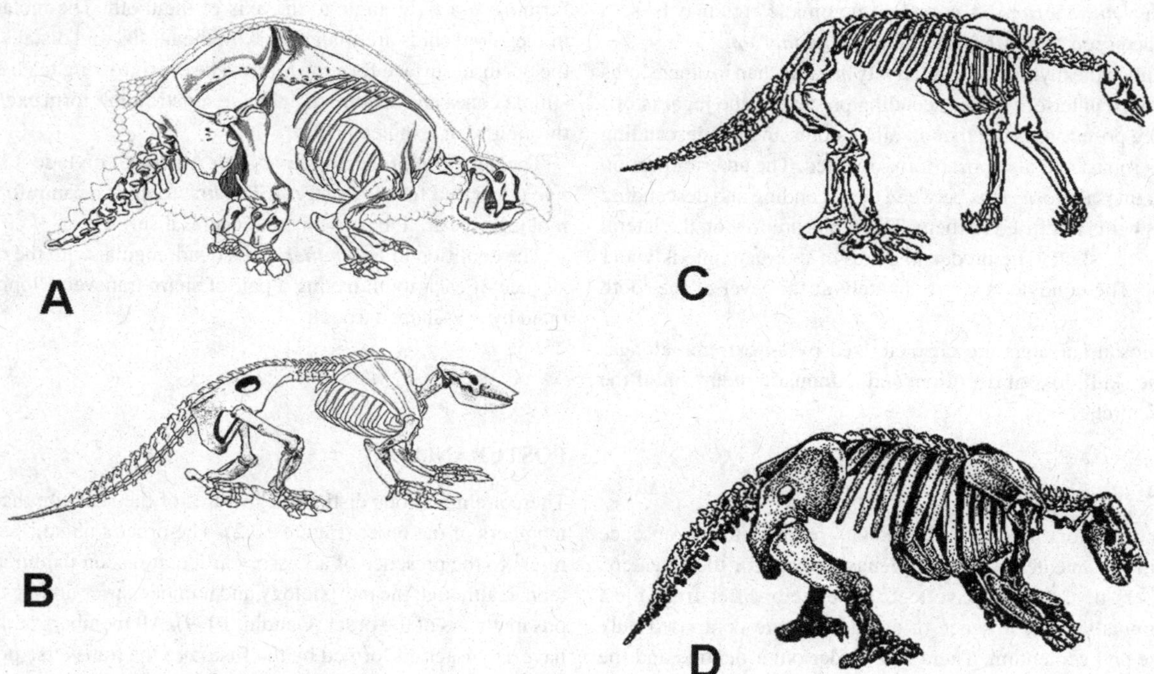

Figure 10.2. Skeletons of Xenarthrans. A. *Glyptotherium arizonae* (after Gillette and Ray, 1981). B. *Holmesina septentrionalis* (after Edmund, 1985b). C. *Nothrotheriops shastensis* (after Stock, 1925). D. *Paramylodon harlani* (after Stock, 1925). (C,D: courtesy of the Carnegie Institution of Washington.)

CINGULATA

The cervicals are fused from the axis to the seventh vertebra in all cingulates to form a cervical tube. In *Glyptotherium*, the second, third, and fourth thoracic vertebrae are fused to form the trivertebral element to which the first rib is fused (Gillette and Ray, 1981). Thoracic vertebrae five through thirteen are fused to form a dorsal tube. The lumbar vertebrae are fused to the synsacrum to form a single structure.

The humerus lacks an entepicondylar foramen, in contrast to some of the South American genera in which it is present. The femur has a prominent third trochanter in all three groups of cingulates. In *Glyptotherium*, it is shifted distally and is continuous with the lateral epicondyle. The third trochanter in pampatheres and armadillos is positioned about midshaft. The tibia and fibula are fused in the adults in all three groups. The unguals on the manus are clawlike, less so in *Glyptotherium* than in *Holmesina* and *Dasypus*. The unguals of the pes in *Glyptotherium* are hooflike but in *Holmesina* the unguals of the manus are clawed while those of the pes are hooflike; the unguals of the manus and pes of *Dasypus* are claws.

All cingulates have a carapace consisting of individual bony osteoderms (Figure 10.1). Armadillos and pampatheres have transverse imbricating bands that allow some movement of the carapace, while in glyptodonts the dermal ossicles fuse to form a single immovable structure. While the presence of dermal armor is characteristic of the cingulates, the construction of the carapace and the caudal tube varies considerably, particularly in the form of the external ornamentation. In living armadillos, the number of movable imbricating bands is variable and this is also true for fossil forms. The extant species of armadillo in North America, *Dasypus novemcinctus*, generally has nine bands, but the lack of any complete carapaces in the extinct species *D. bellus* precludes determining how many bands its carapace possessed. The pampathere *Holmesina* has three movable bands (Edmund, 1985a). The other osteoderms making up the carapace are polygonal, and generally hexagonal, although the number of sides can vary. A casque formed from numerous osteoderms covers the top of the skull in all three groups, and a tube of osteoderms surrounds the tail. In *Glyptotherium*, the caudal armor is composed of ten rings with each ring formed by a double row of osteoderms for each caudal vertebra (Gillette and Ray, 1981). The caudal armor in pampatheres and armadillos is also formed of rings. In *Holmesina*, there is a single row of osteoderms per ring, but in *Dasypus* each ring is composed of a double row of osteoderms.

PILOSA

In the scapula, the acromion process is enlarged so that it fuses with the coracoid process to form the acromiocoracoid arch. There is a coracoid foramen, which is also present in anteaters and is a shared skeletal feature that unites sloths and anteaters in the Pilosa. In the astragalus, pilosans are distinguished from the cingulates by the concave surface of the navicular process. The surface is convex in cingulates as in other mammals. The proximal surface of the navicular in pilosans has a complementary raised mammilary process. The articular surface for the patella is continuous with the distal condyles in mylodonts, the two surfaces are separate in megalonychids, and in megatheres it connects only with the lateral condylar surface.

The humerus in megalonychids retains an entepicondylar foramen, which is lost in the mylodonts and megatheres. The ulna and radius in megalonychids and megatheres are about equal in length to the humerus while in the mylodonts they are about half the length of the humerus. In all sloths, the tibia and fibula are considerably shortened and are about half the length of the femur. All North American sloths lack an ungual on digit five. In megalonychids and mylodonts, there are claws on digits one through four. This is also characteristic of the Blancan megathere *Eremotherium eomigrans*, although the number of claws is reduced in later species. The claws in megalonychids and megatheres are triangular in cross section while those of mylodonts are semi-circular. The pes in mylodonts and megatheres is rotated so the weight of the body is supported by the fifth metatarsal and calcaneum. The astragalus has a prominent odontoid process and the lateral trochlear surface is flat. There is no modification of the astragalus in megalonychids and it more generally resembles that of other mammals.

SYSTEMATICS

SUPRAORDINAL

The monophyly of the Xenarthra has clearly been accepted based on numerous morphological synapomorphies (Gaudin, 2003), many of them skeletal, as described above, and has been independently supported by recent molecular studies (Delsuc *et al.*, 2002). Xenarthrans have long been considered primitive and the sister group to other placental mammals (McKenna, 1975; Shoshani and McKenna, 1998), but this view of their relationship to other mammals has recently been greatly modified by molecular studies. Recent work (Delsuc *et al.*, 2002) divides placental mammals into two Southern Hemisphere clades, Xenarthra and Afrotheria (including Proboscideans, Sirenians, Hyracoids, and Tublidentata) and a Northern Hemisphere clade composed of the Laurasiatheria (Carnivora, Pholidota, Artiodactyla, Perissodactyla, and Chiroptera) and Euarchontoglires (Insectivora, Primates, Lagomorpha, and Rodentia). As such, xenarthrans represent part of the basal radiation of placental mammals, with subsequent diversification and evolution in South America during its isolation as an island continent throughout most of the Tertiary.

INFRAORDINAL

Xenarthrans are readily subdivided into two major groups, the Pilosa (hairy forms) including sloths and anteaters, and the Cingulata (armored forms), that includes living armadillos and the extinct pampatheres and glyptodonts (Figure 10.3). Sloths, often placed in the Phyllophaga (or Tardigrada), are divided into four well-defined families, Megatheriidae, Megalonychidae, Mylodontidae, and Nothrotheriidae, each of which has at least one representative in North America, with the first three present in the Tertiary of North America and the last family, Nothrotheriidae, restricted to the Pleistocene. Anteaters consist of a single family, Myrmecophagidae, with a single Pleistocene (Irvingtonian) record in North America.

Currently two major clades within the Cingulata are recognized: Dasypoda for armadillos and the Glyptodonta, which includes both pampatheres and glyptodonts, although there is still some question about the relationship of pampatheres. Representatives of each major clade of the Cingulata are present in the North American Tertiary and each has members that survived into the Pleistocene. Armadillos are subdivided in the dasypodids and euphractines and only dasypodids are present in late Tertiary and Pleistocene North American faunas (Figure 10.4).

INCLUDED GENERA OF XENARTHRA COPE, 1899

The locality numbers listed for each genus refer to the list of unified localities in Appendix I. The locality numbers may be listed more than one way. The acronyms for museum collections are listed in Appendix III.

Parentheses around the locality (e.g. [CP101]) mean that the taxon in question at that locality is cited as an "aff." or "cf." Parentheses are usually used for individual species, implying that the genus is firmly known from the locality, but the actual species identification may be questionable. Question marks in front of the locality (e.g., ?CP101) mean that the taxon is questionably known from that locality, implying some doubt that the taxon is actually present at that locality, either at the genus or species level. An asterisk (*) indicates the type locality.

CINGULATA ILLIGER, 1811

GLYPTODONTIDAE GRAY, 1869

***Glyptotherium* Osborn, 1903**

Type species: *Glyptotherium texanum* Osborn, 1903.

Type specimen: AMNH 10704 (carapace, caudal vertebrae, caudal armor, pelvis, seven chevrons).

Characteristics: *Glyptotherium* is a medium- to large-sized glyptodont; skull short and truncated with a flat dorsal profile that is anteriorly inclined; nasal aperture trapezoidal in outline; postorbital bar incomplete; manus lacks digit one, digit five is reduced and digits two, three and four are large; caudal vertebrae unfused, each caudal protected by individual movable rings, except the "sacrocaudals" and the last two or three, which may be fused into a short tube; individual osteoderms of the carapace firmly articulated and immovable except in the anterolateral region, each osteoderm polygonal with a rosette sculpting and the central figure larger or equal in size to the peripherals, central figure convex, flat, or weakly concave while peripheral figures are flat (Gillette and Ray, 1981).

Included species: *G. texanum*, (known from localities SB18B, SB49, SB50, SP1H, SP5); *G. arizonae* (localities GC14A, GC15B, C, [GC16], GC17B, GC18IIB, GC18IIIA, SB14E, F*).

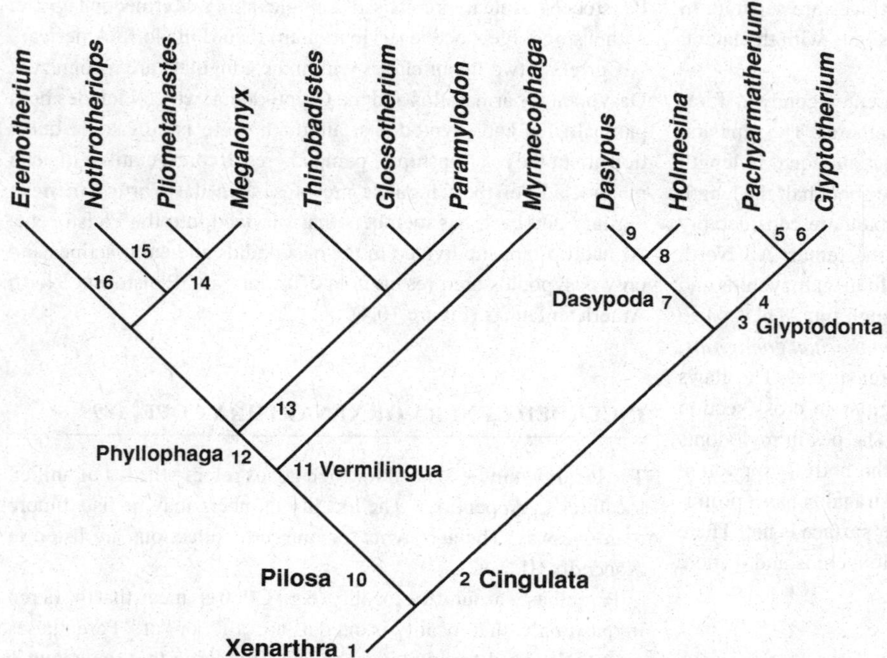

Figure 10.3. Proposed relationships of the North American xenarthrans. This cladogram is restricted to North American xenarthran taxa and is not meant to be a comprehensive overview of all xenarthran relationships. Characters listed are from Engleman (1985), Gaudin (2003) and, research in progress by McDonald and is not meant to be comprehensive. Characteristics at node 1–16 are as Follows. 1. XENARTHRA: xenarthrous vertebrae, synacrum incorporating proximal caudal vertebrae; sternal ribs with synovial articulations with sternum; dentition ever-growing and lacking enamel. 2. CINGULATA: carapace composed of osteoderms, axis fused to one or more cervical vertebrae; tibia and fibula fused in all forms; strong development of the greater trochanter above the head of the femur, third trochanter always strongly developed. 3. GLYPTODONTA: solid carapace with no imbricating bands, osteoderms with rosette patterns; dentition lobate; rostrum short square and not tapering; anteriorly inclined ascending ramus of mandible, deep horizontal ramus of mandible. 4. GLYPTODONTIDAE: dentition trilobate; jugal with anteroposteriorly compressed descending flange; premaxillae greatly reduced or lost. 5. GLYPTOTALINAE: anterior portion of carapace imbricating on posterior portion, osteoderms with rosette offset ventrally, osteoderms thick in proportion to size. 6. GLYPTODONTINAE: rosette centrally placed on osteoderms, osteoderms relatively thin for size; dentition trilobate. 7. DASYPODA: carapace flexible with imbricating bands of osteoderms; dentition simple and ovate to bilobate; triangular anteriorly expanding lachrymal; anteriorly tapering rostrum; shallow horizontal ramus of mandible. 8. PAMPATHERIIDAE: carapace with only three rows of imbricating osteoderms; dentition reniform to bilobate. 9. DASYPODIDAE: three or more rows of imbricating osteoderms in carapace; teeth simple and peglike. 10. PILOSA: coracoid foramen in scapula, navicular process of astragalus with concavity; zygomatic arch incomplete but may form complete arch secondarily. 11. VERMILINGUA: no dentition: skull elongated, jugal reduced and zygomatic arch incomplete. 12. PHYLLOPHAGA: dentition reduced to maximum of five upper and four lower; jugal with mediolaterally compressed descending flange. 13. MYLODONTIDAE: dentition lobate, tooth rows diverge anteriorly; calcaneum broadly expanded, astragalus with odontoid process; unguals semi-circular in cross section; entepicondylar foramen of humerus present in early forms but absent in North American forms. 14. MEGALONYCHIDAE: first tooth modified into large caniniform; astragalus unmodified and lacking odontoid process; calcaneum with tuber calcis expanded and winglike. 15. NOTHROTHERIIDAE: caniniform small, when present in early forms and separated from molariforms by diastema, but absent in later forms; vomer with keel exposed in internal nares; all unguals triangular in cross section except for on digit two of manus, which is semi-circular; astragalus with odontoid process. 16. MEGATHERIIDAE: dentition bilophate and square, no diastema in tooth row; humerus lacking entepicondylar foramen; astragalus with odontoid process.

Glyptotherium sp. (including *G. floridanum*, *G. cylindricum*, *G. mexicanum*) is also known from localities (GC15B), SB49, SB58IIC, SB65, SB65II, and from the Pleistocene of Florida, Oklahoma, Texas, New Mexico, and Mexico.

***Pachyarmatherium* Downing and White, 1995**

Type species: *Pachyarmatherium leiseyi* Downing and White, 1995.

Type specimen: UF 64347 (internal carapace osteoderm).

Characteristics: Currently *Pachyarmatherium* known only from its armor; the internal carapacial osteoderm small but extremely thick with a greater relative thickness for its size than for any other cingulate; central figure of each osteoderm polygonal and convex with no medial depression; central figure displaced toward one edge and larger than the peripheral figures, which vary from three to six in number and align in a single row; hair follicle pits positioned at the intersection of the groove separating the central figure from the peripheral figures and the radial grooves separating the peripheral figures (Downing and White, 1995).

Included species: *P. leiseyi* only (known from localities GC15C, D, GC16, GC17B, GC18IIB, GC18IIIA*, C, GC18IV, EM9).

PAMPATHERIIDAE EDMUND 1987

***Holmesina* Simpson, 1930 (including *Kraglievichia* in part)**

Type species: *Holmesina septentrionalis* (Leidy, 1889) (originally described as *Kraglievichia septentrionalis*).

Type specimen: WFIS 4076 (three osteoderms).

Characteristics: Anterior dentition of *Holmesina* distinguishes it from other pampatheres; second tooth subreniform with its long axis oriented at about 45° to the midline; third tooth reniform in outline and closely aligned with the other alveoli; all other cheek teeth bilobate.

Included species: *H. septentrionalis* (known from locality GC15F, and also known from the later Irvingtonian and the Rancholabrean); *H. floridanus* Robertson, 1976 only (localities GC14A, D, GC15A*–D, GC16, GC17A, B, GC18IIB, GC18IIIA–C, GC18IV).

Holmensina sp. is also known from localities SB68, GC17B. An indeterminate pampathere (noted as *Plaina* in Carranza-Castañeda and Miller, 2004), is also reported from localities SB58C and SP1H.

Comments: Robertson (1976) distinguished *H. floridanus* (at the time placed in *Kraglievichia*) from *Kraglievichia paranensis* of South America by the reniform fourth upper tooth, rather than peglike, and its long axis, oriented anterolingually instead of parallel to the tooth row. *H. floridanus* is much smaller than *H. septentrionalis* and size has been one of the primary criteria to distinguish the two species. As documented by Edmund (1985b) and Hulbert and Morgan (1993), there is a gradual size increase from *H. floridanus* to *H. septentrionalis* that blurs this distinction. Morphological characters that distinguish the two species include a deep depression on the palmar edge of the proximal end of the third metacarpal in *H. septentrionalis* that is absent in *H. floridanus* and the presence of a lip on the medial side of the proximal end of the second metatarsal in *H. floridanus*. This feature is absent in *H. septentrionalis*. In the astragalus, there is a medial projection present in the earlier species that is absent in the Rancholabrean species; the astragalus of the Rancholabrean species has deep concavity that forms on the anterior margin of the trochlea, which is absent in the earlier species. In the earlier species, the astragalocalcanear facet and sustentacular facets are confluent but become separate in *H. septentrionalis* (a similar separation is seen in the complementary facets on the calcaneum).

The buckler osteoderms in *Holmesina* are generally hexagonal but longer than wide. The internal surface is slightly concave, relatively smooth but with a fibrous appearance. The outer surface is perforated by numerous foramina. The ornamentation of the outer surface includes a marginal band, a shallow depression paralleling the outer margin of each osteoderm. The anterior portion of the marginal band is generally the widest part and bears a number of craterlike pits, which are interpreted as the site of hair follicles (Edmund, 1985a). The remainder of the surface of the osteoderm consists of a sharply defined submarginal ridge separated from the central raised keel by a depressed sulcus.

The osteoderm forming the movable or imbricating bands of the carapace has a length–width ratio of 2.5 to 1. The outer surface can be divided into three areas: an exposed ornamented part, a usually depressed rugose area, and a sharply defined raised table at the anterior end (Edmund, 1985a). The ornamented area occupies about half of the length of the osteoderm. A continuation of the submarginal band or at least an area of depression may divide the ornamented area from the rugose band. The smooth anterior part of the imbricating osteoderm fits under the internal surface of the posterior edge of the osteoderm anterior to it.

DASYPODIDAE BONAPARTE, 1838

Dasypus Linnaeus, 1758

Type species: *Dasypus novemcinctus*, the extant long-nosed armadillo.

Type specimen: (*D. bellus*) AMNH 23542 (single imbricating osteoderm).

Characteristics: The diagnosis of the genus *Dasypus* by Wetzel (1985) utilizes the following characters: skull with a long and slender rostrum constituting 55% or more of the length of the head; forelimb with four claws, the largest on the second and third digits; hind foot with five claws, with the longest on digit three; teeth smaller in diameter than other armadillos; palate extending far posterior to the tooth row; mandible longer and more slender than in other armadillos.

Included species: *D. bellus* Simpson, 1929 only (known from localities GC14A, D, GC15A–G, GC16, GC17A, B, GC18IIB, GC18IIIB, GC18IV, [CP121]). *D. bellus* is also known from numerous Pleistocene sites in the eastern United States (Klippel and Parmalee, 1984). *Dasypus* is also known from the late Pliocene or early Pleistocene to Recent of South America (McKenna and Bell, 1997).

Comments: While there are numerous fossil armadillo records in North America referred to the extinct species *D. bellus*, (Klippel and Parmalee, 1984), there has not been a systematic review of this species. While most records are based on isolated osteoderms, there is sufficient cranial and postcranial material to make direct comparisons with extant species of *Dasypus* to confirm the assignment of *D. bellus* to the genus or to permit reassignment, should that be warranted. Recognition of *D. bellus* has traditionally been based on the larger-sized osteoderms and skeletal elements compared with *D. novemcinctus*. While this criterion may be satisfactory for late Pleistocene material, as has been shown by Klippel and Parmalee (1984), Blancan members of the lineage are similar in size to *D. novemcinctus* and there is an increase in size in the lineage until its extinction at the end of the Pleistocene. A thorough reexamination of the North American record of *D. bellus* is needed.

PILOSA FLOWER, 1883

MEGALONYCHIDAE ZITTEL, 1892

***Megalonyx* Harlan 1825 (including *Megatherium*, in part)**

Type species: *Megalonyx jeffersonii* (Demarest, 1822) (originally described as *Megatherium jeffersonii*).

Type specimen: ANSP 12508 (left radius, ulna, second, third, fifth metacarpals and phalanges).

Cranial characteristics: *Megalonyx* skull shorter and stouter than in other sloths, reflecting a short facial region anterior to the orbit; zygomatic arch complete, with the anterior portion of the squamosal process fitting into a slot between the ascending and descending processes of the jugal, both processes gracile, triangular and posteriorly oriented; premaxillae reduced to two small plates fitting between the enlarged caniniforms.

Dental characteristics: Dentition five upper and four lower teeth, with the anterior-most tooth of the skull and mandible separated from the cheek teeth by a prominent diastema; anterior-most tooth greatly enlarged, modified into a caniniform, oval in cross section with a lingual column and the occlusal surface at a right angle to the axis of the tooth; molariforms triangular in cross section with prominent lophs on the mesial and distal edges; predental spout of the mandible greatly reduced in *Megalonyx*; condyle of the jaw above the level of the cheek teeth.

Postcranial characteristics: Humerus with entepicondylar foramen; manus with claws on digits one through four, claws triangular in cross section; third trochanter of the femur small but distinct and positioned midway on the shaft; calcaneum with expanded tuber calcis that is wing-like; astragalus primitive with no modification of medial trochlea into an odontoid process.

Included species: *M. curvidens* Matthew, 1924 (known from localities GC13B, CP115A*, CP116E, F); *M. leptostomus* Cope, 1893 (including *M. leptonyx*, *M. rohrmani*) (localities GC14A, D, GC15A–C, GC16, GC17A, B, GC18IIB, NB13B, SB15C, SB18D, SP1H, SP5, CP97, CP117A, B, CP131, PN3C, PN4, PN23A, C); *M. mathisi* Hirschfeld and Webb, 1968 (locality CC25B); *M. wheatleyi* (localities GC15D, F, GC18IIIA, GC18IV). *M. jeffersonii* is also known from the Rancholabrean.

Megalonyx sp. is also known from localities CC37, CC38B, CC40, CC46II, CC47(C or D), NB13B, SB49, SB50, SB58IIB, SP1D, CP116E, CP121, CP126II, CP128B, C, PN14.

***Pliometanastes* Hirschfeld and Webb, 1968**

Type species: *Pliometanastes protistus* Hirschfeld and Webb, 1968.

Type specimen: UF 9479 (partial skull).

Characteristics: *Pliometanastes* distinguished by a triangular caniniform with the occlusal surface oblique to the long axis of the tooth; predental spout of the mandible better developed than in *Megalonyx*; generally body size smaller than Hemphillian species of *Megalonyx*.

Included species *P. protistus* (known from localities GC11B*–D, GC12IIA, CC25AA); ?*P. galushai* Hirschfeld and Webb, 1968 (locality SB34C*).

Pliometanastes sp. is also known from localities GC12IIB, CC17I, SB34C, SP3A.

Indeterminate megalonychids

Indeterminate megalonychids are reported from localities CC36, CC37, CC40, CP116D, CP123C, PN11A, PN12B, PN14, CE1, (EM8B).

MEGATHERIIDAE OWEN, 1843

***Eremotherium* Spillman, 1948 (including *Schaubia; Megatherium*, in part)**

Type species: *Eremotherium laurillardi* (Lund, 1842) (originally described as *Megatherium laurillardi*).

Type specimen: ZMUC 1130 (isolated molariform tooth).

Cranial characteristics: *Eremotherium* skull robust, zygomatic arch secondarily fused; both ascending and descending rami of the jugal large and vertically oriented, ascending ramus with a second posterior process that probably represents a secondary ossification.

Dental characteristics: Five upper and four lower teeth in a continuous series, all molariform, square in cross section with well-developed transverse lophs on the mesial and distal edges of the occlusal surface, resulting in a greatly increased surface area for the entire row of cheek teeth compared with other sloths; mandible unusually deep with an elongate and narrow predental spout; condyle is elevated above the cheek teeth.

Postcranial characteristics: Humerus lacks an entepicondylar foramen; distal end of the ulna modified into a narrow styloid process; claws present on digits one through four in the manus of *E. eomigrans* but digits one and two reduced and lack claws in the Pleistocene species *E. laurillardi* with claws present on digits three and four; femur lacks a third trochanter, the articular surface for the patella connects with the articular surface of the lateral condyle and medial condyle separated by a prominent groove; tibia and fibula fused in adults; pes rotated so that the weight of the animal is supported by the calcaneum and fifth metatarsal; calcaneum elongate and the entire plantar surface contacts the ground with a secondary ossification on the tuber calcis; medial trochlea of the astragalus modifies into an odontoid process and the lateral trochlea is a flat surface.

Included species: *E. eomigrans* De Iuliis and Cartelle, 1999 only (known from localities GC15B*–F, GC17B, GC18II[A], B, GC18IIIA–C, GC18IV). *E. laurillardi* (including *E. guanajuatense*) is known from the Irvingtonian of Florida and the Rancholabrean in Texas, Florida, Georgia, South Carolina, New Jersey, and Mexico (see Cartelle and De Iuliis, 1995).

Eremotherium sp. is also known from localities GC17B, GC18IIA.

Comments: *Eremotherium eomigrans* is similar in size to *E. laurillardi.* It is indistinguishable from the latter species in its cranial, mandibular, and dental morphology. Despite the similarity in size to the latter species its postcranial skeleton is more gracile. The humerus has a more prominent lesser tubercle and the proximodistal and mediolateral fusion between the tibia and fibula is narrower. Its primary distinguishing character is a pentadactyl manus with unguals on digits one through four. The metacarpal–carpal element may be formed by either the fusion of the first metacarpal and trapezium or these two bones plus the trapezoid. *Eremotherium* is also known from the Pleistocene of South America (McKenna and Bell, 1997).

MYLODONTIDAE AMEGHINO, 1889

***Thinobadistes* Hay, 1919**

Type species: *Thinobadistes segnis* Hay, 1919.

Type specimen: USNM 3333 (left astragalus).

Characteristics: Front tooth modified into a caniniform that is triangular in cross section; third trochanter of the femur is a small thickened area positioned slightly distal to the middle of the shaft; articular surface for the patella continuous with the articular surfaces of the distal condyles; a deep furrow separates the astragalar facets of the calcaneum, in contrast to *Glossotherium* in which the two facets are confluent; second metatarsal fused with the mesocuneiform (Webb, 1989).

Included species: *T. segnis* (known from localities [GC10D], GC11B*); *T. wetzeli* Webb, 1989 (localities [GC11D], GC12IIA*, SP3A).

Thinobadistes sp. is also known from locality SP3B.

***Glossotherium* Owen, 1840**

Type species: *Glossotherium robustus* Owen, 1840.

Type specimen: RCS (destroyed during World War II).

Characteristics: For *Glossotherium chapadmalense* Kraglievich, 1925 (North American species). Front tooth modified into a caniniform that is triangular in cross section; no third trochanter on the femur; astragalar facets of the calcaneum confluent; second metatarsal and mesocuneiform separate; dermal ossicles in the skin.

Included species: *G. chapadmalense* Kraglievich, 1925 (known from localities GC14A, D, GC15A, C, D, GC17A, B, GC18IIA, NB13C, SB18D, SB34IVA, SB37C, [SP5], CP117A, B, CP121, PN23C); *G. garbanii* Montellano-Ballesteros and Carranza-Castañeda, 1986 (locality SB58C).

Glossotherium sp. is also known from localities SB50, SB58B, SP1H, CP80, CP118.

Comments. *Paramylodon harlani*, the descendent of *G. chapadmalense*, is known from the Irvingtonian and Rancholabrean of North America (including localities GC15E, GC18IIIA, B, GC18IV). *Glossotherium* is also known from the late Miocene to Pleistocene of South America (McKenna and Bell, 1997).

BIOLOGY AND EVOLUTIONARY PATTERNS

The dispersal of various groups of xenarthrans from South America into North America took place at different times beginning in the late Miocene and continuing into the Quaternary (Webb, 1985) (Figure 10.4). The first appearance of many of these taxa has served an important biostratigraphic role (Tedford *et al.*, 1987). The lack of study to produce a detailed and exact chronology of morphological change that defines each chronospecies within the different lineages has limited the utility of the group in North American biostratigraphy.

The appearance of two sloths, a megalonychid, *Pliometanastes*, and a mylodont, *Thinobadistes*, in North America is used to aid in defining the Clarendonian–Hemphillian (early late/late late Miocene) boundary. Based on current knowledge of the age of the sites for their earliest appearance, their appearance in North America is essentially simultaneous. A partial skeleton of *Pliometanastes* from the Siphon Canal locality in the Mehrten Formation (locality CC25AA) has been dated at 8.19 ± 0.16 Ma (Hirschfeld, 1981). The earliest appearance of *Thinobadistes* is from the type locality, Mixon's Bone Bed in Florida (locality GC11B), considered to be about 8 Ma. Both genera entered North America prior to the formation of the Panamanian Isthmus and are confined to the Hemphillian.

While *Pliometanastes* is predominately confined to the early Hemphillian, there is a record of the genus from the late Hemphillian Box T locality (locality SP3A) dated at 6.1 Ma. Its descendent, *Megalonyx*, first appears in the late Hemphillian. The earliest record for the genus *Megalonyx* based on the diagnostic morphology of the caniniform is the Lemoyne Fauna (locality CP116C) in Nebraska dated at 6.7 Ma (Leite, 1990). It is difficult to distinguish the postcranial skeleton of Hemphillian *Pliometanastes* and *Megalonyx*; therefore, positive identification to genus requires the preservation of the caniniform tooth. As a consequence, there are numerous Hemphillian faunas that contain an indeterminate megalonychid. *Megalonyx* is represented by five chronospecies in North America and the lineage survived until the end of the Pleistocene. While it appears that *Pliometanastes* gave rise to *Megalonyx*, *Thinobadistes* became extinct without issue. Recent work in Mexico (Flynn *et al.*, 2005) has documented the association of xenarthrans with radiometrically dated ashes in Guanajuato and indicates their presence significantly earlier in tropical North America. At Rancho El Ocote, *Glossotherium* and *Megalonyx* have been dated to 4.8–4.7 Ma. The date is not unexpected for *Megalonyx* but extends the age of *Glossotherium* in North America by over 2 million years. Remains of a pampathere from Rancho El Ocote have been referred to the South American genus *Plaina* and are dated to 4.7–4.6 Ma. At Rancho Viejo and La Pantera, the remains of *Glyptotherium* have been found in early Blancan deposits dated to 3.9–3.1 Ma.

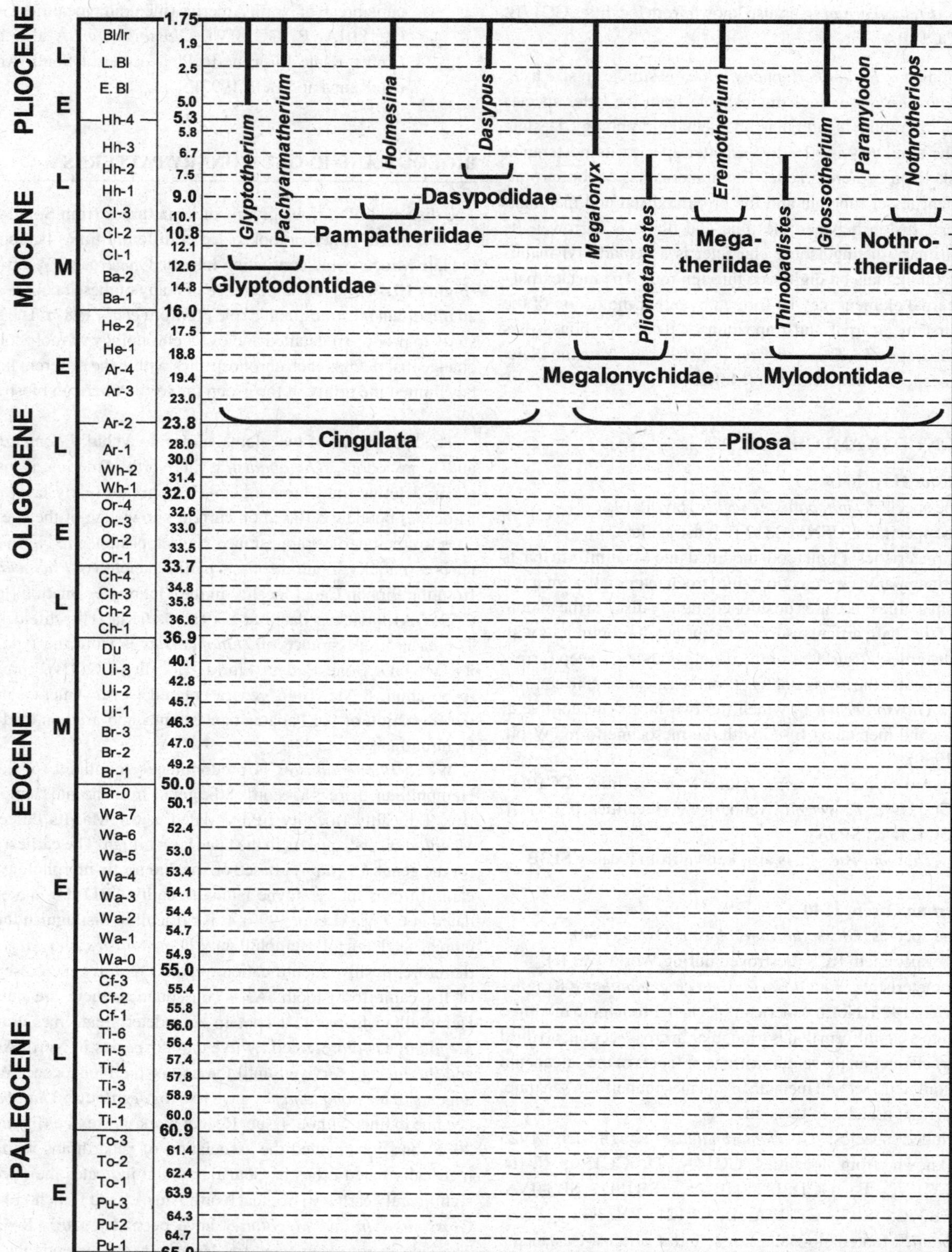

Figure 10.4. Temporal ranges of North American xenarthran genera.

The second influx of xenarthrans into North America did not occur until after the establishment of a land bridge between North and South America in the form of the Panamanian Isthmus. This second dispersal event took place around 2.5 Ma (mid-Blancan). The first part of this dispersal took place during the Gauss chron and included two sloths, a mylodont, "*Glossotherium*" *chapadmalense*, a megathere; *Eremotherium eomigrans*, and four cingulates: the glyptodont *Glyptotherium*, the pampathere *Holmesina*, an armadillo, *Dasypus*, and an unusual cingulate, *Pachyarmatherium*. The appearance of these five taxa allows the Blancan to be conveniently subdivided into an early pre-interchange (approximately 4.8 to 2.5 Ma) and a late post-interchange (2.5–1.9 Ma). This division works best from the southern United States southward, since many of these taxa are geographically restricted and their ranges for the most part did not disperse very far northward. Two exceptions are *Glossotherium chapadmalense*, whose range extended as far north as Idaho and Nebraska in the Blancan, and *Dasypus bellus*, which is present in Nebraska during the Blancan (Voorhies, 1987). All of these genera survived in North America until the Pleistocene extinction except *Pachyarmatherium*, which became extinct in the Irvingtonian. While *Dasypus* is present in North America today, its appearance in the United States appears to be a reinvasion following the extinction of *D. bellus* at the end of the Pleistocene (Taulman and Robbins, 1996).

A third dispersal of xenarthrans occurred in the Irvingtonian. This event is marked by the first appearance of the nothrothere *Nothrotheriops texanum* (known from Florida Irvingtonian localities GC18IIIA and GC18IV). It is also at this time that the giant anteater, *Myrmecophaga*, extended its range into North America (Shaw and McDonald, 1987). *Myrmecophaga* is known from a single site, El Golfo Mexico, so is of limited biostratigraphic use. *Nothrotheriops* survived in North America until the end of the Pleistocene.

Initially the earliest representative of each xenarthran lineage in North America was small in body size compared with its descendents and in most cases there was a general increase in size until the terminal species. This trend in increasing size is best documented in the sloths: *Megalonyx*, *Thinobadistes* and *Glossotherium–Paramylodon*. An increase in size has also been documented in the pampathere *Holmesina* and the armadillo *Dasypus bellus*. In the glyptodont *Glyptotherium*, adults of *G. texanum* in the Blancan are half the size of *G. arizonae* in the Irvingtonian and *G. floridanus* in the Rancholabrean. There is a slight increase in size in the giant sloth *Eremotherium*, but the increase in size is not as pronounced as in the other lineages. The data are insufficient to document any trends in *Pliometanastes* or in *Pachyarmatherium*.

There are distinctive patterns in the distribution of the various genera, both pilosans and cingulates, with many taxa confined to the southeastern United States and southward. In contrast the sloth *Megalonyx* expanded its range so that by the Rancholabrean it had the greatest distribution of all North American xenarthrans, with records as far north as the Old Crow Basin in the Yukon in the Rancholabrean (McDonald, Harington, and De Iuliis, 2000).

The earliest records of the mylodont "*Glossotherium*" *chapadmalense* in the Blancan indicate that it was already widespread, being found as far north as Idaho and Nebraska, so either it was present in North America earlier than the current records indicate or was able to disperse quickly following its entrance into North America. Its descendant, *Paramylodon harlani*, was equally widespread until its extinction at the end of the Pleistocene.

The largest of the sloths in North America was *Eremotherium*. The earliest species in North America, *E. eomigrans*, is currently only known from the Blancan and early Irvingtonian of Florida (De Iuliis and Cartelle, 1999). While the Rancholabrean species *E. laurillardi* was more widespread than *E. eomigrans*, its range in North America was more restricted than other sloths. It is only known from a coastal lowland zone from New Jersey southward through Florida and the Gulf Coast. This seems to reflect a preference for warmer habitats, because its overall range was considerably greater than the other North American sloths as its distribution includes numerous localities in Mexico southward to southern Brazil (Cartelle and De Iuliis, 1995).

All species of the genus *Glyptotherium* are restricted to the southern United States and south into Mexico. Currently there are no records of the genus north of 35° latitude. The initial distribution of the genus in the Blancan and Irvingtonian extended as far west as southeastern Arizona, but by the Rancholabrean its range had contracted eastward to the southeastern United States from Texas to South Carolina and south into Mexico (Gillette and Ray, 1981). The range of the pampathere *Holmesina* extended farther north than *Glyptotherium* and the genus is found as far north as Kansas in the Pleistocene (Hibbard *et al.*, 1978).

The general distribution of *Dasypus bellus* parallels that of the modern species *Dasypus novemcinctus*. The most northern record of the species in the Blancan is Big Springs, Nebraska (locality CP121; Voorhies, 1987), which is comparable to the current northernmost records of *D. novemcinctus* today. Like its modern relative, it probably expanded its range northward during warm periods with the northern part of the population eliminated during climatic cooling.

Based on the proportional abundance of the amino acid hydroxyproline in bone collagen fibers, Ho (1967) was able to establish the core body temperature for a number of extinct species. Using Ho's data, McNab (1985) calculated the core body temperature of *Nothrotheriops shastensis* as 34.4 °C. This is considered very low for an animal with an estimated body weight of 250 kg. Preserved hair of *Nothrotheriops* from dry caves in the southwestern United States has permitted a calculation of its thermal conductance, which is about 2.75 times higher than expected. The combination of low body temperature and high thermal conductance in *Nothrotheriops* suggests that, like modern sloths, it was thermally sensitive and cooler temperatures may have limited the animal's distribution. In comparison, the thermal conductance of preserved hair of the South American late Pleistocene genus *Mylodon*, which is closely related to *Glossotherium* and *Paramylodon*, had a thermal conductance of 1.9 of the expected value. *Mylodon* is comparable in body size to its close relative, the North American late Pleistocene genus *Paramylodon*, and it is reasonable to assume that the hair of *Paramylodon* like that of *Mylodon* provided better insulation than the hair of *Nothrotheriops*. McNab (1985) argued that their larger body size permitted the mylodonts to occupy seasonally cool to cold climates not available to smaller sloths.

Based on this evidence for low basal metabolism in the extinct sloths and other xenarthrans, McNab (1985) felt that, like their modern relatives, the extinct forms were K-selected and limited by the carrying capacity of the environment. He inferred that they would have had small litter sizes, extended parental care, low growth rates, and long gestation periods. Direct evidence for the number of young is unknown for the extinct North American xenarthrans except *D. bellus*. A skeleton from a late Pleistocene site in Florida included four embryonic skeletons, indicating that, like its modern relative *D. novemcinctus*, *D. bellus* was polyembryonic (Auffenberg, 1957).

Modern species of *Dasypus* are considered to be generalist insectivores although with a greater emphasis on termites than previously thought (Redford, 1985). Similar habits have been inferred for *D. bellus*. Fossil fecal pellets attributed to the mound-building termite *Kalotermes* have been reported from Seminole Field, Florida, the type locality for *D. bellus* (Light, 1930).

Vizcaino, De Iuliis, and Bargo (1998) analyzed the masticatory apparatus of *Holmesina* and considered the anatomy of the skull and mandible to be analogous to that of grazing ungulates. They considered the form of the mesial molariforms to indicate that, while it was probably not a strict grazer, the animal was adapted for processing resistant vegetal matter. Striations on the occlusal surface of the teeth indicate a strong labial–lingual component to mastication.

Gillette and Ray (1981) interpreted the habitat of glyptodonts as lowland areas along primary watercourses, with these animals frequenting lake and stream shorelines and perhaps even occupying shallow water. Part of their interpretation is based on the morphology of the feet, which have features suggesting the digits could spread out, thus distributing the weight over a larger area on mud and soft marshy ground. They also note that in many faunas in North America, glyptodonts are associated with the two extinct capybaras. Like the capybaras, the glyptodonts are interpreted as "aquatic grazers" and, like hippos, probably grazed in the savannas adjacent to watercourses.

Among the ground sloths, our knowledge of the diet is best for the sloth *Nothrotheriops shastensis*, thanks to the preservation of its dung in numerous caves in the southwestern United States (Martin, Sabels, and Shutler, 1961; Hansen, 1978). Seventy-two genera of plants have been identified in dung samples from Rampart Cave in Arizona. The major plant taxa identified include desert globe mallow, Nevada Mormon tea, saltbushes, catclaw acacia, common reed, and yucca. Based on the samples available, the species was a browser and fed on desert vegetation similar to that found in the area today. It appears its ecology was similar to that of extant desert herbivores.

Megalonyx is also interpreted as a browser, although this is inferred from its dentition and not based on any direct evidence of diet as for *Nothrotheriops*. Likewise, based on anatomy, *Eremotherium* is considered to have been a browser, but given its size and ability to sit upright it was probably a high browser and filled an ecological niche analogous to a giraffe.

The flat occlusal surface of the molariforms in mylodonts has been used traditionally as a basis to interpret them as grazers. Recently Naples (1989) has suggested that *Paramylodon* may have not been an obligate grazer, but rather was a mixed feeder, and that its low basal metabolism allowed it to feed on high-fiber forage of low nutritional value not utilized by other herbivores. Recent stable isotope studies to reconstruct the diet of members of the Pleistocene fauna of Rancho La Brea (Coltrain *et al.* 2004) included *Paramylodon harlani*. The δ^{15}N value in *P. harlani* was more positive than equids and "true non-ruminants," but less than ruminants, indicating that the sloth could be considered a "foregut fermenting non-ruminant mammal." The general similarity in the dentition of *Glossotherium chapadmalense* and *Thinobadistes* to *Paramylodon* suggests that their diets were roughly similar.

REFERENCES

Akersten, W. A. (1972). Red Light local fauna (Blancan) of the Love Formation, southeastern Hudspeth County, Texas. *Bulletin of the Texas Memorial Museum*, 20, 1–52.

Ameghino, F. (1889). Contribucion al conocimiento de los mamíferos fósiles de la Republica Argentina. *Actas de la Academia Nacional de Ciencias, Córdoba, Buenos Aires*, 6, 1–1027.

Auffenberg, W. (1957). A note on an unusually complete specimen of *Dasypus bellus* (Simpson) from Florida. *Journal of the Florida Academy of Sciences*, 20, 233–7.

Bonaparte, C. L. J. L. (1838). Synopsis vertebratorum systematis. *Nuovi annali di scienze naturali, Bologna*, 2, 105–33.

Burmeister, H. (1879). Description physique de la République Argentine. *Animaux Vertébrés,* Vol. III, Pt. 1: *Mammifères Vivants et Étients*. Buenos Aires, Paul-Émile Coni, 1–555.

Carranza- Castañada, O. and Miller, W. E. (2004). Late Tertiary terrestrial mammals from central Mexico and their relationship to South American immigrants. *Revista Brasileria de Paleontologia*, 7, 249–61.

Cartelle, C. and De Iuliis, G. (1995). *Eremotherium laurillardi*: the Panamerican late Pleistocene megatheriid sloth. *Journal of Vertebrate Paleontology,* 15, 830–41.

Coltrain, J. B., Harris, J. M., Cerling, T. E., *et al.* (2004). Rancho La Brea stable isotope biogeochemistry and its implications for the palaeoecology of late Pleistocene, coastal California. *Palaeogeography, Palaeoclimatology, Palaeoecology*, 205, 199–219.

Cope, E. D. (1893). A preliminary report on the vertebrate paleontology of the Llano Estacado. *4th Annual Report of the Texas Geological Survey*. Washington, DC: US Government Printing Office.

(1899). The *Edentata* of North America. *The American Naturalist*, 23, 657–64.

De Iuliis, G. and Cartelle, C. (1999). A new giant megatheriine ground sloth (Mammalia: Xenarthra: Megatheriidae) from the late Blancan to early Irvingtonian of Florida. *Zoological Journal of the Linnean Society*, 127, 495–515.

Delsuc, F., Scally, M., Madsen, O., *et al.* (2002). Molecular phylogeny of living Xenarthrans and the impact of character and taxon sampling on the placental tree rooting. *Molecular Biology and Evolution*, 19, 1656–71.

Demarest, M. A. G. (1822). Mammalogie ou description des especes de mammifères. *Imprimeur Libbararre due des Poitevins*, 6, 1–555.

Downing, K. F. and White, R. S. (1995). The Cingulates (Xenarthra) of the Leisey Shell Pit local fauna (Irvingtonian), Hillsborough County, Florida. *Bulletin of the Florida Museum of Natural History*, 37, 375–96.

Edmund, A. G. (1985a). The fossil giant armadillos of North America (Pampatheriinae, Xenarthra = Edentata). In *The Evolution and Ecology of Armadillos, Sloths, and Vermilinguas*, ed. G. G. Montgomery, pp. 83–93. Washington, DC: Smithsonian Institution Press.

(1985b). The armor of fossil giant armadillos (Pampatheriidae, Xenarthra, Mammalia). *Pearce-Sellards Series, Texas Memorial Museum*, 40, 1–20.

(1987). Evolution of the genus *Holmesina* (Pampatheriidae, Mammalia). *Pearce-Sellards Series, Texas Memorial Museum*, 45, 1–20.

Engleman, G. F. (1985). The phylogeny of Xenartha, In *The Evolution and Ecology of Armadillos, Sloths and Vermilinguas*, ed. G. C. Montgomery, pp. 51–64. Washington, DC: Smithsonian Institution Press.

Flower, W. J. (1883). On the arrangement of the orders and families of existing Mammalia. *Proceedings of the Zoological Society of London*, 1883, 178–86.

Flynn, J. J., Kowallis, B. J., Nuñez, C., *et al.* (2005). Geochronology of Hemphillian–Blancan aged strata, Guanajuato, Mexico, and implications for timing of the Great American Biotic Interchange. *Journal of Geology*, 113, 287–307.

Gaudin, T. J. (1999). The morphology of xenarthrous vertebrae (Mammalia: Xenarthra). *Fieldiana (Geology)*, 41, 1–38.

(2003). Phylogeny of the Xenarthra (Mammalia). *Senckenbergiana Biologica*, 83, 27–40.

Gazin, C. L. (1957). Exploration for the remains of giant ground sloths in Panama. *Annual Reports of the Smithsonian Institution*, 4272, 341–54.

Gillette, D. D. and Ray, C. E. (1981). Glyptodonts of North America. *Smithsonian Contributions to Paleobiology*, 40, 1–255.

Hansen, R. M. (1978). Shasta ground sloth food habits, Rampart Cave, Arizona. *Paleobiology*, 4, 302–19.

Harlan, R. (1825). *Fauna Americana*. Philadelphia, PA: Anthony Finley.

Hay, O. P. (1919). Descriptions of some mammalian and fish remains from Florida of probably Pleistocene age. *Proceedings of the United States National Museum*, 56, 103–12.

Hibbard, C. W., Zakrzewski, R. J., Eshelman, R. E., *et al.* (1978). Mammals of the Kanopolis local fauna, Pleistocene (Yarmouth) of Ellsworth County, Kansas. *Contributions from the Museum of Paleontology, University of Michigan*, 25, 11–44.

Hirschfeld, S. E. (1981). *Pliometanastes protistus* (Edentata, Megalonychidae) from Knight's Ferry, California with discussion of early Hemphillian megalonychids. *PaleoBios*, 36, 1–16.

Hirschfeld, S. E. and Webb, S. D. (1968). Plio-Pleistocene megalonychid sloths of North America. *Bulletin of the Florida State Museum*, 12, 213–96.

Ho, T.-Y. (1967). Relationship between amino acid contents of mammalian bone collagen and body temperature as a basis for estimation of body temperature of prehistoric animals. *Comparative Biochemistry and Physiology*, 22, 113–19.

Hulbert, R. C., Jr. and Morgan, G. S. (1993). Quantitative and qualitative evolution in the giant armadillo *Holmesina* (Edentata: Pampatheriidae) in Florida. In *Morphological Change in Quaternary Mammals of North America*, ed. R. A. Martin and A. D. Barnosky, pp. 134–77. Cambridge, UK: Cambridge University Press.

Illiger, C. (1811). *Prodromus Systematis Mammalium et Avium Additis Terminis Zoographicis Utrudque Classis*. Berlin: C. Salfeld.

Jefferson, T. (1799). A memoir on the discovery of certain bones of a quadruped of the clawed kind in the western parts of Virginia. *Transactions of the American Philosophical Society*, 4, 246–60.

Klippel, W. and Parmalee, P. W. (1984). Armadillos in North American late Pleistocene contexts. *Carnegie Museum of Natural History Special Publication*, 8, 149–60.

Kraglievich, L. (1925). Cuatro nuevos gravigrados de la fauna Araucana Chapadmalalense. *Anales del Museo Nacional de Historia Natural*, 33, 215–35.

Leidy, J. (1889). Fossil vertebrates from Florida. *Proceedings of the Academy of Natural Science, Philadelphia* 1889, 96–7.

Leite, M. (1990). Stratigraphy and mammalian paleontology of the Ash Hollow Formation (Upper Miocene) on the north shore of Lake McConaughy, Keith County, Nebraska. *University of Wyoming, Contributions to Geology*, 28, 1–29.

Light, S. F. (1930). Fossil termite pellets from the Seminole Pleistocene. *University of California Publications in Geological Sciences*, 19, 75–80.

Linnaeus, C. (1758). *Systema Naturae per Regna tria Naturae, secundum Classes, Ordines, Genera, Species, cum Characteribus, Differentiis, Synonymis, Locis*, 10th edn. Stockholm: Laurentii Salvii.

Lund, P. W. (1842). Blik paa Brasiliens dyreverden för sidste jordomvaeltning. Fjerde Afhandling: Fortsaettelse af Pattedyrene. *Det Kongelige Danske Videnskabernes Selskabs Skrifter Naturvidenskabelige og Mathematiske Afhandlingar*, 9, 137–208.

Marshall, L. G. (1981). The Great American Interchange: an invasion induced crisis for South American mammals. In *Biotic Crises in Ecological and Evolutionary Time*, ed. M. Nitecki, pp 133–229. New York: Academic Press.

Martin, P. S., Sabels, B. E., and Shutler, D., Jr. (1961). Rampart Cave coprolite and ecology of the Shasta ground sloth. *American Journal of Science*, 259, 102–27.

Matthew, W. D. (1924). Third contribution to the Snake Creek fauna. *Bulletin of the American Museum of Natural History*, 50, 149–50.

McDonald, H. G. (1977). Description of the osteology of the extinct gravigrade edentate, *Megalonyx*, with observations on its ontogeny, phylogeny and functional anatomy. M.Sc. Thesis, University of Florida, Gainsville.

McDonald, H. G., Harington, C. R., and De Iuliis, G. (2000). The ground sloth *Megalonyx* from Pleistocene deposits of the Old Crow Basin, Yukon Territory, Canada. *Arctic*, 53, 213–20.

McKenna, M. C. (1975). Toward a phylogenetic classification of the Mammalia. In *Phylogeny of the Primates*, ed. W. P. Luckett and F. S Szalay, pp. 221–46. New York: Plenum Press.

McKenna, M. C. and Bell, S. K. (1997). *Classification of Mammals Above the Species Level*. New York: Columbia University Press.

McNab, B. K. (1985). Energetics, population biology, and distribution of xenarthrans, living and extinct. In *The Evolution and Ecology of Armadillos, Sloths and Vermilinguas*, ed. G. G. Montgomery. pp. 219–32. Washington DC: Smithsonian Institution Press.

Montellano-Ballesteros, M. and Carranza-Castañeda C. (1986). Descripción de un Milodontido del Blancano temprano de la mesa central de Mexico. *Universidad Nacional Autónoma de México Institudo de Geología Revista*, 6, 193–203.

Naples, V. L. (1987). Reconstruction of cranial morphology and analysis of function in the Pleistocene ground sloth *Nothrotheriops shastense* (Mammalia, Megatheriidae). *Contributions in Science, Los Angeles County Natural History Museum*, 389, 1–21.

(1989). The feeding mechanism in the ground sloth, *Glossotherium*. *Contributions in Science, Los Angeles County Natural History Museum*, 415, 1–23.

(1995). The artificial generation of wear patterns on tooth models as a means to infer mandibular movement during feeding in mammals. In *Functional Morphology in Vertebrate Paleontology*, ed. J. J. Thomason, pp. 136–50. Cambridge, UK: Cambridge University Press.

Osborn, H. F. (1903). *Glyptotherium texanum*: a new glyptodont, from the lower Pleistocene of Texas. *Bulletin of the American Museum of Natural History*, 19, 491–4.

Owen, R. (1840). Fossil Mammalia. In *The Zoology of the Voyage of H. M. S. Beagle under the Command of Captain Fitzroy, during the Years 1832–1836*, ed. C Darwin, pp. 1–111. London: Smith, Elder.

(1843). Zoological summary of the extinct and living animals of the order Edentata. *Edinburgh New Philosophical Journal*, 35, 353–61.

Patterson, B. and Pascual, R. (1963). The extinct land mammals of South America. *Proceedings of the 16th International Zoological Congress*, Washington, DC, pp. 138–48.

Redford, K. H. (1985). Food habits of armadillos (Xenarthra: Dasypodidae). In *The Evolution and Ecology of Armadillos, Sloths, and Vermilinguas*, ed. G. G. Montgomery, pp. 429–37. Washington DC: Smithsonian Institution Press.

Robertson, J. A. (1976). Latest Pliocene mammals from Haile XVA, Alachua County, Florida. *Bulletin of the Florida State Museum*, 20, 111–86.

Shaw, C. A. and McDonald, H. G. (1987). First record of giant anteater (Xenarthra, Myrmecophagidae) in North America. *Science*, 236, 186–8.

Shoshani, J. and McKenna, M. C. (1998). Higher taxonomic relationships among extant mammals based on morphology, with selected comparisons of results from molecular data. *Molecular Biology and Evolution*, 9, 572–84.

Simpson, G. G. (1929). Pleistocene mammalian fauna of the Seminole Field, Pinellas County, Florida. *Bulletin of the American Museum of Natural History*, 56, 561–99.

(1930). *Holmesina septentrionalis*, an edentate from the Pleistocene of Florida. *American Museum Novitates*, 442, 1–10.

Spillman, F. (1948). Beiträge zur kenntnis eines neuen gravigraden riesensteppentieres (*Eremotherium carolinense* gen. et. sp. nov.). seines lebebsraumes und seiner lebensweise. *Palaebiologica*, 8, 231–79.

Stock, C. (1925). Cenozoic gravigrade edentates of western North America. *Carnegie Institution of Washington Publication*, 331, 1–206.

Taulman, J. F. and Robbins, L. W. (1996). Recent range expansion and distributional limits of the nine-banded armadillo (*Dasypus novemcinctus*) in the United States. *Journal of Biogeography*, 23, 635–48.

Tedford, R. H., Skinner, M. F., Fields, R. W., *et al.* (1987). Faunal succession and biochronology of the Arikareean through Hemphillian interval (Late Oligocene through earliest Pliocene epochs) in North America. In *Cenozoic Mammals of North America Geochronology and Biostratigraphy*, ed. M. O. Woodburne, pp. 153–210. Berkeley, CA: University of California Press.

Vizcaino, S. F., Iuliis, G., de and Bargo, M. S. (1998). Skull shape, masticatory apparatus, and diet of *Vassallia* and *Holmesina* (Mammalia: Xenarthra: Pampatheriidae): when anatomy constrains destiny. *Journal of Mammalian Evolution*, 5, 291–322.

Voorhies, M. R. (1987). Fossil armadillos in Nebraska: the northernmost record. *The Southwestern Naturalist*, 32, 237–43.

Webb, S. D. (1985). Late Cenozoic mammal dispersal between the Americas. In *The Great American Biotic Interchange*, ed. F. G. Stehli and S. D. Webb, pp. 357–86. New York: Plenum Press.

(1989). Osteology and relationships of *Thinobadistes segnis*, the first mylodont sloth in North America. In *Advances in Neotropical Mammalogy*, ed. K. H. Redford and J. F. Eisenberg, pp. 469–532. Gainesville, FL: The Sandhill Crane Press, Inc.

Wetzel, R. M. (1985). Taxonomy and distribution of armadillos, Dasypodidae. In *The Evolution and Ecology of Armadillos, Sloths, and Vermilinguas*, ed. G. G. Montgomery, pp. 23–46. Washington, DC: Smithsonian Institution Press.

Zittel, K. A., von (1892) *Handbuch der Palaeontologie, Palaeozoologie*, Vol. IV: Vertebrata (Mammalia). Munch: Oldenbourg.

Part IV: Archonta

11 Archonta summary

GREGG F. GUNNELL[1] and MARY T. SILCOX[2]

[1]*Museum of Paleontology, University of Michigan, Ann Arbor, MI, USA*
[2]*Department of Anthropology, University of Winnipeg, Winnipeg, MB, Canada*

INTRODUCTION

For the purposes of this summary chapter, we include the orders Primates, Dermoptera (flying lemurs or colugos), Scandentia (tree shrews), and Chiroptera (bats) in the Grandorder Archonta (McKenna, 1975; Szalay, 1977; Szalay and Drawhorn, 1980; Novacek and Wyss, 1986; Szalay and Lucas, 1993, 1996; McKenna and Bell, 1997). When originally proposed, Gregory (1910) also included elephant shrews (Macroscelidea) in Archonta because he believed that macroscelideans and tree shrews should be grouped together in Menotyphla. Butler (1972) convincingly demonstrated that macroscelideans and scandentians are not sister taxa, however, and that Menotyphla is an unnatural grouping. Modern conceptions of Archonta have, therefore, included tree shrews but not elephant shrews (e.g., McKenna, 1975; Szalay, 1977; Novacek and Wyss, 1986; McKenna and Bell, 1997). Although a monophyletic Archonta is a frequent component of classifications based on morphology (but see Novacek, 1980), molecular evidence has consistently failed to support the inclusion of bats in a clade with primates, tree shrews, and colugos (Pumo *et al.*, 1998; Miyamoto, Porter, and Goodman, 2000; Liu *et al.*, 2001; Murphy *et al.*, 2001a,b; Springer *et al.*, 2003, 2004; Van den Bussche and Hoofer, 2004; Teeling *et al.*, 2005). Molecular studies do often feature a clade with Primates, Scandentia, and Dermoptera, however (e.g., Adkins and Honeycutt, 1991; Waddell, Okada, and Hasegawa, 1999; Liu *et al.*, 2001; Murphy *et al.*, 2001a,b; Springer *et al.*, 2003, 2004), which Waddell, Okada, and Hasegawa (1999) named Euarchonta.

There are several fossil groups that may be referable to Archonta or Euarchonta (reviewed by Silcox *et al.*, 2005), of which the potential dermopterans of the family Plagiomenidae, and the paraphyletic assemblage of possible primates referred to as "plesiadapiforms," are the most convincingly connected to the extant forms. These two groups will be treated as archontans here.

GENERAL CONSIDERATIONS

Archontans are a difficult group to diagnose because they typically exhibit a mosaic of derived, ordinal traits, and relatively primitive mammalian characteristics, with few clear archontan-level features that have not been subsequently altered by evolution (Novacek, 1980). For instance, archontan forelimbs may combine very derived characteristics (e.g., elongation in bats and colugos), with primitive features, such as clavicles and five digits.

Archontans are, perhaps, more easily recognized ecologically. Many are nocturnal (except some primates, tree shrews, and bats), most inhabit forests exclusively (except some bats and primates), and most are restricted to equatorial climates (except some bats and humans). Dermopterans and tree shrews apparently never existed outside of southern Asia and Indo-Pakistan (Jacobs, 1980; Ducrocq *et al.*, 1992), unless North American Paleogene plagiomenids are dermopterans (Silcox *et al.*, 2005). Microchiropteran bats are nearly ubiquitous, existing everywhere except in harsh polar regions (Gunnell and Simmons, 2005), while megachiropterans, like most archontans, are restricted to tropical regions. Geographic ranges of some archontans have fluctuated with changing climates. For example, primates, for the most part, are restricted to the equatorial tropics today but had a nearly complete Laurasian presence during the Eocene when more equable, warmer biomes were available at higher latitudes (Fleagle, 1999).

In light of the fact that the preferred archontan milieu is the tropical forest, it seems likely that the group evolved in this environment.

Evolution of Tertiary Mammals of North America, Vol. 2. ed. C. M. Janis, G. F. Gunnell, and M. D. Uhen. Published by Cambridge University Press.

Figure 11.1. Restoration of the Eocene euprimate *Notharctus* (by Marguette Dongvillo).

Indeed, among primates the invasion of more open environments is a relatively recent phenomenon, with the fossil record documenting a long history of arboreality (Fleagle, 1999).

FEATURES UNITING ARCHONTA

There are a number of anatomical features that may be synapomorphies of Archonta (for details see Smith and Madkour, 1980; Szalay and Drawhorn, 1980; Novacek and Wyss, 1986; Johnson and Kirsch, 1993; Wible and Martin, 1993; Szalay and Lucas, 1996; Hooker, 2001; Sargis, 2002a). These have been summarized by Silcox *et al.* (2005) and include both soft tissue (pendulous penis and retinal specializations) and osteological characteristics. Osteological features include (see Silcox *et al.* [2005] and sources cited therein) entotympanic present, with a close proximity to the tubal cartilage, tegmen tympani, and greater petrosal nerve; petrosal epitympanic wing absent; sustentacular and navicular facets of astragalus in direct contact on medial side of neck; large distal sustentacular facet of the calcaneum articulating with a ventral extension of the navicular facet of the astragalus; ribs craniocaudally expanded; C3–C7 spinous processes weak or absent; lesser tuberosity robust and medially protruding; humeral capitulum spheroidal; radial central fossa circular and well excavated; acetabulum elliptical and buttressed cranially; enlarged, flattened, triangular attachment areas for quadratus femoris; patellar groove shallow and short; distal tibiofibular joint synovial; calcaneal cuboid facet concave; distal facet of entocuneiform wide; ungals deep and narrow.

Almost all of the features included in this list are missing in one or more archontan order, presumably as a result of evolutionary changes occurring since the common origin of the group. The only osteological trait that has not undergone such modifications, and therefore is found in all archontans, is the large distal sustentacular facet of the calcaneus, which articulates with a ventral extension of the navicular facet of the astragalus (Hooker, 2001). However, it is worth noting that the inclusion of any one group in Archonta rests not only on features shared by all archontans, but also on traits shared between particular members of the Grandorder. Van den Bussche and Hoofer (2004) recently suggested that the seemingly paltry morphological support for the monophyly of Archonta was insufficient to counter the apparently weighty molecular support for bat affinities elsewhere. However, the association of bats with other archontans rests not only on traits that unite the entire group but also on the more impressive list of morphological traits that link Chiroptera and Dermoptera in Volitantia (Simmons and Geisler, 1998).

FEATURES DISTINGUISHING CHIROPTERA, DERMOPTERA, SCANDENTIA, AND PRIMATES

BODY SIZE

Presumably, archontans had an arboreal ancestry (Szalay and Drawhorn, 1980; Bloch and Boyer, 2002, 2007) and, therefore, originated at a relatively small body size. Bats have never achieved large body size: the largest living bats (some species of fruit bats) achieve masses of only 1.5 kg, while most are well under 100 g (Nowak, 1999). Dermopterans are only slightly heavier than the largest bats, reaching body weights of 1.0-1.75 kg (Nowak, 1999). Living scandentians typically weigh less than 400 g (Nowak, 1999). Although some primates do reach relatively large body sizes, most are under 10 kg (Fleagle, 1999), and only humans, a few fossil Old World monkeys (e.g., *Theropithecus oswaldi*), some living and fossil apes, and some extinct subfossil forms from Madagascar have body weights that could truly be considered large (between 50 and 200 kg; Fleagle, 1999). Plesiadapiforms also are predominantly a small-bodied radiation with only some species of *Plesiadapis* reaching 2.5 kg in estimated body weight.

LOCOMOTION AND POSTCRANIAL SPECIALIZATIONS

Archontan orders can be most readily differentiated from one another by postcranial and locomotor specializations and exhibit nearly every conceivable variation on arboreality that one could imagine, particularly if one surveys both extant and extinct forms (e.g., including the sloth-like and koala-like subfossil lemurs: Walker, 1974; Jungers *et al.*, 1997). The most distinctive, of course, are bats, the only mammalian group, and one of only three known vertebrate groups (the others being birds and pterosaurs) to have achieved powered flight. Unique bat characteristics include elongate hand digits that provide the frame for a flight membrane or patagium formed of skin stretched across this frame. This patagium encompasses the hindlimbs and often includes the tail as well. Other bat characteristics involve structural and orientational changes to the pectoral girdle and thorax to provide attachment areas and increased leverage for flight muscles, and specialization in the hindlimbs involved in the unique upside-down roosting posture of bats (Hill and Smith, 1984; Simmons, 1995; Gunnell and Simmons, 2005). Fossil microchiropteran bats are represented by good skeletons as far back as the early Eocene (Simmons and Geisler, 1998). Unfortunately, virtually all of the unique synapomorphies of bats were already achieved by that time (Simmons and Geisler, 1998) so the fossil record has not shed much light on the mechanics of bat diversification as yet.

Perhaps only slightly less specialized than bats are dermopterans, who have not achieved powered flight but nonetheless exhibit remarkable aerial gliding skills. In many ways, the forelimbs of dermopterans resemble those of bats: both have an elongate radius, a shortened humerus, a reduced ulna, and similar elbow and wrist structures (Szalay and Lucas, 1996). Dermopterans also have a patagium stretched across the digits of the hand (leading to the common name "mitten-gliders"), and they show some elongation of the manual digits.

Cynocephalid dermopterans have recently been described from the late Eocene and early Oligocene of southern Asia (Ducrocq *et al.*, 1992; Marivaux *et al.*, 2006). However, if plagiomenids are dermopterans then the fossil record extends into the Paleocene in North America. Plagiomenids possess multicusped incisors and the distinctive angled cheek-tooth lophs of living dermopterans. MacPhee, Cartmill, and Rose (1989) interpreted the cranial anatomy from one severely damaged plagiomenid skull as being at odds with a dermopteran attribution for the group. However, the anatomy they described, while very different from that of modern dermopterans, does not clearly link plagiomenids to any other group. As such, the ear region may simply be autapomorphous in plagiomenids, and the dental similarities may still be valid synapomorphies linking plagiomenids to dermopterans. Plagiomenidae is not represented by any postcranial remains as yet.

Unlike bats and colugos, who exhibit highly derived postcranial skeletons, tree shrews seem almost pedestrian by comparison. *Ptilocercus*, the sole living member of the Ptilocercidae, is more arboreal than are most of the numerous species of tupaiid tree shrews, and it probably represents the more primitive condition for scandentians (Le Gros Clark, 1926; Szalay and Drawhorn, 1980; Martin, 1990; Sargis, 2001, 2002a,b, 2004; Olson, Sargis, and Martin, 2005). In general, tree shrews have forelimbs that are somewhat shorter than hindlimbs and digits that are tipped with claws. Tupaiid tree shrews show relatively primitive hand proportions, with none of the extreme elongation of the digits seen in bats or colugos, and hand phalanges that are shorter relative to the metacarpals than in living primates (Bloch and Boyer, 2002). The arboreal *Ptilocercus* has more mobile joints than do more terrestrial tree shrews, which is reflected in the disposition of the proximal and distal humerus, acetabulum, ankles, and wrist (Sargis, 2002b). The fossil record for tree shrews is very poor, with only some fragments of teeth known from before the Miocene (Tong, 1988). The only postcranial fossil known for a tree shrew is a Pliocene-aged rib cage attributed to *Tupaia* (Dutta, 1975), but never formally described, which does not add much to our understanding of the evolution of the group.

Euprimate postcranial specializations are indicative of the active, running and leaping arborealism practiced by most members of the group. Both fore- and hindlimbs are specialized for grasping and feature opposable thumbs and big toes. With a few exceptions (e.g., callitrichids, grooming claws in prosimians), digits are tipped with nails. Hindlimbs may be much longer than forelimbs, especially in the most specialized vertical clinging and leaping taxa. Forelimbs have become elongate in primates that practice suspensory locomotion, including apes and some New World monkeys (e.g., the spider monkey, *Ateles*; Fleagle, 1999).

Until recently, very little was known of plesiadapiform postcranial anatomy. Before 1989, only *Plesiadapis* was represented by skeletal material, mostly isolated elements (but see Gingerich and Gunnell, 1992, 2005). Although one early functional reconstruction of *Plesiadapis* suggested that it was a ground-living, marmot-like form (Gingerich, 1976), most studies since have concluded that plesiadapids were committed arborealists (Szalay, Tattersall, and Decker, 1975; Beard, 1989; Youlatos and Godinot, 2004; Bloch and Boyer, 2007).

Beard (1989) reviewed the postcranial record for plesiadapiforms known at the time, which included fairly well-preserved material not only for plesiadapids, but also for paromomyids and micromomyids. His study of this material led him to suggest that paromomyids (Beard, 1989, 1990, 1993a) and micromomyids (Beard, 1993b) were gliders that may have been directly related to dermopterans. Unfortunately, the material that he studied comprised unassociated postcranial elements, and his conclusions depended in some cases on assumptions and identifications that have not been substantiated by new, associated skeletal specimens (Hamrick, Rosenman, and Brush, 1999; Boyer, Bloch, and Gingerich, 2001).

The discovery of a treasure-trove of plesiadapiform skeletons in freshwater limestones in the Paleocene of Wyoming is beginning to provide evidence that plesiadapiforms exhibited a wide variety of locomotor patterns and skeletal specializations (Bloch and Boyer, 2002, 2007). Better associated remains of paromomyids and micromomyids from these limestones suggest that they were non-gliding, arboreal taxa that can be likened to callitrichids and *Ptilocercus*, respectively (Boyer, Bloch, and Gingerich, 2001; Bloch, Boyer, and Houde, 2003; Bloch and Boyer, 2007). *Carpolestes simpsoni* had an opposable big toe like those of euprimates, and elongate manual digits for effective grasping in the hand (Bloch and Boyer, 2002). It is not clear if these similarities represent synapomorphies with euprimates or convergences, but what is clear is that plesiadapiforms exploited arboreal habitats in manners similar to those of some living euarchontans.

Living archontans have developed anatomical features that have allowed them to exploit the natural habitats available in arboreal environments. Arboreal tree shrews are, in general, above-branch quadrupeds and may represent the morphotype from which all of the other archontans were derived (Bloch, Boyer, and Houde, 2003). Dermopterans adopted a gliding lifestyle while bats developed powered flight, perhaps both from a terminal, under-branch feeding ancestral morphotype (Gunnell and Simmons, 2005).

Primates developed grasping and leaping locomotor repertoires that allowed them to exploit all parts of their arboreal world from the lowest vertical trunks to the tips of the emergent canopy – and everything in between. It seems clear that some of these locomotor adaptations to life in the trees had been experimented with by plesiadapiforms by at least the late Paleocene.

DIET AND CRANIODENTAL SPECIALIZATIONS

Much as archontans developed postcranial adaptations that allowed them access to all parts of their arboreal habitus, they also developed a bewildering array of cranial and dietary specializations as well. Life in the trees brings fruits, leaves, flowers, gums and saps, small invertebrates and vertebrates, and insects onto the menu of possible food items, and archontans developed cranial and dental specializations that allowed them to exploit all of these food resources.

Bats possess a variety of dental and cranial specializations. Fruit bats (Megachiroptera), which feed almost exclusively on fruits and flowers, typically have simplified, flattened cheek teeth that are often reduced in number and may be separated by short diastemata; a few forms do have cuspidate molars (e.g., *Pteralopex*, *Harpyoinycteris* [Nowak, 1999]), but these cusps are atypically arranged suggesting they evolved as neomorphs. Skulls tend to be low and relatively long, eyes are relatively large, external ears simple, premaxillae unspecialized, palates extend posteriorly well beyond the last molars, postorbital processes are well developed, and the cochlea is not expanded. Microchiroptera have more primitive dilambdodont dentitions, shortened skulls with crowded teeth, eyes usually small, external (and internal) ears complex, premaxillae derived, short palates that may not extend posteriorly to the last molars, and enlarged cochlea. The enlarged cochlea and complex external and internal ears are specializations developed for echolocation that are used by microchiropterans for orientation, maneuvering, and hunting (aerial hawking). Microchiropterans are mostly insectivorous and either feed in the air or glean insects from leaves or other surfaces (Simmons and Geisler, 1998).

Dermopteran skulls are broad and flat with a pair of laterally positioned and well-developed temporal ridges extending the length of the neurocranium (Stafford and Szalay, 2000). The braincase is globular; the palate is especially broad and flat with a pair of large, anterior, palatal fenestra; and there is a distinct postorbital process (Stafford and Szalay, 2000). Cheek teeth are generally tribosphenic but have relatively well-developed, distinctly angled lophs, and the upper molars have enlarged and distinct conules (MacPhee, Cartmill, and Rose, 1989). Lower incisors (and the lower canine in *Galeopterus variegatus*) are pectinate with multiple tines, which may be used for grooming (Aimi and Inagaki, 1988). The upper first incisor has been lost, and a premaxillary pad is present, which occludes with the first and second lower incisors (Stafford and Szalay, 2000). Dermopterans are folivores (Wischusen and Richmond, 1998).

Tree shrews have relatively long and low skulls with complete postorbital bars. The orbits are relatively large and either face anteriorly (*Ptilocercus*) or more laterally (tupaiids; Sargis, 2004). Like lemuriform primates, tree shrews possess a toothcomb, but unlike primates the canine is not incorporated into the comb (Butler, 1980). Like most primates, the branches of the internal carotid artery that pass through the middle ear are enclosed in bony tubes, although these tubes are formed in part by the entotympanic, not the petrosal, making their homology with those of primates unlikely (Cartmill and MacPhee, 1980). Tupaiid molars are dilambdodont. However, *Ptilocercus* does not have dilambdodont molars, suggesting that dilambdodonty evolved independently in the lineage leading to Tupaiidae (Butler, 1980). Tree shrews are omnivorous, with fruit and insects making up much of their diet (Nowak, 1999).

Euprimates have a variety of cranial specializations but in general can be characterized as having relatively shorter, higher, and more globular skulls than most other archontans, with enlarged and forward-facing orbits, complete postorbital bars (coupled with a postorbital plate in tarsiers and anthropoids), enlarged brains with reduced olfactory bulbs and an enlarged frontal cortex, an auditory bulla completely formed from the petrosal bone, and bony tubes formed by the petrosal surrounding the branches of the internal carotid artery as they pass through the middle ear. Primate dentitions tend to be either tritubercular or quadritubercular, with relatively low-crowned, bunodont molars. Primates eat a wide variety of items, with most forms consuming some combination of fruit and insects, or fruit and leaves (Kay, 1984).

Cranial morphology of plesiadapiforms is an interesting mixture of primitive and apparently derived features, linking them as possible stem primates. Plesiadapiforms, in general, have long and low skulls with small brains, laterally facing orbits, and no postorbital bars. Plesiadapids and *Carpolestes* apparently possess auditory bullae formed from the petrosal bone (Russell, 1959; Szalay, Rosenberger, and Dagosto, 1987; Boyer *et al.*, 2004; Bloch and Silcox, 2006), like euprimates, while paromomyids include an entotympanic element in the bulla, making them more similar to non-euprimate archontans (Kay, Thewissen, and Yoder, 1992; Bloch and Silcox, 2001; Silcox, 2003). Dentitions are highly variable. Dental reduction is a common trend in multiple families, although the most primitive plesiadapiform, *Purgatorius*, retains a lower dental formula of i3, c1, p4, and m3 (Clemens, 2004). Central incisors are enlarged and procumbent in even the most primitive plesiadapiforms (Clemens, 2004). A common feature of several plesiadapiform families are multicuspate upper central incisors (Rose, Beard, and Houde, 1993); this general morphology likely arose multiple times independently (Silcox and Bloch, 2006). Posterior premolars (or m1 in picrodontids) are often highly specialized, with the most extreme examples being found in carpolestids, where lower fourth premolars form well-developed cutting blades and upper third and fourth premolars feature multiple rows of cuspules (Rose, 1975). As in euprimates, cheek teeth are relatively low crowned and bunodont, particularly when contrasted with those of contemporary insectivores, suggesting an early shift to a more herbivorous diet.

OTHER FEATURES

Many primates have developed complex social systems that involve both vocal communication and subtle facial expression. Unlike other

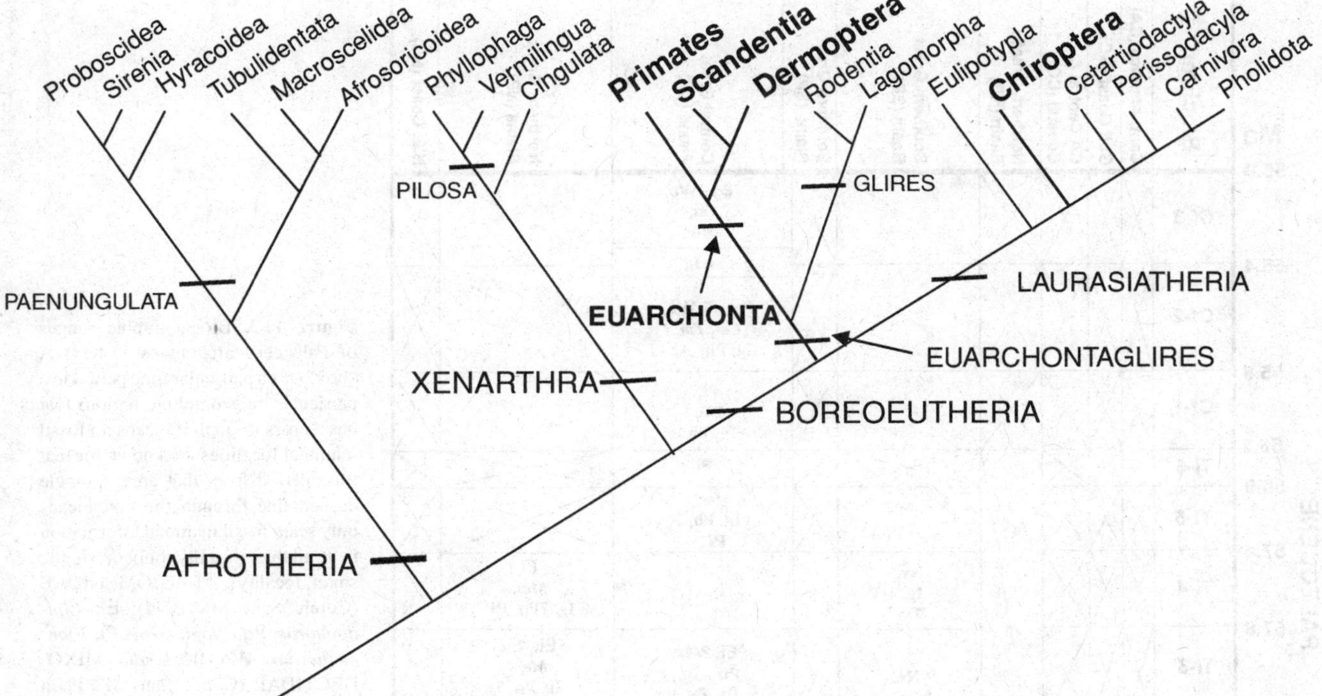

Figure 11.2. Cladogram showing interrelationships of archontans (as the term is used in this volume) with other placental mammals. Archontans are in bold face type.

mammals, primates apparently pass learned behaviors from one generation to the next.

Colugos are amazingly accomplished gliders, making glides that may exceed 100 m (Nowak, 1999). Their patagia encompass the hands as well as the entire length of the body onto the feet and tail (Nowak, 1999). They are similar to some bats (e.g., *Icaronycteris*) in having claws that are unusually high proximally (Szalay and Lucas, 1996), which provide robust support for clinging to tree trunks. Because of the structure of their skeletons and the unique attachments of their patagia to their digits and tail, colugos are not adept at moving on the ground (Nowak, 1999).

SYSTEMATICS

ORIGIN AND AFFINITIES OF ARCHONTA

EXTANT FORMS

Early molecular studies were inconsistent in terms of where the various archontan orders were placed in relation to other mammalian groups (see summary in Allard, McNiff, and Miyamoto [1996]) (Figure 11.2). However, in recent years, a growing consensus has supported a relationship between Euarchonta and Glires in molecular analyses (Waddell, Okada, and Hasegawa, 1999; Murphy *et al.*, 2001a,b; Springer *et al.*, 2003, 2004; Kriegs *et al.*, 2006), which might suggest an Asian origin for the group in light of the Asian location of all basal gliroids (Meng, 2004). However, the earliest archontan or euarchontan fossils are from North America, so an origin from that continent cannot be excluded. The earliest fossils convincingly tied to Archonta or Euarchonta belong to the plesiadapiform *Purgatorius*, and are latest Cretaceous or earliest Paleocene in age (Johnston and Fox, 1984; Lofgren, 1995; Clemens, 2004), implying that the group must have arisen by this time. Molecular dates for the origin of Euarchonta are much earlier. For example, in an analysis that controlled for many of the problems that have plagued previous molecular dating estimates, Springer *et al.* (2003) produced dates for the divergence of Euarchonta that ranged between 80 and 90 million years ago, some 15–25 million years older than the oldest well-documented fossil archontan.

FOSSIL RELATIVES

As noted above, there are no fossil relatives of tree shrews known from North America. The only fossil record of this group consist of isolated occurrences in the Eocene and Miocene of China (Qiu, 1986; Tong, 1988; Ni and Qui, 2002), the Miocene of Thailand (Mein and Ginsberg, 1997), and the Mio-Pliocene of Pakistan and India (Dutta, 1975; Chopra, Kaul, and Vasishat, 1979; Chopra and Vasishat, 1979; Jacobs, 1980).

ARCHONTAN PHYLOGENETICS

ARCHONTAN MONOPHYLY

As noted above, no molecular analysis has ever provided support for the inclusion of bats in a clade with the other supposed archontans. Rather, bats are typically part of a grouping with carnivores,

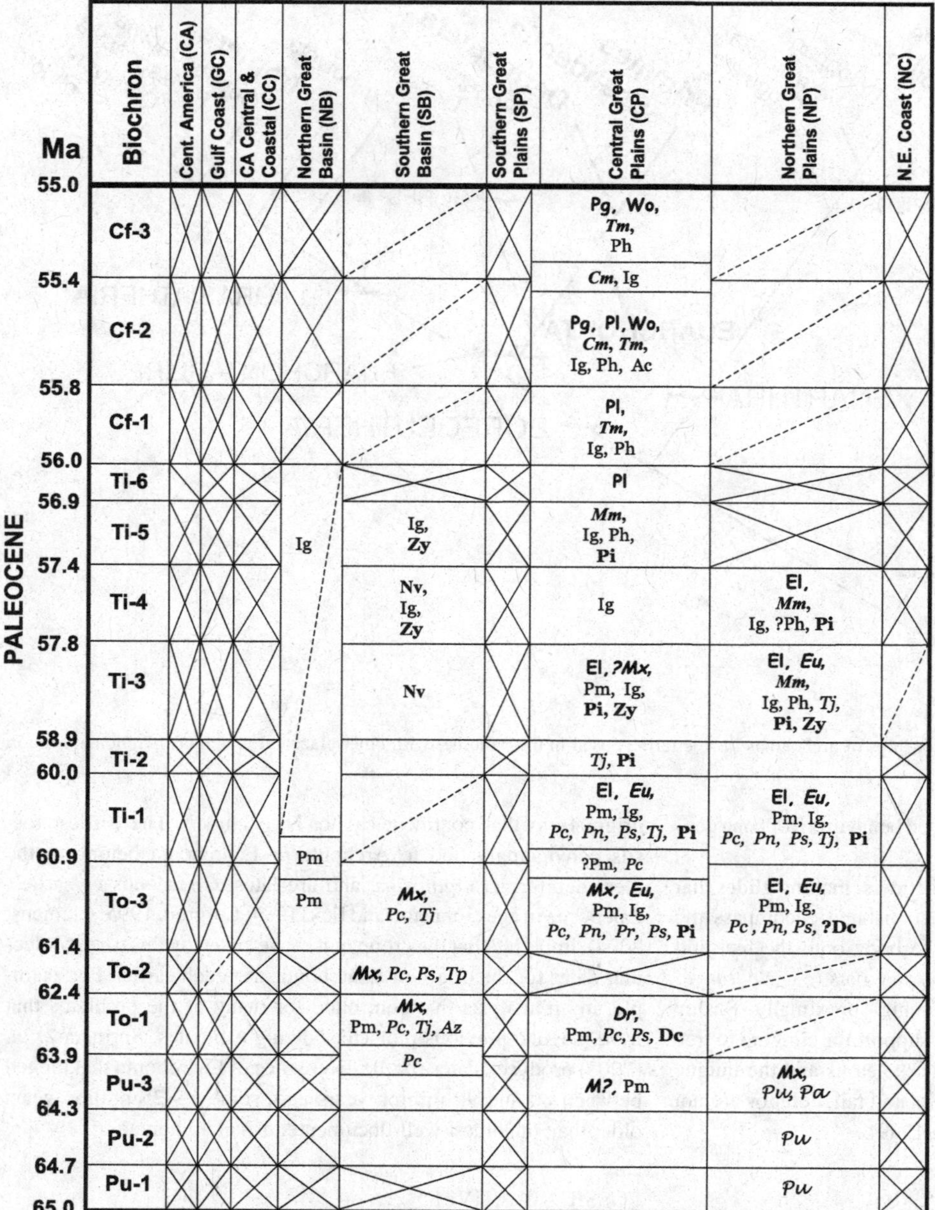

Figure 11.3. Biogeographic ranges of Paleocene archontans 1. Key: A "box" (for a particular time period in a particular biogeographic region) that has a cross through it means no fossil mammal localities are known for that time period from that area; a single dashed line through the box means only scant fossil mammal information is available (usually only a single, small locality). PLAGIOMENIDAE (Comic Sans MS Bold): **El**, *Elpidophorus*; **Pg**, *Plagiomene*; **Pl**, *Planetetherium*; **Wo**, *Worlandia*. MIXODECTIDAE (Comic Sans MS Plain Bold Italics): ***Dr***, *Dracontolestes*; ***Eu***, *Eudaemonema*; ***Mx***, *Mixodectes*; ***M?***, *Mixodectid indet*. PLESIADAPIFORMES: INCERTAE SEDIS (Lucida Handwriting): *Pa*, *Pandemonium*; *Pu*, *Purgatorius*. MICROMOMYIDAE (Times New Roman Bold Italics): ***Cm***, *Chalicomomys*; ***Mm***, *Micromomys*; ***Tm***, *Tinimomys*. PAROMOMYIDAE (Georgia Plain): Ac, *Acidomomys*; Ig, *Ignacius*; Pm, *Paromomys*; Ph, *Phenacolemur*. PALAECHTHONIDAE (Georgia Italics): *Az*, *Anasazia*; *Pc*, *Palaechthon*; *Pn*, *Palenochtha*; *Pr*, *Premnoides*; *Ps*, *Plesiolestes*; *Tj*, *Torrejonia*; *Tp*, *Talpohenach*. PICRODONTIDAE (Georgia Bold); **Dc**, *Draconodus*; **Pi**, *Picrodus*; **Zy**, *Zanycteris*.

ungulates, pholidotans, and eulipotyphlan insectivores (Liu *et al.*, 2001; Murphy *et al.*, 2001a,b; Springer *et al.*, 2003, 2004; Van den Bussche and Hoofer, 2004; Kriegs *et al.*, 2006), which has been named Laurasiatheria. This finding contradicts morphologically based classifications (McKenna, 1975; Szalay, 1977; Novacek and Wyss, 1986; McKenna and Bell, 1997) that included bats in Archonta, and in particular the fairly substantial morphological support for Volitantia, a clade including Dermoptera and Chiroptera (Simmons, 1995; Simmons and Geisler, 1998). It is worth noting, however, that many of the traits shared by bats and colugos are postcranial features that may have evolved independently as a result of the similar locomotor modes in these two groups (Gunnell and Simmons, 2005).

INTERRELATIONSHIPS OF ARCHONTAN ORDERS

One conclusion for which recent studies of both molecules and morphology do agree is that Chiroptera is monophyletic (Bailey, Slightom, and Goodman, 1992; Simmons, 1994; Springer *et al.*, 2004; Kriegs *et al.*, 2006). This contradicts the "flying primate" hypothesis, which suggested that flight originated twice in mammals, with megachiropterans being more closely related to primates

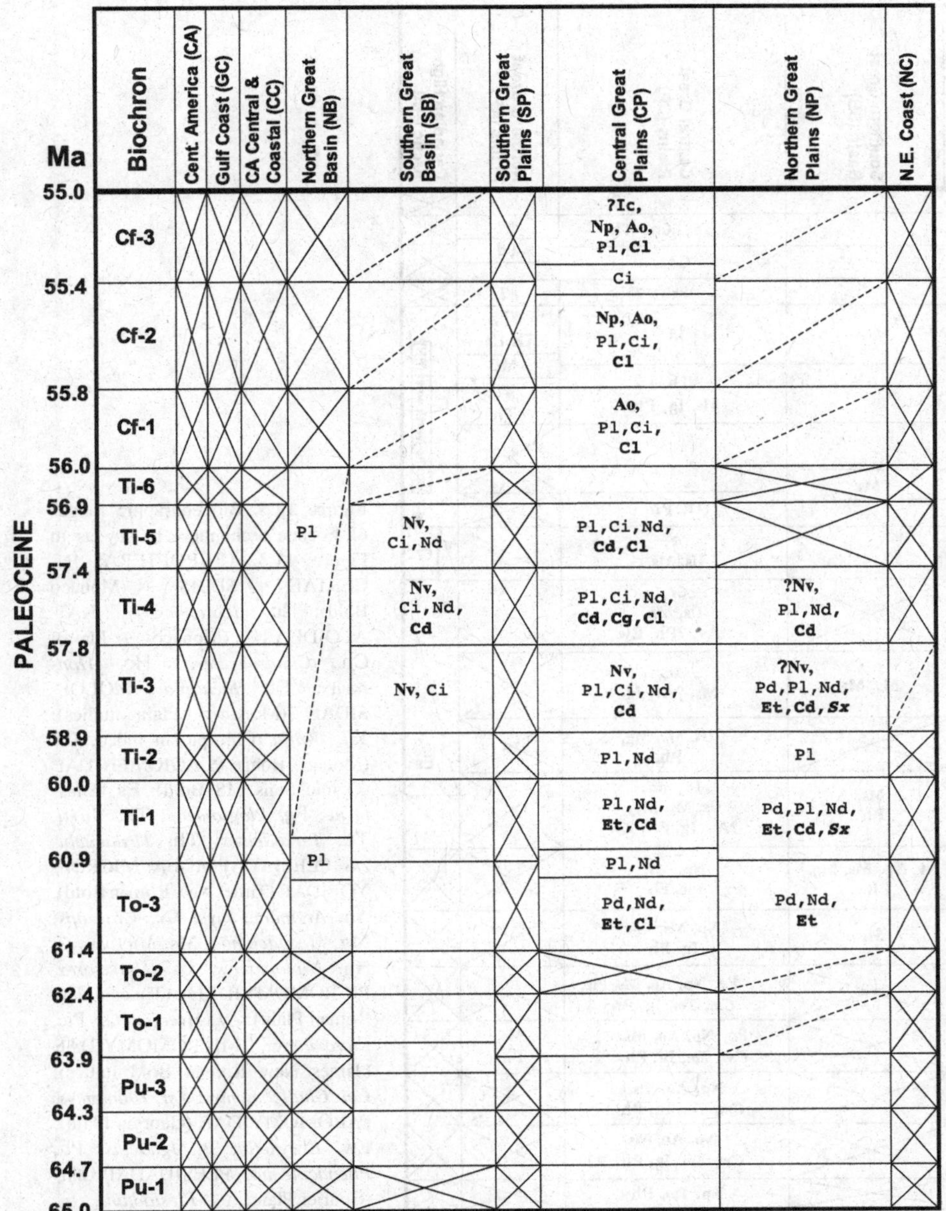

Figure 11.4. Biogeographic ranges of Paleocene archontans 2. Key as in Figure 11.3. CHIROPTERA: INCERTAE SEDIS (Monaco Bold): **Ic**, *Icaronycteris*. PLESIADIFORMES: MICROSYOPIDAE (Times New Roman Bold): **Ao**, *Arctodontomys*; **Nv**, *Navajovius*; **Np**, *Niptomomys*. PLESIDADAPIDAE (Courier Plain): Ci, *Chiromyoides*; Nd, *Nannodectes*; Pd, *Pronothodectes*; Pl, *Plesiadapis*. CARPOLESTIDAE (Courier Bold): **Cd**, *Carpodaptes*; **Cg**, *Carpomegodon*; **Cl**, *Carpolestes*; **Et**, *Elphidotarsius*; SAXONELLIDAE (Courier Bold Italic): ***Sx***, *Saxonella*.

than to microchiropterans (Smith and Madkour, 1980; Pettigrew *et al.*, 1989).

There has been a long history of allying tree shrews with primates, either as a member of the Primates (e.g., Carlsson, 1922; Le Gros Clark, 1925, 1926; Simpson, 1945) or as the sister taxon to Primates (e.g., Wible and Covert, 1987; Kay, Thewissen, and Yoder, 1992; Novacek, 1992). The primary basis for this hypothesis is the shared presence of various cranial features between these groups. However, it is now clear that some of these similarities are likely non-homologous (e.g., bony tubes for the internal carotid artery branches are at best only partly homologous since they are partly formed by the entotympanic in tree shrews [Cartmill and MacPhee, 1980] and/or are also found in other putative primate sister taxa [Silcox, 2003]). Recent molecular studies have most often allied scandentians with dermopterans to the exclusion of primates (Liu and Miyamoto, 1999; Liu *et al.*, 2001; Madsen *et al.*, 2001; Murphy *et al.*, 2001a,b; Springer, 2004), in a clade named Sundatheria by Olson, Sargis, and Martin (2005). This clade has also received some support from morphological studies, when Chiroptera is excluded (Sargis, 2001; Silcox, 2001).

The growing consensus for a sister-group relationship between Scandentia and Dermoptera is contradicted by the "primatomorphan" hypothesis of Beard (1990, 1993a,b), which linked Dermoptera and Primates to the exclusion of Scandentia. Although there

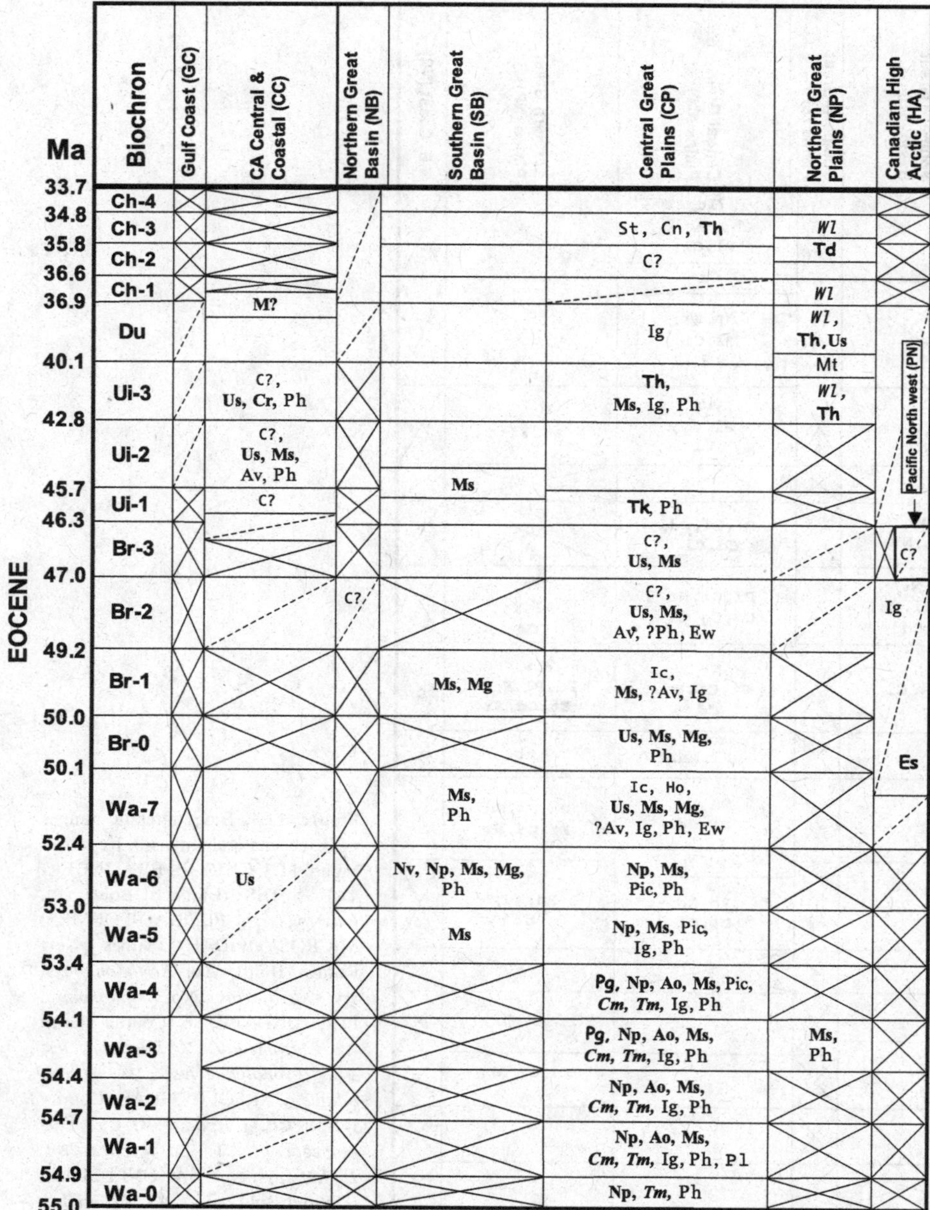

Figure 11.5. Biogeographic ranges of Eocene archontans 1. Key as in Figure 11.3. CHIROPTERA: INCERTAE SEDIS (Monaco Bold): **Ic**, *Icaronycteris*. NATALOIDEA (Monaco Plain): Cn, *Chadronycteris*; Ho, *Honrovits*; St, *Stehlinia*. MOLOSSIDAE (Monanco Plain Italics): ***Wl***, *Wallia*. Chiropteran indet., **C?** (Monaco Bold). PLAGIOMENIDAE (Comic Sans MS Bold): **Es**, *Ellesmene*; **Pg**, *Plagiomene*; **Tk**, *Tarka*; **Td**, *Tarkadectes*; **Th**, *Thylacaelurus*. PLESIADAPOIDEA: MICROSYOPIDAE (Times New Roman Bold). **Ao**, *Arctodontomys*; **Cr**, *Craseops*; **Mg**, *Megadelphus*; **Ms**, *Microsyops*; **Np**, *Niptomomys*; **Us**, *Uintasorex*. PICROMOMYIDAE (Times New Roman Plain): Av, *Alveojunctus*; Pic, *Picromomys*. MICROMOMYIDAE (Times New Roman Bold Italics): ***Cm***, *Chalicomomys*; ***Tm***, *Tinimomys*. PAROMOMYIDAE (Georgia Plain): Ew, *Elwynella*; Ig, *Ignacius*; Ph, *Phenacolemur.* PLESIDADAPIDAE (Courier Plain): Pl, *Plesiadapis*.

are some molecular studies that have provided apparent support for this hypothesis, they often failed to include Scandentia in the analysis, implying that they did not test the monophyly of Sundatheria (e.g., Teeling *et al.*, 2005). Sargis (2002a) found that many of the features that were supposed to link Primates and Dermoptera to the exclusion of Scandentia are actually found in the most primitive living tree shrew, *Ptilocercus lowii*, weakening support for Primatomorpha. Beard's (1990, 1993a,b) hypothesis rested in large part on his interpretation of various unassociated remains of paromomyid and micromomyid plesiadapiforms. Study of better associated, and in some cases more complete, material for these groups (Boyer and Bloch, 2002, 2007) have failed to uphold the Primatomorpha hypothesis, as have analyses including a broader range of characters and taxa (Silcox, 2001; Bloch, Silcox, and Sargis, 2002; Bloch *et al.*, 2004). Therefore, although disagreements abound, the current consensus for the relationships of extant euarchontan groups are that Scandentia and Dermoptera form a clade called Sundatheria, and that Sundatheria is the sister taxon to Primates.

EVOLUTIONARY AND BIOGEOGRAPHIC PATTERNS

GENERAL TRENDS

Figures 11.3 through 11.7 summarize the biogeographic distribution of archontans in the fossil record from North America. Plagiomenids and mixodectids are represented throughout most of the

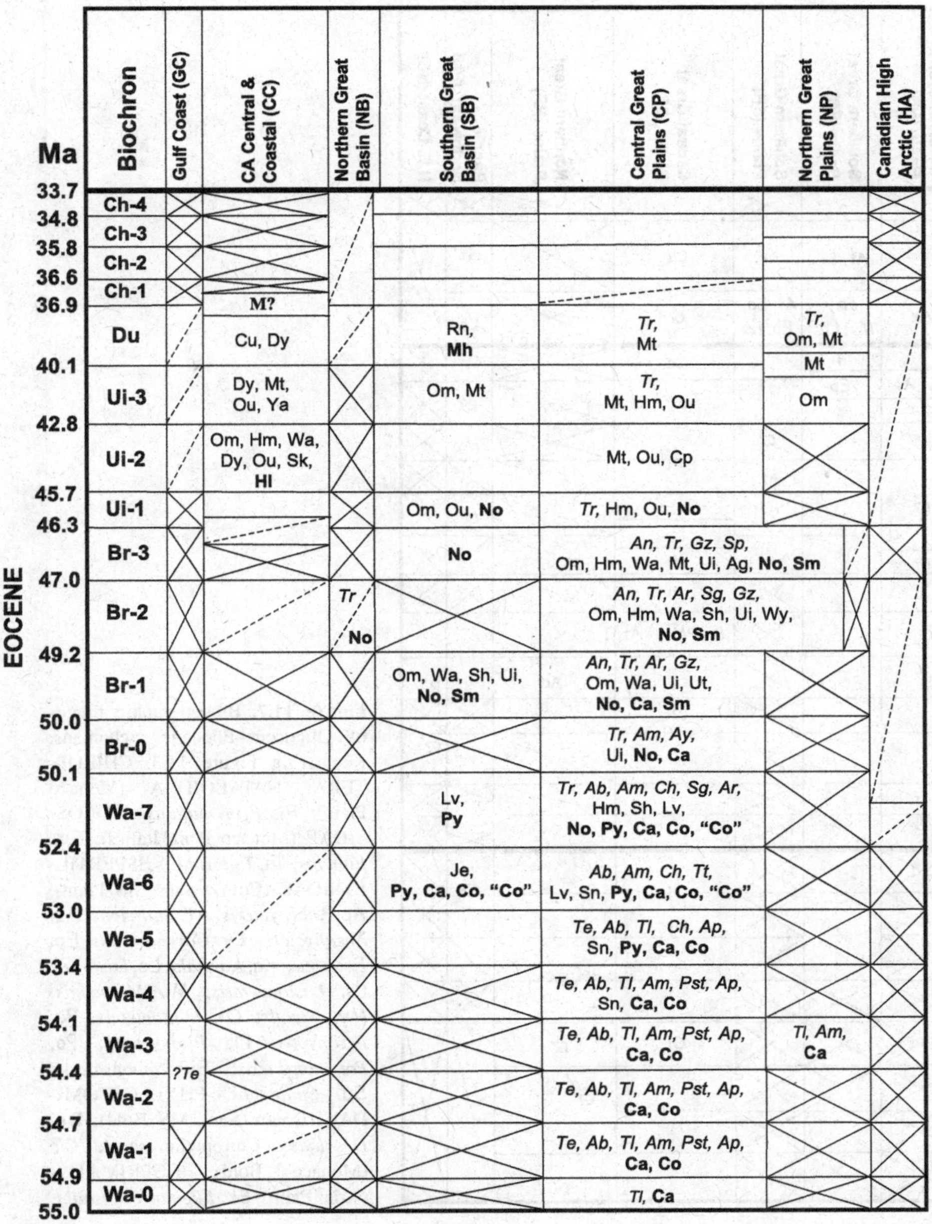

Ma	Biochron	Gulf Coast (GC)	CA Central & Coastal (CC)	Northern Great Basin (NB)	Southern Great Basin (SB)	Central Great Plains (CP)	Northern Great Plains (NP)	Canadian High Arctic (HA)
33.7	Ch-4							
34.8	Ch-3							
35.8	Ch-2							
36.6	Ch-1		M?					
36.9	Du		Cu, Dy		Rn, **Mh**	*Tr*, Mt	*Tr*, Om, Mt / Mt	
40.1	Ui-3		Dy, Mt, Ou, Ya		Om, Mt	*Tr*, Mt, Hm, Ou	Om	
42.8	Ui-2		Om, Hm, Wa, Dy, Ou, Sk, **Hl**			Mt, Ou, Cp		
45.7	Ui-1				Om, Ou, **No**	*Tr*, Hm, Ou, **No**		
46.3	Br-3				**No**	*An, Tr, Gz, Sp*, Om, Hm, Wa, Mt, Ui, Ag, **No, Sm**		
47.0	Br-2			*Tr* / **No**		*An, Tr, Ar, Sg, Gz*, Om, Hm, Wa, Sh, Ui, Wy, **No, Sm**		
49.2	Br-1				Om, Wa, Sh, Ui, **No, Sm**	*An, Tr, Ar, Gz*, Om, Wa, Ui, Ut, **No, Ca, Sm**		
50.0	Br-0					*Tr, Am, Ay*, Ui, **No, Ca**		
50.1	Wa-7				Lv, **Py**	*Tr, Ab, Am, Ch, Sg, Ar*, Hm, Sh, Lv, **No, Py, Ca, Co, "Co"**		
52.4	Wa-6				Je, **Py, Ca, Co, "Co"**	*Ab, Am, Ch, Tt*, Lv, Sn, **Py, Ca, Co, "Co"**		
53.0	Wa-5					*Te, Ab, Tl, Ch, Ap*, Sn, **Py, Ca, Co**		
53.4	Wa-4					*Te, Ab, Tl, Am, Pst, Ap*, Sn, **Ca, Co**		
54.1	Wa-3					*Te, Ab, Tl, Am, Pst, Ap*, **Ca, Co**	*Tl, Am*, **Ca**	
54.4	Wa-2	*?Te*				*Te, Ab, Tl, Am, Pst, Ap*, **Ca, Co**		
54.7	Wa-1					*Te, Ab, Tl, Am, Pst, Ap*, **Ca, Co**		
54.9	Wa-0					*Tl*, **Ca**		
55.0								

EOCENE

Figure 11.6. Biogeographic ranges of Eocene archontans 2: Euprimates. Key as in Figure 11.3. ANAPTOMORPHINAE (Arial Italics): *Ab*, *Absarokius*; *Am*, *Anemorhysis*; *An*, *Anaptomorphus*; *Ar*, *Artimonius*; *Ap*, *Arapahovius*; *Ay*, *Aycrossia*; *Ch*, *Chlororhysis*; *Gz*, *Gazinius*; *Pst*, *Pseudotetonius*; *Sg*, *Strigorhysis*; *Sp*, *Sphacorhysis*; *Te*, *Tetonius*; *Tl*, *Teilhardina*; *Tr*, *Trogolemur*; *Tt*, *Tatmanius*. OMOMYINAE (Arial Plain): Ag, *Ageitodendron*; Cp, *Chipetaia*; Cu, *Chumashius*; Dy, *Dyseolemur*; Hm, *Hemiacodon*; Je, *Jemezius*; Lv, *Loveina*; Mt, *Macrotarsius*; Om, *Omomys*; Ou, *Ourayia*; Rn, *Rooneyia*; Sh, *Shoshonius*; Sk, *Stockia*; Sn, *Steinius*; Ui, *Uintanius*; Ut, *Utahia*; Wa, *Washakius*; Wy, *Wyomomys*; Ya, *Yaquius*. NOTHARCTINAE (Arial Bold): **Ca**, *Cantius*; **Co**, *Copelemur*; **"Co"**, *"Copelemur"*; **Hl**, *Hesperolemur*; **Mh**, *Mahgarita*; **No**, *Notharctus*; **Py**, *Pelycodus*; **Sm**, *Smilodectes*.

Paleocene in the Central Great Plains, first by *Dracontolestes* in the earliest Torrejonian, followed by *Mixodectes, Eudaemonema*, and *Elpidophorus* into the middle Tiffanian. *Mixodectes* is also represented throughout the Torrejonian in the Southern Great Basin and in the late Puercan in the Northern Great Plains. *Elpidophorus* and *Eudaemonema* occur in Northern Great Plains localities beginning in the late Torrejonian through the beginning of the late Tiffanian (Ti4). Curiously, no plagiomenids are recorded from Ti5, one of the best-sampled intervals of the Tiffanian in the Central Great Plains, but *Planetetherium* appears subsequent to that biochron in Ti6 and persists through the middle Clarkforkian (Cf2). *Plagiomene* and *Worlandia* both appear in the middle Clarkforkian in the Central Great Plains, with the former lasting through Wa4 and the latter disappearing at the Paleocene–Eocene boundary. Interestingly, *Plagiomene* is not known from the earliest Wasatchian (Wa0 through Wa2) but does occur before and after this interval of time. The only possible late Wasatchian or Bridgerian record of a plagiomenid is that of *Ellesmene* from the Canadian High Arctic. After that there are scattered records of *Tarka, Tarkadectes*, and *Thylacaelurus* from the early Uintan through the early late Chadronian in the Central Great Plains and Northern Great Plains. The last occurrence of a plagiomenid in North America is *Thylacaelurus* in Ch3 in the Central Great Plains (Figure 11.5).

There are possible occurrences of fossil bats from the latest Paleocene of the Central Great Plains. After that, no bats are known until the late Wasatchian (Wa7) in the Central Great Plains where

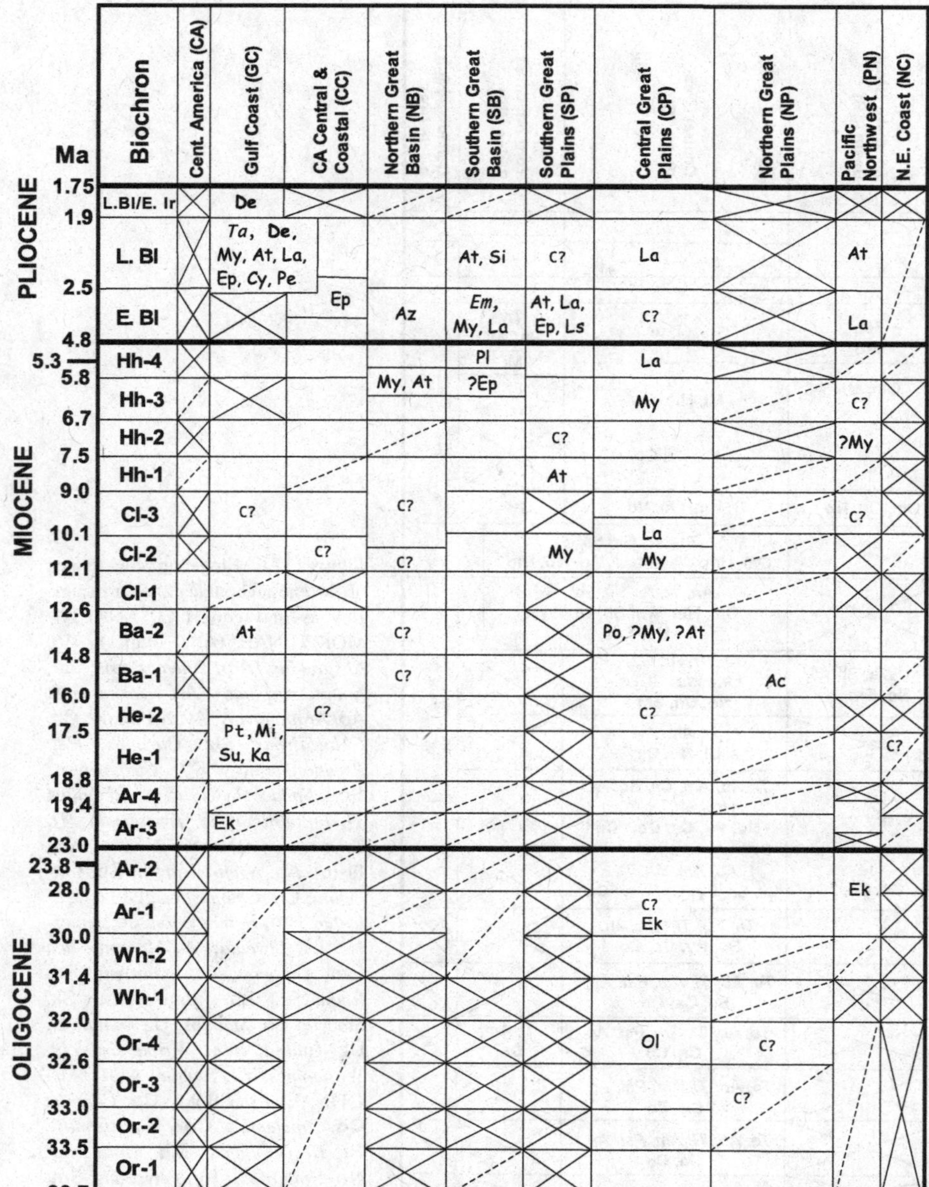

Figure 11.7. Biogeographic ranges of Oligocene–Pliocene archontans. Key as in Figure 11.3. CHIROPTERA: NATALOIDEA (Monaco Plain): Pt, *Primonatalus*. MOLOSSIDAE (Monaco Plain Italics): *Em*, *Eumops*; *Ta*, *Tadarida*. VESPERTILIONIDAE (Comic Sans MS Plain): Ac, *Ancenycteris*; At, *Antrozous*; Az, *Anzanycteris*; Cy, *Corynorhinus*; Ep, *Eptesicus*; Ka, *Karstala*; La, *Lasiurus*; Ls, *Lasionycteris*; Mi, *Miomyotis*; My, *Myotis*; Ol, *Oligomyotis*; Pe, *Perimyotis*; Pl, *Plionycteris*; Po, *Potamonycteris*; Si, *Simonycteris*; Su, *Suaptenos*. PHYLLOSTOMIDAE (Comic Sans MS Bold): **De**, *Desmodus*. Chiropteran indet., **C?** (Monaco Bold). EUPRIMATES (Arial Plain): Ek, *Ekgmowechashala*.

Icaronycteris and *Honrovits* co-occur. *Icaronycteris* is also known from the early Bridgerian in the Central Great Plains. Most of the remainder of the North American Eocene bat record consists of scattered occurrences of possible bats from the Gulf Coast, Central Great Plains, and the Pacific Northwest. By the late middle and late Eocene, other bats begin to appear in the Central Great Plains (*Stehlinia* and *Chadronycteris*) and Northern Great Plains (*Wallia*). A few other records of bats occur in the Central Great Plains and Northern Great Plains in the later Oligocene but bats do not begin to become common elements of North American faunal assemblages until the Hemingfordian in the Gulf Coast region. By the late Miocene on into the Pliocene, bats are persistently represented from localities across the United States (Figure 11.7).

By far the most common archontans in the North American Paleocene are plesiadapiforms. They dominate mammalian faunal samples from the Southern Great Basin, Central Great Plains, and Northern Great Plains beginning in the Puercan through all of the Paleocene. They are also represented by *Plesiadapis* and paromomyids in the Northern Great Basin. The early part of the plesiadapiform radiation is dominated by paromomyids and palaechthonids. Beginning in the Tiffanian, plesiadapids and carpolestids become more common, although paromomyids maintain a relatively steady abundance and diversity throughout the entire temporal distribution of plesiadapiforms.

Plesiadapiforms are as diverse as euprimates during the early Eocene and are especially common in the Central Great Plains and Southern Great Basin. However, compared with the Paleocene, the taxonomic makeup of the Eocene plesiadapiform radiation is much different (Figure 11.5). Instead of being dominated by plesiadapids and carpolestids, Eocene plesiadapiforms are represented by microsyopids, micromomyids, picromomyids, and paromomyids. Micromomyids have disappeared by the end of Wa4 while the other families all hold on until at least the middle Uintan in the case of picromomyids, and the Duchesnean in the case of paromomyids and microsyopids.

Euprimates do not appear until the beginning of the Eocene, being first represented in North America by *Teilhardina* and *Cantius* in the earliest Wasatchian (Wa0) (Figure 11.6). Omomyids are diverse beginning in the early Wasatchian (Wa1) and remain so throughout most of the early and middle Eocene in the Central Great Plains and Southern Great Basin. Omomyids remain relatively diverse and become relatively common in California Central and Coast and the Northern Great Plains in the Uintan and Duchesnean. Adapiform primates are less diverse throughout the Eocene than are omomyids but they are much more abundant, especially in the early Eocene of the Central Great Plains. By the beginning of the middle Eocene, adapiforms are still abundant but their diversity begins to decrease to the point where they last occur in the Uintan of California Central and Coast and the Duchesnean of the Southern Great Basin. The only possible North American primate known after the Duchesnean is the enigmatic *Ekgmowechashala*, which has a relatively broad geographic distribution in North America (Oregon, South Dakota, Nebraska, Florida) but a limited temporal distribution (Arikareean only). *Ekgmowechashala* is not only temporally restricted but is never abundant either (Figure 11.7).

CONCLUSIONS

The Tertiary record of Archonta in North America is of central importance to understanding the group, as it includes the earliest known archontan (or euarchontan) and primate (*Purgatorius*), some of the earliest euprimates (*Teilhardina*, *Cantius*), possibly the earliest bat specimens (Cf. *Icaronycteris* sp.; Gingerich, 1987), and a group that may include the first dermopterans (Plagiomenidae). Therefore, even as the fossil record from other parts of the world improves, the North American record will continue to occupy a central place in debates about the origin and evolution of this group.

REFERENCES

Adkins, R. M. and Honeycutt, R. L. (1991). Molecular phylogeny of the superorder Archonta. *Proceedings of the National Academy of Sciences, USA*, 88, 10 317–21.

Aimi, M. and Inagaki, H. (1988). Grooved lower incisors in flying lemurs. *Journal of Mammalogy*, 69, 138–40.

Allard, M. W., McNiff, B. E., and Miyamoto, M. M. (1996). Support for interordinal eutherian relationships with an emphasis on primates and their archontan relatives. *Molecular Phylogenetics and Evolution*, 5, 78–88.

Bailey, W. J., Slightom, J. L., and Goodman, M. (1992). Rejection of the flying primate hypothesis by phylogenetic evidence from the 1-globin gene. *Science*, 256, 86–9.

Beard, K. C. (1989). Postcranial anatomy, locomotor adaptations, and paleoecology of early Cenozoic Plesiadapidae, Paromomyidae, and Micromomyidae (Eutheria, Dermoptera). Ph.D. Thesis. Johns Hopkins University School of Medicine, Baltimore.

(1990). Gliding behavior and palaeoecology of the alleged primate family Paromomyidae (Mammalia, Dermoptera). *Nature*, 345, 340–1.

(1993a). Phylogenetic systematics of the Primatomorpha, with special reference to Dermoptera. In *Mammal Phylogeny: Placentals*, ed. F. S. Szalay, M. J. Novacek, and M. C. McKenna, pp.129–50. New York: Springer-Verlag.

(1993b). Origin and evolution of gliding in early Cenozoic Dermoptera (Mammalia, Primatomorpha). In *Primates and Their Relatives in Phylogenetic Perspective*, ed. R. D. E. MacPhee, pp. 63–90. New York: Plenum Press.

Bloch, J. I. and Boyer, D. M. (2002). Grasping primate origins. *Science*, 298, 1606–10.

(2007). New skeletons of Paleocene–Eocene Plesiadapiformes: a diversity of arboreal positional behaviors in early primates. In *Primate Origins and Adaptations: A Multidisciplinary Perspective*, ed. M. J. Ravosa and M. Dagosto, pp. 535–81. New York: Plenum Press.

Bloch, J. I. and Silcox, M. T. (2001). New basicrania of Paleocene–Eocene *Ignacius*: re-evaluation of the plesiadapiform–dermopteran link. *American Journal of Physical Anthropology*, 116, 184–98.

(2006). Cranial anatomy of Paleocene plesiadapiform *Carpolestes simpsoni* (Mammalia, Primates) using ultra high-resolution X-ray computed tomography, and the relationships of plesiadapiforms to Euprimates. *Journal of Human Evolution*, 50, 1–35.

Bloch, J. I., Silcox, M. T., and Sargis, E. J. (2002). Origin and relationships of Archonta (Mammalia, Eutheria): re-evaluation of Eudermoptera and Primatomorpha. *Journal of Vertebrate Paleontology*, 22(suppl. to no. 3), p. 37A.

Bloch, J. I., Boyer, D. M., and Houde, P. (2003). Skeletons of Paleocene–Eocene micromomyids (Mammalia, Primates): functional morphology and implications for euarchontan relationships. *Journal of Vertebrate Paleontology*, 23(suppl. to no. 3), p. 35A.

Bloch, J. I., Silcox, M. T., Boyer, D. M., and Sargis, E. J. (2004). New hypothesis of primate supraordinal relationships and its bearing on competing models of primate origins: a test from the fossil record. *American Journal of Physical Anthropology*, 123(suppl. 38), p. 64.

Boyer, D. M., Bloch J. I., and Gingerich, P. D. (2001). New skeletons of Paleocene paromomyids (Mammalia,?Primates): were they mitten gliders? *Journal of Vertebrate Paleontology*, 21(suppl. to no. 3), p. 35A.

Boyer, D. M., Bloch J. I., Silcox, M. T., and Gingerich, P. D. (2004).New observations on the anatomy of *Nannodectes* (Mammalia, Primates) from the Paleocene of Montana and Colorado. *Journal of Vertebrate Paleontology*, 24(suppl. to no. 3), p. 40A.

Butler, P. M. (1972). The problem of insectivore classification. In *Studies in Vertebrate Evolution*, ed. K. A. Joysey and T. S. Kemp, pp. 253–65. Edinburgh: Oliver and Boyd.

(1980). The tupaiid dentition. In *Comparative Biology and Evolutionary Relationships of Tree Shrews*, ed. W. P. Luckett, pp. 171–204. New York: Plenum Press.

Carlsson, A. 1922. Über die Tupaiidae und ihre Beziehungen zu den Insectivora und den Prosimiae. *Acta Zoologica*, 3, 227–70.

Cartmill, M. and MacPhee, R. D. E. (1980). Tupaiid affinities: the evidence of the carotid arteries and cranial skeleton. In *Comparative Biology and Evolutionary Relationships of Tree Shrews*, ed. W. P. Luckett, pp. 95–132. New York: Plenum Press.

Chopra, S. R. K. and Vasishat, R. N. (1979). Sivalik fossil tree shrew from Haritalyangar, India. *Nature*, 281, 214–15.

Chopra, S. R. K., Kaul, S., and Vasishat, R. N. (1979). Miocene tree shrews from the India Sivaliks. *Nature*, 218, 213–14.

Clemens, W. A. (2004). *Purgatorius* (Plesiadapiformes, Primates?, Mammalia), a Paleocene immigrant into Northeastern Montana:

stratigraphic occurrences and incisor proportions. *Bulletin of the Carnegie Museum of Natural History*, 36, 3–13.

Ducrocq, S., Buffetaut, E., Buffetaut-Tong, H., *et al.* (1992). First fossil flying lemur: a dermopteran from the Late Eocene of Thailand. *Palaeontology*, 35, 373–80.

Dutta, A. K. (1975). Micromammals from Siwaliks. *Indian Minerals,* 29, 76–7.

Fleagle, J. G. (1999). *Primate Adaptation and Evolution*, 2nd edn. New York: Academic Press.

Gingerich, P. D. (1976). Cranial anatomy and evolution of early Tertiary Plesiadapidae (Mammalia, Primates). *University of Michigan Papers on Paleontology*, 15, 1–140.

(1987). Early Eocene bats (Mammalia, Chiroptera) and other vertebrates in freshwater limestones of the Willwood Formation, Clarks Fork Basin, Wyoming. *Contributions from the Museum of Paleontology, University of Michigan*, 27, 275–320.

Gingerich, P. D. and Gunnell, G. F. (1992). A new skeleton of *Plesiadapis cookei. The Display Case*, 6, 1–2.

(2005). Brain of *Plesiadapis cookei* (Mammalia, Proprimates): surface morphology and encephalization compared to those of Primates and Dermoptera. *Contributions from the Museum of Paleontology, University of Michigan*, 32, 185–95.

Gregory, W. K. (1910). The orders of mammals. *Bulletin of the American Museum of Natural History*, 27, 1–524.

Gunnell, G. F. and Simmons, N. B. (2005). Fossil evidence and the origin of bats. *Journal of Mammalian Evolution*, 12, 209–46.

Hamrick, M. W., Rosenman, B. A., and Brush, J. A. (1999). Phalangeal morphology of the Paromomyidae (?Primates, Plesiadapiformes): the evidence for gliding behavior reconsidered. *American Journal of Physical Anthropology*, 109, 397–413.

Hill, J. E. and Smith, J. D. (1984). *Bats: a Natural History*. Austin, TX: University of Texas Press.

Hooker, J. J. (2001). Tarsals of the extinct insectivoran family Nyctitheriidae (Mammalia): evidence for archontan relationships. *Zoological Journal of the Linnean Society*, 132, 501–29.

Jacobs, L. L. (1980). Siwalik fossil tree shrews. In *Comparative Biology and Evolutionary Relationships of Tree Shrews*, ed. W. P. Luckett, pp. 205–16. New York: Plenum Press.

Johnson, J. I. and Kirsh, J. A. W. (1993). Phylogeny through brain traits: interordinal relationships among mammals including Primates and Chiroptera. In *Primates and Their Relatives in Phylogenetic Perspective*, ed. R. D. E. MacPhee, pp. 293–331. New York: Plenum Press.

Johnston, P. A. and Fox, R. C. (1984). Paleocene and late Cretaceous mammals from Saskatchewan, Canada. *Palaeontographica Abteilung A*, 186, 163–222.

Jungers, W. L., Godfrey, L. R., Simons, E. L., and Chatrath, P. S. (1997). Phalangeal curvature and positional behavior in extinct sloth lemurs (Primates, Paleopropithecidae). *Proceedings of the National Academy of Sciences, USA*, 34, 11 998–2001.

Kay, R. F. (1984). On the use of anatomical features to infer foraging behavior in extinct primates. In *Adaptation for Foraging in Nonhuman Primates: Contributions to an Organismal Biology of Prosimians, Monkeys and Apes*, ed. P. S. Rodman and J. G. H. Cant, pp. 21–53. New York: Columbia University Press.

Kay, R. F., Thewissen, J. G. M., and Yoder, A. D. (1992). Cranial anatomy of *Ignacius graybullianus* and the affinities of the Plesiadapiformes. *American Journal of Physical Anthropology*, 89, 477–98.

Kriegs, J. O., Churakov, G., Kiefmann, M., *et al.* (2006). Retroposed elements as archives for the evolutionary history of placental mammals. *PLoS Biology*, 34, 537–44.

Le Gros Clark, W. E. (1925). On the skull of *Tupaia. Proceedings of the Zoological Society of London*, 1925, 559–67.

(1926). On the anatomy of the pen-tailed tree shrew (*Ptilocercus lowii*). *Proceedings of the Zoological Society of London*, 1926, 1179–309.

Liu, F.-G. R. and Miyamoto, M. M. (1999). Phylogenetic assessment of molecular and morphological data for eutherian mammals. *Systematic Biology*, 48, 54–64.

Liu, F.-G. R., Miyamoto, M. M., Freire, N. P., *et al.* (2001). Molecular and morphological supertrees for eutherian (placental) mammals. *Science*, 291, 1786–9.

Lofgren, D. L. (1995). The Bug Creek problem and the Cretaceous–Tertiary boundary at McGuire Creek, Montana. *University of California Publications in Geological Science*, 140, 1–185.

MacPhee, R. D. E., Cartmill, M., and Rose, K. D. (1989). Craniodental morphology and relationships of the supposed Eocene dermopteran *Plagiomene* (Mammalia). *Journal of Vertebrate Paleontology*, 9, 329–49.

Madsen, O., Scally, M., Douady, C. J., *et al.* (2001). Parallel adaptive radiations in two major clades of placental mammals. *Nature*, 409, 610–14.

Marivaux, L., Bocat, L., Chaimanee, Y., *et al.* (2006). Cynocephalid dermopterans from the Palaeogene of South Asia (Thailand, Myanmar, and Pakistan): systematic, evolutionary and palaeobiogeographic implications. *Zoologica Scripta*, 35, 395–420.

Martin, R. D. (1990). *Primate Origins and Evolution: A Phylogenetic Approach.* Princeton, NJ: Princeton University Press.

McKenna, M. C. (1975). Toward a phylogenetic classification of the Mammalia. In *Phylogeny of the Primates*, ed. W. P. Luckett and F. S. Szalay, pp. 21–46. New York: Plenum Press.

McKenna, M. C. and Bell, S. K. (1997). *Classification of Mammals Above the Species Level*. New York: Columbia University Press.

Mein, P. and Ginsburg, L. (1997). Les mammifères du gisement Miocène inférieur de Li Mae Long, Thaïlande: systématique, biostratigraphie et paléoenvironnement. *Geodiversitas,* 19, 783–844.

Meng, J. (2004). Phylogeny and divergence of basal Glires. *Bulletin of the American Museum of Natural History*, 285, 93–109.

Miyamoto, M. M., Porter, C. A., and Goodman, M. (2000). c-*myc* gene sequences and the phylogeny of bats and other eutherian mammals. *Systematic Biology*, 49, 501–14.

Murphy, W. J., Eizirik, E., Johnson, W. E., *et al.* (2001a). Molecular phylogenetics and the origins of placental mammals. *Nature*, 409, 614–18.

Murphy, W. J., Eizirik, E., O'Brien, S. J., *et al.* (2001b). Resolution of the early placental mammal radiation using Bayesian phylogenetics. *Science*, 294, 2348–51.

Ni, X. and Qiu, Z. (2002). The micromammalian fauna from the Leilao, Yuanmou hominoid locality: Implications for biochronology and paleoecology. *Journal of Human Evolution*, 42, 535–46.

Novacek, M. J. (1980). Phylogenetic analysis of the chiropteran auditory region. *Proceedings of the Fifth International Bat Research Conference*, ed. D. E. Wilson and A. L. Gardner, pp. 317–30. Lubbock, TX: Texas Technical Press.

(1992). Mammalian phylogeny: shaking the tree. *Nature*, 356, 121–5.

Novacek, M. J. and Wyss, A. R. (1986). Higher-level relationships of the Recent eutherian orders: morphological evidence. *Cladistics*, 2, 257–87.

Nowak, R. M. (1999). *Walker's Mammals of the World*, 6th edn. Baltimore, MD: Johns Hopkins University Press.

Olson, L. E., Sargis, E. J., and Martin, R. D. (2005). Intraordinal phylogenetics of tree shrews (Mammalia: Scandentia) based on evidence from the mitochondrial 12S rRNA gene. *Molecular Phylogenetics and Evolution*, 35, 656–73.

Pettigrew, J. D., Jamieson, B. G. M., Robson S. K., *et al.* (1989). Phylogenetic relations between microbats, megabats and primates (Mammalia: Chiroptera, Primates). *Philosophical Transactions of the Royal Society of London, B*, 325, 489–559.

Pumo, D. E., Finamore, P. S., Franek, W. R., *et al.* (1998). Complete mitochondrial genome of a neotropical fruit bat, *Artibeus jamaicensis*, and a new hypothesis of relationships of bats to other eutherian mammals. *Journal of Molecular Evolution,* 47, 709–17.

Qiu, Z. (1986). Fossil tupaiid from the hominoid locality of Lufeng, Yunnan. *Vertebrata PalAsiatica*, 24, 308–19.

Rose, K. D. (1975). The Carpolestidae: early Tertiary primates from North America. *Bulletin of the Museum of Comparative Zoology*, 147, 1–74.

Rose, K. D., Beard, K. C., and Houde, P. (1993). Exceptional new dentitions of the diminutive plesiadapiforms *Tinimomys* and *Niptomomys* (Mammalia), with comments on the upper incisors of Plesiadapiformes. *Annals of the Carnegie Museum*, 62, 351–61.

Russell, D. E. (1959). Le crâne de *Plesiadapis*. *Bulletin Societe Géologique de France*, 4, 312–4.

Sargis, E. J. (2001). The phylogenetic relationships of archontan mammals: Postcranial evidence. *Journal of Vertebrate Paleontology*, 21(suppl. to no. 3), p. 97A.

(2002a). The postcranial morphology of *Ptilocercus lowii* (Scandentia, Tupaiidae): an analysis of Primatomorphan and Volitantian characters. *Journal of Mammalian Evolution*, 9, 137–60.

(2002b). Functional morphology of the forelimb of tupaiids (Mammalia, Scandentia) and its phylogenetic implications. *Journal of Morphology*, 253, 10–42.

(2004). New views on tree shrews: the role of tupaiids in primate supraordinal relationships. *Evolutionary Anthropology*, 13, 56–66.

Silcox, M. T. (2001). A phylogenetic analysis of Plesiadapiformes and their relationship to Euprimates and other Archontans. Ph.D. Thesis, Johns Hopkins School of Medicine, Baltimore.

(2003). New discoveries on the middle ear anatomy of *Ignacius graybullianus* (Paromomyidae, Primates) from ultra high resolution X-ray computed tomography. *Journal of Human Evolution*, 44, 73–86.

Silcox, M. T. and Bloch, J. I. (2006). Upper incisor evolution in plesiadapiform primates. *American Journal of Physical Anthropology*, supple. 42, p. 165.

Silcox, M. T., Bloch, J. I., Sargis, E. J., and Boyer, D. M. (2005). Euarchonta (Dermoptera, Scandentia, Primates). In *The Rise of Placental Mammals: Origins and Relationships of the Major Extant Clades*, ed. K. D. Rose and J. D. Archibald, pp. 127–44. Baltimore, MD: Johns Hopkins University Press.

Simmons, N. B. (1994). The case for chiropteran monophyly. *American Museum Novitates*, 3103, 1–54.

(1995). Bat relationships and the origin of flight. In *Ecology, Evolution, and Behavior of Bats*, ed. P. A. Racey and S. M. Swift, pp. 27–43. Oxford: Oxford University Press.

Simmons, N. B. and Geisler, G. H. (1998). Phylogenetic relationships of *Icaronycteris*, *Archaeonycteris*, *Hassianycteris*, and *Palaeochiropteryx* to extant bat lineages, with comments on the evolution of echolocation and foraging strategies in Microchiroptera. *Bulletin of the American Museum of Natural History*, 235, 1–182.

Simpson, G. G. (1945). The principles of classification and a classification of mammals. *Bulletin of the American Museum of Natural History*, 85, 1–350.

Smith, J. D. and Madkour, G. (1980). Penial morphology and the question of chiropteran phylogeny. In *Proceedings of the Fifth International Bat Research Conference*, ed. D. E. Wilson and A. L. Gardner, pp. 347–65. Lubbock, TX: Texas Technical Press.

Springer, M. S., Murphy, W. J., Eizirik, E., and O'Brien, S. J. (2003). Placental mammal diversification and the Cretaceous–Tertiary boundary. *Proceedings of the National Academy of Sciences, USA*, 100, 1056–61.

Springer, M. S., Stanhope, M. J., Madsen, O., and de Jong, W. W. (2004). Molecules consolidate the placental mammal tree. *Trends in Ecology and Evolution*, 19, 430–48.

Stafford, B. J. and Szalay, F. S. (2000). Craniodental functional morphology and taxonomy of dermopterans. *Journal of Mammalogy*, 81, 360–85.

Szalay, F. S. (1977). Phylogenetic relationships and a classification of the eutherian Mammalia. In *Major Patterns in Vertebrate Evolution*, ed. M. K. Hecht, P. C. Goody and B. M. Hecht, pp. 315–74. New York: Plenum Press.

Szalay, F. S. and Drawhorn, G. (1980). Evolution and diversification of the Archonta in an arboreal milieu. In *Comparative Biology and Evolutionary Relationships of Tree Shrews*, ed. W. P. Luckett, pp. 133–69. New York: Plenum Press.

Szalay, F. S. and Lucas, S. G. (1993). Cranioskeletal morphology of Archontans, and diagnoses of Chiroptera, Volitantia, and Archonta. In *Primates and their Relatives in Phylogenetic Perspective*, ed. R. D. E. MacPhee, pp. 187–226. New York: Plenum Press.

(1996). The postcranial morphology of Paleocene *Chriacus* and *Mixodectes* and the phylogenetic relationships of archontan mammals. *Bulletin of the New Mexico Museum of Natural History and Science*, 7, 1–47.

Szalay, F. S., Tattersall, I., and Decker, R. L. (1975). Phylogenetic relationships of *Plesiadapis*: postcranial evidence. In *Approaches to Primate Paleobiology*, ed. F. S. Szalay, pp. 136–66. Basel: Karger.

Szalay, F. S., Rosenberger, A. L., and Dagosto, M. (1987). Diagnosis and differentiation of the order Primates. *Yearbook of Physical Anthropology*, 30, 75–105.

Teeling E. C., Springer M. S., Madsen O., Bates P., O'Brien S. J., and Murphy, W. J. (2005). A molecular phylogeny for bats illuminates biogeography and the fossil record. *Science*, 307, 580–4.

Tong, Y. (1988). Fossil tree shrews from the Eocene Hetaoyuan formation of Xichuan, Henan. *Vertebrata PalAsiatica*, 26, 214–20.

Van den Bussche R. A. and Hoofer S. R. (2004). Phylogenetic relationships among recent chiropteran families and the importance of choosing appropriate out-group taxa. *Journal of Mammalogy*, 85, 321–30.

Waddell, P. J., Okada, N., and Hasegawa, M. (1999). Towards resolving the interordinal relationships of placental mammals. *Systematic Biology*, 48, 1–5.

Walker, A. C. (1974). Locomotor adaptations in past and present prosimian primates. In *Primate Locomotion*, ed. F. A. Jenkins, Jr., pp. 349–82. London: Academic Press.

Wible J. R. and Covert, H. H. (1987). Primates: cladistic diagnosis and relationships. *Journal of Human Evolution*, 16, 1–22.

Wible J. R. and Martin, J. R. (1993). Ontogeny of the tympanic floor and roof in archontans. In *Primates and their Relatives in Phylogenetic Perspective*, ed. R. D. E. MacPhee, pp. 111–46. New York: Plenum Press.

Wischusen, E. W. and Richmond, M. E. (1998). Foraging ecology of the Philippine flying lemur (*Cynocephalus volans*). *Journal of Mammalogy*, 79, 1288–95.

Youlatos, D. and Godinot, M. (2004). Locomotor adaptations of *Plesiadapis tricuspidens* and *Plesiadapis* n. sp. (Mammalia, Plesiadapiformes) as reflected on selected parts of the postcranium. *Journal of Anthropological Science*, 82, 103–18.

12 Chiroptera

NICHOLAS J. CZAPLEWSKI,[1] GARY S. MORGAN,[2] and SAMUEL A. McLEOD[3]

[1]*Oklahoma Museum of Natural History, Norman, OK, USA*
[2]*New Mexico Museum of Natural History, Albuquerque, NM, USA*
[3]*Los Angeles County Museum of Natural History, Los Angeles, CA, USA*

INTRODUCTION

With the exception of rodents, bats are the most diverse extant mammalian order. The fossil record of bats is notoriously poor, however. There are several good reasons for the impoverished fossil record of chiropterans. Bats are generally quite small mammals with very light and delicate bones. They are volant (the only mammals to achieve powered flight) and typically roost in trees and caves away from areas conducive to long-term sedimentary accumulation. Their taxonomic diversity increases dramatically toward the equator, especially in tropical rain forests that have a depauperate fossil record for their entire biota. The Pleistocene fossil record for bats is much better than that for the Tertiary, but it is mostly restricted to cave deposits and extant, cave-adapted species.

Prior to the Pleistocene, the fossil record of bats is too incomplete to provide a comprehensive picture of their evolution. With the exception of the exquisite examples from the Quercy Phosphorites, Geiseltal Coal Deposits, Green River Formation, and the Messel Oil Shales (Gunnell, 2001; Storch, 2001), diagnostic remains of bats are more tantalizing than informative. Typical Tertiary bat fossils are isolated teeth, jaw fragments, or portions of long bones. Unfortunately, the primitive and still widespread dilambdodont dentition of insectivorous bats is not highly distinctive and has appeared in a number of mammalian groups. This problem perplexes paleontologists attempting to identify fossils. In particular, the fossil teeth of various Insectivora and Chiroptera are often confused. There are probably good bats represented among the many taxa of Paleocene insectivorous mammals. There may also be non-bats among the taxa listed below.

DEFINING FEATURES OF THE ORDER CHIROPTERA

CRANIAL

In its primitive configuration, the skull is generally long and low with a rather long and narrow rostrum and somewhat inflated braincase. The nasal opening is generally large and may separate the premaxillae. A postorbital bar is sometimes present but usually not complete. Zygomatic arches are well developed. Most cranial sutures are obliterated in adults. The tympanic is composed of an annular ectotympanic and entotympanic parts.

In derived species the configuration of the skull and jaws (Figure 12.1A) varies greatly because these elements are related to strikingly diverse feeding habits, roosting habits, echolocation and flight styles, and other morphological correlates. The premaxillae vary in their structure, development of the nasomaxillary and palatal branches, and articulation or degree of fusion with the adjacent bones. These different features of the premaxillae have been used as characteristics to separate higher taxa of bats. (Jepsen, 1966, 1970; Novacek, 1982, 1985, 1987, 1991; Wible and Novacek, 1988; Simmons, Novacek, and Baker, 1991; Koopman, 1994; Simmons, 1994, 1995, 1998; Simmons and Geisler, 1998).

DENTAL

The maximum dental formula is I2/3, C1/1, P3/3, M3/3, with one less upper incisor and one less upper and lower premolar than most primitive eutherians with an unreduced dental formula (Figure 12.1B). The tooth position for the missing premolar has often been

Evolution of Tertiary Mammals of North America, Vol. 2. ed. C. M. Janis, G. F. Gunnell, and M. D. Uhen. Published by Cambridge University Press.

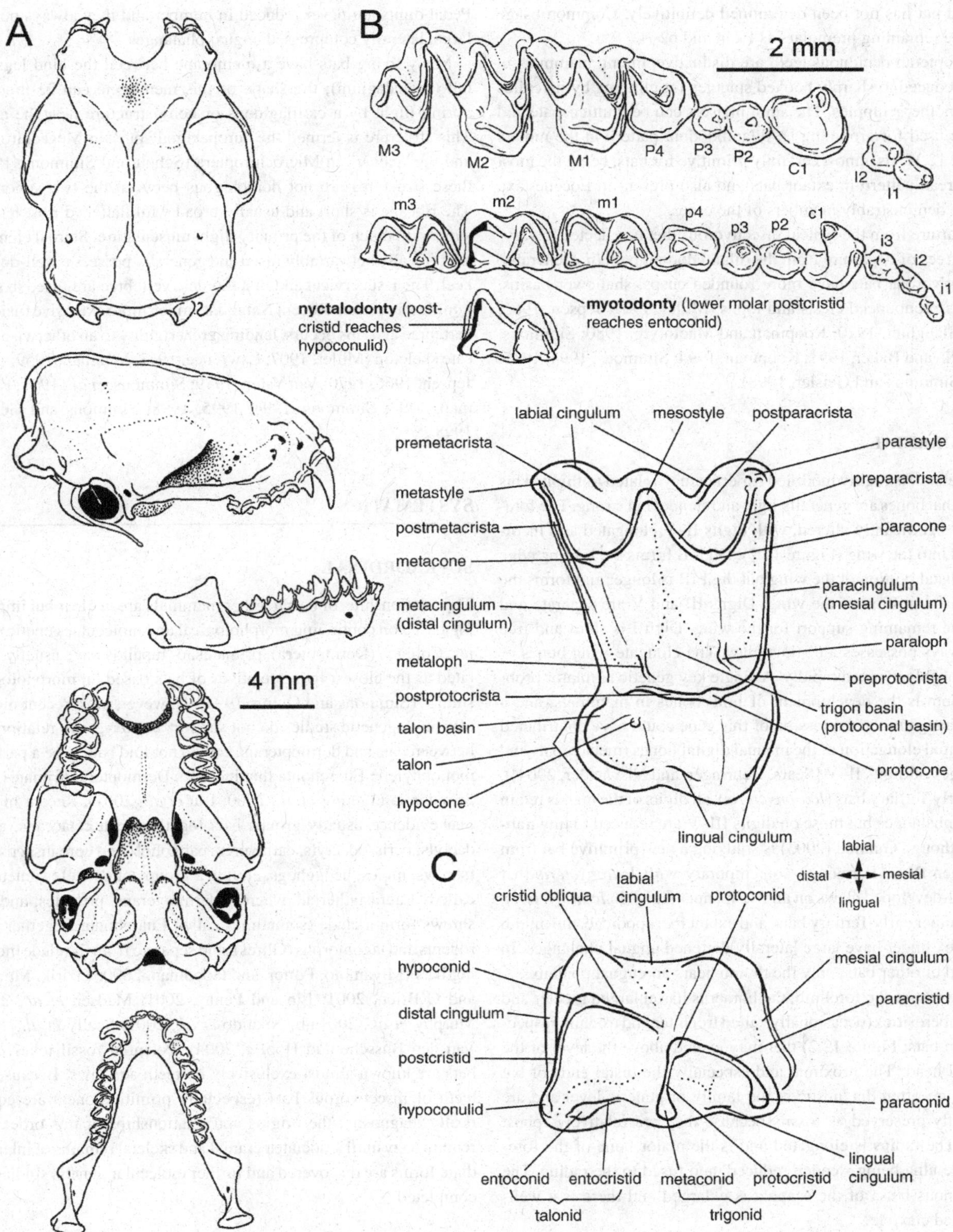

Figure 12.1. A. Skull and mandible of *Lasiurus cinereus*; skull in dorsal, lateral, and palatal views, and mandible in lateral and occlusal views. B. Teeth of *Myotis auriculus*: right upper tooth row in occlusal view; left lower tooth row in occlusal view. Additional lower molar contrasts myotodont and nyctalodont configurations of the talonids (Menu and Sigé, 1971). C. Diagrams of a right upper and left lower molar of a bat. Potential structures are labeled with the terms usually used in recent literature.

debated but has not been determined definitively. Common usage cites the remaining premolars as P2–4 and p2–4.

Chiropteran deciduous teeth are distinctive among mammals in being reduced to sharply hooked spicules for clinging by juveniles to the mother's nipples. The adult incisors can be multicuspate and may be used for grooming. A dilambdodont pattern in the molars (Figure 12.1C) is almost certainly primitive for bats, being the most widespread pattern in extant bats and also present in Eocene taxa that are demonstrably members of the order.

Departure from the typical insectivorous diet is reflected in varying degrees of departure from the dilambdodont pattern. Generally, the frugivorous bats have more rounded cusps, shallower basins, and less-pronounced crests and lophs (Miller, 1907; Jepsen, 1966, 1970; Slaughter, 1970; Koopman and MacIntyre, 1980; Simmons, Novacek, and Baker, 1991; Koopman, 1994; Simmons, 1994, 1995, 1998; Simmons and Geisler, 1998).

POSTCRANIAL

The skeleton is highly modified for efficiency related to flight. This means that bones are generally light and slender but strong. The forelimb is significantly altered, with digits II–V elongated and incorporated into the wing (Figure 12.2). Digit II forms the leading edge of the distal portion of the wing but digit III is longer and forms the most distal portion of the wing. Digits III and V are separate and form the remaining support for the wing. Digit I is short and free and always possesses a distinct claw. The elongate wing bones of bats probably evolved rapidly. A single key genetic regulator probably controls the development of limb bones in mammals, and a small change in the expression of this gene could have contributed to the rapid elongation of the manual digital bones (metacarpals and phalanges of digits II–V [Sears, Behringer, and Niswander, 2004]).

In early Tertiary bats (*Icaronycteris*), all digits of the manus retain ungual phalanges but those on digits III–V are reduced to tiny nubbins, although Gunnell (2003) is studying a new primitive bat from the Green River Formation contemporary with *Icaronycteris* that has well-developed claws on all five manual digits. In *Icaronycteris*, certain other early Tertiary bats, and extant Pteropodidae, the thumb and index finger have large laterally flattened ungual phalanges. In the hand of other bats, only the thumb bears an ungual phalanx.

In the rest of the forelimb, the humerus has enlarged greater and lesser tuberosities (occasionally called trochiter and trochin, respectively, in bats; Figure 12.2) that may extend above the level of the humeral head. The proximal and especially the distal ends of bat humeri are often diagnostic at the family or generic level and are frequently preserved as fossils because they are relatively robust bones. The radius is elongated and is the major bone of the forelimb, the ulna being slender, reduced, and fused to the radius. The infraspinous fossa of the scapula is enlarged and there is a well-developed clavicle.

In the hindlimb, the femoral head is rotated so that the knees point backwards and the palmar surfaces of the hind feet point forward. The femur, tibia, and foot bones are slender and resist tensional rather than compressional stresses because of bats' habit of hanging by the feet instead of standing. The fibula is usually greatly reduced. Pedal digits are never reduced in number and they always possess large, laterally compressed ungual phalanges.

Many living bats have a membrane between the hind legs and tail (uropatagium); the shape of this membrane can be modified during flight by a cartilaginous or bony structure near the ankle. This structure is termed the "uropatagial spur" in Megachiroptera and the "calcar" in Microchiroptera (Schutt and Simmons, 1998); these structures are not homologous between the two suborders. The ribcage is short and usually broad with flattened ribs, lending support to much of the primary flight musculature. Sternal elements are typically but variably fused and generally possess a well-defined keel. The last cervical and first thoracic vertebrae are fused in some families; in certain Recent Natalidae, much more extensive fusion of vertebrae and ribs occurs, lending great rigidity to an otherwise delicate skeleton (Miller, 1907; Lawrence, 1943; Vaughan, 1959, 1970; Jepsen, 1966, 1970; Van Valen, 1979; Simmons, *et al.*, 1991; Koopman, 1994; Simmons, 1994, 1995, 1998; Simmons and Geisler, 1998).

SYSTEMATICS

SUPRAORDINAL

The relationships of bats to other mammals are unclear but improving based on continuing morphological and molecular genetic studies. Colugos (Dermoptera), primates, or tupaiids have usually been cited as the closest living relatives of bats based on morphological studies (Simmons and Quinn, 1994). However, most recent molecular phylogenetic studies do not support a sister-group relationship between bats and dermopterans and do not find bats to be a part of a monophyletic Euarchonta that includes Dermoptera, Primates, and Scandentia (Teeling *et al.*, 2000; Liu *et al.*, 2001). Recent molecular evidence usually groups bats together with cetaceans, artiodactyls, perissodactyls, carnivores, pangolins, and core insectivores (shrews, moles, hedgehogs, etc.) in a group collectively sometimes called "Laurasiatheria," whereas dermopterans, primates, and tree shrews form a clade (sometimes called Euarchonta) together with rodents and lagomorphs (Glires), and separate from the clade including bats (Miyamoto, Porter, and Goodman, 2000; Eizirik, Murphy, and O'Brien, 2001; Lin and Penny, 2001; Madsen *et al.*, 2001; Murphy *et al.*, 2001a,b; Nikaido *et al.*, 2001; Scally *et al.*, 2002; Van den Bussche and Hoofer, 2004). Potential fossil relatives of bats are known almost exclusively by teeth and jaws. Because the teeth of insectivorous bats (especially primitive ones) are equivocally diagnostic, the origins and relationships of this order will remain hazy until associated cranial and skeletal remains of intermediate forms are discovered and further molecular genetic studies are completed.

INFRAORDINAL

Until recently the order Chiroptera was typically split into two suborders, Megachiroptera and Microchiroptera. The order and its two suborders are generally considered to be monophyletic (Simmons,

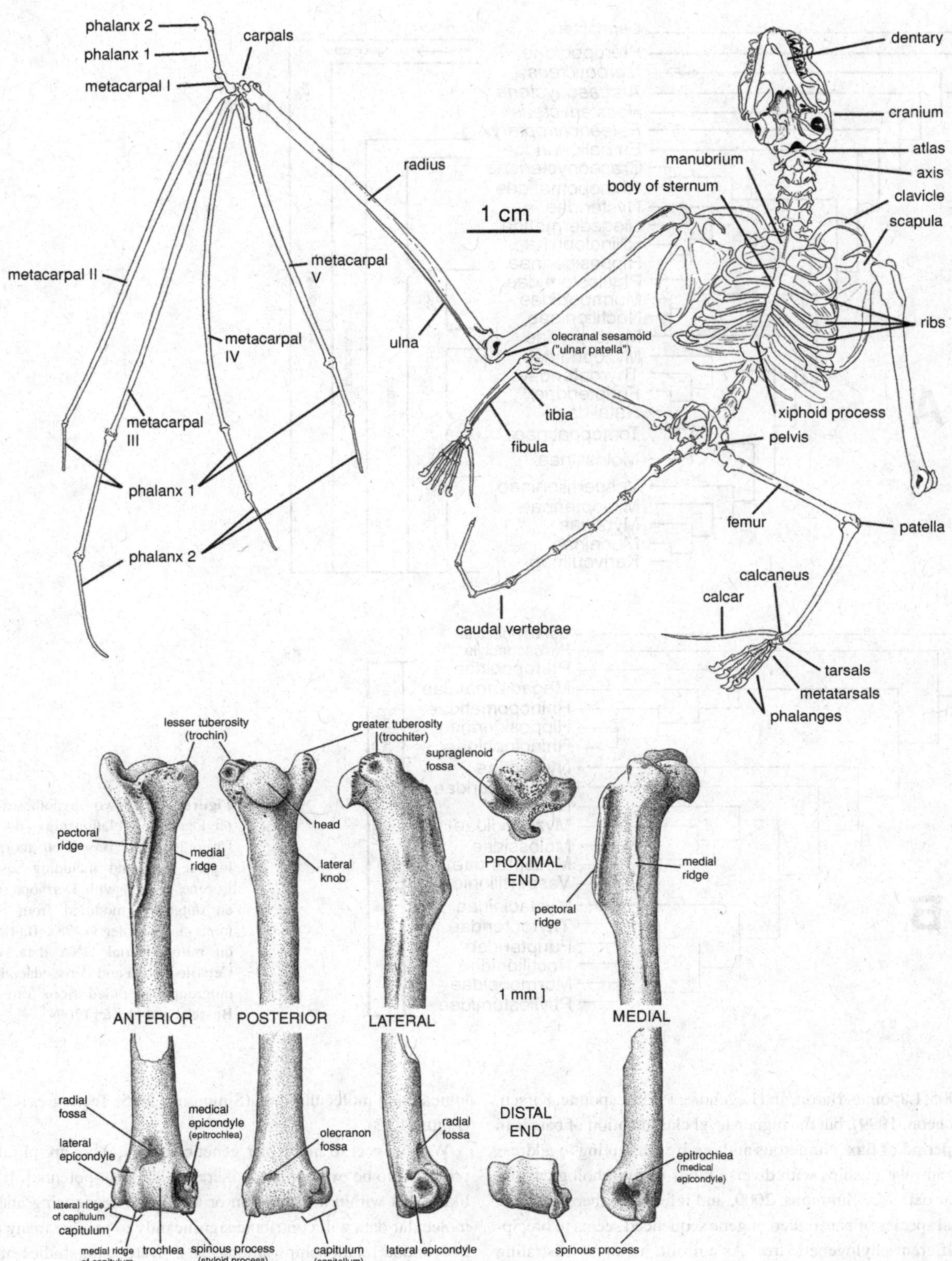

Figure 12.2. Skeleton of modern *Lasiurus cinereus* minus left forearm and manus, ventral aspect, with elements labeled. Proximal and distal portions of the right humerus of a bat (*Karstala silva*, Miocene, Florida). Many anatomical features are labeled with terms used in the literature. Modified from Czaplewski and Morgan (2000a). (Courtesy of the Society of Vertebrate Paleontology.)

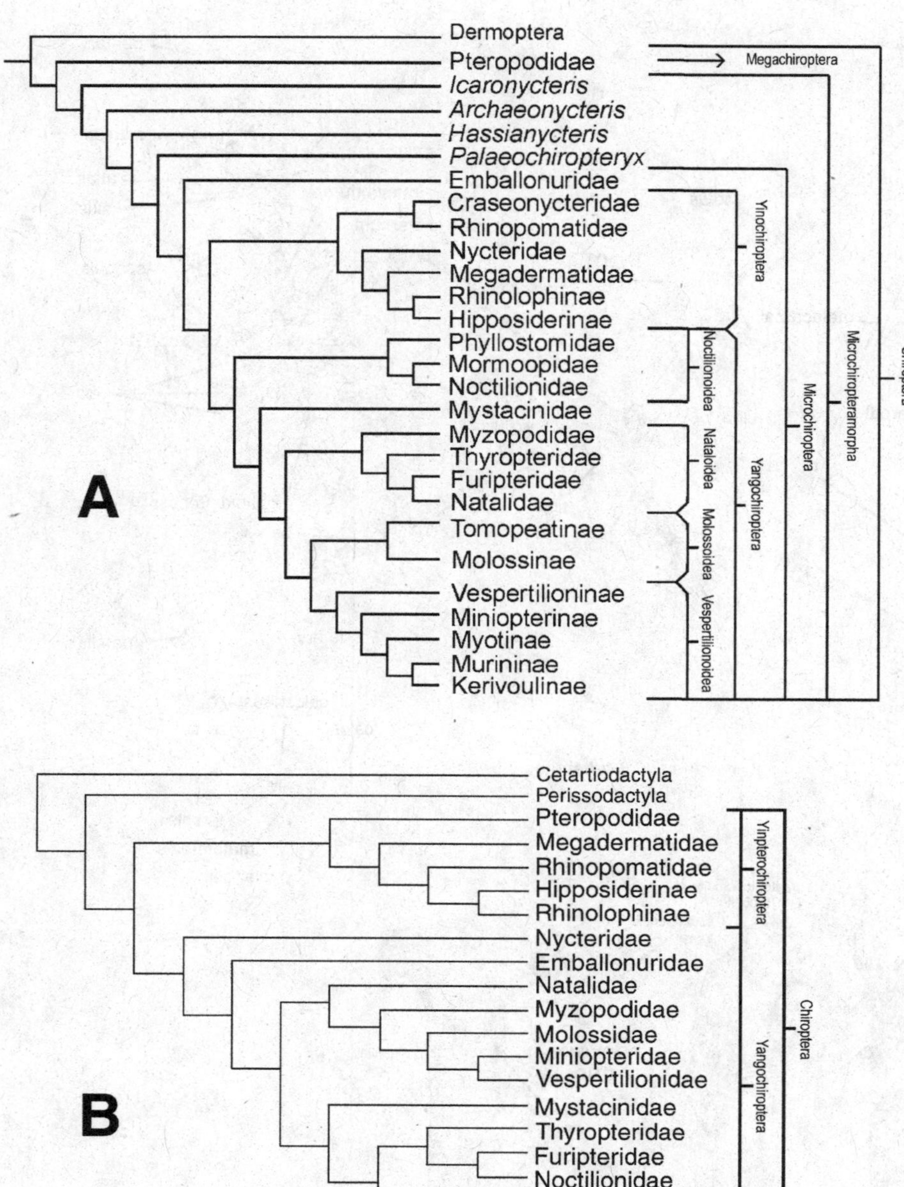

Figure 12.3. Two hypotheses of phylogenetic relationships of the Chiroptera: A. Based on morphological data and including several Eocene genera, with Dermoptera as an outgroup, modified from Simmons and Geisler (1998). B. Based on mitochondrial DNA data, with Cetartiodactyla and Perissodactyla as outgroups, modified from Van den Bussche and Hoofer (2004).

1998, 2005; Lapointe, Baron, and Legendre, 1999; Lapointe, Kirsch, and Hutcheon, 1999), but the higher-level classification of bats is in a major period of flux. Numerous authors are attempting to address chiropteran relationships with diverse kinds of morphological and molecular data (see Simmons, 2000, and references therein). Each additional species of bat studied or gene sequenced seems to precipitate a different phylogenetic tree. As a result, it can be frustrating for the casual observer to struggle with conflicting hypotheses of phylogeny and classification. Yet, at the same time, it is encouraging to see the enthusiasm with which studies of the evolution and diversification of bats are being pursued. Progress is being made, and authors are attempting to harmonize morphological, paleontological, and molecular data (Simmons, 2005; Teeling *et al.*, 2005) (Figure 12.3).

With respect to molecular genetic studies, there are pitfalls and problems to be overcome but there is also great potential. It seems likely that within a generation or two, gene sequencing and other molecular data will contribute significantly to resolve many poorly understood relationships among bats. Systematic studies of fossil bats will be critical to this process in serving to calibrate the molecular clock of bat evolution. It is not possible in this chapter to discuss and reconcile all the many recent papers and hypotheses on the higher-level phylogeny of bats, but we note that some authors, such as Teeling *et al.* (2000, 2002, 2005), Liu *et al.* (2001), Springer,

Teeling, and Stanhope (2001a), Springer *et al.* (2001b), recently suggested that Microchiroptera may not be monophyletic. For the reader wishing to explore the recent literature on the relationships of bats to other eutherian mammals, see Pumo *et al.* (1998); Miyamoto *et al.* (2000); Eizirik, Murphy, and O'Brien (2001); Hooker (2001); Lin and Penny (2001); Murphy *et al.* (2001a,b); Nikaido *et al.* (2001); Scally *et al.* (2002); Van den Bussche and Hoofer (2004). For views on the suprafamilial relationships among bats (especially those including North American groups), see also Simmons (1998, 2000); Simmons and Geisler (1998); Kirsch *et al.* (1998); Kennedy *et al.* (1999); Lapointe, Baron, and Legendre (1999); Lapointe, Kirsch, and Hutcheon (1999); Göbbel (2000); Van den Bussche and Hoofer (2000, 2001, 2004); Springer, Teeling, and Stanhope (2001a); Jones *et al.* (2002); Hutcheon and Kirsch (2004); Teeling *et al.* (2005); and references therein.

The Megachiroptera traditionally includes only one monophyletic family, Pteropodidae (the Old World fruit bats or "flying foxes") (Kirsch *et al.* 1995; Álvarez *et al.*, 1999; Hutcheon and Garland, 2004), that has never occurred in North America. All North American bats belong to the Microchiroptera. Microchiroptera consists of about 16 extant families and at least five extinct families (Simmons and Geisler, 1998).

In their recent systematic treatment of the order that included fossil forms, Simmons and Geisler (1998) proposed a new higher-level classification of bats based on their phylogenetic hypothesis, including two major groupings, the Megachiropteramorpha and Microchiropteramorpha. The traditional suborder Megachiroptera together with the Oligocene genus *Archaeopteropus* were placed in Megachiropteramorpha (*Archaeopteropus* was later removed from Megachiropteramorpha and placed in Microchiropteramorpha by Schutt and Simmons [1998]). Microchiropteramorpha comprises all other known bats.

At the core of this group is the suborder Microchiroptera with two infraorders (Yinochiroptera and Yangochiroptera) plus the Emballonuroidea. (In the molecular studies that find Microchiroptera to be paraphyletic, some families that are usually included in Microchiroptera [Megadermatidae and Rhinolophidae] are placed together with Pteropodidae in Yinpterochiroptera [Springer, Teeling, and Stanhope, 2001a; Springer *et al.*, 2001b].) In addition, Microchiropteramorpha includes several primitive, extinct families: the Icaronycteridae, Archaeonycteridae, Palaeochiropterygidae, and Hassianycteridae. Only the Icaronycteridae, plus several "nataloids" and the extant families Phyllostomidae, Molossidae, and Vespertilionidae occurred in North America during the Tertiary (Figure 12.3). Although the confirmed fossil record of bats does not yet extend back prior to the early Eocene, the presence of a "fully fledged" bat (*Icaronycteris*) at that time suggests that the origin of bats is earlier still. Indeed, *Icaronycteris* itself may have occurred in the late Paleocene; Gingerich (1987: his Fig. 23) illustrated three fragmentary teeth as "Cf. *Icaronycteris* sp." and Bloch and Bowen (2001) listed the same specimens from Clarkforkian 3 in the Clarks Fork Basin, Wyoming (locality CP18B).

Recent molecular phylogenetic studies of eutherian mammals have yet to be successful at resolving whether divergence of the major eutherian clades (orders) occurred prior or subsequent to the end of the Cretaceous (Kumar and Hedges, 1998; Archibald, 1999; Foote *et al.*, 1999a,b; Hedges and Kumar, 1999; Rich, Vickers-Rich, and Flannery, 1999; Eizirik, Murphy, and O'Brien, 2001; Nikaido *et al.*, 2001). However, Springer *et al.* (2001b) examined fossil evidence in addition to molecular evidence and found that "molecular dating suggests that crown-group bats last shared a common ancestor 52 to 54 million years ago," which matches the known fossil record well.

Superfamilies of bats that have representatives in North America include the following constituent families: Emballonuroidea (including Emballonuridae); Noctilionoidea (including Mormoopidae, Noctilionidae, Phyllostomidae, and sometimes Mystacinidae); Nataloidea (including Myzopodidae, Furipteridae, Thyropteridae, and Natalidae); Molossoidea (including Molossidae); and Vespertilionoidea (including Vespertilionidae). Several of these families have no known living representatives or fossil record in North America. Recently, Simmons (1998) placed two genera that are traditionally considered to belong to the Vespertilionidae (*Antrozous* and *Bauerus*) in Molossoidea: Antrozoidae. In addition, Van den Bussche and Hoofer (2001) found that mitochondrial DNA sequences failed to support the monophyly of Nataloidea, calling into question the validity of the superfamily. Instead, mitochondrial DNA evidence suggests Natalidae are basal members of the Vespertilionoidea (Van den Bussche and Hoofer, 2004). Although we anticipate additional future data pertinent to these relationships, for the present paper we tentatively chose to retain the traditional concept of Nataloidea and to retain *Antrozous* and *Bauerus* within Vespertilionidae, as listed above.

INCLUDED GENERA OF CHIROPTERA BLUMENBACH, 1779

The locality numbers listed for each genus refer to the list of unified localities in Appendix I. The acronyms for museum collections are listed in Appendix III.

Brackets around the locality code (e.g., [CP101]) mean that the taxon in question at that locality is cited as an "aff." or "cf." of the taxon in question. Brackets are usually used for individual species, thus implying the genus is firmly known from the locality, but the actual species identification may be questionable. A question mark in front of the locality code (e.g., ?CP101) means that the taxon is questionably known from that locality, either at the genus or the species level. An asterisk (*) indicates the type locality.

Dental terminology primarily follows Legendre (1984). The terms nyctalodont (or nyctalodonty) and myotodont (or myotodonty) refer to alternative conditions of the postcristid in the anterior lower molars of certain bats (Menu and Sigé, 1971; Figure 12.1C). In molars that exhibit nyctalodonty, the postcristid connects the hypoconid with the hypoconulid; this is considered to be the primitive character state. In bats with derived molars that exhibit myotodonty, the postcristid connects the hypoconid with the entoconid, bypassing the hypoconulid. The lower molars of a few species of bats show an intermediate condition called submyotodonty (Legendre, 1984, Fig. 3).

ICARONYCTERIDAE JEPSEN, 1966

Cranial and dental characteristics: Premaxillae are not united at the midline. Maximum dental formula for bats (I2/3, C1/1, P3/3, M3/3) present; dilambdodont upper molars have high cusps with a pronounced ectoloph and distinct cingula; M3 not markedly truncated posteriorly, having only a somewhat abbreviated metastyle and postmetacrista; P3–4 three-rooted and non-molariform but have pronounced cingula, P4 has an expanded lingual shelf; p3–4 two-rooted and non-molariform; lower molars with relatively high cusps with a deep talonid basin and distinct cingula; incisors non-overlapping and faintly multicuspate.

Postcranial characteristics: Humerus relatively long and straight but with a large greater tuberosity that articulates slightly with the scapula and a high flangelike deltopectoral crest; radius long but much stouter than the ulna; a greatly reduced olecranon process on the ulna, which is shorter than the radius; digital formula of the manus 2–3–3–3–3, with a decreasing digital length of 3-4-5–2-1; manus digit II with a distinct sheathed claw and phalanx free of the wing as long as that of digit I, although more slender; femur longer than the tibia, the head and short neck sitting at an angle to the shaft; pes digital formula 2-3-3-3-3, with decreasing digital length of 4-3-2-5-1; all pes digits with claws; no calcar and long tail, presumably free of any uropatagium that might have been present; no fusion or coalescence of the axial elements except for the sacrum; no mesosternal keel. (Taken from Jepsen, 1966, 1970; Novacek, 1985, 1987; Simmons and Geisler, 1998).

Icaronycteris Jepsen, 1966

Type species: *Icaronycteris index* Jepsen, 1966.

Type specimen: PU 18150 (now housed in the YPM collection as YPM-PU 18150), a nearly complete skeleton; late Wasatchian of Wyoming.

Characteristics: Same as for the family.

Average length of m1: 1.5 mm.

Included species: *I. index* only, known from locality CP33A* only.

Comments: Although *Icaronycteris* is known with certainty only from locality CP33A, specimens identified as Cf. *Icaronycteris* are also known from late Paleocene [CP17B, CP18B] localities (Gingerich, 1987; Bloch and Bowen, 2001) and *Icaronycteris* sp. from early Eocene localities (CP27D, E). A second species of *Icaronycteris* might be represented by UW 2244, a partial skeleton from the Green River Formation, Gosiute Lake, from a different level and younger (Bridgerian) than *I. index* (locality CP33C; Jepsen, 1966). This specimen was called *Icaronycteris* cf. *I. index* by Novacek (1987); Jepsen (1966) noted that it is smaller than the type of *I. index* and differs from it in many proportions.

PHYLLOSTOMIDAE GRAY, 1825

This family has an exceedingly poor pre-Pleistocene fossil record and almost no record in North America. Only one highly derived subfamily, Desmodontinae, the vampires, is represented by one genus, *Desmodus*, in the latest Tertiary of North America. Other subfamilies of Phyllostomidae are Phyllostominae, Glossophaginae, Phyllonycterinae, Carolliinae, and Stenodermatinae (Wetterer, Brockman, and Simmons, 2000); Baker *et al.* (2003) gave a different classification, recognizing Macrotinae, Micronycterinae, Desmodontinae, Lonchorhininae, Phyllostominae, Glossophaginae, Lonchophyllinae, Carolliinae, Glyphonycterinae, Rhinophyllinae, and Stenodermatinae.

Dental characteristics: Teeth in many genera of subfamily Phyllostominae are similar to those of most other families of insect-eating bats, but other phyllostomid subfamilies have highly derived dentitions specialized for diets of fruit, nectar, and pollen. Phyllostomid subfamilies other than Desmodontinae are unknown in the Tertiary of North America and are not discussed herein. In Desmodontinae, the first upper incisors are greatly enlarged, flattened, and bladelike. The upper canines are also flattened and bladelike. The cheek teeth are greatly reduced in size and number, such that the cheek tooth row is shorter than the anteroposterior length of the canine.

Postcranial characteristics: The skeleton is robustly built, with the humerus having well-developed secondary articulation between the greater tuberosity and scapula. The greater and lesser tuberosities of the femur are subequal in size and extend beyond the humeral head. The distal articular surface of the humerus is not aligned with the long axis of the shaft. The humeral epitrochlea is broad and bears a small distal spinous process. Digit two of the wing (manus) retains a single phalanx distal to the metacarpal. In the Desmodontinae, the femur, tibia, and fibula are stout and deeply grooved for the accommodation of muscles related to the peculiar terrestrial locomotion (Altenbach, 1979).

Desmodus Wied-Neuwied, 1826

Type species: *Desmodus rotundus* (E. Geoffroy, 1810), the extant New World common vampire.

Type specimen: None designated.

Characteristics: Dental formula reduced to I1/2, C1/1, P1/2, M1–2/1; upper incisors greatly enlarged and projecting anteriorly, with acute tip and long, concave, very sharp posterior cutting edge; lower incisors small and deeply bilobed; canines large and flattened, with posterior bladelike cutting edge; cheek teeth greatly reduced in size; braincase large and rather globose, narrowed anteriorly; interorbital region narrow; rostrum reduced; palate strongly domed (concave).

Average length of m1: Not available.

Average length of humerus: 39.7 mm.

Mastoid breadth; 12.4 mm.

Included species: *D. archaeodaptes* Morgan, Linares, and Ray, 1988 only (known from localities GC15C, D, E*).

Comments: The type locality of *D. archaeodaptes* is Haile 21A, Alachua County, Florida (locality GC15E), which is early Irvingtonian, early Pleistocene. A referred humerus of *D. archaeodaptes* formerly considered to date to the early Pleistocene is now considered to be latest Pliocene in age (from locality GC15C, latest Blancan: Morgan, Linares, and Ray, 1988; Morgan 1991; Morgan and Hulbert, 1995). This is the oldest record of the vampire bats, Desmodontinae.

Desmodus is also known from the Pleistocene (to Recent) of North America, South America, and the West Indies (McKenna and Bell, 1997).

NATALOIDEA OR VESPERTILIONOIDEA

?*Stehlinia* Revilliod, 1919

Type species: *Stehlinia gracilis* Revilliod, 1919.

Type specimen: Naturhistorisches Museum de Bale QP 602, cranium; middle Eocene (MP 13–14) of France.

Characteristics: Nataloid with generalized cranial and dental characters; premaxillary complete; dental formula I2/3, C1/1, P3/3, M3/3; P3 two- or three-rooted; p3 two-rooted, slightly larger than p4; P4 and p4 non-molariform; upper molars lack hypocone; lower molars nyctalodont with large talonid; greater tuberosity of humerus articulates with secondary articular surface on the dorsum of the scapula (Sigé, 1974).

Average length of m1: 1.7 mm.

Included species: "*Stehlinia*? sp." only, known from locality CP98B only.

Comments: *Stehlinia* is known from the middle Eocene to late Oligocene of Europe. The presence of this genus in North America needs to be confirmed or refuted based on restudy of the material on which the queried identification was originally made (Ostrander [1985, 1987] followed Sigé [1974] in placing it in the Kerivoulidae), or else by collecting more and better specimens. Moreover, the genus *Stehlinia* and the Kerivoulinae are not referable to Natalidae sensu stricto as revised by Van Valen (1979). Simmons and Geisler (1998) placed *Stehlinia* in Nataloidea incertae sedis. Sigé (1997) placed it in Vespertilionoidea: Palaeochiropterygidae. The North American material from locality CP98B listed by Ostrander (1985) is Chadronian in age.

Honrovits Beard, Sigé, and Krishtalka, 1992

Type species: *Honrovits tsuwape* Beard, Sigé, and Krishtalka, 1992.

Type specimen: CM 43156, left dentary with m1–3; late Wasatchian (Lostcabinian) of Wyoming.

Characteristics: Dentary bears a high, posteriorly recurved coronoid process; lower molars with nearly median, cuspidate hypoconulid; talonid on m3 unusually elongated by strong and distally projecting hypoconulid; p2 with strong mesial and distal crests; lingual lobe of P4 reduced; p4 low and simple, lacks paraconid and metaconid; upper molars with centrocrista not reaching labial border; postprotocrista joins lingual base of metacone; P3 and p3 strongly reduced.

Average length of m1: 1.6 mm.

Included species: *H. tsuwape* only, known from locality CP27D* only.

Comments: The genus *Honrovits* cannot be definitely referred to the family Natalidae until more complete skeletal material becomes available. Its unusually strong, median, cuspidate hypoconulid is a primitive character shared with many Mesozoic therians. This character is unlike that of most bats, including true natalids, in which the hypoconulids are reduced in size and situated low at the posterolingual corner of the lower molar talonids. We follow Simmons and Geisler (1998) in placing *Honrovits* in Nataloidea incertae sedis but note that this taxon may not represent a chiropteran.

Chadronycteris Ostrander, 1983

Type species: *Chadronycteris rabenae* Ostrander, 1983.

Type specimen: SDSM 09931, left maxilla fragment with P4–M3; middle Chadronian of Nebraska.

Characteristics: M1–2 lack a hypocone and talon and with rounded subtriangular occlusal outline; parastyle on the M1 reduced; M3 with reduced parastyle and mesostyle and no metacone, metastyle, or metacristae; P4 nonmolariform, unicuspate, triangular, and three-rooted.

Average length of M1: 1.5 mm.

Included species: *C. rabenae* only, known from locality CP98B* only.

Comments: In naming *Chadronycteris*, Ostrander (1983) originally compared the type maxilla and a referred M1 only with the proscalopine talpid *Oligoscalops*, the heterosoricine soricid *Domnina*, and the chiropteran *Stehlinia*. He did not compare *Chadronycteris* with any other bats. Because of its similarity to *Stehlinia*, and following Sigé's (1974) placement of that genus, Ostrander placed *Chadronycteris* in Vespertilionoidea, Kerivoulidae (= Kerivoulinae). Van Valen (1979) relegated kerivouline bats to the Natalidae (which he placed in Vespertilionoidea). Nevertheless, most recent authors (e.g., Simmons and Geisler, 1998; Hoofer and Van den Bussche, 2003) have placed Kerivoulinae within the Vespertilionidae rather than Natalidae. Membership of *Chadronycteris* in the Kerivoulinae, Natalidae, or Vespertilionidae is not demonstrable based on the type maxilla; more complete material will be necessary to determine its relationships with other bats. Therefore, we tentatively follow Simmons and Geisler (1998) in placing the genus *Chadronycteris* in Nataloidea incertae sedis, although it may belong in the Vespertilionoidea. McKenna and Bell (1997) placed *Chadronycteris* in Microchiroptera incertae sedis.

NATALIDAE GRAY, 1866

Characteristics: Skeleton delicately built; greater tuberosity of the humerus forms a second articulation with the scapula; distal articular surface of the humerus is out of line with the humeral shaft; humeral epitrochlea broad with short distal spinous process. Second digit of the wing reduced to a metacarpal with no phalanges; third digit of the wing with only two phalanges. Premaxillae with unreduced palatal and nasal branches; premaxillae are fused with one another and with the maxillae in adults. The last cervical and the first thoracic vertebrae fused or unfused; most of the lumbar series (except the last one or two) and sometimes the thoracic vertebrae fused into a rigid bony structure. Three lower premolars present, subequal in size, appearing as nearly equilateral triangles in lateral profile; lower molars with talonids and small carnassial-like notches in the cristid obliquae and postcristids.

Primonatalus Morgan and Czaplewski, 2003

Type species: *Primonatalus prattae* Morgan and Czaplewski, 2003.

Type specimen: UF 108641, partial right dentary with m1–3; early Hemingfordian of Florida.

Characteristics: No other Tertiary natalids known, so by comparison with Recent natalids, the following features characterize *Primonatalus*. Mandibular ramus deep below molars; ventral margin of mandible straight between p4 and mandibular angle; summit of coronoid process of mandible comparatively well developed with triangular-shaped dorsal tip; anterior edge of coronoid process curves posterodorsally from alveolar margin to dorsal tip of coronoid at angle of roughly 70°, coronoid process rises above level of articular condyle; posterior portion of coronoid process slopes ventrally from tip of coronoid down to condyle; distinct mandibular angle present directly ventral to tip of coronoid; base of angular process located halfway between ventral edge of mandible and alveolar margin in vertical dimension, and about halfway between coronoid and condyle in anteroposterior dimension; angular process with little flare laterally; mandibular foramen opens level with alveolar margin. Two features, the comparatively well-developed triangular-shaped coronoid process, which is taller than the articular condyle, and the ventral position of the mandibular foramen and angular process reflect a lesser degree of dorsal cranial flexion in *Primonatalus* than in *Natalus*; posterior mental foramen located in deep concavity near alveolar margin between roots of c1 and p2; p3 larger, longer, and more compressed with distinct concavity posterior to main cusp; metaconid and entoconid same height on m1 and m2; carnassial-like notches on cristid obliqua and postcristid on lower molars present but weak; no lophid associated with carnassial-like notch extending posteriorly from cristid obliqua into talonid basin; talonid notch deeper than in Recent natalids because of taller metaconid and more ventral connection of entocristid to trigonid; cristid obliqua also connects to trigonid more ventrally; labial cingular cusp of P4 low; occlusal outline of P4 with anterior indentation. Talon weakly developed on M1–2; lingual cingulum weak on upper molars. Humerus with broad triangular-shaped epitrochlea and no notch between weak medial process and distal spinous process (Figure 12.2); distal spinous process small, projecting only to edge of trochlea or very slightly distal to it, and separated from trochlea by narrow but distinct notch.

Average length of m1: 1.2 mm.

Included species: *P. prattae* only, known from locality GC8D* only.

MOLOSSIDAE GERVAIS, 1856

Characteristics: Humerus with a greater tuberosity that extends well beyond the humeral head and forms a strong secondary articulation with the scapula; distal end of humerus with a narrow epitrochlea and a long spinous process. Digit two of the wing reduced to an elongate metacarpal and a single vestigial phalanx; greater trochanter of femur bears a hooklike process; last cervical vertebrae fused with first thoracic vertebra; skull lacks postorbital processes; palatal branches of the premaxilla reduced and fused with maxillae in adults; nasal branches of premaxillae present.

Wallia Storer, 1984

Type species: *Wallia scalopidens* Storer, 1984.

Type specimen: SMNH P1654.312, right M1; late Uintan of Saskatchewan.

Characteristics: Upper molars strongly dilambdodont; preprotocrista extends as a paracingulum to parastyle; M1 and M2 with indistinct para- and metaconule and with large, low, cuspate hypocone not fused to posterolingual wall of postprotocrista; M1 and M2 with lingual cingulum at anterior base of protocone; M3 lacks hypocone, postprotocrista, and postmetacrista.

Average length of m1: not available.

Average length of M1: 1.47–1.75 mm.

Included species: *W. scalopidens* only (known from localities NP8*, NP9A).

Wallia sp. is also known from localities NP10B, Bii.

Comments: This taxon was originally placed in Proscalopidae (Insectivora) by Storer (1984) but was moved to Molossidae by Legendre (1985). To date, the genus *Wallia* is known only by isolated upper molars (Storer, 1984, 1995) from the Uintan and Duchesnean; a lower jaw originally referred by Storer (1984) to *Wallia* is not that of a chiropteran. *Wallia* is the earliest known unquestionable molossid, and close study is necessary to determine its relationship to the late Eocene European *Cuvierimops* Legendre and Sigé, 1983 (Legendre, 1985).

A fragmentary mandible with a m2 from the Bridgerian at locality CP34D was assigned by McKenna, Robinson and Taylor (1962) to "Order Chiroptera Family Uncertain Undescribed genus and species." Based on a personal communication from B. Sigé that this specimen appeared to be a vespertilionoid and possibly a molossid, Legendre

(1985) tentatively included it in a list as ?Molossidae. If the specimen is indeed a molossid, then it would be the oldest member of the family.

***Eumops* Miller, 1906 (including *Molossus*, in part)**

Type species: *Eumops californicus* (Merriam, 1890) (originally described as *Molossus californicus*, including *Molossus perotis* Schinz, 1821), the extant California mastiff bat.

Type specimen: None designated.

Characteristics: Skull narrow, with nearly flat dorsal profile and rather tubular area encircling the external nares, lending an overall "cylindrical" appearance to the cranium; basisphenoid pits well developed and moderately to very deep; palate slightly arched; anterior palatal emargination absent; incisive foramina paired. Dental formula I1/2, C1/1, P1–2/2, M3/3; upper incisors slender, with curved crown; I1s in contact with one another and often in contact with canines; anterior upper premolar usually greatly reduced when present; M1 hypocone apex merged into postprotocrista or distinct; reduction of M3 interspecifically variable, but always with third commissure (premetacrista) ranging in length from tiny to greater than half the length of the preparacrista; lower incisors bilobed; lower molars nyctalodont.

Average length of m1: 1.7–3.4 mm.

Included species: *Eumops* cf. *E. perotis* only, known from locality SB14C only.

Comments: *Eumops* is also known from the Pleistocene (and Recent) of North America and South America (McKenna and Bell, 1997).

***Tadarida* Rafinesque, 1814 (including *Nyctinomus* Geoffroy, 1818; *Rhizomops* Legendre, 1984; *Cephalotes*)**

Type species: *Tadarida teniotis* (Rafinesque, 1814) (originally described as *Cephalotes teniotis*), the extant European freetailed bat.

Type specimen: None designated.

Characteristics: Anterior palatal emargination well developed, narrow to wide; basisphenoid pits absent, shallow, or moderate in depth; dental formula I1/2–3, C1/1, P2/2, M3/3; upper incisors convergent, their crowns not hooked; P3 variable in size from small to large; M1 hypocone prominent, its apex merged into postprotocrista or distinct; M1 and M2 with trigon basin closed posteriorly by the postprotocrista–metacingulum; M1 and M2 with extensive talon; upper molars paraloph weak or absent; M3 not greatly reduced, with three commissures always present, the third (premetacrista) at least half the length of the first (preparacrista); p3 two-rooted; lower premolars with roots aligned at an angle oblique to the axis of the lower tooth row; lower molars nyctalodont.

Average length of m1: 1.6–2.6 mm.

Included Tertiary species: *Tadarida* sp. only, known from locality GC17A only.

Comments: The genus *Tadarida* is also known from the Eocene to Recent of Europe; the Miocene to Pleistocene of Africa; the Pleistocene to Recent of North America, Asia, and the West Indies; and the Recent of South America and New Guinea (McKenna and Bell, 1997).

VESPERTILIONIDAE GRAY, 1821

Simmons (1998) noted that there were no unique derived features uniting the members of this family. It may be polyphyletic. Hoofer and Van den Bussche (2003) provided a molecular phylogeny and classification for this family.

Characteristics: Greater tuberosity of the humerus forms a second articulation with the scapula; second digit of the wing with a metacarpal and a single remaining, reduced phalanx; no vertebral fusion; skull lacks postorbital processes; premaxillae lack palatal branches and are fused with maxillae in adults but widely separated from one another.

***Oligomyotis* Galbreath, 1962**

Type species: *Oligomyotis casementi* Galbreath, 1962.

Type specimen: SIU P-418, distal left humerus; Orellan of Colorado.

Characteristics: Humeral trochlea and capitulum not laterally displaced, medial ridge of trochlea medial to midline of humeral shaft; capitulum shallowly rounded and not keeled; medial ridge of trochlea and rounded portion of capitulum approximately the same in anteroposterior dimensions; lateral ridge of capitulum approximately one-half the anteroposterior dimensions of trochlea; spinous process of epitrochlea separated from trochlea by deep but narrow groove and does not project distally beyond level of trochlea; spinous process with broad and flattened base.

Average length of m1: not available.

Distal width of humerus: 3.6 mm.

Included species: *O. casementi* only, known from locality CP68C* only.

Comments: The type (and only) specimen of *O. casementi* may be lost.

***Miomyotis* Lawrence, 1943**

Type species: *Miomyotis floridanus* Lawrence, 1943.

Type specimen: MCZ 4055, left humerus lacking the lesser tuberosity and most of the capitulum; early Hemingfordian of Florida.

Characteristics: Humerus medium-sized with shaft nearly straight when viewed laterally; epitrochlea relatively narrow, spinous process small and extends as far as capitulum; trochlear groove and ridge of capitulum at slight angle to main axis of shaft.

Average length of m1: not available.

Average length of humerus: 24.8 mm.

Included species: *M. floridanus* only, known from locality GC8D* only.

***Suaptenos* Lawrence, 1943**

Type species: *Suaptenos whitei* Lawrence, 1943.

Type specimen: MCZ 4056, left humerus; early Hemingfordian of Florida.

Characteristics: Humerus large with slightly curving shaft when viewed laterally; anteroposteriorly elongated greater tuberosity merges with deltoid crest and possesses a very deep subspinous fossa; epitrochlea broad with deep median fossa; spinous process also broad and not extending below level of capitulum.

Average length of m1: not available.

Average length of humerus: 29.1 mm.

Included species: *S. whitei* only, known from locality GC8D* only.

Comments: This species is by far the most abundant among the diverse paleocommunity of chiropterans in the Thomas Farm Local Fauna, Florida.

***Karstala* Czaplewski and Morgan, 2000a**

Type species: *Karstala silva* Czaplewski and Morgan, 2000a.

Type specimen: UF 108672, left m2; early Hemingfordian of Florida.

Characteristics: M1 and M2 postprotocrista–metaloph extends from protocone to base of metacone, closing off trigon basin posteriorly; M1 with unique short and low but sharp crest that extends from juncture of postprotocrista and metaloph down posterior wall of talon, where it stops abruptly before reaching postcingulum (thus forming a Y-shaped structure with the postprotocrista–metaloph); lower molars myotodont; m3 trigonid lacks lingual cingulum; m3 talonid reduced and lacks hypoconulid. Humerus differs from that known for North American genera in its combination of large size, rounded head, deep supraglenoid pit, lack of a notch between medial side of trochlea and blunt spinous process, and small olecranon fossa (Figure 12.2). Differs from described genera of vespertilionids except *Antrozous* (including *Bauerus* as a subgenus) in presence of relatively strong, steeply inclined lingual cingulum at foot of trigonid valley in m1 and m2; lingual cingulum ("metacristid" of Menu [1985]: Figure 12.1C) as small crest on anterolingual base of metaconid, extending steeply downward and forward to posterolingual base of paraconid. Differs from *Antrozous* in much larger size, relatively longer talonid, more labially situated hypoconulid, and entoconid aligned with paraconid–metaconid, not lingually displaced, and taller, better-developed paraconid.

Average length of m1: 2.2 mm.

Included species: *K. silva* only, known from locality GC8D* only.

***Ancenycteris* Sutton and Genoways, 1974**

Type species: *Ancenycteris rasmusseni* Sutton and Genoways, 1974.

Type specimen: TTU-P 4093, left dentary fragment with I1–M2; early late Barstovian of Montana.

Characteristics: Incisors tri- or quadritubercular and compressed anteroposteriorly so that they overlap; canine with pronounced cingulum on lingual and posterior sides, lingual portion sloping steeply from anterior side at level of incisors to base of tooth; two lower premolars present; p2 relatively small with single, rounded, peglike cusp but pronounced cingular shelf surrounding cusp; p4 two-rooted and laterally compressed while possessing a single large and very high caniniformlike cusp; molars m1–m2 very similar, with open trigonids and hypoconulid twinned with laterally displaced entoconid.

Average length of m1: 1.7 mm.

Included species: *A rasmusseni* only, known from locality NP41B* only.

Comments: The type and only specimen of *A. rasmusseni* is lost, although a mold and cast were made of the specimen prior to its disappearance (B. Henry, personal communication, 1989).

***Potamonycteris* Czaplewski, 1991**

Type species: *Potamonycteris biperforatus* Czaplewski, 1991.

Type specimen: UNSM 52008, left half of rostrum with M2–3; late Barstovian of Nebraska.

Characteristics: Vespertilionid with double infraorbital foramen; dental formula for upper toothrow I1 or 2, C1, P2, M3; upper molars lack hypocone and talon; M3 premetacrista equal in length to postparacrista.

Average length of m1: not available.

Average length of M2: 1.4 mm.

Included species: *P. biperforatus* only, known from locality CP114B* only.

***Myotis* Kaup, 1829 (including *Vespertilio*, in part)**

Type species: *Myotis myotis* (Borkhausen, 1797) (originally described as *Vespertilio myotis*), the extant European mouse-eared bat.

Type specimen: None designated.

Characteristics: Dental formula I2/3, C1/1, P2–3/2–3, M3/3 (Figure 12.1B); M1 and M2 with weak to absent talon and hypocone; M3 with premetacrista and metacone present, postmetacrista absent; lower molars myotodont; humeral head rounded, shaft straight; greater tuberosity extends beyond humeral head; distal articular surface of humerus aligned with shaft; distal spinous process.

Average length of m1: 1.2–2.5 mm.

Included species: *Myotis* cf. *M. yumanensis* only, known from locality SP1A only.

Myotis sp. is also known from localities GC15C, NB35II, SB12, [CP114B], CP116A, CP123E, [PN12B].

Comments: The genus *Myotis* is also known from the Oligocene to Recent of Europe; the Miocene to Recent of Africa and Asia; the Pleistocene to Recent of North and

South America; and the Recent of Madagascar, the West Indies, Australia, and New Guinea (McKenna and Bell, 1997).

Antrozous H. Allen, 1862 (including _Pizonyx_, in part; _Simonycteris_, in part; _Vespertilio_, in part))

Type species: *Antrozous pallidus* (LeConte, 1856) (originally described as *Vespertilio pallidus*), the extant North American pallid bat.

Type specimen: None designated.

Characteristics: Dental formula I1/2–3, C1/1, P1/2, M3/3; M1 and M2 lack hypocones; M1 and M2 preprotocrista extends to anterior base of paracone rather than to parastyle, and postprotocrista extends to base of metacone (Figure 12.1C); M3 reduced to protocone, paracone, and parastyle (postparacrista, mesostyle, and metacone are absent); humerus distal articular surface displaced laterally from longitudinal axis of humeral shaft; central ridge of capitulum rounded and parallel to longitudinal axis of shaft; olecranon fossa and radial fossa absent; medial epitrochlea of humerus broad with short distal spinous process.

Average length of m1: 2.0 mm.

Included species: *A. pallidus* (known from localities NB35II, SB14D, SP1C, F, PN23C).

Antrozous sp. is also known from localities GC4DD, GC15C, GC26, [CP114A].

Comments: The localities listed above include two records of *Antrozous* specimens previously mistakenly identified as *Pizonyx wheeleri* (Dalquest and Patrick, 1989; Czaplewski, 1993a) and as *Simonycteris stocki* (Harrison, 1978; Czaplewski, 1993b).

The genus *Antrozous* is also known from the Pleistocene of North America and Cuba.

Anzanycteris White, 1969

Type species: *Anzanycteris anzensis* White, 1969.

Type specimen: LACM 19300, skull with left and right dentaries in articulation; early Blancan.

Characteristics: Lower incisors crowded between canines, i2 unicuspate, much smaller than i1 and wedged into depression on anteromedial side of canine, but not covered by it; upper incisors relatively long, about half length of canines; coronoid process of dentary oriented vertically; P4 nearly as wide as M1 and contacts that tooth through about half of its width; M1 and M2 without hypocones, with weakly developed cingula, postprotocrista slopes gently from relatively low protocone toward base of metacone but does not close off trigon basin posteriorly; M3 truncated in middle of postparacrista and lacks mesostyle.

Average length of m1: not available.

Average length of m2: 1.7 mm.

Included species: *A. anzensis* only, known from locality NB13B* only.

Lasiurus Gray, 1831 (including _Nycteris_, in part)

Type species: *Lasiurus borealis* (Müller, 1776) (originally described as *Nycteris borealis*), an extant New World red bat.

Type specimen: None designated.

Characteristics: Skull short, with rostrum broad and massive (Figure 12.2A); external nares broadly open and confluent with deep and wide anterior palatal emargination. Dental formula I1/3, C1/1, P1–2/2, M3/3; M1 and M2 preprotocrista extends to parastyle or to base of paracone; M1 and M2 postprotocrista connects directly to base of metacone, closing trigon basin posteriorly; M3 reduced, postmetacrista present but mesostyle essentially absent, metacone, premetacrista and postmetacrista absent; p4 single-rooted or double-rooted; lower molars submyotodont, myotodont, or occasionally nyctalodont. Head of humerus ovoid and canted toward lesser tuberosity; greater tuberosity extends well above head of humerus; distal articular surface of humerus aligned with longitudinal axis of shaft; humeral shaft relatively stout; olecranon fossa deep and groovelike; radial fossa shallow; distal spinous process of humerus long and attached to trochlea.

Average length of m1: 1.3–1.9 mm.

Included Tertiary species: *L. borealis* (known from locality SP1F); *L.* cf. *L. blossevillii* (locality SB12); *L. fossilis* Hibbard, 1950 (localities CP128A*, PN23A).

Lasiurus sp. is also known from localities GC15C, CP116B, CP121.

Comments: The specimen from PN23A in the Glenns Ferry Formation, Idaho (see Thewissen and Smith, 1987), an isolated humerus, was referred to *L. fossilis*, a species that is otherwise known only by the type specimen, a lower jaw fragment.

The genus *Lasiurus* is also known from the Pleistocene (to Recent) of North and South America, and the Recent of the West Indies (McKenna and Bell, 1997).

Eptesicus Rafinesque, 1820

Type species: *Eptesicus melanops* Rafinesque, 1820 (including *Eptesicus fuscus* [Palisot de Beauvois, 1796]), the extant New World big brown bat.

Type specimen: None designated.

Characteristics: Dental formula I2/3, C1/1, P1–2/2, M3/3; P4 with large talon; upper molars with small talon; upper molars lack paraloph and metaloph; M3 metacone and premetacrista reduced; p4 double- or single-rooted; lower molars myotodont, with thick labial cingulum.

Average length of m1: 1.7–2.0 mm.

Included species: *E. fuscus* only (known from localities CC49, ?SP1F).

Eptesicus sp. is also known from localities GC15C, [?SB10].

Comments: In naming and illustrating *Eptesicus hemphillensis*, Dalquest (1983) inadvertently confused the intended type specimen with a specimen of "*Limnoecus*(?)" described in the same publication. A right dentary fragment with m2–3 (TMM 41261-49) of a bat is listed by this number and also illustrated by Dalquest (1983: his Figure 1) as "*Limnoecus* (?)," whereas a left dentary fragment with m1–2 and empty alveoli for m3 (TMM 41261-33) of a soricid is listed in the text and illustrated as the holotype of *E. hemphillensis* (Dalquest, 1983: his Figure 3). In the text, both taxa are described as being represented by a fragmentary lower jaw containing m1 and m2 and the alveoli of m3. Our reexamination of both original specimens confirms this mix-up. Dalquest's diagnosis of *E. hemphillensis* alludes to the "M_1 and M_2," and his description mentions that in the holotype "only the alveoli of M_3 remain." Therefore, the diagnosis of *E. hemphillensis* is clearly based on the soricid specimen and not on the bat specimen.

Moreover, based on our comparisons and measurements, the undescribed bat jaw fragment TMM 41261-49, mistakenly illustrated as the jaw of a soricid, cannot be distinguished from a series of modern specimens of *E. fuscus*, and in fact the jaw fragment does not exhibit characteristics sufficient to allow it to be differentiated from several genera of Vespertilionidae. Thus, *E. hemphillensis* is a nomen dubium for a shrew, and the bat specimen for which that name was actually intended is Vespertilionidae indet. The nomen *E. hemphillensis* should be abandoned.

The genus *Eptesicus* is also known from the Pleistocene (to Recent) of North America, the Miocene to Recent of Europe and Asia, the Pliocene to Recent of Africa and South America, and the Recent of the West Indies and Australia (McKenna and Bell, 1997).

Plionycteris Lindsay and Jacobs, 1985

Type species: *Plionycteris trusselli* Lindsay and Jacobs, 1985.

Type specimen: IGCU 1165, right maxilla fragment with P4–M1; latest Hemphillian of Chihuahua.

Characteristics: P4 has straight posterior margin abutting M1 and pronounced parastylar shelf; M1 trigon basin closed by postprotocrista; M1 with distinct but small and rounded hypocone.

Average length of m1: not available.

Average length of M1: 1.2 mm.

Included species: *P. trusselli* only, known from locality SB60* only.

Simonycteris Stirton, 1931

Type species: *Simonycteris stocki* Stirton, 1931.

Type specimen: LACM (CIT) 394, palate and rostrum with several broken teeth; late Blancan of Arizona.

Characteristics: Upper tooth-row formula I2, C1, P1, M3; upper incisors well developed, I2 larger than I3; I3 separated from C1 by space equal to transverse diameter of root of I3; I3s nearly aligned with I2s or slightly posterior to them; remnants of P4 resemble the P4 of *Eptesicus* except in having more sinuous posterior margin and in having its talon rounded rather than angular; weak talon on M1; preprotocrista of M1 less trenchant than that in *Eptesicus*; M1 longer anteroposteriorly than wide; M2 wider transversely than long; M3 small. Premaxillae fused with maxillae; rostrum short; nasals slightly elevated; forehead rises abruptly above rostrum; orbits rounded in front; palate emarginate anteriorly and domed posteriorly.

Average length of m1: not available.

Average length of M1: 2.0 mm.

Included species: *S. stocki* only, known from locality SB14F* only.

Comments: Based on a personal communication from Karl Koopman, Kurtén and Anderson (1980) indicated that *Simonycteris* is referable to the extant South American genus *Histiotus*; this assignment was followed by McKenna and Bell (1997). However, *Simonycteris* is clearly distinct from *Histiotus* spp. in a number of rostral and dental characters. *Simonycteris* has a wider and relatively shorter and shallower rostrum than *Histiotus*; as a result, the maxillary tooth rows of *Simonycteris* are much farther apart than in *Histiotus*. *Simonycteris* also has a slightly larger infraorbital foramen and much more abrupt facial angle (i.e., a more concave forehead profile) than *Histiotus* spp. The M1 and M2 of the two genera are very similar but those of *Simonycteris* have a more robust talon with less of an indentation on the lingual face of the molar between the protocone and talon than those of *Histiotus*.

Lasionycteris Peters, 1865 (including *Vespertilio*, in part)

Type species: *Lasionycteris noctivagans* (LeConte, 1831) (originally described as *Vespertilio noctivagans*), the extant North American silver-haired bat.

Type specimen: None designated.

Characteristics: Dental formula I2/3, C1/1, P2/3, M3/3; M1 and M2 with moderately developed talon (for a vespertilionid); lower molars submyotodont; rostrum broad and somewhat flattened, with paired lateral concavities on the dorsal surface between the orbits and external nares.

Average length of m1: 1.5 mm.

Included species: *L. noctivagans* only, known from locality SP1F only.

Comments: *Lasionycteris* is also known from the Pleistocene of North America (McKenna and Bell, 1997).

Corynorhinus H. Allen, 1865 (including *Plecotus*)

Type species: *Corynorhinus rafinesquii* (Lesson, 1827) (originally described as *Plecotus rafinesquii*), the extant North American Rafinesque's big-eared bat.

Type specimen: None designated.

Characteristics: Rostrum arched, with median concavity weak or absent; braincase dorsally domed; zygomatic arch expanded in its posterior one-third. Dental formula I2/1,

C1/1, P2/3, M3/3; P4 transversely much wider than long; upper molars lack talon, paraloph, and metaloph; upper molar trigon basin is widely open posteriorly; M3 with metacone and occasionally with postmetacrista; p3 present; p4 double-rooted; basioccipital pits present; lower molars myotodont and with strong cingula.

Average length of m1: 1.3–1.5 mm.

Included species: *Corynorhinus* sp. only, known from locality GC15C only.

Comments: The genus *Corynorhinus* is also known from the Pleistocene (and Recent) of North America, and the Miocene to Recent of Europe and Asia (McKenna and Bell, 1997).

***Perimyotis* Menu, 1984 (including *Vespertilio*, in part)**

Type species: *Perimyotis subflavus* (F. Cuvier, 1832) (originally described as *Vespertilio subflavus*), the extant North American eastern pipistrelle.

Type specimen: None designated.

Characteristics: Dental formula I1–2/3, C1/1, P1–2/2, M3/3; P4 with moderately developed talon; talon of upper molars weak to absent; M3 with strong metacone but little or no postmetacrista; lower molars nyctalodont or myotodont; humerus with relatively deep olecranon fossa.

Average length of m1: 1.0–1.2 mm.

Included species: *Perimyotis* sp. only, known from locality GC15C only.

Comments: Menu (1984) made a strong argument based on dental and humeral characters for separating the bat formerly known as *Pipistrellus subflavus* from all other species of *Pipistrellus* as *Perimyotis subflavus*. Recent molecular genetics studies of living bats have confirmed this distinction (Hoofer and Van den Bussche, 2003; Stadelmann *et al.*, 2004).

INDETERMINATE CHIROPTERANS

Isolated, fragmentary, and imprecisely identifiable remains of bats have also been reported from localities GC11A, CC4, CC6A, B, CC17G, CC19, NB6C, E, NB7 (sublocality uncertain), NB8, NB14, NB35II, SB14C, SB46, SP1A, SP3B, SP5, CP5A, CP31E, CP34D, CP39B, C, CP42A, CP54B, CP98B, CP100III, CP114A, CP128A–C, NP8, NP9A, NP32B, C, NP49II, PN5A, PN10, PN12 (sublocality uncertain), PN14, PN15, and NC2B.

BIOLOGY AND EVOLUTIONARY PATTERNS

Bats have long been recognized as a distinct group, being the only mammals capable of powered flight. Modern forms of bats reach their greatest diversity and abundance in tropical regions around the Earth. Temperate regions in North America have relatively few modern species, most of which are insectivorous, and scarcity of food forces many of these to migrate or hibernate in winter (Figure 12.4). Almost certainly, the lineage leading to bats underwent rapid evolution to achieve powered flight, probably through an intermediate gliding stage. The genetic, morphological, behavioral, and physiological developments to achieve flight are complex (e.g., Maina, 2000), and the sequence in which they might have evolved is controversial (Speakman, 2001).

The oldest known bats come from early Eocene beds in western North America and the middle Eocene in Europe. The remarkably complete skeletal remains of *Icaronycteris* show that these earliest known bats were already highly specialized fliers with fully developed wings and probably the ability to echolocate. Therefore, bats must have differentiated even earlier, probably during the widespread tropical climes of the Paleocene.

Having achieved the key adaptation of flight, bats blossomed into myriad variations on the theme of crepuscular or nocturnal volant insectivores. Even more diversity evolved in the tropical and subtropical regions as they expanded into adaptive zones for nocturnal frugivores (eating fruit), nectarivores (nectar and pollen), piscivores (fish), carnivores (meat), and even sanguivores (blood). The typical bat, however, retained the conservative features of the teeth for insectivory and of the humerus for flight (the two most common skeletal elements preserved in the fossil record). Even today, extant insectivorous bats are more easily distinguished by external characters than by their teeth. Hence, the fossil record of bats, represented by rare isolated teeth and humeri, even more rarely by skull or jaw parts with multiple teeth, probably underrepresents their contribution to mammalian faunas throughout almost all of the Cenozoic. (Most of the North American record is from the temperate regions, where the bats are almost exclusively insectivorous.) Compounding this poor record is the frustrating Achilles heel of vertebrate paleontology, the problem of non-comparable parts.

Several North American fossil genera originally described as chiropterans were subsequently referred to other orders. Late in the nineteenth century, Marsh (1872a,b) referred *Nyctitherium velox*, *Nyctitherium priscus*, and *Nyctilestes serotinus* to the Chiroptera, but these were later shown to be soricoid insectivores. Based on a rostrum from the Wind River Eocene, Cope (1880) described *Vesperugo anemophilus* as a bat, but the specimen was never figured or adequately described and apparently was lost early in the twentieth century or late in the nineteenth (Matthew, 1917). Matthew (1917) named *Zanycteris paleocenus*, based on an incomplete skull from the Tiffany beds of southwestern Colorado. Matthew did not place *Zanycteris* in a family but he noted similarities in its extremely specialized upper molars to extant fruit-eating phyllostomid bats of the Neotropics. *Zanycteris* was later shown to be a picrodontid plesiadapiform (see Szalay, 1968; Schwartz and Krishtalka, 1977; Tomida, 1994). Hall (1930) described the supposed vespertilionid bat *Mystipterus vespertilio* from a worn m3 and associated dentary fragment from the Clarendonian (late Miocene) of Nevada. Hutchison (1968) reidentified *Mystipterus* as a uropsiline mole (Talpidae). *Wyonycteris chalix*, described from a mandible, maxillary fragment, and several isolated teeth from the late Clarkforkian (late Paleocene) of Wyoming, was originally thought to be the oldest known bat (Gingerich, 1987). However, Hand *et al.* (1994) presented convincing dental evidence that *Wyonycteris* is not a bat. Smith (1995) described a new species of *Wyonycteris* from the early Eocene of

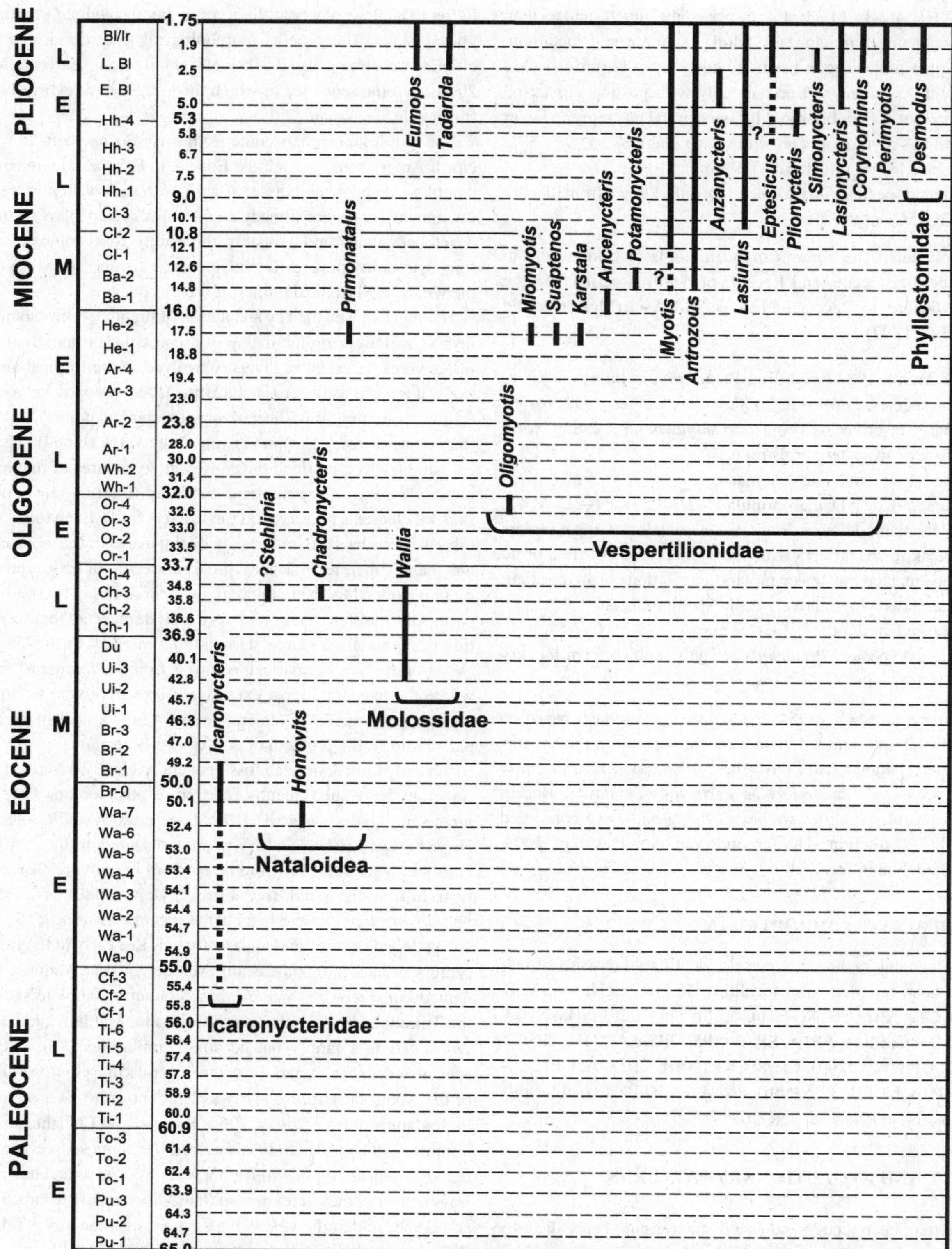

Figure 12.4. Temporal ranges of North American bat genera.

Belgium and tentatively referred that species and *W. chalix* to the insectivoran family Adapisoriculidae. Hooker (1996) and McKenna and Bell (1997) placed *Wyonycteris* in the Nyctitheriidae.

Evolutionary relationships within the order Chiroptera are also uncertain. Although there has been discussion by some authors about megachiropteran bats being more closely related to primates than to microchiropterans, other authors have argued that the shared derived characters of the two suborders of bats overwhelm characters that each might share with some other mammalian order (Novacek, 1982, 1987; Pettigrew, 1986; Wible and Novacek, 1988). Phylogenetic analyses based primarily on morphological data indicate monophyly of the Chiroptera (Simmons, Novacek, and Baker, 1991; Simmons, 1993a,b, 1994, 1995, 1998; Simmons and Geisler, 1998). Progress to date in genomics and other molecular studies also tends to affirm monophyly of the Chiroptera. Some recent studies have placed the Megachiroptera (Pteropodidae) within the Microchiroptera (Hutcheon, Kirsch, and Pettigrew, 1998; Teeling *et al.*, 2000, 2002), thus indicating non-monophyly for the Microchiroptera. A recent molecular "supertree" analysis failed to predict monophyly of the Microchiroptera, whereas morphological and combined supertree analyses of morphological and molecular data found the suborder to be monophyletic (Liu *et al.*, 2001).

Like many other extant orders of mammals, the first firm occurrence of the Chiroptera is in the early Eocene, and extant families begin to appear in the middle to late Eocene and Oligocene (Figure 12.4). Early evolution to the chiropteran body plan coupled with a generally conservative evolution within the group results in most fossil bats back through the Oligocene being easily assignable to living families and, in many cases, even to extant genera. The earlier Eocene taxa, however, are difficult to assign to a family and sometimes to suborder. Novacek (1987) marshaled evidence that *Icaronycteris* should be recognized as a member of the Microchiroptera incertae sedis. More recently, in an overview of Eocene bats worldwide, Simmons and Geisler (1998) found the best-known Eocene genera to fall phylogenetically outside Microchiroptera but closer to them than to Pteropodidae (Figure 12.3). Accordingly, Simmons and Geisler (1998) proposed a classification in which these Eocene genera are separate from the Megachiroptera, and are outside Microchiroptera but included within new higher taxa, the Microchiropteramorpha and Microchiropteriformes. These last two taxa also include Microchiroptera.

The earliest known bat, *Icaronycteris*, constitutes a unique family, Icaronycteridae, the only extinct North American bat family; however, Gunnell (2003) placed *Icaronycteris* and an undescribed new bat from the Green River Formation in Archaeonycteridae. Two other Eocene species, *Honrovits tsuwape* and *Chadronycteris rabenae*, may represent early members of the vespertilionoid or nataloid radiation, but are not members of the Vespertilionidae or Natalidae sensu stricto. A fourth Eocene species, *Wallia scalopidens*, represents the extant family Molossidae. *Wallia* was originally identified as a proscalopine (Insectivora) and gives further testimony to the difficulty of identifying bats based on isolated teeth and differentiating them from insectivorans. One remaining species, *Oligomyotis casementi*, represented by a humerus fragment from the Orellan of Colorado, completes the meager Paleogene record of bats in North America, although large unpublished samples of bats are known from two Oligocene sites in Florida (see discussion below).

In the Neogene, representatives of Vespertilionidae predominate in North American faunas along with a few records of Molossidae, and one highly specialized member (vampire) of the derived family Phyllostomidae that appears in the latest Pliocene. Although the stratigraphic distribution of Tertiary bats in North America is sparse (Figure 12.4), all North American Land-Mammal Ages from the Wasatchian (early Eocene) onward are represented by at least one published record of a bat (but some of the records are of indeterminate material). Of course, the epochs and land-mammal ages of the Tertiary represent unequal periods of time.

It is a generally accepted axiom among vertebrate paleontologists (and repeated above) that bats are one of the least common groups of mammals in the fossil record. There are about 50 Tertiary localities from North America that have produced bats, although many of these sites are represented by specimens currently identified only as indeterminate Chiroptera. Furthermore, most North American Tertiary bat localities have produced only one species, many represented by a single fossil specimen. By comparison with the North American record, the pre-Pleistocene history of the Chiroptera in Europe is rich, with several species represented at many of the localities (Sigé and Legendre, 1983). The published North American Tertiary chiropteran fauna consists of 5 families, 24 genera, and 28 species, distributed chronologically as follows (the numbers do not equal the total for the entire Tertiary because several genera and species occur in both the Miocene and Pliocene): Eocene (five genera and five species), Oligocene (one genus and one species), Miocene (nine genera and nine species), and Pliocene (13 genera and 15 species).

Except for the Oligocene, the number of genera and species of bats from North America gradually increases from the Eocene to the present. Although the Oligocene record of bats from North America is weak at present, two unpublished Oligocene sites from Florida discussed below contain significant samples of bats that will greatly augment the North American middle Tertiary bat fauna. Bats are well represented in Pleistocene deposits in North America, primarily in caves (Kurtén and Anderson, 1980), but virtually all the species represented are extant. As a result, the Pleistocene record tells us little about the evolution, diversification, and historical biogeography of the North American chiropteran fauna.

Karstic deposits (fissure fillings) and fluvio-lacustrine sediments have produced the majority of Tertiary bat remains in western Europe (Sigé and Legendre, 1983). Outside of Florida, fossiliferous karst deposits of Tertiary age are uncommon in North America, and fluvio-lacustrine deposits have produced comparatively few bats. In peninsular Florida, paleocaves represented by sinkhole and fissure deposits formed in Paleogene marine limestones have produced the most diverse Tertiary chiropteran faunas from North America (Lawrence, 1943; Morgan, 1989, 1991; Czaplewski and Morgan, 2000a,b; Morgan and Czaplewski, 2000), including I-75 (locality GC7; Whitneyan, early Oligocene), Brooksville 2 (locality GC8AA; Arikareean, late Oligocene), Thomas Farm (locality GC8D; Hemingfordian, early Miocene), and Inglis 1A (locality GC15C; Blancan, late Pliocene).

Most Tertiary bat fossils from North America consist of fragmentary material, including partial maxillae and dentaries with teeth, isolated teeth, and postcranial elements. Complete or partially articulated skeletons of Tertiary bats have been recovered from lake sediments in the Eocene Green River Formation of Wyoming (locality CP33A) and Colorado (locality CP64IB) (Jepsen, 1966; Grande, 1984; Gunnell, 2003). Very few articulated or associated skeletons of bats are known from other Tertiary sites in North America. Exceptions include a skull and partial skeleton from the Bridgerian (early-middle Eocene) of Nevada (locality NB14) (Emry, 1990) and an associated mandible, scapula, and humerus from the Chadronian (late Eocene) of Oregon (locality PN5A) (Brown, 1959).

The North America Tertiary chiropteran record, although incomplete, does provide a general overall picture of the evolutionary and biogeographic history of several families of bats. The record is limited to the temperate portion of the continent in the United States, Canada, and northern Mexico. There are only four published records of North American Tertiary bats from outside of the United States, the molossid *Wallia scalopidens* from the Uintan and Duchesnean (middle and late Eocene) of Saskatchewan, Canada (Storer, 1984, 1995), two indeterminate bats from the Duchesnean of Saskatchewan (Storer, 1995), and the vespertilionid *Plionycteris trusselli* from the late Hemphillian (early Pliocene) of Chihuahua, Mexico (Lindsay and Jacobs, 1985). The Tertiary record of bats in tropical North America (southern Mexico and Central America) is non-existent. The earliest record of a bat from this region is a partial skeleton tentatively referred to the mormoopid *Pteronotus parnellii* from the Irvingtonian of El Salvador (Webb and Perrigo, 1984). As in temperate North America (e.g., Kurtén and Anderson, 1980), the fossil record of bats in tropical North America improves in the Pleistocene, particularly in Mexico and Belize (Arroyo-Cabrales, 1992; Czaplewski, Krejca, and Millar, 2003a), although the record there is still sparse.

The North American Tertiary record of the Chiroptera includes many important undescribed specimens or faunas not listed above under "Included North American Genera." Some of these specimens have been known for many years, and several are illustrated. We briefly summarize the undescribed taxa mentioned previously in the literature, as a significant chapter of the North American chiropteran fossil record would be undocumented without reference to these samples. Jepsen (1966) mentioned a partial bat skeleton of Bridgerian age from the Green River Formation in Wyoming that is smaller and of different proportions than the type of *Icaronycteris index*. Novacek (1985: Fig. 1; 1987: Fig. 2) illustrated and described the basicranial region of this same specimen, and identified it as *Icaronycteris* cf. *I. index*. Emry (1990: Fig. 5) reported and illustrated a skull, mandible, and partial associated postcranial skeleton of a bat from the Bridgerian Elderberry Canyon Local Fauna from the Sheep Pass Formation in Nevada (locality NB14). Grande (1984: Fig. III.26) illustrated a complete skeleton of an unidentified bat of Uintan age from the Parachute Creek Member of the Green River Formation in Colorado (locality CP64IIIB). Emry (1973) listed a bat from the Chadronian Flagstaff Rim Fauna from the White River Formation in Wyoming (locality CP39F).

Two undescribed chiropteran faunas from the Oligocene of Florida, I-75 (locality GC7; Whitneyan) and Brooksville 2 (locality GC8AA; early Arikareean), are dominated by taxa belonging to tropical families that are now rare or absent in temperate North America, including the Emballonuridae, Mormoopidae, Natalidae, and possibly Phyllostomidae. One species of mormoopid and four phyllostomids occur in the modern fauna of the southwestern United States, but otherwise the Mormoopidae, Phyllostomidae, and Natalidae are endemic to the Neotropics. The Emballonuridae are pantropical but reach their greatest diversity in tropical America. Both I-75 and Brooksville 2 contain a new genus and two new species of emballonurids and a new genus and species of mormoopid. A new genus that may represent the oldest phyllostomid or other noctilionoid is represented by a large species from Brooksville 2 and a smaller species from I-75. The oldest natalid, known from a partial radius, occurs at I-75 (Morgan, 1989; Czaplewski and Morgan, 2000b; Morgan and Czaplewski, 2000, 2003).

The richest Tertiary chiropteran fauna in North America is the early Hemingfordian (early Miocene) Thomas Farm Local Fauna from Florida (locality GC8D), which is composed of at least eight species representing four families, Emballonuridae, Natalidae, Vespertilionidae, and Molossidae (Lawrence, 1943; Morgan, 1989; Czaplewski and Morgan, 2000a,b; Morgan and Czaplewski, 2000). Three vespertilionids known only from Thomas Farm were described as new genera and species, *Miomyotis floridanus*, *Suaptenos whitei*, and *Karstala silva* (Lawrence, 1943; Czaplewski and Morgan, 2000a), and two other new genera of vespertilionids await description. The Thomas Farm bat fauna also contains a natalid, an emballonurid, and two molossids (Czaplewski and Morgan, 2000b; Morgan and Czaplewski, 2000, 2003; Czaplewski, Morgan, and Naeher, 2003b). The Thomas Farm sample represents a transitional period in Florida bat faunas, between late Oligocene faunas dominated by families now restricted to tropical America and middle Miocene and younger faunas that are almost exclusively composed of vespertilionids. Another large undescribed bat fauna, including at least two genera of vespertilionids represented by abundant skeletal material, is known from the late Barstovian (middle Miocene) Myers Farm site in the Valentine Formation of Nebraska (locality CP114B) (Corner, 1976; Czaplewski, Bailey, and Corner, 1999).

The majority of bats previously described from the North American Tertiary belong to the Vespertilionidae, the predominant family of bats in the modern temperate fauna of the continent. The earliest vespertilionid from North America is *Oligomyotis* from the Orellan (early Oligocene) of Colorado (Galbreath, 1962). From the early Oligocene through the middle Miocene, six extinct genera of vespertilionids have been described from North America (*Oligomyotis* from the Oligocene; *Karstala*, *Miomyotis*, and *Suaptenos* from the early Miocene; and *Ancenycteris* and *Potamonycteris* from the middle Miocene). Three other extinct genera of vespertilionids are restricted to the Pliocene (*Anzanycteris*, *Plionycteris*, and *Simonycteris*). Four extant vespertilionid genera first occur in North America in the middle to late Miocene (*Antrozous*, *Eptesicus*, *Lasiurus*, and *Myotis*), and three additional extant genera appear in the Pliocene (*Corynorhinus*, *Lasionycteris*, and *Perimyotis*). Seven of the eleven extant temperate North American genera of vespertilionids, as well as several living species, are known from the Blancan. The only North American vespertilionid genera that currently lack a Tertiary fossil record are *Euderma*, *Idionycteris*, *Pipistrellus* (= *P. hesperus*), and *Nycticeius*. As in North America, vespertilionids first appear in Europe and Asia during the early Oligocene (Sigé and Legendre, 1983; Gabunia and Gabunia, 1987). Apparently, vespertilionids became widespread in the northern continents early in their evolutionary history. Vespertilionids are currently virtually unknown in the South American fossil record before the Pleistocene, being represented by a single tooth of indeterminate genus from the middle Miocene (Czaplewski *et al.*, 2003c).

The Molossidae are primarily a tropical group in the New World, although three genera and six extant species extend their ranges northward into the southern United States. A partial dentary from the late Bridgerian of Wyoming (locality CP34D) (McKenna, Robinson and Taylor, 1962: Fig. 9), tentatively referred to the Molossidae by Legendre (1985), may represent the earliest record of the family. The oldest described species of molossid from North America is *Wallia scalopidens* from the Uintan of Saskatchewan (locality NP8; Storer, 1984). An indeterminate molossid occurs in the Arikareean of Florida (locality GC8AA), and two indeterminate species similar to *Tadarida* and *Mormopterus* occur in the early Hemingfordian (locality GC8D; Czaplewski, Morgan, and Naeher, 2003b). After an absence from the early Miocene through the early Pliocene, molossids reappear in the North American fossil record in the late Pliocene. The extant molossid genera *Tadarida* and *Eumops* are first recorded in North America in the late Blancan localities GC17 and SB14C, respectively (Morgan and Ridgway, 1987; Czaplewski, 1993b; Czaplewski, Morgan, and Naeher, 2003b). The third genus from temperate North America, *Nyctinomops*, has no fossil record in temperate North America before the Pleistocene but has records in tropical Mexico (Arroyo-Cabrales, 1992). The oldest European molossid is *Cuvierimops* from the late Eocene of France (Legendre and Sigé, 1983; Sigé and Legendre, 1983; Simmons and Geisler, 1998). The earliest molossid from South America is an extinct species of the living genus *Mormopterus* from the Oligocene of Brazil (Paula Couto, 1956; Legendre, 1985). The molossid fossils from North America and South America are too incomplete to provide a coherent picture of the evolutionary history of the New World members of the family. However, it may not be coincidental that *Eumops* and *Tadarida* first appear in North America in the late Blancan, after the onset of the Great American Faunal Interchange. An isolated tooth from the Blanco Local Fauna (locality SP5) reported as a "bat, near *Tadarida*" by Dalquest (1975) probably represents an indeterminate vespertilionid (Czaplewski, Morgan, and Naeher, 2003b).

The earliest definite North American member of the primarily neotropical family Phyllostomidae is an extinct species of vampire bat, *Desmodus archaeodaptes*, from the late Blancan (late Pliocene, locality GC15C) and early Irvingtonian (early Pleistocene) of Florida (Morgan, Linares, and Ray, 1988). Bat fossils previously referred to the Phyllostomidae from the Hemingfordian (Hutchison and Lindsay, 1974) and Clarendonian (James, 1963) localities CC17G and CC19 of California are too incomplete to justify referral to this family. *Desmodus* almost certainly immigrated to North America from South America in the late Blancan as a participant in the Great American Faunal Interchange, presumably following the northward migration of its primary "prey" species (large xenarthrans?). The Phyllostomidae presumably evolved in South America. Fossils referred to this family are known from the middle Miocene La Venta Fauna of Colombia, including the phyllostomines *Notonycteris magdalenensis*, *N. sucharadeus*, and *Tonatia* or *Lophostoma* sp., and the nectar-feeding bat *Palynephyllum antimaster* (Czaplewski, 1997; Czaplewski *et al.*, 2003c). Among the three phyllostomid genera now found in the modern fauna of temperate North America (not including *Desmodus*, which is now restricted to the Neotropics), *Macrotus* and *Leptonycteris* are unknown before the late Pleistocene, and *Choeronycteris* has no fossil record.

A new genus and two undescribed species from the Oligocene of Florida localities GC7 and GC8AA represent the oldest New World members of the Emballonuridae. A second undescribed genus of emballonurid occurs in the early Miocene of Florida locality GC8D. The oldest emballonurid is *Tachypteron* from the middle Eocene of Europe (Storch, Sigé, and Habersetzer, 2002). The only previously published Tertiary emballonurids from the New World are the extant genus *Diclidurus* and an indeterminate genus from the middle Miocene of Colombia (Czaplewski, 1997; Czaplewski *et al.*, 2003c). The Florida fossils are the oldest records for the emballonurid tribe Diclidurini, the monophyletic group that includes all living neotropical members of the family. Therefore, New World emballonurids must have branched off from Old World emballonurids sometime prior to the late Oligocene. Emballonurids may have entered South America by overwater dispersal from tropical North America sometime prior to the middle Miocene. The presence of *Diclidurus* in the Miocene of Colombia and the diverse living fauna of emballonurids in the Neotropics, including 9 genera and 19 species, suggests that this family reached South America in the early Neogene, if not the Paleogene. Although emballonurids surely inhabited tropical North America after the early Miocene, they are unknown in the New World fossil record between the middle Miocene and the late Pleistocene.

An undescribed new genus and species of Mormoopidae from the Oligocene of Florida localities GC7 and GC8AA represents the first Tertiary fossil record for this endemic New World family. The oldest previous record of a mormoopid was a partial skeleton tentatively assigned to the living species *Pteronotus parnellii* from the early Pleistocene of El Salvador (Webb and Perrigo, 1984). A partial limb bone referable to the family Natalidae from the Oligocene of Florida (locality GC7) is the oldest record of this family, and the new genus and species *Primonatalus prattae* was recently described from the early Miocene of Florida (locality GC8D; Morgan and Czaplewski, 2003). The Natalidae are an endemic neotropical family of bats that had no previous unequivocal Tertiary fossil record. Other supposed natalids reported previously from North American Tertiary sites cannot be referred to this family. The Mormoopidae and Natalidae probably originated in tropical North America, as recent fossil discoveries establish that both families were present in Florida in the Oligocene (locality GC7). No pre-Pleistocene fossils of these two families are known from South America. The Mormoopidae and Natalidae are currently found throughout Middle America and the West Indies, but their distribution is somewhat restricted in South America where they occur primarily in the northern one-third of the continent (Smith, 1972). Mormoopids and natalids probably evolved in tropical North America prior to the late Oligocene, and then later reached the West Indies by overwater dispersal. Both families underwent extensive speciation in the West Indies, where they reach their greatest diversity in the extant fauna. The restricted distribution, lack of endemism, and absence of a Tertiary fossil record suggest that mormoopids and natalids may not have arrived in South America until after the Great American Interchange.

The Eocene chiropteran fauna of North America mostly consists of archaic groups whose relationships to modern families of bats are unclear (Simmons and Geisler, 1998). The early Eocene record is composed of *Icaronycteris* of the extinct family Icaronycteridae and the possible nataloid *Honrovits* (Beard, Sigé, and Krishtalka, 1992). The middle Eocene fauna includes *Icaronycteris* and the first representatives of an extant family, an indeterminate ?molossid from the Bridgerian of Wyoming (locality CP34D) (McKenna, Robinson and Taylor, 1962; Legendre, 1985) and the molossid *Wallia scalopidens* from the Uintan of Saskatchewan (locality NP8) (Storer, 1984). Late Eocene bats include two genera of possible vespertilionoids or nataloids, *Chadronycteris* and *Stehlinia*?, both from the Chadronian of Nebraska (locality CP98B; Ostrander, 1983, 1985). The Oligocene bat fauna of western North America includes the earliest North American records of the family Vespertilionidae: *Oligomyotis* from the Orellan of Colorado (locality CP68C; Galbreath, 1962) and ?*Oligomyotis* or ?*Myotis* from the early Arikareean of Nebraska (locality CP100III; Czaplewski, Bailey, and Corner, 1999). Oligocene bat faunas from Florida are composed primarily of extinct genera belonging to the tropical families Emballonuridae and Mormoopidae. Emballonurids also occur in the early Miocene of Florida, represented by an undescribed genus, along with *Primonatalus*, a new genus in the endemic neotropical family Natalidae. Other early Miocene bats from Florida include three extinct genera of vespertilionids, *Karstala*, *Miomyotis*, and *Suaptenos*, two undescribed genera of vespertilionids, and two genera of the Molossidae. The dominance of families now restricted to tropical America appears to be limited to Florida late Oligocene and early Miocene faunas, because none of these families are known from faunas elsewhere in North America.

From the middle Miocene until the late Pliocene, the North American chiropteran fauna consists exclusively of species of Vespertilionidae. Most living genera of North American temperate vespertilionids make their first appearance during this time interval. By the early Pliocene (approximately 3–4 Ma), the North American bat fauna had begun to take on a modern aspect. Dalquest (1978) reported four extant species of Vespertilionidae, *Lasionycteris noctivagans*, *Eptesicus fuscus*, *Lasiurus borealis*, and *Antrozous pallidus*, from the early Blancan of Texas (locality SP1F), and Czaplewski (1993b) tentatively identified the extant red bat *Lasiurus blossevillii* from the early Blancan of Arizona (locality SB12). After the beginning of the Great American Faunal Interchange in the late Blancan, several bats with neotropical affinities appear in the North American fossil record. These late Blancan bats may represent immigrants from South America, and include the extant molossid *Eumops* cf. *E. perotis* from Arizona (Czaplewski, 1993b), the molossid *Tadarida* sp. from Florida (Morgan and Ridgway, 1987), and the extinct vampire *Desmodus archaeodaptes* from Florida (Morgan, Linares, and Ray, 1988; Morgan, 1991). Along with *Desmodus*, six species of vespertilionids, including one species each in the genera *Myotis*, *Antrozous*, *Lasiurus*, *Eptesicus*, *Perimyotis*, and *Corynorhinus*, are known from the late Blancan Inglis 1A fauna in Florida (locality GC15C). Not only is Inglis 1A the most diverse Pliocene chiropteran fauna in North America, but it also includes the earliest North American *Perimyotis* and *Corynorhinus*, and the only record of *Antrozous* from the eastern United States (Morgan, 1991). Most living species of North American bats do not appear until the Pleistocene (Kurtén and Anderson, 1980), although study of the Inglis bats may push the origin of several other vespertilionids back into the late Pliocene.

ACKNOWLEDGMENTS

We offer our sincere thanks to the many curators at various institutions who have allowed us to study the collections in their care, loaned us specimens, provided casts, or otherwise supported our work (USNM, UNSM, UF [FLMNH], UCMP, MWSU, TMM, AMNH, DMNH, FM, SDSM, OMNH, CM, SMNH, UALP, IGCU, LACM, MCZ, TTU, SBCM, NAU, MNA). We also appreciate aid from Patrick Fisher and Tiffany Naeher Stephens of the OMNH for help with illustrations. Useful constructive comments by Mary Silcox and an anonymous reviewer improved the manuscript. The work for this chapter was supported in part by a grant from the National Science Foundation (DEB 9981512) to NJC and GSM.

REFERENCES

Allen, H. E. (1862). Descriptions of two species of Vespertilionidae and some remarks on the genus *Antrozous*. *Proceedings of the Academy of Natural Sciences of Philadelphia*, 14, 246–8.

(1865). On a new genus of Vespertilionidae. *Proceedings of the Academy of Natural Sciences of Philadelphia*, 17, 173–5.

Altenbach, J. S. (1979). Locomotor morphology of the vampire bat, *Desmodus rotundus*. *American Society of Mammalogists Special Publication* 6, 1–137. Washington, DC: American Society of Mammalogists.

Álvarez, Y., Juste B. J., Tabares, E., *et al.* (1999). Molecular phylogeny and morphological homoplasy in fruitbats. *Molecular Biology and Evolution*, 16, 1061–7.

Archibald, J. D. (1999). Divergence times of eutherian mammals. *Science*, 285, 2031a.

Arroyo-Cabrales, J. (1992). Sinopsis de los murciélagos fósiles de México. *Revista de la Sociedad Mexicana de Paleontologia*, 5, 1–14.

Baker, R. J., Hoofer, S. R., Porter, C. A., and Van den Bussche, R. A. (2003). Diversification among New World leaf-nosed bats: an evolutionary hypothesis and classification inferred from digenomic congruence of DNA sequence. *Occasional Papers of the Museum of Texas Technical University*, 230, 1–32.

Beard, K. C., Sigé, B., and Krishtalka, L. (1992). A primitive vespertilionoid bat from the early Eocene of central Wyoming. *Comptes Rendus de l'Académie des Sciences, Paris*, 314, 735–41.

Bloch, J. I. and Bowen, G. J. (2001). Paleocene–Eocene microvertebrates in freshwater limestones of the Willwood Formation, Clarks Fork Basin, Wyoming. In *Eocene Biodiversity: Unusual Occurrences and Rarely Sampled Habitats*, ed. G. F. Gunnell, pp. 95–129. New York: Kluwer Academic/Plenum.

Blumenbach, J. F. (1779–1780). *Handbuch der Naturgeschichte*. Göttingen: Johann Christian Dieterich.

Borkhausen, M. B. (1797). *Deutsche Fauna*, Vol. 1: *Säugethiere und Vogel*. Frankfurt.

Brown, R. W. (1959). A bat and some plants from the upper Oligocene of Oregon. *Journal of Paleontology*, 33, 125–9.

Cope, E. D. (1880). The badlands of the Wind River and their fauna. *American Naturalist*, 14, 745–8.

Corner, R. G. (1976). An early Valentinian vertebrate local fauna from southern Webster County, Nebraska. M.Sc. Thesis, University of Nebraska, Lincoln.

Cuvier, F. (1832). Essai de classification naturelle des Vespertilions, et description de plusieurs espèces de ce genre. *Nouvelles Annales du Museum d'Histoire Naturelle, Paris*, 1, 1–20.

Czaplewski, N. J. (1991). Miocene bats from the lower Valentine Formation of northeastern Nebraska. *Journal of Mammalogy*, 72, 715–22.

(1993a). *Pizonyx wheeleri* Dalquest and Patrick (Mammalia: Chiroptera) from the Miocene of Texas referred to the genus *Antrozous* H. Allen. *Journal of Vertebrate Paleontology*, 13, 378–80.

(1993b). Late Tertiary bats (Mammalia, Chiroptera) from the southwestern United States. *The Southwestern Naturalist*, 38, 111–18.

(1997). Chiroptera. In *Vertebrate Paleontology in the Neotropics: The Miocene Fauna of La Venta, Colombia*, ed. R. F. Kay, R. H. Madden, R. L. Cifelli, and J. J. Flynn, pp. 410–31. Washington, DC: Smithsonian Institution Press.

Czaplewski, N. J. and Morgan, G. S. (2000a). A new vespertilionid bat (Mammalia: Chiroptera) from the Early Miocene (Hemingfordian) of Florida, USA. *Journal of Vertebrate Paleontology*, 20, 736–42.

(2000b). Faunal evolution of Tertiary bats in North America, with emphasis on Florida. *Journal of Vertebrate Paleontology*, 20(suppl. to no. 3), p. 37A.

Czaplewski, N. J., Bailey, B. E., and Corner, R. G. (1999). Tertiary bats (Mammalia: Chiroptera) from northern Nebraska. *Transactions of the Nebraska Academy of Sciences*, 25, 83–93.

Czaplewski, N. J., Krejca, J., and Millar, T. E. (2003a). Late Quaternary bats from Cebada Cave, Chiquibul Cave System, Belize. *Caribbean Journal of Science*, 39, 23–33.

Czaplewski, N. J., Morgan, G. S., and Naeher, T. (2003b). Molossid bats from the late Tertiary of Florida with a review of the Tertiary Molossidae of North America. *Acta Chiropterologica*, 5, 61–74.

Czaplewski, N. J., Takai, M., Naeher, T. M., Shigehara, N., and Setoguchi, T. (2003c). Additional bats from the Middle Miocene La Venta fauna of Colombia. *Revista de la Academia Colombiana de Ciencias Exactas, Físicas y Naturales*, 27, 263–82.

Dalquest, W. W. (1975). Vertebrate fossils from the Blanco local fauna of Texas. *Occasional Papers of the Museum of Texas Technical University*, 30, 1–52.

(1978). Early Blancan mammals of the Beck Ranch Local Fauna of Texas. *Journal of Mammalogy*, 59, 269–98.

(1983). Mammals of the Coffee Ranch Local Fauna Hemphillian of Texas. *The Pearce–Sellards Series, Texas Memorial Museum*, 38, 1–41.

Dalquest, W. W. and Patrick, D. B. (1989). Small mammals from the early and medial Hemphillian of Texas, with descriptions of a new bat and gopher. *Journal of Vertebrate Paleontology*, 9, 78–88.

Eizirik, E., Murphy, W. J., and O'Brien, S. J. (2001). Molecular dating and biogeography of the early placental mammal radiation. *Journal of Heredity*, 92, 212–19.

Emry, R. J. (1973). Stratigraphy and preliminary biostratigraphy of the Flagstaff Rim area, Natrona County, Wyoming. *Smithsonian Contributions to Paleobiology*, 18, 1–43.

(1990). Mammals of the Bridgerian (middle Eocene) Elderberry Canyon Local Fauna of eastern Nevada. [In *Dawn of the Age of Mammals in the Northern Part of the Rocky Mountain Interior, North America*, ed. T. M. Bown and K. D. Rose.] *Geological Society of America, Special Paper*, 243, 197–210.

Foote, M., Hunter, J. P., Janis, C. M., and Sepkoski, J. J., Jr. (1999a). Evolutionary and preservational constraints on origins of biologic groups: divergence times of eutherian mammals. *Science*, 283, 1310–13.

(1999b). Divergence times of eutherian mammals. *Science*, 285, 2031a.

Gabunia, L. K. and Gabunia, V. J. (1987). On the first find of fossil bats (Chiroptera) in the Paleogene of the USSR. *Bulletin of the Academy of Sciences of the Georgian SSR*, 126, 197–200.

Galbreath, E. C. (1962). A new myotid bat from the middle Oligocene of northeastern Colorado. *Transactions of the Kansas Academy of Science*, 65, 448–51.

Geoffroy Saint-Hilaire, É. (1810). Sur les Phyllostomes et les Mégadermes, deux genres de la famille des chauve-souris. *Annales Muséum National d'Histoire Naturelle, Paris*, 15, 157–98

(1818). Description de l'Egypte. Histoire naturelle. Description des mammiféres qui se trouvent en Egypte. *Paris*, 2, 99–135.

Gervais, P. (1856). Deuxième mémoire. Documents zoologiques pour servir à la monographie des chéiroptères Sud-Américains. In *Mammifères*, ed. F. L. P. Gervais pp. 25–88. [In *Animaux Nouveaux ou Rares Recueillis pendant L'expédition dans les Parties Centrales de l'Amérique du Sud, de Rio de Janeiro à Lima, et de Lima au Parà; Exécutée par Ordre du Gouvernement Français pendant les Années 1843 à 1847, sous la Direction du Comte Francis de Castelnau*, ed. F. de Castelnau, Vol. 1, pp. 1–116. Paris: P. Bertrand.]

Gingerich, P. D. (1987). Early Eocene bats (Mammalia, Chiroptera) and other vertebrates in freshwater limestones of the Willwood Formation, Clark's Fork Basin, Wyoming. *Contributions from the Museum of Paleontology, University of Michigan*, 27, 275–320.

Göbbel, L. (2000). The external nasal cartilages in Chiroptera: significance for intraordinal relationships. *Journal of Mammalian Evolution*, 7, 167–201.

Grande, L. (1984). Paleontology of the Green River Formation, With a Review of the Fish Fauna (2nd edn). *Geological Survey of Wyoming Bulletin*, 63, 1–333.

Gray, J. E. (1821). On the natural arrangement of vertebrose animals. *London Medical Repository, Monthly Journal, and Review*, 15, 296–311.

(1825). Outline of an attempt at the disposition of the Mammalia into tribes and families with a list of the genera apparently appertaining to each tribe. *Annals of Philosophy, New Series*, 10(26 of whole series), 337–44.

(1831). Descriptions of some new genera and species of bats. *The Zoological Miscellany London*, 1, 37–8.

(1866). Synopsis of the genera of the Vespertilionidae and Noctilionidae. *Annals and Magazine of Natural History, Series* 3, 17, 89–93.

Gunnell, G. F. (ed.) (2001). *Eocene Biodiversity: Unusual Occurrences and Rarely Sampled Habitats*. New York: Kluwer Academic/Plenum.

(2003). New primitive microbat (Chiroptera) from the Green River Formation (upper Lower Eocene), Fossil Basin, southwestern Wyoming. *Journal of Vertebrate Paleontology*, 23(suppl. to no. 3), p. 58A.

Hall, E. R. (1930). A new genus of bat from the later Tertiary of Nevada. *University of California Publications, Bulletin of the Department of Geological Sciences*, 19, 319–20.

Hand, S., Novacek, M., Godthelp, H., and Archer, M. (1994). First Eocene bat from Australia. *Journal of Vertebrate Paleontology*, 14, 375–81.

Harrison, J. A. (1978). Mammals of the Wolf Ranch Local Fauna, Pliocene of the San Pedro Valley, Arizona. *Occasional Papers of the Museum of Natural History, University of Kansas*, 73, 1–18.

Hedges, S. B. and Kumar, S. (1999). Divergence times of eutherian mammals. *Science*, 285, 2031a.

Hibbard, C. W. (1950). Mammals of the Rexroad Formation from Fox Canyon, Meade County, Kansas. *Contributions from the Museum of Paleontology, University of Michigan*, 8, 113–192.

Hoofer, S. R., and Van den Bussche, R. A. (2003). Molecular phylogenetics of the chiropteran family Vespertilionidae. *Acta Chiropterologica*, 5(suppl.), 1–63.

Hooker, J. J. (1996). A primitive emballonurid bat (Chiroptera, Mammalia) from the earliest Eocene of England. [In *Paléobiologie et Évolution des Mammifères Paléogenes: Volume Jubilaire en Hommage à Donald E. Russell*, ed. M. Godinot and P. D. Gingerich.] *Palaeovertebrata*, 25, 287–300.

(2001). Tarsals of the extinct insectivoran family Nyctitheriidae (Mammalia): evidence for archontan relationships. *Zoological Journal of the Linnean Society*, 132, 501–29.

Hutcheon, J. M. and Garland, T., Jr. (2004). Are megabats big? *Journal of Mammalian Evolution*, 11, 257–77.

Hutcheon, J. M. and Kirsch, J. A. W. (2004). Camping in a different tree: results of molecular systematic studies of bats using DNA–DNA hybridization. *Journal of Mammalian Evolution*, 11, 17–47.

Hutcheon, J. M., Kirsch, J. A. W., and Pettigrew, J. D. (1998). Base-compositional biases and the bat problem. III. The question of microchiropteran monophyly. *Philosophical Transactions of the Royal Society of London, B*, 353, 607–17.

Hutchison, J. H. (1968). Fossil Talpidae (Insectivora, Mammalia) from the later Tertiary of Oregon. *Bulletin of the Museum of Natural History, the University of Oregon*, 11, 1–117.

Hutchison, J. H. and E. H. Lindsay. (1974). The Hemingfordian mammal fauna of the Vedder Locality, Branch Canyon Formation, Santa Barbara County, California. Part 1: Insectivora, Chiroptera, Lagomorpha, and Rodentia (Sciuridae). *PaleoBios*, 15, 1–19.

James, G. T. (1963). Paleontology and nonmarine stratigraphy of the Cuyama Valley badlands, California. Part I. Geology, faunal interpretations, and systematic descriptions of Chiroptera, Insectivora, and Rodentia. *University of California Publications in Geological Sciences*, 4, 1–154.

Jepsen, G. L. (1966). Early Eocene bat from Wyoming. *Science*, 154, 1333–9.

(1970). Bat origins and evolution. In *Biology of Bats*, Vol. I, ed. W. A. Wimsatt, pp. 1–64. New York: Academic Press.

Jones, K. E., Purvis, A., MacLarnon, A., Bininda-Emonds, O. R. P., and Simmons, N. B. (2002). A phylogenetic supertree of the bats (Mammalia: Chiroptera). *Biological Reviews*, 77, 223–59.

Kaup, J. J. (1829). *Skizzirte Entwickelungs-Geschichte und Natürliches System der Europäischen Thierwelt*. Darmstadt, Leipzig: C. W. Leske.

Kennedy, M., Paterson, A. M., Morales, J. C., *et al.* (1999). The long and short of it: branch lengths and the problem of placing the New Zealand short-tailed bat, *Mystacina. Molecular Phylogenetics and Evolution*, 13, 405–16.

Kirsch, J. A. W., Flannery, T. F., Springer, M. S., and Lapointe, F.-J. (1995). Phylogeny of the Pteropodidae (Mammalia: Chiroptera) based on DNA hybridization, with evidence for bat monophyly. *Australian Journal of Zoology*, 43, 395–428.

Kirsch, J. A. W., Hutcheon, J. M., Byrnes, D. G. P., and Lloyd, B. D. (1998). Affinities and historical zoogeography of the New Zealand short-tailed bat, *Mystacina tuberculata* Gray 1843, inferred from DNA-hybridization comparisons. *Journal of Mammalian Evolution*, 5, 33–64.

Koopman, K. F. (1994). Chiroptera: systematics. In *Handbook of Zoology*, Vol. VIII: *Mammalia, Part 60*. New York: Walter de Gruyter.

Koopman, K. F. and MacIntyre, G. T. (1980). Phylogenetic analysis of chiropteran dentition. In *Proceedings of the Fifth International Bat Research Conference*, ed. D. E. Wilson and A. L. Gardner, pp. 279–88. Lubbock, TX: Texas Technical Press.

Kumar, S. and Hedges, S. B. (1998). A molecular timescale for vertebrate evolution. *Nature*, 392, 917–20.

Kurtén, B. and Anderson, E. (1980). *The Pleistocene Mammals of North America*. New York: Columbia University Press.

Lapointe, F.-J., Baron, G., and Legendre, P. (1999). Encephalization, adaptation and evolution of Chiroptera: a statistical analysis with further evidence for bat monophyly. *Brain, Behavior and Evolution*, 54, 119–26.

Lapointe, F.-J., Kirsch, J. A. W., and Hutcheon, J. M. (1999). Total evidence, consensus, and bat phylogeny: a distance-based approach. *Molecular Phylogenetics and Evolution*, 11, 55–66.

Lawrence, B. (1943). Miocene bat remains from Florida, with notes on the generic characters of the humerus of bats. *Journal of Mammalogy*, 24, 356–69.

LeConte, J. (1831). *Lasionycteris*. In *The Animal Kingdom Arranged in Conformity with Its Organization by the Baron Cuvier; Translated from the French with Notes and Additions*, Vol. 1, ed. H. McMurtrie, p. 431. New York: Carvill.

(1856). Observations on the North American species of bats. *Proceedings of the Academy of Natural Sciences of Philadelphia*, 7, 431–8.

Legendre, S. (1984). Étude odontologique des représentants actuels du groupe *Tadarida* (Chiroptera, Molossidae). Implications phylogéniques, systématiques et zoogéographiques. *Revue Suisse de Zoologie*, 91, 399–442.

(1985). Molossidés (Mammalia, Chiroptera) cénozoiques de l'Ancien et du Nouveau Monde; statut systématique; intégration phylogénique de données. *Neues Jahrbuch für Geologie und Paläontologie, Abhandlungen*, 170, 205–27.

Legendre, S. and Sigé, B. (1983). La place du "Vespertilion de Montmartre" dans l'histoire des chiroptères molossidés. In *Actes du Symposium Paléontologique Georges Cuvier*, ed. E. Buffetaut, J. M. Mazin, and E. Salmon, pp. 347–61. Montbéliard.

Lesson, R. P. (1827). *Manuel de Mammalogie, ou Histoire Naturelle des Mammifères*. Paris: Roret.

Lin, Y.-H. and Penny, D. (2001). Implications for bat evolution from two new complete mitochondrial genomes. *Molecular Biology and Evolution*, 18, 684–8.

Lindsay, E. H. and Jacobs, L. L. (1985). Pliocene small mammal fossils from Chihuahua, Mexico. *Universidad Nacional Autónoma de México, Instituto de Geología, Paleontología Mexicana*, 51, 1–53.

Liu, F.-G. R., Miyamoto, M. M., Freire, N. P., *et al.* (2001). Molecular and morphological supertrees for eutherian (placental) mammals. *Science*, 291, 1786–9.

Madsen, O., Scally, M., Douady, C. J., *et al.* (2001). Parallel adaptive radiations in two major clades of placental mammals. *Nature*, 409, 610–14.

Maina, J. N. (2000). What it takes to fly: the structural and functional respiratory refinements in birds and bats. *Journal of Experimental Biology*, 203, 3045–64.

Marsh, O. C. (1872a). Preliminary description of new Tertiary mammals. Part I. *American Journal of Science and Arts* (third series), 4, 122–8.

(1872b). Preliminary description of new Tertiary mammals. Parts II, III, and IV. *American Journal of Science and Arts* (third series), 4, 202–24.

Matthew, W. D. (1917). A Paleocene bat. *Bulletin of the American Museum of Natural History*, 37, 569–71.

McKenna, M. C. and Bell, S. K. (1997). *Classification of Mammals Above the Species Level*. New York: Columbia University Press.

McKenna, M. C., Robinson, P., and Taylor, D. W. (1962). Notes on Eocene Mammalia and Mollusca from Tabernacle Butte, Wyoming. *American Museum Novitates*, 2102, 1–33.

Menu, H. (1984). Revision du statut de *Pipistrellus subflavus* (F. Cuvier, 1832). Proposition d'un taxon générique nouveau: *Perimyotis* nov. gen. *Mammalia*, 48, 409–16.

(1985). Morphotypes dentaires actuels et fossiles des chiroptères vespertilioninés Ie partie: Étude des morphologies dentaires. *Palaeovertebrata*, 15, 71–128.

Menu, H. and Sigé, B. (1971). Nyctalodontie et myotodontie, importants caractères de grades évolutifs chez les chiroptères entomophages. *Comptes Rendus de Séances de l'Académie des Sciences*, 272, 1735–8.

Merriam, C. H. (1890). Descriptions of a new species of *Molossus* from California (*Molossus californicus*). *North American Fauna*, 4, 31–2.

Miller, G. S., Jr. (1906). Twelve new genera of bats. *Proceedings of the Biological Society of Washington*, 19, 83–6.

(1907). The families and genera of bats. *Bulletin of the United States National Museum*, 57, 1–282.

Miyamoto, M. M., Porter, C. A., and Goodman, M. (2000). c-*myc* gene sequences and the phylogeny of bats and other eutherian mammals. *Systematic Biology*, 49, 501–14.

Morgan, G. S. (1989). New bats from the Oligocene and Miocene of Florida, and the origins of the Neotropical chiropteran fauna. *Journal of Vertebrate Paleontology*, 9(suppl. to no. 3), p. 33A.

(1991). Neotropical Chiroptera from the Pliocene and Pleistocene of Florida. *Bulletin of the American Museum of Natural History*, 206, 176–213.

Morgan, G. S. and Czaplewski, N. J. (2000). A new bat in the Neotropical family Natalidae from the early Miocene (Hemingfordian) Thomas Farm Local Fauna, Florida. *Journal of Vertebrate Paleontology*, 20(suppl. to no. 3), p. 59A.

Morgan, G. S. and Czaplewski, N. J. (2003). A new bat in the Neotropical family Natalidae from the early Miocene of Florida. *Journal of Mammalogy*, 84, 729–52.

Morgan, G. S. and Hulbert, R. C., Jr. (1995). Overview of the geology and vertebrate biochronology of the Leisey Shell Pit local fauna, Hillsborough County, Florida. *Bulletin of the Florida Museum of Natural History*, 37, 1–92.

Morgan, G. S. and Ridgway, R. B. (1987). Late Pliocene (late Blancan) vertebrates from the St. Petersburg Times site, Pinellas County, Florida, with a brief review of Florida Blancan faunas. *Papers in Florida Paleontology*, 1, 1–22.

Morgan, G. S., Linares, O. J., and Ray, C. E. (1988). New species of fossil vampire bats (Mammalia: Chiroptera: Desmodontidae) from Florida and Venezuela. *Proceedings of the Biological Society of Washington*, 101, 912–28.

Müller, P. L. S. (1776). *Mit Einer Ausführlichen Erklärung Ausgefertiget. Des Ritters Carl von Linné. Vollständigen Natursystems Supplements und Register-ban Üaller Sechs Theile Oder Classen des Thierreichs*, pp. 3–34. Nurnberg: G. N. Raspe.

Murphy, W. J., Eizirik, E., Johnson, W. E., *et al.* (2001a). Molecular phylogenetics and the origins of placental mammals. *Nature*, 409, 614–18.

Murphy, W. J., Eizirik, E., O'Brien, S. J., *et al.* (2001b). Resolution of the early placental radiation using Bayesian phylogenetics. *Science*, 294, 2348–51.

Nikaido, M., Kawai, K., Cao, Y., *et al.* (2001). Maximum likelihood analysis of the complete mitochondrial genomes of eutherians and a reevaluation of the phylogeny of bats and insectivores. *Journal of Molecular Evolution*, 53, 508–16.

Novacek, M. J. (1982). Information for molecular studies from anatomical and fossil evidence on higher eutherian phylogeny. In *Macromolecular Sequences in Systematic and Evolutionary Biology*, ed. M. Goodman, pp. 3–41. New York: Plenum Press.

(1985). Evidence for echolocation in the oldest known bats. *Nature*, 315, 140–1.

(1987). Auditory features and affinities of the Eocene bats *Icaronycteris* and *Palaeochiropteryx* (Microchiroptera, incertae sedis). *American Museum Novitates*, 2877, 1–18.

(1991). Aspects of the morphology of the cochlea in microchiropteran bats: an investigation of character transformation. *Bulletin of the American Museum of Natural History*, 206, 84–99.

Ostrander, G. E. (1983). New early Oligocene (Chadronian) mammals from the Raben Ranch local fauna, northwest Nebraska. *Journal of Paleontology*, 57, 128–39.

(1985). Correlation of the Early Oligocene (Chadronian) in northwestern Nebraska. *Dakoterra. South Dakota School of Mines and Technology*, 2, 205–31.

(1987). The early Oligocene (Chadronian) Raben Ranch local fauna, northwest Nebraska: Marsupialia, Insectivora, Dermoptera, Chiroptera, and Primates. *Dakoterra, South Dakota School of Mines and Technology*, 3, 92–104.

Palisot de Beauvois, A. M. F. J. (1796). *Catalogue Raisonne du Muséum, de Mr. C. W. Peale*. Philadelphia, PA: S. H. Smith, Parent.

Paula Couto, C. de (1956). Une chauve-souris fossile des argiles feuilletées Pléistocènes de Tremembé, État de Sao Paulo (Brésil). *Actes IV Congrès Internationale Quaternaire*, 1, 343–7.

Peters, W. C. H. (1865). Über die zu den *Vampyri* gehörigen Fleder thiere und uber die naturliche stellung der Gattung *Antrozous*. *Monatsberichte der Königlich Preussischen Akademie der Wissenschaften zu Berlin*, 1865, 503–24.

Pettigrew, J. D. (1986). Flying primates? Megabats have the advanced pathway from eye to midbrain. *Science*, 231, 1304–6.

Pumo, D. E., Finamore, P. S., Franek, W. R., *et al.* (1998). Complete mitochondrial genome of a Neotropical fruit bat, *Artibeus jamaicensis*, and a new hypothesis of the relationships of bats to other eutherian mammals. *Journal of Molecular Evolution*, 47, 709–17.

Rafinesque, C. S. (1814). *Précis des Découvertes et Travaux Somiologiques de Mr. C. S. Rafinesque Schmaltz entre 1800 et 1814 ou Choix Raisonné de ses Principales Découvertes en Zoologie et en Botanique, pour Servir d'Introduction à ses Ouvrages Futurs*. Palerme, Royale Typographie Militaire, aux dépens de l'auteur, 1814. [Alternatively, same author and year, *Principes Fondamentaux de Somologie, ou les Loix de la Nomenclature et de la Classification de l'Empire Organique ou des Animaux et des Végétaux: Contenant les Régles Essentielles de l'Art de leur Imposer des Noms Immutables et de les Classer Méthodiquement*. Palerme, De l'imprimerie de Franc. Abate, aux dépens de l'auteur, 1814.]

(1820). *Annals of Nature; or Annual Synopsis of New Genera and Species of Animals, Plants, &c., Discovered in North America*. First annual number, for 1820. Lexington, KY: T. Smith.

Revilliod, P. (1919). L'état actuel de nos connaissances sur les Chiroptères fossiles (Note préliminaire). *Comptes Rendus Société des Physique et Histoire Naturelle Genève*, 36, 93–6.

Rich, T. H., Vickers-Rich, P., and Flannery, T. F. (1999). Divergence times of eutherian mammals. *Science*, 285, 2031a.

Scally, M., Madsen, O., Douady, C. J., *et al.* (2002). Molecular evidence for the major clades of placental mammals. *Journal of Mammalian Evolution*, 8, 239–77.

Schinz, H. R. (1821). In *Das Thierreich Eingetheilt nach dem Bau der Thiere als Grundlage ihrer Naturgeschichte und der Vergleichenden Anatomie von dem Herrn Ritter von Cuvier*. Vol. 1, *Säugethiere und Vogel*. Stuttgart: J. G. Cotta'schen Buchhandlung.

Schutt, W. A., Jr. and Simmons, N. B. (1998). Morphology and homology of the chiropteran calcar, with comments on the phylogenetic relationships of *Archaeopteropus*. *Journal of Mammalian Evolution*, 5, 1–32.

Schwartz, J. H. and Krishtalka, L. (1977). Revision of Picrodontidae (Primates, Plesiadapiformes): dental homologies and relationships. *Annals of the Carnegie Museum*, 46, 55–70.

Sears, K., Behringer, R., and Niswander, L. (2004). The development of powered flight in Chiroptera: the morphological and genetic evolution of bat wing digits. *Journal of Vertebrate Paleontology*, 24(suppl. to no. 3), p. 111A.

Sigé, B. (1974). Données nouvelles sur le genre *Stehlinia* (Vespertilionoidea, Chiroptera) du Paléogène d'Europe. *Palaeovertebrata*, 6, 253–72.

(1997). Les remplissages karstiques polyphasés (Éocène, Oligocène, Pliocène) de Saint-Maximin (phosphorites du Gard) et leur apport à la connaissance des faunes européenes, notamment pour l'Éocène moyen (MP 13). 3: Systématique: euthériens entomophages. [In *Actes du Congrès BiochroM'97*, ed. J.-P Aguilar, S. Legendre, and J. Michaux.] *Mémoires et Travaux de l'Institut de Montpellier de l'École Pratique des Hautes Études*, 21, 737–50.

Sigé, B. and Legendre, S. (1983). L'histoire des peuplements de chiroptères du bassin méditerranéen: l'apport comparé des remplissages karstiques et des dépôts fluviolacustres. *Mémoires de Biospéologie*, 10, 207–24.

Simmons, N. B. (1993a). The importance of methods: archontan phylogeny and cladistic analysis of morphological data. In *Primates and Their Relatives in Phylogenetic Perspective*, ed. R. D. E. MacPhee, pp. 1–61. New York: Plenum Press.

(1993b). Morphology, function, and phylogenetic significance of pubic nipples in bats (Mammalia: Chiroptera). *American Museum Novitates* 3077, 1–37.

(1994). The case for chiropteran monophyly. *American Museum Novitates* 3103, 1–54.

(1995). Bat relationships and the origin of flight. *Symposia of the Zoological Society of London*, 67, 27–43.

(1998). A reappraisal of interfamilial relationships of bats. In *Bat Biology and Conservation*, ed. T. H. Kunz and P. A. Racey, pp. 3–26. Washington, DC: Smithsonian Institution Press.

(2000). Bat phylogeny: an evolutionary context for comparative studies. In *Ontogeny, Functional Ecology, and Evolution of Bats*, ed. R. A. Adams and S. C. Pedersen, pp. 9–58. Cambridge, UK: Cambridge University Press.

(2005). Chiroptera. In *The Rise of Placental Mammals. Origins and Relationships of the Major Extant Clades*, ed. K. D. Rose and J. D. Archibald, pp. 159–74. Baltimore, MD: Johns Hopkins University Press.

Simmons, N. B. and Geisler, J. H. (1998). Phylogenetic relationships of *Icaronycteris*, *Archaeonycteris*, *Hassianycteris*, and *Palaeochiropteryx* to extant bat lineages, with comments on the evolution of echolocation and foraging strategies in Microchiroptera. *Bulletin of the American Museum of Natural History*, 235, 1–182.

Simmons, N. B. and Quinn, T. H. (1994). Evolution of the digital locking mechanism in bats and dermopterans: a phylogenetic perspective. *Journal of Mammalian Evolution*, 2, 231–54.

Simmons, N. B., Novacek, M. J., and Baker, R. J. (1991). Approaches, methods, and the future of the chiropteran monophyly controversy: a reply to J. D. Pettigrew. *Systematic Zoology*, 40, 239–43.

Slaughter, B. H. (1970). Evolutionary trends of chiropteran dentitions. In *About Bats: A Chiropteran Symposium*, ed. B. H. Slaughter and D. W. Walton, pp. 51–83. Dallas, TX: Southern Methodist University Press.

Smith, J. D. (1972). Systematics of the chiropteran family Mormoopidae. *University of Kansas Museum of Natural History, Miscellaneous Publication*, 56, 1–132.

Smith, T. (1995). Présence du genre *Wyonycteris* (Mammalia, Lipotyphla) à la limite Paléocène-Éocène en Europe. *Comptes Rendus de l'Académie des Sciences, Paris*, 321, 923–30.

Speakman, J. R. (2001). The evolution of flight and echolocation in bats: another leap in the dark. *Mammal Review*, 31, 111–30.

Springer, M. S., Teeling, E. C., and Stanhope, M. J. (2001a). External nasal cartilages in bats: evidence for microchiropteran monophyly? *Journal of Mammalian Evolution*, 8, 231–6.

Springer, M. S., Teeling, E. C., Madsen, O., Stanhope, M. J., and de Jong, W. W. (2001b). Integrated fossil and molecular data reconstruct bat echolocation. *Proceedings of the National Academy of Sciences, USA*, 98, 6241–6.

Stadelmann, B., Herrera, L. G., Arroyo-Cabrales, J., *et al.* (2004). Molecular systematics of the fishing bat *Myotis* (*Pizonyx*) *vivesi*. *Journal of Mammalogy*, 85, 133–9.

Stirton, R. A. (1931). A new genus of the family Vespertilionidae from the San Pedro Pliocene of Arizona. *University of California Publications, Bulletin of the Department of Geological Sciences*, 20, 27–30.

Storch, G. (2001). Paleobiological implications of the Messel mammalian assemblage. In *Eocene Biodiversity Unusual Occurrences and Rarely Sampled Habitats*, ed. G. F. Gunnell, pp. 215–35. New York: Kluwer Academic/Plenum.

Storch, G., Sigé, B., and Habersetzer, J. (2002). *Tachypteron franzeni* n. gen., n. sp., earliest emballonurid bat from the Middle Eocene of Messel (Mammalia, Chiroptera). *Paläontologische Zeitschrift*, 76, 189–99.

Storer, J. E. (1984). Mammals of the Swift Current Creek local fauna (Eocene: Uintan), Saskatchewan. *Natural History Contributions (Saskatchewan Culture and Recreation)*, 7, 1–158.

(1995). Small mammals of the Lac Pelletier lower fauna, Duchesnean, of Saskatchewan, Canada: insectivores and insectivore-like groups, a plagiomenid, a microsyopid and Chiroptera. In *Vertebrate Fossils and the Evolution of Scientific Concepts. A Tribute to L. Beverly Halstead*, ed. W. A. S. Sarjeant, pp. 595–615. London: Gordon and Breach.

Sutton, J. F. and Genoways, H. H. (1974). A new vespertilionine bat from the Barstovian deposits of Montana. *Occasional Papers of the Museum of Texas Technical University*, 20, 1–8.

Szalay, F. S. (1968). The Picrodontidae, a family of early primates. *American Museum Novitates*, 2329, 1–55.

Teeling, E. C., Scally, M., Kao, D. J., *et al.* (2000). Molecular evidence regarding the origin of echolocation and flight in bats. *Nature*, 403, 188–92.

Teeling, E., Madsen, O., Van den Bussche, R. A., *et al.* (2002). Microbat paraphyly and the convergent evolution of a key innovation in Old World rhinolophoid microbats. *Proceedings of the National Academy of Sciences, USA*, 99, 1431–6.

Teeling, E. C., Springer, M. S., Madsen, O., *et al.* (2005). A molecular phylogeny for bats illuminates biogeography and the fossil record. *Science*, 307, 580–4.

Thewissen, J. G. M. and Smith, G. R. (1987). Vespertilionid bats (Chiroptera, Mammalia) from the Pliocene of Idaho. *Contributions from the Museum of Paleontology, University of Michigan*, 27, 237–45.

Tomida, Y. (1994). Phylogenetic reconstruction of fossil mammals based on the cheek tooth morphology. *Honyurui Kagaku (Mammalian Science)*, 34, 19–29.

Van den Bussche, R. A. and Hoofer, S. R. (2000). Further evidence for inclusion of the New Zealand short-tailed bat (*Mystacina tuberculata*) within Noctilionoidea. *Journal of Mammalogy*, 81, 865–74.

(2001). Evaluating monophyly of Nataloidea (Chiroptera) with mitochondrial DNA sequences. *Journal of Mammalogy*, 82, 320–7.

(2004). Phylogenetic relationships among Recent chiropteran families and the importance of choosing appropriate out-group taxa. *Journal of Mammalogy*, 85, 321–30.

Van Valen, L. (1979). The evolution of bats. *Evolutionary Theory*, 4, 103–21.

Vaughan, T. A. (1959). Functional morphology of three bats: *Eumops*, *Myotis*, and *Macrotus*. *University of Kansas Publications, Museum of Natural History*, 12, 1–153.

(1970). Adaptations for flight in bats. In *About Bats: A Chiropteran Symposium*, ed. B. H. Slaughter and D. W. Walton, pp. 127–43. Dallas, TX: Southern Methodist University Press.

Webb, S. D. and S. C. Perrigo. (1984). Late Cenozoic vertebrates from Honduras and El Salvador. *Journal of Vertebrate Paleontology*, 4, 237–54.

Wetterer, A. L., Brockman, M. V., and Simmons, N. B. (2000). Phylogeny of phyllostomid bats (Mammalia: Chiroptera): data from diverse morphological systems, sex chromosomes, and restriction sites. *Bulletin of the American Museum of Natural History*, 248, 1–200.

White, J. A. (1969). Late Cenozoic bats (subfamily Nyctophylinae) from the Anza-Borrego Desert of California. *University of Kansas Museum of Natural History Miscellaneous Publications*, 51, 275–82.

Wible, J. R. and Novacek, M. J. (1988). Cranial evidence for the monophyletic origin of bats. *American Museum Novitates*, 2911, 1–19.

Wied-Neuwied, M. A., Prinz zu (1826). Beiträge zur Naturgeschichte von Brasilien. Verzeichniss der Amphibien, Säugethiere, und Vögel, welche auf einer Reise Zwischen dem 13ten und dem 23sten Grade südlicher Breite im östlichen Brasilien beobachtet wurden. II. *Abteilung Mammalia Säugethiere*, *Weimar*, 2, 1–620.

13 Plagiomenidae and Mixodectidae

KENNETH D. ROSE
Johns Hopkins University School of Medicine, Baltimore, MD, USA

INTRODUCTION

Plagiomenids and mixodectids are small early Tertiary insectivore-like mammals that are known mainly from their teeth and jaws. Their broader relationships among Eutheria have been difficult to work out because of the very fragmentary nature of known fossil remains. In the case of plagiomenids, peculiarities of the molar cusp patterns and of the incisor crowns are reminiscent of the teeth of living flying lemurs; consequently plagiomenids have long been considered to be related to or members of the Dermoptera (Matthew, 1918; Simpson, 1945; Jepsen, 1962). Only *Plagiomene* is known from cranial material other than jaws, and these skulls have proven critical in reassessing the affinities of the genus (Dawson *et al.*, 1986; MacPhee, Cartmill, and Rose, 1989). However, no postcrania have been described for any plagiomenid. Mixodectids have been recognized for decades as possible close relatives of plagiomenids, based on dental similarities (e.g., Simpson, 1936; Sloan, 1969). A couple of poorly preserved skull fragments of *Mixodectes* are the only reported cranial remains for the family. More informative are several postcranial elements recently attributed to *Mixodectes* by Szalay and Lucas (1996), which suggest arboreal habits and archontan relationships. In their recent classification of mammals, McKenna and Bell (1997) included both Plagiomenidae and Mixodectidae in the Dermoptera, Grandorder Archonta.

Both plagiomenids and mixodectids seem to have been exclusively North American clades. (Early Eocene *Placentidens* from France, once considered a plagiomenid [Russell, Louis, and Savage, 1973], is regarded as an erinaceomorph lipotyphlan by McKenna and Bell [1997].) In addition, plagiomenids were evidently a northern group, all known fossils coming from central Wyoming (above 42° north latitude) or farther north. Although the group was never diverse or very common, certain species were locally abundant, perhaps related to a specific (and unusual?) environment. For example, Clarkforkian *Planetetherium mirabile* was the most common of 20 species known from the Bear Creek Locality, Montana (locality CP17A) (Rose, 1981), and *Worlandia inusitata* ranked third among 30 species at University of Michigan Locality SC-188 in the Clark's Fork Basin (locality CP17B) (Krause, 1986); yet both species are otherwise extremely scarce. It is particularly curious that plagiomenids are abundant and diverse elements of the Eocene mammalian fauna from Ellesmere and Axel Heiberg Islands, Canada (West and Dawson, 1978; Dawson *et al.*, 1986, 1993), sites that were well within the Arctic Circle even then (McKenna, 1980; Dawson *et al.*, 1993). At least six plagiomenid species were present, but only one has been described so far. This species, *Ellesmene eureka*, represents the northernmost occurrence known for any early Tertiary mammal.

Although evidence suggests that Plagiomenidae and Mixodectidae are closely related and may be sister groups, their broader relationships require further study, and the two families will be considered separately in this chapter.

DEFINING FEATURES OF THE FAMILIES PLAGIOMENIDAE AND MIXODECTIDAE

PLAGIOMENIDAE

The dentition of plagiomenids (Figure 13.1) is characterized by a tendency to molarize posterior premolars and to reduce third molars; in most forms, there is a decrease in molar size from front to back, with m3 being markedly reduced. The lingual cusps of the posterior premolars and molars are anteriorly displaced, and those of the lowers are more elevated than the buccal cusps, giving these teeth a distinct anterolingual cant. The molar trigonids are anteroposteriorly compressed, with low paraconids or paracristids. Strong conules are present on the upper molariform teeth, but the hypocone is typically lost (except in *Elpidophorus* and the questionable plagiomenid

Evolution of Tertiary Mammals of North America, Vol. 2. ed. C. M. Janis, G. F. Gunnell, and M. D. Uhen. Published by Cambridge University Press.

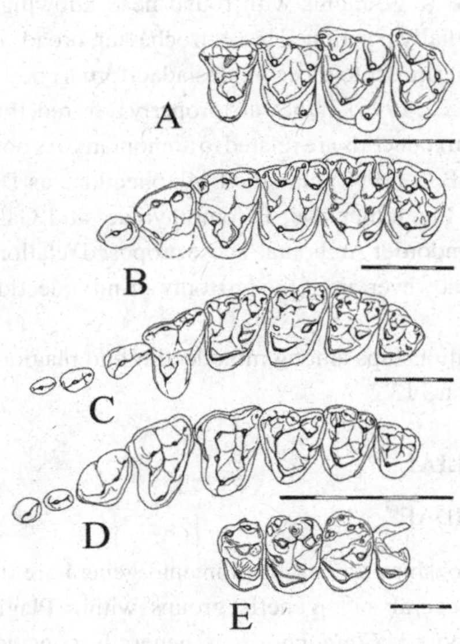

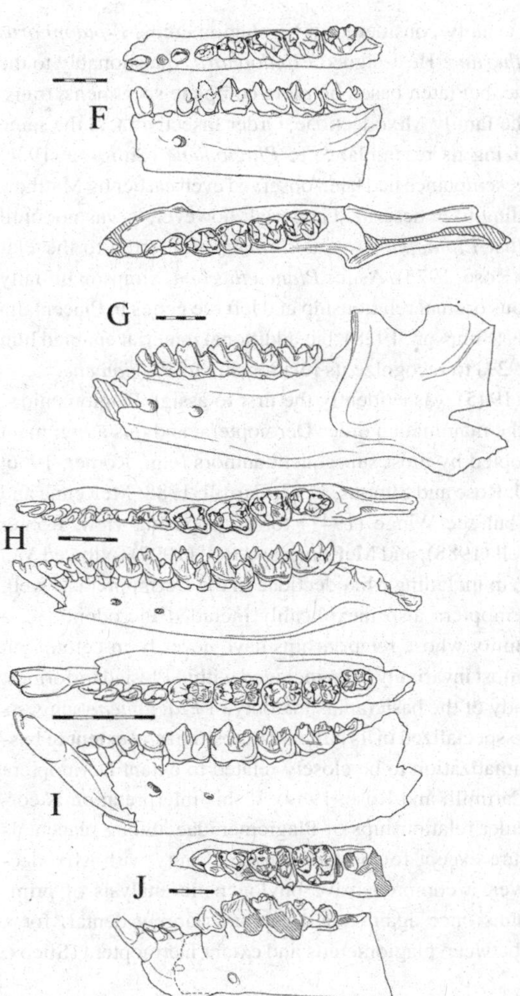

Figure 13.1. Left upper and lower dentitions of mixodectids and plagiomenids. A, F. *Eudaemonema cuspidata* (partly after Simpson, 1937). B, G. *Elpidophorus elegans* (after Simpson, 1936), C, H. *Plagiomene multicuspis* (after Rose and Simons, 1977, and MacPhee, Cartmill, and Rose, 1989), D, I. *Worlandia inusitata* (after Rose, 1982), and E, J, *Tarka stylifera* (after McKenna, 1990). Scale bars = 5 mm.

Thylacaelurus). The upper molariform teeth are further characterized by a V-shaped valley that runs buccally from the buccal side of the protocone and separates the mesial from the distal cusps. Accessory cusps are commonly present (especially stylar cusps, an entoconulid, and a twinned hypoconulid), and the enamel is often crenulated. The medial incisor is always enlarged and the incisor crowns are usually digitate (but not in *Tarka*). (Based mainly on Matthew, 1918; Rose, 1973, 1982.)

MIXODECTIDAE

The dentition of mixodectids is characterized, where known, by two pairs of enlarged incisors, the central pair generally larger than the lateral pair. The upper molars have a broad stylar shelf, distinct mesostyle, and prominent and lingually placed hypocone; the lower molars have a low, crestlike paraconid, a high, cuspate metaconid, and more or less twinned large entoconid and smaller hypoconulid. The molar talonids are broad, basined, and wider than the trigonids, and the basin opens lingually through a deep notch between the entoconid and the metaconid (talonid notch). The lower cheek teeth lack ectocingulids. (These features and generic characters are based principally on Szalay [1969] and Gunnell [1989]).

SYSTEMATICS

SUPRAFAMILIAL

PLAGIOMENIDAE

The first known plagiomenids were recognized to be unusual insectivores or insectivore-like forms. Matthew (1918) remarked on the resemblance of the molars of *Plagiomene* to those of extant Dermoptera (*Cynocephalus*) and less so to desman moles (*Myogale*); nevertheless, he assigned the new family Plagiomenidae to Insectivora. A decade later, Simpson (1927, 1928) described two more genera now usually considered to be plagiomenids, *Elpidophorus* and *Planetetherium*. He assigned *Elpidophorus* questionably to the Oxyclaenidae, but later, based on more complete specimens, transferred it to the family Mixodectidae, Order Insectivora, at the same time recognizing its resemblance to *Plagiomene* (Simpson, 1936, 1937). This resemblance had been observed even earlier by Matthew (1918), alluding to undescribed material; however, it was not until much later that *Elpidophorus* was formally tranferred to the Plagiomenidae (Rose, 1975). As for *Planetetherium*, Simpson initially saw no obvious ordinal relationship and left the genus in Placentalia incertae sedis (Simpson, 1928), but additional material enabled him (Simpson, 1929) to recognize its relationship to *Plagiomene*.

Simpson (1945) was evidently the first to assign Plagiomenidae formally to the mammalian order Dermoptera, and this assignment has been adopted by most subsequent authors (e.g., Romer, 1966; Jepsen, 1970; Rose and Simons, 1977; Carroll, 1988; McKenna and Bell, 1997) but see Winge (1941) for an opposing view. Romer (1966), Carroll (1988), and McKenna and Bell (1997) followed Van Valen (1967) in including Mixodectidae in the Dermoptera as well. Romer's Dermoptera also inexplicably included Picrodontidae, a Paleocene family whose relationships have never been certain but which has almost invariably been included within Plesiadapiformes.

Recent study of the basicranial anatomy of *Plagiomene* suggests that it was too specialized in its bullar composition and extent of basicranial pneumatization to be closely related to extant Dermoptera (MacPhee, Cartmill, and Rose, 1989). If this interpretation is correct, the broader relationships of Plagiomenidae among placentals remain obscure except for probable close affinity with Mixodectidae. However, a comprehensive phylogenetic analysis of primitive archontans once again found support (largely dental) for a relationship between plagiomenids and extant Dermoptera (Silcox, 2001).

Beard (1990) and Kay, Thorington, and Houde (1990) argued that paromomyids (generally recognized as plesiadapiform primates or near-primates) are in fact the oldest known Dermoptera. It should be noted, however, that the dentition of paromomyids bears no special resemblance to that of either plagiomenids or galeopithecids (including late Eocene *Dermotherium*, the oldest known definitive dermopteran). Recent analysis of dental and basicranial remains of paromomyids reaffirms their close relationship to plesiadapiforms rather than to dermopterans (Bloch and Silcox, 2001; Silcox, 2001).

MIXODECTIDAE

It has long been known that mixodectids share various dental traits with microsyopids (usually assigned to Plesiadapiformes), but whether or not these similarities reflect close relationship has been contentious (e.g., Osborn, 1902; Szalay, 1969; Gunnell, 1989). Although earlier vertebrate paleontologists (e.g., Osborn, 1902; Wortman, 1903; Simpson, 1945) usually grouped them together, Szalay (1969) argued that the two families are not specially related. More recently, Gunnell (1989) observed that details of the upper dental formula and molar structure are contrary to a close relationship between mixodectids and microsyopids and are more supportive of a broader clade also containing plesiadapoids, plagiomenids, and apatemyids; nevertheless, he assigned Mixodectidae questionably to the Insectivora.

Several postcranial elements attributed to *Mixodectes* show arboreal adaptations (e.g., radius with round head allowing extensive supination, medially positioned lesser trochanter, broad and shallow patellar groove) similar to those of plesiadapiforms and *Ptilocercus* (Szalay and Lucas, 1996). If they are properly referred, this could be evidence that mixodectids are related to archontans. As noted earlier, McKenna and Bell (1997) classified Mixodectidae as Dermoptera (together with Plagiomenidae, Paromomyidae, and Galeopithecidae), in the Grandorder Archonta. These proposed relationships will not be secure, however, until the anatomy of mixodectids is better understood.

Possible relationships among mixodectids and plagiomenids are depicted in Figure 13.2.

INFRAFAMILIAL

PLAGIOMENIDAE

Probable relationships among plagiomenid genera are depicted in Figure 13.2. Several suprageneric groups within Plagiomenidae have been proposed. *Elpidophorus* is generally considered to be the primitive sister taxon of all other Plagiomenidae (Rose, 1975; Gunnell, 1989; McKenna, 1990). A close relationship between *Planetetherium* and *Worlandia* is probable (based on their small size, shorter dentary, simpler and smaller i2–P2, and relatively broad, low-crowned cheek teeth) and is reflected by recognition of a subfamily, Worlandiinae (Bown and Rose, 1979; Rose, 1982), or tribe, Worlandiini (McKenna, 1990; McKenna and Bell, 1997). As observed by McKenna (1990), although *Plagiomene* is a primitive plagiomenid in some respects (e.g., dental formula), it shares several synapomorphies with Worlandiini, including relatively large and complex (submolariform) P3/p3, p2 with a metacone, and P4/p4 fully molarized and as big as the first molar or bigger. However, these characteristics could have been acquired independently in these genera.

Ellesmene resembles worlandiins in having a foreshortened dentary and a single-rooted p2. However, it retains the primitive lower dental formula of *Plagiomene* (i3, c1, p4, m3) and appears to be more closely related to the latter based on synapomorphous traits, including cheek teeth with very similar crown structure and highly molarized third and fourth premolars (Dawson *et al.*, 1993).

Tarka and *Tarkadectes* appear to be closely related to each other and divergent from other plagiomenids, and they probably represent a separate subfamily. They seem to be more primitive than *Plagiomene* and Worlandiini in retaining simple p3 and i1 (enlarged but apparently without a divided crown), but more derived in having larger and/or additional accessory (stylar) cusps, and in losing the talonid of p4, the metacone of P4, and two anterior teeth, probably i2–3 (McKenna, 1990). Consequently, it is possible that *Tarka* and *Tarkadectes* are the most derived plagiomenids; however, this would require that simplification of the third and fourth premolars is a derived condition.

McKenna (1990) proposed that the Arikareean genus *Ekgmowechashala* (which has generally been considered an omomyid

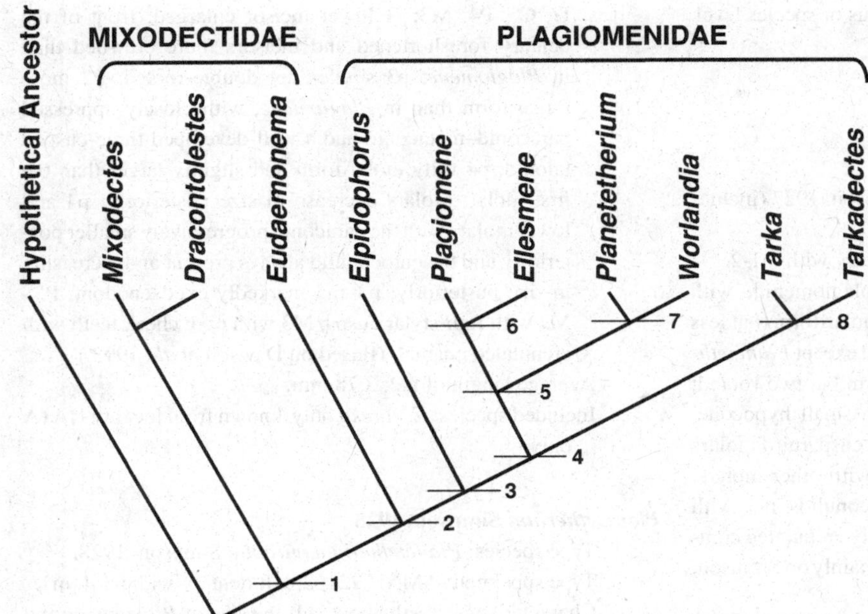

Figure 13.2. Cladogram of relationships of plagiomenids and mixodectids, modified mainly after McKenna (1990), supplemented by Dawson *et al.* (1993). *Thylacaelurus* is too poorly known to be included. Characters at nodes are as follows: 1. I1–2/i1–2 large, lower molars with low trigonids and strong anterobuccal cingulum, molar hypoconids buccally extended, upper molars with five stylar cusps, M1–2 with hypocone. 2. P3–4 with metacone, P4/p4 with all molar cusps, m3 talonid narrower. 3. Hypocone reduced; metaconid and entoconid of lower molars elevated relative to protoconid and hypoconid, paracristid and protocristid of lower molars appressed, compressing trigonid basin; lower molars with strong ectocingulum and slightly to strongly exodaenodont. 4. Lower molars with entoconulid, hypocone lost from upper molars, enamel wrinkled, molars decrease in size from M1/m1-M3/m3. 5. P2 with metacone, P3/p3 large and nearly molariform, P4/p4 as large as or larger than M1/m1. 6. P3–4 more fully molariform, and each with 3–4 stylar cusps. 7. Size of animal smaller, dentary foreshortened and two anterior lower teeth lost (probably i2–3). 8. Secondary simplification of P4/p4: P4 metacone lost, p4 talonid reduced; lower molars with buccal stylar cusp(s), enlarged metastylids, and mesoconid.

primate [Szalay, 1976]) was an aberrant plagiomenid, based in part on reinterpretation of the cusp homologies of the upper teeth. He further concluded that *Tarka* and *Tarkadectes* belong to the same subfamily, Ekgmowechashalinae. The latter two genera are here accepted as plagiomenids (albeit omomyid-like in some respects, particularly crown shape of the i1, p3, and m3), whereas *Ekgmowechashala* is tentatively retained in the Omomyidae. Fossils recently recovered in China support the omomyid affinities of *Ekgmowechashala* (K. C. Beard, personal communication). Until the interrelationships among plagiomenid genera are better resolved, formal use of subfamilial names seems premature and is avoided here.

Thylacaelurus has been considered a didelphid (Russell, 1954), a lipotyphlan (Szalay, 1969), or a plagiomenid (Van Valen, 1967; Krishtalka and Setoguchi, 1977; Bown and Rose, 1979; Storer, 1984; Ostrander, 1987), but the last assignment has never been secure. Recently described (isolated) lower molars referred to the genus (Storer, 1984) appear to strengthen the likelihood that it is correctly included in Plagiomenidae, but until more complete specimens are available its affinities will remain tenuous.

MIXODECTIDAE

Mixodectidae is a small family, consisting of just three North American genera and possibly the European *Remiculus*; all are rather poorly known. Alternatively, *Remiculus* could be related to plagiomenids (Gunnell, 1989). *Elpidophorus* was classified as a mixodectid for many years (e.g., Simpson, 1936; Szalay, 1969) but more recently has been considered a plagiomenid (see above and Rose, 1975; Gunnell, 1989). *Mixodectes*, the type genus, is relatively derived in having lost the canine and p1. As Gunnell pointed out, *Dracontolestes* (known from two very fragmentary lower jaws) differs from *Mixodectes* only in trivial features and the two may be congeneric, whereas *Eudaemonema* is significantly different in ways that approach plagiomenids (perhaps because it is closer to the stem of mixodectids). Gunnell portrayed *Eudaemonema* as the sister group of plagiomenids, although classifying it as a mixodectid. *Eudaemonema* differs from plagiomenids and resembles other mixodectids in having a large, lingual hypocone, but it resembles both groups in having an enlarged, laterally compressed i1, very high metaconids and entoconids, and strong transverse valleys on the upper molars (Gunnell, 1989).

INCLUDED GENERA OF PLAGIOMENIDAE AND MIXODECTIDAE

The locality numbers listed for each genus refer to the list of unified localities in Appendix I. The locality numbers may be listed more than one way. The acronyms for museum collections are listed in Appendix III.

Parentheses around the locality (e.g., [CP101]) mean that the taxon in question at that locality is cited as an "aff." or "cf." the taxon in question. Parentheses are usually used for individual species, implying that the genus is firmly known from the locality, but the actual species identification may be questionable. Question marks in front of the locality (e.g., ?CP101) mean that the taxon is questionably known from that locality, implying some doubt that the taxon is actually present at that locality, either at the genus or species level. An asterisk (*) indicates the type locality.

PLAGIOMENIDAE MATTHEW, 1918

Elpidophorus Simpson, 1927

Type species: *Elpidophorus elegans* Simpson, 1927 (including *E. patratus*, Simpson, 1936).

Type specimen: AMNH 15541, right dentary with m1-2.

Characteristics: The least specialized plagiomenid, with P4/p4 almost fully molariform, P3 submolariform (but less molarized than in any other plagiomenid except *Planetetherium*), and p3 essentially premolariform but two-rooted; p2 simple and single-rooted; M2 with small hypocone; M1 with hypocone shelf but no distinct cusp; third molars only very slightly reduced compared with other molars; molariform lower teeth lacking entoconulids but with well-developed cingulids; enamel weakly crenulated compared with other plagiomenids. (Based mainly on Simpson, 1936; Szalay, 1969; Rose, 1975.)

Average length of m2: 3.10 mm (*E. minor)* to 3.62 mm (*E. elegans*).

Included species: *E. elegans* (known from localities CP13E, [CP26A], [NP1C], NP3A–D*, NP4, NP19IIA, C, NP20D, *E. minor* (localities [CP15A], NP19C*, [NP3A0]).

Undescribed new species of *Elpidophorus* have been reported from localities NP3F and NP20E.

Plagiomene Matthew, 1918

Type species: *Plagiomene multicuspis* Matthew, 1918.

Type specimen: AMNH 15084, upper and lower jaws with damaged teeth.

Characteristics: Dental formula I?3/3, C 1/1, P4/4, M3/3; dentary relatively shallow and elongate; lower incisors with bilobed crowns, decreasing in size posteriad; canine only slightly bigger than i3 and p1; of the premolars, only P1 simple and single-rooted; p2-4 and P2-4 progressively more molariform (P4/p4 fully molariform); posterior premolars and molars with crenulated enamel; uppers with accessory stylar cusps; lowers with entoconulid, twinned hypoconulid; M3/m3 reduced; lingual cusps anteriorly skewed relative to buccal cusps, and in lower teeth higher than buccal cusps; lower cheek teeth exodaenodont, with strong ectocingulids. (Mostly from Rose, 1973.)

Average length of m2: 2.80 mm (*P. accola*) to 3.73 mm (*P. multicuspis*).

Included species: *P. multicuspis* (known from localities CP19B, CP20A*); *P. accola* (localities CP17B, CP18B*).

Ellesmene Dawson *et al.*, 1993

Type species: *Ellesmene eureka* Dawson *et al.*, 1993.

Type specimen: NMC 30860, right dentary fragment with p3–4-m1–2, and roots or alveoli for i1-3, c1, p1-2, m3.

Characteristics: Lower dental formula i3, c1, p4, m3, as in *Plagiomene*; upper dental formula probably similar, I?, C?, P4, M3; i1 lower incisor enlarged; front of the dentary foreshortened and incisors more crowded than in *Plagiomene*; p3 single- not double-rooted; P2 more molariform than in *Plagiomene*, with closely appressed paraconid–metaconid and a well-developed three-cusped talonid; p4 fully molariform and slightly larger than the first molar; molars decrease in size posteriorly; p4 and lower molars with the paraconid progressively smaller posteriorly and the entoconulid always present and increasing in size posteriorly; p3–m2 markedly exodaenodont; P3–M2 with four stylar cusps; M3 with two; cheek teeth with crenulated enamel. (Based on Dawson *et al.*, 1993.)

Average length of m2: 3.78 mm.

Included species: *E. eureka* only, known from locality HA1A only.

Planetetherium Simpson, 1928

Type species: *Planetetherium mirabile* Simpson, 1928.

Type specimen: AMNH 22162, left dentary with p3-4, m1.

Characteristics: Teeth about half the size of *Plagiomene* and generally lower crowned; dental formula unknown (only cheek teeth described); P3 smaller than p3, and with metacone much smaller than and closely appressed to paracone, and no conules; p3 submolariform but not as broad or as molarized as in *Worlandia*; P4/p4 fully molariform; third molars reduced; enamel crenulated; accessory stylar cusps and entoconulids present; lower cheek teeth exodaenodont. (Based on Simpson 1928, 1929; Szalay, 1969.)

Average length of m2: 2.28 mm.

Included species: *P. mirabile* only (known from localities CP17A*, [CP26C]).

Planetetherium new sp. is also known from locality CP14E.

Worlandia Bown and Rose, 1979

Type species: *Worlandia inusitata* Bown and Rose, 1979.

Type specimen: UM 68381, left dentary with p2–m1 and root of i1.

Characteristics: Small plagiomenid, with teeth less than half the size of those in *Plagiomene* and slightly smaller than those of *Planetetherium*; dentary relatively short and deep compared with that of *Plagiomene*; 10 lower teeth; dental formula interpreted as I?/2, C1/1, P4/4, M3/3; lower medial incisor hypertrophied and with bilobed crown; i2–p2 and C1–P2 all small, simple, and single-rooted; third and fourth premolars as large as or larger than molars; P3/p3 semi-molariform (more molarized than in *Planetetherium*) and P4/p4 molariform; upper and lower molars decrease in size posteriorly; cheek teeth low crowned, enamel crenulated on p3–m3 and P3-M3. (Based on Bown and Rose, 1979; Rose, 1981, 1982.)

Average length of m2: 1.63 mm.

Included species: *W. inusitata* only (known from localities [CP17B], CP18B).

***Tarka* McKenna, 1990**

Type species: *Tarka stylifera* McKenna, 1990.

Type specimen: AMNH 113133, left dentary with i1, c, p1, p3–m2.

Characteristics: Dentary short and deep relative to that of *Plagiomene*, with eight teeth; lower dental formula interpreted as i1, c1, p3, m3; i1 enlarged, procumbent, crown apparently not digitate; teeth interpreted as canine and p1 small, simple, and single-rooted; short diastema between p1 and p3, p2 presumably suppressed; p3 small and essentially premolariform; p4 with molariform trigonid and reduced talonid heel; lower molariform teeth with accessory cusps (labial stylar cusps, metastylids, and extra cusps between protoconid and metaconid); presumed P4 not molariform and smaller than M1 or M2; upper molars with strong conules and large stylar cusps; enamel crenulated. (Based on McKenna, 1990.)

Average length of m2: 5.2 mm.

Included species: *T. stylifera* only, known from locality CP36B only.

***Tarkadectes* McKenna, 1990**

Type species: *Tarkadectes montanensis* McKenna, 1990.

Type specimen: CMNH 40818, left dentary with p4–m2.

Characteristics: Generally similar to *Tarka* but smaller; lower cheek teeth not as wide and lacking large labial stylar cusps; p4 with molariform trigonid and reduced talonid; m1 cristid obliqua bearing distinct cusps; enamel highly crenulated. (Based on McKenna, 1990.)

Length of m2: 4.0 mm (holotype).

Included species: *T. montanensis* only, known from locality NP31 only.

? PLAGIOMENIDAE

***Thylacaelurus* Russell, 1954**

Type species: *Thylacaelurus montanus* Russell, 1954.

Type specimen: NMC 8910, right maxilla with P4–M2 (or P3–M1?).

Characteristics: Slightly smaller than *Planetetherium*; first upper molariform tooth (M1 according to Russell [1954] and Szalay [1969]; P4 according Storer [1984]) with small but distinct hypocone; upper molariform teeth with prominent conules, broad basin, and paired stylar cusps buccal to the paracone and metacone; anterior-most preserved upper tooth (Russell's P4) peculiarly hypertrophied, with a tall and very large paracone (twinned with metacone?), a low, anterolingually shifted protocone, and prominent, long buccal roots reminiscent of a canine root; lower molar trigonids anteroposteriorly compressed and oblique (anterolingually skewed), talonid basins broad, entoconid high, hypoconulid isolated behind postcristid; referred lower incisors (Krishtalka and Setoguchi, 1977; Storer, 1984) with digitate crowns having three to five lobes; but these could belong instead to nyctitheriids, which are found in the same deposits. (Based on Szalay, 1969; Krishtalka and Setoguchi, 1977; Storer, 1984).

Average length of m2: 1.76 mm (*T. campester*, m1 or m2).

Included species: *T. montanus* (known from localities [CP29C], CP98B, NP13*); *T. campester* (NP8, NP9A) (Storer, 1995).

Thylacaelurus sp. is also known from locality NP9B (Storer, 1996).

Comments: Russell (1954) initially believed *Thylacaelurus* was a didelphid marsupial, but subsequently it has usually been considered to be a plagiomenid, following Van Valen (1967). The genus remains very poorly known, represented principally by isolated teeth. The homologies of the teeth in the holotype are uncertain, particularly the enlarged and odd-shaped anterior-most tooth (presumably a posterior premolar). McKenna and Bell (1997) assigned the genus to an unassigned subfamily of Dermoptera, exclusive of Plagiomenidae. Fox (1990) reported an unidentified taxon, Cf. *Thylacaelurus*, from the early Tiffanian of Alberta (NP1C).

MIXODECTIDAE COPE 1883a

***Mixodectes* Cope, 1883b (including *Indrodon* Cope, 1883c; *Olbodotes* Osborn, 1902)**

Type species: *Mixodectes pungens* Cope, 1883c (including *M. crassiusculus* Cope, 1883b; *Olbodotes copei* Osborn, 1902).

Type specimen: AMNH 3081, left dentary with p3–m3, roots of i1–2, and alveolus of p2.

Characteristics: Lower dental formula i2, c0, p3, m3; p4 tall and premolariform; p3 also simple and much smaller; lower molars unlike *Eudaemonema* in having a low precingulid, more basined talonids, and much more buccal junction of the cristid obliqua with the back of the trigonid (postvallid); P4 premolariform; hypocone forms on posterior slope of (and only slightly separated from) the protocone.

Average length of m2: 3.48 mm (*M. malaris*) to 4.83 mm (*M. pungens*).

Included species: *M. pungens* (known from locality SB23H); *M. malaris* (including *Indrodon malaris* Cope, 1883a) (localities SB23E-H, CP14A).

Mixodectes sp. is also known from localities CP14A, ?CP26A, NP16C, Mixodectid nr. *Mixodectes* sp. indet. is known from locality SP23C, and mixodectid gen. and sp. indet. is known from locality CP1B.

***Dracontolestes* Gazin, 1941**

Type species: *Dracontolestes aphantus* Gazin, 1941.

Type specimen: USNM 16180, fragment of left dentary with m2 talonid and m3.

Characteristics: Very similar to *Mixodectes*, and approximately the size of *M. malaris*, but with somewhat narrower

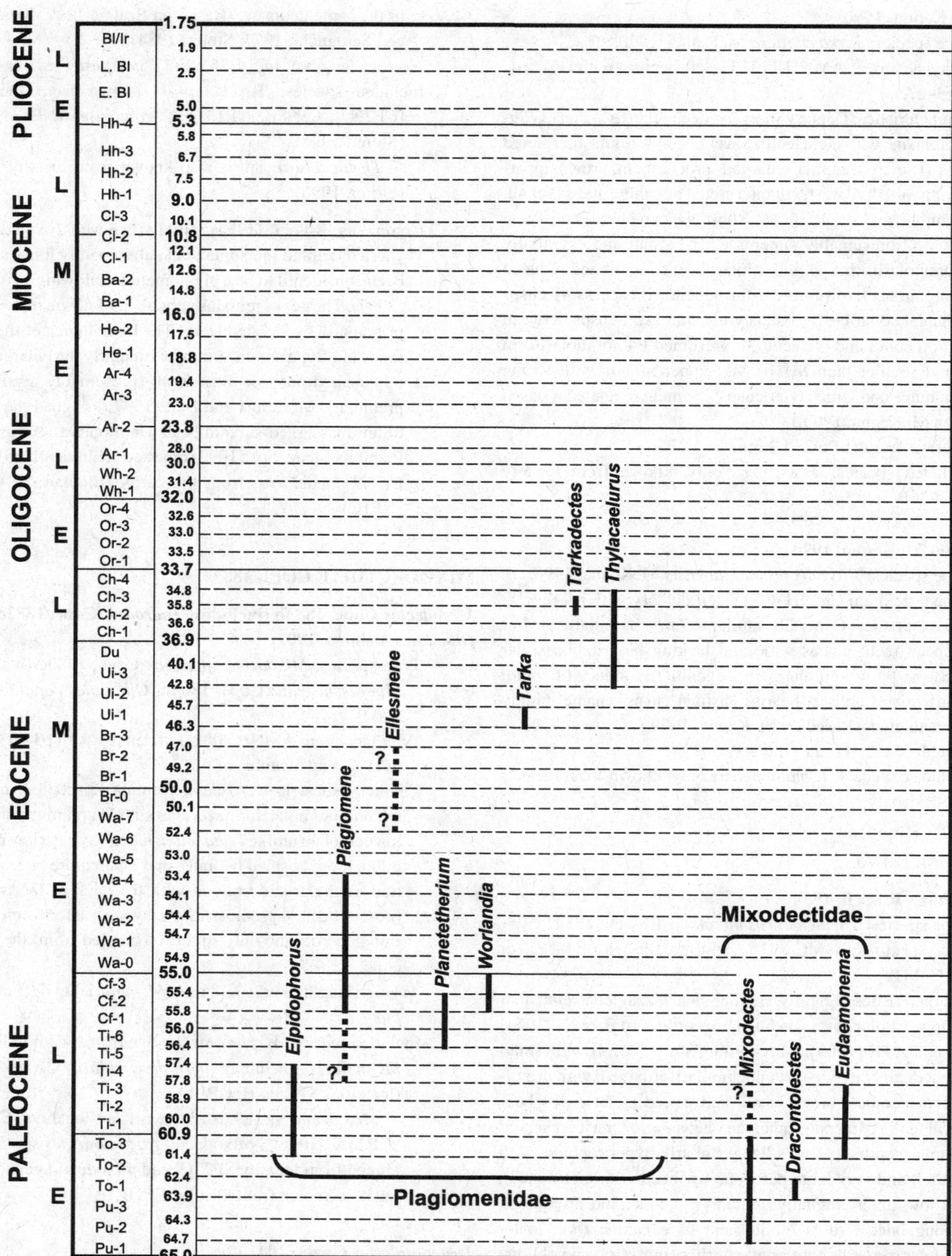

Figure 13.3. Temporal ranges of North American mixodectid and plagiomenid genera.

lower molars, shallower talonid notch (owing to presence of an entocristid), and more centrally placed hypoconulid on m3; as in *Mixodectes*, m2-3 (the only teeth known) with a precingulid and buccal cristid obliqua/postvallid contact.

Length of m2: 2.4 mm (holotype).

Included species: *D. aphantus* only, known from locality CP1C only.

***Eudaemonema* Simpson, 1935**

Type species: *Eudaemonema cuspidata* Simpson, 1935.

Type specimen: USNM 9314, left dentary with canine and p2–m3, and roots of i1-2 and p1.

Characteristics: Lower dental formula i2, c1, p4, m3; P3/p3 premolariform to semi-molariform; P4/p4 semi-molariform to molariform; lower molars with a small mesoconid but lacking precingulids; cristid obliqua crossing talonid toward or joining metaconid; upper molars with well-developed preparacrista, postmetacrista, and pre- and postcingula, and strong conules, the paraconule larger than the metaconule; hypocone large, rising from the posterolingual cingulum, and separated from the protocone by a deep notch.

Average length of m2: 3.33 mm.

Included species: *E. cuspidata* only (known from localities CP13B, C, CP15A, [NP1C], NP19C*).

Eudaemonema new spp. are also known from locality NP3B, C.

BIOLOGY AND EVOLUTIONARY PATTERNS

Because of the limited fossil evidence for plagiomenids, it is not possible to infer much about their paleobiology with confidence. Cheek teeth of the extant dermopteran *Cynocephalus* are superficially insectivoran-like, but crenulated enamel and heavy wear along shearing crests are consistent with their well-documented strictly herbivorous diet of leaves and flowers (Lim, 1967; Wischusen, 1990). The wrinkled enamel, molarized premolars, and similar pattern of shearing-wear facets in plagiomenids – whatever their phylogenetic relationships – suggests that they, too, were primarily herbivorous (Rose and Simons, 1977). Crenulated enamel and presence of multiple accessory cusps in *Tarka* and *Tarkadectes* led McKenna (1990) to postulate that these genera fed on fruit, nectar, and exudates, but most of the closest analogues are extinct forms whose diets are equally speculative. Microwear analysis might provide a more precise assessment (Figure 13.3).

If these inferred dietary preferences are correct, then plagiomenids were probably arboreal, a habitat seemingly corroborated by the occurrence of *Planetetherium* in coal beds at Bear Creek, Montana (locality CP17A). But the presence of a patagium for gliding, as has been portrayed in various artistic renderings of plagiomenids (presumably because of the putative relationship to extant Dermoptera), is purely hypothetical. Confirmation or rejection of either arboreal or gliding capabilities may soon be possible if plagiomenid postcrania can be identified in the collections from Ellesmere Island.

Mixodectids are less well known than plagiomenids, but the fragmentary skeletal remains, if properly referred, suggest relationship to plesiadapiforms and arboreal habits (Szalay and Lucas, 1996).

Elpidophorus, the most primitive known plagiomenid, possesses dental characters in common with both Mixodectidae and Plagiomenidae and supports a close relationship between them; but its more derived traits clearly link it with plagiomenids (Rose, 1975; McKenna, 1990).

REFERENCES

Beard, K. C. (1990). Gliding behaviour and palaeoecology of the alleged primate family Paromomyidae (Mammalia, Dermoptera). *Nature*, 345, 340–1.

Bloch, J. I. and Silcox. M. T. (2001). New basicrania of Paleocene–Eocene *Ignacius*: re-evaluation of the plesiadapiform–dermopteran link. *American Journal of Physical Anthropology*, 116, 184–98.

Bown, T. M. and Rose, K. D. (1979). *Mimoperadectes*, a new marsupial, and *Worlandia*, a new dermopteran, from the lower part of the Willwood Formation (early Eocene), Bighorn Basin, Wyoming. *Contributions from the Museum of Paleontology, University of Michigan*, 25, 89–104.

Carroll, R. L. (1988). *Vertebrate Paleontology and Evolution*. New York: Freeman.

Cope, E. D. (1883a). On the mutual relations of the bunotherian Mammalia. *Proceedings of the Academy of Natural Sciences, Philadelphia*, 35, 77–83.

(1883b). First addition to the fauna of the Puerco Eocene. *Proceedings of the American Philosophical Society*, 20, 545–62.

(1883c). Second addition to the knowledge of the Puerco epoch. *Paleontological Bulletin*, 37, 309–24.

Dawson, M. R., Hickey, L. J., Johnson, K., and Morrow, C. J. (1986). Discovery of a dermopteran skull from the Paleogene of Arctic Canada. *National Geographic Research*, 2, 112–15.

Dawson, M. R., McKenna, M. C., Beard, K. C., and Hutchison, J. H. (1993). An early Eocene plagiomenid mammal from Ellesmere and Axel Heiberg Islands, Arctic Canada. *Kaupia*, 3, 179–92.

Fox, R. C. (1990). The succession of Paleocene mammals in western Canada. [In *Dawn of the Age of Mammals in the Northern Part of the Rocky Mountain Interior, North America*, ed. T. M. Bown and K. D. Rose.] *Geological Society of America Special Paper*, 243, 51–70.

Gazin, C. L. (1941). The mammalian fauna of the Paleocene of central Utah, with notes on the geology. *Proceedings of the United States National Museum*, 91, 1–53.

Gunnell, G. F. (1989). Evolutionary history of Microsyopoidea (Mammalia, ?Primates) and the relationship between Plesiadapiformes and Primates. *University of Michigan Papers on Paleontology*, 27, 1–157.

Jepsen, G. L. (1962). Futures in retrospect. *Annual Report of the Peabody Museum of Natural History*, 3, 1–14.

(1970). Bat origins and evolution. In *Biology of Bats*, Vol. I, ed. W. A. Wimsatt, pp. 1–64. New York: Academic Press.

Kay, R. F., Thorington, R. W., Jr., and Houde, P. (1990). Eocene plesiadapiform shows affinities with flying lemurs not primates. *Nature*, 345, 342–4.

Krause, D. W. (1986). Competitive exclusion and taxonomic displacement in the fossil record: the case of rodents and multituberculates in North America. *Contributions to Geology, University of Wyoming, Special Paper*, 3, 95–117.

Krishtalka, L. and Setoguchi, T. (1977). Paleontology and geology of the Badwater Creek area, central Wyoming. Part 13. The late Eocene

Insectivora and Dermoptera. *Annals of the Carnegie Museum*, 46, 71–99.

Lim, B. L. (1967). Observations on the food habits and ecological habitat of the Malaysian flying lemur. *International Zoo Yearbook*, 7, 196–7.

MacPhee, R. D. E., Cartmill, M., and Rose, K. D. (1989). Craniodental morphology and relationships of the supposed Eocene dermopteran *Plagiomene* (Mammalia). *Journal of Vertebrate Paleontology*, 9, 329–49.

Matthew, W. D. (1918). A revision of the lower Eocene Wasatch and Wind River faunas. Part V. Insectivora (continued), Glires, Edentata. *Bulletin of the American Museum of Natural History*, 38, 565–657.

McKenna, M. C. (1980). Eocene paleolatitude, climate, and mammals of Ellesmere Island. *Palaeogeography, Palaeoclimatology, Palaeoecology*, 30, 349–62.

(1990). Plagiomenids (Mammalia:? Dermoptera) from the Oligocene of Oregon, Montana, and South Dakota, and middle Eocene of northwestern Wyoming. [In *Dawn of the Age of Mammals in the Northern Part of the Rocky Mountain Interior, North America*, ed. T. M. Bown and K. D. Rose.] *Geological Society of America Special Paper*, 243, 211–34.

McKenna, M. C. and Bell, S. K. (1997). *Classification of Mammals Above the Species Level*. New York: Columbia University Press.

Osborn, H. F. (1902). American Eocene primates, and the supposed rodent family Mixodectidae. *Bulletin of the American Museum of Natural History*, 16, 169–214.

Ostrander, G. E. (1987). The early Oligocene (Chadronian) Raben Ranch local fauna, northwest Nebraska: Marsupialia, Insectivora, Dermoptera, Chiroptera, and Primates. *Dakoterra, South Dakota School of Mines and Technology*, 3, 92–104.

Romer, A. S. (1966). *Vertebrate Paleontology*. Chicago, IL: University of Chicago Press.

Rose, K. D. (1973). The mandibular dentition of *Plagiomene* (Dermoptera, Plagiomenidae). *Breviora, Museum of Comparative Zoology*, 411, 1–17.

(1975). *Elpidophorus*, the earliest dermopteran (Dermoptera, Plagiomenidae). *Journal of Mammalogy* 56, 676–9.

(1981). The Clarkforkian Land-Mammal Age and mammalian faunal composition across the Paleocene–Eocene boundary. *University of Michigan Papers on Paleontology*, 26, 1–197.

(1982). Anterior dentition of the early Eocene plagiomenid dermopteran *Worlandia*. *Journal of Mammalogy*, 63, 179–83.

Rose, K. D. and Simons, E. L. (1977). Dental function in the Plagiomenidae: origin and relationships of the mammalian order Dermoptera. *Contributions from the Museum of Paleontology, University of Michigan*, 24, 221–36.

Russell, D. E., Louis, P., and Savage, D. E. (1973). Chiroptera and Dermoptera of the French early Eocene. *University of California Publications in Geological Sciences*, 95, 1–57.

Russell, L. S. (1954). Mammalian fauna of the Kishenehn Formation, southeastern British Columbia. [Annual Report of the National Museum of Canada for 1952–53.] *Bulletin of the National Museum of Canada*, 132, 92–111.

Silcox, M. T. (2001). A phylogenetic analysis of Plesiadapiformes and their relationship to Euprimates and other Archontans. Ph.D. Thesis, Johns Hopkins University, Baltimore.

Simpson, G. G. (1927). Mammalian fauna and correlation of the Paskapoo Formation of Alberta. *American Museum Novitates*, 268, 1–10.

(1928). A new mammalian fauna from the Fort Union of southern Montana. *American Museum Novitates*, 297, 1–15.

(1929). A collection of Paleocene mammals from Bear Creek, Montana. *Annals of the Carnegie Museum*, 19, 115–22.

(1935). New Paleocene mammals from the Fort Union of Montana. *Proceedings of the United States National Museum*, 83, 221–44.

(1936). A new fauna from the Fort Union of Montana. *American Museum Novitates*, 873, 1–27.

(1937). The Fort Union of the Crazy Mountain Field, Montana, and its mammalian faunas. *Bulletin of the United States National Museum*, 169, 1–277.

(1945). The principles of classification and a classification of the mammals. *Bulletin of the American Museum of Natural History*, 85, 1–350.

Sloan, R. E. (1969). Cretaceous and Paleocene terrestrial communities of western North America. *Proceedings of the North American Paleontological Convention*, part E, pp. 427–53.

Storer, J. E. (1984). Mammals of the Swift Current Creek local fauna (Eocene: Uintan, Saskatchewan). *Natural History Contributions from the Saskatchewan Museum of Natural History*, 7, 1–158.

(1995). Small mammals of the Lac Pelletier Lower Fauna, Duchesnean, of Saskatchewan, Canada: insectivores and insectivore-like groups, a plagiomenid, a microsyopid and Chiroptera. In *Vertebrate Fossils and the Evolution of Scientific Concepts*, ed. W. A. S. Sarjeant, pp. 595–615. Australia: Gordon and Breach.

(1996). Eocene–Oligocene faunas of the Cypress Hills Formation, Saskatchewan. In *The Terrestrial Eocene–Oligocene Transition in North America*, ed. D. R. Prothero and R. J. Emry, pp. 240–61. Cambridge, UK: Cambridge University Press.

Szalay, F. S. (1969). Mixodectidae, Microsyopidae, and the insectivore–primate transition. *Bulletin of the American Museum of Natural History*, 140, 193–330.

(1976). Systematics of the Omomyidae (Tarsiiformes, Primates): taxonomy, phylogeny, and adaptations. *Bulletin of the American Museum of Natural History*, 156, 157–450.

Szalay, F. S. and Lucas, S. G. (1996). The postcranial morphology of Paleocene *Chriacus* and *Mixodectes* and the phylogenetic relationships of archontan mammals. *Bulletin of the New Mexico Museum of Natural History and Science*, 7, 1–47.

Van Valen, L. (1967). New Paleocene insectivores and insectivore classification. *Bulletin of the American Museum of Natural History*, 135, 217–84.

West, R. M. and Dawson, M. R. (1978). Vertebrate paleontology and the Cenozoic history of the North Atlantic region. *Polarforschung*, 48, 103–19.

Winge, H. (1941). *The Interrelationships of the Mammalian Genera*, Vol. I. [English translation by E. Deichmann and G. M. Allen.] Copenhagen: C. A. Reitzels Forlag.

Wischusen, E. W. (1990). The foraging ecology and natural history of the Philippine flying lemur (*Cynocephalus volans*). Ph.D. Thesis, Cornell University, Ithaca.

Wortman, J. L. (1903). Classification of the Primates. *American Journal of Science*, 15, 399–414.

14 Plesiadapiformes

MARY T. SILCOX[1] and GREGG F. GUNNELL[2]

[1]*University of Winnipeg, MB, Canada*

[2]*Museum of Paleontology, University of Michigan, Ann Arbor, MI, USA*

INTRODUCTION

Plesiadapiforms (Simons and Tattersall in Simons, 1972) are a geographically widespread and temporally long-lived group of archaic mammals. They are found on all Holarctic continental land masses, first appearing near the Cretaceous–Tertiary boundary and surviving through the middle Eocene. The largest and most diverse radiation is found in North America, while a more moderate radiation of plesiadapiforms is found in Europe. Representatives of the group have only recently been uncovered in Asia (Beard and Wang, 1995; Tong and Wang, 1998; Thewissen, Williams, and Hussain, 2001; Fu, Wang, and Tong, 2002; Smith, Van Itterbeeck, and Missiaen, 2004; Beard *et al.*, 2005). Tabuce *et al.* (2004) have suggested that the African aziibids may also be plesiadapiforms.

Most plesiadapiforms are small bodied, with the largest taxa being comparable in size to an extant marmot (Gingerich, 1976; Gunnell, 1989). Plesiadapiforms include insectivorous, herbivorous, and omnivorous taxa, with most forms probably having quite diverse diets (Kay and Cartmill, 1977). Some taxa may have specialized on tree exudates or larvae and very soft fruits (Szalay, 1968; Rose and Bown, 1996; Silcox, Rose, and Walsh, 2002). Existing evidence indicates that the plesiadapiforms for which postcrania are known were arboreal but lacked the specializations for leaping seen in most early euprimates (Szalay and Decker, 1974; Szalay, Tattersall, and Decker, 1975; Szalay and Dagosto, 1980; Szalay and Drawhorn, 1980; Beard, 1989, 1993a; Gingerich and Gunnell, 1992; Boyer, Bloch, and Gingerich, 2001; Bloch and Boyer, 2002, 2003; Bloch, Boyer, and Houde, 2003). Hypotheses that some plesiadapiforms were gliders (Beard, 1989, 1990a, 1993a) have not been supported by recent analyses and discoveries (Krause, 1991; Runestad and Ruff, 1995; Hamrick, Rosenman, and Brush, 1999; Boyer, Bloch, and Gingerich, 2001; Bloch, Boyer, and Houde, 2003; see below).

Plesiadapiforms reach their highest diversity in the Paleocene and are reduced in diversity by the early Eocene with only a few taxa surviving into the middle Eocene (Maas, Krause, and Strait, 1988; McKenna and Bell, 1997; Silcox, Rose, and Walsh, 2002). Reduction in diversity and the ultimate extinction of plesidapiforms in North America was likely the result of competition, first with rodents appearing in the late Paleocene (Maas, Krause, and Strait, 1988) and then with euprimates appearing in the early Eocene (Gunnell, 1989).

Perhaps the most important recent discoveries of plesiadapiforms have come from the acid preparation of limestone blocks by workers at the University of Michigan, which has yielded a number of well-associated and exquisitely preserved partial skeletons of various plesiadapiforms (Gingerich and Gunnell, 1992; Bloch and Bowen, 2001; Bloch and Boyer, 2001, 2002, 2003, 2007; Bloch and Silcox, 2001, 2003; Boyer, Bloch and Gingerich, 2001; Bloch *et al.*, 2002, 2007; Bloch, Boyer, and Houde, 2003). Many of these new specimens remain unpublished, but the material that has been detailed in print has led to substantial changes to our understanding of the adaptive features of the group. As noted above, these new finds have failed to uphold the gliding hypothesis for paromomyids or micromomyids (Boyer, Bloch, and Gingerich, 2001; Bloch, Boyer, and Houde, 2003; Bloch *et al.*, 2007). The first reasonably complete skeleton of a carpolestid demonstrates a divergent, grasping hallux with a nail (Bloch and Boyer, 2002, 2003), a character complex thought previously to be limited to euprimates (e.g., Cartmill, 1972). Hand remains from various plesiadapiform taxa (with the exception of *Plesiadapis*) demonstrate grasping proportions, providing further evidence of adaptive similarities to euprimates (Bloch and Boyer, 2002, 2003). These finds strengthen the hypothesis that plesiadapiforms should be included in the order Primates (see discussion below) by providing adaptive, as well as phylogenetic, evidence in support of that placement.

Bloch and Boyer (2002) incorporated the new carpolestid postcranial material into a cladistic analysis. The results of this analysis are generally similar to those pictured in Figure 14.3, below (Silcox,

Evolution of Tertiary Mammals of North America, Vol. 2. ed. C. M. Janis, G. F. Gunnell, and M. D. Uhen. Published by Cambridge University Press.

2001), with the exception that carpolestids fell out as the sister taxon to euprimates. This finding does not gain support from the cranial data partition (Silcox, 2001; Bloch and Silcox, 2003, 2006), or from an analysis with more complete anatomical sampling that includes these new finds (Bloch, Silcox, and Sargis, 2002; Bloch and Boyer, 2003; Bloch *et al.*, 2007).

The treatment of plesiadapiforms as a group separated from euprimates here does not indicate support for the formal taxonomic recognition of this group, or for monophyly of this cluster. Rather, it is done as a product of convenience and recognition that the members of this probably paraphyletic cluster of forms do share certain similarities (many of which are likely to be symplesiomorphies).

DEFINING FEATURES OF THE SUBORDER PLESIADAPIFORMES

CRANIAL

Relatively few plesiadapiform taxa are represented by cranial material. In general, the cranium is primitive, being relatively low with a small brain, having a strong postorbital constriction, and lacking a postorbital bar or evidence of convergent orbits (Figure 14.1). Sagittal crests are often developed and zygomatic arches are relatively broad and widely splayed for strong chewing musculature. The rostral portion and the neurocranium are approximately equivalent in length, implying a fairly long snout. The endocasts known for plesiadapiforms exhibit relatively large olfactory lobes (Gingerich and Gunnell, 2005). Known taxa either lack evidence of an ossified auditory bulla (McKenna, 1966; but see MacPhee, Cartmill, and Gingerich, 1983; MacPhee, Novacek, and Storch, 1988; Silcox, 2001) in primitive forms or have well-developed auditory bullae. In some cases these auditory bullae are entotympanic in origin (Kay, Thorington, and Houde, 1990; Kay, Thewissen, and Yoder, 1992; Bloch and Silcox, 2001; Silcox, 2003), in others they are continuous with the petrosal and, therefore, may have been formed by this bone (Russell, 1959, 1964; Szalay, Rosenberger, and Dagosto 1987; Silcox, 2001; Bloch and Silcox, 2003, 2006; but see MacPhee, Cartmill, and Gingerich, 1983; MacPhee and Cartmill, 1986; Beard and MacPhee, 1994).

DENTAL

All plesiadapiforms share enlarged, procumbent pairs of upper and lower central incisors. Lower central incisors tend to be relatively simple with few accessory cusps being developed. Upper incisors are more complex, often with three or more main cuspules developed (see Rose, Beard, and Houde, 1993; Figure 14.3, below). Except for the most primitive species, many taxa have reduced numbers of teeth between the central incisors and the distal two premolars, with long diastemata often present. If these teeth are present, then they are usually reduced in size and may not occlude with their counterparts.

Many plesiadapiforms have enlarged and specialized premolars, especially the upper and lower fourth premolars. These teeth are often larger than m1 and exhibit cusp proliferation and/or crown distention and exodaenodonty in many taxa. Lower molars tend to retain paraconids on all molars and have low, flat, well-developed talonid basins. The m3 hypoconulid is almost always larger than the corresponding cusp on m1 (a feature shared with euprimates), and in many cases becomes substantially expanded and fissured. The upper forth premolar has a postprotocingulum (a feature shared with euprimates). Upper molars vary from being quite primitive in appearance, to distinctly modified with postprotocingula associated with enlarged distolingual basins, mesostyles, and/or profound cusp reduction. Frequently a hypocone is present that is either cingular in development or develops off a postprotocingulum (Figure 14.2).

POSTCRANIAL

As with the cranium, few plesiadapiforms are well known postcranially, and fewer still are known from certainly associated (or published) postcranial and dental remains. Of those that are represented, plesiadapiforms appear generally to have been arboreal quadrupeds, with grasping features of the manus (except *Plesiadapis*) and with claws on most digits that made them capable of locomoting on large diameter vertical supports. Some taxa also show features for suspension under small branches (e.g., micromomyids; Bloch, Boyer, and Houde, 2003; see Bloch and Boyer [2007] for a more detailed overview).

SYSTEMATICS

The phylogenetic history and taxonomic placement of plesiadapiforms is complex and contentious. When originally described, *Plesiadapis* was favorably compared with the European adapiform primate *Adapis* (although as it turns out the "*Plesiadapis*" specimens actually compared with *Adapis* by Gervais [1877] were later placed in a different genus, *Protoadapis* [see Gingerich, 1976]). Gidley (1923), and Simpson (1935a, b) noted dental similarities between plesiadapiforms and euprimates. Inclusion of Plesiadapiformes as a suborder or infraorder in Primates was subsequently reaffirmed by Russell (1959), Simons (1972), Szalay (1973), Gingerich (1976), and Van Valen (1994), among others. Martin (1972), Cartmill (1972), and Wible and Covert (1987) questioned the inclusion of plesiadapiforms within Primates. Gunnell (1989) reviewed much of the evidence available at the time and concluded that plesiadapiforms could not be included within Primates nor could they be viewed as ancestral to euprimates. Gunnell (1989) retained plesiadapiforms in ?Primates pending discovery of more fossil material that could lead to a more complete resolution of their true relationships, and suggested that plesiadapiforms shared their closest relationships with Plagiomenidae, a family at that time considered to be dermopterans (but see MacPhee, Cartmill, and Rose, 1989).

Gingerich (1989) proposed a new, gradistically defined order, Proprimates, for all plesiadapiforms, a concept rejected by Beard (1990a; see also Gingerich, 1990), who suggested (1990a,b, 1991a,

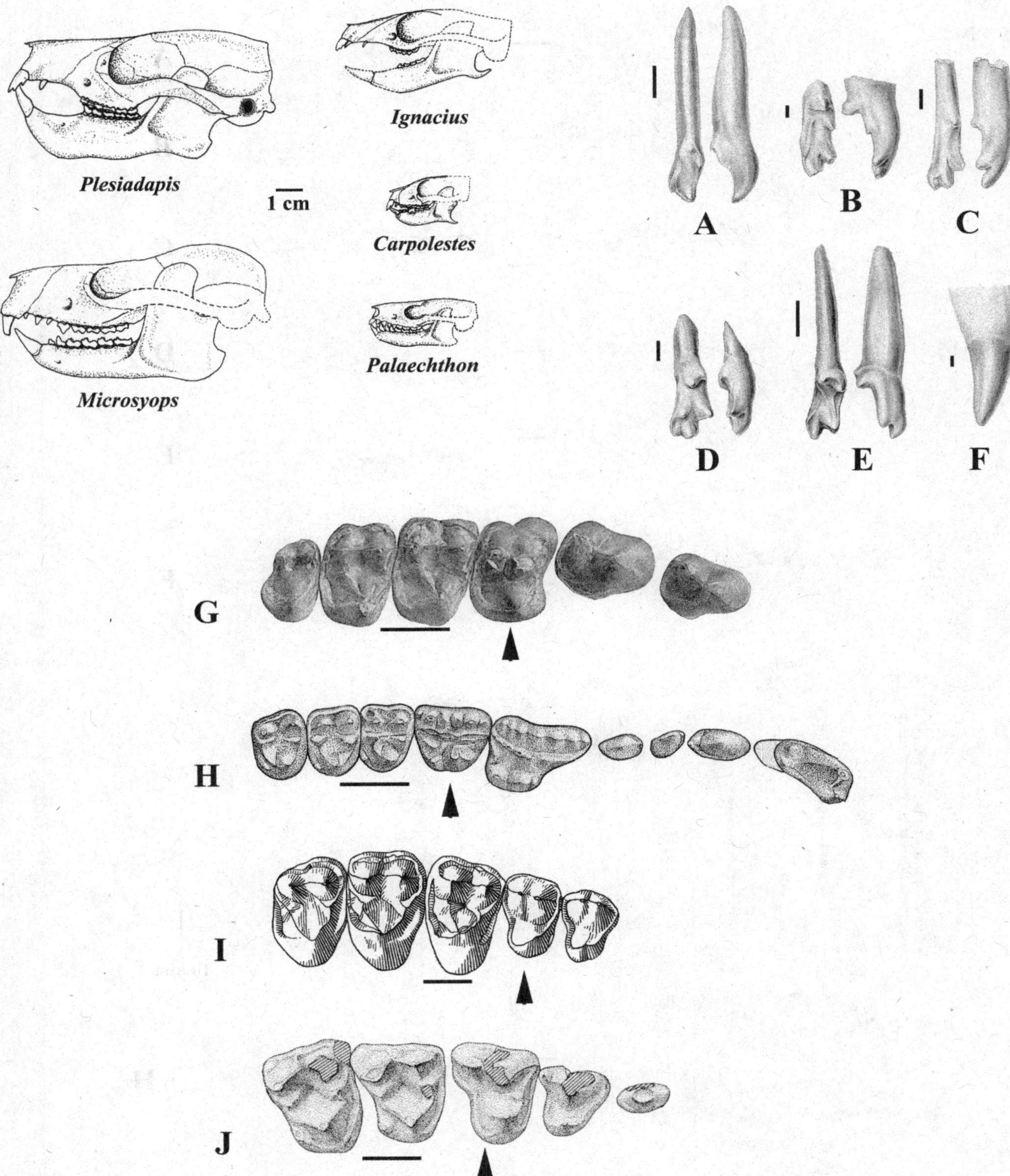

Figure 14.1. Representative skulls of plesiadapiforms (from Fleagle, 1988). A-F. Representative upper central incisors of plesiadapiforms (adapted from Rose, Beard, and Houde, 1993, scale bars = 1 mm): A. Micromomyid *Tinimomys graybulliensis*. B. Plesiadapid *Nannodectes gidleyi*. C. Paromomyid *Phenacolemur simonsi*. D. Saxonellid *Saxonella crepaturae*. E. Carpolestid *Carpodaptes cygneus*. F. Microsyopid *Megadelphus lundeliusi*. G-J. Representative right upper dentitions of plesiadapiforms (Scale bars = 2 mm). G. Paromomyid *Phenacolemur pagei* (reversed, from Simpson, 1955). H. Carpolestid *Carpolestes simpsoni* (from Bloch and Gingerich, 1998). I. Plesiadapid *Plesiadapis anceps* (from Simpson, 1936). J. Picrosyopid *Microsyops latidens* (from Gunnell, 1989).

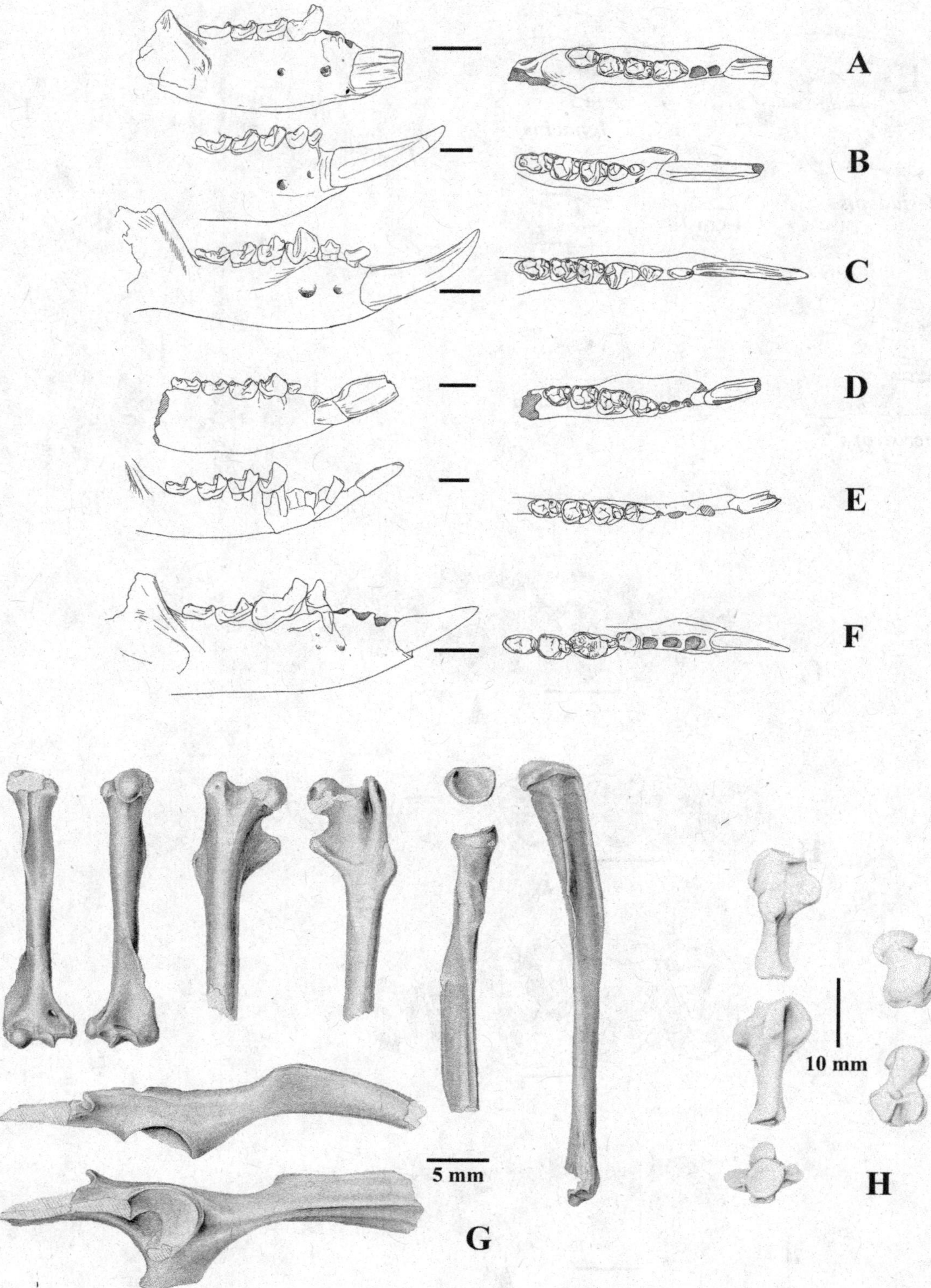

Figure 14.2. A-F. Representative right lower dentitions of plesiadapiforms (adapted from Rose and Bown, 1996, scale bars = 2 mm). A. Picromomyid *Picromomys petersonorum*. B. Paromomyid *Ignacius fremontensis*. C. Micromomyid *Tinimomys graybulliensis*. D. Uintasoricine microsyopid *Niptomomys doreenae*. E. Navajoviine microsyopid *Navajovius kohlhaasae*. F. Picrodontid *Picrodus silberlingi*. G. Representative postcranial elements of plesiadapiforms: includes right humerus of paromomyid *Phenacolemur simonsi*, right proximal femur of *P. simonsi*, right proximal radius of paromomyid *Ignacius graybullianus*, right tibia *P. simonsi*, and right pelvis *I. graybullianus* (adapted from Beard, 1991a). H. Left calcaneum (dorsal, plantar, and distal views) and right astragalus (dorsal and plantar views) of plesiadapid *Plesiadapis tricuspidens* (from Gunnell, 1989).

1993a,b) that Plesiadapiformes are not a monophyletic group (Gingerich [1989] explicitly rejected plesiadapiform monophyly as well) and that some traditionally recognized plesiadapiform taxa are more closely related to modern dermopterans than to any other taxon. Beard (1993a,b) proposed a new classification wherein most plesiadapiforms examined were included as a suborder within Dermoptera while others were included in a new suborder, Eudermoptera, with extant colugos. This reconstituted Order Dermoptera is united with Primates in a new Mirorder Primatomorpha. The arrangement endorsed by McKenna and Bell (1997) was fundamentally compatible with Beard's view – they placed all of the taxa usually grouped in Plesiadapiformes and Dermoptera in Primates, an arrangement since adopted by Beard (Beard *et al.*, 2005). McKenna and Bell (1997) did not include any concept of a united Plesiadapiformes, however, and actually placed some traditional plesiadapiforms (carpolestids) in Euprimates as a member of the parvorder Tarsiiformes. This latter attribution has not received support in subsequent analyses (Beard, 2000; Stafford and Szalay, 2000; Bloch *et al.*, 2001, 2002, 2007; Silcox, 2001; Silcox *et al.*, 2001; Bloch and Boyer, 2002, 2003; Bloch and Silcox, 2006).

Silcox (2001) undertook the most extensive phylogenetic analysis of plesiadapiforms to date. Unlike other cladistic analyses of the group (e.g., Gunnell, 1989; Kay, Thewissen, and Yoder, 1992; Beard, 1993b; Bloch and Boyer, 2002; Bloch and Silcox, 2003), she included members of all 11 well-established families of plesiadapiforms (excluding the more questionably associated aziibids) and sampled characters from the dental, cranial, and postcranial data partitions. The taxonomy used here is modified from Silcox's findings, and the summary tree presented as Figure 14.3 derives from her analysis. The results of Silcox's total evidence analysis failed to uphold the Primatomorpha or Eudermoptera hypotheses. Instead, dermopterans fell out in a clade with chiropterans (Volitantia), which was the sister taxon to Scandentia. Plesiadapiforms did not form a monophyletic group but rather were arranged as a series of stem taxa leading to euprimates. Carpolestids did not form an exclusive relationship with the included tarsiiform euprimates, failing to support McKenna and Bell's (1997) proposed taxonomy, but were part of a clade (Plesiadapoidea) with saxonellids and plesiadapids.

The hypothesis that Scandentia is the sister taxon to Euprimates to the exclusion of plesiadapiforms (e.g., Wible and Covert, 1987; Kay, Thewissen, and Yoder, 1992) was not supported by Silcox (2001), and subsequent discoveries have further weakened support for this idea (Silcox, 2003). Sargis' (2002) work on the previously poorly known scandentian *Ptilocercus lowii* found that it demonstrated many features thought to be primatomorphan synapomorphies, removing much of the support for an exclusive plesiadapiform–dermopteran–euprimate relationship.

Silcox (2001, 2007) has argued that the implication of this result is that the only viable taxonomic positions for plesiadapiforms are as members of the order Primates. This is based first on the apparently paraphyletic nature of the group Plesiadapiformes – modern taxonomic practice frowns on the use of paraphyletic grouping, and classifying all of these taxa together would obfuscate the important point that some plesiadapiforms share a more recent common ancestor with euprimates than they do with other plesiadapiforms. Second, other hypothesized taxonomic positions (e.g., in Dermoptera or Insectivora) are not supported by the phylogenetic results. Finally, as the taxa key to an understanding of euprimate origins, it seems important to position plesiadapiforms in the order Primates to appropriately focus studies of the early evolution of the group.

The superfamilial divisions used here (Paromomyoidea and Plesiadapoidea) follow Silcox (2001). A competing hypothesis would recognize a group including microsyopids and some paromomyoids (palaechthonids) as the Microsyopoidea (e.g., Gunnell, 1989; Rose, Beard, and Houde, 1993).

INCLUDED GENERA OF PLESIADAPIFORMES

The locality numbers listed for each genus refer to the list of unified localities in Appendix I. The locality numbers may be listed in several ways. The acronyms for museum collections are listed in Appendix III.

Parentheses around the locality (e.g., [CP101]) mean that the taxon in question at that locality is cited as an "aff." or "cf." the taxon in question. Parentheses are usually used for individual species, thus implying that the genus is firmly known from the locality, but the actual species identification may be questionable. Question marks in front of the locality (e.g., ?CP101) mean that the taxon is questionably known from that locality, thus implying some doubt that the taxon is actually present at that locality, either at the genus or species level. An asterisk (*) indicates the type locality.

PURGATORIIDAE (VAN VALEN AND SLOAN, 1965) (= PURGATORIINAE)

Characteristics: Very primitive plesiadapiforms with lower dental formula of i3, c1, p4, m3; double-rooted p2, large canine; p4 smaller than m1 and premolariform, with no metaconid; lower molar trigonids tall, weakly inclined mesially; m3 hypoconulids enlarged relative to m1 and m2, but not lobate; upper molars with weak postprotocingulum (at most) (modified from Gunnell, 1989; Buckley, 1997). There is no cranial or postcranial material known for this family.

Comments: Purgatoriidae is included here as a distinct, monotypic family, following Gunnell (1989) and McKenna and Bell (1997), because of its very primitive nature and lack of demonstrated ties to any particular other family. As a plausible ancestor not only for several families of plesiadapiforms, but also for Euprimates (Rose, 1995), this taxon is inherently paraphyletic and was found to be such by Silcox (2001). Accordingly, the family's characteristics are either primitive (e.g., unreduced lower dentition) or shared with more derived forms not included in the family (e.g., mesially inclined trigonids). Van Valen (1994) expanded the

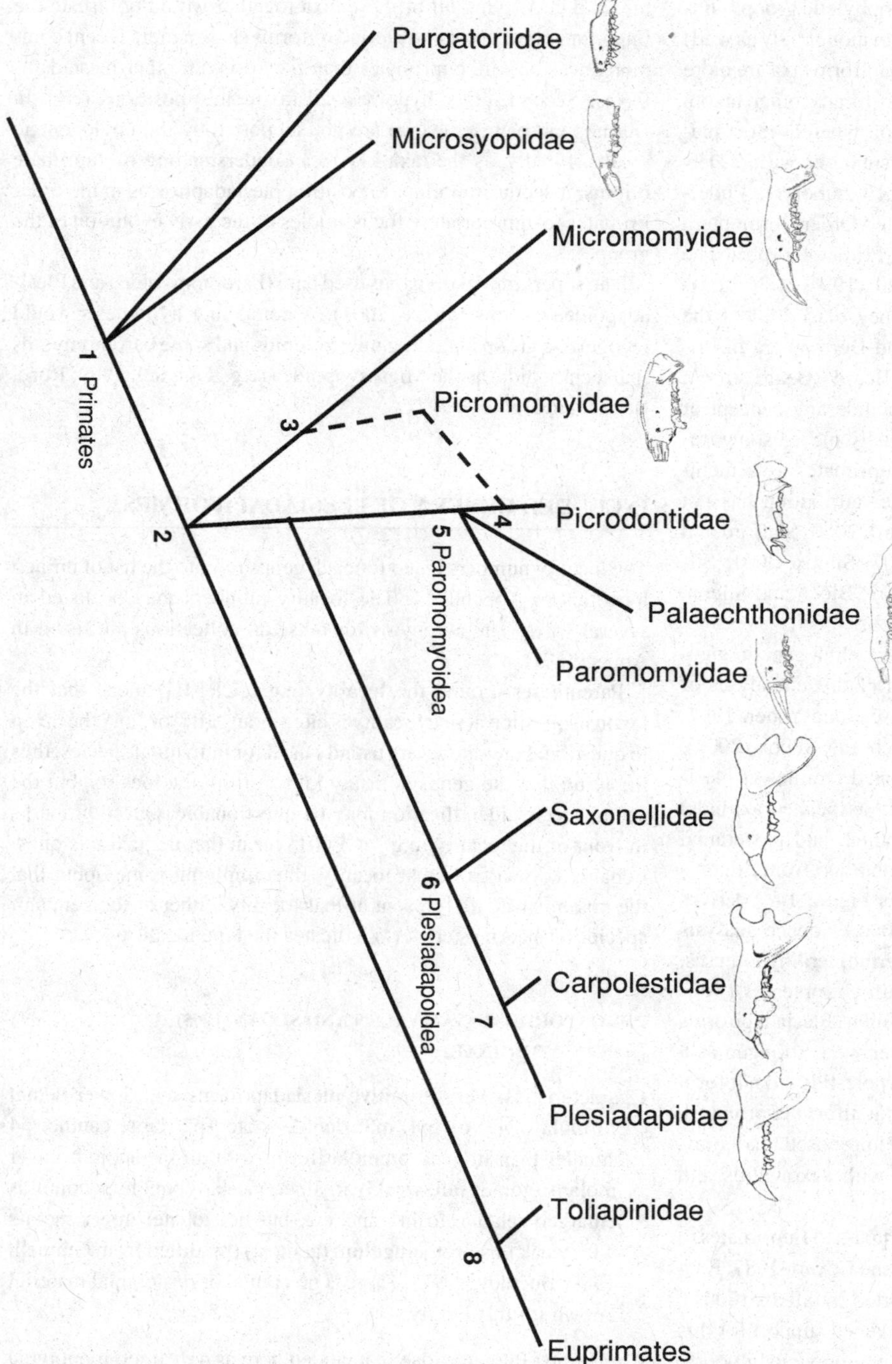

Figure 14.3. Proposed interrelationships among plesiadapiforms. Pattern of relationships is derived from Silcox (2001). Buccal outlines of jaws are modified from Clemens (1974), Fleagle (1988), Gunnell (1989), and Rose and Bown (1996), and are scaled to approximately the same m2 length. Key to characters at nodes on the cladogram (modified from Silcox, 2001; Silcox, Rose, and Walsh, 2002; Bloch and Boyer, 2003): 1. PRIMATES Linnaeus, 1758- postprotocingulum present on p4, m3 hypoconulids enlarged relative to m1–2. 2. Euprimate-like manual grasping, breadth of m1–2 talonids wider than trigonids, narrow central stem of the skull, hypotympanic sinus extensive, fattened area on medial promontorium, contact between lacrimal and palatine in the orbit obscured by maxillofacial contact. 3. Very small size, dorsolateral crest in the root region of i1 becomes re-oriented to lie on the medial side of the crown, p4 larger than m1. 4. Lower canine absent, shallow hypoflexid on p4, crest between protoconid and cristid obliqua absent on p4, molar hypoflexids indistinct, molars very low crowned with very strong mesial inflection of the lower molar trigonids. 5. PAROMOMYOIDEA (Schwartz, 1986) I1 with a posterocone, developing divisions at the tip; M1–2 with pronounced postprotocingulum and large posterointernal basin, protocone skewed mesiodistally; m3 talonid similar in breadth to the talonid or wider. 6. PLESIADAPOIDEA Trouessart, 1897 – unbasined p4 talonid with no entocristid, m1 trigonid swollen at the base, large conule on P4, non-projecting parastylar lobe on P4, protocone on M1–2 that is placed centrally on the tooth, intermediately sized greater tuberosity (note that excluding *Pandemonium* and the Asian taxon *Chronolestes* from this clade allows the addition of the following traits: multicuspate I1, P3 with a conule). 7. i1 broad at the base with a margoconid, m1 with a stepped postvallid (cristid obliqua continues onto metaconid, which is distally offset from protoconid). 8. P4 parastylar lobe small and not projecting, molar hypoconulids shifted buccally, molars moderately low crowned, straight preparacrista on M1–2 (note that the European family Toliapinidae here includes only *Avenius* and *Toliapina*, both European taxa).

Purgatoriidae to include taxa that are here placed in the Micromomyidae and Palaechthonidae. He claimed that these forms "lack sufficient adaptive divergence from the ancestral plesiadapiform to be separated at the family level" (Van Valen, 1994, p. 14), indicating that no derived features actually link the three clusters of genera that he included in the Purgatoriidae. Given that existing hypotheses would place these latter taxa in monophyletic groupings with more derived plesiadapiforms (e.g., Gunnell, 1989; Beard, 1993a; Silcox, 2001), Van Valen's arrangement seems unwarranted.

Purgatorius Van Valen and Sloan, 1965

Type species: *Purgatorius unio* Van Valen and Sloan, 1965.

Type specimen: University of Minnesota V. P. 1597, Puercan of Montana.

Characteristics: as for the family.

Average length of m2: 1.98 mm.

Included species: *P. unio* (including *P. titusi* Buckley, 1997) (known from locality NP17, NP16C*); *P. ceratops* Van Valen and Sloan, 1965 (locality NP15C*); *P. janisae* Van Valen, 1994 (locality NP16B*); *P.* sp. nov. (locality NP7B*).

Purgatorius sp. is also known from locality (NP20E).

Comments: Plesiadapiform specimens from the Garbani Local Fauna (locality NP16B) have generally (e.g., Clemens, 1974; Kielan-Jaworowska, Bown, and Lillegraven, 1979) been considered to be referable to the type species, *P. unio*. Van Valen (1994) named a new taxon for this material, *P. janisae*, restricting *P. unio* to the original Purgatory Hill (locality NP16) sample, an attribution followed recently by Clemens (2004). In naming "*P. titusi*," Buckley (1997) focused his comparisons on the *Purgatorius* material from Garbani. With the recognition that this material represents a species distinct from *P. unio*, it is necessary to reconsider the validity of "*P. titusi*" in the context of the original type collection of *P. unio*. Silcox (2001) considered this question in detail and found that the only potentially systematically informative difference between "*P. titusi*" and *P. unio* was the presence of a paraconid on p4 in some (but not all) specimens of the former. In light of the variable expression of this cusp, and the small sample size known for *P. unio* sensu stricto, she opined that "*P. titusi*" should be considered a junior synonym of *P. unio*, an arrangement that we follow here (see also Clemens, 2004).

Van Valen (1994) also placed the specimens described by Johnston and Fox (1984) from the Ravenscrag Formation of Saskatchewan (locality NP4) into this new taxon. Although Van Valen's distinction of the Garbani and Purgatory Hill specimens is followed here, his attribution of the Ravenscrag material to *P. janisae* does not seem to be supported by the preserved morphology (as Van Valen himself notes in the accompanying discussion; 1994, p. 18); this sample likely pertains to a new species. All of the specimens that have been identified as *Purgatorius* are from the Puercan, with the exception of a referred specimen (an isolated m3) from the Olive locality of Montana (locality NP20E) (Wolberg, 1978, 1979) that is late Tiffanian in age. Given this very recent age, the fact that the specimen is smaller than known species of *Purgatorius*, and the high variability known for m3s (e.g., Gingerich, 1974a), this identification needs to be substantiated with further evidence before the implied range extension for *Purgatorius* can be confidently accepted.

MICROSYOPIDAE OSBORN AND WORTMAN, 1892

Characteristics: Enlarged, bladelike, fully lanceolate lower central incisors that are rotated mesially relative to the condition in comparable forms (e.g., *Plesiolestes*); upper and lower first premolars lacking; lower molars with closely approximated entoconid–hypoconulids; small mesoconid often developed; upper molars relatively simple, either lacking hypocones or having a small cusp developed off the postcingulum; small mesostyles present in more derived genera; lower p2 reduced, single-rooted, and separated from i1 (or canine, where present) and p3 by diastemata; upper canines double-rooted. Cranium with strong sagittal crest; heavy, broadly splayed zygomatic arches; strong postorbital constriction; postorbital process of the frontal but no complete postorbital bar; no evidence of an ossified bulla (if a bulla was present it was not formed from the petrosal; MacPhee, Novacek, and Storch, 1988); grooves for both promontorial and stapedial arteries evident on the promontorium (several unpublished specimens of petrosals are known, and there is no indication of bony tubes around these vessels). Postcrania associated with dental remains were first discussed for the family by Beard (1991b), but have not yet been formally described.

NAVAJOVIINAE (SZALAY AND DELSON, 1979)

Characteristics: p3 double-rooted, smaller than p4 and subequal in size to p2; lower molars with distinct, cuspate paraconids, acute cusps, and relatively deep but narrow talonid basins; M3/m3 reduced; upper canine transversely narrow and double-rooted, as is P3; P4 lacks metacone but may have distinct metastylar cusp; M1-2 with distinct hypocones formed on the postcingulum, relatively deep trigon basins, and lacking mesostyles (modified from Szalay and Delson, 1979; Gunnell, 1989).

Comments: The only North American taxon included in Navajoviinae is *Navajovius* itself, but the European genus *Berruvius* may belong in this subfamily as well (Gunnell, 1989; Silcox, 2001; but see Hooker, Russell, and Phélizon, 1999).

Navajovius Matthew and Granger, 1921

Type species: *Navajovius kohlhaasae* Matthew and Granger, 1921.

Type specimen: AMNH 17390, late Paleocene of southwestern Colorado.

Characteristics: Small taxon; lower central incisor lanceolate; p2–3 equal in size; p3 double-rooted and smaller than p4; small, distinct paraconids present on all lower molars; M3/m3 reduced; P4 metacone absent; upper molars with distinct hypocones; upper canine and P3 double-rooted and buccolingually compressed.

Average length of m1: 1.5 mm.

Included species: *N. kohlhaasae* (known from localities SB20A*, SB39B, C, CP13IIA, CP21B); *N.? mckennai* Szalay, 1969a (locality SB24*).

Navajovius sp. also may be represented at localities NP20D, E.

Comments: This is the most primitive North American microsyopid known, with marked similarities to palaechthonids and to the French *Berruvius*. *Navajovius* is a small microsyopid with an estimated body mass of 80 g. The genus is very rare, and no specimens more complete than the original type material have ever been described.

UINTASORICINAE SZALAY, 1969b

Characteristics: Small microsyopids; p3 either single-rooted or with very mesiodistally compressed double roots and much smaller than p4; lower molars with indistinct paraconids (especially m2–3), bulbous cusps, and flattened, shallow talonid basins; M3/m3 reduced; P4 lacks metacone; upper molars lack hypocones and have relatively wide and shallow trigon basins (modified form Szalay, 1969b; Gunnell, 1989).

Uintasorex Matthew, 1909

Type species: *Uintasorex parvulus* Matthew, 1909.

Type specimen: AMNH 12052, Bridgerian of southwestern Wyoming.

Characteristics: Very small taxon; lower central incisor slender and lanceolate; lower canine absent; lower molars with relatively narrow trigonids and broader talonids; m2–3 with indistinct or absent paraconids; p4 with small paraconid; upper molars lacking hypocones.

Average length of m1: 0.9–1.2 mm.

Included species: *U. parvulus* (known from localities [CP27E], CP31E, CP34C, D*); *U. montezumicus* Lillegraven, 1976 (localities CC4*, [CC9A]).

Uintasorex sp. also is known from localities CC1, CC6, CC7B, CC8, CP27D, NP9A, B. An indeterminate uintasoricine is present at CC8IIB.

Comments: *Uintasorex* is the smallest taxon included within Microsyopidae, with a body mass estimate of approximately 20 g. It is a relatively common taxon in the later Bridgerian of Wyoming and the Uintan of California but is otherwise very rare (Szalay, 1969b).

Niptomomys McKenna, 1960

Type species: *Niptomomys doreenae* McKenna, 1960.

Type specimen: UCMP 44081, Wasatchian of northern Colorado.

Characteristics: Lower central incisors lanceolate; upper and lower fourth premolars enlarged; p4 lacking a paraconid; lower molar talonid and upper molar trigon basins relatively large and deep; upper molar conules buccally placed.

Average length of m1: 1.2–1.5 mm.

Included species: *N. doreenae* (known from localities SB24, CP17B, CP18B, CP19A, [CP19AA], CP20A–C, CP63*); *N. favorum* Strait, 2001 (CP20AA*); *N. thelmae* Gunnell and Gingerich, 1981 (CP20C*).

Comments: This genus shows some notable similarities to the European, Ypresian, microsyopoid *Avenius*, which Russell, Phélizon, and Louis (1992) used to support the presence of a land bridge between Europe and North America in the early Eocene. Their hypothesis requires two immigration events between North America and Europe – the first of a *Navajovius*-like form that gave rise to *Berruvius*, and the second of a *Niptomomys* relative that produced *Avenius*. Hooker, Russell, and Phélizon (1999) considered *Avenius* more comparable to another European taxon, *Toliapina*. Their phylogenetic hypothesis requires only a single immigration event with a *Navajovius*-like form giving rise to an exclusively Old World clade, the Toliapinidae, including all European microsyopoids. If this view is correct, then the similarities between *Avenius* and *Niptomomys* must be parallelisms. Silcox's (2001, 2007) findings suggest another scenario. She found that *Berruvius* fell out as the basal-most microsyopid, while *Avenius* and *Toliapina* formed a clade (Toliapinidae sensu lato) that was the sister taxon to euprimates. This hypothesis would imply that the similarities between *Avenius* and *Niptomomys* are convergences. In any case, the biogeographic implications of these similarities remain unclear.

A serially associated upper central incisor is known from an unpublished cranium of *Niptomomys*. This specimen exhibits a posterocone, as in most other plesiadapiforms (Silcox and Bloch, 2006), but contrasting with microsyopine I1s (Rose, Beard, and Houde, 1993).

MICROSYOPINAE OSBORN AND WORTMAN, 1892

Characteristics: Microsyopines differ from navajoviines and uintasoricines most notably in being much larger; lower central incisors very robust and enlarged; M3/m3 unreduced. Microsyopines differ further from navajoviines in having P3 with a distinct protocone lobe and upper molars with more centrally placed protocones. Microsyopines differ further from uintasoricines in having more acute cusps and crests, paraconids on all lower molars, and in having deeper and more restricted trigon and talonid basins (modified from Gunnell, 1985, 1989).

Arctodontomys Gunnell, 1985 (including *Microsyops*, in part; *Pantolestes*, in part; *Diacodexis*, in part)

Type species: *Arctodontomys simplicidens* (Rose, 1981) (originally described as *Microsyops simplicidens*).

Type specimen: UM 67214, Clarkforkian of northwestern Wyoming.

Characteristics: Lower central incisors lanceolate; p3 smaller than p2 and p4; p3 single-rooted where known; p4 lacking metaconid and having a weakly developed talonid basin; upper molars lack mesostyles; P4 lacking metacone; all teeth with relatively acute cusps.

Average length of m1: 2.7–3.5 mm.

Included species: *A. simplicidens* (known from localities CP14E, CP17B*, CP18B, and possibly CP62B); *A. nuptus* (Cope, 1882; originally described as *Pantolestes nuptus*; including *Diacodexis nuptus*, Gazin, 1952) (locality CP20A*); *A. wilsoni* (Szalay, 1969a; originally described as *Microsyops wilsoni*; including *M. wilsoni* Delson, 1971, Lucas and Froehlich, 1989) (localities SB24, CP19A, CP20A*, CP25B, CP63, CP64A).

A small species of *Arctodontomys* is also present at locality CP14E.

Comments: *Arctodontomys* can be distinguished from *Microsyops* by its relatively simple teeth, especially upper and lower fourth premolars, and by its reduced p3. It is otherwise very similar to *Microsyops* and is almost certainly ancestral to this later-occurring taxon.

Microsyops Leidy, 1872 (including *Bathrodon*; *Chriacus*, in part; *Cynodontomys*, in part; *Limnotherium*; *Notharctus*, in part; *Palaeacodon*; *Pelycodus*, in part)

Type species: *Microsyops elegans* (Marsh, 1871) (originally described as *Limnotherium elegans*).

Type specimen: YPM 11794, Bridgerian of Wyoming.

Characteristics: Upper molars with mesostyles; P4 with distinct to strong metacone; p3 smaller than p4 but not reduced as in *Arctodontomys*; p4 with distinct to strong metaconid and relatively well-developed talonid basin; molars with relatively more bulbous cusps than *Arctodontomys*; cranial features as for family.

Average length of m1: 3.0–4.6 mm.

Included species: *M. elegans* (including *Microsyops gracilis* Leidy, 1872; *Palaeacodon versus* Marsh 1872; *Bathrodon typus* Marsh, 1872; *Mesacodon speciosus* Marsh, 1872; *Microsyops* [*Mesacodon*] *speciosus* Osborn, 1902; *Microsyops* [*Bathrodon*] *typus* Osborn, 1902) (known from localities CP20F, CP25J, CP31A, B, E, CP34A, C*); *M. angustidens* (Matthew, 1915; originally described as *Cynodontomys angustidens*; including *Cynodontomys alfi* McKenna, 1960; *Microsyops alfi* Szalay, 1969a) (localities SB24, CP20B*, CP25A, CP31A, CP63, CP64B, NP49); *M. kratos* Stock, 1938 (locality CC4*); *M. annectens* (Marsh, 1872; originally descsribed as *Bathrodon annectens*; including *Microsyops* [*Bathrodon*] *annectens* Osborn, 1902; *Microsyops schlosseri* Wortman, 1903–1904) (localities CC4, SB43A, [CP20E], CP31E, CP32B, CP34D*); *M. cardiorestes* Gunnell, 1989 (including *Microsyops* sp. A, Gunnell, 1985) (locality CP20B*); *M. knightensis* (Gazin, 1952; originally described as *Cynodontomys knightensis*) (localities SB22AC, CP25G*, GG, J); *M. latidens* (Cope, 1882; originally described as *Cynodontomys latidens*; including *Pelycodus angulatus* Cope, 1882; *Chriacus pelvidens* Cope, 1883; *Notharctus palmeri* Loomis, 1906; *Notharctus cingulatus* Loomis, 1906) (localities SB24, CP20C*, CP25A, CP27B, CP64B); *M. scottianus* Cope, 1881 (Matthew, 1915; originally described as *Cynodontomys scottianus*; including *C. scottianus* White, 1952; Robinson, 1966) (localities SB22B, CP20D, E, CP25H, J, CP27C, D*, E, CP64C).

Microsyops sp. also is known from localities SB25A, CP6B, CP27D, and CP33A1.

Comments: *Microsyops* first appears in the middle Wasatchian and survives into the Uintan. While never particularly species diverse, *Microsyops* is a relatively common taxon, especially in the Bridgerian where it may constitute as much as 5–7% of the mammalian faunal sample in terms of total numbers of specimens. The genus reaches its greatest diversity in the latest early and earliest middle Eocene in which as many as three species may co-occur (Gunnell, 1989, 1998).

Megadelphus Gunnell, 1989 (including *Cynodontomys*, in part; *Microsyops*, in part)

Type species: *Megadelphus lundeliusi* (White, 1952) (originally described as *Cynodontomys lundeliusi*).

Type specimen: USNM 18371, Bridgerian of Wyoming.

Characteristics: Upper and lower central incisors less procumbent than in other microsyopids; i1 very robust; I2 and upper canine reduced and peglike; upper canine single-rooted; larger than all other microsyopid taxa. Cranium with a long, low profile; very strong sagittal crest; small orbits; strong postorbital constriction.

Average length of m1: 5.4–6.5 mm.

Included species: *M. lundeliusi* only (including *Microsyops lundeliusi* McKenna, 1966; Robinson, 1966; Szalay, 1969a; Guthrie, 1971; Lucas and Froehlich, 1989) (known from localities SB22C, SB24, CP27C*, E).

Craseops Stock, 1934

Type species: *Craseops sylvestris* Stock, 1934.

Type specimen: LACM (CIT) 1580, Uintan of California.

Characteristics: Relatively large taxon; upper molars strongly dilambdodont; metaconules lacking; lower molars with hypoconid very buccal to protoconid; p4 and lower molars with strong paracristids and weak, crestiform paraconids.

Average length of m1: 5.2 mm.

Included species: *C. sylvestris* only, known from locality CC9A* only.

Comments: *Craseops* is only known from the Uintan of California and represents the last known occurrence of microsyopids in North America.

PICROMOMYIDAE ROSE AND BOWN, 1996

Characteristics: Very small taxa; p4 larger than m1, with expansive talonid and no hypoflexid; low-crowned lower molars with shallow or absent hypoflexids, trigonids low and mesially inclined, entoconids and hypoconulids close together.

Comments: When this family was originally described, Rose and Bown (1996) suggested a relationship between it and either Microsyopidae or Micromomyidae. These authors considered the similarities to picrodontids (e.g., very-low-crowned molars) to be attributable to homoplasy, likely associated with a similar diet. Silcox (2001) was unsuccessful in determining an appropriate taxonomic placement for the group, suggesting relationships with Microsyopidae, Picrodontidae, or Micromomyidae were all possible. The first complete lower central incisor of a picromomyid (Silcox, Rose, and Walsh, 2002) exhibits features suggestive of a closer relationship with micromomyids, although this hypothesis remains tentative.

Alveojunctus Bown, 1982

Type species: *Alveojunctus minutus* Bown, 1982.

Type specimen: USGS 2005, Bridgerian of Wyoming.

Characteristics: Differs from *Picromomys* in being larger, having a p4 with a basined talonid and more distinct entoconid and hypoconulid cusps.

Average length of m1 or m2: 1.75–2.25 mm.

Included species: *A. minutus* Bown, 1982 (known from locality CP31E*); *A. bowni* Silcox, Rose, and Walsh, 2002 (locality CC4*).

Alveojunctus also may be represented at localities ?CP25J and ?CP27D.

Comments: *Alveojunctus* is very poorly known, being represented with certainty only by six isolated teeth from Bridgerian-aged sediments in Wyoming (Bown, 1982), and a single isolated tooth from the early Uintan of California (Silcox, Rose, and Walsh, 2002). Two additional specimens that may belong to this genus, from the early Bridgerian of the Cathedral Bluff Tongue (locality CP25J; Bown, 1982) and the early Eocene of the Wind River Formation (locality CP27D; Stucky, 1984), appear to have been lost (Silcox, Rose, and Walsh, 2002). *Alveojunctus* is clearly allied with *Picromomys* (Rose and Bown, 1996; Silcox, Rose, and Walsh, 2002; contra Hooker, Russell, and Phélizon, 1999) based on the close similarity in the unusual morphology of p4 in these two genera.

Picromomys Rose and Bown, 1996

Type species: *Picromomys petersonorum* Rose and Bown, 1996.

Type specimen: USNM 487900, Wasatchian of Wyoming.

Characteristics: m1-2 with cuspate expansion mesiobuccal to protoconid; p4 cusps blunt and undifferentiated, talonid wider than trigonid and unbasined, lacking distinct entoconid or hypoconulid.

Average length of m1: 1.1 mm.

Included species: *P. petersonorum* only, known from locality CP20B* only.

Comments: The p4 of this taxon is very similar to the m1 of *Picrodus* in possessing a tall but mesiodistally compressed trigonid, and a broad, unbasined talonid that slopes lingually. The difference in tooth position, and the dissimilarity between the lower molars of *Picromomys* and *Picrodus*, suggests that these similarities are convergent and probably indicative of a similar diet emphasizing soft fruits, larvae, and/or nectar (Szalay, 1968; Rose and Bown, 1996). At an estimated mass of around 10 g, *Picromomys* is the smallest known plesiadapiform, and it may be the smallest known primate (Rose and Bown, 1996; Silcox, Rose, and Walsh, 2002); a euprimate calcaneum reported from southeastern China produces a similar mass estimate (Gebo *et al.*, 1998, 2000).

MICROMOMYIDAE SZALAY, 1974

Characteristics: Very small taxa; enlarged, non-lanceolate central incisors; p4 larger than m1, with a trenchant paracristid; molar trigonids longer than talonids. Humerus with a spherical capitulum, shallow olecranon fossa; radius with a round, well-excavated radial head, robust bicipital tuberosity, strong dorsal ridge on the distal end; ulna with a moderately long, medially flaring olecranon process; intermediate manual phalanges shorter than proximal phalanges; innominate with long, rodlike ilium, cranially buttressed, elliptical acetabulum; femur with robust lesser trochanter projecting beyond the level of the head, wide, shallow patellar groove; fibula wide proximally; astragalus with a short neck, large groove for flexor fibularis muscles; calcaneum short with a distally positioned peroneal tubercle. Modified in part from Szalay, 1974; Beard, 1989; Bloch, Boyer, and Houde, 2003.

Comments: The family is unknown from Europe. A tentatively identified specimen from Asia has been mentioned in print (Tong and Wang, 1998) but not formally described; Silcox considers this specimen not to pertain to a micromomyid. Postcranial material is known for *Tinimomys*, *Chalicomomys*, and *Dryomomys* (Beard, 1989, 1993a; Bloch, Boyer, and Houde, 2003; Bloch *et al.* 2007). Beard (1993a) reconstructed the locomotor mode of micromomyids as involving mitten-gliding in a manner similar to modern dermopterans. As with paromomyids, better-associated material has demonstrated that the key feature supporting this reconstruction (long intermediate manual phalanges) is actually not present in this group (Bloch, Boyer, and Houde, 2003; Bloch *et al.*, 2007). The evidence presented to date demonstrates that micromomyids were committed claw-climbing arborealists capable of moving on large-diameter supports and of taking suspensory postures under small branches, but lacking features for leaping found in most early euprimates (Beard, 1989; Bloch, Boyer, and Houde, 2003: Bloch *et al.* 2007). As noted under

Tinimomys (below), the cranial morphology of this family is extremely controversial.

Micromomys Szalay, 1973

Type species: *Micromomys silvercouleei* Szalay, 1973.

Type specimen: PU 17676, Tiffanian of Wyoming.

Characteristics (modified from Beard and Houde, 1989): Lower canine present; diastemata separate i1 and c1, and c1 and p2; p2 double-rooted; p3 short and high crowned; P4 tall and short with a narrow trigonid; P3 short, low crowned, simple with only a remnant of a protocone lobe; molars with sharply crested trigonids and relatively broad talonids; protoconids taller than metaconids on m1; upper molars lacking hypocone but with distinct conules.

Average length of m1: 1.1–1.3 mm.

Included species: *M. silvercouleei* Szalay, 1973 (known from locality CP13G*); *M. fremdi* Fox, 1984a (locality NP3C*); *M. vossae* Krause, 1978 (locality NP7D*).

Micromomys may be represented by a new species from locality ?CP13G.

Comments: *Micromomys fremdi* is the most primitive member of the family (Fox, 1984a) and may have given rise to the *Chalicomomys–Tinimomys* grouping as well as to later species of *Micromomys*, which would make this genus paraphyletic. The very limited material available for the other species of *Micromomys* (fewer than five specimens for the two species combined and no upper teeth) makes it difficult to test this hypothesis adequately (Beard and Houde, 1989).

Chalicomomys Beard and Houde, 1989

Type species: *Chalicomomys antelucanus* Beard and Houde, 1989.

Type specimen: USNM 42559, Wasatchian of Wyoming.

Characteristics: Lower canine present; p2 single-rooted, not separated from c1 by diastema; p4 trigonid moderately inflated; p4 weakly exodaenodont; P3 without metacone; lower molar cusps sharp, protoconid and metaconid similar in height; upper molars lack lingual cingula, pericones, and hypocones (modified from Beard and Houde, 1989).

Average length of m1: 1.0 mm.

Included species: *C. antelucanus* (known from locality CP19*); *C. willwoodensis* Rose and Bown, 1982 (localities CP19*, CP25B).

Chalicomomys may be represented by new species from localities CP14E (see Wilf *et al.*, 1998) and CP17B (see Bloch and Boyer, 2001).

Comments: Rose and Bown (1996) suggested that this taxon should be congeneric with *Micromomys*, although admitting that it "... strengthens the probability of a close relationship between *Micromomys* and *Tinimomys*" (p. 313). One of us (Silcox) feels that the presence of characters shared by *Tinimomys* and *Chalicomomys*, to the exclusion of *Micromomys* (e.g., single-rooted p2, p4 trigonid relatively wide and p4 at least moderately exodaenodont; Beard and Houde, 1989), necessitates a generic separation. Although *Micromomys* is already likely paraphyletic, adding *Chalicomomys* to the genus would only complicate matters further.

Tinimomys Szalay, 1974 (including *Myrmekomomys*)

Type species: *Tinimomys graybulliensis* Szalay, 1974.

Type specimen: PU 17899, Wasatchian of Wyoming.

Characteristics: Lower canine absent; p4 exodaenodont, relatively long and not as tall as in *Micromomys*; lower molar cusps relatively bunodont; protoconid and metaconid similar in height on all molars; P3 with a metacone and prominent protocone lobe; upper molars with distinct hypocones and often small pericones and lingually continuous cingula, conules relatively weak (modified from Beard and Houde, 1989).

Average length of m1: 1.2 mm.

Included species: *T. graybulliensis* Szalay, 1974 (including *Myrmekomomys loomisi* Robinson, 1994) (known from localities CP14E, CP17A, B, [C], CP19*, CP20A, CP25B).

Tinimomys may be represented by a new species at locality CP14E (see Wilf *et al.*, 1998).

Comments: *T. graybulliensis* is currently the best-known species of micromomyid including nearly complete upper and lower jaws (Rose, Beard, and Houde, 1993) and some well-preserved postcranial elements (Beard, 1989, 1993a,b). Two petrosal fragments have been attributed to *Tinimomys* based on size and association with dental material (Gunnell, 1989; MacPhee *et al.*, 1995). The two specimens differ markedly in morphology, however. The specimen described by Gunnell (1989) is comparable to that of a microsyopid in appearance, with distinct grooves for stapedial and promontorial arteries. The MacPhee *et al.* (1995) specimen is quite similar to the petrosal of *Plesiadapis* in morphology, with no arterial grooves apparent.

Dryomomys Bloch *et al.*, 2007

Type species: *Dryomomys szalayi* Bloch *et al.*, 2007.

Type specimen: UM 41870, Clarkforkian of Wyoming.

Characteristics: P3 lingually expanded with strong protocone and lacking a metacone; P4 very wide and inflated with large protocone crown that is situated close to the paracone and metacone; upper molars lacking continuous lingual cingulum and pericone and having only a very weak hypocone; no diastemata present between i1 and c1; p2 single-rooted; p3 low crowned with open talonid; p4 relatively higher crowned with narrow talonid.

Average length of m1: 1.2 mm.

Included species: *D. szalayi* only, known from locality CP17B* only.

PAROMOMYOIDEA (SIMPSON, 1940)

Characteristics: I1 with a posterocone, developing divisions at the tip; M1–2 with pronounced postprotocingulum and large posterointernal basin, protocone skewed mesiodistally; m3 talonid similar in breadth to the talonid or wider (modified from Silcox, 2001).

Comments: Paromomyoidea as constructed here follows Silcox (2001) and is less inclusive than Schwartz (1986) envisioned, since it excludes *Navajovius* (included in Microsyopidae here). The grouping is likely paraphyletic, including an array of primitive forms grouped in the Palaechthonidae and two more cohesive families that may have originated from an ancestor that would be classified as a palaechthonid.

PAROMOMYIDAE SIMPSON, 1940 (= PAROMOMYINAE [IN PART] SIMPSON, 1940, PHENACOLEMURIDAE [IN PART] SIMPSON, 1955)

Characteristics modified from Bown and Rose (1976); Rose and Gingerich (1976); Gunnell (1989); Kay, Thewissen, and Yoder (1992); Bloch and Silcox (2001); Boyer, Bloch, and Gingerich (2001); Bloch and Boyer (2002, 2003) and Silcox, 2003.

Cranial characteristics: Cranium with long, tapering snout and long nasals that have a broad contact with the frontals; premaxillae do not contact the frontals; orbits small and widely spaced (presumably indicating large olfactory lobes); postorbital bar absent; orbital canal small; lacrimal foramen associated with a lacrimal tubercle, located on the antorbital rim; large infraorbital foramen; strong postorbital constriction; very wide zygomatic arches (the brain was apparently quite small); auditory bulla ossified from the entotympanic and inflated; central stem, between the enlarged bullae, narrow; posterior carotid foramen very small and laterally located; groove or bony tube present at the lateral extreme of the promontorium for internal carotid nerves and/or a small promontorial artery.

Dental characteristics: Small plesiadapiforms; hypertrophied, subhorizontal incisors that are slender, pointed, and not lanceolate; p4 with a tall, pointed, broad-based protoconid, no well-defined paraconid, and a basined talonid; lower molars with paraconids placed lingually and reduced or indistinct from metaconid on m2–3; lower molar trigonids with postvallids that are strongly mesially inclined, and a protoconid/metaconid notch that is obscured by a fold of enamel; m3 with expansion distally in the form of an additional lobe; upper and lower molars low crowned; upper molars with distinct postprotocingula enclosing distolingual basins and weak conules.

Postcranial characteristics: Indicative of arborealism including features for claw-climbing on large vertical supports; manual proportions indicative of strong grasping abilities, but not gliding; features of the pes suggestive of hindlimb suspension.

Comments: Until Bown and Rose's (1976) revision of plesiadapiform relationships, the composition of the Paromomyidae was generally held to be much broader than its make up here, including taxa that are now considered to be microsyopids (*Navajovius*), palaechthonids (*Plesiolestes*, *Palaechthon*, *Torrejonia*, *Palenochtha*), and micromomyids (*Tinimomys*, *Micromomys*). Bown and Rose (following Simpson, 1955; McKenna, 1960) suggested that *Paromomys*, *Phenacolemur*, and *Ignacius* share derived traits that are missing in these other taxa, including a shift from cusps to crests on low-crowned molars, and simple p4s that develop tall, pointed protoconids. The similarities shared by the more inclusive group are considered here to be primitive, or as features of the more inclusive Paromomyoidea. The Paromomyidae is well documented in the fossil record, with fairly extensive dental, postcranial, and cranial material available. The affinities and locomotor behavior of this family have been a matter of intense debate, with some authors suggesting a link to Dermoptera, and a pattern of behavior that would include gliding (Kay, Thorington, and Houde, 1990; Kay, Thewissen, and Yoder, 1992; Beard, 1989, 1990a, 1993a,b) and others challenging elements of this interpretation (e.g., Krause, 1991; Szalay and Lucas, 1993, 1996; Van Valen, 1994; Runestad and Ruff, 1995; Stafford and Thorington, 1998; Hamrick, Rosenman, and, Brush, 1999; Stafford and Szalay, 2000; Bloch and Silcox, 2001; Boyer, Bloch, and Gingerich, 2001; Silcox, 2001, 2003). Novel discoveries and analyses of cranial material suggest that there is no convincing cranial support for a paromomyid–dermopteran relationship (Bloch and Silcox, 2001, 2006; Silcox, 2001, 2003). Better-associated postcranial finds have demonstrated that some features indicative of gliding which were thought to be characteristic of paromomyids (e.g., long intermediate manual phalanges) are actually not present in this family (Bloch and Boyer, 2001, 2007; Boyer, Bloch, and Gingerich, 2001). Comparisons of this new material with dermopteran postcranials also document profound differences in the vertebral column and other regions that suggest paromomyids were not gliders (Boyer, Bloch, and Gingerich, 2001).

Paromomys Gidley, 1923

Type species: *Paromomys maturus* Gidley, 1923.

Type specimen: USNM 9473, Torrejonian of Montana.

Characteristics: Lower dental formula i2, c1, p3, m3; p4 with no metaconid; m1 with distinct paraconid that is placed less proximal to the metaconid than in *Acidomomys*, *Ignacius*, or *Phenacolemur*; lower molars with distinct external cingulids; upper molars with less-expanded distolingual basins than in *Acidomomys*, *Ignacius*, or *Phenacolemur*, and metacone and paracone subequal in size.

Average length of m2: 2.0–3.2 mm.

Included species: *P. maturus* (known from localities CP13C, CP14A, NP19C*, NP19IIC); *P. depressidens* Gidley, 1923 (localities NB1A-B, CP1C, CP13B, CP13IIA, C, CP14A, B, CP15A, CP26A, NP19C*).

Paromomys sp. is also known from localities SB23C, CP1B, NP3A0, NP19IID (see Lofgren, McKenna, and Walsh [1999] for additional locality information).

Comments: Bown and Rose (1976) suggested that *P. depressidens* might have given rise to *Ignacius* and *P. maturus* to *Phenacolemur*. Rigby (1980) favored derivation of *Phenacolemur* from *P. depressidens* and suggested that *P. depressidens* may be generically distinct from *P. maturus*. There are few traits that would support special relationships between either *Paromomys* spp. and particular later forms. Yet, there are many derived features shared by *Phenacolemur*, *Ignacius*, *Acidomomys* (and, where known, *Elwynella*) that are absent in any known *Paromomys*, including no external cingulids on the lower molars, even less distinct paraconids on all the lower molars (especially m1), expanded distolingual basins on the upper molars, and further enlargement of the m3 hypoconulid lobe (Gunnell, 1989; Bloch *et al.*, 2002). This shared constellation of features suggests that *Phenacolemur*, *Ignacius*, *Acidomomys*, and *Elwynella* may have shared an ancestor more recent than known *Paromomys*, although perhaps ultimately derived from a species referable to that genus.

Ignacius Matthew and Granger, 1921 (including *Phenacolemur*, in part)

Type species: *Ignacius frugivorus* Matthew and Granger, 1921.

Type specimen: AMNH 17368, Tiffanian of Colorado.

Characteristics (modified from Bown and Rose, 1976): Very reduced dentition with lower dental formula of i1, c0, p1/2, m3; p4 with tall, pointed protoconid, no paraconid or metaconid, and narrow basin, shorter mesiodistally and narrower than m1 or p4 of *Phenacolemur* and lacking the mesial bulge near the base of the protoconid of *Phenacolemur*; lower molars very low crowned, with no cingulids, and paraconids on m2–3 poorly distinguished from the metaconid; paraconid less distinct on m1 than in *Paromomys*; upper and lower molar basins very shallow, more so than in *Phenacolemur*; metacone on upper molars reduced, somewhat smaller than paracone; upper molar distolingual basins expanded relative to *Paromomys*, less so than in *Phenacolemur*, especially on M3; centrocristae on upper molars invaginated buccally. Cranial and postcranial features as for the family.

Average length of m2: 1.45–2.1 mm.

Included species: *I. frugivorus* (including *Phenacolemur frugivorus* Simpson 1935; Simpson 1955; Gazin, 1971; Schiebout, 1974; Krishtalka, Black, and, Riedel, 1975) (known from localities NB1B, SB20A*, SB39C, CP13E, CP14C, CP15A, B, CP22B, CP24A, CP26A, NP1C, NP3A–F, NP7D, NP19IIC, NP20D, E, NP47B, BB); *I. fremontensis* (including *Phenacolemur fremontensis* Gazin, 1971) (localities CP13B, CP15A*, NP1C, NP3A, [NP3A0]); *I. graybullianus* Bown and Rose, 1976 (localities CP17B, CP19, CP20A*, CP25B); *I. mcgrewi* Robinson, 1968 (including *Phenacolemur frugivorus* Robinson, 1968; Krishtalka, 1978) (localities CP25J, CP29C*, D); *I. clarkforkensis* Bloch *et al.*, 2007 (locality CP17B*).

Ignacius sp. is also known from localities NB1B, ?CP13G, CP14E, CP17B, CP27D, NP4, NP19IIA, NP47C, HA1A (two species) (see Lofgren, McKenna, and Walsh [1999] for additional locality information).

Comments: *Ignacius* is one of the biogeographically most widely spread genera of primates ever known, stretching north to Ellesmere Island (West and Dawson, 1977; McKenna, 1980), south into Texas (Schiebout, 1974), and east to deposits from the early Eocene of Asia (Tong and Wang, 1998). The paromomyid material from Ellesmere Island (locality HA1A) includes two distinct species, the larger of which is the largest known member of this family and may be referable to a different genus (McKenna, 1980). No formal descriptions or measurements of this material have been published, so it is not included in the average m2 size for the genus.

Simpson (1955) questioned the distinctness of *Ignacius* from *Phenacolemur*, and for many years the type species (*I. frugivorus*) was referred to the latter genus. Bown and Rose (1976) argued that it was possible to distinguish the two genera, a distinction which is also followed here, since it seems likely that they are the products of discrete evolutionary lineages.

Phenacolemur Matthew, 1915 (including *Dillerlemur*; *Simpsonlemur*; *Pulverflumen*)

Type species: *Phenacolemur praecox* Matthew, 1915.

Type specimen: AMNH 16102, Wasatchian of Wyoming.

Characteristics: Very reduced lower dentition with lower dental formula i1, c0, p1, m3; p4 premolariform with no paraconid or metaconid, subequal in size to m1, mesiodistally expanded relative to *Ignacius*, particularly near the base, which is mesially swollen; lower molars similar to *Ignacius*, low crowned, paraconid indistinct from metaconid, no cingulids; upper molars low crowned with distolingual basins that are expanded beyond the condition in *Ignacius*, especially on M3, metacone reduced relative to paracone, more so than in *Ignacius* or *Acidomomys*; centrocrista not invaginated (unlike *Ignacius* and *Acidomomys*) (modified from Bown and Rose, 1976; Bloch *et al.*, 2002). Cranial and postcranial features as for the family.

Average length of m2: 1.75–3.1 mm.

Included species: *P. praecox* (known from localities [SB40], CP17B, CP19AA, CP19A*, CP20A, CP25B, CP63, NP20D, NP49); *P. citatus* Matthew, 1915 (including *Simpsonlemur citatus* Robinson and Ivy, 1994) (localities CP20B*, CP27B); *P. fortior* Robinson and Ivy, 1994 (locality CP20*); *P. jepseni* Simpson, 1955 (including *Simpsonlemur jepseni* Robinson and Ivy, 1994) (localities SB24*, CP25I2, CP27B); *P. pagei* Jepsen, 1930 (including *Dillerlemur pagei* Robinson and Ivy, 1994) (localities CP13G*, CP14E, CP17A, B, E, CP26C, NP20E); *P. shifrae* Krishtalka, 1978 (localities CC7B, CC9A*,

CP29C); *P. simonsi* Bown and Rose, 1976 (localities CP14E, CP19A, CP20A*, CP25B, CP63). A new, undescribed species of *Phenacolemur* is also known from CP20C (Silcox *et al.*, 2004).

Phenacolemur sp. (including *Dillerlemur* sp. Robinson and Ivy, 1994; *Pulverflumen* sp. Robinson and Ivy, 1994; *Simpsonlemur* sp. Robinson and Ivy, 1994; *Pulverflumen magnificum* Robinson and Ivy, 1994; *Dillerlemur robinettei* Robinson and Ivy, 1994) is also known from localities CC8, CP17C, CP20A, CP25B, F, CP26C, D, CP27D, E, (CP31E), CP36B, CP63, (NP47C).

Comments: Szalay (1972) described the only known cranial specimen for this genus. Comparison between this description and the cranial characteristics listed above for Paromomyidae reveals some discrepancies. Reappraisal of the relevant specimens suggests that *P. jepseni* was extremely similar in basicranial morphology to *Ignacius* (as described by Kay, Thewissen, and Yoder, 1992; see Bloch and Silcox, 2001; Silcox, 2001, 2003).

***Acidomomys* Bloch *et al.*, 2002**

Type species: *Acidomomys hebeticus* Bloch *et al.*, 2002.

Type specimen: UM 108206, Clarkforkian of Wyoming.

Characteristics: Lower dental formula i2, c0, p2, m3; p4 short relative to m1, with large metaconid, compressed talonid basin with no hypoconulid, and no distinct hypoflexid; lower molars low crowned with no external cingulids, metaconid and paraconid closely appressed, m3 with inflated third lobe and indistinct paraconid. Upper dental formula I2, C1, P3, M3; I1 with mediocone, strong mediocrista, anterocone, laterocone, and a doubled posterocone; P2 single-rooted; M1–2 with metacone somewhat smaller than paracone, and distolingual basins expanded relative to *Paromomys*; centrocristae on upper molars invaginated buccally (modified from Bloch *et al.*, 2002).

Average length of m2: 2.50 mm.

Included species: *A. hebeticus* only, known from locality CP17B* only.

Comments: *Acidomomys* is the only known paromomyid that exhibits a metaconid on its p4. Although this is a feature seen in some other plesiadapiforms (e.g., some palaechthonids and microsyopids), the absence of this cusp from the p4 of *Paromomys* suggests this is likely to be a neomorph (Bloch *et al.*, 2002). *Acidomomys* is also unique for a Clarkforkian paromomyid in retaining both i2 and a double-rooted p3. Therefore, although this taxon shares some derived features with *Ignacius* and *Phenacolemur* relative to *Paromomys* (see above), it seems likely that it branched off quite early and has a substantial ghost lineage. In retaining p3, *Acidomomys* is similar to Bridgerian *Elwynella.* In *Elwynella,* this tooth is single-rooted, however, and the molars are derived beyond the condition in *Acidomomys* in lacking a distinct paraconid on m1 and in having crests in the place of clear cusps. While it is possible that these two taxa are part of the same evolving lineage, this argument is only supported by a symplesiomorphy (shared retention of p3).

The hypodigm of *Acidomomys* is made up entirely of subadult specimens, providing the best evidence available for any plesiadapiform of the morphology of the deciduous dentition and of the dental eruption sequence (Bloch *et al.*, 2002). Postcranial and cranial material of *Acidomomys* has been recovered (Bloch and Boyer, 2001) but is not yet formally described (but see Boyer, Bloch, and Gingerich, 2001).

***Elwynella* Rose and Bown, 1982**

Type species: *Elwynella oreas* Rose and Bown, 1982.

Type specimen: USGS 2351, Bridgerian of Wyoming.

Characteristics: Lower dental formula i1, c0, p2, m3; p3 single-rooted; short diastema between i1 and p3; lower incisor broader than in *Phenacolemur* or *Ignacius*; lower molars with strongly arcuate paracristids, ill-defined or absent paraconids, and well-defined crests in the place of clearly distinct cusps.

Average length of m2: 2.55 mm.

Included species: *E. oreas* only (including Cf. *Phenacolemur* sp. Bown, 1982) (known from localities CP31E*, [CP20E]).

Comments: This relatively late-occurring form is included in a separate genus because the retention of p3 (lost in all other paromomyids known from after the Clarkforkian [Rose and Bown, 1982; Bloch *et al.*, 2002]) suggests it may represent a separate, unsampled line, with a ghost lineage stretching at least the entire length of the Wasatchian. The unusually derived form of the molars and presence of only a single root to p3 also distinguishes this taxon from *Acidomomys*.

PALAECHTHONIDAE (SZALAY, 1969a)

Characteristics: Relatively primitive plesiadapiforms with enlarged, procumbent, semi-lanceolate i1; p2 single-rooted; p4 premolariform or semi-molariform, sub-equal or smaller than m1; lower molar trigonids slightly mesially inclined; lower molar talonids often bearing strong mesoconids; upper molars with strong conules and well-demarcated postprotocingula that enclose a distinct distolingual basin; cranium with fairly long snout; small, widely spaced and laterally directed orbits; lacrimal foramen intraorbital; large infraorbital foramen located above P4 (modified from Kay and Cartmill, 1977; Gunnell, 1989). There are no known postcrania for any member of this family.

Comments: The only cranial material that has been described for this family is a very badly damaged skull of *Plesiolestes nacimienti* (Kay and Cartmill, 1977). This specimen does not preserve any remnants of ear morphology.

Silcox (2001) failed to support the monophyly of the family Palaechthonidae, suggesting that it was a paraphyletic assemblage of primitive forms from which various other groups likely took their origin. In spite of this fact, the family is retained here as a generally similar group of primitive taxa.

Gunnell (1989) divided the family into two subfamilies, the Palaechthoninae and the Plesiolestinae. The primary character distinguishing these two families (particularly in light of the inclusion of *Plesiolestes nacimienti* in the latter) is a difference in the relative proportions of the lower, antemolar teeth. In palaechthonines, the lower canine is smaller or similar in size to p2, whereas in plesiolestines it is relatively larger. These subfamilies are included here, although Silcox (2001) failed to support their monophyly.

PALAECHTHONINAE SZALAY, 1969a

Characteristics: Lower canine smaller or equal in size to p2; p4 with a less-distinct entoconid than in plesiolestines. No cranial or postcranial material is known for this subfamily (modified from Gunnell, 1989).

Palaechthon Gidley, 1923

Type species: *Palaechthon alticuspis* Gidley, 1923.

Type specimen: USNM 9532, Torrejonian of Montana.

Characteristics: Somewhat larger than *Palenochtha*; lower molars more mesiodistally compressed with less-elevated trigonids; upper molars less narrow lingually, with less mesially skewed protocones; paraconids more distinct on m2–3 than in *Premnoides* and retains an i2 in contrast to that genus (modified from Gunnell, 1989).

Average length of m2: 1.9–2.1 mm.

Included species: *P. alticuspis* (known from localities SB39IIB, CP13C, CP13IIA, CP15A, NP1C, NP19C*, [CP13IIB]); *P. woodi* Gazin, 1971 (localities SB23C, F, H, CP15A*, [NP19D], NP19IIB).

Palaechthon sp. is also known from localities CP1C, NP20B.

Comments: As the characteristics given for this genus indicate, the two species included in *Palaechthon* are largely linked by a lack of shared specializations with other genera, rather than any well-supported synapomorphies.

Palenochtha Simpson, 1935b (including *Palaechthon*, in part)

Type species: *Palenochtha minor* (Gidley, 1923) (originally described as *Palaechthon minor*).

Type specimen: USNM 9639, Torrejonian of Montana.

Characteristics: Very small palaechthonids with p4 trigonids much taller than the extremely narrow talonids; lower molars with sharp cusps and an extremely well-demarcated and deep protoconid/metaconid notch; molar trigonids less mesiodistally compressed and less mesially inclined than in most other plesiadapiforms, with paraconids that are not lingually shifted (placed centrally rather than being appressed to the metaconid); lower molars with buccal cingulids and no mesoconids; P4 with a distinct metacone; upper molars very narrow lingually with a protocone that is strongly "twisted" mesially (modified from Gunnell, 1989).

Average length of m2: 1.45 mm.

Included species: *P. minor* (known from localities CP13B, C, CP13IIA, CP14A, CP15A, [NP3A0], NP19C*, NP19IIB); *P. weissae* Rigby, 1980 (locality CP14A*).

Comments: Rigby's (1980) description of *P. weissae* emphasizes the less-reduced dental formula of that species as the primary distinguishing feature from *P. minor* (see discussion in Gunnell, 1989, pp. 19–21). The p4 (AMNH 100355) shares with *P. minor* the derived feature of a very narrow talonid but bears a better-defined paraconid and a distinct metaconid, making the tooth more molariform, and thus more *Palaechthon*-like, than is the case in *P. minor*. The form of the p4 trigonid provides support for a close relationship between *Palenochtha* and *Palaechthon*.

Premnoides Gunnell, 1989

Type species: *Premnoides douglassi* Gunnell, 1989.

Type specimen: YPM-PU 14802, Torrejonian of Wyoming.

Characteristics: p4 similar in size to m1 but premolariform, with no sign of a paraconid and no distinct metaconid; lower molars with mesially compressed trigonids, squared with indistinct, lingually placed paraconids and shallow protoconid–metaconid notch; lower molars with small but distinct mesoconids and lacking distinct buccal cingulids (modified from Gunnell, 1989).

Average length of m2: 2.1 mm.

Included species: *P. douglassi* only, known from locality CP13B* only.

Comments: This taxon shares resemblances with both palaèchthonids (e.g., the overall morphology of p4) and with paromomyids (e.g., the morphology of the lower molar trigonids), making it a possible ancestor for the latter group.

PLESIOLESTINAE GUNNELL, 1989

Characteristics (modified from Gunnell, 1989): Medium- to large-sized palaechthonids; lower canine larger than p2; p4 with deep talonid basin and strong entoconid.

Plesiolestes Jepsen, 1930 (including *Palaechthon*, in part)

Type species: *Plesiolestes problematicus* Jepsen, 1930.

Type specimen: YPM-PU 13291, Torrejonian of Wyoming.

Characteristics: Medium-sized palaechthonid; enlarged, lanceolate lower central incisor, the flattened surface of which faces dorsally (i.e., the tooth is not rotated mesially as in microsyopids); p4 with distinct metaconid; molar and premolar cusps sharply defined; lower molars and p4 frequently with mesoconids; cranial features as for family.

Average length of m2: 2.5 mm.

Included species: *P. problematicus* (including *Palaechthon problematicus* Rigby, 1980; Gingerich, Houde, and Krause, 1983) (known from localities CP13B*, C, CP13IIA, CP14A, CP15A, NP3A0, NP19IIB, NP20A); *P. nacimienti* (including *Palaechthon nacimienti* Wilson and Szalay, 1972) (localities SB23F*, CP1C, CP13IIA, A1).

Plesiolestes sp. is also known from locality CP13IIB.

Comments: Various authors (e.g., Rigby, 1980) have synonymized *Plesiolestes* with *Palaechthon*. This practice is not followed here since the two genera appear to be parts of different subfamilies. The inclusion of *P. nacimienti* in *Plesiolestes* rather than *Palaechthon* follows Gunnell (1989) and is largely based on the relative sizes of the antemolar teeth.

Talpohenach Kay and Cartmill, 1977 (including _Palaechthon_, in part)

Type species: *Talpohenach torrejonius* Kay and Cartmill, 1977 (originally described as ?*Palaechthon nacimienti* Wilson and Szalay, 1972).

Type specimen: UKMNH 7903, Torrejonian of New Mexico.

Characteristics: Medium-sized palaechthonid; upper canine large; P3 and P4 with protocone and strong posterolingual basin; P4 with a paracone, metacone, and parastyle; upper molars with relatively broad stylar shelves and crenulated enamel (modified from Kay and Cartmill, 1977).

Average length of M2 (no m2 known): 2.45 mm.

Included species: *T. torrejonius* only, known from locality SB23E* only.

Comments: The distinctiveness of this taxon from *Palaechthon* at the generic level (Szalay and Delson, 1979; Williamson and Lucas, 1993; Williamson, 1996) and *P. nacimienti* at the specific level (Gunnell, 1989) has been a matter of debate. One of us (Silcox) has recently reexamined the type (and only) specimen. Based on detailed comparisons with the entire hypodigm of *P. nacimienti*, it seems that Kay and Cartmill (1977) actually underemphasized the differences between this taxon and *P. nacimienti* (see discussion in Silcox, 2001). For this reason, the generic and specific distinctions are maintained here. The placement of this taxon in the Plesiolestinae is based more on general similarities to *Plesiolestes nacimienti* than on any strong, character-based argument. The discovery of lower teeth could alter this placement.

Torrejonia Gazin, 1968 (including _Plesiolestes_, in part)

Type species: *Torrejonia wilsoni* Gazin, 1968.

Type specimen: USNM 25255, Torrejonian of New Mexico.

Characteristics: Large palaechthonid, substantially larger than *Plesiolestes*; p4 with indistinct paraconid and no metaconid; molar trigonids relatively low and broad based; premolar and molar cusps bulbous; molar teeth relatively low crowned (compared with *Plesiolestes*).

Average length of m2: 3.3–3.65 mm.

Included species: *T. wilsoni* (including *Plesiolestes wilsoni* Wilson and Szalay, 1972; Szalay, 1973; Szalay and Delson, 1979; Williamson and Lucas, 1993; Williamson, 1996) (known from localities SB23C, [GG], H*); *T. sirokyi* (including *Plesiolestes sirokyi* Szalay, 1973; Szalay and Delson, 1979) (localities CP13IIA, CP15A, CP16A*, CP21A, NP1C, [NP3*A*, B]). See Webb (1996, 1998) for additional locality information.

Comments: The distinctiveness of this genus from *Plesiolestes* has been the subject of debate. Several authors (Wilson and Szalay, 1972; Szalay, 1973; Szalay and Delson, 1979; Williamson, 1996) have argued for placing both *T. sirokyi* and *T. wilsoni* in the genus *Plesiolestes*. Although the taxonomic arrangement followed here does recognize a relationship between *Torrejonia* and *Plesiolestes*, the very large size and distinct p4 morphology shared by *T. wilsoni* and *T. sirokyi* suggests that retaining them in a separate genus is preferable.

Anasazia Van Valen, 1994 (including _Palaechthon_, in part)

Type species: *Anasazia williamsoni* Van Valen, 1994.

Type specimen: NMMNH P-19860, Torrejonian of New Mexico.

Characteristics: Large palaechthonid; p1 retained; canine and p2 relatively large; p4 with broad trigonid region, no metaconid, and small but distinct paraconid placed very low on the tooth.

Length of p4 (no m2 known): 2.9 mm.

Included species: *A. williamsoni* only (including *Palaechthon* sp. Williamson and Lucas, 1993; *Anisazia* Williamson, 1996), known from locality SB23C* only.

Comments: The type specimen of this genus was confiscated from a commercial collector, making its precise provenance unknown. It was apparently found in the Kutz Canyon area of the San Juan Basin, New Mexico (Van Valen, 1994). The only other known specimen, a tentatively referred maxillary fragment (NMMNH P-16191), is from the upper part of the Kimbeto Wash, and is probably somewhat more recent than the holotype (Williamson, 1996).

PICRODONTIDAE SIMPSON, 1937

Characteristics: Lower molar trigonids reduced and constricted, talonids expanded and flattened with papillated enamel; upper and lower first molars enlarged with molars reducing in size from first to third; premolars relatively small and simple; upper molars with low relief and greatly expanded, flattened trigon basins with papillated enamel; cranial material indicates a fairly long and narrow snout; no postcranial material for this family.

Comments: The systematic affinities of this very unusual group were extremely obscure until the detailed analysis by Szalay in 1968. Douglass (1908), in describing the first plesiadapiform discovered in North America, noted similarities between *Picrodus* and marsupials described by Ameghino from Patagonia. Matthew (1917a) considered *Zanycteris* a Paleocene bat of the family Phyllostomatidae. Simpson (1937), in naming the group, asserted that they were likely to be placental mammals and placed them in Insectivora incertae sedis. He also noted similarities to Chiroptera and Primates, although he disagreed with Matthew and considered the similarities with phyllostomatids to be convergent (see also Simpson, 1935b; McGrew and Patterson, 1962). Szalay (1968) convincingly demonstrated that picrodontid teeth could be most plausibly derived from a primitive plesiadapiform model.

***Draconodus* Tomida, 1982 (including *Picrodus*, in part)**

Type species: *Draconodus apertus* Tomida, 1982.

Type specimen: UALP 13217, Torrejonian of Utah.

Characteristics: Upper molars with paracone and metacone relatively close together (retaining a more cuspate appearance than in other picrodonts), relatively narrow and rectangular in outline; M1 with postparacrista and premetacrista separated and of equal length (modified from Tomida, 1982).

Average length of M1: 1.85 mm.

Included species: *D. apertus* Tomida, 1982 (including *Picrodus apertus*, Williams, 1985) only, known from locality CP1C*only.

Comments: This poorly known taxon is the oldest and most primitive genus of picrodontid, retaining more distinct paracone and metacone cusps than in later species. Tomida (1994) has argued for a direct origin for picrodonts from *Purgatorius*, through this taxon (see also Van Valen, 1994; Scott and Fox, 2005). Silcox's (2001) results would argue against this, supporting derivation instead from a somewhat more derived, palaechthonid-like form. The primary basis for this debate are two features (Scott and Fox, 2005): the possibly primitive, small m3 of *Picrodus*, which would suggest an origin from a very primitive plesiadapiform, and the presence of a well-developed postprotocingulum enclosing a large posterointernal basin in all known picrodontids, which would support an origin from a somewhat more derived plesiadapiform. Since the fossil sister taxon to Primates (including plesiadapiforms) is currently unresolved, it is not clear whether or not the small m3 seen in *Picrodus* is primitive or derived. One candidate for the sister taxon to Primates, Mixodectidae, for example, has an m3 morphology that is quite unlike *Picrodus*, including the presence of a large hypoconulid. However, the absence of a well-developed postprotocingulum in *Purgatorius* and *Micromomys fremdi* makes it clear that the development of such a crest is a derived state within plesiadapiforms. Although the form of the postprotocingulum exhibits some evidence of homoplasy among plesiadapiforms (Scott and Fox, 2005), until the wider relationships of plesiadapiforms to other fossil taxa are resolved, derivation of picrodontids from a primitive paromomyoid (or possibly plesiadapoid) with a well-developed postprotocingulum seems more likely than an origin directly out of *Purgatorius*.

***Picrodus* Douglass, 1908 (including *Megopterna*)**

Type species: *Picrodus silberlingi* Douglass, 1908.

Type specimen: CM 1670, Torrejonian of Montana.

Characteristics: Upper molars with distinct stylar shelves, reduced paracone, and incomplete postprotocrista; no parastylar lobe on M1.

Average length of m1: 2.2–2.5 mm.

Included species: *P. silberlingi* Douglass, 1908 (including *Megopterna minuta* Douglass, 1908; *Picrodus* sp. probably *P. silberlingi*, Winterfeld, 1982) (known from localities CP13B, E, CP13IIA, CP14A, D, NP19C*, NP20A, [B]); *P. calgariensis* Scott and Fox, 2005 (localities CP13C, NP3A0*); *P. canpacius* Scott and Fox, 2005 (localities [CP13IIA], CP15A, CP16A, NP1C*, NP19IIB); *P. lepidus* Scott and Fox, 2005 (locality NP3C*).

Picrodus sp. has also been reported from localities CP24A, NP3A, B, C, NP20E (see Scott and Fox, 2002, 2005). Possible *Picrodus* specimens are also known from SB39C, NP1C.

***Zanycteris* Matthew, 1917a (including *Palaeonycteris*)**

Type species: *Zanycteris paleocenus* Matthew, 1917a.

Type specimen: AMNH 17180, Tiffanian of Colorado.

Characteristics: Upper molars with reduced stylar shelves, paracone relatively distinct, postprotocrista extends to apex of protocone; clear parastylar lobe on M1. Cranial characteristics as for the family.

Average length of M1: 2.3 mm.

Included species: *Z. paleocenus* Matthew, 1917a only (including *Palaeonycteris paleocenica* Weber and Abel, 1928) (known from localities SB20A*, CP26A, NP3B, C).

Comments: The only known cranial material for this family is the type specimen of this genus. Unfortunately this specimen is badly crushed, and lacks any remnants of the ear region.

PLESIADAPOIDEA TROUESSART, 1897

Characteristics: Upper and lower central incisors enlarged and procumbent; lower incisors non-lanceolate; upper incisors generally bi- or tricuspate; lateral incisors, canines, and anterior premolars often reduced or lost; p4 with an unbasined talonid and no entocristid; m3 with enlarged hypoconulid; P3 generally with a large conule; P4 with a conule and a reduced, non-projecting parastylar lobe; upper molars with a centrally placed protocone on M1–2; upper and lower third and fourth premolars often enlarged and complex; auditory bullae fully ossified (modified from Beard and Wang, 1995; Silcox *et al.*, 2001).

Comments: The description of the Asian "carpolestid" *Chronolestes simul* (Beard and Wang, 1995) makes it more difficult to find defining characteristics for Plesiadapoidea, since it lacks several traits (e.g., conule on P3, multicuspate upper incisor) that are present in all of the North American representatives of the group (as defined here, so excluding *Pandemonium* and Paromomyidae – see discussion below).

PLESIADAPIDAE TROUESSART, 1897

Characteristics (modified from Russell, 1964; Szalay, Tattersall, and Decker, 1975; Gingerich, 1976; MacPhee, Cartmill, and Gingerich, 1983): Differ from microsyopids in having a tricuspate upper central incisor and in lacking a lanceolate lower central incisor; differ from carpolestids in lacking a bladelike lower fourth premolar and in having simple upper third and fourth premolars; differ from paromomyids and saxonellids in having more robust lower central incisors; differ from picromomyids and picrodontids in lacking fourth premolar or first molar specializations. Cranium with broadly splayed zygomatic arches; strong postorbital constriction; no postorbital bar (Russell, 1959; Szalay, 1971; Gingerich, 1976); endocast with large olfactory lobes (Gingerich and Gunnell, 2005); the internal carotid artery appears to have involuted in plesiadapid ontogeny and grooves or tubes for the promontorial and stapedial arteries are lacking; bulla continuous with the promontorium, which may be indicative of a petrosal origin (but see, MacPhee, Cartmill, and Gingerich, 1983). Postcranial features include very dorsoplantarly deep claws; fore- and hindlimbs similar in length; generally robust bones; rounded, centrally excavated radial head features (suggestive of a generalized arboreal mode of locomotion, perhaps including some suspensory behavior).

Pronothodectes Gidley, 1923

Type species: *Pronothodectes matthewi* Gidley, 1923.

Type specimen: USNM 9547, Torrejonian of Montana.

Characteristics: Small taxon; retains primitive upper and lower dental formula of i2, c1, p3, m3.

Average length of m1: 2.0–3.0 mm.

Included species: *P. matthewi* Gidley, 1923 (known from localities NP3A0, NP19C*, NP20A); *P. gaoi* Fox, 1990a (localities NP3A– C*); *P. jepi* Gingerich, 1975 (including *Pronothodectes* sp. cf. *P. jepi*, Winterfeld, 1982) (localities CP13B*-C, [CP13IIA], CP14B).

Pronothodectes sp. also is represented at localities NP1B, C. See Webb (1996, 1998) for additional locality information.

Plesiadapis Gervais, 1877 (including *Nothodectes*, in part; *Tetonius*, in part)

Type species: *Plesiadapis tricuspidens* Gervais, 1877.

Type specimen: MNHN Crl-16, Thanetian of France.

Characteristics: Lower canines lacking; I1 tricuspate and generally bearing a centroconule; cheek teeth broad; m3 with a squared, lobate hypoconulid that is often fissured; mandible shallower and incisors less robust than *Chiromyoides* (modified from Gingerich, 1976). Cranial and postcranial features as for family.

Average length of m1 (North American species only): 2.6–5.4 mm.

Included species: *P. anceps* Simpson, 1936 (including *P. jepseni* Gazin, 1956) (known from localities [NB1B], CP13D, CP16A,?NP1C, NP19IIC*, NP20C); *P. churchilli* Gingerich, 1975 (including *P. farisi* Krishtalka, Black, and Riedel, 1975; *P.* cf. *P. churchilli* Wolberg, 1979; *P.* sp. probably *P. churchilli*, Winterfeld, 1982) (localities [NB1B], CP13F*, CP14C, CP15B, NP3F, ?NP4, NP7D, NP20E, NP47BB, C); *P. cookei* Jepsen, 1930 (localities ?CP14E, ?CP15C, CP17B*, C, CP24B, CP26C); *P. dubius* (including *Nothodectes dubius* Matthew, 1915; ?*P.* sp. indet. Simpson, 1928; *P.* cf. *fodinatus* Van Houten, 1945; *P ?pearcei* Gazin, 1956) (localities CP14E, CP17A–C, CP18B*, CP24B, CP26B, C, CP62B, C); *P. fodinatus* Jepsen, 1930 (including *P. rubeyi* Gazin, 1942; *P. farisi* Dorr, 1952; *P.* cf. *jepseni* Gazin, 1956) (localities CP13G*, CP14D, CP16C, CP22B, CP24B, [NP47C]); *P. gingerichi* Rose, 1981 (localities CP17A0*, B); *P. praecursor* Gingerich, 1975 (including *P.* sp. Gazin, 1971) (localities CP13D, CP13IIA, CP15A, NP1C, NP19IIA*); *P. rex* (including *Tetonius rex* Gidley, 1923; *Nothodectes* cf. *gidleyi* Simpson, 1927; *P. rex* Simpson, 1937; *P.* cf. *fodinatus* Gazin, 1956; *P. gidleyi* Dorr, 1958; *P. paskapooensis* Russell, 1964; *P.* sp., McKenna [in Love, 1973]; *P.* cf. *P. rex* Wolberg, 1979) (localities CP13E, CP13IIA1, CP16B, CP21B, CP22A, CP24A, CP26A, NP3A–D, NP19IID*, NP20D, NP47B); *P. simonsi* Gingerich, 1975 (locality CP13F*).

Plesiadapis sp. is also known from localities NB1A, CP13IIC, NP3F, NP4, NP47C. See Secord (1998), Higgins (1999), and Lofgren, McKenna, and Walsh (1999) for additional locality information.

Comments: *Plesiadapis* is one of the most common plesiadapiforms present in later Paleocene sediments in North America (Gingerich, 1976; Rose, 1981). As such, *Plesiadapis* is a very useful biostratigraphic indicator. Biochronological zonations for the Tiffanian and Clarkforkian North American Land-Mammal Ages have been developed based on the stratigraphic occurrence of *Plesiadapis* species (Gingerich, 1975, 1976). A single specimen of *Plesiadapis dubius* has been described from the Wasatchian (Rose and Bown, 1982), which is the only documentation of the survival of this family past the

Clarkforkian–Wasatchian boundary. Unlike the supposed Wasatchian carpolestid material, this specimen is known to have been found with clear Wasatchian index fossils (*Hyracotherium* and *Cantius*; Rose, 1981; Rose and Bown, 1982) and is accompanied by an extremely carefully detailed provenance (Rose and Bown, 1982; Rose personal communication). This genus is very well known from cranial and postcranial material from Europe (Russell, 1959, 1964; Szalay, Tattersall, and Decker, 1975; Beard, 1989), and North America (Gingerich and Gunnell, 1992, 2005).

***Chiromyoides* Stehlin, 1916**

Type species: *Chiromyoides campanicus* Stehlin, 1916.

Type specimen: NMB Cy-153, Thanetian of France.

Characteristics: i2 and lower canine absent; cheek teeth small but broad; mandible very deep; lower central incisor very robust.

Average length of m1: 2.0–2.8 mm.

Included species: *C. caesor* Gingerich, 1973 (known from localities SB20A, CP13F*, CP14C, D); *C. major* Gingerich, 1975 (localities CP17B*, C); *C. minor* Gingerich, 1975 (localities SB39B, CP24A*); *C. potior* Gingerich, 1975 (including *C. caesor* Schiebout, 1974) (localities SB20B*, SB39C, CP13G, H, CP17A, CP22B).

An additional species of *Chiromyoides* is known from locality CP14E. *Chiromyoides* may also be represented at CP13IIA1.

Comments: *Chiromyoides* is primarily known from Europe, and relatively rare in North America being represented mostly by isolated upper or lower incisors (Gingerich and Dorr, 1979). The genus is distinct from other plesiadapids (and other plesiadapiforms) in the very deep mandible and extremely enlarged and robust upper and lower first incisors (Gingerich, 1973). The unique nature of this dentition may have been related to a specialization for seed eating (Gingerich, 1976).

***Nannodectes* Gingerich, 1975 (including *Nothodectes*, in part; *Pronothodectes*, in part; *Plesiadapis*, in part)**

Type species: *Nannodectes gazini* Gingerich, 1975.

Type specimen: AMNH 92008, Tiffanian of Wyoming.

Characteristics: Small taxon; i2 lacking; incisors and cheek teeth narrow; P3 triangular; I1 lacking centroconule. Postcranial characteristics as for the family.

Average length of m1: 2.1–3.1 mm.

Included species: *N. gazini* Gingerich, 1975 (Gazin, 1956; originally described as *Pronothodectes* cf. *matthewi*; including *Pronothodectes* cf. *simpsoni* Gazin, 1956; cf. *Pronothodectes matthewi* Gazin, 1969) (known from localities CP16A*, CP21A); *N. gidleyi* (Matthew, 1917b; originally described as *Nothodectes gidleyi*; including *Plesiadapis gidleyi* Simpson, 1928) (localities SB20A*, CP14C, D); *N. simpsoni* (Gazin, 1956; originally described as *Pronothodectes simpsoni*) (localities CP13IIA1, CP16B*, NP3B); *N. intermedius* (Gazin, 1971; originally described as *Pronothodectes intermedius*) (localities CP13D, CP13IIA, CP15A*, NP1C, NP19IIA,B, [NP20B]).

Nannodectes sp. is also known from localities SB39C, NP47B, and may be present at localities NP3E and NP20E.

Comments: This taxon was named by Gingerich to recognize the existence of a distinct, smaller lineage of plesiadapids in North America (Gingerich, 1975, 1976). *Plesiadapis* and *Nannodectes* are very similar morphologically, and both probably evolved from a similar species of *Pronothodectes*. *Nannodectes* stayed comparable in size with *Pronothodectes* for the early part of its evolution, however, and shows an increase in size only after *Plesiadapis* has become substantially enlarged (see Gingerich, 1976: Figure 24). Postcrania are known for two species, *Nannodectes gidleyi* (Simpson, 1935a) and *Nannodectes intermedius* (Beard, 1989), and are generally similar to those described for *Plesiadapis*. Fragmentary cranial material is also known for both of these species (Simpson, 1935a; Gingerich, Houde, and Krause, 1983; Boyer *et al.*, 2004)

Indeterminate plesiadapids

Indeterminate plesiadapids are also known from localities NP3D, NP3F, and CP13IIB (Secord, 1998).

CARPOLESTIDAE SIMPSON, 1935b

Characteristics modified from Rose (1975), Beard and Wang (1995), Silcox *et al.* (2001), Bloch and Boyer (2002), and Bloch and Silcox (2003, 2006).

Cranial characteristics: Cranium with non-convergent orbits; no postorbital bar; long snout; large palatal fenestra(e); unreduced internal carotid circulation with grooves for both promontorial and stapedial arteries running transpromontorially from a posteromedially positioned posterior carotid foramen; two-chambered auditory bulla.

Dental characteristics: Small plesiadapoids; p2 very small and buttonlike or absent; p3 small relative to m1; p4 hypertrophied, very strongly exodaenodont, polycuspidate (including a paraconid and metaconid), and plagiaulacoid; lower molars exodaenodont; m1 with trigonid elongated mesially and talonid that is shortened relative to m2–3; I1 with complex, multicuspate crown; P3 with at least two buccal cusps, a paraconule that is continuous with a median crest, and a protocone; P4 with at least four buccal cusps and a crest that stretches both mesially and distally from the paraconule; hypocones on M1–2 distinct.

Postcranial characteristics: Humerus with a spherical capitulum, strong deltopectoral crest, shallow olecranon fossa, abruptly flaring supinator crest; ulna with a long, anteriorly inflected olecranon process; long manual phalanges; femur with a posteroproximally extended articular surface of the head, distally positioned, medially extended lesser trochanter, shallow patella groove; tibia with wide, ungrooved distal articular surface; astragalus with a short neck, elliptical head; calcaneum short with a distally positioned peroneal process; innominate with a cranially buttressed, elliptical acetabulum and cranially positioned ischial spine; claws on pedal digits II to V; hallux divergent with a nail.

Comments: Many of the dental characteristics listed here for Carpolestidae are not found in the putative Asian carpolestid *Chronolestes simul*. Following analyses by Silcox *et al.* (2001) and Silcox (2001; see also Bloch and Fisher, 1996; Fox, 2002), this taxon is considered here not to be a carpolestid, but rather a basal plesiadapoid.

Carpolestids are unusual for plesiadapiforms in retaining a third upper incisor even in advanced members of the group (Fox, 1993; Bloch and Gingerich, 1998), and a third lower incisor in at least one species (*Elphidotarsius shotgunensis*; Fox, 1993; Silcox, 2001)

All of the postcranial material known for carpolestids can be attributed to *Carpolestes simpsoni* (Bloch and Gingerich, 1994, 1998; Bloch and Boyer, 2001, 2002, 2003) the humerus described by Beard [1989] is probably *C. simpsoni*, not *C. nigridens* [J. I. Bloch]), so this taxon forms the basis for the list of characteristics given above. This sample includes very complete and well-associated material (Bloch and Boyer, 2001, 2002, 2003) which demonstrates that *C. simpsoni* was a committed arborealist, but without the leaping specializations characteristic of most early euprimates (Bloch and Boyer, 2002). Like euprimates, however, this species had features of the hands and feet for grasping, including a divergent hallux with a nail, which had been thought to be a uniquely euprimate character complex within primates (Bloch and Boyer, 2002, 2003). Well-preserved, fairly complete cranial material is only known for *Carpolestes simpsoni* (Gingerich, 1987; Bloch and Gingerich, 1994, 1998; Bloch and Silcox, 2003, 2006).

Elphidotarsius Gidley, 1923

Type species: *Elphidotarsius florencae* Gidley, 1923.

Type specimen: USNM 9411, Torrejonian of Montana.

Characteristics: p3 premolariform and larger both relatively and absolutely than in *Carpodaptes*, or *Carpolestes*; p4 enlarged, exodaenodont, and bladelike, larger than m1 but bearing no more than five longitudinally arranged apical cuspules, followed by a single, strong talonid cusp on a well-demarcated heel; trigonid of m1 extended mesiodistally, so that the protoconid is placed mesial to the metaconid, but arrangement of cusps not as nearly linear as in *Carpodaptes*, *Carpolestes*, or *Carpocristes*; P1 and P2 small, single-rooted, with peglike crowns; P3 enlarged with an intermediate crest bearing a conule, no more than four buccal cusps, and hypocone weakly developed; P4 larger than P3, also with a single intermediate ridge bearing a cusp; P4 with no more than five buccal cusps, hypocone and pericone weak or absent; upper molars generally plesiadapoid-like (modified from Rose, 1975; Sikox *et al.*, 2001).

Average length of m2: 1.38 mm.

Included species: *E. florencae* (known from localities CP13B, NP19C*, NP20A); *E. russelli* Krause, 1978 (localities NP1C*, NP3A, NP19IIA, AA); *E. shotgunensis* Gazin, 1971 (locality CP15A*); *E. wightoni* Fox, 1984b (localities NP3A–C*). See Hartman and Krause (1993) and Webb (1996, 1998) for additional locality information.

Comments: This genus contains the most primitive North American members of Carpolestidae. As an assemblage of primitive forms, the taxon is almost certainly a paraphyletic stem taxon to the rest of the family (Bloch *et al.*, 2001; Silcox *et al.*, 2001). This is reflected in the characteristics of the grouping – *Elphidotarsius* shows the basic features of a carpolestid, but not developed to the extent seen in more advanced forms.

Carpodaptes Matthew and Granger, 1921 (including *Carpolestes*, in part; *Carpocristes*, in part)

Type species: *Carpodaptes aulacodon* Matthew and Granger, 1921.

Type specimen: AMNH 17367, Tiffanian of Colorado.

Characteristics: p2 absent; p3 small and single-rooted; p4 enlarged and bladelike with five or six apical cuspules, intermediate between *Elphidotarsius* and *Carpolestes* in terms of the size relative to m1 and distinctiveness of talonid; m1 trigonid elongated and open, with a more linear arrangement of cusps than *Elphidotarsius*; P3 and P4 subequal in size, both enlarged relative to upper molars, each with three mesiodistally aligned rows of cuspules, intermediate in relative size between *Elphidotarsius* and *Carpolestes*; P3 buccal row with four cuspules, lingual section of the tooth with distinct protocone and hypocone; P4 with five or six buccal cusps, lingual region with protocone, hypocone, and pericone all clearly developed (modified from Rose, 1975; Silcox *et al.*, 2001).

Average length of m2: 1.1–1.4 mm.

Included species: *C. aulacodon* (known from locality SB20A*); *C. cygneus* (Russell, 1967; originally described as *Carpolestes cygneus*; including *Carpocristes cygneus* Beard and Wang, 1995) (localities NP3 [C], F*); *C. hazelae* Simpson, 1936 (localities CP13E, ?CP14C, CP15A, NP1C, NP3A–D, [NP4], NP19IIC*, NP20D, [NP47B]); *C. hobackensis* Dorr, 1952 (including *Carpocristes hobackensis* Beard and Wang, 1995) (localities CP22B*, CP26A, NP3D, E, NP20E, [NP47BB]); *C. rosei* (Beard, 2000; originally described as *Carpocristes rosei*) (localities CP16B*, NP4, NP47B); *C. stonleyi* Fox, 2002 (localities CP13F, [NP3E], NP7D*).

Carpodaptes sp. is known from localities CP24A, ?NP19IID, NP47BB.

Comments: Beard and Wang (1995) proposed the name *Carpocristes* for *C. oriens*, a very autapomorphic carpolestid from the Wutu Formation, Shandong Province of the People's Republic of China. Beard and Wang (1995) and Beard (2000) also referred three North American species to the genus: *C. cygneus*, *C. hobackensis*, and *C. rosei*. This referral was based on a hypothesis of relationships that would put these four species in a distinct clade. Given the aberrant nature of the dentition of *C. oriens*, and its retention of p2 (a tooth lost not only in *C. rosei*, *C. cygneus*, and *C. hobackensis* but also in *Elphidotarsius wightoni* and *E. russelli*; Fox, 1984b, 2000; Silcox *et al.*, 2001), it seems likely that *C. oriens* diverged from the carpolestid tree at a more primitive stage and proceeded on its own unique evolutionary course. If the taxon evolved from some (unknown) *Elphidotarsius* or *Carpodaptes* that retained p2, the similarities between *C. oriens*, *C. rosei*, *C. cygneus*, and *C. hobackensis* would likely be convergent. (See also Fox [2002] for a similar conclusion reached independently).

Almost all of the features in the diagnosis of *Carpocristes* (Beard and Wang, 1995, p. 14; see also Beard, 2000) are present either in *Carpodaptes* (subequal P3–4, no anterolabial spur on P3); some species of *Carpolestes* (pronounced posterolingual excavation on p4, in *C. dubius* and *C. simpsoni*), or are missing in some supposed *Carpocristes* (proliferation of crests in the median row of P3–4, lacking in *C. cygneus* and *C. hobackensis*; posterior apical cusp on p4 very posterior in position, lacking in *C. oriens*; pronounced posterolingual excavation on p4, lacking in *C. rosei*). (See also the discussion in Bloch *et al.* [2001] and Fox [2002].)

For these reasons, the name *Carpocristes* is limited here to the markedly divergent Asian form, which is easily distinguished by its addition of several crests between the buccal and lingual cusps on the upper premolars. Bloch *et al.* (2001) supported another possibility, which would group *C. hobackensis* (but not *C. cygneus*) with *C. oriens* in a possible ancestor–descendent relationship. They recognized this by including *C. hobackensis* in *Carpocristes*. The character support for this relationship is limited, however, and this scenario requires the re-evolution of p2 in *C. oriens* (see also discussion in Fox [2002]). In light of these considerations, *C. hobackensis* is retained here in *Carpodaptes*. In any case, *Carpodaptes* is an inherently paraphyletic group, giving rise to one or more lineage of *Carpolestes*, as well as possibly *Carpocristes* (Bloch *et al.*, 2001; Silcox *et al.*, 2001).

Holtzman (1978) referred the material from the Tongue River Formation to *Carpodaptes cygneus* and *C. hobackensis*. Beard and Wang (1995) argued that the supposed *C. hobackensis* material did not belong to that genus and considered the collection to contain *C. hazelae* instead. Beard (2000) assigned the "*C. cygneus*" material to *C. rosei*. More recently, Fox (2002) assigned the Tongue River Formation material to *Carpodaptes* cf. *hobackensis* and *C.* cf. *hazelae*. Although we have followed Fox's (2002) attributions of this material, we agree with him that a revision of *Carpodaptes hazelae* is needed, which may lead to further revisions to the attribution of this material.

Carpomegodon Bloch _et al._, 2001 (including _Carpodaptes_, in part; _Carpolestes_, in part)

Type species: *Carpomegodon jepseni* (Rose, 1975) (originally described as *Carpodaptes jepseni*)

Type specimen: YPM-PU 20716, Tiffanian of Wyoming.

Characteristics: Largest carpolestid known with higher-crowned p4 than other species; lower dental formula i2, c1, p2, m3; i1 very robust; teeth between i1 and p4 single-rooted and very compact, i2 extremely reduced; p4 with six apical cuspules (one or two additional cuspules variably present), pointed apex; upper dental formula I?, C1, P4, M3; P3 crown larger than that of P4, with antero-external extension, four buccal cusps (incipient fifth); M1–2 smaller than P3 and P4, similar to *Carpolestes dubius* (modified from Bloch *et al.*, 2001).

Average length of m2: 1.71 mm.

Included species: *C. jepseni* only (including *Carpodaptes jepseni* Rose, 1975; Szalay and Delson, 1979; *Carpolestes jepseni* Gingerich, 1980; Bloch and Gingerich, 1998), known from locality CP13F* only.

Comments: The generic placement of the material included here in *Carpomegodon* has been somewhat controversial. *Carpomegodon* exhibits distinctive features of both *Carpodaptes* (e.g., only six distinct apical cusps on p4) and *Carpolestes* (e.g., relative size of p4, anternoexternal extension on P3, P3 > P4). This mosaic of features has led to shifting taxonomic placement between those two genera (e.g., Rose, 1975; Bloch and Gingerich, 1998; Bloch, Rose, and Gingerich, 1998). New specimens described by Bloch *et al.* (2001) document a derived feature (extreme shortening of the anterior lower jaw) not seen in other carpolestids, which suggests this taxon represents a side branch not directly ancestral to *Carpolestes*, which, therefore, requires its own distinct generic designation.

Carpolestes Simpson, 1928 (including _Litotherium_; _Carpodaptes_, in part)

Type species: *Carpolestes nigridens* Simpson, 1928.

Type specimen: AMNH 22159, Clarkforkian of Montana.

Characteristics: p3 very small or absent; p4 very enlarged, hypertrophied and exodaenodont with eight or nine apical cuspules and a talonid that merges totally with the main blade; p4 significantly longer, taller, and more exodaenodont than any other tooth in the lower dentition; trigonid of m1 with cusps arranged in a line, of a similar height to the talonid of p4 and forming part of a continuous blade with that tooth; P3–4 larger than upper molars, with P3 larger than P4, both teeth relatively larger than in *Carpodaptes* or *Elphidotarsius*, although with a similar polycuspate pattern to the former (modified from Rose, 1975).

Average length of m2: 1.48–1.52 mm.

Included species: *C. nigridens* (including *C. aquilae*, *Litotherium complicatum* Simpson, 1929; *Carpodaptes nigridens* Szalay and Delson, 1979) (known from localities CP13H, CP14E, CP17A*, B, CP26C); *C. dubius* Jepsen, 1930 (including *Carpodaptes dubius* Szalay and Delson, 1979) (localities CP13G*, CP24B); *C. simpsoni* Bloch and Gingerich, 1998 (localities CP17B*, CP18B).

Carpolestes sp. is also known from localities CP14B, D, CP15B, CP17C, CP24B.

Comments: There are a few specimens that have been identified as *Carpolestes* sp. from Wasatchian localities (Twisty-Turn Hollow, SC-2, SC-161 [within locality CP19A]; Rose, 1975, 1977, 1981), which would imply a range extension for carpolestids beyond the end of the Paleocene in North America (see also McKenna and Bell, 1997). Rose (1981) indicated that the locality data for the SC-2 and SC-161 specimens were probably erroneous, in light of the absence of any Wasatchian carpolestid material with a better-substantiated provenance. SC-161 and Twisty-Turn Hollow (SC-1) are located very close to numerous Clarkforkian localities (Rose, 1981: p. 10 and Figure 2); consequently, imprecise locality data could have been incorrectly interpreted as Wasatchian, rather than Clarkforkian, sites. The lack of subsequent discoveries of Wasatchian carpolestids, in spite of very intense sampling for this time period (e.g., Bown *et al.*, 1994), supports this view. As noted above, well-preserved cranial and postcranial material is known for *Carpolestes simpsoni* (Bloch and Boyer, 2002; Bloch and Silcox, 2006).

SAXONELLIDAE (RUSSELL, 1964) (= SAXONELLINAE)

Characteristics: Long, narrow, tapering i1 with a root that extends distally to at least m1; lower dentition very reduced (i1, c0, p2, m3); p3 enlarged and bladelike (plagiaulacoid), larger than p4; P2 small, single-rooted and peglike; P3 very enlarged with an expanded buccal border (particularly via an enlarged parastylar lobe and cusp), central conule, and relatively very narrow lingual border with no hypocone or pericone; P4 smaller than P3 and M1, less molariform than P3, with only paracone and metacone cusps buccally, but a similar narrow lingual border lacking hypocone or pericone; molars generally plesiadapoid-like (modified from Russell, 1964; Fox, 1991). There is no cranial material known for this family. Two postcranial specimens referred to *Saxonella crepaturae* (Szalay and Delson, 1979; Szalay and Dagosto, 1980; Szalay and Drawhorn, 1980), although attribution not based on a direct dental association.

***Saxonella* Russell, 1964**

Type species: *Saxonella crepaturae* Russell, 1964.

Type specimen: Geiseltalmuseum Wa/351, Thanetian of Germany.

Characteristics: As for the family.

Average length of m2: 1.65 mm.

Included species: *S. naylori* Fox, 1991 (known from localities NP3B, C*); *S.* sp. nov. (locality NP3B); *S.* sp. (locality NP3A). See Webb (1998) for additional locality information.

Comments: *Saxonella* was considered to be a solely European genus for almost 20 years, stretching from the description by Russell (1964) of the type material from the Walbeck fissure in eastern Germany, to Fox's (1984c) report of a specimen from western Canada, some 10 000 km from Walbeck. This family is unknown from the richer, contemporary deposits to the south, which would suggest its range was restricted by climatic or ecological factors (Fox, 1984c, 1991). The systematic position of the genus has been a subject of debate. Russell (1964) originally placed *Saxonella* in a subfamily of the Carpolestidae, based on the shared possession of plagiaulacoid premolars, an attribution that was followed in some subsequent taxonomies (e.g., Romer, 1966; McKenna, 1967). Rose (1975) pointed out that the tooth that is enlarged is different in carpolestids and *Saxonella* (p3 rather than p4) and suggested that the subfamily be raised to familial rank to recognize the degree of divergence between this group and other plesiadapoids.

Various authors (e.g. Van Valen, 1969, 1994; McKenna and Bell, 1997) have indicated a closer tie to plesiadapids than to carpolestids and have classified *Saxonella* as a subfamily of the Plesiadapidae. Gingerich (1976) dismissed this notion and suggested that the similar incisor morphology shared by *Phenacolemur* and *Saxonella* indicates that the latter should be placed in a subfamily of the Paromomyidae. As noted by Fox (1991), the post-incisal morphology of *Saxonella* does not support this view. Silcox *et al.* (2001) and Silcox (2001) provided evidence for a closer relationship between Plesiadapidae and Carpolestidae than either share with *Saxonella*, which would support the placement of this genus in its own family.

FAMILY OR SUPERFAMILY UNCERTAIN

***Pandemonium* Van Valen, 1994**

Type species: *Pandemonium dis* Van Valen, 1994.

Type specimen: UMVP 1631, Puercan of Montana.

Characteristics: Upper molars with wide stylar shelf, moderate conules, and small hypocone that is differentiated from the postcingulum; P3–4 lack conules; lower molars moderately transverse with relatively low trigonid, paraconid small; lower molar talonids broad and shallow; m3 hypoconulid expanded into incipient third lobe; p3–4 simple and transverse.

Average length of m1: 2.4–2.8 mm.

Included species: *P. dis* only (known from localities NP16B, C*).

Comments: Van Valen (1994) described *Pandemonium* as a plesiadapid and made comparisons with both *Purgatorius* and *Pronothodectes*, apparently to show that it has features that are intermediate between these taxa. Van Valen's own diagram of plesiadapiform phylogeny (1994, p. 61) placed it at the base of Plesiadapoidea (including Plesiadapidae, Carpolestidae, and *Saxonella*), however, rather than being ancestral solely to Plesiadapidae. Although there are general similarities between *Pandemonium* and *Pronothodectes* that might support the inclusion of the former in Plesiadapoidea, this hypothesis was not supported by Silcox (2001), who suggested rather that it could be considered a palaechthonid. In light of these conflicting views the genus is left in an uncertain position here.

BIOLOGY AND EVOLUTIONARY PATTERNS

The North American plesiadapiform radiation is characterized by taxonomic diversity and morphological complexity (Figure 14.4). Paleobiological attributes of plesiadapiforms can be separated into dietary and locomotor categories.

Based on dental evidence, it is clear that the more derived members of most plesiadapiform clades became dietary specialists with adaptations that included loss of anterior teeth, enlargement and enhancement of upper and lower central incisors, and modification and specialization of distal premolars, especially upper and lower fourth premolars. Early and primitive members of all families were small animals with sharply cusped and crested teeth, suggestive of insectivorous habits, although they had teeth that had lower crowns and broader talonid basins than contemporary insectivores, signifying a move towards omnivory. These small, primitive taxa include *Purgatorius*, all palaechthonids except *Torrejonia* and *Anasazia*, the microsyopid *Navajovius*, the plesiadapid *Pronothodectes*, and the paromomyid *Paromomys* (Kay and Cartmill, 1977). Several groups of small plesiadapiforms (e.g., micromomyids, uintasoricine microsyopids, and toliapinids) became more bunodont through time, implying that they probably ate a progressively more omnivorous diet (Hooker, Russell, and Phélizon, 1999). Derived, small-bodied taxa that may have specialized on soft fruit, larvae, nectar, and/or tree exudates include the picromomyids *Alveojunctus* and *Picromomys* (Rose and Bown, 1996), the picrodontids (Szalay, 1968), and the more derived paromomyids (Gingerich, 1974b). These forms presumably used their enlarged basin areas to crush and mash their food.

Biknevicius (1986) studied the dental function in carpolestids. She concluded that their unique pattern of premolar morphology would have functioned in processing food with a soft inside and harder, or more ductile, outside, such as tough fruit, nuts, or insects. Given the similarities in its dental specializations, *Saxonella* probably had a comparable diet. Carpolestids underwent a progressive increase in features such as the size and number of apical cuspules on p4 through the Paleocene (Rose, 1975; Biknevicius, 1986), likely indicating an increase in the amount of fruit consumed by the late Paleocene. This shift may have been associated with a radiation of angiosperms in the Northern Hemisphere, and possibly the evolution of euprimate-like grasping abilities (Bloch and Boyer, 2002, 2003).

Larger-bodied taxa like the more derived microsyopids (*Microsyops* and *Megadelphus*) and *Torrejonia* and *Anasazia* probably consumed a mixed diet of fruits and insects. Only *Craseops* seems to have a dentition that might allow for some folivory in its dietary regimen, but no plesiadapiform developed the combination of large body size and extremely crested teeth needed to be strictly folivorous (Kay, 1975). *Plesiadapis* probably specialized on fruits but may have included other vegetation such as roots or tubers in its diet as well. *Chiromyoides* combines a very deep mandible with extremely stout incisors, a pattern that has been suggested to represent a specialization for granivory (Gingerich, 1976) (Figure 14.5).

Plesiadapiform locomotion is difficult to assess given the lack of postcranial specimens for most taxa. Until recently the larger species of *Plesiadapis* (e.g., *P. tricuspidens*, *P. walbeckensis*, *P. cookei*) were the best-represented plesiadapiform taxa postcranially, and these taxa have long formed the basis for an understanding of plesiadapiform postcranial anatomy. Preliminary indications of some differences between these forms and other plesiadapiforms (e.g., they lack grasping hand proportions found in all other plesiadapiforms studied [Bloch and Boyer, 2002, 2003]) suggests that these larger plesiadapids may not be "typical plesiadapiforms" and should not be treated as such. These large plesiadapids were probably arboreal quadrupeds, based on their large, laterally compressed claws and their body proportions (Szalay, Tattersall, and Decker, 1975; Beard, 1989; Gingerich and Gunnell, 1992). However, some taphonomic evidence suggests that certain smaller species of *Plesiadapis* may have been terrestrial. One small species of *Plesiadapis*, *P. rex*, is very abundant at Cedar Point Quarry, a Tiffanian-aged assemblage from northwestern Wyoming. Out of 1988 mammal specimens known from Cedar Point, 811 represent *P. rex* (41%, see Rose, 1981). It is difficult to reconcile such high abundance with an arboreal lifestyle, so perhaps when postcranial remains become available for *P. rex* these may indicate that small *Plesiadapis* species were more terrestrial than larger species.

Beard (1990a, 1993a) hypothesized that paromomyids and micromomyids had specializations of the digits of the forelimb that were part of an adaptive complex that allowed for gliding locomotion. Krause (1991) cast doubt on the associations of these postcranial elements with identifiable teeth and questioned whether there was any strong evidence to hypothesize gliding as an important behavior of paromomyids. Subsequently, Runestad and Ruff's (1995) examination of limb proportions in extant gliders and non-gliding mammals and Hamrick, Rosenman, and, Brush's (1999) reexamination of phalangeal proportions in paromomyids and other mammals found little support for the gliding hypothesis. The discovery of better-associated postcranial material for paromomyids and micromomyids demonstrates that there is no longer any convincing

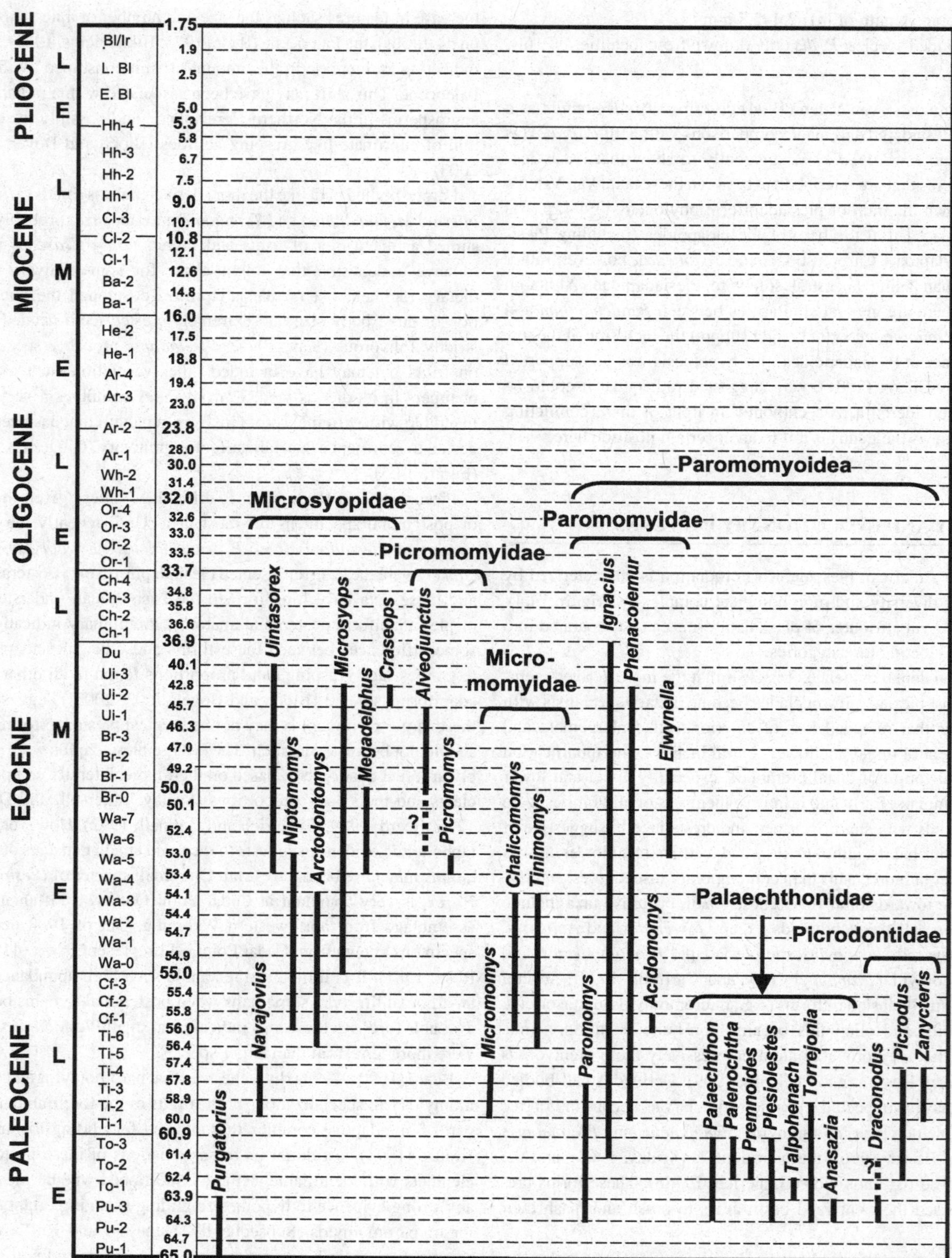

Figure 14.4. Temporal ranges of genera of Microsyopidae, Micromomyidae, Paromomyidae, Palaechthonidae, and Picrodontidae.

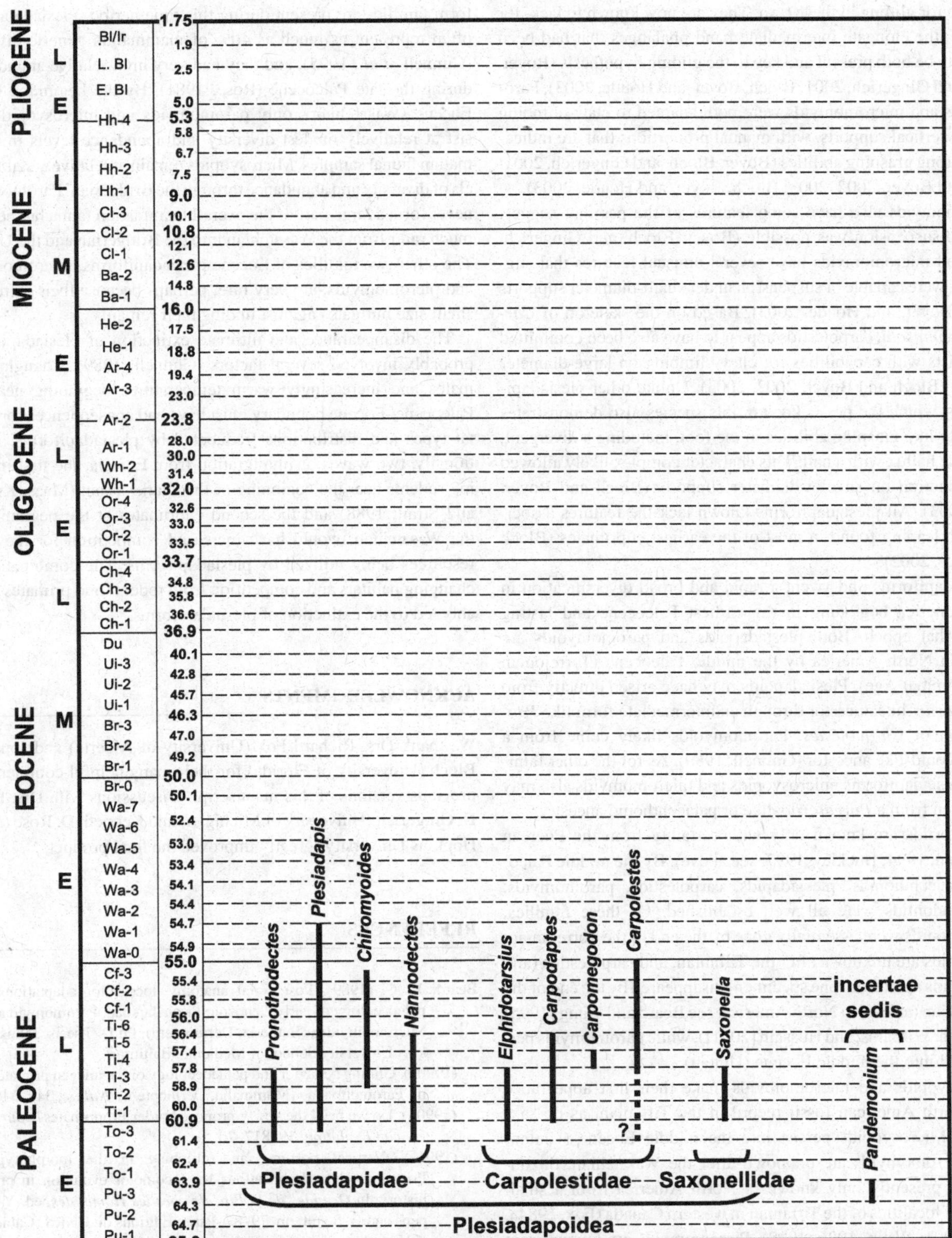

Figure 14.5. Temporal ranges of genera of Plesiadapidae, Carpolestidae, Saxonellidae, and the enigmatic *Pandemonium*.

evidence for gliding in these taxa. They are now known to lack, for example, the elongate intermediate hand phalanges that had been thought to be both present and key to the gliding hypothesis (Boyer, Bloch, and Gingerich, 2001; Bloch, Boyer, and Houde, 2003). Paromomyids and micromomyids were both adapted to claw-climbing on large vertical supports, with manual proportions that are indicative of strong grasping abilities (Boyer, Bloch, and Gingerich, 2001; Bloch and Boyer, 2002, 2003; Bloch, Boyer, and Houde, 2003).

Paromomyids also have some features of the pes that suggest hindlimb suspension was possible (Boyer, Bloch, and Gingerich, 2001) and micromomyids have several unusual features that suggest they were capable of suspension under small-diameter supports (Bloch, Boyer, and Houde, 2003). Based on the skeleton of *Carpolestes simpsoni*, carpolestids appear to have also been committed arborealists with capabilities for claw-climbing on large-diameter supports (Bloch and Boyer, 2002, 2003). Unlike other plesiadapiforms for which the pes is known, this species also demonstrates euprimate-like grasping abilities in the foot, including a divergent, opposable hallux with a nail. This character complex likely allowed for locomotion on smaller-diameter supports (Bloch and Boyer, 2002, 2003). All plesiadapiforms known lack the features associated with leaping found in most of the earliest euprimates (Bloch and Boyer, 2002).

Plesiadapiforms underwent a rapid and broad diversification in North America beginning in the earliest Paleocene and lasting through that epoch. Both plesiadapoids and paromomyoids are present in North America by the middle Paleocene (Torrejonian Land-Mammal Age). Plesiadapoids may have arisen directly from a *Purgatorius*-like ancestor through an intermediate form like *Pandemonium* or *Chronolestes*. Paromomyoids likely come from a palaechthonid-like ancestor (Gunnell, 1989). As for the other families of plesiadapiforms, microsyopids and micromomyids also may have arisen from a *Purgatorius*-like or palaechthonid ancestry.

Purgatoriids are known only with certainty from the Puercan (Van Valen, 1994; Buckley, 1997; see above). By the middle Paleocene, palaechthonids, plesiadapids, carpolestids, paromomyids, and picrodontids were all well established. Of these families, palaechthonids were essentially gone by the end of the Torrejonian; picrodontids are unknown after the Tiffanian, and carpolestids (and plesiadapids apart from one specimen) disappeared by the end of the Clarkforkian (at least in North America; see Beard and Wang [1995] Thewissen, Williams, and Hussain [2001]), while paromomyids persisted well into the middle Eocene (Uintan).

Microsyopids and micromomyids make their first appearance in the North American fossil record in the Tiffanian, as do saxonellids. Microsyopids persist into the middle Eocene (Uintan) while micromomyids are unknown after the Wasatchian. Saxonellids are presently only known in North America from a small handful of localities of the Tiffanian in western Canada (Fox, 1984a, 1990b, 1991; Webb, 1996, 1998). Picromomyids are known from the Wasatchian and Bridgerian of western Wyoming and the Uintan of California, but their record remains very sparse (Silcox, Rose, and Walsh, 2002).

Plesiadapiform diversity reaches its zenith in the Tiffanian and Clarkforkian. Seven of ten recognized North American plesiadapiform families are present during this time period. Plesiadapiforms often represent as much as 40% of mammalian generic diversity (Gunnell *et al.*, 1995), and they had very high relative abundances during the late Paleocene (Rose, 1981). By the beginning of the Eocene (Wasatchian), only paromomyids and microsyopids persist at relatively modest diversity and abundance levels in mammalian faunal samples. Microsyopids remain at relatively stable levels of diversity and abundance through the Bridgerian. Paromomyids never form a large part of the overall mammalian fauna but become much rarer from the Wasatchian into the Bridgerian and the Uintan. The other two families of Eocene plesiadapiforms, micromomyids and picromomyids, are very rare, perhaps because their extremely small size mitigates against finding their remains.

The disappearance and ultimate extinction of plesiadapiforms probably involved several factors (Gunnell, 1989). Changing climates and increasingly warm temperatures beginning near the Paleocene–Eocene boundary must have had consequences for habitat types and distributions frequented by plesiadapiforms. Additionally, two waves of immigration from Eurasia, the first involving rodents near the beginning of the Clarkforkian (Maas, Krause, and Strait, 1988) and the second euprimates at the beginning of the Wasatchian, would have increased competition for the same resources being utilized by plesiadapiforms. The combination of changing habitats and competition from rodents and primates probably led to the extinction of plesiadapiforms.

ACKNOWLEDGMENTS

We thank Drs. Richard Fox (University of Alberta) and Jonathan Bloch (University of Florida) for their many helpful comments on previous versions of this manuscript. Discussions with Drs. Philip D. Gingerich (University of Michigan) and Kenneth D. Rose (Johns Hopkins University) greatly improved the final product.

REFERENCES

Beard, K. C. (1989). Postcranial anatomy, locomotor adaptations, and paleoecology of early Cenozoic Plesiadapidae, Paromomyidae, and Micromomyidae (Eutheria, Dermoptera). Ph.D. Thesis, Johns Hopkins University School of Medicine, Baltimore.

(1990a). Gliding behavior and palaeoecology of the alleged primate family Paromomyidae (Mammalia, Dermoptera). *Nature*, 345, 340–1.

(1990b). Do we need the newly proposed order Proprimates? *Journal of Human Evolution*, 19, 817–20.

(1991a). Vertical postures and climbing in the morphotype of Primatomorpha: Implications for locomotor evolution in primate history. In *Origines de la Bipédie chez les Hominidés*, ed. Y. Coppens and B. Senut, pp. 79–87. Paris: Editions du CNRS (Cahiers de Paléoanthropologie).

(1991b). Postcranial fossils of the archaic primate family Microsyopidae. *American Journal of Physical Anthropology*, suppl 12, pp. 48–9.

(1993a). Origin and evolution of gliding in early Cenozoic Dermoptera (Mammalia, Primatomorpha). In *Primates and Their Relatives in Phylogenetic Perspective*, ed. R. D. E. MacPhee, pp. 63–90. New York: Plenum Press.

(1993b). Phylogenetic systematics of the Primatomorpha, with special reference to Dermoptera. In *Mammal Phylogeny: Placentals*, ed. F. S. Szalay, M. J. Novacek, and M. C. McKenna, pp. 129–50. New York: Springer-Verlag.

(2000). A new species of *Carpocristes* (Mammalia: Primatomorpha) from the middle Tiffanian of the Bison Basin, Wyoming, with notes on carpolestid phylogeny. *Annals of the Carnegie Museum*, 69, 195–208.

Beard, K. C. and Houde, P. (1989). An unusual assemblage of diminutive plesiadapiforms (Mammalia,?Primates) from the early Eocene of the Clark's Fork Basin, Wyoming. *Journal of Vertebrate Paleontology*, 9, 388–99.

Beard, K. C. and MacPhee, R. D. E. (1994). Cranial anatomy of *Shoshonius* and the antiquity of Anthropoidea. In *Anthropoid Origins*, ed. J. G. Fleagle and R. F. Kay, pp. 55–97. New York: Plenum Press.

Beard, K. C. and Wang, J. (1995). The first Asian plesiadapoids (Mammalia: Primatomorpha). *Annals of the Carnegie Museum*, 64, 1–33.

Beard, K. C., Ni, X., Wang, Y., Gebo, D., and Meng, J. (2005). Phylogenetic position and biogeographic significance of *Subengius mengi* (Mammalia, Carpolestidae), the oldest Asian plesiadapiform. *Journal of Vertebrate Paleontology*, 25(suppl. to no. 3), p. 35A.

Biknevicius, A. R. (1986). Dental function and diet in the Carpolestidae (Primates, Plesiadapiformes). *American Journal of Physical Anthropology*, 71, 157–71.

Bloch, J. I. and Bowen G. J. (2001). Paleocene–Eocene microvertebrates in freshwater limestones of the Clarks Fork Basin, Wyoming. In *Eocene Biodiversity: Unusual Occurrences and Rarely Sampled Habitats*, ed. G. F. Gunnell, pp. 94–129. New York: Plenum Press.

Bloch, J. I. and Boyer, D. M. (2001). Taphonomy of small mammals in freshwater limestones from the Paleocene of the Clarks Fork Basin. [In *Paleocene–Eocene Stratigraphy and Biotic Change in the Bighorn and Clarks Fork Basins, Wyoming*, ed. P. D. Gingerich.] *University of Michigan Papers on Paleontology*, 33, 185–98.

(2002). Grasping primate origins. *Science*, 298, 1606–10.

(2003). Response to comment on "Grasping primate origins". *Science*, 300, 741c.

(2007). New skeletons of Paleocene-Eocene Plesiadapiformes: A diversity of arboreal positional behaviors in early primates. In *Primate Origins: Adaptations and Evaluation*, ed. M. J. Ravosa and M. Dagosto, pp. 535–81. New York: Springer.

Bloch, J. I. and Fisher, D. C. (1996). Phylogeny of the Carpolestidae (Mammalia, Proprimates): an application of stratocladistics. *Journal of Vertebrate Paleontology*, 14(suppl. to no. 3), pp. 22A–3A.

Bloch, J. I. and Gingerich, P. D. (1994). New species of *Carpolestes* (Mammalia, Proprimates) from Clarkforkian late Paleocene limestones of the Clarks Fork Basin, Wyoming: teeth, skulls, and femur. *Journal of Vertebrate Paleontology*, 14(suppl. to no. 3), pp. 17A–18A.

(1998). *Carpolestes simpsoni*, new species (Mammalia, Proprimates) from the late Paleocene of the Clarks Fork Basin, Wyoming. *Contributions from the Museum of Paleontology, University of Michigan*, 30, 131–62.

Bloch, J. I., and Silcox, M. T. (2001). New basicrania of Paleocene–Eocene *Ignacius*: re-evaluation of the plesiadapiform-dermopteran link. *American Journal of Physical Anthropology*, 116, 184–98.

(2003). Comparative cranial anatomy and cladistic analysis of Paleocene primate *Carpolestes simpsoni* using ultra high-resolution X-ray computed tomography. *American Journal of Physical Anthropology*, suppl. 36, p. 68.

(2006). Cranial anatomy of Paleocene plesiadapiform *Carpolestes simpsoni* (Mammalia, Primates) using ultra high-resolution X-ray computed tomography, and the relationships of plesiadapiforms to Euprimates. *Journal of Human Evolution*, 50, 1–35.

Bloch, J. I., Rose, K. D., and Gingerich, P. D. (1998). New carpolestids from late Tiffanian Divide Quarry, Bighorn Basin, Wyoming. *Journal of Vertebrate Paleontology*, 18(suppl. to no. 3), p. 28A.

Bloch, J. I., Fisher, D. C., Rose, K. D., and Gingerich, P. D. (2001) Stratocladistic analysis of Paleocene Carpolestidae (Mammalia, Plesiadapiformes) with description of a new late Tiffanian genus. *Journal of Vertebrate Paleontology*, 21, 119–31.

Bloch, J. I., Boyer, D. M., Gingerich, P. D., and Gunnell, G. F. (2002). New primitive paromomyid from the Clarkforkian of Wyoming and dental eruption in Plesiadapiformes. *Journal of Vertebrate Paleontology*, 22, 366–79.

Bloch, J. I., Silcox, M. T., and Sargis, E. J. (2002). Origin and relationships of Archonta (Mammalia, Eutheria): re-evaluation of Eudermoptera and Primatomorpha. *Journal of Vertebrate Paleontology*, 22 (suppl. to no. 3), p. 37A.

Bloch, J. I., Boyer, D. M., and Houde, P. (2003). Skeletons of Paleocene–Eocene micromomyids (Mammalia, Primates): functional morphology and implications for euarchontan relationships. *Journal of Vertebrate Paleontology*, 23(suppl. to no. 3), p. 35A.

Bloch, J. I., Silcox, M. T., Boyer, D. M., and Sargis, E. J. (2007). New Paleocene skeletons and the relationship of plesiadapiforms to crown-clade primates. *Proceedings of the National Academy of Sciences*, 104, 1159–64.

Bown, T. M. (1982). Geology, paleontology, and correlation of Eocene volcaniclastic rocks, southeast Absaroka Range, Hot Springs County, Wyoming. *United States Geological Survey, Professional Paper*, 1201A, A1–A75.

Bown, T. M. and Rose, K. D. (1976). New early Tertiary primates and a reappraisal of some Plesiadapiformes. *Folia Primatologica*, 26, 109–38.

Bown, T. M., Rose, K. D., Simons, E. L., and Wing, S. L. (1994). Distribution and stratigraphic correlation of upper Paleocene and lower Eocene fossil mammal and plant localities of the Fort Union, Willwood, and Tatman formations, southern Bighorn Basin, Wyoming. Washington: *United States Geological Survey Professional Paper*, 1540, 1–103.

Boyer, D. M., Bloch J. I., and Gingerich, P. D. (2001). New skeletons of Paleocene paromomyids (Mammalia,?Primates): were they mitten gliders? *Journal of Vertebrate Paleontology*, 21(suppl. to no. 3), p. 35A.

Boyer, D. M., Bloch J. I., Silcox, M. T., and Gingerich, P. D. (2004). New observations on the anatomy of *Nannodectes* (Mammalia, Primates) from the Paleocene of Montana and Colorado. *Journal of Vertebrate Paleontology*, 24(suppl. to 3), p. 40A.

Buckley, G. A. (1997). A new species of *Purgatorius* (Mammalia: Primatomorpha) from the lower Paleocene Bear Formation, Crazy Mountains Basin, south-central Montana. *Journal of Paleontology*, 71, 149–55.

Cartmill, M. (1972). Arboreal adaptations and the origin of the order Primates. In *The Functional and Evolutionary Biology of Primates*, ed. R. Tuttle, pp. 97–122. Chicago, IL: Aldine-Atherton.

Clemens, W. A. (1974). *Purgatorius*, an early paromomyid primate (Mammalia). *Science*, 184, 903–5.

(2004). *Purgatorius* (Plesiadapiformes, Primates?, Mammalia), a Paleocene immigrant into Northeastern Montana: stratigraphic occurences and incisor proportions. *Bulletin of the Carnegie Museum of Natural History*, 36, 3–13.

Cope, E. D. (1881). On the Vertebrata of the Wind River Eocene beds of Wyoming. *Bulletin of the United States Geologic and Geographic Survey of the Territories* (F. V. Hayden, director), 6, 183–202.

(1882). Contributions to the history of the Vertebrata of the lower Eocene of Wyoming and New Mexico, made during 1881. I. The fauna of the Wasatch beds of the basin of the Big Horn River. II. The fauna of the *Catathlaeus* beds, or lowest Eocene, New Mexico. *Proceedings of the American Philosophical Society*, 10, 139–97. [Also privately published as *Paleontological Bulletin*, no. 37.]

(1883). First addition to the fauna of the Puerco Eocene. *Proceedings of the American Philosophical Society*, 20, 545–63.

Delson, E. (1971). Fossil mammals of the early Wasatchian Powder River Local Fauna, Eocene of northeast Wyoming. *Bulletin of the American Museum of Natural History*, 146, 307–64.

Dorr, J. A., Jr. (1952). Early Cenozoic stratigraphy and vertebrate paleontology of the Hoback Basin, Wyoming. *Bulletin of the Geological Society of America*, 63, 59–94.

(1958). Early Cenozoic vertebrate paleontology, sedimentation, and orogeny in Central Western Wyoming. *Bulletin of the Geological Society of America*, 69, 1217–44.

Douglass, E. (1908). Vertebrate fossils from the Fort Union beds. *Annals of the Carnegie Museum*, 5, 11–26.

Fleagle, J. G. (1988). *Primate Adaptation and Evolution*. New York: Academic Press.

Fox, R. C. (1984a). The dentition and relationships of the Paleocene primate *Micromomys* Szalay, with description of a new species. *Canadian Journal of Earth Sciences*, 21, 1262–7.

(1984b). A new species of the Paleocene primate *Elphidotarsius* Gidley: its stratigraphic position and evolutionary relationships. *Canadian Journal of Earth Sciences*, 21, 1268–77.

(1984c). First North American record of the Paleocene primate *Saxonella*. *Journal of Paleontology*, 58, 892–4.

(1990a). *Pronothodectes gaoi* n. sp., from the late Paleocene of Alberta, Canada, and the early evolution of the Plesiadapidae (Mammalia, Primates). *Journal of Paleontology*, 64, 637–47.

(1990b). The succession of Paleocene mammals in western Canada. [In *Dawn of the Age of Mammals in the Northern Part of the Rocky Mountain Interior, North America*, ed. T. M. Bown and K. D. Rose]. *Geological Society of America, Special Paper*, 243, 51–70

(1991). *Saxonella* (Plesiadapiformes:?Primates) in North America: *S. naylori*, sp. nov., from the late Paleocene of Alberta, Canada. *Journal of Vertebrate Paleontology*, 11, 334–49.

(1993). The primitive dental formula of the Carpolestidae (Plesiadapiformes, Mammalia) and its phylogenetic implications. *Journal of Vertebrate Paleontology*, 13, 516–24.

(2002). The dentition and relationships of *Carpodaptes cygneus* (Russell) (Carpolestidae, Plesiadapiformes, Mammalia) from the Late Paleocene of Alberta, Canada. *Journal of Paleontology*, 76, 864–81.

Fu, J.-F., Wang, J.-W., and Tong, Y.-S. (2002). The new discovery of the Plesiadapiformes from the early Eocene of Wutu Basin, Shandong Province. *Vertebrata PalAsiatica*, 40, 219–27.

Gazin, C. L. (1942). Fossil Mammalia from the Almy Formation in western Wyoming. *Journal of the Washington Academy of Sciences*, 32, 217–20.

(1952). The lower Eocene Knight Formation of western Wyoming and its mammalian faunas. *Smithsonian Miscellaneous Collections*, 117, 1–82.

(1956). The upper Paleocene Mammalia from the Almy Formation in western Wyoming. *Smithsonian Miscellaneous Collections*, 131, 1–18.

(1968). A new primate from the Torrejonian middle Paleocene of the San Juan Basin, New Mexico. *Proceedings of the Biological Society of Washington*, 81, 629–34.

(1969). A new occurrence of Paleocene mammals in the Evanston Formation, southwestern Wyoming. *Smithsonian Contributions to Paleobiology*, 2, 1–17.

(1971). Paleocene primates from the Shotgun Member of the Fort Union Formation in the Wind River Basin, Wyoming. *Proceedings of the Biological Society of Washington*, 84, 13–38.

Gebo, D. L., Dagosto, M., Beard, K. C., and Qi, T. (1998). The smallest primate? *American Journal of Physical Anthropology*, suppl. 26, p. 86.

(2000). The smallest primates. *Journal of Human Evolution*, 38, 585–94.

Gervais, P. (1877). Enumération de quelques ossements d'animaux vertébrés recueillis aux environs de Reims par M. Lemoine. *Journal Zoologie (Paris)*, 6, 74–9.

Gidley, J. W. (1923). Paleocene primates of the Fort Union, with discussion of relationships of Eocene primates. *Proceedings of the United States National Museum*, 63, 1–38.

Gingerich, P. D. (1973). First record of the Paleocene primate *Chiromyoides* from North America. *Nature*, 244, 517–18.

(1974a). Size variability of the teeth in living mammals and the diagnosis of closely related sympatric fossil species. *Journal of Paleontology*, 48, 895–903.

(1974b). Function of pointed premolars in *Phenacolemur* and other mammals. *Journal of Dental Research*, 53, 497.

(1975). New North American Plesiadapidae (Mammalia, Primates) and a biostratigraphic zonation of the middle and upper Paleocene. *Contributions from the Museum of Paleontology, University of Michigan*, 24, 135–48.

(1976). Cranial anatomy and evolution of early Tertiary Plesiadapidae (Mammalia, Primates). *University of Michigan Papers on Paleontology*, 15, 1–140.

(1980). Evolutionary patterns in early Cenozoic mammals. *Annual Reviews of Earth and Planetary Science*, 8, 407–24.

(1987). Early Eocene bats (Mammalia, Chiroptera) and other vertebrates in freshwater limestones of the Willwood Formation, Clarks Fork Basin, Wyoming. *Contributions from the Museum of Paleontology, University of Michigan*, 27, 275–320.

(1989). New earliest Wasatchian mammalian fauna from the Eocene of northwestern Wyoming: composition and diversity in a rarely sampled high-floodplain assemblage. *University of Michigan Papers on Paleontology*, 28, 1–97.

(1990). Mammalian order Proprimates: response to Beard. *Journal of Human Evolution*, 19, 821–2.

Gingerich, P. D. and Dorr, J. A., Jr. (1979). Mandible of *Chiromyoides minor* (Mammalia, Primates) from the upper Paleocene Chappo Member of the Wasatch Formation. *Journal of Paleontology*, 53, 550–2.

Gingerich, P. D. and Gunnell, G. F. (1992). A new skeleton of *Plesiadapis cookei*. *The Display Case*, 6, 1–2.

(2005). Brain of *Plesiadapis cookei* (Mammalia, Proprimates): surface morphology and encephalization compared to those of Primates and Dermoptera. *Contributions from the Museum of Paleontology, University of Michigan*, 32, 185–95.

Gingerich, P. D., Houde, P., and Krause, D. W. (1983). A new earliest Tiffanian (late Paleocene) mammalian fauna from Bangtail Plateau, Western Crazy Mountain Basin, Montana. *Journal of Paleontology*, 57, 957–70.

Gunnell, G. F. (1985). Systematics of early Eocene Microsyopinae (Mammalia, Primates) in the Clark's Fork Basin, Wyoming. *Contributions from the Museum of Paleontology, University of Michigan*, 27, 51–71.

(1989). Evolutionary history of Microsyopoidea (Mammalia, ?Primates) and the relationship between Plesiadapiformes and Primates. *University of Michigan Papers on Paleontology*, 27, 1–157.

(1998). Mammalian fauna from the lower Bridger Formation (Bridger A, early middle Eocene) of the southern Green River Basin, Wyoming. *Contributions from the Museum of Paleontology, University of Michigan*, 30, 83–130.

Gunnell, G. F. and Gingerich, P. D. (1981). A new species of *Niptomomys* (Microsyopidae) from the early Eocene of Wyoming. *Folia Primatologica*, 36, 128–37.

Gunnell, G. F., Morgan, M. E., Maas, M. C., and Gingerich, P. D. (1995). Comparative paleoecology of Paleogene and Neogene mammalian faunas: trophic structure and composition. *Palaeogeography, Palaeoclimatology, Palaeoecology*, 115, 265–86.

Guthrie, D. (1971). The mammalian fauna of the Lost Cabin Member, Wind River Formation (Lower Eocene) of Wyoming. *Annals of the Carnegie Museum*, 43, 47–113.

Hamrick, M. W., Rosenman, B. A., and Brush, J. A. (1999). Phalangeal morphology of the Paromomyidae (?Primates, Plesiadapiformes): the evidence for gliding behavior reconsidered. *American Journal of Physical Anthropology*, 109, 397–413.

Hartman, J. H., and Krause, D. W. (1993). Cretaceous and Paleocene stratigraphy and paleontology of the Shawmut Anticline and the Crazy Mountains Basin, Montana: road log and overview of recent investigations. In *Montana Geological Society, South Central Field Conference Guidebook*, pp. 71–84. Billings, MT: Montana Geological Society.

Higgins, P. (1999). Biostratigraphy of the boundary between Torrejonian and Tiffanian North America Land Mammal Ages. *Journal of Vertebrate Paleontology*, 19 (suppl. to no. 3), p. 51A.

Holtzman, R. C. (1978). Late Paleocene mammals of the Tongue River Formation, western North Dakota. *North Dakota Geological Survey, Report of Investigation* 65, 1–88.

Hooker, J. J., Russell, D. E., and Phélizon, A. (1999). A new family of Plesiadapiformes (Mammalia) from the Old World lower Paleogene. *Palaeontology*, 42, 377–407.

Jepsen, G. L. (1930). Stratigraphy and paleontology of the Paleocene of northeastern Park County, Wyoming. *Proceedings of the American Philosophical Society*, 69, 463–528.

Johnston, P. A. and Fox, R. C. (1984). Paleocene and late Cretaceous mammals from Saskatchewan, Canada. *Palaeontographica Abteilung A*, 186, 163–222.

Kay, R. F. (1975). The functional adaptations of primate molar teeth. *American Journal of Physical Anthropology*, 43, 195–216.

Kay, R. F., and Cartmill, M. (1977). Cranial morphology and adaptations of *Palaechthon nacimienti* and other Paromomyidae (Plesiadapoidea,?Primates), with a description of a new genus and species. *Journal of Human Evolution*, 6, 19–53.

Kay, R. F., Thorington, Jr., R. W., and Houde, P. (1990). Eocene plesiadapiform shows affinities with flying lemurs not primates. *Nature*, 345, 342–4.

Kay, R. F., Thewissen, J. G. M., and Yoder, A. D. (1992). Cranial anatomy of *Ignacius graybullianus* and the affinities of the Plesiadapiformes. *American Journal of Physical Anthropology*, 89, 477–98.

Kielan-Jaworowska, Z., Bown, T. M., and Lillegraven, J. A. (1979). Eutheria. In *Mesozoic Mammals: the First Two-thirds of Mammalian History*, ed. J. A. Lillegraven, Z. Kielan-Jaworowska, and W. A. Clemens, pp. 221–58. Berkeley, CA: University of California Press.

Krause, D. W. (1978). Paleocene primates from western Canada. *Canadian Journal of Earth Sciences*, 15, 1250–71.

(1991). Were paromomyids gliders? Maybe, maybe not. *Journal of Human Evolution*, 21, 177–88.

Krishtalka, L. (1978). Paleontology and geology of the Badwater Creek area, central Wyoming. Part 15: Review of the late Eocene Primates from Wyoming and Utah, and the Plesitarsiiformes. *Annals of the Carnegie Museum*, 47, 335–60.

Krishtalka, L., Black, C. C., and Riedel, D. W. (1975). Paleontology and geology of the Badwater Creek area, central Wyoming, Part 10: A late Paleocene mammal fauna from the Shotgun Member of the Fort Union Formation. *Annals of the Carnegie Museum*, 45, 179–212.

Leidy, J. (1872). On the fossil vertebrates in the early Tertiary formations of Wyoming. *Annual Report of the United States Geological Survey of the Territories* (F. V. Hayden, director), 5, 353–77.

Lillegraven, J. A. (1976). Didelphids (Marsupialia) and *Uintasorex* (?Primates) from later Eocene sediments of San Diego County, California. *Transactions of the San Diego Society of Natural History*, 18, 85–112.

Linnaeus, C. (1758). *Systema Naturae per Regna Tria Naturæ Secundum Classes, Ordines, Genera, Species, cum Characteribus, Differentilis, Synonymis, Locis*. London: The British Museum (Natural History), reprinting 1956.

Lofgren, D. L., McKenna, M. C., and Walsh, S. L. (1999). New records of Torrejonian–Tiffanian mammals from the Paleocene–Eocene Goler Formation, California. *Journal of Vertebrate Paleontology*, 19 (suppl. to no. 3), p. 60A.

Loomis, F. B. (1906). Wasatch and Wind River primates. *American Journal of Science*, 21, 277–85.

Love, J. D. (1973). Harebell Formation (Upper Cretaceous) and Pinyon Conglomerate (uppermost Cretaceous and Paleocene) northwestern Wyoming. *United States Geological Survey Professional Paper*, 734A, 1–54.

Lucas, S. G., and Froehlich, J. W. (1989). Fossil primates from New Mexico and the early adaptive radiation of the Adapidae. *Journal of Anthropological Research*, 45, 67–75.

Maas, M. C., Krause, D. W., and Strait, S. G. (1988). The decline and extinction of Plesiadapiformes (Mammalia: ?Primates) in North America: Displacement or replacement? *Paleobiology*, 14, 410–31.

MacPhee R. D. E. and Cartmill, M. (1986). Basicranial structures and primate systematics. In *Comparative Primate Biology,* Vol. 1: *Systematics, Evolution, and Anatomy*, ed. D. R. Swisher and J. Erwin, pp. 219–75. New York: Alan R. Liss.

MacPhee, R. D. E., Cartmill, M., and Gingerich, P. D. (1983). New Paleogene primate basicrania and the definition of the order Primates. *Nature*, 301, 509–11.

MacPhee, R. D. E., Novacek, M. J., and Storch, G. (1988). Basicranial morphology of early Tertiary erinaceomorphs and the origin of Primates. *American Museum Novitates*, 2921, 1–42.

MacPhee, R. D. E., Cartmill, M., and Rose, K. D. (1989). Craniodental morphology and relationships of the supposed Eocene dermopteran *Plagiomene* (Mammalia). *Journal of Vertebrate Paleontology*, 9, 329–49.

MacPhee, R. D. E., Beard, K. C., Flemming, C., and Houde, P. (1995). Petrosal morphology of *Tinimomys graybulliensis* is plesiadapoid not microsyopoid. *Journal of Vertebrate Paleontology*, 15 (suppl. to no. 3), p. 42A.

Marsh, O. C. (1871). Notice of some fossil mammals from the Tertiary formation. *American Journal of Science*, 2, 34–45.

(1872). Preliminary description of new Tertiary mammals. Parts I–IV. *American Journal of Science*, 4, 122–8, 202–4.

Martin, R. D. (1972). Adaptive radiation and behavior of the Malagasy lemurs. *Philosophical Transactions of the Royal Society of London*, 264, 295–352.

Matthew, W. D. (1909). The Carnivora and Insectivora of the Bridger Basin, middle Eocene. *Memoirs of the American Museum of Natural History*, 9, 291–567.

(1915). A revision of the lower Eocene Wasatch and Wind River faunas: Part IV: Entelonychia, Primates, and Insectivora (Part). *Bulletin of the American Museum of Natural History*, 34, 429–83.

(1917a). A Paleocene bat. *Bulletin of the American Museum of Natural History*, 37, 569–71.

(1917b). The dentition of *Nothodectes*. *Bulletin of the American Museum of Natural History*, 37, 831–9.

Matthew, W. D. and Granger, W. (1921). New genera of Paleocene mammals. *American Museum Novitates*, 13, 1–7.

McGrew, P. O. and Patterson, B. (1962). A picrodontid insectivore(?) from the Paleocene of Wyoming. *Breviora, Museum of Comparative Zoology*, 175, 1–9.

McKenna, M. C. (1960). Fossil Mammalia from the early Wasatchian Four Mile fauna, Eocene of northwest Colorado. *University of California, Publications in Geological Sciences*, 37, 1–130.

(1966). Paleontology and the origin of the primates. *Folia Primatologica*, 4, 1–25.

(1967). Classification, range, and deployment of the prosimian primates. *Problems actuel de Paléontologie*, 163, 603–9.

(1980). Late Cretaceous and early Tertiary vertebrate paleontological reconnaissance, Togwotee Pass area, Northwestern Wyoming. In *Aspects of Vertebrate History: Essays in Honor of Edwin Harris Colbert*, ed. L. L. Jacobs, pp. 321–43. Flagstaff, AZ: Museum of Northern Arizona Press.

McKenna, M. C. and Bell, S. K. (1997). *Classification of Mammals Above the Species Level*. New York: Columbia University Press.

Osborn, H. F. (1902). The American Eocene primates and the supposed rodent family Mixodectidae. *Bulletin of the American Museum of Natural History*, 16, 169–214.

Osborn, H. F. and Wortman, J. L. (1892). Fossil mammals of the Wasatch and Wind River beds. Collected in 1891. *Bulletin of the American Museum of Natural History*, 4, 80–147.

Rigby, J. K., Jr. (1980). Swain Quarry of the Fort Union Formation, middle Paleocene (Torrejonian), Carbon County, Wyoming: geologic setting and mammalian fauna. Chicago: *Evolutionary Monographs*, 3, 1–179.

Robinson, P. (1966). Fossil Mammalia of the Huerfano Formation, Eocene of Colorado. *Bulletin of the Peabody Museum of Natural History, Yale University*, 21, 1–95.

(1968). The paleontology and geology of the Badwater Creek Area, central Wyoming, Part 4: Late Eocene primates from Badwater, Wyoming, with a discussion of material from Utah. *Annals of the Carnegie Museum*, 39, 307–26.

(1994). *Myrmekomomys*, a new genus of micromomyine (Mammalia, ?Microsyopidae) from the lower Eocene rocks of the Powder River Basin, Wyoming. *University of Wyoming Contributions to Geology*, 30, 85–90.

Robinson, P. and Ivy, L. D. (1994). Paromomyidae (?Dermoptera) from the Powder River Basin, Wyoming and a discussion of microevolution in closely related species. *University of Wyoming Contributions to Geology*, 30, 91–116.

Romer, A. S. (1966). *Vertebrate Paleontology*. Chicago, IL: University of Chicago Press.

Rose, K. D. (1975). The Carpolestidae: early Tertiary primates from North America. *Bulletin of the Museum of Comparative Zoology*, 147, 1–74.

(1977). Evolution of carpolestid primates and chronology of the North American middle and late Paleocene. *Journal of Paleontology*, 51, 536–42.

(1981). The Clarkforkian Land-Mammal Age and mammalian faunal composition across the Paleocene–Eocene boundary. *University of Michigan Papers on Paleontology*, 26, 1–197.

(1995). The earliest primates. *Evolutionary Anthropology*, 3, 159–73.

Rose, K. D. and Bown, T. M. (1982). New plesiadapiform primates from the Eocene of Wyoming and Montana. *Journal of Vertebrate Paleontology*, 2, 63–9.

(1996). A new plesiadapiform (Mammalia: Plesiadapiformes) from the early Eocene of the Bighorn Basin, Wyoming. *Annals of the Carnegie Museum*, 65, 305–21.

Rose, K. D. and Gingerich, P. D. (1976). Partial skull of the plesiadapiform primate *Ignacius* from the early Eocene of Wyoming. *Contributions from the Museum of Paleontology, University of Michigan*, 24, 181–9.

Rose, K. D., Beard, K. C., and Houde, P. (1993). Exceptional new dentitions of the diminutive plesiadapiforms *Tinimomys* and *Niptomomys* (Mammalia), with comments on the upper incisors of Plesiadapiformes. *Annals of the Carnegie Museum*, 62, 351–61.

Runestad, J. A. and Ruff, C. B. (1995). Structural adaptations for gliding in mammals with implications for locomotor behavior in paromomyids. *American Journal of Physical Anthropology*, 98, 101–19.

Russell, D. E. (1959). Le crâne de *Plesiadapis*. *Bulletin Societe Géologique de France*, 4, 312–14.

(1964). Les mammifères paléocène d'Europe. *Mémoire Museum National d'Histoire Naturelle*, 13, 1–324.

Russell, D. E., Phélizon, A., and Louis, P. (1992). *Avenius* n. gen. (Mammalia, Primates?, Microsyopidae) de l'Eocène inférieur de France. *Comptes Rendus des Seances de l'Academie des Sciences, Paris*, 314, 243–50.

Russell, L. S. (1967). Palaeontology of the Swan Hills Area, north-central Alberta. *Contributions from the Royal Ontario Museum*, 71, 1–31.

Sargis, E. J. (2002). The postcranial morphology of *Ptilocercus lowii* (Scandentia, Tupaiidae): an analysis of Primatomorphan and Volitantian characters. *Journal of Mammalian Evolution*, 9, 137–60.

Schiebout, J. A. (1974). Vertebrate paleontology and paleoecology of Paleocene Black Peaks Formation, Big Bend National Park, Texas. *Austin: Bulletin of the Texas Memorial Museum*, 24, 1–88.

Schwartz, J. H. (1986). Primate systematics and a classification of the order. In *Comparative Primate Biology: Systematics, Evolution, and Anatomy*, ed. D. R. Swindler and J. Erwin, pp. 1–41. New York: Alan R. Liss.

Scott, C. S. and Fox, R. C. (2002). Picrodontids (Mammalia: Plesiadapiformes) from the Paleocene of Alberta, Canada. *Journal of Vertebrate Paleontology*, 22(suppl. to no. 3), p. 104A.

(2005). Windows on the evolution of *Picrodus* (Plesiadapiformes, Primates): morphology and relationships of a species complex from the Paleocene of Alberta. *Journal of Paleontology*, 79, 634–56.

Secord, R. (1998). Paleocene mammalian biostratigraphy of the Carbon Basin, southeastern Wyoming, and age constraints on local phases of tectonism. *Rocky Mountain Geology*, 33, 119–54.

Silcox, M. T. (2001). A phylogenetic analysis of Plesiadapiformes and their relationship to Euprimates and other Archontans. Ph.D. Thesis Johns Hopkins School of Medicine, Baltimore.

(2003). New discoveries on the middle ear anatomy of *Ignacius graybullianus* (Paromomyidae, Primates) from ultra high resolution X-ray computed tomography. *Journal of Human Evolution*, 44, 73–86.

(2007). Primate taxonomy, plesiadapiforms, and approaches to primate origins. In *Primate Origins and Adaptations: a Multidisciplinary Perspective*, ed. M. J. Ravosa and M. Dagosto, pp. 143–78. New York: Plenum Press.

Silcox, M. T. and Bloch, J. I. (2006). Upper incisor evolution in plesiadapiform primates. *American Journal of Physical Anthropology*, suppl. 42, p. 165.

Silcox, M. T., Krause, D. W., Maas, M. C., and Fox, R. C. (2001). New specimens of *Elphidotarsius russelli* (Mammalia,?Primates, Carpolestidae) and a revision of plesiadapoid relationships. *Journal of Vertebrate Paleontology*, 21, 132–52.

Silcox, M. T., Rose, K. D., and Walsh, S. L. (2002). New specimens of picromomyids (Plesiadapiformes, Primates) with description of a new species of *Alveojunctus*. *Annals of the Carnegie Museum*, 71, 1–11.

Silcox, M. T., Rose, K. D., and Bown, T. M. (2004). Early Eocene Paromomyidae (Mammalia, Primates) from the Southern Bighorn Basin (Willwood Formation, Wasatchian NALMA, Wyoming): taxonomy, variation and evolution. *Journal of Vertebrate Paleontology*, 24 (suppl. to no. 3), p. 113A.

Simons, E. L. (1972). *Primate Evolution: An Introduction to Man's Place in Nature*. New York: Macmillan.

Simpson, G. G. (1927). Mammalian fauna and correlation of the Paskapoo Formation of Alberta. *American Museum Novitates*, 268, 1–10.

(1928). A new mammalian fauna from the Fort Union of southern Montana. *American Museum Novitates*, 297, 1–15.

(1929). Third contribution to the Fort Union Fauna at Bear Creek, Montana. *American Museum Novitates*, 345, 1–12.

(1935a). New Paleocene mammals from the Fort Union of Montana. *Proceedings of the United States National Museum*, 83, 221–44.

(1935b). The Tiffany fauna, upper Paleocene. III. Primates, Carnivora, Condylarthra, and Amblypoda. *American Museum Novitates*, 817, 1–28.

(1935c). The Tiffany fauna, upper Paleocene. II. Structure and relationships of *Plesiadapis*. *American Museum Novitates*, 816, 1–30.

(1936). A new fauna from the Fort Union of Montana. *American Museum Novitates*, 873, 1–27.

(1937). The Fort Union of the Crazy Mountain Field, Montana and its mammalian faunas. *Bulletin of the United States National Museum*, 169, 1–287.

(1940). Studies on the earliest Primates. *Bulletin of the American Museum of Natural History*, 77, 185–212.

(1955). The Phenacolemuridae, new family of early Primates. *Bulletin of the American Museum of Natural History*, 105, 415–41.

Smith, T., Van Itterbeeck, J., and Missiaen, P. (2004) Oldest plesiadapiform (Mammalia, Proprimates) from Asia and its palaeobiogeographical implications for faunal interchange with North America. *Comptes Rendus Palevol*, 3, 43–52.

Stafford, B. and Thorington, Jr., R. W. (1998). Carpal development and morphology in archontan mammals. *Journal of Morphology*, 235, 135–55.

Stafford, B. J. and Szalay, F. S. (2000). Craniodental functional morphology and taxonomy of dermopterans. *Journal of Mammalogy*, 81, 360–85.

Stehlin, H. G. (1916). Die Säugetiere des schweizerischen Eocaens. *Caenopithecus*, etc. *Abhandlungen Schweizerischen Paläontologischen Gesellschaft*, 41, 1299–552.

Stock, C. (1934). Microsyopsinae and Hyopsodontidae in the Sespe upper Eocene, California. *Proceedings of the National Academy of Science, Philadelphia*, 20, 349–54.

(1938). A tarsiid primate and a mixodectid from the Poway Eocene, California. *Proceedings of the National Academy of Science, Philadelphia*, 24, 288–93.

Strait, S. G. (2001). New Wa0 fauna from the southeastern Bighorn Basin. [In *Paleocene–Eocene Stratigraphy and Biotic Change in the Bighorn and Clarks Fork Basins, Wyoming*, ed. P. D. Gingerich.] *University of Michigan Papers on Paleontology*, 33, 127–43.

Stucky, R. K. (1984). Revision of the Wind River Faunas, early Eocene of central Wyoming. Part 5. Geology and biostratigraphy of the upper part of the Wind River Formation, Northeastern Wind River Basin. *Annals of the Carnegie Museum*, 53, 231–94.

Szalay, F. S. (1968). The Picrodontidae, a family of early primates. *American Museum Novitates*, 2329, 1–55.

(1969a). Mixodectidae, Microsyopidae, and the insectivore-primate transition. *Bulletin of the American Museum of Natural History*, 140, 195–330.

(1969b). Uintasoricinae, a new subfamily of early Tertiary mammals (?Primates). *American Museum Novitates*, 2363, 1–36.

(1971). Cranium of the late Paleocene primate *Plesiadapis tricuspidens*. *Nature*, 230, 324–5.

(1972). Cranial morphology of the early Tertiary *Phenacolemur* and its bearing on primate phylogeny. *American Journal of Physical Anthropology*, 36, 59–76.

(1973). New Paleocene primates and a diagnosis of the new suborder Paromomyiformes. *Folia Primatologica*, 22, 243–50.

(1974). A new species and genus of early Eocene primate from North America. *Folia Primatologica*, 22, 243–50.

Szalay, F. S. and Dagosto, M. (1980). Locomotor adaptations as reflected on the humerus of Paleogene primates. *Folia Primatologica*, 34, 1–45.

Szalay, F. S. and Decker, R. L. (1974). Origins, evolution, and function of the tarsus in late Cretaceous Eutheria and Paleocene Primates. In *Primate Locomotion*, ed. F. A. Jenkins, Jr., pp. 223–359. New York: Academic Press.

Szalay, F. S. and Delson, E. (1979). *Evolutionary History of the Primates*. New York: Academic Press.

Szalay, F. S. and Drawhorn, G. (1980). Evolution and diversification of the Archonta in an arboreal milieu. In *Comparative Biology and Evolutionary Relationships of Tree Shrews*, ed. W. P. Luckett, pp. 133–69. New York: Plenum Press.

Szalay, F. S. and Lucas, S. G. (1993). Cranioskeletal morphology of Archontans, and diagnoses of Chiroptera, Volitantia, and Archonta. In *Primates and their Relatives in Phylogenetic Perspective*, ed. R. D. E. MacPhee, pp. 187–226. New York: Plenum Press.

(1996). The postcranial morphology of Paleocene *Chriacus* and *Mixodectes* and the phylogenetic relationships of archontan mammals. *Bulletin of the New Mexico Museum of Natural History and Science*, 7, 1–47.

Szalay, F. S., Tattersall, I., and Decker, R. L. (1975). Phylogenetic relationships of *Plesiadapis*: postcranial evidence. In *Approaches to Primate Paleobiology*, ed. F. S. Szalay, pp. 136–66. Basel: Karger.

Szalay, F. S., Rosenberger, A. L., and Dagosto, M. (1987). Diagnosis and differentiation of the order Primates. *Yearbook of Physical Anthropology*, 30, 75–105.

Tabuce, R., Mahboubi, M., Tafforeau, P., and Sudre, J. (2004) Discovery of a highly-specialized plesiadapiform primate in the early-middle Eocene of northwestern Africa. *Journal of Human Evolution*, 47, 305–21.

Thewissen, J. G. M., Williams, E. M., and Hussain, S. T. (2001). Eocene mammal faunas from northern Indo-Pakistan. *Journal of Vertebrate Paleontology*, 21, 347–66.

Tomida, Y. (1982). A new genus of picrodontid primate from the Paleocene of Utah. *Folia Primatologica*, 37, 37–43.

(1994). Phylogenetic reconstruction of fossil mammals based on cheek tooth morphology. *Honyuri Kagaku*, 34, 19–29.

Tong, Y. and Wang, J. (1998). A preliminary report on the early Eocene mammals of the Wutu fauna, Shandong Province, China. [In *Dawn of the Age of Mammals in Asia*, ed. K. C. Beard and M. R. Dawson.] *Bulletin of the Carnegie Museum of Natural History*, 34, 186–93.

Trouessart, E.-L. (1897). *Catalogus Mammalium*, 1, 1–644.

Van Houten, F. B. (1945). Review of latest Paleocene and early Eocene mammalian faunas. *Journal of Paleontology*, 19, 421–61.

Van Valen, L. M. (1969). A classification of the Primates. *American Journal of Physical Anthropology*, 30, 295–6.

(1994). The origin of the plesiadapid primates and the nature of *Purgatorius*. *Evolutionary Monographs, Chicago*, 15, 1–79.

Van Valen, L. M. and Sloan, R. E. (1965). The earliest primates. *Science*, 150, 743–5.

Webb, M. W. (1996). Late Paleocene mammals from near Drayton Valley, Alberta. M.Sc. Thesis, Edmonton University of Alberta, Edmonton.

(1998). New Saxonellidae, Plagiomenidae, and Pantolestidae from the late Paleocene of Central Alberta, Canada. *Journal of Vertebrate Paleontology* 18 (suppl. to no. 3), pp. 84A–5A.

Weber, M. and Abel, O. (1928). *Die Säugetiere*, Vol. III. Jena: G. Fischer.

West, R. M. and Dawson, M. R. (1977). Mammals from the Palaeogene of the Eureka Sound Formation: Ellesmere Island, Arctic Canada. *Géobios Special Memoir*, 1, 107–24.

White, T. E. (1952). Preliminary analysis of the vertebrate fossil fauna of the Boysen Reservoir area. *Proceedings of the United States National Museum*, 102, 185–207.

Wible J. R. and Covert, H. H. (1987). Primates: cladistic diagnosis and relationships. *Journal of Human Evolution*, 16, 1–22.

Wilf, P., Beard, K. C., Davies-Vollum, K. S., and Norejko, J. W. (1998). Portrait of a Late Paleocene (Early Clarkforkian) terrestrial ecosystem: Big Multi Quarry and associated strata, Washakie Basin, Southwestern Wyoming. *Palaios*, 13, 514–32.

Williams, J. A. (1985). Morphology and variation in the posterior dentition of *Picrodus silberlingi* (Picrodontidae). *Folia Primatologica*, 45, 48–58.

Williamson, T. E. (1996). The beginning of the age of mammals in the San Juan Basin, New Mexico: biostratigraphy and evolution of Paleocene mammals of the Nacimiento Formation. *Bulletin of the New Mexico Museum of Natural History and Science*, 8, 1–141.

Williamson, T. E. and Lucas, S. G. (1993). Paleocene vertebrate paleontology of the San Juan Basin, New Mexico. In *Vertebrate*

Paleontology in New Mexico, ed. S. G. Lucas and J. Zidek, pp. 105–35. Albuquerque, NM: New Mexico Museum of Natural History and Science.

Wilson, R. W. and Szalay, F. S. (1972). New paromomyid primate from middle Paleocene beds, Kutz Canyon area, San Juan Basin, New Mexico. *American Museum Novitates*, 2499, 1–18.

Winterfeld, G. F. (1982). Mammalian paleontology of the Fort Union Formation (Paleocene), eastern Rock Springs Uplift, Sweetwater County, Wyoming. *Contributions to Geology, University of Wyoming*, 21, 73–112.

Wolberg, D. L. (1978). The mammalian paleontology of the Late Paleocene (Tiffanian) Circle and Olive Localities, McCone and Powder River Counties, Montana. Ph.D. Thesis, University of Minnesota, Minneapolis.

(1979). Late Paleocene (Tiffanian) mammalian fauna of two localities in eastern Montana. *Northwest Geology*, 8, 83–93.

Wortman, J. L. (1903–1904). Studies of Eocene Mammalia in the Marsh Collections, Peabody Museum. Part 2. Primates. *American Journal of Science*, 15, 163–76, 399–414, 419–36; 16, 345–68; 17, 23–33, 133–40, 203–14.

15 Euprimates

GREGG F. GUNNELL,[1] KENNETH D. ROSE,[2] and D. TAB RASMUSSEN[3]

[1]*Museum of Paleontology, University of Michigan, Ann Arbor, MI, USA*
[2]*Johns Hopkins University School of Medicine, Baltimore, MD, USA*
[3]*Washington University, St. Louis, MO, USA*

INTRODUCTION

North American fossil primates are known from the earliest Eocene to the earliest Miocene. They are very diverse and abundant in the early and early middle Eocene but decrease in diversity and abundance by the late middle Eocene and make up only a very small part of mammalian faunal samples from then until their disappearance in the early Miocene.

Two suborders of fossil primates are represented in North America: Tarsiiformes, traditionally viewed as the "tarsier- or galago-like" primates of the early Cenozoic, and Adapiformes, their "lemur-like" counterparts. Tarsiiforms, represented in North America by the family Omomyidae, were a relatively small-bodied radiation and were specialized insectivores, frugivores, and faunivores. Adapiforms, represented in North America by the subfamily Notharctinae and possibly by the subfamily Cercamoniinae, were typically larger than tarsiiforms and were more generalized frugivores and folivores. Both groups were arboreal with elongate hindlimb elements, digits with nails, and opposable big toes for grasping limbs and branches of trees. Tarsiiforms had short faces and enlarged orbits and were probably nocturnal while adapiforms had longer, lower skulls with relatively smaller orbits and were probably diurnal. The decline of North American primates was the result of changing climatic conditions as relatively warm, subtropical conditions gave way to more temperate, seasonal climates beginning in the late middle Eocene. These changing climatic conditions resulted in reduction of primate habitats as closed, forested areas were replaced by more open woodlands and, later, by prairie grasslands.

DEFINING FEATURES OF THE SUBORDER EUPRIMATES

CRANIAL

TARSIIFORMS AND ADAPIFORMS

Crania possess a fully ossified auditory bulla formed from the petrosal bone, relatively front-facing and enlarged orbits, an ossified postorbital bar, a large braincase with a relatively large and complex neocortex, and bony tubes enclosing the internal carotid, promontory, and stapedial arteries as they pass through the middle ear.

OMOMYIDAE

Only a few North American omomyid taxa are represented by cranial remains, including *Tetonius* (Cope, 1882; Szalay, 1976), *Omomys* (Alexander and MacPhee, 1999), *Shoshonius* (Beard, Krishtalka, and Stucky 1991; Beard and MacPhee, 1994), and *Rooneyia* (Wilson, 1966; Szalay, 1975, 1976). All known omomyid skulls have large, forward-facing orbits and relatively short and rounded braincases. All except *Shoshonius* have an ectotympanic extended into an auditory tube, and the petromastoid region is inflated in all except *Rooneyia*. *Tetonius* and *Shoshonius* have relatively short rostra, while *Rooneyia* and *Omomys* have somewhat longer rostra but not as long as in known adapiforms. The foramen magnum in all known omomyids except *Rooneyia* is positioned further underneath the skull than in adapiforms, indicating a more vertical habitual posture (Figure 15.1).

Evolution of Tertiary Mammals of North America, Vol. 2. ed. C. M. Janis, G. F. Gunnell, and M. D. Uhen. Published by Cambridge University Press.

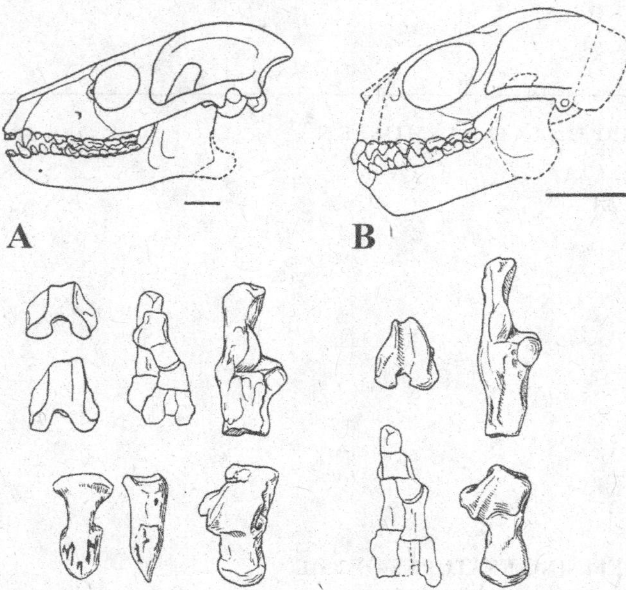

Figure 15.1. Skulls and postcranial elements of North American euprimates. A. *Notharctus* skull, adapiform postcrania. B. *Tetonius* skull, omomyid postcrania. Note elongate skull of adapiform and shorter skull of omomyid, deep knee joint in both groups, somewhat elongate astragalus and calcaneum in adapiforms and more elongate foot elements in omomyids, and flattened nails typical of all euprimates (from Rose, 1995a). Scale bars = 10 mm.

NOTHARCTIDAE

North American adapiform cranial morphology is well represented by good skulls of *Smilodectes* and *Notharctus* (Gazin, 1958), as well as by more poorly preserved skulls of *Cantius* (Rose, MacPhee, and Alexander, 1999) and *Hesperolemur* (Gunnell, 1995a). In general, adapiform crania have relatively elongate rostra, long, low braincases, often with well-developed sagittal and nuchal crests (perhaps reflecting sexual differences), and relatively laterally facing orbits. There is relatively strong postorbital constriction. The lacrimal foramen is positioned at or just inside the anterior orbital margin except in *Cantius*, where it is anterior to the orbital margin (Gunnell, 1995a). As in extant lemurs, the ectotympanic of the middle ear forms a free annular ring fused to the bulla only at the posterior end (except in *Hesperolemur* where both the anterior and posterior crus are fused to the lateral wall of the bulla).

The branches of the internal carotid artery (promontorial and stapedial) in the middle ear cavity of notharctids are typically enclosed in bony tubes as they are in omomyids. In *Cantius abditus*, however, it appears that most of the promontorial artery traveled in an open sulcus rather than a bony tube, a resemblance to *Adapis* and to living lemurs, which have an intact internal carotid system (Rose, MacPhee, and Alexander, 1999). Even so, it is uncertain whether this represents the primitive state for adapiforms (most likely) or a possible synapomorphy of *Cantius* and lemuriforms (less likely). Gunnell (1995a) suggested that *Hesperolemur* also lacked bony tubes for the middle ear arteries. Such bony tubes are extremely delicate and easily broken, so the condition of the arterial pathways in both *Cantius* and *Hesperolemur* should be regarded as tentative, pending discovery of additional well-preserved fossils.

DENTAL

TARSIIFORMS AND ADAPIFORMS

Molars are relatively low crowned with broad basins; lower molars have reduced paraconids and an extended m3 hypoconulid, and upper molars have reduced stylar shelves. Tarsiiforms tend to have pointed, procumbent lower and upper incisors and often have the teeth between the central incisor and the fourth premolar reduced and crowded together, with teeth lost (reduced dental formula) in more derived forms. Adapiforms have spatulate incisors that are generally more vertically implanted. Teeth between the incisors and the fourth premolar are normally retained (unreduced dental formula), often with short diastemata between more anterior teeth. Unlike tarsiiforms, in which the canine is seldom enlarged, adapiforms tend to have larger, overlapping canines and often exhibit some degree of canine size dimorphism between males and females (Figure 15.2).

OMOMYIDAE

Omomyids exhibit a wide range of dental morphologies. Many of the most primitive taxa (including *Teilhardina*, *Steinius*) have simple lower molars with well-developed, marginally placed cusps and little or no development of accessory cuspules. Upper molars are also simple, with small stylar shelves, no stylar cusps, small conules, and weak or absent hypocones and postprotocingula. Upper and lower molar basins are relatively shallow with smooth enamel. P3-4 have single, large, buccal cusps and low protocones; p3-4 have tall protoconids and low talonid heels with only p4 having a metaconid. These primitive forms were almost certainly predominantly frugivorous (Strait, 2001).

There are several dental themes developed through various North American omomyid lineages. Some are typified by dentaries that are shortened and deepened anteriorly with development of a very large, pointed, procumbent central incisor. Anterior teeth between i1 and p4 are either lost or shortened, reduced, and crowded together with reduction in root number. Molars are low crowned with constricted trigonids. Some of these developed specialized premolars, especially P4/p4 that became tall, broad, and exodaenodont. Other taxa have sharply defined cusps with progressive proliferation of accessory cuspules including mesostyles and other stylar cusps, metastylids, pericones, and complex conules. These taxa retain paraconids on all lower molars and have more constricted, deeper talonid basins. Their incisors tend to be small and pointed, but not procumbent.

NOTHARCTIDAE

Unlike omomyids, which exhibit a variety of dental morphologies, North American notharctids are all relatively similar in tooth

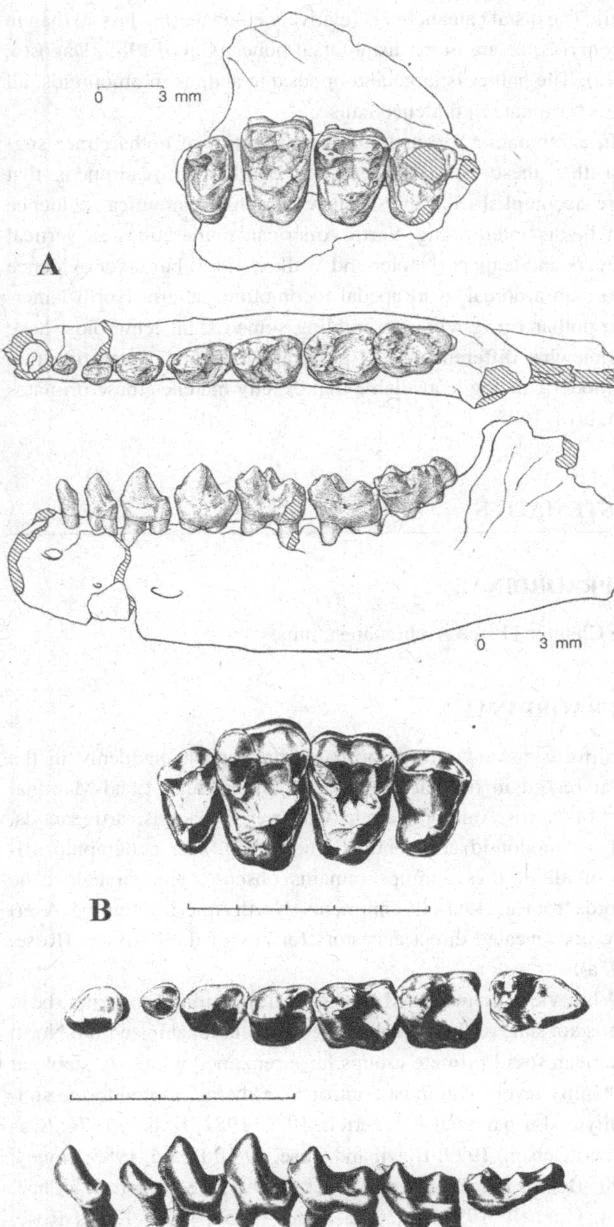

Figure 15.2. Dental comparisons between early Tertiary North American adapiforms and omomyids. A. Adapiform *Cantius torresi* (from Gingerich, 1995). B. Omomyid *Teilhardina americana* (from Bown and Rose, 1987). Scale bar = 3 mm (A): scale bar in 1 mm increments (B). (A: courtesy of the University of Michigan. B: courtesy of the Paleontological Society.)

structure. Notharctids have spatulate lower and upper incisors that are implanted subvertically except in the most primitive genus, *Cantius*, in which they are more anteriorly inclined. Canines are usually enlarged and often overlapping and exhibit some degree of size dimorphism (Gingerich, 1995). Premolars are relatively simple anteriorly but increase in complexity posteriorly with upper and lower fourth premolars often being semi-molariform. Lower molars have relatively low trigonids with a well-developed trigonid fovea (except in *Hesperolemur*) and shallow, broad talonid basins. The lower third molar usually has an extended, broad third lobe, and the enamel of the talonid basin is often crenulated. Upper molars typically have a well-developed postprotocingulum from which there is progressive differentiation and enlargement of a hypocone and often a strong mesostyle in more derived taxa; however, postprotocingula and mesostyles are weak or absent in the most primitive known notharctids.

POSTCRANIAL

TARSIIFORMS AND ADAPIFORMS

The elbow joint is mobile with a rounded humeral capitulum that is extended distally; the radial head is round. The hindlimbs are much longer than forelimbs and the femur is often longer than the tibia. There is an anteroposteriorly deep distal femur, an opposable hallux, and nails on all digits. Tarsiiforms can be differentiated postcranially from adapiforms by being much smaller and in possessing a more cylindrical femoral head, relatively elongate tarsal elements, and close appression or fusion of the distal tibiofibula.

OMOMYIDAE

Several omomyids are represented by at least some postcranial elements (Gebo, 1988; Dagosto, 1993). Among North American forms these include *Teilhardina* (Szalay, 1976; Gebo, 1988), *Tetonius* (Szalay, 1976; Gebo, 1988; Rosenberger and Dagosto, 1992), *Absarokius* (Covert and Hamrick, 1993), *Arapahovius* (Savage and Waters, 1978), *Omomys* (Rosenberger and Dagosto, 1992; Anemone, Covert, and Nachman, 1997; Hamrick, 1999; Anemone and Covert, 2000), *Hemiacodon* (Simpson, 1940; Szalay, 1976; Dagosto, 1985; Gebo, 1988), *Washakius* (Szalay, 1976; Gebo, 1988), and *Shoshonius* (Dagosto, Gebo, and Beard, 1999).

While there is some variability within known postcranial elements of omomyids, overall there is remarkable consistency. Forelimb elements of omomyids are very poorly known, represented only by complete humeri of *Shoshonius* (Dagosto, Gebo, and Beard, 1999), distal humeri of *Omomys* (Anemone, Covert, and Nachman, 1997; Anemone and Covert, 2000) and *Hemiacodon* (Szalay, 1976; Dagosto, 1993), and some carpal elements of *Omomys* (Hamrick, 1999). *Shoshonius* has a relatively higher humerofemoral index (relatively shorter femur) than extant tarsiers, in this regard being more similar to galagos and cheirogaleids. Known omomyid hamates resemble those of extant monkeys, tarsiers, and adapiforms, differing only in having a less well-developed hamulus (Hamrick, 1999). In general, omomyid carpals are consistent with powerful manual grasping during arboreal locomotion.

Hindlimb features of omomyids (see Dagosto, Gebo, and Beard [1999] for summary) include a long, rodlike ilium, relatively long femur with semi-cylindrical head and an anteroposteriorly deep distal end, and a relatively long, bowed tibia that has a long tibiofibular articulation (*Absarokius* and *Shoshonius*), a relatively shorter tibiofibular articulation (*Omomys*), or a distally fused tibiofibula (European *Necrolemur*). Omomyid astragali have relatively long necks, a high astragalar body, a long and relatively narrow trochlea, and a well-developed posterior trochlear shelf (Gebo, 1988). The calcaneum is relatively longer than in adapiforms, with a well-developed cuboid pivot (except perhaps in *Shoshonius* [Dagosto, Gebo, and Beard, 1999]). The cuboid, navicular, and cuneiforms are also elongate and the first metatarsal has a long and prominent peroneal tubercle. Omomyids, like adapiforms, have an opposable hallux. Unlike some adapiforms (e.g. *Europolemur*), it appears that all omomyid digits terminate in nails as there is no unequivocal evidence for the existence of toilet claws among known omomyids.

Judging from postcranial evidence, known omomyids were capable of powerful leaping but were not specialized vertical clingers and leapers like extant tarsiers (Dagosto, Gebo, and Beard, 1999). Omomyids were probably more like extant galagos – active arboreal quadrupeds that included frequent leaping in their locomotor repertoire.

NOTHARCTIDAE

North American notharctines are among the best known postcranially of any fossil primates (Gregory, 1920). Several nearly complete skeletons of *Notharctus* have been described and *Smilodectes* is also known from several relatively good skeletons. *Cantius* is represented by some well-preserved skeletal elements (mostly humeral and tarsal) while the other North American notharctines (*Copelemur*, *Pelycodus*, and *Hesperolemur*) remain unknown postcranially. Postcranial anatomy among the known forms is quite similar. In general, North American notharctines are very similar skeletally to some extant lemuroids, in particular *Lemur* and *Propithecus*. In the forelimb, the humerus has a large, ovoid head that faces posteriorly and projects proximally slightly beyond the tuberosities. The supinator crest is well developed as is the deltopectoral crest. The bicipital groove is shallow; the capitulum is distinct and spherical and separated from the trochlea by a distinct groove; the olecranon fossa is relatively shallow, and an entepicondylar foramen is present. Combined with other features of the scapula, radius, and ulna, the humeral morphology indicates that notharctines had a relatively mobile forelimb that was probably habitually partially supinated (Gregory, 1920; Szalay and Delson, 1979; Gebo, 1987; Dagosto, 1993).

Notharctine hindlimbs are much longer than the forelimbs, resulting in an intermembral index of about 60 (Dagosto, 1993). The pelvis has a relatively broad ilium and a long ischium. The femur has a round head and a greater trochanter that extends proximally as far as the head. The femoral shaft is round in cross section and longer than the tibia. The distal end of the femur is deep anteroposteriorly and has a narrow, elevated patellar groove with a high, rounded lateral rim (Gregory, 1920; Covert, 1985; Rose and Walker, 1985).

Tarsal elements of notharctines also resemble those of extant *Lemur*. The astragalar body is high and short with a shallow trochlea; the neck is relatively long, and there is a large posterior trochlear shelf. The distal calcaneum is relatively elongate (but less so than in omomyids) as are more distal tarsal bones (Gebo, 1988; Dagosto, 1993). The hallux is large and opposable and, as in omomyids, all digits terminate in flattened nails.

In combination, postcranial characteristics of notharctines suggest that these animals were active arboreal quadrupeds that were accomplished leapers. There is some anatomical evidence that favors interpreting North American notharctines as vertical clingers and leapers (Napier and Walker, 1967) but other evidence favors an arboreal quadrupedal locomotion pattern. North American notharctines, while resembling some extant lemuroids, have a somewhat different overall postcranial pattern, suggesting that no modern analog is available that exactly matches these primates (Dagosto, 1993).

SYSTEMATICS

SUPRAORDINAL

See Chapter 11, the Archontan summary.

INFRAORDINAL

Tarsiiform and adapiform primates first appear suddenly in the fossil record in the earliest Eocene (Wasatchian Land-Mammal Age of North America) along with perissodactyls, artiodactyls, and hyaenodontid creodonts (Figure 15.3). The geographic origin of all of these groups remains obscure. Known Paleocene records from all Holarctic continents (North America, Europe, Asia) have not revealed direct ancestors for any of these groups (Rose, 1995a).

While views on intra- and inter-primate relationships seem to be in a constant state of flux, the phylogenetic relationships within North American fossil primate groups have remained relatively stable at the family level, with most controversies being limited to the subfamily and tribal level (Gingerich, 1976, 1981; Szalay, 1976; Szalay and Delson, 1979; Bown and Rose, 1987; Beard, 1988; Honey, 1990; Beard, Krishtalka, and Stucky, 1991; Beard and MacPhee, 1994; Gunnell, 1995a,b, 2002; Rose, 1995a; Gunnell and Rose, 2002; Muldoon and Gunnell, 2002).

The earliest known omomyids are three species of *Teilhardina*: *T. asiatica* (China) *T. belgica* (Europe) and *T. brandti* (North America). All three species are primitive in variably retaining small p1 teeth and relatively unreduced canines. Upper molars of these species are simple with very weak or absent postprotocingula. However, these species seem more derived than the later-occurring North American genus *Steinius* in having somewhat reduced third molars, more basally inflated molar cusps that are set in from tooth margins, and relatively shorter posterior premolars. The presence of the more primitive taxon *Steinius* in sediments stratigraphically younger than those in which *Teilhardina* occurs suggests that even more primitive tarsiiforms must have existed as early as or earlier than *Teilhardina* (Bown and Rose, 1984, 1987; Rose and Bown, 1991; Rose, 1995a).

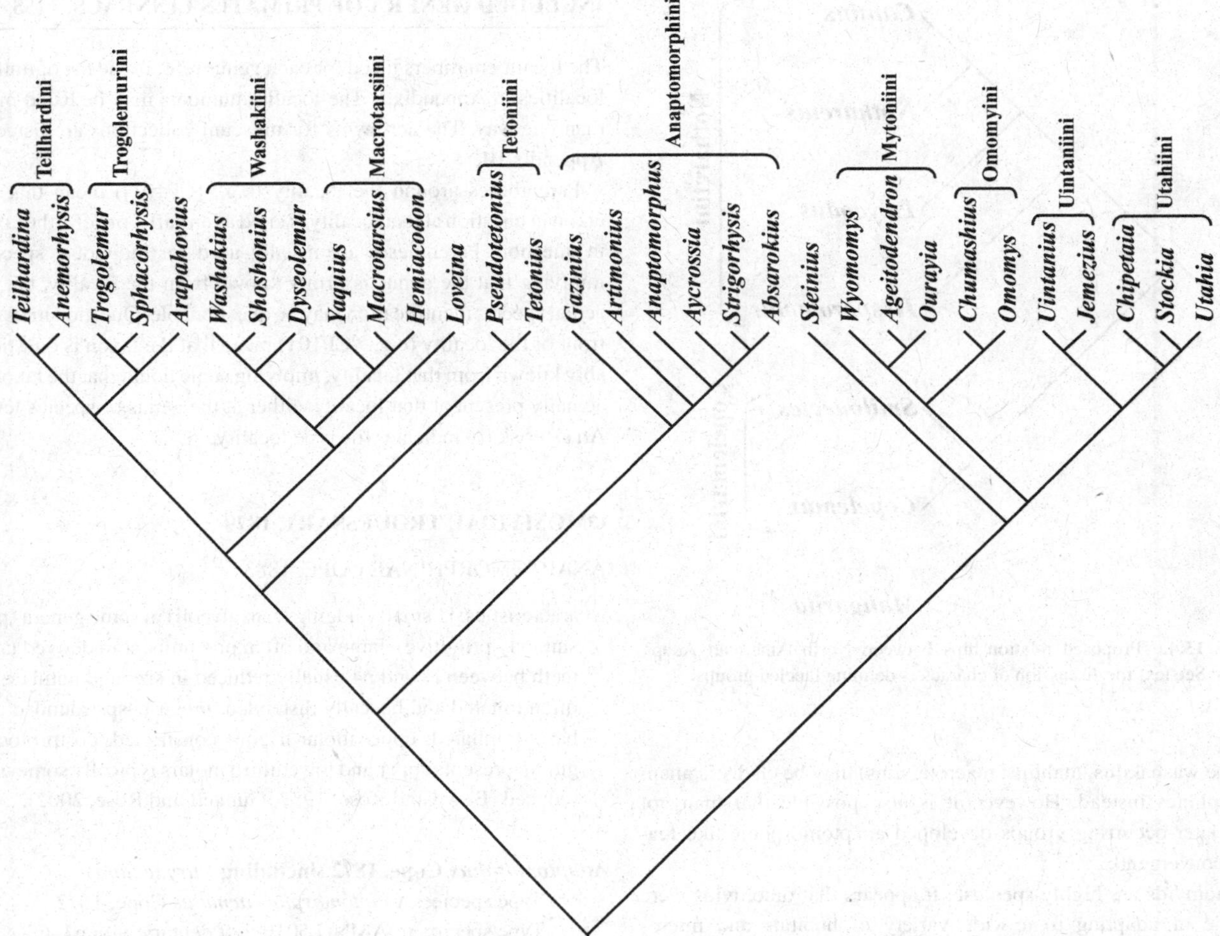

Figure 15.3. Proposed relationships between North American Omomyidae. See text for discussion of characters defining labeled groups.

The earliest known notharctines are two species of *Cantius*: *C. torresi* from the earliest Wasatchian of North America, and *C. eppsi* from the early Ypresian of Europe. *Cantius torresi* is primitive in having squared upper molars that lack mesostyles, a very weak postprotocingulum, and relatively short and broad lower premolars (Gingerich, 1986). The early Eocene adapiform *Donrussellia* from Europe, although slightly younger, may be more primitive than *C. torresi* (Godinot, 1992, 1998) in having a smaller p4 metaconid, a less lingually distended p4 talonid, and shorter and wider upper molars – features that resemble those in early omomyids like *Teilhardina*. This suggests that omomyids and adapiforms are sister taxa that diverged slightly before the first appearances of *Donrussellia*, *Teilhardina*, and *Cantius*, as these earliest known representatives of each clade differ very little from one another dentally (Gingerich, 1986; Rose, 1995a; Godinot, 1998). The presence of what appear to be true tarsiids in the Eocene of Asia (Beard *et al.*, 1994; Beard, 1998a,b) lends support to the idea of a very early split between tarsiids and adapiforms–omomyids.

After their initial occurrence in the earliest Eocene, omomyids undergo an increasingly diverse radiation through the early and middle Eocene in North America, with an ancestry that can be traced essentially from *Teilhardina* and *Steinius* (Bown and Rose, 1987). The interrelationships of early and middle Eocene North American omomyids are not well understood nor are their possible relationships with later-occurring primates. This is because conflicting characters imply that various specialized dental features (such as hypertrophied p4, crenulated enamel, enlarged incisors, and reduction of teeth between the incisors and p4) have evolved in parallel multiple times, and there has been little consensus among experts as to which characteristics are synapomorphic and which are homoplasious.

Szalay (1976) arranged omomyid taxa into several tribes and subtribes based on his interpretations of relationships within Omomyidae. These tribal units have remained relatively stable but the composition of each tribe or subtribe has been a point of contention (Bown and Rose, 1987; Honey, 1990; Gunnell, 1995b; McKenna and Bell, 1997; Gunnell and Rose, 2002; Muldoon and Gunnell, 2002). If the anaptomorphine–omomyine dichotomy can be traced to a split between *Teilhardina* and *Steinius* (Bown and Rose, 1991), then some traditionally recognized omomyine groups (e.g. middle

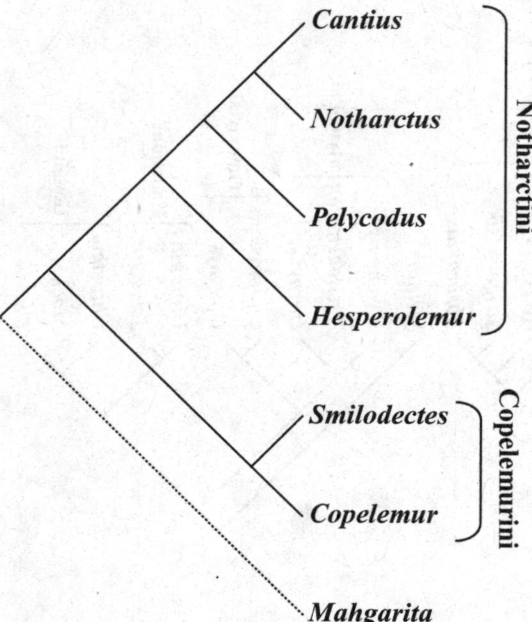

Figure 15.4. Proposed relationships between North American Adapiformes. See text for discussion of characters defining labeled groups.

Eocene washakiins, utahiins, macrotarsiins) may be cladistic anaptomorphines instead. However, it is also possible that many of these later-occurring groups developed anaptomorphine-like features convergently.

Omomyids are highly speciose. It appears that omomyids were capable of adapting to a wide variety of habitats and microenvironments (Gunnell, 1997). The North American record presents an interesting dichotomy (as the subfamilies are recognized here) with anaptomorphines being the most common group in the early Eocene and omomyines being much more common in the middle Eocene. A combination of climatic and paleoecological change along with competition between these two groups result in the distributions present in the early and middle Eocene of North America. Interestingly, recent work in the earliest middle Eocene has suggested that anaptomorphines are displaced by omomyines, rather than replaced (Gunnell, 1997; Muldoon and Gunnell, 2002), with anaptomorphines being more common in distal and upland habitats while omomyines are more common in the much more frequently sampled proximal habitats.

Notharctines (Figure 15.4), in contrast to omomyids, were never especially diverse but they do constitute a very common element in early and early middle Eocene North American faunal samples, often making up as much as 10% of all mammalian specimens in a sample, especially in the early Eocene. There are rarely more than three genera of notharctines present at any one faunal horizon, and even this diversity is uncommon (Gunnell, 2002). Except for the latest early and earliest middle Eocene, no more than two notharctine genera are present at any one time in North American faunal samples, with each genus seldom being represented by more than one species from any horizon.

INCLUDED GENERA OF PRIMATES LINNEAUS, 1758

The locality numbers listed for each genus refer to the list of unified localities in Appendix I. The locality numbers may be listed more than one way. The acronyms for museum collections are listed in Appendix III.

Parentheses around the locality (e.g., [CP101]) mean that the taxon in question at that locality is cited as an "aff." or "cf." the taxon in question. Parentheses are usually used for individual species, implying that the genus is firmly known from the locality, but the actual species identification may be questionable. Question marks in front of the locality (e.g., ?CP101) mean that the taxon is questionably known from that locality, implying some doubt that the taxon is actually present at that locality, either at the genus or species level. An asterisk (*) indicates the type locality.

OMOMYIDAE TROUESSART, 1879

ANAPTOMORPHINAE COPE, 1883

Characteristics: i1 small (judging from alveoli) in some genera (presumably primitive), large and often procumbent in derived taxa; teeth between i1 and p4 usually reduced in size and number; p4 often inflated and buccally distended; molar cusps rounded and basally inflated; upper molar trigons constricted, postprotocingulum present; upper and lower third molars typically somewhat reduced (Bown and Rose, 1987; Gunnell and Rose, 2002).

Anaptomorphus Cope, 1872 (including *Euryacodon*)

Type species: *Anaptomorphus aemulus* Cope, 1872.

Type specimen: AMNH 5010, left dentary with p4–m2.

Characteristics: Lower dental formula i2, c1, p2, m3; lower teeth relatively low crowned with bulbous cusps, paraconids distinct on m1–2, weaker on m3; m1–2 approximately the same length, m2 broader; M3/m3 reduced, m3 hypoconulid weak and short; p3–4 simple, lacking paraconids and metaconids (p4 sometimes with weak enamel fold), talonids short and weak; upper molars lingually extended (especially M1–2), postprotocristae distinct, paraconules weak, metaconules stronger, trigon basins shallow and restricted (Szalay, 1976; Gunnell, 1995b).

Average length of m1: 2.0–3.0 mm.

Included species: *A. aemulus* (known from localities CP5A, CP34C*, D); *A. westi* (localities CP25J, CP34A3, CP34B*).

Anaptomorphus sp. is also known from locality CP32B.

Comments: Despite being the type genus of the subfamily Anaptomorphinae, *Anaptomorphus* is relatively poorly represented and much of its anatomy awaits discovery.

Trogolemur Matthew, 1909

Type species: *Trogolemur myodes* Matthew, 1909.

Type specimen: AMNH 12599, right dentary with p2–m3.

Characteristics: Lower dental formula i1-2, c1, p2, m3; i1 greatly enlarged, very procumbent, i2–p3 (or c1–p3 when c1 present) very small, roots shallow; anterior teeth posterior to i1 crowded and overhanging; p4 small with weak paraconid and metaconid, protoconid not projecting above tooth row, talonid very short; paraconid distinct on m1, small to absent on m2-3, trigonids anteroposteriorly compressed; m3 relatively long and narrow, hypoconulid relatively broad; P4 protocone distinct, low, anterior; paracone high, centrally placed, postprotocrista strong, enclosing trigon basin; upper molar protocones prominent, pre- and postprotocristae well developed, postprotocingulum high and compressed, paraconules weak, metaconules weak to absent, mesostyles absent; mandible very deep anteriorly (Szalay, 1976; Emry, 1990; Gunnell, 1995b).

Average length of m1: 1.5–1.8 mm.

Included species: *T. myodes* (known from localities NB14, CP34C*, D, [CP25F], [CP36B]); *T. amplior* (localities CP27?D, E*, [CP34A3]); *T. fragilis* (locality CP27E*); *T. leonardi* (locality NP9A*).

Trogolemur sp. is also known from localities CP29C–D, CP34E and may be present from NP9B. Previously published records of *Trogolemur* from CP63 and NP20D are in error.

Comments: *Trogolemur* is among the smallest known anaptomorphines with an estimated body weight of 50 g. *Trogolemur* is represented from the earliest Bridgerian through the Duchesnean, making it the latest surviving anaptomorphine known.

Tetonius Matthew, 1915 (including _Paratetonius_; _Anaptomorphus_, in part)

Type species: *Tetonius homunculus* (Cope, 1882) (originally described as *Anaptomorphus homunculus*).

Type specimen: AMNH 4194, skull with left and right C1–M3.

Characteristics: Lower dental formula i2, c1, p2-3, m3; anterior teeth not crowded together; p3 unreduced, double-rooted; i1 always larger than i2, both relatively procumbent; differs from *Teilhardina* in being larger and in having relatively broader cheek teeth than most species of *Teilhardina*, p4 taller, and c1 and p2 more reduced; differs from *Absarokius* in lacking ventrobuccally distended p4 and in having a larger i1 and relatively smaller p2 (Bown and Rose, 1987).

Average length of m1: 2.1–2.3 mm.

Included species: *T. homunculus* (known from locality CP20A*); *T. matthewi* (localities CP19A–B, CP20A*, CP25A, CP27A, CP63); *T. mckennai* (locality CP63*).

Tetonius sp. also is known from localities CP19, CP20, CP25A–B, and may be present at GC21 and GC21II (Beard and Tabrum, 1990; Beard and Dawson, 2001).

Absarokius Matthew, 1915 (including _Anaptomorphus_, in part)

Type species: *Absarokius abbotti* (Loomis, 1906) (originally described as *Anaptomorphus abboti*).

Type specimen: ACM 3479, right dentary p3–m3.

Characteristics: Lower dental formula i2, c1, p3, m3; differs from *Tetonius* and *Pseudotetonius* in having P4 typically larger, narrower, more quadrate, and more buccally distended; M3/m3 relatively more reduced; differs further from *Pseudotetonius* in having i1 much smaller and c1 larger; differs further from *Tetonius* in having i1 smaller, c1 larger, and p2 much larger; differs from *Artimonius* in retaining p2, in lacking extreme p4 enlargement, in having narrower upper molars with relatively small trigon basins that are covered by smooth enamel (Bown and Rose, 1987; Muldoon and Gunnell, 2002).

Average length of m1: 2.3–2.5 mm.

Included species: *A. abbotti* (known from localities CP19C–E, CP20B–C, CP25[A], G, CP27B*, D, CP28A); *A. gazini* (localities CP25C*, D); *A. metoecus* (localities SB22C, CP20B*, C, CP27B).

Teilhardina Simpson, 1940 (including _Protomomys_; _Tetonoides_, in part)

Type species: *Teilhardina belgica* (Teilhard de Chardin, 1927) (originally described as *Protomomys belgica*).

Type specimen: IRSNB 64, left dentary with p3–m3 (lectotype, Simpson, 1940).

Characteristics: Lower dental formula i2, c1, p3–4, m3; p3 simple and unreduced; p4 not enlarged, paraconid and metaconid small; differs from *Anemorhysis* in having a short and unbasined p4 talonid and a more lingually directed hypoflexid; smaller than *Tetonius* and *Absarokius* (Bown and Rose, 1987).

Average length of m1: 1.8–2.0 mm.

Included species: *T. americana* (known from localities CP19A, CP20A*); *T. brandti* (locality CP19AA*); *T. crassidens* (localities CP19A, CP20A*); *T. demissa* (locality CP20B*); *T. tenuicula* (= *Tetonoides tenuiculus*) (locality CP20A*).

Teilhardina sp. is also known from locality NP49.

Comments: *Teilhardina* is one of the most primitive omomyids known, rivaled only by *Steinius* in North America. It appears to lie at or near the base of Anaptomorphinae. *Teilhardina* is the only known anaptomorphine represented outside of North America, with *T. belgica* being present in the early Ypresian in Europe and *T. asiatica* present in early Eocene sediments of China (Ni *et al.*, 2004). Szalay and Delson (1979) rejected the presence of *Teilhardina* in North America and instead subsumed *T. americana*, *T. tenuicula*, and *Anemorhysis pearcei* under the species *Anemorhysis tenuicula*. The American *Teilhardina* species are more similar to *T. belgica*, however, and lack the diagnostic traits of *Anemorhysis*. Nonetheless, it is likely that several of the North American species now assigned to *Teilhardina* will ultimately be placed in a different genus.

***Anemorhysis* Gazin, 1958 (including *Uintalacus*; *Paratetonius*?; *Tetonoides*, in part)**

Type species: *Anemorhysis sublettensis* (Gazin, 1952) (originally described as *Paratetonius*? *sublettensis*).

Type specimen: USNM 19205, left dentary with p4–m2.

Characteristics: Similar in size to *Teilhardina*, smaller than *Tetonius*. Differs from other similar anaptomorphines (*Teilhardina* and *Tetonius*) in having a semi-molariform p4; p4 paraconid prominent, metaconid high and well developed, hypoconid distinct, entoconid smaller, and cristid obliqua short but distinct forming a small talonid basin; molar cusps sharply defined, not basally inflated; m1-2 postcristid straight (Bown and Rose, 1987).

Average length of m1: 1.65–1.95 mm.

Included species: *A. sublettensis* (known from locality CP25G*); *A. natronensis* (locality CP27E); *A. pearcei* (including *Tetonoides pearcei* Gazin, 1962), (localities CP25A*, NP49); *A. pattersoni* (locality CP20C*); *A. savagei* (localities CP25A*, CP27B); *A. wortmani* (localities CP19D, CP20C*).

Comments: Some authors recognize the genus *Tetonoides* for the primitive species *A. pearcei* and *A. savagei*, which differ from the other species of *Anemorhysis* in having smaller talonid basins and weaker paraconid and metaconid cusps on p4.

***Chlororhysis* Gazin, 1958**

Type species: *Chlororhysis knightensis* Gazin, 1958.

Type specimen: USNM 21901, left dentary with c1–p4.

Characteristics: Lower canine unreduced; p2-4 relatively narrow and uninflated; p4 metaconid tiny to absent, talonid unbasined; differs from *Loveina* in premolar proportions and in having p4 metaconid weak and low (Gazin, 1958, 1962; Bown and Rose, 1984, 1987).

Average length of p4: 1.9 mm (lower molars unknown).

Included species: *C. knightensis* (known from localities CP25F, G*); *C. incomptus* (locality CP20B*).

Comments: *Chlororhysis* is a very poorly known anaptomorphine represented by only a few specimens from the Green River, Bighorn, and Washakie Basins in Wyoming.

***Pseudotetonius* Bown, 1974 (including *Mckennamorphus*)**

Type species: *Pseudotetonius ambiguus* (Matthew, 1915) (originally described as *Tetonius*? *ambiguus*).

Type specimen: AMNH 15072, left dentary with p3–m2.

Characteristics: Lower dental formula i2, c1, p2, m3; differs from *Tetonius* in having larger i1, smaller i2 and c1, and more reduced p3 with a mesiodistally compressed single or bilobed root; dentary foreshortened compared with *Tetonius* and teeth between i1 and p4 crowded; p4–m3 and P4–M3 similar to *Tetonius*.

Average length of m1: 2.2 mm.

Included species: *P. ambiguus* only (including *Mckennamorphus ambiguus* Szalay, 1976) (known from localities CP20A*, CP63).

***Arapahovius* Savage and Waters, 1978**

Type species: *Arapahovius gazini* Savage and Waters, 1978.

Type specimen: UCMP 100 000, right maxilla with P3–M3.

Characteristics: Lower dental formula i2, c1, p3, m3; i1 moderately enlarged and procumbent, i2–c1 small; p2 reduced, smaller than c1; p3–4 with distinct paraconids and metaconids, p3 talonid short, p4 talonid somewhat longer; all upper and lower teeth with rugose enamel (Savage and Waters, 1978; Rose, 1995b).

Average length of m1: 1.8–2.4 mm.

Included species: *A. gazini* (known from locality CP25A*); *A. advena* (localities CP20B, CP20BB*).

Comments: Savage and Waters (1978) attributed several isolated tarsal elements and a distal tibia to *Arapahovius*. The astragalus is of a typical omomyid design with a well-developed posterior shelf, an elongate neck, and a continuous sustentacular–navicular facet. Other elements include an elongate calcaneum, navicular, and cuboid. The elongation of all of these elements suggests that *Arapahovius* was an accomplished leaper. *Arapahovius* is well represented in the Washakie Basin of Wyoming but is almost unknown outside of this area being otherwise represented by only a few specimens from the Bighorn and Great Divide Basins of Wyoming.

***Aycrossia* Bown, 1979**

Type species: *Aycrossia lovei* Bown, 1979.

Type specimen: USNM 250561, left maxilla with C1–M3.

Characteristics: Lower dental formula i2?, c1, p2-3, m3; p3 small relative to p4; p4 tall and premolariform, somewhat enlarged and buccally distended; m1 metaconid large, tiny parastylid, located between paraconid and metaconid; lower molar talonids with weakly rugose enamel; upper molars short and broad, lacking mesostyles, pericones and basal cingula, enamel slightly rugose; M3/m3 relatively unreduced compared with M2/m2 (Bown, 1979).

Average length of m1: 2.4 mm.

Included species: *A. lovei* only (known from localities CP27E, CP31E*).

***Strigorhysis* Bown, 1979**

Type species: *Strigorhysis bridgerensis* Bown, 1979.

Type specimen: USNM 250556, palate with left P4–M3 and right P3–M3.

Characteristics: Lower dental formula i2, c1, p2, m3; p3 much smaller than p4; p4 enlarged and buccally distended, lacking a paraconid; upper and lower molars with rugose, complex enamel; upper molars lingually expanded with strong postprotocingula (Bown, 1979; Bown and Rose, 1987).

Average length of m1: 2.2–2.85 mm.

Included species: *S. bridgerensis* (known from localities [CP20D], C P31E*); *S. huerfanensis* (locality SB22C*); *S. rugosus* (locality CP31E*).

Comments: *Strigorhysis* is closely related to *Absarokius* but differs in having relatively larger i1, smaller p3, and more crenulated enamel.

***Gazinius* Bown, 1979**

Type species: *Gazinius amplus* Bown, 1979.

Type specimen: USNM 250554, right maxilla with M1-2.

Characteristics: Differs from most other anaptomorphines in the following features: M2 protocone centrally placed, massive, bulbous; M2 lacking both conules and postprotocingulum (present in *G. bowni*), greatly expanded lingually; m2 trigonid compressed, cusps low and bulbous, distinct cristids (except for paracristid) absent, talonid short, tall hypoconid, low entoconid, hypoconulid absent; m2 talonid basin with weakly rugose enamel (Bown, 1979; Gunnell, 1995b).

Average length of M2: 2.3–2.5 mm.

Included species: *G. amplus* (known from localities CP31E*, [CP34A3]); *G. bowni* (locality CP34D*).

***Tatmanius* Bown and Rose, 1991**

Type species: *Tatmanius szalayi* Bown and Rose, 1991.

Type specimen: USGS 21654, left dentary with p3–m1.

Characteristics: Very small, size of *Anemorhysis*; differs from all other anaptomorphines in having p3 small and single-rooted; p4 high crowned and basally inflated but mesiodistally shorter than m1, with weak paraconid and metaconid; p4 root bilobed, single-rooted buccally, double-rooted lingually; dentary very deep anteriorly, indicating a large i1 (Bown and Rose, 1991).

Length of m1: 1.9 mm.

Included species: *T. szalayi* only, known from locality CP20C* only.

***Sphacorhysis* Gunnell, 1995b**

Type species: *Sphacorhysis burntforkensis* Gunnell, 1995b.

Type specimen: UM 30966, left dentary with p4-m3.

Characteristics: Differs from *Trogolemur* (to which it is otherwise similar) in having i1 relatively less enlarged; i2 and c1 smaller and less vertically implanted; p4 metaconid very small, hypoconid and cristid obliqua lacking; m2 larger than m1; lower molar talonids very shallow with moderately rugose enamel; m3 trigonid very low relative to talonid, hypoconulid short and broad, hypoconid massive and basally inflated, and cristid obliqua short and straight (Gunnell, 1995b).

Average length of m1: 1.9 mm.

Included species: *S. burntforkensis* only, known from locality CP34D* only.

***Artimonius* Muldoon and Gunnell, 2002 (including *Absarokius*, in part)**

Type species: *Artimonius witteri* (Morris, 1954) (originally described as *Absarokius witteri*).

Type specimen: YPM-PU 14972, left dentary with p3-4, m2-3.

Characteristics: Differs from most anaptomorphines in lacking p2 and having p4 extremely inflated; differs from all anaptomorphines except *Pseudotetonius* in the extreme reduction of p3; differs from *Pseudotetonius* in lacking an enlarged i1 and in having a relatively taller and more inflated p4; differs from *Anemorhysis* in having a simple, non-basined p4 talonid; differs from *Absarokius* in being larger, in having more transverse upper molars, molars with relatively larger, crenulated basins, and in having extreme p4 inflation and ventrobuccal distension (Muldoon and Gunnell, 2002).

Average length of m1: 2.60 mm.

Included species: *A. witteri* (known from localities CP25I*, J); *A. australis* (locality SB22C*); *A. nocerai* (localities SB22C*, CP25GG, J).

Comments: These species were all formerly included in *Absarokius*.

OMOMYINAE TROUESSART, 1879

Characteristics: i1 (where known) relatively small, moderately procumbent to vertical; teeth between i1 and p4 not usually reduced in size or number; p4 usually not inflated or buccally distended (*Uintanius* and *Jemezius* being exceptions); molar cusps acute, marginally placed, and not basally inflated; upper molar trigons open with weak to absent postprotocingula; lower molar talonid basins relatively deep; M3/m3 typically not reduced (Szalay, 1976; Bown and Rose, 1987; Gunnell and Rose, 2002).

***Omomys* Leidy, 1869 (including *Euryacodon*; *Palaeacodon*)**

Type species: *Omomys carteri* Leidy, 1869.

Type specimen: ANSP 10335, right dentary with p3-4, m2.

Characteristics: *Omomys* differs from other omomyines in having M1-3 with lingually continuous cingula, pericones well developed, hypocones distinct and cingular, and lacking postprotocingula; p3 relatively tall and unreduced.

Average length of m1: 2.5 mm.

Included species: *O. carteri* (known from localities [CC4], SB22C, SB43AA, B, CP5A, CP25J, CP31E, CP32B, CP34A3, B, C*, D, CP35, CP38B); *O. lloydi* (localities CP5A*, CP25J).

Omomys sp. is also known from NP8, NP9A–B and may also be present at SB42.

Comments: *Omomys* was the first tarsiiform and first primate described from North America (Leidy, 1869).

Omomys was only known from dental remains until very recently, when the first skull and postcranial remains were discovered (Rosenberger and Dagosto, 1992; Anemone, Covert, and Nachman, 1997; Alexander and MacPhee, 1999; Anemone and Covert, 2000; Murphey *et al.*, 2001). Postcranial remains include elongate foot elements and a long, slender, straight-shafted femur. The femur has a cylindrical head and an expansion of the articular surface onto the short, robust neck. The distal femur is deep antero-posteriorly. These features indicate an animal with powerful leaping capabilities (Anemone, Covert, and Nachman, 1997; Anemone and Covert, 2000). The one reported skull of *Omomys* (Alexander and MacPhee, 1999) has enlarged, relatively laterally facing orbits and a more elongate rostrum than either *Shoshonius* or *Tetonius*. Enlarged orbits suggest that *Omomys*, like other known omomyids, was probably nocturnal.

Szalay (1976) synonymized *Omomys sheai* (Gazin, 1962) with *Utahia kayi*. However, *O. sheai* lacks the compressed molar trigonids of utahiin omomyines and retains distinctive paraconids on m2–3. *O. sheai* is of the same size and similar morphology to *Loveina zephyri* and we believe that it can be assigned to that taxon (Stucky, 1984; Honey, 1990). Bown and Rose (1984) transferred *Omomys minutus* (Loomis, 1906) to *Loveina minuta*, a decision with which we concur.

Hemiacodon Marsh, 1872 (including _Omomys_, in part)

Type species: *Hemiacodon gracilis* Marsh, 1872.

Type specimen: YPM 11806, right dentary with p3–m3.

Characteristics: *Hemiacodon* is very similar to *Macrotarsius* (Gunnell, 1995b) but differs in having very rugose enamel; distinct M1-2 postprotocingula, large conules, and more distinct hypocones; p4 trigonid more open and better developed; m1-3 talonids much wider than trigonids.

Average length of m1: 3.6 mm.

Included species: *H. gracilis* (known from localities [CC4], CP34C, D*, CP35, CP37B, CP38B); *H. casamissus* (locality CP27D*).

Hemiacodon sp. is also known from localities CP34E and CP65.

Comments: *Hemiacodon* is one of the best known omomyines dentally and postcranially. Szalay (1976) suggested that *Hemiacodon* is more closely related to washakiins (including *Washakius*, *Loveina*, *Shoshonius*, and *Dyseolemur*) than to *Macrotarsius*. Honey (1990) questioned this allocation and removed *Hemiacodon* from washakiins. Gunnell (1995b) argued for a close relationship between *Hemiacodon* and *Macrotarsius*. *Hemiacodon* is known from the Bridgerian and may be present in the Wasatchian if *H. casamissus* truly does represent a species of *Hemiacodon* (see Gunnell [1995b] for discussion).

Washakius Leidy, 1873 (including _Yumanius_; _Shoshonius_ in part)

Type species: *Washakius insignis* Leidy, 1873.

Type specimen: ANSP 10332, right dentary with m2-3.

Characteristics: Differs from all other washakiins (*Loveina*, *Shoshonius*, *Dyseolemur*) in having well-developed hypocones on M1-2; differs from *Loveina* in having M1-2 second metaconule present, metacone more buccally inflated; p3-4 less posteriorly distended; m1-3 trigonids with metastylid (as in *Shoshonius* and *Dyseolemur*); m1-3 talonids widely open lingually; m3 with double-cusped hypoconulid; differs from *Shoshonius* in having M1-3 with weak to absent mesostyle; m2 paraconid more anterior with shorter paracristid; m1-3 talonid notch always present; differs from *Dyseolemur* in having M1-2 second metaconule present, protocones more peripheral; p4 buccal cingulum weak, paraconid and metaconid stronger; m1-2 talonids relatively wider, cusps more marginal; m3 with a double-cusped hypoconulid (Honey, 1990).

Average length of m1: 2.0–2.2 mm.

Included species: *W. insignis* (known from localities [CP25J], CP31E, CP34A3, B, C*, D); *W. izetti* (localities [SB22C], CP25J, CP64IIIA*); *W. laurae* (localities CP31E, CP34D*); *W. woodringi* (localities CC4*, CC7B).

Washakius sp. is also known from localities CP27E and CP31E.

Shoshonius Granger, 1910

Type species: *Shoshonius cooperi* Granger, 1910.

Type specimen: AMNH 14664, right maxilla with P3–M3.

Characteristics: Differs from other washakiins (*Loveina*, *Washakius*, *Dyseolemur*) in having strong mesostyles on M1–3; differs from *Loveina* in having m1-3 with distinct metastylids; differs from *Washakius* in having m2 paraconid more lingual with longer paracristid, and m1–3 talonid notch usually absent; differs from *Dyseolemur* in having M2 hypocone less inflated, m1-3 paraconid more lingual with longer paracristid, m1-3 talonids relatively wider with more marginally placed cusps (Szalay, 1976).

Average length of m1: 2.1 mm.

Included species: *S. cooperi* (known from localities SB22C, CP27D*); *S. bowni* (locality CP31E*).

Comments: *Shoshonius*, like *Omomys*, was until recently only represented by dental remains. Recently described skulls show that *Shoshonius* had relatively large orbits, lacked postorbital closure, had a reduced rostrum, lacked a tubular ectotympanic, and had a posterior carotid foramen ventrolaterally placed on the auditory bulla (Beard, Krishtalka, and Stucky, 1991; Beard and MacPhee, 1994). Postcranially, *Shoshonius* resembles other known omomyids in having a relatively long femur with proximally placed lesser and third trochanters, a deep distal femur, and elongate tarsal elements (Dagosto, Gebo, and Beard, 1999).

Uintanius Matthew, 1915 (including _Huerfanius_; _Omomys_, in part)

Type species: *Uintanius ameghini* (Wortman, 1904) (originally described as *Omomys ameghini*).

Type specimen: YPM 13241, left dentary with m2-3.

Characteristics: *Uintanius* is distinguished from all other omomyines by its very large upper and lower third and fourth premolars (especially P4/p4); *Uintanius* differs from *Jemezius* in having P3-4 protocones relatively smaller; M1-3 conules stronger; p4 talonid simple, lacking a well-developed basin; m1-3 paraconids shifted buccally; and m3 relatively smaller (Szalay, 1976; Beard, Krishtalka, and Stucky, 1992; Gunnell, 1995b).

Average length of m1: 1.9 mm.

Included species: *U. ameghini* (known from localities CP5A, CP34C*, D); *U. rutherfurdi* (localities SB22C*, CP25J, CP27E, [CP34A3]).

Chumashius Stock, 1933

Type species: *Chumashius balchi* Stock, 1933.

Type specimen: LACM 1391, left dentary with p3–m3.

Characteristics: *Chumashius* is very similar to *Omomys*, differing only in having a relatively larger lower canine, relatively lower p3-4, and in lacking distinct pericones and hypocones on upper molars (Szalay, 1976).

Average length of m1: 2.5 mm

Included species: *C. balchi* only (known from localities CC7E, CC9BB*, B2).

Dyseolemur Stock, 1934

Type species: *Dyseolemur pacificus* Stock, 1934.

Type specimen: LACM 1395, right dentary with p4–m3.

Characteristics: Differs from *Shoshonius* in lacking M1-3 mesostyles and in having an expanded M2 distolingual cingulum; differs from *Washakius* in lacking M1-3 double metaconule, in having buccally shifted m1-3 paraconids, and a relatively smaller m3; differs from *Loveina* in having m1-3 metastylids (Szalay, 1976).

Average length of m1: 2.0 mm.

Included species: *D. pacificus* only (known from localities CC6B, CC7B–C, E, CC8, CC8IIA, CC9A*, AA, B, CC9IVA).

Loveina Simpson, 1940 (including _Tetonius_, in part; _Notharctus_, in part; _Omomys_, in part)

Type species: *Loveina zephyri* Simpson, 1940.

Type specimen: AMNH 32517, left dentary with p3–m1.

Characteristics: *Loveina* is more primitive than other washakiins (*Shoshonius*, *Dyseolemur*, and *Washakius*) in lacking or having only incipient m1–3 metastylids and in lacking hypocones, pericones, double metaconules, and mesostyles on upper molars (one or more of which occur in other genera) (Szalay, 1976; Honey, 1990; Gunnell *et al.*, 1992).

Average length of m1: 2.1 mm.

Included species: *L. zephyri* (known from localities SB22B, CP27D*); *L. minuta* (including *Notharctus minutus* Loomis, 1906) (localities CP25F, CP27B*); *L. sheai* (including *Omomys sheai*, Gazin, 1962) (localities CP25G*, H); *L. wapitiensis* (locality CP20E*).

Macrotarsius Clark, 1941 (including _Hemiacodon_, in part)

Type species: *Macrotarsius montanus* Clark, 1941.

Type specimen: CM 9592, right dentary with c1, p3–m3.

Characteristics: *Macrotarsius* is closely related to *Hemiacodon* but differs in having P3 lingually broader with hypocone shelf better developed; M1–3 conules relatively weaker, stylar shelves moderate, mesostyles well developed, hypocones and pericones somewhat weaker giving upper molars a more squared shape; m1-3 trigonids closed lingually and distally, metaconid marginal and not basally inflated; m3 as broad as m1-2 (Honey, 1990; Gunnell, 1995b).

Average length of m1: 4.0–4.5 mm.

Included species: *M. montanus* (known from localities [NP9A], NP24B*); *M. jepseni* (localities [SB42], CP6A*); *M. roederi* (locality CC9AA*); *M. siegerti* (localities CP29C*, D, [CP37B]).

Comments: *Macrotarsius* is one of the most geographically widespread omomyines known, being represented from deposits all along the Rocky Mountain corridor westward to California as well as from China.

Ourayia Gazin, 1958 (including _Mytonius_; _Microsyops_, in part)

Type species: *Ourayia uintensis* (Osborn, 1895) (originally described as *Microsyops uintensis*).

Type specimen: AMNH 1899, left dentary with p3–m2.

Characteristics: *Ourayia* differs from *Wyomomys* and *Ageitodendron* in being larger; having m2-3 paraconids indistinct; m1-3 molar enamel more rugose and complex; further differs from *Ageitodendron* in lacking m1-2 hypoconulids, having talonid notch with a weaker inflection of the entocristid and postmetacristid, and ectocingulid stronger (Gunnell, 1995b).

Average length of m1: 4.1 mm.

Included species: *O. uintensis* (known from localities SB43AA, CP6A*, B, CP38C); *O. hopsoni* (localities CP6B*, CP29C).

Ourayia sp. is also known from CC4, CC6A–B, CC7B, CC8, CC8IIA.

Comments: *Ourayia* has had a complex taxonomic history (Osborn, 1902; Wortman, 1903–4; Gazin, 1958; Simons, 1961; Robinson, 1968; Szalay, 1976; Krishtalka, 1978; Honey, 1990; Gunnell, 1995b; Rasmussen, 1996). We consider *Ourayia* to be the sister taxon to *Wyomomys* and *Ageitodendron* (Gunnell, 1995b), a clade that forms a sister-group relationship with *Utahia*, *Stockia*, and *Chipetaia*. The former clade can be differentiated from the latter genera by the presence of rounded and relatively shallow lower molar talonids, lower molars with very heavy, notched, buccally distended ectocingulids, and basally inflated protoconids and hypoconids.

***Utahia* Gazin, 1958**

Type species: *Utahia kayi* Gazin, 1958.

Type specimen: CM 6488, right dentary with p3–m3.

Characteristics: *Utahia* seems closely related to *Stockia* and *Chipetaia* but is considerably smaller; less-crenulated enamel; p4 relatively smaller; trigonids of m1–3 higher than in *Stockia. Utahia* differs from *Chipetaia* in p4 lacking a distinct metaconid; smaller and more mesial entoconids on m1-2; more lingual cristid obliquae on m2-3; more lingual paraconid on m2 (Szalay, 1976; Rasmussen, 1996; Muldoon and Gunnell, 2002).

Average length of m1: 1.8 mm.

Included species: *U. kayi* (known from locality CP5A*); *U. carina* (locality CP25J*).

***Stockia* Gazin, 1958**

Type species: *Stockia powayensis* Gazin, 1958.

Type specimen: LACM 2235, right dentary with m1-3.

Characteristics: *Stockia* differs from *Utahia* in having relatively larger p4 and more crenulated molar enamel; differs from *Chipetaia* in m2-3 with less-distinct paraconids, less-inflated metaconids, stronger premetacristids, more buccal cristid obliquae, and shallower talonid notches; differs from *Asiomomys*, in having higher molar trigonids, less distinctly separated protoconids and metaconids, more-distinct cusps and crests, and more crenulated enamel (Szalay, 1976; Beard and Wang, 1991; Rasmussen, 1996).

Average length of m1: 2.9 mm.

Included species: *S. powayensis* only, known from locality CC4* only.

Comments: *Stockia* is quite similar to *Utahia* and *Chipetaia*. *Stockia* also shares many similarities with *Asiomomys* (Beard and Wang, 1991) from China. Lillegraven (1980) suggested that *Stockia* was congeneric with *Omomys*, an idea rejected by Honey (1990) and Beard and Wang (1991). *Stockia*, *Utahia*, *Asiomomys*, and *Chipetaia* appear to be closely related, but none of these taxa is well known and only *Chipetaia* is represented by any upper dental elements (Rasmussen, 1996). All of these taxa must await much better material before certain taxonomic allocation will be possible.

***Ekgmowechashala* Macdonald, 1963**

Type species: *Ekgmowechashala philotau* Macdonald, 1963.

Type specimen: SDSM 5550, left dentary with p3–m2.

Characteristics: Lower dental formula i2, c1, p3, m3. *Ekgmowechashala* differs from all other known omomyids in having lower molars progressively reduced in size from m1 to m3; p4 flattened and molarized with closely joined and enlarged entoconid and metastylid; m1-3 paraconids absent and paracristids weak to absent; P4 paraconule very large, situated between enlarged and bulbous protocone and paracone; all tooth enamel crenulated.

Average length of m1: 4.1 mm.

Included species: *E. philotau* only (known from localities CP85C*, CP101).

Ekgmowechashala sp. is also known from localities GC5A, PN6D and possibly from locality PN6D2.

Comments: The phylogenetic affinities of *Ekgmowechashala* remain vague. Szalay and Delson (1979) suggested that it was most closely related to *Rooneyia* among known North American fossil mammals. Rose and Rensberger (1983) supported Szalay and Delson but did note that *Ekgmowechashala* was also similar to some European microchoerine omomyids (specifically *Necrolemur* and *Microchoerus*). McKenna (1990) suggested that *Ekgmowechashala* may not have been a primate at all, choosing instead to ally it with *Tarka* and *Tarkadectes* in Plagiomenidae, a family that may be related to extant flying lemurs (Dermoptera). If *Ekgmowechashala* is a primate, it is the latest-occurring primate in North America (Arikareean, latest Oligocene, or earliest Miocene).

***Rooneyia* Wilson, 1966**

Type species: *Rooneyia viejaensis* Wilson, 1966.

Type specimen: UTBEG 40688-7, skull.

Characteristics: *Rooneyia* differs from all other omomyids in having a reduced and uninflated petromastoid region, fused frontal bones, and squared orbits. Other cranial characters of *Rooneyia* include the presence of a tubular ectotympanic, a lacrimal foramen opening anterior to the orbital margin, absence of postorbital closure, and a foramen magnum that opens posteriorly. M1-3 low crowned with very bunodont cusps, well-developed conules, and robust, cingular hypocones that are nearly as large as protocones (Wilson, 1966; Szalay, 1976).

Average length of M1: 4.0 mm.

Included species: *R. viejaensis* only, known from locality SB44B* only.

Comments: *Rooneyia* is still represented only by the type skull. The taxonomic position of *Rooneyia* is in question (Wilson, 1966; Simons, 1968; Szalay, 1976; Ross, Williams, and Kay, 1998) and further evidence is required to elucidate its phylogenetic relationships.

***Steinius* Bown and Rose, 1984 (including *Omomys*, in part; *Loveina*, in part; *Uintanius*, in part)**

Type species: *Steinius vespertinus* (Matthew, 1915) (originally described as ?*Omomys vespertinus*).

Type specimen: AMNH 16835, left dentary with m1-3.

Characteristics: Lower dental formula of i2, c1, p4, m3; i1 slightly larger than c1, both larger than i2; p1-2 small, single-rooted; p3-4 simple and uninflated; p3 relatively tall and unreduced; p4 metaconid small and low; upper and lower molars with acute cusps set on the tooth margins, and not basally inflated; differs from *Omomys* (to which it is otherwise quite similar) in retaining p1, in having a less-projecting p3, smaller molar paraconids with m1 paraconid less lingual and m2–3 paraconids more lingual than in *Omomys* (Bown and Rose, 1984, 1987, 1991; Rose and Bown, 1991).

Average length of m1: 2.4–2.9 mm.

Included species: *S. vespertinus* (known from localities CP20B*, BB, CP28A); *S. annectens* (locality CP20C*).

Comments: *Steinius* is one of the most primitive tarsiiforms known even though it is not among the earliest occurring. As such *Steinius* forms a plausible ancestral taxon for many later-occurring omomyines.

Jemezius Beard, 1987

Type species: *Jemezius szalayi* Beard, 1987.

Type specimen: CM 34843, left dentary with p4–m1.

Characteristics: *Jemezius* differs from its probable sister taxon *Uintanius* in having P3-4 protocones relatively larger; upper molar conules weaker; p4 lower and more complex with well-developed talonid basin; lower molar paraconids more lingual; and m3 relatively larger with less-compressed trigonid (Beard, 1987).

Average length of m1: 2.3 mm.

Included species: *J. szalayi* only, known from locality SB24* only.

Comments: *Jemezius* is known only from New Mexico but appears to be closely related to *Uintanius* and *Steinius* (Beard, 1987), both known only from Wyoming (Bown and Rose, 1991; Gunnell, 1995b).

Yaquius Mason, 1990

Type species: *Yaquius travisi* Mason, 1990.

Type specimen: LACM 40201, right m2.

Characteristics: *Yaquius* is very similar to *Macrotarsius*, differing in having strong p3-4 buccal cingulids; a less anteroposteriorly compressed p4 talonid basin; m1-3 protoconids and hypoconids shifted lingually and metaconids twinned; m3 talonid relatively shorter and entoconid higher; M3 with tiny to absent conules. Differs from *Ourayia* in having broader p3–m3 buccal cingulids; more quadrate p3-4; p4 anteroposteriorly compressed; m2-3 talonid notch closed lingually by entocristid; m1-3 paraconids taller and metaconids twinned (Mason, 1990).

Average length of m2: 4.3 mm.

Included species: *Y. travisi* only, known from locality CC9A* only.

Comments: *Yaquius* is only known from a small sample of isolated teeth making its affinities difficult to discern.

Wyomomys Gunnell, 1995b

Type species: *Wyomomys bridgeri* Gunnell, 1995b.

Type specimen: UM 98874, right dentary with m2–3.

Characteristics: Compressed m2-3 trigonids, paraconid distinct and separated from metaconid, paracristids long and relatively tall, hypoconids buccally inflated; m2 cristid obliqua robust and heavy, buccally projecting ectocingulid; m3 entocristid relatively long. Differs from *Ageitodendron* and *Ourayia* (its apparent closest relatives) in having distinct m2-3 paraconids that are separated from metaconids by sharply incised valleys; lacking inflected talonid notches; having m3 as large as or larger than m2; having smooth molar; and being smaller (Gunnell, 1995b).

Average length of m2: 2.4 mm.

Included species: *W. bridgeri* only, known from locality CP34C* only.

Ageitodendron Gunnell, 1995b.

Type species: *Ageitodendron matthewi* Gunnell, 1995b.

Type specimen: UM 30924, right dentary with m1-3.

Characteristics: *Ageitodendron* is larger than *Wyomomys*, differing in having m2-3 paraconids small and undifferentiated; m2 with weakly twinned metaconid cusp; m3 relatively smaller than m2; talonid basin enamel weakly rugose; talonid notch inflected (formed by buccally angled entocristid and postmetacristid); ectocingulid very robust and notched basal to hypoflexid. *Ageitodendron* is smaller than *Ourayia*, differing in having small but distinct m1–2 hypoconulids; m2 with weakly twinned metaconid cusp; m3 hypoconulid relatively longer; molar talonid notch with very well-developed inflection; molar talonid enamel weakly rugose; molar ectocingulids relatively less robust (Gunnell, 1995b).

Average length of m1: 3.3 mm.

Included species: *A. matthewi* only, known from locality CP34D* only.

Chipetaia Rasmussen, 1996

Type species: *Chipetaia lamporea* Rasmussen, 1996.

Type specimen: CM 69800, right m2-3.

Characteristics: *Chipetaia* appears closely related to *Utahia* and *Stockia*. Differs from *Utahia* in having m1-2 entoconids relatively larger and more distally placed; m2-3 cristids obliquae more buccal; m1-3 paraconids shifted buccally; p4 metaconid distinct and with a broader talonid, and enamel more rugose. Differs from *Stockia* in having m2–3 with paraconids more distinct, metaconids more inflated, premetacristid weaker, cristid obliquae more lingually placed, and talonid notches more open. *Chipetaia* differs from most other omomyids in having m2–3 with compressed trigonids and reduced paraconids (Rasmussen, 1996).

Average length of m1: 3.6 mm.

Included species: *C. lamporea* only, known from locality CP6AA* only.

NOTHARCTIDAE TROUESSART, 1879

NOTHARCTINAE TROUESSART, 1879

Notharctines (known from North America and Europe) differ from cercamoniines (present in North America, Europe, and Africa) in having relatively smaller orbits, upper molars with postprotocingula, which in most forms give rise to a hypocone (or "pseudohypocone" instead of the cingular hypocone of cercamoniines), always retaining unreduced upper and lower first and second premolars, having mesostyles on upper molars (except for most primitive taxa), and in having a very large posterior astragalar shelf (Franzen, 1994; Godinot, 1998).

***Notharctus* Leidy, 1870 (including *Cantius*, in part; *Hipposyus*; *Thinolestes*; *Telmatolestes*; *Limnotherium*; *Prosinopa*, in part; *Tomitherium*, in part)**

Type species: *Notharctus tenebrosus* Leidy, 1870.

Type specimen: USNM 3752, right dentary with c1, p2–m3.

Characteristics: *Notharctus* differs from all other notharctines in having relatively long lower molar paracristids that are anteriorly expanded and posteriorly curving; large, distinct, and lobate upper molar hypocones that are separated from protocones by a distinct notch; a buccolingually compressed and bladelike P4 paracone with a metacone present in more derived species; and sharply defined crests on all cheek teeth. Differs from *Cantius* in having strong and well-developed upper molar mesostyles; weakly defined but present lower molar paraconids; well-developed entoconid notches. Differs from *Copelemur* in lacking buccolingually compressed and bladelike postprotocingula on upper molars; having more posteriorly placed lower molar entoconids with longer and straighter entocristids. Differs from *Smilodectes* in having a longer and lower skull with uninflated frontal bones; a flexed cristid obliqua on m3; high and strong entocristid on m1-2; weak but present paraconids on lower molars. Differs from *Pelycodus* and *Hesperolemur* in having more acute cusps and crests and strong mesostyles.

Average length of m1: 5.0–7.5 mm.

Included species: *N. tenebrosus* (known from localities NB14, [CP5A], CP31A, CP34C); *N. pugnax* (locality CP34C*); *N. robinsoni* (SB22C, [CP27E], CP34A3*); *N. robustior* (localities CP34D*, CP35, CP38B–C, CP65); *N. venticolus* (including *Cantius antediluvius*) (localities CP25G–J, CP27D*, E, CP64C).

Notharctus sp. is also known from localities SB43AA, SB26II, CP29A, CP31E, CP32B, CP34B-E, CP35.

Comments: *Notharctus* is among the best known fossil primates, being represented by virtually complete skulls and skeletons (Gregory, 1920; Gazin, 1958; Alexander, 1994). *Notharctus* is a large adapiform with body size estimates ranging from 4 to 8 kg. The combination of relatively large size and sharply crested teeth suggests that *Notharctus* was a folivore.

***Pelycodus* Cope, 1875 (including *Prototomus*, in part)**

Type species: *Pelycodus jarrovii* (Cope, 1874) (originally described as *Prototomus jarrovii*).

Type specimen: Unnumbered dentary with broken m2, complete m3 in Cope Collection, specimen now lost. A topotype, AMNH 16298, left dentary with m1-2, was noted by Gingerich and Simons (1977) and this specimen was later designated as a neotype by Gingerich and Haskin (1981).

Characteristics: *Pelycodus* differs from other notharctines in having molars with low, rounded, bulbous cusps; upper molars with reduced stylar shelf and no development of mesostyles or para- and metastyles; lower molars with broad and shallow talonid basins, anteroposteriorly compressed trigonids, and reduced paraconids on m2-3. Differs from *Hesperolemur* in retaining lower molar paraconids and lacking distinctly inflated cusps.

Average length of m2: 5.6–6.9 mm.

Included species: *P. jarrovii* (known from localities SB22B, SB24*, CP20B, [CP27B]); *P. danielsae* (including *P. schidelerorum*) (localities SB24*, CP64C).

***Smilodectes* Wortman, 1903 (including *Aphanolemur*; *Hyopsodus*, in part)**

Type species: *Smilodectes gracilis* (Marsh, 1871) (originally described as *Hyopsodus gracilis*).

Type specimen: YPM 11800, left dentary with p4-m1.

Characteristics: *Smilodectes* differs from other notharctines in having a skull with a relatively short rostrum, rounded braincase, and inflated frontal bones; lacking a flexed cristid obliqua on m3 (the cristid obliqua does not turn buccally to join the lingually angled postprotocristid as it does in all other notharctines). Differs from all other notharctines except *Copelemur* in having reduced paraconids in combination with an elongate and straight paracristid that leaves molar trigonids widely open lingually. Shares strong mesostyles with *Notharctus* but differs in having more buccolingually compressed postprotocristids with less-distinct hypocones, in being somewhat smaller, and in lacking a fused mandibular symphysis.

Average length of m1: 4.0–5.2 mm.

Included species: *S. gracilis* (known from localities CP5A, CP31E, CP34B, C*, D); *S. mcgrewi* (localities SB22C, CP20F, CP34A3*); *S. sororis* (locality CP25J*).

Comments: *Smilodectes gingerichi* likely does not represent a species of *Smilodectes* (Gunnell, 2002) (see inclusion within *Cantius*, below), thereby limiting the stratigraphic range of *Smilodectes* to the Bridgerian only.

***Cantius* Simons, 1962 (including *Tomitherium*, in part; *Pelycodus*, in part; *Protoadapis*; *Smilodectes*, in part)**

Type species: *Cantius angulatus* (Cope, 1875) (originally described as *Pelycodus angulatus*).

Type specimen: Unnumbered left dentary with m1 in Cope Collection, specimen now lost. A neotype, AMNH 55511, left dentary p3-4, right dentary p4–m2, m3 trigonid, was designated by Beard (1988).

Characteristics: *Cantius* differs from most other notharctines typically in being somewhat smaller (except for some late-occurring and derived species); differs from *Copelemur* and *Smilodectes* in lacking or generally having shallower lower molar entoconid notches and short, curving entocristids, entoconid more posteriorly placed, and in having short paracristids that close trigonids lingually; upper molars lacking distinct, bladelike postprotocingula. Differs from *Notharctus* in having upper molars with weak, undifferentiated or smaller hypocones, absent or small mesostyles; lower molars lacking or having shallower entoconid notches and retaining distinct paraconids and short paracristids that close trigonids lingually. Differs from *Pelycodus* in having more-distinct, less-bulbous cusps and crests; upper molars with well-developed stylar cusps; lower molars with distinct paraconids. Differs from *Hesperolemur* in having more rectangular, lingually narrower upper molars with weakly developed hypocones and weak to absent mesostyles; lower molars retaining paraconids and high paracristids that close trigonid lingually.

Average length of m1: 3.3–5.5 mm.

Included species: *C. angulatus* (including *Smilodectes gingerichi*) (known from locality SB24*); *C. abditus* (localities CP19D, CP20BB, C*, CP25E); *C. frugivorus* (localities SB24*, CP25C, E, G, [CP27A], CP64B–C); *C. mckennai* (localities CP19A*, B, CP20A, CP25A–B); *C. nunienus* (localities CP25F–J, CP27D*, E); *C. ralstoni* (localities CP19A*, CP20A, CP25B); *C. torresi* (localities CP19AA*, CP25B); *C. trigonodus* (localities CP19C, CP20B*, CP25A–B, NP49); *C. simonsi* (locality CP19E*).

Comments: *Cantius* is the only notharctine known from outside North America, being represented by *C. eppsi* and *C. savagei* in Europe (Godinot, 1998). *Cantius* is a paraphyletic taxon, giving rise to *Notharctus* and Copelemurini in North America (Gunnell, 2002) and perhaps at least to some cercamoniine adapiforms. This is not unexpected given the rather primitive nature and relatively ancient age of *Cantius*. *Cantius* spp., especially those that are more derived and later occurring, exhibit a fair degree of dental variability (Gunnell, 2002), often resembling *Notharctus* in some features. For example, some specimens lack entoconid notches completely, while others exhibit more distinct notches. Hypocones vary from poorly developed to quite strong, as do mesostyles. Some specimens with relatively distinct entoconid notches have been assigned to *Copelemur* (Gingerich and Simons, 1977), however, these specimens seem to lack other derived character states of *Copelemur* and could simply represent an independent acquisition of entoconid notches in some *Cantius* spp. (Beard, 1988). The fact that *Notharctus* also has strong entoconid notches favors this latter interpretation.

***Copelemur* Gingerich and Simons, 1977 (including *Tomitherium*, in part)**

Type species: *Copelemur tutus* (Cope, 1877) (originally described as *Tomitherium tutum*).

Type specimen: Unnumbered right dentary with p3–m1 in Cope Collection, specimen now lost. A neotype, UALP 10233, right dentary with p3–m2, was designated by Beard (1988).

Characteristics: *Copelemur* differs from all other notharctines in having upper molars with a well-developed, posteriorly extended, buccolingually compressed postprotocingulum; lower molars with strong anterolingually extending paracristid retaining a distinct paraconid, entoconid anterobuccally placed, and entocristid short and curving. Differs from all other notharctines except *Notharctus*, *Smilodectes*, and some species of *Cantius* in having lower molars with a distinct entoconid notch.

Average length of m1: 3.8–5.2 mm.

Included species: *C. tutus* (known from localities SB24*, CP25E, F); *C. australotutus* (localities CP25C, E*); *C. praetutus* (localities CP25C*, CP64B).

Comments: *Copelemur* appears to have a generally southern distribution being certainly known only from southern Wyoming and New Mexico. Some specimens from the Bighorn Basin in northwestern Wyoming may represent *Copelemur* but are more likely to represent *Cantius* spp. that have converged on *Copelemur* in some dental features. *Copelemur* is restricted to the Wasatchian.

Two other species of "*Copelemur*," "*C.*" *feretutus* (known from localities CP19D, CP20C, CP27B*) and "*C.*" *consortutus* (known from localities SB24, CP20D, CP27D*) have been described (Gingerich and Simons, 1977). Neither species is likely to represent *Copelemur* but it remains uncertain where they do belong.

***Hesperolemur* Gunnell, 1995a**

Type species: *Hesperolemur actius* Gunnell, 1995a.

Type specimen: SDSNH 35233, crushed skull with left and right P4–M3.

Characteristics: *Hesperolemur* differs from all known notharctines (except for one species of *Cantius*, see Rose, MacPhee, and Alexander, 1999) in lacking bony tubes for transmitting middle ear arteries; having the anterior crus of the ectotympanic solidly fused to the lateral wall of the auditory bulla. Further differs from all other notharctines in having upper molars with a combination of a well-developed, lobate hypocone and a small mesostyle with weakly developed stylar cuspules; lower molars lacking paraconids and with very low and short paracristid; lower molars lacking development of a trigonid fovea (Gunnell, 1995b).

Average length of M1: 5.2 mm.

Included species: *H. actius* only, known from locality CC4 only.

Comments: The validity of the genus *Hesperolemur* has been questioned by Rose, MacPhee, and Alexander (1999). These authors prefer to refer the species *H. actius* to the genus *Cantius*.

CERCAMONIINAE GINGERICH, 1975

Characteristics: Cercamoniines, a primarily European radiation, differ from notharctines in having relatively larger orbits; upper molars with true hypocones (derived from postcingulum) and lacking postprotocingula; reduced upper and lower first and second premolars (P1/p1 often lost), lacking mesostyles on upper molars, and in having a smaller posterior astragalar shelf (Franzen, 1994; Godinot, 1998). In most other postcranial attributes, cercamoniines are much more similar to notharctines than to European Adapidae.

***Mahgarita* Wilson and Szalay, 1976**

Type species: *Mahgarita stevensi* Wilson and Szalay, 1976.

Type specimen: TMM 41578-9, crushed skull with left and right C1-M3.

Characteristics: Dental formula I2/2, C1/1, P3/3, M3/3; P2/2 very reduced in size; P3 with a reduced protocone; upper molar metaconules crestiform, hypocones sharply defined; cingula strong and essentially continuous; rostrum relatively short; promontory canal relatively large and stapedial canal reduced or absent; postorbital closure lacking but strong postorbital bar present; mandibular symphysis solidly fused (Wilson and Szalay, 1976; Rasmussen, 1990).

Average length of M1: 3.6 mm.

Included species: *M. stevensi* only, known from locality SB43C* only.

Comments: *Mahgarita* is the only non-notharctine possible adapiform known from North America. It has been variously interpreted as a cercamoniine adapiform (= Adapidae of Wilson and Szalay, 1976; = Protoadapini of Szalay and Delson, 1979) or as a protoanthropoid (Rasmussen, 1990). *Mahgarita* and *Rooneyia* are among the latest surviving North American primates, and the phylogenetic relationships of both are still in dispute.

BIOLOGY AND EVOLUTIONARY PATTERNS

Three principal clades of anaptomorphine omomyids can be recognized in the early Eocene of North America, one comprising *Teilhardina*, *Anemorhysis*, and possibly *Arapahovius*, a second including *Tetonius* and *Pseudotetonius*, and a third consisting of *Absarokius* and probably *Strigorhysis*. All three of these lineages show evidence of anagenetic change through time as documented by dense samples in the Bighorn Basin of Wyoming (Bown and Rose, 1987). The early Eocene radiation also includes more enigmatic or poorly represented taxa such as *Chlororhysis* and *Tatmanius* (Bown and Rose, 1984, 1991; Rose and Bown, 1991; Rose, 1995b) (Figure 15.5).

The *Teilhardina–Anemorhysis* clade spans much of the early Eocene with *Teilhardina* being present from Wasatchian biochron Wa0 through Wa5 (Rose, 1995b) and *Anemorhysis* from Wa4 through Wa7. This clade is characterized by relatively small teeth, progressively shorter, more molariform posterior premolars (especially p4), little anterior tooth loss (except for p1 and occasionally p2) or reduction, and progressive enlargement of central incisors (Bown and Rose, 1987).

The *Tetonius–Pseudotetonius* clade was derived from a *Teilhardina* ancestry and spans the early part of the early Eocene from Wa1 through Wa4. The principal trend in this lineage was toward hypertrophy of i1 and reduction in size and number of teeth between i1 and p4, concomitant with reduction from a two-rooted to one-rooted p3 (Bown and Rose, 1987). *Tatmanius* may be a derivative or sister taxon to the *Tetonius–Pseudotetonius* clade (Bown and Rose, 1991).

The *Absarokius* lineage spans the latter part of the early Eocene from Wasatchian biochron Wa5 through Wa7 (Bown and Rose, 1987; Muldoon and Gunnell, 2002). In general, the *Absarokius* lineage is typified by retention of relatively unreduced canine and p2 and a relatively small central incisor, along with increasing hypertrophy of P4/p4. *Strigorhysis* and other anaptomorphines such as *Anaptomorphus*, *Aycrossia*, *Artimonius*, and *Gazinius* all are probably derived from this *Absarokius* clade. Some taxa such as *Strigorhysis*, *Gazinius*, and *Artimonius* continued enlargement of P4/p4 and crenulation of enamel while others such as *Anaptomorphus* and to a lesser extent *Aycrossia*, retained smaller P4/p4 teeth and did not develop complex, crenulated enamel.

Steinius is the only omomyid besides *Teilhardina* known to retain p1, and the alveolus for this tooth suggests that it is less reduced than in *Teilhardina*. However, *Steinius* apparently has a larger central incisor (i1) than in primitive species of *Teilhardina*, presumably a derived trait (Rose and Bown, 1991). *Steinius* closely resembles *Omomys* in having generalized molars with peripheral cusps, and elongate, pointed posterior premolars with an unreduced p3 and prominent but low metaconids on p4. Both forms have relatively large (unreduced) lower canines. The very primitive nature of *Steinius* and its close resemblance to *Omomys* suggests that *Steinius* is the most primitive known omomyine. *Steinius* may be linked to uintaniins as well through the Wasatchian genus *Jemezius*.

The evolutionary history of middle Eocene omomyid tribes is not as well understood or as well documented as are early Eocene anaptomorphines, but some relationships seem well supported. As noted above, middle Eocene anaptomorphins (*Artimonius*, *Anaptomorphus*, *Aycrossia*, *Strigorhysis*, *Gazinius*) are likely derived from one of the lineages of *Absarokius* documented from the early Eocene (Bown and Rose, 1987). Trogolemurins (*Trogolemur* and *Sphacorhysis*) seem distantly related to *Anemorhysis* but more closely related to *Arapahovius*. Omomyins (*Omomys* and *Chumashius*) may have been derived from a *Steinius*-like form (Bown and Rose, 1984; Honey, 1990; Rose and Bown, 1991; Gunnell, 1995b). Washakiins (*Washakius*, *Shoshonius*, *Dyseolemur*) can be traced to latest early Eocene *Loveina* in Wasatchian Wa7 but their origins from

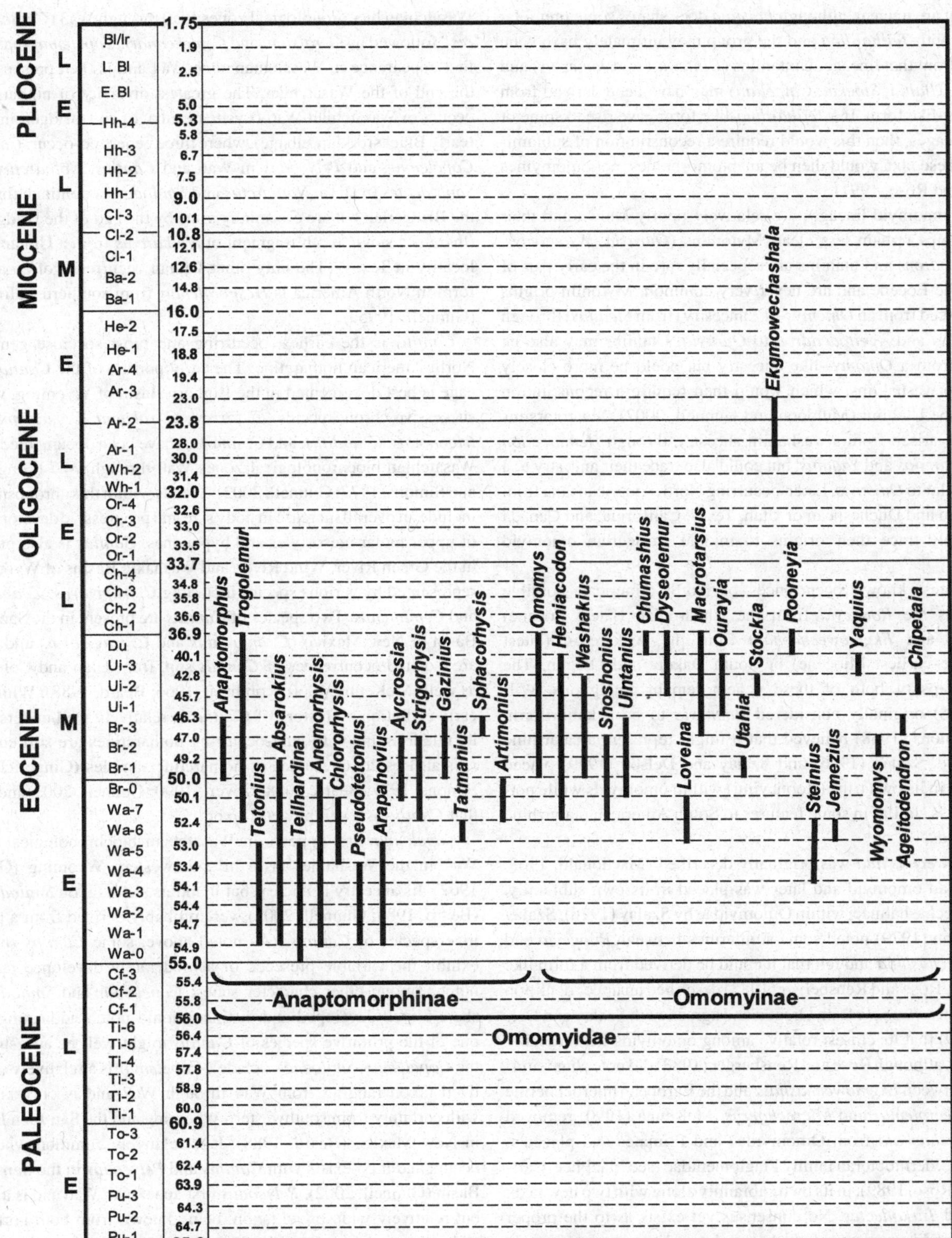

Figure 15.5. Temporal ranges of North American genera of Omomyidae.

that point are unclear, although *Loveina* does shares some premolar features with *Teilhardina* and the group may ultimately have been derived from that lineage. Uintaniins (*Uintanius* and *Jemezius*) and utahiins (*Utahia*, *Stockia*, *Chipetaia*) may have been derived from a *Steinius*-like form. If a *Teilhardina*-like form gave rise to some of these lineages, then this would require a reconstitution of subfamilies, as these taxa would then be anaptomorphines, not omomyines (Bown and Rose, 1991).

Other omomyid lineages are relatively poorly known and their relationships remain less clear. Mytoniins (*Ourayia*, *Wyomomys*, *Ageitodendron*) and utahiins are especially rare in the early part of the middle Eocene and are never very common. Mytoniin origins can be traced from an *Omomys*-like ancestry (from *Steinius*) through *Wyomomys* and *Ageitodendron* to *Ourayia*. Utahiins may also be derived from a *Omomys*-like ancestry but could be more closely related to washakiins, which would then require a reconstitution of the tribe Utahiini (Muldoon and Gunnell, 2002). Macrotarsiins may have arisen from a washakiin ancestry, through *Hemiacodon* to *Macrotarsius* and *Yaquius*, but could also trace their ancestry to a form similar to *Omomys*. Later-occurring North American taxa from the Uintan and Duchesnean of Utah, Texas, California, and Canada presumably trace their origins to middle Bridgerian omomyid tribes.

The latest known occurrences of North American possible omomyids are *Rooneyia* from the Chadronian (latest Eocene) of Texas and *Ekgmowechashala* from the Arikareean (latest Oligocene–earliest Miocene) of South Dakota and Oregon. The relationships of both of these genera remain ambiguous. Wilson (1966) originally considered *Rooneyia* to be a lemuriform, while Simons (1968) believed that it might represent a catarrhine anthropoid. Szalay (1976) and Szalay and Delson (1979) placed *Rooneyia* in its own tribe (Rooneyiini) within omomyids while noting that it is similar in some features to South American platyrrhine anthropoids.

Ekgmowechashala was originally described (Macdonald, 1963, 1970) as an omomyid and later was placed in its own subfamily, Ekgmowechashalinae, within Omomyidae by Szalay (1976). Szalay and Delson (1979) noted some similarities between *Rooneyia* and *Ekgmowechashala* and felt that it could be derived from a form like *Rooneyia*. Rose and Rensberger (1983) described an upper dentition of *Ekgmowechashala* from Oregon and agreed with Szalay and Delson (1979) that its closest relative among omomyids was probably *Rooneyia*, although Rose and Rensberger (1983) did note some similarities between *Ekgmowechashala* and the European microchoerine genera *Necrolemur* and *Microchoerus*. McKenna (1990) removed *Ekgmowechashala* from Omomyidae and Primates and placed it within the ?dermopteran family Plagiomenidae (see MacPhee, Cartmill, and Rose, 1989), in its own subfamily along with two new taxa, *Tarka* and *Tarkadectes*. No consensus yet exists as to the proper affinities of *Ekgmowechashala*, but we have chosen to treat it as an omomyid in this chapter.

Notharctines are much less diverse than omomyids in the North American early and middle Eocene (Figure 15.6). After their initial appearance in the earliest Wasatchian (early Eocene), notharctines are represented by only one species at any given time during Wasatchian biochronological zones Wa0 through Wa3 (*Cantius torresi*, followed by *C. ralstoni* and *C. mckennai*). *Copelemur* makes its first appearance in Wasatchian zone Wa4 and is then present until the end of the Wasatchian. The greatest diversity of notharctines occurs in Wasatchian Wa6 (Lysitean subage) and Bridgerian Br1a (early Blacksforkian subage), where three genera co-occur: *Cantius*, *Copelemur*, and *Pelycodus* in Wa6 and *Cantius*, *Notharctus*, and *Smilodectes* in Br1a. *Notharctus* and *Smilodectes* continue through the Bridgerian but are essentially gone by the end of the Bridgerian (there is a single tooth fragment of *Notharctus* from a Uintan-aged locality in Texas). The only other Uintan occurrence of an adapiform in North America is *Hesperolemur* from southern California (Gunnell, 1995a).

Cantius is the earliest occurring and most speciose genus of North American notharctine. The early portion of the *Cantius* lineage is best documented in the Bighorn Basin of Wyoming, where successive chronospecies (*C. torresi*, *C. ralstoni*, *C. mckennai*, *C. trigonodus*, *C. abditus*, and *C. simonsi*) have been documented from Wasatchian biochronological zones Wa0 through Wa7 (Gingerich and Simons, 1977; Gunnell, 2002). Trends within this chronospecies include an overall increase in body size and progressive development of upper molar mesostyles and hypocones. *Cantius* is also present in the Green River, Wind River, and Washakie Basins of Wyoming, represented by various species including *C. frugivorus*, *C. abditus*, and *C. nunienus*. Two species of *Cantius* are present in the San Juan Basin of New Mexico (*C. angulatus* and *C. frugivorus*) and there are isolated occurrences of *Cantius* spp. from Utah and Colorado as well (McKenna, 1960; Robinson, 1966; Beard, 1988). While the early evolutionary history of *Cantius* is relatively well understood, its relationships with later-occurring notharctines are still controversial. Results of studies of notharctine samples (Gingerich and Simons, 1977; Beard, 1988; Covert, 1990; Gunnell, 2002) indicate that *Cantius* is a paraphyletic taxon.

Copelemur first appears in Wasatchian biochronological zone Wa4 in the Washakie Basin in southwestern Wyoming (Gazin, 1962). Its ancestry is unclear but it and its sister taxon *Smilodectes* (Beard, 1988; Gunnell, 2002) were probably derived from a primitive species of *Cantius*. (As noted above, some *Cantius* species exhibit the variable presence of a moderately developed talonid notch, among other character states; *Copelemur* and *Smilodectes* share a well-developed and distinct talonid notch, indicating that one of the primitive species of *Cantius* might well be ancestral to the *Copelemur–Smilodectes* clade). *Copelemur* is a relatively short-lived taxon ranging from Wa4 through Wa7 and is only known with certainty from southwestern Wyoming and the San Juan Basin in New Mexico (Beard, 1988). It is relatively common where it occurs and it co-exists with *Cantius* and *Pelycodus* in the San Juan Basin (Gunnell, 2002). *Pelycodus* first appears in Wa6 and is a rare, but relatively widespread taxon, being known from both northern and southern Wyoming as well as New Mexico (Rose and Bown, 1984).

Notharctus first appears in the latest Wasatchian (Wa7) while *Smilodectes* makes its first appearance in the latest early Eocene (Bridgerian zone Br1a). These taxa are relatively common elements of mammalian faunal samples through the early and middle

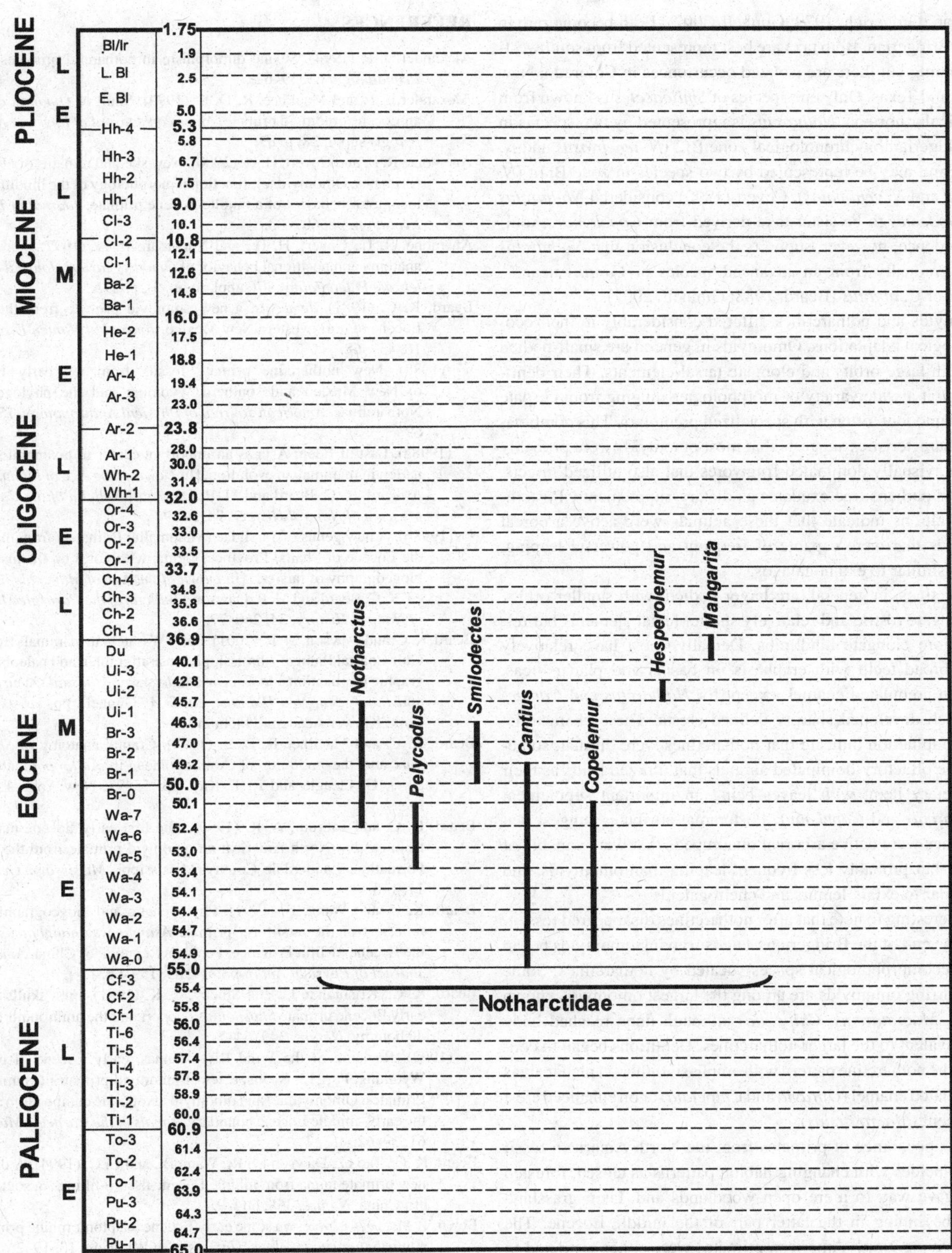

Figure 15.6. Temporal ranges of North American genera of Adapiformes.

Bridgerian (Gingerich, 1979; Gunnell, 2002). Both become rare in the later Bridgerian. Both taxa are best represented from southwestern Wyoming but there are isolated occurrences in Colorado, New Mexico, and Texas. Only one species of *Smilodectes* is known from any particular horizon. *Notharctus* is represented by two species in early Bridgerian biochronological zone Br2 (*N. tenebrosus* and *N. pugnax*) and may be represented by two species in zone Br1a (*N. robinsoni* and *N. venticolus*). Covert (1990) considered *Notharctus* and *Smilodectes* to be sister taxa derived from *Copelemur*. Additional evidence, however, supports the conclusion that *Notharctus* evolved gradually from an advanced species of *Cantius* near *C. nunienus* or *C. abditus* (Beard, 1988; Gunnell, 2002).

Omomyids and notharctines differed considerably in their ecomorphological adaptations. Omomyids in general are small-bodied forms with large orbits and elongate tarsal elements. Their dentitions exhibit a wide variety of morphologies ranging from cuspate to more bunodont, often with specialized premolars. This combination of characteristics suggests that most omomyids were probably nocturnal, visually dominated frugivores that also utilized insects, seeds, and perhaps tree exudates as dietary supplements. Postcranial adaptations indicate that these animals were active arboreal quadrupeds that were capable of frequent and powerful leaping, probably similar to extant galagos.

Notharctines, in general, are larger bodied, with smaller orbits, more elongate rostra, and relatively shorter tarsal elements but relatively more elongate hindlimbs. Dentally, most have relatively flat and broad teeth with emphasis on basins and planar areas, often with crenulated enamel, except for *Notharctus* and *Copelemur*, which developed sharply defined crests. These characteristics in combination indicate that notharctines were diurnal, somewhat more olfactory-dominated animals that utilized fruits as their major dietary item, with leaves being an important supplement for *Notharctus* and *Copelemur*. Postcranial attributes suggest that notharctines were active arboreal quadrupeds that were capable of powerful, but probably less frequent leaping than omomyids, and were similar to extant lemurs and cheirogaleids.

It is interesting to note that after notharctines disappeared (essentially at the end of the Bridgerian), later surviving omomyids began to fill in ecomorphological spaces vacated by notharctines. Some later-occurring omomyids are among the largest omomyids known (*Ourayia*, *Macrotarsius*, *Chipetaia*) although none reached body sizes equivalent to the larger notharctines. Dentitions began to converge on the notharctine pattern with emphasis either on planar areas and crenulated enamel (*Ourayia* and *Chipetaia*) or on enhanced crest development (*Macrotarsius*).

The disappearance of primates from the North American fossil record coincides with changing habitat patterns as closed, forested habitats gave way to more open woodlands and, later, grassland prairies, beginning in the latter part of the middle Eocene. The latest surviving North American primates were either restricted to southern or coastal refugia (Texas, California, Oregon) or developed bizarre dental adaptations (and presumably cranial and postcranial adaptations as well) that allowed them to exploit some specific habitats available during these changing environmental conditions (*Ekgmowechashala* in South Dakota).

REFERENCES

Alexander, J. P. (1994). Sexual dimorphism in notharctid primates. *Folia Primatologia* 63, 59–62.

Alexander, J. P. and MacPhee, R. D. E. (1999). Skull of *Omomys carteri*, an Eocene omomyid primate. *American Journal of Physical Anthropology*, suppl. 28, p. 83.

Anemone, R. L. and Covert, H. H. (2000). New skeletal remains of *Omomys* (Primates, Omomyidae): functional morphology of the hindlimb and locomotor behavior of a middle Eocene primate. *Journal of Human Evolution*, 38, 607–33.

Anemone, R. L., Covert, H. H., and Nachman, B. A. (1997). Functional anatomy and positional behavior of *Omomys carteri*. *Journal of Vertebrate Paleontology* 17(suppl. to no. 3), p. 29A.

Beard, K. C. (1987). *Jemezius*, a new omomyid primate from the early Eocene of northwestern New Mexico. *Journal of Human Evolution*, 16, 457–68.

(1988). New notharctine primate fossils from the early Eocene of New Mexico and southern Wyoming and the phylogeny of Notharctinae. *American Journal of Physical Anthropology*, 75, 439–69.

(1998a). East of Eden: Asia as an important center of taxonomic origination in mammalian evolution. [In *Dawn of the Age of Mammals in Asia*, ed. K. C. Beard and M. R. Dawson.] *Bulletin of the Carnegie Museum of Natural History*, 34, 5–39.

(1998b). A new genus of Tarsiidae (Mammalia: Primates) from the middle Eocene of Shanxi Province, China, with notes on the historical biogeography of tarsiers. [In *Dawn of the Age of Mammals in Asia*, ed. K. C. Beard and M. R. Dawson.] *Bulletin of the Carnegie Museum of Natural History*, 34, 260–77.

Beard, K. C. and Dawson, M. R. (2001). Early Wasatchian mammals from the Gulf Coastal Plain of Mississippi: biostratigraphic and paleobiogeographic implications. In *Eocene Biodiversity: Unusual Occurrences and Rarely Sampled Habitats*, ed. G. F. Gunnell, pp. 75–94. New York: Kluwer Academic/Plenum.

Beard, K. C. and MacPhee R. D. E. (1994). Cranial anatomy of *Shoshonius* and the antiquity of Anthropoidea. In *Anthropoid Origins*, ed. J. G. Fleagle and R. F. Kay, pp. 55–97. New York: Plenum Press.

Beard, K. C. and Tabrum, A. R. (1990). The first early Eocene mammal from eastern North America: an omomyid primate from the Bashi Formation, Lauderdale County, Mississippi. *Mississippi Geology*, 11, 1–6.

Beard, K. C. and Wang, B. (1991). Phylogenetic and biogeographic significance of the tarsiiform primate *Asiomomys changbaicus* from the Eocene of Jilin Province, People's Republic of China. *American Journal of Physical Anthropology*, 85, 159–66.

Beard, K. C., Krishtalka, L., and Stucky, R. K. (1991). First skulls of the early Eocene primate *Shoshonius cooperi* and the anthropoid-tarsier dichotomy. *Nature*, 349, 64–7.

(1992). Revision of the Wind River faunas, early Eocene of central Wyoming. Part 12. New species of omomyid primates (Mammalia: Primates: Omomyidae) and omomyid taxonomic composition across the early–middle Eocene boundary. *Annals of the Carnegie Museum*, 61, 39–62.

Beard, K. C., Tao Q., Dawson M. R., Wang B., and Li C. (1994). A diverse new primate fauna from middle Eocene fissure-fillings in southeastern China. *Nature*, 368, 604–9.

Bown, T. M. (1974). Notes on some early Eocene anaptomorphine primates. *Contributions to Geology, University of Wyoming*, 13, 19–26.

(1979). New omomyid primates (Haplorhini, Tarsiiformes) from middle Eocene rocks of west-central Hot Springs County, Wyoming. *Folia Primatologia*, 31, 48–73.

Bown, T. M. and Rose, K. D. (1984). Reassessment of some early Eocene Omomyidae, with description of a new genus and three new species. *Folia Primatologia*, 43, 97–112.

(1987). Patterns of dental evolution in early Eocene anaptomorphine primates (Omomyidae) from the Bighorn Basin, Wyoming. *Paleontological Society Memoir*, 23, 1–162.

(1991). Evolutionary relationships of a new genus and three new species of omomyid primates (Willwood Formation, lower Eocene, Bighorn Basin, Wyoming). *Journal of Human Evolution*, 20, 465–80.

Clark, J. (1941). An anaptomorphid primate from the Oligocene of Montana. *Journal of Paleontology*, 15, 562–3.

Cope, E. D. (1872). On a new vertebrate genus from the northern part of the Tertiary basin of Green River. *Proceedings of the American Philosophical Society*, 12, 554.

(1874). Report upon vertebrate fossils discovered in New Mexico, with descriptions of new species. In *Annual Report of the Chief Engineers, Appendix FF3*, pp. 588–606. Washington, DC: US Government Printing Office.

(1875). Systematic catalogue of Vertebrata of the Eocene of New Mexico, collected in 1874. In *United States Army, Engineer Department, Geographical Explorations and Surveys West of 100th Meridian* (G. Wheeler, in charge), pp. 5–37. Washington, DC: US Government Printing Office.

(1877). Report upon the extinct Vertebrata obtained in New Mexico by parties of the expedition of 1874. In *United States Army, Engineer Department, Geographical Explorations and Surveys West of 100th Meridian*, Vol. 4 (G. Wheeler, in charge), pp. 1–370. Washington, DC: US Government Printing Office.

(1882). Contributions to the history of the Vertebrata of the lower Eocene of Wyoming and New Mexico, made during 1881. I. The fauna of the Wasatch beds of the basin of the Bighorn River. II. The fauna of the *Catathlaeus* beds, or lowest Eocene, New Mexico. *Proceedings of the American Philosophical Society*, 20, 139–97.

(1883). On the mutual relations of the Bunotherian Mammalia. *Proceedings of the Academy of Natural Sciences, Philadelphia*, 35, 77–83.

Covert, H. H. (1985). Adaptations and evolutionary relationships of the Eocene Primate Family Notharctidae. PhD Thesis, Duke University, Durham.

(1990). Phylogenetic relationships among Notharctinae of North America. *American Journal of Physical Anthropology*, 81, 381–97.

Covert, H. H. and Hamrick M. W. (1993). Description of new skeletal remains of the early Eocene anaptomorphine primate *Absarokius* (Omomyidae) and a discussion about its adaptive profile. *Journal of Human Evolution*, 25, 351–62.

Dagosto, M. (1985). The distal tibia of primates with special reference to the Omomyidae. *International Journal of Primatology*, 6, 45–75.

(1993). Postcranial anatomy and locomotor behavior in Eocene primates. In *Postcranial Adaptation in Nonhuman Primates*, ed. D. L. Gebo, pp. 199–219. DeKalb, IL: Northern Illinois Press.

Dagosto, M., Gebo, D. L., and Beard, K. C. (1999). Revision of the Wind River faunas, early Eocene of central Wyoming. Part 14. Postcranium of *Shoshonius cooperi* (Mammalia: Primates). *Annals of the Carnegie Museum*, 68, 175–211.

Emry, R. J. (1990). Mammals of the Bridgerian (middle Eocene) Elderberry Canyon Local Fauna of eastern Nevada. [In *Dawn of the Age of Mammals in the Northern Part of the Rocky Mountain Interior, North America*, ed. T. M. Bown and K. D. Rose.] *Geological Society of America Special Paper*, 243, 187–210.

Franzen, J. L. (1994). The Messel primates and anthropoid origins. In *Anthropoid Origins*, ed. J. G. Fleagle and R. F. Kay, pp. 99–122. New York: Plenum Press.

Gazin, C. L. (1952). The lower Eocene Knight Formation of western Wyoming and its mammalian faunas. *Smithsonian Miscellaneous Collections*, 117, 1–82.

(1958). A review of the middle and upper Eocene Primates of North America. *Smithsonian Miscellaneous Collections*, 136, 1–112.

(1962). A further study of the lower Eocene mammalian faunas of southwestern Wyoming. *Smithsonian Miscellaneous Collections*, 144, 1–98.

Gebo, D. L. (1987). Humeral morphology of *Cantius*, an early Eocene adapid. *Folia Primatologia*, 49, 52–6.

(1988). Foot morphology and locomotor adaptation in Eocene primates. *Folia Primatologia*, 50, 3–41.

Gingerich, P. D. (1975). A new genus of Adapidae (Mammalia, Primates) from the late Eocene of southern France, and its significance for the origin of higher primates. *Contributions from the Museum of Paleontology, University of Michigan*, 24, 163–70.

(1976). Cranial anatomy and evolution of early Tertiary Plesiadapidae (Mammalia, Primates). *University of Michigan Papers in Paleontology*, 15, 1–141.

(1979). Phylogeny of middle Eocene Adapidae (Mammalia, Primates) in North America: *Smilodectes* and *Notharctus*. *Journal of Paleontology*, 53, 153–63.

(1981). Early Cenozoic Omomyidae and the evolutionary history of tarsiiform primates. *Journal of Human Evolution*, 10, 345–74.

(1986). Early Eocene *Cantius torresi*: oldest primate of modern aspect from North America. *Nature*, 319, 319–21.

(1995). Sexual dimorphism in earliest Eocene *Cantius torresi* (Mammalia, Primates, Adapoidea). *Contributions from the Museum of Paleontology, University of Michigan*, 29, 185–99.

Gingerich, P. D. and Haskin, R. A. (1981). Dentition of early Eocene *Pelycodus jarrovii* (Mammalia, Primates) and the generic attribution of species formerly referred to *Pelycodus*. *Contributions from the Museum of Paleontology, University of Michigan*, 25, 327–37.

Gingerich, P. D. and Simons, E. L. (1977). Systematics, phylogeny, and evolution of early Eocene Adapidae (Mammalia, Primates) in North America. *Contributions from the Museum of Paleontology, University of Michigan*, 24, 245–79.

Godinot, M. (1992). Apport à la systématique de quatre genres d'Adapiformes (Primates, Eocène). *Comptes Rendus de l'Académie des Séances, Paris*, 314, 237–42.

(1998). A summary of adapiform systematics and phylogeny. *Folia Primatologia*, 69, 218–49.

Granger, W. (1910). Tertiary faunal horizons in the Wind River Basin, Wyoming, with descriptions of new Eocene mammals. *Bulletin of the American Museum of Natural History*, 28, 235–51.

Gregory, W. K. (1920). On the structure and relations of *Notharctus*, an American Eocene primate. *Memoirs of the American Museum of Natural History*, 3, 49–243.

Gunnell, G. F. (1995a). New notharctine (Primates, Adapiformes) skull from the Uintan (middle Eocene) of San Diego County, California. *American Journal of Physical Anthropology*, 98, 447–70.

(1995b). Omomyid primates (Tarsiiformes) from the Bridger Formation, middle Eocene, southern Green River Basin, Wyoming. *Journal of Human Evolution*, 28, 147–87.

(1997). Wasatchian–Bridgerian (Eocene) paleoecology of the western interior of North America: changing paleoenvironments and taxonomic composition of omomyid (Tarsiiformes) primates. *Journal of Human Evolution*, 32, 105–32.

(2002). Notharctines primates (Adapiformes) from the early to middle Eocene (Wasatchian–Bridgerian) of Wyoming: transitional species and the origins of *Notharctus* and *Smilodectes*. *Journal of Human Evolution*, 43, 353–80.

Gunnell, G. F. and Rose, K. D. (2002). Tarsiiformes: evolutionary history and adaptation. In *The Primate Fossil Record*, ed. W. C. Hartwig, pp. 45–82. Cambridge, UK: Cambridge University Press.

Gunnell, G. F., Bartels, W. S., Gingerich, P. D., and Torres, V. (1992). Wapiti Valley faunas: early and middle Eocene fossil vertebrates from the North Fork of the Shoshone River, Park County, Wyoming. *Contributions from the Museum of Paleontology, University of Michigan*, 29, 247–87.

Hamrick, M. W. (1999). First carpals of the Eocene primate family Omomyidae. *Contributions from the Museum of Paleontology, University of Michigan*, 30, 191–8.

Honey, J. G. (1990). New washakiin primates (Omomyidae) from the Eocene of Wyoming and Colorado, and comments on the evolution of the Washakiini. *Journal of Vertebrate Paleontology*, 10, 206–21.

Krishtalka, L. (1978). Paleontology and geology of the Badwater Creek area, central Wyoming. Part 15: Review of the late Eocene primates from Wyoming and Utah, and the Plesitarsiiformes. *Annals of the Carnegie Museum*, 47, 335–60.

Leidy, J. (1869). Notice of some extinct vertebrates from Wyoming and Dakota. *Proceedings of the Academy of Natural Sciences, Philadelphia*, 21, 63–7.

(1870). Descriptions of *Palaeosyops paludosus*, *Microsus cuspidatus*, and *Notharctus tenebrosus*. *Proceedings of the Academy of Natural Sciences, Philadelphia*, 22, 113–14.

(1873). *United States Geological Survey of the Territories, Report*: *Contribution to the Extinct Vertebrate Fauna of the Western territories*. Part 1, pp. 14–358. Washington, DC: US Geological Survey.

Lillegraven, J. A. (1980). Primates from the later Eocene rocks of southern California. *Journal of Mammalogy*, 61, 181–204.

Linnaeus, C. (1758). *Systema Naturae per Regna tria Naturae, secundum Classes, Ordines, Genera, Species, cum Characteribus, Differentiis, Synonymis, Locis*, 10th edn. Stockholm: Laurentii Salvii.

Loomis, F. B. (1906). Wasatch and Wind River Primates. *American Journal of Science*, 21, 277–85.

Macdonald, J. R. (1963). The Miocene faunas from the Wounded Knee area of western South Dakota. *Bulletin of the American Museum of Natural History*, 125, 141–238.

(1970). Review of the Miocene Wounded Knee faunas of southwestern South Dakota. *Bulletin of the Los Angeles County Museum*, 8, 1–82.

MacPhee, R. D. E., Cartmill, M., and Rose, K. D. (1989). Craniodental morphology and relationships of the supposed Eocene dermopteran *Plagiomene* (Mammalia). *Journal of Vertebrate Paleontology*, 9, 329–49.

Marsh, O. C. (1871). Notice of some fossil mammals from the Tertiary formation. *American Journal of Science*, 2, 35–44, 120–7.

(1872). Preliminary description of new Tertiary mammals. Parts I through IV. *American Journal of Science*, 4, 22–8, 202–24.

Mason, M. A. (1990). New fossil primate from the Uintan (Eocene) of southern California. *PaleoBios*, 13, 1–7.

Matthew, W. D. (1909). The Carnivora and Insectivora of the Bridger Basin, middle Eocene. *Memoirs of the American Museum of Natural History*, 9, 291–567.

(1915). A revision of the lower Eocene Wasatch and Wind River faunas. Part IV. Entelonychia, Primates, Insectivora (part). *Bulletin of the American Museum of Natural History*, 34, 429–83.

McKenna, M. C. (1960). Fossil Mammalia from the early Wasatchian Four Mile Fauna, Eocene of Northwest Colorado. *University of California Publications in Geological Sciences*, 37, 1–130.

(1990). Plagiomenids (Mammalia:? Dermoptera) from the Oligocene of Oregon, Montana, and South Dakota, and middle Eocene of northwestern Wyoming. [In *Dawn of the Age of Mammals in the Northern Part of the Rocky Mountain Interior, North America*, ed. T. M. Bown and K. D. Rose.] *Geological Society of America Special Paper*, 243, 211–34.

McKenna, M. C. and Bell, S. K. (1997). *Classification of Mammals Above the Species Level*. New York: Columbia University Press.

Morris, W. J. (1954). An Eocene fauna from the Cathedral Bluffs Tongue of the Washakie Basin, Wyoming. *Journal of Paleontology*, 28, 195–203.

Muldoon, K. M. and Gunnell, G. F. (2002). Omomyid primates (Tarsiiformes) from the early middle Eocene at South Pass, Greater Green River Basin, Wyoming. *Journal of Human Evolution*, 43, 479–511.

Murphey, P. C., Torick, L., Bray, E., Chandler, R., and Evanoff, E. (2001). Taphonomy, fauna, and depositional environment of the *Omomys* Quarry, an unusual accumulation from the Bridger Formation (Middle Eocene) of southwestern Wyoming. In *Eocene Biodiversity: Unusual Occurrences and Rarely Sampled Habitats*, ed. G. F. Gunnell, pp. 361–402. New York: Kluwer Academic/Plenum.

Napier, J. H. and Walker, A. (1967). Vertical clinging and leaping: a newly recognized category of locomotor behaviour of primates. *Folia Primatologia*, 6, 204–19.

Ni, X., Wang, Y., Hu, Y., and Li, C. (2004). A euprimate skull from the early Eocene of China. *Nature*, 427, 65–8.

Osborn, H. F. (1895). Fossil mammals of the Uinta Basin. Expedition of 1894. *Bulletin of the American Museum of Natural History*, 7, 71–105.

(1902). The American Eocene Primates and the supposed rodent family Mixodectidae. *Bulletin of the American Museum of Natural History*, 16, 169–214.

Rasmussen, D. T. (1990). The phylogenetic position of *Mahgarita stevensi*: protoanthropoid or lemuroid? *International Journal of Primatology*, 11, 439–69.

(1996). A new middle Eocene omomyine primate from the Uinta Basin, Utah. *Journal of Human Evolution*, 31, 75–87.

Robinson, P. (1966). Fossil Mammalia of the Huerfano Formation, Eocene, of Colorado. *Bulletin of the Peabody Museum of Natural History*, 21, 1–95.

(1968). The paleontology and geology of the Badwater Creek area, central Wyoming. Part 4. Late Eocene primates from Badwater, Wyoming, with a discussion of material from Utah. *Annals of the Carnegie Museum*, 39, 307–26.

Rose, K. D. (1995a). The earliest Primates. *Evolutionary Anthropology*, 3, 159–73.

(1995b). Anterior dentition and relationships of the early Eocene omomyids *Arapahovius advena* and *Teilhardina demissa*, sp. nov. *Journal of Human Evolution*, 28, 231–44.

Rose, K. D. and Bown, T. M. (1984). Early Eocene *Pelycodus jarrovii* (Primates: Adapidae) from Wyoming: phylogenetic and biostratigraphic implications. *Journal of Paleontology*, 58, 1532–5.

(1991). Additional fossil evidence on the differentiation of the earliest euprimates. *Proceedings of the National Academy of Sciences, USA*, 88, 98–101.

Rose, K. D. and Rensberger, J. M. (1983). Upper dentition of *Ekgmowechashala* (omomyid primate) from the John Day Formation, Oligo-Miocene of Oregon. *Folia Primatologia*, 41, 102–11.

Rose, K. D. and Walker, A. (1985). The skeleton of early Eocene *Cantius*, oldest lemuriform primate. *American Journal of Physical Anthropology*, 66, 73–89.

Rose, K. D., MacPhee, R. D. E., and Alexander, J. P. (1999). Skull of early Eocene *Cantius abditus* (Primates: Adapiformes) and its phylogenetic implications, with a reevaluation of "*Hesperolemur*" *actius*. *American Journal of Physical Anthropology*, 109, 523–39.

Rosenberger, A. L. and Dagosto, M. (1992). New craniodental and postcranial evidence of fossil tarsiiforms. In *Topics in Primatology,* Vol. 3, ed. R. H. Tuttle, H. Ishida, and M. Goodman, pp. 37–51. Kyoto: University of Kyoto Press.

Ross, C., Williams, B., and Kay, R. F. (1998). Phylogenetic analysis of anthropoid relationships. *Journal of Human Evolution*, 35, 221–306.

Savage, D. E. and Waters B. T. (1978). A new omomyid primate from the Wasatch Formation of southern Wyoming. *Folia Primatologia*, 30, 1–29.

Simons, E. L. (1961). The dentition of *Ourayia*: its bearing on relationships of omomyid primates. *Postilla, Yale Peabody Museum*, 54, 1–20.

(1962). A new Eocene primate genus, *Cantius*, and a revision of some allied European lemuroids. *Bulletin of the British Museum of Natural History, Geology*, 7, 3–36.

(1968). Early Cenozoic mammalian faunas, Fayum Province, Egypt, Introduction. *Bulletin of the Peabody Museum of Natural History*, 28, 1–21.

Simpson, G. G. (1940). Studies on the earliest primates. *Bulletin of the American Museum of Natural History*, 77, 185–212.

Stock, C. (1933). An Eocene primate from California. *Proceedings of the National Academy of Sciences, Philadelphia,* 19, 954–9.

(1934). A second Eocene primate from California. *Proceedings of the National Academy of Sciences, Philadelphia*, 20, 150–4.

Strait, S. G. (2001). Dietary reconstruction of small-bodied omomyoid primates. *Journal of Vertebrate Paleontology*, 21, 322–34.

Stucky, R. K. (1984). The Wasatchian–Bridgerian Land Mammal Age boundary (early to middle Eocene) in western North America. *Annals of the Carnegie Museum*, 53, 347–82.

Szalay, F. S. (1975). Phylogeny of primate higher taxa: the basicranial evidence. In *Phylogeny of the Primates: A Multidisciplinary Approach*, ed. W. P. Luckett and F. S. Szalay, pp. 91–125. New York: Plenum Press.

(1976). Systematics of the Omomyidae (Tarsiiformes, Primates): taxonomy, phylogeny, and adaptations. *Bulletin of the American Museum of Natural History*, 156, 163–449.

Szalay, F. S. and Delson, E. (1979). *Evolutionary History of the Primates*. New York: Academic Press.

Teilhard de Chardin, P. (1927). Les mammifères de l'Éocène inferieur de la Belgique. *Memoire, Museum Royal Histoire Naturelle Belgique*, 36, 1–33.

Trouessart, E. L. (1879). Catalogue des mammifères vivants et fossiles. *Revue et Magasin de Zoologie Pure et Applique*, 7, 223–30.

Wilson, J. A. (1966). A new primate from the earliest Oligocene, west Texas, preliminary report. *Folia Primatologia*, 4, 227–48.

Wilson, J. A. and Szalay, F. S. (1976). New adapid primate of European affinities from Texas. *Folia Primatologia*, 25, 294–312.

Wortman, J. L. (1903–4). Studies of Eocene Mammalia in the Marsh Collection, Peabody Museum. *American Journal of Science*, 15, 163–76, 399–414, 419–36; 16, 345–68; 17, 23–33, 33–140, 203–14.

Part V: Glires

16 Glires summary

CHRISTINE M. JANIS,[1] MARY R. DAWSON,[2] and LAWRENCE J. FLYNN[3]

[1]*Brown University, Providence, RI, USA*

[2]*Carnegie Museum of Natural History, Pittsburgh PA, USA*

[3]*Peabody Museum, Harvard University, Cambridge, MA, USA*

INTRODUCTION

GENERAL CONSIDERATIONS

Rodents and lagomorphs are usually grouped in the higher taxon Glires, and paleontological, morphological, and molecular studies generally bear out this relationship (see discussion below). The first fossil relatives of Glires are known from the late Paleocene of Asia (Li and Ting, 1985; Li *et al.*, 1987), approximately correlative with the first appearance of rodents in North America.

Rodents constitute over half the species of living mammals and are also extremely diverse as fossil forms. Extant rodents comprise 29 families, 468 genera, and 2052 species (Nowak, 1999), with an additional 743 known extinct genera (McKenna and Bell, 1997), and are found worldwide except for Antarctica and some oceanic islands. Extant lagomorphs are much less diverse than rodents, comprising two families, 13 genera, and 81 species (Nowak, 1999), with an additional 56 known extinct genera (McKenna and Bell, 1997). Although lagomorphs have a broad geographic distribution (and are generally individually numerous), they are absent, except for recent introductions, in Australia and most of the southeast Asian islands, southern South America, Madagascar, the West Indies, and Antarctica (Nowak, 1999).

FEATURES UNITING RODENTS AND LAGOMORPHS

Rodents and lagomorphs share a number of dental features, many related to common ancestry (McKenna, 1975; Novacek, 1985) but also certainly adaptive for a small herbivore or omnivore. Both rodents and lagomorphs have an essentially diprotodont dentition, with a large pair of procumbent, ever-growing incisors with thickened enamel anteriorly; the enamel is highly structured, with decussating prisms. These incisors are the retained deciduous second incisors (Luckett, 1985). The other incisors are lost in rodents, but lagomorphs retain a small pair of upper incisors located behind the first pair; the canines, and the anterior premolars are also lost, resulting in a prominent post-incisor diastema. The cheek teeth tend to be squared up (quadritubercular): uppers do this with a posterior shelf, often supporting a hypocone; lowers do this with a lowered trigonid that lacks a paraconid (although some early rodents retain a small paraconid). In both orders, the lower jaw is shorter than the upper jaw, with the procumbent lower incisors making up some of the difference in length (see Figure 16.2A, below), and the capability of propalinal extension of the dentary along a large glenoid making up the rest (see Jacobs, 1984).

Meng and Wyss (2005) list other synapomorphies of Glires: lack of centrocrista on the upper molars; a prominent angular process; mandibular condyle oriented anteroposteriorly; longtitudinally elongate and anteriodorsally shifted glenoid fossa and reduced postglenoid process (these features together allowing for propalinal jaw motion to engage the incisors); somewhat elongate incisive foramina; short infraorbital canals; narrow premaxilla–frontal contact; anteriorly shifted posterior edge of anterior zygomatic root; anteriorly shifted orbits; and an ectotympanic bulla. Martinez (1985) argued that many of these cranial synapomorphies have also appeared in other mammals and so do not provide strong support for a sister group association between rodents and lagomorphs.

FEATURES DISTINGUISHING RODENTS AND LAGOMORPHS (FROM OTHER MAMMALS AND FROM EACH OTHER)

BODY SIZE

Despite their great species diversity, rodents are in general of small body size, most less than around 0.5 kg in body mass. The median body size of extant terrestrial non-volant mammals is approximately

Evolution of Tertiary Mammals of North America, Vol. 2. ed. C. M. Janis, G. F. Gunnell, and M. D. Uhen. Published by Cambridge University Press.

Figure 16.1. Restoration of the Neogene archaeolagine *Hypolagus* (by Marguette Dongvillo)

0.25 kg, at least in part because of the large numbers of small rodents. The exception to this size limit appears to be among the hystricognaths (although a few other large rodents also exist, such as beavers and some gophers). The capybara (*Hydrochaeris hydrochaeris*) is the largest extant rodent, with a body mass of approximately 50 kg. Other large extant caviomorphs include the pacarana (*Dinomys branickii*, ~15 kg), the paca (*Agouti paca*, ~10 kg), and the mara or Patagonian cavy (*Dolichotis patagonium*, ~13 kg) (Nowak, 1999). Some extinct caviomorphs were truly large, with body masses of over 100 kg. These include the Pliocene capybara *Protohydrochaerus*, with an estimated mass of over 200 kg (Vaughan, Ryan, and Czaplewski, 2000), and the recently discovered late Miocene *Phoberomys pattersoni*, a dinomyid with an estimated mass of 700 kg (Sánchez-Villagra, Aguillera, and Horovitz, 2003). The largest known sciuromorph is the Pleistocene giant beaver, *Castoroides ohioensis*, with a body mass estimated to be around 220 kg. The smallest extant rodents have body masses in the region of 10 g (e.g., small species of the silky pocket mouse, *Perognathus flavus*). Lagomorphs exhibit a more narrow range of body sizes than rodents, ranging from around 125 g in the smallest pikas (e.g., *Ochotona princeps*) to around 7 kg in the largest hares (e.g., *Lepus alleni*) (Nowak, 1999). No significantly larger lagomorphs are known from the fossil record.

LOCOMOTION AND POSTCRANIAL SPECIALIZATIONS

Rodents exhibit a number of different lifestyles and locomotor behaviors. While the generalized rodent adaptation is generally terrestrial/scansorial, various forms are more specialized for climbing, gliding, swimming, or digging. The general small size of rodents means that few exhibit any true cursorial specializations, except for ones associated with ricochetal (= saltatorial) locomotion (bipedal hopping). However, some caviomorphs (e.g., the mara) have limb proportions suggestive of more cursorial adaptations and share some derived features of the scapula seen in lagomorphs (Elissamburu and Vizcaíno, 2004; see below). Rodents that have adopted a subterranean mode of life to the extent that they are committed to life underground at least 95% of the time can be termed fossorial. Truly fossorial rodents show burrowing adaptations in the skull and postcrania. This lifestyle is adopted by gophers (Geomyidae) and, in the past, by some beavers (Castoridae) and the extinct Mylagaulidae.

In contrast with the range of locomotor and substrate adaptations seen in rodents, all lagomorphs are terrestrial, with some forms (hares and some rabbits) being more specialized for rapid quadrupedal bounding. Szalay (1985) noted that, while the ancestral form of the rodent foot appears to be fairly generalized for a broad range of locomotor behaviors, the lagomorph ancestral foot is more cursorially specialized. All lagomorphs have an elbow joint with movement limited to the parasaggital plane, a reduced fibula with distal fusion of the tibia and fibula, a narrow and elongated astragalus, and a calcaneum with a broad fibular facet and a distinct heel. The more cursorial leporids have relatively elongated limbs, with especially elongated hindlimbs and metatarsals. The Arctic hare (*Lepus arcticus*) can hop bipedally (Vaughan, Ryan, and Czaplewski, 2000).

Leporids are also noted for their long ears (although the pika has short pinnae). Bramble (1989) has proposed that these ears are associated with a unique cranial specialization. The occipital portion of the skull is rather loosely joined to the main body of the skull, and often becomes disarticulated from it in museum specimens. Bramble (1989) proposes that this line of weakness represents an adaptation for the absorption of force impacted on the forelimbs during rapid locomotion, in the absence of the type of shock-absorbing tendons seen in cursorial ungulates: the elongated ears act to "re-set" the skull after impact absorption. Specializations of the lagomorph scapula, with a long, caudally directed metacromion process, may represent muscular adaptations for counteracting such forces (Janis and Seckel, 2008). Note that an animal the size of a rabbit would be too small to benefit from ungulate-like tendons, which only function as shock absorbers and energy storers under conditions of greater impact forces (Biewener, 1998). Likewise, ricochetal rodents are too small to benefit from the type of "pogo stick" energy-storing effect seen with the tendons of kangaroos (Biewener, 1989). (Note, however, that the springhare [*Pedetes capensis*], with body mass of around 5 kg, may gain more tendon energy storage during hopping than the small ricochetal rodents such as jerboas and kangaroo rats. Personal observation of video footage of a springhare on a treadmill by the senior author certainly suggests that energy storage may be occurring in the Achilles tendon.)

DIET AND CRANIODENTAL SPECIALIZATIONS

Although both rodents and lagomorphs are generally regarded as herbivorous, rodents exhibit a broader range of diets than lagomorphs. Lagomorphs are primarily herbivorous, relying mainly on grass but also eating leaves, juice and pulp of stems, tubers, some bark and seeds. Pikas are unique in building haystacks, curing and storing up a supply of a variety of vegetation for winter use. The smaller rodents (under ~700 g) are too small to subsist exclusively on herbage and are more omnivorous (or, at least, gramnivorous, depending on grass seeds rather than leaves), and many rodents also take a certain amount of animal material. Many rodents have diets that vary seasonally, for example during certain months of the year consuming large amounts of their favorite insects. Lagomorphs and many rodents deal with the cellulose in their diet by being cecal fermenters, relying on coprophagy to obtain sufficient nutrients from the food. However, some small, gramnivorous rodents, such as voles and lemmings, have enlarged forestomachs with a degree of foregut fermentation.

Although both rodents and lagomorphs have incisors with thickened enamel on their anterior surface (extending onto the lateral surface in lagomorphs), the posterior incisors of lagomorphs are completely covered by thin enamel. The enamel on the main incisors is highly structured at the microstructural level, with decussating bands of hydroxyapatite prisms. In leporids, there is one layer of enamel, and in ochotonids the enamel has two or three layers (Martin, 2004). In rodents, the enamel has a two-layered microstructure. Rodent incisors are laterally compressed, the longer cross-sectional axis having an anteroposterior direction. This differential is much less in lagomorphs and even reversed. Rodent incisors are also longer and more deeply rooted than those of lagomorphs, with the base extending well into the maxilla and mandible. From their first appearance in the Eocene, lagomorphs have a longitudinal groove on the anterior surface of dI2.

Both rodents and lagomorphs have the ability to disengage the lower jaw from molar to molar occlusion, and project it forward so that the incisor tips occlude. This is very important in cropping and ingesting food. Additionally, incisor–incisor attrition is necessary for sharpening and offsetting the continual growth of the incisors; if a lower incisor breaks, nothing stops the upper from growing along its tight arc and recurving back and up into the palate. This capability of movement is incorporated to some degree into the cheek tooth occlusion in many rodents. An orthal (aft–fore, proal) component to the power stroke is used in the chewing of many kinds of rodent; some, like arvicolines, are almost exclusively proal. Lagomorphs generally retain a mainly transverse direction of motion in cheek tooth occlusion.

The cheek teeth of rodents and lagomorphs are rather different in form. Occlusal morphologies are modified from the general tribosphenic condition to the bunodont, bunolophodont, or multilophed condition (see Janis and Fortelius, 1988). The basic cheek tooth pattern in early rodents is readily derivable from the tribosphenic plan. In lagomorphs, however, homologies for structures in the upper molariform teeth remain unclear. Early rodents and some squirrels have two upper premolars, but most later rodents have a cheek tooth formula of three molars and a single (fourth) premolar, or a reduced state in which all premolars are absent and the cheek tooth formula is 3/3. Lagomorphs also have a reduced cheek tooth dentition, but retain three upper and two lower premolars, and some ochotonids lack third molars. Among the lagomorphs, leporids retain a non-molariform P2/p3, and ochotonids have non-molariform P2-3 and p3. The premolars (when retained) are molarized in most rodents. The cheek teeth of many rodents are either hypsodont (high-crowned) or hypselodont (ever-growing); rodent teeth are often termed "rootless," but in fact rodents exhibit crown hypselodonty, with continually growing crowns, thus retaining at least a rim of enamel around the tooth. All lagomorphs have unilaterally hypsodont upper cheek teeth and hypsodont lower cheek teeth.

Rodents and lagomorphs also differ in their chewing mechanisms. In rodents, the masseter and temporalis muscles are important in mastication, whereas in lagomorphs the masseter and pterygoid muscles are dominant (Turnbull, 1970; Martinez, 1985). The greatly reduced coronoid process of the lagomorphs is related to the reduction in importance of the temparalis muscle in mastication. Lagomorphs have the maxillary tooth rows more widely separated than the mandibular, a factor in their transverse mastication. In contrast, many rodents have the more derived type of proal mastication, moving the lower jaw forwards during tooth occlusion. Accordingly, major features of the rodent skull reflect the modification of the masseter muscle to effect this proal motion. These modifications have arisen convergently, and somewhat differently, in different rodent groups (see below). Rodents also have a rounded (rather than mediolaterally elongated) jaw condyle to permit this motion. The skull of lagomorphs is usually highly fenestrated in the lateral side of the maxilla. This is especially pronounced in leporids but is also present in some extinct ochotonids. The extant *Ochotona* has a single opening in this region.

OTHER ANATOMICAL FEATURES

Many rodents have long tails, while all lagomorphs have very short tails (the tail of pikas is not visible externally). Lagomorphs are unique among placental mammals in the males possessing a prepenile scrotum, like the condition in marsupials. (This is actually true only of leporids: the testes of pikas are abdominal, as is true also for many rodents.) Male rodents also usually possess a baculum (penis bone), whereas male leporids do not. A number of features of the fetal membranes and placentation are similar, and apparently homologous, between rodents and lagomorphs (Luckett, 1985). Cheek pouches for food storage are common among rodents but are never seen in lagomorphs.

CRANIAL SPECIALIZATIONS OF RODENTS

As mentioned above, different types of rodent exhibit different types of cranial specialization in association with modifications of the zygomasseteric musculature. These differences have been important to hypotheses of rodent interrelationships, but we will discuss the differences in anatomy in this section, rather than in the section on systematics.

There are four different types of rodent cranial morphology: protrogomorphous, sciuromorphous, hystricomorphous, and myomorphous (Figure 16.2). These different types of skull morphology were originally assumed to be correlated with systematic groupings. However, as discussed in the following section, the latter three morphologies appear to have evolved convergently on several occasions from the primitive protrogomorphous condition.

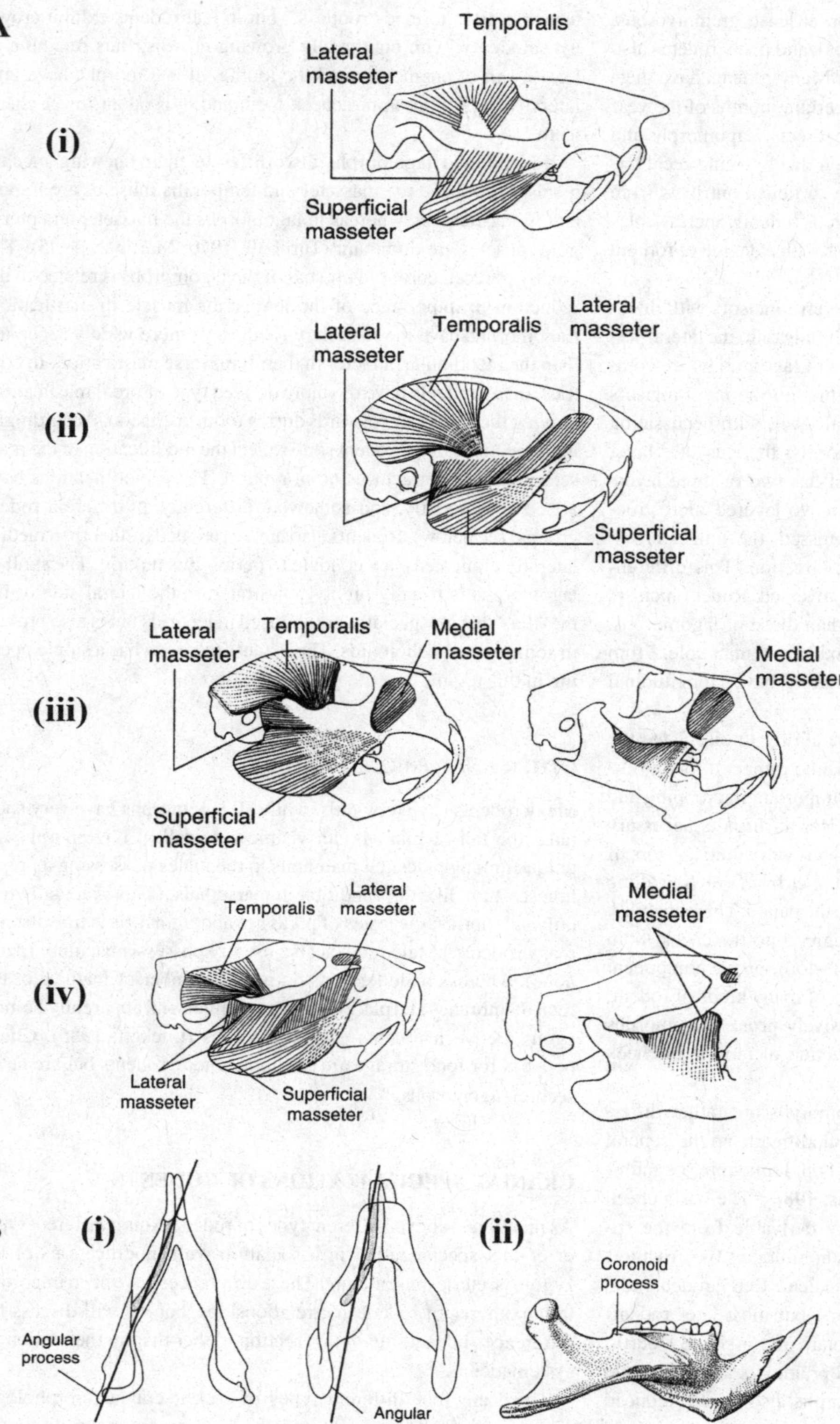

Figure 16.2. Rodent cranial morphology. A. Skull and zygomasseteric musculature: (i) Protrogomorphous condition; (ii) sciuromorphous condition; (iii) hystricomorphous condition; (iv) myomorphous condition. B. Lower jaws: (i) sciurognathus condition; (ii) hystricognathous condition (ventral lateral view of the lower jaw). (Adapted from Vaughan, Ryan and Czaplewski [2000] with permission of the authors.)

1. The protrogomorphous type (Figure 16.2Ai), seen in many primitive rodents (e.g., ischyromyids) and in the mountain beaver (*Aplodontia rufa*) among extant rodents, is basically unmodified from the general mammalian condition with respect to the accommodation of anterior projections of the masseter muscle. As in most mammals, the superficial masseter originates from the masseteric prominence on the anterior end of the zygomatic arch, below the infraorbital foramen; the lateral masseter (= zygomaticomandibularis) originates from the length of the zygomatic arch; and the medial masseter (= deep masseter) originates from the anterior portion of the base of the zygomatic arch.
2. The sciuromorphous type (Figure 16.2Aii) is typical of sciurids, but also seen in castoroids and geomyoids. Here the lateral masseter has an anterior portion that originates from anterior to the orbit. Accordingly, there is a deep trough for this muscle on the rostrum anterior to the orbit. Both protrogomorphous and sciuromorphous rodents also usually retain a moderately sized temporalis muscle and a coronoid process that is not as reduced as in some other rodents.
3. The hystricomorphous type (Figure 16.2Aiii) is seen in hystricognathous rodents (the South American caviomorphs and several Old World groups) and also among some other types of rodent such as some primitive cricetids, dipodids, ctenodactylids, pedetids, and anomalurids. Here it is the medial masseter that has an anterior projection, originating from the rostrum and passing through a greatly enlarged infraorbital canal.
4. The myomorphous type (Figure 16.2Aiv) combines the hystricomorphous condition of the masseter muscle with expansion of the lateral masseter anteriorly onto a plate of the anterior root of the zygoma. This zygomatic plate constricts the infraorbital foramen in keyhole fashion. The skull shows both a sculpted plate (although not as pronounced as in sciuromorphs) and an enlarged infraorbital foramen (although not as enlarged as in hystricomorphs). This condition is seen in modern muroids, and in less derived state among the glirids (dormice) which, despite their name, are probably more closely related to squirrels than to true mice (see following section).

Another difference in craniodental morphology among different types of rodent is in the form of the lower jaw, which can be divided into two types: sciurognathous and hystricognathous (see Figure 16.2B). The sciurognathous jaw type is typical of most rodents and resembles the primitive mammalian condition. Here the angular process of the dentary is in the same plane as the main ramus of the dentary. The hystricognathous jaw type is the more derived condition, seen only among hystricomorphous forms (except in the protrogomorphous bathyergids). Here the angular process is deflected laterally with respect to the plane of the dentary ramus. In hystricognaths, there is a more prominent projection at the base of the incisor root and the coronoid process is usually vestigial (in association with an extremely reduced temporalis muscle). The hystricognathous condition is used by most systematists to define a large division of Rodentia, the Hystricognathi, often considered a suborder or infraorder.

SYSTEMATICS

ORIGIN AND AFFINITIES OF GLIRES

EXTANT FORMS

The notion of a taxonomic grouping of "Glires" was originally proposed by Linnaeus (1735), although a variety of other mammalian taxa, including soricids and rhinocerotids were variously included in his Glires (Linnaeus, 1758) as his classification evolved. Later the Glires were divided into the Duplicidentata Illiger, 1811 (lagomorphs) and Simplicidentata Lilljeborg, 1866 (rodents). Gidley (1912) recognized the lagomorphs as being distinct from the rodents and considered them to be separate orders, and Simpson (1945) united the orders Rodentia and Lagomorpha within the cohort Glires. Many workers in the mid twentieth century (especially Wood, 1957) sought to unite rodents and lagomorphs with different eutherian groups. As reviewed by Meng and Wyss (2005), rodents have been considered at various times to be related to multituberculates, mixodectids, tillodonts, taeniodonts, and leptictids, while lagomorphs have been grouped with triconodontids, condylarths, uintatheres, zalambdodont insectivores, zalambdalestids, and macroscelidids.

Most morphologists in the later part of the twentieth century have regarded rodents and lagomorphs as sister taxa. Early molecular work claimed that rodents and rabbits do not form a monophyletic grouping but presently Glires is supported by a large number of molecular studies (e.g., Honeycutt and Adkins, 1993; Stanhope *et al.*, 1996; Reyes, Pesole, and Saccone, 1998; Adkins *et al.*, 2001; DeBry and Sagel, 2001; Huchon *et al.*, 2002; Adkins, Walton, and Honeycutt, 2003; Misawi and Janke, 2003; Douzery and Huchon, 2004; Reyes *et al.*, 2004). Meng and Wyss (2005) provide a more extensive review of this issue.

The "standard" cladistic classification to emerge from morphologists in the 1990s (e.g., Novacek, 1993) considered the Rodentia, Lagomorpha, and Macroscelidea (elephant shrews) to represent a single clade (grandorder Glires), with rodents and lagomorphs being sister taxa. The relationship of Glires to other eutherians within the cohort Epitheria was uncertain. Molecular phylogenies emerging in the late 1990s and in the twenty-first century (e.g., Eizirik, Murphy, and O'Brien, 2001; Murphy *et al.*, 2001; de Jong *et al.*, 2003, Reyes *et al.*, 2004: see Springer *et al.* [2004, 2005] for a summary) have maintained the sister-group relationship of rodents and lagomorphs but have removed macroscelidids from this grouping, uniting them with other endemic African mammalian orders in the Afrotheria. The Glires (rodents plus lagomorphs) are now united with the Euarchonta (primates, bats, and dermopterans) in the Euarchontoglires, and this grouping is, in turn, united with the Laurasiatheria (all other eutherians with the exception of Afrotheria and Xenarthra) in the Boreoeutheria.

FOSSIL RELATIVES

During the mid twentieth century, extensive material of rodent-like eutherians was collected from the Paleocene and Eocene of eastern

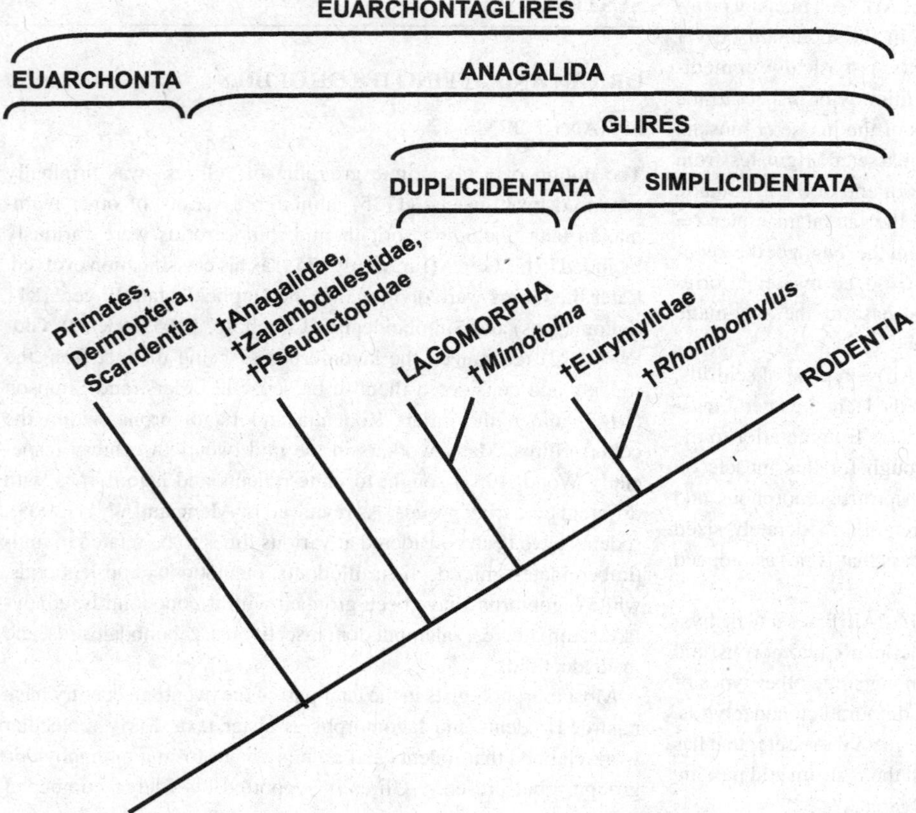

Figure 16.3. Position of the Glires within the Eutheria (extinct taxa indicated by daggers).

Asia. One group, the eurymylids, appears to be closely related to rodents. Other genera, sometimes grouped as family Mimotonidae, have been proposed as closely related to lagomorphs (Li and Ting, 1985). Li *et al.* (1987) proposed the following classification: cohort Glires: superorders Simplicidentata (orders Mixodontia and Rodentia), and Duplicidentata (orders Mimotonida and Lagomorpha). The Mixodontia would include the extinct families Eurymylidae and Rhombomylidae. Currently, *Rhombomylus* is seen as a eurymylid and Mimotonidae is restricted, essentially, to *Mimotoma* (Meng, Hu, and Li, 2003).

McKenna and Bell (1997) considered rodents and lagomorphs to be in the superorder Preptotheria, grandorder Anagalida, which comprised some extinct families (anagalids, pseudictopids, and zalambdalestids), and the mirorders Macroscelidea (elephant shrews), Duplicidentata (lagomorphs), and Simplicidentata (rodents and the extinct Asian eurymylids). Archibald, Averimov and Ekdale (2001) have argued for zalambdalestids as the sister taxon to Glires, but see Meng and Wyss (2005) and Wible, Rougier and Novacek (2005a) for other views.

Fossil evidence for early members of the order Rodentia, and their possible interrelationships, is discussed in more detail by Anderson (in Chapter 18). Figure 16.3 shows the interrelationships within the Glires, based on fossil evidence, and of the Glires to other mammals, based on molecular evidence.

The divergence times of rodents from lagomorphs, or of the rodent subgroups from each other, was originally estimated from molecular evidence to be very distant (e.g., 110 Ma for the split between sciurognaths and hystricognaths; Kumar and Hedges, 1998). However, the molecular analysis by Adkins, Walton, and Honeycutt (2003) has produced divergence times including an estimate of 94–100 Ma for the origin of rodents. All of these greatly exceed expectations based on fossils. Note that Jacobs and Flynn (2005) showed that the molecular estimates for splitting times of rodent families greatly predate the fossil record evidence.

RODENT PHYLOGENETICS

RODENT MONOPHYLY

The monophyly of rodents had never been in doubt until the early molecular phylogenetic studies of the 1990s, where various workers took considerable glee in announcing that "the guinea pig is not a rodent" (e.g., Graur, Hide, and Li, 1991; Graur *et al.*, 1992; Li *et al.*, 1992). These conclusions were vigorously debunked by Luckett and Hartenberger (1993), who discussed both morphological and molecular studies and concluded that results suggesting rodent paraphyly could be attributed to poor sampling of both taxon numbers and sequence data. Phillippe (1997) and Sullivan and Swofford (1997) (both reacting in particular to the paper claiming rodent paraphyly by D'Erichia *et al.* [1996]) also discussed methodological problems in molecular phylogenetic analyses and showed that different modes of analysis can yield different topologies.

Some more recent papers have also argued for rodent paraphyly, again showing the separation of the guinea pig (*Cavia porcellus*) from other rodents (Ma *et al.*, 1993; Noguchi *et al.*, 1994; D'Erchia *et al.*, 1996; Janke, Xu, and Arnason, 1997; Reyes, Pesole, and Saccone, 1998), although other papers published around the same time argued for rodent monophyly (e.g., Frye and Hedges, 1995; Cao, Okada, and Hasewaga, 1997). Note, however, that these papers mostly sample a small number of rodent species, even if a large number of mammalian taxa are sampled overall. For example, Reyes, Pesole, and Saccone (1998) claimed a paraphylic Rodentia when only four rodent species had been sampled among an array of 19 other mammals: but later these same authors (Reyes *et al.*, 2004) found that a larger sampling of taxa in general, including nine rodents, resulted in a monophyletic Rodentia. Recent molecular papers that have included a large number of rodent species (see Table 16.2, below) all support rodent monophyly (e.g., Catzeflis *et al.*, 1995; Nedbal, Honeycutt, and Schlitter, 1996; Huchon, Catzeflis, and Douzery, 1999; Adkins *et al.*, 2001; DeBry and Sagel, 2001; Huchon *et al.*, 2002; Adkins, Walton, and Honeycutt, 2003). Meng and Wyss (2005) also provide extensive discussion of this issue.

INTERRELATIONSHIPS OF RODENT FAMILIES

Blainville (1816) perceived rodents as "grimpeurs," "fouisseurs," and "marcheurs," groups broadly equivalent to the later designations, based on zygomasseteric muscle structure, of Sciuromorpha, Myomorpha, and Hystricomorpha (see Figure 16.2A) (Waterhouse, 1839; Brandt, 1855). A fourth category, Protrogomorpha (for the primitive unspecialized condition seen in many Eocene rodents and retained in aplodontids), was coined by A. E. Wood (Wood, 1937). Tullberg (1899) emphasized the alternative classification of rodents based on the jaw structure as the Sciurognathi and the Hystricognathi (see Figure 16.2B).

This classification was retained, more or less, by Simpson (1945) (see Table 16.1). However, during the later part of the twentieth century, it became increasingly apparent that the zygomasseteric morphologies were subject to much convergence and were not a suitable basis for classification (e.g., Wilson [1949] noted that the sciuromorphous geomyoids were in other respects more like the myomorphous muroids). Note also that the morphologies typical of extant forms may not be the condition of the earlier members of the group. For example, as detailed elsewhere in this volume, the earliest squirrels were protogomorphous rather than sciuromorphous, and early muroids were hystricomorphous rather than myomorphous.

With the advent of cladistic methodologies, it also became apparent that, while the derived hystricognathous condition likely described a monophyletic group (the South American caviomorphs and various Old World groups), the "Sciurognathi" was a paraphyletic waste-basket (Luckett and Hartenberger, 1985). As shown in Table 16.1, the classification of Nowak (1999), considering only extant rodent families, presented an unresolved Sciurognathi, while those by Korth (1994) and McKenna and Bell (1997), also incorporating extinct families, retained the notion of the groupings of Sciuromorpha and Myomorpha but no longer based these groupings solely on the form of the zygomassetic musculature. Both Korth (1994) and McKenna and Bell (1997) grouped the protrogomorphous aplodontids with the Sciuromorpha, and the sciuromorphous geomyoids and hystricomorphous dipodoids with the Myomorpha, but retained the sciuromorphous castoroids within the Sciuromorpha.

The literature on the history of rodent subordinal classification based on morphological features is extensive. A. E. Wood published extensively on this subject in the second half of the twentieth century (e.g., Wood, 1955, 1965, 1975, 1985), changing his ideas on the subject a number of times. This history of rodent systematics is summarized in Korth (1994) and Landry (1999) and will not be extensively reviewed here. Rather, we will focus on the ideas that have emerged since the mid 1990s as the result of numerous molecular studies. Again, the literature here is extensive, and we have attempted to summarize some of the main ideas in Table 16.2, with a consensus presented in Figure 16.4, where the placement of fossil taxa follows Korth (1994) and the various authors in this volume. Note that the most primitive rodent family is probably the extinct Alagomyidae (Dawson and Beard, 1996); this family is predominantly Asian, although the genus *Alagomys* makes a brief appearance in the late Paleocene of North America (see also Chapter 29).

Several groupings of rodents have almost universal support; aplodontids are united with sciurids, dipodoids with muroids (= Myodonta), geomyids with heteromyids (= Geomyoidea), and caviomorphs with a few Old World rodents (= Hystricognathi). There is considerable debate about the relationships within the Hystricognathi, although as few of these taxa are represented in the North American Tertiary this will not be elaborated on here. Note that while molecular studies unite the Erethezionidae (New World porcupines) with the other caviomorphs, several authors would place them as originating separately from the phiomorphs (African hystricognaths) on morphological grounds (e.g., Bugge, 1985; McKenna and Bell, 1997; Landry, 1999). The monophyly or paraphyly of South American rodents, of course, affects the notion of whether or not they arrived on that continent by a single invasion. The relationship of the Hystricognathi (or of the Ctenohystricia, see below) to other rodents remains equivocal: molecular studies that combine multiple genes and a large number of taxa favor an association with myomorphs, but fossil record evidence supports a position of the Ctenohystricia as the sister taxon to other major rodent groupings (Marivaux *et al.*, 2004; Wible *et al.*, 2005b).

Other groupings of rodents have moderately strong, if not universal, support. With respect to the North American groups, geomyoids usually cluster with the myodonts, reflecting the views of Wilson (1949), and also Wood (1955). A possibly unexpected grouping is that of geomyoids and castorids, but their common sciuromorphy may be a synapomorphy. Note, however, that various molecular studies place castorids in different positions among the rodents (even if a geomyoid/castorid grouping is the most common), and many morphologists have considered castorids to be a group with unresolved affinities (e.g., Wood, 1955; Bugge, 1985; Wahlert, 1985). This grouping would receive some support from paleontological studies, as both taxa probably have a North American origin (Vianey-Liaud, 1985) (Figure 16.4).

Table 16.1. *Classifications of the order Rodentia*

Simpson (1945) classification: extant families not found in North America in brackets; not all extinct families shown

Suborder	Superfamily	Family
Sciuromorpha	Aplodontoidea	†Ischyromyidae (inc. †Sciuravinae, †Cylindrodontinae)
		Aplodontidae
		Mylagaulidae
		†Protoptychidae
		†Eomyidae
	Sciuroidea	Sciuridae
	Geomyoidea	Geomyidae
		Heteromyidae (inc. †Florentiamyidae)
	Castoroidea	Castoridae
		†Eutypomyidae
	Anomaluroidea	[Anomaluridae, Pedetidae]
Myomorpha	Muroidea	Cricetidae (inc. Lemmini and Microtini)
		[Spalacidae, Rhizomyidae, Muridae]
	Gliroidea	Gliridae
	Dipodoidea	Zapodidae
		Dipodidae
Hystricomorpha	Hystricoidea	[Hystricidae]
	Bathyergoidea	[Bathyergidae]
	Erethizodontioidea	Erethizontidae
	Cavoidea	Hydrochaeridae
		[Caviidae, Dinomyidae, Dasyproctidae]
	Chinchilloidea	[Chinchillidae]
	Octodontoidea	[Octodontidae and several others]
?Hystricomorpha (or ?Myomorpha)	Ctenodactyloidea	[Ctenodactylidae]

†Extinct taxa.

McKenna and Bell (1997) classification: families not found in North America in brackets; not all extinct non-North American taxa shown

Suborder	Infraorder	Parvorder	Superfamily	Family
Sciuromorpha				
			†Ischyromyoidea	†Ischyromyidae
			Aplodontoidea	†Allomyidae
				Aplodontidae
				†Mylagaulidae
	†Theridomyomorpha			[†Theridomyidae]
	Sciurida			†Reithroparamyidae
				Sciuridae
	Castorimorpha			†Eutypomyidae
				Castoridae
Myomorpha				
	Myodonta		Dipodoidea	†Armintomyidae
				Dipodidae (inc. Zapodinae)
			Muroidea	Muridae (inc. Cricetinae, Arvicolinae)
	Glirimorpha			[Gliridae]
	Geomorpha		Geomyoidea	Geomyidae (inc. Heteromyinae)
				†Florentiamyidae
			†Eomyoidea	†Eomyidae
	Incertae sedis			†Protoptychidae

Table 16.1. *(cont.)*

McKenna and Bell (1997) classification: families not found in North America in brackets; not all extinct non-North American taxa shown

Suborder	Infraorder	Parvorder	Superfamily	Family
Anomaluromorpha			Pedetoidea	[Pedetidae]
			Anomaluroidea	[Anomaluridae]
Sciuravida				†Sciuravidae
				†Cylindrodontidae
				[Ctenodactylidae]
				[†Chappattimyidae]
Hystricognatha				
	Hystricognathi			Erethizontidae
				[Hystricidae, Petromuridae, Thyronomyidae]
		Bathyergomorphi		[Bathyergidae]
		Caviida	Cavoidea	Hydrochaeridae
				[Agoutidae, Caviidae, Dasyproctidae, Dinomyidae]
			Chinchilloidea	[Chinchillidae, Abrocomidae]
			Octodontoidea	[Octodontidae, plus others]
Uncertain				†Alagomyidae
				†Laredomyidae

†Extinct taxa.

Nowak (1999) classification: extant families only: families not found in North America in brackets

Suborder	Infraorder	Superfamily	Family
Sciurognathi	Protrogomorpha		Aplodontidae
	Sciuromorpha		Sciuridae
	Castorimorpha		Castoridae
	Myomorpha	Geomyoidea	Geomyidae
			Heteromyidae
		Dipodoidea	Dipodidae
		Muroidea	Muridae (inc. Cricentinae and Arvicolinae)
	Anomaluromorpha		[Anomaluridae, Pedetidae]
	Ctenodactylomorpha		[Ctenodactylidae]
	Gliromorpha		[Gliridae]
Hystricognathi	Bathyergomorpha		[Bathyergidae]
	Hystricomorpha		[Hystricidae]
	Phiomorpha		[Petromuridae, Thyronomyidae]
	Caviomorpha	Erethizodonta	Erethizontidae
		Chinchilloidea	[Chinchillidae]
		Cavoidea	Hydrochaeridae
			[Agoutidae. Caviidae, Dasyproctidae, Dinomyidae]
			Octodontoidea
			[Octodontidae and others]

Some interesting groupings have also emerged among rodent groups that are not native to North America. The ctenodactylids (gundis) are usually found to group with the Hystricognathi (in a grouping termed the Ctenohystricia), although this very early-appearing family has also been considered as basal to other rodents (Dawson, Li, and Qi, 1984). New fossil record evidence from the Paleogene of Asia strongly supports the validity of the Ctenohystricia (Mariveaux *et al.*, 2004; Wible *et al.*, 2005b). The Gliridae (the dormice, also known as the Myoxidae or the Muscardinidae) clusters with the sciuroids in most analyses. Their zygomasseteric structure is apparently myomorphous, but modified by a strong origin of the lateral masseter on an abbreviated zygomatic plate; it could also be interpreted as derived sciuromorphy. Two groups that remain enigmatic in their placement, both hystricomorphous African endemics,

Table 16.1. *(cont.)*

Korth (1994) classification: North American Tertiary taxa only, excluding caviomorphs. Italic text in brackets indicates classification followed in this volume

Suborder	Infraorder	Superfamily	Family
Sciuromorpha		†Ischyromyoidea	†Ischyromyidae (inc. †Reithroparamyinae)
		Aplodontoidea	Aplodontidae (inc. †Allomyinae)
			†Mylagaulidae
		Sciuroidea	Sciuridae
		Castoroidea	Castoridae
			†Eutypomyidae
		Incertae sedis	†Cylindrodontidae
Myomorpha	Myodonta	Muroidea	Cricetidae (inc. Arvicolinae)
			[*Arvicolidae as separate family*]
		Dipodoidea	Zapodidae
			†Simimyidae
			†Armintomyidae
			[*all within Dipodidae*]
	Geomorpha	Geomyoidea	Geomyidae[a]
			Heteromyidae[a]
			†Florentiamyidae
			†Heliscomyidae
			[†*Jimomyidae*]
[?Myomorpha]		†Eomyoidea	†Eomyidae
			†Sciuravidae
?Myomorpha			
Uncertain			†Protoptychidae [*possibly a myomorph*]
			†Laredomyidae [*possibly a glirid*]
			[†*Alagomyidae, basal to all other rodents*]

[a]These groups united into the Parvorder Geomyina.
†Extinct taxa.

are the Anomaluridae (the scaly-tailed flying squirrels) and the Pedetidae (the springhare) (these two families are commonly united in the Anomaluroidea). Huchon *et al.* (2002) weakly associated *Anomalurus* with the castorid–geomyoid cluster, and somewhat more distantly from Myomorpha; Adkins *et al.* (2001) placed *Pedetes* near Myomorpha. It would be instructive to analyze both genera simultaneously by multiple genes.

EVOLUTIONARY AND BIOGEOGRAPHIC PATTERNS

GENERAL TRENDS

We have followed the pattern of chronofaunas detailed in Volume 1. These are as follows: Paleocene Fauna, early to middle Eocene Fauna (Wasatachian–Uintan), White River Chronofauna (Duchesnean–early Arikareean), Runningwater Chronofauna (late Arikareean–early Hemingfordian), Clarendonian Chronofauna (late Hemingfordian–Clarendonian), and Mio-Pliocene Chronofauna (Hemphillian–Blancan). Table 16.3 shows the distribution of various paleobiological features of Glires at various "snapshots" through the Tertiary, usually around the middle of each chronofauna, but with the White River and Mio-Pliocene Chronofaunas divided into middle and late portions. These diagrams include every genus known from all or part of the particular North American Land-Mammal Age subdivision: they are intended to provide an overall picture of evolutionary trends only, not to represent precise quantification.

The paleobiological attributes are designated as follows, from the information provided in the original chapters. Body size, loosely divided into "small," "medium," and "large" is determined by the dental measurements provided. "Small" taxa have an m2 length of less than 1 mm: this corresponds to an animal the size of a chipmunk or less (around 100 g, or less). "Large" taxa have an m2 length of 4 mm or greater: this corresponds to an animal the size of a muskrat or greater (around 2 kg, or greater). Dietary categories are determined by tooth crown height: brachydont, mesodont, hypsodont, and hypselodont. Brachydont forms are presumed to be browsers: increasing levels of hypsodonty reflect increasing levels of abrasive material in the diet, relating either to silica contained in grasses or to the accumulation of dust and grit on the food (see Janis, 1988). We realize that there might be some difference in determination of these features between chapters: brachydont and hypselodont (rootless, ever-growing molars) are discrete categories, but the boundary between mesodont and hypsodont may be idiosyncratic depending on the author.

Table 16.2. *Molecular (and additional morphological) support for suprafamilial groupings in rodents*

Groupings	Sources
1. Universally supported clusterings	
Caviomorphs and Old World hystricomorphs (= Hystricognathi)	Catzeflis *et al.*, 1995;[a] Nedbal, Honeycutt, and Schiltter, 1996;[b] Huchon, Catzeflis, and Douzery, 1999;[c] Douady *et al.*, 2000;[g] Adkins *et al.*, 2001;[h] Huchon and Douzery, 2001;[d] Huchon *et al.*, 2002;[e] Adkins, Walton, and Honeycutt, 2003;[i] (also supported by Lavocat and Parent, 1985;[o] Luckett, 1985;[o] Landry, 1999;[r])
Sciurids + aplodontids (= sciuromorphs)	Nedbal, Honeycutt, and Schiltter, 1996;[b] Huchon, Catzeflis, and Douzery, 1999;[c] Douady *et al.*, 2000;[g] Adkins *et al.*, 2001 ;[h] DeBry and Sagel, 2001;[j] Huchon *et al.*, 2002;[e] Adkins, Walton, and Honeycutt, 2003;[i] (also supported by Bugge, 1985;[l] Lavocat and Parent, 1985;[n] Wahlert, 1985;[p] Landry, 1999;[r])
Geomyids + heteromyids (= geomyoids)	Nedbal, Honeycutt, and Schiltter, 1996;[b] Douady *et al.*, 2000;[g] Adkins *et al.*, 2001 ;[h] DeBry and Sagel, 2001;[j] Huchon *et al.*, 2002:[e] Montgelard *et al.*, 2002;[k] Adkins, Walton, and Honeycutt, 2003;[i] (also supported by Bugge, 1985;[l] Lavocat and Parent, 1985;[n])
Muroids + dipodoids (= Myodonta)	Nedbal, Honeycutt, and Schiltter, 1996;[b] Huchon, Catzeflis, and Douzery 1999;[c] Adkins *et al.*, 2001;[h] DeBry and Sagel, 2001;[j] Huchon *et al.*, 2002;[e] Montgelard *et al.*, 2002;[k] Adkins, Walton, and Honeycutt, 2003;[i] (also supported by Bugge, 1985 ;[l] Luckett, 1985;[o])
2. Position of geomyoids:	
Geomyoids + myodonts	Nedbal, Honeycutt, and Schiltter, 1996;[b] Douady *et al.*, 2000;[g] Huchon *et al.*, 2000 ;[e] Adkins *et al.*, 2001;[h] Montgelard *et al.*, 2002;[k] Adkins, Walton, and Honeycutt, 2003;[i] (also supported by Bugge, 1985 ;[l] Luckett, 1985;[o] Wahlert, 1985;[p] Landry, 1999;[r])
Geomyoids (+ castorids) as separate grouping	Montgelard *et al.*, 2002;[k]
Geomyoids + hystricognaths	DeBry and Sagel, 2001;[j]
3. Position of castoroids	
Castorids + geomyoids (included with myodonts)	Douady *et al.*, 2000;[g] Huchon *et al.*, 2000;[e] Adkins *et al.*, 2001;[h] Montgelard *et al.*, 2002;[k] Adkins, Walton, and Honeycutt, 2003;[i]
Castorids + sciuromorphs	Nedbal, Honeycutt, and Schiltter, 1996;[b] DeBry and Sagel, 2001;[j] (also supported by Landry, 1999 ;[r])
4. Position of Hystricognathi	
Hystricognaths + sciuromorphs	Huchon, Catzeflis, and Douzery 1999;[c] Montgelard *et al.*, 2002;[k]
Hystricognaths + myomorphs	Adkins *et al.*, 2001;[h] Huchon *et al.*, 2002;[e] Adkins, Walton, and Honeycutt, 2003;[i]
5. Position of Old World hystricomorph groups	
Ctenodactylids + hystricognaths (= Ctenohystricia)	Huchon, Catzeflis, and Douzery 1999;[c] Douady *et al.*, 2000;[g] Adkins *et al.*, 2001;[h] Huchon and Douzery 2001;[d] Huchon *et al.*, 2002;[d] Montgelard *et al.*, 2002;[k] Adkins, Walton, and Honeycutt, 2003;[i] (also supported by George, 1985;[m] Lavocat and Parent, 1985;[n] Luckett 1985;[o] Martin, 1993 ;[q] Landry, 1999;[r])
Ctenodactylids + sciuromorphs	Nedbal, Honeycutt, and Schiltter, 1996;[b]
Glirids + sciuromorphs	Nedbal, Honeycutt, and Schiltter, 1996;[b] Kramereov, Vassetzky, and Serdobova, 1999;[f] Huchon *et al.*, 2000;[e] Adkins *et al.*, 2001 ;[h] DeBry and Sagel, 2001;[j] Montgelard *et al.*, 2002;[k] Adkins, Walton, and Honeycutt, 2003;[i] (also supported by Bugge, 1985 ;[l] Lavocat and Parent, 1985;[n] Wahlert, 1985;[p])
Glirids + myomorphs	Douady *et al.*, 2000;[g] (also supported by Wahlert, 1985;[p])
Glirids as sister taxon to other rodents	Huchon *et al.*, 1999;[c]
Anomaluroids + sciuromorphs	Montgelard *et al.*, 2002;[k]
Anomaluroids + myomorphs	Nedbal, Honeycutt, and Schiltter, 1996;[b] Adkins *et al.*, 2001;[h] Huchon *et al.*, 2002;[e] Adkins, Walton, and Honeycutt, 2003;[i]
Anomaluroids + hystricognaths	Landry 1999;[r]
Anomaluroids undetermined	Douady *et al.*, 2000;[g] (also supported by Bugge, 1985 [l])

[a] Mitochondrial gene (12S rRNA) 35, spp. (mainly hystricognaths). [b] One mitochondrial gene (12S RNA), 53 spp. [c] One nuclear gene (*vWF*), 15 spp. (no geomyoids, *Castor*, anomalurids). [d] One nuclear gene (*vWF*), 27 spp. (mainly hystricognaths). [e] Three nuclear genes (*vWF*, *IRBP*, *A2AB*), 21 spp. [f] Two SINES (B1, B1-dID), 15 spp. [g] DNA CsCl profiles, 49 spp. (50% muroids, no *Castor*). [h] One mitochondrial gene 12S rRNA, three nuclear genes (*GHR*, *BRCA*1, *vWF*), 26 spp. [i] As note *h*, but additional information (from *GHR* + *BRCA*1), 35 spp. [j] Nuclear gene *IRBP*, 22 spp. (no ctenodactylids or anomalourids). [k] Two mitochondrial genes (for cytochrome *b* and 12S RNA), 28 spp. [l] Carotid arterial patterns. [m] Reproductive and chromosomal characters. [n] Middle ear anatomy. [o] Dental and placentation features. [p] Cranial foramina. [q] Incisor enamel. [r] Multiple cranial, postcranial, and soft anatomy characters (excluding zygomasseteric structure).

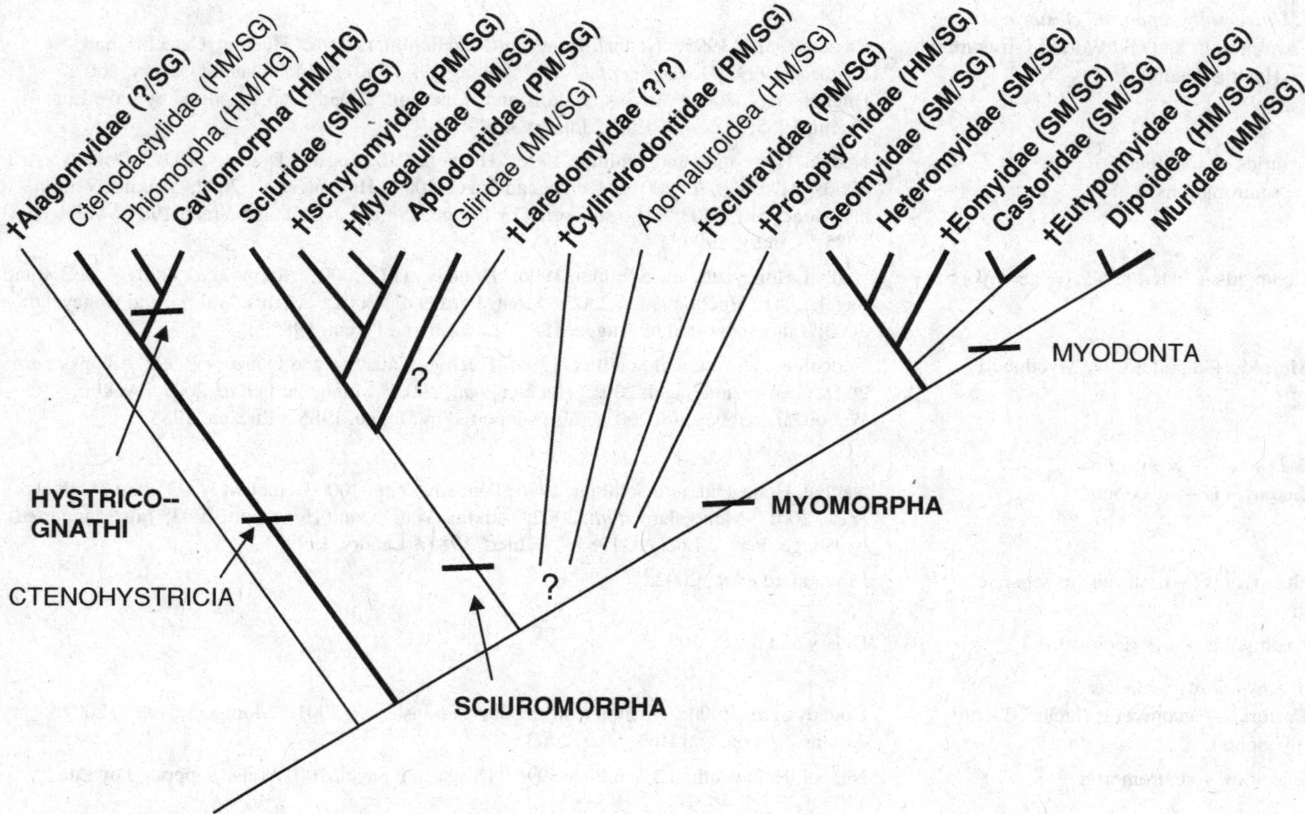

Figure 16.4. Probable interrelationships of rodent families and superfamilies. A thick line represents a greater degree of confidence in the association. Families found in the North American Tertiary in bold face (extinct families indicated by a dagger). Extant families never found in North America are in plain type (extinct families not shown). The letters in parentheses after the taxon names indicate the form of the jaw and the zygomasseteric musculature. HG, hystricognathus; HM, hystricomorphous; MM, myomorphous; PM, protrogomorphous; SG, sciurognathus; SM, sciuromorphous.

Table 16.3. *Distribution of Glires ecomorphological attributes*

	Percentage at each time						
Attributes:	Br2 ($n = 22$)	Du ($n = 41$)	Or ($n = 35$)	Ar4 ($n = 45$)	Ba2 ($n = 55$)	Hh1 ($n = 41$)	L.Bl ($n = 54$)
Body size							
Small	36	49	43	42	45	24	22
Medium	54	44	56	54	40	59	59
Large	10	7	3	4	15	17	19
Cheek tooth height							
Brachydont	86	66	60	32	27	29	20
Mesodont	14	24	29	24	25	15	13
Hypsodont	0	10	11	40	44	39	41
Hypselodont	0	0	0	2	5	17	26
Locomotor specialization							
Generalized	95	91	86	54	44	49	50
Arboreal	0	7	14	9	13	7	5
Fossorial	5	2	0	22	18	17	11
Cursorial	0	0	0	15	25	25	27
Aquatic	0	0	0	0	0	2	5

Locomotory behavior is determined by direct report from the chapters, based on postcranial information: deviations from the general scansorial form are noted as arboreal, burrowing, aquatic, and cursorial (the last representing quadrupedal bounding in lagomorphs and bipedal hopping, or ricochetal behavior, in rodents). In some cases, all members of a clade are considered to have the locomotor specializations that had been determined for only a few members: for example, among the squirrels, all members of the Petaurisdinae (flying squirrels) and the Sciurini (tree squirrels) are considered to be arboreal, but members of other subfamilies are not. Among the Eomyidae, members of the tribe Eomyini are considered to be arboreal (one European eomyid is known to be a glider [Storch, Engesser, and Wuttke, 1996]). Sicistine dipodids are also considered to be arboreal, following the habits of the extant members of the group. The burrowing forms are primarily the palaeocastorine beavers, the mylagauline aplodontids, the geomyids, and a few marmotine squirrels (the extant taxa *Marmota* and *Cynomys* are considered to be burrowers here). Among the leporids, archaeolagines and leporines are considered to be cursorial, but palaeolagines are not. Cursorial rodents include the ricochetal heteromyids and zapodine dipodids. We probably err on the side of caution here and also note that the numbers of scansorial (i.e., unspecialized) forms are probably overestimated owing to the paucity of good postcranial evidence, and the general lack of postcranial specializations in smaller mammals for habits such as arboreality or swimming.

In the Eocene fauna (there are too few Paleocene Glires for consideration) during the "snapshot" of the mid Bridgerian, the majority of Glires (all rodents at this time) are of medium size (only 36% were small and 10% were large). Only a few (14%) are mesodont, and none is hypsodont. Most of these taxa have generalized postcrania, although it is possible that some ischyromyids were arboreal (see Rose and Chinnery, 2004). A large, probably burrowing, ischyromyid, *Pseudotomus*, was present, and the cylindrodontines (two genera present during this time interval) may also have been fossorial, judging from cranial evidence. No taxa are cursorial, although note the presence of the ricochetal *Protoptychus* in the middle Eocene (see Figure 16.5, below).

In the early White River Chronofauna (Duchesnean), a greater proportion of Glires are small (49%), and there are three large taxa (7%, one ischyromyid and two eomyids). The large ischyromyid, *Pseudotomus*, is the only burrower. There are three eomyid genera that are likely aborealists. No Glires from this time are cursorial. However, significant changes in dietary behavior seem to have occurred since the early Eocene: now mesodont cheek teeth are seen in 24% of the fauna, and hypsodont teeth in 10% (leporids and dipodids).

In the later White River Chronofauna (Orellan), the size distribution in the fauna is broadly similar, although there is only one large-sized taxon (the ischyromyid rodent *Manitsha*), as are the distribution of tooth crown heights. Fourteen percent of the fauna is apparently arboreal (three eomyids and one sciurid), and no other locomotory specializations are evident.

In the Runningwater Chronofauna (latest Arikareean, Ar4), the proportion of small rodents in the fauna, 42%, is similar to that in the White River Chronofauna, although a couple of large taxa are now present (one aplodontid and one geomyid). With regard to diet, the number of mesodont taxa remains similar at 24%, but the number of hypsodont taxa increases considerably to 40% of the fauna (representing geomyoids, dipodids, and leporids), and one geomyid (*Entoptychus*) is hypselodont. Burrowers (palaeocastoroines and geomyids) now make up 22% of the fauna, arboreal taxa (eomyids and dipodids) make up 9% of the fauna, and cursorial taxa are now in evidence (archaeolagines and heteromyids), constituting 15% of the fauna.

These trends of increasing numbers of large rodents and increasing locomotor specializations continue into the Clarendonian Chronofauna, at the late Barstovian (Ba2) interval. Now there are seven large taxa (aplodontids and castorids), representing 15% of the fauna, although the percentage of small Glires is similar to before. The percentages of dietary types (25% mesodont, 44% hypsodont, 5% hypselodont) are similar to those of the late Arikareean. However, there is an increase in the numbers of cursorial forms (again archaeolagines and heteromyids), now making up 25% of the fauna. Other locomotor types, burrowers (mylagaulids and geomyids) and arboreal forms (eomyids, dipodids, and sciurids) are at similar proportions to the late Arikareean (18% and 13%, respectively).

During the Mio-Pliocene Chronofauna, the number of small rodents decreases to around 25% of the fauna. Although some cricetids and arvicolids are predominant small rodents during this time, many of them are more medium sized, and other small forms (eomyids, primitive geomyoids) are lost. The proportion of large Glires (around 18%) is similar to that of the late Barstovian, represented now mainly by leporines, castorids, and caviomorphs. The castorids and caviomorphs also include aquatic forms, making this locomotor category apparent for the first time (at around 6% of the fauna, although smaller aquatic forms, with less specialized postcrania, may have been around during earlier times). The numbers of other locomotor types are similar to those in the Clarendonian Chronofauna, with cursorial forms continuing to represent around 25% of the fauna, although the percentage of burrowers decreases in the Pliocene (from 17% to 11%) with the extinction of the mylagaulids. However, significant changes to cheek tooth height are now seen: in the early part of the chronofauna (earliest Hemphillian, Hh1) hypsodont teeth occur in 39% of taxa and hypselodont teeth in 17%, while in the late part of the chronofauna (late Blancan and earliest Irvingtonian), 43% is hypsodont and 26% is hypselodont. This increase reflects, in part, the radiation of the cricetids and arvicolines but also reflects a general increase in hypsodonty among most rodent groups.

In summary, with regard to body size, small members of the Glires make up from a half to a quarter of the fauna throughout the Tertiary: large rodents are not prominent until the middle Miocene, but never make up more than 20% of the fauna. With regard to tooth crown height, reflecting either dietary preference and/or changing habitats, a certain degree of hypsodonty is seen by the middle Eocene; by the Oligocene more than a third of the Glires fauna is mesodont or hypsodont, and by the early Miocene (late Arikareean) this is true of more than half the fauna. This contrasts with the condition

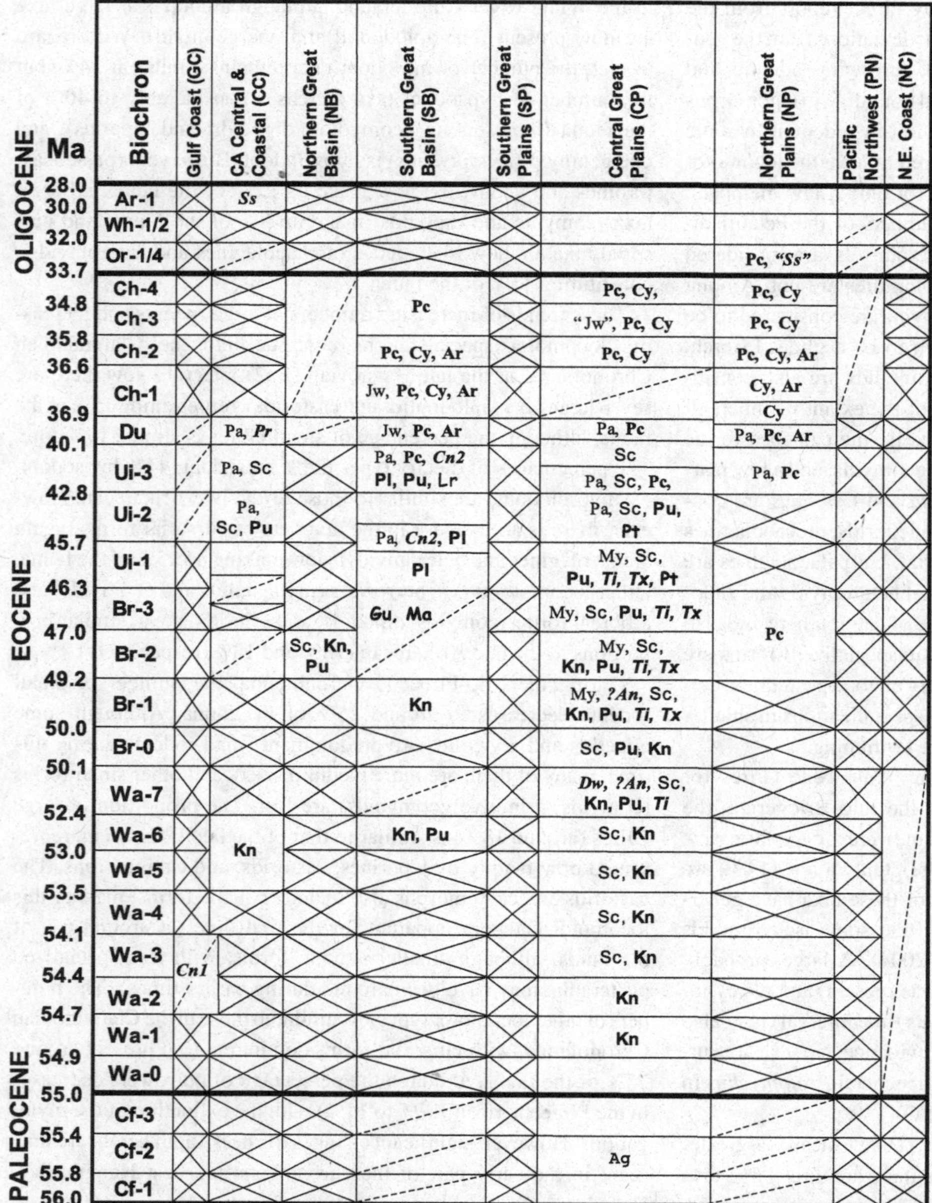

Figure 16.5. Biogeographic ranges of sciuriavids, cylindrodontids, and other Paleogene rodents (excluding ischyromyids). Key: A "box" (for a particular time period in a particular biogeographic region) that has a cross through it means no fossil mammal localities are known for that time period from that area; a single dashed line through the box means only scant fossil mammal information is available (usually only a single, small locality). CYLINDRODONTIDAE: JAYWILSONOMYINAE (Time New Roman Plain): Jw, *Jaywilsonomys*; My, *Mysops*; Pa, *Pareumys*. CYLINDRODONTINAE (Times New Roman Bold): **Ar**, *Ardynomys*; **Cy**, *Cylindrodon*; **Pc**, *Pseudocylindrodon*. ?CYLINDRODONTIDAE (Times New Roman Bold Italics): ***An***, *Anomoemys*; ***Cn1***, Cylindrodontid new genus 1; ***Cn2***, Cylindrodontid new genus 2; ***Dw***, *Dawsonomys*; ***Pr***, *Presbymys*; ***Ss***, *Sespemys*. SCIURAVIDAE: LINEAGE A (Arial Plain): Sc, *Sciuravus*. LINEAGE B (Arial Bold): **Kn**, *Knightomys*; **Pl**, *Prolapsus*; **Pu**, *Pauromys*. SCIURAVIDS INDET. (Arial Bold Italics): ***Ti***, *Tillomys*; ***Tx***, *Taxymys*. ALAGOMYIDAE (Courier Plain): Ag, *Alagomys*. LAREDOMYIDAE (Comic Sans MS Plain): Lr, *Laredomys*. PROTYPTYCHIDAE (Comic Sans MS Bold): **Gu**, *Guanajuatomys*; **Ma**, *Marfilomys*; **Pt**, *Protyptychus*.

in ungulates (see Janis, Damuth, and Theodor, 2000), where a rise in hypsodonty among ungulates is not apparent until the late early Miocene. It may be that rodents and rabbits are more sensitive indicators to environmental changes than ungulates, as open habitats are thought to have been prevalent in North America since the late Eocene (Retallack, 2001), and palynological evidence shows that grasslands spread by the start of the Miocene (Strömberg, 2006). The further increase in the percentage of hypsodont Glires in the Mio-Pliocene (now around two-thirds of the fauna, including many hypselodont taxa) is also paralleled in the ungulates. Note, however, that the numbers of ungulates decline at this point, so the increase in the percentage of the hypsodont forms represents the decline of the brachydont ones rather than an absolute increase in numbers of hypsodont taxa (Janis, Damuth, and Theodor, 2000). This is not true for the Glires, whose numbers hold constant at between 40 to 50 genera since the late Eocene, and although the numbers of brachydont taxa do decline over time, there is also a rise in the absolute numbers of higher-crowned taxa.

The increase in hypsodonty among the Glires is coincident with the appearance of burrowing adaptations (although it is not only burrowing rodents that are hypsodont), again suggesting a change to more open habitats at this time. The appearance of cursorial rodents, in the late Oligocene to early Miocene, is of a similar timing to that of the appearance of cursorial ungulates (Janis and Wilhelm, 1993; Janis, Theodor, and Boisvert, 2002), again suggesting correlation with the spread of open habitats. However, unlike the situation in

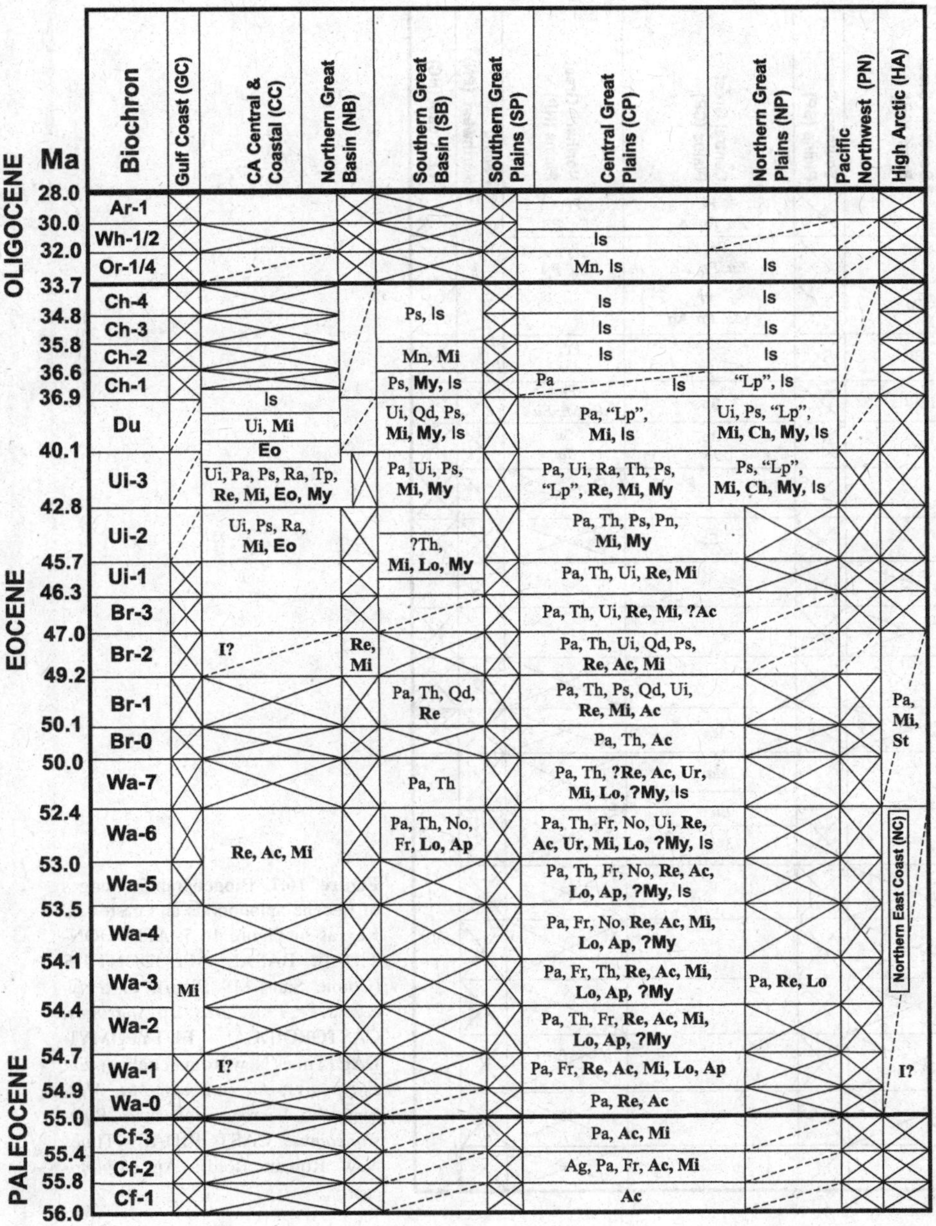

Epoch	Ma	Biochron	Gulf Coast (GC)	CA Central & Coastal (CC)	Northern Great Basin (NB)	Southern Great Basin (SB)	Southern Great Plains (SP)	Central Great Plains (CP)	Northern Great Plains (NP)	Pacific Northwest (PN)	High Arctic (HA)
OLIGOCENE	28.0	Ar-1									
	30.0	Wh-1/2						Is			
	32.0	Or-1/4						Mn, Is	Is		
EOCENE	33.7	Ch-4				Ps, Is		Is	Is		
	34.8	Ch-3						Is	Is		
	35.8	Ch-2				Mn, **Mi**		Is	Is		
	36.6	Ch-1				Ps, **My**, Is		Pa Is	"Lp", Is		
	36.9	Du		Is / Ui, **Mi** / **Eo**		Ui, Qd, Ps, **Mi, My**, Is		Pa, "Lp", **Mi**, Is	Ui, Ps, "Lp", **Mi, Ch, My**, Is		
	40.1	Ui-3		Ui, Pa, Ps, Ra, Tp, **Re, Mi, Eo, My**		Pa, Ui, Ps, **Mi, My**		Pa, Ui, Ra, Th, Ps, "Lp", **Re, Mi, My**	Ps, "Lp", **Mi, Ch, My**, Is		
	42.8	Ui-2		Ui, Ps, Ra, **Mi, Eo**		?Th, **Mi, Lo, My**		Pa, Th, Ps, Pn, **Mi, My**			
	45.7	Ui-1						Pa, Th, Ui, **Re, Mi**			
	46.3	Br-3						Pa, Th, Ui, **Re, Mi, ?Ac**			
	47.0	Br-2		I?	**Re, Mi**			Pa, Th, Ui, Qd, Ps, **Re, Ac, Mi**			
	49.2	Br-1				Pa, Th, Qd, **Re**		Pa, Th, Ps, Qd, Ui, **Re, Mi, Ac**			Pa, **Mi, St**
	50.1	Br-0						Pa, Th, **Ac**			
	50.0	Wa-7				Pa, Th		Pa, Th, **?Re, Ac, Ur, Mi, Lo, ?My**, Is			
	52.4	Wa-6		**Re, Ac, Mi**		Pa, Th, No, Fr, **Lo, Ap**		Pa, Th, Fr, No, Ui, **Re, Ac, Ur, Mi, Lo, ?My**, Is			Northern East Coast (NC)
	53.0	Wa-5						Pa, Th, Fr, No, **Re, Ac, Lo, Ap, ?My**, Is			
	53.5	Wa-4						Pa, Fr, No, **Re, Ac, Mi, Lo, Ap, ?My**			
	54.1	Wa-3	**Mi**					Pa, Fr, Th, **Re, Ac, Mi, Lo, Ap, ?My**	Pa, **Re, Lo**		
	54.4	Wa-2						Pa, Th, Fr, **Re, Ac, Mi, Lo, Ap, ?My**			
PALEOCENE	54.7	Wa-1		I?				Pa, Fr, **Re, Ac, Mi, Lo, Ap**			I?
	54.9	Wa-0						Pa, **Re, Ac**			
	55.0	Cf-3						Pa, **Ac, Mi**			
	55.4	Cf-2						Ag, Pa, Fr, **Ac, Mi**			
	55.8	Cf-1						**Ac**			
	56.0										

Figure 16.6. Biogeographic ranges of ischyromyid. Key as in Figure 16.5. PARAMYINAE (Times New Roman Plain): Fr, *Franimys*; "Lp", "*Leptotomus*"; Mn, *Manitsha*; No, *Notoparamys*; Pa, *Paramys*; Pn, new paramyine genus; Ps, *Pseudotomus*; Qd, *Quadratomus*; Ra, *Rapamys*; Th, *Thisbemys*; Tp, *Tapomys*; Ui, *Uintaparamys*. REITHROPARAMYINAE (Times New Roman Bold): **Ac**, *Acritoparamys*; **Ap**, *Apatosciuravus*; **Ch**, *Churcheria*; **Lo**, *Lophiparamys*; **Mi**, *Microparamys*; **Re**, *Reithroparamys*; **St**, *Strathcona*; **Ur**, *Uriscus*. AILURAVINAE (Arial Bold): **Eo**, *Eohaplomys*; **My**, *Mytonomys*. ISCHYROMYINAE (Arial Plain): Is, *Ischyromys*.

ungulates, where few non-cursorial taxa remain in the Neogene, the percentage of cursorial Glires never exceeds 25%. The retention of generalized, non-cursorial forms probably represents the use of microhabitats by these smaller mammals.

CHRONOFAUNAL CHANGES

EARLY-MID EOCENE FAUNA

Rodents first appear in the latest Paleocene (Clarkforkian), with *Acritoparamys* in the early Clarkforkian, and *Paramys, Franimys, Microparamys*, and *Alagomys* in the middle Clarkforkian (Figure 16.5). All but *Alagomys*, which is the sole representative of the primitive Asian family Alagomyidae and is known only from a single locality, were members of the Ischyromyidae, a family of mainly medium-sized generalized rodents that were prominent in the early and middle Eocene and persisted into the Oligocene. Ischyromyid diversity is at its maximum in the late Wasatchian to early Bridgerian, with around a dozen genera (Figure 16.6).

Other rodents in this time interval were the generalized sciuravids, and the possibly fossorial cylindrodontids. Sciuravids first appear in the early Wasatchian, and are at their greatest generic diversity (five genera) in the late Wasatchian through Bridgerian. They become reduced in diversity in the Uintan (and are also now rare in individual abundance: Gregg Gunnell, personal communication) and are extinct by the end of this interval (with the exception of

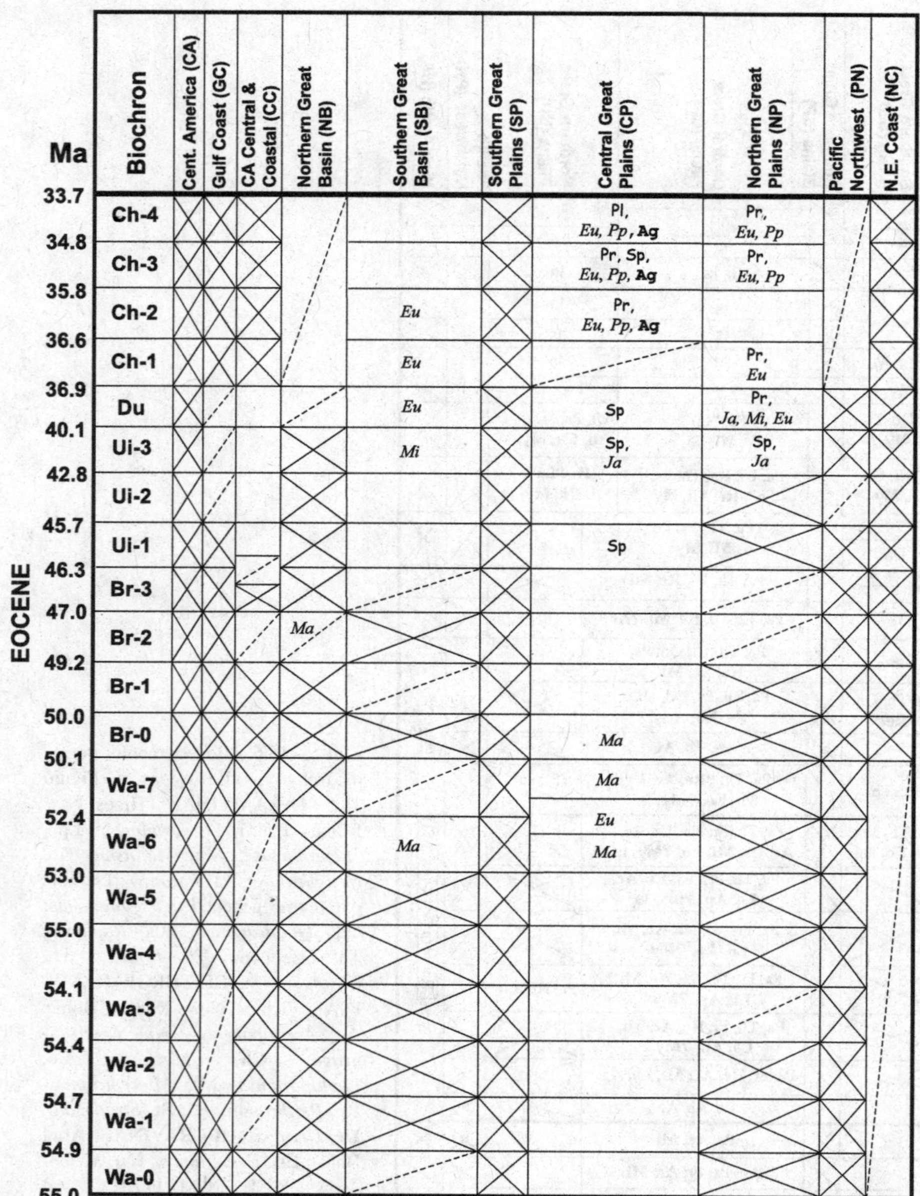

Figure 16.7. Biogeographic ranges of Eocene aplodontids and castorids. Key as in Figure 16.5. APLODONTIDAE: BASAL APLODONTIDS (Comic Sans MS Plain): Pl, *Pelycomys*; Pr, *Prosciurus*; Sp, *Spurimus*. CASTOROIDEA: EUTYPOMYIDAE (Times New Roman Italics): *Eu*, *Eutypomys*; *Ja*, *Janimus*; *Ma*, *Mattimys*; *Mi*, *Microeutypomys*; *Pp*, *Pipestoneomys*. CASTORIDAE (Times New Roman Bold): **Ag**, *Agnotocastor*.

a possible Duchesnean occurrence of *Sciuravus*). Sciuravids were apparently absent from the Northern Great Plains. Cylindrodontids first appear in the early Wasatchian but do not diversify until the latest Wasatchian and Bridgerian. Their generic diversity during this time period was greatest (three genera) in the later Uintan (Figure 16.5).

A few other odd rodents are known from (usually) single occurrences the Bridgerian and Uintan: *Guanajuatomys* and *Marfilomys* from Mexico, which are possibly related to the specialized ricochetal *Protoptychus* (known from a couple of Central Great Plains sites), and *Laredomys* from Texas (the only member of the family Laredomyidae) (Figure 16.5). A few members of more derived rodent families also have a first appearance in the early or early middle Eocene, such as the eutypomyid castoroid *Mattimys*, and primitive dipodids such as *Armintomys* and *Elymys*. By the Uintan, more members of more derived rodent families make their first appearance, such as the aplodontid *Spurimus*, a couple more eutypomyids, eomyine eomyids such as *Adjidaumo* and *Paradjidaumo*, the dipodid *Simimys*, and geomorphs such as *Griphomys* and *Floresomys* (Figures 16.7 and 16.8; see also Figure 16.12, below). Primitive lagomorphs also make a first appearance in the late Uintan (see Figure 16.15, below).

WHITE RIVER CHRONOFAUNA

Some members of the primitive rodent families persist into this time interval. Ischyromyines are still quite diverse (10 genera) in the Duchesnean, but only a few of these persist into the Chadronian,

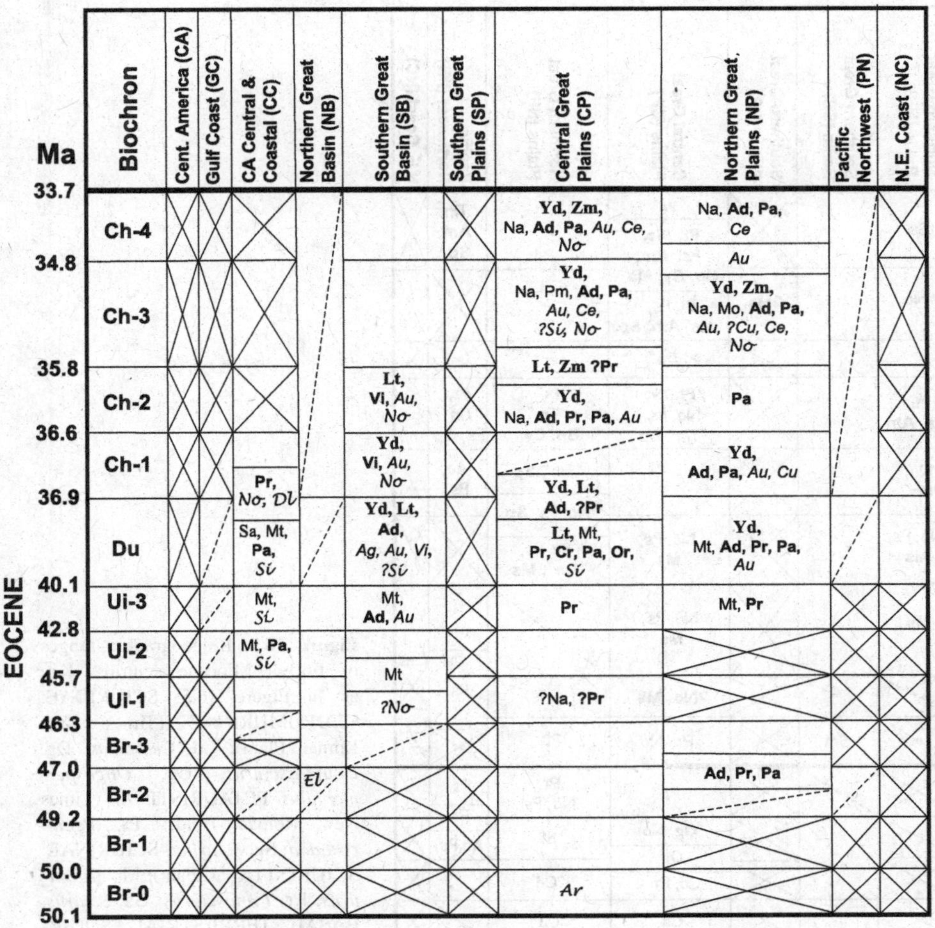

Ma	Biochron	Cent. America (CA)	Gulf Coast (GC)	CA Central & Coastal (CC)	Northern Great Basin (NB)	Southern Great Basin (SB)	Southern Great Plains (SP)	Central Great Plains (CP)	Northern Great Plains (NP)	Pacific Northwest (PN)	N.E. Coast (NC)
33.7	Ch-4							**Yd, Zm,** Na, **Ad, Pa,** *Au, Ce,* No	Na, **Ad, Pa,** *Ce*; *Au*		
34.8	Ch-3							**Yd,** Na, Pm, **Ad, Pa,** *Au, Ce,* ?Si, No; **Lt, Zm ?Pr**	**Yd, Zm,** Na, Mo, **Ad, Pa,** *Au, ?Cu, Ce,* No		
35.8	Ch-2					**Lt, Vi,** *Au,* No		**Yd,** Na, **Ad, Pr, Pa,** *Au*	**Pa**		
36.6	Ch-1			**Pr,** No, Dl		**Yd, Vi,** *Au,* No		**Yd, Lt, Ad, ?Pr**	**Yd, Ad, Pa,** *Au, Cu*		
36.9	Du			Sa, Mt, **Pa,** Si		**Yd, Lt, Ad,** *Ag, Au, Vi,* ?Si		**Lt,** Mt, **Pr, Cr, Pa, Or,** Si	**Yd,** Mt, **Ad, Pr, Pa,** *Au*		
40.1	Ui-3			Mt, Si		Mt, **Ad,** *Au*		**Pr**	Mt, **Pr**		
42.8	Ui-2			Mt, **Pa,** Si		Mt					
45.7	Ui-1					?No		?Na, ?Pr			
46.3	Br-3										
47.0	Br-2				El				**Ad, Pr, Pa**		
49.2	Br-1										
50.0	Br-0							Ar			
50.1											

EOCENE

Figure 16.8. Biogeographic ranges of Eocene eomyids and dipodids. Key as in Figure 16.5. EOMYIDAE: INCERTAE SEDIS (Times New Roman Plain): Sa, *Simiacritomys*. YODERMYINAE (Times New Roman Bold): **Lt**, *Litoyoderomys*; **Yd**, *Yoderimys*; **Zm**, *Zemiodontomys*. EOMYINAE: NAMATOMYINI (Arial Plain): Mo, *Montanamus*; Mt, *Metanoiamys*; Na, *Namatomys*; Pm, *Paranamatomys*. EOMYINI (Arial Bold): **Ad**, *Adjidaumo*; **Cr**, *Cristadjidaumo*; **Or**, *Orelladjidaumo*; **Pa**, *Paradjidaumo*; **Pr**, *Protadjidaumo*; **Vi**, *Viejadjidaumo*. EOMYINAE INCERTAE SEDIS (Arial Italics): *Ag*, *Aguafriamys*; *Au*, *Aulolithomys*; *Ce*, *Centimanomys*; *Cu*, *Cupressimus*. DEPODIDAE: ARMINTOMYINAE (Lucida Handwriting Plain): Ar, *Armintomys*. SUBFAMILY UNCERTAIN (Lucida Handwriting Bold): **Dl**, *Diplolophus*; **El**, *Elymys*; **No**, *Nonomys*; **Si**, *Simimys*.

including the large burrowing taxon *Pseudotomus*. The widespread *Ischyromys*, present since the mid Wasatchian, survives through the Whitneyan (Figure 16.6). Cylindrodontids remain fairly diverse through the Chadronian, but only one genus, *Sespemys*, which is confined to the California Central and Coast region, is reported (there are more but not yet published, M. R. Dawson) from the Oligocene.

Members of more modern Glires families diversify at this point. Squirrels first appear in the Chadronian with the primitive family Cedromurinae, which survives until near the end of the Oligocene (Figure 16.9). The primitive sciurine *Protosciurus* also appears at the start of the Oligocene. At least some of these squirrels were probably arboreal specialists (e.g., *Douglassciurus* and *Protosciurus*). Other likely arboreal rodents in this time interval are members of the Eomyini (Eomyinae), some of which were present in the middle Eocene, and which further diversity in the late Eocene and Oligocene. Other eomyids, such as the Yodermyinae and the Namatomyini (Eomyinae), also diversify at this time: in fact, the maximal eomyid generic diversity (nine genera) is in the mid Chadronian. Dipodids are fairly rare through this time interval, with only a couple of genera present. Both eomyids and dipodids are rarely found in the Gulf Coast and California Central and Coast regions, and dipodids are absent from the Northern Great Plains (Figure 16.8).

Basal aplodontids, and members of the Allomyinae, diversify in the Oligocene. Eutypomyid castoroids, and basal members of the Castoridae, diversify in the Chadronian and Orellan, respectively, and the burrowing palaeocastorines first appear in the Whitneyan. During the Chadronian, at least, the castorids and aplodontids are confined to the Central and Northern Great Plains regions, while the eutypomyids are also present in the Southern Great Basin. During the later Oligocene, castorids are found in the Gulf Coast and California Central and Coast regions, while aplodontids are absent from these regions but are diverse in the Pacific Northwest (including some endemic meniscomyines) (Figure 16.10).

Other burrowers at this time, although of smaller size than the beavers, are the entoptychine geomyids, first appearing in the Arikareean. Other geomorphs in this time interval include members of the Heliscomyidae (mainly confined to the Orellan) and the Florentiamyidae (present in the Whitneyan and Arikareean), plus some geomyoids of uncertain affinities, such as *Tenudomys* and *Harrymys* (see Figure 16.12, below). Geomorphs, in general, are rare from the Northern Great Plains region until the very end of the Oligocene. The first heteromyids, representing the first radiation of ricochetal rodents (apart from the middle Eocene *Protoptychus*), first appear in the Whitneyan, and four genera are present by the end of the Oligocene. The first cursorial lagomorphs, represented by the

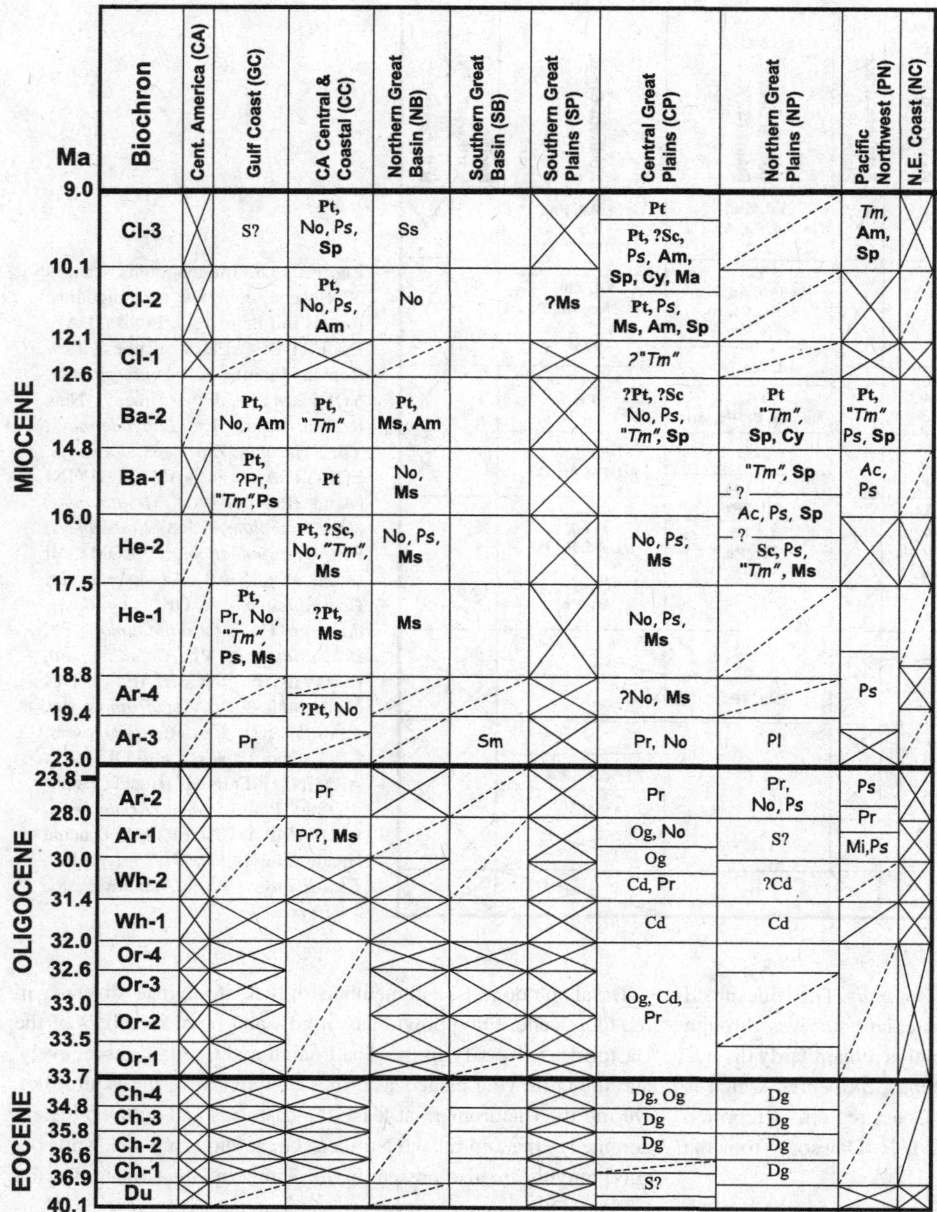

Ma	Biochron	Cent. America (CA)	Gulf Coast (GC)	CA Central & Coastal (CC)	Northern Great Basin (NB)	Southern Great Basin (SB)	Southern Great Plains (SP)	Central Great Plains (CP)	Northern Great Plains (NP)	Pacific Northwest (PN)	N.E. Coast (NC)
9.0	Cl-3		S?	Pt, No, Ps, Sp	Ss			Pt; Pt, ?Sc, Ps, Am, Sp, Cy, Ma		*Tm*, Am, Sp	
10.1	Cl-2			Pt, No, Ps, Am	No		?Ms	Pt, Ps, Ms, Am, Sp			
12.1	Cl-1							?"*Tm*"			
12.6	Ba-2		Pt, No, Am	Pt, "*Tm*"	Pt, Ms, Am			?Pt, ?Sc No, Ps, "*Tm*", Sp	Pt "*Tm*", Sp, Cy	Pt, "*Tm*" Ps, Sp	
14.8	Ba-1		Pt, ?Pr, "*Tm*", Ps	Pt	No, Ms				"*Tm*", Sp ?	Ac, Ps	
16.0	He-2			Pt, ?Sc, No, "*Tm*", Ms	No, Ps, Ms			No, Ps, Ms	Ac, Ps, Sp ? Sc, Ps, "*Tm*", Ms		
17.5	He-1		Pt, Pr, No, "*Tm*", Ps, Ms	?Pt, Ms	Ms			No, Ps, Ms			
18.8	Ar-4							?No, Ms		Ps	
19.4	Ar-3		Pr	?Pt, No		Sm		Pr, No	Pl		
23.0 / 23.8	Ar-2			Pr				Pr	Pr, No, Ps	Ps	
28.0	Ar-1			Pr?, Ms				Og, No	S?	Pr; Mi,Ps	
30.0	Wh-2							Og; Cd, Pr	?Cd		
31.4	Wh-1							Cd	Cd		
32.0	Or-4										
32.6	Or-3							Og, Cd, Pr			
33.0	Or-2										
33.5	Or-1										
33.7	Ch-4							Dg, Og	Dg		
34.8	Ch-3							Dg	Dg		
35.8	Ch-2							Dg	Dg		
36.6	Ch-1								Dg		
36.9–40.1	Du							S?			

Epochs: MIOCENE (9.0–23.8), OLIGOCENE (23.8–33.7), EOCENE (33.7–40.1).

Figure 16.9. Biogeographic ranges of Eocene–Miocene sciurids. Key as in Figure 16.5. SCIURIDAE: CEDROMURINAE (Times New Roman Plain): Cd, *Cedromus*; Dg, *Douglassciurus*; Og, *Oligospermophilus*. PETAURISTINAE (Times New Roman Bold): **Pt**, *Petauristodon*; **Sc**, *Sciurion*. SCIURINAE: SCIURINI (Arial Plain): Mi, *Miosciurus*; Pr, *Protosciurus*; Ss, *Sciurus*. BASAL TERRESTRIAL SQUIRRELS (Comic Sans MS Plain): Ac, *Arctomyoides*; No, *Nototamias*; Pl, *Palaearctomys*; Ps, *Protospermophilus*; Sm, *Similisciurus*; "Tm", "*Tamias*." TAMIINI (Comic Sans MS Bold): **Tm**, *Tamias*. MARMOTINI (Arial Bold): **Am**, *Ammospermophilus*; **Cy**, *Cynomyoides*; **Ma**, *Marmota*; **Ms**, *Miospermophilus*; **Sp**, *Spermophilus*. S?, Sciurid indet (Times New Roman Plain).

archeolagine *Archaeolagus*, also make a first (definite) appearance in the Arikareean, although this time interval is mainly dominated by the less cursorially adapted palaeolagines, which were mainly confined to the Central and Northern Great Plains regions (see Figure 16.15, below). Finally, cricetids make their first appearance in the White River Chronofauna, with the eumyines appearing in the Chadronian, and the eucricetodontines in the Orellan, with further diversification in the Arikareean. During this time the cricetodontines are mainly found in the Central Great Plains, while the eucricetodontines are also found in the California Central and Coast, Northern Great Plains, and Pacific Northwest regions (Figure 16.14, below).

RUNNINGWATER CHRONOFAUNA

This short time period, representing the earliest Miocene (late Arikareean through early Hemingfordian), was noted as a distinctive faunal unit for large mammals in Volume 1 (see also Webb and Opdyke, 1995). Among the carnivores, this represents the time of the "cat gap" (Van Valkenburgh, 1991), when many non-feloid carnivorans evolved adaptations for hypercarnivory (see also Hunt and Tedford, 1993). Among the ungulates, this marks a rather curious time when animals of an open habitat type were present, such as parahippine horses and stenomyline camelids, although oreodonts remained abundant prior to the later (late Hemingfordian) diversification of

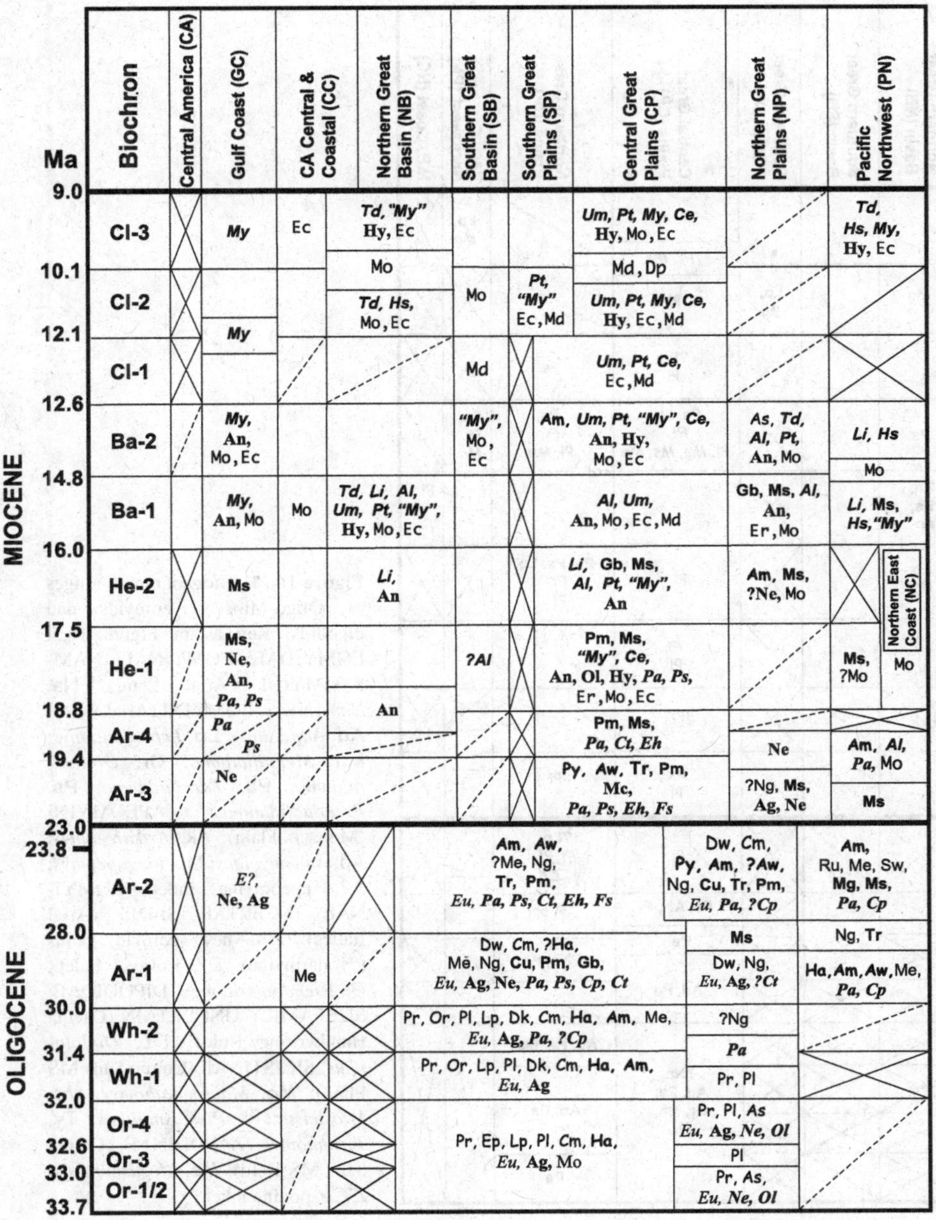

Ma	Biochron	Central America (CA)	Gulf Coast (GC)	CA Central & Coastal (CC)	Northern Great Basin (NB)	Southern Great Basin (SB)	Southern Great Plains (SP)	Central Great Plains (CP)	Northern Great Plains (NP)	Pacific Northwest (PN)
9.0	Cl-3		My	Ec	Td, "My" Hy, Ec			Um, Pt, My, Ce, Hy, Mo, Ec		Td, Hs, My, Hy, Ec
10.1	Cl-2		My		Mo; Td, Hs, Mo, Ec	Mo	Pt, "My" Ec, Md	Md, Dp; Um, Pt, My, Ce, Hy, Ec, Md		
12.1	Cl-1					Md		Um, Pt, Ce, Ec, Md		
12.6	Ba-2		My, An, Mo, Ec			"My", Mo, Ec		Am, Um, Pt, "My", Ce, An, Hy, Mo, Ec	As, Td, Al, Pt, An, Mo	Li, Hs; Mo
14.8	Ba-1		My, An, Mo	Mo	Td, Li, Al, Um, Pt, "My", Hy, Mo, Ec			Al, Um, An, Mo, Ec, Md	Gb, Ms, Al, An, Er, Mo	Li, Ms, Hs, "My"
16.0	He-2		Ms		Li, An			Li, Gb, Ms, Al, Pt, "My", An	Am, Ms, ?Ne, Mo	Northern East Coast (NC)
17.5	He-1		Ms, Ne, An, Pa, Ps			?Al		Pm, Ms, "My", Ce, An, Ol, Hy, Pa, Ps, Er, Mo, Ec		Ms, ?Mo; NC: Mo
18.8	Ar-4		Pa; Ps		An			Pm, Ms, Pa, Ct, Eh	Ne	Am, Al, Pa, Mo
19.4	Ar-3		Ne					Py, Aw, Tr, Pm, Mc, Pa, Ps, Eh, Fs	?Ng, Ms, Ag, Ne	Ms
23.0 / 23.8	Ar-2		E?, Ne, Ag			Am, Aw, ?Me, Ng, Tr, Pm, Eu, Pa, Ps, Ct, Eh, Fs			Dw, Cm, Py, Am, ?Aw, Ng, Cu, Tr, Pm, Eu, Pa, ?Cp	Am, Ru, Me, Sw, Mg, Ms, Pa, Cp
28.0	Ar-1			Me		Dw, Cm, ?Ha, Me, Ng, Cu, Pm, Gb, Eu, Ag, Ne, Pa, Ps, Cp, Ct			Ms; Dw, Ng, Eu, Ag, ?Ct	Ng, Tr; Ha, Am, Aw, Me, Pa, Cp
30.0	Wh-2					Pr, Or, Pl, Lp, Dk, Cm, Ha, Am, Me, Eu, Ag, Pa, ?Cp			?Ng; Pa	
31.4	Wh-1					Pr, Or, Lp, Pl, Dk, Cm, Ha, Am, Eu, Ag			Pr, Pl	
32.0	Or-4					Pr, Ep, Lp, Pl, Cm, Ha, Eu, Ag, Mo			Pr, Pl, As, Eu, Ag, Ne, Ol	
32.6	Or-3								Pl	
33.0–33.7	Or-1/2								Pr, As, Eu, Ne, Ol	

MIOCENE (9.0–23.0 Ma); OLIGOCENE (23.0–33.7 Ma)

Figure 16.10. Biogeographic ranges of Oligo-Miocene aplodontids and castorids. Key as in Figure 16.5. APLODONTIDAE: BASAL APLODONTIDS (Comic Sans Ms Plain): As, *Ansomys*; Cm, *Campestrallomys*; Dk, *Dakotallomys*; Dw, *Downsimus*; Ep, *Epeiromys*; Lp, *Leptoromys*; Or, *Oropyctis*; Pl, *Pelycomys*; Pr, *Prosciurus*. ALLOMYINAE (Comic Sans MS Bold): **Am**, *Allomys*; **Aw**, *Alwoodia*; **Ha**, *Haplomys*; **Py**, *Parallomys*. APLODONTINAE (Comic Sans MS Bold Italics): ***Li***, *Liodontia*; ***Td***, *Tardontia*. MENISCOMYINAE (Arial Plain): Me, *Meniscomys*; Ng, *Niglarodon*; Ru, *Rudiomys*; Sw, *Sewelleladon*. MYLAGAULIDAE: PROMYLAGAULINAE (Arial Bold): **Cu**, *Crucimys*; **Gb**, *Galbreathia*; **Mg**, *Mylagaulodon*; **Ms**, *Mesogaulus*; **Pm**, *Promylagaulus*; **Tr**, *Trilaccogaulus*. MYLAGAULINAE (Arial Bold Italics): ***Al***, *Alphagaulus*; ***Ce***, *Ceratogaulus*; ***Hs***, *Hesperogaulus*; ***My***, *Mylagaulus*; ***"My"***, "*Mylagaulus*"; ***Pt***, *Pterogaulus*; ***Um***, *Umbogaulus*. CASTOROIDAE: EUTYPOMYIDAE (Times New Roman Italics): *Eu*, *Eutypomys*; *E*, Eutypomid indet. CASTORIDAE: BASAL CASTORIDAE (Times New Roman Bold): **Ag**, *Agnotocastor*; **An**, *Anchitheriomys*; **Hy**, *Hystricops*; **Ol**, *Oligotheriomys*; **Mc**, *Migmacastor*; **Ne**, *Neatocastor*. PALAEOCASTORINAE (Times New Roman Bold Italics): ***Cp***, *Capacikala*; ***Ct***, *Capatanka*; ***Eh***, *Euhapsis*; ***Fs***, *Fossorcastor*; ***Pa***, *Palaeocastor*; ***Ps***, *Pseudopalaeocastor*. CASTOROIDINAE (Monaco Plain): Dp, *Dipoides*; Ec, *Eucastor*; Er, *Euroxenomys*; Md, *Microdipoides*; Mo, *Monosaulax*.

hypsodont ungulates such as merychippine horses and antilocaprid ruminants. Strömberg (2006) noted a paradox in paleoenvironmental evidence at this time: grasslands were present, but hypsodont, grazing ungulates had yet to radiate (although stenomyline camels were hypsodont, their narrow muzzles clearly show that they were not grazers). As noted previously, hypsodont Glires were already diverse by this time. Whether or not the "Runningwater" interval represents a clear chronofauna for small mammals remains to be determined.

Among the squirrels, the cedromurines are now extinct, and there is now a diversification of basal ground squirrels, plus the appearance of the marmotine *Miospermophilus*, but this radiation commences in the late Oligocene rather than the early Miocene (Figure 16.9). Basal aplodontids and meniscomyines are now extinct, and there is a diversification of allomyine and promylagauline aplodontids in this general time interval, but this also commences in the late Oligocene, and none of these taxa persists past the early Barstovian. The beavers hold more promise for defining this time interval: the fossorial palaeocastorines, which mainly first appear in the early Arikareean, survive until the end of the Arikareean or into the early Hemingfordian but do not persist into the late Hemingfordian. Basal castorids, such as *Agnotocastor*, *Neatocastor*, and *Oligotheriomys*, also become extinct around this time, and although a few castoroidines are present in this time interval, their radiation is mainly in the later Miocene (Figure 16.10). Only one dipodid, *Plesiosminthus*, is known from this time interval, and the eomyids are mainly represented by the continuation of earlier eomyines, although

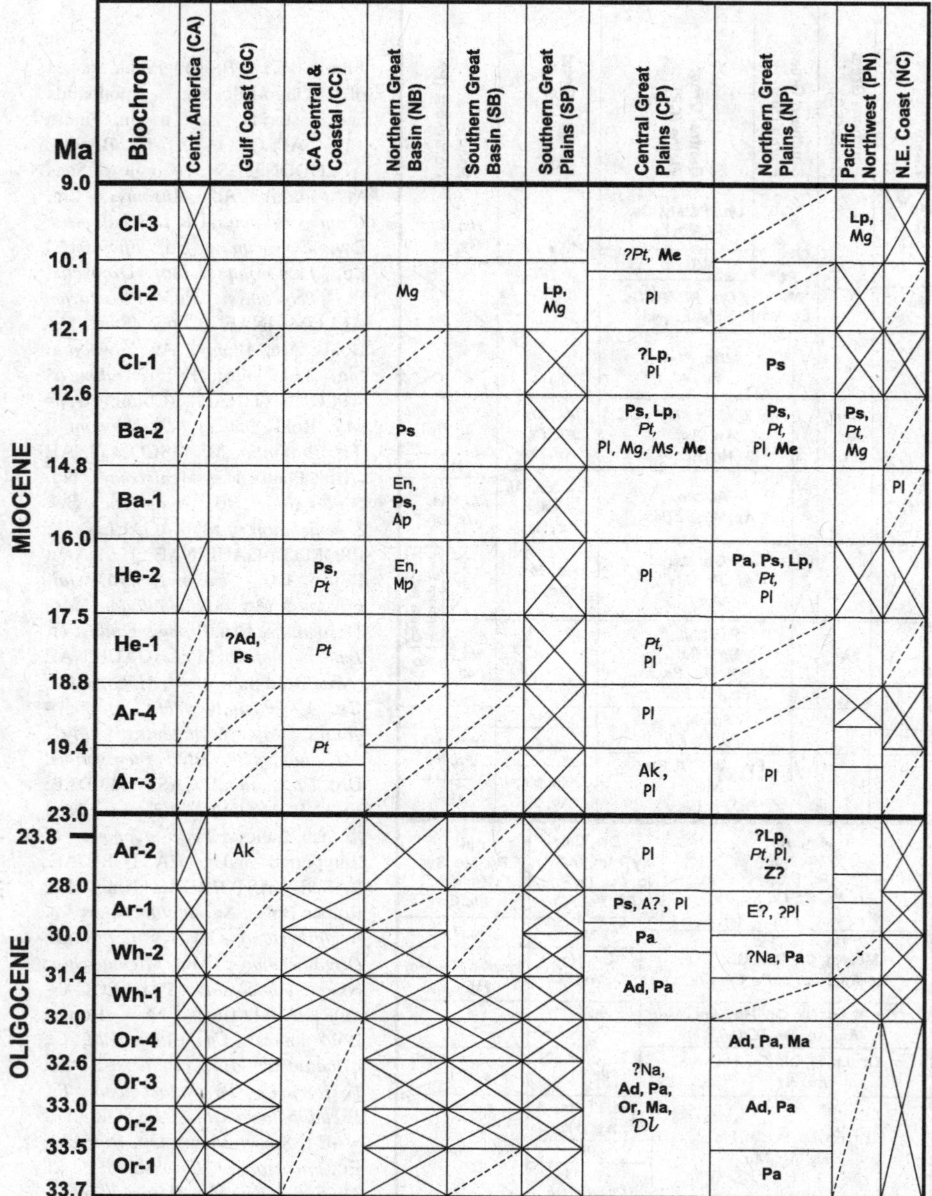

Figure 16.11. Biogeographic ranges of Oligo-Miocene eomyids and dipodids. Key as in Figure 16.5. EOMYIDAE: EOMYINAE: NAMATOMYINI (Arial Plain): Na, *Namatomys*. EOMYINI (Arial Bold): **Ad**, *Adjidaumo*; **Lp**, *Leptodontomys*; **Ma**, *Metadjidaumo*; **Or**, *Orelladjidaumo*; **Pa**, *Paradjidaumo*; **Ps**, *Pseudadjidaumo*. APEOMYINI (Monaco Plain): Ak, *Arikareeomys*; Ap, *Apeomyoides*; Mp, *Megapeomys*; A?, apeomyine indet. EOMYINAE INCERTAE SEDIS (Arial Italics): *En*, new eomyid genus (?yodermyine); *E?*, Eomyid indet.; *Pt*, *Pseudotheridomys*. DIPODIDAE: SUBFAMILY UNCERTAIN (Lucida Handwriting Bold): DL, *Diplolophus*. SICISTINAE (Comic Sans MS Plain): Mg, *Macrognathomys*; Ms, *Miosicista*; Pl, *Plesiosminthus*; Ty, *Tyrannomys*. ZAPODINAE (Comic Sans MS Bold): **Me**, *Megasminthus*; **Z?**, zapodine indet.

Pseudotheridomys appears in the late Arikareean (but continues into the Barstovian) (Figure 16.11).

Among the Geomorpha, several florentiamyids and the genus *Dikkomys* are characteristic of this interval and become extinct at the end of the early Hemingfordian. This is also true for the entoptychine geomyids, with the exception of *Gregorymys*, which persists through the Barstovian (Figure 16.12). Among the heteromyids, primitive taxa and mioheteromyines continue their Oligocene diversification and range through this interval, mainly becoming extinct at the end of the Barstovian (Figure 16.13). Several cricetids, such as the eucricetodontines *Leidymys, Paciculus*, and *Yatkolamys*, and the cricetodontine "*Deperetomys*," span this interval, and the last two taxa are confined to it (Figure 16.14). Among the lagomorphs, this interval is characterized by the continuation of palaeolagines such as *Megalagus*, and the primitive archaeolagine *Archaeolagus*, but precedes the radiation of more derived archaeolagines and leporids, although *Hypolagus* first appears in the early Hemingfordian. Ochotonids first appear in the early Arikareean: a couple of genera span the Runningwater interval, but further diversification occurs later (Figure 16.15).

CLARENDONIAN CHRONOFAUNA

Among ungulates, this time interval represents the radiation of the great "savanna fauna" (Webb), and first commences in the late Hemingfordian (see Janis, Damuth, and Theodor, 2000), ending

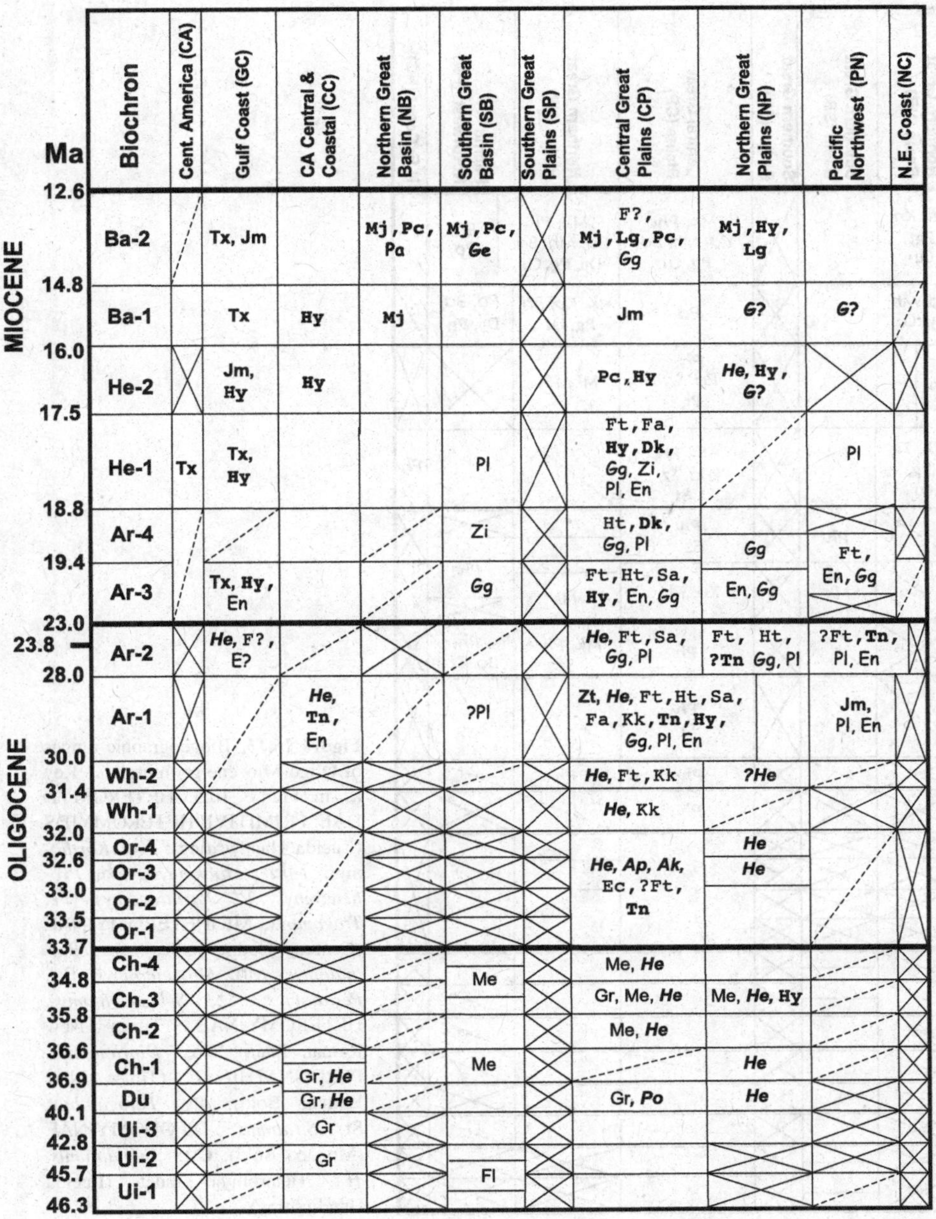

Epoch	Ma	Biochron	Cent. America (CA)	Gulf Coast (GC)	CA Central & Coastal (CC)	Northern Great Basin (NB)	Southern Great Basin (SB)	Southern Great Plains (SP)	Central Great Plains (CP)	Northern Great Plains (NP)	Pacific Northwest (PN)	N.E. Coast (NC)
MIOCENE	12.6–14.8	Ba-2		Tx, Jm		Mj, Pc, Pa	Mj, Pc, Ge		F?, Mj, Lg, Pc, Gg	Mj, Hy, Lg		
	14.8–16.0	Ba-1		Tx	Hy	Mj			Jm	G?	G?	
	16.0–17.5	He-2		Jm, Hy	Hy				Pc, Hy	He, Hy, G?		
	17.5–18.8	He-1	Tx	Tx, Hy			Pl		Ft, Fa, Hy, Dk, Gg, Zi, Pl, En		Pl	
	18.8–19.4	Ar-4					Zi		Ht, Dk, Gg, Pl	Gg	Ft, En, Gg	
	19.4–23.0	Ar-3		Tx, Hy, En			Gg		Ft, Ht, Sa, Hy, En, Gg	En, Gg		
OLIGOCENE	23.0–28.0 (23.8)	Ar-2		He, F?, E?					He, Ft, Sa, Gg, Pl	Ft, Ht, ?Tn Gg, Pl	?Ft, Tn, Pl, En	
	28.0–30.0	Ar-1			He, Tn, En		?Pl		Zt, He, Ft, Ht, Sa, Fa, Kk, Tn, Hy, Gg, Pl, En		Jm, Pl, En	
	30.0–31.4	Wh-2							He, Ft, Kk	?He		
	31.4–32.0	Wh-1							He, Kk			
	32.0–32.6	Or-4								He		
	32.6–33.0	Or-3							He, Ap, Ak, Ec, ?Ft, Tn	He		
	33.0–33.5	Or-2										
	33.5–33.7	Or-1										
	33.7–34.8	Ch-4					Me		Me, He			
	34.8–35.8	Ch-3							Gr, Me, He	Me, He, Hy		
	35.8–36.6	Ch-2							Me, He			
	36.6–36.9	Ch-1			Gr, He		Me			He		
	36.9–40.1	Du			Gr, He				Gr, Po	He		
	40.1–42.8	Ui-3			Gr							
	42.8–45.7	Ui-2			Gr		Fl					
	45.7–46.3	Ui-1										

Figure 16.12. Biogeographic ranges of Eocene–Miocene geomorphs 1. Key as in Figure 16.5. GEOMORPHA: INCERTAE SEDIS (Arial Plain): Gr, *Griphomys*; Fl, *Floresomys*; Me, *Meliakrouniomys*. JIMOMYIDAE (Arial Bold): **Jm**, *Jimomys*; **Tx**, *Texomys*; **Zt**, *Zetamys*. HELISCOMYIDAE (Arial Bold Italics): ***Ak***, *Akmaiomys*: ***Ap***, *Apletotomeus*; ***He***, *Heliscomys*; ***Po***, *Passaliscomys*. FLORENTIAMYIDAE (Courier Plain): Ec, *Ecclesimus*; Fa, *Fanimus*; Ft, *Florentiamys*; F?, Florentiamyid indet.; Ht, *Hitonkala*; Kk, *Kirkomys*; Sa, *Sanctimus*. GEOMYOIDEA INCERTAE SEDIS (Courier Bold): **Dk**, *Dikkomys*; **Hy**, *Harrymys*; **Lg**, *Lignimus*; **Mj**, *Mojavemys*; **Pc**, *Phelosaccomys*; **Tn**, *Tenudomys*. GEOMYIDAE: ENTOPTYCHINAE (Comic Sans MS Plain): En, *Entoptychus*; E?, Entoptychine indet.; *Gg*, *Gregorymys*; Pl, *Pleurolicus*; Zi, *Ziamys*. GEOMYINAE (Comic Sans MS Bold): **Ge**, *Geomys*; **G?**, Geomyid indet.; **Pa**, *Parapliosaccomys*.

with the end of the Clarendonian and the extinction of many browsing taxa.

Among the squirrels, there is a continuation of the radiation of basal terrestrial squirrels, which does not survive the end of the Clarendonian. *Tamias, Sciurus*, and several marmotines also make their first appearance during this time, although they are scarce or absent from California Central and Coast and Southern Great Basin Faunas (Figure 16.9). This interval marks the radiation of the aplodontine and mylagauline aplodontids, some of which survive into the Hemphillian, but their maximum diversity is in the Barstovian and Clarendonian. Mylagaulines are at their maximum diversity in the Clarendonian, with five genera. The allomyines and promylagaulines are now mainly extinct (none survives past the early Barstovian). The mylagaulines are large-sized rodents, and they now take over the burrowing niche from the palaeocastorine beavers. The castoroidine beavers, however, diversify during this interval: *Euroxenomys, Monosaulax*, and *Microdipoides* (the last also fossorial) span this interval and become extinct by the end of the Clarendonian (Figure 16.10). Eomyids are rare during this time interval, comprising the eomyine *Leptodontomys* (mainly confined to the Northern Great Plains and Pacific Northwest) and the apeomyines *Megapeomys* and *Apeomyoides* (confined to the Northern Great Basin). Dipodids are also rare, mainly comprising sicistines such as *Plesiosminthus* and *Macrognathomys*, which are mainly restricted

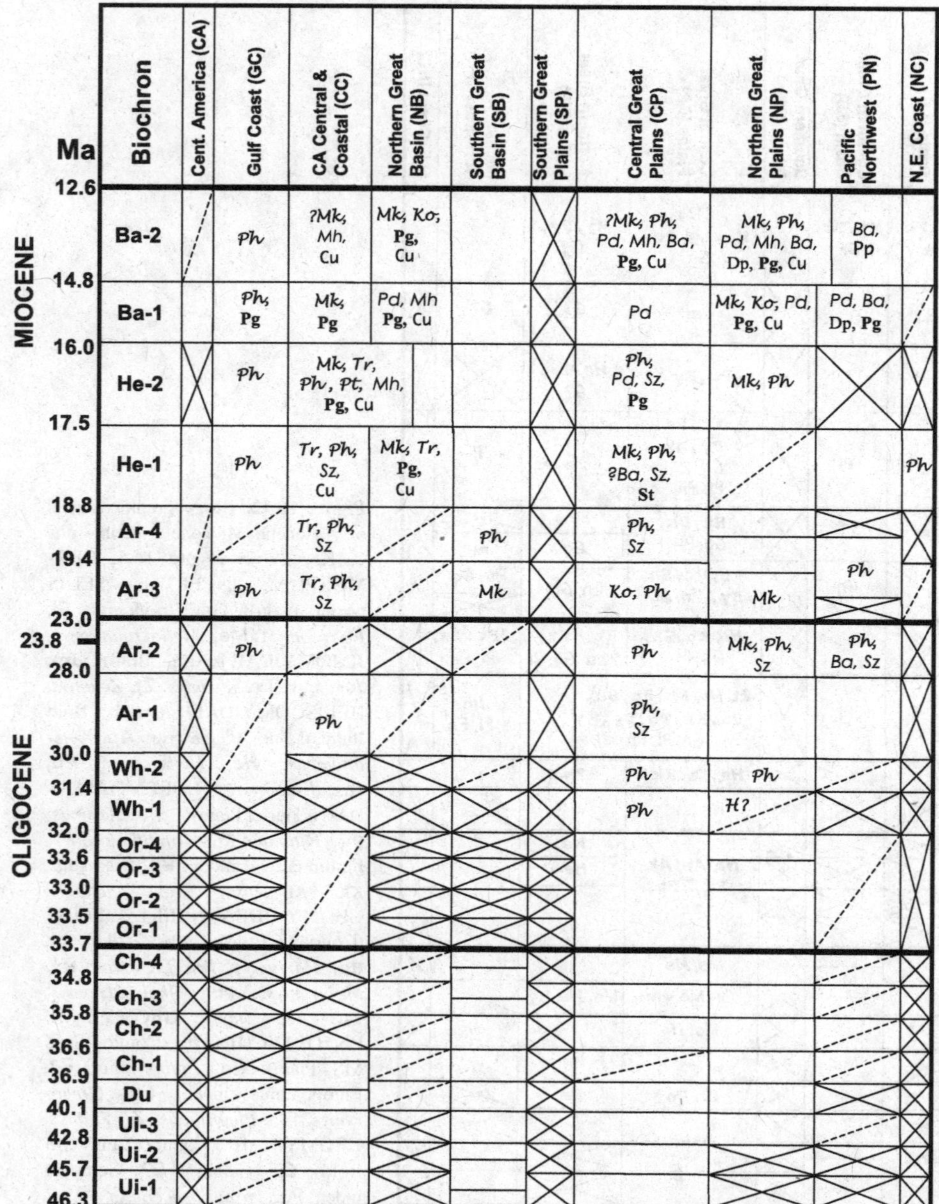

Figure 16.13. Biogeographic ranges of Oligo-Miocene geomorphs 2. Key as in Figure 16.5. HETEROMYIDAE: PRIMITIVE HETEROMYIDS (Lucida Handwriting): Ko, *Korthomys*; Mk, *Mookomys*; Ph, *Proheteromys*; Pt, *Paratrogomys*; Tr, *Trogomoys*. MIOHETEROMYINAE (Century Gothic italics): Ba, *Balantiomys*; Mh, *Mioheteromys*; Pd, *Peridiomys*; Sz, *Schizodontomys*. HETEROMYINAE (Times New Roman Plain): Dp, *Diprionomys*. PEROGNATHINAE (Times New Roman Bold): **Pg**, *Perognathus*; **St**, *Stratimus*. DIPODOMYINAE (Monaco Bold): **Cu**, *Cupidinimus*. H?, Heteromyid indet. (Lucida Handwriting).

to the Central Great Plains. However, the first member of the ricochetal zapodines, *Megasminthus*, appears in the late Barstovian (Figure 16.11).

More primitive geomorphs are now mainly extinct, although some jimomyids persist until the end of the Barstovian (mainly confined to the Gulf Coast region), and some taxa of uncertain affinity, such as *Phelosaccomys* and *Lignimus*, extend through most of this interval. Entoptychine geomyids are now mainly extinct, and some more primitive members of the geomyines, such as *Geomys*, first appear in the late Barstovian. Note that geomyids are absent from California Central and Coast, and largely absent from the Gulf Coast region (Figure 16.12 and Figure 16.19, below). Among the heteromyids, primitive taxa and mioheteromyines, which are been abundant in Gulf Coast and Southern Great Basin regions, become largely extinct by the end of the Barstovian. However, some more derived taxa, such as *Diprionomys*, *Perognathus*, and *Cupidinimus*, appear during this time (Figure 16.13 and Figure 16.19, below). Cricetids are not very diverse during this time: the cricetine *Copemys* (first appearing in the late Arikareean) is common, although it is absent from Gulf Coast Faunas until the late Clarendonian. Some more derived taxa, such as *Peromyscus* and *Prosigmodon*, first appear in the Clarendonian but are mainly confined to the Central Great Plains during this time (Figure 16.14). The archaeolagine *Hypolagus* is the most common lagomorph during this time interval: *Archaeolagus* is extinct by the end of the Hemingfordian, but some leporids, such as *Pronotolagus* and the hare-sized *Alilepus*, appear

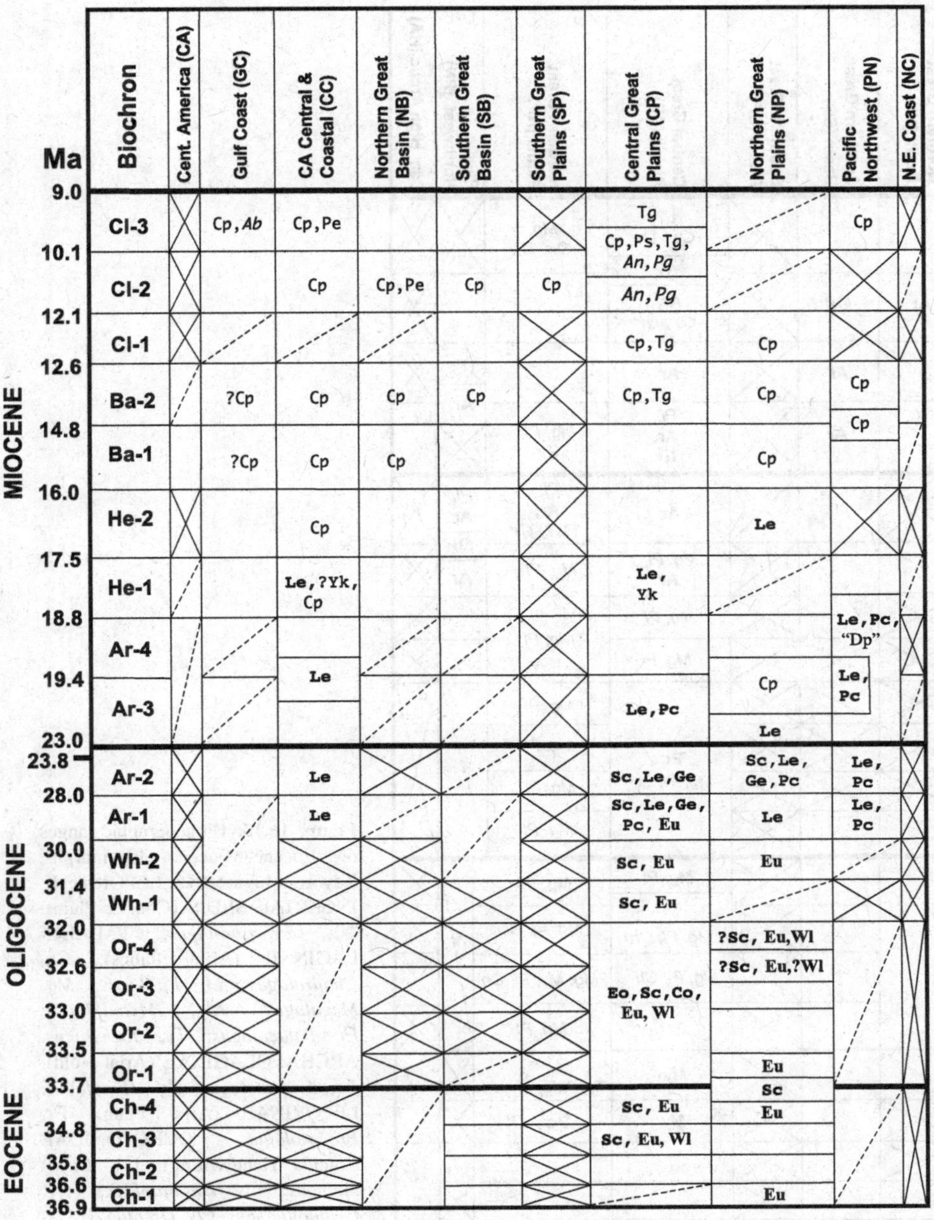

Epoch	Ma	Biochron	Cent. America (CA)	Gulf Coast (GC)	CA Central & Coastal (CC)	Northern Great Basin (NB)	Southern Great Basin (SB)	Southern Great Plains (SP)	Central Great Plains (CP)	Northern Great Plains (NP)	Pacific Northwest (PN)	N.E. Coast (NC)
MIOCENE	9.0	Cl-3		Cp, *Ab*	Cp, Pe				Tg Cp, Ps, Tg, *An*, *Pg*		Cp	
	10.1	Cl-2			Cp	Cp, Pe	Cp	Cp	*An*, *Pg*			
	12.1	Cl-1							Cp, Tg	Cp		
	12.6	Ba-2		?Cp	Cp	Cp	Cp		Cp, Tg	Cp	Cp	
	14.8	Ba-1		?Cp	Cp	Cp				Cp	Cp	
	16.0	He-2			Cp					Le		
	17.5	He-1			Le, ?Yk, Cp				Le, Yk			
	18.8	Ar-4			Le						Le, Pc, "Dp"	
	19.4	Ar-3							Le, Pc	Cp Le	Le, Pc	
OLIGOCENE	23.0 23.8	Ar-2			Le				Sc, Le, Ge	Sc, Le, Ge, Pc	Le, Pc	
	28.0	Ar-1			Le				Sc, Le, Ge, Pc, Eu	Le	Le, Pc	
	30.0	Wh-2							Sc, Eu	Eu		
	31.4	Wh-1							Sc, Eu			
	32.0	Or-4								?Sc, Eu, Wl		
	32.6	Or-3							Eo, Sc, Co, Eu, Wl	?Sc, Eu, ?Wl		
	33.0	Or-2										
	33.5	Or-1								Eu		
EOCENE	33.7	Ch-4							Sc, Eu	Sc Eu		
	34.8	Ch-3							Sc, Eu, Wl			
	35.8	Ch-2										
	36.6	Ch-1								Eu		
	36.9											

Figure 16.14. Biogeographic ranges of Eocene–Miocene mice. Key as in Figure 16.5. CRICETIDAE: EUCRICETODONTINAE: EUCRICETODONINI (Courier Bold): **Co**, *Coloradoeumys*; **Eo**, *Eoeumys*; **Ge**, *Geringia*; **Le**, *Leidymys*; **Pc**, *Paciculus*; **Sc**, *Scottimus*. INCERTAE SEDIS (Courier Plain); Yk, *Yatkolamys*. CRICETODONTINAE: EUMYINI (Times New Roman Bold): **Eu**, *Eumys*; **Wl**, *Wilsoneumys*. CRICETODONTINI (Times New Roman Plain): "Dp", "*Deperetomys*." CRICETINAE: DEMOCRICETODONTINI (Monaco Bold): **Cp**, *Copemys*; **Pe**, *Peromyscus*; **Ps**, *Pseudomyscus*. MEGACRICETODONTINI (Monaco Plain): Tg, *Tregomys*. SIGMODONTINI (Monaco Bold Italics): ***Ab***, *Abelmoschomys*; ***An***, *Antecalomys*; ***Pg***, *Prosigmodon*.

in the Barstovian, as do several ochotonid genera. Ochotonids are at their most diverse during this time interval, with four genera in the late Hemingfordian, and three in the mid to late Clarendonian. Note that the ochotonids are absent from faunas in California Central and Coast (except for the endemic Hemingfordian *Cuyamalagus*), and the Gulf Coast, and they are rarely found in the Southern Great Basin (Figure 16.15 and Figure 16.21, below).

MIO-PLIOCENE CHRONOFAUNA

Among the squirrels, this time interval represents the maximum diversity of the marmotines, with four to five genera present. The flying squirrel (petauristine) *Cryptopterus* is also apparent in Gulf Coast faunas. Other squirrels are absent from this region and are only sparsely represented in the California Central and Coast region (Figure 16.16). Aplodontids (both aplodontines and mylagaulines) persist into the Hemphillian but are extinct by the end of the Miocene (with the exception of the extant taxon *Aplodontia*, the mountain beaver, known only from the Pacific Northwest). The mountain beaver is limited in its habitat by its dependence on water (McNab, 2002): if this was true for aplodontids in general, then their disappearance from the rest of North America in the Pliocene may have relevance for the interpretation of environmental conditions. The beavers are now represented by the modern castorine *Castor* and four castoroidine genera, three of which were of large size (including the extremely large *Procastoroides* and *Castoroides*). The

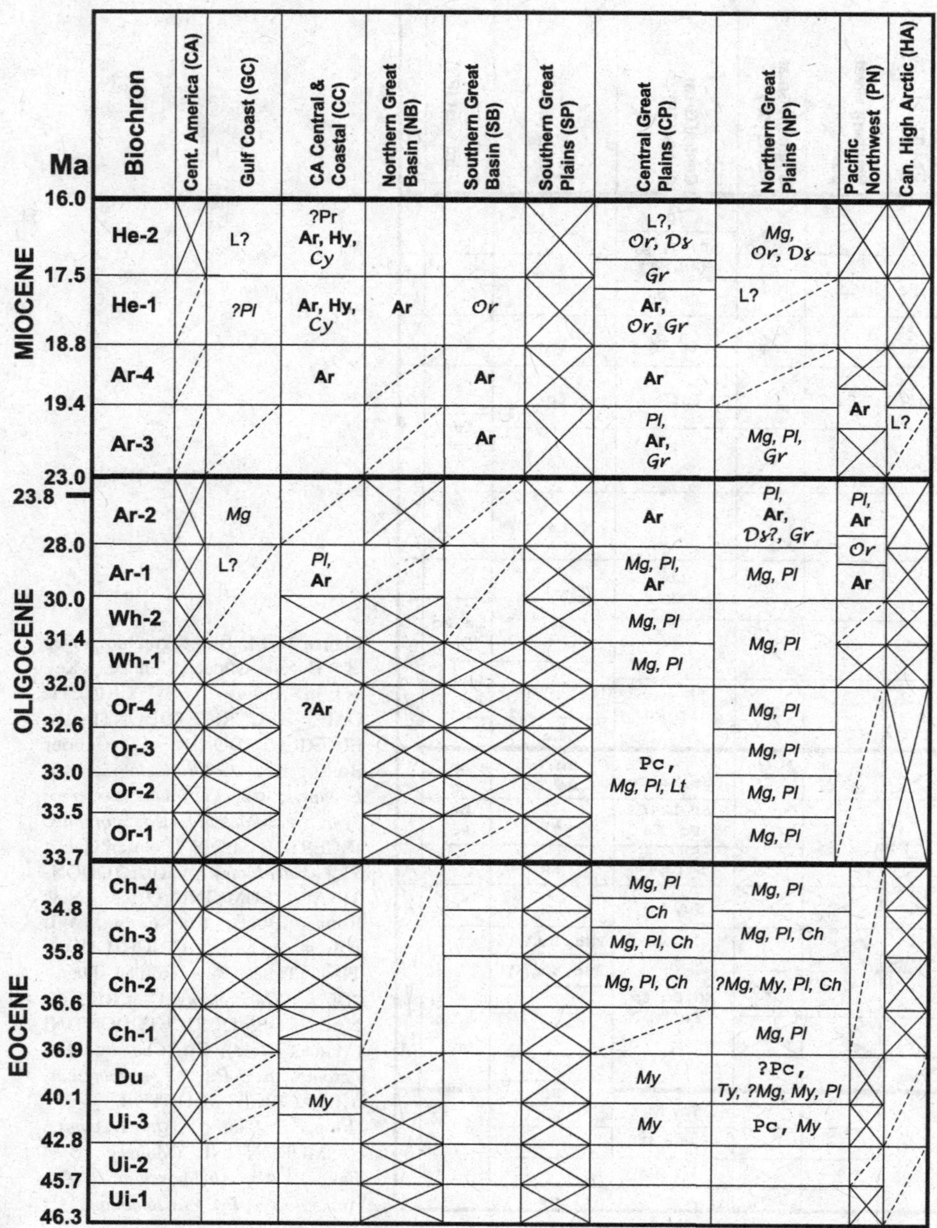

Figure 16.15. Biogeographic ranges of Eocene–Miocene lagomorphs. Key as in Figure 16.5. LEPORIDAE: INCERTAE SEDIS (Courier Plain): Pc, *Procaprolagus*. PALAEOLAGINAE (Arial Italics): *Ch*, *Chadrolagus*; *Lt*, *Litolagus*; *Mg*, *Megalagus*; *My*, *Mytonolagus*; *Pl*, *Palaeolagus*; *Ty*, *Tachylagus*. ARCHAEOLAGINAE (Arial Bold): **Ar**, *Archaeolagus*; **Hy**, *Hypolagus*. LEPORINAE (Arial Plain): Pr, *Pronotolagus*. OCHOTONIDAE (Lucida Handwriting): Cy, *Cuyamalagus*; Ds, *Desmatolagus*; Gr, *Gripholagomys*; Or, *Oreolagus*.

beaver-sized *Dipoides* also seems to have been aquatic (Tedford and Harington, 2003; Rybczynski, 2004). Castorids are absent from the Gulf Coast, and Northern and Southern Great Basins until the late Blancan, and only have a single appearance in California Central and Coast faunas (Figure 16.17).

A few eomyids survive into the Hemphillian, including the eomyine *Leptodontomys*, and the Hemphillian is actually the only time of the appearance of the eomyine tribe Ronquilomyini, found mainly in Central and Southern Great Plains Faunas. However, none survives into the Pliocene. The dipodids during this interval are primarily the ricochetal zapodines, and a couple of sicistines, and are mainly found in Central Great Plains and the Pacific Northwest (Figure 16.18). The more primitive geomorphs are now all extinct: Geomorpha during this time interval are represented by the ricochetal heteromyine, perognathine, and dipodomyine heteromyids, and by geomyine geomyids. Heteromyids are absent from the Gulf Coast region, and geomyids are not apparent here until the Pliocene (Figure 16.19).

The cricetids are represented primarily by the cricetines and a new tribe of cricetodonines, the galushamyines, which range from the early Hemphillian to the mid Blancan. Cricetids are virtually absent from the Gulf Coast region until the Pliocene, and sigmodontine cricetines are not found in the California Central and Coast and Northern Great Basin regions during the Hemphillian. Cricetids are at their maximum diversity in the early Blancan with 12 genera, half of which are of small size. A new rodent family, the

Epoch	Ma	Biochron	Cent. America (CA)	Gulf Coast (GC)	CA Central & Coastal (CC)	Northern Great Basin (NB)	Southern Great Basin (SB)	Southern Great Plains (SP)	Central Great Plains (CP)	Northern Great Plains (NP)	Pacific Northwest (PN)	C.E. America (CE)
PLIOCENE	1.75–1.9	L. Bl/ E. Ir		Er, ?Ne, Hy		Sp, Er	Sp, Ma		Sp, Cn			
PLIOCENE	1.9–2.5	L. Bl		Cr, Er, Ne	Am, Er	Sp, Ma	Sp, Cn, Pa, Er, Ne	Sp, Pa	Sp, Cn		Er	
PLIOCENE	2.5–4.8	E. Bl			?Er	"Tm", Sp	Sp, Pa, Pp, Ne	Sp	"Tm", Sp, Pa		Am, Sp, Pa	
MIOCENE	5.3–5.8	Hh-4		Cr	Sp		Sp, Pa, Ne	Sp	Sp, Ma, Pa			Sp
MIOCENE	5.8–6.7	Hh-3					Sp	Ma	Sp, Pa		Tm, Sp, Pp	
MIOCENE	6.7–7.5	Hh-2			Sp			Sp	Sp		Tm, Am, Sp	
MIOCENE	7.5–9.0	Hh-1				Sp, Ma, ?Pa	Sp	Sp	Sp	Sp	Sp	

Figure 16.16. Biogeographic ranges of Mio-Pliocene sciurids and caviomorphs. Key as in Figure 16.5. SCIURIDAE: PETAURISTINAE (Times New Roman Bold): **Cr**, *Cryptopterus*. SCIURINAE: TAMIINI (Comic Sans MS Bold): **Tm**, *Tamias*; **"Tm"**, "*Tamias*." MARMOTINI (Arial Bold): **Am**, *Ammospermophilus*; **Cn**, *Cynomys*; **Ma**, *Marmota*; **Pa**, *Paenemarmota*; **Pp**, *Parapaenemarmota*; **Sp**, *Spermophilus*. CAVIOMORPHA (Lucida Handwriting): ERETHIZONTIDAE: *Er*, *Erethizon*. HYDROCHAERIDAE: *Hy*, *Hydrochaeris*; *Ne*, *Neochoerus*.

Epoch	Ma	Biochron	Cent. America (CA)	Gulf Coast (GC)	CA Central & Coastal (CC)	Northern Great Basin (NB)	Southern Great Basin (SB)	Southern Great Plains (SP)	Central Great Plains (CP)	Central East America (CE)	Northern Great Plains (NP)	Pacific Northwest (PN)	High Arctic (HA)
PLIOCENE	1.75–1.9	L. Bl/ E. Ir		Cd		Ca	Ca						
PLIOCENE	1.9–2.5	L. Bl		Ca			Ca		Ca, Pc			Ca, Pc	
PLIOCENE	2.5–4.8	E. Bl						Pc	Dp, Pc	Ca		Ca, Dp, Pc	Dp, ?Pc
MIOCENE	5.3–5.8	Hh-4		Dp			Um, Dp	"My"	Eg, Ca, Dp				
MIOCENE	5.8–6.7	Hh-3						"My", Dp	My, Eg, Um, Ca, Ec, Dp			Li, Hs. Ca, Dp	
MIOCENE	6.7–7.5	Hh-2			Ca		"My", Dp	Um, Dp	Um, Pt, Ca, Ec, Dp			Td, Li, Hs, "My", Hy, Dp	
MIOCENE	7.5–9.0	Hh-1		My		Li, Hs, Dp	Dp	Um, "My"	Eg, ?Ec, Dp		Dp	Li	

Figure 16.17. Biogeographic ranges of Mio-Pliocene aplodontids and castorids. Key as in Figure 16.5. APLODONTIDAE: APLODONTINAE (Comic Sans MS Bold Italics): ***Li***, *Liodontia*; ***Td***, *Tardontia*. MYLAGAULINAE (Arial Bold Italics): ***Eg***, *Epigaulus*; ***Hs***, *Hesperogaulus*; ***My***, *Mylagaulus*; ***"My"***, "*Mylagaulus*"; ***Pt***, *Pterogaulus*; ***Um***, *Umbogaulus*. CASTORIDAE: BASAL CASTORIDAE (Times New Roman Bold): **Hy**, *Hystricops*. CASTORINAE (Monaco Italics): *Ca*, *Castor*. CASTOROIDINAE (Monaco Bold): **Cd**, *Castoroides*; **Dp**, *Dipoides*; **Ec**, *Eucastor*; **Pc**, *Procastoroides*.

Epoch	Ma	Biochron	Cent. America (CA)	Gulf Coast (GC)	CA Central & Coastal (CC)	Northern Great Basin (NB)	Southern Great Basin (SB)	Southern Great Plains (SP)	Central Great Plains (CP)	Northern Great Plains (NP)	Pacific Northwest (PN)	N.E. Coast (NC)
PLIOCENE	1.75–1.9	L. Bl/ E. Iv		Zp					Ty, Zp			
PLIOCENE	1.9–2.5	L. Bl						Zp	Jz, Zp			
PLIOCENE	2.5–4.8	E. Bl							Zp			
MIOCENE	5.3–5.8	Hh-4					Ro					
MIOCENE	5.8–6.7	Hh-3							Ka, Mg, Me, Pz		Lp, Pz	
MIOCENE	6.7–7.5	Hh-2						Co			Lp, Mg, Pz	
MIOCENE	7.5–9.0	Hh-1				Pz		Ka	Ka			

Figure 16.18. Biogeographic ranges of Mio-Pliocene eomyids and dipodids. Key as in Figure 16.5. EOMYIDAE: EOMYINAE: EOMYINI (Arial Bold): **Lp**, *Leptodontomys*. RONQUILLOMYINI (Monaco Bold): **Co**, *Comancheomys*; **Ka**, *Kansasimys*; **Ro**, *Ronquillomys*. DIPODIDAE: SICISTINAE (Comic Sans MS Plain): Mg, *Macrognathomys*. ZAPODINAE (Comic Sans MS Bold): **Jz**, *Javazapus*; **Pz**, *Pliozapus*; **Zp**, *Zapus*.

Ma		Biochron	W.C. Marine (WM)	Gulf Coast (GC)	CA Central & Coastal (CC)	Northern Great Basin (NB)	Southern Great Basin (SB)	Southern Great Plains (SP)	Central Great Plains (CP)	Northern Great Plains (NP)	Pacific Northwest (PN)	Central East America (CE)
1.75–1.9	PLIOCENE	L. Bl/ E. Ir		Ge, Og, Th	Dd; Th	Pg, Dd, Mp, Ge, Th			Pg, Pp, Dd, Ge			
1.9–2.5	PLIOCENE	L. Bl	Th	Og	Pg, Dd, Th	Ge, Th	Pg, Pp, Dd, Ge, Cr	Pg, Pp, Ge	Pg, Pp, Ge		Dd, Th	
2.5–4.8	PLIOCENE	E. Bl			Pg, Pp, Dd, Re, Th	Or, Pg, ?Cu, Pp, Dd, Mp, Pe, Ge	Pg, Pp, Dd, Ge, Cr	Pg, Pp, Ge	Pg, Pp, Ge		Or, Pg, Pp, Pe, Th	Ge
4.8–5.8 (5.3)	MIOCENE	Hh-4				?Dd	Pg, Pp, Ge	Pg, Pp, Pe, Ge	Pc, Pg, Pp, Pe, Th			
5.8–6.7	MIOCENE	Hh-3			Pg, Cu, Ps	Ps	Pg, Cu, Pp		Pc, Pg, Pp, Ps, Pe		Dp, Or, Pg, Ps	
6.7–7.5	MIOCENE	Hh-2			H?			Pg, Cu, Pp, Pe	Ps		Dp, Or, Pg, Ps	
7.5–9.0	MIOCENE	Hh-1				Pg, Dp, Cu, Ps, ?Th	?Pg	Pg, Ps	Pg, Ps		Or, ?Dd, Ps	
9.0–10.1	MIOCENE	Cl-3			Mj	Pg, Cu, ?Th			Pc, Lg, Mh, Pg, Cu, Ep, G?		Dp, Pg, Ps	
10.1–12.1	MIOCENE	Cl-2			Pg, Cu	Pc, Dp, Pg, Cu, ?Ps, ?Th		Pg	Pc, Mh, Pg, Cu, G?			
12.1–12.6	MIOCENE	Cl-1							Pc, Lg, Mh, ?Pg, Cu			

Figure 16.19. Biogeographic ranges of Mio-Pliocene geomorphs. Key as in Figure 16.5. GEOMORPHA: GEOMYOIDEA INCERTAE SEDIS (Courier Bold): **Lg**, *Lignimus*; **Mj**, *Mojavemys*; **Pc**, *Phelosaccomys*. HETEROMYIDAE: MIOHETEROMYINAE (Century Gothic Italics): *Mh*, *Mioheteromys*. HETEROMYINAE (Times New Roman Plain): Dp, *Diprionomys*: PEROGNATHINAE (Times New Roman Bold): **Pg**, *Perognathus*; **Or**, *Oregonomys*. DIPODOMYINAE (Monaco Bold): **Cu**, *Cupidinimus*; **Dd**, *Dipodomys*; **Ep**, *Eodipodomys*; **Mp**, *Microdipodops*; **Pp**, *Prodipodomys*. **H?**, Heteromyid indet. (Lucida Handwriting). GEOMYIDAE: GEOMYINAE (Comic Sans MS Bold): ***Cr***, *Cratogeomys*; ***Ge***, *Geomys*; ***G?***, Geomyid indet.; ***Og***, *Orthogeomys*; **Pe**, *Pliogeomys*; **Ps**, *Pliosaccomys*; **Re**, *Reynoldsomys*; **Th**, *Thomomys*.

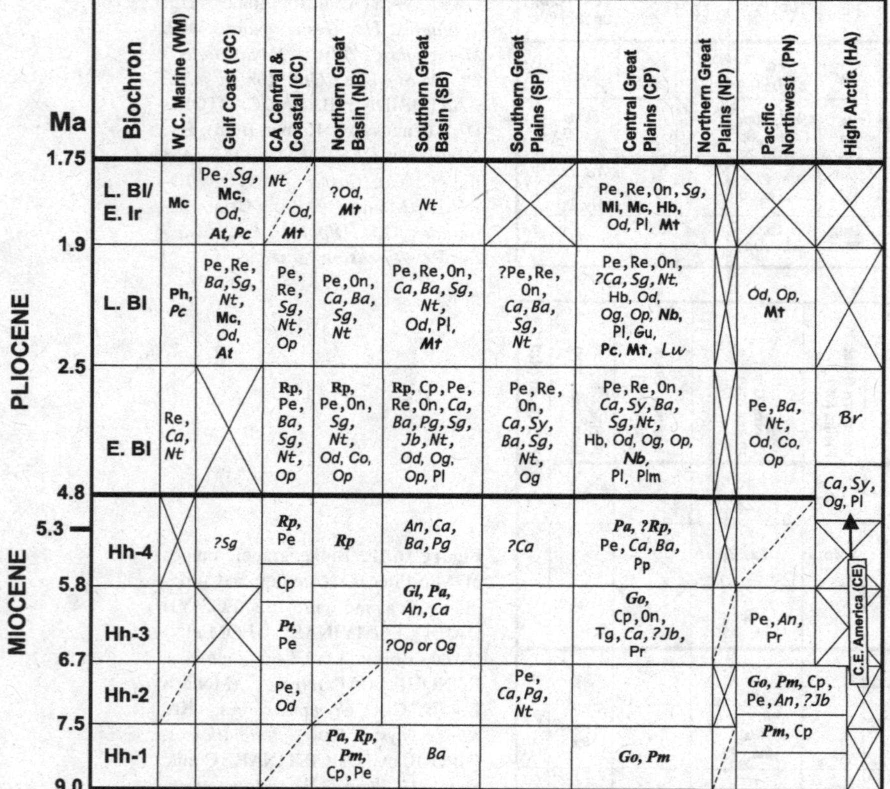

Ma		Biochron	W.C. Marine (WM)	Gulf Coast (GC)	CA Central & Coastal (CC)	Northern Great Basin (NB)	Southern Great Basin (SB)	Southern Great Plains (SP)	Central Great Plains (CP)	Northern Great Plains (NP)	Pacific Northwest (PN)	High Arctic (HA)
1.75–1.9	PLIOCENE	L. Bl/ E. Ir	Mc	Pe, Sg, Mc, Od, At, Pc	Nt; Od, Mt	?Od, Mt	Nt		Pe, Re, On, Sg, Mi, Mc, Hb, Od, Pl, Mt			
1.9–2.5	PLIOCENE	L. Bl	Ph, Pc	Pe, Re, Ba, Sg, Nt, Mc, Od, At	Pe, Re, Sg, Nt, Op	Pe, On, Ca, Ba, Sg, Nt	Pe, Re, On, Ca, Ba, Sg, Nt, Od, Pl, Mt	?Pe, Re, On, Ca, Ba, Sg, Nt	Pe, Re, On, ?Ca, Sg, Nt, Hb, Od, Og, Op, Nb, Pl, Gu, Pc, Mt, Lu		Od, Op, Mt	
2.5–4.8	PLIOCENE	E. Bl	Re, Ca, Nt		Rp, Pe, Ba, Sg, Nt, Op	Rp, Pe, On, Sg, Nt, Od, Co, Op	Rp, Cp, Pe, Re, On, Ca, Ba, Pg, Sg, Jb, Nt, Od, Og, Op, Pl	Pe, Re, On, Ca, Sy, Ba, Sg, Nt, Og	Pe, Re, On, Ca, Sy, Ba, Sg, Nt, Hb, Od, Og, Op, Nb, Pl, Plm		Pe, Ba, Nt, Od, Co, Op	Br; Ca, Sy, Og, Pl
4.8–5.8 (5.3)	MIOCENE	Hh-4		?Sg	Rp, Pe	Rp	An, Ca, Ba, Pg	?Ca	Pa, ?Rp, Pe, Ca, Ba, Pp			C.E. America (CE)
5.8–6.7	MIOCENE	Hh-3			Cp; Pt, Pe		Gl, Pa, An, Ca; ?Op or Og		Go, Cp, On, Tg, Ca, ?Jb, Pr		Pe, An, Pr	
6.7–7.5	MIOCENE	Hh-2			Pe, Od			Pe, Ca, Pg, Nt			Go, Pm, Cp, Pe, An, ?Jb	
7.5–9.0	MIOCENE	Hh-1				Pa, Rp, Pm, Cp, Pe	Ba		Go, Pm		Pm, Cp	

Figure 16.20. Biogeographic ranges of Mio-Pliocene mice. Key as in Figure 16.5. CRICETIDAE: GALUSHAMYINI (Times New Roman Bold Italics): ***Gl***, *Galushamys*; ***Go***, *Goniodontomys*; ***Pa***, *Paronychomys*; ***Pm***, *Paramicrotoscoptes*; ***Pt***, *Pliotomodon* ***Rp***, *Repomys*. CRICETINAE: DEMOCRICETODONTINI (Monaco Bold): **Cp**, *Copemys*; **On**, *Onychomys*; **Pe**, *Peromyscus*; **Re**, *Reithrodontomys*. MEGACRICETODONTINI (Monaco Plain); Tg, *Tregomys*. SIGMODONTINI (Monaco Bold Italics): ***An***, *Antecalomys*; ***Ba***, *Baiomys*; ***Ca***, *Calomys*; ***Jb***, *Jacobsomys*; ***Pg***, *Prosigmodon*; ***Sg***, *Sigmodon*; ***Sy***, *Symmetrodontomys*. NEOTOMINI (Monaco Plain Italics): *Nt*, *Neotoma*. ARVICOLIDAE: PROMIMOMYINAE (Arial Plain): Pr, *Promimomys*. ARVICOLINAE (Arial Bold): **Hb**, *Hibbardomys*; **Mc**, *Microtus*; **Mi**, *Mimomys*; **Ph**, *Phenacomys*. ONDATRINAE (Arial Plain Italics): *Co*, *Cosomys*; *Od*, *Ondatra*; *Og*, *Ogmodontomys*; *Op*, *Ophiomys*. NEBRASKOMYINAE (Arial Bold Italics): ***At***, *Atopomys*; ***Nb***, *Nebraskomys*. PLIOPHENACOMYINAE (Comic Sans MS Plain); Gu, *Guildayomys*; Pl, *Pliophenacomys*; Plm, *Pliolemmus*; Pp, *Protopliophenacomys*. LEMMINAE (Comic Sans MS Bold); **Mt**, *Mictomys*; **Pc**, *Pliophenacomys*. INCERTAE SEDIS (Lucida Handwriting): *Br*, *Baranomys*; *Lu*, *Luopomys*.

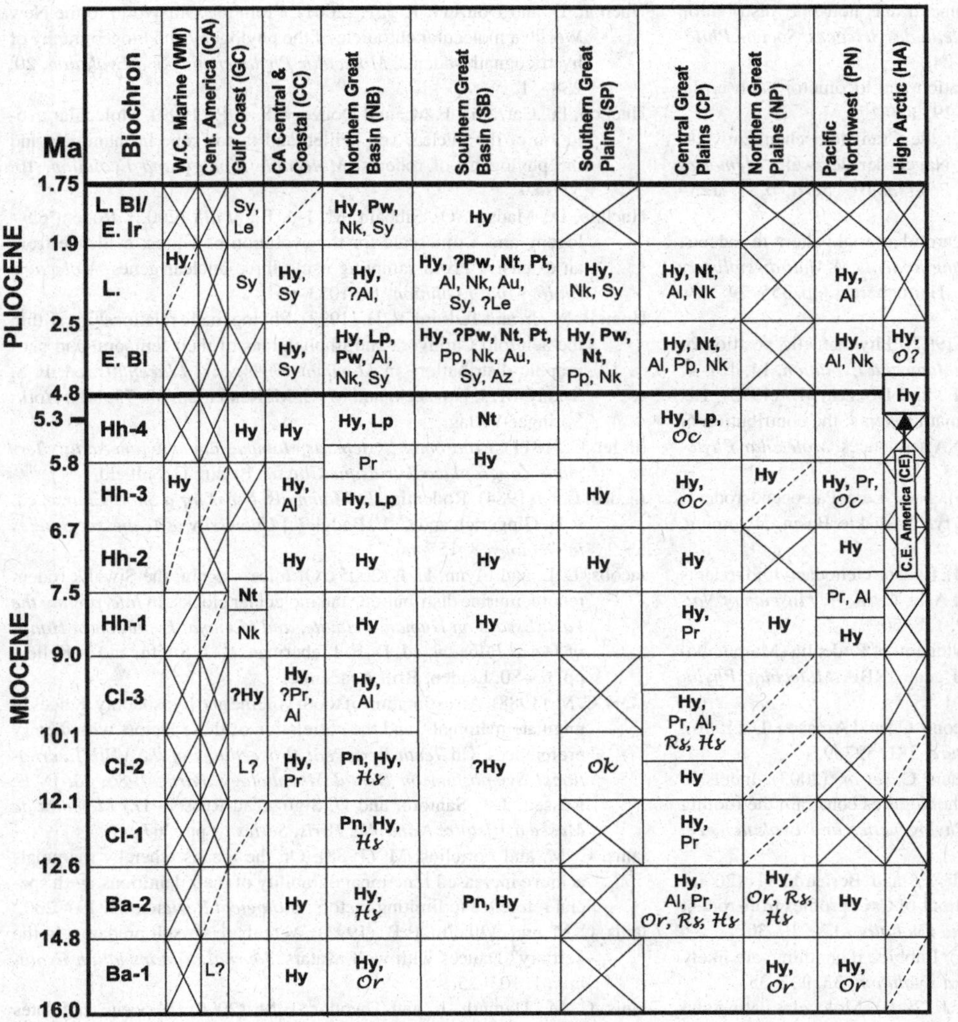

Epoch	Ma	Biochron	W.C. Marine (WM)	Cent. America (CA)	Gulf Coast (GC)	CA Central & Coastal (CC)	Northern Great Basin (NB)	Southern Great Basin (SB)	Southern Great Plains (SP)	Central Great Plains (CP)	Northern Great Plains (NP)	Pacific Northwest (PN)	High Arctic (HA)
PLIOCENE	1.75–1.9	L. Bl/ E. Ir			Sy, Le		Hy, Pw, Nk, Sy	Hy					
PLIOCENE	1.9–2.5	L.	Hy		Sy	Hy, Sy	Hy, ?Al,	Hy, ?Pw, Nt, Pt, Al, Nk, Au, Sy, ?Le	Hy, Nk, Sy	Hy, Nt, Al, Nk		Hy, Al	
PLIOCENE	2.5–4.8	E. Bl				Hy, Sy	Hy, Lp, Pw, Al, Nk, Sy	Hy, ?Pw, Nt, Pt, Pp, Nk, Au, Sy, Az	Hy, Pw, Nt, Pp, Nk	Hy, Nt, Al, Pp, Nk		Hy, Al, Nk	Hy, O? / Hy
MIOCENE	4.8–5.8 (5.3)	Hh-4			Hy	Hy	Hy, Lp / Pr	Nt / Hy		Hy, Lp, Oc			
MIOCENE	5.8–6.7	Hh-3	Hy			Hy, Al	Hy, Lp		Hy	Hy, Oc	Hy	Hy, Pr, Oc	C.E. America (CE)
MIOCENE	6.7–7.5	Hh-2				Hy	Hy	Hy	Hy	Hy		Hy	
MIOCENE	7.5–9.0	Hh-1			Nt / Nk		Hy	Hy		Hy, Pr	Hy	Pr, Al / Hy	
MIOCENE	9.0–10.1	Cl-3			?Hy	Hy, ?Pr, Al	Hy, Al			Hy, Pr, Al, Rs, Hs			
MIOCENE	10.1–12.1	Cl-2			L?	Hy, Pr	Pn, Hy, Hs	?Hy	Ok	Hy			
MIOCENE	12.1–12.6	Cl-1					Pn, Hy, Hs			Hy, Pr			
MIOCENE	12.6–14.8	Ba-2				Hy	Hy, Hs	Pn, Hy		Hy, Al, Pr, Or, Rs, Hs	Hy, Or, Rs, Hs	Hy	
MIOCENE	14.8–16.0	Ba-1		L?		Hy	Hy, Or				Hy, Or	Hy, Or	

Figure 16.21. Biogeographic ranges of Mio-Pliocene lagomorphs. Key as in Figure 16.5. LEPORIDAE: ARCHAEOLAGINAE (Arial Bold): **Hy**, *Hypolagus*; **Lp**, *Lepoides*; **Nt**, *Notolagus*; **Pn**, *Panolax*; **Pt**, *Paranotolagus*; **Pw**, *Pewelagus*. LEPORINAE (Arial Plain): Al, *Alilepus*; Au, *Aluralagus*; Az, *Aztlanolagus*; Le, *Lepus*; L?, leporid indet.; Nk, *Nekrolagus*; Pp, *Pratilepus*; Pr, *Pronotolagus*; Sy, *Sylvilagus*. OCHOTONIDAE (Lucida Handwriting): Hs, *Hesperolagomys*; Oc, *Ochotona*; O?, ochotonid indet.; Ok, *Oklahomalagus*; Or, *Oreolagus*; Rs, *Russellagus*.

Arvicolidae (sometimes considered as a subfamily of cricetids), makes its first appearance at this time, immigrating from Eurasia. Arvicolids first appear in the mid Hemphillian but do not diversify until the Blancan. One early-appearing form is the muskrat, *Ondatra*, which adds another large aquatic rodent to the list of Mio-Pliocene ecomorphological types. Arvicolid maximum diversity is in the late Blancan (including the earliest Irvingtonian), with 13 genera (Figure 16.20).

Another immigrant rodent group in this time interval, this time from South America, is the caviomorphs; these are all large rodents, including the arboreal porcupines and the aquatic capybaras. Capybaras are confined to the Gulf Coast and Southern Great Basin; porcupines are found in the Gulf Coast, but also in the California Central and Coast, Northern Great Basin, and Pacific Northwest (Figure 16.16). Finally, the Mio-Pliocene represents the apogee of lagomorph radiation, with 12 genera present in the early Blancan. Almost half of these genera belong to the subfamily Archaeolaginae, although archaeolagine occurrences are fairly rare during the Pliocene (and archeolagines are absent from the Gulf Coast region during this time), and the subfamily does not survive past the Irvingtonian. The only ochotonid to survive into this time interval is the extant pika, *Ochotona*, confined to Central Great Plains and Pacific Northwest (Figure 16.21).

REFERENCES

Adkins, R. M., Gelke, E. L., Rowe, D., and Honeycutt, R. L. (2001). Molecular phylogeny and divergence time estimates for major rodent groups: evidence from multiple genes. *Molecular Biology and Evolution*, 18, 777–91.

Adkins, R. M., Walton, A. H., and Honeycutt, R. L. (2003). Higher-level systematics of rodents and divergence time estimates based on two congruent nuclear genes. *Molecular Phylogenetics and Evolution*, 26, 409–20.

Archibald, J. D., Averianov, A. O., and Ekdale, E. G. (2001). Late Cretaceous relatives of rabbits, rodents, and other extant eutherian mammals. *Nature*, 414, 62–5.

Biewener, A. A. (1989). Scaling body support in mammals: limb posture and muscle mechanics. *Science*, 245, 45–8.

(1998). Muscle–tendon stresses and elastic energy storage during locomotion in the horse. *Comparative Biochemistry and Physiology B*, 120, 73–87.

Blainville, H. M. D. de (1816). Prodrome d'une nouvelle distribution systématique du règne animal. *Bulletin des Sciences, Société Philomathique de Paris, Série* 3, 3, 105–24.

Bramble, D. M. (1989). Cranial specialization and locomotor habit in the Lagomorpha. *American Zoologist*, 29, 303–17.

Brandt, J. F. (1855). Untersuchungen über die Cranialogischen Entwicklungsufen und Classification der Nager der Jetzwelt. *Mémoires de l'Acadamie Impériale de St. Pétersbourg, Série 6*, 7, 125–336.

Bugge, J. (1985). Systematic value of the carotid arterial pattern in rodents. In *Evolutionary Relationships Among Rodents: A Multidisciplinary Analysis*, ed. W. P. Luckett and J.-L. Hartenberger, pp. 355–79. New York: Plenum Press.

Cao, Y., Okada, N., and Hasegawa, M. (1997). Phylogenetic position of guinea pigs revisited. *Molecular Biology and Evolution*, 14, 461–4.

Catzeflis, F. M., Hänni, C., Sourrouille, F., and Douzery, E. (1995). Re: molecular systematics of hystricognath rodents: the contribution of sciurognath mitochondrial 12S rRNA sequences. *Molecular Phylogenetics and Evolution*, 4, 357–60.

Dawson, M. R. and Beard, K. C. (1996). New Late Paleocene rodents (Mammalia) from Big Multi Quarry, Washakie Basin, Wyoming. *Palaeovertebrata*, 25, 301–21.

Dawson, M. R., Li, C.-K., and Qui, T. (1984). Eocene ctenodactyloid rodents (Mammalia) of Eastern and Central Asia. *Carnegie Museum of Natural History Special Publication*, 9, 138–50.

DeBry, R. W. and Sagel, R. M. (2001). Phylogeny of Rodentia (Mammalia) inferred from the nuclear-encoded gene IRBP. *Molecular Phylogenetics and Evolution*, 19, 290–301.

D'Erchia, A. M., Gissi, C., Pesole, G., Saccone, C., and Arnason, U. (1996). The guinea-pig is not a rodent. *Nature*, 381, 597–9.

de Jong, W. W., van Dijk, M. A. M., Poux, C., *et al.* (2003). Indels in protein-coding sequences of Euarchontoglires constrain the rooting of the eutherian tree. *Molecular Phylogenetics and Evolution*, 28, 328–40.

Douady, C., Carels, N., Clay, O, Catzefli, F., and Bernardi, G. (2000). Diversity and phylogenetic implications of CsCl profiles from rodent DNAs. *Molecular Phylogenetics and Evolution*, 17, 219–30.

Douzery, E. J. P. and Huchon, D, D. (2004). Rabbits, if anything, are likely Glires. *Molecular Phylogenetics and Evolution*, 33, 922–35.

Eizirik, E., Murphy, W. J., and O'Brien, S. J. (2001). Molecular dating and biogeography of the early placental mammal radiation. *Journal of Heredity*, 92, 212–19.

Elissamburu, A., and Vizcaíno, S. F. (2004). Limb proportions and adaptations in caviomorph rodents (Rodentia: Caviomorpha). *Journal of Zoology, London*, 262, 145–59.

Frye, M. S. and Hedges, S. B. (1995). Monophyly of the order Rodentia inferred from mitochondrial DNA sequences of the genes for 12S rRNA, 16S rRNA, and tRNA-valine. *Molecular Biology and Evolution*, 12, 168–76,

George, W. (1985). Reproductive and chromosomal characters of ctenodactylids as a key to their evolutionary relationships. In *Evolutionary Relationships Among Rodents: A Multidisciplinary Analysis*, ed. W. P. Luckett and J.-L. Hartenberger, pp. 453–74. New York: Plenum Press.

Gidley, J. W. (1912). The lagomorphs, an independent order. *Science*, 36, 285–6.

Graur, D., Hide, W. A., and Li, W.-H. (1991). Is the guinea pig a rodent? *Nature*, 351, 649–52.

Graur, D., Hide, W. A., Zarkikh, A., and Li, W.-H. (1992). The biochemical phylogeny of guinea pigs and gundis and the paraphyly of the order Rodentia. *Comparative Biochemistry and Physiology, B*, 101, 495–8.

Honeycutt, R. L. and Adkins, R. M. (1993). Higher level systematics of eutherian mammals: An assessment of molecular characters and phylogenetic hypotheses. *Annual Review of Ecology and Systematics*, 24, 279–305.

Huchon, D. and Douzery, E. J. P. (2001). From the Old World to the New World: a molecular chronicle of the phylogeny and biogeography of hystricognath rodents. *Molecular Phylogenetics and Evolution*, 20, 238–51.

Huchon, D., Catzeflis, F. M., and Douzery, E. J. P. (1999). Molecular evolution of the nuclear von Willebrand factor gene in mammals and the phylogeny of rodents. *Molecular Biology and Evolution*, 16, 577–89.

Huchon, D., Madsen, O., Sibbald, M. J. J. B., *et al.* (2002). Rodent phylogeny and a timescale for the evolution of Glires: evidence from an extensive taxon sampling using three nuclear genes. *Molecular Biology and Evolution*, 19, 1053–65.

Hunt, R. M., Jr. and Tedford, R. H. (1993). Phylogenetic relationships within the aeluroid Carnivora and implications of their temporal and geographic distribution. In *Mammal Phylogeny: Placentals*, ed. F. S. Szalay, M. J. Novacek, and M. C. McKenna, pp. 53–73. New York: Springer-Verlag.

Illiger, C. (1811). *Prodromus Systematis Mammalium et Avium Additis Terminis Zoographicis Utruisque Classis*. Berlin: C. Salfield.

Jacobs, L. L. (1984). Rodentia. [In *Mammals, Notes for a Short Course*, ed. P. D. Gingerich and C. E. Badgley.] *University of Tennessee Studies in Geology*, 8, 155–66.

Jacobs, L. L. and Flynn, L. J. (2005). Of mice – again: the Siwalik rodent record, murine distribution, and molecular clocks. In *Interpreting the Past: Essays on Human, Primate, and Mammal Evolution in Honor of David Pilbeam*, ed. D. E. Lieberman, R. J. Smith, and J. Kelley, pp. 63–80. Leiden: Brill Academic.

Janis, C. M. (1988). An estimation of tooth volume and hypsodonty indices in ungulate mammals, and the correlation of these factors with dietary preferences. [In *Teeth Revisited: Proceedings of the VIIth International Symposium on Dental Morphology, Paris, 1986*, ed. D. E. Russell, J.-P. Santoro, and D. Sigogneau-Russell, D.] *Mémoirs de Musée d'Histoire Naturelle, Paris, Series C*, pp. 367–87.

Janis, C. M. and Fortelius, M. (1988). On the means whereby mammals achieve increased functional durability of their dentitions, with special reference to limiting factors. *Biological Reviews*, 63, 197–230.

Janis, C. M. and Wilhelm, P. B. (1993). Were there pursuit predators in the Tertiary? Dances with wolf avatars. *Journal of Mammalian Evolution*, 1, 103–25.

Janis, C. M., Damuth, J., and Theodor, J. M. (2000). Miocene ungulates and terrestrial primary productivity: Where have all the browsers gone? *Proceedings of the National Academy of Sciences, USA*, 97, 7899–904.

Janis, C, M., Theodor, J. M., and Boisvert, B. (2002). Locomotor evolution in camels revisited: a quantitative analysis of pedal anatomy and the acquisition of the pacing gait. *Journal of Vertebrate Paleontology*, 22, 110–21.

Janke, A., Xu, X., and Arnason, U. (1997). The complete mitochondrial genome of the wallaroo (*Macropus robustus*) and the phylogenetic history among Monotremata, Marsupialia and Eutheria. *Proceedings of the National Academy of Sciences, USA*, 94, 1276–81.

Korth, W. W. (1994). *The Tertiary Record of Rodents in North America*. New York: Plenum Press.

Kramereov, D., Vassetzky, N., and Serdobova, I. (1999). The evolutionary position of dormice (Gliridae) in Rodentia determined by a novel short retroposon. *Molecular Biology and Evolution*, 16, 715–17.

Kumar, S. and Hedges, S. B. (1998). A molecular timescale for vertebrate evolution. *Nature*, 392, 917–20.

Landry, S. O., Jr. (1999). A proposal for a new classification and nomenclature for the Glires (Lagomorpha and Rodentia). *Mitteilungen aus dem Museum für Naturkunde in Berlin, Zoologische Reihe*, 75, 283–316.

Lavocat, R. and Parent, J.-P. (1985). Phylogenetic analysis of middle ear features in fossil and living rodents. In *Evolutionary Relationships Among Rodents: A Multidisciplinary Analysis*, ed. W. P. Luckett and J.-L. Hartenberger, pp. 333–54. New York: Plenum Press.

Li, C.-K. and Ting, S.-Y. (1985). Possible phylogenetic relationship of Asiatic eurymylids and rodents, with comments on mimotomids. In *Evolutionary Relationships Among Rodents: A Multidisciplinary Analysis*, ed. W. P. Luckett and J.-L. Hartenberger, pp. 35–58. New York: Plenum Press.

Li, C.-K., Wilson, R. W., Dawson, M. R., and Krishtalka, L. (1987). The origins of rodents and lagomorphs. *Current Mammalogy*, 1, 97–108.

Li, W.-H., Hide, W. A., Zharkika, A., Ma, D. P., and Graur, D. (1992). The molecular taxonomy and evolution of the guinea pig. *Journal of Heredity*, 83, 174–81.

Lilljeborg, W. (1866). *Systematisk Öfveresigt af de Gnagande Däggdjuren, Glires*. Uppsala: Kungliga Akademi. Boktryckeribt.

Linnaeus, C. (1735). *Systema Naturae, sive Regna Tria Naturae Systematice Proposita Classes, Ordines, Genera, and Species*. Leiden: Theodorum Haak.

(1758). *Systema Naturae per Regna Tria Naturae, Secundum Classes, Ordines, Genera, Species cum Characteribus, Differentiis, Synonymis, Locis*. Vol. 1. *Regnum Animale*. Edito decima, reformata. Stockholm: Laurentii Salvii.

Luckett, W. P. (1985). Superordinal and intraordinal affinities of rodents: developmental evidence from the dentition and placentation. In *Evolutionary Relationships Among Rodents: A Multidisciplinary Analysis*, ed. W. P. Luckett and J.-L. Hartenberger, pp. 227–76. New York: Plenum Press.

Luckett, W. P. and Hartenberger, J.-P. (1985). Evolutionary relationships among rodents: comments and conclusions. In *Evolutionary Relationships Among Rodents: A Multidisciplinary Analysis*, ed. W. P. Luckett and J.-L. Hartenberger, pp. 685–712. New York: Plenum Press.

(1993). Monotype or polyphyly of the order Rodentia: possible conflict between morphological and molecular interpretation. *Journal of Mammalian Evolution*, 1, 127–47

Ma, D.-P., Zharkikh, A., Graur, D., Vanderberg, J. L., and Li, W.-H. (1993). Structure and evolution of opossum, guinea pig, and porcupine cytochrome *b* genes. *Journal of Molecular Evolution*, 36, 327–34.

Marivaux, L., Vianey-Liaud, M., and Jaeger, J.-J. (2004). High-level phylogeny of early Tertiary rodents: dental evidence. *Zoological Journal of the Linnean Society*, 142, 105–34.

Martin, T. (1993). Early rodent incisor enamel evolution: phylogenetic implications. *Journal of Mammalian Evolution*, 1, 227–54.

(2004). Evolution of incisor enamel microstructure in Lagomorpha. *Journal of Vertebrate Paleontology*, 24, 411–26.

Martinez, N. L. (1985). Reconstruction of ancestral cranioskeletal features in the order Lagomorpha. In *Evolutionary Relationships Among Rodents: A Multidisciplinary Analysis*, ed. W. P. Luckett and J.-L. Hartenberger, pp. 151–89. New York: Plenum Press.

McKenna, M. C. (1975). Toward a phylogenetic classification of the Mammalia. In *Phylogeny of the Primates*, ed. W. P. Luckett and F. S. Szalay, pp. 221–46. New York: Plenum Press.

McKenna, M. C. and Bell, S. K. (1997). *Classification of Mammals Above the Species Level*. New York: Columbia University Press.

McNab, B. K. (2002). *The Physiological Ecology of Vertebrates: A View from Energetics*. Ithaca, NY: Comstock.

Meng, J. and Wyss, A. R. (2005). Glires (Lagomorpha, Rodentia). In *The Rise of Placental Mammals: Origins and Relationships of the Major Clades*, ed. K. D. Rose and J. D. Archibald, pp. 144–58. Baltimore, MD: Johns Hopkins University Press.

Meng, J., Hu, Y., and Li, C. (2003). The osteology of *Rhombomylus* (Mammalia, Glires): implications for phylogeny and evolution of Glires. *Bulletin of the American Museum of Natural History*, 275, 1–247.

Misawa, K. and Janke, A. (2003). Revising the Glires concept: phylogenetic analysis of molecular sequences. *Molecular Phylogenetics and Evolution*, 28, 320–7.

Montgelard, C., Bentz, S., Tirard, C., Verneau, O., and Catzeflis, F. M. (2002). Molecular systematics of Sciurognathi (Rodentia): the mitochondrial cytochrome *b* and 12S rRNA genes support the Anomaluroidea (Pedetidae and Anomaluridae). *Molecular Phylogenetics and Evolution*, 22, 220–33.

Murphy, W. J., Eizirik, E., Johnson, W. E., *et al.* (2001). Molecular phylogenetics and the origins of placental mammals. *Nature*, 409, 614–18.

Nedbal, M. A., Honeycutt, R. L., and Schiltter, D. A. (1996). Higher-level systematics of rodents (Mammalia: Rodentia): evidence from the mitochondrial 12S rRNA gene. *Journal of Mammalian Evolution*, 3, 201–37.

Noguchi, T., Fujiwara, S., Hayashi, S, and Sakuraha, H. (1994). Is the guinea pig (*Cavia porcellus*) a rodent? *Comparative Biochemistry and Physiology, B*, 107, 179–82.

Novacek, M. J. (1985). Cranial evidence for rodent affinities. In *Evolutionary Relationships Among Rodents: A Multidisciplinary Analysis*, ed. W. P. Luckett and J.-L. Hartenberger, pp. 59–81. New York: Plenum Press.

(1993). Reflection on higher mammalian phylogenetics. *Journal of Mammalian Evolution*, 1, 3–30.

Nowak, R. M. (1999). *Walker's Mammals of the World*, 6th edn. Baltimore, MD: Johns Hopkins University Press.

Phillippe, H. (1997). Rodent phylogeny: pitfalls of molecular phylogenies. *Journal of Molecular Evolution*, 45, 712–25.

Retallack, G. J. (2001). Cenozoic expansion of grasslands and climatic cooling. *Journal of Geology*, 109, 407–26.

Reyes, A., Pesole, G., and Saccone, C. (1998). Complete mitochondrial DNA sequence of the fat dormouse, *Glis glis*; further evidence of rodent paraphyly. *Molecular Biology and Evolution*, 15, 499–505.

Reyes, A., Gissi, C., Catzefliz, F., *et al.* (2004). Congruent mammalian trees from mitochondrial and nuclear genes using Bayesian methods. *Molecular Biology and Evolution*, 21, 397–403.

Rose, K. D. and Chinnery, B. J. (2004). The postcranial skeleton of early Eocene rodents. [In *Fanfare for an Uncommon Paleontologist. Essays in Honor of Malcolm C. McKenna*, ed. M. R. Dawson and J. A. Lillegraven.] *Bulletin of the Carnegie Museum of Natural History*, 36, 211–44.

Rybczynski, N. (2004). Effect of incisor shape on woodcutting performance in two beavers (Castoridae, Rodentia). *Journal of Vertebrate Paleontology*, 24(suppl. to no. 3), p. 107A.

Sánchez-Villagra, M. R., Aguillera, O., and Horovitz, I. (2003). The anatomy of the world's largest extinct rodent. *Science*, 301, 1708–10.

Seckel, L. and Janis, C. M. (2008). Convergences in scapula morphology among small cursorial mammals: an osteological correlate for locomotory specialization. *Journal of Mammalian Evolution*, in press.

Simpson, G. G. (1945). The principles of classification and a classification of mammals. *Bulletin of the American Museum of Natural History*, 85, 1–350.

Springer, M. S., Stanhope, M. J., Madsen, O., and de Jong, W. W. (2004). Molecules consolidate the placental mammal tree. *Trends in Ecology and Evolution*, 19, 430–8.

Springer, M. S., Murphy, W. J., Eizirik, E., and O'Brien, S. J. (2005). Molecular evidence for major placental clades. In *The Rise of Placental Mammals: Origins and Relationships of the Major Clades*, ed. K. D. Rose and J. D. Archibald, pp. 37–49. Baltimore, MD: Johns Hopkins University Press.

Stanhope, M. J., Smith, M. A., Waddell, V. G., *et al.* (1996). Mammalian evolution and the interphotoreceptor retinoid binding protein (IRPB) gene: convincing evidence for several superordinal clades. *Journal of Molecular Evolution*, 43, 83–92.

Storch, G., Engesser, B., and Wuttke, M. (1996). Oldest fossil record of gliding in rodents. *Nature*, 379, 439–41.

Strömberg, C. A. E. (2006). Evolution of hypsodonty in equids: testing a hypothesis of adaptation. *Paleobiology*, 32, 236–58.

Sullivan, J. and Swofford, D. L. (1997). Are guinea pigs rodents? The importance of adequate models in molecular phylogenetics. *Journal of Mammalian Evolution*, 4, 77–86.

Szalay, F. S. (1985). Rodent and lagomorph morphotype adaptations, origins, and relationships: some postcranial attributes analyzed. In *Evolutionary Relationships Among Rodents: A Multidisciplinary Analysis*, ed. W. P. Luckett and J.-L. Hartenberger, pp. 83–132. New York: Plenum Press.

Tedford, R. H. and Harington, C. R. (2003). An Arctic mammal fauna from the Early Pliocene of North America. *Nature*, 425, 388–90.

Tullberg, T. (1899). Über das System der Nagetiere. Eine phlogenetische Studie. *Nova Acta Regiae Societatis Scientiarium Uppsaliensis, Series 3*, 18, 1–514.

Turnbull, W. D. (1970). Mammalian masticatory apparatus. *Fieldiana, Geology*, 18, 149–356.

Van Valkenburgh, B. (1991). Iterative evolution of hypercarnivory in canids (Mammalia: Carnivora): evolutionary interactions among sympatric predators. *Paleobiology*, 17, 340–62.

Vaughan, T. A., Ryan, J. M., and Czaplewski, N. J. (2000). *Mammalogy*, 4th edn. Fort Worth, TX: Saunders College Press.

Vianey-Liaud, V. (1985). Possible evolutionary relationships among Eocene and Lower Oligocene rodents of Asia, Europe, and North America. In *Evolutionary Relationships Among Rodents: A Multidisciplinary Analysis*, ed. W. P. Luckett and J.-L. Hartenberger, pp. 277–309. New York: Plenum Press.

Wahlert, J. H. (1985). Cranial foramina of rodents. In *Evolutionary Relationships Among Rodents: A Multidisciplinary Analysis*, ed. W. P. Luckett and J.-L. Hartenberger, pp. 311–32. New York: Plenum Press.

Waterhouse, G. R. (1839). The distribution of Rodentia. *Proceedings of the Zoological Society of London*, 1839, 172–4.

Webb, S. D. and Opdyke, N. D. (1995). Global climatic influence on Cenozoic land mammal faunas. In *Studies in Geophysics: Effects of Past Global Change on Life*, pp. 184–208. Washington, D.C.: National Academy Press for the Board on Earth Sciences and Resources Commission on Geosciences, Environment, and Resources National Research Council.

Wible, J. R., Rougier, G. W., and Novacek, M. J. (2005a). Anatomical evidence for supraordinal/ordinal eutherian taxa in the Cretaceous. In *The Rise of Placental Mammals: Origins and Relationships of the Major Clades*, ed. K. D. Rose and J. D. Archibald, pp. 15–36. Baltimore. MD: Johns Hopkins University Press.

Wible, J. R., Wang, Y., Chuamkui, L., and Dawson, M. R. (2005b). Cranial anatomy and relationships of a new ctenodactyloid (Mammalia, Rodentia) from the Early Eocene of Hubei Province, China. *Annals of the Carnegie Museum*, 74, 91–150.

Wilson, R. W. (1949). Early Tertiary rodents of North America. *Carnegie Institution of Washington Publication*, 584, 67–164.

Wood, A. E. (1937). The mammalian fauna of the White River Oligocene. Part II. Rodentia. *Transactions of the American Philosophical Society*, 28, 155–269.

(1955). A revised classification of the rodents. *Journal of Mammalogy*, 36, 165–87.

(1957). What, if anything, is a rabbit? *Evolution*, 11, 417–25.

(1965). Grades and clades among rodents. *Evolution*, 19, 115–30.

(1975). The problem of hystricognathous rodents. *Papers on Paleontology, University of Michigan*, 12, 75–80.

(1985). The relationships, origins, and dispersal of the hystricognathous rodents. In *Evolutionary Relationships Among Rodents: A Multidisciplinary Analysis*, ed. W. P. Luckett and J.-L. Hartenberger, pp. 475–513. New York: Plenum Press.

17. Lagomorpha

MARY R. DAWSON

Carnegie Museum of Natural History, Pittsburgh, PA, USA

INTRODUCTION

Lagomorphs – the rabbits, hares, and pikas – were Eocene immigrants into North America from Asia, where their origins can be traced to the Paleocene family Mimotonidae, and they have formed an important component of many North American faunas over the past 40 million years. Members of this successful, but morphologically rather conservative, order have populated a wide variety of habitats, including Arctic and alpine tundra, taiga, grassland, woodland, marsh, swamp, and desert. The ultimate success of members of the order can be traced, insofar as osteological and dental characters are concerned, to early acquisition of hypsodonty, both of gliriform incisors and cheek teeth, presumably a legacy of anagalid relationships.

The typical dental features of lagomorphs were acquired early in their evolutionary history. The order has been characterized by morphological stability with relatively few subsequent dental changes, except for various patterns of increased hypsodonty. This conservatism is reflected in low taxonomic diversity. Only two families are recognized: the Leporidae, known in North America from the Uintan (middle Eocene) to the Recent, and the Ochotonidae, known in North America from the late early Arikareean (latest Oligocene) to the Recent.

The Leporidae underwent most of their mid Tertiary evolution in North America. The Ochotonidae achieved their greatest diversity and widest geographic distribution during the Miocene, evolving along distinctly different endemic paths in Europe, Asia, Africa, and North America. The low diversity of lagomorphs – only two families since the Uintan – contrasts strongly with the record of the much more diverse Rodentia, the other major group of gliriform mammals, which includes 27 extinct and 25 extant families since the late Paleocene.

An unfilled morphological gap is present between the Paleocene mimotonids and the true lagomorphs, which have a first appearance in Asia in the Irdinmanhan (Tong, 1997). The first appearance of lagomorphs in North America is slightly later, in the late Uintan. One genus of Ochotonidae (*Ochotona*) and four of Leporidae (*Lepus*, *Sylvilagus*, *Brachylagus*, *Romerolagus*) still inhabit North America. Ochotonids are currently restricted to Eurasia and North America, but the natural distribution of leporids is nearly worldwide, except for Antarctica, Australia, New Zealand, Madagascar, and some oceanic islands.

DEFINING FEATURES OF THE ORDER LAGOMORPHA

The following list of characters includes, so far as possible, features that can be confirmed in the known fossil record.

Characteristics: Small- to medium-sized gliriform mammals; dental formula I2/1, C0/0, P3/2, M2–3/2–3; anterior upper and lower incisors enlarged, ever growing, and rootless (embryological evidence indicates that the enlarged incisors are dI2/di2); small, peglike upper incisor behind anterior upper incisor; upper cheek teeth with at least some unilateral hypsodonty (upper and lower cheek teeth hypsodont and become rootless in many fossil and all living members of the order); anterior upper incisor grooved; upper cheek teeth wider than lower and maxillary tooth rows more widely separated than mandibular, with masticatory movements vertical and transverse; trigonids higher than talonids; facial portion of maxilla with single fenestra or numerous small fenestrae; incisive foramina usually elongate, bony palate short; orbitosphenoid large; optic foramina confluent; ectotympanic auditory bulla; glenoid fossa transverse oval; coronoid process of mandible rudimentary; quadrupedal with five digits on manus, four on pes; clavicle present; humerus lacking epicondylar foramen; tibia and fibula united distally; fibula articulates with calcaneum; tail short or absent.

Evolution of Tertiary Mammals of North America, Vol. 2. ed. C. M. Janis, G. F. Gunnell, and M. D. Uhen. Published by Cambridge University Press.

LEPORIDAE

Cranial characteristics: Lateral surface of maxilla fenestrated into a lacework of bone; snout tapering in width anteriorly; lateral process of nasal extending further posteriorly than remainder; supraorbital process of frontal present; tendency toward reduction of palatine component of palate; zygoma vertically expanded plate; tendency toward loss of posterior mental foramen (Figure 17.1).

Dental characteristics: Dental formula I2/1, C0/0, P3/2, M2–3/3; tendency toward molarization of P3–4; trigonid higher than talonid; tendency toward union of trigonid and talonid by lingual bridge of enamel and dentine; shafts of upper cheek teeth relatively straight, extending up medial side of orbit; premolar foramen absent; m3 comprising two columns; enamel of upper and lower incisors single layer with Hunter–Schreger bands.

Postcranial characteristics: Vertebrarterial canal in seventh cervical vertebra; 19 (12:7) thoracolumbar vertebrae; four fused sacral vertebrae; pelves fused at pubic symphysis. Most evolutionary changes in the skeleton have been gradual improvements in features associated with rapid quadrupedal locomotion, either for short bursts of speed (rabbits) or rapid, sustained bounding locomotion (hares).

OCHOTONIDAE

No North American Tertiary ochotonids are well known from skulls or postcranial remains. Accordingly, the defining features given here are limited and contain plausible extensions from remains of better-known ochotonid taxa (e.g, *Prolagus, Ochotona*).

Cranial characteristics: Nasals narrow posteriorly; lateral surface of maxilla with well-developed area of fenestrated bone, single opening, or a combination of both; zygoma sturdily constructed anteriorly and contacted by maxillary tuberosity; distinct groove on posterior zygomatic root suggesting well-developed temporalis muscle relative to that in leporids; one or two pairs of incisive foramina; palate with premolar foramen and shortened maxillary component of bony palatal bridge; supraorbital processes absent; symphysis of jaw without posterior buttress that occurs in leporids; tendency toward enlargement and posterior shift of posterior mental foramen.

Dental characteristics: Dental formula I2/1, C0/0, P3/2, M2–3/2–3; P3 persistently non-molariform; enamel of lingual reentrant of upper molariform teeth not crenulated; shafts of upper cheek teeth more or less curved, extending toward outer side of orbit; trigonid and talonid of p4–m2 not united by lingual bridge; m3, if present, with talonid reduced or absent; incisor enamel, where known, of several layers and mostly composed of radial enamel; enamel in upper and lower incisors dissimilar in composition (Koenigswald, 1996).

Postcranial characteristics: Twenty two (sometimes 21) thoracolumbar vertebrae (16:6); anteroposteriorly wide transverse processes on the lumbar vertebrae; fused spines and processes on sacral vertebrae; pubic symphysis unfused; second cuneiform fused to metatarsal II.

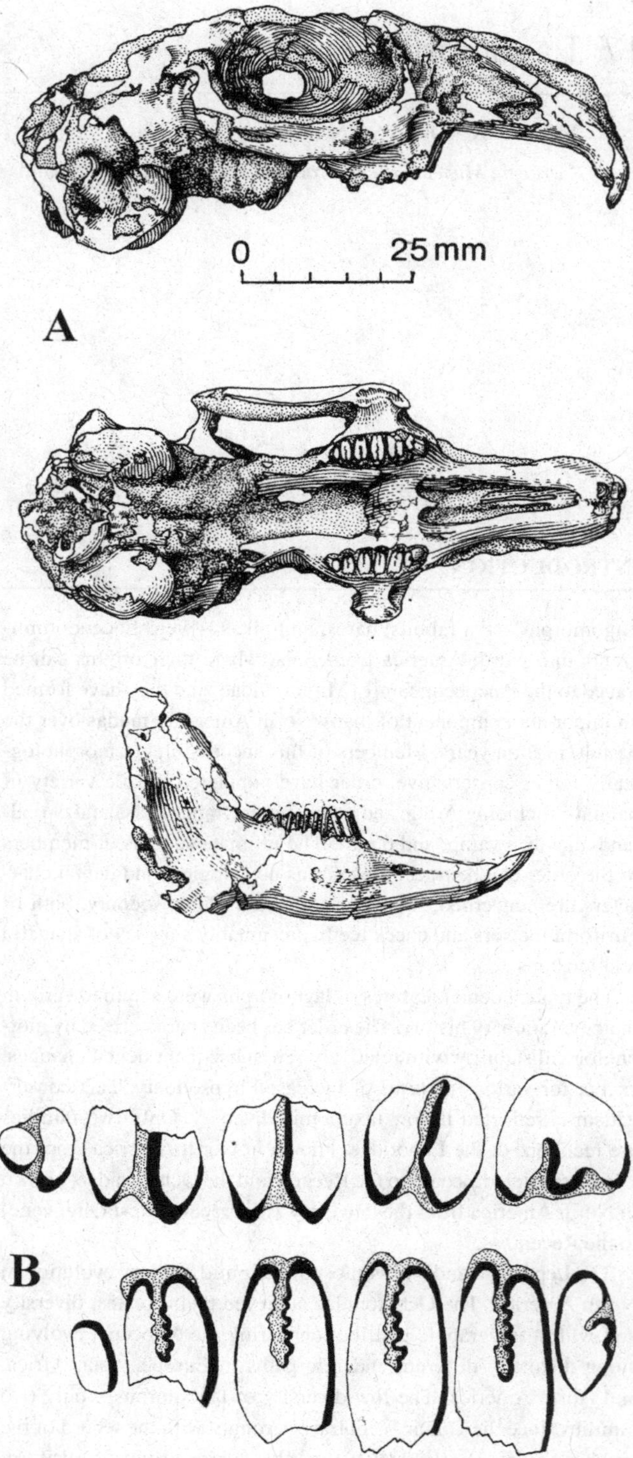

Figure 17.1. A. Skull and dentary of *Hypolagus vetus*. B. Upper and lower cheek tooth enamel patterns of *Hypolagus vetus*. (From White, 1984; courtesy of the Carnegie Museum of Natural History.)

SYSTEMATICS

SUPRAORDINAL

The careful morphological analysis done by Tullberg (1899) strongly supported the concept of the order Lagomorpha (called Duplicidentati) as one branch of the larger taxonomic unit Glires, which includes both the rodents and lagomorphs. The content of the Glires has expanded greatly since 1899, principally as the result of discoveries of Paleocene and Eocene fossils in Asia. The Lagomorpha, however, remain as a discrete unit not greatly different from that outlined by Tullberg.

The mimotonids and lagomorphs, characterized by their two pairs of upper incisors, have been united as a superorder Duplicidentata within the larger cohort Glires, which includes mammals having enlarged, ever-growing anterior incisors. In contrast, the Simplicidentata, which includes the mixodonts and rodents, have a single pair of upper incisors. Mimotonids are characterized by prismatic, unilaterally hypsodont upper cheek teeth, enlarged anterior incisors with a single layer of enamel, non-molariform upper premolars, unfused tibia and fibula, and no fibular facet on the calcaneum (Li and Ting, 1985). The Duplicidentata has also been interpreted, using a crown-clade-restricted phylogenetic definition, as all gliriform eutherians sharing a more recent common ancestor with Lagomorpha than with Rodentia, and Lagomorpha as the crown clade stemming from the most recent common ancestor of *Ochotona* and Leporidae (Wyss and Meng, 1996). A diverse array of primitive Glires existed in Asia during the Paleocene and early Eocene (Dashzeveg *et al.*, 1998). If broader relationships of the Glires can be found (Wood, 1957), they appear to be with the primarily Paleocene anagalids, an Asian group in which the cheek teeth are prismatic and unilateral hypsodont but without loss of incisors or development of a diastema (Novacek, 1984).

Several attempts at classification within the Lagomorpha (Gureev, 1964; Erbajeva, 1988) have led to recognition of some new families (e.g., Palaeolagidae) that are united only by shared primitive features and others (e.g., Prolagidae) that are of very restricted scope and are better regarded as subfamilies. On the whole, these attempts at subdivision have contributed neither to phylogenetic nor to paleobiogeographic understanding of the order. The recognition of two families of Lagomorpha is followed here.

In a morphologically homogeneous order like the Lagomorpha, in which there have been many parallel morphological changes in both dentition and limb structure, determination of significant characters on which to establish relationships becomes difficult. The lagomorphs are primitive in the structure of the bulla, many aspects of the soft anatomy, and the reproductive system, but are highly derived in dental and skull characters. The family Leporidae, which has a fossil record dating back to the middle Eocene, is the more primitive of the two families in dental formula and palatal morphology but becomes the more progressive of the two in hypsodonty, occlusal pattern of the cheek teeth, and limb development. The family Ochotonidae, first definitely recorded from the early Oligocene, has one derived character in the palate, the presence of a premolar foramen, and is characterized by greater dental reduction than the leporids. However, the ochotonids retain such features as rooted cheek teeth and a primitive occlusal pattern well into the Miocene, when the leporids have more derived dental features.

The polarities of some leporid and ochotonid characters are not yet firmly established. In a very important study, Koenigswald (1996) examined a variety of lagomorph incisors and cheek teeth. He reaffirmed that leporids have a single layer of incisor enamel with Hunter–Schreger bands in I2/i2. This had previously been considered to be the universal lagomorph condition, but Koenigswald's new investigation of ochotonid enamel uncovered a much more complex situation. In ochotonids, the upper and lower incisors of each taxon studied differed in number of layers of enamel: in the lower incisors, two or even three layers are present, with Hunter–Schreger bands in at least one layer. The upper incisor I2, however, has two or three layers but lacks any development of Hunter–Schreger bands. Cheek teeth of both leporids and ochotonids have enamel composed of an inner layer of radial enamel and an outer layer having Hunter–Schreger bands, basically similar to the condition in the lower incisors of ochotonids.

Koenigswald's (1996) interpretation of these structures is that the ochotonid condition, which is more similar to that of the cheek teeth, is primitive, and that the leporid condition, having a single layer of enamel with Hunter–Schreger bands, is derived. The logic of this argument is excellent but, unfortunately, this interpretation does not correspond well with the geological record. This shows the single-layered, or leporid, condition to be much older, dating to the middle Eocene, whereas the oldest examined ochotonid is of Miocene age. (Note that the so-called middle Eocene lagomorph from Split Rock, Wyoming, cited by Koenigswald, is actually a middle Miocene ochotonid.) Until some older ochotonids, including *Desmatolagus gobiensis*, and such Eocene lagomorph genera as *Lushilagus*, *Shamolagus*, and *Procaprolagus* have been evaluated for enamel structure, the question of polarity for incisor enamel should remain open. The question is an important one for determining both lagomorph origins and the relationships between the two families.

Martin (2004) has recently considered the incisor enamel microstructure of a selected set of lagomorphs. He showed that the more complex enamel construction is present in the eurymylids and in most mimotonids. He accordingly considered the double-layered enamel schmelzmuster to be the primitive condition for lagomorphs, with the more simple condition in most leporids representing the derived condition. While *Desmatolagus gobiensis*, here considered to be the oldest known definite ochotonid, has a single layer of incisor enamel, Martin (2004) saw this feature as sufficient reason to refer this taxon to the Leporidae.

INFRAORDINAL

LEPORIDAE

Interrelationships within the Leporidae at the subfamily level have proven difficult to establish, largely because of the few characters that differentiate genera within the family. One solution has been to disregard subfamilies entirely (McKenna and Bell, 1997).

A more constructive approach has been the recognition of various subfamilial arrangements, the first of which was proposed by Dice (1929), who named the subfamilies Palaeolaginae, Archaeolaginae, and Leporinae based mostly on the pattern of enamel folding on p3. Some subsequent specialists on North American leporids (Dawson, 1958; White, 1984, 1987, 1991a) have also utilized this subfamilial classification.

While other morphological characters have been assigned to the subfamilies, the occlusal pattern of p3 and, to a lesser extent, of P2 remain the most useful basis for this arrangement. Dice regarded each of his three subfamilies as an independent, and thus monophyletic, phylogenetic series. Although the subfamily Archaeolaginae was established with some question, Dice finally concluded that *Archaeolagus* and *Hypolagus* represent a distinct line of evolution. Wood (1940) did not recognize a separate Archaeolaginae, noting that in very late stages of wear some of the palaeolagines attained the p3 pattern characteristic of *Archaeolagus*. In Wood's interpretation, the palaeolagines (including archaeolagines as used here) were not ancestral to the leporines.

Burke (1941) added the subfamilies Mytonolaginae and Desmatolaginae for relatively primitive Asian and North American leporids, and Megalaginae for the genus *Megalagus*. Burke also retained Palaeolaginae and Archaeolaginae as distinct subfamilies, which he considered to be "evolutionary stages." Dice's subfamilies were retained by Dawson (1958), with added importance placed on the development of the anteroexternal fold on p3 as a derived feature shared by archaeolagines and leporines. Retention of juvenile characters (paedomorphy) in the dentition was mentioned as important in the development of the internal folds on p3 in the leporines. White (1987, 1991a) reviewed the later Tertiary archaeolagines and leporines and concluded that the Leporinae arose from the Archaeolaginae sometime in the Clarendonian (early late Miocene).

As presently constituted, the palaeolagines are a broadly based, primitive grouping. Studies of the numerous Asian Eocene lagomorph genera will be necessary to determine the interrelationships within this group and also to establish characters that might explain the origins of North American *Procaprolagus* and *Mytonolagus*. Archaeolagines and leporines are both derived, and possibly united, in having added an external fold on the trigonid of p3. In archaeolagines, any internal folding on p3 has been suppressed, whereas in leporines these folds, which are basically juvenile characters, are retained into the adult. The fossil record does not at present reveal whether these two groups were derived separately from palaeolagines or if one group was derived from the other. As interpreted here (Figure 17.2), palaeolagines and archaeolagines contain closely related genera that can be arranged in two more or less horizontal groups, with leporines as a monophyletic derivative of the archaeolagines.

OCHOTONIDAE

Although the Ochotonidae have long been recognized in the North American Pleistocene (Gidley and Gazin, 1938; Guilday, 1979), McGrew (1941) was the first to suggest that ochotonids might be present in the North American Tertiary. In his description of a new species of the lagomorph *Oreolagus*, McGrew suggested the possibility that it might represent an ochotonid but left its familial assignment unresolved. Wilson (1949) confirmed that *Oreolagus* is an ochotonid. The immigration event that brought *Oreolagus* to North America appears to have been distinct from those of other immigrant ochotonids, including *Hesperolagomys* and *Russellagus*. The relatively incompletely known *Cuyamalagus*, *Gripholagomys*, and *Oklahomalagus* may also represent distinct lineages and immigration events.

Ochotonids originated in Asia, and *Desmatolagus gobiensis* is the oldest known lagomorph to have the derived character of a premolar foramen. Attempts to establish a subfamily arrangement within the Ochotonidae have not been successful, possibly because of the great amount of endemism in Asia and Europe and later in Africa and North America when ochotonids expanded their range during the Miocene. The most distinct lineages are those of *Piezodus–Prolagus* and *Lagopsis–Paludotona* in Europe, and a series of species of *Oreolagus* in North America (Dawson, 1965, 1967).

INCLUDED GENERA OF LAGOMORPHA BRANDT, 1855

The locality numbers listed for each genus refer to the list of unified localities in Appendix I. The locality numbers may be listed in several alternative ways. The acronyms for museum collections are listed in Appendix III.

Parentheses around the locality (e.g., [CP101]) mean that the taxon in question at that locality is cited as an "aff." or "cf." the taxon in question. Parentheses are usually used for individual species, thus implying that the genus is firmly known from the locality, but the actual species identification may be questionable. Question marks in front of the locality (e.g., ?CP101) mean that the taxon is questionably known from the locality, thus implying some doubt that the taxon is actually present at the locality, either at the genus or the species level. An asterix (*) indicates the type locality.

LEPORIDAE GRAY, 1821

***Procaprolagus* Gureev, 1960 (including *Desmatolagus*, in part)**

Type species: *Procaprolagus vetustus* (Burke, 1941) (originally described as *Desmatolagus vetustus*).

Type specimen: AMNH 26089, Jhama Obo, East Mesa, Ulan Gochu Formation, Inner Mongolia, Ergilian (approximately Chadronian).

Characteristics: Cheek teeth rooted, lower crowned than in *Mytonolagus*; P3–M2 with persistent hypostria, crescentic valley, and buccal pattern; p3 and M3/m3 reduced; lower incisor shaft extending to below posterior end of m2; p3 with anteroposteriorly short talonid, persistent lingual valley, and anterolingual fold; p4–m2 with anteroposteriorly compressed trigonid, and a longer, rounded talonid (Storer, 1984).

Average length of lower molariform tooth: 2.0–2.8 mm.

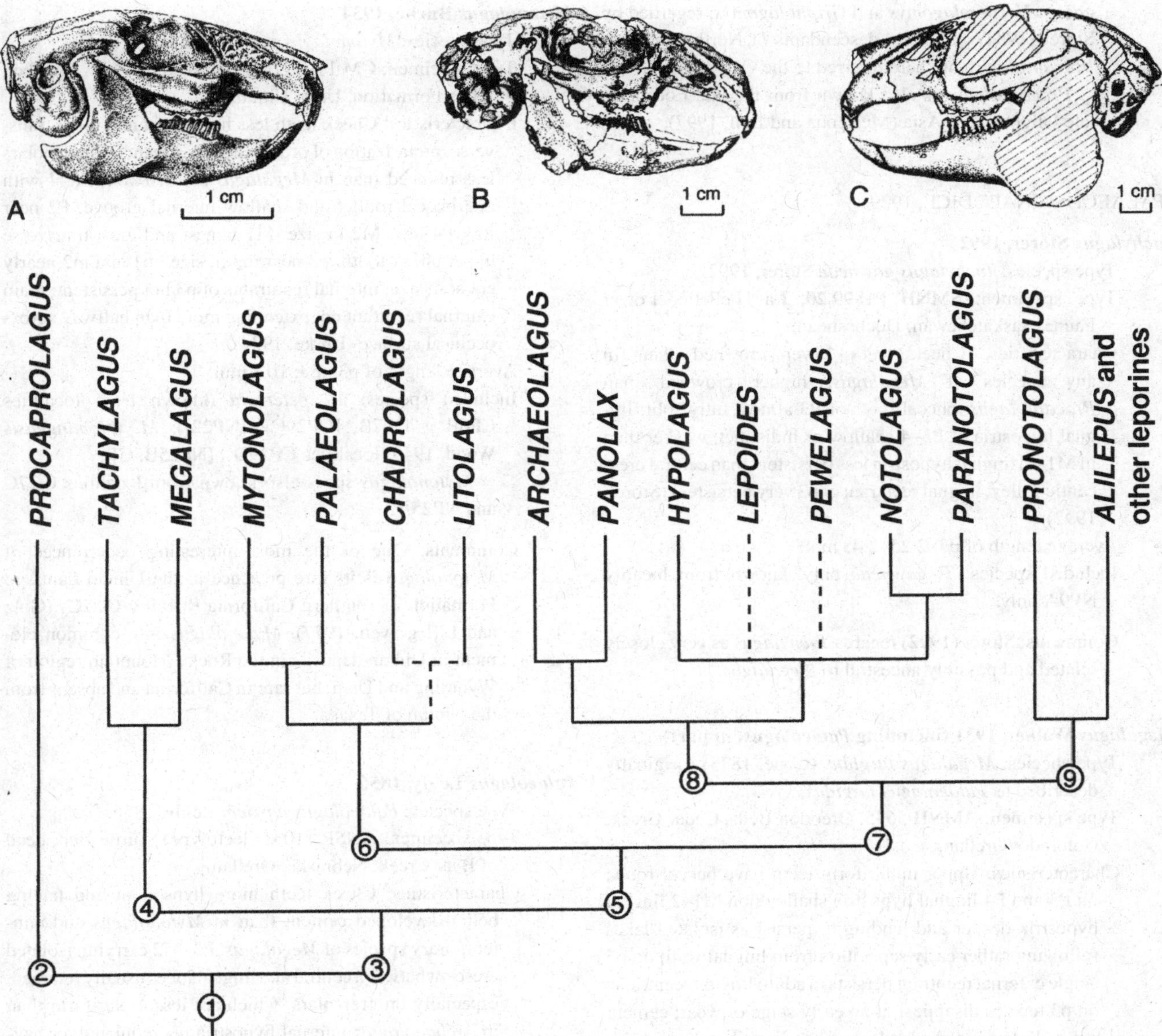

Figure 17.2. Interrelationships among genera of North American leporids. Characteristics of nodes 1–9 as follows. 1. Lateral surface of maxilla fenestrated into lacework of bone; lateral process of nasal extending further posteriorly than remainder of bone; supraorbital process of frontal present; incisor enamel single layer with Hunter–Schreger bands; p3 pattern based on one buccal fold and one lingual fold between trigonid and talonid. 2. P2, p3, M3, and m3 strongly reduced. 3. Persistent lingual bridge between trigonid and talonid of p4–m2; P2 and p3 increased in size. 4. p3 trigonid rotated posterolinguad; posterolophid on p4–m2 persistent. 5. Buccal folds on upper molars reduced or absent. 6. p3 trigonid parallel to talonid. 7. P3–4 completely molariform in pattern; p3 with persistent anteroexternal fold. 8. Lingual side of p3 lacking posterointernal fold. 9. On p3 posterointernal fold deep and persistent as fold, isolated lake, or united with posteroexternal fold. Note: *Tachylagus* was synonymized with *Megalagus* by McKenna and Bell (1997). (A: courtesy of the American Philosophical Society. B: courtesy of the Carnegie Museum of Natural History. C: courtesy of the Society of Vertebrate Paleontology.)

Included species: *P. gazini* Burke, 1936 (known from locality CP99A*); *P. vusillus* Storer, 1984 (locality NP8*).

Procaprolagus sp. is also known from locality NP9D.

Comments: The oldest North American species of *Procaprolagus*, *P. vusillus*, is both geologically older and morphologically more primitive (lower-crowned teeth, greater persistence of buccal pattern of upper molariform teeth) than the Asian *P. vetustus* (Storer, 1984). Reference of the North American species to the Asian genus must be viewed with caution; however, it is apparent that *P. vusillus* represents a separate immigration from Asia than that of the approximately contemporaneous *Mytonolagus* (Dawson, 1970; Storer, 1984). *Procaprolagus* does not appear to be a long-lived North American lineage, and no close relatives are known after the Orellan (early Oligocene) *P.*

gazini. *Hesperolagomys* and *Gripholagomys*, regarded by Storer (1984) as leporid descendants of North American *Procaprolagus*, are here referred to the Ochotonidae.

Procaprolagus is also known from the late Eocene to late Oligocene of Asia (McKenna and Bell, 1997).

"PALAEOLAGINAE" DICE, 1929

Tachylagus Storer, 1992

Type species: *Tachylagus gawneae* Storer, 1992.

Type specimen: SMNH P1899.20, Lac Pelletier Lower Fauna, Saskatchewan, Duchesnean.

Characteristics: Cheek teeth lower crowned than in any species of *Megalagus*, higher crowned than *Procaprolagus*; buccal roots of P3 almost fully split; lingual hypostria of P3–4 shallow or indistinct, weaker than in M1–2; lingual hypostria less persistent than central crescentic valley; lingual reentrant of p3 very persistent (Storer, 1992).

Average length of p3: 2.25–2.43 mm.

Included species: *T. gawneae* only, known from locality NP9A only.

Comments: Storer (1992) regards *Tachylagus* as very closely related and possibly ancestral to *Megalagus*.

Megalagus Walker, 1931 (including *Palaeolagus*, in part)

Type species: *Megalagus turgidus* (Cope, 1873) (originally described as *Palaeolagus turgidus*).

Type specimen: AMNH 5635, Oreodon Beds, Cedar Creek, Colorado, Orellan.

Characteristics: Upper molariform teeth have buccal roots; on P3 and P4 lingual hypostria shallow; on M1–2 lingual hypostria deeper and tending to persist as isolated lakes following rather early separation from lingual wall; on p3 single external reentrant persists in adult; lingual reentrants on p3 tend to disappear at an early stage of wear; cement not well developed; size larger than in all but one species of *Palaeolagus*, *P. intermedius* (Dawson, 1958).

Average length of p2–m2: 8.9–9.8 mm.

Included species: *M. turgidus* (known from localities CP83C, CP84A–C, CP99A–C, NP10[B2], BB, NP29C, [NP32B, C]); *M. abaconis* Hayes 2000 (locality GC8AA); *M. brachyodon* (Matthew, 1903) (localities CP39C, F, [CP98C], NP10[B], Bii, NP24C*–E, NP27C, D; NP49II); *M. dawsoni* Black, 1961 (localities [NP10E], NP34C*); *M. primitivus* (Schlaikjer, 1935) (localities CP42D*, [?CP85C], NP10C2).

Megalagus sp. is also known from localities GC8AA, AB, CP39D, E, ?CP43, NP10Bi, ?NP25B, C, NP36A.

Comments: Evolutionary trends in *Megalagus* from the late Eocene to early Miocene parallel similar trends in *Palaeolagus*, but at any given time *Megalagus* tends to lag behind *Palaeolagus* in most dental characters.

Mytonolagus Burke, 1934

Type species: *Mytonolagus petersoni* Burke, 1934.

Type specimen: CM 11937, Myton Pocket, Myton Member, Uinta Formation, Utah, Uintan.

Characteristics: Cheek teeth less hypsodont and more transverse, molarization of premolars less advanced, and molars less reduced than in *Megalagus brachyodon*; P3–4 with one buccal rootlet and shallow internal groove. P2 near M3, P4 near M2 in size; M1 largest and most transverse upper cheek tooth; p3 near m3 in size, m1 and m2 nearly equal in size; internal reentrants of p3 not persistent; main external reentrant not extending more than halfway across occlusal surface (Burke, 1934).

Average length of p3–m3: 10.8 mm.

Included species: *M. petersoni* (known from localities CP6B*, ?CP7B, [CP29C], [NP22]); *M. wyomingensis* Wood, 1949 (localities CP29D*, [NP25B, C]).

Mytonolagus sp. is also known from localities CC7C and NP25A.

Comments: One of the most interesting occurrences of *Mytonolagus* is its rare presence in the Uintan Santiago Formation of southern California (locality CC7C) (Golz and Lillegraven, 1977). *Mytonolagus* is a common element in Uintan deposits in the Rocky Mountain region of Wyoming and Utah, but rare in California and absent from the Uintan of Texas.

Palaeolagus Leidy, 1856

Type species: *Palaeolagus haydeni* Leidy, 1856.

Type specimen: ANSP 11031 (lectotype), Turtle Bed, head of Bear Creek, Nebraska, Orellan.

Characteristics: Cheek teeth more hypsodont and having better-developed cement than in *Mytonolagus* and contemporary species of *Megalagus*; P3–M2 carrying isolated crescent between central and lingual lobes usually retained, especially on premolars, which are less molariform than in *Archaeolagus*; internal hypostria uncrenulated, or having a few small irregularities, usually shorter transversely and worn away more rapidly on premolars than on molars; lower incisor extends posteriorly along ventromedial edge of lower jaw to below m1 in primitive species and to below talonid of p3 in more derived species; on p3 anteroexternal groove of trigonid usually lacking, or, if present, shallow and associated with internal reentrant between trigonid and talonid; anterior root of zygomatic arch situated slightly more posteriorly and extending laterally less abruptly than in *Archaeolagus*; in known skulls, angle between basicranial axis and palate smaller than in *Archaeolagus* (Dawson, 1958).

Average length of lower tooth row, p4–m2: 5.0–8.0 mm.

Included species; *P. haydeni* (including *P. temnodon*, in part) (known from localities CP39C, CP40B, CP41A, B, CP68B, CP84A, B, CP98C, CP99A–C, NP10[Bi], B2, BB, B3, NP27C, NP27C1, [NP32B, C], NP50B, C); *P.*

burkei Wood, 1940 (localities CP41C, CP46, CP68C, D, CP84A–C, CP99A, B, NP10BB, [C], [C2], NP27C1, D, NP29D, NP32B, C, NP50B, C); *P. hemirhizis* Korth and Hageman, 1988 (localities CP99A*, [NP10B2]); *P. hypsodus* Schlaikjer, 1935 (localities CP42D*, CP46, CP85A–C, CP101, NP34C); *P. primus* Emry and Gawne, 1986 (locality CP39B); *P. intermedius* Matthew, 1899 (localities [CP46], CP68C, ?CP84A–C, CP99A, B, NP27C1, D, NP29D, NP50B, C, NP51A); *P. philoi* Dawson, 1958 (localities CP85A, C*, CP86A, [C], CP101); *P. temnodon* Douglass, 1901 (localities CP43, CP98C, NP10B, [Bii], NP24C*, D, E, NP27C, NP29C, NP49II).

Palaeolagus sp. is also known from localities CC12, CP39D, NP10CC, NP25A–C, NP32B, NP36A, PN6D2, D3, G, and questionable records are known from localities GC3A, CC9C.

Comments: This long-ranging genus of leporid was a common member of many Oligocene mammalian assemblages. The morphological limits of *Palaeolagus* have been set broadly, and within the genus several species groups are recognized. These include the *P. primus–P. temnodon–P. haydeni* group; the *P. intermedius–P. philoi* group, and the *P. burkei–P. hypsodus* group. The morphologically most distinctive of the species is *Palaeolagus hypsodus*, in which the auditory bullae are greatly enlarged (Dawson, 1958). "The temporal and morphological series *Mytonolagus petersoni*, *M. wyomingensis*, *P. primus*, and *P. temnodon* can be interpreted as a phylogenetic series" (Emry and Gawne, 1986).

***Chadrolagus* Gawne, 1978**

Type species: *Chadrolagus emryi* Gawne, 1978.

Type specimen: F:AM 99106, 9 m below Ash G, Flagstaff Rim area, Wyoming, Chadronian.

Characteristics: Distinguished from *Palaeolagus* by morphology of premolars: P2 narrow, buccal lobe absent or greatly reduced, central lobe projecting well anterior to lingual lobe; p3 relatively short, protoconid smaller and located further posteriorly than metaconid and lacking anterior projection (Gawne, 1978).

Average length of p4–m2: 5.6–6.6 mm.

Included species: *C. emryi* only (known from localities CP39C, F*, NP24C, D).

Comments: While *Chadrolagus* is closely related to *Palaeolagus temnodon*, it appears not to be related to any later leporids (Gawne, 1978).

***Litolagus* Dawson, 1958**

Type species: *Litolagus molidens* Dawson, 1958.

Type specimen: LACM(CIT) 1568, lower nodular layer, Oreodon Beds, near Douglas, Wyoming, Orellan.

Characteristics: Cheek teeth hypsodont, having well-developed cement. On P3–M2 relatively long, straight-walled hypostria; shaft of lower incisor extends posteriorly to below p4; p3 with two lobes, separated by single buccal reentrant between trigonid and talonid; internal reentrant lost at early stage of wear; trigonid and talonid joined by lingual bridge on p4–m2; palatine component of palate shorter, anterior root of zygoma extends laterally less abruptly than in *Palaeolagus haydeni* (Dawson, 1958).

Average length of p3–m3: 9.8 mm.

Included species: *L. molidens* only, known from locality CP40B only.

Comments: *Litolagus* has an interesting combination of derived dental characters with primitive structure of the zygomatic region. It parallels *Palaeolagus* in postcranial characters.

ARCHAEOLAGINAE DICE, 1929

***Archaeolagus* Dice, 1917 (including *Lepus*, in part)**

Type species: *Archaeolagus ennisianus* (Cope, 1881) (originally described as *Lepus ennisianus*).

Type specimen: AMNH 7190, John Day Formation, Oregon, Arikareean.

Characteristics: In adult, single shallow anterior reentrant on P2; internal hypostria on P3–M2 usually straight walled; P3–4 less molariform than molars and also less than in corresponding premolars of *Hypolagus*; on p3, two external grooves with anterior groove shallow and usually lacking cement and posterior groove between trigonid and talonid cement filled; anterior root of zygoma situated more anteriorly and extending outward more abruptly than in *Palaeolagus* or *Megalagus* (Dawson, 1958).

Average length of p3–m3: 9.6–15.7 mm.

Included species; *A. ennisianus* (known from localities [CP48], PN6C–F, [G]); *A. acaricolus* Dawson, 1958 (locality CC15*); *A. buangulus* Dawson, 1969 (in Stevens, Stevens, and Dawson, 1969) (localities SB46*, [SB47]); *A. emeraldensis* Barnosky, 1986 (locality CP45B*); *A. macrocephalus* (Matthew, 1907) (localities [SB29A], CP86B*, D, CP104A, [NP10CC]); *A. sonaranus* Alvarez, 1963 (locality SB53*); *A. primigenius* (Matthew, 1907) (localities CP86B*, D, CP103B, CP104A, [CP105]), CC15, CC19, [NP10CC]).

Archaeolagus sp. is also known from localities CC9IIIA, CC9IVD, NB4, NP36B, NP37, and questionably from localities CC9D and CC10II.

Comments: Although widely distributed geographically in western North America, *Archaeolagus* is seldom a common element in any fauna.

***Panolax* Cope, 1874**

Type species: *Panolax sanctaefidei* Cope, 1874.

Type specimen: USNM 1095, Tesuque Formation, New Mexico, Barstovian.

Characteristics: Single anterior reentrant on P2; on upper molariform teeth internal reentrant straight walled, cement filled, and crossing half or less of occlusal surface (Dawson, 1958).

Average length of upper molariform tooth: 2.5 mm.

Included species: *P. sanctaefidei* only (known from localities NB8, SB32D*).

Comments: This poorly known genus seems to be more closely allied to *Archaeolagus* than to *Hypolagus* (Dawson, 1958).

***Hypolagus* Dice, 1917 (including *Lepus*, in part)**

Type species: *Hypolagus vetus* (Kellogg, 1910) (originally decribed as *Lepus vetus*).

Type specimen: UCMP 12565, Thousand Creek Formation, Nevada, Hemphillian.

Characteristics: On anterior surface of P2, one well-developed reentrant and shallower, more external groove; internal hypostria on P3-M2 crenulated at some stage of wear, crenulations usually more pronounced and persistent on premolars than on molars; anteroexternal fold on p3 deeper than in *Archaeolagus*, usually cement filled (Dawson, 1958).

Average alveolar length p3–m3: 11.9–17.5 mm.

Included species: *H. vetus* (known from localities CC35A, ?CC52II, NB13A–C, NB28, NB31*, SB10, SB34C, SB35A, [CP116B], ?NP5B, [NP11 including ?*H. gidleyi*], NP45, [PN7], PN11A, B, PN13, PN21B), [HA3]; *H. arizonensis* Downey, 1962 (locality SB18B*); *H. edensis* Frick, 1921 (localities CC40*, CC47D, NB13B, NB33A, PN4); *H. fontinalis* Dawson, 1958 (localities ?NB6E, NB23C, SB33II, CP90A*, CP114[A], B, CP116C, PN7, [CE2]); *H. furlongi* Gazin, 1934 (localities NB35IIIC, SP1G, CP121, [CP131], PN23C*); *H. gidleyi* White, 1987 (localities NB35IIIA, B, [SB34, sublocality uncertain], SB38IIB, [C], SP3B, CP116B, [NP11], PN4, PN23A*, PN23A*, B, C); *H. limnetus* Gazin, 1934 (localities WM26, [CC32B], CC38A, NB12 [sublocality uncertain], PN23A*); *H. parviplicatus* Dawson, 1958 (localities NB6C, D, NB18A*, [NB23B], CP56, CP114A, B); *H. oregonensis* Shotwell, 1956 (localities PN14*, PN15); *H. regalis* Hibbard, 1939 (localities CP116F, CP128C*); *H. ringoldensis* Gustafson, 1978 (localities GC13B, NB29B, SB11, SB14A, [SB34 (sublocality uncertain)], [SB58 (sublocality uncertain)], CP116F, CP128A, PN3C*); *H. tedfordi* White, 1987 (localities [GC13B], CC17G*, NB29B, SB9, SB10, SP1F, [SP1H (or could be *Sylvilagus*)]); *H. voorhiesi* White, 1987 (localities SP5, CP117B, CP121*, CP128C, PN23C).

Hypolagus sp. is also known from localities WM20, [WM26], [GC11A], CC17B–F, I, CC19, CC21B, C, CC25B, CC32A, CC36, CC37, CC53, NB7 (sublocality uncertain), NB8, NB10, NB13B, NB19A, NP29B, SB12, SB17, SB18A, SB31B, SB32D, SB48, SB64, SP1D, SP5, CP45E, CP54C, CP56IIB, CP76B, CP114C, CP116A, B, D, E, CP118, NP40A, B, NP42, PN8B, PN9B, PN12B.

Comments: *Hypolagus* is a wide-ranging genus, known from California to Florida, and from Mexico to Ellesmere Island (N. Rybczynski, personal communication) and northern Greenland (Funder *et al.*, 1985). It also has a long stratigraphic range, Barstovian through Blancan, but became extinct in North America by the end of the Blancan (White, 1987). The very primitive Barstovian species *Hypolagus parviplicatus* may be near the origin of the oldest known leporine *Pronotolagus*, a transition that occurred by at least 14 Ma (Voorhries and Timperley, 1997). *Hypolagus* is also known from the late Miocene to Pleistocene of Europe and Asia (McKenna and Bell, 1997).

***Lepoides* White, 1987**

Type species: *Lepoides lepoides* White, 1987.

Type specimen: UNSM 91196, Santee Local Fauna, Nebraska, Hemphillian.

Characteristics: Larger than all other archaeolagines; P2 with three anterior reentrants; on p3 anterior reentrant present, posterior external reentrant strongly deflected posteriorly (White, 1987).

Average alveolar length of p3–m3: 18.7–19.9 mm.

Included species: *L. lepoides* only (known from localities [NB29B], NB33A, NB35IIIA, CP116F*).

Comments: This largest archaeolagine is close in size to the living Arctic hare, *Lepus arcticus*. It is also distinctive in the complexity of folds on the anterior surface of P2 (White, 1987).

***Pewelagus* White, 1984**

Type species: *Pewelagus dawsonae* White, 1984.

Type specimen: LACM 24644, Upper Tapiado Wash, Arroyo Seco Local Fauna, Palm Spring Formation, Blancan.

Characteristics: Tympanic bulla larger proportionally and actually than in any other known leporid; internal nares narrow; P2 ovoid with medial anterior reentrant inflected at medial half and extending posteriad; external anterior reentrant shallow or absent; on p3 crenulations present on walls of anterior and posterior external reentrants (White, 1987).

Average alveolar length of p3–m3: 11.2 mm.

Included species: *P. dawsonae* (known from localities NB12B, NB13A, B*, C, NB33A, SP1F); ?*P. mexicanus* (locality SB65*).

Comments: This relatively small archaeolagine is characterized by its enlarged auditory bullae, which are larger than in any other leporid (White, 1987). *Hypolagus mexicanus* Miller and Carranza-Casta, 1982 was referred to ?*Pewelagus mexicanus* by White (1987), a tentative reference until better material of the species indicates whether it possesses the characteristically large auditory bulla of *Pewelagus*.

Notolagus Wilson, 1938

Type species: *Notolagus velox* Wilson, 1938.

Type specimen: LACM(CIT) 2133, Rincon Local Fauna, Yepomera, Chihuahua, Mexico, Hemphillian.

Characteristics: Two to three anterior reentrants on P2; P3–M2 with crenulated internal hypostriae that extend approximately two-thirds of occlusal width; anterior upper incisor with cement-filled narrow groove on anterior surface; p3 with posterior external reentrant extending across from 40 to 60% of occlusal width, anteroexternal reentrant deep and complexly folded, crossing from 16 to 35% of occlusal width or completely cutting across tooth to unite with anterior internal reentrant and isolate anterior portion of tooth as a separate column (Wilson, 1938; White, 1991a).

Average length of p3: 2.2–3.4 mm.

Included species: *N. velox* (known from localities SB58B, SB60*, SB62); *N. lepusculus* (Hibbard, 1939) (localities SP1F, CP128B, C*).

Notolagus sp. is also known from localities GC11D, SB14A, D.

Comments: *Notolagus* was originally described as an archaeolagine, based on the fact that the internal reentrant on p3 is anterointernal, not posterointernal. White (1991a) reassigned the genus to the Leporinae because of the anterointernal reentrant on p3. However, the internal fold that characterizes p3 in the leporines is the posterointernal fold or its remnant that fuses with the posteroexternal reentrant (Hibbard, 1963). Accordingly, *Notolagus* is regarded here as a highly derived archaeolagine, not a leporine.

Paranotolagus Miller and Carranza-Castañeda 1982

Type species: *Paranotolagus complicatus* Miller and Carranza-Castañeda 1982.

Type specimen: IGCU 3957, Rancho Viejo, Guanajuato, Mexico, Blancan.

Characteristics: Anterior upper incisor with three cement-filled grooves; P3 with one to two anterior reentrants, anterior internal reentrant deep with crenulated enamel, anterior external reentrant shallow, posterior external reentrant with enamel strongly folded on posterior side; lower molariform teeth with trigonid and talonid separated on lingual side (Miller and Carranza-Castañeda, 1982).

Average length of p3: 2.1–2.4 mm.

Included species: *P. complicatus* only, known from locality SB65* only.

Paranotolagus sp. is also known from locality SB58C.

Comments: *Paranotolagus* exhibits even more complicated enamel folding than its relative *Notolagus* or other archaeolagines. The folding even extends to multiple grooves on the incisor.

Indeterminate archaeolagines

Indeterminate archaeolagines are reported from localities CC47A, B, CC48B, and NP9B.

LEPORINAE

A variety of small differences between p3 distinguish a number of mostly Blancan (Pliocene) genera, which are known mainly from incomplete material. Where known, skulls are essentially in a *Lepus–Sylvilagus* grade of development. Whether the differences in p3 warrant generic separation may be questioned, in that living species of *Lepus* show as much variation in p3 as do many of these genera. The 10 genera and 14 species recognized by White (1991a) share a temporal range from Clarendonian (early late Miocene) through Blancan (11 to 3 Ma). The diversity in occlusal pattern represented among these taxa documents some of the morphological trends leading to the establishment of the leporine pattern of p3, while not clearly establishing the precise lineage(s) to *Sylvilagus* and *Lepus*.

Pronotolagus White, 1991a (including _Hypolagus_, in part)

Type species: *Pronotolagus apachensis* (Gazin, 1930) (originally described as *Hypolagus apachensis* Wood, 1937; including *Hypolagus*? *apachensis*).

Type specimen: LACM (CIT) 36, Apache Canyon, Ventura County, California, early Clarendonian.

Characteristics: p3 lacking anterior reentrant, anterior internal reentrant more deeply incised than posterior internal reentrant, anterior external reentrant shallow and wide, posterior external reentrant crosses less than half occlusal width (White, 1991a).

Average length of p3: 2.2–2.8 mm.

Included species: *P. apachensis* (known from localities CC17G*, [CC19], CP116C); *P. albus* Voorhies and Timberley, 1997 (localities CP56IIB, CP114B*); *P. nevadensis* Kelly, 2000 (locality NB34II); *P. whitei* Korth, 1998 (locality CP116B).

Pronotolagus sp. is also known from localities CC31, PN15, PN21B.

Comments: As the varied taxonomic references imply, this leporid is close to the archaeolagines but may be regarded as the most primitive leporine on the basis of the posterointernal reentrant on p3 (White, 1991a), a character that it shares also with *Alilepus*. The Barstovian *P. albus* is the more primitive of the two known species (Voorhies and Timberley, 1997).

Alilepus Dice, 1931 (including _Lepus_, in part)

Type species: *Alilepus annectens* (Schlosser, 1924) (originally described as *Lepus annectens*).

Type specimen: Lagrelius Collection, Uppsala, Ertemte, Nei Mongol, China, latest Miocene.

Characteristics: Medium to large leporine with skull and mandible modernized; P2 with deep median anterior reentrant and shallow external anterior reentrant; p3 with posterior internal reentrant often forming isolated lake, anterior internal reentrant shallower than posterior internal reentrant, when present, anterior external reentrant shallow (White, 1991a).

Average alveolar length of p3–m3: 11.7–18.0 mm.

Included species: *A. hibbardi* White, 1991a (known from localities CC26B, NB29A, PN22II*); *A. vagus* Gazin, 1934 (localities NB35IIIB, CP116F, PN4, PN23A*, C); *A. wilsoni* White, 1991a (localities SB31D*, CP132B).

Alilepus sp. is also known from localities CP114B, CP116B, CP128C.

Comments: The late Clarendonian *A. hibbardi* and Hemphillian to Blancan *A. vagus* retain primitive features of p3. *A. wilsoni* is clearly very close to the genus *Alurolagus*. *Alilepus* is also known from the Miocene and Pliocene of Europe and Asia (McKenna and Bell, 1997).

***Pratilepus* Hibbard, 1939**

Type species: *Pratilepus kansasensis* Hibbard, 1939.

Type specimen: KUVP 4582, Rexroad Locality 3, Meade Formation, Kansas, Blancan.

Characteristics: On p3 anterior internal reentrant more deeply incised than in any other leporid except *Auralagus*; enamel on posterior wall of anterior and posterior external reentrants highly crenulated; posterior internal reentrant variable, connecting to lingual wall, isolated as lake, or united with posterior external reentrant; three anterior reentrants on P2, the medial one deep and the internal and external ones shallower (White, 1991a).

Average alveolar length of p3–m3: 13.2–13.8 mm.

Included species: *P. kansasensis* only (known from localities SP1F, CP128C*).

Pratilepus sp. is also known from locality SB58C.

Comments: *Pratilepus* and *Alurolagus* must be very closely allied, if not congeneric.

***Nekrolagus* Hibbard, 1939 (including *Pediolagus*)**

Type species: *Nekrolagus progressus* Hibbard, 1939 (originally described as *Pediolagus progressus*).

Type specimen: KU 4570, Rexroad Fauna, Meade Formation, Kansas, Blancan.

Characteristics: Leporine very similar in dental pattern and jaws to *Lepus* and *Sylvilagus*, differing mainly in posterior reentrant on p3, in which the posterior internal reentrant forms a lake that is either isolated or united with the posterior external reentrant (Hibbard, 1963; White, 1991a).

Average alveolar length of p3–m3: 15.2 mm.

Included species: *N. progressus* only (known from localities GC13B, NB13A, NB33A, SP1F, [SP5], CP128C*, PN3C).

Nekrolagus sp. is also known from localities NB13C, SB14A, D, CP132B.

Comments: Hibbard's (1963) perceptive analysis of the dental pattern of the Blancan leporines *Pratilepus* and *Nekrolagus* led to understanding of the origin of the p3 pattern in *Lepus* and *Sylvilagus*. *Nekrolagus* is also possibly known from the early Pliocene of Asia (McKenna and Bell, 1997).

***Aluralagus* Downey, 1968 (including *Sylvilagus*, in part)**

Type species: *Aluralagus bensonensis* (Gazin, 1942) (originally described as *Sylvilagus*? *bensonensis*).

Type specimen: USNM 16595, south of Benson, Arizona, Saint David Formation, Blancan.

Characteristics: p3 having anterior external reentrant strongly inflected and with thin, well-crenulated enamel, lake from posterior internal reentrant isolated or united with posterior external reentrant (White, 1991a).

Average length of p3; 2.2–3.2 mm.

Included species: *A. bensonensis* (known from locality SB14A*); *A. virginiae* Downey, 1970 (localities SB14F*, SB18 [sublocality uncertain], SB37D).

Aluralagus sp. is also known from localities SB12, SB14A, [SB68].

Comments: There are numerous "genera" in the Blancan approaching the *Lepus–Sylvilagus* dental pattern but retaining some vestiges of the *Nekrolagus–Pratilepus* condition (White, 1991a). *Aluralagus* is one such genus, and could equally well be included in *Pratilepus* or *Sylvilagus*.

***Sylvilagus* Gray, 1867**

Type species: *Sylvilagus floridanus* Lyon, 1904 (originally described as *Lepus sylvaticus* Bachman, 1837; including *Sylvilagus sylvaticus* Gray, 1867), the extant North American rabbit.

Type specimen: None designated.

Characteristics: Medium to large leporine; interparietal bone separate; P2 with two or three anterior reentrants; p3 with *Lepus*-pattern of enamel folds, including shallow anteroexternal fold and posterointernal fold extending to lingual side of tooth (White, 1991a).

Average alveolar length of p3–m3: 11.6–15.1 mm.

Included species: *S. hibbardi* White, 1984 (known from localities CC47D, NB13C, SB14B, SP1H); *S. webbi* White, 1991b (localities GC15 ?A,?B, C, D, GC17A*, ?B, GC18IIB).

Sylvilagus sp. is also known from localities GC15A, GC16, CC48B, NB13B, C, SB14E, F, SB37D (could be *Lepus*). The genus is also known from most US Pleistocene localities (including *S. floridanus* at localities GC15E, GC18IIIA, GC18IV, and *S. palustris* at locality GC18IV).

Comments: It is difficult on dental evidence alone to differentiate several leporids that have Blancan records: *Alurolagus*, *Pratilepus*, *Sylvilagus*, and *Lepus* share many dental features. Both *Lepus* and *Sylvilagus* have been recorded from Blancan localities but these generic assignments must be considered tentative. *Sylvilagus* is also known from the Pleistocene and Recent of South America (McKenna and Bell, 1997).

***Aztlanolagus* Russell and Harris, 1986**

Type species: *Aztlanolagus agilis* Russell and Harris, 1986.

Type specimen: UTEP 1-1202, Dry Cave, Eddy County, New Mexico, Rancholabrean.

Characteristics: Lower incisor root ends anterior to p3; p3 with anterior external reentrant, anterior reentrant, and anterior internal reentrant, posterior external reentrant occasionally unites with lake from posterior internal reentrant (Russell and Harris, 1986; White, 1991a).

Average length of p3: 2.0–2.7 mm.

Included Tertiary species: *A. agilis* only, known from locality SB15C only (also known from the Pleistocene of New Mexico, Arizona, and Texas).

Comments: This small leporid is very incompletely known; only a single lower jaw and a few lower cheek teeth are represented from a few Blancan and Pleistocene localities in New Mexico, Arizona, and Texas (Russell and Harris, 1986; Winkler and Tomida, 1988). *Aztlanolagus* has a highly distinctive pattern of p3 with deep anterior and anterointernal folds and an isolated posterointernal lake. This might be an immigrant from Eurasia with affinities to *Pliopentalagus*. However, the distinctive p3 structure may have developed in parallel among different descendants of *Alilepus* across the Holarctic, and *Aztlanolagus* may be traceable to a North American *Alilepus*.

Indeterminate leporids

Indeterminate leporids have been reported from the following localities: CA1, GC5B, GC6B, GC7, GC9C, CC35A, CC41, NB27A, B, NB36IIIA, B, SB37A, B, D, SB49, SB50, CP107, CP123D, CP128AA, NP9C, NP10D, NP34B, HA2.

Lepus sp. (the extant North American hare, or jackrabbit) is known from latest Blancan/early Irvingtonian localities GC15C, F, GC18IIIA, GC18IV.

OCHOTONIDAE THOMAS, 1897

Oreolagus Dice, 1917 (including *Palaeolagus*, in part)

Type species: *Oreolagus nevadensis* (Kellogg, 1910) (originally described as *Palaeolagus nevadensis*).

Type specimen: UCMP 12575, Virgin Valley, Nevada, Barstovian.

Characteristics: Ochotonids near Recent *Ochotona* (the extant pika) in size; dental formula I(2)/1, C0/0, P3/2, M2/2; on P3 anterior loph crossing half or less of tooth width, crescentic valley retains anteroexternal connection, lingual hypostria short; on P4, hypostria crosses less than half tooth width, directed anteriorly toward but not extending past crescentic valley; M1 primitively with crescentic valley and longer hypostria than P4, valley progressively absent in adults and hypostria lengthening until it crosses approximately three-fourths of tooth width; M2 smaller than M1, shallow lingual fold, lingual part of posterior loph swinging posteriorly; lower incisor terminating below p4 or m1; on p3, buccal fold crossing approximately half tooth width between trigonid and wider talonid, shallow grooves in some specimens occurring anteriorly and lingually; p4, m1, and m2 having trigonid and slightly narrower to equally wide talonid, talonid with narrow anterior protrusion, column of each tooth joined by cement; two mental foramina on lateral surface of lower jaw, one anterior to p3, second more ventral, below molars (Dawson, 1965).

Average length of p4–m2: 5.0–6.4 mm.

Included species: *O. nevadensis* (known from localities NB18A*, NP41B); *O. colteri* Barnosky, 1986 (locality CP45E*); *O. nebrascensis* McGrew, 1941 (localities [SB29B], CP54B, CP105*, [NP10E]); *O. wallacei* Dawson, 1965 (locality PN8A*); *O. wilsoni* Dawson, 1965 (locality CP71*).

Oreolagus sp. is also known from localities NP40?A, B, NP42, PN16.

Comments: *Oreolagus* was widely distributed during the Miocene and has been reported from localities in Nebraska, Wyoming, Montana, New Mexico, Nevada, and Oregon. The species of *Oreolagus* illustrate gradual changes in dental and jaw morphology. *Oreolagus* is more derived in dental characters than *Desmatolagus gobiensis* but could be related to *D. gobiensis*. The *Oreolagus* lineage is distinct from those of the other North American ochotonids (Dawson, 1965).

Desmatolagus Matthew and Granger, 1923

Type species: *Desmatolagus gobiensis* Matthew and Granger, 1923.

Type specimen: AMNH 19103, Hsanda Gol Formation, Red Beds, Loh, Mongolia, Shandgolian (approximately Orellan).

Characteristics of ?*D. schizopetrus*, Dawson, 1965, the only North American species that might represent *Desmatolagus*: cheek teeth hypsodont but rooted; concave buccal and convex lingual walls indicate unilateral hypsodonty of upper cheek teeth; anterior loph of P3 crossing approximately two-thirds width of tooth, crescentic valley connecting to anteroexternal wall, lingual hypostria shallow; upper molars with distinct crescentic valley, hypostria crossing more than one-third occlusal width in early wear, pattern reducing with wear to remnant of valley and lingual notch; lower molariform teeth with trigonid having relatively straight posterior wall and narrower talonid with anteriorly directed protrusion and curved posterior wall (Dawson, 1965).

Length upper molar of ?*D. schizopetrus*: 1.7–1.8 mm.

Included species: ?*D. schizopetrus* only (known from localities CP54B*, NP10E, ?NP36B).

Comments: *Desmatolagus* is primarily an Old World genus, known from the late Eocene to late Miocene of Asia, and the early Oligocene of Europe (McKenna and Bell, 1997). With considerable perception, Matthew and Granger (1923) recognized that *Desmatolagus gobiensis* was related to the Ochotonidae. Indeed, presence of a premolar foramen in the maxilla is the most reliable shared derived feature of early members of the Ochotonidae (a premolar foramen is absent in leporids). *D. gobiensis* is thus the oldest known definite ochotonid. "*Desmatolagus*" *robustus* from the same deposits in Mongolia is a leporid whose real generic affinities remain to be determined. The only North American lagomorph that might represent *Desmatolagus* is ?*D. schizopetrus*, known only from isolated teeth. Reference to *Desmatolagus* is, however, highly questionable. Voorhies (1990) suggested that the ochotonid *Russellagus vonhofi* might be a junior synonym of ?*D. schizopetrus*, in which case the appropriate name would be *Russellagus schizopetrus*.

Note, however, that Martin (2004) has recently considered *Desmatolagus* to be a leporid, based on the fact that its incisor enamel contains a single-layered shmelzmuster with Hunter–Schreger bands (see p. 295).

Cuyamalagus Hutchison and Lindsay, 1974

Type species: *Cuyamalagus dawsoni* Hutchison and Lindsay, 1974 (in Hutchison and Lindsay, 1974).

Type specimen: UCMP 82151, Vedder Locality, Branch Canyon Formation, California, Hemingfordian.

Characteristics: Cheek teeth hypsodont but rooted; upper molariform teeth unilaterally hypsodont, with distinct crescentic valleys and hypostriae crossing one-third or less of occlusal width in early wear; lower molars with talonid and trigonid separate to near base of crown, talonid nearly as wide transversely as trigonid (Hutchison and Lindsay, 1974).

Average length of lower molariform tooth: 2.2–2.3 mm.

Included species: *C. dawsoni* only (known from localities CC9IIIB, CC9IVD, CC19*).

Comments: Although only known by six teeth, *Cuyamalagus dawsoni* shows a relatively advanced condition of the lower molariform teeth, with the trigonid and talonid nearly equal in width. No clear affinities with other North American Miocene ochotonids can be positively determined based on the limited available material. Hutchison and Lindsay (1974) suggested possible relationships to ?*Desmatolagus schizopetrus* or *Gripholagomys*.

Russellagus Storer, 1970

Type species: *Russellagus vonhofi* Storer, 1970.

Type specimen: ROM 7384, Kleinfelder Farm Locality, Wood Mountain Formation, Saskatchewan, Barstovian.

Characteristics: Cheek teeth hypsodont but rooted; lower jaw with large mental foramen beneath trigonid of m1, smaller foramen beneath trigonid of p4; below cheek teeth ventral border of horizontal ramus straight, curving ventrally to angle; talonid of lower molariform teeth narrower, longer anteroposteriorly than trigonid, sending forward wide, buccally plicate processes to meet trigonid; upper cheek teeth unilaterally hypsodont, columns convex lingually with two small buccal roots and one large lingual root, persistent, cement-filled, crescentic valley; short lingual hypostria traversing one-fourth or less width of crown and ending lingual to crescentic valley (Storer, 1970).

Average length of p3–m3: 8.95 mm.

Included species: *R. vonhofi* only (known from localities CP114A, B, NP11*).

Russellagus sp is also known from locality CP116B.

Comments: Bair (1998) has shown that *Russellagus vonhofi* has a premolar foramen, a diagnostic character of ochotonids. Voorhies (1990) suggested that *R. vonhofi* is a junior synonym of ?*Desmatolagus schizopetrus*. *Russellagus*, *Hesperolagomys*, and ?*Desmatolagus schizopetrus* are closely related, and may represent divergence following a single immigration event from Asia.

Oklahomalagus Dalquest, Baskin and Schultz, 1996

Type species: *Oklahomalagus whisenhunti* Dalquest, Baskin, and Schultz, 1996.

Type specimen: MSUCFV 12604, Whisenhunt Quarry, Laverne Formation, Oklahoma, Clarendonian.

Characteristics: Upper molariform teeth with strong roots; P3 with elaborate folding of anterior reentrant; internal crescentic valleys U- or V-shaped; lower molariform teeth with anteroposteriorly compressed trigonids with pinched buccal ends (Dalquest, Baskin, and Schultz, 1996).

Average length of P3: 2.00 mm.

Included species: *O. whisenhunti* only, known from locality SP1A only.

Comments: Known from less than 20 isolated cheek teeth, the broader affinities of this taxon are not yet possible to determine.

Hesperolagomys Clark, Dawson and Wood, 1964

Type species: *Hesperolagomys galbreathi* Clark, Dawson and Wood, 1964.

Type specimen: MCZ 17890, Fish Lake Valley, Esmeralda Formation, Nevada, Clarendonian.

Characteristics: Cheek teeth hypsodont but rooted; occlusal surface of P4–M1 with persistent crescentic valley, hypostria extending almost to crescent, and anteroloph transversely wider than posteroloph; p3 with buccal fold between trigonid and talonid, anterointernal groove in trigonid, lingual wall short anteroposteriorly; trigonid of p4–m2 wider and shorter than talonid; large mental foramen below p3, smaller mental foramina on ramus ventral and anterior to p3 and below m1–2; m3 present (Clark, Dawson, and Wood, 1964).

Average length of p3–m2: 5.8 mm.

Included species: *H. galbreathi* (known from localities NB23C*, [CP116B]); *H. fluviatilis* Storer, 1970 (localities CP114A–?C, NP11*).

Hesperolagomys sp. is also known from locality NB7B.

Comments: Although Storer (1984) questioned the assignment of *Hesperolagomys* to the Ochotonidae, Bair (1998) reaffirmed this familial assignment, demonstrating that the genus possessed a premolar foramen in the maxilla, a shared derived feature of ochotonids.

***Gripholagomys* Green, 1972**

Type species: *Gripholagomys lavocati* Green, 1972.

Type specimen: SDSM 62294, Black Bear Quarry II, Rosebud Formation, South Dakota, Hemingfordian.

Characteristics: P3 with short anterior loph; P4 rectangular in outline; posterior arms of crescents of upper cheek teeth long; p3 with slight anterolingual groove (Green, 1972).

Average length of lower molariform tooth: 1.8–2.2 mm.

Included species: *G. lavocati* only (known from localities CP86E, F*, CP103B, NP10CC, NP36D).

Comments: Storer (1984) considered *Gripholagomys* to be a leporid related to *Procaprolagus vusillus*. Assignment here to the Ochotonidae is based on dental characters alone More complete material will be required to establish familial relationships. Relationship to *Cuyamalagus* has also been suggested (Hutchison and Lindsay, 1974).

***Ochotona* Link, 1795 (including *Lepus*, in part)**

Type species: *Ochotona daurica* (Pallas, 1778) (originally described as *Lepus daurica*), the extant North American pika.

Type specimen: None designated.

Characteristics: Skull not arched, constricted between orbits; lateral side of maxilla with single large opening, finer fenestration present or absent; mental foramen below m2; P3 non-molariform; hypostria of M1–2 not crenulated; M2 with posterolingual projection; M3 absent; p3 with two buccal and one lingual fold; m3 single column; clavicle well developed; pubic symphysis absent.

Included Tertiary species: *O. spanglei* Shotwell, 1956 only (known from localities [CP116E, F], PN14*).

Ochotona is also known from the Pleistocene of West Virginia, Pennsylvania, Maryland, Colorado, Wyoming, Idaho, Alaska, and the Yukon.

Comments: The only definitive Tertiary record of *Ochotona* in North America is from the McKay Reservoir Local Fauna, Shutler Formation, Hemphillian, of Oregon. The genus is clearly an immigrant from Asia, as both the lineage and the genus itself can be traced to older horizons there (Dawson, 1961; Qiu, Wu, and Qiu, 1999). During the Pleistocene, *Ochotona* was widely distributed across North America, ranging as far east as Maryland (Guilday, 1979). The present geographic range of *Ochotona*, limited to two species in the western United States and Canada, is a relict distribution of this formerly widespread lagomorph. The geographic restriction of *Ochotona* may be related to competitive exclusion, possibly by microtine rodents.

The genus *Ochotona* is also known from the late Miocene to late Pleistocene of Europe, the late Miocene to Recent of Asia, and possibly from the Pleistocene of Africa (McKenna and Bell, 1997).

Indeterminate ochotonids

An indeterminate ochotonid is reported from locality HA3.

BIOLOGY AND EVOLUTIONARY PATTERNS

Lagomorphs have formed an important component of many North American faunas over the past 40 million years.

While Eocene lagomorphs may have been adapted to a variety of habitats, with the cooling and drying that characterized the Eocene–Oligocene transition (Retallack, 1992) they became more hypsodont, probably in relation to floral changes promoted by changing environmental conditions.

The evolutionary and paleobiogeographic histories of the Leporidae and Ochotonidae are markedly different (Figures 17.3 and 17.4). The Leporidae entered North America much earlier than the Ochotonidae and have a North American record extending back to the middle Eocene (late Uintan). It appears that two separate leporid lineages arrived approximately simultaneously from Asia and gave rise to *Mytonolagus* and North American "*Procaprolagus*." The more successful lineage was that of *Mytonolagus*. During the late Eocene and Oligocene, the *Mytonolagus* lineage produced several branches, all retaining the palaeolagine level of evolutionary development. The early and middle Miocene were times of low diversity for leporids in many localities. The archaeolagine *Archaeolagus* was replaced by *Hypolagus* in the Hemingfordian. The greatest diversity known for the family from one locality is three genera from the Barstovian Stewart Quarry in Nebraska (within locality 114B), which has been interpreted as a preserved woodland habitat and which records the first presence of the more derived leporines (Voorhies and Timperley, 1997). The peak of leporid generic diversity was reached in the Pliocene. While part of this diversity is real, it should also be recognized that some of the Pliocene genera are very closely related and separated by very small morphological distances. The four Recent North American leporid genera, *Romerolagus*, *Brachylagus*, *Sylvilagus*, and *Lepus*, were derived from this diversity of Pliocene genera. A high rate of turnover occurred among leporids during the Blancan, when 11 species of archaeolagines and 10 species of leporines became extinct (White, 1987).

Dental evolution in the leporids was accompanied by changes in the posture, skull, and locomotion (Dawson, 1958; White, 1984). Many of these changes occurred in parallel in the palaeolagines, archaeolagines, and leporines. An interesting parallelism is the development of greatly enlarged auditory bullae in the Arikareean palaeolagine *Palaeolagus hypsodus* (Dawson, 1958) and in the Blancan archaeolagine *Pewelagus dawsonae* (White, 1984).

Following a middle Eocene (late Uintan, *c.* 43 Ma) immigration from Asia, North America became the principal site of leporid

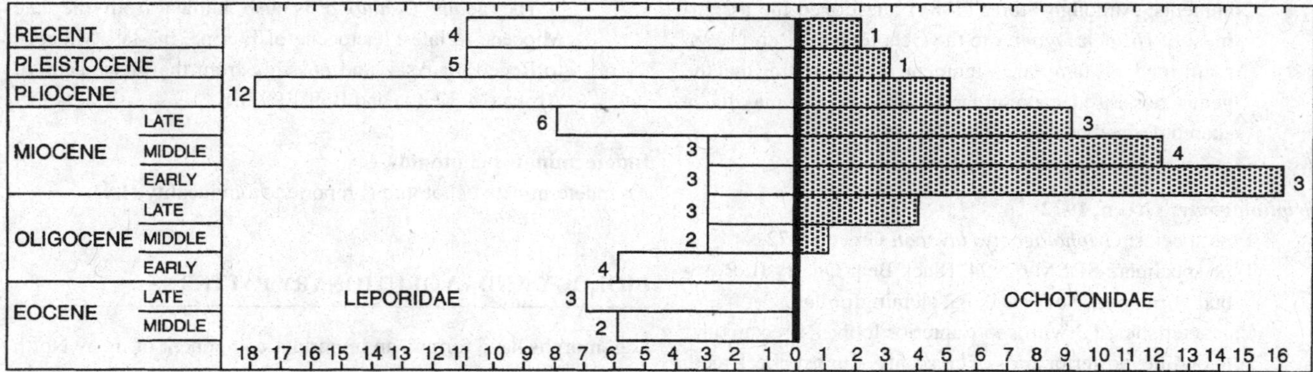

Figure 17.3. Diversity of leporids and ochotonids through the Cenozoic, shown by plotting numbers of genera known (each horizontal unit equals one genus) for each time interval indicated. Number at the end of a bar indicates North American genera out of the total.

evolution until the late Miocene (*c.* 8 Ma), when leporids reappeared in Eurasia (Dawson, 1967). The oldest known North American leporids are the late Uintan (middle Eocene) genera *Mytonolagus* and *Procaprolagus*. *Mytonolagus* led to the main line of leporid evolution. By the late Eocene (Chadronian), three genera were present: *Palaeolagus*, *Chadrolagus*, and *Megalagus*. While *Chadrolagus* appears to have been a short-lived lineage, species of *Palaeolagus* and *Megalagus*, both characterized by gradual morphological changes, persist into the Arikareean (late Oligocene; *Palaeolagus hypsodus*) and Hemingfordian (early Miocene; *Megalagus* cf. *dawsoni*). *Procaprolagus* is a dentally more primitive, primarily Asian, leporid that survives only into the early Oligocene (Orellan) in North America (Storer, 1984).

The early and middle Miocene record of the Leporidae is exclusively North American. *Archaeolagus* is a relatively rare element in faunas of the Great Plains. The younger, more cursorially adapted *Hypolagus* appears to have been more abundant and successful, especially in localities of the Great Basin, where drier, more open habitats can be inferred.

Considerable morphological, particularly dental, diversity is shown by a number of leporid genera from the late Miocene and Pliocene. Many of these genera show dental variants leading to the occlusal pattern, mostly on P2 and p3, characteristic of the most successful modern leporids: *Lepus*, *Sylvilagus*, and *Oryctolagus*. The origin of the modern dental pattern in these leporids was thoroughly documented by Hibbard (1963). More recent additions (especially by White, 1984, 1991a) have confirmed this scenario.

The ochotonids were relatively late arrivals in North America, first recorded in the late early Arikareean (latest Oligocene). Of the North American Miocene ochotonids, only *Oreolagus* has a fairly extensive record. The North American ochotonid record appears to reflect at least three, and possibly four, separate immigration events from Asia or Beringia (Dawson, 1999) leading to *Oreolagus*, *Hesperolagomys–Russellagus–?Desmatolagus*, and *Ochotona*, with perhaps a separate event for *Cuyamalagus*. Unlike the leporids, which occur only in North America during the middle Tertiary, ochotonids reach their widest geographic distribution in the Miocene. Several distinct lineages are represented, with endemic members in Europe, others in Asia, and probably two in Africa in addition to those that developed in North America.

The Ochotonidae first appear relatively late in the fossil record. The first well-documented species is the Asian *Desmatolagus gobiensis* of Shandgolian age (early Oligocene, approximately equivalent to the North American Orellan). By the late Oligocene, ochotonids had expanded into western Europe, where early diversification led to several morphologically distinct lineages. Geographic extension from Asia into North America occurred somewhat later, with the first North American appearance of *Oreolagus* in the late Arikareean. *Oreolagus* is abundant at several localities in the western United States and is morphologically well known from jaws and dentitions. It has been interpreted as an inhabitant of stream-border communities, where leporids are relatively rare (Wilson, 1960).

Several other North American ochotonids appear to represent distinct Miocene immigrations from Asia. The best known of these are *Russellagus* and *Hesperolagomys*, two relatively closely related ochotonids that range from the Great Plains west to Nevada and north to Saskatchewan. *Cuyamalagus* is distinct from these but is known only from a few isolated teeth. The Clarendonian *Oklahomalagus*, also known from isolated teeth, is of uncertain affinities.

Ochotonids became rare in the Pliocene of North America. *Ochotona* appears as an immigrant from Asia in the Hemphillian and persists in the modern North American fauna.

ACKNOWLEDGEMENTS

It is a pleasure to recognize here my past masters in lagomorphology, Robert W. Wilson, Albert E. Wood, John J. Burke, Claude W. Hibbard, and John A. White. Discussing lagomorphs with these savants has always been a worthwhile experience. Present masters Mike Voorhies and John Storer have contributed extensively to the understanding of these animals in the North American Cenozoic.

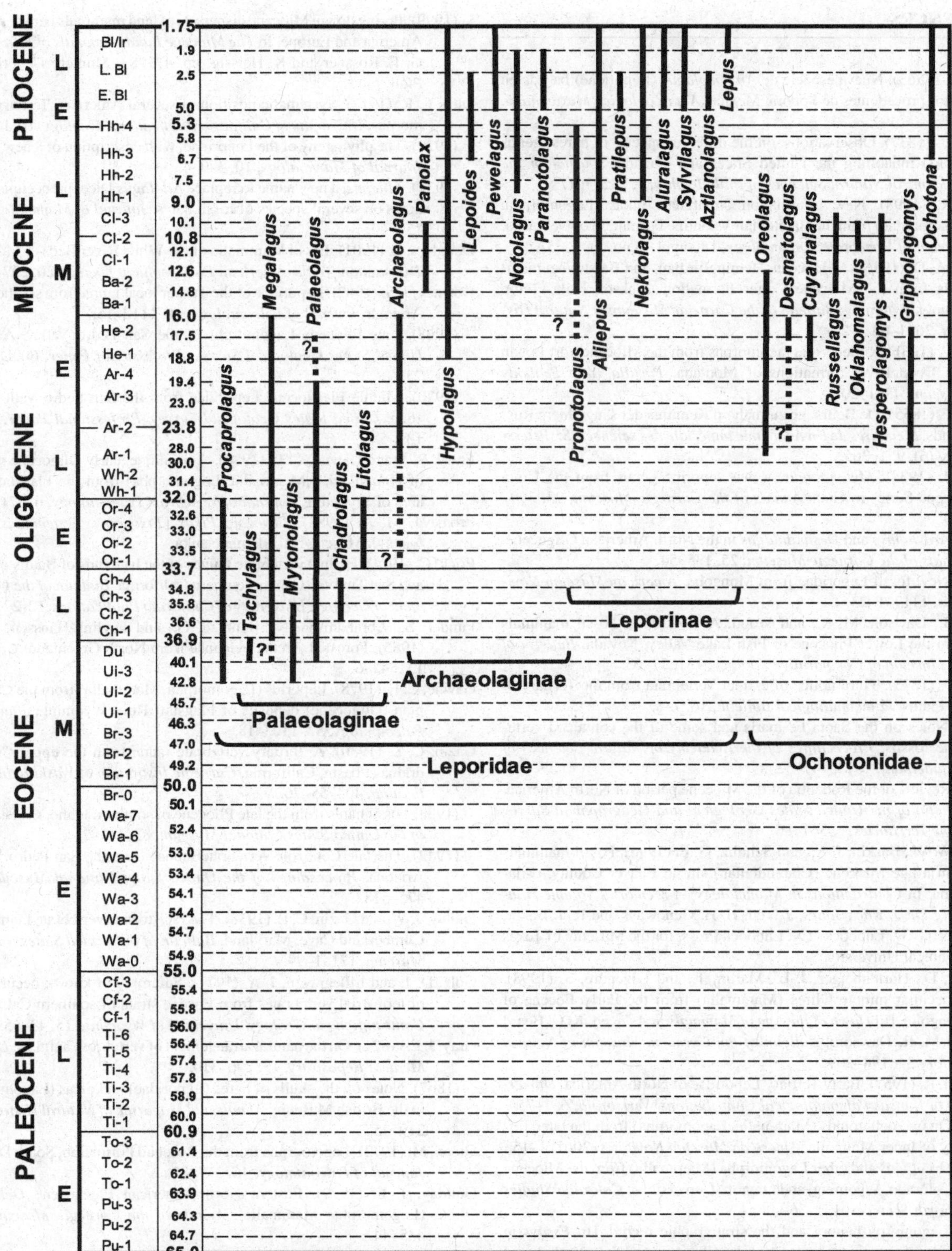

Figure 17.4. Temporal ranges of North American lagomorph genera.

REFERENCES

Alvarez, T. (1963). Nueva especie de *Archaeolagus* (Leporidae) basada en restos procedentes de Sonora, Mexico. *Acta Zoologica Mexicana*, 6, 1–4.

Bachman, J. (1837). Observations, on the different species of hares (genus *Lepus*) inhabiting the United States and Canada. *Journal of the Academy of Natural Sciences of Philadelphia*, 7, 282–361.

Bair, A. R. (1998). New information on archaic pikas (Lagomorpha, Ochotonidae) from the Barstovian Valentine Formation of Nebraska. *Journal of Vertebrate Paleontology*, 18(suppl. to no. 3), p. 25A.

Barnosky, A. D. (1986). Arikareean, Hemingfordian, and Barstovian mammals from the Miocene Colter Formation, Jackson Hole, Teton County, Wyoming. *Bulletin of the Carnegie Museum of Natural History*, 26, 1–69.

Black, C. C. (1961). Rodents and lagomorphs from the Miocene Fort Logan and Deep River formations of Montana. *Postilla, Yale Peabody Museum*, 48, 1–20.

Brandt, J. F. (1855). De Beitraege zur nähern Kenntnis der säugethiere Russlands. *Mémoires de l'n Académie impériale des sciences, St. Petersburg* (6), 9, 1–365.

Burke, J. J. (1934). *Mytonolagus*, a new leporine genus from the Uinta Eocene Series in Utah. *Annals of the Carnegie Museum*, 23, 399–420.

(1936). *Ardynomys* and *Desmatolagus* in the North American Oligocene. *Annals of the Carnegie Museum*, 25, 135–54.

(1941). New fossil Leporidae from Mongolia. *American Museum Novitates*, 1117, 1–23.

Clark, J. B., Dawson, M. R., and Wood, A. E. (1964). Fossil mammals from the Lower Pliocene of Fish Lake Valley, Nevada. *Bulletin of the Museum of Comparative Zoology*, 131, 27–63.

Cope, E. D. (1873). Third notice of extinct Vertebrata from the Tertiary of the Plains. *Palaeontological Bulletin*, 16, 1–8.

(1874). Notes on the Santa Fe marls and some of the contained vertebrate fossils. *Proceedings of the Academy of Natural Sciences of Philadelphia*, 1874, 147–52.

(1881). Review of the Rodentia of the Miocene period of North America. *Bulletin of the United States Geological and Geographical Survey of the Territories*, 6, 361–86.

Dalquest, W. W., Baskin, J. A., and Schultz, G. E. (1996). Fossil mammals from a late Miocene (Clarendonian) site in Beaver County, Oklahoma. In *Contributions in Mammalogy: A Memorial Volume Honoring Dr. J. Knox Jones, Jr*,, ed. H. H. Genoways and R. J. Baker, pp. 107–37. Lubbock. TX: Lubbock Press, for the Museum of Texas Technical University.

Dashzeveg, D., Hartenberger, J.-L., Martin, T., and Legendre, S. (1998). A peculiar minute Glires (Mammalia) from the Early Eocene of Mongolia. [In *Dawn of the Age of Mammals in Asia*, ed. K. C. Beard and M. R. Dawson.] *Bulletin of the Carnegie Museum of Natural History* 34, 194–209.

Dawson, M. R. (1958). Early Tertiary Leporidae of North America. *University of Kansas Paleontological Contributions, Vertebrata*, 6, 1–75.

(1961). On two ochotonids (Mammalia, Lagomorpha) from the later Tertiary of Inner Mongolia. *American Museum Novitates*, 2061, 1–15.

(1965). *Oreolagus* and other Lagomorpha (Mammalia) from the Miocene of Colorado, Wyoming, and Oregon. *University of Colorado Studies in Earth Sciences*, 1, 1–36.

(1967). Lagomorph history and the stratigraphic record. [In *Essays in Paleontology and Stratigraphy*, ed. C. Teichert and E. L. Yochelson.] *Department of Geology, University of Kansas, Special Publication*, 2, 287–316.

(1970). Paleontology and geology of the Badwater Creek area, central Wyoming. Part 6. The leporid *Mytonolagus* (Mammalia, Lagomorpha). *Annals of the Carnegie Museum*, 41, 215–30.

(1999). Bering down: Miocene dispersals of land mammals between North America and Europe. In *The Miocene Land Mammals of Europe*, ed. G. E. Rössner and K. Heissig, pp. 473–83. Munich: Dr. Friedrich Pfeil.

Dice, L. R. (1917). Systematic position of several American Tertiary lagomorphs. *University of California Publications, Geology*, 10, 179–83.

(1929). The phylogeny of the Leporidae, with description of a new genus. *Journal of Mammalogy*, 10, 340–4.

(1931). *Alilepus*, a new name to replace *Allolagus* Dice, preoccupied, and notes on several species of fossil hares. *Journal of Mammalogy*, 12, 159–60.

Douglass, E. (1901). Fossil Mammalia of the White River Beds of Montana. *Transactions of the American Philosophical Society*, 20, 237–79.

Downey, J. S. (1962). Leporidae of the Tusker local fauna from southeastern Arizona. *Journal of Paleontology*, 36, 1112–15.

(1968). Late Pliocene Lagomorphs of the San Pedro Valley, Arizona. *United States Geological Survey, Professional Paper*, 600D, 169–73.

(1970). Middle Pleistocene Leporidae from the San Pedro Valley, Arizona. *United States Geological Survey, Professional Paper*, 700B, 131–6.

Emry, R. J. and Gawne, C. E. (1986). A primitive, early Oligocene species of *Palaeolagus* (Mammalia, Lagomorpha) from the Flagstaff Rim area of Wyoming. *Journal of Vertebrate Paleontology*, 6, 271–80.

Erbajeva, M. A. (1988). *Cenozoic Pikas (Taxonomy, Systematics, Phylogeny)*. Moscow: Academia Nauka.

Frick, C. (1921). Extinct vertebrate faunas of the badlands of Bautista Creek and San Timoteo Cañon, southern California. *Bulletin of the Department of Geology, University of California Publications*, 12, 277–424.

Funder, S., Abrahamsen, N., Bennike, O., and Feyling-Hanssen, R. W. (1985). Forested Arctic: evidence from North Greenland. *Geology*, 13, 542–6.

Gawne, C. E. (1978). Leporids (Lagomorpha, Mammalia) from the Chadronian (Oligocene) deposits of Flagstaff Rim, Wyoming. *Journal of Paleontology*, 52, 1103–18.

Gazin, C. L. (1930). A Tertiary vertebrate fauna from the upper Cuyama drainage basin, California. *Carnegie Institution of Washington Publication*, 404, 55–76.

(1934) Fossil hares from the late Pliocene of southern Idaho. *Proceedings of the United States National Museum*, 83, 111–21.

(1942). The late Cenozoic vertebrate faunas from the San Pedro Valley, Arizona. *Proceedings of the United States National Museum*, 92, 475–518.

Gidley, J. W. and Gazin, C. L. (1938). The Pleistocene vertebrate fauna from Cumberland Cave, Maryland. *Bulletin of the United States National Museum*, 171, 1–99.

Golz, D. J. and Lillegraven, J. A. (1977). Summary of known occurrences of terrestrial vertebrates from Eocene strata of southern California. *Contributions to Geology, University of Wyoming*, 15, 43–65.

Gray, J. E. (1821). On the natural arrangement of vertebrose animals. *London Medical Repository*, 15, 296–310.

(1867). Notes on the skulls of hares (Leporidae) and picas (Lagomyidae) in the British Museum. *Annals and Magazine of Natural History*, 20, 219–25.

Green, M. (1972). Lagomorpha from the Rosebud Formation, South Dakota. *Journal of Paleontology*, 46, 377–85.

Guilday, J. E. (1979). Eastern North American Pleistocene *Ochotona* (Lagomorpha: Mammalia). *Annals of the Carnegie Museum*, 48, 435–44.

Gureev, A. A. (1960). [Oligocene Lagomorpha of Mongolia and Kazakhstan]. *Trudy, Paleontologicheskii Institut, Akademia Nauk SSSR*, 77, 5–34.

(1964). Zaitzeobraznye [Lagomorpha]. Fauna USSR, Mammalia, 3. *Academy of Science SSSR, Zoological Institute*, 87, 1–276 [in Russian.]

Gustafson, E. P. (1978). The vertebrate faunas of the Pliocene Ringold Formation, south-central Washington. *Bulletin of the Museum of Natural History, University of Washington*, 23, 1–62.

Hayes, F. G. (2000). The Brooksville 2 local fauna (Arikareean, latest Oligocene): Hernando County, Florida. *Bulletin of the Florida Museum of Natural History*, 43, 1–47.

Hibbard, C. W. (1939). Four new rabbits from the upper Pliocene of Kansas. *The American Midland Naturalist*, 21, 506–13.

(1963). The origin of the p3 pattern of *Sylvilagus, Oryctolagus*, and *Lepus*. *Journal of Mammalogy*, 44, 1–15.

Hutchison, J. H. and Lindsay, E. H. (1974). The Hemingfordian mammal fauna of the Vedder locality, Branch Canyon Formation, Santa Barbara County, California. Part 1: Insectivora, Chiroptera, Lagomorpha, and Rodentia (Sciuridae). *PaleoBios*, 15, 1–19.

Kellogg, L. (1910). Rodent fauna of the late Tertiary beds of Virgin Valley and Thousand Creek, Nevada. *Bulletin of the Department of Geology, University of California Publications*, 5, 421–37.

Kelly, T. S. (2000). A new Hemphillian (late Miocene) mammalian fauna from Hoye Canyon, west central Nevada. *Contributions in Science, Los Angeles County Museum of Natural History*, 481, 1–21.

Koenigswald, W. von (1996). Die Zahl der Schmelzschichten in den Inzisiven bei den Lagomorpha und ihre systematische Bedeutung. *Bonner zoologische Beiträge*, 46, 33–57.

Korth, W. W. (1998). Rodents and lagomorphs (Mammalia) from the late Clarendonian (Miocene) Ash Hollow Formation, Brown County, Nebraska. *Annals of the Carnegie Museum*, 67, 299–348.

Korth, W. W. and Hageman, J. (1988). Lagomorphs (Mammalia) from the Oligocene (Orellan and Whitneyan) Brule Formation, Nebraska. *Transactions of the Nebraska Academy of Sciences*, 16, 141–52.

Leidy, J. (1856). Notices of remains of extinct Mammalia, discovered by Dr. F. V. Hayden in Nebraska Territory. *Proceedings of the Academy of Natural Sciences of Philadelphia*, 8, 88–90.

Li, C.-K. and Ting, S.-Y. (1985). Possible phylogenetic relationship of Asiatic eurymylids and rodents, with comments on mimotonids. In *Evolutionary Relationships Among Rodents*, ed. W. P. Luckett and J.-L. Hartenberger, pp. 35–58. New York: Plenum Press.

Link, H. F. (1795). *Beitraege zur Naturgeschichte* 1794–1801, 1(2), 74.

Lyon, M. W. (1904). Classification of the hares and their allies. *Smithsonian Miscellaneous Collections*, 45, 321–447.

Martin, T. (2004). Evolution of incisor enamel microstructure in Lagomorpha. *Journal of Vertebrate Paleontology*, 24, 411–26.

Matthew, W. D. (1899). A provisional classification of the fresh water Tertiary of the West. *Bulletin of the American Museum of Natural History*, 12, 19–75.

(1903). The fauna of the *Titanotherium* beds at Pipestone Springs, Montana. *Bulletin of the American Museum of Natural History*, 19, 197–226.

(1907). A lower Miocene fauna from South Dakota. *Bulletin of the American Museum of Natural History*, 23, 169–219.

Matthew, W. D. and Granger, W. (1923). Nine new rodents from the Oligocene of Mongolia. *American Museum Novitates*, 102, 1–10.

McGrew, P. O. (1941). A new Miocene lagomorph. *Field Museum of Natural History, Geological Series*, 8, 37–41.

McKenna, M. C. and Bell, S. K. (1997). *Classification of Mammals Above the Species Level*. New York: Columbia University Press.

Miller, W. E. and Carranza-Castañeda, O. (1982). New lagomorphs from the Pliocene of central Mexico. *Journal of Vertebrate Paleontology*, 2, 95–107.

Novacek, M. J. (1984). Cranial evidence for rodent affinities. In *Evolutionary Relationships Among Rodents*, ed. W. P. Luckett and J.-L. Hartenberger, pp. 59–81. New York: Plenum Press.

Pallas, P. S. (1778). Novae species quadrupedum e glirium ordine. *Erlangae*, 1–388.

Qiu, Z.-X., Wu, W., and Qiu, Z.-D. (1999). Miocene mammal faunal sequence of China: palaeozoogeography and Eurasian relationships. In *The Miocene Land Mammals of Europe*, ed. G. E. Rössner and K. Heissig, pp. 443–55. Munich: Dr. Friedrich Pfeil.

Retallack, G. J. (1992). Paleosols and changes in climate and vegetation across the Eocene/Oligocene boundary. In *Eocene–Oligocene Climatic and Biotic Evolution*, ed. D. R. Prothero and W. A. Berggren, pp. 382–98. Princeton, PA: Princeton University Press.

Russell, B. D. and Harris, A. H. (1986). A new leporine (Lagomorpha: Leporidae) from Wisconsinan deposits of the Chihuahuan desert. *Journal of Mammalogy*, 67, 632–9.

Schlaikjer, E. M. (1935). Contributions to the stratigraphy and palaeontology of the Goshen Hole area, Wyoming. IV. New vertebrates and the stratigraphy of the Oligocene and early Miocene. *Bulletin of the Museum of Comparative Zoology*, 76, 97–189.

Schlosser, M. (1924). Tertiary vertebrates from Mongolia. *Palaeontologia Sinica, C*, 1, 1–132.

Shotwell, J. A. (1956). Hemphillian mammalian assemblage from northeastern Oregon. *Bulletin of the Geological Society of America*, 67, 717–38.

Stevens, M. S., Stevens, J. B., and Dawson, M. R. (1969). New early Miocene Formation and vertebrate Local Fauna, Big Bend National Park, Brewster County, Texas. *Pearce-Sellards Series, Texas Memorial Museum*, 15, 1–53.

Storer, J. E. (1970). New rodents and lagomorphs from the Upper Miocene Wood Mountain Formation of southern Saskatchewan. *Canadian Journal of Earth Sciences*, 7, 1125–9.

(1984). Mammals of the Swift Current Creek local fauna (Eocene: Uintan, Saskatchewan). *Saskatchewan Museum of Natural History, Natural History Contributions*, 7, 1–158.

(1992). *Tachylagus*, a new lagomorph from the Lac Pelletier lower fauna (Eocene: Duchesnean) of Saskatchewan. *Journal of Vertebrate Paleontology*, 12, 230–5.

Thomas, O. (1897). On the genera of rodents: and an attempt to bring up to date the current arrangements of the Order. *Proceedings of the Zoological Society of London*, 10, 12–28.

Tong, Y.-S. (1997). Middle Eocene small mammals from Liguanqiao Basin of Henan Province and Yuanqu Basin of Shanxi Province, central China. *Palaeontologia Sinica*, 18, (n.s. C, 26), 1–256 [Chinese, English summary.]

Tullberg, T. (1899). Ueber das System der Nagethiere: eine Phylogenetische Studie. *Nova Acta Regiae Societatis Scientiarum Upsaliensis*, ser. 3, 18, 1–514.

Voorhies, M. R. (1990). *Technical Report 82–09: Vertebrate Paleontology of the Proposed Norden Reservoir Area, Brown, Cherry, and Keya Paha Counties, Nebraska*, pp. 1–138, A1–A593. Lincoln, NE: Division of Archeological Research, University of Nebraska.

Voorhies, M. R. and Timperley, C. L. (1997). A new *Pronotolagus* (Lagomorpha: Leporidae) and other leporids from the Valentine Railway quarries (Barstovian, Nebraska), and the archaeolagine–leporine transition. *Journal of Vertebrate Paleontology*, 17, 725–37.

Walker, M. V. (1931). Notes on North American fossil lagomorphs. *The Aerend*, 2, 227–40.

White, J. A. (1984). Late Cenozoic Leporidae (Mammalia, Lagomorpha) from the Anza-Borrego Desert, southern California. *Carnegie Museum of Natural History, Special Publication*, 9, 41–57.

(1987). The Archaeolaginae (Mammalia, Lagomorpha) of North America, excluding *Archaeolagus* and *Panolax*. *Journal of Vertebrate Paleontology*, 7, 425–50.

(1991a). North American Leporinae (Mammalia: Lagomorpha) from Late Miocene (Clarendonian) to latest Pliocene (Blancan). *Journal of Vertebrate Paleontology*, 11, 67–89.

(1991b). A new *Sylvilagus* (Mammalia: Lagomorpha) from the Blancan (Pliocene) and Irvingtonian (Pleistocene) of Florida. *Journal of Vertebrate Paleontology*, 11, 243–6.

Wilson, R. W. (1938). A new genus of lagomorph from the Pliocene of Mexico. *Bulletin of the Southern California Academy of Science*, 36, 98–104.

(1949). Rodents and lagomorphs of the Upper Sespe. *Carnegie Institution of Washington Publication*, 584, 51–65.

(1960). Early Miocene rodents and insectivores from northeastern Colorado. *University of Kansas Paleontological Contributions, Vertebrata*, 7, 1–92.

Winkler, A. J. and Tomida, Y. (1988). New records of the small leporid *Aztlanolagus agilis* Russell and Harris (Leporidae, Leporinae). *The Southwestern Naturalist*, 33, 391–6.

Wood, A. E. (1937). Additional material from the Tertiary of the Cuyama Basin of California. *American Journal of Science*, 33, 29–43.

(1940). The mammalian fauna of the White River Oligocene. Part III. Lagomorpha. *Transactions of the American Philosophical Society, New Series*, 28, 271–362.

(1949). Small mammals from the uppermost Eocene (Duchesnean) from Badwater, Wyoming. *Journal of Paleontology*, 23, 556–65.

(1957). What, if anything, is a rabbit? *Evolution*, 11, 417–25.

Wyss, A. R. and Meng, J. (1996). Application of phylogenetic taxonomy to poorly resolved crown clades: a stem-modified node-based definition of Rodentia. *Systematic Biology*, 45, 559–68.

18 Ischyromyidae

DEBORAH ANDERSON
St. Norbert College, De Pere, WI, USA

INTRODUCTION

Ischyromyids are among the first rodents to appear in the fossil record. Fossils are recorded from the late Paleocene to early Eocene of North America, early to late Eocene of Europe, and the early Oligocene of East Asia. Molar crown patterns of members of the family are strikingly conservative, many species being originally described based on significant differences in size. In general, cheek teeth exhibit a characteristic rhomboid shape in occlusal outline, making these rodents easy to identify to family. Upper molars, while less rhomboid overall, retain the feature of two distinct widths, both of which exceed the anteroposterior molar dimension.

Currently Ischyromyidae (Alston, 1876) is recognized as the major family with Paramyinae as a subfamily. Wood (1962, 1977) and others recognized Paramyidae as a family distinct from the Ischyromyidae based on the inferred location of the deep masseter muscle, the development of lophs in the cheek teeth, and other aspects of the cranial anatomy. However, Black (1968, 1971) considered that these differences are not significant enough to justify two separate families because the differences between members of other recognized subfamilies of paramyids are equally extensive. Following Korth (1984, 1994), Black (1968, 1971) and others, use of Ischyromyidae to include the family Paramyidae (now as the subfamily Paramyinae) seems most appropriate given that the former has priority.

Size range estimates and habits for ischyromyids are based on comparisons with recent rodents. Larger species of ischyromyids, such as *Manitsha*, were approximately one and a half times larger than a beaver and were probably terrestrial and subfossorial, based on cranial and postcranial morphology (Wood, 1962). Species of *Microparamys*, one of the smallest members of the family, were the size of a small vole and likely scampering in habit. The family includes a large range of sizes and probable habits, which could lead to the suggestion that the current group may include more than one family. As more specimens are recovered, in particular more postcranial material, taxonomic groupings will certainly need to be reevaluated. Ischyromyids were most diverse taxonomically during the early to middle Eocene. Diversity dropped off dramatically from the early Oligocene to extinction of the group in the late Oligocene (Korth, 1984).

DEFINING FEATURES OF THE FAMILY ISCHYROMYIDAE

CRANIAL

The skull is relatively large and robust with a long snout and long basicranial region (versus the short snout and short basicranium of Recent rodents). The skull has a well-developed zygoma, zygomatic arches that extend parallel to the skull their entire length, a narrowing of the skull in the postorbtial region, and long nasals that taper rapidly at their posterior margin (Figure 18.1A). The zygomasteric structure is protrogomorphous, as evidenced by the limitation of the masseter to the ventral surface of the zygoma with no relation to the relatively large, circular infraorbital foramen. In lateral view, the skull is low dorsoventrally with the dorsal and ventral margins essentially parallel, not unlike Recent rodents. Other distinguishing features include the suture between the maxillae and premaxillae, which is crenulated and arches forward. The auditory bullae are ossified in some species and may be separate or attached to the skull. A distinct sagittal crest varies in height.

The mandible is relatively large and thick dorsoventrally, more slender in smaller species. The ascending ramus is broad; the well-developed coronoid process is well above the plane of lower cheek teeth and there are one or two mental foraminae, the anterior-most located below the diastema. The anterior edge of the masseteric fossa extends to a position below m1 or m2, and the diastema is

Evolution of Tertiary Mammals of North America, Vol. 2. ed. C. M. Janis, G. F. Gunnell, and M. D. Uhen. Published by Cambridge University Press.

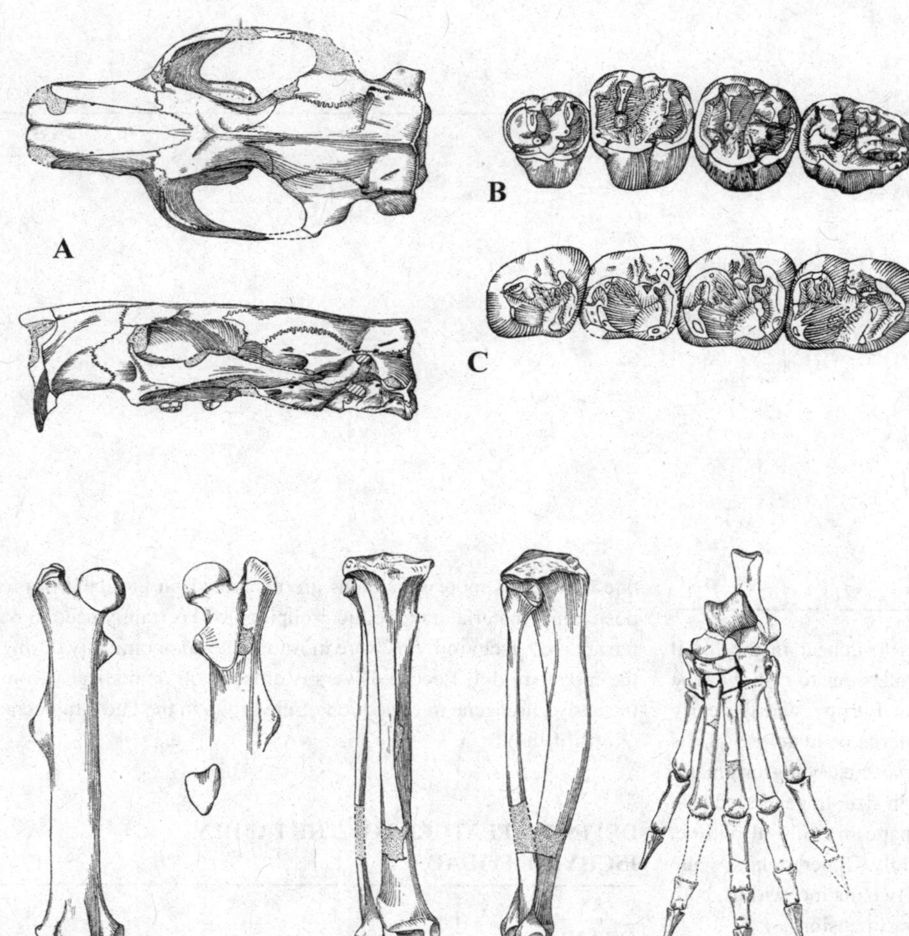

Figure 18.1. A. Skull of *Paramys delicatus* in lateral and ventral views (AMNH 12506: anterior part of nasals restored from YPM 13381). Note narrowing of skull in postorbital region, tapering of nasals posteriorly, and parallel dorsal and ventral margin. B. Upper cheek teeth of *Paramys*. C. Lower cheek teeth of *Paramys*. D. Hindlimb elements of *Paramys*. (All after Wood, 1962; courtesy of the American Philosophical Society.)

short. The mandibular angle is primarily sciurognathus with some indications of hystricognathy.

DENTAL

The dental formula is I1/1, C0/0, P2/1, M3/3. The lower cheek teeth exhibit a characteristic rhomboid shape with the anteroposterior dimension exceeding the narrower anterior width and posterior width (Figure 18.1C). Four distinct cusps include a protoconid, a metaconid, an entoconid, and a hypoconid, with the metaconid being the highest and anterior-most cusp. Mesoconids are common among species but vary in distinctness. Anterior and posterior cingulae form the outermost boundaries of the reduced trigonid and broad talonid basins. In most species the basin enamel is smooth. Distinct enamel crenulations are characteristic of some genera such as *Thisbemys* and *Lophiparamys*. The first and second lower molars are similar in size and morphology, while premolar four and molar three are more variable.

The upper teeth are more molariform in outline, and the anterior width exceeds the posterior width and anteroposterior measure, the latter being the smallest dimension. Protocone, hypocone, paracone, and metacone are the most prominent cusps. Anterior and posterior cingulae are well developed and meet up with the protocone at a sharp angle forming two distinctive internal notches (Figure 18.1B). The basin enamel is generally smooth, but may be crenulated.

POSTCRANIAL

Postcranial material is best known for *Paramys*, *Reithroparamys*, *Pseudotomus*, and *Uintaparamys* (Figure 18.1D). Most taxa have a generalized scampering (scansorial) type of skeleton, while some genera have incipient adaptations for burrowing. In the appendicular skeleton, the femur is longer than skull; the limbs range from slender to robust according to the size of the taxon; the humeral and femoral shafts are nearly straight; there is a partial spreading of manus digits and the pes is massive and powerful. In the axial skeleton, there are large vertebrarterial and atlantal foraminae. The cervical vertebrae may have ventral keels or ridges. The thoracic vertebrae have long transverse processes and small ventral keels. The lumbar vertebrae have prominent ventral foraminae and keels. There is one sacral and two pseudosacral vertebrae and the tail is long and slender (Wood, 1962).

SYSTEMATICS

SUPRAFAMILIAL

It is still a point of debate as to whether the ischyromyids represent the basal stock from which all other rodent groups evolved. One of the earliest known rodents in the fossil record is an ischyromyid, *Acritoparamys atavus*. The fossil includes an isolated lower molar and incisor fragments from the late Paleocene of Bear Creek, Montana (locality CP17A) (Wood, 1962; Carroll, 1988). More recently Dawson and Beard (1996) have reported a new species of paramyid, *Paramys adamus*, which they propose is more primitive than *A. atavus*. Discovered in the Big Multi Quarry, northern Washakie Basin of Wyoming (locality CP14E), *P. adamus* includes nearly complete upper and lower dentitions. This quarry also includes the first North American record of a species of alagomyid, *Alagomys russelli* (Dawson and Beard, 1996); previously members of this family of rodents were known only from the late Paleocene–early Eocene of Mongolia and China.

Dawson and colleagues (Dawson, Li, and Qui, 1984) first suggested that the Asian genus *Cocomys* represents the basal rodent, based on the primitive lower molars that include cusps and a non-molariform premolar. Other ancestral features of *Cocomys*, also found in early ischyromyids, include upper molars with large conules and a hypocone, and lower molars exhibiting well-developed hypoconulids and isolated entoconids (Korth, 1984). Since then, a new rodent from the earliest Eocene of Asia has been discovered, *Exmus mini*, classified as a member of the family Ctenodactyloidea (Wible *et al.* 2005). Recent phylogenetic analyses indicate that *Exmus* and *Cocomys* are sister taxa that share many dental features with members of the Ischyromyidae but exhibit many cranial features not found in the North American ischyromyids.

A recent phylogenetic analysis completed by Marivaux, Vianey-Liaud, and Jaeger (2004) reveals two distinct clades among early Tertiary rodents: (1) the earliest ischyromyid rodents and their close relatives (Muroidea, Dipodoidea, Geomyoidea, Anomaluroidea, Castoroidea, Sciuravidae, Gliroidea, Sciuroidea, Aplodontoidea, and Theridomorpha), which is referred to the clade Ischyromyiformes, and (2) the earliest ctenodactyloids (Ctenodactyloidea, Chapattimyidea, Yuomyidae, Diatomyidae) and hystrignathous rodents (Tsaganomyidae, Baluchimyinae, phiomorphs, caviomorphs), referred to the clade Ctenohystrica. The analysis is based on 108 characters, which are primarily molar features with the inclusion of two cranial and six incisor characters. Significant diagnostic characters for differentiating between the two clades using molars is the presence of an anterior crest, the paralophid, which links the protoconid to the paraconid in the Ctenohystrica. An anterolophid connecting the protoconid to the metaconid (the paraconid is lost) essentially replaces the ancestral crest in the Ischyromyiformes. In addition, the reduction or loss of the hypoconulid occurs in members of the Ischyromyiformes, while this feature is retained in the Ctenohystrica.

Cocomys, *Exmus*, and *Paramys* all have brachydont, cuspidate molars, and the possession of molar cusps versus lophs or crests continues to be accepted as a primitive condition in rodents (Butler, 1985; Korth, 1994; Meng, Hu, and Li, 2003; Marivaux, Vianey-Liaud, and Jaeger, 2004; Wible *et al.*, 2005). Likewise, a non-molariform upper premolar, characteristic of *Cocomys*, continues to be recognized as a taxonomically significant feature, because the acquisition of a hypocone is recognized as advanced (Dawson, Li, and Qui, 1984; Carroll, 1988; Meng and Wyss, 2001; Meng, Hu, and Li, 2003; Marivaux, Vianey-Liaud, and Jaeger 2004; Wible *et al.*, 2005). Wood (1962) hypothesized that the ancestors of rodents had a lower premolar with a single cusp in the trigonid and a two-cusped talonid, while the upper premolar had two cusps, one lingual and one buccal. This idea stems from Wood's proposal that primates were ancestral to rodents, an idea no longer well accepted. Currently, the Tertiary Eurymyloidea of Asia are considered ancestral to the order Rodentia, a hypothesis proposed by Hartenberger (1985) and supported by Marivaux, Vianey-Liaud, and Jaeger (2004). While the alagomyids appear simultaneously in North America with the paramyids, the absence of a hypocone in the former sets them apart. The presence of a hypocone (a derived feature) has evolved in several different lineages of mammal (Hunter and Jernvall, 1995).

An analysis of ischyromyids as ancestral rodents based on jaw anatomy and the reconstruction of the arrangement of the masseter musculature conflicts with evolutionary relationships derived primarily from dental features. The extension of the masseteric fossa on the dentary to below m2 is a character that unites *Alagomys* and *Paramys*, leaving out *Cocomys* with an extension only to m3 (Dawson and Beard, 1996). The position of the mandibular angle relative to the horizontal ramus suggests that ischyromyids are sciurognathus, a primitive condition in rodents shared by *Cocomys*, *Paramys*, and *Alagomys*. However, there are differences in jaw musculature: the masseter pattern for ischyromyids is protrogomorphous while the ctenodactyloids are hystricomorphous (see discussion in Chapter 16); cranial anatomy is unknown for the alagomyids. Figure 18.2 shows the relationships among these genera.

INFRAFAMILIAL

A preliminary phylogenetic analysis of the taxa currently included in the family Ischyromyidae is included in this chapter as a starting point for determining the evolutionary relationships of this very diverse albeit morphologically conservative group of "rodents." The term rodents used in this chapter refers to the group formerly known as Rodentia, most recently defined by Meng and Wyss (2001) as the taxon Rodentiaformes. Using a crown-clade definition of rodents, Meng and Wyss (2001) introduced the new taxon to encompass the alagomyids (found to be excluded from the group in their study) and the group formerly known as Rodentia. A total of 45 characters including dental and mandibular characters were scored from original specimens, casts, and from the literature.

Individual characters were polarized based on a comparison of the outgroups (Watrous and Wheeler, 1981). *Tribosphenomys* was chosen as an outgroup because it is very closely related to the Rodentia but had been shown to be outside of the clade as defined by Meng

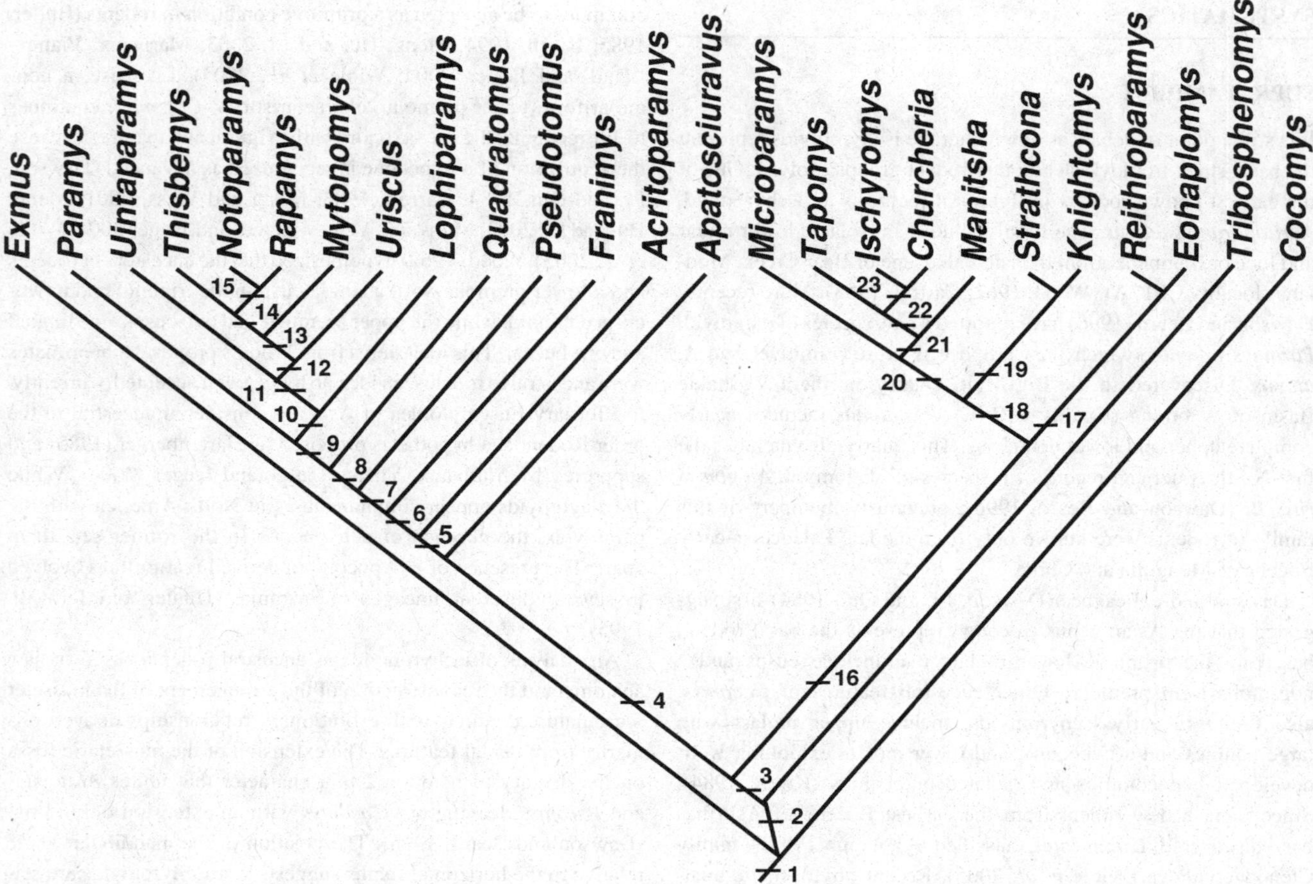

Figure 18.2. Ischyromyid relationships: the single most parsimonious tree resulting from a heuristic search of 5000 trees. Characteristics at nodes 1–23 as follows. 1. Upper molar shape triangular; P4 hypocone absent, labial cusps single, metaconule present, protoconule indistinct; M1–2 protoconule present; p4 hypoconid present, hypoconulid absent; m1–2 hypoconulid distinct; m3 hypoconulid forming third narrow lobe; anterior edge of masseteric fossa below m2; anterior cingulum not well developed. 2. ISCHYROMYIDAE: P4 labial cusps double, with metacone and paracone equal in size; metalophulid II of m2 absent; m1–2 ectolophid complete. 3. Endoloph present; hypoconulid of m3 indistinct; mental foramen below diastema (anterior to p4); anterior cingulum uniting protoconid and metaconid; m2 posterior cingulum distinct from entoconid. 4. Ectolophid absent on m1–2. 5. Upper molar hypocone weak or absent. 6. m1–2 hypoconulid indistinct; angular process lateral to plane of mandible body; m2 posterior cingulum joining with entoconid. 7. P4 protoconule present; ectolophid of m1–2 incomplete. 8. Upper molar hypocone present. 9. Upper molar shape rectangular; angular process originating in plane of mandibular body. 10. Mesoloph present; metalophulid II of m2 short; ectolophid of m2 complete. 11. Upper molar hypoflexus shallow. 12. Ectolophid of m1–2 anteriorly incomplete; hypoconulid m1–2 distinct; double metaconules. 13. p4 hypoconulid present; lower molar hypoflexid deep. 14. Upper molar shape triangular; hypolophid of lower molar incomplete; anterior mental foramen below p4. 15. p4 hypolophulid present; mesoconid absent; entoconid anteriorly shifted; ectolophid of m1–2 complete; one mental foramen; M2 posterior cingulum distinct from entoconid. 16. P4 protoconulę present; metalophulid II of m2 well developed; hypolophid of lower molar incomplete. 17. P4 protoconule present; metalophulid II of m2 absent. 18. Anterior cingulum separate from protoconid. 19. P4 hypocone small, P4 protoconule indistinct; p4 hypoconulid present; metalophulid II of m2 short. 20. p4 hypolophid present; anterior edge of masseteric fossa below anterior half of m1 or p4; one mental foramen. 21. Mesoconid absent; hypolophid of lower molar complete; anterior cingulum not well developed; 22. P4 metaconule indistinct; upper molar shape rectangular. 23. P4 hypocone small, P4 protoconule indistinct; M1–2 protoconule absent; ectolophid of m1–2 incomplete anteriorly.

and Wyss (2001). Sister taxa, *Cocomys* and *Exmus*, were included as primitive rodent outgroups, currently classified outside clade Ischyromyiformes. *Knightomys*, a sciuravid thought to be closely related to the ischyromyids (and previously classified in the family; covered in Chapter 19), was included in the analysis to facilitate comparison with other phylogenetic analyses, which have included *Paramys* and the chosen outgroups. Multistate characters were treated as unordered. A heuristic search of 5000 trees using phylogenetic analysis using parsimony, PAUP* (Swofford, 2000) resulted in a single most parsimonious tree (Figure 18.2). Few clades were resolved to a bootstrap level of 70%; most clades are yet to be resolved.

Inferred relationships among *Paramys*, *Reithroparamys*, *Knightomys*, and the outgroups are consistent with the results of previous phylogenetic analyses that include these genera (Meng *et al.* 1994; Dawson and Beard, 1996; Wyss and Meng, 1996; Meng and Wyss, 2001; Meng, Hu, and Li, 2003; Marivaux, Vianey-Liaud, and Jaeger, 2004; Wible *et al.*, 2005). New insights include the well-supported (bootstrap value, 75%) *Knightomys–Strathcona* clade formed by the presence of a P4 hypocone, indistinct-to-absent P4 protoconule, p4 hypoconulid, and short metalophulid II. The *Mytonomys–Rapamys* clade is supported (bootstrap value, 72%) by possession of a p4 hypolophid, entoconid anteriorly shifted and distinct from the posterolophid, a complete ectolophid crest without mesoconid, and a single mental foramen. The *Ischyromys–Churcheria* clade is supported (bootstrap value, 75%) by possession of a P4 hypocone, lack of protoconules on upper cheek teeth, and an anteriorly incomplete ectolophid crest. Several members of the subfamily Paramyinae (*Paramys, Uintaparamys, Thisbemys, Notoparamys*, and *Rapamys*) occur within a common clade, supporting their classification based on cranial anatomy. Other subfamilies are not well supported by classification based on dental and mandibular characters. Little is known about cranial anatomy for most of these taxa and it is likely that they are currently classified in one subfamily primarily because of their small size. Detailed study of the auditory region of *Reithroparamys* by Meng (1990) revealed the possession of eight derived features making this taxon more closely related to sciurids and aplodontids than to the primitive condition found in members of the subfamily Paramyinae. It may represent a unique family, supported by the phylogeny presented in this chapter. Classification of *Uriscus* is challenged by the fact that upper molars and cranial material are unknown and there are very few specimens. Based on dental pattern of mandibular teeth, this taxon fits well with other Paramyinae. The basal position of *Acritoparamys* relative to many Paramyinae taxa fits previous hypothesis that this represents one of earliest ischryomyids in the fossil record and is consistent with its early stratigraphic record (late Paleocene). Well-supported synapomorphies (consistency index, 1.00) in this phylogeny include the crown anatomy of P4, and the development of lophs and crests. This is consistent with recognition of members of the family Ischyromyidae as distinct from the closely related members of Sciuravidae in the retention of cusps versus the development of lophs and crests.

INCLUDED GENERA OF ISCHYROMYIDAE ALSTON, 1876

The locality numbers listed for each genus refer to the list of unified localities in Appendix I. The acronyms for museum collections are listed in Appendix III.

The locality numbers may be listed in several alternative ways. Parentheses around the locality (e.g., [CP101] mean that the taxon in question at that locality is cited as an "aff." or "cf." the taxon in question. Parentheses are usually used for individual species, thus implying that the genus is firmly known from the locality but that the species identification may be questionable. Question marks in front of the locality (e.g.,?CP101) mean that the taxon is questionably known from that locality, thus implying some doubt that the taxon is actually present at that locality, at either the genus or the species level.

PARAMYINAE HAECKEL, 1895

Characteristics: Medium to large size; long basicranial region; M3 near middle of skull; small orbit; skull large with a heavy snout; short frontals; distinct sagittal crest; heavy parallel-sided antorbital region; auditory bullae unossified. Cheek tooth enamel ranging from smooth to minor pitting to distinct crenulations; metaconid highest cusp; distinctly rhomboid-shaped cheek teeth; anterior cingulum continuous from metaconid to protoconid; size increase from p4 to m3; length exceeding width in lower molars; width exceeding length in upper molars (Wood 1962; Korth 1984, 1994).

PARAMYINI HAECKEL, 1895

***Paramys* Leidy, 1871 (including *Ischyrotomus*, in part)**

Type species: *Paramys delicatus* Leidy, 1871.

Type specimen: ANSP 10273.

Characteristics: Medium to large size; heavy mandibular ramus, uniform depth; pitting of crown surface in *P. delicatus*, relatively smooth enamel in other species; no cresting; marginal, rounded cusps; P4 molariform, no hypocone; M1, M2 with hypocone separated from protocone by distinct lingual groove; posterior displacement of metacone on M3; protoconule, metaconule, and single mesostyle on P4–M3; lower incisors oval in outline, smooth, covered anterolaterally with enamel; anterior cingulum tying metaconid and protoconid together; trigonid basins opening posteriorly; posterolophid continuous with hypoconid; distinct mesoconid.

Average length of m1: 2.73–4.27 mm.

Included species: *P. delicatus* (known from localities ?CP19C, CP20F, CP25J, [CP32B], CP34A, B, D, CP38B); *P. adamus* Dawson and Beard, 1996 (locality CP14E); *P. compressidens* (Peterson, 1919) (localities [CC9AA], CP6AA, A, B, CP38C, D, CP39A); *P. copei* (including *P. copei bicuspis* and *P. copei major*) Loomis, 1907 (localities SB22C, SB24, CP19A–C, CP20[AA], B, C, CP23B, CP25F, G, H, I, [I2, J2], CP27A, C, E, CP31E, [CP32B], [CP37A, B], [CP63], CP64A–C); *P. delicatior* Leidy, 1871 (localities [SB43B], [CP5A], CP25J, CP34A, C, D); *P. excavatus* Loomis, 1907 (localities SB22B, C, SB24, SB40, CP20A, [B, C], CP25C–E, G, CP27B, D, E, CP32B, [CP34D], [CP35], CP64A–C, [NP49]); *P. hunti* (Dawson, 2001) (locality HA1A); *P. nini* (Wood, 1962) (localities SB22C, SB24, [CP32B]); *P. pycnus* Ivy, 1990 (localities CP19A, B); "*P.*" *simpsoni* (locality SB24) Korth, 1984; *P. taurus* Wood, 1962 (localities

SB24, CP17B, CP18, CP19AA-C, CP20AA, A, B, C, CP63).

Paramys sp. is also known from localities CP20E, CP25?B, GG, J, CP26C, CP28B, CP31(A), D, CP34E, CP36B, CP37B, CP38A, B, CP64A.

Comments: Korth (1988a) noted that "*Paramys simpsoni*" is known from a single skull that differs in a number of cranial features from other ischyromyids and considered that it should belong to a new genus. *Paramys* is also known from the early to middle Eocene of Europe and Asia (McKenna and Bell, 1997).

***Uintaparamys* (formerly *Leptotomus* Matthew, 1910; including *Paramys*, in part)**

Type species: *Uintaparamys leptodus* (Cope, 1873) (originally described as *Paramys leptodus*).

Type specimen: AMNH 5026.

Characteristics: Medium to large size; distinctly rhomboid-shaped cheek teeth; relatively smooth enamel; protoloph and metaloph essentially parallel on P4 and M1 but at more of an angle on M2 and M3; protoloph and metaloph approximately equal in length for P4–M3; metaconule of P4–M3 present and equal in size to the metacone; protoconule of P4–M3 half the size of the protocone; anteroloph well developed and longer than posterolophs, both ending lingually at the protocone; shelflike cingulum buccal to single mesostyle; lower incisors narrow, convex anteriorly and laterally compressed, no groove; protoconid arm distinct and doubled in m1–3; entoconid isolated from posterolophid by distinct groove; trigonid basin open posteriorly; mesoconid small, not well rounded, lingually displaced; nearly straight, well-developed ectolophid crest; no trace of a hypolophid; metastylids inconsistently present.

Average length of m1: 3.70–4.86 mm.

Included species: *U. leptodus* (known from localities CP6B, [CC9AA], SB44A, CP38C); *U. bridgerensis* Wood, 1962 (localities CP34A, C, D, CP35, CP38B, CP65); ?*U. caryophilus* Wilson, 1940 (localities CC4, CC9AA, CP34A, C); ?*U. coelumensis* Wilson and Stevens, 1986 (locality SB43C); ?*U. parvus* Wood, 1959 (localities SB26II, CP25J2, CP31E, CP32B, CP34A, C, CP35).

Uintaparamys sp. is also known from localities CC7C, CC9BB, SB43B, C, CP20C, (CP31C), CP31E (two spp.), CP34C, CP65, NP9A.

Comments: The original name of this genus, *Leptotomus*, is preoccupied by a beetle of the same generic name. McKenna and Bell (1997) claimed that the correct name is *Uintaparamys* Kretzoi 1968, as followed here, although they have this taxon subsumed under *Tapomys* following Black 1971. However, there are generic-level distinctions that warrant a separate genus for *Uintaparamys*.

"*Leptotomus guildayi*" Black 1971 is most likely not a species of *Uintaparamys*, as suggested by the large trigonid basins, wide buccal valley on m1–3, and complete hypolophid on p4–m3; however, at present there is no agreement on a new generic designation. It is known from localities CP29(C), D, (NP8), (NP9A), (NP10B).

***Thisbemys* Wood, 1959 (including *Leptotomus*, in part)**

Type species: *Thisbemys corrugatus* Wood, 1959.

Type specimen: USNM 17165.

Characteristics: Medium size; teeth highly corrugated, irregularities are part of crown pattern; distinct lingual gorge between metaconid and entoconid; molars exhibit characteristic rhomboid shape.

Average length of m1: 2.65–4.27 mm.

Included species: *T. corrugatus* (known from localities CP31E, CP32B, CP34C, D, CP35, [CP37A]); *T. elachistos* Korth, 1984 (locality SB22B); *T. medius* (Peterson, 1919) (locality CP6B); *T. perditus* (Wood, 1962) (including *Leptotomus huerfanensis, Paramys gardneri*) (localities SB22A, B,?C, CP25I, J, CP27B, CP38B, CP64B, C); *T. plicatus* Wood, 1962 (localities CP31E, CP34A, C); *T. uintensis* (Osborn, 1895) (locality CP6B).

Thisbemys sp. is also known from localities ?SB43A, CP19B, CP27E, CP31C, [E], CP34E, CP65.

***Notoparamys* Korth, 1984 (including *Leptotomus*, in part)**

Type species: *Notoparamys costilloi* (Wood, 1962) (originally described as *Leptotomus costilloi*).

Type specimen: AMNH 55111.

Characteristics: Medium size; posterior aspect of anterior root of zygoma in line with anterior cingulum of M1 buccally; distinct, buccally positioned hypocone adjacent to metaconule on M1–2, absent on P4; single mesostyle with shelf buccal to this on M1–2; distinct hypoconulid; entoconid crest extending to hyoconulid and posterolophid continuous with hypoconid; distinct mesoconid; ectolophid not well developed; single protoconid arm closing trigonid basin posteriorly on m1–3.

Average length of m1: 3.02–3.84 mm.

Included species: *N. costilloi* (known from localities SB22A, SB24); *N. arctios* Korth, 1984 (locality CP64B).

Notoparamys sp. is also known from locality CP19C.

***Quadratomus* Korth, 1984 (including *Leptotomus*, in part; *Paramys*, in part; *Pseudotomus*, in part)**

Type species: *Quadratomus grandis* (Wood, 1962) (originally described as *Leptotomus grandis*).

Type specimen: USNM 20137

Characteristics: Large size; quadrate outline with distinct hypocone and mesostyle buccolingually elongate on M1–2; no hypocone on P4; protoloph and metaloph essentially parallel for P4–M2; protoconule and metaconule present, metaconules sometimes double; trigonid basin anteroposteriorly compressed, slightly open posteriorly; minute mesoconids; ectolophid frequently absent; pitting of enamel common; no entoconid crest or hypolophid; posterolophid essentially continuous with entoconid.

Average length of m1: 4.88–5.30 mm.

Included species: *Q. grandis* (known from localities SB22C, CP34A, C); ?*Q. gigans* (Wood, 1974) (locality SB44B); *Q. grossus* (including *Pseudotomus robustus*) Korth, 1984 (localities CP34B, C); *Q. sundelli* (Eaton, 1982) (locality CP32B).

Quadratomus sp. is also known from locality SB22C.

***Rapamys* Wood, 1962 (including *Leptotomus*, in part)**

Type species: *Rapamys fricki* (Wilson, 1940) (originally described as *Leptotomus fricki*).

Type specimen: LACM 2181.

Characteristics: Hypocone distinct on M1–2, lacking on P4; double or triple metaconules on M1–2; external cingula located buccal to elongate mesostyle on P4–M2; lingual metaconule distinctly larger than metacone on P4–M2; anterior portion of M3 molariform; lower incisor rounded anteriorly with enamel extending laterally and medially; trigonid basin opening posteriorly owing to short metalophid; isolated entoconid, markedly separate from cuspate, transverse posterolophid; entoconid crest well developed; mesoconid indistinct; nearly straight ectolophid.

Average length of m1: 3.31–3.68 mm.

Included species: *R. fricki* (known from localities CC7B, CC9AA); *R. wilsoni* Black, 1971 (locality CP29C).

Rapamys sp. is also known from locality CC8.

***Tapomys* Wood, 1962 (including *Leptotomus*, in part)**

Type species: *Tapomys tapensis* Wilson, 1940 (originally described as *Leptotomus tapensis*).

Type specimen: LACM 2157.

Characteristics: Large size; P4 with no hypocone; hypocone present on M1–2; large metaconule, small protoconule on P4–M3; anteroloph well developed and larger than posteroloph; protoloph and metaloph essentially parallel on P4–M2; compressed and grooved lower incisors; entoconid isolated from posterolophid by distinct groove; no distinct entoconid crest; indistinct mesoconid; small trigonid basin closing posteriorly by complete metalophid.

Average length of m1: 4.32–4.92 mm.

Included species: *T. tapensis* only, known from locality CC9AA only.

Comments: *Tapomys* is also possibly known from the late Eocene of Asia (McKenna and Bell, 1997).

MANITSHINI SIMPSON, 1941

***Manitsha* Simpson, 1941**

Type species: *Manitsha tanka* Simpson, 1941.

Type specimen: AMNH 39081.

Characteristics: Large size, with teeth, skull, and jaws massive for an ischyromyid; nasals short; inflated snout; unique essentially straight line suture of premaxilla–maxilla; infraorbital foramen relatively small; posterior margin of anterior root of zygoma in line with middle of P4; molar crown pattern simple, like *Paramys*; well-developed protoloph and metaloph on upper molars; elongate buccal crest present medial to mesostyle; triangular p4; hypoconid indistinct; trigonid basin very reduced anteroposteriorly; complete metalophid closing trigonid posteriorly; no mesoconid, weak ectolophid; well-developed hypolophid; posterolophid continuous with entoconid.

Average length of p1: 9.11 mm.

Included species: *M. tanka* only, known from locality CP84A only.

Manitsha sp. is also known from locality SB44D (see Wood, 1962).

***Pseudotomus* Cope, 1872 (including *Ischyrotomus*; *Manitsha*, in part; *Paramys*, in part)**

Type species: *Pseudotomus hians* Cope, 1872.

Type specimen: AMNH 5025.

Characteristics: Large size; short snout; robust skull; infraorbital foramen relatively small; prominent chin process extends depth of mandible inferior to m1; incisors flattened anteriorly; no hypocone on P4; protoconules and metaconules single or doubled; mesostyles often doubled; hypocones on M1–2, but not well developed; no entoconid crest; small trigonid basin opening to rear; deep lingual gorge between metaconid and entoconid; entoconid distinct from posterolophid; mesoconid distinct; lower cheek tooth width exceeding length, unique for an ischyromyid.

Average length of m1: 4.80–7.65 mm.

Included species: *P. hians* (including *Ischyrotomus superbus*) (known from locality CP34C); *P. californicus* (Wilson, 1940) (locality CC4); *P. eugenei* (Burke, 1935) (originally described as *Ischyrotomus eugini*) (localities CP6B, NP8); *P. horribilis* (Wood, 1962) (originally described as *Ischyrotomus horribilus*) (locality CP34C); *P. johanniculi* (Wood, 1974) (originally described as *Manitsha johanniculi*) (localities SB44A–C); *P. littoralis* (Wilson, 1940) (locality CC4); *P. petersoni* (Matthew, 1910) (originally described as *Ischyrotomus petersoni*) (localities SB44B, CP6A, B); *P. robustus* (Marsh, 1872) (originally described as *Paramys robustus*, including *Ischyrotomus oweni*) (localities [CP5A], [CP31E], CP32B, [CP34B], CP34C, [CP25J]); *P. timmys* Storer, 1988 (locality NP9A).

Pseudotomus sp. is also known from localities CC7C, SB43A–C, E (including "*Manitsha*"), CP6A, CP29C, (CP31D, E).

PSEUDOPARAMYINI MICHAUX, 1964

***Franimys* Wood, 1962**

Type species: *Franimys amherstensis* Wood, 1962.

Type specimen: ACM 10524.

Characteristics: Small size; snout short, but long, slender nasals extend past posterior margin of zygoma anterior; quadrate-shaped frontal; unossified bulla; P3 large; P4 non-molariform, no hypocone; hypocone on M1–2; single metaconule, protoconule not well developed; low anterior cingulum; protoconid of p4 practically nonexistent; hypocone looking like a cusp of cingulum; distinct metaconid arm, but trigonid basin open posteriorly; mesoconid distinct; posterolophid meets up with entoconid.

Average length of m1: 2.38 mm.

Included species: *F. amherstensis* (known from localities CP17B, [CP19B, C], [CP20A]); *F. ambos* Korth, 1984 (including *F. lysitensis*) (locality CP27B); *F. buccatus* (Cope, 1877) (locality SB24).

Comments: *Franimys* is also questionably known from the middle Eocene of South Asia (McKenna and Bell, 1997).

REITHROPARAMYINAE WOOD, 1962

Characteristics: Small to medium sized; unique among the ischyromyids in possession of tympanic bullae co-ossified with the skull; posterior aspect of nasal bones at same position as anterior aspect of zygomatic arch and aligned with premaxillaries; suture between maxilla and palatine relatively straight; rostrum narrowed anteriorly; double metaconules common on upper molars; well-developed hypocones; molariform P4; M2 nearly equal in size and shape to M1; mandible slender; incisors narrowed and anteroposteriorly compressed; lower molars with distinct hypoconulids and buccal aspect of posterolophid posterior to entoconid (Wood, 1962; Korth, 1984, 1994).

REITHROPARAMYINI WOOD, 1962

Reithroparamys Matthew, 1920 (including *Paramys*, in part)

Type species: *Reithroparamys delicatissimus* (Leidy, 1871) (originally described as *Paramys delicatissimus*).

Type specimen: ANSP 10277.

Characteristics: Medium size; bulla co-ossified with skull; interorbital constriction of skull; nasals not as tapered as in other ischryomyids; low anterior and posterior cingula on upper molars; protoloph and metaloph well developed; metaconule sometimes doubled, protoconule single; hypocone on M1–2 small; slender lower jaw; incisors oval in cross section, some with anterior sulcus; distinct mesoconid associated with nearly straight ectolophid; double protoconid arm; entoconid crest; entoconid isolated from posterolophid.

Average length of m1: 2.34–3.20 mm.

Included species: *R. delicatissimus* (including *R. matthewi*) (known from localities NB14, CP25I, J, [CP31E], [CP34B–D], [CP37B]); *R. ctenodactylops* Korth, 1984 (localities CP20AA, CP25B); *R. debequensis* Wood, 1962 (localities [CP19C)], ?CP20A, CP25E, CP27D, CP63, CP64B); *R. huerfanensis* Wood, 1962 (localities [NB14], SB22C, SB26II, CP25I, J, CP34A, B); *R. sciuroides* (Scott and Osborn, 1887) (originally described as *Paramys sciuroides*) (including *R. gidleyi*) (locality CP6B).

Reithroparamys sp. is also known from localities CC1, CC9AA, CP17B, CP25J, (CP31A, B), C, CP32B, ?CP36B, NP49.

Acritoparamys Korth, 1984 (including *Paramys*, in part; *Reithroparamys*, in part)

Type species: *Acritoparamys francesi* (Wood, 1962) (originally described as *Paramys francesi*).

Type specimen: AMNH 14724.

Characteristics: Tapering snout, nasals extend past posterior aspect of anterior zygoma; posterior margin of anterior root of zygoma in line with the center of P4; moderate size infraorbital foramen; incisors oval in cross section, enamel limited to anterior surface; hypocone distinct on M1–2, derived from cingulum; strong protoloph and metaloph directed toward protocone; single metaconule and protoconule; large mesostyle; low anterior cingulum joining protoconid and metaconid; trigonid basin open posteriorly, reduced anteroposteriorly; well-developed hypoconulid; posterolophid separated from entoconid by shallow valley; distinct, lingually displaced mesoconid.

Average length of m1: 1.88–3.22 mm.

Included species: *A. francesi* (known from localities [CP18], [CP19], [CP20A], CP25G, CP27D, E, CP64B); *A. atavus* (Jepsen, 1937) (localities CP17A, [CP19A], [CP25B]); *A. atwateri* (Loomis, 1907) (localities CP14F, CP17B, CP18, CP19A-C, [CP20A, AA, BB], CP25A, CP27B, [CP64A]); *A. pattersoni* (Wood, 1962) (localities [CP19A, C], CP64B, C); *A. wyomingensis* (Wood, 1959) (localities CP25I, J, CP31E, CP32A, ?CP34D).

Acritoparamys sp. is also known from locality CC1.

Comments: *Acritoparamys* is also questionably known from the early Eocene of Asia (McKenna and Bell, 1997).

Uriscus Wilson, 1940

Type species: *Uriscus californicus* Wilson, 1940.

Type specimen: LACM 2194.

Characteristics: Small size; upper teeth unknown; trigonid basin minute, open posteriorly; mesoconid distinct, lingually displaced; mesostylids present; single protoconid arm; posterior cingulum continuous.

Average length of m1: 2.1 mm.

Included species: *U. californicus* only, known from locality CP4 only.

MICROPARAMYINI WOOD, 1962

***Microparamys* Wood, 1959 (including *Paramys*, in part)**

Type species: *Microparamys minutus* (Wilson, 1937) (originally described as *Paramys minutus*).

Type specimen: YPM 10730.

Characteristics: Very small ischyromyids; posterior margin of anterior zygoma in line with P4 anteroloph; no hypocone on P4; hypocone present on M1–2; metaloph and protoloph well developed, masseteric fossa ending beneath m1 instead of m2; anterior cingulum distinct from protoconid; distinct metastylid and mesoconid; single protoconid arm; entoconid distinctly separate from posterolophid.

Average length of m1: 1.31–1.72 mm.

Included species: *M. minutus* (known from localities [CC4], [CC6A], [CC7A], SB43A, CP5A, [CP17B], [CP19A, B], CP25J, CP34[A], B, C, CP35); *M. bayi* Dawson 2001 (locality HA1A); *M. cheradius* Ivy, 1990 (localities CP17B, CP18, CP19B); *M. dubius* (Wood, 1949) (localities CP6AA, A, CP29C, D, [CP36B], NP8); *M. hunterae* Ivy, 1990 (locality CP19A); *M. nimius* Storer, 1988 (locality NP9A); *M. perfossus* Wood, 1974 (localities SB44B, D); *M. sambucus* Emry and Korth, 1989 (locality NB14); *M. solidus* Storer, 1984 (locality NP8); *M. tricus* (Wilson, 1940) (locality CC7C, E, CC9BB); *M. woodi* (localities [CC4], CC6B, CC7B, CC8, CC8IIA, CC9IVA).

Microparamys sp. is also known from localities GC21II, CC1, CC7D, CC9AA, CC10, SB42, SB43B, C, CP20A, BB, CP25GG, J, CP27B (including *M. "lysitensis"*), CP27D, CP31C, E (two spp.), CP34A, B, ?D, CP64B.

Comments: *Microparamys* is also known from the early to middle Eocene of Europe (McKenna and Bell, 1997).

***Lophiparamys* Wood, 1962 (including *Paramys*, in part)**

Type species: *Lophiparamys murinus* (Matthew, 1918) (originally described as *Paramys murinus*).

Type specimen: AMNH 15131.

Characteristics: Very small ischyromyids; cusps of upper molars difficult to distinguish among cresting; slender mandible, anterior end of masseteric fossa inferior to m2; extensive crenulations of cheek teeth, numerous accessory ridges and crestlets; anterior cingulum distinct from protoconid, the latter quite reduced in p4 as in *Franimys*; distinct metastylid; trigonid basin very reduced anteroposteriorly, filled in part with metaconid ridge; entoconid distinctly separate from posterolophid; well-developed metalophid, curved in center.

Average length of m1: 1.47–2.01 mm.

Included species: *L. murinus* (known from localities CP19A, C, CP20A, CP27C, [NP49]); *L. debequensis* Wood, 1962 (localities CP27B, CP64B); *L. woodi* Guthrie, 1971 (localities SB24, CP27D).

Lophiparamys sp. is also known from locality SB43A.

***Churcheria* Storer, 2002**

Type species: *Churcheria baroni* (Storer, 1988) (originally described as *Anonymus baroni*, but replaced by *Churcheria* because *Anonymus* was already in use for another organism).

Type specimen: SMNH P1899.936.

Characteristics (from Storer, 1988): Basins and ridges of cheek teeth very rugose; hypocone of upper cheek teeth well separated from protocone, uniting with metaloph; protoloph and metaloph of upper teeth low, interrupted or weak centrally, without distinct protoconule and metaconule; protoloph and metaloph slightly flexed anterad, converging towards protocone; lower cheek teeth with strong lingual crest extending posterad from metaconid, ending in distinct metastylid; mesoconid extending distinctly buccad; mesolophid of moderate length.

Length of m2: 1.58 mm.

Included species: *C. baroni* only, known from locality NP9A only.

Churcheria sp. is also known from locality NP8.

Comments: Storer (1988) considered *Anonymus* to be a microparamyine distinguished by very rugose enamel and abundant accessory cuspules and ridges, but apart from this rugosity the teeth bear a strong resemblance to *Microparamys perfossus*.

***Strathcona* Dawson, 2001**

Type species: *Strathcona minor* Dawson, 2001.

Type specimen: CM 323667.

Characteristics (from Dawson, 2001): Hypocone on M1–2; metaloph and protoloph oriented toward protocone; single metaconule and protoconule; mesostyle small; lower molars transversely short, anteriorly protruding anterior cingulum that is nearly cuspate buccally, metaconid crest reaching buccally into trigonid basin, which opens posteriorly; protoconid distinct from anterior cingulum; short protoconid arm; distinct mesoconid, entoconid isolated from posterolophid, metastylid well developed. Differs from other microparamyines (except for *Microparamys cheradius*, a much smaller species), in having the metaconid crest extending into the trigonid basin.

Average length of m1: 1.95 mm (*S. minor*); 2.31 mm (*S. major*).

Included species. *S. minor* (known from locality HA1A); *S. major* Dawson 2001 (locality HA1A).

Comments: *S. minor* is the smallest rodent known from the Arctic Eocene, but it is still bigger than other microparamyids. A number of dental features (described in Dawson, 2001) support the inclusion of this genus in the Microparamyinae. The similarity of the metaconid crest to *Microparamys cheradius* (see above) suggests that it might have been an early derivative of *Microparamys*) (Dawson, 2001).

REITHROPARAMYINAE?

***Apatosciuravus* Korth, 1984**

Type species: *Apatosciuravus bifax* Korth, 1984.

Type specimen: CM 38765.

Characteristics: Small size; posterior margin of anterior zygoma in line with P3; upper molars with cusps and transverse lophs; large hypocone on M1–2; distinct mesostyle; anterior cingulum joins metaconid to protoconid; distinct mesoconid; distinct hypoconulid; narrow trigonid basin; entoconid isolated from posterolophid.

Average length of m1: 1.33–1.60 mm (*A. bifax*).

Included species: *A. bifax* (known from localities CP20A, BB, CP63); *A. jacobsi* Flanagan, 1986 (locality SB24).

AILURAVINAE MICHAUX, 1968

These are large ischyromyids. For a description of the skull, refer to Weitzel (1949). Features unique to the upper cheek teeth include P3 present; M3 molariform; hypocone present on P4–M3, best developed on M1–2; prominent protostyle on M1–2. The mandible is slender with a somewhat flattened ventral side. Lower incisors are oval in cross section and narrowed anteriorly, similar to those of *Uintaparamys*, unique in that they are short. Lower molars include an entoconid distinct from the posterolophid, well-developed hypoconulid, large mesoconid, curved ectolophid, protoconid lingually displaced, p4 relatively large, protoconid arm on m1–2. Cheek teeth are characterized by progressive development of deep valleys and cusplets (Wood, 1976).

***Mytonomys* Wood, 1959 (including *Leptotomus*, in part; *Prosciurus*, in part; *Pseudotomus*, in part)**

Type species: *Mytonomys robustus* (Peterson, 1919) (originally described as *Prosciurus*? or *Leptotomus robustus*).

Type specimen: CM 2925.

Characteristics: Large size; anterior maxillary root of zygoma level with P4; no hypocone on P4; hypocone present on M1–2; single protoconule and metaconule; large mesostyle; protoloph and metaloph essentially parallel on P4-M2; talonid basin very reduced; entoconid ridge extends toward ectolophid; metastylid; lower incisor nearly equidimensional; trigonid basin defined posteriorly by crest from protoconid and metaconid, but remaining open to talonid basin; mesoconid indistinct, lingually displaced, associated with straight ectolophid; entoconid isolated from posterolophid.

Average length of m1: 3.84–5.20 mm.

Included species: *M. robustus* (including *Leptotomus kayi*) (known from localities SB43C, CP6A, B, CP7A, B, NP22); *M. burkei* (Wilson, 1940) (locality CC9AA); ?*M. coloradensis* (Wood, 1962) (originally described as *Pseudotomus coloradensis*) (localities [CP20B–D], CP27B, D, CP64A–C); *M. gaitania* Ferrusquía and Wood, 1969 (locality SB52); *M. mytonensis* (Wood, 1962) (locality CP6B); *M. wortmani* (Wood, 1962) (locality CP27D).

Mytonomys sp. is also known from localities SB42, SB43A, C.

***Eohaplomys* Stock, 1935**

Type species: *Eohaplomys serus* Stock, 1935.

Type specimen: CIT 1572.

Characteristics: Posterior margin of anterior root of zygoma at level of posteroloph of P4; prominent parastyle and mesostyle; hypocone present on M1–2; single protoconule and metaconule; protoloph and metaloph well developed; protoconid arm forming strong crest; trigonid opening posteriorly; entoconid distinctly separated from posterolophid; metastylid and hypoconulid present; basin enamel with wrinkled appearance.

Average length of m1: 2.25 mm.

Included species: *E. serus* (known from localities [CC7D], CC9AA); *E. matutinus* Stock, 1935 (locality CC9AA); *E. tradux* Stock, 1935 (localities [CC7B, C], [CC8], CC9AA).

Eohaplomys sp. is also known from localities CC6A, B, CC8, CC9AA.

ISCHYROMYINAE ALSTON, 1876

These medium- to large-sized ischyromyids have auditory bullae co-ossified with the skull. The infraorbital foramen is moderate sized and inclined. Nasals and premaxillaries form a straight line ending in front of the orbit. The posterior margin of the anterior root of zygoma is level with dP3 in smaller species and level with middle of P4 in larger ones. Upper and lower cheek teeth are partially to completely lophate with a hypocone distinct on M1–2, no mesoconid, no metastylid, and hypolophid well developed (Black, 1968, 1971; Wahlert, 1974; Wood, 1976).

***Ischyromys* Leidy, 1856 (including *Titanotheriomys*)**

Type species: *Ischyromys typus* Leidy, 1856.

Type specimen: ANSP 11015.

Characteristics: Medium size; auditory bulla co-ossified with skull; upper incisors oval in cross section; metaconule more distinct than protoconule; posterior border of anterior root of zygomatic arch level with anterior aspect of P4; lingual notch between protocone and hypocone sometimes shallow; protoloph and metaloph essentially parallel on P4–M2; lower incisors oval in cross section, tapered anteriorly; lower molar crown pattern crestlike; length of molars slightly greater than width; no mesoconid; entoconid isolated from posterlophid; hypolophid present.

Average length of m1: 3.18–3.70 mm

Included species: *I. typus* (known from localities [CP39IIA], CP68C, CP84A, B [including "*I. pliacus*"], CP98BB [including "*I. pliacus*"], C [including "*I. veterior*"], CP99A, B); *I. blacki* Wood, 1974 (localities SB44B, C); *I. douglassi* Black, 1968 (localities CP98B, [NP22], NP25 [A, B], C); *I. junctus* Russell, 1972 (locality NP10B); *I. parvidens* Miller and Gidley, 1920 (localities CP84A, CP99A); *I. pliacus* Troxell, 1922 (localities CP40B, CP41B, CP99A, ?B, NP29C); *I. veterior* Matthew, 1902 (localities SB44E, CP39B–F, CP39IIC, CP43, CP45A, [CP68B], CP98B, BB (including "*Titanotheriomys wyomingensis*"), CP99A, NP10Bi, NP24C–E, NP27C, NP29C).

Ischyromys sp. is also known from localities CC8IIB, SB43C, E, CP39A-D (large sp.), CP68B ("*I. troxelli*"), C, CP83A, NP9A, NP24C–E, NP27C, NP29C, NP30A, NP32C, NP50B.

INDETERMINATE ISCHYROMYIDS

Unidentified ischyromyids are also reported from localities GC21II, CC2, CC4, CC7B, CC8, CC8IIA, CC9AA, BB, ?CC10, CC50, NB14, SB25A, CP20BB (several spp.), CP25A, CP31B (two small spp.), CP34E, CP38C, CP65, NP9A (two spp.), NP9B, NC0II.

Note contributed by Rachel Dunn (see Dunn and Rasmussen, 2005). A new ischryomyid specimen, possibly representing a new taxon of the subfamily Paramyinae, has recently been recovered from the Uinta Basin, middle Eocene of Utah (locality CP6A). Cranial and postcranial materials have been recovered including a manus, pes and associated limbs, pelvis, and vertebral column. The specimen closely resembles *Manitsha* (Simpson, 1941) and differs from *Pseudotomus* (and *Ischyrotomus*) in some key dental traits (Wood, 1962). Characteristics of the lower molars include a low but unmistakable hypolophulid extending from the entoconid to the ectolophid; the ectolophid crest is only slightly developed, there is a clean split into mesial and distal halves by a deep, straight groove, and the teeth are robust in build and heavily enamelled. Average M1 length is 7.1 mm. The upper molars bear four smooth lophs (anteroloph, paraloph, metaloph, posteroloph) lacking conules; the simple paracones and metacones are separated by a deep notch.

The new rodent differs from *Manitsha* in having more triangular upper molars (rather than squarish), in lacking the anterior extension of the scar for the masseter muscle, and in being significantly smaller than the Oligocene species of that genus. The new rodent is very robust including prominent crests on the proximal limb bones, a large and rugose olecranon process, and long, deep ungula phalanges. The medial epicondyle of the humerus is robust, and the ungual phalanges have large attachment points for the digital flexors. The new rodent may be conspecific with *Ischyrotomus eugenei*, which is based on a woefully worn lower jaw.

BIOLOGY AND EVOLUTIONARY PATTERNS

From a paleobiological and evolutionary perspective, rodents of the family Ischyromyidae are an interesting group. It is possible that members of this family, *Acritoparamys atavus* perhaps, represent some of the earliest rodents in the fossil record worldwide (Figure 18.3). Even if they do not represent the actual basal stock, these rodents may have been ancestral to later Tertiary rodent families such as Sciuridae, Aplodontidae, Old World Gliridae, and possibly Castoridae (Carroll, 1988). The transition from ischyromyids to the sciurids is one of the best supported by the fossil record. The earliest sciurid, the Chadronian (late Eocene) *Protosciurus jeffersoni*, shares a number of derived features with *Reithroparamys*. According to Korth (1994) these features include: a protocone crest, two crests from the protoconid to metaconid, a double metaconule, presence of a hypolophid, and the presence of relatively large ossified bullae. In general, crests and lophs characterize the molar occlusal pattern of modern lineages replacing the distinct cusps of early rodents such as ischyromyids.

Ischyromyid cranial and mandibular anatomy reveals a combination of primitive and derived features, further evidence to suggest that this group is ancestral to later Tertiary rodent families. Ossified auditory bullae, a derived feature in rodents, occurs in some genera of ischyromyids but is absent in others. While an ossified bulla is found in *Reithroparamys* and *Pseudotomus* (*Ischyrotomus*) *petersoni*, *Paramys* lacks ossified bullae. The bullae of *Pseudotomus* may represent an intermediate stage because they are not co-ossified with the skull and are not enlarged as are the bullae of *Reithroparamys*. Crania with preserved bullae for many other genera of ischyromyids have yet to be found; in particular data for small genera of ischyromyids are lacking. As more fossilized crania are recovered, evidence of a transitional series of bullae among members of the family Ischyromyidae would provide further support to the above hypotheses.

Another derived skull feature shared by ischyromyids and recent rodents is the low dorsoventral aspect with dorsal and ventral margins essentially parallel in lateral view (Wood, 1962; Meng, Hu, and Li, 2003). Primitive characters found in ischyromyids include a basioccipital with a strong median keel, a feature absent in recent rodents (Wood, 1962; Meng, Hu, and Li, 2003; Wible *et al.*, 2005), and nasals long and tapering rapidly at their posterior margin (lacking in *R. sciuridoides*), which is relatively uncommon for recent rodents. Primitive characters of the auditory region of *Paramys* include a low promontorium, narrow and shallow epitympanic recesses, and separate foraminae for the stapedial artery and facial nerve (Wahlert, 2000). A significant derived feature uniting rodents and lagomorphs is that the squamosal does not form part of the epitympanic recess in the auditory region (Wahlert, 2000). Examples of cranial anatomy characteristics that are intermediate include the anterior portion of the frontals oriented lateral to the nasals in *Paramys*, *Pseudotomus*, and in some species of *Uintaparamys*, a condition not present in *Reithroparamys*. This characteristic is found in some recent rodents but is absent in others.

Significant primitive features of mandibular anatomy include sciurognathy; the condyle of the mandible above the tooth row (Korth, 1994); a masseteric fossa in the dentary that extends anteriorly to below m2–3; a well-marked fossa on the side of the coronoid process, for the anterior part of the temporalis (Wood, 1962); and a broad ascending ramus of the dentary. Intermediate to derived features include lengthening of the diastema and a less pronounced masseteric fossa in later species of *Uintaparamys*, anteroposterior shortening of the ascending ramus in *U. bridgerensis*, single versus double mental foraminae, and a knob at the end of the masseteric fossa. In summary, this mix of ancestral, derived, and intermediate features of mandibular and cranial anatomy are grounds for

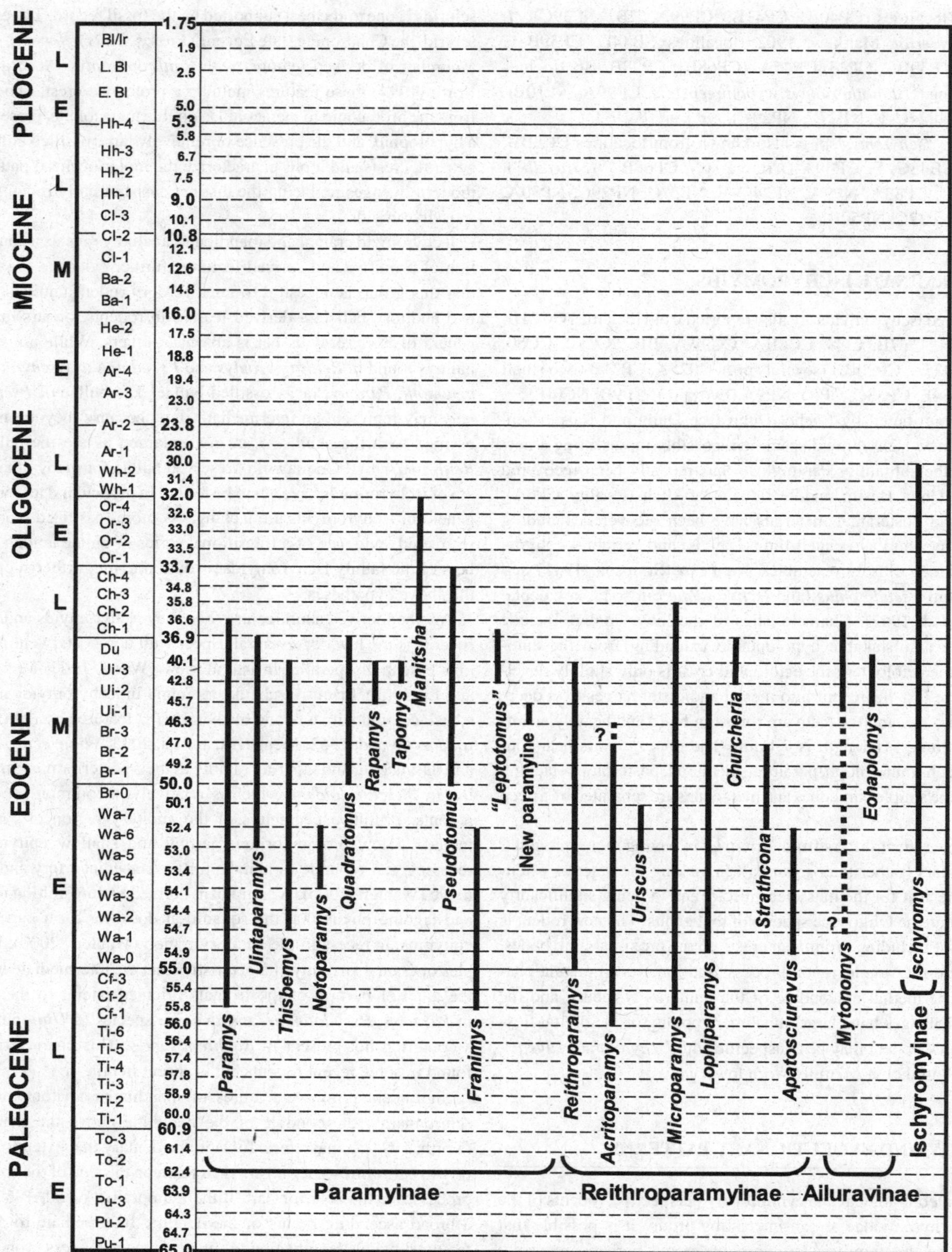

Figure 18.3. Temporal ranges of North American ischyromyid genera.

considering the ischyromyids as an ancestral group to later Tertiary rodents.

Another aspect of ischyromyid anatomy that makes them likely candidates as an ancestral rodent group is the hypothesized wide range of locomotor adaptations. Ischyromyids appear to range from scampering to subfossorial to possibly burrowing based on their skeletal anatomy in comparison with living rodents. *Paramys* skeletons do not show any particular specializations with respect to locomotion, and members of this genus were probably ground scamperers based on the limb proportions compared with living scamperers such as *Rattus*, *Mus*, and *Peromyscus* (Wood, 1962). However, the very generalized skeleton of *Paramys* is distinct from digging forms such as *Castor* and *Marmota*. One primitive skeletal feature of *Paramys* is that femur length exceeds skull length, the opposite of the condition in extant rodents. *Uintaparamys* has a skeleton very similar to that of *Paramys* except that it is more slender in some respects (e.g., femur, pelvis). Members of this genus were also most likely scamperers.

Pseudotomus exhibits a striking difference in hindlimb length compared with forelimb length with the hindlimb twice the length of the forelimb. These proportions are similar to those found in living saltatorial rodents. However, given the heavy bone construction of *Pseudotomus*, it is not likely that this animal was saltatorial. More than likely, *Pseudotomus* used its massive stout bones, including heavy feet and claws, for digging shallow burrows and spent most of its time foraging on the ground. Other adaptations of the skeleton associated with burrowing include heavy phalanges, stout bones, short claws, prominent crests and fossae for muscle attachment, and a dorsally rounded skull with a broad, short snout (Wood, 1962). The unusual hindlimb to forelimb ratio, which is inconsistent with extant fossorial forms, is evidence that *Pseudotomus* retained generalist features and was probably able to exploit more than one niche. This conclusion is consistent with the findings of Rose and Chinnery (2004) who studied the postcranial skeleton of other paramyid rodents.

Rodent classification is more challenging and evolutionary patterns more interesting owing to the preponderance of instances of parallelism and convergence among rodents, and the ischyromyids are no exception. Most examples of these two phenomena result in confusion at the family and generic level of taxonomy. For example, Korth (1984) recognized the development of a massive postcranial skeleton occurring in three distinct lineages of ischyromyids (Manitishini, Paramyini, and the European genus *Plesiarctomys*), a case of parallel evolution. As a result of this observation, *Quadratomus* was removed from the Manitishini and placed in the tribe Paramyini.

Another example of convergence occurs between ischyromyids such as *Pseudotomus* and members of the suborder Palaeanodonta. Both are early Tertiary mammals that evolved skeletal adaptations for digging and/or burrowing, such as relatively short, clawed forelimbs (Rose, 1990). Dental characters such as lophs and crests have evolved many times in unrelated lineages in rodent evolution: specifically, Marivaux, Vianey-Liaud, and Jaeger (2004) recognized the acquisition of a transverse crest in the talonid and development of a protoconid arm as convergent evolution in the clades Ctenohystrica and Ischyromyiformes. Dawson and Beard (1996) suggested that the loss of a paracone occurring before development of the hypocone might be an example of mosaic evolution in rodents.

There are examples of directional and stabilizing selection among the ischyromyids, and some evidence that one species developed as a product of both. Within the family, there are several species that exhibit the same molar crown pattern but are significantly different in size. For example, *Paramys* (*excavatus*) *taurus* has the same molar crown pattern as *Paramys copei*, four cusps, small mesoconid, no hypolophid, trigonid basin open to the rear, low anterior cingulum, posterolophid continuous with the hypoconid, no entoconid crest, and smooth enamel. But the two are very different in size, the first lower molar of *P. taurus* averages 2.75 mm anteroposteriorly while the same tooth in *P. copei* is 3.30 mm. These two species are found to occur together in the same time period of early Eocene, but *P. copei* originated later in time; therefore, this represents a case of directional selection with a trend towards large size. Other species of ischyromyids are approximately the same size (based on molar length and widths), but differ in the molar crown pattern. Consider *Franimys* sp., which averages 2.38 mm anteroposteriorly and *Acritoparamys francesi*, which averages 2.32 mm. One feature unique to *Franimys* is the absence of a protoconid on p4, a cusp that is present and well defined in *A. francesi*. Stabilizing selection was maintaining the small size first seen in *Franimys* while other selective pressures were acting on the crown pattern, differences that were most likely the result of changing environmental conditions.

Taxa from two subfamilies, Paramyinae and Reithroparamyinae, two distinct genera, occur as early as Clarkforkian (latest Paleocene) time in North America (Korth, 1994), reflecting morphological diversity in the family from its time of origin (Figure 18.3). The earliest paramyines have the morphologically simplest crown patterns among the ischyromyids and are small- to medium-sized rodents. Late early Eocene paramyinae include larger genera and more variations in crown pattern, including enamel irregularities, development of a hypolophid, isolated entoconids, and reduced mesoconids. Diversity measured in number of genera peaks around Bridgerian (early middle Eocene), falling dramatically at the end of the Uintan (late middle Eocene) (Wood, 1962; Korth, 1984).

Reithroparamyinae show peak generic diversity in the Wasatchian (early Eocene) and following a slight decline, the number of genera falls until the subfamily becomes extinct in the Orellan (early Oligocene) (Wood, 1962; Korth, 1994). This subfamily includes small genera of ischyromyids, each displaying a unique, more complex crown pattern than seen in most paramyinae. Larger hypocones, isolation of the anterior cingulum, and presence of a distinctive hypolophid are some of the significant molar crown pattern characters seen in this group. The ossified auditory bullae are a unique cranial feature found in only one other subfamily, Ailuravinae. This character is thought to be derived but appears in the first members of the subfamily Reithroparamyinae in the Clarkforkian.

The other two subfamilies, Ailuravinae and Ischyromyinae, are relatively small with a total of three genera. Both genera of Ailuravinae are large in size and occurred during the Uintan. Their crown patterns are more complex than those of the Paramyinae, including large hypocones, complex crown patterns for P3, elongated conules on upper molars, and crests and lophs on the lower cheek teeth (Korth, 1988b). The Ischyromyinae with one genus represent the most recently appearing ischyromyids. The group was most diverse in number of species during the Chadronian (late Eocene) (Korth, 1984). Not surprisingly, this genus includes a number of derived cranial and dental features not seen in older ischyromyids.

REFERENCES

Alston, E. R. (1876). On the classification of the order Glires. *Proceedings of the Zoological Society of London*, 1876, 61–98.

Black, C. C. (1968). The Oligocene rodent *Ischyromys* and discussion of the family Ischyromyidae. *Annals of the Carnegie Museum*, 39, 273–305.

(1971). Paleontology and geology of the Badwater Creek area, central Wyoming. Part 7: Rodents of the family Ischyromyidae. *Annals of the Carnegie Museum*, 43, 179–217.

Burke, J. J. (1935). Fossil rodents from the Uinta Eocene series. *Annals of the Carnegie Museum*, 25, 5–12.

Butler, P. M. (1985). Homologies of molar cusps and crests, and their bearing on assessments of rodent phylogeny. In *Evolutionary Relationships Among Rodents*, ed. W. P. Luckett and J.-L. Hartenberger, pp. 381–401. New York: Plenum Press.

Carroll, R. L. (1988). *Vertebrate Paleontology and Evolution*. New York: Freeman.

Cope, E. D. (1872). Second account of new Vertebrata from the Bridger Eocene. *Proceedings of the American Philosophical Society*, 12, 466–8.

(1873). On the extinct Vertebrata of the Eocene of Wyoming, observed by the expedition of 1872, with notes on the geology. *Sixth Annual Report of the US Geological Survey of the Territories* (F. V. Hayden in charge), pp. 545–649. Washington, DC: US Government Printing Office.

(1877). Report upon the extinct Vertebrata obtained in New Mexico by parties of the expedition of 1874. *Report of the US Geographical Survey West of the 100th Meridian (G. M. Wheeler in charge)*, Vol. 4, pp. 1–370. Washington, DC: US Government Printing Office.

Dawson, M. R. (2001). Early Eocene rodents (Mammalia) from the Eureka Sound Group of Ellesmere Island, Canada. *Canadian Journal of Earth Sciences* 38, 1107–16.

Dawson, M. R. and Beard, K. C. (1996). New Late Paleocene rodents (Mammalia) from Big Multi Quarry, Washakie Basin, Wyoming. *Palaeovertebrata*, 25, 301–21.

Dawson, M. R., Li, C. K., and Qui, T. (1984). Eocene ctenodactyloid rodents (Mammalia) of eastern and central Asia, [In *Papers in Vertebrate Paleontology honoring Robert Warren Wilson*, ed. R. M. Mengel.] *Carnegie Museum of Natural History Special Publication*, 9, 138–50.

Dunn, R. and Rasmussen, T. (2005). Large rodent from the middle Eocene of Utah. *Journal of Vertebrate Paleontology*, 24(suppl. to no. 3), p. 53A.

Eaton, J. G. (1982). Paleontology and correlation of Eocene volcanic rocks in the Carter Mountain area, Park County, southeastern Absaroka Range, Wyoming, *Contributions to Geology, University of Wyoming*, 21, 153–94.

Emry, R. J. and Korth, W. W. (1989). Rodents of the Bridgerian (Middle Eocene) Elderberry Canyon Local Fauna of Eastern Nevada. *Smithsonian Contributions to Paleobiology*, 67, 1–14.

Ferrusquía-Villafranca, I. and Wood, A. E. (1969). New fossil rodents from the early Oligocene Rancho Gaitan Local Fauna, northeastern Chihuahua, Mexico. *Texas Memorial Museum, Pearce-Sellards Series* 16, 1–13.

Flanagan, K. M. (1986). Early Eocene rodents from the San Jose formation, San Juan Basin, New Mexico. [In *Vertebrates, Phylogeny, and Philosophy*, ed. K. M. Flanagan and J. A. Lillegraven.] *University of Wyoming Contributions to Geology Special Paper* 3, 197–220.

Guthrie, R. D. (1971). Factors regulating the evolution of microtine tooth complexity, *Zeitschrift für Säugetierkunde*, 36, 37–54.

Haeckel, E. (1895). *Systematische Phylogenie der Wirbelthiere (Vertebrata). Dritter Theil des Entwurfs einer systematischen Phylogenie*. Berlin.

Hartenberger, J.-L. (1985). The order Rodentia: major questions on their evolutionary origin, relationships, and suprafamilial systematics. In *Evolutionary Relationships Among Rodents: A Multidisciplinary Analysis*, ed. W. P. Luckett, and J.-L. Hartenberger, pp. 1–34. New York: Plenum Press.

Hunter, J. P. and Jernvall, J. (1995). The hypocone as a key innovation in mammalian evolution. *Proceedings of the National Academy of Sciences. USA*, 92, 10 718–22.

Jepsen, G. L. (1937). A Paleocene rodent, *Paramys atavus*. *Proceedings of the American Philosophical Society*, 78, 291–301.

Ivy, L. D. (1990). Systematics of late Paleocene and early Eocene Rodentia (Mammalia) from the Clark's Fork Basin, Wyoming. *Contributions from the Museum of Paleontology, University of Michigan*, 28, 21–71.

Korth, W. W. (1984). Earliest Tertiary evolution and radiation of rodents in North America. *Bulletin of the Carnegie Museum of Natural History*, 24, 1–71.

(1988a). *Paramys compressidens* Peterson and the systematic relationships of the species of *Paramys* (Paramyinae, Ischyromyidae). *Journal of Paleontology*, 62, 468–71.

(1988b). The rodent *Mytonomys* from the Uintan and Duchesnean (Eocene) of Utah, and the content of the Ailurvinae (Ischyromyidae, Rodentia). *Journal of Vertebrate Paleontology*, 8, 290–4.

(1994). *The Tertiary Record of Rodents in North America*. New York: Plenum Press.

Kretzoi, M. (1968). New generic names for homonyms. *Vertebrata Hungarica*, 10L, 163–6.

Leidy, J. (1856). Notices of remains of extinct Mammalia discovered by Dr. F. V. Hayden in Nebraska Territory. *Proceedings of the Academy of Natural Sciences, Philadelphia*, 8, 88–90.

(1871). Notice of some extinct rodents. *Proceedings of the Academy of Natural Sciences, Philadelphia*, 22, 230–2.

Loomis, F. B. (1907). Wasatch and Wind River rodents. *American Journal of Science, Series 4*, 23, 123–30.

Marivaux, L., Vianey-Liaud, M., and Jaeger, J.-J. (2004). High-level phylogeny of early Tertiary rodents: dental evidence. *Zoological Journal of the Linnean Society*, 142, 105–34.

Marsh, O. C. (1872). Preliminary description of new Teritary mammals. *American Journal of Science, Series* 3, 4 202–24.

Matthew, W. D. (1902). A horned rodent from the Colorado Miocene, with a revision of the Mylagauli, beavers, and hares of the American Tertiary. *Bulletin of the American Museum of Natural History*, 16, 291–310.

(1910). On the osteology and relationships of *Paramys* and the affinities of the Ischyromyidae. *Bulletin of the American Museum of Natural History*, 28, 43–71.

(1918). Insectivora (continued), Glires, Edentata. [In *A Revision of the Lower Eocene Wasatch and Wind River Faunas*, Part 5, ed. W. D. Matthew and W. Granger.] *Bulletin of the American Museum of Natural History*, 38, 565–657.

(1920). A new genus of rodents from the middle Eocene. *Journal of Mammalogy*, 1, 168–9.

McKenna, M. C. and Bell, S. K. (1997). *Classification of Mammals Above the Species Level*. New York: Columbia University Press.

Meng, J. (1990). The auditory region of *Reithroparamys delicatissimus* (Mammalia, Rodentia) and its systematic implications. *American Museum Novitates*, 2972, 1–35.

Meng, J. and Wyss, A. R. (2001). The morphology of *Tribosphenomys* (Rodentiaformes, Mammalia): phylogenetic implications for basal Glires. *Journal of Mammalian Evolution*, 8, 1–71.

Meng, J., Wyss, A. R., Dawson, M. R., and Zhai, R. (1994). Primitive fossil rodent from Inner Mongolia and its implications for mammalian phylogeny. *Nature*, 370, 134–6.

Meng, J., Hu, Y., and Li, C. (2003). The osteology of *Rhombomylus* (Mammalia, Glires): implications for phylogeny and evolution of Glires. *Bulletin of the American Museum of Natural History*, 275, 1–247.

Michaux, J. (1964). Diagnoses de quelques Paramyidés de l'Eocene inférieur de France. *Comptes Rendus Sommaires de la Societé Géologique de France*, 4, 153.

Miller, G. S. and Gidley, J. W. (1920). A new fossil rodent from the Oligocene of South Dakota. *Journal of Mammalogy*, 1, 73–4.

Osborn, H. F. (1895). Perissodactyls of the Lower Miocene White River Beds. *Bulletin of the American Museum of Natural History*, 7, 343–75.

Peterson, O. A. (1919). Report upon the material discovered in the upper Eocene of the Unita Basin by Earl Douglass in the years 1908–1909 and by O. A. Peterson in 1912. *Annals of the Carnegie Museum*, 12, 40–168.

Rose, K. D. (1990). Postcranial skeletal remains and adaptations in early Eocene mammals from the Willwood Formation, Bighorn Basin, Wyoming. [In *Dawn of the Age of Mammals in the Northern Part of the Rocky Mountain Interior, North America*, ed. T M. Bown and K. D. Rose.] *Special Paper of the Geological Society of America*, 243: 107–34.

Rose, K. D. and Chinnery, B. J. (2004). The postcranial skeleton of early Eocene rodents. [In *Fanfare for an Uncommon Paleontologist. Essays in Honor of Malcolm C. McKenna*, ed. M. R. Dawson and J. A. Lillegraven.] *Bulletin of the Carnegie Museum of Natural History*, 36, 211–44.

Russell, L. S. (1972). Evolution and classification of the pocket gophers of the subfamily Geomyinae. *University of Kansas, Publications of the Museum of Natural History*, 16, 473–579.

Scott, W. B. and Osborn, H. F. (1887). Preliminary report on the vertebrate fossils of the Uinta Formation, collected by the Princeton expedition of 1886. *American Philosophical Society Proceedings*, 24, 255–64.

Simpson, G. G. (1941). A giant rodent from the Oligocene of South Dakota. *American Museum Novitates*, 1149, 1–16.

Stock, C. (1935). A new genus of rodent from the Sespe Eocene. *Bulletin of the Geological Society of America*, 46, 61–8.

Storer, J. E. (1984). Mammals of the Swift Current Creek Local Fauna (Eocene: Uintan), Saskatchewan. *Natural History Contributions, Saskatchewan Museum of Natural History*, 7, 1–158.

(1988). The rodents of the Lac Pellatier lower fauna, late Eocene (Duchesnean) of Saskatchewan. *Journal of Vertebrate Paleontology*, 8, 84–101.

(2002). *Churcheria*, new name for *Anonymus storer*, 1988 (Vertebrata, Mammalia, Rodentia), not *Anonymus* Land, 1884 (Platyhelminthes,Turbellaria, Polycladida). *Journal of Vertebrate Paleontology*, 22, 734.

Swofford, D. L. (2000). PAUP*. *Phylogenetic Analysis Using Parsimony (*and Other Methods)*, Version 4. Sunderland, MA: Sinauer.

Troxell, E. L. (1922). Oligocene rodents of the genus *Ischyromys*. *American Journal of Science, Series 5*, 3, 123–30.

Wahlert, J. H. (1974). The cranial foramina of protrogomorphous rodents; an anatomical and phyologenetic study. *Bulletin of Museum of Comparative Zoology*, 146, 363–410.

(2000). Morphology of the auditory region in *Paramys copei* and other Eocene rodents of North America. *American Museum Novitates*, 3307, 1–16.

Watrous, L. E. and Wheeler, Q. D. (1981). The out-group comparison method of character analysis. *Systematic Zoology*, 30, 1–11.

Weitzel, K. (1949). Neue Wirbeltiere (Rodentia, Insectivora, Testudinata) aus dem Mitteleozän von Messel bei Darmstadt. *Abhandlung der Sencken-bergischen Naturforschenden Gesellschaft*, 480, 1–24.

Wible, J. R., Wang, Y., Chuamkui, L., and Dawson, M. R. (2005). Cranial anatomy and relationships of a new ctenodactyloid (Mammalia, Rodentia) from the Early Eocene of Hubei Province, China. *Annals of the Carnegie Museum*, 74, 91–150.

Wilson, J. A. and Stevens, M. S. (1986). Fossil vertebrates from the latest Eocene, Skyline channels, Trans-Pecos Texas. [In *Vertebrates, Phylogeny and Philosophy*, ed. K. M. Flanagan and J. A. Lillegraven.] *Contributions to Geology, University of Wyoming Special Paper* 3, 221–36.

Wilson, R. W. (1937). Two new Eocene rodents from the Green River Basin, Wyoming. *American Journal of Science*, 34, 447–56.

(1940). Californian paramyid rodents. *Carnegie Institution of Washington Publications*, 514, 59–83.

Wood, A. E. (1949). A new Oligocene rodent genus from Patagonia, *American Museum Novitates*, 1435, 1–54.

(1959). Eocene radiation and phylogeny of the rodents. *Evolution*, 13, 354–61.

(1962). The early Tertiary rodents of the family Paramyidae. *Transactions of the American Philosophical Society*, 52, 1–261.

(1974). Early Tertiary Vertebrate Faunas Vieja Group Trans-Pecos Texas: Rodentia. *Texas Memorial Museum, Bulletin*, 21, 1–112.

(1976). The paramyid rodent *Ailuravus* from the middle and late Eocene of Europe, and its relationships. *Palaeovertebrata*, 7, 117–49.

(1977). The evolution of the rodent family Ctenodactylidae. *Journal of the Palaeontological Society of India*, 20, 120–37.

Wyss, A. R. and Meng, J. (1996). Application of phylogenetic taxonomy to poorly resolved crown clades: a stem-modified node-based definition of Rodentia. *Systematic Biology*, 45, 559–68.

19 Sciuravidae

ANNE H. WALTON and READ M. PORTER

Pratt Museum of Natural History, Amherst College, MA, USA

INTRODUCTION

Sciuravids are a paraphyletic group of primitive rodents composed of at least three branches at the base of the speciose suborder Myomorpha. They are known only from the early to middle Eocene of North America (Wasatchian–Uintan). They are small (pygmy mouse to hamster size) and were widespread and locally abundant. Most species are represented by isolated teeth or jaw fragments. Their diets and ecological roles are unknown, but they probably played an important role near the base of the terrestrial food chain in subtropical and temperate habitats. The three lineages recognized here include one composed of certain species of *Sciuravus*, one of *Knightomys* and the controversial hystricognathous genus *Prolapsus*, and one for species of *Pauromys*, which may be sister to the Eomyidae.

For purposes of this chapter, the informal term "sciuravids" refers collectively to the three lineages listed above, and other poorly represented taxa that might be related to them. The formal family term "Sciuravidae" is confined to the *Sciuravus* lineage, a crown clade composed of the type species and two other species demonstrably sister to it.

DEFINING FEATURES OF THE FAMILY SCIURAVIDAE

CRANIAL

The family Sciuravidae was defined by Miller and Gidley (1918) using mainly primitive features: sciurognathy, a small infraorbital foramen (protrogomorphy), a primitive rodent dental formula with two upper premolars (I1/1, C0/0, P2/1, M3/3), and brachydont cheek teeth. The small infraorbital foramen is visible in lateral view, indicating little modification of the zygoma. Matthew (1910) and Korth (1994) further distinguished Sciuravidae from contemporaneous Paramyidae by the more gracile skull with a less pronounced postorbital constriction, a large interparietal bone, ossified bullae weakly attached to the skull, and a more anterior termination of the masseteric scar (at the level of m1–2).

Well-preserved skull material has been described only for the Bridgerian type species, *Sciuravus nitidus* (Figure 19.1A). Wahlert (1974) reported that a medial internal carotid artery is absent from the middle ear of *S. nitidus* but that a promontory artery is present, unlike the arrangement of vessels in Paramyidae (Lavocat and Parent, 1985). In most respects the middle ear of *Sciuravus* is primitive.

DENTAL

The quadritubercular upper molars with a well-developed hypocone, the single derived character listed by Miller and Gidley (1918), distinguished Sciuravidae from Paramyidae, with their more primitive tritubercular molars. The hypocone is separated from the protocone by a pronounced intervening valley, and conules are reduced or absent (Wilson, 1949). Korth (1994) listed the following additional dental characters for sciuravids: P3 is small and peglike; P4 is smaller than molars but submolariform, with a distinct hypocone; upper molars are square, with four well-developed transverse crests, derived from anterior and posterior cingula (anteroloph and posteroloph), protocone–protoconule (protoloph), and metacone (metaloph); the metaloph typically contacts the hypocone; lower molars are rectangular with three or four transverse crests formed by the anterior cingulid (anterolophid), posterior arm of the protoconid (metalophid II), entoconid (hypolophid), and a posterior cingulid (posterolophid), isolated from the entoconid but in contact with the hypolophid; the mesoconid is typically distinct, weakly attached to the hypoconid and often elongate to form a fifth crest in the central basin or talonid (mesolophid); the lower premolar is smaller than m1 and, in some species, reduced.

Evolution of Tertiary Mammals of North America, Vol. 2. ed. C. M. Janis, G. F. Gunnell, and M. D. Uhen. Published by Cambridge University Press.

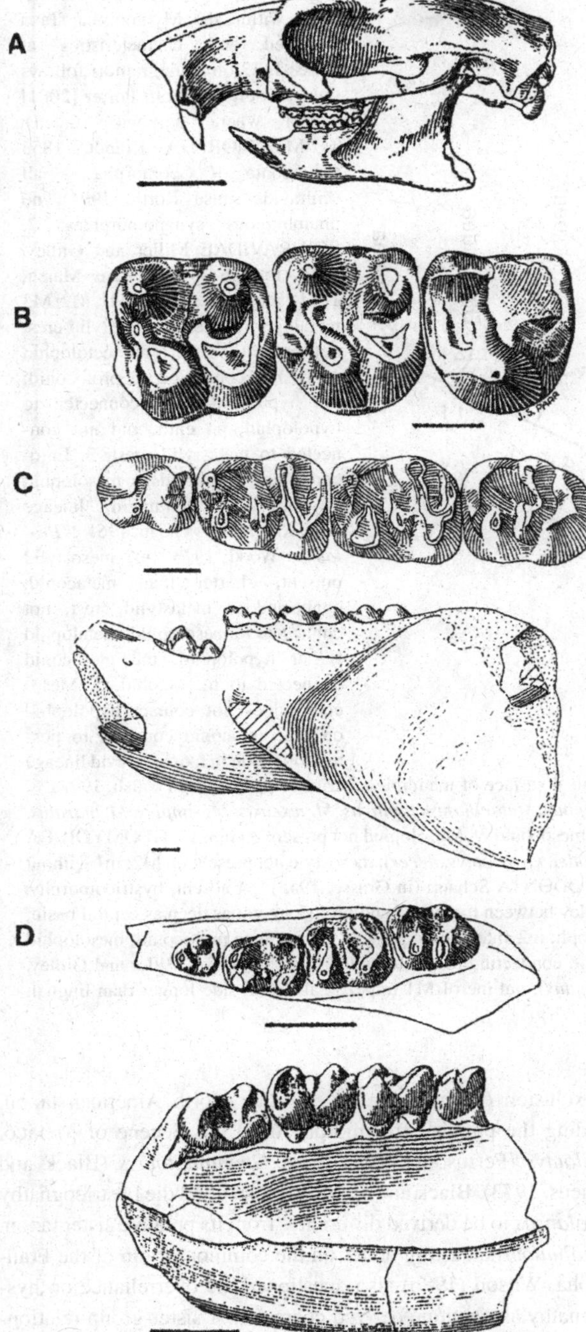

Figure 19.1. A. Skull of *Sciuravus nitidus* (from Matthew, 1910, his Fig. 13 [illustration of AMNH 12551]). Scale bar = 1 cm. B. Upper molars (left M1–3, anterior to left) of *Sciuravus nitidus*; type specimen (from Wilson, 1938a, his Fig. 1; [illustration of YPM 13333]). Scale bar = 1 mm. C. Lower dentition (left p4–m3, anterior to left) and mandible of *Prolapsus sibilatorius* (from Wilson and Runkel, 1991, their Fig. 1 [illustration of TMM 41672-106]). Scale bars = 1 mm. D. Lower dentition (left p4–m3, anterior to left) and mandible (lingual view) of *Pauromys perditus*; type specimen (from Troxell, 1923a, his Fig. 1 [illustration of YPM 13601]). Scale bars = 1 mm. (A: courtesy of the American Museum of Natural History. B,D: reprinted by permission of the American Journal of Science. C: courtesy of the University of Texas.)

POSTCRANIAL

Postcranial material has been described only for *Sciuravus nitidus* by Matthew (1910), who noted no specializations (i.e., generalized scansorial). Comparisons rely mainly on dental characters.

SYSTEMATICS

SUPRAFAMILIAL

The first description of a sciuravid (*Sciuravus nitidus* Marsh, 1871) was based on a partial skull from the Bridger Eocene. Loomis (1907) listed the characters that distinguish the genus from *Paramys* and described a new species, *Sciuravus depressus*, later assigned to *Knightomys* by Gazin (1961). Miller and Gidley (1918) erected the family Sciuravidae for primitive rodent genera with quadritubercular molars, listing only *Sciuravus*. Wilson (1949), in his comprehensive review of North American Tertiary rodents, followed the classification of Simpson (1945); for family Ischyromyidae, subfamily Sciuravinae, he listed *Sciuravus*, *"S." depressus*, *Tillomys*, *Taxymys*, and possibly *Pauromys*.

Because of their generalized morphology, sciuravids have been proposed as ancestral to several major branches of the Rodentia. Wood (1955, p. 171) observed that a revision of the group "may well establish them as an important intermediate link in rodent relationships." A cladistic analysis of mainly dental characters (Porter 2001; Figure 19.2) confirms that at least some sciuravids are basal members of the diverse and abundant suborder Myomorpha (Phaneraulata of Landry [1999]), including the extinct Eomyidae and the living superfamilies Geomyoidea, Dipodoidea, and Muroidea.

That sciuravids are primitive members of the Myomorpha has been proposed many times (Matthew, 1910; Wood, 1937, 1959; Wilson, 1938a, 1949; Black, 1965; Lindsay, 1968; Dawson, 1977; Flynn, Jacobs, and Lindsay, 1985). Specifically, sciuravids have been compared repeatedly to primitive members of the Eomyidae (Fahlbusch, 1973, 1979; Storer, 1987). Lavocat and Parent (1985) observed auditory similarities between fossil Eomyidae and a group including *Sciuravus* and ancient Muroidea. Fahlbusch (1985) argued for an origin of the Geomorpha (Eomyidae + Geomyoidea) from some sciuravid, specifically *Pauromys*. The tendency towards bilophodonty in *Pauromys* molars suggested a relationship with Geomorpha to Vianey-Liaud (1985). Chiment and Korth (1996) accepted a sciuravid origin for Eomyidae.

The earliest hystricomorphous rodent, the possible dipodoid *Armintomys* (Dawson, Krishtalka, and Stucky, 1990), has sciuravid-like molars. Walsh (1996) emphasized the traits shared by *Pauromys* and the intriguing dipodoid *Simimys*. The reduced p4 of *Pauromys* is a key character in some phylogenetic reconstructions leading to Muroidea ("Cricetidae"), in which P4/p4 is lost. Schaub (1925) considered *Pauromys* an ancestral myomorph, a view elaborated by Stehlin and Schaub (1951). In contrast, Hartenberger (1998) derived Geomorpha and Dipodoidea from Sciuravidae (his suborder 2B), but excluded the Muroidea.

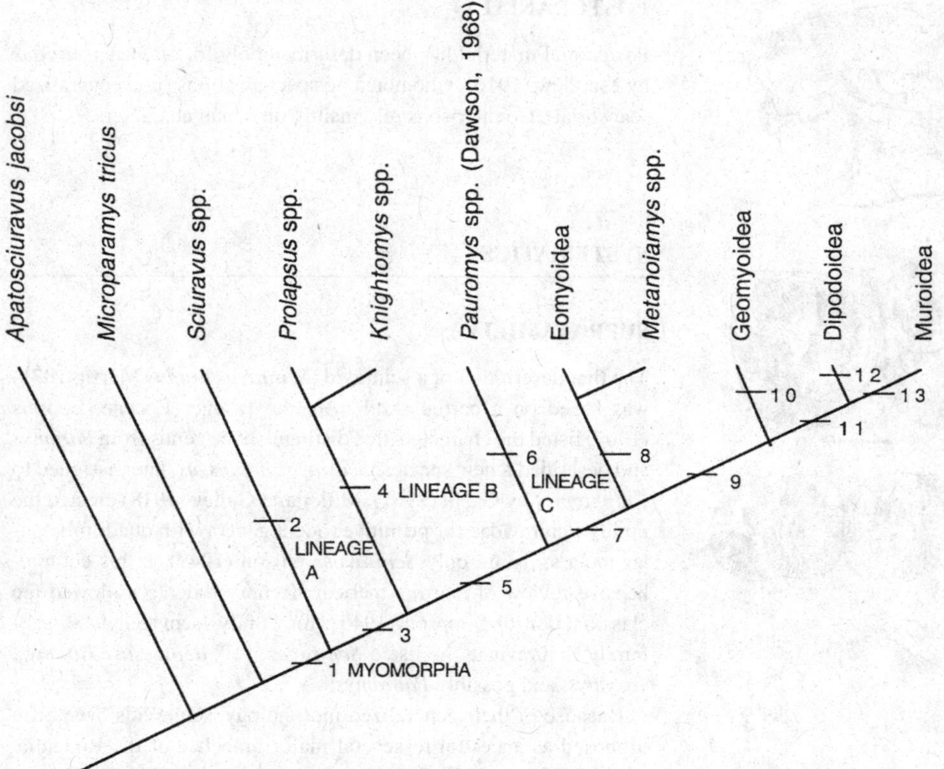

Figure 19.2. Proposed interrelationships within the Myomorpha. Taxa analyzed and characteristics at nodes 1–13 on cladogram as follows (synapomorphies from Porter [2001] except where otherwise noted). 1. MYOMORPHA Brandt, 1855 (Myodonta + Geomorpha + all sciuravids sensu Korth, 1994): no unambiguous synapomorphies. 2. SCIURAVIDAE Miller and Gidley, 1918, lineage A (*Sciuravus* Marsh, 1871): metaloph short on M1, M2; M3 metaloph transverse; metastylid crest not present on m1; m2 ectolophid short, not connected to protoconid; m2 hypoconulid not connected to hypolophid; p4 entoconid not connected to metastylid crest. 3. Entoconid not connected to mesolophid on m1. 4. Sciuravid lineage B (= *Knightomys* Gazin, 1961 + *Prolapsus* Wood, 1973): m2 mesostylid present, shorter than metaconid, connected to metastylid crest, not connected to mesolophid, metalophid II, or hypolophid; m2 mesoconid connected to mesolophid. 5. Mesoconid does not contact hypolophid on m2; entoconid connects to posterolophid on m3. 6. Sciuravid lineage C (= *Pauromys* Troxell, 1923a): p4 reduced. 7. Uniserial enamel microstructure; dorsal surface of mandible anterior to p4 concave (Walsh, 1997). 8. EOMYOIDEA Depéret and Douxami, 1902 (*Namatomys lacus, Protadjidaumo altilophus, Metanoiamys fugitivus, M. texensis, M. simplex, M. marinus, M. agorus*): p4 with short mesolophid that does not contact hypoconid, metaconid, or mesoconid. 9. Mesolophid not present on m3. 10. GEOMYOIDEA Bonaparte, 1845 (*Floresomys guanajuatoensis, Proheteromys sulculus, Tenudomys bodei, Griphomys alecer*): mesostyle not present on M2; m1 without mesoconid or mesolophid; m2 with straight posterolophid, no mesolophid. 11. MYODONTA Schaub (in Grassé, 1955): p4 absent, hystricomorphy. 12. DIPODOIDEA Weber, 1904 (*Schaubemys grangeri, Elymys complexus*): M2 valley between hypocone and protocone not as deep as central basin, metacone does not connect to posteroloph; M3 paracone does not connect to anteroloph; m2 anteroconid present, shorter than protoconid; mesolophid shorter than mesoconid and transverse, metalophid II directed posteriorly, mesoconid connecting to ectolophid. 13. MUROIDEA Miller and Gidley, 1918 (*Eumys elegans*; *Copemys longidens, Simimys simplex, Pappocricetodon antiquus*): outline of M1 trapezoidal, labial side longer than lingual, with hypocone as tall as protocone.

In addition to the well-documented basal position within Myomorpha, other, more distant, relationships have been proposed. According to Wortman (1903, p. 368) and Taylor (1914), *Sciuravus* is part of the phylogenetic series leading to Castoridae. Walton (1993a) observed that the dental characteristics of the most primitive Eutypomyidae (possibly basal castoroids; Wahlert, 1974, 1977) are shared with sciuravids, a similarity also noted by Lavocat and Parent (1985) for auditory characters. These observations do not demonstrate a close relationship between sciuravids and beavers but instead emphasize that sciuravid characters are primitive and widespread.

Prolapsus, from the Uintan of Texas, has been classified as a primitive member of the Hystricognathi (Franimorpha) by Wood (1973, 1985) and Patterson and Wood (1982) based on its hystricognathous mandible, though these authors recognized its sciuravid molar pattern. The study of Porter (2001) recovered no groupings of earliest caviomorphs with any sciuravid. The South American genera *Migraveramus* (Deseadan), *Platypittamys* (Deseadan), and *Acarechimys* (Santacrucian) consistently grouped together to the exclusion of any sciuravid or other North American taxon, including the putative franimorphs from the Eocene of Mexico, *Marfilomys* (Ferrusquía, 1989), and *Guanajuatomys* (Black and Stephens, 1973). Black and Sutton (1984) found the hystricognathy of *Prolapsus* to be derived differently from its putative sister taxon, *Guanajuatomys*, casting doubt on the common origin of the Franimorpha. Wilson (1986) also questioned the overreliance on hystricognathy as a unique derived character. A sister-group relationship between any sciuravid and the South American Caviomorpha is not supported by the available evidence.

The origin of sciuravids, and the Myomorpha, remains unclear. According to Ivy (1990), the most likely sister taxon is *Microparamys* (Reithroparamyinae), based on its antiquity (Clarkforkian first appearance), large hypocones, and apparent retention of deciduous premolars late into life, a trait shared with *Sciuravus* (and some modern sciurids [Lillegraven, 1977]). In the cladistic study of Porter (2001; Figure 19.2), the relatively advanced Uintan species *Microparamys tricus* falls basal to the sciuravid and myomorph clade. Bootstrap support is high, but nodes at this depth should be viewed with caution in the absence of more non-myomorph taxa, especially reithroparamyines (see Kelly and Whistler, 1994). The Clarkforkian reithroparamyine *Apatosciuravus* has also been proposed as near the ancestry of sciuravids (Korth, 1984, 1994) or at least *Pauromys* (Flanagan, 1986); but this genus is more remote, falling basal to the *Microparamys* + Myomorpha clade in the study by Porter (2001) (Figure 19.2). Material of *Microparamys hunterae*, suggested by Ivy (1990) to be a potential ancestor to *Pauromys*, was too incomplete to allow Porter (2001) to test the proposed relationship. For the same reason, "*Microparamys*" *scopaiodon* Korth, 1984 (early Wasatchian) and "*Sciuravus*" *rarus* Wilson, 1938a (Bridgerian), which share a reduced p4 with *Pauromys* (Korth, 1994), could not be tested.

Though the results presented here suggest the earliest radiation of the Myomorpha took place in North America, other interpretations are possible. Myomorphs appear early in the Asian fossil record and may have originated there. *Ivanantonia*, from the early Eocene of Mongolia, may be a primitive myodont (Hartenberger, Dashzeveg, and Martin, 1997). The earliest unambiguous muroid, *Pappocricetodon*, comes from the middle Eocene of China (Tong, 1992; Tong and Dawson, 1995; Dawson and Tong, 1998). Wang and Dawson (1994) favored the hypothesis that sciuravids, rather than Alagomyidae or Ctenodactyloidea, were sister to Myomorpha. Sciuravids, however, are not known from Asia. There have been reports of sciuravids from the Eocene and Oligocene of Asia (Li, 1963; Dawson, 1964; Shevyreva, 1971, 1972; Wang and Li, 1990), but these fossils were later reidentified as members of other groups (Dawson, 1977; Wood, 1977; Korth, 1994; Wang and Dawson, 1994; Wang, 1997).

INFRAFAMILIAL

There has been no attempt to describe sciuravid subfamilies formally. At most, Wilson (1949) and Korth (1994) pointed out that *Pauromys* is separate from *Knightomys* and all other sciuravid genera. The three lineages presented in Figure 19.2 represent the first attempt to define subdivisions among the most basal myomorphs.

The most recent and comprehensive review of Tertiary North American rodents (Korth, 1994) lists the following genera in the Sciuravidae: *Sciuravus*, *Tillomys*, *Taxymys*, *Knightomys*, *Prolapsus*, *Floresomys*, and *Pauromys*. Of these, a cladistic analysis of tooth characters (Porter, 2001) suggests that *Floresomys* should be classified in Geomyoidea (Figure 19.2), as identified elsewhere (e.g., Vianey-Liaud, 1985), whereas there is not enough identified material to classify *Tillomys* or *Taxymys* with more confidence than Myomorpha incertae sedis.

INCLUDED GENERA OF SCIURAVIDAE MILLER AND GIDLEY, 1918

The locality numbers listed for each genus refer to the list of unified localities in Appendix I. The locality numbers may be listed more than one way. The acronyms for museum collections are listed in Appendix III.

Parentheses around the locality (e.g., [CP101]) mean that the taxon in question at that locality is cited as an "aff." or "cf." the taxon in question. Parentheses are usually used for individual species, implying that the genus is firmly known from the locality, but the actual species identification may be questionable. Question marks in front of the locality (e.g., ? CP101) mean that the taxon is questionably known from that locality, implying some doubt that the taxon is actually present at that locality, either at the genus or species level. An asterisk (*) indicates the type locality.

LINEAGE A: *SCIURAVUS* SPP.

Characteristics: P4 without hypocone; short metaloph on M1–2, oriented transversely on M3; m1 with no metastylid crest, entoconid connecting to mesolophid; m2 ectolophid short, not connected to protoconid, hypoconulid not connected to hypolophid; p4 entoconid not connected to metastylid crest.

Sciuravus Marsh, 1871 (including *Colonomys*)

Type species: *Sciuravus nitidus* Marsh, 1871.

Type specimen: YPM 13333, early Bridgerian of Wyoming.

Characteristics: As above.

Length of tooth row (p4–m3): 9.7 mm.

Included species: *S. nitidus* (including *S. undans* Marsh, 1871, *Colonomys celer* Marsh 1872, see Wilson, 1938a, p. 129–130) (known from localities CP25I, J, J2, CP33B, CP34A, B, C*, D, CP36A [CP38B]); *S. eucristadens* Burke, 1937 (locality CP5A*); *S. powayensis* (Wilson, 1940; emended by Lillegraven, 1977) (localities CC4*, CC6A, CC7A, B).

Species tentatively included: *S. altidens* (Peterson, 1919) (localities CP6AA, A*); *S. bridgeri* Wilson, 1938a (localities CP25J, CP34A, B*, C); *S. popi* Dawson, 1966 (localities CP6A, [CP32B]); *S.*? *rarus* Wilson, 1938a (locality CP34C); *S. wilsoni* Gazin, 1961 (localities CP25H, CP25I).

Sciuravus sp. is also known from localities CC7C, NB14, CP20C, CP25J, CP27A, E, CP29C, D, CP31A, B, CP32B, CP34A, B, CP36B, CP38B, CP65.

Comments: The generic description of *Sciuravus* was emended by Wilson (1938a), who noted that the type species (*S. nitidus*) is highly variable. The monophyly of the *Sciuravus* spp. listed above is not strongly supported, and future analyses or discoveries may lead to revision of the genus. Those species only tentatively included in *Sciuravus* are known from material that is too incomplete to confirm their identity. *S.*? *rarus* Wilson 1938a, though represented by p4–m1 only (YPM 10729), is significant for the extreme reduction of p4, a character shared with *Pauromys* (Korth, 1994). *S. powayensis* is a complex taxon that may represent two species (Lillegraven, 1977).

LINEAGE B: *KNIGHTOMYS* AND *PROLAPSUS*

Characteristics: low mesostylid on m2 (shorter than metaconid), connected to metastylid crest but not to mesolophid, metalophid II, or hypolophid; m2 mesoconid connected to mesolophid; entoconid not connected to mesolophid on m1.

Knightomys Gazin, 1961 (including *Dawsonomys*, in part; *Microparamys*, in part; *Paramys*, in part; *Sciuravus*, in part; *Taxymys*, in part; *Tillomys*, in part)

Type species: *Knightomys senior* (Gazin, 1952) (originally described as *Tillomys senior*).

Type specimen: USNM 19308, mandible with p4–m2, Wasatchian of Wyoming.

Characteristics: Smaller but deeper-jawed than *Tillomys*; skull longer, less inflated than *Sciuravus*; cheek teeth more cuspate, less lophate, than *Sciuravus*; minute second mental foramen variably present posterior to larger foramen; masseteric fossa terminating inferior to m1; hypolophid on m1–2 short, entoconid isolated, mesoconid distinct and entering into talonid.

Approximate length of tooth row (p4–m3): 6.6 mm.

Included species: *K. senior* (known from localities [SB24], CP25G*, [I], J, CP27E); *K. cuspidatus* (Bown, 1982) (originally described as *Taxymys cuspidatus*) (locality CP31E*); *K. depressus* (Loomis, 1907) (originally described as *Sciuravus depressus*; emended by Wood, 1965) (including *Microparamys lysitensis* Wood, 1962; *Microparamys cathedralis* Wood, 1962; see Guthrie, 1971) (localities SB22C, SB24, [CP20C, D], CP25G, I, I2, CP27B*, D, E, CP64C); *K. reginensis* (Korth, 1984) (originally described as *Microparamys reginensis*; emended by Flanagan, 1986) (locality SB24*).

Species probably included: *K. huerfanensis* (Wood, 1962) (originally described as *Paramys huerfanensis*; emended by Korth, 1984) (localities SB22C, CP25I, CP27D, E); *K. minor* (Wood, 1965) (originally described as *Dawsonomys minor*; emended by Korth 1984) (localities [SB24], CP27B*, CP64B).

Species tentatively included: *K. cremneus* (Ivy, 1990) (localities CP19A-C*).

Knightomys sp. is also reported from localities (CC1), NB14, CP20BB, CP25J (see Emry and Korth, 1989; Walsh, 1991; Silcox and Rose, 2001).

Comments: Specific descriptions for *K. depressus*, *K. minor*, *K. senior*, and *K. huerfanensis* were emended by Korth (1984). Flanagan (1986) emended the diagnosis of *K. reginensis* and discussed the generic characters of *Knightomys*. The generic diagnosis was emended by Ivy (1990). *K. depressus* is the most common sciuravid in the early Eocene (Korth, 1984). Korth (1994) observed that *K. cuspidatus* Bown, 1982 is the same size as *K. depressus*, and the two species may be synonymous.

Prolapsus Wood, 1973

Type species: *Prolapsus sibilatorius* Wood, 1973.

Type specimen: TMM 41372-179, mandible with m1–3, Uintan of Texas.

Characteristics: Robust mandible with deep masseteric fossa; angle originating lateral and ventral to incisor alveolus (hystricognathous); protrogomorphous snout; mesoconid on m1–2 transversely elongate; trigonid basin enclosed posteriorly by metalophid; anterior arm of hypocone on M1 and M2 widely separate from protocone and protoloph, extending into central basin almost to paracone (pseudomesoloph of Wood, 1973).

Approximate length of tooth row (p4–m3): 10.4 mm.

Included species: *P. sibilatorius* (known from localities SB41, SB43A*, SB43B); *P. junctionis* Wood, 1973 (localities SB43A*, SB43B).

Comments: Wood's (1973) diagnoses of genus *Prolapsus* and both species were emended by Wilson and Runkel (1991), who described the mandible as "questionably hystricognathous." The analysis by Porter (2001) confirmed their conclusion that the basal sister taxon to *Prolapsus* is *Knightomys huerfanensis* (Figure 19.2), forming a clade with a southern Rockies occurrence. Wilson and Runkel (1991) also concluded that the closest relative of *Prolapsus* is the eomyid *Aguafriamys*, from the Duchesnean of trans-Pecos Texas.

LINEAGE 3: *PAUROMYS* SPP.

Characteristics: p4 reduced (< 60% length of m1); entoconid not connected to mesolophid on m1; m2 mesoconid does not contact hypolophid; m3 entoconid connects to posterolophid.

Pauromys Troxell, 1923a

Type species: *Pauromys perditus* Troxell, 1923a.

Type specimen: YPM 13601, mandible with p4–m3, Bridgerian of Wyoming.

Characteristics: Small rodent with very small p4; molars longer than wide; well-developed anterior cingulids and distinct mesoconids on lower molars; P3 absent from at least *P. lillegraveni*; other dental characters as above.

Length of tooth row (p4–m3): 3.6 mm.

Included species: *P. perditus* (known from localities CP34C, D*); *P. lillegraveni* Walsh, 1997 (locality CC6A*).

Species tentatively included: *P. schaubi* (Wood, 1959) (localities CP25J, CP34D*); *P. exallos* (Emry and Korth, 1989) (locality NB14*).

Pauromys sp. is also known from localities SB24, SB44A, CP5A, (CP6A), CP27D, E, CP34E, CP35, CP38C (see Dawson, 1968; Nelson, 1974; Korth, 1984; Flanagan, 1986; Walton, 1993a,b; Walsh, 1996).

Comments: An emended diagnosis for *Pauromys*, with a description of the best-preserved material available, is provided by Walsh (1997); the implications of his analysis of early myomorph evolution are discussed below. Most of the teeth from the Uintan of Texas ascribed by Walton (1993b) to *Pauromys* have been reidentified as *Metanoiamys* (Chiment and Korth, 1996), an eomyid, in part because of their unreduced premolars (Walsh, 1997). However, a close relationship may exist between some species referred to *Pauromys* and Eomyidae of the tribe Namatomyini (Korth, 1992); analyses of subsets of dental character data (Porter, 2001), though inconclusive, frequently showed groupings of the *Pauromys* spp. from Powder Wash (locality CP5A) (Dawson, 1968) or the Lost Cabin localities (localities CP27D, E) (Korth, 1984) with *Metanoiamys*. New discoveries of more complete material may alter the branching pattern shown in this part of the cladogram (Figure 19.2).

TAXONOMIC POSITION UNCERTAIN

***Tillomys* Marsh, 1872 (including *Sciuravus*, in part)**

Type species: *Tillomys senex* Marsh, 1872.

Type specimen: YPM 11788, right mandible with m1, Bridgerian of Wyoming.

Characteristics: Known only from mandibles; trigonids higher and more massive than in *Sciuravus*, with well-developed metastylid crests; entoconid–hypolophid oriented obliquely posteriad, unlike more transverse orientation in *Sciuravus*, contacts posterolophid with moderate wear; hypolophid more developed than in *Knightomys*.

Alveolar length of tooth row (p4–m3): 8.1 mm.

Included species: *T. senex* (known from localities CP34D*, [CP65]);? *Tillomys parvidens* Wilson, 1938b (including *Sciuravus parvidens* Marsh, 1872) (localities CP34B, C*).

Tillomys sp. is also known from localities CP25J, CP31B, CP34A, B, CP35, CP38B.

Comments: *Taxymys* (below) and *Tillomys* are Bridgerian taxa of roughly the same size and from the same area. *Taxymys* is known only from maxillary material and *Tillomys* only from mandibles, a coincidence that led Troxell (1923b) to synonymize the genera under *Tillomys*. Wilson (1938b) disagreed with the synonymy on morphological grounds and recommended retention of both genera (separate from *Sciuravus*) until associated upper and lower dental material is recovered. Associated material was still unknown at the time of Korth's (1994) review. New, little-worn material will also be necessary to establish the relationship between *Sciuravus* and *Tillomys*, which Wilson (1938b) suggested may be cogeneric.

***Taxymys* Marsh, 1872**

Type species: *Taxymys lucaris* Marsh, 1872.

Type specimen: YPM 10020, maxillary fragment with P3–4, Bridgerian of Wyoming.

Characteristics: Known only from maxillary material; molars more crested than in *Sciuravus*; no connection between protocone and hypocone; protoconule does not form separate spur from protoloph; P4 narrow, without distinct hypocone or metaconule, but molariform in ?*T. progressus* (Wilson, 1938c).

Transverse width of P4: 1.9 mm.

Included species: *T. lucaris* (Marsh, 1872) (known from localities CP34B, C, D*, [CP38B]); ?*T. progressus* (Wilson, 1938c) (locality uncertain, from the Bridger Basin, Wyoming).

Taxymys sp. is also known from localities CP31C, CP34E, CP38B.

Comments: As for *Tillomys* above. The taxon *T. lucaris* may include more than one species (Wilson, 1938c).

INDETERMINATE SCIURAVIDS

Indeterminate sciuravids are also reported from localities? CP7A, CP25GG, CP25J, CP29C, CP34E, CP63, CP64A.

BIOLOGY AND EVOLUTIONARY PATTERNS

Sciuravids are known from the Wasatchian (early Eocene) through Uintan (late middle Eocene) Land-Mammal Ages, but they are quite rare in the Wasatchian, and extremely abundant in the Bridgerian and Uintan, following the start of the higher latitude climatic changes to lower mean annual temperatures and a greater extent of seasonality (Figure 19.3). However, they did not survive the greater extent of climatic change that took place later in the middle Eocene.

Little can be inferred about the biology of these early mice. The identified material is so sparse (mainly isolated teeth and jaw fragments) and generalized it offers few clues to style of life. For one of the few well-supported branches of the phylogeny in Figure 19.2, it is reasonable to postulate that *Prolapsus* is derived from *Knightomys*, persisting through the Uintan in the warm, damp climate of trans-Pecos Texas (Westgate and Gee, 1990) long after *Knightomys* had disappeared from the increasingly seasonal and less-densely forested Rocky Mountain areas.

The *Knightomys*–*Prolapsus* and *Sciuravus* lineages both show an increase in size over time (Korth 1994). A tendency towards increasing crown height, crest height, and length, and a bipartite tooth form can be detected in the progression from basal myomorphs of the Wasatchian, such as *Knightomys*, to Uintan forms such as *Pauromys lillegraveni* or *Sciuravus powayensis*. Similar trends characterize many rodent lineages. For some authors, however, these tendencies foreshadowed developments in geomorphs, specifically the Eomyidae (Lindsay, 1968; Wilson and Runkel, 1991), Heteromyidae (Black and Sutton, 1984), and Geomyidae (Vianey-Liaud, 1985).

The reduced p4 of *Pauromys perditus* (also seen in *P. lillegraveni* and *Sciuravus rarus*) suggested to several authors an ancestral relationship to the Myodonta (Stehlin and Schaub, 1951; Dawson, 1968,

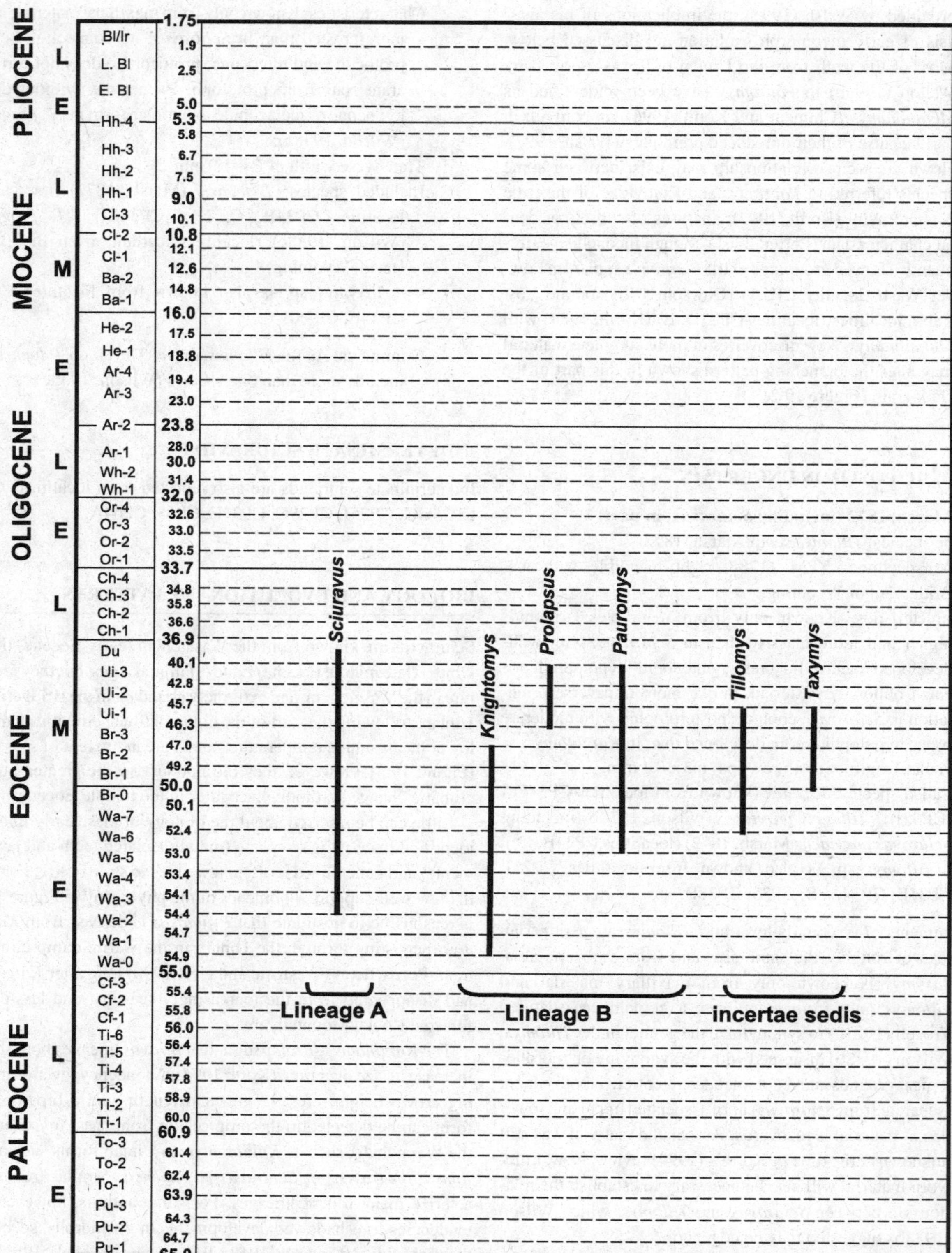

Figure 19.3. Temporal ranges of sciuravid genera.

1977; Lindsay, 1968). Lower premolars are absent in all living Myodonta, whereas upper premolars are highly reduced in Dipodoidea and absent in Muroidea. The earliest unambiguous myodont (*Elymys* Emry and Korth, 1989) is contemporaneous with *Pauromys* (Bridgerian). Walsh (1997) used selected characters to propose relationships among Paleogene myomorphs (his Fig. 10) in which *Pauromys* is nested within the Myodonta, more derived than Eomyidae and distant from other sciuravids, based mainly on premolar reduction. Walsh (1997) prudently did not attempt a numerical cladistic analysis as did Porter (2001), whose results (Figure 19.2) show the *Pauromys* lineage basal to Eomyidae and its premolar reduction to be acquired independently from the Myodonta. (*P. lillegraveni* was not included in this analysis, which required complete dental material.) Curiously, the derived character of a concave mandibular diastema is shown to be acquired once in Porter's cladogram (after *Pauromys*, which presumably has the primitive condition) but more than once in Walsh's cladogram (in the Eomyidae and again in advanced Myodonta). Truly, the collection of more complete material and a revision of the Sciuravidae has much to offer toward resolving the early evolution of the most diverse mammalian clade.

ACKNOWLEDGEMENTS

We are grateful for conversations with John Storer (our éminence grise), William Korth, and Steve Walsh. Many thanks go to Tekla Harms of the Amherst College Geology Department, co-advisor for the senior thesis on which the systematic portion was based, and Linda Thomas, conservator at the Pratt Museum.

REFERENCES

Black, C. C. (1965). Fossil mammals from Montana. Part 2. Rodents from the early Oligocene Pipestone Springs local fauna. *Annals of the Carnegie Museum*, 38, 1–48.

Black, C. C. and Stephens, J. R., III (1973). Rodents from the Paleogene of Guanajuato, Mexico. *Occasional Papers of the Museum of Texas Technical University*, 14, 1–10.

Black, C. C. and Sutton, J. F. (1984). Paleocene and Eocene rodents of North America. *Carnegie Museum of Natural History Special Publication*, 9, 67–84.

Bonaparte, C. L. (1845). Specchio generale dei sistemi erpetelogico ed anfibiologico. *Atti Congresso Scienzia Italia*, 6, 376–8.

Bown, T. M. (1982). Geology, paleontology, and correlation of Eocene volcaniclastic rocks, Southeast Absaroka Range, Hot Springs County, Wyoming. *United States Geological Survey, Professional Paper*, 1201–A, 1–75.

Brandt, J. F. (1855). Beitrage zur nähern Kenntniss der Säugethiere Russlands. *Mémoires de l'Academie des Sciences Impériale de St. Pétersbourg*, 6, 1–365.

Burke, J. J. (1937). A new *Sciuravus* from Utah. *Annals of the Carnegie Museum*, 27, 1–9.

Chiment, J. J. and Korth, W. W. (1996). A new genus of eomyid rodent (Mammalia) from the Eocene (Uintan–Duchesnean) of southern California. *Journal of Vertebrate Paleontology*, 16, 116–24.

Dawson, M. R. (1964). Late Eocene rodents (Mammalia) from Inner Mongolia. *American Museum Novitates*, 2191, 1–15.

(1966). Additional Late Eocene rodents (Mammalia) from the Uinta Basin, Utah. *Annals of the Carnegie Museum*, 38, 97–114.

(1968). Middle Eocene rodents (Mammalia) from northeastern Utah. *Annals of the Carnegie Museum*, 39, 327–70.

(1977). Late Eocene rodent radiations: North America, Europe and Asia. *Géobios, Mémoire Spécial*, 1, 195–209.

Dawson, M. R. and Tong, Y.-S. (1998). New material of *Pappocricetodon schaubi*, an Eocene rodent (Mammalia: Cricetidae) from the Yuanqu Basin, Shanxi Province, China. *Bulletin of the Carnegie Museum of Natural History*, 34, 278–85.

Dawson, M. R., Krishtalka, L., and Stucky, R. K. (1990). Revision of the Wind River faunas, early Eocene of central Wyoming. Part 9. The oldest known hystricomorphous rodent (Mammalia: Rodentia). *Annals of the Carnegie Museum*, 59, 135–47.

Depéret C. and Douxami, H. (1902). Les vertébrés oligocènes de Pyrimont-Challonges (Savoie). *Abhandlungen Schweizerischen Paläontologischen Gesellschaft*, 29, 1–91.

Emry, R. J. and Korth, W. W. (1989). Rodents of the Bridgerian Elderberry Canyon local fauna of eastern Nevada. *Smithsonian Contributions to Paleobiology*, 67, 1–14.

Fahlbusch, V. (1973). Die stammesgeschichtlichen Beziehungen zwischen dem Eomyiden (Mammalia, Rodentia) Nordamerikas und Europas. *Mitteilungen Bayerische Staatsammlung für Palaeontolgie und Historische Geologie, München*, 13, 141–75.

(1979). Eomyidae–Geschichte einer Säugetierfamilie. *Paläontologische Zeitschrift*, 53, 88–97.

(1985). Origin and evolutionary relationships among Geomyoids. In *Evolutionary Relationships Among Rodents: A Multidisciplinary Analysis*, ed. W. P. Luckett and J.-L. Hartenberger, pp. 617–29. New York: Plenum Press.

Ferrusquía-Villafranca, I. (1989). A new rodent genus from central Mexico and its bearing on the origin of the Caviomorpha. In *Papers on Fossil Rodents in Honor of Albert Elmer Wood*, ed. C. C. Black and M. R. Dawson, pp. 91–117. Los Angeles, CA: Natural History Museum of Los Angeles County.

Flanagan, K. M. (1986). Early Eocene rodents from the San Jose Formation, San Juan Basin, New Mexico. [In *Vertebrates, Phylogeny, and Philosophy*, ed. K. M. Flanagan and J. A. Lillegraven.] *Contributions to Geology, University of Wyoming Special Paper*, 3, 197–220.

Flynn, L. J., Jacobs, L. L., and Lindsay, E. H. (1985). Problems in muroid phylogeny: relationship to other rodents and origin of major groups. In *Evolutionary Relationships Among Rodents: A Multidisciplinary Analysis*, ed. W. P. Luckett and J.-L. Hartenberger, pp. 589–616. New York: Plenum Press.

Gazin, C. L. (1952). The lower Eocene Knight Formation of western Wyoming and its mammalian faunas. *Smithsonian Miscellaneous Collections*, 117, 1–49.

(1961). New sciuravid rodents from the Lower Eocene Knight Formation of western Wyoming. *Proceedings of the Biological Society of Washington*, 74, 193–4.

Grassé, P.-P. (ed.). (1955). *Traité de Zoologie.* Vol. XVII, 1–2: *Mammifères. Les Ordres: Anatomie, Éthologie, Systématique*. Paris: Masson.

Guthrie, D. A. (1971). The mammalian fauna of the Lost Cabin Member, Wind River Formation (Lower Eocene) of Wyoming. *Annals of the Carnegie Museum*, 43, 47–113.

Hartenberger, J.-L. (1998). Déscription de la radiation des Rodentia (Mammalia) du Paléocène supérieur au Miocène; incidences phylogénétiques. *Comptes Rendus de l'Académie des Sciences, Paris, Sciences de la Terre et des Planètes*, 326, 439–44.

Hartenberger, J.-L., Dashzeveg, D., and Martin, T. (1997). What is *Ivantonia efremovi* (Rodentia, Mammalia)? *Paläontologische Zeitschrift*, 71, 135–43.

Ivy, L. D. (1990). Systematics of late Paleocene and early Eocene Rodentia (Mammalia) from the Clarks Fork Basin, Wyoming. *Contributions from the Museum of Paleontology, University of Michigan*, 28, 21–70.

Kelly, T. S. and Whistler, D. P. (1994). Additional Uintan and Duchesnean (Middle and Late Eocene) mammals from the Sepse Formation, Simi Valley, California. *Contributions in Science, Los Angeles County Museum of Natural History*, 439, 1–29.

Korth, W. W. (1984). Earliest Tertiary evolution and radiation of rodents in North America. *Bulletin of the Carnegie Museum of Natural History*, 24, 1–71.

(1992). Cylindrodonts (Cylindrodontidae, Rodentia) and a new genus of eomyid, *Paranamatomys* (Eomyidae, Rodentia), from the Chadronian of Sioux County, Nebraska. *Transactions of the Nebraska Academy of Sciences*, 19, 75–82.

(1994). *The Tertiary Record of Rodents in North America*. New York: Plenum Press.

Landry, S. O. (1999). A proposal for a new classification and nomenclature for the Glires (Lagomorpha and Rodentia). *Mitteilungen des Museum für Naturkunde Berlin, Zoologische Reihe*, 75, 283–316.

Lavocat, R. and Parent, J.-P. (1985). Phylogenetic analysis of middle ear features in fossil and living rodents. In *Evolutionary Relationships Among Rodents: A Multidisciplinary Analysis*, ed. W. P. Luckett and J.-L. Hartenberger, pp. 333–54. New York: Plenum Press.

Li, C.-K. (1963). Paramyid and sciuravids from North China. *Vertebrata Palasiatica*, 7, 151–60.

Lillegraven, J. A. (1977). Small rodents (Mammalia) from Eocene deposits of San Diego County, California. *Bulletin of the American Museum of Natural History*, 158, 225–61.

Lindsay, E. (1968). Rodents from the Hartman Ranch Local Fauna. California. *Contributions from the University of California Museum of Paleontology*, 6, 1–22.

Loomis, F. B. (1907). Wasatch and Wind River rodents. *American Journal of Science*, 4, 123–30.

Marsh, O. C. (1871). Notice of some new fossil mammals and birds from the Tertiary formations of the West. *American Journal of Science*, 2, 120–7.

(1872). Preliminary description of new Tertiary mammals. *American Journal of Science*, 4, 202–24.

Matthew, W. D. (1910). On the osteology and relationships of *Paramys*, and the affinities of Ischyromyidae. *Bulletin of the American Museum of Natural History*, 28, 43–72.

Miller, G. S. and Gidley, J. W. (1918). Synopsis of the supergeneric groups of Rodents. *Journal of the Washington Academy of Sciences*, 8, 431–48.

Nelson, M. E. (1974). Middle Eocene rodents (Mammalia) from southwestern Wyoming. *Contributions to Geology, University of Wyoming*, 13, 1–10.

Patterson, B. and Wood, A. E. (1982). Rodents from the Deseadan Oligocene of Bolivia and the relationships of the Caviomorpha. *Bulletin of the Museum of Comparative Zoology*, 149, 371–543.

Peterson, O. A. (1919). Report upon the material discovered in the Upper Eocene of the Uinta Basin by Earl Douglas in the years 1908–1909 and by O. A. Peterson in 1912. *Annals of the Carnegie Museum*, 12, 40–168.

Porter, R. (2001). A cladistic analysis of the Sciuravidae (Mammalia: Rodentia) and implications for rodent phylogeny. Senior Thesis, Amherst College, Amherst.

Schaub, S. (1925). Die hamsterartigen Nagetiere des Tertiärs und ihre lebenden Verwandten. *Abhandlungen Schweizerischen Paläontologischen Gesellschaft*, 45, 1–114.

Shevyreva, N. S. (1971). The first find of Eocene rodents in the USSR. *Bulletin of the Academy of Sciences of the Georgian SSR, Tblisi*, 61, 745–7. [In Russian.]

(1972). New rodents in the Paleogene of Mongolia and Kazakhstan. *Paleontological Journal of Moscow*, 3, 134–45. [In Russian.]

Silcox, M. T. and Rose, K. D. (2001). Unusual vertebrate microfaunas from the Willwood Formation, early Eocene of the Bighorn Basin, Wyoming. In *Eocene Biodiversity: Unusual Occurrences and Rarely Sampled Habitats*, ed. G. F. Gunnell, pp. 131–64. New York: Kluwer Academic/Plenum.

Simpson, G. G. (1945). The principles of classification and a classification of mammals. *Bulletin of the American Museum of Natural History*, 85, 1–350.

Stehlin, H. G. and Schaub, S. (1951). Die Trigonodontie der simplicidentaten Nager. *Schweizerische Paläontologische Abhandlungun*, 67, 1–385.

Storer, J. E. (1987). Dental evolution and radiation of Eocene and early Oligocene Eomyidae (Mammalia, Rodentia) of North America, with new material from the Duchesnean of Saskatchewan. [In *Papers in Vertebrate Paleontology in Honor of Morton Green*, ed. J. E. Martin and G. E. Ostrander.] *Dakoterra*, 3, 108–17.

Taylor, W. P. (1914). Outline of the history of the Castoridae. *Proceedings Fifth Annual Meeting of the Pacific Coast Section of the Paleontological Society*, p. 167.

Tong, Y.-S. (1992). *Pappocricetodon*, a pre-Oligocene cricetid genus (Rodentia) from central China. *Vertebrata Palasiatica*, 30, 1–16.

Tong, Y-S. and Dawson, M. R. (1995). Early Eocene rodents (Mammalia) from Shandong Province, People's Republic of China. *Annals of the Carnegie Museum*, 64, 51–63.

Troxell, E. L. (1923a). *Pauromys perditus*, a small rodent. *American Journal of Science*, 5, 155–6.

(1923b). The Eocene rodents *Sciuravus* and *Tillomys*. *American Journal of Science*, 5, 383–96.

Vianey-Liaud, M. (1985). Possible evolutionary relationships among Eocene and lower Oligocene rodents of Asia, Europe and North America. In *Evolutionary Relationships Among Rodents: A Multidisciplinary Analysis*, ed. W. P. Luckett and J.-L. Hartenberger, pp. 277–309. New York: Plenum Press.

Wahlert, J. H. (1974). The cranial foramina of protrogomorphous rodents: an anatomical and phylogenetic study. *Bulletin of the Museum of Comparative Zoology*, 146, 363–410.

(1977). Cranial foramina and relationships of *Eutypomys* (Rodentia, Eutypomyidae). *American Museum Novitates*, 2626, 1–8.

Walsh, S. L. (1991). Eocene mammal faunas of San Diego County. In *Eocene Geologic History San Diego Region*, ed. P. L. Abbott and J. A. May, pp. 161–78. Fullerton, CA: Pacific Section of the Society for Sedimentary Geology.

(1996). Middle Eocene mammal faunas of San Diego County, California. In *The Terrestrial Eocene–Oligocene Transition in North America*, ed. D. R. Prothero and R. J. Emry, pp. 75–119. Cambridge, UK: Cambridge University Press.

(1997). New specimens of *Metanoiamys*, *Pauromys*, and *Simimys* (Rodentia: Myomorpha) from the Uintan (middle Eocene) of San Diego County, California, and comments on the relationships of selected Paleogene Myomorpha. *Proceedings of the San Diego Society of Natural History*, 32, 1–20.

Walton, A. H. (1993a). A new genus of eutypomyid (Mammalia: Rodentia) from the middle Eocene of the Texas Gulf Coast. *Journal of Vertebrate Paleontology*, 13, 262–6.

(1993b). *Pauromys* and other small Sciuravidae (Mammalia: Rodentia) from the middle Eocene of Texas. *Journal of Vertebrate Paleontology*, 13, 243–61.

Wang, B.-Y. (1997). The mid-Tertiary Ctenodactylidae (Rodentia, Mammalia) of eastern and central Asia. *Bulletin of the American Museum of Natural History*, 234, 1–88.

Wang, B.-Y. and Dawson, M. R. (1994). A primitive cricetid (Mammalia: Rodentia) from the middle Eocene of Jiangsu Province, China. *Annals of the Carnegie Museum*, 63, 239–56.

Wang, B.-Y. and Li, C.-K. (1990). First Paleogene mammalian fauna from northeast China. *Vertebrata Palasiatica*, 28, 165–205.

Weber, M. C. W. (1904). *Die Säugetiere. Einführing in die Anatomie und Systematik der Recenten und Fossilen Mammalia*. Jena.

Westgate, J. W. and Gee, C. T. (1990). Paleoecology of a middle Eocene mangrove biota (vertebrates, plants, and invertebrates) from southwest Texas. *Palaeogeography, Palaeoclimatology, Palaeoecology*, 78, 163–77.

Wilson, J. A. and Runkel, A. C. (1991). *Prolapsus*, a large sciuravid rodent, and new eomyids from the late Eocene of Trans-Pecos Texas. *Pearce-Sellards Series, Texas Memorial Museum*, 48, 1–30.

Wilson, R. W. (1938a). Review of some rodent genera from the Bridger Eocene. Part I. *American Journal of Science*, 35, 123–37.

(1938b). Review of some rodent genera from the Bridger Eocene. Part II. *American Journal of Science*, 35, 207–22.

(1938c). Review of some rodent genera from the Bridger Eocene. Part III. *American Journal of Science*, 35, 297–304.

(1940). Two new Eocene rodents from California. *American Journal of Science*, 34, 85–95.

(1949). Early Tertiary rodents of North America. *Carnegie Institution of Washington, Contributions to Paleontology*, 584, 59–83.

(1986). The Paleogene record of the rodents: fact and interpretation. [In *Vertebrates, Phylogeny, and Philosophy*, ed. K. M. Flanagan and J. A. Lillegraven.] *Contributions to Geology, University of Wyoming, Special Paper*, 3, 163–75.

Wood, A. E. (1937). The mammalian fauna of the White River Oligocene. Part II, Rodentia. *Transactions of the American Philosophical Society*, 28, 155–269.

(1955). A revised classification of the rodents. *Journal of Mammalogy*, 36, 165–87.

(1959). Eocene radiation and phylogeny of the rodents. *Evolution*, 13, 354–61.

(1962). The early Tertiary rodents of the Family Paramyidae. *Transactions of the American Philosophical Society*, 52, 1–261.

(1965). Small rodents from the early Eocene Lysite Member, Wind River Formation of Wyoming. *Journal of Paleontology*, 39, 124–34.

(1973). Eocene rodents, Pruett Formation, southwest Texas; their pertinence to the origin of the South American Caviomorpha. *Pearce-Sellards Series, Texas Memorial Museum*, 20, 1–41.

(1977). The evolution of the rodent family Ctenodactylidae. *Journal of the Paleontological Society of India*, 20, 120–37.

(1985). The relationships, origin and dispersal of the hystricognathous rodents. In *Evolutionary Relationships among Rodents: A Multidisciplinary Analysis*, ed. W. P. Luckett and J.-L. Hartenberger, pp. 475–514. New York: Plenum Press.

Wortman, J. L. (1903). Studies of Eocene Mammalia in the Marsh Collection. *American Journal of Science*, 4, 345–68.

20 Cylindrodontidae

STEPHEN L. WALSH[1] and JOHN E. STORER[2]

[1]*Deceased; late of San Diego Natural History Museum, San Diego, CA, USA*
[2]*Sechelt, BC, Canada*

INTRODUCTION

The cylindrodontids are a group of generalized to possibly fossorial rodents, possessing broad, somewhat flattened skulls and slightly inflated auditory bullae (known mainly from later taxa). The postcranial skeleton is unknown with the exception of some undescribed vertebrae of *Cylindrodon* and some isolated tarsals and a proximal radius and ulna of *Pareumys*.

The geologically older cylindrodontid taxa are small in body size (mouse to rat sized), while one of the younger forms (*Jaywilsonomys ojinagaensis*) is as large as a modern ground squirrel. Until recently, the known record of cylindrodontids is mainly from North America, but several additional taxa have now been reported from the Eocene and Oligocene of Asia (Tong, 1997; Tyut'kova, 1997; Dashzeveg and Meng, 1998), and the geographic origin and patterns of dispersal of the family are less clear than previously believed.

The known temporal range of cylindrodontids extends from possibly the early Wasatchian (earliest Eocene) to possibly the late Orellan (early Oligocene) of North America, and from the middle Eocene to the early Oligocene of Asia. Flynn, Jacobs, and Cheema (1986) regarded the Asian early Miocene genus *Downsimys* as a possible cylindrodontid (see also McKenna and Bell, 1997; Dashzeveg and Meng, 1998), but this genus has recently been referred to the Anomaluroidea and assigned an Oligocene age (Marivaux and Welcomme, 2003, their Table 1). Interestingly, cylindrodontids have not been discovered in the European Paleogene (Beard and Dawson, 2001). Recent reviews by Korth (1994) and Emry and Korth (1996) have brought about a better understanding of the North American taxa known to date.

DEFINING FEATURES OF THE FAMILY CYLINDRODONTIDAE

CRANIAL

Fragmentary skulls are known for *Mysops* (locality CP34), *Pareumys* (localities CC4, CC7B, CC8IIA), and *Jaywilsonomys* (locality SB52), which provide only limited information. Complete skulls are known for the younger and more derived genera *Pseudocylindrodon* (Figure 20.1A), *Cylindrodon*, and *Ardynomys*. The skulls of these genera are somewhat flattened dorsoventrally (but no more so than in primitive ischyromyids), with auditory bullae that are more inflated than in ischyromyids but much less so than in *Protoptychus*. The skull is shortened anteroposteriorly, especially the rostrum, which is broad and dorsoventrally deep. Zygomatic arches extend widely and are nearly anteroposterior in orientation. Weak sagittal crests are present. Zygomasseteric structure is protrogomorphous (Emry and Korth, 1996), and the infraorbital foramen is relatively small in all cylindrodontids. The anterior zygomatic root is situated somewhat more posteriorly relative to the cheek teeth than in sciuravids but is similar in position to that of most primitive ischyromyids.

Wahlert (1974, p. 397) described a pattern of cranial foramina distinctive to the Cylindrodontidae, and similar in some features to the advanced ischyromyid subfamily Ischyromyinae, noting that "The shift in emphasis of the venous system away from the postglenoid foramen, the presence of a single temporal foramen below the parietal-squamosal suture, and the opening-up of a squamosomastoid foramen are all changes from the paramyid condition that

Evolution of Tertiary Mammals of North America, Vol. 2. ed. C. M. Janis, G. F. Gunnell, and M. D. Uhen. Published by Cambridge University Press.

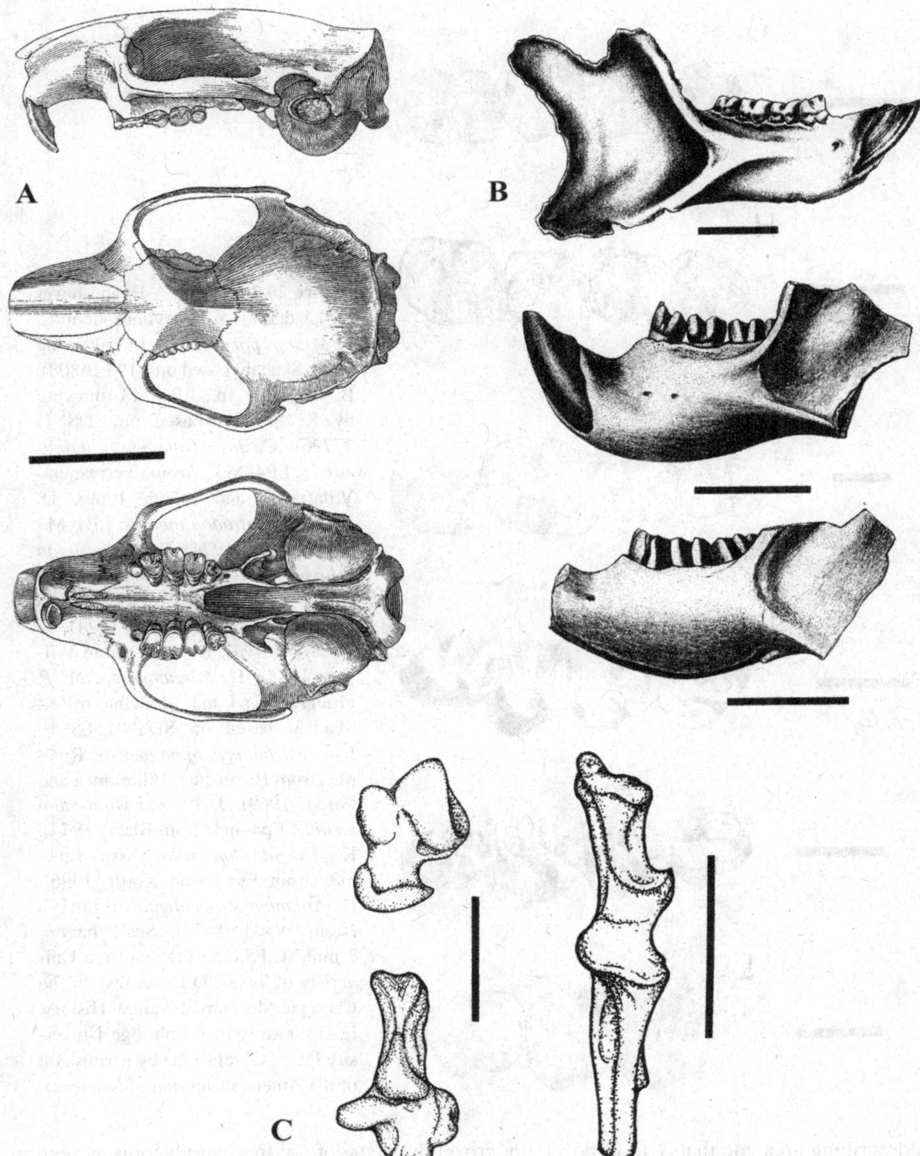

Figure 20.1. A. Skull of *Pseudocylindrodon medius* (from Burke, 1938). Scale bar = 1 cm. B. Lateral views of mandibles of cylindrodontids: top, *Pareumys* sp. (drawing by K. Skadahl based on UCMP 85198); middle, *Pseudocylindrodon neglectus* (from Burke, 1935a); bottom, *Cylindrodon fontis* (from Burke, 1935a). Scale bars = 5 mm. C. Postcrania of *Pareumys*: top left, *Pareumys* sp., dorsal view of left astragalus (SDSNH 46924); bottom left, *Pareumys* sp., dorsal view of left calcaneus (based on SDSNH 47684 and 54721); right, *Pareumys* sp. n. *P. grangeri*; anterior view of left proximal ulna (SDSNH 87142). Scale bars = 5 mm. (Postcranial drawings by R. A. Cerutti. A,B (in part): courtesy of the Carnegie Museum of Natural History.)

occur in [ischromyine] ischyromyids." These characters are known only for the younger subfamily Cylindrodontinae.

Relative to younger cylindrodontids, the mandibles of *Mysops* and *Pareumys* tend to be less robust and shallow dorsoventrally while mandibles of *Pseudocylindrodon*, *Cylindrodon*, and *Ardynomys* tend to be more robust and deeper dorsoventrally, and with more upright incisors (Figure 20.1B). The diastema between i1 and p4 is usually horizontal in older genera (*Mysops* and *Pareumys*), becoming slightly concave in younger forms (Figure 20.1B). The masseteric fossa extends anteriorly to as far as the m2 trigonid in older forms, and as far as the m1–2 contact in younger forms. The pocket for the pterygoideus internus muscle on the medial side of the jaw is very pronounced in cylindrodontines (Emry and Korth, 1996), but not as deep in *Pareumys*. The condyle occurs slightly above the level of the cheek teeth in all cylindrodontids where this character is known (*Pareumys* and cylindrodontines).

Wood (1975, 1980, 1984), from examination of *Cylindrodon* mandibles with incomplete angular processes, thought that the angle showed some lateral extension and was thus "subhystricognathous" or "incipiently hystricognathous." However, in a study of better-preserved material Emry and Korth (1996, p. 400) interpreted the mandible of *Cylindrodon* as "sciurognathous, differing little from those of Eocene ischyromyids and sciuravids." These disagreements are partly semantic in nature, with some of the confusion being caused by the phylogenetically loaded terms "subhystricognathous" and "incipiently hystricognathous." Even the term "hystricognathous" is ambiguous, with Wood (1984) defining it as describing an angle that originates lateral to the plane of the incisor,

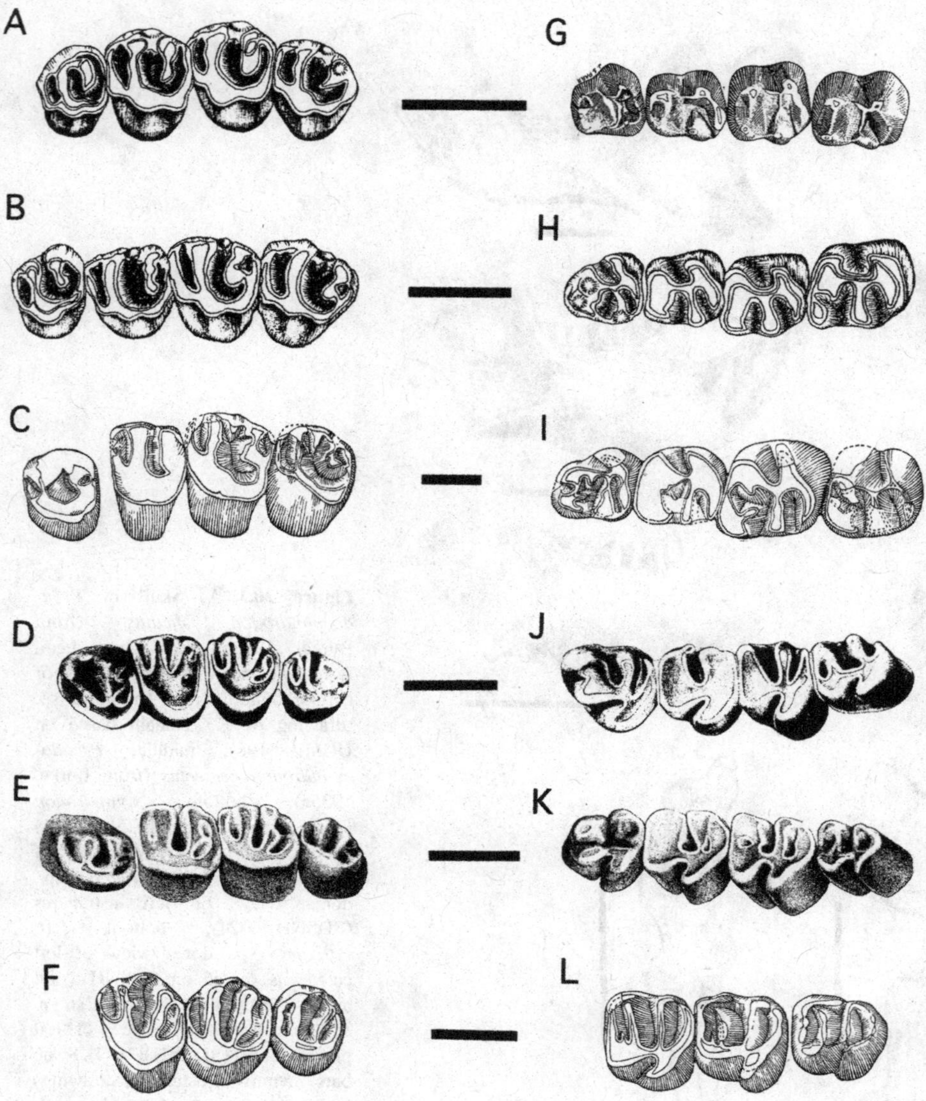

Figure 20.2. Upper (A-F) and lower (G-L) dentitions of cylindrodontids. A. *Mysops parvus*; LP4–M3 (drawing by K. Skadahl based on YPM 16808). B. *Pareumys* sp., LP4–M3 (drawing by K. Skadahl based on SDSNH 47796). C. *Jaywilsonomys ojinagaensis*; LP4–M3 (from Ferrusquía-Villafranca and Wood, 1969). D. *Pseudocylindrodon medius*; LP4–M3 (from Black, 1974). E. *Cylindrodon solarborus*; LP4–M3 (from Emry and Korth, 1996). F. *Ardynomys occidentalis*; LM1–3 (from Wood, 1974). G. *Mysops parvus*; Rp4–m3 (from Wilson, 1938). H. *Pareumys* sp. nr. *P. grangeri*; Rp4–m3 (drawing by K. Skadahl based on SDSNH 50583). I. *Jaywilsonomys ojinagaensis*; Rp4–m3 (from Ferrusquía–Villafranca and Wood, 1969). J. *Pseudocylindrodon medius*. Lp4–m3 (from Black, 1974). K. *Cylindrodon solarborus*; Lp4–m3 (from Emry and Korth, 1996). L. *Ardynomys occidentalis*; Lm1–3 (from Wood, 1974). Scale bars = 2 mm. (C,F,I,L: courtesy of the University of Texas. D,J: courtesy of the Carnegie Museum of Natural History. E,K: courtesy of Cambridge University Press. G: reprinted by permission of the American Journal of Science.)

and Emry and Korth (1996) defining it as describing an angle that originates lateral to the body of the horizontal ramus.

Interestingly, Landry (1999) noted that paramyid jaws are not as fully sciurognathous as in sciurids, because the middle of the angular process is not pushed inward and the posterior end is not turned upward. Landry (1999, p. 295), therefore, coined the descriptive term "protrognathous" for the condition in which "the origin of the angular process with respect to the incisor alveolus is equivocal and in which the edge of the angular process is straight, never pushed inward at its anterior end." This condition is apparently characteristic of many Eocene rodents including paramyines and sciuravids, and is depicted in the illustrations by Wood of *Reithroparamys* (1962, Fig. 41E) and *Cylindrodon* (1984, Fig. 1). The same condition is present in at least some mandibles of *Pareumys* (e.g., SDSNH 31327), in which most of the angle originates immediately lateral to the incisive alveolus as a posteroventral extension of the ridge forming the inferior border of the masseteric fossa, and it then either continues directly posteriad or curves slightly medially to join (but not cross) the plane of the incisive alveolus as seen in ventral view.

DENTAL

The dental formula is I1/1, C0/0, P2/1, M3/3. Upper and lower cheek teeth of several cylindrodontid genera are illustrated in Figure 20.2. The cheek teeth are always rooted, primitively brachydont, and unilaterally hypsodont, becoming progressively higher-crowned and ultimately hypsodont in *Cylindrodon*. The upper incisors are usually strongly recurved, although the incisors of *Cylindrodon* and *Ardynomys* tend to be slightly less recurved. The P4 is primitively smaller than the upper molars and lacks a hypocone but is subequal in size (i.e., approximately the same size) to M1–2 in some younger forms. The upper molars are "sciuroid" in general crested pattern, with relatively weak hypocones and with strong anterior cingula, protolophs, and posterolophs, but with variably developed metalophs. Mesolophs are absent, but mesostyles are variably present.

The lower incisors are relatively slender in older genera, becoming broader and often with a triangular cross section in younger forms. The enamel microstructure is pauciserial in *Mysops* and *Pareumys* (see Wahlert, 1968; D. Kalthoff, personal communication 2004), and uniserial in *Cylindrodon* and *Ardynomys* (see Wahlert, 1968; Martin, 1992; Dashzeveg and Meng, 1998). The p4 is primitively smaller than the lower molars, becoming subequal in size to m1–2 in younger forms. The lower molars usually have complete ectolophids and complete, anteriorly positioned hypolophids, resulting in two distinct subequal basins on the lingual side of the teeth. Mesolophids and mesoconids are absent, and metastylids are weak or absent. The m3 is primitively subequal to m2, becoming smaller than m2 in younger forms.

The major evolutionary trends in the cheek teeth are increased crown height and enamel thickness in all lineages. Unilateral hypsodonty in the upper cheek teeth (greater crown height lingually than labially) was present in *Mysops* and further emphasized in *Pareumys*. Most cylindrodontids retained a single-rooted, peglike P3 in the adult dentition, although Wood (1974) reported this tooth to be absent in *Jaywilsonomys*. According to Emry and Korth (1996), P3 was lost in *Cylindrodon*, with some individuals retaining DP3 into adulthood.

POSTCRANIAL

Vertebrae associated with lower jaws of *Cylindrodon fontis* were noted by Wood (1984), but neither these nor any other postcranial remains of cylindrodontids have been described. Therefore, a few fragmentary postcrania referable to *Pareumys* are described and illustrated here because of their relevance in inferring the habitus of an early member of the family and in the hope that they may aid in the recognition of postcranial samples of other cylindrodontids.

Several isolated astragali and calcanea from the late Uintan Jeff's Discovery Local Fauna (Walsh, 1996; part of composite locality CC7B) and the late Uintan Mission Valley Formation (locality CC8) are referred to *Pareumys* on the basis of their size and relative abundance at sites where teeth and jaws of the genus are common and other rodents of similar body size are rare. The astragali (e.g., SDSNH 46924; Figure 20.1C) have a deep, highly asymmetrical tibial trochlea typical of most rodents, with a lateral crest much larger than the medial crest (terminology after Szalay [1985]). This asymmetry is more pronounced than in *Sciuravus* (numerous referred specimens in SDSNH collections), *Paramys*, *Thisbemys*, and *Reithroparamys* (see Wood, 1962 [Figs. 7F, 11E, 39F, 43K]; Szalay, 1985 [Fig. 6]) but is similar to the condition in *Microparamys* (numerous referred specimens in SDSNH collections). The calcaneoastragalar facet is very wide and deep. The head is smaller relative to the entire astragalus than in *Paramys*, is rounded distally, moderately expanded medially with a prominent facet for the medial tarsal (Szalay, 1985), and has a hooklike lateral process as in *Microparamys* and *Rattus*. The neck is highly divergent from the rotational axis of the trochlea and somewhat variable in length, but it is always quite short, being distinctly shorter than in the astragali of *Microparamys*, *Sciuravus*, *Neotoma*, and *Ischyromys* (see Wood, 1937: Plate XXV), and equal in length to or slightly shorter than the astragali of *Paramys*, *Thisbemys*, and *Reithroparamys*. The neck is distinctly thicker than the relatively thin neck present in *Microparamys*. The astragali of *Pareumys* are similar in overall proportions to those of the middle Eocene paramyine *Leptotomus* (e.g., Wood, 1962: Fig. 29D), the late Eocene sciurid *Protosciurus* (see Emry and Thorington, 1982: Fig. 15C) and the late Oligocene geomyid *Entoptychus* (see Rensberger, 1971: Plate 22), but again with a slightly shorter neck. However, the *Pareumys* astragali do not have the degree of neck reduction seen in the modern fossorial geomyid rodent *Thomomys*.

A few distinctive calcanea almost certainly referable to *Pareumys* also occur at localities CC7B and CC8 (e.g., SDSNH 47684 and 54721; Figure 20.1C). These articulate well with the astragali described above and are similar in overall proportions to the calcaneus of *Neotoma* (e.g., Stains, 1959: Fig. 4). The body is short but robust, with a grooved proximal end for the tendon of Achilles. The sustentacular facet is flat and more elongated medially than that of *Paramys*, *Sciuravus*, *Thomomys*, and *Neotoma* but is similar to the condition in *Microparamys*. The peroneal process is proximodistally elongated as in *Neotoma* and *Microparamys*, with a distinct groove for the tendon of the peroneus longus muscle. Unlike the condition in *Thomomys*, the calcaneoastragalar facet is very long proximodistally, extending for nearly a third of the length of the entire bone, and is similar in this respect to the calcaneus of *Titanotheriomys* figured by Wood (1937: Plate XXVII). The calcaneoastragalar facet is also very tall dorsally as in *Titanotheriomys*. Unlike the distinctly inclined calcaneocuboid facet in *Sciuravus*, this facet in *Pareumys* is nearly perpendicular to the long axis of the calcaneus, slightly concave, and in distal view is not circular as in *Paramys* and *Microparamys* but is oval in outline (compressed in a dorsomedial–ventrolateral direction), with a weak plantar tubercle, again similar to the condition in *Titanotheriomys*.

A proximal ulna and radius are associated with a partial skull of *Pareumys* sp. cf. *P. grangeri* from locality CC4 (SDSNH 87142; Figure 20.1C). The olecranon process of the ulna is not significantly enlarged in comparison with presumed or known fossorial rodents such as *Ischyromys* (e.g., Wood, 1937: Plate XXV), *Mylagaulus* (e.g., Fagan, 1960: Fig. 13), and *Thomomys* (e.g., Hill, 1937) and is comparable in relative length and proportions to those of *Paramys* (e.g., Wood, 1962: Fig. 6E,F), the middle Eocene ricochetal rodent *Protoptychus* (Turnbull, 1991: Fig. 5B), the late Oligocene moderately fossorial geomyid *Pleurolicus* (e.g., Rensberger, 1973: Plate 8e), and the non-fossorial *Neotoma*. However, the *Pareumys* olecranon is relatively thicker mediolaterally than that of *Neotoma* and *Thomomys*. A prominent process occurs on the medial side of the proximal end of the olecranon, forming the medial margin of the groove for the tendon of the triceps longus. This process and groove are stronger than in *Paramys* and *Neotoma*, but similar to the condition in *Sciurus* (see Emry and Thorington, 1982: Fig. 8A). A posteromedial tuberosity is present but much weaker than in *Ischyromys*, *Entoptychus*, and *Thomomys*, and the olecranon is only slightly curved medially when seen in anterior view, similar to the condition reported by Wood (1962) for *Paramys*, as well as in *Neotoma* and *Spermophilus*. The anconeal and coronoid processes seem to be more pronounced than in *Paramys*, are slightly weaker than those of *Neotoma*, and much weaker than those of *Thomomys*. The radial facet is continuous with the adjacent surface of the semilunar notch as in *Neotoma* and *Sciurus* (see Emry and Thorington, 1982: Fig. 8A) and does not form a distinct angle to the latter as in *Paramys*, *Spermophilus*, and especially *Thomomys*. The sharp ridge and accompanying groove just distal to the coronoid process (forming the insertion of the brachialis) are stronger than in *Thomomys* and comparable to the conditions in *Paramys* and *Neotoma*. The lateral fossa containing the abductor pollicis longus muscle is more pronounced than in *Paramys* and comparable to that of *Pleurolicus* and *Neotoma*, but shallower and shorter in proximal extent than in *Entoptychus* and *Thomomys*.

The head of the radius of SDSNH 87142 is elliptical in proximal view (ratio of major to minor axis, 1.38), but slightly less elliptical than in *Neotoma* and much less elliptical than in *Thomomys*. This suggests that *Pareumys* had the ability to pronate and supinate its forelimb to a greater degree than in *Thomomys* (Holliger, 1916). A bicipital tuberosity is present on the posteromedial side of the shaft and is similar in development to this tuberosity in *Neotoma*.

The few available postcrania of *Pareumys* show a mixture of characters seen in a variety of rodents and seem to suggest a fairly generalized habitus. While having a relatively short proximal tarsus is consistent with fossoriality, it also does not imply it, as this condition is present in a number of rodents with rather different lifestyles. Most importantly, the relatively unmodified proximal ulna of *Pareumys* suggests that this genus was not a fully fossorial, specialized claw-digger (Lessa and Thaeler, 1989). Recognition of postcranial material of the later cylindrodontid taxa is eagerly awaited and will be necessary to document any changes in habitus of the family over time.

SYSTEMATICS

SUPRAFAMILIAL

The family Cylindrodontidae was for much of its early history placed within the superfamily Ischyromyoidea, as were all primitive members of the order Rodentia (Miller and Gidley, 1918; Wood, 1937, 1955; Black and Sutton, 1984). Ischyromyoidea were defined (Wood, 1955, p. 170) as "either having the origin of the masseter limited to the ventral surface of the zygoma, or advancing up the anterior face of the arch . . . Dental formula primitive for the order."

Wahlert (1974, p. 408) demoted the cylindrodontids to the rank of a subfamily Cylindrodontinae within a redefined family Ischyromyidae, on the basis of a suite of cranial characters that are more derived than those of the Paramyidae of his usage (sensu Wood, 1980) and shared with *Ischyromys* and *Titanotheriomys*. Wood (1975, 1980, p. 6) included the Cylindrodontidae in a "subhystricognathous to hystricognathous; protrogomorphous to . . . hystricomorphous" infraorder Franimorpha, within the suborder Hystricognathi. Emry and Korth (1996), however, found no evidence of "subhystricognathy" in *Cylindrodon* (but see above discussion) and considered the Cylindrodontidae to be of uncertain position within the order Rodentia.

Korth (1994) assigned the Cylindrodontidae to the suborder Sciuromorpha incertae sedis, and included only the Ischyromyidae (including the Paramyinae and Ischyromyinae as subfamilies) within the superfamily Ischyromyoidea. Korth (1994) also suggested that cylindrodontids could have originated directly from the basal ctenodactyloid stock. McKenna and Bell (1997, p. 186) assigned the Cylindrodontidae to their new suborder Sciuravida, which was defined as "the most recent common ancestor of Ivanantoniidae, Sciuravidae, Chapattimyidae, Cylindrodontidae, Ctenodactylidae, and all of its descendants." Dashzeveg and Meng (1998) rejected the sister-group relationship between Cylindrodontidae and Ctenodactylidae proposed by Averianov (1996) and speculated on the possibility of a relationship between cylindrodontids, sciurids, and aplodontids. As the sister taxon and nearest outgroup of the Cylindrodontidae are not confidently known, identifying character polarities to be used in determining cladistic structure within the family is difficult, a matter discussed in more detail below.

INFRAFAMILIAL

Wood (1974) recognized two subfamilies of cylindrodontids. The Cylindrodontinae was considered to include the Bridgerian (early middle Eocene) species of *Mysops* and the mainly Chadronian (late Eocene) genera *Cylindrodon*, *Pseudocylindrodon*, and *Ardynomys*, while the Jaywilsonomyinae was considered to include the Uintan–Duchesnean (middle middle Eocene to late middle Eocene) *Pareumys* and the Chadronian *Jaywilsonomys*. However, like McKenna and Bell (1997), we consider *Mysops* to be a jaywilsonomyine on the basis of its primitive mandibular and skull characters. Revised diagnoses of Jaywilsonomyinae and Cylindrodontinae are given below, and a hypothesis of cladistic relationships for undoubted North American members of the Cylindrodontidae is shown in Figure 20.3.

Mysops is known from the early Bridgerian, and so the basal subfamily Jaywilsonomyinae may have arisen in North America during the early Eocene (Wasatchian) from an ancestor near, but perhaps not within, the subfamily Paramyinae (but see discussion of certain Asian taxa below). The more derived subfamily of cylindrodontids, the Cylindrodontinae, is first known from the late Uintan (*Pseudocylindrodon*), although members of this subfamily are much more common in the Chadronian. We also questionably include *Dawsonomys*, *Anomoemys*, *Presbymys*, and *Sespemys* in the Cylindrodontidae but do not assign them to either subfamily.

INCLUDED GENERA OF CYLINDRODONTIDAE MILLER AND GIDLEY, 1918

Acronyms for museum collections are listed in Appendix III. The locality numbers listed for each genus refer to the list of unified localities in Appendix I.

The locality numbers may be listed in several ways. Square brackets around the locality (e.g., [CP101]) mean that the taxon in question at that locality is cited using the modifiers "aff.", "cf.," or "nr." Square brackets are usually used for individual species, thus

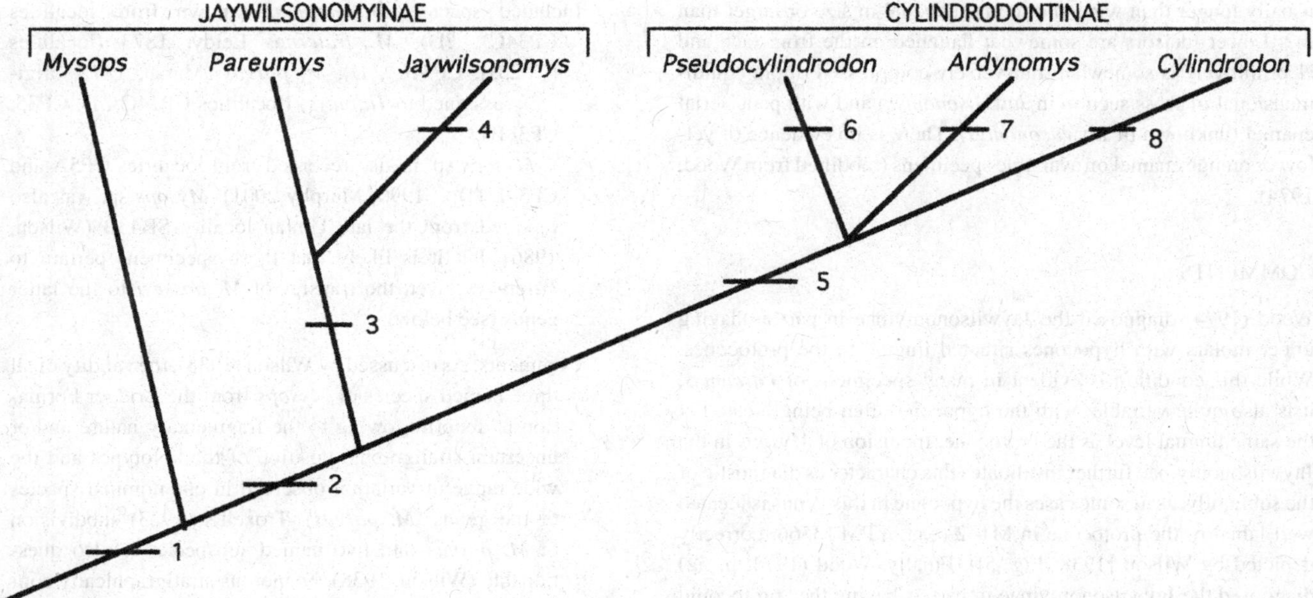

Figure 20.3. Cladogram showing hypothesis of relationships of North American cylindrodontid genera. Characteristics at nodes 1–8 as follows. 1. Incisive foramina greater than 50% of the length of the upper diastema; P4 asymmetrical; moderate unilateral hypsodonty of upper cheek teeth; complete hypolophid and ectolophid on lower molars, forming two distinct lingual basins; incisors slightly broadened, with medial side of anteroventral face flattened. 2. Increased body size; greater unilateral hypsodonty; stronger hypolophids and ectolophids; development of peneplaned occlusal surface on cheek teeth with moderate wear. 3. Tendency for metaloph or metaconule to fuse to posteroloph. 4. Even larger body size; P3 apparently absent; upper molar metaconules usually larger than metacones; strong marginal anterolophid on lower molars, metalophulid II interrupted between protoconid and metaconid; lower incisor with nearly equidimensional, subtriangular cross section. 5. Incisive foramina less than 50% of the length of the diastema; weak sagittal crest present; increased crown height of cheek teeth; P4 symmetrical or only slightly asymmetrical and subequal to M1; metalophs complete to lingual crest; M3 smaller than M2; mandible generally more robust and diastema slightly concave; lower incisor enamel uniserial; p4 subequal to m1; m3 smaller than m2. 6. Two mental foramina usually present on mandible. 7. Skull more robust and shorter anteroposteriorly, with upper incisor exposed in anterior wall of orbit; upper incisors very large; chin process prominent on mandible. 8. Cheek teeth fully hypsodont, with floors of basins not visible in unworn cheek teeth; DP3 present but not replaced by permanent P3; anteroposterior length of P4 greater than transverse width.

implying that the genus is firmly known from the locality, but the actual species identification may be questionable. Question marks in front of the locality (e.g., ?CP101) mean that the taxon is questionably known from that locality, thus implying some doubt that the taxon is actually present there, either at the genus or the species level. An asterisk (*) indicates the type locality. See Emry and Korth (1996) for detailed synonymies for various cylindrodontid species.

JAYWILSONOMYINAE WOOD, 1974

CHARACTERISTICS

If the condition in *Pareumys* can be taken as representative of the subfamily, the rostrum is slightly longer relatively to that in *Pseudocylindrodon* and distinctly longer than in *Cylindrodon* and *Ardynomys*. The supraorbital temporal crests are very weak and a sagittal crest is absent. The incisive foramina are relatively long (more than half the length of the diastema in the upper dentition) and extend posteriorly past the maxillary–premaxillary suture (unknown in *Jaywilsonomys*).

Upper incisors never invade the orbit and are strongly recurved. Cheek teeth are always rooted, primitively low crowned, becoming higher crowned in derived species of *Pareumys* and in *Jaywilsonomys*, but never hypsodont. P4 and p4 generally are distinctly smaller than M1/m1; P4 is not fully molariform and has the labial face of the metacone extending much further labially than the labial face of the paracone (less marked in *Jaywilsonomys*). The M1–2 are usually wider than long, with strong anterior cingulum, protoloph, and posteroloph; the hypocone of upper molars is present but much weaker than protocone; metaloph sometimes completes to hypocone on P4 but are usually incomplete or contacting posterior cingulum on M1–3 (sometimes directed toward lingual wall in *Mysops*, as in the cylindrodontines). The M3 is roughly circular in occlusal outline and subequal in size to or larger than M2.

The mandible is primitively slender and shallow (more robust and deeper in late Uintan and younger species of *Pareumys* and in *Jaywilsonomys*); the dorsal surface of the mandibular diastema is essentially horizontal, with a distinct ridge between i1 and p4. The mental foramen is usually single and the anterior-most part of the masseteric fossa extends to below the m2 trigonid. The medial surface of the mandible is rugose, with numerous tiny nutritive foramina and a distinct pocket for the pterygoideus internus, but not as deep as in cylindrodontines. The p4 has a complete hypolophid and sometimes a complete metalophulid II (term of Wood and Wilson, 1936); the trigonid basin is usually opening anteriorly; m1–3 are usually longer than wide and m3 is subequal in size or larger than m2. Lower incisors are somewhat flattened on the front face and i1 primitively is somewhat transversely compressed (more equidimensional in cross section in *Jaywilsonomys*) and with pauciserial enamel (unknown in *Jaywilsonomys*). There is no evidence of yellow or orange enamel on available specimens (modified from Wood, 1974).

COMMENTS

Wood (1974) diagnosed the Jaywilsonomyinae in part as having upper molars with hypocones situated lingual to the protocones. While this condition is evident in many specimens of *Pareumys*, it is also quite variable, with the hypocone often being located at the same lingual level as the protocone. Inclusion of *Mysops* in the Jaywilsonomyinae further invalidates this character as diagnostic of the subfamily, as in some cases the hypocone in this genus is located well labial to the protocone in M1–2 (e.g., YPM 13566, correctly depicted by Wilson [1938: Fig. 5]). Finally, Wood (1974, p. 52) diagnosed the Jaywilsonomyinae in part as having the "protoconid and metaconid of the lower molars connected by way of anterolophid and never via metalophulid II, so that trigonid basin drains freely posterad." This character is valid only for the genus *Jaywilsonomys*, however, because a complete metalophulid II is clearly present in the lower molars of some specimens of *Mysops* and *Pareumys* (e.g., Wilson, 1940: Plate 1), even though other specimens from the same localities have no metalophulid II.

Mysops **Leidy, 1871 (including *Syllophodus* Cope, 1881)**

Type species: *Mysops minimus* Leidy, 1871.

Type specimen: ANSP 10266, early Bridgerian of Wyoming.

Characteristics: Small cylindrodontid (teeth smaller than any known species of *Pareumys*), with cheek teeth relatively low crowned; upper cheek teeth moderately unilaterally hypsodont; P4 asymmetrical, with labial face of metacone extending much further labially than paracone; metaloph directed toward protocone and sometimes complete with moderate wear; metaloph directed toward hypocone in M1–2 but usually contacting the lingual crest only after heavy wear. Mandible slender, with distinct ridge present atop the horizontal diastema between i1 and p4; masseteric fossa often showing two distinct depressions, the posterior depression ending anteriorly below m2–3 contact, the anterior depression ending below m2 trigonid; single mental foramen present below and slightly anterior to p4 (rare accessory mental foramen sometimes present below p4); ascending ramus crossing alveolar border behind m3; i1 somewhat transversely compressed and subtriangular in cross section, with medial edge of the anteroventral face flattened; trigonid slightly higher than talonid in unworn lower cheek teeth; hypolophids and ectolophids usually complete but relatively weak, reaching a maximum height well below that of the major cusps.

Average length of m1: 1.5 mm.

Included species: *M. minimus* (known from localities CP34C*, ?D); *M. fraternus* Leidy, 1873 (localities [CP25J], CP34C*, D); *M. parvus* (Marsh, 1872, originally assigned to *Tillomys*) (localities CP34C*, D, CP35, CP38B).

Mysops sp. is also recorded from localities CP5A and CP34E (Doi, 1990; Murphy 2001). *Mysops* sp. was also reported from the late Uintan locality SB43B (Wilson, 1986), but it is likely that these specimens pertain to *Pareumys* given the transfer of *M. boskeyi* to the latter genus (see below).

Comments: As discussed by Wilson (1938), the validity of all three named species of *Mysops* from the Bridger Formation is doubtful, owing to the fragmentary nature and/or uncertain stratigraphic position of the holotypes and the wide range of variation observed in one nominal species of the genus (*M. parvus*). Troxell's (1923) subdivision of *M. parvus* into two named subspecies is also questionable (Wilson, 1938). Numerous stratigraphically controlled specimens (mostly isolated teeth) of *Mysops* have recently been collected from the Bridger and Green River Formations by personnel from the University of Colorado Museum (P. Robinson and T. Culver, personal communication). These specimens could aid in a revision of the genus, which is beyond the scope of this chapter.

Microparamys wilsoni Wood 1962 was referred to *Mysops* by Korth (1984), but this species was included neither in *Microparamys* nor *Mysops* by Korth (1994). Wood's (1962: Fig. 55E) illustration of the m1 in the holotype of *M. wilsoni* (YPM 13449) showed a strong metastylid, a feature absent in all m1s of *Mysops* that we have seen. Wood (1962) reported that the masseteric fossa in YPM 13449 extends to below the middle of m1 (consistent with the condition in *Microparamys*; e.g., Walsh, 1997: Fig. 2D), but this fossa extends only to below the m2 trigonid in the mandibles of *Mysops* that we have seen. Wood's (1962, Fig. 55F) illustration of the P4 in a maxilla fragment referred to *M. wilsoni* (YPM 13451) showed the paracone extending further labially than the metacone, a distinct hypocone, and a metaloph connected to the hypocone rather than to the protocone. We have not observed these features in any P4 of *Mysops*. *M. wilsoni* is retained in *Microparamys* by Anderson (Chapter 18) and we likewise exclude the species from *Mysops*.

Wahlert (1973, p. 9) hesitated to include *Mysops* in the family Cylindrodontidae "because, in the one partial skull of the genus (USNM 18043) [specimen not located for this study], the incisive foramina are considerably longer relative to the diastemal length than in *Cylindrodon*, *Pseudocylindrodon*, and *Ardynomys*." However, the relative lengths of the incisive foramina reported by Wahlert (1973) for *Mysops* appear to be approximately the same as those in *Pareumys* (see below), which suggests that this is the primitive cylindrodontid condition, thus excluding *Mysops* from the subfamily Cylindrodontinae. The Bridgerian species of *Mysops* were implicitly included in the subfamily Cylindrodontinae by Wood (1974) and explicitly included in the Cylindrodontinae by Korth (1994). In contrast, McKenna and Bell (1997) assigned *Mysops* to the Jaywilsonomyinae. We agree with the latter interpretation, as this genus lacks almost all of the derived characters of *Cylindrodon*, *Pseudocylindrodon*, and *Ardynomys* and is much more similar to *Pareumys* in most respects.

***Pareumys* Peterson, 1919**

Type species: *Pareumys milleri* Peterson, 1919.

Type specimen: CM 2938, late Uintan of Utah.

Characteristics: Rostrum slightly longer than in *Pseudocylindrodon* and distinctly longer than in *Cylindrodon* and *Ardynomys*, similar in relative length to rostrum of *Sciuravus* but shorter than that of *Paramys*; distinct knob of bone and associated ridge for origin of masseter lateralis on ventral surface of anterior zygomatic root anterolateral to P3, similar in position to that of *Sciuravus*, but usually weaker; infraorbital foramen relatively small, as in cylindrodontines, vertical in lateral view, and elliptical in anterior view, with dorsoventral height twice the mediolateral width, and with major axis slanted outward (dorsal part more lateral than ventral part); distinct notch in frontals at the postorbital constriction; unlike the condition in cylindrodontines, supraorbital temporal crests very weak, converging only slightly posteriad, and not meeting to form a sagittal crest on the frontal, remaining separate to at least the frontal-parietal suture; length of incisive foramina more than half that of the upper diastema, always extending posteriorly beyond the maxillary–premaxillary suture, similar to the condition in *Mysops*; maxillary–premaxillary suture on side of rostrum vertical as in cylindrodontines, *Sciuravus*, and ischyromyines, and not anteriorly convex as in paramyines (Wood, 1976); anteriormost extent of maxillary–palatine suture occur opposite P4–M1 contact; posterior palatine foramina in the palatines medial to m2 and the m1–m2 contact; maxillary bone very thin and closely appressed to pterygoid extension of the palatine around lingual base of M3; posterior maxillary foramen absent. Upper cheek tooth pattern very similar to that of *Mysops*, but with larger, higher-crowned, and more unilaterally hypsodont teeth, with metalophs either free-ending or directed toward and often contacting posteroloph; P4 asymmetrical, with labial face of metacone extending much further labially than paracone; distinct ridge atop the horizontal mandibular diastema between i1 and p4; as in *Mysops*, masseteric fossa often showing two distinct depressions, the posterior depression ending anteriorly below m2–3 contact, the anterior depression ending anteriorly below m2 trigonid; single mental foramen below and slightly anterior to p4 (rare accessory mental foramen sometimes present below p4); ascending ramus crossing alveolar border behind m3; pocket for pterygoideus internus on medial side of mandible not as deep as in cylindrodontines; condition of angle somewhat variable but at least some specimens protrognathous sensu Landry (1999); as in *Mysops*, lower incisor transversely compressed relative to cylindrodontines, but becoming progressively broader in younger species. Lower cheek tooth pattern very similar to *Mysops*, but with larger, higher-crowned teeth, stronger hypolophids and ectolophids, and progressive development of peneplaned occlusal surface after moderate wear. Astragalus with deep, asymmetrical tibial trochlea and divergent, relatively short neck; calcaneus with short, robust body, elongated sustentacular process, and relatively large calcaneoastragalar facet; ulna with relatively unspecialized olecranon process.

Average length of m1: 1.60–2.30 mm.

Included North American species: *P. milleri* (known from localities [CC9AA, BB], CP6AA–B*, [CP7A]); *P. boskeyi* (Wood, 1973, originally assigned to *Mysops*) (locality SB43A); *P. grangeri* Burke, 1935b (localities [CC4], CP6AA, A*, CP38D); *P. guensbergi* Black, 1970a (localities CP7C*, [CP29D]); *P. troxelli* Burke, 1935b (locality CP6A*).

Pareumys sp. is also recorded from localities CC6A, CC7B-E, CC8, CC8IIA, CC9A, AA, CC9II, CC9IVA, CC10, ?SB43B, CP29D, NP22, NP24B, NP25B.

Comments: Wilson (1940) described a mandible fragment with m2 of *Pareumys* (?) sp. from locality CC9AA (LACM[CIT] 2221) and noted that the tooth was unusually high-crowned, flat-topped, and had an enclosed posterior basin. Black (1970b) suggested that this specimen might represent a species of *Pseudocylindrodon*, but we posit individual variation in the Brea Canyon population of *Pareumys* as an alternative explanation. The incisor remnant in this jaw has a flattened medial edge to the anteroventral face as in *Pareumys*, and the squared outline, high crown height, and enclosed posterior basin are nearly matched in m2 teeth of *Pareumys* sp. from the late Uintan of San Diego County (locality CC7B).

The cranial characters of *Pareumys* noted above are evident in several recently collected partial skulls from southern California (SDSNH 31327, 50584, 56172, 87142), none of which preserves the basicranium. Most of the characters of the orbital foramina discussed by Wahlert (1974) cannot be determined from available specimens, although the optic foramen is located dorsal to M3 as in cylindrodontines. The supraorbital temporal crests of *Pareumys* are similar to those of *Reithroparamys* as illustrated by Wood (1962, Fig. 41A), but weaker. It is unclear whether this condition is present in other jaywilsonomyines.

Wood (1974) noted that the presence of P3 in *Pareumys* was uncertain. Lillegraven (1977) documented the presence of a single-rooted P3 in one specimen of the early Uintan species *Pareumys* sp. nr. *P. grangeri*, and a partial skull (SDSNH 31327) of *Pareumys* sp. from locality CC7B has a single-rooted P3 in place, possessing an anteroposteriorly compressed, peglike crown. Incisor growth rates of *Pareumys* are discussed later in this chapter. Specimens of *Pareumys* from localities CC4 and CC7B (early and late Uintan, respectively) have a lower incisor microstructure that is pauciserial, but somewhat more derived than that of *Mysops* (D. Kalthoff, personal communication 2004).

Mysops boskeyi Wood 1973 was transferred to *Pareumys* by Korth (1994), an assignment with which we agree (see also Black and Sutton, 1984). *P. troxelli* was only doubtfully assigned to *Pareumys* by Burke (1935b) but clearly pertains to this genus, although its separation from *P. grangeri* is questionable. Fortunately, numerous stratigraphically controlled specimens of *Pareumys* have recently been collected from the Uinta Formation (Rasmussen *et al.*, 1999).

Specific taxonomy of *Pareumys* from southern California is also difficult. Although early Uintan specimens of *P.* sp. nr. *P. grangeri* from locality CC4 are consistently small in size and have a free-ending metaloph, late Uintan and early Duchesnean populations of the genus generally lack persistent morphological characters needed to diagnose distinct species confidently. Early late Uintan specimens (e.g., from locality CC7B) commonly have cheek teeth similar in standard linear dimensions to those of *P.* sp. nr. *P. grangeri* but are slightly higher crowned, and the termination of the metaloph is variable. Latest Uintan and earliest Duchesnean specimens tend to have slightly larger, slightly higher-crowned teeth, with the metaloph contacting the posterior cingulum (Wilson, 1940; Lillegraven, 1977; Kelly, 1990). Kelly (1990) reasonably suggested that most of the specimens described by Wilson (1940) from localities CC9AA and CC9BB and by Lillegraven (1977) from locality CC7E may be conspecific or represent a group of closely related species. Direct comparison of large samples from southern California and the Western Interior will be necessary for a valid species-level revision of the genus. We would also tend to agree with Lillegraven's (1977, p. 253) view that "Any generic distinction between an advanced species of *Mysops* and a primitive species of *Pareumys* is probably quite arbitrary."

Jaywilsonomys **Ferrusquía-Villafranca and Wood, 1969**

Type species: *Jaywilsonomys ojinagaensis* Ferrusquía-Villafranca and Wood, 1969.

Type specimen: IGM 65-24, early Chadronian of Chihuahua, Mexico.

Characteristics: Upper molars with strong anterior cingulum; central basin weakly closed labially in P4–M1, strongly closed-in M2, and strongly closed in M3 by large mesostyle; metaconules of upper cheek teeth connecting with posteroloph; hypocones of M1–3 considerably linguad of protocones from which they are only faintly separated, so that protocone and hypocone form a diagonal crest; P3 reportedly absent; P4 ovoid in occlusal outline, with labial face of metacone extending only slightly further labially than labial face of paracone; upper molar metaconules usually larger than metacones. Masseteric fossa ending anteriorly below m2 trigonid; ascending ramus crossing alveolar border near rear of m3; strong marginal anterolophid connecting protoconid and metaconid of lower molars; metalophulid II between protoconid and metaconid interrupted at its middle, so trigonid basin drains posteriorly; the two talonid basins of the lower teeth wide open lingually; hypolophid usually with a cusplike enlargement buccad of entoconid; lower incisor with an almost flat, gently rounded anterior face and nearly equidimensional, subtriangular cross section (modified from Ferrusquía-Villafranca and Wood, 1969; Wood, 1974).

Average length of m1: 3.30 mm.

Included North American species: *J. ojinagaensis* only (known from localities SB27B, SB52*).

Jaywilsonomys sp. is known from localities SB27A, SB43C and E, and "*Jaywilsonomys* n. sp." was also recorded from CP98B by Ostrander (1985).

Comments: The species *Jaywilsonomys pintoensis* was originally named by Ferrusquía-Villafranca and Wood (1969) from locality SB52 and later transferred to *Pseudocylindrodon* by Ferrusquía-Villafranca (1984). As concluded by Ferrusquía-Villafranca and Wood (1969), *Jaywilsonomys ojinagaensis* appears to be a very close relative of *Pareumys* and is possibly descended from some species of the latter genus.

CYLINDRODONTINAE MILLER AND GIDLEY, 1918

CHARACTERISTICS

The rostrum is shorter than in jaywilsonomyines, and the incisive foramina are relatively short, almost always extending less than half the length of the diastema in the upper dentition. Supraorbital temporal ridges are distinct and converge posteriorly to form a sagittal crest. The upper incisors tend to be slightly less recurved than in jaywilsonomyines (*Pseudocylindrodon* excepted). The P4 is nearly molariform, with the labial faces of the metacone and paracone extending roughly the same distance labially. The P4 and p4 are generally subequal to or larger than the M1/m1 (*Ardynomys* excluded). The M3 and m3 are generally distinctly smaller than the M2 and m2. The upper cheek teeth may or may not have well-developed hypocones, and the metaloph is usually complete to the lingual crest after slight wear (occasionally incomplete in *Pseudocylindrodon*). The mandible is robust; the dorsal surface of the mandibular diastema between i1 and p4 is slightly concave. The masseteric fossa extends anteriorly to about the m1–m2 contact; the medial side of mandible has a deep pocket for the insertion of the pterygoideus internus. The metalophulid II of the lower molars is complete, isolating the trigonid basin from the talonid basin (modified from Wood, 1980). Lower incisor enamel is uniserial.

COMMENTS

Pseudocylindrodon, *Cylindrodon*, and *Ardynomys* are chiefly Chadronian (late Eocene) genera, and along with *Jaywilsonomys* account for the early and middle Chadronian maximum in cylindrodontid generic diversity. *Cylindrodon* is the most common fossil rodent in some faunas and is almost always abundant in early and middle Chadronian assemblages.

Simpson (1945) was the first to use the name Cylindrodontinae for a subfamily-ranked taxon identical to Miller and Gidley's (1918) concept of Cylindrodontidae. Wood (1974) later used the name Cylindrodontinae in a more restricted sense, including only *Cylindrodon*, *Pseudocylindrodon*, *Ardynomys*, and possibly some species of *Mysops*. Even though the concept of Cylindrodontinae used here is very similar to that of Wood (1974), the rules of zoological nomenclature require that Miller and Gidley (1918) be regarded as the author of this subfamily name. See McKenna and Bell (1997, p. 2 and Appendix A) and the International Commission on Zoological Nomenclature (1999, article 36) for discussion of this point.

***Pseudocylindrodon* Burke, 1935a**

Type species: *Pseudocylindrodon neglectus* Burke, 1935a.

Type specimen: USNM 13758, middle Chadronian of Montana.

Characteristics: Rostrum slightly longer than in *Cylindrodon* and *Ardynomys*. Upper incisors strongly recurved; cheek teeth much lower crowned than in *Cylindrodon*, similar in relative crown height to *Ardynomys*; permanent P3 present, often minute; occlusal pattern of cheek teeth retained until very late wear. Premaxillary–maxillary suture across the palate intersecting the incisive foramina near their center; anterior-most extent of maxillary–palatine suture occuring opposite M1 or the M1–M2 contact; diastema relatively short; two mental foramina usually present, located below p4 rather than below the diastema as in *Cylindrodon* and *Ardynomys*; ascending ramus crossing alveolar border by m3 (Wood, 1980; Emry and Korth, 1996).

Average length of m1 (North American species): 1.60–2.05 mm.

Included species: *P. neglectus* (known from localities SB44B–D, CP68B, CP98B, C, NP9B, NP10Bi, B2, NP24C*); *P. citofluminis* Storer, 1984 (locality NP8*); *P. lateriviae* Storer, 1988 (locality NP9A*); *P. medius* Burke, 1938 (localities CP44, NP25C*); *P. pintoensis* (Ferrusquía-Villafranca and Wood, 1969, originally assigned to *Jaywilsonomys*) (localities [SB44D], SB52*); *P. texanus* Wood, 1974 (localities SB43E, SB44B, C*, [E]); *P. tobeyi* Black, 1970b (locality CP29D*); *P.* new species (locality NP32C).

Pseudocylindrodon sp. is also known from localities SB43C, CP29C, CP39E, NP9B, NP10Bii, NP13, NP22, NP25B, NP32C.

Comments: *Jaywilsonomys pintoensis* Ferrusquía-Villafranca and Wood, 1969 was transferred to *Pseudocylindrodon* without discussion by Ferrusquía-Villafranca (1984). This recombination does not modify the temporal range of either genus. *Pseudocylindrodon* may also be known from the Oligocene of Mongolia, represented by the species *P. mongolicus* Kowalski 1974 (McKenna and Bell, 1997; Dashzeveg and Meng, 1998).

Wood (1980) stated that two mental foramina were present on the mandible in all reported material of *Pseudocylindrodon*. In contrast, Black (1974) stated that only one mental foramen was present in specimens of *P. medius* and *P.* sp. near *P. tobeyi*. However, Burke (1938) originally reported that two mental foramina were present in *P. medius*. Two mental foramina are clearly present in *P. neglectus* (see Burke, 1935a: Fig. 1) and *P. texanus* (see Wood, 1974: Fig. 18C), while only one mental foramen is evident in a mandible referred to *P. pintoensis* (see Wood, 1974: Fig. 24C). The status of this character is unknown for other species of the genus. The value of a doubled mental foramen as an autapomorphy of *Pseudocylindrodon* is questionable in any case, as this condition is occasionally present in *Mysops* and *Pareumys*. The record of *Pseudocylindrodon* new species from the Cook Ranch Local Fauna (locality NP32C, Tabrum, Prothero, and Garcia, 1996), is apparently of late Orellan (early Oligocene) age, and if verified would be the youngest known cylindrodontid from North America. The Kealey Springs West Local Fauna (locality NP10B2) contains *Pseudocylindrodon neglectus* (see Storer, 1994) and is placed by some authors in the earliest Orellan rather than the latest Chadronian, thus representing another possible extension of North American Cylindrodontidae into the Oligocene.

Incisors of *Pseudocylindrodon medius* from McCarty's Mountain (locality NP25C, Chadronian) have been determined to have uniserial enamel as in other cylindrodontines (D. Kalthoff, personal communication 2004). It will be interesting to learn whether a transition from pauciserial to uniserial enamel can be documented in the late Uintan to early Duchesnean species *P. citofluminus*, *P. tobeyi*, and *P. lateriviae*.

***Cylindrodon* Douglass, 1901**

Type species: *Cylindrodon fontis* Douglass, 1901.

Type specimen (lectotype): CM 738a, middle Chadronian of Montana.

Characteristics: Skull similar in shape to *Pseudocylindrodon* but with shorter rostrum; maxillary–premaxillary suture intersecting posterior ends of the incisive foramina. Cheek teeth much higher crowned than in *Pseudocylindrodon* and *Ardynomys*; floors of basins not visible in unworn cheek teeth; despite elevated crown height and deepened crown pattern, the occlusal pattern disappearing in late wear; cheek teeth tapering toward base, so outline of occlusal surface changes from longer than wide to wider than long with wear; hypocone probably not present in upper cheek teeth; DP3 present but not a permanent P3; anteroposterior length of P4 greater than transverse width. Mandible relatively robust and deep; ascending ramus crossing alveolar border by or in front of m3; deep pocket for insertion of pterygoideus internus on medial side of mandible; p4 without hypolophid; upper incisors oval in cross section, slightly less recurved than in *Pseudocylindrodon*; lower incisors triangular in cross section; incisor enamel yellow and uniserial in microstructure (Wahlert, 1968; Wood, 1980; Emry and Korth, 1996).

Average length of m1: 1.81 – 2.39 mm.

Included species: *C. fontis* (known from localities SB44C, D, CP45A, CP98C, NP24C*, D, [NP25C]); *C. collinus* Russell, 1972 (localities NP10B*, [NP49II]); *C. natronensis* Emry and Korth, 1996 (localities CP39B, C*, E); *C. nebraskensis* Hough and Alf, 1956 (localities CP68B, CP98B*); *C. solarborus* Emry and Korth, 1996 (locality CP39F*).

Cylindrodon sp. is also known from localities ?CP39IIB, CP43, CP44, CP98C, NP9C, NP10B, Bii, NP25C, and NP30A.

Comments: Emry and Korth (1996) opined that *C. fontis* was restricted to the Pipestone Springs Fauna (localities NP24C, D). Although we agree with Emry and Korth (1996) that the specimens referred by Galbreath (1969) to *C. fontis* should be reassigned to *C. nebraskensis*, we continue to list *C. fontis* from several other localities reported in the literature (Wood, 1974, 1980).

The holotype lower jaw of *Cylindrodon nebraskensis* was variably assigned the specimen numbers WSM 108 and 1908 by Hough and Alf (1956). The number "WSM 108" is actually written on the jaw. However, this holotype has subsequently been assigned the specimen number RAM 7009, and the locality number is RAM 200130 (D. Lofgren, personal communication 2002). *C. nebraskensis* was regarded as a junior synonym of *C. fontis* by Wood (1980) but was retained as the largest known species of *Cylindrodon* by Emry and Korth (1996). Korth (1992) had previously made *C. galbreathi* Ostrander 1983 a junior synonym of *C. nebraskensis*.

Ardynomys Matthew and Granger, 1925

Type species: *Ardynomys olseni* Matthew and Granger, 1925.

Type specimen: AMNH 20368, late Eocene of Mongolia.

Characteristics: Skull larger, relatively more robust, and shorter anteroposteriorly than in *Cylindrodon* and *Pseudocylindrodon*; apparently causing the base of the upper incisor to create a distinct bulge in the anterior wall of the orbit; maxillary–premaxillary suture just posterior to the incisive foramina or crossing the incisive foramina near their posterior ends. Upper incisors very large relative to upper cheek teeth and slightly less recurved than in *Pseudoylindrodon*, with subtriangular to nearly circular cross sections; anteroventral face of lower incisor flat or slightly concave; single large mental foramen present just anterior to p4; "chin" process prominent on the mandible; permanent P3 is present but small; cheek teeth similar in general morphology to and about as relatively high crowned as those of *Pseudocylindrodon* but relatively broader buccolingually and larger in overall size; ascending ramus crossing alveolar border near front of m3 (Burke, 1936; Wood, 1980; Martin, 1992; Emry and Korth, 1996; Dashzeveg and Meng, 1998).

Average length of m1 (North American species): 2.30 mm.

Included North American species: *A. occidentalis* Burke, 1936 (known from localities SB44B–D, NP25[A], B, C*); *A. saskatchewanensis* (Lambe, 1908, originally assigned to *Sciurus*?) (localities NP9B, NP10B).

Ardynomys sp. is also known from SB43E.

Comments: Storer's (1978) transfer of *Sciurus*? *saskatchewanensis* to *Ardynomys* was questioned by Wood (1980), who tentatively assigned this species to *Prosciurus*. However, the holotype of *S.*? *saskatchewanensis* "is clearly referable to *Ardynomys*" in the opinion of Emry and Korth (1996, p. 402), and we agree.

Ardynomys is also known from the late Eocene to late Oligocene of Asia (McKenna and Bell, 1997). *Ardynomys* has long been assumed to have originated in North America and migrated to Asia (Wood, 1970), but the recent discovery of *Proardynomys* from the middle Eocene of Mongolia suggests the reverse (Dashzeveg and Meng, 1998).

?CYLINDRODONTIDAE

Dawsonomys Gazin, 1961

Type species: *Dawsonomys woodi* Gazin, 1961.

Type specimen: USNM 19309, Lostcabinian (late early Eocene) of Wyoming.

Characteristics: Lower incisor relatively broad, rounded anteriorly and laterally; lower p4 and m1 low crowned, trigonids anteroposteriorly compressed, and hypolophids relatively posterior; lower p4 nearly rectangular in occlusal outline and only slightly smaller than m1, with metaconid anterior in position, incomplete hypolophid, and incomplete ectolophid; lower m1 rectangular in occlusal outline, with complete hypolophid and a mesoconid on the larger ectolophid (modified from Gazin, 1962 and Korth, 1984).

Average length of m1: 2.0 mm.

Included North American species: *D. woodi* only, known from locality CP25G only.

Comments: *Dawsonomys* was originally regarded as a sciuravid by Gazin (1961, 1962). Korth (1984) tentatively assigned *Dawsonomys* to the Cylindrodontidae on the basis of its incisor morphology and several characters of the m1, such as the anteroposteriorly compressed and elevated trigonid and complete hypolophid. As in sciuravids, however, the hypolophid in p4 and m1 is situated more posteriorly than in most undoubted cylindrodontids, such that the anterior lingual valley is much larger than the posterior lingual valley.

Dashzeveg and Meng (1998) pointed out several features of the m1 of *D. woodi* that are unusual for cylindrodontids, such as the presence of a mesoconid, the lack of an anterolabially projecting hypoconid, and the posterolabial trend of the hypolophid. Definite inclusion of *Dawsonomys* in the Cylindrodontidae should be based on additional material, as the only cheek teeth known for the genus are p4 and m1. Martin (1992) reported that the incisor enamel of *Dawsonomys* was pauciserial, but the species investigated (*D. minor* Wood, 1965) was transferred to *Knightomys* by Korth (1984), a decision with which we agree.

Anomoemys Wang, 1986

Type species: *Anomoemys lohiculus* (Matthew and Granger, 1923) (originally described as *Prosciurus lohiculus*).

Type specimen: AMNH 19100, early Oligocene of Mongolia.

Characteristics: Size similar to larger species of *Pareumys*. Wang (1986, p. 293) gave the characteristics as: "Cheek teeth unilaterally hypsodont, four transverse lophs developed and almost equal to each other in height, protocone extends anteroposteriorly, wear surface of protocone–hypocone area of upper teeth aligned antero-posteriorly; no free anterior arm of protocone; paracone and metacone indistinct; protoconule and metaconule distinct; metaloph incomplete; the presence of a longitudinal crest connecting metaconule with protoloph and posteroloph; P4 with parastyle; p4 with metaloph[ul]id I; lower cheek teeth trigonid broad and closed, metastylid developed, metaloph[ul]id II extends posterointernally and is complete, ectolophid straight, hypoconid hypsodont and extends anteroexternally, hypoconulid distinct."

Average length of m1 (North American species): 2.55 mm.

Included species: *Anomoemys? lewisi* (Black, 1974 [originally assigned to *Pareumys*]) only (known from localities CP29C*, D; see Black, 1970b, 1974; Maas, 1985).

Comments: A possible representative of *Anomoemys* in North America is *A.*? *lewisi*, from the late Uintan and earliest Duchesnean of Badwater, Wyoming (Wang, 1986; Korth, 1994). This species is only doubtfully assignable to *Anomoemys* in our view. The type species of the genus, *A. lohiculus*, was described by Wang (1986) as having lower molars with metastylids, hypsodont hypoconids, and distinct hypoconulids. In contrast, Black (1970b) described the isolated lower molars that Black (1974) later assigned to *Pareumys lewisi* as lacking metastylids, lacking swollen hypoconids, and having indistinct hypoconulids. *A. lohiculus* is from the Hsanda Gol Formation of Mongolia. This unit is now assigned to the early Oligocene (Bryant and McKenna, 1995; Dashzeveg and Meng, 1998), and the type locality of *A. lohiculus* could be as much as 10 Ma younger than the type locality of *A.? lewisi* (see radiometric dates on the Hsanda Gol Formation discussed by Meng and McKenna [1998] and their Fig. 2). Better-preserved, associated upper and lower dentitions of *A.? lewisi* will be necessary to corroborate the generic allocation of this species.

Wood (1937, his Fig. 11) assigned an isolated M3 in his personal collection from the "Lower Oligocene of the Cypress Hills of Saskatchewan" to *Prosciurus* cf. *lohiculus*. Wood (1980) later referred this tooth to *Plesiospermophilus*? *altidens* (Russell, 1972), but this species had been made a junior synonym of *Cylindrodon collinus* by Storer (1978), a decision with which Emry and Korth (1996) agreed.

Presbymys Wilson, 1949a

Type species: *Presbymys lophatus* Wilson, 1949a.

Type specimen: LACM (CIT) 3506, early Duchesnean (late middle Eocene) of southern California.

Characteristics: Cheek teeth relatively high crowned but rooted; enamel thick, wrinkled; presence of P3 uncertain; labial face of P4 metacone extends further labially than labial face of paracone; upper cheek teeth rectangular in occlusal outline and unilaterally hypsodont; pattern consisting of one main median inflection opening externally, but becoming an isolated enamel basin with continued wear; teeth apparently quadritubercular, but protocone–hypocone not well separated and forming a continuous crest with wear; mesostyles present; lower cheek teeth more or less bilophate, with ectolophid absent or so weak as to not become evident until after considerable wear; metastylids present; unworn p4 talonid with hypolophid, post-hypolophid valley, and angulate posterolophid (modified from Wilson, 1949a).

Average length of m1: 2.05 mm.

Included species: *P. lophatus* only (known from locality CC9BB and a new locality correlated with locality CC9BB by Lander *et al.* [2000]).

Presbymys sp. is also known from locality CC10 (Walsh, 1991).

Comments: *Presbymys* remains poorly known and does not fit comfortably into any recognized family. The holotype lower jaw contains only p4–m1, the cheek teeth in the paratype maxilla fragment are heavily worn, and the new specimens reported by Lander *et al.* (2000) duplicate the tooth positions in the holotype. Although the P4 is asymmetrical as in *Pareumys* and the upper cheek teeth are unilaterally hypsodont as in other cylindrodontids, the bilophate lower cheek teeth are unlike those of cylindrodontids (which always have a strong ectolophid), being more similar to the geomyoid pattern (Wilson, 1949b).

Presbymys was tentatively assigned to the Cylindrodontidae (then ranked as a subfamily) by Wilson (1949a,b), although several resemblances to the Protoptychidae in the upper teeth were noted. Korth (1994) assigned *Presbymys* to the Protoptychidae without giving explicit reasons, and McKenna and Bell (1997) did the same. It may be noted that the P4 is essentially symmetrical in *Protoptychus* (e.g., Wahlert, 1973: Fig. 1; Turnbull, 1991: Fig. 8), unlike the condition in *Mysops*, *Pareumys*, and *Presbymys*. Any familial assignment of *Presbymys* must be regarded as tentative until more complete material is discovered.

***Sespemys* Wilson, 1934**

Type species: *Sespemys thurstoni* Wilson, 1934.

Type specimen: LACM (CIT) 1397, early Arikareean (late early Oligocene) of southern California.

Characteristics: Single-rooted P3 present; P4 without pronounced unilateral hypsodonty, with distinct parastyle, paracone and metacone equal in labial extent, metaloph strong but just barely contacts labial base of protocone, distinct metaconule present, hypocone weak or absent; upper molars unknown; mandibular incisor broad, with flat anteroventral face; anterior border of masseteric fossa prominent and ridged, terminating beneath the posterior border of m1; ascending ramus crossing alveolar border behind m3; anterior marginal ridge of lower molars forming the principle element in the anterior crest; protolophid a minor element in the anterior crest and consisting of a short spur only; anterior lingual basin much larger than posterior lingual basin; both basins opening lingually and separated from one another by a complete hypolophid connecting labially with a complete ectolophid; weak mesoconid and metastylid present in cheek teeth (modified from Wilson, 1934, 1949c).

Average length of m1: 2.5 mm.

Included species: *S. thurstoni* only, known from locality CC9C only.

"*Sespemys*? new species" is also recorded from locality NP32C.

Comments: The relationships of *Sespemys* have been discussed by Wilson (1934, 1949c), Burke (1936), Wood (1937, 1974, 1980), Korth (1994), and Dashzeveg and Meng (1998). The genus has generally been regarded as either the youngest known cylindrodontid or as an aberrant prosciurine. The lower cheek teeth are similar in gross morphology to those of cylindrodontids and ischyromyines. The presence of a mesoconid on the ectolophid is unusual for cylindrodontids, as is the presence of a P4 that is not unilaterally hypsodont.

Korth (1994) regarded *Sespemys* as a prosciurine aplodontid convergent on cylindrodontids in its massive cheek teeth and large lower incisor. Discovery of the upper molars of this genus should clarify its taxonomic position. A lower jaw from locality CP99B referred to Cf. *Sespemys* by Wood (1937, 1980) was regarded as an aplodontid of uncertain identification by Korth (1989), and as a new prosciurine genus by Korth (1994). If verified, the listing of "*Sespemys*? new species" reported by Tabrum, Prothero, and Garcia (1996) from the late Orellan Cook Ranch Local Fauna would constitute the oldest known record of the genus.

ADDITIONAL POTENTIAL CYLINDRODONTIDS

A new genus and two new species of a primitive cylindrodontid are recorded from locality GC21II (earliest Eocene; Beard and Dawson, 2001; Dawson, 2004). These taxa have yet to be described but would represent the oldest known record of cylindrodontids in the world.

Several isolated teeth from locality SB42 were listed as representing a new genus and species of cylindrodontid by Westgate (1988, 2001). Additional mandibular and maxillary specimens from localities SB42 and SB43A, B are described in an unpublished manuscript by A. Runkel, A. Walton, J. W. Westgate, and R. W. Wilson. These specimens possess many characters that are atypical for cylindrodontids, including cheek teeth that are not unilaterally hypsodont and have very low cusps; upper molars that have a mesoloph and other tiny ridges in the central basin; and lower molars that have a complete but very weak hypolophid and ectolophid, and occasionally a doubled metalophulid II. This new genus is known only from the Uintan of Texas, and in our view it is not a cylindrodontid. We speculate that it may be more closely related to forms like *Janimus* Dawson, 1966 and *Mattimys* Korth, 1984, but include it in the present discussion for completeness.

A few comments on the genus *Spurimus* are relevant here. Two species of *Spurimus* were named by Black (1971) on the basis of isolated teeth from the late to latest Uintan of Badwater, Wyoming. Additional isolated teeth from the early Chadronian Pilgrim Creek Fauna (locality CP44) were assigned to *Spurimus* sp. by Sutton and Black (1975). Black (1971) originally included *Spurimus* in the Prosciurinae. However, Rensberger (1975) only tentatively accepted *Spurimus* as a prosciurine. He noted several characters of *Spurimus* that were atypical of the subfamily and concluded that more data were needed to establish a relationship between *Spurimus* and the prosciurines. Korth (1994) included *Spurimus* in the Prosciurinae without discussion, and Flynn (Chapter 22) also includes *Spurimus* in the Prosciurinae. Alternatively, similarities of the teeth of *Spurimus* to those of *Pareumys* sp. nr. *P. grangeri* were noted by Lillegraven (1977), and a tacit assignment of *Spurimus* to the Cylindrodontidae was made by Rasmussen *et al.* (1999). Although *Spurimus* does possess several dental characters similar to those of jaywilsonomyine cylindrodontids, the referred P4 teeth have a paracone that extends much further labially than the metacone (Black, 1971: Figs. 56 and 65), which is the reverse of the condition in *Mysops* and *Pareumys*. As usual, better material of *Spurimus* should clarify matters.

Finally, other unidentified cylindrodontids are recorded from localities CC10, CP25J, and CP29D (Black, 1970b; Walsh, 1991; Gunnell and Bartels, 2001).

BIOLOGY AND EVOLUTIONARY PATTERNS

The known temporal ranges of North American cylindrodontids are shown in Figure 20.4. Cylindrodontids were relatively small rodents, ranging from the size of large mice and small rats during the Bridgerian and Uintan (middle Eocene) (*Mysops* and *Pareumys*) to the size of a ground squirrel during the Chadronian (late Eocene)

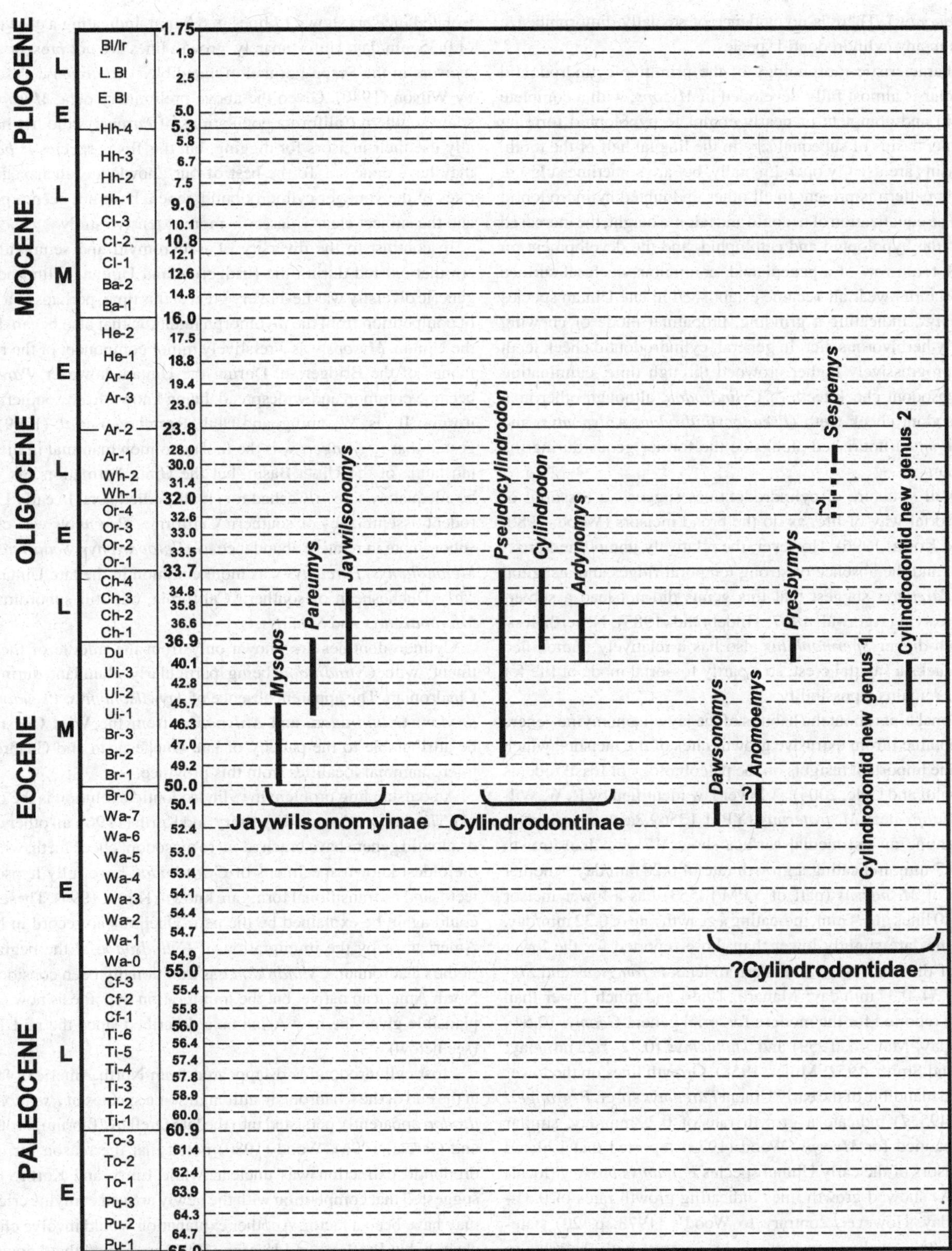

Figure 20.4. Temporal ranges of cylindrodontid genera.

(*Jaywilsonomys*). There is no evidence of sexually dimorphic size variation in any cylindrodontid species.

The characteristic morphology of the primitive cylindrodontid lower molar is almost fully developed in *Mysops*, with a complete ectolophid and complete or nearly complete hypolophid forming two distinct basins of subequal size in the lingual half of the tooth. These basins are usually open lingually, but are sometimes closed. This basic pattern is present in all other undoubted cylindrodontid genera. The equalization of trigonid and talonid height, the increased height of the hypolophid and ectolophid, and the development on the cheek tooth row of a peneplaned, continuous occlusal surface after substantial wear all became established in late Uintan species of *Pareumys*, indicating a grinding, propalinal mode of chewing and a fully herbivorous diet. In general, cylindrodontid cheek teeth became progressively higher crowned through time, culminating in the hypsodont cheek teeth of *Cylindrodon*, although other taxa with mesodont cheek teeth (*Pseudocylindrodon*, *Ardynomys*, and *Jaywilsonomys*) flourished alongside the former genus during the Chadronian.

The skull structure of cylindrodontines suggests a fossorial or semi-fossorial way of life, as do the broad incisors (Wood, 1980; Emry and Korth, 1996). However, the relatively unmodified proximal ulna and the absence of strong temporal ridges and a sagittal crest in *Pareumys* suggest that this genus did not lead a subterranean lifestyle (Agrawal, 1967; Hildebrand, 1985). Nevertheless, the scratch-digger *Spermophilus* also has a relatively unmodified ulna and lacks a sagittal crest, so a partly fossorial mode of life for *Pareumys* remains a possibility.

Additional hints about the habitus of the jaywilsonomyine genera may be obtained from extrusive growth rates of the incisors, which can provide important insights on the paleobiology of fossil rodents (e.g., Rinaldi and Cole, 2004). A lower jaw identified by R. W. Wilson as *Mysops* near *M. fraternus* (YPM 13569.2a) has an incisor whose dentine on the medial surface shows 17 circadian growth lines in 4.7 mm, indicating a growth rate of 0.28 mm/day. Another mandible of *M. parvus* (part of YPM 13553) has a lower incisor showing 20 lines in 6.4 mm, indicating a growth rate of 0.32 mm/day. These values are slightly lower than those reported for the lower incisors of the non-fossorial modern rodents *Peromyscus* and *Sigmodon* (0.34–0.37 mm/day; Manaro, 1959) and much lower than the values reported for the modern fossorial genera *Geomys* (0.64–0.70 mm/day; Manaro, 1959) and *Thomomys* (0.72–1.22 mm/day; Howard and Smith, 1952; Miller, 1958). Growth lines on the lower incisor in a mandible of the early Uintan *Pareumys* sp. cf. *P. grangeri* (SDSNH 49537) indicate a growth rate of 0.29 mm/day, similar to that obtained for *Mysops*. Wood (1973) reported that isolated upper incisors of the early Uintan species *Pareumys boskeyi* (locality SB43A) showed growth lines indicating growth rates of 0.31–0.36 mm/day. However, contrary to Wood's (1973, p. 20) statement that "these values are low . . . by contrast with the rates in *Geomys* . . .", the values he reported for *P. boskeyi* are nearly equal to those obtained by Manaro (1959) for upper incisors of *Geomys*. Growth lines on incisors from southern California late Uintan populations of *Pareumys* seem to be highly variable and have been difficult to count. The most unambiguous specimen (SDSNH 97895, an isolated incisor) shows 17 lines in 6.0 mm, indicating a growth rate of 0.35 mm/day. Unfortunately, growth lines are not present on the incisors in the Sespe Formation mandibles of *Pareumys* described by Wilson (1940). Given the above preliminary data, *Mysops* and some southern California populations of *Pareumys* did not habitually use their incisors for digging, but the Texas species *P. boskeyi* may have done so. To the best of our knowledge, incisor growth rates of the younger cylindrodontid genera have not been reported, and the subject clearly deserves more extensive study.

In contrast to the diversity of ischyromyid and sciuravid genera that existed during the Bridgerian and Uintan, cylindrodontid generic diversity was never very great at this time, perhaps as a result of competition from the myomorph radiation that also began during the Uintan. *Mysops* was a relatively minor component of the rodent faunas of the Bridgerian. During the Uintan, however, *Pareumys* became common and widespread, being known from southern California, Texas, Wyoming, and Utah. Rasmussen *et al.* (1999) suggested that *P. grangeri* was the most common mammal in the Uintan fauna of the Uinta Basin, but the Uinta Formation has yet to be adequately screen-washed for small-bodied taxa. In early Uintan rodent assemblages of southern California, *Pareumys* was clearly subordinate in relative abundance to *Microparamys*, *Sciuravus*, and *Metanoiamys*. *Pareumys* was more common in the late Uintan and early Duchesnean of southern California, but still subordinate to *Microparamys* and *Simimys*.

Cylindrodontines are known only from the middle of the continent, with *Cylindrodon* being particularly abundant during the Chadronian. The apparent absence of *Jaywilsonomys, Pseudocylindrodon*, *Cylindrodon*, and *Ardynomys* from the West Coast may be attributable to the paucity of late Duchesnean and Chadronian micromammal localities from this province.

An outstanding problem in cylindrodontid evolution is the origin of *Cylindrodon*. As noted by Emry and Korth (1996), all other cylindrodontid genera have brachydont to mesodont cheek teeth, whereas the oldest known specimens of *Cylindrodon* have fully hypsodont teeth, and no transitional forms are known (Korth, 1994). These facts could again be explained by the poor Duchesnean record in North America, or by the immigration of *Cylindrodon* at the beginning of the Chadronian. *Cylindrodon* has traditionally been considered a North American native, but the immigration hypothesis now seems plausible given the new Asian taxa described since the mid 1990s (see below).

Most cylindrodontids disappeared from North American faunas at the end of the Chadronian, although some species of *Pseudocylindrodon* apparently persisted into the late Orellan (Tabrum, Prothero, and Garcia, 1996). Wood (1980) noted that the reason for cylindrodontid extinction was unclear, while Emry and Korth (1996) suggested that competition with the newly arrived eumyine cricetids may have been a factor. Another explanation would involve climate change, but Prothero and Heaton (1996) suggested that known climate changes around the Eocene/Oligocene boundary were poorly correlated with changes in the North American mammalian fauna (the extinction of brontotheres being a prominent exception). The introduction of new predators is another possible cause for cylindrodontid extinction. Until recently, the oldest known record of gopher snakes (Colubridae) in North America was from the Orellan (Holman, 2000), but a Chadronian record of the family was reported by Parmley and Holman (2003). However, these early colubrids were apparently too small to be significant rodent predators (J. A. Holman, personal communication 2004). Other possible agents of cylindrodontid extinction are mammalian carnivores, with one candidate being *Palaeogale*, whose oldest known record in North America is from the middle Chadronian (NP24C; Baskin and Tedford, 1996; Tabrum, Prothero, and Garcia, 1996), which was the time of greatest cylindrodontid abundance and diversity.

As noted above, our inability to confidently identify the sister taxon and nearest outcrop of the Cylindrodontidae leads to uncertainty about character polarity and cladistic structure within the family. The conservative hypothesis of relationships shown in Figure 20.3 is based on the assumption that the characters found in most early and middle Eocene ischyromyids and sciuravids are primitive for Cylindrodontidae. Given this assumption, several trends in cylindrodontid evolution are evident, in part reflected by the subdivision of this family into the Jaywilsonomyinae and Cylindrodontinae as diagnosed here. In *Mysops* and *Pareumys*, the incisive foramina are relatively long, being greater than 50% of the length of the diastema (character unknown for *Jaywilsonomys*). The labial face of the metacone on the P4 of jaywilsonomyines extends further labially than that of the paracone; P4 is usually distinctly smaller than M1; the upper molars usually have a free-ending metaloph or a metaloph that contacts the posteroloph rather than the lingual crest; the upper molars are usually transversely wider than long; and the M3 is subequal in size to M2. The mandibular diastema is essentially horizontal, with a distinct ridge on its dorsal surface (the primitive protrogomorph condition). As in most ischyromyids and sciuravids, the lower p4 of jaywilsonomyines is usually distinctly smaller than m1, m3 is subequal in size to m2, and all lower molars are longer than wide (Black, 1970a).

In contrast, in cylindrodontines, the incisive foramina are generally shorter than 50% of the length of the diastema. The labial face of the metacone on P4 usually extends no further or only slightly further labially than that of the paracone; P4 tends to be subequal in size to M1; the upper molars almost always have a complete metaloph that contacts the lingual crest, and they are sometimes longer than wide; and the M3 is usually distinctly smaller than M2. The mandibular diastema in cylindrodontines is slightly concave, and the dorsal ridge is either weak or absent. The lower p4 is usually subequal in size to m1; the m3 is usually smaller than m2, and the lower molars are sometimes wider than long (e.g., *Cylindrodon*, *Ardynomys*). Wilson (1949b) believed that *Cylindrodon* and *Pseudocylindrodon* were more closely related to one another than either was to *Ardynomys*, whereas Wood (1974, his Fig. 13) viewed *Pseudocylindrodon* and *Ardynomys* as being more closely related to one another than either was to *Cylindrodon*. The recent discovery of *Proardynomys* in Mongolia may strengthen Wilson's view (see below). The three genera are shown as forming an unresolved group in Figure 20.3.

Other trends in North American cylindrodontid evolution that do not precisely reflect the subfamilial classification include the broadening and increased relative size of the incisors (*Jaywilsonomys* and Cylindrodontinae), a modest increase in body size within some genera (*Pareumys*, *Pseudocylindrodon*, *Cylindrodon*), and a general increase in hypsodonty of the cheek teeth in various lineages (especially developed in *Cylindrodon*). If the cladistic relationships depicted in Figure 20.3 are correct, then two important characters underwent reversal in cylindrodontid evolution. These are the relative length of the incisive foramina (short in ischyromyids and sciuravids, long in *Mysops* and *Pareumys*, and short again in cylindrodontines), and the structure of P4 (essentially symmetrical in ischyromyids and sciuravids, asymmetrical in *Mysops* and *Pareumys*, and essentially symmetrical again in cylindrodontines).

Finally, the relationships of the North American cylindrodontids to the increasing numbers of known Asian forms are unclear. The alliance by McKenna and Bell (1997) of cylindrodontids, sciuravids, chapattimyids, and ctenodactylids in their new suborder Sciuravida may imply an Asian origin for the cylindrodontids. Indeed, Tong (1997) reported isolated teeth from the middle Eocene of China that were similar to but less specialized than those of *Mysops* (see Dashzeveg and Meng, 1998). Several genera currently classified as chapattimyids are very similar in dental morphology to cylindrodontids, one example being *Petrokozlovia* from the early and middle Eocene of Mongolia, Kazakhstan, and Kyrgyzstan (Shevyreva, 1972; Averianov, 1996). Interestingly, *Hulgana* Dawson, 1968, from the "Ulan Gochu" beds of Mongolia (late Eocene; see Meng and McKenna 1998: Fig. 2), has an asymmetrical P4 strikingly like that of *Pareumys* (see Dawson, 1968: Fig. 1). Originally assigned to the Ischyromyidae, *Hulgana* was regarded as a cylindrodontid by Flynn, Jacobs, and Cheema (1986) and Averianov (1996), but this allocation was questioned by Korth (1994) and Dashzeveg and Meng (1998).

Proardynomys is an important taxon described from the middle Eocene of Mongolia by Dashzeveg and Meng (1998). It has several dental characters that are primitive for cylindrodontids, such as relatively low-crowned cheek teeth with low lophids. Traditionally, *Ardynomys* was thought to have originated in North America and later migrated to Asia (Wood, 1970). However, Dashzeveg and Meng (1998) derived *Ardynomys* from *Proardynomys* in Asia, thus making *Ardynomys* an immigrant to North America. Dashzeveg and Meng (1998, p. 14) suggested that "*Proardynomys* and *Mysops* may represent two evolutionary lineages within the Cylindrodontinae. These lineages may have evolved from a morphotype that had a *Mysops* type of enamel microstructure and molar patterns similar to those of *Proardynomys*." Clearly, much more remains to be learned about cylindrodontid evolution.

ACKNOWLEDGEMENTS

We thank Denny Diveley (AMNH), Alan Tabrum (CM), Sam McLeod (LACM), Pamela Owen (TMM), Pat Holroyd (UCMP), Toni Culver (UCM), and Daniel Brinkman, Lyn Murray, and Mary Ann Turner (YPM) for the loan of specimens and casts. Deborah Anderson, Larry Flynn, J. Alan Holman, Daniela Kalthoff, Stuart Landry, Don Lofgren, Laurent Marivaux, Caroline Rinaldi, Peter Robinson, Thomas Martin, and Hugh Wagner provided helpful information. James Westgate provided a copy of an unpublished manuscript that describes a new rodent genus from Texas. John Alroy's Paleobiology Database was helpful in checking taxon and locality occurrences, and Kai Skadahl and Richard Cerutti (SDSNH) provided original drawings of several specimens. William Korth reviewed this manuscript and provided helpful suggestions.

REFERENCES

Agrawal, V. C. (1967). Skull adaptations in fossorial rodents. *Mammalia*, 31, 300–12.

Averianov, A. (1996). Early Eocene Rodentia of Kyrgyzstan. *Bulletin of the Museum of Natural History, Paris, 4th séries*, 18C, 629–62.

Baskin, J. A. and Tedford, R. H. (1996). Small arctoid and feliform carnivorans. In *The Terrestrial Eocene–Oligocene Transition in North America*, ed. D. R. Prothero and R. J. Emry, pp. 486–97. Cambridge, UK: Cambridge University Press.

Beard, K. C. and Dawson, M. R. (2001). Early Wasatchian mammals from the Gulf Coastal Plain of Mississippi: biostratigraphic and paleobiogeographic implications. In *Eocene Biodiversity: Unusual Occurrences and Rarely Sampled Habitats*, ed. G. F. Gunnell, pp. 75–94. New York: Kluwer Academic/Plenum.

Black, C. C. (1970a). A new *Pareumys* (Rodentia: Cylindrodontidae) from the Duchesne River Formation, Utah. *Fieldiana, Geology*, 16, 453–9.

(1970b). Paleontology and geology of the Badwater Creek area, central Wyoming. Part 5. The cylindrodont rodents. *Annals of the Carnegie Museum*, 41, 201–14.

(1971). Paleontology and geology of the Badwater Creek area, central Wyoming. Part 7. Rodents of the Family Ischyromyidae. *Annals of the Carnegie Museum*, 43, 179–217.

(1974). Paleontology and geology of the Badwater Creek area, central Wyoming. Part 9. Additions to the cylindrodont rodents from the late Eocene. *Annals of the Carnegie Museum*, 45, 151–60.

Black, C. C. and Sutton, J. F. (1984). Paleocene and Eocene rodents of North America. [In *Papers in Vertebrate Paleontology Honoring Robert Warren Wilson*, ed. R. M. Mengel.] *Carnegie Museum of Natural History Special Publication*, 9, 67–84.

Bryant, J. D. and McKenna, M. C. (1995). Cranial anatomy and phylogenetic position of *Tsaganomys altaicus* (Mammalia: Rodentia) from the Hsanda Gol Formation (Oligocene), Mongolia. *American Museum Novitates*, 3156, 1–42.

Burke, J. J. (1935a). *Pseudocylindrodon*, a new rodent genus from the Pipestone Springs Oligocene of Montana. *Annals of the Carnegie Museum*, 25, 1–4.

(1935b). Fossil rodents from the Uinta Eocene Series. *Annals of the Carnegie Museum*, 25, 5–12.

(1936). *Ardynomys* and *Desmatolagus* in the North Amercian Oligocene. *Annals of the Carnegie Museum*, 25, 135–54.

(1938). A new cylindrodont rodent from the Oligocene of Montana. *Annals of the Carnegie Museum*, 27, 255–75.

Cope, E. D. (1881). Review of the Rodentia of the Miocene Period of North America. *Bulletin of the United States Geological and Geographical Survey of the Territories*, 6, 361–86.

Dashzeveg, D. and Meng, J. (1998). A new Eocene cylindrodontid rodent (Mammalia, Rodentia) from the eastern Gobi of Mongolia. *American Museum Novitates*, 3253, 1–18.

Dawson, M. R. (1966). Additional late Eocene rodents (Mammalia) from the Uinta Basin, Utah. *Annals of the Carnegie Museum*, 38, 97–114.

(1968). Oligocene rodents (Mammalia) from East Mesa, Inner Mongolia. *American Museum Novitates*, 2324, 1–12.

(2004). Early Wasatchian cylindrodontid rodents: evolution in the Gulf Coastal Plain. *Journal of Vertebrate Paleontology*, 24(suppl. to no. 3), p. 51A.

Doi, K. (1990). Geology and paleontology of two primate families of the Raven Ridge, Northwestern Colorado and Northeastern Utah. M. Sc. Thesis, University of Colorado, Boulder.

Douglass, E. (1901). Fossil Mammalia of the White River beds of Montana. *Transactions of the American Philosophical Society*, 20, 237–79.

Emry, R. J. and Korth, W. W. (1996). Cylindrodontidae. In *The Terrestrial Eocene–Oligocene Transition in North America*, ed. D. R. Prothero and R. J. Emry, pp. 399–416. Cambridge, UK: Cambridge University Press.

Emry, R. J. and Thorington, R. W. Jr. (1982). Descriptive and comparative osteology of the oldest fossil squirrel, *Protosciurus* (Rodentia: Sciuridae). *Smithsonian Contributions to Paleobiology*, 47, 1–35.

Fagan, S. R. (1960). Osteology of *Mylagaulus laevis*, a fossorial rodent from the upper Miocene of Colorado. *University of Kansas Paleontological Contributions, Vertebrata*, 9, 1–32.

Ferrusquía-Villafranca, I. (1984). A review of the early and middle Tertiary mammal faunas of Mexico. *Journal of Vertebrate Paleontology*, 4, 187–198.

Ferrusquía-Villafranca, I. and Wood, A. E. (1969). New fossil rodents from the early Oligocene Rancho Gaitan local fauna, northeastern Chihuahua, Mexico. *The Pearce-Sellards Series, Texas Memorial Museum*, 16, 1–13.

Flynn, L. J., Jacobs, L. L., and Cheema, I. U. (1986). Baluchimyinae, a new ctenodactyloid rodent subfamily from the Miocene of Baluchistan. *American Museum Novitates*, 2841, 1–58.

Galbreath, E. C. (1969). Cylindrodont rodents from the Lower Oligocene of northwestern Colorado. *Transactions of the Illinois Academy of Science*, 62, 94–7.

Gazin, C. L. (1961). New sciuravid rodents from the lower Eocene Knight Formation of western Wyoming. *Proceedings of the Biological Society of Washington*, 74, 193–4.

(1962). A further study of the lower Eocene mammalian faunas of southwestern Wyoming. *Smithsonian Miscellaneous Collections*, 144, 1–98.

Gunnell, G. F. and Bartels, W. S. (2001). Basin margins, biodiversity, evolutionary innovation, and the origin of new taxa. In *Eocene Biodiversity: Unusual Occurrences and Rarely Sampled Habitats*, ed. G. F. Gunnell, pp. 403–40. New York: Kluwer Academic/Plenum.

Hildebrand, M. (1985). Digging of quadrupeds. In *Functional Vertebrate Morphology*, ed. M. Hildebrand, D. M. Bramble, K. F. Liem and D. B. Wake, pp. 89–109. Cambridge, MA: Belknap Press.

Hill, J. E. (1937). Morphology of the pocket gopher, mammalian genus *Thomomys*. *University of California Publications in Zoology*, 42, 81–174.

Holliger, C. D. (1916). Anatomical adaptations in the thoracic limb of the California pocket gopher and other rodents. *University of California Publications in Zoology*, 13, 447–94.

Holman, J. A. (2000). *Fossil Snakes of North America*. Bloomington, IN: Indiana University Press.

Hough, J. and Alf, R. (1956). A Chadron mammalian fauna from Nebraska. *Journal of Paleontology*, 30, 132–40.

Howard, W. E. and Smith, M. E. (1952). Rates of extrusive growth of incisors of pocket gophers. *Journal of Mammalogy*, 38, 485–7.

International Commission on Zoological Nomenclature (1999). *International Code of Zoological Nomenclature*, 4th edn. London: International Trust for Zoological Nomenclature.

Kelly, T. S. (1990). Biostratigraphy of the Uintan and Duchesnean land mammal assemblages from the middle member of the Sespe Formation, Simi Valley, California. *Contributions in Science, Los Angeles County Natural History Museum*, 419, 1–42.

Korth, W. W. (1984). Earliest Tertiary evolution and radiation of rodents in North America. *Bulletin of the Carnegie Museum of Natural History*, 24, 1–71.

(1989). Aplodontid rodents (Mammalia) from the Oligocene (Orellan and Whitneyan) Brule Formation, Nebraska. *Journal of Vertebrate Paleontology*, 9, 400–14.

(1992). Cylindrodonts (Cylindrodontidae, Rodentia) and a new genus of eomyid, *Paranamatomys*, (Eomyidae, Rodentia) from the Chadronian of Sioux County, Nebraska. *Transactions of the Nebraska Academy of Sciences*, XIX, 75–82.

(1994). *Tertiary Record of Rodents in North America.* New York: Plenum Press.

Kowalski, K. (1974). Middle Oligocene rodents from Mongolia. *Acta Palaeontologica Polonica*, 30, 147–78.

Lambe, L. M. (1908). The Vertebrata of the Oligocene of the Cypress Hills, Saskatchewan. *Geological Survey of Canada, Contributions to Paleontology*, 3, 1–65.

Lander, E. B., Whistler, D. P., Anderson, E. S. and Kennedy, C. L. (2000). *Project 97–17: Big Sky Country Club, LLC, Lost Canyons Golf Club (Tapo and Dry Canyon Portions, Whiteface Specific Plan Area), Simi Valley, Ventura County, California Paleontologic Resource Impact Mitigation Program Final Technical Report of Results and Findings.* Altadena, CA: Paleo Environmental Associates.

Landry, S. O. (1999). A proposal for a new classification and nomenclature for the Glires (Lagomorpha and Rodentia). *Mitteilungen aus dem Museum für Naturkunde in Berlin, Zoologische Reihe*, 75, 283–316.

Leidy, J. (1871). Notice of some extinct rodents. *Proceedings of the Academy of Natural Sciences of Philadelphia*, 1871, 230–2.

(1873). Contributions to the extinct vertebrate fauna of the western territories. *Report of the United States Geological Survey of the Territories*, 1, 7–358.

Lessa, E. P. and Thaeler, C. S. Jr. (1989). A reassessment of morphological specializations for digging in pocket gophers. *Journal of Mammalogy* 70, 689–700.

Lillegraven, J. A. (1977). Small rodents (Mammalia) from Eocene deposits of San Diego County, California. *Bulletin of the American Museum of Natural History*, 158, 221–62.

Maas, M. C. (1985). Taphonomy of a late Eocene microvertebrate locality, Wind River Basin, Wyoming (USA). *Palaeogeography, Palaeoclimatology, Palaeoecology*, 52, 123–42.

Manaro, A. J. (1959). Extrusive incisor growth in the rodent genera *Geomys*, *Peromyscus*, and *Sigmodon*. *Quarterly Journal of the Florida Academy of Sciences*, 22, 25–31.

Marivaux, L. and Welcomme, J.-L. (2003). New diatomyid and baluchimyine rodents from the Oligocene of Pakistan (Bugti Hills, Balochistan): systematic and paleobiogeographic implications. *Journal of Vertebrate Paleontology*, 23, 420–34.

Marsh, O. C. (1872). Preliminary description of new Tertiary mammals. Part II. *American Journal of Science, Third Series*, 4, 202–4.

Martin, T. (1992). Schmelzmikrostructur in den Inzisiven alt- und neuweltlicher hystricognather Nagetiere. *Palaeovertebrata, Mémoire Extraordinaire*, 1–168.

Matthew, W. D. and Granger, W. (1923). Nine new rodents from the Oligocene of Mongolia. *American Museum Novitates*, 102, 1–10.

(1925). New creodonts and rodents from the Ardyn Obo formation of Mongolia. *American Museum Novitates*, 193, 1–7.

McKenna, M. C. and Bell, S. K. (1997). *Classification of Mammals Above the Species Level.* New York: Columbia University Press.

Meng, J. and McKenna, M. C. (1998). Faunal turnovers of Palaeogene mammals from the Mongolian Plateau. *Nature*, 394, 364–7.

Miller, G. S. and Gidley, J. W. (1918). Synopsis of the supergeneric groups of rodents. *Journal of the Washington Academy of Science*, 8, 431–48.

Miller, R. S. (1958). Rates of incisor growth in the mountain pocket gopher. *Journal of Mammalogy*, 39, 380–85.

Murphey, P. C. (2001). Stratigraphy, fossil distribution, and depositional environments of the upper Bridger Formation (middle Eocene) of southwestern Wyoming, and taphonomy of an unusual Bridger microfossil assemblage. Ph.D. Thesis, University of Colorado, Boulder.

Ostrander, G. E. (1983). New early Oligocene (Chadronian) mammals from the Raben Ranch local fauna, northwest Nebraska. *Journal of Paleontology*, 57, 128–39.

(1985). Correlation of the early Oligocene (Chadronian) in northwestern Nebraska. [In *Fossiliferous Cenozoic Deposits of Western South Dakota and Northwestern Nebraska*, ed. J. E. Martin.] *Dakoterra*, 2, 203–51.

Parmley, D. and Holman, J. A. (2003). *Nebraskophis* Holman from the late Eocene of Georgia (USA), the oldest known North American colubrid snake. *Acta Zoologica Cracoviensia*, 46, 1–8.

Peterson, O. A. (1919). A report upon the material discovered in the upper Eocene of the Uinta Basin by Earl Douglass in the years 1908–1909, and by O. A. Peterson in 1912. *Annals of the Carnegie Museum*, 12, 40–168.

Prothero, D. R. and Heaton, R. H. (1996). Faunal stability during the early Oligocene climate crash. *Palaeogeography, Palaeoclimatology, Palaeoecology*, 127, 257–83.

Rasmussen, D. T., Conroy, G. C., Friscia, A. R., Townsend, K. E., and Kinkel, M. D. (1999). Mammals of the middle Eocene Uinta Formation. [In *Vertebrate Paleontology in Utah*, ed. D. D. Gillette.] *Utah Geological Survey Miscellaneous Publication*, 99–1, 401–20.

Rensberger, J. M. (1971). Entoptychine pocket gophers (Mammalia, Geomyoidea) of the early Miocene John Day Formation, Oregon. *University of California Publications in Geological Sciences*, 90, 1–163.

(1973). Pleurolicine rodents (Geomyoidea) of the John Day Formation, Oregon. *University of California Publications in Geological Sciences*, 102, 1–95.

(1975). *Haplomys* and its bearing on the origin of the aplodontid rodents. *Journal of Mammalogy*, 56, 1–14.

Rinaldi, C. and Cole, T. M. III. (2004). Environmental seasonality and incremental growth rates of beaver (*Castor canadensis*) incisors: implications for paleobiology. *Palaeogeography, Palaeoclimatology, Palaeoecology*, 206, 289–301.

Russell, L. S. (1972). Tertiary Mammals of Saskatchewan, Part II: The Oligocene fauna, non-ungulate orders. *Contributions of the Royal Ontario Museum, Life Sciences*, 84, 1–63.

Shevyreva, N. S. (1972). New rodents from the Paleogene of Mongolia and Kazakhstan. *Paleontological Journal*, 3, 399–408.

Simpson, G. G. (1945). The principles of classification and a classifcation of mammals. *Bulletin of the American Museum of Natural History*, 85, 1–350.

Stains, H. J. (1959). Use of the calcaneum in studies of taxonomy and food habits. *Journal of Mammalogy*, 40, 392–401.

Storer, J. E. (1978). Rodents of the Calf Creek local fauna (Cypress Hills Formation, Oligocene, Chadronian), Saskatchewan. *Contributions of the Saskatchewan Museum of Natural History*, 1, 1–54.

(1984). Mammals of the Swift Current Creek Local Fauna (Eocene, Uintan, Saskatchewan). *Contributions of the Saskatchewan Museum of Natural History*, 7, 1–158.

(1988). The rodents of the Lac Pelletier lower fauna, late Eocene (Duchesnean) of Saskatchewan. *Journal of Vertebrate Paleontology*, 8, 84–101.

(1994). A latest Chadronian (late Eocene) mammalian fauna from the Cypress Hills, Saskatchewan. *Canadian Journal of Earth Sciences*, 31, 1335–41.

Sutton, J. F. and Black, C. C. (1975). Paleontology of the earliest Oligocene deposits in Jackson Hole, Wyoming. Part 1. Rodents exclusive of the Family Eomyidae. *Annals of the Carnegie Museum*, 45, 299–315.

Szalay, F. S. (1985). Rodent and lagomorph morphotype adaptations, origins, and relationships: some postcranial attributes analyzed. In *Evolutionary Relationships Among Rodents: A Multidisciplinary Analysis*, ed. W. P. Luckett and J.-L. Hartenberger, pp. 83–132. New York: Plenum Press.

Tabrum, A. R., Prothero, D. R., and Garcia, D. (1996). Magnetostratigraphy and biostratigraphy of the Eocene–Oligocene transition, southwestern Montana. In *The Terrestrial Eocene–Oligocene Transition in North America*, ed. D. R. Prothero and R. J. Emry, pp. 278–311. Cambridge, UK: Cambridge University Press.

Tong, Y. (1997). Middle Eocene small mammals from Liguanqiao Basin of Henan Province and Yuanqu Basin of Shanxi Province, Central China. *Paleontologia Sinica Series C*, 26, 1–256. [In Chinese, English summary.]

Troxell, E. L. (1923). The Eocene rodents *Sciuravus* and *Tillomys*. *American Journal of Science (Series 5)*, 5, 383–96.

Turnbull, W. D. (1991). *Protoptychus hatcheri* Scott, 1895. The mammalian faunas of the Washakie Formation, Eocene age, of southern Wyoming. Part II. The Adobetown Member, middle division (= Washakie B), Twka/2 (in part). *Fieldiana Geology New Series*, 21, 1–33.

Tyut'kova, L. A. (1997). A new cylindrodontid (Rodentia, Mammalia) from the *Indricotherium* fauna. *Paleontological Journal*, 31, 662–6.

Wahlert, J. H. (1968). Variability of rodent incisor enamel as viewed in thin section, and the microstructure of the enamel in fossil and recent rodent groups. *Breviora*, 309, 1–18.

(1973). *Protoptychus*, a hystricomorphous rodent from the late Eocene of North America. *Breviora*, 419, 1–14.

(1974). The cranial foramina of protrogomorphous rodents: an anatomical and phylogenetic study. *Bulletin of the Museum of Comparative Zoology*, 146, 363–410.

Walsh, S. L. (1991). Late Eocene mammals from the Sweetwater Formation, San Diego County, California. [In *Eocene Geologic History San Diego Region*, ed. P. L. Abbott and J. A. May]. *Pacific Section, Society of Economic Paleontologists and Mineralogists*, 68, 149–59.

(1996). Middle Eocene mammal faunas of San Diego County, California. In *The Terrestrial Eocene–Oligocene Transition in North America*, ed. D. R. Prothero and R. J. Emry, pp. 75–119. Cambridge, UK: Cambridge University Press.

(1997). New specimens of *Metanoiamys*, *Pauromys*, and *Simimys* (Rodentia: Myomorpha) from the Uintan (middle Eocene) of San Diego County, California, and comments on the relationships of selected Paleogene Myomorpha. *Proceedings of the San Diego Society of Natural History*, 32, 1–20.

Wang, B. Y. (1986). On the systematic position of *Prosciurus lohiculus*. *Vertebrata PalAsiatica*, 24, 285–94.

Westgate, J. W. (1988). Biostratigraphic implications of the first Eocene land mammal fauna from the North American coastal plain. *Geology*, 16, 995–8.

(2001). Paleoecology and biostratigraphy of marginal marine Gulf Coast Eocene vertebrate localities. In *Eocene Biodiversity: Unusual Occurrences and Rarely Sampled Habitats*, ed. G. F. Gunnell, pp. 263–97. New York: Kluwer Academic/Plenum.

Wilson, J. A. (1986). Stratigraphic occurrence and correlation of early Tertiary vertebrate faunas, Trans-Pecos Texas: Aguna Fria–Green Valley areas. *Journal of Vertebrate Paleontology*, 6, 350–73.

Wilson, R. W. (1934). Two rodents and a lagomorph from the Sespe of the Las Posas Hills, California. *Publication of the Carnegie Institution of Washington*, 453, 11–17.

(1938). Review of some rodent genera from the Bridger Eocene. Part II. *American Journal of Science*, 35, 207–22.

(1940). *Pareumys* remains from the later Eocene of California. *Publication of the Carnegie Institution of Washington*, 514, 97–108.

(1949a). Additional Eocene rodent material from southern California. *Publication of the Carnegie Institution of Washington*, 584, 1–25.

(1949b). Early Tertiary rodents of North America. *Publication of the Carnegie Institution of Washington*, 584, 66–164.

(1949c). Rodents and lagomorphs of the upper Sespe. *Publication of the Carnegie Institution of Washington*, 584, 51–65.

Wood, A. E. (1937). The mammalian fauna of the White River Oligocene. Part II. Rodentia. *Transactions of the American Philosophical Society*, 28, 157–269.

(1955). A revised classification of the rodents. *Journal of Mammalogy*, 36, 165–87.

(1962). The early Tertiary rodents of the Family Paramyidae. *Transactions of the American Philosophical Society, New Series*, 52 (Part 1), 1–261.

(1965). Small rodents from the early Eocene Lysite member, Wind River Formation of Wyoming. *Journal of Paleontology*, 39, 124–34.

(1970). The early Oligocene rodent *Ardynomys* (Family Cylindrodontidae) from Mongolia and Montana. *American Museum Novitates*, 2418, 1–18.

(1973). Eocene rodents, Pruett Formation, southwest Texas; their pertinence to the origin of the South American Caviomorpha. *The Pearce-Sellards Series, Texas Memorial Museum*, 20, 1–40.

(1974). Early Tertiary vertebrate faunas Vieja Group Trans-Pecos Texas: Rodentia. *Bulletin of the Texas Memorial Museum*, 21, 1–112.

(1975). The problem of the hystricognathous rodents. *University of Michigan Papers in Paleontology*, 12, 75–80.

(1976). The Oligocene rodents *Ischyromys* and *Titanotheriomys* and the content of the Family Ischyromyidae. In *Athlon: Essays in Paleontology in Honour of Loris Shano Russell*, ed. C. S. Churcher, pp. 244–77. [*Royal Ontario Museum Life Science Miscellaneous Publication*.] Toronto: University of Toronto Press.

(1980). The Oligocene rodents of North America. *Transactions of the American Philosophical Society*, 70, 5 (Part), 1–68.

(1984). Hystricognathy in the North American Oligocene rodent *Cylindrodon* and the origin of the Caviomorpha. [In *Papers in Vertebrate Paleontology Honoring Robert Warren Wilson*, ed. R. M. Mengel.] *Special Publication of the Carnegie Museum of Natural History*, 9, 151–60.

Wood, A. E. and Wilson, R. W. (1936). A suggested nomenclature for the cusps of the cheek teeth of rodents. *Journal of Paleontology*, 10, 388–91.

21 Sciuridae

H. THOMAS GOODWIN
Andrews University, Berrien Springs, MI, USA

INTRODUCTION

The family Sciuridae comprises about 275 living and numerous extinct species of squirrels – the fourth most diverse extant mammal family (after Muridae, Vespertilionidae, Soricidae; Vaughan, Ryan, and Czaplewski, 2000) – and squirrels inhabit all continents with land mammals except Australia (Nowak, 1999). The family includes nocturnal gliders and a diverse assemblage of diurnal species occupying arboreal and terrestrial niches, with sizes of extant species ranging from approximately 10 g (tropical pygmy squirrels) to approximately 7500 g (some marmots) (Nowak, 1999); the late Neogene *Paenemarmota* was larger still. Extant generic diversity is centered in southern and eastern Asia and adjacent islands (Nowak, 1999), but a speciose terrestrial clade of chipmunks and the tribe Marmotini (ground squirrels, marmots, prairie dogs, and extinct relatives) is most diverse in North America. The Marmotini exhibit remarkable annual cycles and diverse patterns of social behavior, and they have been intensively studied (e.g., Murie and Michener, 1984). The North American fossil record for the family extends from the late Eocene (Chadronian), documenting over 20 Tertiary genera as reviewed below.

DEFINING FEATURES OF THE FAMILY SCIURIDAE

CRANIAL

Characterizations of Sciuridae provided by Vianey-Liaud (1985), Korth and Emry (1991), and Korth (1994) were primary sources for the cranial, dental, and skeletal descriptions which follow. The skull is relatively broad and variably convex in dorsal profile. The rostrum is typically short and bears small, paired incisive foramina ventrally.

All living sciurids are sciuromorphous (see Chapter 16) with a relatively broad zygomatic arch that angles anterodorsally from its ventral origin; a small infraorbital foramen in a bony canal (except *Tamias*), positioned low on the side of the rostrum; and a distinct masseteric tubercle ventral or ventrolateral to the infraorbital foramen for insertion of a slip of the lateral masseter. However, the earliest sciurids are either protrogomorphous (*Douglassciurus*; Emry and Thorington, 1982) or exhibit a unique zygomasseteric architecture, possibly reflecting a modified myomorphous condition (Cedromurinae; Korth and Emry, 1991) (Figure 21.1).

The interorbital region is characteristically broad and bounded posteriorly by prominent postorbital processes extending laterally and posteriorly from the frontals. The postglenoid foramen is large and exposes the petrosal beneath. The auditory bulla is inflated and septate, retains the stapedial artery (primitive) within a bony canal (derived), and exhibits a bony meatocochlear bridge (also in aplodontids; Lavocat and Parent, 1985). The pterygoid flange is elongate, and the basisphenoid–basiocciptial region is angled relative to the plane of the palate. The palate is usually broad.

DENTAL

Incisor enamel is uniserial in microstructure and variably ornamented. The cheek teeth vary from brachydont (primitive) to hypsodont. Most species have P3–M3 with P3 small in early and many extant taxa, lost in a few taxa, and substantially enlarged in advanced ground-dwelling squirrels.

In most species, the anterior cingulum of P4 is broad and rounded buccally at the parastyle but tapers lingually, often giving the tooth a vaguely subtriangular appearance. In primitive taxa (and to a certain extent in flying squirrels), M1 and M2 bear a distinct hypocone, a prominent metaconule strongly constricted at protocone, and in some cases additional conules on the protoloph and metaloph; however, derived taxa tend to simplify upper cheek teeth in various

Evolution of Tertiary Mammals of North America, Vol. 2. ed. C. M. Janis, G. F. Gunnell, and M. D. Uhen. Published by Cambridge University Press.

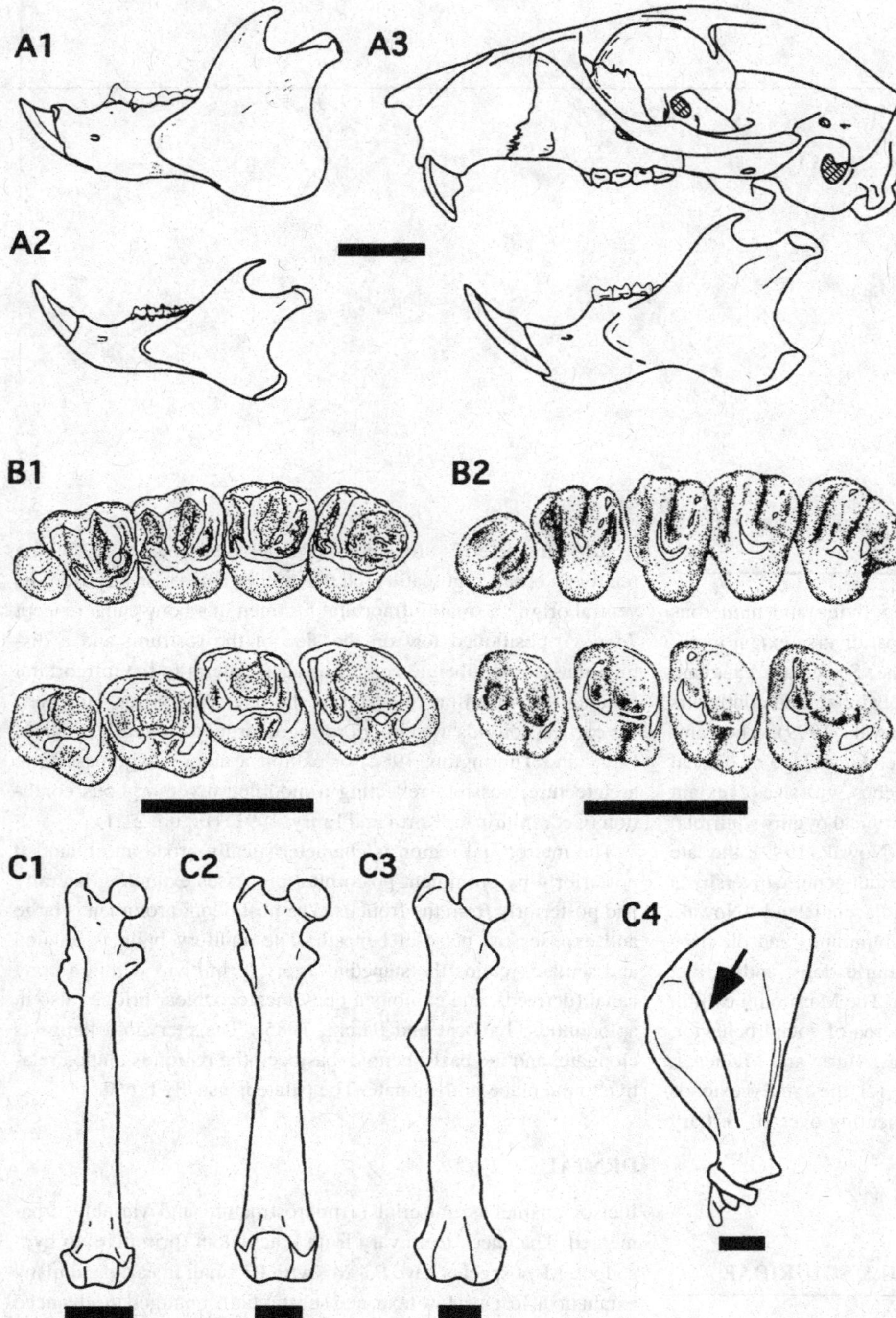

Figure 21.1. A. Jaws and skull. A1. Left lower jaw of Chadronian *Douglassciurus jeffersoni* (USNM 214936; redrawn from Emry and Korth, 1996). A2. Right lower jaw (reversed) of Hemingfordian *Miospermophilus bryanti* (KU 10149; redrawn from Black, 1963). A3. Skull and jaw of Recent *Sciurus carolinensis* (KU [mammal collection] 9637; redrawn from Hall, 1981). Scale bar = 10 mm. B1. Upper and lower cheek dentition *Douglassciurus jeffersoni* (right uppers [reversed]) of FAM 79307, left lowers a composite of FAM 79305 [p4-m2] and FAM 79301 [m3]; redrawn from Emry and Korth, 1996). B2. Late Blancan derived ground squirrel, ? *Spermophilus cragini* (left uppers a composite of KUVP 113207 [P3], KUVP 113217 [P4], and KUVP 6168 [M1–3]; right lowers [reversed] a composite of UMMP 50139 [p4], UMMP 50144 [m1–2], and UMMP 50157 [m3]; redrawn from Goodwin and Hayes, 1994). Scale bars = 5 mm. C. Femur. C1. *Douglassciurus jeffersoni* (left [reversed], USNM 243981; redrawn from Emry and Thorington, 1982). C2. Extant *Marmota monax* (right, AU 7975); C3. For comparison, *Paramys delicatus* (right, AMNH 12506; redrawn from Wood, 1962). C4. Right scapula of *Marmota monax*; medial view (AU 7975), with prominent, ridgelike subscapular spine marked by arrow. Scale bars = 10 mm. (A1,B1,B2: courtesy of the Society of Vertebrate Paleontology. C3: courtesy of the American Philosophical Society.)

ways. The M3 is subtriangular in occlusal outline, with variably expanded posterior cingulum and primitively lacking a distinct metacone or metaloph (the latter well developed in some advanced terrestrial taxa).

The lower cheek teeth of early squirrels display a distinct entoconid with variably developed hypolophid extending towards the hypoconid, and a distinct mesoconid positioned between protoconid and hypoconid, features that tend to be reduced or lost in many subsequent lineages (although flying squirrels tend to retain complex lower cheek teeth). The p4 of most sciurids is substantially narrower across the trigonid than across the talonid, with protoconid and metaconid adjacent; a distinct anteroconid is present in many. The m1–2 primitively exhibit a low trigonid commonly bounded distally by a complete metalophid coursing from protoconid to metaconid, but the trigonid is higher and the metalophid more variably developed in more derived taxa. The m3 is roughly triangular in shape because of distal expansion in the hypoconid area, with a low posterior ridge coursing lingually and mesially towards the metaconid, usually with only a minor entoconid if any.

POSTCRANIAL

The sciurid postcranial skeleton resembles the general pattern for Eocene ischyromyids (Korth, 1994), but the earliest known squirrel has a more gracile appendicular skeleton (Figure 21.1C1, compare with Figure 21.1C3; Emry and Thorington, 1982). Among extant squirrels, tree and flying squirrels (not known from postcranial material in the North American Tertiary) retain the gracile limb elements whereas terrestrial squirrels (especially the tribe Marmotini) have limbs and digits that are more robust (Figure 21.1C2). The tibia and fibula are never fused. The scaphoid and lunate of the carpus are fused in all taxa known from appropriate postcranial material except *Douglassciurus* (Emry and Thorington, 1982). The scapula bears a prominent, ridgelike subscapular spine (Figure 21.1C4), known otherwise only in the anomalurids (Emry and Thorington, 1982), and the iliac wings of the pelvic girdle are usually strongly curved outward.

SYSTEMATICS

Much work on suprafamily and broad infrafamily relationships of Sciuridae has been based on morphological and molecular evidence from modern taxa (e.g., Bryant, 1945; Moore, 1959; Hight, Goodman, and Prychodko, 1974; Hafner, 1984; Nedbal, Honeycutt, and Schlitter, 1996; Huchon, Catzeflis, and Douzery, 1999; Montgelard *et al.*, 2002). Paleontological studies have contributed in important ways by documenting extinct taxa, exploring relationships among fossil and recent lineages, and revealing stratigraphic patterns in morphology and diversity (e.g., Bryant, 1945; Black, 1963; Emry and Korth, 1996).

SUPRAFAMILIAL

Matthew (1910) noted strong resemblance between sciurids and *Prosciurus*, an enigmatic genus variously placed in Ischyromyidae (e.g., Matthew, 1910; Wilson, 1949a), Sciuridae (Matthew, 1903), or Aplodontidae (Rensberger, 1975; Korth and Emry, 1991; Korth, 1994; see also Chapter 22). Wilson (1949a) supported the relationship between Sciuridae, Aplodontidae (and Mylagaulidae), and Ischyromyidae (especially his subfamilies Prosciurinae and Paramyinae) but did not propose a specific phylogenetic scenario because of insufficient evidence. More recent evidence for Aplodontidae plus Sciuridae includes close similarity of early aplodontid and sciurid fossils (e.g., Vianey-Liaud, 1985; Korth and Emry, 1991), patterns of cranial foramina (Wahlert, 1985), middle ear morphology (Lavocat and Parent, 1985; Meng, 1990), patterns of implantation and placental and embryonic membrane structure (Luckett, 1985), and molecular similarity (Nedbal, Honeycutt, and Schlitter, 1996; Huchon, Catzeflis, and Douzery, 1999; Montgelard *et al.*, 2002). A possible relationship was proposed between Sciuridae and Gliridae (dormice) based on the pattern of cranial arteries (Bugge, 1985); the morphology of the auditory region (Meng, 1990) and mitochondrial DNA sequence data (Montgelard *et al.*, 2002) provide additional support for glirids as the extant sister group of sciurids + aplodontids (see discussion in Chapter 16).

The relationship between aplodontids + sciurids and certain ischyromyids has likewise been supported, although some doubt the hypothesis (e.g., Lavocat and Parent, 1985). Wilson (1949a) reported that *Reithroparamys*, unique among ischyromyids known to him, shared with sciurids ossified auditory bullae. Others have confirmed shared features between reithroparamyines, as presently understood, and aplodontids + sciurids (e.g., Wood, 1962; Meng, 1990), and a reithroparamyine-like ischyromyid is a plausible ancestor for Sciuridae + Aplodontidae (Emry and Korth, 1996).

Rodent classification has not clearly reflected these relationships. Most classifications placed ischyromyids, aplodontids, and sciurids within the suborder Sciuromorpha (e.g., Matthew, 1910; Simpson, 1945; Wilson, 1949a; Wood, 1955) but have positioned these families either in separate superfamilies (Ischyromyoidea, Aplodontoidea, Sciuroidea [Wood, 1955; Korth, 1994]), or allied ischyromyids and aplodontids (Aplodontoidea) but located sciurids in a separate superfamily, Sciuroidea (e.g., Matthew, 1910; Simpson, 1945; Wilson, 1949a). Matthew (1910) constituted Sciuroidea to include Castoridae, Geomyidae, and Heteromyidae along with Sciuridae, but Simpson (1945) removed all families but Sciuridae, a solution followed by most subsequent workers (e.g., Wilson, 1949a; Wood, 1955; Black, 1963; Korth, 1994). Hartenberger (1985) included Sciuridae, Aplodontidae, and the extinct Ischyromyidae and Mylagaulidae within Sciuroidea, appropriately reflecting the close relationship among these families. Some recent molecular systematists have followed this (e.g., Huchon, Catzeflis, and Douzery, 1999; Montgelard *et al.*, 2002).

INFRAFAMILIAL

Early work on Tertiary North American squirrels was descriptive, and conservatively assigned most fossil taxa (e.g., 12 of 13 recognized prior to 1930) to the extant genera *Sciurus, Marmota* (known as *Arctomys*), and *Spermophilus* (known as *Citellus*) (Marsh, 1871; Cope, 1879, 1881, 1883; Douglass, 1901, 1903; Kellogg, 1910; Hay, 1921; Gidley, 1922; Matthew, 1924; Merriam, Stock, and Moody, 1925). Descriptive work has continued unabated to the present (about 70 species of sciurids are currently recognized from the North American Tertiary), but the majority of species are now assigned to extinct genera.

Bryant (1945) made the first attempt to comprehensively assess relationships between fossil and extant Nearctic sciurids. He did not name formal suprageneric taxa but recognized three informal divisions: terrestrial squirrels and chipmunks, tree squirrels, and flying squirrels. Within the first of these divisions, he recognized terrestrial squirrels and chipmunks as separate sections, and further subdivided the terrestrial squirrels into the marmot group and the ground squirrel–prairie dog group. Simpson (1945) included flying squirrels as a subfamily of squirrels, Petauristinae, and grouped

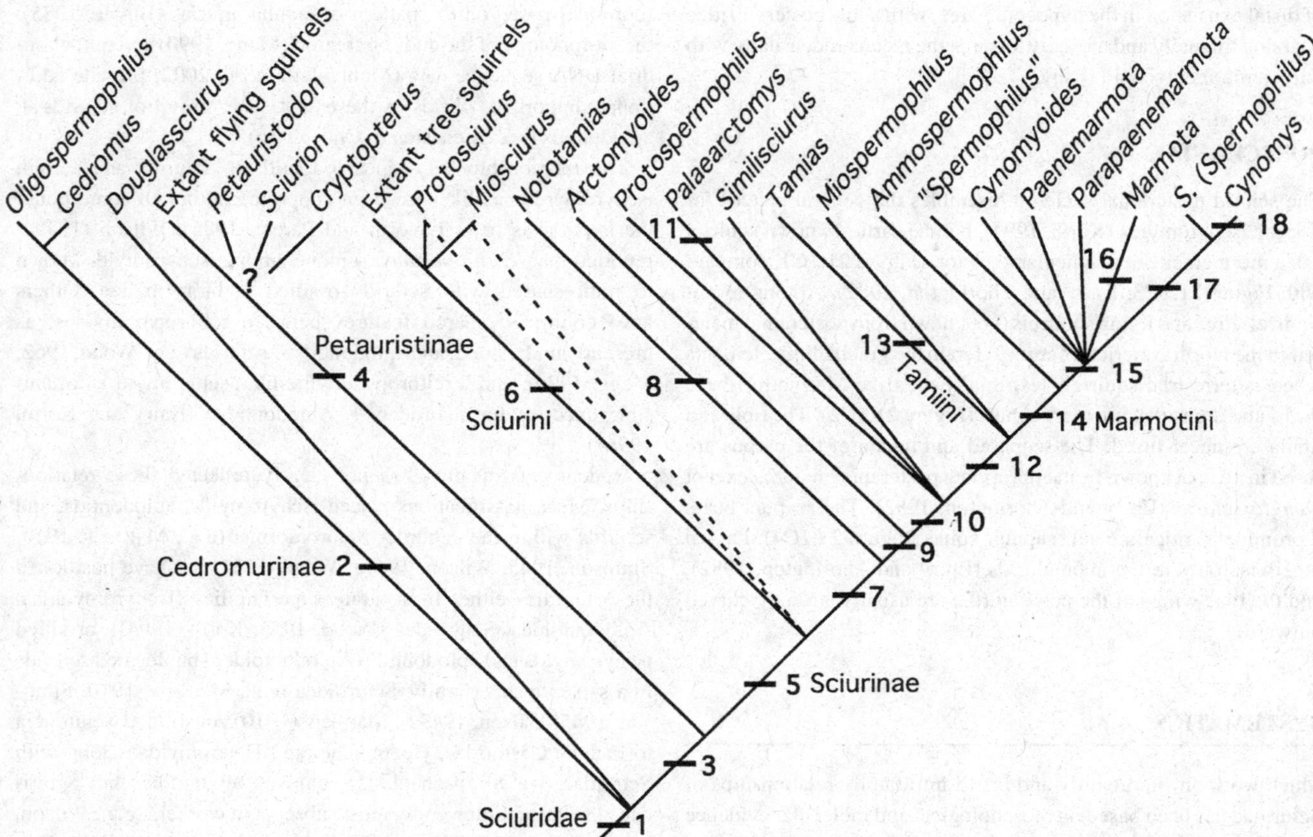

Figure 21.2. Proposed interrelationships within the Sciuridae. Characteristics supporting nodes 1–18 on the cladogram as follows. 1. SCIURIDAE Gray, 1821: elongate pterygoid flanges; stapedial artery enclosed in bony tube across promontorium; basioccipital tilted ventrally; postorbital process with posterior extension; postglenoid foramen enlarged (from Korth and Emry, 1991). 2. CEDROMURINAE Korth and Emry, 1991: zygoma anteriorly broadened and tilted lateral to infraorbital foramen. 3. At least partial sciuromorphy developed. 4. PETAURISTINAE Simpson, 1945: complex character set of wrist associated with gliding (Thorington, 1984). 5. SCIURINAE Baird, 1857: interrelationships within Sciurinae based on preliminary phylogenetic analyses of unpublished dataset (derived from Bryant [1945] and personal observation for Recent taxa; literature and personal observation for extinct taxa); simplified cheek teeth (reduction/loss of hypocone, accessory conules on proto- and metaloph, hypolophid). 6. SCIURINI Burmeister, 1854: angular process does not extend posteriorly as far as condylar process. (Extant North American Sciurini distinguished by other features not observed on fossils.) 7. Plane of alveolar row at or below level of distal end of jaw. 8. *Nototamias* Pratt and Morgan, 1989: mesoconid lost on m1–3. 9. Occlusal outline of M1–2 subquadrate (versus quadrate); diastema more elongate relative to jaw depth. 10. Metalophid incomplete on m2, entoconid relatively distinct but entoconid corner rounded. 11. *Palaearctomys* Douglass, 1903: post–palatal vacuities; deep and strongly arced diastemal region of lower jaw. 12. Masseteric fossa extending to level of p4–m1 contact or more anteriorly, rounded entoconid corner with entoconid less distinct. 13. TAMIINI Black, 1963: small size infraorbital foramen piercing zygomatic plate and not in canal (unknown in fossil material). 14. MARMOTINI Simpson, 1945: masseteric tubercle well developed; P3 of moderate size; mesoconid absent on m1–2. 15. Squamosal root of zygomatic arch projecting more laterally and less anteroventrally (molecular evidence also confirms that *Ammospermophilus* is outside *Spermophilus* [Hafner, 1984]). 16. *Marmota* Blumenbach, 1779: numerous unique characters such as highly flattened skull; paroccipital process extending well below auditory bullae; mesial and distolingual cingula on P3 confluent with each other; p4 protolophid complete but attached low on protoconid and metaconid. 17. Strongly molariform p4; inturned angle of lower jaw; hypsodont cheek teeth. 18. *Cynomys* Rafinesque, 1817: extreme hypsodonty (by sciurid standards); accessory lophule characteristically present behind paracone on M1–3; complete protolophid on p4.

other North American genera into the tribes Sciurini (tree squirrels) and Marmotini (terrestrial squirrels), both in the subfamily Sciurinae. Black (1963) followed Simpson (1945) in most respects, but erected the tribe Tamiini for chipmunks.

Korth and Emry (1991) erected a third sciurid subfamily, Cedromurinae, to incorporate the problematic genera *Cedromus* (interpreted as sciurid based on well-preserved cranial evidence) and *Oligospermophilus*. Korth's (1994) survey of Tertiary squirrels followed the suprageneric classification of Black (1963) with the addition of Cedromurinae (Korth and Emry, 1991). Emry and Korth (1996) did not discuss Cedromurinae in a treatment of early sciurids.

Relationships among and within these major sciurid clades have been difficult to establish because of the fragmentary fossil material and the conservative morphology of most sciurids. A tentative phylogenetic model (Figure 21.2) exhibits a basal polytomy with Cedromurinae, *Douglassciurus*, and the lineage leading to the crown clade Sciurinae + Petauristinae. Cedromurines, if sciurid as accepted here (after Korth and Emry, 1991), clearly originated near the base of the sciurid clade. *Douglassciurus* has been interpreted elsewhere as an early tree squirrel (tribe Sciurini, subfamily Sciurinae; Emry and Korth, 1996), primarily based on detailed postcranial similarities with modern tree squirrels (Emry and Thorington, 1982); as closely related to the cedromurine *Oligospermophilus douglassi* (Korth, 1981); or as not a sciurid at all (e.g., Wood, 1937, 1980; Vianey-Liaud, 1985). Placement outside Sciurinae + Petauristinae, as suggested here, implies that postcranial features shared with tree squirrels are primitive for the family.

The systematic position of petauristines (flying squirrels) has been controversial. Petauristines have been treated as a separate family (e.g., Pocock, 1923; Bruijn and Ünay, 1989) or as a subfamily within Sciuridae (e.g., Bryant, 1945; Simpson, 1945; Black, 1963; Korth, 1994), despite persistent worries about monophyly of the group. Black (1963) suggested independent origins for Old and New World Recent flying squirrels from within tree squirrels (tribe Sciurini); diphyletic derivation from within tree squirrels was likewise supported with immunological evidence (Hight, Goodman, and Prychodko, 1974). Engesser (1979) doubted James' (1963) assignment of North American Tertiary flying squirrels to the Old World genus *Sciuropterus*. He erected *Petauristodon* to accommodate North American Tertiary fossils, and argued that Old and New World Tertiary flying squirrels may have evolved independently from early Tertiary rodents, with North American *Petauristodon* plausibly derivable from *Prosciurus* and possibly not a sciurid at all.

Based on this uncertainty, Skwara (1986) described *Sciurion* only as a "flying squirrel" and did not allocate it to subfamily. In contrast, Thorington (1984) argued strongly for monophyly of at least Recent flying squirrels based on a detailed character complex of the wrist shared across Old and New World Recent flying squirrels and not shared by other gliders, but these features are unknown for Tertiary fossils in North America. Pratt and Morgan (1989) extended this argument by showing that some dental features alleged to distinguish Old and New World Tertiary species were shared between members of both assemblages. Assignment of Tertiary "flying squirrels" from North America (*Petauristodon, Sciurion, Cryptopterus*) to Petauristinae is tentative.

Three tribes are recognized within North American Sciurinae (e.g., Black, 1963; Korth, 1994): Sciurini (tree squirrels), Tamiini (chipmunks), and Marmotini (ground-dwelling squirrels). Evidence equivocally supports inclusion of *Miosciurus* and *Protosciurus* within Sciurini, with tree squirrels evolving before chipmunks and Marmotini (Figure 21.2), in line with phylogenetic models based on molecular evidence (e.g., Hafner, 1984; Giboulet *et al.*, 1997), but contra Black (1963), who argued that chipmunks were the primitive stem group in Sciurinae. The accepted composition of Tamiini (*Nototamias, Tamias*; Pratt and Morgan, 1989; Korth, 1994) and Marmotini (all the remaining Tertiary and extant North American terrestrial squirrels; Black, 1963; Korth, 1994) falls apart when examined phylogenetically (Figure 21.2). To reflect these relationships accurately, *Nototamias* and other basal "terrestrial squirrels" are here left unassigned to tribe, Tamiini is restricted to *Tamias* and Marmotini is restricted to the crown group above *Tamias* (*Ammospermophilus* and above, and possibly *Miospermophilus*).

The relationships among early "terrestrial squirrels" are poorly resolved (Figure 21.2). The chipmunk-like *Nototamias* is the most primitive, followed by the large, vaguely marmot-like *Arctomyoides*. The enigmatic *Similisciurus*, initially considered a tree squirrel (Stevens, 1977), is placed here in an unresolved polytomy with *Protospermophilus* and *Palaearctomys*. The latter two genera are not closely related to *Marmota* (contra Black, 1963). *Miospermophilus* clearly belongs in the crown clade of terrestrial Sciurinae, probably as the earliest member of Marmotini although it could be allied with *Tamias* (Wilson, 1960; Black, 1963). The most primitive extant genus of Marmotini is *Ammospermophilus*, consistent with molecular evidence for its early divergence relative to other ground squirrels (e.g., Hafner, 1984).

Systematics of Marmotini above *Ammospermophilus* are confused at best. *Spermophilus* is a large, unwieldy genus with numerous extant subgenera (Hall, 1981) that are difficult to recognize based on derived dental evidence with the exception of the morphologically advanced *S.* (*Spermophilus*) (e.g., Goodwin and Hayes, 1994). As presently constituted, the genus is almost certainly paraphyletic relative to both *Marmota* and *Cynomys*, with the latter sharing a recent common ancestor with the subgenus *S.* (*Spermophilus*) (e.g., Bryant, 1945; Black, 1963; Goodwin and Hayes, 1994). Mitochondrial cytochrome *b* data provide additional support for the evolution of these genera from within *Spermophilus* (*Marmota* [Thomas and Martin, 1993], *Cynomys* [Harrison *et al.*, 2003]). These observations are difficult to harmonize with Korth's (1996) scenario that *Spermophilus, Marmota*, and *Cynomys* represent different subtribes (Spermophilina, Marmotina, Cynomyina) separate since at least the Barstovian, based in part on remarkable similarity between Miocene *Cynomyoides* and extant *Cynomys. Cynomyoides* may also be related to *Paenemarmota* and *Parapaenemarmota*, large squirrels from the Hemphillian and Blancan previously allied with *Marmota* (e.g., Black, 1963; Zakrzewski, 1969) or advanced *Spermophilus* (e.g., Repenning, 1962). In the hypothesis tentatively suggested here, these three extinct genera likewise arose from within *Spermophilus*.

INCLUDED GENERA OF SCIURIDAE GRAY, 1821

Black (1963) provided the last major review of Tertiary North American Sciuridae, although many taxa have been described since his work and other taxa are now better known. A comprehensive review is again needed, especially of the diverse Marmotini. Traditional taxa are retained here, even if certainly paraphyletic (e.g., *Spermophilus*), until such a revision can be accomplished. The sequence of taxa below follows presumed phylogenetic sequence (Figure 21.2), and thus differs considerably from previous classifications (e.g., Black, 1963; Korth, 1994).

The locality numbers listed for each genus refer to the list of unified localities in Appendix I. The locality numbers may be listed in several ways. The acronyms for museum collections are listed in Appendix III.

Parentheses around the locality (e.g., [CP101]) mean that the taxon in question at that locality is cited as an "aff." or "cf." the taxon in question. Parentheses are usually used for individual species, thus implying that the genus is firmly known from the locality, but the actual species identification may be questionable. Question marks in front of the locality (e.g., ?CP101) mean that the taxon is questionably known from the locality, thus implying some doubt that the taxon is actually present at that locality, either at the genus or the species level. An asterisk (*) indicates the type locality.

PRIMITIVE SCIURIDS

Douglassciurus Emry and Korth, 2001 (including *Douglassia*; *Sciurus*, in part; *Prosciurus*, in part; *Cedromus*, in part; *Protosciurus*, in part)

Type species: *Douglassciurus jeffersoni* (originally described as *Sciurus jeffersoni*, Douglass, 1901) (including *Prosciurus jeffersoni* Osborn and Matthew, 1909; ?*Prosciurus jeffersoni* Wood, 1937; *Cedromus jeffersoni* Wood, 1962, 1980; ?*Protosciurus jeffersoni* Black, 1965; *Protosciurus jeffersoni* Storer, 1978; *Douglassia jeffersoni* Emry and Korth, 1996).

Type specimen: CM 736, middle Chadronian of Montana.

Characteristics: Skull protrogomorphous and bearing a relatively short diastema between incisors and cheek teeth; upper cheek teeth complex when unworn, with a doubled metaconule, an enlarged hypocone (for sciurids), and an accessory lophule extending from the protocone into the anterior valley (protocone crest of Emry and Korth, 1996); mandible robust with masseteric fossa below the posterior half of m1; lower cheek teeth bearing a large entoconid (for sciurids) and a distinct hypolophid. Postcranial skeleton closely similar to that of extant tree squirrels, genus *Sciurus* (modified from Emry and Korth, 1996).

Average length of m2: 2.9 mm.

Included species: *D. jeffersoni* only (known from localities CP39B–F, CP98B, NP10B, [Bii], B2, NP24C*).

Comments: *D. jeffersoni* is both the earliest and perhaps the most completely known Tertiary squirrel from North America, thanks to an excellent comparative analysis of a well-preserved skeleton (Emry and Thorington, 1982). The complex taxonomic history of *D. jeffersoni* (see account of synonymy above) reflects conflicting interpretations of primitive craniodental and essentially modern postcranial morphology (see Emry and Thorington, 1982; Emry and Korth, 1996).

CEDROMURINAE KORTH AND EMRY, 1991

Characteristics: Low, broad skull, short rostrum, and unique zygomasseteric structure – the masseteric scar on the mandible extending anteriorly beneath the posterior one-half of m1, as in other primitive sciurids, but the zygoma broadening and tilting anterodorsally lateral to the infraorbital foramen; with upper molars a low ridge ("ectoloph") from paracone to mesostyle; ectoloph–mesostyle complex partially or completely blocking central valley buccally; metaloph constricted at its contact with the protocone; hypocone commonly evident on M1–2; lower molars with variably developed hypolophid extending into the talonid basin from the entoconid and an enclosed trigonid basin; mesoconid well developed (modified from Korth and Emry, 1991).

Cedromus Wilson, 1949b

Type species: *Cedromus wardi* Wilson, 1949b.

Type specimen: UCM 19808, Orellan of Colorado.

Characteristics: Cedromurines differ from *Oligospermophilus* in larger size, smaller hypocone, and larger metaconules on upper molars; ectoloph connecting with a distally placed mesostyle (adjacent to the metacone), thus fully blocking the buccal end of the central valley; lower molars with posterior cingulum confluent with the entoconid (not separated by a notch), and mesostylids having greater buccal expansion (modified from Korth and Emry, 1991).

Average length of m2: 2.3–2.7 mm.

Included species: *C. wardi* (known from localities CP46, CP68C*, CP84A, CP99A); *C. wilsoni* Korth and Emry, 1991 (localities CP41B*, [CP99A]).

Cedromus sp. is also known from localities CP68C, CP99B, ?NP10C2.

Comments: Wilson (1949b) interpreted *Cedromus* as an ischyromyid but left open the possibility that it was a prosciurid evolving towards squirrels, or perhaps a primitive squirrel (see Korth and Emry [1991] for the taxonomic history of *Cedromus*). He also noted similarity with ?*Prosciurus jeffersoni* (now *Douglassciurus*). *Cedromus* was most recently allied with sciurids on the basis of well-preserved skull material (Korth and Emry, 1991).

Oligospermophilus Korth, 1987 (including *Protosciurus*, in part)

Type species: *Oligospermophilus douglassi* (Korth, 1981) (originally described as ?*Protosciurus douglassi*).

Type specimen: CM 38659, Orellan of Nebraska.

Characteristics: Cedromurines smaller than *Cedromus*; upper molars with smaller metaconule and enlarged hypocone, tending to square off the occlusal outline; lower molars distinguished from *Cedromus* by presence of narrow valley separating the posterolophid from the entoconid (modified from Korth and Emry, 1991).

Average length of m2: 2.1 mm.

Included species: *O. douglassi* only (known from localities CP98C, CP99A*).

Oligospermophilus sp. is also known from localities CP41B, CP42D, CP85C.

Comments: *Oligospermophilus* was originally tentatively allied with Marmotini (Korth, 1987) but subsequently recognized as a cedromurine (Korth and Emry, 1991). In the initial description of the type species, Korth (1981) treated it as congeneric with what is now *Douglassciurus*.

SCIURINAE BAIRD, 1857

Characteristics: Simplified tooth pattern with reduction or loss of the hypocone on M1–2 and hypolophid on m1–2; protoloph and metaloph on uppers tending to be less conulate than in *Douglassciurus* (a strong metaconule is present in some, but not the multiple conules observed in *Douglassciurus* and some "flying squirrels"). Tooth shape and the presence and development of mesostyles, mesostylids, and mesoconids highly variable; all members lack gliding membranes; at least incipiently sciuromorphous (all extant North American taxa are completely sciuromorphous).

SCIURINI BURMEISTER, 1854

Characteristics: Skull broad interorbitally; extant North American taxa with more strongly convex dorsal skull profile than terrestrial squirrels, but extinct genera (*Protosciurus, Miosciurus*) with flattened profile; zygomatic plate angling more steeply (greater than ~60° [Black, 1963]) relative to the basicranial axis than in terrestrial squirrels; diastemal region of lower jaw characteristically short, anterior end of diastema below plane of the alveolar border. Cheek teeth brachydont with low lophs/lophids; M1 and M2 roughly quadrate shape with mesiodistally expanded protocone and complete lophs with the metaconule reduced or absent; p4–m2 typically exhibit distinct entoconid at an angled linguodistal corner; metalophid usually complete on m1–2, fully closing a low trigonid basin; appendicular skeleton is gracile with long, slender elements, retaining the primitive morphology of *Douglassciurus* (modified from Black, 1963)

Protosciurus Black, 1963 (including *Sciurus*, in part)

Type species: *Protosciurus condoni* Black, 1963.

Type specimen: UO F-5171, early Arikareean of Oregon.

Characteristics: Skull not fully sciuromorphous but advanced over the condition in *Douglassciurus*; masseter inserting on masseteric tubercle, positioned ventrolateral to infraorbital foramen, and on lateral margin of the zygoma but not passing over the infraorbital foramen onto the rostrum; masseteric fossa on lower jaw characteristically falling beneath m1, diastemal depression deep. Upper cheek teeth typically with reduced conules and hypocone (both conules and hypocone better developed in the earliest species, *P. mengi*; see Korth, 1987); lower cheek teeth likewise low crowned with prominent entoconid, broad buccal valley, distinct mesoconid, and (usually) a constriction at the junction of the protolophid and protoconid (modified from Black, 1963).

Average length of m2: 2.8–3.6 mm.

Included species: *P. condoni* (known from locality PN6D3*); *P. mengi* Black, 1963 (locality CP99A*); *P. rachelae* Black, 1963 (locality PN6* [exact placement not known]); *P. tecuyensis* (Bryant, 1945, including *Sciurus tecuyensis*) (locality CC13*).

Protosciurus sp. is also known from localities GC3B, ?GC4C, GC8B, CC12, CP85A, CP86B, CP103A, NP10CC.

Comments: Black (1963) recognized four species of *Protosciurus*, two of which he placed with confidence (*P. condoni* and *P. mengi*) and the others as a matter of convenience (*P. tecuyensis* and *P. rachelae*). Only *P. mengi* is known from more than a couple of specimens (Korth, 1987), thus variability within species is essentially undocumented. None of the named species with known stratigraphic location (but see *P. rachelae* above) occur later than the early Arikareean (late Oligocene), but records of *Protosciurus* sp. have been questionably reported as late as early Barstovian (middle Miocene).

Miosciurus Black, 1963 (including *Sciurus*, in part; *Prosciurus*, in part)

Type species: *Miosciurus ballovianus* (Cope, 1881) (originally described as *Sciurus ballovianus*, also Bryant, 1945) (including *Prosciurus ballovianus* Matthew, 1909).

Type specimen: AMNH 6901, early Arikareean of Oregon.

Characteristics: Approximately size of a chipmunk; skull not fully sciuromorphous, resembling *Protosciurus* in zygomasseteric form; M1 and M2 exhibiting lesser mesiodistal expansion of the protocone than in *Protosciurus*; diastemal region of lower jaw with shallow depression (unlike deep depression characteristic of *Protosciurus* and extant *Sciurus*); masseteric fossa beneath m1; m1 and m2 with expanded anteroconid on the protolophid having a clear constriction at its contact with the protoconid (modified from Black, 1963).

Average length of m2: 1.6 mm.

Included species: *M. ballovianus* only, known from locality PN6C2* only.

Comments: Known only from the type specimen; *M. ballovianus* documents presence of a sciurid substantially smaller than *Protosciurus* by the late Oligocene. Black (1963) considered *Miosciurus* intermediate between chipmunks and tree squirrels, perhaps close to the divergence of arboreal and terrestrial lineages.

Sciurus Linnaeus, 1758

Type species: *Sciurus vulgaris* Linnaeus, 1758, the extant Eurasian red squirrel.

Type specimen: None designated.

Characteristics: Small- to moderate-sized squirrels; skull sciuromorphous, strongly convex dorsally, broad interorbitally with slender postorbital processes, lacking pronounced sagittal or occipital crests. Upper cheek teeth with relatively simple lophs, lacking the conules characteristic of *Protosciurus*; mesostyle of upper cheek teeth the anteroconid, mesostylid, and metastylid of lower cheek teeth variably developed but tending to be less pronounced than in *Protosciurus* (Emry, Korth, and Bell, 2005).

Average length of m2: 1.4 mm (for Tertiary taxon; Emry, Korth, and Bell, 2005). Extant North American species much larger (~ 2.8 mm in *Sciurus carolinensis* [Eastern gray squirrel] and *Sciurus niger* [Eastern fox squirrel]).

Included species: *S. olsoni* Emry, Korth, and Bell (2005) only, known from locality NB29A only. *Sciurus* is also known from Pleistocene sites ranging from Pennsylvania to Florida in eastern North America, and south and west through Texas to northern Mexico and New Mexico (Kurtén and Anderson, 1980).

Comments: The single Tertiary fossil assigned to *Sciurus* is substantially smaller than extant members of the genus. In addition, it has a slightly more-prominent anteroconid on p4, and somewhat more-developed mesostylids and metastylids on lower cheek teeth, than is typical of modern species. In these respects, it may be intermediate between *Protosciurus* and *Sciurus* (Emry, Korth, and Bell, 2005).

This important record helps to fill a gap between early Tertiary and extant sciurines from North America and significantly extends the known North American range of the genus *Sciurus*, otherwise known only from the Pleistocene and Holocene. The virtual absence of tree squirrels between the Hemingfordian and Pleistocene may be a sampling artifact owing to underrepresentation of their preferred forest habitat in fossil sites and perhaps low original population densities in these habitats (Emry, Korth, and Bell, 2005). The genus *Sciurus* is also known from the Pleistocene and Recent of Europe and Asia, and the Recent of South America (McKenna and Bell, 1997).

Indeterminate sciurinines

Indeterminant sciurines are also known from localities PN6, PN8A. Black (1963) reported *Sciurus* from the early Barstovian (locality PN8A), but his description suggests a taxon rather unlike the modern genus. This fossil and another indeterminate find reported by Black (1963) are listed here.

BASAL "TERRESTRIAL SQUIRRELS"

Placement of taxa as basal "terrestrial squirrels" differs from previous classifications (see p. 359 for discussion and Figure 21.2).

Nototamias Pratt and Morgan, 1989 (including *Tamias*, in part; *Eutamias*, in part)

Type species: *Nototamias hulberti* Pratt and Morgan, 1989.

Type specimen: UF 3873, early Hemingfordian of Florida.

Characteristics: Chipmunk-sized squirrels exhibiting a characteristically short, shallow diastemal region of the lower jaw with the mental foramen positioned beneath the anterior root of the p4; masseteric fossa terminating beneath the anterior part of m1; upper cheek teeth not typically separable from those of *Tamias* (true chipmunks), but lower cheek teeth with fused anterior and posterior roots and no mesoconid; m1 and m2 with more reduced (or absent) proximobuccal groove at the junction of the protolophid and protoconid relative to *Tamias* (modified from Pratt and Morgon, 1989).

Average length of m2: 1.2–1.5 mm.

Included species: *N. hulberti* (known from locality GC8D*); *N. quadratus* Korth, 1992 (localities CP103A*, NP10CC); *N. ateles* (Hall, 1930) (including *Eutamias ateles, Tamias* [*Neotamias*] *ateles* Bryant, 1945, *Tamias ateles* Black, 1963) (localities [CC17G, H], NB6B*, C, E, NB23B, C, [CP45E], [CP114A]).

Nototamias sp. is also known from localities GC4DD, CC9IVC, CC19, [CP54B], CP71, CP85C, [CP86D].

Comments: Pratt and Morgan (1989) regarded *Nototamias* as a monophyletic sister taxon to chipmunks of the genera *Tamias* and *Eutamias* (now included within a single genus, *Tamias*; Wilson and Reeder, 1993) and included *N. ateles* within the genus. They diagnosed *Nototamias* on presumably derived characters, including the fusion of roots on lower cheek teeth. Sutton and Korth (1995) returned *N. ateles* to *Tamias* on the argument that the possession of unfused roots on lower cheek teeth is derived for sciurids, with *N. ateles* having unfused roots. However, *N. ateles* exhibits roots that are mostly fused (Pratt and Morgan, 1989) and shares with *Nototamias* other features that are presumably derived, including the absence of the mesoconid on p4–m3 and a shortened diastema. I follow Pratt and Morgan (1989) and leave the species in *Nototamias*. However, the genus does not appear to be closely related to chipmunks when all Tertiary genera are included in phylogenetic reconstruction (Figure 21.2), and it is here treated as probably near the base of the clade, eventually giving rise to chipmunks and ground squirrels.

The problem of assigning small, chipmunk-like isolated cheek teeth, especially uppers (Pratt and Morgan, 1989) will be treated in the account of *Tamias* (p. 364).

Arctomyoides Bryant, 1945 (including *Sciurus*, in part)

Type species: *Arctomyoides arctomyoides* (Douglass, 1903) (originally described as *Sciurus arctomyoides*).

Type specimen: CM 741, probably early or middle Miocene of Montana.

Characteristics: Marmot-sized squirrel with robust lower jaw beneath the cheek teeth and a long, shallow diastemal depression; upper incisors with median groove and fine striations, like *Palaearctomys*; lower incisors mediolaterally compressed, finely striated, but lacking median groove; M1 with broad protocone and consequent subquadrate form and low but complete lophs; m1–2 likewise roughly quadrate in outline, bearing prominent entoconid and mesoconid, but weak metalophids incompletely closing the trigonid basin of each tooth; m3 with prominent, crenulated posterolophid (modified from Black, 1963).

Average length of m2: 4.20 mm.

Included species: *Arctomyoides arctomyoides* only, known from locality NP41* only (exact horizon unknown).

Arctomyoides sp. is also known from locality PN7.

Comments: Known from a single specimen, *Arctomyoides arctomyoides* represents a distinctive, large sciurid. It is probably early or middle Miocene in age, although its exact age remains uncertain. The type was reported only from the Lower Madison Valley Beds (Douglass, 1903) but with inadequate stratigraphic data (see Sutton and Korth, 1995). It does not appear to have any close affinity with later marmots (contra Bryant, 1945).

Palaearctomys Douglass, 1903

Type species: *Palaearctomys montanus* Douglass, 1903 (including *Palaearctomys macrorhinus* Douglass, 1903).

Type specimen: CM 740, probably early or middle Miocene of Montana.

Characteristics: Distinctive, marmot-sized squirrel with broad, deep rostrum; zygomatic plate failing to reach dorsal surface of skull; infraorbital foramen slitlike with masseteric tubercle set well ventral to it; palate broad; zygomatic arch more twisted than in Sciurini but less than in Marmotini (in which the primitive medial surface faces dorsomedially); most distinctive cranial feature a pair of deep postpalatal vacuities, otherwise unknown in Sciuridae; lower jaw deep, massive, and strongly curved in the diastemal region, which housed a large incisor. Upper incisor with single median groove and fine striations; lower incisors strongly compressed, flattened medially and laterally, and with finely striated enamel; M1 and M2 with low, complete lophs lacking conules, a small parastyle, and a large mesostyle; M3 lacking metaloph or metaconule; lower molars with distinct entoconid and an incomplete metalophid, leaving the trigonid basin open distally (modified from Black, 1963).

Average length of m2: 3.60 mm.

Included species: *Palaearctomys montanus* only, known from locality NP41* only (exact horizon unknown).

Comments: Known from only two specimens (initially described as separate species; Douglass, 1903), *Palaearctomys* represents one of the most distinctive squirrels from the North American Tertiary with features found in no other North American member of the family (e.g., postpalatal vacuities; deep and strongly curved diastemal region of the lower jaw). Unfortunately, *Palaearctomys* shares the same problem of stratigraphic placement as does *Arctomyoides* (see preceding account of that genus). It is unlikely to be related to marmots (contrary to the opinion of Black [1963]).

Similisciurus Stevens, 1977

Type species: *Similisciurus maxwelli* Stevens, 1977.

Type specimen: TMM 40635-90, early late Arikareean of Texas.

Characteristics: Lower jaw deep with the masseteric fossa, terminating beneath m1; lower cheek teeth much more hypsodont and with narrower buccal valleys than in other Arikareean or earlier squirrels; m1 and m2 longer than wide, with high, narrow trigonid, a distinct groove separating swollen protolophid from metaconid, a distinct entoconid, and possibly a trace of a hypolophid; talonid basin deep and pinched centrally; m3 elongate, lacking a distinct entoconid (from Stevens, 1977).

Average length of m2: 2.9 mm.

Included species: *S. maxwelli* only, known from locality SB46* only.

Comments: Known from a single mandible with m1–3 and an isolated p4, this enigmatic taxon was overlooked in subsequent reviews of Sciuridae (e.g., Hafner, 1984; Korth, 1994). It has a primitive position of the masseteric fossa and was judged more similar to *Protosciurus* than to any other early squirrel (Stevens, 1977), with greatest similarity to the Orellan *P. mengi*. Stevens (1977) also noted similarity to *Cedromus* in the morphology of the masseteric fossa, the possible presence of hypolophids (also shared with *Douglassciurus*), and prominent entoconids. However, the hypsodonty of the cheek teeth is extreme by Miocene standards. It appears to be one of the basal "terrestrial squirrels" rather than an aberrant member of Sciurini (Figure 21.2).

Protospermophilus Gazin, 1930 (including *Sciurus*, in part; *Citellus*, in part; *Prosciurus* in part; *Arctomyoides*, in part)

Type species: *Protospermophilus quatalensis* (Gazin, 1930) (originally described as *Citellus* (*Protospermophilus*) *quatalensis*) (including *Sciurus venturus* Bryant, 1945).

Type specimen: LACM (CIT) 30, probably Clarendonian of California.

Characteristics: Medium-sized squirrels with skull roof flattened to gently convex; masseteric fossa extending onto the rostrum well beyond the infraorbital foramen, but dorsal limit of zygomatic plate on side of rostrum and not reaching its dorsal border; infraorbital foramen small, typically slitlike, and bearing a ventrally placed masseteric tubercle; distinct pits immediately behind each upper incisor presumably for insertion of muscles associated with a cheek pouch (as in extant Marmotini); masseteric fossa of lower jaw terminating anteriorly beneath the anterior one-half of m1; cheek teeth low crowned and robust; upper cheek teeth with a metaloph constricted at its junction with the protocone; lingual end of the posterior cingulum on M1 and M2 characteristically expanded adjacent to the protocone, sometimes forming an incipient hypocone and distally expanding the lingual margin of the tooth; lower cheek teeth variable in form but typically exhibiting distinct entoconid and slightly rugose talonid floor (lost with wear) (modified from Black, 1963).

Average length of m2: 2.3–3.1 mm.

Included species: *P. quatalensis* (known from localities GC4C, CC17* [exact horizon of type uncertain, but probably Clarendonian; James, 1963], CC17G, H, [CC26B], [CP116B]); *P. vortmani* (Cope, 1879) (including *Sciurus vortmani, Prosciurus wortmani* Matthew, 1909, *Sciurus vortmani* Bryant, 1945) (localities PN6* [exact sublocality uncertain], PN6D1, PN6D2, PN6H); *P. kelloggi* Black, 1963 (localities CP54B*, NP10E); *P. angusticeps* (Matthew and Mook, 1933) (including *Sciurus angusticeps*) (localities NB17, NP34C*); *P. oregonensis* (Downs, 1956) (including *Arctomyoides oregonensis*) (locality PN7*); *P. malheurensis* (Gazin, 1932) (including *Sciurus malheurensis*) (localities PN8A, B*).

Protospermophilus sp. is known from localities GC5B, [GC9C], NB17, CP71, CP90A, CP114B, NP10CC, E.

Comments: Most species in *Protospermophilus* were initially assigned to other genera, with the current generic content assembled by Bryant (1945) and Black (1963). The genus does not appear to be ancestral or even closely related to *Marmota* (contrary to the opinion of Black [1963]).

TAMIINI BLACK, 1963

Characteristics: Same as for the one included genus, *Tamias*

Tamias Illiger, 1811 (including *Sciurus*, in part; *Eutamias*, in part)

Type species: *Tamias striatus* (Linnaeus, 1758) (originally described as *Sciurus striatus*), the extant Eastern chipmunk.

Type specimen: None designated; based on extant specimen in collection of King Frederic Adolphus of Sweden (Wilson and Reeder, 1993).

Characteristics (sources: Bryant, 1945; Black, 1963; Pratt and Morgan, 1989): Small sciurids with a fully sciuromorphous zygomasseteric structure: infraorbital foramen in extant members a simple hole in the zygomatic plate, lacking the bony canal typical of other sciurids (this feature is not preserved on any Tertiary material); lower jaw bearing a relatively long, slender diastema with masseteric fossa terminating anteriorly beneath p4, and mental foramen positioned anterior to p4 alveolus. Upper molars subquadrate with low, complete lophs, metaloph typically only slightly constricted at its contact with the protocone and showing only a slight metaconule swelling. Lower cheek teeth with a mesoconid and a proximobuccal groove at the junction of the protolophid and protoconid (may be lost with wear); entoconid not usually distinct, but distolingual corner of the tooth more angular than typical of Marmotini; m2 and m3 characteristically with an incomplete metalophid and, consequently, a distally open trigonid valley. Limb elements intermediately proportioned between those of tree and ground squirrels.

Average length of m2: 1.2–1.7 mm.

Included Tertiary fossil species: *Tamias malloryi* (Martin, 1998) (including *Eutamias malloryi*) only, known from locality PN14* only.

Tamias sp. is also known from Tertiary localities NB13B, PN10, PN11, PN14. *Tamias* is widespread across North America in the Pleistocene, with fossils known as far east as Maryland and Virginia, as far south as Texas, and as far west as California (Kurtén and Anderson, 1980).

Comments: North American chipmunks were previously grouped in two genera, *Tamias* (including the eastern chipmunk, *T. striatus*) and *Eutamias* (24 species, mostly of western North America; Wilson and Reeder, 1993), which can be distinguished with adequate material (*Eutamias* exhibits grooved incisors and a P3, both absent in *Tamias* [Pratt and Morgan, 1989]). Both names have been used in identification of small, chipmunk-like sciurids of the Tertiary (reviewed in Martin, 1998), although *T. malloryi* (Martin, 1998) and the few other late Tertiary specimens that can be assigned with confidence represent the *Eutamias* morphology. However, recent taxonomic revision includes all extant species of chipmunks in *Tamias* (Wilson and Reeder, 1993). This convention is here applied to fossil species as well. The genus *Tamias* is also known from the Miocene to Recent of Europe and Asia (McKenna and Bell, 1997).

The status of many Tertiary "chipmunks" is confused at best and is likely to remain so lacking better material. Much of the isolated material from the Arikareean to Barstovian (early to middle Miocene) may be assignable to *Nototamias*. However, the upper teeth are not diagnostic; most descriptions of the lowers do not include characteristics that clearly distinguish the genera (e.g., root fusion; Pratt and Morgan, 1989), and some of the putative distinguishing characters appear mosaically in fossils (e.g., Sutton and Korth [1995] reported material lacking mesoconids [as in *Nototamias*] but with unfused roots [as in *Tamias*]). The jaws of *Nototamias* and *Tamias* are distinctly different, but the degree of dental similarity seems surprising given the apparent phylogenetic separation of the two (Figure 21.2).

Pratt and Morgan recommended that unassignable Tertiary "chipmunks" be identified only as "*Tamias* sp." (indicating a chipmunk-like form, maybe *Tamias*, maybe *Nototamias*); this practice is followed here. "*Tamias* sp." is known from localities GC3B, GC4C, CC21C, CC22A, CP56, CP89, CP114B, CP128C, NP10E, NP11, NP40A, B, NP41B, PN8B.

MARMOTINI SIMPSON, 1945

Characteristics: Skull roof varying from flattened to moderately convex with a relatively narrow interorbital region, not as strongly convex nor as broadened interorbitally as in extant Sciurini; small to well-developed pits behind each upper incisor for origins of muscles associated with a cheek pouch (also known in some basal "terrestrial squirrels"); infraorbital foramen oval to triangular with well-developed masseteric tubercle positioned ventrally or ventrolaterally to it, never slitlike as is typical of Sciurini and some early "terrestrial squirrels" (e.g., *Palaearctomys*); zygomatic arch (not often preserved on fossils) twisted such that the primitive medial surface faces mediodorsally; diastemal region of the lower jaw characteristically long, with a shallow diastemal depression in many forms (deepened in some derived taxa); anterior end of the lower jaw typically at or above the plane of the alveolar border. Cheek teeth more high crowned than in Sciurini with more pronounced lophs/lophids; upper molars subquadrate to triangular, each typically bearing a distinct metaconule, variably constricted at its contact with the protocone; lower p4–m2 each bearing an indistinct entoconid incorporated into the posterolophid, the distolingual (entoconid) corner usually rounded; m1 and m2 rhomboidal with lingual half of tooth more constricted mesiodistally than the buccal half (modified from Black, 1963).

Miospermophilus Black, 1963 (including *Palaearctomys*, in part; *Sciurus*, in part)

Type species: *Miospermophilus bryanti* (Wilson, 1960) (originally described as *Palaearctomys*? *bryanti*), (including *Sciurus* sp. Galbreath, 1953).

Type specimen: KU 10149, early Hemingfordian of Colorado.

Characteristics: Small squirrel with brachydont teeth; skull unknown; lower jaw with long diastemal region and shallow diastemal depression; masseteric fossa terminating beneath posterior half of p4. Incisors with finely striated enamel; upper M1–2 with complete but slightly constricted metaloph, generally a small metaconule, and a somewhat mesiodistally compressed protocone, resulting in a more triangular shape than is characteristic of other early Miocene squirrels; distobuccal corner of M3 expanded distally; p4 much more compressed buccolingually across the trigonid than talonid, with adjacent protoconid and metaconid; lower molars clearly rhomboidal with the lingual half narrower owing to anterior displacement of a small, inconspicuous entoconid; junction between protolophid and protoconid on m1–2 commonly marked by a distinct notch; metalophid sometimes enclosing a low, small trigonid pit (modified from Black, 1963).

Average length of m2: 1.75–1.90 mm.

Included species: *M. bryanti* (known from locality CP71*); *M. wyomingensis* Black, 1963 (localities [CC22A], NB17, CP54B*, NP10E); ?*M. lavertyi* Dalquest, Baskin, and Schultz, 1996 (this query is the author's) (locality SP1A*).

Miospermophilus sp. is also known from localities [GC8D], ?CC9IIIA, CC9IVD, CC19, [NB4], NB6B, C, E, NB17, [CP86D], CP114A.

Comments: *Miospermophilus* shows resemblance both to chipmunks and to ground squirrels (Wilson, 1960; Black, 1963; Dalquest, Baskin, and Schultz, 1996) but does not seem to have any special relationship with *Arctomyoides* or *Palaearctomys* (Black, 1963; contra Wilson, 1960). The genus is placed here within Marmotini, but this assignment is tentative. The record of *M. lavertyi* is much later than other described species of the genus and differs in having simpler cheek teeth (Dalquest, Baskin, and Schultz, 1996), but it also differs from known *Tamias* (e.g., lack of mesoconid). Its generic placement seems uncertain and is queried here.

Ammospermophilus Merriam, 1892 (including *Spermophilus*, in part; *Citellus*, in part)

Type species: *Ammospermophilus leucurus* (Merriam, 1889) (originally described as *Tamias leucurus*), the extant white-tailed antelope squirrel.

Type specimen: USNM 1108/1660.

Characteristics: Small ground squirrel with low-crowned teeth; extant species very similar cranially and dentally to small species of *Spermophilus*; skull exhibiting convex dorsal profile, usually narrower interorbitally than postorbitally; infraorbital foramen oval with lateral wall sloping ventromedially to small but distinct masseteric tubercle, located ventral to the foramen zygomatic arch twisted such that the lateral surface of the jugal faces ventrolaterally (as in all extant terrestrial sciurids), squamosal root of arch directed anteroventrally. Cheek teeth relatively small; P3 not much enlarged; P4–M2 each with distinct metaconule separate from the protocone until advanced wear and a variably developed mesostyle; M3 not as elongate relative to M2 as in most *Spermophilus*; p4 non-molariform, with closely appressed protoconid and metaconid; first and second lower molars typically with a complete metalophid, the trigonid basin small; entoconid on p4–m2 usually indistinct; m3 more triangular in outline than in m1–2, typically with incomplete metalophid.

Average length of m2 in Tertiary fossils: 1.8–1.9 mm.

The characterization above applies to extant species and the Blancan taxon *A. hanfordi*. Other fossils assigned to *Ammospermophilus* differ in important respects. The earliest species assigned to the genus, *A. fossilis*, and the Blancan *A. jeffriesi* both lack the twisted zygomatic arch characteristic of all extant terrestrial sciurids. In addition, *A. fossilis* has a humerus to skull ratio more like that of flying squirrels and lacks a small, anteroventrally projecting process on the lateral ridge bounding the masseteric fossa, which is present on all terrestrial squirrels today.

Included Tertiary species: *A. fossilis* James, 1963 (known from localities CC17G*, NB6E); *A. junturensis* (Shotwell and Russell, 1963) (including *Citellus junturensis, Spermophilus junturensis* Shotwell, 1970, including ?*Ammospermophilus* sp. Black, 1963) (localities CP116[A], B, PN10*, [PN11]); *A. jeffriesi* Miller, 1980 (locality CC53*); A *hanfordi* Gustafson, 1978 (localities PN3C*, [PN23A]).

Ammospermophilus sp. is also known from localities [GC4DD], PN4.

Ammospermophilus is known from Pleistocene sites ranging from southern California northeast to northern Utah, and southeast to Texas (FAUNMAP, 1994).

Comments: It is difficult to identify fragmentary fossils confidently to *Ammospermophilus* even in late Quaternary contexts because of strong phenetic similarity to small species of *Spermophilus* (personal observations). Nevertheless, Blancan material of *A. hanfordi* seems phenetically close to *Ammospermophilus* and is probably appropriately assigned. The other Blancan taxon, *A. jeffriesi*, was placed in the genus with confidence (Miller, 1980) but differs significantly in cranial features, including the untwisted zygomatic arch (like tree squirrels). Placement of two Clarendonian species, *A. junturensis* and *A. fossilis*, seems even less certain. The latter is dentally like a ground squirrel but cranially and postcranially resembles tree or flying squirrels. Such mosaic morphology is not surprising if ground squirrels evolved from tree squirrel stock, but it compromises clear taxonomic placement given the paucity of diagnostic generic characters otherwise.

Lindsay (1972) interpreted *A. fossilis* to be derived from *Nototamias ateles*, and proposed removing *Ammospermophilus* from the tribe Marmotini to the tribe Tamiini. This proposal has not been followed by subsequent authors and is not supported by molecular data (e.g., Hafner, 1984). The genus needs to be reassessed.

Spermophilus Cuvier, 1825 (including *Otospermophilus*; *Citellus*; *Protospermophilus*, in part; *Sciurus*, in part)

Type species: *Spermophilus citellus* (Linnaeus, 1766) (originally described as *Mus citellus*), the extant European ground squirrel.

Type specimen: None designated.

Characteristics: A variable, paraphyletic genus exhibiting characteristics of Marmotini but difficult to differentially characterize; some primitive, small species closely resembling *Ammospermophilus* but typically having more expanded infraorbital foramen and ventrolaterally projecting squamosal root of the zygomatic arch (projects more ventrally and anteriorly in *Ammospermophilus*); all known species smaller than *Marmota* and lacking the peculiar features of that taxon (see account of that genus). The most morphologically derived species (placed in the subgenus *S.* [*Spermophilus*]) resembling prairie dogs in the strong medial deflection of the angle of the lower jaw (90°); greater hypsodonty and mesiodistal compression of cheek teeth; enlarged and lophate P3; high and complete metaloph on P4–M2; development (in some species) of a distinct metaloph on M3; molariform p4 with transversely expanded trigonid; elongate M3 and m3; but lack other derived features of prairie dogs (see account of that genus [*Cynomys*]).

Average length of m2 (for Tertiary North American fossils): 1.8–3.5 mm.

Included species: Fossils not assignable to subgenus: *S. argonautus* (Stirton and Goeriz, 1942) (including *Otospermophilus argonautus; Citellus* sp. Kellogg, 1910, Bryant, 1945, Wilson, 1937; *Citellus* [*Otospermophilus*] *argonautus* Black, 1963) (localities CC25B*, CC32B, CC38A, NB27B, NB31); *S. bensoni* (Gidley, 1922) (including *Citellus bensoni*) (localities SB14A*, C–E, [SB18, sublocality unknown], SB34IIB); *S. boothi* Hibbard, 1972 (localities [CP117A], CP118*, CP131II); *S. cyanocittus* Korth, 1997 (localities CP116A, B*); *S. dotti* (Hibbard, 1954) (including *Citellus dotti*) (locality SP4B*); *S. finlayensis* (Strain, 1966) (including *Citellus finlayensis*) (locality SB49*); *S. fricki* (Hibbard, 1942a) (including *Citellus fricki*) (locality CP123C*); *S. gidleyi* (Merriam, Stock, and Moody, 1925) (including *Otospermophilus gidleyi, Citellus* [*Otospermophilus*] *gidleyi* Bryant, 1945, Black, 1963) (locality PN12B*); *S. howelli* (Hibbard, 1941a) (including *Citellus howelli*) (localities NB35IIIC, [SP1H], [SP5], CP128A, C*, [PN4], [PN23A], [CE1]); *S. jerae* Sutton and Korth, 1995 (locality NP41B*); *S. johnsoni* Hibbard, 1972 (locality CP118*); *S. matachicensis* (Wilson, 1949c) (including *Citellus matachicensis*, also Black, 1963) (locality SB61*); *S. matthewi* (Black, 1963) (including *Citellus* [*Otospermophilus*] *matthewi, Sciurus* cf. *aberti* Matthew, 1924, Bryant, 1945) (localities CP115C*, [CP116F]); *S. meadensis* (Hibbard, 1941b) (including *Citellus meadensis*) (localities CP132B*, [C, D]); *S. mcgheei* (Strain, 1966) (including *Citellus mcgheei*) (locality SB49*); *S. meltoni* Hibbard, 1972 (locality CP118*); *S. primitivus* (Bryant, 1945) (including *Citellus primitivus, Sciurus* sp. Douglass, 1903, *Citellus* [*Otospermophilus*] *primitivus* Black, 1961, 1963) (localities [CP45E], NP41* [sublocality unknown], NP42); *S. rexroadensis* (Hibbard, 1941a) (including *Citellus rexroadensis*) (localities SP4B, CP128A, C*); ?*S. russelli* Gustafson, 1978 (localities PN3C*, [PN4], PN23A); *S. shotwelli* (Black, 1963) (including *Citellus* [*Otospermophilus*] *shotwelli*) (localities [SB57], [CP116F], PN14*); *S. tephrus* (Gazin, 1932) (including *Sciurus tephrus, Citellus ridgwayi* Gazin, 1932, *Citellus longirostris* Scharf, 1935, *Protospermophilus tephrus* Bryant, 1945, *Citellus* [*Otospermophilus*] *tephrus* Black, 1963) (localities PN8B*, [PN9B]); *S. tuitus* (Hay, 1921) (including *Citellus tuitus*) (locality SB17*); *S. wellingtonensis* Kelly, 1997 (localities NB34II, NB36IIIA*); *S. wilsoni* (Shotwell, 1956) (including

Citellus [*Otospermophilus*] *wilsoni*, also Black, 1963) (localities PN2 [sublocality unknown], [PN11], [PN13], PN14*).

Fossils assignable to the subgenus *S.* (*Spermophilus*): *S. cochisei* (Gidley, 1922) (including *Citellus cochisei*) (locality SB14F*); ?*S. cragini* (Hibbard, 1941b) (including *Citellus cragini, S. cragini* Kurtén and Anderson, 1980) (locality CP132B*); *S. mckayensis* (Shotwell, 1956) (including *Citellus* [*Citellus*] *mckayensis*, also Black, 1963) (locality PN14*).

Spermophilus sp. is also known from localities CC17H, CC25B, CC26B, CC31, NB13C, [NB35IIIB], SB10, SB11, SB12, SB18, SB37D, SB38IIA, C, SB48, SB58C, SB60, SB61, SB65, SP1C, F, H, SP3B, SP5, CP80, CP89, CP90A, CP114A, CP116C–E, CP117A, CP128AA–C, CP131II, NP11, NP40A, B, NP45, PN14, PN15, PN21B, CE1.

Pleistocene fossils of *Spermophilus* are widely distributed in North America, from California to Virginia, and from Alaska to Texas (Kurtén and Anderson, 1980). The genus *Spermophilus* is also known from the Pleistocene to Recent of Europe and Asia (McKenna and Bell, 1997).

Comments: The current taxonomy of *Spermophilus* recognizes six North American subgenera (two additional subgenera restricted to Eurasia; Nowak, 1999), reflecting significant morphological variability. These subgenera can be characterized phenetically, but most are difficult to diagnose phylogenetically on morphological features available for fossils, perhaps because the radiation of *Spermophilus* is a relatively recent phenomenon (Bryant, 1945). Most Tertiary fossils assigned to *Spermophilus* exhibit a conservative dental morphology typical of the subgenus *S.* (*Otospermophilus*). On this basis, Black (1963) assigned 10 of 13 Tertiary species of *Spermophilus* then known to *S.* (*Otospermophilus*). Korth (1994) likewise followed this approach. However, it is unclear which if any of these fossil species actually fall within the lineage of extant *S.* (*Otospermophilus*) since they are recognized based on generalized, presumably primitive morphology. Furthermore, early species of *Spermophilus* differ from extant *S.* (*Otospermophilus*) in some respects (e.g., compressed upper incisors of *S. tephrus* [Bryant, 1945] and lower incisors of *S. primitivus* [Bryant, 1945]; developed mesoconid on m1–2 of *S. primitivus* [Black, 1963, his Plate 18 and Fig. 1] and *S. jerae* [Sutton and Korth, 1995, their Fig. 5]; distinct notch at contact between protoconid and protolophid on m1–2 of *S. jerae* [Sutton and Korth, 1995, their Fig. 5]; complete metaloph on M1–2 of *S. tephrus* [Bryant, 1945]). Assignments to *S.* (*Otospermophilus*) are best avoided at present. In contrast, assignments of a few species to *S.* (*Spermophilus*) are based on a number of presumably derived features (e.g., greater hypsodonty, enlarged P3, molariform p4, etc.; Goodwin and Hayes, 1994) and seem valid. Given growing evidence that *Spermophilus* as currently conceived is ill defined and paraphyletic (see discussion on the previous page), and the substantial Tertiary diversity of the genus, this group badly needs comprehensive revision.

Cynomys Rafinesque, 1817 (including *Arctomys*, in part)

Type species: *Cynomys ludovicianus* (Ord, 1815) (originally described as *Arctomys ludoviciana*) (including *Cynomys socialis* Rafinesque, 1817), the extant black-tailed prairie dog.

Type specimen: None designated.

Characteristics: Moderately large ground squirrels with distinctive, high crowned, transversely expanded cheek teeth; infraorbital foramen strongly triangular with a prominent, laterally extending masseteric tubercle; maxillary tooth rows typically converging strongly posteriorly; zygomatic arches showing strong lateral expansion; lower jaw robust with strong medial deflection of the angle (about 90°). Teeth large for animal's size; P3 large and lophate; P4–M3 each bearing prominent lophs; metaloph usually complete, with a weakly developed metaconule; M1–3 often with accessory lophule along the distal surface of the paracone; M3 strongly expanded distally, with well-developed metaloph; p4 molariform, with the trigonid subequal in transverse width to the talonid, a complete protolophid uniting protoconid and metaconid without interruption, fully enclosing the trigonid pit; m1–3 with basin trench, especially developed on m3 and coursing in an arc from distal base of protoconid, along the ectolophid, to the confluence of the hypoconid and posterolophid; m1–3 often with strong rugosity on the talonid floor when unworn and prominent metalophid enclosing a high, transversely elongate trigonid basin (modified from Goodwin, 1995a).

Average length of m2: 2.9–3.4 mm.

Included Tertiary species: *C. hibbardi* Eshelman, 1975 (localities [SB37II], CP131II*); *C. vetus* Hibbard, 1942b (locality CP132II). *Cynomys* sp. is also known from localities CP131II, CP132D.

Cynomys is also known from the Pleistocene, ranging from southern Alberta to central Mexico, and from Oregon to westernmost Iowa (Goodwin, 1995b; Martin, 1996).

Comments: At least one late Blancan (?*S. cragini*; Goodwin and Hayes, 1994) and several undescribed early Irvingtonian squirrels (personal observation) are either large, *Cynomys*-like members of *Spermophilus* or small, primitive members of *Cynomys*, indicating an adaptive radiation of prairie-dog-like squirrels in the late Blancan and early Irvingtonian (late Pliocene–early Pleistocene).

Marmota Blumenbach, 1779 (including *Arctomys*, in part)

Type species: *Marmota marmota* (Linnaeus, 1758) (originally described as *Mus marmota*), the extant Alpine marmot.

Type specimen: None designated.

Characteristics: Large ground-dwelling squirrels; modern species with distinctive, flattened dorsal profile of the skull and expanded zygomatic arches (skull material unknown for most Tertiary taxa); lower jaw bearing strong crest along the dorsal surface of the masseteric fossa, unlike *Paenemarmota*, and lower incisor terminating less posteriorly than in *Paenemarmota* (see Voorhies, 1988); diastemal region long, characteristically falling sharply anterior to p4. Upper cheek teeth somewhat cuspate instead of lophate; P3 large, oblique loph not extending to the mesiolingual margin of the tooth, allowing the mesial and distolingual cingula to merge; P4–M2 bearing prominent metaconules; M3 lacking the complete metaloph of *Paenemarmota, Cynomys*, and some *Spermophilus*, typically bearing a swollen metaconule; p4 transversely compressed across the trigonid with a distinct, strongly curved and mesially bulging protolophid. Lower molars with variably developed metalophid, typically incomplete on m2 and m3; upper incisors varying from smooth to shallowly grooved; lower incisors varying from smooth to finely striated, but not deeply striated as in *Paenemarmota* and *Parapaenemarmota*.

Average length of m1 (m2 unknown for some fossil taxa): 3.6–4.7 mm for Tertiary *Marmota*; 5.2–5.7 mm for extant North American taxa (Gordon and Czaplewski, 2000).

Included species: *M. arizonae* Hay, 1921 (known from locality SB17*); *M. korthi* Kelly, 2000 (locality NB34II); *M. minor* (Kellogg, 1910) (including *Arctomys minor*) (locality NB31*); *M. vetus* (Marsh, 1871) (including *Arctomys vetus*) (localities CP116[B], F; type locality* known only as "Loup Fork Beds" of Nebraska).

Marmota sp. is also known from localities NB36IIIA, [SB12], SP1D. In the late Pleistocene, *Marmota* is common from sites within and around the Great Basin and central Rocky Mountains of western North America. It also occurs as far south as Texas and is known across eastern North America in a belt extending from Missouri to Virginia (FAUNMAP, 1994). The genus *Marmota* is also known from the Pleistocene and Recent of Europe and Asia (McKenna and Bell, 1997).

Comments: The Tertiary record of *Marmota* in North America is extremely poor after removal of numerous taxa once assigned to the genus (see accounts of *Paenemarmota* and *Parapaenemarmota*). Tertiary marmots on average were smaller than their modern counterparts (Gordon and Czaplewski, 2000), although the unmeasured Blancan form, *M. arizonae*, was reported as similar to extant *M. flaviventris* (Hay, 1921).

Cynomyoides Korth, 1996

Type species: *Cynomyoides vatis* Korth, 1996.

Type specimen: UNSM 101817, late Clarendonian of Nebraska.

Characteristics: Lower cheek teeth lophate, mesiodistally compressed (except m3, which is strongly elongate mesiodistally), with an elevated trigonid pit; lower molars with basin trench that extends in an arc from the base of the trigonid, along the ectolophid, terminating adjacent to the posterolophid; floor of talonid basin with slight rugosity, more developed on m3 along the margin of the arcing basin trench.

Average length of m2: 2.6 mm.

Included species: *Cynomyoides vatis* only, known from locality CP116B* only.

Cynomyoides sp. is also known from locality NP41B.

In these features, *Cynomyoides* resembles late Cenozoic prairie dogs (*Cynomys*), and, in some features, *Paenemarmota* and *Parapaenemarmota* (especially the basin trench). It differs from prairie dogs in the small size and oblique (versus transverse) orientation of the floor of the trigonid pit on m1–3, the lack of a strong metalophid bounding this pit distally on m1–3, the prominently incomplete protolophid and strong buccal expansion of the hypoconid on p4, and the lesser tooth crown height. It is also smaller than all described species of *Cynomys*.

Comments: Resemblance to *Cynomys* is striking (Korth, 1996) and surprising given the Miocene age of the fossils (*Cynomys* is unknown before the late Blancan). If *Cynomys* is derived from within derived *Spermophilus*, as supported here, this similarity represents iterative evolution of a prairie-dog-like squirrel rather than the close affinity proposed by Korth (1996). More complete material would assist in systematic interpretation (the genus is unknown from cranial or upper dental material).

Paenemarmota Hibbard and Schultz, 1948 (including *Marmota*, in part; *Arctomys*, in part)

Type species: *Paenemarmota barbouri* Hibbard and Schultz, 1948.

Type specimen: KU 6994, early Blancan of Kansas.

Characteristics: Largest known Tertiary squirrel from North America. Upper and lower incisors robust and markedly striated, lower incisor extending well behind m3; P3 relatively large; P4 as large or larger than M1 and bearing a well-developed metaconule that is absent or poorly developed on molars; P4–M2 each bearing prominent posterior cingula; M3 with well-developed metaloph roughly parallel to the protoloph and separated from the protocone when unworn; lower cheek teeth bearing strongly rugose talonid basin and deep, crescentic trench along margins of the talonid basin adjacent to the metalophid, ectolophid, and buccal portion of the posterolophid; metalophid strong on lower cheek teeth, sometimes enclosing the trigonid basin (modified from Voorhies, 1988).

Average length of m2: 6.8–7.8 mm.

Included species: *P. barbouri* (known from localities SB16, SB33C, SP1H, SP5, CP117A, CP128A, B*, PN23A); *P.*

mexicana (Wilson, 1949c) (including *Marmota mexicana, P. barbouri* Repenning, 1962 in part) (localities SB58B, SB60*, SB64), *P. sawrockensis* (Hibbard, 1964) (including *Marmota sawrockensis, Marmota*? sp. Hibbard, 1953) (localities CP10II, CP116F, CP128AA*); ?*Paenemarmota nevadensis* (Kellogg, 1910) (including *Arctomys nevadensis, Marmota nevadensis* Wilson, 1937, Bryant, 1945, Black, 1963) (localities NB31*, [CP116E]).

Paenemarmota sp. is also known from localities SB38IIC, SB63, SB65, PN3C.

Comments: Repenning (1962) synonymized *P. barbouri* and *Marmota mexicana*, but Dalquest and Mooser (1980) resurrected the latter based on a divided posterior valley on M3 in Hemphillian forms from Mexico. Voorhies (1988) provided an updated diagnosis of the genus after including the slightly smaller *Marmota sawrockensis* within it. Voorhies (1988), Nelson and Miller (1990), and Martin (1998) reviewed relationships among large, terrestrial sciurids of the late Tertiary and noted, in particular, similarity between *Marmota nevadensis* (Kellogg, 1910) and *Paenemarmota*, with Martin (1998) assigning it to ?*Paenemarmota*.

***Parapaenemarmota* Martin, 1998 (including *Marmota*, in part; *Citellus*, in part)**

Type species: *Parapaenemarmota oregonensis* (Shotwell, 1956) (originally described as *Marmota oregonensis*).

Type specimen: UO F-3625, Hemphillian of Oregon.

Characteristics: Resembles *Paenemarmota* in the presence of a prominent metaloph on M3 and lower cheek teeth each with rugose talonid basin and deep trench along metalophid and ectolophid; differs in smaller size, large metaconule on M2, greater posterior expansion of M3, and closely appressed metaconid and protoconid on p4. Metalophid on lower molars enclosing complete trigonid basin, resembling *P. sawrockensis* (Voorhies, 1988; Nelson and Miller, 1990) but not most *Marmota* (modified from Martin, 1998).

Average length of M2 (m2 unknown for included species): 4.5–4.8 mm.

Included species: *P. oregonensis* (known from locality PN14*); *P. pattersoni* (Wilson, 1949c) (including *Citellus pattersoni*) (locality SB61*).

Comments: Both included species are poorly known. *P. pattersoni* is known only from a maxillary fragment with P4–M3 (Wilson, 1949c); consequently, nothing can be said of its lower dental morphology. With current evidence, the genus seems more closely allied with *Paenemarmota* than with *Marmota*, which appears to have been a separate but perhaps related lineage (Voorhies, 1988).

PETAURISTINAE SIMPSON, 1945

Characteristics: Extant flying squirrels with gliding membrane and a unique bony complex of the wrist (Thorington, 1984); these unknown for Tertiary North American fossils assigned to the Petauristinae. Teeth of petauristines with some or all of the following attributes (modified from James, 1963): heavily crenulated occlusal surface of cheek teeth in early wear stages (sometimes more lophulate rather than crenulate; see Robertson, 1976); hypocone a distinct cusp on upper M1–2, and lophs of upper cheek teeth complicated; accessory lophules arising from protocone and extending into the central and sometimes anterior valley (termed protocone crest by Emry and Korth, 1996); lower teeth with a distinct mesoconid, prominent entoconid, and clear groove marking the connection of the protolophid and protoconid. (Note that a number of these features likewise characterize primitive squirrels such as *Douglassciurus*.)

***Petauristodon* Engesser, 1979 (including *Sciuropterus*, in part)**

Type species: *Petauristodon mathewsi* (James, 1963) (originally described as *Sciuropterus mathewsi*).

Type specimen: UCMP 50429, late Clarendonian of California.

Characteristics: Cheek teeth with crenulate enamel; upper cheek teeth usually with distinct conules within the protoloph and metaloph, sometimes giving rise to spur-like accessory lophs; a distinct accessory loph extending buccally from the protocone between the protoloph and metaloph, and often another extending into the anterior valley (protocone crest of Emry and Korth [1996]); an evident hypocone separated from the protocone by a lingual notch (especially on M1 and M2); lower cheek teeth with an isolated mesostylid, transversely expanded mesoconid reaching nearly to the buccal margin of the tooth, an entoconid that is submerged within the posterolophid, and strongly crenulated enamel; lower cheek teeth lacking the mesiobuccal cingulum evident on the m3 of *Cryptopterus* (see Engesser, 1979; Pratt and Morgan, 1989).

Average length of m1–2: 1.00–2.73 mm. (*P. minimus* unknown from m1 or m2, but known from P4 and M3. P4 measures 0.96 mm; size relationships among other species suggests M2 size given as ~ 1.00 mm. Other length estimates from Pratt and Morgan 1989].)

Included species: *P. jamesi* (Lindsay, 1972, including *Sciuropterus jamesi*) (known from localities CC17G, NB6E* [probably the type locality, UCMP locality V-6448, *Copemys russelli* zone of Lindsay, 1972]); *P. mathewsi* (localities CC17E, G, H*, [CC22A]); *P. minimus* (Lindsay, 1972, including *Sciuropterus minimus*) (localities [GC3B], NB6E*); *P. pattersoni* Pratt and Morgan, 1989 (locality GC8D*); *P. uphami* (James, 1963, including *Sciuropterus uphami*) (localities CC17D*, F).

Petauristodon sp. is also known from localities GC3B, GC4C, GC4DD, ?CC9IVC, D, CC17G, ?CP45E, CP114A, B, [CP116B], NP11, NP40B, PN8B.

Comments: All but one species now assigned to *Petauristodon* (*P. pattersoni*) were originally assigned to the Old World genus *Sciuropterus* (James, 1963; Lindsay,

1972). Engesser (1979) pointed out that features uniting North American specimens with *Sciuropterus* – specifically the presence of distinct conules in the protoloph and metaloph, accessory lophules, and hypocones – were also present in North American paramyids and he erected *Petauristodon* for the North American forms. Emry and Thorington (1982) pointed out that these features were also present in the early sciurid *Douglassciurus jeffersoni* and took this as support for Engesser's hypothesis.

***Sciurion* Skwara, 1986 (including *Blackia*, in part)**

Type species: *Sciurion campestre* Skwara, 1986.

Type specimen: SMNH P1593.123, late Hemingfordian of Saskatchewan.

Characteristics: Small "flying squirrel" with finely crenulated enamel in basins of unworn cheek teeth; anteroconid present on m1 to m3 separated distinctly from protoconid; mesoconid present on each lower cheek tooth; M1 or M2 with paracone distinctly offset lingually and bearing a partial, buccally placed cingulum; transverse lophs of upper cheek teeth relatively high (for "flying squirrels"), simple, lacking the distinct conules and accessory lophs present in *Petauristodon* (modified from Skwara, 1986).

Average length of m2: 1.3 mm.

Included species: *Sciurion campestre* only, known from locality NP10E* only (but note that Bell, Meyer, and Bryant [2003] reported that three other species of *Sciurion* have been recognized from Orellan, Whitneyan, and early Arikarreean localities within the Cypress Hills Formation [locality NP10]).

?*Sciurion* sp. is also reported from locality CP116B. Specimens previously assigned to *Blackia* sp. but here tentatively assigned to *Sciurion* sp. are reported from localities CC19, CP114A, B.

Comments: Skwara (1986) explicitly avoided referring *Sciurion* to Petauristinae because of contention over systematics of the group. She found the closest comparison among fossil petauristines to be with *Blackia miocaenica* from France but noted several distinct differences. Hutchison and Lindsay (1974) referred two isolated teeth from locality CC19 (Vedder Local Fauna, California) to *Blackia* sp. They are similar in size to *Sciurion* but differ in a few features (P4 with small accessory loph and lacking buccal cingulum; m3 lacking mesoconid, faintly present in *Sciurion*). I tentatively refer them to *Sciurion*, pending more information, and do not recognize *Blackia* in the North American record on available evidence. Specimens of *Blackia* from localities CP114A, B (Valentine Formation, Nebraska) were assigned without morphological comment; they are treated as *Sciurion* sp. as a matter of convenience.

***Cryptopterus* Mein, 1970 (including *Petauria*)**

Type species: *Cryptopterus crusafonti* Mein, 1970.

Type specimen: Catalog number not given in description (Mein, 1970), listed only as coming from the collection of the Museo de la Ciudad de Sabadell, middle Miocene of Spain.

Characteristics of North American member (modified from Robertson, 1976): The single North American species of the genus (several others known from the Old World [from the Miocene and Pliocene of Europe and the Pliocene of Asia, McKenna and Bell, 1997]) is largest "flying squirrel" from the North American Tertiary; m3 distinguished by a distinct, broad, mesiobuccal cingulum adjacent to the protoconid; talonid of m3 bearing numerous, low lophulids but not finely crenulated; mesostylid nearly submerged in a crest extending distally from the metaconid; indistinct hypolophid extending into the talonid basin from the entoconid; distinct mesoconid lying between the protoconid and hypoconid, and an anteroconid present along the protolophid.

Average length of m3 (m2 unknown for North American forms): 4.91 mm.

Included species: *C. webbi* Robertson, 1976 (including *Petauria* sp. Webb, 1974) only (known from localities GC13B, GC15A*).

Comments: If the generic assignment is correct, the presence of this Old World flying squirrel in the North American Hemphillian and Blancan is significant biogeographically (Robertson, 1976).

INDETERMINATE SCIURIDS

Sciurids gen. and sp. indet. are known from localities GC11A, CC9C, CC32A, NB6C, E, NB13A, NB18A, NB23B, SB11, CP39A, CP45E, CP56, CP56IIB, CP84B, CP114D, NP5A, NP34C, NP36A, B, PN16.

BIOLOGY AND EVOLUTIONARY PATTERNS

Overviews of historical pattern in the Tertiary record of North American Sciuridae were provided by several previous workers (e.g., Wilson, 1960; Repenning, 1962; Black, 1963, 1972; Korth, 1994). The synopsis that follows traces the family and its North American clades (Cedromurinae [extinct group], Petauristinae [flying squirrels], Sciurini [tree squirrels], various basal "terrestrial squirrels," Tamiini [chipmunks], and Marmotini [advanced ground-dwelling squirrels]) in temporal sequence (Figure 21.3).

The family Sciuridae first appear in the Chadronian of North America (*Oligospermophilus, Douglassciurus*) and in roughly contemporaneous deposits in Europe (*Palaeosciurus*; Vianey-Liaud, 1985), with a reithroparamyine-like ischyromyid a plausible ancestor (Emry and Korth, 1996). At least five North American species are present during the Chadronian–Whitneyan (latest Eocene to mid Oligocene) interval: three cedromurines (two species of *Cedromus, Oligospermophilus*), a subfamily of squirrels restricted to this

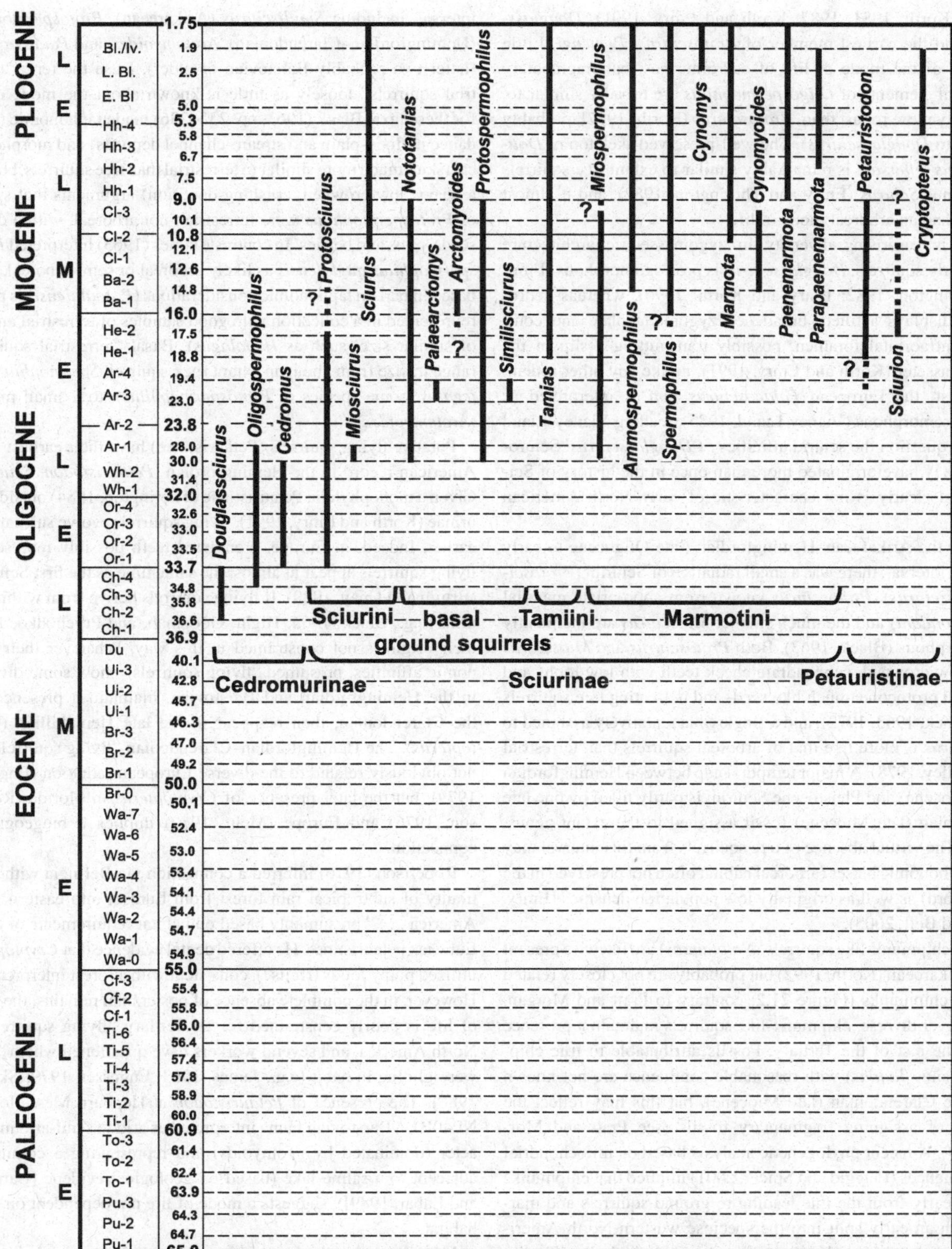

Figure 21.3. Temporal ranges of North American sciurid genera.

interval (Korth, 1981, 1987; Korth and Emry, 1991), *Douglassciurus*, and the earliest member of *Protosciurus, P. mengi*. Little is known of the mode of life of cedromurines, but fragmentary postcranial elements of *Oligospermophilus* are broadly similar to, but slightly more robust than, *Protosciurus* (Korth [1987], probably referring to *Douglassciurus*). The well-preserved skeleton of *Douglassciurus jeffersoni* is remarkably similar to extant tree squirrels of the genus *Sciurus* (Emry and Thorington, 1982), and it almost certainly had an arboreal mode of life.

There is surprising diversity in zygomasseteric architecture among early squirrels. *Douglassciurus* is protrogomorphous (Emry and Thorington, 1982; Emry and Korth, 1996), whereas cedromurines displays a tilted, broadened zygomatic plate and compressed infraorbital foramen, possibly transmitting a slip of the medial masseter (Korth and Emry, 1991), unlike any other rodent. In contrast, the European *Palaeosciurus goti* is interpreted as fully sciuromorphous (Vianey-Liaud, 1985), leading Vianey-Liaud (1985) to question the sciurid affinities of *Douglassciurus*. Sciuromorphy may have originated more than once in the history of Sciuridae and clearly is not characteristic of early North American forms.

During the Arikareean–Hemingfordian (late Oligocene to early Miocene) interval, there was a small radiation of Sciurini (*Protosciurus, Miosciurus*). *Protosciurus* known from appropriate material (e.g., *P. condoni*) and the much smaller *Miosciurus* are incipiently sciuromorphous (Black, 1963). Both *Protosciurus* and *Miosciurus* exhibit low crowned, subquadrate cheek teeth with low lophs and broadened protocone, much like seed- and nut-eating tree squirrels today (Black, 1963, 1972), and a single tibia tentatively assigned to *Protosciurus* is more like that of arboreal squirrels than terrestrial ones (Frailey, 1978). A major temporal gap between Hemingfordian (early Miocene) and Pleistocene Sciurini is partly filled by one late Clarendonian (late Miocene) fossil assigned to the extant genus, *Sciurus*. The virtual absence of tree squirrels from this interval may reflect taphonomic biases (arboreal habitat often not preserved in the fossil record) as well as originally low population densities (Emry, Korth, and Bell, 2005).

Small, chipmunk-like squirrels (*Nototamias*) are likewise present by the Arikareean (Korth, 1992) but probably are not closely related to extant chipmunks (Figure 21.2; contrary to Pratt and Morgan, 1989). Never diverse, chipmunk-like squirrels maintain a presence through the rest of the Tertiary. Fossils attributable to true chipmunks (genus *Tamias*) with reasonable confidence are not known before the Clarendonian (late Miocene), but this may reflect the difficulty of assigning fragmentary fossils (see Pratt and Morgan, 1989). A recent phylogenetic analysis based on mitochondrial DNA sequences (Piaggio and Spicer, 2001) implied that chipmunks diverged early from the line leading to ground squirrels and marmots, with an early split into the speciose western North American clade *Eutamias* (or *Neotamias*) and a clade including *Tamias striatus* (eastern North America) and *Tamias sibiricus* (Asia). This interpretation suggests a deeper history for *Tamias* than is clearly documented with evidence from the fossil record.

A cluster of distinctive, basal "terrestrial squirrels" radiate in the late Arikareean–Barstovian (earliest to middle Miocene) interval, including *Similisciurus* (Arikareean), *Protospermophilus* (Hemingfordian–Clarendonian), *Arctomyoides*, and *Palaearctomys* (latter two probably Barstovian or older). I use the term "terrestrial squirrels" loosely as little is known about the mode of life for these taxa. Black (1963, pp. 235–236) made taphonomic (abundance in flood-plain and stream-channel deposits) and morphological (skull much more similar to terrestrial than tree squirrels, but with a heavy, low crowned, crushing dentition) arguments that species of *Protospermophilus* were terrestrial, not arboreal, with a diet of seeds, nuts, and berries. In contrast, James (1963) interpreted *Protospermophilus quatalensis* as likely arboreal or semi-arboreal, again based in part on taphonomic considerations (*P. quatalensis* is poorly represented in a collection with good samples of terrestrial animals of similar size, such as *Hypolagus*). Basal "terrestrial squirrels" range in size from about an extant rock squirrel, *Spermophilus variegatus* (some species of *Protospermophilus*), to a small marmot (*Arctomyoides*).

Putative flying squirrels (Petauristinae) have their earliest North American record in the Hemingfordian (*Petauristodon, Sciurion*). Given a monophyletic Petauristinae (Thorington, 1984) outside Sciurinae (Korth and Emry, 1991), flying squirrels evolve substantially earlier. Indeed, in Europe, crenulated teeth possibly representing flying squirrels appear at almost the same time as the first Sciurinae (Bruijn and Ünay, 1989). If flying squirrels derive from within Sciurini (e.g., Black, 1963; Hight, Goodman, and Prychodko, 1974), their origin is not constrained in this way. Whatever their taxonomic affinities, presumed "flying squirrels" show some diversity in the Hemingfordian and Barstovian, maintain a presence into the Clarendonian, then reappear in the late Hemphillian (*Cryptopterus*). The Hemingfordian–Clarendonian "flying squirrels" are not obviously related to the diverse European radiation (Engesser, 1979), but the later presence of *Cryptopterus* in Florida (Robertson, 1976) and Europe (Mein, 1970) implies a biogeographic connection.

Robertson (1976) inferred a connection via Beringia with "continuity of subtropical rain forest from Eurasia into eastern North America . . .," presumably based on habitat requirements of extant Eurasian petauristines. He offered dental evidence that *Cryptopterus* utilized pulpy foods (fruits?), consistent with a forest interpretation. However, in the complete absence of postcranial remains, the mode of life is poorly constrained for all Tertiary "flying squirrels" in North America, and several workers have questioned whether they were gliding in habit (e.g., James, 1963; Engesser, 1979; Skwara, 1986). The presence of *Petauristodon* at Hepburn Mesa (locality NP40B), a Barstovian fauna interpreted as arid to semi-arid in character (dominated by geomyoids) and representing a community adjacent to a saline lake (based on geological evidenc [Barnosky and Labar, 1989]), suggests a mode of life not dependent on forest habitat.

If *Miospermophilus* is a member of the tribe Marmotini, this clade likewise first appears in North America during the Hemingfordian (late early Miocene). *Miospermophilus* shares dental features with later *Spermophilus* (e.g., relatively narrow protocone, rhomboidal lower molars) as well as with chipmunks (low molar lophs, small size) (Black, 1963). The Marmotini do not diversify greatly from the Hemingfordian through the Clarendonian (late early to early late Miocene). Most early species are primitive dentally and cranially, but the prairie-dog-like *Cynomyoides* is remarkably advanced dentally for its age (the genus first appears in the Barstovian; Korth, 1996). *Ammospermophilus* may have appeared in the Barstovian (*A. fossilis*), although assignment of early species seems dubious (see generic account). Skeletal proportions published by James (1963) indicate closer resemblance to small tree and flying squirrels than to terrestrial squirrels.

The Marmotini experience a major evolutionary radiation in the Hemphillian and Blancan (latest Miocene and Pliocene). The initial phase of this radiation (within the Hemphillian) is roughly coincident with the expansion of the C4 grassland ecosystems reported between approximately 7–5 Ma (e.g., Cerling, Wang, and Quade, 1993) although much greater diversity was achieved in the Blancan. The clade clearly was responding to the expansion of open, grass-dominated habitats during this time (e.g., Potts *et al.*, 1992). Notable events included: (1) explosive diversification of "*Spermophilus*;" (2) a small radiation of very large, ground-dwelling forms either related to marmots or advanced ground squirrels (*Parapaenemarmota, Paenemarmota*), and possibly to the earlier *Cynomyoides*; and (3) the first appearance of the specialized, extant genera *Marmota* (in the Hemphillian, possibly the Clarendonian) and *Cynomys* (in the Blancan), lineages derived from within "*Spermophilus*" (e.g., Bryant, 1945; Black, 1963; Thomas and Martin, 1993; Goodwin and Hayes, 1994; Piaggio and Spicer, 2001). Most of the Hemphillian and Blancan species of *Spermophilus* are generalized dentally and cannot be assigned with confidence to modern subgenera (Martin, 1989), but *S. mckayensis* from the Hemphillian is clearly referable to the dentally advanced subgenus *S.* (*Spermophilus*). Postcranial material is available for several of these ground squirrels (e.g., *S. shotwelli* [Black, 1963], *S. fricki* [Hibbard, 1942a], and an undescribed species from locality CP117A [the Broadwater Formation of Nebraska; personal observation]), thus firmly demonstrating their terrestrial habits. The *Parapaenemarmota–Paenemarmota* radiation is restricted to North America, but *Marmota* is also found in Eurasia. The time of the appearance of the genus in North America (Hemphillian to latest Miocene) and molecular phylogenies placing North American species most basal in the extant radiation of *Marmota* (Kruckenhauser *et al.*, 1999; Steppan *et al.*, 1999) both support a North American origin.

ACKNOWLEDGEMENTS

The author thanks W. Korth for a helpful review of the manuscript.

REFERENCES

Baird, S. F. (1857). *Explorations and Surveys for a Railroad Route from the Mississippi River to the Pacific Ocean*. Washington, DC: Nicholson.

Barnosky, A. D. and Labar, W. J. (1989). Mid-Miocene (Barstovian) environmental and tectonic setting near Yellowstone Park, Wyoming and Montana. *Bulletin of the Geological Society of America*, 101, 1448–56.

Bell, S., Meyer, T., and Bryant, H. (2003). "Flying squirrel" (Mammalia: Rodentia) diversity in the Oligocene of the Cypress Hill Formation (southwest Saskatchewan, Canada). *Journal of Vertebrate Paleontology*, 23(suppl. to no. 3), p. 33A.

Black, C. C. (1961). Fossil mammals from Montana. Pt. 1. Additions to the late Miocene Flint Creek Local Fauna. *Annals of the Carnegie Museum*, 36, 69–76.

(1963). A review of North American Tertiary Sciuridae. *Bulletin of the Museum of Comparative Zoology*, 130, 111–248.

(1965). Fossil mammals from Montana. Pt. 2. Rodents from the early Oligocene Pipestone Springs Local Fauna. *Annals of the Carnegie Museum*, 38, 1–48.

(1972). Holarctic evolution and dispersal of squirrels (Rodentia: Sciuridae). In *Evolutionary Biology*, ed. T. Dobzhansky, M. K. Hecht, and W. C. Steere, pp. 305–22. New York: Appleton-Century-Crofts.

Blumenbach, J. F. (1779). *Handbuch der Naturgeschichte*. Gottingen: Dieterich.

Bruijn, H. de and Ünay, E. (1989). Petauristinae (Mammalia, Rodentia) from the Oligocene of Spain, Belgium, and Turkish Thrace. *Contributions in Science, Los Angeles County Museum of Natural History*, 33, 139–45.

Bryant, M. D. (1945). Phylogeny of Nearctic Sciuridae. *The American Midland Naturalist*, 33, 257–390.

Bugge, J. (1985). Systematic value of the carotid arterial pattern in rodents. In *Evolutionary Relationships Among Rodents: A Multidisciplinary Analysis*, ed. W. P. Luckett and J.-L. Hartenberger, pp. 355–79. New York: Plenum Press.

Burmeister, H. (1854). *Systematische Uebersicht der Thiere Brasiliens, welche während einer Reise durch die Provinzen von Rio de Janeiro und Minas Geraës. gesammelt oder beobachtet. Erster Teil, Säugethiere (Mammalia)*. Berlin: Georg Reimer.

Cerling, T. E., Wang, Y., and Quade, J. (1993). Expansion of C4 ecosystems as an indicator of global ecological change in the late Miocene. *Nature*, 361, 344–5.

Cope, E. D. (1879). Second contribution to a knowledge of the Miocene fauna of Oregon. *Paleontological Bulletin*, 31, 1–7.

(1881). On the Nimravidae and Canidae of the Miocene period of North America. *Bulletin of the United States Geological and Geographic Survey of the Territories*, 6, 165–81.

(1883). The extinct Rodentia of North America. *The American Naturalist*, 17, 43–57.

Cuvier, F. (1825). *Des Dent des Mammifères*. Strasbourg: F. G. Levrault.

Dalquest, W. W. and Mooser, O. (1980). Late Hemphillian mammals of the Ocote local fauna, Guanajuato, Mexico. *The Pearce-Sellards Series of the Texas Memorial Museum*, 32, 1–25.

Dalquest, W. W., Baskin, J. A., and Schultz, G. E. (1996). Fossil mammals from a late Miocene (Clarendonian) site in Beaver County, Oklahoma. In *Contributions in Mammalogy: a Memorial Volume honoring Dr. J. Knox Jones, Jr.*, ed. H. H. Genoways and R. J. Baker, pp. 107–37. Lubbock, TX: Museum of Texas Technical University.

Douglass, E. (1901). Fossil Mammalia of the White River Beds of Montana. *Transactions of the American Philosophical Society*, 20, 237–79.

(1903). New vertebrates from the Montana Territory. *Annals of the Carnegie Museum*, 2, 145–99.

Downs, T. (1956). The Mascall fauna from the Miocene of Oregon. *University of California Publications in Geological Sciences*, 31, 199–354.

Emry, R. J. and Korth, W. W. (1996). The Chadronian squirrel "*Sciurus*" *jeffersoni* Douglass, 1901: a new generic name, new material, and its bearing on the early evolution of Sciuridae (Rodentia). *Journal of Vertebrate Paleontology*, 16, 775–80.

(2001). *Douglassciurus*, a new name for *Douglassia* Emry and Korth, 1996, not *Douglassia* Bartsch, 1934. *Journal of Vertebrate Paleontology*, 21, 401.

Emry, R. J. and Thorington, R. W. Jr. (1982). Descriptive and comparative osteology of the oldest fossil squirrel, *Protosciurus* (Rodentia: Sciuridae). *Smithsonian Contributions to Paleobiology*, 47, 1–35.

Emry, R. J., Korth, W. W., and Bell, M. A. (2005). A tree squirrel (Rodentia, Sciuridae, Sciurini) from the late Miocene (Clarendonian) of Nevada. *Journal of Vertebrate Paleontology*, 25, 228–35.

Engesser, B. (1979). Relationships of some insectivores and rodents from the Miocene of North America and Europe. *Bulletin of the Carnegie Museum of Natural History*, 14, 1–68.

Eshelman, R. E. (1975). Geology and paleontology of the early Pleistocene (late Blancan) White Rock fauna from north-central Kansas. [In *Studies on Cenozoic Paleontology and Stratigraphy. Claude W. Hibbard Memorial Volume 4.*] *University of Michigan Papers on Paleontology*, 13, 1–60.

FAUNMAP Working Group (1994). FAUNMAP: a database documenting late Quaternary distributions of mammal species in the United States. *Illinois State Museum Scientific Papers*, 25, 1–690.

Frailey, D. (1978). An early Miocene (Arikareean) fauna from northcentral Florida (the SB-1A Local Fauna). *Occasional Papers of the Museum of Natural History, University of Kansas*, 75, 1–20.

Galbreath, E. C. (1953). A contribution to the Tertiary geology and paleontology of northeastern Colorado. *University of Kansas Paleontological Contributions, Vertebrata*, 4, 1–119.

Gazin, C. L. (1930). A Tertiary vertebrate fauna from the upper Cuyama drainage basin, California. *Carnegie Institution of Washington Publication*, 404, 55–76.

(1932). A Miocene mammalian fauna from southeastern Oregon. *Carnegie Institution of Washington Publication*, 418, 36–86.

Giboulet, O., Chevret, P, Ramousse, R., and Catzeflis, F. (1997). DNA-DNA hybridization evidence for the recent origin of marmots and ground squirrels (Rodentia: Sciuridae). *Journal of Mammalian Evolution*, 4, 271–84.

Gidley, J. W. (1922). Preliminary report on fossil vertebrates of the San Pedro Valley, Arizona, with descriptions of new species of Rodentia and Lagomorpha. *United States Geological Survey Professional Paper*, 131E, 119–31.

Goodwin, H. T. (1995a). Systematic revision of fossil prairie dogs with descriptions of two new species. *University of Kansas Museum of Natural History Miscellaneous Publications*, 86, 1–38.

(1995b). Pliocene-Pleistocene biogeographic history of prairie dogs, genus *Cynomys* (Sciuridae). *Journal of Mammalogy*, 76, 100–22.

Goodwin, H. T. and Hayes, F. E. (1994). Morphologically derived ground squirrels from the Borchers Local Fauna, Meade County, Kansas, with a redescription of ?*Spermophilus cragini*. *Journal of Vertebrate Paleontology*, 14, 278–91.

Gordon, C. L. and Czaplewski, N. J. (2000). A fossil marmot from the late Miocene of western Oklahoma. *Oklahoma Geology Notes*, 60, 28–32.

Gray, J. E. (1821). On the natural arrangement of vertebrate animals. *London Medical Repository*, 15, 296–310,

Gustafson, E. P. (1978). The vertebrate faunas of the Pliocene Ringold Formation, south-central Washington. *Bulletin of the Museum of Natural History, University of Oregon*, 23, 1–62.

Hafner, D. J. (1984). Evolutionary relationships of the Nearctic Sciuridae. In *The Biology of Ground-Dwelling Squirrels*, ed. J. O. Murie and G. R. Michener, pp. 3–23. Lincoln, NB: University of Nebraska Press.

Hall, E. R. (1930). Rodents and lagomorphs from the Barstow beds of southern California. *University of California Publications in Geological Sciences*, 19, 313–18.

(1981). *Mammals of North America.* Vol. 1. New York: John Wiley.

Harrison, R. G., Bogdanowicz, S. M., Hoffmann, R. S., Yensen, E., and Sherman, P. W. (2003). Phylogeny and evolutionary history of the ground squirrels (Rodentia: Marmotinae). *Journal of Mammalian Evolution*, 10, 249–76.

Hartenberger, J.-L. (1985). The order Rodentia: major questions on their evolutionary origin, relationships and suprafamilial systematics. In *Evolutionary Relationships Among Rodents: A Multidisciplinary Analysis*, ed. W. P. Luckett and J.-L. Hartenberger, pp. 1–33. New York: Plenum Press.

Hay, O. P. (1921). Descriptions of species of Pleistocene Vertebrata, types or specimens of most of which are preserved in the United States National Museum. *Proceedings of the United States National Museum*, 59, 599–642.

Hibbard, C. W. (1941a). New mammals from the Rexroad fauna, upper Pliocene of Kansas. *The American Midland Naturalist*, 26, 337–68.

(1941b). The Borchers fauna, a new Pleistocene interglacial fauna from Meade County, Kansas. *Geological Survey of Kansas Bulletin*, 38, 197–220.

(1942a). A new fossil ground squirrel *Citellus* (*Pliocitellus*) *fricki* from the Pliocene of Clark County, Kansas. *Transactions of the Kansas Academy of Sciences*, 45, 253–7.

(1942b). Pleistocene mammals from Kansas. *Geological Survey of Kansas Bulletin*, 41, 261–9.

(1953). The Saw Rock Canyon fauna and its stratigraphic significance. *Papers of the Michigan Academy of Science, Arts, and Letters*, 38, 387–411.

(1954). A new Pliocene vertebrate fauna from Oklahoma. *Papers of the Michigan Academy of Science, Arts, and Letters*, 39, 339–59.

(1964). A contribution to the Saw Rock Canyon Local Fauna of Kansas. *Papers of the Michigan Academy of Science, Arts, and Letters*, 49, 115–27.

(1972). Class Mammalia. *Bulletin of the American Museum of Natural History*, 148, 77–130.

Hibbard, C. W. and Schultz, C. B. (1948). A new sciurid of Blancan age from Kansas and Nebraska. *Bulletin of the University of Nebraska State Museum*, 3, 19–29.

Hight, M., Goodman, E. M., and Prychodko, W. (1974). Immunological studies of the Sciuridae. *Systematic Zoology*, 23, 12–25.

Huchon, D., Catzeflis, F., and Douzery, E. (1999). Molecular evolution of the nuclear von Willebrand factor gene in mammals and the phylogeny of rodents. *Molecular Biology and Evolution*, 16, 577–89.

Hutchison, J. H. and Lindsay, E. H. (1974). The Hemingfordian mammal fauna of the Vedder locality, Branch Canyon Formation, Santa Barbara County, California. *PaleoBios*, 15, 1–19.

Illiger, J. K. W. (1811). *Prodromus Systematis Mammalium et Avium.* Berolini: Sumptibus C. Salfield.

James, G. T. (1963). Paleontology and nonmarine stratigraphy of the Cuyama Valley Badlands, California. *University of California Publications in Geological Sciences*, 45, 1–154.

Kellogg, L. (1910). Rodent fauna of the late Tertiary beds at Virgin Valley and Thousand Creek, Nevada. *University of California Publications in Geological Sciences*, 5, 421–37.

Kelly, T. S. (1997). Additional late Cenozoic (latest Hemphillian to earliest Irvingtonian) mammals from Douglas County, Nevada. *PaleoBios*, 18, 1–31.

(2000). A new Hemphillian (late Miocene) mammalian fauna from Hoye Canyon, west central Nevada. *Contributions in Science, Los Angeles Museum of Natural History*, 481, 1–21.

Korth, W. W. (1981). New Oligocene rodents from western North America. *Annals of the Carnegie Museum*, 50, 289–318.

(1987). Sciurid rodents (Mammalia) from the Chadronian and Orellan (Oligocene) of Nebraska. *Journal of Paleontology*, 61, 1247–55.

(1992). Fossil small mammals from the Harrison Formation (late Arikareean: earliest Miocene), Cherry County, Nebraska. *Annals of the Carnegie Museum*, 61, 69–131.

(1994). *The Tertiary Record of Rodents in North America.* New York: Plenum Press.

(1996). A new genus of prairie dog (Sciuridae, Rodentia) from the Miocene (Barstovian of Montana and Clarendonian of Nebraska) and the classification of Nearctic ground squirrels (Marmotini). *Transactions of the Nebraska Academy of Sciences*, 23, 109–13.

(1997). Additional rodents (Mammalia) from the Clarendonian (Miocene) of northcentral Nebraska and a review of Clarendonian rodent biostratigraphy of that area. *Paludicola*, 1, 97–111.

Korth, W. W. and Emry, R. J. (1991). The skull of *Cedromus* and a review of the Cedromurinae (Rodentia, Sciuridae). *Journal of Paleontology*, 65, 984–94.

Kruckenhauser, L., Pinsker, W., Haring, E., and Arnold, W. (1999). Marmot phylogeny revisited: molecular evidence for a diphyletic origin of sociality. *Journal of Zoological Systematics and Evolution Research*, 37, 49–56.

Kurtén, B. and Anderson, E. (1980). *Pleistocene Mammals of North America*. New York: Columbia University Press.

Lavocat, R. and Parent, J.-P. (1985). Phylogenetic analysis of middle ear features in fossil and living rodents. In *Evolutionary Relationships Among Rodents: A Multidisciplinary Analysis*, ed. W. P. Luckett and J.-L. Hartenberger, pp. 333–54. New York: Plenum Press.

Lindsay, E. H. (1972). Small mammal fossils from the Barstow Formation, California. *University of California Publications in Geological Sciences*, 93, 1–104.

Linnaeus, C. (1758). *Systema Naturae per Regna Tria Naturae, Secundum Classes, Ordines, Genera, Species, cum Characteribus and Differentiis, 10th edn., Reformata, Vol.* I. Holmiae: Laurentii Salvii.

(1766). *Systema Naturae per Regna Tria Naturae, Secundum Classes, Ordines, Genera, Species, cum Characteribus and Differentiis. 12th edn., Reformata, Vol.* I. Holmiae: Laurentii Salvii.

Luckett, W. P. (1985). Superordinal and intraordinal affinities of rodents: developmental evidence from the dentition and placentation. In *Evolutionary Relationships Among Rodents: A Multidisciplinary Analysis*, ed. W. P. Luckett and J.-L. Hartenberger, pp. 227–76. New York: Plenum Press.

Marsh, O. C. (1871). Notice of some new fossil mammals and birds from the Tertiary formation of the west. *American Journal of Science*, 2, 120–7.

Martin, J. E. (1996). First occurrence of *Cynomys* from west of the Rocky Mountains. *Journal of Vertebrate Paleontology*, 16(suppl. to no. 3), p. 51A.

(1998). Two new sciurids, *Eutamias malloryi* and *Parapaenemarmota* (Rodentia), from the late Miocene (Hemphillian) of northern Oregon. [In *Contributions to the Paleontology and Geology of the West Coast. In Honor of V. Standish Mallory*, ed. J. E. Martin.] *Thomas Burke Memorial Washington State Museum Research Report*, 6, 31–42.

Martin, L. D. (1989). Plio-Pleistocene rodents in North America. *Contributions in Science, Los Angles County Museum of Natural History*, 33, 47–58.

Matthew, W. D. (1903). The fauna of the *Titanotherium* beds at Pipestone Springs, Montana. *Bulletin of the American Museum of Natural History*, 19, 197–226.

(1909). Faunal lists of the Tertiary Mammalia of the West. *Bulletin of the United States Geological Survey*, 361, 91–120.

(1910). On the osteology and relationships of *Paramys* and the affinities of the Ischyromyidae. *Bulletin of the American Museum of Natural History*, 28, 43–72.

(1924). Third contribution to the Snake Creek fauna. *Bulletin of the American Museum of Natural History*, 50, 59–210.

Matthew, W. D. and Mook, C. C. (1933). New fossil mammals from the Deep River beds of Montana. *American Museum Novitates*, 601, 1–7.

McKenna, M. C. and Bell, S. K. (1997). *Classification of Mammals Above the Species Level*. New York: Columbia University Press.

Mein, P. (1970). Les Sciuropteres (Mammalia, Rodentia) Neogenes d'Europe Occidentale. *Geobios*, 3, 7–77.

Meng, J. (1990). The auditory region of *Reithroparamys delicatissimus* (Mammalia, Rodentia) and its systematic implications. *American Museum Novitates*, 2972, 1–35.

Merriam, C. H. (1889). Description of fourteen new species and one new genus of North American mammals. *North American Fauna*, 2, 1–52.

(1892). The geographic distribution of life in North America with special reference to the Mammalia. *Proceedings of the Biological Society of Washington*, 7, 1–64.

Merriam, J., Stock, C. C., and Moody, C. L. (1925). The Pliocene Rattlesnake Formation and fauna of eastern Oregon, with notes on the geology of the Rattlesnake and Mascall deposits. *Carnegie Institution of Washington Publication*, 347, 43–92.

Miller, W. E. (1980). The late Pliocene Las Tunas local fauna from southernmost Baja California, Mexico. *Journal of Paleontology*, 54, 762–805.

Montgelard, C., Bentz, S., Tirard, C., Verneau, O, and Catzeflis, F. (2002). Molecular systematics of Sciurognathi (Rodentia): the mitochondrial cytochrome *b* and 12S rRNA genes support the Anomaluroidea (Pedetidae and Anomaluridae). *Molecular Phylogenetics and Evolution*, 22, 220–33.

Moore, J. C. (1959). Relationship among the living squirrels of the Sciurinae. *Bulletin of the American Museum of Natural History*, 118, 159–206.

Murie, J. A. and Michener, G. R. (ed.) (1984). *The Biology of Ground-Dwelling Squirrels*. Lincoln, NB: University of Nebraska Press.

Nedbal, M. A., Honeycutt, R. L., and Schlitter, D. A. (1996). Higher-level systematics of rodents (Mammalia, Rodentia): evidence from the mitochondrial 12S rRNA gene. *Journal of Mammalian Evolution*, 3, 201–37.

Nelson, M. E. and Miller, D. M. (1990). A Pliocene record of the giant marmot, *Paenemarmota sawrockensis*, in northern Utah. *Contributions to Geology, University of Wyoming*, 28, 31–7.

Nowak, R. M. (1999). *Walker's Mammals of the World*, 6th edn. Baltimore, MD: Johns Hopkins University Press.

Ord, G. (1815). In *A New Geographical, Historical and Commercial Grammar and Present State of the Several Kingdoms of the World*, ed. W. Guthrie, J. Ferguson, and J. Knox. Philadelphia, PA: Johnson and Warner.

Osborn, H. F. and Matthew, W. D. (1909). Cenozoic mammal horizons of western North America, with faunal lists of the Tertiary Mammalia of the West. *United States Geological Survey Bulletin*, 361, 1–138.

Piaggio, A. J. and Spicer, G. S. (2001). Molecular phylogeny of the chipmunks inferred from mitochondrial cytochrome *b* and cytochrome oxidase II gene sequence. *Molecular Phylogenetics and Evolution*, 20, 335–50.

Pocock, R. I. (1923). The classification of the Sciuridae. *Proceedings of the Zoological Society of London*, 1923, 209–46.

Potts, R., Behrensmeyer, A. K., Taggart, R. E. *et al.* (1992). Late Cenozoic terrestrial ecosystems. In *Terrestrial Ecosystems Through Time: Evolutionary Paleoecology of Terrestrial Plants and Animals*, ed. A. K. Behrensmeyer, J. D. Damuth, W. A. DiMichele *et al.*, pp. 419–541. Chicago, IL: University of Chicago Press.

Pratt, A. E. and Morgan, G. S. (1989). New Sciuridae (Mammalia: Rodentia) from the early Miocene Thomas Farm local fauna, Florida. *Journal of Vertebrate Paleontology*, 9, 89–100.

Rafinesque. C. (1817). Description of seven new genera of North American quadrupeds. *American Monthly Magazine*, 2, 44–6.

Rensberger, J. M. (1975). *Haplomys* and its bearing on the origin of the aplodontoid rodents. *Journal of Mammalogy*, 56, 1–14.

Repenning, C. A. (1962). The giant ground squirrel *Paenemarmota*. *Journal of Paleontology*, 36, 540–56.

Robertson, J. S. (1976). Latest Pliocene mammals from Haile XV A, Alachua County, Florida. *Bulletin of the Florida State Museum, Biological Sciences*, 20, 111–86.

Scharf, D. W. (1935). A Miocene mammalian fauna from Sucker Creek, southeastern Oregon. *Carnegie Institution of Washington Publication*, 453, 97–118.

Shotwell, J. A. (1956). Hemphillian mammalian assemblage from northeastern Oregon. *Bulletin of the Geological Society of America*, 67, 717–38.

(1970). Pliocene mammals of southeast Oregon and adjacent Idaho. *Museum of Natural History University of Oregon Bulletin*, 17, 1–30.

Shotwell, J. A. and Russell, D. E. (1963). Mammalian fauna of the upper Juntura Formation, the Black Butte local fauna. *Transactions of the American Philosophical Society*, 53, 42–69.

Simpson, G. G. (1945). The principles of classification and a classification of mammals. *Bulletin of the American Museum of Natural History*, 85, 1–350.

Skwara, T. (1986). A new "flying squirrel" (Rodentia: Sciuridae) from the Early Miocene of southwestern Saskatchewan. *Journal of Vertebrate Paleontology*, 6, 290–4.

Steppan, S. J., Akhverdyan, M. R., Lyapunova, E. A. *et al.* (1999). Molecular phylogeny of the marmots (Rodentia: Sciuridae): tests of evolutionary and biogeographic hypotheses. *Systematic Biology*, 48, 715–34.

Stevens, M. S. (1977). Further study of Castolon local fauna (Early Miocene) Big Bend National Park, Texas. *The Pearce-Sellards Series, Texas Memorial Museum*, 28, 1–69.

Stirton, R. A. and Goeriz, H. F. (1942). Fossil vertebrates from the superjacent deposits near Knights Ferry, California. *University of California Publications in Geological Sciences*, 26, 447–72.

Storer, J. E. (1978). Rodents of the Calf Creek local fauna (Cypress Hills Formation, Oligocene, Chadronian), Saskatchewan. *Saskatchewan Museum of Natural History Contributions*, 1, 1–54.

Strain, W. S. (1966). Blancan mammalian fauna and Pleistocene formations, Hudspeth County, Texas. *Bulletin of the Texas Memorial Museum*, 10, 1–55.

Sutton, J. F. and Korth, W. W. (1995). Rodents (Mammalia) from the Barstovian (Miocene) Anceney Local Fauna, Montana. *Annals of the Carnegie Museum*, 64, 267–314.

Thomas, W. K. and Martin, S. L. (1993). A recent origin of marmots. *Molecular Phylogenetics and Evolution*, 2, 330–6.

Thorington, R. W., Jr. (1984). Flying squirrels are monophyletic. *Science*, 225, 1048–50.

Vaughan, T. A., Ryan, J. M., and Czaplewski, N. J. (2000). *Mammalogy*, 4th edn. Fort Worth, TX: Saunders College Press.

Vianey-Liaud, M. (1985). Possible evolutionary relationships among Eocene and lower Oligocene rodents of Asia, Europe and North America. In *Evolutionary Relationships Among Rodents: A Multidisciplinary Analysis*, ed. W. P. Luckett and J.-L. Hartenberger, pp. 277–309. New York: Plenum Press.

Voorhies, M. R. (1988). The giant marmot *Paenemarmota sawrockensis* (new combination) in Hemphillian deposits of northeastern Nebraska. *Transactions of the Nebraska Academy of Sciences*, 16, 165–72.

Wahlert, J. H. (1985). Cranial foramina of rodents. In *Evolutionary Relationships Among Rodents: A Multidisciplinary Analysis*, ed. W. P. Luckett and J.-L. Hartenberger, pp. 311–32. New York: Plenum Press.

Webb, S. D. (1974). Chronology of Pleistocene land mammals in Florida. In *Pleistocene Mammals of Florida*, ed. S. D. Webb, pp. 5–31. Gainesville, FL: University of Florida Press.

Wilson, D. E. and Reeder, D. M. (eds.) (1993). *Mammal Species of the World*. Washington, DC: Smithsonian Institution Press.

Wilson, R. W. (1937). New middle Pliocene rodent and lagomorph faunas from Oregon and California. *Carnegie Institution of Washington Publication*, 551, 113–34.

(1949a). Early Tertiary rodents of North America. *Carnegie Institution of Washington Publications*, 584, 67–164.

(1949b). On some White River fossil rodents. *Carnegie Institution of Washington Publications*, 584, 27–50.

(1949c). Rodents of the Rincon fauna, Western Chihuahua, Mexico. *Carnegie Institution of Washington Publications*, 584, 165–76.

(1960). Early Miocene rodents and insectivores from northeastern Colorado. *University of Kansas Paleontological Contributions, Vertebrata*, 7, 1–92.

Wood, A. E. (1937). The mammalian fauna of the White River Oligocene. Part II. Rodentia. *Transactions of the American Philosophical Society*, 28, 155–269.

(1955). A revised classification of the rodents. *Journal of Mammalogy*, 36, 165–87.

(1962). The early Tertiary rodents of the family Paramyidae. *Transactions of the American Philosophical Society*, 52, 1–261.

(1980). The Oligocene rodents of North America. *Transactions of the American Philosophical Society*, 70, 3–68.

Zakrzewski, R. J. (1969). The rodents from the Hagerman local fauna, upper Pliocene of Idaho. *Contributions from the Museum of Paleontology, University of Michigan*, 23, 1–36.

22 Aplodontoidea

LAWRENCE J. FLYNN[1] and LOUIS L. JACOBS[2]

[1]*Peabody Museum, Harvard University, Cambridge, MA, USA*

[2]*Southern Methodist University, Dallas, TX, USA*

INTRODUCTION

The Aplodontoidea is a Holarctic radiation of small- to medium-sized, sometimes large, rodents, with but a single surviving species, *Aplodontia rufa*, the mountain beaver of the Pacific Northwest. Aplodontoids have been viewed traditionally as a North American group, although some European members have been known for half a century. That view has been tempered further by finds stemming from the growing Asian record that reveal diverse Paleogene and Neogene members of the group on that continent. However, a center of diversity in the mid Cenozoic of North America is still apparent. Consequently, the origin and major radiation of the taxon were probably North American, but repeated dispersal of aplodontoid rodents followed by cladogenesis of higher taxa punctuated the Eurasian record.

Aplodontoids are distinguished by protrogomorphy, a condition of the zygomasseteric musculature in which the masseter origin is restricted primarily to the ventral part of the zygoma (see Chapter 16). The degree of protrogomorphy seen in these rodents is of unclear polarity; it is usually taken as a primitive condition but is possibly secondarily derived.

Aplodontoids were diverse in the western part of North America during the Oligocene and Miocene epochs. The late middle Eocene/early Oligocene aplodontoid radiation produced two distinctive mid-Tertiary groups. The Allomyinae–Aplodontinae appeared in the later early Oligocene, diversified, and survive today in *Aplodontia*. The Mylagaulidae originated in the late Oligocene and were diverse through the Miocene. Aplodontoids are small- to medium-sized rodents that show dental specializations, including extra cusps on cheek teeth, development of an ectoloph on the upper cheek teeth, and presence of molar styles. Derived aplodontines attain large body size for rodents.

The Mylagaulidae are characterized by a greatly enlarged fourth premolar, and in derived forms by bony protuberances of the nasal bones; these are the bizarre horned rodents. Mylagaulids are medium- to large-sized rodents with broad, flattened skulls that show numerous adaptations for burrowing. The lower jaw is deep, especially at the diastema, to accommodate a fourth premolar that becomes progressively larger and higher crowned as an evolutionary trend. In derived species, the premolar of the adult crowds out the first or even second molar during the process of its eruption.

Some authors have used the family Prosciuridae as a basal taxon, but current work has aligned most of its members with Aplodontidae. Wood (1980) and Korth (1994a) use Prosciurinae as a subfamily, but we do not in this study. McKenna and Bell (1997) use Allomyidae Marsh, 1877 for primitive aplodontids plus allomyines. Meng (1990) conducted a thorough analysis of the ear region of the Eocene genus *Reithroparamys*, arguing for a special relationship with aplodontids, squirrels, and dormice. Possibly this Eocene genus will emerge as the oldest known aplodontoid.

DEFINING FEATURES OF THE SUPERFAMILY APLODONTOIDEA

CRANIAL

The zygomasseteric musculature is protrogomorphous, although presence of some fibers of the masseter muscle in the infraorbital foramen has been noted in *Aplodontia rufa* (Eastman, 1982). This may be consistent with a view that confinement of the masseter to the ventral part of the zygomatic arch is derived, rather than primitive. The vertical infraorbital foramen is small, and the diastema is relatively short. Aplodontoids have a small interpremaxillary foramen or pit in this position; the incisive foramina are short; the posterior palatine foramina are shifted rather far posteriorly; the ethmoid foramen occurs in the frontal bone (Wahlert, 1974).

Evolution of Tertiary Mammals of North America, Vol. 2. ed. C. M. Janis, G. F. Gunnell, and M. D. Uhen. Published by Cambridge University Press.

The Mylagaulidae are renowned among the Rodentia in developing horns from nasal bones in terminal sister genera *Epigaulus* and *Ceratogaulus*. The cranium is low and broad, with nearly vertical occiput. There are strong lambdoidal and sagittal crests. The rostrum is short and heavy, and there are prominent postorbital processes. The coronoid process is high and broad. The auditory bulla is somewhat inflated, and there is a bony, elongate external auditory meatus (see Korth 1994a) (Figure 22.1).

DENTAL

The incisor enamel microstructure shows the derived uniserial pattern. Cheek teeth become high crowned and complexly crested in later members of the superfamily. The fourth premolar, upper and lower, tends to be the largest tooth of the cheek tooth row and dominates the molars in later forms. The molars tend to have high, unbroken walls, with extra styles and crests added in most forms. The upper cheek teeth usually lack a developed hypocone but show conules; lower cheek teeth have a mesoconid and usually a mesostylid.

Within the Aplodontidae, the cheek tooth formula is P2/1, M3/3. The upper molars show little or no development of the hypocone, conules are strong, the paracone and metacone are subtriangular. An ectoloph with prominent mesostyle, and sometimes parastyle, may be developed in the upper cheek teeth, while the lower cheek teeth show mesostylids and crested mesoconids. Crests isolate enamel lakes in the cheek teeth and the incisor enamel is uniserial.

Within the Mylagaulidae, the incisors are heavy and gently convex anteriorly. The upper cheek tooth rows converge posteriorly. The P3 is present primitively, lost in later forms, and later forms lose molars, as noted above. The fourth premolars are the largest cheek teeth. Numerous infoldings of enamel on cheek teeth are manifest after wear as isolated fossettes (-ids) on a relatively flat occlusal surface, a means of increasing cutting surfaces during mastication. The nature of these "lakes" and ontogenetic change in them caused by wear should be understood in terms of their origin as infoldings, not as entities in themselves.

POSTCRANIAL

Aplodontoidea in general tend toward moderate and large size and show terrestrial and fossorial skull and postcranial adaptations. The postcranial skeleton of mylagaulids is replete with fossorial adaptations (see Fagan, 1960).

Fagan (1960: Fig. 1) performed an analysis of fossorial features in mylagaulids from the Barstovian (middle Miocene). She noted that the forelimb shows typical burrowing features: heavy muscle scars, broad condyles, expanded forefoot and claws. However, the skull shows that this animal was a chisel tooth-digger that used the forelimbs to move dirt, and Fagan goes on to suggest that nasal ornamentation and, ultimately, horn development in later mylagaulids could have been related to the subterranean mode of life.

SYSTEMATICS

SUPRA-SUPRAFAMILIAL

Aplodontoidea Brandt, 1855 are distinguished from extant rodent groups by their protrogomorphous zygomasseteric musculature. They are presumed to be derived from ischyromyoid rodents of the North American early Tertiary, which were also protrogomorphous. Lack of advanced jaw musculature would seem to rule out close relationship with any other extant rodent group, because these are characterized by sciuromorphy, hystricomorphy, or myomorphy. However, it is possible that the protrogomorphous condition is secondarily simplified, which could cloud relationships. A recent molecular study (Adkins, Walton, and Honeycutt, 2002) supported phylogenetic relationship to squirrels (Sciuridae). This hypothesis has been suggested by morphologists who noted that early aplodontoid molars resemble squirrel molars in arrangement of cusps and lack of a hypocone.

INFRA-SUPRAFAMILIAL

The content of the Aplodontidae Brandt, 1855 has fluctuated somewhat as researchers have attempted to redefine the group in a way that includes all primitive members of the clade. In particular, the prosciurines vary in number of derived features of the family and do not hold together as a subfamily. Wang (1987) supported prosciurine affinity of Asian *Anomoemys lohiculus* (formerly a species of *Prosciurus*), but later workers consider this a cylindrodontid (Korth, 1994a). *Cedromus wardi* Wilson, 1949, for years accepted as a prosciurine, was argued by Wood (1980) to be a squirrel related to *Protosciurus jeffersoni*, and Korth and Emry (1991) erected Cedromurinae within Sciuridae for it. Note, however, that Wood's observations could be used to argue that *Protosciurus* is not a squirrel, a contention long held by Vianey-Liaud (1985).

The array of late Paleogene taxa that are more clearly allied with advanced aplodontids and mylagaulids have been grouped in higher taxa such as Prosciuridae. The group is not monophyletic in that it includes the origins of both Aplodontidae and Mylagaulidae, so it is not used here. Some interrelationships of genera are evident, as noted below, but naming higher taxa at this point does not seem useful. Among more derived aplodontoids, Allomyinae, Meniscomyinae, and Ansomyinae (see below) are distinctive groups. Allomyines appear close to Aplodontinae. At present, Meniscomyinae appear to be related to the mylagaulids, implying that meniscomyines should perhaps be transferred to Mylagaulidae Cope, 1881 (Figure 22.2).

The Mylagaulidae are closely related to Aplodontidae, sharing such features as uniserial incisor enamel and similarly arrayed cranial foramina (Wahlert, 1974). The family is in sore need of revision due to apparently greater morphological variation in dental pattern and sexual dimorphism than is seen in most rodents. Korth's (2000a) study recognized suites of cranial and dental features that are useful in unraveling the systematics of the group. The mylagualid subfamily Promylagaulinae is distinguished by a number of characteristic traits, but most appear to be primitive. Korth (2000a) noted derived

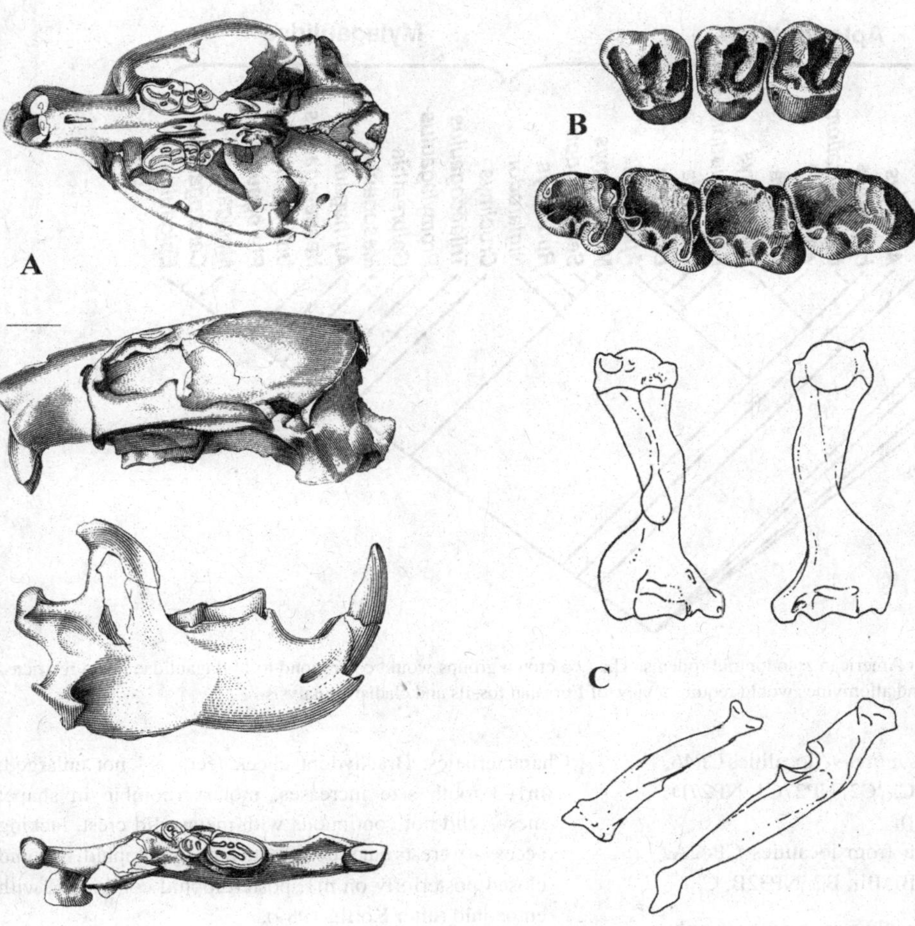

Figure 22.1. A. Skull of *Galbreathia novellus* in ventral and lateral views, and the right dentary in lateral and dorsal views. Scale bar = 1 cm. B. Dentition of the primitive aplodontoid, *Prosciurus*: left P4 M1–2 above; left p4 m1–3 below; anterior to left for both (McGrew, 1941). C. Forelimb anatomy of *Pterogaulus laevis*: anterior and posterior views of right humerus above; lateral view of left radius and ulna below (from Fagan, 1960). (B: courtesy of the Field Museum of Natural History. C: courtesy of the University of Kansas.)

features of *Mesogaulus* relative to promylagaulines and created subfamily Mesogaulinae for that genus.

Note that Hopkins (2005a) provided a reanalysis of the phylogenetic relationships of the genera within the family Mylagaulidae.

INCLUDED GENERA OF APLODONTOIDEA BRANDT, 1855

The locality numbers listed for each genus refer to the list of unified localities in Appendix I. The locality numbers may be listed more than one way. The acronyms for museum collections are listed in Appendix III.

Parentheses around the locality (e.g., [CP101]) mean that the taxon in question at that locality is cited as an "aff." or "cf." the taxon in question. Parentheses are usually used for individual species, implying that the genus is firmly known from the locality, but the actual species identification may be questionable. Question marks in front of the locality (e.g., ?CP101) mean that the taxon is questionably known from that locality, implying some doubt that the taxon is actually present at that locality, either at the genus or species level. An asterisk (*) indicates the type locality.

APLODONTIDAE BRANDT, 1855

BASAL APLODONTIDS

Prosciurus Matthew, 1903 (including *Paramys*, in part)

Type species: *Prosciurus vetustus* Matthew, 1903.

Type specimen: AMNH 9626, partial left maxilla with P3–M3.

Characteristics: Cheek teeth brachydont; upper molars with continuous anteroloph–protocone–posteroloph crest; hypocone absent or minute; incomplete metaloph, sometimes double metaconules; parastyle on P4; rhombic lower molars with broad, shallow basins, strong mesoconid, posterior arm of protoconid (metalqphulid II) approaching metaconid, entoconid on hypolophid (Figure 22.1).

Average length of m2: 1.88 mm (average of five specimens of *P. vetustus* from Pipestone Springs; Black, 1965).

Included species: *P. vetustus* (including *P. minor* Russell, 1972, fide Storer, 1978) (known from localities CP39C, D, E, CP44, CP98B, NP10B, NP24C, D); *P. albiclivus* Korth, 1994b (localities CP99A, NP50B); *P. magnus* Korth, 1989 (localities [CP99A, B], [NP10C2]); *P. parvus* Korth, 1989 (localities CP99A, [NP32B, C]); *P. relictus* (Cope, 1873)

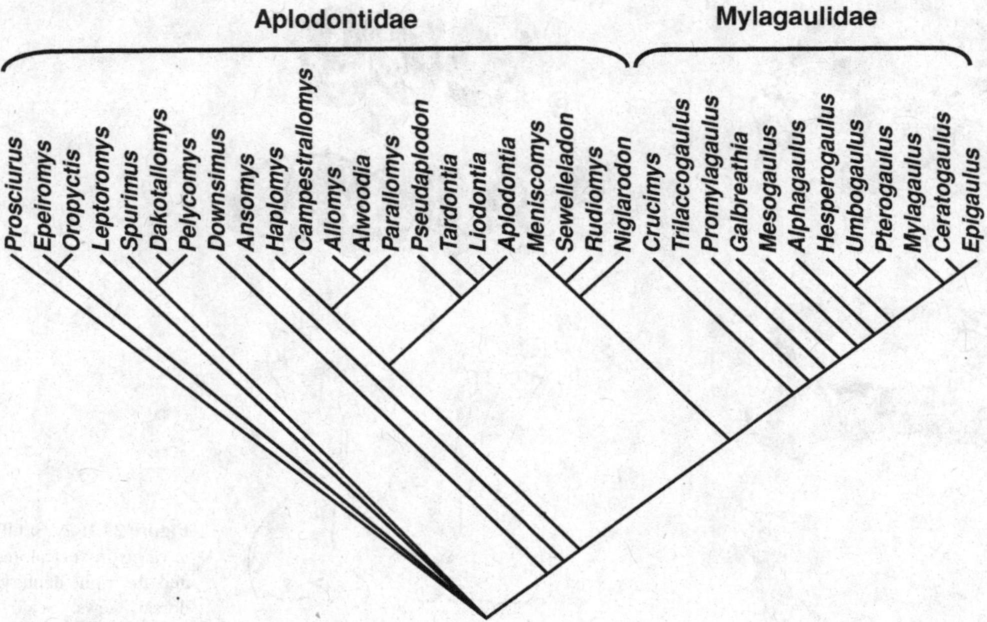

Figure 22.2. Proposed relationships of North American aplodontoid rodents. The two crown groups would correspond to Mylagaulidae and a restricted Aplodontidae. Resolution of "prosciurines" and allomyines would require review of Eurasian fossils and cladistic analysis.

(originally described as *Paramys relictus*) (localities CP46, CP68C, CP84A, CP99A, NP10C, ?C2, NP27C1, NP27D, [NP29D] [see also Wood, 1980]).

Prosciurus sp. is also known from localities CP42A, CP84B, CP85A, ?NP9B, NP10Bi, Bii, B2, NP32B, C.

Comments: *Prosciurus* was originally proposed as a subgenus of *Sciurus* (Matthew, 1903). Storer (1988) saw prosciurine affinity with material from locality NP9A, and (1978) referred Cypress Hills material to *Ardynomys saskatchewanensis*. Asian *P. lohiculus* was removed from *Prosciurus* and made the type species of *Anomoemys* by Wang (1987), but she cited *P. ordosicus* from Inner Mongolia, so the genus still has members on two continents.

Oropyctis Korth, 1989

Type species: *Oropyctis pediasius* Korth, 1989.

Type specimen: USNM 56397, left dentary with dp4, p4, m1–3.

Characteristics: Brachydont, non-lophate cheek teeth; smooth enamel; p4 longer than m1; incomplete hypolophid in p4, m2–3 (weak on m1), weak mesoconid and buccal mesolophid, mesostylid connected to metastylid crest (after Korth, 1989).

Average length of m2: 2.45 mm.

Included species: *O. pediasius* only, known from locality CP99B only.

Epeiromys Korth, 1989

Type species: *Epeiromys spanius* Korth, 1989.

Type specimen: USNM 56399, right dentary with m1–3.

Characteristics: Brachydont cheek teeth; p4 not enlarged; m1–3 tooth size increases; molars rhombic in shape; mesostylid not continuous with metastylid crest, lacking accessory crests but pitted enamel; no hypolophid, trigonid closed posteriorly on m1, posterolophid continuous with entoconid (after Korth, 1989).

Average length of m2: 2.30 mm.

Included species: *E. spanius* only, known from locality CP99A only.

Leptoromys Tedrow and Korth, 1997

Type species: *Leptoromys wilsoni* Tedrow and Korth, 1997.

Type specimen: SDSM 63629, left dentary fragment with p4m1–3.

Characteristics: Brachydont cheek teeth with broad enamel irregularities in basins; multiple mesostylids, broad hypolophids uniting with ectolophids, curving mesostylid loph running into talonid basin (after Tedrow and Korth, 1997).

Average length of m1: 2.1 mm.

Included species: *L. wilsoni* only, known from locality CP84A only.

Leptoromys sp. is also known from locality CP84B.

Spurimus Black, 1971

Type species: *Spurimus scottii* Black, 1971.

Type specimen: CM 16088, right M1 or 2.

Characteristics: Cheek teeth brachydont; narrow anterior cingulum low on upper cheek teeth; large hypocone on M1–2; incomplete metaloph ending in small metaconule; small trigonid basin enclosed on lower molars; notch between anterior cingulum and protoconid; complete hypolophid, entoconid separate from posterolophid, mesoconid weak or absent.

Average length of M1 or M2: 2.06 mm (average of five hypodigm; Black, 1971).

Included species: *S. scottii* and *S. selbyi* Black, 1971 (known from localities CP29C, D, CP36B).

Spurimus sp. is also known from localities CP44, ?NP8.

***Pelycomys* Galbreath, 1953**

Type species: *Pelycomys rugosus* Galbreath, 1953.

Type specimen: KU 8343, right dentary fragment with m1–3.

Characteristics: Large, brachydont, cheek teeth subtriangular to subrhombic, round and inflated cusps; large trigonids with metalophulid II nearly complete; strong, separate hypolophid; small metaconid, entoconid separated anteriorly and posteriorly by notches.

Average length of m2: 3.20 mm (holotype).

Included species: *P. rugosus* (known from locality CP68B); *P. brulanus* Korth, 1986 (locality CP99A); *P. placidus* Galbreath, 1953 (localities CP46, CP68C, CP99A [see Korth, 1989]).

Pelycomys sp. is also known from localities CP85A, CP99B, NP32B, C.

***Dakotallomys* Tedrow and Korth, 1999 (including *Dakotamys* Tedrow and Korth, 1997, not *Dakotamys* Eaton, 1995)**

Type species: *Dakotallomys pelycomyoides* (Tedrow and Korth, 1997) (originally described as *Dakotamys pelycomyoides*).

Type specimen: UCMP 82834, right mandible with p3–m3.

Characteristics: Large size, nearly equal to *Pelycomys*; cheek teeth brachydont; mesostylid on lower cheek teeth small and attached to metastylid crest as in some species of *Pelycomys*; differs in presence of a unique labial cingulum on protoconid and hypoconid of lower cheek teeth; hypolophid complete on m1-m2 as in *Pelycomys*; lower incisor wider in cross section relative to length than *Pelycomys* and other prosciurines (from Tedrow and Korth, 1997).

Average length of m2: 3.18 mm (holotype).

Included species: *D. pelycomyoides* (known from locality CP84B); *D. lillegraveni* (Tedrow and Korth, 1997) (originally described as *Dakotamys lillegraveni*) (locality CP84B).

Comments: Tedrow and Korth (1997) suggest the possibility of a clade comprising *Spurimus, Pelycomys*, and *Dakotallomys*.

***Downsimus* Macdonald, 1970 (including *Allomys*, in part)**

Type species: *Downsimus chadwicki* Macdonald, 1970 (originally described as *Allomys sharpi*, Macdonald, 1970).

Type specimen: LACM 17031, right dentary fragment with p4, m1–2.

Characteristics: Rhomboidal molars, trigonid basin not enclosed; incomplete metalophulid II, strong mesostylid and mesoconid, hypoconulid with anterior crest on midline; entoconid isolated; short hypolophid; p4 with protoconid, metaconid, mesoconid, and mesostylid.

Average length of m2: 2.41 mm.

Included species: *D. chadwicki* (including "*Allomys sharpi* Macdonald, 1970", fide Rensberger, 1983) only (known from localities CP85C, CP101, NP10CC).

Downsimus sp. is also known from locality NP36A.

***Campestrallomys* Korth, 1989 (including *Prosciurus*, in part)**

Type species: *Campestrallomys dawsonae* (Macdonald, 1963) (originally described as *Prosciurus dawsonae*).

Type specimen: SDSM 56112, left dentary with incisor, p4m1–3.

Characteristics: Cheek teeth brachydont; moderately crescentic buccal cusps on upper molars with ectoloph and large mesostyle; protoconule moderate in size and metaconule single; double (triple on m3) mesostylids on lower cheek teeth connecting to the metastylid crest; hypolophids very weak to absent; lower incisor flattened anteriorly (after Korth, 1989; see also his comments on the auditory bulla).

Average length of m2: 1.9 mm (*C. annectens*).

Included species: *C. dawsonae* (known from localities CP85C, CP101, [NP10CC]); *C. annectens* Korth, 1989 (locality CP99A); *C. siouxensis* Korth, 1989 (localities CP99B, C).

Cf. *Campestrallomys* sp. (including "cf. *Sespemys*" from Wood, 1937) is also known from locality CP99C.

***Ansomys* Qiu, 1987 (including *Pseudallomys* Korth, 1992a)**

Type species: *Ansomys orientalis* Qiu, 1987.

Type specimen of *A. nexodens*: CM 11898, right dentary with m1–3.

Characteristics: Brachydont cheek teeth with high cusps and crests; large and bifid mesostyle; ectoloph with straight components (flat labial walls of paracone and metacone, straight wall bridging mesostyles); protocone convex dorsoventrally; single metaconule; complex of small accessory crests in lower molar basins; molars increasing in size from m1 to m3; main cusps anteroposteriorly compressed; cingulum anterior to protoconid; complete hypolophid; broad shelf at base of crown buccal to large mesoconid; double mesostylid on m2–3 with buccally directed crests; large hypoconulid (after Korth, 1992a; Hopkins, 1999).

Length of m2 of *A. nexodens*: 1.99 mm.

Average length of m2 for *A. hepburnensis*: 1.52 mm.

Included species: *A. hepburnensis* Hopkins, 1999 (known from locality NP40B); *A. nexodens* (Korth, 1992a) (originally described as *Pseudallomys nexodens*) (locality NP30B).

Comments: Korth (1992a) recognized a new taxon for North America and referred it to Old World *Pseudallomys*, based on a single lower jaw. Hopkins (1999) found that her excellent sample included the features of Korth's *P. nexodens*, and saw, based on complete dentitions, that the North American fossils demonstrate occurrence of the Asiatic *Ansomys* on this continent. *Ansomys* is a derived aplodontid, but Hopkins noted conditions precluding affinity with Allomyinae, including low crown height and low lower molar crests, curvature of the protocone, and compressed metaconid. *Ansomys* lacks the allomyine double metaconule.

Qiu (1987) similarly saw *Ansomys* (known from the middle Miocene of Asia) as representing a primarily Asiatic aplodontid radiation and defined Ansomyinae. Characteristics include the bifid to lophate mesostyle, often not closing off the central transverse valley; ectoloph flat at paracone; single metaconule; no hypocone; basined lower cheek teeth with metaconid–metastylid crest; metaconid pinched anteroposteriorly; anteroconid crest absent; mesostylid with long crest; transverse metalophulid II; entoconid anterior to hypoconid, large hypoconulid and mesoconid; and p4 with metaconid crest running posteriorly.

Qiu's (1987) analysis of character polarity led to a cladogram of relationships in which he saw Ansomyinae as more closely related to allomyines plus aplodontines, than to Meniscomyinae. Meniscomyinae appear to be the sister taxon to Mylagaulidae, with the implication that mylagaulids (below) are a subset of the aplodontids, derived in many features, but perhaps properly seen as submerged in Meniscomyinae.

ALLOMYINAE MARSH, 1877

Rensberger (1983) redefined this group as brachydont aplodontids, but with crested cheek teeth.

Characteristics: Upper molars with mesostyle closing a transverse valley; ectoloph crestlike (not cuspate) and concave at the paracone; double metaconule; small hypocone developed on cingulum joining protocone. Lower cheek teeth with strong mesostylid and a crest joining it to the metaconid; mesostylid running lingual to a line joining the metaconid–entoconid, lacking labially directed crest; posteriorly directed crest from the small anteroconid; small mesoconid in strongly labial position, hypolophid incomplete, leaving a large posterior fossettid; p4 is only slightly larger than m1 and with posteriorly directed internal metaconid crest; incisors relatively narrow.

Haplomys Miller and Gidley, 1918 (including *Meniscomys*, in part)

Type species: *Haplomys liolophus* (Cope, 1881) (originally described as *Meniscomys liolophus*).

Type specimen: AMNH 6987, left maxilla fragment with P4M1–2.

Characteristics: Larger and with higher-crowned cheek teeth than most prosciurines; strongly expanded parastyle on P4; triangular paracone and metacone; distinct mesostylid and metastylid; triangular mesoconid with buccal crest converging toward hypoconid; hypolophid joins ectolophid posterior to mesoconid (see Rensberger, 1975).

Average length of m2: 1.6 mm (*H. galbreathi*).

Included species: *H. liolophus* (known from localities PN6C1, C2); *H. galbreathi* Tedrow and Korth 1997 (locality CP84A).

Haplomys sp. is also known from localities CP84B, C, CP85A, [CP101].

Comments: Rensberger (1975) showed that *Haplomys* presents many allomyine features and argued that it revealed Prosciurinae as a subfamily of Aplodontidae. Further, his analysis could be used to support inclusion of the genus in Allomyinae. Wang (1987) agreed, associating *H. arboraptus* from the late Oliogcene of Inner Mongolia with allomyine aplodontids.

Parallomys Rensberger, 1983 (including *Plesispermophilus*, in part)

Type species: *Parallomys ernii* (Stehlin and Schaub, 1951) (originally described as *Plesispermophilus ernii*).

Type specimen: Museum Basel Coderet 2180, left dentary with p4-m1, early Miocene of Europe.

Characteristics: Cheek teeth with relatively weak lophs and no accessory crests; upper molars with broad transverse valley; enamel smooth to faintly crenulated; labial faces of paracone and metacone sloping strongly; metastylar crest trending labially; mesoconid not connected to hypoconid (after Rensberger, 1983).

Average length of M1 or M2: 2.2 mm (holotype of *P. americanus*).

Included species: *P. americanus* only (known from localities CP103A, [NP10CC]; see Korth, 1992b).

Comments: This genus is known mainly from late Oligocene material from Europe.

Allomys Marsh, 1877 (including *Meniscomys*, in part)

Type species: *Allomys nitens* Marsh, 1877.

Type specimen: YPM 13604, palate with left P3–M3, right P4–M2.

Characteristics: Upper cheek teeth wide and with narrow central transverse valley; weak metaloph and protoloph crests, other crests variably developed; lower cheek teeth with tendency to develop accessory crests; posterointernal crest from mesoconid joins hypoconid, forming fossettid (after Rensberger, 1983; his Fig. 3).

Length of M1: 2.4 mm (holotype).

Included species: *A. nitens* (known from localities PN6E, F); *A. cavatus* (Cope, 1881) (originally described as *Meniscomys cavatus*) (locality PN6C2 [see Rensberger, 1983]); *A. cristabrevis*, Barnosky, 1986 (locality CP45B); *A.*

reticulatus Rensberger, 1983 (locality PN6F); *A. simplicidens* Rensberger, 1983 (locality PN6D1); *A. stirtoni* Klingener, 1968 (localities CP114A, [B]); *A. storeri* Tedrow and Korth, 1997 (locality CP84B); *A. tesselatus* Rensberger, 1983 (locality PN6G).

Allomys sp. is also known from localities ?CP49B, ?CP85A, NP10CC, E, NP36B, PN6D2, PN16.

***Alwoodia* Rensberger, 1983 (including *Allomys*, in part)**

Type species: *Alwoodia magna* Rensberger, 1983.

Type specimen: UCMP 76941, right maxilla fragment with P3–M3.

Characteristics: Large allomyine with narrow incisors; crests on cheek teeth high, heavy, elongate, and without accessory processes; upper cheek teeth with constricted central valleys; posterolabial fossettid as in *Allomys*.

Length of M1: 2.4 mm (holotype).

Included species: *A. magna* (known from locality PN6D1); *A. harkseni* (Macdonald, 1963) (originally described as *Allomys harkseni*) (localities CP86B, CP102, CP103A); *A.* sp. (locality ?NP10CC).

Comments: We follow Korth (1994a) in placing *A. harkseni* here. The species was not formally placed in this genus by Rensberger (1983). *Alwoodia* is stratigraphically low, but morphologically derived for the John Day allomyines.

APLODONTINAE BRANDT, 1855

The large, hypsodont aplodontid genera are a Neogene radiation of the last 15 million years, mainly North American, but also Asian.

Characteristics: Most enamel lakes in cheek teeth disappear during wear; upper cheek teeth with very prominent mesostyles, other styles reduced, including the parastyle on P4; trigonids reduced in the lower molars (see Rensberger, 1975).

***Tardontia* Shotwell, 1958 (including *Pseudaplodon*, in part)**

Type species: *Tardontia occidentale* (Macdonald, 1956) (originally described as *Pseudaplodon occidentale*).

Type specimen: UCMP 38662, left dentary fragment with p4m1–3.

Characteristics: High-crowned, rooted cheek teeth; lower cheek teeth with anterolabial fold that persists despite wear; m3 unreduced, with length greater than that of other molars; no mesostylids on molars; weak mesostylid on p4; anterior reentrant on p4 extending to base of crown, posterior reentrant on m3 opening to base of crown.

Length of m2: 2.5 mm (holotype).

Included species: *T. occidentale* (known from localities NB25B, NB29A, and probably PN10 and Montana, according to Shotwell, 1958); *T. nevadans* Shotwell, 1958 (locality NB23B).

Tardontia sp. *A* is also known from locality PN11B and is advanced in having rootless cheek teeth (Shotwell, 1970); *Tardontia* sp. *B* is also known from locality NP40B (Barnosky and Labar, 1989).

Comments: Macdonald (1956) originally assigned the genotypic species to *Pseudaplodon* Miller, 1927, comparing it to *P. asiatica* (Schlosser, 1924). The latter species, from the terminal Miocene of Inner Mongolia, also has rooted teeth, lack of mesostylids on lower molars, presence of anterolabial reentrants on lower teeth, posterior reentrant on m3, and anterior reentrant on p4. *P. asiatica* differs in its reduced m3, lower crown height, and retention of enamel lakes into late wear. Shotwell (1958) reviewed the controversial history of *P. asiatica* and showed that it is considerably less derived than American *Tardontia*, despite its younger age as we now know it.

***Liodontia* Miller and Gidley, 1918 (including *Aplodontia*, in part)**

Type species: *Liodontia alexandrae* (Furlong, 1910) (originally described as *Aplodontia alexandrae*).

Type specimen: UC 11325, palate.

Characteristics: Smaller size and less flattened skull roof than in *Aplodontia* Richardson, 1829; strong masseteric crest, short lower diastema; cheek teeth hypsodont, tooth growth at least until old age; mesostylids weak or absent, anterolabial fold in lower cheek teeth lost in early wear, but enamel lakes persisting through moderate wear; p4 with anterior reentrant nearly to base of crown, m3 lacks posterior reentrant.

Average length of m2: 2.65 mm (Gazin, 1932).

Included species: *L. alexandrae* (known from localities NB18A, PN7, PN8A, B); *L. furlongi* Gazin, 1932 (localities NB31, PN11A, B).

Liodontia sp. is also known from localities NB17, NB23B, CP10, PN14.

MENISCOMYINAE RENSBERGER, 1981

Meniscomyines have a thin-walled tympanic bulla. The group exhibits a trend of increasing tooth crown height, all taxa being basically hypsodont.

Characteristics: Upper molars with strong mesostyle closing off central valley; ectoloph concave at the paracone and metacone; single metaconule; no hypocone. Lower molars with strong mesostylid isolated from the metaconid; large central mesoconid; entoconid positioned opposite hypoconid; p4 much larger than m1; incisors are broad and flattened.

***Rudiomys* Rensberger, 1983**

Type species: *Rudiomys mcgrewi* Rensberger, 1983.

Type specimen: UCMP 105122, left dentary with incisor, m1–2, and associated left periotic and femur epiphysis.

Characteristics: Cheek teeth relatively low crowned; mesostylids bladelike, mesostylid crest arched; wide m1 trigonid; closed central fossettid on m2; mesostylid and metaconid widely separated, metaconid not inflated, prominent metastylid crest.

Length of m2: 2.2 mm (holotype).

Included species: *R. mcgrewi* only, known from locality PN6E only.

***Meniscomys* Cope, 1879 (including *Allomys*, in part)**

Type species: *Meniscomys hippodus* Cope, 1879.

Type specimen: AMNH 6962, skull with mandible.

Characteristics: Small meniscomyine; lower cheek teeth with small posterolabial fossettid; p4 narrow anteriorly, anterior inflection forms basin with wear, open anterolabial inflection, no anterolingual inflection; molar mesostylid bulbous, metastylid and metaconid closely set and without accessory crests; entoconid anterolabially elongated and may be expanded; relatively long m2–3, weak posterior inflection on m3; P4 relatively long with separate fossettes; small posterolabial fossette on M1; upper molars relatively long (Rensberger, 1983).

Length of m2: 1.7 mm (holotype).

Included species: *M. hippodus* (known from localities CP85C, PN6C2, D2, D3, E); *M. editus* Rensberger, 1983 (PN6E, but above *M. hippodus* level); *M. uhtoffi* Rensberger, 1983 (localities PN6C2, D1).

Meniscomys sp. is also known from localities CC12, CP85 ?A, D, CP88.

***Sewelleladon* Shotwell, 1958**

Type species: *Sewelleladon predontia* Shotwell, 1958.

Type specimen: UOMNH F-4734, left dentary with incisor, p4m1–3.

Characteristics: Largest known meniscomyine; p4 with open anterior and closed anterolabial inflection, anterolingual fossettid present, rounded mesoconid touched by anterolabial process from hypoconid; lower molars with flat lingual walls; m1–2 with cylindrical entoconid and small posterolingual inflection, metaconid expanded.

Length of m2: 2.4 mm (holotype).

Included species: *S. predontia* only, known from locality PN6G only.

***Niglarodon* Black, 1961 (including *Horatiomys; Meniscomys* in part)**

Type species: *Niglarodon koreneri* Black, 1961.

Type specimen: YPM 14024, right dentary with p4, m1–3.

Characteristics: Lower cheek teeth with large posterolabial fossettid; p4 mesostylid crest when present lacks distinct mesostylid; molar mesostylid compressed and poorly differentiated from its crest, elongated mesostylid crest and adjacent inflection, straight entoconid crest; m2 lacking central fossettid, labial end of posterolingual inflection unexpanded in molars, but inflection strong in m3; P4 with posterolabial fossette joining central valley, central juncture of metaloph and protoloph, protoconule joining anteroloph, posterolingual fossette deeper than anterolingual inflection; metaloph of M1–2 joining paracone and protoloph, anterolingual fossette smaller than anterolabial fossette, posterolabial and posterolingual fossettes equal and elongated, posterolabial and posterolingual inflections on M3 broadly open until late wear (from Rensberger, 1981, 1983: Fig. 4).

Length of m2: 2.0 mm (holotype).

Included species: *N. koerneri* (known from localities CP85C, NP34B, NP36A); *N. blacki* Rensberger, 1981 (localities CP45B, NP34B); *N. loneyi* Rensberger, 1981 (locality NP34B); *N petersonensis* Nichols, 1976 (locality PN16); *N. progressus* Rensberger, 1981 (locality NP34B); *N. yeariani* Nichols, 1976 (locality PN16).

Niglarodon sp. is also known from localities ?CP49B, NP36A, B, ?C, D.

Comments: *N. koerneri, N. progressus, N. blacki*, and *N. loneyi* are a time-successive sequence within NP34B; *N. yeariani* and *N. petersonensis* are superposed in PN16. *Meniscomys* appears to be a western radiation, confined to Oregon, and parallel to the montane radiation of *Niglarodon*. A meniscomyine from South Dakota (Macdonald, 1963) is listed here as *N. koerneri. N. blacki* may extend into Wyoming (Rensberger, 1981); Barnosky (1986) referred a specimen from locality CP45B to this species. *Niglarodon* includes two lineages, one in Montana and one in Idaho, which span much of Arikareean time. *N. koerneri* includes *Horatiomys montanus* Wood, 1935.

INDETERMINATE APLODONTIDS

Indeterminate aplodontids have been reported from localities NB2, NB27B, NP10C2, NP29C, NP32C.

MYLAGAULIDAE COPE 1881

PROMYLAGAULINAE RENSBERGER, 1980

Characteristics: P3 present, unmodified and unincorporated into dental battery; P4 with lingually expanded anterocone, small labial styles, elongated cusps flattened mediolaterally; M1, M2 with equal anterior and posterior cingula and anterior part of ectoloph not invaginated more than posterior part, most fossettes lost by wear, posterolingual one being most persistent; p4 with minute mesoconid expanded by wear, labial reentrant without strong anterior branch, but partially closed by expanded hypoconid; m1, m2 with persistent central fossettid, metastylid crest from metastylid to metaconid, posterolabially compressed hypoconid, strong convex posterolophid, anterolabially directed posterior fossettid; m3 reduced with posteriorly flattened hypoconid and reduced labial inflection (after Rensberger, 1980).

***Crucimys* Rensberger, 1980 (including *Meniscomys*, in part)**

Type species: *Crucimys milleri* (Macdonald, 1970) (originally described as *Meniscomys milleri*).

Type specimen: SDSM 6272, left dentary with incisor, p4, m1–3.

Characteristics: Monotypic genus known only from lower jaws; small lower jaw; distinct mesostylid on p4, no anterolingual reentrant (fossettid); m1–2 with distinct mesostylid and transversely oriented mesostylid crest, but incomplete metastylid crest (after Rensberger, 1980).

Average length of m2: 1.6 mm (Rensberger, 1980).

Included species: *C. milleri* only (known from localities CP85C, NP10CC).

Trilaccogaulus Korth, 1992b (including *Promylagaulus*, in part)

Type species: *Trilaccogaulus lemhiensis* (Nichols, 1976) (originally described as *Promylagaulus lemhiensis*).

Type specimen: UM 4038, anterior portion of skull with rostrum, palate, left P3–M2, right incisor root.

Characteristics: Size and crown height as in *Promylagaulus*; P4 oval and with only three persistent fossettes; p4 with three major fossettids, lower molars lacking mesoconid and spurring directed posteriorly from the mesoconid (after Korth, 1992b).

Length of P4: 4.16 mm (holotype, Nichols, 1976).

Included species: *T. lemhiensis* (known from locality PN16); *T. montanensis* (Rensberger, 1979) (locality NP34B [Korth, 1992b]); *T. ovatus* (Rensberger, 1979) (localities CP86B, CP103A, NP10CC).

Promylagaulus McGrew, 1941

Type species: *Promylagaulus riggsi* McGrew, 1941.

Type specimen: FMNH 26256, facial region of skull with nearly complete upper dentition with incisors and P3–M3, lacking left M3.

Characteristics: Increased cheek tooth crown height; fossettids persistent, premolars relatively large; P4 squared anteriorly, with expanded anterocone, small mesostyle, four persistent enamel lakes; labial inflection on p4, two small fossettids persist; lower molars with dentine tracts, lacking mesoconids (after Korth, 1992b).

Length of M2: 1.7–2.0 mm (holotype, after Rensberger, 1979).

Included species: *P. riggsi* only (known from localities CP47, CP86[B], D, CP103A, [NP10CC]).

Promylagaulus sp. is also known from localities CP74A, CP101.

Galbreathia Korth, 1999 (including *Mesogaulus*, in part)

Type species: *Galbreathia novellus* (Matthew, 1924) (originally described as *Mesogaulus novellus*)

Type specimen: AMNH 18911, left dentary with dp4, p4, m1–2.

Characteristics: Large promylagauline, with skull not highly modified for fossorial life, and vertical occiput; five enamel lakes on premolars; mesostyle and parastyle evident on P4; buccal position of anteroconid on p4, mesoconid crest on p4 but not on lower molars, undivided anterior fossettid on lower molars; higher crowned than *Promylagaulus*; lacking P3, unlike *Promylagaulus* and the early mylagauline *Mesogaulus*.

Average length of M2: 2.5 mm (Munthe, 1988).

Included species: *G. novellus* (known from localities CP54B, CP86A, CP108B); *G. bettae* Sutton and Korth, 1995 (locality NP41B).

Comments: Galbreath (1984), Munthe (1988), and Korth, (1992b) all have noted that "*M.*" *novellus* should not be considered congeneric with other species assigned to *Mesogaulus*. Munthe's (1988; Fig. 5) fine sample allowed understanding of the species for Korth (1992b, 1999) to explicitly tie this Hemingfordian taxon to older *Promylagaulus*, to the exclusion of close relationship with later mylagaulines.

Mylagaulodon Sinclair, 1903

Type species: *Mylagaulodon angulatus* Sinclair, 1903.

Type specimen: UCMP 1652, poorly preserved cranium with P3–4.

Characteristics: Similar to *Promylagaulus*, but with larger parastyle and mesostyle on P4.

Included species: *M. angulatus* only, known from locality PN6D3 only.

Comments: Korth (1994a, 1998) restricted the record of this species to its type specimen from Oregon. He pointed out the transitional resemblance to meniscomyine aplodontids.

Mesogaulus Riggs, 1899

Type species: *Mesogaulus ballensis* Riggs, 1899.

Type specimen: Field Museum specimen P25223, left dentary with incisor, p4, m1–2.

Characteristics: Fossorial features of the humerus and manus, but hornless; cheek teeth higher crowned than in *Promylagaulus*, with larger premolars, but not as large as in *Mylagaulus*; premolars not oval, showing mesostyle on P4 and metastylid on p4; adult dentition (described by Black and Wood 1956) dominated by large fourth premolar, followed by low crowned first molar that is shed early in life, and two hypsodont molars; single sagittal crest and vertical occiput. This taxon is the unique genus of subfamily Mesogaulinae created by Korth (2000a).

Average length of m2: 3.9–4.0 mm.

Included species: *M. ballensis* (known from locality NP34B); *M. paniensis* (Matthew, 1902) (localities CP68C, CP71, [CP86D], [NP10E], [PN6G]); *M. praecursor* Cook and Gregory, 1941 (locality CP104B).

Mesogaulus sp. is also known from localities GC8D, DB, GC9B, CP107, NP34C, PN6H, PN19A.

MYLAGAULINAE COPE, 1881

Characteristics: Large aplodontids showing strong fossorial adaptations and typically with cranial ornamentation, at least thickened bone at anterior end of nasals; P3 lacking; fourth premolar more than twice molar in size; first molar lost when permanent fourth premolar erupts; premolars lacking roots; cheek teeth high crowned and premolars oval in outline (exception, P4 of *A. vetus*); upper incisor shallowly grooved; posterior skull width nearly equal to length, with twinned parasaggital crests; sloping occipital (nearly vertical in *Alphagaulus*) (after Korth, 2000a).

***Alphagaulus* Korth, 2000a (including *Mylagaulus*, in part)**

Type species: *Alphagaulus vetus* (Matthew, 1924) (originally described as *Mylagaulus vetus*).

Type specimen: AMNH 18905, right dentary with incisor, p4, m2.

Characteristics: Small mylagaulines with skull moderately expanded posteriorly; few fossettes in premolars; low rugosities at anterior ends of nasals.

Length of p4: 8.1 mm (topotypic material, Korth, 2000a).

Included species: *A. vetus* (known from localities ?SB29C, CP54B, CP108B, CP110); *A. douglassi* (McKenna, 1955) (originally described as *Mylagaulus douglassi*) (localities NP40A, [B]); *A. pristinus* (Douglass, 1903) (originally described as *Mylagaulus pristinus*) (localities NB18A, NP41B, NP42, PN18); *A. tedfordi* Korth, 2000a (locality CP111).

***Hesperogaulus* Korth, 1999 (including *Mylagaulus*, in part)**

Type species: *Hesperogaulus wilsoni* Korth, 1999.

Type specimen: LACM 142506, skull with incisors and right P4.

Characteristics: Large mylagaulid with elongated oval premolar; posterior paired horns lacking, but thickened bone present at anterior end of nasals; low sloping occipital.

Average length of P4: 11.8 mm (Shotwell, 1958).

Included species: *H. wilsoni* (known from localities PN11B, PN14); *H. gazini* Korth, 1999 (localities PN8A, B, PN9A).

Hesperogaulus sp. is also known from localities NB23C, NB27II, NB31, PN10.

***Umbogaulus* Korth, 2000a (including *Mylagaulus*, in part)**

Type species: *Umbogaulus galushai* Korth, 2000a.

Type specimen: FAMNH 65576, left p4.

Characteristics: Larger than *Ceratogaulus*; paired bosses at ends of nasals; large number of fossettes in premolars.

Length of p4: 12.0 mm (topotypic material, Korth, 2000a).

Included species: *U. galushai* (known from locality CP111); *U. monodon* (Cope, 1881, including *Mylagaulus*) (localities NB18A, [SB11], SP1C, [SP3B], CP114B, D, CP115C, CP116A, B, CP123D).

***Pterogaulus* Korth, 2000a (including *Mylagaulus*, in part)**

Type species: *Pterogaulus laevis* (Matthew, 1902) (originally described as *Mylagaulus laevus*).

Type specimen: FAMNH 65576, left p4.

Characteristics: Large mylagaulines lacking nasal horns or thickening; postorbital processes expanded; premolar elongating anteroposteriorly, fossettes pinched and elongated; second molar (as well as first) lost when permanent premolar erupts.

Average length of p4: 8.2 mm (*P. laevis*, Korth, 2000a).

Included species: *P. laevis* (known from localities [NB18A], [SP1A], [CP56], CP75B, CP76B, CP108B, [CP114C], [NP11]); *P. barbarellae* Korth, 2000a, (locality CP116B); *P. cambridgensis* (Korth, 2000b) (originally described as "*Mylagaulus*" *cambridgensis*) (locality CP116D).

***Mylagaulus* Cope, 1878 (including *Epigaulus*, in part)**

Type species: *Mylagaulus sesquipedalis* Cope, 1878.

Type specimen: AMNH 8329, upper molar.

Characteristics: Premolars simpler than in other mylagaulines, with fewer fossettes; P4 with C-shaped posterobuccal fossette, parafossette with lingual branch separated in early wear; presence of horns unknown (if *M. minor* belongs here, as listed below, then horns occur in the genus).

Average length of p4: 8.4 mm (*M. elassos*, Baskin, 1980).

Included species: *M. sesquipedalis* (known from locality CP114B); *M. elassos* Baskin, 1980 (locality GC11A); *M. kinseyi* Webb, 1966 (locality GC11B); *M. minor* (Hibbard and Phillis, 1945) (originally described as *Epigaulus minor*) (localities CP116B, CP123E, PN10).

"*Mylagaulus*" sp. is also known from localities GC4D, E, GC10C, NB23B, NB29A, SB32D, SB34C, SP1A, C, D, SP4B, CP88, CP89, CP90A, CP107, CP114A, PN7, PN12B.

***Ceratogaulus* Matthew, 1902**

Type species: *Ceratogaulus rhinocerus* Matthew, 1902.

Type specimen: AMNH 9456, nearly complete cranium and left dentary.

Characteristics: Fossorial features; cheek teeth high crowned; paired nasal horn cores.

Average length of p4: 11.0 mm (Korth, 2000a).

Included species: *C. rhinocerus* (known from localities CP76B, CP105, CP114B, D [Korth 2000a]); *C. anecdotus* Korth, 2000a (localities CP116A, B).

***Epigaulus* Gidley, 1907**

Type species: *Epigaulus hatcheri* Gidley, 1907.

Type specimen: USNM 5485, nearly complete skeleton.

Characteristics: Mylagaulid of largest body size; cheek teeth high crowned; paired nasal horns; reduced postorbital processes; numerous premolar fossettes. Korth (2000a) sees this species as an end member of a broadly conceived

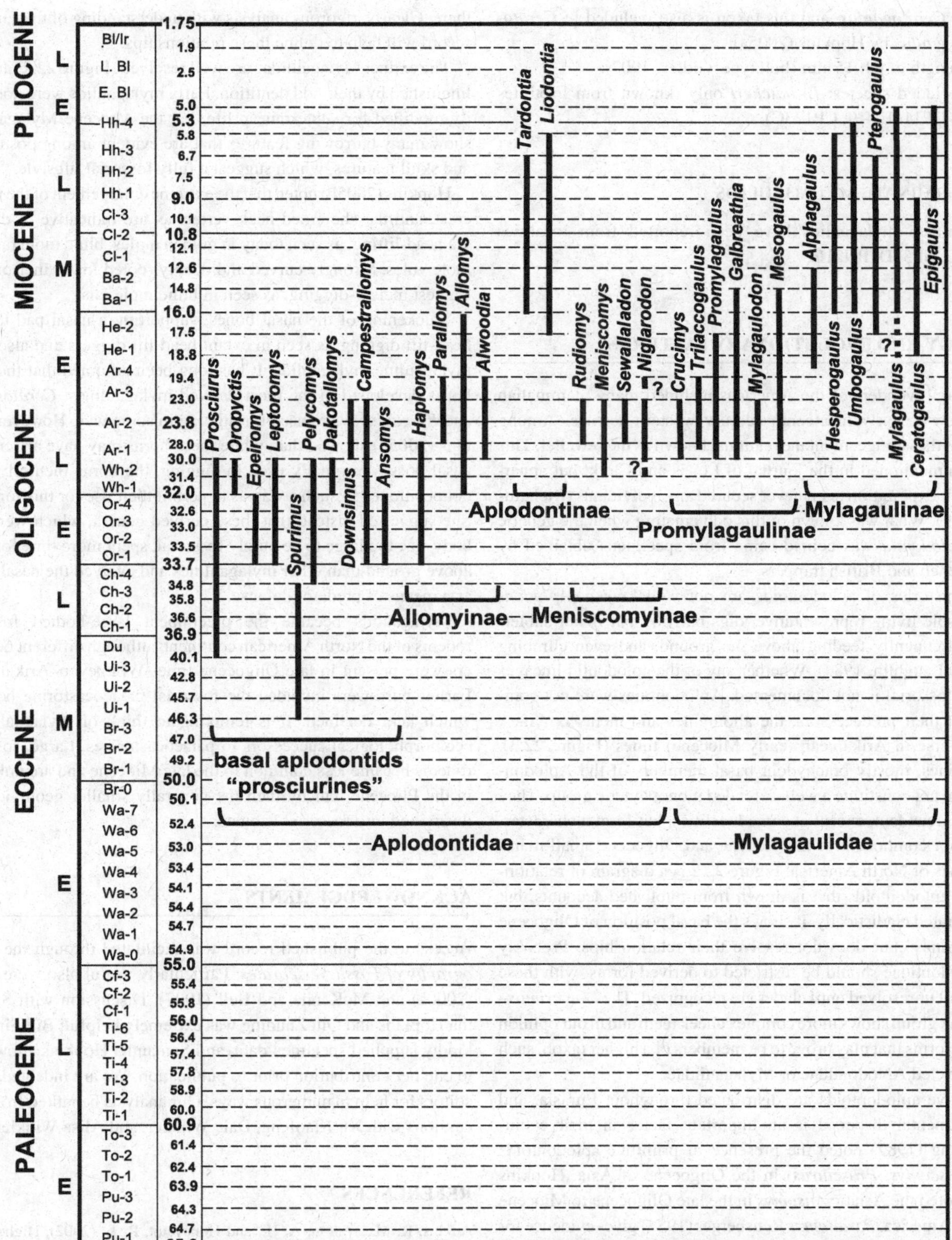

Figure 22.3. Temporal ranges of North American aplodontoid genera.

Ceratogaulus, and this taxon is also included in *Ceratogaulus* by Hopkins (2005a).

Length of p4: 15 mm (holotype, Gidley, 1907).

Included species: *E. hatcheri* only (known from localities CP116E, [F], CP123C).

INDETERMINATE MYLAGAULIDS

Indeterminate mylagaulids have been reported from localities NP10CC, NP34D, PN21B.

BIOLOGY AND EVOLUTIONARY PATTERNS

The natural wonders of the Americas included many mammalian forms new to early nineteenth century western science. Among these was the strange, montane mammal known as the sewellel. This rodent is mentioned in the journal of Lewis and Clark, but apparently their exploration team never secured any individuals (Richardson, 1829). What was known of this odd creature when the generic name *Aplodontia* was coined came from specimens obtained by early French and British trappers.

Life histories of aplodontoids are not well known. *Aplodontia*, the sole living representative, digs burrows but is not subterranean, frequently feeding above the ground, and even climbing trees (McLaughlin, 1984). Whether any of the aplodontid lineages became fossorial is not documented. Aplodontines were not common, but their predecessors, the allomyines and meniscomyines, were diverse in Arikareean (early Miocene) times (Figure 22.3). Prosciurines, mostly brachydont basal members of the Aplodontoidea, do not constitute a well-defended monophyletic group. They were not abundant, except at some localities, but form a characteristic minor component of late Eocene and Oligocene small mammal faunas of North America. Figure 22.2 is a diagram of relationships of aplodontoids that is drawn from published accounts, but not generated cladistically. It shows the basal position of Oligocene "prosciurines" but does not resolve their relationships. Probably the Aplodontidae should be restricted to derived forms, with these genera left unresolved until clades are recognized. The *Downsimus–Haplomys* group shows more complex cheek teeth and in our opinion includes forms that may prove to be members of a higher taxon, such as a restricted Aplodontidae or Mylagaulidae.

Primitive aplodontoids are distributed throughout Eurasia, and intercontinental dispersal is an important factor in their evolution. Wang (1987) noted the presence of primitive aplodontoids (*Promeniscomys, Prosciurus*) in the Oligocene of Asia. Hopkins (1999) noted the Asiatic *Ansomys* in the late Oligocene to Miocene of North America. *Parallomys* Rensberger (1983) gives evidence for dispersal of more advanced aplodontids, and *Pseudaplodon* Miller (1927) shows that aplodontines crossed the Bering Land Bridge. Even mylagaulids appear to have Asiatic records (Wu, 1988; but see McKenna and Bell [1997] for classification of *Sinomylagaulus*). Given the substantial Eurasian record of some aplodontoid subgroups, such as Allomyinae, those higher taxa possibly originated there. Clearly, rigorous analysis with understanding of the Eurasian record will help to define these relationships.

The extinct Mylagaulidae are well resolved (Figure 22.2) and distinguished by their odd dentition. Early mylagaulids were not heavily modified for subterranean life, but the Miocene Mylagaulinae show many burrowing features that are evident among postcranial and skull features, which suggest a fully fossorial lifestyle.

Hopkins (2005b) noted that the extreme enlargement of the nuchal crest, and the shortened neck vertebrae, are indicative of digging via head-lifting, as seen today in golden moles, blind mole rats, and mole voles. Strongly curved and deeply rooted lower incisors also suggest incisor-digging, as seen in blind mole rats.

Thickening of the nasal bones, supporting a nasal pad used in head-lift digging, is seen in extant head-lift diggers and also in all mylagauline mylagaulids. It has long been assumed that the nasal horns developed in the most derived mylagaulines, *Ceratogaulus* and *Epigaulus*, were used for digging in some way. However Hopkins (2005b) showed that, while these horns may have arisen from nasal bosses originally used in digging, the horns themselves are not positioned in such a way as to permit their use for this function. She suggested instead that these derived genera, which were of a body size too large to be purely fossorial, spent more time foraging above ground than other mylagaulines and evolved the nasal horns as a means of predator defense.

Mylagaulids became the preeminent large-bodied fossorial rodents of the North American continent, although different burrowers were present in late Oligocene time (Whitneyan–Arikareean). Earlier burrowers included the fossorial palaeocastorine beavers, which later declined. It is tempting to think of mylagualids as ecomorphological successors to palaeocastorines. Large, fossorial rodents became less common in the late Miocene and are unknown in the Pliocene, but by then the generally smaller geomyids had diversified into fossorial habitats.

ACKNOWLEDGEMENTS

Access to the published record was facilitated through the *Bibliography of Fossil Vertebrates*. Particularly useful also were Korth (2000a) and McKenna and Bell (1997). Discussion with Samantha Hopkins and Qiu Zhuding was extremely helpful; Bill Simpson kindly supplied specimen data, and Samantha Hopkins allowed us to cite her contribution prior to publication. We are indebted to the editors for help in numerous ways. Our analysis benefited from discussions with Xu Xiaofeng, Dale Winkler, and Alisa Winkler.

REFERENCES

Adkins, R. M., Walton, A. H., and Honeycutt, R. L. (2002). Higher level systematics of rodents and divergence time estimates based on two congruent nuclear genes. *Molecular Phylogenetics and Evolution*, 26, 409–20.

Barnosky, A. D. (1986). Arikareean, Hemingfordian, and Barstovian mammals from the Miocene Colter Formation, Jackson Hole, Teton County, Wyoming. *Bulletin of the Carnegie Museum of Natural History*, 26, 1–69.

Barnosky, A. D. and Labar, W. J. (1989). Mid-Miocene (Barstovian) environmental and tectonic setting near Yellowstone Park, Wyoming and Montana. *Bulletin of the Geological Society of America*, 101, 1448–56.

Baskin, J. A. (1980). Evolutionary reversal in *Mylagaulus* (Mammalia, Rodentia) from the late Miocene of Florida. *The American Midland Naturalist*, 104, 155–62.

Black, C. C. (1961). Rodents and lagomorphs from the Miocene Fort Logan and Deep River Formations of Montana. *Postilla, Yale Peabody Museum*, 48, 1–20.

(1965). Fossil mammals from Montana. Part 2: Rodents from the early Oligocene Pipestone Springs Local Fauna. *Annals of the Carnegie Museum*, 38, 1–48.

(1971). Paleontology and geology of the Badwater Creek area, Central Wyoming. Part 7: Rodents of the family Ischyromyidae. *Annals of the Carnegie Museum*, 43, 179–217.

Black, C. C. and Wood, A. E. (1956). Variation and tooth-replacement in a Miocene mylagaulid rodent. *Journal of Paleontology*, 30, 672–84.

Brandt, J. F. (1855). Beitrage zur nähern Kenntnis der Säugethiere Russlands. *Mémoires de l'Academie imperiale des Sciences de St. Petersbourg Series 6*, 9, 1–375.

Cook, H. J. and Gregory, J. T. (1941). *Mesogaulus praecursor*, a new rodent from the Miocene of Nebraska. *Journal of Paleontology*, 15, 549–52.

Cope, E. D. (1873). Synopsis of new Vertebrata from the Tertiary of Colorado. *United States Geological and Survey of the Territories*, pp. 1–19. Washington, DC: Government Printing Office.

(1878). Description of new extinct vertebrates from the upper Tertiary and Dakota formations. *Bulletin of the United States Geological and Geographical Survey*, 4, 379–96.

(1879). On some characters of the Miocene fauna of Oregon. *Proceedings of the American Philosophical Society*, 18, 63–78.

(1881). Review of the Rodentia of the Miocene period of North America. *Bulletin of the United States Geological and Geographical Survey*, 6, 361–86.

Douglass, E. (1903). New vertebrates from the Montana Tertiary. *Annals of the Carnegie Museum*, 2, 145–99.

Eastman, C. B. (1982). Histricomorphy as the primitive condition of the rodent masticatory apparatus. *Evolutionary Theory*, 6, 163–5.

Eaton, J. G. (1995). Cenomanian and Turonian (early Late Cretaceous) multituberculate mammals from southwestern Utah. *Journal of Vertebrate Paleontology*, 15, 761–84.

Fagan, S. R. (1960). Osteology of *Mylagaulus laevis*, a fossorial rodent from the Miocene of Colorado. *University of Kansas Paleontological Contributions, Vertebrata*, 9, 1–32.

Furlong, E. (1910). An aplodont rodent from the Tertiary of Nevada. *University of California Publications, Bulletin of the Department of Geological Sciences*, 3, 397–403.

Galbreath, E. C. (1953). A contribution to the Tertiary geology and paleontology of northeastern Colorado. *University of Kansas Paleontological Contributions. Vertebrata*, 4, 1–120.

(1984). On *Mesogaulus paniensis* (Rodentia) from Hemingfordian (middle Miocene) deposits in northeastern Colorado. *Carnegie Museum of Natural History Special Publication*, 9, 85–9.

Gazin, L. (1932). A new Miocene mammalian fauna from southeastern Oregon. *Carnegie Institution of Washington, Contributions in Paleontology*, 418, 39–86.

Gidley, J. W. (1907). A new horned rodent from the Miocene of Kansas. *Proceedings of the United States National Museum*, 32, 627–36.

Hibbard, C. W. and Phillis, C. F. (1945). The occurrence of *Eucastor* and *Epigaulus* in the lower Pliocene of Trego County, Kansas. *University of Kansas Science Bulletin*, 30, 549–55.

Hopkins, S. B. (1999). Phylogeny and biogeography of the genus *Ansomys* Qiu, 1987 (Mammalia: Rodentia: Aplodontidae) and description of a new species from the Barstovian (mid-Miocene) of Montana. *Journal of Paleontology*, 78, 731–40.

(2005a). Evolutionary history and paleoecology of aplodontid rodents. Ph.D. Thesis. University of California, Berkeley.

(2005b). The evolution of fossoriality and the adaptive role of horns in the Mylagaulidae (Mammalia: Rodentia). *Proceedings of the Royal Society Biological Sciences B*, 272, 1705–13.

Klingener, D. (1968). Rodents of the Mio-Pliocene Norden Bridge local fauna, Nebraska. *The American Midland Naturalist*, 80, 65–74.

Korth, W. W. (1986). Aplodontid rodents of the genus *Pelycomys* Galbreath from the Orellan (middle Oligocene) Brule Formation, Nebraska. *Journal of Mammalogy*, 67, 545–50.

(1989). Aplodontid rodents (Mammalia) from the Oligocene (Orellan and Whitneyan) Brule Formation, Nebraska. *Journal of Vertebrate Paleontology*, 9, 400–14.

(1992a). A new genus of prosciurine rodent (Mammalia: Rodentia: Aplodontidae) from the Oligocene (Orellan) of Montana. *Annals of the Carnegie Museum*, 61, 171–5.

(1992b). Fossil small mammals from the Harrison Formation (Late Arikareean: earliest Miocene), Cherry County, Nebraska. *Annals of the Carnegie Museum*, 61, 69–131.

(1994a). *The Tertiary Record of Rodents in North America*. New York: Plenum Press.

(1994b). A new species of the rodent *Prosciurus* (Aplodontidae, Prosciurinae) from the Orellan (Oligocene) of North Dakota and Nebraska. *Journal of Mammalogy*, 75, 478–82.

(1998). Rodents and lagomorphs (Mammalia) from the late Clarendonian (Miocene) Ash Hollow Formation, Brown County, Nebraska. *Annals of the Carnegie Museum*, 67, 299–348.

(1999). *Hesperogaulus*, a new genus of mylagaulid rodent (Mammalia) from the Miocene (Barstovian to Hemphillian) of the Great Basin. *Journal of Paleontology*, 73, 945–51.

(2000a). Review of Miocene (Hemingfordian to Clarendonian) mylagaulid rodents (Mammalia) from Nebraska. *Annals of the Carnegie Museum*, 69, 227–80.

(2000b). A new species of mylagulid rodent (Mammalia) from the Hemphillian (Late Miocene) of Nebraska. *Paludicola*, 2, 265–8.

Korth, W. W. and Emry, R. J. (1991). The skull of *Cedromus* and a review of the Cedromurinae (Rodentia, Sciuridae). *Journal of Paleontology*, 65, 984–94.

Macdonald, J. R. (1956). A new Clarendonian mammalian fauna from the Truckee Formation of western Nevada. *Journal of Paleontology*, 30, 186–202.

(1963). The Miocene Wounded Knee faunas of southwestern South Dakota. *Bulletin of the American Museum of Natural History*, 125, 141–238.

(1970). Review of the Miocene faunas from the Wounded Knee area of western South Dakota. *Bulletin of the Los Angeles County Museum of Natural History, Science*, 8, 1–82.

Marsh, O. C. (1877). Notice of some new vertebrate fossils. *American Journal of Science, Series 3*, 14, 249–56.

Matthew, W. D. (1902). A horned rodent from the Colorado Miocene, with a revision of the Mylagauli, beavers, and hares of the American Tertiary. *Bulletin of the American Museum of Natural History*, 16, 291–310.

(1903). The fauna of the *Titanotherium* beds at Pipestone Springs, Montana. *Bulletin of the American Museum of Natural History*, 19, 197–226.

(1924). Third contribution to the Snake Creek fauna. *Bulletin of the American Museum of Natural History*, 50, 59–210.

McGrew, P. O. (1941). The Aplodontoidea. *Field Museum of Natural History, Geological Series*, 9, 1–30.

McKenna, M. C. (1955). A new species of mylagaulid from the Chalk Cliffs local fauna, Montana. *Journal of the Washington Academy of Science*, 45, 107–10.

McKenna, M. C. and Bell, S. K. (1997). *Classification of Mammals Above the Species Level*. New York: Columbia University Press.

McLaughlin, C. A. (1984). Protrogomorph, sciuromorph, castorimorph, myomorph, geomyoid, anomaluroid, pedetoid, and ctenodacyloid rodents. In *Orders and Families of Recent Mammals of the World*, ed. S. Anderson and J. K. Jones, Jr., pp. 267–88. New York: John Wiley.

Meng, J. (1990). The auditory region of *Reithroparamys delicatissimus* (Mammalia, Rodentia) and its systematic implications. *American Museum Novitates*, 2972, 1–35.

Miller, G. S. (1927). Revised determinations of some Tertiary mammals from Mongolia. *Palaeontologia Sinica, Series C*, 5, 1–20.

Miller, G. S. and Gidley, J. W. (1918). Synopsis of the supergeneric groups of rodents. *Journal of the Washington Academy of Science*, 8, 431–48.

Munthe, J. (1988). Miocene mammals of the Split Rock area, Granite Mountains Basin, Central Wyoming. *University of California Publications in Geological Sciences*, 126, 1–136.

Nichols, R. (1976). Early Miocene mammals from the Lemhi Valley of Idaho. *Tebiwa*, 18, 9–47.

Qiu, Z.-D. (1987). The Aragonian vertebrate fauna of Xiacaowan, Jiangsu. 7. Aplodontidae (Rodentia, Mammalia). *Vertebrata PalAsiatica*, 25, 283–96.

Rensberger, J. M. (1975). *Haplomys* and its bearing on the origin of the aplodontid rodents. *Journal of Mammalogy*, 56, 1–14.

(1979). *Promylagaulus*, progressive aplodontoid rodents of the early Miocene. *Contributions in Science, Los Angeles County Museum of Natural History*, 312, 1–18.

(1980). A primitive promylagauline rodent from the Sharps Formation, South Dakota. *Journal of Paleontology*, 54, 1267–77.

(1981). Evolution in a late Oligocene-early Miocene succession of meniscomyine rodents in the Deep River Formation, Montana. *Journal of Vertebrate Paleontology*, 1, 185–209.

(1983). Successions of meniscomyine and allomyine rodents (Aplodontidae) in the Oligo-Miocene John Day Formation, Oregon. *University of California Publications in Geological Sciences*, 124, 1–157.

Richardson, J. (1829). On *Aplodontia*, a new genus of the Order Rodentia, constituted for the reception of the Sewellel, a burrowing animal which inhabits the North Western Coast of America. *Zoological Journal*, 4, 333–7.

Riggs, E. S. (1899). The Mylagaulidae: an extinct family of sciuromorph rodents. *Field Columbian Museum, Geological Series*, 1, 181–7.

Russell, L. S. (1972). Tertiary mammals of Saskatchewan. Part II: the Oligocene fauna, non-ungulate orders. *Life Sciences Contributions, Royal Ontario Museum*, 84, 1–97.

Schlosser, M. (1924). Tertiary vertebrates from Mongolia. *Palaeontologia Sinica, Series C*, 1, 1–119.

Shotwell, J. A. (1958). Evolution and biogeography of the aplodontid and mylagaulid rodents. *Evolution*, 12, 451–84.

(1970). Pliocene mammals of southeast Oregon and adjacent Idaho. *Bulletin of the Museum of Natural History, University of Oregon*, 17, 1–103.

Sinclair, W. S. (1903). *Mylagaulodon*, a new rodent from the upper John Day of Oregon. *American Journal of Science*, 15, 143–4.

Stehlin, H. G. and Schaub, S. (1951). Die Trigoonodontie der simplicidentaten Nager. *Schweizerische Palaeontologie Abhandlungen*, 67, 1–385.

Storer, J. E. (1978). Rodents of the Calf Creek local fauna (Cypress Hills Formation, Oligocene, Chadronian) Saskatchewan. *Natural History Contributions, Museum of Natural History, Regina*, 1, 1–54.

(1988). The rodents of the Lac Pelletier lower fauna, late Eocene (Duchesnean) of Saskatchewan. *Journal of Vertebrate Paleontology*, 8, 84–101.

Sutton, J. F. and Korth, W. W. (1995). Rodents (Mammalia) from the Barstovian (Miocene) Anceny local fauna, Montana. *Annals of the Carnegie Museum*, 64, 267–314.

Tedrow, A. R. and Korth, W. W. (1997). New aplodontid rodents (Mammalia) from the Oligocene (Orellan and Whitneyan) of Slim Buttes, South Dakota. *Paludicola*, 1, 80–90.

(1999). *Dakotallomys* (Rodentia, Aplodontidae) a replacement name for *Dakotamys* Tedrow and Korth, 1997 not *Dakotamys* Eaton, 1995. *Paludicola*, 2, 257.

Vianey-Liaud, M. (1985). Possible evolutionary relationships among Eocene and lower Oligocene rodents of Asia, Europe and North America. In *Evolutionary Relationships among Rodents: A Multidisciplinary Analysis*, ed. W. P. Luckett and J. L. Hartenberger, pp. 277–309. New York: Plenum Press.

Wahlert, J. H. (1974). The cranial foramina of protrogomorphous rodents; an anatomical and phylogenetic study. *Bulletin of the Museum of Comparative Zoology*, 146, 363–410.

Wang, B. (1987). Discovery of Aplodontidae (Rodentia, Mammalia) from middle Oligocene of Nei Mongol, China. *Vertebrata PalAsiatica*, 11, 370–7.

Webb, S. D. (1966). A relict species of the burrowing rodent, *Mylagaulus*, from the Pliocene of Florida. *Journal of Mammalogy*, 47, 401–12.

Wilson, R. W. (1949). Early Tertiary rodents of North America. *Carnegie Institution of Washington Publication*, 584, 67–164.

Wood, A. E. (1935). Two new genera of cricetid rodents from the Miocene of western United States. *American Museum Novitates*, 789, 1–8.

(1937). The mammalian fauna of the White River Oligocene. Part 2: Rodentia. *Transactions of the American Philosophical Society*, 28, 155–269.

(1980). The Oligocene rodents of North America. *Transactions of the American Philosophical Society*, 70, 1–68.

Wu, W.-Y. (1988). The first discovery of middle Miocene rodents from the northern Junggar Basin, China. *Vertebrata PalAsiatica* 26, 250–64.

23 Castoroidea

LAWRENCE J. FLYNN[1] and LOUIS L. JACOBS[2]

[1] *Peabody Museum, Harvard University, Cambridge, MA, USA*
[2] *Southern Methodist University, Dallas, TX, USA*

INTRODUCTION

Castoroidea include the familiar beavers, which became diverse and abundant in the middle Tertiary of Holarctica, North America in particular. The superfamily also includes Eocene and younger eutypomyids, which have a distinctive dentition and are allied with the beavers based on cranial and other features. Some individuals of extant *Castor* approach 40 kg in body mass, but earlier forms included species less than 1 kg. The castoroidine lineage attained great size, including the Pleistocene giant beaver of surprising body mass, up to 100 kg (Reynolds, 2002). Whereas modern beavers are semi-aquatic, many forms were terrestrial, and some show clear burrowing adaptations. Oligocene beavers were responsible for constructing the famous *Daemonelix* helical burrow fills of the White River Group (Peterson, 1905).

The castorids are a late Eocene to Recent group of medium (muskrat size) to large rodents, homogeneous in dentition but variably adapted to fossorial and aquatic habitats. Postcrania are known for some taxa, and these should be surveyed for evolutionary patterns. The group has two modern species of beaver, which are both semi-aquatic but quite distinct. Palaearctic *Castor fiber* differs from Nearctic *Castor canadensis* in a number of cranial and dental features. Given that usually both species are placed in the same genus *Castor*, consistency requires that similar fossil forms, e.g., *Sinocastor*, be subsumed in this genus. Other specialists (e.g., Mylan Stout) have stressed that it is logical to separate these species generically.

In contrast to the Recent occurrence of only two living forms, the fossil record shows a rich diversity of Oligocene–Miocene beavers, many of which were clearly terrestrial. This diversity is only partly reflected by current taxonomy yet has been long recognized by researchers including L. Martin, J. Martin, and W. Korth, who are currently advancing our understanding of beaver evolution. Incisor morphology has been used in beaver systematics, but distributions of incisor features are inconsistent with currently recognized groups; this intriguing feature should be resurveyed. The following synopsis of beaver diversity is based mainly on formally recognized taxa but is influenced by our realization that much work remains to be done. Consistent with this approach, our survey also benefits from the fresh analysis of castorid systematics undertaken by Xu Xiaofeng (1995), who has made important strides toward discerning clades based on morphometrics.

DEFINING FEATURES OF THE SUPERFAMILY CASTOROIDEA

Castoroidea presents features that occur in other rodents but in combination distinguish the group. Members are sciuromorphous rodents, achieving anterior extension of the origin of the masseter muscle by expanding it anterodorsally on to the snout. The anterior part of the zygomatic plate is deeply sculpted by the muscle and the infraorbital foramen is constricted ventrally. Castoroids have heavy incisors, used variously for digging or gnawing wood, and the enamel is uniserial in microstructure. The dental battery is high crowned, lophate, with large fourth premolars, and constructed for anteromedial shear.

Castoroidea also present autapomorphies. Wahlert (1977) reviewed derived cranial conditions that distinguish superfamily Castoroidea, including enclosed posterior maxillary foramen, sciuromorphy and the constricted infraorbital canal, sphenopalatine foramen within maxilla, separate dorsal palatine foramen in the maxilla or shared by maxilla and orbitosphenoid, and presence of an interorbital foramen posterior to the optic foramen. These features unite Eutypomyidae and Castoridae as superfamily Castoroidea, to the exclusion of other rodents.

Evolution of Tertiary Mammals of North America, Vol. 2. ed. C. M. Janis, G. F. Gunnell, and M. D. Uhen. Published by Cambridge University Press.

CRANIAL

EUTYPOMYIDAE

The skull is unknown in the earliest species of the group, but all known material demonstrates a fully sciuromorphous condition; that is, a pronounced origin for the superficial masseter on the side of the snout, which displaces downward a small infraorbital foramen with a short infraorbital canal (see Wahlert, 1977). Korth (1994) noted that the rostrum (and diastemal portion of the mandible) are rather elongate where known. The anterior insertion of the masseter muscle on the mandible is in the vicinity of the anterior margin of the first molar.

CASTORIDAE

The sciuromorphous skull is low and robust, with deep rostrum and mandible, broad zygomatic arches, and expanded occiput. A prominent tubercle below the infraorbital foramen is the insertion point for the superficial part of the masseter muscle. Castorids have short incisive foramina, display long and low infraorbital canals, possess an interorbital foramen, and present mandibular eminences. Tooth rows generally diverge posteriorly, and the bulla, not inflated, has a prominent auditory meatus (Korth, 1994). Incisors are heavy, with the lower incisor terminating in a lateral capsule.

DENTAL

EUTYPOMYIDAE

The dentition includes two upper premolars: dental formula I1/1, C0/0, P2/1, M3/3. The quadritubercular molar pattern is modified by five transverse crests, and a tendency toward a complicated pattern resulting from accessory enamel connections, which are most exaggerated in *Eutypomys* (Figure 23.1A). The fourth premolars are large, and are replaced.

CASTORIDAE

The dentition comprises 4/4 cheek teeth (except in early taxa, e.g., *Agnotocastor*, which retains P3). The cheek teeth are medium to high crowned with strong lophodonty and progressive increase in length of hypostria (hypostriid). The cheek teeth generally have four simple crests, modified in some taxa, and with crowns tapering toward their bases. Later forms have diverging maxillary cheek tooth rows. The incisors are heavy with uniserial enamel, rounded to flattened and with ornamentation in some forms.

POSTCRANIAL

Korth (1994) reviewed known castoroid postcrania (see also Stirton, 1935; Macdonald, 1963, 1970; Wood, 1980; Korth 2001a). According to Wood (1937), eutypomyid bones are conservative, reminiscent of ischyromyoid postcrania (see Chapter 18), slender as in squirrel bones, but with more prominent muscle scars. Castorids also present an ischyromyoid-like skeleton. Postcrania are robust; the forelimb is longer than the hindlimb, and the tibia and fibula

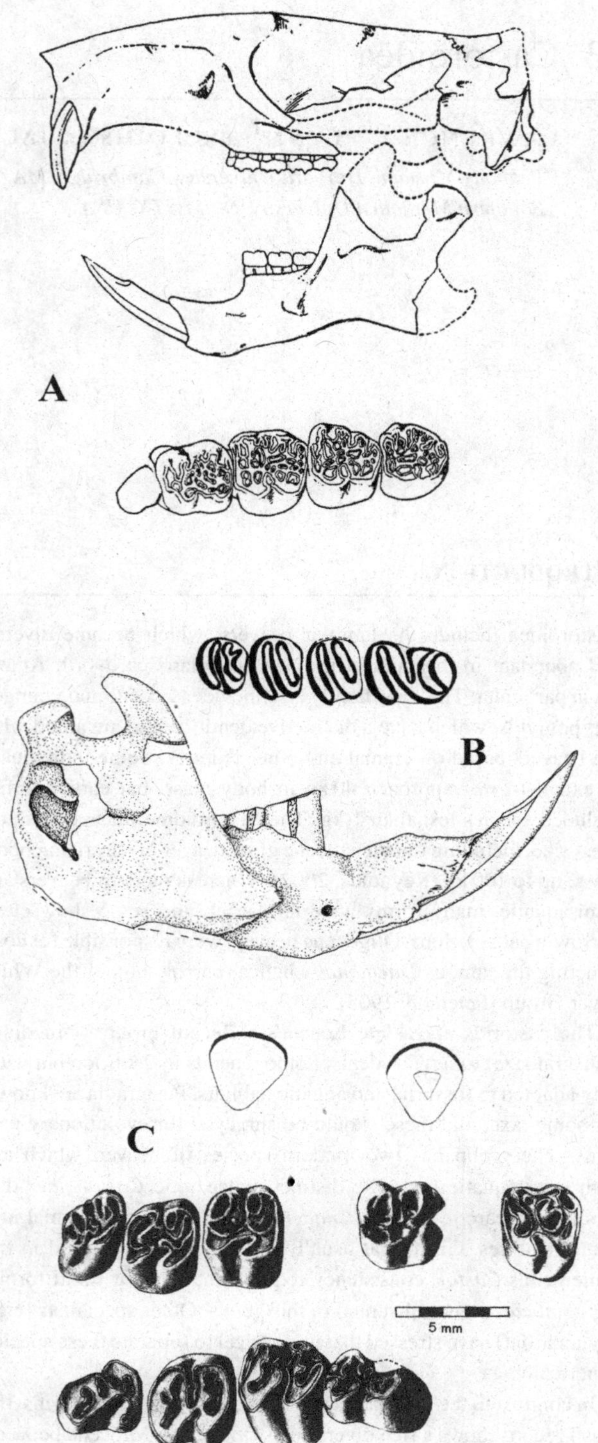

Figure 23.1. A. Upper dentition of *Eutypomys thomsoni* (from Matthew, 1905), showing the five cheek teeth and myriad of enamel lakes characteristic of Eutypomyidae. B. Dentary and cheek teeth of *Eucastor phillisi*; a castoroidine beaver showing the S-pattern of enamel loops on molars (from Wilson, 1968). C. Dentition of *Neatocastor hesperus* (from Korth, 1996a): top are cross sections from two lower incisors of different individual age; middle is upper right M1–3 (anterior to right), P4 and DP4; and bottom is lower left dentition, anterior to left. (A: courtesy of the American Museum of Natural History. B: courtesy of the University of Michigan. C: courtesy of the Carnegie Museum of Natural History.)

are unfused. Mid Cenozoic terrestrial to fossorial beavers present digging adaptations, including massive forelimb and manus, elongate scapula, and large skull with shortened rostrum. Semi-aquatic adaptations in castorines and castoroidines include elongate distal portions of limbs, an elongate pes, and (in *Castor*) dorsoventrally flattened caudal vertebrae.

SYSTEMATICS

SUPRA-SUPERFAMILIAL

Castoroidea are sciuromorphous rodents and, by consensus, are placed in the suborder Sciuromorpha. However, they are not particularly close to the squirrels phylogenetically. Both dentitions and skeletal features differ fundamentally. Current molecular studies (see Adkins, Walton, and Honeycutt, 2003) suggest relationship to geomyoids and, by inference, to the extinct sciuromorphous Eomyidae. (See further discussion in Ch. 16.)

The several genera attributed to the family Eutypomyidae agree in some features with *Eutypomys*, but the resemblance is not compelling in all cases. Korth (1994) reviewed the group and many of the published opinions on its affinities. The clear consensus based on dental and cranial features (Wahlert, 1977) is to ally Eutypomyidae with Castoridae.

INFRA-SUPRAFAMILIAL

Figure 23.2 shows hypothetical generic relationships. It is a diagram generated to summarize major features of castoroid evolution. It is not a cladogram built by analysis of trait distribution, although it draws on the cladogram of Xu (1995), in part. It is modified by current findings, including new fossils, and ongoing research, and reflects mainly the American record of castorids. Consequently, the relationships of genera are shown with respect to each other, minus close Eurasian relatives. Groups like the Castorinae, being mainly a Palaearctic group, are poorly represented in North America until the late Miocene when *Castor* appears.

Figure 23.2 shows Palaeocastorinae, with content corresponding approximately with the cladogram of Xu (1996). *Euhapsis, Fossorcastor*, and *Capatanka* are closely linked, and near *Pseudopalaeocastor* (Xu's *Nannasfiber*). The Castoroidinae are as Xu (1995) portrayed them, although *Microdipoides* is injected into the subfamily, and with the exception that *Monosaulax* and *Eucastor* can now be distinguished, thanks to the work of Korth (2002a). We depict a group related to *Agnotocastor* and reflecting increased size and molar complexity but do not feel confident that it is monophyletic. The group includes Agnotocastorinae of Korth and Emry (1997) and Anchitheriomyini of Korth (2001b). Agnotocastorinae differs from that of Korth and Emry (1997) in excluding *Anchitheriomys* but including *Hystricops*.

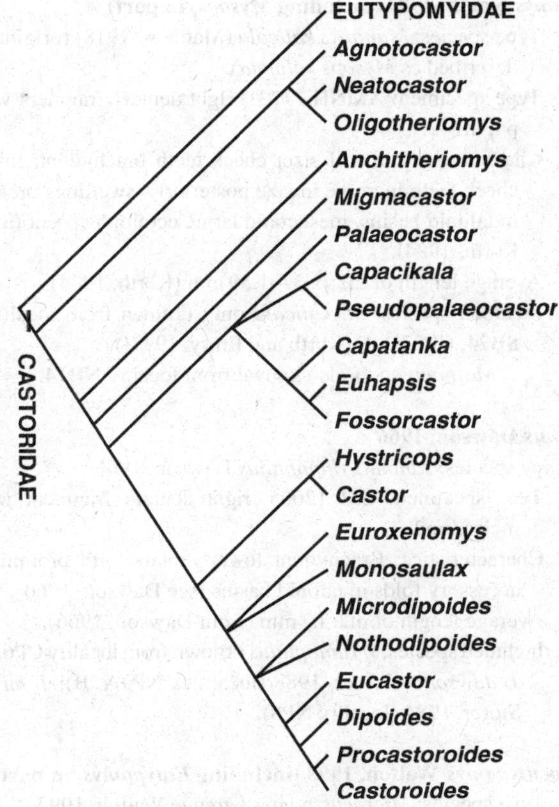

Figure 23.2. Proposed relationships of North American castoroid rodents based on analysis of previous work as discussed in the text. It is a blend of approaches, yet itself not generated by cladistic analysis. Inclusion of Eurasian forms would help to test the tree.

INCLUDED GENERA OF CASTOROIDEA GILL, 1872

The locality numbers listed for each genus refer to the list of unified localities in Appendix I. The locality numbers may be listed more than one way. The acronyms for museum collections are listed in Appendix III.

Parentheses around the locality (e.g., [CP101]) mean that the taxon in question at that locality is cited as an "aff." or "cf." the taxon in question. Parentheses are usually used for individual species, implying that the genus is firmly known from the locality but the actual species identification may be questionable. Question marks in front of the locality (e.g., ?CP101) mean that the taxon is questionably known from that locality, implying some doubt that the taxon is actually present at that locality, either at the genus or species level. An asterisk (*) indicates the type locality.

EUTYPOMYIDAE MILLER AND GIDLEY, 1918

Eutypomyidae was first distinguished as a family rank group by Miller and Gidley (1918), who noted that its taxa tend to build high-crowned teeth based on a four-cusped pattern. This is one feature of the group that is also advanced in Castoridae. In addition, Miller and Gidley were drawn to the complicated crown pattern of *Eutypomys*, namely a complex of secondary loops that result in isolated enamel islets (see Figure 23.1A). Thus, the complex eutypomyid tooth morphology is built upon a fundamentally primitive dentition, including 5/4 cheek tooth formula.

***Mattimys* Korth, 1984 (including *Mysops*, in part)**

Type species: *Mattimys kalicola* (Matthew, 1918) (originally described as *Mysops kalicola*).

Type specimen: AMNH 14731, right dentary fragment with p4, m1–2.

Characteristics: Small size; cheek teeth brachydont; lower cheek teeth increase in size posteriorly, swellings present in talonid basins, mesoconid large, ectoloph absent (after Korth, 1984).

Average length of m2: 1.37–1.59 mm (Korth, 1984).

Included species: *M. kalicola* only (known from localities SB24, CP27D, E [Korth and Emry, 1997]).

Mattimys sp. is also known from locality NB14.

***Janimus* Dawson, 1966**

Type species: *Janimus rhinophilus* Dawson, 1966.

Type specimen: CM 12005, right dentary fragment with incisor, m2–3.

Characteristics: Brachydont lower molars with prominent accessory folds in talonid basins (see Dawson, 1966).

Average length of m2: 1.7 mm (from Dawson, 1966).

Included species: *J. rhinophilus* (known from locality CP6B); *J. dawsonae* Storer, 1988 (localities NP9A, B); *J. mirus* Storer, 1984 (locality NP8).

***Microeutypomys* Walton, 1993 (including *Eutypomys*, in part)**

Type species: *Microeutypomys karenae* Walton, 1993.

Type specimen: TMM 42486-451, left M1.

Characteristics: Cheek teeth low crowned, but strongly lophate and crenulated; mesoloph and mesolophid not well developed, complete lingual wall on lower molars; molars with transverse lophs and valleys but few cross connections (after Walton, 1993).

Average length of m2: 1.14 mm (Walton, 1993).

Included species: *M. karenae* (known from locality SB42); *M. tilliei* (Storer, 1988) (originally described as *Eutypomys tilliei*) (locality NP9A).

***Eutypomys* Matthew, 1905**

Type species: *Eutypomys thomsoni* Matthew, 1905.

Type specimen: AMNH 12254, skull with mandible and partial skeleton.

Characteristics: Large eutypomyid with high-crowned cheek teeth and complex enamel pattern formed by cross connections of five transverse lophs (including complete mesoloph and mesolophid).

Average length of m2: 3.4–3.68 mm (from Wood, 1937)

Included species: *E. thomsoni* (known from localities [CP83C], CP84A, CP99A, B, NP24D, NP50B); *E. acares* Storer, 1988 (locality NP9A); *E. hybernodus* Korth, 2000a (localities CP84A, NP50B); *E. inexpectatus* Wood, 1974 (localities SB44B, C, D); *E. magnus* Wood, 1937 (locality CP84A); *E. montanensis* Wood and Konizeski, 1965 (localities CP85C, [CP86B], [NP10CC], NP36A); *E. obliquidens* Storer, 1988 (locality NP9A); *E. parvus* Lambe, 1908 (localities NP10B, NP24C).

Eutypomys sp. is also known from localities SB43C, CP39B, CP84B, CP98B, NP9B, C, NP10Bi, Bii, CC, NP25B, NP30A, B.

***Pipestoneomys* Donohoe, 1956**

Type species: *Pipestoneomys bisulcatus* Donohoe, 1956.

Type specimen: CNHM UM409, right maxilla fragment with M1–3.

Characteristics: High-crowned cheek teeth, possibly lacking P3; cheek teeth decreasing in size posteriorly, strong internal and anteroexternal reentrants on upper molars meet in early wear; isolated posterointernal enamel lake on M1–2; M1 wider anteriorly than posteriorly; M2 similar but smaller; M3 reduced and rounded; lower teeth with strong mesoconid, no mesoloph–mesolophid.

Average length of m1: 1.83 mm (Donohoe, 1956).

Included species: *P. bisulcatus* (known from localities ?CP44, CP98B, C, NP24C, D); *P. pattersoni* Alf, 1962 (locality CP98BB).

Pipestoneomys sp. is also known from locality CP99A (West and Korth, 1994).

Comments: *P. pattersoni* is somewhat smaller than the genotypic species, but Wood (1980) considered it a synonym of the latter. Often referred with question to Castoridae, familial assignment of *Pipestoneomys* remains uncertain. Aplodontid affinity has been suggested (e.g., Donohoe, 1956), but Wood (1980) and Alf (1962) consider relationship to *Eohaplomys*. Black (1965) described the Pipestone Springs dentary, showing that the masseteric crest extends below p4. This anterior position is consistent with the sciuromorphous condition of castoroids, as opposed to the protrogomorphy of *Eohaplomys* and aplodontids.

Indeterminate eutypomyids

Eutypomyids of indeterminate genus and species are reported from localities ?GC8AA, SB44A, NP10C2.

CASTORIDAE HEMPRICH, 1820

Castoridae display great diversity, but hypothesized subfamily level groupings are not fully tested. Herein, we recognize affinities of certain genera but focus on the better-defended subfamilies as formal entities. For example, we treat early members of the family loosely as "basal castorids" rather than assign them variously to Agnotocastorinae or Asiacastorinae. We utilize three subfamilies to highlight the beaver radiations: Palaeocastorinae, Castorinae, and Castoroidinae.

BASAL CASTORIDS

***Agnotocastor* Stirton, 1935**

Type species: *Agnotocastor praetereadens* Stirton, 1935.

Type specimen: AMNH 1428; crushed cranium lacking zygomatic arches and incisors, left P3 and right P3–4.

Characteristics: Relatively low-crowned teeth, narrow rostrum and skull, kidney-shaped bulla; short hypostria, rounded lower incisors, rounded incisor faces; P3 present; P4 square and molars rectangular (wider than long), wide fossettes and flexi persisting to late wear, mesostria and hypostria of subequal length.

Average length of m2: 3.7–3.8 mm (*A. coloradensis*, from Wilson, 1949; Galbreath, 1953).

Included species: *A. praetereadens* (known from locality CP84B); *A. coloradensis* Wilson, 1949 (locality CP68C); *A. galushai* Emry, 1972 (localities CP39C–F); *A. readingi* Korth, 1988 (locality CP99A).

Agnotocastor sp. is also known from localities GC8AA, CP83C, CP84A, B, CP98B, NP32B, C, NP34C, NP36A.

Comments: Most authors consider *Agnotocastor* to be so generalized that allocation to any major beaver clade is not secure. Korth (1996b) proposed arguments that *Agnotocastor* is derivable from the dental pattern shown by *Eutypomys magnus*. In his current treatment of the systematics of the Castoridae, Korth (2001a) endorsed subfamily Agnotocastorinae (Korth and Emry, 1997), but the group remains weakly supported at present.

Agnotocastor is also known from the Oligocene of Asia (McKenna and Bell, 1997).

***Neatocastor* Korth, 1996a (including *Steneofiber*; *Monosaulax*, in part)**

Type species: *Neatocastor hesperus* (Douglass, 1901) (originally described as *Steneofiber hesperus*).

Type specimen: CM 71, left dentary with dentition.

Characteristics: DP3 present (possibly replaced?); cheek teeth high crowned and with accessory enamel lakes; incisor faces broadly convex; upper cheek tooth rows parallel; premolars subequal in size to molars (before wear); see Korth (1996a) for other cranial features.

Average length of m2: 4.00 mm.

Included species: *N. hesperus* (including "*Steneofiber*" *complexus* Douglass, 1901) (known from localities GC5A, [B], CP101, NP33B, ["Madison River," see Korth, 1996a], NP41A, NP50B); "*S. montanus*" (Scott, 1893) (originally described as *Steneofiber*, could be referred to *Neatocastor*, see Stirton, 1935) (locality NP34D).

Neatocastor sp. is also possibly known from locality GC8AB.

Comments: Korth (1996a) named this genus to accommodate early Miocene North American beavers that are clearly advanced over *Agnotocastor* and are not part of the Castoroidinae clade. Although this taxon is quite primitive (Figure 23.2), Korth saw ties to the *Anchitheriomys* group and attempted to show that these early Neogene beavers are not allied to Old World *Steneofiber*, in disagreement with the findings of Xu (1995). Martin (1973) also saw generic distinctiveness of the skull of "*S.*" *complexus*. While remaining neutral on affinities, we agree with taxonomic distinction of *Neatocastor* from *Steneofiber*.

***Oligotheriomys* Korth, 1998a (including *Miotheriomys*)**

Type species: *Oligotheriomys primus* Korth, 1998a.

Type specimen: F:AM 64016, partial right maxilla with M1–2.

Characteristics: Posterior maxillary notch instead of foramen; slightly divergent maxillary tooth rows; retained P3; upper molars with moderately complex enamel pattern and shallow mesoflexus (after Korth, 1998a), long diastema; lower premolar larger than molars (after Korth, 2004).

Average length of M1: 4.6 mm; for *O. stenodon*.

Length of M1: 4.4 mm.

Length of m1: 4.6 mm.

Length of p4: 7.2 mm.

Included species: *O. primus* (known from locality NP50B); *O. stenodon* (Korth, 2004) (originally described as *Miotheriomys stenodon*) (locality CP106).

Oligotheriomys sp. is also known from locality CP105.

Comments: Increase in size and complication of molar patterns seems to characterize the *Agnotocastor*–*Neatocastor*–*Oligotheriomys* lineage. Korth (1998a) linked *O. primus* to *Anchitheriomys*, yet *O. primus* is poorly known and 15 Ma older than *A. fluminis*. The late early Miocene species *A. senrudi* may help to bridge this gap to some extent. Korth (2000a) considered that *Eutypomys magnus* could be transferred to *Oligotheriomys*. *O. stenodon* was considered to represent a distinct genus *Miotheriomys* by Korth (2004), but characteristics of known material do not appear to distinguish it at the generic level from *O. primus*. *O. stenodon* is much younger, contemporaneous with early *Anchitheriomys*, but it is represented by material not readily comparable to known specimens of *O. primus*.

***Anchitheriomys* Roger, 1898 (including *Amblycastor*; *Monosaulax*, in part)**

Type species: *Anchitheriomys suevicus* (Schlosser, 1884) (originally described as *Hystrix suevicus*).

American species: *Anchitheriomys fluminus* (Matthew, 1918) (originally described as *Amblycastor fluminus*).

Type specimen: AMNH 17213, right dentary fragment with incisor and p4.

Characteristics: Large beaver with rounded incisor bearing longitudinal ornamentation; long diastema; lower molars with four roots; crescentic parafossetid on p4; short striae on upper teeth.

Average length of m1–2 for *A. fluminus*: 6.5–7.3 mm (Storer, 1975).

Included species: *A. fluminus* (known from localities CP108B, CP110, CP114A); *A. nanus* Korth, 2004 (locality CP105); *A. senrudi* (Wood, 1945) (originally described as *Monosaulax senrudi*) (localities CP51B, CP108A, NP40A); *A. stouti* Korth, 2001b (locality CP105).

Anchitheriomys sp. is also known from localities GC4C, E, GC5B, GC10B, [NB3D], CP75B (= ?*Hystricops* sp.), NP11.

Comments: *Anchitheriomys* is represented by a large middle Miocene beaver from Tunggur, Inner Mongolia and allied forms found elsewhere in Eurasia. *Anchitheriomys tungurensis* (Stirton, 1934) has flattened postcrania reminiscent of modern, aquatic beavers. Old World species are becoming fairly well known, but until recently the status of related North American relatives remained enigmatic. It was thought that similar North American species had simpler enamel patterns, and they were consequently retained under the genus *Amblycastor* Matthew, 1918.

Until recently, the scant remains in Europe allowed suspicion of hystricid affinity for *Anchitheriomys* to persist. Korth (1994) offered the surprising suggestion that North American species are not beavers, but large eutypomyids. However, the morphometric work of Xu (1995) placed *Anchitheriomys* solidly in Castoridae, and Koenigswald and Mörs (2001) specified enamel microstructure evidence for castorid assignment; in fact, the derived condition of *Anchitheriomys* enamel argues for its grouping with castoroidines. Note also that, in gross morphology, the rounded lower incisor with ornamentation is shared with castoroidines.

That American forms are primitive and assignable to *Anchitheriomys* Roger, 1898 is reaffirmed recently by analysis of a well-preserved skull from Norden Bridge, Nebraska (locality CP114A) (Korth and Emry [1997] citing Voorhies). Here, *A. fluminis* is seen to resemble Eurasian *Anchitheriomys* (making *Amblycastor* a junior synonym) and to share features with *Agnotocastor* and *Neatocastor*, which led Korth and Emry (1997) to establish subfamily Agnotocastorinae for the three genera. Because the subfamily is defended only by presence of a long rostrum with procumbent incisors, we await further analysis of new specimens before using this group. This school of thought sees *Anchitheriomys* as a derivative of North American beaver evolution and implies westward migration into Asia.

Xu (1995) offer an alternative. Cranial evidence supports an Asian radiation of beavers that may contain the origin of *Anchitheriomys*. He endorsed Asiacastorinae Lytschev, 1983 as a subfamily that includes *Anchitheriomys* and *Trogontherium*. This fresh viewpoint sees American *Anchitheriomys* as an immigrant antedating a later radiation that produced *Castor. Trogontherium* is also thought to be a castoroidine (see, e.g., Korth, 2001c). Affinity with castoroidines may be supported by enamel microstructure (Koenigswald and Mörs, 2001).

***Hystricops* Leidy, 1858**

Type species: *Hystricops venustus* Leidy, 1858 (originally proposed as a subgenus of *Hystrix*).

Type specimen: USNM 1180, p4.

Characteristics: Large size, smooth incisor enamel; poorly known genus represented mainly by isolated cheek teeth; premolars much larger than molars (P4 about 1.5 times longer than M1 in *H. browni*); hypostria (-id) the longest reentrant, but terminating well above base of enamel; parastria nearly as long, but metastria (-id) short; hypostriid long and mesostriid short.

Size of p4 of *H. venustus* holotype: 11.5 × 8.4 mm (given by Stirton [1935]; calculation from Leidy's [1869] figure suggests larger size).

Included species: *H. venustus* (known from localities CP105, [CP114B], CP116A, B); *H. browni* Shotwell, 1963 (locality PN11B).

Hystricops sp. is also known from localities CP115B, NB19A, NB29A, PN10; an unnamed species (="*Anchitheriomys*" n. sp., Wilson, 1960) from CP71 is placed here, following Korth (2004).

Comments: *Hystricops* is also possibly known from the middle Miocene of Asia (McKenna and Bell, 1997).

***Migmacastor* Korth and Rybczynski, 2003**

Type species: *Migmacastor procumbodens* Korth and Rybczynski, 2003.

Type specimen: FAM 65723, skull with mandible.

Characteristics: Palate deeply grooved with posteriorly diverging tooth rows; incisive foramina < 30% diastema; P3 absent; premolars largest cheek teeth; cheek teeth mesodont and rooted, with smooth enamel lakes (one open flexus/flexid plus three fossettes), oblique parafossette; procumbent incisors with rounded enamel surface; lower incisor rooted in large lateral capsule.

Length of m2: 3.30 mm.

Width of m2: 3.70 mm.

Included species: *M. procumbodens* only, known from locality CP103A only.

PALAEOCASTORINAE MARTIN, 1987

Martin (1987) and Korth (1994, 2001a) built an argument for recognition of a clade of beavers as a higher taxon (subfamily), which is characterized by adaptations consistent with radiation into subterranean habitat. Members of this group show numerous fossorial features, including flattening of the enamel surface of incisors to form chisellike digging tools. Alternative classifications (e.g., Xu, 1995) have utilized osteological features that do not indicate the same content of this subfamily. Xu's Fossorcastorinae excludes early palaeocastorines, such as *Palaeocastor* and *Capacikala*. Below, we follow Korth's most recent review (2001a).

Characteristics: Skull with shortened rostrum and stout postorbital constriction, broad posteriorly; robust forelimbs and short tails; chisel-shaped incisors; coronoid process placed anteriorly (all features correlating with other fossorial adaptations). Other advanced traits shared with later beavers: P3 lost; shortened incisive foramina; small posterior palatine foramina in maxilla–palatine suture; P4/p4 the largest cheek teeth and maxillary rows diverging posteriorly; stapedial branch of carotid artery and its foramen lost; nasal bones posteriorly attenuated (after Korth, 2001a).

***Palaeocastor* Leidy, 1869 (including *Steneofiber*, in part)**

Type species: *Palaeocastor nebrascensis* (Leidy, 1856) (originally described as *Steneofiber nebrascensis*).

Type specimen: Of the syntypes, the unnumbered fragmentary skull (see Matthew, 1907) is generally taken as the name bearer.

Characteristics: High-crowned cheek teeth; large, rounded bullae, convergent supraorbital ridges, moderate lambdoidal crest; incisors somewhat rounded; premolars close to molars in size; premolars with short mesostria (-id) that does not touch the base of the enamel, para- and metafossettid present in early wear.

Average length of lower cheek tooth row: 12.4–14.0 mm (Macdonald, 1963).

Included species: *P. nebrascensis* (known from localities CP41 [sublocality unknown], CP85A, B, CP85C, CP86B, C, D, CP101, NP51A, PN6C [sublocality unknown]); *P. fossor* (Peterson, 1905, originally described as *Steneofiber fossor*) (localities CP52, CP103A, PN6C [sublocality unknown]); *P. peninsulatus* (Cope, 1881) (originally described as *Steneofiber peninsulatus*) (localities PN6C2, [PN16]); *P. simplicidens* (Matthew, 1907) (originally described as *Steneofiber simplicidens*) (localities GC3A, [GC5B], CP85C?, CP86B, D, CP103A).

Palaeocastor sp. is also known from localities CP51B, CP103A, CP104A, [B], NP10CC, NP36D, PN6E, F.

Comments: Wood (1937, 1980) noted that an Orellan (early Oligocene) specimen with somewhat rounded incisors may be an early record of the genus. Reynolds (2002) estimated the body mass of *Palaeocastor* at approximately 1 kg but does not specify what species.

The fact that the type specimen of *P. nebrascensis* remains unnumbered after a century and a half is inconsistent with normal practice. Leidy (1869) did not designate a type among the three specimens that he had in hand. Today the type series should be considered as a whole and numbered as syntypes.

"*P*". *peninsulatus* (perhaps including AMNH 7000) has distinctly flattened incisors and consequently may not belong in this genus. It is characteristic of the Arikareean (late Oligocene to early Miocene) of the Oregon John Day Formation (locality PN6; see Stirton, 1935; Macdonald, 1963; Fisher and Rensberger, 1972; Fremd, Bestland, and Retallack, 1997), and possibly also of Idaho (locality PN16; see Nichols, 1979).

T. M. Stout (AMNH notes), based on his curation of the castorids of the American Museum of Natural History, does not recognize *Agnotocastor* but includes that material in *Palaeocastor*. He placed the holotype of the genotypic species *A. praetereadens* in his *P. complexus*. We do not follow such a broad concept of *Palaeocastor* or of the species *complexus*. We follow Korth (1996a) in distinguishing *N. complexus*, which is consistent with Xu (1995), who considered the species an American *Steneofiber*, and with Stirton (1935), who considered it "*Monosaulax*."

Martin (1987) included *Capatanka* and its species as a subgenus of *Palaeocastor*. He further noted that both *P. fossor* and *C. magnus* were recovered from *Daemonelix* burrows, showing that *Palaeocastor* dug these amazing spiral features. However, the analysis of Xu (1995) firmly linked *Capatanka* with *Euhapsis*. These forms are all fossorial and likely to be recovered from burrows.

Palaeocastor is also known from the early Miocene of Asia (McKenna and Bell, 1997).

***Capacikala* Macdonald, 1963 (including *Steneofiber*, in part)**

Type species: *Capacikala gradatus* (Cope, 1878) (originally described as *Steneofiber gradatus*).

Type specimen: AMNH 7008, damaged cranium.

Characteristics: Small beaver; short, deep skull with orthodont incisors that have flattened anterior faces, double temporalis muscle scars not meeting in single sagittal crest; large bullae expanded ventrally; cheek teeth decrease in size posteriorly.

Average length of M2: 2.8 mm (from Cope, 1878).

Included species: *C. gradatus* (known from localities CP85C, ?NP37, PN6C2, D1, D3); *C. parvus* Xu, 1996 (localities CP48, ?CP85A).

Capacikala sp. is also known from locality PN6G.

Comments: *Capacikala* is also known from the early Miocene of Asia (McKenna and Bell, 1997).

***Pseudopalaeocastor* Martin, 1987 (including *Nannasfiber*; *Steneofiber*, in part)**

Type species: *Pseudopalaeocastor barbouri* (Peterson, 1905) (originally described as *Steneofiber barbouri*).

Type specimen: CM 1210, skull and partial skeleton.

Characteristics: Small beaver; broad skull; incisors broad and very flattened; temporal muscle scars merging posteriorly as sagittal crest.

Average length of m2: 2.9 mm.

Included species: *P. barbouri* (including *Steneofiber sciuroides, Palaeocastor milleri*) (known from localities GC3A, [GC5B], CP51A, B, CP52, CP69, CP86C, CP102, CP103A); *P. osmagnus* (Xu, 1996) (originally described as *Nannasfiber ostellatus*) (localities CP51A, B, CP52).

Comments: Synonyms of *Pseudopaleocastor* are discussed in Korth (2001a).

***Capatanka* Macdonald, 1963 (including *Palaeocastor*, in part)**

Type species: *Capatanka cankpeopi*, Macdonald, 1963.

Type specimen: SDSM 53421, skull with complete dentition.

Characteristics: Massive skull with strong sagittal and lambdoidal crests; large bullae expanded anteroposteriorly; premolars somewhat bigger than molars; incisors flat to slightly rounded; posterior glenoid pocket on skull; roughened parietals.

Average length of lower cheek tooth row: 15.4 mm.

Included species: *C. cankpeopi*, (known from localities CP85C, CP86B); *C. magnus* (Romer and McCormack, 1928) (originally described as *Palaeocastor magnus*) localities CP51A, CP86C); *C. minor* Xu, 1996 (locality CP86A).

Capatanka sp. is also questionably known from locality NP36B.

Comments: As Macdonald (1963, 1970) suspected, "*C*". *brachyceps* does not belong in this genus, and is placed in *Fossorcastor*.

***Euhapsis* Peterson, 1905**

Type species: *Euhapsis platyceps* Peterson, 1905.

Type specimen: CM 1220, cranium lacking cheek teeth, plus left dentary fragment with p4–m2.

Characteristics: Medium-sized beaver with short snout and broad occiput and zygomas; origin of superficial masseter restricted downward so that infraorbital foramen expands as elongate slit; glenoid as a pocket in occipital region; enlarged premolars; incisor not completely flat; strongly fossorial (adapted from Martin, 1987; his Fig. 4).

Average length of m2: 2.5 mm.

Included species: *E. platyceps* (known from locality CP102); *E. breugerorum* Martin, 1987 (locality CP103A); *E. ellicottae* Martin, 1987 (localities CP51A, CP103A); *E. luskensis* Xu, 1996 (localities CP51A, CP52).

Euhapsis sp is also known from locality ?NP36B.

Comments: Martin (1987) advanced understanding of the previously enigmatic *Euhapsis*. He united this genus with *Fossorcastor* in a clade of subterranean beavers but noted that these have not been found in spiral burrows, which suggests different life-history characteristics than those probable for *Palaeocastor*. Martin (1987) further showed *E. ellicottae* to be one of the most fossorial rodents of all time. It displays many burrowing adaptations in the skeleton, probably including a horny pad on the snout.

***Fossorcastor* Màrtin, 1987 (including *Steneofiber*, in part)**

Type species: *Fossorcastor greeni* Martin, 1987.

Type specimen: KUVP 80845, skull with mandible from lower part of the Harrison Formation, Arikareean of Wyoming.

Characteristics: Short, deep skull with vertical occiput; glenoid fossa not extending into occipital region; superficial masseter running under infraorbital foramen displacing it dorsally; deep jaw modified with vertical ascending ramus and condyle above angular process; incisors strongly proodont (after Martin, 1987).

Average length of m2: 3.0 mm (Martin, 1987).

Included species: *F. greeni* (known from localities CP52, CP103A); *F. brachyceps* (Matthew, 1907) (originally described as *Stenofiber brachyceps*) (locality CP86B).

Comments: *F. greeni* is a somewhat smaller species than *F. brachyceps*, which has been known through the past century. *Steneofiber brachyceps* Matthew, 1907 (including Matthew's "*Euhapsis gaulodon*") was submerged in *Palaeocastor simplicidens* by Stirton (1935). This unsatisfactory assignment was recognized by Macdonald (1963), who considered the species to have affinity with *Capatanka*. On clear morphometric grounds, *Fossorcastor* is distinct from these other beavers (Xu, 1995).

CASTORINAE HEMPRICH, 1820

Characteristics: Hypsodonty with reentrants persisting along length of crown (three buccal flexi and three lingual flexids extending to near crown base; long striae); reentrants extending just past midway into crown; aquatic adaptations include flattened limb elements (after Korth, 2001a).

***Castor* Linnaeus, 1758**

Type species: *Castor fiber* Linnaeus, 1758, the extant North American beaver.

Type specimen: None designated.

Characteristics: Large beavers with hypsodont cheek teeth and slightly rounded lower incisors; three external striae on upper cheek teeth, three internal on lowers (longer in Recent species); no distinct palatal groove; dorsally rounded rostrum; nasal bones rounded in outline, lacrymal not expanded dorsally; wide interorbital region.

Average length of m2: 7.1 mm (*Castor californicus*, from Shotwell, 1970).

Included Tertiary species: *C. californicus* Kellogg, 1911 (including "*C. accessor*" Hay, 1927) (known from localities CC39B, CP115C, CP116E, F, PN3C, PN23A, C); *C. canadensis* (localities GC15A, SB37E). *Castor* sp. is also known from localities NB13C, SB18E, SB19, CP131, PN14, PN15, CE2.

Comments: The living beavers *Castor canadensis* and *Castor fiber* are morphometrically quite distinct. Considering them as congeneric requires a broad interpretation to the content of *Castor*, and classic fossil beavers such as *Sinocastor* from Yushe and other late Miocene red bed sequences of China have to be subsumed at the subgeneric level. A Miocene origin of *Castor* is consistent with the fact that some late Miocene and Pliocene beavers from North America are assigned to the genus. Although the origin of *Castor* is not yet clear, it quite possibly is a late Miocene immigrant from Asia. The genus *Castor* is also known from the late Miocene to Recent of Europe and Asia (McKenna and Bell, 1997).

CASTOROIDINAE ALLEN, 1877

Characteristics: Jugal not contacting lacrimal; mandibular condyle medially placed relative to coronoid and angle; increasing crown height (Xu, 1995); palate with diverging tooth rows and deep grooves; American forms with cheek tooth crown "S" pattern (dominated by single deep buccal and lingual reentrants); some achieve large size (mass up to 100 kg).

Euroxenomys Samson and Radulesco, 1973 (including *Chalicomus*; *Monosaulax*, in part)

Type species: *Euroxenomys minutum* (Meyer, 1838) (originally described as *Chalicomus minutus*).

Type specimen of *E. inconnexus*: CM 8925, partial skull with right mandible.

Characteristics: Small beavers with rounded incisors; rooted cheek teeth; large premolars; elongate M3; persistent parastria on P4; p4 with one lingual reentrant (mesostriid).

Average length of m1: 2.8 mm (*E. inconnexus*).

Included species: *E. inconnexus* Sutton and Korth, 1995 (known from locality NP41B); *E. wilsoni* Korth, 2001c (originally described as *Monosaulax* n. sp. of Wilson, 1960) (localities CP71, CP105).

Comments: Sutton and Korth (1995) studied material from the Anceney Local Fauna (locality NP41B) previously called "*Monosaulax pansus*." They found it to differ significantly from *Eucastor* and referred the species to *Euroxenomys*, a genus previously known only from Europe. Hugueney (1999) considered *Euroxenomys* a subgenus of *Trogontherium*. Consistent with affinity to *Trogontherium*, *Euroxenomys* is considered a primitive castoroidine of Asian origin. Possibly *Anchitheriomys* is a member of this group, rather than a North American derivative of *Agnotocastor*.

Monosaulax Stirton, 1935 (including *Steneofiber*, in part; *Dipoides*, in part)

Type species: *Monosaulax pansus* (Cope, 1874) (originally described as *Steneofiber pansus*).

Type specimen: YPM-PU 10575, right dentary with all teeth, plus right maxilla with P4-M3.

Characteristics: Relatively short rostrum, upper diastema approximately 50% greater than upper cheek tooth row; cheek teeth mesodont, showing accessory lakes occasionally, and with roots; rounded incisor enamel; premolars with short striids; parasagittal crests not known to meet at the midline.

Size of m2 of holotype: 3.66 × 3.9 mm.

Length p4–m3: 17.3 mm.

Length P4–M3: 14.8 mm (Korth, 2000b).

Included species: *M. pansus* (known from localities GC4DD, CC23A, NB6C, NB23B, C, SB32D, F, SB34A, CP110, CP114A, CP115B, NP11, NP42); *M. baileyi* Korth, 2004 (locality CP105); *M. curtus* (Matthew and Cook, 1909) (originally described as *Dipoides curtus*) (localities [CP45E], CP75B, BB, CP110, [CP114A, B]); *M. progressus* Shotwell, 1968 (locality PN9B); *M. skinneri* Evander, 1999 (localities NB18A, CP114B); *M. typicus* Shotwell, 1968 (locality PN9B); *M. valentinensis* Evander, 1999 (locality CP114B).

Monosaulax sp. is also known from localities GC4D, NB7B, C, NB18A, NB23B, CP68C, CP88, NP10E, PN6H, PN9B, NC2B.

Comments: Many authors, following Stout (1967), tentatively considered *Monosaulax* a junior synonym of *Eucastor*, because the holotype of *Monosaulax pansus* was lost for many years and attributed specimens were much like typical *Eucastor tortus*. Korth (2000b) made a significant breakthrough by relocating the holotype of *M. pansus*, which allowed redescription and rediagnosis of the species, and comparison of cranial features with *E. tortus* (Korth, 2002a). The genus is resurrected as a useful taxon illustrating early evolution of subfamily Castoroidinae. *M. pansus* co-occurs with the higher-crowned *E. tortus*, but in general, *Monosaulax* is seen as an early (Hemingfordian to Barstovian age, late early to middle Miocene) castoroidine; more derived and younger (usually Clarendonian age, early late Miocene) species are assigned to *Eucastor*. Species attribution of the abundant material frequently referred variously to *M. pansus* or *E. tortus* should be testable by recomparison with holotypes. Many citations in the literature need this reevaluation.

A morphometric analysis of castorid skulls and dentitions demonstrated for Xu (1995) the homogeneity of the *Steneofiber–Monosaulax* group. He, therefore, considered *Steneofiber* the appropriate name and submerged many of the species discussed here (plus Eurasian forms) as *Steneofiber hesperus*. However, Korth's (1996a) careful review of the holotype of *S. hesperus* shows it to represent the distinct genus *Neatocastor*.

Eucastor Leidy, 1858 (including *Sigmophius*)

Type species: *Eucastor tortus* Leidy, 1858.

Type specimen: USNM 11020, rostrum and palate with broken incisors, left and right P4-M2.

Characteristics: Cheek teeth high crowned with roots formed after wear; rounded incisor enamel; worn molars showing two lakes (hypoflexus [-id] and mesofossette [-id]) in early wear and especially in more primitive species; hypostria (-id) long; mesostria (-id) short; para- and metafossettes (-ids) present (two or three external flexi and internal flexids in early wear); lower premolars higher crowned with longer striids than in *Monosaulax*; "S" pattern (Figure 23.1B) possibly developing in late wear in advanced species (after Stirton, 1935). *E. tortus* smaller in size, with greater postorbital constriction, and with rostrum elongated relative to *Monosaulax pansus* (upper diastema more than twice length of upper cheek tooth row; Korth [2002a]).

Length of P4-M3 alveolus of holotype: 12 mm.

Lower cheek tooth row (AMNH 10821): 19 mm.

Included species: *E. tortus* (known from localities?SB32F, SP1A, CP90A, CP105, CP110, CP114A, B, CP115B, [C]); *E. dividerus* Stirton, 1935 (localities NB23C, CP90A, CP110, CP114B, D); *E. lecontei* (Merriam, 1896) (originally described as *Sigmophius lecontei*) (localities [CC26B], CC35A, NB18A, NB19C, NB25B, [NB29A]); *E. malheurensis* Shotwell and Russell, 1963 (locality PN10); *E. phillisi*, Wilson, 1968 (locality CP123E); *E. tedi* Korth, 1999 (locality CP111).

Eucastor sp. is also known from localities GC4DD, GC10B, CP114D, CP116B, [CP123C].

Comments: Miocene beavers are surprisingly diverse, for example in the Harrison through Valentine Formation sequence of Nebraska (curation of T. M. Stout). Many of these species belong to the *Eucastor* clade, in which observed temporal trends are increase in crown height, suppression of accessory flexi/fossettes, and development of the "S" pattern on the occlusal surface of cheek teeth. Several species of *Eucastor* have been named from the Miocene of Eurasia. For some of these, *Steneofiber* is the appropriate name, but large, high-crowned forms from the Pliocene of Yushe Basin, China, for example, are called *Eucastor* (Teilhard de Chardin, 1942; Xu, 1994).

Microdipoides Korth and Stout, 2002 (including _Nothodipoides_; _Eucastor_, in part)

Type species: *Microdipoides eximius* Korth and Stout, 2002.

Type specimen: F:AM 65295, right dentary with incisor, p4, m1–2.

Characteristics: Small castoroidines exhibiting size decrease despite increasing relative crown height; hypsodont molars with persistent "S" pattern on lower molars and p4; short diastema; incisor flattened anteriorly.

Average length of m1: 3.1 mm.

Included species: *M. eximius* (known from locality CP90A); *M. planus* (Stirton, 1935) (originally described as *Eucastor planus*) (localities [SB32FF], SP1A, CP116A, B); *M. stirtoni* (Korth, 2002b) (originally described as *Nothodipoides stirtoni*) (localities CP110, CP114D).

Comments: The flattened incisor is usually considered a hallmark of a tooth-digging burrower. Korth and Stout (2002) drew close affinity of their new species with Stirton's *Eucastor planus*, and transferred the latter to *Microdipoides*. Korth (2002b) recognized more species showing affinity to *Microdipoides* and presumed diversity in this group, including his *Nothodipoides stirtoni*. Being presumed transitional to *Microdipoides planus*, we consider *Nothodipoides* a junior synonym.

Dipoides Schlosser, 1902

Type species: *Dipoides problematicus* Schlosser, 1902.

Type specimen of *Dipoides stirtoni* Wilson, 1934, first named New World species of the genus: CIT 1662, left dentary with all cheek teeth.

Characteristics: Cheek teeth hypsodont (roots closing only in old age); round incisors; molars with "S" pattern (hypostria [-id] not opposite mesostria [-id], these striae extending to bases of crowns; parastria [-id] present to adult stage; metastria [-id] shorter, fossettes [-ids] in young animals; molars nearly as large as premolars (after Shotwell, 1955).

Average length of p4: 6.35 mm (*D. stirtoni*, from Wilson, 1934).

Included species: *D. stirtoni* Repenning, 1987 (known from localities NB27B, NB30, NB31, CP58, CP116D, NP45, PN11B, PN12B, PN13, PN15, PN22); *D. rexroadensis* Hibbard, 1949 (localities CP128B, C, PN3C); *D. smithi* Shotwell, 1955 (localities PN14, PN15); *D. tanneri* Korth, 1998b (locality CP116B); *D. vallicula* Shotwell, 1970 (localities [SB33II], PN15); *D. williamsi* Stirton, 1936 (localities SB11, SB34C); *D. wilsoni* Hibbard, 1949 (locality CP128AA).

Dipoides sp. is also known from localities GC13B, NB31, SP1D, SP3B, CP115C, CP116C, E, F; the *Dipoides* cf. *D. intermedius* record from the Pliocene of Ellesmere Island (locality HA3) is listed here.

Comments: *Dipoides* is a Holarctic taxon first named from the late Miocene and Pliocene of the Old World. It is distributed throughout Eurasia and across North America, into Ellesmere Island (Tedford and Harrington, 2003). North American *Dipoides* is so well represented that its many named species document the rich evolution of this group rather well into the early Pleistocene. *Paradipoides stovalli* Rinker and Hibbard, 1952 was named for a "post-Pearlette" jaw from Beaver County, Oklahoma, larger than *D. intermedius* (transferred to *Procastoroides*, below). This rather large *Castor*-size form resembles older *Dipoides* and perhaps should be assigned to that genus (Zakrzewski, 1969).

Flynn (1997) saw *Dipoides* as a late immigrant into China at approximately 7 Ma and considered the genus as abruptly dispersing from North America. Hugueney (1999) documented an array of late Miocene fossils of *Eucastor* and *Dipoides* affinity, one record perhaps as old as 12 Ma. Her hypothesis called for parallel evolution of *Dipoides* in North America and Europe, as a vicariant event from a *Steneofiber–Monosaulax* stock.

***Procastoroides* Barbour and Schultz, 1937 (including *Eocastoroides*)**

Type species: *Procastoroides sweeti* Barbour and Schultz, 1937.

Type specimen: NSM 100-12-6–36S. P.

Characteristics: Large beaver 75% the size of *Castoroides ohioensis* (exceeding 50 kg), but with flatter, less broad skull (see Woodburne, 1961); 9–10 longitudinal grooves in incisor of *P. idahoensis*, but absent in *P. sweeti*; cheek tooth reentrants filling; with cementum "S" pattern modifying to oblique cross-lophs joined only by thin enamel or isolated; narrow dentine tracts in some specimens.

Average length of p4: 15 mm (Barbour and Schultz, 1937).

Included species: *P. sweeti* (known from localities SP1G, CP94, CP117B, CP118, CP128B, C, CP131, CP131II); *P. idahoensis* (locality PN23C); *P. intermedius* (Zakrzewski, 1969) (localities PN23A, ?HA3; this species transferred from *Dipoides*, following Xu [1995]).

Procastoroides sp. is also known from locality PN3C.

Comments: Woodburne (1961) reviewed *P. sweeti* and its synonym *Eocastoroides lanei* Hibbard, 1938. *Castoroides* Foster, 1838 appears in the Irvingtonian and has widely been perceived as closely related to *Procastoroides*. However, Woodburne (1961) noted features (e.g., its plain incisor) that made *P. sweeti* an unlikely ancestor. When Shotwell (1970) broadened the content of *Procastoroides* by naming *P. idahoensis*, he reopened the idea that *Castoroides* evolved from *Procastoroides*. As an alternative view, we suggest that *P. idahoensis* could be regarded as an early species of *Castoroides* (i.e., late Blancan age, 2.5 to 2 Ma Neville, Opdyke, and Lindsay, 1979; Lundelius *et al.*, [1987]) (note that *Castoroides* is actually reported from localities GC18IIIA, ?B).

INDETERMINATE CASTORIDS

Castorids of indeterminate genus and species are reported from localities NP10C2, CC.

BIOLOGY AND EVOLUTIONARY PATTERNS

The evolution of Castoridae is a Holarctic phenomenon. The group appears to have originated in North America, as its older sister taxon, Eutypomyidae, is North American, as are the Oligocene *Agnotocastor* and allies. Although European beavers are among the oldest known, *Asteneofiber* appears abruptly in the early Oligocene as a fully derived beaver and without apparent clue to its ancestry (Hugueney, 1999). We consider early Eurasian beavers as immigrants from North America. Although Xu (1995) used selected cranial features to ally *Steneofiber* with North American *Monosaulax* (or early *Eucastor*), Korth's (2001a) arguments that *Steneofiber* is simply primitive, but at the base of the Castorinae, are also based on good evidence. Under this scenario, Castorinae is a Eurasian radiation of beavers, absent from North America until the late Miocene. Independent research led to the recognition of subfamily Asiacastorinae Lytschev (Xu, 1995) to accommodate genera of apparent Asian origin. We suggest that Asiacastorinae are contained in Castoroidinae and this group is of Eurasian origin. Sutton and Korth (1995) recognized *Euroxenomys* in North America, representing an early Miocene dispersion of Eurasian Castoriodinae into North America. Korth (2004) sees *Anchitheriomys* as a large beaver of North American origin, but we suggest that it may be another early Miocene immigrant.

In North America, two castorid clades diversified and dominated the middle Cenozoic record: the Palaeocastorinae and the Castoroidinae (Figure 23.3). Palaeocastorinae are a terrestrial group of North American origin and without known record in Eurasia. Unlike the aquatic lineages, palaeocastorines demonstrate radiation into subterranean habitats, with highly fossorial features developed in some forms, such as *Euhapsis*. The North American group of Castoroidines, the endemic Castoroidini of Korth (2001a), evolved hypsodont, grinding dentitions and included a lineage that achieved great mass, up to 100 kg, in the Pleistocene. *Dipoides* appears to have dispersed into Asia during the late Miocene. In addition, subsequent vicariant evolution of castoroidines in North America and Eurasia appears likely. The alternative that Eurasian "*Dipoides*" species evolved independently from Asian castoroidines remains to be tested. The successful castoroidines spread throughout North America and are recorded in Ellesmere Island, where they were aquatic (Tedford and Harington, 2003; Rybczynski, 2004). To what degree earlier castoroidines were aquatic is unknown.

The generalized lifestyle and aquatic preferences of beavers appear to have established the group as a Holarctic faunal element predisposed to take advantage of cross-continent dispersal opportunities. More dispersal events are likely to emerge, documented by the growing fossil record, and these will be highly useful for intercontinental correlation. However, the Castoridae are a group that demonstrate how perilous are conclusions about habitat based on extant species alone. The modern aquatic beavers are not a good biological model for the highly successful terrestrial to fossorial Palaeocastorinae. The underground niche so successfully exploited by this subfamily likely contributed to confinement of the group to North America.

Further information on life histories of different beavers could derive from their best-represented fossil element: the incisors. It is possible to calculate growth rates from periradicular lines in *Castor* incisors (Rinaldi and Cole, 2004). This also presents a good possibility of interpreting season of death in fossil beavers. One might predict and test different patterns of birth, growth, and death in palaeocastorine and castoroidine beavers.

ACKNOWLEDGEMENTS

Access to the published record was facilitated through the *Bibliography of Fossil Vertebrates*. Particularly useful also were Korth (2001a) and McKenna and Bell (1997). We are indebted to Thomas Mörs, Christine Janis, and Julia Keller for help in running down

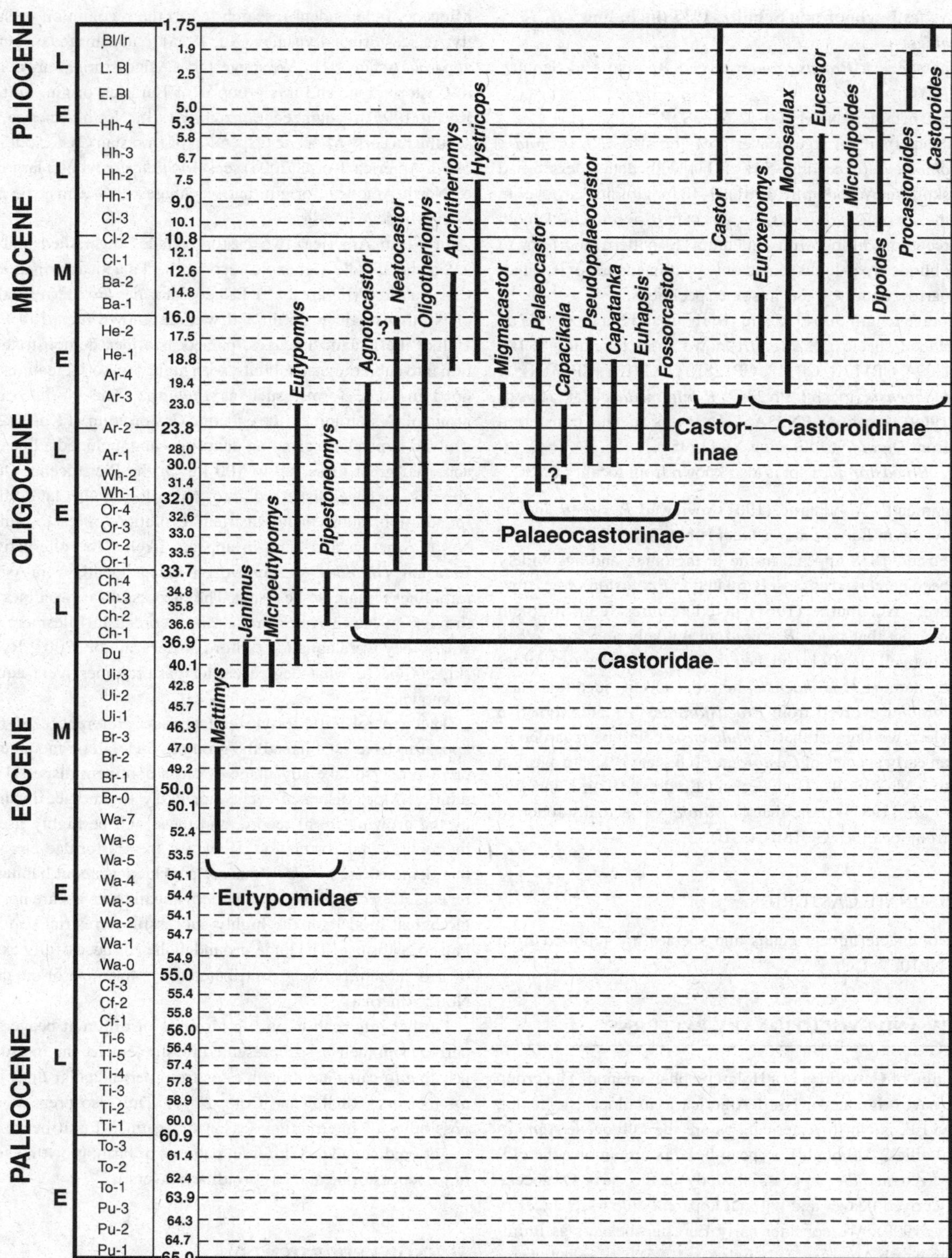

Figure 23.3. Temporal ranges of North American castoroid genera.

citations. The thesis work of Xu Xiaofeng formed a fundamental basis for our review. Alisa Winkler and Dale Winkler provided crucial feedback on earlier drafts.

REFERENCES

Adkins, R. M., Walton, A. H., and Honeycutt, R. L. (2003). Higher level systematics of rodents and divergence time estimates based on two congruent nuclear genes. *Molecular Phylogenetics and Evolution*, 26, 409–20.

Alf, R. (1962). A new species of the rodent *Pipestoneomys* from the Oligocene of Nebraska. *Breviora, Museum of Comparative Zoology*, 172, 1–7.

Allen, J. A. (1877). Castoroididae. In *Monographs of North American Rodentia*, ed. E. Coues and J. A. Allen, pp. 415–42. Washington, DC: US Government Printing Office for the US Geological Survey of the Territories.

Barbour, E. H. and Schultz, C. B. (1937). An early Pleistocene fauna from Nebraska. *American Museum Novitates*, 942, 1–10.

(1965). Fossil mammals from Montana. Part 2: Rodents from the early Oligocene Pipestone Springs Local Fauna. *Annals of the Carnegie Museum*, 38, 1–48.

Cope, E. D. (1874). On a new mastodon and a rodent. *Proceedings Academy of Natural Science, Philadelphia*, 1874, 221–3.

(1878). On some characters of the Miocene fauna of Oregon. *Proceedings of the American Philosophical Society*, 18, 63–78.

(1881). Review of the Rodentia of the Miocene period of North America. *Bulletin of the United States Geological and Geographic Survey of the Territories*, 6, 361–86.

Dawson, M. R. (1966). Additional Late Eocene rodents from the Uinta Basin, Utah. *Annals of the Carnegie Museum*, 38, 97–114.

Donohoe, J. C. (1956). New aplodontid rodent from Montana Oligocene. *Journal of Mammalogy*, 37, 264–8.

Douglass, E. (1901). Fossil Mammalia of the White River beds of Montana. *Transactions of the American Philosophical Society*, 20, 243–9.

Emry, R. J. (1972). A new species of *Agnotocastor* (Rodentia, Castoridae). *American Museum Novitates*, 2485, 1–7.

Evander, R. L. (1999). Rodents and lagomorphs (Mammalia) from the Railway Quarries local fauna (Miocene, Barstovian) of Nebraska. *Paludicola*, 2, 240–57.

Fisher, R. V. and Rensberger, J. M. (1972). Physical stratigraphy of the John Day Formation, central Oregon. *University of California Publications in Geological Sciences*, 101, 1–45.

Flynn, L. J. (1997). Late Neogene mammalian events in North China. *Actes du Congrès BioChrom '97. Mémoire Travaux E.P.H.E., Institut Montpellier*, 21, 1183–92.

Foster, J. W. (1838). Organic remains. In *Ohio Geological Survey Second Annual Report*, pp. 79–83. Columbus, OH: Division of Geological Survey.

Fremd, T., Bestland, E. A., and Retallack, G. J. 1997. *John Day Basin Paleontology: Field Trip Guide and Road Log, for 1994 Society of Vertebrate Paleontology Annual Meeting*. Seattle, WA: Northwest Interpretive Association (in association with John Day Fossil Beds National Monument, Kimberly, OR).

Galbreath, E. C. (1953). A contribution to the Tertiary geology and paleontology of northeastern Colorado. *University of Kansas Paleontological Contributions. Vertebrata*, 4, 1–120.

Gill, T. (1872). Arrangement of the families of mammals with analytical tables. *Smithsonian Miscellaneous Collections*, 11, 1–98.

Hay, O. P. (1927). *The Pleistocene of the Western Region of North America and its Vertebrated Animals*. Washington, DC: The Carnegie Institute.

Hemprich, W. (1820). *Grundriss der Naturgeschichte für höhere Lehranstalten Entworfen von Dr. W. Hemprich*. Berlin: August Rucker; Vienna: Friedrich Volke.

Hibbard, C. W. (1938). An upper Pliocene fauna from Meade County, Kansas. *Transactions of the Kansas Academy of Science*, 40, 239–65.

(1949). Pliocene Saw Rock Canyon fauna in Kansas. *Contributions from the Museum of Paleontology, University of Michigan*, 7, 91–105.

Hugueney, M. (1999). Family Castoridae. In *The Miocene Land Mammals of Europe*, ed. G. E. Rössner and K. Heissig, pp. 281–300. Munich: Dr. Friedrich Pfeil.

Kellogg, L. (1911). A fossil beaver from Kettleman Hills, California. *University of California Publications in Geological Sciences*, 6, 401–2.

Koenigswald, W., von and Mörs, T. (2001). The enamel microstructure of *Anchitheriomys* (Rodentia, Mammalia) in comparison with that of other beavers and porcupines. *Paläontologische Zeitschrift*, 74, 601–12.

Korth, W. W. (1984). Earliest Tertiary evolution and radiation of rodents in North America. *Bulletin of the Carnegie Museum of Natural History*, 24, 1–71.

(1988). A new species of beaver (Rodentia, Castoridae) from the middle Oligocene (Orellan). *Journal of Paleontology*, 62, 965–7.

(1994). *The Tertiary Record of Rodents in North America*. New York: Plenum Press.

(1996a). A new genus of beaver (Mammalia: Castoridae: Rodentia) from the Arikareean (Oligocene) of Montana and its bearing on castorid phylogeny. *Annals of the Carnegie Museum*, 65, 167–79.

(1996b). Additional specimens of *Agnotocastor readingi* (Rodentia, Castoridae) from the Orellan (Oligocene) of Nebraska and a possible origin of the beavers. *Paludicola*, 1, 16–20.

(1998a). A new beaver (Rodentia, Castoridae) from the Orellan (Oligocene) of North Dakota. *Paludicola*, 1, 127–31.

(1998b). Rodents and lagomorphs (Mammalia) from the late Clarendonian (Miocene) Ash Hollow Formation, Brown County, Nebraska. *Annals of the Carnegie Museum*, 67, 299–348.

(1999). A new species of beaver (Rodentia, Castoridae) from the earliest Barstovian (Miocene) of Nebraska and the phylogeny of *Monosaulax* Stirton. *Paludicola*, 2, 258–64.

(2000a). A new species of *Eutypomys* Matthew (Rodentia, Eutypomyidae) from the Orellan (Oligocene) and reevaluation of "*Eutypomys*" *magnus* Wood. *Paludicola*, 2, 273–8.

(2000b). Rediscovery of lost holotype of *Monosaulax pansus* (Rodentia, Castoridae). *Paludicola*, 2, 279–81.

(2001a). Comments on the systematics and classification of beavers (Rodentia, Castoridae). *Journal of Mammalian Evolution*, 8, 279–96.

(2001b). A new species of *Anchitheriomys* (Rodentia: Castoridae) and a review of the anchitheriomyine beavers from North America. *Paludicola*, 3, 51–5.

(2001c). Occurrence of the European genus of beaver *Euroxenomys* (Rodentia: Castoridae) in North America. *Paludicola*, 3, 73–9.

(2002a). Topotypic cranial material of the beaver *Monosaulax pansus* Cope (Rodentia, Castoridae). *Paludicola*, 4, 1–5.

(2002b). Review of the castoroidine beavers (Rodentia, Castoridae) from the Clarendonian (Miocene) of northcentral Nebraska. *Paludicola*, 4, 15–24.

(2004). Beavers (Rodentia, Castoridae) form the Runningwater Formation (Early Miocene, Early Hemingfordian) of western Nebraska. *Annals of the Carnegie Museum*, 73, 2, 61–71.

Korth, W. W. and Emry, R. J. (1997). The skull of *Anchitheriomys* and a new subfamily of beavers (Castoridae, Rodentia). *Journal of Paleontology*, 71, 343–7.

Korth, W. W. and Rybczynski, N. (2003). A new, unusual castorid (Rodentia) from the earliest Miocene of Nebraska. *Journal of Vertebrate Paleontology*, 23, 667–75.

Korth, W. W. and Stout, T. M. (2002). A diminutive beaver (Rodentia, Castoridae) from the late Clarendonian (Miocene) of South Dakota. *Paludicola*, 3, 134–8.

Lambe, L. M. (1908). The Vertebrata of the Oligocene of the Cypress Hills, Saskatchewan. *Contributions to Canadian Palaeontology*, 3, 56–7.

Leidy, J. (1856). Notices of remains of extinct Mammalia discovered by Dr. F. V. Hayden in Nebraska Territory. *Proceedings of the Academy of Natural Sciences, Philadelphia*, 8, 1–89.

(1858). Notices of some remains of extinct Vertebrata from the valley of the Niobrara River, collected during the exploring expedition of 1857 in Nebraska (under the command of Lieut. G. K. Warren, U.S. Top. Eng.). *Proceedings of the Academy of Natural Sciences, Philadelphia*, 10, 1–23.

(1869). The extinct mammalian fauna of Dakota and Nebraska, including an account of some allied forms from other localities, together with a synopsis of the mammalian remains of North America. *Journal of the Academy of Natural Sciences, Philadelphia*, 2, 1–472.

Linnaeus, C. (1758). *Systema Naturae Per Regna Tria Naturae, Secundum Classes, Ordines, Genera, Species Cum Characteribus, Differentiis, Synonymis, Locis*. Vol. 1: *Regnuma Animale*, 10th edn, revised. Stockholm: Laurentii Salvi.

Lundelius, E. L., Downs, T., Lindsay, E. H., *et al.* (1987). The North American Quaternary sequence. In *Cenozoic Mammals of North America, Geochronology and Biostratigraphy*, ed. M. O. Woodburne, pp. 211–35. Berkeley, CA: University of California Press.

Lytschev, G. F. (1983). Osnovnyye napravleniya evolyutsii v semeystve Castoridae. [The basic directions in the evolution of the family Castoridae.] In *Istoriya I Evolyutsiya Sovremennoy Fauny Gryzunov SSSR (Neogen-Sovreennost')*, ed. I. M. Gromov, pp. 179–203. [In Russian.]

Macdonald, J. R. (1963). The Miocene Wounded Knee faunas of southwestern South Dakota. *Bulletin of the American Museum of Natural History*, 125, 141–238.

(1970). Review of the Miocene faunas from the Wounded Knee area of western South Dakota. *Bulletin of the Los Angeles County Museum of Natural History, Science*, 8, 1–82.

Martin, L. D. (1973). The mammalian fauna of the Lower Miocene Gering Formation and the early evolution of the North American Cricetidae. Ph.D. Thesis, University of Kansas, Lawrence.

(1987). Beavers from the Harrison Formation (early Miocene) with a revision of *Euhapsis*. *Dakoterra, South Dakota School of Mines and Technology*, 3, 73–91.

Matthew, W. D. (1905). Notice of two new genera of mammals from the Oligocene of South Dakota. *Bulletin of the American Museum of Natural History*, 21, 21–6.

(1907). A lower Miocene fauna from South Dakota. *Bulletin of the American Museum of Natural History*, 23, 169–219.

(1918). Contributions to the Snake Creek fauna, with notes upon the Pleistocene of western Nebraska. *Bulletin of the American Museum of Natural History*, 38, 197–9.

Matthew, W. D. and Cook, H. J. (1909). A Pliocene fauna from western Nebraska. *Bulletin of the American Museum of Natural History*, 26, 380–1.

McKenna, M. C. and Bell, S. K. (1997). *Classification of Mammals Above the Species Level*. New York: Columbia University Press.

Merriam, J. C. (1896). *Sigmophius Le Contei*, a new castoroid rodent from the Pliocene near Berkeley, California. *University of California Publications in Geological Sciences*, 1, 363–70.

Meyer, H., von (1838). Letter on Swiss fossils. *Magazine of Natural History New Series*, 2, 583–8.

Miller, G. S. and Gidley, J. W. (1918). Synopsis of the supergeneric groups of rodents. *Journal of the Washington Academy of Science*, 8, 431–48.

Neville, C., Opdyke, N. D. and Lindsay, E. H. (1979). Magnetic stratigraphy of Pliocene deposits of the Glenns Ferry Formation, Idaho, and its implications for North American mammalian biostratigraphy. *American Journal of Science*, 279, 503–26.

Nichols, R. (1979). Additional early Miocene mammals from the Lemhi Valley of Idaho. *Tebiwa, Miscellaneous Papers*, 17, 1–12.

Peterson, O. A. (1905). Description of new rodents and discussion of the origin of *Daemonelix*. *Memoirs of the Carnegie Museum*, 2, 139–200.

Repenning, C. A. (1987). Biochronology of the microtine rodents of the United States. In *Cenozoic Mammals of North America, Geochronology and Biostratigraphy*, ed. M. O. Woodburne, pp. 236–68. Berkeley, CA: University of California Press.

Reynolds, P. S. (2002). How big is a giant? The importance of method in estimating body size of extinct mammals. *Journal of Mammalogy*, 83, 321–32.

Rinaldi, C. and Cole, T. M., III (2004). Environmental seasonality and incremental growth rates of beaver (*Castor canadensis*) incisors: implications for paleobiology. *Palaeogeography, Palaeoclimatology, Palaeoecology*, 206, 289–301.

Rinker, G. C. and Hibbard, C. W. (1952). A new beaver, and associated vertebrates, from the Pleistocene of Oklahoma. *Journal of Mammalogy*, 33, 98–101.

Roger, O. (1898). Wirbeltierreste aus dem Dinotheriensande der bayerisch-schwäbischen Hochebene. *Bericht des Naturwissenschaftlichen Vereins für Schwaben und Neuburg*, 33, 1–46.

Romer, A. S. and McCormack, J. T. (1928). A large *Palaeocastor* from the lower Miocene. *American Journal of Science*, 15, 59–60.

Rybczynski, N. (2004). Effect of incisor shape on woodcutting performance in two beavers (Castoridae, Rodentia). *Journal of Vertebrate Paleontology*, 24(suppl. to no. 3), p. 107A.

Samson, P. and Radulesco, C. (1973). Remarques sur l'evolution des Castorides (Rodentia, Mammalia). In *Livre du Cinquantenaire de l'Institut de Spéléologie "Emile Racovitza"*, ed. T. Orghidan, pp 437–49. Bucharest: Academiei Republicii Socialiste Romania.

Schlosser, M. (1884). Die Nager des europäischen Tertiärs nebst Betrachtungen über die Organisation und die geschichtliche Entwicklung der Nager überhaupt. *Palaeontographica*, 31, 19–162.

(1902). Beiträge zur Kenntniss der Säugetierreste aus den süddeutschen Bohnerzen. *Geologie und Paläontologie Abhandlungen Jena*, 5, 21–3.

Scott, W. B. (1893). The mammals of the Deep River beds. *The American Naturalist*, 27, 680.

Shotwell, J. A. (1955). Review of the Pliocene beaver *Dipoides*. *Journal of Paleontology*, 29, 129–44.

(1963). Mammalian fauna of the Drewsey Formation, Bartlett Mountrain, Drinkwater and Otis Basin local faunas. *Transactions of the American Philosophical Society*, 53, 70–7.

(1968). Miocene mammals of southeast Oregon. *Bulletin of the Museum of Natural History, University of Oregon*, 14, 1–67.

(1970). Pliocene mammals of southeast Oregon and adjacent Idaho. *Bulletin of the Museum of Natural History, University of Oregon*, 17, 1–103.

Shotwell, J. A. and Russell, D. E. (1963). Juntura Basin: Studies in earth history and paleontology. *Transactions of the American Philosophical Society*, 53, 42–69.

Stirton, R. A. (1934). A new species of *Amblycastor* from the *Platybelodon* beds, Tung Gur Formation, of Mongolia. *American Museum Novitates*, 694, 1–4.

(1935). A review of the Tertiary beavers. *University of California Publications in Geological Sciences*, 23, 391–458.

(1936). A new beaver from the Pliocene of Arizona with notes on the species of *Dipoides*. *Journal of Mammalogy*, 17, 279–81.

Storer, J. E. (1975). Tertiary mammals of Saskatchewan. Part III: The Miocene fauna. *Life Sciences Contributions, Royal Ontario Museum*, 103, 1–134.

(1984). Mammals of the Swift Current Creek local fauna (Eocene: Uintan, Saskatchewan). *Natural History Contributions, Museum of Natural History, Regina*, 7, 1–158.

(1988). The rodents of the Lac Pelletier lower fauna, late Eocene (Duchesnean) of Saskatchewan. *Journal of Vertebrate Paleontology*, 8, 84–101.

Stout, T. M. (1967). [Addendum in M. F. Skinner and B. E. Taylor] A revision of the geology and paleontology of the Bijou Hills, South Dakota. *American Museum Novitates*, 2300, 1–53.

Sutton, J. F. and Korth, W. W. (1995). Rodents (Mammalia) from the Barstovian (Miocene) Anceny local fauna, Montana. *Annals of the Carnegie Museum*, 64, 267–314.

Tedford, R. H. and Harington, C. R. (2003). An Arctic mammal fauna from the Early Pliocene of North America. *Nature*, 425, 388–90.

Teilhard de Chardin, P. (1942). New rodents of the Pliocene and lower Pleistocene of North China. *Publications de l'Institut Géo-Biologie, Pekin*, 9, 1–101.

Wahlert, J. H. (1977). Cranial foramina and relationships of *Eutypomys* (Rodentia, Eutypomyidae). *American Museum Novitates*, 2626, 1–8.

Walton, A. H. (1993). A new genus of eutypomyid (Mammalia: Rodentia) from the middle Eocene of the Texas Gulf Coast. *Journal of Vertebrate Paleontology*, 13, 262–6.

Wilson, R. L. (1968). Systematics and faunal analysis of a lower Pliocene vertebrate assemblage from Trego County, Kansas. *Contributions from the Museum of Paleontology, University of Michigan*, 22, 75–126.

Wilson, R. W. (1934). A new species of *Dipoides* from the Pliocene of eastern Oregon. *Carnegie Institution of Washington Publication*, 453, 19–28.

(1949). Early Tertiary rodents of North America. *Carnegie Institution of Washington Publication*, 584, 67–164.

(1960). Early Miocene rodents and insectivores from northeastern Colorado. *University of Kansas Paleontological Contributions, Vertebrata*, 7, 1–92.

Wood, A. E. (1937). The mammalian fauna of the White River Oligocene. Part 2: Rodentia. *Transactions of the American Philosophical Society*, 28, 155–269.

(1974). Early Tertiary vertebrate faunas, Vieja Group, Trans-Pecos Texas: Rodentia. *Bulletin of the Texas Memorial Museum*, 21, 1–112.

(1980). The Oligocene rodents of North America. *Transactions of the American Philosophical Society*, 70, 1–68.

Wood, A. E. and Konizeski, R. L. (1965). A new eutypomyid rodent from the Arikareean (Miocene) of Montana. *Journal of Paleontology*, 39, 492–6.

Wood, H. E. (1945). Late Miocene beaver from southeastern Montana. *American Museum Novitates*, 1299, 1–6.

Woodburne, M. O. (1961). Upper Pliocene geology and vertebrate paleontology of part of the Meade Basin, Kansas. *Papers of the Michigan Academy of Science, Arts, and Letters, Paleontology*, 46, 61–95.

Xu, X. F. (1994). Evolution of Chinese Castoridae. In *Rodent and Lagomorph Families of Asian Origins and Diversification*, ed. Y. Tomida, C. K. Li, and T. Setoguchi. *National Science Museum of Tokyo Monograph*, 8, 77–98.

(1995). Phylogeny of beavers (Family Castoridae): applications to faunal dynamics and biochronology since the Eocene. Ph.D. Thesis, Southern Methodist University, Dallas.

(1996). Castoridae. In *The Terrestrial Eocene-Oligocene Transition in North America*, ed. D. R. Prothero and R. J. Emry, pp. 417–32. Cambridge, UK: Cambridge University Press.

Zakrzewski, R. J. (1969). The rodents from the Hagerman local fauna, upper Pliocene of Idaho. *Contributions from the Museum of Paleontology, University of Michigan*, 23, 1–36.

24 Dipodidae

LAWRENCE J. FLYNN

Peabody Museum, Harvard University, Cambridge, MA, USA

INTRODUCTION

Dipodidae are small- to medium-size rodents that include the extant jumping mice and birch mice of North America and Asia, both of small size, and the bigger, more widely distributed Old World jerboas. Both jumping mice and jerboas show saltatorial adaptations, especially in the hindlimb. The group is not species rich although jerboas show some diversity. Many authors refer to the generally small-bodied fossil members of this group as "zapodids," a term with misleading higher taxonomic connotation (below). Dipodids are rarely common in the fossil record but are present in many Recent small mammal communities of North America. They appear to be useful as paleoclimatic indicators, given their preference for moist conditions, the arboreal habits of birch mice, and the ability to hibernate for jumping mice.

Mid Cenozoic genera are usually grouped with extant *Sicista* in the subfamily Sicistinae. Sicistines do not show saltatorial skeletal modifications but are good climbers with prehensile tails (Klingener, 1984). Late Cenozoic zapodines hibernate and are saltatorial, although the postcrania are not as derived as in some Asian groups. As noted above, the family name Dipodidae is employed here, rather than Zapodidae, because the small genera of the mid Cenozoic fossil record do not constitute a monophyletic group with respect to *Zapus*. Fossil evidence suggests multiple dispersal events between Asia and North America. Clearly, use of Zapodidae for some genera would mandate other family rankings (e.g., Sicistidae), and would require thorough revision of the group.

In the North American fossil record, some dipodid species have apparently long temporal ranges, as from the Arikareean through the Barstovian Land-Mammal Ages (late Oligocene to middle Miocene), an extraordinary longevity for a rodent species. This is perhaps artificial and a result of limited study of poor samples of variable and small rodents (see also the discussion in Skwara, 1988).

DEFINING FEATURES OF THE FAMILY DIPODIDAE

CRANIAL

The skull has a hystricomorphous zygomasseteric structure, and a neurovascular passageway separated as a foramen ventromedial to the infraorbital foramen (Figure 24.1D). Extant forms show elongated incisive foramina, a slender mandible, and a masseteric fossa extending below the first molar (noted by Korth, 1994). The foramen in the coronoid fossa for the mandibular canal is large.

DENTAL

Cheek tooth dental formula is typically P1/0, M3/3, with a small P4, but oldest forms preserve a peglike P3 as well, and extant *Napeozapus* lacks both premolars. First and second molars (including uppers) are usually elongated anteroposteriorly. Upper incisors bear a longitudinal groove in some genera; incisor enamel microstructure is uniserial.

Molars show the "cricetid plan," and opposite cusps are joined by transverse crests. In the M1 of all but the earliest forms, there is a mesocone and associated mesoloph, and a strong anterior cingulum, usually with anterostyle. Typically, an endoloph (including the mure or longitudinal portion joining major cusps) joins the hypocone, mesocone, and protocone. The m1 has an anteroconid. The lower molars have a strong posterolophid, and the anteroposterior mure that joins cusps is often labial in location (in this case termed an ectolophid). A distinct mesoconid usually bears a mesolophid. There is a strong anterior cingulum on m2 and m3. Typical evolutionary trends are toward increasing lophodonty and emphasis of shearing crests.

Evolution of Tertiary Mammals of North America, Vol. 2. ed. C. M. Janis, G. F. Gunnell, and M. D. Uhen. Published by Cambridge University Press.

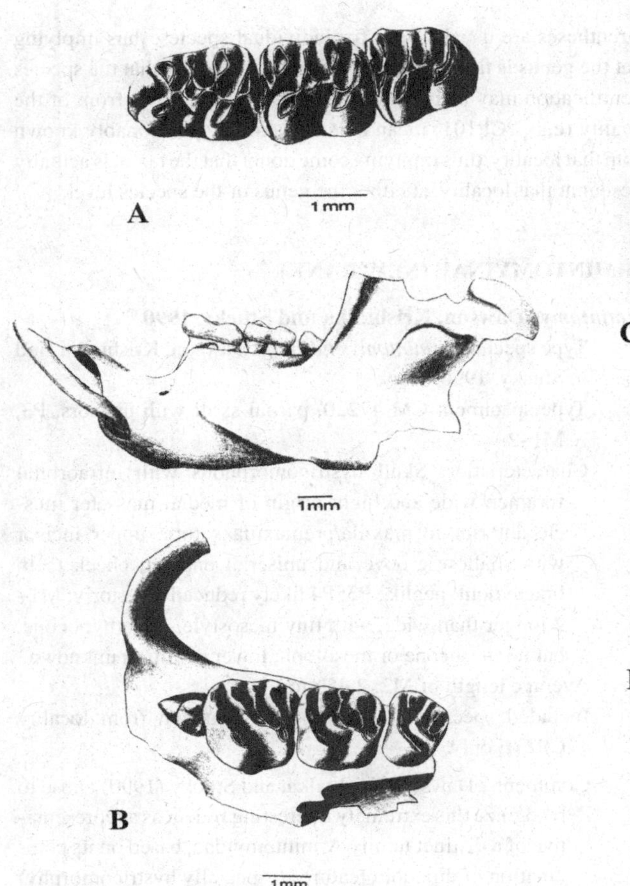

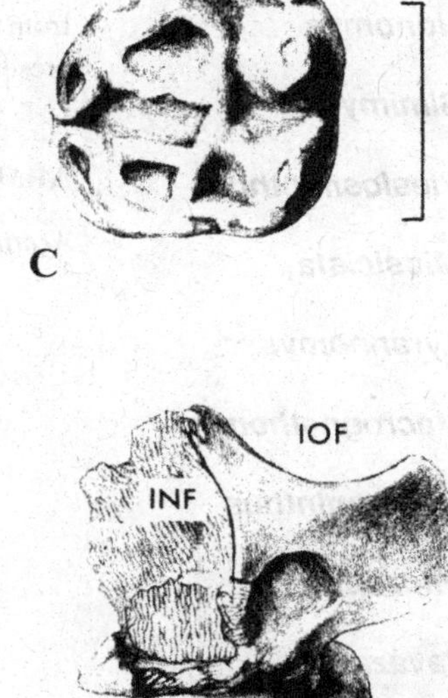

Figure 24.1. A,B. Upper and lower dentitions of *Megasminthus tiheni*, from Korth (1980a). C,D. Upper molar (C) and anterior view of zygomatic structure (D) of *Nonomys* (after Emry (1981). The enlarged infraorbital foramen transmits the medial masseter muscle in the traditional hystricomorphous fashion. The ventromedial opening is a separate neurovascular channel. This dipodoid feature is present early in the evolution of the group. (A,B: courtesy of the Carnegie Museum of Natural History. C,D: courtesy of Robert J. Emry.)

POSTCRANIAL

Postcrania are rare in the fossil record, but extant zapodines and sicistines are gracile, with hindlimbs sometimes modified for saltatorial locomotion. Modifications are extreme, including fused metatarsals, in Old World jerboas.

SYSTEMATICS

SUPRAFAMILIAL

Dipodoids are members of the rodent suborder Myomorpha and considered to be closely related to muroids, the familiar hamsters, mice, and relatives of global distribution. A host of features including cranial morphology and soft anatomy clearly ally the Dipodoidea with the Muroidea (see, e.g., Bugge, 1985; Vianey-Liaud, 1985). For most systematists, the more inclusive taxon Myodonta is used to unite these two groups. McKenna and Bell (1997) utilized Myodonta at the infraorder taxonomic level. The sister taxa Muroidea and Dipodoidea both have fossil records extending into the early Eocene; their time of divergence can be estimated in excess of 50 Ma. The origin of suborder Myomorpha would be perhaps as much as 55 Ma.

Despite the name, most myomorphs are not myomorphous. Primitively, myodonts show an enlarged infraorbital foramen through which part of the deep masseter passes. This condition is hystricomorphy. Early muroids retain hystricomorphy, but later muroid fossils demonstrate derivation of myomorphy from hystricomorphy (see Chapter 16). Derived muroids become myomorphous progressively, with part of the masseter passing lateral to the enlarged infraorbital foramen. This lateral masseter constricts the ventral portion of the infraorbital foramen, lending a keyhole shape to the opening.

INFRAFAMILIAL

Dipodidae, although not numerically dominant in genera, are a diverse group with center of radiation in Asia and divergence of several clades. The North American record includes primitive, bunodont sicistines plus the zapodines, which are strongly lophodont, multicrested species similar to *Zapus hudsonius* (Figure 24.2). Some authors regard the American dipodoid radiation, at least the surviving ones, as a distinctive family, Zapodidae. Obviously extant *Zapus* and *Napeozapus* are strongly derived in dentition and monophyletic. Several North American forms show affinity with them, but other primitive North American fossil genera represent separate subfamily-level groupings, including the Asiatic Sicistinae. Furthermore, Zapodinae, even when restricted to the allies of *Zapus*, usually includes the Asian *Eozapus*, with first record in the terminal Miocene Ertemte Fauna, Inner Mongolia (Fahlbusch, 1992). Therefore,

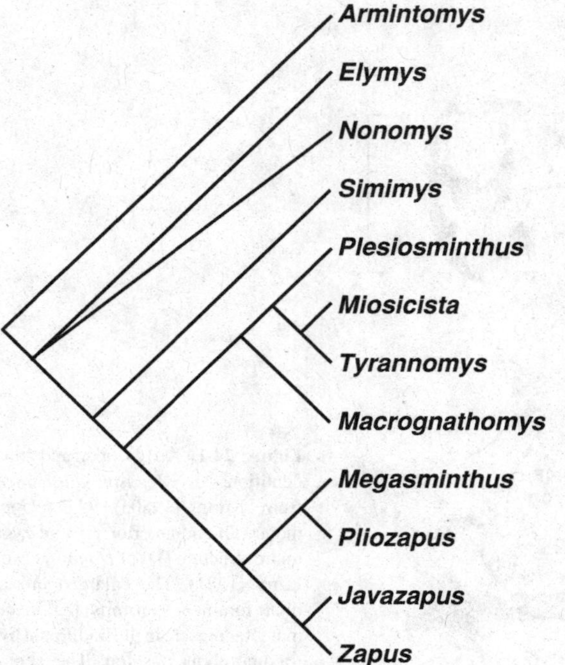

Figure 24.2. Relationships of fossil dipodid genera of North America, including basal taxa, Sicistinae, and Zapodinae.

Zapodinae as well as the older Sicistinae may be of Asian origin, and both are distinctive subfamilies of Dipodidae on a par with other Old World groups, such as the jerboas. The family rank Dipodidae is applied here to include primitive and derived forms, some exhibiting scansorial, and others saltatorial, locomotion. This usage is parallel in content with the diverse Heteromyidae, which also include scansorial and saltatorial genera (see Klingener, 1984).

Dipodids have an Eocene origin, possibly in North America, but the heyday of the group on this continent was a Neogene event. The Neogene record of Dipodidae includes both birch mice (Sicistinae) and jumping mice (Zapodinae). The early history of the group includes the aberrant early *Armintomys* Dawson, Krishtalka, and Stucky (1990), which has been proposed as a distinct family-level taxon. Herein it is considered a dipodid, but monotypic subfamily Armintomyinae Dawson, Krishtalka, and Stucky (1990) is utilized. Other Paleogene genera are not assigned here to subfamily. Some small Tertiary forms have occasionally been proposed as primitive dipodoids. These are detailed elsewhere, for example *Simiacritomys* Kelly, 1992, here considered an eomyid (see Chapter 25).

INCLUDED GENERA OF DIPODIDAE FISCHER VON WALDHEIM, 1817

The locality numbers listed for each genus refer to the list of unified localities in Appendix I. The acronyms for museum collections are listed in Appendix III. The localities are listed in several ways. Parentheses around the locality (e.g., [CP101] mean that the taxon in question at that locality is cited as an "aff." or "cf." to the taxon name. Parentheses are usually used for individual species, thus implying that the genus is firmly known from the locality but that the species identification may be questionable. Question marks in front of the locality (e.g., ?CP101) mean that the taxon is questionably known from that locality, thus implying some doubt that the taxon is actually present at that locality, at either the genus or the species level.

ARMINTOMYINAE (NEW RANK)

Armintomys Dawson, Krishtalka and Stucky, 1990

Type species: *Armintomys tullbergi* Dawson, Krishtalka, and Stucky, 1990.

Type specimen: CM 47220, partial skull with incisors, P3, M1–2.

Characteristics: Skull hystricomorphous with infraorbital foramen wide and high; origin of medial masseter muscle anterior to maxilla/premaxilla suture; upper incisor with shallow groove and uniserial enamel; cheek teeth brachydont; peglike P3; P4 likely reduced anteriorly; M1–2 longer than wide, with tiny mesostyle, large hypocone, but no mesocone or mesoloph; lower dentition unknown.

Average length of M2: 2.45 mm.

Included species: *A. tullbergi* only, known from locality CP27E only.

Comments: Dawson, Krishtalka, and Stucky (1990) chose to recognize this extremely interesting rodent as a representative of a distinct family Armintomyidae, based on its combination of dipodoid features (especially hystricomorphy) and retention of primitive characters (such as P3). They also considered the incisor enamel to be derived, describing it as transitional to uniserial. I consider its derived condition synapomorphic, an advanced (uniserial) microstructure shared with all Myodonta. I treat *Armintomys* as a basal dipodid, rather than a distinct family, in keeping with my conservative treatment of zapodines within Dipodidae. *Armintomys* is a surprisingly large rodent for a basal dipodoid position. Consequently, subfamily Armintomyinae is warranted.

GENERA NOT ASSIGNED TO SUBFAMILY

Elymys Emry and Korth, 1989

Type species: *Elymys complexus* Emry and Korth, 1989.

Type specimen: USNM 404720, right maxilla with P4, M1–3.

Characteristics: Small size, brachydont molars with mesostyle but no mesocone or mesoloph; no P3; P4 peglike; molars longer than wide, with low crests; upper molars with large hypocone (small in M3), but conules evident only on M1 (as a protoconule), with endoloph; lower molars with ectolophid, including a variable mesoconid (no mesolophid).

Average length of M2: 0.91 mm.

Included species: *E. complexus* only, known from locality NB14 only.

Comments: This rodent is quite small, much smaller than *Armintomys*, and appropriate for the plesiomorphic condition of early Dipodidae. The molars of the two genera are similar in morphology. A major difference from *Armintomys* appears to be the greatly reduced premolars of *Elymys*, which is the modern condition. Larger samples for both genera would show the relative sizes of deciduous and replacement premolars and permit understanding of the role played by premolar replacement, which could alter the systematic interpretation and allocation of these early Eocene forms.

***Simimys* Wilson, 1935a (including *Eumysops*)**

Type species: *Simimys simplex* (Wilson, 1935b) (originally described as *Eumyops simplex*).

Type specimen: CIT 1759, left dentary fragment with m1–3.

Characteristics: Bunodont, brachydont molars with crests moderately developed including incomplete mesoloph (-id); small anterocone and anteroconid; mesocone small, but mesoconid prominent and with labial spur; hypoconulid on posterolophid; third molars relatively large; P4 variably retained.

Average length of m2: 1.42 mm (after Lillegraven and Wilson, 1975).

Included species: *S. simplex* (including *S. murinus* and *S. vetus*, see Lillegraven and Wilson, 1975) (known from localities CC7E, CC8, CC9A–B2, CC9VA, CC10); *S. landeri* Kelly, 1992 (locality CC9B2).

Simimys sp. is also known from localities CC6B, CC7B–D, CC8IIA, CC9BB, CC9II, ?SB43C, (SB44B), (CP5II), ?CP98B, (CP99B).

Comments: This primitive myodont has been considered close to the ancestry of both "zapodids" and cricetids. Because it shows many dipodoid features including zygomasseteric structure with a separate neurovascular foramen ventral to the infraorbital foramen (Emry, 1981) and prominent mesoconid. Here it is placed in Dipodidae. Some authors remain concerned that it apparently had lost P4 and therefore "could not be ancestral to" later dipodids. This does not seem to warrant separate family status (Simimyidae Wood, 1980), and indeed Walsh (1997) described a specimen preserving a small fourth premolar, thereby removing the need for an exclusive higher taxon.

***Nonomys* Emry and Dawson, 1973 (including *Nanomys; Subsumus*)**

Type species: *Nonomys simplicidens* (Emry and Dawson, 1972) (originally described as *Nanomys simplicidens*).

Type specimen: F:AM 79304, right dentary with m1–3, broken incisor.

Characteristics: Emry (1981) clearly showed the isolated foramen for the infraorbital nerve and blood vessels that characterizes this small rodent as a dipodoid (this opening is ventral to the large, hystricomorphous infraorbital foramen of early myomorphs); cheek teeth brachydont; P4 lost, molars cuspate with indistinct crests; prominent cingula (lingual on uppers, buccal on lowers); large anterocone on M1 on prominent shelf; strong anteroconid on m1; hypoconulids distinct; no mesocone or mesoconid.

Length of m2: 1.23 mm (holotype).

Included species: *N. simplicidens* (including *Subsumus candelariae* Wood, 1974) only (known from localities SB44D, CP39C–E).

Nonomys sp. is also known from localities CC8IIB, (SB42), SB43E, NP10Bi, Bii.

Comments: Martin (1980) saw *Nonomys* as a primitive cricetid and utilized Nonomyinae as a higher taxon for it. Emry (1981) compared *Nonomys* and *Simimys* and noted that both show the dipodoid condition of neurovascular canal separate from the hystricomorphous infraorbital foramen. He elected to refer both to primitive Muroidea without further affiliation, in recognition that they lack derived characters of extant families. The genera do not show compelling evidence of sister-taxon relationship. Although the cheek tooth formula of *Nonomys* is 3/3, a shared muroid condition, *Simimys* variably retains P4, as do dipodoids. The taxonomy used here implies loss of P4 in *Nonomys*, in parallel with muroids.

In this interpretation, *Nonomys* represents primitive dipodoids in the late Eocene of North America. However, the continent of origin of the group remains enigmatic, given the possibility that *Ivanantonia*, from the early Eocene of Mongolia (Hartenberger, Dashzeveg, and Martin, 1997), could be a *Nonomys* relative.

As noted in Chapter 27, the taxon identified as *Eumys* sp. by Storer (1988) from locality NP9A is considered a species of *Nonomys*. If these Duchesnean age fossils represent *Eumys*, they would be the earliest North American record of the Cricetidae.

***Diplolophus* Troxell, 1923 (including *Gidleumys*)**

Type species: *Diplolophus insolens*, Troxell, 1923.

Type specimen: YPM 10368, right dentary fragment with m1–3.

Characteristics: Large size; mesodont, strongly bilophodont dentition; longitudinal connections in cheek teeth weak in upper molars, absent in the lowers; cheek teeth reduced in number to 3/3; strong anterocone on shelf in first upper tooth; anterior portion of first lower tooth narrow and bipartite; second molars wider than long, with evidence of hypoconulid; strong masseteric crest anterior to first lower tooth.

Average length of m2: 2.1 mm (from Barbour and Stout, 1939).

Included species: *D. insolens* only (known from localities CP68C, CP84A, CP99A [see Barbour and Stout, 1939]).

Diplolophus sp. is also known from locality CC8IIB.

Comments: *Diplolophus* is an early rodent with three cheek teeth, not obviously muroid and, therefore, plausibly dipodoid by default. I list it here accordingly, without conviction. Korth (1994) concisely discussed this interesting rodent, and its possible affinities with geomyoids; it could also represent a derived eomyid. It would be an early record for a strongly bilophodont geomyoid. Some resemblance to the Old World eomyid *Rittenaria* is apparent, yet the latter is a rather small rodent. Wood (1980) discussed the possibility that if *Diplolophus* is related to Geomorpha then the cheek teeth could represent fourth premolars through first and second molars.

SICISTINAE

The Sicistinae is characterized by small size, usually smooth upper incisors, and hind legs unmodified for saltatorial locomotion. These primitive features do not define the group well. Survey of features to test its cohesiveness would have to include fossil Asian forms (e.g., López-Antoñanzas and Sen, 2006). The sole extant genus is *Sicista*, the birch mouse. Birch mice are small, 6–14 g (Nowak and Paradiso, 1983). They excavate shallow burrows and climb readily in environments ranging from forest to steppe. Martin (1994) judged North American fossil *Macrognathomys* to be close to *Sicista*, implying possibly more than one migration of sicistines across Beringia.

Plesiosminthus Viret, 1926 (including *Schaubemys*)

Type species: *Plesiosminthus schaubi* from St.-Gerand-Le-Puy, France.

Type specimen (North American *P. clivosus* Galbreath, 1953): KU 9279, left dentary with m1–3.

Characteristics: Small, generalized sicistine; cheek teeth brachydont; M1 usually with strong anterocone, upper molars with long mesoloph; reduced M3; m1 usually with prominent mesoconid, variable mesolophid, low and tiny anteroconid; upper incisor grooved (unlike *Parasminthus*).

Average length of lower molar row: 2.85 mm in *P. clivosus*, 3.7 mm in *P. sabrae*, and 4.1 mm in *P. grangeri*.

Included species: *P. cartomylos* (Korth, 1987) (originally described as *Schaubeumys cartomylos*) (known from localities CP87B, CP114A, B, D, NP11); *P. clivosus* Galbreath, 1953 (including *P. geringensis* Martin, 1974) (localities CP75B, CP86B-D, CP88, CP101, NP10CC, E); *P. galbreathi* Wilson, 1960 (possible junior synonym of *P. grangeri*) (localities CP75B, CP88); *P. grangeri* (Wood, 1935) (localities CP86B–D); *P. sabrae* (Black, 1958) (originally described as *Schaubeumys sabrae*) (locality CP54B).

Plesiosminthus sp. is also known from localities CP90A, NP10E, NP34C, ?D, NP36?A, B, D, NP40B, NC2B.

Comments: *Plesiosminthus* is also known from the Oligocene and Miocene of Europe and Asia (McKenna and Bell, 1997). Skwara (1988) noted that many earlier Miocene sites at which *Plesiosminthus* is recorded yield a large and small form. While many authors have applied the generic nomen *Shaubeumys* for small North American early sicistines (Engesser, 1979; Korth, 1980; Skwara, 1988), recent reviewers (Green, 1992; McKenna and Bell, 1997) have failed to find synapomorphies that distinguish the American species and have supported Wilson's (1960) view that they are not generically distinct from Old World *Plesiosminthus*. This hypothesis invokes dispersal of one or more species eastward across Beringia. Some authors (see Green, 1992) have also suggested that the Asiatic *Parasminthus* (which is distinct) occurs in North America. I agree with use of *Plesiosminthus* for early North American sicistines but see no compelling evidence for *Parasminthus* on this continent.

Miosicista Korth, 1993

Type species: *Miosicista angulus* Korth, 1993.

Type specimen: UNSM 45424, right dentary fragment with m1–3.

Characteristics: Cheek teeth brachydont; molars with oblique (alternating) setting of cusps, with lingual cusps anterior to buccal neighbors; weak ectolophid and mesolophid; accessory crest internal to hypoconid on m2–3; masseteric fossa extending only to the back of m1; m3 somewhat reduced (after Korth, 1993).

Length of m2: 1.23 mm (holotype).

Included species: *M. angulus* only, known from locality CP114B only.

Macrognathomys Hall, 1930

Type species: *Macrognathomys nanus* Hall, 1930.

Type specimen: UCMP 29634, left dentary with incisor, m1–3.

Characteristics: Cheek teeth brachydont; large anteroconid on m1 usually joining protoconid–metalophid; m1–2 approximately equal in size, third molar only somewhat smaller.

Length of lower molar row: 3.2 mm (holotype).

Included species: *M. nanus* (known from localities NB23C, CP123E, and probably PN10, PN11B); *M. gemmacollis* (Green, 1977) (localities SP1A, CP89).

Macrognathomys sp. is also known from localities CP114A, PN9B.

Comments: Martin (1994) judged this genus to be close to *Sicista*, the extant Asian birch mouse. The earliest records of the genus *Sicista* are relatively younger (late Miocene), for example approximately 6.5 Ma in the Mahui Formation of Yushe Basin, China.

Tyrannomys Martin, 1989

Type species: *Tyrannomys harkseni* Martin, 1989.

Type specimen: SDSM 8078, left m2.

Characteristics: Small sicistine with brachydont cheek teeth; molar cusps anteroposteriorly compressed, and with sharp connecting crests; mesocone (-id) not inflated; longitudinal crests rather central in position.

Length of m2: 1.02 mm (holotype).

Included species: *T. harkseni* only, known from locality CP97II only.

Comments: This interesting taxon is included for completeness. It records a late-surviving record of sicistines in North America. It is quite small (attesting to hard work in its recovery) and has alternate cusps, reminiscent of the condition in *Miosicista*.

Indeterminate sicistines

A sicistine of indeterminate genus and species is reported from locality CP123E.

ZAPODINAE

The Zapodinae are restricted here to relatives of *Zapus hudsonius*, the extant, widespread jumping mice of North America. Zapodines present derived dental conditions, which can be found in a number of extinct taxa. In general, lophodonty and crown height of the cheek teeth are accentuated, with loss of distinct cusps. Also, the lophs are oblique, rather than transverse. Korth (1994) noted several useful features, including absence of anteroposterior connections between dental crests, especially with M1 showing an isolated anteroloph consisting of lobate protocone continuous with the anterior cingulum (no connection to the mesocone; Figure 24.1A). The anteroconid of m1 is often modified (or lost). In addition, zapodines have grooved upper incisors, and elongated hindlimbs. Dipodid relationships are presented in Figure 24.2.

Among extant genera, *Napeozapus* is very close to *Zapus* but lacks P4. *Eozapus* is less derived dentally and is restricted to China. The body mass of these small rodents is 13 to 26 g. They have long tails, and hindlimbs modified for hopping. They inhabit cool forest to meadow settings, and range into high altitudes. Extant forms are proficient hibernators.

***Megasminthus* Klingener, 1966**

Type species: *Megasminthus tiheni* Klingener, 1966.

Type specimen: UMMP 52874, fragmentary palate with partial left and right toothrows.

Characteristics: Cheek teeth brachydont; P4 with two small cusps; anteroloph usually separate on M1–2, strong mesocone, mesostyle, mesoloph; m1 with mesostylid, strong ectostylid, usually on a crest, and with a median but isolated anteroconid; ectolophid weak between protoconid and mesoconid; m2 lacks ectostylid.

Average length of m2: 1.84 mm (*M. tiheni* from Korth, 1980).

Included species: *M. tiheni* (known from localities CP87B, CP89, CP114A, B, NP11); *M. gladiofex* Green, 1977 (localities CP87B, CP114A).

Megasminthus sp. is also known from localities CP116B, CP123E.

***Pliozapus* Wilson, 1936**

Type species: *Pliozapus solus* Wilson, 1936.

Type specimen: CIT 1811, right dentary fragment with incisor, m1–3.

Characteristics: Small zapodine with brachydont cheek teeth; m1 and m2 about equal in size, m3 smaller, as in *Megasminthus*; molars apparently more lophodont than in *Megasminthus*, although the type specimen is worn; m1 and m3 three-rooted; m2 four-rooted; m1 anteroconid reduced, variable; prominent, flared angular process of the dentary as in *Zapus*.

Length of lower molar row: 3.9 mm (holotype).

Included species: *P. solus* only (known from localities NB27B, [CP116E], PN11B, PN14).

***Javazapus* Martin, 1989**

Type species: *Javazapus weeksi* Martin, 1989.

Type specimen: SDSM 8071, right m1.

Characteristics: Resembles *Pliozapus* in size and absence of anteroconid on m1; cheek teeth mesodont; lower molars all with two roots (fewer than *Pliozapus*); considered by Martin (1989) to be higher crowned and more lophodont than *Pliozapus*.

Average length of m2: 1.33 mm.

Included species: *J. weeksi* only, known from locality CP97II only.

Javazapus sp. is also known from locality CP131 (Martin, 1989).

***Zapus* Coues, 1875**

Type species: *Zapus hudsonius*, the extant North American meadow jumping mouse.

Type specimen: UMMP 27036, right dentary with incisor, m1–3.

Characteristics: Molars mesodont and lophodont, with planed occlusal surfaces; large (often bilobed) anteroconid on m1, which has no posterior connection; angled, pinched protoconids and hypoconids, protoconid continuous with an oblique strong crest running to entoconid and separated from posterior lobe (the hypoconid–posterolophid with small posterior lake); strong mesostylid on mesolophid, no ectostylid; reduced third molars; strongly angled protocones with apices elongated posteriorly (and joining hypocone on M2).

Average length of lower molar row: 4.3 mm (*Z. rinkeri)*.

Included species: *Z. burti* Hibbard, 1941 (known from localities CP132B, D); *Z. rinkeri* (localities CP128A, [B]); *Z. sandersi* Hibbard, 1956 (localities CP128C, D, CP130, CP131).

Zapus sp. is also known from localities GC15D, SP1H, CP97II, CP117B, and (sometimes with *Napeozapus)* from many Pleistocene sites.

Indeterminate zapodines

A zapodine of indeterminate genus and species is reported from locality NP10CC.

BIOLOGY AND EVOLUTIONARY PATTERNS

Dipodid history in North America includes tantalizing glimpses into the origin of the group and the Neogene radiations of the sicistine and zapodine clades (Figure 24.3). Early Eocene *Armintomys* provides clues about the structure of the cheek teeth in early hystricomorphs. Herein, this genus is considered a dipodid with sciuravid-like cheek teeth. The slightly younger *Elymys* is dentally more typical of dipodids. These Eocene taxa point to a complex origin for the group, complexity that encompasses early Asian rodents as well. Later Eocene *Simimys* and *Nonomys* also show primitive dentitions that may help to indicate dipodoid roots among sciuravid-like predecessors.

Given the antiquity of *Armintomys* and *Elymys*, approximately 50 Ma, it appears that Myodonta had evolved by this time. Myomorpha would have differentiated from sciuravid-like rodents earlier, probably shortly after the earliest Eocene appearance of the order Rodentia in North America.

Dipodids are, typically, a small component of North American small mammal faunas throughout the Cenozoic, yet a characteristic of the North American Neogene is the presence of both sicistines and zapodines. Some localities, particularly in Asia, contain rich dipodid assemblages. Extant Sicistinae (birch mice) are good climbers, suggestive of vegetation cover, probably trees, and presumably relatively moist conditions. It is reasonable to presume this habitat also for Neogene forms, and their spotty fossil record is consistent with this; tree squirrels, like birch mice, are uncommon elements in many faunas, presumably those where fossilization of arboreal taxa was not optimal. Zapodinae (jumping mice) are primarily quadrupedal but tend toward saltatorial locomotion and are nocturnal (Nowak and Paradiso, 1983). They exploit wooded to grassy and alpine habitat. Zapodines diversified in the late Neogene especially, a time of increasing seasonality, including colder winters, and their lifestyle is consistent with this; they hibernate.

Dipodids show an array of interesting features that can be surveyed in fossil forms in the future. Hindlimb elongation with fusion of metatarsals and loss of side toes is progressive across lineages. Highly saltatorial rodents generally have modified cervical vertebrae and frequently develop greatly inflated auditory bullae for detecting predators in open habitat. Unfortunately, these features are rarely preserved among fossils. A comprehensive molecular study of Dipodoidea would provide an exciting opportunity to map functional morphology across phylogeny. One feature of the hard anatomy that can be reevaluated now is incisor grooving. The distribution of this feature across known fossil species potentially will be important in tracking relationships.

Dipodids, like other rodents, attest to intervals of intercontinental dispersal during the Cenozoic, well in advance of the extraordinary global climate change late in that era. *Plesiosminthus* had invaded the North American land mass by the early Miocene, and early species are indistinguishable from Eurasian forms at the generic level, although some authors use *Schaubeumys* for North American species. Late in the Neogene, apparently both modern sicistines, like *Macrognathomys*, and zapodines, possibly comparable to *Eozapus*, participated in dispersal across the Bering Land Bridge.

ACKNOWLEDGEMENTS

I wish to thank the paleontological community for the wealth of information on which this survey is based. The publication record is a crucial resource, and access to it was facilitated through the Bibliography of Fossil Vertebrates. Particularly useful in checking details are Korth (1994) and McKenna and Bell (1997). I am indebted to Julia Keller and Christine Janis for help in running down citations. Ev Lindsay and Bob Martin kindly helped in evaluating some of the records.

REFERENCES

Barbour, E. H. and Stout, T. M. (1939). The White River Oligocene rodent *Diplolophus*. *Bulletin of the University of Nebraska State Museum*, 2, 29–36.

Black, C. C. (1958). A new sicistine rodent from the Miocene of Wyoming. *Breviora, Museum of Comparative Zoology*, 86, 1–7.

Bugge, J. (1985). Systematic value of the carotid arterial pattern in rodents. In *Evolutionary Relationships Among Rodents: A Multidisciplinary Analysis*, ed. W. P. Luckett and J.-L. Hartenberger, pp. 355–79. New York: Plenum Press.

Coues, E. (1875). Some account, critical, descriptive, and historical of *Zapus hudsonius*. *Bulletin of the United States Geological and Geographical Survey of the Territories*, Vol. 1, Series 2, pp. 253–62. Washington, DC: Government Printing Office.

Dawson, M. R., Krishtalka, L., and Stucky, R. K. (1990). Revision of the Wind River faunas, early Eocene of central Wyoming. Part 9: The oldest known hystricomorphous rodent (Mammalia: Rodentia). *Annals of the Carnegie Museum*, 59, 135–47.

Emry, R. J. (1981). New material of the Oligocene muroid rodent *Nonomys*, and its bearing on muroid origins. *American Museum Novitates*, 2712, 1–14.

Emry, R. J. and Dawson, M. R. (1972). A unique cricetid (Mammalia, Rodentia) from the early Oligocene of Natrona County, Wyoming. *American Museum Novitates*, 2508, 1–14.

(1973). *Nonomys*, new name for the cricetid (Rodentia, Mammalia) genus *Nanomys* Emry and Dawson. *Journal of Paleontology*, 47, 103.

Emry, R. J. and Korth, W. W. (1989). Rodents of the Bridgerian (middle Eocene) Elderberry Canyon local fauna of eastern Nevada. *Smithsonian Contributions to Paleobiology*, 67, 1–14.

Engesser, B. (1979). Relationships of some insectivores and rodents from the Miocene of North America and Europe. *Bulletin of the Carnegie Museum of Natural History*, 14, 1–68.

Fahlbusch, V. (1992). The Neogene mammalian faunas of Ertemte and Harr Obo in Inner Mongolia (Nei Mongol), China. 10. *Eozapus* (Rodentia). *Senckenbergiana Lethaea*, 72, 199–217.

Fischer von Waldheim, G. (1817). Adversaria zoologica. *Memoires de la Société Impériale des Naturalistes du Moscou*, 5, 357–428.

Galbreath, E. C. (1953). A contribution to the Tertiary geology and paleontology of northeastern Colorado. *University of Kansas Paleontological Contributions, Vertebrata*, 4, 1–120.

Green, M. (1977). Neogene Zapodidae (Mammalia, Rodentia) from South Dakota. *Journal of Paleontology*, 51, 996–1015.

(1992). Comments on North American fossil Zapodidae (Mammalia, Rodentia) with reference to *Megasminthus, Plesiosminthus*, and *Schaubeumys*. *Occasional Papers of the Museum of Natural History, University of Kansas*, 148, 1–11.

Hall, E. R. (1930). Rodents and lagomorphs from the later Tertiary of Fish Lake Valley, Nevada. *University of California Publications in Geological Sciences*, 19, 295–312.

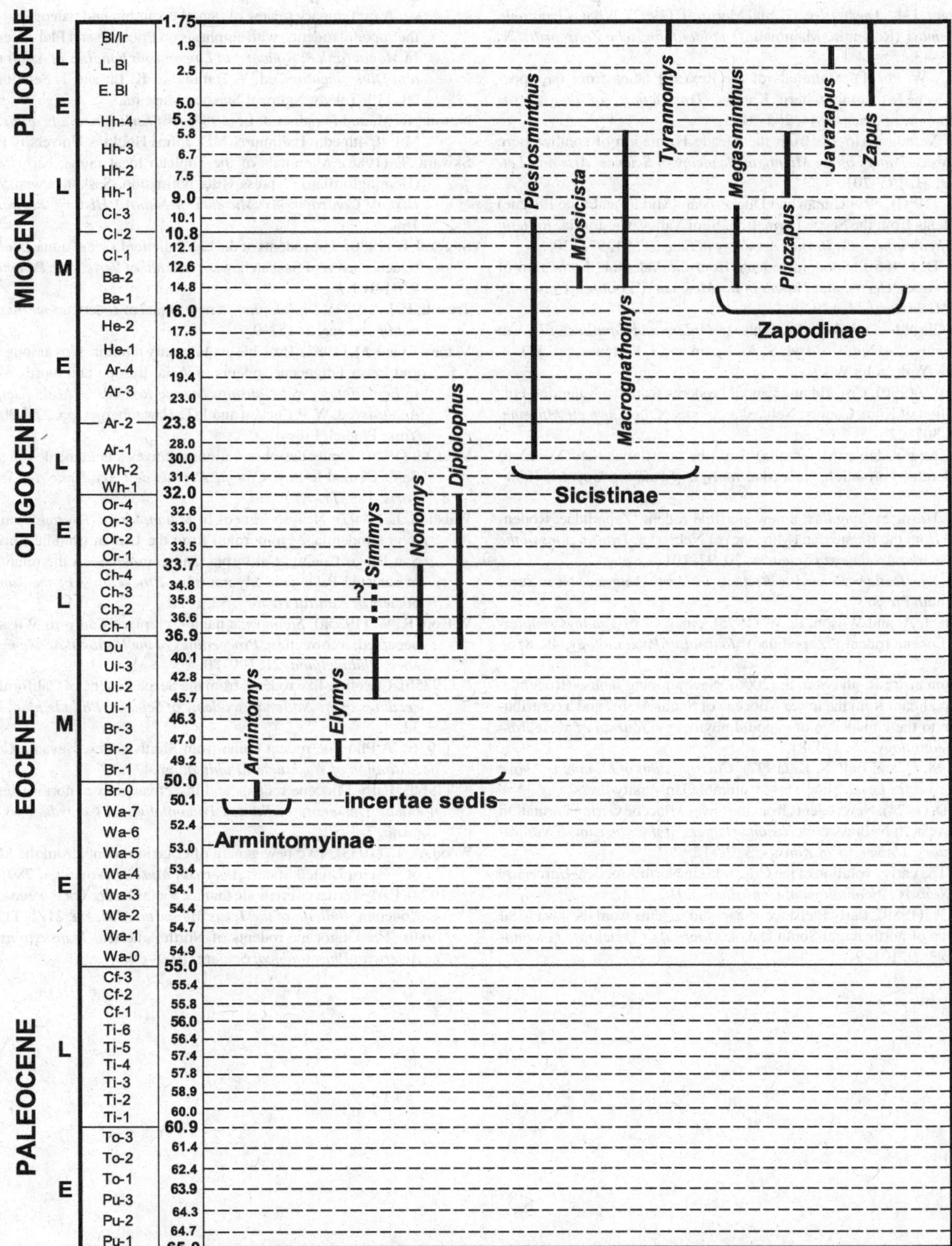

Figure 24.3. Temporal ranges of North American dipodid genera.

Hartenberger, J.-L., Dashzeveg, G., and Martin, T. (1997). What is *Ivantonia efremovi* (Rodentia, Mammalia)? *Paläontologische Zeitschrift*, 71, 135–43.

Hibbard, C. W. (1941). Mammals of the Rexroad fauna from the upper Pliocene of southwestern Kansas. *Transactions of the Kansas Academy of Science*, 44, 265–313.

(1956). Vertebrate fossils from the Meade Formation of southwestern Kansas. *Papers of the Michigan Academy of Science, Arts and Letters*, 41, 145–201.

Kelly, T. S. (1992). New Uintan and Duchesnean (Middle and Late Eocene) rodents from the Sespe Formation, Simi Valley, California. *Bulletin of the Southern California Academy of Science*, 91, 97–120.

Klingener, D. (1966). Dipodoid rodents from the Valentine Formation of Nebraska. *Occasional Papers of the Museum of Zoology, University of Michigan*, 644, 1–9.

(1984). Gliroid and dipodoid rodents. In *Orders and Families of Recent Mammals of the World* ed. S. Anderson and J. K. Jones, pp. 381–8. New York: John Wiley.

Korth, W. W. (1980). Cricetid and zapodid rodents from the Valentine Formation of Knox County, Nebraska. *Annals of the Carnegie Museum*, 49, 307–22.

(1987). New rodents (Mammalia) from the late Barstovian (Miocene) Valentine Formation, Nebraska. *Journal of Paleontology*, 61, 1058–64.

(1993). *Miosicista angulus*, a new sicistine rodent (Zapodidae, Rodentia) from the Barstovian (Miocene) of Nebraska. *Transactions of the Nebraska Academy of Sciences*, 20, 97–101.

(1994). *The Tertiary Record of Rodents in North America*. New York: Plenum Press.

Lillegraven, J. A. and Wilson, R. W. (1975). Analysis of *Simimys simplex*, an Eocene rodent (?Zapodidae). *Journal of Paleontology*, 49, 856–74.

López-Antoñanzas, R. and Sen, S. (2006). New jumping mouse (Rodentia, Zapodidae) from the lower Miocene of Saudi Arabia and a contribution to the knowledge of zapodid phylogeny. *Journal of Vertebrate Paleontology*, 26, 170–81.

McKenna, M. C. and Bell, S. K. (1997). *Classification of Mammals Above the Species Level*. New York: Columbia University Press.

Martin, L. D. (1974). New rodents from the lower Miocene Gering Formation of western Nebraska. *Occasional Papers of the Museum of Natural History, University of Kansas*, 32, 1–12.

(1980). The early evolution of the Cricetidae in North America. *University of Kansas Paleontological Contribution*, 102, 1–42.

Martin, R. A. (1989). Early Pleistocene zapodid rodents from the Java local fauna of northcentral South Dakota. *Journal of Vertebrate Paleontology*, 9, 101–9.

(1994). A preliminary review of dental evolution and paleogeography in the zapodid rodents, with emphasis on Pliocene and Pleistocene taxa. In *Monograph 8: Rodent and Lagomorph Families of Asian Origins and Diversification*, ed. Y. Tomida, C. K. Li, and T. Setoguchi, pp. 99–113. Tokyo: National Science Museum.

Nowak, R. M. and Paradiso, J. L. (1983). *Walker's Mammals of the World*, Vol. II, 4th edn. Baltimore, MD: Johns Hopkins University Press.

Skwara, T. (1988). Mammals of the Topham local fauna: early Miocene (Hemingfordian) Cypress Hills Formation, Saskatchewan. *Natural History Contributions, Museum of Natural History, Regina*, 9, 1–168.

Storer, J. E. (1988). The rodents of the Lac Pelletier Lower Fauna, late Eocene (Duchesnean) of Saskatchewan. *Journal of Vertebrate Paleontology*, 8, 84–101.

Troxell, E. L. (1923). *Diplolophus*, a new genus of rodents. *American Journal of Science*, 5, 157–9.

Vianey-Liaud, M. (1985). Possible evolutionary relationships among Eocene and lower Oligocene rodents of Asia, Europe and North America. In *Evolutionary Relationships Among Rodents: A Multidisciplinary Analysis*, ed. W. P. Luckett and J.-L. Hartenberger, pp. 277–309. New York: Plenum Press.

Viret, G. (1926). Nouvelles observations relatives à la faune de rongeurs de Saint-Gerand-le-Puy, *Compte Rendus de l'Académie des Sciences, Paris*, 183, 71–2.

Walsh, S. L. (1997). New specimens of *Metanoiamys, Pauromys*, and *Simimys* (Rodentia, Myomorpha) from the Uintan (middle Eocene) of San Diego County, California, and comments on the relationships of selected Paleogene Myomorpha. *Proceedings of the San Diego Society of Natural History*, 32, 1–20.

Wilson, R. W. (1935a). *Simimys*, a name to replace *Eumyops* Wilson, preoccupied: a correction. *Proceedings of the National Academy of Science, Philadelphia*, 21, 179–80.

(1935b). Cricetine-like rodents from the Sespe Eocene of California. *Proceedings of the National Academy of Science, Philadelphia*, 21, 26–32.

(1936). A Pliocene rodent fauna from Smith Valley, Nevada. *Carnegie Institution of Washington Publication*, 473, 17–34.

(1960). Early Miocene rodents and insectivores from northeastern Colorado. *University of Kansas Paleontological Contributions, Vertebrata*, 7, 1–92.

Wood, A. E. (1935). Two new genera of cricetid rodents from the Miocene of western United States. *American Museum Novitates*, 789, 1–8.

(1974). Early Tertiary vertebrate faunas, Vieja Group, Trans-Pecos, Texas: Rodentia. *Bulletin of the Texas Memorial Museum*, 21, 1–112.

(1980). The Oligocene rodents of North America. *Transactions of the American Philosophical Society*, 70, 1–68.

25 Eomyidae

LAWRENCE J. FLYNN
Peabody Museum, Harvard University, Cambridge, MA, USA

INTRODUCTION

The family Eomyidae constitutes an extinct radiation of superficially squirrel-like rodents. Eomyids were both diverse and species rich at times during the middle Cenozoic and dominated some small-mammal fossil assemblages. Eomyids were small to medium size, occasionally large, and Holarctic in distribution. Because primitive genera occur in the late middle Eocene of North America and appear during the Oligocene of Asia and Europe, the family is considered to have a Nearctic origination. Old World members of this extinct family were reviewed by Fahlbusch (1973) and Engesser (1999). Eomyids persisted to the end of the Pliocene in Europe and to the end of the Miocene in North America.

Eomyidae are squirrel-like in body proportions and, presumably, ground-dwelling to arboreal in habits. Eomyids resemble squirrels in jaw musculature, having the masseter originate on the side of the snout in sciuromorphous fashion. They are universally considered related to Geomorpha, but placement of the group Eomyidae + Geomorpha lacks consensus. Often these rodents have been nestled in suborder Myomorpha (but see below). Eomyids do share with many myomorphs the "cricetid plan" of molars: four major cusps joined by thin to strong lophs, longitudinal connection (mure) between protocone (-id) and hypocone (-id), and variably developed anterocone (-id) on first molars. They share with a broader group of rodents simple dentary structure (sciurognathy) and uniserial incisor enamel microstructure, but eomyid enamel shows additional unique character states (Wahlert and Koenigswald, 1985).

In recent years, thanks to wide application of wet screening, the group has become known as quite diverse in the fossil record. Numerous localities throughout Europe and North America show many taxa to be broadly distributed, some demonstrating intercontinental dispersal. With an emerging fossil record in Asia, the Eomyidae will benefit from thorough phylogenetic analysis.

DEFINING FEATURES OF THE FAMILY EOMYIDAE

CRANIAL

The skull has sciuromorphous zygoma (masseter originating on the side of the zygoma, and the infraorbital foramen small and restricted ventrally) and is similar in both primitive and derived features to muroids (Wahlert, 1978). The mandible is slender with a long diastema and a masseteric crest usually extending far anteriorly to below p4; (see Wahlert, 1978; Engesser, 1979; Wood, 1980; Wahlert and Koenigswald, 1985) (Figure 25.1).

The eomyid skull and neck do not show fossorial modification (known material lacks heavy bone, strong contacts, sloping occiputs, posterior skull breadth) or saltatorial adaptations (bullae not greatly inflated, no modified cervical vertebrae). Eomyid skulls are rather generalized and gracile, reminiscent of underived muroids. No taxa possess a sagittal crest, and only one has a postorbital process (Korth, 1994).

DENTAL

The cheek tooth formula is P1/1, M3/3 (except in early forms that retain P3). The premolars are submolariform and nearly rectangular. The molars are often wider than long and there is a strong anterior cingula on upper and lower molars. The cheek teeth are bunodont to pentalophodont, with a central mure, a mesocone (-id) developed as a transverse crest or mesoloph (-id), and hypoconulids present in bunodont forms. The incisor enamel is uniserial, but with lamellae oriented longitudinally.

POSTCRANIAL

The known slender postcrania possibly resemble those of squirrels (Wood, 1937). Given how little we know of the habitus of eomyids,

Evolution of Tertiary Mammals of North America, Vol. 2. ed. C. M. Janis, G. F. Gunnell, and M. D. Uhen. Published by Cambridge University Press.

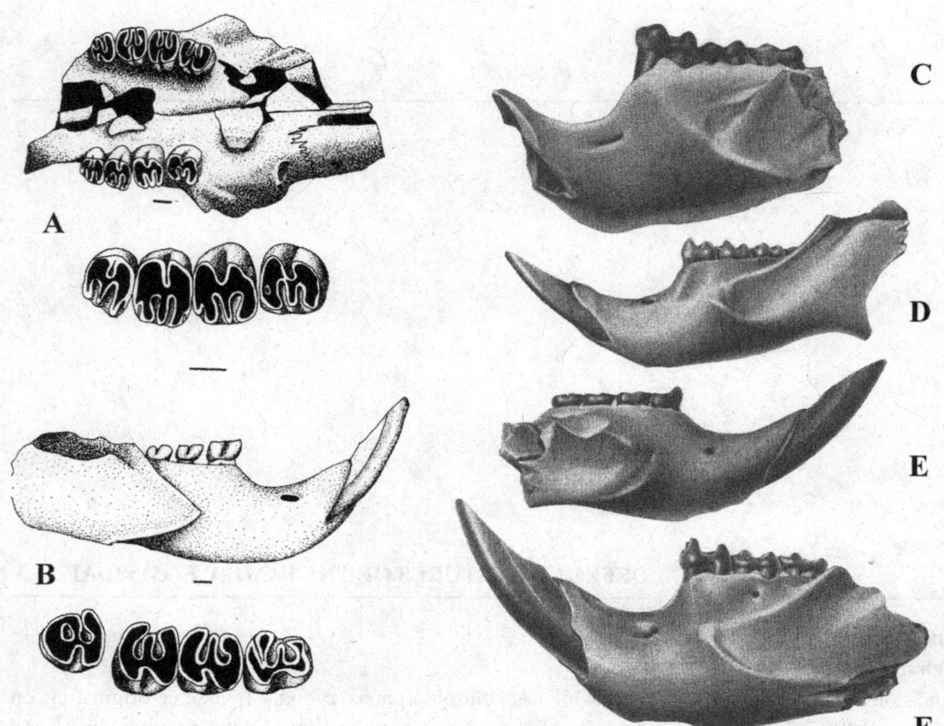

Figure 25.1. A,B. Cranial, mandibular, and dental features of the early eomyid. *Paradjidaumo* (after Korth, 1980). Scale bars = 1 mm. C–F. Dentaries C. *Yoderimys*. D. *Adjidaumo*. E. *Aulolithomys*. F. *Namatomys*. (All from Black, 1965); A,B: courtesy of the Society for Sedimentary Geology. C–F: courtesy of the Carnegie Museum of Natural History.)

it is with great interest that Storch, Engesser, and Wuttke (1996) documented gliding membranes in *Eomys quercyi*. This is in harmony with "squirrel" features of the skeleton, although Storch, Engesser, and Wuttke (1996) cautioned against assuming volant locomotion for all eomyids.

SYSTEMATICS

SUPRAFAMILIAL

Eomyidae have often been placed with Geomorpha in suborder Myomorpha. The name Myomorpha derives from classical division of Rodentia into suborders based on jaw musculature morphology (see Chapter 16). Basal Myomorpha are hystricomorphous; the myomorphous condition is a derived set of characters arising within the group. In contrast, all known Eomyidae (and the Geomorpha) are sciuromorphous, which appears to contradict classification within suborder Myomorpha. Do all "sciuromorphous" rodents share the same detailed autapomorphous character states? Is this an argument for inclusion in the same suborder? Whether the eomyid condition is identical in detail with the sciuromorphy of squirrels has not been investigated adequately from the perspective of anatomy or development of extant sciuromorphous rodents. Neither is the sciuromorphy of extant squirrels and beavers adequately compared. Still, the jaw musculature seems at odds with classification of eomyids as primitive Myomorpha.

Wahlert (1978) surveyed other features of the cranial anatomy in eomyids, geomyoids, and myomorphs. He showed that eomyids share with geomyoids several special features: sciuromorphy, short incisive foramen, long infraorbital canal, anterior position of sphenopalatine foramen, and occurrence of a sphenopterygoid foramen. Along with Myomorpha, eomyids share the following features: entrance to transverse canal separated from alisphenoid canal, short carotid canal, and temporal foramina usually absent. Are these last features sufficient to defend inclusion of Eomyidae in Myomorpha? Limited cranial data support linkage of Eomyidae with suborder Myomorpha, but not without the contradiction of jaw musculature morphology.

Recent molecular data appear to resolve the issue. Nuclear genes demonstrate that extant Geomorpha group more closely with Castoridae than with Myomorpha (Adkins, Walton, and Honeycutt, 2002; Huchon *et al.*, 2002). However, molecular data do not support Sciuromorpha monophyly; the squirrels cluster with *Aplodontia* and perhaps dormice (Gliridae). By inference, then, and accepting affinity of Eomyidae with Geomorpha, eomyids are no longer considered basal myomorphs but are part of a Nearctic radiation of rodents showing both sciurognathy and sciuromorphy, and with uniserial incisor enamel. This radiation did not necessarily give rise to both beavers and squirrels; "sciuromorphy" may have evolved twice. (See further discussion in Chapter 16.)

INFRAFAMILIAL

Considering the entire Holarctic radiation of the family, there is little consensus regarding systematics below the family level for Eomyidae. At present only the Yoderimyinae Wood, 1955, are stripped

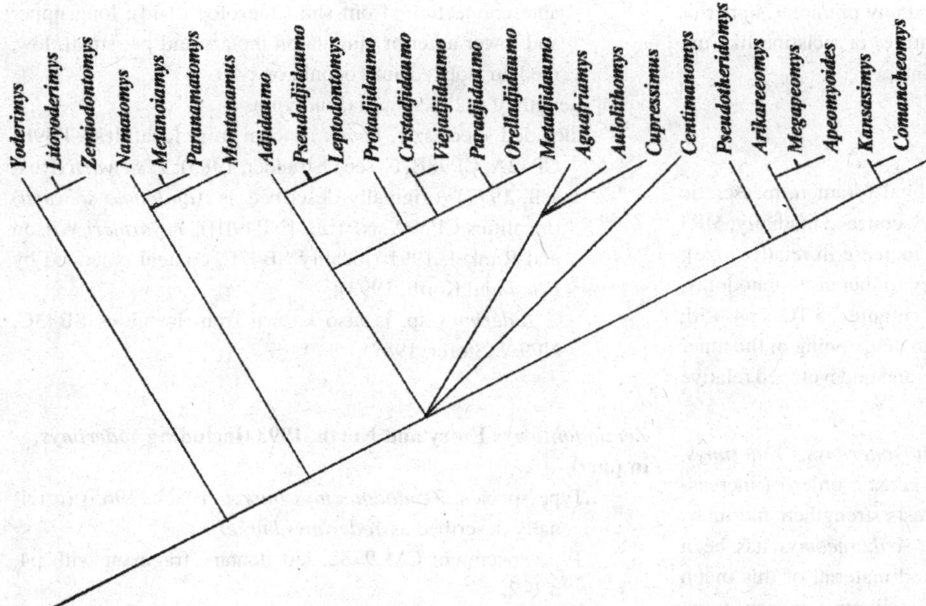

Figure 25.2. Relationships of eomyid genera of North America. The Yoderimyinae form a basal cluster of uncertain relationship to the namatomyines. *Adjidaumo* and derivatives are closely allied with *Eomys* and Eurasiatic derivatives (not shown). *Megapeomys* and *Pseudotheridomys* are later immigrants to North America. Remaining eomyids are a North American radiation terminating in the *Kansasimys* cluster of large, bunodont forms.

from the constellation of other eomyid genera as a named subfamily. These are rather large early forms that show a trend of increasing lophodonty (Figure 25.1C). The Yoderimyinae have been thought of as primitive counterparts of the rest of the family, the Eomyinae, which display derived conditions of presence of the interorbital foramen and loss of DP3 and P3 (but note retention of P3 in *Metanoiamys*). A cluster of small, bunodont forms among eomyines, including *Eomys*, *Adjidaumo*, *Pseudadjidaumo*, and *Leptodontomys*, likely are the core of a tribe-level grouping. Certainly the late Miocene large-body size terrestrial genera *Kansasimys*, *Ronquillomys*, and *Comancheomys* are closely related. Korth (1992a) proposed a tribe-level group of primitive genera, the Namatomyini. At least some of these appear linked by a trend of increasing cusp height, but the taxon is weakly supported at present. Fahlbusch (1985) pointed out that a number of groups are apparent in Europe, but diagnosis of these depends on thorough analysis of eomyid features in both Old and New World genera. Recently, Apeomyini (first known from Eurasia) have been recognized in North America. The following is a contribution in tabulating character distribution. Based on this review, a hypothesis of interrelationships of groups of genera is developed (Figure 25.2).

INCLUDED GENERA OF EOMYIDAE WINGE, 1887

The locality numbers listed for each genus refer to the list of unified localities in Appendix I. The acronyms for museum collections are listed in Appendix III.

The localities are listed in several ways. Parentheses around the locality (e.g., [CP101]) mean that the taxon in question at that locality is cited as an "aff." or "cf." to the taxon name. Parentheses are usually used for individual species, thus implying that the genus is firmly known from the locality but that the species identification may be questionable. Question marks in front of the locality (e.g., ?CP101) mean that the taxon is questionably known from that locality, thus implying some doubt that the taxon is actually present at that locality, at either the genus or the species level.

EOMYIDAE INCERTAE SEDIS

?*Simiacritomys* Kelly, 1992

Type species: *Simiacritomys whistleri* Kelly, 1992.

Type specimen: LACM 131462, left m2.

Characteristics: Small rodent with brachydont cheek teeth, squared upper molars with high crests; broken entoloph and ectolophid, five crests include mesoloph (-id); tiny anterocone on anterior crest (cingulum) and four major cusps on upper molars, no mesostyle, protoloph usually incomplete; lower molars with metalophid usually incomplete, distinct hypoconulid (from Kelly, 1992).

Average length of M2: 1.73 mm.

Included species: *S. whistleri* only, known from locality CC9B2 only.

Comments: *Simiacritomys* is a primitive rodent showing the "cricetid plan" molar morphology and is consistent with classification as primitive Eomyidae. I am not convinced (nor was Kelly [1992]) that assignment to Dipodoidea is more probable, since the molars are squared, precociously lophodont, and with incomplete longitudinal crests. The genus is poorly represented, lacking diagnostic anterior cheek teeth (certain M1, m1, and any premolar, upper or lower). A lower p4 (eomyid feature) or sicistine-like m1 would likely settle the assignment.

YODERIMYINAE WOOD, 1955

Characteristics: Eomyids lacking interorbital foramen, masseteric crest scar prominent below m1; dentary concave medially; DP3 present; P4 and p4 large (tendency to increase in relative size); cheek teeth low crowned with tendency to increase lophodonty, and with prominent anterior cingula (Figure 25.1C); p4 with distinct anteroconid; incisor enamel showing zoning of the inner enamel with the innermost zone thinner and undeveloped relative to eomyines.

Included genera in this subfamily: *Litoyoderimys*, *Yoderimys*, *Zemiodontomys* are included in phylogenetic order of increasing lophodonty and reduced cusps (crests strengthen and intersect at distinct angles progressively). *Arikareeomys* has been placed in this subfamily, but the limited material of this much younger genus prevents its secure assignment; it appears to be derived with respect to *Litoyoderimys*, but along different lines than *Zemiodontomys*. I follow Smith, Cifelli, and Czaplewski (2006) in transferring the genus to Apeomyini.

Comments: Wood (1955) defined the subfamily, but it was Emry and Korth (1993) who demonstrated its utility to accommodate an early radiation of large-bodied, lophodont eomyids. They recognized the evolutionary trends within the group and the useful characters in its diagnosis.

Litoyoderimys Emry and Korth, 1993 (including *Yoderimys*, in part)

Type species: *Litoyoderimys lustrorum* (Wood, 1974) (originally described as *Yoderimys lustrorum*).

Type specimen: FMNH PM 431, partial cranium with left cheek teeth and incisors.

Characteristics: Cheek teeth mesodont; P3 present, inflated cusps dominate low lophs; short mesolophs (-ids); no ectoloph; p4 smaller than m1; anterior cingula not prominent on upper molars; small, enclosed trigonid on p4.

Average length of m2: 1.85 mm (*L. lustrorum*, from Wood, 1974).

Included species: *L. lustrorum* (known from localities SB44B, D); *L. auogoleus* Emry and Korth, 1993 (localities CP39A, B).

An undescribed species of *Litoyoderimys* is also reported from locality CP5II.

Yoderimys Wood, 1955

Type species: *Yoderimys bumpi* Wood, 1955.

Type specimen: SDSM 5330, left dentary with incisor, p4, m1–3.

Characteristics: Cheek teeth brachydont; P3 present, lophodont but with inflated cusps; paracone and metacone joined by ectoloph, metacone with oblique outer wall; variable connections from short mesoloph (-id); long upper and lower anterior cingula on molars and p4; small, low, median isolated anteroconid on p4.

Length of m2: 1.94 mm (holotype).

Included species: *Y. bumpi* (known from localities CP39B, CP42A, CP98B, C, see Ostrander, 1985); *Y. stewarti* (Russell, 1972) (originally described as *Adjidaumo stewarti*) (localities CP39A, NP10B, [NP49II]); *Y. yarmeri* Wilson and Runkel, 1991 (locality SB43E, content emended by Emry and Korth, 1993);

Yoderimys sp. is also known from localities SB43C, NP9A (Storer, 1987).

Zemiodontomys Emry and Korth, 1993 (including *Yoderimys*, in part)

Type species: *Zemiodontomys burkei* (Black, 1965) (originally described as *Yoderimys burkei*).

Type specimen: CM 9782, left dentary fragment with p4, m1–2.

Characteristics: Cheek teeth mesodont; P3 lost, but DP3 present; P4 larger than molars; lophodont without evident cusps; protoloph and mesoloph join at buccal ends, as do metaloph and posteroloph (ectoloph incomplete); heavy mandible.

Length of m2: 2.3 mm (holotype).

Included species: *Z. burkei* only (known from localities CP39B, CP98C, NP24C).

EOMYINAE WINGE, 1887

It is likely that this subfamily will be subdivided in the future. Its basal tribe Namatomyini perhaps will have to be removed as an independent subfamily. Tribe Eomyini appears to represent a mid Tertiary radiation including a North American lineage, and origin of Eurasian clades that have no North American record. As an exception, the Apeomyini has at least one North American representative. The interrelationships of these groups remain untested.

NAMATOMYINI KORTH, 1992a

Namatomys Black, 1965

Type species: *Namatomys lloydi* Black, 1965.

Type specimen: CM 8976, left dentary with incisor, p4, m1–3.

Characteristics: Small eomyid with short mandibular diastema; lower molars with short metalophulid I and II and short mesolophid; p4 with small anteroconid in addition to metaconid and protoconid; strong anterior cingula; molars higher crowned than *Adjidaumo*.

Length of m2: 1.4 mm (holotype).

Included species: *N. lloydi* only (known from localities CP39B–E, CP98B, (NP10B2), NP24C).

Cf. *Namatomys* sp. is also known from localities CP34E, CP46, CP65, CP99A, NP10Bi, Bii, C2.

Metanoiamys **Chiment and Korth, 1996 (including *Namatomys*, in part)**

Type species: *Metanoiamys agorus* Chiment and Korth, 1996.

Type specimen: UCMP 106432, right p4.

Characteristics: Small size; molars brachydont, bunodont; P3 retained (Walsh, 1997), anteroconid on p4 (not dp4); protoconid joining to metaconid by metalophulid II or through the anteroconid; ectolophid weak, mesoconid transversely elongate; molariform P4 with strong anterior cingulum; endoloph of upper molars weak; masseteric fossa extending to below front of m1.

Average length of m2: 1.18 mm (*M. agorus*), 1.29 mm (*M. fantasma*) (Lindsay, 1968).

Included species: *M. agorus* (known from localities CC4, CC6A, CC7A); *M. fantasma* (Lindsay, 1968) (originally described as *Namatomys fantasma*) (locality CC9II); *M. fugitivus* (Storer, 1984) (locality NP8); *M. korthi* Kelly and Whistler, 1998 (locality CC9B2); *M. lacus* (Storer, 1987, = "*Namatomys*" quotes was then meant to reflect differences from *Namatomys*) (localities [CP5II], NP9A); *M. marinus* Chiment and Korth, 1996 (localities CC7E, CC9A, AA).

Metanoiamys sp. is also known from localities CC7B, D, CC8IIB, SB43A, B, SB44A, NP9B.

Comments: Dawson (1977) and Storer (1984, 1987) noted that several early eomyid species cluster together and represent a genus similar to but distinct from *Namatomys*. Chiment and Korth (1996) reviewed these and named *Metanoiamys*, building on observations by Korth (1992a), who had defined tribe Namatomyini. This tribe may be a cohesive grouping for *Namatomys*, *Metanoiamys*, and *Paranamatomys*, but *Montanamus* (treated below; the specimen is referred to Namatomyini in Korth [1994]) is not consistent with them. At least one of the features of the tribe may need review: presence of a small anteroconid may be a retained primitive feature, seen also in Yoderimyinae, and not a synapomorphy.

Paranamatomys **Korth, 1992a (including *Cupressimus*, in part)**

Type species: *Paranamatomys storeri* (Ostrander, 1983a) (originally described as *Cupressimus storeri*).

Type specimen: SDSM 10139, left dentary fragment with incisor, p4, m1.

Characteristics: Cheek teeth brachydont; masseteric scar extending anterior to p4 on dentary; molars brachydont with cusps uninflated; small anteroconid on p4 connecting with protoconid; anterior cingulum on lower molars joining protoconid; slightly oblique and off-center mesolophid.

Length of m2: 1.00 mm.

Included species: *P. storeri* only, known from locality CP98B only.

Montanamus **Ostrander, 1983b**

Type species: *Montanamus bjorki* Ostrander, 1983b.

Type specimen: SDSM 10540, left dentary fragment with incisor, p4, m1–2.

Characteristics: Cheek teeth brachydont; p4 lacking metaconid, anteroconid, and mesolophid; molars cuspate with weak lophs; anterior cingula confluent with metaconids; short mesolophid and small mesoconid on lower molars; reduced posterior cingula; laterally compressed incisor with rounded enamel.

Length of m2: 1.22 mm (holotype).

Included species: *M. bjorki* only, known from locality NP24C only.

EOMYINI WINGE, 1887

Adjidaumo **Hay, 1899 (including *Gymnoptychus*, in part)**

Type species: *Adjidaumo minutus* (Cope, 1873) (originally described as *Gymnoptychus minutus*).

Type specimen: AMNH 5362, left dentary fragment with p4, m1–3.

Characteristics: Small eomyid, cheek teeth brachydont; molars cuspate with low crests; lower molars with mesolophids of moderate length, mesoconid uninflated, ectolophid often incomplete; strong anterior cingulum joined centrally; incisor enamel smooth and rounded.

Length of lower cheek tooth row: 4.7 mm (type species).

Included species: *A. minutus* (known from localities [SB44B, D], CP39[A, B], C–E, [CP42A], CP68C, CP83C, CP84A, CP98B, C, CP99A); *A. craigi* Storer, 1987 (locality NP9A); *A. intermedius* Korth, 1989 (locality CP99A); *A. maximus* Korth, 1989 (localities CP99A, NP10B2); *A. minimus* (Matthew, 1903) (originally described as *Gymnoptychus minimus*, including *A. burkei* [Russell, 1954], [originally described as *Protadjidaumo burkei*]) (localities CP39C–E, CP98B, C, NP10B, NP13, NP24C, D, [NP25B], NP32C).

Adjidaumo sp. is also known from localities ?GC5B, SB43C, D, CP46, CP68B, CP83B, CP84B, NP9B, C, NP10BB.

Comments: Isolated specimens of *Adjidaumo* can be confused with European *Eomys* Schlosser, 1884. Korth (1994) reviewed the history of views on validity of the genus and, in agreement with Engesser (1979), elected to utilize *Adjidaumo* because features evident in large samples do consistently provide separation. *Eomys* is the older name, yet European species apparently immigrated from North America. McKenna and Bell (1997) submerged *Adjidaumo* in *Eomys*, thereby defining the entire group of species as monophyletic. At present I accept that the consistent, albeit minor, differences between Eurasian and North American species groups validate *Adjidaumo* and reflect vicariant evolution of these conservative eomyids on the two land masses. *Eomys* continues to the Miocene boundary in Europe (Fahlbusch, 1973), and simple forms occur late in North America as well. Lindsay (1974) referred a specimen from the Hemingfordian Vedder Locality (locality

CC19) to *Eomys*; Korth and Bailey (1992) transfer this to *Leptodontomys*, but I list it below under *Pseudadjidaumo*.

***Protadjidaumo* Burke, 1934 (including *Adjidaumo*, in part)**

Type species: *Protadjidaumo typus* Burke, 1934.

Type specimen: CM 11931, left dentary fragment with p4, m1–2.

Characteristics: Small eomyid with molars with weakly developed lophs, cusps still evident, and crown height greater than *Adjidaumo*; elongated lower molars; short cingula, short mesolophs (-ids).

Average length of m2: 1.2 mm.

Included species: *P. typus* (known from localities CP7A, B, ?CP36B); *P. altilophus* Storer, 1984 (locality NP8); *P. pauli* Storer, 1987 (locality NP9A).

Protadjidaumo sp. is also known from localities CC8IIB, CP5II, ?CP39A, B (R. J. Emry, personal communication, 1983), NP13.

***Cristadjidaumo* Korth and Eaton, 2004**

Type species: *Cristadjidaumo mckennai* Korth and Eaton, 2004.

Type specimen: UMNH VP5656, left m2.

Characteristics: Small, low-crowned eomyid with molars having moderately developed lophs, two lophs in both upper and lower cheek teeth dominating others on the crowns, cusps uninflated.

Average length of m2: 1.15 mm.

Average width of m2: 1.15 mm.

Included species: *C. mckennai* only, known from locality CP5II only.

***Viejadjidaumo* Wood, 1974**

Type species: *Viejadjidaumo magniscopuli* Wood, 1974.

Type specimen: TMM 40492–2B, skull and left dentary.

Characteristics: Cheek teeth brachydont; skull lacking knob for superficial masseter on sciuromorphous snout; molars bunodont; short anterior cingulum on P4; anterior cingulum on lower molars joining protoconid buccal to midline; weak posterior cingulum, weak mesolophs (-ids); prominent mesostyles; rounded incisor.

Length of m2: 1.27 mm (holotype).

Included species: *V. magniscopuli* only (known from localities SB44B, D).

Viejadjidaumo sp. is also known from localities SB43C, E, (SB44B).

***Paradjidaumo* Burke, 1934 (including *Gymnoptychus*, in part; *Adjidaumo*, in part)**

Type species: *Paradjidaumo trilophus* (Cope, 1873) (originally described as *Gymnoptychus trilophus*).

Type specimen: AMNH 5401, right dentary fragment with dp4, m1–2.

Characteristics: Small eomyid with molars with developed lophs and moderately high crowns; valleys, especially central ones, narrow and deep; lower incisor flattened anteriorly and bearing a longitudinal ridge (after Black, 1965).

Average length of m2: 1.4 mm (*P. trilophus*).

Included species: *P. trilophus* (including *P. minor*) (known from localities CP9, [CP39B], CP40B, CP41B, CP42A, CP43, CP45A, CP68B, C, CP83C, CP84A, B, CP98B, C, CP99A, NP24C, D, [NP25B, C], NP27C, NP29C, [D], NP30B, [NP49II]); *P. alberti* Russell, 1954 (localities [CP5II], NP13); *P. hansonorum* (Russell, 1972) (originally described as *Adjidaumo hansonorum*) (localities NP10B, [NP49II]); *P. hypsodus* Setoguchi, 1978 (locality CP46); *P. spokanensis* White, 1954 (locality NP29D); *P. reynoldsi* Kelly, 1992 (locality CC9B2); *P. validus* Korth, 1980, (locality CP99A).

Paradjidaumo sp. is also known from localities CC7B, [CC8IIB], CP39C–E, CP83C, CP84C, ?NP9B, NP10Bii, B2, BB, C2, E, NP25C, NP32C.

***Orelladjidaumo* Korth, 1989**

Type species: *Orelladjidaumo xylodes* Korth, 1989.

Type specimen: UNSM 81068, partial maxilla with left DP4, M1–2.

Characteristics: Small size; lophodont molars, moderately high crowned with some unilateral hypsodonty; short mesoloph; no lingual extension of anterior cingulum.

Length of M2: 1.12 mm (holotype).

Included species: *O. xylodes* only, known from locality CP99A only.

Orelladjidaumo sp. is also tentatively known from locality CP5II.

***Metadjidaumo* Setoguchi, 1978**

Type species: *Metadjidaumo hendryi* Setoguchi, 1978.

Type specimen: CM 33786, left m2.

Characteristics: Small size, moderately high crowned and lophodont molars; elevated trigonid; free posterior cingulum on lingual side of m1; no posterior cingulum on m2; weak mesolophs and mesolophids; buccal attachment of anterior cingulum on lower molars; lower incisor flattened and lacking ridge.

Average length of m2: 0.99 mm (Setoguchi, 1978).

Included species: *M. hendryi* (known from locality CP46); *M. cedrus* Korth, 1981 (localities ?CP46, CP68C).

Metadjidaumo sp. is also known from locality NP32C.

Comments: Korth (1994) suggested reallocation of *M. cedrus* to *Orelladjidaumo*.

***Pseudadjidaumo* Lindsay, 1972 (including *Adjidaumo*, in part; *Leptodontomys*, in part; *Eomys*, in part)**

Type species: *Pseudadjidaumo stirtoni* Lindsay, 1972.

Type specimen: UCMP 78468, right m1.

Characteristics: Small, low-crowned eomyid with molar cusps evident in moderately developed lophs; anterior and posterior cingula in upper and lower molars enclosing shallow valleys; deep central valleys with short mesolophs (-ids); mesocone not prominent and mesostyle rare; anterior cingulum of lower molars connecting near the anterior arm of the protoconid and ending without strong labial extension.

Length of m1: 0.96 mm.

Width of m1: 0.93 mm (holotype).

Included species: *P. stirtoni* (known from localities NB6D, E); *P. douglassi* (Burke, 1934) (originally described as *Adjidaumo douglassi*) (localities CP101, NP51B); *P. quartzi* (Shotwell, 1967) (originally described as *Adjidaumo quartzi*) (locality PN9B); *P. russelli* Storer, 1970 (localities [NP5A], NP11); *P.* sp. A (locality CC19, see Lindsay, 1974); *P.* sp. B (locality CP114A, see Klingener, 1968).

Pseudadjidaumo sp. is also known from locality NP40B.

Comments: The histories of small eomyids on this continent and Eurasia are separate and vicarious. The small North American eomyids assigned to *Adjidaumo* appear to continue as *Pseudadjidaumo* and *Leptodontomys*. Korth and Bailey (1992) noted the minute differences separating *Pseudadjidaumo* and *Leptodontomys* and elected to submerge them into one genus (the latter, since it is the older name). This approach may gain support, pending description of suites of Barstovian "*Leptodontomys*" from the Great Plains. Presently, I retain *Pseudadjidaumo* and recognize its earliest members as *P. douglassi* (an Arikareean species transferred from *Adjidaumo*, following Korth and Bailey [1992]), and *P.* "sp. A" from the Hemingfordian Vedder Locality, California (CC19) (originally assigned to *Eomys*). Bailey (2005) noted variation in the small, bunodont eomyine from locality CP101 and suggest allocation of some specimens to the European genera *Eomys* and *Pentabuneomys*; at present, this hypothesis needs further testing.

***Leptodontomys* Shotwell, 1956**

Type species: *Leptodontomys oregonensis* Shotwell, 1956.

Type specimen: UOMNH F-3633, right dentary fragment with incisor and p4.

Characteristics: Small eomyid, with molar cusps joined by narrow lophs; p4 nearly rectangular and lacking anteroconid; small mesoconids, short mesolophs (-ids); anterior cingula spaning both upper and lower molars (lingual portion in uppers and labial portion in lowers developed); lower molars with short posterolingual cingula; rounded incisor with smooth enamel.

Length of cheek tooth row: 2.9 mm (alveolar length of holotype).

Included species: *L. oregonensis* (known from locality PN14); *L.* sp. A (localities PN10, PN11B); *L.* sp. B (locality PN11B); *L.* sp. C (locality NP10E) (see Shotwell, 1967; Skwara, 1988).

Leptodontomys sp. is also known from localities SP1A, CP114A, ?D (Korth, 1997; Smith 2005), NP10CC.

Comments: *Leptodontomys*, in this restricted sense, is represented by only one named species on this continent, although perceived diversity in samples is growing. The genus is also known from the middle Miocene to Pliocene of China, and the middle Miocene to early Pleistocene of Europe. *Plesiosminthus* is also known from the Oligocene and Miocene of Europe and Asia (McKenna and Bell, 1997). Engesser (1979) argued for distinct separation of these late Tertiary small eomyids from European counterparts, for which he erected the genus *Eomyops*. Qiu (1994) transferred the Eurasian forms to *Leptodontomys*, given recent finds of intermediate species from China, and saw corresponding intercontinental dispersal. Currently, Engesser ([1999], who agreed with Korth and Bailey [1992] to include *Pseudadjidaumo* with *Leptodontomys*), noted the distinctness of the Eurasian and American lineages after the early Miocene.

APEOMYINI FEJFAR, RUMMEL AND TOMIDA, 1998

Characteristics: Eomyids of medium to large body size; large premolars; long diastema; cheek teeth lophodont and bilobed; cheek tooth morphology with anterior and posterior moieties, each with a pair of transverse crests enclosing an interior enamel lake, moieties isolated or weakly joined; molars markedly wider than long.

***Arikareeomys* Korth, 1992b**

Type species: *Arikareeomys skinneri* Korth, 1992a.

Type specimen: UNSM 81031, right maxilla with P4, M1.

Characteristics: Cheek teeth mesodont; P3 absent, P4 quite large, longer than wide and bigger than M1; lophodont, but mesoloph abbreviated and ectoloph incomplete; protoloph and anteroloph joining buccally as do metaloph-posteroloph, thus forming two lakes on M1; anterior lake opening bucally on P4.

Length of M1: 1.9 mm.

Width of M1: 2.13 mm (holotype).

Included species: *A. skinneri* only, known from locality CP103A only.

Arikareeomys sp. is also known from locality GC8AB.

***Megapeomys* Fejfar, Rummel and Tomida, 1998**

Type species: *Megapeomys lavocati* Fejfar, Rummel and Tomida, 1998.

Type specimen of *M. bobwilsoni* Morea and Korth, 2002: UCR 15412, broken skull and mandible with left P4, M1–3, right M1–3, right p4, m1–3, and left m2–3.

Characteristics: Large eomyid with mesodont and lophodont teeth; four transverse crests, with central mure weak in lower teeth, such that crown pattern makes isolated anterior and posterior moieties; p4 largest of lower cheek teeth; lower molars with four roots; extremely large diastema; incisors not proodont; skull robust with strong lambdoid crest.

Length of m2: 2.5 mm (*M. bobwilsoni*).

Width of m2: 3.2 mm (*M. bobwilsoni*).

Alveolar length of P4, M1–3: 10.4 mm.

Included North American species: *M. bobwilsoni* only, known from locality NB17 only.

Apeomyoides Smith, Cifelli and Czaplewski, 2006

Type species: *Apeomyoides savagei* Smith, Cifelli and Czaplewski, 2006.

Type specimen: UCMP 109300, right dentary with incisor, p4, m1–3.

Characteristics: Large eomyid with mesodont, rectangular, cheek teeth with four lophs; p4 large with only two roots; m1–3 with three roots; long diastema, masseteric scar ending under m1; incomplete entoloph and ectolophid, lower cheek teeth bilobed with anterior and posterior moieties bearing transversely elongated enamel lakes.

Average length of m2: 2.00 mm.

Average width of m2: 2.23 mm.

Included species: *A. savagei* only, known from locality NB22A only.

Comments: *Apeomyoides* is the latest, most derived apeomyine. *Megapeomys lavocati* and *M. lindsayi* Fejfar, Rummel and Tomida (1998) are Eurasian species that represent the apeomyine radiation. Never common as fossils, the group became widespread in the late early Miocene, dispersing across the Bering Land Bridge to North America (as *M. bobwilsoni*). Apeomyines were generally large, which is unusual for eomyids, and Smith, Cifelli and Czaplewski (2006) have added interesting comparisons with the equally large *Arikareeomys*, arguing phylogenetic affinity with Apeomyini rather than Yoderimyinae.

Indeterminate apeomyini

An indeterminate apeomyine is reported from locality CP101.

RONQUILLOMYINI (NEW)

Characteristics: Eomyids of medium to large body size; molars with bulbous cusps; upper molars with anterior and posterior cingula; lower molars with anterior cingula, prominent mesocone and mesoconid present, mesoloph and mesolophid absent to weakly developed; fourth premolars large; p4 wider than m1; anterior extension of masseteric crest reaches below p4 on the dentary.

Kansasimys Wood, 1936

Type species: *Kansasimys dubius* Wood, 1936.

Type specimen: KU 3582, left dentary with incisor, p4, m1–3.

Characteristics: Medium-size eomyid, with moderately high-crowned molars with rounded cusps and weak lophs; molariform premolars; p4 with small anteroconid crest joined to protoconid; central mesocones (-ids); weak mesolophs (-ids); no mesostyle; narrow anterior cingula spanning molars and connected at midline to metalophids; masseteric crest ending with anterior extension below p4; flattened incisors.

Length of m2: 1.94 mm (holotype).

Included species: *K. dubius* only (known from localities SP1C, CP123D).

Kansasimys sp. is also known from locality CP116C.

Comancheomys Dalquest, 1983

Type species: *Comancheomys rogersi* Dalquest, 1983.

Type specimen: TMM 41261–41 (an upper first or second molar; see Korth, 1994).

Characteristics: Eomyid similar to *Kansasimys*, but judged to have molars more lophodont and slightly higher crowned (Dalquest, 1983).

Length of holotype (M1 or M2): 2.2 mm.

Width of holotype (M1 or M2): 2.35 mm.

Included species: *C. rogersi* only, known from locality SP3B only.

Ronquillomys Jacobs, 1977

Type species: *Ronquillomys wilsoni* Jacobs, 1977.

Type specimen: F:AM 98589, right dentary fragment with incisor, p4, m1–3, plus left maxillary fragment with P4, M1.

Characteristics: Large eomyid with molars having bulbous cusps on strong lophs; low anterior and posterior cingula on upper molars; mesocone present, but no mesoloph; strong mesoconid with short mesolophid; entolophulid directed posterolabially from entoconid to short posterior cingulum, not touching hypoconid; long, low anterior cingula on lower molars; p4 without anteroconid; deep dentary with incisor capsule; masseteric crest extending beyond p4.

Length of m2: 2.5 mm (holotype).

Included species: *R. wilsoni* only, known from locality SB10 only.

Comments: The last three genera are a clade of late Miocene eomyids in North America characterized by molars with retention of bulbous cusps despite their large size. Only *Centimanomys* is larger than *Ronquillomys*. The strong resemblance of the ronquillomyines to each other led Korth (1994) to prefer synonymy, at least at the generic level. Indeed, *Comancheomys rogersi* is close to *Kansasimys dubius*, although some differences exist among known specimens. The larger *Ronquillomys wilsoni* is more distinct, but still closely related. All occur in the 7 to 6 Ma range (see Dalquest, 1983; Harrison, 1983; Lindsay, Opdyke, and Johnson, 1984).

OTHER GENERA NOT ASSIGNED TO TRIBE

Pseudotheridomys Schlosser, 1926 (including *Theridomys*, in part)

Type species: *Pseudotheridomys parvulus* (Schlosser, 1884) (originally described as *Theridomys parvulus*).

Type specimen of *P. hesperus* Wilson, 1960: KU 10195, right dentary fragment with incisor, p4, m1–2.

Characteristics: Molars moderately high crowned and lophodont with mesolophs (-ids) usually complete; mesocone (-id) undeveloped; upper molars with large anterior valleys closed labially, posterior valleys also usually closed; mesostyles present; mure may be incomplete in early wear; lower molars elongated and with metastylids.

Length of cheek tooth row: 3.7 mm (alveolar length of lower jaw of *P. hesperus*, Wilson, 1960).

Included species: *P. cuyamensis* Lindsay, 1974 (known from localities CC9IVC, D, CC19, [CP45E]); *P. hesperus* Wilson, 1960 (localities CP71, NP10CC); *P. pagei* Shotwell, 1967 (localities PN8B, PN9B).

Pseudotheridomys sp. is also known from localities GC3B, (CC9IVD), CP88, (CP116B), NP10E, NP40B.

Comments: This genus, like the dipodid *Plesiosminthus*, occurs in the Miocene of North America but is known more broadly from the Oligocene and Miocene of Europe and Asia (McKenna and Bell, 1997). Both genera are represented in North America by several species. Consensus (e.g., Engesser, 1979; Qiu, 1994) agrees that the North American forms demonstrate at least one dispersal event across the Bering Land Bridge during the late Oligocene. Engesser's (1999) chart implies dispersal from east to west.

Aguafriamys Wilson and Runkel, 1991

Type species: *Aguafriamys raineyi* Wilson and Runkel, 1991.

Type specimen: TMM 41580-32, right dentary fragment with p4, m1–2.

Characteristics: Large eomyid with unenlarged p4 that has closely positioned protoconid and metaconid; molars brachydont and not strongly lophodont; anterior cingulum on molars not prominent; mesoconid with labial spur and short mesolophid.

Length of m2: 2.60 mm (holotype).

Included species: *A. raineyi* only, known from locality SB43C only.

Aulolithomys Black, 1965

Type species: *Aulolithomys bounites* Black, 1965.

Type specimen: USNM 20974, left dentary with incisor, p4, m1–2 (Figure 25.1E).

Characteristics: Molars brachydont, short p4 lacking anteroconid; molars short with entoconid close to metaconid; short anterior and posterior cingula; short and low mesolophs (-ids); minute anterior cingulum on P4; thick enamel; incisor flattened anteriorly (Figure 25.1E; see also Wood, 1974).

Length of m2: 1.6 mm (holotype).

Included species: *A. bounites* (known from localities SB44B, [C], D, CP39B, CP98B, C, NP10B, NP24C, D [Storer, 1978]); *A. vexilliames* Korth and Emry, 1997 (localities CP39C, [D, E]).

Aulolithomys sp. is also known from localities SB43B, C, E, CP83B, NP9B, NP10Bii.

Cupressimus Storer, 1978

Type species: *Cupressimus barbarae* Storer, 1978.

Type specimen: P661.101, left dentary with p4, m1.

Characteristics: Molars brachydont; square molars with weak crests; short, low mesolophid; lingual, short posterior cingulum; anterior cingulum free at both ends; p4 lacking posterior cingulum, hypolophid weak, triangular metaconid, may have anteroconid, masseteric fossa extending beyond p4; compressed, rounded incisor.

Length of m1: 1.2 mm (holotype).

Included species: *C. barbarae* only, known from locality NP10B only.

Cupressimus sp. is also known from locality ?NP24C.

Centimanomys Galbreath, 1955

Type species: *Centimanomys major* Galbreath, 1955.

Type specimen: KU 9902, left dentary with incisor, p4, m1–3.

Characteristics: Large size; brachydont molars, with strong transverse crests and mesolophids; p4 lacks anteroconid and metaconid; anterior cingulum on molars joining protoconid labially; incisor with longitudinal ridge.

Length of m2: 2.6 mm (holotype).

Included species: *C. major* (known from localities CP68B, CP98B, see Ostrander, 1985); *C. galbreathi* Martin and Ostrander, 1986 (locality CP68B).

Centimanomys sp. is also known from localities NP10B2, NP49II.

INDETERMINATE EOMYIDS

Fragmentary material ascribed to eomyids are reported from localities GC5A (might be a cricetid), GC8AA, AB, CC6B, NB17, SP1C, CP29C, CP36B, NP10B, C2, E, NP36A.

An undescribed new genus of eomyid is reported from localities GC8AA and GC8B.

BIOLOGY AND EVOLUTIONARY PATTERNS

Eomyids have no living representatives but are universally considered the sister taxon to Geomorpha (living gophers and pocket mice). The family has been placed with geomorphs in suborder Myomorpha, but this is poorly defended by morphological evidence. On the contrary, new molecular analyses associate geomorphs with beavers (Castoridae). Geomorphs and beavers share sciuromorphous zygomatic morphology – but they also share this condition with squirrels. Presently, molecular data do not suggest that squirrels cluster with

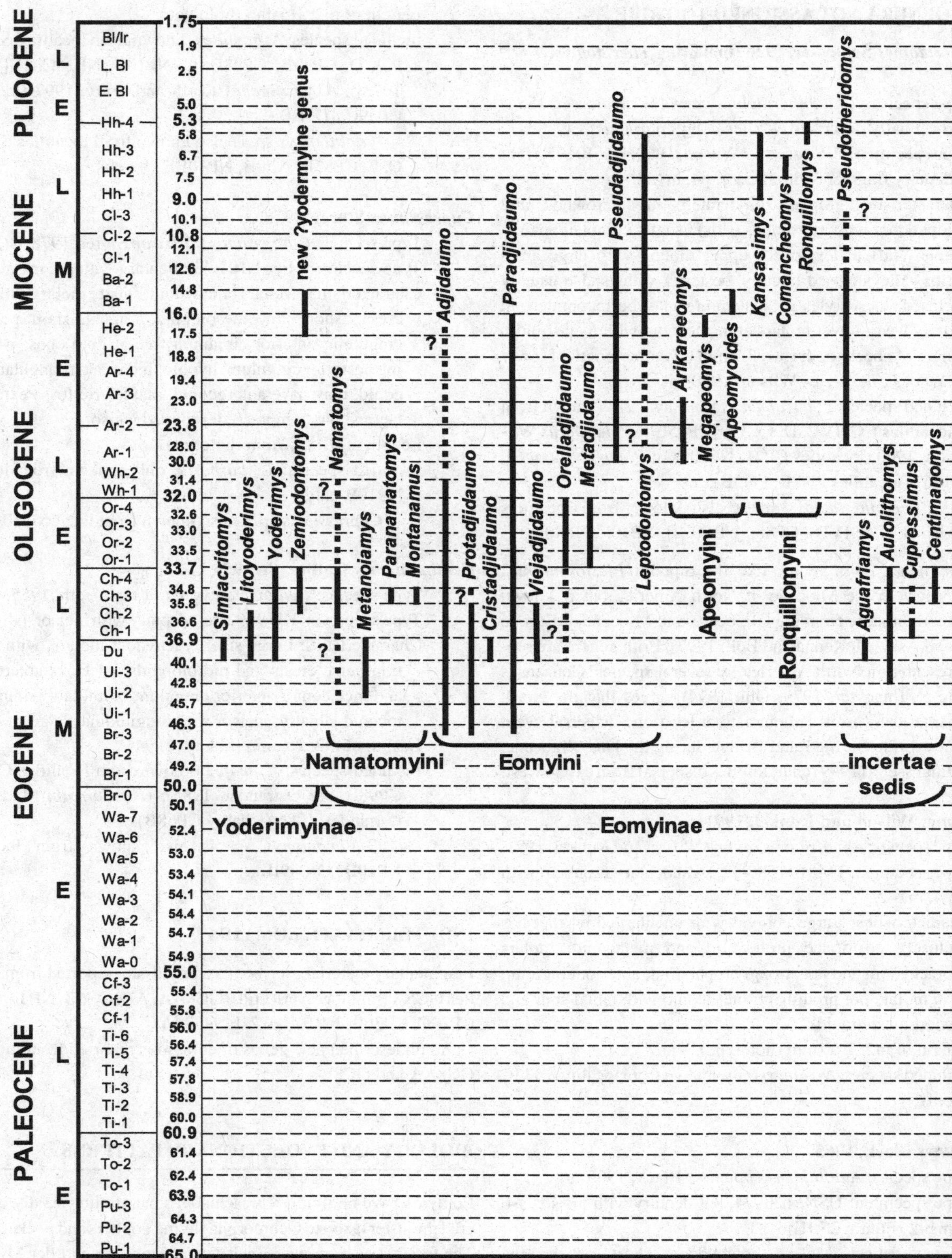

Figure 25.3. Temporal ranges of North American eomyid genera.

geomorphs and beavers. All of these relationships require further testing and review of character states in morphological features used to sort them (see further discussion in Chapter 16).

Primitive species are problematic in assignment to higher taxon. Some have been assigned variously to Eomyidae, Geomorpha, or Sciuridae. Early taxa grouped here or by other studies within Eomyidae may prove to belong to other families of rodents. Those "eomyid" genera that have been considered transitional with geomorphs are placed therein in this volume. Consequently, the remaining early Eomyidae are primitive in many features. Taking known eomyids as closely related to geomorphs, the origin of Eomyidae and Geomorpha is constrained to the middle Eocene, probably before 45 Ma. By the middle Cenozoic, several eomyid clades had diverged separately in North America and Eurasia, and these can form the basis for diagnosis of subfamilies or tribes. Namatomyines, eomyines, and various genera diversified in the late Eocene and dominate some small mammal assemblages through the Oligocene and early Miocene (Figure 25.3).

Ecomorphological counterparts to Eomyidae might be living small- to middle-sized squirrels, although many eomyids are like conservative muroids in dentition. In some fossil assemblages, eomyids are well represented by relatively complete material, suggesting that some genera utilized burrows. Postcrania have been compared with those of squirrels. Little can be concluded for the paleoecology of the family as a whole beyond the concise comments of Engesser (1999), where he concluded a mesic and forest habitat for most eomyids and cited the gliding adaptation of a fossil form that is indistinguishable from *Eomys quercyi*. This supports arboreality for at least some eomyids, but Engesser (1999) echoed the note of caution of Storch, Engesser, and Wuttke (1996) against presuming such habitat for all members of the family.

Eomyids demonstrate intercontinental range expansion. A classic case of nomenclature driven by continent of occurrence involves the late Eocene genera *Eomys* (Europe) and *Adjidaumo* (North America), which are exceedingly difficult to differentiate based alone on morphology of isolated teeth. McKenna and Bell (1997) submerged the latter into *Eomys*. I follow Korth (1994) in accepting minor features (correlated with vicariant biogeography) as distinctions. There are other taxa that indicate long-distance dispersal. Engesser (1979) argued strongly for Miocene dispersal to North America of *Leptodontomys*, which was confirmed by Qiu (1994). *Pseudotheridomys* is another immigrant form, and to this list is now added the distinctive *Megapeomys*, a genus now known from all northern continents (Morea and Korth, 2002).

Eomyidae are a complex and rich group that at times dominated the small-mammal faunas of North America. The group apparently originated in North America in the middle Eocene and subsequently spread throughout the Holarctic. Taken as squirrel analogs, this extinct group probably included both ground-dwelling and arboreal forms, at least one of these being volant. Eomyids became uncommon in the late Neogene. The last forms in North America (late Hemphillian age) were large and presumably terrestrial; in Europe, two lineages survived to the end of the Pliocene (Engesser, 1999). Eomyidae demonstrates intercontinental dispersal at many times since the late Eocene. A thorough phylogenetic analysis, global in scale, would provide a basis for definition of tribes and subfamilies and would test evident vicariant biogeographic events, thereby constraining time of dispersal and increasing the usefulness of the Eomyidae in long-distance correlation.

ACKNOWLEDGEMENTS

I wish to thank the paleontological community for the wealth of information on which this survey is based. The publication record is a crucial resource, and access to it was facilitated through the Bibliography of Fossil Vertebrates. Particularly useful in checking details are Korth (1994) and McKenna and Bell (1997). I am indebted to Julia Keller, Everett Lindsay, and Christine Janis for help in running down citations. Bill Korth's careful review encouraged rethinking higher taxon relationships and developing some of the ideas presented here.

REFERENCES

Adkins, R. M., Walton, A. H., and Honeycutt, R. L. (2002). Higher-level systematics of rodents and divergence time estimates based on two congruent nuclear genes. *Molecular Phylogenetics and Evolution*, 26, 409–20.

Bailey, B. (2005). Eomyids from the early Arikareean Ridgeview local fauna of western Nebraska and their biogeographic significance. *Journal of Vertebrate Paleontology*, 25(suppl. to no. 3), p. 33A.

Black, C. C. (1965). Fossil mammals from Montana. Part 2: Rodents from the early Oligocene Pipestone Springs Local Fauna. *Annals of the Carnegie Museum*, 38, 1–48.

Burke, J. J. (1934). New Duchesne River rodents and a preliminary survey of the Adjidaumidae. *Annals of the Carnegie Museum*, 23, 391–8.

Chiment, J. J. and Korth, W. W. (1996). A new genus of eomyid rodent (Mammalia) from the Eocene (Uintan–Duchesnean) of southern California. *Journal of Vertebrate Paleontology*, 16, 116–24.

Cope E. D. (1873). Third notice of extinct Vertebrata from the Tertiary of the Plains. *Paleontological Bulletin*, 16, 1–8.

Dalquest, W. W. (1983). Mammals of the Coffee Ranch Local Fauna, Hemphillian of Texas. *The Pearce-Sellards Series, Texas Memorial Museum*, 38, 1–41.

Dawson, M. R. (1977). Late Eocene rodent radiations: North America, Europe, and Asia. *Géobios, Memoire Speciale*, 1, 195–209.

Emry, R. J. and Korth W. W. (1993). Evolution in Yoderimyinae (Eomyidae: Rodentia), with new material from the White River Formation (Chadronian) at Flagstaff Rim, Wyoming. *Journal of Paleontology*, 67, 1047–57.

Engesser, B. (1979). Relationships of some insectivores and rodents from the Miocene of North America and Europe. *Bulletin of the Carnegie Museum of Natural History*, 14, 1–68.

(1999). Family Eomyidae. In *The Miocene Land Mammals of Europe*, ed. G. Rössner and K. Heissig, pp. 319–35. München; Dr. F. Pfeil.

Fahlbusch, V. (1973). Die Stammegeschichtlichen Beziehungen zwischen den Eomyiden (Mammalia, Rodentia) Nordamerikas und Europas. *Mitteilungen Bayerische Staatssammlung für Paläontologie und Historisches Geologie*, 13, 141–75.

(1985). Origin and evolutionary relationships among geomyoids. In *Evolutionary Relationships Among Rodents: A Multidisciplinary Analysis*, ed. W. P. Luckett and J.-L. Hartenberger, pp. 617–30. New York: Plenum Press.

Fejfar, O., Rummel, M., and Tomida, Y. (1998). New eomyid genus and species from the early Miocene (MN zones 3–4) of Europe and Japan related to *Apeomys* (Eomyidae, Rodentia, Mammalia). In

Monograph 14: Advances in Vertebrate Paleontology and Geochronology, ed. Y. Tomida, L. J. Flynn, and L. L. Jacobs, pp. 123–43. Tokyo: National Science Museum.

Galbreath, E. C. (1955). A new eomyid rodent from the lower Oligocene of northeastern Colorado. *Transactions of the Kansas Academy of Science*, 58, 75–8.

Harrison, J. A. (1983). The Carnivora of the Edson Local Fauna (Late Hemphillian), Kansas. *Smithsonian Contributions to Paleobiology*, 54, 1–42.

Hay, O. P. (1899). Notes on the nomenclature of some North American fossil vertebrates. *Science*, 10, 253–4.

Huchon, D., Madsen, O., Sibbald, M. *et al.* (2002). Rodent phylogeny and a timescale for the evolution of Glires: Evidence from an extensive taxon sampling using three nuclear genes. *Molecular Biology and Evolution*, 19, 1053–65.

Jacobs, L. L. (1977). Rodents of the Hemphillian age Redington local fauna, San Pedro Valley, Arizona. *Journal of Paleontology*, 51, 505–19.

Kelly, T. S. (1992). New Uintan and Duchesnean (Middle and Late Eocene) rodents from the Sespe Formation, Simi Valley, California. *Bulletin of the Southern California Academy of Science*, 9, 97–120.

Kelly, T. S. and Whistler, D. P. (1998). A new eomyid rodent from the Sespe Formation of southern California. *Journal of Vertebrate Paleontology*, 18, 440–3.

Klingener, D. (1968). Rodents of the Mio-Pliocene Norden Bridge local fauna, Nebraska. *The American Midland Naturalist*, 80, 65–74.

Korth, W. W. (1980). *Paradjidaumo* (Eomyidae, Rodentia) from the Brule Formation, Nebraska. *Journal of Paleontology*, 54, 933–41.

(1981). *Metadjidaumo* (Eomyidae, Rodentia) from Colorado and Wyoming. *Journal of Paleontology*, 55, 598–602.

(1989). Geomyoid rodents from the Orellan (middle Oligocene) of Nebraska. [In *Papers on Fossil Rodents in Honor of Albert Elmer Wood*, ed. C. C. Black and M. R. Dawson.] *Natural History Museum of Los Angeles County Science Series*, 33, 31–46.

(1992a). Cylindrodonts (Cylindrodontidae, Rodentia) and a new genus of eomyid, *Paranatomys* (Eomyidae, Rodentia) from the Chadronian of Sioux County Nebraska. *Transactions of the Nebraska Academy of Sciences*, 19, 75–82.

(1992b). Fossil small mammals from the Harrison Formation (Late Arikareean: earliest Miocene), Cherry County, Nebraska. *Annals of the Carnegie Museum*, 61, 69–131.

(1994). *The Tertiary Record of Rodents in North America*. New York: Plenum Press.

(1997). Additional rodents (Mammalia) from the Clarendonian (Miocene) of northcentral Nebraska and a review of Clarendonian rodent biostratigraphy of that area. *Paludicola*, 1, 97–111.

Korth, W. W. and Bailey, B. E. (1992). Additional specimens of *Leptodontomys douglassi* (Eomyidae, Rodentia) from the Arikareean (Late Oligocene) of Nebraska. *Journal of Mammalogy*, 73, 651–62.

Korth, W. W. and Eaton, J. G. (2004). Rodents and a marsupial (Mammalia) from the Duchesnean (Eocene) Turtle Basin local fauna, Sevier Plateau, Utah. [In *Fanfare for an Uncommon Paleontologist: Essays in Honor of Malcolm C. McKenna*, ed. M. R. Dawson and J. A. Lillegraven.] *Bulletin of the Carnegie Museum of Natural History*, 36, 109–119.

Korth, W. W. and Emry, R. J. (1997). A new species of *Aulolithomys* (Rodentia, Eomyidae) from the Chadronian (Late Eocene) of Wyoming. *Paludicola*, 1, 112–16.

Lindsay, E. H. (1968). Rodents from the Hartman Ranch local fauna, California. *PaleoBios*, 6, 1–22.

(1972). Small mammal fossils from the Barstow Formation, California. *University of California Publications in Geological Sciences*. 93, 1–104.

(1974). The Hemingfordian mammal fauna of the Vedder locality, Branch Canyon Formation, Santa Barbara County, California. Part II: Rodentia (Eomyidae and Heteromyidae). *PaleoBios*, 16, 1–20.

Lindsay, E. H., Opdyke, N. D., and Johnson, N. M. (1984). Blancan–Hemphillian Land Mammal Ages and Late Cenozoic mammal dispersal events. *Annual Reviews of Earth and Planetary Sciences*, 12, 445–88.

McKenna, M. C. and Bell, S. K. (1997). *Classification of Mammals Above the Species Level*. New York: Columbia University Press.

Martin, L. D. and Ostrander, G. E. (1986). A new species of the Oligocene eomyid rodent *Centimanomys*. *Transactions of the Kansas Academy of Science*, 89, 119–23.

Matthew, W. D. (1903). The fauna of the *Titanotherium* beds at Pipestone Springs, Montana. *Bulletin of the American Museum of Natural History*, 19, 197–226.

Morea, M. F. and Korth, W. W. (2002). A new eomyid rodent (Mammalia) from the Hemingfordian (early Miocene) of Nevada and its relationship to Eurasian Apeomyinae (Eomyidae). *Paludicola*, 4, 10–14.

Ostrander, G. E. (1983a). New early Oligocene (Chadronian) mammals from the Raben Ranch local fauna, northwest Nebraska. *Journal of Paleontology*, 57, 140–4.

(1983b). A new genus of eomyid (Mammalia, Rodentia) from the early Oligocene (Chadronian), Pipestone Springs, Montana. *Jounal of Paleontology*, 57, 128–39.

(1985). Correlation of the early Oligocene (Chadronian) in northwestern Nebraska. [In *Fossiliferous Cenozoic Deposits of Western South Dakota and Northwestern Nebraska*, ed. J. E. Martin.] *Dakoterra, South Dakota School of Mines and Technology*, 2, 205–31.

Qiu Z.-D. (1994). Eomyidae in China. In *Monograph 8: Rodent and Lagomorph Families of Asian Origins and Diversification*, ed. Y. Tomida, C. K. Li, and T. Setoguchi, pp. 49–55. Tokyo: National Science Museum.

Russell, L. S. (1954). Mammalian fauna of the Kishenehn Formation, southeastern British Columbia. *Bulletin of the National Museum of Canada*, 132, 92–111.

(1972). Tertiary mammals of Saskatchewan. Part II: the Oligocene fauna, non-ungulate orders. *Life Sciences Contributions, Royal Ontario Museum*, 84, 1–97.

Schlösser, M. (1884). Die Nager des europäischen Tertiärs nebst Betrachtungen über die Organisation und die geschichtliche Entwicklung der Nager überhaupt. *Palaeontographica*, 31, 1–143.

(1926). Die Säugetierfauna von Peublanc (Dep. Allier). *Societas Scientiarum Naturalium Croatica*, 38/39, 372–94.

Setoguchi, T. (1978). Paleontology and geology of the Badwater Creek Area, Central Wyoming. Part 16: The Cedar Ridge local fauna (Late Oligocene). *Bulletin of the Carnegie Museum of Natural History*, 9, 1–61.

Shotwell, J. A. (1956). Hemphillian mammalian assemblage from northeastern Oregon. *Bulletin of the Geological Society of America*, 67, 717–38.

(1967). Late Tertiary geomyoid rodents of Oregon. *Bulletin of the Museum of Natural History, University of Oregon*, 9, 1–51.

Skwara, T. (1988). Mammals of the Topham Local Fauna: early Miocene (Hemingfordian) Cypress Hills Formation, Saskatchewan. *Natural History Contributions, Museum of Natural History, Regina*, 9, 1–168.

Smith, K. (2005). First record of *Leptodontomys* (Eomyidae: Rodentia) in the Clarendonian (Miocene) of the Southern Great Plains. *Proceedings of the Oklahoma Academy of Science*, 85, 47–53.

Smith, K., Cifelli, R., and Czaplewski, N. (2006). A new genus of eomyid rodent from the Miocene of Nevada. *Acta Palaeontologica Polonica*, 51, 385–92.

Storch, G., Engesser, B., and Wuttke, M. (1996). Oldest fossil record of gliding in rodents. *Nature*, 379, 439–41.

Storer, J. E. (1970). New rodents and lagomorphs from the Upper Miocene Wood Mountain Formation, of southern Saskatchewan. *Canadian Journal of Earth Sciences*, 7, 1125–9.

(1978). Rodents of the Calf Creek local fauna (Cypress Hills Formation, Oligocene, Chadronian) Saskatchewan. *Natural History Contributions, Museum of Natural History, Regina*, 1, 1–54.

(1984). Mammals of the Swift Current Creek local fauna (Eocene: Uintan, Saskatchewan). *Natural History Contributions, Museum of Natural History, Regina*, 7, 1–158.

(1987). Dental evolution and radiation of Eocene and early Oligocene Eomyidae (Mammalia: Rodentia) of North America, with new material from the Duchesnean of Saskatchewan. *Dakoterra, South Dakota School of Mines and Technology*, 3, 108–17.

Wahlert, J. H. (1978). Cranial foramina and relationships of the Eomyoidea (Rodentia, Geomorpha). Skull and upper teeth of *Kansasimys*. *American Museum Novitates*, 2645, 1–16.

Wahlert, J. H. and Koenigswald, W., von (1985). Specialized enamel in incisors of eomyid rodents. *American Museum Novitates*, 2832, 1–12.

Walsh, S. L. (1997). New specimens of *Metanoiamys, Pauromys*, and *Simimys* (Rodentia, Myomorpha) from the Uintan (middle Eocene) of San Diego County, California, and comments on the relationships of selected Paleogene Myomorpha. *Proceedings of the San Diego Society of Natural History*, 32, 1–20.

White, T. E. (1954). Preliminary analysis of the fossil vertebrates of the Canyon Ferry Reservoir area. *United States National Museum, Proceedings*, 103, 395–438.

Wilson, J. A. and Runkel, A. C. (1991). *Prolapsus*, a large sciuravid rodent, and new eomyids from the late Eocene of Trans-Pecos Texas. *The Pearce-Sellards Series, Texas Memorial Museum*, 48, 1–30.

Wilson, R. W. (1960). Early Miocene rodents and insectivores from northeastern Colorado. *University of Kansas Paleontological Contributions, Vertebrata*, 7, 1–92.

Winge, H. (1887). Jordfundne og nulevende Gnavere (Rodentia) fra Lagoa Santa, Minas Geraes, Brasilien. *E Museo Lundii, University of Copenhagen*, 1, 1–178.

Wood, A. E. (1936). A new rodent from the Pliocene of Kansas. *Journal of Paleontology*, 10, 392–4.

(1937). The mammalian fauna of the White River Oligocene, Part 2, Rodentia. *Transactions of the American Philosophical Society*, 28, 155–269.

(1955). Rodents from the Lower Oligocene Yoder Formation of Wyoming. *Journal of Paleontology*, 29, 519–24.

(1974). Early Tertiary vertebrate faunas, Vieja Group, Trans-Pecos Texas: Rodentia. *Bulletin of the Texas Memorial Museum*, 21, 1–112.

(1980). The Oligocene rodents of North America. *Transactions of the American Philosophical Society*, 70, 1–68.

26 Geomorpha

LAWRENCE J. FLYNN,[1] EVERETT H. LINDSAY[2] and ROBERT A. MARTIN[3]

[1] *Peabody Museum, Harvard University, Cambridge, MA, USA*
[2] *University of Arizona, Tucson, AZ, USA*
[3] *Murray State University, Murray, KY, USA*

INTRODUCTION

Geomorpha are small- to medium-size rodents, including the extant pocket gophers and pocket mice plus the allied kangaroo rats, traditionally classified in the suborder Myomorpha. Geomorphs include the extant families Heteromyidae and Geomyidae, which together with close relatives are termed Geomyoidea herein. The fossil record reveals rich diversity and radiations of extinct lineages, such as the entoptychine gophers and the florentiamyids.

Throughout the Tertiary, geomorph rodents are an enduring component characteristic of evolving North American ecosystems. With no clear exceptions (occasionally, Asiatic genera have been referred with question to the group), the Geomorpha are restricted to the New World, and few of them managed to penetrate into the South American land mass. Geomorphs are especially prominent in the fossil record during the Oligocene and Miocene, particularly in the Great Plains.

Modern geomorph diversity is considerable, including small, scansorial forms, ricochetal or saltatorial forms (i.e., bipedal hoppers), and large burrowers. Past diversity was also great, including several extinct families named to encompass these. Study of Geomorpha is progressing from a phase of alpha taxonomy to preliminary attempts to tease apart the interrelationships of the diverse fossil forms by using both shared derived morphology and molecular data.

DEFINING FEATURES OF THE INFRAORDER GEOMORPHA

Geomorphs exhibit considerable diversity: small, brachydont forms, high-crowned saltatorial genera, and robust burrowers. These rodents have four upper and four lower cheek teeth and the crown pattern is usually bilophodont.

The parvorder Geomyina includes the families Heliscomyidae, Florentiamyidae, and the Geomyoidea. This taxon is characterized by the following features: skull sciurognathous and sciuromorphous, 4/4 cheek teeth with cusps arranged in transverse rows and teeth widened with addition of styles (protostyle, entostyle, hypostyle above; protostylid and hypostylid below), and uniserial incisor enamel (see Wahlert, 1983, 1985) (Figure 26.1).

The superfamily Geomyoidea unites the diverse extant Heteromyidae (pocket mice) and the Geomyidae (pocket gophers). Characteristics include shortened incisive foramina (less than half diastema length); posterior maxillary foramen closed; cranially, the premaxillae extend further posteriorly than the nasal bones; there is a parapterygoid fossa and a bulla at the back of foramen ovale; the cheek teeth tend to be high crowned and lophodont, with stylar cusps well developed.

This list of features is short, allowing accommodation of the several affiliated genera below. If Geomyoidea were further restricted to contain only Heteromyidae plus Geomyidae (including entoptychines), then the list of shared derived features expands considerably (Wahlert and Sousa, 1988): length of incisive foramina reduced to less than 30% diastema length, deep pterygoid fossae opening into sphenopterygoid canal, laterally flared superior angular processes of dentary, bone of auditory bulla with spongy internal texture (smooth in dipodomyines), alisphenoid and maxilla in broad contact, and premolars at least as large as molars. Pocket gophers and pocket mice also share numerous features of the soft anatomy (Ryan, 1989).

SYSTEMATICS

SUPRA-INFRAORDINAL

The taxon Geomorpha is utilized at the infraorder level within Myomorpha by McKenna and Bell (1997) to contain geomyoids and their outgroup, the extinct Eomyidae. Myomorphs share the broadly

Evolution of Tertiary Mammals of North America, Vol. 2. ed. C. M. Janis, G. F. Gunnell, and M. D. Uhen. Published by Cambridge University Press.

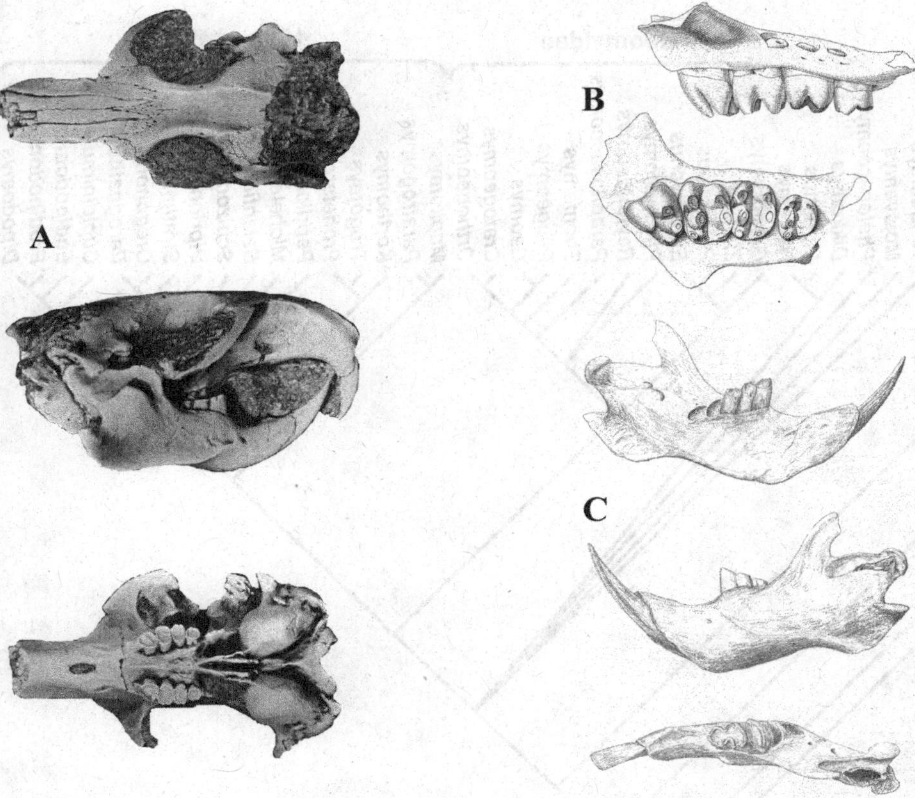

Figure 26.1. A. Skull of the fossorial entoptychine geomyid *Pleurolicus sulcifrons* (from Rensberger, 1973). B. Maxilla of the heteromyid *Perognathus furlongi* (from Lindsay, 1972). C. Dentary of the primitive geomyine *Parapliosaccomys oregonensis* (from Shotwell, 1967). Geomyids have simplified molars that are hypsodont to hypselodont. (A,B: courtesy of the University of California. C: courtesy of the University of Oregon.)

conceived condition of uniserial incisor microstructure. However, a survey of detailed microstructural features (Kalthoff, 2000) is beginning to distinguish important variation with the potential to distinguish phylogenetic groups among myomorphs. Both the Eomyidae and the Geomorpha share the broadly conceived cranial musculature feature of sciuromorphy (see Chapter 16). Future careful analysis could test whether the resemblance in sciuromorphy to that of squirrels is superficial and independently acquired. Eomyidae are described in Chapter 25.

The assignment of Geomorpha to the suborder Myomorpha has been questioned from time to time, but Wahlert (1985) listed a number of cranial characters to support it (short carotid canal near the basisphenoid–basioccipital junction, transverse canal entrance separate from alisphenoid canal, slender zygomatic arch, and uniserial incisor enamel microstructure). This suborder assignment has little to do with the cranial anatomy associated with zygomasseteric musculature (see Chapter 16). The name Myomorpha comes from those derived members of this group that display myomorphy. But all known geomorphs are distinctive in having a sciuromorphous type of anatomy, the condition where medial fibers of the masseteric musculature extend anterior to the zygomatic plate and originate on the lateral side of the snout. This condition was probably acquired independently, perhaps by a different pathway relative to the sciuromorphy of squirrels. The presumed convergence of sciuromorphy can perhaps be tested by careful survey of osteological features of the skull and dentary. The analysis may be assisted by examining the style of origin of superficial fibers of the masseter, the insertion points on the dentary, or the morphology of neighboring cranial features (such as the infraorbital foramen). Fossil genera should be resurveyed to determine the distribution of true sciuromorphy and test whether early Myomorpha display the feature, and whether Geomorpha belong in that suborder.

Recent molecular studies (Adkins, Walton and Honeycutt, 2002; Huchon *et al.*, 2002) offer an alternative phylogenetic interpretation. They see an association of Geomorpha with Castoroidea, which would suggest that Geomorpha do not belong in Myomorpha but in an independent higher-level taxon. This higher-level taxon would be characterized by sciuromorphy, again likely acquired independently from squirrels, but perhaps as a synapomorphy of castoroids and geomorphs (see further discussion in Chapter 16).

SUB-INFRAORDINAL

Perhaps it is the long history (most of the Cenozoic) and the high diversity among earliest records of the group that obscure the phylogeny of the Geomorpha. Despite great advances in taxonomy and recognition of clades not allied to extant crown groups, geomorph evolution remains poorly understood. We attempt a tree here, but it is not a cladogram, rather a summary of relationships as we understand them, and drawing on the works of others. The tree is based on derived morphology, but character distributions are not completely integrated into a phylogeny based on parsimony (Figure 26.2). We hope that this hypothesis is heuristic in the sense of stimulating further research.

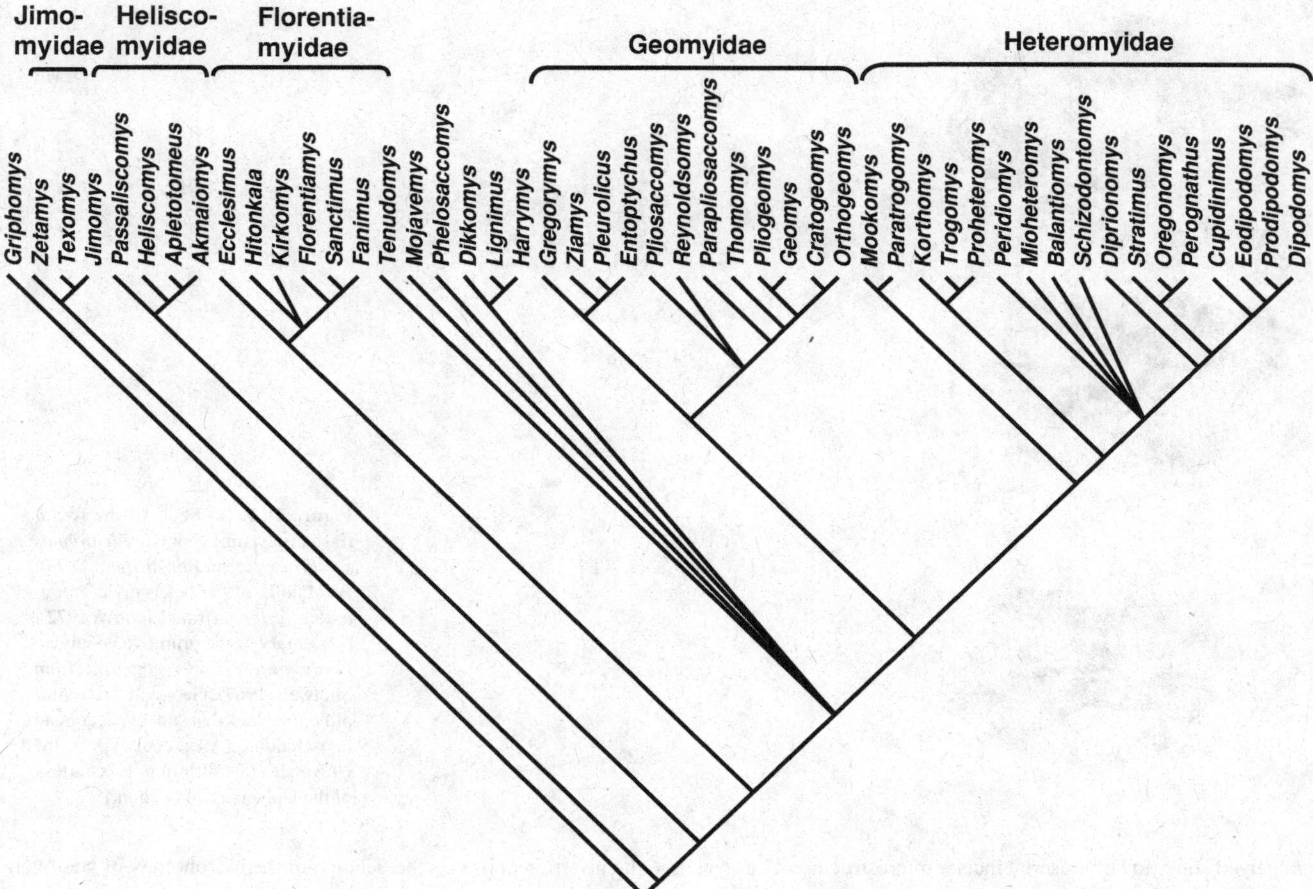

Figure 26.2. Proposed relationships of North American Tertiary Geomorpha. Only genera with discussed fossil records are included. This tree is based on the pattern of evolution as perceived herein. It is not independently generated, which would require a new survey of character states in many fossils. Geomorpha contains the large radiation of sciuromorphous, bilophodont North American rodents. *Griphomys* is a basal form, but its relationships to other early forms are unknown (except that *Floresomys* is not sciuromorphous and, therefore, presumed to be an outgroup). Early clades that originated during the late Eocene and Oligocene are recognized as extinct families. Geomyoids originated in the early Miocene, but earliest forms did not display defining features of extant groups and are not assigned to a family. Geomyid and heteromyid radiations are distinct, but early heteromyids are not assigned to living subfamilies. Proposed groupings of early genera remain to be fully tested.

Formerly, diverse mid Tertiary fossil taxa with incomplete suites of derived features were attributed variously to Geomyidae or Heteromyidae, making familial assignment insecure. Some genera display "incongruent" (apparently derived) conditions characteristic of both families, which confounds classification. Not surprisingly, the Heteromyidae, a family of generally smaller and lower-crowned forms, came to be used as a wastebasket for conservative fossil geomorphs. Reinterpretation of character distributions and better fossils have led to recognition of separate monophyletic groups, such as the families Heliscomyidae and Florentiamyidae.

In our approach, a few genera are placed within Geomyoidea without assignment to a family. Nevertheless, many fossil geomorphs lie external to our concept of Geomyoidea. Of these, some genera share derived features with geomyoids while others do not, and hence the taxon Geomyina is selected to contain a larger grouping of geomyoids plus relatives, but not all basal Geomorpha. Heliscomyidae and Florentiamyidae are examples of primitive Geomyina. Still more primitive are early forms, which are considered Geomorpha incertae sedis. The taxon Jimomyidae (new family) is placed there.

Our analysis attempts to recognize the successive and incremental modernization of geomorphs by utilizing a nested hierarchy (more to less inclusive): infraorder Geomorpha, parvorder Geomyina, superfamily Geomyoidea containing the extant families. The higher groups contain taxa (other than the families Geomyidae and Heteromyidae) that present some but not all derived conditions. Unfortunately, our brief survey below lacks thorough comparisons (owing in many cases to unknown character states) and suggests that the polarity of some traits should be reviewed. For example, what is the significance of rostral fenestration? It is derived at some level, but the level might be that of Geomyoidea (as used here) and secondarily lost (walled over) among Geomyidae. "More and better fossils," as rodent paleontologist Oldrich Fejfar would say, will provide the raw data to test such hypotheses.

One tool that may prove very useful in discerning supraspecific taxonomy of mid Tertiary forms was developed by Korth (1997). He showed (his Table 2) that crown height (e.g., at the base of the transverse valley on the labial side of p4 or m1), divided by posterior tooth width defines an index that can be applied across diverse species. Prior to Korth's analysis, height of crown among geomorphs was never successfully quantified. There is great promise that this "Korth index" will standardize "height of crown" in heteromyid rodents. However, preliminary results suggest that premolars are subject to excessive variation and may not provide useful comparisons. An example with greater promise is the Korth index for m1: height of the transverse valley above the base of the crown (labial view), divided by the transverse width of the posterior loph. We believe that this approach will help to clarify alpha taxonomy and generic content.

Another tool that may prove useful is quantification and distribution of minute accessory cuspules on teeth, especially on premolars. One problem that has limited this tool is that accessory cuspules can be identified only in unworn or slightly worn teeth. Another problem is that polarity of individual cuspules (i.e., are they new or vestigial?) is unexplored. However, modern collecting techniques yield large samples, which should clarify the basis for judging polarities and permit statistical treatment to enhance interpretations.

INCLUDED GENERA OF GEOMORPHA THALER, 1966

The locality numbers listed for each genus refer to the list of unified localities in Appendix I. The locality numbers may be listed more than one way. The acronyms for museum collections are listed in Appendix III.

Parentheses around the locality (e.g., [CP101]) mean that the taxon at that locality is cited as "aff." or "cf." the taxon in question. Parentheses are usually used for individual species, implying that the genus is firmly known from the locality, but the species identification may be questionable. Question marks in front of the locality (e.g., ?CP101) mean that the taxon is questionably known from that locality, implying some doubt at either the genus or species level that the taxon is actually present there. An asterisk (*) indicates the type locality.

GEOMORPHA INCERTAE SEDIS

The following taxa have been considered related to Geomyoidea yet do not share enough advanced features with main groups to demonstrate particular affinity, although they are considered here to be more closely related to Geomyoidea than are the Eomyidae. *Griphomys* shows a precocious tendency toward bilophodonty and is sciuromorphous. These features plus others (Wilson, 1940) strongly argue for geomyoid affinity. *Meliakrouniomys*, also problematical in its assignment to Eomyidae or Geomyoidea (Harris and Wood, 1969; Emry, 1972), is placed here with *Griphomys*, based on shared features (Wood, 1974, 1980). *Floresomys* resembles the two, but its protrogomorphous zygoma and retention of P3 make its assignment to Geomorpha more questionable. Although this taxon is considered to be a sciuravid by Korth (1994), it is treated in this chapter, following the phylogenetic analysis of Walton and Porter (Chapter 19).

Griphomys Wilson, 1940

Type species: *Griphomys alecer* Wilson, 1940.

Type specimen: CIT 2522, right dentary fragment with p4, m1–2.

Characteristics: Inclined zygomatic plate indicating sciuromorphy, masseteric scar extending below premolar; cheek teeth brachydont and bilophodont, 4/4 cheek teeth; upper teeth quadritubercular with anterior and posterior cingula; lower teeth quadritubercular with reduced mesoconid and anterior cingulum.

Length of m2: 1.3 mm (holotype).

Included species: *G. alecer* (known from localities CC7B, E, CC9AA, [BB], CC9IVA); *G. toltecus* Lillegraven, 1977 (locality CC7E).

Unidentified species of *Griphomys* are known from a number of localities. *Griphomys* sp. A (locality CC9II); *Griphomys* sp. B (locality CP44); *Griphomys* sp. C (localities CC7C, D, CC8, [CC8IIB], CC9BB). Korth and Eaton (2004) note an additional record at CP5II.

Meliakrouniomys Harris and Wood, 1969

Type species: *Meliakrouniomys wilsoni* Harris and Wood, 1969.

Type specimen: UTBEG 40283-80, left dentary fragment with incisor, p4, m1–3.

Characteristics: Sciuromorphous zygomasseteric structure like that of geomyoids; low crowned, bilophodont quadrate molars; large p4 and m3, p4 bearing metaconid and protoconid, narrowing anteriorly with straight labial and lingual walls; short anterior and posterior cingula, hypoconulids present; incisors laterally compressed, long diastema.

Length of m2: 1.83 mm (holotype).

Included species: *M. wilsoni* (known from locality SB44E); *M. skinneri* Emry, 1972 (localities CP39C-F).

Meliakrouniomys sp. is also known from localities SB43E, CP98B, NP10Bi.

Floresomys Fries, Hibbard, and Dunkle, 1955

Type species: *Floresomys guanajuatoensis* Fries, Hibbard, and Dunkle, 1955.

Type specimen: 52-137, Instituto de Geologia de la Universidad Nacional Autonoma de Mexico, mandible with right and left p4, m1–3 and left maxilla fragment with P3–4/M1–3.

Characteristics: Skull protrogomorphous, rostrum long and narrow; cheek teeth 5/4, with premolars reduced; molars brachydont but higher crowned than in sciuravids, with central basins of cheek teeth broad and deep; upper molars cuspate with small metaconule and strong anterior cingulum; mesostyle not present on M2; lower molars with hypoconulid, open transverse valley; no mesolophid or mesoconid on m1; straight posterolophid and no mesolophid on m2; lower incisor compressed laterally.

Average length of m2: 1.4 mm (mean from Black and Stephens, 1973).

Included species: *F. guanajuatoensis* only, known from locality SB51 only.

JIMOMYIDAE (NEW)

Jimomyidae is a small, monophyletic group, basal to the Geomyoidea and within Geomorpha. Whether family Jimomyidae stands alone as a unique higher taxon within Geomorpha or falls within Geomyina depends on finding more complete material, especially cranial elements. The group is characterized by strong lophodonty in which anterior and posterior moieties of the molars are bisected by a deep, central, transverse valley. The clade includes the mesodont genera, *Jimomys*, *Texomys*, and *Zetamys*.

Jimomyidae is diagnosed by 4/4 cheek teeth, with large, nearly molariform premolars, and derived crown pattern as follows. Lophodont cheek teeth lack cingula. Upper cheek teeth have a deep, lingual, transverse valley nearly to completely bisecting the tooth; this valley is oblique, separates protocone from hypocone, and has a complete wall, identified as the mesoloph, as its posterior border. In some species, the metaloph isolates a small posterolabial enamel lake and the anteroloph isolates an anterolabial enamel lake. The lower cheek teeth have a similar, oblique labial valley, separating the protoconid and hypoconid and nearly to completely bisecting the tooth. The posterior moiety of the tooth has a complete hypolophid and complete posterior arm of the hypoconid enclosing a small posterolingual enamel lake. Anteriorly, there is a deep lingual reentrant anterior to the metalophid. The large lower premolar is expanded by an anterolophid complex that, with wear, defines an additional anterior enamel lake. Based on *Jimomys* (see Wahlert, 1976) the incisor enamel is uniserial with interprismatic crystallites strongly inclined with respect to prisms.

Jimomys Wahlert, 1976 (including *Florentiamys*, in part)

Type species: *Jimomys labaughi* Wahlert, 1976.

Type specimen: F:AM 97816, left dentary fragment with incisor, p4, m1–3.

Characteristics: Cheek teeth low crowned; premolar largest of four lower cheek teeth; molars wider than long; two deep valleys on premolar, one on molars, only slightly oblique; anterior and posterior portions of premolar and molars with small enamel lakes; molars with multiple roots (after Wahlert, 1976).

Average length of m2: 1.6 mm (Wahlert, 1976).

Included species: *J. labaughi* (known from localities GC4E, CP111); *J. lulli* (Wood, 1936a) (originally described as *Florentiamys lulli*) (localities PN6D1, PN6E).

Jimomys sp. is also possibly known from locality GC8DB.

Texomys Slaughter, 1981

Type species: *Texomys ritchei* Slaughter, 1981.

Type specimen: F:AM 97816, left dentary fragments with incisor, p4, m1–3.

Characteristics: Cheek teeth high crowned; oblique transverse valleys completely crossing teeth; other reentrants completely closed; upper molars with three roots, lower molars with only two.

Average length of m2: 2.1 mm (see Slaughter, 1981).

Included species: *T. ritchiei* (known from localities GC4C, DD); *T. stewarti* Slaughter, 1981 (locality CA7).

Texomys sp. is also known from localities GC4C, E, GC5A, B, GC8B.

Zetamys Martin, 1974

Type species: *Zetamys nebraskensis* Martin, 1974.

Type specimen: UNSM 11514, right maxilla fragment with M1–2.

Characteristics: Cheek teeth lophodont and of moderate crown height; simplified crown morphology making Z-pattern (oblique transverse valley not completely bisecting teeth).

Average length of m2: 1.86 mm.

Included species: *Z. nebraskensis* only, known from locality CP101 only.

Zetamys sp. is also known from locality CP85B.

GEOMYINA, BONAPARTE, 1845

The taxonomic level Parvorder Geomyina (Bonaparte, 1845) is utilized here to reflect the diversity of mid Tertiary geomorphs. In other classifications, Heliscomyidae and Florentiamyidae (below) are considered primitive geomyoids. While they have attained some advanced characteristics of extant groups, they lack many of the distinctive features of extant pocket gophers and pocket mice, but do show unique features. Herein, the superfamily Geomyoidea is reserved for extant crown groups plus closely related fossil forms.

HELISCOMYIDAE KORTH, WAHLERT, AND EMRY, 1991

Green and Bjork (1980) made apparent that the simplistic view of *Heliscomys* as a primitive heteromyid does not do justice to the complex evolution of such interesting North American rodents. Additionally, the many species assigned to the genus did not make a cohesive group. Quite different species had been assembled as closely related and were somehow thought to be primitive heteromyids, yet they were characterized by divergent features. This problem has been dealt with in two ways: first, expansion of Florentiamyidae to accommodate a distinct radiation (see below), and second, the recognition of Heliscomyidae by Korth, Wahlert, and Emry (1991). These authors (plus Korth, 1994) are followed here in recognizing a restricted *Heliscomys* and retaining *Akmaiomys* and *Apletotomeus* as variant heliscomyids.

Characteristics (modified after Korth, Wahlert, and Emry, 1991): Elongated and depressed incisive foramina; long extension of crest on mandible for insertion of the masseter muscle, with mental foramen anterodorsal to its tip; nasals extending far posteriorly, between orbits; imperforate rostrum; posterior palatine foramina within palatines; sphenofrontal foramen and accessory foramen ovale present; no sphenopterygoid canal; back of mandible not deeply sculpted, processes not gracile; brachydont teeth with weak bilophodonty and small stylar cusps.

***Heliscomys* Cope, 1873**

Type species: *Heliscomys vetus* Cope, 1873.

Type specimen: AMNH 5461, left dentary fragment with m1–2.

Characteristics: Small geomyine with brachydont cheek teeth showing isolated cusps and strong cingulae bearing accessory cusps (labial on lower teeth, lingual on uppers); upper premolar triangular with one main anterior cusp and two posterior cusps plus style; lower premolar with four main cusps, narrow anteriorly; tubercle for superfical fibers of masseter muscle; masseteric fossa terminates far forward on dentary (see Green and Bjork, 1980).

Average length of cheek tooth row: 2.8–3.1 mm (Wood, 1980).

Included species: *H. vetus* (including *H. senex*, *H. gregoryi*, and *H. hatcheri*) (known from localities CP44, CP46, CP68C, CP84A, B, CP98B, C, CP99A, NP10B2, NP32B, C); *H. mcgrewi* Korth, 1989 (localities CP46, CP99A, [NP32C]); *H. ostranderi* Korth, Wahlert, and Emry 1991 (localities CP39B–E, [NP10B], NP10Bi, NP24C); *H. woodi* McGrew, 1941 (localities CP86B, CP99C).

Heliscomys sp. is also known from localities GC8AB, CC8IIB, CC9B2, CC9IVB, NP9A, ?NP10C2, NP10E.

Comments: Korth (1995a) treated these species differently, with *H. ostranderi*, *H. gregoryi*, and *H. hatcheri* constituting a morphologically distinct array that he recognized as subgenus *Syphyriomys*.

***Passaliscomys* Korth and Eaton, 2004**

Type species: *Passaliscomys priscus* Korth and Eaton, 2004.

Type specimen: UMNH VP5657, left P4.

Characteristics: Four-cusped molars lacking stylar cusps; P4 with large (unreduced) paracone, but endostyle small or absent.

Average length of m1 or m2: 0.94 mm.

Included species: *P. priscus* only, known from locality CP5II.

***Apletotomeus* Reeder, 1960a**

Type species: *Apletotomeus crassus* Reeder, 1960a.

Type specimen: UMMP 25893, left dentary fragment with incisor, p4, m1–3.

Characteristics: p4 quadrate and unreduced; broad and strongly curved lower incisor with flattened anterior face; dentary with short diastema and heavily built.

Length of m2: 0.97 mm (holotype).

Included species: *A. crassus* only (known from localities CP68C, CP99A).

***Akmaiomys* Reeder, 1960a**

Type species: *Akmaiomys incohatus* Reeder, 1960a.

Type specimen: CNHM-PM 381, right dentary fragment with incisor, p4, m1.

Characteristics: Large p4 with fifth prominent anterocentral cusp; procumbent and narrow lower incisor with flattened anterior surface.

Length of m1: 0.95 mm (holotype).

Included species: *A. incohatus* only, known from locality CP68C only.

Comments: Wahlert (1993), as a reviewer, considered this taxon to be a species of *Proheteromys*. W. W. Korth (personal communication, 2004) sees *Akmaiomys* and *Apletotomeus* species as distinct from *Heliscomys*, and congeneric.

Indeterminate heliscomyids

A heliscomyid of indeterminate genus and species is reported from locality GC8AA.

FLORENTIAMYIDAE WOOD, 1936A

Characteristics: Alisphenoid plus palatine process forming part of margin of anterior alar fissure; large optic foramen; reduced incisive foramina; masticatory and buccinator foramina united with accessory foramen ovale; entostyle elongated anteroposteriorly to block lingual end of transverse valley; lingual styles in lower molars; cheek teeth brachydont (after Wahlert, 1983). Korth's (1993a) study of *Hitonkala* cranial anatomy clarifies distinctiveness of the group and demonstrates outgroup relationship to geomyoids (as defined below).

***Ecclesimus* Korth, 1989 (including *Heliscomys*, in part)**

Type species: *Ecclesimus tenuiceps* (Galbreath, 1948) (originally described as *Heliscomys tenuiceps*).

Type specimen: KU 7702, partial skull with left P4, M1–3.

Characteristics: Small florentiamyid, known only from upper dental elements at present; paracone sometimes present on P4; relatively small M3; incisive foramina unreduced.

Average length of M2: 0.93 mm (Galbreath, 1948).

Included species: *E. tenuiceps* only (known from localities CP68C, CP99A).

***Florentiamys* Wood, 1936a (including *Sanctimus*, in part)**

Type species: *Florentiamys loomisi* Wood, 1936a.

Type specimen: Amherst No. 27-126, skull, mandible, and partial skeleton.

Characteristics: Large florentiamyid; small protostyle continuous with entostyle of P4; double protostylid of p4 (from Wahlert, 1983).

Average length of m2: 2.3–2.6 mm (Wahlert, 1983).

Included species: *F. loomisi* (known from localities CP42D, PN6F, ?G); *F. agnewi* Macdonald, 1963 (locality CP85C); *F. kennethi* Wahlert, 1983 (localities CP48, CP50);

F. kingi Wahlert, 1983 (locality CP48) *F. kinseyi* Wahlert, 1983 (locality CP48); *F. tiptoni* (Macdonald, 1970) (originally described as *Sanctimus tiptoni*) (localities CP86B, NP10CC).

Florentiamys sp. is also known from localities ?CP84A, CP103A, CP105.

Kirkomys Wahlert, 1984 (including *Heliscomys*, in part)

Type species: *Kirkomys milleri* Wahlert, 1984.

Type specimen: F:AM 105337, partial skull with incisors and all upper cheek teeth.

Characteristics: Small florentiamyid; P4 with protocone and strong anterior root, but no paracone or protostyle; incisive foramina reduced; non-molariform p4 (after Wahlert, 1984).

Average length of M2: 1.1 mm.

Included species: *K. milleri* (known from locality CP99B); *K. schlaikjeri* (Black, 1961) (originally described as *Heliscomys schlaikjeri*) (localities CP42D, CP85C, CP99C, CP101).

Comments: Korth (1997) noted that *Kirkomys milleri*, known only from upper cheek teeth, may be a junior synonym of *Proheteromys matthewi* (known from lower dentition, and herein accepted as a heteromyid); both taxa come from the Whitney Member of the Brule Formation in Nebraska.

Hitonkala Macdonald, 1963

Type species: *Hitonkala andersontau* Macdonald, 1963.

Type specimen: SDSM 56120, skull with complete dentition.

Characteristics: Small florentiamyid; single anterior cusp on P4; p4 highly molariform; M1 with separated protostyle and hypostyle.

Length of m2: 1.27 mm (holotype).

Included species: *H. andersontau* (known from localities CP51A, CP85C); *H. macdonaldtau* Korth, 1992 (localities CP103A, NP10CC).

Sanctimus Macdonald, 1970

Type species: *Sanctimus stuartae* Macdonald, 1970.

Type specimen: LACM 15292, nearly complete skull.

Characteristics: Large florentiamyid; protostyle large on P4, connected to protocone but divided from entostyle; single protostylid on p4.

Length of m2: 2.0 mm (holotype).

Included species: *S. stuartae* (known from localities CP85C, CP99C, CP101); *S. falkenbachi* Wahlert, 1983 (locality CP103A); *S. simonisi* Wahlert, 1983 (locality CP50); *S. stouti* Wahlert, 1983 (locality CP50).

Fanimus Korth, Bailey, and Hunt, 1990 (including *Pleurolicus*, in part; *Sanctimus*, in part)

Type species: *Fanimus ultimus* Korth, Bailey, and Hunt, 1990.

Type specimen: UNSM 26504, left dentary fragment with incisor stub, p4, m1–2.

Characteristics: Large florentiamyid with molars much wider than long; P4 without protostyle; p4 with single protostylid and large anteroconid.

Average length of m2: 1.9 mm.

Included species: *F. ultimus* (known from locality CP105); *F. clasoni* (Macdonald [1963]; originally *Pleurolicus*, then *Sanctimus* in Macdonald [1970]) (locality CP85C; also CP101 if including "*Tenudomys titanus*;" see Korth *et al.* 1990, and discussion below).

Indeterminate florentiamyids

Florentiamyids of indeterminate genus and species are reported from localities GC8AA, CP114A, NP10C2.

GEOMYOIDEA, BONAPARTE, 1845

Geomyoids are mainly North American groups, although both geomyids and heteromyids range throughout Central America and Heteromyidae reach northern South America. The usage of the superfamily here is reserved for these extant crown groups and their close fossil relatives. Several genera treated below have uncertain familial attributes. They converge on Heteromyidae or Geomyidae in various features of crown height and dentine tract development, and species have been taken as early pocket mice or true gophers. Here, they are shown as distinct, and listed as Geomyoidea incertae sedis.

GEOMYOIDEA INCERTAE SEDIS

Tenudomys Rensberger, 1973 (including *Pleurolicus*, in part; *Proheteromys*, in part; *Grangerimus*, in part)

Type species: *Tenudomys macdonaldi* Rensberger, 1973.

Type specimen: SDSM 53380, right dentary fragment with incisor, p4, m1–3.

Characteristics: Features of the superfamily plus skull deepened dorsoventrally; masseteric scar extends below mental foramen; small p4; cheek teeth mesodont; narrow incisors (Korth, 1993b).

Average length of m1: 1.65 mm.

Included species: *T. macdonaldi* (known from localities CP85C [type incorrectly assigned to *Pleurolicus leptophrys* by Macdonald, 1963], PN6E); *T. basilaris* Korth, 1989 (locality CP99A); *T. bodei* (Wilson, 1949) (originally described as *Mookomys? bodei*) (locality CC9D); *T. dakotensis* (Macdonald, 1963) (originally described as *Grangerimus dakotensis*) (locality CP85C).

T. titanus Martin (1974) (locality CP101) is a dubious record of the genus; Korth *et al.* (1990) transfer it with question to *Fanimus clasoni* (see the latter above).

Tenudomys sp. is also questionably known from locality NP36B.

Mojavemys Lindsay, 1972

Type species: *Mojavemys alexandrae* Lindsay, 1972.

Type specimen: UCMP 78155, right maxilla fragment with P4, M1–3.

Characteristics: Relatively large size; cheek teeth mesodont with thick enamel (weak dentine tracts evident at base of crown), high cusps with shallow transverse valleys; P4 with oval protocone; dp4 simple; no superior masseteric crest; rostrum fenestrated (Lindsay, 1972; Korth and Chaney, 1999).

Average length of M1: 1.55 mm.

Included species: *M. alexandrae* (known from locality NB6C); *M. galushai* Korth and Chaney, 1999 (localities SB32D, E); *M. lophatus* Lindsay, 1972 (localities NB6C, E); *M. magnumarcus* Barnosky, 1986a (locality CP45E); *M. wilsoni* (James, 1963) (originally described as *Pliosaccomys wilsoni*) (locality CC17H).

Mojavemys sp. is also known from localities NP40B, NP42.

***Phelosaccomys* Korth and Reynolds, 1994 (including *Parapliosaccomys*, in part; *Lignimus*, in part)**

Type species: *Phelosaccomys annae* (Korth, 1987) (originally described as *Parapliosaccomys annae*).

Type specimen: UNSM 56377, left dentary fragment with incisor, p4, m1–3.

Characteristics: Small perforation of rostrum; masseteric crest with superior branch and terminating anterior to p4; small and hypselodont teeth with dentine tracts; p4 with trigonid basin; protoloph and metaloph on P4 uniting lingually, wide incisors (Korth and Chaney, 1999).

Length of m1: 1.25 mm (*P. annae*).

Included species: *P. annae* (known from locality CP114B); *P. hibbardi* (Storer, 1973) (originally described as *Lignimus hibbardi*) (localities CP56IIB, CP114D, CP116A, B, CP123E, Korth, 1997); *P. shotwelli* Korth and Reynolds, 1994 (locality NB8); *P. neomexicanus* Korth and Chaney, 1999 (locality SB32D).

Phelosaccomys sp. is also known from localities NB6E (= *Parapliosaccomys* of Lindsay, 1972), NB7A, CP108A, CP114A, CP116E, F.

Comments: Korth (1987) perceived that the Great Plains species formerly placed in *Lignimus* and *Parapliosaccomys* represented another primitive geomorph. Korth (1994) concurred with Akersten (1988) that these species perhaps represent yet another genus. This led to the definition of *Phelosaccomys*. Korth and Chaney (1999) later proposed a higher taxon (subfamily) for the two genera *Mojavemys* and *Phelosaccomys*. This higher taxon, based mainly on primitive characters and a mosaic of derived conditions (some shared with heteromyids and others with geomyids), is weakly supported given the poor state of knowledge of early Miocene geomyoids. One key character is fenestration of the snout. This is considered an autapomorphy of Heteromyidae, but it appears in *Mojavemys* and to a lesser degree in *Phelosaccomys*. The presence of rostral fenestrae in *Tenudomys* is apparently unknown. It seems parsimonious to consider the presence of snout fenestration as derived for the Geomyoidea (as used here) and reversed or refilled in the Geomyidae. Similarly, the polarity of many other features surveyed is open to question, and we feel it is premature to erect higher taxa for these basal geomyoids. However, a clue to their affinity may lie in tracing the development of dentine tracts among geomyoids.

***Harrymys* Munthe, 1988 (including *Dikkomys*, in part; *Proheteromys*, in part)**

Type species: *Harrymys irvini* Munthe, 1988.

Type specimen: UCMP 12004, skull and mandible with complete dentition.

Characteristics: Primitive geomyoid features, rostral fenestra, incisive foramina unreduced; stapedial canal encased in bone; auditory bullae inflated and meeting anteromedially, mastoid chambers enlarged; short interparietal suture, large interparietal; cheek teeth high crowned, but rooted; uppers with U-pattern open buccally, lowers with ectolophid forming H-pattern (protostylid–hypostylid ridge produces R-pattern), hypolophid V-shaped; smooth incisors (after Wahlert, 1991).

Length of m1: 1.8 mm (holotype).

Included species: *H. irvini* (known from localities CP49A, CP54B); *H. canadensis* Korth, 1995b (locality NP11); *H. magnus* (Wood, 1932) (originally described as *Proheteromys magnus*) (localities GC5B, [GC8A, B], GC9B, CC19, ?CP105); *H. maximus* (James, 1963) (originally described as *Proheteromys maximus*) (locality CC17D); *H. woodi* (Black, 1961) (originally described as *Dikkomys woodi*) (localities CP49A, NP24C, NP34D).

Harrymys sp. is also known from localities CP74A or B, CP103II, CP108A.

Comments: This large genus has been considered a primitive heteromyid, incertae sedis, or as a member of a basal heteromyid subfamily Harrymyinae (Wahlert, 1991). Its features are mainly primitive, lacking some of the "geomyid" features of *Mojavemys*. *Harrymys* has autapomorphies as Wahlert (1991) indicated, including the H-pattern of loph connection in lower molars. Few characters suggest special heteromyid affinity, and the polarity of these should be tested. *Harrymys* does not definitively demonstrate classification within Heteromyidae. We agree with Bailey (1999) that *Proheteromys magnus* should be transferred to *Harrymys*, as should *P. maximus*.

We place *Harrymys* among genera such as *Tenudomys* and *Mojavemys* as taxa basal to Heteromyidae and Geomyidae; these are ripe for phylogenetic analysis, which would be necessary to establish special relationship to either extant family. Possibly future studies will show that there does exist a taxon "Harrymyinae," including certain poorly represented species (*Proheteromys magnus* and *P. maximus*), and possibly *Stratimus* (here listed as a heteromyid). The genera *Dikkomys* and *Lignimus*, represented only by dentitions and jaw fragments, also resemble *Harrymys*. Because both Wahlert (1991) and Korth (1994) proposed relationship of these genera, discussion of *Dikkomys* and *Lignimus* follows here.

Dikkomys Wood, 1936b

Type species: *Dikkomys matthewi* Wood, 1936b.

Type specimen: AMNH 22720, right p4.

Characteristics: High-crowned, rooted cheek teeth; premolars large; teeth bilophodont after moderate wear; lophs of worn upper cheek teeth joining lingually; lophs of lower teeth joining centrally (this is the sagicristid of Green and Bjork [1980]); masseteric crest terminating beyond p4, above the mental foramen.

Average length of m1: 1.6 mm (Galbreath, 1948).

Included species: *Dikkomys matthewi* only (known from localities CP74A, CP86D, CP88).

Dikkomys sp. is also reported from locality CP103II (Bailey, 1999).

Lignimus Storer, 1970

Type species: *Lignimus montis* Storer, 1970.

Type specimen: ROM 7108, left p4.

Characteristics: High-crowned teeth, but rooted and without dentine tracts; p4 with strong anterior cingulum; lower molars with mesially directed V-shaped ridge joining entoconid and hypoconid, protostylid and metalophid isolate anterobuccal lake; upper molars with central enamel lakes and anterior cingula with variable anterocone (after Storer, 1970; note that dentine tract elongation is not part of the diagnosis).

Average length of m1–2: 1.53 mm (Storer, 1975).

Included species: *L. montis* (known from localities [CP114A], NP11); *L. austridakotensis* Korth, 1996a (locality CP89); *L. transversus* Barnosky, 1986a (localities CP45E, NP40B, NP42).

Lignimus sp. is also known from localities CP114D, CP116B.

HETEROMYIDAE GRAY, 1868

Heteromyidae are a diverse group. Although represented today by only six genera, there are over 300 species. They show a range of locomotor adaptations from scansorial to fully ricochetal. The fossil record is rich in heteromyids. Later Tertiary forms can be attributed to extant subfamilies, but many early Miocene representatives are of undetermined affinities and are not classified to subfamily herein. As widely acknowledged, the group is in sore need of revision.

Characteristics: Generally small- to medium-size sciuromorphous rodents, infraorbital foramen anterior, on the side of the snout, with the inclined zygomatic plate behind it; side of snout perforated; reduced incisive foramina; thin skull bones, showing minor inflation; nasal bones anteriorly projecting; masseteric crest extending anteriorly, well beyond premolar and without a superior branch; mental foramen anterior in position, near the middle of the diastema; angular process distinct and well separated from coronoid process, coronoid small and low; molars generally six-cusped with large styles (-ids); p4 small and narrow anteriorly; incisors laterally compressed, often sulcate; postcrania gracile; with elongated hindlimb and fused tibiofibula (modified from Korth *et al.*, 1991).

BASAL HETEROMYIDS

Mookomys Wood, 1931 (including *Proheteromys*, in part)

Type species: *Mookomys altifluminus* Wood, 1931.

Type specimen: AMNH 21360, broken mandible with incisors, left p4, m1–2, and right p4, m1; skull fragments and partial skeleton.

Characteristics: Cheek teeth low crowned (but see comment below); upper incisors grooved, four-cusped p4; upper molar lophs joining non-centrally with wear; P4 lophs joining lingually.

Average length of m1: 1.1 mm.

Included species: *M. altifluminus* (known from localities ?CC19, NP34D); *M. subtilis* Lindsay, 1972 (locality NB6E); *M. sulculus* (Wilson, 1960) (originally described as *Proheteromys sulculus*) (localities CC17C, CC19, CC22A, CP71); *M. thrinax* Sutton and Korth, 1995 (locality NP41B).

Mookomys sp. is cited for localities (CC21C), NB4, SB3, ?CP114B, NP36B, C, NP40B, but generic level allocation is uncertain.

Comments: Korth (1994, 1997) pointed out that numerous species have been assigned to *Mookomys* in error. Wood (1931, 1935a) characterized *Mookomys* as a small, low-crowned heteromyid rodent, lower crowned than the modern heteromyid *Perognathus*, with a simple four-cusped p4 and a grooved upper incisor. Subsequently, virtually any small, low-crowned, middle Miocene heteromyid that resembled *Perognathus* was assigned to the genus *Mookomys*, regardless of other considerations. Korth (1997: Table 2) noted that the type of *Mookomys*, *M. altifluminus*, is actually mesodont rather than brachydont (Korth index for m1 of *M. altifluminus* is 0.34–0.45; *Perognathus* species range is 0.27–0.39) and transferred low-crowned species that had been assigned to *Mookomys* (*M. formicarum* Wood, 1935a; *M. bodei* Wilson, 1949; *M. subtilis* Lindsay, 1972; *Mookomys* cf. *M. altifluminus* Lindsay, 1974; *M.* sp. Whistler, 1984) to other genera. Korth (1997) recognized only two species (*M. altifluminus* and *M. thrinax*) in the genus *Mookomys*.

We concur that species previously assigned to *Mookomys* should be reexamined, but do not accept all of Korth's (1997) arguments for new generic assignments. Specifically, we disagee with transfer of *M. subtilis* to

Heliscomys, which was based (in part) on its low-crowned dentition with weakly developed cusps. Korth (1997) also mistakenly noted a posterior cingulum on M1 of *M. subtilis*, which is known only from a single dentary with no associated upper teeth. Korth (1997) suggested that the masseteric scar on the dentary of *M. subtilis* is either even with or slightly ventral to the mental foramen on the side of the dentary. The mental foramen on the dentary of *M. subtilis* is located very low, ventral to the masseteric scar (Lindsay, 1972: Fig. 21) much below the dorsal border of the diastema, which is strikingly different relative to the mental foramen and masseteric scar seen in *Heliscomys ostranderi* (Korth *et al.* 1991: Fig. 2), which is located on the upper part of the dentary. Also, cheek tooth cusps in *M. subtilis* are generally more robust and larger than those of *Heliscomys*, and the anterior cingulum in lower molars of *M. subtilis* is shorter, not as distinct as the anterior cingulum in lower molars of *Heliscomys* species. We reject the assignment of *M. subtilis* to *Heliscomys*, and retain it in *Mookomys*, until it is better known.

We agree that *M. fomicarum* Wood (1935a) does not match the revised generic diagnosis for *Mookomys*, and this species is treated separately below. Green (2002) documented wide intraspecific variation in a large sample of *Mookomys*, which led him to question whether the named species can be distinguished morphologically. His analysis further led to transfer of *Proheteromys sulculus* to *Mookomys*, based in large part on its sulcate incisor. *Mookomys* is one of the first geomyoids to show grooving of upper incisors (another grooved genus is *Mioheteromys*). Grooved and sulcate incisors, like dentinal tracts, are derived states that should be plotted on dendrograms to evaluate congruence.

***Paratrogomys* Lindsay and Reynolds, 2007**

Type species: *Paratrogomys whistleri* Lindsay and Reynolds, 2007.

Type specimen: LACM (CIT) 5184, skull fragment with complete upper dentition.

Characteristics: Mesodont cheek teeth and sulcate upper incisors; p4 lacking central pit, subparallel lophids joining centrally after moderate wear; bilophate lower molars joining labially and centrally in late wear (the metalophid becoming J-shaped); P4 with both lingual and medial union of protoloph and metaloph in late wear; upper molars bilophate with lingual union after moderate wear (metaloph J-shaped), discontinuous anterior cingulum.

Average length of m1: 1.04 mm.

Included species: *P. whistleri* only, known from locality CC22A only.

***Korthomys* Lindsay and Reynolds, 2006 (including *Mookomys*, in part)**

Type species: *Korthomys formicarum* (Wood, 1935a) (originally described as *Mookomys formicorum*).

Type specimen: CM 10177, right M1 (lost); neotype: SBCM L1816-3087, left P4.

Characteristics: Small heteromyid, with low-crowned, bunodont cheek teeth; P4 with large protocone and small paracone, metaloph J-shaped; upper molars U-shaped with lophs joined lingually; p4 with central pit and subparallel protolophid and metalophid; lower molars with weak cingula and lophs joining centrally and labially with wear.

Average length of m1: 0.94 mm (m2 slightly smaller).

Included species: *K. formicarum* only (known from localities NB6E [including *Mookomys* cf. *formicorum*, Lindsay, 1972], CP103A, NP41B).

***Trogomys* Reeder, 1960b**

Type species: *Trogomys rupimenthae* Reeder, 1960b.

Type specimen: LACM (CIT) 5184, skull fragment with complete upper dentition.

Characteristics: Cheek teeth low crowned but strongly bilophodont; molars wide; protostylid joined to protoconid by anterior cingulum; minute posterior cingulum in lower molars; four-cusped p4 nearly as broad anteriorly as posteriorly; short diastema; upper incisor asulcate.

Average length of m2: 0.89 mm.

Included species: *T. rupimenthae* only (known from localities CC9IIIB, CC9IVC, D, CC15, CC18A, [NB4]).

***Proheteromys* Wood, 1932**

Type species: *Proheteromys floridanus* Wood, 1932.

Type specimen: FSGS V5329, left dentary fragment with p4, m1.

Characteristics: Smooth, asulcate upper incisor enamel; dentition mesodont with lophs higher than in *Mookomys*; four-cusped premolars; anterior cusps on p4 close and joining early in wear to isolate a central basin; weak anterior and posterior cingula on lower molars.

Average length of m1: 0.85 mm.

Included species: *P. floridanus* (known from localities GC5B, GC8[A], DB, GC9B, ?SB29A); *P. cejanus* Gawne, 1975 (locality SB29A); *P. bumpi* Macdonald, 1963 (locality CP85C); *P. fedti* Macdonald, 1963 (localities CP85C, CP101); *P. gremmelsi* Macdonald, 1963 (locality CP85C); *P. ironcloudi* Macdonald, 1970 (localities CP86B, CP101, NP10CC); *P. magnus* Wood, 1932 (localities GC5B, [GC8A, B], GC9B, CC19, ?CP105); *P. matthewi* Wood, 1935a (localities CP86D, [NC2B]); *P. maximus* James, 1963 (locality CC17D); *P. nebraskensis* Wood, 1937 (localities CP84C, CP99B); *P. parvus* (Troxell, 1923, including *Diplolophus*, locality unknown); *P. sabinensis* Albright, 1996 (locality GC5A); *P. thorpei* Wood, 1935b (localities PN6D3, F); *P. toledoensis* Albright, 1996 (locality GC5A).

Proheteromys sp. is also known from localities GC4DD, GC5B, GC8AA, AB, A, B, DB, GC9C, CC9IVB–D,

(CC12), CC19, CP45E, CP54B, CP76B, CP84B, CP86E, CP88, CP99C, CP103A, NP10C2, E, NP40B.

Comments: Green (2002) argued that the crown height of *Proheteromys* is not demonstrably greater than that of *Mookomys*. Revisers fortunate enough to have large samples of *Proheteromys* recognize considerable variation and that some named species are likely to represent variant individuals of single populations. It is usually apparent that more than one species is present at a single locality; sometimes three size classes are discernable (sexual dimorphism is unlikely in these small rodents). It seems unlikely, however, that the same sympatric taxa occur at all localities, especially those as far separated as California and Florida. A species-rich primitive heteromyid genus is not unexpected. Still, the 14 named species of *Proheteromys*, which span much of the early to middle Miocene, suggest oversplitting and call for revision. Larger species of *Proheteromys* (e.g., *P. magnus* and *P. maximus*) are possibly related to *Harrymys*.

"MIOHETEROMYINAE" KORTH, 1997

Korth (1997) created his new subfamily Mioheteromyinae to accommodate mid Miocene diversity among heteromyids. Hemingfordian and Barstovian heteromyids do not show clear affinity with extant heteromyid subfamilies and the poor fossil material has made it difficult to evaluate the scattered dental remains. Even as cranial material has emerged, relationships have not become much clearer. Middle Miocene forms did attain a stage of evolution approaching modern conditions (notably higher crown height, shorter incisive foramina, anteriorly projecting masseteric crests, sulcate incisors in some genera), but these are not as derived as in extant subfamilies.

Korth (1997) hypothesized a monophyletic taxon (subfamily) for many of these forms. Whether it will withstand future testing seems dubious, given that few unique features support Mioheteromyinae, and of these features, interpretation and parallelism are in question. Korth cited the broad, low neurocranium, lack of the sphenofrontal foramen, and cheek teeth moderately high crowned. This taxonomic analysis is an important step in clarifying the early heteromyid radiation and presents testable hypotheses to stimulate further work. Quite possibly a core Mioheteromyinae will emerge, but its contents are not obvious. Here, we suggest that certain forms (above) do not belong in the subfamily.

Peridiomys Matthew, 1924

Type species: *Peridiomys rusticus* Matthew, 1924.

Type specimen: AMNH 18894, right dentary with incisor, p4, m1–2.

Characteristics: Cheek teeth of mesodont crown height and with bulbous cusps; metaconid much larger than protostylid on p4; incisors broad and flattened; dentary robust (modified from Korth, 1997).

Average length of m1: 1.55 mm (Korth, 1997; note that widths are much greater than lengths).

Included species: *P. rusticus* (known from localities CP110, CP111, CP114A); *P. halis* Sutton and Korth, 1995 (locality NP41B).

Peridiomys sp. is also known from localities NB6C, CP45E, CP54B, CP76B, NP40A, B, NP41B, PN7.

Mioheteromys Korth, 1997 (including *Diprionomys*, in part; *Oregonomys*, in part)

Type species: *Mioheteromys amplissimus* Korth, 1997.

Type specimen: UCMP 37134, partial skull with all cheek teeth and associated right dentary with all teeth.

Characteristics: Upper incisor sulcate, lower incisor narrow; cheek teeth mesodont, slightly inflated cusps, with roots; p4 often with a minute anterostylid, and protostylid and metaconid joining anteriorly after moderate wear; P4 large, protocone often with a small accessory cuspule (modified after Korth, 1997). Wood (1935a) interpreted *M. agrarius* to exhibit saltatorial adaptations.

Average length of m1: 1.55 mm.

Included species: *M. amplissimus* (known from localities CP45E, CP114A–C); *M. agrarius* (Wood, 1935a) (originally described as *Diprionomys agrarius*) (localities CP114D, CP116A, B); *M. arcarius* (Korth, 1996a) (originally described as "*Diprionomys*" *arcarius*) (localities CP89, NP11); *M. crowderensis* Lindsay and Reynolds, 2006 (localities CC22A, B, NB6D).

Balantiomys Korth, 1997 (including *Diprionomys*, in part; *Peridiomys*, in part)

Type species: *Balantiomys oregonensis* (Gazin, 1932) (originally described as *Diprionomys oregonensis*).

Type specimen: LACM(CIT) 371, partial skull with left cheek teeth and associated right dentary with all teeth.

Characteristics: Smooth, asulcate, narrow upper incisors; mesodont cheek teeth; p4 with subequal metaconid and protostylid that are wide medially (D-shaped), yielding a narrow central depression or pit, occasionally with a minute anterostylid; P4 large with lingual union of lophs, hypostyle placed anteriorly relative to hypocone (modified after Korth, 1997).

Average length of m1: 1.25 mm.

Included species: *B. oregonensis* (known from localities PN6G, PN7, PN8B, PN9B); *B. borealis* (Storer, 1970) (originally described as *Peridiomys borealis*) (localities CP114B, NP11); *B. nebraskensis* Korth, 1997 (localities ?CP106, CP113A).

Schizodontomys Rensberger, 1973 (including *Grangerimus*, in part)

Type species: *Schizodontomys greeni* Rensberger, 1973.

Type specimen: UCMP 39435, left dentary with incisor, p4, m1.

Characteristics: Large size, with thick enamel; smooth, asulcate incisors; rooted and mesodont cheek teeth; p4 metaconid slightly larger than protostylid; bullae inflated as in *Harrymys*; maxilla–premaxilla suture in palate confluent with posterior end of incisive foramen (modified after Korth, 1997).

Average length of m1: 1.54 mm (Rensberger, 1973).

Included species: *S. greeni* (known from locality PN6G); *S. amnicolus* Korth, Bailey, and Hunt, 1990 (locality CP105; note spelling error in publication); *S. harkseni* (Macdonald, 1970) (originally described as *Grangerimus harkseni*) (localities CP54B, CP86D, NP10CC); *S. sulcidens* Rensberger, 1973 (localities CP51A, CP86A, ?CP105).

Schizodontomys sp. is also known from localities CC9IVC, D.

HETEROMYINAE GRAY, 1868

Characteristics of the subfamily were enumerated by Williams, Genoways and Braun (1993). Briefly, and as far as limited fossil material goes, heteromyines are the primitive group of modern Heteromyidae. They lack hypertrophy of the bulla and are mostly scansorial in locomotion. Molars are mesodont to hysodont, with the cusp pattern submerged into lophs after moderate wear; heteromyines never develop hypselodonty (cheek teeth remain rooted). The upper incisor is smooth or with shallow sulcus. Cheek tooth lophs unite lingually first, then labially, often isolating a central enamel basin, especially in M1.

Diprionomys Kellogg, 1910 (including *Entoptychus*, in part)

Type species: *Diprionomys parvus* Kellogg, 1910.

Type specimen: UCMP 12566, right dentary fragment with incisor, p4-m1.

Characteristics: Cheek teeth high crowned and strongly bilophodont; p4 equaling m1 in size, with large anteroconid; transverse lophs uniting with wear, lingually first then labially, thus isolating a central basin; molars somewhat wider than long.

Average length of m1: 2.5 mm (Hall, 1930).

Included species: *D. parvus* (known from localities [NB23C], NB31, [PN11], PN13, PN15); *D. minimus* (Wood, 1936b) (originally described as *Entoptychus minimus*) (locality NB31).

Diprionomys sp. is also known from localities NP40B, PN8A, PN10.

Comments: *Diprionomys* is not well represented, but may be an early record of subfamily Heteromyinae (see Korth, 1997) although all *Diprionomys* could possibly be nothing more than worn *Cupidinimus* specimens.

PEROGNATHINAE COUES, 1875

Characteristics of the subfamily were listed by Williams, Genoways, and Braun (1993). Relevant to fossils, note that perognathines are generally small, mesodont heteromyids. The anterior loph of P4 bears one cusp, the protocone, and the p4 has four cusps that usually join in an X-pattern. The transverse lophs of upper molars usually unite lingually first, then buccally; lophs of lower molars usually unite buccally first, sometimes medially and forming an H-pattern. The third molars are reduced, and the upper incisor is sulcate. The auditory bullae are expanded ventrally. The postcrania are generally delicate, locomotor specializations include scansorial to moderately saltatorial, with the hindlimbs relatively long for muroids.

Stratimus Korth, Bailey, and Hunt, 1990 (including *Mookomys*, in part)

Type species: *Stratimus strobelli* Korth, Bailey, and Hunt, 1990.

Type specimen: UNSM 26688, left dentary with incisor, p4, m1–2.

Characteristics: Quadritubercular p4 with median loph directed from metalophid to separate the cusps of the protolophid (protoconid and metaconid); transverse lophs of lower molars joining centrally; short anterior cingulum on upper molars; P4 with robust, transversely elongated protocone that joins both the hypocone (centrally) and the hypostyle (lingually) in late wear, hypostyle relatively large and located anterior relative to hypocone (modified after Korth, Bailey, and Hunt, 1990).

Average length of m1: 1.2 mm.

Included species: *S. strobelli* only (known from localities CP71 [including ?*Mookomys formicorum* of Wilson, 1960], CP105).

Comments: Korth, Bailey, and Hunt (1990) interpreted the lower molar crown pattern of *Stratimus* as having small lophules from the hypoconid and entoconid that converge anteriorly at the midline to form a V-shaped hypolophid. The key point is central convergence of transverse lophs to form (later in wear) the H-pattern seen in *Harrymys* (geomyoid incertae sedis, above).

Oregonomys Martin, 1984 (including *Perognathus*, in part)

Type species: *Oregonomys pebblespringensis* Martin, 1984.

Type specimen: UWBM 57116, left dentary with incisor, p4, m1–3.

Characteristics: Perognathine with sulcate upper incisor and inflated auditory bulla; wide palate with tooth size decreasing posteriorly; bilophodont upper molars with lingual union; P4 with protocone and accessory protostyle, three roots, initial union of lophs usually medial; dentary with low mental foramen, coronoid process deflected posteriorly, strong angular process; p4 with metalophid complex of at least three cusps, and with medial to buccal union of lophs (lingual valley shallow); transverse lophs of lower molars usually joining buccally initially, sometimes medially (modified from Martin, 1984).

Average length of m1: 1.3 mm.

Included species: *O. pebblespringensis* (known from localities PN14, [PN21B]); *O. magnus* (Zakrzewski, 1969) (originally described as *Perognathus magnus*) (localities

PN14, PN23A); *O. sargenti* (Shotwell, 1956) (originally described as *Perognathus sargenti*) (locality PN14).

Oregonomys sp. is also known from locality NB33A.

Comments: Martin (1984) proposed this genus as a perognathine and included Wood's *Diprionomys agrarius* (which is represented by a skeleton). Subsequent workers have disputed the position of the latter species, with Korth (1997) most recently placing it in *Mioheteromys*. Wahlert (1993) interpreted the inflated mastoid of *O. pebblespringensis* with thin (non-cancellous), simple walls as evidence for inclusion of the genus with Dipodomyinae. Dipodomyinae are generally characterized by thin-bone bullae, but Korth (1998) reviewed the distribution of this character, noting that thin bullae are widespread in early heteromyids. He interpreted this structure as primitive and the spongy bullae typical of later (not all) perognathines as derived. This difference in interpretation of polarity tends to support placement of *Oregonomys* in Perognathinae.

***Perognathus* Wied-Neuwied, 1839**

Type species: *Perognathus fasciatus* Wied-Neuwied, 1839, the extant olive-backed pocket mouse.

Type specimen: *P. furlongi* Gazin, 1930, first named fossil species: CIT 35, with incisors and all molars (premolars broken).

Characteristics: Cheek teeth brachylophodont to mesodont, lower part of crown slightly inflated (crown height increasing somewhat through time); cusps uniting after early wear, roots long and separate; posterior molars reduced in size; anterior cingulum continuous on M1–2 but broken on M3; lower molars with H-pattern; p4 four-cusped, cusps uniting medially with wear; P4 with single anterior cusp (a buccal cuspule may occur in *P. furlongi*) centrally joined with the three-cusped metaloph; upper incisors sulcate (modified after Lindsay, 1972; Martin, 1984).

Average length of m2: 0.88 mm (*P. furlongi*).

Included species: *P. ancenensis* Sutton and Korth, 1995 (known from locality NP41B); *P. brevidens* Korth, 1987 (localities CP114A, B); *P. carpenteri* Dalquest, 1978 (locality SP1F); *P. dunklei* Hibbard, 1939 (locality CP123D); *P. furlongi* Gazin, 1930 (localities CC17E, G–I, NB6C, E, CP45E); *P. gidleyi* Hibbard, 1941 (localities SB14D, SB15A, SB18A, CP128C); *P. henryredfieldi* Jacobs, 1977 (locality SB10); *P. maldei* Zakrzewski, 1969 (locality PN23A); *P. mclaughlini* Hibbard, 1949 (localities NB33A, SB10, [SP4B], CP128AA); *P. minutus* James, 1963 (localities CC17G, NB6C, E, NB7A–E, NB33A); *P. pearlettensis* Hibbard, 1941 (localities SB14D, [SP1A, F, H], SP5, CP128A, C, [CP130], CP131, CP132B); *P. rexroadensis* Hibbard, 1950 (localities [SP1F, H], SP5, CP128A); *P. stevei* Martin, 1984 (locality PN14); *P. trojectioansrum* Korth, 1979 (localities CP114A, B, NP11).

Perognathus sp. is also known from localities GC9C, CC19, CC21B, CC47C, D, NB4, NB13A–C, SB11, SB12, SB14B, C, E, F, SB18B, ?SB48, SB50, SB58A, SB60, SP1C, H, SP3B, CP54B, CP56, CP90A, CP116B, C, F, CP123E, CP128AA, B–D, CP131, CP132C, D, NP40B, NP41B, PN4, PN10, PN11, PN14, PN15.

Comments: Known widely from Barstovian age and younger deposits, and with a Hemingfordian record at Split Rock, *Perognathus* is one of the longest-ranging rodent genera. It is also geographically widespread, spanning the North American continent. Pleistocene records of the genus are distributed from California to Texas to Kansas, and north through Colorado to Idaho and Washington. Some of these records likely include *Chaetodipus*, which is distinguished from *Perognathus* based on features of the pelage, pinna, and pes, as well as the bulla (Williams, Genoways, and Braun, 1993). Korth (1998) shows potential to distinguish fossil species of these genera by bullae, if preserved.

DIPODOMYINAE COUES, 1875

Characteristics of the subfamily were listed by Williams, Genoways, and Braun (1993). Body mass of some taxa exceeds 100 g. Hindlimbs tend to be elongated and adapted to ricochetal locomotion. Incisors usually sulcate. Cheek teeth are hypsolophodont to hypselodont, bilophodont with cusps evident only in early wear; wear results in U-shaped lingual union of lophs of upper teeth and H-pattern medial union of lophs in lower molars. The P4 usually has one cusp in protoloph with the lophs uniting centrally. Lophs unite with an X-pattern in p4. The third molars are small; dentine tracts develop progressively. Auditory bullae are expanded dorsally and an inflated mastoid is exposed on the skull surface. Bullae are mostly hollow, without spongy trabeculae, and expand caudally beyond the occiput in extant genera. Dalquest and Carpenter (1986) have pointed out that the variable features of the premolars limit their utility in diagnoses.

***Cupidinimus* Wood, 1935a (including *Perognathoides*; *Diprionomys*, in part)**

Type species: *Cupidinimus nebraskensis* Wood, 1935a.

Type specimen: CM 10193, partial skull and skeleton with left P4 and M3, left incisor, p4, m1–3.

Characteristics: Cheek teeth bilophodont with near-vertical sides and well-developed roots; molars with six cusps and prominent cingula; upper molar lophs not fusing labially until late wear; lower molars joining centrally to form H-pattern; weak dentine tracts occasionally in lower molars; P4 occasionally with an accessory protostyle; p4 with central union of protolophid and metalophid and occasional accessory anterior cusp; upper incisor asulcate.

Average length of m1: 0.89 mm (*C. nebraskensis*, from Korth, 1979).

Included species: *C. nebraskensis* (known from localities [CC21C], CP114A, B); *C. avawatzensis* Barnosky, 1986b (localities NB7B, NB8); *C. bidahochiensis* (Baskin, 1979) (originally described as *Perognathoides bidahochiensis*) (locality SB11); *C. boronensis* Whistler, 1984 (localities CC19, NB4); *C. cuyamensis* (Wood, 1937) (originally described as *Perognathoides cuyamensis*) (locality CC17G); *C. eurekensis* (Lindsay, 1972) (originally described as *Perognathoides eurekensis*) (localities NB6E, NP41B); *C. halli* (Wood, 1936c) (originally described as *Perognathoides halli*) (localities NB6C, E); *C. kleinfelderi* (Storer, 1970) (originally described as *Perognathoides kleinfelderi*) (locality NP11); *C. lindsayi* Barnosky, 1986b (localities [CC9IVD], NB6C, E); *C. madisonensis* (Dorr, 1956) (originally described as *Perognathoides madisonensis*) (locality NP41B); *C. prattensis* Korth, 1998 (locality CP116B); *C. quartus* (Hall, 1930) (originally described as *Diprionomys quartus*) (locality NB23C); *C. saskatchewanensis* (Storer, 1970) (originally described as *Perognathus saskatchewanensis*) (locality NP11); *C. tertius* (Hall, 1930) (originally described as *Diprionomys tertius*) (localities NB7A–C, NB8, NB23C); *C. whitlocki* Barnosky, 1986b (locality CP45E).

Cupidinimus sp. is also known from localities CC36, NB7D–E, NB20A, ?NB33A, SP3B, CP90A, CP114D, CP116A, B, NP40B.

Comments: This genus contains diverse primitive dipodomyines distributed from Saskatchewan to Arizona, and California to Texas. *Cupidinimus* is recognized as the preferred taxon name for *Perognathoides* Wood (1935a) (see review by Wahlert, 1993) and accommodates some species at one time placed in *Diprionomys* (*C. quartus*, *C. tertius*). The genus had been considered a perognathine, but recent work that incorporated cranial data argued for inclusion in Dipodomyinae (see Korth, 1998). The presence of asulcate upper incisors in *Cupidinimus* has led researchers to conclude that Dipodomyinae and Perognathinae achieved sulcate incisors independently. However, reversal (loss) of this character is also possible.

***Prodipodomys* Hibbard, 1939 (including *Liomys; Etadonomys; Perognathus*, in part)**

Type species: *Prodipodomys kansensis* (Hibbard, 1937) (originally described as *Dipodomys kansensis*).

Type specimen: KUMNH VP 3945, left dentary with incisor, p4.

Characteristics: Dipodomyine sharing with *Dipodomys* a small foramen in the trough between m3 and the coronoid process; sulcate upper incisors; cheek teeth high crowned but with roots and low, weak dentine tracts; P4 with separate anteroloph and three roots; p4 with X-pattern and two roots; subequal m1 and m2 with H-pattern early in wear; small m3.

Average length of p4: 1.15 mm (from Zakrzewski, 1981).

Included species: *P. kansensis* (known from locality CP123D); *P. centralis* (Hibbard, 1941) (originally described as *Liomys centralis*, and inc. "*P. rexroadensis*") (localities SB14A, SP1F, H, CP80, CP118, [CP128A], CP130); *P. coquorum* (Wood 1935a) (originally described as *Perognathus coquorum*) (locality CP115D); *P. griggsorum* Zakrzewski, 1970 (localities SP4B, CP128AA); *P. idahoensis* Hibbard, 1962 (localities CC47C, SB14D, SB60, CP97II, CP128?B, C, PN4, PN23A [Martin *et al.*, 2002]); *P. mascallensis* Downs, 1956 (including "*Mojavemys*" *mascallensis*, Barnosky, 1986a) (localities PN7, PN9B); *P. minor* (Gidley, 1922) (originally described as *Dipodomys minor*) (localities NB33A, SB14B–E, SB60, PN23A); *P. riversidensis* Albright, 2000 (localities CC47A, CC48B); *P. tiheni* (Hibbard, 1943) (originally described as *Etadonomys tiheni*) (localities NB33A, CP132B, [D]); *P. timoteoensis* Albright, 2000 (localities CC47A, C).

Prodipodomys sp. is also known from localities ?CC47B, CC48A, B, NB13B, SB10, SB18B, SB50, SP3B, CP116E, F, CP128AA, B–D, CP130, CP131.

***Eodipodomys* Voorhies, 1975**

Type species: *Eodipodomys celtiservator* Voorhies, 1975.

Type specimen: UGV 109, partial skull and skeleton with complete lower dentition, upper incisors, premolars, and right M1.

Characteristics: Large dipodomyine sharing with *Dipodomys* a small foramen in the trough between m3 and the coronoid process of the dentary; rooted cheek teeth hypsodont and with well-developed dentine tracts; p4 longer than wide, with oval anterior loop without reentrant; asulcate upper incisors.

Average length of p4: 2.7 mm.

Included species: *E. celtiservator* only, known from locality CP116B only.

Comments: This interesting fossil of the Clarendonian suggests that *Eodipodomys* is sister to *Dipodomys*, and that *Prodipodomys* is more remotely related. Zakrzewski (1981) interpreted *Eodipodomys* as an independent entry in the "*Dipodomys*" niche. It is very like a kangaroo rat in its inflated bullae and elongated hind legs. It was recovered from a burrow, demonstrably within the Ash Hollow Formation since it is overlain by continuous sandstone. Wahlert (1993) noted that the asulcate incisors are incongruent with the condition in *Dipodomys* and *Prodipodomys*, where the incisors are sulcate, but *Cupidinimus* also has asulcate incisors.

***Dipodomys* Gray, 1841 (including *Perodipus*; *Dipodops*; *Macocolus*, in part; *Cricetodipus*)**

Type species: *Dipodomys phillipsii* Gray, 1841, the extant Phillips kangaroo rat.

Type specimen: None designated.

(Type specimen of *D. hibbardi* Zakrzewski, 1981: UM 72614, right P4.)

Characteristics: Cheek teeth with crown height greater than root length, roots generally single and open, dentine tracts well developed; p4 with anterior reentrant continuous to base of crown; upper incisors sulcate.

Average length of P4: 1.14 mm (*D. hibbardi*).

Included Tertiary species: *D. gidleyi* Wood, 1935a, (known from localities NB35IV, SB14F, SB18A); *D. hibbardi* Zakrzewski, 1981 (localities NB13A–C, SB15A, SB18A, B, CP132B); *D. pattersoni* Dalquest and Carpenter, 1986 (Irvingtonian of Texas). The extant *D. compactus* True, 1888, is noted from localities NB13B, C.

Dipodomys sp. is also known from localities CC47D, CC48A–C, ?NB10, NB13A–C, SB37D, PN11, PN23C. Pleistocene records of the genus are scattered throughout California, Nevada, Arizona, and New Mexico.

Comments: *Dipodomys* is characterized by rootless (hypselodont) cheek teeth with the development of dentine tracts. The tracts are short in early fossil and some extant species. *Prodipodomys* ranges through most of Blancan NALMA. *Dipodomys* becomes widespread late in this land-mammal age. Wahlert (1993) included in this genus Wilson's (1939) dipodomyine from the Avawatz Fauna (locality NB8). The p4 in this taxon is very slightly indented and wider than long, which seems to rules out assignment to *Eodipodomys*. This very early (Clarendonian) record for *Dipodomys* seems unlikely and we suspect that it is not truly from deposits yielding the Avawatz Fauna, or that it represents another hypsodont geomyoid (Shotwell, 1967, suggested *Pliosaccomys*).

Microdipodops Merriam, 1891

The extant kangaroo mouse genus is recognized from the Pleistocene of Nevada and has a possible Pliocene record from California (localities NB13B, C).

INDETERMINATE HETEROMYIDS

Heteromyids of indeterminate genus and species are reported from localities GC8AA, AB, CC12, CC32A, CC38A, CP5II, NB8, NB32, SB11, SB46, SB48, CP88, NP10CC, E, PN19A.

GEOMYIDAE, BONAPARTE, 1845

Geomyids are the familiar North and Central American pocket gophers. Extant forms are grouped as subfamily Geomyinae and distributed in two tribes. The extant *Thomomys* is assigned to Thomomyini, whereas the extant *Geomys*, *Pappogeomys* (including *Cratogeomys*), *Orthogeomys*, and *Zygogeomys* make up the Geomyini. Geomyinae comprises medium-sized rodents, larger than most heteromyids the subgenera of *Orthogeomys* (*Heterogeomys* and *Macrogeomys*) are large for rodents, some specimens exceeding 1 kg in body mass. Whereas *Thomomys* is the smallest of the extant genera and has rounded, smooth incisor enamel, geomyines are larger, with chisellike heavy incisors, and a deep groove running the length of the upper incisor. Some forms have one or even two additional shallow grooves parallel to the deep groove. Fossil taxa show some degree of grooving, suggesting that this feature is primitive for the group. The skulls are heavily built, and the postcrania show modifications for burrowing. The Neogene is the acme of geomyids diversity and abundance. Geomyines are fairly widely known in the fossil record, particularly the later Neogene, but thomomyines are relatively less common.

Several fossil genera included in the Geomyidae are not attributed to extant subfamilies. The Miocene Entoptychinae is distinctive and well documented by Rensberger (1971) and by Wahlert and Souza (1988). The latter authors reaffirmed that entoptychines are geomyids and not an independent basal geomyoid taxon. *Pleurolicus* (appearing in the late Oligocene) and other forms have been placed in a distinct subfamily, but the current practice is to submerge *Pleurolicus* in Entoptychinae (McKenna and Bell, 1997; see discussion in Korth, 1994).

Characteristics: Skull progressively shortened; diastema sometimes strongly arched; greatly reduced incisive foramina; palate deeply furrowed; optic foramen small and anterior; region dorsal to orbitosphenoid usually not ossified; masticatory foramen with anteromedial opening of alisphenoid canal; no sphenofrontal or stapedial foramen; internal pterygoid muscle originating far anteriorly (through sphenopterygoid canal, sometimes beyond optic foramen); boss on squamosal redirecting pull of posterior temporal muscle, deep anterior temporal fibers arising in broad vertical fossa in orbit; alisphenoid extending far dorsally; mandibular angle reducing to a ridge on the dentary, and a robust superior angular process projecting laterally below condyloid process; temporal muscle inserting in broad mandibular pit between coronoid and posterior cheek teeth; alveoli of upper and lower incisors attenuating posteriorly; premolars larger than molars. This list paraphrases the wording of Wahlert (1985) and is selected based on subsequent work, including that of Wahlert and Souza (1988).

ENTOPTYCHINAE MILLER AND GIDLEY, 1918

Characteristics: The geomyid characteristics indicated above are generally expressed to a moderate degree in known entoptychines, but Wahlert and Souza (1988) point out other derived conditions for the subfamily. Anteromedial bullar processes meet ventrally at the basicranium midline, the mastoid is inflated and exposed dorsally in the skull roof, and correspondingly the squamosal is emarginated posteriorly. The entoptychine diastema is not as strongly arched as in geomyines, and the upper incisors, not as elongated as in geomyines, have a faint to strong groove, typically on the medial corner of the enamel. Having a lower-crowned dentition than geomyines, entoptychines retain a strong transverse valley in cheek teeth. Burrowing adaptations of the postcrania are generally less developed than in geomyines (Rensberger, 1971).

Gregorymys Wood, 1936b (including *Entoptychus*, in part)

Type species: *Gregorymys formosus* (Matthew, 1907) (originally described as *Entoptychus formosus*).

Type specimen: AMNH 12887, cranium with complete upper dentition.

Characteristics: Medium-size geomyids, with rooted cheek teeth of moderate crown height and undulating base of enamel; groove on upper incisor faint or obvious; lower incisor flattened anteriorly; p4 larger than m1, with variable anterior cingulum and hypolophid with three cusps; transverse lophs of lower cheek teeth join initially on the labial side; P4 larger than M1, with multicusped protoloph; transverse lophs of upper cheek teeth joining initially on the lingual sides.

Length of M1: 1.5 mm (holotype).

Included species: *G. formosus* (known from localities CP52, CP85C, CP86B–D, CP103A); *G. curtus* (Matthew, 1907) (originally described as *Entoptychus curtus*) (localities CP86B–D); *G. douglassi* Wood, 1936b (locality NP39); *G. kayi* Wood, 1950 (locality NP38A); *G. larsoni* Munthe, 1977 (locality CP76II); *G. riggsi* Wood, 1936b (localities CP52, CP86D); *G. riograndensis* Stevens, 1977 (localities SB46, SB47).

Gregorymys sp. is also known from localities CP88, NP34C, NP36B, C, PN6F.

Comments: Korth (1994) commented on the diversity of *Gregorymys*, noting the morphologically advanced *G. kayi* from Montana and similar specimens from Wyoming. The late-occurring *G. larsoni* is less anomalous given occurrence of *Gregorymys* sp. in the Hemingfordian (locality CP88; Martin, 1976). Closely related *Ziamys* brings more Hemingfordian species to this array.

Ziamys Gawne, 1975

Type species: *Ziamys tedfordi* Gawne, 1975.

Type specimen: F:AM 51264, partial cranium with upper dentition lacking only right P4.

Characteristics: Geomyid smaller than *Gregorymys*, with cusps of low-crowned cheek teeth merging late in wear; rostrum short and broad; upper incisor greatest dimension width, with medial groove and additional deep central groove; lower incisor flattened anteriorly; transverse lophs of lower cheek teeth joining centrally late in wear; P4 with unicusped protoloph; transverse lophs of upper molars join variably late in wear.

Length of M1: 1.4 mm (holotype).

Included species: *Z. tedfordi* (known from locality SB29A); *Z. hugeni* Korth, Bailey, and Hunt (1990) (locality CP105).

Pleurolicus Cope, 1878 (including *Grangerimus*)

Type species: *Pleurolicus sulcifrons* Cope, 1878.

Type specimen: AMNH 7165, partial skull with right P4, M1–2, left m1.

Characteristics: Medium-size geomyid; mesodont cheek teeth with cusps merging after early wear, enamel base undulating, but not developing dentine tracts; rostrum elongated; incisor enamel gently rounded and showing weak longitudinal grooves (upper and lower), shallow upper incisor groove in place of the medial groove of *Entoptychus*; transverse lophs of P4 joining in late wear, p4 relatively short, without anterior cingulum, but a minute accessory cusp may be present.

Length of M1: 1.6 mm (holotype, from Rensberger, 1973).

Included species: *P. sulcifrons* (including *P. diplophysus* Cope, 1881, *P. copei*, and *Grangerimus oregonensis*) (known from localities CP86D, [CP101], PN6D1, E, H); *P. dakotensis* Wood, 1936b (locality CP86B); *P. exiguus* Korth, 1996b (localities CP104A, CP105); *P. hemingfordensis* Korth, Bailey, and Hunt, 1990 (locality CP105); *P. leptophrys* Cope, 1881 (localities PN6D3, G, H); *P. selardsi* (Hibbard and Wilson, 1950) (originally described as *Grangerimus selardsi*) (locality ?SB45).

Pleurolicus sp. is also known from localities SB29B, CP47, NP36B.

Comments: A junior synonym of *P. sulcifrons* is *Grangerimus oregonensis* Wood, 1936b. Rensberger (1973) recognized distinction of *Pleurolicus* from *Entoptychus*. He considered two other primitive geomyoids (*Tenudomys* and *Schizodontomys*) as related to *Pleurolicus*, and he constructed the subfamily Pleurolicinae to contain the three. Presently, *Tenudomys* is considered a basal geomyoid, while *Schizodontomys* is seen as a heteromyid, leaving *Pleurolicus* as an early entoptychine, and Pleurolicinae a redundant taxon.

Entoptychus Cope, 1878 (including *Palustrimus; Gregorymys*, in part)

Type species: *Entoptychus cavifrons* Cope, 1878.

Type specimen: AMNH 7052, partial skull with upper dentition.

Characteristics: Medium-size geomyids; cheek teeth high crowned with cusps merging early in wear, dentine tracts well developed, tendency to root loss; rostrum elongated; upper incisor with medial groove only; transverse lophs of P4 joining lingually; p4 with weak anterior cingulum; molar lophs tending to join laterally to form transversely elongated enamel lakes.

Average length of M1: 1.5 mm (Wood, 1936b).

Included species: *E. cavifrons* (known from localities PN6D3, F); *E. basilaris* Rensberger, 1971 (localities CC9D, PN6D2, F); *E. fieldsi* Nichols, 1976 (locality PN16); *E. germanorum* Wood, 1936b (locality PN6F); *E. grandiplanus* Korth, 1992 (locality CP103A); *E. individens* Rensberger, 1971 (localities PN6D3, G); *E. minor* Cope, 1881 (localities CP85C, PN6D3, F); *E. montanensis* (Hibbard and Keenmon, 1950) (originally described as *Gregorymys montanensis*) (locality NP33B); *E. planifrons* Cope, 1878 (including *E. sperryi* and *E. rostratus*, both named by Sinclair, 1905) (localities PN6D3, F, G); *E. productidens* Rensberger, 1971 (localities PN6D3, F); *E. sheppardi* Nichols, 1976 (locality PN16); *E. transitorius*

Rensberger, 1971 (localities PN6D3, F); *E. wheelerensis* Rensberger, 1971 (localities PN6D2, D3, F).

Entoptychus sp. is also known from localities CP47, CP51B.

Comments: *Palustrimus lewisi* (Wood, 1935b) is a DP4 of *Entoptychus*. Rensberger (1971) considered *E. lambdoideus* and *E. crassiramus* (from locality PN6H) based by Cope (1878) on insufficient material as useless taxonomic names. The types of each are fragmentary with heavily worn dentitions and could be attributed to any of several species.

Indeterminate entoptychines

Entoptychines of indeterminate genus and species are from localities GC8AA, A, B, ?NP34C.

GEOMYINAE BONAPARTE, 1845

Characteristics: Medium- to large-sized geomyids with strongly arched diastema and narrow, grooved palate; optic foramen small and (with the anterior alar fissure) located anteriorly; internal pterygoid originating far anteriorly, through the sphenopterygoid canal, into the orbit; stapedial and sphenofrontal foramina lost; temporal muscle inserting on dentary in prominent mandibular pit between cheek teeth and coronoid; extended incisor roots; fossorial adaptations (after Wahlert and Souza, 1988).

***Pliosaccomys* Wilson, 1936 (including *Diprionomys*, in part; *Cupininimus*, in part)**

Type species: *Pliosaccomys dubius* Wilson, 1936.

Type specimen: LACM(CIT) 1796, right dentary with incisor, p4, m1–3.

Characteristics: Small geomyids with high-crowned cheek teeth that have weak dentine tracts and roots fused proximally, separate distally; p4 longer than m1, with incomplete anterior cingulum and wide metalophid and hypolophid joining in early wear; large P4 with narrow anteroloph flattened anteriorly and joining wide metaloph lingually in early wear; molar lophs appressed (join lingually in uppers, labially in lowers); upper incisors wide, rounded, lacking grooves; lower incisor enamel flattened medially (after Shotwell, 1967).

Average length of p4: 1.3 mm (Wilson, 1936).

Included species: *P. dubius* (known from localities NB27B, PN14); *P. higginsensis* Dalquest and Patrick, 1989 (locality SP1C); *P. magnus* (Kellogg, 1910) (originally described as *Diprionomys magnus*, later *Cupininimus magnus* of Wood, 1935a) (localities NB31, PN11, [PN15]).

Pliosaccomys sp. is also known from localities CC36, ?NB8, CP116C, D, PN10, PN11.

***Reynoldsomys* Albright, 2000**

Type species: *Reynoldsomys timoteoensis* Albright, 2000.

Type specimen: UCR 22416, right P4.

Characteristics: Small geomyine with rooted premolars having dentine tracts present lingually and labially on metaloph (no tracts on protoloph).

Length of P4: 1.28 mm (holotype).

Included species: *R. timoteoensis* only, known from locality CC47B only.

***Parapliosaccomys* Shotwell, 1967**

Type species: *Parapliosaccomys oregonensis* Shotwell, 1967.

Type specimen: UO 3631, left dentary with incisor, p4, m1.

Characteristics: Medium-size geomyid with high-crowned cheek teeth and roots forming in late wear; molars with one long root; dentine tracts developed halfway up crown; p4 longer than m1, with vestige of anterior cingulum, narrow metalophid joining hypolophid in early wear; large P4 with wide anteroloph flattened anteriorly and joining wider metaloph lingually in early wear; molar lophs appressed (join lingually in uppers, labially in lowers); upper incisors flattened (no grooves); lower incisor enamel flattened; dentary with mental foramen anteroventral to anterior limit of masseteric crest; pit absent between molars and ascending ramus (modified after Shotwell, 1967).

Average length of lower cheek teeth row: 6 mm (calculated from Shotwell figure of alveoli).

Included species: *P. oregonensis* only (known from localities NB34II, PN14).

Parapliosaccomys sp. is also known from locality NB6E.

***Thomomys* Wied-Neuwied, 1839 (including *Plesiothomomys*)**

Type species: *Thomomys talpoides* (Richardson, 1828), the extant Western pocket gopher.

Type specimen: None designated. (Type specimen of *T. gidleyi* Wilson, 1933: USNM 12651, left dentary with p4, m1.)

Characteristics: Small geomyids; cheek teeth hypsodont and rootless but lacking cementum; tight constriction (fold) in lateral enamel on labial side of upper cheek teeth and lingual side of lower cheek teeth; dentinal tracts well developed; enamel thickness reduced on posterior plates of upper cheek teeth and on anterior plates of lower molars (with wear, thin enamel plates are lost but full anterior enamel plate retained on upper cheek teeth and full posterior enamel plate retained on lower cheek teeth); p4 with anteroposteriorly extending metalophid joining to wide hypolophid at midline; P4 with small protoloph and wide metaloph joining slightly lingual to midline; upper incisors wide and gently rounded, smooth or with minute groove lingually; mental foramen anterior to masseteric crest; dentary with shallow to deep temporal fossa between m3 and capsular process of incisor.

Average length of m1:1.2 mm (average over considerable variation).

Included species: *T. gidleyi* (known from localities WM26, CC47A–D, CC48B, [PN3C], PN23A); *T. carsonensis* Kelly, 1994 (localities NB35IIIC, NB36IIIB).

Thomomys sp. is also known from localities CC48C, ?NB7B–D, NB13C, ?NB36IIIA, CP116F, PN23C. Extant *Thomomys orientalis* is reported from the Pleistocene of Florida (including locality GC15D).

Orthogeomys **Merriam, 1895**

Type species: *Orthogeomys scalops* (Thomas, 1894) (including *Geomys scalops*), the extant southern Mexican taltuza.

Type specimen: None designated. Type specimen of *Orthogeomys propinetis* (Wilkins, 1984): UF 46001, partial cranium with complete dentition, lacking only right M3.

Characteristics (for fossil *Orthogeomys*): Upper incisors with single groove; skull dolichocephalic; shallow retromolar fossa in dentary; cheek teeth hypselodont, with strong dentine tracts and no enamel plate on posterior surface of P4, or on anterior walls of m1 and m2; M3 bicolumnar.

Average maximum length of p4: 2 mm (*O. propinetis*).

Included species: *O. propinetis* only (known from localities GC15A, C, D, GC17A, GC18IIB).

Comments: *Geomys propinetis* Wilkins, 1984 was referred to *Orthogeomys* by Ruez (2001) after reanalysis of original and newer material. Wilkins (1984) had rejected this assignment. This record represents a considerable range extension for the genus.

Cratogeomys **Merriam, 1895 (including *Geomys*, in part; *Pappogeomys*, in part)**

Type species: *Cratogeomys merriami* (Thomas, 1893).

Type specimen: not specified, probably from the Valley of Mexico. (Type specimen of *Pappogeomys* (*Cratogeomys*) *sansimonensis* Tomida, 1987: UALP 10344, partial skeleton including anterior portion of skull with incisors, right P4, M1–2 and left P4; right dentary with all teeth, left incisor).

Characteristics: Upper incisors with a single groove; sagittal crest on skull; open-rooted, hypselodont cheek teeth; dentinal tracts present but enamel plate persisting on posterolingual corner of metaloph of P4.

Average length of occlusal surface of lower cheek tooth row: 8 mm (*C. bensoni*); 9 mm (*C. sansimonensis*) (from Tomida, 1987).

Included species: *Cratogeomys bensoni* Gidley, 1922 (known from localities SB14A–C); *C. sansimonensis* Tomida, 1987 (locality SB15A).

Cratogeomys sp. is also known from localities SB18A, SB50, SB68.

Comments: Tomida (1987) named his species at a time when *Cratogeomys* was generally considered a subgenus of *Pappogeomys*. *Pappogeomys* has page priority in Merriam (1895). He further noted that *C. sansimonensis* shows *Cratogeomys* characters, but that the type material of *C. bensoni* does not preserve those features. We retain *C. bensoni* here, pending review of undescribed material from the type locality of the latter. Interrelationships of living geomyid species and genera was greatly clarified by Spradling *et al.* (2004), who showed that *Pappogeomys bulleri* (Thomas, 1892) is distinct from all species of *Cratogeomys*. In some analyses it is basal to *Cratogeomys*, but in others it associates with *Orthogeomys*. Spradling *et al.* (2004) presented their Bayesian analysis on living genera, including those without a fossil record, and we follow them in seeing *Cratogeomys* as a separate late Blancan radiation.

Pliogeomys **Hibbard, 1954 (including *Progeomys*)**

Type species: *Pliogeomys buisi* Hibbard, 1954.

Type specimen: UMMP 29147, right dentary with incisor, p4, m1–2.

Characteristics: Medium-size geomyid; cheek teeth high crowned; roots fused but open in molars, separate distally in premolars; dentine tracts usually well developed; premolar long, with rounded metalophid joining wider hypolophid near midline after early wear; large P4 with enamel persisting on posterior side, rounded protoloph joining wider metaloph lingual to midline in early wear; enamel persisting anteriorly on lower molars until late wear; p4 generally with entire enamel, V-shaped reentrant fold lacking cement; wide lower incisors; upper incisors with deep median groove, plus shallow lingual groove (some authors finding two grooves); presence of shallow pit in mandible between m3 and ascending ramus (in part after Hibbard, 1954).

Length of cheek teeth alveolus: 7.3 mm (holotype).

Included species: *P. buisi* (known from locality SP4B); *P. parvus* Zakrzewski, 1969 (localities NB33A, PN23A); *P. russelli* Korth, 1995c (locality CP115D); *P. sulcatus* (Dalquest, 1983) (originally described as *Progeomys sulcatus*) (locality SP3B).

Pliogeomys sp. is also known from localities ?NB33A, CP116E, F.

Comments: Our usage of the genus is meant to contain a radiation of rooted forms close to *Geomys*. Dalquest (1983) named *Progeomys sulcatus* for a primitive member of this group that shows no dentine tracts but has doubled shallow incisor grooves. Given the few primitive features, we include this early geomyine in *Pliogeomys*. We also note that *Ziamys* has double ornamentation on its incisor, suggesting that double grooving may be primitive for modern geomyines. Furthermore, Rensberger (1971) has reported ornamentation in juvenile *Thomomys*.

Geomys **Rafinesque, 1817 (including *Zygogeomys*; *Pliogeomys*, in part)**

Type species: *Geomys pinetis* Rafinesque, 1817, the extant Eastern pocket gopher.

Type specimen: None designated.

Characteristics: Derived geomyine with hypselodont cheek teeth with cementum and fossorial postcranial adaptations;

large and strongly inclined upper diastema; two longitudinal grooves in upper incisors; lower incisors wide and chisel shaped; heavy masseteric scar on dentary; lower premolars with a figure-of-eight occlusal outline; hypsodont, ever-growing molars, anteroposteriorly compressed, with bilobed pattern submerged by wear to produce simple oval column; variation in dentinal tracts resulting in incomplete enamel around occlusal outline of cheek teeth. Subgenus *Nerterogeomys* Gazin (1942) was defined as having generally complete enamel around the back of P4, in contrast to the subgenus *Geomys* with interrupted enamel (but workers should note that this is dependent on wear). Hibbard (1967) noted that the mental foramen usually is more posterior in *Nerterogeomys* (below masseteric scar).

Average length of p4: 3.3–3.6 mm (Franzen, 1947).

Included species: *Geomys* (*Geomys*) *adamsi* Hibbard, 1967 (known from localities CP128A, [B] [CE1] [Martin, Goodwin, and Farlow, 2002]); *G.* (*G.*) *caranzai* (Lindsay and Jacobs, 1985) (originally described as *Pliogeomys caranzai)* (locality SB60); *G.* (*G.*) *garbanii* White and Downs, 1961 (localities NB13B,C); *G.* (*G.*) *pinetis* (localities GC15A, C–E, GC17A, GC18IIIA, B, GC18IV); *G.* (*G.*) *quinni* McGrew, 1944 (including *G. jacobi*) (localities CP117B, CP118, CP128C, CP131, [CP132B]) (see Hibbard and Riggs, 1949; Skinner and Hibbard, 1972); *G.* (*G.*) *tobinensis* (localities CP130, CP132B).

Geomys (*Nerterogeomys*) *smithi* (including *G.* (*N.*) *minor* Gidley, 1922) (known from localities NB13A–C, SB14A, B, E, SB32C, SB35A, B, SB38IIA, C, SB62, SP1F, CP128[AA]–C); *Geomys* (*N.*) *anzensis* Becker and White, 1981 (localities NB13A–C); *G.* (*N.*) *paenebursarius* strain, 1966 (localities SB37B, [SB38IIC], SB49, SB50); *G.* (*N.*) *persimilis* (Hay, 1927) (originally described as *Zygogeomys persimilis*) (localities SB14D–F, SB15A, SB18A, B, [SB34IVA]).

G. (*N.*) sp. is also known from localities NB36IIIA, SP1F, H, CP128C, 129, 130.

Geomys sp. is also known from localities NB13A–C, SB12, SB15A, SB18A, SB31C, SB35D, SB37D, SP4B, SP5, CP80, CP121, CP128AA (including "*Pliogeomys*"), B–D, CP130, CP131, CP132B–D.

Comments: Fossil *Geomys* of Pleistocene age are widespread from Florida westward.

INDETERMINATE GEOMYIDS

Geomyids of indeterminate genus and species are reported from localities NB32, CP114A, CP116B, NP10C2, E, NP42, PN19A.

BIOLOGY AND EVOLUTIONARY PATTERNS

ECOMORPHOLOGY

Geomorpha is a distinctly North American group of Rodentia. Relationship with certain Asiatic fossils (e.g., *Diatomys*) has been suspected, but these taxa are now seen as separate (Flynn, 2006). Geomorpha are well represented among the Cenozoic faunas everywhere across the vast North American landmass at every time interval since the late middle Eocene epoch (Figure 26.3). The theme of this complex mammalian radiation is sciuromorphy combined with bilophodont molar morphology. Individuals are generally small to medium sized, and a few forms achieved a large body size for rodents (mass greater than 500 g). Some extant geomyids (e.g., the taltuzas of Central America, *Orthogeomys*) are sexually dimorphic with males near 1 kg in mass (Nowak and Paradiso, 1983).

Much as the biology of extant heteromyids displays diversity, fossil geomorphs also were diverse (Figure 26.4). Evidence from fossil forms shows a range of cheek tooth occlusal patterns from bunobrachydont to lophohypsodont. Some taxa show inflated auditory bullae (extreme in Dipodomyinae). Inflated auditory bullae house the middle ear and apparently create a resonating chamber that is efficient in sensing predator approach. Such bullae are, therefore, often taken as indicators of life in open habitats.

Heteromyids, dipodomyines especially, also display adaptation to saltatorial locomotion, typically elongated hindlimbs. This is another feature consistent with living in open habitats, insofar as hopping can be a good escape strategy. Postcrania of extinct heteromyids are not common, but later Tertiary fossils usually show greater hindlimb to forelimb length ratios or altered muscle attachments for leaping. Skeletal evidence indicates saltatorial locomotion for the fossil dipodomyines *Cupidinimus* and *Eodipodomys* (Wood, 1935a; Voorhies, 1975). Kangaroo rats (Dipodomyinae) show extreme saltatorial adaptation. Their leaping ability is sometimes classified as "ricochetal locomotion." One hallmark of ricochetal locomotion is fusion of the cervical part of the vertebral column (Hatt, 1932). Commitment to open habitats and saltation precludes commitment to subterranean life. Although many heteromyids dig burrows, they are not truly subterranean.

The contrasting strategy adopted by gophers (Geomyidae) for life in open habitats, as in the plains and adjacent mountain regions, is the strategy of life underground. Geomyidae are fully subterranean (Figure 26.5). The term "fossorial" can be applied and implies life primarily under the ground, with only up to 5% of the time spent on the surface. Under such conditions, the skull and postcrania show considerable subterranean modification. The skull is heavily built, often wedge shaped with broad, sloping occiput and heavy muscle attachments; the lower incisor protrudes and is utilized as a chisel in digging. The forelimb is usually developed as a powerful soil transport mechanism – a shovel. Muscle attachments are distal to supply power, the manus is relatively broad, sometimes heavily clawed.

The extent to which many fossil forms dug burrows is not known owing to incomplete material. The distribution of digging traits is virtually unknown among non-geomyid geomorphs. Among extinct geomyids, *Pleurolicus* and *Entoptychus* display skulls adapted for tooth-digging burrowing (Rensberger, 1971, 1973). Rensberger (1973) interpreted remains of *Schizodontomys* to indicate fossorial adaptations, but these are not as derived as in geomyids, suggesting incomplete commitment to subterranean life. Only Geomyidae, therefore, indicate full exploitation of the subterranean niche, and an implied diet including underground plant storage organs. This

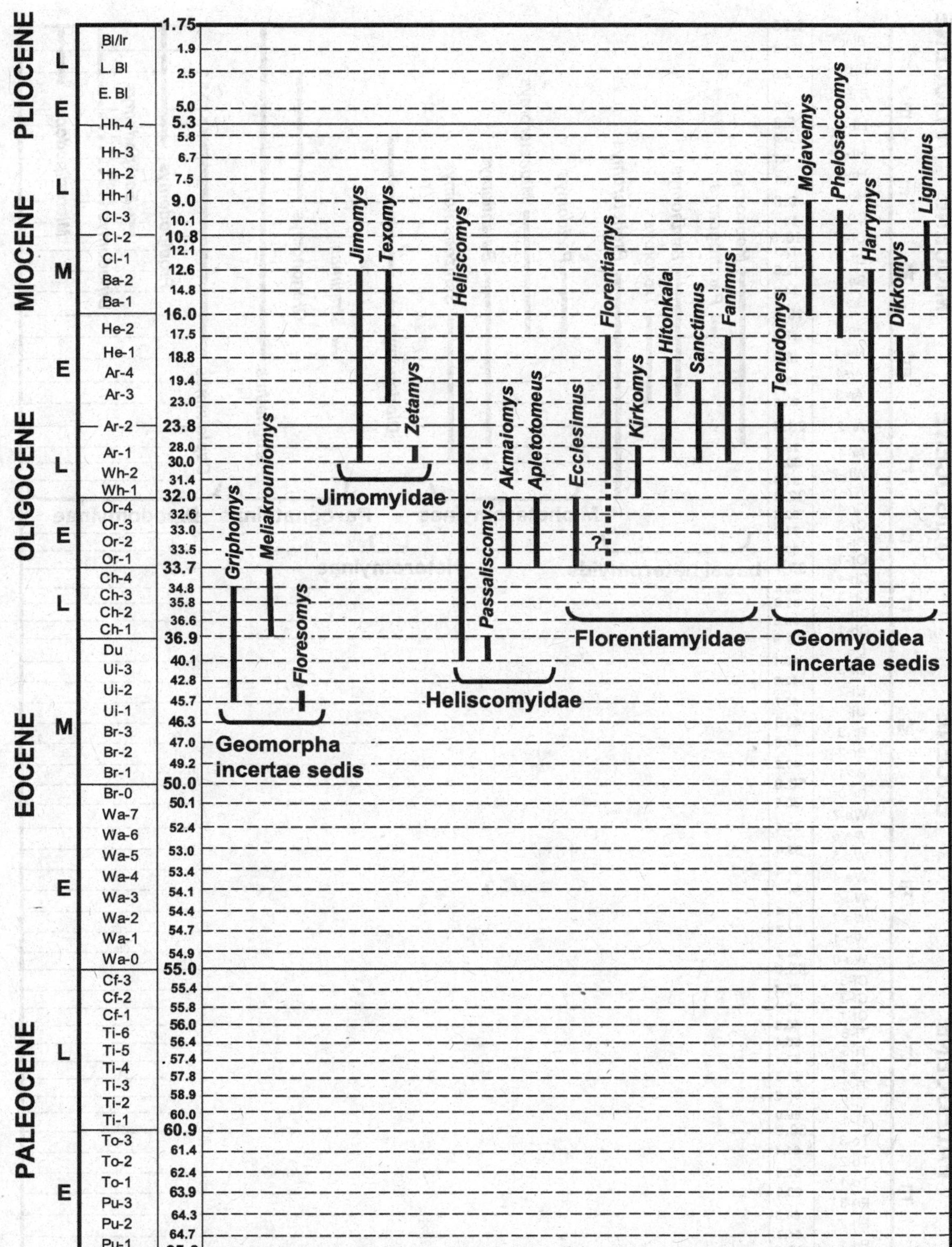

Figure 26.3. Temporal ranges of Geomorpha 1: Primtive geomyoids.

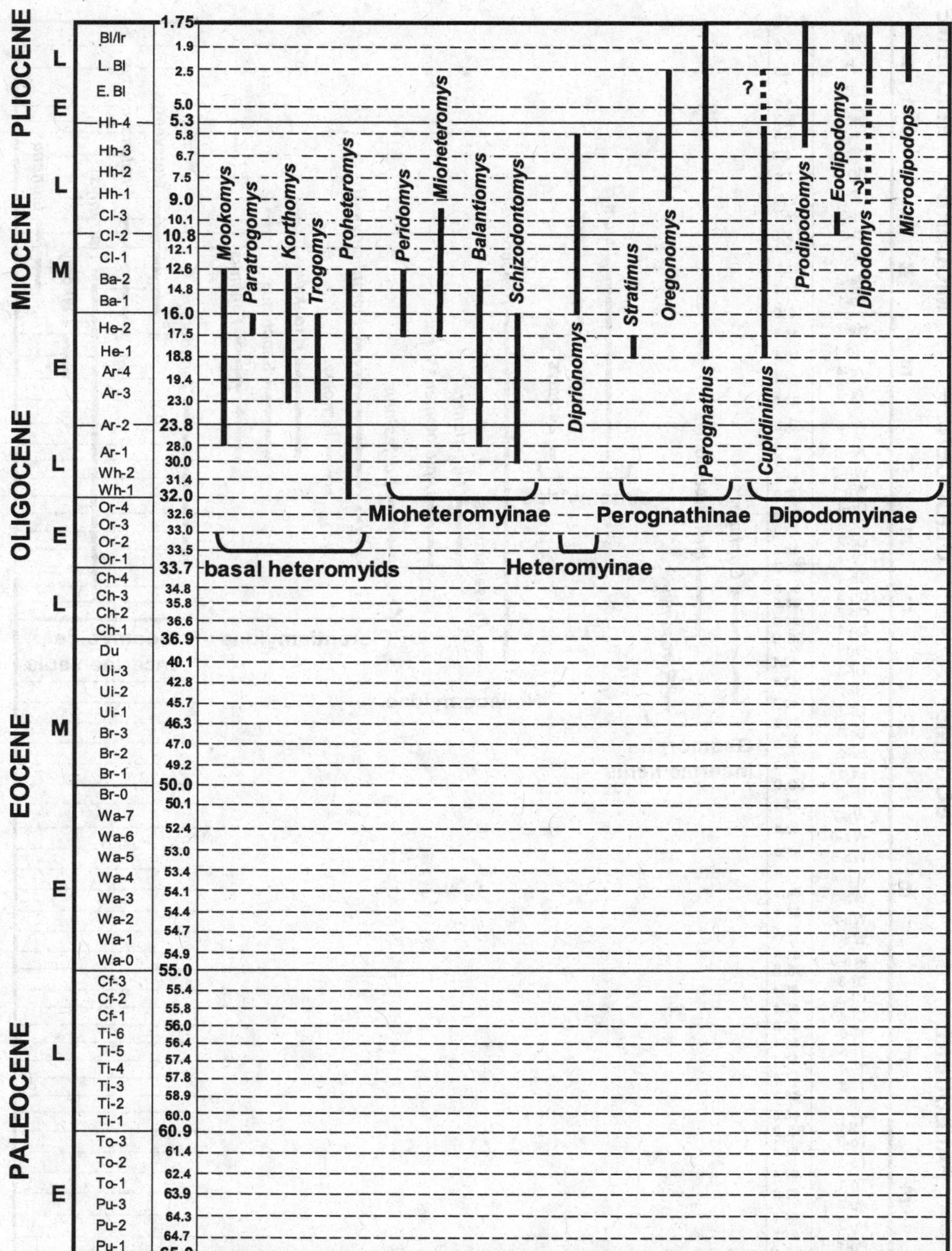

Figure 26.4. Temporal ranges of Geomorpha 2: Heteromyids.

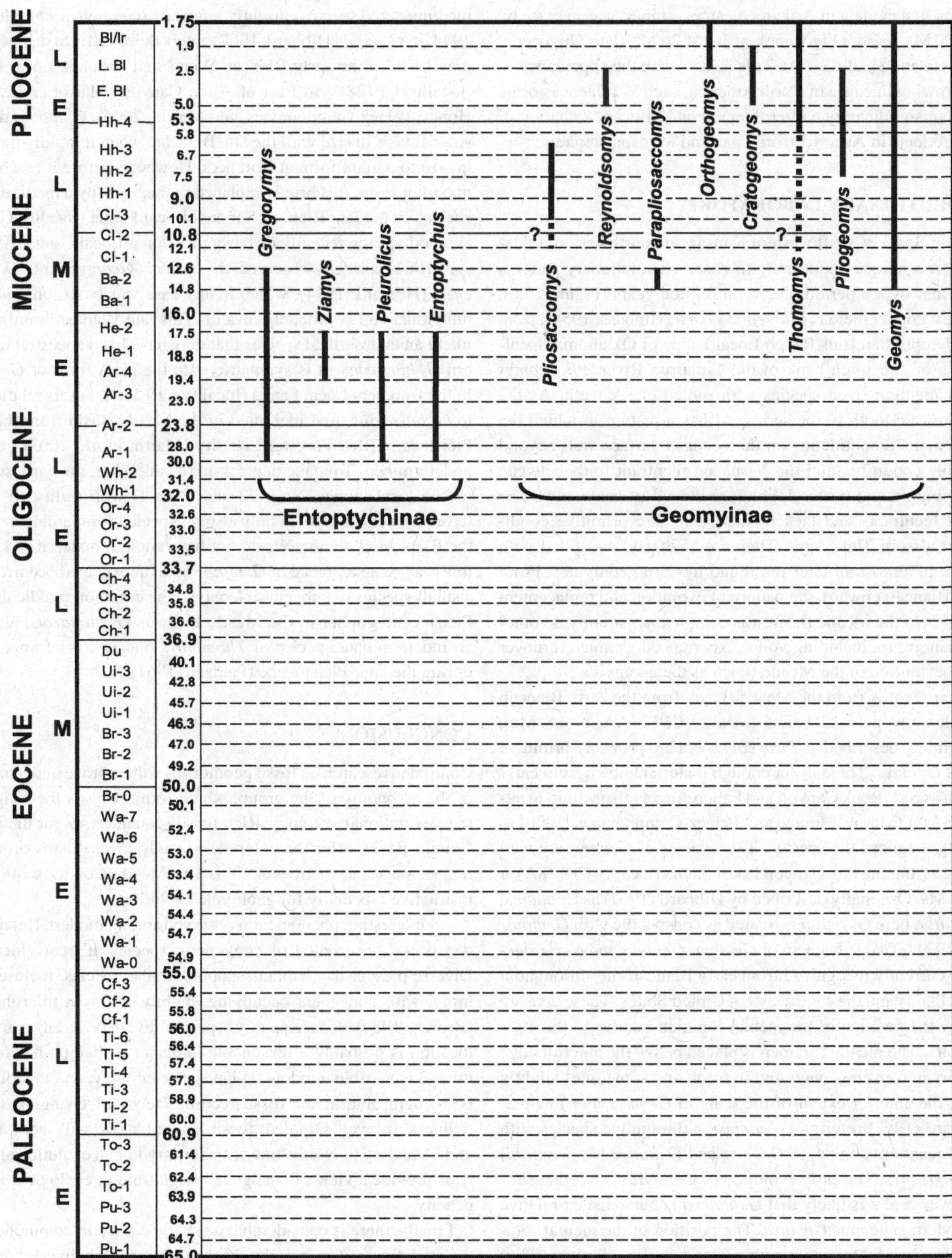

Figure 26.5. Temporal ranges of Geomorpha 3: Geomyids.

adaptation, first evident in Arikareen entoptychines, was in place by perhaps 30 Ma (early Oligocene), at latest 26 Ma (late Oligocene; Cook, Lessa and Hadly, 2000). This is later than the appearance of other burrowing lineages in North America, such as palaeocastorine beavers, but significantly earlier than the late Miocene radiation of fossorial rodents in Asia (Rhizomyinae and Myospalacinae).

AN EVOLUTIONARY LABORATORY

The Meade Basin of southwestern Kansas and northwestern Oklahoma offers a unique opportunity to observe evolutionary patterns in geomyines over a period of several million years (Figure 26.6). The earliest gopher there is *Pliogeomys buisi* (Hibbard, 1954), from the late Hemphillian Buis Ranch Local Fauna of Oklahoma (locality SP4B), on the south bank of the Cimarron River. *Pliogeomys buisi* is a medium-sized species with rooted cheek teeth. As the most diagnostic tooth in the series, p4 has a pattern in which the enamel remained continuous on the occlusal surface well beyond the juvenile condition, and the V-shaped reentrant folds between the anterolophid and posterolophid are devoid of cementum. The V-shaped reentrants and lack of cementum are primitive conditions seen also in *Thomomys*. They are U-shaped and filled with cementum in almost all later fossil and modern geomyines. From the early Blancan onward, the pattern of evolution and replacement of geomyids in the Meade Basin was complex; apparently no other mammalian group, including voles, experienced as much turnover and microevolution in the Meade Basin as the geomyids.

The next records from the Meade Basin, from the early Blancan Saw Rock Canyon Local Fauna (locality CP128AA, *c.* 4.8 Ma), were originally described as *Pliogeomys* but are probably primitive species of *Geomys*. There is not enough material known from early sites, such as Saw Rock Canyon and Fallen Angel (also within locality CP128AA) (Martin, Honey, and Peláez-Campomanes, 2000) to adequately diagnose the species. Two species of *Geomys* are well represented from the Fox Canyon Local Fauna (locality CP128A) at about 4.3 Ma. Originally described by Hibbard (1967) as *G. adamsi* and *G. smithi*, here *G. smithi* is treated as conspecific with *G. minor* (Gidley, 1922). Two subgenera of *Geomys*, *Geomys* and *Nerterogeomys*, are currently recognized from early Blancan sites throughout the Great Plains and the southwestern United States. These taxa are defined by the position of the mental foramen relative to the masseteric scars. The mental foramen is placed below the anterior edge of the scars in *Nerterogeomys* and in front, and sometimes slightly dorsal to, the anterior extent of the scars in *Geomys*. For this reason Hibbard (1967) referred *G. adamsi*, a diminutive species with V-shaped reentrants on p4, to *Geomys* and *G. minor* (= *G. smithi*) to *Nerterogeomys*. The early evolution of geomyids is not that simple, however, and it is likely that *G. adamsi* is but a distant relative of the modern subgenus *Geomys*. The position of the mental foramen, while a helpful taxonomic character, may have limited utility for higher-level taxonomy in *Geomys*. *G. adamsi* is also tentatively identified in the early Blancan Pipe Creek Sinkhole Local Fauna (locality CE2) from Indiana (Martin, Goodwin, and Farlow, 2002).

Throughout the Blancan, over a period of about two million years, *G. minor* expresses a dwarfing trend. During the early Blancan, a medium-sized species currently referred to *Geomys quinni* McGrew, 1944 (= *G. jacobi* Hibbard, 1967) enters the area. Its first appearance may be from the early Blancan Wiens and Vasquez Local Faunas (locality CP128B) in East of Alien Canyon (Martin *et al.*, 2002; Honey, Peláez-Campomanes, and Martin, 2005). *G. quinni* demonstrates stasis in size until the late Blancan, when it rapidly increases in size to a giant form at Borchers. However, before it reaches this maximum size, it is briefly replaced in the basin by a medium-sized species in the late Blancan Sanders Local Fauna (locality CP130), and perhaps the recently discovered Paloma Local Fauna. As Paulson (1961) correctly observed, the Sanders *Geomys* is not *G. tobinensis* (Hibbard, 1944), which in any case was based on a juvenile individual and is probably invalid. This late Blancan immigrant is likely an undescribed species that may have been ancestral to modern *G. bursarius*. It is sympatric with the giant form of *G. quinni* in the Borchers Local Fauna (locality CP132B), but its relationship to *G. bursarius*, first identified in the Meade Basin from the early Pleistocene Rick Forester Locality (Martin *et al.*, 2003), remains undetermined. To make matters more confusing, the earliest Pleistocene local faunas, Nash 72 and Short Haul (locality CP132C), have produced molars of only a small species comparable in size to the Blancan *G. minor*. We do not have enough material to know if this is a last appearance of *G. minor* or an unexpected occurrence of a small species of subgenus *Geomys*. The common middle and late Pleistocene gopher in the Meade Basin is *G. bursarius*, although an indeterminate species of *Thomomys* makes a brief appearance during the later Pleistocene (Paulson, 1961).

CONCLUSION

Continued research on fossil geomorphs will improve understanding of the taxonomy of the group, which forms a basis for insights to their evolutionary biology. Rich fossil sequences, as for the Meade County Basin, offer the potential to study mechanisms of species replacement and microevolution. The Meade County sequence is instructive especially for geomyine evolution.

An interesting phenomenon observed among modern Heteromyidae is the occurrence of congeneric species at many locations. Diverse present-day habitats support similar species in close proximity, with congeners occupying slightly different microhabitats (Brown, 1984). Consequently, species diversity at any particular location is probably a reasonable indicator of habitat heterogeneity and a possible window on landscape ecology, and this ought to be recognizable in the fossil record. Trends of changing species richness in well-sampled fossil assemblages will complement understanding of North American Tertiary faunal evolution and perhaps provide a view of changes through time in ecological heterogeneity.

Finally, there is considerable work to be done in comprehending the evolution of Geomorpha. Molecular studies help considerably, from recognizing relationship with other rodent groups to mapping interrelationships of living genera. However, the pattern of evolution over 40 million years is highly complex and undeduced from extant members. Living families Heteromyidae and Geomyidae, as diagnosed morphologically, do not extend to the origin of the

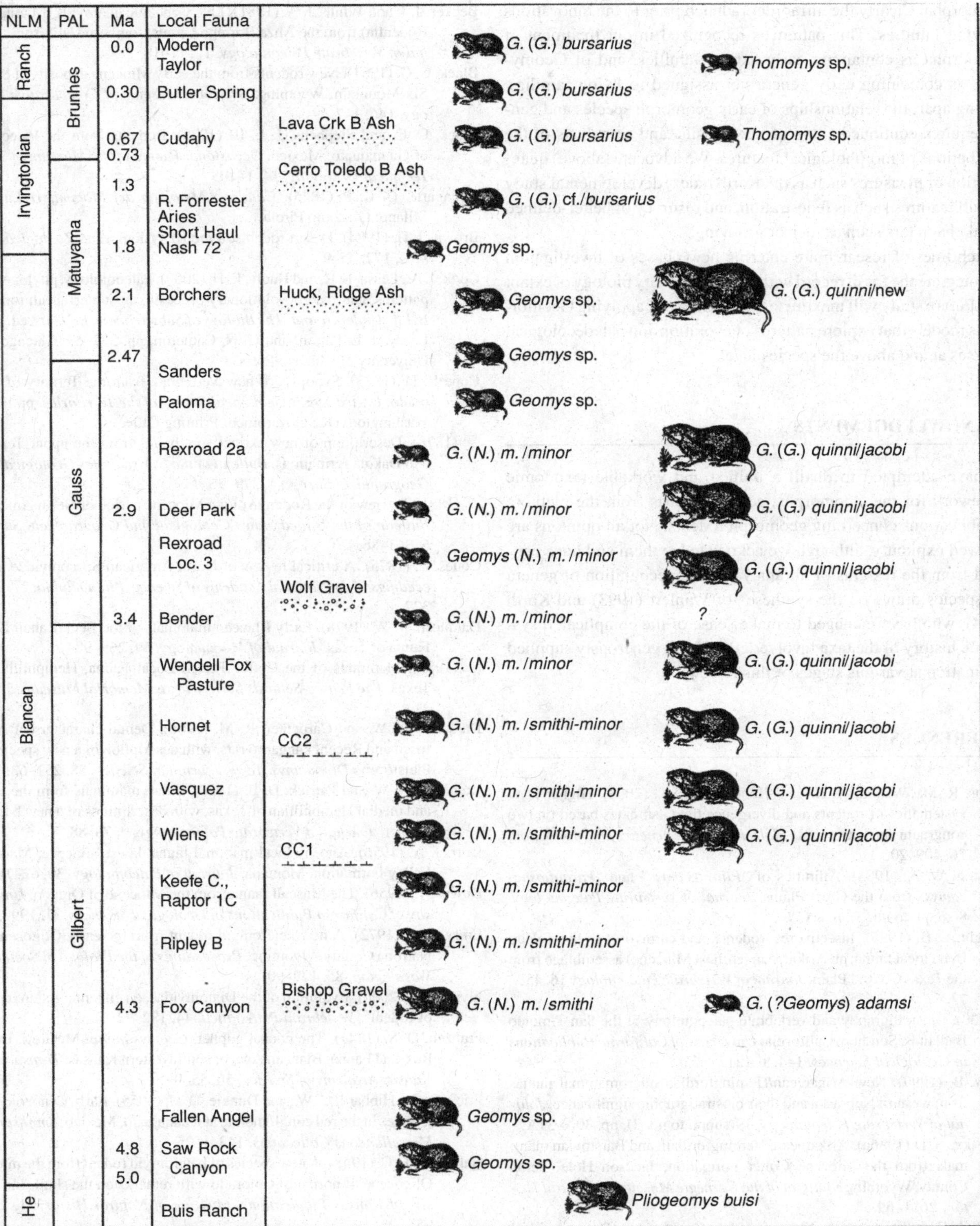

Figure 26.6. Preliminary replacement and evolutionary scenario for geomyids in the Meade Basin of southwestern Kansas and northwestern Oklahoma. *N.*, *Nerterogeomys*; He, Hemphillian; Keefe C., Keefe Canyon; R. Forrester, Rick Forrester; Lava Crk, Lava Creek; CC1 and CC2, calcium carbonate marker intervals, Bishop and Wolf gravels, informal sand and gravel units replacing older units such as the Angell and Meade gravels; Huck, Huckleberry; NLM, North American Land-Mammal Age; PAL, paleomagnetic stratigraphy. Sizes of gophers reflect preliminary statistical analyses, and may change with current studies. This summary is a contribution from the Meade Basin Rodent Project supported by NSF grant EAR-0207582.

Geomorpha. Clearly the infraorder radiated prior to the innovations of living families. This pattern is recognized in our treatment of Geomorpha as containing early extinct families, and of Geomyoidea as containing early genera not assigned to living families. Teasing apart the relationships of early geomorph species and genera requires continued study of new fossils, and reanalysis of the distributions of morphological features. We advocate (above) quantification of measures such as the Korth index, developmental study of skull features such as fenestration, and resurvey of better-defined dental characters such as incisor grooving.

Such lines of research are entering new phases of investigation that integrate the fossil record with the evolutionary biology of extant taxa. Future study will maximize the potential for applying Geomorpha as models that explore patterns of evolution and paleoecological analyses at and above the species level.

ACKNOWLEDGEMENTS

We have attempted to distill a unified and workable taxonomic framework for the Geomorpha, selecting ideas from the plethora of publications concerning geomorph rodents. Not all opinions are followed explicitly, although we acknowledge them and have benefited from the insights of all analyses. Our recognition of genera and species draws on the syntheses of Wahlert (1993) and Korth (1994), who have managed to make sense of the complicated systematic history of the taxa involved. Bill Korth generously supplied information at various stages of this review.

REFERENCES

Adkins, R. M., Walton, A. H., and Honeycutt, R. L. (2002). Higher level systematics of rodents and divergence time estimates based on two congruent nuclear genes. *Molecular Phylogenetics and Evolution*, 26, 409–20.

Akersten, W. A. (1988). Affinities of "*Pliosaccomys*" and "*Parapliosaccomys*" from the Great Plains. *Journal of Vertebrate Paleontology*, 8(suppl. to no. 3), p. 8A.

Albright, L. B. (1996). Insectivores, rodents, and carnivores of the Toledo Bend local fauna: an Arikareean (earliest Miocene) assemblage from the Texas Coastal Plain. *Journal of Vertebrate Paleontology* 16, 458–73.

(2000). Biostratigraphy and vertebrate paleontology of the San Timoteo Badlands, Southern California. *University of California Publications in Geological Sciences*, 144, 1–121.

Bailey, B. (1999). New Arikareean/Hemingfordian micromammal faunas from western Nebraska and their biostratigraphic significance. *Journal of Vertebrate Paleontology*, 19(suppl. to no. 3), pp. 30A–31A.

Barnosky, A. D.(1986a). Arikareean, Hemingfordian, and Barstovian mammals from the Miocene Colter Formation, Jackson Hole, Teton County, Wyoming. *Bulletin of the Carnegie Museum of Natural History* 26, 1–69.

(1986b). New species of the Miocene rodent *Cupidinimus* (Heteromyidae) and some evolutionary relationships within the genus. *Journal of Vertebrate Paleontology*, 6, 46–64.

Baskin, J. A. (1979). Small mammals of the Hemphillian age White Cone local fauna, northeastern Arizona. *Journal of Paleontology*, 53, 695–708.

Becker, J. J. and White, J. A. (1981). Late Cenozoic geomyids (Mammalia, Rodentia) from the Anza Borrego Desert, southern California. *Journal of Vertebrate Paleontology*, 1, 211–18.

Black, C. C. (1961). New rodents from the early Miocene deposits of Sixty-Six Mountain, Wyoming. *Breviora, Museum of Comparative Zoology*, 146, 1–7.

Black, C. C. and Stephens, J. J., III (1973). Rodents from the Paleogene of Guanajuato, Mexico. *Occasional Papers of the Museum of Texas Technical University*, 14, 1–10.

Bonaparte, C. L. P. (1845). *Catologo Metodico dei Mammiferi Europe*. Milano: Giacomo Pirola.

Brown, J. H. (1984). Desert rodents: a model system. *Acta Zoologica Fennica*, 172, 45–9.

Cook, J. A., Lessa, E. P., and Hadly, E. H. (2000). Paleontology, phylogenetic patterns, and macroevolutionary processes in subterranean rodents. In *Life Underground: The Biology of Subterranean Rodents*, ed. E. A. Lacey, J. L. Patton, and G. N. Cameron, pp. 332–69. Chicago, IL: University of Chicago Press.

Cope E. D. (1873). Synopsis of new Vertebrata from the Tertiary of Colorado. *United States Geological Survey of the Territories*, pp. 1–19. Washington, DC: Government Printing Office.

(1878). Description of new extinct vertebrates from the upper Tertiary and Dakota formations. *Bulletin of the United States Geological and Geographical Survey*, 4, 379–96.

(1881). Review of the Rodentia of the Miocene period of North America. *Bulletin of the United States Geological and Geographical Survey*, 6, 361–86.

Coues, E. (1875). A critical review of North American Saccomyidae. *Proceedings of the National Academy of Science, Philadelphia*, 27, 272–327.

Dalquest, W. W. (1978). Early Blancan mammals of the Beck Ranch Local Fauna of Texas. *Journal of Mammalogy*, 59, 269–98.

(1983). Mammals of the Coffee Ranch Local Fauna, Hemphillian of Texas. *The Pearce-Sellards Series, Texas Memorial Museum*, 38, 1–41.

Dalquest, W. W. and Carpenter, R. M. (1986). Dental characters of some fossil and Recent kangaroo rats, with description of a new species of Pleistocene *Dipodomys*. *Texas Journal of Science*, 38, 251–63.

Dalquest, W. W. and Patrick, D. B. (1989). Small mammals from the early and medial Hemphillian of Texas, with descriptions of a new bat and gopher. *Journal of Vertebrate Paleontology*, 9, 78–88.

Dorr, J. A. (1956). Anceny local mammal fauna, latest Miocene, Madison Valley Formation, Montana. *Journal of Paleontology*, 30, 62–74.

Downs, T. (1956). The Mascall fauna from the Miocene of Oregon. *University of California Publications in Geological Sciences*, 31, 199–354.

Emry, R. J. (1972). A new heteromyid rodent from the early Oligocene of Natrona County, Wyoming. *Proceedings of the Biological Society of Washington*, 85, 179–90.

Flynn, L. J. (2006). Evolution of the Diatomyidae, an endemic Asian family of rodents. *Vertebrata PalAsiatica*, 44, 182–92.

Franzen, D. S. (1947). The pocket gopher, *Geomys quinni* McGrew, in the Rexroad Fauna, Blancan Age, of southwestern Kansas. *Transactions Kansas Academy of Science*, 50, 55–9.

Fries, C., Jr., Hibbard, C. W., and Dunkle, D. H. (1955). Early Cenozoic vertebrates in the red conglomerate at Guanajuato, Mexico. *Smithsonian Miscellaneous Collections*, 123, 1–25.

Galbreath, E. C. (1948). A new species of heteromyid rodent from the middle Oligocene of northeast Colorado with remarks on the skull. *University of Kansas Publications, Museum of Natural History*, 1, 285–95.

Gawne, C. E. (1975). Rodents from the Zia Sand Miocene of New Mexico. *American Museum Novitates*, 2586, 1–25.

Gazin, L. (1930). A Tertiary vertebrate fauna from the upper Cuyama drainage basin, California. *Carnegie Institution of Washington, Contributions in Paleontology*, 404, 55–76.

(1932). A new Miocene mammalian fauna from southeastern Oregon. *Carnegie Institution of Washington, Contributions in Paleontology*, 418, 39–86.

(1942). The late Cenozoic vertebrate faunas from the San Pedro Valley, Arizona. *Proceedings of the United States National Museum*, 92, 475–518.

Gidley, J. W. (1922). Preliminary report on fossil vertebrates of the San Pedro Valley, Arizona, with descriptions of new species of rodents and lagomorphs. *United States Geological Survey Professional Paper*, 131, 119–31.

Gray, J. E. (1841). A new genus of Mexican glirine Mammalia. *Annals and Magazine of Natural History, Series 1*, 7, 521–2.

(1868). Synopsis of the species of Saccomyinae, or pouched mice, in the collection of the British Museum. *Proceedings of the Zoological Society of London*, 1868, 199–206.

Green, M. (2002). Variation in Black Bear Quarry II (Hemingfordian, South Dakota) *Proheteromys* (Geomyoidea: Mammalia) with comments on *Mookomys*. *Institute for Tertiary–Quaternary Studies, TER-QUA, Symposium Series*, 3, 187–94.

Green, M. and Bjork, P. R. (1980). On the genus *Dikkomys* (Geomyoidea, Mammalia). *Palaeovertebrata, Mémoire Jubilaire, Réné Lavocat*, 1, 343–53.

Hall, E. R. (1930). Rodents and lagomorphs from the later Tertiary of Fish Lake Valley, Nevada. *University of California Publications in Geological Sciences*, 19, 295–312.

Harris, J. M. and Wood, A. E. (1969). A new genus of eomyid rodent from the Oligocene Ash Spring local fauna of Trans-Pecos Texas. *The Pearce-Sellards Series, Texas Memorial Museum*, 14, 1–7.

Hatt, R. T. (1932). The vertebral column of ricochetal rodents. *Bulletin of American Museum of Natural History* 63, 599–738.

Hay, O. P. (1927). The Pleistocene of the western region of North America and its vertebrate animals. *Carnegie Institution of Washington, Contributions in Paleontology*, 322B, 1–346.

Hibbard, C. W. (1937). Additional fauna of Edson Quarry of the Middle Pliocene of Kansas. *The American Midland Naturalist*, 18, 460–4.

(1939). Notes on additional fauna of Edson Quarry of the middle Pliocene of Kansas. *Transactions of the Kansas Academy of Science*, 42, 457–62.

(1941). New mammals from the Rexroad fauna, upper Pliocene of Kansas. *The American Midland Naturalist*, 26, 337–68.

(1943). *Etadonomys*, a new Pleistocene heteromyid rodent, and notes on other Kansas mammals. *Transactions of the Kansas Academy of Science*, 46, 185–91.

(1944). Stratigraphy and vertebrate paleontology of Pleistocene deposits of southwestern Kansas. *Bulletin of the Geological Society of America*, 55, 707–54.

(1949). Pliocene Saw Rock Canyon fauna in Kansas. *Contributions from the Museum of Paleontology, University of Michigan*, 7, 91–105.

(1950). Mammals of the Rexroad Formation from Fox Canyon, Kansas. *Contributions from the Museum of Paleontology, University of Michigan*, 8, 113–92.

(1954). A new Pliocene vertebrate fauna from Oklahoma. *Papers of the Michigan Academy of Science, Arts, and Letters*, 39, 339–59.

(1962). Two new rodents from the early Pleistocene of Idaho. *Journal of Mammalogy*, 43, 482–5.

(1967). New rodents from the late Cenozoic of Kansas. *Papers of the Michigan Academy of Science, Arts, and Letters*, 52, 115–31.

Hibbard, C. W. and Keenmon, K. A. (1950). New evidence of the lower Miocene age of the Blacktail Deer Creek Formation in Montana. *Contributions from the Museum of Paleontology, University of Michigan*, 8, 193–204.

Hibbard, C. W. and Riggs, E. S. (1949). Upper Pliocene vertebrates from Keefe Canyon, Meade County, Kansas. *Bulletin of the Geological Society of America*, 60, 829–60.

Hibbard, C. W. and Wilson, J. A. (1950). A new rodent from subsurface stratus in Bee County, Texas. *Journal of Paleontology*, 24, 621–3.

Honey, J. G., Peláez-Campomanes, P., and Martin, R. A. (2005). Stratigraphic framework of Early Pliocene fossil localities along the North bank of the Cimarron River, Meade County, Kansas. *Ameghiniana* 42, 461–72.

Huchon, D., Madsen, O., Sibbald, M. J. J. B., *et al.* (2002). Rodent phylogeny and a timescale for the evolution of Glires: evidence from an extensive taxon sampling using three nuclear genes. *Molecular Biology and Evolution*, 19, 1053–65.

Jacobs, L. L. (1977). Rodents of the Hemphillian age Redington local fauna, San Pedro Valley, Arizona. *Journal of Paleontology*, 51, 505–19.

James, G. T. (1963). Paleontology and nonmarine stratigraphy of the Cuyama Valley Badlands, California. *University of California Publications in Geological Sciences*, 45, 1–145.

Kalthoff, D. (2000). Die Schmelzmikrostruktur in den Incisiven der hamsterartigen Nagetiere und anderer Myomorpha (Rodentia, Mammalia). *Palaeontographica A*, 259, 1–193.

Kellogg, L. (1910). Rodent fauna of the late Tertiary beds at Virgin Valley and Thousand Creek, Nevada. *University of California Publications in Geological Sciences*, 5, 421–37.

Kelly, T. S. (1992). New Uintan and Duchesnean (Middle and Late Eocene) rodents from the Sespe Formation, Simi Valley, California. *Bulletin of the Southern California Academy of Science*, 91, 97–120.

(1994). Two Pliocene (Blancan) vertebrate faunas from Douglas County, Nevada. *PaleoBios*, 16, 1–23.

Korth, W. W. (1979). Geomyoid rodents from the Valentine Formation of Knox County, Nebraska. *Annals of the Carnegie Museum*, 48, 287–310.

(1987). New rodents (Mammalia) from the late Barstovian (Miocene) Valentine Formation, Nebraska. *Journal of Paleontology*, 61, 1058–64.

(1989). Geomyoid rodents from the Orellan (middle Oligocene) of Nebraska. [In *Papers on Fossil Rodents in Honor of Albert Elmer Wood*, ed. C. C. Black and M. R. Dawson.] *Los Angeles: Science Series, Natural History Museum of Los Angeles County*, 33, 31–46.

(1992). Fossil small mammals from the Harrison Formation (Late Arikareean: earliest Miocene), Cherry County, Nebraska. *Annals of the Carnegie Museum*, 61, 69–131.

(1993a). The skull of *Hitonkala* (Florentiamyidae, Rodentia) and relationships within Geomyoidea. *Journal of Mammalogy*, 74, 168–74.

(1993b). A review of the Oligocene (Orellan and Arikareean) genus *Tenudomys* Rensberger (Geomyoidea: Rodentia). *Journal of Vertebrate Paleontology*, 13, 335–41.

(1994). *The Tertiary Record of Rodents in North America*. New York: Plenum Press.

(1995a). The skull and upper dentition of *Heliscomys senex* Wood (Heliscomyidae: Rodentia). *Journal of Paleontology*, 69, 191–4.

(1995b). A new heteromyid rodent (Mammalia) from the Barstovian (Miocene) of Saskatchewan. *Canadian Journal of Earth Science*, 32, 21–3.

(1995c). Rodents from the late Hemphillian (latest Miocene), Sioux County, Nebraska. *Transactions of the Nebraska Academy of Sciences*, 22, 87–92.

(1996a). Geomyoid rodents (Mammalia) from the Bijou Hills, South Dakota (Barstovian, Miocene). *Contributions to Geology, University of Wyoming*, 31, 49–55.

(1996b). A new species of *Pleurolicus* (Rodentia, Geomyidae) from the early Miocene (Arikareean) of Nebraska. *Journal of Vertebrate Paleontology*, 16, 781–4.

(1997). A new subfamily of primitive pocket mice (Rodentia, Heteromyidae) from the middle Tertiary of North America. *Paludicola*, 1, 33–66.

(1998). Cranial morphology of two Tertiary pocket mice, *Perognathus* and *Cupidinimus* (Rodentia, Heteromyidae). *Paludicola* 1, 132–42.

Korth, W. W. and Chaney, D. S. (1999). A new subfamily of geomyoid rodents (Mammalia) and a possible origin of Geomyidae. *Journal of Paleontology*, 73, 1191–200.

Korth, W. W. and Eaton, J. G. (2004). Rodents and a marsupial (Mammalia) from the Duchesnean (Eocene) Turtle Basin local fauna, Sevier Plateau, Utah. *Bulletin of the Carnegie Museum of Natural History*, 36, 109–19.

Korth, W. W. and Reynolds, R. E. (1994). A hypsodont gopher (Rodentia, Geomyidae) from the Clarendonian (Miocene) of California. *San Bernardino County Museum Association, Special Publication*, 94, 91–5.

Korth, W. W., Bailey, B. E., and Hunt, R. M., Jr. (1990). Geomyoid rodents from the early Hemingfordian (Miocene) of Nebraska. *Annals of the Carnegie Museum*, 59, 2–47.

Korth, W. W., Wahlert, J. H., and Emry, R. J. (1991). A new species of *Heliscomys* and recognition of the family Heliscomyidae (Geomyoidea: Rodentia). *Journal of Vertebrate Paleontology*, 11, 247–56.

Lillegraven, J. A. (1977). Small rodents (Mammalia) from Eocene deposits of San Diego County, California. *Bulletin of the American Museum of Natural History*, 158, 221–62.

Lindsay, E. H. (1972). Small mammal fossils from the Barstow Formation, California. *University of California Publications in Geological Sciences*, 93, 1–104.

(1974). The Hemingfordian mammal fauna of the Vedder Locality, Branch Canyon Formation, Santa Barbara County, California. Part II: Rodentia (Eomyidae and Heteromyidae). *Geobios*, 16, 1–20.

Lindsay, E. H. and Jacobs, L. L. (1985). Pliocene small mammal fossils from Chihuahua, Mexico. *Paleontologia Mexicana*, 51, 1–53.

Lindsay, E. H. and Reynolds, R. E. (2007). Heteromyid rodents from Miocene faunas of Southern California. [In *Geology and Vertebrate Paleontology of Western and Southern North America: Contributions in Honor of David P. Whistler*, ed. X. Wang and L. Barnes.] *Contributions in Science, Los Angeles County Museum of Natural History*, in press.

Macdonald, J. R. (1963). The Miocene Wounded Knee faunas of southwestern South Dakota. *Bulletin of the American Museum of Natural History*, 125, 141–238.

(1970). Review of the Miocene faunas from the Wounded Knee area of western South Dakota. *Bulletin of the Los Angeles County Museum of Natural History, Science*, 8, 1–82.

McGrew, P. O. (1941). Heteromyids from the Miocene and lower Oligocene. *Geological Series of the Field Museum of Natural History*, 8, 55–7.

(1944). An early Pleistocene (Blancan) fauna from Nebraska. *Geological Series of the Field Museum of Natural History*, 9, 33–66.

McKenna, M. C. and Bell, S. K. (1997). *Classification of Mammals Above the Species Level.* New York: Columbia University Press.

Martin, J. E. (1976). Small mammals from the Miocene Batesland Formation of South Dakota. *Contributions to Geology, University of Wyoming*, 14, 69–98.

(1984). A survey of Tertiary species of *Perognathus* (Perognathinae) and a description of a new genus of Heteromyinae. [In *Papers in Vertebrate Paleontology Honoring Robert Warren Wilson*, ed. R. M. Mengel.] *Carnegie Museum of Natural History, Special Publication*, 9, 90–121.

Martin, L. D. (1974). New rodents from the lower Miocene Gering Formation of western Nebraska. *Occasional Papers of the Museum of Natural History, University of Kansas*, 32, 1–12.

Martin, R. A., Honey, J. G., and Peláez-Campomanes, P. (2000). The Meade Basin rodent project: a progress report. *Paludicola*, 3, 1–32.

Martin, R. A., Goodwin, H. T., and Farlow, J. L. (2002). Late Neogene (Late Hemphillian) rodents from the Pipe Creek Sinkhole, Grant County. Indiana. *Journal of Vertebrate Paleontology*, 22, 137–51.

Martin, R. A., Honey, J. G., Peláez-Campomanes, P., *et al.* (2002). Blancan lagomorphs and rodents of the Deer Park assemblages, Meade County, Kansas. *Journal of Paleonotology*, 76, 1072–90.

Martin, R. A., Hurt, R. T., Honey, J. G., and Peláez-Campomanes, P. (2003). Late Pliocene and early Pleistocene rodents from the northern Borchers Badlands (Meade County, Kansas), with comments on the Blancan–Irvingtonian boundary in the Meade Basin. *Journal of Paleontology*, 77, 985–1001.

Matthew, W. D. (1907). A lower Miocene fauna from South Dakota. *Bulletin of the American Museum of Natural History*, 23, 169–219.

(1924). Third contribution to the Snake Creek fauna. *Bulletin of the American Museum of Natural History*, 50, 59–210.

Merriam, C. H. (1891). Description of a new genus and species of dwarf kangaroo rat from Nevada (*Microdipodops megacephalus*). *North American Fauna*, 5, 115–17.

(1895). Monographic revision of the pocket gophers, Family Geomyidae (exclusive of the species of *Thomomys*). *North American Fauna*, 8, 1–259.

Miller, G. S. and Gidley, J. W. (1918). Synopsis of the supergeneric groups of rodents. *Journal of the Washington Academy of Science*, 8, 431–48.

Munthe, J. (1977). A new species of *Gregorymys* (Rodentia, Geomyidae) from the Miocene of Colorado. *PaleoBios*, 26, 1–12.

(1988). Miocene mammals of the Split Rock area, Granit Mountains Basin, Central Wyoming. *University of California Publications in Geological Sciences*, 126, 1–136.

Nichols, R. (1976). Early Miocene mammals from the Lemhi Valley of Idaho. *Tebiwa*, 18, 9–47.

Nowak, R. M. and Paradiso, J. L. (1983). *Walker's Mammals of the World*, Vol. II, 4th edn. Baltimore, MD: Johns Hopkins University Press.

Paulson, G. R. (1961). The mammals of the Cudahy fauna. *Papers of the Michigan Academy of Science, Arts and Letters*, 46, 127–53.

Rafinesque, C. S. (1817). Description of seven new genera of North American quadrupeds. *American Monthly Magazine, New York*, 2, 44–6.

Reeder, W. G. (1960a). Two new rodent genera from the Oligocene Whiter River Formation. *Fieldiana, Geology*, 10, 511–24.

(1960b). A new rodent genus (Family Heteromyidae) from the Tick Canyon Formation of California. *Bulletin of the Southern California Academy of Sciences*, 59, 121–32.

Rensberger, J. M. (1971). Entoptychine pocket gophers (Mammalia, Geomyoidea) of the early Miocene John Day Formation, Oregon. *University of California Publications in Geological Sciences*, 90, 1–163.

(1973). Pleurolicine rodents (Geomyoidea) of the John Day Formation, Oregon. *University of California Publications in Geological Sciences*, 102, 1–95.

Richardson, J. (1828). Short characters of a few quadrupeds procured on Capt. Franklin's last expedition. *Zoological Journal*, 3, 516–20.

Ruez, D. R., Jr. (2001). Early Irvingtonian (latest Pliocene) rodents from Inglis 1C, Citrus County, Florida. *Journal of Vertebrate Paleontology*, 21, 53–71.

Ryan, J. M. (1989). Comparative myology and phylogenetic systematics of the Heteromyidae (Mammalia, Rodentia). *Miscellaneous Publications, Museum of Zoology, University of Michigan*, 176, 1–103.

Shotwell, J. A. (1956). Hemphillian mammalian assemblage from northeastern Oregon. *Bulletin of the Geological Society of America*, 67, 717–38.

(1967). Late Tertiary geomyoid rodents of Oregon. *Bulletin of the Museum of Natural History, University of Oregon*, 9, 1–51.

Skinner, M. F. and Hibbard, C. W. (1972). Pleistocene pre-glacial and glacial rocks and faunas of North-Central Nebraska. *Bulletin of the American Museum of Natural History*, 148, 1–148.

Sinclair, W. J. (1905). New or imperfectly known rodents and ungulates from the John Day series. *Bulletin of the University of California Department of Geology*, 4, 125–43.

Slaughter, B. H. (1981). A new genus of Geomyoid rodent from the Miocene of Texas and Panama. *Journal of Vertebrate Paleontology*, 1, 111–15.

Spradling, T. A., Brant, S. V., Hafner, M. S., and Dickerson, C. J. (2004). DNA data support a rapid radiation of pocket gopher genera (Rodentia: Geomyidae). *Journal of Mammalian Evolution*, 11, 105–25.

Stevens, M. S. (1977). Further study of the Castolon local fauna (early Miocene) Big Bend National Park, Texas. *The Pearce-Sellards Series, Texas Memorial Museum*, 28, 1–69.

Storer, J. E. (1970). New rodents and lagomorphs from the upper Miocene Wood Mountain Formation of southern Saskatchewan. *Canadian Journal of Earth Sciences*, 7, 1125–9.

(1973). The entoptychine geomyid *Lignimus* (Mammalia: Rodentia) from Kansas and Nebraska. *Canadian Journal of Earth Sciences*, 10, 72–83.

(1975). Tertiary mammals of Saskatchewan. Part III: The Miocene fauna. *Life Sciences Contributions, Royal Ontario Museum*, 103, 1–134.

Strain, W. S. (1966). Blancan mammalian fauna and Pleistocene formations, Hudspeth County, Texas. *Bulletin of the Texas Memorial Museum*, 10, 1–55.

Sutton, J. F. and Korth, W. W. (1995). Rodents (Mammalia) from the Barstovian (Miocene) Anceny local fauna, Montana. *Annals of the Carnegie Museum*, 64, 267–314.

Thaler, L. (1966). Les rongeurs fossiles du Bas-Languedoc dans leurs rapports avec l'histoire des faunes et la stratigraphie du Tertiare d'Europe. *Mémoires du Muséum National d'Histoire Naturelle, Nouvelle Série C*, 17, 1–295.

Thomas, O. (1892). Diagnosis of a new Mexican *Geomys*. *Annals and Magazine of Natural History, Series 6*, 10, 196.

(1893). On some larger species of *Geomys*. *Annals and Magazine of Natural History, Series 6*, 12, 269–73.

(1894). On two new Neotropical mammals. *Annals and Magazine of Natural History, Series 6*, 13, 436–9.

Tomida, Y. (1987). *Monograph 3: Small Mammal Fossils and Correlation of Continental Deposits, Safford and Duncan Basins, Arizona, USA*. Tokyo: National Science Museum.

Troxell, E. L. (1923). *Diplolophus*, a new genus of rodents. *American Journal of Science*, 5, 157–9.

True, F. W. (1888). Description of *Geomys personatus* and *Dipodomys compactus*, two new species of rodents from Padre Island, Texas. *Proceedings of United States National Museum*, 11, 159–60.

Voorhies, M. R. (1975). A new genus and species of fossil kangaroo rat and its burrow. *Journal of Mammalogy*, 56, 160–76.

Wahlert, J. H. (1976). *Jimomys labaughi*, a new geomyoid rodent from the early Barstovian of North America. *American Museum Novitates*, 2591, 1–6.

(1983). Relationships of the Florentiamyidae (Rodentia, Geomyoidea) based on cranial and dental morphology. *American Museum Novitates*, 2769, 1–23.

(1984). *Kirkomys*, a new florentiamyid (Rodentia, Geomyoidea) from the Whitneyan of Sioux County, Nebraska. *American Museum Novitates*, 2793, 1–8.

(1985). Skull morphology and relationships of geomyoid rodents. *American Museum Novitates*, 2812, 1–20.

(1991). The Harrymyinae, a new heteromyid subfamily (Rodentia, Geomorpha) based on cranial and dental morphology of *Harrymys* Munthe, 1988. *American Museum Novitates*, 3013, 1–23.

(1993). The fossil record. [In *Biology of the Heteromyidae*, ed. H. H. Genoways and J. H. Brown.] *American Society of Mammalogists Special Publication*, 10, 1–37.

Wahlert, J. H. and Souza, R. A. (1988). Skull morphology of *Gregorymys* and relationships of the Entoptychinae (Rodentia, Geomyidae). *American Museum Novitates*, 2922, 1–13.

Whistler, D. P. (1984). An early Hemingfordian (early Miocene) fossil vertebrate fauna from Boron, western Mojave Desert, California. *Contributions in Science, Los Angeles County Museum of Natural History*, 355, 1–36.

White, J. A. and Downs, T. (1961). A new *Geomys* from the Vallecito Creek Pleistocene of California, with notes on variation in Recent and fossil species. *Contributions in Science, Los Angeles County Museum of Natural History*, 42, 1–34.

Wied-Neuwied, Prinz Maximilian zu (1839). Ber einige Nager mit äusseren Backentaschen aus dem westlichen Nord-America. I. Uber ein paar neue Gattungen der Nagethiere mit äusseren Backentaschen. *Nova Acta Physico-Medica, Academiae Caesareae Leopoldino-Carolinae, Naturae Curiosoorum*, 19, 365–74.

Wilkins, K. T. (1984). Evolutionary trends in Florida Pleistocene pocket gophers (genus *Geomys*), with description of a new species. *Journal of Vertebrate Paleontology*, 3, 166–81.

Williams, D. F., Genoways, H. H., and Braun, J. K. (1993). Taxonomy. [In *Biology of the Heteromyidae*, ed. H. H. Genoways and J. H. Brown.] *American Society of Mammalogists, Special Publication*, 10, 38–196.

Wilson, R. W. (1933). A rodent fauna from the later Cenozoic beds of southwestern Idaho. *Carnegie Institution of Washington Publication*, 440, 117–35.

(1936). A Pliocene rodent fauna from Smith Valley, Nevada. *Carnegie Institution of Washington Publication*, 473, 17–34.

(1939). Rodents and lagomorphs of the late Teriary Avawatz fauna, California. *Carnegie Institution of Washington Publication*, 514, 31–8.

(1940). Two new Eocene rodents from California. *Carnegie Institution of Washington Publication*, 514, 85–95.

(1949). Early Tertiary rodents of North America. *Carnegie Institution of Washington Publication*, 584, 67–164.

(1960). Early Miocene rodents and insectivores from northeastern Colorado. *University of Kansas Paleontological Contributions, Vertebrata*, 7, 1–92.

Wood, A. E. (1931). Phylogeny of the heteromyid rodents. *American Museum Novitates*, 501, 1–19.

(1932). New heteromyid rodents from the Miocene of Florida. *Florida State Geological Survey Bulletin*, 10, 43–51.

(1935a). Evolution and relationship of the heteromyid rodents. *Annals of the Carnegie Museum*, 24, 73–262.

(1935b). Two new rodents from the John Day Miocene. *American Journal of Science*, 30, 368–72.

(1936a). A new subfamily of heteromyid rodents from the Miocene of western United States. *American Journal of Science*, 31, 41–9.

(1936b). Geomyoid rodents from the middle Tertiary. *American Museum Novitates*, 866, 1–31.

(1936c). Fossil heteromyid rodents in the collections of the University of California. *American Journal of Science*, 32, 112–19.

(1937). Additional material from the Tertiary of the Cuyama Basin of California. *American Journal of Science*, 33, 29–43.

(1950). A new geomyid rodent from the Miocene of Montana. *Annals of the Carnegie Museum*, 31, 335–8.

(1974). Early Tertiary vertebrate faunas, Vieja Group, Trans-Pecos Texas: Rodentia. *Bulletin of the Texas Memorial Museum*, 21, 1–112.

(1980). The Oligocene rodents of North America. *Transactions of the American Philosophical Society*, 70, 1–68.

Zakrzewski, R J. (1969). The rodents from the Hagerman local fauna, upper Pliocene of Idaho. *Contributions from the Museum of Paleontology, University of Michigan*, 23, 1–36.

(1970). Notes on kangaroo rats from the Pliocene of southwestern Kansas, with the description of a new species. *Journal of Paleontology*, 44, 474–7.

(1981). Kangaroo rats from the Borchers local fauna, Blancan, Meade County, Kansas. *Transactions of the Kansas Academy of Sciences*, 84, 78–88.

27 Cricetidae

EVERETT H. LINDSAY
University of Arizona, Tucson, AZ, USA

INTRODUCTION

The Cricetidae is a very successful group of rodents, having established a record in North America from the late Eocene to the Recent. Zoologists generally recognize 8 genera and 35 species of Cricetidae (plus 9 genera and 30 species of Arvicolidae, derived from Cricetidae; see Chapter 28) presently living in North America (Burt and Grossenheider, 1976). If one were to include both North and Central America (e.g., using the data of Hall and Kelson [1959]), those numbers swell to 22 genera and 188 species currently inhabiting these areas. These expanded numbers reflect a dynamic radiation of Cricetidae from North America into the Neotropics during the Pliocene.

During the late Cenozoic (i.e., Blancan, Irvingtonian, and Rancholabrean NALMAs), there were 32 genera and 134 species of cricetid and arvicolid rodents recorded from North America (Kurtén and Anderson, 1980). Within that late Cenozoic group, 12 genera and 98 species are either extinct or no longer inhabit North America, reflecting a significant taxonomic turnover during the last 5 Ma.

Cricetid rodents are not known from South America prior to the Pliocene (i.e., Montehermosan SALMA). It is difficult to obtain consistent counts of cricetid species living in South America: Webb and Marshall (1982) listed 47 genera of cricetids, Marshall (1988) listed 54 and Marshall and Cifelli (1990) recognized 64. In spite of this uncertainty, there is little question that cricetid rodents have undergone an explosive radiation in Central and South America during the last 5 Ma.

DEFINING FEATURES OF THE FAMILY CRICETIDAE

The Cricetidae are characteristically small rodents whose adult length, excluding the tail, is usually between 100 and 250 mm. Groups within the North American Cricetidae are well adapted to xeric, boreal, aquatic and semi-fossorial environments. Most of the cricetids are herbivorous, although the common North America genus *Onychomys* is strongly insectivorous. Among cricetid groups outside North America, the Myospalacinae of Asia are fossorial, and the Ichthyomyinae of Central and South America are semi-aquatic, feeding mostly on aquatic arthropods and fishes (Voss, 1988).

CRANIAL

Perhaps the most distinctive feature of the cricetid cranium is the transition from a hystricomorphous to a myomorphous zygoma. The earliest known cricetid is *Pappocricetodon*, reported from Asia during the middle and late Eocene (Tong, 1992; Wang and Dawson, 1994); it is hystricomorphous. No suitable ancestor for cricetids is presently known from Asia, although primitive dipodids (e.g., *Sinosminthus, Heosminthus*, and *Allosminthus*) with dental morphology close to that of *Pappocricetodon* are recorded from late Eocene of Asia (Wang, 1985). Dipodids are all hystricomorphous; they differ from cricetids in lacking a prominent anterocone on M1, in addition to having a peglike P4.

Other possible ancestors to North American cricetids (and likely immigrants from Asia) include primitive North American Dipodidae such as *Armintomys*, *Elymys*, *Simimys*, *Nonomys*, and *Simiacritomys*. *Armintomys* is known from Bridgerian NALMA (early Eocene) deposits of Wyoming (Dawson, Kristhalka, and Stucky, 1990), and *Elymys* from Bridgerian (early Eocene) deposits of Nevada (Emry and Korth, 1989; see also Chapter 24). *Armintomys* has an hystricomorphous zygoma but it retains P3, a tooth that has been lost in all known Cricetidae. *Elymys* has lost its P3 but retains a reduced P4 (like the Asian cricetid *Pappocricetodon* and the North American dipodid *Simimys*); the zygoma of *Elymys* is not known. The transition from hystricomorphy to myomorphy in cricetids is documented in Europe during the Oligocene (Vianey-Liaud, 1985). The earliest cricetid skull known in North America, *Eoeumys* of Martin (1980)

Evolution of Tertiary Mammals of North America, Vol. 2. ed. C. M. Janis, G. F. Gunnell, and M. D. Uhen. Published by Cambridge University Press.

from the Chadronian (= late Eocene), had a "transitional" myomorphous zygoma, with the ventral infraorbital foramen weakly constricted and the zygomatic plate not fully vertical. Martin (1980, his Fig. 25) illustrated the zygoma of many early North American cricetids, showing much variation in ventral closure of the infraorbital canal; within this group of illustrated cricetids he considered *Leidymys* (and probably *Cricetops*) hystricomorphous but apparently considered all the other genera myomorphous.

Many of the skulls of early North American cricetids illustrated by Martin (1980) should be evaluated relative to hystricomorphy. *Coloradoeumys galbreathi* (his Fig. 6), *Eoeumys vetus* (his Fig. 9), *Leidymys lockingtonianus* (his Fig. 15), and *Geringia magregori* (his Fig. 19) appear to be subhystricomorphous (Martin, 1980); only *Eumys elegans* (his Figs. 2 and 25H) are weakly myomorphous. This was noted by Martin (1980, p. 39) but his statement "The other tendency is for the [infraorbital] foramen to enlarge until it is essentially hystricomorphous as in *Leidymys*" suggests that Martin believed hystricomorphy in cricetids developed from myomorphy. This is contrary to all of the evidence generated since the mid 1980s.

The middle and late Eocene North American dipodid rodents *Simimys* and *Nonomys* have often been considered Cricetidae; they are now placed in Dipodidae (see Chapter 24). Both genera are clearly hystricomorphous and the dental morphology of *Simimys* is very similar to that of *Pappocricetodon* and other early cricetids. Both *Simimys* and *Nonomys* had a neurovascular foramen ventral to the infraorbital foramen, which is known in some fossil and recent dipodids but is unknown in any cricetids. Walsh (1997) showed that some specimens securely identified as species of *Simimys* had a single-rooted P4, like that of *Pappocricetodon*. In the only previously known skull of *Simimys* (LACM[CIT] 3529), the P4 is absent (Wilson, 1949a).

Kelly (1992) described *Simiacritomys whistleri*, as a ?zapodid from late middle Eocene (Duchnesnean) deposits of southern California. This genus is also placed in the Dipodidae; it has well-developed cross-lophs in its cheek teeth and a small anterocone on M1, suggesting a relationship with *Simimys*; see Chapter 23).

Other distinctive cranial features of Cricetidae are a smooth, hour-glass-shaped interorbital area; interparietal present, sometimes large; jugal reduced; incisive foramina generally long, projecting near to level of first molar in derived genera; auditory bulla inflated, usually small; the mandible is slender, usually with a prominent or distinct masseteric crest (Figure 27.1).

DENTAL

The most characteristic feature of the cricetid dentition (which is shared by all but the most primitive of the Muroidea) is the cheek tooth formula, restricted to three molars on each side, upper and lower dentitions. Only *Pappocricetodon* (along with *Simimys* and *Elymys*, discussed above) retain a reduced single-rooted P4, as seen in the Zapodidae. In primitive genera, the third molar is large, approaching the size of the first molar. In some derived genera, the posterior molar is reduced or may even be absent. The anterior molar always has an anterior cusp (anterocone or anteroconid), which may be bilobed. Middle and posterior upper molars lack an anterocone and have a prominent paracone and protocone. The posterior upper molars have reduced metacone and hypocone, either of which may be indistinct or absent, especially in derived genera. The middle and posterior lower molars lack an anteroconid and have a prominent metaconid and protoconid, and usually a smaller hypoconid. In addition, the posterior molar usually has a reduced or indistinct entoconid, which may be absent, especially in derived genera. All of these cusps are usually joined by narrow arms or lophs that are directed toward the midline of the tooth, forming a "wall" or longitudinal mure.

Primitive cricetids have three prominent roots in the upper molars and two prominent roots in the lower molars. Accessory rootlets, which may become prominent roots in more derived genera, occur in sigmodontine cricetids. Many of the South American cricetids with large, rooted cheek teeth may have well-developed accessory roots.

When crown height is increased, accessory cuspules and lophs are reduced and a single loph joins alternating cusps. This results in the expression of cusps as tall, alternating prisms; hence, the term *prismatic* or *microtine* teeth. Primitively, prismatic teeth are rooted, the base of the crown is relatively straight, and cementum is absent from the narrow medial reentrants where prisms are joined. As the height of crown increases, the roots are lost, and the base of the crown becomes irregular, ultimately resulting in absence of enamel (e.g., dentinal tracts) with dentine exposed on the labial and lingual basal sides of teeth. Cementum penetrates into the alveolus and also develops in reentrants to help to stabilize the tooth. The resulting cheek teeth are ever-growing and rootless, comparable in this respect to rodent incisors.

POSTCRANIAL

Postcranial morphology of cricetids is highly variable, as expected from the broad range of habitats exploited by these mammals. Also, few studies have rigorously examined variation in postcranial "hard" (i.e., skeletal) morphology among extant members of the cricetids with the goal of identifying either generic or specific distinctions. The most common method of collecting small mammal fossils is underwater screening of previously dried and soaked sediment. Unfortunately, this process destroys skeletal associations, which would make identification of isolated postcranial elements from such a site suspect unless the site was monotypic, and this virtually never occurs.

Perhaps the postcranial skeletal anatomy most rigorously studied among cricetids is the male glans penis. Hooper and Musser (1964) studied the cricetid phallus, concluding that sigmodontine cricetids (i.e., the South American radiation) possess a complex penis, in comparison with the simple or straight penis of peromyscine–neotomine cricetids (i.e., the North American radiation). Moreover, the generalized and most common type of phallus among members of the superfamily Muroidea is a complex penis, suggesting that the penis of peromyscine–neotomine cricetids is derived relative to that of Old World cricetids and South American sigmodontine cricetids.

Carleton (1980) made an exhaustive analysis of hard and soft anatomy of peromyscine–neotomine cricetids, identifying 79

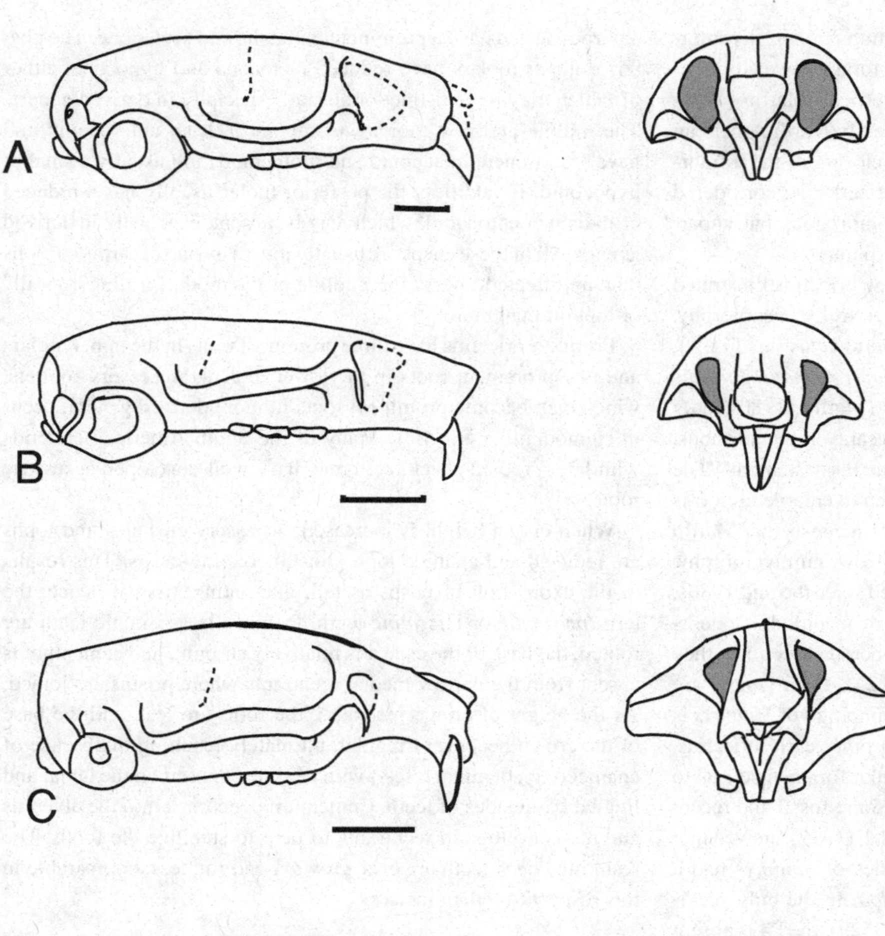

Figure 27.1. Lateral and anterior views of cricetid skulls showing zygoma. A. *Leidymys* sp. (modified from Korth, 1994) with hystricomorphous zygoma. B. *Scottimus* sp. 1 (AMNH 97815) with transitional myomorphous zygoma. C. *Eumys* sp. (composite from AMNH 38924 and 39599) with myomorphous zygoma. D. *Peromyscus maniculatus* (Recent) with myomorphous zygoma. Scale bars = 5 mm. (A: reproduced with kind permission of Springer Science and Business Media.)

anatomical (including 16 phallic) characters among 49 species of peromyscine–neotomine cricetids and 26 species of other muroids. Regarding the polarity and distinction between types of penis, Carleton (1980, p. 128) concluded "there is a structural continuum between the traditionally defined 'complex' and 'simple' penes," and "the hypothesis of the simple penis as ancestral is an equally plausible, if not more harmonious, interpretation than the hypothesis that 'complex' type is primitive for Muroidea."

Carleton and Musser (1984) did not list any postcranial skeletal characters restricted to the Cricetidae (their Sigmodontinae).

SYSTEMATICS

SUPRAFAMILIAL

Controversy and conflicting nomenclature have always been a factor of rodent systematics. That is not to say that progress in systematics of Rodentia has been absent, as reflected in the 1984 *International Conference* proceedings on "Evolutionary Relationships among Rodents", edited by Luckett and Hartenberger (1985). Rodents with a myomorphous zygoma and cheek teeth restricted to three teeth in each division of the jaw are a large and distinctive group that has commonly been recognized as the family Cricetidae Rochebrune, 1883 (Simpson, 1945), or the family Muridae Illiger, 1811 (McKenna and Bell, 1997). This familial group is universally recognized as a member of the larger superfamily Muroidea Illiger, 1811. As with other rodent specialists, I am biased, and prefer to call this taxon the family Cricetidae rather than the family Muridae. The rationale for this preference is based on our current knowledge that distinctive morphological differences separate restricted cricetids and restricted murids (except when dealing with primitive restricted murids), and that evolution of early restricted murids occurred in the mid Miocene of southern Asia, derived from endemic restricted cricetids (Jacobs, 1977a; Flynn, Jacobs, and Lindsay, 1985).

The taxa Muroidea and Dipodoidea are usually united (with the Gliroidea) in the suborder Myomorpha (Simpson, 1945), implying a phyletic relationship, although not all fossil members of the Myomorpha possess a myomorphous zygoma. The Gliroidea, a completely Old World group of rodents, are now considered to have a separate derivation, relative to the Muroidea and Dipodoidea. The leading candidate for ancestry of the Gliroidea is a microparamyine ischyromoid rodent, following Hartenberger (1971) and Vianey-Liaud (1985). As noted above, we now recognize that primitive members of the Myomorpha had an hystricomorphous zygoma. (See also discussion in Ch. 16.)

Clearly, all members of the taxon Cricetidae and the taxon Muridae are united in the taxon Muroidea, as used in the sense of Simpson (1945) and McKenna and Bell (1997). Currently, many neontologists and some paleontologists place all of these taxa in the family Muridae (and superfamily Muroidea), as subfamilies (e.g., Cricetinae and Murinae). As noted by Carleton and Musser (1984, p. 300), the Cricetinae-and-Murinae-within-Muridae taxonomic hierarchy reflects "uncertainty rather than conviction" of intergroup relationships and accepts a high incidence of parallelism within the Muroidea.

Fossil Cricetidae and Muridae are recognized primarily on the basis of their teeth, which are predominantly low crowned and cuspate or high crowned and prismatic in Cricetidae. Muridae are distinguished from Cricetidae by the addition of internal cusps on upper molars, and particularly by new connections between cusps, especially the enterocone to the protocone (Jacobs, 1977a, 1978). In addition, the murid cheek teeth are characteristically lophate or chevron-shaped and are usually low crowned. As with all rules, there are exceptions, in that dentition of the high-crowned but nonprismatic cricetid genus *Sigmodon* is strongly lophate. Also, arvicoline rodents are characterized by having prismatic cheek teeth with a planed occlusal surface, but some non-arvicoline cricetids, such as *Neotoma*, have similar prismatic cheek teeth.

The murid dentition is derived, and the transition from cricetid to murid dental morphology is well documented in Miocene deposits of southern Asia (Jacobs, 1978; Flynn, Jacobs, and Lindsay, 1985; Lindsay, 1988). Additionally, there is no known record of the restricted Muridae (in the sense of Simpson [1945]) in North America prior to their introduction by humans (e.g., post-Columbian). Hence, the restricted family Muridae is not included in this review of North American terrestrial mammals.

The Muridae (and Muroidea) of most neontologists usually has 15 (Carleton and Musser, 1984) or 16 (Nowak, 1999) subfamilies including Sigmodontinae as the New World rats and mice (i.e., Cricetidae as used here), and the Murinae as the Old World rats and mice (i.e., Muridae as used here). Virtually all of the remaining subfamilies of this broader group Muridae would be found in Simpson's (1945) Cricetidae. Nowak (1999) listed 90 genera in the Sigmodontinae and 119 genera in the Murinae. The neontological Muridae is one of the most successful families of mammals, with 301 genera and 1336 species (Nowak, 1999).

The Simimyidae was erected by Wood (1980) for the small genus *Simimys* from the middle to late Eocene (Uintan, Duschesnean and ?Chadronian NALMAs), which had previously been allied (with question) to the Cricetidae. As discussed above, *Simimys* species have an hystricomorphous zygoma and dental formula with either three or four upper cheek teeth and three lower cheek teeth. The first upper molar of *Simimys* lacks a prominent anterocone, which is virtually always present in Cricetidae, and the first lower molar usually has a minute anteroconid in *Simimys*. Lillegraven and Wilson (1975) showed that *Simimys* also has a small neurovascular foramen ventral to the infraorbital foramen, which occurs in some Dipodoidea but is not present in any known cricetid. Consequently, *Simimys* is now placed in the Dipodidae rather than the Muroidea

The late Eocene (Chadronian) genus *Nonomys* has also been included in the Cricetidae, primarily because it has a hystricomorphous zygoma and three upper and lower cheek teeth. Emry (1981) showed that *Nonomys* has an isolated neurovascular foramen ventral to the infraorbital foramen, similar to *Simimys*. *Nonomys* also has a prominent "parastyle" (rather than an anterocone) on M1, plus medial stylar cusps on M1 and M2; it also lacks well-developed cross-lophs on both upper and lower cheek teeth. Therefore, *Nonomys* and *Simimys* are excluded from the Cricetidae because they possess characters that are unknown in cricetids, and lack characters that are considered primitive for cricetids. However, *Simimys* and *Nonomys* share characters that are considered primitive for both cricetids and dipodids, suggesting possible ancestral affinity of dipodids and cricetids.

One passing note; I do not consider that the specimen (SMNH P1899.995) illustrated by Storer (1988) as *Eumys* sp. from locality NP9A (the Lac Pelletier Lower Fauna in Saskatchewan) is a cricetid rodent. The illustration of M1 shows a greatly enlarged parastyle anterior to the paracone that is directed transversely and posteriorly, terminating anterior to the protocone. This morphology suggests relationship (if not identity) with *Nonomys* or some related dipodid rather than to any known cricetid, especially the genus *Eumys*, as determined by Storer. A second specimen, identified as a left M2, was recorded by Storer (1988) but not illustrated. If these specimens were truly *Eumys*, they would represent the earliest record of the Cricetidae in North America. As currently understood, the oldest record of Cricetidae in North America is the genus *Eumys*, known from several Chadronian NALMA sites in North America.

INFRAFAMILIAL

Some cricetid genera have evolved high-crowned, prismatic teeth. This group experienced an explosive radiation during the Pleistocene, resulting in recognition of a taxonomic group identified as microtines or arvicolines (e.g., Microtinae or Arvicolinae). This group is presented in Chapter 28 as the Arvicolidae. Two relatively high-crowned genera, *Goniodontomys* and *Paramicrotoscoptes*, have primitive prismatic teeth but are excluded from the Arvicolidae; they are presented in this chapter.

Thirty one genera and 124 species of North American fossil Cricetidae are recognized herein; a rigorous phylogenetic analysis of the Cricetidae is beyond the limits of this study. In its place, I present a review of some previous phylogenetic interpretations of North American Cricetidae and give my biased (and poorly justified) summation of the classification used in Figure 27.2.

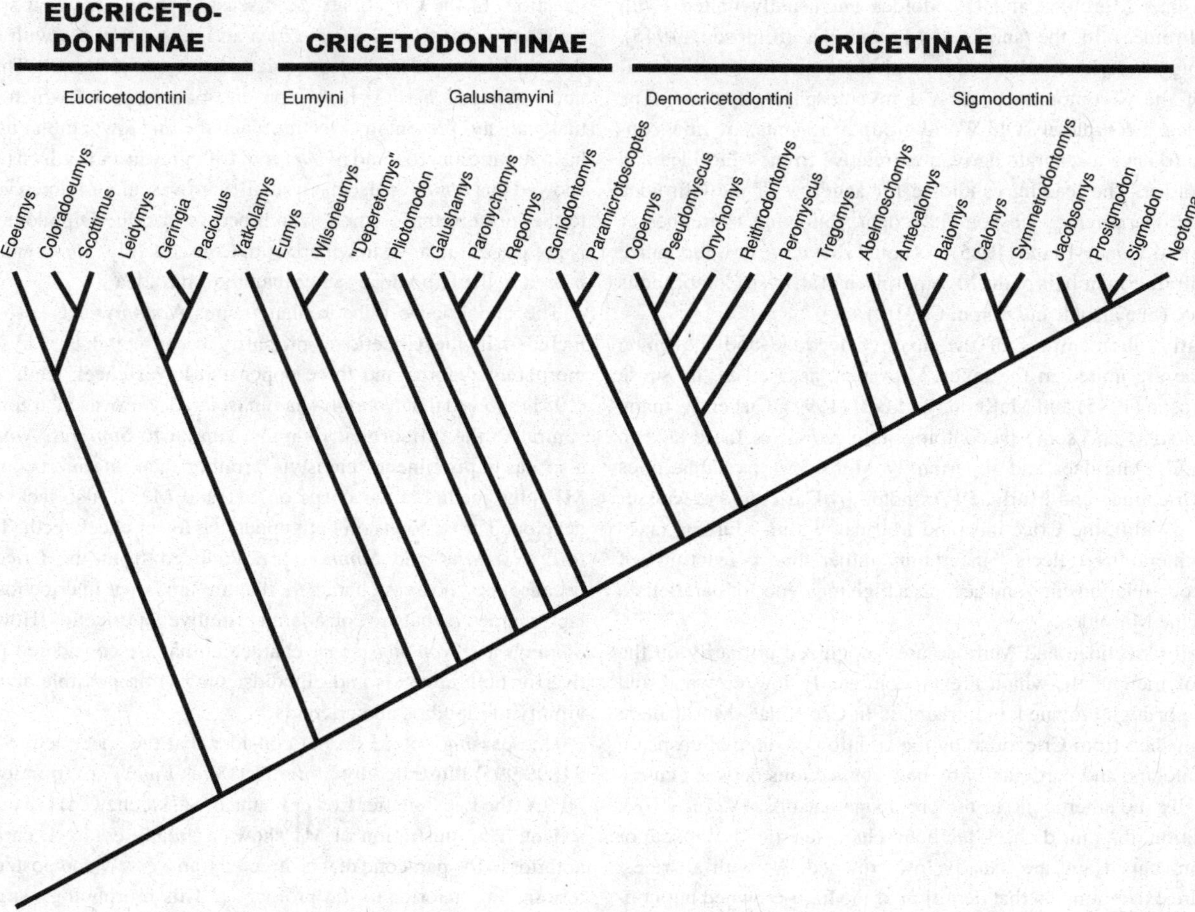

Figure 27.2. Relationships of North American cricetid genera. See text for discussion of characters supporting clades.

Simpson (1945) provided a stable taxonomic framework and discussion for study of cricetid phylogenetic patterns. He placed all of the New World (= tribe Eumyini and tribe Hesperomyini) along with the Eurasian (= tribes Cricetopini, Cricetodontini, Cricetini, and Myospalacini) taxa in the subfamily Cricetinae, in contrast to the subfamilies Nesomyinae (e.g., Madagascar cricetids), Lophomyinae (African cricetids), and Microtinae (or Arvicolinae; Holarctic arvicolids). Simpson also placed the genera *Paciculus* and *Copemys* in "tribe Eumyini or Hesperomyinae incertae sedis." This taxonomic framework is primarily biogeographic, reflecting the classification of living mammals, with a strong biogeographic foundation. Thus, North American Cricetidae were organized by Simpson (1945) primarily into a primitive tribe Eumyini and a more derived tribe Hesperomyini, which also included the South American cricetids.

The most recent comprehensive review of North American Cricetidae is by Korth (1994), who recognized three subfamilies of North American Cricetidae. Korth (1994) raised the tribe Eumyini of Simpson (1945) to subfamily rank (Eumyinae), including the genera *Eoeumys*, *Coloradoeumys*, and *Wilsoneumys*. Korth added the subfamily Eucricetodontinae of Mein and Freudenthal (1971) to include the genera *Paciculus*, *Leidymys*, *Scottimus*, *Geringia*, and *Yatkolomys*. He additionally raised the Nearctic cricetid genera of Simpson's tribe Hesperomyini to the subfamily Sigmodontinae Wagner, 1843, with *Copemys* the stem of the subfamily Sigmodontinae (and *Paciculus* the stem of the subfamily Eucricetodontinae). Korth (1994) recognized three tribes in the Sigmodontinae; Peromyscini (with *Copemys*, *Poamys*, *Tregomys*, *Peromyscus*, *Reithrodontomys*, *Paronychomys*, and *Onychomys*), Neotominae (with *Repomys*, *Galushamys*, and *Neotoma*), and Sigmodontini (with *Abelmoschomys*, *Calomys*, *Prosigmodon*, *Symmetrodontomys*, *Sigmodon*, and *Orozomys*), plus subfamily incertae sedis (with *Pliotomodon*). The classification that I have adopted (Figure 27.2) differs little from that of Korth (1994).

Similarity of North American and Eurasian cricetid genera has been a common theme of rodent researchers for many years, with controversy over interpreting these resulting from "convergence" (i.e., lacking phyletic significance) or "dispersal" (i.e., with phyletic significance). Many of these controversies can and will be resolved by the addition of new material, and by scrutiny of earlier interpretations.

In summary, I have utilized the systematics of previous authors wherever I found it reliable and consistent with my understanding of current knowledge of relationships among the Cricetidae. This classification is reflected in Figure 27.2. The classification provided herein is not the "final word" regarding relationships, as it should (and will) be revised by correction of my errors and additional insight.

INCLUDED GENERA OF CRICETIDAE ROCHEBRUNE, 1883

The "characteristics" of taxa are taken from various sources, which are identified (or emended); if not identified, characterizations are the biased observations of the author. Localities listed for each species are coded in reference to the unified localities in Appendix I. The acronyms for museum collections are listed in Appendix III. Brackets surrounding a locality (e.g., [CP101]) indicate that the taxon was cited in a published faunal list with identity open to question (as "aff." or "cf."). A question mark in front of a locality (e.g., ?CP101) means that the taxon is questionably known from that locality or that the reference lacks adequate description of the taxon, implying some doubt that this record is reliable. In a number of instances, "records" were added by the senior editor, in the interest of completeness; some of those "records" were deleted by the author, in the interest of veracity. The serious reader should use these records with caution, verifying any that are questionable. An asterisk (*) indicates the type locality.

When possible, measurement of the length and width of the m1 are provided, usually as an average of the hypodigm; alternatively, measurement of a single specimen from that locality or fauna is provided as a common parameter to aid comparisons and size evaluations. Some of the older references did not give the dimensions of m1, in which case measurements of casts (occasionally the type) are given, if available.

Characteristics: Zygoma hystricomorphous, subhystricomorphous (lower infraorbital canal not constricted and zygomatic plate more horizontal than vertical) or myomorphous (lower infraorbital canal strongly constricted and zygomatic plate more vertical than horizontal); incisor ornamentation common in primitive taxa, usually absent in derived taxa; dentition brachydont to mesodont until the Pliocene, when several lineages become higher crowned; M1 anterocone medium to large size, commonly bilobed; M3 relatively large to small size; m1 anteroconid medium to large size; m1–2 with union of protoconid and metaconid either posterior or anterior, or cusps separate; m3 large to small size; longitudinal ridge (mure) well developed to weakly developed, or absent; incisive foramina long or short, usually with posterior termination slightly anterior to position of M1; accessory rootlets absent in primitive taxa, present in some derived taxa.

EUCRICETODONTINAE MEIN AND FREUDENTHAL, 1971

Characteristics: Zygoma subhystricomorphous to myomorphous; incisor ornamentation common; dentition brachydont to mesodont; M1 anterocone medium size; M3 relatively large; m1 with posterior arm of protoconid long, directed toward and usually contacting metaconid; m3 comparatively large, usually about equal to size of m1; accessory rootlets absent.

EUCRICETODONTINI MCKENNA AND BELL, 1997

Characteristics: Zygoma subhystricomorphous; dentition brachydont, except for *Geringia* and *Paciculus*; m1 anteroconid usually medium size; m3 large relative to m2.

Eoeumys Martin, 1980 (including *Leidymys*, in part; *Eumys*, in part; *Scottimus*, in part)

Type species: *Eoeumys vetus* (Wood, 1937) (originally described as *Leidymys vetus*).

Type specimen: AMNH 8742, partial skull (with upper dentition) and left dentary.

Characteristics: Small and brachydont rodent; skull with short, slightly tapering snout, presence of parasagittal crests, short incisive foramina, and large infraorbital foramina that are weakly constricted ventrally; upper and lower incisors lacking ornamentation (note that lower incisors of *Scottimus exiguus*, described as *Eoeumys exiguus* by Martin [1980], characterized as having faint, non-parallel raised ridges that radiate pinnately; that species is placed in *Scottimus* in this study, following Korth [1981]); cheek teeth *Eoeumys* virtually indistinguishable from *Eumys* (modified from Martin, 1980). Lower dentition of *Eoeumys vetus* not securely identified.

Length of M1 (m1 is usually slightly smaller) = 1.20 mm (average of right and left M1).

Included species: *E. vetus* only, known from locality CP68C* only.

Comments: Martin (1980) described *Eoeumys* based on a skull (AMNH 8742) that had been described as *Leidymys vetus* by Wood (1937) and a partial palate (plus dentary, AMNH 12261) that had been described as *Eumys exiguus* by Wood (1937). He also referred numerous unlisted specimens that had been grouped with the above taxa (as *Scottimus exiguus* or *Paracricetodon exiguus*) to *Eoeumys*. Korth (1981) pointed out the complicated history of *Leidymys vetus* and *Scottimus exiguus* and reassigned all the unlisted specimens of *Eoeumys* (by Martin, 1980) to *Scottimus*. Korth (1981) also redefined the genus *Scottimus*, named three new species of *Scottimus*, and suggested that *Eoeumys* should be restricted to the type specimen of *E. vetus* (i.e., AMNH 8742).

This classification follows Korth (1981) as the most recent reviser. One of the salient characters ascribed to *Eoeumys* by Martin (1980), that of "pinnate" ridge ornamentation on the lower incisors, is thereby restricted to *Scottimus*. The "pinnate" ornamentation described and illustrated by Martin (1980: Fig. 26) should be termed a fine "pinnafoliant" ornamentation because it lacks a central ramus or "petiole" from which the minute ridges disperse. Similar incisor ornamentation is seen in some early European cricetids assigned to the genus *Pseudocricetodon* Thaler, 1969.

Scottimus **Wood, 1937**

Type species: *Scottimus lophatus* Wood, 1937.

Type specimen: MCZ 5064, right maxilla with M1–3.

Characteristics: Small, brachydont rodent with parallel-sided snout, parasagittal crests, long incisive foramina on the palate, and large infraorbital foramina with ventral constriction; cheek teeth developing lophs directed anteroposteriorly between the labial cusps and on the lingual side of lingual cusps in upper molars, and between lingual cusps in lower molars; upper molars decreasing in size posteriorly; M1 with large anterocone centered labial to the midline; M3 small metacone and with indistinct hypocone; anterior cingulum on M2–3 long, extending from the labial to the lingual margin; m1 with relatively large anteroconid, frequently isolated from posterior cusps; incisors with fine pinnafoliant (= pinnate of Martin [1980]) raised ridges as ornamentation, which appear to be absent in the lower incisors of advanced species; masseteric ridge prominent on the lateral side of the dentary, terminating anteriorly below M1; mental foramen on the lateral surface of the dentary below the diastema, near termination of the masseteric ridge (modified from Korth 1981).

Average length of m1: 1.97 mm (*S. ambiguus*) to 2.51 mm (*S. longiquus*) (Korth, 1981).

Included species: *S. lophatus* (known from localities CP84B, CP99B*); *S. ambiguus* (localities [CP46], CP68C*); *S. exiguus* (localities CP68C, CP84B*); *S. kellamorum* (locality CP42D*); *S. longiquus* (localities CP84B, NP29D*); *S. viduus* (localities CP44, CP84B, CP98BB, CP99A*, NP32B2–?C).

Unnamed species include *Scottimus* sp. 1 (locality CP40B); *S.* sp. 2 (locality NP32B2–?C); *S.* sp. 3 (locality CP84B); *S.* sp. 4 (localities CP85D, CP86B); *S.* sp. 5 (locality NP10CC).

Comments: The specimen discussed with *Eomys vetus* by Martin (1980) is the unnamed species called *Scottimus* sp. 1 herein.

Coloradoeumys **Martin, 1980**

Type species: *Coloradoeumys galbreathi* Martin, 1980.

Type specimen: KUVP 11132, skull and left dentary.

Characteristics: Small, brachydont rodent; skull with short narrow snout, single sagittal crest, short incisive foramina, large sphenoidal fissure in the palate, infraorbital foramina that are weakly constricted ventrally (almost hystricomorphous); M1 with large anterocone centered labial to the midline and joined centrally by the anterior arm of the protocone; m1 with small anteroconid joined by both paraconid and protoconid; incisors lack ornamentation (modified from Martin, 1980).

Average length of m1: 2.48 mm (Martin, 1980).

Included species: *C. galbreathi* only, known from locality CP68C* only.

Comments: Martin (1980) characterized this taxon as a primitive eumyine. Apparently, it differs from *Eumys* in having less constriction of the infraorbital foramen, and differs from *Eoeumys* in having a single sagittal crest.

Leidymys **Wood, 1936 (including *Cotimus* in part; *Eumys*, in part; *Hesperomys*, in part; *Paciculus*, in part; *Peromyscus*, in part)**

Type species: *Leidymys nematodon* (Cope, 1879) (originally described as *Hesperomys nematodon*).

Type specimen: AMNH 7018, partial skull.

Characteristics: Medium- to large-size brachydont rodent; skull with broad parallel-sided snout, parasagittal crests, palate long with wide incisive foramina, infraorbital foramina large with weak (to strong?) ventral constriction (e.g., hystricomorphous or transitional myomorphous); dentition weakly lophodont with cusps slightly inflated; upper molars with mesoloph well developed; M1 with relatively narrow anterocone centered slightly labial to the midline; M3 with small metacone and hypocone; lower molars having long posterior arm of protoconid directed transversely across the midline and commonly joining the metaconid on m1 and m2; mesolophid usually long in m1, short or absent in m2 and m3; a minute mesoconid common in m1, absent in m2 and m3; m1 usually with small wide anteroconid; m3 large, with distinct entoconid; lower incisors commonly with ornamentation including one raised ridge on or near the midline and two or three minute, parallel raised ridges labial to the midline; masseteric ridge terminating anteriorly beneath or anterior to m1; mental foramen located on the side of the dentary below the posterior limit of the diastema at the level of the masseteric ridge termination (modified by Wood, 1936).

Average length of m1: 1.36 mm (*L. parvus*) to 2.50 mm (*L. eliensis*, Black, 1961b).

Included species: *L. nematodon* (known from localities CC9IIIA, CC9IVB, C, PN6D3*, E, F); *L. alicae* (locality NP36B*); *L. blacki* (formerly *Eumys*) (localities CP85C*, CP86B, CP99C, CP101, ?NP10E); *L. cerasus* (locality CP103B*); *L. eliensis* (formerly *Eumys*) (locality NP34C*); *L. korthi* (locality NP10CC*); *L. lockingtonianus* (locality PN6H*); *L. parvus* (localities PN6C, PN6D*); *L. montanus* (formerly *Paciculus*) (localities NP34D*, PN16); *L. woodi* (formerly *Eumys*) (localities CP85C*, CP86B).

Unnamed *Leidymys* species are also known from the following localities: *L.* sp. 1 (localities CC12, CC13); *L.* sp. 2 (localities NP36A, B); *L.* sp. 3 (locality NP36D); *L.* sp. 4 (very small size) (locality CP88); *L.* sp. 5 (small) (localities CP9IVC, D).

Comments: The genus *Cotimus* was described by Black (1961a) as a cricetid rodent (*C. alicae*) from the Miocene Flint Creek Fauna in Montana (locality NP42). This genus was subsequently recognized and described from middle Miocene faunas of Europe by Fahlbusch (1964). Later, Martin (1980) transferred *Cotimus alicae* to *Leidymys alicae*, and European species of *Cotimus* were transferred to the genus *Eumyarian* Thaler, 1966 (Mein and Freudenthal, 1971). *Leidymys alicae* is listed below; its relationship with European specimens of *Eumyarian* has never been adequately resolved. The genus *Leidymys* has never been reported in the Old World.

Skwara (1988) described the Hemingfordian Topham Fauna in Saskatchewan (locality NP10E) with both large and small mammals. She identified three species (*Leidymys* sp. cf. *L. blacki*, *Yatkolamys edwardsii*) and Cricetidae genus indet. based on 23 isolated teeth, with measurements on 13 of the teeth. After plotting these measurements and examining her illustrations, I am not convinced that these specimens represent three taxa; I have included Skwara's entire cricetid sample as a single taxon, *Leidymys* sp. cf. *L. blacki*.

With respect to localities within the locality designation PN6, early collections from the John Day Formation lacked good stratigraphic resolution. One of the productive collecting areas at that time was designated as "The Cove." This is currently interpreted as the general vicinity of Turtle Cove, which is centered in sec. 20, T.11S, R.26E in Grant County on the Picture Cliffs Quadrangle. Several hundred feet of the Turtle Cove Member of the John Day Formation are exposed in Turtle Cove and fossils are more commonly found there from strata that overlie the Deep Creek Tuff, dated 27.5 Ma (which overlies the Picture Gorge Ignimbrite, considered about 28.7 Ma [Woodburne and Swisher, 1995]). The overlying Kimberly and Haystack Mountain Members of the John Day Formation are not well developed in the vicinity of Turtle Cove. Fossils occur below the Deep Creek Tuff, but it is much more likely that the fossils collected from the Turtle Cove area during early collecting were also from above the Deep Creek Tuff (Fisher and Rensberger, 1972; Rensberger, 1983). The interval between the Deep Creek Tuff and the Picture Gorge Ignimbrite in the Turtle Cove Member of the John Day Formation marks the lower part of the *Meniscomys* Concurrent Range Zone of Rensberger (1983); it corresponds to the stratigraphic interval of NP6D in the locality framework of Appendix I. Therefore, fossil localities designated "The Cove" and "Turtle Cove" in the John Day Formation are listed here as locality PN6D.

The teeth of *Leidymys* are significantly lower crowned than the teeth of *Paciculus*. This is very evident in the material from the John Day beds in Oregon, although crown height was never utilized in characterization of these genera. Cope (1879) described fossil skulls of *Hesperomys nematodon* and *Paciculus insolitus* from the John Day beds, without giving a specimen number to either taxon. *Peromyscus parvus* was described from the John Day beds by Sinclair (1905), who noted that *P. parvus* is smaller than *Peromyscus nematodon* (*Hesperomys* had been abandoned by that time, in place of *Peromyscus*, the most common modern cricetid). Sinclair (1905, p. 26) also noted that "The lower incisor is delicately grooved on the outer face" a feature (ornamentation) that is now recognized characteristic of *Leidymys* and of *Paciculus*.

Cope (1881a) also named *Eumys lockingtonianus* from the John Day beds, designated *Paciculus lockingtonianus* by Cope (1881b) and *Leidymys lockingtonianus* by Wood (1936). In Wood (1936), *Leidymys nematodon* was designated the genotype for *Leidymys* with AMNH specimen 7108, the skull believed to be that described by Cope (1879) as the genoholotype. Wood (1936) also designated *Paciculus insolitus* the genotype for *Paciculus* with AMNH specimen 7022, a palate lacking M3, believed to be the skull described by Cope (1879) as the genoholotype. Thus, Wood (1936) recognized four cricetid rodents (*Leidymys nematodon*, *Leidymys lockingtonianus*, *Paciculus insolitus*, and *Paciculus parvus*) from the John Day beds.

Wood (1936) used size and dental cusp morphology as primary features to distinguish these four species; he did not utilize cheek tooth crown height to distinguish *Leidymys* from *Paciculus*. Crown height is difficult to measure in these genera. The height of the transverse valley on the labial side of lower and lingual side of upper cheek teeth in these genera is very similar, the cusps of *Leidymys* are more slanted and not as tall as the more vertical and higher cusps of *Paciculus*, which accounts for the main difference in crown height of these genera. Note that the current dogma recognizes incisor ornamentation in both *Leidymys* and *Paciculus* but with slight differences (fide Martin, 1980: Fig. 26).

Black (1961b) described *Paciculus montanus*, which was reviewed by Martin (1980), as *Paciculus montanus*; this taxon appears to be lower crowned than *Paciculus insolitus* (the holotype) and is, therefore, assigned to *Leidymys* in this study (e.g., *L. montanus*). Black (1961b) described *Eumys eliensis*, which appears to be more lophate than material assigned to *Eumys* and is, therefore, transferred to *Leidymys* (e.g., *L. eliensis*) in this study. Macdonald (1963) described *Eumys blacki* and *Eumys woodi*, which were both transferred to *Leidymys* by Martin (1980) (e.g., *L. blacki* and *L. woodi*); these changes are followed in this study. Note that topotypic material of *L. montanus*, *L. woodi*, and *L. eliensis* were not examined during this study; their taxonomic placement herein is based primarily on illustrations in Black (1961b) and Martin (1980). I have wondered whether *P. nebraskensis* should also be moved to *Leidymys* (e.g., *L. nebraskensis*) but Fig. 21A in Martin (1980) appears relatively high crowned, so I am leaving *P. nebraskensis* as *Paciculus*.

***Geringia* Martin, 1980 (including *Eumys*, in part; *Paciculus*, in part)**

Type species: *Geringia mcgregori* (Macdonald, 1970) (originally described as *Paciculus mcgregori*).

Type specimen: LACM 9271, partial skull.

Characteristics: Medium size; cheek teeth relatively high crowned (compared with *Leidymys*), with planar wear; M1 with very reduced anterocone and nearly straight anterior margin; lower incisor broad with one ventral ridge, upper incisor undescribed (modified from Martin, 1980).

Average length of m1: 1.55–1.62 mm (Macdonald, 1970; Williams and Storer, 1998).

Included species: *G. mcgregori* (known from localities CP85C*, CP99C, CP101); *G. gloveri* (localities CP86B*, NP10CC).

Comments: The incidence of *G. mcgregori* at locality CP101 includes UNSM 11527 identified as *Paciculus nebraskensis* by Martin (1980: Fig. 21) fide Korth (1992). Martin (1980) characterized *Geringia* as smaller and less advanced (e.g., M1 anterocone more reduced) than *Paciculus*. It is questionable whether *Geringia* and *Paciculus* should be separated. *Geringia* is separated here because the ornamentation of the lower (single raised line) incisors seems distinctive. Upper incisors of *Paciculus* are undescribed; lower incisors have ornamentation similar to that of *Leidymys*. Apparently, *Geringia* is higher crowned than *Leidymys*, and *Paciculus* is higher crowned than *Geringia*. Note that specimens of *Geringia* were not examined in this study.

***Paciculus* Cope, 1879**

Type species: *Paciculus insolitus* Cope, 1879.

Type specimen: AMNH 7022, partial palate with left M1–2.

Characteristics: Medium- to large-size rodent with relatively high-crowned and weakly lophodont rooted cheek teeth; skull with broad parallel-sided snout, with parasagittal crests, long and narrow incisive foramina in the palate, infraorbital foramina with a weak ventral constriction; upper molars with mesoloph of variable length; M1 with small anterocone joining the anterior arm of the protocone in early wear; M3 with small hypocone and metacone; lower molars having long posterior arm of the protoconid that is directed transversely across the midline and commonly joining the metaconid on m1; mesolophid usually long on m1 and shorter than the posterior arm of the protoconid on m2 and m3; m1 with small and wide anteroconid; m3 relatively large with small or indistinct entoconid; masseteric ridge terminating anteriorly below m1; mental foramina located just below the posterior limit of the diastema at the level of the masseteric ridge termination; upper incisors inadequately described; lower incisors with one large lingual and three smaller labial raised lines that are parallel to the long axis of the incisor, as in *Leidymys* (modified from Martin, 1980). Lower incisors from Oregon identified as *Paciculus* have ornamentation comparable to that of *Leidymys*. Lower incisors of *P*. cf *P. insolitus* from Nebraska (below) described as smooth (Alker, 1969). It is questionable whether this material should be assigned to *Paciculus*.

Average length of m1: 2.16 mm (*P. insolatus*, cast of UCMP 75543) to 2.28 mm (*P. nebraskensis*, Korth, 1992).

Included species: *P. insolitus* (known from localities PN6D*–F, H); *P. nebraskensis* (localities CP101*, CP103B, NP10CC).

Paciculus sp. is also known from localities NP36B, D.

EUCRICETODONTINAE, INCERTAE SEDIS

Yatkolamys cannot be assigned with confidence to a tribe until the zygoma and other critical features are reported and described.

***Yatkolamys* Martin and Corner, 1980**

Type species: *Yatkolamys edwardsi* Martin and Corner, 1980.

Type species: UNSM 45390.

Characteristics: Distinguished from all other North American cricetids except *Eoeumys* by having a large number of minute raised lines on the lower incisor; differs from *Eoeumys* in that the minute raised lines are parallel rather than branching (as in pinnae) (from Martin and Corner, 1980).

Length of m1: 2.04 mm (Martin and and Corner, 1980).

Included species: *Yatkolamys edwardsi* only, known from locality CP104C* only.

Yatkolamys sp. is also reported from localities ?CC9IIIB, ?C9IVD.

Comments: The incisor ornamentation of *Yatkolamys* is similar to that of the European Oligocene genus *Pseudocricetodon*. *Yatkolamys*, known from a single specimen, has a lower dentition very similar to *Leidymys*. A better sample is needed before relationships of this taxon relative to *Leidymys* and other cricetids can be established.

CRICETODONTINAE STEHLIN AND SCHAUB, 1951

Characteristics: Zygoma myomorphous; incisor ornamentation common in primitive taxa, rare in derived taxa; dentitions brachydont to mesodont in primitive taxa, becoming hypsodont and prismatic in advanced taxa; M1 anterocone medium to large size, commonly with both anterior and posterior union of paracone and protocone or with only posterior union; M3 large to medium size, with reduced metacone and hypocone; m1 anteroconid medium size, with posterior arm of protoconid long, directed toward metaconid and usually crossing midline, commonly with both anterior and posterior union of protoconid and metaconid; m3 large to medium size with reduced but distinct entoconid and hypoconid; accessory rootlets rare or absent.

EUMYINI SIMPSON, 1945

Characteristics: Dentition brachydont to mesodont; M1 anterocone medium size; M3 large size; m1 anteroconid medium size; M3 large size; m1 with posterior union of protoconid and metaconid; m3 large size.

Eumys Leidy, 1856 (including *Cricetodon*, in part)

Type species: *Eumys elegans* Leidy, 1856.

Type specimen: ANSP 11027, partial left dentary with m1 and m2 fragment.

Characteristics: Small- to medium-size brachydont rodent; skull with broad rostrum, median sagittal crest, large infraorbital foramina slightly constricted ventrally; dentition slightly lophodont, with robust cusps; upper cheek teeth with labial cusps opposite lingual cusps, joined by transverse lophs; the mesoloph usually short, rarely absent; M1 with large, single-cusped anterocone centered labial to the midline, joined medially or lingually with the anterior arm of the protocone; M3 hypocone reduced or absent; M3 and m3 relatively large; lower cheek teeth with lingual cusps placed slightly anterior relative to labial cusps, with lophs from lingual cusps directing toward and joining the anterior side of labial cusps; posterior arm of the protoconids long, directed transversely across the midline and usually joining the metaconid; mesolophid usually short, rarely absent; m1 with small medial anteroconid, usually joining the anterior arm of the protoconid near the midline; m3 entoconid is usually indistinct or absent; masseteric ridge prominent on the lateral side of the dentary, terminating anteriorly below the anterior or posterior root of m1; mental foramen on the lateral side of the dentary below the diastema, near the level of the masseteric ridge; upper and lower incisors may have very fine, wavy, raised ridge ornamentation (modified from Martin, 1980).

Average length of m1: 1.995 (*E. brachyodus*, Setoguchi, 1978) to 2.40 mm (*E. parvidens*).

Included species: *E. elegans* (known from localities CP46, CP68C, CP84A*, CP98B, ?C, CP99A, NP10B2, NP29D, NP32?B, [C], NP50B); *E. brachyodus* (localities CP68D, CP84B, CP85A, CP99B, CP101, NP10C2, NP?32B, [C]); *E. cricetodontoides* White 1954 (localities NP27[C1], ?D, NP29D*, NP32?B, [C]); *E. latidens* White 1954 (locality NP29D); *E. nebraskensis* Wood 1937 (locality CP99A); *E. obliquidens* Wood 1937 (localities CP68C, CP84A, ?NP27C1, D); *E. parvidens* (localities CP84A, CP99A, [NP32B, C]); *E. pristinus* (locality NP10B); *E. spokanensis* White 1954 (locality NP29D).

Eumys sp. is also known from locality NP32B.

Comments: Wood (1980) assigned to *E. parvidens* some of the material referred to *Paracricetodon exiguus* (= *Scottimus exiguus*) by Alker (1968), along with some of the material assigned to *Scottimus viduus* by Korth (1981).

Eumys is generally regarded as the most common, best-known Oligocene cricetid. Martin (1980) synonymized the following species (*E. obliquidens, E. cricetodontoides, E. latidens, E. spokanensis*, and *Cricetodon nebraskensis*) with *E. elegans*. *C. nebraskensis* was described by Wood (1937); it is now considered a species of *Eumys* by most authorities (Wood, 1980).

Wilsoneumys Martin, 1980 (including *Eumys*, in part)

Type species: *Wilsoneumys planidens* Martin, 1980.

Type specimen: UCM 19810, partial left dentary with m2–3.

Characteristics: Medium-size, brachydont rodent with uninflated cusps expanded laterally, producing "squarish" lateral margins; crests of upper and lower molars narrow, leaving occlusal surfaces nearly planar; M1 (illustrated by Setoguchi [1978]) with large and wide anterocone that joins the anterior arm of the protocone labial to the midline; lower cheek teeth with minute mesoconid, mesolophid short or absent; m1 with broad anteroconid that usually joins the metaconid before joining the protoconid; m3 large; lower incisors small, flat, with smooth enamel; masseteric ridges terminating anteriorly below m2 (modified from Martin, 1980).

Average length of m1: 2.156 mm (Setoguchi, 1978).

Included species: *W. planidens* only (known from localities CP46, CP68C*, CP98B, CP99A).

Wilsoneumys sp. is also known from localities NP32B, ?C.

Comments: Galbreath (1953) reported two specimens (KU 8472 and 8467) that he assigned to *Eumys planidens*, which were more lophate, heavier, slightly larger and collected from higher stratigraphic levels than the type of *W. planidens*; he suggested they might be more advanced than the type. Other authors have included these specimens in *W. planidens*. This species (*W. planidens*) apparently (type material not examined) is higher crowned relative to most species of *Eumys*. Wilson (1949b) suggested that *W. planidens* (as *Eumys planidens*) might be ancestral to *Paciculus*; Martin (1980) disagreed, stating that *Paciculus* was probably derived from *Leidymys*, and that possibly *Geringia* paralleled *Paciculus*.

CRICETODONTINI SIMPSON, 1945

Characteristics: Dentition mesodont; incisor ornamentation common; M1 anterocone medium to large size, with both anterior and posterior union of paracone and protocone; M3 medium size; m1 anteroconid medium size, with posterior union of protoconid and metaconid (primitive), or both posterior and anterior union, or anterior union of these cusps (derived); m3 medium size.

"*Deperetomys*" Mein and Freudenthal, 1971 (including *Cricetodon*, in part)

Type species: *Deperetomys hagni* Fahlbusch, 1964.

Type specimen: München 1952 XIV 259 (middle Miocene of Europe).

Characteristics: Cuspate low-crowned cheek teeth with dimensions very similar to those of European *Cricetodon*, *Eumyarion* and *Deperetomys*; m1 with small central anteroconid, plus weak posterior union of protoconid and metaconid; m3 with prominent entoconid and hypoconid; lower incisor with two parallel minute raised lines labial to the midline and a more prominent single raised line lingual to the midline. Upper dentition of the Warm Springs cricetid is unknown.

Length of m1: 2.48 mm (UCR 16883).

Included species: "*Deperetomys*" sp. only, known from locality PN6H* only.

Comments: The "Warm Springs cricetid" was described by Dingus (1978) as *Eumyarion*, a European genus believed very similar to *Leidymys* from North America. Dingus made extensive comparisons with *Leidymys*, the most abundant cricetid from the John Day Formation and concluded that the "Warm Springs cricetid" is larger and more advanced than any known *Leidymys* from that area. Later, after comparison of casts and specimens of other cricetids from Europe, he concluded that the "Warm Springs cricetid" is closer to *Deperetomys*, a European contemporary of *Eumyarion*. The lower incisor of European *Deperetomys* has two minute parallel raised lines labial to the midline; the lower incisor of the "Warm Springs cricetid" has a prominent raised line on the midline plus two less-prominent parallel raised lines labial to the midline.

GALUSHAMYINI (NEW)

Characteristics: Dentition subhypsodont to hypsodont, tending toward prismatic (with cusps elongated into enamel prisms); lophs and lophids strongly aligned; accessory lophs and lophids absent; M3 medium or small size; m1 anteroconid medium size, with anterior union of protoconid and metaconid; accessory rootlets absent.

Galushamys Jacobs, 1977b

Type species: *Galushamys redingtonensis* Jacobs, 1977b.

Type specimen: F:AM 98598, right maxillary fragment with M1–2.

Characteristics: Medium-size cricetine with slender, very high-crowned cheek teeth with well-developed roots; upper molars joining lingually, partly joining labially by anteroposterior lophs; lower molars poorly known (probably cusps not joined lingually in lower molars); M1 with relatively small anterocone; m1 with small anteroconid that joins the protoconid and metaconid in early wear; M3 small, subcircular in outline with indistinct cusps. *Galushamys* higher crowned than *Paronychomys*, crown height approximating that of *Repomys*, but is derived relative to *Repomys* in having greater closure of labial reentrants in upper cheek teeth. As presently understood, an isolated enamel islet is never developed in M1 or M2 of *Repomys* (modified from Jacobs, 1977b).

Length of m1: 2.23 mm (Jacobs, 1977b).

Included species: *G. redingtonensis* only, known from locality SB10* only.

Comments: Jacobs (1977b) argued that *Galushamys* was derived from a New World *Copemys*-like cricetine rather than from *Pliotomodon*, which is generally believed an immigrant from Eurasia. A major factor of that argument is the small size M3 in *Galushamys* compared with the medium size M3 in *Pliotomodon*. However, the similarity of these two genera in crown height, shape of M1 and m1, and development of posterolabial lophs in upper molars is striking.

Paronychomys Jacobs, 1977b

Type species: *Paronychomys lemredfieldi* Jacobs, 1977b.

Type specimen: F:AM 98592, right maxillary fragment with M1–3.

Characteristics: Medium- to small-size cricetine with relatively high-crowned check teeth, especially between the base of cusps and base of crown, but also with well-developed roots; molar cusps with steep walls and alternate in position; M1 with wide anterocone; m1 with large anteroconid; M3 and m3 both slightly reduced in size (modified from Jacobs, 1977b).

Average length of m1: 1.49 mm (*P. alticuspis*: Baskin, 1979) to 1.93 mm (*P. tuttlei*; Jacobs, 1977b).

Included species: *P. lemredfieldi* (known from locality SB10*); *P. alticuspis* (locality SB11*); *P. tuttlei* (locality SB10*).

Paronychomys sp. is also known from localities NB7E, CP116F.

Repomys May, 1981 (including *Neotomodon*, in part)

Type species: *Repomys gustelyi* May, 1981.

Type specimen: UCR 20178, isolated m3.

Characteristics: Slender, very high-crowned cricetine with strongly lophate cheek teeth having thick enamel and well-developed roots; cusps distinctive only in unworn teeth; cheek teeth lophate although fossettes and enamel-surrounded islets not developed on M1 and M2; anterior enamel islet may be present on m3; most diagnostic tooth is the small M3, displaying an E-shaped occlusal outline when unworn or slightly worn (resulting from two deep labial reentrants) or an F-shaped occlusal outline when moderately worn (i.e., the posterior reentrant less deep and removed with wear), ephemeral posterior enamel islet may form with moderate wear, or (when well worn) only an anterior enamel islet may be displayed; M1 with anterocone well separated from protocone and paracone; m1 with small anteroconid joining protoconid and metaconid in early wear; m3 reduced, keyhole-shaped, with a minute anterolingual reentrant and a short anterolabial cingulum, both of which are lost in early wear (modified from Mou, 1999).

Average length of m1: 2.02 mm (*R. panacaenensis*; Mou, 1999) to 2.45 mm (*R. gustelyi*; May, 1981).

Included species: *R. gustelyi* (known from localities CC40, NB10*); *R. arizonensis* (locality SB18B*); *R. maxumi* (localities CC26C*, CC40); *R. panacaenensis* (localities NB33A*, SB12, [SB34IIB]).

Repomys sp. is also known from localities NB7E, ?NB9, NB13B, NB33A, ?CP116F.

Comments: The taxon *Repomys* was listed as "Cf. *Neotomodon*" by Tedford *et al.* (1987). The Recent volcano mouse of Mexico (*Neotomodon*) has no known fossil record but prior to the description of *Repomys* many of the taxa assigned to *Repomys* by May (1981) had been informally referred to Cf. *Neotomodon*. Also, as noted above, the species described as *Neotoma minutus* from Coffee Ranch, Texas (locality SP3B), by Dalquest (1983) should be compared to species of *Repomys*.

***Pliotomodon* Hoffmeister, 1945**

Type species: *Pliotomodon primitivus* Hoffmeister, 1945.

Type specimen: UCMP 36030.

Characteristics: Cricetine with slender, very high-crowned, semi-lophate cheek teeth that enclose deep fossettes and have well-developed roots; enamel relatively thick but not as thick as in *Repomys*; M3 and m3 comparatively large although smaller than anterior teeth; upper molars with distinctive lingual anteroposterior loph, joining the anterocone (M1) or anterior cingulum (M2 and M3) with the protocone, hypocone and posterior cingulum; lingual loph with two inflections (M1), one inflection (M2), or no inflections (M3); bilobed anterocone of M1 continuing labially to enclose a large anterior fossette, joining the paracone by a short and narrow loph, the posterior cingulum continuing labially to join the metacone (indistinct on M3) and partially enclosing a large posterior fossette, which completes with high narrow labial loph joining the paracone and metacone; anterior and posterior fossettes separated by high transverse loph (= mesoloph) joining mure and base of the paracone; posterior fossette on M3 divided further (into three unequal fossettes) by a second transverse loph (= metalophule) joining the hypocone and metacone; lower molars more cuspate, only in m3 do fossettids close after moderate wear; anteroconid on m1 narrow and single-cusped, joining the protoconid labially and the metaconid lingually; minute anterior fossettid may be closed posteriorly by narrow protolophulid; posterior arm of protoconid on m1 and m2 nearly aligned with entolophulid, separating a deep, narrow, obliquely oriented anterior lingual valley from a shorter, wider obliquely oriented posterior lingual valley; m3 with continuous lingual loph joining posterior metaconid with a short and transverse mesolophid and entolophulid, and with the posterior cingulum, to enclose three narrow fossettids (modified from Hoffmeister, 1945).

Length of m1: 2.75 mm (Hoffmeister, 1945).

Included species: *P. primitivus* only, known from locality CC36* only.

Comments: *Pliotomodon* is generally considered an immigrant, probably derived from the European genus *Ruscinomys*. Mein and Freudenthal (1971) placed *Ruscinomys* in the subfamily Cricetodontinae, tribe Cricetodontini, along with *Deperetomys*. The "gap" between middle Miocene "*Deperetomys*" and Pliocene *Pliotomodon* in North America is filled with abundant records of brachydont cricetid rodents bearing small M3 and m3, none of which resembles *Ruscinomys* or *Pliotomodon*.

***Goniodontomys* Wilson, 1937 (including *Microtoscoptes*, in part)**

Type species: *Goniodontomys disjunctus* Wilson, 1937.

Type specimen: LACM (CIT) 1959.

Characteristics: Small cricetid rodent with hypsodont, prismatic cheek teeth lacking cementum in reentrants and with well-developed roots; reentrant angles of enamel prisms generally opposed rather than alternate, and the apices of reentrants commonly touch; enamel relatively thin, not differentiated into thick and thin segments; m1 with posterior loop, three opposing prisms, and a complex anterior loop; second reentrant angle (T_2) opposite the third reentrant angle (T_3) in m1 (modified from Wilson, 1937).

Length of m1: 3.00 mm (Wilson, 1937).

Included species: *G. disjunctus* only (known from localities CP58, CP116III, PN11B*).

Comments: Following Schaub (1940), *Goniodontomys* has generally been synonymized with *Microtoscoptes*, described from Asia (Schaub, 1934). New material collected from the topotype locality of *Microtoscoptes* in Asia reported by Fahlbusch (1987) indicates that *Goniodontomys disjunctus* is generically distinct from *Microtoscoptes praetermissus*.

***Paramicrotoscoptes* Martin, 1975 (including *Microtoscoptes*, in part)**

Type species: *Paramicrotoscoptes hibbardi* Martin, 1975.

Type specimen: UNSM 47504, right maxilla fragment with M1–2.

Characteristics: Small cricetid with hypsodont, prismatic cheek teeth lacking cementum or dentine tracts on the reentrants; reentrant angles in lower molars and M1 tending to be directly opposite one another and confluent; M2 with single large lingual reentrant; M3 with two lingual reentrants and a posterior enamel pit (from Martin, 1975).

Average length of m1: 2.04 mm (UO 27030) to 2.38 mm (Martin, 1975).

Included species: *P. hibbardi* only (known from localities NB30, CP116C*, PN11B, PN13, PN22).

Comments: Fahlbusch (1987) pointed out that *Microtoscoptes praetermissus* from Asia shares the character of a single lingual reentrant on M2, which suggests that *Paramicrotoscoptes* is closer to Asian species of *Microtoscoptes* than to North American *Goniodontomys*.

CRICETINAE STEHLIN AND SCHAUB, 1951

Characteristics: Overall body size usually small; zygoma myomorphous; dentitions usually mesodont; incisor ornamentation rare or absent; M1 anterocone large, with paracone united posteriorly (and sometimes anteriorly) to protocone; M3 medium size to small, with metacone and hypocone reduced; m1 with anterior union of protoconid and metaconid; m3 medium size to small, usually with hypoconid reduced and entoconid indistinct or absent; accessory rootlets developed in some taxa.

DEMOCRICETODONTINI (NEW)

Characteristics: Incisor ornamentation rare; upper molars with tendency toward alignment of metalophule II and anterior arm of the hypocone; lower molars with tendency toward alignment of entolophulid and posterior arm of the protoconid; M1 becoming bilobed in some taxa; m1 anteroconid wide and asymmetrical in early taxa; accessory rootlets rare or absent.

Copemys Wood, 1936 (including *Hesperomys; Miochomys; Poamys; Peromyscus*, in part)

Type species: *Copemys loxodon* (Cope, 1874) (originally described as *Hesperomys loxodon*).

Type specimen: USNM 1204.

Characteristics: Small, brachydont cricetid in which the alternation of labial and lingual cusps on cheek teeth has begun developing but is incomplete; protolophules on M1 and M2 weakly developed, protolophule II not aligned with the anterior arm of the hypocone, entolophulids of m1 and m2 weakly developed, not aligned with the posterior arm of the protoconid; posterior arm of the protoconid on m1 and m2 short, not directed transversely across the midline; M3 small with the hypocone reduced or absent; incisors lacking ridged ornamentation (modified from Lindsay, 1972).

Average length of m1: 1.35 mm (*C. lindsayi*, Sutton and Korth, 1995) to 1.92 mm (*C. esmeraldensis*, Lindsay, 1972).

Included species: *C. loxodon* (known from localities SB32D*, SB34A); *C. barstowensis* (localities NB6C*–E, PN9A, B); *C. dentalis* (localities [CC17D], NB7B–E, NB8, NB23C*; CP123E, PN9A, PN10); *C. esmeraldensis* (localities NB23C*, [PN9A], PN10, ?PN11B); *C. lindsayi* (locality NP41B*); *C. longidens* (localities NB6C–E*, NB7A, NB23B); *C. mariae* (localities SP1A*, CP116B); *C. niobrariensis* (including *C. kelloggae*) (localities CP45E, CP89, CP114A, B*, C, NP11); *C. pagei* (localities GC4C, ?GC9C, NB6B–D, PN8B*, PN9A, B); *C. russelli* (localities CC16, CC17G*, H, CC21C, NB6E, NB7A); *C. tenuis* (localities NB6D, E*).

Copemys sp. is also known from localities?GC4DD, ?GC9C, GC11A, CC17I, CC18A, CC21C, SB12, CP114D, CP116B, NP5A, NP40B, NP41B, PN9B, PN10, PN21.

Comments: Matthew (1924) described *Poamys rivicola* based on a right dentary with worn m2 from the lower Snake Creek Formation (now recognized as the Olcott Formation [locality CP110, Skinner, Skinner, and Gooris, 1977]) of Nebraska. The m2 (AMNH 18892) was subsequently lost but its illustration (Matthew, 1924: Fig. 10) and morphology of the dentary are consistent with the tooth and dentary of either a cricetid or a zapodid rodent. Cricetids are much more common in these deposits than zapodids, so it is more likely that *P. rivicola* is a cricetid, and if so *Poamys* Matthew, 1924 should be the senior synonym of *Copemys* Wood, 1936. However, because of this uncertainty, *P. rivicola* is considered a nomen nudum, or a nomen oblitum. Hoffmeister (1959) described *Miochomys niobrarensis* (an isolated M1) and *Peromyscus kelloggae* (a dentary with m1) from UCMP locality V3218 (locality CP114B) in Cherry County, Nebraska. This site was considered part of the Niobrara River Fauna by collectors at UCMP and is now placed within the Crookston Bridge Member of the Valentine Formation by Skinner and Johnson (1984). *Miochomys* is considered synonymous with *Copemys* (as *C. niobrarensis*), and all Miocene species referred to *Peromyscus* were placed in *Copemys* by Lindsay (1972). *C. kelloggae* is considered a junior synonym of *C. niobrariensis* (see comment below). Several of these species are very similar and have never been adequately differentiated, including *C. loxodon, C. niobrariensis, C. longidens*, and *C. dentalis*. It is likely that some of these species will be synonymized upon close scrutiny.

C. loxodon, the genotype, had been described inaccurately (Wood, 1937; Clark, Dawson, and Wood, 1964) and was later redescribed by Fahlbusch (1967), who pointed out the similarity of *Copemys* to the European cricetid genus *Democricetodon*. Fahlbusch suggested that these two taxa are probably congeneric with the North American (e.g., *Copemys* [*Copemys*]) and European (e.g., *Copemys* [*Democricetodon*]) forms and distinct at the subgeneric level. Some European and Asian species of *Democricetodon* have ornamentation (raised minute ridges) of the lower incisor; ornamentation has never been observed on lower incisors of North American specimens of *Copemys*. Lindsay and Qiu (personal communication, 1991) have suggested that the earliest and most primitive North American species (*Copemys pagei*) might more appropriately be assigned to the genus *Democricetodon*, in recognition of its greater similarity to *Democricetodon* from China than with all the later North American species of *Copemys*, which are probably derived from it.

Provenance of the genotypic species, *C. loxodon*, is problematic. The type was collected from the Espanola Basin in New Mexico by Cope (1874), prior to the establishment of accurate mapping and stratigraphic resolution of the Santa Fe Group. Later geological work in that area by Galusha and Blick (1971) placed the type imprecisely in the Pojoaque Member of the Tesuque Formation. Tedford (1981) has identified a productive Barstovian collection, the "Santa Cruz sites" from the lower Pojoaque Member in that area. Also, Dan Chaney (personal communication, 2002) collected a large sample of *Copemys loxodon* from the Pojoaque Member in the general vicinity of the "Santa Cruz sites" for his thesis; hence the location is provisionally assigned to locality SB32D based on the material collected by Chaney.

Skinner and Taylor (1967) listed *Peromycus* (*Copemys*) sp. (identified by T. M. Stout) from the Bijou Hills Fauna of South Dakota (locality CP89). This identification is wrong, verified by correspondence with the late M. F. Skinner; the illustrated taxon is a dipodid, probably *Plesiosminthus*. Later, Green (1985) again identified a cricetid (*Copemys* sp. cf. *C. kelloggae*) from the Bijou Hills Fauna but did not illustrate or diagnose the 25 specimens he assigned to that taxon. The presence of a cricetid in the important Bijou Hills Fauna (locality CP89) is questionable, until that material is adequately described and illustrated.

***Pseudomyscus* Korth, 1997**

Type species: *Pseudomyscus bathygnathus* Korth, 1997.

Type specimen: UNSM 101818, right dentary with m1–3.

Characteristics: Small brachydont cricetid with deep mandible and steep posterior border of the diastema; mandible with two mental foramina below the posterior base of the diastema and anterior to the masseteric ridge; lower molars with lingual cusps placed slightly anterior relative to labial cusps. Known only from mandible with lower dentition and incisor (modified from Korth, 1997).

Average length of m1: 1.55 mm (Korth, 1997).

Included species: *P. bathygnathus* only, known from locality CP116B only.

Comments: *P. bathygnathus* is distinctive in having two mental foramina, a morphology that is unknown in any other North American cricetid rodent, and having a steep posterior diastema in a deep mandible. As noted by Korth (1987), the steep posterior diastema is present in arvicoline rodents; it is also present in some North American sigmodontine rodents, such as *Prosigmodon* and *Sigmodon*. The ecological significance of a steep posterior diastema is not clearly known; the phyletic significance is suggestive of a relationship with either arvicolines or sigmodontines but is equivocal.

***Peromyscus* Gloger, 1841 (including *Hesperomys*, in part; *Podomys*, in part)**

Type species: *Peromyscus leucopus novaeboracensis*, the extant white-footed mouse.

Type specimen: None designated.

Characteristics: Small- to medium-size rodent, white feet, white belly, large ears, long, usually hairy, tail; skull with long narrow rostrum; cheek teeth small, brachydont and usually tuberculate; upper molars with protolophule aligned with the anterior arm of the hypocone; lower molars have entolophid aligned with the posterior arm of the protocone; incisors neither grooved nor ornamented (modified from Kurtén and Anderson, 1980).

Average length of m1: 1.58 mm (*P. hagermanensis*, Tomida, 1987) to 2.36 mm (*P. pliocenicus*, USNM 23567).

Included species: *P. antiquus* (known from localities NB19C, NB27B, NB31*, PN11B, PN15); *P. baumgartneri* (localities CC47C, CC48A, CP128A*); *P. complexus* (locality CC47C*); *P. cragini* (localities CP131*, [CP132D]); *P. hagermanensis* (localities CC47A, C, D, CC48A, NB33A, SB12, SB15A, SB18B, CP128, PN23A*); *P. kansasensis* (localities SP1F, ?SP1H, ?SP5, CP118, CP128AA*); *P. maximus* (localities CC47D*, CC48A, B); *P. nosher* (locality PN3C*); *P. pliocenicus* (localities CC32B, CC38A*, PN11B, ?PN13, ?PN15); *P. polionotus* (locality GC15C); *P. sarmocophinus* (locality GC15C).

Peromyscus sp. is also known from localities GC15C, E, GC18IIA, B, GC18IV, CC36, CC37, NB13B, C, SB14D, SB15A, C, SB34IIB, SP3B, CP128AA, B, CP130, CP132B, C, PN12B.

Peromyscus spp. are also widely distributed in the Pleistocene, represented from various Irvingtonian and Rancholabrean localities in Arkansas, California, Colorado, Florida, Georgia, Idaho, Kansas, Missouri, Montana, New Mexico, Oklahoma, Pennsylvania, Tennessee, Texas, Utah, Virginia, West Virginia, and Wyoming.

Comments: Nothing in the description or illustrations of *P. nosher* distinguishes it from the genus *Copemys*. Material of *P. nosher* was not examined in this study, however, so it is retained in the genus *Peromyscus* with reservation and this query.

Fossil species of *Peromyscus* are distinguished from those of *Copemys* by diagonal alignment of the protolophule relative to the anterior arm of the hypocone in upper molars, and with the entolophid aligned diagonally relative to the posterior arm of the protoconid in lower molars. Based on those criteria, at least one modern species of *Peromyscus*, *P. eremicus*, should be considered a species of the genus *Copemys*. *P. nosher* also resembles *Prosigmodon* and *Antecalomys* but *P. nosher* has a mesoloph (on M1) plus a short mesolophid and long entolophid (on m1), which are vestigial or absent in sigmodontines. Also, sigmodontines tend to develop accessory rootlets on upper and lower molars, and roots on *P. nosher* were undescribed by Gustafson (1978). *Peromyscus kelloggae* of Hoffmeister (1959), based on (UCMP 36105) a left dentary with m1 from the Niobrara River Fauna, is considered a junior synonym of *Copemys niobrariensis*: Voorhies (1990), as first revisor, selected *C. niobrariensis* as the senior synonym.

Reithrodontomys Giglioli, 1874

Type species: *Reithrodontomys megalotis*, the extant Western harvest mouse.

Type specimen: None designated.

Characteristics: Small brachydont cricetine; both M3 and m3 very reduced; deep medial groove on upper incisors; anterocone of M1 broad, usually single-cusped; anteroconid on m1 single- or bicuspid; masseteric ridge high, terminating anteriorly below anterior root of m1, usually above level of diastema; mental foramina located on the posterior dorsal rim of the diastema.

Average length of m1: 1.25 mm (*R. wetmorei*; Ruez, 2001) to 1.43 mm (*R. wetmorei*; Czaplewski, 1990).

Included species: *R. galushai* (known from locality SB18B*); *R. humulis* (locality GC15C); *R. moorei* (localities CP132C, D); *R. pratincola* (CP131, CP132B*); *R. rexroadensis* (localities SB15A, SB18B, SP1F, CP128AA*); *R. wetmorei* (localities GC15C, SB12, CP128AA*, [D]).

Reithrodontomys sp. is also known from localities WM26, GC15C, CC47D, SB12, SP1H, CP128AA.

Comments: *Reithrodontomys* is represented by five extant species, widely distributed over central and southern North America and extending into Central America. Four of these species (*R. humulis*, *R. megalotis*, *R. montanus*, and *R. fulvescens*) are recorded from Rancholabrean NALMA, and two of these species (*R. humulis* and *R. fulvescens*), plus an extinct species (*R. moorei*), are recorded from Irvingtonian NALMA. *R. humulis*, with an early Pleistocene record in Florida, is presently distributed along the northern Gulf Coast to Florida and on the Atlantic Coast from Florida to Chesapeake Bay; it may have the longest temporal range on any extant rodent species.

Onychomys Baird, 1857

Type species: *Onychomys leucogaster*, the extant Northern grasshopper mouse.

Type specimen: None designated.

Characteristics: Relatively small, brachydont cricetine; high narrow cusps on rooted cheek teeth, slightly alternate in upper molars and strongly alternate in lower molars; M1 with narrow anterocone displaced to labial side of the tooth; m1 with narrow single-cusped anteroconid; M3 and m3 reduced in size, especially posteriorly; masseteric ridge high, terminating anteriorly below root of m1; mental foramina on the posterolateral surface of the diastema, below level of masseteric ridge. *Onychomys* is omnivorous, feeding primarily on insects, larvae, and worms.

Average length of m1: 1.62 mm (*O. bensoni*; Czaplewski, 1990) to 1.92 mm (*O. pedroensis*; 1.92 mm USNM 10506).

Included species: *O. leucogaster* (known from locality [CP132C]); *O. bensoni* (localities SB12, SB14A*, D); *O. gidleyi* (including O. *larrabeei*) (localities SP1F, ?CP116E, CP128AA*, A); *O. hollisteri* (localities CP128C*, CP131, CP132B); *O. martini* (locality CP123D*); *O. pedroensis* (including *O. fossilis*) (localities SB14F*, SB18B, CP131, CP132B).

Onychomys sp. is also known from localities NB13B, C, NB33A, SB50, SP1H, SP5, CP128AA, B.

Comments: Fossil species of *Onychomys* are very similar, separated with great difficulty, and samples of *Onychomys* are usually small. Carleton and Eshelman (1979) reviewed the fossil record of *Onychomys*, noting that there were frequently two species of slightly different size recorded from well-sampled sites, possibly reflecting the presence of two slightly different extant species of *Onychomys* (*O. leucogaster*, the large species, and *O. torridus*, the small species). Sexual dimorphism has never been observed in any species of rodent. Two species of *Onychomys* with slightly different size are frequently found in Pliocene faunas. Both *O. leucogaster* and *O. torridus*, along with the extinct species *O. jinglebobensis*, are recorded from Rancholabrean NALMA; two extinct species, *O. pedroensis* and *O. jinglebobensis*, are recorded from Irvingtonian NALMA.

MEGACRICETODONTINI MEIN AND FREUDENTHAL, 1971

Characteristics: Incisor ornamentation absent; dentition brachydont to mesodont with tendency to reduce accessory lophs and lophids plus reduction of longitudinal mure; m1 anteroconid narrow and symmetrical in primitive taxa.

Tregomys Wilson, 1968 (including *Gnomomys; Copemys*, in part)

Type species: *Tregomys shotwelli* Wilson, 1968.

Type specimen: UMMP V55787.

Characteristics: Small cricetid; simple, low-crowned cheek teeth; m1 with narrow, symmetrical median anteroconid not closely appressed to metaconid; metalophulid directed toward protoconid, not anterior to it (as in *Copemys*) and with subequal anterior cingula; M1 with relatively narrow, single-cusped anterocone; M2 with prominent protolophule II but usually lacking protolophule I and accessory lophs; posterior half of m1 (and m2) noticeably wider than the anterior half (modified from Wilson, 1968). Note that *Gnomomys saltus* Wilson (1968) is considered a junior synonym of *T. shotwelli*, following Voorhies (1990a). *T. shotwelli* and *G. saltus* are virtually identical in size, which is smaller than any known species of *Copemys*.

Average length of m1: 1.19 mm (*T. shotwelli*) to 1.40 mm (*T. pisinnus*) (Wilson, 1968).

Included species: *T. shotwelli* (including *Gnonomys saltus*) (known from localities CP114A, CP116B, CP123B, E); *T. pisinnus* (originally described as *Copemys pisinnus*) (localities CP56IIB, CP116B, CP123B, E).

Tregomys sp. is also known from localities CP114A, B.

Comments: Korth (1998) considered *T. pisinnus* as a species of *Copemys* (e.g., *Copemys pisinnus). Tregomys* looks suspiciously like the Eurasian genus *Megacricetodon*, but it does appear to be distinct, based primarily on a single-cusped anterocone in M1 and the small size of *T. shotwelli*. *Tregomys* seems closest to *C. tenuis* and *C. russelli*, both of which have a narrow, slightly asymmetrical anteroconid on m1; it differs from these species in being smaller, the metaconid is not close to the anteroconid, and the metalophulid is directed toward the protoconid, not anterior to it. *Tregomys* probably represents a second Barstovian immigrant (relative to *Copemys*) from Eurasia that became established in the eastern part of North America; it could very well represent the ancestor of the sigmodont lineage of cricetids.

SIGMODONTINI WAGNER, 1843

Characteristics: Incisor ornamentation absent; dentition brachydont to mesodont; near alignment of lophs and lophids in primitive taxa, with progressive alignment of lophs and lophids in advanced taxa; general reduction of accessory lophs and lophids; M1 anterocone weakly bilobed (primitive taxa) becoming strongly bilobed (advanced taxa); accessory rootlets poorly developed in primitive taxa, well developed in derived taxa.

Abelmoschomys Baskin, 1986

Type species: *Abelmoschomys simpsoni* Baskin, 1986.

Type specimen: UF 61326, isolated right M1.

Characteristics: Small, brachydont sigmodontine; M1 with weakly to moderately bifurcated anterocone; m1 with unicusped to incipiently bifurcated anteroconid that is not close to or appressed to the metaconid; accessory rootlets usually present on M1, occasionally present on m1; upper molar protolophule II not aligned with anterior arm of hypocone; lower molar entolophid not aligned with posterior arm of protoconid; alternation of labial and lingual cusps poorly developed. (Modified from Baskin [1986], who did not describe medial and posterior molars.)

Average length of m1: 1.48 mm (Baskin, 1986).

Included species: *A. simpsoni* only, known from locality GC11A* only

Comments: *Abelmoschomys* is distinguished from *Tregomys* by having a wide, weakly bilobed anterocone on M1; it is distinguished from *Copemys* based on the development of accessory rootlets in *Abelmoschomys*. Roots of cheek teeth in *Copemys* are well documented, but rootlets are virtually unknown. Baskin (1986) suggested that the incipient development of a bifurcated anterocone and anteroconid on the anterior molars of *Abelmoschomys* marks the initiation of the sigmodont line of cricetids in North America.

Antecalomys Korth, 1998 (including *Peromyscus*, in part)

Type species: *Antecalomys phthanus* Korth, 1998.

Type specimen: UNSM 101543, isolated left M1.

Characteristics: Small brachydont cricetid lacking accessory lophs and styles; alignment (or near-alignment) of posterior arm of protoconid and metalophid in lower molars; non-alignment of protolophule II with anterior arm of hypocone in upper molars; M1 with asymmetrically bilobed anterocone; m1 with transversely broad, slightly asymmetrical anteroconid; M3 and m3 small, lacking metacone (in M3) and entoconid (in m3); accessory rootlets variably present on M1 (modified from Korth, 1998).

Average length of m1: 1.34 mm (*A. vasquezi*; Jacobs, 1977b) to 1.59 mm (*A. phthanus*; Korth, 1998).

Included species: *A. phthanus* (known from locality CP116B); *A. valensis* (originally described as *Peromyscus valensis* (localities NB35IV, SB60, PN11B, PN14, PN15); *A. vasquezi* (locality SB10).

Antecalomys sp. is also known from localities SB11, CP116A.

Calomys (Bensonomys) Waterhouse, 1837 (including *Cimmaronomys*)

Type species: *Calomys bimaculatus*, the extant vesper mouse.

Type specimen: None designated.

Characteristics: Small, very brachydont cricetines (primitively), bilobed anterocone on M1; bilobed anteroconid on m1; M3 and m3 reduced, comparable to Recent species of *Peromyscus*; anterior end of heavy masseteric crest situated below the anterior root of m1; mental foramina very high, on dorsal side of dentary anterior to the end of the masseteric crest (after Baskin, 1978).

Average length of m1: 1.25 mm (*C. elachys*) to 1.89 mm. (*C. baskini*) (Lindsay and Jacobs, 1985).

Included species: *C. arizonae* (known from localities SB12, SB14A*, B, D, E, SB18B); *C. baskini* (localities SB60, SB62*); *C. coffeyi* (locality SP3B*); *C. elachys* (localities SB12, SB60, SB62*); *C. eliasi* (localities SP1F, CP128AA–?C); *C. gidleyi* (locality SB11*); *C. hershkovitzi* (locality CE1*); *C. meadensis* (localities WM26, ?CP118, CP128C*, D, CP130B); *C. stirtoni* (locality CP128AA*); *C. yazhi* (locality SB11*).

Calomys (Bensonomys) sp. is also known from localities NB13C, SB12, SB15A, ?SP4B, SP5, CP116E, F, CP128AA, B.

Comments: Hibbard (1953) described *Cimarronomys stirtoni* from the Saw Rock Canyon Fauna of Kansas.

Cimarronomys has no features that separate it from *Calomys* (*Bensonomys*); Martin, Goodwin, and Farlow, 2002). *Cimarronomys* is herein considered a junior synonym of *Calomys* and is identified as *C. stirtoni*, following Martin, Goodwin, and Farlow, (2002). Several of the species of *Calomys* are very similar (e.g., *C. stirtoni, C. baskini, C. arizonae*, and *C. eliasi*); some of them may be synonymized when placed under close scrutiny.

Symmetrodontomys Hibbard, 1941 (including _Peromyscus_, in part)

Type species: *Symmetrodontomys simplicidens* Hibbard, 1941.

Type specimen: KU 4601.

Characteristics: Small brachydont cricetine; large, very wide anterocone on M1; bilobed anteroconid on m1; cusps slightly alternate in position, with strong transverse lophs joining cusps on the opposite side of the tooth; lophs directed anteroposteriorly and weakly developed; M3 and m3 relatively large; masseteric ridge terminating below anterior root of m1; mental foramina high, anterior to end of masseteric ridge, on upper margin of diastema (modified from Hibbard, 1941).

Average length of m1: 1.55 mm (*S. daamsi*) to 1.75 mm (*S. simplicidens*) (Martin, Goodwin, and Farlow, 2002).

Included species: *S. simplicidens* (known from localities SP1F, CP128AA*, A, [D]); *S. beckensis* (originally described as *Peromyscus beckensis*) (locality SP1F*); *S. daamsi* (locality CE1*).

Comments: *Symmetrodontomys* is very similar to *Calomys* (*Bensonomys*), differing primarily in having a very large and wide anterocone on M1 of *Symmetrodontomys* and greater tooth crown height. Martin (2000) recognized that *Peromyscus sawrockensis* Hibbard (1964) is inseparable from *Symmetrodontomys simplicidens* Hibbard (1941). Martin, Goodwin, and Farlow (2002) placed *P. sawrockensis* Hibbard (1953) as a junior synonym of *S. simplicidens*, which is followed here. When preparing this manuscript, I considered the possibility that *Peromyscus beckensis* of Dalquest (1978), with a bilobed anteroconid and mental foramen on the dorsal surface of the diastema, is a synonym of *Prosigmodon oroscoi* (Jacobs and Lindsay, 1981). However, R. A. Martin (personal communication, 2002) pointed out that the species described by Dalquest is better assigned to *Symmetrodontomys* Hibbard than to *Prosigmodon* Jacobs and Lindsay, and I agree. R. A. Martin suggested that the appearance of *Symmetrodontomys* approximates the Blancan/Hemphillian NALMA boundary.

Baiomys True, 1894

Type species: *Baiomys taylori*, the extant Northern pygmy mouse.

Type specimen: None designated.

Characteristics: Small brachydont cricetines; small narrow incisors; cheek teeth with alternate cusps; upper cheek teeth protolophule II not aligned with anterior arm of hypocone; M1 usually carrying weakly bilobed anterocone; lower cheek teeth; entolophulid not aligned with posterior arm of protoconid; m1 anteroconid weakly to strongly bilobed; M3 and m3 reduced in size; masseteric ridge terminating anteriorly below anterior root of m1; mental foramina high, on posterolateral rim of diastema.

Average length of m1: 1.12 mm (*B. aquilonius*; UMMP 55055) to 1.70 mm (*B. sawrockensis*; Hibbard, 1964).

Included species: *B. aquilonius* (known from locality PN23A*); *B. brachygnathus* (localities SB14D, F*, SB18); *B. kolbi* (localities SB60, SB62, CP128A*); *B. minimus* (localities SB14A*, C, D, SB15A); *B. mowi* (localities CC47C*, SB15A); *B. rexroadi* (localities SP1F, CP128AA*–C); *B. sawrockensis* (locality CP128AA*).

Baiomys sp. is also known from localities GC15C, NB13C, SB18B, SB34IIB, SP1H, SP5, CP116F, CP128B, PN4, PN23A. The genus is also known from the Pleistocene.

Prosigmodon Jacobs and Lindsay, 1981 (including _Sigmodon_, in part)

Type species: *Prosigmodon oroscoi* Jacobs and Lindsay, 1981.

Type specimen: IGCU 1217, isolated left M1.

Characteristics: Medium- to large-size cricetid; mesodont, rooted teeth bearing large robust cusps; accessory rootlets often developed, especially on M1; M1 with wide, bilobed anterocone; m1 with wide, strongly bilobed anteroconid, anterior arm of protoconid joining the lingual lobe of anteroconid; M3 with indistinct metacone; both M3 and m3 little reduced; masseteric ridge terminating below anterior root of m1; mental foramina high, on dorsal rim of posterior diastema (modified from Jacobs and Lindsay, 1981).

Average length of m1: 2.24 mm (*P. chihuahuensis*; Lindsay and Jacobs, 1985) to 2.60 mm (*P. holocuspis*; Czaplewski, 1987).

Included species: *P. oroscoi* (known from localities SB60, SB62, SP3B); *P. chihuahuensis* (localities SB60, SB62, SB64); *P. holocuspis* (locality SB12*).

Prosigmodon sp. is also known from localities SB35A, CP123A.

Comments: Repenning and May (1986) identified two sigmodontines (*Prosigmodon* sp. aff. *Sigmodon intermedius* and Cf. *Oryzomys*) and illustrated only the latter. The illustration of Cf. *Oryzomys* does not distinguish it from *Prosigmodon*, nor does the brief description. These two taxa are grouped here, as *Prosigmodon* sp. The Truth or Consequences (locality SB35A) material should also be compared with *Jacobsomys* from the Verde Fauna of Arizona (locality SB12), and with ?*Oryzomys* sp. from the Bartlett Mountain Fauna of Oregon (locality PN11B).

Dalquest (1983) described *Paronychomys*(?), an m2 (as M2), and *Copemys*(?), an M1 (as m1), from Coffee

Ranch (locality SP3B); based on illustrations and descriptions therein, both specimens seem better assigned to *Prosigmodon oroscoi* (author's notes).

***Sigmodon* Say and Ord, 1825**

Type species: *Sigmodon hispidus*, the extant hispid cotton rat.
Type specimen: None designated.
Characteristics: Medium- to large-size cricetid; robust, mesodont, strongly lophate, rooted cheek teeth, usually with accessory rootlets; M1 with broad, single-cusped anterocone joined medially by anterior arm of protocone; m1 with large, single-cusped anteroconid; M3 and m3 relatively large; masseteric ridge terminating anteriorly below anterior root of m1; mental foramina high, on dorsal surface of posterior diastema.
Average length of m1: 2.03 mm (*S. medius*; Tomida, 1987) to 2.88 mm (*S. lindsayi*; Martin and Prince, 1989).
Included species: *S. bakeri* (known from locality GC15F); *S. curtisi* (localities GC15C, GC18IIB, SB14D–F, CP80, CP132C, D); *S. hudspethensis* (localities SB49*, SB50); *S. libitinus* (localities GC15D*, E, GC18IIIA, GC18IV); *S. lindsayi* (locality NB13C*); *S. medius* (localities ?GC13B, GC15A, GC17A, NB13A, B, SB12, SB14B*, SB15A, C, SB35[A], D, SP1F, H, SP5, CP118, CP128AA, C, CP130, CP131); *S. minor* (localities GC18IIB, CC47B–D, CC48A, SB14D–F*, CP80, CP128D, CP132B).

Sigmodon sp. is also known from localities ?NB13A–C, SB14C, SB15B, C, SB35D, CP128B.

Sigmodon spp. are well represented in Pleistocene faunas of Arizona, Florida, Georgia, Kansas, New Mexico, Oklahoma, Texas, and Mexico, and the genus is also known from the Recent of South America.

Comments: The morphological transition between *S. medius* and *S. minor* is well documented (Cantwell, 1969; Martin, 1970, 1974, 1979; Harrison, 1978) and uncontested. Some authors (e.g., Harrison 1978; Martin, 1979) interpreted these as a single polymorphic species, *S. minor*, based on the significant morphological continuity of specimens (especially m1 size) of this taxon when represented by large samples. I recognize this as an important morphological and chronological cline preserved in the fossil record, but I prefer to distinguish these as two separate species primarily because (1) the early Blancan morphs (e.g., *S. medius*) are distinct from the later Irvingtonian morphs (e.g., *S. minor*), and (2) the morphocline crosses a temporal boundary (Blancan–Irvingtonian boundary).

It is virtually impossible (and impractical) to separate these species when recorded in faunas that approximate the Blancan–Irvingtonian boundary. In those instances, and in this compilation, I follow a practical "rule of thumb" to assign the morphological cline to *S. minor* when (and only when) the larger and morphologically distinct species *Sigmodon curtisi* also occurs in the fauna. When small samples and/or undiagnostic measurements of the *Sigmodon* morphocline are represented, the taxon should be designated *Sigmodon* sp. With more testing, we might find that the appearance of *S. curtisi* is a convenient small mammal boundary–taxon for the Blancan–Irvingtonian NALMA boundary.

***Jacobsomys* Czaplewski, 1987 (including *Oryzomys*, in part)**

Type species: *Jacobsomys verdensis* Czaplewski, 1987.
Type specimen: MNA V4849, isolated left M1.
Characteristics (from Czaplewski, 1987): Medium-size cricetid mesodont; cheek teeth with distinctly bilobed anterocone on M1; major cusps of upper molars opposite rather than alternate in position; anterior mures of M1, M2, and m1 oriented anteroposteriorly and situated along midline; accessory lophs well coalesced to bases of the major cusps.
Length of m1: 2.04 mm (holotype).
Included species: *Jacobsomys verdensis* only, known from locality SB12* only.

Jacobsomys sp. is also reported from locality SB35A. Questionable species of *Jacobsomys* are ?*Oryzomys pliocaenicus* Hibbard (1937) (locality CP123D), plus ?*Oryzomys* sp. of Shotwell (1970) (locality PN11B), and ?*Oryzomys* sp. of Repenning and May (1986) (locality SB35A).

Comments: *Orozomys palustris*, the extant marsh rice rat, inhabits the Atlantic Coast as far north as Pennsylvania and waterways along the Gulf Coast and Mississippi River. The genus *Orozomys* is well represented in modern faunas of Mexico, Central America, and South America; it was a major contributor to the radiation of Sigmodontini in Central and South America. *Orozomys* is known only from the three sites (listed in the paragraph above) in latitudes north of Mexico; these records are markedly different from those of *Sigmodon*, another major contributor to the radiation of Sigmodontini in Central and South America. This material, from Edson Quarry (locality CP123D) Kansas, from Bartlett Mountain (locality PN11B) Oregon, and Truth or Consequences (locality SB35A) New Mexico, may represent species of *Jacobsomys*, the most likely ancestor of *Orozomys*. The Bartlett Mountain material is approximately the same size as *J. verdensis* but has less-robust cusps and lacks some of the accessory lophs of *J. verdensis*; the Edson Quarry material and the Truth or Consequences material have never been adequately described and compared with *Jacobsomys*.

NEOTOMINI MCKENNA AND BELL, 1997

Characteristics: Dentitions high crowned, tending toward hypsodonty but with strong roots; opposing lophs and lophids strongly aligned; accessory lophs and lophids absent; primitively m3 S-shaped in occlusal outline; accessory rootlets present primitively, with reduction of accessory roots and incipient fusion of primary roots in derived taxa.

Note: *Neotoma* is placed here in the Cricetinae, in a separate tribe near the tribe Sigmodontini, primarily because early records of the genus have accessory rootlets in M1 (Tomida, 1987 [Fig. 19A]; Zakrzewski, 1991); a feature present in primitive Sigmodontini. Accessory rootlets in extant species of *Neotoma* are rare, to my knowledge. Many other studies have placed *Neotoma* along with other high-crowned cricetids (the tribe Galushamyini of this study) as the foundation of the tribe Neotomini. By implication, *Neotoma* is not closely related to Galushamyini; it should be placed near the Sigmodontini, its presumed closest relatives.

***Neotoma* Say and Ord, 1825 (including *Parahodomys*)**

Type species: *Neotoma floridana*, the extant Eastern wood rat.

Type specimen: None designated.

Characteristics: Medium- to large-size cricetine; very high-crowned but rooted cheek teeth (including variable development of rootlets), lacking cementum in reentrants; cusps strongly lophate, usually joining by a mure lingual to midline in upper molars and labial to midline in lower molars; anterocone on M1 and anteroconid on m1 usually located slightly labial to midline; M3 and m3 are relatively small, compared with other cheek teeth, metacone and entoconid indistinct or absent; masseteric ridge terminating anteriorly below anterior root of m1, below level of diastema; mental foramina located on the lateral side of the dentary anterior to the masseteric ridge.

***Neotoma (Paraneotoma)* Hibbard 1967**

Characteristics (from Tomida, 1987): Relatively high-crowned cheek teeth with thick enamel; upper cheek teeth labial flexi and lower cheek teeth lingual flexids directed more or less perpendicular to main axis of tooth; m3 wearing to a distinctive S-pattern.

Average length of m1: 2.80 mm (*N. vaughani*; MNA V-4882) to 3.82/2.24 mm (*N. leucopetrica*; Zakrzewski, 1991).

Included species: *N. fossilis* (known from localities [WM26], CC47B– D, CC48A, SB14A*, B, D, SB15A); *N. leucopetrica* (locality CP131*); *N. minutus* (locality SP3B*); *N. quadriplicata* (localities SB15A, C, SB35A, SP1F, ?H, ?SP5, CP128AA*–C, CP130, ?PN3C, ?PN4, PN23A); *N. sawrockensis* (localities SP1F, CP128AA*); *N. taylori* (localities SB18B, CP131, CP132B*); *N. vaughani* (locality SB12*, ?CP128AA).

***Neotoma (Parahodomys)* Hibbard, 1967**

Characteristics (from Tomida, 1987): Relatively high-crowned cheek teeth with thick enamel; upper cheek teeth labial flexi and lower cheek teeth lingual flexids directed more oblique relative to the long axis of the tooth; m3 lacking a posteroflexid.

Included species: *Neotoma (Parahodomys) primaevus* only, known from locality CP128A* only.

Neotoma (Parahodomys) sp. is also known from locality CP117A.

Neotoma sp. (subgenus uncertain) is also known from localities GC15C, CC47D, CC48C, CC53, NB13A-C, SB12, SB14C, D, F, SB17, SB64 (author's notes). *Neotoma* is well represented in Irvingtonian NALMA and Rancholabrean NALMA, with seven extant species living primarily in the western and southern parts of North America.

Comments: Based only on the description and illustrations (material was not examined), *Neotoma minutus* from Coffee Ranch (locality SP3B) appears as though it should be assigned to *Repomys* May, 1981. The Coffee Ranch material is based on two isolated specimens (M2 and m3); the size of M2 from Coffee Ranch is well within the range of *Repomys* material illustrated by Tomida (1987: Fig. 22). Voorhies (1990b) described *Neotoma* cf. *N. sawrockensis* from the Mailbox (locality CP116E), Santee, and Devil's Nest Airstrip (locality CP116F) localities, all Hemphillian NALMA, in Nebraska. My colleague Dr. Rick Zakrzewski later examined the material from Santee and Devil's Nest Airport localities and concluded that these specimens are from an unnamed dipodid rather than *Neotoma*. Voorhies assured Zakrzewski that the material from the Mailbox Locality is similar to that from the Santee and Devil's Nest Airport localities. Therefore, the appearance of *Neotoma* prior to the Blancan NALMA (e.g., at localities CP116E and CP116F) is seriously questioned; it is widespread and well represented in early Blancan localities.

BIOLOGY AND EVOLUTIONARY PATTERNS

Cricetidae have been an important part of the North American small mammal fauna for much of the middle and late Cenozoic. There were two or three major pulses of North American cricetid diversity: an Oligocene pulse that began in the Chadronian NALMA (with *Eumys*), an expansion during the Orellan NALMA (with *Eoeumys*, *Scottimus*, and *Eumys*), and a climax during the Arikareean NALMA (with *Leidymys* and *Paciculus*). Cricetids declined during the latter part of the Arikareean NALMA and during the Hemingfordian NALMA, with the second cricetid pulse beginning in the Barstovian NALMA, with *Copemys*. The Blancan NALMA recorded a rejuvenation (or third pulse) of North American cricetid diversity, with many of the modern genera (e.g., *Peromyscus*, *Baiomys*, *Neotoma*, *Sigmodon*, *Onychomys*, and *Reithrodontomys*) appearing, or with ancestors well established in faunas of the Hemphillian NALMA (Figure 27.3).

The success of modern Cricetidae (and their descendants, the Muridae) resides primarily in their high fecundity and broad ecological tolerance. We often think of rabbits as highly productive mammals, but cricetid rodents represent greater biomass than rabbits, both in modern and in fossil faunas, based on their greater fecundity and diversity. Reproductive capacity within cricetids is highly variable, but it is very reasonable to propose that a small

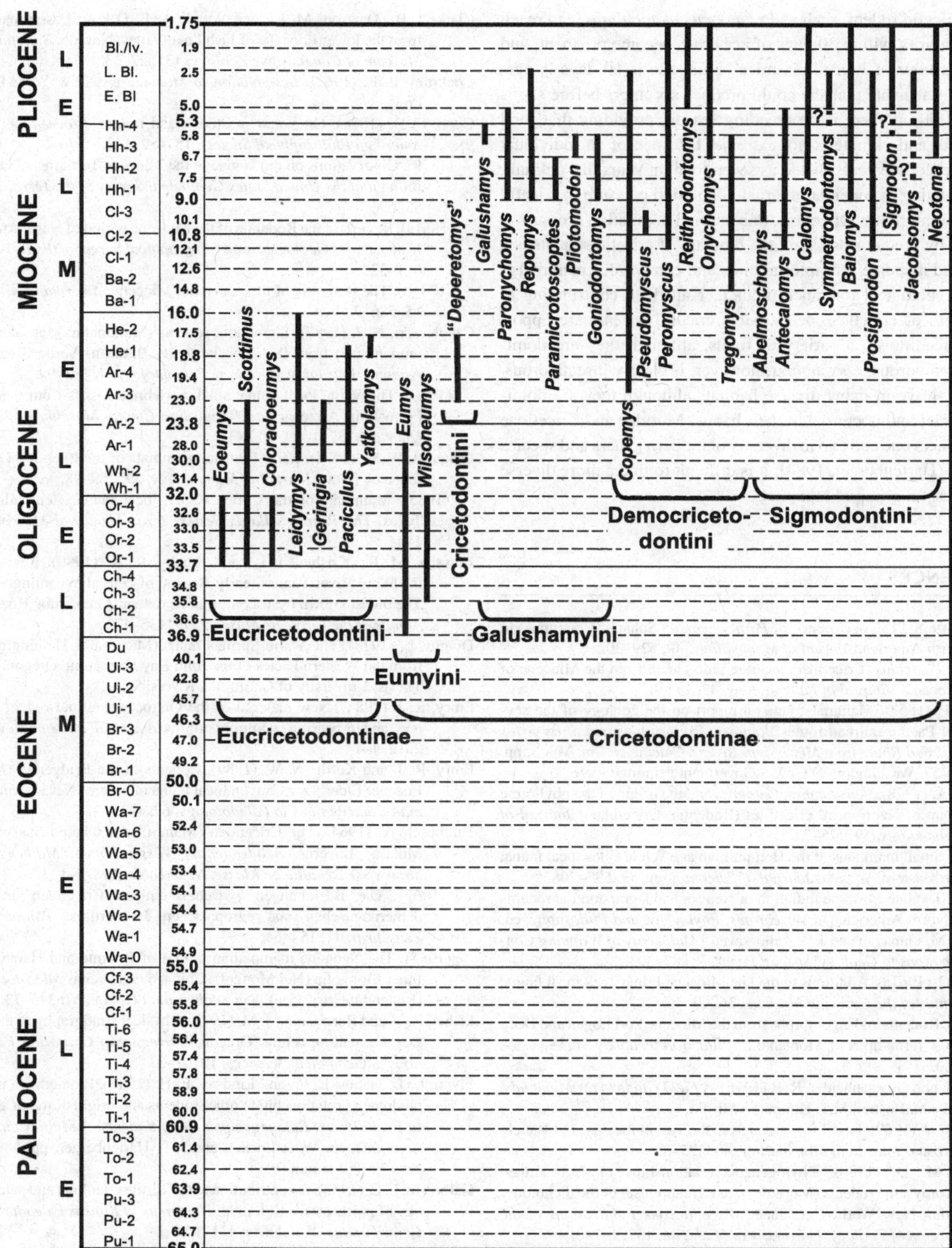

Figure 27.3. Temporal ranges of North American cricetid genera.

female cricetid rodent, such as *Peromyscus maniculatus*, can reach sexual maturity within 30 days of her birth, become pregnant, and produce a litter of four to six mice before she is 60 days in age. It is highly feasible that she could produce six litters before she is one year old, as there is no breeding season, providing that food is plentiful and climate is not extréme. Lifespan of an individual might exceed five years (listed as seven to eight years in Cockrum, 1962), creating exponential internal population pressure in cricetid rodent communities. Of course, this internal population pressure is balanced by predator pressure; the biomass of cricetid rodents is a prime food resource for raptorial birds and carnivorous mammals.

Broad ecological tolerance is another attribute contributing to the success of cricetid rodents. Most cricetid rodents are opportunistic, feeding on a variety of foods, although they are dominantly granivorous; *Onychomys*, however, is clearly insectivorous. Cricetids thrive in many diverse habitats although they are dominantly secret and nocturnal in their habits. Muridae and Cricetidae rank numbers one and two in diversity of modern genera and species of rodent (Hartenberger, 1984); it is difficult to find a more diverse and successful group of mammals.

REFERENCES

Alker, J. (1968). The occurrence of *Paracricetodon* Schaub (Cricetidae) in North America. *Journal of Mammalogy*, 49, 529–30.

(1969). *Paciculus* (Cricetinae, incertae sedis) teeth from the Miocene of Nebraska. *Journal of Paleontology*, 43, 171–4.

Baird, S. F. (1857). Mammals: general report on the zoology of the several Pacific railroad routes. *Reports, Explorations and Surveys for Railroad Route from Mississippi River to Pacific Ocean*, Vol. 8, pp. 1–757. Washington, D.C.: US Goverment Printing Office.

Baskin, J. A. (1978). *Bensonomys, Calomys*, and the origin of the phyllotine group of Neotropical cricetines (Rodentia: Cricetidae). *Journal of Mammalogy*, 59, 125–35.

(1979). Small mammals of the Hemphillian age White Cone local fauna, northeastern Arizona. *Journal of Paleontology*, 53, 695–708.

(1986). The late Miocene radiation of Neotropical sigmodontine rodents in North America. [In *Vertebrates, Phylogeny, and Philosophy*, ed. K. M. Flanagan and J. A. Lillegraven.] *University of Wyoming Contributions to Geology, Special Paper*, 3, 287–303.

Black, C. C. (1961a). Additions to the late Miocene Flint Creek local fauna. *Annals of the Carnegie Museum*, 36, 69–76.

(1961b). Rodents and lagomorphs from the Miocene Fort Logan and Deep River formations of Montana: *Postilla. Yale Peabody Museum*, 48, 1–20.

Burt, W. H. and Grossenheider, R. P. (1976). *A Field Guide to the Mammals*, 3rd edn. Boston, MA: Houghton Mifflin.

Cantwell, R. J. (1969). Fossil *Sigmodon* from the Tusker locality, 111 Ranch, Arizona. *Journal of Mammalogy*, 50, 375–8.

Carleton, M. D. (1980). Phylogenetic relationships of Neotomine–Peromyscine rodents (Muroidea) and a reappraisal of the dichotomy within New World Cricetinae. *Miscellaneous Publications of the Museum of Zoology, University of Michigan*, 157, 1–146.

Carleton, M. D. and Eshelman, R. E. (1979). A synopsis of fossil grasshopper mice, genus *Onychomys*, and their relationships to Recent species. [*Claude W. Hibbard Memorial Volume 7.*] *University of Michigan, Papers on Paleontology*, 21, 1–63.

Carleton, M. D. and Musser, G. G. (1984). Muroid Rodents. In *Orders and Families of Recent Mammals of the World*, ed. S. Anderson and J. K. Jones, pp. 289–379. New York: Wiley-Interscience.

Clark, J. B., Dawson, M. R., and Wood, A. E. (1964). Fossil mammals from the lower Pliocene of Fish Lake Valley, Nevada. *Bulletin of the Museum of Comparative Zoology*, 131, 27–63.

Cockrum, E. L. (1962). *Introduction to Mammalogy*. New York: Ronald Press.

Cope, E. D. (1874). On a new mastodon and rodent. *Proceedings of the American Philosophical Society*, 1874, 221–3.

(1879). Observations on the faunae of the Miocene Tertiaries of Oregon. *Bulletin of the United States Geological Survey of the Territories*, 5, 55–69.

(1881a). Review of the Rodentia of the Miocene period of North America. *United States Geological and Geographical Survey of the Territories*, 6, 1–370.

(1881b). The Rodentia of the American Miocene. *American Naturalist*, 15, 586–7.

Czaplewski, N. J. (1987). Sigmodont rodents (Mammalia: Muroidea: Sigmodontinae) from the Pliocene (early Blancan) Verde Formation, Arizona. *Journal of Vertebrate Paleontology*, 7, 183–99.

(1990). The Verde local fauna: small vertebrate fossils from the Verde Formation, Arizona. *San Bernardino County Museum Associates, Quarterly*, 37, 1–39.

Dalquest, W. W. (1978). Early Blancan mammals of the Beck Ranch local fauna of Texas. *Journal of Mammalogy*, 59, 269–98.

(1983). Mammals of the Coffee Ranch local fauna, Hemphillian of Texas. *The Pearce-Sellards Series, Texas Memorial Museum*, 38, 1–41.

Dawson, M. R., Krishtalka, L., and Stucky, R. K. (1990). Revision of the Wind River Faunas, early Eocene of central Wyoming. Part 9: The oldest known hystricomorphous rodent (Mammalia: Rodentia). *Annals of the Carnegie Museum*, 59, 135–47.

Dingus, L. (1978). The Warm Springs fauna (Mammalia, Hemingfordian) from the Western Facies of the John Day Formation, Oregon. M.Sc. Thesis, University of California, Riverside.

Emry, R. J. (1981). New material of the Oligocene muroid rodent *Nonomys*, and its bearing on muroid origins. *American Museum Novitates*, 2712, 1–14.

Emry, R. J. and Korth, W. W. (1989). Rodents of the Bridgerian (Middle Eocene) Elderberry Canyon local fauna of eastern Nevada. *Smithsonian Contributions to Paleobiology*, 67, 1–14.

Fahlbusch, V. (1964). Die Cricetiden (Mamm.) der Oberen Susswasser–Molasse Bayerns. *Akademie der Wissenshaften, Mathematisch-naturwissenschaftliche Klasse, Neue Folge*, 118, 1–134.

(1967). Die Beziehungen zwischen einigen Cricetiden des nordamerikanischen und europaischen Jungtertiars. *Palaontologie Zeitschrift*, 41, 154–64.

(1987). The Neogene mammalian faunas of Ertemte and Harr Obo in Inner Mongolia (Nei Mongol), China, 5. The genus *Microtoscoptes* (Rodentia: Cricetidae). *Senckenbergiana Lethae*, 67, 345–73.

Fisher, R. V. and Rensberger, J. M. (1972). Physical stratigraphy of the John Day Formation, central Oregon. *University of California Publications in Geological Sciences*, 101, 1–33.

Flynn, L. J., Jacobs, L. L., and Lindsay, E. H. (1985). Problems in muroid phylogeny: relationship to other rodents and origin of major groups. In *Evolutionary Relationships Among Rodents: A Multidisciplinary Approach*, ed. W. P. Luckett and J. L. Hartenberger, pp. 589–616. New York: Plenum Press.

Galbreath, E. C. (1953). A contribution to the Tertiary geology and paleontology of northeastern Colorado. *University of Kansas Paleontological Contributions, Vertebrata*, 4, 1–120.

Galusha, T. and Blick, J. C. (1971). Stratigraphy of the Santa Fe Group, New Mexico. *Bulletin of the American Museum of Natural History*, 144, 1–128.

Giglioli, X. (1874). *Bolletino Societe Geographica Italia, Roma*, 11, 326.

Gloger, X. (1841). *Gemeinutziges Hand-und Hilfs buch der Naturgeschichte*, 1, 95.

Green, M. (1985). Micromammals from the Miocene Bijou Hills local fauna, South Dakota. [*Guidebook for 45th Annual Meeting, Society Vertebrate Paleontology.*] *Dakoterra, South Dakota School of Mines and Technology*, 2, 141–54.

Gustafson, E. P. (1978). The vertebrate faunas of the Pliocene Ringold Formation, south-central Washington. *Bulletin of the Natural History Museum, University of Oregon*, 23, 1–62.

Hall, E. R. and Kelson, K. R. (1959). *Mammals of North America*. New York: Ronald Press.

Harrison, J. A. (1978). Mammals of the Wolf Ranch local fauna, Pliocene of the San Pedro Valley, Arizona. *Occasional Papers of the Museum of Natural History, University of Kansas*, 73, 1–18.

Hartenberger, J.-L. (1971). Contribution a l'etude des genres *Gliravus* et *Microparamys* (Rodentia) de l'Eocene d'Europe. *Palaeovertebrata*, 4, 97–135.

(1984). The Order Rodentia: major questions on their evolutionary origin, relationships, and suprafamilial systematics. In *Evolutionary Relationships Among Rodents*, ed. W. P. Luckett and J.-L. Hartenberger, pp. 1–33. New York: Plenum Press.

Hibbard, C. W. (1937). Additional fauna of Edson Quarry of the middle Pliocene of Kansas. *The American Midland Naturalist*, 18, 460–4.

(1941). New mammals from the Rexroad fauna, upper Pliocene of Kansas. *The American Midland Naturalist*, 26, 337–68.

(1953). The Saw Rock Canyon fauna and its stratigraphic significance. *Papers of the Michigan Academy of Science, Arts and Letters*, 38, 387–411.

(1964). A contribution to the Saw Rock Canyon local fauna of Kansas. *Papers of the Michigan Academy of Science, Arts and Letters*, 44, 115–27.

(1967). New rodents from the late Cenozoic of Kansas. *Papers of the Michigan Academy of Science, Arts, and Letters*, 52, 115–32.

Hoffmeister, D. F. (1945). Cricetine rodents of the middle Pliocene of the Mulholland fauna, California. *Journal of Mammalogy*, 26, 186–91.

(1959). New cricetid rodents from the Niobrara River fauna, Nebraska. *Journal of Paleontology*, 33, 696–9.

Hooper, E. T. and Musser, G. G. (1964). The glans penis in Neotropical cricetines (Family Muridae) with comments on the classification of muroid rodents. *Miscellaneous Publications of the Museum of Zoology, University of Michigan*, 123, 1–57.

Illiger, C. (1811). *Prodromus Systematis Mammalilum et Avium Additis Terminis Zoographicis Utriudque Classis*. Berlin: C. Salfeld.

Jacobs, L. L. (1977a). A new genus of murid rodent from the Miocene of Pakistan and comments on the origin of Muridae. *PaleoBios*, 25, 1–11.

(1977b). Rodents of the Hemphillian age Redington local fauna, San Pedro Valley, Arizona. *Journal of Paleontology*, 51, 505–19.

(1978). Fossil rodents (Rhizomyidae and Muridae) from Neogene Siwalik deposits, Pakistan. *Bulletin Series, Museum of Northern Arizona*, 52, 1–103.

Jacobs, L. L. and Lindsay, E. H. (1981). *Prosigmodon oroscoi*, a new sigmodont rodent from the late Tertiary of Mexico. *Journal of Paleontology*, 55, 425–30.

Kelly, T. S. (1992). New Uintan and Duchesnean (middle and late Eocene) rodents from the Sespe Formation, Simi Valley, California. *Bulletin of the Southern California Academy of Sciences*, 91, 97–120.

Korth, W. W. (1981). New Oligocene rodents from western North America. *Annals of the Carnegie Museum*, 50, 289–318.

(1987). New rodents (Mammalia) from the late Barstovian (Miocene) Valentine Formation, Nebraska. *Journal of Paleontology*, 61, 1058–64.

(1992). Fossil small mammals from the Harrison Formation (late Arikareean: earliest Miocene), Cherry County, Nebraska. *Annals of the Carnegie Museum*, 61, 69–131.

(1994). *The Tertiary Record of Rodents in North America*. New York: Plenum Press.

(1997). Additional rodents (Mammalia) from the Clarendonian (Miocene) of northcentral Nebraska, and a review of Clarendonian rodent biostratigraphy of that area. *Paludicola*, 1, 97–111.

(1998). Rodents and lagomorphs (Mammalia) from the late Clarendonian (Miocene) Ash Hollow Formation, Brown County, Nebraska. *Annals of the Carnegie Museum*, 67, 299–348.

Kurtén, B. and Anderson, E. (1980). *Pleistocene Mammals of North America*. New York: Columbia University Press.

Leidy, J. (1856). Notices of remains of extinct Mammalia discovered by Dr. F. V. Hayden in Nebraska Territory. *Academy Natural Sciences, Proceedings*, 8, 88–90.

Lillegraven, J. A. and Wilson, R. W. (1975). Analysis of *Simimys simplex*, an Eocene rodent (?Zapodidae). *Journal of Paleontology*, 49, 856–74.

Lindsay, E. H. (1972). Small mammal fossils from the Barstow Formation, California. *University of California Publications in Geological Sciences*, 93, 1–104.

(1988). Cricetid rodents from Siwalik deposits near Chinji Village, part I. Megacricetodontinae, Myocricetodontinae and Dendromurinae. *Palaeovertebrata*, 18, 95–154.

Lindsay, E. H. and Jacobs, L. L. (1985). Pliocene small mammal fossils from Chihuahua, Mexico. *Paleontologia Mexicana, Universita Nacional Autonomica Mexico, Instituto de Geologia*, 51, 1–50.

Luckett, W. P. and Hartenberger, J.-L. (eds.) (1985). *Evolutionary Relationships among Rodents: A Multidisciplinary Analysis*. New York: Plenum Press.

Macdonald, J. R. (1963). The Miocene faunas from the Wounded Knee area of western South Dakota. *Bulletin of the American Museum of Natural History*, 125, 139–238.

(1970). Review of the Miocene Wounded Knee faunas of southwestern South Dakota. *Bulletin of the Los Angeles County Museum of Natural Science*, 8, 1–82.

Marshall, L. G. (1988). Land Mammals and the Great American Interchange. *American Scientist*, 76, 380–8.

Marshall, L. G. and Cifelli, R. L. (1990). Analysis of changing diversity patterns in Cenozoic land mammal age faunas, South America. *Palaeovertebrata*, 19, 169–210.

Martin, L. D. (1975). Microtine rodents from the Ogallala Pliocene of Nebraska and the early evolution of the Microtinae in North America. [In *Claude W. Hibbard Memorial Volume 3: Studies on Cenozoic Paleontology and Stratigraphy*, ed. G. R. Smith and N. E. Friedland.] *Papers on Paleontology, University of Michigan*, 12, 101–10.

(1980). The early evolution of the Cricetidae in North America. *University of Kansas Paleontological Contributions*, 102, 1–42.

Martin, L. D. and Corner, R. G. (1980). A new genus of cricetid rodent from the Hemingfordian (Miocene) of Nebraska. *University of Kansas Paleontological Contributions*, 103, 1–6.

Martin, R. A. (1970). Line and grade in the extinct *medius* species group of *Sigmodon*. *Science*, 167, 1504–6.

(1974). Fossil mammals from the Coleman IIA fauna, Sumter County. In *Pleistocene Mammals of Florida*, ed. S. D. Webb, pp. 35–99. Gainesville, FL: University of Florida Press.

(1979). Fossil history of the rodent genus *Sigmodon*. *Evolutionary Monographs*, 2, 1–36.

(2000). The taxonomic status of *Peromyscus sawrockensis* Hibbard, 1964 (Rodentia, Cricetidae). *Paludicola*, 2, 269–72.

Martin, R. A. and Prince, R. H. (1989). A new species of early Pleistocene cotton rat from the Anza-Borrego Desert of southern California. *Bulletin of the Southern California Academy of Sciences*, 88, 80–7.

Martin, R. A., Goodwin, H. T., and Farlow, J. O. (2002). Late Neogene (late Hemphilllian) rodents from Pipe Creek Sinkhole, Grant County, Indiana. *Journal of Vertebrate Paleontology*, 22, 137–51.

Matthew, W. D. (1924). Third contribution to the Snake Creek fauna. *Bulletin of the American Museum Natural History*, 50, 61–210.

May, S. R. (1981). *Repomys* (Mammalia: Rodentia) from the late Neogene of California and Nevada. *Journal of Vertebrate Paleontology*, 1, 219–30.

McKenna, M. C. and Bell, S. K. (1997). *Classification of Mammals Above the Species Level*. New York: Columbia University Press.

Mein, P. and Freudenthal, M. (1971). Une nouvelle classification des Cricetidae (Mammalia, Rodentia) du Tertiaire de l'Europe. *Scripta Geologica*, 2, 1–37.

Mou, Y. (1999). Biochronology and magnetostratigraphy of the Pliocene Panaca Formation, Southeast Nevada. Ph.D. Thesis, University of Arizona, Tucson.

Nowak, R. M. (1999). *Walker's Mammals of the World*, 6th edn. Baltimore, MD: Johns Hopkins University Press.

Rensberger, J. M. (1983). Successions of Meniscomyine and Allomyine Rodents (Aplodontidae) in the Oligo-Miocene John Day Formation, Oregon. *University of California Publications in Geological Sciences*, 124, 1–157.

Repenning, C. A. and May, S. R. (1986). New evidence for the age of the Palomas Formation, Truth or Consequences, New Mexico. In *New Mexico Geological Society Guidebook for the 37th Field Conference, Truth or Consequences*, pp 257–60. Socorro, NM: New Mexico Geological Society.

Rochebrune, A. T., de (1883). Faune de la Sénégambie. Mammifères. *Actea Societe Linnéenne* (Bordeaux), 37, 49–203.

Ruez, D. R. (2001). Early Irvingtonian (latest Pliocene) rodents from Inglis IC, Citrus County, Florida. *Journal of Vertebrate Paleontology*, 21, 172–85.

Say, T. and Ord, G. (1825). Description of a new species of Mammalia, whereon a new genus is proposed to be found. *Journal of the Academy of Natural Sciences, Philadelphia*, 4, 352–5.

Schaub, S. (1934). Uber einige fossile Simplicidentaten aus China und der Mongolei. *Abhhandlungen Schweizerische Palaeontologische Gesellschaft*, 54, 1–40.

(1940). Zur Revision des Genus *Trilophomys* Deperet. *Verhandlugen Naturforschen Gesellschaft, Basel*, 51, 65–75.

Setoguchi, T. (1978). Paleontology and geology of the Badwater Creek area, central Wyoming. Part 16: The Cedar Ridge local fauna (late Oligocene). *Bulletin of the Carnegie Museum of Natural History*, 9, 1–61.

Shotwell, J. A. (1970). Pliocene mammals of southeast Oregon and adjacent Idaho. *Bulletin of the Museum of Natural History, University of Oregon*, 17, 1–103.

Simpson, G. G. (1945). Principles of classification and a classification of mammals. *Bulletin of the American Museum of Natural History*, 85, 1–350.

Sinclair, W. J. (1905). New or imperfectly known rodents and ungulates from the John Day series. *University of California Bulletin of the Geology Department*, 4, 125–43.

Skinner, M. F. and Johnson, F. W. (1984). Tertiary stratigraphy and Frick Collection of fossil vertebrates from North-Central Nebraska. *Bulletin of the American Museum of Natural History*, 178, 215–368.

Skinner, M. F. and Taylor, B. E. (1967). A revision of the geology and paleontology of the Bijou Hills, South Dakota. *American Museum Novitates*, 2300, 1–53.

Skinner, M. F., Skinner, S. M., and Gooris, R. J. (1977). Stratigraphy and biostratigraphy of late Cenozoic deposits in central Sioux County, Western Nebraska. *Bulletin of the American Museum of Natural History*, 158, 263–370.

Skwara, T. (1988). Mammals of the Topham local fauna: Early Miocene (Hemingfordian), Cypress Hills Formation, Saskatchewan. *Natural History Contributions, Saskatchewan Museum of Natural History*, 9, 1–169.

Stehlin, H. G. and Schaub, S. (1951). Die Trigonodontie der simplicidentaten Nager. *Abhandlungen Schweizerische Palaeontologische, Basel*, 67, 1–384.

Storer, J. E. (1988). The rodents of the Lac Pelletier lower fauna, late Eocene (Duchesnean) of Saskatchewan. *Journal of Vertebrate Paleontology*, 8, 84–101.

Sutton, J. F. and Korth, W. W. (1995). Rodents (Mammalia) from the Barstovian (Miocene) Anceney local fauna, Montana. *Annals of the Carnegie Museum*, 64, 267–314.

Tedford, R. H. (1981). Mammalian biochronology of the late Cenozoic basins of New Mexico. *Bulletin of the Geological Society of America*, 92, 1008–22.

Tedford, R. H., Skinner, M. S., Fields, R. S., *et al.* (1987). Faunal succession and biochronology of the Arikareean through Hemphilllian interval (late Oligocene through earliest Pliocene epochs) in North America. In *Cenozoic Mammals of North America*, ed. M. O. Woodburne, pp. 153–210. Berkeley, CA: University of California Press.

Thaler, L. (1966). Les rongeurs fossiles du Bas-Languedoc dans leurs rapports avec l'histoire de faunes et la stratigraphie d'Europe. *Mémoires du Muséum National d'Histoire Naturelle, Nouvelle Séries C*, 17, 1–295.

(1969). Rongeurs nouveaux de l'Oligocene moyen d'Espagne. *Palaeovertebrata*, 2, 191–207.

Tomida, Y. (1987). *Monograph 3: Small Mammal Fossils and Correlation of Continental Deposits, Safford and Duncan Basins, Arizona, USA*. Tokyo: National Science Museum.

Tong, Y. (1992). *Pappocricetodon*, a pre-Oligocene cricetid genus (Rodentia) from central China. *Vertebrata PalAsiatica*, 30, 1–16.

True, F. W. (1894). On the relationships of Taylor's mouse, *Sitomys taylori*. *Proceedings of United States National Museum*, 16, 757–8.

Vianey-Liaud, M. (1985). Possible evolutionary relationships among Eocene and lower Oligocene rodents of Asia, Europe and North America. In *Evolutionary Relationships Among Rodents: A Multidisciplinary Analysis*, ed. W. P. Luckett and J. L. Hartenberger, pp. 277–309. New York: Plenum Press.

Voorhies, M. R. (1990a). *Technical Report 82-09: Vertebrate Paleontology of the Proposed Norden Reservoir Area, Brown, Cherry and Keya Paha Counties, Nebraska*. Denver, CO: US Bureau of Land Reclaimation Division of Archeological Research.

(1990b). Vertebrate biostratigraphy of the Ogallala Group in Nebraska. In *Geologic Framework and Regional Hydrology: Upper Cenozoic Blackwater Draw and Ogallala Formations, Great Plains*, ed. T. C. Gustavson, pp. 115–51. Austin, TX: Bureau of Economic Geology, University of Texas.

Voss, R. S. (1988). Systematics and ecology of Ichthyomyine rodents (Muroidea): patterns of morphological evolution in a small adaptive radiation. *Bulletin of the American Museum of Natural History*, 188, 259–493.

Wagner, J. A. (1843). *Archiv für Naturgeschellschaft*, 9, 52.

Walsh, S. L. (1997). New specimens of *Metanoiamys*, *Pauromys*, and *Simimys* (Rodentia: Myomorpha) from the Uintan (middle Eocene) of San Diego County, California, and comments on the relationships of selected Paleogene Myomorpha. *Proceedings of the San Diego Society of Natural History*, 32, 1–20.

Wang, B. Y. (1985). Zapodidae (Rodentia, Mammalia) from the lower Oligocene of Qujing, Yunnan, China. *Mainzer Geowissenschaftliche Mitteilungen*, 14, 345–67.

Wang, B. Y. and Dawson, M. R. (1994). A primitive cricetid (Mammalia: Rodentia) from the middle Eocene of Jiangsu Province, China. *Annals of the Carnegie Museum*, 63, 239–56.

Waterhouse, G. R. (1837). *Proceedings of the Zoological Society of London*, 1837, 15–17, 27–32.

Webb, S. D. and Marshall, L. G. (1982). Historical biogeography of Recent South American land mammals, In *Special Publication Series 6, Pymatuning Laboratory of Ecology*, ed. M. A. Mares and H. H. Genoways, pp. 39–52. Pittsburg, PA: University of Pittsburg Press.

White, T. H. (1954). Preliminary analysis of the fossil vertebrates of the Canyon Ferry Reservoir area. *Proceedings of the United States National Museum*, 103, 395–438.

Williams, M. R. and Storer, J. E. (1998). Cricetid rodents of the Kealey Springs local fauna (early Arikareean, late Oligocene) of Saskatchewan. *Paludicola*, 1, 143–9.

Wilson, R. L. (1968). Systematics and faunal analysis of a lower Pliocene vertebrate assemblage from Trego County, Kansas. *Contributions from the Museum of Paleontology, University of Michigan*, 22, 75–126.

Wilson, R. W. (1937). New middle Pliocene rodent and lagomorph faunas from Oregon and California. *Carnegie Institution of Washington Publications*, 487, 1–19.

(1949a). Additional Eocene rodent material from southern California. *Carnegie Institution of Washington Publications*, 584, 1–25.

(1949b). On some White River fossil rodents. *Carnegie Institution of Washington Publications*, 584, 27–50.

Wood, A. E. (1936). Cricetid rodents described by Leidy and Cope from the Tertiary of North America. *American Museum Novitates*, 822, 1–8.

(1937). Rodentia. [In *The Mammalian Fauna of the White River Oligocene*, ed. B. Scott, G. L. Jepsen and A. E. Wood.] *Transactions of the American Philosophical Society, New Series*, 28, 155–269.

(1980). The Oligocene rodents of North America. *Transactions of the American Philosophical Society*, 70, 1–68.

Woodburne, M. O. and Swisher, C. C. III (1995). Land mammal high-resolution geochronology, intercontinental overland dispersals, sea level, climate, and vicariance. [In *Geochronology, Time Scales and Global Stratigraphic Correlation: Unified Temporal Framework for an Historical Geology,* ed. W. A. Berggren, D. V. Kent, M.-P. Aubry, and J. Hardenbol.] *Society of Economic and Petroleum Mineralogists Special Publication*, 54, 335–64.

Zakrzewski, R. J. (1991). New species of Blancan woodrat (Cricetidae) from north-central Kansas. *Journal of Mammalogy*, 72, 104–9.

28 Arvicolidae

ROBERT A. MARTIN
Murray State University, Murray, KY, USA

INTRODUCTION

The Arvicolidae is a cosmopolitan, mostly Holarctic, family of rodents that first appear in the fossil record during the late Miocene. The earliest records of this group may be from North America, although it is not likely that the family originated here. The arvicolids dispersed and explosively radiated at the higher latitudes during the late Tertiary and are represented throughout Eurasia and North America from that time onwards. Depending on the classification, there are approximately 10 extant genera in North America (see Appendix to this chapter). They range in size from less than 20 g (e.g., *Clethrionomys gapperi*) to over 1 kg (*Ondatra zibethicus*). While most modern *Microtus* are pastoral, found in relatively open, prairie or tundra environments, some North American *Microtus* and other genera can be found in forested (e.g., *M. pinetorum*, *Clethrionomys*, *Phenacomys*), bog (e.g., *Synaptomys*), and aquatic (e.g., *Ondatra*, *Neofiber*) habitats. *Microtus*, in particular, has very high reproductive rates in comparison with rodents of comparable size in other families, and the oscillating population cycles of arvicolids and their predators are well known. In many parts of the world, arvicolids represent the dominant prey species for carnivorous animals.

There is an extensive literature on the systematics of fossil arvicolids. Summaries have been published by Miller (1896), Hinton (1926), Ognev (1950), Kretzoi (1955a, 1969), Rabeder (1981), Carleton and Musser (1984), Chaline (1987), Repenning, Fejfar, and Heinrich (1990), and Gromov and Polyakov (1992), and additional taxonomic changes have been made that apply to North American arvicolids in a variety of publications focused on specific groups (e.g., Koenigswald and Martin, 1984a; Repenning and Grady, 1988; Martin, 1989, 1995; Repenning, 1992; Repenning *et al.*, 1995; Fejfar and Repenning, 1998).

DEFINING FEATURES OF THE FAMILY ARVICOLIDAE

Skull material of arvicolids is rarely found, and has been used minimally in species descriptions and diagnoses. A full discussion of arvicolid skull features can be found in Hinton (1926) and Gromov and Polyakov (1992). A combination of dental and mandibular features is most commonly used in the classification of fossil arvicolids. Schematic illustrations of the first lower molar of most arvicolid genera are shown in Figure 28.1.

Repenning (1968) provided a useful examination of mandibular musculature in the arvicolids and showed how these muscles defined a mandibular form characteristic of the Arvicolidae. In particular, he identified a muscle scar characteristic of arvicolids, representing the insertion area for muscularis masseter medialis, pars anterior.

The occlusal dentition of all arvicolids is planed and prismatic; a detailed terminology has developed to allow for precision of comparison. I follow the terminology of van der Meulen (1973, 1978), though others exist (e.g., Fejfar and Repenning, 1998). Enamel edges often define triangular (T) features on arvicolid molars, and are so named. Triangle apices are also referred to as anticlines, and the valleys between them as reentrant folds or synclines. Partly following Rabeder (1981), reentrant folds may be horizontal (basically perpendicular to midline of tooth; internal apices not bent anterior or posterior), provergent (internal apices distinctly bent anteriorly), or postvergent (internal apices bent posteriorly; predominantly in upper molars). Rabeder (1981) also provided a lexicon of terms for undulations in the lateral dentine–enamel junction, the linea sinuosa. Extensions of enamel-free (dentine) areas up the molar sides are referred to in the literature as dentine tracts. Characters of the occlusal enamel edges at both the macroscopic and microscopic scales are also important. Definition of the macroscopic conditions follows Martin (1987). Molars in which the posterior edges of the

Evolution of Tertiary Mammals of North America, Vol. 2. ed. C. M. Janis, G. F. Gunnell, and M. D. Uhen. Published by Cambridge University Press.

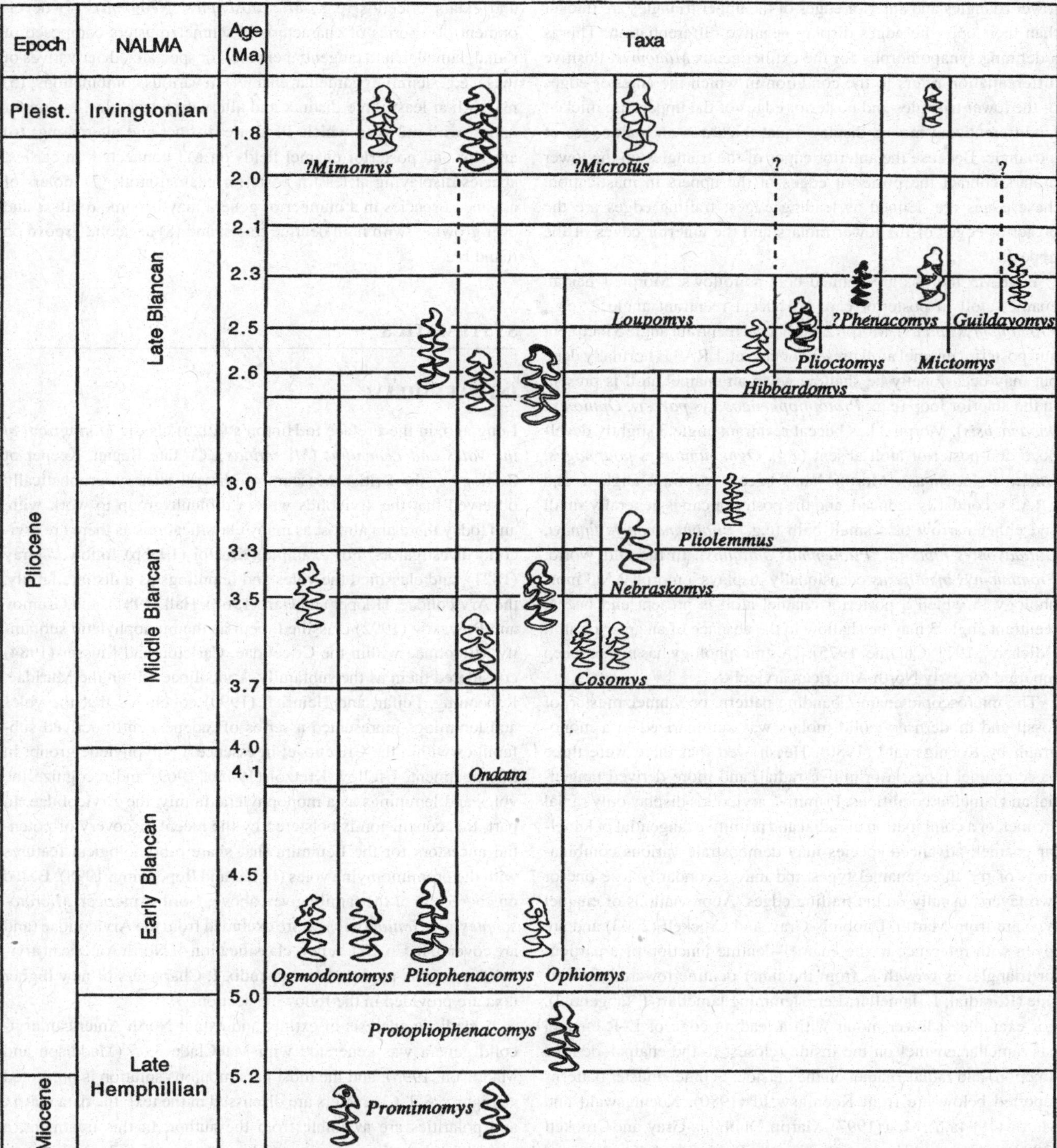

Figure 28.1. Replacement chronology for North American arvicolid rodents. Pleist., Pleistocene; NALMA, North American Land-Mammal Age. Illustrations of typical first lower molars are provided for most genera. The Blancan is arbitratrily broken into three units of approximately equal duration. Illustrations not to scale.

lower triangles and anterior edges of the upper triangles are thicker than their opposite edges display negative differentiation. This is a defining synapomorphy for the extinct genus *Mimomys*. Positive differentiation refers to the condition in which the anterior edges of the lower triangles and posterior edges of the uppers are thicker. Undifferentiated molars display equal thickness on both edges of a triangle. Because the anterior edges of the triangles on the lower molars contact the posterior edges of the uppers in mastication, these edges are defined as leading edges; trailing edges are the posterior edges of the lower molars and the anterior edges of the uppers.

Four M3 morphs are defined here as follows. Morph 1 has an enamel atoll in posterior cap, no buccal reentrant angle 3 (e.g., *Promimomys mimus*); Morph 2 has buccal reentrant angle 3 incipient and posterior enamel atoll present or absent. LRA3 is normally deep but may occasionally be shallow when an enamel atoll is present in the anterior loop (e.g., *Protopliophenacomys parkeri, Ophiomys panacaensis*). Morph 3 has buccal reentrant angle 3 slightly developed and posterior atoll absent (e.g., *Ogmodontomys poaphagus, Ondatra zibethicus*). Morph 4 has buccal reentrant angle 1 and LRA3 secondarily reduced, and the posterior cap is generally small and either narrow or a small bulb (e.g., *Pliophenacomys finneyi, Guildayomys hibbardi, Pliolemmus antiquus*). In the Old World, *Promimomys insuliferus* occasionally displays a morph 2 M3 morphology in which a posterior enamel atoll is present and buccal reentrant angle 3 may be shallow in the absence of an anterior atoll (Michaux, 1971; Chaline, 1975). This morphology has not yet been reported for early North American arvicolids.

The microscopic enamel banding pattern, or schmelzmuster, of fossil and modern arvicolid molars was summarized in a monograph by Koenigswald (1980). He showed that there were three basic enamel types: a primitive radial, and more derived tangential and lamellar conditions. Primitive arvicolids display only radial enamel, or a combination of radial and primitive tangential or lamellar enamel; advanced species may demonstrate various combinations of the three enamel types and may secondarily lose one or two layers, usually on the trailing edges. Abbreviations of enamel type are from Martin, Duobinis-Gray, and Crockett (2003) and are given with reference to the enamel–dentine junction of a particular triangle, as growth is from the inner dentine towards the outside (R, radial; L, lamellar; Lem, lemming lamellar; T, tangential). For example, a lower molar with a leading edge of L–R enamel has lamellar enamel on the inside (closest to the enamel–dentine junction) and radial enamel on the outside. Schmelzmuster patterns reported below are from Koenigswald (1980), Koenigswald and Martin (1984a,b), Mou (1997), Martin, Duobinis-Gray, and Crockett (2003), and unpublished observations by R. Martin, A. Tesakov, and C. Crockett.

Defining dental features for the Arvicolidae include the following: (1) three upper and lower molars, separated from incisors by a diastema; (2) molars planed, and occlusal surface arranged in distinct enamel prisms surrounding dentine cores (the triangle geometry mentioned above); (3) relative to the Cricetidae (or cricetine murids; Carleton and Musser, 1984), there is a general tendency towards hypsodonty, though this is also seen in a number of extinct and extant cricetids (e.g., *Microtoscoptes, Neotoma*); (4) development of a series of characteristic schmelzmusters composed of radial, lamellar, and tangential enamel (or specialized derivatives of these, e.g., lemming lamellar enamel) in various combinations; (5) m1 with at least three distinct and alternating triangles, in combination with m2–3 in which T1–2 are distinct and alternating; (6) anterior and posterior enamel fields on M3 connected, in earliest species displaying at least a posterior enamel atoll; (7) molars of advanced species in a number of genera may become rootless and ever growing, with high dentine tracts; and (8) arvicolid groove on mandible.

SYSTEMATICS

INFRAFAMILIAL

Long ago, in the Preface to Hinton's (1926) classic *Monograph of the Voles and Lemmings (Microtinae)*, C. Tate Regan, Keeper of Zoology at the British Museum of Natural History, prophetically observed that the arvicolids were a difficult group to work with, and today there are almost as many classifications as there are arvicolid investigators. For example, Kretzoi (1955b) followed Gray (1821), and classified the voles and lemmings as a distinct family, the Arvicolidae. Hooper and Hart (1962), Hall (1981), and Gromov and Polyakov (1992) classified them as the monophyletic subfamily Microtinae within the Cricetidae. Carleton and Musser (1984) considered them as the subfamily Arvicolinae within the Muridae. Repenning, Fejfar, and Heinrich (1990) concluded that the voles and lemmings represented a series of independently derived subfamilies within the Cricetidae; in essence a polyphyletic group. In this treatment, I follow Kretzoi (1955b, 1969) and recognize the voles and lemmings as a monophyletic family, the Arvicolidae. In part, this conclusion is bolstered by the recent discovery of potential ancestors for the Lemmini that share morphological features with the promimomyine voles (Fejfar and Repenning, 1998). Based on characters of the family given above, North American *Microtoscoptes* and *Goniodontomys* are excluded from the Arvicolidae (and are covered in Ch. 27). A new classification of North American arvicolid genera is provided in Appendix I. Characters of new higher taxa are provided in the following section.

A cladistic analysis of extinct and extant North American arvicolid genera was generated with MacClade 3.05 (Maddison and Maddison, 1995), and the most parsimonious solution is presented as Figure 28.2. Characters are discussed in the text; the data matrix and polarities are available from the author. In this treatment, a species is a lineage, and no matter how much morphological change occurs within a lineage only one species is recognized (R. A. Martin, 1993, 1996). The temporal distribution of arvicolid rodents during the Blancan does not allow recognition of distinct "early" and "late" subdivisions based on faunal content. Consequently, for this study, the Blancan is arbitrarily broken into three zones (early, middle, late) of basically equal duration, ending with the Blancan–Irvingtonian/Pliocene–Pleistocene boundary (considered coeval here) at approximately 2.0 Ma (Figure 28.1).

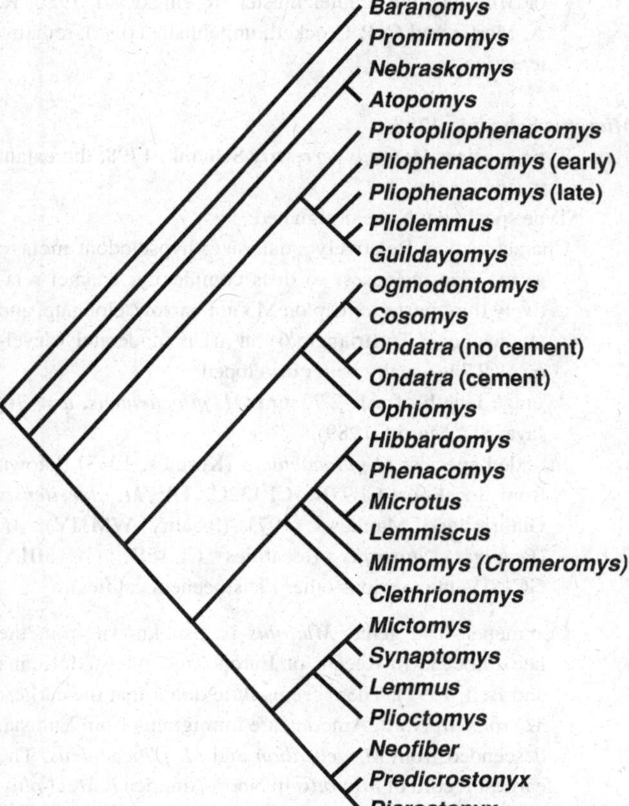

Figure 28.2. Cladogram representing relationships of extinct and extant North American arvicolid genera. *Baranomys*; an extinct European arvicolid-like cricetid, is the outgroup. The positions of *Lemniscus* and *Neofiber* are highly tentative. Characters are to be found in the text.

INCLUDED GENERA OF ARVICOLIDAE GRAY, 1821

The locality numbers listed for each genus refer to the list of unified localities in Appendix I. The acronyms for museum collections are listed in Appendix III.

The locality numbers may be listed in several ways. Parentheses around the locality (e.g., [CP101] mean that the taxon in question at that locality is cited as an "aff." or "cf." the taxon in question. Parentheses are usually used for individual species, thus implying that the genus is firmly known from the locality but that the species identification may be questionable. Question marks in front of the locality (e.g., ?CP101) mean that the taxon is questionably known from that locality, thus implying some doubt that the taxon is actually present at that locality, at either the genus or the species level.

PROMIMOMYINAE (NEW)

Characteristics: Thick, undifferentiated enamel; dentine tracts absent to very low; m1 with three triangles, a relatively simple anteroconid and often an enamel atoll; no *Mimomys*-kante or prism fold on m1 in early species, may be developed somewhat in later ones; M3 morph 1 or 2, with three roots; schmelzmuster (histological banding pattern of enamel prisms): all radial enamel or (1) radial on leading edges and radial plus primitive tangential sections on trailing edges or (2) radial on leading edges and slight expression of lamellar enamel at tips of triangle apices (see Martin, 2003a,b for detailed definitions and descriptions of arvicolid dental morphology).

Promimomys Kretzoi, 1955b (including *Prosomys*)

Type species: *Promimomys cor* Kretzoi, 1955b.

Type specimen: VT 31, Hungarian Geological Institute.

Characteristics: Small size; m1 with three triangles and simple anteroconid; relatively brachydont molars with little development of dentine tracts; enamel atoll often present in m1; reentrant folds on m1 usually horizontal, occasionally provergent; M3 usually morph 1; lingual reentrant folds on lower molars mostly horizontal; triangles on molars alternate; schmelzmuster (Koenigswald, 1980) with leading edges composed only of radial enamel, trailing edges composed of inner radial and outer radial plus some primitive tangential enamel.

Average length of m1: 2.9 mm (*P. cor*, holotype m1); 2.5 mm (P. mimus) ($n = 49$, Christmas Valley; Repenning, 1968).

Included species: *P. mimus* only (known from localities CP116E, CP116III, PN14, PN15).

Promimomys sp. is also known from locality CP116E.

ARVICOLINAE BONAPARTE, 1837

Characteristics: with early evidence of, or descended from, the *Mimomys* schmelzmuster; early species with three triangles on m1 plus a *Mimomys*-kante, later species possibly five or more triangles on m1 and the *Mimomys*-kante lost; advanced species in some genera (e.g., *Mimomys, Microtus, Lemmiscus*) with cementum in reentrant folds; lingual reentrants on lower molars distinctly provergent; labial (buccal) triangles on m1 at least incipiently smaller than lingual triangles; M3 may be morphs 2–4.

ARVICOLINI KRETZOI, 1955B

Mimomys Forsyth Major, 1902 (including *Cosomys*, in part)

Type species: *Mimomys pliocaenicus* Forsyth Major, 1902.

Type specimen: VT 31, Hungarian Geological Institute.

Characteristics: Schmelzmuster usually L–R on leading and R–T on trailing edges; relatively high-crowned, rooted molars; m1 with three triangles plus a variable anteroconid; negative enamel differentiation under light microscope; enamel atolls often present on m1 and M3; *Mimomys*-kante and prism fold often present. Advanced species may lose the atolls, gain cementum in the reentrant folds, and develop high dentine tracts.

Average length of m1: 2.76 mm (estimate from one *M. dakotaensis* specimen; Martin [1989]).

Included species: *M. dakotaensis* only, known from locality CP97II only.

Comments: Arvicolids from the Java Local Fauna (locality CP97II) were described by Martin (1989) and represent predominantly early Pleistocene species with a few species more typical of late Pliocene (late Blancan) time. Whether this assemblage includes reworked older specimens or the last appearance of some archaic forms remains to be determined.

The genus *Mimomys* is also known from the late Miocene and Pleistocene of Europe, and the Pliocene and Pleistocene of Asia (McKenna and Bell, 1997). The presence of *Mimomys* in North America has been debated for many years. For example, although initially described as *Cosomys primus* (Wilson, 1932), this species has also been considered as a member of *Mimomys* by Hinton (1926) and more recently by Repenning (1987) and Bell (2000). Originally described as *Mimomys monohani* (Martin, 1972), an arvicolid from the Irvingtonian Mullen Assemblage of Nebraska was later referred to the new genus *Loupomys* by Koenigswald and Martin (1984b). Repenning (1987) included *Ogmodontomys*, *Ophiomys* and *Cosomys* as subgenera of *Mimomys*. Mou (1997) described a new species of *Mimomys*, *M. panacaensis*, from an early Blancan locality near Panaca, Nevada. Martin (2003b) examined the features of the type species of *Mimomys*, *M. pliocaenicus* Forsyth Major 1902 and concluded that the crucial aggregate synapomorphy for *Mimomys* included negative enamel differentiation and a *Mimomys* schmelzmuster, in combination with rooted molars. In North America, this combination is found only in *Mimomys dakotaensis* from the Java Local Fauna of South Dakota (Martin, 1989) and, presumably, *M. virginianus* from the Cheetah Room Local Fauna of Hamilton Cave, West Virginia (Repenning and Grady, 1988) and the Pit Locality in Porcupine Cave, Colorado (Bell and Barnosky, 2000), regarded as of early Pleistocene (Irvingtonian) age. *Ogmodontomys*, *Ophiomys*, and *Cosomys* were considered by Koenigswald and Martin (1984b) and Martin (2003a) to represent independent North American radiations, though probably descended from Asian immigrants.

Advanced species of the Arvicolini have ever-growing molars with cementum in the reentrant angles, an m1 with very high dentine tracts on both sides of the posterior loop and on the anteroconid, and lack a *Mimomys*-kante on m1. In North America, this advanced group is represented by *Microtus* and, perhaps, by *Lemmiscus*. Although *Lemmiscus* has been previously classified within the genus *Lagurus* (e.g., Hall, 1981), the dental pattern and schmelzmuster of *Lemmiscus* appears to have arisen independently of the Old World Lagurini (Koenigswald and Tesakov, 1997). Nevertheless, there are enough distinctive dental characters of *Lemmiscus* that its assignment to the Arvicolini, despite what appears to be a derived *Mimomys*- or *Microtus*-like schmelzmuster (Koenigswald, 1980; R. A. Martin and C. P. Crockett, unpublished data), remains tentative.

***Microtus* Schrank, 1798**

Type species: *Microtus terrestris* Schrank, 1798, the extant meadow vole.

Type specimen: None designated.

Characteristics: Relatively small size; hypselodont molars; as m1 size increases, so does complexity; enamel relatively thin; posterior cap on M3 not narrow, elongate, and attenuated; if T6 (triangle 6) on m1 is moderately developed, T7 is usually better developed.

Average length of m1: 2.73 mm (*M. pliocaenicus*, $n = 76$, Java, SD; Martin, 1989).

Included species: *M. pliocaenicus* (Kormos, 1933) (known from localities CP97II, CP132C, D); *M. deceitensis* Guthrie and Matthews, 1971 (locality WM1IV); *M.* (*Pedomys*) *australis* (localities GC15D, GC18IIIA, GC18IV, plus various other Pleistocene localities).

Comments: The genus *Microtus* is also known from the late Pliocene to Recent of Europe and Asia (McKenna and Bell, 1997). There seems little doubt that the earliest *Microtus* in North America are immigrants from Eurasia, descended from *M. deucalion* and *M. pliocaenicus*. The earliest record of *Microtus* in North America is *M.* cf *pliocaenicus* from the Short Haul Local Fauna of the Meade Basin of Kansas (locality CP132C), which lies above the Huckleberry Ridge ash dated at 2.10 Ma and below sediments recording the Olduvai Normal subchron, the base of which is 1.95 Ma (R. Martin, E. Lindsay, and N. Opdyke, unpublished data). *Microtus pliocaenicus* has also been recovered from the Aries A and Aries NE Local Faunas (localities CP132C and D), which occur in superposition above Short Haul and also beneath the Olduvai event (Martin *et al.*, 2003). The Nash 72 Local Fauna (locality CP132D) was recovered in sediments of reversed polarity above the Olduvai and is, therefore, of Pleistocene age. On the Central Great Plains, we use the *Microtus* immigration event in the Meade Basin to define the Blancan–Irvingtonian transition. Thus, Short Haul, Aries A, and Aries NE are from the Irvingtonian, but by definition must also be Pliocene in age. The latest Blancan local fauna in the Meade Basin is Borchers (locality CP132B), which is considered to be coeval with the Huckleberry Ridge ashfall at 2.10 Ma.

PHENACOMYINI (NEW)

Characteristics: The Phenacomyini demonstrate the defining character of the Arvicolinae, namely a *Mimomys*-type schmelzmuster, plus the following features: very hypsodont molars; absence of enamel atolls on m1 and M3; early evidence of asymmetry of m1, with lingual triangles usually distinctly larger than buccal ones;

three to five triangles on m1 (usually five) plus an anteroconid; dentine tracts moderately to well developed; positive enamel differentiation in most species; lingual reentrants on lower molars provergent; minimal development to loss of *Mimomys*-kante in most species; M3 morph 3. This tribe includes the modern genera *Phenacomys* (= *Arborimus*) and the extinct *Hibbardomys*.

***Phenacomys* Merriam, 1889**

Type species: *Phenacomys intermedius* Merriam, 1889.

Type specimen: Museum of Geology, and Natural History Museum of Canada 700.

Characteristics: Dentine tract at T6 ("mimosinuid" of Rabeder [1981]) never high on m1 (as in *Hibbardomys*), but hyposinuid often very high.

Average length of m1: 2.5–2.8 mm (*P. gryci*; Repenning *et al.*, 1987).

Included species: *P. gryci* Repenning *et al.* 1987 only, known from locality WM1III only.

Comments: Specimens reported as *Hibbardomys zakrzewskii* by Martin (1989) from the latest Blancan or early Irvingtonian Java Local Fauna of South Dakota (locality CP97II) may also include some *Phenacomys* (personal observation). *Phenacomys* is also known from the Pleistsocene.

***Hibbardomys* Zakrzewski, 1984**

Type species: *Hibbardomys marthae* Zakrzewski, 1984.

Type specimen: UNSM 51279.

Characteristics: m1 with three to five triangles; lingual reentrants narrow; very high dentine tract at T6.

Average length of m1: 2.92 mm (*H. fayae*, $n = 8$); 2.93 mm (*H. skinneri*, $n = 2$); 2.89 mm (*H. voorhiesi*, $n = 15$) (Zakrzewski, 1984).

Included species: *H. marthae* (known from localities [CP97II], CP121, CP131); *H. fayae* Zakrzewski, 1984 (locality CP131II); *H. skinneri* Zakrzewski, 1984 (locality CP118); *H. voorhiesi* Zakrzewski, 1984 (locality CP131); *H. zakrzewskii* Martin, 1989 (locality CP97II).

ONDATRINAE

Characteristics: Rooted molars without cement in all but the most advanced *Ondatra* populations; enamel undifferentiated in occlusal view with light microscope; variable schmelzmuster, advanced over the Promimomyinae but not as in *Mimomys*; most early species with enamel atolls on m1 and/or M3 (both absent in *Ogmodontomys pipecreekensis*; rare in *Ondatra zibethicus*); M3 morph 2 or 3; *Mimomys*-kante and prism fold present on m1 in most species, may be lost in later ones; most species with three triangles on m1, tendency towards five or more triangles in advanced species; lingual reentrants at least incipiently provergent, distinctly provergent in most species of *Ophiomys*, *Cosomys*, and *Ondatra*; low to moderately low dentine tracts, never breaking through at occlusal surface in teeth with moderate wear; anteroconid on juvenile m1 often highly crenulated.

This subfamily is not the same, in either concept or content, as the Ondatrinae of Repenning (1982) and Repenning, Fejfar, and Heinrich (1990). They envisioned the ondatrines to represent a monophyletic group of "muskrats," including taxa that developed five triangles on m1 during the Pliocene (e.g., *Ondatra, Dolomys, Pliomys, Dinaromys, Kislangia*, and *Neofiber*). This is a diverse group of taxa, which are not all closely related to each other. Most *Dolomys*, for example, have the negative enamel differentiation of *Mimomys*, as well as a *Mimomys*-type schmelzmuster (Koenigswald, 1980). For similar reasons, *Kislangia* also seems to be closely related to *Mimomys* (see Agustí, Galobart, and Martín Suárez, 1993). *Pliomys* has a schmelzmuster reminiscent of *Ondatra* but never develops the advanced lamellar edges of *Ondatra* and, in later species (e.g., *P. lenki*), evolves positive enamel differentiation. *Ondatra* retains very thick, undifferentiated enamel throughout its history. According to Rabeder (1981), *Pliomys* may be more closely related to *Clethrionomys* than to other arvicolids, but *Clethrionomys* displays a *Mimomys*-like schmelzmuster, and a close relationship with *Pliomys* is, therefore, doubtful. The schmelzmuster of the North American *Neofiber* (known in the early Irvingtonian of Florida from locality GC15F) is very unusual, composed of a derived lamellar-like enamel on both leading and trailing edges that is more structurally similar to that of the lemmings than to any other arvicolids (Koenigswald, 1980; author's personal observation). An Irvingtonian species with rooted molars (*Proneofiber guildayi*) is the earliest known relative.

The extant Balkan high-montane vole *Dinaromys bogdanovi* is not likely related to the North American *Ondatra zibethicus*. *D. bogdanovi* is a terrestrial species, living in rocky talus slopes. Unlike the thick, undifferentiated enamel of *Ondatra*, the dentition of *D. bogdanovi* is positively differentiated. The schmelzmuster of *Dinaromys* is somewhat similar to that of *Pliomys*, and this may indicate a close relationship, as suggested by Chaline (1975) and Gromov and Polyakov (1992). As in *Pliomys*, there may be an extension of primitive lamellar enamel into the trailing edges of the lower molar triangles; on other lower molar triangles, while there may be well-developed lamellar enamel on the leading edge, the trailing edge may have mostly radial enamel with a bit of primitive tangential enamel (Koenigswald, 1980).

***Ondatra* Link, 1795 (including *Fiber*; *Pliopotamys*)**

Type species: *Ondatra zibethicus* (Linnaeus, 1766) (originally described as *Fiber zibethicus*), the extant common muskrat.

Type specimen: None designated.

Characteristics: Very thick, undifferentiated enamel; tendency in early species to develop five triangles plus anteroconid on m1; enamel atolls very rare on m1 and M3, mostly absent; M3 morph 3; *Mimomys*-kante on m1 lost early in development and in evolutionary history; schmelzmuster radial in early populations, adding lamellar enamel first at the triangle apices and then throughout both the leading and trailing edges in descendant populations; cementum absent in early populations, added in later ones.

Average length of m1: approximately 4.1 mm (early populations) to 9.0 mm (advanced, late Wisconsinan samples) (see R. A. Martin, 1993: Table 1).

Included species: *O. z. /annectens* Brown, 1908 (originally described as *O. annectens)* (localities GC18IIIA, GC18IV, CP97II); *O. z. /idahoensis* Wilson, 1933 (originally described as *O. idahoensis*) (localities GC15C, GC18IIB, CC48C, ?NB13C, [NB36III], SB14E, CP131, CP132B, PN23C); *O. z. /meadensis* (Hibbard, 1938) (originally described as *Pliopotamys meadensis*) (localities NB33A, CP118, CP128C); *O. z. /minor* (Wilson, 1933) (originally described as *Pliopotamys minor*) (known from localities PN4, PN23A, B).

Ondatra sp. is also known from localities CC39B, SB15A, B, CP121. The genus is also known from the Pleistocene.

Comments: Martin (1996) concluded that the fossil history of the North American muskrat included only a single species that had changed somewhat during the past 3.75 million years. In order to represent this example of phyletic evolution, likely the result of an aquatic lifestyle, he subsumed the named fossil genera and species of muskrats (not including the "round-tailed" muskrats, *Neofiber*) under the extant *Ondatra zibethicus* Linnaeus, recognizing five chronomorphs (see R. A. Martin, 1993, 1996 for discussion of chronomorph concept).

Ogmodontomys Hibbard, 1941

Type species: *Ogmodontomys poaphagus* Hibbard, 1941.

Type specimen: KUMNH 4594.

Characteristics: Thick enamel; three triangles on m1, with some tendency to express T4-5 in advanced populations; LRA4 and buccal reentrant angle 3 on m1 alternate, and the anteroconid complex is distinctly asymmetrical, especially in advanced populations; schmelzmuster composed, in primitive species, of mostly radial enamel, but changing to radial on the leading triangle edges and tangential and radial on the trailing edges in later populations; lamellar enamel poorly developed or absent, undifferentiated under light microscope but may be slightly negatively differentiated when measured under a scanning electron microscope, never develops distinct negative differentiation as in European *Mimomys*; reentrant angles deep, narrow and at least incipiently provergent, but rarely as deeply provergent as in *Ophiomys* and *Cosomys*; usually a *Mimomys*-kante and prism fold at some point in development; M3 without enamel atoll (though a species combining other characters here, but with an enamel atoll, could still be included within the genus); M3, morph 3, with three roots in most populations but tendency towards reduction to two roots in advanced populations; relatively low dentine tracts; upper incisor with faint lateral groove in *O. poaphagus*.

Average length of m1: 3.28 mm (*O. poaphagus*, $n = 318$; Rexroad Loc. 3 [locality CP128E]), (Zakrzewski, 1967).

Included species: *O. poaphagus* Hibbard, 1941 (known from localities SP1F, CP118, CP128A–D); *O. pipecreekensis* Martin, Goodwin, and Farlow, 2002 (locality CE2); *O. sawrockensis* Hibbard, 1957 (locality CP128AA).

Ogmodontomys sp. is also known from localities ?NB11 (could be *Ophiomys*), SB12, CP131.

Comments: As with *Cosomys* (see below), *Ogmodontomys* has been considered at various times as a member of the genus *Mimomys* (e.g., Wilson, 1932; Repenning, 1987; Bell, 2000). However, *Ogmodontomys* does not display distinct negative enamel diffentiation and lamellar enamel edges have not been found in any samples of *Ogmodontomys poaphagus* examined to date (Koenigswald, 1980; Martin, 2001; Martin Crockett, and Marcolini, 2006; R. Martin and C. Crockett, unpublished data).

Czaplewski (1990) reported *Ogmodontomys poaphagus* from deposits in the Verde Formation (SB12), but the illustrations of referred m1 teeth suggest that the taxon represented is not *O. poaphagus* (see Martin, Goodwin, and Farlow [2002] for discussion), though it is an *Ogmodontomys*.

Cosomys Wilson, 1932

Type species: *Cosomys primus* Wilson, 1932.

Type specimen: CIT 500.

Characteristics: Very hypsodont m1 with three triangles and anteroconid; anteroconid on m1 relatively elongate, with very deep LRA4, and with distinct crenulations and enamel atoll through moderate wear; lingual triangles on lower molars highly provergent; M3 morph 3 and without enamel atolls; M3 with two roots; schmelzmuster, leading edges R–L–R, trailing edges R–T–R; upper incisors without grooves.

Average length of m1: 3.2 mm ($n = 3$).

Included species: *C. primus* only (known from localities NB12B, PN23A).

Comments: *Cosomys* has been recently considered as a member of the genus *Mimomys* (Repenning, 1998; Bell, 2000), but as discussed above, *Cosomys* lacks the critical character of *Mimomys*, negative enamel differentiation. It has a schmelzmuster similar to that of *Mimomys*, but with an extra layer of radial enamel in both leading and trailing edges. Koenigswald and Martin (1984b) concluded that this schmelzmuster developed independently of European *Mimomys*. *Cosomys* is considered here as an endemic form, and its ancestry is currently undetermined.

Ophiomys Hibbard and Zakrzewski, 1967

Type species: *Ophiomys parvus* (Wilson, 1933).

Type specimen: CIT 1369.

Characteristics: m1 with three to five alternating triangles plus anteroconid; LRA3 and buccal reentrant angle 2 almost directly opposite in most species; when distinct, LRA4 relatively shallow and almost opposite buccal reentrant angle 3, expressing a symmetrical and relatively broad anteroconid complex (in contrast to the asymmetrical pattern in *Ogmodontomys*); dentine tracts poorly developed; *Mimomys*-kante poorly developed to absent in early species, absent in later ones; enamel atoll present on m1 in early species, lost in later ones; M3 morph 2 and 3, rarely with anterior or posterior atoll in early species, lost in later ones; M3 approximately 50% three-rooted and 50% two-rooted in early species, later species two-roots; schmelzmuster: leading edges (1) radial outer, some primitive lamellar inner, advanced lamellar only at triangle apices (*O. taylori*, *O. parvus*) or (2) only radial (*O. meadensis*, Sanders Local Fauna [locality CP130B]); trailing edges mostly radial enamel, occasionally with some primitive lamellar enamel near triangle apices (all species).

Average length of m1: 2.74 mm (*O. panacaensis*, $n = 38$); 2.7 mm (*O. mcknighti*, $n = 2$); 2.6 mm (*O. taylori*, $n = 27$; USGS Cenozoic Locality 19216, Glenns Ferry Formation [PN23A]).

Included species: *O. parvus* (known from locality PN23C); *O. meadensis* Hibbard, 1956 (localities CP130, CP131); *O. magilli* Hibbard, 1972 (locality CP118); *O. mcknighti* Gustafson, 1978 (locality PN3C); *O. panacaensis* Mou, 1997 (locality NB33A); *O. taylori* (Hibbard, 1959) (localities [SB15A], PN23A).

Ophiomys sp. is also known from localities CC26C, CC47D, ?NB11 (could be *Ogmodontomys*), PN4.

Comments: *Ophiomys fricki* dental materials described and illustrated by Hibbard (1972) have none of the distinguishing features of *Ophiomys* and are currently left incertae sedis.

Mou (1997) has recently shown that the schmelzmuster of "*Mimomys*" *panacaensis* is characterized by two-layered radial enamel on the leading and very primitive tangential plus radial enamel on the trailing edges. There are also occasional small sections of discrete (primitive) lamellar enamel on some leading edges and at the apices of some triangles. Mou pointed out that this pattern is similar to the earliest *Mimomys* of Europe, but this is a pattern seen in a number of vole taxa as lamellar enamel first begins to appear (e.g., *Ondatra*, *Ophiomys*; Koenigswald, 1980). Because "*M.*" *panacaensis* does not possess negatively differentiated enamel and does not display a typical *Mimomys* schmelzmuster, it should not be included within *Mimomys*. However, proper placement is problematic, as there are a number of early arvicolid species in North America that display similar dental morphologies and were probably ancestral to both *Ophiomys* and *Ogmodontomys*. Some of these include *Ogmodontomys sawrockensis* from the early Blancan Saw Rock Canyon Local Fauna of Kansas (locality CP128A0) (Hibbard, 1957), *Ophiomys mcknighti* of the early Blancan Ringold Formation (locality CP120A0) in Washington state (Gustafson, 1978), *Ophiomys magilli* of the late Blancan Sand Draw Local Fauna (locality CP118) from Nebraska (Hibbard, 1972), and a series of undescribed arvicolid samples from the western United States (e.g., Maxum, Alturas of California [locality CC26C]) briefly listed or discussed by Repenning (1987) and Bell (2000). These taxa are closely related and represent the early radiation of a clade including *Ophiomys*, *Ogmodontomys*, *Cosomys*, and probably *Ondatra*, but the features that distinguish these genera become obscured as we move back in time, into the "metaregion" (defined by Martin and Tesakov [1998]) for the group.

At least with regard to the species I have examined, *Ogmodontomys* dentitions have thicker enamel than those of *Ophiomys*, and there are usually three roots on M3 in early species of the former genus. The anteroconid complex on m1 in *Ophiomys* apparently remains simple and broad even in the most advanced species, with little development of buccal reentrant angle 3 and LRA4. Even in *Ogmodontomys sawrockensis* from early deposits in the Meade Basin (including the original Saw Rock Canyon Quarry and new early Blancan localities in southwestern Kansas such as Argonaut and Fallen Angel [locality CP128AA] [Martin, Honey, and Peláez-Campomanes, 2000]) there is modest development of these reentrants and a constriction that tends to define T4–5 on the anterior cap.

The source sediments of "*M.*" *panacaensis* in the Panaca Formation (locality NB33A) are now considered to lie close to the Hemphillian–Blancan boundary, from approximately 4.98–4.60 Ma (Lindsay *et al.*, 2002). They were presumably deposited at about the same time or slightly later than the Saw Rock Canyon Local Fauna (Martin, Honey, and Peláez-Campomanes, 2000), though age estimates for the latter locality independent of biochronologic data are currently lacking. At least based on available information, it seems likely that "*M.*" *panacaensis* was not ancestral to *O. sawrockensis*, despite retaining a simple dental pattern and an enamel atoll on M3 as well as m1. That "*M.*" *panacaensis* has begun to develop a unique trajectory can be seen in the reduction of roots on M3. Furthermore, the enamel atoll of M3 was lost in *O. sawrockensis* and the Blancan *O. poaphagus*, a derived feature for that clade. Thus, *O. sawrockensis* was also somewhat derived in certain regards. In summary, the features that characterize "*M.*" *panacaensis* do not preclude it from membership in *Ophiomys*, and it is here tentatively referred to that genus, as *Ophiomys panacaensis*.

NEBRASKOMYINAE (NEW)

Characteristics: Small size; thick, undifferentiated enamel; m1 simple, buccal reentrant angle 3 undeveloped, *Mimomys*-kante usually present at some point in development; T1–2 highly confluent on m1–2; M3 morph 3 and lacking atolls; simple schmelzmuster composed mostly of radial enamel (*Nebraskomys*; unknown in *Atopomys*).

Nebraskomys (Hibbard, 1957) and its Pleistocene descendant, *Atopomys* (Patton, 1965) represent a distinct North American group that evolved independently from a promimomyine ancestor. This can be seen in the simple dentition and underived schmelzmuster, composed even in the late Blancan *Nebraskomys* of mostly radial enamel (Koenigswald, 1980; R. Martin and A. Tesakov, unpublished data). Zakrzewski (1975) and Winkler and Grady (1990) reviewed the fossil *Atopomys* and confirmed Hibbard's (1970) initial speculation that *Atopomys* and *Nebraskomys* were closely related. Based on the characters illustrated and discussed in these studies, there is very little morphological difference between *Nebraskomys* and *Atopomys*, and further study may determine they should be considered as members of a single genus.

Nebraskomys Hibbard, 1957

Type species: *Nebraskomys mcgrewi* Hibbard, 1957.

Type specimen: UMMP 25610.

Characteristics: thick, undifferentiated enamel; low dentine tracts; relatively less hypsodont compared with *Atopomys*; M3 large and with distinct T3–4 (in contrast to *Atopomys*, in which the posterior cap is reduced and T4 is vestigial).

Average length of m1: 2.15 mm (*N. mcgrewi*, $n = 4$, Sand Draw; Hibbard, 1972).

Included species: *N. mcgrewi* (known from localities CP118, CP131); *N. rexroadensis* Hibbard, 1970 (locality CP128C).

Comments: *Atopomys salvelinus*, the Pleistocene descendent of *Nebraskomys*, is known from latest Blancan/earliest Irvingtonian localities GC15D and GC18IIB.

Nebraskomys retains a predominantly radial schmelzmuster and very thick, undifferentiated enamel, with an absence of atolls from m1 and M3, plus a gain of complete confluency of T1–2 on m1–2 and, in *N. mcgrewi*, more-developed dentine tracts. The ancestral schmelzmuster cannot be used by itself to define *Nebraskomys*, but in combination with the other characters it makes *Nebraskomys* very distinct. Because the earliest records of *Nebraskomys* are from late Blancan sediments, it is not clear if it is descended directly from a promimomyine without a *Mimomys*-kante or from one with this feature. There are a number of Eurasian taxa that bear a similar relationship to Eurasian *Promimomys* (e.g., *Ungaromys*, *Stachomys*, *Germanomys*).

PLIOPHENACOMYINAE REPENNING, FEJFAR, AND HEINRICH, 1990 NEW RANK

Characteristics: No enamel atoll on m1; *Mimomys*-kante weakly developed in ancestral species, absent in later species; enamel atolls usually absent on M3 (present in *Protopliophenacomys parkeri*); M3 morph 2 or 4; lingual reentrants on m1 generally horizontal, becoming distinctly provergent only in the most advanced species, *P. osborni*; most genera with rooted molars, one derived genus (*Guildayomys*) without roots; most species with at least five triangles plus anteroconid on m1 (*P. parkeri* with three triangles); anterior enamel border of buccal reentrant angle 3 on m1 often at 90° or greater relative to anteroposterior midline of tooth; schmelzmuster with radial enamel only in early species (*P. finneyi*), adding lamellar enamel on leading and either primitive lamellar (*Pliophenacomys* cf. *P. primaevus*; Sand Draw [locality CP118]) or primitive tangential (*P. osborni*; White Rock [locality CP131]) enamel on the trailing edges of advanced species (Koenigswald, 1980; R. Martin and A. Tesakov, unpublished data).

Protopliophenacomys Martin, 1994 (including *Propliophenacomys*)

Type species: *Protopliophenacomys parkeri* (Martin, 1994) (originally described as *Propliophenacomys parkeri*).

Type specimen: UNSM 47597.

Characteristics: Enamel undifferentiated; m1 with three triangles and anteroconid, no enamel atoll; slight *Mimomys*-kante present on m1 with little wear; dentine tracts low; M3 morph 2, with posterior enamel atoll.

Average length of m1: 2.52 mm ($n = 2$).

Included species: *P. parkeri* only, known from locality CP116F only.

Comments: *Protopliophenacomys parkeri* was initially described as a species of the genus *Propliophenacomys* (Martin, 1975). The holotype specimen for the type species of this genus was shown to belong to an intrusive *Phenacomys* specimen (Martin, 1994). Later, Martin (1994) provided the new generic name *Protopliophenacomys* for *P. parkeri*.

Pliophenacomys Hibbard, 1938

Type species: *Pliophenacomys primaevus* Hibbard, 1938.

Type specimen: KUMNH 3905.

Characteristics: Enamel undifferentiated in early species, differentiated in later ones; dentine tracts moderately to well developed; five triangles on m1; lingual reentrant folds on lower molars often horizontal; *Mimomys*-kante absent; schmelzmuster variable (see above); M3 morph 4.

Average length of m1: 2.86–3.12 mm (range from *P. primaevus* through *P. osborni* [from Zakrzewski, 1984]).

Included species: *P. primaevus* (known from localities SB18B, [CP118], CP128C, [CP131]); *P. dixonensis* Zakrzewski, 1984 (locality CP131II); *P. finneyi* Hibbard and Zakrzewski, 1972 (locality CP128A); *P. koenigswaldi* Martin, Goodwin, and Farlow, 2002 (locality CE2); *P. osborni* Zakrzewski, 1984 (locality CP131); *P. wilsoni* Lindsay and Jacobs, 1985 (locality SB62).

Pliophenacomys sp. is also known from localities SB14A–C, E, CP97II.

***Pliolemmus* Hibbard, 1938**

Type species: *Pliolemmus antiquus* Hibbard, 1938.

Type specimen: KUMNH 3889.

Characteristics: Large and very hypsodont; high dentine tracts at buccal and lingual edges of posterior loop and all triangles; M3 morph 4.

Average length of m1: 3.11 mm ($n = 13$, Sand Draw [locality CP118]; Hibbard, 1972).

Included species: *P. antiquus* only (known from localities CP118, CP128C, CP130).

***Guildayomys* Zakrzewski, 1984**

Type species: *Guildayomys hibbardi* Zakrzewski, 1984.

Type specimen: UNSM 51839.

Characteristics: Ever-growing molars; dentine tracts may be very high, especially on the posterior loop and at T6 and T7 on m1; M3 morph 4.

Average length of m1: 2.99 mm ($n = 23$, Big Springs [locality CP121]; Zakrzewski, 1984).

Included species: *G. hibbardi* only (known from localities CP121, CP131).

LEMMINAE GRAY, 1825

Characteristics: m1 relatively short compared with m3; buccal reentrants slightly to significantly deeper than lingual reentrants in lower molars, resulting in larger buccal than lingual triangles; cementum in reentrant folds; dentine tracts moderately to well developed; schmelzmuster, all species with some expression of lemming lamellar enamel, tangential enamel as a full edge only in *Tobienia kretzoii*; M3 morph 3 or highly derived from that condition.

***Pliomys* Suchov, 1976 (including *Synaptomys*, in part)**

Type species: *Pliomys mimomiformis* Suchov, 1976.

Type specimen: Zoological Institute, Russian Academy of Sciences, St. Petersburg, ZIN S. 61636 (right m1).

Characteristics: Negatively differentiated (e.g., *P. mimomiformis*) or undifferentiated (*P. rinkeri*) enamel; T1–2 closed on m1; m1 axis only slightly shifted towards lingual tooth border.

Average length of m1: 3.08 mm (*P. rinkeri*, $n = 1$, Dixon [locality CP131II]; Hibbard, 1956).

Included species: *P. rinkeri* (Hibbard, 1956) (originally described as *Synaptomys rinkeri*) only, known from locality CP131II only.

Pliomys sp. is also known from localities WM1III, GC15D.

Comments: The proper assignment of early lemmines in North America is difficult, in part because of recent, conflicting opinions on the relationships of the earliest lemmines in both Europe and North America. Despite earlier references of European lemmings to *Synaptomys* (e.g., Kowalski, 1977), experts working with European arvicolids now suggest that bog lemmings are absent from Europe, and that all fossil European lemmines with rootless molars are likely to be members of the genus *Lemmus* (Carls and Rabeder, 1988; Fejfar and Repenning, 1998; Abramson and Nadachowski, 2001). *Mictomys* and *Synaptomys* are, therefore, considered to be endemic North American taxa. According to Repenning and Grady (1988), *Synaptomys* evolved from a *Mictomys* ancestor, and the late Blancan *Synaptomys rinkeri* (Hibbard, 1956) should be classified as a *Pliomys*. However, Martin, Duobinis-Gray, and Crockett (2003) showed that modern *Synaptomys* could have evolved from either a *Mictomys* species or *Pliomys rinkeri*.

Reference of North American specimens to *Pliomys* must remain tentative. Despite superficial similarities, most notably in the closure of T1–2 on m1, North American *P. rinkeri* molars have undifferentiated enamel (Martin, Duobinis-Gray, and Crockett, 2003), whereas the type species of *Pliomys*, *P. mimomiformis*, has negative differentiation (Fejfar and Repenning, 1998: Fig. 15). Repenning *et al.* (1995) referred a *Pliomys* m1 from the late Blancan Fish Creek Local Fauna of Alaska (locality WM1III) to *P. mimomiformis*, but the specimen does not have negative enamel differentiation and is, therefore, not *P. mimomiformis*. Consequently, there are currently no lemmine molars in North America with negative enamel differentiation.

According to Abramson and Nadachowski (2001), there is a continuity of dental morphological evolution in Europe from the Pliocene *Pliomys mimomiformis* to later Pleistocene *Lemmus* that appears to be independent of lemmine evolution in the New World. They classify *Pliomys* as a subgenus of *Lemmus*. Consequently, it may be that "*Pliomys*" *rinkeri* is a New World species unrelated to European *Lemmus*, in which case it may eventually revert to *Synaptomys*. However, no research has conclusively linked *P. rinkeri* with either *Mictomys* or later *Synaptomys* species, and until a detailed phylogenetic analysis is available for North American bog lemmings, I will tentatively follow Fejfar and Repenning (1998) and consider "*Synaptomys*" *rinkeri* to be a North American *Pliomys* with undifferentiated enamel.

***Mictomys* True, 1894 (including *Synaptomys*, in part)**

Type species: *Mictomys innuitus* True, 1894.

Type specimen: None designated.

Characteristics: T1–2 of m1 confluent; m1 axis strongly to extremely shifted towards lingual side, resulting in very reduced lingual and enlarged labial triangles on lower molars.

Average length of m1: 2.82 mm (*Mictomys kansasensis*, $n = 25$; Martin, 1989).

Included species: *M. anzaensis* (locality NB13C); *M. kansasensis* (known from localities CC48C, CP132C, D); *M. landesi* (Hibbard, 1954) (originally described as

Synaptomys landesi) (locality CP132B); *M. vetus* (Wilson, 1932) (originally described as *Synaptomys vetus*) (localities SB17E, SB18B, CP119, PN23C).

Comments: *Tobienia kretzoii*, from late Ruscinian (early Pliocene) deposits at Wölfersheim, Germany, was recently proposed as ancestral to *Mictomys* and *Synaptomys* (Fejfar and Repenning, 1998). This species has a number of features suggesting alliance with the Lemminae, plus confluence of T1–2 on m1, as in *Mictomys*. In general structure, it makes a good ancestor for the earliest North American *Mictomys vetus*, but its relationship to *Synaptomys* is less secure, for reasons noted above.

SUBFAMILY INCERTAE SEDIS

Loupomys Koenigswald and Martin, 1984b (including *Mimomys*, in part)

Type species: *Loupomys monahani* (Martin, 1972) (originally described as *Mimomys monahani*).

Type specimen: UNSM 39216.

Characteristics: Molars hypsodont, with very thick, undifferentiated enamel; three closed triangles plus anteroconid on m1; *Mimomys*-kante (perhaps not homologous to the same condition in European *Mimomys*) present on m1; cement in reentrant angles; schmelzmuster comprising mostly radial enamel.

Average length of m1: 2.6 mm ($n = 6$; Martin, 1972).

Included species: *L. monahani* only, known from locality CP120 only.

Comments: The genus *Loupomys* combines a primitive schmelzmuster with the advanced trait of cementum in the reentrant angles. The thick enamel of *Loupomys* is reminiscent of the same condition in *Nebraskomys* and the Old World *Ungaromys*. Koenigswald and Martin (1984b) reported a small amount of discrete (primitive) lamellar enamel at the apices of some triangles in a *Loupomys* molar, but this condition offers no help in phylogenetic determination because it heralds later development in a number of arvicolid lineages (Koenigswald, 1980).

OTHER ARVICOLID OCCURRENCES

A specimen resembling the European taxon *Baranomys* (which is more likely to be an arvicolid-like cricetid rather than a member of the Arvicolidae) has been reported from the early Pliocene of Ellesmere Island in the Canadian High Arctic (locality HA3) (Zakrzewski and Harrington, 2001).

BIOGEOGRAPHY AND EVOLUTIONARY PATTERNS

The North American arvicolid fossil record preserves a history replete with intercontinental dispersal events, speciation events representing various "experiments" in morphology, and a number of microevolutionary patterns within lineages representing both stasis and considerable phyletic change.

We have few definitive radiometric dates for the earliest North American arvicolids by which to calibrate our arvicolid evolutionary clock. Sediments that produced the Santee Local Fauna (locality CP116F) of Nebraska with *Protopliophenacomys parvus* appear to lie beneath an unnamed tuff dated at 5.0 Ma (Voorhies, 1988), but the fossiliferous sediments are 4.3 km from this unit. Specimens that may be either *Ophiomys* or *Ogmodontomys* have been recovered with the Upper Alturas Local Fauna of California (locality NB11), taken 21 m below a tuff provisionally correlated with a nearby marine ash dated at 4.8 Ma (Repenning, 1987). An *Ogmodontomys* sp. record (originally referred to *O. poaphagus*, but it is not that species; see above) is from 55 m above an ash in the Verde Formation of Arizona (locality SB12) dated at 5.7 Ma. According to Czaplewski (personal communication, 2002), this record may correspond to a time period around 4.3–4.5 Ma (Figure 28.3).

The earliest muskrat in North America (*Ondatra* sp.) comes from the Etchegoin Formation of California (locality CC39), below the Lawlor Tuff (4.1 Ma) and, according to Repenning (1998), below the Nunivak event of the Gilbert Chron. *Cosomys primus*, *Ondatra zibethicus/minor*, and *Ophiomys taylori* have been recovered from about 100 m of sediments in the Glenns Ferry Formation (locality PN23) that possibly range between 3.75 and 3.20 Ma (with radiometric dates of 3.2, 3.3, 3.48 Ma; Zakrzewski, 1969; Lich, 1990). Radiometric dates bracketing the Coso Mountain Locality of California (locality NB12B) provide about 3.0 Ma as the highest stratigraphic datum for *Cosomys primus* (Repenning, 1987). *Mictomys landesi* and *Ondatra zibethicus/idahoensis* are associated with an ash dated at 2.10 Ma from the Borchers Locality (locality CP132B) in the Meade Basin of Kansas (Izett and Honey, 1995), and the lowest stratigraphic datum for the Pleistocene *Microtus pliocaenicus* in the Meade Basin is from the Short Haul Local Fauna, in paleomagnetically reversed sediments below the base of the Olduvai event at 1.95 Ma (R. Martin, E. Lindsay, and N. Opdyke, unpublished data) and above the Huckleberry Ridge ash at 2.10 Ma. Further refinement of potential ages for North American arvicolids (Figure 28.1) is based on a combination of correlation of sediments with the magnetic polarity time scale of Berggren *et al.* (1995) and presumed replacement chronologies for a variety of North American mammal taxa (see Lindsay, Johnson, and Opdyke, 1975; Lindsay, Opdyke and Johnson, 1984; Tomida, 1987; Woodburne, 1987 [and papers therein]; Janis, Scott and Jacobs, 1998; Albright, 1999; Cassiliano, 1999; Martin, Honey, and Peláez-Campomanes, 2000).

The evolutionary history of North American arvicolids is dominated by a series of immigration events from Eurasia (see Repenning [1987, 1998] for interpretation of dispersal routes). Interestingly, there is no verified record of reciprocal exchange in this group, although the current distribution of the collared lemmings, *Dicrostonyx*, and the presence of an ancestral taxon *Predicrostonyx* (Guthrie and Matthews, 1971) in Alaska suggests at least one east–west dispersal event. The epicenter of arvicolid evolution and dispersal appears to have been in central Eurasia, with various waves eventually penetrating the North American continent. The earliest arvicolid-like cricetids to reach North America include

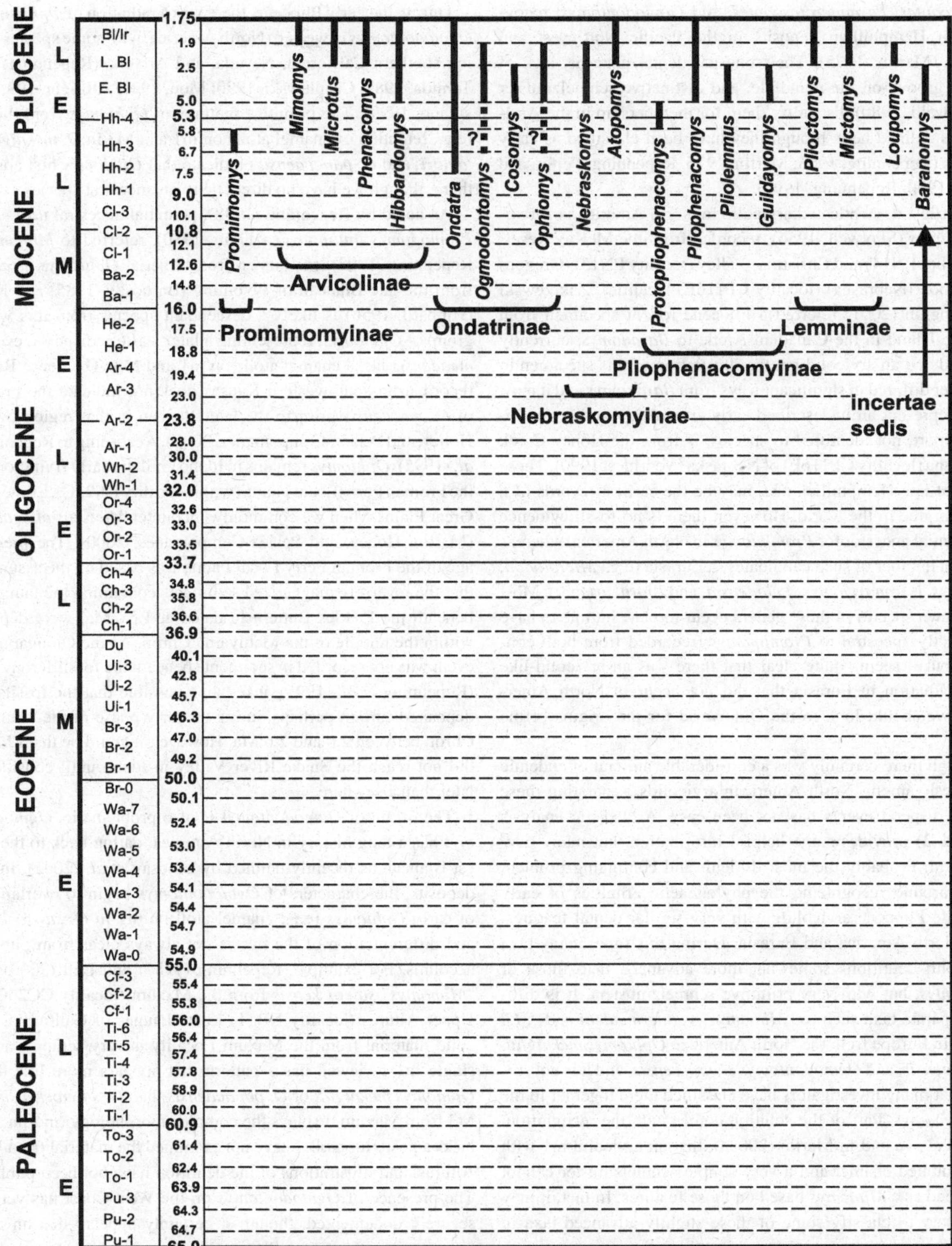

Figure 28.3. Temporal ranges of North American arvicolid genera.

Microtoscoptes, *Paramicrotoscoptes*, and *Goniodontomys*, recovered from Hemphillian deposits in the Pacific Northwest and Nebraska (Martin, 1975). Their unusual dental patterns, lack of arvicolid groove on the mandible, and distinctive schmelzmuster (Koenigswald, 1980) disqualify them for membership in the Arvicolidae as defined here, though they have been classified as early arvicolids (Repenning, 1968; Martin, 1975, Repenning, Fejfar, and Heinrich 1990; Repenning, 1998).

The earliest authenticated arvicolid in North America is *Promimomys mimus* (Shotwell, 1956), recorded from the McKay Reservoir (locality PN14) and Christmas Valley (locality PN15) (Oregon) and Mailbox (Nebraska) (locality CP116E) localities. Zakrzewski and Harington (2001) referred a Pliocene rodent specimen from Ellesmere Island in the Canadian Arctic to *Baranomys*, currently not considered an arvicolid but the illustration of this specimen in their poster differed in significant ways from *Baranomys* and it may, instead, represent an undescribed early arvicolid. Another *Promimomys* report, not identified to species, is from the Honey Creek Local Fauna (locality CP116E) of Nebraska (Voorhies, 1990). These records, of late Hemphillian age, may be the earliest records of a promimomyine in the world. However, there is no fossil evidence for a potential ancestor for *Promimomys* in North America, whereas there are a number of such candidates in Eurasia (e.g., *Microtodon*, *Baranomys*, *Baranarviomys*, *Celadensia*, and *Bjornkurtenia*). Most of the known species of these genera occur too late in time to have been directly ancestral to *Promimomys* (recorded from both continents), but it seems quite clear that there was an arvicolid-like cricetid radiation in Eurasia that did not occur in North America, and we should look to the Old World for the origins of the family.

Although there certainly was a considerable amount of endemic cladogenesis among North American arvicolids, separating these out from disperal events has not been easy. A cladistic analysis (Figure 28.2) provides some helpful information, but many puzzles remain. Probably the most difficult and continuing problem revolves around recognizing the phylogenetic affinities of early and middle Pliocene arvicolids with very similar dental features, both in North America and Eurasia. During this time, many taxa appear with dentitions somewhat more advanced than those of *Promimomys*, but with very primitive schmelzmusters. It is difficult to separate taxa such as "*Mimomys*" *vandermeuleni* and "*M.*" *davakosi* in Europe from the North American *Ophiomys mcknighti*, *O. panacaensis*, and *Ogmodontomys sawrockensis*, and it is not surprising that many investigators have classified them together in the genus *Mimomys*. Part of the solution comes with the recognition, discussed above and in Martin (2003a), that an arvicolid m1 with undifferentiated enamel and a very simple schmelzmuster cannot be classified as a *Mimomys* based on those features. In fact, it may be necessary to classify some of these slightly advanced taxa in the Promimomyinae, recognizing this subfamily as an ancestral, natural group, the arvicolid metaregion (see Martin and Tesakov, 1998), from which later taxa would descend. However, at least with North American taxa, features provided in the generic accounts tentatively serve to distinguish many of our early arvicolids from the Promimomyinae.

During the early Pliocene, there was a radiation of *Ophiomys* and *Ogmodontomys* in western North America, with some species reaching southern California, Nevada, and Arizona (Repenning, 1987; Tomida, 1987; Czaplewski, 1990; Mou, 1997; Albright, 1999; Cassiliano, 1999). The primitive features of *Ophiomys* from this area (e.g., retention of enamel atolls on m1 and M3 in *O. mcknighti*, *O. taylori*, and *O. panacaensis*) indicate that *Ophiomys* first appeared there, though we have no direct tie to an ancestral species.

As noted by Repenning (1998), potential ancestral taxa such as *Promimomys antiquus* (Zazhigin, 1980), referred to *Mimomys* by Repenning, 1998), are known from Siberia. *Ophiomys meadensis* from the late Blancan of Wyoming (Barnosky, 1985) and Kansas (Hibbard, 1956) is likely a descendant species from this western group of *Ophiomys*, representing a later, eastern dispersal event. *O. meadensis* lacks enamel atolls on m1 and M3. The dense Blancan record from southwestern Kansas does not indicate the presence of *O. meadensis* prior to the late Blancan in that region (Martin, Honey, and Peláez-Campomanes, 2000). According to Repenning *et al.* (1995), *Ophiomys* remains in Idaho until the early Irvingtonian in the Froman Ferry Local Fauna (within locality PN23C), a time on the Great Plains when we commonly encounter *Microtus pliocaenicus* (Martin, Honey, and Peláez-Campomanes, 2000). The presumed age of the Froman Ferry Local Fauna was based on the assumption that the reversely magnetized sediments containing the fauna from beneath the Pickles Butte ash, dated at 1.58 Ma, were deposited within the middle of the Matuyama Chron. As the Olduvai normal event was not recorded in sediments beneath the fossiliferous layers (Repenning *et al.*, 1995), it remains possible that the fossils were deposited earlier, perhaps during an early phase of the Matuyama Chron between 2.0 and 2.6 Ma. However, it could be that *Microtus* did not reach the Snake River Valley in Idaho until considerably later than elsewhere.

The origin of *Ogmodontomys* is also problematic. *Ogmodontomys* has a long history in North America, dating back to the earliest Blancan as recently defined by Lindsay *et al.* (2002). In older deposits, the characters of *Ogmodontomys* begin to overlap those of early *Ophiomys* (e.g., enamel atoll on m1 in *O. sawrockensis*) and differentiation of the two is not always clear from published accounts. For example, Repenning (1987) and Bell (2000) listed "*Mimomys*" *sawrockensis* from the Maxum (locality CC26C) and Upper Alturas (locality NB11) Local Faunas of California. Arvicolid material from the Maxum Locality is very scrappy, and the single m1 I viewed has a pattern that appears more like that of *Ophiomys mcknighti* or *O. panacaensis* than *O. sawrockensis*. One M3 from Maxum displays the complex morphology (morph 3) and lacks a posterior atoll. I have not examined the material from Upper Alturas, and illustrations of the dentition have not been published. The presence of *Ogmodontomys* on the West Coast has yet to be securely documented, though it certainly appears that an ancestral taxon to the Great Plains *O. sawrockrensis*, *O. poaphagus*, and *O. pipecreekensis* would have its origins in that region of the country if these taxa did not descend directly from Great Plains *Promimomys*.

As shown to some extent by Zakrzewski (1967), and confirmed in a preliminary fashion by examination of hundreds of new *Ogmodontomys* molars from many new localities in the Meade Basin (Martin, Honey, and Peláez-Campomanes, 2000), molars of *O. poaphagus* remain in morphological stasis through much of Blancan time. Whether they evolved rapidly in a speciation event associated with origin from *O. sawrockensis* or more gradually through phyletic change from that species is currently uncertain. Preliminary study indicates that the average size of *Ogmodontomys* from the Sand Draw Local Fauna of Nebraska (locality CP118) is greater than that of *O. poaphagus* from the Meade Basin, and it remains to be seen if this population represents a distinct species, an expression of a Bergmann's response, or an endpoint in an *O. poaphagus* phyletic series.

The advanced features of *Cosomys primus* (from California and Idaho; Wilson, 1932; Zakrzewski, 1969), most notably the considerable hypsodonty and high dentine tracts, double-rooted M3, and complex schmelzmuster (Koenigswald, 1980), are coupled in that taxon with retention of the *Mimomys*-kante and an enamel atoll on m1 through all but late wear stages (Zakrzewski, 1969). A potential ancestor is the undescribed arvicolid from the Upper Alturas Local Fauna of California noted above, tentatively dated at more than 4.8 Ma (Repenning, 1987, 1998). Lich (1990) showed that size and morphology of *C. primus* m1 teeth remained stable though part of the Glenns Ferry Formation in Idaho (locality PN23), a period that could range from 45 000 to 164 000 years.

At one point, the evolutionary history of the modern muskrat was represented by at least two genera and six species, although by the 1960s it was assumed that these taxa represented a single lineage (Semken, 1966). R. A. Martin (1993) subsumed them under the modern *O. zibethicus*, and in 1996 he published a more complete history of phyletic change in this species, recognizing a series of intermediate, informally named, populations (see above). Repenning (1998) mentioned the presence of an early muskrat from the Etchegoin Formation of California (locality CC39B), from beneath a tuff dated at 4.1 Ma, but this material remains undescribed. If, as Repenning (1998) suggests, it is referable to "*Pliopotamys*," then it may represent the earliest record of *O. zibethicus* in North America. Repenning (1998; Repenning, Fejfar, and Heinrich 1990) considered *Ondatra*, *Pliomys*, *Dinaromys*, and *Dolomys* as related taxa, but as discussed above in the generic accounts, this is highly unlikely on purely morphological grounds. The origin of *Dolomys* from *Mimomys* seems well established (see Maul, 1996; Radulescu and Samson, 1996), and an ancestral species of *Dinaromys* has been recognized in Italy (Sala, 1996). The North American muskrat fossil record remains as one of the best examples of phyletic evolution among terrestrial small mammals, spanning at least 3.75 Ma. Dental trends for the m1 include increased size (body size estimates suggest a change from about 100 g in *O. zibethicus/minor* to over 1 kg in *O. zibethicus/zibethicus*), increase in hypsodonty and dentine tract height, increase in number of triangles, addition of cement in reentrant folds, and increase in complexity of the schmelzmuster (Semken, 1966; Koenigswald, 1980; Martin, 1996). These changes were not added monotonically or simultaneously; as shown by L. D. Martin (1993) and R. A. Martin (1996), the push for large size dramatically increased subsequent to 1.0 Ma, while cementum was added almost a million years earlier.

The Pliophenacomyinae is an extinct and wholly endemic North American clade. The earliest record of a pliophenacomyine vole is *Protopliophenacomys* (Martin, 1994) from the Santee and Devil's Nest Local Faunas (locality CP116F) of Nebraska. An ash presumed to overlie the Santee locality has been fission-track dated at 5.0 Ma (Voorhies, 1988). *Protopliophenacomys* displays features intermediate between *Promimomys mimus* and the earliest *Pliophenacomys*, *P. koenigswaldi* (Martin, Goodwin, and Farlow, 2002). The Pliophenacomyinae is recognized by its horizontal lingual reentrants on the lower molars, distinctively wide buccal reentrant angle 3, and secondarily simple (morph 4) M3. Evolutionary trends in the tribe include moderate gain of complexity of the schmelzmuster, increase in number of triangles and dentine tracts on m1, loss of roots (*Guildayomys*) and, in one species (*Pliophenacomys osborni*), development of provergent lingual reentrant angles. The latest records of pliophenacomyinines may be the Java Local Fauna (locality CP97II) (Martin, 1989), unless this is a mixed Blancan and Irvingtonian fauna. Until the contemporaneity of the Java arvicolids is determined, it will not be known if the pliophenacomyinines became extinct prior to the close of the Blancan or during the early Irvingtonian.

In North America, the subfamily Arvicolinae includes a number of taxa assigned to the Arvicolini, Clethrionomyini, and Phenacomyini. These tribes seem to be Eurasian immigrants, as ancestral taxa are unknown in North America, and none appears in the New World before the latest Blancan or possibly earliest Irvingtonian (e.g., Java, Fish Creek). As Martin (1989) noted, there is a distinct similarity in the dentitions of *Hibbardomys* and *Phenacomys*, and these taxa are combined here in a new tribe, the Phenacomyini. This distinction is based on the observation that some of the defining features of the Phenacomyini appear in very primitive m1 teeth, such as those of *Phenacomys gryci* from Alaska and Idaho (Repenning *et al.*, 1987, 1995). A *Mimomys*-kante is present on m1 in *P. gryci*, but an atoll is absent, and the lingual triangles are distinctly elongate relative to the buccal side. The schmelzmuster of modern *Phenacomys* demonstrates the arvicoline condition (Koenigswald, 1980). These observations lead to a tentative hypothesis that the Phenacomyini split off early from an ancestral arvicolid and retained the genetic predisposition to develop the arvicoline schmelzmuster. However, a piece from an m1 of *Hibbardomys voorhiesi* from the late Blancan White Rock Local Fauna of northcentral Kansas (locality CP131) (Eshelman, 1975) has well-developed lamellar enamel on the leading triangle edges and only radial on the trailing edges (R. A. Martin and C. P. Crockett, unpublished data). If further study fails to reveal tangential enamel on the trailing edges in *Hibbardomys* molars, then this genus may have to be removed from the Phenacomyini and the similarities of *Hibbardomys* and *Phenacomys* would then be considered an example of parallelism. Also, we must consider the possibility that the L–R (leading edges), R–T (trailing edges) schmelzmuster of the Arvicolinae evolved independently in some groups, in which case *Phenacomys* and *Clethrionomys* could represent distinct subfamilies, and the Arvicolinae would be restricted to *Mimomys* (as defined here), *Microtus* (as defined here), *Arvicola* (not including the North American water vole, *M. richardsoni* [see Hoffman and Pattie, 1968]), *Loupomys* (perhaps), and any Old World taxa shown to be derived from *Mimomys*.

Both early *Microtus* (sometimes referred to the genus or subgenus *Allophaiomys* [see Martin and Tesakov, 1998]) and *Mimomys* (*Cromeromys*) have been reported in North American faunas that lie very close to the Blancan–Irvingtonian and Pliocene–Pleistocene boundaries (Repenning and Grady, 1988; Martin, 1989; Repenning, 1998). (See Martin, Duobinis-Gray, and Crockett [2003] for a revised date of the Cheetah Room Local Fauna from Hamilton Cave.) Neither is reported from faunas prior to that time, and it seems possible that a single dispersal event was responsible for their simultaneous arrival in the New World. While additional immigration and subsequent diversification characterizes *Microtus* evolution in North America (Martin, 1989, 1998; Conroy and Cook, 1999), *Mimomys* appears to have been restricted mostly to colder climates or more northern dispersal routes, and its temporal range in North America is currently unresolved.

The red-backed voles, *Clethrionomys*, have no Tertiary record in North America. A number of late Pliocene through modern species are known throughout Eurasia (Tesakov, 1998) and ancestry of the Clethrionomyini may be with a small species of *Mimomys*-like arvicolid during the Pliocene. Although the enamel of *Clethrionomys* molars is very thick, its schmelzmuster is distinctly of the *Mimomys* type (Koenigswald, 1980).

From their first appearance in the North American fossil record at over 2.5 Ma, the lemmings (including the modern genera *Lemmus*, *Plioctomys*, *Mictomys*, and *Synaptomys*) have demonstrated advanced dental traits, with ever-growing molars, cementum in the reentrants, and loss of enamel atolls from m1 and M3. Despite this, they have retained a very simple m1, with only three triangles and slight development of T4–5 on the anteroconid. Repenning and Grady (1988) and Martin, Duobinis-Gray, and Crockett (2003) presented their respective views on evolution of North American *Synaptomys*. The origin of our northern (*Mictomys*) and southern (*Synaptomys*) bog lemmings may both trace to an ancestral species of *Mictomys* such as *M. vetus* (Wilson, 1932), or *Synaptomys* may have descended separately from *Plioctomys*. Fejfar and Repenning (1998) recently identified two taxa of ancient lemmings from Wölfersheim, Germany, both with rooted molars, which they hypothesize as ancestors to later lemmines. As noted above, the modern collared lemmings evolved from *Predicrostonyx hopkinsi* of the Cape Deceit Locality of Alaska (Repenning *et al.*, 1987). A prior ancestral taxon is not known.

Molars of the round-tailed "muskrat" *Neofiber alleni* superficially resemble those of the modern *Ondatra zibethicus*, but this is likely an example of convergence. An ancestor with rooted molars, *Proneofiber guildayi*, has been reported from Irvingtonian deposits in Texas (Hibbard and Dalquest, 1973). As Koenigswald (1980) reported, the schmelzmuster of *N. alleni* is unique, with well-developed lemming lamellar enamel on both leading and trailing edges. Based on this and other features, in the cladogram of Figure 28.2, *Neofiber* falls out with the lemmings. Its relationships remain to be determined, but it seems unlikely that it is phylogenetically close either to the Ondatrinae or the Pliophenacomyinae (contra Chaline *et al.*, 1999).

REFERENCES

Abramson, N. and Nadachowski, A. (2001). Revision of fossil lemmings (Lemminae) from Poland with special reference to the occurrence of *Synaptomys* in Eurasia. *Acta Zoologica Cracoviensia*, 44, 65–77.

Agustí, J., Galobart, A., and Martín Suárez, E. (1993). *Kislangia gusii* sp. nov., a new arvicolid (Rodentia) from the Late Pliocene of Spain. *Scripta Geologica*, 103, 119–34.

Albright, L. B., III (1999). Biostratigraphy and vertebrate paleontology of the San Timeteo Badlands, southern California. *University of California Publications in Geological Sciences*, 144, 1–121.

Baird, S. F. (1858). *Reports of Exploration and Surveys to Ascertain the Most Practicable and Economical Route for a Railroad from the Mississippi River to the Pacific Ocean*, Part 1: *Mammalia*, p. 558. [Executive Document for the 33rd Congress.] Washington, DC: Nicholson.

Barnosky, A. D. (1985). Late Blancan (Pliocene) microtine rodents from Jackson Hole, Wyoming: biostratigraphy and biogeography. *Journal of Vertebrate Paleontology*, 5, 255–71.

Bell, C. J. (2000). Biochronology of North American microtine rodents. In *Quaternary Geochronology: Methods and Applications*, ed. J. S. Noller, J. M. Sowers, and W. R. Lettis, pp. 379–405. Washington, DC: American Geophysical Union.

Bell, C. J. and Barnosky, A. D. (2000). The microtine rodents from the Pit locality in Porcupine Cave, Park County, Colorado. *Annals of the Carnegie Museum*, 69, 93–134.

Berggren, W. A., Kent, D. V., Swisher, C. C., and Aubry, M. P. (1995). A revised Cenozoic geochronology and chronostratigraphy. [In *Geochronology, Time Scales, and Global Stratigraphic Correlation*, ed. W. A. Berggren, D. V. Kent, M. P. Aubry, and J. Hardenbol.] *Society for Sedimentary Geology Special Publication*, 54, 129–212.

Bonaparte, C. L. J. L. (1837). New systematic arrangement of vertebrated animals. *Transactions of the Linnaean Society of London*, 18, 247–304.

Brown, B. (1908). The Conard fissure, a Pleistocene bone deposit in northern Arkansas. *Memoirs of the American Museum of Natural History*, 9, 157–208.

Carleton, M. D. and Musser, G. G. (1984). Muroid rodents. In *Orders and Families of Recent Mammals of the World*, ed. S. Anderson and J. Knox Jones, Jr., pp. 289–379. New York: John Wiley.

Carls, N. and Rabeder, G. (1988). Die Arvicoliden (Rodentia, Mammalia) aus dem Ältest-Pleistozän von Schernfeld (Bayern). *Beiträge zur Paläontologie von Österreich*, 14, 123–237.

Cassiliano, M. L. (1999). Biostratigraphy of Blancan and Irvingtonian mammals in the Fish Creek–Vallecito Creek section, southern California, and a review of the Blancan–Irvingtonian boundary. *Journal of Vertebrate Paleontology*, 19, 169–86.

Chaline, J. (1975). Evolution et rapports phylétiques des Campagnols (Arvicolidae, Rodentia) apparentés à *Dolomys* et *Pliomys* dans l'hémisphère Nord. *Comptes Rendu Académie Science de Paris, Series D*, 28, 33–6.

(1987). Arvicolid data (Arvicolidae, Rodentia) and evolutionary concepts. *Evolutionary Biology*, 21, 237–310.

Chaline, J., Brunet-Lecomte, P., Montuire, S., Viriot, L., and Courant, F. (1999). Anatomy of the arvicoline radiation (Rodentia): palaeogeographical, palaeoecological history and evolutionary data. *Annals Zoologica Fennica*, 36, 239–67.

Conroy, C. J. and Cook, J. A. (1999). MtDNA evidence for repeated pulses of speciation within arvicoline and murid rodents. *Journal of Mammalian Evolution*, 6, 221–45.

Czaplewski, N. J. (1990). The Verde local fauna: small vertebrate fossils from the Verde Formation, Arizona. *Quarterly Journal of the San Bernardino County Museum Association*, 37, 1–39.

Eshelman, R. (1975). Geology and paleontology of the early Pleistocene (late Blancan) White Rock fauna from north-central Kansas. [In *Claude W. Hibbard Memorial Volume 4*.] *University of Michigan Papers on Paleontology*, 13, 1–60.

Fejfar, O. and Repenning, C. A. (1998). The ancestors of the lemmings (Lemmini, Arvicolinae, Cricetidae, Rodentia) in the early Pliocene of Wölfersheim near Frankfurt am Main, Germany. *Senckenbergiana Lethaea*, 77, 161–93.

Forsyth Major, C. I. (1902). Exhibition of, and remarks upon, some jaws and teeth of Pliocene voles (*Mimomys* gen. nov.). *Proceedings of the Zoological Society*, 1, 102–7.

Gloger, C. W. L. (1841). Naturgesischichte. *Gemeinutziges Hand-und Hilfsbuch der Naturgeschichte*, 1, 97.

Gray, J. E. (1821). On the natural arrangement of vertebrate animals. *London Medical Repository*, 15, 296–310.

(1825). Outline of an attempt at the distribution of the Mammalia into tribes and families with a list of the genera apparently appertaining to each tribe. *Annals of Philosophy New Series*, 10, 337–44.

Gromov, I. M. and Polyakov, I. Y. (1992). Voles (Microtinae). In *Fauna of the USSR: Mammals*, Vol. 3. Washington, DC: Smithsonian Institution Libraries. [Originally published in 1977; translated from the Russian.]

Gustafson, E. P. (1978). The vertebrate faunas of the Pliocene Ringold Formation, south-central Washington. *Bulletin of the Museum of Natural History, University of Oregon*, 23, 1–62.

Guthrie, R. D. and Matthews, J. V., Jr. (1971). The Cape Deceit fauna: early Pleistocene mammalian assemblage from the Alaska Arctic. *Quaternary Research*, 1, 474–510.

Hall, E. R. (1981). *The Mammals of North America*. New York: John Wiley.

Hibbard, C. W. (1938). An upper Pliocene fauna from Meade County, Kansas. *Transactions of the Kansas Academy of Sciences*, 40, 239–65.

(1941). New mammals from the Rexroad fauna, Upper Pliocene of Kansas. *American Midland Naturalist*, 26, 337–68.

(1954). A new *Synaptomys*, an addition to the Borchers interglacial (Yarmouth?) fauna. *Journal of Mammalogy*, 35, 249–52.

(1956). Vertebrate fossils from the Meade formation of southwestern Kansas. *Papers of the Michigan Academy of Science, Arts, and Letters*, 41, 145–203.

(1957). Two new Cenozoic microtine rodents. *Journal of Mammalogy*, 38, 39–44.

(1959). Late Cenozoic microtine rodents from Wyoming and Idaho. *Papers of the Michigan Academy of Science, Arts, and Letters*, 44, 3–40.

(1970). A new microtine rodent from the upper Pliocene of Kansas. *Contributions to the Museum of Paleontology, University of Michigan*, 23, 99–103.

(1972). Class Mammalia. [In *Early Pleistocene Preglacial Rocks and Faunas of North-central Nebraska*, ed. M. S. Skinner and C. W. Hibbard.] *Bulletin of the American Museum of Natural History*, 148, 77–130.

Hibbard, C. W. and Dalquest, W. W. (1973). *Proneofiber*, a new genus of vole (Cricetidae, Rodentia) from the Pleistocene Seymour Formation of Texas, and its evolutionary and stratigraphic significance. *Quaternary Research*, 3, 269–74.

Hibbard, C. W. and Zakrzewski, R. J. (1967). Phyletic trends in the late Cenozoic microtine *Ophiomys* gen. nov. from Idaho. *Contributions to the Museum of Paleontology, University of Michigan*, 21, 255–71.

(1972). A new species of microtine from the late Pliocene of Kansas. *Journal of Mammalogy*, 53, 834–9.

Hinton, M. A. C. (1926). *Monograph of the Voles and Lemmings (Microtinae) Living and Extinct*. London: British Museum of Natural History.

Hoffman, R. S. and Pattie, D. L. (1968). *A Guide to Montana Mammals*. Missoula, MT: University of Montana Printing Services.

Hooper, E. T. and Hart, B. S. (1962). A synopsis of Recent North American microtine rodents. *Miscellaneous Publications of the Museum of Zoology, University of Michigan*, 120, 1–68.

Izett, G. A. and Honey, J. G. (1995). Geologic map of the Irish Flats quadrangle, Meade County, Kansas. *United States Geological Survey Miscellaneous Investigations Series Map* I-2498 (scale 1:24 000).

Janis, C. M., Scott, K. M., and Jacobs, L. L. (eds.) (1998). *Evolution of Tertiary Mammals of North America.* Vol. 1: *Terrestrial Carnivores, Ungulates, and Ungulatelike Mammals*. Cambridge, UK: Cambridge University Press.

Koenigswald, W., von (1980). Schmelzstruktur und morphologie in den molaren der Arvicolidae (Rodentia). *Abhandlungen der Senckenberischen Naturforschenden Gesellschaft*, 539, 1–129.

Koenigswald, W., von and Martin, L. D. (1984a). Revision of the fossil and Recent Lemminae (Rodentia, Mammalia). [In *Papers in Vertebrate Paleontology Honoring Robert Warren Wilson*, ed. R. M. Mengel.] *Carnegie Museum of Natural History Special Publication*, 9, 122–37.

(1984b). The status of the genus *Mimomys* (Arvicolidae, Rodentia, Mamm.) in North America. *Neues Jahrbuch Geologie und Paläontologie*, 168, 108–24.

Koenigswald, W. von and Tesakov, A. (1997). The evolution of the schmelzmuster in Lagurini (Arvicolinae, Rodentia). *Palaeontographica B*, 245, 45–61.

Kormos, T. (1933). Neues wühlmäuse aus dem Oberpliocän von Püspökfürdö. *Neues Jahrbuch für Mineralogie, Geologie und Paläontologie. Abteilung B*, 69, 323–46.

Kowalski, K. (1977). Fossil lemmings (Mammalia, Rodentia) from the Pliocene and early Pleistocene of Poland. *Acta Zoologica Cracoviensia*, 22, 297–317.

Kretzoi, M. (1955a). *Dolomys* and *Ondatra*. *Acta Geologica*, 3, 347–55.

(1955b). *Promimomys cor* n. g. n. sp., ein altertümichler Arvicolide aus dem ungarishen Unterpleistozän. *Acta Geologica*, 3, 1–3.

(1969). Skizze einer arvicoliden-phylogenie: stand 1969. *Vertebrata Hungarica*, 11, 155–93.

Lich, D. (1990). *Cosomys primus*, a case for stasis. *Paleobiology*, 16, 384–95.

Lindsay, E. L. and Jacobs, L. L. (1985). Small mammal fossils from Chihuahua, Mexico. *Paleontologia Mexicana*, 51, 1–45.

Lindsay, E. L., Johnson, N. M., and Opdyke, N. D. (1975). Preliminary correlation of North American Land Mammal Ages and geomagnetic polarity. [In *Studies on Cenozoic Paleontology and Stratigraphy, C. W. Hibbard Memorial Volume*, ed. G. R. Smith and N. E. Friedland.] *Contributions from the Museum of Paleontology, the University of Michigan*, 3, 111–19.

Lindsay, E. L., Opdyke, N. D., and Johnson, N. M. (1984). Blancan–Hemphillian land mammal ages and late Cenozoic dispersal events. *Annual Reviews of Earth and Planetary Science*, 12, 445–88.

Lindsay, E. H., Mou, Y., Downs, W. *et al.* (2002). Recognition of the Hemphillian/Blancan boundary in Nevada. *Journal of Vertebrate Paleontology*, 22, 429–42.

Link, H. F. (1795). *Beiträge zur Naturgeschichte*, 1, 76.

Linnaeus, C. (1766). *Systema Naturae per Regna Tria Naturae, Secundum Classes, Ordines, Genera, Species, cum Characteribus and Differentiis.* 12th edn, *Reformata*, Vol. I. Holmiae: Laurentii Salvii.

Maddison, W. and Maddison, D. (1995). *MacClade 3.03*. Cambridge, MA: Harvard University Press.

Martin, L. D. (1972). The microtine rodents of the Mullen assemblage from the Pleistocene of north central Nebraska. *Bulletin of the Nebraska State Museum*, 9, 173–82.

(1975). Microtine rodents from the Ogallala Pliocene of Nebraska and the early evolution of the Microtinae in North America. [In *Studies on Cenozoic Paleontology and Stratigraphy, C. W. Hibbard Memorial Volume*, ed. G. R. Smith and N. E. Friedland.] *Contributions to the Museum of Paleontology, University of Michigan*, 3, 101–11.

(1993). Evolution of hypsodonty and enamel structure in Plio-Pleistocene rodents. In *Morphological Change in Quaternary Mammals of North America*, ed. R. A. Martin and A. D. Barnosky, pp. 205–25. Cambridge, UK: Cambridge University Press.

(1994). A new genus of Miocene vole possibly related to *Phenacomys*. *Nebraska Academy of Sciences, Ter-Qua Symposium Series*, 2, 9129–30.

Martin, R. A. (1987). Notes on the classification and evolution of some North American fossil *Microtus*. *Journal of Vertebrate Paleontology*, 7, 270–83.

(1989). Arvicolid rodents of the early Pleistocene Java local fauna from north-central South Dakota. *Journal of Vertebrate Paleontology*, 9, 438–50.

(1993). Patterns of variation and speciation in Quaternary rodents. In *Morphological Change in Quaternary Mammals of North America*, ed. R. A. Martin and A. D. Barnosky, pp. 226–80. Cambridge, UK: Cambridge University Press.

(1995). A new middle Pleistocene species of *Microtus* (*Pedomys*) from the southern United States, with comments on the taxonomy and early evolution of *Pedomys* and *Pitymys* in North America. *Journal of Vertebrate Paleontology*, 15, 171–86.

(1996). Dental evolution and size change in the North American muskrat: classification and tempo of a presumed phyletic sequence. In *Palaeoecology and Palaeoenvironments of Late Cenozoic Mammals: Tributes to the Career of C.S. (Rufus) Churcher*, ed. K, Stewart and K. Seymour, pp. 431–57. Toronto: University of Toronto Press.

(1998). Time's arrow and the evolutionary position of *Orthriomys* and *Herpetomys*. [In *The Early Evolution of Microtus*, ed. R. A. Martin and A. Tesakov.] *Paludicola*, 2, 70–3.

(2003a). Biochronology of latest Miocene through Pleistocene arvicolid rodents from the Central Great Plains of North America. *Colloquios de Paleontología, Special Issue*, 1, 373–83.

(2003b). The status of *Mimomys* in North America revisited. *Deinsia*, 10, 399–406.

Martin, R. A. and Tesakov, A. (1998). Introductory remarks. Does *Allophaiomys* exist? [In *The Early Evolution of Microtus*, ed. R. A. Martin and A. Tesakov.] *Paludicola*, 2, 1–7.

Martin, R. A., Honey, J. G., and Peláez-Campomanes, P. (2000). The Meade Basin rodent project: a progress report. *Paludicola*, 3, 1–32.

Martin, R. A., Goodwin, H. T., and Farlow, J. O. (2002). Late Neogene (late Hemphillian) rodents from the Pipe Creek Sinkhole, Grant County, Indiana. *Journal of Vertebrate Paleontology*, 22, 137–51.

Martin, R. A., Duobinis-Gray, L., and Crockett, C. P. (2003). A new species of early Pleistocene *Synaptomys* from Florida and its relevance to southern bog lemming origins. *Journal of Vertebrate Paleontology*, 23, 917–36.

Martin, R. A., Hurt, R. T., Honey, J. G., and Peláez-Campomanes, P. (2003). Late Pliocene and early Pleistocene rodents from the northern Borchers Badlands (Meade County, Kansas), with comments on the Blancan–Irvingtonian boundary in the Meade Basin. *Journal of Paleontology*, 77, 985–1001.

Martin, R. A., Crockett, C. P., and Marcolini, F. (2006). Variation of the schmelzmuster and other enamel characters in molars of the primitive Pliocene vole *Ogmodontomys* from Kansas. *Journal of Mammalian Evolution*, 12, 223–41.

Maul, L. (1996). A discussion of the referral of *Mimomys occitanus* Thaler, 1955 (Rodentia: Arvicolidae) to the genus *Mimomys*. [In *Neogene and Quaternary Mammals of the Palaearctic*, ed. A. Nadachowski and L. Werdelin.] *Acta Zoologica Cracoviensia*, 39, 343–8.

McKenna, M. C. and Bell, S. K. (1997). *Classification of Mammals Above the Species Level*. New York: Columbia University Press.

Merriam, J. (1889). *North American Fauna*, 2. Washington, DC: US Department of Agriculture.

Michaux, J. (1971). Arvicolinae (Rodentia) du Pliocène terminal at du Quaternaire ancien de France et d'Espagne. *Palaeovertebrata*, 4, 138–214.

Miller, G. S., Jr. (1896). The genera and subgenera of voles and lemmings. *North American Fauna*, 12, 1–84.

Mou, Y. (1997). A new arvicoline species (Rodentia: Cricetidae) from the Pliocene Panca Formation, southeast Nevada. *Journal of Vertebrate Paleontology*, 17, 376–83.

Ognev, S. I. (1950). *The Mammals of the USSR. and Adjacent Countries*, Vol. 7. Moscow: Academy of Science.

Patton, T. H. (1965). A new genus of fossil microtine from Texas. *Journal of Mammalogy*, 46, 466–71.

Rabeder, G. (1981). Die Arvicoliden (Rodentia, Mammalia) aus dem Pliozän und dem alteren Pleistozän von Niederösterreich. *Beiträge Paläontologie und Geologie Österreich*, 8, 1–373.

Radulescu, C. and Samson, P. (1996). Contributions to the knowledge of the mammalian faunas from Malusteni and Beresti (Romania). *Traveaux Institute Speologie Émile Racovitza*, 28, 43–56.

Repenning, C. A. (1968). Mandibular musculature and the origin of the subfamily Arvicolinae (Rodentia). *Acta Zoologica Cracoviensia*, 13, 29–72.

(1982). Classification notes. In *Mammal Species of the World: A Taxonomic and Geographic Reference*, ed. J. H. Honacki, K. E. Kinman and J. W. Koeppl, p. 484. Lawrence, KS: Allen Press.

(1987). Biochronology of the microtine rodents of the United States. In *Cenozoic Mammals of North America*, ed. M. O. Woodburne, pp. 236–68. Berkeley, CA: University of California Press.

(1992). *Allophaiomys* and the age of the Olyor Suite, Krestovka sections, Yakutia. *Bulletin of the United States Geological Survey*, 2037, 1–98.

(1998). North American mammalian dispersal routes: rapid evolution and dispersal constrain precise biochronology. In *Monograph 14: Advances in Vertebrate Paleontology and Geochronology*, ed. Y. Tomida, L. J. Flynn, and L. L. Jacobs, pp. 39–780. Tokyo: National Science Museum.

Repenning, C. W. and Grady, F. (1988). The microtine rodents of the Cheetah Room fauna, Hamilton Cave, West Virginia, and the spontaneous origin of *Synaptomys*. *Bulletin of the United States Geological Survey*, 1853, 1–32.

Repenning, C. W., Brouwers, E. M., Carter, L. D., Marincovich, Jr., L., and Ager, T. A. (1987). The Beringian ancestry of *Phenacomys* (Rodentia: Cricetidae) and the beginning of the modern Arctic Ocean borderland biota. *Bulletin of the United States Geological Survey*, 1687, 1–31.

Repenning, C. A., Fejfar, O., and Heinrich, W.-D. (1990). Arvicolid rodent biochronology of the Northern Hemisphere. In *International Symposium: Evolution, Phylogeny, and Biostratigraphy of Arvicolids (Rodentia, Mammalia)*, ed. O. Fejfar and W.-D. Heinrich, pp. 385–418. Prague: Geological Survey.

Repenning, C. A., Weasma, T. R., and Scott, G. R. (1995). The early Pleistocene (latest Blancan–earliest Irvingtonian) Froman Ferry fauna and history of the Glenns Ferry Formation, southwestern Idaho. *Bulletin of the United States Geological Survey*, 2105, 1–86.

Sala, B. (1996). *Dinaromys allegranzii* n. sp. (Mammalia, Rodentia) from Rivoli Veronese (northeastern Italy) in a Villanyian association. [In *Neogene and Quaternary Mammals of the Palaearctic*, ed. A. Nadachowski and L. Werdelin.] *Acta Zoologica Cracoviensia*, 39, 469–72.

Schrank, (1798). *Fauna Boica* 1 (Abth, 1), 1–72.

Semken, H. A., Jr. (1966). Stratigraphy and paleontology of the McPherson *Equus* beds (Sandahl local fauna), McPherson County, Kansas. *Contributions from the Museum of Paleontology, University of Michigan*, 20, 121–78.

Shotwell, J. A. (1956). Hemphillian mammal assemblage from northeastern Oregon. *Bulletin of the Geological Society of America*, 67, 717–38.

Suchov, V. P. (1976). Remains of lemmings in the Bashkirian Pliocene deposits. [In *Rodents: Evolution and History of their Recent Fauna*, ed. I. Gromov.] *Proceedings of the Zoological Institute Russian Academy of Sciences*, 66, p. 117.

Tesakov, A. (1998). Voles of the Tegelen fauna. *Mededelingen Nederlands Instituut voor Toegepaste Geowettenschappen TNO*, Nr60, 71–134.

Tilesius, W. G., von (1850). *Isis* (Oken), p. 228.

Tomida, Y. (1987). *Small Mammal Fossils and Correlation of Continental Deposits, Safford and Duncan Basins, Arizona*. Tokyo: National Science Museum.

True F. W. (1884). *Babirussa* tusks from an Indian grave in British Columbia. *Science*, 4, 34.

(1894). Diagnoses of new North American mammals. *Proceedings of the United States National Museum*, 17, 241–3.

van der Meulen, A. J. (1973). Middle Pleistocene smaller mammals from the Monte Peglia (Orvieto, Italy) with special reference to the phylogeny of *Microtus* (Arvicolidae: Rodentia). *Quaternaria*, 17, 1–144.

(1978). *Microtus* and *Pitymys* (Arvicolidae) from Cumberland Cave, Maryland, with a comparison of some New and Old World species. *Annals of the Carnegie Museum of Natural History*, 47, 101–45.

Voorhies, M. R. (1988). The giant marmot *Paenemarmota sawrockensis* (new combination) in Hemphillian deposits of northeastern Nebraska. *Transactions of the Nebraska Academy of Sciences*, 16, 165–72.

(1990). Vertebrate biostratigraphy of the Ogallala Group in Nebraska. In *Geologic Framework and Regional Hydrology: Upper Cenozoic Blackwater Draw and Ogallala Formations, Great Plains*, ed. T. C. Gustafson, pp. 115–51. Austin, TX: Bureau of Economic Geology, University of Texas.

Wilson, R. W. (1932). *Cosomys*, a new genus of vole from the Pliocene of California. *Journal of Mammalogy*, 13, 150–4.

(1933). A rodent fauna from later Cenozoic beds of southwestern Idaho. *Carnegie Institute of Washington, Contributions to Paleontology*, 440, 117–35.

Winkler, A. J. and Grady, F. (1990). The middle Pleistocene rodent *Atopomys* (Cricetidae: Arvicolinae) from the eastern and south-central United States. *Journal of Vertebrate Paleontology*, 10, 484–90.

Woodburne, M. W. (ed.) (1987). *Cenozoic Mammals of North America*. Berkeley, CA: University of California Press.

Zakrzewski, R. J. (1967). The primitive vole, *Ogmodontomys*, from the late Cenozoic of Kansas and Nebraska. *Papers of the Michigan Academy of Science, Arts, and Letters*, 52, 133–50.

(1969). The rodents of the Hagerman local fauna, Upper Pliocene of Idaho. *Contributions from the Museum of Paleontology, University of Michigan*, 23, 1–36.

(1975). The late Pleistocene arvicoline rodent *Atopomys*. *Annals of the Carnegie Museum*, 45, 255–61.

(1984). New arvicolines (Mammalia: Rodentia) from the Blancan of Kansas and Nebraska. [In *Contributions in Quaternary Vertebrate Paleontology: A Volume in Memorial to John E. Guilday*, ed. H. H. Genoways and M. R. Dawson.] *Special Publication of the Carnegie Museum of Natural History*, 8, 200–17.

Zakrzewski, R. J. and Harington, C. R. (2001). Unusual Pliocene rodent from the Canadian Arctic islands. *Journal of Vertebrate Paleontology*, 21(suppl. to no. 3), pp. 116A–17A.

Zazhigin, V. (1980). Late Pliocene and Anthropogene rodents of the south of western Siberia. *Academy of Sciences USSR, Geological Institute, Moscow*, Transaction 339, 1–156.

APPENDIX: CLASSIFICATION OF NORTH AMERICAN ARVICOLID GENERA

Family Arvicolidae Gray, 1821

Subfamily Promimomyinae, new

Promimomys Kretzoi, 1955b; synonym *Prosomys* Shotwell, 1956

Subfamily Arvicolinae Bonaparte, 1837

Tribe Arvicolini Kretzoi, 1955b

Mimomys Forsyth Major, 1902

Microtus Schrank, 1798

Lemmiscus Thomas, 1912

Tribe Clethrionomyini, Hooper and Hart, 1962

Clethrionomys Tilesius, 1850

Tribe Phenacomyini, new

Phenacomys Merriam, 1889

Hibbardomys Zakrzewski, 1984

Subfamily Ondatrinae Repenning, 1982

Tribe Ondatrini Kretzoi, 1955a

Ondatra Link, 1795; synonym *Pliopotamys* Hibbard, 1938

Tribe Ogmodontomyini, new

Ogmodontomys Hibbard, 1941

Cosomys Wilson, 1932

Ophiomys Hibbard and Zakrzewski, 1967

Subfamily Nebraskomyinae, new

Nebraskomys Hibbard, 1957

Atopomys Patton, 1965

Subfamily Pliophenacomyinae Repenning, Fejfar, and Heinrich, 1990; new rank

Protopliophenacomys Martin, 1994

Pliophenacomys Hibbard, 1938

Pliolemmus Hibbard, 1938

Guildayomys Zakrzewski, 1984

Subfamily Lemminae Gray, 1825

Lemmus Link, 1795

Plioctomys Suchov, 1976

Mictomys True, 1894

Synaptomys Baird, 1858

Subfamily Discrostonychinae Kretzoi, 1955b

Predicrostonyx Guthrie and Matthews, 1971

Dicrostonyx Gloger, 1841

North American arvicolid genera of uncertain taxonomic affinity

Loupomys Koenigswald and Martin, 1984b; *Neofiber* True, 1884

Proneofiber Hibbard and Dalquest, 1973.

29 Hystricognathi and Rodentia incertae sedis

LAWRENCE J. FLYNN
Peabody Museum, Harvard University, Cambridge, MA, USA

INTRODUCTION

This chapter treats the Tertiary records of some of the most primitive and most derived rodents of the North American fossil record. Its focus is two tiny rodents, three larger genera implicated in the origin of South American "caviomorphs," and members of two quite different South American rodent families, the New World porcupines and the capybaras. The species of minute body size are an alagomyid, which is an early, basal form, and the enigmatic monotypic family Laredomyidae, which has derived dentition despite its early age. The larger taxa are *Protoptychus*, a saltatorial rodent from Wyoming and Utah, and two genera from the Paleogene of Mexico. The rodents assigned to extant families of South American origin are modern in aspect and represent a late Neogene dispersal event into North America, part of the Great American Biotic Interchange.

This chapter considers some of the smallest and largest rodents in the fossil record of North America. Simpson (1928) studied the fossil capybara *Hydrochaeris*, the present species of which is the largest living rodent.

PRIMITIVE ENIGMATIC RODENTS

ALAGOMYIDAE DASHZEVEG, 1990

DEFINING FEATURES

Alagomyidae are among the earliest and most primitive of Rodentia. Two Asian genera are becoming increasingly well known. *Alagomys* was first described from the earliest Eocene (Bumbanian) of Mongolia (Dashzeveg, 1990). *Tribosphenomys*, another alagomyid from the late Paleocene(?) Bayan Ulan beds of Inner Mongolia (Meng *et al.*, 1994), apparently antedates *Alagomys*. A species of *Alagomys* from the latest Paleocene of Wyoming attests to intercontinental dispersal of the family with other rodents at the beginning of the Clarkforkian. The genera share interesting features that generally are primitive, including small body size, a short diastema, upper molars of much greater width than length and lacking an expanded hypocone shelf, and the presence of a hypoconulid on lower molars. Dawson and Beard (1996) noted the synapomorphy for the family of a widely spaced paracone and metacone on P4. In addition, alagomyids have cheek teeth with thin shearing crests and tribosphenic arrangement of cusps (Figure 29.1A).

SYSTEMATICS

Alagomyids indicate no close familial relationship, falling in most analyses as an outgroup to other rodents. Some analyses define Rodentia as excluding Alagomyidae, a nomenclatural choice. Dawson and Beard (1996) saw Paramyidae plus Ctenodactylidae as closer to each other than either is to Alagomyidae. Meng, Hu, and Li (2003) placed the alagomyid *Tribospenomys* outside the primitive Asian Eocene rodent *Cocomys* and extant rodents.

Most rodents tend to have cheek teeth that are square in shape, attenuating the protolophid in lower molars and adding a hypocone (or hypocone shelf) in upper molars. Dawson and Beard (1996) noted that the alagomyid condition (transverse upper molars without a hypocone plus thin crests joining high cusps) may be interpreted as a retained, primitive state reflecting the original rodent morphotype. It is also possible that the conditions of transverse upper molars with shearing crests and no hypocones actually reflect a functional adaptation to an insectivorous diet and are, therefore, derived.

INCLUDED GENUS OF ALAGOMYIDAE DASHZEVEG, 1990

***Alagomys* Dashzeveg, 1990**

Type species: *Alagomys inopinatus* Dashzeveg, 1990.

Type specimen of American species, *A. russelli* Dawson and

Evolution of Tertiary Mammals of North America, Vol. 2. ed. C. M. Janis, G. F. Gunnell, and M. D. Uhen. Published by Cambridge University Press.

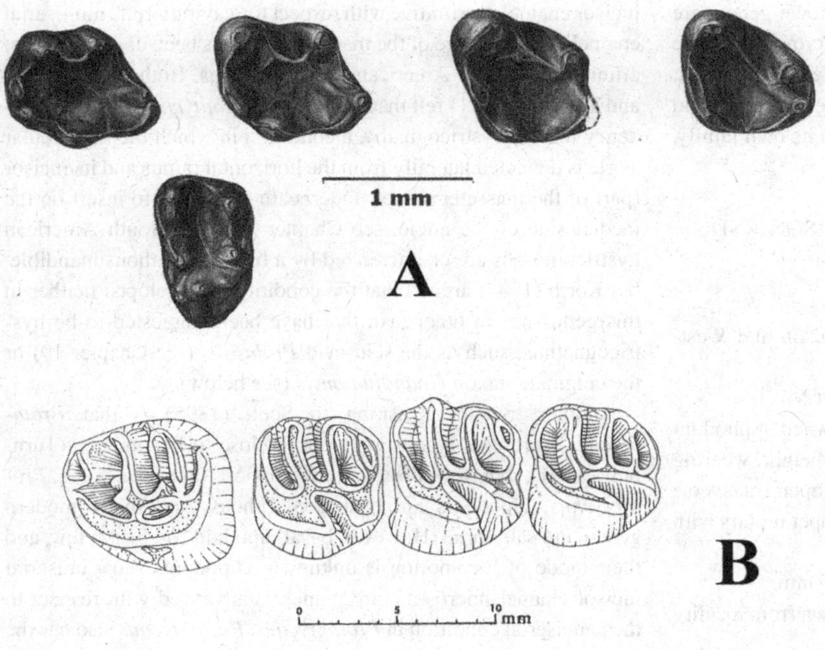

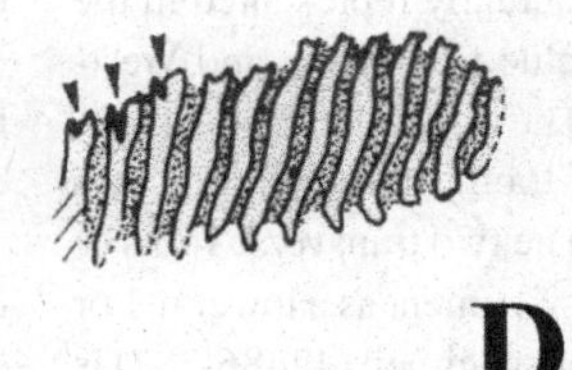

C

20 mm

D

Figure 29.1. A. Cheek teeth of *Alagomys* (from Dawson and Beard, 1996). B. Upper dentition of *Erethizon cascoensis* (from White, 1970). C,D. Upper dentition and incomplete left M3, showing labial reentrants in several laminae, of *Neochoerus dichroplax* (after Mones, 1980). (A: courtesy of the University of Montpellier II.)

Beard, 1996: CM 68692, right dentary with incisor, p4, m1–2.

Characteristics: Small rodent with short diastema; pauciserial incisor enamel in *A. russelli*; cheek teeth brachydont, basined with well-defined cusps joined by sharp crests; paracone and metacone on buccal margin of upper molars, with low crest between them; high protocone isolated from conules; pericone but no hypocone; lower cheek teeth rhombic, lacking any hint of paraconid, with transversely attenuated hypoconulid; m3 without expanded heel.

Length of m2: 0.91 mm (*A. russelli*).

Included species: *A. russelli* only, known from locality CP14E only.

Comments: *Alagomys* is also known from the early Eocene of Mongolia. Regardless of whether *Alagomys* is derived towards insectivory in some features, this taxon provides evidence of a very early branch in rodent evolution, and Alagomyidae participated in the earliest invasion of Rodentia into North America from Asia. Dawson and Beard (1996) suggested that the Clarkforkian presence of *Alagomys* in North America implies that the Bumbanian in Asia may be Paleocene. The incisor enamel of *A. russelli* is pauciserial, with outer layer in evidence; the outer layer is apparently lacking in the Asian *A. inopinatus* (Martin, 1993).

LAREDOMYIDAE WILSON AND WESTGATE, 1991

DEFINING FEATURES

Laredomys, the only member of this monotypic family, is known from the Uintan (late middle Eocene) of Texas. It is of minute body size as indicated by cheek tooth occlusal area less than 1 mm^2. Osteology and skull structure are unknown. Cheek teeth are low crowned and lophate, with crests dominating cusps. and transverse in orientation, sometimes bifurcating. Cheek teeth have a median transverse valley, open at both ends in specimens considered as lower cheek teeth and closed lingually in an upper tooth.

SYSTEMATICS

This exciting new taxon is represented by four published teeth. Its affinities are completely unknown, based in large part on lack of comparable material. The molar structures are so autapomorphous that primitive similarities are obscured. Nothing is known for North America that shows such derived conditions at such an early age (middle Eocene). *Laredomys* invokes comparisons with dormice

(Gliridae), but Gliridae are Old World and well-known genera are younger. Recently, late Eocene glirids of similar crown structure have become known from western Europe (Vianey-Liaud, 1994), reopening the impression that *Laredomys* could be related to Old World glirids. At present, this genus is classified in its own family, pending closer comparison to the oldest Gliridae.

INCLUDED GENUS OF LAREDOMYIDAE WILSON AND WESTGATE, 1991

***Laredomys* Wilson and Westgate, 1991**

Type species: *Laredomys riograndensis* Wilson and Westgate, 1991.

Type specimen: TMM 42486-709, left M1 or M2.

Characteristics: Minute rodent with low-crowned, lophodont cheek teeth; transverse crests of uniform height, wearing evenly; lower molars with complete and open transverse valley bounded by crests that bifurcate; upper molars with transverse valley blocked lingually.

Length of M1 or M2 (the largest tooth): 0.85 mm.

Included species: *L. riograndensis* only, known from locality SB42 only.

Comments: As this taxon is so poorly represented, it is difficult to identify which teeth are actually represented in the fossil material. One important clue for Wilson and Westgate (1991) was root structure. The three roots of the holotype suggest that it is an upper tooth. Its shape indicates that it is not a premolar or M3. The two transverse roots of other teeth are consistent with assignment as a lower m1 or m2. The root structure of another specimen (42486-387) is enigmatic and appears weak; possibly this is a milk tooth.

PROTOPTYCHIDAE SCHLOSSER, 1911

DEFINING FEATURES

This is a monotypic genus and family comprising rodents of moderate body size with hystricomorphous zygomasseteric structure, large incisive foramina, inflated auditory bulla and mastoid, and postcranial adaptations for saltatorial locomotion, notably an extremely elongated hindlimb. The cheek tooth dental formula is P2/0, M3/3, with molariform premolars and a peglike P3. The molars are cuspate with a four-crested pattern that becomes mainly bilophodont with wear. The upper incisors are ungrooved and the incisor enamel is pauciserial.

SYSTEMATICS

Wood (1955) considered the family, known from the Uintan (late middle Eocene) of Utah and Wyoming, to represent an independently specialized ischyromyoid group. Wahlert (1973) advanced understanding of cranial features of the holotype and, noting the hystricomorphy and pauciserial enamel, suggested that *Protoptychus* could represent the stem group of South American hystricomorphs, or "caviomorphs." The molar structure of *Protoptychus* does resemble that of some South American genera, and pauciserial incisor enamel is primitive with respect to "caviomorph" multiserial enamel. The structure of the mandible also has been used to suggest affinity with South American hystricomorphs. Both Wood (1975) and Turnbull (1991) felt that the jaw of *Protoptychus* showed a tendency toward hystricognathy, a condition in which the mandibular angle is deflected laterally from the horizontal ramus and its incisor (part of the masseter passes underneath the ramus to insert on the medial side of the angle; see Chapter 16). The South American hystricomorphs are characterized by a hystricognathous mandible, but Korth (1984) argued that the condition is developed neither in this genus nor in other taxa that have been suggested to be hystricognathus, such as the sciuravid *Prolapsus* (see Chapter 19) or the enigmatic taxon *Guanajuatomys* (see below).

Another hypothesis dating to Scott (1895) is that *Protoptychus* represents a basal myomorph close to Dipodoidea (Turnbull, 1991), but few observations support this idea. Like *Protoptychus*, dipodoids are hystricomorphous, and many modern genera are saltatorial. However, basal dipodoid fossils are few, and their mode of locomotion is unknown. Dipodoids show uniserial incisor enamel microstructure, which is advanced with respect to the pauciserial condition in *Protoptychus*. *Protoptychus* also has the S-type of molar schmelzmuster (internal enamel structure), which is apparently the primitive condition, and unlike the derived C-type typical of myomorph rodents (dipodoids, geomyoids, and muroids) (Koenigswald, 2004). The molar structure is inconsistent with a hypothesis of dipodoid relationship, being unlike that of any myodonts, including muroids.

Morphological characters do not support a special relationship of *Protoptychus* with either South American hystricomorphs or dipodoids. The genus remains a candidate, however, as a basal myomorph or as an advanced ischyromyoid. Koenigswald (2004) argued for a relationship of this taxon with the ischyromyids, but this was based primarily on primitive characters. McKenna and Bell (1997) classified *Protoptychus* as a myomorph incertae sedis. Synapomorphies with higher rodent groups are not apparent.

INCLUDED GENERA OF PROTOPTYCHIDAE SCHLOSSER, 1911

***Protoptychus* Scott, 1895**

Type species: *Protoptychus hatcheri* Scott, 1895.

Type specimen: PU 11235, skull lacking mandible.

Characteristics: Hystricomorphous with infraorbital foramen high but not flaring laterally, origin of medial masseter muscle reaching premaxilla suture; cheek teeth mesodont; peglike P3 present; P4 molariform; p4 submolariform and considerably smaller than m1; molars cuspate, wearing lophodont; first molars largest in tooth rows; upper molars with hypocone, small mesostyle; lower molars with anterior protolophid; m1–2 with complete hypolophid; postcranial morphology indicative of saltatorial (ricochetal) locomotion.

Length of m2: 1.91 mm.

Length of cheek tooth row: approximately 8 mm (Turnbull, 1991: Table 1).

Included species: *P. hatcheri* only (including *P. smithi* Wilson, 1937) (known from localities CP6A, CP38C).

Comments: All specimens appear to be early Uintan in age. The type of *P. hatcheri* is from Utah; the type of *P. smithi* is from beds probably equivalent to the localities that yielded the fine Washakie Basin specimens of Turnbull (1991). Based on the Washakie skeletons, the family is characterized more fully than it could be when Wood (1955) diagnosed it.

?PROTOPTYCHIDAE

Guanajuatomys Black and Stephens, 1973

Type species: *Guanajuatomys hibbardi* Black and Stephens, 1973.

Type specimen: TTU-P 1140, left mandible with incisor, p4, m1–3.

Characteristics: Deep mandible with short, flattened incisor terminating below and medial to molars; convex lower border of ramus; weak masseteric crest ending below m2; brachydont, cuspate cheek teeth; m1 and m2 nearly square; large hypoconulids; short trigonid basins.

Length of m2: 2.3 mm.

Included species: *G. hibbardi* only, known from locality SB51 only.

Comments: While Black and Stephens (1973) noted that familial assignment of *Guanajuatomys* is unclear, they saw general similarity to the Ischyromyidae and possible special relationship to South American "caviomorphs." They were impressed by a suggestion in the dentary of hystricognathy, the condition in which its angle is strongly deflected laterally from the incisor and horizontal ramus. Subsequent workers disagree with this finding (Korth, 1984; Wilson, 1986), removing any apparent synapomorphy with any hystricognaths. Korth (1994) suggested that *Guanajuatomys* could be a protoptychid, although the dentary shape is quite different from that of *Protoptychus*; he also noted that cranial material from the same locality named *Marfilomys* Ferrusquía-Villafranca (1989) could represent the upper dentition of *Guanajuatomys*.

Marfilomys Ferrusquía-Villafranca, 1989

Type species: *Marfilomys aewoodi* Ferrusquía-Villafranca, 1989.

Type specimen: IGM-4032, partial skull with complete upper dentition.

Characteristics: Medium-sized rodent with arched diastema; anterior root of zygomatic arch at level of premolars; infraorbital foramen somewhat enlarged; incisors relatively small, and rounded anteriorly; P3 present, with three cusps; P4 submolariform and about the size of M2; M1–3 decreasing in size posteriorly; bunodont, brachydont, cheek teeth, somewhat higher crowned lingually; P4 and M1–3 with distinct conules; molars with hypocones.

Length of M2: 2.0 mm.

Cheek tooth row length: 9.2 mm. Diastema length; 6.8 mm

Included species: *M. aewoodi* only, known from locality SB51 only.

Comments: The body size and the bunodont condition of the teeth are consistent with the referral of *Marfilomys aewoodi* to *Guanajuatomys hibbardi* (see above), but Ferrusquía-Villafranca (1989) felt that the specimens could not represent closely related species. He made exhaustive comparisons with other taxa, ruling out many but retaining the idea that *Marfilomys* shows a relationship to ischyromyoids, cylindrodontids, and protoptychids. He further felt that affinity to early South American hystricomorphs was supported by jaw structure in related taxa, at least in *Protoptychus*, and "incipient" hystricomorphy in IGM-4032, the specimen assigned to *Marfilomys*. The incipient qualifier is irrelevant. IGM-4032 does have an inflated infraorbital foramen, but whether the masseter muscle penetrated it is unclear. *M. aewoodi* is listed here under ?Protoptychidae, for want of a more convincing assignment. Affinity to South American rodents is unsupported.

SUBORDER HYSTRICOGNATHI TULLBERG, 1899

DEFINING FEATURES

Hystricognathi are extremely diverse today, with many genera and species occurring in South America and through parts of Asia and Africa. One member, the Old World porcupine (*Hystrix*), occurs throughout Africa, across South Asia into Indonesia, and in Italy. The suborder name derives from the feature that unites them, the hystricognathous mandible, which is characterized by the lateral flare of the angle beyond the plane of the incisor and horizontal ramus; a portion of the masseter muscle passes beneath the dentary in a groove anterior to the angle, and inserts on the medial side of the angular process. This complex feature groups diverse South American families with Old World porcupines (*Hystrix*) and African/South Asian groups such as the cane rats (Thryonomyidae). The offset of the ramus and angle of the dentary is pronounced in hystricognaths, unlike the slight offset sometimes noted in *Protoptychus* or *Prolapsus*.

Other than the hystricognathous dentary, few features appear to distinguish the entire suborder. Two traits are characteristic and universal but occur in other rodents as well. One is hystricomorphy, the condition in which the infraorbital foramen is greatly expanded to accommodate passage of part of the masseter muscle. The masseter muscle extends anteriorly through it and originates on the side of the snout (see Chapter 16). Other rodents such as Ctenodactylidae display this feature, and the latter family probably represents the sister taxon to the hystricognaths among living rodents (see Luckett and Hartenberger, 1985). However, there are other hystricomorphous rodents, such as Dipodoidea, which probably derived the condition independently. A second characteristic of Hystricognathi is multiserial incisor enamel. This feature also occurs outside the suborder, for example in ctenodactylids. It does not occur in the hystricomorphous dipodoids, which have uniserial incisor enamel.

A feature shared by many, but not all, hystricognaths in both the Old World and the New World, is the retention of the deciduous premolar in the adult (lack of replacement). This derived condition would appear to unite most South American forms with certain African taxa, but it is important to note that members of the Erethizontidae, for example, do not have this feature, and that early hystricognaths from the Fayum of Africa show premolar replacement. The implication is that a single immigrant from Africa did not account for the entire South American radiation, or that retention of the deciduous premolar developed more than once, probably independently on the two continents.

HYSTRICOGNATHI AND THE ORIGIN OF THE SOUTH AMERICAN GROUPS

Hystricognathi are monophyletic based on diverse sources of information from morphology and molecules (see Bugge, 1985; Luckett and Hartenberger, 1985). Hystricognaths appear in the Eocene of northern Africa and dominate the Oligocene Fayum Faunas of Egypt. Hystricognathi in the present rodent fauna of Africa include Hystricidae (Old World porcupines), Thryonomyidae (cane rats), Petromuridae (dassie rats), and Bathyergidae (mole rats). The Hystricidae have a broad Old World distribution, and several hystricognaths (including thryonomyids) occurred in the mid Tertiary of southern Asia. The South American hystricognaths appeared in the latest Eocene (Wyss *et al.*, 1993); no other rodents preceded them on that continent.

Evolving in isolation from other rodents, the South American hystricognaths diversified and have been thought of as a cohesive group called Caviomorpha. However, there has been considerable discussion as to whether or not the assemblage of New World hystricognathous rodents is indeed monophyletic. Landry (1957, 1999), for example, found morphological similarities to unite *Erethizon* and *Hystrix*, and others that group *Chinchilla* with the African *Petromus*. McKenna and Bell (1997) also grouped the family Erethizontidae outside of their parvorder Caviida, which contains the other South American hystricognaths. The appearance of erethizontids and other hystricognaths in the New World might require two invasions. The monophyly of Caviomorpha remains controversial, but recent molecular studies (Adkins, Walton, and Honeycut, 2002; Huchon *et al.*, 2002) have failed to demonstrate otherwise.

Believing the South American forms to be a natural group, Wood (1975) endorsed Caviomorpha and sought their ancestry in a loosely grouped array of North American rodents. He dubbed this assemblage the Franimorpha and saw within its members generalized ancestors of caviomorphs, including some that showed "incipient hystricognathy" of the dentary. Wood (1975, 1985) called for caviomorph colonization of South America by dispersal from North America, a hypothesis making close relationship to Old World hystricognaths unlikely. Lavocat (1969, 1980) championed the competing hypothesis of dispersal of hystricognaths from Africa to South America. This view was consistent with hystricognath monophyly and the disjunct distribution of the group. It is discussed further below.

INCLUDED GENERA OF HYSTRICOGNATHI TULBERG, 1899

ERETHIZONTIDAE BONAPARTE, 1845

Characteristics: Large rodents (*Erethizon* males exceed 10 kg) with hystricognathus mandible and hystricomorphous zygomasseteric structure; cheek tooth dental formula P1/1, M3/3; premolars are replaced; permanent premolars molariform and large (Figure 29.1B); cheek teeth highly lophodont, all with four crests, except the deciduous premolar, which may have five; crests elevated, but teeth not hypsodont; incisor enamel microstructure multiserial; carotid foramen and separate internal carotid artery present; second and third cervical vertebrae fused; specialized hairs modified as quills; short, non-prehensile tail; feet heavily clawed and modified for climbing trees; reduced hallux and pollux.

Erethizon Cuvier, 1822 (including *Hystrix*, in part; *Coendu*, in part)

Type species: *Erethizon dorsatum* (Linnaeus 1758) (originally described as *Hystrix dorsata*), the extant North American porcupine.

Type specimen of *E. bathygnathum* Wilson, 1935: USNM 13684, left dentary with broken incisor, p4, m1–2.

Characteristics: Largest body size for the family (males of *E. dorsatum* weighing up to 18 kg [Nowak and Paradiso, 1983]); cheek teeth hypsodont; skull and zygomatic arches flaring widely to the posterior; maxillary tooth rows diverging posteriorly; P4 larger than molars in living species.

Average length of m2: 6.8 mm (*E. bathygnathum*).

Included species: *E. dorsatum* (including Pleistocene *Coendu cumberlandicus* White, 1970) (known from localities GC15D, E, GC18IIIA); *E. bathygnathum* Wilson 1935 (including *Coendu stirtoni* White, 1968) (localities NB13C, SB14D, PN23C); *E. cascoensis* (White, 1970; originally described as *Coendu cascoensis*) (localities [CC47D], CC48C); *E. kleini* Frazier, 1981 (locality GC15C); *E. poyeri* Hulbert, 1997 (locality GC15A).

Erethizon sp. is also known from locality GC15B.

Comments: *Erethizion* is also known from the Pleistocene of North America. Extant *Coendu* Lacépède, 1799, the prehensile-tailed porcupine, is known as far north as central Mexico. Frazier (1981) considered no fossils from further north to represent the genus.

HYDROCHAERIDAE GRAY, 1825

Characteristics: Very large rodents (*Hydrochaeris* is the largest living rodent, with a body mass of 50–60 kg), with modified hystricognathus mandible and hystricomorphous zygomasseteric structure; mandible modified with high and heavy masseteric crest, separated from tooth row by deep groove; cheek tooth dental formula P1/1, M3/3, premolars not replaced; cheek teeth hypselodont; multilaminate third upper molar extraordinarily elongated, longer than other cheek teeth combined; lower cheek teeth with accessory laminae, so two tooth rows are about the same length; tooth rows converge anteriorly; maxillary premolars nearly touch; lower incisors short, both upper and lower incisors with shallow anterior furrow; incisor enamel microstructure multiserial; long paroccipital processes.

***Neochoerus* Hay, 1926 (including *Hydrochaeris*, in part)**

Type species: *Neochoerus pinckneyi* (Hay, 1923) (originally described as *Hydrochaeris pinckneyi*).

Type specimen: (*N. dichroplax*) F:AM 107691, left M2 and M3.

Characteristics: Largest body size for the family (larger than *Hydrochaeris*); coronoid process greatly reduced; upper third molar with more laminae (14–17) than in *Hydrochaeris* (12–13) some laminae showing labial bifurcations (Figure 29.1C,D); masseteric ridge extending anteriorly to back of premolar (see Ahearn and Lance, 1980).

Average length of M2: 8 mm *(N. dichroplax).*

Average length of M3: 34 mm (*N. dichroplax*).

Included species: *N. cordobai* Carranza-Castañeda and Miller, 1993 (known from localities SB59, SB65); *N. dichroplax* Ahern and Lance, 1908 (including *N. lancei* Mones, 1980) (localities [GC18IIIA], GC17A, B, SB14D, SB18A, B).

Comments: Hay (1923) based his new species *Hydrochaeris pinckneyi* on a single M3 (unnumbered) from the Pleistocene of South Carolina, near Charleston. He later created the genus *Neochoerus*, being impressed by better material from the Pleistocene of Texas (near Sinton on the Arkansas River). *N. pinckneyi* is a large capybara of late Pleistocene age (Ahearn and Lance, 1980). Blancan localities from Arizona and Florida document earlier (late Pliocene) spread of *N. dichroplax*. However, the genus was represented by an even larger species in central Mexico in the late Hemphillian (Carranza-Castañeda and Miller, 1993), implying earlier immigration than previously suspected. *Neochoerus* is also known from the Pleistocene of North and South America.

***Hydrochaeris* Brünnich, 1772**

Type species: *Hydrochaeris hydrochaeris* (Linnaeus, 1766) (originally described as *Sus hydrochaeris*), the extant capybara of South and Central America.

Type specimen of earliest North American fossil, *H. holmesi*: AMNH 23434, lower incisor, m1 and m3 (holotype from Saber Tooth Cave, Citrus County, Florida).

Characteristics: Large body size (females average 61 kg, males average 50 kg [Nowak and Paradiso, 1983]); M3 laminae with no bifurcation; masseteric crest extending to middle of p4, further than in *Neochoerus*; lower incisor shorter than in *Neochoerus*; vestigial tail.

Length of m3 (no m2 length published): 20.5 mm (*H. holmesi*).

Included species: *H. holmesi* Simpson, 1928 only (known from localities GC15A, C [Morgan and White, 1995]).

Comments: *Hydrochaeris* is also known from the Pleistocene of North and South America.

BIOLOGY AND EVOLUTIONARY PATTERNS

This chapter groups together North American fossil records of enigmatic and derived rodents (Figure 29.2). *Alagomys* represents the basal rodent radiation. Only relatively recently recognized outside Asia, this diminutive rodent presents the primitive rodent morphotype for most features but may possibly be derived in features reflecting insectivory. *Laredomys* is also diminutive, but contrasts with *Alagomys* in its highly derived molar crown pattern. Its crown pattern of folds making thin, sinuous lophs is an adaptation for gum eating or frugivory, and it invokes the dentition of early dormice (Gliridae). If *Laredomys* is indeed a glirid, then its late middle Eocene record in Texas stands as an extraordinary example of rodent dispersal.

The interesting rodent *Protoptychus* is derived in its skeletal adaptations to saltatorial (ricochetal) locomotion. Such adaptation implies open habitats in the late Eocene of Utah and Wyoming. This genus presents the interesting feature of hystricomorphy of the zygoma. Hystricomorphy is a derived condition that evolved at least twice among Rodentia, once in a group more inclusive than suborder Hystricognathi, and once within Myomorpha (it is characteristic of early Myodonta). Whether *Protoptychus* is hystricomorphous independently or shares its ancestry with either suborder is not demonstrated, and it is treated as representing the distinct family Protoptychidae. Two other enigmatic taxa, *Guanajuatomys* and *Marfilomys*, resemble *Protoptychus* in dentition, but a special relationship is not well supported. *Marfilomys*, like *Protoptychus*, appears to have been hystricomorphous; claims of hystricognathy in *Guanajuatomys* (and *Protoptychus*) are not substantiated. Do protoptychids have anything to do with South American hystricognaths? We do not know, but the implication has been that protoptychids and other genera are North American vestiges of the origin of the South American hystricognaths.

Rodentia are absent from the South American record until the late Eocene. Their appearance is a clear case of intercontinental dispersal. Also remarkable is that all of the mid-Cenozoic South American rodents are hystricognaths, and these taxa show affinity with Old World lineages. How the Hystricognathi dispersed to South America remains a mystery. Accepting the Hystricognathi as monophyletic, and struck by their disjunct distribution, some workers have invoked waif dispersal across the Atlantic Ocean from Africa. Lavocat (1969) noted that the distance between Africa and South America in the late Eocene, when rodents would have crossed the Atlantic, was much less than it is today. Rafting early hystricognaths from Africa to South America seems only remotely possible, but this scenario could also explain the presence of platyrrhine primates there (Lavocat, 1980).

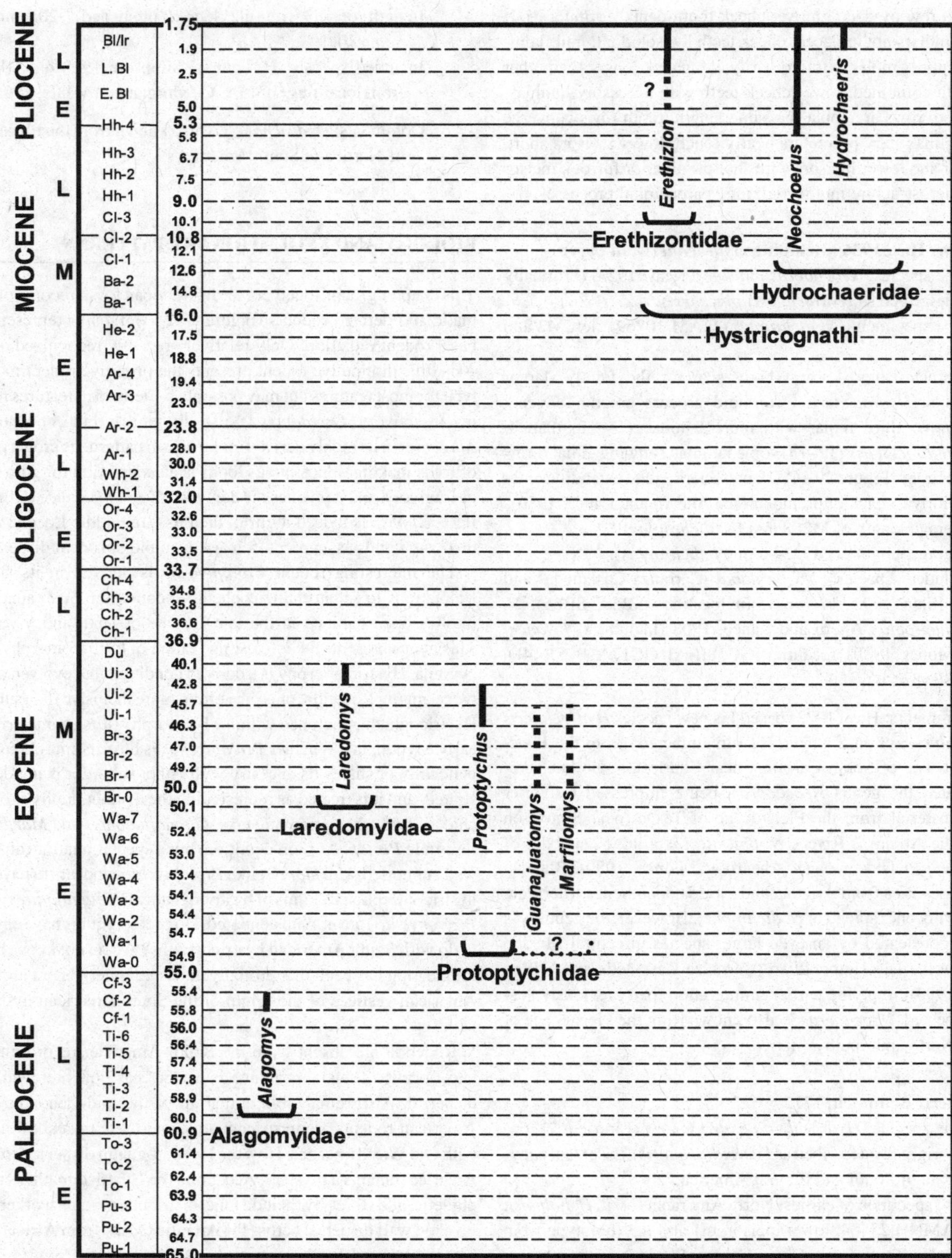

Figure 29.2. Temporal ranges of North American Hystricognathi and Rodentia incertae sedis.

Wood (1975) preferred an independent "caviomorph" origin from the loosely defined North American "franimorphs" (including *Protoptychus*), which are probably not monophyletic. He felt that the South American lineages evolved independently and in parallel with Old World hystricognaths. The dispersal distance (north–south) is much shorter, but morphological ties to any franimorphs are not convincing. There would still have been a water barrier between the American continents.

Flynn, Jacobs, and Cheema (1986) used elements of the arguments of Lavocat and Wood (morphological similarity with African forms, proximity to North America) to propose a scenario of dispersal eastward from Asia. Recognizing potential hystricognaths in the mid Tertiary of southern Asia (the Baluchimyinae), Flynn, Jacobs, and Cheema (1986) proposed dispersal to the New World via Beringia, and then to South America by island hopping. The weakness of this scenario is a lack of hystricognaths in the late Eocene record of North America. Huchon and Douzery (2001) have advocated dispersal of caviomorphs to South America by a southern route from southeast Asia via Australia and Antarctica. A fresh cladistic analysis of diverse fossil taxa (Marivaux *et al.*, 2002) concurs in perceiving a special relationship between South Asian ctenodactyloids and both African and South American hystricognaths. Such dispersalist scenarios remain conjectural, and all are subject to reevaluation by new fossil finds and molecular research on living hystricognaths.

Hystricognaths diversified in South America, and some genera expanded into North America in the Late Tertiary. Capybaras illustrate a complex history. Given that the living *Hydrochaeris* is not as derived as *Neochoerus*, its later presence in North America indicates range extension separate from that of *Neochoerus*. *Hydrochaeris* appears to have arrived from Central America at the end of the Pliocene, about 2 Ma, several million years after the arrival of *Neochoerus*. Indeed, the Hemphillian age of *N. cordobai* shows that *Neochoerus* was present in southern North America quite early, as early as any other late Miocene invader from South America. Porcupines entered the North American fossil record in the late Pliocene, conceivably about the time that *Hydrochaeris* immigrated. The earliest porcupine records are few but widespread, and late Blancan in age; *Erethizon*, of course, was quite successful and is widespread today.

ACKNOWLEDGEMENTS

Pathways to the diverse publication record were facilitated through the Bibliography of Fossil Vertebrates, Korth (1994), and McKenna and Bell (1997). Nowak and Paradiso (1983), Woods (1984), and the collections of the Museum of Comparative Zoology were crucial in developing the accounts on hystricognaths. Dennis Ruez generously supplied information on Florida *Hydrochaeris*. Understanding the interesting rodents of this chapter is a product of the groundbreaking research of individual researchers cited here. I appreciate the privilege to use their work for this synthesis.

REFERENCES

Adkins, R. M., Walton, A. H., and Honeycutt, R. L. (2002). Higher level systematics of rodents and divergence time estimates based on two congruent nuclear genes. *Molecular Phylogenetics and Evolution*, 26, 409–20.

Ahearn, M. E. and Lance, J. F. (1980). A new species of *Neochoerus* (Rodentia: Hydrochoeridae) from the Blancan (Late Pliocene) of North America. *Proceedings of the Biological Society of Washington*, 93, 435–42.

Black, C. C. and Stephens, J. J., III. (1973). Rodents from the Paleogene of Guanajuato, Mexico. *Occasional Papers of the Museum of Texas Technical University*, 14, 1–10.

Bonaparte, C. L. P. (1845). *Catologo Metodico dei Mammiferi Europe*. Milano: Giacomo Pirola.

Brünnich, M. T. (1772). *Zoologiae Fundamenta Praelectionibus Academicis Accomodata*. Grunde i Dyrelaeren: Hasniae et Lipsiae.

Bugge, J. (1985). Systematic value of the carotid arterial pattern in rodents. In *Evolutionary Relationships Among Rodents: A Multidisciplinary Analysis*, ed. W. P. Luckett and J.-L. Hartenberger, pp. 355–79. New York: Plenum Press.

Carranza-Castañeda, O. and Miller, W. E. (1993). Roedores caviomorfos de la Mesa Central de México, Blancano temprano (Plioceno tardío) de la fauna local Rancho Viejo, Estado de Guanajuato. *Revista Instituto de Geologia, Universidad Nacional Autónoma de México*, 7, 182–99.

Cuvier, F. (1822). Examen des éspèces du genre porc-épic, et formation des genres ou sous-genres *Acanthion*, *Eréthizon*, *Sinéthère* et *Sphiggure*. *Mémoires du Museum d'Histoire Naturelle*, 9, 413–456.

Dashzeveg, D. (1990). New trends in adaptive radiation of Early Tertiary rodents (Rodentia, Mammalia). *Acta Zoologica Cracoviensia*, 33, 11–35.

Dawson, M. R. and Beard, K. C. (1996). New Late Paleocene rodents (Mammalia) from Big Multi Quarry, Washakie Basin, Wyoming. *Palaeovertebrata*, 25, 301–21.

Ferrusquía-Villafranca, I. (1989). A new rodent genus from Central Mexico and its bearing on the origin of the Caviomorpha. [In *Papers on Fossil Rodents in Honor of Albert Elmer Wood*, ed. C. C. Black and M. R. Dawson.] *Natural History Museum of Los Angeles County. Science Series* 33, 91–117.

Flynn, L. J., Jacobs, L. L., and Cheema, I. U. (1986). Baluchimyinae, a new ctenodactyloid rodent subfamily from the Miocene of Baluchistan. *American Museum Novitates*, 2841, 1–58.

Frazier, M. K. (1981). A revision of the fossil Erethizontidae of North America. *Bulletin of the Florida State Museum, Biological Sciences*, 27, 1–76.

Gray, J. E. (1825). Outline of an attempt at the disposition of the Mammalia into tribes and families with a list of the genera apparently appertaining to each tribe. *Annals of Philosopy, new series*, 10, 337–344.

Hay, O. P. (1923). The Pleistocene of North America and its vertebrated animals from the states east of the Mississippi River and from the Canadian Provinces east of longitude 95°. *Carnegie Institution of Washington Publication*, 322, 1–499.

(1926). A collection of Pleistocene vertebrates from southwestern Texas. *Proceedings of the National Museum of the United States*, 68, 1–18.

Huchon, D. and Douzery, E. J. P. (2001). From the Old World to the New World: a molecular chronicle of phylogeny and biogeography of hystricognath rodents. *Molecular Phylogenetics and Evolution*, 20, 238–51.

Huchon, D., Madsen, O., Sibbald, M. J. J. B., *et al.* (2002). Rodent phylogeny and a timescale for the evolution of Glires: evidence from an extensive taxon sampling using three nuclear genes. *Molecular Biology and Evolution*, 19, 1053–65.

Hulbert, R. C., Jr. (1997). A new Late Pliocene porcupine (Rodentia: Erethizontidae) from Florida. *Journal of Vertebrate Paleontology*, 17, 623–6.

Koenigswald, W., von (2004). Enamel microstructure of rodents, classification, and parallelisms, with a note on the systematic affiliation of the enigmatic Eocene rodent *Protoptychus*. *Journal of Mammalian Evolution*, 11, 127–42.

Korth, W. W. (1984). Earliest Tertiary evolution and radiation of rodents in North America. *Bulletin of Carnegie Museum of Natural History*, 24, 1–71.

(1994). *The Tertiary Record of Rodents in North America*. New York: Plenum Press.

Lacépède, M. le comte de (1799). Mémiore sur une nouvelle table méthodique des animaux à mamelles. *Mémoires de Mathématique et de Physique*, Year 7 of the Republic, 469–502

Landry, S. O., Jr. (1957). The interrelationships of the New and Old World hystricomorph rodents. *University of California Publications in Zoology*, 56, 1–118.

(1999). A proposal for a new classification and nomenclature for the Glires (Lagomorpha and Rodentia). *Mitteilungen aus dem Museum fur Naturkunde in Berlin, Zoologische Reihe*, 75, 283–316.

Lavocat, R. (1969). La systématique des rongeurs hystricomorphes et la dérive des continents. *Comptes Rendus des Seances de l'Académie des Sciences, Paris*, 269, 1496–7.

(1980). The implications of rodent paleontology and biogeography to the geographical sources and origin of Platyrrhine primates. In *Evolutionary Biology of the New World Monkeys and Continental Drift*, ed. R. L. Ciochon and A. B. Chiarelli, pp. 93–102. New York: Plenum Press.

Linneaus, C. (1758). *Systema Naturae per Regna Tria Naturae, Secundum Classes, Ordines, Genera, Species, cum Characteribus, Differentiis, Synonymis, Locis*. Vol. 1: *Regnum Animale*, 10th edn. Stockholm: Laurentii Salvii.

(1766). *Systema Naturae per Regna Tria Naturae, Secundum Classes, Ordines, Genera, Species, cum Characteribus, Differentiis, Synonymis, locis*. Vol. 1: *Regnum Animale*, 12th edn. Stockholm: Laurentii Salvii.

Luckett, W. P. and Hartenberger, J.-L. (1985). Evolutionary relationships among rodents: comments and conclusions. In *Evolutionary Relationships Among Rodents: A Multidisciplinary Analysis*, ed. W. P. Luckett and J.-L. Hartenberger, pp. 685–712. New York: Plenum Press.

Marivaux, L., Vianey-Liaud, M., Welcomme, J.-L. and Jaeger, J. J. (2002). The role of Asia in the origin and diversification of hystricognathous rodents. *Zoologica Scripta*, 31, 225–39.

Martin, T. (1993). Early rodent incisor enamel evolution: phylogenetic implications. *Journal of Mammalian Evolution*, 1, 227–54.

McKenna, M. C. and Bell, S. K. (1997). *Classification of Mammals Above the Species Level*. New York: Columbia University Press.

Meng, J., Wyss, A. R., Dawson, M. R., and Zhai R.-J. (1994). Primitive fossil rodent from Inner Mongolia and its implications for mammalian phylogeny. *Nature*, 370, 134–6.

Meng, J., Hu, Y., and Li, C. (2003). The osteology of *Rhombomylus* (Mammalia, Glires): implications for phylogeny and evolution of Glires. *Bulletin of the American Museum of Natural History*, 275, 1–247.

Mones, A. (1980). Estudios sobre la Familia Hydrochaeridae (Rodentia), IX. *Neochoerus lancei*, nueva especie del Plioceno de Norteamerica. *Communicaciones Paleontologicas del Museo de Historia Natural de Montevideo*, 1, 171–81.

Morgan, G. S. and White, J. A. (1995). Small mammals (Insectivora, Lagomorpha, and Rodentia) from the early Pleistocene (early Irvingtonian) Leisey Shell Pit Local Fauna, Hillsborough County, Florida. *Bulletin of the Florida Museum of Natural History*, 37(Part 2), 397–461.

Nowak, R. M. and Paradiso, J. L. (1983). *Walker's Mammals of the World*, Vol. II, 4th ed. Baltimore, MD: Johns Hopkins University Press.

Schlosser, M. (1911). Mammalia Säugetiere. In *Grundzüge der Paläontologie*, vol. II: *Vertebrata*, neubearbeitet von F. Broili, E. Koken, M. Schlosser, ed. K. A. von Zittel, pp. 325–585. Munich: R. Oldenbourg.

Scott, W. B. (1895). *Protoptychus hatcheri*, a new rodent from the Uinta Eocene. *Proceedings of the Academy of Natural Sciences, Philadelphia*, 1895, 269–86.

Simpson, G. G. (1928). Pleistocene mammals from a cave in Citrus County, Florida. *American Museum Novitates*, 328, 1–16.

(1930). Rodent Giants. *Natural History*, 30, 305–13.

Tullberg, T. (1899). Ueber das System der Nagethiere, eine phylogenetische Studie. *Nova Acta Regiae Societatis Scientiarium Upsaliensis*, 18(Series 3, Section 2), 1–514.

Turnbull, W. D. (1991). *Protoptychus hatcheri* Scott, 1895. The Mammalian faunas of the Washakie Formation, Eocene Age, of southern Wyoming. Part II. The Adobetown Member, Middle Division (= Washakie B, Twka/2 In Part). *Fieldiana, Geology New Series*, 21, 1–33.

Vianey-Liaud, M. (1994). La radiation des Gliridae (Rodentia) à l'Eocène supérieur en Europe Occidentale, et sa descendance Oligocène. *Münchener Geowissenschaftliche Abhandlungen A*, 26, 117–60.

Wahlert, J. H. (1973). *Protoptychus*, a hystricomorphous rodent from the late Eocene of North America. *Breviora, Museum of Comparative Zoology*, 419, 1–14.

White, J. A. (1968). A new porcupine from the middle Pleistocene of the Anza-Borrego Desert of California. With notes on mastication in *Coendu* and *Erethizon*. *Contributions in Science, Los Angeles County Natural History Museum*, 136, 1–15.

(1970). Late Cenozoic porcupines (Mammalia, Erethizontidae) of North America. *American Museum Novitates*, 2421, 1–15.

Wilson, J. A. and Westgate, J. M. (1991). A lophodont rodent from the middle Eocene of the Gulf Coastal Plain. *Journal of Vertebrate Paleontology*, 11, 257–60.

Wilson, R. W. (1935). A new species of porcupine from the later Cenozoic of Idaho. *Journal of Mammalogy*, 16, 220–2.

(1937). Two new Eocene rodents from the Green River Basin, Wyoming. *American Journal of Science, Series 5*, 34, 447–56.

(1986). The Paleogene record of rodents: fact and interpretation. [In *Vertebrates, Phylogeny, and Philosophy*, ed. K. M. Flanagan and J. A. Lillegraven.] *Contributions to Geology, University of Wyoming Special Paper*, 3, 163–76.

Wood, A. E. (1955). A revised classification of the rodents. *Journal of Mammalogy*, 36, 165–87.

(1975). The problem of hystricognathous rodents. *Papers on Paleontology, University of Michigan*, 12, 75–80.

(1985). Northern waif primates and rodents. In *The Great American Biotic Interchange*, ed. F. G. Stehli and S. D. Webb, pp. 267–82. New York: Plenum Press.

Woods, C. A. (1984). Hystricognath rodents. In *Orders and Families of Recent Mammals of the World*, ed. S. Anderson and J. K. Jones, Jr., pp. 389–446. New York: John Wiley.

Wyss, A. R., Flynn, J. J., Norell, M. A., *et al.* (1993). South America's earliest rodent and recognition of a new interval of mammalian evolution. *Nature*, 365, 434–7.

Part VI: Marine mammals

30 Marine mammals summary

MARK D. UHEN
Smithsonian Institution, Washington, DC, USA

INTRODUCTION

Tetrapods have invaded the aquatic habitat on numerous occasions. They have completely returned to the water in at least six cases: ichthyosaurs, plesiosaurs, mosasaurs, some snakes, cetaceans, and sirenians. Many other groups live almost exclusively in the water but return to the terrestrial environment for birth of their young. These include many turtles, many crocodilians, some lacertilians, some snakes, penguins, pinnipeds, and probably placodonts, nothosaurs, thalattosaurs, desmostylians, and some Jurassic sphenodonts (Mazin, 2001). In addition, there are many tetrapods that spend a great deal of time in and around water but still retain their abilities to locomote on land or in the air, such as marine iguanas, otters, polar bears, aquatic sloths, and various sea birds. It is interesting to note that virtually all partially aquatic tetrapods feed in the water (hippopotamus are one of the rare counterexamples, since they spend a great deal of time in the water but feed on grass on land), and that evolutionary scenarios regarding the evolution of aquatic tetrapods often involve feeding in the water as an early part of the evolutionary scenario. Additionally, it appears that giving birth and sometimes mating are the activities that are not able to be performed in the water, preventing full commitment to an aquatic existence among the diverse groups of semi-aquatic tetrapods, be they reptiles, birds, or mammals.

Among mammals, only Cetacea and Sirenia have completely adopted an aquatic existence. Both groups originated in the early Eocene and proceeded along somewhat parallel evolutionary pathways. Fossils of both groups demonstrate that they each went through a semi-aquatic stage where all four limbs were present, and some form of limb-based aquatic locomotion was used for propulsion. By the late Eocene, both Cetacea and Sirenia had reduced their hindlimbs to the point where they were no longer functional on land, and sirenians no longer ventured out of the water. Both cetaceans and sirenians convergently evolved tail flukes and associated tail-based styles of locomotion.

Pinnipeds are first known from the fossil record in the late Oligocene. Pinnipeds evolved very differently from either Cetacea or Sirenia in that they originally used limb-based locomotion when they first entered the aquatic environment and they still do today. That said, there are three quite different styles of locomotion found in the three groups of modern pinnipeds (odobenids, otariids, and phocids), indicating that their locomotor styles have diverged a great deal since the Oligocene.

Desmostylians are somewhat enigmatic creatures that arose in the late Oligocene and became extinct by the end of the Miocene. Desmostylian fossils are usually found in nearshore marine deposits, which suggests that they lived in this type of environment. Their skeletons have been reconstructed in various ways, but no matter how they are put together, the bones themselves do not particularly suggest much adaptation to an aquatic existence (Domning, 2002). Recent studies have shown that desmostylians probably ate marine algae, which supports the idea that they lived, or at least fed, in nearshore aquatic environments (Clementz, Hoppe, and Koch, 2003).

SYSTEMATICS

CETACEA

There are two groups of living Cetacea: Odontoceti and Mysticeti (variously termed crown-group Cetacea, or Neoceti, or Autoceta [Fordyce and Muizon, 2001]), and one paraphyletic group of exclusively fossil Cetacea from which Neoceti were derived: Archaeoceti (see Figure 30.3). Archaeocetes (and thus Cetacea) originated in the late early Eocene (Bajpai and Gingerich, 1998). Currently it is

Evolution of Tertiary Mammals of North America, Vol. 2. ed. C. M. Janis, G. F. Gunnell, and M. D. Uhen. Published by Cambridge University Press.

Figure 30.1. Restoration of the Miocene eurhinodelphid odontocete *Xiphiacetus cristatus* (by Marguette Dongvillo).

Figure 30.2. Restoration of the Miocene hydrodamaline sirenian *Dusisiren jordani* (by Marguette Dongvillo).

thought that cetaceans arose from within the Artiodactyla (Geisler and Uhen, 2003, 2005), or perhaps are sister taxa with Artiodactyla (Thewissen *et al.*, 2001), sharing a common stem lineage from somewhere within early ungulates. Archaeocetes range from the late early Eocene to the late Eocene and are thought to be ancestral to Neoceti (see Figure 30.4, below; but see Fordyce [2002] for a possible extension of the range of archaeocetes into the late Oligocene). Odontoceti and Mysticeti are thought to be sister taxa, united in a monophyletic Neoceti, which is thought to have been derived from among the basilosaurid archaeocetes, most likely among the dorudontines (Uhen and Gingerich, 2001; Uhen, 2004), a view not supported by Barnes, Goedert, and Furusawa (2001). Synapomorphies for Cetacea include a pachyosteosclerotic tympanic bulla, an elongated rostrum with anterior teeth in line with the cheek teeth, and a prominent falcate process of the basioccipital or basioccipital crest.

Cetaceans have a substantial fossil record. Fossil archaeocetes show multiple diversifications of semi-aquatic forms (middle Eocene Pakicetidae, Ambulocetidae, Remingtonocetidae, Protocetidae) and fully aquatic forms (late middle and late Eocene Basilosauridae). The oldest reported and named species of Neoceti is the toothed mysticete *Llanocetus denticrenatus* (Mitchell, 1989), which is thought to be latest Eocene in age. The oldest odontocete is an unnamed taxon (Barnes, Goedert, and Furusawa, 2001) from the Lincoln Creek Formation, which is thought to be earliest Oligocene in age (Prothero, Jaquette, and Armetrout, 2001). There is a sparse published record of early Oligocene Cetacea, mostly presumed mysticetes (Fordyce, 2003), but the record of late Oligocene and younger Neoceti is excellent.

DEFINING FEATURES OF NEOCETI

Fordyce and Muizon (2001) have given an excellent overview (paraphrased below) of the characteristics that both characterize and delimit Neoceti as a monophyletic group. As was noted by Fordyce and Muizon (2001), characteristics such as "telescoping" do not help to delimit Neoceti, because the style of cranial bone rearrangement differs significantly between Odontoceti and Mysticeti. Those characteristics that are shared by both groups are noted here. The posterior portion of the maxilla is at least slightly concave, rather than smoothly convex, and presents one or more dorsal infraorbital foramina. An open mesorostral groove extends far anteriorly on the rostrum; consequently the premaxillae have little or no midline contact. The posterior-most teeth in odontocetes and toothed mysticetes lie anterior to the antorbital notch. The zygomatic process of the squamosal is robust, anteriorly produced, and the jugal is delicate. Finally, odontocetes and mysticetes are amastoid, with the posterior (mastoid) process of the periotic not exposed laterally on the skull wall. Relationships among neocetans are shown in Figure 30.4.

Most (not all) basal species are polydont, with more than the usual mammalian number of cheek teeth; tooth succession is unknown (Fordyce and Muizon, 2001). Monophyodonty, a potential synapomorphy of Neoceti, is potentially shared with at least one dorudontine archaeocete, *Chrysocetus* (Uhen and Gingerich, 2001). All early odontocetes and mysticetes have cheek teeth with multiple accessory denticles and two roots, but this is a feature that is shared with basilosaurid archaeocetes, rather than a synapomorphy of Neoceti.

Barnes (1984) listed three postcranial characteristics of Neoceti that are lacking in Archaeoceti and are potential synapomorphies of Neoceti: elbow joint non-rotational with anteroposterior position of the radius and ulna, olecranon fossa of the humerus lost, and hyperphalangy. The elbow joint in basilosaurid archaeocetes is still somewhat mobile in flexion and extension but lacks the ability to pronate and supinate (Uhen, 1998, 2004). Basilosaurids also retain an olecranon fossa and lack hyperphalangy (Uhen, 2004). Another potential synapomorphy of Neoceti is the presence of transverse processes on the lumbar vertebrae at nearly 90°

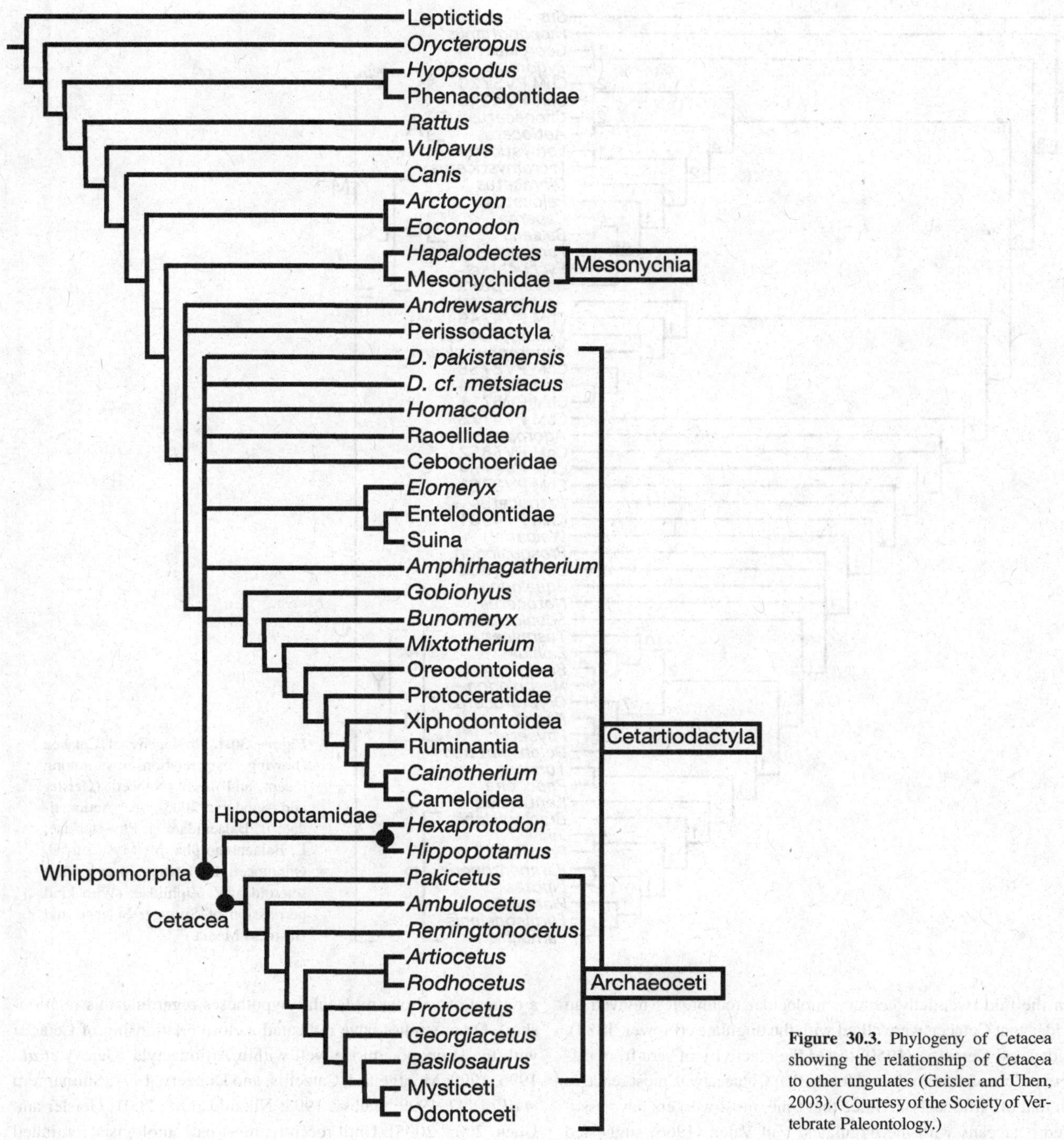

Figure 30.3. Phylogeny of Cetacea showing the relationship of Cetacea to other ungulates (Geisler and Uhen, 2003). (Courtesy of the Society of Vertebrate Paleontology.)

to the neural spine, which in archaeocetes angle distinctly ventrally. All of these postcranial characteristics remain undescribed in most early odontocetes and mysticetes. Once more of the postcranial material of these early neocetes is described, it will become clear when and where these characteristics evolved in the cetacean phylogeny and whether or not they represent synapomorphies of Neoceti.

Cetacea have been phylogenetically allied with almost every group of living mammals at one time or another. Kellogg (1936 [and references therein]) listed many of the groups from which cetaceans have been said to have been derived: Triconodonta, Marsupialia, Insectivora, Creodonta, Pinnipedia, Edentata, Artiodactyla, Perissodactyla, and Sirenia. Linnaeus ([1758], as followed by Simpson [1945]) placed cetaceans in their own group, Mutica, to highlight their strange, or maimed mammalian appearance. By placing them in their own cohort, Simpson (1945) indicated his complete lack of confidence in associating cetaceans with any other order or orders of living mammals.

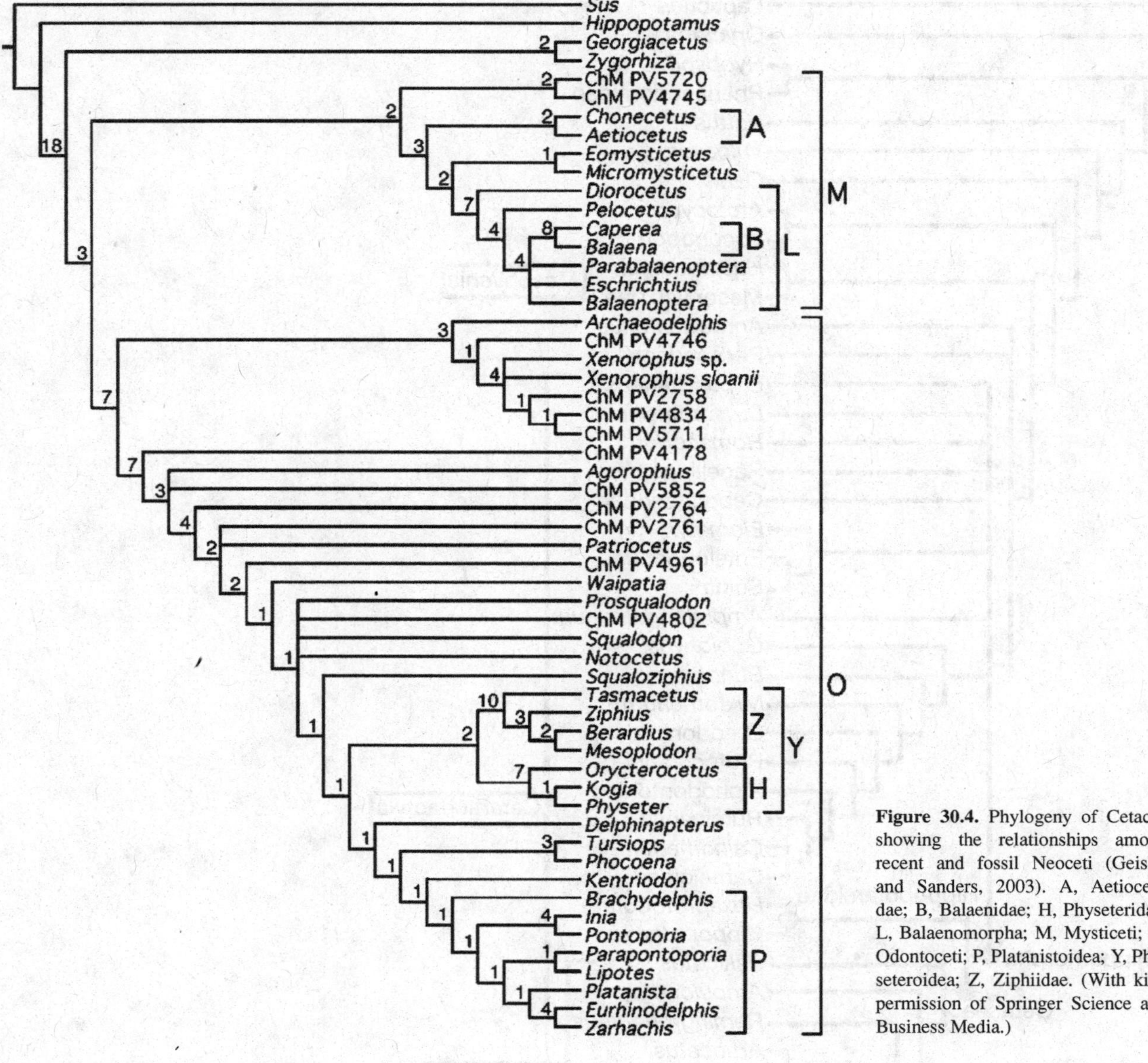

Figure 30.4. Phylogeny of Cetacea showing the relationships among recent and fossil Neoceti (Geisler and Sanders, 2003). A, Aetiocetidae; B, Balaenidae; H, Physeteridae; L, Balaenomorpha; M, Mysticeti; O, Odontoceti; P, Platanistoidea; Y, Physeteroidea; Z, Ziphiidae. (With kind permission of Springer Science and Business Media.)

In the mid twentieth century, molecular techniques revived an old idea that Cetacea were allied with the ungulates (Flower, 1883). Boyden and Gemeroy (1950) tested the reactivity of sera from different orders of mammals and found that Cetacea was most reactive with that of Artiodactyla. Since that time, most workers have associated cetaceans with the ungulates. Van Valen (1966) suggested that Mesonychia (early ungulates thought to have been omnivorous or carnivorous [O'Leary, 2002]) were ancestral to Archaeoceti, and thus to Cetacea as a whole. The more basal relationship of Mesonychia + Cetacea (also known as Cete, but see below) to other groups of ungulates has been unclear at best. Prothero, Manning, and Fischer (1988) suggested that Cete was actually the sister taxon to some early ungulates + Paenungulata (= Uintatheriidae + (miscellaneous taxa + (Tethytheria + Perissodactyla))).

Studies using nuclear and mitochondrial DNA have supported the relationship of Cetacea to Artiodactyla (see Gatesy [1998] for a review of various molecular hypotheses regarding this relationship). Other studies have indicated a close relationship of Cetacea with the Hippopotamidae well within Artiodactyla (Gatesy *et al.*, 1996, 1999; Montgelard, Catzeflis, and Douzery, 1997; Shimamura *et al.*, 1997, 1999; Gatesy, 1998; Nikaido *et al.*, 2001; Geisler and Uhen, 2003, 2005). Until recently, most paleontologists continued to support a relationship with Mesonychia, but this could not be directly addressed by molecular techniques since mesonychians are extinct and thus could not be sampled for molecular studies (Geisler and O'Leary, 1997; Uhen, 1999; Geisler, 2001). Recent fossil discoveries (Gingerich *et al.*, 2001; Thewissen *et al.*, 2001) clearly indicate that early cetaceans share a suite of synapomorphies with artiodactyls to the exclusion of mesonychians (Thewissen *et al.*, 2001; Geisler and Uhen, 2003, 2005). Figure 30.3 illustrates a recent hypothesis of relationships of Cetacea to the rest of Mammalia. This at least indicates that Cetacea + Artiodactyla form the clade

Cetartiodactyla, although it remains an open question what forms the sister taxon to this clade. Perhaps, as shown by Thewissen *et al.* (2001), Mesonychia are the sister taxa to Cetartiodactyla. This hypothesis, however, cannot be correct if the molecular studies indicating that Cetacea and Hippopotamidae are sister taxa is correct. The single analysis incorporating relevant morphological, fossil, and molecular evidence strongly supports inclusion of Cetacea within a traditionally delimited Artiodactyla – and weakly supports a sister-group relationship of Cetacea and Hippopotamidae, Whippomorpha (Geisler and Uhen, 2003). Additional data, including stratigraphy, continue to support this hypothesis (Geisler and Uhen, 2005). Cetartiodactyla as originally conceived included Whippomorpha (Waddell, Okada, and Hasegawa, 1999), so both taxa may still be appropriate.

PINNIPEDIA

RELATIONSHIPS TO OTHER CARNIVORA

Since the 1960s, there has been a great deal of interest in pinniped phylogenetic relationships, which, in part, has resulted from the impressive expansion in the numbers of fossil pinniped specimens available in museum collections together with the advent of molecular phylogenetic methods (Árnason and Widegren, 1986; Vrana *et al.*, 1994; Flynn and Nedbal, 1998).

Some early investigators postulated a non-carnivoran origin for pinnipeds. For example, Wortman (1894, 1906) proposed that they evolved from oxyaenid creodonts. Most researchers, however, have concluded that pinnipeds evolved from arctoid fissiped carnivorans, based on dental and osteological characters (Matthew, 1909), soft anatomical features (Weber, 1904), brains (Fish, 1903), karyotypes (Árnason, 1977), dentitions and cranial anatomy (Tedford, 1976), basicranial circulation (Hunt and Barnes, 1994), mitochondrial DNA (Árnason and Widegren, 1986); cytochrome *b* and 12S genes (Vrana *et al.*, 1994; Flynn and Nedbal, 1998).

However, there has been continuing controversy (Kellogg, 1922; Howell, 1930; Mitchell, 1967) over whether pinnipeds evolved from a single aquatic ancestor (monophyletic origin) or had two independent origins (diphyletic origin). The monophyletic viewpoint contends that all pinnipeds shared a single common aquatic ancestor (Simpson, 1945; Davies, 1958; Scheffer, 1958; Árnason, 1977; King, 1983) and were derived from terrestrial arctoids, usually with ursids being the most likely sister group (Wyss, 1987; Flynn, Neff, and Tedford, 1988; Berta, Ray, and Wyss, 1989). The diphyletic hypothesis proposes that true seals (Phocidae) are most closely related to mustelids, whereas sea lions and walruses (Otariidae and Odobenidae) are related to ursids (McLaren, 1960; Mitchell, 1967; Tedford, 1976; Muizon, 1982).

MONOPHYLY

Wyss (1987) reviewed osteological evidence for walrus relationships and concluded that pinnipeds are monophyletic. He proposed that phocids have their closest relationships with the most derived of the animals that have been traditionally classified as Otarioidea, the allodesmines and the walruses. This is in direct contrast to proposals of monophyly in which the shared common ancestry of Phocidae and Otariidae is very ancient and involved very primitive carnivorans (e.g., Davies, 1958; Scheffer, 1958; King, 1983).

In support of pinniped monophyly Wyss (1988, 1994), Berta *et al.* (1989), and Wyss and Flynn (1993) used skeletal features of various fossil and living pinnipeds to support pinniped monophyly and to support the taxon Pinnipedimorpha, containing the fossil *Enaliarctos* and all other pinnipeds, including phocids, otariids, and odobenids. Deméré, Berta, and Adam (2003) recently constructed a composite tree based on numerous studies of pinniped relationships, depicting a monophyletic Pinnipedimorpha and Pinnipedia (Figure 30.5). Features of the forelimb that are synapomorphies of Pinnipedimorpha include elongation of digit I; decrease in the emphasis of digits I–V, including the reduction of intermediate phalanx V; flattened phalanges with non-trochleated, hingelike interphalangeal articulations and reduction of the medial keel on the distoplantar surfaces of the metacarpal heads; enlarged greater and lesser tuberosities of the humerus; and strongly developed deltopectoral crest of the humerus; and short and robust humerus (Wyss, 1988; Berta and Sumich, 1999). Features of the hindlimb that are synapomorphies of Pinnipedimorpha include elongation of metatarsal I and the first proximal phalanx, well-developed major elements of digit V, and weak digit III (Wyss, 1988; Berta and Sumich, 1999). Features of the skull that are synapomorphies of Pinnipedimorpha include a large infraorbital foramen, a large contribution of the maxilla to the orbital wall, and a lacrimal that is absent or fused early in ontogeny and that does not contact the jugal (Berta and Sumich, 1999).

Numerous molecular studies also support pinniped monophyly within Arctoidea. Using highly repetitive DNA components of the mitochondrial DNA of pinnipeds, Árnason and Widegren (1986) concluded that pinnipeds were monophyletic and that they were the sister taxon to Mustelidae. This study did not include many other groups of Carnivora, much less Arctoidea as possible sister taxa to Pinnipedia. Vrana *et al.* (1994) used a total evidence approach combining cytochrome *b* and ribosomal 12S gene data with morphologic data from Wyss and Flynn (1993) to conclude that again pinnipeds were monophyletic and that they were the sister taxon to Ursidae, and that the red panda (*Aliurus*) was the sister taxon to Ursidae + Pinnipedia. Flynn and Nedbal (1998) used a similar approach combing data from transthyretin intron I, cytochrome *b*, and morphology to conclude again that pinnipeds were monophyletic and that they formed the sister taxon to (Ursidae + ((Procyonidae + Mustelidae) + *Aliurus*)). Despite the disagreement among these studies as to the sister taxon of Pinnipedia (or more inclusively Pinnipedimorpha) within Arctoidea, all agree that pinnipeds are a monophyletic group of arctoid carnivorans.

DIPHYLY

McLaren (1960), Mitchell (1967), Tedford (1976), Barnes (1987), and Wozencraft (1989) have argued that Pinnipedia is an artificial taxon. Wozencraft (1989), in a phylogenetic analysis of Recent Carnivora, followed Tedford (1976) in supporting a close relationship

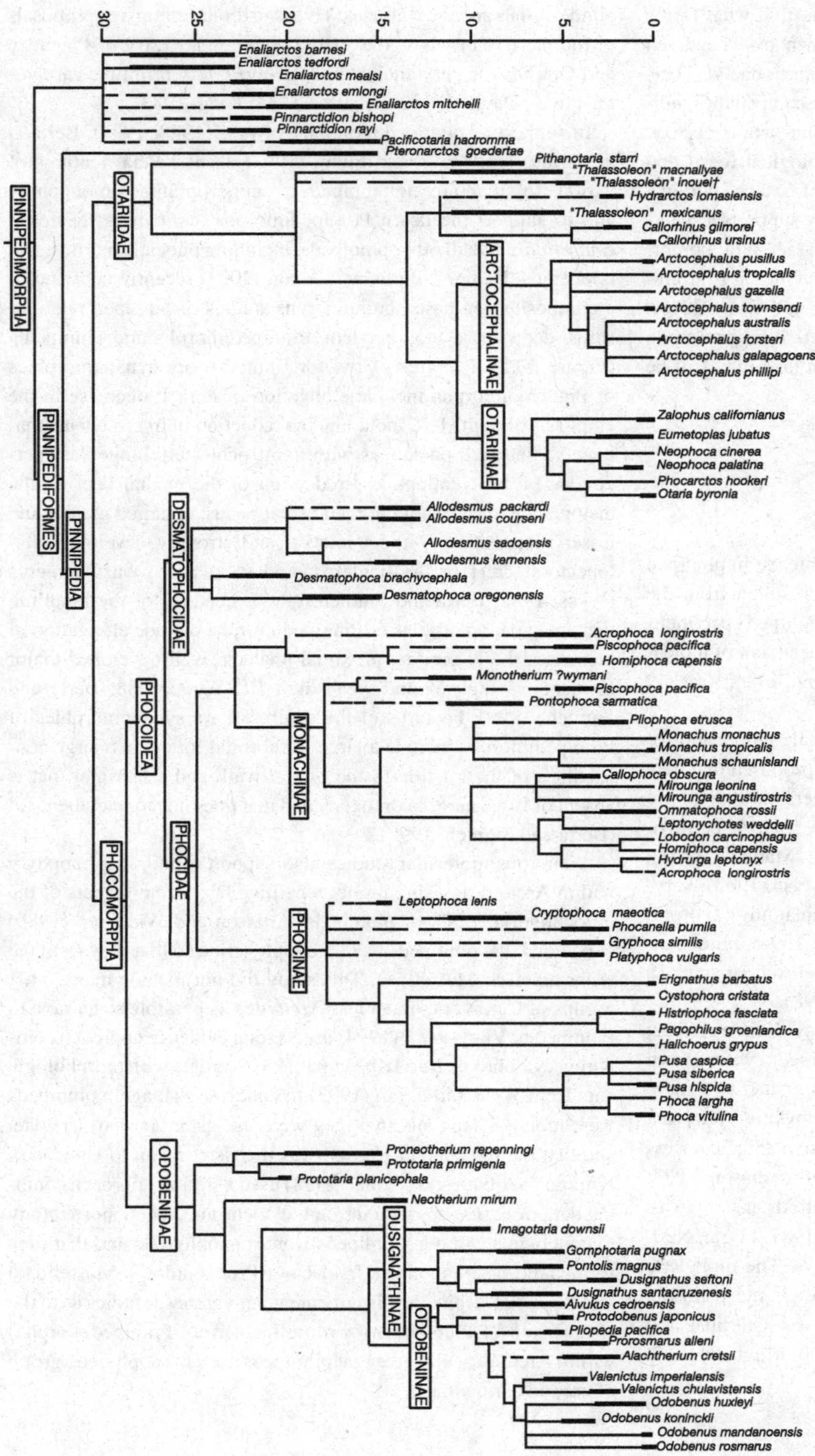

Figure 30.5. Monophyletic hypothesis of pinniped origins (after Deméré, Berta, and Adam, 2003).

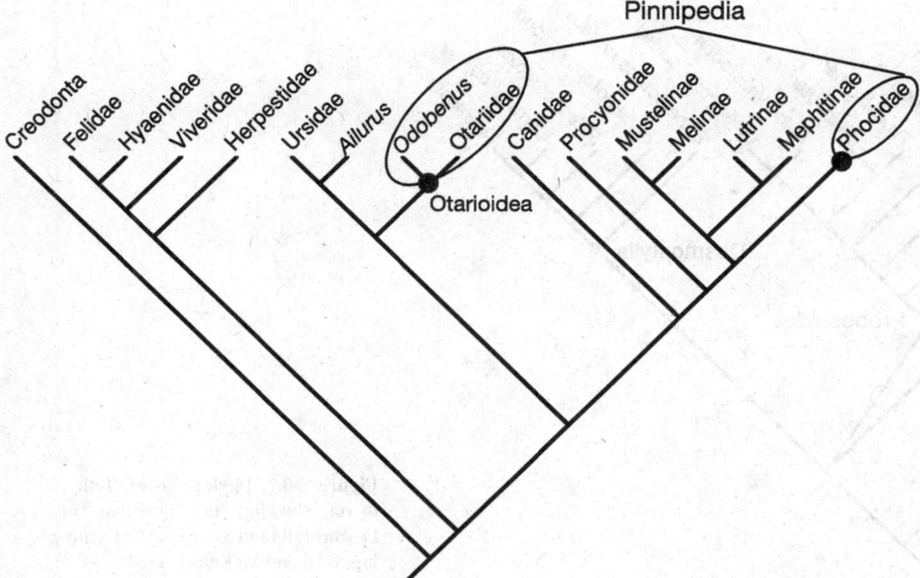

Figure 30.6. Diphyletic hypothesis of pinniped origins (after Wozencraft, 1989).

between phocids and musteloids, although he differed from these authors in details of the interrelationships of these groups. In addition, Wozencraft (1989) followed the traditional practice of uniting otarioids and ursids as a monophyletic group (Figure 30.6).

The evidence for diphyly comes from these synapomorphies uniting phocids and musteloids: the alisphenoid canal is absent in phocids and mustelids (McLaren, 1960), the molars are greatly reduced when present (Tedford, 1976), and the postscapular fossa is absent (Tedford, 1976). Also, additional synapomorphies unite the otarioids and ursids: the presence of a groove between the mastoid and the tympanic on the auditory bullae (McLaren, 1960), the caudal entotympanic is small (Tedford, 1976), the hypotympanic cavity is not significantly extended behind the promontorium (Tedford, 1976), the mastoid process projects ventrally (McLaren, 1960), and the paroccipital and mastoid processes are united by a ridge of bone (McLaren, 1960). Additional support for diphyly comes from the biogeographic patterns seen in living and fossil taxa. Based on the distribution of fossil finds, otarioids originated and evolved in the North Pacific Ocean, while phocids originated and evolved in the North Atlantic–Paratethyan region (Ray, 1977; Koretsky and Sanders, 2002). Given the separation of these two oceanic basins by the continent of North America, it is difficult to see how these two groups could have had a common ancestor that postdated the invasion of the sea, as is suggested by the hypothesis of pinniped monophyly (but see Deméré, Berta, and Adam, 2003).

The differences between the proposed hypotheses for pinniped relationships reflect differences in the interpretations of the polarity of characters, the level of analysis, and the choice of characters included in the analysis (Howell, 1930; Barnes 1972, 1989; Mitchell, 1975; Berta *et al.* 1989; Repenning, Berta, and Wyss, 1990). Proponents of pinniped diphyly have yet to produce a comprehensive phylogenetic analysis including all relevant pinnipedimorphs and potential sister taxa to support their viewpoint.

SIRENIA AND DESMOSTYLIA

Sirenia and Desmostylia are both members of the Tethytheria. This group consists of Proboscidea, Sirenia, and Desmostylia (McKenna and Bell, 1997). Domning, Ray, and McKenna (1986) reviewed the systematics of Tethytheria and determined that desmostylians are more closely related to Proboscidea than either are to Sirenia (Figure 30.7). It is unclear from this study what taxa were used to root their phylogeny. Possible outgroups to the Tethytheria include Perissodactyla, where the two groups are united as Pantomesaxonia (Prothero, Manning, and Fischer, 1988) or Hyracoidea (Gheerbrant *et al.*, 2001); some phenacodontid (*Minchenella*, e.g., Domning, Ray, and McKenna, 1986); or possibly Embrithopoda (McKenna and Bell, 1997). The earliest members of the Tethytheria are all known from Africa (as are many of the possible sister taxa to Tethytheria), and the center of diversity of this group today is in Africa, suggesting a possible African origin for this group. While many of the earliest fossils of Sirenia are known from Africa (Fayum region, Egypt) and Indo-Pakistan, the very earliest sirenian is *Prorastomus* from Jamaica (Savage, Domning, and Thewissen, 1994). Also, all known desmostylians are known exclusively from the North Pacific Ocean Basin, although they are not known before the Oligocene. Presently, the biogeography of desmostylian origins is poorly understood, and there is no consensus opinion on the sister-group relationship to the tethytheres.

EVOLUTIONARY AND BIOGEOGRAPHICAL PATTERNS

EOCENE

CETACEA

Cetacea originated in Indo-Pakistan in the late early Eocene (Bajpai and Gingerich, 1998). From these early beginnings, archaeocetes

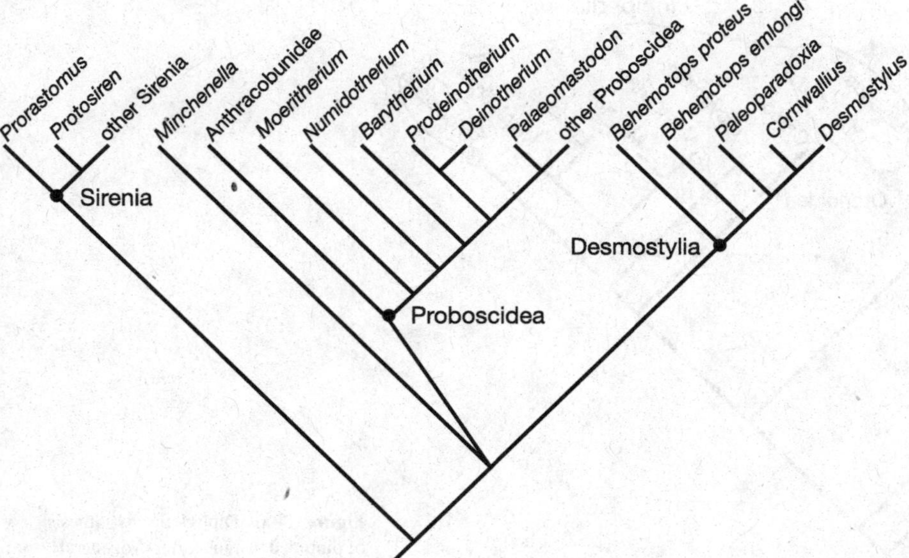

Figure 30.7. Phylogeny of Tethytheria showing the positions of Desmostylia and Sirenia (after Domning, Ray, and McKenna, 1986).

diversified into many semi-aquatic forms during the middle Eocene. During the middle Eocene, semi-aquatic protocetids spread from Indo-Pakistan, across west Africa, and all the way to the Atlantic and Gulf Coasts of North America. It is unclear what route these immigrant taxa took, either following the shorelines across the North Atlantic or directly across the Atlantic Ocean (Uhen, 1999; Figure 30.8).

By the end of the middle Eocene, basilosaurid archaeocetes had evolved from among the most aquatic forms of protocetids. In most phylogenetic analyses, either *Georgiacetus* or *Eocetus* is the protocetid sister taxon to Basilosauridae + Neoceti (Uhen and Gingerich, 2001). Since both of these taxa are known from North America, it suggests that basilosaurids may have originated in North America, although that is far from certain. As basilosaurids diversified and spread around the world in the late middle and early late Eocene, protocetids disappear. There are no known protocetids from the late Eocene or later. Basilosaurids, and thus archaeocetes as a whole, disappear from the fossil record at the end of the Eocene (but see below).

While the origin of Neoceti is firmly rooted in the dorudontine basilosaurids, the exact nature of the origin is unclear at best. There may be separate dorudontine roots for Odontoceti and Mysticeti, or there may be a single root for stem Neoceti within the Dorudontinae. A single described late Eocene taxon, *Chrysocetus healyorum*, might represent the latter. *Chrysocetus* may be monophyodont, a characteristic that would unite it with Neoceti to the exclusion of all other dorudontines (Uhen and Gingerich, 2001). Additional archaeocete-like fossils are known from the late Oligocene of New Zealand that are thought to be close to the base of Neoceti (Fordyce, 2002). Until these fossils are fully described and placed in a phylogenetic context, the origin of Neoceti will remain in question.

The earliest known Mysticeti, from the latest Eocene of Antarctica (Mitchell, 1989), predates the earliest known Odontoceti, from the early Oligocene of North America (Barnes, Goedert, and Furusawa, 2001) (Figure 30.9). Although the original description of the earliest known mysticete, *Llanocetus denticrenatus*, was based on a fragmentary specimen, additional material has been found representing the skull, jaw, and much of the skeleton (Fordyce, 2003). This whale had large, widely spaced posterior teeth whose crowns have prominent accessory cusps. Mitchell (1989) postulated that *L. denticrenatus* filter-fed on planktonic Crustacea, in a manner analogous to crabeater seals. It may also be the case that early mysticetes filter-fed on schools of small fish, which would be more similar to the hypothesized feeding mechanism of their archaeocete forebears. Further study on the more complete material of *Llanocetus* and other toothed mysticetes will hopefully elucidate the origins of mysticete feeding mechanisms.

PINNIPEDIA

Pinnipeds are unknown from the Eocene, although other arctoid carnivorans are known from North America during the Eocene.

DESMOSTYLIA

Desmostylians are unknown from the Eocene.

SIRENIA

Although Sirenia are considered to belong to the Tethytheria, which are often thought to have an African origin, and despite the fact that many of the earliest records of Sirenia are from the Old World, the earliest known sirenian, *Prorastomus sirenoides* is known from the late early Eocene of Jamaica (Savage, Domning, and Thewissen, 1994; Figure 30.11, below). Another Jamaican sirenian, *Pezosiren portelli*, from the early middle Eocene, shows that early sirenians were semi-aquatic quadrupeds (Domning, 2001a). A possible

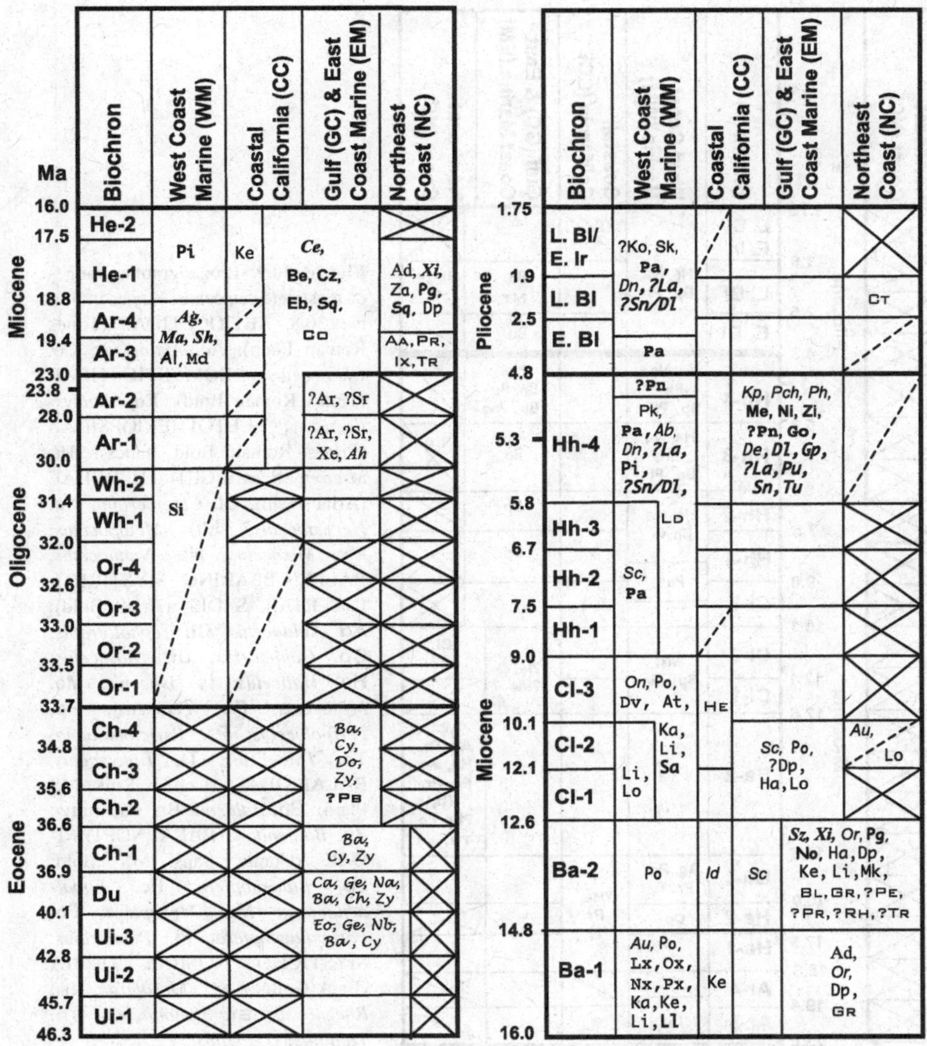

Figure 30.8. Biogeographic ranges of archaeocete and odontocete cetaceans. Key: A "box" (for a particular time period in a particular biogeographic region) that has a cross through it means no fossil marine mammal localities are known for that time period from that area; a single dashed line through the box means only scant fossil marine mammal information is available (usually only a single, small locality). ARCHAEOCETI: PROTOCETIDAE (Lucida Handwriting Plain): Ca, *Carolinacetus*; Eo, *Eocetus*; Ge, *Georgiacetus*; Na, *Natchitochia*; Nb, New Bern Whale. BASILOSAURIDAE (Lucida handwriting Bold): Ba, *Basilosaurus*; Ch, *Chrysocetus*; Cy, *Cynthiacetus*; Do, *Dorudon*; Zy, *Zygorhiza*. ARCHAEOCETI NOMINA DUBIA (Bank Gothic Bold): PB, *Pontobasileus*. ODONTOCETI: INSERTAE SEDIS (Times New Roman Plain): Ad, *Araeodelphis*; Ar, *Archaeodelphis*; Sr, *Saurocetus*; Xe, *Xenorophus*. AGOROPHIIDAE (Times New Roman Italics): *Ah*, *Agorophius*. SIMOCETIDAE (Times New Roman Bold): **Si**, *Simocetus*. EURHINODELPHINIDAE (Times New Roman Bold Italics): ***Ag***, *Argryocetus*; ***Ce***, *Ceterhinops*; ***Ma***, *Macrodelphinus*; ***Sh***, *Squaloziphius*; ***Sz***, *Schizodelphis*; ***Xi***, *Xiphiacetus*. ODONTOCETI: PHYSETEROIDEA: KOGIIDAE (Arial Plain): Ko, *Kogia*; Pk, *Praekogia*; Sk, *Scaphokogia*. PHYSETERIDAE (Arial Italics): *Au*, *Aulophyseter*; *Id*, *Idiophyseter*; *Kp*, *Kogiopsis*; *On*, "*Ontocetus*"; *Or*, *Orycterocetus*; *Pc*, *Paleophoca*; *Ph*, *Physeterula*; *Sc*, *Scaldicetus*. ZIPHIOIDEA: ZIPHIDAE (Arial Bold): **An**, *Anoplonassa*; **Be**, *Belemnoziphius*; **Cz**, *Choneziphius*; **Eb**, *Eboroziphius*; **Me**, *Mesoplodon*; **Ni**, *Ninoziphius*; **Py**, *Pelycorhamphus*; **Pz**, *Proroziphius*; **Zi**, *Ziphius*. PLATANISTOIDEA: PLATANISTIDAE (Comic Sans MS Plain): Al, *Allodelphis*; Po, *Pomatodelphis*; Za, *Zarhachis*. SQUALODELPHINIDAE (Comic Sans MS Bold): **Pg**, *Phocageneus*; **No**, *Notocetus*; **Sq**, *Squalodon*. ODONTOCETI: DELPHINIDA: INCERTAE SEDIS (Courier Plain): Dv, *Delphinavus*; Lx, *Lamprolithax*; Md, *Miodelphis*; Nx, *Nannolithax*; Ox, *Oedolithax*; Px, *Platylithax*. LIPOTIDAE (Courier Bold): **Pa**, *Parapontoporia*. PONTOPORIIDAE (Courier Bold): **Pn**, *Pontoporia*. INIIDAE (Courier Bold): **Go**, *Goniodelphis*. DELPHINOIDEA: KENTRIODONTIDAE (Monaco Plain): At, *Atocetus*; Dp, *Delphinodon*; Ha, *Hadrodelphis*; Ka, *Kampholophos*; Ke, *Kentriodon*; Li, *Liolithax*; Lo, *Lophocetus*; Mk, *Macrokentriodon*. ALBERIONIDAE (Monaco Italics): *Ab*, *Albireo*. MONODONTIDAE (Monaco Italics): *De*, *Delphinapterus*; *Dn*, *Denebola*. PHOCOENIDAE (Monaco Bold): **Ll**, *Loxolithax*; **Pi**, *Piscolithax*, **Sa**, *Salumiphocoena*. DELPHINIDAE (Monaco Bold Italics): ***Dl***, *Delphinus*; ***Gp***, *Globicephala*; ***La***, *Lagenorhynchus*; ***Pu***, *Pseudorca*; ***Sn***, *Stenella*; ***Tu***, *Tursiops*. ODONTOCETI NOMINA DUBIA (Bank Gothic): AA, *Agabelus*; BL, *Belosphys*; CT, *Cetophis*; CO, *Colophonodon*; GR, *Graphiodon*; HE, *Hesperocetus*; IX, *Ixacanthus*; LD, *Lonchodelphis*; PE, *Pelodelphis*; PR, *Priscodelphinus*; RH, *Rhabdosteus*; TR, *Tretosphys*.

prorastomid has also been noted from the early middle Eocene of Israel (Goodwin, Domning, and Lipps, 1998). Many additional middle and late Eocene localities from North America, Europe, North Africa, East Africa, and Indo-Pakistan show that sirenians spread throughout the Atlantic and Tethyan region during their early history (Domning, Morgan, and Ray, 1982).

Many of these Old World Eocene sirenians retain well-developed hindlimbs, and they were most likely still amphibious (Domning and Gingerich, 1994). By the end of the Eocene, however, sirenians are thought to have become fully aquatic (Domning, 2001b). Both family Prorastomidae and family Protosirenidae are extinct by the end of the Eocene. The modern paraphyletic family Dugongidae originated

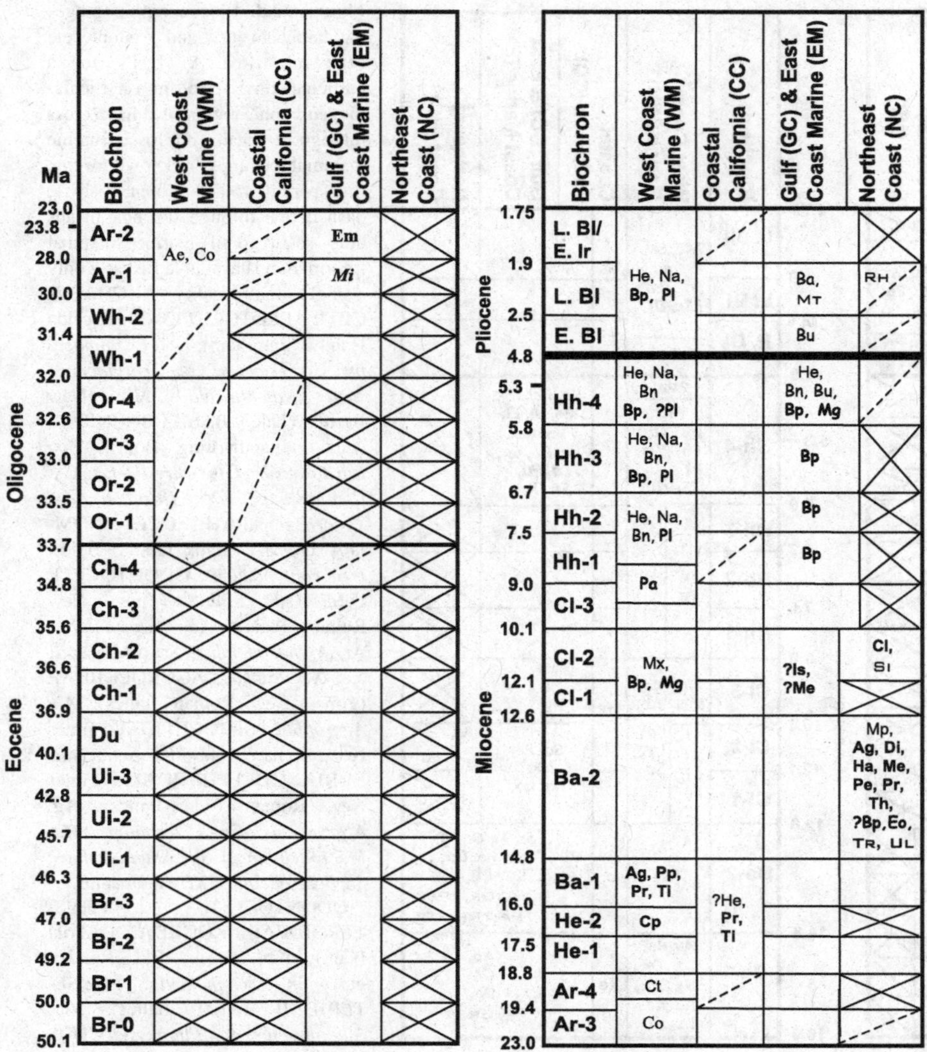

Figure 30.9. Biogeographic ranges of mysticete cetaceans. Key as in Figure 30.8. AETIOCETIDAE (Times Roman Plain): Ae, *Aetiocetus*; Co, *Chonecetus*. EOMYSTICETIDAE (Times Roman Bold): **Em**, *Eomysticetus*. CETOTHERIOPSIDAE (Times Roman Bold Italics): ***Mi***, *Micromysticetus*. CETOTHERIIDAE (Arial Plain): Ct, *Cetotherium*; He, *Herpetocetus*; Mt, *Metopocetus*; Mx, *Mixocetus*; Na, *Nannocetus*. BALEEN-BEARING MYSTICETI: INCERTAE SEDIS (Arial Bold): **Ag**, *Aglaocetus*; **Cl**, *Cephalotropis*; **Cp**, *Cophocetus*; **Di**, *Diorocetus*; **Ha**, *Halicetus*; **Is**, *Isocetus*; **Me**, *Mesocetus*; **Pe**, *Pelocetus*; **Pp**, *Peripolocetus*; **Pr**, *Parietobalaena*; **Th**, *Thinocetus*; **Ti**, *Tiphyocetus*. BALAENIDAE (Comic Sans MS Plain): Ba, *Balaena*; Bn, *Balenulua*; Bu, *Balaenotus*. BALAENOPTERIDAE (Comic Sans MS Bold: **Bp**, *Balaenoptera*; **Eo**, *Eobalaenoptera*; **Mg**, *Megaptera*; **Pa**, *Parabalaenoptera*; **Pl**, *Plesiocetus*. MYSTICETI NOMINA DUBIA (Bank Gothic): MT, *Mesoteras*; RH, *Rhegnopsis*; SI, *Siphonocetas*; TR, *Tretulias*; UL, *Ulias*.

in the Eocene and is represented by the genera *Eotheroides*, *Prototherium*, and *Eosiren* (Domning, 1994).

OLIGOCENE

CETACEA

Cetacean fossils are extremely rare during the early Oligocene (Figures 30.8 and 30.9). This is, in part, because of the major fall in sea level during the late early Oligocene, which created a hiatus and probably destroyed some previously deposited early Oligocene sediments. During this time, it is thought that both mysticetes and odontocetes radiated and became globally distributed. Cetaceans are conspicuously absent from the Gulf Coast of North America during the early Oligocene, despite their relative abundance in the late Eocene, and a continuous record of sedimentation in the Gulf during this interval (Dockery and Lozouet, 2003). Although odontocetes are quite diverse by the late Oligocene, indicating a prior period of evolution and diversification, there are no demonstrably late Eocene or early Oligocene named odontocetes in the published record.

A variety of toothed mysticetes are known from the late Oligocene from around the world. Some of these had large teeth rather like those of archaeocetes (*Kekenodon*, "archaeomysticetes" [Barnes and Sanders, 1996a,b; Uhen, 2002]). Others had small, closely appressed teeth that sustained substantial wear during life (Mammalodontidae), while still others had diminutive teeth that may not have had any function at all (Aetiocetidae), possibly in conjunction with proto-baleen of some form (Emlong, 1966). By the late Oligocene, baleen-bearing mysticetes were also known from the east coast of North America (Sanders and Barnes, 2002), the west coast of North America (Crowley and Barnes, 1996), Australasia (Fordyce and Muizon, 2001), and Japan (Okazaki, 1995). No modern families of mysticetes are known from the Oligocene.

Many types of odontocete are also known from the late Oligocene. In North America, they are found in the Pacific Northwest from

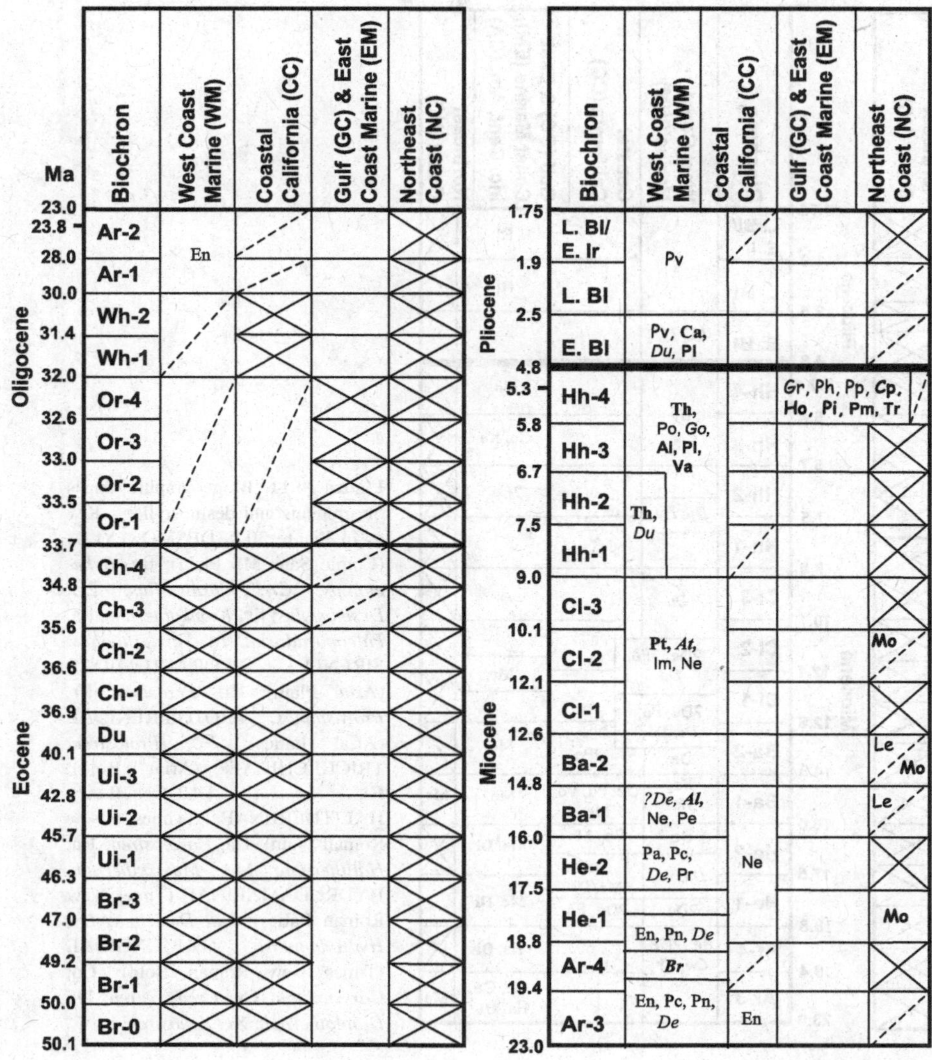

Figure 30.10. Biogeographic ranges of pinnipeds. Key as in Figure 30.8. PHOCIDAE: PHOCINAE (Comic Sans MS Plain): Gr, *Gryphoca*; Le, *Leptophoca*; Ph, *Phocanella*; Pp, *Platyphoca*; Pv, *Phoca*. MONACHINAE (Comic Sans MS Bold): **Cp**, *Callophoca*; **Ho**, *Homiphoca*; **Mo**, *Monotherium*; **Pi**, *Pliophoca*. OTARIOIDEA: OTARIDAE: ENALIARCTINAE (Times New Roman Plain): En, *Enaliarctos*; Pa, *Pacificotaria*; Pc, *Pteronarctos*; Pn, *Pinnarctidion*. OTARIINAE (Times New Roman Bold): **Ca**, *Callorhinus*; **Pt**, *Pithanotaria*; **Th**, *Thalassoleon*. DESMATOPHOCIDAE: DESMATOPHOCINAE (Times New Roman Italics): *De*, *Desmatophoca*. ALLODESMINAE (Times New Roman Bold Italics): ***Al***, *Allodesmus*; ***At***, *Atopotarus*; ***Br***, *Brachyallodesmus*. ODOBENIDAE: IMAGOTARIINAE (Arial Plain). Im, *Imagotaria*; Ne, *Neotherium*; Pe, *Pelagiarctos*; Po, *Pontolis*; Pr, *Proneotherium*. DUSIGNATHINAE (Arial Italics): *Du*, *Dusignathus*; *Go*, *Gomphotaria*. ODOBENINAE (Arial Bold): **Ai**, *Aivukus*; **Pl**, *Pliopedia*; **Pm**, *Prorosmarus*; **Tr**, *Trichecodon*; **Va**, *Valenictus*.

British Columbia to northern Californian, Baja California, and on the east coast in the vicinity of Charleston, South Carolina. All early odontocetes are thought to be echolocators, although a wide variety of skull forms indicate a diversity of feeding habits. No modern families of odontocetes are known from North America during the Oligocene, although the Physeteridae (sperm whales) are known from the late Oligocene of Georgia (Eurasia, not North America; Mchedlidze, 1984).

PINNIPEDIA

The earliest named pinnipeds (*Enaliarctos*) are from the late Oligocene of the Pacific Northwest coast of North America (Berta, 1991; Figure 30.10). Recent biogeographic analysis places the center of origin for pinnipeds in the eastern North Pacific in the late Oligocene at the onset of Antarctic glaciation (Deméré, Berta, and Adam, 2003). Berta and Wyss (1994) concluded that *Enaliarctos* was the sister taxon to all other pinnipeds. The late Oligocene Ashley and Chandler Bridge Formations of South Carolina (localities EM4 and EM5) have produced proximal femora that have been ascribed to Phocidae (Koretsky and Sanders, 2002), but these specimens have not been placed in formally named species to date.

DESMOSTYLIA

The earliest known desmostylians (Figure 30.11) are from the late Oligocene deposits of Oregon and Washington (*Behemotops*; Domning, Ray, and McKenna, 1986), western coastal Canada and Alaska (*Cornwallius*; Domning, 2002), and Japan (*Behemotops* and *Ashoroa*; Inuzuka, 2000). The entire known fossil record of Desmostylia is restricted to the margins of the North Pacific Ocean Basin. While there is no known Eocene or early Oligocene records of desmostylians, Domning (2001b) speculated that desmostylians evolved in Asia from their common ancestor with other tethytheres, possibly something like the phenacolophid *Minchenella* (Domning, 2002).

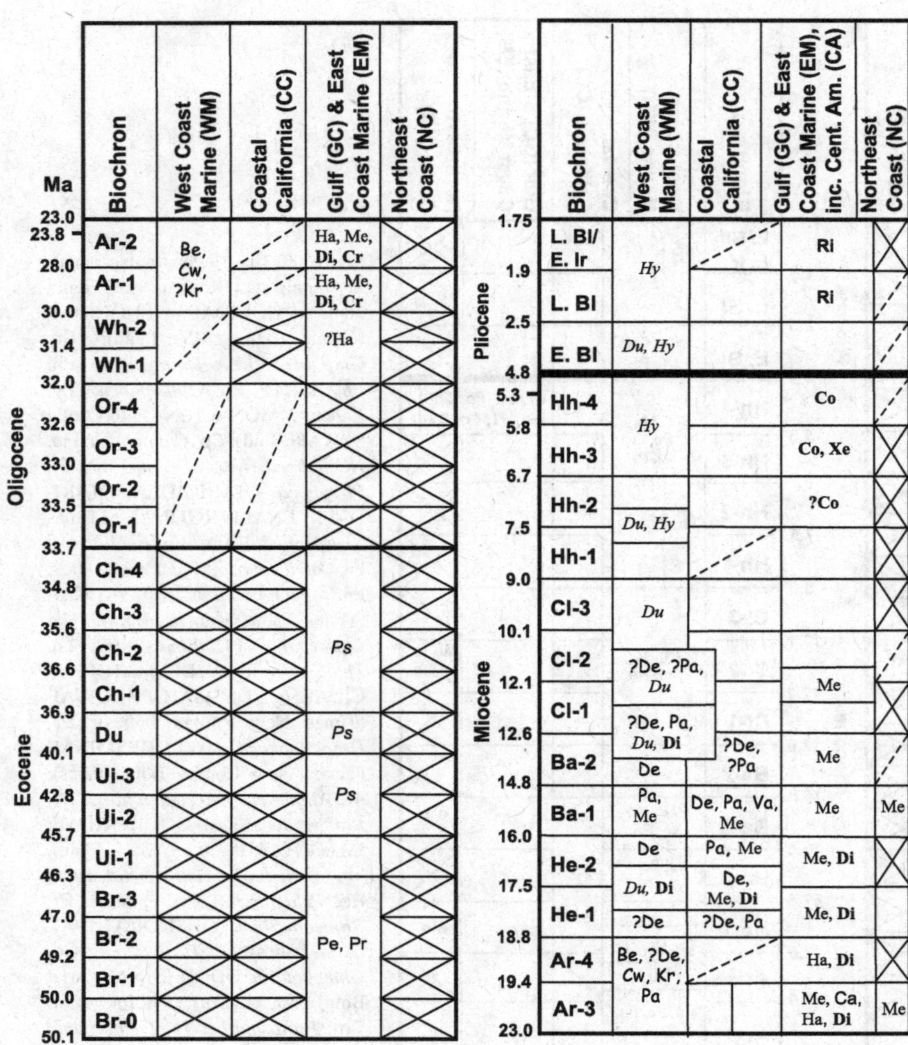

Figure 30.11. Biogeographic ranges of sirenians and desmostylians. Key as in Figure 30.8. DESMOSTYLIA (Comic Sans MS Plain): Be, *Behemotops*; Cw, *Cornwallius*; De, *Desmostylus*; Kr, *Kronkotherium*; Pa, *Paleoparadoxia*; Va, *Vanderhoofius*. SIRENIA: PRORASTOMIDAE (Arial Plain): Pe, *Pezosiren*; Pr, *Prorastomus*. PROTOSIRENIDAE (Arial Italics): *Ps*, *Protosiren*. TRICHECHIDAE (Arial Bold): **Ri**, *Ribodon*. DUGONGIDAE: HALITHERIINAE (Times New Roman Plain): Ca, *Caribosiren*; Ha, *Halitherium*; Me, *Metaxytherium*. HYDRODAMALINAE (Times New Roman Italics): *Du*, *Dusisiren*; *Hy*, *Hydrodamalis*. DUGONGINAE (Times New Roman Bold): **Co**, *Corystosiren*; **Cr**, *Crenatosiren*; **Di**, *Dioplotherium*; **Xe**, *Xenosiren*.

SIRENIA

The early Oligocene sees the last of the "Eocene" taxa in the species *Eosiren imenti*, which is known from the Gebel Qatrani Formation of the Fayum, Egypt (Domning *et al.*, 1994; Figure 30.11). The earliest members of the modern family Trichechidae make their first appearance in the taxon *Annomotherium* from the late Oligocene of Europe (Domning *et al.*, 1994). Dugongids continue to diversify in the Oligocene, with the first appearance of the subfamilies Halitheriinae (*Halitherium*) and Dugonginae (*Crenatosiren*) (McKenna and Bell, 1997).

MIOCENE

CETACEA

Cetaceans diversify wildly in the Miocene, reaching a global diversity peak during the middle Miocene that is unmatched even in the Recent cetacean fauna (Figures 30.8 and 30.9). Stem baleen-bearing mysticetes are abundant and globally distributed during the Miocene. Both the Balaenidae and Balaenopteridae originate in the Miocene, with the earliest known Balaenidae near the Oligocene/Miocene boundary and the earliest known Balaenopteridae from the middle Miocene. Toothed mysticetes are gone from the global cetacean fauna by the early Miocene.

Archaic families of odontocetes diversify and spread globally during the early and middle Miocene, and most modern families of odontocetes originated by the end of the Miocene. Archaic families are either extinct (Squalodontidae, Dalpiazinidae, Eurhinodelphinidae, Eoplatanistidae) or of low diversity (Platanistidae) by the end of the Miocene. Kentriodontidae (the earliest representatives of the modern group Delphinoidea) originates in the Oligocene and further diversifies in the early and middle Miocene, before becoming extinct by the end of the Miocene. Modern families of Delphinoidea (Delphinidae, Phocoenidae, Monodontidae, Pontoporiidae, Iniidae, and Lipotidae) originate in the Miocene and range to the Recent. The delphinoid family Albireonidae is known only from the late Miocene and early Pliocene of western North America (Barnes, 1984).

PINNIPEDIA

Pinnipeds diversify greatly and spread globally during the Miocene (Figure 30.10). All of the living groups of pinnipeds (Otariidae, Odobenidae, and Phocidae) are definitively present by the end of the middle Miocene (Berta and Sumich, 1999). Otariids are first known from and restricted to the North Pacific during the Miocene (Deméré, Berta, and Adam, 2003). Odobenids are first known from the late early Miocene of the North Pacific and undergo a rapid diversification in the late Miocene (Deméré, Berta, and Adam, 2003). The earliest named phocids are known from the middle Miocene of the North Atlantic, although an extinct group of archaic phocoids, the desmatophocids, are known only from the Miocene of the North Pacific (Deméré, Berta, and Adam, 2003).

DESMOSTYLIA

Miocene desmostylians are more highly derived than their Oligocene antecedents. *Desmostylus* developed very distinctive molars, with circular bundles of thick enamel columns that may be an adaptation for consumption of a more abrasive diet (Domning, 2002). *Desmostylus*, *Kronokotherium*, and *Paleoparadoxia* are known from both the western coast of North America and Asia, while *Vanderhoofius* is known only from North America. Desmostylia are extinct by the end of the Miocene (Figure 30.11).

SIRENIA

Sirenia, particularly dugongids, continue to diversify though the Miocene (Figure 30.11). It appears that these often sympatric taxa developed different and complimentary feeding strategies within the same ecosystem (Domning, 2001c). Trichechidae is not known from North America in the Miocene (McKenna and Bell, 1997). Domning and Ray (1986) noted the presence of a dugongid from the early Miocene Nye Formation of Oregon (locality WM6) as the earliest sirenian known from the North American West Coast. Hydrodamalines make their first appearance in the late early Miocene (*Dusisiren*), and an abundance of hydrodamalines then becomes evident on the West Coast (Domning, 1978). The *Dusisiren* lineage evolves anagenetically (rather than cladogenetically) over the remainder of the Miocene (Domning, 1978).

PLIOCENE–RECENT

CETACEA

The earliest records of both Eschrichtiidae and Neobalaenidae are essentially subfossil, although both families probably originated earlier. The very last of the stem baleen-bearing mysticetes are known from the Pliocene. Odobenocetopsidae, a family of delphinoid odontocetes thought to be closely related to the Monodontidae, is known only from the early Pliocene of Peru. Odobenocetopsids are quite a divergent form among cetaceans, possessing an abbreviated rostrum and posteriorly projecting tusks rather like a walrus.

As a whole, Pliocene and Pleistocene cetacean faunas are similar to Recent faunas (Figures 30.8 and 30.9). The Pleistocene contraction toward the equator and re-expansion toward the poles of cetacean ranges is thought to have given rise to geminate species pairs in the Northern and Southern Hemispheres (Davies, 1963).

PINNIPEDIA

During the Pliocene to Recent interval, pinnipeds take on their modern biogeographic patterns (Figure 30.10). The diversification of odobenids comes to an abrupt halt, with only a single species (*Odobenus rosmarus*) surviving to the Recent. Otariids, by comparison, expand their range into the Southern Ocean during the late Pliocene (Deméré, Berta, and Adam, 2003). Phocids continue to diversify and spread globally during the Pliocene to Recent (Deméré, Berta, and Adam, 2003).

DESMOSTYLIA

No Pliocene to Recent desmostylians are currently known.

SIRENIA

North American dugongids become extinct on the East Coast by the end of the Pliocene (Figure 30.11). A late Pliocene skull from Florida, however, appears to be very closely related to, if not congeneric with, the modern *Dugong dugon* from the Indo-Pacific (Domning, 2001b). This suggests that *Dugong* evolved in the Caribbean and dispersed to the Indo-Pacific realm prior to the closure of the Isthmus of Panama (Domning, 2001b). The modern sea cow, *Trichechus manatus*, first appears in North America in the early Pleistocene fossil record of Florida (Morgan, 1994).

On the North American West Coast, the genus *Hydrodamalis* evolves from *Dusisiren* in the early Pliocene. This single lineage of large, kelp-eating sirenians continues to expand into much cooler waters than any other group of sirenians. The final species of the lineage, *Hydrodamalis gigas*, was hunted to extinction by the close of the eighteenth century (Domning, 1978).

SUMMARY

Although marine mammals are linked by their occupation of similar habitats, their evolutionary histories have remained quite separated with only minimal exceptions. The only known marine mammal predator that preys on other marine mammals is *Orcinus orca*, the modern killer whale. Some groups of killer whales are known to prey on pinnipeds, much larger cetaceans, and sea otters (Estes *et al.*, 1998). The only other possible ecological interaction between groups of marine mammals is the possible filling of the ecological niche abandoned by desmostylians when they became extinct by the evolution of kelp-eating hydrodamaline sirenians (Domning, 1978). Most other potential evolutionary responses to ecological pressures appear to have been from within the groups of marine mammals.

Currently, many populations and even entire species of marine mammals are threatened by loss of habitat, by catch of the fishing industry, and pollution. Some species such as Steller's sea cow,

Hydrodamalis gigas, and the Caribbean monk seal, *Monachus tropicalis*, have become extinct in historical times as the direct result of human actions; others are almost certain to follow. The Chinese river dolphin or baiji is currently at very low numbers and its habitat is rapidly shrinking owing to human encroachment and the building of dams along the rivers in which it once lived. So, while many once-threatened species of marine mammals have been saved from extinction through banning of most forms of whaling, the indirect effects of human activity are continuing to cause a significant fall in marine mammal diversity.

REFERENCES

Árnason, U. (1977). The relationship between the four principal pinniped karyotypes. *Hereditas*, 87, 227–42.

Árnason, Ú. and Widegren, B. (1986). Pinniped phylogeny enlightened by molecular hybridizations using highly repetitive DNA. *Molecular Biology and Evolution*, 3, 356–65.

Bajpai, S. and Gingerich, P. D. (1998). A new Eocene archaeocete (Mammalia, Cetacea) from India and the time of origin of whales. *Proceedings of the National Academy of Science, USA*, 95, 15 464–8.

Barnes, L. G. (1972). Miocene Desmatophocinae (Mammalia: Carnivora) from California. *University of California, Publications in Geological Sciences*, 89, 1–76.

(1984). Whales, dolphins and porpoises: origin and evolution of the Cetacea. *University of Tennessee Department of Geological Sciences Studies in Geology*, 8, 139–54.

(1987). An early Miocene pinniped of the genus *Desmatophoca* (Mammalia: Otariidae) from Washington. *Contributions in Science, Los Angeles County Museum of Natural History*, 382, 1–20.

(1989). A new enaliarctine pinniped from the Astoria Formation, Oregon, and a classification of the Otariidae (Mammalia: Carnivora). *Contributions in Science, Los Angeles County Museum of Natural History*, 403, 1–26.

Barnes, L. G. and Sanders, A. E. (1996a). The transition from archaeocetes to mysticetes: late Oligocene toothed mysticetes from near Charleston, South Carolina. [*Proceedings of the Sixth North American Paleontological Convention.*] *Paleontological Society Special Publication*, 8, p. 24. Baltimore, MD: Paleontological Society.

(1996b). The transition from Archaeoceti to Mysticeti: late Oligocene toothed mysticetes from South Carolina, USA. *Journal of Vertebrate Paleontology*, 16(suppl. to no. 3), p. 21A.

Barnes, L. G., Goedert, J. L., and Furusawa, H. (2001). The earliest known echolocating toothed whales (Mammalia; Odontoceti): preliminary observations of fossils from Washington State. *Mesa Southwest Museum Bulletin*, 8, 91–100.

Berta, A. (1991). New *Enaliarctos** (Pinnipedimorpha) from the Oligocene and Miocene of Oregon and the role of "enaliarctids" in pinniped phylogeny. *Smithsonian Contributions to Paleobiology*, 69, 1–33.

Berta A. and Sumich, J. L. (1999). *Marine Mammals: Evolutionary Biology*. London: Academic Press.

Berta, A. and Wyss, A. R. (1994). Pinniped phylogeny. *Proceedings of the San Diego Society of Natural History*, 29, 33–56.

Berta, A., Ray, C. E., and Wyss, A. R. (1989). Skeleton of the oldest known pinniped, *Enaliarctos mealsi. Science*, 244, 60–2.

Boyden, A. and Gemeroy, D. (1950). The relative position of the Cetacea among the orders of Mammalia as indicated by precipitin tests. *Zoologica*, 35, 145–51.

Clementz, M. T., Hoppe, K. A., and Koch, P. L. (2003). A paleoecological paradox: the habit and dietary preferences of the extinct tethythere *Desmostylus*, inferred from stable isotope analysis. *Paleobiology*, 29, 506–19.

Crowley, B. E. and Barnes, L. G. (1996). A new late Oligocene mysticete from Washington State. *Paleontological Society Special Publication*, 8, 90.

Davies, J. L. (1958). Pleistocene geography and the distribution of northern pinnipeds. *Ecology*, 39, 97–113.

(1963). The antitropical factor in cetacean speciation. *Evolution*, 17, 101–16.

Deméré, T. A., Berta, A., and Adam, P. J. (2003). Pinnipedimorph evolutionary biogeography. *Bulletin of the American Museum of Natural History*, 279, 33–76.

Dockery, D. T., III and Lozouet, P. (2003). Molluscan faunas across the Eocene/Oligocene boundary in the North American Gulf Coastal Plain, with comparisons to those of the Eocene and Oligocene of Paris. In *From Greenhouse to Icehouse: The Marine Eocene-Oligocene Transition*, ed. D. R. Prothero, L. C. Ivany, and E. A. Nesbitt, pp. 303–40. New York: Columbia University Press.

Domning, D. P. (1978). Sirenian evolution in the North Pacific Ocean. *University of California, Publications in Geological Sciences*, 118, 1–176.

(1994). A phylogenetic analysis of the Sirenia. *Proceedings of the San Diego Society of Natural History*, 29, 177–89.

(2001a). The earliest known fully quadrupedal sirenian. *Nature*, 413, 62–7.

(2001b). Evolution of the Sirenia and Desmostylia. In *Secondary Adaptation of Tetrapods to Life in Water*, ed. J.-M. Mazin and V. de Buffrénil, pp. 151–68. Munich: Dr. Friedrich Pfeil.

(2001c). Sirenians, seagrasses, and Cenozoic ecological change in the Caribbean. *Palaeogeography, Palaeoclimatology, Palaeoecology*, 166, 27–50.

(2002). The terrestrial posture of Desmostylians. *Smithsonian Contributions to Paleobiology*, 93, 99–111.

Domning, D. P. and Gingerich, P. D. (1994). *Protosiren smithae*, new species (Mammalia, Sirenia), from the late middle Eocene of Wadi Hitan, Egypt. *Contributions from the Museum of Paleontology, University of Michigan*, 29, 69–87.

Domning, D. P. and Ray, C. E. (1986). The earliest sirenian (Mammalia: Dugongidae) from the eastern Pacific Ocean. *Marine Mammal Science*, 2, 263–76.

Domning, D. P., Morgan, G. S., and Ray, C. E. (1982). North American Eocene sea cows (Mammalia, Sirenia). *Smithsonian Contributions to Paleobiology*, 52, 1–69.

Domning, D. P., Ray, C. E., and McKenna, M. C. (1986). Two new Oligocene desmostylians and a discussion of tethytherian systematics. *Smithsonian Contributions to Paleobiology*, 59, 1–55.

Domning, D. P., Gingerich, P. D., Simons, E. L., and Ankel-Simons, F. A. (1994). A new early Oligocene Dugongid (Mammalia, Sirenia) from the Fayum Province, Egypt. *Contributions from the Museum of Paleontology, University of Michigan*, 29, 89–108.

Emlong, D. R. (1966). A new archaic cetacean from the Oligocene of northwest Oregon. *Bulletin of the Oregon University Museum of Natural History*, 3, 1–51.

Estes, J. A., Tinker, M. T., Williams, T. M., and Doak, D. F. (1998). Killer whale predation on sea otters linking oceanic and nearshore ecosystems. *Science*, 282, 473–6.

Fish, P. A. (1903). The cerebral fissures of the Atlantic walrus. *Proceedings of the United States National Museum*, 26, 675–88.

Flower, W. H. (1883). On the arrangement of the orders and families of existing Mammalia. *Proceedings of the Zoological Society of London*, 1883, 178–86.

Flynn, J. J. and Nedbal, M. A. (1998). Phylogeny of the Carnivora (Mammalia): congruence vs. incompatibility among multiple data sets. *Molecular Phylogenetics and Evolution*, 9, 414–26.

Flynn, J. J., Neff, N. A., and Tedford, R. H. (1988). Phylogeny of the Carnivora. In *The Phylogeny and Classification of the Tetrapods*, Vol. 2:

Mammals. ed. M. J. Benton, pp. 73–115. [*Systematics Association Special Volume* No. 35B.] Oxford: Clarendon Press.

Fordyce, R. E. (2002). Oligocene archaeocetes and toothed mysticetes: Cetacea from times of transition. *Geological Society of New Zealand Miscellaneous Publication, Secondary Adaptation to Life in Water*, pp. 16–17. Wellington: Geological Society of New Zealand.

(2003). Cetacean evolution and Eocène–Oligocene oceans revisited. In *From Greenhouse to Icehouse: The Marine Eocene–Oligocene Transition*, ed. D. R. Prothero, L. C. Ivany, and E. A. Nesbitt, pp. 154–70. New York: Columbia University Press.

Fordyce, R. E. and Muizon, C., de (2001). Evolutionary history of cetaceans: a review. In *Secondary Adaptation of Tetrapods to Life in Water*, ed. J.-M. Mazin and V. Buffrénil de, pp. 169–223. Munich: Dr. Friedrich Pfeil.

Gatesy, J. (1998). Molecular evidence for the phylogenetic affinities of Cetacea. In *The Emergence of Whales*, ed. J. G. M. Thewissen, pp. 63–112. New York: Plenum Press.

Gatesy, J., Hayashi, C., Cronin, M., and Arctander, P. (1996). Evidence from milk casein genes that cetaceans are close relatives of hippopotamid artiodactyls. *Molecular Biology and Evolution*, 13, 954–63.

Gatesy, J., Milinkovitch, M., Waddell, V., and Stanhope, M. (1999). Stability of cladistic relationships between Cetacea and higher-level artiodactyl taxa. *Systematic Biology*, 48, 6–20.

Geisler, J. H. (2001). New morphological evidence for the phylogeny of Artiodactyla, Cetacea, and Mesonychidae. *American Museum Novitates*, 3344, 1–53.

Geisler, J. H. and O'Leary, M. A. (1997). A phylogeny of Cetacea, Artiodactyla, Perissodactyla, and archaic ungulates: the morphological evidence. *Journal of Vertebrate Paleontology*, 17(suppl. to no. 3), p. 48A.

Geisler, J. H. and Sanders, A. E. (2003). Morphological evidence for the phylogeny of Cetacea. *Journal of Mammalian Evolution*, 10, 23–129.

Geisler, J. H. and Uhen, M. D. (2003). Morphological support for a close relationship between hippos and whales. *Journal of Vertebrate Paleontology*, 23, 991–6.

(2005). Phylogenetic relationships of extinct Cetartiodactyls: results of simultaneous analyses of molecular, morphological, and stratigraphic data. *Journal of Mammalian Evolution*, 12, 145–60.

Gheerbrant, E., Sudre, J., Iarochené, M., and Moumni, A. (2001). First ascertained African "condylarth" mammals (primitive ungulates: cf. Bulbulodentata and cf. Phenacodonta) from the earliest Ypresian of the Ouled Abdoun Basin, Morocco. *Journal of Vertebrate Paleontology*, 21, 107–18.

Gingerich, P. D., Haq, M. U., Zalmout, I. S., Khan, I. H., and Malakani, M. S. (2001). Origin of whales from early artiodactyls: hands and feet of Eocene Protocetidae from Pakistan. *Science*, 293, 2239–42.

Goodwin, M. B., Domning, D. P., and Lipps, J. H. (1998). The first record of an Eocene (Lutetian) marine mammal from Israel. *Journal of Vertebrate Paleontology*, 18, 813–15.

Howell, A. B. (1930). *Aquatic Mammals: Their Adaptations to Life in the Water*. Springfield, IL: Charles C. Thomas.

Hunt, R. M. Jr. and Barnes, L. G. (1994). Basicranial evidence for ursid affinity of the oldest pinnipeds. *Proceedings of the San Diego Society of Natural History*, 29, 57–67.

Inuzuka, N. (2000). Primitive late Oligocene desmostylians from Japan and the phylogeny of the Desmostylia. *Bulletin of the Ashoro Museum of Paleontology*, 1; 91–123.

Kellogg, R. (1922). Pinnipeds from the Miocene and Pleistocene deposits of California. *University of California, Publications in Geology*, 13, 23–132.

(1936). A Review of the Archaeoceti. *Carnegie Institute of Washington Special Publication*, 482, 1–366.

King, J. E. (1983). *Seals of the World*. Ithaca: Comstock.

Koretsky, I. A. and Sanders, A. E. (2002). Paleontology of the late Oligocene Ashley and Chandler Bridge Formations of South Carolina, 1: Paleogene pinniped remains; the oldest known seal (Carnivora: Phocidae). *Smithsonian Contributions to Paleobiology*, 93, 179–83.

Matthew, W. D. (1909). The Carnivora and Insectivora of the Bridger Basin, Middle Eocene. *Memoirs of the American Museum of Natural History*, 6, 291–567.

Mazin, J.-M. (2001). Mesozoic marine reptiles: an overview. In *Secondary Adaptation of Tetrapods to Life in Water*, ed. J.-M. Mazin and V. D. Buffrénil, pp. 95–117. Munich: Dr. Friedrich Pfeil.

Mchedlidze, G. A. (1984). *General Features of Paleobiological Evolution of Cetacea*. New Delhi: Amerind.

McKenna, M. C. and Bell, S. K. (1997). *Classification of Mammals Above the Species Level*. New York: Columbia University Press.

McLaren, I. A. (1960). Are the Pinnipedia biphyletic? *Systematic Zoology*, 9, 18–28.

Mitchell, E. D. (1967). Controversy over diphyly in pinnipeds. *Systematic Zoology*, 16, 350–1.

(1975). Parallelism and convergence in the evolution of the Otariidae and Phocidae. *Rapports et Procès-Verbaux des Réunions*, 169, 12–26.

(1989). A new cetacean from the late Eocene La Meseta Formation, Seymour Island, Antarctic Peninsula. *Canadian Journal of Fisheries and Aquatic Science*, 46, 2219–35.

Montgelard, C., Catzeflis, F. M., and Douzery, E. (1997). Phylogenetic relationships of artiodactyls and cetaceans as deduced from the comparison of cytochrome *b* and 12S rRNA mitochondrial sequences. *Molecular Biology and Evolution*, 14, 550–9.

Morgan, G. S. (1994). Miocene and Pliocene marine mammal faunas from the Bone Valley Formation of central Florida. *Proceedings of the San Diego Society of Natural History*, 29, 239–68.

Muizon, C., de (1982). Phocid phylogeny and dispersal. *Annals of the South African Museum*, 89, 175–213.

Nikaido, M., Matsuno, F., Hamilton, H., *et al.* (2001). Retroposon analysis of major cetacean lineages: the monophyly of toothed whales and the paraphyly of river dolphins. *Proceedings of the National Academy of Science, USA*, 98, 7384–9.

O'Leary, M. A. (2002). Mesonychia. In *Encyclopedia of Marine Mammals*, ed. W. F. Perrin, B. Würsig, and Thewissen J. G. M., pp. 735–7. London: Academic Press.

Okazaki, Y. (1995). A new type of primitive baleen whale (Cetacea; Mysticeti) from Kyushu, Japan. *The Island Arc*, 3, 432–5.

Prothero, D. R., Manning, E. M., and Fischer, M. (1988). The phylogeny of the ungulates. In *The Phylogeny and Classification of the Tetrapods*, Vol. 2, ed. M. J. Benton, pp. 201–34. Oxford: Clarendon Press.

Prothero, D. R., Jaquette, C. D., and Armentrout, J. M. (2001). Magnetic stratigraphy of the upper Eocene–upper Oligocene Lincoln Creek Formation, Porter Bluffs, Washington. In *Magnetic Stratigraphy of the Pacific Coast Cenozoic*, ed. D. Prothero, pp. 169–78. Santa Fe Springs, CA: Pacific Section of the Society for Sedimentary Geology.

Ray, C. E. (1977). Geography of phocid evolution. *Systematic Zoology*, 25, 391–406.

Repenning, C. A., Berta, A., and Wyss, A. R. (1990). Oldest pinniped. *Science*, 248, 499–500.

Sanders, A. E. and Barnes, L. G. (2002). Paleontology of the late Oligocene Ashley and Chandler Bridge Formations of South Carolina, 3: Eomysticetidae, a new family of primitive mysticetes (Mammalia: Cetacea). *Smithsonian Contributions to Paleobiology*, 93, 313–56.

Savage, R. J. G., Domning, D. P., and Thewissen, J. G. M. (1994). Fossil Sirenian of the west Atlantic and Caribbean region. V. The most primitive known sirenian, *Prorastomus sirenoides* Owen, 1855. *Journal of Vertebrate Paleontology*, 14, 427–49.

Scheffer, V. B. (1958). *Seals, Sea Lions, and Walruses*. Stanford, CA: Stanford University Press.

Shimamura, M., Yasue, H., Ohshima, K., *et al.* (1997). Molecular evidence from retroposons that whales form a clade within even-toed ungulates. *Nature*, 388, 666–70.

Shimamura, M., Abe, H., Nikaido, M., Ohshima, K., and Okada, N. (1999). Genealogy of families of SINEs in cetaceans and artiodactyls: the presence of a huge superfamily of $tRNA^{Glu}$-derived families of SINEs. *Molecular Biology and Evolution*, 16, 1046–60.

Simpson, G. G. (1945). The principles of classification and a classification of mammals. *Bulletin of the American Museum of Natural History*, 85, 1–350.

Tedford, R. H. (1976). Relationship of pinnipeds to other carnivores (Mammalia). *Systematic Zoology*, 25, 363–74.

Thewissen, J. G. M., Williams, E. M., Roe, L. J., and Hussain, S. T. (2001). Skeletons of terrestrial cetaceans and the relationship of whales to artiodactyls. *Nature*, 413, 277–81.

Uhen, M. D. (1998). Middle to late Eocene basilosaurines and dorudontines. In *The Emergence of Whales*, ed. J. G. M. Thewissen, pp. 29–61. New York: Plenum Press.

(1999). New species of protocetid archaeocete whale, *Eocetus wardii* (Mammalia, Cetacea), from the middle Eocene of North Carolina. *Journal of Paleontology*, 73, 512–28.

(2002). Evolution of dental morphology (cetacean). In *Encyclopedia of Marine Mammals*, ed. W. F. Perrin, B. Würsig, and J. G. M. Thewissen, pp. 316–19. London: Academic Press.

(2004). Form, function, and anatomy of *Dorudon atrox* (Mammalia, Cetacea): an archaeocete from the middle to late Eocene of Egypt. *University of Michigan Museum of Paleontology Papers on Paleontology*, 34, 1–222.

Uhen, M. D. and Gingerich, P. D. (2001). New genus of dorudontine archaeocete (Cetacea) from the middle-to-late Eocene of South Carolina. *Marine Mammal Science*, 17, 1–34.

Van Valen, L. M. (1966). Deltatheridia, a new order of mammals. *Bulletin of the American Museum of Natural History*, 132, 1–126.

Vrana, P. B., Milinkovitch, M. C., Powell, J. R., and Wheeler, W. C. (1994). Higher level relationships of the arctoid Carnivora based on sequence data and "total evidence". *Molecular Phylogenetics and Evolution*, 3, 47–58.

Waddell, P. J., Okada, N., and Hasegawa, M. (1999). Towards resolving the interordinal relationships of placental mammals. *Systematic Biology*, 48, 1–5.

Weber, M. (1904). *Die Säugetiere*. Jena: G. Fischer.

Wortman, J. L. (1894). Osteology of *Patriofelis*, a middle Eocene creodont. *Bulletin of the American Museum of Natural History*, 6, 129–64.

(1906). A new fossil seal from the marine Miocene of the Oregon coast. *Science*, 24, 89–92.

Wozencraft, W. C. (1989). The phylogeny of the Recent Carnivora. In *Carnivore Behavior, Ecology, and Evolution*, ed. J. L. Gittleman, pp. 495–535, Ithaca NY: Cornell University Press.

Wyss, A. R. (1987). The walrus auditory region and the monophyly of pinnipeds. *American Museum Novitates*, 2871, 1–31.

(1988). Evidence from flipper structure for a single origin of pinnipeds. *Nature*, 334, 427–8.

(1994). The evolution of body size in phocids: some ontogenetic and phylogenetic observations. *Proceedings of the San Diego Society of Natural History*, 29, 69–76.

Wyss, A. R. and Flynn, J. J. (1993). A phylogenetic analysis and definition of the Carnivora. In *Mammal Phylogeny: Placentals*, ed. F. S. Szalay, M. J. Novacek, and M. C. McKenna, pp. 32–52. New York: Springer-Verlag.

31 Otarioidea

LAWRENCE G. BARNES
Natural History Museum of Los Angeles County, Los Angeles, CA, USA

INTRODUCTION

The carnivoran superfamily Otarioidea is used here as it was by Tedford (1976), Repenning (1976), Repenning and Tedford (1977), Muizon (1978), and King (1983) and includes the sea lions, fur seals, walruses, and their extinct fossil relatives. The superfamily Otarioidea is thus equal in its content to the family Otariidae as it was used in the broad sense by Mitchell (1968, 1975), Barnes (1972, 1979, 1987), and Barnes, Domning and Ray (1985: Table 1), and it includes the classically used families Otariidae (sensu Gray, 1866a,b) and Odobenidae (Allen, 1880).

The fossil record, going back to the latest Oligocene, documents a surprising diversity of extinct members of this group, which affects the classification of the extant species and interpretations of their phylogeny. The work of Mitchell (1966, 1968, 1975) showed us that a great number of extinct animals in different lineages acquired an habitus like that of the sea lion. The work of Repenning and Tedford (1977) demonstrated that an animal does not need to have tusks to be a walrus, or to be related to the walruses.

A comprehensive statement of the diversity and distribution of characters, mostly cranial and dental, among the various fossil and living otarioid subfamilies can be found in the work of Mitchell (1968). Although perhaps not at first easy to sort out, the critical diagnostic characters of the major groups of otarioid pinnipeds are shown to be made up of partially overlapping suites of characters. For example, the extinct and relatively derived species in the desmatophocid subfamily Allodesminae share significant derived cranial characters with the Odobenidae (walruses and extinct relatives), but they also share some convergent limb characters with the Otariinae (sea lions and fur seals). Additionally, the extinct species of the Imagotariinae ("pseudo-sea lions") share phylogenetically significant basicranial and postcranial characters with the Odobeninae (true walruses) but have some dental and mandibular features that are convergent with the Recent Otariinae (sea lions and fur seals). Greatly enlarged canine tusks and the accompanying extreme rostral and mandibular modifications are important derived characters of the living walruses (*Odobenus rosmarus*). However, these are relatively recently acquired features within the odobenine lineage, as proven by the discovery of fossil true walruses of early Pliocene that exhibit markedly lesser modifications.

Most of the described fossil otarioid genera are monotypic or include, at the most, only two or three species. This is partly because the study of fossil pinnipeds is still in its relative infancy, and partly because few fossil pinnipeds from stratigraphically superimposed faunal assemblages or rock units have yet been collected and/or described. Repenning and Tedford (1977) outlined a preliminary chronological sequence of occurrences of fossil pinnipeds from the Pacific margin of North America, and they summarized the fossil associations of taxa. Many of these are still known by specimens from a single locality, or at best only a few localities, and it is presently not possible to recreate extensive chronologic, stratigraphic, and geographic ranges for most taxa. Miyazaki *et al.* (1995) provided a more recent review of North Pacific fossil otarioid taxa and occurrences, and this includes all the then-named North American Tertiary taxa.

Most of the recently named fossil species were based on specimens that include at least the major part of a cranium or mandible, but this has not always been the case. Earlier designations of disparate, non-comparable skeletal elements as holotype specimens have in some situations hindered subsequent objective comparisons. For example, *Pliopedia pacifica* Kellogg, 1921, *Neotherium mirum* Kellogg, 1931, and *Valenictus imperialensis* Mitchell, 1961 were each named on a few limb elements (in some instances noncomparable), and only later were other bones assigned to them.

Evolution of Tertiary Mammals of North America, Vol. 2. ed. C. M. Janis, G. F. Gunnell, and M. D. Uhen. Published by Cambridge University Press.

Most other fossil otarioids have been named on the basis of at least crania and mandibles, in most cases of male individuals that show secondary sex characters that help to define otarioids at the genus and species level.

There is ample evidence at present to demonstrate that otarioid pinnipeds are a monophyletic group (Mitchell, 1968; Sarich, 1969; Repenning and Tedford, 1977; Wiig, 1983; Repenning, 1990; Barnes and Hirota, 1995; Flynn *et al.*, 2005), and otarioid clades (which are here recognized as the subfamilies Enaliarctinae, Otariinae, Desmatophocinae, Allodesminae, Imagotariinae, Dusignathinae, and Odobeninae) share a suite of derived characters that supports their classification in a single superfamily. Three families, not recognized in all published works, can be recognized within the Otarioidea to contain these subfamilies. The superfamily Otarioidea is similar in its morphologic and taxonomic diversity to some other suprafamilial groups of mammals and is more diverse than the family Phocidae (true seals). The extinct otariid subfamily Enaliarctinae contains the earliest and most primitive known species of the Otarioidea (Mitchell and Tedford, 1973; Barnes, 1979, 1990; Berta, Ray, and Wyss, 1989; Berta, 1991). The other families and subfamilies represent geologically more recent and morphologically more derived lineages that acquired various types of specializations, and that derive from the Enaliarctinae.

DEFINING FEATURES OF THE SUPERFAMILY OTARIOIDEA

CRANIAL

All otarioids have the following cranial characteristics (Figure 31.1):

Cranium with mastoid process large and salient (especially developed in adult males); distinct alisphenoid canal; hamular process of pterygoid bent medially and extending posteriorly; basal whorl of cochlea directed posterolaterally; anterior process of malleus present; no head on incus; internal acoustic meatus on petrosal; adult tympanic bulla composed of about three-fourths ectotympanic ossification of the tympanic, entotympanic ossification largely confined to the formation of the carotid canal; slightly inflated to little-inflated tympanic bulla; no fossa for the origin of the tensor tympani muscle (except for enaliarctines [see Repenning and Tedford, 1977]); bulla fused to postglenoid process, mastoid process and exoccipital, but separated by a fissure from the basioccipital; paroccipital process fused to the mastoid in maturity, except in the odobenines, where fusion, if it occurs, takes place in old age; posterior lacerate foramen large.

DENTAL

Except for the enaliarctines, otarioids have homodont or nearly homodont cheek teeth. The early enaliarctines retain carnassial cheek teeth inherited from their terrestrial actoid ancestors (Mitchell and Tedford, 1973; Repenning and Tedford, 1977; Hunt and Barnes, 1994), and these carnassial structures are reduced in geochronologically later enaliarctines (Barnes, 1992). The presence of some degree of remaining carnassial structures, the primitive character state, arbitrarily defines the admittedly paraphyletic Enaliarctinae (see Repenning and Tedford, 1977).

All later and derivative otarioid clades possess one or more derived character states in their dentitions. The Otariinae retain relatively primitive cheek tooth cusps, and have various degrees of coalescing of the cheek tooth roots, with the ultimate condition being one root on most cheek teeth, near homodonty of the cheek tooth row, and loss of the last upper and lower molars. The I3 is greatly enlarged in the sea lions, but not in the fur seals. The Desmatophocidae similarly show progressive coalescence of cheek tooth roots, simplification of the cheek tooth crowns, and enlargement of the I3. In the desmatophocid subfamily Allodesminae, the cheek tooth crowns are exceptionally bulbous and have relatively smooth enamel, I3 is very large, and the canines have lost the posterior and lingual cristae. Members of the odobenid subfamily Imagotariinae also have progressively coalesced roots on the cheek teeth, with the ultimate condition of the cheek tooth series being near homodonty, and an enlargement of the part of the posterior crista on the upper canine near the apex of the crown. Many imagotariines have elaborately cuspate lingual cingula on their cheek teeth.

Members of the odobenid subfamily Dusignathinae develop coalesced cheek tooth roots, ultimately producing a single root that is round in cross section, and bulbous crowns with smooth enamel. Both the upper and the lower canines are enlarged, in some species becoming tusklike. The incisors are concomittantly reduced in size, and in the more highly derived taxa the incisors are lost, with the exception of an enlarged I3. Derived dusignathines lack their last molars. Odobenine odobenids, the true walruses, progressively develop enlarged upper canine tusks with osteodentine in the pulp cavities, smaller lower canines that become premolariform and join the cheek tooth row, and single roots of the cheek teeth that are round in cross section. The more derived odobenines lose, either evolutionarily or ontogenetically, both the upper and lower incisors, and one or both molars in each upper and lower cheek tooth row.

POSTCRANIAL

All otarioids have hind feet that are capable of being turned forward and used in terrestrial locomotion; the astragalus and calcaneum are of the normal carnivore type (not elongated promimodistally, and articular facets not modified to accommodate permanent extension). The neck is lengthened, the humerus lacks an entepicondylar foramen, and the ilium is not bent laterally. Sexual dimorphism is great in all members, with the adult males differing from the adult females by being larger, and by having skulls with canines, sagittal crests, and other cranial crests and processes that are proportionally larger.

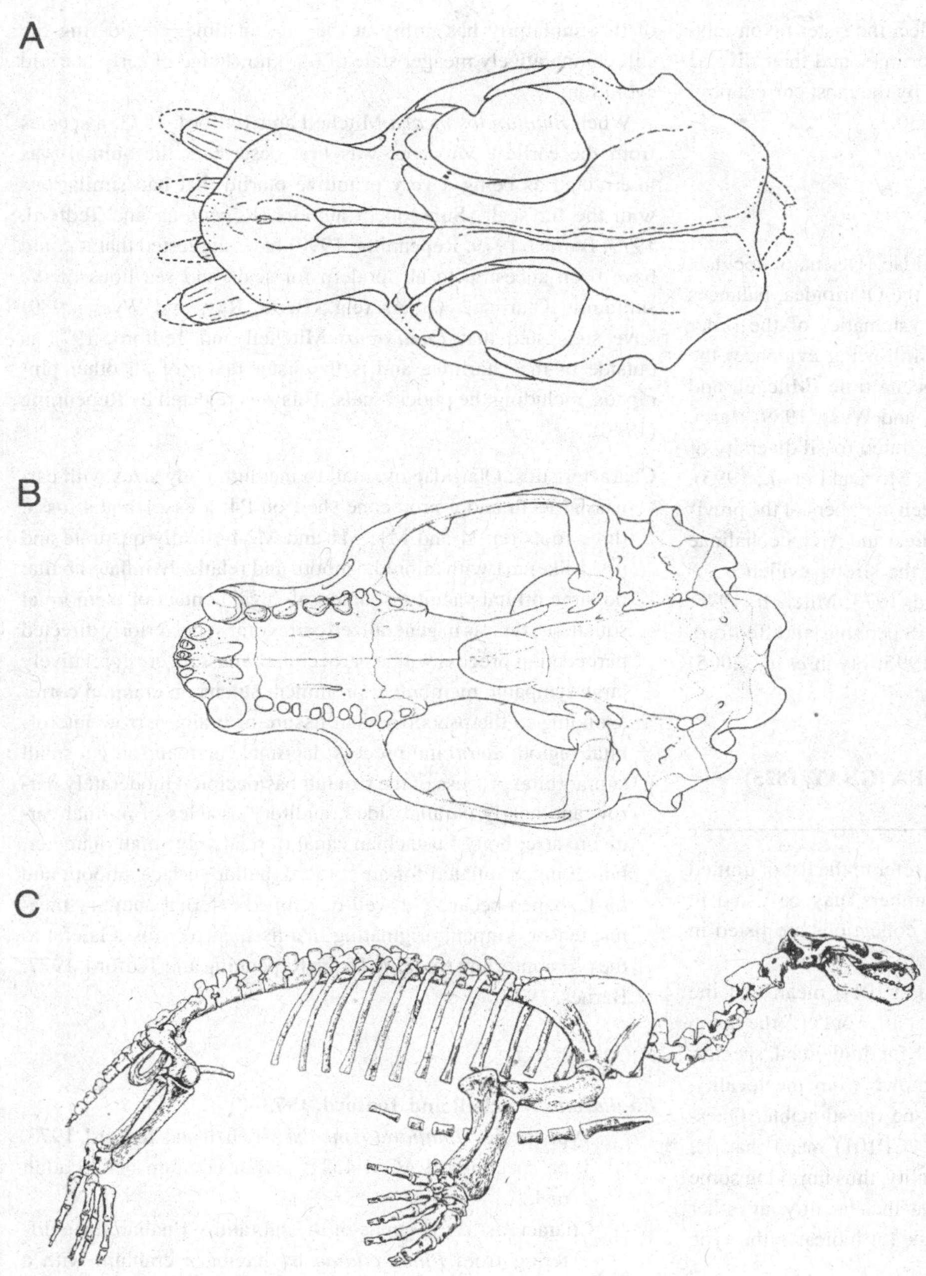

Figure 31.1. Examples of Tertiary otarioid pinnipeds. A. *Gomphotaria pugnax* Barnes and Raschke, 1991, a derived dusignathine odobenid of latest Miocene age (Hemphillian correlative) LACM 121508, the holotype cranium, is shown in dorsal view. This is from an old adult male, as demonstrated, among other characters, by the presence of the baculum, from the Capistrano Formation, southern Coastal California; note the sagittal crest and the large and procumbent canines and lateral incisors. B. *Pacificotaria hadromma* Barnes, 1992, a derived enaliarctine otariid of latest early Miocene age (Hemingfordian correlative). LACM 127973, the holotype cranium, is shown in ventral view. This is from an adult male, as demonstrated by the relatively large canine teeth, from the Astoria Formation, coastal Oregon; note the remnant carnassial teeth. C. *Allodesmus kelloggi* Mitchell, 1966, a derived allodesmine desmatophocid of middle Miocene age (Barstovian correlative). LACM 4320, the holotype skeleton, is shown in lateral view. This is from an old adult male, as demonstrated, among other characters, by the presence of the baculum, from the Sharktooth Hill Local Fauna, central California (after Mitchell, 1979, illustration on p. 5); note the long neck, which is characteristic of the Otarioidea.

SYSTEMATICS

SUPRAFAMILIAL

Classifications of otarioid pinnipeds have varied among authors (see Barnes, Domning, and Ray, 1985; Berta, 1991). The classification that is used in this chapter is based on the one that was proposed by Mitchell (1968, 1975) and the one that was used by Barnes (1987), and as modified by Repenning and Tedford (1977). A single family Otariidae was used by Hall (1981), but the family was placed within the order Pinnipedia, not within the order Carnivora. The use of only one family Otariidae by Hall contrasts with two (the Otariidae and Odobenidae [e.g., by Scheffer, 1958; King, 1983]; or the Enaliarctidae and Otariidae [including Odobeninae] [Tedford, 1976]) or four (the Enaliarctidae, Otariidae, Desmatophocidae, and Odobenidae [by Repenning, 1976; Repenning and Tedford, 1977]). The differences between these classifications are principally ones of rank and hierarchy, however, rather than of perceived phylogenetic relationships. The several alternative phyogenetic analyses of the Carnivora that were presented by Flynn *et al.* (2005) all showed the Otariinae (sea lions and fur seals) as sister to the Odobeninae (walruses), to the exclusion of Phocidae. Therefore, the superfamily Otarioidea is used here to reflect the sister-taxon relationship between the Odobeninae, the Otariinae, and their diverse fossil relatives, and this move is supported by the most current phylogenetic analyses. (See discussion in Ch. 30.)

INFRAFAMILIAL

The recognition of three families (Otariidae, Desmatophocidae, Odobenidae) within a single superfamily, the Otarioidea, balances the classification in the context of the systematics of the order Carnivora (Tedford, 1976) based on the following evidence: the apparent origin of otarioids in late Oligocene time (Mitchell and Tedford, 1973; Barnes, 1979; Berta, Ray, and Wyss, 1989; Berta, 1991; Hunt and Barnes, 1994), the demonstrated fossil diversity of this group (Repenning and Tedford, 1977; Miyazaki *et al.*, 1995), the cases of successful hybridization between members of the previously recognized otariid subfamilies Otariinae and Arctocephalinae (Mitchell, 1968; Van Gelder, 1977), and the strong evidence for otarioid monophyly (Mitchell and Tedford, 1973; Mitchell, 1975; Repenning, 1976, 1990; Tedford, 1976; Repenning and Tedford, 1977; Barnes, 1979; Barnes and Hirota, 1995; Flynn *et al.*, 2005) (Figure 31.2).

INCLUDED GENERA OF OTARIOIDEA (GRAY, 1825) SMIRNOV, 1908

The locality numbers listed for each genus refer to the list of unified localities in Appendix I. The locality numbers may be listed in several ways. The acronyms for museum collections are listed in Appendix III.

Parentheses around the locality (e.g., [CP101]) mean that the taxon in question at that locality is cited as "aff." or "cf." the taxon in question. Parentheses are usually used for individual species, thus implying that the genus is firmly known from the locality, but the actual species identification may be questionable. Question marks in front of the locality (e.g., ?CP101) mean that the taxon is questionably known from that locality, thus implying some doubt that the taxon is actually present at that locality, at either the genus or the species level. An asterisk (*) indicates the type locality.

OTARIIDAE GILL, 1866

ENALIARCTINAE MITCHELL AND TEDFORD, 1973

The primitive (paraphyletic) subfamily Enaliarctinae is here retained as a horizontal group, admittedly a grade rather than a clade, because it includes taxa from which all other otarioid lineages apparently evolved (Repenning and Tedford, 1977; Barnes, 1979; Barnes, Domning, and Ray, 1985). Repenning and Tedford (1977) defined the Enaliarctinae, in part, using an arbitrary morphological feature: the retention of the primitive carnassial structure of the cheek teeth (P4 and m1), which is inherited from their terrestrial fissiped carnivoran ancestors (Hunt and Barnes, 1994). Retention of this subfamily has utility at the present time, considering the still comparatively meager state of our knowledge of early otarioid evolution.

When *Enaliarctos mealsi* Mitchell and Tedford, 1973, a species from the earliest Miocene, was first described, the animal was interpreted as being a very primitive otariid that had similarities with the fur seals. Subsequent authors (Repenning and Tedford, 1977; Barnes, 1979; Repenning, 1990) have suggested that it could have been ancestral to all modern fur seals and sea lions of the subfamily Otariinae. Others (e.g., Berta, Ray, and Wyss, 1989) have suggested that *Enaliarctos* Mitchell and Tedford, 1973 is outside of the Otariidae and is the sister taxon of all other pinnipeds, including the phocid seals. This was rejected by Repenning (1990).

Characteristics: Otariidae of small to medium body sizes with carnassial teeth and a protocone shelf on P4; a nasolabialis fossa; three roots on P4 and M1; M1 and M2 basically quadrate and ursid-like; m1 with talonid; smooth and relatively inflated bulla; no large orbital vacuity; squamosal–jugal contact of the normal squamous type as in generalized carnivorans; posteriorly directed paroccipital process with a paroccipital–mastoid crest; relatively large tympanic membrane; prominent sulcus on cranium corresponding to the pseudosylvian fissure of brain; narrow interorbital region; antorbital process; lacrimal foramen; only a small supraorbital process of the frontal; basioccipital moderately narrow and nearly parallel sided; auditory ossicles of normal carnivore size; bony Eustachian canal of relatively small diameter; bullae much inflated for an otarioid, bullar surface smooth and flask shaped because of well-developed external auditory meatus; tensor tympani originating in a fissiped-like fossa lateral to the promontorium (modified from Repenning and Tedford, 1977; Barnes, 1979).

Enaliarctos Mitchell and Tedford, 1973

Type species: *Enaliarctos mealsi* Mitchell and Tedford, 1973.

Type specimen: LACM 4321, partial cranium of old adult male.

Characteristics: A genus of the subfamily Enaliarctinae differing from *Pinnarctidion* by having a cranium with a wider interorbital region; smaller orbit; smaller antorbital processes; external openings of optic foramina located relatively higher on the anterior wall of the braincase; narrower and more arched palate both anteroposteriorly and transversely; smaller pterygoid process of maxilla ventral to orbit. Cheek teeth proportionally larger and more closely spaced; deep embrasure pit for m1 on the palate between P4 and M1; protocone shelf of P4 larger and positioned more anteriorly on the tooth; anterolabial corner of M1 large; internal narial opening higher and narrower; strut between palate and braincase less concave lateral to pterygoid hamulus; tympanic cavity smaller; paroccipital process smaller and joined to mastoid process by only a low paroccipital–mastoid crest; occipital condyles nearly

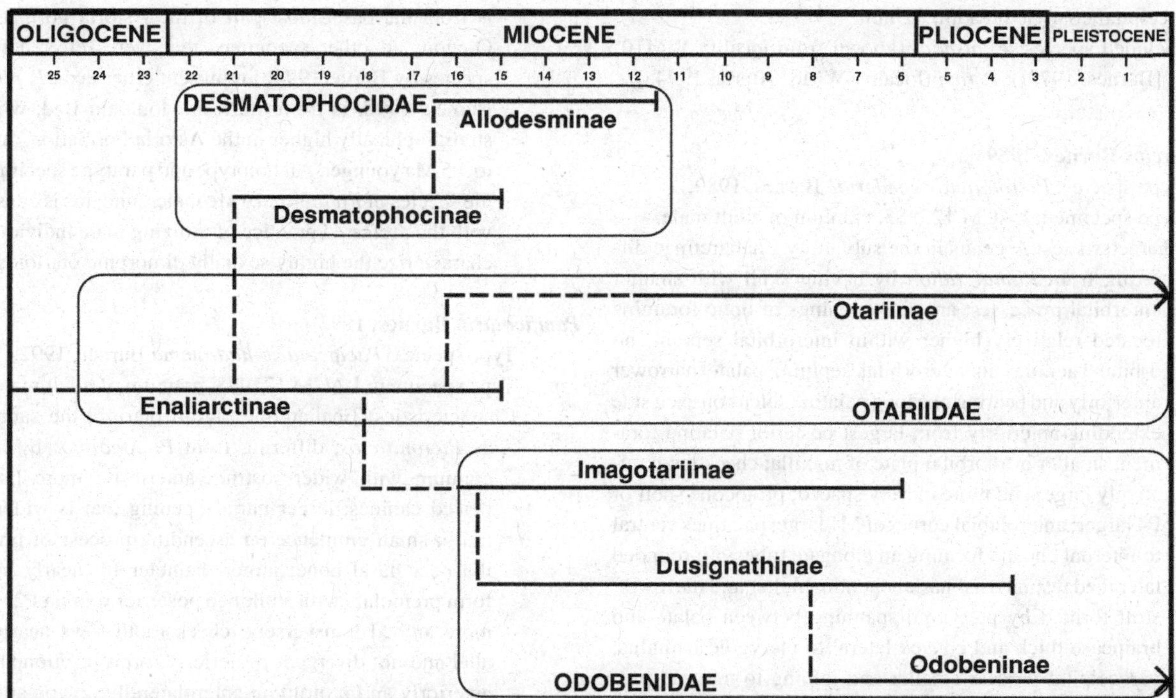

Figure 31.2. Known geochronologic ranges and proposed interrelationships of the subfamilies of North American otarioid pinnipeds. Solid lines represent known ranges; dashed lines that extend from these represent implied ghost lineages. The most widely accepted evolutionary relationships between the subfamilies are shown. These are essentially as they were proposed by Mitchell (1968) and by Repenning and Tedford (1977). Further details of these relationships were demonstrated between the Enaliarctinae and the Otariinae by Mitchell and Tedford (1973), Barnes (1979), Berta (1991), and Berta *et al.* (1989), between the Desmatophocinae and the Allodesminae by Mitchell (1966, 1968), Barnes (1972), and Deméré and Berta (2002), between the Enaliarctinae and the Imagotariinae by Kohno, Barnes, and Hirota (1995), between the Imagotariinae and the Dusignathinae by Mitchell (1968) and Barnes and Raschke (1991), and between the Dusignathinae and the Odobeninae by Deméré (1994). All clades terminate with extinction during the Tertiary, with the exception of the Otariinae and Odobeninae, which have Pleistocene fossil records, and which are extant. The first and last known occurrences of each subfamily are as follows. Enaliarctinae, a late Oligocene specimen from Washington (Hunt and Barnes, 1994) and an un-named middle Miocene specimen from the Sharktooth Hill Bonebed in California. Otariinae, late Miocene *Pithanotaria starri* Kellogg, 1927 from California and extant species Desmatophocinae (early Miocene *Desmatophoca brachycephala* Barnes, 1987) from Washington, and early middle Miocene specimens from the Astoria Formation in Oregon (Barnes, 1987; Deméré and Berta, 2002). Allodesminae, a latest early Miocene specimen from Oregon (Barnes and Hirota, 1995) and an early late Miocene species of *Allodesmus* from Washington (Bigelow, 1994). Imagotariinae, latest early Miocene *Proneotherium repenningi* Barnes, 1995 (in Kohno, Barnes, and Hirota, 1995) from Oregon and latest Miocene specimens from California (Repenning and Tedford, 1977). Dusignathinae, a middle Miocene specimen from Baja California (Barnes, Aranda-Manteca, and Tellez, 1992) and Pliocene *Dusignathus seftoni* Deméré, 1994, from San Diego. Odobeninae, latest Miocene *Aivukus cedrosensis* Repenning and Tedford, 1977, from Baja California, and the extant species.

parallel in posterior view rather than diverging dorsally (modified from Barnes, 1979).

Occipital condyle breadth: 48 mm (extrapolating from the complete left occipital condyle).

Included species: *E. mealsi* (known from locality WM10* [Mitchell and Tedford, 1973]); *E. barnesi* (locality WM5B* [Berta, 1991]); *E. emlongi* (locality WM6* [Berta, 1991]); *E. mitchelli* (localities WM6, CC14* [Barnes, 1979]); *E. tedfordi* (locality WM5B* [Berta, 1991]).

Comments: *Enaliactos* is also known from the early Miocene of Japan (McKenna and Bell, 1997).

***Pinnarctidion* Barnes, 1979**

Type species: *Pinnarctidion bishopi* Barnes, 1979.

Type specimen: UCMP 86334, partial cranium of old adult male.

Characteristics: A genus of the subfamily Enaliarctinae differing from *Enaliarctos* by having a cranium with narrower interorbital region; larger orbit; larger antorbital processes; external openings of optic foramina located more ventrally and posteriorly; wider and flatter palate; larger pterygoid process of maxilla ventral to orbit; cheek teeth proportionally smaller and more widely spaced; no embrasure pit for m1 on palate between P4 and M1; protocone shelf of P4 smaller and more posteriorly on tooth; anterolabial corner M1 reduced; internal narial opening wider and lower; strut between palate and braincase deeply concave lateral to ptyergoid hamulus; tympanic cavity larger; paroccipital process larger and joining to mastoid process by large paroccipital–mastoid crest; broader basicranium; occipital condyles more widely separated dorsally (modified from Barnes, 1979).

Occipital condyle breadth: 52 mm.

Included species: *P. bishopi* (known from locality WM10* [Barnes, 1979]); *P. rayi* (locality WM6* [Berta, 1994a]).

***Pteronarctos* Barnes, 1989**

Type species: *Pteronarctos goedertae* Barnes, 1989.

Type specimen: LACM 123883, cranium of adult male.

Characteristics: A genus in the subfamily Enaliarctinae differing from *Pinnarctidion* by having skull with smaller antorbital processes; anterior openings of optic foramina located relatively higher within interorbital septum; no orbital vacuities in interorbital septum; palate narrower anteriorly and bearing one long palatine sulcus on each side extending anteriorly from largest posterior palatine foramen; smaller infraorbital plate of maxilla; cheek teeth relatively larger and more closely spaced; protocone shelf of P4 larger; anterolabial corner of M1 large; palatines ventral to internal choana forming an elongate tube with rounded lateral edges; internal narial opening higher and narrower; strut formed by pterygoid spanning between palate and braincase thick and convex lateral to pterygoid hamulus; paroccipital process smaller and joining to mastoid process by thinner and narrower crest; occipital condyles more prominent and separated ventrally by deeper intercondylar notch. Differing from *Enaliarctos* by having skull with more nearly parallel cheek tooth rows; roots of P2–M1 more closely appressed; P3 with bilobed posterior root; lesser carnassial function of P4 and m1 as indicated by fusion of protocone root of P4 to posterolateral root and by shallower embrasure pit for m1 on palate between P4 and M1; smaller M1 with bilobed instead of trilobed root; larger and double-rooted M2; larger supraorbital process of frontal; narrower posterior part of interorbital area; no preglenoid process on lateral part of glenoid process; more prominent anterolateral corner of braincase; shallower fossa on lateral side of braincase corresponding to pseudosylvian sulcus of brain; narrower squamosal fossa between braincase and zygomatic arch; incisive foramina (palatine fissures) entering external nares less vertically; smaller orbit; smaller external nares; greater facial angle (Barnes, 1989).

Occipital condyle breadth: 52 mm (holotype, LACM 123883 [Barnes, 1989]).

Included species: *P. goedertae* (known from localities WM6II; WM7* [Barnes, 1989]); *P. piersoni* (locality WM7* [Barnes, 1990]).

Comments: Berta (1994b) concluded that *Pteronarctos piersoni* (Barnes, 1990) and *Pacificotaria hadromma* (Barnes, 1992) were junior synonyms of *Pteronarctos goedertae*, and that the type specimens of all were females. However, these specimens have well-formed sagittal crests and relatively large canines, demonstrating that they are males, but of relatively small species. *P. goedertae* is known only by one specimen, the holotype adult male cranium, which is from the basal-most part of the Astoria Formation in Oregon; all other specimens that were referred to this species by Berta (1994b) represent other taxa. *P. piersoni* Barnes, 1990, is from the Iron Mountain Bed, which is stratigraphically higher in the Astoria Formation, and is 2 to 3.5 Ma younger. All holotype and paratype specimens of the species of *Pteronarctos* are males, and this is consistent with the preferred practice of utilizing male individuals to characterize the highly sexually dimorphic otarioids.

***Pacificotaria* Barnes, 1992**

Type species: *Pacificotaria hadromma* Barnes, 1992.

Type specimen: LACM 127973, cranium of adult male.

Characteristics: Enaliarctine otariid otarioid the same size as *Pteronarctos*; differing from *Pteronarctos* by having cranium with wider rostrum anteriorly; more laterally placed canines; larger narial opening that is wider ventrally; small eminence on ascending process of premaxilla near nasal bone; larger diameter I3; nearly molariform premolars with a bilobed posterior root on P2; palate more arched transversely; cheek tooth rows nearly parallel and not divergent posteriorly; orbit positioned more anteriorly and protruding anterolaterally; zygomatic arch deeper dorsoventrally and departing more abruptly from the sagittal plane of the cranium to accommodate the larger orbit; pharyngeal tubercle on basicranium larger and protruding further ventrally; mastoid process more curved rather than cubic; dorsal surface of braincase more convoluting and rugose; sagittal and nuchal crests larger.

Occipital condyle breadth: 48 mm.

Included species: *P. hadromma* only, known from locality WM7* only.

OTARIINAE GRAY, 1824

Characteristics: Small- to large-size otariid otarioids; supraorbital processes present on cranium and very large in adult males; occipital condyles high on the occipital shield and relatively close together, being parallel or nearly so; paroccipital process knoblike, usually thick and fused to mastoid in adulthood; basioccipital moderately narrow and parallel sided to trapezium shaped (modified from Repenning and Tedford, 1977); cheek tooth crowns with lingual cingulum, often with small cingular cuspules, no single major cingular cusp; upper and lower canines not greatly elongated, caniniform with thick enamel; optic foramina positioned relatively dorsally on the anterior surface of the braincase; optic chiasma narrow so that optic nerve tracts are close together, or in some cases joined; internal acoustic meatus circular and tracts for vestibulocochlear and facial nerves not separated; paroccipital and mastoid process broadly continuous, directed ventrally or posteroventrally; paroccipital process reduced; postglenoid foramen small; ventral surface of bulla usually rugose, with strong projections (often smooth and flat in Arctocephalini); posterior edge of bulla extending markedly down from basicranial surface with great relief; apical bullar spur and fossa absent; vagina processus styloidei present but variable, shallow, small to medium size, posterolateral to bulla, not connected with stylomastoid foramen by a distinct groove; posterior bullar projection absent; posterior lacerate foramen large; auditory cavity long, narrow, relatively deep; hypotympanicum large, inflated medially, posteriorly, and ventrally well beyond promontorium, with small eminentia vagina processus styloidei on posterior wall; recessus meatus acoustics externi small to medium size; crista tympani projected into auditory cavity, may have overhanging, free ventral edge; bulge of carotid canal touching promontorium, underlain by hypotympanicum posteriorly; tympanohyal enclosed within vagina processus styloidei at or below level of membrane tympanica; tympanic membrane relatively small, generally ovate, with long axis dorsoventrally, oriented ventrointernally; fossa incudis not usually extending posteriorly beyond level of posterolateral corner of promontorium; promontorium low, not hanging far down within auditory chamber, inflated anteriorly and posteriorly; fossa cerebellaris not deeper than transverse width of entrance; tentorium not appressed to eminence containing semicircular canals (modified from Mitchell, 1968); internal acoustic meatus oval to circular, with little or no separation of canals for facial and vestibulocochlear nerves; petrosal apex pointed or bluntly pointed and relatively little enlarged; bony Eustachian tube not conspicuously enlarged; palate flat to arched; hypophyseal fossa deep and globose; upper cheek teeth lacking any persistent posterolingual cusp; lower lateral incisor anteromedial to lower canine; mandibular symphysis shallow and weak, broadly oval in shape, usually not articulating over the entire depth of the chin, and never fused; angle of the dentary, where the digastricus inserts, small and anteriorly located; humerus short and stout, more slender in the fur seals, deltoid tuberosity located on the deltopectoral crest; anteroposterior diameter of the trochlear lip approximately the same as that of the distal capitulum; distal termination of the pectoral crest approximately in line with the midpoint of distal articulation, except in the fur seals; radius short, with pronator origin proximal to midpoint of shaft; ulna with flat anterior surface of olecranon above semilunar notch (narrower in more primitive taxa), distal radial articulation inconspicuous and united with distal articulation; scapholunar with facet for articulation with magnum little or no deeper than adjacent facet for unciform; femoral head approximately as high as greater trochanter; calcaneum with medial tuberosity on calcaneal tuber no larger than that on lateral side; astragalus with low fibular articulation flaring distally and curving toward a paraplantar plane onto an enlarged lateral process (modified from Repenning and Tedford, 1977).

Callorhinus Gray, 1859

Type species: *Callorhinus ursinus* Linnaeus, 1758, the extant northern fur seal.

Type specimen of the only fossil species, *Callorhinus gilmorei*: SDSNH 25176, partial young adult female skeleton including left I3 and P2; right C1, I1 or I2, and P1; right i1; right dentary with p2–3 and m1; left i1; ribs and rib fragments; lumbar, thoracic, and caudal vertebrae and fragments; all associated and from the same individual.

Characteristics: *Callorhinus gilmorei* with long and shallow pterygoid process on the dentary that does not form a medial shelf; double-rooted lower third and fourth premolars; closely spaced M1–2; small I3 with oval cross section; dorsally directed anterior margin of the mandibular foramen (Berta and Deméré, 1986).

Occipital condyle breadth: not available on the holotype specimen.

Included species: *C. gilmorei* Berta and Deméré, 1986, only, known from locality WM26* only.

Comments: *Callorhinus* is also known from the Pleistocene and Recent of North America, and the Recent of Asia (McKenna and Bell, 1997).

Pithanotaria Kellogg, 1925

Type species: *Pithanotaria starri* Kellogg, 1925.

Type specimen: Stanford University Museum No. 11, now CAS 13665 (Firby, 1972), partial skeletal impression of an adult female on diatomite slab.

Characteristics: Small otariine otariid without M2; P4 to M1 diastema; double-rooted cheek teeth except p1; cheek tooth crowns simple, with sharply pointed crowns having very weak internal cingula and no secondary cuspules; I3 with a distinct posterolingual cuspule rather than being simple and nearly conical as in most modern otariids; dental formula I3/2, C1/1, P4/4, M1/1–2.

Occipital condyle breadth: not available on the holotype specimen.

Included species with localites: *P. starri* only (known from localities WM16A, WM17B*).

Comments: Because the holotype is only the impression of bones, comparisons are difficult with three-dimensional bones from other localities. A rubber mold was made of the impressions, and a reversed made to reveal the three-dimensional surfaces of the bones, and this is curated with the holotype.

Thalassoleon Repenning and Tedford, 1977

Type species: *Thalassoleon mexicanus* Repenning and Tedford, 1977.

Type specimen: IGCU 902, cranium, mandible, and some postcranial bones, all of an adult male individual.

Characteristics: Medium body size with short rostrum; very broad nasal bones anteriorly; facial angle of about 150° in adult males; m2 lost; M2 present; all cheek teeth (except P1) double-rooted; diastema present between M1 and M2; lesser and variably present diastema between P4 and M1; cheek tooth crowns forming a single lanceolate cusp, with moderate and rounded internal cingulum, and no accessory cusps; trapezium-shaped basioccipital that is very broad posteriorly (for an otariinae); coronoid process of dentary very broad, its posterior margin not overhanging the mandibular condyle; pterygoid process of dentary shallow, with very little shelflike medial protrusion; vertebral foramina of all vertebrae small relative to living species of Otariinae; dental formula I3/2, C1/1, P4/4, M2/1; baculum very long, without forked apex, dorsal process of apex small, base flattened dorsoventrally.

Occipital condyle breadth: 59.8 mm (but may not be accurate because the condyles are slightly distorted).

Included species: *T. mexicanus* (known from locality WM30A*); *T. macnallyae* (locality WM20* [Repenning and Tedford, 1977]).

DESMATOPHOCIDAE HAY, 1930

DESMATOPHOCINAE (HAY, 1930) MITCHELL, 1966

Characteristics: Otarioids of moderate size. Differing from Enaliarctinae by having crania without an embayment in the lateral edge of the basioccipital for a loop of the median branch of the internal carotid artery and by lacking carnassial teeth. Differing from all other subfamilies except Allodesminae by having tympanic crest not projecting into tympanic cavity; nasal bones anteroposteriorly elongate, tapering posteriorly, and inserting between frontals. Differing from Otariinae, Dusignathinae, Imagotariinae, and Odobeninae by having large, elongate, posterolaterally directed paroccipital process that is separate from mastoid process and not joined to it by a crest. Differing from most Enaliarctinae, all Otariinae, Imagotariinae, Dusignathinae, and Odobeninae by having pterygoid process of maxilla enlarged so palate is expanded ventral to orbit. Differing from Allodesminae, Dusignathinae, and Odobeninae by retaining vertical posterior and medial cristae on canine crowns. Differing from most Imagotariinae and from all Odobeninae by having hyoid fossa separated from stylomastoid foramen by only a thin bridge of bone, not joined by an elongate, narrow sulcus. Differing from Otariinae by having posterior lacerate foramen not greatly elongate anteroposteriorly. Differing from Allodesminae by having more inflated tympanic bulla; posterior lacerate foramen not expanded transversely; smaller orbit with anterior margin flared anterodorsally instead of being retracted posteriorly; unreduced incisive foramina; squamosal–jugal contact not greatly expanded dorsoventrally. Differing further from Odobeninae by having canines that are not modified as tusks (modified from Barnes, 1987); palate relatively flat; dental formula I3/3, C1/1, P4/4, M2/2; thick enamel on canines; lower canine not incorporated into crushing cheek tooth row; cheek teeth with smooth lingual cingulum; postglenoid foramen small; auditory bulla smooth, slightly inflated; apical bullar spur and fossa absent; inframental rim smooth rather than spinose; hyoid fossa located posterolateral to bulla, not connected with stylomastoid foramen via groove; posterior bullar projection absent; basioccipital not contributing to wall of carotid canal posteromedially; posterior carotid foramen distinctly separate from posterior lacerate foramen; auditory cavity long, narrow, and deep; hypotympanicum of medium size, inflated anteriorly and laterally well beyond promontorium, with small eminentia vagina processus styloidei on posteroexternal wall; crista tympani not projecting into auditory cavity, almost round, oriented slightly ventrolaterally; bulge of carotid canal touching promontorium, underlain slightly by hypotympanicum; promontorium globous, hanging in marked relief from dorsal wall of auditory cavity, inflated mainly posteriorly; fenestra cochlea medium to small (modified from Mitchell, 1968).

***Desmatophoca* Condon, 1906**

Type species: *Desmatophoca oregonensis* Condon, 1906.

Type specimen: UO F-735, cranium, partial dentary, various postcranial bones of an adult male individual.

Characteristics: The same as for the subfamily, as long as it is monotypic.

Occipital condyle breadth: 65 mm (holotype of *Desmatophoca oregonensis*), 69.8 mm (adult male holotype of *Desmatophoca brachycephala*, LACM 120199) (Barnes, 1987, p. 3).

Included species: *D. oregonensis* (known from localities WM6II, WM7*); *D. brachycephala* (locality WM4* [Barnes, 1987]).

Comments: The holotype is from the base of the Astoria Formation, and specimens from higher stratigraphic levels within the Astoria Formation (locality WM7) that were referred to this species by Barnes (1987) and by Deméré and Berta (2002) probably do not represent this species.

ALLODESMINAE (KELLOGG, 1931) MITCHELL, 1966

Characteristics: Medium- to large-size otarioids; cranium with incisive foramina very reduced; orbits very large relative to Desmatophocinae, occupying 20–25% of condylobasal length (modified from Repenning and Tedford, 1977); anterior part of zygomatic arch dorsal to infraorbital foramen retracting posteriorly; anterolateral margin of orbit, therefore, not cuplike, not overhanging infraorbital foramen; upper and lower canines caniniform, not greatly elongated, but deeply rooted; enamel on canines relatively thick; crowns of cheek teeth bulbous and smooth; lingual cingulae of cheek teeth reduced and smooth; dental formula I3/2, C1/1, P4/4, M2/2; mandibular symphysis elongate, unfused; angle of dentary expanded and smoothly convex; supraorbital process of frontal small; antorbital process of lacrymal absent; anterior part of zygomatic process of squamosal vertically expanded; postorbital process of jugal extended dorsally and applied to nearly vertical anterior face of expanded zygomatic process of squamosal; mastoid process separate from large paroccipital process, which is formed mainly by exoccipital; auditory bulla nearly flat, relatively smooth; postglenoid foramen absent; optic foramina and anterior lacerate foramina located posteriorly beneath anterior end of braincase; sagittal crest reduced; paroccipital process not thin and platelike, but large, elongate, directed posterolaterally, not connected with mastoid process by a crest; apical bullar spur and fossa absent; stylomastoid foramen located posterior to bulla, not directly between bulla and mastoid process; tympanohyal facet large, flattened, and located within hyoid fossa; hyoid fossa large, shallow, posterior to bulla, not connected with stylomastoid foramen via a groove around bulla; posterior bullar projection absent; labrum vagina processus styloidei absent; posterior carotid foramen separate from posterior lacerate foramen; posterior lacerate foramen small; auditory cavity long, deep, bowed laterally; hypotympanicum large, inflated externally and centrally well beyond promontorium, with eminentia vagina processus styloidei small to absent on posterior wall; recessus meatus acousticus externi small to absent; tympanic crest not projected into auditory cavity, without overhanging free ventral edge, almost round, oriented almost vertically; bulge of carotid canal vascularized, appressed to promontorium, not extensively underlain by hypotympanicum; epitympanic recess large, globous, not well delineated from rest of tympanic cavity, with wide fossa incudis; promontorium hanging far into auditory cavity, globous, inflated mainly posteriorly; fossa cerebellaris deeper than transverse width of entrance; tentorium appressed to eminence containing semicircular canals; internal acoustic meatus bilobed, with tracts for vestibulocochlear and facial nerves well separated (modified from Mitchell, 1968); baculum recurved as in *Odobenus*, circular in cross section, except for slight ventral flattening, one ventral and one larger and slightly bilobed dorsal process on the apex (modified from Repenning and Tedford, 1977); humerus relatively long, with long deltopectoral crest, deltoid tuberosity incorporated within the deltopectoral crest; olecranon process of ulna thick and massive; trochanteric fossa present on femur; podial bones relatively thick and expanded.

Allodesmus Kellogg, 1922

Type species: *Allodesmus kernensis* Kellogg, 1922.

Type specimen: CAS 2472, partial dentary of old adult male.

Characteristics: Medium to large body size; cranium with exceptionally large eyes, prenarial shelf present in both sexes; convoluted dorsal surface of braincase that is pierced by small vascular foramina; tympanic bulla finely wrinkled; canines cylindrical in cross section, having crowns that lack both lingual and posterior cristae; M1–2 greatly reduced and positioned on the ventral root of the zygomatic arch; lateral wall of the alisphenoid canal frequently atrophied or absent, but alisphenoid canal present none the less; paroccipital process large and protruding posteriorly; angle of dentary large and posteroventrally expanded.

Occipital condyle breadth (not known for *A. kernensis*): 97 mm (old adult male holotype of *A. kelloggi*, LACM 4320), 106 mm (young adult male holotype of *A. gracilis*, UCMP 81708).

Included species: *A. kernensis* (known from locality WM13C*); *A. kelloggi* (locality WM13A* [Mitchell, 1966]); *A. gracilis* (locality WM13A* [Barnes, 1995, in Barnes and Hirota, 1995; see also Barnes, 1972]).

Comments: *Allodesmus* is also known from the Miocene of Japan (Barnes and Hirota, 1995).

Atopotarus Downs, 1956

Type species: *Atopotarus courseni* Downs, 1956.

Type specimen: LACM 1376, nearly complete semiarticulated partial skeleton of adult male individual, lacking the hindquarters.

Characteristics: Elongate skull; cranium not bulbuous; lambdoidal crests large, prenarial shelf absent; mastoid process elongate and protruding ventral to level of the postglenoid process; ventral border of dentary nearly straight rather than curved dorsally; p2 through m1 with two widely divergent roots; m2 absent; basal portion of crowns of cheek teeth inflated. Differing from *Brachyallodesmus* by having more elongate cranium; large sagittal and lambdoidal crests; smaller canines; double-rooted rather than single-rooted P2–4 and p2–4. Differing from *Allodesmus* by lacking prenarial platform; having greatly expanded angle of dentary; having p2 through p4 with two widely divergent roots; lacking m2 (Barnes and Hirota, 1995).

Occipital condyle breadth: cannot be determined from the holotype, in which the condyles are still embedded in rock.

Included species: *A. courseni* only, known from locality WM17H* only.

Brachyallodesmus Barnes and Hirota, 1995

Type species: *Brachyallodesmus packardi* (Barnes, 1972) (originally described as *Allodesmus packardi*).

Type specimen: CAS 4371A, incomplete cranium.

Characteristics: Allodesmine with short and broad cranium with bulbous braincase, lacking sagittal and lambdoidal crests; orbits enlarged, interorbital region correspondingly narrow; posterior part of palate very wide, with secondary and centrally located arch; all premolars single-rooted; P1 located posteromedial to canine. Differing from all other genera of Allodesminae by having a cranium that is wider with more rounded braincase that lacks sagittal and lambdoidal crests; palate flatter, and wider posteriorly. Differing further from *Atopotarus* and *Allodesmus* by having sulcus on lateral side of braincase marking the presence of a pseudosylvian sulcus on the brain. Differing further from *Atopotarus* by having single-rooted cheek teeth. Differing further from *Megagomphos* Hirota and Barnes, 1995, by having smaller teeth; canines oval rather than round in cross section; cheek teeth retaining vertical sulci, marking the previously double-rooted condition. Differing further from *Allodesmus* by having relatively larger orbits and concomitantly narrower interorbital region; complete alisphenoid canal laterally; less cubic mastoid process; retaining a well-developed inferior petrosal venous sinus.

Occipital condyle breadth: not preserved in the holotype specimen.

Included species: *B. packardi* only, known from locality WM11* only.

ODOBENIDAE ORLOV, 1906

Characteristics: Otarioids with transversely thick pterygoid struts between the palate and alisphenoid bone; optic foramina located relatively ventrally on the anterior surface of the braincase; optic chiasma widely separated with optic nerve tracts not conjoined; internal acoustic meatus bilobed and tracts for nerves well separated; deltoid tubercle on the humerus moved to the external side of the deltopectoral crest; calcaneal tuber elongate.

IMAGOTARIINAE MITCHELL, 1968

Characteristics: Small- to large-sized otarioid; very small supraorbital processes of the frontals; small saggital crest; palate flat to arched transversely; cheek tooth crowns with prominent lingual cingulae; upper and lower canines not greatly elongated, caniniform, with thick enamel; an enlargement of the part of the posterior crista on the upper canine near the apex of the crown; lower canines not incorporated into crushing cheek tooth row; dental formula I3/2, C1/1, P4/4, M2/1–2; squamosal–jugal contact of the normal, squamous type, as in generalized carnivorans; anterior border of orbit overhanging infraorbital foramen, but not flaring so much anterolaterally as in the Otariinae; paroccipital process small, thin, and platelike; paroccipital and mastoid processes joined into a continuous crest, directed ventrally; postglenoid foramen large; auditory bulla smooth, may be sightly inflated or nearly flat; stylomastoid foramen between ectotympanic and mastoid process; hyoid fossa present, large, and posterior to bulla; basioccipital contributing to wall of carotid canal posteromedially, which is otherwise enclosed in entotympanic; posterior carotid foramen separate from posterior lacerate foramen; posterior lacerate foramen small; auditory cavity long, narrow, deep; mastoid not contributing to wall internal to eminentia vagina processus stylodei; hypotympanicum large, inflated medially and ventrally well beyond promontorium, with pronounced eminentia vagina processus styloidei on posterior wall; recessus meatus acousticus externi small; tympanic crest projected into auditory cavity, without overhanging, free ventral edge, almost round, longest diameter anteroposteriorly, oriented ventromedially 60° from axis of external acoustic meatus; bulge of carotid canal almost touching promontorium, underlain by hypotympanicum posteriorly; tympanohyal enclosed within hyoid fossa at level of tympanic membrane; epitympanic recess pear shaped, well delineated, with narrow fossa incudis; septum canalis musculotubarii a cylindrical shell almost touching dorsal roof of auditory cavity; promontorium low, inflated only posteriorly; fenestra cochlea huge; groove on promontorium ventral to fenestra vestibuli; nervus vagus ramus auricularis passing posterodorsal to tip of tympanohyal; pars petrosa not excavated for canalis petrobasilaris; tympanic wall of carotid canal possibly not contributing to wall of canalis petrobasilaris; pars petrosa dorsal to internal acoustic meatus flattened, striated slightly, defined from ventral part by sulcus; fossa cerebellaris deeper than transverse width of entrance; tentorium appressing to eminence containing semicircular canals (modified from Mitchell, 1968); humerus relatively long, with narrow but well-formed deltopectoral crest; olecranon process of ulna large and with curved posterior border; femur relatively short and transversely wide.

***Imagotaria* Mitchell, 1968**

Type species: *Imagotaria downsi* Mitchell, 1968.

Type specimen: SBMNH 342, partial cranium and dentary, various postcranial bones.

Characteristics: Otarioid of medium to large body size; hyoid fossa connected with stylomastoid foramen via groove at posterolateral side of bulla; posterior bullar projection present; cheek teeth with prominent central cusp, cuspidate lingual cingulae; approximate area ratio of oval window to tympanic membrane 1:10, comparable to living sea lions and deep-diving phocids; mandibular symphysis not ankylosed; horizontal ramus of dentary relatively short, having a sinusoidal profile in lateral view; angle of dentary prominent; coronoid process not high, broadly triangular, with narrow apex; mandibular foramen positioned high on dentary, at level of cheek tooth row.

Occipital condyle breadth: unknown.

Included species: *I. downsi* only (known from localities WM6A, WM16B [Barnes, 1971], WM17B* [Mitchell, 1968]).

***Neotherium* Kellogg, 1931**

Type species: *Neotherium mirum* Kellogg, 1931.

Type specimen: USNM 11542 (lectotype), right calcaneum of apparent adult female individual.

Characteristics: Otarioid of small to medium body size; palate relatively flat; incisive foramina large; P2–4 with main cusp and additional smaller anterior and posterior cusps, P3–4 with protocone root; making the premolars nearly molariform; m1 with talonid; all cheek teeth with prominent, smooth lingual cingulae; no groove between hyoid fossa and stylomastoid foramen; posterior bullar projection absent; mandibular symphysis not ankylosed; horizontal ramus of dentary elongate, with dorsal and ventral margins nearly parallel; angle of dentary not large; pterygoid process prominent, projecting medially; coronoid process broad and high; mandibular foramen low on dentary, positioned below level of cheek tooth row.

Occipital condyle breadth: 60 mm (old adult male cranium, LACM 131950; not present on the type specimen).

Included species: *N. mirum* only, known from locality WM13A* only.

Neotherium sp. is also known from localities (WM28II), (CC52) (Kohno, Barnes, and Hirota, 1995; Barnes, 1998).

Comments: The co-type specimens of Kellogg (1931) include a navicular, cuboid, and an astragalus, all representing different individuals. Repenning and Tedford (1977) referred a radius to this species. A dentary of an adult female was referred to this species by Barnes (1988), and a cranium of an old adult male was referred to the species by Kohno, Barnes and Hirota (1995). Many additional specimens are now available in the LACM and UCMP, and all known specimens are from the Sharktooth Hill Local Fauna.

***Pelagiarctos* Barnes, 1988**

Type species: *Pelagiarctos thomasi* Barnes, 1988.

Type specimen: LACM 121501, partial mandible of an old adult male individual.

Characteristics: Otarioid of relatively large body size (estimated to be ~2.5–3 m in body length for adult males); body of dentary thick transversely and deep dorsoventrally (somewhat as in *Allodesmus* Kellogg, 1922); mandibular symphysis ankylosed, elongate, and sloping posteroventrally; canine large and procumbent, with longitudinal sulcus on each side of root, creating bilobed cross section; cheek teeth with prominent conical cusps, and cuspidate lingual cingulae; roots of p2–3 double; tooth enamel thick and slightly crenulated. Differing from both *Neotherium* Kellogg, 1931 and *Imagotaria* Mitchell, 1968 by having ankylosed mandibular symphysis. Differing further from *Neotherium* by being much larger; having cheek tooth crowns bearing more prominent lingual cingulae with numerous small cuspules; roots of two-rooted cheek teeth closer together. Differing further from *Imagotaria* by having cheek teeth with larger and more irregular cuspules on the lingual cingulae; more prominent labial cingulum; one secondary cusp (metacone and metaconid) on the posterior crista of the crowns; more widely spread roots on lower cheek teeth posterior to p1.

Occipital condyle breadth: unknown.

Included species: *P. thomasi* only, known from locality WM13A* only.

Comments: Isolated premolars that fit within the empty alveoli of the holotype mandible have been referred to the species, but no other specimens are known of the species. All known specimens are apparently from adult male individuals.

***Pontolis* True, 1905b (including *Pontoleon*)**

Type species: *Pontolis magnus* (True, 1905b) (originally described as *Pontoleon magnus*).

Type specimen: USNM 3792, partial braincase.

Characteristics: Otarioid of very large body size; cranium with large, thick, and vertically extended nuchal crest; narrow squamosal fossa; thick and arcuate combined mastoid and paroccipital processes; zygomatic process of squamosal canted medially; wide basioccipital; transversely very thick pterygoid strut; relatively flat tympanic bulla with ventrally deflected lateral border.

Occipital condyle breadth: 113 mm (holotype, USNM 3792 [True, 1909, p. 144]).

Included species: *P. magnus* only, known from locality WM8* only.

Comments: True (1905a) originally placed *Pontolis magnus* in a new genus (*Pontoleon*), but later it was transferred to *Pontolis* (True, 1905b). The species is known at this time only by the holotype partial cranium of a very large adult male individual.

***Proneotherium* Barnes, 1995, in Kohno, Barnes, and Hirota, 1995**

Type species: *Proneotherium repenningi* Barnes, 1995, in Kohno, Barnes, and Hirota, 1995.

Type specimen: USNM 205334, nearly complete cranium with all teeth except left I1, right P1, left P4, and right and left M1–2; missing part of the right zygomatic arch (Kohno, Barnes and Hirota, 1995).

Characteristics: Otarioid of small to medium body size; elongate cranium with deep fossa on rostrum dorsal to canine; cheek region convex; antorbital process prominent; nasals not transversely expanded anteriorly; lacrimal foramen present; nasolabialis fossa deep, broad, with tubercle at its dorsal margin; anterior border of orbit (zygomatic root) cupped as in the Otariinae; supraorbital process only a small bump; low sagittal crest, extending as far anterior as posterior parts of orbits; braincase surface finely wrinkled and pitted; pterygoid strut moderately thickened transversely, convex, not expanded posterolaterally; anterior end of zygomatic process of squamosal uniformly tapered; elongate groove on each side of palate emanating from posterior palatine foramen; postglenoid process large, projecting anteriorly; paroccipital/mastoid crest large, arcuate in shape; optic nerves not widely separated anteriorly; basioccipital surface level with medial side of bulla; bulla with fossa posterior to postglenoid process; median lacerate foramen and anterior end of carotid canal not covered by tympanic bulla; crowns of I1-2 transversely bifid; P1 not enlarged; P2-4 double rooted; P3-4 with posterior root transversely bilobed and large posterolingual protocone shelf; M1 three-rooted; M2 with a bilobed single root (see Kohno, Barnes and Hirota, 1995).

Occipital condyle breadth: 60 mm (extrapolated from the undistorted left side on the holotype, but not listed by Kohno, Barnes and Hirota [1995]).

Included species: *P. repenningi* only, known from locality WM7* only.

Comments: The paratype is very similar in size to the holotype, and both specimens appear to represent similar-age adult males. Both are from the same stratgraphic level of the Astoria Formation.

DUSIGNATHINAE MITCHELL, 1968

Characteristics: Otarioids of medium to very large body size; cranium with anterior end of frontal thickened and extending anterolaterally over the ascending process of the maxilla dorsal to the anterior part of the orbit; squamosal–jugal contact of the normal, squamous type, as in generalized carnivorans; dorsal edge of zygomatic process of squamosal tilting medially; squamosal fossa narrow; mastoid process large and crescent shaped, joined to the paroccipital process, which is much reduced, and applied as a thin plate to the posterior side of the mastoid process, a deep cleft housing the stylomastoid foramen between the tympanic bulla and the mastoid process; angle of dentary flaring laterally; both upper and lower canines enlarged, in some taxa developed as tusks; upper and lower incisors reduced in size and numbers, some incisors absent in some taxa; bulbous-crowned cheek teeth; m2 lost in some taxa; relatively long and stout humerus, with thick, long deltoid crest, small greater tubercle, large entepicondyle, and distal trochlea inclined medially.

***Dusignathus* Kellogg, 1927**

Type species: *Dusignathus santacruzensis* Kellogg, 1927.

Type specimen: Holotype UCMP 27131, partial cranium and mandible of adult, sex unknown, but probably a male.

Characteristics: Otarioid of medium body size; elongate but slender upper and lower canine crowns and roots; palate slightly transversely arched; lower canines very close together because of an extremely narrow mandibular symphyseal region and apparently occluding only with the elongate I3; lower incisors small or absent; cheek teeth closely spaced, with thick, long peglike roots and simple, nearly conical crowns including a fully developed and peg-rooted p4 and m1; cheek tooth crowns capped by thin, smooth enamel, showing wear on their anterior and posterior surfaces and on proximal part of lingual sides; braincase with low sagittal and lambdoidal crests; paroccipital process continuous with mastoid process; postglenoid process reduced; mandibular condyle directed mostly dorsally into glenoid fossa; stylomastoid foramen not connected via a groove with hyoid fossa; hyoid fossa near posterior end of bulla, opening into auditory cavity; hypotympanicum not greatly inflated; surface of bulla relatively smooth and flat; eminentia vagina processus styloidei small to absent; recessus meatus acousticus externi small; tympanic crest slightly projected into auditory cavity, without overhanging, free ventral edge, subquadrate in shape, longest side oriented dorsoventrally, slanting anteroventromedially about 60° from axis of external acoustic meatus; epitympanic recess well delineated, with medium to narrow fossa incudis; mandiublar symphysis elongate, narrow, not fused in adult; genial tuberoisity pronounced; enamel thin; lower canine not premolariform and not incorporated into crushing cheek tooth row, moderately tusklike; horizontal ramus of dentary very deep, but relatively thin transversely, with concave ventral margin and dorsally turned posterior end; dental formula I3/0, C1/1, P4/4, M2/1 (modified from Mitchell, 1968; Repenning and Tedford, 1977).

Occipital condyle breadth: not preserved on holotypes of either *D. santacruzensis* or *D. seftoni*.

Included species: *D. santacruzensis* (known from locality WM20*); *D. seftoni* (locality WM26* [Deméré, 1994]).

Comments: The limb bones that Repenning and Tedford (1977) referred to the species cannot be objectively demonstrated to be conspecific. These specimens were originally identified as an odobenid by Mitchell (1962) and are from the same stratigraphic horizon, and they might belong to one or more other odobenids from the Purisima Formation (locality WC20) whose limb bones have not yet been identified.

***Gomphotaria* Barnes and Raschke, 1991**

Type species: *Gomphotaria pugnax* Barnes and Raschke, 1991.

Type specimen: LACM 121508, virtually complete skeleton, including cranium, and left and right dentaries (Barnes and Raschke, 1991).

Characteristics: Genus of Dusignathinae differing from *Dusignathus* by having cranium with high sagittal crest; rostrum laterally expanded distally to accommodate enlarged tusks; postglenoid process of squamosal relatively larger; mastoid–paroccipital crest more compressed anteroposteriorly and expanded dorsoventrally; relatively larger horizontal shelf projecting laterally over external acoustic meatus; external aperture of stylomastoid foramen directed anterolaterally rather than ventrolaterally; dentary relatively shallower dorsoventrally but relatively thicker transversely, with more recumbent coronoid process; more prominent and laterally projecting angle of dentary; relatively larger and more rugose symphyseal surface; pterygoid process tabular and directed medially rather than posteromedially; lesser degree of angular divergence from mandibular symphysis at midline; larger upper and lower canines developing as more procumbent tusks, with fluted roots covered with thick cementum and exposed outside of alveoli; cheek teeth with crowns more bulbous; P1 and p1 relatively shallow-rooted (Barnes and Raschke, 1991).

Occipital condyle breadth: not preserved on the holotype.

Included species: *G. pugnax* only, known from locality WM24A* only.

ODOBENINAE ALLEN, 1880

Characteristics: Supraorbital process reduced or absent; sagittal crest reduced or absent; occipital condyles widely flaring; basioccipital very wide and approximately pentagonal in shape; squamosal appressed to posterodorsal surface of jugal in simple squamous contact; nasals extend further back than maxillae on top of cranium; palate arched both transversely and anteroposteriorly; cheek teeth all single-rooted, having very slight lingual cingulum, or no lingual cingulum, crowns wearing with maturity; upper canines may or may not be enlarged and tusklike; lower canines reduced, in derived taxa incorporated into cheek tooth row; enamel reduced to absent; paroccipital process of exoccipital thin, platelike, appressed to paroccipital–mastoid flange; paroccipital and mastoid processes continuous, massive, directed ventrally; postglenoid foramen medium size; auditory bulla flat, wrinkled; bony Eustachian tube of large diameter; apical bullar spur and fossa small to absent; inframeatal rim spinose, major spine situated far posterodorsally; stylomastoid foramen between ectotympanic and mastoid process; hyoid fossa present, small, and located posteroexternal to bulla, connected with stylomastoid foramen via groove between bulla and mastoid; posterior bullar projection absent; basioccipital may contribute to wall of carotid canal; posterior carotid foramen separate from posterior lacerate foramen; posterior lacerate foramen medium size; auditory cavity long, narrow, and deep; mastoid may contribute to wall internal to eminentia vagina processus stylodei; hypotympanicum large, inflated posteriorly, ventral, and anterior to promontorium, with small eminentia vagina processus styloidei on posterior wall; recessus meatus acousticus externi small to absent; tympanic crest projected into auditory cavity, without overhanging, free edge ventrally; bulge of carotid canal almost touching promontorium, not underlain by hypotympanicum posteriorly; tympanohyal enclosed within hyoid fossa at level of tympanic membrane, contacts and may fuse with petrosal in or near wall of facial canal; tympanic membrane almost round, large, oriented ventromedially; tympanic membrane to: oval window ratio approximately 20:1; auditory ossicles very large; recessus epitympanicus pear shaped, well defined, with wide fossa incudis; septum canalis musculotubarii a simple bony shelf, not touching roof of auditory cavity; promontorium low, broad, inflated only posteriorly, bearing major flange on posterolateral edge; fenestra cochlea wide; petrosal excavated for canalis petrobasilaris; tympanic wall of carotid canal may contribute to wall of canalis petrobasilaris; petrosal dorsal to internal acoustic meatus flattened, slightly striated, defined from ventral part by flange; internal acoustic meatus very wide, shallow, with widely spaced canals for vestibulocochlear and facial nerves nearly separated by a septum; petrosal apex large, rounded, flat; fossa cerebellaris deeper than transverse width of entrance; tentorium appressed to eminence containing semicircular canals; mandibular symphysis deep, large, narrowly oval in shape, articulating over the entire depth of the prominent chin; mandibular symphysis fused in derived taxa; angle of dentary, where digastricus muscle inserts, weak and anteriorly located ventral to a point between the last cheek tooth and the coronoid process; humerus long for a pinniped and relatively slender; distal termination of pectoral crest approximately in line with the medial lip of the trochlea of the distal articulation; anteroposterior diameter of the trochlear lip much greater than that of the distal capitulum; radius with pronator origin at or distal to the midpoint of shaft; ulna without flat anterior surface of olecranon above semilunar notch; distal radial articulation distinct and elevated; scapholunar with deep pit for articulation with magnum; femoral head distinctly higher than greater trochanter; proximal end of fibula usually not fused to the tibia, commonly so even in individuals of great age; calcaneum with most prominent tuberosity on medial side of calcaneal tuber; astragalus with high and essentially vertical fibular articulation, not flaring distally onto extended lateral process (modified from Mitchell, 1968; Repenning and Tedford, 1977).

Aivukus Repenning and Tedford, 1977

Type species: *Aivukus cedrosensis* Repenning and Tedford, 1977.

Type specimen: IGCU 901, partial cranium and a dentary, various forelimb bones, latest Miocene, Hemphillian correlative, Lower Member of the Almejas Formation, Baja California, Mexico.

Characteristics: Otarioid of moderate size; skull slightly modified to accommodate small canine tusks; rostrum in adult elongate and not massive; upper canines elongate and long-growing, but growth stopping when pulp cavity is filled with annular dentine; canines without globular dentine in pulp cavity; lower canines reduced in size but still 50% larger in diameter than p1; mandibular symphysis not ankylosed and much less sloping than in *Prorosmarus*; posterior end of mandibular symphysis terminating approximately beneath p1 and p2; dental formula I2/?, C1/1, P4/4, M1/1 (modified from Repenning and Tedford, 1977).

Occipital condyle breadth: not preserved on the holotype.

Included species: *A. cedrosensis* only, known from locality WM30A* only.

Pliopedia Kellogg, 1921

Type species: *Pliopedia pacifica* Kellogg, 1921.

Type specimen: USNM 13627 (previously Stanford University Museum No. 537), various bones of a forelimb.

Characteristics: Otarioid for which dentition and facial portion of skull are unknown; sagittal and lambdoidal crest lacking; supraoccipital area for insertion of neck muscles low and broad; carotid canal very large; petrosal and bulla as in *Odobenus*; humerus with greater tubercle no higher than head, lesser tubercle large, curved medially, and sloping distally; bicipital groove deep and narrow; deltoid tubercle not on pectoral crest but on lateral surface of shaft; entepicondyle directed posteriorly as well as medially; shaft of humerus straight and not shortened or curved; ulna short and thick with moderately deep olecranon process; radius short and thick with radial process medially positioned; first metacarpal having a depression on its dorsal surface distal to proximal articulation, and more slender than that of *Imagotaria* (modified from Repenning and Tedford, 1977).

Occipital condyle breadth: unknown.

Included species: *P. pacifica* only, known from locality WM23* only.

Prorosmarus Berry and Gregory, 1906

Type species: *Prorosmarus alleni* Berry and Gregory, 1906.

Type specimen: USNM 9343, partial dentary, possibly of an adult male.

Characteristics: Otarioid with moderate to large body size; mandible somewhat like that of *Odobenus*, not dorsally turned in the anterior part as in *Alachtherium* (which implies a shorter facial region); cheek teeth closely spaced as in *Odobenus*; mandibular symphysis not ankylosed, elongate, not extending nearly as far posteriorly as in *Alachtherium* but terminating beneath p2; two well-developed incisors; canine reduced but retaining its primitive position, not in line with the cheek teeth, and caniniform in shape; cranium probably retained functional incisors in the adult, judged by wear on the medial side of the lower canine; inferring from the small canine and mandibular tip, the upper canine may have been nearly as large as in extant *Odobenus*.

Occipital condyle breadth: not available on the holotype.

Included species: *P. alleni* only, known from locality EM8A* only.

***Trichecodon* Lankester, 1865**

Type species: *Trichecodon huxleyi* Lankester, 1865.

Type specimen: None designated. Figured in Lankester (1865: Plates 10 and 11).

Characteristics: Large canine tusks resembling those of living *Trichechus* in curvature, varying lateral compression, large surface furrows, short and wide pulp cavity having globular 'osteo-dentine," and in every detail of minute structure. Canine tusks differing from living *Trichechus* in greater curvature at the point of the tusk; greater lateral compression; minor development of cement (Lankester, 1865).

Occipital condyle breadth: unknown.

Included species: *T. huxleyi* only (known from localities GC13B, EM8B [Morgan, 1994]).

Comments: *Trichecodon* is also known from the Pliocene of western Europe (Deméré, 1994), and from the Pleistocene of Europe and North America (McKenna and Bell, 1997).

***Valenictus* Mitchell, 1961**

Type species: *Valenictus imperialensis* Mitchell, 1961.

Type specimen: LACM (CIT) 3926, humerus of old adult.

Characteristics: Large body size; major limb bones both osteosclerotic and pachyostotic; humerus with a relatively small head, elongate but low and broad deltopectoral crest having a large deltoid tuberosity on its lateral margin; broad but anteroposteriorly flattened distal end; greater tubercle equal in height to head; lesser tubercle lower in height than head; deltopectoral crest broad; bicipital groove narrow; entepicondyle large, thick, projecting very prominently medially; distal end of humerus inclined medially at 70° from vertical axis of shaft (modified from Mitchell, 1961).

Occipital condyle breadth: 110 mm (based on SDSNH 36786, *Valenictus chulavistensis* [Deméré, 1994]).

Included species: *V. imperialensis* (known from locality WM22*); *V. chulavistensis* (locality WM26* [Deméré, 1994]).

BIOLOGY AND EVOLUTIONARY PATTERNS

GENERAL MORPHOLOGY

All fossil otarioid pinnipeds for which sufficiently complete fossils have been obtained indicate a body with general features that are similar to those of the living representatives: elongate neck, anterolaterally directed eyes, strong jaws, short limbs and tail, rotational hip joint, flattened manus and pes with elongate digits, and a rotational ankle joint that allows them to walk on land.

Mitchell (1966, 1979) and Mitchell and Tedford (1973) have presented life reconstructions of some otarioids from fossil osteological evidence. Most fossil otarioids, however, are known by too little material to attempt such reconstructions. It is yet another step to postulate behavior from the available fossil record. One of the most completely known fossil otarioids is *Allodesmus kelloggi*, and Mitchell (1966) discussed possible sexual dimorphism, polygyny, and feeding behavior for this species.

LOCOMOTION

Both terrestrial and aquatic locomotion have been accomplished by different methods in the three major Recent groups of pinnipeds. These methods have been summarized often (King, 1964, 1983; Gordon, 1983). Living sea lions and fur seals of the subfamily Otariinae swim by propelling themselves with rapid, repeated strokes of the forelimbs. The forelimb is streamlined and has a flattened anterior (radial) edge. The flattened hind flippers are held with their plantar surfaces appressed, and they trail behind the body, moving occasionally while swimming and functioning mostly as rudders. The propulsive stroke is caused by adduction of the forelimb, accompanied by medial forelimb rotation to produce forward thrust (English, 1976), virtually "flying" through the water in much the way that penguins do. To reduce drag during the recovery stroke, the forelimb is rotated laterally. The rapid swimming and darting of the neck by sea lions and fur seals enables them to capture fast-swimming prey (Mitchell, 1966; English, 1976; King, 1983). The walrus *Odobenus rosmarus* uses both the fore- and hindlimbs in swimming: the posterior end of the body moves laterally, and the forelimbs accomplish a paddling movement (King, 1964; Gordon, 1983). In contrast, the true seals of the family Phocidae all use their hindlimbs for aquatic propulsion.

Otarioid locomotion on land is constrained, however, by some of the same anatomical specializations that have evolved for efficient swimming. Sea lions, fur seals, and walruses all place their fore and hind feet flat on the ground, with at least the lower limb elements vertical in order to support the body. The forelimbs, with the manus pointed laterally away from the body, step alternately. The hindlimbs, bound tightly to the pelvic area, step alternately or synchronously when moving rapidly (King 1964, 1983; English 1976), with the pelvis held in a near-vertical position (see Howell 1929; King 1964). English (1976) has interpreted the transverse position of the manus as a transverse lever, unlike the "use of the manus as a sagittal plane lever" in fissipeds, and this is a compromise with a manus that has been modified in ways to make it valuable for swimming. Otarioid terrestrial locomotion is a more primitive method than that used by the Phocidae, which cannot "walk" on land, but rather hunch and/or wriggle along, and which have permanently extended hindlimbs as a result of extreme modifications of the ankle joints.

The bones of the forelimb of all fossil and Recent pinnipeds are shortened and flattened. This flattening of the radius and ulna, while an indication of aquatic adaptation, does not necessarily indicate that an animal's primary propulsion is by the forelimb. Howell (1930) has pointed out that the flattening of these bones does not serve a similar function in the otariids as it does in the phocids. In sea lions, the flattening has a mechanical basis: the creation of a streamlined hydrodynamic paddle (see English, 1976), of which the manus is a continuation. Howell (1930) saw few logical reasons, however, for flattening of the radius and ulna in the true seals (Phocidae), because paddle formation is not now, and presumably has never been, a major factor in the development of the antebrachium. Phocids do, however, hold the forelimbs close to the side of the body while swimming and they do use them as rudders to guide themselves; therefore, the flattening may merely be to reduce drag.

Another osteological feature of the forelimb, however, is an indication of how it was used in aquatic locomotion. In terrestrial carnivores, the shafts of the radius and the ulna are aligned so that in the usual position for walking or running, the manus points forward. In the highly derived Otariinae, the shafts of the radius and the ulna are twisted 90° outward (supination). This agrees with the lateral projection of the manus on land and with the flipper being held in a plane parallel to the body in swimming. The laterally rotated position of the manus relative to the elbow joint can be discerned from the forelimb bones of otariines.

In contrast, the degree of lateral rotation of the forelimbs of phocids and of fossil walruses is less, varying between 0° (standard for fissipeds) and 90° (standard for otariines). The phocids, with less rotation, do not use the forelimbs for power strokes. In the Recent walrus *Odobenus rosmarus*, the degree of rotation is not quite as much as in the Otariinae. They do a sort of "dog paddle" with the forelimbs.

The derived condition of 90° rotation is found in all members of the Otariinae since the latest Miocene, but is also fully developed by the rather distantly related desmatophocines *Allodesmus kelloggi* and *A. gracilis* by as early as the middle Miocene. In contrast, the distal end of the radius of *Neotherium mirum*, like that of the slightly later imagotariine *Imagotaria downsi*, is rotated almost exactly halfway (~45°) between the primitive fissiped condition and the derived condition that is found in the Otariinae. The manus of both *N. mirum* and *I. downsi* in terrestrial posture undoubtedly pointed anterolaterally. In swimming, these animals probably held their forelimbs oblique to the anteroposterior axis of the body. During propulsive strokes, this would place the plantar surface of the manus at a 45° angle to either anteroposterior or medial adduction–abduction movements of the forelimb. Thus, the manus could be used more like an expanded paddle in a "crawl" stroke rather than the way that the flattened flipper is used by the modern species of Otariinae. Such a function of the manus in swimming seems to correlate with the structure of a complete manus that has been referred to *I. downsi* by Repenning and Tedford (1977) in which the metacarpals spread out more than they do in species of the Otariinae (see Howell, 1929). In the Imagotariines, the hindlimbs must have also been used for propulsion, probably in a fashion similar to that of living walruses.

DIVING AND FEEDING

Depth of diving by fossil otariids can be inferred from oval to round window ratios, size of the tympanic membrane, and size of the tympanic cavity (Repenning, 1972). Recent pinnipeds, which have relatively large tympanic membranes as in terrestrial animals, are not deep divers, while the otariines, with small tympanic membranes, are deep divers.

There is more dental diversity among the fossil otarioids than there is among the living taxa. Some fossil odobenines (e.g., *Prorosmarus alleni*) apparently had tusks that were nearly as large as those of the living walrus. Walruses that have large tusks tend to develop a fused mandibular symphysis. In contrast, *Aivukus cedrosensis*, which has a cranium that is in many ways similar to Recent *Odobenus rosmarus*, has small tusks and a non-ankylosed mandibular symphysis. These and other fossil otarioids that had blunt, peg-like, single-rooted teeth might have fed on benthic invertebrates as do living walruses. The piston-tongue method of suction feeding in walruses may even be extended by analogy to the fossil *Aivukus cedrosensis*, and to the more distantly related members of the subfamily Dusignathine, because members of both groups have relatively highly arched palates.

Most fossil otarioids have dentitions that are generally similar to those of the Recent otariines, and it may be inferred that they fed primarily on fish. This group of fish feeders would also include the enaliarctines, the desmatophocines, and *Neotherium* and *Imagotaria* among the Imagotariinae, all of which have the full complement of incisors, normal-sized canines, and a full (or nearly full) complement of cheek teeth that have small- to moderate-sized crowns with a prominent central cusp and/or additional cusps and a cingulum. Fish must have been the primitive diet of otariines.

The large, relatively specialized imagotariine *Pelagiarctos thomasi* might have been a predator of large marine vertebrates (Barnes, 1988), based on the size and shape of its mandible and teeth. A very different type of feeding, however, was probably utilized by various fossil pinnipeds, which have a bulbous, smooth type of cheek tooth (e.g., *Allodesmus*, *Desmatophoca*, and possibly *Dusignathus*). There is no modern analog for these. The first two groups were apparently deep divers and fast swimmers, and therefore had not acquired a walrus-like habitus. Their teeth are somewhat like those of squid-eating extant toothed whales, and these fossil otarioids were possibly also teuthophagous. No otarioid has yet been found that appears to have evolved a crustacean-catching dentition like that of the Recent crabeater seal (*Lobodon carcinophagus*). We do not yet have information about the diets of these diverse otarioids with various types of dentition, but it is clear that they represent a greater diversity of morphological types than exist now. Barnes and Raschke (1991) argued that the bizarre dusignathine *Gomphotaria pugnax* ate hard-shelled intertidal marine invertebrates, based on the degree and type of wear on its teeth, coupled with its apparent lack of deep-diving adaptations.

The most direct line of evidence for fossil pinniped diets would be preserved stomach contents in the visceral area of a skeleton. Two such fossil occurrences are known, but they have yet to be recorded in the scientific literature.

POLYGYNY

Sexual dimorphism appeared early in the history of otarioids (Mitchell, 1966; Bartholomew, 1970). It is known in the enaliarctines (Barnes, 1979), in the middle Miocene *Allodesmus* (Mitchell, 1966; Barnes, 1972), and in the contemporaneous and geochronologically more recent imagotariines, *Neotherium* and *Imagotaria*. Such size dimorphism is a characteristic of polygynous pinnipeds (see Bartholomew, 1970), and is clearly demonstrable by the fossil record. Fossils of male otarioids show the same types of secondary sex characters as are present in living species: larger body size, relatively larger canine teeth, rugose bone texture, and larger crests and tuberosities, particularly on the skull. Many incidences of broken teeth, especially canines, in the fossil record may have resulted from fights for dominance and territory among males. From this observation, other non-paleontological attributes of the species may be inferred. The members of the subfamily Otariinae are the only group of otarioids to have developed a large and tabular supraorbital process, and this is especially well developed in male otariines. The males of living otariine species are highly combative and territorial, and this large supraorbital process is likely a form of protection for the eye from the jabs of the large canine teeth of other males. Terrestrial parturition probably occurred among all fossil species because it is the primitive trait for carnivorans, and because it is universal among Recent otarioids (Bartholomew, 1970).

ZOOGEOGRAPHY

Recent otarioids range from the equatorial latitudes (one species, the Galapagos sea lion) to the Arctic, but most species are cool–temperate to polar animals. This sort of distribution among various marine organisms has been termed antitropical (or bitemperate or bipolar), and it is variably demonstrated by extant otarioids. The classic examples among marine mammals involve pairs of genera, species, subspecies, or populations that are distributed in the Northern and Southern Hemispheres and are separated by equatorial and/or temperate latitudes (Figure 31.3).

The walrus inhabits the Arctic, but it has no Southern Hemisphere counterpart. It is antitropical, but not bipolar. The sea lion *Zalophus* has an antitropical, circum-North Pacific distribution (including the recently extinct Japanese sea lion, *Z. japonicus*, and excepting the disjunct population of *Z. wollebaeki* that inhabits the Galapagos Islands). In the broad sense, however, the northern sea lions *Zalophus* and *Eumetopias jubatus* can be regarded as the Northern Hemisphere antitropical counterparts of the southern sea lions *Phocarctos hookeri*, *Neophoca cinerea*, and *Otaria byronia*. Davies (1958a) even identified possible taxon pairs among these (e.g., *Eumetopias* and *Otaria; Zalophus* and *Neophoca*). Mitchell (1968) offered the alternative suggestion that the Southern Hemisphere sea lions are more closely related to each other than they are to *Zalophus* or to *Eumetopias*. The fur seals offer better examples of north–south taxon pairs, and they seem to fit the antitropical distribution model better .

The most frequently postulated place of origin of the Otarioidea is the Northern Hemisphere, either in the Arctic (Davies, 1958b) or the North Pacific Ocean (McLaren, 1960). The latter remains the favored location, because the rocks around the North Pacific margin have yielded the earliest and most primitive members of the family, as well as the greatest taxonomic diversity.

Among Recent North Pacific otarioids, particularly the California sea lion *(Zalophus californianus)*, the sexes segregate during non-breeding seasons. Males occupy the northern parts of each species range, and the females and young the southern part.

Fossil deposits that contain mostly bones of adult males, and few or none of females and juveniles, may be interpreted as having been part of the non-breeding range of the species.

NICHES AND COMPETITION

At least two, and as many as five, species of otarioid pinnipeds (Repenning and Tedford, 1977) have now been discovered associated in several fossil-bearing deposits around the Pacific margin. These various otarioids may not have been in competition with one another. They might have fed on different food items, or they might have occupied the area of the fossil localities at different times of the year owing to different seasonal migration cycles. The result of this would be extensive niche division and/or seasonal species allopatry or sexual allopatry. Up to six species of pinnipeds, otarioids as well as phocids, regularly live at some time of the year along the California coast today, and at times three pinniped species may even simultaneously occupy the same beach (Orr and Poulter, 1965; Odell, 1981). Some of the fossil otarioids, such as *Pelagiarctos*, the desmatophocids, the dusignathines, and some of the extinct odobenines have no apparent living analogs, and clearly they had diets and/or feeding methods that differed from those of any of the Recent species. The fossil evidence indicates, therefore, that during the Tertiary the otarioid pinnipeds had much greater taxonomic and ecologic diversity than they do today.

ACKNOWLEDGEMENTS

I thank Francisco J. Aranda-Manteca, Annalisa Berta, Thomas A. Deméré, Kiyoharu Hirota, Naoki Kohno, Edward D. Mitchell, Rodney E. Raschke, Clayton E. Ray, and Richard H. Tedford for many years of dialogue and free exchange of information about fossil otarioid pinnipeds. I am especially grateful to the late Charles A. Repenning, who mentored me while I was preparing my Master's Thesis dealing with allodesmine pinnipeds. We lost a good friend when Charles Repenning was killed in 2005, and I dedicate this work to his memory. Our approaches to the systematics and classification of the otarioid pinnipeds have varied greatly, both through the years and between ourselves, but elucidation of the wonderful fossil record of the otarioid pinnipeds, and the morphology that is documented by it, is the most important aspect of our work. I thank Daniel Gabai for assistance with some of the graphics, and the late Robert L. Clark, Jr., Michael D. Quarles, Michael J. Stokes, and Howell W. Thomas for detailed specimen preparation and replication. I thank William A. Clemens and the late Donald E. Savage, who served on my Master's Thesis Committee while I was studying the Allodesminae as a graduate student. Without the input of the persons mentioned above, and many others too numerous to list here, this chapter would not have been possible. I also thank the University of California Museum of Paleontology and the Natural History Museum of Los Angeles County and its staff for support during many years of study of fossil pinnipeds and the production of publishable results.

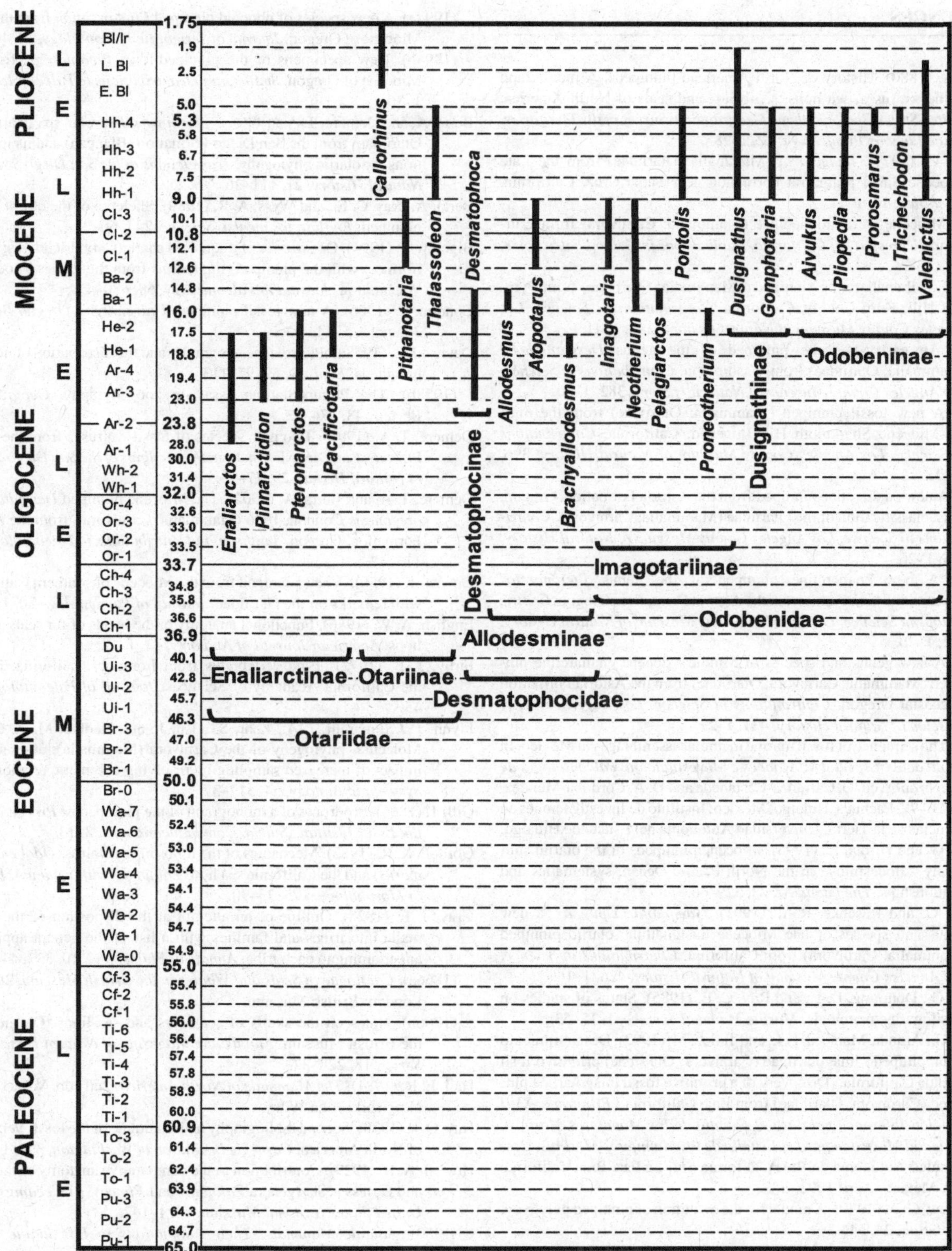

Figure 31.3. Temporal ranges of North American otarioid genera.

REFERENCES

Allen, J. A. (1880). History of North American pinnipeds. A monograph of the walruses, sea-lions, sea-bears, and seals of North America. *United States Geological and Geographical Survey of the Territories, Miscellaneous Publication*, 12, 1–785.

Barnes, L. G. (1971). *Imagotaria* (Mammalia: Otariidae) from the Late Miocene Santa Margarita Formation near Santa Cruz, California. *PaleoBios*, 11, 1–11.

(1972). Miocene Desmatophocinae (Mammalia: Carnivora) from California. *University of California, Publications in Geological Science*, 89, 1–68.

(1979). Fossil enaliarctine pinnipeds (Mammalia: Otariidae) from Pyramid Hill, Kern County, California. *Contributions in Science, Los Angeles County Museum of Natural History*, 318, 1–41.

(1987). An early Miocene pinniped of the genus *Desmatophoca* (Mammalia: Otariidae) from Washington. *Contributions in Science, Los Angeles County Museum of Natural History*, 382, 1–20.

(1988). A new fossil pinniped (Mammalia: Otariidae) from the middle Miocene Sharktooth Hill Bonebed, California. *Contributions in Science, Los Angeles County Museum of Natural History*, 396, 1–11.

(1989). A new enaliarctine pinniped from the Astoria Formation, Oregon, and a classification of the Otariidae (Mammalia: Carnivora). *Contributions in Science, Los Angeles County Museum of Natural History*, 403, 1–26.

(1990). A new enaliarctine pinniped of the genus *Pteronarctos* (Mammalia: Carnivora) from the Astoria Formation, Oregon. *Contributions in Science, Los Angeles County Museum of Natural History*, 422, 1–20.

(1992). A new genus and species of Middle Miocene enaliarctine pinniped (Mammalia, Carnivora, Otariidae) from the Astoria Formation in coastal Oregon. *Contributions in Science, Los Angeles County Museum of Natural History*, 431, 1–27.

(1998). The sequence of fossil marine mammal assemblages in México. In *Publicación Especial* 1: *Avances en Investigación. Paleontología de Vertebrados*, ed. O. Carranza-Castañeda and D. A. Córdoba-Méndez, pp. 26–79. Pachuca, Hidalgo, México: Instituto de Investigaciones en Ciencias de la Tierra, Universidad Autónoma del Estado de Hidalgo.

Barnes, L. G. and Hirota, K. (1995). Miocene pinnipeds of the otariid subfamily Allodesminae in the North Pacific Ocean: systematics and relationships. *The Island Arc*, 3, 329–60.

Barnes, L. G. and Raschke, R. E. (1991). *Gomphotaria pugnax*, a new genus and species of late Miocene dusignathine otariid pinniped (Mammalia: Carnivora) from California. *Contributions in Science, Los Angeles County Museum of Natural History*, 426, 1–16.

Barnes, L. G., Domning, D. P., and Ray, C. E. (1985). Status of studies on fossil marine mammals. *Marine Mammal Science*, 1, 15–53.

Barnes, L. G., Aranda-Manteca, F. J., and Tellez, M. (1992). Descubrimiento de un pinípedo Imagotariinae (Carnivora; Otariidae) primitivo fósil en Baja California. [Discovery of a primitive fossil imagotariine pinniped (Carnivora; Otariidae) from Baja California.] *Programa, XVII Reunión Internacional para el Estudio de los Mamíferos Marinos, Sociedad Mexicana para el Estudio de los Mamíferos Marinos*, Universidad Autónoma de Baja California Sur, La Paz, Baja California Sur, México, April 1992.

Bartholomew, G. A. (1970). A model for the evolution of pinniped polygyny. *Evolution*, 24, 546–9

Berry, E. W. and Gregory, W. K. (1906). *Prorosmarus alleni*, a new genus and species of walrus from the upper Miocene of Yorktown, Virginia. *American Journal of Science*, 21, 444–50.

Berta, A. (1991). New *Enaliarctos** (Pinnipedimorpha) from the Oligocene and Miocene of Oregon and the role of "enaliarctids" in pinniped phylogeny. *Smithsonian Contributions to Paleobiology*, 69, 1–33.

(1994a). A new species of phocoid pinniped *Pinnarctidion* from the early Miocene of Oregon. *Journal of Vertebrate Paleontology*, 14, 405–13.

(1994b). New specimens of the pinnipediform *Pteronarctos* from the Miocene of Oregon. *Smithsonian Contributions to Paleobiology*, 78, 1–30.

Berta, A. and Deméré, T. A. (1986). *Callorhinus gilmorei* n. sp., (Carnivora: Otariidae) from the San Diego Formation (Blancan) and its implications for otariid phylogeny. *Transactions of the San Diego Society of Natural History*, 21, 111–26.

Berta, A., Ray. C. E., and Wyss, A. R. (1989). Skeleton of the oldest known pinniped, *Enaliarctos mealsi. Science*, 244, 60–2.

Bigelow, P. (1994). Occurrence of a squaloid shark (Chondrichthyes: Squaliformes) with the pinniped *Allodesmus* from the Upper Miocene of Washington. *Journal of Paleontology*, 68, 680–4.

Condon, T. (1906). A new fossil pinniped. *University of Oregon Bulletin*, 3(suppl.), 5–14.

Davies, J. L. (1958a). Pleistocene geography and the distribution of northern pinnipeds. *Ecology*, 39, 97–113.

(1958b). The Pinnipedia: an essay in zoogeography. *Geographical Review*, 48, 474–93.

Deméré, T. A. (1994). Two new species of fossil walruses from the Upper Pliocene San Diego Formation. *Proceedings of the San Diego Society of Natural History* 29, 77–98.

Deméré, T. A. and Berta, A. (2002). The Miocene pinniped *Desmatophoca oregonesis* Condon, 1906 (Mammalia: Carnivora), from the Astoria Formation, Oregon. *Smithsonian Contributions to Paleobiology*, 93, 113–47.

Downs, T. (1956). A new pinniped from the Miocene of southern California: with remarks on the Otariidae. *Journal of Paleontology* 30, 115–31.

English, A. W. (1976). Functional anatomy of the hands of fur seals and sea lions. *American Journal of Anatomy*, 47, 1–18.

Firby, J. B. (1972). Type vertebrates from Lompoc, California, now at the California Academy of Sciences. *Journal of Paleontology*, 46, 450.

Flynn, J. J., Finarelli, J. A., Zehr, S., Hsu, J., and Nedbal, M. A. (2005). Molecular phylogeny of the Carnivora (Mammalia): assessing the impact of increased sampling on resolving enigmatic relationships. *Systematic Biology*, 54, 317–37.

Gill, T. (1866). Prodrome of a monograph of the pinnipedes. *Proceedings of the Essex Institute, Salem, Communications*, 5, 3–13.

Gordon, K. R. (1983). Mechanics of the limbs of the walrus (*Odobenus rosmarus*) and the California sea lion (*Zalophus californianus*). *Journal of Morphology*, 175, 73–90.

Gray, J. E. (1825). Outline of an attempt at the dispostion of the Mammalia into tribes and families with a list of the genera apparently appertaining to each tribe. *Annals of Philosophy*, 10, 337–44.

(1866a). *Catalogue of Seals and Whales in the British Museum*, 2nd edn. London: British Museum.

(1866b). Notes on the skulls of sea-bears and sea-lions (Otariidae) in the British Museum. *Annals and Magazine of Natural History, 3rd Series*, 18, 228–37.

Hall, E. R. (1981). *The Mammals of North America*, 2nd edn, Vols. 1 and 2. New York: John Wiley.

Hay, O. P. (1930). Second bibliography and catalogue of the fossil Vertebrata of North America. *Carnegie Institution of Washington*, 390, 1–1074.

Howell, A. B. (1929). Contribution to the comparative anatomy of the eared and earless seals (genera *Zalophus* and *Phoca*). *Proceedings of the United States National Museum*, 73, 1–142.

(1930). *Aquatic Mammals: Their Adaptations to Life in the Water*. Springfield, IL: Charles C. Thomas.

Hunt, R. M., Jr. and Barnes, L. G. (1994). Basicranial evidence for ursid affinity of the oldest pinnipeds. *Proceedings of the San Diego Society of Natural History* 29, 57–67.

Kellogg, A. R. (1921). A new pinniped from the Upper Pliocene of California. *Journal of Mammalogy*, 2, 212–26.

(1922). Pinnipeds from Miocene and Pleistocene deposits of California. A description of a new genus and species of sea lion from the Temblor Formation together with seal remains from the Santa Margarita and San Pedro Formations and a résumé of current theories regarding origin of Pinnipedia. *University of California Publications, Bulletin of the Department of Geological Sciences*, 13, 23–132.

(1925). Additions to the Tertiary history of the pelagic mammals on the Pacific coast of North America. IV. New pinnipeds from the Miocene diatomaceous earth near Lompoc, California. *Carnegie Institution of Washington Publication*, 348, 71–96.

(1927). Fossil pinnipeds from California. *Carnegie Institution of Washington Publication*, 346, 27–37.

(1931). Pelagic mammals from the Temblor Formation of the Kern River region, California. *Proceedings of the California Academy of Sciences, Series 4*, 19, 217–397.

King, J. E. (1964). *Seals of the World*. London: British Museum (Natural History).

(1983). *Seals of the World*, 2nd edn. Ithaca, NY: Comstock.

Kohno, N., Barnes, L. G., and Hirota, K. (1995). Miocene fossil pinnipeds of the genera *Prototaria* and *Neotherium* (Carnivora; Otariidae; Imagotariinae) in the North Pacific Ocean: evolution, relationships, and distribution. *The Island Arc*, 3, 285–308.

Lankester, E. R. (1865). On the sources of the mammalian fossils of the Red Crag, and on the discovery of a new mammal in that deposit, allied to the walrus. *Quarterly Journal of the Geological Society*, 21, 221–32.

Linnaeus, C. (1758). *Systema Naturae per Regna Tria Naturae, Secundum Classes, Ordines, Genera, Species, cum Characteribus and Differentiis,* 10th edn, *Reformata,* Vol. I. Holmiae: Laurentii Salvii.

McKenna, M. C. and Bell, S. K. (1997). *Classification of Mammals Above the Species Level*. New York: Columbia University Press.

McLaren, I. A. (1960). Are the Pinnipedia biphyletic? *Systematic Zoology*, 9, 18–28.

Mitchell, E. D. (1961). A new walrus from the Imperial Pliocene of southern California: with notes on odobenid and otariid humeri. *Contributions in Science, Los Angeles County Museum of Natural History*, 44, 1–28.

(1962). A walrus and a sea lion from the Pliocene Purisima Formation at Santa Cruz, California: with remarks on the type locality and geologic age of the sea lion *Dusignathus santacruzensis* Kellogg. *Contributions in Science, Los Angeles County Museum of Natural History*, 56, 1–24.

(1966). The Miocene pinniped *Allodesmus*. *University of California, Publications in Geological Sciences*, 61, 1–46.

(1968). The Mio-Pliocene pinniped *Imagotaria*. *Journal of the Fisheries Research Board of Canada*, 25, 1843–900.

(1975). Parallelism and convergence in the evolution of Otariidae and Phocidae. *Rapports et Procés-verbaux des Réunions, Conseil International pour l'Exploration de la Mer*, 169, 12–26.

(1979). Origins of eastern North Pacific sea mammal fauna. *Whalewatcher, Journal of the American Cetacean Society*, 13, 1–7.

Mitchell, E. D. and Tedford, R. H. (1973). The Enaliarctinae. A new group of extinct aquatic Carnivora and a consideration of the origin of the Otariidae. *Bulletin of the American Museum of Natural History*, 151, 201–84.

Miyazaki, S., Horikawa, H., Kohno, N., *et al.* (1995). Summary of the fossil record of pinnipeds of Japan, and comparisons with that from the eastern North Pacific. *The Island Arc*, 3, 361–72.

Morgan, G. S. (1994). Miocene and Pliocene marine mammal faunas from the Bone Valley Formation of central Florida. *Proceedings of the San Diego Society of Natural History*, 29, 239–68.

Muizon, C., de (1978). *Arctocephalus* (*Hydrarctos*) *lomasiensis*, subgen. nov. et nov. sp., un nouvel Otariidae du mio-pliocène de Sacaco (Pérou). *Bulletin de l'Institut Français d'Études Andines*, 7, 169–89.

Odell, D. K. (1981). California sea lion *Zalophus californianus* (Lesson, 1828). In *Handbook of Marine Mammals,* Vol. I: *The Walrus, Sea Lions, Fur Seals and Sea Otter*, ed. S. H. Ridgway and R. J. Harrison, pp. 67–97. London: Academic Press.

Orr, R. T. and Poulter, T. C. (1965). The pinniped population of Ano Nuevo Island, California. *Proceedings of the California Academy of Sciences*, 32, 377–404.

Repenning, C. A. (1972). Underwater hearing in seals: functional morphology. In *Functional Anatomy of Marine Mammals*, ed. R. J. Harrison. pp. 307–31. London: Academic Press.

(1976). Adaptive evolution of sea lions and walruses. *Systemic Zoology*, 25, 375–90.

(1990). Technical comments. Oldest pinniped. *Science*, 248, 499.

Repenning, C. A. and Tedford, R. H. (1977). Otarioid seals of the Neogene. *United States Geological Survey Professional Paper*, 992, 1–93.

Sarich, V. M. (1969). Pinniped phylogeny. *Systematic Zoology*, 18, 416–22.

Scheffer, V. B. (1958). *Seals, Sea Lions, and Walruses. A Review of the Pinnipedia*. Stanford, CA: Stanford University Press.

Smirnov, N. A. (1908). [Review of Russian pinnipeds.] *Mémoires del'Académie Impériale des Sciences de St.-Pétersbourg, 8 série, Sciences, Mathématiques et Physiques*, 23, 1–75.

Tedford, R. H. (1976). Relationship of pinnipeds to other carnivores (Mammalia). *Systematic Zoology*, 25, 363–74.

True, F W. (1905a). Diagnosis of a new genus and species of fossil sea-lion from the Miocene of Oregon. *Smithsonian Miscellaneous Collections*, 48, 47–9.

(1905b). New name for *Pontoleon*. *Proceedings of the Biological Society of Washington*, 18, 253.

(1909). A further account of the fossil sea lion *Pontolis magnus*, from the Miocene of Oregon. *United States Geological Survey Professional Paper*, 59, 143–8.

Van Gelder, R. G. (1977). Mammalian hybrids and generic limits. *American Museum Novitates*, 2635, 1–25.

Wiig, O. (1983). On the relationship of pinnipeds to other carnivores. *Zoologica Scripta*, 12, 225–7.

32 Phocidae

IRINA A. KORETSKY[1] and LAWRENCE G. BARNES[2]

[1] *Howard University, Washington, DC, USA*

[2] *Natural History Museum of Los Angeles County, Los Angeles, CA, USA*

INTRODUCTION

The family Phocidae includes four subfamilies (Phocinae, Monachinae, Cystophorinae, and Devinophocinae), is extremely diverse in size, ranging from the smallest to the large-sized mammals in the superfamily Phocoidea. Phocids are morphologically well distinguished from other carnivorans in their extreme adaptations to an aquatic life, and they are least adapted to movement on hard substrates because of their inability to raise the trunk on the hindlimbs, and also because of their shortened and weak fore-flippers.

Phocidae separated from ancient Carnivora probably in the early Oligocene or before, then became widely distributed in the middle Miocene, and practically ceased to exist in Eastern North America in the early Pliocene. Phocidae includes 34 genera, of which 21 are extinct and 13 extant. The total number of extinct species is 25, which compares with (23) of the Recent species of the family.

The range of size differences among the various species of Phocidae is not particularly large. The smallest of them, some Phocinae, are rarely longer than 100 cm as adults and weigh up to 20 kg. The largest of the Phocidae (*Mirounga* in Cystophorinae) are up to 600 cm long and weigh 4060–5080 kg (4–5 tons). Sexual dimorphism is not obvious in most Phocidae (except *Mirounga*) and is mainly visible in coloration, body and skull dimensions, and other features; in general, males are slightly larger than the females. In the males some features of dimorphism are: processes of the nasal cavity form a trunk or a hoodlike swelling (Cystophorinae); more massive skull; more powerful dental apparatus, in particular the much larger canines; and some difference in color, size, and proportions of the body and postcranial skeleton (Koretsky, 1987a, 2001; Van Bree and Erdbrink, 1987).

Seals live in diverse climatic zones: Arctic and Antarctic (circumpolar), boreal and austral, and even subtropical. Depending on the season, they live in herds (often large and dense), in small isolated groups, or singly or in pairs (for short intervals). At present, seals are most numerous in the boreal and arctic regions of the Northern Hemisphere and in the temperate zone of the Southern Hemisphere, especially in the zones of convergence of cold and warm waters. A majority of the species is isolated in the regions of cold and moderately cold waters in which the surface temperature does not exceed 20 °C. The exceptions are primarily thermophilic seals of the genus *Monachus*, which are isolated in three regions of the subtropical belt (Mediterranean Sea, Caribbean Sea, and in the Hawaiian Islands). The seals are not found in the Indian Ocean, the Malayan archipelago, southwestern Pacific Ocean and the central parts of the Pacific Ocean, south and north of the Equator. The only exception is the small area around the Hawaiian Islands occupied by the relict small population of the Hawaiian monk seal. Many subpolar species living at the edge of ice have developed migratory patterns in response to seasonal benefits in food and protection.

Despite their superb adaptations to life in water, true seals do require a solid substrate on which to give birth, suckle the pups, molt, and simply rest. Some use only ice for these purposes (pagophilic species), while others select beaches, mostly of islands (pagophobic species). As a rule, the more pagophobic species use land (subtropical seals Monachinae and some west European Phocinae). Some pagophilic species (Phocinae) make round openings in the ice cover for breathing and ventilation, and for crawling onto the ice. Some burrow holes in the ice for themselves and for camouflaging their pups under the snow cover.

The different types of diet are reflected in the general organization of seals (Reidman, 1990). Many species of Phocidae are chiefly or partly dependent on fish, and the teeth in these aquatic predators are characterized by the sharpness of their sawlike cusps. In *Erignathus* (Phocinae), which feeds chiefly on benthic crustaceans, worms, and some shelled molluscs, the teeth become worn early and fall out.

The feeding habits of pinnipeds are not clearly differentiated. Some are mainly benthic feeders, among them some species of Phocinae; some are hunters of fish and cephalopods, and some eat

Evolution of Tertiary Mammals of North America, Vol. 2. ed. C. M. Janis, G. F. Gunnell, and M. D. Uhen. Published by Cambridge University Press.

planktonic crustaceans (Monachinae: *Lobodon*). The leopard seal (Monachinae: *Hydrurga*) feeds also on large fish and warm-blooded animals, including birds (penguins). Nevertheless, most of the seals survive on a mixed diet.

DEFINING FEATURES OF THE FAMILY PHOCIDAE

CRANIAL

The skull is similar in shape and size in all phocids. The calvarium of the skull is usually more or less flattened, and the sagittal crest is poorly developed or absent. The cerebral hemispheres are large and well developed. The facial portion of the skull is usually shorter, narrower, and lower than the cerebral portion. The eyes are large, well adapted for use at night and in deep or murky water, and situated well forward. The large orbits greatly reduce the dorsal skull surface in the interorbital region, and are joined widely with the temporal fossa.

With rare exceptions, the zygomatic arches are shifted laterally. The palate includes the anterior and posterior palatine fossae. The auditory bullae are large and relatively thick walled, highly swollen and rounded, but more flattened and complex than other pinnipeds. The bullae are composed of ento- and endotympanic, the latter forming the external auditory meatus. In contrast to most mammals, the lacrimal bone is absent, the fossa pterygoidea is not developed, and the alisphenoid canal is absent. The ethmoturbinal bones are small and arranged in five folds; the bones of the nasal concha (maxilloturbinal) are usually highly developed, large, and fill most of the nasal cavity.

The mandible is usually weak and thin, with a small angular process and long coronoid process; the symphysis is unfused. The mandibular condyle is semi-cylindrical with a saddle-shaped concavity in the middle. The mouth, jaws, teeth, and associated structures are developed for grasping and tearing, as opposed to chewing. The temporal and masseter muscles are reduced, with corresponding reduction of sagittal crest and postorbital processes. This tendency is less prominent in some Monachinae due to the development of strong jaws for combat purposes (Figure 32.1).

DENTAL

Phocids are heterodont but lack a carnassial notch, except in rare cases where the tooth in the carnassial position might have a rudimentary notch. The number of teeth varies from 30 (Cystophorinae) to 34 (Phocinae). The mandibular incisors are relatively small, sometimes very weak; the lateral incisors in the maxilla are more developed. The canines, especially the upper ones, are quite large. The cheek teeth are less differentiated than in other carnivores and usually flattened laterally. The structure of the crown is extremely diverse, in most cases with many cusps, but the main cusp is usually raised much above the others located mesially and distally. Some species have monocuspid teeth. The premolars (except p1, P1) and the molars usually have two roots.

POSTCRANIAL

The body is spindle shaped, narrowed toward both ends, with a poorly developed or undeveloped tail. The neck is thick, not sharply marked off from head or trunk, more muscular, and more flexible than in most Carnivora. The external genitalia and mammary teats are usually withdrawn beneath the contour of the body; the bases of the limbs, to the elbows and knees, are deeply enclosed with the body. The pinniped body is enveloped in thick subcutaneous fat (blubber). Blubber provides thermal insulation and reserve energy.

The fore- and hindlimbs are modified into flippers adapted for swimming, with only their distal portions projecting from the cylindrical trunk. The fore- and hindlimbs each have five digits. In some species, claws are well developed on the fore-flippers and, usually, on the hind-flippers. The phalanges bearing the claws are often slightly broadened and elongated, with distinct gaps for nail beds. The highly elongated digits of the fore and hind extremities are not separated externally into individual rays; the digits are covered with a web of skin, like a sheath, and set close together. The elasticity of the web assists in the powerful movement of the digits. When stretched, the web forms a broad fan-shaped surface resembling the marginate caudal fin of a fish. The first digit and the fifth are particularly elongated, and the fifth digit is also considerably broadened.

The scaphoid, lunate, and centrale bones are fused in the carpus. The ulna and radius are shortened but quite independent and not fused. The humerus is even more shortened but maintains a highly developed deltoid crest. The clavicle is not developed. The articular facet of the astragalus is a sharp crest. The tibia is relatively long; the fibula is independent for most of its length but fuses with the tibia in the proximal epiphysis. The femur is highly shortened, flattened, and broadened distally; the third trochanter is absent; the lesser trochanter is usually not developed. The vertebral column is extremely flexible with highly developed intervertebral discs. The left and right innominata are unfused, and connected by ligaments over a very short distance. Unlike the condition in most mammals, the acetabular fossa in pinnipeds faces laterally.

SYSTEMATICS

SUPERFAMILIAL

There has been continuing controversy (e.g., Kellogg, 1922; Howell, 1930; Mitchell, 1967; Berta, Ray, and Wyss, 1989; Repenning, 1990; Árnason *et al.*, 2006) over whether pinnipeds evolved from a single terrestrial ancestor (monophyletic origin) or had two independent terrestrial origins (diphyletic origin). This controversy is in some cases irrespective of any arguments about the validity of using an order, Pinnipedia. The "classic" theory of the diphyletic origin of pinnipeds proposes that true seals (Phocoidea or Phocidae) had a North Atlantic origin and are most closely related to musteloids, whereas sea lions and walruses (Otarioidea or Otariidae, sensu lato) had a North Pacific origin and are most closely related to ursids (Mivart, 1885; McLaren, 1960a; Mitchell, 1967; Tedford, 1976;

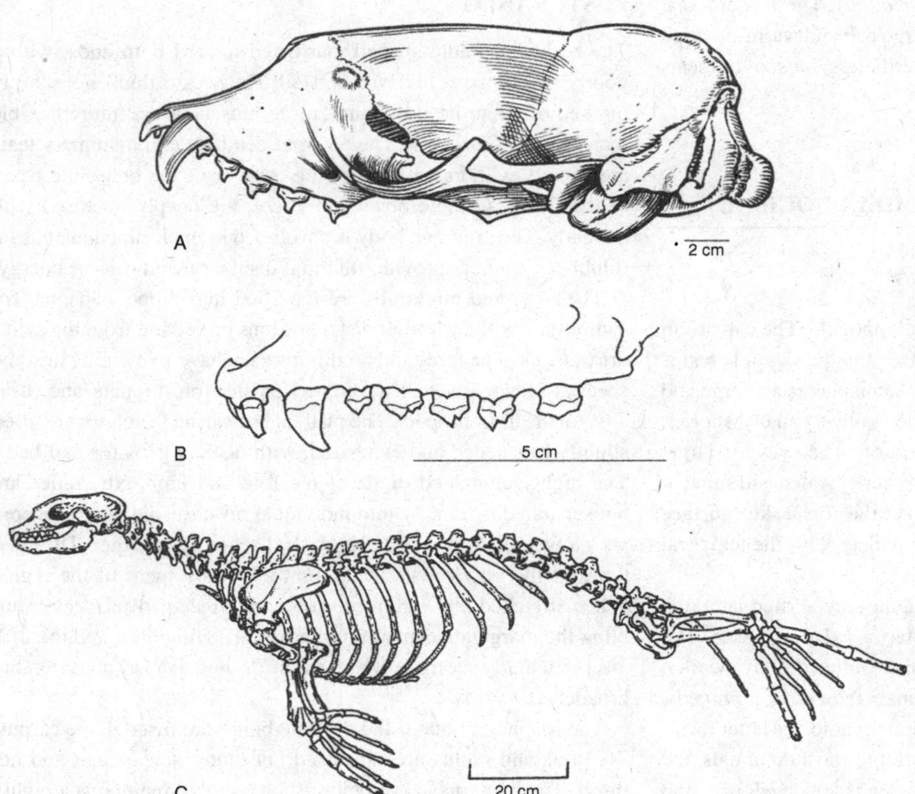

Figure 32.1. Skull, teeth and skeleton of phocids. A. Skull (*Leptonychotes weddelli*), in left lateral view. B. Upper dentition of *Phoca vitulina* in left lateral view (modified from King, 1983). C. Skeleton in lateral view (*Monachus tropicalis*).

Muizon, 1982a,b). Both major pinniped groups are recognized as being derived from among the terrestrial arctoid carnivorans, not from among the groups containing cats, viverrids, or dogs (Tedford, 1976; Árnason, 1977; Árnason *et al.*, 2006).

The monophyletic hypothesis contends that all pinnipeds derive from a single terrestrial arctoid (Simpson, 1945; Davies, 1958a,b; Scheffer, 1958; Árnason, 1977; King, 1983; Wiig, 1983), usually with ursids being the most likely sister group (Wyss, 1987; Flynn, Neff, and Tedford, 1988; Berta, Ray, and Wyss, 1989). Proponents of this view usually recognize a suborder or other higher taxon called the Pinnipedia or Pinnipedimorpha.

McLaren (1960a), Tedford (1976), and Repenning and Tedford (1977) reviewed the evidence for relationships of both otariid and phocid seals within the context of the order Carnivora. They summarized evidence that otariids had an origin from terrestrial fissiped carnivores separate from that of phocids. Tedford (1976) classified both phocids and otariids in the infraorder Arctoidea, but accepted pinniped diphyly. He classified the Otariidae (including Odobeninae) in the parvorder Ursida with bears, and the Phocidae with Mustelidae in the superfamily Musteloidea, in the parvorder Mustela. Later Wiig (1983) rejected Tedford's hypothesis (1976) concerning a sister-group relationship of the otariids with the ursids and the phocids with the mustelids. Tedford also recognized the sea lions and walruses in a pinniped group, with variously related fossil relatives, and it is now commonly accepted that the group is monophyletic. (See discussion in Ch. 30.)

INFRAFAMILIAL

For many years, there was discussion about the relationships between the Phocinae, Monachinae, and Cystophorinae, but a skull from Bratislava of *Devinophoca*, which is the most primitive, least specialized, and oldest Phocinae from Central Paratethys (Vienna Basin), helped to solve this problem (Koretsky and Holec, 2002: Fig. 9).

Regrettably, up to the present there has been no clear conception of the relationships within subfamily Monachinae. In the past, in accordance with the classification of Trouessart (1897), the subfamily contained the genera *Lobodon* Gray, 1844; *Ommatophoca* Gray, 1844; *Hydrurga* Gistel, 1848; and *Leptonychotes* Gill, 1872. However, in the opinion of Simpson (1945), *Lobodon* and all the other genera mentioned above should be placed in the subfamily Lobodontinae. This is still a controversial problem. One group of investigators considers that these genera should be assigned to one subfamily (Ognev, 1935; Grassé, 1955; King, 1964, 1983; Muizon, 1982a,b) while others (e.g., Hendey, 1972) separate them into two subfamilies. Finally some investigators (Sokolov, 1979; Pavlinov and Rossolimo, 1987; Wozencraft, 1989) do not separate true seals (Phocidae) into subfamilies.

The prevailing division of the family into subfamilies is extremely simple. Four subfamilies that include Recent and extinct genera are usually recognized, based on the number of incisors: (1) Phocinae, 10 incisors (I 3/2); (2) Monachinae, eight incisors (I 2/2); (3) Cystophorinae, six incisors (I 1/2); and (4) extinct subfamily

Devinophocinae (I 3/?), with mixture of the characters of three previously mentioned subfamilies. Chapskii (1955, 1961, 1971, 1974) provided a comprehensive analysis of suprageneric systematics. He clearly described diagnostic cranial characters forming the basis of the separation of true seals into three subfamilies: Phocinae, Monachinae, and Cystophorinae, which he, in turn, divided into tribes and subtribes. We refer to Chapskii (1974) and Repenning and Ray (1977), where a detailed analysis is presented; King's (1966) hypothesis proven untenable and is not followed in King (1983).

According to King's (1966) hypothesis, the genus *Cystophora* should be transferred from the subfamily Cystophorinae into the subfamily Phocinae, and the genus *Mirounga* into the subfamily Monachinae. It may be assumed that Muizon (1982a) was unaware of the study of Chapskii (1974), because he accepted the systematics of King (1964, 1966) without any reservations. Consequently, he returned to the concept of subdivision of the subfamily Cystophorinae. Evidence of the correctness of Chapskii's concept was provided in a comparative study of the pinniped calcaneum (Robinette and Stains, 1970). We support Chapskii's view on the subfamily Cystophorinae as being in its classical position.

We have included in our matrix character–state data for 25 species of fossil and modern phocids and data for three outgroup taxa. These outgroup taxa are the fossil otarioids *Allodesmus kelloggi* and *Enaliarctos emlongi*, and the Recent mustelid *Lutra canadensis*, reflecting the competing hypotheses of pinniped relationships: monophyly (Wyss, 1987, 1988a,b; Flynn, Neff, and Tedford, 1988; Berta, Ray, and Wyss, 1989; Berta and Wyss, 1990; 1994; Wyss and Flynn, 1993; Berta and Sumich, 1999) or diphyly (McLaren, 1960a; Mitchell, 1966, 1968, 1975; Repenning, 1975, 1976; Tedford, 1976; Hunt and Barnes, 1994; Bininda-Emonds and Russell, 1996; Koretsky and Sanders, 2002). Points where the nodes of the present tree correspond to traditionally recognized phocid taxa are indicated (Figure 32.2).

INCLUDED GENERA IN THE PHOCIDAE GRAY, 1821

The locality numbers listed for each genus refer to the list of unified localities in Appendix I. The locality numbers may be listed more than one way. The acronyms for museum collections are listed in Appendix III.

Parentheses around the locality (e.g., [CP101]) mean that the taxon in question at that locality is cited as an "aff." or "cf." the taxon in question. Parentheses are usually used for individual species, implying that the genus is firmly known from the locality, but the actual species identification may be questionable. Question marks in front of the locality (e.g., ?CP101) mean that the taxon is questionably known from that locality, implying some doubt that the taxon is actually present at that locality, either at the genus or species level. An asterisk (*) indicates the type locality.

PHOCINAE GILL, 1866

Characteristics: Seals of small and medium size; 10 incisors (I3/2); mastoid very pronounced, narrow, cylindrical, width not greater than half the length of the tympanic bulla; bulla directed sharply downward behind the mastoid process; maxilla immediately swollen in front of orbit, lateral contour convex; anterior palatal foramen well developed, with a more or less pronounced groove-like shape (from Chapskii, 1974; Koretsky, 2001).

PHOCINI CHAPSKII, 1955

The species in the tribe Phocini are small- or medium-sized Northern Hemisphere seals related to *Phoca* (the extant Harbor seal). Barnes and Mitchell (1975) reported various isolated latest Pliocene and Pleistocene phocid bones from North America West Coast that closely resemble those of Harbor seals and identified them as *Phoca*, cf. *P. vitulina*. These bones are rare and, if correctly identified, constitute the only phocine fossils from the eastern North Pacific. The only fossil Phocidae from the western North Pacific, identified as *Phoca*, were found in late Pleistocene cave deposits near the northern tip of Honshu, Japan (Hasegawa *et al.*, 1988). Such records indicate that Harbor seals entered the North Pacific in latest Pliocene time and probably not before. The phocid record is much older in the Atlantic and is consistent with theories that this is the area of origin and primary evolution of the group.

Gryphoca Van Beneden 1876 (including *Halichoerus*, in part)

Type species: *Gryphoca similis* Van Beneden 1876 (including *Halichoerus* [*Gryphoca*] *similis* Trouessart, 1897, 1904; Friant, 1947).

Type specimen: IRSNB 1081, Ct. M. 217 (original), USNM 10359 (cast); left humerus (see Van Beneden, 1877).

Characteristics: Medium size. Bones of postcranial skeleton of *Gryphoca similis* very similar in size and dimensions to those of extant *Halichoerus grypus* (gray seal); deltoid crest of humerus narrow and short, terminating hardly below half the length of the bone; thin in the parasaggital plane, lesser tubercle located slightly above head and greater tubercle, head round, diaphysis swelling on inclined almost parallel to the deltoid crest; scapular tuberosity well developed, curving only a short way below glenoid cavity, glenoid cavity broad but abraded medially, neck of the scapula wide not clearly defined, supraspinous fossa deeper and narrower than infraspinous fossa; caudal border of scapula very thick; ilium thickened and only slightly excavated on exterior surface; greater ischiatic notch shallowly concave, caudal dorsal iliac spine strongly developed, forming a robust articular surface, iliopectineal eminence elongated caudally, extending approximately half the length of the body of the ilium, edges of the acetabular fossa rise sharply above the body of the bone, acetabulum with a weakly marked cotyloid notch and acetabular notch, eminence for the gluteus medius muscles well developed and ending nearly at the same level as the iliopectineal eminence but degree of development of this eminence varies individually; ischium and pubis; not preserved greater trochanter of femur is higher than

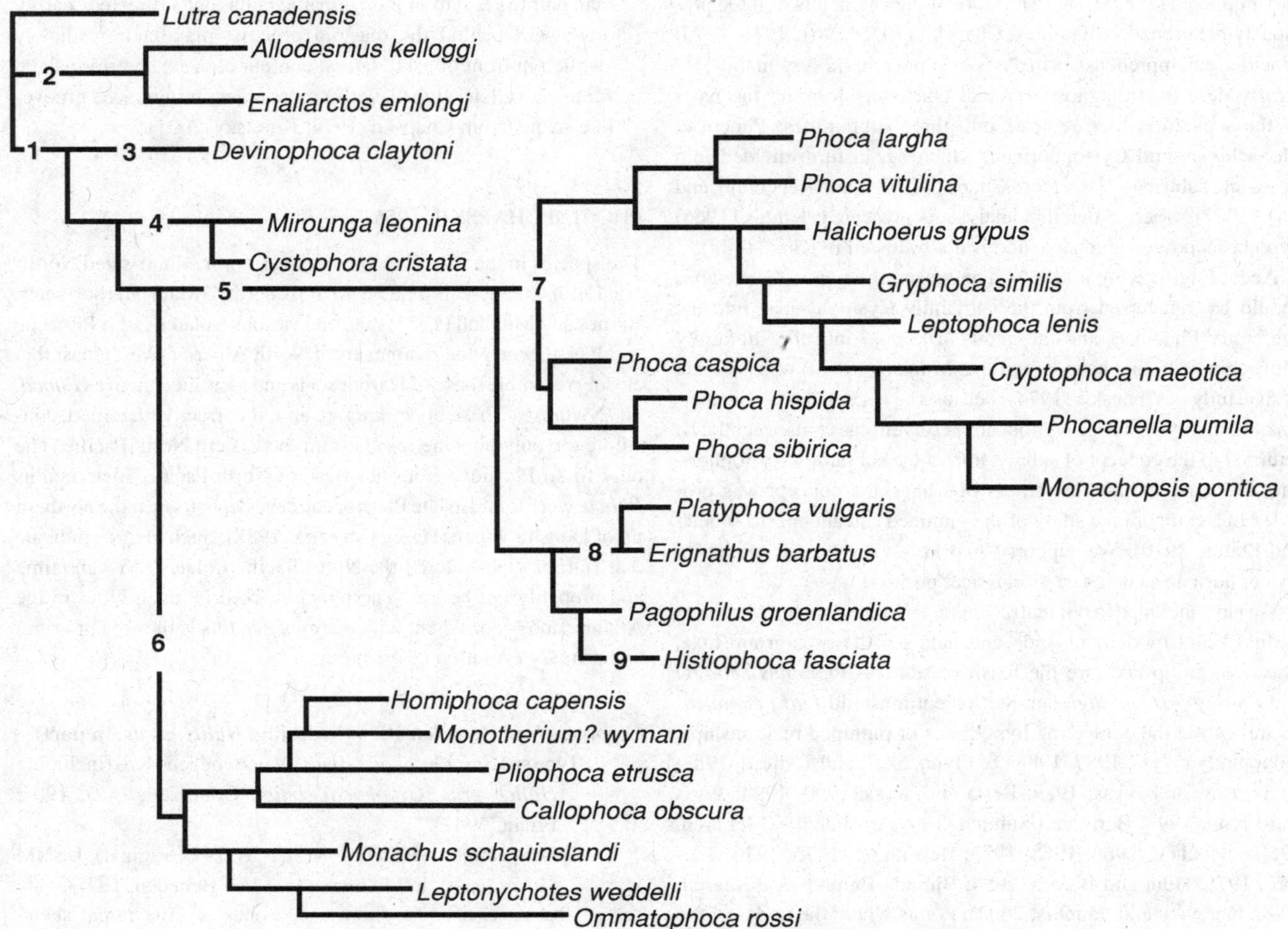

Figure 32.2. The mh*; routine in Hennig86 (Farris, 1988) produced two maximally parsimonious trees, each 295 steps long with a consistency index of 0.55 and a retention index of 0.48. Use of the successive-weighting option in Henning 86 reduced the number of trees from two to one, leaving part of the tree much better resolved. The nodes of the cladogram are supported by the following character transformations. 1. PHOCIDAE: relative dimensions of postcanine teeth compared with the size and massivity of the skull; absence of lesser trochanter of femur; width of proximal and distal epiphyses about equal, or distal wider than proximal by 20–25 %. 2. OTARIIDAE: one branch forms the possibly paraphyletic subfamilies. 3. DEVINOPHOCINAE (paraphyletic): maxilla with long concavity anterior to orbit (shared with Monachinae); width of interorbital area equal or greater than 25% of mastoid width (shared with Cystophorinae); absence of basal cingulum on postcanine teeth (shared with Cystophorinae and Phocinae). 4. CYSTOPHORINAE (paraphyletic): symphyseal part of the mandible reaching only to the alveolus of p1 (autapomorphy); lesser tubercle of humerus oval-shaped (shared with *Halichoerus* and *Erignathus*). 5. PHOCINAE plus PHOCINI: diameter of the infraorbital foramen equal to or greater than diameter of alveolus of upper canine; lesser tubercle of humerus pronounced; coronoid fossa shallow; greater trochanter of femur enlarged and triangular. 6. MONACHINAE (paraphyletic): maxilla with long concavity (shared with Devinophocinae); anterior palatine foramina indistinctly marked; middle of internal crest of the humeral trochlea raised archlike over coronoid fossa; intertrochanteric crest of femur absent. 7. PHOCINI, in part (paraphyletic): anterior palatal foramina round and deep; alveoli of p4 larger than alveoli of m1; upper incisors arranged in a straight line (shared with *Cystophora*). 8. ERIGNATHINI; lesser tubercle of humerus oval shaped (shared with *Halichoerus* and *Cystophora*). 9. Genus *Histriophoca*: alveoli shallow; diastemata unequal (shared with *Praepusa* and *Monachopsis*).

head, smallest width of diaphysis located in middle part of bone, two condyles well developed and relatively large, intercondylar fossa very broad, not deep, two epicondyles considerably thinned (proximal part of *Gryphoca similis* femur, hence no description available); fragmentary, rolled and abraded; two condyles of tibia and fibula elongated and ellipsoidal, intercondyloid eminence very high, with well-developed borders, popliteal notch very narrow, tibial crest flattened dorsomedially, muscular groove flattened and broad, distal articular surface almost round, not deep, groove for tendon of muscularis flexor digitorum longus caudal to medial malleolus, deep and narrow with elevated medial border forming a distinct crest.

Included species: *G. similis* only, known from locality (EM8B) only (Ray, 1976a).

Comments: *G. similis* is a late Pliocene phocine from Europe (Scaldisian, first found near Antwerp, Belgium) and the eastern coast of North America. Van Beneden (1877) allied it with *Halichoerus*; Simpson (1945) classified it in the

Phocinae, but Ray (1976a) regarded it as closely related to *Phocanella pumila* Van Beneden, 1877.

This rare taxon from Van Beneden's collection is represented only by a few fragmentary and rolled bones, but with very distinctive morphological features. The situation is similar for the remains of this species at the Lee Creek Mine (locality EM8B). Nevertheless, we are reasonably sure about the assignment of these bones from eastern United States to *Gryphoca similis*, and that this name represents an identifiable species.

Leptophoca True, 1906

Type species: *Leptophoca lenis* True, 1906.

Type specimen: USNM 5359; right humerus (see True, 1906).

Characteristics: Phocine of medium size, with total skull length near 235 mm; upper incisors forming U-shaped arcade; P2–M1 double-rooted with posterior alveoli larger than anterior; p4, P4 larger than m1, M1; cheek teeth except p1, P1 with three or more cusps; diastemata present between teeth; preorbital part of maxilla with long, pronounced convexity; small antorbital process present on anterior margin of orbit; frontal contact of nasal bones twice as long as maxillary contact; interorbital space narrowing between anterior part of orbits and much narrowed in most posterior part of interorbital area; interorbital width 11.4% of width of skull at mastoid processes; sagittal crest beginning at posterior end of orbit and becoming deeper at middle of braincase; infraorbital foramen not visible in dorsal view; diameter of infraorbital foramen equal to diameter of alveolus of upper canine; palatal process of maxilla flat; anterior palatal foramina oval and deep; palatal groove shallow but well defined; anteroposterior length of tympanic bulla less than smallest distance between bullae; jugular process well developed; width of mastoid process less than half length of tympanic bulla; mastoid convexity not turned down behind mastoid process; connection present between zygomatic process of squamosal and mastoid process; body of mandible swollen and thick, symphyseal part not pronounced; chin prominence absent; ramus of mandible thin and low; cheek teeth aligned parallel to axis of mandible with equal diastemata; alveoli of p4 equal in size to alveoli of m1. Lesser tubercle of humerus distal to proximal part of deltoid crest and head; head compressed craniocaudally; deltoid crest extending less than two-thirds of humeral length; maximum width of deltoid crest in its proximal end; lateral epicondyle reaching distal part of deltoid crest; greater trochanter of femur higher than head, proximal part wider than distal, distinct lesser trochanter located far below distal border of greater trochanter; trochanteric fossa deep, wide, and covered medioproximally; head large, seated on narrow, short neck; minimum width of diaphysis shifted proximally; greatest breadth across condyles 65.0–66.0% of bone's length.

Included species: *L. lenis* only (known from localities NC3A*, B–D, [True, 1906; Bohaska, 1998]).

Comments: The name *Leptophoca* is widely known to researchers (Koretsky, 2001), but material of this genus has heretofore been insufficient, except for the humerus, radius, and a fragment of the innominate. Dr. Clayton Ray has assembled additional material collected over the last 30 years. As a result, the USNM collection now includes an almost complete associated skeleton of this genus, which is very rare for any fossil seal taxon.

The species that was originally called *Prophoca proxima* Van Beneden, 1876, from middle Miocene deposits in Belgium, was considered by Ray (1976a) to be closely related to *Leptophoca*. Van Beneden (1876) had recognized the very primitive nature of *P. proxima* when he named it.

Phoca Linnaeus, 1758

Type species: *Phoca vitulina* Linnaeus, 1758, the extant harbor seal.

Type specimen: None designated.

Characteristics: Condylobasal length of adult skull exceeding 200 mm; interorbital width twice the diameter of infraorbital foramen; longitudinal diameter of alveolus of maxillary canine 1.5–2.0 times more than maximal width of infraorbital foramen; total width of nasal bones (at level of frontomaxillary suture) not less than 20.0% of total length of skull; length of tympanic bulla less than distance between them; foramen ovale not covered by tympanic bulla; teeth with three or four cusps, these additional cusps weakly developed; symphyseal part of mandible blunt and rounded, alveolar part massive and swollen; chin prominence weakly pronounced, extending from anterior alveolus p3 to posterior alveolus p4; tooth alveoli rounded, with equal diastemata between them. Deltoid crest of humerus longer than half length of bone, maximal width at proximal end; lesser tubercles considerably higher than head and proximal part of deltoid crest; index of head's height (ratio of head's width to its height), 0.83; lateral supracondylar crest strongly developed, reaching level of distal end of deltoid crest; greater trochanter of femur considerably higher than head, proximal end slightly wider than distal; trochanteric fossa opening medially, not deep; head of femur not bending in distal direction, attached to short, wide neck; maximal distance between epicondyles 50.0–54.0% of bone's length (Koretsky, 2001).

Included species: *Phoca* sp. (known from locality WM26IV [Barnes and Mitchell, 1975]); *P.* cf. *vitulina* (locality [WM8II] [Barnes and Mitchell, 1975]).

Comments: *Phoca* is also known from Pleistocene deposits of the East and West Coasts, as well as the St. Lawrence seaway. Fossil *Phoca* are also known from localities around the North Atlantic. The genus *Phoca* is also known from the Recent of Europe (possibly also from the Pleistocene),

Asia (Lake Baikial and the Caspian Sea), and the Arctic (McKenna and Bell, 1997).

***Phocanella* Van Beneden, 1877**

Type species: *Phocanella pumila* Van Beneden, 1877.

Type specimen: IRSNB 1080, Ct. M. 227 (original), USNM 10358 (cast); right humerus (see Van Beneden, 1877).

Characteristics: Humerus and femur of *Phocanella pumila* similar to those of modern *Phoca largha* (spotted seal); intertubercular groove of humerus very deep, rather than narrow; deltoid crest terminating at half of length of bone; maximal width of deltoid crest near middle of its length, below the greater tubercle; lesser tubercle well developed and located considerably proximal to the humeral head and greater tubercle; head compressed proximodistally; radial groove well expressed; medial epicondyle well developed and extending a little bit further distally than the lateral; proximal part of lateral epicondyle flat, tending to a well-developed supinator crest; entepicondylar foramen small, with a wide bridge over it; olecranon fossa not deep, but flat; coronoid fossa not deep and forming a rounded triangular depression extending somewhat more proximal than the medial epicondyle; radial tuberosity relatively prominent, big, and almost round; articular circumference very small; neck relatively wide; grooves for all tendons shallow; insertion of muscularis pronator teres weak, mostly absent; groove for tendon of muscularis abductor pollicis longus narrow and deep; groove for extensor digitorum communis shallow. Ilium thin; iliac crest only slightly averted and excavated on exterior surface; iliac tuberosity and caudal dorsal iliac spine very well developed compared with size of bone; greater ischiatic notch very short and concave; a deep fovea occurring at the level of caudal dorsal iliac spine on the body of the ilium varies in its degree of development; edges of acetabular fossa rising above surface of ilium body; acetabulum deep, circular, with well-marked cotyloid notch; eminence for gluteus medius muscle triangular in shape, poorly developed and terminating more posteriorly than the iliopectoral eminence. Ischium and pubis are not preserved. Femur greater trochanter height greatly exceeds height of femoral head; greater trochanter triangular with distal part narrower than proximal; trochanteric fossa large and open; intertrochanteric crest relatively strongly developed, and prolonged almost to the medial border of the bone distal to neck of femoral head; femoral head seated on a narrow, long neck; smallest width of the diaphysis located in middle of bone; medial condyle extending considerably distal to lateral condyle. Popliteal notch and muscular groove of tibia and fibula very flattened and broad; popliteal notch on dorsal side of tibia very broad but well marked and considerably deepened; tibial crest not strongly expressed; two condyles flattened and round; intercondyloid eminence weak and only slightly raised above the surfaces of the two condyles.

Included species: *P. pumila* only (known from localities GC13B [Morgan, 1994], EM8B [Ray, 1976a].)

Comments: *Phocanella* originally included two named species, *P. pumila* Van Beneden, 1877, and *P. minor* Van Beneden, 1877, from middle(?) Pliocene deposits of Belgium. Ray (1976a) reported *P. pumila* from the earliest Pliocene Yorktown Formation on the East Coast of North America. Van Beneden (1877) had allied both species with *Pusa hispida*. Simpson (1945) placed them in the Phocinae, and Ray (1976a) considered them to be generalized phocines near *Phoca*. *P. minor* is a synonym of *P. pumila*, and is closely related to *Phoca largha*. In the same manuscript, another species, *Phocanella couffoni* Friant, 1947, was interpreted as a nomen dubium. It had been named based on a very poorly preserved femur from France. Savage and Russell (1983) included in the genus *Phocanella* the species *Phoca vitulinoides* but did not provide an explanation for the transfer of this taxon. In our opinion, there is no basis for placing *Phoca vitulinoides* in the genus *Phocanella*.

ERIGNATHINI CHAPSKII, 1955

McLaren (1960b) considered the extant bearded seal, *Erignathus barbatus* (Erxleben, 1777), to represent a highly specialized phocine lineage that had become separated very early in its evolutionary history from the other Phocinae. King (1966) concluded that *Erignathus* occupies a primitive position among the Phocinae and shares certain similarities with the Monachinae, especially with *Monachus*. Repenning (1983) reported a few bones of this genus from Pleistocene deposits in Alaska. It is unique among the Phocinae in not having the anterior end of the ilium bent laterally (a derived character of all other Phocinae [King, 1966; Hendey and Repenning, 1972]). Chapskii (1955) and later King (1964) recognized the unique nature of *Erignathus* by classifying it in its own tribe Erignathini, as we do here.

***Platyphoca* Van Beneden, 1877 (including *Phoca*, in part)**

Type species: *Platyphoca vulgaris* Van Beneden, 1877 (including *Phoca* [*Platyphoca*] *vulgaris*, Trouessart, 1904; Friant, 1947).

Type specimen: IRSNB 1117, Ct. M. 210 (original), USNM 10354 (cast); left humerus (see Van Beneden, 1877).

Characteristics: Phocinae of large size. Intertubercular groove of humerus very shallow and not well defined, undoubtedly owing in part to absence of the proximal epiphysis; deltoid crest very short, terminating at less than half of the length of the bone (extrapolating for the missing proximal epiphysis), appearing less prominent than it would in a complete, mature humerus, owing to loss of the traction epiphysis extending along its crest distally from the proximal epiphysis; crest passing from base of lesser tubercle, parallel to deltoid crest, along medial surface of humerus only slightly shorter than deltoid crest;

radial groove absent; two epicondyles strongly developed and very wide, imparting a spatulate character to the distal end of the bone, lateral epicondyle extending further proximally than medial; distal parts of both epicondyles flat; entepicondylar foramen large, almost round, with a broad, flat bridge over it; entire bone unusually long, relatively slender, and straight, not bowed medially as is typical of Phocinae; entepicondyle with isosceles triangle shape, apex at its medial extremity (in typical Phocinae entepicondyle is an irregular polygon with medial extremity situated more proximally); coronoid fossa very big, extending proximally further than the medial epicondyle, almost as far as the lateral epicondyle; olecranon fossa situated transversely, very deep throughout width; trochlea of humerus relatively high and prominent. Ilium body very broad; iliac crest very well marked; ischiatic notch considerably shortened and expanded laterally; iliopectineal eminence extended medially and elongated, to approximately one-third length of ilium body; edges of acetabular fossa well expressed, very sharp; acetabulum not deep, the lunate surface not being perpendicular to acetabular fossa; cotyloid notch and acetabular notch well marked and strongly developed; iliac fossa shallow, wide, not high. Ischium and pubis have not been identified in the collections.

Included species: *P. vulgaris* only, known from locality EM8B only (Ray, 1976a).

Comments: The lectotype of *P. vulgaris* is from the Pliocene (Scaldisian), Borgerhout Third Section, Antwerp Basin, Belgium. Besides the type locality, *Platyphoca* is also known from the Pliocene eastern United States and was classified by Simpson (1945) in the Phocinae. Ray (1976a) considered it to be a large, aberrant phocine and agreed with Van Beneden's (1877) conclusion that it is related to *Erignathus*. This intriguing species has been virtually ignored in the literature except for listing in a few comprehensive works, where *P. vulgaris* is listed among species from the Pliocene of Europe (e.g., Trouessart, 1904; Kellogg, 1922; Simpson, 1945; King, 1964; Muizon, 1992).

MONACHINAE GRAY, 1869

MONACHINAE INCERTAE SEDIS

Monotherium Van Beneden, 1876 (including *Phoca*, in part)

Type species: *Monotherium delognii* Van Beneden, 1876.

Type specimen: None designated.

Characteristics: Particularly remarkable in the length of the bodies of the vertebrae; *Monotherium delognii* somewhat similar to *Phoca barbata* (Van Beneden, 1876).

Included species: *M. aberratum* (known from locality NC2A [Ray, 1976b]); *M. affine* (locality NC5 [Ray, 1976b]; *Monotherium*? *wymani* (locality NC3E [Leidy, 1853]).

Comments: Ray (1976b) tentatively moved *Phoca wymani* to the genus *Monotherium*. King (1956) emphasized the generally primitive structure of the genus *Monachus* relative to the Phocinae. *Monotherium*? *wymani* (Leidy, 1853) is the most primitive monachine. The species may belong in a different genus and its occurrence suggests that the subfamily originated in the Northern Hemisphere from Phocinae. *Monotherium* is also known from the Miocene of Europe (McKenna and Bell, 1997).

MONACHINI (GRAY, 1869) SCHEFFER, 1958

The tribe Monachini has been recognized by several authors and appears to be a natural grouping among the Monachinae. Hendey and Repenning (1972) concluded that *Monachus* has a primitive ear structure compared with that of other phocids. The genus *Monachus* has three species. King (1956) concluded that the recently extinct Caribbean *Monachus tropicalis* Gray, 1850 and the extant Hawaiian *Monachus schauinslandi* Matschie, 1905 are more closely related to each other than either is to *Monachus monachus* Hermann, 1779. Repenning and Ray (1977) concluded that *M. schauinslandi* is the most primitive species of *Monachus*.

Pliophoca Tavani, 1941–2 (including *Pristiphoca*; *Phoca*, in part; *Monachus*, in part)

Type species: *Pliophoca etrusca* Tavani, 1941–2 (="Scheletro fossile di Foca"; Ugolini, 1900, p. 147) (including *Phoca occitana*, Gervais, 1853, 1859; *Pristiphoca occitana*, Gervais, 1859; Kellogg, 1922; Tavani, 1941–2; King, 1960; Muizon, 1992; *Pristiphoca occitanica*, Toula, 1897; Trouessart, 1897; *Pristiphoca occitanicus*, Trouessart, 1904, 1905; *Monachus albiventer*, Ugolini, 1902; Muizon, 1992).

Type specimen: Museo Geopaleontologico della Università di Pisa; Italy; major part of associated skeleton. Holotype, plus referred specimens from same beds in Museo Geo-Paleontologico di Firenze (Tavani, 1942–3).

Characteristics: Bones of the skull and extremities of *Pliophoca etrusca* are very similar in their dimensions to those of modern *Monachus monachus*. Palatal process of maxillary bone of skull flattened; infraorbital foramen visible from above, compressed dorsoventrally; palatal groove well pronounced along total length, reaching lateral notch of palatal bone; tympanic bulla mostly broken away; opening of auditory tube (meatus acousticus) round, relatively large, with small obliquely extended ventrorostrum by a flat wide groove. Mandibular fossa very broad and flattened; I1 shortened, narrower than I2; i1 and i2 seated on same level in mandible (as in *Callophoca obscura*), body of mandible swollen but not high; retromolar space elongated; ramus very strong and high, especially condyloid process; condyloid crest very well developed, as the masseteric fossa; mandibular notch flattened, not deep; symphyseal part of mandible very large, reaching the anterior alveoli of p3; chin prominence usually located between p2 and p3, but sometimes failing to reach posterior alveoli of p2; diastemata between teeth absent in immature individuals, in contrast to mature individuals; largest diastemata P1–2 and P4–M1; crowns of teeth broad in immature but worn flat with age in mature individuals; lower canines not large, smaller than upper canines, round in cross section; cheek tooth row oriented slightly oblique to axis of symphyseal part of mandible body; alveoli of posterior roots of cheek teeth slightly compressed anteriorly; widest dimension at more oblique angle to the axis of the dentary than that of anterior roots; p2–m1 five cuspid, double-rooted; basal cingula very well developed, especially on lingual side; m1 alveolus shorter in length than that of p4; alveoli of p1 and P1 very large relative to other cheek teeth. Glenoid cavity of scapula oval, only posterocranial margin of the cavity slightly flattened; coracoid process not developed, a small scapular tuberosity replacing it; infraarticular tuberosity concave, not reaching upper angle of glenoid cavity; angle between scapular spine and glenoid cavity less than 90° (caudal part of scapula unknown as our material is very fragmentary). Proximal part of deltoid crest of humerus located below the lesser tubercle but higher than the head; lesser tubercle spherical; intertubercular groove width and slightly concave; head relatively large and spherical; deltoid crest well developed but relatively short, overhanging humerus shaft and terminating a little below the midshaft; lateral epicondylus broad and shorter than medial epicondylus; trochlea humeri very small; radial groove and olecranon fossa flat; radial tuberosity relatively small and flattened, oval, and pulled out of the axis of the bone; articular circumference higher on radial aspect than on lateral one; grooves for all tendons very shallow; insertion of muscularis pronator teres weak, mostly absent. Radial notch of ulna absent, but ridge of ulnar tuberosity well developed; anconeal process relatively wide, high, and square; coronoid process extending with the angle to axis of the bone; ulna of *Pliophoca etrusca* with a concavity of the trochlear notch on medial aspect of the bone, but lateral aspect is prominent; interosseous crest very large and prominent; olecranon not preserved. Epiphysis of femur well developed; diaphysis relatively short and narrow; greater trochanter rectangular and on same level as head; head considerably smaller than bony mass, strongly bent in a cranial direction, and on a wide, short neck; intertrochanteric crest not developed; trochanteric fossa shallow and wide; minimum width of diaphysis located in middle part of bone; condyles relatively large and widely spaced.

Included species: *P. etrusca* only, known from locality EM8B only (Ray, 1976a).

Comments: *Pliophoca etrusca* has not been widely cited in the literature, in spite of the excellent material and good description available. *Pliophoca* is also known from the Pliocene of Italy (Tavani, 1942–1943). Only a few authors (Simpson, 1945; Hendey, 1972; Ray, 1976a; Muizon and Hendey, 1980) have discussed the phylogenetic relations of this species or its taxonomic status. Abundant material of this species from the Yorktown Formation in North Carolina (locality EMB8) includes cranial and postcranial bones.

A seemingly related species is *Messiphoca mauretanica* Muizon, 1981a, from the late Miocene of Algeria. According to Muizon (1981a), this species is close to the origin of *Pliophoca* and *Monachus*.

***Callophoca* Van Beneden, 1877 (including *Mesotaria; Paleophoca*; *Platyphoca*, in part; *Phoca*, in part)**

Type species: *Callophoca obscura* Van Beneden, 1877 (*Mesotaria ambigua*, Van Beneden, 1876 [nomen nudum]; *Platyphoca vulgaris*, Van Beneden, 1876 [nomen nudum]; *Paleophoca nystii*, Van Beneden, 1876 [van Beneden, 1877; Kellogg, 1922; King, 1964]; *Palaeophoca nystii*, Hendey, 1972 [Hendey and Repenning, 1972]; *Paläophoca nystii*, Toula, 1897; *Palaeophoca nysti*, Allen, 1880 [Tavani, 1941–2]; Misonne, 1958; *Paloeophoca nysti*, Allen, 1880; *Callophoca obscura*, Van Beneden, 1877 [Dollo, 1909; Kellogg, 1922; Misonne, 1958; King, 1964; Ray, 1976a; Repenning, Ray, and Grigorescu, 1979; Muizon, 1981a, 1982a, 1992; Savage and Russell, 1983; Barnes, Domning, and Ray, 1985]; *Phoca* (*Callophoca*) *obscura*, Trouessart, 1897 [Troussart, 1904; Friant, 1947]; *Mesotaria ambigua*, Van Beneden, 1877 [Trouessart, 1897, 1904, 1905; Dollo, 1909; Kellogg, 1922; Friant, 1947; King, 1964; Misonne, 1958; Ray, 1976bb; Muizon, 1982a, 1992]; *Callophoca ambigua*. Ray, 1976a [Repenning, Ray, and Grigorescu, 1979; Muizon, 1982a, 1992; Savage and Russell, 1983]; *Platyphoca vulgaris*, Van Beneden, 1877 [Dollo, 1909; Kellogg, 1922; Misonne, 1958; King, 1964; Ray, 1976b; Muizon, 1982a, 1992; Savage and Russell, 1983]; *Phoca* (*Platyphoca*) *vulgaris*, Trouessart, 1897, [Trouessart, 1904, 1905; Friant, 1947]).

Type specimen: IRSNB 1198, Ct. M. 203 (original), USNM 10434 (cast); right humerus (from Deurne), deltoid crest absent (see Van Beneden, 1877; identified as *Callophoca obscura*).

Characteristics: Bones of *Callophoca obscura* are similar to those modern *Mirounga leonina* (Southern elephant seal) in dimensions. The fragment of skull belongs to a juvenile individual, as the teeth are not worn, and some sutures are still open. Maxilla with very well-developed concavity anterior to the orbits; nasal bones separated; incisors flattened; alveoli of I2 considerably wider than those of I1; incisors arranged in a shallow semi-circle; tooth row forming a straight line and considerably shortened; canines very large; roots of most teeth exhibiting a tendency to fusion; P1, p1 single-rooted; other teeth double-rooted; diastemata between teeth unequal; alveoli all rounded, and of same size, except posterior alveoli of m1, which are considerably smaller; tooth roots very large, but crowns become obliterated with age; cusps of teeth strongly demarcated; tooth crown of immature individuals carrying one big middle cusp (paraconid), one front and one rear cusp equal size and lying on the same level; third cusp seated on the basal cingulum, relatively smaller than other two, very small on m1; cingular cusp smaller on upper than on lower teeth. Anterior palatine foramina flattened and oval; palatal grooves weakly pronounced along entire length, starting from posterior part of p1; maxillary process of zygomatic bone clearly visible; lacrimal process of zygomatic bone reaching infraorbital foramen; jugular process failing to approach border closely to mastoid process, and with a convexity on its anterolateral part; symphyseal part very large, reaching anterior alveoli of p3; chin prominence absent; retromolar space shortened; I1 somewhat shortened and narrower than I2; i2 lies behind i1, near the canine in the mandible. Glenoid cavity of scapula oval triangular in shape; infra-articular tuberosity high and flat, reaching upper angle of glenoid cavity; well-developed acromion perpendicular to scapular spine; angle between scapular spine and glenoid cavity less than 90°; no recognizable coracoid process present. Humerus with a weak scapular tuberosity proximal part of deltoid crest located below lesser tubercle and below or on same level as head; lesser tubercle resembling a flattened square; head compressed in craniocaudal direction, considerably larger in males; ratio of width of head to height is 0.698–0.809; intertubercular groove wide and flat; deltoid crest strongly developed, its distal part reaching the coronoid fossa, considerably larger in males; maximum width of deltoid crest reached from its distal part to deltoid tuberosity; deltoid tuberosity well developed, overhanging the bone, considerably larger in males; medial epicondyle elongated, reaching distal part of deltoid crest; lateral epicondyle half length of medial epicondyle; radial groove and olecranon fossa flat and wide; trochlea humeri very large, especially in males. Radius tuberosity relatively large, prominent, and oval; articular circumference on lateral aspect of bone very small, almost reaching level of radial tuberosity, high and flattened on radial surface (medial aspect); deep groove on distolateral aspect for tendon of muscularis abductor pollices longus; shallow groove for tendon of muscularis extensor digitorum communis; well-marked insertion for muscularis pronator teres; radial notch of ulna absent, replaced by sharpened ridge; anconeal process relatively high, large, and wide; coronoid process extending along axis of bone; trochlear notch concave on medial aspect of radius, prominent on lateral aspect; head of ulna and olecranon not preserved. Ilium very thick and flattened; greater ischiatic notch slightly curved with well-developed caudal dorsal ischial spine; iliopectineal eminence well expressed, situated on same level as proximal border of acetabular fossa; edges of acetabular fossa raised slightly above plane surface of ilium; acetabulum, shallow cone shaped, with well-marked cotyloid notch and acetabular notch; eminence for psoas minor muscle not well marked; ischium flattened; sexual dimorphism strongly pronounced in development of body of ilium and in size of acetabular fossa. Femoral head height exceeds height of greater trochanter (9.9 mm in males [$n = 10$]; 5.0 mm in females [$n = 15$], which is not significant); proximal part of trochanter wide and elongated; femoral head considerably smaller in females than in males; head strongly bent in a distal direction, seating on a narrow, short neck; intertrochanteric crest very big and wide, not reaching midline of head; trochanteric fossa deep and long, distal border shallow; minimum width of diaphysis on proximal part of femur; condyles prominent, widely spaced. Popliteal notch of tibia and fibula deep, quite narrow, but well marked; two condyles weakly concave in centers, and oval; tibial crest not developed; intercondyloid eminence weak, only slightly rising above two lateral borders of condyles; tibial tuberosity flattened on dorsal side of tibia, extending along the axis of the bone, and triangular in shape. Sexual dimorphism in bones of the mandible and extremities is described in detail in Koretsky (1987a) and Van Bree and Erdbrink (1987).

Included species: *C. obscura* only (known from localities GC13B [Morgan, 1994], EM8B [Ray, 1976a]).

Comments: Two well-known species from the early Pliocene of Belgium, *Callophoca obscura* and *Mesotaria ambigua*, were described and illustrated by Van Beneden in 1877. Ray (1976a) recognized these species as being synonyms, differentiated only by sexual dimorphism. Additional material, recently collected from Lee Creek, North Carolina (locality EM8B), supports this notion. About 60% of the phocid material from this locality belongs to *Callophoca obscura*.

LOBODONTINI GRAY, 1869

This tribe has been used by authors to unite the Antarctic monachine seals (*Lobodon*, *Hydrurga*, *Ommatophoca*, *Leptonychotes*). The maxillary process is not distinct from the body of the zygomatic bone. The lower border of the zygomatic bone is straight, or with a slight gentle arch only in its posterior part. The anterodorsal end (lacrimal process) of the zygomatic bone does not reach the infraorbital foramen but terminates behind and at the same level as the foramen. The greatest downward flexure of the upper edge (masseteric margin) of the zygomatic bone is lower than the infraorbital foramen. The nasal bones are usually fused, and their frontal part is longer than the maxillary part. The double fontanels in the presphenoid region narrow in their anterior and posterior parts and are slitlike (Chapskii, 1971).

A unique derived character of the group is the lack of an entepicondylar foramen on the humerus. While the presence or absence of the entepicondylar foramen has been used for phocid systematics, this requires some qualification. The foramen is always absent in Monachini, always present in Cystophorinae, and mostly present, but variable (and in some individuals, even present on one side and absent on the other) in Phocinae. We consider the phocine and cystophorine conditions as being primitive and the monachine condition (absence of the foramen) to be derived. This is in opposition to the conclusion of Wyss (1994) and results in the opposite polarity of the character.

***Homiphoca* Muizon and Hendey, 1980**

Type species: *Homiphoca capensis* (Hendey and Repenning, 1972) (originally described as *Prionodelphis capensis*).

Type specimen: SAM-PQ-L1 5695; incomplete and partly restored skull with left C and P4, and right P3 (Hendey and Repenning, 1972; Muizon and Bond, 1982).

Characteristics: Monachine phocid; skull similar to *Monachus*; large rostrum, wide posteriorly and narrow anteriorly; premaxilla with prominent tuberosity; m1 largest of cheek teeth, with principal cusp oblique posteriorly, and often with small accessory cusp; interorbital region broad; tympanic bulla covering petrosal; mastoid forming a lip overlapping posterior border of the bulla; humerus with entepicondylar foramen; tibia and fibula fused proximally.

Included species: *H. capensis* only, known from locality EM8B only (Ray, 1976a).

Comments: *Homiphoca capensis* from the Pliocene of South Africa was first described as *Prionodelphis capensis*. The age of this species is uncertain. For example, according to Ray (1976a) this species is middle Pliocene in age, but Hendey (1978) concluded that the specimens are from the late Pliocene. Muizon (1981b) simply stated that this deposit is Pliocene, although earlier (Muizon and Hendey, 1980) he had stated that it was late Miocene/early Pliocene. The latter opinion was also expressed by Berta and Wyss (1994). According to Muizon and Hendey (1980), *H. capensis* is morphologically intermediate between the extant Monachini and Lobodontini, but more closely related to *Lobodon* than to any other living seal. The Monachinae appear to have a range from the middle Miocene to the Recent, and Ray (1976a) reported bones resembling those of *H. capensis* from the latest Miocene Yorktown Formation in Virginia (locality EM8B). Most elements of the postcranial skeleton, skull, and mandible are diagnosed and illustrated in detail by Muizon and Hendey (1980) and other authors (Hendey and Repenning, 1972; Ray, 1976a; Repenning and Ray, 1977).

We are very tentative about our reference of the listed material to *Homiphoca*. We also have other specimens possibly representing the same taxon or similar taxa. We are at present unable to separate these specimens from South African *Homiphoca*, and we wish to call attention to the similarities while recognizing that our reference is inconclusive but with possibly profound biogeographic implications.

BIOLOGY AND EVOLUTIONARY PATTERNS

As we discussed above, the subfamilies Phocinae, Monachinae, Cystophorinae, and Devinophocinae should be considered as separate phylogenetic branches of ancient Phocidae, which separated from ancient Carnivora probably in the early Oligocene or before, then became widely distributed in the middle Miocene, and practically ceased to exist in the eastern United States in the early Pliocene (Figure 32.3). These fossil animals were members of the subfamilies to which modern phocids belong.

Thus, true seals are no doubt descendants of some Oligocene pinnipeds. Indeed, as we mentioned above, specimens referred to the family Phocidae have already been found in the late Oligocene of South Carolina (Koretsky and Sanders, 2002). Unfortunately, seal-bearing Oligocene marine deposits were not found in the Old World, not because sea did not exist there then, but more likely because some geological event destroyed it (perhaps the Alpinian and Karpathian orogenic events). Members of the family Phocidae do not appear in the North Pacific until latest Pliocene time (Barnes and Mitchell, 1975), and their evolutionary history is separate from that of the Otariidae at least since late Oligocene time (Koretsky, 2001).

Because Wyss considered the otaroid *Allodesmus* as being close to Phocidae, he concluded that the origin of the Phocidae was in the North Pacific and that primitive phocids were large animals (Wyss, 1994; Berta and Sumich, 1999; see also Koretsky 1987a,b; 2001). On the contrary, the North Pacific record of fossil Phocidae is relatively late, represents only a few genera, and provides no evidence about the earlier evolution of the family. Thus, the suggestion of Flynn (1988) that " . . .even the idea of phocid origin in the Atlantic may be incorrect" is a slight overstatement. The close correspondence of successive early Pliocene seal faunas on both sides of the North Atlantic indicates relatively easy interchange, probably with amphi-Atlantic distribution of a large number of species. Perhaps only with (Pleistocene) climatic deterioration were monachines restricted to lower latitudes, and transatlantic gene flow interrupted to the extent that speciation ensued. *Monachus* might have been pushed southward by cold-loving phocines from the north (Walsh and Naish, 2002). Perhaps the ancestors of *Monachus schauinslandi* did not spread on the west coast of middle America owing to preemption of the habitat by otariids. The record from the Paratethyan region agrees with earlier conclusions (Chapskii, 1971; Hendey, 1972; Repenning, 1975; Ray, 1976a; Repenning and Ray, 1977; Muizon, 1981c, 1982c; Arnason *et al.*, 2006) that early phocid evolution was primarily in the Atlantic basin, including Tethys. It is in the Atlantic and Tethyan areas that the earliest occurrences and greatest diversity of mid Tertiary fossil phocids are known.

Phocids, otariids, and odobenids now seem to have equally early origins, in the late Oligocene, but the evidence suggests that they have separate ancestries, with phocids in the North Atlantic and otariids in the North Pacific. The morphologies of the two groups show a relationship between sea lions (and walruses) and bear-like animals, and between phocids and otter-like animals. The paleozoogeography of "pinnipeds" indicates that the Otariidae developed in the North Pacific and the Phocidae in the North Atlantic, and their early geographical separation was a consequence of separate origins.

Tertiary seals are not very similar morphologically to any known terrestrial or semi-aquatic carnivorans that could have been the original ancestors of phocids. They also do not differ very much from

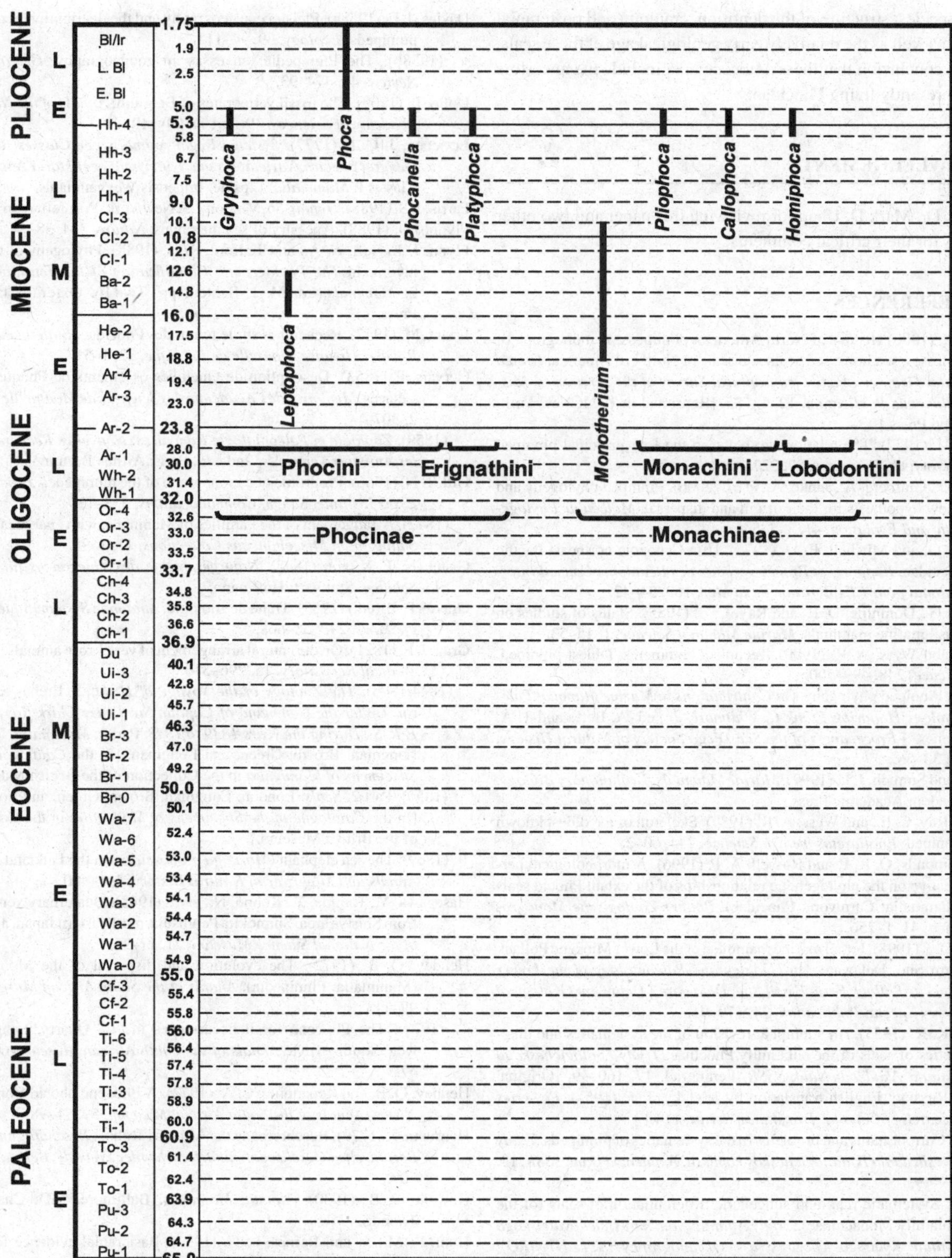

Figure 32.3. Temporal ranges of North American phocid genera.

modern species; structure of the dentition, cranium, and postcranial skeleton, as well as the relatively early geological age of these seals, allow the conclusion that these Miocene taxa include the ancestors of some presently living Phocidae.

ACKNOWLEDGEMENTS

We thank Dr. Mark D. Uhen for reviewing this paper and two other reviewers for their critical comments.

32.4 REFERENCES

Allen, J. A. (1880) History of North American pinnipeds. A monograph of walruses, sea lions, sea bears and seals of Northern America. *United States Geological and Geographical Survey of the Territories, Miscellaneous Publication*, 12, 1–785. [Reprinted in 1974; New York: Arno Press.]

Árnason, U. (1977). The relationship between the four principal pinniped karyotypes. *Hereditas*, 87, 227–42.

Árnason, U., Gullberg, A., Janke, A., *et al.* (2006). Pinniped phylogeny and a new hypothesis for their origin and dispersal. *Molecular Phylogenetic and Evolution*, 41, 345–54.

Barnes, L. G. and Mitchell, E. D. (1975). Late Cenozoic northeast Pacific Phocidae. *Rapports et Procés-verbaux des Réunions, Conseil International pour l'Exploration de la Mer*, 169, 34–42.

Barnes, L. G., Domning, D. P., and Ray, C. E. (1985). Status of studies on fossil marine mammals. *Marine Mammal Science*, 1, 15–53.

Berta, A. and Wyss, A. R. (1990). Technical comments. Oldest pinniped. *Science*, 248, 499–500.

(1994). Pinniped phylogeny. [In *Contributions in Marine Mammal Paleontology Honoring Frank C. Whitmore, Jr.*, ed. A. Berta and T. A. Deméré.] *Proceedings of the San Diego Society of Natural History*, 29, 33–56.

Berta, A. and Sumich, J. L. (1999). *Marine Mammals. Evolutionary Biology*. London: Academic Press.

Berta, A., Ray, C. E., and Wyss, A. R. (1989). Skeleton of the oldest known pinniped, *Enaliarctos mealsi*. *Science*, 244, 60–2.

Bininda-Emonds, O. R. P. and Russell, A. P. (1996). A morphological perspective on the phylogenetic relationships of the extant Phocid seals (Mammalia: Carnivora: Phocidae). *Bonner Zoologische Monographien*, 41, 1–256.

Bohaska, D. J. (1998). Fossil marine mammals of the Lower Miocene Pollack Farm Site, Delaware. [In *Geology and Paleontology of the Lower Miocene Pollack Farm Fossil Site, Delaware*.] *Delaware Geological Survey Special Publication*, 21, 179–91.

Chapskii, K. K. (1955). An attempt at revision of the systematics and diagnostics of seals of the subfamily Phocinae. *Trudy Zoologicheskogo Instituta Akademii Nauk SSSR*, Leningrad, 17, 160–99. [Original in Russian; English translation by Jeletzky, T. F. (1957). *Fisheries Research of Canada, Translation Series*, 114.]

(1961). Current status and problems in the systematics of pinnipeds. *Trudy Soveshchania Ikhtiologicheskoy komissii. Akademya Nauk SSSR*, 12, 138–49.

(1971). Systematic rank and subgeneric differentiation of seals (of the subfamily Monachinae). *Trudy Atlantic Nauchno-Issledovatelskogo Institute Rybnogo Khoziaistva e Okeanography (ATLANTNIRO)*, Kaliningrad, 39, 305–16. [Original in Russian; English translation (1974). *Fisheries Research Board of Canada, Translation Series*, 3185.]

(1974). In defense of classical taxonomy of the seals of the family Phocidae. In *Theoretical Questions on the Systematics and Phylogeny of Animals*, pp. 282–334. Leningrad: Nauka. [In Russian.]

Davies, J. L. (1958a). Pleistocene geography and the distribution of northern pinnipeds. *Ecology*, 39, 97–113.

(1958b). The Pinnipedia: an essay in zoogeography. *Geographical Review*, 48, 474–93.

Dollo, L. (1909). The fossil vertebrates of Belgium. *Annals of the New York Academy of Sciences*, 19(Part 1), 99–119.

Erxleben, J. C. P. (1777). *Systema Regni Animalis per Classes, Ordines, Genera, Species, Varietates cum Synonymia et Historia Animalivm*, Classis I: *Mammalia*. Lipsiae: Impensis Weygandianis.

Farris, J. S. (1988). *Hennig 86*, Version 1.5. New York: Port Jefferson Station.

Flynn, J. J. (1988). Ancestry of sea mammals. *Nature*, 334, 383–4.

Flynn, J. J., Neff, N. A., and Tedford, R. H. (1988). Phylogeny of the Carnivora. In *The Phylogenetic Classification of the Tetrapods*, Vol. 2: *Mammals*, ed. M. J. Benton, pp. 73–116. Oxford: Clarendon Press.

Friant, M. (1947). Recherches sur le fémur des Phocidae. *Bulletin du Musée Royal d'Histoire Naturelle de Belgique*, 23, 1–51.

Gervais, P. (1853). Description de Quelques ossements de Phoques et de Cétacés. *Mémoires de l'Académie des Sciences de Montpellier, Paris*, 2, 307–14.

(1859). *Zoologie et Paleontologie françaises: Nouvelles Recherches sur les Animaux Vertébrés*, 2nd edn. Paris: Arthus Bertrand.

Gill, T. N. (1866). Prodrome of a monograph of the pinnipedes. *Proceedings Essex Institute, Salem, Communications*, 5, 3–13.

(1872). Arrangement of the Families of Mammals, with analytical tables. *Smithsonian Miscellaneous Collections*, 11, 1–98.

Gistel, N. F. X. von (1848). *Naturgeschichte des Thierreichs für höhere Schulen*. Stuttgart: Hoffmann.

Grassé, P.-P. (ed.) (1955). Traité de zoologie. *Anatomie Systematique, Biologie Paris*, 27, 568–668.

Gray, J. E. (1821). On the natural arrangement of vertebrose animals. *London Medical Repository*, 15, 296–310.

(1844/1875). *The Zoology of the Voyage of H. M. S.* Erebus *and* Terror, *Under the Command of Captain Sir James Clark Ross, N. N., F. R. S., During the years 1839 to 1843*, Part 1; *Mammalia*. London: Longman, Brown, Green, and Longman. [In the *Catalogue of the Specimens of Mammalia* in the Collection of the British Museum.]

(1850). Part 2: *Seals*. London: Longman, Brown, Green, and Longman. [In the *Catalogue of the Specimens of Mammalia* in the Collection of the British Museum.]

(1869). The sea elephant (*Morunga proboscidea*) at the Falkland Islands. *Annals and Magazine of Natural History*, 4, 1–400.

Hasegawa, Y., Tomida, Y., Kohno, N., *et al.* (1988). Quaternary vertebrates from Shiriya area, Shimokita Peninsula, northeastern Japan. *Memoirs of the National Science Museum*, 21, 17–36.

Hendey, Q. B. (1972). The evolution and dispersal of the Monachinae (Mammalia: Pinnipedia). *Annals of the South African Museum*, 59, 99–113.

(1978). The Pliocene fossil occurrences in "E" Quarry, Langebaanweg, South Africa. *Annals of the South African Museum*, 69, 215–47.

Hendey, Q. B. and Repenning, C. A. (1972). A Pliocene phocid from South Africa. *Annals of the South African Museum*, 59, 71–98.

Hermann, J. (1779). Beschreibung der Münchs-Robbe. *Beschäftigungen der Berlinischen Gesellschaft naturforschender Freunde, Berlin*, 4, 456–509.

Howell, A. B. (1930). *Aquatic Mammals*. Baltimore, MD: Charles C. Thomas.

Hunt, R. M., Jr. and Barnes, L. G. (1994). Basicranial evidence for ursid affinity of the oldest pinnipeds. *Proceedings of the San Diego Society of Natural History*, 29, 57–67.

Kellogg, A. R. (1922). Pinnipeds from Miocene and Pleistocene deposits of California. A description of a new genus and species of sea lion from the Temblor together with seal remains from the Santa Margarita and San Pedro Formations and a resume of current theories regarding origin of Pinnipedia. *University of California Publications, Bulletin of the Department of Geological Sciences*, 13, 23–132.

King, J. E. (1956). The monk seals (genus *Monachus*). *Bulletin of the British Museum of Natural History, Zoology*, 3, 201–56.

(1960). Sea-lions of the genera *Neophoca* and *Phocarctos*. *Mammalia*, 24, 445–56.

(1964). *Seals of the World*. London: British Museum (Natural History) Publications.

(1966). Relationships of the hooded and elephant seal (genera *Cystophora* and *Mirounga*). *Zoological Journal*, 148, 385–98.

(1983). *Seals of the World*, 2nd edn. Ithaca, NY: Comstock.

Koretsky, (=Koretskaya) I. A. (1987a). Sexual dimorphism in the structure of the humerus and femur of *Monachopsis pontica* (Pinnipedia: Phocinae). *Vestnik Zoologii*, 4, 77–82. [In Russian.]

(1987b). The position of the genus *Praepusa* in the Phocinae system. *Praeprint/Akademii Nauk Ukrainian SSR. Institute of Zoology*, 87, 3–7. [In Russian.]

(2001). Morphology and systematics of Miocene Phocinae (Mammalia: Carnivora) from Paratethys and the North Atlantic region. *Geologica Hungarica, Budapest*, 54, 1–109.

Koretsky, I. A. and Holec, P. (2002). A primitive seal (Mammalia: Phocidae) from the early middle Miocene of Central Paratethys. *Smithsonian Contributions to Paleobiology*, 93, 163–78.

Koretsky, I. A. and Sanders, A. (2002). Paleontology of the Late Oligocene Ashley and Chandler Bridge Formations of South Carolina, 1: Paleogene pinniped remains; the oldest known seal (Carnivora: Phocidae). *Smithsonian Contributions to Paleobiology*, 93, 179–84.

Leidy, J. (1853). The ancient fauna of Nebraska. *Smithsonian Contributions to Knowledge*, 6, 1–126.

Linnaeus, C. (1758). *Systema Naturae per Regna Tria Naturae, Secundum Classes, Ordines, Genera, Species, cum Characteribus and Differentiis*, 10th edn, *Reformata*, Vol. I. Holmiae: Laurentii Salvii.

Matschie, P. (1905). Eine Robbe von Laysan. *Beschäftigungen Gessellschaft Naturforschender Freunde, Berlin*, 254–62.

McKenna, M. C. and Bell, S. K. (1997). *Classification of Mammals Above the Species Level*. New York: Columbia University Press.

McLaren, I. A. (1960a). Are the Pinnipedia biphyletic? *Systematic Zoology*, 9, 18–28.

(1960b). On the origin of the Caspian and Baikal seals and the paleoclimatological implications. *American Journal of Science*, 258, 47–65.

Misonne, X. (1958). Faune du Tertiaire et du Pleistocene inferieur de Belgique. (Oiseaux et Mammiferes). *Bulletin du Institut Royal des Sciences Naturelles de Belgique*, 34, 1–36.

Mitchell, E. D. (1966). The Miocene pinniped *Allodesmus*. *University of California, Publications in Geological Sciences*, 61, 1–105.

(1967). Controversy over diphyly in pinnipeds. *Systematic Zoology*, 16, 350–1.

(1968). The Mio-Pliocene pinniped *Imagotaria*. *Journal of the Fisheries Research Board of Canada*, 25, 1843–900.

(1975). Parallelism and convergence in the evolution of Otariidae and Phocidae. *Rapports et Proces-verbaux des Reunions, Conseil International pour l'Exploration de la Mer*, 169, 12–26.

Mivart, St. G. (1885). Notes on the Pinnipedia. *Proceedings of the Zoological Society, London*, 1885, 484–500.

Morgan, G. S. (1994). Miocene and Pliocene marine mammal faunas from the Bone Valley Formation of Central Florida. *Proceedings of the San Diego Society of Natural History*, 29, 239–68.

Muizon, C., de (1981a). Premier signalement de Monachinae (Phocidae: Mammalia) dans le Sahèlien (Miocène Supérieue) d'Oran (Algérie). *Palaeovertebrata*, 11, 181–94.

(1981b). Les vertébrés fossiles de la formation Pisco (Pérou). Première partie. Deux nouveaux Monachinae (Phocidae, Mammalia) du Pliocène du Sud-Sacaco. *Institut Français d'Études Andines*, 6, 1–150.

(1981c). Le grand voyage des Phoques. *La Recherche Paris*, 12, 750–2.

(1982a). Phocid phylogeny and dispersal. *Annals of the South African Museum*, 89, 175–213.

(1982b). Les relations phylogenetiques des Lutrinae (Mustelidae, Mammalia). *Geobios, Mémoire Special*, 6, 259–72.

(1982c). Dispersion des Monachinae (Phocidae, Mammalia) dans L'Hemisphere Sud. 9–eme *Réunion Annuelle Sciences de la Terre*, pp. 1–463. Paris: Société Géologique de France.

(1992). Paläontologie. In *Handbuch der Säugetiere Europas*, ed. R. Duguy and D. Robineau, Vol. 6: Meeressäuger. Part II: Robben – *Pinnipedia*. pp. 34–41. Wiesbaden: AULA-Verlag.

Muizon, C., de and Bond, M. (1982). Le Phocidae (Mammalia) Miocène de la formation Parana (Entre Rios, Argentina). *Bulletin du Muséum National d'Histoire Naturelle, Paris*, (Série 4, section C) 3–4, 165–207.

Muizon, C., de and Hendey. Q. B. (1980). Late Tertiary seals of the South Atlantic Ocean. *Annals of the South African Museum*, 82, 91–128.

Ognev, S. I. (1935). *Mammals of the USSR and Adjacent Countries*, Vol. 3: *Carnivora*, pp. 1–752. Moscow: Glavpushnina. [In Russian; English translations by A. Birron, and Z. S. Coles (1962), the *Israel Program for Scientific translations*.]

Pavlinov, I. J. and Rossolimo, O. L. (1987). *Systematics of Mammalia of the Soviet Union*. Moscow: Moscow University [In Russian.]

Ray, C. E. (1976a). Geography of phocid evolution. *Systematic Zoology*, 25, 391–406.

(1976b). *Phoca wymani* and other Tertiary seals (Mammalia: Phocidae) described from the eastern seaboard of North America. *Smithsonian Contributions to Paleobiology*, 28, 1–36.

Reidman, M. (1990). *The Pinnipeds. Seals, Sea Lions, and Walruses*. Berkeley, CA: University of California Press.

Repenning, C. A. (1975). Otarioid evolution. *Rapports et Proces-verbaux des Reunions, Conseil International pour l'Exploration de la Mer*, 169, 27–33.

(1976). Adaptive evolution of sea lions and walruses. *Systematic Zoology*, 25, 375–90.

(1983). Faunal exchanges between Siberia and North America. *Proceedings of the American Quaternary Association, Conference*, 5, 40–55.

(1990). Technical comments. Oldest pinniped. *Science*, 248, 499.

Repenning, C. A. and Ray, C. E. (1977). The origin of the Hawaiian Monk seal. *Proceedings of the Biological Society of Washington*, 89, 667–88.

Repenning, C. A. and Tedford, R. H. (1977). Otarioid seals of the Neogene. *Geological Survey Professional Paper*, 992, 1–93.

Repenning, C. A., Ray, C. E., and Grigorescu, D. (1979). Pinniped biogeography. In *Historical Biogeography, Plate Tectonics, and the Changing Environment*, ed. J. Gray and A. J. Boucot, pp. 357–69. Corvallis, OR: Oregon State University Press.

Robinette, H. and Stains, H. J. (1970). Comparative study of the calcanea of the Pinnipedia. *Journal of Mammalogy*, 51, 1–527.

Savage, D. E. and Russell, D. E. (1983). *Mammalian Paleofaunas of the World*. Reading, MA: Addison-Wesley.

Scheffer, V. B. (1958). *Seals, Sea Lions, and Walruses. A Review of the Pinnipedia*. Stanford, CA: Stanford University Press.

Simpson, G. G. (1945). The principles of classification and a classification of mammals. *Bulletin of the American Museum of Natural History*, 85, 1–350.

Sokolov, V. E. (1979). *Systematika Mlekopitauzchich* [*Systematics of Mammals*]*: Cetacea, Pinnipedia, Carnivora, Tublidentata, Tylopoda, Perissodactyla*. Moscow: High School. [In Russian.]

Tavani, G. (1941–2). Revisione dei resti del Pinnipede conservato nel Museo di Geologia di Pisa. *Palaeontographica Italica, Memorie di Paleontologia*, 40, 97–113.

(1942–3). Revisione dei resti di Pinnipedi conservati nel Museo geopaleontologica di Firenze. *Atti della Societá Toscana di Scienze Naturali*, 5, 3–11.

Tedford, R. H. (1976). Relationship of pinnipeds to other carnivores (Mammalia). *Systematic Zoology*, 25, 363–74.

Toula, F. (1897). *Phoca vindobonensis* n. sp. von Nussdorf. *Beiträge zur Paläontologie und Geologie Österreich-Ungarns und des Orients, Mittheilungen des Paläontologischen Institutes der Universität Wien*, 11, 47–70.

Trouessart, E.-L. (1897). Carnivora, Pinnipedia, Rodentia I. Fascicle II. In *Catalogus Mammalium tam Viventium quam Fossilium*, new edn, pp. 219–452. Berlin: R. Friedländer.

(1904). Primates, Prosimiae, Chiroptera, Insectivora, Carnivora, Pinnipedia. Fascicle I. In *Catalogus Mammalium tam Viventium quam Fossilium Quinquennale Supplementum* (1904), p. 288. Berlin: R. Friedländer.

(1905). Cetacea, Edentata, Marsupialia, Allotheria, Monotremata. Fascicle IV. In *Catalogus Mammalium tam Viventium quam Fossilium. Quinquennale Supplementum* (1899–1904), pp. 753–929. Berlin: R. Friedländer.

True, F. W. (1906). Diagnosis of a new genus and species of fossil sea-lion from the Miocene of Oregon. *Smithsonian Miscellaneous Collections*, 48, 47–9.

Ugolini, R. (1900). Di uno scheletro fossile di Foca trovato ad Orciano. *Atti della Società Toscana di Scienze Naturali, Pisa*, 12, 146–8.

(1902). Il *Monachus albiventer* Bodd. del Pliocene di Orciano Paleontographica Italica. *Memorie de Paleontologia, Pisa*, 8, 1–21.

Van Beneden, P.-J. (1876). Les Phoques Fosiles du basin d'Anvers. *Bulletin de l'Académie Royale des Sciences, des Lettres et des Baux-Arts de Belgique* 41, 783–802.

(1877). Description des ossements fossiles des environs d'Anvers, première partie. Pinnipèdes ou Amphithèriens. *Musée Royal d'Histoire Naturelle de Belgique, Annales*, 1, 1–88.

Van Bree, P. J. H. and Erdbrink, D. P. B. (1987). Fossil Phocidae in some Dutch collections (Mammalia, Carnivora). *Beaufortia*, 37, 43–66.

Walsh, S. and Naish, D. (2002). Fossil seals from late Neogene deposits in South America: a new pinniped (Carnivora, Mammalia) assemblage from Chile. *Paleontology*, 45, 821–42.

Wiig, Ø. (1983). On the relationships of Pinnipeds to other Carnivores. *Zoologica Scripta*, 12, 225–7.

Wozencraft, C. (1989). The phylogeny of the Recent Carnivora. In *Carnivore Behavior, Ecology, and Evolution*, ed. J. L. Gittleman, pp. 495–535. New York: Cornell University Press.

Wyss, A. R. (1987). The walrus auditory region and monophyly of pinnipeds. *American Museum Novitates*, 2871, 1–31.

(1988a). Evidence from flipper structure for a single origin of pinnipeds. *Nature*, 334, 427–8.

(1988b). On "retrogression" in the evolution of the Phocinae and phylogenetic affinities of the monk seals. *American Museum Novitates*, 2924, 1–38.

(1994). The evolution of body size in phocids: some ontogenetic and phylogenetic observations. *Proceedings of the San Diego Society of Natural History*. 29, 69–77.

Wyss, A. R. and Flynn, J. J. (1993). A phylogenetic analysis and definition of the Carnivora. In *Mammal Phylogeny: Placentals*, ed. F. S. Szalay, M. J. Novacek, and M. C. McKenna, pp. 32–52. New York: Springer-Verlag.

33 Archaeoceti

MARK D. UHEN
Smithsonian Institution, Washington, DC, USA

INTRODUCTION

Archaeoceti are Eocene cetaceans that originated in the Old World (probably Indo-Pakistan) and dispersed around the world in the late middle and late Eocene. North American archaeocetes are found in Gulf Coast and Atlantic Coast deposits from Texas to North Carolina. The earliest North American archaeocetes in the family Protocetidae were semi-aquatic, possessing substantial hindlimbs and the ability to locomote on land. Archaeocetes probably came to North America from the western Tethys where similar whales are known. Later North American archaeocetes in the family Basilosauridae (or other non-North American basilosaurids) probably gave rise to the modern cetaceans, Mysticeti and Odontoceti. Like modern odontocetes, archaeocetes were marine carnivores that probably consumed individual prey items rather than bulk feeding like mysticetes. Stomach contents indicate that archaeocetes ate a variety of fish. There are four named species of protocetids and five named species of basilosaurid currently known from North America, although knowledge of additional, unpublished material suggests that the diversity of both groups was higher. All confirmed North American archaeocetes are known from the southeastern United States.

DEFINING FEATURES OF THE SUBORDER ARCHAEOCETI

CRANIAL

All cetaceans, including archaeocetes, can be characterized by the presence of a pachyosteosclerotic involucrum of the auditory bulla (Luo, 1998). In addition, cetaceans exhibit a rotation of the ossicular chain within the middle ear region, and this rotation is at least partially present in the earliest archaeocetes (Lancaster, 1990; Thewissen and Hussain, 1993).

In addition to characters of the ear region, all cetaceans exhibit elongate premaxillae, which results in a relative retraction of the external nares. The external nares are retracted even further toward the cranial vertex in odontocetes and mysticetes, but the nares are retained on the face in archaeocetes. Cetaceans are also characterized by the presence of a basioccipital crest (also known as the falcate process of the basioccipital). Even the earliest cetaceans for which the basicranium is known (*Pakicetus*) have some development of the basioccipital crest (Figure 33.1).

DENTAL

O'Leary (1998) listed three dental synapomorphies for Cetacea, which can also be used to delimit the Archaeoceti. They are lack of a protocone on P4, P4 paracone twice the height of the paracone of M1, and presence of a well-developed lingual cingulid on the lower molars. In addition, the presence of vertically elongate wear facets on the buccal faces of the lower molars is characteristic of archaeocetes (O'Leary and Uhen, 1999). While not a morphological character per se, it is indicative of a particular mode of oral processing that is not seen in cetacean relatives or the modern groups of cetaceans. Archaeocetes also possess reentrant grooves on the mesial faces of their lower molars that are lacking in cetacean relatives and Neoceti.

In addition to characters of the teeth themselves, archaeocetes can be distinguished from their relatives by the position of the incisors, which are in line with the cheek teeth rather than in an arc across the anterior end of the rostrum. This feature is shared with later Neoceti.

POSTCRANIAL

Since many of the postcranial synapomorphies of modern cetaceans developed within archaeocetes, it is difficult to characterize the postcrania of archaeocetes as a whole. The earliest archaeocetes

Evolution of Tertiary Mammals of North America, Vol. 2. ed. C. M. Janis, G. F. Gunnell, and M. D. Uhen. Published by Cambridge University Press.

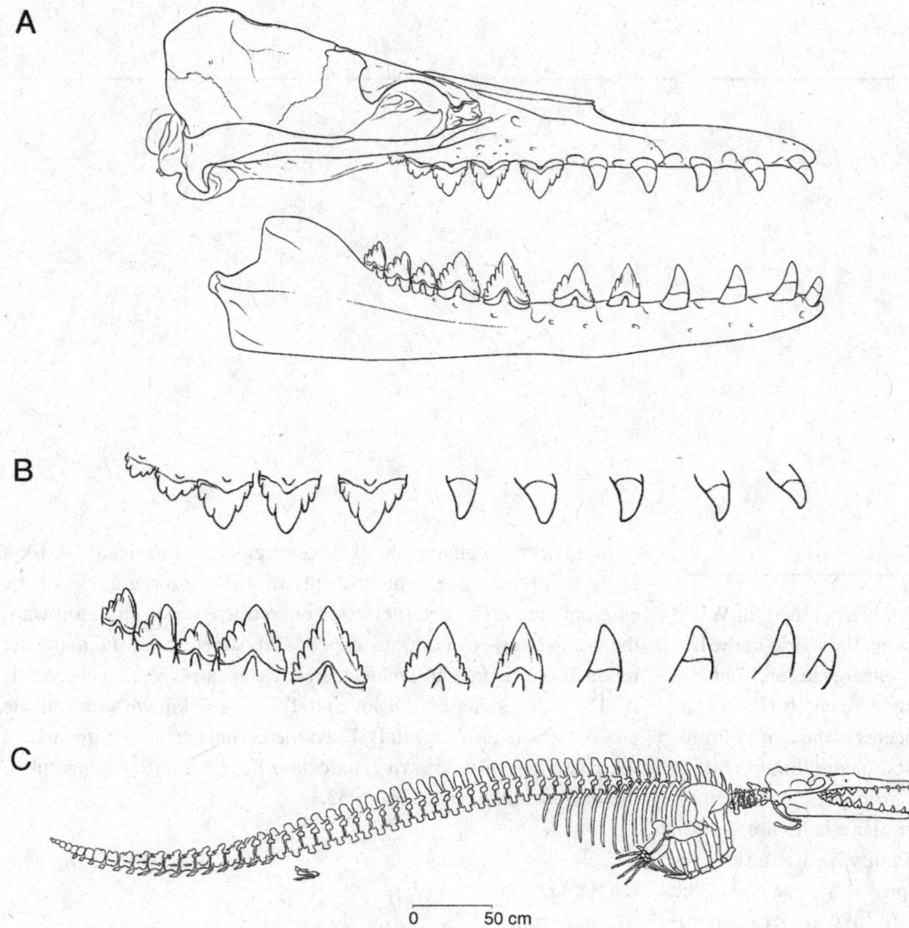

Figure 33.1. Skull and mandible, dentition, and skeleton of a typical archaeocete cetacean, *Dorudon atrox*. These drawings are reconstructions based on multiple specimens. A. Skull and mandible in lateral view. B. Dentition in lateral view. Note that *D. atrox* lacks M3 and has multiple accessory denticles on the cheek teeth. These features are typical of basilosaurids, but not more primitive archaeocetes. C. Skeleton in lateral view. *D. atrox* has a small hindlimb that is not attached to the vertebral column. Protocetids have a larger hindlimb attached to the vertebral column at the sacrum. (Drawings by Bonnie Miljour.)

are quite terrestrial, while later archaeocetes are fully aquatic. The trend toward a more aquatic lifestyle is indicated by reduction of the hindlimbs and detachment of the hindlimbs from the vertebral column, shortening of the cervical vertebral bodies, increase in the number of vertebrae in the posterior thoracic and lumbar regions, reduction of the sacrum, and development of forelimb flippers (Uhen, 1998a).

SYSTEMATICS

SUPRAORDINAL

Much recent work has examined the relationship of Cetacea to other mammals. This has included studies of molecular data from recent cetaceans (see Gatsey [1998] for a thorough review), morphological data from recent cetaceans (Novacek, Wyss, and McKenna, 1988), morphological data from fossil and recent cetaceans (Thewissen, 1994), and morphological data from fossil and recent cetaceans and molecular data from recent cetaceans (O'Leary and Geisler, 1999; Geisler and Uhen, 2003, 2005). Most of these analyses agree that the living group of mammals most closely related to Cetacea is Artiodactyla (but see Prothero, Manning, and Fischer [1988] for a different view). Beyond that, there is little agreement among the different hypotheses of relationships.

Most molecular studies indicate that cetaceans originate within Artiodactyla, which would make Artiodactyla paraphyletic. Many of the most recent studies indicate that cetaceans are the sister taxon to ruminant artiodactyls (Gatesy, 1998), but others have placed cetaceans as the sister taxon to Hippopotamidae (e.g., Milinkovitch, Bérubé, and Palsbøll, 1998), Suidae (Philippe and Douzery, 1994), or Camelidae (Irwin, Kocher, and Wilson, 1991). Phylogenetic analyses combining molecular and morphological data from recent taxa with morphological data from fossil taxa show that hippopotamids are the sister group to Cetacea, creating a paraphyletic Artiodactyla if Cetacea are excluded from that group (Geisler and Uhen, 2003, 2005) (Figure 33.2). (See discussion in Ch 30.)

INFRAORDINAL

Within Archaeoceti, there are five families: Pakicetidae, Ambulocetidae, Remingtonocetidae, Protocetidae, and Basilosauridae (Thewissen, Madar, and Hussain, 1996; Uhen, 1998a). Pakicetids, ambulocetids, and remingtonocetids are all early, semi-aquatic cetaceans that are restricted to the early and middle Eocene of Indo-Pakistan.

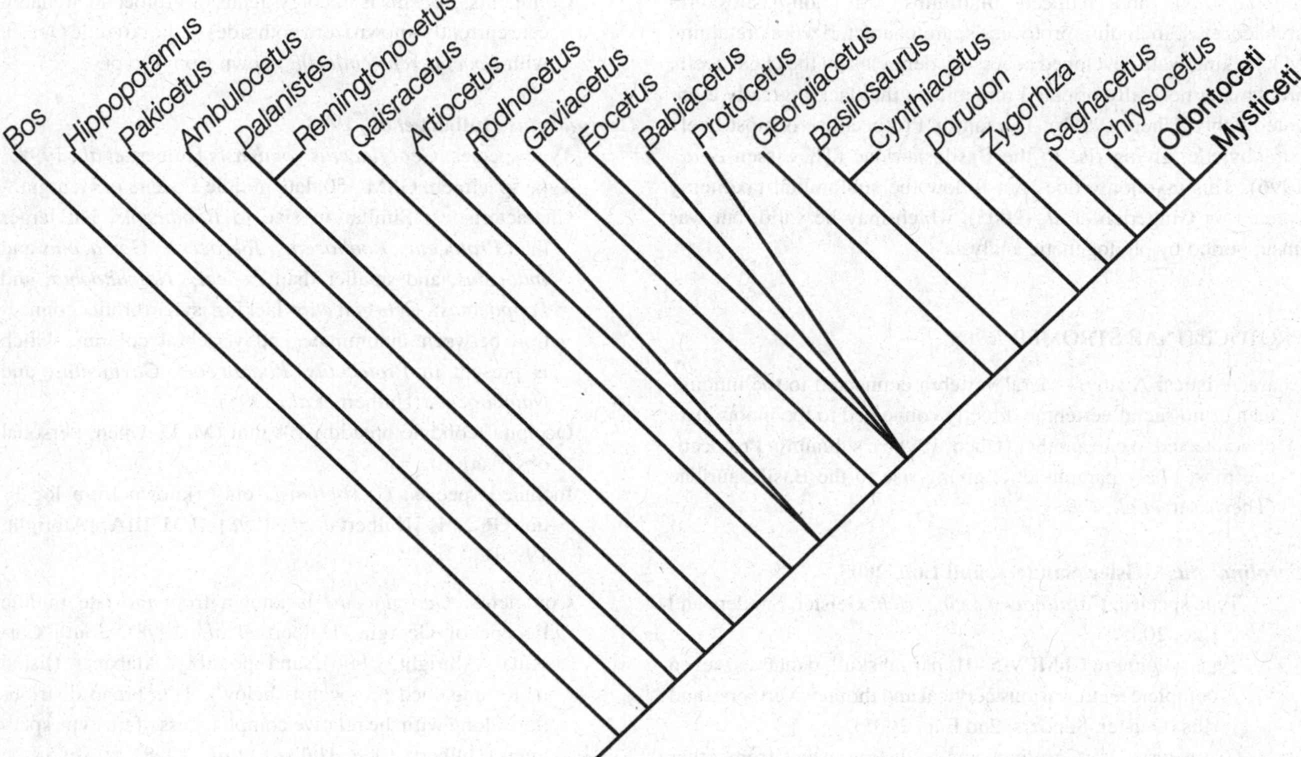

Figure 33.2. Strict consensus cladogram of six equally overall most parsimonious cladograms resulting from the procedure outlined in Uhen (2005). The stratigraphic character was not included in the morphological cladistic analysis. Basilosaurinae is represented by *Basilosaurus*. Basilosaurines have elongate posterior thoracic, lumbar, and anterior caudal vertebrae. Dorudontinae are a paraphyletic assemblage, including *Cynthiacetus*, that give rise to Neoceti and Basilosaurinae. Note that *Cynthiacetus* is the sister taxon to Basilosaurinae (*Basilosaurus*). Neoceti (Odontoceti + Mysticeti) arises from within the Dorudontinae. Early neocetines share a number of derived features with basilosaurids and lack the autapomorphic features of basilosaurines.

Protocetids are the earliest cetaceans found outside of Indo-Pakistan, reaching across Africa (*Pappocetus lugardi* Andrews, 1920) to the western Atlantic. North American *Georgiacetus*, which has a large innominate, *Natchitochia*, having a large single sacral vertebra (and presumably an innominate as well), and *Eocetus*, which has a well-developed innominate, demonstrate that the protocetids dispersed widely prior to their adoption of the fully aquatic existence seen in basilosaurid archaeocetes (Uhen, 1998b). All North American representatives of the family Protocetidae are from the subfamily Protocetinae. The family Protocetidae is most likely paraphyletic, giving rise to the Basilosauridae (Thewissen, Madar, and Hussain, 1996).

Basilosaurids are currently known from every continent except Antarctica and South America. Basilosaurids are thought to be fully aquatic based on the great reduction of the hindlimb, the detachment of the hindlimb from the vertebral column, and the increase in the number of trunk vertebrae (Buchholtz, 1998; Uhen, 1998a). Basilosauridae includes two subfamilies, Basilosaurinae and Dorudontinae. Basilosaurines are specialized in having elongated trunk vertebrae, a character that dorudontines lack. Dorudontinae is probably paraphyletic, giving rise to both basilosaurines and Neoceti (Uhen, 1998a).

INCLUDED GENERA IN THE ARCHAEOCETI FLOWER, 1883

The locality numbers listed for each genus refer to the list of unified localities in Appendix I. The locality numbers may be listed more than one way. The acronyms for museum collections are listed in Appendix III.

Parentheses around the locality (e.g., [CP101]) mean that the taxon in question at that locality is cited as an "aff." or "cf." the taxon in question. Parentheses are usually used for individual species, implying that the genus is firmly known from the locality, but the actual species identification may be questionable. Question marks in front of the locality (e.g., ?CP101) mean that the taxon is questionably known from that locality, implying some doubt that the taxon is actually present at that locality, either at the genus or species level. An asterisk (*) indicates the type locality.

PROTOCETIDAE STROMER, 1908

Thewissen *et al.* (1996) characterized members of the family Protocetidae as having a supraorbital shield, greatly enlarged mandibular foramina, a reduced nasopharyngeal duct (compared with

Ambulocetus), and reduced hindlimbs. All non-basilosaurid archaeocetes, including protocetids, are characterized as retaining M3, lacking well-developed accessory denticles on the cheek teeth, and having normally oriented innominata that lack a greatly elongated pubis (Uhen, 1998b). The family Protocetidae is most likely paraphyletic, giving rise to the Basilosauridae (Thewissen *et al.*, 1996). This taxonomy does not follow the subfamilial taxonomy erected by Gingerich *et al.* (2005), which may be valid, but was unsupported by phylogenetic analysis.

PROTOCETINAE STROMER, 1908

Characteristics: A single sacral vertebra connected to the innominata or no sacral vertebrae directly connected to the innominata or connected to one another (Uhen, 1998b); subfamily Protocetinae most likely paraphyletic, giving rise to the Basilosauridae (Thewissen *et al.*, 1996).

***Carolinacetus* Geisler, Sanders, and Luo, 2005**

Type species: *Carolinacetus gingerichi* Geisler, Sanders and Luo, 2005.

Type specimen: ChMPV 5401, partial skull, dentaries, seven complete teeth, various cervical and thoracic vertebrae and ribs (Geisler, Sanders, and Luo, 2005).

Characteristics: *Carolinacetus* is distinguished from other protocetids by presence of a posterodorsal tongue of the petrosal that is exposed between the exoccipital and the squamosal with the skull in posterior view; mandible with a steep ascending process and a deeply descending ventral margin posteriorly (Geisler, Sanders, and Luo, 2005).

Occipital condyle breadth: The occipital condyle is not known from *Carolinacetus*.

Included species: *C. gingerichi* only, known from locality EM3IIIA* only (Geisler, Sanders, and Luo, 2005).

Comments: This animal has been referred to previously in the literature as the "Cross Whale" (Geisler, Sanders, and Luo, 1996; Williams, 1998).

***Eocetus* Fraas, 1904**

Type species: *Eocetus schweinfurthi* Fraas, 1904.

Type specimen: SMNS 10986, late middle Eocene of Egypt.

Characteristics: Larger than all other protocetines, except possibly *Pappocetus* and *Natchitochia* vertebrae of *Eocetus* differing from vertebrae assigned to *Pappocetus* and *Natchitochia* (Halstead and Middleton, 1974; Uhen, 1998b) in having distinctive pock-marks, small vascular channels penetrating deeply into the bone; centra, neural arches, neural spines, and transverse processes of lumbar vertebrae elongated anteroposteriorly.

Occipital condyle breadth: occipital condyles are unknown from *Eocetus*.

Included species: *E. wardii* only (known from localities EM3B, C* [Uhen, 1999]).

Comments: *Eocetus* is the only genus of protocetid archaeocete currently known from both sides of the Atlantic Ocean, with *Eocetus schweinfurthi* known from Egypt.

***Georgiacetus* Hulbert *et al.*, 1998**

Type species: *Georgiacetus vogtlensis* Hulbert *et al.*, 1998.

Type specimen: GSM 350, late middle Eocene of Georgia.

Characteristics: Similar in size to *Babiacetus*, but larger than *Protocetus*, *Rodhocetus*, *Takracetus*, *Gaviacetus* and *Indocetus*, and smaller than *Eocetus*, *Natchitochia*, and *Pappocetus*; *Georgiacetus* lacking synarthrotic connection between innominate and vertebral column, which is present in *Protocetus*, *Rodhocetus*, *Gaviacetus*, and *Natchitochia* (Hulbert *et al.*, 1998).

Occipital condyle breadth: 109 mm (M. D. Uhen, personal observation).

Included species: *G. vogtlensis* only (known from localities GC22II* [Hulbert *et al.*, 1998], [EM3IIIA] [Albright, 1996]).

Comments: *Georgiacetus* is known from the late middle Eocene of Georgia (Hulbert *et al.*, 1998), South Carolina (Albright, 1996), and possibly Alabama (listed under unnamed protocetids below). This broad distribution, along with the relative completeness of the type specimen (Hulbert, 1998; Hulbert *et al.*, 1998), makes it the best-known North American protocetid.

***Natchitochia* Uhen, 1998b**

Type species: *Natchitochia jonesi* Uhen, 1998b.

Type specimen: USNM 16805, late middle Eocene of Louisiana.

Characteristics: Significantly larger than most other protocetids, with the exception of *Eocetus* and *Pappocetus*; vertebrae lacking elongation of lumbar centra of *Eocetus* (Uhen, 1999) and ventral keel seen in vertebrae assigned to *Pappocetus* (Halstead and Middleton, 1974); well-developed articular surface on the transverse process of the single sacral vertebra (Uhen, 1998b).

Occipital condyle breadth: the skull is not known from *Natchitochia*.

Included species: *N. jonesi* only, known from locality GC23IIB* only (Uhen, 1998b).

Comments: The type specimen of *N. jonesi* was discovered in 1943 and was briefly mentioned by Maher and Jones (1949), citing a letter from Remington Kellogg, but was not formally described until 1998 (Uhen, 1998b). It was also referred to as "Protocetid whale of Uhen" by Williams (1998).

Indeterminate protocetids

Several specimens that have been assigned to the Protocetidae have been discussed in the literature but not described formally. These include the lower jaw of a *Pappocetus*-like whale from North Carolina (the "New Bern whale"), locality EM3II (McLeod and Barnes, 1996) and vertebrae from Texas (GC1B: Kellogg, 1936), Louisiana (GC23IIA), Mississippi (GC21V), and Alabama (GC21IVB, GC23IVC) (M. D. Uhen, unpublished data).

Kellogg (1936) identified a single vertebra from Texas (GC1B) as belonging to *Protocetus* sp. The thoracic vertebra is similar in vertebral body width and height to those of *Protocetus*, but the body is considerably longer, suggesting that it may belong to a different taxon. McLeod and Barnes (1996) stated that the *Pappocetus*-like whale from North Carolina was the "first named species of protocetid from the western North Atlantic," yet failed to name the species in that or any subsequent publication to date. Since 1996, four other species of North American protocetids have been named (*Georgiacetus vogtlensis*, *Natchitochia jonesi*, *Eocetus wardii*, and *Carolinacetus gingerichi*). At least two different species of protocetid are known from the Upper Lisbon Formation of Alabama (locality GC21IVB), based on the disparate sizes of the vertebra from the four specimens known (M. D. Uhen, unpublished data).

BASILOSAURIDAE COPE, 1868

Basilosauridae are distinguished from all other archaeocetes by the loss of M3, the presence of multiple accessory denticles on the cheek teeth, loss of the sacrum, reduction of the hindlimb, and rotation of the pelvis (Uhen, 1998a).

BASILOSAURINAE COPE, 1868

Characteristics: Basilosaurinae had all the above characters distinguishing it from other archaeocetes; additionally, greatly elongated posterior thoracic, lumbar, and anterior caudal vertebrae.

Basilosaurus Harlan, 1834

Type species: *Basilosaurus cetoides* Harlan, 1834.

Type specimen: ANSP 12944A early late Eocene of Louisiana.

Characteristics: Distinguished from other basilosaurids by large body size; elongate posterior thoracic, lumbar, and anterior caudal vertebrae. Lacks the vertically oriented metapophyses seen in *Basiloterus* (Gingerich *et al.*, 1997).

Occipital condyle breadth: 145 mm (*Basilosaurus cetoides*, USNM 4674 [Kellogg, 1936]).

Included species: *B. cetoides* only (known from localities GC22B, C [Kellogg, 1936; Westgate, Gilette, and Rolater, 1994], GC23A–C* [Kellogg, 1936; Westgate, 2001], GC23III [Daly, 1999], GC23IVA–J [Kellogg, 1936; Dockery and Johnston, 1986; Lancaster, 1986], GC23VB, C [Kellogg, 1936], GC23VIA [Palmer, 1939; Westgate, 2001], GC23VIIB [Kellogg, 1936], GC23VIIIA, P [UF Collections], GC23IX [Corgan, 1976], GC23X [Randazzo *et al.*, 1990], EM3IIIB [Uhen and Gingerich, 2001]).

Comments: *Basilosaurus* is also represented in the Old World by *Basilosaurus isis* from Egypt (Gingerich, Smith, and Simons, 1990) and *Basilosaurus drazindai* from Pakistan (Gingerich *et al.*, 1997).

DORUDONTINAE MILLER, 1923

Characteristics: With the derived characteristics of the Basilosauridae; lacking elongate posterior thoracic, lumbar, and anterior caudal vertebrae. Dorudontine genera are very similar to each other, each having a few autapomorphies for differentiation.

Chrysocetus Uhen and Gingerich, 2001

Type species: *Chrysocetus healyorum* Uhen and Gingerich, 2001.

Type specimen: SCSM 87.195, from the early late Eocene of South Carolina.

Characteristics: Much smaller than *Basilosaurus* or *Cynthiacetus*; differs from *Dorudon* and *Zygorhiza* in having smooth enamel on crowns of premolars.

Occipital condyle breadth: the occipital condyles are not known from *Chrysocetus*.

Included species: *C. healyorum* only, known from locality EM3IIIB* only (Uhen and Gingerich, 2001).

Comments: The type specimen of *C. healyorum* is a skeletally juvenile individual with adult teeth. *C. healyorum* may be the sister taxon or an ancestor to odontocetes and mysticetes (Uhen and Gingerich, 2001).

Cynthiacetus Uhen, 2005

Type species: *Cynthiacetus maxwelli* Uhen, 2005.

Type specimen: MMNS VP 445.

Characteristics: Differs from other dorudontines in having skull, cervical vertebrae, and anterior thoracic vertebrae similar in size and morphology to those of *Basilosaurus*; differs from *Basilosaurus* in having posterior thoracic vertebrae, lumbar vertebrae, and anterior caudal vertebrae not elongate, but proportioned similar to other dorudontines.

Occipital condyle breadth: approximately 142 mm (MMNS VP 445 [Uhen, 2005]).

Included species: *C. maxwelli* only (known from localities GC22B [Domning, Morgan, and Ray, 1982], GC23IVA, B [Kellogg, 1936; Uhen, 2005], GC23VC [Uhen, 2005], EM3A [Uhen, 2001]).

Comments: Specimens assigned to *Cynthiacetus* have in the past often been referred to *Pontogeneus*. *Pontogeneus* is known only from the type specimen, a cervical vertebral centrum from locality GC23 (Leidy, 1852; Uhen, 2005). This material is incomplete and probably not diagnostic; other material cannot be referred to the genus with confidence, and the family-level relationships are uncertain. The name *Pontogeneus priscus* should be restricted to the type specimen (Uhen, 2005). Specimens of *Cynthiacetus* that lack posterior trunk vertebrae may be erroneously assigned to *Basilosaurus* because of the similarity in size and morphology of their skulls.

***Dorudon* Gibbes, 1845**

Type species: *Dorudon serratus* Gibbes, 1845.

Type specimen: MCZ 8763, late Eocene of South Carolina.

Characteristics: Smaller than both *Basilosaurus* and *Cynthiacetus*; lacks the elongate trunk vertebrae of *Basilosaurus*; differs from *Zygorhiza* in lacking tiny denticles on the cingula of the upper premolars (Köhler and Fordyce, 1997); premolars relatively lower crowned than those of *Chrysocetus*.

Occipital condyle breadth: occipital condyle is not known from *Dorudon serratus*; approximately 123 mm (mean of three adult individuals [Uhen, 2004]) for *Dorudon atrox* from Egypt.

Included species: *D. serratus* only, known from locality EM3IV* only (Gibbes, 1845; Uhen, 2004).

Comments: *Durodon* is also known from the Eocene of Africa, Europe, and New Zealand. Much of the additional material of *D. serratus* described by Kellogg (1936) is now missing. It is difficult to determine how *D. serratus* differs from *Zygorhiza kochii* based on the poor type specimen, which is a juvenile individual with deciduous teeth.

***Zygorhiza* True, 1908**

Type species: *Zygorhiza kochii* Reichenbach, 1847.

Type specimen: MNB 15324a-b (m. 44).

Characteristics: Smaller than both *Basilosaurus* and *Cynthiacetus*; lacking elongate trunk vertebrae of *Basilosaurus*; differs from *Dorudon* and *Chrysocetus* in having accessory denticles on the cingula of the upper premolars (Köhler and Fordyce, 1997).

Occipital condyle breadth: 125 mm (USNM 16638; Millsaps College Museum adult specimen [Kellogg, 1936]).

Included species: *Z. kochii* only (known from localities GC22C [Westgate, Gillette, and Rolater, 1994], GC23III [Daly, 1999], GC23IVA–E, G [Kellogg, 1936], GC23VA [Kellogg, 1936], GC23VIIC, D [Kellogg, 1936], GC23VIIIC, F, K–O [UF Collections]).

Comments: *Zygorhiza kochii* is the most common archaeocete in North America. It is known from both juvenile and adult specimens, which include parts of most of the skeleton except the hindlimb. *Zygorhiza* is also known from the Eocene of Europe and New Zealand.

ARCHAEOCETI NOMINA DUBIA

***Pontobasileus* Leidy, 1873**

Type species: *Pontobasileus tuberculatus* Leidy 1873.

Type specimen: ANSP 11216 Eocene of Alabama? (Leidy, 1873).

Characteristics: Back part of crown forming a wide, thick, heel, extending over more than half the width of the posterior root; enamel tuberculate (Leidy, 1873).

Occipital condyle breadth: occipital condyles are not known from *Pontobasileus*.

Included species: *Pontobasileus tuberculatus* is known only from the type specimen, which was discovered in the collections of the ANSP along with *Basilosaurus* material from Alabama (Leidy, 1873).

Comments: Material is incomplete and probably not diagnostic. Other material cannot be referred to the genus with confidence, and the family-level relationships are uncertain. Hulbert *et al.* (1998) suggested that *Pontobasileus*, which is known only from a single partial tooth, might represent the same taxon as *Georgiacetus*, but they did not synonymize the two taxa.

BIOLOGY AND EVOLUTIONARY PATTERNS

Archaeoceti includes species that document the transition from fully terrestrial, quadrupedal mammals to fully aquatic, bipinnate mammals. Many of the most significant changes in cetacean anatomy happen in the pakicetids, ambulocetids, and early protocetids, which are known only from the Old World. These changes affected feeding (elongation of the snout, separation of the anterior teeth, increase in relative cheek tooth crown height), hearing (increase in size and density of the auditory bulla, decrease in contact of the bulla from the base of the skull, rotation of the ear ossicles, enlargement of the mandibular canal and foramen), locomotion (reduction of the hindlimbs, unfusing of the sacral vertebrae), and streamlining (shortening of the neck). Protocetids arrived in North America at a time when these changes were already apparent in their anatomy but they still retained large hindlimbs (Hulbert *et al.*, 1998; Uhen, 1998b, 1999) (Figure 33.3).

Subsequently, the origin of basilosaurids involved the development of accessory denticles on the cheek teeth, loss of M3, loss of the protocone and third root on upper molars, further shortening of the neck, an increase in the number of trunk vertebrae, loss of the sacrum (as an attachment for the innominate), development of the tail fluke, development of the forelimb as a flipper, and further reduction of the hindlimb. All of these features together indicate that basilosaurids are the earliest fully aquatic cetaceans (Uhen, 1998a).

The development of an elongate snout and possibly accessory denticles on the cheek teeth appear to be adaptations for a piscivorous feeding strategy. Modern toothed cetaceans feed mainly on fish and squid, with little or no oral processing. Many teuthophagus (squid-eating) cetaceans have blunt snouts and are thought to use suction feeding for prey acquisition (Werth, 2000), while more piscivorous or mixed-prey feeding cetaceans have more elongate snouts. Archaeocetes also have well-developed wear facets on the buccal surfaces of their lower cheek teeth and lingual surfaces of their upper cheek teeth, which are indicative of a high degree of vertical motion of the mandible (O'Leary and Uhen, 1999). In addition, apical wear on the cusps and accessory denticles of basilosaurids are indicative of a high degree of tooth–food contact, as some of these wear facets could not physically be made by tooth–tooth contact. Both of these sets of wear facets indicate that archaeocetes retained

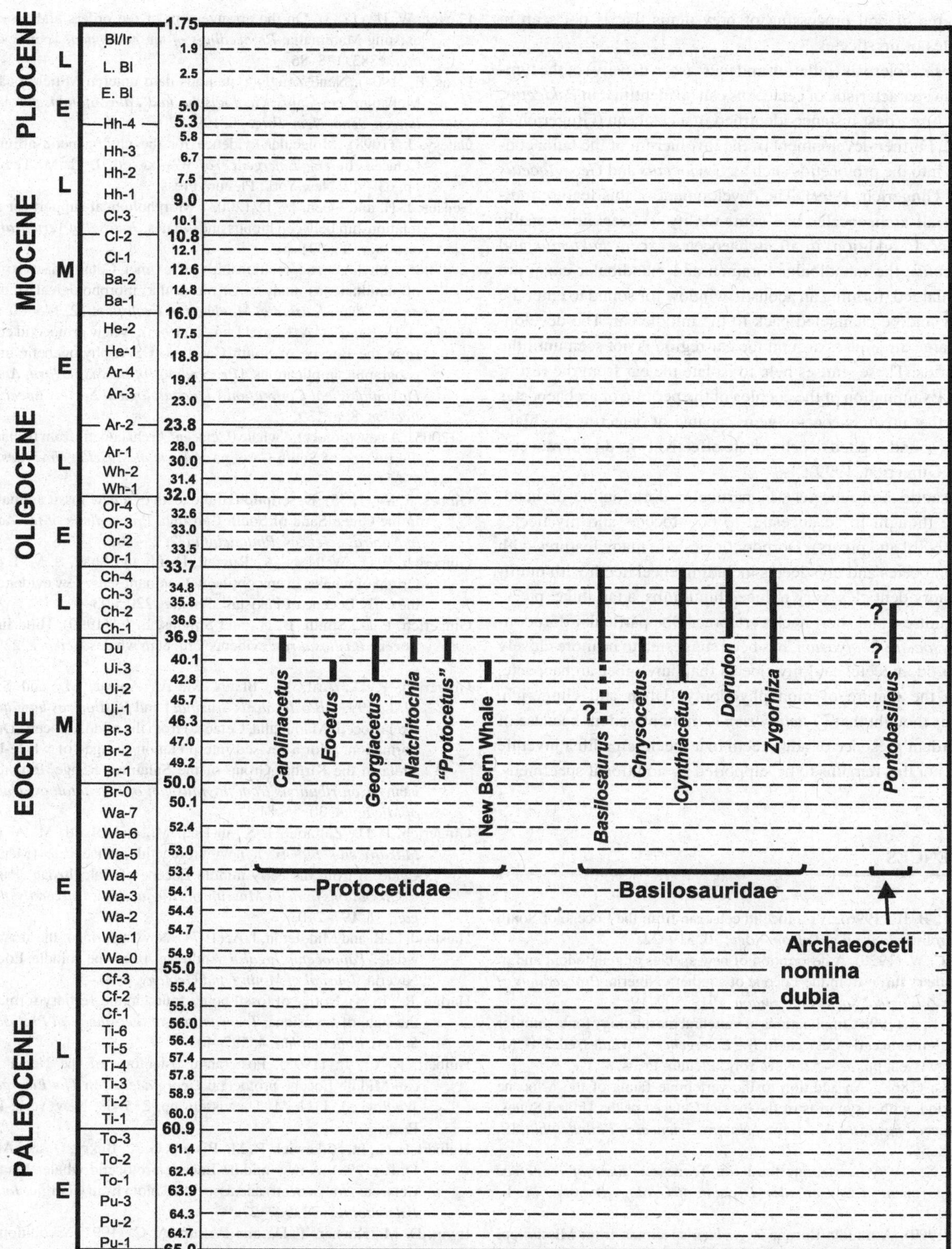

Figure 33.3. Temporal ranges of North American archaeocete genera.

some degree of oral processing of prey items that is not seen in odontocetes or mysticetes.

Pachyosteosclerosis (inflation and increase in density) of the tympanic bulla characteristic of cetaceans can be identified in *Pakicetus*, which is the earliest instance identified in a cetacean (Gingerich *et al.*, 1983). Further development of the involucrum of the bulla continued on into the protocetids such as *Gaviacetus* and *Georgiacetus* (Luo and Gingerich, 1999). The development of this feature indicates that all archaeocetes had some ability to hear directionally underwater. In addition, in all archaeocetes except *Pakicetus* and *Himalyacetus*, the mandibular foramen and mandibular canal are greatly enlarged, forming an acoustic window for sound to enter the lower jaw and be channeled back to the middle ear. The development of large air sinuses around the ear region is not seen until the basilosaurids. These sinuses help to isolate the ear from the rest of the skull. Examination of the cochlea of the periotic of archaeocetes indicates that no archaeocetes were capable of detecting the high-frequency sound needed to echolocate effectively as do odontocetes (Luo and Gingerich, 1999).

Basilosaurid archaeocetes (in particular dorudontine archaeocetes) are thought to be ancestral to odontocetes and mysticetes (Uhen [1998b] and others). Dorudontines share many features with early odontocetes and mysticetes including cheek teeth with multiple accessory denticles, very reduced hindlimbs, a tail fluke, pterygoid air sinuses, and short neck vertebrae. One particular dorudontine, *Chrysocetus healyorum*, has been suggested to be more closely related to odontocetes and mysticetes than any other archaeocete, based on the feature of monophyodonty (Uhen and Gingerich, 2001). It is thought that *C. healyorum* is monophyodont because of the discordant presence of adult teeth in a specimen with a juvenile skeleton, but this remains to be supported by additional specimens.

REFERENCES

Albright, L. B. III (1996). A protocetid cetacean from the Eocene of South Carolina. *Journal of Paleontology*, 70, 519–22.

Andrews, C. W. (1920). A description of new species of zeuglodont and of leathery turtle from the Eocene of southern Nigeria. *Proceedings of the Zoological Society of London*, 1919, 309–19.

Buchholtz, E. A. (1998). Implications of vertebral morphology for locomotor evolution in early Cetacea. In *The Emergence of Whales*, ed. J. G. M. Thewissen, pp. 325–52. New York: Plenum Press.

Cope, E. D. (1868). An addition to the vertebrate fauna of the Miocene period, with a synopsis of the extinct Cetacea of the United States. *Proceedings of the Academy of Natural Sciences of Philadelphia*, 19, 138–56.

Corgan, J. X. (1976). Vertebrate fossils of Tennessee. *State of Tennessee Department of Conservation Division of Geology Bulletin*, 77, 1–100.

Daly, E. (1999). A middle Eocene *Zygorhiza* specimen from Mississippi (Cetacea, Archaeoceti). *Mississippi Geology*, 20, 21–31.

Dockery, D. T., III and Johnston, J. E. (1986). Excavation of an archaeocete whale, *Basilosaurus cetoides* (Owen), from Madison, Mississippi. *Mississippi Geology*, 6, 1–10.

Domning, D. P., Morgan, G. S., and Ray, C. E. (1982). North American Eocene sea cows (Mammalia, Sirenia). *Smithsonian Contributions to Paleobiology*, 52, 1–69.

Flower, W. H. (1883). On the arrangement of the orders and families of existing Mammalia. *Proceedings of the Zoological Society of London*, 1883, 178–86.

Fraas, E. (1904). Neue Zeuglodonten aus dem unteren Mittelecocän vom Mokattam bei Cairo. *Geologische und Palæontologische Abhandlungen, Jena, Neue Folge*, 6, 199–220.

Gatesy, J. (1998). Molecular evidence for the phylogenetic affinities of Cetacea. In *The Emergence of Whales*, ed. J. G. M. Thewissen, pp. 63–112. New York: Plenum Press.

Geisler, J. H. and Uhen, M. D. (2003). Morphological support for a close relationship between hippos and whales. *Journal of Vertebrate Paleontology*, 23, 991–6.

(2005). Phylogenetic relationships of extinct Cetartiodactyls: results of simultaneous analyses of molecular, morphological, and stratigraphic data. *Journal of Mammalian Evolution*, 12, 145–60.

Geisler, J. H., Sanders, A. E., and Luo, Z. (1996). A new protocetid cetacean from the Eocene of South Carolina, USA: phylogenetic and biogeographic implications. [*Proceedings of the Sixth North American Paleontological Convention.*] *Paleontological Society Special Publication*, 8, p. 139.

(2005). A new protocetid whale (Cetacea: Archaeoceti) from the late middle Eocene of South Carolina. *American Museum Novitates*, 3480, 1–65.

Gibbes, R. W. (1845). Description of the teeth of a new fossil animal found in the Green Sand of South Carolina. *Proceedings of the Academy of Natural Sciences, Philadelphia*, 2, 254–6.

Gingerich, P. D., Wells, N. A., Russell, D. E., and Shah, S. M. I. (1983). Origin of whales in epicontinental remnant seas: new evidence from the early Eocene of Pakistan. *Science*, 220, 403–6.

Gingerich, P. D., Smith, B. H., and Simons, E. L. (1990). Hind limbs of Eocene *Basilosaurus*: evidence of feet in whales. *Science*, 229, 154–7.

Gingerich, P. D., Arif, M., Bhatti, M. A., Anwar, M., and Sanders, W. J. (1997). *Basilosaurus drazindai* and *Basiloterus hussaini*, new Archaeoceti (Mammalia, Cetacea) from the middle Eocene Drazinda Formation, with a revised interpretation of ages of whale-bearing strata in the Kirthar Group of the Sulaiman Range, Punjab (Pakistan). *Contributions from the Museum of Paleontology, University of Michigan*, 30, 55–81.

Gingerich, P. D., Zalmout, I. S., ul-Haq, M., and Bhatti, M. A. (2005). *Makaracetus bidens*, a new protocetid archaeocete (Mammalia, Cetacea) from the early middle Eocene of Balochistan (Pakistan). *Contributions from the Museum of Paleontology, University of Michigan*, 31, 197–210.

Halstead, L. B. and Middleton, J. A. (1974). New material of the archaeocete whale, *Pappocetus lugardi* Andrews, from the middle Eocene of Nigeria. *Journal of Mining and Geology*, 8, 81–5.

Harlan, R. (1834). Notice of fossil bones found in the Tertiary formation of the state of Louisiana. *Transactions of the American Philosophical Society Philadelphia*, 4, 397–403.

Hulbert, R. C., Jr. (1998). Postcranial osteology of the North American Middle Eocene protocetid *Georgiacetus*. In *The Emergence of Whales*, ed. J. G. M. Thewissen, pp. 235–68. New York: Plenum Press.

Hulbert, R. C., Jr., Petkewich, R. M., Bishop, G. A., Bukry, D., and Aleshire, D. P. (1998). A new middle Eocene protocetid whale (Mammalia: Cetacea: Archaeoceti) and associated biota from Georgia. *Journal of Paleontology*, 72, 907–27.

Irwin, D. M., Kocher, T. D., and Wilson, A. C. (1991). Evolution of the cytochrome *b* gene of mammals. *Journal of Molecular Evolution*, 2, 37–55.

Kellogg, R. (1936). A Review of the Archaeoceti. *Carnegie Institution of Washington Special Publication*, 482, 1–366.

Köhler, R. and Fordyce, R. E. (1997). An archaeocete whale (Cetacea: Archaeoceti) from the Eocene Waihao Greensand, New Zealand. *Journal of Vertebrate Paleontology*, 17, 574–83.

Lancaster, W. C. (1986). The taphonomy of an archaeocete skeleton and its associated fauna. In *Montgomery Landing Site, Marine Eocene (Jackson) of Central Louisiana: Proceedings of a Symposium at the 36th Annual GCAGS*, ed. J. A. Schiebout and W. van den Bold, pp. 119–31. Baton Rouge, LA: Gulf Coast Association of Geological Societies.

(1990). The middle ear of the Archaeoceti. *Journal of Vertebrate Paleontology*, 10, 117–27.

Leidy, J. (1852). [Description of *Pontogeneus priscus*.] *Proceedings of the Academy of Natural Sciences of Philadelphia*, 6, 52.

(1873). *Report of the United States Geological Survey of the Territories: Contributions to the Extinct Vertebrate Fauna of the Western Territories*, pp. 1–358. Washington, DC: US Government Printing Office.

Luo, Z. (1998). Homology and transformation of cetacean ectotympanic structures. In *The Emergence of Whales*, ed. J. G. M. Thewissen, pp. 269–302. New York: Plenum Press.

Luo, Z. and Gingerich, P. D. (1999). Terrestrial Mesonychia to aquatic Cetacea: transformation of the basicranium and evolution of hearing in whales. *University of Michigan Papers on Paleontology*, 31, 1–98.

Maher, J. C. and Jones, P. H. (1949). Ground-water exploration in the Natchitoches area Louisiana. *United States Geological Survey Water Supply Paper*, 968-D, 159–211.

McLeod, S. A. and Barnes, L. G. (1996). The systematic position of *Pappocetus lugardi* and a new taxon from North America (Archaeoceti: Protocetidae). [*Proceedings of the Sixth North American Paleontological Convention*.] *Paleontological Society Special Publication*, 8, p. 270.

Milinkovitch, M. C., Bérubé, M., and Palsbøll, P. J. (1998). Cetaceans are highly derived artiodactyls. In *The Emergence of Whales*, ed. J. G. M. Thewissen, pp. 113–32. New York: Plenum Press.

Miller, G. S., Jr. (1923). The telescoping of the cetacean skull. *Smithsonian Miscellaneous Collections*, 76, 1–71.

Novacek, M. J., Wyss, A. R., and McKenna, M. C. (1988). The major groups of eutherian mammals. In *The Phylogeny and Classification of the Tetrapods*, Vol. 2, ed. M. J. Benton, pp. 31–71. Oxford: Clarendon Press.

O'Leary, M. A. (1998). Phylogenetic and morphometric reassessment of the dental evidence for a mesonychian and cetacean clade. In *The Emergence of Whales*, ed. J. G. M. Thewissen, pp. 133–62. New York: Plenum Press.

O'Leary, M. A. and Geisler, J. H. (1999). The position of Cetacea within Mammalia: phylogenetic analysis of morphological data from extinct and extant taxa. *Systematic Biology*, 48, 455–90.

O'Leary, M. A. and Uhen, M. D. (1999). The time of origin of whales and the role of behavioral changes in the terrestrial-aquatic transition. *Paleobiology*, 25, 534–56.

Palmer, K. V. W. (1939). *Basilosaurus* in Arkansas. *Bulletin of the American Association of Petroleum Geologists*, 23, 1228–9.

Philippe, H. and Douzery, E. (1994). The pitfalls of molecular phylogeny based on four species, as illustrated by the Cetacea/Artiodactyla relationships. *Journal of Mammalian Evolution*, 2, 133–52.

Prothero, D. R., Manning, E. M., and Fischer, M. (1988). The phylogeny of the ungulates. In *The Phylogeny and Classification of the Tetrapods*, Vol. 2, ed. M. J. Benton, pp. 201–34. Oxford: Clarendon Press.

Randazzo, A. F., Kosters, M., Jones, D. S., and Portell, R. W. (1990). Paleoecology of shallow-marine carbonate environments, middle Eocene of Peninsular Florida. *Sedimentary Geology*, 66, 1–11.

Reichenbach, L. (1847). Systematisches. In *Resultate geologischer, anatomischer und zoologischer untersuchungen über das unter den Namen Hydrarchos von Dr. A. C. Koch zuerst nach Europa gebrachte und in Dresden augestelte grofse fossile Skelett*, ed. C. G. Carus, pp. 1–15. Dresden: Arnoldische Buchhandlung.

Stromer, E. (1908). Die Urwale (Archaeoceti). *Anatomischer Anzeiger*, 33, 81–8.

Thewissen, J. G. M. (1994). Phylogenetic aspects of cetacean origins: a morphological perspective. *Journal of Mammalian Evolution*, 2, 157–84.

Thewissen, J. G. M. and Hussain, S. T. (1993). Origin of underwater hearing in whales. *Nature*, 361, 444–5.

Thewissen, J. G. M., Madar, S. I., and Hussain, S. T. (1996). *Ambulocetus natans*, an Eocene cetacean (Mammalia) from Pakistan. *Courier Forschungsinstitut Senckenberg*, 191, 1–86.

True, F. W. (1908). The fossil cetacean, *Dorudon serratus* GIBBES. *Bulletin of the Museum of Comparative Zoology*, 52, 5–78.

Uhen, M. D. (1998a). Middle to Late Eocene basilosaurines and dorudontines. In *The Emergence of Whales*, ed. J. G. M. Thewissen, pp. 29–61. New York: Plenum Press.

(1998b). New protocetid (Mammalia, Cetacea) from the late middle Eocene Cook Mountain Formation of Louisiana. *Journal of Vertebrate Paleontology*, 18, 664–8.

(1999). New species of protocetid archaeocete whale, *Eocetus wardii* (Mammalia, Cetacea), from the middle Eocene of North Carolina. *Journal of Paleontology*, 73, 512–28.

(2001). New material of *Eocetus wardii* (Mammalia, Cetacea), from the middle Eocene of North Carolina. *Southeastern Geology*, 40, 135–48.

(2004). Form, function, and anatomy of *Dorudon atrox* (Mammalia, Cetacea): an archaeocete from the middle to late Eocene of Egypt. *University of Michigan Papers on Paleontology*, 34, 1–222.

(2005). A new genus and species of archaeocete whale from Mississippi. *Southeastern Geology*, 43, 157–72.

Uhen, M. D. and Gingerich, P. D. (2001). New genus of dorudontine archaeocete (Cetacea) from the middle-to-late Eocene of South Carolina. *Marine Mammal Science*, 17, 1–34.

Werth, A. (2000). A kinematic study of suction feeding and associated behavior in the long-finned pilot whale, *Globicephala melas* (Traill). *Marine Mammal Science*, 16, 299–314.

Westgate, J. W. (2001). Paleoecology and biostratigraphy of marginal marine Gulf Coast Eocene vertebrate localities. In *Eocene Biodiversity: Unusual Occurrences and Rarely Sampled Habitats*, ed. G. F. Gunnell, pp. 263–97. New York: Kluwer Academic/Plenum.

Westgate, J. W., Gillette, C. N., and Rolater, E. (1994). Paleoecology of an Eocene coastal community from Georgia. *Journal of Vertebrate Paleontology*, 14(suppl. to no. 3), p. 52A.

Williams, E. M. (1998). Synopsis of the earliest cetaceans: Pakicetidae, Ambulocetidae, Remingtonocetidae, and Protocetidae. In *The Emergence of Whales*, ed. J. G. M. Thewissen, pp. 1–28. New York: Plenum Press.

34 Odontoceti

MARK D. UHEN,[1] R. EWAN FORDYCE,[2] and LAWRENCE G. BARNES[3]

[1]*Smithsonian Institution, Washington, DC, USA*
[2]*University of Otago, Dunedin, New Zealand*
[3]*Natural History Museum of Los Angeles County, Los Angeles, CA, USA*

INTRODUCTION

Odontoceti are the toothed whales, dolphins, and porpoises. These animals are markedly more diverse and disparate than the mysticetes or baleen whales. Most species are significantly smaller than mysticetes, and no odontocete has baleen. Most, but not all, adult odontocetes have erupted teeth; those without erupted teeth usually have vestigial teeth. Further, the rostrum lacks the flat-based maxilla from which baleen arises in mysticetes. Living odontocetes are predators that generally capture and swallow prey whole. Odontocetes use what teeth they have for prey capture and in some cases sexual display, but not for chewing.

The earliest named odontocete is *Simocetus rayi* Fordyce, 2002, from near Yaquina in Oregon, which was reported initially as late Oligocene; according to dating by Prothero *et al.* (2002); however, the type horizon is probably early Oligocene. Other unnamed odontocetes have also been noted from early Oligocene deposits of Washington State (Barnes and Goedert, 2000; Barnes, Goedert, and Furusawa, 2001). Many well-documented odontocetes are known from late Oligocene deposits around the world (Fordyce and Muizon, 2001). Despite the strong support for odontocete monophyly (Geisler and Sanders, 2003), the phylogenetic relationships among early odontocetes, as well as the relationship of Odontoceti and Mysticeti to archaeocetes has yet to be determined conclusively.

DEFINING FEATURES OF THE SUBORDER ODONTOCETI

CRANIAL

Living odontocetes can be recognized by key soft anatomical features, including a single blowhole, hypertrophied nasofacial muscles, a well-developed melon, and extensive narial diverticula in tissues distal to the external bony nares. These and other soft tissue features in the skull base apparently help to generate the high-frequency sounds used in echolocation (Mead, 1975a; Heyning, 1989a; Cranford, Amundin, and Norris, 1996). Osteological characters of the skull are diagnostic: each maxilla expands dorsally and posteriorly to form a supraorbital plate that is the origin for the enlarged nasofacial muscles. Multiple dorsal infraorbital foramina supply nerves and blood vessels; there is a vertical antorbital notch for passage of the facial nerve. Other osteological and soft-tissue features, including the basicranial sinuses and details of the ear region, have also been cited as diagnostic (Figure 34.1). Winge (1921) and Miller (1923) provided early discussion on such features in odontocetes, as they did for mysticetes, but the main understanding of key functional complexes of odontocetes came much later, for example through the work of Fraser and Purves (1960), Kasuya (1973), Mead (1975a), Heyning (1989a), Cranford, Amundin, and Norris (1996) and others.

In fossil odontocetes, the supraorbital plate of the maxilla is generally similar to that of living species, implying the presence of enlarged nasofacial muscles. This structure of the maxilla and accompanying features have been used to infer echolocation in odontocetes as old as early Oligocene (e.g., *Simocetus* of Fordyce, 2002). In one fossil family, the Odobenocetopsidae, maxillary structure appears to have reversed as part of extreme specialization for suction feeding (Muizon, 1993).

DENTAL

There are no clear dental synapomorphies of Odontoceti that exclude Archaeoceti and Mysticeti. Some modern odontocetes are extremely polydont, while others have reduced numbers of teeth, or lack teeth altogether. The teeth of most living odontocetes are simple conical structures with single roots, but there is a great variety of form among the exceptions to this generalization (Uhen, 2002a). Most

Evolution of Tertiary Mammals of North America, Vol. 2. ed. C. M. Janis, G. F. Gunnell, and M. D. Uhen. Published by Cambridge University Press.

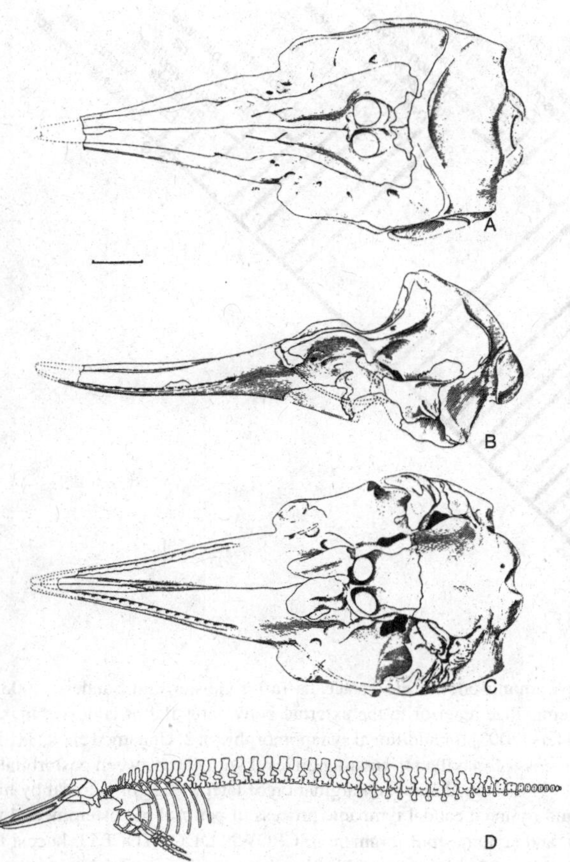

Figure 34.1. Representative odontocete cranial morphology. A. *Albireo whisleri*, dorsal view. B. *Albireo whistleri*, lateral view. C. *Albireo whistleri*, ventral view. (After Barnes 1984; courtesy of the University of California.) D. Skeleton of *Kentriodon pernix*, left lateral view (after Kellogg, 1927b).

early odontocetes are polydont (as are most early toothed mysticetes), except *Simocetus rayi* (Fordyce, 2002), which leaves open the question of whether polydonty is a characteristic of Neoceti, Odontoceti, or some subgroup of Odontoceti.

POSTCRANIAL

Odontocetes generally have a single, rather straight innominate and lack the added process and remnant femur seen in mysticetes. Most other postcranial characteristics either unite some subgroups of odontocetes or are also found in archaeocetes; consequently, they fail to distinguish odontocetes from all other cetaceans.

SYSTEMATICS

SUPRASUBORDINAL

Odontocetes are thought to have been derived from basilosaurid archaeocetes, based on numerous shared derived features of basilosaurids, mysticetes, and odontocetes (Uhen and Gingerich, 2001; Uhen, 2002b). Currently, it is thought that Mysticeti and Odontoceti share a single common ancestor within the Basilosauridae, but this remains to be demonstrated by a thorough cladistic analysis including all relevant basilosaurids, early mysticetes, and early odontocetes.

INFRASUBORDINAL

CROWN- AND STEM-ODONTOCETI

The currently recognized living families of odontocete (Rice, 1998) are Physeteridae, Kogiidae, Platanistidae, Ziphiidae, Delphinidae, Phocoenidae, Monodontidae, Iniidae, Pontoporiidae, and Lipotidae. These families, along with their most recent common ancestor and all of its descendants, form the crown-Odontoceti. The Physeteridae is the most-basal crown-family. Fossil species have been reported for all of these groups, although family-level allocations for many of the fossils might usefully be clarified using crown- and stem-group concepts.

Fossil odontocetes add significantly to the family-level diversity of the Odontoceti. *Archaeodelphis* and *Xenorophus* represent extinct odontocetes not formally placed in families yet. Extinct families of odontocetes include Agorophiidae, and four apparent platanistoids: Squalodelphinidae, Squalodontidae, Waipatiidae, and Dalpiazinidae. Extinct delphinoids include Kentriodontidae, Albireonidae, and Odobenocetopsidae. Higher-level groups (above family level) that are discussed below include the Physeteroidea, Platanistoidea, Delphinoidea, and Delphinida. This is a conservative approach using traditional names. Recent cladistic analyses identify alternative higher classifications; however, these have not yet been accepted widely amongst cetologists, and deserve more study (Geisler and Sanders, 2003).

Many cladograms have been published showing odontocete relationships; some are based on explicit character matrices with formally described methods of analysis, while others are branching diagrams in which nodes are supported by characters, but for which no formal data matrix or analysis is available. Published accounts of diagnostic characters for different families include lists of both general characters and synapomorphies. Cladistic approaches to odontocete higher relationships include those of Muizon (1987, 1988a,b, 1991), Barnes (1990), Heyning (1989a, 1997), Fordyce (1994), Messenger (1994), Luo and Marsh (1996), Messenger and McGuire (1998), and Geisler and Sanders (2003) (Figure 34.2). Living species were listed by Rice (1998); see also Perrin and Brownell (2001). Useful guides to living species include Nowak (2003) and contributions in Ridgway and Harrison (1989, 1994, 1999) and in Perrin, Würsig, and Thewissen (2002).

INCLUDED GENERA OF ODONTOCETI FLOWER, 1867

The locality numbers listed for each genus refer to the list of unified localities in Appendix I. The locality numbers may be listed in more

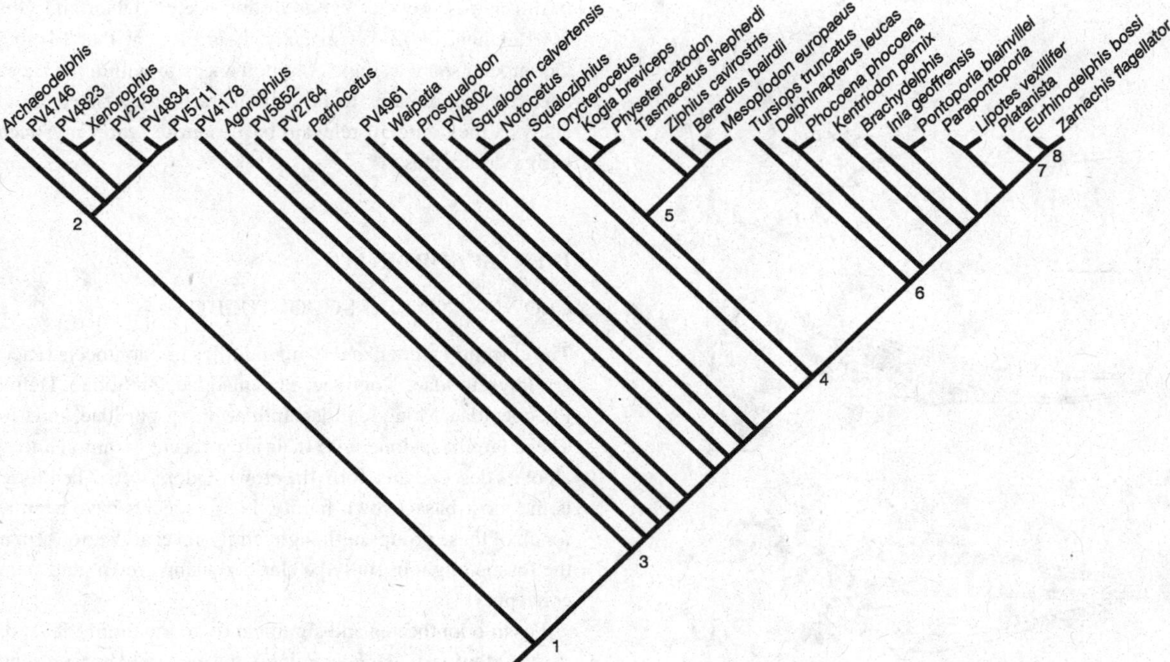

Figure 34.2. Taxa and characteristics at nodes 1–8 are as follows. Relationships among odontocete cetaceans (after Geisler and Sanders, 2003). 1. ODONTOCETI: lacrimal and jugal forming posterior wall of antorbital notch; premaxillae anterior to the external bony nares flat or concave; maxillae covering almost entire surface of supraorbital process of frontal (see Geisler and Sanders [2003] for additional synapomorphies). 2. Unnamed clade: lacrimal excluded from the lateral side of the skull; maxillary foramen present; ascending process of maxilla terminating on level with gap between postorbital and zygomatic process (see Geisler and Sanders [2003] for additional synapomorphies). 3. Unnamed clade: emargination of lateral edge of petrosal by hiatus epitampanicus wide and approximately in line with gap between fenestrae ovalis and rotunda; caudal tympanic process of petrosal in posteromedial view forming a narrow separation or contact; clear separation of stapedial muscle fossa and stylomastoid foramen. 4. CROWN ODONTOCETI: buccal teeth without accessory cusps; premaxillae immediately anterior to external bony nares, gap < 28% of nares width (derived from data in Geisler and Sanders [2003]). 5. PHYSETEROIDEA: lacrimal abuting against and slightly overlapping the anterior edge of supraorbital process of the frontal; zygomatic process of the squamosal directed anterolaterally; basioccipital crests forming angle of 74–90° (see Geisler and Sanders [2003] for additional synapomorphies). 6. Unnamed clade: proximal ethmoid region exposed dorsally; right posterior nasal sac reduced, short, and never reaching osteological vertex; palatine fossa well developed, dividing palatine into medial and lateral laminae (derived from data in Geisler and Sanders [2003]). 7. "PLATANISTIDS" (Geisler and Sanders [2003] include eurhinodelphinids in the family Platanistidae, which does not follow traditional usage and is not followed here): clade including *Platanista* and Eurhinodelphinidae has the following synapomorphies; lateral edge of the rostrum adjacent to the antorbital notch straight; postorbital ridge; three or more facial infraorbital foramina (see Geisler and Sanders [2003] for additional synapomorphies). 8. EURHINODELPHINIDAE: elongate rostrum; premaxillae contact each other along most of the rostrum; 50–60 upper teeth per quadrant (see Geisler and Sanders [2003] for additional synapomorphies).

than one way. The acronyms for museum collections are listed in Appendix III.

Parentheses around the locality (e.g., [CP101]) mean that the taxon in question at that locality is cited as an "aff." or "cf." the taxon in question. Parentheses are usually used for individual species, implying that the genus is firmly known from the locality, but the actual species identification may be questionable. Question marks in front of the locality (e.g., ?CP101) mean that the taxon is questionably known from that locality, implying some doubt that the taxon is actually present at that locality, either at the genus or species level. An asterisk (*) indicates the type locality.

FAMILY INCERTAE SEDIS, INCLUDING ARCHAIC ODONTOCETES

Archaic odontocetes (skull with a marked intertemporal region and heterodont teeth) have long been placed in the Agorophiidae, in turn widely used as a grade taxon (see review of taxa by Fordyce [2002]). Fordyce (1981) suggested that the Agorophiidae be recognized as the clade including the late Oligocene *Agorophius pygmaeus* of the Atlantic Coastal Plain. No other named odontocetes clearly belong to such a group, although see comment under *Agorophius pygmaeus*. The sister taxon of Agorophiidae is uncertain, and the family is not demonstrably related to Squalodontidae. The holotype skull of *Agorophius pygmaeus* is lost, and synapomorphies for the family have not been identified.

The relationships of other described archaic odontocetes from North America are equally uncertain. Of these, the late Oligocene *Archaeodelphis patrius* Allen appears to be the most basal named odontocete (Geisler and Sanders, 2003).

Xenorophus sloani Kellogg 1923a and *Xenorophus* spp. of Whitmore and Sanders (1977; see also Sanders 1996) are here excluded from the Agorophiidae. These peculiar late Oligocene odontocetes from South Carolina show a mix of archaic skull features (prominent intertemporal constriction, bulk of nasofacial muscles probably lying anterior to orbit) and apomorphies (greatly enlarged lacrimals) that make it hard to judge their relationships within the Odontoceti. Luo and Gingerich (1999: Fig. 26) and Marino, McShea, and Uhen (2004) used the new name Xenorophidae without defining or diagnosing the family, making Xenorophidae currently a nomen nudum. The results of the phylogenetic analysis of Geisler and Sanders (2003) includes a clade containing *Archaeodelphis*, *Xenorophus sloanii*, *Xenorophus* sp., and other unnamed taxa, indicating that these taxa may appropriately be grouped into a family-level taxon.

Araeodelphis **Kellogg, 1957**

Type species: *Araeodelphis natator* Kellogg, 1957.

Type specimen: USNM 10478, essentially complete rostrum except for portion immediately in front of antorbital notches, symphyseal portion of mandibles, and 12 detached teeth (Kellogg, 1957).

Characteristics: Rostrum slender, attenuated toward anterior extremity; mesorostral trough roofed over by close approximation of premaxillaries; narrow median longitudinal groove on palatal surface extends forward from 45th tooth (counting backward from anterior-most alveolus) to extremity; more than 45 teeth located on each side of rostrum; mandibles slender with symphysis firmly ankylosed, elongated and tapered toward anterior extremity; width of symphysis greater than depth; median longitudinal groove on dorsal surface of symphysis indistinctly divided longitudinally for approximately two-thirds its length by the raised thin ridge formed at line of contact of mandibles; opposite free posterior portions of mandibles coming together at a blunt angle (55°) at symphysis; external surface of each mandible convex dorsally above the deep channel or groove, commencing about 20 mm behind level of posterior end of symphysis and extending forward almost to anterior end; these channels or grooves on the opposite rami bound the median and somewhat convex longitudinal strip on ventral surface of symphysis; 36 to 37 teeth located on each side of symphysis; roots of teeth implanted in the alveoli, which slope more inward than backward; roots of teeth noticeably swollen below base of crown and attenuated near extremity; crowns of most of the teeth on symphysis and rostrum slender, pointed, and curved inward; enamel on crowns of majority of teeth indistinctly wrinkled in a vertical direction on internal surface; posterior teeth on symphysis have enamel more coarsely sculptured internally and possess a small tubercle on posterior margin of internal face; posterior teeth (41–45) on rostrum possessing one small tubercle on the posterior surface about half way up height of crown and varying number of minute rugosities on internal surface; interval between teeth at base of crowns varies from 3 to 6 mm (Kellogg, 1957).

Occipital condyle breadth: unknown.

Included species: *A. natator* only (known from localities NC2D*, NC3A).

Comments: Material is incomplete and debatably diagnostic, but the concept of the genus could probably be firmed up if more-complete topotypic specimens were found. For now, family-level relationships are uncertain.

Archaeodelphis **Allen, 1921a**

Type species: *Archaeodelphis patrius* Allen, 1921a.

Type specimen: MCZ 15749, posterior portion of a skull with an incomplete braincase.

Characteristics: Teeth unknown, but apparently long-rooted, probably resembling those of *Agorophius* and *Prosqualodon*; nasals long, narrow, and flattened dorsally; maxillae covering the anterior three-fourths of the orbital portion of the frontals; orbit large, with thickened rim and prominent postorbital process; parietal processes meeting across the vertex of the skull behind the orbits; zygomatic process of squamosal relatively small, with small and nearly horizontal glenoid fossa; mastoid region thickened and produced obliquely downward and backward to or beyond the posterior edge of the condyles, which are small and protuberant; palatals large, expanded anteriorly, separated medially for more than half their length at the back end by a deep notch at the front end of their combined margin; pterygoids widely sundered, free margins partly overarching the narial passage; well-marked nasal chamber present above the anterior end of the passage; vomer forming a cylinder that completely encloses the basal end of the mesethmoid cartilage (Allen, 1921a).

Occipital condyle breadth: 57 mm (MCZ 15749; Allen, 1921a, p. 11).

Included species: *A. patrius* only, locality unknown.

Comments: The type specimen of *Archaeodelphis patrius* was discovered in the collections of the Museum of Comparative Zoology, Harvard by Allen with no label of any kind. Allen stated that "The likelihood is that it was sent to Louis Agassiz in the early days of the Museum, possibly from some locality in the southeastern United States, at the time when he was planning a memoir on '*Phocodon*'." Allen sent some of the adhering matrix to Dr. Joseph A. Cushman who reported that, "there are a few Foraminifera contained in it, most of which are not well preserved. A few, however, seem to show that the material is probably Upper Eocene (Jackson) in age, and its general appearance would seem to indicate that it came from the Gulf Coastal Plain of the United States, probably from Alabama." The specimen is quite different from any others known from the Jackson (age or group). Recently, one of us (M. D. Uhen) sent another sediment sample from inside the braincase of MCZ 15749 to the US Geological Survey, where Lucy Edwards analyzed the sample and reported that, "The sediment from the *Archaeodelphis* skull yielded dinoflagellate cysts of Oligocene age. Although the preservation is poor to fair and there is a lot of amorphous debris." The only beds of Oligocene age in the Atlantic and Gulf Coastal regions that have been reported to produce cetacean fossils are the Ashley Formation (locality EM4), Chandler Bridge Formation (locality EM5), and Tiger Leap Formation (not included in this volume). Robert Weems of the US Geological Survey examined the sediment sample and concluded that, "of these, only the Tiger Leap lithology matches what you pulled from the braincase [of *Archaeodelphis*]." While it is possible that the fossil came from some other deposit, we think it likely that *Archaeodelphis* is late Oligocene and not late Eocene in age and that it is more likely to be from the Charleston area of South Carolina than from Alabama.

***Saurocetus* Agassiz, 1848**

Type species: *Saurocetus gibbesii* Agassiz, 1848.

Type specimen: MCZ 8760, isolated tooth.

Characteristics: Distinguished from *Dorudon* by its "great flatness and acute serrated edge" (Agassiz, 1848).

Occipital condyle breadth: unknown.

Included species: *S. gibbesi* only (locality unknown, probably EM4 or EM5 [Agassiz, 1848]).

Comments: It is unclear where the isolated holotype tooth of *Saurocetus gibbesi* was found. The genus was named in a letter from Professor Agassiz to Dr. Gibbes (who found the specimen and transmitted it to Agassiz) that was printed in the *Proceedings of the Academy of Natural Sciences of Philadelphia*. Allen (1924a) reviewed the history of the holotype.

Dooley (2003) excluded *Saurocetus* from the Squalodontidae because the tooth lacks cristae rugosae, which Dooley regarded as characteristic of the family. Dooley (2003) suggested that *Saurocetus* may be related to genus Y of Whitmore and Sanders (1977). We have seen comparably smooth-crowned teeth in Oligocene Squalodontidae from New Zealand and so cannot rule out squalodontid relationships for *Saurocetus*.

***Xenorophus* Kellogg, 1923a**

Type species: *Xenorophus sloanii* Kellogg, 1923a.

Type specimen: USNM 11049, partial skull.

Characteristics: Extraordinary widening of the proximal end of the premaxillary, extending outward under the maxilla, and the overspreading of the supraorbital process by the lacrimal, which also sheaths much of the orbit; nasals situated on a level with the postorbital extension of the supraorbital process; nasal passages sloping backward; maxilla sloping abruptly downwards in front of the orbit; intertemporal constriction present (Kellogg, 1923a; Whitmore and Sanders, 1977).

Occipital condyle breadth: not preserved.

Included species: *X. sloanii* only, known from locality EM4* only.

Comments: *Xenorophus* was placed in the family Agorophiidae by Miller (1923); however, both Whitmore and Sanders (1977) and Fordyce (1981) agreed that *Xenorophus* and *Agorophius* are not closely related and belong in separate families. Luo and Gingerich (1999) included the family-level name Xenorophidae in a figure and Marino, McShea, and Uhen (2004) included it in a table, but there is not yet a formal published description of the Xenorophidae to list diagnostic features and constituent taxa. The Xenorophidae have not been defined or diagnosed, and mention here should not be construed as validation.

AGOROPHIIDAE ABEL, 1913

The family Agorophiidae is based solely on the monospecific genus *Agorophius*, described below.

***Agorophius* Cope, 1895 (including *Zeuglodon*, in part)**

Type species: *Agorophius pygmaeus* (Müller, 1849) (originally described as *Zeuglodon pygmaeus*).

Type specimen: MCZ 8761, partial skull and cheek tooth. Current whereabouts of the skull is unknown, but the cheek tooth is still housed at the MCZ (Fordyce, 1981).

Characteristics: Incomplete rostrum triangular and straight sided, dorsoventrally narrow in lateral view, with cheek teeth inserted apparently vertically along margin; anterior details of rostrum unknown; cheek teeth tall crowned, relatively smooth, with multiple small denticles; supraorbital process of maxilla with multiple posterior dorsal infraorbital foramina, expanding laterally over supraorbital process of frontal, posteriorly toward orbitotemporal crest, and rising medially to form a shallow fossa for nasofacial muscles; premaxilla bifurcated posteriorly, premaxillary foramen and premaxillary sac fossa present, other details of premaxilla lost; short and broad intertemporal constriction present, formed by parietals and associated with large open temporal fossae; supraoccipital not inflated posteriorly, with a rounded anterior nuchal crest; lateral walls of braincase somewhat inflated; zygomatic process of squamosal with robust base; occipital condyles prominent and rounded (based on illustration published by True, 1907a).

Occipital condyle breadth: approximately 80 mm (estimated from the lithograph of the holotype skull published by True [1907a]).

Included species: *A. pygmaeus* only, known from locality EM4* only (Tuomey, 1848).

Comments: Cheek teeth that Ravn (1926) attributed to the east Atlantic Oligocene species *Squalodon (Microzeuglodon) wingei* may be conspecific with *A. pygmaeus*.

SIMOCETIDAE FORDYCE, 2002

The extinct family Simocetidae is based on *Simocetus emlongi* Fordyce, 2002, known from a skull, mandible, and a few postcranial fragments of early Oligocene age from Oregon. Although the skull is of archaic ("agorophiid") grade, it shows some highly specialized features: toothless premaxillae, apex of rostrum dorsoventrally thin with shallow mesorostral groove, and anterior of rostrum and mandible downturned. When the species was described, its broader relationships were uncertain. Fordyce now considers that the toothless premaxillae and elongate conical pterygoid hamulus place *Simocetus* and Simocetidae closer to Eurhinodelphinidae than to other odontocetes.

Simocetus Fordyce, 2002

Type species: *Simocetus rayi* Fordyce, 2002.

Type specimen: USNM 256697, skull and lower jaw with teeth, fragmentary postcrania.

Characteristics: Thin spatulate edentulous rostral apex, formed by edentulous premaxillae, deflected ventrally, resulting in a down-turned rostrum; relatively short rostrum; supraorbital process of maxilla not expanded laterally over supraorbital process of frontal; deep, narrow optic infundibulum; complex fine sutures between alisphenoid and squamosal; posterior maxillary cheek teeth occluding into facing diastemata between mandibular cheek teeth, and vestigial lower I1 (Fordyce, 2002).

Occipital condyle breadth: 83 mm (estimated from the single preserved condyle on USNM 256697).

Included species: *S. rayi* only, known from locality WM4II* only.

Comments: The type specimen is the only described individual of *Simocetus rayi*.

EURHINODELPHINIDAE (ABEL, 1901) RICE, 1998

The Eurhinodelphinidae includes extinct medium- to long-beaked small odontocetes from widely scattered early to middle Miocene localities (e.g., Abel, 1909; Kellogg, 1932), with a few possible records of late Oligocene age. They take their identity from the Belgian late early Miocene *Eurhinodelphis cocheteuxi* Du Bus, 1867. The many reported bony diagnostic features (Kellogg 1925a, 1928, 1932; Muizon 1988b, 1991; Lambert, 2004, 2005a) include the following: rostrum very long; premaxillae toothless and extending beyond the mandible; premaxillary–maxillary suture on rostrum forming a prominent lateral fissure; mesorostral groove roofed by premaxillae; premaxillary fossae with a marked slope upwards to vertex; zygomatic process and postglenoid process of squamosal robust; posterior part of pterygoid hamulus long and conical. This group has previously been called Eurhinodelphidae and Rhabdosteidae (for the latter see Fordyce, 1983, who followed Myrick, 1979).

Because the skull structure is quite specialized, and few details of the pterygoid sinuses and tympanoperiotics are described, the relationships of eurhinodelphinids are uncertain. Muizon (1988a, 1991) placed them close to the extinct *Eoplatanista* (early Miocene, Europe; Eoplatanistidae) and, in turn, regarded the Eurhinodelphinoidea as a possible sister taxon to Delphinida. Geisler and Sanders (2003) placed *Eurhinodelphis* in a paraphyletic Platanistidae but included no other eurhinodelphinids in their analysis.

Argyrocetus Lydekker, 1894 (including *Doliodelphis; Eurhinodelphis*, in part)

Type species: *Argyrocetus patagonicus* Lydekker, 1894.

Type specimen: Museo de La Plata specimen number unknown, figured in Lydekker (1894: Plate 5, Figs. 1, 1a, 2, and 3).

Characteristics: Differing from all living Odontoceti in having well-developed nasals overhanging the narial cavity; nasals large squared bones articulated by a broad base with the frontals, projecting over the narial aperture, where they terminate in a nearly straight transverse edge; inferior surface of nasals regularly beveled from the thickened base to the sharp anterior edge (Lydekker, 1894).

Occipital condyle breadth: 74 mm (USNM 11996, type specimen of *Argyrocetus joaquinensis*, Kellogg, 1932, p. 11).

Included species: *A. joaquinensis* (known from locality WM10* [Kellogg, 1932]); *A. bakersfieldensis* (locality WM10* [Wilson, 1935]).

Comments: The genus *Argyrocetus* is known primarily from the Miocene of South America (McKenna and Bell, 1997). *Argyrocetus bakersfieldensis* was named by Wilson (1935) as *Acrodelphis bakersfieldensis* (Barnes 1977). *Argyrocetus joaquinensis* includes *Doliodelphis littlei* Wilson, 1935 and *Eurhinodelphis extensus* Wilson, 1935 (Barnes, 1977). Lambert (2004) implied that the Californian species might not be congeneric with *Argyrocetus patagonicus* Lydekker, 1894.

Ceterhinops Leidy, 1877

Type species: *Ceterhinops longifrons* Leidy, 1877.

Type specimen: ANSP 11420, facial portion of cranium.

Characteristics: Differs from *Argyrocetus* by having larger temporal fossae, more prominent lambdoid crests, lambdoid crests that curve posteromedially at the posterior edges of the temporal fossae, and larger roots of teeth. Posterior portion of the vomer exhibiting a comparatively capacious groove for the mesethmoid cartilage; mesethmoid bone forming a thick partition separating the external nares and ending in a stout tuberosity at the commencement of the supra vomerine canal; frontal remarkably long, forming a tranversely concave line at the occipital boundary, approximately 2.5 cm in thickness; surface of forehead nearly flat, but slightly convex laterally and towards the fore part; posterior portions of premaxillaries comparatively narrow where they bound the nares, ending in a point extending three-fourths the length of the forehead between the frontal and the expanded supra orbital portion of the maxillary (Leidy, 1877).

Occipital condyle breadth: not preserved in the type.

Included species: *C. longifrons* only, known from locality EM6A* only.

Comments: This genus was originally named by Leidy (1877) on the basis of a fragment of the posterodorsal part of the facial region of the skull. According to Gillette (1975), the type specimen is from the Ashley River Phosphate Beds, South Carolina. Kellogg (1923b) specified the formation as the late Miocene Edisto Marl and stated that it "resembles *Eurhinodelphis* so closely that there appears to be no valid reason for separating this form from the latter genus." Later, however, Kellogg (1928: Fig. 22) labeled an illustration of this same specimen as *Squalodon crassus*.

Macrodelphinus Wilson, 1935

Type species: *Macrodelphinus kelloggi* Wilson, 1935.

Type specimen: YPM 13402, incomplete skull in which most of the basicranium and lower occipital region are lacking, as are the right supraorbital process of the frontal, the postero-external borders of both maxillae, and part of the rostrum; several of preserved rostral parts have teeth, or parts of teeth, in situ (Wilson, 1935).

Characteristics: Large size of skull; wide exposure of the frontals on the vertex; short, wide nasals; hinder ends of premaxillaries not extending posteriorly behind nasals; concave supraoccipital; heavy, elongated rostrum (Wilson, 1935).

Occipital condyle breadth: unknown from the holotype.

Included species: *M. kelloggi* only, known from locality WM10* only.

Comments: Wilson (1935) included *Macrodelphinus* in the family Delphinidae, but Barnes (1977) recognized it as a eurhinodelphinid based on its close similarities with the genus *Argyrocetus*. *Macrodelphinus* is also possibly known from the Miocene of Europe (McKenna and Bell, 1997).

Schizodelphis Gervais, 1861 (including *Delphinorhynchus*)

Type species: *Schizodelphis sulcatus* (Gervais, 1853) (originally described as *Delphinorhynchus sulcatus*).

Type specimen: MNHN RL 12, skull (Muizon, 1988a).

Characteristics: Symphysis long, occupying approximately two-thirds length of jaw; outer side of lower jaw with long vascular impressions, often uniting to form a groove that is broad and shallow at first and afterward narrow and deep; angle of symphysis rounded; lower borders of rami concave, upper bent inward, if seen from above, the rami with the angle of the symphysis assume an elongated oval form; alveoli numerous, reaching up the rami; teeth small; those on the rami short cone shaped, with the crowns bent backward and swollen above the base; premaxillae approximated, ankylosed together in age, depressed in the nasal region, with many grooves in the broader part that converge anteriorly; nasals small; interparietal enclosed between the two frontals and the occipital; frontals mostly free, covered by the maxillae only on the sides (True, 1908a [translating Abel, 1900]).

Occipital condyle breadth: 80 mm (*Schizodelphis sulcatus* from Belgium, IRSNB 3235-M.343; O. Lambert, personal communication 2003).

Included species: See comments below. The genus *Schizodelphis* is at least known from locality NC3AA (Muizon, 1988a).

Comments: *Schizodelphis* is a predominantely Old World genus, known from the Miocene of North Africa, Europe, and western Asia (McKenna and Bell, 1997). Muizon (1988a) stated that the type specimen of *Rhabdosteus* (*R. latiradix* Cope, 1868a) is too incomplete for taxonomic use, and thus the name should be restricted to the type specimen. He referred all supposed *Rhabdosteus* specimens from the Calvert Formation to *Schizodelphis* (see Muizon, 1988a). Muizon (1988a) also clearly stated that, because of the poor quality of the type specimen, *S.* [= *Rhabdosteus*] *latiradix* is Eurhinodelphidae incertae sedis. Based on Muizon (1988a) and discussions with another researcher (O. Lambert, 2003), there are probably at least two distinct species in the Calvert Formation (*E. longirostris*, and *Rhabdosteus barnesi* sensu Myrick, 1979). Both belong in *Schizodelphis* (see Muizon, 1988a) but there are no combinations of names that have been correctly applied to these taxa. Clearly, more work needs to be done to sort out the taxonomy of *Schizodelphis* in North America. *Schizodelphis* (= *Priscodelphinus*) *crassangulum* Case, 1904 (True, 1908a) is probably a platanistid (O. Lambert, personal communication 2003; see *Pomatodelphis* for other nominal species of *Schizodelphis* now considered to be platanistids).

Squaloziphius Muizon, 1991

Type species: *Squaloziphius emlongi* Muizon, 1991.

Type specimen: USNM 181528, skull lacking the major part of the rostrum and the auditory bones (Muizon, 1991).

Characteristics: Close in size to *Mesoplodon bidens*; rostrum wide at its base and dorsoventrally flattened; narial fossa very elongated and passing continuously to the mesorostral gutter; premaxillae with spiracular plates narrow and straight in lateral view, showing the typical transverse premaxillary crest of ziphiids on the vertex, and overhanging the maxillae laterally; nasofrontal platform elongated anteroposteriorly; temporal fossa visible dorsally and not totally covered by maxillae and parietals; jugal and lacrimal not fused; preorbital process prominent and antorbital notch deep; supraorbital process thin; zygomatic process very large and stout, having an enormous rounded postglenoid process; pterygoid bones with long and unexcavated hamular apices; wide and large hamular fossa for the hamular lobe of the pterygoid sinus; dorsal border of the pterygoid sinus not reaching the roof of the orbit and temporal fossa; palatines large; alar process of the basioccipital excavated by a fossa for a medial lobe of pterygoid sinus; basioccipital basin extremely large and wide; lambdoid crests not very long; supraoccipital wide and flat; cerebral skull not as developed as in living ziphiids (Muizon, 1991).

Occipital condyle breadth: 96 mm (USNM 181528, M. D. Uhen, personal observation).

Included species: *S. emlongi* only, known from locality WM3II* only.

Comments: *Squaloziphius* is only known from the type specimen of *S. emlongi*. *Squaloziphius* was originally described as a ziphiid, but it is provisionally identified here as a eurhinodelphinid.

Xiphiacetus Lambert, 2005b

Type species: *Xiphiacetus cristatus* (Du Bus, 1872) (originally described as *Priscodelphinus cristatus*).

Type specimen: IRSNB 3234-M.361 (Lambert, 2005b).

Characteristics: Differs from *Schizodelphis* in more robust skull with more progressive elevation of the premaxillae toward vertex; medial plate of maxilla along the vertex less concave and less erected; thicker supraorbital process; rostrum generally relatively wider at its base. Differs from *Ziphiodelphis* in narrower but thicker triangular part of the premaxilla medially to premaxillary foramen; lacking the more regular flatness and lateral slope seen in *Ziphiodelphis*; mesorostral groove widely open at that level; posterolaterally shorter plate of the maxilla along the vertex giving the posterodorsal outline of the skull a more rounded aspect in lateral view. Differs from *Eurhinodelphis* in maxillary part of the rostrum relatively shorter; dorsomedial portion of the supraoccipital shield concave; less-elevated and wider paroccipital process of the exoccipital with lower occipital condyles. Differs from *Argyrocetus* in relatively wider face; more elevated vertex with nasal longer than wide, at the same level or lower than the frontal; more erected supraoccipital shield. Differs from *Macrodelphinus* in more elevated and more transversly compressed vertex with nasal as long as wide, or longer than wide; frontal as long as, or shorter than, nasal: supraoccipital shield closer to the vertical. Differs from *Mycteriacetus* in relatively wider cranium; more elevated vertex with nasal at the same level or lower than the frontal; frontal as long as, or shorter than the nasal; more erected supraoccipital shield (Lambert, 2005b).

Occipital condyle breadth: 90 mm (IRSNB 3241-M.4893, Lambert, 2005b).

Included species: *X. cristatus* (known from locality NC3AA [Myrick, 1979]); *X. bossi* (locality NC2D* [Kellogg, 1925a]).

Comments: Myrick (1979) named four new species of *Eurhinodelphis* (*E. vaughni*, *E. ashbyi*, *E. morrisi*, and *E. whitmorei*) in his doctoral dissertation. Mention here should not be taken to constitute validation; the Myrick names were discussed by Muizon (1988a). Lambert (2004) indicated that *E. cristatus* (Du Bus, 1872) includes the species *E. ashbyi* and *E. whitmorei* sensu Myrick (1979) and that *E. bossi* Kellogg, 1925a includes the species *E. vaughni* and *E. morrisi* sensu Myrick (1979). Lambert (2004) also noted, however, that both *E. cristatus* and *E. bossi* do not belong in the genus *Eurhinodelphis*, and he later placed them in the genus *Xiphiacetus* (Lambert, 2005b), leaving no representatives of the genus *Eurhinodelphis* from the Calvert Formation.

PHYSETEROIDEA GILL, 1872

These living sperm whales encompass the Physeteridae and Kogiidae, which are classified together in the Physeteroidea (e.g., Simpson, 1945; Fraser and Purves, 1960). Alternative classifications have placed Physeterinae and Kogiinae in one family, Physeteridae (e.g., Simpson, 1945; Luo and Marsh, 1996; Kazár, 2002). Some systematists also include the Ziphiidae (below) in the Physeteroidea.

Physeteroids differ from the other odontocetes by having a very large dorsal supracranial basin (e.g., Raven and Gregory, 1933), which holds the highly modified melon and/or spermaceti organ; facial soft tissues vary in detail between *Kogia* and *Physeter* (e.g., Heyning, 1989a; Cranford, Amundin, and Norris, 1996; Clarke, 2003).

Characteristics: Diagnostic bony features for crown-Physeteroidea include extreme left-skew cranial asymmetry involving the external nares and mesethmoid, with enlarged left naris and reduced right naris; nasal bones reduced or absent; marked asymmetry of premaxillary sac fossae and foramina, with reduction in structures on the left side; antorbital notch a deep slit; bony supracranial basin formed by crests involving maxillae and bones that normally form the postnarial vertex in other odontocetes; lacrimo jugal robust and enlarged; pterygoid sinus greatly enlarged but lacking anterior or orbital extensions, with only the dorsal bony lamina well developed; falciform process absent; posterior process of tympanic bulla enlarged; foramen ovale in an extreme anterior position; dentition homodont; centra of thoracic vertebrae not anteroposteriorly compressed; humerus retaining relatively large deltopectoral crest; proximal end of ulna with radial facet on anterior border (e.g., Flower, 1868; Kellogg, 1928; Fraser and Purves, 1960; Muizon, 1991; Kazár, 2002).

Detailed study is needed to confirm which characters are physeteroid synapomorphies.

KOGIIDAE GILL, 1871

There are two relatively uncommon warm-temperate to tropical species of living Kogiidae: *Kogia breviceps*, the pygmy sperm whale, and *K. simus*, the dwarf sperm whale. Together these represent the crown-Kogiidae. Some authors have regarded the species of *Kogia* as forming a subfamily within the Physeteridae (e.g., Simpson, 1945; Luo and Marsh, 1996; Kazár, 2002). Muizon (1991) recognized two subfamilies, Kogiinae (living *Kogia*, fossil *Praekogia*), and Scaphokogiinae (*Scaphokogia*). Bianucci and Landini (1999) described *Kogia pusilla*, which they placed basal to living *Kogia*. The fossils *Kogia pusilla*, *Praekogia* and *Scaphokogia* are presumed to be stem-Kogiidae.

Kogiids are relatively rare as fossils, and those that have been described (e.g., *Scaphokogia, Praekogia*) are quite disparate from other odontocetes. The fossil record is too poor to say whether the Kogiidae has always been a low-diversity clade, but fossils show a greater disparity than seen in the two living species of *Kogia*. Kogiids of Miocene and Pliocene age have been reported from Mexico, Europe, and New Zealand, and of apparent Pleistocene age from Japan.

Kogiids differ from *Physeter* and stem-physeterids in, for example, body size, body form (shark-like profile in *Kogia* spp.), and different arrangement of soft tissues associated with the nasofacial muscles (e.g., Heyning, 1989a; Cranford, Amundin, and Norris, 1996; Cranford, 1999; Clarke, 2003).

Characteristics: Diagnostic bony features for living *Kogia* (for crown-Kogiidae) include massive maxillary walls that laterally bound the melon and lie lateral to antorbital notches; facial sagittal crest formed by both premaxillae, orbit shifted posteroventrally; blunt squared anterior end of the atrophied zygomatic process of the squamosal; flat platelike posterior process on the periotic; inflated posterior process of the tympanic bulla (e.g., Muizon, 1991; Luo and Marsh, 1996; Bianucci and Landini, 1999; Kazár, 2002).

Rostral shape is variable in the kogiid lineage, as indicated by the long rostrum in the fossil *Scaphokogia* Muizon, 1988b. Some of the characters of the living species in the genus *Kogia* are pedomorphic (see Raven and Gregory, 1933).

Kogia Gray, 1846 (including _Physeter_, in part)

Type species: *Kogia breviceps* (Blainville, 1838) (originally described as *Physeter breviceps*).

Type specimen: Skull only, MNHN; collected by M. Verreaux (Hershkovitz, 1966, Robineau 1989).

Characteristics: Skull very broad and deep dorsoventrally; bone texture very porous and oil-filled; rostrum very short and very rapidly tapering; supracranial basin deep, with an elevated cup formed by the premaxillae in its center; orbit placed posteriorly within the temporal fossa; mandibular rami diverging posteriorly; symphyseal region narrow; teeth in the mandible with elongate crowns, diverging dorsolaterally; teeth number 14 or 15; narrow, slender, conical, acute and bent medially.

Occipital condyle breadth: 95.1 mm (average of adult *Kogia breviceps* specimens USNM 550072, 550103, and 550114).

Included species: *Kogia* sp. only (known from localities [WM32], [WM33] [Barnes, 1998]).

Comments: Some Mexican records of *Kogia*-like specimens consist of isolated periotics that are morphologically very similar to those of the living species of *Kogia*. It is assumed that they do not represent *Praekogia*, for which there are no known periotics, and whose skull is more archaic than that in living *Kogia*.

Praekogia Barnes, 1973

Type species: *Praekogia cedrosensis* Barnes, 1973.

Type specimen: UCR 15229, incomplete cranium lacking the rostrum occipital crest, pterygoids, and with the left zygomatic process of the squamosal separated from the braincase (Barnes, 1973).

Characteristics: *Praekogia* differs from Recent *Kogia* spp. by having dense rather than spongy cranial bone; lower cranial crests; orbit situated anterior to the end of the zygomatic process of the squamosal, not posteriorly within the temporal fossa; small skull, slightly more than one-half the size of that of adult *Kogia breviceps*; cranium less foreshortened; lateral maxillary crests high and with sharp margins, not rounded and inflated, forming nearly straight lateral margins of the supracranial basin; left premaxilla wrapping around posterior side of the left naris and contributing to a sagittal facial crest in the supracranial basin that is less developed than in *Kogia*; right margin of right premaxilla failing to form a *Kogia*-like crest but with deep fossa; temporal fossa large; lateral surface of jugal very slender (Barnes, 1973).

Occipital condyle breadth: 71 mm.

Included species: *P. cedrosensis* only, known from locality WM30A* only.

Scaphokogia Muizon, 1988b

Type species: *Scaphokogia cochlearis* Muizon, 1988b.

Type specimen: MNHN PPI 229, partial skull.

Characteristics: Kogiid close to *Kogia* in size, but cranium longer and less broad; rostrum long, high and narrow, flattened like that of *Kogia*; anterior end round and broad as in *Kogia*; base of rostrum contacts the nasal fossae abruptly and unevenly; antorbital notches very narrow and parallel to the edges of the rostrum, not penetrating in supracranial basin; occiput stretched toward the back (in lateral view), and not showing hyperostosis of the alveolar gutter as in *Kogia*; temporal fossa much larger than that of *Kogia* (Muizon, 1988b).

Occipital condyle breadth: unknown.

Included species: *Scaphokogia* sp. only, known from locality (WM32) only (Barnes, 1998).

Comments: *Scaphokogia* is primarily known from the late Miocene of Peru (McKenna and Bell, 1997).

PHYSETERIDAE GRAY, 1821

The near-cosmopolitan living sperm whale *Physeter catodon* Linnaeus, 1758, or *Physeter macrocephalus* Linnaeus, 1758 (see Rice [1998] for nomenclature), is sufficiently disparate from the related living species of *Kogia* (Kogiidae) to be placed in its own family: Physeteridae or crown-Physeteridae. Many fossil species have been allied to *Physeter*, but nomenclature is confusing especially above genus level; crown-stem terminology has yet to be applied to the *Physeter* lineage. The fragmentary and enigmatic late Oligocene *Ferecetotherium kelloggi* Mchedlidze, 1970, from Azerbaidjan, is the oldest known sperm whale (Barnes, 1985a), making sperm whales the oldest surviving family-level clade of Odontoceti.

The distinctive skull form of *Physeter macrocephalus* reflects the unique morphology of cranial soft tissues, as documented by Raven and Gregory (1933), Fraser and Purves (1960), Heyning (1989a), Cranford (1999), Cranford, Amundin, and Norris (1996), and others.

Characteristics: Diagnostic bony features for *Physeter* (namely, for crown-Physeteridae) include large absolute size; long flattened triangular rostrum with generally vestigial teeth; large supracranial basin with elevated maxillary walls that lie medial to antorbital notches and large slitlike dorsal infraorbital foramina; no facial sagittal crest or elevated postnarial vertex within the supracranial basin; parabolic posterior part of maxillary crest confluent with nuchal crest of supraoccipital; teeth homodont; periotic with enlarged accessory ossicle; cervical vertebrae two to seven fused.

Many fossils show some of these features, or others seen in *Physeter* but not detailed above, and there is little doubt that such fossils are in the *Physeter* lineage. For example, the early Miocene *Diaphorocetus poucheti* (South Atlantic) has a supracranial basin but a more plesiomorphic braincase, which links it firmly with other odontocetes. Many fossil species, for example in the genus *Scaldicetus*, retain the plesiomorphic condition of large functional teeth in both the upper and lower jaws (e.g., Hirota and Barnes, 1995), while others have small or vestigial upper teeth (e.g., *Aulophyseter* and *Orycterocetus*), as does living *P. macrocephalus*. The more archaic fossil sperm whales have enamel-covered tooth crowns, but in various later Tertiary fossil sperm whales the crowns of the teeth lack enamel, as they do in *P. macrocephalus*. The homodonty and enlarged supracranial basin in fossil Physeteridae have led, by analogy with the living *P. macrocephalus*, to the idea that sperm whales have mainly been deep-diving squid eaters (e.g., Cozzuol, 1996).

Miocene and Pliocene fossil sperm whales include many named species worldwide. Many taxa have been based on isolated teeth of uncertain taxonomic value (e.g., Kellogg, 1925b), thus creating problems in understanding the taxonomy and history of the family. Subfamily groupings used within the Physeteridae include Physeterinae, Scaldicetinae, Hoplocetinae, and Aulophyseterinae (e.g., Simpson, 1945; Kazár, 2002).

The living *P. macrocephalus* is not known from the fossil record of North America. Despite reports cited by Dorr and Eschman (1970) and Holman (1995) of *Physeter* from Great Lakes deposits in Michigan, these specimens have been determined to be of historical (less than 1000 years) age and unlikely to have been naturally deposited (Harington, 1988).

***Aulophyseter* Kellogg, 1927a (including *Paracetus*)**

Type species: *Aulophyseter morricei* Kellogg, 1927a.

Type specimen: USNM 11230, partial skull, lacking extremity of rostrum and apex of occipital crest, supraorbital process of the left frontal and overlying maxilla, the left lacrimal and jugal.

Characteristics: Teeth of *A. morricei*, as evidenced by specimens found associated (cataloged in the LACM) very similar to those of *Orycterocetus crocodilinus*, being of small diameter, curved, fluted on the surface, with annular rings around the roots, having no enamel on the crowns. Dorsal cranial elements in the large supracranial basin markedly asymmetrical; left narial passage much larger than the right; shallow alveolar grooves that probably had palatal teeth in ligaments; no trace of distinct or vestigial alveoli for palatal teeth; large maxillary incisure; smaller posterior maxillary foramen; more or less flattened, broad, premaxillae forming the dorsal surface of the median portion of rostrum and a deep supracranial basin; right premaxilla expanding behind narial passages into a broad median thin plate, applying to upper surface of frontal and overlapping maxilla along its internal margin; left premaxillary turning out of its course by enlargement of the corresponding narial passage and apparently terminating near posterior margin of this passage; premaxillae forming the extremity of rostrum; maxilla relatively thick posterior to antorbital notch, limited posteriorly by transverse crest of supraoccipital, continuing laterally to elevated portion of maxillae; postnarial portion of right maxillary not meeting the left in middle line; tapering zygomatic processes rather far anteriorly; small temporal fossa; parietal excluded from vertex of skull; two orifices for the infraorbital system on ventral face of maxilla; deep jugular incisure; alisphenoid thin, large, expanded horizontally, bounded anteriorly by supraorbital process of frontal, suturally united posteriorly with squamosal, and in contact with anterior surface of exoccipital; no tympano-periotic recess; optic canal confluent with sphenoidal fissure; foramen ovale piercing basal portion of alisphenoid; large jugulo-acoustic canal (modified from Kellogg, 1927a). Also, premaxillae dominating the dorsal surface of anterior part of rostrum; supraoccipital shield low; mesethmoid forming a protruding plate between premaxillae (mesethmoid crest); vomer exposed as a long strip on the palatal surface; palatine not covered by pterygoid; separate infraorbital foramina (Kazár, 2002).

Occipital condyle breadth: 248 mm (USNM 11230 [Kellogg, 1927a, p. 22]).

Included species: *A. morricei* (known from locality WM13A*); *A. mediatlanticus* (locality NC5* [Kellogg, 1925b]).

Comments: Kazár (2002) established Aulophyseterinae for two species of *Aulophyseter*, namely *A. morricei* Kellogg, 1927a and *A. mediatlanticus* (Cope, 1895; originally described as a new species of *Paracetus*); the latter was transferred by Kazár from *Orycterocetus*. Kellogg (1927a) referred to *A. morricei* several enamel-crowned teeth and periotics that were found near the holotype skull, but Barnes doubted that the teeth illustrated by Kellogg belong to *A. morricei*. One periotic illustrated by Kellogg (1927a, Plate. 8, Fig. 5; and Plate 9, Fig. 1) probably also does not belong to *A. morricei*. *Aulophyseter* sp. reported by Okazaki (1992) from Japan is not a member of this genus (Hirota and Barnes, 1995), and the South American *Aulophyseter rionegrensis* Gondar (1975) may also belong elsewhere (Kazár, 2002).

Idiophyseter Kellogg, 1925b

Type species: *Idiophyseter merriami* Kellogg, 1925b.

Type specimen: UCMP 24287, incomplete skull, missing the jugals, periotics, tympanic bullae, extremity of the rostrum, and the zygomatic processes; currently lost.

Characteristics: Small-size sperm whale, with a *Physeter*-like skull; well-developed supracranial basin bounded by thick crests and having elevated nuchal crest; rostrum possibly short; maxilla with single-rooted alveoli; lacrimal lacking ventral internal process of the sort present in *Physeter*; occipital shield nearly vertically oriented; condyles large relative to skull size; temporal fossa relatively small (based on Kellogg, 1925b).

Occipital condyle breadth: 216.5 mm (Kellogg, 1925b, p. 29)

Included species: *I. merriami* only, known from locality CC23B*.

Comments: The holotype was stated by Kellogg (1925b) to have been collected from the Temblor Formation (locality CC23B), but it is also possibly from the Salinas Shale Member of the Monterey Formation (see also Barnes, 1977); it is now lost (Barnes, 1977). No other specimens have been referred to this species. Many of the projections and tuberosities of the skull were broken off by erosion, and much of the morphology that is shown in Kellogg's illustrations is represented by clay reconstruction. Barnes (1977) concluded that *Idiophyseter* may be close to *Orycterocetus* and *Aulophyseter*, but Kazár (2002) placed the former two genera in Physeterinae and the latter in Aulophyseterinae.

Kogiopsis Kellogg, 1929

Type species: *Kogiopsis floridana* Kellogg, 1929.

Type specimen: AMNH 20470, symphysial portion of lower jaws, with 11 teeth in place.

Characteristics: Similar to corresponding portions of mandibles of *Kogia*; differing in having broader symphysis, six instead of eight teeth on the symphysial portion of ramus, alveoli not extending over so far on lateral surface of ramus, much larger. Rami firmly ankylosed in region of symphysis; dorsal surface of symphysis rather broad and flat; lateral surfaces of rami slope to midline of ventral face of symphysis; large alveoli located on upper outer edge or angle of mandible; 10 teeth in each mandible, six are lodged in symphysis; teeth long (7.5 – 10 cm [3 – 4 inches] in length); somewhat curved, with small crowns, no distinct neck, rather closely spaced; outer surfaces of cement wrinkled and ornamented with coarse longitudinal grooves; outer surface of internal cone of dentine encircled by fine ridges (Kellogg, 1929).

Occipital condyle breadth: unknown.

Included species: *K. floridana* only, known from locality GC13B* only (Morgan, 1994).

Comments: Kellogg (1929) believed *Kogiopsis* to be a large relative of the pygmy sperm whales (Kogiidae). There are, however, no characters of the holotype mandible that indicate unequivocally that it is a kogiid, and following Morgan (1994) we refer *Kogiopsis floridana* to the Physeteridae.

"Ontocetus" Leidy, 1859

Type species: *Ontocetus emmonsi* Leidy, 1859.

Type specimen: ANSP 11255, plaster cast of holotype tooth present (Gillette, 1975): location of the original is unknown.

Characteristics: Tooth curved conical, compressed and fluted laterally; in perfect condition was over 25 cm (10 inches) in length, approximately 16 cm (4 inches) in greatest diameter, and 16 cm (2.5 inches) wide (Leidy, 1859).

Occipital condyle breadth: not preserved in the holotype.

Included species: The type locality for *Ontocetus emmonsi* is unclear. Leidy (1869) stated that the specimen was from the Miocene of North Carolina; "*Ontocetus*" *oxymycterus* is known from locality WM17B (Kellogg, 1925c).

Comments: The type species of *Ontocetus* is *Ontocetus emmonsi* Leidy, 1859, whose holotype is a large broken tooth of apparent Miocene age from North Carolina. This species has been referred to the Physeteridae, to the Sirenia (by Cope, 1869a), and ultimately to the Odobenidae (walruses; by Leidy, 1869; Ray, 1975). There is no objective basis for assigning the large fossil sperm whale from California (*Ontocetus oxymycterus*; USNM 10923 [Kellogg, 1925a; see also Barnes, 1977]) to the genus *Ontocetus*; therefore, we use the genus in quotes. It is beyond the scope of the present work to decide what the animal is and what generic name is appropriate. In its large size, elongate mandible and rostrum, and large diameter teeth it resembles the Recent *Physeter*, and so we have tentatively classified the species in the subfamily Physeterinae.

Orycterocetus Leidy, 1853

Type species: *Orycterocetus quadratidens* Leidy, 1853.

Type specimen: ANSP 9065 through ANSP 9069, two large teeth, fragments of both dentaries, rib (fide Leidy, 1853; Gillette, 1975), two teeth and upper jaw (fide Leidy, 1869; Kellogg, 1925a, 1965).

Characteristics: Small- to medium-sized sperm whale; 20 teeth in each upper jaw (17 in maxilla; 3 in premaxilla); dentine core of slender curved teeth often with open funnel-like pulp cavity; fine annular lines of growth and longitudinal fluting characterizing dentine core of larger teeth; outer layer of cementum may completely cover dentine on undamaged teeth; conical tip or crown of teeth occasionally black and polished, but lacking enamel; no perceptible distinction between crown and root or visible constriction to form neck below crown; vertex of cranium with large supracranial basin bounded laterally on right side by elevated border of right maxillary, on left side by left premaxillary and elevated crest of underlying left maxillary, and posteriorly by posterior borders of both maxillae, which override medially the posterior ends of the frontals and abut against the dorsal crest of the supraoccipital; right nasal bone either lost or greatly reduced; left nasal bone flattened against the frontal behind greatly enlarged left nasal passage and partially concealed by squamous overlap of markedly expanded posterior portion of right premaxillary (Kellogg, 1965).

Occipital condyle breadth: 122–150 mm (Kellogg, 1965, p. 57).

Included species: *O. quadratidens* (locality unclear, "Miocene of Virginia" according to Kellogg [1965]); *O. crocodilinus* (known from localities NC3A, B*, C [Kellogg, 1965]).

Comments: The precise identities of all species within this genus, and therefore the characterization of the genus, are questionable because all of the species were originally based upon teeth, and teeth are questionably diagnostic at the species level for physeterid systematics. Kazár (2002) transferred *O. mediatlanticus* (Cope, 1895) to *Aulophyseter*. *Orycterocetus* is also known from the Miocene of Europe (McKenna and Bell, 1997).

***Paleophoca* Van Beneden, 1859**

Type species: *Paleophoca nystii* Van Beneden, 1859.

Type specimen: Described from a large canine, but none designated by Van Beneden (1859).

Characteristics: Originally considered to be a seal, differentiated from other seals by enormous type canine (Van Beneden, 1859); phocid canine recognized as actually a tooth of a sperm whale (Ray 1977), but no characteristics given for this reidentification.

Occipital condyle breadth: unknown.

Included species: *P. nystii* only, known from locality EM8B only (Ray, 1977).

Comments: Ray (1977) stated that *Paleophoca nystii* is an "odontocete cetacean, near *Scaldicetus*," while Kazár (2002) identified it as Physeteridae incertae sedis. Van Beneden (1876) referred additional material to *Paleophoca nystii* (cranial material, teeth, limb bones), which are clearly pinniped in nature and presumably do not belong in *Paleophoca* if it is actually a physeterid cetacean rather than a phocid pinniped. *Paleophoca* is also known from the Pliocene and Pleistocene of Europe (McKenna and Bell, 1997).

***Physeterula* Van Beneden, 1877**

Type species: *Physeterula dubusii* Van Beneden, 1877.

Type specimen: MRHN 3192, skull and mandible, MRHN 3191, mandible and skull.

Characteristics: Jawbone slender, lower edge lightly corrugated; two branches deviating strongly behind compared with living *Physeter*, while ahead uniting over a considerable length to form a symphysis, which occupies approximately one-third the length of the jaw; living sperm whale has a much longer mandibular symphysis (Van Beneden, 1877).

Occipital condyle breadth: unknown.

Included species: *Physeterula* sp. only, known from locality GC13B only (Morgan, 1994).

Comments: The type specimen is from the late Miocene (Bolderien) near Antwerp, Belgium; Kazár (2002) classified it in the Physeterinae. This sole North American record is based on isolated teeth.

***Scaldicetus* Du Bus, 1867 (including *Dinoziphius*)**

Type species: *Scaldicetus caretti* Du Bus, 1867.

Type specimen: 45 teeth, assumed to be of one individual.

Characteristics: Low skull profile with anteriorly sloping occipital shield; large and anteroposteriorly elongated temporal fossae; strong and laterally flaring zygomatic processes; functional upper dentition; teeth retaining enamel on crowns (Kazár, 2002).

Occipital condyle breadth: not preserved in the type material.

Included species: *S. grandis* only, known from locality (WM21B) only (Barnes *et al.*, 1981).

Scaldicetus sp. is also known from localities WM20, WM21B (Barnes, 1977), GC10C, EM7 (Morgan, 1994).

Comments: The type specimen of the type species, *Scaldicetus caretti*, is from late Miocene (Bolderian age) sediments, near Antwerp, Belgium (see also Kellogg, 1925c). *Scaldicetus* includes *Dinoziphius carolinensis* (Leidy, 1877), which was referred to *Scaldicetus caretti* by Abel (1905). Allen (1926), citing Hay (1902), placed *D. carolinensis* in *Physeter*. The genus is also known from the early Miocene to early Pliocene of Europe and the late Miocene of Australia (McKenna and Bell, 1997).

Identification of *Scaldicetus*-like specimens can be problematic; Miocene and Pliocene specimens and species attributable to this genus have been reported from various parts of the world (e.g., Hirota and Barnes, 1995; Kazár, 2002). Typically such identifications are based on teeth with corrugated enamel on their crowns, and the indication, either by skull parts or by wear facets on the teeth, that both upper and lower functional teeth were present; the genus is, in practice, a grade. Some specimens of *Scaldicetus* are medium sized for sperm whales, and some are nearly as large as Recent adult sperm whales.

ZIPHIOIDEA (GRAY, 1868)

The Ziphioidea comprises only the family Ziphiidae, characterized below.

ZIPHIIDAE GRAY, 1865

The beaked whales represent a single living family Ziphiidae that takes its identity from the living Cuvier's beaked whale, *Ziphius cavirostris* Cuvier, 1823 (Heyning, 1989b). The alternative family name, Hyperoodontidae, is not followed here (e.g., Moore, 1968; McKenna and Bell, 1997; Rice, 1998). Crown-group Ziphiidae comprises living species in the genera *Ziphius*, *Mesoplodon*, *Indopacetus*, *Berardius*, *Hyperoodon*, and *Tasmacetus*, and all descendants of their most-recent common ancestor. Relationships between the living genera, and especially within *Mesoplodon*, have long been vexing (compare Moore, 1968; Muizon, 1991; Dalebout *et al.*, 1998, 2002). Opinions vary on whether Ziphiidae belong close to *Physeter* and *Kogia* in the Physeteroidea or whether they are closer to Delphinida and other more-crownward odontocetes. Ziphiids are structurally quite disparate from other odontocetes, which explains their placement by some authors in a separate superfamily (Fraser and Purves, 1960).

There is much literature on the supposed diagnostic features of living Ziphiidae (e.g., Flower, 1872; True, 1910a; Kernan, 1919; Fraser and Purves, 1960; Heyning, 1989a; Ridgway and Harrison 1989), but limited discussion of osteological synapomorphies.

Characteristics: Diagnostic bony features include rostrum long, distally usually narrow and subcylindrical; pterygoid sinus and associated fossa on basicranium voluminous, long and deep, extending forward on rostrum; apical or symphysial mandibular teeth enlarged, with post-apical or post-symphyseal teeth small or not erupted; teeth markedly sexually dimorphic (reduced in females); posterior process of tympanic bulla thickened distally and underlying paroccipital process of exoccipital; extreme reduction of the tuberculum of the malleus; body in the living animal markedly deeper than wide, with concomitantly tall neural spines on thoracic and lumbar vertebrae (based on Heyning [1989a], Muizon [1991], and direct observations). In most genera, the skull is deeply basined at the face, so will have an elevated vertex and generally the consequences of a transverse premaxillary crest (Heyning 1989a); the rostrum may be densely ossified (Zioupos *et al.*, 1997).

Ziphiids have an extensive late Miocene and younger record. The oldest well-preserved skull of a named species appears to be that of the late Miocene *Messapicetus longirostris* (Italy; Bianucci, Landini, and Varola, 1992). Worldwide, most of the fossil ziphiids described are dense secondarily ossified jaws – rosta or mandibles – without associated crania (e.g., Owen, 1889; Glaessner, 1947; Whitmore, Morejohn, and Mullins; 1986). Mead (1975b) cautioned that such material cannot be reliably referred to genus. Further, such jaws often occur as fossils in sediment-starved sequences associated with unconformities, settings that might indicate reworking from older rocks.

Anoplonassa Cope, 1869b

Type species: *Anoplonassa forcipata* Cope, 1869b.

Type specimen: MCZ 8766, anterior portion of a mandible.

Characteristics: Slender; shallow symphysis in proportion to length, with strong convexity of sides; upturned, expanded termination of the symphysis; large pair of nearly round and very slightly depressed terminal alveoli; rudimentary alveolar groove, with a pair of small and shallow elliptical alveoli not far distant from the terminal pair; large size and peculiar disposition of the inferior terminal foramina (True, 1907b).

Occipital condyle breadth: unknown

Included species: *A. forcipata* (locality unclear, stated as phosphate deposits near Savannah, Georgia; Cope, 1869b).

Comments: Nothing resembling the type specimen of *Anoplonassa forcipata* has ever been recovered, and, therefore, nothing has been referred to this genus. The genus is presumed to represent a ziphiid.

Belemnoziphius Huxley, 1864

Type species: *Belemnoziphius compressus* Huxley, 1864.

Type specimen: No specimen number listed. Type specimen, a rostrum, (Huxley, 1864, his Plate XIX; see also Owen 1889)

Characteristics: Vomer occupying fully a third of width of upper face of rostrum; in the few instances when extremity of rostrum complete, it is not bifid, but sharply pointed, almost like the end of the guard of a belemnite; vomer and premaxillae seeming to coalesce into one solid terminal cone (Huxley, 1864).

Occipital condyle breadth: unknown.

Included species: *B. prorops* only, known from locality EM6A* only (Leidy 1876a).

Comments: Since this genus is based solely on a rostral fragment, it is best considered Ziphiidae incertae sedis. Leidy (1877) moved his species *Belemnoziphius prorops* to the genus *Dioplodon* based on the "complete solidity of the rostrum," which he felt it shared with both *Dioplodon* and *B. compressus*. *Dioplodon* is now considered a junior synonym of *Mesoplodon* (Hershkovitz, 1966). *Belemnoziphius* is also known from the late Miocene and/or Pliocene of Europe (McKenna and Bell, 1997).

Choneziphius Duvernoy, 1851 (including *Proroziphius* and *Ziphius*, in part)

Type species: *Choneziphius planirostris* (Cuvier, 1823) (originally described as *Ziphius planirostris*).

Type specimen: None designated.

Characteristics: Primary feature two funnel-shaped cavities in incisive bones at base of rostrum immediately in front of nostrils, line being much stronger on the left; intermaxillaires joining over entire length of rostrum, in top, not revealing vomer (Duvernoy, 1851).

Occipital condyle breadth: unknown.

Included species: *C. liops* (known from locality EM6A*); *C. trachops* (locality EM6A*) (Leidy, 1876a); *C. macrops* (locality EM6A*); *C. chonops* (locality EM6A*) (Lambert, 2005c).

Comments: As with many other genera of ziphiids defined on the basis of isolated parts of the rostrum, the named material is incomplete and debatably diagnostic below family level. *Choneziphius* is also known from the late Miocene and/or Pliocene of Europe (McKenna and Bell, 1997).

***Eboroziphius* Leidy, 1876a**

Type species: *Eboroziphius coelops* Leidy, 1876a.

Type specimen: The type specimen was figured by Leidy (1877: Plate 30, Fig. 5 and Plate 31, Fig. 3).

Characteristics: Rostrum above forming a broad gutter as in *Hyperoodon*; not divided by intermaxillary crest; maxillaries with prominent later crests at base, convex inwardly; right prenarial fossa occupied by thick osseous disk; intermaxillaries co-ossified; supravomerian canal open (Leidy, 1876a).

Occipital condyle breadth: unknown

Included species: *E. coelops* only, known from locality EM6A* only.

Comments: The type specimen could not be definitively located in the ANSP, MCZ, or USNM catalogs. Cope (1890) included the species "*coelops*" in a list of taxa in the genus *Choneziphius*. It is unclear whether this was a mistake, or he actually intended to synonymize the genus *Eboroziphius* with *Choneziphius*. Again, the named material is incomplete and debatably diagnostic below family level.

***Mesoplodon* Gervais, 1850 (including *Physeter*, in part)**

Type species: *Mesoplodon bidens* (Sowerby, 1804) (originally described as *Physeter bidens*), the extant beaked whale.

Type specimen: Skull only, originally in Sowerby's museum, now in the museum of the University of Oxford; collected in 1800 by Mr. James Brodie (Hershkovitz, 1966).

Characteristics: Head and body length 3–7 m; pectoral fin length 20–70 cm; dorsal fin height approximately 15–20 cm; expanse of tail flukes is approximately 100 cm; only two teeth become well developed in males, one on each side of the lower jaw, these may be lost in old age; functional teeth of females much smaller than those of males, often, not erupting above the gums; small non-functional teeth may occur both jaws (Nowak and Paradiso, 1983). True (1910a, 1913), Heyning (1989a), Mead (1989), and others gave detailed guides to osteology in living species of *Mesoplodon*.

Occipital condyle breadth: 104.4 mm (average of adult *Mesoplodon bidens* specimens USNM 504146, 550204, and 550414).

Included species: *M. longirostris* only, known from locality EM8B only (Whitmore, 1994).

Mesoplodon sp. is also known from locality GC13B (Morgan, 1994).

Comments: *Mesoplodon* is also known from the Miocene of Japan, the Miocene and/or Pliocene of Europe, the Pliocene of Australia, and the Recent of the Indian, Atlantic, and Pacific Oceans (McKenna and Bell, 1997).

***Ninoziphius* Muizon, 1983b**

Type species: *Ninoziphius platyrostris* Muizon, 1983b.

Type specimen: SAS 941, incomplete skeleton including the skull, mandible, six cervical vertebra, five thoracic vertebrae, three lumbar vertebrae, three caudal vertebrae with chevrons, six ribs, sternum, and assorted metapodials (Muizon, 1983b).

Characteristics: Ziphiid close in size to *Mesoplodon*; many conical teeth in an alveolar gutter (at least 40 teeth per maxilla and approximately 45 per dentary); cranium with long and flat rostrum without pachyostosis, mesorostral gutter open dorsally; fossae for pterygoid sinuses more developed on ventral face than in living Ziphiidae, presenting with vertical bony pillars; periotic spindly; edges of internal acoustic meatus and cochlear and vestibular aqueducts emiting an osseous spine medially; mandible with long symphysis carrying three pairs of large apical and subapical teeth; atlas and axis fused; other cervical vertebrae free (Muizon, 1983a).

Occipital condyle breadth: unknown.

Included species: *N. platyrostris* only (known from localities GC13B [Morgan, 1994], EM8B [Muizon and DeVries, 1985]).

Comments: *Ninoziphius* is also known from the Pliocene of Peru (McKenna and Bell, 1997).

***Pelycorhamphus* Cope, 1895**

Type species: *Pelycorhamphus pertortus* Cope, 1895.

Type specimen: USNM 13796, rostrum with nothing preserved posterior to the nares and the edge of the left maxillary; anterior tip is broken (Cope, 1895).

Characteristics: Vomer bifurcating posteriorly, embracing basin taking place of maxillary basin of right side, reducing maxillary basin of left side to very small dimensions; blow-holes very unsymmetrical (Cope, 1895).

Occipital condyle breadth: unknown

Included species: *P. pertortus* only (see Comments below).

Comments: Cope (1895) was very unclear about the provenance of the type specimen of *P. pertortus*. He stated that all of the specimens described come from "the marine deposits of middle Neocene age of the Atlantic coastal region, and more exactly, in the Yorktown formation of Dana, or the Chesapeake formation of Darton and Dall." The named material is incomplete and debatably diagnostic below family level.

***Proroziphius* Leidy, 1876**

Type species: *Proroziphius macrops* Leidy, 1876.

Type specimen: The type specimen was figured by Leidy (1877: Plate 32, Figs. 1 and 2) but cannot currently be located.

Characteristics: Larger than any other specimen types from the Ashley phosphate beds; also one of the largest of its kind with the exception of the *Hyperoodon*; bones completely co-ossified, leaving barely a trace of their original separation; specimen of usual ivory-like density; supravomerian canal open throughout but exposed only for a few centimeters at the end; rostrum long and narrow, the sides more nearly parallel than in *Choneziphius*, *Eboroziphius*, and *Belemnoziphius*; basal part of rostrum prismoid, upper surface remarkably flat, anterior part with more conical form; no crest extending along the middle above as in the two rostra referred to *Choneziphius*; no visible trace of median groove indicating the original separation of the intermaxillaries (Leidy, 1877).

Occipital condyle breadth: unknown.

Included species: *P. macrops* (known from locality EM6A* [Leidy, 1876b]), *P. chonops* (locality EM6A* [Leidy, 1876c]).

Comments: The type specimen of *Proroziphius chonops* (USNM 16689) is in the US National Museum, but the type specimen of *P. macrops* could not be located in either the ANSP or USNM catalogs. The named material is incomplete and debatably diagnostic below family level. Lambert (2005c) considered *Proroziphius* a subjective synonym of *Choneziphius*.

***Ziphius* Cuvier, 1823**

Type species: *Ziphius cavirostris* Cuvier, 1823, the extant goose-beaked whale.

Type specimen: Fossilized skull, MNHN, Paris, collected in 1804 by Raymond Gorsse (Hershkovitz, 1966; Robineau, 1989).

Characteristics: Broad, deep cranium with high vertex; broad-based and relatively short toothless rostrum, narrowing rapidly forward in posterior part; rostrum densely ossified, with deep prenarial basin, in adult males but not females; large nasals overhang nares; single apical pair of mandibular teeth (based on True, 1910a, 1913; Heyning, 1989b).

Occipital condyle breadth: 175 mm (*Ziphius cavirostris*, USNM 288019).

Included species: *Z. cavirostris* only, known from locality (EM8B) only.

Comments: The genus *Ziphius* is also known from the Pliocene of Australia, and from the Recent of the Mediterranean Sea, and the Indian, Atlantic, and Pacific Oceans (McKenna and Bell, 1997).

PLATANISTOIDEA SIMPSON, 1945

There has been more than a century of debate about the relationships of the Ganges and Indus dolphins (susu, bulhan), *Platanista gangetica*, to other living "river dolphins" (e.g., Flower, 1867; True, 1908a; Kellogg, 1924, 1926, 1928; Simpson, 1945; Barnes, Domning, and Ray, 1985; Muizon, 1987, 1988a, 1991, 1994; Fordyce, 1994; Messenger, 1994; Fordyce and Muizon, 2001). *P. gangetica* provides the basis for the family Platanistidae Gray, 1863, and for Simpson's (1945) superfamily Platanistoidea, a group linked with many taxonomic problems until identified as polyphyletic and revised cladistically by Muizon (1987, 1991, 1994; see also Messenger, 1994). Muizon excluded other extant "river dolphins" – *Inia*, *Pontoporia*, and *Lipotes* – from the Platanistoidea and instead placed them in the Delphinida together with Delphinoidea; this move for "river dolphins" other than *Platanista* is supported by molecular analyses (e.g. Cassens *et al.*, 2000; Nikaido *et al.*, 2001; Yang *et al.*, 2002). Muizon expanded a redefined clade Platanistoidea by including other hitherto problematic extinct groups: Squalodelphinidae (spelling after Rice [1998]), Squalodontidae, and provisionally Dalpiazinidae. Muizon (1988c) recommended that the Acrodelphidae not be used. Fordyce (1994) placed the extinct Waipatiidae in the Platanistoidea, noting that *Waipatia*-like fossils were previously regarded as small species of Squalodontidae. *Platanista* (or Platanistidae, or Platanistoidea) has generally been placed crownward from the Ziphiidae as a sister group to the Delphinida, but two recent studies (Nikaido *et al.*, 2001; Yang *et al.*, 2002) put *Platanista* basal to Ziphiidae – a position in odontocete phylogeny not yet corroborated by anatomical cladistic studies. Barnes (2002), in a review of Platanistidae, provisionally recognized a family Acrodelphidae, in contrast to Muizon (1988c). Recently, Geisler and Sanders' (2003) anatomical cladistic study proposed a concept of Platanistoidea close to that of Simpson (1945). Further comment on nomenclature and diagnostic features of the *Platanista* clade is made below, under Platanistidae and *Platanista*.

PLATANISTIDAE GRAY, 1863

The family Platanistidae has been used very broadly in the past, and family concepts need more clarification. For a start, Platanistidae must be based around the monotypic living genus, *Platanista* Wagler, 1830, in turn represented by one species (*P. gangetica*, with two subspecies [Rice, 1998]) – the freshwater susu of the Indian subcontinent. Beyond that, our views differ on best action and here we do not necessarily offer a consensus. To consider some concepts of Platanistidae, Simpson (1945) used the family broadly to encompass the four monotypic living "river dolphin" genera (*Platanista*,

Inia, Pontoporia, Lipotes) and relatives (see also Geisler and Sanders 2003). Muizon (1987, 1991, 1994) used Platanistidae more narrowly as a clade for *Platanista* and its closest relatives; Muizon placed other putative platanistoids into clades based around *Squalodelphis*, *Squalodon*, and *Dalpiazina*, and also *Waipatia* (see Fordyce, 1994). Recently, one of us (Barnes, 2002) revised the Platanistidae in the traditional sense, establishing a new subfamily Pomatodelphininae for a range of marine early platanistids (including *Zarhachis* and *Pomatodelphis* from the Atlantic Coastal Plain and Gulf Coast [Kellogg, 1926; Case, 1934; Morgan, 1994]) and placing *Platanista* in a redefined subfamily Platanistinae. Barnes (1977) saw the fossil record of Platanistidae as long, including *Allodelphis pratti* Wilson, 1935 and "*Squalodon*" *errabundus* Kellogg, 1931 from West Coast Miocene sequences. An alternative view defines Platanistidae as strictly the crown-taxon containing the living species of *Platanista* Wagler, 1830. Crown-Platanistidae and crown-Platanistoidea are thus monotypic, and all presently known fossil "Platanistidae" in the *Platanista* clade are stem-Platanistidae; whether the latter is strictly equivalent to the subfamily Pomatodelphininae of Barnes (2002) is uncertain. Given these points, and the uncertainties about relationships of *Platanista* to other odontocetes (e.g., Muizon, 1991, 1994; Messenger, 1994; Nikaido *et al.*, 2001; Geisler and Sanders, 2003), previously cited osteological synapomorphies for the Platanistidae and related clades face ongoing revision.

Platanista is remarkably disparate amongst the odontocetes, differing markedly from other living odontocetes and from stem-platanistids. Autapomorphic skull features include, for example, the rostrum and mandibles are laterally compressed; maxillary crests arise from lateral margins of the face and over-arch the nasofacial region, pneumatized by dorsal extensions of the basicranial pterygoid sinuses; pterygoid sinuses have extensive lateral walls of pterygoid in the basicranium; nasals are flattened and appressed into the frontals; temporal fossae are widely open to dorsal view, with large elongate zygomatic processes of squamosals; orbits are anteroposteriorly short; the jugal is short and robust; and the skull is relatively dense (e.g., Anderson, 1878; Fraser and Purves, 1960; Reeves and Brownell, 1989).

Stem-platanistids, which are perhaps equivalent to the Pomatodelphininae of Barnes, share distinctive bony features with *Platanista* (e.g., Muizon, 1987, 1991, 1994; Barnes, 2002). For example, all have crania with elongate and narrow rostra; premaxillae and maxillae both reach the anterior rostral extremity; premaxillae and maxillae are fused distally; anteroposteriorly elongated grooves are present on the lateral sides of the rostrum approximately following the maxilla/premaxilla suture; the rostrum and symphysial part of mandible are more or less mirror images of each other; teeth are single-rooted; anterior teeth have higher crowns than posterior teeth; tooth roots are elongate, transversely compressed, and recurved posteriorly; the supraorbital crest may be present laterally on maxilla, but unlike *Platanista* the crest is low and not heavily pneumatized (although pterygoid sinuses may invade the orbital roof); the posterior dorsal infraorbital foramen lies very close to the posterior end of the premaxilla, positioned at the posterior margin of the premaxilla, and in some taxa is overhung by the margin of that bone; a hooklike peg is present on ridge on articular process of the periotic; the tympanic bulla has an elongate and pointed anterior process or spine and an anterolateral convexity; the atlas has double (upper and lower) transverse processes; the thoracic and caudal vertebral bodies are not shortened anteroposteriorly; transverse and neural processes of the lumbar and anterior caudal vertebrae are large and expanded anteroposteriorly; the scapula lacks a coracoid process, and the acromion lies on the anterior edge of the scapula with loss of the supraspinous fossa (based on Muizon, 1987, 1991, 1994; Barnes, 2002; and personal observations).

Allodelphis Wilson, 1935

Type species: *Allodelphis pratti* Wilson, 1935.

Type specimen: YMP 13408, incomplete skull with major portion of the rostrum missing, bones enclosing the internal choanae are imperfectly preserved, and the right exoccipital is incomplete. Both tympanic bullae and the left periotic are missing. Right periotic is complete and no teeth are preserved. Also includes the right humerus, left half of atlas, 30 incomplete vertebrae and a few rib fragments (Wilson, 1935).

Characteristics: Resembling *Argyrocetus joaquinensis* in presence of elongated nasals. Differing from it in reduced exposure of frontals on vertex; relatively large occipital condyles; conformation of the occiput; rostrum elongated; vertex elevated; postero internal angles of maxillaries rather closely approximated on vertex; nasals long and relatively narrow; premaxillaries narrowed; lambdoidal crests prominent; occiput flattened (Wilson, 1935).

Occipital condyle breadth: 96 mm (YPM 13408, Wilson, 1935, p. 20).

Included species: *A. pratti* only, known from locality WM10* only; *Allodelphis* sp. is also known from locality WM1II.

Comments: Wilson (1935) included *Allodelphis* in the family Delphinidae, but Fordyce and Muizon (2001) included it in Platanistoidea incertae sedis. We cannot reach a consensus on whether the genus belongs strictly in the stem-Platanistidae.

Pomatodelphis Allen, 1921a (including *Schizodelphis*, in part)

Type species: *Pomatodelphis inaequalis* Allen, 1921a.

Type specimen: MCZ 15750, posterior part of right maxilla.

Characteristics: Long-beaked dolphins resembling *Schizodelphis* in general form of skull except rostrum has a convex expansion of the maxillary outline at the proximal end of the tooth rows. Combined width of lower jaws narrower than upper, so lower tooth rows close inside the upper; teeth of lower jaw directed upward into the maxillary, tips of more posterior received in shallow pits (in *Schizodelphis*, these teeth directed outward and interlocking with the maxillary teeth outside the tooth rows [Allen, 1921b]); rostrum dorsoventrally flattened; mandibular symphyses dorsoventrally flattened.

Occipital condyle breadth: 91.5 mm (*Pomatodelphis inaequalis* [Allen, 1921b, p. 151]).

Included species: *P. inaequalis* (known from localities GC10C* [Morgan, 1994], GC27 (Hulbert and Whitmore, 2006); *P. bobengi* (locality GC10C* [Morgan, 1994]).

Pomatodelphis sp. is also known from locality (WM28II) (Barnes, 1998).

Comments: *Pomatodelphis bobengi* was originally described and named by Case (1934) as *Schizodelphis bobengi*, while *Pomatodelphis inaequalis* Allen, 1921a has as a synonym *Schizodelphis depressus* Allen, 1921b. Muizon (1988a) showed that *Schizodelphis* is a eurhinodelphinid. Later, Morgan (1994) pointed out that both *S. bobengi* and *S. depressus* have teeth at the rostral apex – a feature which precludes them from belonging to *Schizodelphis* or to the Eurhinodelphinidae.

When Allen (1921b) named *Pomatodelphis inaequalis*, he designated as its holotype (MCZ 15750) a fragment of the proximal part of the right maxilla, representing the posterior part of the rostrum, but he designated as a referred specimen a much more complete fossil (UF/FGS 2343) consisting of the rostrum and cranial parts of one individual. Kellogg (1959) referred a nearly complete skull to this species. Because *P. inaequalis* has become relatively well known, Morgan (1994) declared it to be a senior synonym of *Schizodelphis depressus* Allen, 1921b, a species which, however, has page priority over *P. inaequalis*. Both species, *Schizodelphis depressus* and *Pomatodelphis inaequalis*, have holotypes that are partial rostra. Morgan (1994) reported that *P. inaequalis* is the most abundant cetacean in the early Clarendonian (*c.* 11 Ma) Agricola Fauna of the Bone Valley Formation in central Florida (locality GC10C). *Pomatodelphis* is also known from the Miocene of Europe (McKenna and Bell, 1997).

Zarhachis Cope, 1868b

Type species: *Zarhachis flagellator* Cope, 1868b.

Type specimen: ANSP 11231, two lumbar vertebrae (Gillette, 1975).

Characteristics: Lumbar vertebrae with broadly obtuse medial line, offering distinct traces of the two keels; dorsoventrally compressed rostra and symphyseal regions of the mandibles (Cope, 1868b).

Occipital condyle breadth: 105 mm (USNM 10911, Kellogg, 1926, p. 17).

Included species: *Z. flagellator* only (known from localities NC2B, C* [Kellogg, 1926; Bohaska, 1998]).

Comments: Cope's exact words about the type specimen of *Zarhachis flagellator* are as follows: "This species is represented by only two lumbar and two caudal vertebrae, which belonged to at least three different individuals, none of them adult." Kellogg (1924) stated that: "The caudal vertebra in the Academy of Natural Sciences of Philadelphia which appears to be the type, and is so labeled, is much worn at both ends, and the anterior epiphysis is missing." Gillette (1975) stated: "[ANSP] 11231 as the holotype, [ANSP] 11232? paratype, 2 caudal vertebrae and 2 lumbar vertebrae, three present only." This fits with Kellogg's implication that there was only a single caudal at the time of his observations, but it is unclear which specimen numbers apply to which specimens. Despite the poor nature and quality of the type specimen of *Z. flagellator*, Kellogg (1924, 1926) felt that there were sufficient "structural peculiarities" of the type caudal vertebra that he could confidently assign his new specimens to *Z. flagellator*.

SQUALODELPHINIDAE DAL PIAZ, 1917

Dal Piaz (1917) proposed the Squalodelphinidae (originally spelled Squalodelphidae) for the Italian early Miocene species *Squalodelphis fabianii*. The family was mostly ignored, and often subsumed under Ziphiidae, until Muizon (1987, 1991) identified it as an extinct clade including *Squalodelphis*, *Notocetus*, and the North American *Phocageneus* (see Kellogg, 1957), which forms the sister taxon for the Platanistidae sensu Muizon, and has these synapomorphies: periotic has square-shaped pars cochlearis and large thin-edged aperture for the cochlear aqueduct; tympanic bulla has a ventral groove along its whole length; the manubrium of the malleus has an anterior extension; the dorsal transverse process of the atlas is strongly developed but the ventral process is extremely reduced. The subcircular fossa of the basicranium may be characteristic, although the homology of such structures in platanistoids is not clear (Fordyce, 1994; Muizon, 1994). As in *Platanista* and the stem-Platanistidae, the scapula lacks the coracoid process, and the acromion lies on the anterior edge of the scapula with loss of the supraspinous fossa.

Phocageneus Leidy, 1869

Type species: *Phocageneus venustus* Leidy, 1869.

Type specimen: ANSP 11227, isolated tooth.

Characteristics: Crown conical, compressed, oval in section at base, and moderately curved, forming an acute ridge before and behind, with an acute point; tooth base conspicuously swollen internally, contracting all around toward the neck; anterior acute border of crown expanding in triangular surface of swollen base; posterior border embraced by an attempt to form a basal cingulum; enamel of crown nearly uniformly corrugated, wrinkles much interrupted; root, broken at its point, has been about twice length of crown; root conical, slightly curved, feebly gibbous (Leidy, 1869).

Occipital condyle breadth: unknown.

Included species: *P. venustus* only (known from localities NC2B [Bohaska, 1998], NC3AA, E* [Kellogg, 1957]).

Comments: Kellogg (1957) referred a specimen (USNM 21039) that included lower jaws, teeth, a periotic, and a partial skeleton to *P. venustus* based on the similarity of the teeth to the type tooth described by Leidy

(1869). The material provides the best record of the family Squalodelphinidae in North America. *Phocageneus* is also known from the Miocene of Europe (McKenna and Bell, 1997).

***Notocetus* Moreno, 1892**

Type species: *Notocetus vanbenedeni* Moreno, 1892.

Type specimen: Museo de La Plata specimen number MLP 5–5 (Muizon, 1987), figured by Moreno (1892, Plate 11).

Characteristics: Longirostral; teeth single-rooted, posterior teeth with complex crowns; skull asymmetrical, involving bones around nares and anterior of supraoccipital; premaxillary sac fossae short and broad, nasals and frontals nodular with prominent medial sutures; thickened supraorbital prominences present, associated with apparently asymmetrical antorbital notches; zygomatic process robust, blunt ended (from Moreno, 1892; True, 1910b; Muizon, 1987; Cozzuol, 1996).

Occipital condyle breadth: unknown.

Included species: *Notocetus* sp. only, known from locality NC3AA only (Muizon, 1987).

Comments: *Notocetus* is represented by a tympanic bulla and periotic, USNM 206286, from the Calvert Formation (Gottfried, Bohaska, and Whitmore, 1994); Muizon (1987) regarded the material as close to *N. vanbenedeni* (also known as *Diochotichus vanbenedeni* [True, 1910b]) of the Miocene of Patagonia.

SQUALODONTIDAE BRANDT, 1873

The family Squalodontidae (shark-toothed dolphins) has traditionally been a repository for heterodont fossil odontocetes in which the teeth have multiple denticles (e.g., Kellogg, 1923b; Simpson, 1945) including many from North America. Traditionally, the loss of the intertemporal constriction separated squalodontids from the more-archaic agorophiids. Rothausen (1968) identified the Squalodontidae as a grade, for which he used two subfamilies, Squalodontinae and Patriocetinae. It is clear now that some supposed squalodontids represent other cetaceans (e.g., Dooley, 2003). Muizon (1987, 1991, 1994) recognized that features of the scapula (acromion on anterior edge, loss of supraspinous fossa, loss of coracoid process) identify the Squalodontidae as a basal clade of Platanistoidea, and (Muizon, 1994) further listed squalodontid synapomorphies, including long rostrum with dilated apex; increased area of vomer exposed on ventral surface of rostrum; and periotic with low, wide regularly convex or flat dorsal process [sic]. Fordyce (1994) listed other ear features when differentiating Squalodontidae from Waipatiidae. Cozzuol (1996) suggested that the Australian genus *Prosqualodon* is not a squalodontid but represents a separate family Prosqualodontidae; indeed, Muizon (1994) earlier excluded *Prosqualodon* from the Squalodontidae. Further study is indicated, and, for now, we do not use Prosqualodontidae.

Squalodon calvertensis Kellogg, 1923b is one of the best known of the Squalodontidae, well represented by late early to early middle Miocene fossils from the Calvert Formation of the Atlantic Coastal Plain (locality NC3). Other species (Dooley, 1996, 2003) from the Chesapeake region include *Squalodon tiedemani* and *S. atlanticus*. Skulls of unnamed late Oligocene squalodontids occur in South Carolina (Whitmore and Sanders, 1977). Squalodontids are rare in Pacific Coast sequences (e.g., Barnes, 1977), with no material yet named formally.

***Squalodon* Grateloup, 1840 (including *Arionius*; *Colophonodon; Delphinoides*; *Phocodon*; *Rhytisodon; Trirhizodon*)**

Type species: *Squalodon gratelupi* (Pedroni, 1845) (originally described as *Delphinoides gratelupi*).

Type specimen: "rostral fragment described by Grateloup as *Squalodon*" (Kellogg 1923b, p. 19); whereabouts unknown.

Characteristics: Rostrum elongated, equaling nearly twice the length of that portion of the skull posterior to maxillary notches; external nasal openings situated far posteriorly; braincase telescoped; no intertemporal construction; supraorbital process of frontal expanded laterally, constricted proximally; frontals in contact posteriorly with supraoccipital, receiving ascending processes of premaxillae in paired grooves along internal margins of maxillae; nasals abbreviated, apparently synostosing with frontals in old adults; mesorostral channel open; mesethmoid filling large frontal fontanelle, forming thick partition separating nasal passages externally, providing support for abbreviated nasals, pierced by a second pair of passages (foramina for the nasal or ophthalmic branches of the trigeminal nerve); cheek teeth increased above typical number 44, variably 58 to 60; molars not exceeding seven; premolars five in either jaw (Kellogg, 1923b).

Occipital condyle breadth: 95 mm (*Squalodon calvertensis*, USNM 10484).

Included species: *S. calvertensis* (known from localities NC2B, C* [Kellogg, 1923b; Bohaska, 1998]); *S. tiedmani* (locality EM6B* [Allen, 1887]); *S. whitmorei* (locality NC2E* [Dooley, 2005]).

Comments: See discussion by Kellogg (1923b: p. 19) about the complex typology for *Squalodon* and the proposed type-species *Squalodon typicus* Kellogg, 1923. Dooley (2003, 2005) considered *Squalodon tiedmani* to be similar to genus Y of Whitmore and Sanders (1977), and that, therefore, it should be placed in Odontoceti incertae sedis.

Grateloup (1840) originally described *Squalodon* as a marine reptile based on a piece of lower jaw. Kellogg (1923b) reviewed many species of *Squalodon* from North America, but he did not make clear statements about the status of different species. Dooley (2003) reviewed all of the North American cetaceans attributed to the

Squalodontidae. The status of additional named species of *Squalodon* from North America is as follows. *Squalodon (= Macrophoca) atlanticus* (Leidy, 1856) is probably a squalodontid (Kellogg, 1923b; Dooley, 2003); *Squalodon crassus* (?EM6A; Allen, 1926) is a non-diagnostic portion of a mandible (Dooley, 2003); *Squalodon debilis* (Leidy, 1856) is not a squalodontid (Kellogg, 1923b); "*Squalodon*" *errabundus* (Kellogg, 1931) is a platanistid related to *Zarhachis* (Barnes [1977 [citing Barnes and Mitchell unpublished data], although a species "*errabundus*" was not listed by Barnes [2002] under his concept of Platanistidae); *Squalodon* (= *Colophonodon*) *holmesii* (Leidy, 1853) is too incomplete for definite allocation (Kellogg, 1923b); *Squalodon* (= *Phoca*) *modesta* (Leidy, 1869) is probably a pinniped tooth (Kellogg, 1923b); *Squalodon pelagius* (Leidy, 1869), is unclear (Kellogg, 1923b) or Odontoceti incertae sedis (Dooley, 2003); *Squalodon protervus* (Cope, 1868a) is now *Dicotyles protervus*, a peccary (Dooley, 2003); *Squalodon wymanii* (Cope 1868a) is not a squalodontid according to Dooley (2003). McKenna and Bell (1997) have listed *Phocodon*, *Arionius*, *Rhytisidon*, *Trirhizodon*, and *Rhytisodon* as junior synonyms of *Squalodon*; on genera of Squalodontidae, see also Fordyce and Muizon (2001), which includes a review of the McKenna and Bell classification.

Squalodon is also known from the late Oligocene of Japan, the late Oligoene to Pliocene of Europe, and possibly also from the late Oligocene of Australia and New Zealand (McKenna and Bell, 1997).

DELPHINIDA MUIZON, 1984

The Delphinida of Muizon (1984, 1988a) encompasses the clade Delphinoidea – living and extinct ocean dolphins – and three monotypic families of extant "river dolphins" (Iniidae, Pontoporiidae, and Lipotidae). The delphinoids include the Delphinidae (ultimately based on *Delphinus delphis*), Phocoenidae, and Monodontidae, and the extinct Kentriodontidae, Albireonidae, and Odobenocetopsidae. Molecular studies support Muizon's (1988a) concept of Delphinida (e.g., Cassens *et al.*, 2000; Nikaido *et al.*, 2001; Yang *et al.*, 2002), although the recent anatomical cladistic analysis of Geisler and Sanders (2003) placed the "river dolphins" with *Platanista* rather than with the Delphinoidea, making the Delphinida of Muizon paraphyletic. Within the Delphinida, the "river dolphins" *Inia geoffrensis*, *Pontoporia blainvillei*, and *Lipotes vexillifer* show some superficial similarities yet are skeletally rather disparate from each other; the disparity explains their placement in three separate genera and families.

Characteristics: Diagnostic bony features proposed for the Delphinida include presence of a lateral lamina on palatine; virtual loss of posterior region of lateral lamina of pterygoid (with occasional secondary occurrences of the lamina); depressed or excavated posterodorsal region of involucrum of tympanic; reduction of anterior process of periotic; development of processus muscularis of malleus; transverse process of lumbar vertebrae enlarged in triangular blades; anterior inflexion of anteroventral angle of sigmoid process of periotic; frontals narrower (or as wide as) than nasal on vertex (from Muizon, 1988a).

DELPHINIDA FAMILY INCERTAE SEDIS

Delphinavus Lull, 1914

Type species: *Delphinavus newhalli* Lull, 1914.

Type specimen: YPM 10040, partial skull and skeleton.

Characteristics: Differs from *Delphinus* in having more nearly vertical occiput; exoccipitals taller than wide; large, heavy squamosal; maxilla that is not laterally or posteriorly as expanded (Lull, 1914).

Occipital condyle breadth: not preserved.

Included species: *D. newhalli* only, known from locality WM17E* only,

Comments: Because of the nature of preservation and preparation of the specimen, it is difficult to assess enough morphology to place *Delphinavus* confidently in any particular family of Delphinida. Material is quite complete and should be diagnostic upon preparation. For now, family-level relationships are uncertain.

Lamprolithax Kellogg, 1931

Type species: *Lamprolithax simulans* Kellogg, 1931.

Type specimen: USNM 11566, right periotic.

Characteristics: Apex of depressed subtriangular area behind internal acoustic meatus situated considerably beyond fossa inclosing cerebral orifice of the aqueduct of the vestibule (Kellogg, 1931).

Occipital condyle breadth: unknown.

Included species: *L. simulans*, (known from locality WM13A*); *L. annectens*, (locality WM13A* [Kellogg, 1931]).

Comments: *Lamprolithax* periotics were reported from the late Miocene of Pietra leccese, Salento Peninsula, Italy, by Bianucci and Varola (1994), who considered *Lamprolithax* a probable kentriodontid. Because periotics are generally quite diagnostic, it should be possible to identify the family if a skull is found with a periotic in place. For now, family-level relationships are uncertain.

Miodelphis Wilson, 1935

Type species: *Miodelphis californicus* Wilson, 1935.

Type specimen: YPM 13407, incomplete skull, lacking the rostrum from the maxillary notches forward, the nasals, the right supraorbital process of the frontal, most of the left zygomatic process of the squamosal, and most of the pterygoids. Also includes two hyoid bones and two vertebrae (Wilson, 1935).

Characteristics: Close approximation of the maxillae at posterior extremities; oblique truncation of maxilla above temporal fossa; large, rectangular exposures of frontals on vertex; occipital condyles small (Wilson, 1935).

Occipital condyle breadth: 63.5 mm (YPM 13407 [Wilson, 1935, p. 80]).

Included species: *M. californicus* only, known from locality WM10* only.

Comments: Material is incomplete and debatably diagnostic, but the concept of the genus could probably be firmed up if more-complete topotypic specimens were found. For now, family-level relationships are uncertain.

Nannolithax Kellogg, 1931

Type species: *Nannolithax gracilis* Kellogg, 1931.

Type specimen: USNM 11569, right periotic (Kellogg, 1931).

Characteristics: Diminutive ear bone; decidedly angular; anterior process strongly compressed from side to side distally. Important characters: compressed outline of internal acoustic meatus; attenuation of anterior process; relatively large size of epitympanic orifice of aquaeductus Fallopii; absence of pit on posterior face above stapedial fossa (Kellogg, 1931).

Occipital condyle breadth: unknown.

Included species: *N. gracilis* only, known from locality WM13A* only.

Comment: Because periotics are generally quite diagnostic, it should be possible to identify the family if a skull is found with a periotic in place. For now, family-level relationships are uncertain. *Nannolithax* is also known from the Miocene of Europe (McKenna and Bell, 1997).

Oedolithax Kellogg, 1931

Type species: *Oedolithax mira* Kellogg, 1931.

Type specimen: USNM 11572, right periotic (Kellogg, 1931).

Characteristics: Periotic in some respects larger replica of those referred to *Lamprolithax simulans*. Attention is directed to: pronounced inflation of pars cochlearis posterior to cochlear aqueduct; depression with apex formed by fossa inclosing orifice of vestibular aqueduct, base formed by outer rim of internal acoustic meatus; rather deep pit on posterior face above stapedial fossa; anteroposterior diameter of the pars cochlearis (Kellogg, 1931).

Occipital condyle breadth: unknown.

Included species: *O. mira* only, known from locality WM13A* only.

Comments: Because periotics are generally quite diagnostic, it should be possible to identify the family if a skull is found with a periotic in place. For now, family-level relationships are uncertain.

Platylithax Kellogg, 1931

Type species: *Platylithax robusta* Kellogg, 1931.

Type specimen: CAS 4339, right periotic (Kellogg, 1931).

Characteristics: Flattened and depressed area external to the internal acoustic meatus; cerebral and external faces nearly at right angles to each other; swelling behind cerebral orifice of cochlea (Kellogg, 1931).

Occipital condyle breadth: unknown.

Included species: *P. robusta* only, known from locality WM13A* only.

Comments: *Platylithax* sp. was reported from the middle Miocene of the Esvres syncline, Val-de-Loire, France (Ginsburg, 1990), but we are unconvinced of the identification. Because periotics are generally quite diagnostic, it should be possible to identify the family if a skull is found with a periotic in place that matches the type of *P. robusta*. For now, family-level relationships are uncertain.

LIPOTIDAE ZHOU, QIAN, AND LI, 1979

The Lipotidae encompasses one living species, the highly endangered Chinese river dolphin or baiji *Lipotes vexillifer* Miller, 1918 (e.g., Brownell and Herald, 1972; Zhou, Quian, and Li, 1978, 1979; Chen, 1989). At first, the dolphin was considered a species of Iniidae, but later, because of its distinctive structure, it was placed in the new family Lipotidae Zhou, Quian, and Li, 1978. Muizon (1988a) proposed a new superfamily Lipotoidea for *L. vexillifer*, although this superfamily has been cited little.

Characteristics: Diagnostic bony features for *Lipotes* (for crown-Lipotidae) include narrow slightly upturned rostrum with constriction of rostrum at base; maxilla deeply grooved leading to antorbital notch, producing supraorbital crest; premaxillary sac fossa lying far posteriorly, short anteroposteriorly; perinasal area markedly asymmetrical; nasals large, prolonged vertically; frontals small, nodular, elevated to form vertex; maxillae extending relatively far back on skull, closely approximating supraoccipital, pinching medially behind vertex; periotic with short-cupped bullar facet on posterior process, lacking prominent parabullary ridge, retaining anterior bullar facet on anterior process (summary of features illustrated or elaborated by Miller [1918], Kasuya [1973], Fraser and Purves [1960], Barnes [1985b], Geisler and Sanders [2003], and personal observations).

Two fossil genera have been allied with *Lipotes* in the Lipotidae. One, *Prolipotes* Zhou, Zhou, and Zhao (1984), is based on a fragment of mandible from Chinese freshwater strata; it is probably not diagnostic. Species of marine long-beaked dolphins in *Parapontoporia* Barnes 1984 were initially placed with Pontoporiidae, but Muizon (1988a) put *Parapontoporia* closer to *Lipotes* than to *Pontoporia*; here, following Muizon, *Parapontoporia* is regarded as possibly in the stem-Lipotidae.

Parapontoporia Barnes, 1984 (including *Stenodelphis*)

Type species: *Parapontoporia sternbergi* (Gregory and Kellogg, 1927) (originally described as *Stenodelphis sternbergi*).

Type specimen: AMNH 21905, rostral fragment bearing 24 teeth with three vacant alveoli (type of *Stenodelphis sternbergi* Gregory and Kellogg, 1927).

Characteristics: Cranial vertex offset asymmetrically to left; premaxillary sac fossae on premaxillary surfaces flat, not elevated, convex; posterior terminations of premaxillae not widely separated from anterolateral corners of nasals; deep squamosal fossa between zygomatic arch and braincase; lateral laminae of pterygoid joining with posterior plate of palatine to form a bony wall within the orbit not connecting posteriorly with the basisphenoid and vomer exposed on the palate (Barnes, 1984).

Occipital condyle breadth: unknown.

Included species: *P. sternbergi* (known from locality WM26); *P. pacifica* (localities WM21Al [Barnes *et al.*, 1981]), WM30A* [Barnes, 1984]); *P. wilsoni* (locality WM20* [Barnes, 1985b]).

Parapontoporia sp. is also known from localities WM30B, (WM32) (Barnes, 1998).

Comments: Fordyce and Muizon (2001) questionably listed *Parapontoporia* as a member of Lipotidae. More-complete material is needed to confirm relationships.

Pontoporia Gray, 1846 (including _Delphinus_, in part)

Type species: *Pontoporia blainvillei* (Gervais and d'Orbigny, 1844) (originally described as *Delphinus blainvillei*), the extant La Plata river dolphin.

Type specimen: Skull only, MNHN; collected by M. de Fréminville (Hershkovitz, 1966; Robineau, 1989).

Characteristics: Skull roundish; rostrum very long, compressed, with a strong groove on each side above; eyebrow with a long cylindrical crest; lower jaw compressed, with a deep groove on each side; symphysis very long; teeth small, subcylindrical, smooth rather hooked, and acute (Gray, 1846). See also diagnostic features for crown-Pontoporiidae, above.

Occipital condyle breadth: 62 mm (average of 19 specimens of mature individuals from the USNM collection).

Included species: *Pontoporia* sp. only (known from localities [WM31], [EM8B] [Whitmore, 1994; Barnes, 1998]).

Comments: *Pontoporia* is also known from the Pleistocene of Argentina (possibly also from the late Miocene) and the Recent of the Atlantic Ocean (the La Plata River estuary and coastal waters of southeast South America) (McKenna and Bell, 1997).

PONTOPORIIDAE GRAY, 1870

The Pontoporiidae contains a single living species, the franciscana *Pontoporia blainvillei* (Gervais and d'Orbigny, 1844). In spite of its wide recognition as a "river dolphin," the franciscana is a nearshore marine to estuarine species found along the South American East Coast (Brownell, 1989). *Stenodelphis* is a former generic name (Rice, 1998). *P. blainvillei* was initially placed in the Iniidae (e.g., Flower, 1867), sometimes in a subfamily Stenodelphininae Miller, 1923 (e.g., Simpson, 1945), but in recent decades it has been put in the family Pontoporiidae. The family has been used both narrowly and widely; here we differentiate the crown-Pontoporiidae containing only *P. blainvillei* (which has a negligible fossil record) from the stem-Pontoporiidae, which includes the fossil genera *Brachydelphis*, *Pontistes*, and *Pliopontos* (see Muizon, 1988a). Muizon (1988a) omitted *Parapontoporia*, which he placed in the Lipotidae; earlier, Barnes (1984), placed *Parapontoporia* new genus in the new subfamily Parapontoporiinae.

Characteristics: Diagnostic bony features for *Pontoporia* (namely, for crown-Pontoporiidae) include relatively small skull; rostrum and mandible long and narrow, with over 50 teeth per quadrant; relatively long narrow nasals; secondary facial symmetry; prominent lateral lamina of pterygoid; periotic with markedly short anterior and posterior processes (e.g., Fraser and Purves, 1960; Muizon, 1988a; Brownell, 1989).

North American pontoporiids are fragmentary (e.g., Morgan, 1994; Whitmore, 1994), in contrast to better-preserved South American specimens.

INIIDAE GRAY, 1846

The Iniidae Gray, 1846 contains a single living species, the bouto, *Inia geoffrensis* Gervais and d'Orbigny, 1844 (e.g., Best and da Silva, 1989). The bouto is widespread in the Amazon and other river systems in northeastern South America. *Inia* is the type genus for the Iniidae, a family that has been used very widely for fossils in the past (e.g., True, 1908b; Simpson, 1945).

Characteristics: Diagnostic bony features for *Inia* (namely, for crown-Iniidae) include rostrum long and narrow, not waisted at the base; teeth differentiated, posterior teeth tuberculate with rough ornament; tooth row closely approaching posterior of rostrum; orbit short, with long obliquely descending postorbital process; ascending or supraorbital process of maxilla narrow with more or less straight crest along lateral edge (dorsal view), facial borders narrowing posteriorly, no marked crest above orbit; perinasal area and frontals on vertex showing minor asymmetry; nasals showing marked asymmetry; vertex forming prominent frontal hump; temporal fossa large and well exposed to dorsal view; squamosal base elongate; laminae of pterygoid reduced; extreme reduction in length of posterior process of periotic (summary of features illustrated or noted by Miller [1918], Kasuya [1973], and Muizon [1988a]).

Since the late 1800s, many fossils have been identified as iniids, but only a few apparently belong in the stem-Iniidae: the South American genera *Ischyrorhynchus* and *Saurodelphis* (see Cozzuol, 1985, 1996), and possibly the poorly known North American *Goniodelphis*.

***Goniodelphis* Allen, 1941**

Type species: *Goniodelphis hudsoni* Allen, 1941.

Type specimen: MCZ 3920, portion of a skull.

Characteristics: Larger than *Inia*; beak very narrow; tooth rows separated in the distal portion by a knifelike ridge, from where they suddenly diverge proximally to form an inverted Y; in hinder part of palate two depressions between vertical wing of palatal and midline, partially roofed over by the pterygoids when intact, form a pair of shallow gutters that are parallel instead of converging forward as in most small odontocetes, floor of each gutter smoothly continuous with inner wall of the posterior nares, as if it were in life open to the nasal passage; nasal passage sloping strongly upward and backward instead of being nearly vertical, forming an angle of some 30° with the plane of the palate; posterior teeth with roots circular in section below the crown; tooth at the angle of the Y on each side triangular in section, with the base of the triangle formed by the outer side (Allen, 1941).

Occipital condyle breadth: unknown.

Included species: *G. hudsoni* only, known from locality GC13B* only (Morgan, 1994).

Comments: *G. hudsoni* belongs uncertainly in the stem-Iniidae (Muizon, 1988a); a complete skull is needed to be sure of family relationships. Kellogg (1944) and Morgan (1994) provide details of Florida material.

DELPHINOIDEA FLOWER, 1864

The Delphinoidea encompasses the dolphins (family Delphinidae) and their relatives (Phocoenidae, Monodontidae, and the extinct Kentriodontidae, Albireonidae and Odobenocetopsidae), forming the most speciose group of living Cetacea. Living delphinoids are somewhat variable in external form but are unified by features of the skull, including asymmetrical premaxillae, style of development of extensive basicranial sinuses, and details of the tympanoperiotics (Fraser and Purves, 1960; Kasuya, 1973; Barnes, 1978, 1990; Muizon, 1988a; Geisler and Sanders, 2003). Most living delphinoids have basicranial pterygoid sinuses that invade the orbits and have asymmetrical bones around the external nares, but some lack orbital sinuses, and some have almost symmetrical facial bones. Anatomical cladistic studies have identified some synapomorphies (e.g., Muizon, 1988a; Heyning, 1989a; Barnes, 1990; Messenger and McGuire, 1998), with an extended list offered by Geisler and Sanders (2003). However, a detailed computer-aided cladistic analysis of crown-delphinoids is sorely needed to unravel patterns. Molecular studies support the morphologically based concept of Delphinoidea (e.g., Cassens *et al.*, 2000; Nikaido *et al.*, 2001).

KENTRIODONTIDAE (SLIJPER, 1936)

Kentriodontids are relatively small archaic delphinoids, typified by *Kentriodon pernix* Kellogg, 1927b from the middle Miocene Calvert Formation of Maryland and Virginia (locality NC3). Kentriodontids have been reported widely from North American localities and beyond, including Belgium, Romania, Azerbaidjan, New Zealand, Japan, Oregon, Washington, California, Mexico, and Peru, with a stratigraphic range of late Oligocene to late Miocene (Ichishima *et al.*, 1995).

Concepts of Kentridontidae have changed significantly since Slijper (1936) proposed a subfamily Kentriodontinae within the Delphinidae. Barnes (1978) elevated the taxon to family rank and used it as a grade for delphinoids with a more or less symmetrical vertex of the skull, in contrast with the asymmetrical cranial vertex of living dolphins, Delphinidae. Barnes viewed kentriodontids as broadly ancestral to delphinids. Later work identified other species of *Kentriodon* Kellogg, 1927b, such as *Kentriodon obscurus* Kellogg, 1931 (Barnes and Mitchell, 1984) and *Kentriodon hobetsu* Ichishima, 1995; all are small, with a body that is probably less than 2 m long. Many other genera and species, including long-established taxa, have been attributed to the family, for example, *Delphinodon dividum* True, 1912a, *Hadrodelphis calvertense* Kellogg, 1966, *Liolithax pappus* (Kellogg, 1955), and *Macrokentriodon morani* Dawson, 1996a.

Characteristics: Diagnostic features reported for Kentridontidae include procumbent large first upper incisor; combination of more or less symmetrical facial bones (plesiomorphy) associated with pterygoid sinuses that invade the orbits.

Muizon's (1988a) cladistic study found it difficult to demonstrate kentriodontid monophyly. No particular kentriodontid has yet been reported as the sister taxon to the Delphinidae. The proposed use of crown and stem concepts for Delphinidae, and the revision of Kentriodontidae as the clade for *Kentriodon*, would clarify relationships.

***Atocetus* Muizon, 1988b (including *Pithanodelphis*, in part)**

Type species: *Atocetus nasalis* (Barnes, 1985c) (originally described as *Pithanodelphis nasalis*).

Type specimen: LACM 30093, relatively complete but deformed skull, with periotics, auditory bullae, ear ossicles, and mandible (Barnes, 1985c; Muizon, 1988b).

Characteristics: Differs from *Pithanodelphis* primarily by nasals much more developed (thicker, wider, longer); premaxilla/nasal suture located on anterobasal side of nasal; absence of clear V-shaped notch in the anterior end of the nasal (Cf. *Pithanodelphis* and other Kentriodontidae); posteromedial ends of the maxillae each excavated by a small cavity extending laterally while skirting the occipital crest fossae for the middle and posterior sinuses less developed; the basioccipital basin broader alar processes; thicker. Wing of exoccipital and paroccipital process facing posteriorly and in a plane roughly perpendicular to the cranial axis, rather than directed posterolaterally and strongly inflected anteriorly; temporal fossa relatively open dorsally; lateral and posterior faces of zygomatic process of squamosal forming an almost right angle (from Muizon, 1988b).

Occipital condyle breadth: unknown.

Included species: *A. nasalis* only, known from locality WM17G* only (Barnes, 1985c, 1988).

Comments: Originally described as *Pithanodelphis nasalis*; the species was moved to *Atocetus* by Muizon (1988b). All other occurrences of *Atocetus* are restricted to South America.

Delphinodon Leidy, 1869 (including _Squalodon_, in part)

Type species: *Delphinodon mento* (Cope, 1868a) (originally described as *Squalodon mento*).

Type specimen: ANSP 11228-11230, four premolars (one currently missing [Gillette 1975]).

Characteristics: Crown of tooth subtrihedral conical, as broad as it is long, ovoid in section at base, and with a slight twist inwardly; inner and outer surfaces very unequal and separated by linear rugose ridges; back of crown forming, at its basal half, a thick convex tubercle, crossed by posterior dividing ridge, and bounded near base by a short embracing ridge; anterior dividing ridge of crown pursuing sigmoid course from the summit posterointernally to the base anteroexternally; inner and outer surfaces of crown conspicuously wrinkled, inner surface with an irregular curved ridge, terminating in a basal tubercle and dividing off the anterior more wrinkled third of the inner surface from the posterior two-thirds of the same surface; root more than three times the length of the crown, strongly curved backward, slightly gibbous near the crown and compressed near the point (Leidy, 1869).

Notable characteristics of *D. dividum* True, 1912a (the species normally taken as representing *Delphinodon*): face large relative to rostral length, broad with rounded margins; orbit short; nasal with deeply excavated anteromedial angle; tabular vertex (Kellogg, 1928; Muizon, 1988a).

Occipital condyle breadth: occipital condyle breadth of the type specimen of *D. dividum* not listed by True 1912a for USNM 7278, but is 70.6 mm (D. Bohaska, personal communication); the skull is significantly crushed.

Included species: *Delphinodon division* (known from localities NC3AA* [True, 1912a]; NC2A [USNM collections]), *D. mento* (locality NC3A [Leidy, 1869]), *Delphinodon* cf *D. mento* (locality [GC10C] [Morgan, 1994]).

Comments: Based on comparisons with the teeth, True (1912a) referred a much more complete specimen (USNM 7278) to the genus *Delphinodon*, placing it in the new species *Delphinodon dividum*. Dawson (1996b) considered at least one of the teeth referred to *Delphinodon mento* as similar to *Hadrodelphis calvertense*, while Dooley (2003) considered *Delphinodon mento* to be Odontoceti incertae sedis.

Hadrodelphis Kellogg, 1966

Type species: *Hadrodelphis calvertense* Kellogg, 1966.

Type specimen: USNM 23408, portion of right mandible ankylosed anteriorly to a shorter portion of the left mandible at symphysis (Kellogg, 1966).

Characteristics: Mandibles thick, robust, ankylosed anteriorly by symphyseal fusion; mandibular alveoli large, anteroposterior diameter 18–23 mm separated by 5–8 mm septa or interspaces; mandibular teeth with black enamel crowns; crowns of posterior teeth with anastomosing fine striae and with apical portion of subconical crown bent inward, overhanging broad internal basal shelf; enamel crown of more anterior mandibular teeth nearly conical and with internal shelf progressively reduced (Kellogg, 1966). See also Dawson (1996b).

Occipital condyle breadth: unknown.

Included species: *H. calvertense* only (known from localities NC3B*, D [Kellogg, 1966; Dooley, 1993; Dooley, Fraser, and Luo, 2004]),

Hadrodelphis sp. is also known from locality GC10C (Morgan, 1994).

Comments: *Hadrodelphis* is also known from the Miocene of Europe (McKenna and Bell, 1997).

Kampholophos Rensberger, 1969

Type species: *Kampholophos serrulus* Rensberger, 1969.

Type specimen: UCMP 35045, skull with all but symphysial portion of rostrum and mandibles, two postsymphysial teeth represented in each maxillary, all mandibular teeth posterior to break at symphysis, three unattached teeth, presumably from maxillary, seven cervical vertebrae.

Characteristics: Wings of lambdoidal crest extremely prominent, lacking angularity, converging and joining at relatively lower vertex in a broadly rounded ridge, displaying constriction of posterior-most flanges, exposing slightly swollen lateral walls of braincase in posterior view; posterior border of skull in lateral view formed by prominently projecting, evenly rounded flanges of lambdoidal crest; supraoccipital extending forward almost three-fifths of the distance between occipital surface of skull and posterior border of nasals, forming almost horizontal, dorsal boundary of braincase; vertex small and roughly square, composed of posterointernal extremities of premaxillaries, nasals exposed posterointernal angles of frontals and crest of supraoccipital; approximately same height reached by all dorsal elements of skull except frontal; left nasal highest medially and anteriorly, sloping down laterally at rather steep angle; zygomatic processes of squamosal strongly tapered in both vertical and transverse dimensions, quite slender anteriorly (Rensberger, 1969).

Occipital condyle breadth: 89 mm (Rensberger, 1969, p. 8).

Included species: *K. serrulus* only, known from locality WM17A* only.

Kampholophos sp. is also known from locality WM13C (Barnes, 1977).

Comments: *K. serrulus* was initially assigned to the family Iniidae but has subsequently been recognized as an archaic member of the family Kentriodontidae (Barnes, 1978; Ichishima *et al.*, 1995). Reputedly iniid dental characters, such as the ornamented (wrinkled) enamel and broad crowns on the posterior teeth, are merely primitive for Kentriodontidae, as they are also found among other basal Odontoceti.

***Kentriodon* Kellogg, 1927b (including *Grypolithax*, in part)**

Type species: *Kentriodon pernix* Kellogg, 1927b.

Type specimen: USNM 8060, skull, mandibles, cervical and dorsal vertebrae, ribs (Kellogg, 1927b).

Characteristics: Braincase short and narrow, about five-eighths as long as rostrum; rostrum relatively long, slender, slightly constricted at base; vertex small, more or less pentagonal; nasals relatively large, anterior margin deeply notched, elevated anteriorly; apophysis large, conspicuously produced; antorbital notch deep and narrow; maxillary foramina situated posterior to antorbital notches; horizontally expanded cranial plate of maxilla wider than premaxilla at level of anterior margin of respiratory passages; premaxillae not noticeably expanded posterior to antorbital notches; curvature of transverse crest of supraoccipital irregular (Kellogg, 1927b).

Occipital condyle breadth: 48 mm (USNM 8060 [Kellogg, 1927b, p. 22]).

Included species: *K. pernix* (known from locality NC3AA*); *K. obscurus* (locality WM13A* [Barnes and Mitchell, 1984]).

Kentriodon sp. is also known from localities (WM13C), (CC52) (Barnes, 1998).

Comments: *Kentriodon obscurus* includes *Grypolithax pavida* Kellogg 1931 (Barnes and Mitchell, 1984). *Kentriodon* is also possibly known from the Miocene of Europe and Japan (McKenna and Bell, 1997).

***Liolithax* Kellogg, 1931 (including *Lophocetus*, in part)**

Type species: *Liolithax kernensis* Kellogg, 1931.

Type specimen: CAS 4340, left periotic.

Characteristics: Prolongation of anteroventral angle of anterior process to support outer lip of bulla; transverse crease on ventral surface of anterior process marking anterior limit of tuberosity; deep incisure on posterior face above and internal to stapedial fossa; constriction of entrance to the aquaeductus Fallopii (Kellogg, 1931). Barnes (1978) noted: nasals more elongated anteroposteriorly than in *Kampholophos*, nearly parallel sided, depressed medially; shape of premaxillae around narial region nearly circular and not as parallel sided as in *Kampholophos* or *Lophocetus*; periotic with relatively long anterior process and short posterior process that is deep dorsoventrally and with small articular facet for the auditory bulla compared with *Kampholophos*; cerebral or dorsal surface of lateral part of periotic featureless and nearly flat; teeth with conical crowns and no accessory tubercles such as are on *Kampholophos*; striae on enamel only on lingual side of teeth.

Occipital condyle breadth: unknown.

Included species: *L. kernensis* (known from localities WM13A* [Kellogg, 1931], WM17G [Barnes, 1978], WM28II [Barnes, 1998], CC52 [Barnes, 1998]); *L. pappus* (locality NC3C [Kellogg, 1955]).

Liolithax, sp. is also known from locality WM16A (Barnes, 1977, 1978).

Comments: *Liolithax pappus* was originally described by Kellogg (1955) as *Lophocetus pappus*, but it was moved to the genus *Liolithax* by Barnes (1978) based on the similarity of the type periotic of *Liolithax kernensis* to the type periotic of "*Lophocetus*" *pappus*. *Liolithax* is also known from the Miocene of Europe (McKenna and Bell, 1997).

***Lophocetus* Cope, 1868a (including *Delphinus*, in part)**

Type species: *Lophocetus calvertensis* (Harlan, 1842) (originally described as *Delphinus calvertensis*).

Type specimen: USNM 16314, partial skull with left periotic, cervical and first two thoracic vertebrae.

Characteristics: Temporal fossa truncated by horizontal crest above, prolonged backwards and bounded by a projecting crest, rendering occipital plane concave; same crest prolonged upwards and thickened, each not meeting that of the opposite side, but continuing on inner margins of maxillary bones, turning outwards and ceasing opposite nares; front, therefore, deeply grooved; premaxillaries separated by a deep groove; teeth with cylindrical roots (Cope, 1868a). Barnes (1978) noted: skull with wide postorbital part of facial surface with convex lateral margin of frontal between postorbital process and lambdoid crest; relatively large nasals; prominent tuberosities on lateral part of occipital shield; periotic with relatively large pars cochlearis without deep fissure separating it from anterior process; anterior process of periotic short and globular.

Occipital condyle breadth: 93.1 mm (D. Bohaska, personal communication, for type specimen; not reported by Cope [1868a]).

Included species: *L. calvertensis* (known from locality NC5* [Kellogg, 1955]); *L. repenningi* (locality WM16C [Barnes, 1978]).

Lophocetus sp. is also known from locality (GC10C) (Morgan, 1994).

***Macrokentriodon* Dawson, 1996a**

Type species: *Macrokentriodon morani* Dawson, 1996a.

Type specimen: CMM-V-15

Characteristics: Differs from other kentriodontids in having longer nasals than frontals; nasal–frontal suture perpendicular to long axis of the skull; premaxillae constricted at the level of the antorbital notch; lambdoidal crests prominent; fossae for the pterygoid sinuses long, tapered, extending rostrally to antorbital notch; fossa for middle sinus prominent and deep (Dawson, 1996a).

Occipital condyle breadth: Occipital condyle not preserved in the type and only known specimen.

Included species: *M. morani* only, known from locality NC4A* only.

ALBIREONIDAE BARNES, 1984

The porpoise-like extinct *Albireo whistleri* Barnes, 1984 (latest Miocene), from Isla Cedros, is the sole named species within the Albireonidae. Barnes (1984, 1990) gave many diagnostic characters, including synapomorphies, listed under the genus below. Muizon (1988a) placed the family close to Phocoenidae, while Barnes (1990) placed it between Monodontidae and Kentriodontidae.

Albireo Barnes, 1984

Type species: *Albireo whistleri* Barnes, 1984.

Type specimen: UCR 14589, cranium, periotics, tympanic bullae, ossicles, mandible with teeth, isolated teeth, complete postcranial skeleton lacking only terminal caudal vertebrae, pelvic bones, and some phalanges and metacarpals (Barnes, 1984).

Characteristics: Skull with cranial vertex elevated and symmetrical; facial region steeply sloping posterior to nares; occipital condyles and zygomatic processes of squamosals large, rostrum of intermediate length, tapered and curved dorsally at distal end; teeth large with conical crowns covered by smooth enamel; premaxillary eminences anterior to nares, low, relatively narrow, smoothly rounded and highest near midline; posterior terminations of premaxillae wide and flat, extending further posteriorly than posterior margins of nasal openings; nasal bones large and flat; no fossae for air sinuses developed in lateral side of basisphenoid nor in the orbit between the frontal and maxilla; zygomatic process of jugal relatively large; narial passages large, circular, passing vertically through the skull; pterygoid sinuses large and totally floored ventrally by pterygoid bones; periotic large, globose with round and pointed anterior process; cochlear portion large, elongate anteroposteriorly; posterior process large with smooth articular facet for bulla; dentary elongate with upturned anterior end; mandibular symphysis moderately short (approximately 27% length of dentary); atlas large, not fused to axis; thoracic and lumbar vertebrae with anteroposteriorly compressed centra and with elongate transverse and neural spines; scapula high; forelimb elements short and expanded anteroposteriorly (slightly modified from Barnes, 1984).

Occipital condyle breadth: 93 mm, UCR 14589.

Included species: *A. whistleri* only (known from localities WM30A*, [B] [Barnes, 1984, 1998]).

Comments: *Albireo whistleri* is the only species in *Albireo*, and *Albireo* is the only genus currently known from the family Albireonidae. Barnes (1988) mentioned that other as yet unnamed material is known.

MONODONTIDAE GRAY, 1821

Crown-Monodontidae includes two species, *Monodon monoceros* (the narwhal) and *Delphinapterus leucas* (the beluga) of Arctic and cold North American waters.

Characteristics: Cranium relatively wide; rostrum short, broad at base, tapering rapidly to middle, with distal margins roughly parallel; distal end of premaxillae expanded transversely; maxillae broadly exposed medial to premaxillae at posterior end of mesorostral groove; narrow cranial vertex conspicuously elevated, but adjacent bones of face not elevated, so most of dorsal surface of skull is in the one plane; pterygoid hamuli separated; lateral lamina of palatine strongly developed ventral to the optic canal, with pterygoid sinuses thus restricted to the basicranium and not invading orbits; cervical vertebrae unfused (True, 1889; Fraser and Purves, 1960; Muizon, 1988a; Barnes, 1990).

There is no comprehensive anatomically based published cladistic analysis of monodontid relationships. *Monodon* and *Delphinapterus* are reported widely as North American Quaternary fossils, but older material is sparse. Whitmore (1994) reported early Pliocene *Delphinapterus* sp. from the Yorktown Formation (locality EM8), while Barnes (1977) noted undescribed late Miocene monodontids from California. The one extinct monodontid described from a skull is *Denebola brachycephala* Barnes, 1984 (late or latest Miocene to early Pliocene, Isla Cedros).

Species in the extinct Odobenocetopsidae are related closely to Monodontidae, according to Muizon (1993; Muizon, Domning, and Parish, 1999). There are no reported records from North America.

Delphinapterus Lacépède, 1804 (including *Delphinus*, in part)

Type species: *Delphinapterus leucas* (Pallas, 1776) (originally described as *Delphinus leucas*), the extant beluga.

Type specimen: None designated (see Hershkovitz, 1966).

Characteristics: No prominent tusk (cf. narwhal, *Monodon monoceros*); generally nine erupted homodont teeth per quadrant; skull more symmetrical than *Monodon* but otherwise similar (True 1889; Reeves and Tracey, 1980; Stewart and Stewart, 1989).

Occipital condyle breadth: 133.9 mm (average of adult *Delphinapterus leucas* specimens USNM 305071, 485826, and 504673).

Included species: *Delphinapterus* sp. only, known from locality EM8B only (Whitmore, 1994).

Comments: The genus *Delphinapterus* is also known from the middle Miocene of western Asia, the Pleistocene of Europe, and the Recent of the Arctic, North Atlantic and Pacific Oceans (McKenna and Bell, 1997).

Denebola Barnes, 1984

Type species: *Denebola brachycephala* Barnes, 1984.

Type specimen: UCR 21245, nearly complete but badly shattered skull, periotics, tympanic bullae, isolated teeth, parts of both mandibles, several vertebrae, ribs, parts of the pectoral girdle and forelimbs (Barnes, 1984).

Characteristics: Skull relatively shorter and wider than that of *Delphinapterus*, with rostrum broad and flat; occipital crest more prominent; cranial vertex moderately elevated; nasals longer; premaxillae laterally expanded but not arched along medial margins; 15 or 16 teeth present on each side of rostrum with one in each premaxilla (Barnes, 1984).

Occipital condyle breadth: 110 mm (estimated from a half width on UCR 21245).

Included species: *D. brachycephala* (known from locality WM30A [Barnes, 1984]); *Denebola* n. sp. (locality WM30A [Aranda-Manteca and Barnes, 1993]).

Denebola sp. is also known from locality (WM33) (Barnes, 1998).

PHOCOENIDAE GRAY, 1825

Crown-Phocoenidae includes living species of porpoises in the genera *Phocoena*, *Phocoenoides*, and *Neophocaena* (genera follow Rosel, Haygood, and Perrin [1995]). For the crown-Phocoenidae,

Characteristics: Widely recognized diagnostic characters (some synapomorphic) include blunt-snouted external profile; premaxillary sac fossae elevated to form premaxillary eminences; reduced and ventrally displaced nasals; "nasal hump" present dorsal to nasals; reduced ascending processes of premaxillae, not contacting nasals; preorbital lobe of pterygoid sinus expanded dorsally; tooth crowns spatulate (laterally flattened, elongate) rather than conical; posterior process of bulla pachyostotic and lacking spines (see Fraser and Purves, 1960; Kasuya, 1973; Barnes 1984, 1985d, 1990; Muizon, 1988a).

There is no comprehensive, anatomically based, published cladistic analysis of phocoenid relationships. There appear to be no firm fossil records for extinct species of crown-group porpoises; the described fossils presumably represent stem-Phocoenidae. The globally oldest phocoenid is the North American *Salumiphocoena stocktoni* (Wilson, 1973), of late Miocene age, reportedly >10 Ma (Barnes, 1985d).

Loxolithax Kellogg, 1931

Type species: *Loxolithax sinuosa* Kellogg, 1931.

Type specimen: CAS 4352.

Characteristics: Highly angular processes; broad facet on the ventral face of the posterior process; relatively wide fossa incudis; elongated internal acoustic meatus; large fossa surrounding cerebral orifice of the aqueduct of vestibule; anterior process with a laterally compressed extremity (Kellogg, 1931).

Occipital condyle breadth: unknown.

Included species: *L. sinuosa* only, known from locality WM13A* only.

Comments: Kellogg (1931) placed *Loxolithax* within the Delphinidae, but in the past such family assignments were commonly made uncritically for many small, generalized, mid Tertiary fossil odontocetes. Wilson (1973) assigned his new late Miocene species *Loxolithax stocktoni* to this genus, but the periotics of the two species (*L. sinuosa, L. stocktoni*), the only basis for any comparison between the two taxa, are not markedly similar. Barnes (1985d) referred *Loxolithax sinuosa* to the family Phocoenidae and transferred the late Miocene species to the new genus *Salumiphocoena*.

Piscolithax Muizon, 1983b

Type species: *Piscolithax longirostris* Muizon, 1983b.

Type specimen: SAS 933, nearly complete skull and mandible with an associated atlas and additional cervical and dorsal vertebrae (Muizon, 1983b).

Characteristics: Larger than other phocoenids, but similar to *Loxolithax stocktoni*; rostrum long and broad at base and pointed at its end; maxilla carrying approximately 37 slightly transversely flattened conical teeth; toothed premaxilla and mandible carrying approximately 38 teeth each; antorbital notch deep and near to the anterior maxillary foramen; posterior premaxillae smooth, convex, narrow, side edge shows a well-marked gutter; ascending processes of premaxillae not exceeding the posterior edge of the nasal opening; occipital covering the frontals; broad and flat palate; together pterygoid and palatine forming a block narrower than the posterior part of the palate; pterygoids moved further back than in *Phocoena*; tympanic with smooth, thick pedicle, much smaller than in *Phocoena*; periotic with transversely flattened cochlea and more robust processes than in *Phocoena*; long mandible for Phocoenidae with coronoid process more advanced than in *Phocoena* and *Loxolithax*; opening of the mandibular canal further back than in *Phocoena* (Muizon, 1983b).

Occipital condyle breadth: unknown.

Included species: *P. boreios* (known from locality WM30A [Barnes, 1984]); *P. tedfordi* (locality WM30A [Barnes, 1984]).

Piscolithax sp. is also known from locality (WM28) (Barnes, 1998).

Comments: *Piscolithax* is also known from the Pliocene (and possibly also the late Miocene) of Peru (McKenna and Bell, 1997). The geographic distribution of *Piscolithax* could be interpreted as antitropical, similar to that of living phocoenids (Barnes, 1985d), but because there is negligible fossil record from tropical latitudes it is not possible to tell whether *Piscolithax* was truly absent from the Tropics.

Salumiphocoena Barnes, 1985 (including *Loxolithax*, in part)

Type species: *Salumiphocoena stocktoni* (Wilson, 1973) (originally described as *Loxolithax stocktoni*).

Type specimen: UCMP 34576, a mature skull with tympanic bullae, petrosals and auditory ossicles, mandibles, 29 vertebrae, a left forelimb lacking phalanges, and several ribs (Wilson, 1973).

Characteristics: A phocoenid about the size of extant *Phocoenoides dalli*; rostrum triangular; cranium longer than wide; braincase somewhat inflated, with cranial vertex not significantly elevated; posterior apices of premaxillae extending only a little beyond midpoint of external nares; zygomatic processes of the squamosals slender and short; temporal fossa short; teeth numerous, crowns more cylindrical than spatulate, with a tiny cusp both anterior and posterior (summarized from Wilson. 1973; Barnes, 1985d; Ichishima and Kimura, 2000).

Occipital condyle breadth: 76.3 mm (UCMP 34576 [Wilson, 1973, p. 6]).

Included species: *S. stocktoni* only, known from locality WM17D* only.

Comments: Wilson (1973) originally assigned his new late Miocene species to *Loxolithax* Kellogg, 1931, using the combination *Loxolithax stocktoni*. However, the periotic of *Loxolithax stocktoni* differs sufficiently from that of *L. sinuosa* Kellogg, 1931 to suggest that the two species are probably not congeneric. Barnes (1985d) referred Kellogg's *L. sinuosa* to the family Phocoenidae, and transferred Wilson's late Miocene species to the new genus *Salumiphocoena*.

DELPHINIDAE GRAY, 1821

The true dolphins or ocean dolphins (family Delphinidae) form the most speciose group of living Cetacea (Rice, 1998). The family is based on *Delphinus*, for which the type species is *Delphinus delphis* Linnaeus. Living delphinids are variable in external form but are easily recognized on the basis of osteological and soft tissue characters. The literature on delphinid systematics is confused; until recently, the Delphinidae was used widely as a grade family for small odontocetes (e.g., True, 1889; Kellogg, 1928; Simpson, 1945) including the porpoises (now accorded separate family status as Phocoenidae). Crown-group Delphinidae contains the extant species of *Delphinus*, *Cephalorhynchus*, *Feresa*, *Globicephala*, *Grampus*, *Lagenodelphis*, *Lagenorhynchus*, *Lissodelphis*, *Orcaella*, *Orcinus*, *Peponocephala*, *Pseudorca*, *Sotalia*, *Sousa*, *Stenella*, *Steno*, and *Tursiops*, plus all descendants of their most recent common ancestor. Subfamilies used within the crown-Delphinidae in recent decades include Delphininae, Steninae, Lissodelphininae, Cephalorhynchinae, and Globicephalinae (e.g., Fraser and Purves, 1960; Mead, 1975a; Muizon, 1988a; Bianucci, 1996). Results of recent molecular analyses (e.g., LeDuc, Perrin, and Dizon, 1999) indicate that such groups should probably be revised, and that some genera (e.g., the speciose *Lagenorhynchus*) are probably polyphyletic.

Extensive published comment on the diagnostic features of Delphinidae includes that of Fraser and Purves (1960), Kasuya (1973), Barnes (1978, 1990), Heyning (1989a), and Muizon (1988a, 1991). Some contributions are framed in terms of apomorphies, but we are not aware of any comprehensive matrix-based analysis of relationships. The prime diagnostic feature of Delphinidae is the synapomorphy of asymmetrical premaxillae, with the left narrower and shorter than the right, and retracted further forward from nasal. In soft tissues, the preorbital and postorbital lobes of the pterygoid sinuses are diagnostic (Fraser and Purves, 1960); these structures are substantially reflected in skull topography. Literature on fossil Delphinidae has not previously discriminated between crown- and stem-group; here, the stem-Delphinidae is presumed to include some species previously referred to the Kentriodontidae (below).

Globally, fossils with delphinid cranial features can be traced back to the late Miocene, >10 Ma (Barnes, 1977), but there are no reliable records from the middle Miocene or older. The literature contains many references to fossil "dolphins" from North America (e.g., *Hesperocetus californicus* True, 1912b, *Delphinavus newhalli* Lull, 1914), but there appear to be no distinct fossil species of Delphinidae based on North American types. According to Barnes (1977) and Whitmore (1994), latest Miocene and younger fossils from sequences along the Pacific and Atlantic are likely to represent *Delphinus*, *Globicephala*, *Lagenorhynchus*, *Pseudorca*, *Stenella*, and *Tursiops*.

Delphinus Linnaeus, 1758

Type species: *Delphinus delphis* Linnaeus, 1758, the extant common dolphin.

Type specimen: None designated (see Hershkovitz, 1966).

Characteristics: The most salient character by which this genus, in its present restricted limits, is distinguished is the presence of two broad and deep lateral grooves in the palate. This is, indeed, the only character by which in the present state of our knowledge it is distinguishable from *Prodelphinus* [= *Stenella*], its nearest ally (True, 1889).

Occipital condyle breadth: 87 mm (average of measurements from 88 adult *D. delphis* skulls in the USNM collection).

Included species: *Delphinus* sp. only, known from locality (EM8B) only (Whitmore, 1994). Barnes (1998) listed *Stenella* sp. or *Delphinus* sp. from localities WM30B and WM32.

Comments: The genus *Delphinus* is also known from the Miocene to Recent of Asia, the Miocene to Recent (the Black Sea) of Europe, the Pliocene of Africa, the Pliocene to Pleistocene of South America, the Pliocene of New Zealand, and the Recent of the Mediterranean Sea and the Indian, Atlantic, and Pacific Oceans (McKenna and Bell, 1997).

Globicephala Lesson, 1828

Type species: *Globicephala melas* (Traill), the extant long-finned pilot whale (originally described as *Delphinus melas* [according to Hershkovitz, 1966]).

Type specimen: Lectotype, as noted by Hershkovitz (1966) is a skull, BMNH 363a-44.12.3.2.

Characteristics: Head and body length 360–850 cm; pectoral fin length approximately one-fifth head and body length; dorsal fin height approximately 30 cm; width of tail flukes approximately 130 cm; coloration black throughout, except for a white area often present below the chin; head swollen so the forehead bulges above the upper jaw; pectoral fin narrow and tapering; dorsal fin located just in front of the middle of the back; 7 to 11 teeth on each side of each jaw (Nowak and Paradiso, 1983); skull relatively wide (cranium wider than long); rostrum short, broad and dorsoventrally thin, with teeth in anterior part; rostral portion of premaxillae flat and broad; mandibular symphysis short; antorbital notch deep; perinasal area strongly asymmetrical; nasals nodular and elevated and, with adjacent facial elements, positioned far posteriorly; pterygoids large and in contact (based on True, 1889; Arnold and Heinsohn, 1996).

Occipital condyle breadth: 154 mm (average of measurements of two adult *Globicephala melas* skulls in the USNM collection).

Included species: *Globicephala* sp. only, known from locality EM8B only (Whitmore, 1994).

Comments: The genus *Globicephala* is also known from the Pliocene of Europe, and the Recent of the Mediterranean Sea, and the Indian, Atlantic, and Pacific Oceans (McKenna and Bell, 1997).

Lagenorhynchus Gray, 1846

Type species: *Lagenorhynchus albirostris* Gray 1846, the extant white-beaked dolphin.

Type specimen: Skeleton, BMNH 916a-48.7.12.12 (Hershkovitz, 1966).

Characteristics: Head rather convex, gradually sloping into the beak in front; beak short, tapering in front; lower jaw rather long; body elongate, tapering behind, largest at the pectoral fins; pectoral fins rather far back, rather elongate and slightly falcate; dorsal fin high, falcate, rather behind the middle of the back; back with a lower, rounded, finlike ridge near the teal; tail-lobes rather narrow, elongate; skull rather depressed, hinder ends of maxillary bones expanded, horizontal, and rather thickened on the edge; nose short, broad, flat above, rather narrow in front, scarcely longer than the length of the brain cavity; triangle in front of blowers elongate, reaching beyond the middle of the nose of the skull; intermaxillaries separated by a deep groove filled with cartilage (Gray, 1846).

Occipital condyle breadth: 86 mm (average of measurements of 17 adult *Lagenorhynchus acutus* skulls in the USNM collection).

Included species: *Lagenorhynchus* sp. only (known from localities [WM30B], [WM32], [EM8B] [Whitmore, 1994; Barnes, 1998]).

Comments: Molecular studies (e.g., LeDuc, Perrin, and Dizon, 1999) indicate that *Lagenorhynchus* of traditional use is a form genus. At present there is no satisfactory diagnosis based on skeletons. The genus *Lagenorhynchus* is also known from the Pleistocene of Europe, possibly from the Pliocene of North Africa, and from the Recent of the Indian, Atlantic, and Pacific Oceans (McKenna and Bell, 1997).

Pseudorca Reinhardt, 1862 (including *Phocoena*, in part)

Type species: *Pseudorca crassidens* (Owen, 1846) the extant false killer whale, (originally described as *Phocoena crassidens*).

Type specimen: Subfossil skull, once in the Cambridge University Museum, now lost (Hershkovitz, 1966).

Characteristics: Differs from the skull of the beluga by its concave profile; its greater breadth in proportion to its length, especially across the maxillary portion: shorter temporal fossae; more numerous teeth. The general resemblance of the fossil to the skulls of the *Grampus* and round-headed porpoise is much closer, and its distinctive character requires more detailed comparison with these for its demonstration (Owen, 1846). Other characters include: skull massive; rostrum short, broad; rostral portion of premaxillae broad, roughened anteriorly, and abruptly truncated distally; pterygoids short, nearly or quite in contact medially; palatines prolonged laterally across the optic canal; up to 10 large teeth per quadrant (True, 1889; Stacey, Leatherwood, and Baird, 1994).

Occipital condyle breadth: 149.8 mm (average of adult *Pseudorca crassidens* specimens USNM 484982, 219325, and 485827).

Included species: *Pseudorca* sp. only, known from locality EM8B only (Whitmore, 1994).

Comments: The genus *Pseudorca* is also known from the Pleistocene of Japan and Europe, possibly from the Pliocene of New Zealand, and from the Recent of the Mediterranean Sea and the Indian, Atlantic, and Pacific Oceans (McKenna and Bell, 1997).

Stenella Gray, 1866 (including *Steno* in part)

Type species: *Stenella attenuata* (Gray, 1846) the extant pantropical spotted dolphin (originally described as *Steno attenuata*).

Type specimen: Skull figured by Gray (1846, p. 44, Plate 28) (Hershkovitz, 1966).

Characteristics: Differs from *Delphinus* in absence of deep lateral palatine grooves (shallow grooves present posteriorly on the rostrum); differs from *Tursiops* in its smaller and less numerous teeth, (generally) more numerous vertebrae (True, 1889). Other characters include: relatively small cranium; long dorsoventrally flattened rostrum; large number of small teeth; medially convergent premaxillae; sigmoid ramus; small temporal fossae (Perrin, 1998).

Occipital condyle breadth: 88 mm (average of measurements of 36 adult *Stenella coeruleoalba* skulls in the USNM collection).

Included species: *Stenella* n. sp. only, known from locality EM8B only (Whitmore, 1994). Barnes (1998) listed

Stenella sp. or *Delphinus* sp. from localities WM30B and WM32.

Comments: The genus *Stenella* is also known from the Recent of the Mediterranean Sea, and the Indian, Atlantic, and Pacific Oceans (McKenna and Bell, 1997).

Tursiops **Gervais, 1855 (including *Delphinus*, in part)**

Type species: *Tursiops truncatus* (Montagu, 1821) the extant bottlenose dolphin (originally described as *Delphinus truncatus*).

Type specimen: Skull only, BMNH 353h-62.7.18.15 (type specimen of *Tursiops truncatus* as noted by Hershkovitz [1966; see also Rice, 1998]).

Characteristics: Broad cranium; rostrum moderately long, gradually tapering from broad base; width significantly greater across postorbital process than across preorbital processes; postorbital process triangular; anterior sinus developed and extending onto base of rostrum; total number of vertebrae between 61 and 66 (True, 1889; Fraser and Purves, 1960; Barnes, 1990).

Occipital condyle breadth: 191 mm (average of measurements of 31 adult *Tursiops truncatus* skulls in the USNM collection).

Included species: *Tursiops* sp. only, known from locality EM8B only (Whitmore, 1994).

Comments: Bianucci (1996) discussed Italian Pliocene *Tursiops* (*T. ossenae*, *T. cortesii*, *T. astensis*, and *T. capellinii*) previously mentioned by Barnes (1990); only *T. ossenae* was retained in *Tursiops*. Reports of *T. truncatus* from the Pleistocene to Recent globally (Barnes, 1990) should be reappraised in light of recognition of two or more living species (e.g., Rice, 1998; Turner and Worthy, 2003).

The genus *Tursiops* is also known from the Miocene to Recent of Europe (the Black Sea), the Pliocene and Pleistocene of South America, possibly from the Pliocene of New Zealand, and the Recent of the Mediterranean Sea, and the Indian, Atlantic, and Pacific Oceans (McKenna and Bell, 1997).

ODONTOCETI NOMINA DUBIA

Many genera and species of odontocete cetacean have been named based on very incomplete material. Often, these type specimens cannot be distinguished among other named species, genera, families, or even confidently assigned to a suborder. Occasionally, subsequent authors have assigned much better specimens to taxa with these poor type specimens, and the newly referred material has become specimens of reference for subsequent work rather than the type specimens. For others, the names have been around since their inception, but little or nothing has been referred to them because of their poor quality. These taxa should best be considered nomina dubia, and the names restricted to the type specimen.

Agabelus **Cope, 1875**

Type species: *Agabelus porcatus* Cope, 1875.

Type specimen: USNM 10756, rostrum.

Characteristics: Resembles the elongate muzzle of a *Priscodelphinus* without teeth, but with the alveolar lines excavating into a deep groove on each side; superior surfaces possessing a shallow median groove as in most delphinoid cetaceans; supposed palatal face plane and sharply defined by the lateral grooves. These grooves bounded above by a thin overhanging border on each side, their fundus marked by a series of nutritious foramina of small size, apparently corresponding to the positions of teeth of other genera (Cope, 1875).

Occipital condyle breadth: unknown.

Included species: *A. porcatus* only, known from locality NC1B* only.

Comments: Material is incomplete and probably not diagnostic; other material cannot be referred to the genus with confidence, and the family-level relationships are uncertain.

Belosphys **Cope, 1875 (including *Priscodelphinus*, in part)**

Type species: *Belosphys spinosus* (Cope, 1868c) (originally described as *Priscodelphinus spinosus*).

Type specimen: ANSP 11246, originally 13 vertebrae, of which 11 are present (Gillette, 1975).

Characteristics: Posterior lumbars and caudals spinous; dorsals with flat diapophyses (Cope, 1868c).

Occipital condyle breadth: unknown.

Included species: *B. spinosus* only, known from locality ?NC3AA* only (Leidy, 1869).

Comments: Material is incomplete and probably not diagnostic; other material cannot be referred to the genus with confidence, and the family-level relationships are uncertain.

Cetophis **Cope, 1868b**

Type species: *Cetophis heteroclitus* Cope, 1868b.

Type specimen: ANSP 11253, four caudal vertebrae (Gillette, 1975).

Characteristics: Caudal vertebrae with very thick epiphyses (Cope, 1868b).

Occipital condyle breadth: unknown.

Included species: *C. heteroclitus* only, known from locality NC6* only.

Comments: Material is incomplete and probably not diagnostic; other material cannot be referred to the genus with confidence, and the family-level relationships are uncertain.

Colophonodon **Leidy, 1853**

Type species: *Colophonodon holmesii* Leidy, 1853.

Type specimen: It is unclear where the type specimen is or where it was ever curated; one complete tooth with fragments of five others (Leidy, 1853).

Characteristics: Tooth resembles in general form the teeth of the dolphins; approximately 7.5 cm long; curved; conical crown capped with enamel, forming a salient ridge on two sides; root long and concoidal, gibbous just beyond the crown; enamel smooth on some specimens, corrugated on others (Leidy, 1853).

Occipital condyle breadth: unknown.

Included species: *C. holmesii* only (possibly known from locality EM6A, see Leidy [1853]).

Comments: Kellogg (1923b) noted that *Squalodon* [= *Colophonodon*] *holmesii* was too incomplete for definite allocation; see also Dooley (2003).

Graphiodon Leidy, 1870

Type species: *Graphiodon vinearius* Leidy, 1870.

Type specimen: USNM 875, isolated tooth.

Characteristics: Tooth crown curved, conical; without divisional planes; inner surface only feebly defined from the outer, by a single imperfectly developed ridge posterointernally; enamel singularly roughened, owing to short vermicular, somewhat ramifying and more or less interrupted ridges, giving it a fretted or lettered appearance; transverse section of crown circular (Leidy, 1870).

Occipital condyle breadth: unknown.

Included species: *G. vinearius* only, known from locality NC2II* only.

Comments: No additional material has been referred to *Graphiodon*. Kellogg (1923b) speculated that the material may represent a physeterid.

Hesperocetus True, 1912b

Type species: *Hesperocetus californicus* True, 1912b.

Type specimen: UCMP 1352, mandibular fragment with widely spaced teeth bearing crenulated enamel.

Characteristics: Mandibles with rounded upper margins and sinuous lateral borders of the median ridge; sides of mandibles convex; short, conical teeth with elliptical cross sections; rugose enamel (True, 1912b).

Occipital condyle breadth: unknown.

Included species: *H. californicus*, only known from locality CC31* only.

Comments: *Hesperocetus* has been said to be in the family Iniidae (True, 1912b; Simpson, 1945), but it is difficult to place this genus properly because of the poor type specimen. Material is incomplete and debatably diagnostic, but the concept of the genus could probably be firmed up if more-complete topotypic specimens were found. For now, family-level relationships are uncertain.

Ixacanthus Cope, 1868c (including *Priscodelphinus*, in part)

Type species: *Ixacanthus coelospondylus* Cope, 1868c.

Type specimen: ANSP 11254, 13 vertebrae (3 thoracic, 9 lumbar, 1 caudal [Cope, 1868c]), (Gillette, 1975).

Characteristics: Differs from all other Delphinidae in the manner of attachment of the epiphyses of the vertebrae; instead of being nearly plane and thin discs, epiphyses have two oblique faces above, capped by a projecting roof formed by the floor of the neural canal; central portion of epiphyses forming a knob that fits a corresponding shallow pit of the centrum (Cope, 1868c).

Occipital condyle breadth: unknown

Included species: *I. coelospondylus* only, known from locality NC1A* only.

Comments: It is unclear whether these vertebrae are from a single or multiple individuals. Hay (1902) listed several of the species of *Priscodelphinus* (*P. atropius*, *P. conradi*, *P. stenus*) as species of *Ixacanthus*.

Lonchodelphis Allen, 1924b

Type species: *Lonchodelphis occiduus* (Leidy, 1869) (originally described as *Delphinus occiduus*).

Type specimen: MCZ 8765, fragment from the posterior portion of the palate (Allen, 1924b).

Characteristics: As preserved, the type specimen has 19 closely set, circular alveoli, rather over two lines in diameter. (Leidy [1869] used the designation of "line" in his original description); jaw measures little more than 5 cm as the back of the fragment gradually tapers for half its length, before proceeding with parallel sides to the fore end, where it is ten and one half lines wide; palate behind nearly planar or slightly convex in fore part presenting a deep median groove, closely by the apposition of the maxillaries, groove separated only by a narrow ridge from the alveoli; sides of the maxillaries slightly concave longitudinally, convex transversely; intermaxillaries broken away, leaving a wide, angular gutter between the remains of the maxillaries (Leidy, 1869).

Occipital condyle breadth: unknown.

Included species: *L. occiduus* only (known from localities WM24A, WM30A [Barnes, 1977]).

Comments: *Lonchodelphis occiduus* was the first fossil cetacean described from the west coast of North America (Allen, 1924b). Allen (1924b) erected the genus *Lonchodelphis* for the specimen because it was clearly not a species of *Delphinus*. Material is incomplete and debatably diagnostic, but the concept of the genus could probably be established if more-complete topotypic specimens were found. For now, family-level relationships are uncertain.

Pelodelphis Kellogg, 1955

Type species: *Pelodelphis gracilis* Kellogg, 1955.

Type specimen: USNM 13471, ankylosed mandibles lacking distal ends of both rami; posterior portion of right mandible incomplete.

Characteristics: Mandibles slender with elongated symphysis equivalent to less than one-half the length of each ramus when complete; symphysis ankylosed, tapering toward anterior extremity, approximately U-shaped in cross section about half way of its length, with anterior half of its length bent upward; opposite free hinder portions of mandibles forming a blunt angle at level where they ankylose as symphysis; probably 36 teeth in each mandible; 12 alveoli on each mandible posterior to hinder end of symphysis; 21 alveoli on preserved portion of left mandible anterior to posterior end of symphysis; roots of teeth on symphysis implanted obliquely in alveoli, which slope more backward than inward; anterioposterior diameter of each of two alveoli at hinder end of symphysis in right mandible, 9 mm; transverse diameter of same alveoli, 9 mm; diameters of anterior alveoli approximately 1 mm smaller; opposite alveoli at hinder end of symphysis separated by an interval of 22.5 mm; corresponding interval between 16th alveoli (counting forward from hindmost at posterior end of symphysis), 5.5 mm (Kellogg, 1955).

Occipital condyle breadth: unknown.

Included species: *P. gracilis* only, known from locality NC3AA only.

Comments: Muizon (1988a) listed *Pelodelphis* as a "probable" delphinoid. Fordyce and Muizon (2001) suggested that *Pelodelphis* is possibly an iniid. Material is incomplete and debatably diagnostic, but the concept of the genus could probably be established if more-complete topotypic specimens were found. For now, family-level relationships are uncertain.

***Priscodelphinus* Leidy, 1851**

Type species: *Priscodelphinus harlani* Leidy, 1851.

Type specimen: ANSP 11242, vertebra (Gillette, 1975).

Characteristics: Length of body of vertebra little over 5 cm; posterior articular surface is 20 lines wide and 17 in depth; spinal arch is lost; width of the spinal canal just within its fore part nine lines; length of a transverse process, with an articular facet for a rib, 17 lines (Leidy, 1869).

Occipital condyle breadth: unknown.

Included species: *P. harlani* (known from locality NC1B); *P. acutidens* (locality ?NC3AA), *P. atropius* (locality ?NC3AA); *P. conradi* (locality ?NC3AA); *P. stenus* (locality ?NC3AA) (Leidy, 1869).

Comments: Leidy (1869) listed many species of *Priscodelphinus*, each based on either isolated vertebrae or teeth. Material is incomplete and probably not diagnostic; other material cannot be referred to the genus with confidence, and the family-level relationships are uncertain. Another nominal species of *Priscodelphinus* (*P. squalodontoides*) was removed from the genus and made the type species of the new genus *Rudicetus* by Bianucci (2001).

***Rhabdosteus* Cope, 1868a**

Type species: *Rhabdosteus latiradix* Cope, 1868a.

Type specimen: ANSP 11260, rostrum. Gillette (1975) was unsure of the identity of the holotype of *Rhabdosteus latiradix* but was relatively confident it is either ANSP 11260 or ANSP 11262, which he listed as the paratype.

Characteristics: Premaxillary and maxillary bones forming a cylinder, bearing teeth on its proximal portion, and prolonged in its distal portion into a slender straight beak; teeth with enlarged crown separated from the root by a constriction (Cope, 1868a).

Occipital condyle breadth: unknown.

Included species: *R. latiradix* only, known from locality ?NC3AA only (Gillette, 1975).

Comments: Muizon (1988c) clearly stated: "The family Rhabdosteidae and the genus *Rhabdosteus* are based on a type-species [*R. latiradix*] whose type specimen is too uncomplete [sic] for taxonomic purposes and they are here regarded as incertae sedis, the nomina being restricted to that specimen. All specimens from the Calvert Formation hitherto referred to the genus *Rhabdosteus* are in fact representative of *Schizodelphis*." Despite this clear and decisive action by Muizon, other authors (e.g., McKenna and Bell, 1997) persisted in referring to *Rhabdosteus* without comment on its status. Fordyce (1983) used Rhabdosteidae formally as the senior synonym for Eurhinodelphinidae, following the recommendation of Myrick (1979), but here we follow the traditional use of Eurhinodelphinidae.

***Tretosphys* Cope, 1868c (including *Delphinapterus*, in part)**

Type species: *Tretosphys gabbii* (Cope, 1868c) (described as both *Tretosphys gabbii* and *Delphinapterus gabbii* in the same paper).

Type specimen: ANSP 11234, caudal vertebra (Gillette, 1975).

Characteristics: Caudal vertebra in *T. gabbii* not more than half the length of those in *T. grandaevus*, less strongly constricted everywhere and especially below; ridges and chevron articular surfaces much more elevated, especially those on the anterior part of the centrum, embracing a very deep groove in *T. grandaevus*, a shallow one in *T. gabbii*; an additional longitudinal ridge on each side the inferiors in front *T. grandaevus* is wanting in *T. gabbii*; there is no posterior zygapophysis in *T. gabbii* (Cope, 1868c).

Occipital condyle breadth: unknown.

Included species: *T. gabbii* (known from locality ?NC3AA [Kellogg, 1955]); *T. grandaevus* (locality NC1 [Leidy, 1869]); *T. lacertosus* (locality NC1 [Cope, 1868c]); *T. ruschenbergeri* (locality ?NC3AA [Leidy, 1869]); *T. uraeus* (locality NC1 [Leidy, 1869]).

Comments: The type specimen of *T. gabbii* and the type specimens of all other species referred to *Tretosphys* are based on isolated vertebrae. There is a great deal of confusion surrounding this genus because of Cope's (1868c) reference of some species to *Delphinapterus* and others to *Tretosphys*, and apparently some species to both in the

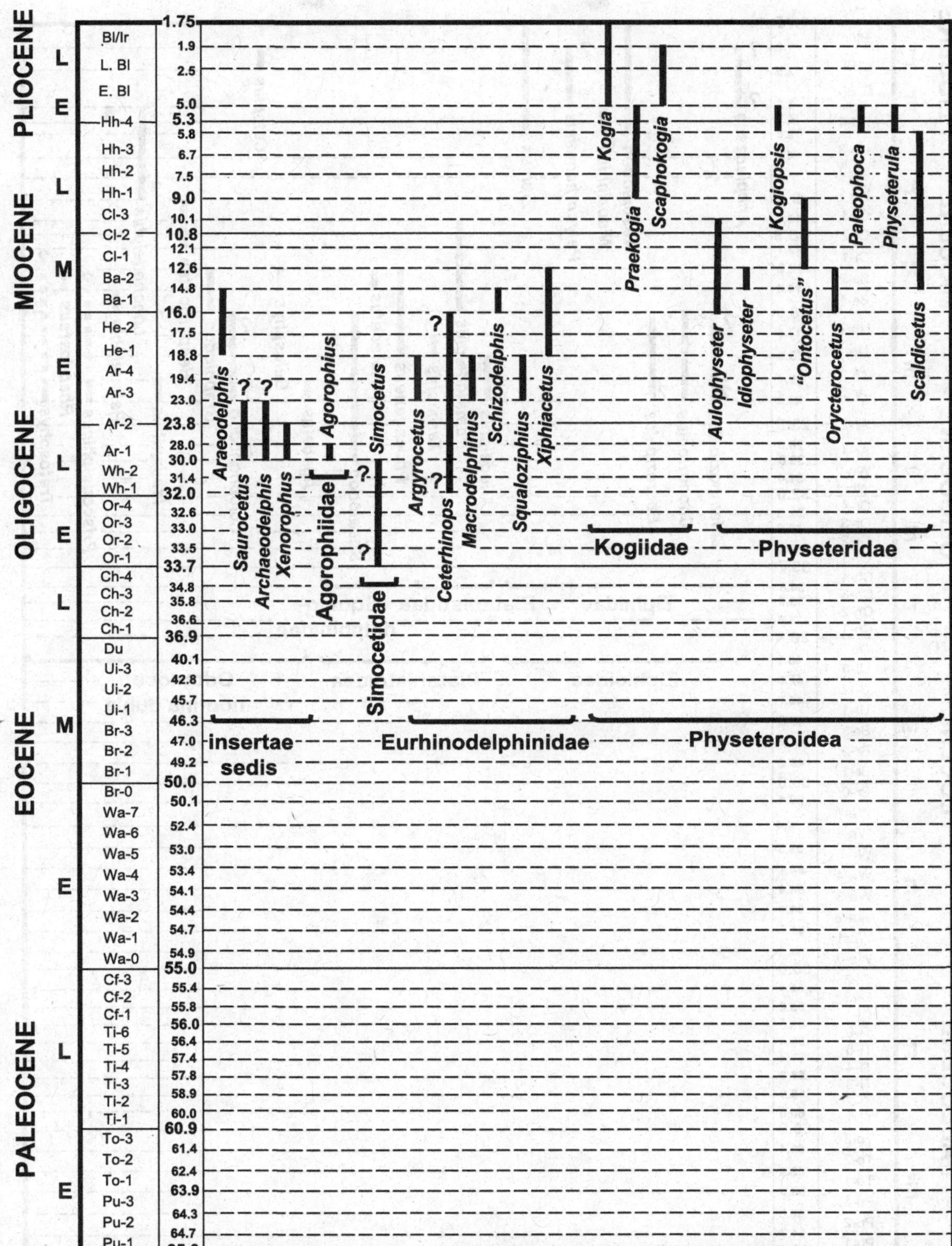

Figure 34.3. Temporal ranges of North American odontocete genera 1: incertae sedis, Eurhinodelphinidae, and Physeteroidea.

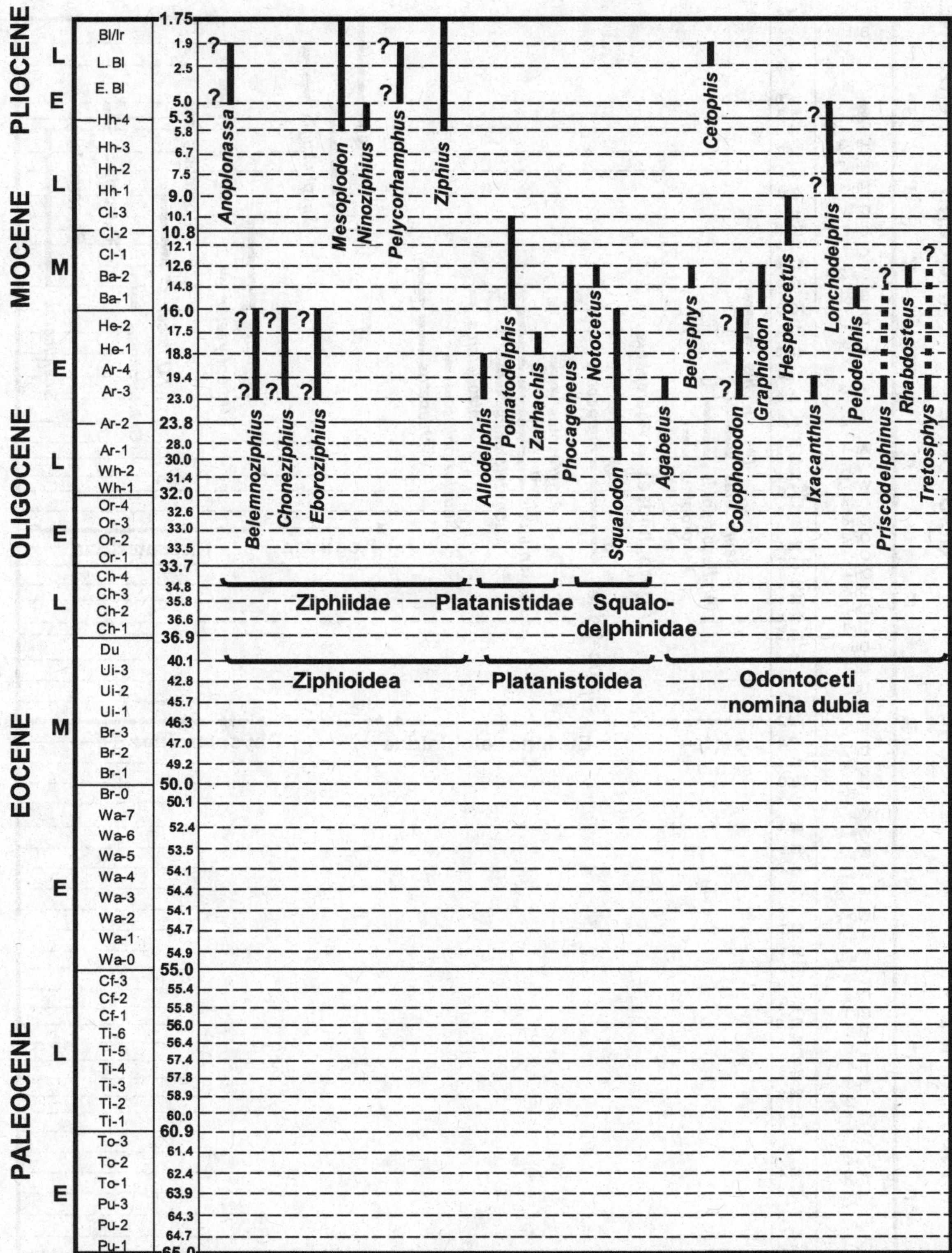

Figure 34.4. Temporal ranges of North American odontocete genera 2: Ziphioidae, Platanistoidae, and Odontoceti nomina dubia.

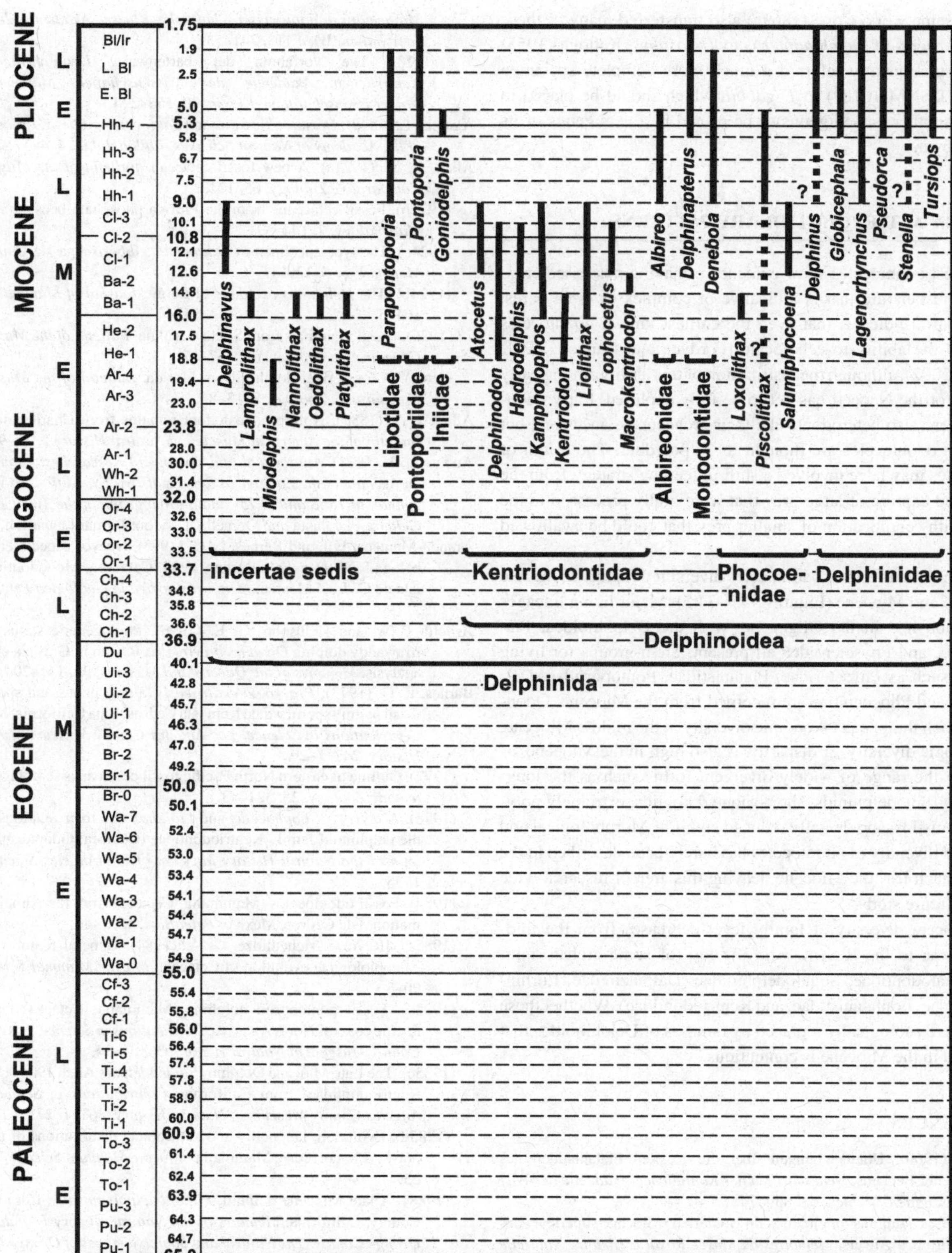

Figure 34.5. Temporal ranges of North American odontocete genera 3: Delphinida.

same work. Cope (1869c) also transferred many of these species of *Delphinapterus* to *Tretosphys*. Kellogg (1955) subsequently referred a much more complete specimen (USNM 10709) to *T. gabbii*, which should be placed in another named genus or be placed in a new genus of its own.

BIOLOGY AND EVOLUTIONARY PATTERNS

The earliest odontocetes show evidence of having possessed craniofacial musculature that is indicative of complex external nares. This, in turn, indicates that even the earliest known odontocetes possessed the ability to echolocate (Fordyce and Muizon, 2001; Fordyce, 2002), although the distribution of this ability among early members of the Neoceti has yet to be fully explored. Early odontocetes were also heterodont, like their archaeocete ancestors, but all known odontocetes are thought to be polydont. The origin of odontocetes may have involved a shift in feeding strategy from the pursuit and capture of large prey that were orally processed to the pursuit with echolocation of smaller prey that could be swallowed whole.

Following their origin, odontocetes diversified rapidly in the late Oligocene and Miocene (Figures 34.3–34.5). Major living lineages were established in the Oligocene, with the Delphinoidea, Platanistoidea, and Physeteroidea all present. Stem-groups for living families, such as Physeteridae, Platanistidae, Pontoporiidae, Delphinidae, and Phocoenidae are recorded from the Miocene. Odontocetes reach their peak taxonomic diversity in the middle Miocene. Morphologic diversity, or disparity, is also high in the Miocene, as shown by the range of widely divergent forms such as the long-snouted eurhinodelphinids, short-snouted kogiids, extremely polydont platanistids, and dentally reduced ziphiids. Morphological and ecological diversity of odontocetes has barely been described in the Recent, much less the Miocene, leaving this area of research wide open for future study.

Odontocete diversity at family level decreases from the middle Miocene to the Recent, with the loss of some archaic families (Squalodontidae, Squalodelphinidae, Dalpiazinidae, Eurhinodelphinidae, Eoplatanistidae, and Kentriodontidae). Whether these groups were ecologically replaced by families of Delphinoidea that originated in the Miocene is contentious.

REFERENCES

Abel, O. (1900). Untersuchungen über die fossilen Platanistiden des Wiener Beckens. *Denkschriften der Koniglichen Akademie der Wissenschaften, Munich*, 68, 839–74.

(1901). Les dauphins longirostres du Boldérien (Miocène supérieur) des environs d'Anvers. *Mémoires du Musée Royal d'Histoire Naturelle de Belgique*, 1, 1–95.

(1905). Les Odontocètes du Boldérien (Miocène supérieur) D'Anvers. *Mémoires du Musée Royal d'Histoire Naturelle de Belgique*, 3, 1–155.

(1909). Cetaceenstudien. I. Mitteilung: Das Skelett von Eurhinodelphis cocheteuxi aus dem Ober-miozän von Antwerpen. *Sitzungsberichte Mathematisch-naturwissenschaftliche Klasse, Akademie der Wissenschaften, Wien* 118, 241–53.

(1913). Die vorfahren der bartenwale. *Denkschriften der Kaiserlichen Akademie der Wissenschaften Mathematisch-Naturwissenschaftliche Klasse*, 90, 155–224.

Agassiz, L. (1848). *Dorudon serratus* and *Saurocetus gibbesii. Proceedings of the Academy of Natural Sciences, Philadelphia*, 4, 4–5.

Allen, G. M. (1921a). A new fossil cetacean. *Bulletin of the Museum of Comparative Zoology*, 65, 1–13.

(1921b). Fossil cetaceans from the Florida phosphate beds. *Journal of Mammalogy*, 2, 144–59.

(1924a). The type-specimen of *Saurocetus gibbesii* Agassiz. *Journal of Mammalogy* 5, 120–1.

(1924b). The *Delphinus occiduus* of Leidy. *Journal of Mammalogy*, 5, 194–5.

(1926). Fossil mammals from South Carolina. *Bulletin of the Museum of Comparative Zoology*, 67, 447–67.

(1941). A fossil river dolphin from Florida. *Bulletin of the Museum of Comparative Zoology*, 89, 3–8.

Allen, J. A. (1887). Note on squalodont remains from Charleston, S. C. *Bulletin of the American Museum of Natural History* 2, 35–9.

Anderson, J. (1878). *Anatomical and Zoological Researches: Comprising an Account of the Zoological Results of Two Expeditions to Western Yunnan in 1868 and 1875; and a Monograph of the Two Cetacean Genera,* Platanista *and* Orcaella. London: Bernard Quaritch.

Aranda-Manteca, F. J. and Barnes, L. G. (1993). Nuevos especimens de la beluga fosil, *Denebola* Barnes, 1984 (Cetacaea, Monodontidae) de Isla de Cedros, Mexico. *Memorias Sociedad Geológica Peninsular*, 2, 19.

Arnold, P. W. and Heinsohn, G. E. (1996). Phylogenetic status of the Irrawaddy dolphin *Orcaella brevirostris* (Owen in Gray): a cladistic analysis. *Memoirs of the Queensland Museum*, 39, 141–204.

Barnes, L. G. (1973). *Praekogia cedrosensis*, a new genus and species of fossil pygmy sperm whale from Isla Cedros, Baja California, Mexico. *Contributions in Science, Los Angeles County Museum of Natural History*, 247, 1–20.

(1977). Outline of eastern North Pacific fossil cetacean assemblages. *Systematic Zoology*, 25, 321–43.

(1978). A review of *Lophocetus* and *Liolithax* and their relationships to the delphinoid family Kentriodontidae (Cetacea: Odontoceti). *Bulletin of the Natural History Museum of Los Angeles Science*, 28, 1–35.

(1984). Fossil odontocetes (Mammalia: Cetacea) from the Almejas Formation, Isla Cedros, Mexico. *PaleoBios*, 42, 1–46.

(1985a). [Review: Mchedlidze, G. A. (1984). General features of the paleobiological evolution of Cetacea.] *Marine Mammal Science*, 1, 90–3.

(1985b). Fossil pontoporiid dolphins (Mammalia: Cetacea) from the Pacific coast of North America *Contributions in Science, Los Angeles County Museum of Natural History*, 363, 1–34.

(1985c). The Late Miocene Dolphin *Pithanodelphis* Abel, 1905 (Cetacea: Kentriodontidae) from California. *Contributions in Science, Los Angeles County Museum of Natural History*, 367, 1–27.

(1985d). Evolution, taxonomy and antitropical distributions of the porpoises (Phocoenidae, Mammalia). *Marine Mammal Science*, 1, 149–65.

(1988). A late Miocene dolphin, *Pithanodelphis nasalis*, from Orange County, California. *Memoirs of the Natural History Foundation of Orange County: The Natural and Social Sciences of Orange County*, 2, 7–21.

(1990). The fossil record and evolutionary relationships of the genus *Tursiops*. In *The Bottlenose Dolphin*, ed. S. Leatherwood and R. R. Reeves, pp. 3–26. London: Academic Press.

(1998). The sequence of fossil marine mammal assemblages in Mexico. *Advances en Investigación Special Publication*, 1, 26–79.

(2002). An Early Miocene long-snouted marine platanistid dolphin (Mammalia, Cetacea, Odontoceti) from the Korneuburg Basin (Austria). *Beiträge Paläontologie*, 27, 407–18.

Barnes, L. G. and Goedert, J. L. (2000). The world's oldest known odontocete (Mammalia, Cetacea). *Journal of Vertebrate Paleontology*, 20(suppl. to no. 3), p. 28A.

Barnes, L. G. and Mitchell, E. D. (1984). *Kentriodon obscurus* (Kellogg, 1931), a fossil dolphin (Mammalia: Kentriodontidae) from the Miocene Sharktooth Hill Bonebed in California. *Contributions in Science, Los Angeles County Museum of Natural History*, 353, 1–23.

Barnes, L. G., Howard, H., Hutchinson, J. H., and Welton, B. J. (1981). The vertebrate fossils of the marine Cenozoic San Mateo Formation at Oceanside, California. In *Geologic Investigations of the Coastal Plain; San Diego County, California*, ed. P. L. Abbott and S. O'Dunn, pp. 53–70. San Diego, CA: San Diego Association of Geologists.

Barnes, L. G., Domning, D. P., and Ray, C. E. (1985). Status of studies on fossil marine mammals. *Marine Mammal Science*, 1, 15–53.

Barnes, L. G., Goedert, J. L., and Furusawa, H. (2001). The earliest known echolocating toothed whales (Mammalia; Odontoceti): preliminary observations of fossils from Washington State. *Mesa Southwest Museum Bulletin*, 8, 91–100.

Best, R. C. and da Silva, V. M. F. 1989. Amazon River dolphin, boto. *Inia geoffrensis* (de Blainville, 1817). In *Handbook of Marine Mammals*, Vol. 4: *River Dolphins and the Larger Toothed Whales*, ed. S. H. Ridgway and R. J. Harrison, pp. 1–23. London: Academic Press.

Bianucci, G. (1996). The Odontoceti (Mammalia, Cetacea) from Italian Pliocene. Systematics and phylogenesis of Delphinidae. *Palaeontographia Italia*, 83, 73–167.

(2001). A new genus of kentriodontid (Cetacea: Odontoceti) from the Miocene of South Italy. *Journal of Vertebrate Paleontology*, 21, 573–7.

Bianucci, G. and Landini, W. (1999). *Kogia pusilla* from the middle Pliocene of Tuscany (Italy) and a phylogenetic analysis of the family Kogiidae (Odontoceti, Cetacea). *Rivista Italiana di Paleontologia e Stratigrafia*, 105, 445–53.

Bianucci, G. and Varola, A. (1994). Kentriodontidae (Odontoceti, Cetacea) from Miocene sediments of the Pietra Leccese (Apulia, Italy). *Atti della Società Toscana di Scienze Naturali Residente in Pisa. Serie A*, 101, 201–12.

Bianucci, G., Landini, W., and Varola, A. (1992). *Messapicetus longirostris*, a new genus and species of Ziphiidae (Cetacea) from the late Miocene of "Pietra leccese" (Apulia, Italy). *Bollettino della Società Paleontologica Italiana* 31, 261–4.

Blainville, H. M. D., de (1838). Sur les cachalots. *Annales Françaises et Etrangeres d'Anatomie et de Physiologie*, 2, 335–7.

Bohaska, D. J. (1998). Fossil marine mammals of the lower Miocene Pollack Farm Site, Delaware. [In *Geology and Paleontology of the Lower Miocene Pollack Farm Fossil Site, Delaware*, ed. R. N. Benson.] *Special Publication of the Delaware Geological Survey*, 21, 178–91.

Brandt, J. F. (1873). Untersuchungen über die fossilen und subfossilen cetaceen Europa's. *Mémoires de L'Académie Impériale des Sciences de St.-Petersbourg, Series 7*, 20, 1–372.

Brownell, R. L. (1989). Franciscana, *Pontoporia blainvillei* (Gervais and d'Orbigny 1844). In *Handbook of Marine Mammals*, Vol. 4: *River Dolphins and the Larger Toothed Whales*, ed. S. H., Ridgway and R. J. Harrison, pp. 45–67. London: Academic Press.

Brownell, R. L. and Herald, E. S. (1972). *Lipotes vexillifer. Mammalian Species*, 10, 1–4.

Case, E. C. (1904). Mammalia. In *Maryland Geological Survey, Miocene*, ed. W. B. Clark, pp. 3–58. Baltimore, MD: Johns Hopkins Press.

(1934). A specimen of a long-nose dolphin from the Bone Valley gravels of Polk County, Florida. *Contributions from the Museum of Paleontology, University of Michigan*, 4, 105–13.

Cassens, I., Vicario, S., Waddell, V. G., *et al.* (2000). Independent adaptation to riverine habitats allowed survival of ancient cetacean lineages. *Proceedings of the National Academy of Sciences, USA*, 97, 11 343–7.

Chen, P. 1989. Baiji. *Lipotes vexillifer* Miller, 1918. In *Handbook of Marine Mammals*, Vol. 4: *River Dolphins and the Larger Toothed Whales*, ed. S. H., Ridgway and R. J. Harrison, pp. 25–43. London: Academic Press.

Clarke, M. R. (2003). Production and control of sound by the small sperm whales, *Kogia breviceps* and *K. sima* and their implications for other Cetacea. *Journal of the Marine Biological Association UK*, 83, 241–63.

Cope, E. D. (1868a). An addition to the vertebrate fauna of the Miocene period, with a synopsis of the extinct Cetacea of the United States. *Proceedings of the Academy of Natural Sciences, Philadelphia*, 19, 138–56.

(1868b). Second contribution to the history of the Vertebrata of the Miocene period of the United States. *Proceedings of the Academy of Natural Sciences, Philadelphia*, 20, 184–94.

(1868c). [Extinct Cetacea from the Miocene bed of Maryland.] *Proceedings of the Academy of Natural Sciences, Philadelphia*, 20, 150–60.

(1869a). Synopsis of the extinct Mammalia of the cave formations in the United States, with observations on some Myriopoda found in and near the same, and on some extinct mammals of the caves of Anguilla, W. I., and of other localities. *Proceedings of the American Philosophical Society*, 11, 171–92.

(1869b). On two new genera of extinct Cetacea. *The American Naturalist*, 3, 444–5.

(1869c). Third contribution to the fauna of the Miocene period of the United States. *Proceedings of the Academy of Natural Sciences, Philadelphia*, 21, 6–12.

(1875). Synopsis of the vertebrata of the Miocene of Cumberland County, New Jersey. *Proceedings of the American Philosophical Society*, 14, 361–4.

(1890). The Cetacea. *The American Naturalist*, 24, 599–616.

(1895). Fourth contribution to the marine fauna of the Miocene period of the United States. *Proceedings of the American Philosophical Society*, 34, 135–55.

Cozzuol, M. A. (1985). The Odontoceti of the "Mesopotamiense" of the Parana River ravines. Systematic review. *Investigations on Cetacea*, 7, 39–53.

(1996). The record of the aquatic mammals in southern South America. *Münchner Geowissenschaftliche Abhandlungen (A)*, 30, 321–42.

Cranford, T. W. (1999). The sperm whale's nose: sexual selection on a grand scale? *Marine Mammal Science*, 15, 1133–57.

Cranford, T. W., Amundin, M., and Norris, K. S. (1996). Functional morphology and homology in the odontocete nasal complex: implications for sound generation. *Journal of Morphology*, 228, 223–85.

Cuvier, G. (1823). Des Ossemens fossiles de Narvals et de Cetacés voisins des Hyperoodons et des Cachalots. *Recherches sur les Ossemens Fossiles*, 5, 349–57.

Dalebout, M. L., Van Helden, A., Van Waerebeek, K., and Baker, C. S. (1998). Molecular genetic identification of Southern Hemisphere beaked whales (Cetacea: Ziphiidae). *Molecular Ecology*, 7, 687–94.

(2002). A new species of beaked whale *Mesoplodon perrini* sp. n. (Cetacea: Ziphiidae) discovered through phylogenetic analyses of mitochondrial DNA sequences. *Marine Mammal Science*, 18, 577–608.

Dal Piaz, G. (1917). Gli Odontoceti del Miocene Bellunese, Parte Terza: *Squalodelphis fabianii. Memoirie Dell'Instituto Geologico Della R. Università Di Padova*, 5, 1–34.

Dawson, S. D. (1996a). A new kentriodontid dolphin (Cetacea; Delphinoidea) from the middle Miocene Choptank Formation, Maryland. *Journal of Vertebrate Paleontology*, 16, 135–40.

(1996b). A description of the skull and postcrania of *Hadrodelphis calvertense* Kellogg 1966 and its position within the Kentriodontidae (Cetacea; Delphinoidea). *Journal of Vertebrate Paleontology*, 16, 125–34.

Dooley, A. C., Jr. (1993). The vertebrate fauna of the Calvert Formation (middle Miocene) at the Caroline Stone Quarry, Caroline County, VA. *Journal of Vertebrate Paleontology*, 13(suppl. to no. 3), p. 33A.

(1996). A newly recognized species of *Squalodon* (Mammalia, Cetacea) from the Miocene of the middle Atlantic Coastal Plain. *Paleontological Society, Special Publication*, 8, 107.

(2003). A review of the eastern North American Squalodontidae (Mammalia: Cetacea). *Jeffersoniana*, 11, 1–26.

(2005). A new species of *Squalodon* (Mammalia, Cetacea) from the Middle Miocene of Virginia. *Virginia Museum of Natural History Memoir*, 8, 1–14.

Dooley, A. C., Jr., Fraser, N. C., and Luo, Z. (2004). The earliest known member of the rorqual-gray whale clade (Mammalia, Cetacea). *Journal of Vertebrate Paleontology*, 24, 453–63.

Dorr, J. A., Jr. and Eschleman, D. F. (1970). *Geology of Michigan*. Ann Arbor, MI: University of Michigan Press.

Du Bus, B. A. L. (1867). Sur quelques Mammifères du crag d'Anvers. *Bulletins de l'Académie Royal des Sciences, des Lettres et des Beaux-Arts de Belgique*, 24, 562–77.

(1872). Mammifères nouveaux du Crag d'Anvers. *Bulletin de l'Académie des Sciences de Belgique*, 34, 491–509.

Duvernoy, G.-L. (1851). Caractères ostéologiques des genres nouveaux ou des espèces nouvelles de cétacés vivants ou fossiles. *Annales des Sciences Naturelles*, 15, 5–71.

Flower, W. H. (1864). Notes on the skeletons of whales in the principal museums of Holland and Belgium, with descriptions of two species apparently new to science. *Proceedings of the Zoological Society of London*, 1864, 384–420.

(1867). Description of the skeleton of *Inia geoffrensis* and the skull of *Pontoporia blainvillii*. *Transactions of the Zoological Society of London*, 6, 87–116.

(1868). On the osteology of the cachalot or sperm-whale (*Physeter macrocephalus*). *Transactions of the Zoological Society of London*, 6, 309–72.

(1872). On the recent ziphioid whales, with a description of the skeleton of *Berardius arnouxi*. *Transactions of the Zoological Society of London*, 8, 203–34.

Fordyce, R. E. (1981). Systematics of the odontocete whale *Agorophius pygmaeus* and the Family Agorophiidae (Mammalia: Cetacea). *Journal of Paleontology*, 55, 1028–45.

(1983). Rhabdosteid dolphins (Mammalia: Cetacea) from the middle Miocene, Lake Frome area, South Australia. *Alcheringa*, 7, 27–40.

(1994). *Waipatia maerewhenua*, new genus and new species (Waipatiidae, new family), an archaic late Oligocene dolphin (Cetacea: Odontoceti: Platanistoidea) from New Zealand. *Proceedings of the San Diego Society of Natural History*, 29, 147–76.

(2002). *Simocetus rayi* (Odontoceti: Simocetidae, New Family): a bizarre new archaic Oligocene dolphin from the eastern North Pacific. *Smithsonian Contributions to Paleobiology*, 93, 185–222.

Fordyce, R. E. and Muizon, C., de (2001). Evolutionary history of cetaceans: a review. In *Secondary Adaptation of Tetrapods to Life in Water*, ed. J.-M. Mazin and V. de Buffrénil, pp. 169–223. Munich: Dr. Friedrich Pfeil.

Fraser, F. C. and Purves, P. E. (1960). Hearing in cetaceans: evolution of the accessory air sacs and the structure of the outer and middle ear in recent cetaceans. *Bulletin of the British Museum (Natural History), Zoology*, 7, 1–140.

Geisler, J. H. and Sanders, A. E. (2003). Morphological evidence for the phylogeny of Cetacea. *Journal of Mammalian Evolution*, 10, 23–129.

Gervais, P. (1850). Sur la famille des cétacés ziphioides, et plus particulierment sur le *Ziphius cavirostris* de la Méditerranée. *Annales des Sciences Naturelles, Partie Zoologique*, 24, 5–17.

(1853). Description de quelques ossements fossiles de phoques et de cétacés. *Mémoires de l'Académie des Sciences et Lettres de Montpellier*, 2, 307–14.

(1855). [*Tursiops*, new name for *Tursio*.] *Histoire Naturelle des Mammifères, avec l'Indication de leurs Moeurs, et de leurs Rapports avec les Arts, le Commerce et l'Agriculture*. Paris: L. Curmer.

(1861). Sur différentes espèces de vertébrés fossils observées pour la plupart dans le midi de la France. *Mémoires de la Section des Sciences Académie des Sciences et Lettres de Montpellier*, 5, 117–32.

Gervais, P. and d'Orbigny, A. (1844). Mammalogie. *Société Philomatique de Paris. Extraits des Procès-verbaux des Séances*, 1844, 38–40.

Gill, T. (1871). The sperm whales, giant and pygmy. *The American Naturalist*, 4, 725–43.

(1872). Arrangement of the families of mammals and synoptical tables of characters of the subdivisions of mammals. *Smithsonian Miscellaneous Collections*, 11, 1–98.

Gillette, D. D. (1975). Catalogue of the type specimens of fossil vertebrates, Academy of Natural Sciences, Philadelphia. Introduction and Part I: marine mammals. *Proceedings of the Academy of Natural Sciences, Philadelphia*, 127, 63–6.

Ginsburg, L. (1990). Les quatre faunes de mammifères Miocènes des faluns du synclinal d'Esvres (Val-de-Loire, France). *Comptes Rendus de L'Académie des Sciences, Série II*, 310, 89–93.

Glaessner, M. F. (1947). A fossil beaked whale from Lakes Entrance, Victoria. *Proceedings of the Royal Society of Victoria*, 58, 25–35.

Gondar, D. (1975). La presencia de cetaceos (Physeteridae) en el Terciario superior (Rionegrense) de la Prov. de Rio Negro. *Actas del Congreso Argentino de Paleontologia y Bioestratigrafia*, 2, 349–56.

Gottfried, M. D., Bohaska, D. J., and Whitmore, F. C., Jr. (1994). Miocene cetaceans of the Chesapeake group. *Proceedings of the San Diego Society of Natural History*, 29, 229–38.

Grateloup, J. P. S. (1840). Description d'un fragment de mâchoire fossile, d'un genre nouveau de reptile (saurien), de taille gigantesque, voisin de *l'Iguanodon*, trouvé dans le Grès Marin, à Léognan, près Bordeaux. *Actes de l'Académie National des Sciences, Belles-lettres et Artes de Bordeaux*, 2, 201–10.

Gray, J. E. (1821). On the natural arrangement of vertebrose animals. *London Medical Repository*, 15, 296–310.

(1825). Outline of an attempt at the disposition of the Mammalia into tribes and families with a list of genera apparently appertaining to each tribe. *Philosophical Annals*, 26, 337–44.

(1846). On the cetaceous animals. In *The Zoology of the Voyage of HMS* Erebus *and* Terror, *under the Command of Capt. Sir J. C. Ross, RN, FRS, during the years 1839 to 1843*, ed. J. Richardson and J. E. Gray, pp. 13–53. London: E. W. Janson.

(1863). On the arrangement of the cetaceans. *Proceedings of the Zoological Society of London*, 1863, 197–202.

(1865). Notice of a new genus of delphinoid whales from the Cape of Good Hope, and of other cetaceans from the same seas. *Proceedings of the Zoological Society, London*, 1865, 522–9.

(1866). Notes on the skulls of dolphins, or bottlenose whales, in the British Museum. *Proceedings of the Zoological Society of London*. 1866, 211–16.

(1868). *Synopsis of the Species of Whales and Dolphins in the Collection of the British Museum*. London: Bernard Quaritch.

(1870). Notes on the arrangement of the genera of delphinoid whales. *Proceedings of the Zoological Society of London*, 1870, 772–3.

Gregory, W. K. and Kellogg, R. (1927). A fossil porpoise from California. *American Museum Novitates*, 269, 1–7.

Harington, C. R. (1988). The Late Quaternary Development of the Champlain Sea Basin. In *Geological Association of Canada Special Paper*, No. 35: *Marine Mammals of the Champlain Sea, and the Problem of*

Whales in Michigan, ed. N. R. Gadd, pp. 225–40. Ottawa: Geological Association of Canada.

Harlan, R. (1842). Description of a new extinct species of dolphin from Maryland. *Bulletin and Proceedings of the National Institution for the Promotion of Science, Washington*, 2, 95–196.

Hay, O. P. (1902). *Bibliography and Catalogue of the Fossil Vertebrata of North America*. Washington, DC: Government Printing Office.

Hershkovitz, P. (1966). Catalog of living whales. *Bulletin of the United States National Museum*, 246, 1–259.

Heyning, J. E. (1989a). Comparative facial anatomy of beaked whales (Ziphiidae) and a systematic revision among the families of extant Odontoceti. *Contributions in Science, Los Angeles County Museum of Natural History*, 405, 1–64.

(1989b). Cuvier's beaked whale *Ziphius cavirostris* G. Cuvier, 1823. In *Handbook of Marine Mammals*, Vol. 4: *River Dolphins and the Larger Toothed Whales*, ed. S. H. Ridgway and R. J. Harrison, pp. 289–308. London: Academic Press.

(1997). Sperm whale phylogeny revisited: analysis of the morphological evidence. *Marine Mammal Science*, 13, 596–613.

Hirota, K. and Barnes, L. G. (1995). A new species of middle Miocene sperm whale of the genus *Scaldicetus* (Cetacea; Physeteridae) from Shiga-mura, Japan. *The Island Arc*, 3, 453–72.

Holman, J. A. (1995). *Ancient Life of the Great Lakes Basin*. Ann Arbor, MI: University of Michigan Press.

Hulbert, R. C., Jr. and Whitmore, R. C., Jr. (2006). Late Miocene mammals from the Mauvilla Local Fauna, Alabama. *Bulletin of the Florida Museum of Natural History*, 46, 1–28.

Huxley, T. H. (1864). On the cetacean fossils termed "*Ziphius*" by Cuvier, with a notice of a new species (*Belemnoziphius compressus*) from the Red Crag. *Quarterly Journal of the Geological Society*, 20, 388–96.

Ichishima, H. (1995). A new fossil kentriodontid dolphin (Cetacea: Kentriodontidae): from the Middle Miocene Takinoue Formation, Hokkaido, Japan. *The Island Arc*, 3, 473–85.

Ichishima, H. and Kimura, M. (2000). A new fossil porpoise (Cetacea; Delphinoidea; Phocoenidae) from the early Pliocene Horokaoshirarika Formation, Hokkaido, Japan. *Journal of Vertebrate Paleontology*, 20, 561–76.

Ichishima, H., Barnes, L. G., Fordyce, R. E., Kimura, M., and Bohaska, D. J. (1995). A review of kentriodontine dolphins (Cetacea; Delphinoidea; Kentriodontidae): systematics and biogeography. *The Island Arc*, 3, 486–92.

Kasuya, T. (1973). Systematic consideration of recent toothed whales based on the morphology of tympano-periotic bone. *Scientific Reports of the Whales Research Institute*, 25, 1–103.

Kazár, E. (2002). Revised phylogeny of the Physeteridae (Mammalia: Cetacea) in the light of *Placoziphius* Van Beneden, 1869 and *Aulophyseter* Kellogg, 1927. *Bulletin de l'Institut Royal des Sciences Naturelles de Belgique*, 72, 150–70.

Kellogg, R. (1923a). Description of an apparently new toothed cetacean from South Carolina. *Smithsonian Miscellaneous Collections*, 76, 1–7.

(1923b). Description of two squalodonts recently discovered in the Calvert Cliffs, Maryland, and notes on the shark-toothed cetaceans. *Proceedings of the United States National Museum*, 62, 1–69.

(1924). A fossil porpoise from the Calvert Formation of Maryland. *Proceedings of the United States National Museum*, 63, 1–39.

(1925a). On the occurrence of remains of fossil porpoises of the genus *Eurhinodelphis* in North America. *Proceedings of the United States National Museum*, 66, 1–40.

(1925b). Additions to the Tertiary history of the pelagic mammals on the Pacific coast of North America. I. Two fossil physeteroid whales from California. *Carnegie Institution of Washington Publication*, 348, 1–34.

(1925c). A fossil physteroid cetacean form Santa Barbara County California. *Proceedings of the United States National Museum*, 66, 1–8.

(1926). Supplementary observations on the skull of the fossil porpoise *Zarhachis flagellator* Cope. *Proceedings of the United States National Museum*, 67, 1–18.

(1927a). Study of the skull of a fossil sperm-whale from the Temblor Miocene of Southern California. *Carnegie Institution of Washington Publication*, 346, 1–24.

(1927b). *Kentriodon pernix*, a Miocene porpoise from Maryland. *Proceedings of the United States National Museum*, 69, 1–55.

(1928). The history of whales: their adaptation to life in the water. *Quarterly Review of Biology*, 3, 29–76.

(1929). A new fossil toothed whale from Florida. *American Museum Novitates*, 389, 1–9.

(1931). Pelagic mammals from the Temblor Formation of the Kern River region, California. *Proceedings of the California Academy of Sciences*, 19, 217–397.

(1932). A Miocene long-beaked porpoise from California. *Smithsonian Miscellaneous Collections*, 87, 1–11.

(1944). Fossil cetaceans from the Florida Tertiary. *Bulletin of the Museum of Comparative Zoology*, 94, 433–71.

(1955). Three Miocene porpoises from the Calvert Cliffs, Maryland. *Proceedings of the United States National Museum*, 105, 101–54.

(1957). Two additional Miocene porpoises from the Calvert Cliffs Maryland. *Proceedings of the United States National Museum*, 107, 279–337.

(1959). Description of the skull of *Pomatodelphis inaequalis* Allen. *Bulletin of the Museum of Comparative Zoology*, 121, 1–26.

(1965). Fossil marine mammals from the Miocene Calvert Formation of Maryland and Virginia. Part 2. The Miocene Calvert sperm whale *Orycterocetus*. *Bulletin of the Smithsonian Institution*, 247, 47–63.

(1966). Fossil marine mammals from the Miocene Calvert formation of Maryland and Virginia. Part 4. A new odontocete from the Calvert Miocene of Maryland. *Bulletin of the United States National Museum*, 247, 90–101.

Kernan, J. D. (1919). The skull of *Ziphius cavirostris*. *Bulletin of the American Museum of Natural History*, 28, 349–93.

Lacépede, B. G. E. (1804). *Histoire Naturelle des Cetaces*. Paris: Chez Plassan.

Lambert, O. (2004). Systematic revision of the Miocene long-snouted dolphin *Eurhinodelphis longirostris* Du Bus, 1872 (Cetacea, Odontoceti, Eurhinodelphinidae). *Bulletin de l'Institute Royal des Sciences Naturelles de Belgique, Sciences de la Terre*, 74, 147–74.

(2005a). Phylogenetic affinities of the long-snouted dolphin *Eurhinodelphis* (Cetacea, Odontoceti) from the Miocene of Antwerp, Belgium. *Palaeontology*, 48, 653–79.

(2005b). Review of the Miocene long-snouted dolphin *Priscodelphinus cristatus* DuBus, 1872 (Cetacea, Odontoceti) and phylogeny among eurhinodelphinids. *Bulletin de l'Institut Royal des Sciences Naturelles de Belgiques, Sciences de la Terre*, 75, 211–35.

(2005c). Systematics and phylogeny of the fossil beaked whale *Z. phirostrum* du Bus, 1868 and *Choneziphius* Duvernay 1851 (Mammalia, Cetacea, Odontoceti), from the Neogene of Antwerp (North Belgium). *Geodiversitas*, 27, 443–93.

LeDuc, R. G., Perrin, W. F., and Dizon, A. E. (1999). Phylogenetic relationships among the delphinid cetaceans based on full cytochrome *b* sequences. *Marine Mammal Science*, 15, 619–48.

Leidy, J. (1851). [Descriptions of a number of fossil reptiles and mammals.] *Proceedings of the Academy of Natural Sciences, Philadelphia*, 5, 325–82.

(1853). [Observations on extinct Cetacea.] *Proceedings of the Academy of Natural Sciences, Philadelphia*, 6, 377–8.

(1856). Notices of remains of extinct vertebrated animals of New Jersey, collected by Prof. Cook of the State Geological Survey under the direction of Dr. W. Kitchell. *Proceedings of the Academy of Natural Sciences, Philadelphia*, 8, 220–21.

(1859). [Remarks on *Dromatherium sylvestre* and *Ontocetus emmonsi*.] *Proceedings of the Academy of Natural Sciences, Philadelphia*, 11, 162.

(1869). The extinct mammalian fauna of Dakota and Nebraska, including an account of some allied forms from other localities, together with a synopsis of the mammalian remains of North America. *Journal of the Academy of Natural Sciences, Philadelphia*, 2, 1–472.

(1870). [Description of *Graphiodon vinearis* and remarks on *Crocodilus elliotti*.] *Proceedings of the Academy of Natural Sciences, Philadelphia*, 1870, 122.

(1876a). Remarks on fossils from the Ashley Phosphate Beds. *Proceedings of the Academy of Natural Sciences, Philadelphia*, 1876, 80–1.

(1876b). Remarks on fossils of the Ashley Phosphate Beds. *Proceedings of the Academy of Natural Sciences, Philadelphia*, 1876, 114–15.

(1876c). Remarks on vertebrate fossils from the phosphate beds of South Carolina. *Proceedings of the Academy of Natural Sciences, Philadelphia*, 1876, 86–7.

(1877). Description of vertebrate remains, chiefly from the phosphate beds of South Carolina. *Journal of the Academy of Natural Sciences, Philadelphia*, 8, 209–61.

Lesson, R.-P. (1828). In *Complément des oeuvres de Buffon ou Histoire Naturelle des animaux rares découverts par les naturalists et les voyageurs depuis la mort de Buffon*, Vol. 1: *Cétacés*. Paris.

Linnaeus, C. (1758). *Systema Naturae per Regna Tria Naturae, Secundum Classes, Ordines, Genera, Species, cum Characteribus and Differentiis*, 10th edn, *Reformata*, Vol. I. Holmiae: Laurentii Salvii.

Lull, R. S. (1914). Fossil dolphin from California. *American Journal of Science*, 37, 209–20.

Luo, Z. and Gingerich, P. D. (1999). Terrestrial Mesonychia to aquatic Cetacea: transformation of the basicranium and evolution of hearing in whales. *University of Michigan Papers on Paleontology*, 31, 1–98.

Luo, Z. and Marsh, K. (1996). Petrosal (periotic) and inner ear of a Pliocene kogiine whale (Kogiinae, Odontoceti): implications on relationships and hearing evolution of toothed whales. *Journal of Vertebrate Paleontology*, 16, 328–48.

Lydekker, R. (1894). Cetacean skulls from Patagonia. *Annales del Museo de la Plata, Paleontología Argentina*, II, 1–13.

Marino, L., McShea, D. W., and Uhen, M. D. (2004). The origin(s) and evolution of large brains in toothed whales. *Anatomical Record*, 281A, 1247–55.

Mchedlidze, G. A. (1970). *Nekotorye Obschie Cherty Istorii Kitoobraznykh*, Vol. 1. Tiblishi, Georgia: Metsniyereba.

McKenna, M. C. and Bell, S. K. (1997). *Classification of Mammals Above the Species Level*. New York: Columbia University Press.

Mead, J. G. (1975a). Anatomy of the external nasal passages and facial complex in the Delphinidae (Mammalia: Cetacea). *Smithsonian Contributions to Zoology*, 207, 1–72.

(1975b). A fossil beaked whale (Cetacea: Ziphiidae) from the Miocene of Kenya. *Journal of Paleontology*, 49, 743–51.

(1989). Beaked whales of the genus *Mesoplodon*. In *Handbook of Marine Mammals*, Vol. 4: *River Dolphins and the Larger Toothed Whales*, ed. S. H, Ridgway and R. J. Harrison, pp. 349–430. London: Academic Press.

Messenger, S. L. (1994). Phylogenetic relationships of platanistoid river dolphins (Odontoceti, Cetacea): Assessing the significance of fossil taxa. *Proceedings of the San Diego Society of Natural History*, 29, 125–33.

Messenger, S. L. and McGuire, J. A. (1998). Morphology, molecules and the phylogenetics of cetaceans. *Systematic Biology*, 47, 90–124.

Miller, G. S., Jr. (1918). A new river-dolphin from China. *Smithsonian Miscellaneous Collections*, 68, 1–12.

(1923). The telescoping of the cetacean skull. *Smithsonian Miscellaneous Collections*, 76, 1–71.

Montagu, G. (1821). Description of a species of *Delphinus*, which appears to be new. *Memoirs of the Wernerian Natural History Society*, 3, 75–82.

Moore, J. C. (1968). Relationships among the living genera of beaked whales with classifications, diagnoses and keys. *Fieldiana: Zoology*, 53, 209–398.

Moreno, F. P. (1892). Lijeros apuntes sobre dos géneros de cetáceos fosiles de la República Argentina. *Revista Museo de La Plata*, 3, 381–400.

Morgan, G. S. (1994). Miocene and Pliocene marine mammal faunas from the Bone Valley Formation of central Florida. *Proceedings of the San Diego Society of Natural History*, 29, 239–68.

Muizon, C., de (1983a). A new Ziphiidae (Cetacea) from the Lower Pliocene of Peru. *Comptes Rendus Hebdomadaires des Séances de l'Académie des Sciences*, 297, 85–8.

(1983b). A new Phocoenidae (Cetacea) from the lower Pliocene of Peru. *Comptes Rendus Hebdomadaires des Séances de l'Académie des Sciences*, 296, 1203–6.

(1984). Les vertébrés fossiles de la Formation Pisco (Pérou), Part 2: les Odontocetes (Cetacea, Mammalia) du Pliocene inferieur de Sus-Sacaco. *Institut Francais d'Études Andines*, 50, 9–175.

(1987). The affinities of *Notocetus vanbenedeni*, an early Miocene platanistoid (Cetacea, Mammalia) from Patagonia, southern Argentina. *American Museum Novitiates*, 2904, 1–27.

(1988a). Les relations phylogénétiques des Delphinida (Cetacea, Mammalia). *Annales de Paléontologie*, 74, 159–227.

(1988b). Les vertébrés fossils de la Formation Pisco (Pérou): les odontocétes (Cetacea, Mammalia) du Miocene. *Éditions Recherche sur les Civilisations, Mémoire*, 78, 1–244.

(1988c). Le polyphylétisme des Acrodelphidae, Odontocètes longirostres du Miocène européen. *Bulletin du Muséum Nationale d'Histoire Naturelle, Paris, Série 4*, 10C, 31–88.

(1991). A new Ziphiidae (Cetacea) from the Early Miocene of Washington State (USA) and phylogenetic analysis of the major groups of odontocetes. *Bulletin du Muséum Nationale d'Histoire Naturelle, Paris*, 12, 279–326.

(1993). Walrus-like feeding adaptation in a new cetacean from the Pliocene of Peru. *Nature*, 365, 745–8.

(1994). Are the squalodonts related to the platanistoids? *Proceedings of the San Diego Society of Natural History*, 29, 135–46.

Muizon, C., de and DeVries, T. J. (1985). Geology and paleontology of the Cenozoic marine deposits in the Sacaco area (Peru). *Geologische Rundschau*, 74, 547–63.

Muizon, C., de, Domning, D. P., and Parish, M. (1999). Dimorphic tusks and adaptive strategies in a new species of walrus-like dolphin (Odobenocetopsidae) from the Pliocene of Peru. *Comptes Rendus Adadémie des Sciences*, 329, 449–55.

Müller, J. (1849). *Über die fossilen Reste der Zeuglodonten von Nordamerica*. Berlin: von G. Reimer.

Myrick, A. C., Jr. (1979). Variation, taphonomy, and adaptation of the Rhabdosteidae (= Eurhinodelphidae) (Odontoceti, Mammalia) from the Calvert Formation of Maryland and Virginia. Ph.D. Thesis, University of California, Los Angeles.

Nikaido, M., Matsuno, F., Hamilton, H., *et al.* (2001). Retroposon analysis of major cetacean lineages: the monophyly of toothed whales and the paraphyly of river dolphins. *Proceedings of the National Academy of Science, USA*, 98, 7384–9.

Nowak, R. M. (2003). *Walker's Marine Mammals of the World*. Baltimore, MD: Johns Hopkins University Press.

Nowak, R. M. and Paradiso, J. L. (1983). *Walker's Mammals of the World*, 4th edn. Baltimore, MD: Johns Hopkins University Press.

Okazaki, Y. (1992). An occurrence of a fossil sperm whale from the Miocene Mizunami Group, central Japan. *Mizunami-shi Kaseki Hakubutsukan kenkyu hokoku*, 19, 295–9.

Owen, R. (1846). *A History of British Fossil Mammals and Birds*. London: Van Voorst.

(1889). Monograph on the British Fossil Cetacea from the Red Crag. No. 1, containing genus *Ziphius*. *Palaeontographical Society*, 1–40.

Pallas, P. S. (1776). *Reise durch verschiedene Provinzen de Russischen Reichs*. [*Voyage dans differentes provinces de l'empire de Russie*.] St. Petersburg: St. Petersburg Academy of Sciences.

Pedroni, P. M. (1845). Note sur une espèce du genre Dauphin. *Comptes Rendus Hebdomadaires des Séances de l'Académie des Sciences*, 21, 1181.

Perrin, W. F. (1998). *Stenella longirostris*. *Mammalian Species*, 599, 1–7.

Perrin, W. F. and Brownell, R. L., Jr. (2001). Update of the list of recognized species of cetaceans. *Journal of Cetacean Research and Management*, 3, 364–6.

Perrin, W. F., Würsig, B., and Thewissen, J. G. M. (eds.) (2002). *Encyclopedia of Marine Mammals*. London: Academic Press.

Prothero, D. R., Bitboul, C. Z., Moore, G. W., and Niem, A. R. (2002). Magnetic stratigraphy and tectonic rotation of the Oligocene Alsea, Yaquina, and Nye Formations, Lincoln County, Oregon. In *Magnetic Stratigraphy of the Pacific Coast Cenozoic*, Book 9, ed. D. R. Prothero, pp. 184–94, Tulsa, OK: Pacific Section of the Society for Sedimentary Geology.

Raven, H. C. and Gregory, W. K. (1933). The spermaceti organ and nasal passages of the sperm whale (*Physeter catodon*) and other odontocetes. *American Museum Novitates*, 677, 1–18.

Ravn, J. P. J. (1926). On a cetacean, *Squalodon* (*Microzeuglodon*?) *wingei* nov. sp., from the Oligocene of Jutland. *Meddelelser fra Dansk Geologisk Forening*, 7, 45–54.

Ray, C. E. (1975). The relationships of *Hemicaulodon effodiens* Cope 1869 (Mammalia: Odobenidae). *Proceedings of the Biological Society of Washington*, 88, 281–304.

(1977). Geography of phocid evolution. *Systematic Zoology*, 25, 391–406.

Reeves, R. R. and Brownell, R. L. (1989). *Susu, Platanista gangetica* (Roxburgh, 1801) and *Platanista minor* Owen, 1853. In *Handbook of Marine Mammals*, Vol. 4: *River Dolphins and the Larger Toothed Whales*, ed. S. H. Ridgway and R. J. Harrison, pp. 69–99. London: Academic Press.

Reeves, R. R. and Tracey, S. (1980). *Monodon monoceros*. *Mammalian Species*, 127, 1–7.

Reinhardt, J. (1862). [on the genus *Pseudorca*.] *Oversigt over det Kgl. Danske videnskabernes selskabs forhandlinger og dets medlemmers arbeider i aaret*, 151, 103–52.

Rensberger, J. M. (1969). A new iniid cetacean from the Miocene of California. *University of California, Publications in Geological Sciences*, 82, 1–36.

Rice, D. W. (1998). Marine mammals of the world. *Society for Marine Mammalogy Special Publication*, 4, 1–231.

Ridgway, S. H. and Harrison, R. J. (1989). *Handbook of Marine Mammals*, Vol. 4: *River Dolphins and the Larger Toothed Whales*. London: Academic Press.

(1994). *Handbook of Marine Mammals*, Vol. 5: *The First Book of Dolphins*. London: Academic Press.

(1999). *Handbook of Marine Mammals*, Vol. 6: *The Second Book of Dolphins and the Porpoises*. London: Academic Press.

Robineau D. (1989). Les types de cetaces actuels du Muséum national d'Histoire naturelle. 1. Balaenidae, Balaenopteridae, Kogiidae, Ziphiidae, Iniidae, Pontoporiidae. *Bulletin du Muséum National d'Histoire Naturelle, Paris*, 11, 271–89.

Rosel, P. E., Haygood, M. G., and Perrin, W. F. (1995). Phylogenetic relationships among the true porpoises (Cetacea: Phocoenidae). *Molecular Phylogenetics and Evolution*, 4, 463–74.

Rothausen, K. (1968). Die systematisch Stellung der europäischen Squalodontidae (Odontoceti, Mamm.). *Paläontologische Zeitschrift*, 42, 83–104.

Sanders, A. E. (1996). The systematic position of the primitive odontocete *Xenorophus sloanii* (Mammalia: Cetacea) and two new taxa from the Late Oligocene of South Carolina, USA. [*Proceedings of the 6th North American Paleontological Congress*.] *Paleontological Society Special Publication*, 8, p. 338.

Simpson, G. G. (1945). The principles of classification and a classification of mammals. *Bulletin of the American Museum of Natural History*, 85, 1–350.

Slijper, E. J. (1936). Die Cetaceen. *Capita Zoologica*, 7, 1–590.

Sowerby, J. (1804). Extracts from the Minute-Book of the Linnean Society [note on *Physeter bidens*]. *Transactions of the Linnaean Society London*, 7, 310.

Stacey, P. J., Leatherwood, S., and Baird, R. W. (1994). *Pseudorca crassidens*. *Mammalian Species*, 456, 1–6.

Stewart, B. E. and Stewart, R. E. A. (1989). *Delphinapterus leucas*. *Mammalian Species*, 336, 1–8.

True, F. W. (1889). Contributions to the natural history of the cetaceans, a review of the family Delphinidae. *Bulletin of the United States National Museum*, 36, 1–191.

(1907a). Remarks on the type of the fossil cetacean *Agorophius pygmaeus* (Müller). *Smithsonian Miscellaneous Collections*, 1694, 1–8.

(1907b). Observations on the type specimen of the fossil cetacean *Anoplonassa forcipata* Cope. *Bulletin of the Museum of Comparative Zoology*, 51, 97–106.

(1908a). On the occurrence of remains of fossil cetaceans of the genus *Schizodelphis* in the United States, and on *Priscodelphinus* (?) *crassangulum* Case. *Smithsonian Miscellaneous Collections*, 50, 449–60.

(1908b). On the classification of the Cetacea. *Proceedings of the American Philosophical Society*, 47, 385–91.

(1910a). An account of the beaked whales of the Family Ziphiidae in the collection of the United States National Museum, with remarks on some specimens in other American museums. *Bulletin of the United States National Museum*, 73, 1–89.

(1910b). Description of a skull and some vertebrae of the fossil cetacean *Diochoticus vanbenedeni* from Santa Cruz, Patagonia. *Bulletin of American Museum of Natural History*, 28, 19–32.

(1912a). Description of a new fossil porpoise of the genus *Delphinodon* from the Miocene Formation of Maryland. *Journal of the Academy of Natural Sciences, Philadelphia*, 15, 165–93.

(1912b). A fossil toothed cetacean from California, representing a new genus and species. *Smithsonian Miscellaneous Collections*, 60, 1–7.

(1913). Diagnosis of a new beaked whale of the genus *Mesoplodon* from the coast of North Carolina. *Smithsonian Miscellaneous Collections*, 60, 1–2.

Tuomey, M. (1848). Notice of the discovery of a cranium of the *Zeuglodon* (*Basilosaurus*). *Proceedings of the Academy of Natural Sciences, Philadelphia*, 1 2 *Series*, 16–17.

Turner, J. P. and Worthy, G. A. J. (2003). Skull morphometry of bottlenose dolphins (*Tursiops truncatus*) from the Gulf of Mexico. *Journal of Mammalogy*, 84, 665–72.

Uhen, M. D. (2002a). Dental morphology (cetacean), Evolution of. In *Encyclopedia of Marine Mammals*, ed. W. F. Perrin, B. Würsig, and J. G. M. Thewissen, pp. 316–19. London: Academic Press.

(2002b). Basilosaurids. In *Encyclopedia of Marine Mammals*, ed. W. F. Perrin, B. Würsig, and J. G. M. Thewissen, pp. 78–81. London: Academic Press.

Uhen, M. D. and Gingerich, P. D. (2001). New genus of dorudontine archaeocete (Cetacea) from the middle-to-late Eocene of South Carolina. *Marine Mammal Science*, 17, 1–34.

Van Beneden, P. J. (1859). Rapport de M. Van Beneden. *Bulletins de l'Académie Royal des Sciences, des Lettres et des Beaux-Arts de Belgique*, 8, 123–46.

(1876). Les Phoques fossiles du bassin d'Anvers. *Journal de Zoologie*, 5, 188–205.

(1877). Note sur un Cachalot main du crag d'Anvers, *Physeterula dubussi*. *Bulletins de l'Académie Royal des Sciences, des Lettres et des Beaux-Arts*, 64, 851–6.

Wagler, J. (1830). *Natürliches System der Amphibien, mit verangehender Classification der Säugthiere un Vögel*. Munich: J. G. Cotta'schen Buchhandlung.

Whitmore, F. C., Jr. (1994). Neogene climate change and the emergence of the modern whale fauna of the North Atlantic Ocean. *Proceedings of the San Diego Society of Natural History*, 29, 223–7.

Whitmore, F. C., Jr. and Sanders, A. E. (1977). Review of the Oligocene Cetacea. *Systematic Zoology*, 25, 304–20.

Whitmore, F. C., Jr., Morejohn, V., and Mullins, H. T. (1986). Fossil beaked whales: *Mesoplodon longirstrostris* dredged from the ocean bottom. *National Geographic Research*, 2, 47–56.

Wilson, L. E. (1935). Miocene marine mammals from the Bakersfield region, California. *Peabody Museum of Natural History Bulletin*, 4, 1–143.

(1973). A delphinid (Mammalia, Cetacea) from the Miocene of Palos Verdes Hills, California. *University of California, Publications in Geological Sciences*, 103, 1–33.

Winge, H. (1921). A review of the interrelationships of the Cetacea. *Smithsonian Miscellaneous Collections*, 72, 1–97.

Yang, G., Zhou, K., Ren, W., *et al.* (2002). Molecular systematics of river dolphins inferred from complete mitochondrial cytochrome *b* gene sequences. *Marine Mammal Science*, 18, 20–9.

Zhou, K., Qian, W., and Li, Y. (1978). Recent advances in the study of the baiji, *Lipotes vexillifer*. *Journal of the Nanjing Normal College*, 1, 8–13.

(1979). The osteology and the systematic position of the baiji, *Lipotes vexillifer*. *Acta Zoologica Sinica*, 25, 58–74.

Zhou, K., Zhou, M., and Zhao, Z. (1984). First discovery of a Tertiary platanistoid fossil from Asia. *Scientific Reports of the Whales Research Institute, Tokyo*, 35, 173–81.

Zioupos, P., Currey, J. D., Casinos, A., and Buffrénil, V., de (1997). Mechanical properties of the rostrum of the whale *Mesoplodon densirostris*, a remarkably dense bony tissue. *Journal of Zoology*, 241, 725–37.

35 Mysticeti

MARK D. UHEN,[1] R. EWAN FORDYCE,[2] and LAWRENCE G. BARNES[3]

[1]*Smithsonian Institution, Washington, DC, USA*

[2]*University of Otago, Dunedin, New Zealand*

[3]*Natural History Museum of Los Angeles County, Los Angeles, CA, USA*

INTRODUCTION

The Mysticeti are the baleen whales and their archaic tooth-bearing relatives. Another vernacular name for the group is mysticetes. Mysticeti are small to very large whales, and their geochronological range is from late Eocene to the Recent, while their geographical range is throughout all of the world's oceans. Mysticetes are notably less diverse and disparate than odontocetes. Currently recognized living species were listed by Rice (1998) and Perrin and Brownell (2001).

Mysticetes are whales that bulk feed and are not known to echolocate. They have a loose mandibular symphysis and a type of jaw articulation that, in some groups, allows the mouth to open widely for bulk feeding; therefore, their feeding adaptations separate them from both the Archaeoceti and the Odontoceti. Mysticeti are also distinguished from the Archaeoceti by the phenomenon that is termed telescoping (see Miller, 1923), a process whereby, unlike the situation in Archaeoceti, the cranial bones of the rostrum extend posteriorly over the facial region, and the bones of the occiput thrust anteriorly toward the cranial vertex. The Mysticeti are further distinguished from the Odontoceti by their distinct type of telescoping, in which the rostral bones dorsally move posteriorly only as far as the interorbital region, and ventrally move posteriorly into the orbit; in all but the most archaic groups, the apex of the occipital shield thrusts anteriorly at least as far as the mid part of the temporal fossae.

The Mysticeti are a monophyletic group that originated from within the paraphyletic Archaeoceti. The oldest named mysticete, *Llanocetus denticrenatus*, is late Eocene in age, and the group has been represented throughout most of its history by up to five or six contemporaneous family-level groups. The baleen-bearing Mysticeti are bulk filter-feeders and have a smooth ventral surface of the maxilla (the palate), which has grooves that hold the blood vessels that nourish the baleen plates.

Tooth-bearing mysticetes have heterodont dentitions, in which incisors, canines, premolars, and molars can be recognized. Such heterodont dentitions are clearly derived, via the Archaeoceti, from the generalized eutherian mammalian dentition. Tooth-bearing mysticetes are now classified in the families Llanocetidae, Mammalodontidae, and Aetiocetidae, and there have been recent reports of at least two other apparently undescribed families (Barnes and Sanders, 1996a,b). Of these, the Aetiocetidae and the as yet undescribed families are known from North America, while the Llanocetidae and Mammalodontidae are from the margins of the Southern Ocean. Some tooth-bearing mysticetes retain skull characters that are very much like those of archaeocetes, and all living species of baleen whales have vestigial teeth that are reabsorbed prior to birth (Karlsen, 1962). Thus, toothed mysticetes form a morphological and presumably phylogenetic link between archaeocetes and baleen-bearing mysticetes.

The geochronologically earliest and structurally most archaic named toothed mysticete is *Llanocetus denticrenatus* from Antarctica. Although the species was based on a fragmentary specimen (Mitchell, 1989), this whale is known from a skull with large, widely spaced posterior teeth whose crowns had prominent accessory cusps (Fordyce, 2003). Mitchell postulated (1989) that *L. denticrenatus* filter-fed on planktonic Crustacea, in a manner analogous to the crab-eating seals, and that this feeding method was a forerunner to filter-feeding by the more crown-ward baleen-bearing Mysticeti.

DEFINING FEATURES OF THE SUBORDER MYSTICETI

CRANIAL

The Mysticeti are the clade of Neoceti that includes crown-Mysticeti (all extant species of *Balaena, Eubalaena, Caperea, Eschrichtius, Balaenoptera*, and *Megaptera*, and all descendants of

Evolution of Tertiary Mammals of North America, Vol. 2. ed. C. M. Janis, G. F. Gunnell, and M. D. Uhen. Published by Cambridge University Press.

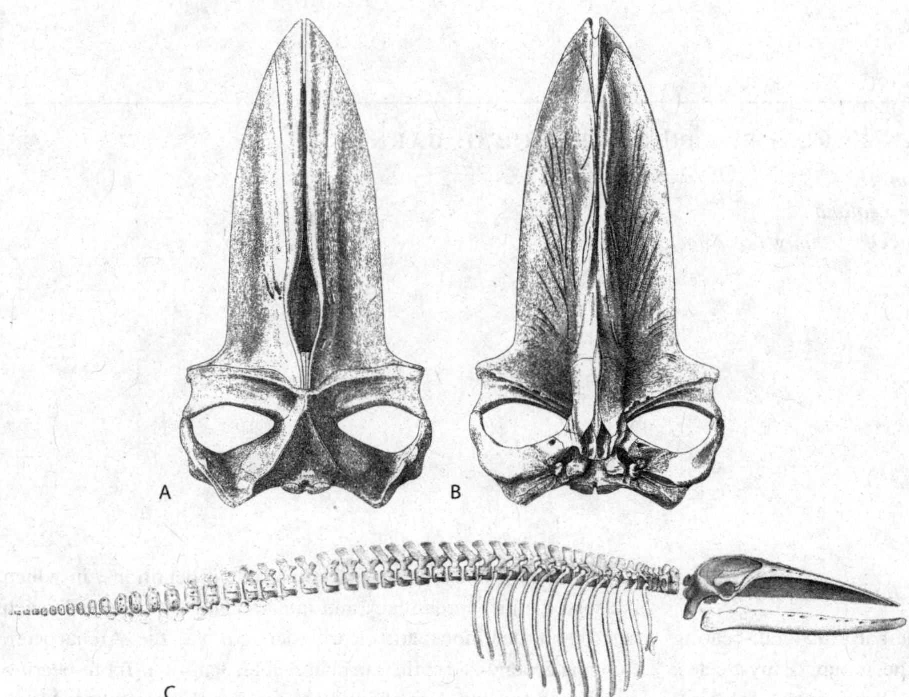

Figure 35.1. Representative mysticete cranial morphology. A. *Pelocetus calvertensis*, dorsal view. B. *Pelocetus calvertensis*, ventral view (A,B after Kellogg, 1965). C. *Megaptera novaeangliae* skeleton, right lateral view, lacking forelimb (after Van Beneden and Gervais, 1880).

their most recent common ancestor) plus the stem-Mysticeti, which are the extinct forms closer to crown-Mysticeti than to Odontoceti. Characteristic features cited in past studies include skulls with (primitively) elongate nasal bones; the anterior bony nares are single and continuous with the mesorostral groove; the maxilla has a lateral process appressed to the anterolateral surface of antorbital process of frontal; the maxilla has an infraorbital plate; the temporal musculature is attached to the temporal crest on the posterior part of supraorbital process of the frontal; the basioccipital crest has a tubercle posteriorly; the peribullary sinus is limited to the area surrounding and immediately anterior to the tympanic bulla; a foramen pseudo-ovale is present and perforates the falciform process of the squamosal; there is no middle sinus, and the pterygoid sinus is subspherical and restricted to the basicranium; and the dentary lacks a symphyseal articulation, instead has a flexible mandibular symphysis with a symphyseal ligament associated with a horizontal groove that curves upward at anterior end (Figure 35.1). This list of features includes plesiomorphies and apomorphies; cladistic studies, discussed below, should continue to clarify diagnostic features. Particular patterns of telescoping (sensu Winge, 1921; Miller, 1923) are widely cited as diagnostic for Mysticeti. In the mysticete form of telescoping, the emphasis is on the rostral bones that are positioned only as far posterior as the dorsal interorbital region, without any platelike ascending supraorbital process on the maxilla; most or perhaps all mysticetes have a posteriorly produced infraorbital process of the maxilla. Further, in all but the most basal groups, the apex of the occipital shield is thrust anteriorly at least as far as the mid part of the temporal fossae. Modern mysticetes have rostral bones that are loosely sutured with the cranium, and this seems to be the case for most fossil mysticetes, but whether the pattern is diagnostic for Mysticeti is uncertain.

DENTAL

Modern mysticetes lack teeth, but they probably evolved from toothed ancestors, in part because as embryos they develop tooth buds that are later resorbed (Karlsen, 1962). Many tooth-bearing early mysticetes are known from the Oligocene (Fordyce and Barnes, 1994). One group of toothed mysticetes includes species with archaeocete-like teeth (Fordyce, 1989; Barnes and Sanders, 1996a,b). This group could be delimited to include *Llanocetus denticrenatus* from the late Eocene of Antarctica (Mitchell, 1989). *Kekenodon onamata* and *Phococetus vasconum* were formerly thought to be members of this group (Fordyce, 1992) but are more likely specialized relict "archaeocetes" currently in taxonomic limbo between Archaeoceti and Neoceti (Fordyce, 2004). The teeth in some of the basal mysticetes are similar in size to those of archaeocetes and the cheek teeth are similarly double-rooted and have accessory denticles. Some species have an extra molar in each quadrant over the count of basilosaurids, yielding three upper and four lower molars (Barnes and Sanders, 1996a,b). Another group of toothed mysticetes, the Aetiocetidae, includes species that all have tiny teeth that are conical anteriorly and have tiny denticles posteriorly, but are all single-rooted. There are no dental characteristics that unite toothed mysticetes and exclude archaeocetes and early odontocetes.

POSTCRANIAL

Mysticetes generally have an innominate bone with a laterally projecting process in the middle of the bone. They also often have a second bone attached to the innominate that may represent the femur (Slijper, 1962). Living mysticetes all have a single bony element in the sternum, but the sternum is incompletely known in many

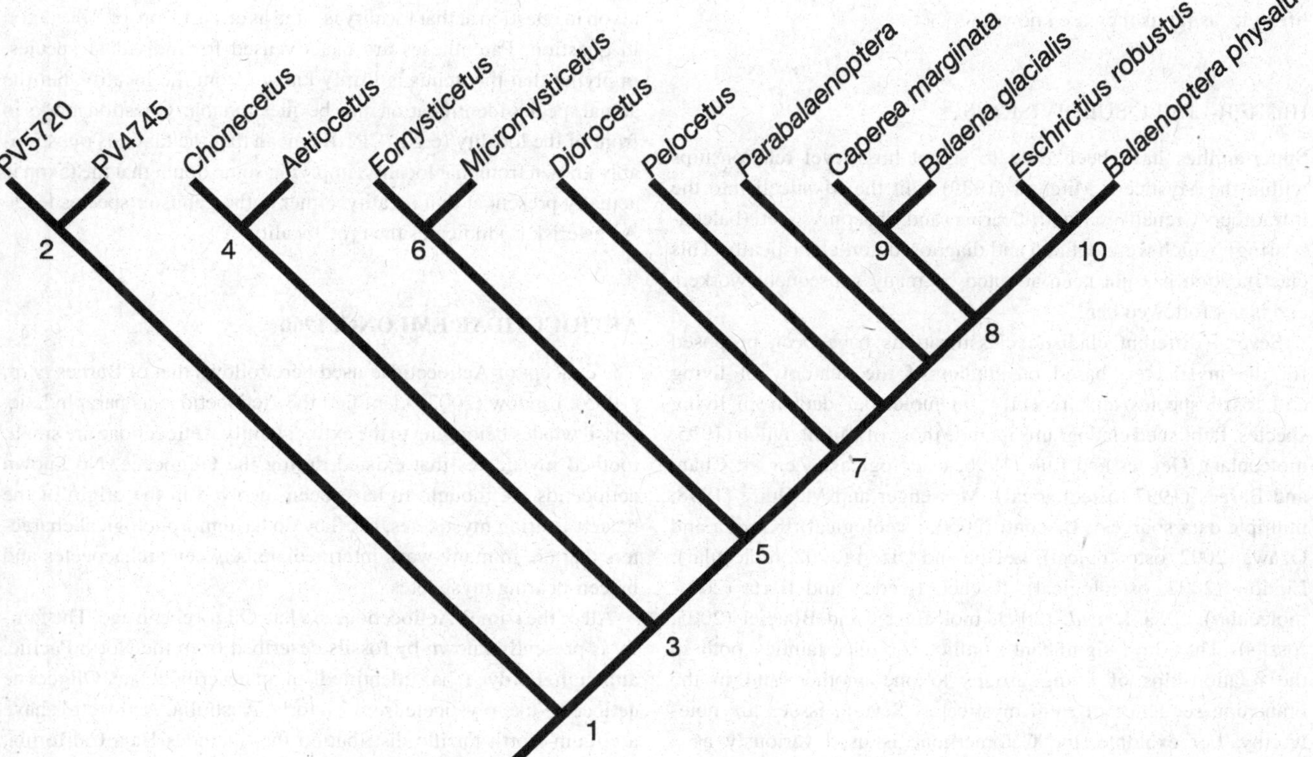

Figure 35.2. Relationships among mysticete cetaceans (after Geisler and Sanders, 2003). Taxa and characteristics at nodes 1–10 as follows. 1. MYSTICETI: zygomatic process of the maxilla bearing a steep face that clearly separates the rostrum from the antorbital process; long postorbital process; wide and bulbous basioccipital crest (Geisler and Sanders, 2003). 2. ARCHAEOMYSTICETES of Barnes and Sanders (1996a,b): large, archaeocete-like teeth. 3. Unnamed clade: anterior half of maxilla forming a highly acute angle with flattened maxilla, mandibular symphysis not sutured, connected by ligaments (derived from data in Geisler and Sanders [2003]). 4. AETIOCETIDAE (sensu Barnes *et al.*, 1995): two facial infraorbital foramina; premaxillae overhang maxillae; cross section of intertemporal region ovoid without a sagittal crest; small, heterodont teeth. 5. CHAEOMYSTICETI (Mitchell, 1989): teeth absent; mandible slightly bowed laterally; frontals same height as nasals (see Geisler and Sanders [2003] for additional synapomorphies). 6. Unnamed clade: fundus of acoustic meatus funnellike, smaller at the blind end and wider near the rim (derived from data in Geisler and Sanders [2003]). 7. Unnamed clade: palate with pronounced, median keel; small mandibular fossa; ascending process of premaxilla terminating on anterior half of supraorbital process of frontal (see Geisler and Sanders [2003] for additional synapomorphies). 8. CROWN-MYSTICETI: width of rostrum at antorbital notch very narrow; dorsoventral thickness of anterior edge of nasal very thin; cross section through intertemporal region, including parietals, pinched ventrally and dorsal part expanded laterally, expanded part rounded over in cross section (derived from data in Geisler and Sanders [2003]). 9. BALAENOIDEA: posterodorsal portion of maxilla sutured to the frontal; palatine/maxilla suture V-shaped; ascending process of premaxilla terminating at anterior edge of supraorbital process of frontal (see Geisler and Sanders [2003] for additional synapomorphies). 10. BALAENOPTEROIDEA.

fossil mysticetes, and this feature is unlikely to unite all mysticetes. Most other postcranial characteristics either unite some subgroups of mysticetes or are also found in archaeocetes as well, and thus fail to distinguish mysticetes from all other cetaceans.

SYSTEMATICS

SUPRASUBORDINAL

Mysticeti are thought to have evolved from basilosaurid archaeocetes, given the numerous shared derived features of basilosaurids, mysticetes, and odontocetes – the last two collectively forming the Neoceti (Uhen and Gingerich, 2001; Fordyce, 2002a; Uhen, 2002). Currently, it is thought that Mysticeti and Odontoceti share a single common ancestor within the Basilosauridae, but this remains to be demonstrated by a thorough cladistic analysis including all relevant basilosaurids, early mysticetes, and early odontocetes.

INFRASUBORDINAL

GROUPS OF STEM-MYSTICETES

Stem-mysticetes pose problems because of their archaic structure. Stem-mysticetes are those Cetacea that are related more closely to crown-group Mysticeti (namely, Balaenopteridae + Eschrichtiidae + Balaenidae + Neobalaenidae) than to Odontoceti or any member of the Archaeoceti (Figure 35.2). The families of stem-group Mysticeti recognized here are Llanocetidae, Mammalodontidae, Aetiocetidae, Eomysticetidae, and an as yet unnamed group from South Carolina (Barnes and Sanders, 1996a,b); all are toothed. Species in these families are recognized as mysticetes because their rostral and mandibular structure is consistent with a filter-feeding lifestyle, as far as they are known.

HIGHER-LEVEL SUBDIVISIONS

Superfamilies have been used to signal high-level relationships within the Mysticeti. Mitchell (1989) split the Mysticeti into the infraorders Crenaticeti (tooth-bearing) and Chaeomysticeti (baleen-bearing), which were defined and diagnosed non-cladistically. This classification has not been adopted by many subsequent workers and is not followed here.

Several different cladistic classifications have been proposed for the mysticetes, based on studies of the anatomy of living and fossil species and, recently, on molecular data from living species. Published cladograms include those of Milinkovitch (1995, molecular); Geisler and Luo (1996, osteological); Zeigler, Chan, and Barnes (1997, osteological); Messenger and McGuire (1998, multiple data sources); Bisconti (2000, osteological); Kimura and Ozawa (2002, osteological); LeDuc and Dizon (2002, molecular); Lindow (2002, osteological); Rychel, Reeder, and Berta (2004, molecular); Sasaki *et al.* (2005, molecular); and Bouetel (2005, fossils). There are significant conflicts or uncertainties both in the relationships of living groups to one another, and in the branching sequence of stem mysticetes. Several issues are noteworthy. For example, the Cetotheriidae is used variously as a grade or as a clade, of uncertain relationship to crown-group families of Mysticeti. Probably, many so-called cetotheres do not belong in the same clade as *Cetotherium rathkii*; many may be stem-Balaenopteridae. Several molecular studies have placed the Neobalaenidae (pygmy right whale) closer to Balaenopteridae (rorquals, humpbacks) than to Balaenidae (right whales), which are basal amongst the crown-Mysticeti. Recent molecular analyses have placed *Eschrichtius* (long placed in the Eschrichtiidae) and *Megaptera* (in the monotypic Megapterinae) with the genus *Balaenoptera*, making the Balaenopteridae and Balaenopterinae of traditional use (e.g., Mitchell, 1989) paraphyletic.

Since there is no agreement among the fossil and recent cetacean workers as to how to group families of mysticetes, we list the genera of North American Mysticeti grouped by family but not in higher groupings such as superfamilies. The mysticete families present in the Tertiary of North America are Aetiocetidae, Cetotheriidae, Balaenopteridae, and Balaenidae. Gray whales, Eschrichtiidae, are known only from the North American Quaternary (Barnes and McLeod, 1984). The extant Neobalaenidae, and extinct Llanocetidae and Mammalodontidae have not been reported from North America.

INCLUDED GENERA OF MYSTICETI COPE, 1891

The locality numbers listed for each genus refer to the list of unified localities in Appendix I. The locality numbers may be listed in more than one way. The acronyms for museum collections are listed in Appendix III.

Parentheses around the locality (e.g., [CP101]) mean that the taxon in question at that locality is cited as an "aff." or "cf." the taxon in question. Parentheses are usually used for individual species, implying that the genus is firmly known from the locality, but the actual species identification may be questionable. Question marks in front of the locality (e.g., ?CP101) mean that the taxon is questionably known from that locality, implying some doubt that the taxon is actually present at that locality, either at the genus or species level. An asterisk (*) indicates the type locality.

AETIOCETIDAE EMLONG, 1966

The concept of Aetiocetidae used here follows that of Barnes *et al.* (1995). Lindow (2002) identified the Aetiocetidae as paraphyletic. Fossil whales belonging to the extinct family Aetiocetidae are small, toothed mysticetes that existed during the Oligocene. No known aetiocetids are thought to have been involved in the origin of the baleen-bearing mysticetes, but they do have morphological characters that are in many ways intermediate between archaeocetes and baleen-bearing mysticetes.

All of the named Aetiocetidae are late Oligocene in age. The family is presently known by fossils described from the North Pacific, although Fordyce has identified an undescribed late Oligocene aetiocetid-like mysticete from Victoria, Australia. Aetiocetids have a circum-North Pacific distribution that includes Baja California, Oregon, Washington, British Columbia, and Japan.

The Aetiocetidae are Mysticeti with functional teeth in the adult stage in both the palate (premaxilla, maxilla) and dentary. Their heterodont dentitions are differentiated as incisors, canines, premolars, and molars, and there are small accessory cusps on the comparatively small molar and premolar crowns. The medial margin of the premaxilla is elevated at the anterolateral corner of the nasal; the parietals are widely exposed in the intertemporal region; there is a sharp subtemporal ridge present within the temporal fossa at the anterior margin of the squamosal fossa; a tubercle is present at the posterior margin of temporal fossa and is continuous with the lambdoidal crest; a large squamosal fossa is present; the zygomatic process of the squamosal is elongate and uninflated; there is an elliptical notch in the posterior margin of the palatine bone; the anterior process of the periotic is closely appressed to the medial side of the falciform process of the squamosal; a relatively large carotid foramen perforates the basioccipital; the fossa for the peribullary air sinus is large; the jugular notch is shallow; and the coronoid process of the dentary is relatively large and its apex is lobate.

Aetiocetus Emlong, 1966

Type species: *Aetiocetus cotylaveus* Emlong, 1966.

Type specimen: USNM 25210 (formerly UO 26351), associated parts of one individual skeleton, including nearly complete skull, lacking parts of the right squamosal and jugal, with both tympanic bullae and periotics, cervical, thoracic and lumbar vertebrae, and some ribs (Emlong 1966).

Characteristics: Relatively wide, anteriorly tapered, relatively flat rostrum; anteriorly thrust apex to supraoccipital; broad, relatively anteroposteriorly short intertemporal region; notch between anterior margins of the nasals; elongate zygomatic process of the squamosal with dorsal curvature and a rounded anterior extremity; large and curved coronoid process of the dentary; heterodont dentition having in each dentary and on each side of the palate three incisors, one canine, four premolars, and three or more molars, the teeth being relatively small (Barnes *et al.* 1995).

Occipital condyle breadth: 94 mm (Emlong, 1966, p. 17).

Included species: *A. cotylaveus* (known from locality WM5B*); *A. weltoni* (locality WM5A [Barnes *et al.*, 1995]).

Aetiocetus sp. is also known from locality (WM27IIA) (Barnes, 1998).

Comments: *Aetiocetus* was originally assigned by Emlong (1966) to the Archaeoceti because it has teeth, but Van Valen (1968) and Barnes (1987, 1989) pointed out the cranial characters that it shares with the Mysticeti and transferred it to that suborder. Barnes *et al.* (1995) rediagnosed and redefined the family Aetiocetidae, named additional species in the genus *Aetiocetus*, named subfamilies and additional genera and species in the Aetiocetidae, and discussed phylogeny.

Chonecetus Russell, 1968

Type species: *Chonecetus sookensis* Russell, 1968.

Type specimen: NMC VP64443, braincase, ventral part of atlas vertebra, centra of three vertebrae.

Characteristics: Differing from *Ashorocetus* by having less anteriorly inclined occipital shield; larger occipital condyles; less prominent vertical crest on occipital shield near its apex; less prominent lambdoidal crest; basioccipital crests thick transversely and not divergent posteriorly; smaller peribullary sinus anterior to tympanic bulla; larger cranial hiatus; short and thick mastoid process that is not expanded dorsally onto lateral surface of braincase. Differing from *Aetiocetus* by having a small, triangular occipital shield; prominent occipital condyles; short and thick postorbital processes of the frontals; small and rectangular zygomatic processes of the squamosals; short posterior processes of the periotic (Barnes *et al.*, 1995).

Occipital condyle breadth: 69 mm (M. D. Uhen, personal observation).

Included species: *C. sookensis* (known from locality WM2II*); *C. goedertorum* (locality WM3* [Barnes *et al.*, 1995]).

Comments: The type locality of *C. sookensis* was originally stated by Russell (1968) to be within the Sooke Formation. Since Russell's publication, the nomenclature of the marine Oligocene rocks on Vancouver Island has been revised. The type locality of *C. sookensis* now falls within the mapped area of the Hesquiat Formation. The age of the Hesquiat Formation, and therefore of the holotype of *C. sookensis*, remains as early late Oligocene.

C. sookensis is known only by the holotype, which has been additionally prepared since it was originally studied by Russell (1968). Although Russell (1968) was equivocal on its familial and subordinal assignment, *C. sookensis* is clearly a mysticete, albeit a very archaic one.

EOMYSTICETIDAE SANDERS AND BARNES, 2002a

The family Eomysticetidae differs from all other mysticete families by having an elongate and narrow intertemporal region. It differs from members of the families of toothed mysticetes (Llanocetidae, Mammalodontidae, and Aetiocetidae) by the absence of teeth in adulthood. It differs from the families Cetotheriidae, Balaenopteridae, Balaenidae, Neobalaenidea, and Eschrichtiidae by having elongate anterolateral processes of nasal bones; extremely long zygomatic processes of squamosals; small posterior process of periotic with a well-defined facet for articulation with tympanic bulla; and a humerus that is longer than the radius and ulna. Eomysticetidae differ from Balaenopteridae, Balaenidae, Neobalaenidae, and Eschrichtiidae by having the dorsal naris located near the midlength of the rostrum; exceptionally elongate nasal bones; parietals exposed on the intertemporal region between frontals and the apex of occipital shield; the nuchal crest directed dorsolaterally; the periotic with transversely compressed anterior process and no dorsal process on cochlear portion; a large coronoid process of the dentary; and the posterior edge of the coronoid elevated above the anterior edge (Sanders and Barnes, 2002a). See also discussion of Cetotheriidae, below.

Eomysticetus Sanders and Barnes, 2002a

Type species: *Eomysticetus whitmorei* Sanders and Barnes, 2002a.

Type specimen: ChM PV4253: skull, both periotics, both tympanic bullae, both dentaries, seven cervical vertebrae, seven thoracic vertebrae, two lumbar vertebrae, and one possible caudal vertebra, parts of at least 17 ribs, a possible sternal rib, right scapula, humerus, radius, and ulna.

Characteristics: Same as for the family Eomysticetidae.

Occipital condyle breadth: 142 mm (*E. whitmorei*, Sanders and Barnes, 2002a: Table 1).

Included species: *E. whitmorei* (known from locality EM5C*); *E. carolinensis* (locality EM5B* [Sanders and Barnes 2002a]).

Comments: Phylogenetic analysis by Geisler and Sanders (2003) indicated that Eomysticetoidea (Eomysticetidae + Cetotheriopsidae) is a monophyletic group of basal baleen-bearing mysticetes.

CETOTHERIOPSIDAE BRANDT, 1872 SENSU GEISLER AND SANDERS, 2003

The subfamily Cetotheriopsinae was proposed by Brandt (1872) (raised to family level by Geisler and Sanders [2003]) and includes late Oligocene mysticetes that lack teeth, have relatively small occipital shields with large lambdoidal crests, elongate and dorsally arching, curved zygomatic processes of the frontals, and a secondary fossa within the squamosal fossa. Named genera are *Cetotheriopsis* and *Micromysticetus*. See also discussion of Cetotheriidae, below.

Micromysticetus Sanders and Barnes, 2002b (including *Cetotheriopsis*, in part)

Type species: *Micromysticetus tobieni* (Rothausen, 1971) (originally described as *Cetotheriopsis tobieni*).

Type specimen: JGU P1289, partial braincase.

Characteristics: Differing from *Cetotheriopsis* by being smaller; having cranium with its occipital shield shaped like a broad equilateral triangle, not an anteroposteriorly elongate triangle; a less prominent medial crest on midline of occipital shield; no prominent sulcus dorsal to the occipital condyles on the occipital shield; an anteroposteriorly shorter squamosal fossa that is shallow and interrupted by a protuberance, the squamosal prominence, extending posterolaterally to the posterior margin of the squamosal fossa; exoccipital thicker anteroposteriorly and narrower transversely; zygomatic process of squamosal elongate, deep, arched, extending anteriorly to a point that is beyond the anterior apex of the supraoccipital; glenoid fossa broad transversely, with a rounded posterointernal margin; basioccipital between the basioccipital crests flat, or slightly convex; sulcus for the external acoustic meatus short (Sanders and Barnes, 2002b).

Occipital condyle breadth: 150 mm (Rothausen, 1971).

Included species: *M. rothauseni* only, known from locality EM4* only.

Comments: The holotype specimen of *M. tobieni* was previously in the private collection of Fritz von der Hocht, Krefeld, Germany and was collected from Kiesgrube Wilhelm Frangen, Lank-Latum, WNW Dusseldorf-Kaiserwerth, Nordrhein-Westfalen, Germany; Meeressande, upper "Chattian A" beds, nannoplankton zone MP24 (Martini and Müller, 1975), Lower Chattian, late Oligocene. The only known North American specimen is *Micromysticetus rothauseni*, ChM PV 4844, a braincase lacking the entire rostrum and the right zygomatic process of the squamosal, with both periotics and the axis vertebra.

CETOTHERIIDAE BRANDT, 1872

The concept of Cetotheriidae used here is the clade that includes the late Miocene mysticete *Cetotherium rathkii* Brandt and those other species that are closer to *C. rathkii* than to other baleen whales. We provisionally accept the conclusion of Bouetel (2005) that the Cetotheriidae sensu stricto probably includes the North American Tertiary species *Metopocetus durinasus, Mixocetus elysius*, and *Nannocetus eremus* as well as *Cetotherium rathkii, Piscobalaena nana*, and *Herpetocetus sendaicus*. Bouetel's (2005) conclusions have yet to be justified by a detailed computer-aided cladistic analysis, but our experience with the relevant specimens and literature suggests that the above cluster is reasonable. Other literature shows widely varying opinions on the content of the family, and on diagnostic features. It is not clear how the clade Cetotheriidae is related to any extant family. It is likely that many "cetotheres" of past use are stem-Balaenopteridae or are stem species in other mysticete crown clades such as Balaenidae, and some may belong in recently recognized groups such as Eomysticetidae and Cetotheriopsidae. Previously, the term cetothere has been widely applied to fossil baleen (toothless) whales that lack the defining features of the four extant families (Balaenopteridae, Eschrichtiidae, Balaenidae, and Neobalaenidae) for example see Kellogg (1931, 1934a, 1965, 1968a, 1969), Simpson (1945), Rothausen (1971), Czyzewska and Ryziewicz (1976), and Barnes and McLeod (1984). Research since 1990 has seen an interest in determining cladistic relationships (e.g., McLeod *et al.*, 1993), the start of computer-aided cladistics of reputed cetotheres (Geisler and Luo, 1996), and the recommendation to define cetotheres as the clade including *Cetotherium rathkii* (e.g., Fordyce and Muizon, 2001). Recent cladistic studies include those of Kimura and Ozawa (2002), Geisler and Sanders (2003) and Bouetel (2005).

Cetotherium Brandt, 1843

Type species: *Cetotherium rathkii* Brandt, 1843.

Type specimen: PIN 1840/1 (see Pilleri, 1986).

Characteristics: Posterior ends of the rostral bones (the nasals, premaxillae, and maxillae) penetrating deeply posteriorly as a pointed wedge between the frontal bones; occipital shield relatively small, anterior apex not extending very far anteriorly, not as far as the end of the zygomatic process of the squamosals; temporal fossae wide open dorsally; paroccipital processes large and thick anteroposteriorly.

Occipital condyle breadth: occipital condyles not preserved in the type specimen (Pilleri 1986, Plate 25).

Included species: *C. furlongi* only, known from locality WM11* only (Kellogg, 1925).

Comments: Some authors cite different dates for the first publication of *Cetotherium*, for example Capellini (1876) cited Brandt (1872). Capellini (1901) used *Cetotherium* as a subgenus of *Aulocetus*. Several species named within the genus *Cetotherium*, or subsequently transferred to it, are of dubious taxonomic position or status. Only one species of *Cetotherium* sensu stricto is reported from North America, and that is *Cetotherium furlongi* Kellogg, 1925, from early Miocene, Vaqueros Formation, Monterrey County, California. The holotype USNM 26521 has not been seen since the time when Kellogg was studying the specimen (Barnes, 1977).

Cetotherium is also known from the late Oligocene to late Miocene of Europe (possibly also the early Pliocene) (McKenna and Bell, 1997).

Herpetocetus Van Beneden, 1872

Type species: *Herpetocetus scaldiensis* Van Beneden, 1872.

Type specimen: None designated.

Characteristics: Differs from *Nannocetus* in being larger; having a cleft or postparietal foramen in the dorsoposterior part of the side wall of the braincase.

Occipital condyle breadth: 98 mm (NSMT PV 19540).

Included species: *Herpetocetus*. n. sp. only, known from locality EM8B only (Whitmore, 1994).

Herpetocetus sp. is also known from localities WM20, WM24A, WM26, (WM30A), (CC52) (Barnes, 1977, 1998).

Comments: Family placement follows Bouetel (2005). *Herpetocetus* is also known from the middle and/or late Miocene of Europe (McKenna and Bell, 1997).

***Metopocetus* Cope, 1896**

Type species: *Metopocetus durinasus* Cope, 1896.

Type specimen: USNM 8518, incomplete skull with right periotic.

Characteristics: Lateral occipital crests continuous with anterior temporal crests that diverge forward; frontal bone elongate, not covered posteriorly by the maxillary, co-ossified with nasals; nasals short, co-ossified with each other, not projecting anterior to frontals (Cope, 1896).

Occipital condyle breadth: 150 mm (Kellogg, 1968b, p. 123).

Included species: *M. durinasus* only, known from locality NC4A* only.

Comments: Kellogg (1968b) stated that the periotic and construction of the skull of *Metopocetus* suggest an "affinity if not identity with the genus *Mesocetus*." Here, family placement follows Bouetel (2005). *Metopocetus* is also known from the late Miocene of Europe (McKenna and Bell, 1997).

***Mixocetus* Kellogg, 1934b**

Type species: *Mixocetus elysius* Kellogg, 1934b.

Type specimen: LACM 882, skull, mandibles, and partial skeleton.

Characteristics: Large size; long, tapered rostrum; heavily built braincase; as in *Cetotherium*, narial opening extending posteriorly to a point behind the antorbital processes; the posterior ends of the rostral bones (nasals, premaxillae, and maxillae) tapering and extending posteriorly to a point between the posterior parts of the supraorbital processes of the frontals; temporal fossa opening dorsally, not being overthrust by the occipital shield. It was described by Kellogg (1934b) as having the "apex of supraoccipital shield not thrust forward beyond level of hinder parietal margin of temporal fossa; nasals located almost if not wholly behind level of preorbital angles of supraorbital processes of frontals; thin anterior process of parietal, which overrides basal portion of supraorbital process, extended forward beyond hinder ends of median rostral elements (ascending processes of maxillaries and of premaxillaries and nasals); backward thrust of rostrum has carried hinder ends of median rostral elements beyond level of anterior-most portion of hinder edge of supraorbital process; and rostrum exhibits a rather gradual distal attenuation". Differs from *Cetotherium* by having very large antorbital process of the maxilla; parallel-sided lateral margins of the supraorbital processes of the frontals; protruding lateral wall of the braincase; very thick and posteriorly protruding exoccipital (see Barnes 1977).

Occipital condyle breadth: approximately 268 mm (Kellogg, 1934b, p. 96).

Included species: *M. elysius* only, known from locality WM18* only.

Comments: Family placement follows Bouetel (2005).

***Nannocetus* Kellogg, 1929**

Type species: *Nannocetus eremus* Kellogg, 1929.

Type specimen: UCMP 26502, imperfectly preserved cranium; upper surface of the braincase, nasals, lacrimals, jugals, maxillae, premaxillae, and vomer are missing. Tympanic bullae and periotics were in place when the specimen was prepared (Kellogg, 1929).

Characteristics: Diminutive mysticete with large orbits; mesially constricted supraorbital process with prominent postorbital projection; large opening on the fore wall of braincase for exit of blood vessels and olfactory nerves; zygomatic process of squamosal proportionately large, transversely thickened, rather bulbous in shape, abruptly produced downward, with rapidly attenuated incurved extremity, with swollen postglenoid projection; exoccipitals thin and produced obliquely downward and backward, not attaining level of posterior surfaces of condyles, which are narrow and widely spaced but not protuberant; basioccipital with large lateral protuberances; pterygoid fossa small, without roof; tympanic bulla relatively large; periotic with rather large, swollen apophysis, compressed anterior process, and rather narrow pars cochlearis (Kellogg, 1929).

Occipital condyle breadth: 90 mm (UCMP 26502 [Kellogg, 1929, p. 457]).

Included species: *N. eremus* only, known from locality WM26III* only.

Nannocetus sp. is also known from localities WM21A, WM21B (Barnes, Hutchinson, and Welton, 1981), (WM30A) (Barnes, 1998).

Comments: Family placement follows Bouetel (2005).

BALEEN-BEARING MYSTICETI FAMILY INCERTAE SEDIS

***Aglaocetus* Kellogg, 1934 (including *Cetotherium*, in part)**

Type species: *Aglaocetus moreni* (Lydekker, 1894) (originally described as *Cetotherium moreni*).

Type specimen: MDLP 5-14, skull, bulla, and vertebrae.

Characteristics: Large size, with a wide rostrum; relatively wide exposure of the parietals across the cranial vertex;

nasals long and slender, parallel not tapered; posterior premaxillary terminations not tapered; ascending process of maxilla wide, triangular; dorsal rostral surfaces of maxillae sloping laterally; large and dorsally flaring lambdoidal crests bordering the occipital shield; wide temporal fossa; narrow zygomatic process of squamosal with sharp dorsal margin; wide, flat, relatively convex glenoid fossa; small postglenoid process; anteroposteriorly thick exoccipital; large and thick basioccipital crest; dentary with deep dorsoventrally, transversely compressed anterior end; high and laterally bent coronoid process; high and narrow mandibular condyle.

Occipital condyle breadth: 167 mm (Kellogg, 1968c, p. 168).

Included species: *A. patulus* only (known from localities NC3AA*, D [Kellogg, 1968c]).

Aglaocetus sp. is also known from locality (WM13A) (Barnes, 1977).

Comments: Lydekker (1894) originally described *Cetotherium moreni* based on a specimen from Argentina, and Kellogg (1934a) erected a new genus (*Aglaocetus*) based on additional Argentinean specimens. Kellogg added the species *A. patulus* in 1968 based on specimens from the Calvert Formation of Virginia. Kimura and Ozawa (2002) showed a cladogram that excluded *Aglaocetus* from a clade of *Cetotherium rathkii* and *Mixocetus elysius. Aglaocetus* is also known from the Miocene of South America, and possibly from the early Miocene of Australia (McKenna and Bell, 1997).

Cephalotropis Cope, 1896

Type species: *Cephalotropis coronatus* Cope, 1896.

Type specimen: USNM 9352, incomplete cranium.

Characteristics: Cranium with a relatively elongate intertemporal region, having relatively wide exposure of the parietals across the top between frontals and apex of supraoccipital; posterior ends of nasals and premaxillae slightly tapering, not wide. Additional information from *C. nectus* (given by Kellogg, 1940) is that *Cephalotropis* differs from *Metopocetus* and *Aulocetus* by the greater length of the exposure of the parietals and the frontals on the cranial vertex; lesser posterior extension of the posterior ends of the median rostral bones (nasals and premaxillae), and by a narrower rostrum.

Occipital condyle breadth: 150 mm (Cope, 1896, p. 143).

Included species: *C. coronatus* only, known from locality NC5* only.

Comments: Winge (1910) suggested that *Cephalotropis* was a synonym of *Plesiocetus*, which is, however, an archaic member of the Balaenopteridae. True (1912a) provided a diagnosis and considered the taxon to be distinct, as did Kellogg (1931). Kellogg (1940) also named *Cephalotropis nectus* from the middle to late Miocene (Tortonian) of Portugal.

Cophocetus Packard and Kellogg, 1934

Type species: *Cophocetus oregonensis* Packard and Kellogg, 1934.

Type specimen: UO 305, skull.

Characteristics: Apex of supraoccipital shield thrusting forward beyond level of anterior ends of zygomatic processes; nasals located in part anterior to level of preorbital angles of supraorbital processes; thin anterior process of parietal, overriding basal portion of supraorbital process, not extending forward to level of hinder ends of median rostral elements (ascending thrust of rostrum carrying hinder ends of median rostral elements to or beyond level of center of orbit, but not to level of anterior-most portion of hinder edge of supraorbital processes of frontals); short, rather narrow intertemporal construction formed by parietals; exposure of frontals in median interorbital region reduced to a narrow strip; rostrum relatively broad, not displaying any pronounced attenuation distally; zygomatic processes robust and directed forward; alisphenoid present in temporal wall of braincase; condyle of mandible broad, not laterally compressed, possessing a deep furrow on internal face near ventral border; periotic prolongation upward into cerebral cavity, short anterior process with outwardly projected posteroventral angle, unusually large and broad fossa on cerebral face in region of orifice of vestibular aqueduct (Packard and Kellogg, 1934).

Occipital condyle breadth: 124 mm (Packard and Kellogg, 1934, p. 35),

Included species: *C. oregonensis* only, known from locality WM7A* only.

Comments: This genus has remained monotypic since its description (see Barnes, 1977). It is apparently most closely related to the middle Miocene species *Agalocetus patulus* from the Calvert Formation (locality NC3) and *Peripolocetus vexillifer* Kellogg (1931), from the Sharktooth Hill Bonebed in California (locality WM13A). Slijper (1936) presented data on the morphology of the vertebrae. Kimura and Ozawa (2002) showed a cladogram that excluded *Cophocetus* from a clade of *Cetotherium rathkii* and *Mixocetus elysius.*

Diorocetus Kellogg, 1968a

Type species: *Diorocetus hiatus* Kellogg, 1968a.

Type specimen: USNM 16783, skull, lower jaws and partial skeleton.

Characteristics: Rostrum strongly tapered anteriorly; incisure of variable length, commencing near the posterior end of each maxillary internal to the base of its posteroexternal process, extending obliquely forward toward the maxillary–premaxillary contact along the mesorostral trough and separating triangular area behind it into a dorsal and ventral plate; backward thrust of rostrum limited, median rostral elements (ascending processes of maxillaries, premaxillaries, nasals) not carried backward beyond the level of the posteroexternal processes of the maxillaries, which project laterally beyond the preorbital angles of the supraorbital processes on the immature type skull but to level of center of orbit on more mature referred skull; no transverse temporal crest developed on supraorbital process; elongated nasals located for most part anterior to level of preorbital angle of supraorbital processes; apex of supraoccipital shield thrusting forward to or slightly behind level of anterior ends of zygomatic processes; palatines elongated; lateral descending processes of basioccipital knoblike, smaller than pterygoid fossa; anterior process of periotic compressed transversely; deep lengthwise groove for facial nerve on ventral surface of posterior process; groove behind stapedial fossa on posterior face of pars labyrinthica extending from posterointernal angle of posterior process to cerebral faces of pars cochlearis; horizontal ramus of mandible robust, depth anteriorly about one-fifteenth of its length; coronoid process small and low; condyle expanded from side to side, with deep groove above angle on internal surface for attachment of internal pterygoid muscle (Kellogg, 1968a)

Occipital condyle breadth: 138–144 mm (Kellogg, 1968a, p. 142).

Included species: *D. hiatus* only (known from localities NC3C*, D [Dooley, 1993; Dooley, Fraser, and Luo, 2004]).

Comments: Kimura and Ozawa (2002) showed a cladogram that clustered *Diorocetus* and *Parietobalaena*, and excluded *Diorocetus* from a clade of *Cetotherium rathkii* and *Mixocetus elysius*.

Halicetus Kellogg, 1969

Type species: *Halicetus ignotus* Kellogg, 1969.

Type specimen: USNM 23636, fractured skull with periotics and tympanic bullae and partial skeleton.

Characteristics: Atlas not unusually thickened, vestigial hypopysial process present; odontoid process of axis short, acutely pointed; pedicles of neural arch of third to seventh cervicals short and rather wide; neural canal relatively high, not unusually widened; roof of neural canal arched; neural spines of dorsals progressively increasing in anteroposterior width and height toward posterior end of series, almost vertical on anterior dorsals in contrast to slight backward inclination of posterior dorsals; neural spine of 11th lumbar shorter but broader than on first lumbar; metapophyses of lumbars and posterior dorsals thin, deep vertically; transverse processes of first to fourth caudals anteroposteriorly widened toward extremity; posterior process of periotic greatly enlarged anteroposteriorly; bulbous anterior process rugose and porous internally, elsewhere irregularly creased or wrinkled longitudinally; dorsal rim of circular internal acoustic meatus projecting internally (cerebrally) beyond slitlike depression for aperture of vestibular aqueduct and the cochlear aqueduct orifice; transverse and vertical diameters of excavation behind stapedial fossa for extension of air sac system approximately equivalent (Kellogg, 1969).

Occipital condyle breadth: not known from the type specimen.

Included species: *H. ignotus* only, known from locality NC4A* only.

Comments: No additional specimens have been referred to *Halicetus* besides the type specimen of *H. ignotus*.

Isocetus Van Beneden, 1880

Type species: *Isocetus depauwii* Van Beneden, 1880.

Type specimen: none designated by Van Beneden (1880).

Characteristics: Condyle of mandible round; opening of dental channel broad, furrowed at its base (Van Beneden, 1880).

Occipital condyle breadth: unknown.

Included species: ?*Isocetus* only, known from locality GC10C only (Kellogg, 1944; Morgan, 1994).

Comments: *Isocetus* was originally described based on specimens from western Europe (it is known in Europe from the middle Miocene and possibly the late Miocene [McKenna and Bell, 1997]). Kellogg (1968b) indicated that Cope's (1895) *Mesocetus siphunculus* probably belongs in *Isocetus* (but see *Mesocetus* below). Kellogg's (1944) identification of *Isocetus* from Florida was based on two auditory bullae (MCZ 17884 [Morgan, 1994]). Morgan (1994) reported that a skull with similar bullae has been found (UF 130000) but it has not been compared with European specimens identified as *Isocetus*. Consequently, the presence of *Isocetus* in North America is still questionable.

Mesocetus Van Beneden, 1880

Type species: *Mesocetus longirostris* Van Beneden, 1880.

Type specimen: Van Beneden (1880) did not designate any specimens as types for the four listed species of *Mesocetus*. They are thought to come from Tortonian deposits near Antwerp, Belgium (Roth, 1978).

Characteristics: Mandibular condyle transversely narrow, high dorsoventrally; parietals tall; long space between the anterior edge of occipital and frontal (Van Beneden, 1880).

Occipital condyle breadth: unknown.

Included species: *M. siphunculus* only, known from locality NC3AA* only (Cope 1895).

Mesocetus sp. is also known from locality ?GC10C (Kellogg, 1944; Morgan, 1994).

Comments: *Mesocetus* was originally described based on specimens from western Europe (it is known in Europe from the middle Miocene to the Pliocene [McKenna and Bell, 1997]). Cope (1895) had originally based *Mesocetus siphunculus* on a nearly complete dentary and parts of three others. Kellogg (1968b) transferred *Mesocetus siphunculus* Cope, 1895 to *Isocetus*, creating the new binomen

Isocetus siphunculus (Cope, 1895). The holotype of the type species of *Isocetus, Isocetus depauwi*, includes a partial postcranial skeleton but no skull, so generic identity is difficult to assess.

When Kellogg (1968b) transferred *M. siphunculus* to *Isocetus*, he redefined and illustrated the holotype *M. siphunculus*, closely compared it with *Isocetus depauwi*, and restricted it to the holotype right dentary, originally in the Johns Hopkins University Museum but now cataloged as AMNH 22665, which is from the Middle "Miocene marl of Pamunky River at the base of the Calvert Formation." Kellogg (1968b) decided that the humerus (AMNH 22669) that Cope had referred to *M. siphunculus* is not diagnostic. *Mesocetus siphunculus* was recognized by Gottfried, Bohaska, and Whitmore (1994) and apparently is different from all species of mysticete from the Calvert Formation (locality NC3) for which a dentary is known: *Aglaocetus patulus* and *Metopocetus durinasus* being the only two for which no dentary has been published. The holotype dentary of *M. siphunculus* is the correct size to fit on the skull of *A. patulus*. Kellogg's (1944) identification of *Mesocetus* from Florida was based on two auditory bullae (MCZ 17885 [see also Morgan, 1994]), but these bullae are not likely to be diagnostic for *Mesocetus*. Consequently, the presence of *Mesocetus* in North America is questionable.

Parietobalaena Kellogg, 1924

Type species: *Parietobalaena palmeri* Kellogg, 1924.

Type specimen: USNM 10668, skull including periotics.

Characteristics: Medium-size, relatively archaic cetotheriid; skull with relatively narrow and anteriorly tapered rostrum; narial opening in the posterior one-third of the rostrum; rostral bones barely reaching the level of the antorbital processes; zygomatic processes of the squamosals lightly built and flaring anterolaterally.

Occipital condyle breadth: 86.5 mm (USNM 10668 [Kellogg, 1924, p. 9]).

Included species: *P. palmeri* (known from localities NC3A*, B–D); *P. securis* (locality WM13A* [Kellogg, 1931]).

Parietobalaena sp. is also known from locality (CC52) (Barnes, 1998).

Comments: Kellogg (1924) based the type species, *P. palmeri*, on a single braincase from the middle Miocene Calvert Formation in Maryland. Later he discussed the species and subsequently (1968d) referred additional specimens, including skulls, to the species. It is the most common mysticete in the Calvert Formation (Goffried, Bohaska, and Whitmore, 1994).

When Kellogg (1931) named *Parietobalaena securis* from the middle Miocene Sharktooth Hill Bonebed in California, he selected as the holotype of this species an isolated periotic, CAS 4371. Subsequently collected specimens from the Sharktooth Hill Bonebed (see Barnes, 1977) include skulls associated with earbones that confirm *Parietobalaena securis* is in fact congeneric with *P. palmeri*. Similar also to its Atlantic congener, *P. securis* is the most abundant mysticete in the Sharktooth Hill Local Fauna.

Kimura and Ozawa (2002) showed a cladogram that clustered *Parietobalaena* and *Diorocetus*, and excluded *Parietobalaena* from a clade of *Cetotherium rathkii* and *Mixocetus elysius*.

Pelocetus Kellogg, 1965

Type species: *Pelocetus calvertensis* Kellogg, 1965.

Type specimen: USNM 11970, skull with tympanic bullae and periotics, both mandibles, and a partial skeleton.

Characteristics: Apex of supraoccipital shield thrust forward slightly beyond level of anterior ends of zygomatic processes; elongated nasals located in part anterior to level of preorbital angles of supraorbital processes of frontals; strong forward overthrust carrying anterior borders of parietals to median interorbital region, frontals overriding but not extending forward to level of posterior ends of median rostral elements (ascending processes of maxillaries, premaxillaries, nasals); backward thrust of rostrum limited, median rostral elements (ascending processes of maxillaries, premaxillaries, nasals) not carrying backward beyond level of middle of orbit; exposure of frontals in median interorbital region reduced to a narrow strip; thin temporal crest on each supraorbital process; short, pinched-in intertemporal constriction formed by opposite parietals meeting on midline; wide temporal fossae; slender, bowed outward zygomatic processes; alisphenoid present in temporal wall of cranium; rostrum broad, sides nearly parallel on basal half and then rather strongly curved to distal end and equivalent to 68% of total length of skull; narial fossae elongated; palatines elongated, diverging posteriorly; lateral protuberances of basioccipital massive, larger than pterygoid fossae; posterior process of periotic elongated and expanded distally; horizontal ramus of mandible relatively deep, and thick, the condyle large and spherical; articular facet for capitulum of following rib situated on first to eighth dorsal vertebrae, inclusive, below level of floor of neural canal and adjacent to edge of posterior face of centrum; lumbar vertebrae with relatively broad and high neural spines, elongated transverse processes, thin lamina-like metapophyses; second to ninth ribs, inclusive, having capitulum at end of elongated neck; scapula fan shaped, exhibiting no vestige of acromion although possessing a coracoid process, having prescapular fossa relatively broad and flat; humerus with anterior or radial face of shaft markedly rugose; distal epiphyses of radius and ulna detached and not completely ossified (Kellogg, 1965).

Occipital condyle breadth: 183 mm (Kellogg, 1965, p. 11).

Included species: *P. calvertensis* only (known from localities NC3B*, C, D [Kellogg, 1965; Dooley, 1993]).

Comments: *Pelocetus* should be considered to include only *P. calvertensis*, because *P. mirabilis* (Ginsburg and Janvier,

1971) from the Miocene of France is based on a mandible that cannot be compared with *P. calvertensis*. Kimura and Ozawa (2002) showed a cladogram that clustered *Pelocetus* in a polytomy that did not include the clade of *Cetotherium rathkii* and *Mixocetus elysius*.

***Peripolocetus* Kellogg, 1931**

Type species: *Peripolocetus vexillifer* Kellogg, 1931.

Type specimen: CAS 4370, squamosal region on the right side, corresponding portion of exoccipital, periotic and bulla, right occipital condyle, and a portion of the basioccipital.

Characteristics: Medium-size mysticete. Closely resembles *Cophocetus oregonensis*, in having skull with massive squamosal with a blunt anterior end of the zygomatic process of the squamosal and a thick postglenoid process; thick paroccipital; posterior process of the periotic elongate; large and swollen anterior process of the periotic, bending medially almost at right angles to the body of the periotic. Differs from *Tiphyocetus* in being larger; having a lambdoid crest that rises more abruptly; braincase swelling out more noticeably above the level of the anterior process of the periotic. (Kellogg, 1931).

Occipital condyle breadth: unknown.

Included species: *P. vexillifer* only, known from locality WM13A* only.

Comments: This species is now represented by several additional specimens in UCMP and LACM, and it appears to be one of the more abundant mysticetes in the Sharktooth Hill Local Fauna.

***Thinocetus* Kellogg, 1969**

Type species: *Thinocetus arthritus* Kellogg 1969.

Type specimen: USNM 23794, posterior portion of a skull with periotics and incomplete left and right tympanic bullae, and partial skeleton.

Characteristics: Atlas massive, thick; short rugose, transversely widened hypophyseal process; odontoid process of axis slender, unusually elongated; pedicles of neural arch of third to sixth cervicals short, widened; neural canal low, unusually wide; roof of neural canal slightly arched, almost horizontal; neural spines of first to fourth dorsals short, nearly vertical, not noticeably slanting backward, their height equivalent to less than half vertical diameter of corresponding vertebra; ninth to twelfth dorsals with somewhat longer backward slanting neural spines, height equivalent to or slightly more than half vertical diameter of corresponding vertebra; neural canals of first to eighth dorsals low, very wide; roof of neural canal not arched, nearly horizontal; diapophyses stout, broad, dorsoventrally compressed; elongated, backward slanting neural spines present on first to ninth lumbars; metapophyses rather slender; transverse process of first to fourth caudals short and broad, not widened distally; scapula wide, acromion broad, coracoid process short and stout, prescapular fossa very narrow; length of humerus approximately 67% length of radius; posterior process of periotic elongated, rather slender, increasing in diameter distally; anterior process compressed from side to side, deeply concave internally, convex externally; posterior profile from a ventral view deeply indented between pars cochlearis and posterior process (Kellogg, 1969).

Occipital condyle breadth: 176 mm (Kellogg, 1969, p. 4).

Included species: *T. arthritis* only (known from localities NC4A*, B).

***Tiphyocetus* Kellogg, 1931**

Type species: *Tiphyocetus temblorensis* Kellogg, 1931.

Type specimen: CAS 4355, partial skull including an incomplete braincase, portions of the rostrum, and right tympanic bulla and periotic are attached to the skull (Kellogg, 1931).

Characteristics: Medium-size mysticete; apex of supraoccipital extending anteriorly slightly beyond the anterior extremities of the zygomatic process of the squamosals; sagittal crest at apex of occipital shield; zygomatic process of squamosal not inflated, with narrow and tapered anterior end; transversely broad postglenoid process; wide temporal fossa; shortened anteroposterior diameter of squamoso-exoccipital region; laterally placed foramen pseudo-ovale; reduced exposure of alisphenoid in lateral wall of braincase; frontals only narrowly exposed in interorbital region between anterior extremities of parietals and posterior extremities of nasals, premaxillae and maxillae; anterior temporal crests weak or absent (modified from Kellogg, 1931).

Occipital condyle breadth: 161.5 mm (CAS 4357, Kellogg, 1931, p. 326).

Included species: *T. temblorensis* only, known from locality WM13A* only.

Tiphyocetus sp. is also known from locality (CC52) (Barnes, 1998).

Comments: Kellogg (1931) stated that *Tiphyocetus* was "intermediate" between *Plesiocetopsis occidentalis* and "*Idiocetus*" *longifrons*. This species appears to be similar to *Diorocetus hiatus*. The holotype was for many years incorrectly numbered as CAS 4370, but on 8 April 1977, the number was changed to CAS 4355, the number under which it was first described by Kellogg (1931). The right tympanic bulla has been broken off and reattached with plaster. As a result, the periotic in the holotype skull is not visible, and it is not possible to confirm that the periotics that Kellogg (1931) referred to this species, CAS 4353, 4354, and 4356, are in fact the same species.

BALAENIDAE GRAY, 1868

The slow-moving, skim-feeding right whales, family Balaenidae, are of uncertain origin. Extant right whales are represented by one living species of *Balaena* and two or three species of *Eubalaena*, with exact number still debated (Rice, 1998; Rosenbaum *et al.*, 2000; Perrin and Brownell, 2001). The living right whales (species of *Balaena* and *Eubalaena*, plus all descendants of their most recent common ancestor) form the crown-Balaenidae, and most or all fossil right whales form the stem-Balaenidae. The oldest described right whale, *Morenocetus* from Patagonia, Argentina, is approximately 20 to 22 million years old (Cabrera, 1926). Bisconti (2000), who produced the most detailed cladistic analysis of fossil balaenids to date, indicated that *Morenocetus* could lie within the crown-Balaenidae, namely in the clade of *Balaena* + *Eubalaena*, although exact position is equivocal. If correctly placed, this fossil occurrence would make the crown-Balaenidae the longest surviving family of the Mysticeti. The partly preserved skull of *Morenocetus* indicates that, like other balaenids, the rostrum was probably narrow and arched. This suggests that it had long baleen, and that it fed in the specialized skim-feeding manner, as do the extant right whales and bowheads (Pivorunas, 1979). A notable older occurrence of putative balaenid is a late Oligocene specimen from New Zealand (Fordyce, 2002b).

McLeod, Whitmore, and Barnes (1993) and Bisconti (2000, 2003) reviewed the evolutionary history of the family Balaenidae. The origin of the family is unknown. Apart from the New Zealand specimen (Fordyce, 2002b), none of the other known baleen-bearing Oligocene Mysticeti has any particularly close similarities with balaenids. Subsequent to the early Miocene occurrence of *Morenocetus*, the Miocene record of the family is almost non-existent, represented mainly by fragmentary fossils. The Pliocene and Pleistocene fossil record of balaenids is, in contrast, much better and includes some well-preserved partial skeletons from Europe and North America, and several named extinct genera and species.

Balaena Linnaeus, 1758

Type species: *Balaena mysticetus* Linnaeus, 1758, the extant bowhead whale.

Type specimen: None designated (see Hershkovitz, 1966).

Characteristics: Large balaenids with an exceedingly narrow rostrum that is arched dorsally in a uniform curve; nasal bones relatively very small and rectangular in dorsal view; supraorbital process of the squamosal extremely elongate, narrow, angled posterolaterally at an extreme angle; supraoccipital with a large, rounded notch emarginating its lateral border; posterolateral corner of cranium projecting posterolaterally with very large exoccipitals and squamosals; dentary bowed far laterally, very much tapered anteriorly, and with extreme torsion in its anterior part.

Occipital condyle breadth: 400 mm. (*Balaena mysticetus*).

Included species: *Balaena ricei* only (including *Balaena* close to *B. mysticetus* of [Whitmore 1994]), known from locality EM8F* only (Westgate and Whitmore, 2002).

Comments: Whitmore (1994) mentioned the nearly complete skeleton, from the Atlantic Coastal Plain of the United States, of a large early Pliocene balaenid that is closely related to *B. mysticetus*. Subsequently, Westgate and Whitmore (2002) put this specimen in a new species, *B. ricei*, thus providing an early record for crown-Balaenidae. *Balaena* spp. of Dorr and Eschman (1971) and Holman (1995), reported from Michigan, have been determined to be of historical (less than 1000 years) age and unlikely to have deposited naturally (Harington, 1988).

The genus *Balaena* is also known from the Pliocene and Pleistocene of Europe, possibly from the Pleistocene of Asia, the Pleistocene of South America, and the Recent of the Indian, Arctic, Atlantic, and Pacific Oceans (McKenna and Bell, 1997).

Balaenotus Van Beneden, 1872

Type species: *Balaenotus insignis* Van Beneden, 1872.

Type specimen: Composite materials described by Van Beneden (1880) as follows: Institut Royal de Sciences Naturelle de Belgique (IRSN) CtM 832, periotic; CtM 833a-b, periotics; CtM 834, tympanic bulla; CtM 835, tympanic bulla; CtM 836a-c, partial frontal, premaxilla, and maxilla; CtM 837, left squamosal; CtM 838, right supraorbital process of frontal; CtM 840, CtM 841a-b, CtM 842a-b, CtM 843, CtM 844a-b, several blocks containing vertebrae of different individuals; CtM 850, partial scapula; CtM 856, proximal part of left dentary.

Characteristics: Narrow supraorbital process of frontal; lacking torsion in anterior part of dentary fusion between atlas and other cervical vertebrae (Bisconti, 2003).

Occipital condyle breadth: unknown.

Included species: *Balaenotus* sp. only, known from locality EM8C only (Baum and Wheeler, 1977).

Comments: The type material of *Balaenotus* is from the Scaldisian, middle Pliocene (De Meuter and Laga, 1976), near Antwerp, Belgium. This genus was considered distinct by Abel (1941), who noted that the seventh cervical vertebra remains distinct, unfused to the sixth cervical, until old age. Simpson (1945), also recognized it as being distinguished by a differently shaped neural canal and the peculiar shape of the thoracic and lumbar vertebrae. The distinctiveness of *Balaenotus* was challenged by Flower and Lydekker (1891), who designated it as a subgenus of *Balaena*. The genus was reviewed by Van Beneden (1872), by Abel (1938, 1941), and by Bisconti (2003), who regarded it as distinct but represented by only the type species, *Balaenotus insignis* Van Beneden, 1872.

Balaenula Van Beneden, 1872

Type species: *Balaenula balaenopsis* Van Beneden, 1872.

Type specimen: IRSNB Ct.M.858a-b, 853d, 859, 860, 861, 862, 863, 865a-b, 867a-c, 868, 869a-b.

Characteristics: Diagnosis of genus *Balaenula* based on the observation that the holotype belongs to a little adult individual and it has proportionally very large and long nasals (Bisconti, 2000).

Occipital condyle breadth: unknown.

Included species: *Balenula* sp. only (known from localities WM21B [Barnes *et al.*, 1981], WM23II [Domning, 1978], EM8B [Whitmore, 1994]).

Comments: *Balenula* is also known from the late Miocene and Pliocene of Europe (McKenna and Bell, 1997) (could refer to *Balaenotus*).

BALAENOPTERIDAE GRAY, 1868

Balaenopterid-like mysticetes are first known in the late Miocene and survive to the present time as the most diverse living group of Mysticeti. The crown-group Balaenopteridae contains living species of *Balaenoptera* and *Megaptera*, and all descendants of their most recent common ancestor. Relationships of balaenopterids, living and fossil, are volatile. Molecular studies (e.g., Rychel, Reeder, and Berta, 2004; Sasaki *et al.*, 2005) suggest that the extant gray whale *Eschrichtius robustus* (nominally Eschrichtiidae) and extant humpback whale *Megaptera novaeangliae* (nominally Balaenopteridae: Megapterinae) nest within crown-*Balaenoptera*, making the rorquals (Balaenopteridae of traditional use) paraphyletic. Balaenopterids are often arranged in several subfamilies, which are not used here because of disagreement about the monophyly of these groups, and disagreement about which genera are crown- or stem-Balaenopteridae. Fossil genera attributed to the Balaenopteridae, and here considered as possible stem-Balaenopteridae, include *Plesiocetus* (generic concept varying with different authors), *Parabalaenoptera* Zeigler, Chan, and Barnes, 1997, and *Eobalaenoptera* Dooley, Fraser, and Luo, 2004.

Characteristics: A distinctive skull structure and other anatomical adaptations allowing both gulp-feeding and near-surface skimming (Lambertson, 1983; Bouetel, 2005); frontal bone above eye abruptly depressed from vertex to form origin for large muscles that assist in closing the lower jaw; deep transverse sulcus between rostral portion of maxilla and supraorbital process of the frontal; cranial vertex; elevated relative to supraorbital process of the frontal; more-specialized taxa with tapered, somewhat twisted horizontal ramus of the dentary which has a large, spherical mandibular condyle.

Older literature emphasizes two key differences from the Cetotheriidae: the abruptly depressed supraorbital processes of the frontals, and the lack of major exposure of the parietals medially at the apex of the skull posterior to the nasal bones. Other differences will become apparent as the cladistic identity of the Cetotheriidae is clarified.

Some balaenopterids, such as the blue and fin whales and their extinct fossil relatives, became very large. Balaenopteridae sensu lato became more abundant in the fossil record as other more archaic baleen-bearing mysticetes became less abundant in latest Miocene and earliest Pliocene time, approximately 5 to 9 Ma.

Balaenoptera Lacépède, 1804 (including *Burtinopsis*; *Balaena*, in part)

Type species: *Balaenoptera physalus*, the extant fin whale (*Balaenoptera gibbar* Lacépède, 1804, which is a junior synonym of *Balaena physalus* of Linnaeus [1758]; see Hershkovitz [1966]).

Type specimen: None designated (see Hershkovitz, 1966).

Characteristics: Supraorbital process of frontal on a lower level than cranial vertex; maxilla extending posteriorly in a narrow ascending process on lateral side of nasal; proximal end of premaxilla narrow, scarcely reaching superior surface, ending opposite distal end of nasals or as a narrow slip of bone on each side of them; temporal ridge absent; parietals barely exposed on cranial vertex; frontals very short on side; alveolar groove and dental canal of mandible as in *Plesiocetus*; proximal end of mandible solid; orifice of mandibular canal small; alveolar groove roofed (modified from True, 1912b).

Occipital condyle breadth: 300 cm (*Balaenoptera physalus*).

Included species: *B. acutorostrata* (known from locality EM8B [Whitmore, 1994]); *B. borealina* (locality EM8B [Whitmore, 1994]); *B.? cephalus* (locality? NC3AA* [Kellogg, 1968b]); *B. davidsonii* (locality WM26* [Déméré, 1986]); *B. floridana* (locality GC13B* [Morgan, 1994]); *B. ryani* (locality WM17F* [Hanna and McLellan, 1924]).

Balaenoptera sp. is also known from localities WM30A (Barnes, 1991, 1998), GC13B (Morgan, 1994).

Comments: Kellogg (1968b) indicated that Cope's (1868a) species *Eschrichtius cephalus* probably belonged in the genus *Balaenoptera*, but that his species *Eschrichtius pusillus* (Cope, 1868b) should be ignored because the type specimen is non-diagnostic. The various fossils attributed to this genus from throughout the world include some that are probably not diagnostic at the species level and others that are well preserved and quite likely highly diagnostic. The problem stems from the fact that no review has been made of the group. Déméré, Berta, and McGowen (2005) stated that the small fossil balaenopterid *B. floridana*, from Florida, might be conspecific with *B. cuvieri* from Europe and that they belong in a new genus, but they did not provide a generic name.

Balaenoptera sp. of Dorr and Eschman (1971) and Holman (1995) reported from Michigan have been determined to be of historical (less than 1000 years) age and unlikely to have naturally deposited (Harington, 1988).

Déméré (1986) and Déméré, Berta, and McGowen (2005) considered *Burtinopsis* to be a junior synonym of *Balaenoptera* and this opinion is followed here. Déméré, Berta, and McGowen (2005) also concluded that *Balaenoptera ryani* belongs in a new genus, but they do not provide a new generic name for the species.

The genus *Balaenoptera* is also known from the Miocene to Pliocene (possibly the Pleistocene) of Europe, the middle Miocene to Pleistocene of Asia, the late Miocene to Pleistocene of South America, and the Recent of the Mediterreanean Sea and the Indian, Arctic, Atlantic, and Pacific Oceans (McKenna and Bell, 1997).

Eobalaenoptera **Dooley, Fraser, and Luo, 2004**

Type species: *Eobalenoptera harrisoni* Dooley, Fraser, and Luo, 2004.

Type specimen: VMNH 742; partial skeleton comprising skull fragments, including most of the supraoccipital, with the basioccipital, occipital condyles, and portions of the squamosals, portions of both periotics, a portion of the right tympanic bulla, the first 28 vertebrae in articulation, part of the glenoid region of one scapula, both forelimbs proximal to the phalanges, and numerous rib fragments (Dooley, Fraser, and Luo, 2004).

Characteristics: Differs from all known "cetotheres" in having the pars cochlearis transversely elongate with a tubular internal auditory meatus; greater petrosal nerve foramen on tympanic side of pars cochlearis; stylomastoid fossa extending onto posterior process, no medial groove on pars cochlearis; long lateral projection of anterior process, directed anterolaterally; cervical vertebrae contributing less than 11% of length of precaudal vertebral column; capitulum and tuberculum on first rib joined by bony lamina. Differs from all known Balaenopteridae and Eschrichtiidae in having a massive posterior process of the periotic, which is elongate and tapering distally; dorsoventrally deep tegmen tympani; lateral process projection of anterior process of periotic directed anterolaterally; tympanic bulla with a swollen dorsal posterior prominence and a pronounced keel (Dooley, Fraser, and Luo, 2004).

Occipital condyle breadth: 169 mm (VMNH 742; Dooley, Fraser, and Luo, 2004).

Included species: *E. harrisoni* only, known from locality NC3D* only.

Comments: Dooley, Fraser, and Luo (2004) noted that they only tentatively place *Eobalenoptera* in the Balaenopteridae until more comprehensive cladistic analyses can be performed.

Megaptera **Gray, 1846**

Type species: *Megaptera novaeangliae* (including *Megaptera longipinna* [sic] Gray, 1846), the extant humpback whale.

Type specimen: None designated (see Hershkovitz, 1966).

Characteristics: Head broad, moderate in size, flattened; throat and chest with deep longitudinal folds; dorsal fins low or tuberous, rather behind the middle of the body; pectoral fin very large, one-third to one-fifth the entire length of the animal, as long as the head, with only four fingers; eyes above the angle of the mouth; navel before the front edge, the male organs under back edge of dorsal fin, with vent nearer the tail; female organs behind back edge of dorsal fin, with vent at its hinder end; nose narrow, broad behind, and contracted in front; temporal bone broad; interorbital space wide; lower jaw much arched (Gray, 1846).

Occipital condyle breadth: 400 mm (*Megaptera novaeangliae*).

Included species: *M. miocaena* only, known from locality WM17B* only (Kellogg, 1922).

Megaptera sp. is also known from locality EM8B (Whitmore, 1994).

Comments: *Megaptera novaeangliae* is a lapsus for *Balaena longimana*, both of which are junior synonyms of *Balaena novaeangliae* Borowski 1781 (see Hershkovitz, 1966). Gray (1846) listed four species of *Megaptera* (*M. poeskop, M. longimana, M. americana*, and *M. antarctica*), but only a single living species (*M. novaeangliae*) is currently recognized. Dooley, Fraser, and Luo (2004) and Déméré, Berta, and McGowen (2005) noted that *Megaptera miocaena* does not belong in the genus *Megaptera*, but neither provides a new genus for the species.

The genus *Megaptera* is also known from the Pliocene of Europe, the Pleistocene of South America (possibly also from the Pliocene) and Asia, and the Recent of the Mediterranean Sea and the Indian, Arctic, Atlantic, and Pacific Oceans (McKenna and Bell, 1997).

Parabalaenoptera **Zeigler, Chan, and Barnes, 1997**

Type species: *Parabalaenoptera baulinensis* Zeigler, Chan, and Barnes, 1997.

Type specimen: CASG 66660, skull and mandible with periotics and bullae and a partial skeleton.

Characteristics: Differs from other balaenopterids in having a transversely narrower intertemporal region; elongate nasal bones (length to width ratio near 4:1) that are narrow, nearly parallel sided, terminating anterior to level of antorbital notch, distally wide; ascending process of maxilla tapering distally.

Occipital condyle breadth: 198 mm (Zeigler, Chan, and Barnes, 1997, p. 118).

Included species: *P. baulinensis* only, known from locality WM18II* only.

Comments: Zeigler, Chan, and Barnes (1997) erected the new subfamily Parabalaenopterinae along with existing subfamilies Megapterinae and Balaenopterinae within the Balaenopteridae to emphasize the morphological differences between *Parabalaenoptera* and other fossil and living balaenopterids, including the elongate and narrow nasals, narrow intertemporal region, long and posteriorly tapering ascending process of the maxillae, conspicuously elevated or swollen anterior portions of the parietals, short postglenoid process, and long and sloping coronoid crest of the dentary. The broader relationships of *Parabalaenoptera* need to be revisited by computer-aided cladistic analysis, and in the light of crown/stem concepts and the possible paraphyly of Balaenopterinae.

Plesiocetus **Van Beneden, 1859**

Type species: *Plesiocetus garopii* [Van Beneden 1859] (fixed as the type by Van Beneden, 1872; later Kellogg [1925] fixed the type species of *Plesiocetus* as *Plesiocetus hupschii* Van Beneden, 1859).

Type specimen: None designated.

Characteristics: Differs from other mysticetes by free, proportionally thick vertebrae; scapula with rudimentary coracoid process, well-developed acromion, located very high and in an oblique direction; bones of the skull indicate a more robust head (Van Beneden, 1859).

Occipital condyle breadth: unknown.

Included species: *Plesiocetus* sp. (only known from localities WM20, WM26, [WM30A] [Barnes 1973, 1977, 1991, 1998]).

Comments: *Plesiocetus* is also known from the Miocene and Pliocene of Europe, and the Miocene and/or early Pliocene of South America (McKenna and Bell, 1997).

ESCHRICHTIIDAE ELLERMAN AND MORRISON-SCOTT, 1951

The gray whales are distinctive in their body form and in skeletal anatomy. The only named genus in the family, *Eschrichtius*, is characterized by a relatively narrow and dorsally arched rostrum with the large nares located at the apex of the curved snout, two large tuberosities for muscle attachment located on the braincase near the nuchal crest, and by a massive mandible that has a low coronoid process and a large, rounded mandibular condyle.

The fossil record of this family is very poor. A reputed gray whale fossil of Pliocene age has been reported from Japan, but the skull is not preserved, so confirmation of the family assignment is not certain. Unfortunately, the poor fossil record provides no certain indication of the origin of the Eschrichtiidae. Morphological studies place the family apparently closest to the Balaenopteridae (Barnes and McLeod, 1984), while molecular studies nest *Eschrichtius robusta* amongst species of Balaenoptera (e.g., Rychel, Reeder, and Berta, 2004; Sasaki *et al.*, 2005).

Eschrichtius Gray, 1864 (including *Balaenoptera*, in part)

Type species: *Eschrichtius robusta* (Lilljeborg, 1861), the extant gray whale (originally described as *Balaenoptera robusta*).

Type specimen: Subfossil skeleton in the University Museum of Uppsala (after Barnes and McLeod, 1984, following Gray, 1864).

Characteristics: Supraorbital process of frontal abruptly depressed at base to a level noticeably below that of dorsal surface of interorbital region; parietals not coming in contact, or nearly contacting, on vertex between occipital shield and frontal; nasals greatly enlarged, their combined dorsal area equal to more than half that of supraorbital portion of frontal; rostrum tending toward depth rather than breadth; mandible heavy, slightly bowed outward (Miller, 1923).

Occipital condyle breadth: 296 mm (average of six adult specimens: USNM 13803, 364969, 364970, 364976, 364971, 364979).

Included species: There are no Tertiary localities in North America that have produced fossils of the genus *Eschrichtius* (Barnes and McLeod, 1984).

Comments: There are many Tertiary fossils from North American that have at one time or another been placed in the genus *Eschrichtius*. These include *E. cephalus* and *E. davidsonii*, which are now both placed in the genus *Balaenoptera* (see above); as well as *E. expansus, E. leptocentrus, E. mysticetoides, E. polyporus, E. priscus*, and *E. pusillus*. All of these others are considered to be either nomina dubia or nomina vana. See Barnes and MacLeod (1984 and references therein) for a complete review.

The genus *Eschrichtius* is also known from the Pleistocene to Recent of North America, and from the Recent of the Atlantic and North Pacific Oceans (McKenna and Bell, 1997).

MYSTICETI NOMINA DUBIA

Many genera and species of mysticete cetacean have been named based on very incomplete material. Often, these type specimens are inadequate for comparisons with other named species, genera, and families, and some cannot confidently be assigned to suborder. Occasionally, later authors have assigned much better specimens to taxa based on poor type specimens, and the newly referred material has become the standard of reference for subsequent work rather than the type specimens. For others, the names have been around since their inception, but little or nothing has been referred to them because of their poor quality. These taxa should best be considered nomina dubia, and the names restricted to the type specimen.

Mesoteras Cope, 1870a

Type species: *Mesoteras kerrianus* Cope, 1870a.

Type specimen: USNM 1633, skull.

Characteristics: Orbital process of frontal narrowed, exceedingly thick and massive at the extremity; posterior lumbars and anterior caudals with short anteroposterior diameter; premaxillary and maxillary bones depressed, the latter thin, horizontal, narrow; otic bulla compressed (Cope, 1870a).

Occipital condyle breadth: not available from holotype specimen.

Included species: *Mesoteras kerrianus* only, known from locality EM8E* only (Cope 1870b).

Comments: Cope (1895) further mentioned this taxon, and a tympanic bulla. Kellogg (1931) provided data on this species. Winge (1910) and True (1912a) doubted that the genus is distinct, but Simpson (1945) listed it as a distinct genus within the family Balaenopteridae.

Rhegnopsis Cope, 1896 (including *Balaena*, in part; *Protobalena*, in part)

Type species: *Rhegnopsis palaeatlantica* (Leidy, 1851) (originally described as *Balaena palaeatlantica*).

Type specimen: ANSP 12919, thoracic vertebra and jaw fragment (Gillette, 1975).

Characteristics: Meckelian fissure present, extending deeply into the mandibular ramus (Cope, 1896).

Occipital condyle breadth: unknown

Included species: *R. palaeatlantica* only, known from locality ?EM8G* only.

Comments: Leidy (1851) originally coined the name *Balaena palaeatlantica* for this animal, which he later (Leidy, 1869) changed to *Protobalaena*, because he thought it was generically distinct from *Balaena*. Cope (1896) pointed out that the genus *Protobalaena* was preoccupied by an animal named by Van Beneden (1867) and coined the name *Rhegnopsis* (see also Kellogg, 1931).

Siphonocetus Cope, 1895 (including *Balaena*, in part)

Type species: *Siphonocetus priscus* (Leidy, 1851) (originally described as *Balaena prisca*).

Type specimen: ANSP 12915, jaw fragment (Gillette, 1975).

Characteristics: Alveolar groove roofed over and perforate (Cope, 1895)

Occipital condyle breadth: unknown

Included species: *S. priscus* (known from locality NC5* [Gillette, 1975]); *S. expansus* (locality uncertain); *S. clarkianus* (locality uncertain).

Siphonocetus sp. is also known from locality NC7 (Baum and Wheeler, 1977).

Comments: The type specimens of all of the species of *Siphonocetus*, including the type species, are mandibular fragments. Kellogg (1968b) considered the type specimen of *Siphoncetus priscus* to be non-diagnostic. Cope (1868b) named a species *Megaptera expansa*, which he later moved to *Eschrichtius* (*E. expansus* [Cope, 1869]), then to *Cetotherium* (*C. expansus* [Cope 1890]), and finally to *Siphonocetus* (*S. expansus* [Cope, 1895]). See Barnes and McLeod (1984) for a review. Kellogg (1968b) and Barnes and McLeod (1984) considered the type material of this species to be non-diagnostic, and thus all of these assignments should be ignored.

Tretulias Cope, 1895

Type species: *Tretulias buccatus* Cope, 1895.

Type specimen: USNM 9345, parts of the mandibular ramus.

Characteristics: Dental canal obliterated, and dental groove without osseous roof; gingival canals and foramina present at one side of the alveolar groove (Cope, 1895).

Occipital condyle breadth: unknown.

Included species: *T. buccatus* only, known from locality ?NC3AA* only (Hay, 1902).

Comments: No other species of *Tretulias* have been named. Kellogg (1968b) considered the type specimen of *Tretulias buccatus* to be non-diagnostic.

Ulias Cope, 1895

Type species: *Ulias moratus* Cope, 1895.

Type specimen: USNM 10595, mandible.

Characteristics: Gingivodental canal open throughout most of its length, closed only near its apex; gingival foramina represented by a few orifices on the alveolar border near the distal extremity (Cope, 1895).

Occipital condyle breadth: unknown.

Included species: *U. moratus* only, known from locality ?NC3AA* only (Hay, 1902).

Comments: No other species of *Ulias* have been named. Kellogg (1968b) considered the type specimen of *Ulias moratus* to be non-diagnostic.

BIOLOGY AND EVOLUTIONARY PATTERNS

The evolutionary history of the Mysticeti can be summarized by four major evolutionary events. The first of these was the origination from within the Archaeoceti, now known to have happened during the Eocene. This resulted in the basic adaptations of the "typical" mysticete cranium. Many of these characters, the apomorphies that link all of the Mysticeti, were prerequisites for later development of bulk filter-feeding. These include an enlarged oval-shaped external bony naris located in the middle to posterior part of the rostrum, a lateral process of the maxilla that is closely appressed to the antorbital process of the frontal, a posteriorly extending infraorbital plate of the maxilla ventral to the orbit, and a flexible mandibular symphysis. Many of the primitive features that are possessed by all fossil and Recent Mysticeti are characters that were simply retained from their (as yet unknown) sister taxon within the Archaeoceti: single bony narial opening, large and tabular supraorbital process of the frontal, large temporal fossa, peribullary sinus that extends only a short distance anteriorly into the pterygoid bone, a thick-walled and heart- to oval-shaped tympanic bulla and, in the earliest Mysticeti, heterodonty and an elongate intertemporal region that is capped by a sagittal crest.

The second major evolutionary event among the Mysticeti was the relatively rapid diversification of the tooth-bearing Mysticeti during the late Eocene and the Oligocene. This resulted in the evolution of many family-level clades: Llanocetidae, Aetiocetidae, Mammalodontidae, Balaenidae, Eomysticetidae, Cetotheriiopsidae, perhaps the Cetotheriidae, and perhaps the Balaenopteridae (Figures 35.3 and 35.4). The fossil record is not complete enough to say whether any of these groups really were endemic to particular ocean basins.

The third major evolutionary event among the Mysticeti was the appearance of baleen. Discovery of early Oligocene toothless Mysticeti with flat palates implies that baleen evolved early in mysticete history, and that baleen-bearing mysticetes were contemporaries of virtually all of the presently known tooth-bearing mysticetes. It is most parsimonious to assume that baleen originated only once in cetacean evolution, and this supports the recognition of a clade Chaeomysticeti.

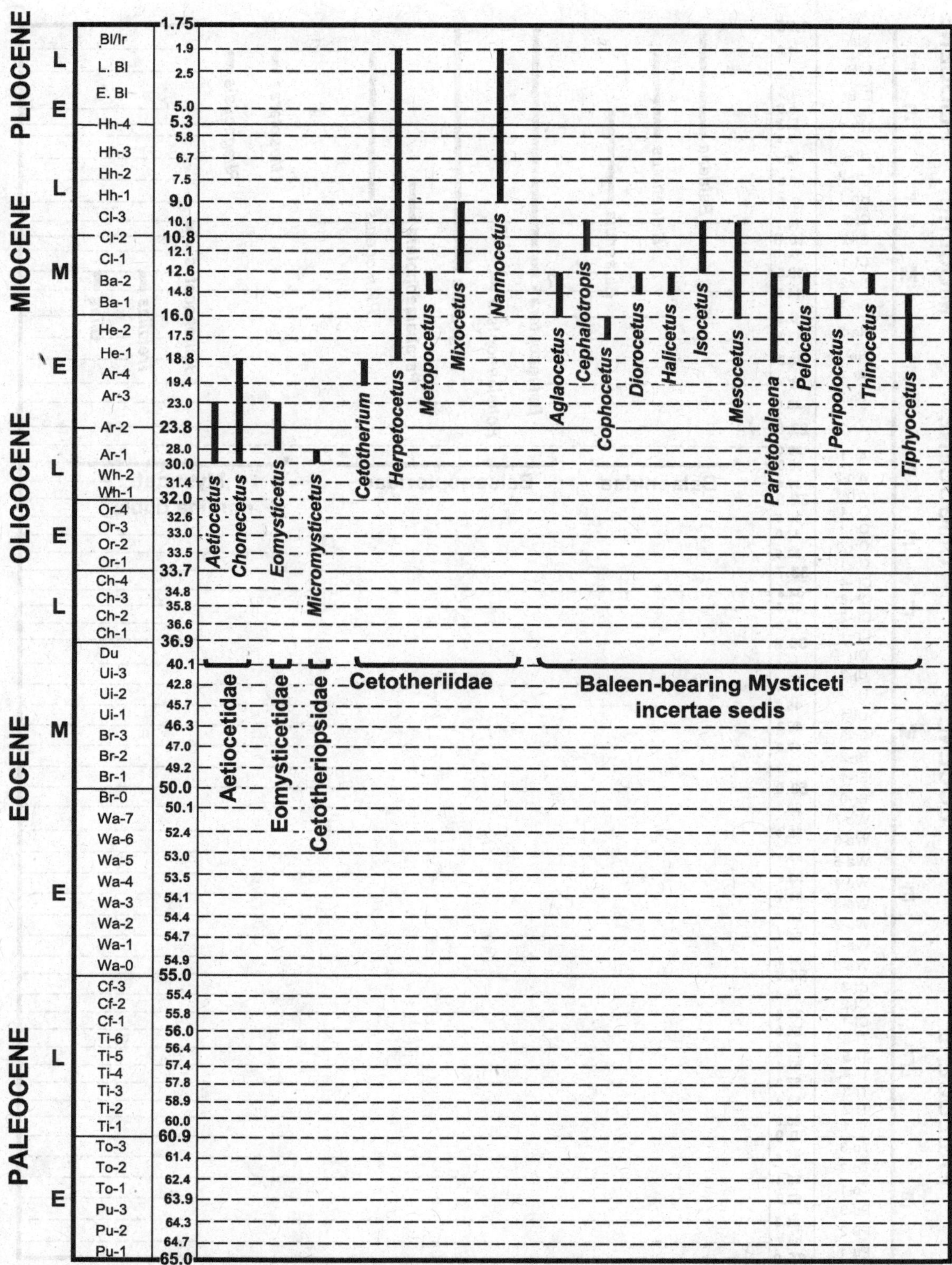

Figure 35.3. Temporal ranges of mysticete genera 1. Aetiocetidae, Eomysticetidae, Cetotheriopsidae, Cetotheriidae, and baleen-bearing Mysticeti incertae sedis.

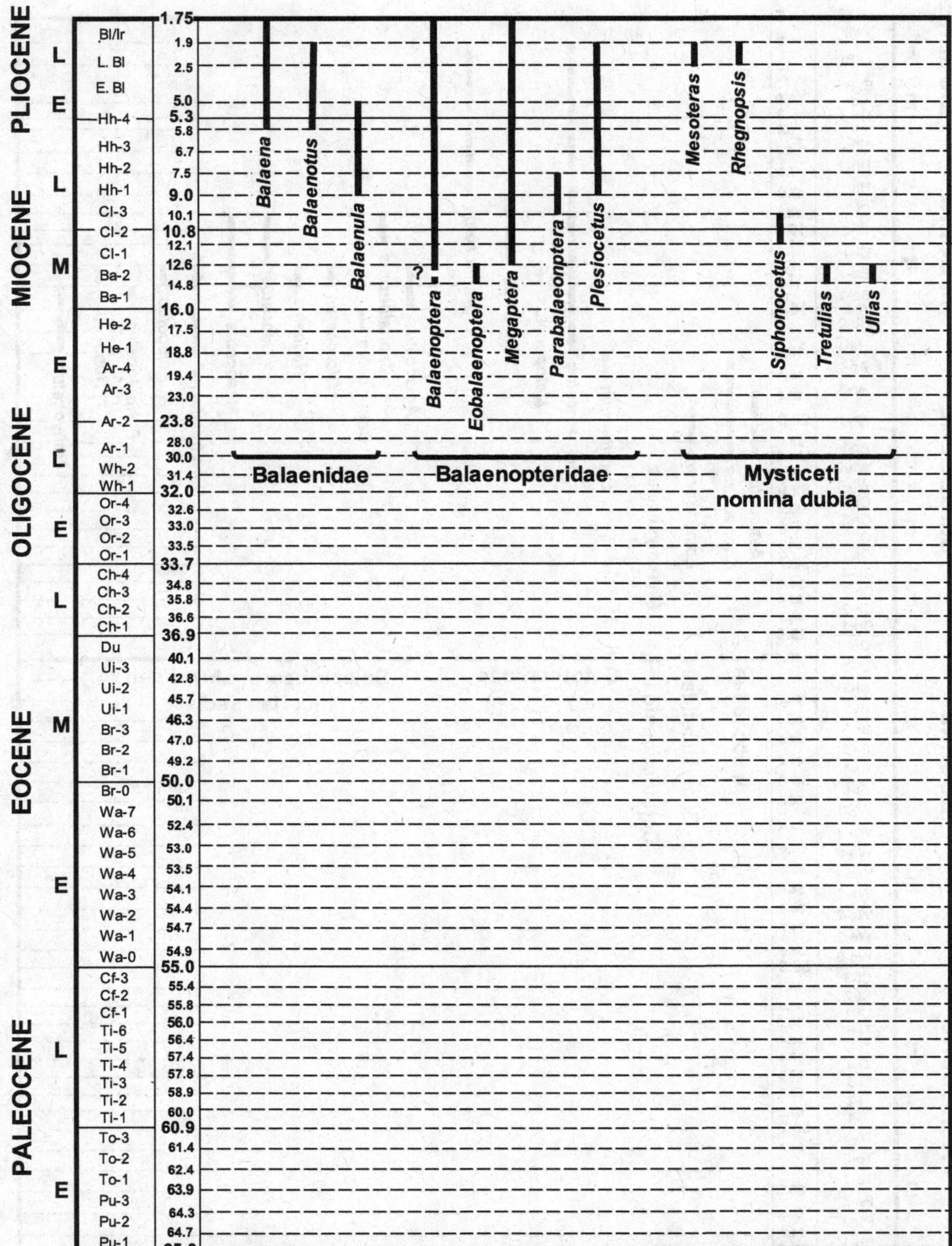

Figure 35.4. Temporal ranges of mysticete genera 2. Balaenidae, Balaenopteridae, and Mysticeti nomina dubia.

The baleen comprises rows of cornified plates, epithelial in origin, that grow downward from the palates of all extant mysticetes. Some rare occurrences of fossilized baleen plates have been reported (Packard, 1947; Barnes, Raschke, and Brown, 1984, 1987; Brand *et al.*, 2004). The extant Dall's porpoise, *Phocoenoides dalli*, an odontocete of the North Pacific, provides a possible model for the evolutionary mechanism for the origin of baleen plates in the Mysticeti. Both the roots and crowns of the teeth of Dall's porpoise are small, indeed they are vestigial, and they are rarely visible on the gums of the animal. The flesh between the teeth forms cornified excrescences that extend apically from the gums between the teeth. These cornified structures in effect act as "false teeth," and they can aid in holding fish. Such cornified flesh might have also formed between the teeth of some groups of tooth-bearing mysticetes. It was mentioned above that the most highly specialized aetiocetids, although not regarded as being closely related to the most archaic baleen-bearing Mysticeti, had small teeth, as would be expected in an ancestor of the baleen-bearing Mysticeti. Reduction in size of the teeth, combined with elaboration of cornified epithelial material between the teeth, might have been the first step in the development of baleen. This could have been followed by formation of the cornified epithelial material into transverse ridges on the palate, and this could have been the inception of the baleen, which grows from the palate in transverse rows. A possibly unanswerable question is which toothed mysticete taxon was the first to have rudimentary baleen plates between its teeth.

The fourth major evolutionary event in mysticete history is the late Oligocene to Recent diversification of the baleen-bearing mysticetes. As the family-level diversity of tooth-bearing mysticetes decreased, that of the baleen-bearing mysticetes increased.

General evolutionary trends among the later Mysticeti include the acquisition of large body size, shortening of the intertemporal region as anterior and posterior parts of the skull "telescope" over and under each other, reduction of the coronoid process of the dentary and mandibular foramen, lateral bowing of the mandibles, and shortening of the neck. Mysticetes are virtually cosmopolitan in their distribution, being found both in the past and now from polar to equatorial latitudes, with fossils known from all ocean basins. Great increase in size among baleen-bearing mysticete whales is a phenomenon of middle Miocene and later time, with the largest known species being among the living species.

REFERENCES

Abel, O. (1938). Vorläufige mitteilungen uber die revision der Fossilen Mystococeten aus dem Tertiär Belgines I. *Bulletin du Musée Royal d'Histoire Naturelle de Belgique*, 14, 1–34.

(1941). Vorläufige mitteilungen uber die revision der Fossilen Mystococeten aus dem Tertiär Belgines V. *Bulletin du Musée Royal d'Histoire Naturelle de Belgique*, 17, 1–29.

Barnes, L. G. (1973). Pliocene cetaceans of the San Diego Formation, San Diego, California. In *Studies on the Geology and Geologic Hazards of the Greater San Diego Area, California*, ed. A. Ross and R. J. Dowlen, pp. 37–42. San Diego, CA: San Diego Association of Geologists.

(1977). Outline of eastern North Pacific fossil cetacean assemblages. *Systematic Zoology*, 25, 321–43.

(1987). *Aetiocetus* and *Chonecetus*, primitive Oligocene toothed mysticetes and the origin of baleen whales. *Journal of Vertebrate Paleontology*, 7 (suppl. to no. 3), p. 10A.

(1989). *Aetiocetus* and *Chonecetus* (Mammalia: Cetacea): primitive Oligocene toothed mysticetes and the origin of baleen whales. In *Proceedings of the Fifth International Theriological Congress*, Vol. 1, p. 479.

(1991). The fossil marine vertebrate fauna of the latest Miocene Almejas formation, Isla Cedros, Baja California, Mexico. *Memorias Universidad Autónoma de Baja California Sur, Primera Reunion Internacional Sobre Geologia de la Peninsula de Baja California*, pp. 147–66.

(1998). The sequence of fossil marine mammal assemblages in Mexico. *Advances en Investigación Special Publication*, 1, 26–79.

Barnes, L. G. and McLeod, S. A. (1984). The fossil record and phyletic relationships of gray whales. In *The Gray Whale:* Eschrichtius robustus, ed. M. L. Jones, S. L. Swartz, and S. Leatherwood, pp. 3–32. London: Academic Press.

Barnes, L. G. and Sanders, A. E. (1996a). The transition from archaeocetes to mysticetes: late Oligocene toothed mysticetes from near Charleston, South Carolina. [*Proceedings of the Sixth North American Paleontological Convention*] *Paleontological Society Special Publication*, 8, p. 24.

(1996b). The transition from Archaeoceti to Mysticeti: late Oligocene toothed mysticetes from South Carolina, USA. *Journal of Vertebrate Paleontology*, 16(suppl. to no. 3), p. 21A.

Barnes, L. G., Hutchinson, J. H., and Welton, B. J. (1981). The vertebrate fossils of the marine Cenozoic San Mateo Formation at Oceanside, California. In *Geologic Investigations of the Coastal Plain, San Diego County, California*, ed. P. L. Abbott and S. O'Dunn, pp. 53–70. San Diego, CA: San Diego Association of Geologists.

Barnes, L. G., Raschke, R. E., and Brown, J. C. (1984). A fossil baleen whale from the Capistrano Formation in Laguna Hills, California. *Memoirs of the Natural History Foundation of Orange County*, 1, 11–18.

(1987). A fossil baleen whale. *Whalewatcher*, 21, 7–10.

Barnes, L. G., Kimura, M., Furusawa, H., and Sawamura, H. (1995). Classification and distribution of Oligocene Aetiocetidae (Mammalia; Cetacea; Mysticeti) from western North America and Japan. *The Island Arc*, 3, 392–431.

Baum, G. R. and Wheeler, W. H. (1977). Cetaceans from the St. Marys and Yorktown Formations, Surry County, Virginia. *Journal of Paleontology*, 51, 492–504.

Bisconti, M. (2000). New description, character analysis and preliminary phyletic assessment of two Balaenidae skulls from the Italian Pliocene. *Palaeontolographia Italica*, 87, 37–66.

(2003). Evolutionary history of Balaenidae. *Cranium (Tijdschrift van de Werkgroep Pleistocene Zoogdieren)*, 20, 9–50.

Borowski, G. H. (1781). *Gemeinnüzzige Naturgeschichte des Thierreichs*, Vol. 2, p. 21. Berlin and Stalsund.

Bouetel, V. (2005). Phylogenetic implications of skull structure and feeding behavior in balaenopterids (Cetacea, Mysticeti). *Journal of Mammalogy*, 86, 139–46.

Brand, L. R., Esperante, R., Chadwick, A. V., Porras, O. P., and Alomía, M. (2004). Fossil whale preservation implies high diatom accumulation rate in the Miocene–Pliocene Pisco Formation of Peru. *Geology*, 32, 165–8.

Brandt, J. F. (1843). De cetotherio, novo balaenarum familiae genre in Rossia Meridionali ante aliquot annos effoso. *Bulletin de La Classe Physico-Mathématique de L'Académie Impérial des Sciences de St. Pétersberg*, 1, 145–8.

(1872). Über eine neue Classification der Bartenwale (Balaenoidea) mit Berücksichtigung der untergegangenen Gattungen derselben.

Bulletin de l'Académie Impériale des Sciences de St. Petersberg, 17, 113–29.

Cabrera, A. (1926). Cétácéos fossiles del Museo de la Plata. *Revista Museo de La Plata* 29, 363–411.

Capellini, G. (1876). Sulle balene fossili toscane. *Atti della Royal Accademia (nazionale) dei Lincei, Rome, Memoire della Classe di Scienze Fisiche, Mathematiche e Naturale, series* 2, 3, 9–14.

(1901). *Balaenottera miocenica* del Monte Titano Repubblica di S. Marino. *Memorie della Royal Accademia delle Scienze dell'Instiuto di Bologna*, 9, 237–60.

Cope, E. D. (1868a). [Extinct Cetacea from the Miocene bed of Maryland.]. *Proceedings of the Academy of Natural Sciences of Philadelphia*, 20, 150–60.

(1868b). Second contribution to the history of the Vertebrata of the Miocene period of the United States. *Proceedings of the Academy of Natural Sciences of Philadelphia*, 20, 184–94.

(1869). Third contribution to the fauna of the Miocene period of the United States. *Proceedings of the Academy of Natural Sciences of Philadelphia*, 11, 6–12.

(1870a). Discovery of a huge whale in North Carolina. *The American Naturalist*, 4, 128.

(1870b). Fourth contribution to the history of the fauna of the Miocene and Eocene periods of the United States. *Proceedings of the American Philosophical Society*, 11, 285–94.

(1890). The Cetacea. *The American Naturalist*, 24, 599–616.

(1891). *Syllabus of Lectures on Geology and Paleontology*, Part 3: *Paleontology of the Vertebrata*. Philadelphia, PA: University of Pennsylvania.

(1895). Fourth contribution to the marine fauna of the Miocene period of the United States. *Proceedings of the American Philosophical Society*, 34, 135–55.

(1896). Sixth contribution to the knowledge of the marine Miocene fauna of North America. *Proceedings of the Philadelphia Society*, 35, 139–46.

Czyzewska, T. and Ryziewicz, Z. (1976). *Pinocetus polonicus* gen. n., sp. n. (Cetacea) from the Miocene limestones of Pinczów, Poland. *Acta Paläeontologia Polonica* 21, 259–91.

De Meuter, F. J. and Laga, P. G. (1976). Lithostratigraphy and biostratigraphy based on benthonic foraminifera of the Neogene deposits of northern Belgium. *Bulletin de la Société Belge de Géologie*, 85, 133–52.

Déméré, T. A. (1986). The fossil whale, *Balaenoptera davidsonii* (Cope 1872), with a review of other Neogene species of *Balaenoptera* (Cetacea: Mysticeti). *Marine Mammal Science*, 2, 277–98.

Déméré, T. A., Berta, A., and McGowen, M. R. (2005). The taxonomic and evolutionary history of fossil and modern balaenopteroid mysticetes. *Journal of Mammalian Evolution*, 12, 99–143.

Domning, D. P. (1978). Sirenian evolution in the North Pacific Ocean. *University of California, Publications in Geological Sciences*, 118, 1–176.

Dooley, A. C., Jr. (1993). The vertebrate fauna of the Calvert Formation (Middle Miocene) at the Caroline Stone Quarry, Caroline County, VA. *Journal of Vertebrate Paleontology*, 13(suppl. to no. 3), p. 33A.

Dooley, A. C., Jr., Fraser, N. C., and Luo, Z. (2004). The earliest known member of the rorqual-gray whale clade (Mammalia, Cetacea). *Journal of Vertebrate Paleontology*, 24, 453–63.

Dorr, J. A., Jr. and Eschman, D. F. (1971). *Geology of Michigan*. Ann Arbor, MI: University of Michigan Press.

Ellerman, J. R. and Morrisson-Scott, T. C. S. (1951). *Checklist of Palaearctic and Indian Mammals: 1758 to 1946*. London: British Museum (Natural History).

Emlong, D. R. (1966). A new archaic cetacean from the Oligocene of northwest Oregon. *Oregon University Museum of Natural History Bulletin*, 3, 1–51.

Flower, W. H. and Lydekker, R. (1891). *An Introduction to the Study of Mammals, Living and Extinct*. London: Adam and Charles Black.

Fordyce, R. E. (1989). Problematic early Oligocene toothed whale (Cetacea, ?Mysticeti) from Waikari, north Canterbury, New Zealand. *New Zealand Journal of Geology and Geophysics*, 32, 395–400.

(1992). Cetacean evolution and Eocene–Oligocene environments. In *Eocene-Oligocene Climatic and Biotic Evolution*, ed. D. R. Prothero and W. A. Berggren, pp. 368–81. Princeton, NJ: Princeton University Press.

(2002a). Neoceti. In *Encyclopedia of Marine Mammals*, ed. W. F. Perrin, B. Würsig, and J. G. M. Thewissen, pp. 787–91. London: Academic Press.

(2002b). Oligocene origins of skim-feeding right whales: a small archaic balaenid from New Zealand. *Journal of Vertebrate Paleontology*, 22(suppl. to no. 3), p. 54A.

(2003). The toothed stem-mysticete *Llanocetus* in the Latest Eocene of the southern Ocean. *Journal of Vertebrate Paleontology*, 23(suppl. to no. 3), p. 50A–1A.

(2004). The transition from Archaeoceti to Neoceti: Oligocene archaeocetes in the Southwest Pacific. *Journal of Vertebrate Paleontology*, 24(suppl. to no. 3), p. 59A.

Fordyce, R. E. and Barnes, L. G. (1994). The evolutionary history of whales and dolphins. *Annual Review of Earth and Planetary Sciences*, 22, 419–55.

Fordyce, R. E. and Muizon, C., de (2001). Evolutionary history of cetaceans: a review. In *Secondary Adaptations of Tetrapods to Life in Water*, ed. J.-M. Mazin and V. de Buffrénil, pp. 169–223. Munich: Friedrich Pfeil.

Geisler, J. H. and Luo, Z. (1996). The petrosal and inner ear of *Herpetocetus* sp. (Mammalia: Cetacea) and their implications for the phylogeny and hearing of archaic mysticetes. *Journal of Paleontology*, 70, 1045–66.

Geisler, J. H. and Sanders, A. E. (2003). Morphological evidence for the phylogeny of Cetacea. *Journal of Mammalian Evolution*, 10, 23–129.

Gillette, D. D. (1975). Catalogue of the type specimens of fossil vertebrates, Academy of Natural Sciences, Philadelphia. Introduction and Part I: marine mammals. *Proceedings of the Academy of Natural Sciences of Philadelphia*, 127, 63–6.

Ginsburg, L. and Janvier, P. (1971). Les mammifères marins des faluns Miocenes de la Touraine et de l'Anjou. *Bulletin du Museum National d'Histoire Naturelle, Sciences de la Terre Série*, 6, 22, 161–95.

Gottfried, M. D., Bohaska, D. J., and Whitmore, F. C., Jr. (1994). Miocene cetaceans of the Chesapeake Group. *Proceedings of the San Diego Society of Natural History*, 29, 229–38.

Gray, J. E. (1846). On the cetaceous animals. In *The Zoology of the Voyage of HMS* Erebus *and* Terror, *under the Command of Capt. Sir J. C. Ross, RN, FRS, During the Years 1839 to 1843*, ed. J. Richardson and J. E. Gray, pp. 13–53. London: E. W. Janson.

(1864). Notes on the whalebone-whales; with a synopsis of the species. *Annals and Magazine of Natural History*, 14, 345–53.

(1868). *Synopsis of the Species of Whales and Dolphins in the Collection of the British Museum*. London: Bernard Quaritch.

Hanna, G. D. and McLellan, M. E. (1924). A new species of whale from the type locality of the Monterey Group. *Proceedings of the California Academy of Sciences*, 13, 237–41.

Harington, C. R. (1988). The late Quaternary development of the Champlain Sea Basin: Geological Association of Canada Special Paper 35. In *Marine Mammals of the Champlain Sea, and the Problem of Whales in Michigan*, ed. N. R. Gadd, pp. 225–40. Ottawa: Geological Association of Canada.

Hay, O. P. (1902). *Bibliography and Catalogue of the Fossil Vertebrata of North America*. Washington, DC: Government Printing Office.

Hershkovitz, P. (1966). Catalog of Living Whales. *Bulletin of the United States National Museum*, 246, 1–259.

Holman, J. A. (1995). *Ancient Life of the Great Lakes Basin*. Ann Arbor, MI: University of Michigan Press.

Karlsen, K. (1962). Development of tooth germs and adjacent structures in the whalebone whale (*Balaenoptera physalus* (L.)). *Hvalrdets Skrifter: Scientific Results of Marine Biological Research*, 45, 1–56.

Kellogg, R. (1922). Description of the skull of *Megaptera miocaena*, a fossil humpback whale from the Miocene diatomaceous earth of Lompoc, California. *Proceedings of the United States National Museum*, 61, 1–18.

(1924). Description of a new genus and species of whalebone whale from the Calvert Cliffs, Maryland. *Proceedings of the United States National Museum*, 63, 1–14.

(1925). Additions to the Tertiary history of the pelagic mammals on the Pacific coast of North America. II. Fossil cetotheres from California. *Carnegie Institution of Washington Publication*, 348, 35–56.

(1929). A new cetothere from Southern California. *Bulletin of the Department of Geological Sciences, University of California Publications*, 18, 449–57.

(1931). Pelagic mammals from the Temblor Formation of the Kern River Region, California. *Proceedings of the California Academy of Sciences*, 19, 217–397.

(1934a). The Patagonian fossil whalebone whale, *Cetotherium moreni* (Lydekker). *Carnegie Institution of Washington Publication*, 447, 65–81.

(1934b). A new cetothere from the Modelo Formation at Los Angeles, California. *Carnegie Institution of Washington Publication*, 447, 83–104.

(1940). Whales, giants of the sea. *National Geographic Magazine*, 67, 35–90.

(1944). Fossil cetaceans from the Florida Tertiary. *Bulletin of the Museum of Comparative Zoology*, 94, 433–71.

(1965). Fossil marine mammals from the Miocene Calvert Formation of Maryland and Virginia. Part 1: A new whalebone whale from the Miocene Calvert Formation. *United States National Museum Bulletin*, 247, 1–45.

(1968a). Fossil marine mammals from the Miocene Calvert Formation of Maryland and Virginia. Part 6: A hitherto unrecognised Calvert cetothere. *United States National Museum Bulletin*, 247, 133–61.

(1968b). Fossil marine mammals from the Miocene Calvert Formation of Maryland and Virginia. Part 5: Miocene Calvert mysticetes described by Cope. *United States National Museum Bulletin*, 247, 103–32.

(1968c). Fossil marine mammals from the Miocene Calvert Formation of Maryland and Virginia. Part 7: A sharp-nosed cetothere from the Miocene Calvert. *United States National Museum Bulletin*, 247, 163–73.

(1968d). Fossil marine mammals from the Miocene Calvert Formation of Maryland and Virginia. Part 8: Supplement to description of *Parietobalaena palmeri. United States National Museum Bulletin*, 247, 175–95.

(1969). Cetothere skeletons from the Miocene Choptank Formation of Maryland and Virginia. *United States National Museum Bulletin*, 294, 1–40.

Kimura, T. and Ozawa, T. (2002). A new cetothere (Cetacea: Mysticeti) from the early Miocene of Japan. *Journal of Vertebrate Paleontology*, 22, 684–702.

Lacépède, B. G. E. (1804). *Histoire Naturelle des Cetaces*. Paris: Chez Plassan.

Lambertson, R. H. (1983). Internal mechanism of rorqual feeding. *Journal of Mammalogy*, 64, 76–88.

LeDuc, R. G. and Dizon, A. E. (2002). Reconstructing the rorqual phylogeny: with comments on the use of molecular and morphological data for systematic study. In *Molecular and Cell Biology of Marine Mammals*, ed. C. J. Pfieffer, pp. 100–10. Melbourne, FL: Kreiger.

Leidy, J. (1851). [Descriptions of two fossil species of *Balaena, B. palaeatlantica* and *B. prisca*.] *Proceedings of the Academy of Natural Sciences, Philadelphia*, 5, 308–9.

(1869). The extinct mammalian fauna of Dakota and Nebraska, including an account of some allied forms from other localities, together with a synopsis of the mammalian remains of North America. *Journal of the Academy of Natural Sciences, Philadelphia*, 2, 1–472.

Lilljeborg, W. (1861). Hvalben, funna i jorden p Gräsön i Roslagen i Sverige. *Föredrag vid Naturforskaremötet i Kopenhamn*, 1860, 599–616.

Lindow, B. E. K. (2002). Bardehvalernes indbyrdes slægtskabsforhold: en forelebig analyse. [The internal relationships of the baleen whales: a preliminary analysis.] In *Resumé-hæfte. Hvaldag 2002*, ed. B. E. K. Lindow, pp. 12–19. Gram, Denmark: Midtsønderjyllands Museum.

Linnaeus, C. (1758). *Systema Naturae per Regna Tria Naturae, Secundum Classes, Ordines, Genera, Species, cum Characteribus and Differentiis*, 10th edn, *Reformata*, Vol. I. Holmiae: Laurentii Salvii.

Lydekker, R. (1894). Cetacean skulls from Patagonia. *Annales del Museo de la Plata, Paleontología Argentina* II, 1–13.

Martini, E. and Müller, C. (1975). Calcareous nannoplankton from the type Chattian. *Proceedings of the Sixth Congress on Neogene Mediterranean Stratigraphy*, Bratislava, Yugoslavia, 1975, Vol. 1, pp. 37–41.

McKenna, M. C. and Bell, S. K. (1997). *Classification of Mammals Above the Species Level*. New York: Columbia University Press.

McLeod, S. A., Whitmore, F. C., Jr., and Barnes L. G. (1993). Evolutionary relationships and classification. In *The Bowhead Whale*, ed. J. J. Burnes, J. J. Montague, and C. J. Cowles, pp. 45–70. Lawrence, KS: Society for Marine Mammalogy.

Messenger, S. L. and McGuire, J. A. (1998). Morphology, molecules and the phylogenetics of cetaceans. *Systematic Biology*, 47, 90–124.

Milinkovitch, M. C. (1995). Molecular phylogeny of cetaceans prompts revision of morphological transformations. *Trends in Ecology and Evolution*, 10, 328–34.

Miller, G. S., Jr. (1923). The telescoping of the cetacean skull. *Smithsonian Miscellaneous Collections*, 76, 1–71.

Mitchell, E. D. (1989). A new cetacean from the late Eocene La Meseta Formation, Seymour Island, Antarctic Peninsula. *Canadian Journal of Fisheries and Aquatic Science*, 46, 2219–35.

Morgan, G. S. (1994). Miocene and Pliocene marine mammal faunas from the Bone Valley Formation of central Florida. *Proceedings of the San Diego Society of Natural History*, 29, 239–68.

Packard, E. L. (1947). Fossil baleen from the Pliocene of Cape Blanco, Oregon. *Oregon State Monographs, Studies in Geology*, 5, 3–11.

Packard, E. L. and Kellogg, R. (1934). A new cetothere from the Miocene Astoria Formation of Newport, Oregon. *Carnegie Institution of Washington Publication*, 447, 1–62.

Perrin, W. F. and Brownell, R. L. (2001). Update of the list of recognized species of cetaceans. *Journal of Cetacean Research Management*, 3(suppl.), pp. 364–7.

Pilleri, G. (1986). *Beobachtungen an den fossilen Cetaceen des Kaukasus*. Ostermundigen, Switzerland: Hirnanatomisches Institut.

Pivorunas, A. (1979). The feeding mechanisms of baleen whales. *American Scientist*, 67, 432–40.

Rice, D. W. (1998). *Marine Mammals of the World. Systematics and Distribution*. Lawrence, KS: Society for Marine Mammalogy.

Rosenbaum, H. C., Brownell, R. L., Brown, M. W. *et al.* (2000). Worldwide genetic differentiation of *Eubalaena*: questioning the number of right whale species. *Molecular Ecology*, 9, 1793–802.

Roth, F. (1978). *Mesocetus argillarius* sp. n. (Cetacea, Mysticeti) from Upper Miocene of Denmark, with remarks on the lower jaw and the echolocation system in whale phylogeny. *Zoologica Scripta*, 7, 63–79.

Rothausen, K. (1971). *Cetotheriopsis tobieni* n. sp., der erste paläogene Bartenwal (Cetotheriidae, Mysticeti, Mamm.) nördlich des Tethysraumes. *Abhandlungen des Hessischen Landesamtes für Bodenforschung*, 60, 131–47.

Russell, L. S. (1968). A new cetacean from the Oligocene Sooke Formation of Vancouver Island, British Columbia. *Canadian Journal of Earth Sciences*, 5, 929–33.

Rychel, A. L., Reeder, T. W., and Berta, A. (2004). Phylogeny of mysticete whales based on mitochondrial and nuclear data. *Molecular Phylogenetics and Evolution*, 32, 892–901.

Sanders, A. E., and Barnes, L. G. (2002a). Paleontology of the Late Oligocene Ashley and Chandler Bridge Formations of South Carolina. 3: Eomysticetidae, a new family of primitive mysticetes (Mammalia: Cetacea). *Smithsonian Contributions to Paleobiology*, 93, 313–56.

(2002b). Paleontology of the Late Oligocene Ashley and Chandler Bridge Formations of South Carolina. 2: *Micromysticetus rothauseni*, a primitive cetotheriid mysticete (Mammalia, Cetacea). *Smithsonian Contributions to Paleobiology*, 93, 271–93.

Sasaki, T., Nikaido, M., Hamilton, H. *et al.* (2005). Mitochondrial phylogenetics and evolution of mysticete whales. *Systematic Biology*, 54, 77–90.

Simpson, G. G. (1945). The principles of classification and a classification of mammals. *Bulletin of the American Museum of Natural History*, 85, 1–350.

Slijper, E. J. (1936). Die Cetaceen. *Capta Zoologica*, 7, 1–590.

(1962). *Whales*. Ithaca, NY: Cornell University Press.

True, F. W. (1912a). The genera of fossil whalebone whales allied to *Balaenoptera*. *Smithsonian Miscellaneous Collections*, 59, 1–8.

(1912b). On the correlation of North American and European genera of fossil cetaceans. *Proceedings of the Seventh International Zoological Congress*, 7, 779–81.

Uhen, M. D. (2002). Dental morphology (cetacean), evolution of. In *Encyclopedia of Marine Mammals*, ed. W. F. Perrin, B. Würsig, and J. G. M. Thewissen, pp. 316–19. London: Academic Press.

Uhen, M. D. and Gingerich, P. D. (2001). New genus of dorudontine archaeocete (Cetacea) from the middle-to-late Eocene of South Carolina. *Marine Mammal Science*, 17, 1–34.

Van Beneden, P. J. (1859). Rapport de M. Van Beneden. *Bulletins de l'Académie Royal des Sciences, des Lettres et des Beaux-Arts de Belgique*, 8, 123–46.

(1867). *Rapport sur les Collections Paléontologiques de l'Université de Louvain*. Louvain: Université de Louvain.

(1872). Les Balaenidés fossiles d'Anvers. *Journal de Zoologie*, 1, 405–19.

(1880). Les Mysticètes à courts fanons des sables des environs d'Anvers. *Bulletins de l'Académie Royale des Sciences, des Lettres et des Beaux-Arts de Belgique*, 50, 11–26.

Van Beneden, P. J. and Gervais, P. (1880). *Ostéographie des Cétacés Vivants et Fossiles*. Paris: Arthus Bertrand.

Van Valen, L. M. (1968). Monophyly or diphyly in the origin of whales. *Evolution*, 22, 37–41.

Westgate, J. W. and Whitmore, F. C., Jr. (2002). *Balaena ricei*, a new species of bowhead whale from the Yorktown Formation (Pliocene) of Hampton, Virginia. *Smithsonian Contributions to Paleobiology*, 93, 295–312.

Whitmore, F. C., Jr. (1994). Neogene climate change and the emergence of the modern whale fauna of the North Atlantic Ocean. *Proceedings of the San Diego Society of Natural History*, 29, 223–7.

Winge, H. (1910). Om *Plesiocetus* og *Squalodon* fra Danmark. *Videnskabelige Meddelelser Dansk Naturhistorisk Foreng, Copenhagen, Series* 7, 1, 1–38.

(1921). A review of the interrelationships of the Cetacea. *Smithsonian Miscellaneous Collections*, 72, 1–97.

Zeigler, C. V., Chan, G. L., and Barnes, L. G. (1997). A new Late Miocene balaenopterid whale (Cetacea: Mysticeti), *Parabalaenoptera baulinensis*, (new genus and species) from the Santa Cruz mudstone, Point Reyes Peninsula, California. *Proceedings of the California Academy of Sciences*, 50, 115–38.

36 Sirenia

DARYL P. DOMNING
Howard University, Washington, DC, USA

INTRODUCTION

The Sirenia (sea cows: manatees and dugongs) are herbivorous marine mammals distributed worldwide in tropical and subtropical waters. They are members of the Tethytheria–Proboscidea and Desmostylia, which together probably constitute their immediate sister group. Their fossil record begins in the early middle Eocene, and they appear to have originated in the Old World along the shores of the former Tethys Sea (Domning, Morgan, and Ray, 1982). They spread rapidly, however, and had already arrived in the New World before the middle Eocene.

Sirenians are medium-sized to very large, thick-skinned, nearly hairless mammals with fusiform bodies, paddlelike forelimbs, and (except in the most primitive forms) no external vestiges of hindlimbs. They feed primarily on marine angiosperms (seagrasses: Hydrocharitaceae and Potamogetonaceae) in warm, shallow, protected waters. Most species have been marine, but trichechids (manatees) favor brackish water or freshwater and eat a wide variety of aquatic plants (especially true grasses: Gramineae) in addition to seagrasses. One lineage, the hydrodamaline dugongids, progressively adapted to cold North Pacific waters and a diet of marine algae (kelps: Laminariales) and attained very large size.

Sirenians seem to have reached their peak diversity during the Miocene. Four families are presently recognized; two are restricted to the Eocene and two are represented today by a total of two living genera and four species. In North America and the West Indies, they are known from the early middle Eocene to the Recent and from the Atlantic, Gulf, Caribbean, and Pacific coasts. For a comprehensive basis to the literature on fossil and living sirenians, see Domning (1996).

DEFINING FEATURES OF THE ORDER SIRENIA

CRANIAL

The most useful taxonomic characters in this order are structural details of the skull (Figure 36.1). The premaxillae are enlarged to form a more or less downturned rostrum; its palatal surface and the opposing mandibular surface are covered by tough masticating pads. The nasal opening is very large and retracted, and the premaxillae are in contact with the frontals. The nasal bones are reduced in the more derived forms, but may be fused with the frontals or absent. There are no paranasal air sinuses. The supraorbital processes are prominent, but there is no postorbital bar, except incipiently in some *Trichechus manatus*. The parietals are fused with each other and with the supraoccipital to form a more or less thick, massive skullcap. The zygomatic process of the squamosal is thick and vertically broad and, together with the large jugal, forms a robust zygomatic arch. The pterygoid processes are large and stout, except in prorastomids.

The external auditory canal of the squamosal is open ventrally. The tympanic is ring shaped. The auditory ossicles are the largest and most massive of any mammal. The periotic is not sutured to the skull except in *Prorastomus*; its mastoid portion is partially exposed between the squamosal and exoccipital, which however contact each other more ventrally. The infraorbital and mental foramina tend to be large, as they carry nerve and blood supply to the large, fleshy, mobile lips. The mandible is heavy and (except in prorastomids) downturned anteriorly, corresponding to the rostral deflection, and it has a long, deep symphysis (usually fused in adults).

Evolution of Tertiary Mammals of North America, Vol. 2. ed. C. M. Janis, G. F. Gunnell, and M. D. Uhen. Published by Cambridge University Press.

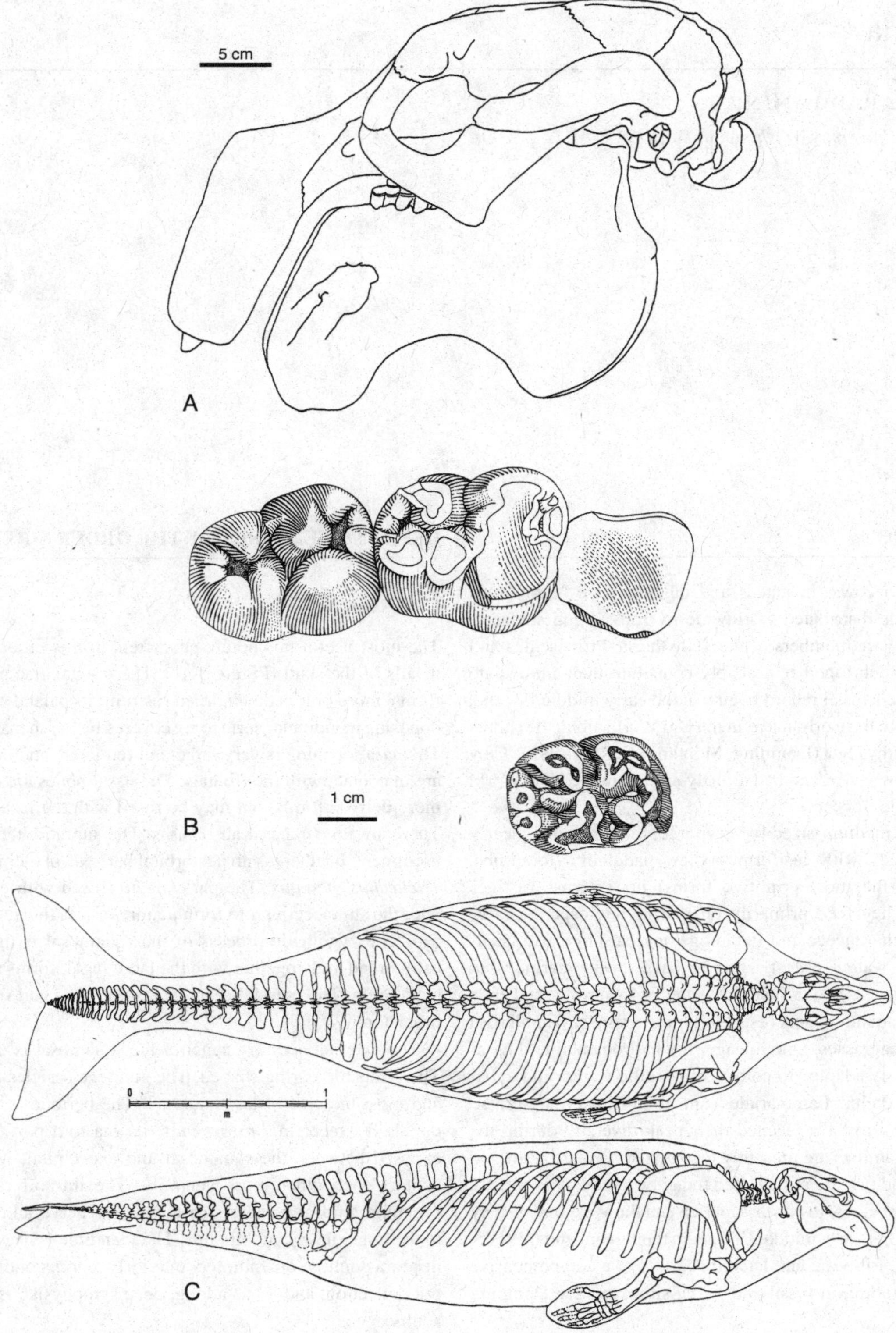

Figure 36.1. A. Lateral view of skull and mandible of *Metaxytherium floridanum*. Length approximately 43 cm. B. Occlusal views of right M1–3 and left m2 of *Dusisiren jordani* (after Domning, 1978). C. Dorsal and lateral views of the skeleton of *Dusisiren jordani*. Total length 4.32 m (after Domning, 1978).

DENTAL

The primitive sirenian dental formula is I3/3, C1/1, P5/5, M3/3 (Domning, Morgan, and Ray, 1982). Where present, the left and right incisor arcades are parallel and aligned longitudinally. Incisors, canines, permanent premolars, and anterior deciduous premolars were for the most part lost by the Miocene, except that many genera retained enlarged first upper incisor tusks. The dp4 is trilobed as in several other ungulate groups – including desmostylians, proboscideans, and artiodactyls. The molariform teeth (DP/dp5 and M/m1–3) in nearly all forms are brachydont and bunobilophodont, and so conservative in cusp pattern that they are of comparatively little taxonomic use. The morphology of the tusk (if any) and the dental formula itself, however, are important.

Upper molars usually bear two rows or groups of three cusps each, together with anterior and posterior cingula; the anterior row of cusps is relatively straight and transverse, the posterior row more or less convex forward. Lower molars typically bear two rows of two cusps each, plus a posterior cingulum or hypoconulid lophule and (sometimes) a small crista obliqua. Accessory cuspules and spurs of cusps are variably present in different taxa.

The three Recent genera are all dentally aberrant: *Hydrodamalis* is completely toothless; *Dugong* has vestigial cheek-tooth crowns, root hypsodonty of M/m2–3, and sexual dimorphism of the tusks (Marsh, 1980); and *Trichechus* has endless horizontal replacement of the cheek teeth by supernumerary molars (Domning and Hayek, 1984).

POSTCRANIAL

The neck is short and the seven cervical vertebrae are strongly compressed (but normally not fused); in *Trichechus* only six cervical are present vertebrae. Ribs are carried by 14 to 21 vertebrae. Beyond these (in dugongids) there are three lumbar, one sacral, and in excess of 20 caudal vertebrae. In the quadrupedal prorastomid *Pezosiren*, there are four lumbar and four sacral vertebrae. In *Trichechus*, the lumbosacral region may be reduced to one vertebra, and a discrete ligamentary attachment of the pelvis to the vertebral column is lost. Except in prorastomids, the anterior caudals support large chevron bones, and the tail bears a horizontally expanded fin and is the main locomotor organ. No dorsal fin is present.

Two to five pairs of ribs are attached to the sternum. A clavicle is absent. The radius and ulna are typically fused. The manus is typically paddlelike. The carpals are serially arranged. The first digit is reduced, the fifth somewhat enlarged and divergent. Hyperphalangy is not normally seen. Pelvic bones are always present, though reduced and (in all but the earliest forms) connected to the vertebral column only by ligaments. In all but the most primitive forms, there is no functional hindlimb, though a vestigial femur is sometimes found. The ribs, neural arches, limb bones, mandibles, and some bones of the skull are osteosclerotic (composed almost entirely of compact bone); the ribs of most forms, and sometimes other bones, are also pachyostotic (swollen in appearance). The ribs are, therefore, highly distinctive and commonly fossilized. The most taxonomically useful postcranial element of post-Eocene sirenians is the humerus.

SYSTEMATICS

SUPRAFAMILIAL

The present concept of the Sirenia evolved from the Linnaean genus *Trichechus*, earlier grouped with the cetaceans but by the 10th edition of the *Systema Naturae* placed within the Bruta immediately after *Elephas* (Linnaeus, 1758). Nonetheless, many early zoologists continued to regard the sea cows as "herbivorous cetaceans," while others tended to associate them with pinnipeds. Illiger (1811) gave them the name Sirenia as a family within his order Natantia, which also included cetaceans. The Sirenia were then raised to ordinal rank by Goldfuss (1820). In the twentieth century, a consensus was reached that they are indeed closely related to the Proboscidea after all. McKenna and Bell (1997) demoted the Sirenia to the status of an infraorder within an order Uranotheria, equivalent to Simpson's (1945) Paenungulata; this usage is not followed here.

By the end of the eighteenth century, all three Recent genera were known to science, and fossil sirenians began to be named in the 1820s. Many taxonomists thereafter placed *Trichechus* (= *Manatus*), *Dugong* (= *Halicore*), and *Hydrodamalis* (= *Rytina*) each in a separate family, and sometimes maintained additional families for certain fossil forms. Hay (1923) applied the subordinal name Trichechiformes to sirenians proper, while including desmostylians within the order as the coordinate suborder Desmostyliformes. The present arrangement took shape in the mid twentieth century, particularly through the contributions of Simpson (1932, 1945), Sickenberg (1934), and Reinhart (1953, 1959). The desmostylians were removed to their own order; *Hydrodamalis* was placed within the Dugongidae; the origin of the Trichechidae, albeit undocumented, was recognized as ancient and probably Eocene; and the Eocene genera *Prorastomus* and *Protosiren* were placed in monotypic (and probably paraphyletic) families. The only suprageneric categories now in general use within the Sirenia are the family and subfamily. A phylogenetic analysis of the order was provided by Domning (1994).

INFRAFAMILIAL

Several subfamilies have been proposed within the Dugongidae. Of those still in use, the paraphyletic Halitheriinae include the earliest and the most conservative dugongids, of which the other subfamilies are evidently offshoots. Within the Halitheriinae, at least some of the European and Mediterranean species of the genera *Eosiren, Halitherium*, and *Metaxytherium* appear to constitute a phyletic series, with representatives in the New World. The Hydrodamalinae (*Dusisiren* and *Hydrodamalis)* are a well-documented lineage derived from *Metaxytherium* and endemic to the North Pacific (Domning, 1978; Muizon and Domning, 1985; Takahashi, Domning, and Saito, 1986; Aranda-Manteca, Domning, and Barnes, 1994;

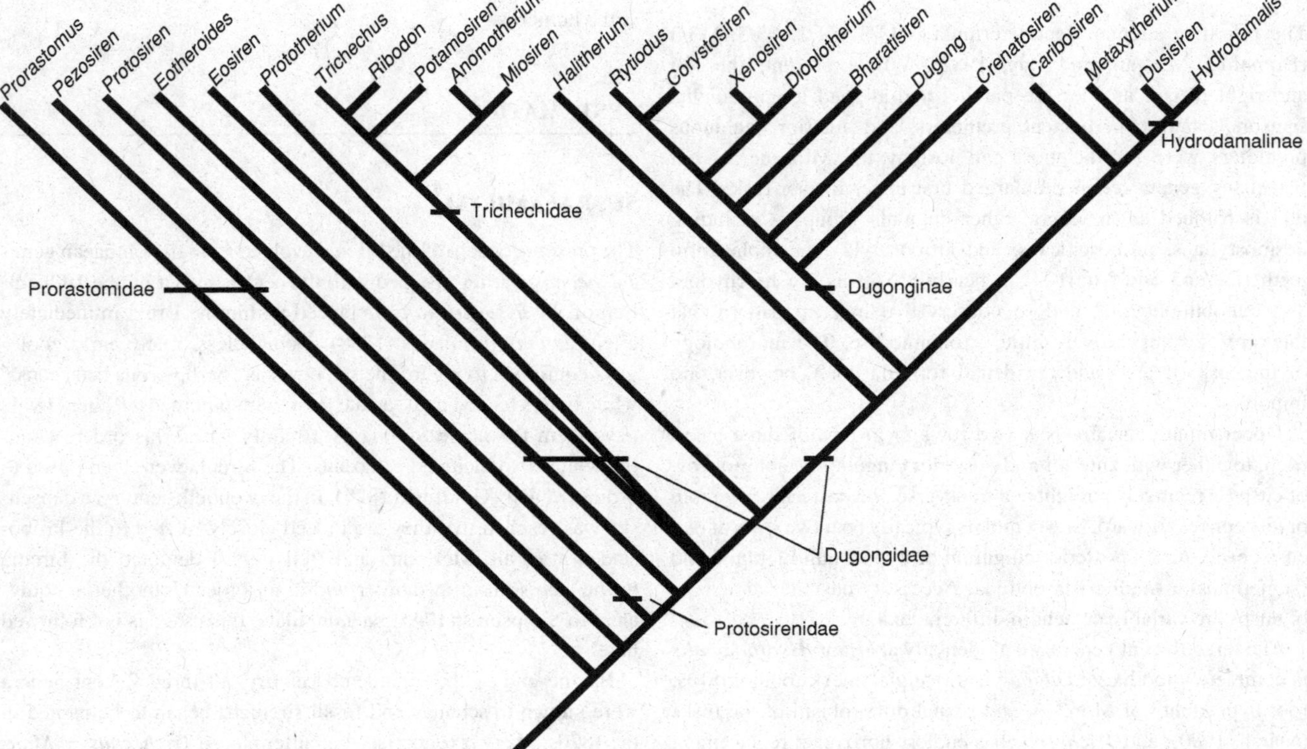

Figure 36.2. Cladogram of sirenian genera.

Domning and Furusawa, 1995). The subfamily Dugonginae, including the Recent *Dugong* and the Old World Miocene genera *Rytiodus* and *Bharatisiren*, appears to have its origin and greatest diversity in the New World, where it includes *Crenatosiren, Dioplotherium, Xenosiren*, and *Corystosiren*. These taxa and others yet to be described constituted a significant adaptive radiation of dugongines centered in the West Atlantic–Caribbean region, and only recently discovered (Domning, 1989a, b, 1990, 1997). Finally, the Miosireninae (*Miosiren* and *Anomotherium*) of northwestern Europe seem to have closer affinities to the Trichechidae than to the Dugongidae; their placement in the former family requires erection of a subfamily Trichechinae for the other trichechids (Domning, 1994) (Figure 36.2).

Apart from the Hydrodamalinae and, increasingly, the Dugonginae, the North American fossil record does not by any means provide an adequate sample or coherent picture of sirenian evolution; the history of this group has been played out on a much larger stage. The scope of this volume encompasses some endemic American sirenians as well as representatives of cosmopolitan genera such as *Protosiren, Halitherium*, and *Metaxytherium*, but the phyletic links between these forms, if known at all, are recorded mainly outside North America. Moreover, the fossil Sirenia of the West Atlantic and Caribbean are still very inadequately known and under active study (cf. Domning, 2001a). The "*Protosiren*" material from Florida (Domning, Morgan, and Ray, 1982) very likely represents a new genus, and other new taxa are known from the Caribbean region but not yet described. The modern manatees (*Trichechus* spp.) are known only from the Pleistocene and Recent but may have reached North America during the Pliocene. Hence, the following synopsis of North American sirenians is no more than an incomplete progress report.

INCLUDED GENERA OF SIRENIA ILLIGER, 1811

The locality numbers listed for each genus refer to the list of unified localities in Appendix I. The locality numbers may be listed in several alternative ways. The acronyms for museum collections are listed in Appendix III.

Parentheses around the locality (e.g., [CP101]) mean that the taxon in question at that locality is cited as "aff." or "cf." the taxon in question. Parentheses are usually used for individual species, thus implying that the genus is firmly known from the locality, but the actual species identification may be questionable. Question marks in front of the locality (e.g., ?CP101) mean that the taxon is questionably known from that locality, thus implying some doubt that the taxon is actually present at that locality, at either the genus or the species level. An asterisk (*) indicates the type locality.

PRORASTOMIDAE COPE, 1889

Characteristics: Dentition complete (I3/3, C1/1, P5/5, M3/3); rostrum deflected little if at all; pterygoid process not enlarged; dental capsule of mandible not exposed; atlas with nearly flat posterior cotyle and long, laterally projecting wing with a prominent ventrolateral process; pelvis retaining bony articulation with sacrum; caudal vertebrae without enlarged transverse processes; hindlimb complete.

***Pezosiren* Domning, 2001b**

Type species: *Pezosiren portelli* Domning, 2001b.

Type specimen: USNM 511925, pair of partial mandibles, early middle Eocene (Bridgerian?) of Jamaica.

Characteristics: Dentition probably complete (I3/3, C1/1, P5/5, M3/3); I1 enlarged into small subconical tusk; canines apparently single-rooted; rostrum very slightly deflected; weak sagittal crest; periotic not sutured to skull; mandibular symphysis narrow, ventral border downturned; dental capsule completely enclosed by bone of mandible; thoracic, lumbar, and sacral neural spines bear prominent horizontal flanges at tips; sacrum comprising four vertebrae, rigidly articulated but usually not fused; transverse processes of caudal vertebrae not enlarged; hindlimb capable of supporting body on land.

Average length of m2: 13 mm (estimated).

Included species: *P. portelli* only, known from locality EM12B* only (Domning, 2001b).

Comments: This site has also produced a late Wasatchian or Bridgerian rhinocerotoid, *Hyrachyus* sp. (Domning *et al.*, 1997) and a possible primate (MacPhee *et al.*, 1999), as well as a more derived prorastomid sirenian (Domning, 2001b).

***Prorastomus* Owen, 1855**

Type species: *Prorastomus sirenoides* Owen, 1855.

Type specimen: BM(NH) 44897, skull, mandible, and atlas, early middle Eocene (Bridgerian?) of Jamaica.

Characteristics: Dentition complete (I3/3, C1/1, P5/5, M3/3); I1 possibly enlarged into tusk; canines double-rooted; rostrum very slightly deflected (about 6°); periotic sutured to skull; mandibular symphysis narrow, ventral border not downturned; dental capsule completely enclosed by bone of mandible.

Average length of m2: Unknown.

Average length of M2: 17 mm.

Included species: *P. sirenoides* only (known from localities EM12A*, [B] [Savage, Domning, and Thewissen, 1994]).

Comments: Fragmentary remains of possible prorastomids are also known from middle and late Eocene sites in Florida (Savage, Domning, and Thewissen, 1994).

PROTOSIRENIDAE SICKENBERG, 1934

Characteristics of the family Protosirenidae are as those for *Protosiren*.

***Protosiren* Abel, 1907**

Type species: *Protosiren fraasi* Abel, 1907.

Type specimen: CGM C.10171, early middle Eocene of Egypt.

Characteristics: Dentition complete (I3/3, C1/1, P5/5, M3/3); canines single-rooted; I1 enlarged; rostral deflection moderate, 35–40°; periotic not sutured to skull; mandibular symphysis broadened; mandibular dental capsule exposed posteroventrally; pelvis well developed, with ligamentary connection to sacrum and large acetabulum; hindlimb possibly retaining some function.

Average length of m2: 15 mm.

Included species: *Protosiren* sp. is known from localities GC23VIIIP, GC23X, EM3A (Domning, Morgan, and Ray, 1982).

Comments: *Protosiren* is also known from the middle Eocene of Egypt, South Asia, and France (McKenna and Bell, 1997).

TRICHECHIDAE GILL, 1872 (1821)

Characteristics (of *Trichechus* and possibly of earlier forms): Incisors and canines vestigial or absent; rostrum small compared with cranium, slightly to moderately deflected; external auditory meatus of squamosal very broad and shallow; periotic not sutured to skull; mandibular symphysis broad; mandibular dental capsule exposed posteroventrally; cervical vertebrae reduced to six; acromion of scapula enlarged, directed forward; bicipital groove of humerus reduced or absent; pelvis reduced to remnant consisting mainly of ischium; tail fin horizontally expanded and paddle shaped, with rounded posterior margin; *Trichechus* is so far known only from the Pleistocene.

***Ribodon* Ameghino, 1883**

Type species: *Ribodon limbatus* Ameghino, 1883.

Type specimen: from late Miocene or early Pliocene of Argentina; deposited in Museo de Paraná, Argentina (Pascual, 1953).

Characteristics: Supernumerary molars present and horizontally replaced, but larger and less lophodont than those of *Trichechus*.

Average length of m2: m2 not identifiable as such.

Average length of a typical lower molar: 22 mm.

Included species: *R. limbatus* is known from the late Miocene or early Pliocene of Argentina. A specimen of *Ribodon* sp. is known from Holden Beach, North Carolina, but it was recovered as beach float, and it is unclear what geological unit and thus what age it can be assigned to.

DUGONGIDAE GRAY, 1821

Characteristics: Rostrum large compared with cranium; processus retroversus of squamosal present; external auditory meatus of squamosal inverted-U-shape, about as wide parasagittally as high; periotic not sutured to skull; mandibular symphysis broad except in the earliest forms; mandibular dental capsule exposed posteroventrally; pelvis reduced to rodlike remnant consisting mostly of ilium and ischium; tail with horizontal, laterally pointed flukes like those of cetaceans.

A handful of specimens previously identified as *Protosiren* sp. (Domning, Morgan, and Ray, 1982: Table 1 [Localities 18 and 21]) from the Castle Hayne Formation (locality EM3) are now thought to represent family Dugongidae gen et. sp. indet. based on comparisons with material of a similar age from Egypt (Gingerich *et al.*, 1994).

HALITHERIINAE (CARUS, 1868)

Characteristics: First incisor tusk may or may not be present; other incisors and canines vestigial or absent; cheek teeth enameled, usually with closed roots; M/m3 not reduced; rostrum moderately to strongly deflected; processus retroversus of squamosal slightly inflected.

Caribosiren Reinhart, 1959

Type species: *Caribosiren turneri* Reinhart, 1959.

Type specimen: UCMP 38722, late Oligocene of Puerto Rico.

Characteristics: Rostrum strongly deflected, about 75°; tusks absent; thoracic neural spines relatively short, anteroposteriorly constricted at bases.

Average length of m2: Unknown.

Average length of M2: 17 mm.

Included species: *C. turneri* only, known from locality EM13* only.

Halitherium Kaup, 1838 (including *Pugmeodon*)

Type species: *Halitherium schinzii* (Kaup, 1838) (originally described as *Pugmeodon schinzii*).

Type specimen: HLMD Az 48, early Oligocene of Germany.

Characteristics: Tusk moderately large, with small subconical enamel crown; other incisors and canines lost, but P/p2–4 still present together with DP/dp5 and M/m1–3; rostrum moderately to strongly deflected, about 45–60°; nasals fairly large, usually in contact in the midline; supraorbital process well developed and dorsoventrally flattened.

Average length of m2: 23 mm (*H. schinzii*).

Included species: ?*H. antillense* Matthew, 1916 (known from locality EM14*); *H. alleni* Simpson, 1932 (locality EM6A*).

Comments: *Halitherium* is also known from the early Oligocene to early Miocene of Europe, the late Oligocene to early Miocene of south Asia, and the Oligocene of Madagascar (McKenna and Bell, 1997).

Metaxytherium Christol, 1840 (including *Felsinotherium* Capellini, 1872; *Halianassa* Studer, 1887 [not Meyer, 1838]; *Hesperosiren* Simpson, 1932).

Type species: *Metaxytherium medium* (Desmarest, 1822) (originally described as *Hippopotamus medius*) see Domning (1996).

Type specimen: MNHN Fs 2706, middle Miocene of France.

Characteristics: Tusk small in North American species, with small conical enamel crown; permanent premolars absent; rostrum strongly deflected, 55–80°; nasals reduced and usually separated in the midline; supraorbital process more or less reduced compared with *Halitherium*.

Average length of m2: 26 mm (*M. floridanum*; see Domning, 1988).

Included species: *M. crataegense* (Simpson, 1932) (probably includes *M. calvertense* Kellogg, 1966 and *M. riveroi* Varona, 1972) (known from localities GC9*, [GC24III], ?NC1, NC3A, [EM4], EM7, [EM15]); *M. floridanum* Hay, 1922 (localities GC8A, B, GC9C, GC10C*); *M. arctodites* Aranda-Manteca, Domning, and Barnes, 1994 (localities WM13A, CC18, CC52*).

Comments: *Metaxytherium* is also known from the early Miocene to Pliocene of Europe, the Miocene (possibly early Pliocene) of Africa, the early Miocene of Asia, the early Miocene to early Pliocene of South America, and the early Pliocene of New Zealand (McKenna and Bell, 1997).

HYDRODAMALINAE (PALMER, 1895 [BRANDT, 1833])

Characteristics: Tusks vestigial or absent; permanent premolars absent; rostral deflection 45° or less; nasals reduced and usually separated in the midline; supraorbital process reduced often blunt or knoblike; lacrimal reduced; jugal in contact with premaxilla; horizontal mandibular ramus slender, ventral border gently concave; processus retroversus of squamosal not inflected; large, body 3 m or more in length; endemic to the North Pacific Ocean.

Dusisiren Domning, 1978

Type species: *Dusisiren jordani* (Kellogg, 1925) (originally described as *Metaxytherium jordani*).

Type specimen: USNM 11051, early Clarendonian of California.

Characteristics: Tusks very reduced or absent; functional molariform teeth (DP5–M3); rostrum moderately deflected, about 40°; size large, length 3–5 m.

Average length of m2: 32 mm.

Included species: *D. jordani* (known from localities WM15, WM16A, WM17A, B*, WM18, WM19 [Domning, 1978]); *D. reinharti* (locality WM28* [Domning, 1978]); *D. dewana* (localities [WM15], [WM20], WM25 [Takahashi, Domning, and Saito, 1986]).

Comments: *Dusisiren* is also known from the late Miocene of Japan (McKenna and Bell, 1997).

Hydrodamalis Retzius, 1794 (including *Rytina* Illiger, 1811)

Type species: *Hydrodamalis gigas* (Zimmermann, 1780) (originally described as *Manati gigas*), the recently extinct Steller's sea cow.

Type specimen: None ever designated; type locality Bering Island, Russia in the Recent.

Characteristics: Teeth absent, at least in adults; rostrum moderately deflected, 45° or less; manus hooklike; carpals highly modified and phalanges extremely reduced or absent; size very large, possibly up to 10 m in length, estimated mass 10 000 kg (Scheffer, 1972)

Average length of m2: no molars available.

Included species: *H. gigas* (Pleistocene to Recent; extinct since 1768 A.D.); *H. cuestae* (known from localities WM21B, WM24A, WM25*, WM26, WM29 [Domning, 1978]).

Comments: *Hydrodamalis* is also known from the Pleistocene of Japan and the Bering Sea (McKenna and Bell, 1997).

DUGONGINAE (GRAY, 1821)

Characteristics: First incisor tusks usually moderate to large in size, often flattened and bladelike; rostrum moderately to strongly downturned; supraorbital processes usually appearing downturned in lateral view, separated by a deep nasal incisure.

***Corystosiren* Domning, 1990**

Type species: *Corystosiren varguezi* Domning, 1990.

Type specimen: IGM 4569, late Miocene or early Pliocene, Yucatan, Mexico.

Characteristics: Tusk very large, flattened and bladelike, with paper-thin enamel on medial side; frontal roof very broad and concave; parietal roof exceptionally thick (4.5 cm), with prominent, upraised temporal crests.

Average length of m2: unknown.

Average length of M2: 28 mm.

Included species: *C. varguezi* only (known from localities CA6A*, ?GC12, GC13B [Domning, 1990]).

***Crenatosiren* Domning, 1991**

Type species: *Crenatosiren olseni* (Reinhart, 1976) (originally described as *Halitherium olseni*).

Type specimen: UF/FGS V6094, late early Arikareean, Florida.

Characteristics: Tusk moderately large, with small subconical enamel crown; rostrum rather strongly deflected, approximately 55°; frontals elongated relative to parietals; nasal incisure very deep; nasals fused with frontals; (see Domning, 1997).

Average length of m2: 16 mm.

Included species: *C. olseni* only (known from localities GC24III*, EM4, EM5 [Domning, 1991]).

***Dioplotherium* Cope, 1883**

Type species: *Dioplotherium manigaulti* Cope, 1883.

Type specimen: ChM PV2896, ?late Oligocene, South Carolina.

Characteristics: Tusk very large, with lozenge-shaped cross section; rostrum strongly deflected, 50–70°; zygomatic process of jugal shorter than diameter of orbit; frontal roof deeply concave; nasals lost or fused with frontals (see Domning, 1989a).

Average length of m2: 23 mm (*D. allisoni* [Domning, 1978]).

Included species: *D. manigaulti* (known from localities GC24III, EM6B* [Domning, 1989a]); *D. allisoni* (Kilmer, 1965) (localities WM15, WM28*, CC18 [Domning, 1978]).

Comments: *Diplotherium* is also known from the early Miocene of Brazil (McKenna and Bell, 1997).

***Xenosiren* Domning, 1989b**

Type species: *Xenosiren yucateca* Domning, 1989b.

Type specimen: IGM 4190, late Miocene or early Pliocene, Yucatan, Mexico.

Characteristics: Tusk very large, flattened, with paper-thin enamel on medial side; frontal roof relatively narrow, deeply concave, with medially overhanging temporal crests; nasals lost or fused with frontals; zygomatic–orbital bridge of maxilla very short anteroposteriorly, continuous with a transverse bony wall on its dorsal side; preorbital process of jugal broad, thin, medially concave.

Average length of m2: unknown.

Average length of M2: 25 mm (estimated).

Included species: *X. yucateca* only (known from localities CA6A, B* [Domning, 1989b]).

INDETERMINATE SIRENIANS

Indeterminate sirenian fossils, mostly fragments of ribs, are found at numerous localities in the eastern and southeastern United States, California, and Mexico, ranging in age from middle Eocene through Pleistocene. Perhaps the most significant of these is a partial skull and mandible of an indeterminate halitheriine dugongid from the early Miocene Nye Mudstone in Oregon (locality WM6 [Domning and Ray, 1986]). This is both the earliest and the northern-most definite record of Tertiary sirenians on the Pacific coast of the Americas to date.

BIOLOGY AND EVOLUTIONARY PATTERNS

In the course of their return to the water, sirenians had to pass through amphibious and probably riverine or estuarine evolutionary stages. These latter stages of adaptation seem to have persisted in the lifestyles of the living manatees, which prefer brackish water and freshwater to the open sea. But the typical habitat of most sirenians for most of their history seems to have been the seagrass beds that flourished in the warm, shallow waters of the tropical Tethys Sea and its successors. Only the hydrodamalines (and possibly the miosirenines) ever abandoned this habitat for cooler climates and a non-angiosperm diet (Figure 36.3).

Seagrasses are not very diverse taxonomically, and marine mammals can cross water barriers more easily than most organisms.

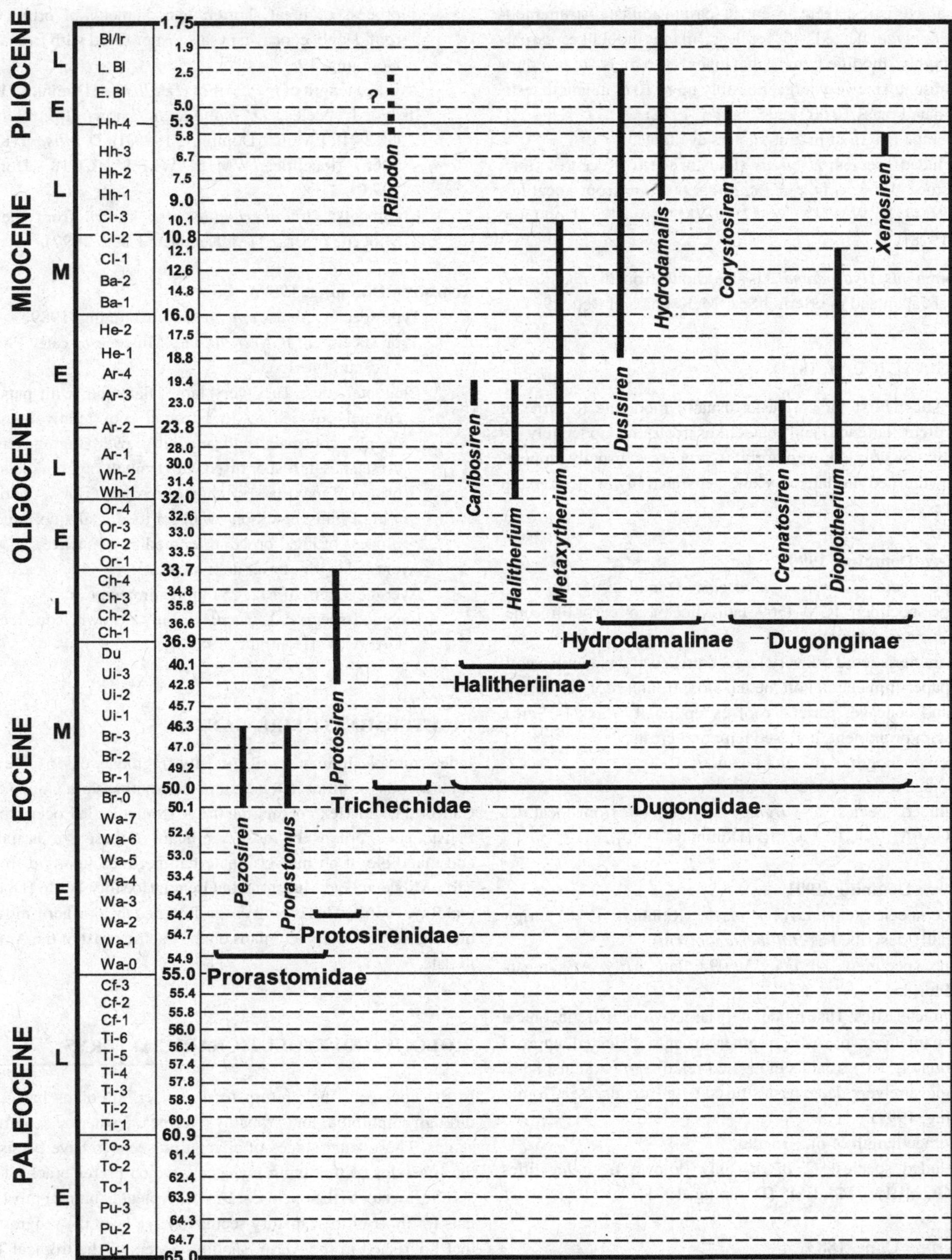

Figure 36.3. Temporal ranges of North American sirenian genera.

Hence it is not hard to explain the lack of diversity among sea cows themselves. On the contrary, it is surprising how much allopatric speciation and niche partitioning actually took place. New species, genera, and even entire adaptive radiations (such as the New World dugongines) are continually being discovered: in one small area of northern Florida, at least three late Oligocene genera seem to have lived sympatrically. Even greater numbers of sympatric species are likely to have been the norm during much of the Tertiary, though how they avoided competition is not well understood (Domning, 2001a).

As early eutherian offshoots and as tropical herbivores, sirenians may have inherited (and subsequently maintained) a low metabolic rate and limited powers of thermoregulation. As shallow divers, they needed a heavy skeleton to submerge easily with large lungs full of air. They also seem to have found it economical of energy to avoid frequent changes in body position (e.g., while surfacing and submerging). Hence the body axis tends to be kept horizontal, and the position of the mouth opening is adjusted to maximize feeding efficiency by evolutionary changes in deflection of the rostrum. Species that feed on floating or emergent plants have only slightly downturned snouts; specialists on bottom plants (predominantly dugongids) have repeatedly and independently evolved sharply deflected ones. Many dugongids probably used their tusks (some with self-sharpening edges) to help to uproot the nutritious rhizomes of seagrasses.

Manatees are apparently a conservative group of sirenians that were isolated in South American rivers and estuaries for most of the Tertiary. When aquatic species of true grasses invaded these rivers, the superior nutritional value of this new food source led them to adopt it as a major component of their diet. Their evolution of unlimited horizontal tooth replacement was probably a response to the tooth abrasion caused by the siliceous phytoliths characteristic of true grasses. Following the extinction of Caribbean dugongids in the late Pliocene, manatees expanded their range northward and also dispersed across the Atlantic to Africa (Domning, 1982).

Viewed as a whole, sirenians are a modestly diverse and successful but quite conservative group of mammals. The tropical seagrass ecosystems that have provided their major habitats have changed little since the Eocene; hence the sirenians, which adapted to these habitats very early and effectively, have likewise had little occasion for major subsequent change. They do not, however, provide many examples of evolutionary stasis. Much of their history has been marked by slow, gradual, and subtle change, particularly in the Halitheriinae (Domning and Thomas, 1987). Anagenesis, well exemplified by the halitheriines and hydrodamalines, has been at least as important as cladogenesis. And sirenians have in no way lost their capacity for rapid evolution or dramatic innovation: witness the extreme specializations of *Hydrodamalis*, and the aberrant dental adaptations of all three Recent genera. Worldwide cooling in the Late Tertiary diminished their geographical range and diversity, but only the rapacity of human activity casts doubt on their future prospects.

REFERENCES

Abel, O. (1907). Die Morphologie der Hüftbeinrudimente der Cetaceen. *Denkschriften der mathematisch-naturwissenschaftlichen Klasse der Kaiserlichen Akademie der Wissenschaften (Wien)*, 81, 139–95.

Ameghino, F. (1883). Sobre una coleccion de mamiferos fosiles del Piso Mesopotamico de la formacion Patagonica. *Boletin de la Academia Nacional de Ciencias, Córdoba*, 5, 101–16.

Aranda-Manteca, F. J., Domning, D. P., and Barnes, L. G. (1994). A new middle Miocene sirenian of the genus *Metaxytherium* from Baja California and California: relationships and paleobiogeographic implications. *Proceedings of the San Diego Society of Natural History*, 29, 191–204.

Brandt, J. F. (1833). Ueber den Zahnbau der Stellerschen Seekuh (Rytina Stelleri) nebst Bemerkungen zur Charakteristik der in zwei Unterfamilien zu zerfaellenden Familie der pflanzenfressenden Cetaceen. *Memoires de l'Academie des Sciences de St.-Petersbourg*, 6, 103–118.

Capellini, G. (1872). Sul Felsinotherio, sirenoide halicoreforme dei depositi littorali pliocenici dell'antico Bacino del Mediterraneo e del Mar Nero. *Memorie della Reale Accademia delle Scienze dell'Istituto di Bologna*, 3, 605–46.

Carus, J. V. (1868). *Handbuch der Zoologie*, Vol. 1: *Wirbelthiere, Mollusken und Molluscoiden*. Leipzig: Wilhelm Engelmann.

Christol, J., de (1840). Recherches sur divers ossements fossiles attribués par Cuvier à deux phoques, au lamantin, et à deux espèces d'hippopotames, et rapportés au *Metaxytherium*, nouveau genre de cétacé de la famille des dugongs. *L'Institut, Annales* 8, Section 1(no. 352), pp. 322–3.

Cope, E. D. (1883). On a new extinct genus of Sirenia from South Carolina. *Proceedings of the Academy of Natural Sciences of Philadelphia*, 1883, 52–4.

(1889). Synopsis of the families of Vertebrata. *The American Naturalist*, 23, 849–77.

Desmarest, A. G. (1822). *Mammalogie ou Description des Espèces de Mammifères*. Paris: Veuve Agasse.

Domning, D. P. (1978). Sirenian evolution in the North Pacific Ocean. *University of California, Publications in Geological Science*, 118, 1–176.

(1982). Evolution of manatees: a speculative history. *Journal of Paleontology*, 56, 599–619.

(1988). Fossil Sirenia of the West Atlantic and Caribbean region. I. *Metaxytherium floridanum* Hay, 1922. *Journal of Vertebrate Paleontology*, 8, 395–426.

(1989a). Fossil Sirenia of the West Atlantic and Caribbean region. II. *Dioplotherium manigaulti* Cope, 1883. *Journal of Vertebrate Paleontology*, 9, 415–28.

(1989b). Fossil Sirenia of the West Atlantic and Caribbean region. III. *Xenosiren yucateca*, gen. et sp. nov. *Journal of Vertebrate Paleontology*, 9, 429–37.

(1990). Fossil Sirenia of the West Atlantic and Caribbean Region. IV. *Corystosiren varguezi*, gen. et sp. nov. *Journal of Vertebrate Paleontology*, 10, 361–71.

(1991). A new genus for *Halitherium olseni* Reinhart, 1976 (Mammalia: Sirenia). *Journal of Vertebrate Paleontology*, 11, 398.

(1994). A phylogenetic analysis of the Sirenia. [In *Contributions in Marine Mammal Paleontology Honoring Frank C. Whitmore, Jr.*, ed. A. Berta and T. A. Deméré.] *Proceedings of the San Diego Society of Natural History*, 29, 177–89.

(1996). Bibliography and index of the Sirenia and Desmostylia. *Smithsonian Contributions to Paleobiology*, 80, 1–611.

(1997). Fossil Sirenia of the West Atlantic and Caribbean Region. VI. *Crenatosiren olseni* (Reinhart, 1976). *Journal of Vertebrate Paleontology*, 17, 397–412.

(2001a). Sirenians, seagrasses, and Cenozoic ecological change in the Caribbean. *Palaeogeography, Palaeoclimatology, Palaeoecology*, 166, 27–50.

(2001b). The earliest known fully quadrupedal sirenian. *Nature*, 413, 625–7.

Domning, D. P. and Furusawa, H. (1995). Summary of taxa and distribution of Sirenia in the North Pacific Ocean. *The Island Arc*, 3, 506–12.

Domning, D. P. and Hayek, L. (1984). Horizontal tooth replacement in the Amazonian manatee (*Trichechus inunguis*). *Mammalia*, 48, 105–27.

Domning, D. P. and Ray, C. E. (1986). The earliest sirenian (Mammalia: Dugongidae) from the eastern Pacific Ocean. *Marine Mammal Science*, 2, 263–76.

Domning, D. P. and Thomas, H. (1987). *Metaxytherium serresii* (Mammalia: Sirenia) from the Lower Pliocene of Libya and France: a reevaluation of its morphology, phyletic position, and biostratigraphic and paleoecological significance. In *Neogene Paleontology and Geology of Sahabi*, ed. N. Boaz, A. El-Arnauti, A. W. Gaziry, J. de Heinzelin and D. D. Boaz, pp. 205–32. New York: Alan R. Liss.

Domning, D. P., Morgan, G. S., and Ray, C. E. (1982). North American Eocene sea cows (Mammalia: Sirenia). *Smithsonian Contributions to Paleobiology*, 52, 1–69.

Domning, D. P., Emry, R. J., Portell, R. W., Donovan, S. K., and Schindler, K. S. (1997). Oldest West Indian land mammal: rhinocerotoid ungulate from the Eocene of Jamaica. *Journal of Vertebrate Paleontology*, 17, 638–41.

Gill, T. N. (1872). Arrangement of the families of mammals. *Smithsonian Miscellaneous Collections*, 11, 1–98.

Gingerich, P. D., Domning, D. P., Blane, C. E., and Uhen, M. D. (1994). Cranial morphology of *Protosiren fraasi* (Mammalia, Sirenia) from the middle Eocene of Egypt: a new study using computed tomography. *Contributions from the Museum of Paleontology, University of Michigan*, 29, 41–67.

Goldfuss, G. A. (1820). *Handbuch der Zoologie*, Part 2. Nürnberg: J. L. Schrag.

Gray, J. E. (1821). On the natural arrangement of vertebrose animals. *London Medical Repository*, 15, 296–310.

Hay, O. P. (1922). Description of a new fossil sea cow from Florida, *Metaxytherium floridanum*. *Proceedings of the United States National Museum*, 61, 1–4.

(1923). Characters of sundry fossil vertebrates. *Pan-American Geologist*, 39, 101–20.

Illiger, C. (1811). *Prodromus Systematis Mammalium et Avium.* Berlin: C. Salfeld.

Kaup, J. J. (1838). [Ueber Zähnen von *Halytherium* und *Pugmeodon* aus Flonheim.] *Neues Jahrbuch für Mineralogie, Geognosie, Geologie, und Petrefaktenkunde*, 1838, 318–20.

Kellogg, R. (1925). A new fossil sirenian from Santa Barbara County, California. *Carnegie Institution of Washington Publications*, 348, 57–70.

(1966). Fossil marine mammals from the Miocene Calvert Formation of Maryland and Virginia. 3. New species of extinct Miocene Sirenia. *United States National Museum Bulletin*, 247, 65–98.

Kilmer, F. H. (1965). A Miocene dugongid from Baja California, Mexico. *Southern California Academy of Sciences Bulletin*, 64, 57–74.

Linnaeus, C. (1758). *Systema Naturae per Regna Tria Naturae, Secundum Classes, Ordines, Genera, Species, cum Characteribus and Differentiis*, 10th edn, *Reformata,* Vol. I. Holmiae: Laurentii Salvii.

MacPhee, R. D. E., Flemming, C., Domning, D. P., Portell, R. W., and Beatty, B. L. (1999). Eocene ?primate petrosal from Jamaica: morphology and biogeographical implications. *Journal of Vertebrate Paleontology* 19(suppl. to no. 3), p. 61A.

Marsh, H. (1980). Age determination of the dugong (*Dugong dugon* (Müller)) in northern Australia and its biological implications. In *Reports of the International Whaling Commission, Special Issue 3*, pp. 181–201. Cambridge, UK: International Whaling Commission.

Matthew, W. D. (1916). New sirenian from the Tertiary of Porto Rico, West Indies. *Annals of the New York Academy of Sciences*, 27, 23–9.

McKenna, M. C. and Bell, S. K. (1997). *Classification of Mammals Above the Species Level*. New York: Columbia University Press.

Meyer, H. von (1838). [Letter to H. G. Bronn.] *Neues Jahrbuch für Mineralogie, Geognosie, Geologie, und Petrefaktenkunde*, 1838, 667–9.

Muizon, C., de and Domning, D. P. (1985). The first records of fossil sirenians in the southeastern Pacific Ocean. *Bulletin du Muséum National d'Histoire Naturelle* 2 Paris, *Series 4*, 7(Sect. C, no. 3), 189–213.

Owen, R. (1855). On the fossil skull of a mammal (*Prorastomus sirenoides*, Owen) from the island of Jamaica. *Quarterly Journal of the Geological Society of London*, 11, 541–3.

Palmer, T. S. (1895). The earliest name for Steller's sea cow and dugong. *Science*, 2, 449–50.

Pascual, R. (1953). Sobre nuevos restos de sirenidos del Mesopotamiense. *Revista de la Asociacion Geologica Argentina*, 8, 163–81.

Reinhart, R. H. (1953). Diagnosis of the new mammalian order, Desmostylia. *Journal of Geology*, 61, 187.

(1959). A review of the Sirenia and Desmostylia. *University of California, Publications in Geological Sciences*, 36, 1–146.

(1976). Fossil sirenians and desmostylids from Florida and elsewhere. *Bulletin of the Florida State Museum, Biological Sciences*, 20, 187–300.

Retzius, A. J. (1794). Ánmärkningar vid genus Trichechi. *Konglig Svenska Vetenskapsakademiens Nya Handlingar*, 15, 286–300.

Savage, R. J. G., Domning, D. P., and Thewissen, J. G. M. (1994). Fossil Sirenia of the West Atlantic and Caribbean Region. V. The most primitive known sirenian, *Prorastomus sirenoides* Owen, 1855. *Journal of Vertebrate Paleontology*, 14, 427–49.

Scheffer, V. B. (1972). The weight of the Steller sea cow. *Journal of Mammalogy*, 53, 912–14.

Sickenberg, O. (1934). Beiträge zur Kenntnis tertiärer Sirenen. *Mémoires du Musée Royal d'Histoire Naturelle de Belgique*, 63, 1–352.

Simpson, G. G. (1932). Fossil Sirenia of Florida and the evolution of the Sirenia. *Bulletin of the American Museum of Natural History*, 59, 419–503.

(1945). The principles of classification and a classification of mammals. *Bulletin of the American Museum of Natural History*, 85, 1–350.

Studer, T. (1887). Ueber den Steinkern des Gehirnraumes einer Sirenoide aus dem Muschelsandstein von Würenlos (Kanton Aargau), nebst Bemerkungen über die Gattung *Halianassa* H. von Meyer und die Bildung des Muschelsandsteins. *Abhandlungen der Schweizerische Paläontologische Gesellschaft*, 14, 1–20.

Takahashi, S., Domning, D. P., and Saito, T. (1986). *Dusisiren dewana*, n. sp. (Mammalia: Sirenia), a new ancestor of Steller's sea cow from the Upper Miocene of Yamagata Prefecture, northeastern Japan. *Transactions and Proceedings of the Palaeontological Society of Japan, New Series*, 141, 296–321.

Varona, L. S. (1972). Un dugongido del Mioceno de Cuba (Mammalia: Sirenia). *Memoria de la Sociedad de Ciencias Naturales La Salle* (Caracas), 91(32), 5–19.

Zimmermann, E. A. W. (1780). *Geographische Geschichte des Menschen und der vierfüssigen Thiere*, Vol. 2. Leipzig: Weygandschen Buchhandlung.

37 Desmostylia

DARYL P. DOMNING
Howard University, Washington DC, USA

INTRODUCTION

Desmostylians are the only wholly extinct order of marine mammals. They are known only from Oligocene and Miocene rocks of the North Pacific littoral, and seem to have arisen from Paleocene or Eocene Asiatic tethytheres; their sister group is evidently the Proboscidea. Desmostylians were large, hippopotamus-like animals with procumbent tusks and massive limbs. Though not visibly modified for swimming, they appear to have been rather awkward on land (with locomotor abilities comparable to those of ground sloths) and seem not to have ventured far from the water. They were probably amphibious browsers on intertidal and subtidal marine algae and seagrasses in subtropical to temperate or cool climates. For a comprehensive overview of the literature on desmostylians, see Domning (1996).

DEFINING FEATURES OF THE ORDER DESMOSTYLIA

CRANIAL

The maxillae and premaxillae are greatly elongated to form a more or less narrow and little-deflected rostrum. The paroccipital process is long and stout. The cranial portion of the squamosal encloses a large epitympanic sinus and is pierced by several foramina, including a large one that communicates with the external auditory canal. This canal is nearly enclosed ventrally by contact of the squamosal posttympanic and postglenoid processes. A sagittal crest is present in more primitive forms (Figure 37.1).

DENTAL

The primitive desmostylian dental formula is I3/3, C1/1, P4/4, M3/3. The incisors and canines are procumbent and more or less tusklike; they may be reduced in number. A long postcanine diastema is seen in the more derived forms, and the permanent premolars are reduced in number. The dp4 has three lobes (more in *Desmostylus*). The molars are primitively brachydont and bunodont, closely resembling those of anthracobunids (primitive proboscideans or tethytheres). Later forms develop thick enamel, closely appressed cusps, and several accessory cusps; *Desmostylus* (meaning "bundle of columns"), the most derived genus, has hypsodont molar crowns consisting of tall, tightly packed enamel cylinders. The homologies of these cusps are explained by Inuzuka, Domning, and Ray (1995).

POSTCRANIAL

The body is stout and compact, with a relatively short neck and a deep thorax. The lumbar spine is strongly arched, and the sacrum and pelvis would be held almost vertically. The tail is very reduced. The ribs show some development of osteosclerosis (increased proportion of compact bone), but not in a degree comparable to that seen in the Sirenia; considerable cancellous bone remains. Neither does the skeleton exhibit sirenian-like pachyostosis (increased volume of bones).

The sternum comprises two parallel rows of broad, flat plates. A clavicle is lacking. The limbs are robust. The radius and ulna are separate but with little or no ability to supinate or pronate. The radius bears a massive styloid process distally. The carpals have a unique semi-alternating arrangement, and the wrist joint holds the manus in a "semi-pronated" position. The first digits of both manus and pes are vestigial or absent, and the phalanges are short and flattened. The innominate is long, and the ilium relatively narrow. The shaft of the femur is strongly compressed anteroposteriorly. The patella is very large. The tibia is twisted on its long axis so that the knee is strongly abducted; the axis of the ankle joint lies at almost 45° to the axis of the tibial shaft and at a similar angle to the long axis of the calcaneum, which points inward. The metatarsals are

Evolution of Tertiary Mammals of North America, Vol. 2. ed. C. M. Janis, G. F. Gunnell, and M. D. Uhen. Published by Cambridge University Press.

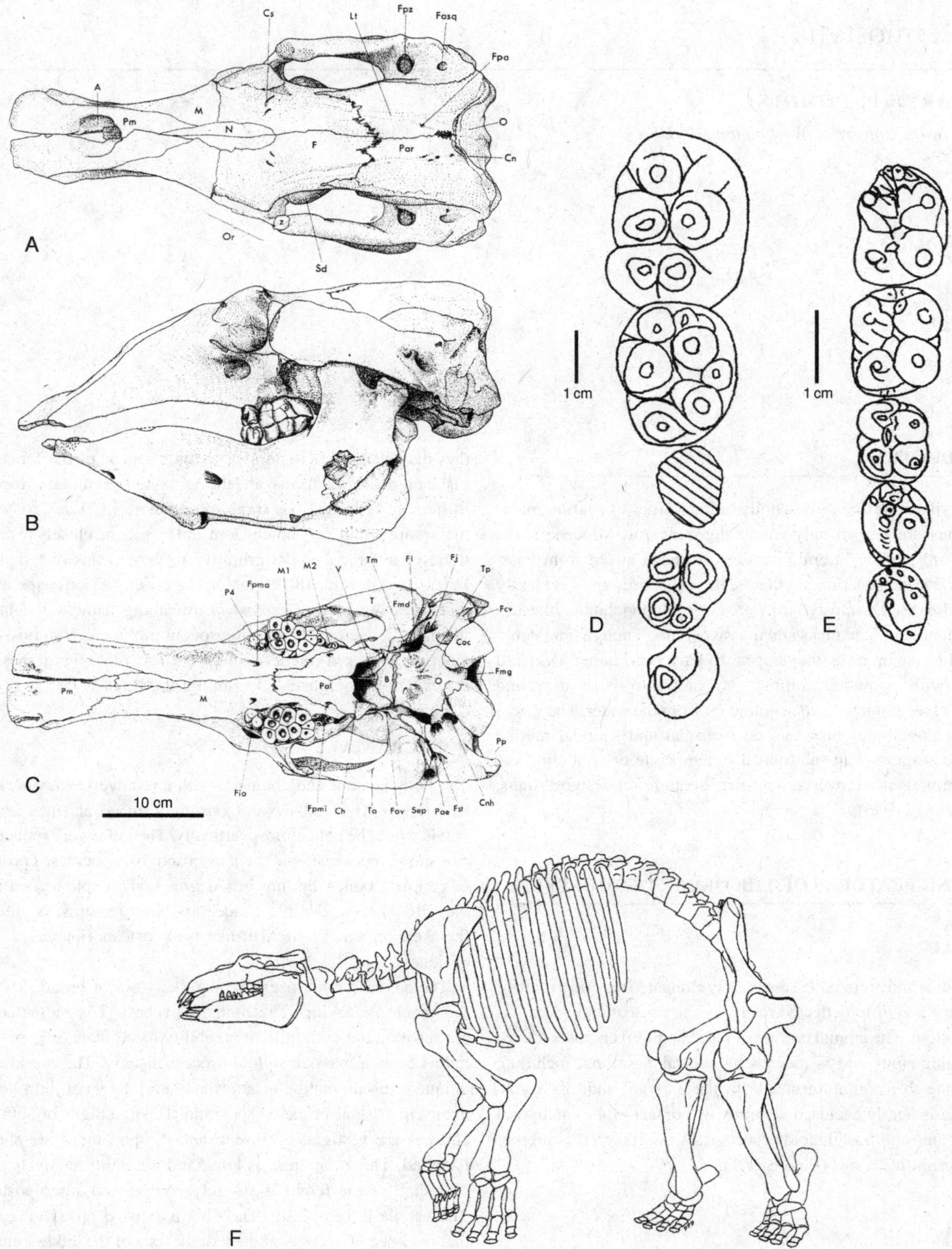

Figure 37.1. A–C. Skull of *Desmostylus hesperus* in dorsal (A), lateral (B), and ventral (C) views (Inuzuka, 1988). Scale = 10 cm. D. Occlusal views of left p3–m3 of *Paleoparadoxia tabatai* (m1 broken). Scale = 1 cm. E. Occlusal views of right p3–m3 of *Behemotops proteus* (after Domning *et al.*, 1986). Scale = ~ 2 cm. F. Skeleton of *Paleoparadoxia* in terrestrial pose, after Domning (2002). Total length ~ 2.2 m. (Courtesy of the Geological Society of Japan.)

shorter than the metacarpals and decrease in length from lateral to medial; the medial metatarsophalangeal joints appear to have been more strongly dorsiflexed than the lateral when standing on level ground. Evidently the weight was borne more by the medial portion of the pes; otherwise the foot posture was rather conventionally digitigrade, and a proboscidean-like pad under the metatarsals may have been present.

SYSTEMATICS

SUPRAORDINAL

Desmostylus was described by Marsh in 1888 on the basis of fragmentary teeth from California and was regarded as a sirenian. A skull was described in Japan in 1902, but was at first considered to represent a proboscidean. Desmostylians were subsequently assigned to several other orders by various workers but were most often put in the Sirenia (e.g., as the suborder Desmostyliformes of Hay, 1923) until the discovery of complete skeletons showed that they had hindlimbs, leading Reinhart to raise them to ordinal rank in 1953. The history of their systematics is summarized by Domning, Ray, and McKenna (1986) and Inuzuka, Domning, and Ray (1995). McKenna and Bell (1997) have since demoted the Desmostylia to a "parvorder" within their order Uranotheria, which is equivalent to Simpson's (1945) Paenungulata; this usage is not followed here.

INFRAORDINAL

The family-level classification of desmostylians has not been stabilized. The known morphological diversity within the order justifies at least two families, and the nominal families Desmostylidae Osborn, 1905, Cornwalliidae Shikama, 1957, Paleoparadoxiidae Reinhart, 1959, and Behemotopsidae Inuzuka, 1987 have been proposed, albeit with partly overlapping membership. However, consensus on which set of these familial names to adopt has not been achieved. Cladistic analyses of the order have been published by Domning *et al.* (1986), Clark (1991), and Ray, Domning, and McKenna (1994); these agreed that the relationships of the then-known genera are as follows: (*Behemotops* (*Paleoparadoxia* (*Cornwallius* (*Desmostylus*, including *Kronokotherium* and *Vanderhoofius*)))). The relationships among the latter three genera have yet to be clarified (Figure 37.2).

INCLUDED GENERA OF DESMOSTYLIA REINHART, 1953

The locality numbers listed for each genus refer to the list of unified localities in Appendix I. The locality numbers may be listed in several alternative ways. The acronyms for museum collections are listed in Appendix III.

Parentheses around the locality (e.g., [CP101]) mean that the taxon in question at that locality is cited as "aff." or "cf." the taxon in question. Parentheses are usually used for individual species, thus implying that the genus is firmly known from the locality, but the actual species identification may be questionable. Question marks in front of the locality (e.g., ?CP101) mean that the taxon is questionably known from that locality, thus implying some doubt that the taxon is actually present at that locality, at either the genus or the species level. An asterisk (*) indicates the type locality.

***Behemotops* Domning, Ray, and McKenna, 1986**

Type species: *Behemotops proteus* Domning, Ray, and McKenna, 1986 (including *B. emlongi* Domning, Ray, and McKenna, 1986; see Ray *et al.*, 1994).

Type specimen: USNM 244035, early Oligocene (Whitneyan?) of Washington (Barnes and Goedert, 2001).

Characteristics: Dental formula I3/3, C1/1, P4/4, M3/3, without marked diastemata; lower canine forming large tusk with subrectangular cross section; premolars large, high, not molariform; molars brachydont, bunodont, with principal cusps neither cylindrical nor appressed, and nearly surrounded by a basal cingulum; symphyseal region of mandible broad, shovellike; lingual surface of mandible lacking swelling at rear of tooth row.

Average length of m2: 31 mm.

Included species: *B. proteus* only (including *B. emlongi*) (known from localities WM3*, WM5B [Domning *et al.*, 1986]).

Comments: *Behemotops* is also known from the late Oligocene of Japan.

***Cornwallius* Hay, 1923**

Type species: *Cornwallius sookensis*, originally described as *Desmostylus sookensis* (Cornwall, 1922).

Type specimen: BCPM 486, late Oligocene, British Columbia, Canada.

Characteristics: Adult dental formula I2/1, C1/1, P2?/0, M3/3 (Beatty, 2002), with long postcanine diastema; upper canine tusks downcurved; molars brachydont, with cylindrical, appressed cusps (six or seven major ones on uppers, five on lowers), slight or no cingulum; rostrum relatively narrow; lingual surface of mandible lacks swelling at rear of tooth row; nasal opening moderate in size, not retracted; supraorbital processes large; zygomatic process of squamosal slender; sagittal crest present.

Average length of m2: 39 mm.

Included species: *C. sookensis* only (known from localities ?WM1, WM2*, WM5A–C, WM27).

***Desmostylus* Marsh, 1888**

Type species: *Desmostylus hesperus* Marsh, 1888.

Type specimen: YPM 11900, late Barstovian of California.

Characteristics: Adult dental formula I0/0–1, C0–1/1, P0/0, M3/3, with very long postcanine diastema; molars hypsodont, with cylindrical, appressed cusps (eight major ones on uppers, seven on lowers), no cingulum; rostrum long and narrow; premaxillary region spatulate; lingual surface of mandible bearing thick swelling at rear

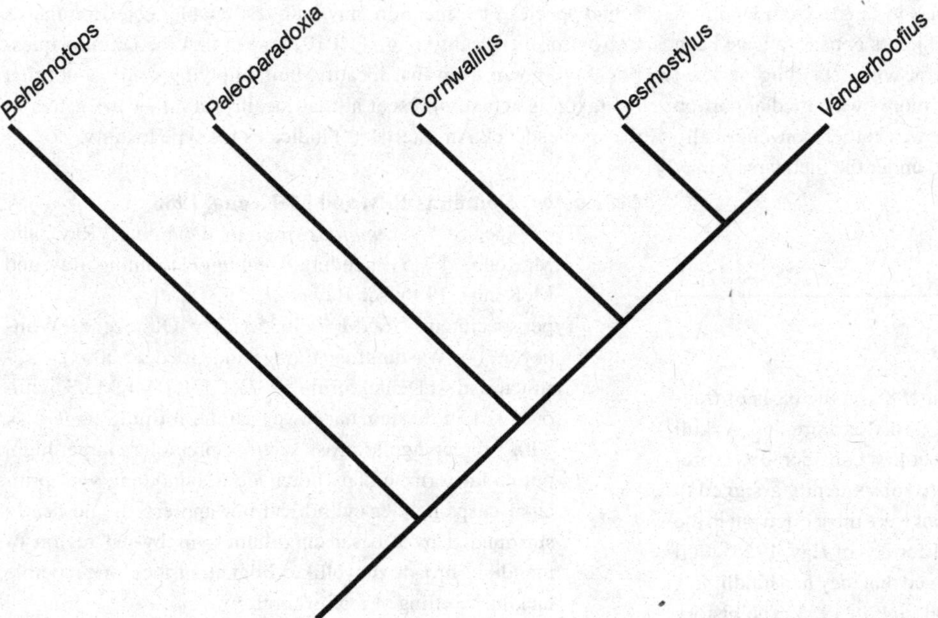

Figure 37.2. Cladogram showing the relationships of desmostylian genera.

of tooth row in adult; nasal opening moderate in size, not retracted; supraorbital processes reduced; zygomatic process of squamosal very broad vertically; sagittal crest present or absent.

Average length of m2: 65 mm.

Included species: *D. hesperus* only (known from localities ?WM3II, ?WM4, ?WM6, WM7, ?WM9, ?WM11, ?WM13, WM14*, ?WM15, ?WM16, ?WM28, CC18, ?CC19, ?CC23, ?CC52).

Comments: *Desmostylus* is also found in the Miocene of Japan, Sakhalin, and Kamchatka. Reports from Texas, Florida, Java, and the New Siberian Islands (Arctic Ocean) are now considered erroneous.

***Kronokotherium* Pronina, 1957**

Type species: *Kronokotherium brevimaxillare* Pronina, 1957.

Type specimen: from early Miocene of Kamchatka; deposited in Zoological Institute, Academy of Sciences, St. Petersburg, Russia.

Characteristics: Molars like those of *Desmostylus*; horizontal ramus of mandible short, deep, mediolaterally thin, with ventral border downturned anterior to m2.

Average length of m2: 42 mm.

Included species: *K. brevimaxillare* (early Miocene of Kamchatka).

Kronokotherium sp. is reported from localities ?WM3 (Barnes and Goedert, 2001), ?[WM3II].

***Paleoparadoxia* Reinhart, 1959 (including *Cornwallius*, in part)**

Type species: *Paleoparadoxia tabatai* (Tokunaga, 1939) (originally described as *Cornwallius tabatai*).

Type specimen: NSMT 5601 (neotype), middle Miocene, Japan.

Characteristics: Dental formula I3/3, C1/1, P3/3, M3/3, with long postcanine diastema; lower canine forms large cylindrical tusk; molars brachydont, bunodont, with cylindrical, appressed cusps, thick cingulum surrounding base of crown; symphyseal region of mandible broad, shovellike; lingual surface of mandible lacking swelling at rear of tooth row; nasal opening large, retracted; supraorbital processes large; zygomatic process of squamosal slender; sagittal crest present.

Average length of m2: 28 mm.

Included species: *P. tabatai* (known from localities [WM13A–C], ?WM16A–C) [CC23A–B]); *P. weltoni* (locality WM9* [Clark, 1991]).

Paleoparadoxia sp. is also known from localities CC18A–B, WM15, WM15II).

Comments: *Paleoparadoxia* is also found in the Miocene of Japan.

***Vanderhoofius* Reinhart, 1959**

Type species: *Vanderhoofius coalingensis* Reinhart, 1959.

Type specimen: UCMP 39989, Barstovian of California.

Characteristics: Adult dental formula I0/0, C0/1, P0/0, M3/3; very long postcanine diastema; molars like those of *Desmostylus*; rostrum long and narrow; premaxillary region spatulate; lingual surface of mandible bearing thick swelling at rear of tooth row in adult; nasal opening moderate in size, not retracted; supraorbital processes reduced; zygomatic process of squamosal somewhat broadened vertically; sagittal crest present.

Average length of m2: 54 mm.

Included species: *V. coalingensis* only, known from locality CC23A* only (Reinhart, 1959).

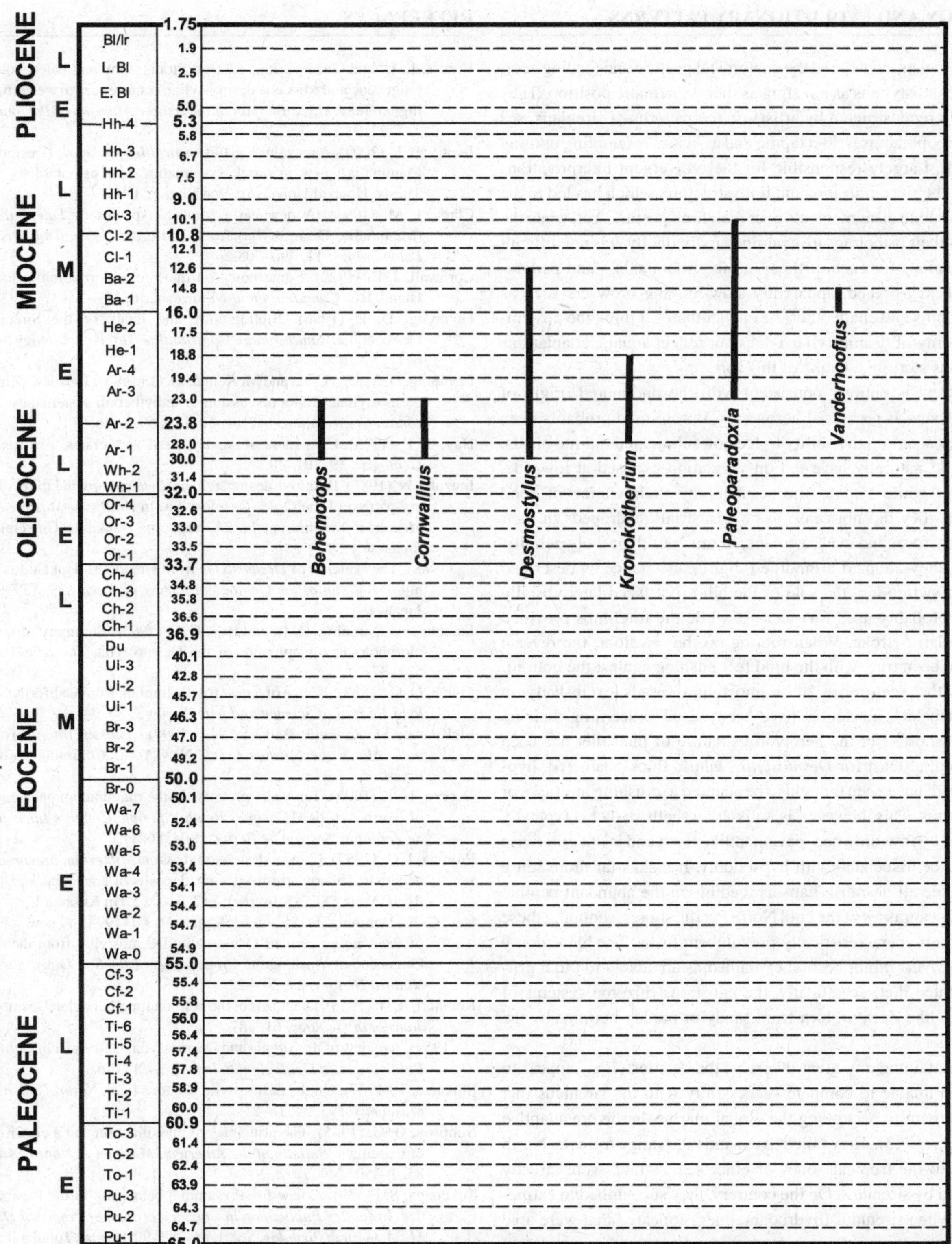

Figure 37.3. Temporal ranges of North American desmostylian genera.

BIOLOGY AND EVOLUTIONARY PATTERNS

Paleontologists have found the anatomical posture and feeding ecology of desmostylians as puzzling as their systematic position. They have been reconstructed by artists on the patterns of sirenians, sea lions, hippopotamuses, and tapirs, and as beasts resembling nothing else at all. Largely responsible for these divergent interpretations has been their peculiar foot and limb structure, which has led to the widest possible disagreements over their limb posture. Yet, when the details of their osteology are examined individually, it is not difficult to find analogs for each of them in other mammals – and particularly in heavy-bodied, apparently slow-moving browsers such as ground sloths and chalicotheriine chalicotheres. I think the postcranial anatomy of desmostylians does not reflect aquatic adaptations so much as sloth-like habits of this sort.

This view is entirely consistent with the undoubted origin of desmostylians as terrestrial herbivores. Before (and probably even after) they acquired the ability to feed on submerged marine plants, they would naturally have fed on vegetation exposed at low tide. Therefore, while tied to the seashore by their choice of food resources, they did not cease to be terrestrial quadrupeds in their basic body form and even, perhaps, in much of their daily activity. On land, they adopted a modified digitigrade stance as described above. In swimming, the sole of the hind foot faced laterodistally and was probably used for steering, while the forelimbs provided the propulsive stroke. When rooting on the sea floor, the reverse may have been true, with the hind feet pushing against the bottom.

Given the occurrence of desmostylian fossils exclusively in marine deposits, no doubts have been raised concerning their tie to the seashore. But the herbivorous nature of their diet has been questioned; at least for *Desmostylus*, whose thick-enameled, hypsodont teeth have been thought to be adapted to crushing mollusks or echinoderms. This, however, is unlikely; a shellfish diet is typically processed by low-crowned, pavementlike teeth and is not otherwise known to be associated with hypsodonty. It makes far more sense to me to see all desmostylians as feeding on the abundant benthic algae and seagrasses of the cool North Pacific shores, scooping these up with their characteristically procumbent tusks. The hypsodonty of *Desmostylus* might best be explained as an adaptation to a grit-contaminated diet: specifically, the nutritious rhizome systems of a seagrass like *Zostera*, torn from sandy or muddy bottoms by the sturdy tusks.

Never attaining any great diversity, the desmostylians appear to have been unable to compete successfully with the sirenians that preceded them in occupying the global marine–herbivore adaptive zone (Figure 37.3). They failed to expand their range from the North Pacific into the tropical zones of other seas, which were already preempted by sirenians. On the contrary, they succumbed to extinction just when sirenians (hydrodamalines) appeared that were able to exploit the cooler-climate plant resources of the desmostylians' own home waters.

REFERENCES

Barnes, L. G. and Goedert, J. L. (2001). Stratigraphy and paleoecology of Oligocene and Miocene desmostylian occurrences in western Washington State, USA. *Bulletin of the Ashoro Museum of Paleontology*, 2, 7–22.

Beatty, B. L. (2002). A reevaluation of *Cornwallius sookensis* (Desmostylia; Mammalia): new material, systematics, and paleobiology. M.Sc. Thesis, Howard University, Washington, DC.

Clark, J. M. (1991). A new early Miocene species of *Paleoparadoxia* (Mammalia: Desmostylia) from California. *Journal of Vertebrate Paleontology*, 11, 490–508.

Cornwall, I. E. (1922). Some notes on the Sooke Formation, Vancouver Island, BC. *Canadian Field-Naturalist*, 36, 121–3.

Domning, D. P. (1996). Bibliography and index of the Sirenia and Desmostylia. *Smithsonian Contributions to Paleobiology*, 80, 1–611.

Domning, D. P., Ray, C. E., and McKenna, M. C. (1986). Two new Oligocene desmostylians and a discussion of tethytherian systematics. *Smithsonian Contributions to Paleobiology*, 59, 1–56.

Hay, O. P. (1923). Characters of sundry fossil vertebrates. *Pan-American Geologist*, 39, 101–20.

Inuzuka, N. (1987). Primitive desmostylians, *Behemotops* and the evolutionary pattern of the Order Desmostylia. In *Professor Masaru Matsui Memorial Volume*, pp. 13–25. Sapporo: Hokkaido University. [In Japanese.]

(1988). The skeleton of *Desmostylus* from Utanobori, Hokkaido. I. Cranium. *Bulletin of the Geological Survey of Japan*, 39, 139–90. [In Japanese.]

Inuzuka, N., Domning, D. P., and Ray, C. E. (1995). Summary of taxa and morphological adaptations of the Desmostylia. *The Island Arc*, 3, 522–37.

Marsh, O. C. (1888). Notice of a new fossil sirenian, from California. *American Journal of Science and Arts, Series 3*, 35, 94–6.

McKenna, M. C. and Bell, S. K. (1997). *Classification of Mammals Above the Species Level.* New York: Columbia University Press.

Osborn, H. F. (1905). Ten years progress in the mammalian palaeontology of North America. *Comptes Rendus du 6me Congrès International de Zoologie, Session de Berne*, 1904, 86–113.

Pronina, I. G. (1957). [A new desmostylid, *Kronokotherium brevimaxillare* gen. nov., sp. nov., from Miocene deposits of Kamchatka.] *Doklady Akad. Nauk SSSR* (Moscow), 117, 310–12. [In Russian.]

Ray, C. E., Domning, D. P., and McKenna, M. C. (1994). A new specimen of *Behemotops proteus* (Mammalia: Desmostylia) from the marine Oligocene of Washington. *Proceedings of the San Diego Society of Natural History*, 29, 205–22.

Reinhart, R. H. (1953). Diagnosis of the new mammalian order, Desmostylia. *Journal of Geology*, 61, 187.

(1959). A review of the Sirenia and Desmostylia. *University of California, Publications in Geological Sciences*, 36, 1–146.

Shikama, T. (1957). On the desmostylid skeltons [sic]. *Natural Science and Museum Tokyo*, 24, 16–21.

Simpson, G. G. (1945). The principles of classification and a classification of mammals. *Bulletin of the American Museum of Natural History*, 85, 1–350.

Tokunaga, S. (1939). A new fossil mammal belonging to the Desmostylidae. In *Jubilee Publication in Commemoration of Professor H. Yabe, M.I.A. Sixtieth Birthday*, Vol. 1, pp. 289–99. Sendai: Tohuku Imperial University.

Addendum

CHRISTINE M. JANIS,[1] RICHARD C. HULBERT, JR.,[2] and MATTHEW C. MIHLBACHLER[3]

[1] *Brown University, Providence, RI, USA*
[2] *Florida Museum of Natural History, University of Florida, Gainesville, FL, USA*
[3] *New York College of Osteopathic Medicine, Old Westbury, NY, USA*

INTRODUCTION

This addendum was conceived as a means of both correcting errors in Vol. 1 and of adding additional information about the species covered in Vol. 1 that has been published in the past decade. We have not attempted to do an exhaustive search of the literature, and we note that much new information is published in obscure publications that are difficult to access. However, even with the limited, rather "happenstance," search that was performed, the amount of new information is immense.

The taxa below are ordered in the same fashion as in Vol. 1, and in each case three main items are considered: corrections to Vol. 1 (indicated with bold); systematic revisions, new taxa, and important new locality or paleobiological information; and additions from new localities. In some cases, more extensive revisions were noted: these included the borophagine canids (from the work of Wang, Tedford, and Taylor, 1999), the merycoidodontid oreodonts (from the work of Stevens and Stevens, 2007), the brontotheres (by Matt Mihlbachler, following his dissertation research), the equids (in part following Froehlich [2002] and also extensive revision by Richard Hulbert), and the rhinocerotids (from Prothero, 2005a). We thank Margaret Stevens and Don Prothero for providing prepublication copies of their work to aid in this addendum. New temporal range charts and biogeographic charts were constructed for these taxa, with the exception of the oreodonts, which just proved too difficult to revise in this fashion. Caitlyn Thompson played an important role in collating the data for these new figures.

The locality numbers listed for each taxon refer to the list of unified localities in Appendix I, and the references can be found in Appendix II. The localities are listed in a several ways. Parentheses around the locality (e.g., [CP101]) mean that the taxon in question at that locality is cited as an "aff." or "cf." to the taxon name. Parentheses are usually used for individual species, thus implying that the genus is firmly known from the locality but that the species identification may be questionable. Question marks in front of the locality (e.g., ?CP101) mean that the taxon is questionably known from that locality, thus implying some doubt that the taxon is actually present at that locality, at either the genus or the species level.

CREODONTA

CORRECTIONS TO VOLUME 1

Prototomus sp. is known from locality CP25A.
Hyaenodon sp. is known from locality CP45A.

SYSTEMATIC REVISIONS, NEW TAXA, AND IMPORTANT NEW LOCALITY OR PALEOBIOLOGICAL INFORMATION

A new diminutive saber-toothed carnivore from the late Uintan of California (locality CC7B) is tentatively assigned to the Hyaenodontidae. This taxon is about half the size of the oxyaenid ?*Machairodus* and resembles the saber-toothed *Apataelurus kayi* (Wagner, 1999).

A new species of *Limnocyon*, *L. cuspidens*, resembling *L. verus* but with dental adaptations suggesting a slightly more carnivorous diet, is reported from the middle Bridgerian of Wyoming (locality CP34C; Morlo and Gunnell, 2005).

Morlo, Gunnell, and Alexander (2000) examined new material of the hyaenodontid *Thinocyon* from the Bridgerian of Wyoming

Evolution of Tertiary Mammals of North America, Vol. 2. ed. C. M. Janis, G. F. Gunnell, and M. D. Uhen. Published by Cambridge University Press.

and concluded that it occupied a weasel-like niche, its morphology suggesting a hypercarnivorous, semi-fossorial existence.

ADDITIONS FROM NEW LOCALITIES (SEE APPENDIX I)

Oxyaena sp. (locality CP20BB).
Patriofelis ulta (locality [CP34A]); *P.* sp. (locality CP25J).
Cf. *Apataelurus* (new gen. and sp.) (locality CC7B).
Arfia junnei (locality [CP20AA]).
Prototomus phobus (locality [GC21II]); *P. robustus* (locality CP20BB); *P.* sp. (locality CP20BB).
Sinopa minor (locality CP34A); *S. rapax* (locality CP34A); *S.* sp. (locality CP25J).
Tritemnodon agilis (locality CP34C); *T.* sp. (localities CP20BB, CP25J, CP34C).
Acarictis ryani (locality CP20AA).
Hyaenodon cruciens (locality NP24D); *H. microdon* (locality NP10B); *H.* sp. (localities CC8IIB, NP10B).
Hemipsalodon sp. (locality PN5B).
Limnocyon sp. (localities CC6A, CC7B, CP34A, C).
Thinocyon velox (locality [CP34A]); *T.* sp. (locality CP25J).
Prolimnocyon sp. (localities CP20BB, CP25J, HA1A2).
Hyaenodontidae indet. (localities CP25I2, CP34A).
Creodont indet. (localities CC4, CC9IVA).

"MIACOIDEA"

SYSTEMATIC REVISIONS, NEW TAXA, AND IMPORTANT NEW LOCALITY OR PALEOBIOLOGICAL INFORMATION

New species of viverravids, *Protictis simpsoni, P. minor*, and *Bryanictis paulus*, are described from the middle Torrejonian of New Mexico (locality SB23; Meehan and Wilson, 2002).

New species of *Viverravus* are reported from the middle Wasatchian of Wyoming (locality CP20BB; Silcox and Rose, 2001).

A new species of *Vulpavus, V. farsonensis*, is reported from the early Bridgerian of Wyoming (locality CP34A; Gunnell, 1998).

Viverravus sp., Cf. *Vulpavus* sp., and *Miacis* sp. are all reported from the Wasatchian of Ellesmere Island (locality HA1A2; Eberle, 2001).

A new postcranial skeleton of *Viverravus acutus* from the Eocene of Wyoming shows that this taxon is scansorial, like *Miacis*. The body mass was estimated as 250 g (Heinrich and Houde, 2006).

Two new species of *Miacis* are reported from the late Duchesnean and early Chadronian of Montana, from localities NP25B (new species A) and NP25C (new species B) (Tabrum, Prothero, and Garcia, 1996).

Wesley and Flynn (2003) have revised the taxonomy of *Tapocyon*, synonymizing *T. occidentalis* with *T. robustus*, and erecting a new species, *T. dawsonae* (known from locality CC7C). New postcranial material shows that *Tapocyon* had retractable claws.

A new small miacid genus is reported from the early Uintan of Wyoming (locality CP6AA; Thornton and Rasmussen, 2001).

ADDITIONS FROM NEW LOCALITIES (SEE APPENDIX I)

Didymictus protenus (locality CP20BB): *D. altidens* (locality CP25J); *D. vancleveae* (locality [CP20E?]); *D.* sp. (localities CP20AA, BB);
Intyrictis vanvaleni (locality CP13IIB).
Protictis paralus (localities NP3A, NP47BB); *P.* sp. (localities NB1B, NP3A0, NP3G).
Raphictis gausion (locality [NP3A]).
Simpsonictis jaynanneae (locality [NP3A0]).
Viverravus actutus (locality CP20BB); *V. gracilis* (localities CP20E, CP25J, CP31E, CCP34A); *V. minutus* (localities CP25I2, J, CP34A): *V. politus* (locality [CP20AA]); *V. sicarius* (locality CP25J); *V.* sp. (localities GC21II, SB51, CP17A, CP25GG).
Viverravid indet. (localities CP47BB [two species previously reported as miacid sp.]).
Miacis deutschi (locality CP20AA); *M. gracilis* (locality [CP6AA]); *M. latidans* (species not in Vol. 1) (locality CP25I2); *M. parvivorus* (species not in Vol. 1) (locality CP34A); *M. petilus* (locality CP20BB); *M.* sp. (localities GC21II, [CC4], CC7B, CP25J, CP34C).
Oödectes sp. (locality CP25J).
Procynodictis progressus (localities CC7B, CC8); *P. vulpiceps* (locality [SB42]).
Tapocyon sp. (localities CC4, CC7B).
Uintacyon rudis (locality CP20BB); *U. vorax* (locality CP34A); *U.* sp. (localities CP25J, CP31E).
Vassocyon promicrodon (locality CP20BB).
Vulpavus australis (localities CP20BB, [E], [CP25GG, J]); *V. acutus* (locality CP20BB); *V. canavus* (localities CP20BB, [CP25J]); *V. palustris* (locality CP34A); *V. politus* (locality [CP20BB]); *V. profectus* (locality CP24I2); *V.* sp. (localities GC21III).
Miocyon scotti (localities [CC7B], [CC8]); *M.* sp. (locality CP6AA).
Miacid indet. (locality CP47BB).

CANIDAE

CORRECTIONS TO VOLUME 1

"*Nothocyon*" *annectens* is known from locality CP104**A**.
Epicyon littoralis is known from localities **CC**32A and B (not **CP**).

SYSTEMATIC REVISIONS, NEW TAXA, AND IMPORTANT NEW LOCALITY OR PALEOBIOLOGICAL INFORMATION

Two new canid species are described by Hayes [2000] from the Brooksville Local Fauna (locality GC8AA) in the early Arikareean of Florida. The hesperocyonine *Osbornodon wangi* is larger than Whitneyan species of this genus, at the small end of the size range of Hemingfordian species, and is less robust than *O. iamonensis* from the Hemingfordian of Florida. The borophagine *Phlaocyon taylori* is the smallest species of the genus; it is also known from the Cow House Slough Local Fauna (also within locality GC8AA).

Wang (2003) described a new species of the large, probably bone-crushing, hesperocyonine *Osbornodon* from the early Hemingfordian of Nebraska and Florida. This material was previously ascribed to *Tephrocyon scitulus, Cynodesmus scitulus,* and *C. iamonensis.* (*Tephrocyon* was synonymized with *Tomarctus* [Borophagini] in Vol. 1, but the species *scitulus* was not included. *Cynodesmus* was included with the hesperocyonines, but the species *scitulus* and *iamonensis* were not included.) *Osbornodon situlus* is known from localities GC3B, GC9IV, CP105.

Possible species of *Canis, Vulpes* sp, and *Borophagus hilli* are reported from the early Pliocene Pipe Creek Sinkhole Site in Indiana (locality CE2) (Farlow *et al.*, 2001).

Baskin (2005) reported Cf. *Urocyon* sp. from the Love Bone Bed, late Claredonian of Florida (locality GC11A). This species was listed in Webb, MacFadden, and Baskin (1981) as *Proturocyon* cf. *macdonaldi*, a nomen nudum.

The early Pliocene Arctic fauna from Ellesmere Island (locality HA3; Tedford and Harington, 2003) has yielded the small canine *Eucyon* sp. (?= *Canis* in Vol. 1).

The borophagine *Tomarctus brevirostris* is reported from the Gaillard Cut Local Fauna, early Hemingfordian of Panama (locality CA7) (MacFadden, 2006). This represents the first report of a carnivoran in this fauna, and there are also fragmentary remains of an unidentified arctocyonid or hemicyonid.

A new species of *Aelurodon, A. montanensis*, is reported from the Barstovian of Montana (locality NP42II; Wang *et al.*, 2004). This taxon is notably hypercarnivorous and is the only representative of the genus in Montana.

Tucker (2003) noted the presence of *Carpocyon limnosus* from the late Hemphillian of Nebraska (locality CP116III), which extends the range of this genus (the previous youngest record in this region was late Clarendonian).

An unidentified canid is reported from the late Miocene/early Pliocene Gray Fossil Site in Tennessee (locality CE1; Wallace and Wang, 2004).

Van Valkenburgh, Sacco, and Wang (2003) reviewed aspects of craniodental morphology and body size in borophagine canids. They concluded that many genera were hypercarnivorous and must have consumed large prey, which led them to propose that at least some borophagines must have been pack hunters. Van Valkenburgh, Wang, and Damuth (2004) discussed macroevolutionary patterns of hypercarnivory in canids, within the subfamilies Hesperocyoninae and Borophaginae. They concluded that energetic constraints and selection for larger body size lead to overspecialization for large, hypercarnivorous taxa, which can be shown to have shorter species durations, indicating that they are more liable to become extinct than are smaller, more generalized ones.

UPDATES FROM WANG, TEDFORD, AND TAYLOR, 1999

Wang, Tedford, and Taylor (1999) have revised the systematics and occurrences of borophagine canids. They divided the subfamily into two tribes, Phlaocyonini (somewhat resembling the "*Cynarctoides* group" of Munthe, Vol. 1) and Borophagini; the Borophagini is further subdivided into subtribes Cynarctina, Aelurodontina, and Borophagina. Eleven new genera are introduced (see below), several genera that were in Munthe have been abandoned (*Aletocyon, Bassariscops, "Nothocyon," Strobodon*, and *Osteoborus*), and there is extensive revision of existing (and new) species among existing and new genera. Wang, Tedford, and Taylor (1999) also provided a discussion of evolutionary patterns in the group.

It is not our intention to repeat this extensive revision and review of borophagines in this addendum. Rather, we list below their new compilation of genera and higher taxa (listed according to phylogenetic rank) and note how these relate to the compilation in Vol. 1; new taxa and type localities are noted with an asterisk, and synonomies are primarily noted for new taxa. We also translate Wang, Tedford, and Taylor's (1999) description of locality occurrences into the locality numbers in this volume (note that there is a considerable addition of locality occurrences). Borophagine occurrences obtained from other sources (including those listed above and below) are added in italics. Descriptions of characteristics (condensed and summarized) are reserved for the higher taxa.

Figure A.1 shows the temporal ranges of borophagines and Figure A.2 shows their biogeographical ranges.

BASAL BOROPHAGINES

Characteristics: Borophagines and canines are distinguished from hesperocyonines by the key innovation of a basined, bicuspid talonid on m1, allowing the initial exploitation of hypocarnivorous niches. Borophagines lack the synapomorphies of canines, including a long slender mandible and long premolars. Borophagines progressively acquire a number of derived features, including upper incisors with lateral cusps and more robust premolars, plus a reversal to more trenchant talonids in hypercarnivorous species.

Archaeocyon * **(including *Pseudocynodictis*, in part; *Hesperocyon*, in part; *Cormocyon*, in part; *Nothocyon*, in part)**

Included species: *A. pavidus* (originally described as *Pseudocynodictis pavidus* Stock, 1933; "*Hesperocyon" pavidus* in Vol. 1) (known from localities CC9C*, CP84B, CP85B, CP99B, PN6C); *A. leptodus* (originally described as *Nothocyon leptodus* Schlaikjer, 1935, including *Hesperocyon leptodus* Macdonald, 1963; equivalent in Vol. 1 not clear) (localities CP41C, CP42D*, CP48, CP49 [sublocality uncertain, A or B], CP50, CP85C, D, CP99B, D, CP101, NP37, NP50II); *A. falkenbachi** (locality CP50*).

Oxetocyon

Included species: *O. cuspidatus* (same as in Vol. 1) (known from localities CP84B*, CP94B, CP101).

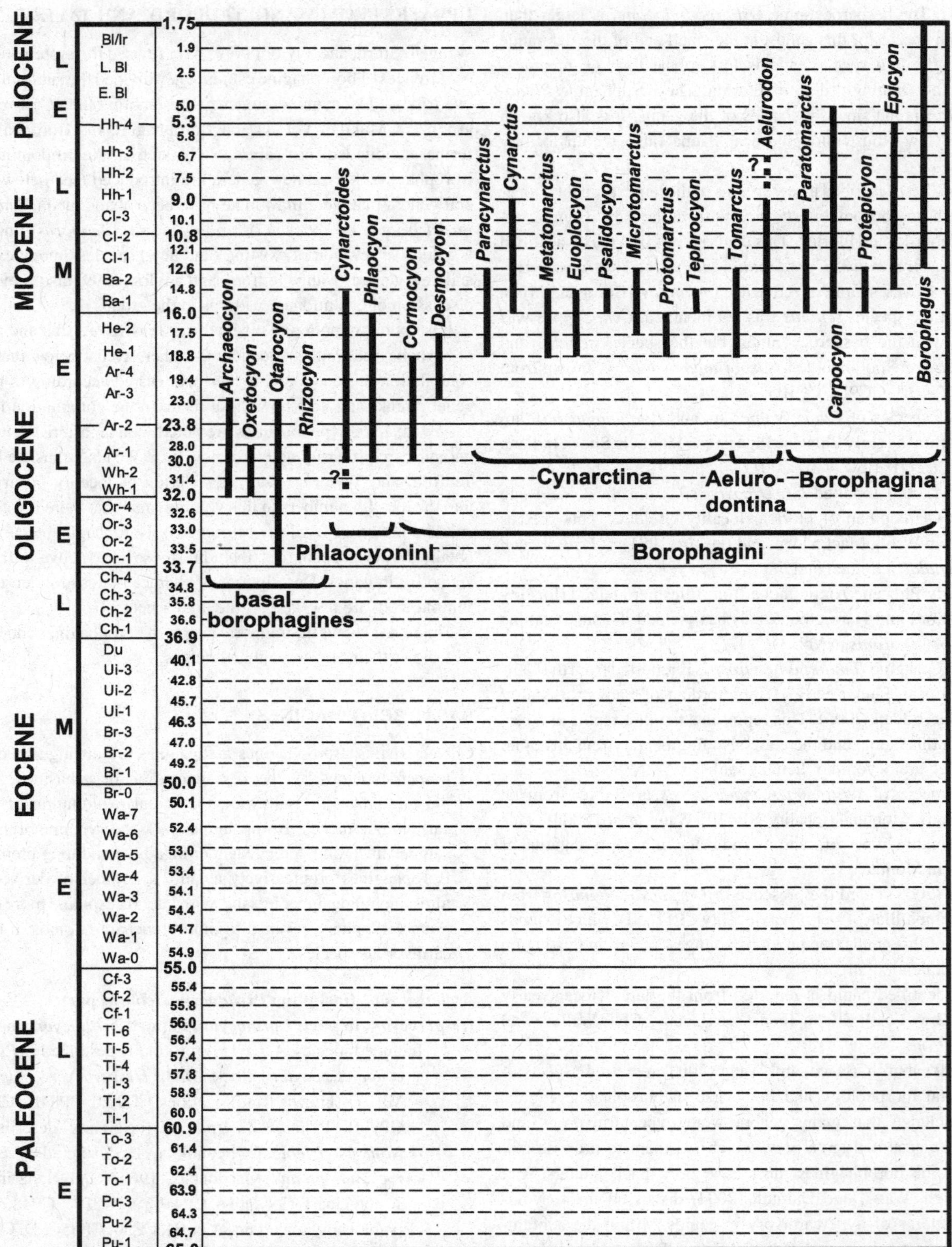

Figure A.1. Temporal ranges of borhyaenid canids.

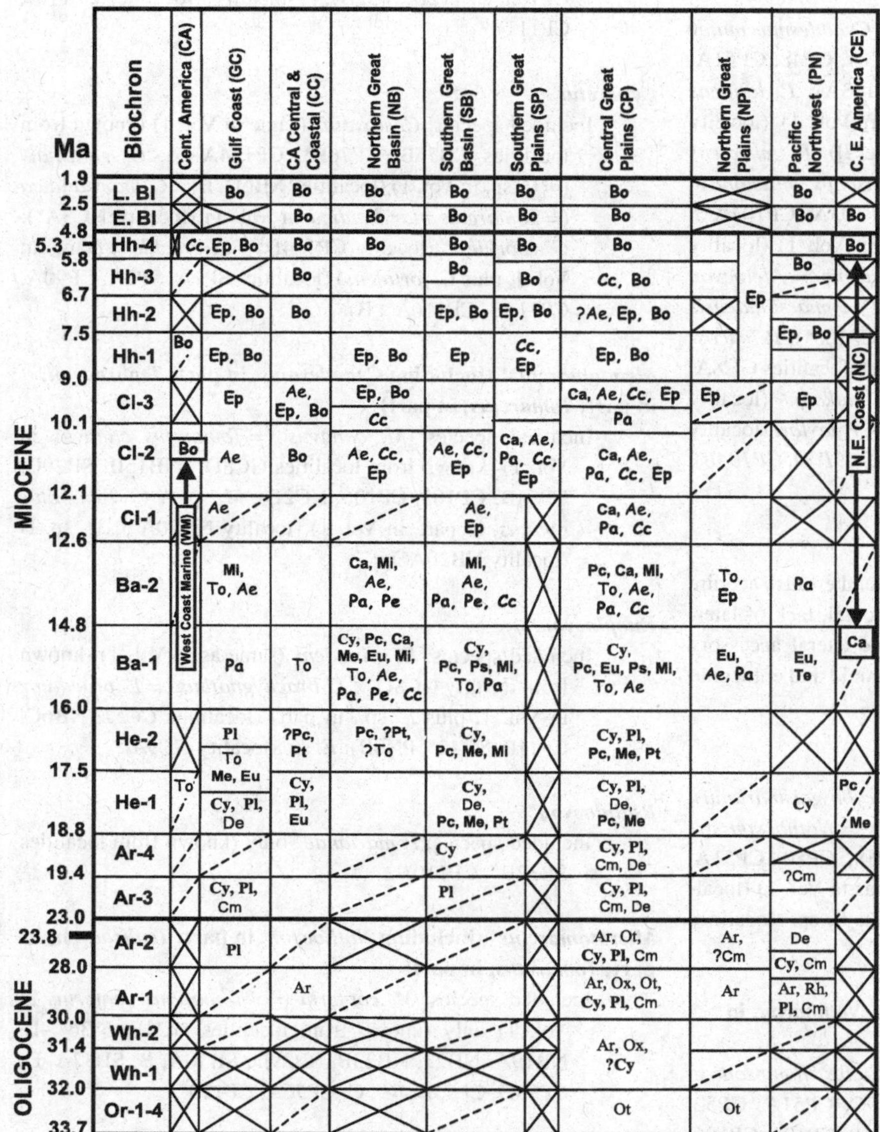

Figure A.2. Biogeographic ranges of borhyaenid canids. Key: A "box" (for a particular time period in a particular biogeographic region) that has a cross through it means no fossil marine mammal localities are known for that time period from that area; a single dashed line through the box means only Scant fossil marine mammal information is available (usually only a single, small locality). BASAL BOROPHAGINES (Times Roman Plain): Ar, *Archaeocyon*; Ot, *Otarocyon*; Ox, *Oxetocyon*; Rh, *Rhizocyon*. PHLAOCYONINI (Times Roman Bold): **Cy**, *Cynarctoides*; **Pl**, *Phlaocyon*. BOROPHAGINI (Arial Plain): Cm, *Cormocyon*; De, *Desmocyon*. CYNARCTINA (Arial Bold): **Ca**, *Cynarctus*; **Eu**, *Euoplocyon*; **Me**, *Metatomarctus*; **Mi**, *Microtomarctus*; **Pc**, *Paracynarctus*; **Ps**, *Psalidocyon*; **Pt**, *Protomarctus*; **Te**, *Tephrocyon*. AELURODONTINA (Comic Sans MS Plain): Ae, *Aelurodon*; To, *Tomarctus*. BOROPHAGINA (Comic Sans MS Bold): **Bo**, *Borophagus*; **Cc**, *Carpocyon*; **Ep**, *Epicyon*; **Pa**, *Paratomarctus*; **Pe**, *Protepicyon*.

***Otarocyon** (including *Cynodesmus*, in part)**

Included species: *O. macdonaldi** (known from localities CP84A*, NP30B); *O. cooki* (mentioned as "*Cormocyon*" *cooki* in Vol. 1) (localities CP48, CP49B?, CP50, CP85C*, D?).

***Rhizocyon** (including *Cynodictis*, in part; *Galecynus*, in part; *Hesperocyon*, in part; *Nothocyon*, in part)**

Included species: *R. oregonensis* (? = *Hesperocyon gregarius* in Vol. 1) only, known from locality PN6C* only.

PHLAOCYONINI (NEW)

Characteristics: Distinguished from basal borophagines by a large M1 metaconule and the presence of a protostylid on m1. Distinguished from Borophagini by the retention of a short, closed m1 trigonid.

Cynarctoides

Included species: *C. lemur* (= *Nothocyon lemur* in Vol. 1) (known from localities GC8B [= GC8A in Vol. 1], CP73A, CP84B, CP85D, PN6D* [and other PN6D sublocalities?]); *C. roii* (equivalence in Vol. 1 unclear) (localities CP85C*, CP101); *C. harlowi* (equivalence in Vol. 1 unclear) (locality CP103A*); *C. luskensis** (localities CP51A*, CP53); *C. gawnae** (localities *CC9IVD*, SB29B, C*); *C. acridens* (same as in Vol. 1, including *C. mustelinus*) (localities GC3B, NB6C, D, SB29A, B, SB32A, B, CP51A*, CP54B, CP71, CP86B, CP103A, CP104A, CP105, CP106, CP107, CP108A, B, CP109A, CP110, CP111, PN17III); *C. emryi** (localities CP104A, CP105*); *C.* sp. (locality *CP104B*).

Phlaocyon

Included species: *P. minor* (mentioned as *Cynodesmus minor* in Vol. 1) (known from localities GC3A, CP48, CP51A, CP52, CP53, CP86C*, CP87A, CP103A); *P. latidens* (mentioned as *Nothocyon latidens* in Vol. 1) (locality PN6C* [plus other PN6C sublocalities?]); *P. annectens* (= "*Nothcyon" annectens* and *Bassariscops willistoni* in Vol. 1) (localities SB46, CP73A, CP103A, CP104A*); *P. achoros* (= *Bassariscops achoros* in Vol. 1) (locality GC8B* [= GC8A in Vol. 1]); *P. multicuspis* (= *Aletocyon multicuspis* in Vol. 1) (locality CP51A*); *P. marslandensis* (= *Phlaocyon* sp. in Vol. 1) (locality CP105*); *P. leucosteus* (same in Vol. 1, plus *P.* sp., in part) (localities GC8A, CP71*, CP103A, CP105, CP107); *P. yatkolai** (locality CP105*); *P. mariae** (locality CP105*); *P. taylori* (locality *GC8AA**); *P.* sp. (localities [*GC8DB*], *CC9IVD, CP104B*).

BOROPHAGINI (NEW)

Characteristics: Elongated m1 trigonid. Most members also acquire the features of premaxillary contact with frontal, lack of laterally flared rim of zygomatic arch, elaborate lateral accessory cusps on I3, and a cristid between the hypoconid and entoconid on m1.

***Cormocyon* (including *Nothocyon*, in part)**

Included species: *C. haydeni** (= *Nothocyon geismarianus*, equivalent in Vol. 1 unclear, probably = *Nothocyon* sp., in part) (known from localities CP48, CP50, CP51A, CP85B, CP86B, C*); *C. copei* (same as in Vol. 1) (localities GC8B, CP74A0, PN6C*, D, *?F*); *C.* sp. (localities *?NP36D, ?NP37*).

***Desmocyon** (including *Cynodesmus*, in part; *Nothocyon*, in part; *Tomarctus*, in part)**

Included species: *D. thomsoni* (= *Tomarctus thompsoni* in Vol. 1) (known from localities SB29C, CP51A, CP52, CP53, CP86C*, *E*, CP103A, CP104A, *B*, CP105, CP106, PN6?G); *D. matthewi**(localities GC9III, CP105*).

CYNARCTINA (NEW RANK)

Characteristics: Distinguished from more primitive Borophagini by the presence of a lateral accessory cusp on I3. Distinguished from more derived Borophagini by the primitive lack of a distinct parastyle crest on P4, and by derived hypocarnivorous features, including a small subangular lobe on the mandible, a high mandibular condyle, and enlarged and elongated molars.

***Paracynarctus** (including *Cynodesmus*, in part; *Tephrocyon*, in part; *Tomarctus*, in part)**

Included species: *P. kelloggi* (= *Tomarctus? kelloggi* in Vol. 1, in part) (known from localities CC21A, NB6C, NB17, NB18A*, NB23B, SB29C, D, SB32A–C, P54B, CP72B, CP105, CP112II, NC2B); *P. sinclairi** (localities CP110*, CP111).

Cynarctus

Included species: *C. saxatilis* (same in Vol. 1) (known from localities CB75B, CP76B*, CP114A–C); *C. galushai** (= *C.* sp. in Vol. 1) (localities NB6D, E); ?*C. marylandica* (= *Tomarctus marylandica* in Vol. 1) (locality NC3A*); *C. voorhiesi** (locality CP114D*); *C. crucidens* (same in Vol. 1, plus *C. fortidens*) (localities SP1A, SP2A, CP90A, CP115B, CP116A*, B).

***Metatomarctus** (including *Cynodesmus*, in part; *Tephrocyon*, in part; *Tomarctus*, in part)**

Included species: *M. canavus* (= *Tomarctus canavus* in Vol. 1) (known from localities GC8D*, NB15II, SB29C, CP54B, CP105, CP106, NC2B); *M.* sp. A (= *Tomarctus optatus*, in part, in Vol. 1) (locality NB20A*); *M.* sp. B (locality NB20A*).

Euoplocyon

Included species: *E. spissidens* (same as in Vol. 1) (known from locality GC8D*); *E. brachygnathus* (= *E. praedator* in Vol. 1, plus *E.* sp., in part) (localities *CC51*, NB6C, CP110, NP42*, PN8B); *E.* sp. (locality *GC9A*).

Psalidocyon*

Included species: *P. marianae** only (known from localities SB32B*, CP110).

***Microtomarctus** (including *Nothocyon*, in part; *Tephrocyon*, in part; *Tomarctus*, in part)**

Included species: *M. conferta* (= ?*Tomarctus confertus* in Vol. 1) only (known from localities GC4E, NB6C–E, NB19A, NB21, NB23B, SB29D, SB30A, B, SB32A–D, CP73D, CP75B, *BB*, C, CP76A, CP110*).

***Protomarctus** (including *Tephrocyon*, in part; *Tomarctus*, in part)**

Included species: *P. optatus* (= *Tomarctus optatus* in Vol. 1) only (known from localities *CC19, CC20*, NB6B, SB29B, CP54B, CP71, CP107, CP108A, B*, CP109A).

Tephrocyon

Included species: *T. rurestris* (= *Tomarctus rurestris* in Vol. 1) only (known from localities PN7*, PN8B).

AELURODONTINA (NEW)

Characteristics: All members have uniformly robust premolars with distinct accessory cusplets. More derived members acquire features such as a broadened palate, an elongated tip of the paraoccipital process, a high sagittal crest, and a nuchal crest that is laterally compressed and overhangs the occipital condyles.

Generally lack the derived features of the Borophagina such as the domed forehead and shortened rostrum.

Tomarctus

Included species: *T. hippophaga* (= *T. hippophagus* in Vol. 1) (known from localities CC17D, NB6C, D, SB32B, CP110*, CP111, *NP11*); *T. brevirostris* (same as in Vol. 1) (localities CA7, GC4E, CC17D, NB6D, SB32B, CP75B*, *BB*, C, CP110); "*T.*" sp. (could be almost anything!) (localities *GC9A, CC23A, NB3D, CP56IIA, CP88, CP90B*).

***Aelurodon* (including *Strobodon*)**

Included species: *A. aesthenostylus* (equivalence in Vol. 1 not clear, = *A.* sp. in part) (known from localities NB6D, E, NB23B*, CP75B–C, CP111, CP113A0); *A. mcgrewi** (localities CP114B, C*); *A. stirtoni* (= *Strobodon stirtoni* in Vol. 1) (localities SB32C, D, CP114D*); *A. ferox* (same in Vol. 1) (localities GC4E, SB30B, SB32C, D, *F*, FF, G, ?H, SB34B, CP113A0, A1, CP114*A–D, *NP41B*); *A. taxoides* (same as in Vol. 1) (localities GC6A, B, GC10C, CC26*A*, B, CC30A, NB23C, SB32EE, SB34A, SP1A0, A, SP2A, CP90A, *B*, *CP112*, CP113A2, CP115B, C [possibly reworked from B], CP116A*, B, CP123B); *A. montanensis* (locality *NP42II*); *A.* sp. (locality *CP54*).

BOROPHAGINA (NEW)

Characteristics: Dorsally enlarged frontal sinus, producing a domed forehead, premolar cingular cusplets reduced or absent, P4 protocone reduced or absent. Generally lack derived features of Aelurodontina such as high sagittal crest and posteriorly extended nuchal crest.

***Paratomarctus** (including *Canis*, in part; *Cynodesmus*, in part; *Tephrocyon*, in part; *Tomarctus*, in part)**

Included species: *P. temerarius* (= *Tomarctus temerarius* in Vol. 1) (known from localities GC4B, C, NB6D, E, NB23B, SB30B, SB32C, D, FF, CP56, CP114*A–D, NP34E, *NP41B*, NP42, PN8B); *P. euthos* (= *Tomarctus euthos* and *T.* sp. in Vol. 1) (localities SP2A, CP90A, CP113A2, CP114D*, CP116A, B).

***Carpocyon* (including *Tomarctus*, in part)**

Included species: *C. compressus* (= *C. cuspidatus* in Vol. 1) (known from localities NB23B, CP75*BB*, C, CP114*A, B, C, CP123A0); *C. webbi** (localities SB32D, FF, SB34A, CP114D*, CP116A); *C. robustus* (= *Tomarctus robustus* in Vol. 1) (localities NB7C*, SB6, SP2A, CP73E, CP90A, CP113A2, CP116A, B); *C. limnosus* (same in Vol. 1) (localities GC13B*, SP1B, *CP116III*).

Protepicyon*

Included species: *P. raki** only (known from localities NB6D, E*, SB30A).

Epicyon

Included species: *E. aelurodontoides** (known from locality CP123C*); *E. saevus* (same in Vol. 1, plus *E. haydeni*, in part) (localities GC11A, B, *GC12*, [*GC27*], NB7*A*, *B*, C, NB12A*0*, NB23C, *NB27A*, *NB28*, NB29A, NB31, SB6, SB30B2, SB32G, SB34A, SP1B, C, SP2A, SP3A, CP90A, *B*, CP113A2, B2, CP115B, C, CP116A*, B, *CP123B*, *NP11*, PN10II); *E. haydeni* (same in Vol. 1, in part, plus *E. validus*) (localities *GC10D*, GC11A, B, GC13B, CC32B, NB7*B*, C, D, SB11A, *SB33II*, SB34A–*C*, *SB48*, SP1A–C, SP2A, SP3A, CP78, CP113A2, CP115B, C, CP116AA, A, B*, C, CP123C, *D*, NP41D, NP45II, PN11A, PN21B; *E.* sp. (localities *PN2A*, *PN10*).

***Borophagus* (including *Osteoborus*; *Epicyon* in part)**

Included species: *B. littoralis* (= *Epicyon littoralis* and *E. diabloensis* in Vol. 1) (known from localities WM17II*, CC26B, CC31, CC32A, B, NB7C, *D*, *E*, NB9, *NB10*, *NB29*); *B. pugnator* (= *Osteoborus pugnator* in Vol. 1) (localities GC11B, GC12IIA, GC13B, SP3A, CP78*, CP113B2, CP123C, *PN2B*, PN11A, PN22II); *B. orc* (= *Osteoborus orc* in Vol. 1) (locality GC12IIA*); *B. parvus** (localities CC25B, CC37, SB9, SB10*); *B. secundus* (= *Osteoborus secundus*, *O. direptor*, and *O. cyonoides* in Vol. 1) (localities CA9, CA10, CC25B, CC40, SB34C, *SB58A*, *B*, *SB58IIB*, SB65, SP1D, SP3B, SP4A, CP113B3, CP115C, D, CP123D*); *B. hilli* (= *Osteoborus hilli* in Vol. 1) (localities GC13B, CC53, SB35*A*, B, SP4B*, *CP128E*, *CE2*, PN3C, PN23A); *B. dudleyi* (= *Osteoborus dudleyi* in Vol. 1) (locality GC13B); *B. diversidens* (same in Vol. 1, plus *B. pachyodon*) (localities GC13A*, GC18IIIC, CC43B, CC45, CC45II, NB12C, NB33B, SB14G, SB18D, SB31E, *SB50*, SB65, SB67, SP1F, G, H, SP5, CP117AA, *A*, *B*, *C*, CP118, CP121, CP128B, PN3D, PN23C); *B.* sp. (localities *GC11D*, *CC6*, *CC38*, *SB37A*, *D*, *SB60*, *CP116D–F*, *CP127*, *PN14*, *PN15*, *EM8B*).

ADDITIONS FROM NEW LOCALITIES (SEE APPENDIX I)

Hesperocyon gregarius (localities NP10BB, [NP49II]); *H.* sp. (localities NP9A, NP10Bi, ?NP10C, ?NP10C2, NP25B, ?NP27C1, NP32C).

Mesocyon sp. (locality ?GC8A).

Enhydrocyon pahinsintewakpa (locality [GC8AA] [= GC8B in Vol. 1]).

Leptocyon sp. (localities ?CC9IVC, CP75B).

Vulpes stenognathus (locality GC13B).

Urocyon cineroargentus (locality GC18IIIA); *U. minicephalus* (species not in Vol. 1) (locality CP15F); *U.* n. sp. (locality GC15C).

Canis armbrusteri (species not in Vol. 1) (localities GC15E, F, GC18IIIA, SB37E); *C. davisi* (locality CP116III); *C. edwardi* (species not in Vol. 1) (localities GC15C, F,

GC18IIB, GC18IIIA–C, CC48C); *C. lepophagus* (localities SB34IVB, SB37A, D); *C.* sp. (localities CC52II, SB37E, SB58IIA).

Eucyon davisi (= *Canis davisi* in Vol. 1) (locality GC13B).

Vulpes stenognathus (locality CP116II); *V.* sp. (locality SB48).

Cynarctoides gawnae (species not in Vol. 1) (locality CC9IVD).

Phlaocyon sp. (localities GC8A, [DB], CC9IVD).

Tomarctus brevirostris (localities CA7, CP75BB): *T.* sp. (locality [CP56IIA]).

Euoplocyon brachygnathus (= *E. predator* in Vol. 1) (locality CC51).

Carpocyon limnosus (locality GC13B).

Epicyon haydeni (localities GC10D, [GC27], SB48).

Osteoborus cyonoides (= *Borophagus parvus* in Wang, Tedford, and Taylor [1999]; locality [SB58IIB]).

Borophagus hilli (= *Osteoborus hilli* in Vol. 1) (localities GC13B, SB35[A], B); *B. pugnator* (= *O. pugnator* in Vol. 1) (locality GC13B); *B.* sp. (localities GC11D, SB37A, [D]).

Canidae indet. (localities CA2II [*Aelurodon* sized], GC8AB, NB35IIIC, NP10Bii, NP10CC, CP56IIB).

PROCYONIDAE

SYSTEMATIC REVISIONS, NEW TAXA, AND IMPORTANT NEW LOCALITY OR PALEOBIOLOGICAL INFORMATION

Baskin (2003) described a number of new procyonid taxa from the mid Miocene of Florida and Texas, including the first North American records of the tribe Potosini (which includes the extant *Bassaricyon* and *Potos*), indicating that the Potosini and Procyonini had diverged from each other by the early Miocene.

New genera within the Potosini include *Bassicyonoides*, and *Parapotos*.

Bassicyonoides is a primitive potosin, intermediate in size between Recent *Bassariscus astutus* and *Procyon lotor*, and with a dentition that shows fewer adaptations for frugivory than extant potosins.

Included species are *B. stewartae* (known from localiy NB17 [= ?mustelid indet. in Vol. 1]) and *B. phyllismillerae* (locality GC9IV).

Parapotos is distinguished by large size and the possession of a massive mandible with large canines, but relatively small molars. Included species is *Parapotos tedfordi* (known from locality GC5E [Stephen Prairie]).

Within the Procyonini, Baskin (2003) also described a new species of *Edaphocyon*, ?*E. palmeri* (known from locality GC9IV).

Wallace and Wang (2004) described a new species of lesser panda of Eurasian affinities, *Pristinailurus bristoli*, from the late Miocene/early Pliocene Gray Fossil Site in Tennessee (locality CE1).

ADDITIONS FROM NEW LOCALITIES (SEE APPENDIX I)

Bassariscus sp. (locality SB35B).

Procyon lotor (species not in Vol. 1) (localities GC15F); *P.* n. sp. (localities GC15C, D, GC17?A, ?B, GC18IIB, GC18IIIA, B).

MUSTELIDAE

CORRECTIONS TO VOLUME 1

Potamotherium sp. is known from locality CP114**A**.

Palaeogale sectoria is known from locality NP24**C**.

Pliotaxidea nevadensis is known from locality PN11**B** (new locality subdivision, emended dating of Hh2).

SYSTEMATIC REVISIONS, NEW TAXA, AND IMPORTANT NEW LOCALITY OR PALEOBIOLOGICAL INFORMATION

A new species of *Oligobunis* is reported from the early Miocene of Florida (locality GC8D) (Labs, 2000).

Two new genera of mustelids are described by Hayes (2000) from the Brooksville 2 Local Fauna from the early late Arikareean of Florida (locality GC8AA), showing that the early diversity of musteloids in North America is greater than that previously known and closing the gap between the late Eocene occurrence of *Mustelavus* and the late Eocene appearance of *Promartes* and ?*Plesictis*. These taxa may represent part of an unknown autochthonous radiation derived from *Mustelavus*, or may be of independent Asian origin.

Acheronictis webbi shares dental characteristics with *Mustelavus* and the European genus *Mustelictis*, but it is 20% smaller than *Mustelavus* and differs from it in a variety of dental characteristics, including a more quadrate M1 and a P4 with a less distinct protocone. *Acheronictis* may be equivalent to the record of ?*Plesictis* from the Monroe Creek Fauna (locality CP102).

Arikarictis chapini is a larger taxon, 30–35% larger than *Mustelavus*, and is more derived than *Mustelavus* and *Mustelictis* in the reduction of postprotocrista and the enlargement of the parastyle on M1. Overall, the teeth are most similar to the European taxa *Angustictis* and *Pseudobassaris*, which some authors have included as primitive members of the Procyonidae. *Arikarictis* is also known from locality GC8A.

Several new species of *Leptarctus* have been described, including *L. martini*, from the late Barstovian of South Dakota (locality CP114C) (Lim and Miao, 2000); *L. kansasensis*, from the late Clarendonian of Kansas (locality CP123B) (Lim and Martin, 2001a) (Baskin [2005] considered this taxon to be conspecific with *L. wortmani*); *L. desuii* (Lim and Martin, 2001b) from Carlson Quarry in the early Hemphillian of Nebraksa (probably within locality CP113B2); *L. supremus* (Lim, Martin, and Wilson, 2001), from the early Hemphillian of Texas (locality SP3A); and *L. webbi*, from the late Clarendonian of Florida (locality GC11A) (Baskin [2005] and see this paper for a discussion of all these new taxa). Lim and Martin (2002) also named *Hypsoparia timmi* from the Barstovian of South Dakota (?locality CP87B). Baskin (in Vol. 1) considered the genus *Hypsoparia* to be synonymous with *Lepartcus primus*, but Lim and Martin (2002) considered it to be distinct from *Leptarctus*. Baskin (2005) discussed this issue of synonomy in some detail, noting that sexual dimorphism within leptarctines may lead to perceived taxonomic differences, and he also considered *H. timmi* to be indistinguishable from *H. bozemanensis* (from locality NP41B).

Lim (1999) provided postcranial evidence for arboreal habits in *Leptarctus* (previously considered to be fossorial) and also considered this taxon to have been frugivorous rather than omnivorous.

A new species of the skunk *Buisnictis, B. chisoensis*, is reported from the earliest Hemphillian of Texas (locality SB48; Stevens and Stevens, 2003).

A new, primitive species of the skunk *Martinogale, M. faulli*, is reported from the late Miocene of California (locality NB7C; Wang, Whistler, and Takeuchi, 2005). This represents one of the earliest known New World mephitines, is the first representaive of this genus on the West Coast of North America, and is also the smallest skunk known. The authors postulated that the New World mephitines represent a single immigration event (contra Baskin, Vol. 1), and that this taxon is close to the origin of the group.

New records of primitive badgers (Taxidiinae), smaller than the Hemphillian *Pliotaxidea*, are reported from the late Clarendonian of Nebraska (locality CP116B) (Owen, 2002). Owen (2006) also reported a new genus of badger, *Chamitataxus avitus*, from the early Hemphillian of New Mexico (locality SB34).

Plionictis oaxacaenis (species not in Vol. 1) is known from the early Barstovian of Mexico (locality CA2II; Ferrusquía-Villafranca, 2003), which represents the southern-most Tertiary carnivoran record in North America.

Tucker (2003) describes a new species of *Mustela* from the late Hemphillian of Nebraska (locality CP116III) that extends back the range of this genus to the Miocene (and also reports on a new "*Meles*"-like taxon at this site).

The early Pliocene Arctic fauna from Ellesmere Island (locality HA3; Tedford and Harington, 2003) has yielded three mustelids: Cf. *Plesiogulo, Martes* sp., and a Eurasian taxon, *Arctomeles sotnikovae* (n. sp.) (a meline badger).

Wallace and Wang (2004) describe a new species of badger of Eurasian affinities, *Arctomeles dimolodontus*, from the late Miocene/early Pliocene Gray Fossil Site in Tennessee (locality CE1).

ADDITIONS FROM NEW LOCALITIES (SEE APPENDIX I)

Brachypsalis modicus (localities CP75B, BB).
Leptarctus primus (locality CP75B).
Mustelavus priscus (locality NP24C).
Leptarctus sp. (locality [CA2III]).
Pliogale furlongi (locality CP116III).
Martinogale alveodens (locality CP116III).
Spilogale putorius (localities GC15F, GC18IIIA); *S.* sp. (localities GC15C, SB37D).
Conepatus leuconotus (genus not in Vol. 1) (locality GC15F).
Pliotaxidea nevadensis (locality CP116III).
Mionictis sp. (locality CP75BB).
Taxidea sp. (localities SN18F, SB34IIB, SB37A, B, D).
Satherium ingens (locality NB35IIIC); *S. piscinarium* (localities GC15A (= *S.* sp. in Vol. 1), GC18IIB).
Lutra canadenis (genus not in Vol. 1) (localities GC18IIIA, B).
Martes sp. (locality SB48).
Plionictis ogygia (locality CP56IIA); *P. oaxaquensis* (species not in Vol. 1) (locality CA1); *P.* sp. (locality GC11A).
?*Sthenictis* nr. ?*S. lacota* (locality GC11A).
Mustela parvus (species not in Vol. 1) (locality CP97II).
Trigonictis cooki (locality GC15D); *T. macrodon* (localities GC15C, GC17A, GC18IIB).
Palaeogale minutus (species not in Vol. 1) (localities GC8AA, A); *P.* sp. (localities ?GC5A, NP10BB, NP32C).
Mustelidae indet. (localities GC8DB, CC8IIB, CC26B, NP10C2, CC, CP56IIB).

URSIDAE

CORRECTIONS TO VOLUME 1

Parictis sp. is known from locality PN6**B**.
Agriotherium schneideri is known from locality GC13**C** (= GC13A in Vol. 1).

SYSTEMATIC REVISIONS, NEW TAXA, AND IMPORTANT NEW LOCALITY OR PALEOBIOLOGICAL INFORMATION

Plionarctos sp. is reported from the late Miocene/early Pliocene Gray Fossil Site in Tennessee (locality CE1; Wallace and Wang, 2004). *Plionarctos edensis* is reported from the early Pliocene Pipe Creek Sinkhole Site in Indiana (locality CE2; Farlow *et al.*, 2001).

McCullough *et al.* (2002) reported on new material at the 111 Ranch Site from the late Blancan of Arizona (locality SB18) and noted the new additions of the ursid Cf. *Tremartcos* (sublocality within SB18 unclear).

The early Pliocene Arctic fauna from Ellesmere Island (locality HA3; Tedford and Harington, 2003) has yielded the ursid *Ursus abtrusus*.

ADDITIONS FROM NEW LOCALITIES (SEE APPENDIX I)

Parictis parvus (locality NP10B); *P.* sp. (locality NP10B).
Cephalogale sp. (locality CC9IVD).
Plithocyon sp. (locality GC11A).
Agriotherium schneideri (locality SB55IIAA).
Ursus abstrusus (locality NB35IIIB).
Arctodus pristinus (genus not in Vol. 1) (localities GC14A, GC15C, D, F, GC17B, GC18IIIA–C, GC18IV).
Ursid indet. (locality SB34IIB)

AMPHICYONIDAE

CORRECTIONS TO VOLUME 1

Cynelos sp. is known from locality CP104**A**.
Daphoenus sp. is known from localities NP9**A**, PN6**C**, **D**.
Paradaphoenus cuspigerus is known from locality PN6**C**.
Paradaphoenus minimus is known from locality CP85**C**.
Ischyrocyon gidleyi is known from locality SP2**A**.

SYSTEMATIC REVISIONS, NEW TAXA, AND IMPORTANT NEW LOCALITY OR PALEOBIOLOGICAL INFORMATION

Hunt (2001) reviewed the occurrences and morphology of the Oligocene genus *Paradaphoenus*, and described a new species, *P. tooheyi*, from localities CP85C, CP99B, and CP99C (Wagner Quarry). *Paradaphoenus minimus* is now considered to be restricted to the Orellan.

Hunt (2002) described new daphoenine taxa from the early Miocene of Wyoming (locality CP51A). These include a new genus, *Adilophontes brackykolos*, closely related to *Daphoenodon*, and two new species of *Daphoenodon*: *D. falkenbachi* and *D. skinneri.*

A new early Arikareean temnocyonine has been found preserved in a burrow in Wildcat Hills, Nebraska (locality CP101; Bailey, Hunt, and Stepleton, 2004).

Hunt (2003) described a new species of *Amphicyon, A. galushai* (known from localities CP74A, CP105). This is the smallest North American species of *Amphicyon* (average length of m2 = 19.6 mm); this specimen extends the known range of the genus (the earliest previously known specimen being from the late Hemingfordian), and is the only *Amphicyon* to overlap with the amphicyonid *Daphoenodon.*

ADDITIONS FROM NEW LOCALITIES (SEE APPENDIX I)

Brachyrhynchocyon dodgei (locality NP24D); *B.* n. sp. (locality NP24C).
Cf. *Daphoenictis* (n. sp.) (locality NP25C).
Daphoenus demilo (species not in Vol. 1) (locality NP9A); *D.* sp. (localities NP10B, BB, C2, NP49II).
Daphoenodon notionastes (species not in Vol. 1) (localities GC5A, GC8A [= GC8B in Vol. 1, = *D.* sp. in Vol. 1], GC8B [= GC8C in Vol. 1, = *D.* sp. in Vol. 1]).
Temnocyon sp. (locality GC8B [= GC8C in Vol. 1]).
Cynelos sp. (locality CC9IVD).
Amphicyon sinapius (species not in Vol. 1) (localities CP75B, BB); *A. ingens* (localities CP75B, BB); *A. reinheimeri* (species not in Vol. 1) (locality CP75B).
Amphicyonid indet. (localities ?CP65, NP10Bi).

NIMRAVIDAE

CORRECTIONS TO VOLUME 1

Hoplophoneus primaevus is known from locality **CP**84A (not **CPP**84A).

SYSTEMATIC REVISIONS, NEW TAXA, AND IMPORTANT NEW LOCALITY OR PALEOBIOLOGICAL INFORMATION

Bryant (1996) reviewed the North American Eocene to Oligocene Nimravidae. A cladistic analysis revealed a number of differences from the phylogeny presented by Martin (Vol. 1). *Nimravus* and *Dinaelurus* are sister taxa (in the Nimravini), separated from the other nimravids and lacking the more extreme development of sabertooth morphologies seen in the other nimravids. *Hoplophoneus* (now seen as a paraphyletic genus) and *Eusmilus* are sister taxa (as in Vol. 1) in the Hoplophoneini, but *Pogonodon* is now placed as the sister taxon to this tribe, and *Dinictis* (also probably a paraphyletic taxon) is the sister taxon to *Pogonodon* plus the Hoplophoneini.

Peigné (2003) provided a review of the systematics phylogenetics of the Nimravidae. Although this paper deals mainly with European taxa, the phylogenetic analysis included all Paleogene genera and reached a different conclusion from the cladogram presented by Martin (Vol. 1), bearing a greater resemblance to the cladogram presented by Bryant (1996) discussed above. This phylogeny shows a basal divergence between the Nimravini (*Nimravus* and *Dinaelurus*) and other taxa, a sister relationship between *Dinictis* and *Pogonodon*, and a clade comprising a *Eusmilus* and a paraphyletic *Hoplophoneus*. The phylogeny also suggests an Old World origin for the group, with most likely three separate immigrations of Paleogene taxa into North America.

Morlo, Peigné, and Nagel (2004) provided evidence to show that the Barbourofelinae are not closely related to the Nimravinae: both should be considered as separate families, with barbourofelids more closely related to felids.

Bryant and Fremd (1998) reviewed the stratigraphic occurrences of the Nimravidae in the John Day Formation of Oregon (locality PN6). *Nimravus brachyops* is known from a number of sites, as noted in Vol. 1. Not noted in Vol. 1 is the occurrence in the John Day of *Hoplophoneus* cf. *H. primaevus* at locality PN6C1 (new), and *Dinictis cyclops* (sublocality uncertain). A new species of *Pogonodon* is noted from locality PN6D2 (new), which is also the sublocality for the type of *Pogonodon davisi* (synonymized with *P. platycopsis* in Vol. 1). The type species of *Eusmilus cerebralis* (noted as PN6 sublocality uncertain in Vol. 1) is from Logan Butte (within locality PN6D).

Baskin (2005) discussed the morphology of *Barbourofelis loveorum* (= *B. lovei* in Vol. 1) from Love Bone Bed, late Clarendonian of Florida (locality GC11A), concluding that its posture was subdigitigrade (as were the Oligocene nimravids), and that it was probably a specialized ambush predator taking prey larger than its own body size.

ADDITIONS FROM NEW LOCALITIES (SEE APPENDIX I)

Dinictis sp. (localities ?NP10B, NP10C2, NP24C).
Hoplophoneus sp. (locality ?NP10B).
Nimravidae indet. (localities GC8B [was GC8A], NP10CC, PN5B).

FELIDAE

CORRECTIONS TO VOLUME 1

Machairodus sp. is known from locality SB58A.
Homotherium sp. is known from locality PN3C.

SYSTEMATIC REVISIONS, NEW TAXA, AND IMPORTANT NEW LOCALITY OR PALEOBIOLOGICAL INFORMATION

Rothwell (2003) has provided a revision of the genus *Pseudaelurus*. Below we list updates to the taxonomy and locality information presented by Martin in Vol. 1 (an asterisk notes localities not reported in Vol. 1). *P. validus* (not recognized in Vol. 1, possibly = *P. sinclairi*, not recognized by Rothwell) (known from localities CP108A, B [= *P.* sp. in Vol. 1], CP110). *P. skinneri* (new) (localities CP108A, B). *P. intrepidus* (localities GC4C*, NB6C*, D*, E*, NB23B, CP75B, CP110, CP114B, C*) (not reported from localities NB8, CP56, CPC76, CP90A, CP115B, as in Vol. 1). *P. stouti* (previously *Lynx stouti*, listed in Vol. 1 as "*P.*" *stouti*) (localities SB30B*, SB32B*, CP76B, CP110*). *P. marshi* (localities GC4C*, NB6C*, D*, E*, SB30B*, SB32D*, CP69*, CP75C*, CP110, CP111*, CP116A*) (not reported from localities SB32A and CP114B, as in Vol. 1). *P. aeluroides* (may be synonymous with *P. marshi*) (locality CP110) (not reported from locality NP11, as in Vol. 1). Rothwell (2003) did not recognize Martin's *P. hibbardi*.

Browne (2005) argued that the Barstow Formation (locality NB6) *Pseudaleurus marshi* should be reassigned to the genus *Nimravides*.

Baskin (2005) discussed the morphology of *Nimravides galiani* from Love Bone Bed, late Clarendonian of Florida (locality GC11A) (recorded as *N.* sp. in Vol. 1), concluding that its postcranial adaptations were similar to the extant jaguar *Panthera onca*.

Machairodus cf. *M. coloradensis* has been confirmed from the Blancan of Mexico (locality SB60; Carranza-Casteñeda, 2001), (previously registered as a queried occurrence in Vol. 1).

Machairodus sp. is reported from the late Miocene/early Pliocene Gray Fossil Site in Tennessee (locality CE1; Wallace and Wang, 2004).

Naples, Martin, and Babiarz (1998) reported on a new type of saber-toothed felid from Florida, related to *Homotherium*. This felid was extremely robustly built, combining the scimitar-toothed morphology of *Homotherium* with the shorter, more robust limb morphology of dirk-toothed felids.

ADDITIONS FROM NEW LOCALITIES (SEE APPENDIX I)

Pseudaelurus sp. (localities [CA2III], ?[GC13B], [SB48], [CP56IIA]).
Nimravides catacopsis (locality [SB48]).
Lynx rexroadensis (localities WM26, GC13B); *L. rufus* (species not in Vol. 1) (localities GC15D, E, GC18IIIA, SB37[D], E); *L.* sp. (localities GC15C, GC17A, [SB37B]).
Miracinonyx inexpectatus (species not in Vol. 1) (localities GC15C, D, GC18IIIA).
Panthera onca (genus not in Vol. 1) (locality GC15F).
Felis sp. (genus not in Vol. 1) (localities CC52II, SB34IIB [small], SB34IVA [small]).
Machairodus coloradense (localities [GC11D], [GC13B], [SB58IIB]); *M.* sp. (localities [GC10D], SB56II, CP116III).
Homotherium n. sp. (localities ?GC14A (= *H.* sp. in Vol. 1), GC15C–E, ?GC17B, GC18IIIA).
Smilodon gracilis (taxon not in Vol. 1) (localities GC14A, GC15A, C, F, GC18IIIA, B, SB37D).
Felidae indet. (localities CA2II (large), NB35IIIA, B, NB36IIIA, B)
Machairodontine indet. (localities GC13B, SB34IIB, SB34IVA).

HYAENIDAE

ADDITIONS FROM NEW LOCALITIES (SEE APPENDIX I)

Chasmaporthetes ossifragus (localities GC15A, C, GC18IIB).

CARNIVORA INDET.

Reported from localities CC6B, CP5II

TAENIODONTA

ADDITIONS FROM NEW LOCALITIES (SEE APPENDIX I)

Ectoganus sp. (localities CP17B, CP20AA, BB).
Stylinodont indet. (locality CP25J).

TILLODONTA

SYSTEMATIC REVISIONS, NEW TAXA, AND IMPORTANT NEW LOCALITY OR PALEOBIOLOGICAL INFORMATION

A new basal tillodont, *Nacimientotherium silvestri*, is reported from the earliest Tiffanian of New Mexico (probably from locality SB23H; Libed, 2004). Libed claimed that this animal is transitional between more derived tillodonts and *Deltatherium* (described as the sister taxon to Pantodonta in Lucas, Vol. 1).

A new specimen of *Megalesthonyx hopsoni* is reported from the early Eocene of the Wind River Basin, Wyoming (locality CP27E) (Williamson, Lucas, and Stucky, 1996); this extends the known range of this taxon into the earliest Bridgerian.

ADDITIONS FROM NEW LOCALITIES (SEE APPENDIX I)

Esthonyx bisculcatus (locality CP20BB); *E.* sp. (locality CP25J).
Trogosus sp. (localities CP25J, CP31E, CP34C).

PANTODONTA

CORRECTIONS TO VOLUME 1

Titanoides primaevus is known from locality NP48**A**.

SYSTEMATIC REVISIONS, NEW TAXA, AND IMPORTANT NEW LOCALITY OR PALEOBIOLOGICAL INFORMATION

A new species of *Deltatherium, D. dandreai*, of larger size than *D. fundaminis*, is reported from the late early Palocene (Torrejonian) of New Mexico (within locality SB23) (Lucas and Kondrashove, 2004a). These authors also considered that the genus *Deltatherium*, designated as the sister taxon to Pantodonta in Vol. 1, should be placed in the family Deltatheriidae within the order Tillodonta.

A new species of *Cyriacotherium* is reported from the late Torrejonian (To3, locality NP3A0) and mid Tiffanian (Ti3, locality NP3B) of Alberta (Scott, 2005). These records represent the earliest known occurrences of this genus.

ADDITIONS FROM NEW LOCALITIES (SEE APPENDIX I)

?*Titanoides primaevus* (localities NP47B, BB) (noted as ?*Pantolambda* sp. in Vol. 1).
Coryphodon eocaenus (locality [GC21II]); *C. lobatus* (locality [CP20BB]); *C.* sp. (localities GC20II, CP20AA, BB).
Pantodont indet. (locality CP13IIC).

DINOCERATA

ADDITIONS FROM NEW LOCALITIES (SEE APPENDIX I)

Cyriacotherium sp. (locality NP3A).
Titanoides sp. (locality NP3A).
Prodinoceras (= *Probathyopsis*) sp. (locality CP17B).
Bathyopsis sp. (locality CP25J).
Uintatherium sp. (locality CC7A).
Uintatheriine indet. (locality CC4).

ARCHAIC UNGULATES

CORRECTIONS TO VOLUME 1

Hyopsodus paulus and *H. minusculus* are known from locality CP5**A**.

Phenacodus trilobatus is known from localities CP5**A**, CP20**A**, CP25**A**, CP27**A**, and CP64**A**.
Ectocion sp. is also known from locality NP48B.

SYSTEMATIC REVISIONS, NEW TAXA, AND IMPORTANT NEW LOCALITY OR PALEOBIOLOGICAL INFORMATION

New taxa of archaic ungulates are reported from the Puercan of Wyoming (Eberle and Lillegraven, 1998a,b): *Loxolopus faulkneri* (localities CP11IIE, F); *Ectoconus cavigelli*, (locality CP11IID); *Baioconodon middletoni* (locality CP11IID); *Ampliconus browni* (genus not in Vol. 1) (locality [CP11IIE]); *A. antoni* (locality CP11IID); and *Alticonus* (? = *Tinuvial*) *gazini* (n. gen. and sp.) (locality CP11IID).

Eberle *et al.* (2002) described new occurrences of archaic ungulates in the Puercan of Colorado (locality CP61). The occurrence in Pu 1 strata (locality CP61A) of *Protungulatum donnae* marks the southern-most occurrence of this taxon, and the occurrence of *Oxyclaenus simplex* marks both a temporal and geographical range extension for this species. New findings in the Pu2 strata (locality CP61B) include *Promioclaenus* and *Loxolophus faulkneri*.

McKenna and Lofgren (2003) described a new species of arctocyonid, *Mimotricentes tedfordi*, from Laudate Locality in the late Paleocene of California (locality NB1A) and suggested that this local fauna may be Tiffanian in age rather than Torrejonian (recorded in this volume as To3–Ti1).

Penkrot *et al.* (2003) described new postcranial material of the hyopsodontid *Apheliscus* and showed that it was very different from *Hyopsodus*. While *Hyopsodus* shows scansorial features, *Apheliscus* shows cursorial or saltatorial postcranial adaptations (resembling leptictids and macroscelidids). Tarsals tentatively assigned to *Haplomylus* show morphological similarities to those of *Apheliscus*, implying similar locomotor adaptations.

Lucas and Kondrashov (2004b) revised the periptychid genus *Ectoconus*. They concluded that *E. majusculus* (species not listed in Vol. 1) is homologous with *E. ditrigonus*, and that *E. cavigellii* (described in 1998) is homologous with *E. symbolus*. Some additional localities are listed from those in Vol. 1: *E. ditrogonius* (localities CP11IIF, H); *E. symbolus* (localities CP1C, CP11IIA, D).

Zack *et al.* (2005) described a new species of apheliscine "condylarth," *Gingerichia*, from the late Paleocene (early Tiffanian, Ti1). *G. geoteretes* is known from Montana (locality NP19II), and *G. hystrix* from Alberta (locality NP1C). They also consider that *Hyopsodus* and mioclaenids form a monophyletic group to the exclusion of other "hyopsodontids," while these other hyopsodontids form a monophyletic clade with the apheliscines. In addition, *Hapaladectes serior* is recognized as the lower dentition of *Utemylus latomius*, or a close relative.

A new species of *Eoconodon, E. ginibitohia*, is reported from early Paleocene of New Mexico (locality SB23B; Clemens and Williamson, 2005).

Based on new material of *Microclaenodon assurgens*, Williamson (2004) claimed that *Oxyclaenus* is a junior synonym of *Microclaenodon*. Phylogenetic analysis also posits *Microclaenodon assurgens* as the sister taxon to the Mesonychia.

Pachyaena is reported from the Wasatchian of Ellesmere Island, Canada (locality HA1A2; Eberle 2001).

Dissacus sp. is now definitely known from locality CP17A (previously a queried occurrence) (Gunnell and Bartels, 2001).

Zack (2004a) described new material of *Wyolestes*, listed in Vol. 1 as a didymoconid, possibly included within the Mesonychoidea. New postcranial material reveals scansorial habits, like the extant coatimundi (*Nasua*). There is a long, narrow rostrum, and a strong, horizontal nuchal plate, as in *Solenodon*. These features weaken the proposed relationships of this taxon to didymoconids or mesonychids, and the skeletal morphology is more consistent with creodont affinities.

Chester *et al.* (2005) discussed the occurrences of archaic ungulates, especially the genera *Phenacodus, Ectocion*, and *Copecion* in the latest Paleocene (Cf3) and earliest Eocene (Wa0) of the Bighorn Basin, Wyoming, and argued against the notion of transient dwarfing previously reported for the latter two genera.

Scott, Webb, and Fox (2006) described a new mammal, *Horolodectes sunae*, from the earliest Tiffanian of the Paskapoo Formation, Alberta (within locality NP3A). This animal bears a resemblance to various "ungulatomorph" eutherians, but the teeth are sufficiently different to prevent unequivocal referral to any known eutherian group.

ADDITIONS FROM NEW LOCALITIES (SEE APPENDIX I)

Protungulatum donnae (localities [CP11IIA, B]); *P. sloani* (locality CP11IIB); *P. gorgun* (localities CP11IIC, NP16A).

Carcinodon sp. (locality ?CP12C).

Princetonia yalensis (locality CP20AA).

Prothryptacodon albertensis (locality CP3A0).

Chriacus badgleyi (locality CP20AA); *C. oconostotae* (locality NP3A); *C.* sp. (localities CP20AA, BB, ?NP3A0).

Loxolophus priscus (localities CP11IID. G); *L. hyattianus* (localities [CP11IID, E, G]); *L. schizophrenus* (locality CP12C); *L.* sp. (localities [CP11IIE], ?CP12C).

Mimotricentes subtrigonus (localities CP13IIA, ?B).

Arctocyon ferox (localities CP13IIA, B, NP3A); *A. mumak* (locality CP13IIC); *A. corrugatus* (species not in Vol. 1) (locality NP3A); *A.* sp. (locality NP47BB).

Colpoclaenus procyonoides (locality NP3A0).

Thryptacon antiquus (locality CP20BB); *T. australis* (localities NP3A, NP47BB); *T. barae* (locality CP20AA); *T.* sp. (locality CP13IIB).

Arctocyonid indet. (localities NP3G, NC0).

Dorraletes diminutivus (locality NP47BB).

Haplaletes disceptatrix (locality CP13IIB).

Haplomylus speirianus (locality GC21II); *H. palustris* (species not in Vol. 1) (locality CP17A).

Hyopsodus uintensis (localities [SB42], CP6); *H. ?simplex* (locality CP20BB); *H. miticulus* (localities CP20BB, E); *H. minusculus* (localities CP20AA, CP25I2, CP34A, C); *H. paulus* (locality CP34C); *H. uintensis* (= *H. "fastigatus"*) (localities CP6AA, NP8); *H. wortmani* (localities CP25I1, I2); *H.* sp. (localities SB51, CP20BB, CP25GG, J, CP34A, C, E).

Litomylus dissentaneus (= *L. "scapheus"*) (locality CP13IIA); *L.* n. sp. (locality NP3A0).

Oxyprimus galadriele (locality CP11IIB); *O.* sp. (localities CP11IIA, B).

Litaletes manitiensis (locality CP12D).

Promioclaenus wilsoni (localities CP11IIF, G); *P. lemuroides* (localities [CP13IIA, B]); *P. acolytus* (localities CP13IIB, [NP3A0]).

Mioclaenus turgidus (locality CP13IIA).

Protoselene sp. (localities ?CP11IIG, [CP13IIB]).

Apheliscus insidiosus (locality CP20BB).

Tinuviel eurydice (locality CP12C).

Maiorana ferrisensis (n. sp.) (locality CP11IIA, D).

Mimatuta sp. (localities [CP11IIA], NP16A).

Ectoconus ditrigonus (localities CP11IIF, H); *E.* sp. (locality CP11IIC).

"Carsioptychus" (= *"Periptychus"*) *coarctatus* (localities CP11IIF, G).

Periptychus carinidens (locality [CP13IIA]); *P.* sp. (locality CP12D).

Periptychid indet. (localities CP11IIC, D, F, G).

Gillisoncus (= *"Mithrandir"*) *gillianus* (localities CP11IIA, D–G).

Conacodon delphae (species not in Vol. 1) (locality CP11IIC); *C. harbourae* (species not in Vol. 1) (locality CP11IIC); *C. cophater* (localities [CP11IIE, F, G]).

Oxyacodon agapetillus (localities CP11IIA, B, D, [E], F); *O. priscilla* (localities CP11IID, F, G); *O. ferronensis* (locality CP12C); *O. apiculatus* (locality CP12C).

Baioconodon denverensis (locality [CP11IIG]); *B. nordicum* (species not in Vol. 1) (locality [CP12C]); *B.* sp. (locality CP11IIF).

Ragnarok nordicum (locality NP15C).

Oxyclaenus cuspidatus (localities CP11IID, G); *O.* sp. (localities [CP11IIF], ?CP12C).

Eoconodon nidhoggi (locality CP12C); *E.* sp. (localities CP11IIA, E).

Hapalodectes sp. (locality CP25GG).

Dissacus navajovius (locality [CP13IIA]); *D.* sp. (localities CP17A, B, CP20AA, NP3A).

Harpagolestes sp. (localities [CC4], NP8).

Hapalorestes lovei (hapalodectid genus, not in Vol. 1) (locality CP31E).

Mesonyx obtusidens (localities [CP33C], CP34C); ?*M.* sp. (localities CP25J, CP65).

Mesonychid indet. (localities CC7B, SB42).

Tetraclaenodon puercensis (locality CP13IIA).

Phenacodus intermedius (localities CP17A, CP20E); *P. primaevus* is *not* present at locality NP47B (now = NP47BB) (this is *Dorraletes diminutivus*), but *P. grangeri is* present at locality NP47BB (formerly described as *Ectocion wyomingensis*); *P. trilobatus* (locality CP20BB); *P. vortmani* (localities [CP20AA, BB, E]); *P.* sp. (localities NB1A, B, CP17B, CP20AA, BB, ?NC0).

Copecion davisi (locality CP20AA).

Ectocion collinus (locality CP47B [not *E. wyomingensis*, as in Vol. 1]); *E. parvus* (localities GC21II, CP20AA); *E. cedrus* (locality [NP3A]); *E. osbornianus* (locality CP20BB); *E.* sp. (localities CP13IIA, NP47BB).

Meniscotherium chamense (localities CP25I1, I2); *M. tapiacitum* (locality [CC1]); *M.* sp. (locality CP20AA).

ARCTOSTYLOPIDA

SYSTEMATIC REVISIONS, NEW TAXA, AND IMPORTANT NEW LOCALITY OR PALEOBIOLOGICAL INFORMATION

An arctostylopid indet. is reported from the Wasatchian of Wyoming (locality CP25C; Zack, 2004b).

Kondrashov and Lucas (2002) described new arctostylopids from Mongolia and claimed, contra to Cifelli, Vol. 1, that arctostylopids are in fact notoungulates, and that the term Arctostylopida should be abandoned. Bloch (1999) also concluded that arctostylopids had affinities with notoungulates, based on new postcranial material, which also indicates that these animals were cursorial.

GENERAL ARTIODACTYLA

SYSTEMATIC REVISIONS, NEW TAXA, AND IMPORTANT NEW LOCALITY OR PALEOBIOLOGICAL INFORMATION

Two papers that reviewed recent developments in molecular phylogenies, both using the approach of supertrees, are Fernández and Vrba (2005) and Price, Bininda-Edmonds, and Gittleman (2005). Both provide a comprehensive review of the recent literature in this field. Both papers show moschids as the sister taxon of cervids, bovids and cervids (plus moschids) as sister taxa, but antilocaprids clustering with giraffids outside of this grouping. Price, Bininda-Edmonds, and Gittleman (2005) have Hippopotamidae (plus Cetacea) as the sister taxon to the Ruminantia, and camelids as the sister taxon to all other artiodactyls.

ADDITIONS FROM NEW LOCALITIES (SEE APPENDIX I)

Indeterminate pecorans are reported from localities CA2II (two species), CA2III (two species). Indeterminate artiodactyls are reported from localities GC5A (small) and CC3.

"DICHOBUNIDS"

SYSTEMATIC REVISIONS, NEW TAXA, AND IMPORTANT NEW LOCALITY OR PALEOBIOLOGICAL INFORMATION

New species of *Ibarus* are reported from the Uintan of California (localities CC7B, CC8; Walsh, 1998).

Stibarus sp. from locality NP24D is now identified as *Stibarus montanus* (Tabrum, Prothero, and Garcia, 1996).

A new species of *Microsus* is reported from the Bridgerian of Wyoming (locality CP25J; Zonneveld, Gunnell, and Bartels, 2000).

A new genus and species of helohyid is reported from the Uintan of Colorado (locality CP65; Stucky *et al.*, 1996).

A new helohyid, *Simojovelhyus pocitosense*, is described from the late Oligocene of Mexico (locality CA0; Ferrusquía-Villafranca, 2007). It differs from other known genera in having a tiny paraconid and a unique combination of cristids and cuspulids. This taxon extends both the geochronological and geographical age of the Helohyidae (the youngest previously known was from the late Eocene, and helohyids have not been previously described from Central America).

Norris (1999) examined crania of the homacodontid *Bunomeryx* and suggested that this taxon had affinities with the Tylopoda (see further comments below, in the section on Protoceratidae).

ADDITIONS FROM NEW LOCALITIES (SEE APPENDIX I)

Diacodexis metsiacus (species not in Vol. 1) (locality CP20BB); *D. secans* (locality CP25I2); *D.* sp. (localities GC21II, CP25GG, CP34A).

Bunophorus sinclairi (locality [CP25GG]); *B.* sp. (localities CP20BB, CP25I1, J, [CP34A]).

Stibarus montanus (localities NP10B, Bi, NP49II).

Leptochoerus sp. (localities NP10BB, C2, NP24C).

Tapochoerus mcmillini (species not in Vol. 1) (localities CC7B, CC8).

Microsus cuspidatus (locality CP34A).

Antiacodon venustus (locality CC4); *A. pygmaeus* (localities CP34[A3], C); *A.* spp. (small and large) (locality CP25J).

Hexacodus sp. (localities CP25GG, J, CP34A).

Homacodon sp. (locality CP34C).

Homacodontid n. spp. A and B (locality NP22).

Mesomeryx grangeri (locality CP6AA).

Achaenodon robustus (localities [CC4], [CP38C]).

Parahyus sp. (locality CC4).

ENTELODONTIDAE

SYSTEMATIC REVISIONS, NEW TAXA, AND IMPORTANT NEW LOCALITY OR PALEOBIOLOGICAL INFORMATION

Foss (2000) performed a phylogenetic analysis of the Entelodontidae, and noted that the taxon *Archaeotherium coarctatum* is separate not only from other species of *Archaeotherium*, but also from other Oligo-Miocene entelodonts, and forms the sister taxon to the Asian *Eoentelodon* and the North American *Brachyhyops*.

Lucas, Foss and Mihlbachler (2004) reported a new species of *Achaenodon, A. fremdi*, from Hancock Quarry, late Uintan of Oregon (locality PN6B).

A new specimen of *Dinohyus* (= *Daeodon*) *hollandi* from the early Miocene of California (locality CC16, previously identified as late late Arikareean in age) is now designated as early

Hemingfordian (see Woodburne, 2004). This extends the biogeographic range of the genus (Lucas, Whistler and Wagner, 1997).

Lucas and Emry (2004) discussed the enigmatic taxon *Brachyhyops*, which they considered to firmly belong within the Entelodontidae (in contrast to Vol. 1), and describe new specimens from the early Chadronian Wyoming (locality CP39A), as well as discussing the stratigraphic distribution of the genus (known additionally from localities SB25B [*B. viensis*] and CP39IIB [type locality of *B. wyomingensis*]).

ADDITIONS FROM NEW LOCALITIES (SEE APPENDIX I)

Archaeotherium coarctatum (locality NP10Biii); *A. mortoni* (locality NP10B3); *A. zygomaticum* (species not in Vol. 1) (locality [NP10C]); *A.* sp. (locality NP10Bii).
Dinohyus sp. (locality ?GC5A).
Entelodontidae indet. (locality NP10Bi).

ANTHRACOTHERIIDAE

CORRECTIONS TO VOLUME 1

Elomeryx armatus is known from locality CC9**C**.

ADDITIONS FROM NEW LOCALITIES (SEE APPENDIX I)

Heptacodon sp. (localities CP5II, PN5B).
Bothriodon sp. (locality NP10Bi).
Elomeryx armatus (locality NP10BB); *E.* sp. (localities NP10B3, C3, ?NP32C).
Aepinacodon sp. (locality CP5II).
Arretotherium acridens (locality GC5A); *A.* sp. (locality NP10C3).
Anthracotheriidae indet. (localities NP10Biii, NP49II).

TAYASSUIDAE

CORRECTIONS TO VOLUME 1

Cf. *Perchoerus proprius* is known from localities CP84**A**, **B**.
Hesperhys sp. is known from locality NC1**A**.
Thinohyus lentus is known from locality PN6**C**.
Prosthennops serus is known from locality PN11**B** (new locality subdivision, emended dating of Hh2).
Platygonus sp. is known from localities SB31**C**, SB68.

SYSTEMATIC REVISIONS, NEW TAXA, AND IMPORTANT NEW LOCALITY OR PALEOBIOLOGICAL INFORMATION

An unidentified tayassuid is reported from the late Miocene/early Pliocene Gray Fossil Site in Tennessee (locality CE1; Wallace and Wang, 2004), and from the early Pliocene Pipe Creek Sinkhole Site in Indiana (locality CE2; Farlow *et al.*, 2001).

ADDITIONS FROM NEW LOCALITIES (SEE APPENDIX I)

Hesperhys sp. (locality ?GC5A).
Floridachoerus olseni (locality ?GC5A).
"*Cynorca*" *sociale* (locality GC5A); *C.* sp. (locality ?CC9IVC).
Dyseohyus or "*Cynorca*" sp. (localities GC4DD, GC8AA [= GC7 in Vol. 1], GC8B [= GC8A in Vol. 1]).
Dyseohyus sp. (locality ?CP75B).
"*Prosthennops*" *xiphidonticus* (localities [SB55], CP75BB, NC3D).
Prosthennops serus (localities GC11D, GC13B, [GC27]); *P.* sp. (localities [CA2II], SB55IIAA, SB55V, SB56II, SB58IIB).
Love Bone Bed species (locality GC10D).
Platygonus pearcei (locality [NB35IIIB]); *P. bicalcaratus* (species not in Vol. 1) (localities GC14A [= *P.* sp. in Vol. 1], GC15A, ?D, GC17A [= *P.* sp. in Vol. 1], B, SB34IVA, [SB35A], [SB36B], SB37A, D); *P. vetus* (species not in Vol. 1) (localities GC15C, E, GC18IIIA, GC18IV); *P.* sp. (locality SB58IIC).
Mylohyus fossilis (localities ?GC15D, GC18IIIA); *M. floridanus* (locality GC14A [= *M.* sp. in Vol. 1], GC15A [= *M.* sp. in Vol. 1], GC17A [= *M.* sp. in Vol. 1], GC18IV).
Tayassuidae indet. (localities SB35D, CP56IIB, NP10CC).

OREODONTOIDEA

CORRECTIONS TO VOLUME 1

Eporeodon occidentalis is known from locality NP28**A**.
Eporeodon major is known from locality NP28**A**.
Merycochoerus chelydra chelydra is questionably known from localities CP52 (not CP52A), and CP64?**A**.
Mercoidodontid indet. is known from locality CP72**A**.

SYSTEMATIC REVISIONS, NEW TAXA, AND IMPORTANT NEW LOCALITY OR PALEOBIOLOGICAL INFORMATION

A new species of *Protoreodon*, *P. walshi* (previously *P.* cf. *P. parvus*), is reported from the late middle Eocene of California (locality CC7B) (a second, unnamed, species may also be present at this locality) (Theodor, 1999).

Stevens and Stevens (1996, 2007) provided a revision of some subfamilies within the Merycoidodontidae (see below). Cobabe (1996) provided a revision of the Lepatauchiniinae.

Specimens of *Diplobunops* from the late Eocene of Oregon (locality PN5B) have been reassigned to *Agriochoerus* (Foss *et al.*, 2004).

A new species of *Merycoidodon* (= *Prodesmatochoerus* in Vol. 1), *M. presidioensis*, is described by Stevens and Stevens (1996) from the Chadronian of Texas and Wyoming (localities SB44D [type] and CP39B).

A new species of oreodont, *Mesoreodon* (= *Eporeodon* in Vol. 1) *floridanus*, distinguished by a large auditory bulla, is reported from the late early Arikareean of Florida (locality GC8AB) (MacFadden and Morgan, 2003).

The species of *Merycochoerus* in the Gaillard Cut Local Fauna (locality CA7; early Hemingfordian of Panama), identified as *M. proprius* in Vol. 1, is identified by MacFadden (2006) as *M. matthewi*.

A new late late Arikareean (Ar4) formation from New Mexico (Seventyfour Draw, locality SB28II) contains two oreodont taxa. *Desmatochoerus* cf. *D. megalodon* (? = *Eporeodon* in Vol. 1) and *Megoreodon* cf. *M.* grandis (= *Merychochoerus superbus* in Vol. 1) previously identified as coming from younger strata (early Barstovian) in the Gila Group (Morgan and Lucas, 2003).

SUMMARY OF REVISION OF STEVENS AND STEVENS, 2007

Here we summarize some of the major points in this paper, especially to show how their classification differs from that of Lander in Vol. 1. The subfamily Aclistomycterinae Lander, 1998 is recognized, but Bathygeniinae Lander, 1998 is not. *Bathygenus* is now included along with *Oreonetes* in the Oreonetinae (this subfamily was synonymized with Miniochoerinae in Vol. 1).

Leptauchiniinae is recognized, as in Vol. 1. This subfamily includes *Leptauchenia* and *Sespia* (as in Vol. 1), and also *Limnetes* (*L. platyceps* was synonymized with *Leptauchinia platyceps* in Vol. 1; "*Limnetes*" *anceps* is considered to be the same taxon as *Oreonetes anceps* both in Vol. 1 and here). The Miniochoerinae is recognized, containing only *Miniochoerus* (*Miniochoerus* was synonymized with *Oreonetes*, in part, in Vol. 1); species of *Miniochoerus* recognized here are *?M. forsythae, M. chadronensis, M. affinis, M. starkensis*, and *M. gracilis*. "*Prodesmatochoerus*" (= *Merycoidodon*) was included in the Miniochoerinae in Vol. 1, but not here, where the Merycoidodontidae is recognized for *Merycoidodon* and *Mesoreodon*. *Merycoidodon* was synonmyized with *Prodesmatochoerus* in Vol. 1, but *Prodesmatochoerus* is not recognized here. *Merycoidon* here includes *M. presidioensis, M. culbertsoni, M. bullatus*, and *M. major*. *Mesoreodon* was synonymized with *Eporeodon* (in part), *Blickohyus* (in part), and *Merycoides* (in part) in Vol. 1. *Merycoidodon major* is considered here to be the ancestor of *Mesoreodon*, which includes *M. chelonyx* (= *Eporeodon occidentalis* in Vol. 1) and *M. minor* (? = *Eporeodon occidentalis minor* in Vol. 1).

The subfamily Promerycochoerinae is recognized (not recognized in Vol. 1). This includes *Desmatochoerus*, *Promerycochoerus*, and *Megoreodon*, none of which were recognized in Vol. 1. *Desmatochoerus* was synonymized with *Eporeodon* (in part) in Vol. 1, and *Eporeodon* was placed in the subfamily Merycochoerinae (but here is placed in the family Eporeodontidae, see below). The only species of *Desmatochoerus* considered valid here is *D. megalodon* (? = *Eporeodon occidentalis major* in Vol. 1). *Promerycochoerus* was synonymized with *Eporeodon* (in part), and with *Merycochoerus* (in part). *Promerycochoerus superbus* (= *Merycochoerus superbus* in Vol. 1), *P. carrikeri* (= *M. carrikeri* in Vol. 1), and *P. chelydra* (= *M. chelydra* in Vol. 1) are recognized here. The only species of *Megoreodon* recognized is *M. grandis* (synonymized with *Merycochoerus superbus* in Vol. 1).

The subfamily Merycochoerinae is recognized, but its composition differs from the Merycochoerinae of Vol. 1, which included many of the species now placed in the Promerycochoerinae (see above). *Hypsiops*, *Submerycochoerus*, and *Merycochoerus* are included here. The species of *Hypsiops* considered valid, *H. latidens* and *H. breviceps*, are probably equivalent to *Hypsiops* as used in Vol. 1, with the exception of *H. bannackensis*, which is considered here to belong to *Submerycochoerus* (not recognized in Vol. 1). Note that in Vol. 1, *Hypsiops* is placed within the Phenacocoelinae, not the Merycochoerinae. The genera of *Merycochoerus* considered valid are *M. matthewi* and *M. proprius* (both included in *Merycochoerus* in Vol. 1).

The subfamily Eporeodontinae is recognized (not recognized in Vol. 1), including *Eporeodon* (placed in the Merycochoerinae in Vol. 1) and *Merycoides* (placed in the Phenacocoelinae in Vol. 1). The species of *Eporeodon* considered valid here are *E. occidentalis, E. trigonocephalus, E. pacificus*, and *E. thurstoni*. Phenacocoelinae of Vol. 1 is not recognized here: the other genera included in this subfamily in Vol. 1 are *Hypsiops* (here placed in the Merycochoerinae, see above), *Oreodontoides* (here placed in the Merychyinae), *Phenacocoelus*, and *Paroreodon* (here both placed in the Ticholeptinae). The Ticholeptinae here also includes *Ticholeptus*. The Ticholeptinae of Vol. 1 also included *Merychyus* (including *Ustatochoerus*), *Mediochoerus*, and *Brachycrus*, which are discussed below. In Vol. 1, *Ticholeptus* only contained the genotypic species *T. zygomaticus*. Here it also contains *T. bluei* (? = *Merychyus elegans bluei* in Vol. 1) and *T. calimontanus* (= *Merychyus elegans ?smithi* in Vol. 1).

The subfamily Merychyinae here includes *Merychyus* (not incorporating *Ustatochoerus* as in Vol. 1, see below), *Oreodontoides*, and *Paramerychyus*. The only recognized species of *Paramerychyus* is *P. harrisonensis* (= *Merycoides harrisonensis* in Vol. 1). *Oreodontoides* has two species: *O. oregonensis* and *O. curtus* (both considered to be *O. oregonensis* in Vol. 1). The species of *Merychyus* recognized here are *M. calaminthus, M. arenrum, M. verrucomalus, M. elegans*, and *M. relictus*.

The subfamily Ustatochoerinae, informal, is being proposed for the genera *Ustatochoerus* and *Mediochoerus*. *Ustatochoerus* was synonymized with *Merychyus* in Vol. 1: species recognized here (all much larger than the species of *Merychyus*) are *U. leptoscelos, U. medius, U. major*, and *U. californicus*. The composition of *Mediochoerus* is more or less the same as in Vol. 1. A new subfamily is also being informally proposed for the genus *Brachycrus*, another large-sized, hypsodont form that appears to have evolved independently of other similar taxa in the Ticholeptinae and Ustatochoerinae. The species recognized (*B. rusticus, B. wilsoni, B. siouense*, and *B. laticeps*) are similar to the situation in Vol. 1.

ADDITIONS FROM NEW LOCALITIES (SEE APPENDIX I)

Protoreodon walshi (? = *P. annectans tardus* in Vol. 1) (localities CC7B, CC8, CC8IIA); *P. annectans* (CC7D); *P.*

parvus (localities SB42, NP8); *P. petersoni* (= "new genus B" in Vol. 1) (locality SB42); *P. minimus* (= "new genus C" in Vol. 1) (locality NP25C); *P.* sp. (localities CC7E, CP6AA, NP22 [small]).

Agriochoerous antiquus (locality NP10Biii); *A.* (= "*Diplobunops*") *matthewi* (locality [GC23A]); *A. pumilus* (= *Protoreodon*) (locality NP22); *A.* (= "*Diplobunops*") n. sp. (locality PN5B); *A.* sp. (localities NP10B, Bi, BB, C, C2).

Agriochoeriidae n. gen. and sp. (locality CP65).

Agriochoeriidae indet. (locality CC8IIB).

Bathygenus sp. (locality NP10B).

Leptauchenia platyceps (= *Limnenetes*) (locality NP25C).

Sespia nitida (locality CC9IVB).

Oreonetes anceps (locality NP25C); *O.* sp. (= *Miniochoerus*) (locality ?NP32C).

Prodesmatochoerus periculorum periculorum (= *Merycoidodon culbertsoni*) (locality NP32C); *P.* (= *Merycoidodon*) sp. (locality NP10Bii).

Eucrotaphus sp. (= *Merycoidodon* [*Anomerycoidodon*]) (locality NP10C).

Phenacocoeline indet. (locality GC8B [= GC8C in Vol. 1]).

Merycochoerus matthewi (locality CA1); *M.* (= *Promerycochoerus*) sp. (locality NP10CC).

Merychyus elegans (locality CC9IVD) *M. medius medius* (= *Ustatochoerus medius*) (localities CP75B, BB); *M.* sp. (locality [GC4DD]).

Ticholeptus zygomaticus (localities CP75B, BB).

Brachycrus laticeps siouxense (= *B. siouxense*) (locality CP75BB).

Merycoidodontidae indet. (localities GC7, NP10Bi, C2, NP25C [large sp.]).

Fragmentary indeterminate oreodonts are also known from the following Florida localities (see MacFadden and Morgan, 2003): GC7, GC8AA, GC8B (= cf. *Mesoreodon floridanus*), GC8C (= cf. *Phenacocoelus luskensis* [species not listed in Vol. 1]), GC8D (= cf. *Merychyus minimus*).

OROMERYCIDAE

SYSTEMATIC REVISIONS, NEW TAXA, AND IMPORTANT NEW LOCALITY OR PALEOBIOLOGICAL INFORMATION

Merycobunodon littoralis (previously = ?*Lophiohyus* [= *Helohyus*]) and a new small species of *Merycobunodon* are reported from the Uintan of California (locality CC4; Walsh, 2001).

ADDITIONS FROM NEW LOCALITIES (SEE APPENDIX I)

Malaquiferus sp. (locality NP10B).

Protylopus stocki (locality CC7B); *P. robustus* (locality [CC7C]); *P.* sp. (localities CC4, ?CC6A, CC8, CC8IIA, CP34E).

Oromeryx sp. (locality CP65).

Eotylopus reedi (locality [NP10B]).

PROTOCERATIDAE

CORRECTIONS TO VOLUME 1

Leptotragulus sp. is known from locality CC7E.

SYSTEMATIC REVISIONS, NEW TAXA, AND IMPORTANT NEW LOCALITY OR PALEOBIOLOGICAL INFORMATION

Leptoreodon sp. in Vol. 1 at locality NP22 is now identified as *L. marshi* (Tabrum, Prothero, and Garcia, 1996).

A new species of *Leptoreodon, L. golzi*, is reported from locality CC7B (Ludtke and Prothero, 2004).

The *Paratoceras* sp. from locality CA7 in Vol. 1. is now identified as *P. wardi* (MacFadden, 2006).

A new species of *Paratoceras, P. tedfordi*, is reported from the late Arikareean of Mexico (locality CA0; Webb, Beatty, and Poinar, 2003), which is the earliest known record of this genus. This skull lacks horns but has rugosities on the maxillary bones and in the supraorbital region and is most likely that of a female.

Two fragmentary protoceratid species are reported from the early Barstovian of Mexico (locality CA1; Ferrusquía–Villafranca, 2003). One is referred to *Paratoceras* sp.; the other is a new genus of kyptoceratine, highly hypsodont and larger than *Kyptoceras amatorum* from the Hemphillian of Florida. These specimens, and the one described below, represent the only definitive protoceratids known from Central America.

A new species of *Prosynthetoceras, P. orthrionanus*, is reported from the late Arikareean of the Texas Gulf Coast (locality GC5A; Albright, 1999).

A new species of *Synthetoceras, S. davisorum*, is reported from the Mauvilla Local Fauna of Alabama (locality GC27). Biochronological analysis of the mammals indicates a younger age than previously considered for this locality (late early Hemphillian rather than latest Clarendonian/earliest Hemphillian) (Hulbert and Whitmore, 2006).

Jiménez-Hidalgo and Carrenza-Casteñeda (2005) reported a possible record of *Kyptoceras* from the latest Hemphillian of central Mexico (locality SB57III). This is the first record of this taxon in the Pliocene of Mexico.

Webb, Beatty, and Poinar (2003) have speculated on the biogeography and adaptive morphology of protoceratids. They noted that, especially with new discoveries in Central America, the group has a primarily tropical distribution and, consequently, their survival in the Gulf Coast region in the Miocene may be an important paleoenvironmental indicator. They also considered that protoceratids may have had a tapir-like proboscis (evidenced by retracted nasals) (although note that retracted nasals may not always be indicative of a proboscis [Clifford and Witmer, 2002; Andrew Clifford personal communication]). Isotopic analysis of *Kyptoceras amatorum* reveals values of oxygen and carbon isotopes consistent with a habit of deep forest browsing, and the limb proportions (long radius and tibia, hindlimbs proportionally longer than forelimbs) resemble those of forest-dwelling tragelaphine bovids such as the bushbuck (*Tragelaphus scriptus*). They also noted that the Synthetoceratinae are unusual in having a knee morphology that would greatly restrict the rotation of the tibia around the long axis of the hindlimb, and they speculated that it might be an adapation to steady the limb while raising the body over the hindlimb (perhaps for high-level browsing), and/or that it might be an adaptation for a leaping style of locomotion in closed woodland.

Considerable doubt has been cast since the mid 1990s concerning the relationships of the Protoceratidae as the sister taxon of the Oromerycidae plus Camelidae, as supported in Vol. 1. Joeckel and Stavas (1996) found little similarity between the braincases of protoceratids and camelids and suggested that camelids might be better considered as basal ruminants. Norris (1999) supported excluding protoceratids from a clade consisting of camelids, bunomerycid "dichobunids," oreodonts, and various European "tylopods," and later (Norris, 2000) also noted that protoceratids were as likely to be basal ruminants as basal tylopods. Webb, Beatty, and Poinar (2003) also noted a possible dental synapomorphy between protoceratids and tylopods: the condiiton of "lingual radicozygy," where the anterolingual and posterolingual roots of each molar are fused to form an irregular anteroposterior dike. More recent analyses of the relationships of all artiodactyls have not resulted in a consensus about protoceratids (see Prothero and Ludtke, 2007). Geisler (2001), using morphological data, placed protoceratids plus oreodonts in a clade together with two European taxa: amphimerycids (usually considered as basal ruminants) and xiphodonts (usually considered as tylopods). This clade is then the sister taxon to a grouping of Ruminantia plus Cameloidea. The total evidence approach of Geisler and Uhen (2005), using both morphological and molecular data, resulted in no strict consensus for the position of protoceratids: many trees placed them as the sister group of Ruminantia, but the addition of stratigraphic data added support for their position as basal tylopods. The analysis of deciduous dental characters by Theodor and Foss (2005) supported the placement of protoceratids as basal selenodonts but did not further clarify their relative position to ruminants versus camelids. The systematic position of protoceratids, whether with the Tylopoda or the Ruminantia, remains in question.

ADDITIONS FROM NEW LOCALITIES (SEE APPENDIX I)

Leptotragulus proavus (locality C6AA); *L.* (= "*Trigenicus*") *profectus* (localities NP10Bi, Bii).

Leptoreodon major (locality CC6A); *L.* sp. (localities CC8, IIA, CC9IVA).

Leptotraguline indet. (locality CC7E).

Pseudoprotoceras seminictus (locality NP10Biii); *P.* sp. (localities NP10Bi, Bii).

Protoceras celer (localities NP10BB, C); *P.* sp. (localities NP10B3, C).

Prosynthetoceras francisi (locality GC4DD); *P. texanus* (localities GC5A, GC9II).

Lambdoceras sp. (locality ?CP1114A: AMNH collections).

CAMELIDAE

CORRECTIONS TO VOLUME 1

Miotylopus gibbi is known from locality CP87**A**.

The taxon "Miolabine sp." should have been reported as being equivalent to "*Homocamelus*."

Procamelus sp. and *Megatylopus* sp. are known from locality PN11**A** and **B** (new locality subdivision, PN11B with emended dating of Hh2).

SYSTEMATIC REVISIONS, NEW TAXA, AND IMPORTANT NEW LOCALITY OR PALEOBIOLOGICAL INFORMATION

A new species of goat-like miolabine, *Capricamelus gettyi*, is reported from the latest Pliocene of California (locality NB12II; Whistler and Webb, 2005). This species not only greatly extends the known range of the miolabines but also has extremely short distal limb proportions, resembling that of mountain goats, unique among camelids. This new specimen suggests a sister-group relationship between the Miolabinae and the Stenomylinae.

Paramiolabis "minutus," a dwarfed miolabine related to *P. singularis*, is reported from the Barstovian of the Barstow Formation, California (localities NB6B, D), and is also known from New Mexico (locality SB32B) (these occurrences were reported as *Paramiolabis* sp. by Honey *et al.* in Vol. 1). A new species of *Miolabis*, intermediate between *M. agatensis* and *M. yavapaiensis*, is also reported from the Barstow Formation (localities NB6A–C) (reported as *Miolabis* sp. by Honey *et al.* in Vol. 1) (Pagnac, 2004).

Cf. *Pliauchenia* sp. is reported from the early Barstovian of Mexico (locality CA2II; Ferrusquía-Villafranca, 2003), extending the biogeographic range of this genus. (Cf. *Protolabis* sp. and Cf. *Procamelus* sp. are also known from this locality.)

A new species of lamine, *Hemiauchenia gracilis*, is reported from the Blancan and Irvingtonian of Florida (localities GC14A, C, GC15C, GC18IIB [type locality] and possibly at SB18; Meachen, 2005). This species has teeth similar to *H. macrocephala* and *H. edensis* but has very long and slender postcranial elements.

The first definitive Pliocene record of *Hemiauchenia (H. blancoensis)* in Mexico (locality SB65) is reported by Jiménez-Hildago and Carranza- Casteñeda (2002).

Jiménez-Hidalgo and Carrenza-Casteñeda (2005) reported a number of camelids from the late early to latest Hemphillian of central Mexico (locality SB57III), including *Hemiauchenia vera* (Hh2–Hh4), a small species of *Hemiauchenia* (Hh3–Hh4), *Alforjas* sp. (Hh3–Hh4), and *Megatylopus matthewi* (Hh3–Hh4). The records for *Alforjas* and *Megatylopus* are the youngest for North America, as well as the southern-most records. They also noted that previous studies have indicated that these taxa were browsers or browser-like mixed feeders.

Megatylopus sp. is reported from the late Miocene/early Pliocene Gray Fossil Site in Tennessee (locality CE1; Wallace and Wang, 2004). Possible species of *Aepycamelus, Gigantocamelus (*possibly = *Titanotylopus),* and *Hemiauchenia* are reported from the early Pliocene Pipe Creek Sinkhole Site in Indiana (locality CE2; Farlow *et al.*, 2001).

Additional camelids have been reported from the late Pliocene/early Pleistocene of Calilfornia (locality NB13C). In addition to *Hemiauchenia, Camelops* and *Titanotylopus* (as noted in Vol. 1), the taxa *Blancocamelus, Gigantocamelus,* and *Palaeolam*a may also have been present (Randall and Jefferson, 2002).

Corner and Voorhies (1998) noted some important differences between *Titanotylopus* and *Gigantocamelus. Titanotylopus* is brachydont, with a pointed muzzle, non-spatulate incisors, narrow and elongated proximal phalanges, and proportionally long and slender metapodials; *Gigantocamelus* is more hypsodont (with cement-filled fossettes), a broader muzzle with spatulate incisors, short and broad proximal phalanges, and proportionally short and broad metapodials.

Ruez (2005) recorded the occurrence of *Palaeolama* at Inglis IC (locality GC15C) from the late Pliocene of Florida, noting that this is the earliest occurrence of this genus. He also noted that *Palaeolama guanajuatensis*, from the latest Hemphillian of Mexico (locality SB58B), probably represents a small specimen of *Hemiauchenia.*

Janis, Theodor, and Boisvert (2002) examined the pedal morphology of camelids to ascertain the time of the acquisition of the pacing gait. They showed that *Poëbrotherium* had no apparent modifications as in those of modern camelids, and while all other camelids showed a degree of modification, a more highly modified foot had evolved independently several times. The evolution of a foot completely modified like that of an extant camel occurred within the Stenomylinae (in the genus *Stenomylus*), in the Protolabinae (in the genus *Protolabis*), and in the common ancestor of the Llamini and Camelini. The time of origin of all of these events was in the late Oligocene to early Miocene, coincident with palynological evidence for the spread of open grassland habitats (see Strömberg, 2006).

ADDITIONS FROM NEW LOCALITIES (SEE APPENDIX I)

Poëbrotherium sp. (localities NP10B, Biii, B3, C, NP32B, C, NP49II).

Paratylopus sp. (locality ?NP25C).

Stenomylus sp. (locality NP10C3).

Nothokemas waldropi (localities GC8AA [= GC8B in Vol. 1], [GC8AB], GC8C [= GC8B in Vol. 1]); *N.* sp. (localities GC5A, GC8AA [= GC7 in Vol. 1], GC8B [= GC8A in Vol. 1], ?GC8C [GC8B in Vol. 1]).

Gentilicamelus sp. (locality [GC8AB]).

Miolabis fissidens (localities CP75B, BB).

Nothotylopus sp. (localities CP75B, BB).

Tanymykter sp. (localities ?CC9IVC, D, NB4)

Protolabis heterodontus (localities CP75B, BB); *P. gracilis* (locality CP56IIB); *P.* sp. (localities [CA2II], [CA2III]).

Michenia agatensis (localities CC9IVC, D); *M.* n. sp. (localities CP75B, BB).

Oxydactylus sp. (locality [GC8AB]).

Aepycamelus giraffinus (localities [CP56IIA], CP75B, BB); *A. major* (localities GC11D, GC13B): *A.* sp. (localities CC18A, ?CC26C, CP56IIA [could be *Procamelus*]).

Blancocamelus meadi (locality SB37D).

Hemiauchenia blancoensis (localities [NB35IIIB], SB34IIB, SB34IVA, SB35B, SB37A–D, [SB38IIC]); *H. vera* (localities [GC27 = *Pleiolama*], NB34II); *H. macrocephala* (localities GC15C, ?GC17B, GC18IIB, GC18IIIA–C, GC18IV); *H. minima* (localities GC10D, GC11D, GC13B, GC15D, GC18IIB); *H.* n. sp, (locality GC15C); *H.* sp. (localities ?CC52II, NB35IIIA, NB36IIIB, SB18F, SB34IIB, SB34IVB, SB35B, SB37A, D, SB38IIC, SB58C, SB58IIB).

Palaeolama mirifica (localities GC15C–F, GC18IIIA, GC18IV).

Alforjas sp. (locality SB58IIB).

Camelops hesternus (locality [SB37E]); *C.* sp. (localities [CC52II], CC52III, SB18E, SB34IIB, SB34IVB, SB35B–D, SB36C, SB37A, D, SP37II [could be *Gigantocamelus*], SB38IIC, SB58IIC, SB64II).

Pliauchenia humphresiana (locality CP116A: USNM collections).

Procamelus grandis (localities GC10D, GC11D, GC13B); *P. occidentalis* (CP56IIB); *P.* sp. (locality CC26B).

Megatylopus sp. (locality CC26B).

Gigantocamelus spatula (localities NB35IIIC, [SB37D]).

Camelidae indet. (localities WM26, CA1, GC8B [= GC8C in Vol. 1] [small], DB, F, CC8IIB, NB35IIIC, NB36IIIA, B, SB18AD, NP10CC).

TYLOPODA INDET

Reported from locality CA1 (?new family).

HORNLESS RUMINANTS

CORRECTIONS TO VOLUME 1

Leptomeryx blacki is known from localities NP10A, B (not NB10A, B).

SYSTEMATIC REVISIONS, NEW TAXA, AND IMPORTANT NEW LOCALITY OR PALEOBIOLOGICAL INFORMATION

Hypertragulus heikeni, from the early Chadronian of Chihuahua, Mexico, is considered by Walsh (2005) to represent a species of *Simimeryx*, thus representing the youngest known species of this genus.

Hypisodus minimus is reported from the late Chadronian/early Orellan of Montana (locality NP27C0; Tabrum, 1998). This represents the first occurrence of this species, and of the Hypertragulidae, in the Chadronian of western Montana.

New species of *Hendryomeryx* (localities NP24C, D) and *Leptomeryx* (localities NP24C, D, NP25C, NP27C, NP32C) are reported from the Chadronian of Montana (Tabrum, Prothero, and Garcia, 1996).

The early Miocene Arctic Fauna from Devon Island, Nunavet (locality HA2) has yielded a highly derived traguloid of Eurasian affinities (Dawson, 2003). The early Pliocene Arctic Fauna from Ellesmere Island (locality HA3) has yielded another probably Eurasian ruminant, likely a primitive cervoid. This was originally described as Cf. *Moschus* (Tedford and Harington, 2003) and has now been named as the new genus and species *Boreameryx braskerudi* (Dawson and Harington, 2007).

The systematics of the Blastomerycinae was revised by Prothero (2007, 2008), who used dental dimensions to synonymize many of the species. All named species of *Problastomeryx* are now synonymized with *Problastomeryx primus*. All named species of *Pseudoblastomeryx* are now synonymized with *Pseudoblastomeryx advena*. Most named species of *Blastomeryx* are now synonymized with *Blastomeryx gemmifer*. "*Blastomeryx*" *vigoratus* is poorly diagnosed and may be a dromomerycid. "*Blastomeryx*" *pristinus* (taxon not listed in Vol. 1) is known from fragmentary material and may be referable to the dromomerycid *Aletomeryx*. "*Blastomeryx*" *cursor* (taxon not listed in Vol. 1) is probably the dromomerycid *Barbouromeryx trigonocorneus*.

Parablastomeryx gregorii and *P. floridanus* are valid species, although *P. floridanus* might possibly be a large *Blastomeryx gemmifer*, but "*Parablastomeryx*" *galushai* is shown to be the dromomerycid *Aletomeryx*. *Machaeromeryx tragulus* and *M. gilcristensis* are both valid species, as in Vol. 1. *Longirostromeryx wellsi* and *L. clarendonensis* are both valid species, but "*Longirostromeryx*" *novomexicanus* is a nomen nudum of unclear affinities. Note that an undescribed skeleton of *Longirostromeryx wellesi*, from the Clarendonian Ashfall Fossil Beds in Nebraska (locality CP116A), shows this taxon to be extremely gracile with a very long neck and long legs, reminiscent of a (smaller) gerenuk gazelle (*Litocranius walleri*) in proportions (Prothero, 2007a; Janis, personal observations).

The relationships of moschids to other ruminants have been the subject of recent controversy. Volume 1 supported the phylogenetic position of the moschids being cervoids, but molecular analysis by Hassanin and Douzery (2003) shows them to be the sister taxon of bovids. Note, however, that the combined dataset of molecular, morphological, and ethological information of Fernández and Vrba (2005) supports cervoid affinities.

ADDITIONS FROM NEW LOCALITIES (SEE APPENDIX I)

Hypertragulus sp. (locality [CP8IIB]).

Nanotragulus loomisi (localities GC8AA [= GC8B in Vol. 1], [CC9IIIA]); *N. ordinatus* (locality [CC9IVC]); *N.* sp. (locality GC5A).

Hendryomeryx defordi (locality [NP25B]); *H.* sp. (localities NP9B, ?NP32C).

Leptomeryx evansi (localities NP10BB, B3, NP27C0, [C1], [NP32C]); *L. speciosus* (localities NP10Bi, Bii, Biii, NP27C0); *L. yoderi* (locality NP49II); *L.* sp. (localities CP67, NP9A, NP10C, C2, NP49II).

Pronodens sp. (locality NP10C2).

Pseudoparablastomeryx advena (locality CC9IVD); *P.* sp. (localities [CA2III], [GC4DD]).

Blastomeryx gemmifer (localities CP75B, BB); *B.* sp. (localities [GC8DB, F]).

Machaeromeryx tragulus (localities CC9IVC, D); *M.* sp. (locality [GC8DB]).

Pseudoceras sp. (localities GC8F, GC11D, GC13B, [GC27], CP116F [USNM collections]).

DROMOMERYCIDAE

CORRECTIONS TO VOLUME 1

Rakomeryx sp. is known from locality CP109**A**.

Pediomeryx (*Yumaceras* sp.) is known from localities **CP**115C, and **CP**116C (not CC115C and CC116C).

SYSTEMATIC REVISIONS, NEW TAXA, AND IMPORTANT NEW LOCALITY OR PALEOBIOLOGICAL INFORMATION

Prothero and Liter (2008) provided a revision of dromomerycids, which they designated as the subfamily Dromomerycinae within the Palaeomerycidae. Many of the species listed by Frick (1937), and repeated by Janis and Manning in Vol. 1, have been grouped together. *Aletomeryx* now includes only the species *A. gracilis* (including *A. lugni, A. marshi*, and *A. scotti*), *A. marslandensis*, and *A. occidentalis*. *Sinclairomeryx* is now a monotypic genus, with *S. riparius* being the only valid species. *Drepanomeryx (Matthomeryx) matthewi* is now subsumed into *Drepanomeryx falciformis*. *Rakomeryx sinclairi* and *Dromomeryx borealis* are the only species recognized for these two genera. *Subdromomeryx* is promoted to the rank of genus, including the species *S. scotti* and *S. antilopinus*, as in Vol. 1. *Barbouromeryx (Protobarbouromeryx) sweeti* and *Barbouromeryx (Probarbouromeryx) marslandensis* can be synonymized with *Barbouromeryx trigonocorneus*, and this taxon now also includes the supposed moschids *Blastomeryx (Parablastomeryx) galushi*, "*Blastomeryx*" *cursor*, and possibly also *Blastomeryx vigoratus*. "*Blastomeryx*" *pristinus* (taxon not listed in Vol. 1) is known from fragmentary material and may be referable to *Aletomeryx* (see Prothero, 2007b). *Bouromeryx americanus* also includes the species *B. milleri, parvus, madisonensis, pawniensis, supernebraskensis* and *pseudonebraskensis* (although *B. submilleri* is distinct). *Cranioceras unicornis* (including *C. granti* and *C. mefferdi*) and *C. teres* (including *C. clarendonensis* and *C. dakotensis*) are the only species recognized for this genus.

Yumaceras is recognized as a separate genus, rather than a subgenus of *Pediomeryx*: the recognized species of *Yumaceras* and *Pediomeryx* remain the same as in Vol. 1.

Semprebon, Janis, and Solounias (2004) examined the dental mesowear and microwear of dromomerycids. Members of the Aletomerycinae were likely mixed feeders, while early to middle Miocene members of Dromomerycinae (Dromomerycini, and Cranioceratini genera *Barbouromeryx, Bouromeryx*, and *Procranioceras*) were likely browsers. However, later members of the Cranioceratini, *Cranioceras* and *Pediomeryx*, had dental wear suggestive of mixed feeding, possibly a response to the cooling and drying environmental conditions of the late Mioene.

ADDITIONS FROM NEW LOCALITIES (SEE APPENDIX I)

Dromomeryx pawniensis (locality CP75BB).
Bouromeryx pawniensis (= *Procranioceras pawniensis* in Tedford, 2004) (locality CP75BB).
Cranioceras sp. (locality CP114B: AMNH collections).
Pediomeryx (*Yumaceras*) *hamiltoni* (localities GC8F, GC10D, GC11D); *P. (Y.)* sp. (CP115C: AMNH collections).
Pediomeryx (*Pediomeryx*) sp. (localities GC27, CP116E: AMNH collections)
Dromomerycid indet. (locality CC18A [large]).

ANTILOCAPRIDAE

CORRECTIONS TO VOLUME 1

Sphenophalus nevadanus is known from locality PN11**A** and **B** (new locality subdivision, PN11B with emended dating of Hh2).

SYSTEMATIC REVISIONS, NEW TAXA, AND IMPORTANT NEW LOCALITY OR PALEOBIOLOGICAL INFORMATION

Merycodus sabulonis is reported in the early Barstovian of Mexico (locality CA2II; Ferrusquía-Villafranca, 2003), extending the biogeographic range of this genus.

Capromeryx has been identified in the latest Hemphillian of Mexico (locality SB59; Jiménez-Hidalgo and Carrenza-Casteñeda, 2001). This is the earliest specimen of the genus known and extends the temporal range. *Capromeryx tauntonensis* has been identified from the early Blancan of the same region (locality SB57III; Jiménez-Hidalgo, Carranza-Casteñeda, and Montellano-Ballesteros, 2004).

Semprebon and Rivals (2007) analyzed antilocaprid microwear and mesowear and concluded that antilocaprines had a more abrasive diet than merycodontines, corresponding with their increased level of hypsodonty. Pleistocene and Recent antilocaprines appear to have reverted to a less abrasive diet.

ADDITIONS FROM NEW LOCALITIES (SEE APPENDIX I)

Paracosoryx pawniensis (species not in Vol. 1) (localities CP75B, BB).
Merycodus sp. (locality ?CC26C).
Cosoryx furcatus (localities CP75B, BB).
Hexobelomeryx fricki (localities SB55IIAA, SB55IV); *H.* sp. (locality SB58IIB).
Capromeryx arizonensis (localities GC15C, GC17B, C18IIB); *C.* sp. (localities SB35A, C, SB37A, B).
Tetrameryx sp. (locality CP117B: USNM collections).
Antilocapridae indet. (localities ?CA1, ?NB34II, SB38IIC [small sp.], CP56IIA, B [cosorycine]).

CERVIDAE AND BOVIDAE

CORRECTIONS TO VOLUME 1

Odocoileus sp. is known from locality GC13**B**.

SYSTEMATIC REVISIONS, NEW TAXA, AND IMPORTANT NEW LOCALITY OR PALEOBIOLOGICAL INFORMATION

Webb (2000) described a new cervid genus, *Eocoileus gentryorum*, from the late Hemphillian of Florida (locality GC13B), making it the earliest known New World cervid. This is a medium-sized cervid, most nearly comparable to the South American cervids *Ozotoceros* and *Mazama* among living cervids, and is the basal member of the tribe Odocoileini. Webb also noted that the genus *Navahoceros* is known from the latest Tertiary, from the Anza-Borrego Desert (locality NB13C). He recognized *Navahoceros* as a member of the tribe Rangiferini, with the implication that the odocoiline/rangiferine split in the New World cervids had occurred by 3 Ma.

Wheatly and Ruez (2006) examined cervid material from the Hagerman Fossil Beds (Hemphillian of Idaho, locality PN23A) and concluded that *Odocoileus* cannot be distinguished from *Bretzia* from the dentition alone, nor can *Odocoileus brachyodontus* (= *O. brachydontus* in Vol. 1) be distinguished from modern species of *Odocoileus*. They, therefore, considered that *O. brachyodontus* is a nomen dubium.

McCullough *et al.* (2002) reported on new material at the 111 Ranch Site from the late Blancan of Arizona (locality SB18) and noted the new additions of the cervids Cf. *Bretzia* and *Cervus* (at SB18E).

A possible occurrence of *Bretzia pseudalces* (represented by portion of an antler) is reported from the late Hemphillian of Nebraska (locality CP116F; = "cervid indet." in Vol. 1; Voorhies and Perkins, 1998).

A rupicaprine-like bovid horncore, together with a partial skeleton, is reported from the early Blancan of Nevada (locality NB33), and another horncore has been found in the late Blancan of Arizona (locality SB18D; Mead, 1997). It is not certain if this material is assignable to *Neotragocerus* or represents an unknown rupicaprine bovid.

ADDITIONS FROM NEW LOCALITIES (SEE APPENDIX I)

Odocoileus brachyodontus (locality SB35A); *O. virgninanus* (species not in Vol. 1) (localities GC14A [= *O.* sp. in Vol. 1], GC15A [= *O.* sp. in Vol. 1], B, C, E, F, GC17A, B, GC18IIB, GC18IIIA–C, GC18IV); *O.* sp. (localities NB35IIIB, SB37D, E, ?SB38IIC).
Eocoileus sp. (genus not in Vol. 1) (locality GC13B).
Navahoceras lacruensis (genus not in Vol. 1) (locality SB37E).
Cervid indet. (localities CC47II, ?NB35IIIA, NB36IIIB [= "*Euceratherium collinum*"], SB18E, SB34IIB).
Bovid indet. (Cf. *Bison*) (localities GC14C, GC17A).

GENERAL PERISSODACTYLA AND PROBOSCIDEA

SYSTEMATIC REVISIONS, NEW TAXA, AND IMPORTANT NEW LOCALITY OR PALEOBIOLOGICAL INFORMATION

The phylogenetics of basal perissodactyls are revised in Froehlich (1999). The notion of the Hippomorpha (Equidae, Palaeotheriidae, and Brontheriidae) is reestablished, and brontotheres may be derived from early palaeotheres. However, Lucas and Holbrook (2004) ran an analysis of basal perissodactyl phylogeny to test the phylogenetic positions of *Lambdotherium* and the early brontotheriid *Eotitanops*. *Lambdotherium* again nested with palaeotheres, although *Eotitanops* occupied a more basal position in the tree, well removed from palaeotheres. In Froehlich's (1999) analysis, the genus *Hyracotherium* emerged as a paraphyletic taxon, containing within it the European taxon *Pliolophus vulpiceps*, suggesting a complex pattern of biogeographical dispersals during the early Eocene.

Hooker and Dashzeveg (2004) revised the supraordinal classification of perissodactyls in their paper discussing the origin of chalicotheres. In their scheme, the "tapiroid" family Lophiodontidae is now included with the Chalicotheriidae in the Chalicotherioidea, with the various genera of the Isectolophidae ("tapiroids") representing the stem lineage to this group, and the entire grouping comprising the infraorder Ancylopoda. The Equoidea (equids and palaeotheres) are now included with the ceratomorphs in the new infraorder Euperissodactyla. Ancylopoda and Euperissodactyla together form the suborder Lophodontomorpha, with the Brontotheriidae as the sister taxon to this grouping.

ADDITIONS FROM NEW LOCALITIES (SEE APPENDIX I)

An unidentified (large) ceratomorph is reported from the earliest Eocene of Wyoming (locality CP20AA), representing a new first appearance for this group (Strait, 2003).

A new "hippomorph" perissodactyl from the Bridgerian of Wyoming (locality uncertain), much larger than *Orohippus* but much smaller than contemporaneous brontotheres, represents an unnamed new genus and species (Holbrook, 2000).

BRONTOTHERIIDAE

CORRECTIONS TO VOLUME 1

Duchesneodus uintensis is known from localities SB27**A**, NP10**A**.
Brontops sp. is known from localities CP41**A**, NP10**B**, NP29**C** (note that this taxon is probably congeneric with *Megacerops*, see revision by Mihlbachler, below).
Menops sp. is known from localities NP10**B**, NP29**C** (note that this taxon is probably congeneric with *Megacerops*, see revision by Mihlbachler, below).
Megacerops sp. is known from locality NP29**C**.
Brontotherid indet. is known from locality CP42**A**.

SYSTEMATIC REVISIONS, NEW TAXA, AND IMPORTANT NEW LOCALITY OR PALEOBIOLOGICAL INFORMATION

New species occurrences are reported from the following localities (late Wasatchian to middle Bridgerian of Wyoming) (Gunnell and Yarborough, 2000): *Palaeosyops paludosus* (locality CP34A); *P. laticeps* (locality CP34D); *P. robustus* (locality CP34C); *P. fontinalis* (localities SB22C, CP20F, CP25J, J2, CP27C, CP31D, CP34A, ?NP21); *P. laevidens* (locality CP34C); *Eotitanops borealis* (localities SB22C, CP20F); *E. minimus* (localities SB22C, CC25GG, CP31D). Gunnell and Yarborough (2000) also provided a number of species synonomies for *Palaeosyops* and *Eotitanops*.

Various brontotheres are reported from the Canadian High Arctic. Cf. *Eotitanops* and a brontotheriid gen. and sp. indet are reported from the early Eocene of Ellesmere Island (locality HA1A0; Eberle, 2005), and a new taxon of brontotheriid is reported from the late middle Eocene of Ellesmere Island (locality HA1C; Eberle and Storer, 1999).

Mihlbachler and Solounias (2002) examined brontothere microwear and concluded that their diet was exclusively folivorous, despite a gross dental morphology that would suggest a folivorous/frugivous diet.

ADDITIONS FROM NEW LOCALITIES (SEE APPENDIX I)

Lambdotherium popoagicum (locality CP25I1); *L.* sp. (locality CP33A).
Eotitanops gregoryi (= *E.* "*minimus*") (localities CP25J, CP31E).
Palaeosyops fontinalis (localities CP25J, CP31E, CP34A).

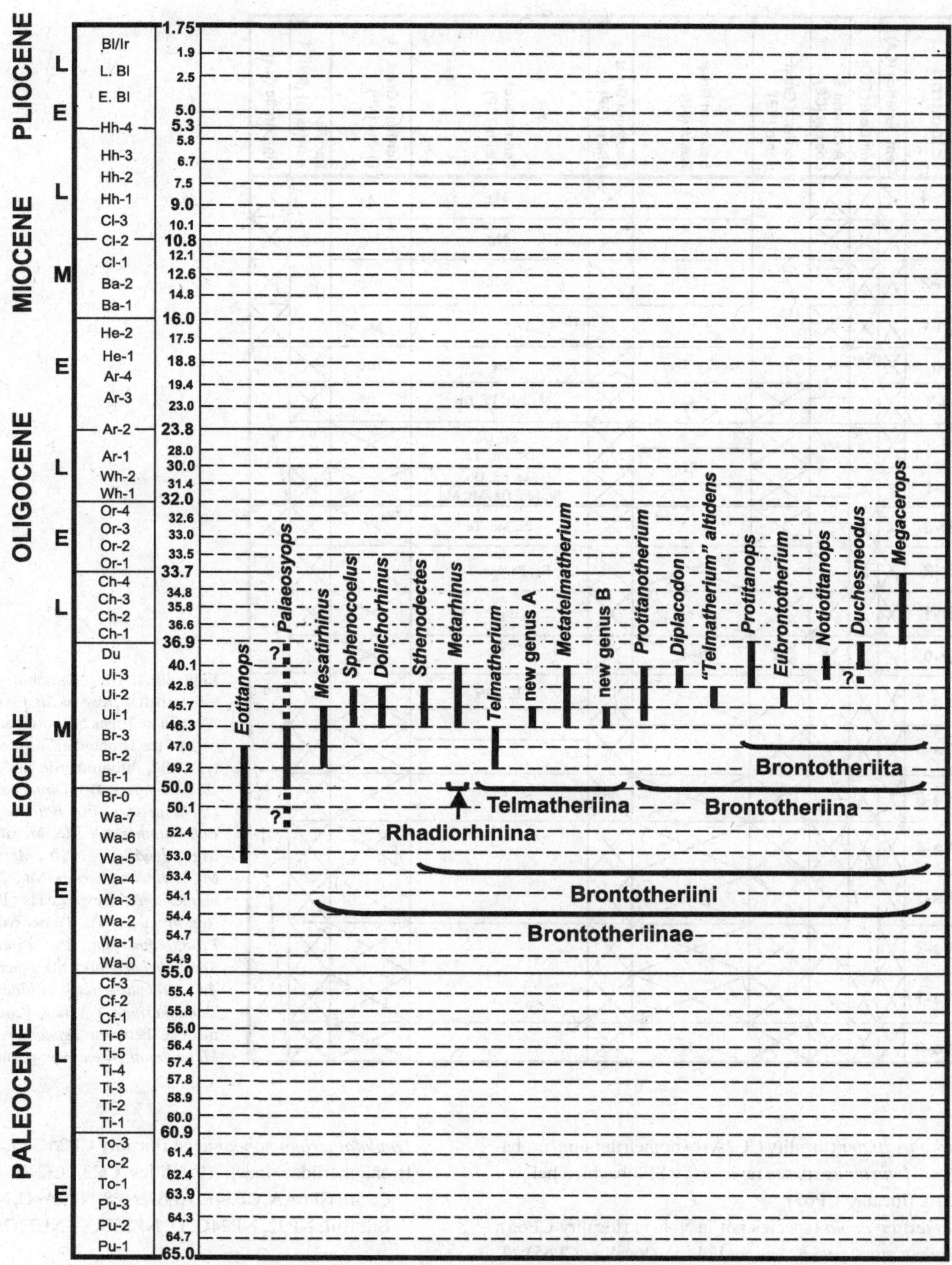

Figure A.3. Temporal ranges of brontotheres. Subfamily divisions follow Mihlbachler (2007). (See caption to Figure A.4 for identity of "new genera" A and B)

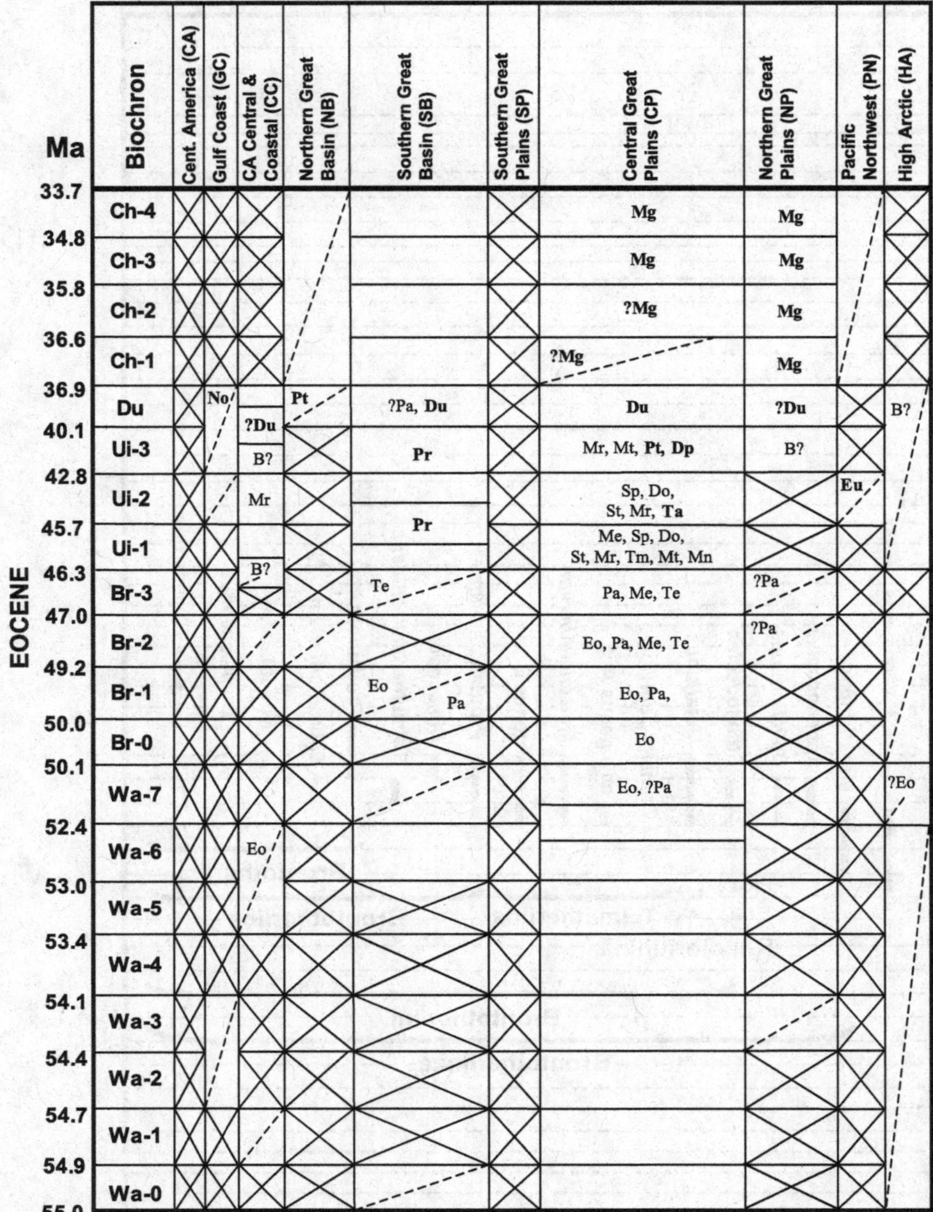

Figure A.4. Biogeographic ranges of brontotheres. Key as in Figure A2. All taxa in Times New Roman, members of the Brontotheriina are in bold type. B?, brontotheriid indet.; Do, *Dolichorhinus*; **Dp**, *Diplacodon*; **Du**, *Duchesneodus*; Eo, *Eotitanops*; **Eu**, *Eubrontotherium*; Me, *Mesatirhinus*; **Mg**, *Megacerops*; Mr, *Metarhinus*; Mt, *Metatelmatherium*; Mn, "*Metatelmatherium*" (new genus B); No, *Notiotitanops*; Pa, *Palaeosyops*; **Pr**, *Protitanotherium*; **Pt**, *Protitanops*; Sp, *Sphenacoelus*; St, *Sthenodects*; **Ta**, "*Telmatherium*" *altidens*; Te, *Telmatherium*; Tm, new genus intermediate between *Telmatherium* and *Metatelmatherium* (new genus A).

Metarhinus ?pater (locality CC7A) (a nomen dubium, probably = *Metarhinus* sp. see revision by Mihlbachler, below); *M*. sp. (locality CP65).

Telmatherium acola (species not in Vol. 1) (locality CP65); *T. advocata* (species not in Vol. 1) (locality CP65); *T*. sp. (locality SB26II). (*Telmatherium accola* and *T. advocata* are both nomina dubia, see revision by Mihlbachler, below.)

Protitanops new sp. (locality PN5B).

Duchesneodus uintensis (locality CC7D) (material may belong to *Protitanops* or *Notiotitanops*, see revision by Mihlbachler, below).

Lambdotherium popoagicum (locality CP25GG).

Brontotheriidae indet. (localities CC3, CC7B, C, SB42, CP5II, CP6AA, CP34E, CP67, NP8, NP9A–C, NP10A, Bi, Bii, Biii, NP22, NP24C, D, NP25B, C, NP27C, NP49II,).

REVISIONS BY MATTHEW MIHLBACHLER

Mihlbachler (2005a, 2008) and Mihlbachler, Lucas, and Emry, (2004) have extensively revised brontothere taxonomy and occurrences. Below we provide a summary of some key features of this work. Figure A.3 shows the temporal range of brontotheres and Figure A.4 shows their biogeographical ranges.

REVISIONS OF SPECIFIC BRONTOTHERE GENERA

Mesatirhinus Osborn, 1908

In Vol. 1, *Mesatirhinus megarhinus* (Earle, 1891) (and its synonym *Mesatirhinus petersoni* Osborn, 1908) was listed as the type and only valid species of *Mesatirhinus*. *Mesatirhinus junius* (Leidy, 1872a) was listed as a nomen dubium. Gunnell and Yarborough (2000) considered *M. junius* valid but mistakenly included it with *Palaeosyops*. Mihlbachler (2005a) returned *M. junius* to *Mesatirhinus* and considered it a valid senior synonym of both *Mesatirhinus megarhinus* and *Mesatirhinus petersoni*. *Mesatirhinus junius* is the type species and only valid species of *Mesatirhinus*. *Mesatirhinus junius* is known from localities CP34D, CP38A, B, ?C, CP65.

Sphenocoelus Osborn, 1895

In Vol. 1, three genera, *Dolichorhinus* Hatcher, 1895, *Tanyorhinus* Cook, 1926, and *Dolichorhinoides* Granger and Gregory, 1943 (an Asian taxon), were considered synonyms of *Sphenocoelus*. However, because of significant structural differences in the crania of these taxa, some of these taxa have been removed from *Sphenocoelus*. *Sphenocoelus uintensis* Osborn, 1895 is the type species and only valid species of *Sphenocoelus*. *Tanyorhinus* remains a synonym of *Sphenocoelus*. *Tanyorhinus blairi* Cook, 1926 and *Tanyorhinus bridgeri* Cook, 1926 are synonyms of *S. uintensis*. *Tanyorhinus harundivorax* Cook, 1926 is a nomen dubium. *Sphenocoelus uintensis* occurs in localities CP6A, CP38C, and CP65. *Dolichorhinus* is again a valid genus (see below). *Dolichorhinoides* is a junior synonym of *Epimanteoceras* Granger and Gregory, 1943 (another Asian taxon).

Dolichorhinus Hatcher, 1895

Although this genus was synonomized with *Sphenocoelus* Osborn, 1895 in Vol. 1, *Dolichorhinus* has been reinstituted as a distinct genus (Mihlbachler, 2005a) because of numerous specializations, particularly in the rostrum and posterior nares. The type species and only valid species of *Dolichorhinus* is *D. hyognathus* (Osborn, 1889). Other species attributed to *Dolichorhinus* (but assigned to *Sphenocoelus* in Vol. 1) include *D. cornutum* (Osborn, 1895), *D. intermedius* (Osborn, 1908), *D. heterodon* Douglass, 1909, *D. longiceps* Douglass, 1909, *D. superior* (Riggs, 1912), and *D. fluminalis* Riggs, 1912). However, none of the holotypes or specimens previously referred to any of these species is morphologically distinct, and all appear to be synonomous with *D. hyognathus*. *Dolichorhinus hyognathus* is known from localilties CP6A and CP65. *Dolichorhinus vallidens* (Cope, 1872) is a nomen dubium.

Metarhinus Osborn, 1908

In Vol. 1, two species were assigned to *Metarhinus*: *M. fluviatilis* Osborn, 1908) (and its synonym *M. riparius* Riggs, 1912) and *M. diploconus* Osborn, 1895 (and its synonym *M. abbotti* [Riggs, 1912]), plus ?*M. pater* Stock, 1937. Recently discovered autapomorphic features in the rostrum of the holotype of *M. diploconus* necessitate removal of this species from *Metarhinus*. "*Metarhinus*" *diploconus* belongs to a new genus whose name is not yet published. *Metarhinus fluviatilis* (= *M. riparius*) remains the type species of *Metarhinus*, while *M. abbotti* is the only additional valid species. The two species of *Metarhinus* can only be distinguished by differing nasal shapes. Other species previously attributed to *Metarhinus*, *M. earlei* Osborn, 1908, *M. cristatus* Riggs, 1912, and ?*M. pater* Stock 1937, are nomina dubia because the diagnostic nasal bone is not preserved. *Heterotitanops parvus* Peterson, 1914a is a nomen dubium based on a juvenile specimen that probably belongs to *Metarhinus* sp. *Metarhinus fluviatilis* is known from localities CP6A and CP38C; *M. abbotti* from locality CP6A; *Metarhinus* sp. from localities CC4, CP6A, and CP38C.

Sthenodectes Gregory, 1912

Sthenodectes incisivum (Douglass, 1909) is the only valid species of this genus. *Sthenodectes priscum* Peterson, 1934 (not mentioned in Vol. 1) is a junior synonym of *S. incisivum*. *Sthenodectes incisivum* is known from localities CP6A and CP38C.

"*Telmatherium*(?)" *altidens* Osborn, 1908

"*Telmatherium*(?)" *altidens*, not mentioned in Vol. 1, is a valid species but does not belong with *Telmatherium*. It represents a new genus whose name is not yet published. It is known from locality CP6A.

Metatelmatherium Granger and Gregory, 1938

Metatelmatherium cristatum Granger and Gregory, 1938 (from Asia) is a junior synonym of *Metatelmatherium ultimum*. Thus *M. ultimum* Osborn, 1908 is the type species. *Metatelmatherium ultimum* is known from localities CP6B and CP38C. In Vol. 1, it was reported from locality CP65. However, the material from CP65, previously assigned to *Metatelmatherium*, represents a new genus and species whose name is not yet published.

A new genus and species that is morphologically intermediate between *Telmatherium validus* and *Metatelmatherium ultimum* is known from localities CP65 and CP38C.

Telmatherium Marsh 1872

Telmatherium validus is the only valid species of *Telmatherium*, known from localities CP31E, CP34D, CP38B. *Telmatherium cultridens* (Osborn, Scott, and Speir, 1878), *Manteoceras manteoceras* Hay, 1901, and *M. washakiensis* Osborn, 1908 are synonyms of *T. validus*. Other species, not mentioned in Vol. 1, including *T. accola* Cook, 1926, *T. advocata* Cook, 1926, *M. foris* Cook, 1926, and *M. pratensis* Cook, 1926, all known from locality CP65, are invalid nomina dubia.

Protitanotherium Hatcher, 1895

Neither *Protitanotherium superbum* Osborn, 1908 nor *Sthenodectes australis* Wilson, 1977 is mentioned in Vol. 1. They are both junior synonyms of *Protitanotherium emarginatum* (Hatcher, 1895). *Protitanotherium emarginatum* is conclusively known from localities SB43A and CP6B. In Vol. 1, a specimen from locality CP37B is reported as possibly belonging to *Protitanotherium*, but this specimen is more likely *Diplacodon* (see below).

Diplacodon

Mader (1989) has considered *Diplacodon elatus* Marsh, 1875 to be a nomen dubium because of the incomplete nature of the holotype (i.e., it lacks horns). More complete specimens belonging to this species were previously assigned to two other genera: *Eotitanotherium* Peterson, 1914b and *Pseudodiplacodon* Mader, 2000. These genera were distinguished by the size and thickness of the horns and nasal bone. However, these aspects of cranial variation are consistent with intraspecific sexual dimorphism and specimens assigned to these genera seem to represent the same, probably sexually dimorphic, species. Thus, *Pseudodiplacodon progressum* (Peterson, 1934) and *Eotitanotherium osborni* Peterson, 1914b are recognized as junior synonyms of *Diplacodon elatus*. *Diplacodon elatus* is known only from locality CP6B. A specimen very similar to *Diplacodon elatus*, but possibly representing a new species, is known from locality CP37B.

***Pseudodiplacodon* Mader, 2000**

See *Diplacodon*.

***Eotitanotherium* Peterson, 1914b**

See *Diplacodon*.

***Protitanops* Stock, 1936**

The brontothere material known from locality PN5B, previously referred to *Protitanops* (Lucas, 1992; Hanson, 1996; Mihlbachler, Lucas, and Emry, 2004), represents a new genus and species whose name is not yet published. (Note added at proof: this taxon has just been named as *Eubrontotherium clarnoensis* [Mihlbachler, 2007]).

***Duchesneodus* Lucas and Schoch, 1982**

Previously, Duchesnian brontothere material was indiscriminantly referred to *Duchesneodus uintensis*. However, without complete skulls (or nearly complete skulls), it is almost impossible to distinguish *Duchesneodus uintensis* from *Protitanops curryi* or *Notiotitanops mississippiensis*. *Duchesneodus uintensis*, known from numerous diagnostic skulls, is the type species and only valid species of *Duchesneodus*. Other species previously assigned to this genus, *D. primitivis* (Lambe, 1908), *D. californicus* (Stock, 1935), and *D. thyboi* (Bjork, 1967), are known from fragmentary material and are nomina dubia. The occurrences of *D. uintensis* at localities SB25B, SB44B, and CP7 are substantiated by adequate fossil material. Reported occurrences of *Duchesneodus* at other localities are not substantiated by adequate fossil material and could represent other species.

***Megacerops* Leidy, 1870a**

Brontotheres from the classic White River deposits have historically been split into a large number of taxa (37 species according to Osborn [1929]). It is universally accepted that Chadronian brontotheres are over split, although a revision does not exist. Mader (Vol. 1) accepted three genera, *Brontops* Marsh, 1887, *Menops* Marsh, 1887, and *Megacerops* Leidy, 1870a, within which a total of 15 species were accepted. These three genera were distinguished based on the thickness of the zygomatic arches and cross-sectional shape of the horns. However, these aspects of cranial variation are entirely continuous, thus preventing the partitioning of the many hundreds of available skulls into discrete morphological units. The conspicuous variation in horn size and shape, and in zygomatic thickness, is consistent with intraspecific sexual dimorphism. Therefore, the number of true species is likely to be fewer than these previous estimates. Only two unambiguously diagnosable species exist: *Megacerops coloradensis* Leidy, 1870a and *Megacerops kuwagatarhinus* Mader and Alexander, 1995. The former is diagnosed by the possession of unbifurcated horns, while the latter is diagnosed by the possession of strongly bifurcated horns (Mihlbachler, Lucas, and Emry, 2004). The many additional names for brontotheres from classic Chadronian deposits that are based on holotypes with unbifurcated horns should be considered synonyms of *Megacerops coloradensis* unless a more thorough analysis demonstrates otherwise. Names for Chadronian brontotheres that are represented by holotypes that lack preserved horns are all nomina dubia. *Megacerops coloradensis* is known from localities CP41A, CP68B, CP83A–C?, CP98A–C?, NP10B, and NP29C. *Megacerops kuwagatarhinus* is known from localities CP67, NP10B? and NP29II.

COMPREHENSIVE LIST OF NORTH AMERICAN BRONTOTHERE SPECIES, SYNONYMS AND NOMINA DUBIA

The following list of valid brontothere species and their synonyms is based on revisions by Gunnell and Yarborough (2000) for *Eotitanops* and *Palaeosyops*, revisions by Mihlbachler, Lucas, and Emry (2004) for *Megacerops* (and synonyms), and revision by Mihlbachler (2005a) for the remaining species.

Valid Taxa

Eotitanops borealis (Cope, 1880) (originally described as *Palaeosyops borealis* Cope, 1880). Synonyms: *E. princeps* Osborn 1913; *E. brownianus* Osborn, 1908; *E. gregoryi* Osborn, 1913; *E. major* Osborn, 1913.

Eotitanops minimus Osborn, 1919.

Palaeosyops paludosis Leidy, 1870b. Synonyms: *P. major* Leidy, 1871; *P. minor* Earle, 1891; *P. longirostris*, Earle, 1891; *Canis montanis* Marsh, 1871; *Canis ?marshi*, Hay, 1899; *Limnohyops matthewi* Osborn, 1908.

Palaeosyops laticeps Marsh, 1872 (also referred to as *Limnohyops laticeps* [Marsh, 1872]).

Palaeosyops robustus Marsh, 1872 (originally described as *Limnohyus robustus* Marsh, 1872). Synonyms: *P. humilis* Leidy, 1872b; *P. diaconus* Cope, 1873a; *P. leidyi* Osborn, 1908; *P. grangeri* Osborn, 1908; *P. copei* Osborn, 1908.

Palaeosyops fontinalis (Cope, 1873b). Synonym: *Eometarhinus huerfanensis* Osborn, 1919.

Palaeosyops laevidens (Cope, 1873b) (originally described as *Limnohyus laevidens* Cope, 1873b). Synonyms: *Limnohyops priscus* Osborn, 1908; *L. monoconus* Osborn, 1908.

Mesatirhinus junius (Leidy 1872a) (originally described as *Palaeosyops junius* Leidy, 1872a). Synonyms: *M. megarhinus* (Earle, 1891) originally described as *Palaeosyops megarhinus* Earle, 1891); *M. petersoni* Osborn, 1908.

Sphenocoelus uintensis Osborn 1895. Synonyms: *Tanyorhinus blairi* Cook, 1926; *T. bridgeri* (Cook, 1926).

Dolichorhinus hyognathus (Osborn, 1889) (originally described as *Palaeosyops hyognathus* Osborn, 1889; and also referred to as *Telmatotherium hyognathus* [Osborn, 1889]). Synonyms: *D. cornutum* (Osborn, 1895) (originally described as *Telmatotherium cornutum* Osborn, 1895); *D. intermedius* (Osborn, 1908); *D. heterodon* Douglass, 1909; *D. longiceps* Douglass, 1909; *D. superior* (Riggs, 1912) (originally described as *Mesatirhinus superior* Riggs, 1912); *D. fluminalis* Riggs, 1912.

Sthenodectes incisivum (Douglass, 1909) (originally described as *Telmatherium*? *incisivum* Douglass, 1909). Synonyms: *S. priscus* Peterson, 1934; *"Metarhinus" diploconus* (Osborn, 1895) (originally described as *Telmathotherium diploconus* Osborn, 1895; also referred to as *Rhadinorhinus diploconus* [Osborn, 1895]).

Metarhinus fluviatilis Osborn, 1908. Synonym: *M. riparius* Riggs, 1912.

Metarhinus abbotti (Riggs, 1912) (originally described as *Rhadinorhinus abbotti* Riggs, 1912).

Telmatherium validus Marsh, 1872. Synonyms: *Telmatherium cultridens* (Osborn, Scott, and Speir, 1878) (originally described as *Leurocephalus cultridens* Osborn, Scott, and Speir, 1878); *Manteoceras manteoceras* Hay, 1901; *M. washakiensis* Osborn, 1908.

Metatelmatherium ultimum (Osborn, 1908) (originally described as *Telmatherium ultimum* Osborn, 1908). Synonyms: *Manteoceras uintensis* Douglass, 1909; *Metatelmatherium cristatum* Granger and Gregory, 1938 (Asian).

Protitanotherium emarginatum Hatcher, 1895. Synonyms: *P. superbum* Osborn 1908; *Sthenodectes australis* Wilson, 1977.

Diplacodon elatus Marsh (1875). Synonyms: *Eotitanotherium osborni* (Peterson, 1914b) (originally described as *Diploceras osborni* Peterson, 1914b [*Diploceras* is preoccupied]); *Pseudodiplacodon progressus* (Peterson, 1934) (originally described as *Diplacodon progressus* Peterson, 1934); *Duchesneodus uintensis* (Peterson, 1931) (originally described as *Teleodus uintensis* Peterson, 1931).

"Telmatherium (?)" altidens Osborn, 1908.

Protitanops curryi (Stock, 1936).

Notiotitanops mississippiensis Gazin and Sullivan, 1942.

Duchesneodus (Teleodus) uintensis (Peterson, 1931).

Megacerops coloradensis Leidy, 1870a. Synonyms: *Megacerops acer* (Cope, 1873c) (originally described as *Megceratops acer* Cope, 1873c); *Menops heloceras* (Cope, 1873c) (originally described as *Megaceratops heloceras* Cope, 1873c; also previously referred to as *Menodus heloceras* Cope, 1873c); *Megacerops bucco* (Cope, 1873d) (originally described as *Symborodon bucco* Cope, 1873d); *Symborodon altirostris* Cope, 1873d; *Menops trigonoceras* (Cope, 1873d) (originally described as *Symborodon trigonoceras* Cope, 1873d); *Brontotherium ingens* Marsh, 1873; *Brontotherium hypoceras* (Cope, 1874) (originally described as *Symborodon hypoceras* Cope, 1874); *Brontotherium dolichoceras* (Scott and Osborn, 1887) (originally described as *Menodus dolichoceras* Scott and Osborn, 1887); *Megacerops platyceras* (Scott and Osborn, 1887) (originally described as *Menodus platyceras* Scott and Osborn, 1887; also previously referred to as *Brontotherium platyceras* [Scott and Osborn, 1887]); *Brontops robustus* Marsh, 1887; *Menops varians* Marsh, 1887 (also previously referred to as *Menodus varians* [Marsh, 1887]); *Megacerops curtus* (Marsh, 1887) (originally described as *Titanops curtus* Marsh, 1887; also previously referred to as *Brontotherium curtus* [Marsh, 1887]); *Brontotherium elatus* (Marsh, 1887) (originally described as *Titanops elatus* Marsh, 1887); *Menops serotinus* (Marsh, 1887) (originally described as *Allops serotinus* Marsh, 1887); *Megacerops syceras* (Cope, 1889) (originally described as *Menodus syceras* Cope, 1889); *Brontops amplus* (Marsh, 1890) (originally described as *Diploclonus amplus* Marsh 1890); *Menops crassicornis* (Marsh, 1891) (originally described as *Allops crassicornis* Marsh, 1891); *Brontops validus* (Marsh, 1891); *Brontotherium medium* (Marsh, 1891) (originally described as *Titanops medius* Marsh, 1891); *Menodus peltoceras* (Cope, 1891); *Brontotherium ramosum* (Osborn, 1896) (originally described as *Titanotherium ramosum* Osborn, 1896); *Brontops brachycephalus* (Osborn, 1902) (originally described as *Megacerops brachycephalus* Osborn, 1902); *Brontops bicornutus* (Osborn, 1902) (originally described as *Megacerops bicornutus* Osborn, 1902; also previously referred to as *Diploclonus bicornutus* [Osborn, 1902]); *Menops marshi* (Osborn, 1902) (originally described as *Megacerops marshi* Osborn, 1902; also referred to as *Allops marshi* [Osborn, 1902]); *Brontotherium leidyi* Osborn (1902); *Brontops tyleri* (Lull, 1905) (originally described as *Megacerops tyleri* Lull, 1905; also previously referred to as *Diploclonus tyleri* [Lull, 1905]); *Brontotherium hatcheri* (Osborn, 1908); *Megacerops copei* (Osborn, 1908) (originally described as *Symborodon copei* Osborn, 1908); *Menops walcotti* (Osborn, 1916) (originally referred to as *Allops walcotti* Osborn, 1916); *Menodus cutleri* Russell, 1934.

Megacerops kuwagatarhinus Mader and Alexander (1995)

Nomina dubia

Menodus giganteus Pomel, 1849; *Menodus prouti*, Owen Norwood and Evans, 1850 (originally described as ?*Palaeotherium prouti* Owen Norwood and Evans, 1850; also referred to as *Titanotherium prouti*); *Palaeotherium maximum* Leidy, 1852; *Eotherium americanus* (Leidy, 1852) (originally described as *Rhinoceros americanus* Leidy, 1852); *Palaeotherium giganteum* Leidy, 1854; *Leidyotherium* (Prout, 1860); *Dolichorhinus vallidens* (Cope, 1872) (originally described as *Palaeosyops vallidens* Cope, 1872; also previously described as *Telmatherium vallidens* [Cope, 1872] and *Manteoceras vallidens* [Cope, 1872]); *Brontotherium gigas* (Marsh, 1873);

Menodus torvus (Cope, 1873c) (originally described as *Symborodon torvus* Cope, 1873c); *Miobasileus ophryas* Cope, 1873c nomen nudum; *Diconodon montanus* (Marsh, 1875) (originally described as *Ansiacodon montanus* Marsh, 1875 [*Ansiacodon* is preoccupied]); *Megacerops angustigenis* (Cope, 1886) (originally described as *Menodus angustigenis* Cope, 1886; also referred to as *Haplacodon angustigenis* [Cope, 1886]); *Brontotherium tichoceras* (Scott and Osborn, 1887) (originally described as *Menodus tichoceras* Scott and Osborn, 1887); *Brontops dispar* (Marsh, 1887); *Brontops selwynianus* (Cope, 1889) (originally described as *Menodus selwynianus* Cope, 1889; also referred to as *Diploclonus selwynianus* [Cope, 1889]); *Teleodus avus* (Marsh, 1890); *Duchesneodus primitivus* (Lambe, 1908) (originally described as *Megacerops primitivus* Lambe, 1908; also referred to as *Teleodus primitivus* [Lambe, 1908]); *Megacerops assiniboiensis* (Lambe, 1908); *Metarhinus earlei* Osborn, 1908; *Metarhinus cristatus* Riggs, 1912; *Heterotitanops parvus* Peterson, 1914a; *Megacerops riggsi* (Osborn, 1916); *Tanyorhinus harundivorax* Cook, 1926; *Telmatherium accola* Cook, 1926; *Telmatherium advocata* Cook, 1926; *Manteoceras foris* Cook, 1926; *Manteoceras pratensis* Cook, 1926; *Brontops canadensis* (Russell, 1934); *Menodus lambei* (Russell, 1934); *Ateleodon osborni* (Schlaikjer, 1935); *Duchesneodus californicus* (Stock, 1935) (originally described as *Teleodus californicus* Stock, 1935); *Metarhinus ?pater* Stock, 1937; *Menodus bakeri* (Stovall, 1948); *Duchesneodus thyboi* (Bjork, 1967) (originally described as *Teleodus thyboi* Bjork, 1967).

EQUIDAE

CORRECTIONS TO VOLUME 1

Mesohippus latidens is known from locality NP23**C**.

Mesohippus sp. is known from locality SB44**B**, but not from locality GC7.

Kalobatippus avus is not present at locality NB6E; *K.* sp. is known from locality CP109**A** but is not present at localities WM13, CA7 (=*Anchitherium clarencei*), CP104A, CP114A, NB17, NC2.

In Vol. 1, *Kalobatippus* sp. was recorded at Hemingfordian localities CP107, CP108A, B, CP109 and Barstovian localities CP110, CP111, and CP114B (mainly from AMNH and NSM collection records). We note (R. Hulbert) that all confirmed records of *Kalobatippus* are Arikareean; Hemingfordian records are likely to represent *Anchitherium* (a taxon synomized with *Kalobatippus* in Vol. 1: see comments in section below), and Barstovian records may represent *Hypohippus* or *Megahippus*. As the Anchitheriinae is in severe need of revision, these locality occurrences are best noted as "large anchithere, indet."

Hypohippus sp. is not present at localities SB32A, CP106.

Archaeohippus blackbergi is not present at localities GC3A, GC5.

Parahippus pawniensis is not present at locality CP75B.

Merychippus gunteri is not present at localities GC3B, GC4D. *M. primus* is known from locality GC9**C**; *M. insignis* is not present at localities GC4B, C, GC10A, NB6C, CP108B, CP114B, D, CP116A, NNP38E; *M. brevidontus* is not present at locality CC17D and is present at locality NB20 (not NB20A); *M. californicus* is questionably present at locality GC10B and is present at locality NB20 (not NB20A); *M. sphenodus* is not present at localities GC10B, CC27, CC31, SB32H, CP76, CP114A–C; *M. republicanus* is not present at localities GC10A, CP114B; *M. goorisi* is not present at localities GC4D, E; *M. seversus* is not present at locality CC17B; *M. sumani* is not present at localities CC17C–F, CC21C; *M. intermontanus* is definitely present at localities NB41B, SB32B, NP41B; *M.* sp. is not present at locality GC9B.

Pseudhipparion sp. is not present at localities GC4E, GC6B, D, CP75C.

Hipparion shirleyae is not present at locality GC4E; *H. tehonense* is not present at locality CP114C; *H.* sp. is not present at localities GC10A, CP114A.

Neohipparion affine is not present at localities GC6D, SB32D; *N. trampasense* is not present at localities GC11C, CP126; *N. leptode* is not present at locality SP3A; *N. eurystyle* is not present at localities CA4A, GC10D, [CC32B], CC36, CP115C, CP116C; *N. gidleyi* is not present at localities SP3A, SB34B; *N.* sp. is not present at localities NB19C, CP78, CP114C, PN14, PN15.

Nannippus peninsulatus is known from locality CP128**C** but is not present at localities SB14F, SP1F; *N. lenticularis* is not present at localities SP2A, NP45; *N. aztecus* is not present at localities GC11B, C, CP118; *N. beckensis* is not present at locality GC6D; *N.* sp. is not present at localities GC4E, GC6D (=*N. lenticularis*), CP123D, PN14.

Cormohipparion occidentale is not present at localities GC6B, CC26B, SP1A; *C. plicatile* is not present at localities CA9, GC13B; *C. emsliei* is not present at locality GC10D; *C.* sp. is not present at localities NB27A (= cf. *C. occidentale*), SB32D, F (= cf. *C. quinni*).

Protohippus perditus is not present at locality GC6B; *P. supremus* is known from locality **CP**90A (not GC90A), and is definitely known from locality GC6B; *P. gidleyi* is not present at locality CP116B; *P.* sp. is not present at localities GC10A, NB21, SB29D, SB32A, B, CP126.

Calippus martini is not present at localities GC13B, SP1A, B, CP90A; *C. cerasinus* is not present at locality GC13B.

Pliohippus pernix is not present at localities SB32D, CP90A; *P. mirabilis* is not present at localities NB7C, CP75C; *P. fossilatus* is not present at locality SP1A; *P.* sp. is not present at localities NB7B–D (= *P. tantalus* and *P. tehonense*), SP1B, C (= *P. nobilis*), SP3A (= *P. nobilis*).

Astrohippus ansae is not present at locality CP116D; *A. stockii* is known from locality SB13 (not SB13B) but is not present at localities GC13C, SP4B; *A.* sp. is not present at localities CC40, SB48, SP3A, CP90A.

Dinohippus leidyanus is not present at locality CP116A; *D. interpolatus* is not present at localities CP17I, NP45; *D. mexicanus* is known from locality SP4**B** but is not present at locality GC13C; *D. spectans* (= *Pliohippus spectans* in Vol. 1) is not present at locality CC17H and is present at locality PN11A and B (new locality subdivision, PN11B with emended dating of Hh 2); *D.* sp. is known from localities NB13**A**, **B**, but is not present at localities NB14, SB10 (= *D. mexicanus*), GC11C (= *D. spectans*).

Equus simplicidens is not present at localities SB31C, SB38, SB66, PN23; *E.* sp. is not present at localities CC53 (= *E. simplicidens*), SB14E, F (= *E. anguinus*), CP121 (= *E. simplicidens*).

Figures A.5 and A.6 show the temporal ranges of equids, and Figures A.7–A.9 show their biographical ranges.

SYSTEMATIC REVISIONS, NEW TAXA, AND IMPORTANT NEW LOCALITY OR PALEOBIOLOGICAL INFORMATION

The North American genus *Hyracotherium* has been extensively revised by Froehlich (2002). Following Hooker (1989, 1994), Froehlich restricted the name "*Hyracotherium*" to the European *H. leporinum*, which is actually a basal palaeothere.

Froehlich (2002) also considered "*Hyracotherium*" *tapirinum* and "*H.*" *cristatum* to be the tapiromorph *Systemodon tapirinus* and listed this taxon as occurring at the following localities: SB24, CP19C, CP20C, E, CP64B, C.

The North American taxa previously referred to as "*Hyracotherium*" are now designated as follows (ordered from the most primitive to the most derived). Localities for taxa cited in the literature, but outside of the range designated by Froelich (and thus of questionable assignment), are given in italics.

Sifrihippus sandrae (known from localities CP19AA and [new] *GC21II*, CP20A).

Minippus index (known from [new] localities CP25E, G, H, CP27B, D, [CP64B, C]; *not* found at localities SB24, CP20C, CP27B); *M. jicarillai* Froehlich, 2002 (localities SB24, CP64A).

Arenahippus grangeri (known from localities CP19AA, and [new] CP20A, AA, CP25A, [B], [CP64A, ?B]); *A. aemulor* (localities CP19A, CP20A); *A. pernix* (localities CP19II, CP20B, BB [no localities listed in Vol. 1]).

Xenicohippus grangeri (same taxon as in Vol. 1) (known from localities CP20C, CP27B, *not* found at localities CP20A, CP64B): *X.* (originally *Hyracotherium*) *craspedotum* ([new] localities *GC21II*, SB21, SB25A, SB40, CP5A, CP20E, F, *CP23A*, *CP24C*, *CP25C, D*, *CP26D*, CP27A, C, E, CP28A–C, CP31A, *CP63*, *NP49*, HA1); *X. osborni* (localities CP64B and [new] SB22A, SB24).

Eohippus angustidens (known from [new] localities [CC50] [= *H. seekinsi*"], SB22A, SB24, CP20C, [CP64B, ?C]).

Protorohippus venticolum (species not recognized in Vol. 1) (known from localities [SB22C], [CP25F, G], CP27B, D, [CP64C]); *P. montanum* (species not recognized in Vol. 1) (locality CP20C).

"*Hyracotherium vacacciense*" (taxon is not mentioned by Froehlich) (also recorded at [new] localities CP25I1, I2).

Unidentified "eohippines" known from the following localities (new from those listed as "*Hyracotherium*" sp. in Vol. 1): SB21, SB40, CP5A, CP20BB, CP23A, CP25C, D, GG, J, CP26D, CP27A, E, CP63).

The taxa *Kalobatippus clarencei* and *K. navasotae* (as listed in Vol. 1) are here considered (by R. Hulbert, following MacFadden, 2001) as belonging to the genus *Anchitherium*. *A. clarencei* is additionally known from localities CA7, GC8D, GC9B, [CP71], CP88, and *A. navasotae* is not present at localities GC8D, GC9B. *A.* sp. is known from localities CC16, NB17, CP105, CP106, CP107, NC2.

A new species of *Archaeohippus*, *A. mannulus* O'Sullivan, 2003, is known from locality GC8AC. O'Sullivan (2000) has also used stable isotopes to study age of maturity in *Archaeohippus* and concluded that this genus matured earlier than *Miohippus*, indicating that its dwarf size was a function of the heterochronic feature of progenesis.

The taxon *Desmatippus texanus* (as listed by MacFadden, Vol. 1) is here considered (by R. Hulbert, following Albright, 1999) as belonging to the genus *Anchippus*. In addition to the localities listed by MacFadden, it is also known from localities GC5A, GC8AB, B, but is not present at locality GC3B.

The taxon *Parahippus coloradensis* (as listed in Vol. 1) is here considered (by R. Hulbert) as belonging to the genus *Desmatippus*. It is additionally known from locality CP75B but is not present at locality CP86C.

The taxon *Merychippus carrizoensis* (as listed in Vol. 1) is here considered (by R. Hulbert, following Kelly, 1995) as belonging to the genus *Parapliohippus*. In addition to the localities listed by MacFadden it is also known from locality CC18, but is not present at locality CC17D.

The taxa *Merychippus tertius*. *M. isonesus*, and *M. stylodontus* (as listed in Vol. 1) are here considered (by R. Hulbert, following Kelly, 1995, 1998) as belonging to the genus *Acritohippus*, as is *M. quinni* Kelly (not listed in Vol. 1). In addition to the localities listed in Vol. 1, *A. tertius* is also known from locality [CC17B]; *A. isonesus* from locality [GC10AA]; *A stylodontus* from locality NP41B; and *A. quinni* from localities CC17C–F, CC21C.

A number of equids are reported from the early Barstovian of Mexico (locality CA2II; Ferrusquía-Villafranca, 2003), representing a biogeographic range extension for all of these taxa. These include *Merychippus* cf. *M. primus*, *M.* cf. *M. sejunctus*, *M.* cf. *M. californicus*, *Neohipparion* aff. *N. trampasense*, *Calippus* sp., and *Pliohippus* aff. *P. pernix*. *Cormohipparion* sp. is further reported from this locality by Bravo-Cuevas (2005).

Woodburne (2003) has analyzed craniodental material of *Merychippus insignis* and *Cormohipparion goorisi* (considered as *Merychippus goorisi* in Vol. 1) and concluded that *Merychippus* and *Cormohipparion* were distinctly different taxa.

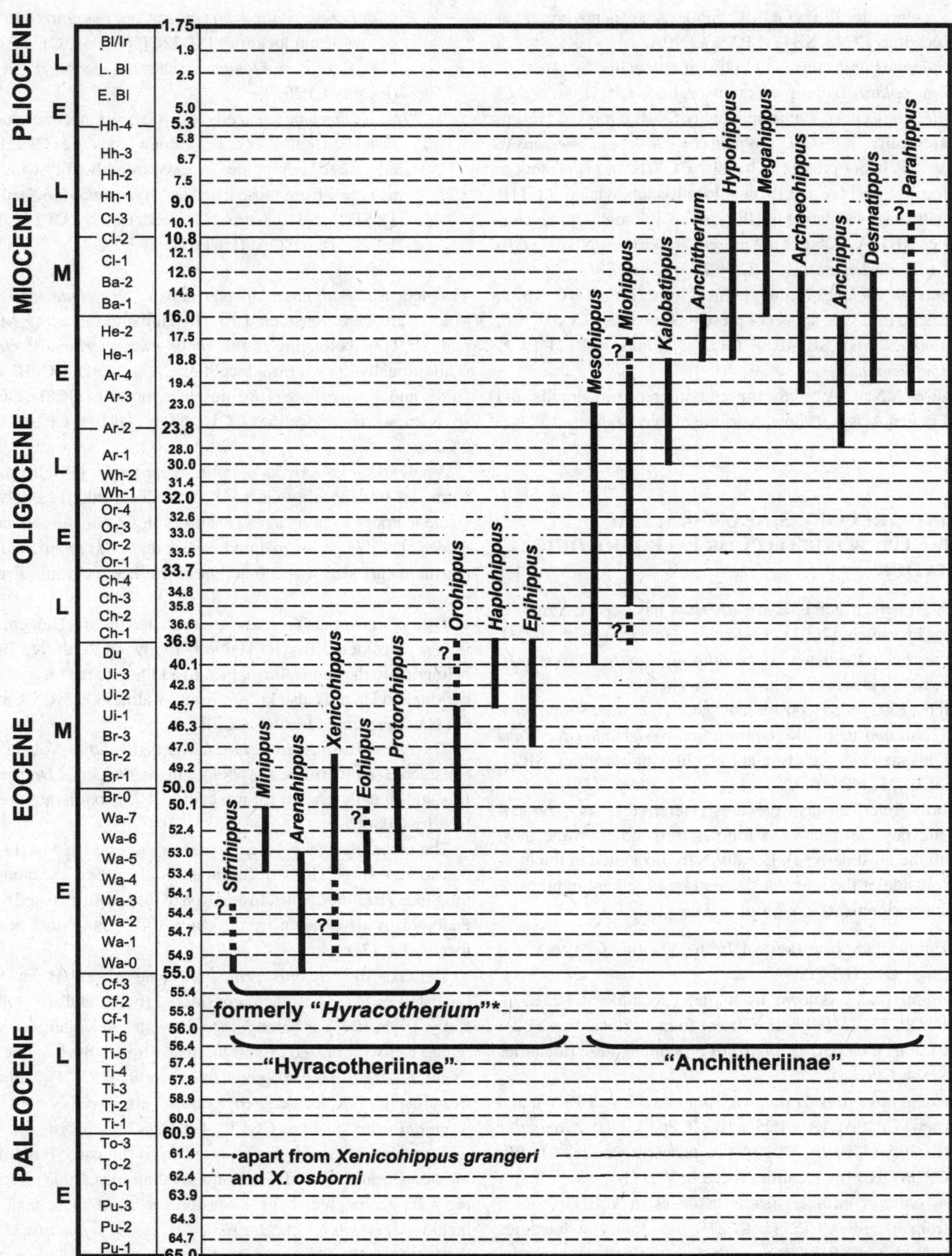

Figure A.5. Temporal ranges of hyracotheriine and anchitheriine equids.

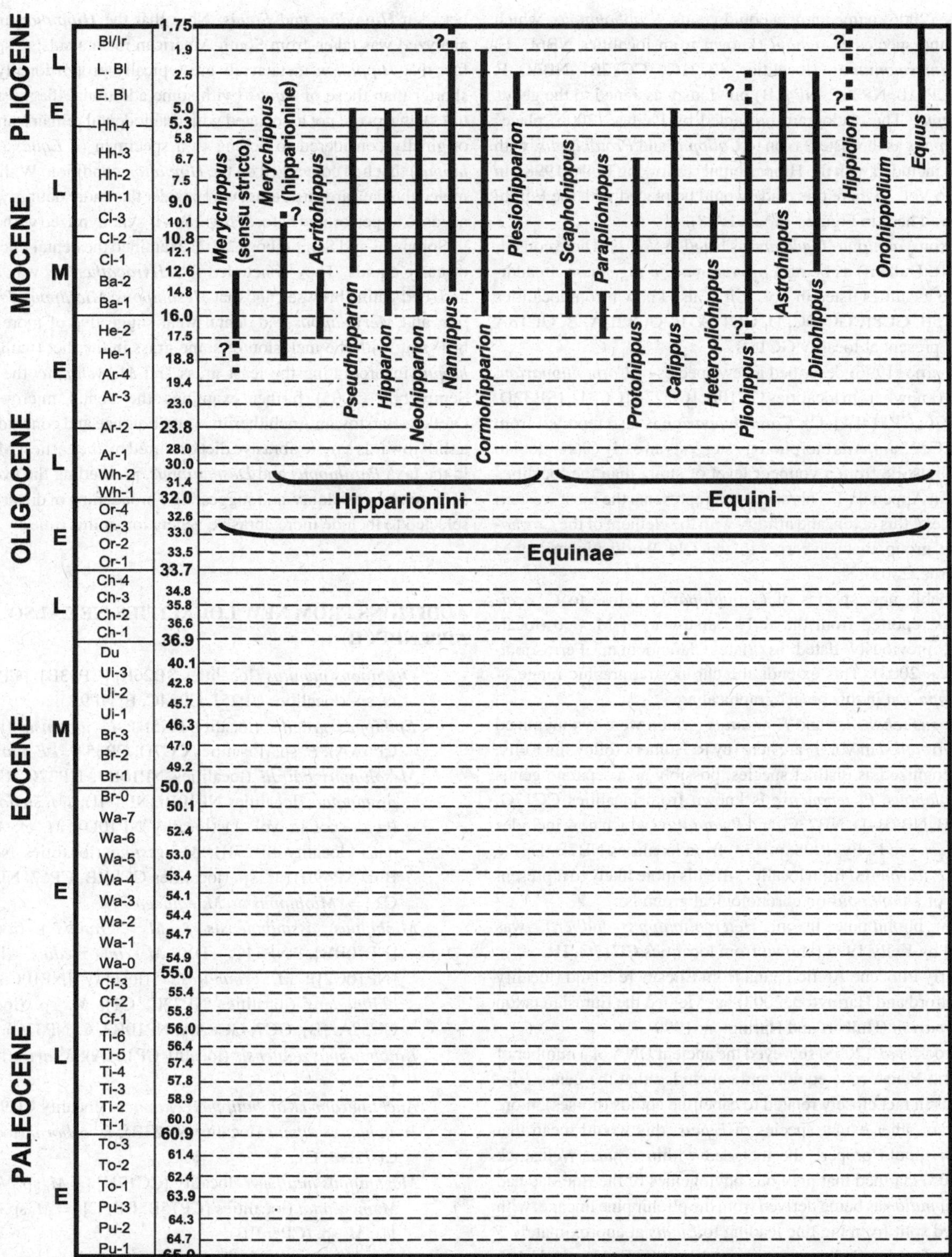

Figure A.6. Temporal ranges of equine equids.

Pagnac (2006) named a new equid genus, *Scaphohippus*, which includes the species *S. sumani* (known from localities NB6E, F) and *S. intermontanus* (localities CC21C, CC22B, NB6D–F, SB32B, CP110, NP38C, NP41B) previously assigned to the genus *Merychippus*. The cladogram presented by Pagnac (2006) places *Scaphohippus* as the sister taxon to *Calippus* and *Protohippus*, with this clade included with the Hipparionini (following Kelly, 1998), in contrast to Vol. 1 where this clade would be placed with the Equini (as followed here in Figures A.6 and A.8).

The taxon *Nannippus ingenuus* (as listed in Vol. 1) is here considered (by R. Hulbert) as being *Cormohipparion ingenuum*. In addition to the localities listed in Vol. 1, it is also known from localities CA9, GC6B, GC8F, GC10C, D, GC11A, B, GC12IIA, B, GC13A but is not present at locality GC13B.

Woodburne (1996) described a new species of *Cormohipparion, C. quinni* (known from localities GC10B, [CC27], [CC31], [SB32D, F, H], CP76, CP114[A]–C). *Cormohipparion* is also reported from locality CC22, and as its morphology suggests an early Clarendonian age, this must be from a younger level of strata than the localities recorded in Appendix I. This specimen represents the most western occurrence of this taxon, and affinity with the element of the *Cormohipparion* group that gave rise to the Old World hipparionines (Woodburne, 2005).

A probable new species of *Cormhipparion*, close to *C. occidentale*, is reported from the early Barstovian of Mexico (locality CA3, previously dated as ?late Clarendonian; Ferrusquía-Villafranca, 2003). This extends the chronostratigraphic range of *Cormohipparion* in this biogeographical area.

Pliohippus tehonense and *P. tantalus* (which were synonymized with *P. fairbanksi* in Vol. 1) are here (by R. Hulbert, following Kelly, 1998) recognized as distinct species, possibly as a separate genus from *Pliohippus*. *P. tehonense* is known from localities CC17G, CC32A, B, NB7B–D, NB23C, and *P. tantalus* (which now includes *P. fairbanks*, see Kelly, 1998, p. 473) from localities NB7B–D (the record of *P. fairbanksi* from locality NB31 is more likely to represent a species of *Dinohippus* on chronological grounds).

A new pliohippine taxon, *Heteropliohippus hulberti*, was described by Kelly 1995 (known from localities CC17G, H).

The early Pliocene Arctic Fauna from Ellesmere Island (locality HA3; Tedford and Harington, 2003) has yielded the Eurasian taxon *Plesiohipparion* (Hulbert and Harington, 1999).

Weinstock *et al.* (2005) surveyed the ancient DNA of a number of extinct Plio-Pleistocene equids and concluded that the genus *Hippidion* was in fact closely related to cabelline equids (horses), more closely than other extant species of *Equus*; this would mean that *Hippidion* should properly be included within *Equus*. Weinstock *et al.* (2005) claimed that previous phylogenies of the Equidae had placed *Hippidion* as being derived from the pliohippine lineage, with a proposed split from the line leading to *Equus* at approximately 5 Ma. Note that the cladogram presented by MacFadden in Vol. 1 shows the position of *Hippidion* and related genera (*Onohippidium* and *Parahippidion*) as being somewhat undetermined within the Equini, and certainly closer to *Equus* (plus *Dinohippus*) than to *Pliohippus*. The earliest known specimen of *Hippidion* (from North America) is from approximately 7 Ma, creating some problems for the molecular estimate of approximately 3.5 Ma of the split between *Hippidion* and *Equus*. Note that the *Hippidion* material analyzed was taken from South American metatarsals: despite the fact that *Hippidion* metatarsals are typically proportionally much shorter than those of *Equus* (with some additional diagnostic features), they were not associated with craniodental remains and were originally considered to belong to a specimen of *Equus (Amerihippus)* that had converged on the *Hippidion* condition. While these molecular data are intriguing, we consider that more definitive study needs to be performed before *Hippidion* is synonymized with *Equus*.

Solounias and Semprebon (2002) examined the dental microwear of early equids. They concluded that *Hyracotherium* was a fruit- and seed-eating browser, and that *Mesohippus, Miohippus, Parahippus*, and *Merychippus* had dental wear suggestive of more coarse browsing with the inclusion of some grass in the diet (with *Mesohippus* incorporating the least grass and *Merychippus* the most). Semprebon (2005) further examined the dental microwear of Orellan–Barstovian Anchitheriinae and Equinae and concluded that a shift towards a more abrasive diet occurred by the earliest Miocene in the taxa *Parahippus* and *Desmatippus*, preceeding the evolution of extreme hypsodonty, and suggesting a broadening of dietary items selected to include more abrasive vegetation at this time.

ADDITIONS FROM NEW LOCALITIES (SEE ALSO APPENDIX II)

Orohippus pumilus (localities SB26II, [CP33B], [CP34A]); *O.* sp. (localities CP25J, CP34C, E, NP9B).

Epihippus gracilis (locality PN5B); *E. uintensis* (locality CP6AA); *E.* sp, (localities CC7D, CP65 [= "*E. parvus*"]).

Mesohippus bairdii (localities NP10B3, NP27C, C1); *M. propinquus* (localities NP10Bi, NP49II); ?*M. stenolophus* (taxon not in Vol. 1) (locality NP10D); *M.* cf. *M. texanus* (locality SB27II); *M. westoni* (localities NP10Bi, Biii, NP49II); *M.* sp. (localities CC8IIB, CP67, NP10Bii, C2 [= "*Miohippus* nr. *M. equiceps*"]).

Miohippus assiniboiensis (= *M.* "*grandis*") (localities [NP10BB], NP24C, D); *M. intermedius* (locality [NP10C2]); *M. gemmarosae* (locality [NP10C3]); *M. obliquidens* (localities NP27C, C1); *M.* sp. (localities GC5?A, [B], GC7, GC8AA, NP10B3, C, NP32C).

Kalobatippus agatensis (locality CP104A); *K. avus* (locality CP52).

Anchitherium (Kalobatippus) clarencei (locality CC9IVD).

Hypohippus affinus (locality NB23B); *H. osborni* (localities CP75BB, C).

Megahippus matthewi (locality [CC17H] [= *M.* sp. Vol. 1]); *M. mckennai* (localities [CP75B, BB, C] [= *M.* sp. in Vol. 1]); *M.* sp. (CP56IIB).

Archaeohippus ultimus (locality NP42); *A. blackbergi* (locality GC8DB).

Desmatippus nebraskensis (= "*Parahippus*") (locality [NP10C3]).

Parahippus pawniensis (locality CC9IVD; *P. wyomingensis* (locality CP51A); *P.* sp. (localities GC8DB, GC10A [could be *Desmatippus*]).

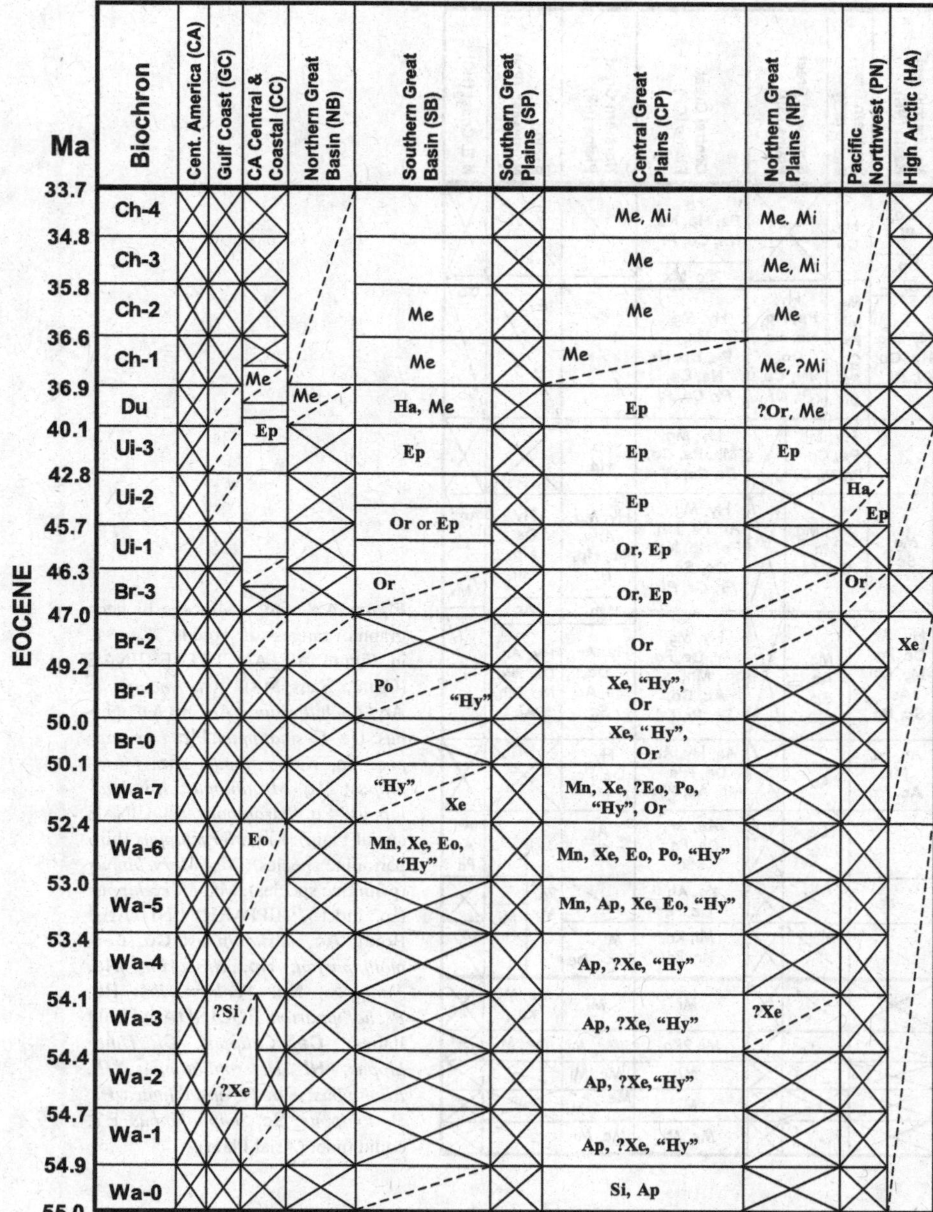

Figure A.7. Eocene biogeographic ranges of equids. Key as in Figure A2. HYRACOTHERIINAE (Times New Roman bold): **Ap**, *Arenahippus*; **Eo**, *Eohippus*; **Ep**, *Ephippus*; **Ha**, *Haplohippus*; **"Hy"**, "*Hyracotherium*" sp.; **Mn**, *Minippus*; **Or**, *Orohippus*; **Po**, *Protorohippus*; **Si**, *Sifrihippus*; **Xe**, *Xenicohippus*. ANCHITHERIINAE (Comic Sans MS): Me, *Mesohippus*; Mi, *Miohippus*.

Merychippus insignis (locality CP75BB); *M. californicus* (locality [CA2II]); *M. goorisi* (locality [GC4DD]); *M. gunteri* (localities GC4DD, GC8DB, [GC26]); *M. brevidontus* (localities CC17E, NB20); *M. primus* (locality [CA2II]); *M. sphenodus* (localities CP75B, BB, C); *M. sejunctus* (localities [CA1II], [CA2II]); *M.* sp. (localities GC4B, C, CC18A, NB6C, CP75B, BB, CP114B–D, CP116A, NP38E, NC3D).

Pseudhipparion curtivallum (locality GC8F); *P. hessei* (localities SB47IIA, B); *P. skinneri* (localities GC10D [= ?*P. simpsoni* in Vol. 1], GC11A, B, GC13A); *P. simpsoni* (localities G6D [= *P.* sp. in Vol. 1], SP4B, [EM8B]).

Hipparion tehonense (localities GC6A, [GC10D], [GC11D], [GC13B], CC26A, CC33); *H.* sp. (localities GC10A, CP114A).

Neohipparion trampasense (localities [CA2II], GC10D, CC26A, CC32B, CP115C); *N. leptode* (localities [CC225B], SP3A, CP78, CP116C, CP126); *N. eurystyle* (localities SB56II, GC6D, GC11C, GC12IIB, GC13C, GC27, [CC36], SB48III, SB58IIA, B, CP123D [AMNH collections], CP127); *N. gidleyi* (localities SB34C, SP1D); *N.* sp. (locality CA4A); *N.* sp. (locality SB55V).

Nannippus peninsulatus (localities GC17A, B, SB18A0, SB34IIB, SB34IVA, SB35D, SB36B, [SB37D], SB38,

Epoch	Ma	Biochron	Cent. America (CA)	Gulf Coast (GC)	CA Central & Coastal (CC)	Northern Great Basin (NB)	Southern Great Basin (SB)	Southern Great Plains (SP)	Central Great Plains (CP)	Northern Great Plains (NP)	Pacific Northwest (PN)	N.E. Coast (NC)
MIOCENE	9.0	Cl-3		Ps, Hp, Ne, Na, Co, Pr, Ca	Mg, Hp, Ne, Co, Hl, Pl, Di	Co, Pl; Hp, Di	Hy, Co		Hy, Ps, Hp, Ne, Na, Co, Pr; ?Pa, Mx		E?	
	10.1	Cl-2		Hy, Ps, Ne, Na, Hp, Co, Pr, Ca, Pl, Di; Ps, Co, Pr, Ca, Pl	Mg, Hp, Co, Hl, Pl, Di	Hy, Mg, Hp, Co, Pl, Di	Mg, Mh, Ps, Co, Pl, Di	Hy, Ps, Hp, Ne, Na, Co, Pr, Ca, Pl, Di	Hy, Mg, Mx, Ps, Hp, Ne, Na, Co, Pr, Ca, Pl		E?	Co
	12.1	Cl-1					Hy, Mh, Ps, Co, Pr, Pl, Di		Hy, Mg, Mh, Ps, Co, Pr, Ca, Pl	Mx, Hp		
	12.6	Ba-2	Pl	Hy, Mg, Mp, Mh, Ps, Na, Co, Pr, Ca, Pl	Ar, Mh, Ac, Hp, Sc	Mg, Sc	Hy, Mg, Mh, Co, Pl		Hy, Mg, Ar, Pa, Mh, Ps, Hp, Ne, Co, Sc, Pr, Ca, Pl	Hy, Mg, Ar, Ac, Hp, Pl; Mh	Hy, De, Mp, Mh; Ac	Ar; Mx
	14.8	Ba-1	Mp, Mh, Ne, ?Co, ?Ca, Pl	Hy, De, Pa, Mp, Mh, Ac, Hp, Co, Pr, Ca	Hy, Ar, De, Pa, Mh, Ac, Pp	Hy, Ar, De, Pa Mp, Mh, Ac, Sc, Pp	Mg, Ac, Sc		Hy, Mg, Ar, De, Pa, Mh, Ac, Co, Sc, Pr, Ca	Hy, Ar, De, Mp, Ac, Sc	Hy, Ar, De, ?Pa, Mp, Mh, Ac	
	16.0	He-2		An, Ar, Pa, Mp	Hy, Ar, Pa, Mx, Ac, Pp	An, Pa, Mx, Ac, Pp	Ar, Ac		An, Hy, Ar, De, ?Pa, Mp, Ac, ?Pl	Hy, De, Pa, Mx		
	17.5	He-1	An, Ar, Mh	?Mi, An, Hy, Ar, Ah, Pa	An, ?Ar Pa, ?Mx; An, Pa		De, Mx		An, Ar, De, Pa, Mx	Ar, Pa	Ar, Pa, Mx	An, Ar, Pa
	18.8	Ar-4		Ka, Ar, Ah, Pa					Ka, Ah, De, Pa		Mi	
	19.4	Ar-3		?Mi Ah	Ka		Pa		Mi, Ka, Ar, Pa	Mi De		Ka
OLIGOCENE	23.0 / 23.8	Ar-2		Ar, Pa, ?Mx; Mi, Ah					Mi	Mi	Me, Mi, Ka	
	28.0	Ar-1		Mi	Mi				Mi, ?Ka	?Me, Mi	Me, Mi	
	30.0	Wh-2							Mi	Me, Mi		
	31.4	Wh-1							Mi	Me Mi		
	32.0	Or							Me, Mi	Me, Mi		
	33.7											

Figure A.8. Oligo-Miocene biogeographic ranges of equids. Key as in Figure A2. ANCHITHERIINAE (Comic Sans MS): Ah, *Anchippus*; An, *Anchitherium*; Ar, *Archaeohippus*; De, *Desmatippus*; Hh, *Hypohippus*; Ka, *Kalobatippus*; Me, *Mesohippus*; Mg, *Megahippus*; Mi, *Miohippus*; Pa, *Parahippus*. EQUINAE (Arial Plain): Mh, *Merychippus* (hipparionine species); Mp, *Merychippus* (primitive species); Mx, *Merychippus* (sp. indet.). HIPPARIONINI (Arial Bold): **Ac**, *Acritohippus*; **Co**, *Cormohipparion*; **Hp**, *Hipparion*; **Na**, *Nanippus*; **Ne**, *Neohipparion*; **Ps**, *Pseudhipparion*. EQUINI (Arial Bold Italics): ***Ca***, *Calippus*; ***Di***, *Dinohippus*; ***Hl***, *Heteropliohippus*; ***Pl***, *Pliohippus*; ***Pp***, *Parapliohippus*; ***Pr***, *Protohippus*; ***Sc***, *Scaphohippus*. E?, equid indet (Arial Plain).

SB50B, CP118); *N. lenticularis* (localities GC6D, SB48II); *N. minor* (=*aztecus*) (localities [GC11C, D], GC13B, GC12IIB, GC27, GC31, SB48II, SB55IV, SB57, SP4B, EM7II); *N. major* (species not in Vol. 1) (locality SB58IIA); *N. westoni* (locality GC10D); *N. morgani* (locality GC13A); *N.* sp. (localities GC8F, SB18E, ?SB64II CP117B, [NP45]).

Cormohipparion occidentale (localities GC6B, GC8F, [GC10C], CC17H, [NB27A], SB47IIA, B, SB48II, SB48III, SP1A, [PN14]); *C. plicatile* (localities GC10D, GC11B, D, GC12IIB, GC13B, A); *C. emsliei* (locality GC27); *C. ingenuum* (localities GC8F, GC10D); *C.* sp. (localities GC27, [SB55IIAA]); *C. paniense* (species not in Vol. 1) (locality CP75BB).

Protohippus gidleyi (localities GC27 [MacFadden and Dobie, 1998], CP126); *P. proparvulus* (species not in Vol. 1) (locality CP75B); *P. eohipparion* (species not in Vol. 1) (locality CP75B).

Calippus small sp. (of Hulbert) (locality GC8F); *C. castilli* (= *C. hondurensis* in Vol. 1) (locality SB55III); *C. elachistus* (locality GC10D); *C. martini* (localities GC8F, SP1A, B, CP90A); *C. hondurensis* (localities GC10D, GC11D, GC13B, SB55IIAA); *C. maccartyi* (locality [GC13C]); *C.* sp. of Breyer (undescribed large species, not included in Vol. 1) (localities SP1B, C, SP3A, CP78, CP116C, D, CP126); *C.* sp. (localities [CA2II], GC8F [spp. A and B], CP75C); *C. labrosus* (species not in Vol. 1) (localities CP75B, BB).

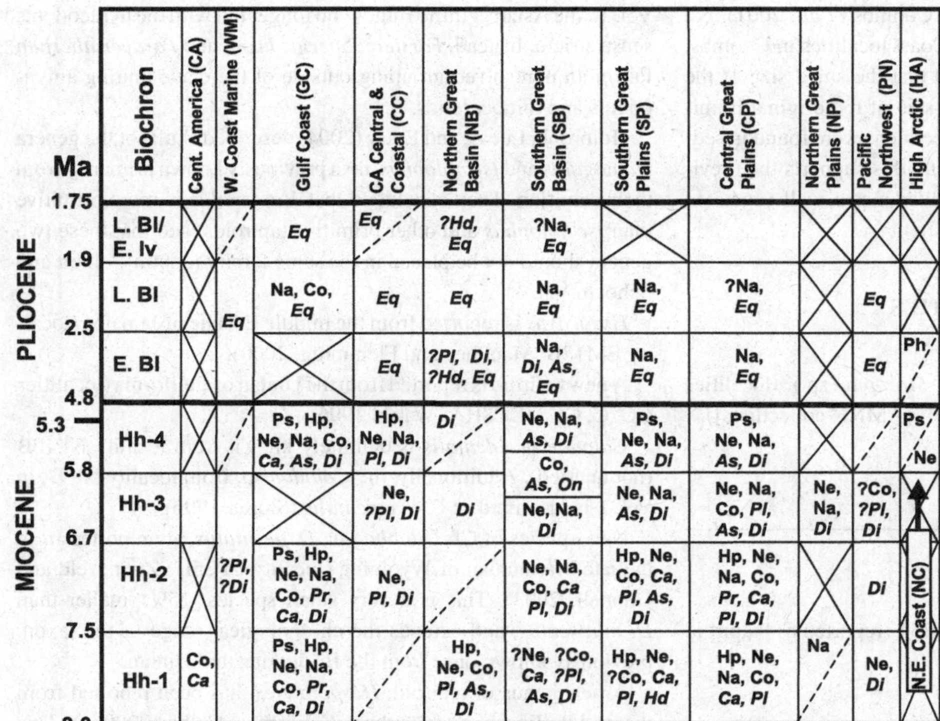

Figure A.9. Mio-Pliocene biogeographic ranges of equids. Key as in Figure A2. EQUINAE: HIPPARIONINI (Arial Bold): **Co**, *Cormohipparion*; **Hp**, *Hipparion*; **Na**, *Nanippus*; **Ne**, *Neohipparion*; **Ph**, *Plesiohipparion*; **Ps**, *Pseudhipparion*. EQUINI (Arial Bold Italic): ***As***, *Astrohippus*; ***Ca***, *Calippus*; ***Di***, *Dinohippus*; ***Eq***, *Equus*; ***Hd***, *Hippidion*; ***On***, *Onohippidium*; ***Pl***, *Pliohippus*; ***Pp***, *Parapliohippus*; ***Pr***, *Protohippus*.

Pliohippus pernix (localities CA2II, CA2III, GC6B, GC8F, SB47IIA, B, CP90A); *P. nobilis* (localities SP1B, C, SP3A, CP116C); *P. mirabilis* (locality SB32D); *P. fossilatus* (locality [CC33]); *P.* sp. (localities SB48II, SP1A, CP78).

Astrohippus ansae (locality SB48III); *A. stockii* (localities SB34IIA, SP4); *A.* sp. (locality CP127).

Dinohippus (most likely) or *Pliohippus* sp. (locality NB35IIIA).

Dinohippus leidyanus (localities CC40, NB10); *D. interpolatus* (localities [CC17I], SB48III, SB55IIAA, III, VI, CP127); *D. mexicanus* (localities SB57II (could be *E. simplicidens*), SB10, SB34IIA, SB56II, SB58IIA, B); *D. leardi* (localities CC26B, CC32B); *D. spectans* (localities GC10D, GC11C); *D.* sp. (localities GC13C, CC25B, CC37, CC39B, CC41, NB7E, NB9, NB13A, B, NB34II, NB35IIIA, SB47IIB, SB66).

Equus (Plesippus) simplicidens (localities WM26, [CC52II], CC53, CC52III, [NB13C], NB35IIIB, SB31C, SB34IIB, SB34IVA, SB35A–D, SB36IIA, SB37A, SB58IIC, ?SB64II, SB65, SB66, CP121); *E. (P.) idahoensis* Merriam (taxon not in Vol. 1) (localities NB35IIIC, NB36IIIA, [SB14F], SB49, SB50, PN12C); *E. (P.) enormis* Downs and Miller (taxon not in Vol. 1) (locality NB13C); *E. (P.) anguinus* Azzaroli and Voorhies (taxon not in Vol. 1) (localities SB14E, F, PN23C); *E. (Hemionus) calobatus* Troxell (species not in Vol. 1) (localities SB37C, D, [E], [SB38], SB38IIC, CP80, CP118); *E. (Hemionus)* n. sp. (localities GC18IIIA, GC18IV); *E. conversidens* (species not in Vol. 1) (locality SB68); *E. scotti* (localities SB31C, SB34IVB, SB35A, [C], SB36B, C, SB37A, D, E, SB38IIC); *E. cumminsi* Cope (species not in Vol. 1) (localities CC47, NB36IIIA, [SB35A, C, D], SB36B, SB38IIB, SB49, SB50, [SPIF, H], SP5, [CP81], CP128C, [CP131II]); *E. giganteus* (species not in Vol. 1) (locality NB35IIIC); *E.* sp. (inc. *E. "fraternus"* and *E. "leidyi"*) (localities GC14A, GC15A–F, GC17A, B, GC18IIA, B, G18IIIA–C, G18IV, F, CC47D (= "*Plesippus*" *idahoensis*), CC47II [= "?*Dolichohippus*"], CC48C (= "*Plesippus*" *francesana*, *E. "bautistensis"*), NB13B, SB15C, SB18A0, SB35A, SB36C, SB37B, E, F.

Equid indet. (localities GC8A, GC8B [=was GC8A in Vol. 1], CP56IIA).

CHALICOTHEROIDEA

SYSTEMATIC REVISIONS, NEW TAXA, AND IMPORTANT NEW LOCALITY OR PALEOBIOLOGICAL INFORMATION

New specimens of *Moropus oregonensis* (now synonymized with *M. distans*) are reported from the late early Arikareean of the John Day Formation, Oregon (locality PN6D3, previously listed only as PN6 in Vol. 1; Coombs *et al.*, 2001).

New small chalicotheres (*Moropus* sp. cf. *M. oregonensis*) are also reported from the late Arikareean of Florida and Texas Gulf

Coast (localities GC5A, GC7II, GC8B; Coombs *et al.*, 2001). No chalicotheres were reported from Gulf Coast localities in Coombs, Vol. 1. Coombs *et al.* (2001) concluded that the small size of the Gulf Coast chalicotheres represents the size of the original immigrants, rather than dwarfing. As these specimens were found in sediments representing well-watered habitats, this supports the previously held notion that chalicotheres inhabited mesic, well-vegetated environments.

ADDITIONS FROM NEW LOCALITIES (SEE APPENDIX I)

Indeterminate chalicotheres (probably *Moropus* sp.) (localities CP114A, B [UNSM collections], CP115B [AMNH collections]).

TAPIROIDEA

CORRECTIONS TO VOLUME 1

Helaletes sp. is known from localities SB51 (reported as "tapiroid indet." in Vol. 1) and CP32**B**.

SYSTEMATIC REVISIONS, NEW TAXA, AND IMPORTANT NEW LOCALITY OR PALEOBIOLOGICAL INFORMATION

Holbrook (1999) performed a cladistic analysis on tapiromorph perissodactyls. *Homogalax* and *Cardiolophus* group outside of other ceratomorphs (but do not cluster together, falling with equal status as chalicotheres, equids and brontotheres). Tapiroids and rhinocerotids then cluster together as two separate groups. Within the tapiroids, *Isectolophus* is the most primitive taxon (contra Colbert and Schoch, Vol. 1, where it was clustered with *Homogalax* and *Cardiolophus* [in the family Isectolophidae] outside of other moropomorphs [i.e., ceratomorphs plus chalicotheres]). *Heptodon* then falls as the sister taxon to other tapiroids (contra Colbert and Schoch, Vol. 1, where it is placed in an unresolved trichotomy with higher tapiroids and rhinocerotids). The next division within the tapiroids leads to *Protapirus* plus *Tapirus* in one branch, and more primitive tapiroids in the other. This branch of more primitive tapiroids groups together the North American taxa *Colodon, Helaletes*, and *Plesiocolophus* in one clade, and three Eurasian taxa (*Lophialetes*, *Schlosseria*, and *Deperetella*) in another. This differs from the scheme in Colbert and Schoch (Vol. 1) where *Plesiocolophus* was placed in the Tapiridae with *Protapirus* and *Tapirus* (the relationship of *Plesiocolophus* and *Protapirus* with respect to *Tapirus* is unresolved), and *Colodon* and *Helaletes* then form successive sister taxa to the Tapiridae.

Within the rhinocerotids, *Hyrachyus* is the most primitive taxon (similar to Colbert and Schoch, Vol. 1). The next branching splits the North American hyracodonts from a group containing the amnynodonts and the rhinocerotids (as sister taxa to one another) (the interrelationships of the three families of rhinocerotoids were not specifically considered in Vol. 1). However, contra the cladogram in Vol. 1., the Asian "giraffe rhinos" no longer fall with the hyracodonts sensu stricto. Instead, *Forstercooperia*, *Juxia*, and *Paraceratherium* fall in an unresolved grouping outside of the clade uniting amynodonts and rhinocerotids.

Holbrook, Lucas, and Emry (2004) described skulls of the genera *Homogalax* and *Isectolophus*, taxa previously known primarily from their dentition. They concluded that *Homogalax* is more primitive than *Isectolophus* and other primitive tapiroids, and that these two genera should not be placed in the same family (contra Colbert and Schoch, Vol. 1).

Hyrachyus is reported from the middle Eocene of Jamaica (locality EM12B; MacPhee and Flemming, 2000).

A new tapiroid is reported from the Uintan of California (localities CC7C, CC8, CC8IIA; Walsh, 1996).

Colodon occidentalis is definitely known from locality NP10B (not queried). Additionally, the *Colodon* sp. from locality NP9A in Vol. 1 is identified as *C. occidentalis* (Storer, 1996).

New species of Cf. *Dilophodon, D. destitutus*, are reported from the late Wasatchian of Wyoming (locality CP25I1; Zonneveld and Gunnell, 2003). This is a very small species, 15% smaller than *D. miniscukus*, and extends the chronological range of the taxon, previously known only from the Bridgerian and Uintan.

A new genus of tapiroid, *Hesperaletes*, has been reported from the middle Eocene of southern California (Colbert, 2006a). This taxon belongs to a clade including *Proapirus* and *Plesiocolopirus* and is characterized by a deeply retracted narial incision, indicative of a proboscis. *H. borineyi* (= "unnamed genus and species A", in part, in Vol. 1) is known from locality CC7B and other associated localities, extending into the Duchesnean. *H. walshi* (= "unnamed genus and species B", in Vol. 1) is known from locality CC4 (previously CC5) and other associated localities. Colbert (2006b) also discussed aspects of variation and species recognition in these tapiroids.

A new genus of tapir *Nexuotapirus marslandensis* (= *Miotapirus marslandensis* and *M. harrisonensis*) is reported from the late Arikareean of Florida, and the early Arikareean of Nebraska (localities GC5A and CP101 [= *Miotapirus harrisonensis* in Vol. 1]; Albright, 1998). This taxon differs from *Miotapirus* (and *Protapirus*) in having molars with a pattern of distinct interloph structures, rather than simple bilophodonty. A separate, larger species, *N. robustus*, is also reported from the late Arikareean of Oregon (localities PN6G, H).

Protapirus validus is a junior synonym of *P. simplex* (Albright, 1998).

Eberle (2004) reported a tapiroid with a relatively enlarged narial incision, possibly of Eurasian affinities, from the early Eocene of the Canadian Arctic (locality HA1A2).

Tapiravus polkensis is reported from the late Miocene/early Pliocene Gray Fossil Site in Tennessee (locality CE1; Wallace and Wang, 2004). Note that Hulbert (1999) considered *Tapiravus polkensis* to be a small species of the genus *Tapirus*, unrelated to the middle Miocene *Tapiravus*.

Hulbert (2005a) provided a phylogenetic analysis of the six extinct species of *Tapirus* from the late Cenozoic of North America (including the four extant species). He noted that *T. johnsoni* is the sister taxon to the other species, and that other Cenozoic species group either into a clade containing the extant species *T. terrestris* and *T. pinchaque* or into a clade containing the extant species *T. bairdii* and *T. indicus*. Hulbert also (2005b) described a new species of *Tapirus, T. webbi*, from the late Clarendonian/early Hemphillian of Florida (localities GC11A, GC11B [type locality]) (originally described as *T. simpsoni* from these localities). This taxon is about the same size as the extant Malaysian tapir (*T. indicus*), but with longer limbs. It is larger than *T. johnstoni* and differs from *T. simpsoni* in having narrower upper premolars and weaker transverse lophs on P1 and P2.

Hulbert (2005b) also discussed the evolutionary history of the genus *Tapirus* in North America and the Old World. In addition to the localities listed in Vol. 1, he noted fragmentary material of *Tapirus* at localities GC11D, GC13A. The early Hemphillian Withlacoochee River 4A Locality (locality GC12II) has produced eight specimens of a *Tapirus* sp. smaller than *T. webbi* or *T. simpsoni*, comparable in size to the extant *T. terrestris*. A smaller tapirid was also recorded within this locality (from Withlacoochee River 5E) and from the latest Hemphillian (locality GC13B), where it was originally referred to as *T. polkensis* (= *Tapiravus polkensis* in Vol. 1).

Graham (2003) reviewed the systematics and paleoecology of Plio-Pleistocene North American tapirs. This work is mainly on Pleistocene taxa, but he makes the interesting observation that, while modern tapirs are tropical forest specialists, their presence at high altitude in the Pleistocene shows that absolute temperature is not a factor in determining their biogeographical distribution. The present-day restricted distribution of tapirs is probably related to the habitat destruction brought about by climatic changes at the end of the Pleistocene.

ADDITIONS FROM NEW LOCALITIES (SEE APPENDIX I)

Homogalax protapirinus (locality [CP20BB]); *H.* sp. (locality CP20BB).
Isectolophus annectens (locality CP65).
Isectolophid indet. (locality CP20BB, CP25J).
Dilophodon sp. (localities CP65, NP8).
Hyrachyus modestus (locality CP31E); *H.* sp. (localities ?CC4, CP25J, CP33B, CP34A, [EM12B]).
Toxotherium hunteri (locality NP10Bii); *T.* sp. (localities ?NP10B3, NP49II).
Helaletes nanus (locality CP34A); *H.* sp. (localites CP25J, CP34A).
Colodon woodi (locality [NP22]); *C.* sp. (locality NP10B3).
"*Dilophodon*" *leotanus* (? = unnamed new genus in Vol. 1) (locality NP22).
Protapirus sp. (locality PN5B).
Tapiravus sp. (localities CP75B, BB).
Tapirus haysii (localities GC15D–F, GC18IIIA, B, GC18IV, SB37D); *T.* n. sp. (localities ?GC14A [= *T.* sp. in Vol. 1], GC15A, ?B, C, GC17B, GC18IIB); *T.* sp. (localities GC11D, GC13B, GC27, CC48C, SB18E, SB35B. CP116C: AMNH collections).
Tapiriidae indet. (localities GC8C [= GC8B in Vol. 1], CP5II [helaletid]).

AMYNODONTIDAE

CORRECTIONS TO VOLUME 1

Metamynodon chadronensis is known from locality SB44**B**.

ADDITIONS FROM NEW LOCALITIES (SEE APPENDIX I)

Amynodon advenus (localities CC6B, SB42); *A. reedi* (species not in Vol. 1) (locality CC4).
Amynodontopsis bodei (localities CC7C, D); *A.* sp. (locality CC7E).

HYRACODONTIDAE

CORRECTIONS TO VOLUME 1

Epitriplopus uintensis is known from locality CP36**B**.
Hyracodontidae indet. is known from locality NP9**A**.

SYSTEMATIC REVISIONS, NEW TAXA, AND IMPORTANT NEW LOCALITY OR PALEOBIOLOGICAL INFORMATION

Cf. *Hyracodon* sp. is reported from the Chadronian of New Mexico (locality SB27II; Lucas, Estep, and Froehlich, 1997).

ADDITIONS FROM NEW LOCALITIES (SEE APPENDIX I)

Triplopus implicatus (locality CP65); *T.*? sp. (localities CC7B, SB42, CP34E).
Hyracodon eximus (species not in Vol. 1, could be *"Forstercooperia grandis"* = *"Uintaceras radinskyi"*) (locality CP65); *H. priscidens* (locality NP10Biii); *H.* sp. (locality ?NP32C).
Hyracodontidae indet. (locality CC7D).

RHINOCEROTIDAE

CORRECTIONS TO VOLUME 1

Trigonias sp. is known from localities **NP**24C (not NB24C), and N**P**27D (not NB27D).
Subhyracodon mitis is only questionably known from locality CP98. *Aphelops megalodus* is known from locality NB29**A**. *Aphelops* sp. is known from locality SB**34**A (now = *Peraceras*, see below) (not SB43A).

SYSTEMATIC REVISIONS, NEW TAXA, AND IMPORTANT NEW LOCALITY OR PALEOBIOLOGICAL INFORMATION

The specimen of *Aphelops* sp., from the late Hemingfordian of Mexico (locality SB55), has been reassigned to Cf. *Menoceras* sp. (Ferrusquía-Villafranca, 2003).

A Cf. *Teleoceras* sp. is reported from the early Barstovian of Mexico (locality CA3, previously dated as ?late Clarendonian; Ferrusquía-Villafranca, 2003). This extends the chronostratigraphic range of this genus in this biogeographical area. An unidentified rhinocerotid is also reported from locality CA2II (Ferrusquía-Villafranca, 2003).

Teleoceras sp. is reported from the late Miocene/early Pliocene Gray Fossil Site in Tennessee (locality CE1; Wallace and Wang, 2004).

Mihlbachler (2003) reviewed the demography of the rhinos *Teleoceras proterum* and *Aphelops malacorhinus* from the late Miocene of Florida. He showed that *Teleoceras* had disproportionally high rates of mortality among young adult males, a mortality pattern more similar to that of the living black rhino than to the hippopotamus, with which *Teleoceras* is often compared. Mortality patterns in *Aphelops* did not resemble those of any extant rhino. Mihlbachler (2005b) showed that there is no correlation between sexual dimorphism in the form of lower incisor tusks and body size dimorphism in *Teleoceras*, and that the balanced sex-specific mortality rates in *Aphelops*, of a similar degree of tusk and size dimorphism as *Teleoceras*, suggests less male–male competition in this genus than in *Teleoceras* or extant rhinos.

Figure A.10 shows the temporal ranges of rhinos, and Figures A.11 and A.12 show their biogeographical ranges.

UPDATES FROM PROTHERO, 2005

Prothero (2005) has compiled an extensive revision of the North American Rhinocerotidae. We note here the following revisions to Vol. 1, including the addition of new taxa (an asterix following a locality number indicates a locality that has been added to Appendix I in this Vol. 2). The rhinoceros taxa are listed below in phylogenetic order, as they appeared in Vol. 1.

Uintaceras

As noted in Vol. 1 (in the comments for the genus *Forstercooperia*), the North American specimens assigned to *Forstercooperia grandis* (an indricotheriid hyracodont) were reassigned by Holbrook and Lucas (1997) to a new genus, *Uintaceras*. Here Prothero includes this taxon, as *Uintaceras radinskyi*, with the Rhinocerotidae. The identity of "*Uintaceras minuta*" (from Vol. 1) is not clear.

Teletaceras

T. radinskyi, as in Vol. 1. *T. mortivallis* is additionally reported from locality SB44B.

Penetrigonias

In Vol. 1, Prothero listed *P. hudsoni* as the type species for this genus, but in Prothero (2005) he listed *P. dakotensis* (Peterson, 1920 = *Caenopus*) as the type species and included *P. hudsoni* within this taxon.

In addition to the localities mentioned in Vol. 1, *P. hudsoni* is also known from localities CP7C, CP68B (*not* CP68D), and NP25C. *P. dakotensis* is additionally reported from localities CP7C, CP39IIC, CP42A, CP68B (*not* D), CP98 (sublocality uncertain), and NP25C. *P. sagittatus* is the same as in Vol. 1.

Trigonias

T. osborni is listed as additionally occurring at localities NB2II, CP98C and NP24C (= *T.* sp. in Vol. 1); *T. wellsi* is listed from localities CP83C (*not* CP83A), and CP98C (*not* CP98A).

Amphicaenopus

A. platycephalus is listed as occurring at localities CP83B (*not* CP83A), CP84C (*not* CP84B), and NP51AA* (*not* NP51A).

Subhyracodon

S. occidentalis is listed as additionally occurring at localities GC24, CP83C, NP32C, and NP50B (*not* NP50C). *S. mitis* is listed as additionally occurring at localities CP84A and NP24C. *S. kewi* is additionally reported from locality CC10III*.

Diceratherium

D. armatum is listed as additionally occurring at localities GC5A (new), CP87A, CP103A, and NP41A. *D. tridactylum* is listed as additionally occurring at localities CP68D and CP84C. *D. annectens* is listed as additionally occurring at localities GC3A, GC5A*, CP85D, CP87A, CP103A, CP104B, and NP36II*. *D. niobrarense* is listed as additionally occurring at localities CP105 and PN19A (*not* PN19B).

Skinneroceras (new genus) (subfamily Diceratheriinae)

Skinneroceras is smaller than *Diceratherium* and most species of *Subhyracodon*, with well-developed parasagittal temporal crests.

Included species is *S. manningi* only, known from locality CP99C only (= *Diceratherium* n. sp. in Vol. 1).

Menoceras

M. arikarense is additionally listed as possibly occurring at localities GC8C (Martin–Anthony site; = GC8B in Vol. 1) and NC1A. *M. barbouri* is listed as additionally occurring at localities CA7 (identified as *Diceratherium* sp. in Vol. 1), GC3B, SB29C, SB55, CP53II*, and CP104B.

Floridaceras

F. whitei is listed as additionally occurring at locality CA7.

Galushaceras (new genus) (Aceratheriinae)

This taxon was listed as "New genus (Prothero, in prep.)" in Vol. 1. There is a single species, *G. levellorum*, known from locality CP107.

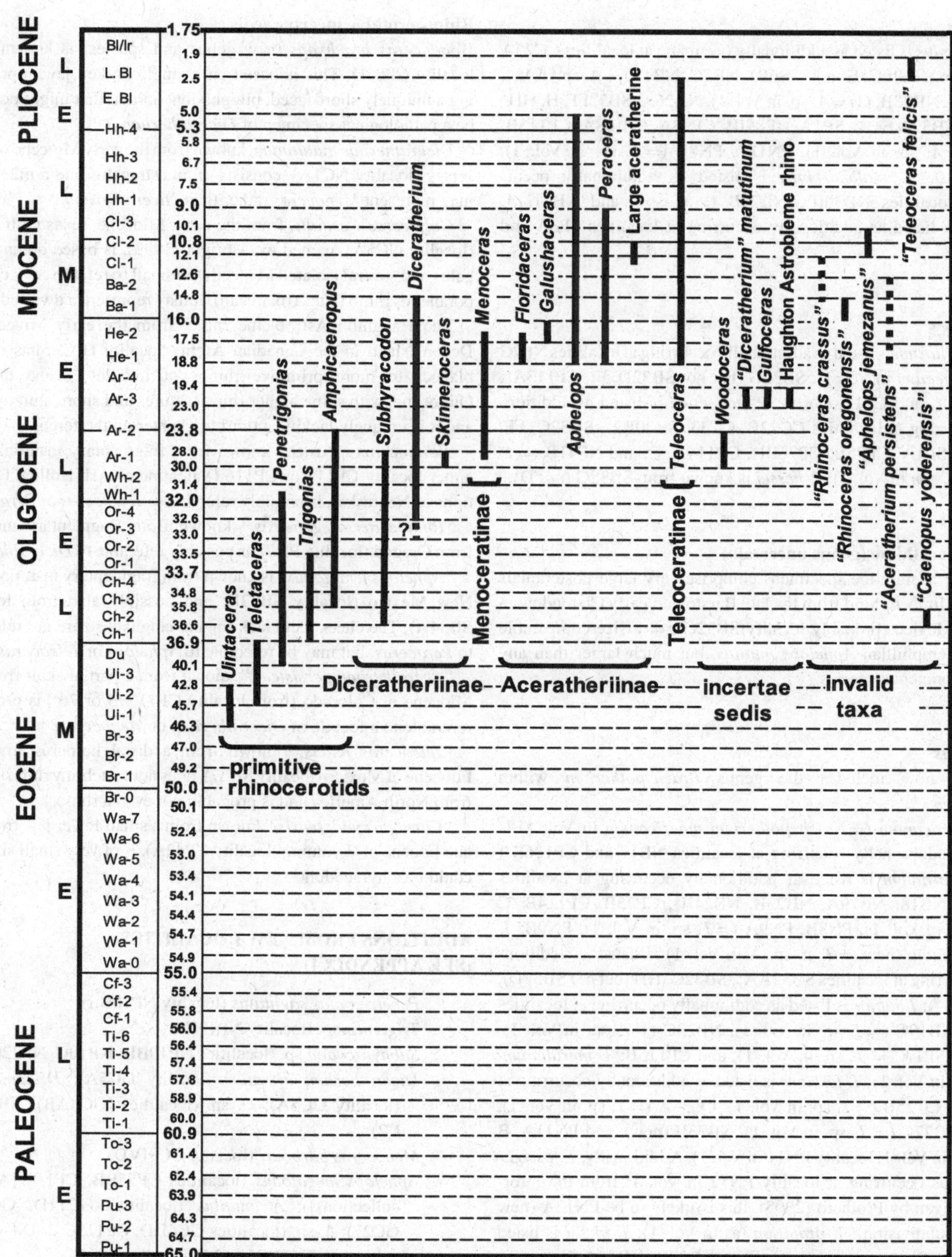

Figure A.10. Temporal ranges of rhinocerotids.

Aphelops

A. megalodus is listed as additionally occurring at localities CC23A (was *A.* sp.), NB6C, E (not NB6B), NB7C, NB17 (= *A.* sp. in Vol. 1), NB18, NB23B, C (= *A.* sp. in Vol. 1), NB26A, SB32FF, H, HH* (but *not* SB32A or F), SP1A, CP89II*, CP93A, CP108A, CP115B, NP11 (= *A.* sp. in Vol. 1), PN1II*, PN8B (= *A.* sp. in Vol. 1), and PN10. *A. malacorhinus* is listed as additionally occurring at localities NB30II*, NB27B (= *A.* sp.), and NB31. *A. mutilus* is listed as additionally occurring at localities SP3A and PN15.

Peraceras

P. superciliosum is listed as additionally occurring at localities NP7C (= *P. profectum* in Vol. 1), SB32G (but not SB32D, F), CP113A*, CP116A0*, and CP116AA, B; *P. profectum* is listed as additionally occurring at localities CC22B, CC33, SB30B2*, SB32C, FF, H, HH*, SP1A, CP75B, CP76III*, CP114A, C, and NP41B (= *P. superciliosum* in Vol. 1); *P. hessei* is known from SB32G (*not* D).

Large acerathiine (genus uncertain)

The large acerathiine specimens comprise very large postcranials of a long-limbed rhino from the late Barstovian/early Clarendonian of New Mexico (probably locality SB32G), of a size comparable to late Hemphillian *Aphelops mutilus*, but much larger than any contemporaneous species.

Teleoceras

Prothero now includes the genus *Brachypotherium* within *Teleoceras*.

T. americanum (= *Brachypotherium americanum* in Vol. 1) is listed as additionally occurring at localities NB17 and CP106II*; *T. medicornutum* is listed as additionally occurring at localities CC22B, NB18, NB19A, NB23B, NB24II*, CP75B, CP114B, C (= *T.* sp. in Vol. 1), PN8B, PN9A (= *T.* sp. in Vol. 1), PN9B; *T. brachyrhinum* (new, = *T.* n. sp. A in Vol. 1) is listed as additionally occurring at localities SB32AA*, SB32G, HH (but *not* SB32D), and SB34A; *T. major* is listed as additionally occurring at localities GC6A, NB19C (= *T.* sp. in Vol. 1), NB23C (= *T.* sp. in Vol. 1), NB29A, SP1A (= *T.* sp. in Vol. 1), and CP115B; *T. meridianum* (same as in Vol. 1); *T. fossiger* is listed as additionally occurring at localities CC25B (= *T.* sp. in Vol. 1), CC38A (= *T.* sp. in Vol. 1), NB11, NB27C (= *T.* sp. in Vol. 1), NB33II (new), and PN11A, B (= *T.* sp. in Vol. 1; locality PN11 in Vol. 1). Additionally, *T. fossiger* is listed as occurring at locality PN12 in Vol. 1: from the information given by Prothero (2005), this is likely to be PN12A (new locality subdivision). *T. proterum* (as in Vol. 1); *T. hicksi* is listed as additionally occurring at localities NB27B, NB33A (note, this locality is early Blancan, not late Hemphillian, according to Lindsay *et al.*, 2002), SB60 (= *T.* sp. in Vol. 1), SP1C (*not* SP1D), SP3A, B, CP115C, PN3(A or B), PN14 (= *T.* sp. in Vol. 1), PN14II (new), PN15 (= *T.* sp. in Vol. 1), and possibly at locality CE2*; *T. guymonensis* (new species = *T.* n. sp. B in Vol. 1) is listed as additionally occurring at locality SB34D*.

Rhinocerotidae incertae sedis

Woodoceras brachyops (new genus and species) is known from locality CP85D. This is based on a single broken jaw, representing a uniquely short-faced, bilophodont animal that might possibly be a pathological specimen of *Diceratherium*.

Diceratherium matutinum, known from the early Miocene of New Jersey (locality NC1A), consists of an astragalus and a m2, and it may represent *Menoceras* rather than *Diceratherium*.

Gulfoceras westfalli, from the early Miocene Texas Gulf Coast (locality GC5A), named by Albright (1999), is based on an astragalus and several isolated M3 teeth too small to refer to any contemporaneous rhino (late Arikareean). It may represent a dwarfed form.

The Haughton Astrobleme rhino, from the early Miocene of Devon Island in the Canadian Arctic (locality HA2), has resemblances to more primitive rhinos of middle Eocene through Oligocene, with some higher rhino features and short, stumpy legs, and it is uniquely lacking a third trochanter on the femur.

"*Rhinoceros crassus*," known from fragmentary material from either locality CP114 or CP116 (Valentine or Ash Hollow Formations of Nebraska), is possibly referable to *Teleoceras* or *Peraceras*.

"*Rhinoceros oregonensis*," known from fragmentary material from Oregon (locality PN7), is possibly referable to *Diceratherium*.

"*Aphelops jemezanus*" is known from fragmentary material from New Mexico (locality SB32H, and possibly also from locality SB34A). The cheek teeth lack lingual cingulae so are not referable to *Peraceras*, but may be referable to *Aphelops* or *Teleoceras*.

"*Aceratherium persistens*," known from a partial skull from the Miocene of Colorado (from locality CP71, 75 or 76), is probably referable to a female of *Diceratherium* or *Menoceras*.

"*Teleoceras felicis*," known from a distal humerus from the Pliocene of Mexico (locality SB67), does not match any rhino known from North America and is probably not even a rhino.

"*Caenopus yoderensis*," known from a small lower jaw from the late Eocene of Wyoming (locality CP42A), is of very small size and could be a hyracodont.

ADDITIONS FROM NEW LOCALITIES (SEE APPENDIX I)

Penetrigonias sagittatus (locality NP10Bii).

Trigonias sp. (locality NP49II).

Subhyracodon sp. (localities NP10BB, NP10C, NP32C).

Diceratherium armatum (locality GC5A); *D. annectans* (locality GC5A); *D.* sp. (localities [GC8AB], NP10B3, C2).

Menoceras barbouri (locality CC9IVD).

Aphelops megalodus (localities CP75BB, CP111 [AMNH collections]); *A. mutilus* (localities GC11D, GC13B, GC27); *A.* sp. (localities GC4DD, CP115C or D [AMNH collections], CP126 [AMNH collections]).

Peraceras superciliosum (locality CP56IIA).

Teleoceras fossiger (localities SB55IIAA, CP126 [AMNH collections]); *T. medicornutum* (localities CP56IIA, CP75BB, CP108B [AMNH collections]); *T.* sp. (localities NB35IIIA, SB58IIA, B, CP56IIA, CP114B [UNSM collections]).

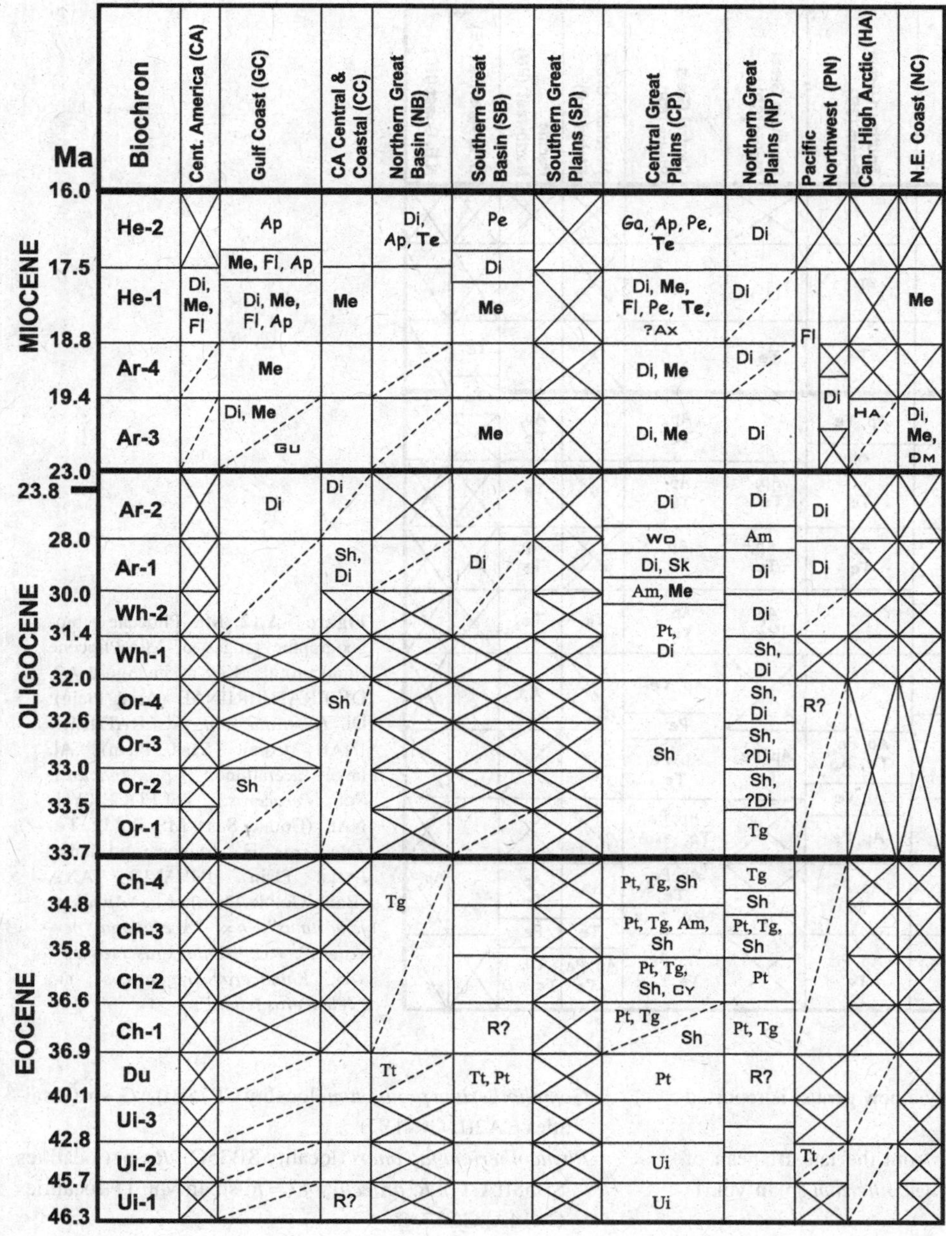

Figure A.11. Biogeographic ranges of Eocene–Miocene rhinocerotids. Key as in Figure A.2. PRIMITIVE RHINOS (Times New Roman): Am, *Amphicaenopus*; Pt, *Penetrigonias*; Tg, *Trigonias*; Tt, *Teleteceras*; Ui, *Uintaceras*. DICERATHERIINAE (Arial Plain): Di, *Diceratherium*; Sh, *Subhyracodon*; Sk, *Skinnoceras*. MENOCERATINAE (Arial Bold): **Me**, *Menoceras*. ACERATHERIINAE (Comic Sans MS Plain): Ap, *Aphelops*; Fl, *Floridaceras*; Ga, *Galushaceras*; Pe, *Peraceras*. TELEOCERATINAE (Comic Sans MS Bold): **Te**, *Teleoceras*. R?, rhinocerotid indet. (Arial Plain): INCERTAE SEDIS (Bank Gothic bold): or INVALID TAXA (Bank Gothic plain): AX, "*Acerathium persistens*"; CY, "*Caenopus yoderensis*"; DM, "*Diceratherium*" *matutinum*; GU, *Gulfoceras*; HA, Haughton Astrobleme rhino; WO, *Woodoceras*.

Indeterminate Rhinocerotidae are known from localities GC8B (= new gen. and sp., Albright 1998), GC8F, NB34II, NP10Bi, B2, NP24C, D.

Indeterminate Rhinocerotoiodea are known from localities CC3, CP67.

PROBOSCIDEA

CORRECTIONS TO VOLUME 1

Mammut americanum is known from locality CP115**C** (not CP115D) and is not known from locality GC14A (= just GC14 in Vol. 1).

Mammutidae indet. is known from locality GC6**A**.

Amebelodon sp. is known from locality PN11**A** and **B** (new locality subdivision, PN11B with emended dating of Hh2).

Amebelodon n. sp. is known from locality GC10**D**.

Gnathobelodon thorpei is known from locality CP123 ?**D**.

SYSTEMATIC REVISIONS, NEW TAXA, AND IMPORTANT NEW LOCALITY OR PALEOBIOLOGICAL INFORMATION

The record of *Gomphotherium* sp. from the Clarendonian of Mexico (locality CA3) has been redated as early Barstovian (Ferrusquía-Villafranca, 2003), restricting the chronostratigraphic range of this genus in biogeographic area.

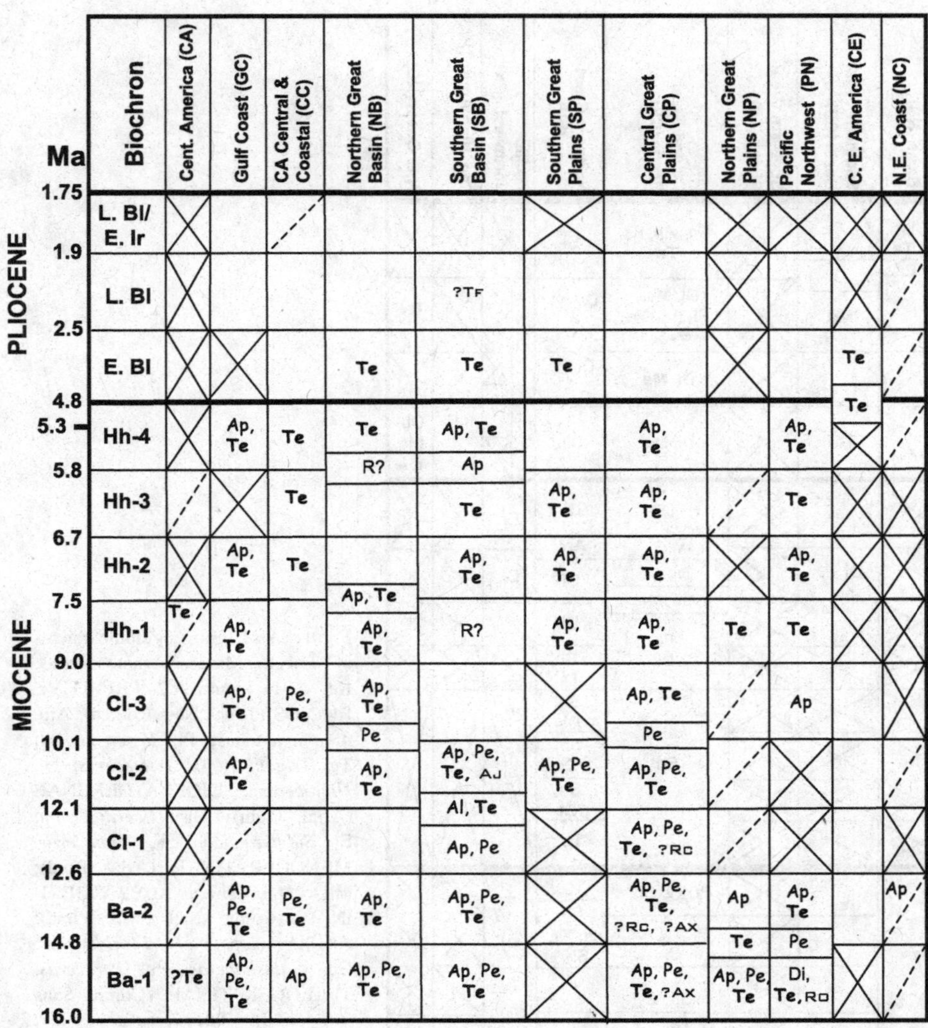

Figure A.12. Mio-Pliocene biogeographic ranges of Mio-Pliocene rhinocerotids. Key as in Figure A2. DICERATHERIINAE (Arial Plain): Di, *Diceratherium*. ACERATHERIINAE (Comic Sans Plain): Al, large acerathine; Ap, *Aphelops*; Pe, *Peraceras*. TELEOCERATINAE (Comic Sans MS Bold): **Te**, *Teleoceras*. R?, rhinocerotid indet. (Arial Plain). INVALID TAXA (Bank Gothic Plain): AJ, "*Aphelops jemazanus*"; AX, "*Acerathium persistens*"; RC, "*Rhinoceras crassus*"; RO, "*Rhinoceras oregonensis*"; TF, "*Teleoceras felicis*".

Cf. *Tetralophodon campester* (possibly a new genus) is reported from locality CP116B (Lambert, 2001).

Rhyncotherium falconeri is reported from the late Blancan of New Mexico (locality SB36, noted as *Rhyncotherium* sp. in Vol. 1) (Lucas and Morgan, 1996.)

Rhyncotherium and *Cuverionius* are reported from the late Miocene or Pliocene of Mexico (from localities SB60II and SB69, respectively; Corona-M and Alberdi, 2006). These occurrences suggest that these gomphotheriids occupied habitats that were mainly subtropical or tropical lowlands.

ADDITIONS FROM NEW LOCALITIES (SEE APPENDIX I)

Zygolophodon sp. (locality CP75BB).

Mammut americanum (localities GC15C, GC18IIB, GC18IIIA, GC18IV, NB35IIIC); *M. hayi* (species not in Vol. 1) (localities GC18IIIA, GC18IV); *M. raki* (species not in Vol. 1) (locality SB35D).

Gomphotherium productum (locality CP75BB); *G.* sp. (localities CA2II, GC4DD).

Rhyncotherium falconeri (locality SB35C); *R.* sp. (localities SB58IIA, C); *R. praecursor* (= *R.* sp. in Vol. 1) (localities GC14A, GC17A).

Cuvieronius tropicus (species not in Vol. 1) (localities GC15B, E, GC17B, GC18IIB, C18IIIA–C, GC18IV, SB37E, F); *S.* sp. (locality SB37D).

Stegomastodon mirificus (localities SB37F, SB38IIB, C); *S. rexroadensis* (= *S. primitivus* in Vol. 1) (localities SB34IVA, [SB35B]); *S.* sp. (localities SB34IIB, SB35A, C, CP121II).

Amebelodon floridanus (localities GC11D, GC13B).

Tetralophodon sp. (?locality SB55IV).

Mammuthus imperator (taxon not in Vol. 1) (localities SB36C, SB37F).

Gomphotheriidae indet. (localities WM26, GC8F, GC27, SB55IIBB [close to *Rhynchotherium* or *Gomphotherium*]).

Indeterminate proboscidean are reported from localities NB35IIIA, B, NB36IIIA, B, CP56IIA.

REFERENCES

Albright, L. B. (1998). A new genus of tapir (Mammalia: Perissodactyla) from the Arikareean (earliest Miocene) of the Texas Coastal Plain. *Journal of Vertebrate Paleontology*, 18, 200–17.

(1999). Ungulates of the Toledo Bend local fauna (late Arikareean, early Miocene), Texas Coastal Plain. *Bulletin of the Florida Museum of Natural History*, 42, 1–80.

Bailey, B., Hunt, R., Jr., and Stepleton, E. (2004). A temnocyonine beardog preserved in a lower Arikareean pumiceous horizon, Wildcat Hills, Western Nebraska. *Journal of Vertebrate Paleontology*, 24(suppl. to no. 3), p. 36A.

Baskin, J. A. (2003). New procyonines from the Hemingfordian and Barstovian of the Gulf Coast and Nevada, including the first fossil record of the Potosini. *Bulletin of the American Museum of Natural History*, 279, 125–46.

(2005). Carnivora from the late Miocene Love Bone Bed of Florida. *Bulletin of the Florida Museum of Natural History*, 45, 413–34.

Bjork, P. R. (1967). Latest Eocene vertebrates from northwestern South Dakota. *Journal of Paleontology*, 41, 227–36.

Bloch, J. I. (1999). Partial skeleton of *Arctostylops* from the Paleocene of Wyoming: arctostylopid–notoungulate relationship revisited. *Journal of Vertebrate Paleontology*, 19(suppl. to no. 3), p. 35A.

Bravo-Cuevas, V. (2005). On the presence of the hipparioninine horse (*Cormohipparion*) from the middle Miocene of the state of Oaxaca, southeastern Mexico. *Journal of Vertebrate Paleontology*, 25(suppl. to no. 3), p. 38A.

Brown, I. (2005). A case for the reassignment of the Barstovian felid *Pseudaelurus marshi* to the genus *Nimravides*. *Journal of Vertebrate Paleontology*, 25(suppl. to no. 3), p. 40A.

Bryant, H. N. (1996). Nimravidae. In *The Terrestrial Eocene–Oligocene Transition in North America*, ed. D. R. Prothero and R. J. Emry, pp. 453–75. Cambridge, UK: Cambridge University Press.

Bryant, H. N. and Fremd, T. J. (1998). Revised biostratigraphy of the Nimravidae (Carnivora) from the John Day Basin of Oregon. *Journal of Vertebrate Paleontology*, 18(suppl. to no. 3), p. 30A.

Carranza-Casteñeda, O. (2001). *Machairodus*, recorded in the Blancan of Guanajuato, Mexico. *Journal of Vertebrate Paleontology*, 21(suppl. to no. 3), p. 38A.

Chester, S., Bloch, J., Boyer, D., and Wing, S. (2005). Anachronistic occurrences of phenacodontid species in the Clarkforkian of the southern Bighorn Basin: possible evidence against transient dwarfing in *Ectocion* and *Copecion* during the CIE-PETM interval. *Journal of Vertebrate Paleontology*, 25(suppl. to no. 3), p. 44A.

Clifford, A. and Witmer, L. (2002). Not all noses are hoses: an appraisal of proboscis evolution in mammals. *Journal of Vertebrate Paleontology*, 22(suppl. to no. 3), p. 45A.

Clemens, W. A. and Williamson, T. E. (2005). A new species of *Eoconodon* (Triisodontidae, Mammalia) from the San Juan Basin, New Mexico. *Journal of Vertebrate Paleontology*, 25, 208–13.

Cobabe, E. A. (1996). Leptaucheniinae. In *The Terrestrial Eocene–Oligocene Transition in North America*, ed. D. R. Prothero and R. J. Emry, pp. 574–80. Cambridge, UK: Cambridge University Press.

Colbert, M. W. (2006a). *Hesperaletes* (Mammalia: Perissodactyla), a new tapiroid from the middle Eocene of California. *Journal of Vertebrate Paleontology*, 26, 697–711.

(2006b). Variation and species recognition in Eocene tapirs from southern California. *Journal of Vertebrate Paleontology*, 26, 712–19.

Cook, H. J. (1926). New Eocene titanotheres from Moffat County, Colorado. *Proceedings of the Colorado Museum of Natural History*, 7, 12–18.

Coombs, M. C., Hunt, R. M., Jr., Stepleton, E., Albright, L. B., III, and Fremd, T. J. (2001). Stratigraphy, chronology, biogeography, and taxonomy of early Miocene small chalicotheres in North America. *Journal of Vertebrate Paleontology*, 21, 607–20.

Cope, E. D. (1872). Second notice of extinct vertebrates from Bitter Creek, Wyoming. *Proceedings of the American Philosophical Society*, 12, 487–8.

(1873a). On some Eocene mammals, obtained by Hayden's Geological Survey of 1872. *Palaeontological Bulletin*, 12, 1–6.

(1873b). On the new perissodactyls from the Bridger Eocene. *Proceedings of the Philadelphia Academy of Natural Sciences*, 13, 35–6.

(1873c). Second notice of extinct vertebrata from the Tertiary of the Plains. *Palaeontological Bulletin*, 15, 1–6.

(1873d). *Synopsis of New Vertebrata from the Tertiary of Colorado, Obtained during the Summer of 1873*. Washington, DC: Government Printing Office.

(1874). Report on the vertebrate palaeontology of Colorado. In *Annual Report of the Geological and Geographical Survey of the Territories for 1873*, pp. 427–533. Washington, DC: Government Printing Office.

(1880). The Bad Lands of the Wind River and their Fauna. *The American Naturalist*, 14, 745–8.

(1886). The Vertebrata of the Swift Current Creek region of the Cypress Hills. In *Annual Report of the Geological and Natural History Survey of Canada for 1885*, Appendix pp. 79–85. Washington, DC: Government Printing Office.

(1889). Vertebrata of Swift Current River. No. III. *The American Naturalist*, 23, 628–9.

(1891). On two new Perissodactyls from the White River Neocene. *The American Naturalist*, 25, 47–9.

Corona-M., E. and Alberdi, M. T. (2006). Two new records of Gompthotheriidae (Mammalia: Proboscidea) in southern México and some biogeographic implications. *Journal of Paleontology*, 80, 357–66.

Corner, R. G. and Voorhies, M. R. (1998). *Titanotylopus* (*Gigantocamelus*): a gracile giant camelid of late Blancan–early Irvingtonian age in North America. *Journal of Vertebrate Paleontology*, 18(suppl. to no. 3), p. 36A.

Dawson, M. (2003). Phylogenetic and geographic affinities of the early Miocene vertebrate fauna of Devon Island, Nunavet, Canada. *Journal of Vertebrate Paleontology*, 23(suppl. to no. 3), p. 44A.

Dawson, M. R. and Harington, C. R. (2007). *Boreameryx*, an unusual new artiodactyl (Mammalia) from the Pliocene of Arctic Canada and endemism in fossil mammals. *Canadian Journal of Earth Sciences*, 44, 585–92.

Douglass, E. (1909). Preliminary descriptions of some new titanotheres from the Uinta deposits. *Annals of the Carnegie Museum*, 6, 304–13.

Earle, C. (1891). *Palaeosyops* and allied genera. *Proceedings of the Academy of Natural Sciences of Philadelphia*, 43, 106–17.

Eberle, J. J. (2001). Early Eocene Leptictida, Pantolesta, Creodonta, Carnivora, and Cete from Ellesmere Island: Arctic links to Europe and Asia. *Journal of Vertebrate Paleontology*, 21(suppl. to no. 3), p. 46A.

(2004). An Arctic "tapir" and implications for northern high-latitude paleobiogeography. *Journal of Vertebrate Paleontology*, 24(suppl. to no. 3), p. 53A.

(2005). Early Eocene Brontotheriidae (Perissodactyla) from the Eureka Sound Group, Ellesmere Island, Canadian High Arctic: implications for brontothere origins and high-latitude dispersal. *Journal of Vertebrate Paleontology*, 25(suppl. to no. 3), p. 52A.

Eberle, J. J. and Lillegraven, J. A. (1998a). A new important record of earliest Cenozoic mammalian history: geologic setting, Multituberculata, and Peradectia. *Rocky Mountain Geology*, 33, 3–47.

(1998b). A new important record of earliest Cenozoic mammalian history: Eutheria and paleogeographic/biostratigraphic summaries. *Rocky Mountain Geology*, 33, 49–117.

Eberle, J. J. and Storer, J. E. (1999). Northernmost record of brontotheres, Axel Heiberg Island, Canada: implications for age of the Buchanan Lake Formation and brontothere paleobiology. *Journal of Paleontology*, 73, 979–83.

Eberle, J. J., Johnson, K., Raynolds, R., Hicks, J., and Nichols, D. (2002). New discoveries of Puercan mammals in the Denver Basin, Colorado:

revisions to local Puercan mammalian biostratigraphy that incorporate paleomagnetic and palynological data. *Journal of Vertebrate Paleontology*, 22(suppl. to no. 3), p. 50A.

Farlow, J. O., Sunderman, J. A., Havens, J. J., *et al.* (2001). The Pipe Creek Sinkhole biota, a diverse late Tertiary continental fossil assemblage from Grant County, Indiana. *American Midland Naturalist*, 145, 367–78.

Fernández, M. H. and Vrba, E. S. (2005). A complete estimate of the phylogenetic relationships in the Ruminatia: a dated species-level supertree of extant ruminants. *Biological Reviews*, 80, 269–302.

Ferrusquía-Villafranca, I. (2003). Mexico's middle Miocene mammalian assemblages: an overview. *Bulletin of the American Museum of Natural History*, 279, 321–47.

(2007). The first Paleogene mammal record of middle America: *Simojovelhyus pocitosense* (Helohyidae: Artiodactyla). *Journal of Vertebrate Paleontology*, 26, 989–1101.

Foss, S. E. (2000). Systematic analysis of the Entelodontidae (Mammalia, Artiodactyla). *Journal of Vertebrate Paleontology*, 20(suppl. to no. 3), p. 42A.

Foss, S. E., Lucas, S., Fremd, T., *et al.* (2004). Reanalysis of the Eocene Hancock Mammal Quarry Local Fauna (Clarno Formation, North-Central Oregon). *Journal of Vertebrate Paleontology*, 24(suppl. to no. 3), p. 59A.

Frick, C. (1937). Horned ruminants of North America. *Bulletin of the American Museum of Natural History*, 69, 1–669.

Froehlich, D. J. (1999). Phylogenetic systematics of basal perissodactyls. *Journal of Vertebrate Paleontology*, 19, 140–59.

(2002). Quo vadis eohippus? The systematics and taxonomy of early Eocene equids (Perissodactyla). *Zoological Journal of the Linnean Society*, 134, 141–256.

Gazin, C. L. and J. M. Sullivan. (1942). A new titanothere from the Eocene of Mississippi with notes on the correlation between the marine Eocene of the Gulf Coastal Plain and continental Eocene of the Rocky Mountain region. *Smithsonian Miscellaneous Collections*, 101, 1–13.

Geisler, J. H. (2001). New morphological evidence for the phylogeny of Artiodactyla, Cetacea, and Mesonychidae. *American Museum Novitates*, 3344, 1–53.

Geisler, J. H. and Uhen, M. D. (2005). Morphological support for a close relationship between hippos and whales. *Journal of Mammalian Evolution*, 12, 145–60.

Graham, R. W. (2003). Pleistocene tapir from Hill Top Cave, Trigg County, Kentucky and a review of Plio-Pleistocene tapirs of North America and their paleoecology. In *Vertebrate Paleontology of Caves*, ed. B. Schubert, J. I. Mead, and R. W. Graham. Bloomington, IN: Indiana University Press.

Granger, W. and Gregory, W. K. (1938). A new titanothere genus from the upper Eocene of Mongolia and North America. *Bulletin of the American Museum of Natural History*, 74, 435–6.

(1943). A revision of the Mongolian titanotheres. *Bulletin of the American Museum of Natural History*, 80, 349–89.

Gregory, W. K. (1912). Note on the upper Eocene titanotheroid *Telmatherium*(?) *incisivum* Douglass from the Uinta Basin. *Science*, 35, 545.

Gunnell, G. F. (1998). Mammalian fauna from the lower Bridger Formation (Bridger A, early middle Eocene) of the southern Green River Basin, Wyoming. *Contributions from the Museum of Paleontology, University of Michigan*, 30, 83–130.

Gunnell, G. F. and Bartels, W. S. (2001). Basin-Margin vertebrate faunas on the Western flank of the Bighorn and Clarks Fork Basins. [In *Paleocene–Eocene Stratigraphy and Biotic Change in the Bighorn and Clarks Fork Basins, Wyoming*, ed. P. D. Gingerich.] *University of Michigan Papers on Paleontology*, 33, 145–55.

Gunnell, G. F. and Yarborough, V. (2000). Brontotheriidae (Perissodactyla) from the late early and middle Eocene (Bridgerian), Wasatch and Bridger Formations, southern Green River Basin, southwestern Wyoming. *Journal of Vertebrate Paleontology*, 20, 349–68.

Hanson, C. B. (1996). Stratigraphy and vertebrate faunas of the Bridgerian–Duchesnean Clarno Formation, North-Central Oregon. In *The Terrestrial Eocene–Oligocene Transition in North America*, ed. D. R. Prothero and R. J. Emry, pp. 206–39. Cambridge, UK: Cambridge University Press.

Hassinin, A. and Douzery, E. J. P. (2003). Molecular and morphological phylogenetics of the Ruminantia and the alternative position of the Moschidae. *Systematic Biology*, 52, 206–28.

Hatcher, J. B. (1895). On a new species of *Diplacodon*, with a discussion of the relations of that genus to *Telmatotherium. The American Naturalist*, 29, 1084–90.

Hay, O. P. (1899). Notes on the nomenclature of some North American vertebrates. Science, 2, 253–4.

(1901). Bibliography and catalogue of the fossil vertebrates of North America. Washington, DC: Government Printing Office.

Hayes, F. G. 2000. The Brooksville 2 local fauna (Arikareean, latest Oligocene): Hernando County, Florida. *Bulletin of the Florida Museum of Natural History*, 43, 1–47.

Heinrich, R. E. and Houde, P. (2006). Postcranial anatomy of *Viverravus* (Mammalia, Carnivora) and implications for substrate use in basal Carnivora. *Journal of Vertebrate Paleontology*, 26, 422–35.

Holbrook, L. T. (1999). The phylogeny and classification of tapiromorph perissodactyls (Mammalia). *Cladistics*, 15, 331–50.

(2000). A new "hippomorph" perissodactyl (Mammalia) from the middle Eocene of Wyoming. *Journal of Vertebrate Paleontology*, 20(suppl. to no. 3), p. 49A.

Holbrook, L. T. and Lucas, S. G. (1997). A new genus of rhinocerotoid from the Eocene of Utah and the status of North American "*Forstercooperia*". *Journal of Vertebrate Paleontology*, 17, 384–96.

Holbrook, L. T., Lucas, S. G., and Emry, R. J. (2004). Skulls of the Eocene perissodactyls (Mammalia) *Homogalax* and *Isectolophus*. *Journal of Vertebrate Paleontology*, 24, 951–6.

Hooker, J. J. (1989). Character polarities in early perissodactyls and their significance for *Hyracotherium* and infraordinal relationships. In *The Evolution of Perissodactyls*, ed. D. R. Prothero and R. M. Schoch, pp. 79–101. Oxford: Oxford University Press.

(1994). The beginning of the equoid radiation. *Zoological Journal of the Linnean Society*, 112, 29–63.

Hooker, J. J. and Dashzeveg, D. (2004). The origin of chalicotheres (Perissodactyla, Mammalia). *Palaeontology*, 47, 1363–86.

Hulbert, R. C., Jr. (1999). Nine million years of *Tapirus* (Mammalia: Perissodactyla) from Florida. *Journal of Vertebrate Paleontology*, 19(suppl. to no. 3), p. 53A.

(2005a). Phylogenetic analysis of late Cenozoic *Tapirus* (Mammalia, Perissodactyla). *Journal of Vertebrate Paleontology*, 25(suppl. to no. 3), p. 72A.

(2005b). Late Miocene *Tapirus* (Mammalia: Perissodactyla) from Florida, with description of a new species, *Tapirus webbi*. *Bulletin of the Florida Museum of Natural History*, 45, 465–94.

Hulbert, R. C., Jr. and Harington, C. R. (1999). An early Pliocene hipparionine horse from the Canadian Arctic. *Palaeontology*, 42, 1017–25.

Hulbert, R. C., Jr. and Whitmore, F. C., Jr. (2006). Late Miocene mammals from the Mauvilla Local Fauna, Alabama. *Bulletin of the Florida Museum of Natural History*, 46, 1–28.

Hunt, R. M., Jr. (2001). Small Oligocene amphicyonids from North America (*Paradaphoenus*, Mammalia, Carnivora). *American Museum Novitates*, 3331, 1–20.

(2002). New amphicyonid carnivorans (Mammalia, Daphoeninae) from then early Miocene of southeastern Wyoming. *American Museum Novitates*, 3385, 1–28.

(2003). Intercontinental migration of large mammalian carnivores: earliest occurrence of the Old World beardog *Amphicyon* (Carnivora,

Amphicyonidae) in North America, *Bulletin of the American Museum of Natural History*, 279, 77–115.

Janis, C. M., Theodor, J. M., and Boisvert, B. (2002). Locomotor evolution in camels revisited: a quantitative analysis of pedal anatomy and the acquisition of the pacing gait. *Journal of Vertebrate Paleontology*, 22, 110–21.

Jiménez-Hidalgo, E. and Carrenza-Casteñeda, O. (2001). The earliest record of the genus *Capromeryx* (Mammalia: Artiodactyla) in Central Mexico. *Journal of Vertebrate Paleontology*, 21(suppl. to no. 3), pp. 65A–6A.

(2002). First Pliocene record of *Hemiauchenia blancoensis* (Mammalia: Camelidae) in Mexico. *Journal of Vertebrate Paleontology*, 22(suppl. to no. 3), pp. 71A–2A.

(2005). Hemphillian camelids and protoceratids from San Miguel de Allende, Guanajuato State, Central Mexico. *Journal of Vertebrate Paleontology*, 25(suppl. to no. 3), p. 75A.

Jiménez-Hidalgo, E., Carranza-Casteñeda, O., and Montellano-Ballesteros, M. (2004). A Pliocene record of *Capromeryx* (Mammalia: Antilocapridae) in Mexico. *Journal of Paleontology*, 78, 1179–86.

Joeckel, R. M. and Stavas, J. M. (1996). Basicranial anatomy of *Syndyoceras cooki* (Artiodactyla, Protoceratidae) and the need for a reappraisal of tylopod relationships. *Journal of Vertebrate Paleontology*, 16, 320–37.

Kelly, T. S. (1995). New Miocene horses from the Caliente Formation, Cuyama Valley Badlands, California. *Contributions in Science, Los Angeles County Museum of Natural History*, 455, 1–33.

(1998). New middle Miocene equid crania from California and their implications for the phylogeny of the Equini. *Contributions in Science, Los Angeles County Museum of Natural History*, 473, 1–44.

Konrashov, P. and Lucas, S. (2002). Late Paleocene Arctostylopidae (Mammalia, Notoungulata) from Monglia. *Journal of Vertebrate Paleontology*, 22(suppl. to no. 3), p. 75A.

Labs, J. (2000). A new mustelid from Thomas Farm, Florida. *Journal of Vertebrate Paleontology*, 20(suppl. to no. 3), p. 53A.

Lambe, L. M. (1908). Part 4: The vertebrata of the Oligocene of the Cypress Hills, Saskatchewan. *Contributions to Canadian Paleontology*, 3, 1–64.

Lambert, W. D. (2001). New tetralophodont gomphothere material from the Clarendonian of Nebraska, and its implication for the status of *Tetralophodon* in North America. *Journal of Vertebrate Paleontology*, 21(suppl. to no. 3), pp. 70A–71A.

Leidy, J. (1852). Description of the remains of extinct Mammalia and Chelonia, from Nebraska territory, collected during the Geological Survey under the direction of Dr. D. D. Owen. In *Report of a Geological Survey of Wisconsin, Iowa, and Minnesota and incidentally a portion of Nebraska Territory*, ed. D. D. Owen, pp. 534–72. Philadelphia, PA: Lippincott, Grambo and Co.

(1854). The ancient fauna of Nebraska, or a description of remains of extinct Mammalia and Chelonia from the Mauvaises Terres of Nebraska. *Smithsonian Contributions to Knowledge*, 6, 1–26.

(1870a). [Description of a new genus and species, *Megacerops coloradensis*.] *Proceedings of the Philadelphia Academy of Natural Sciences*, 22, 1–2.

(1870b). [Descriptions of *Palaeosyops paludosus*, *Microsus cuspidatus*, and *Notharctus tenebrosus*.] *Proceedings of the Philadelphia Academy of Natural Sciences*, 22, 113–14.

(1871). Remarks on fossil vertebrates from Wyoming. *Proceedings of the Philadelphia Academy of Natural Sciences*, 23, 228–9.

(1872a). Remarks on fossils from Wyoming. *Proceedings of the Philadelphia Academy of Natural Sciences*, 24, 277.

(1872b). On some new species of Mammalia from Wyoming. *Proceedings of the Philadelphia Academy of Natural Sciences*, 24, 167–9.

Libed, S. (2004). Late Paleocene *Nacimientotherium silvestrei* (Mammalia) from the San Juan Basin of New Mexico, and the origin of tillodonts. *Journal of Vertebrate Paleontology*, 24(suppl. to no. 3), p. 83A.

Lim, J.-D. (1999). Evidence for *Leptarctus* as an arboreal frugivorous mustelid. *Journal of Vertebrate Paleontology*, 19(suppl. to no. 3), p. 59A.

Lim, J.-D. and Martin, L. D. (2001a). A new species of *Leptarctus* from Kansas, USA. *Neues Jahrbuch für Geologie und Paläontologie Monatshefte*, 10, 633–40.

(2001b). New evidence for plant-eating in a Miocene mustelid. *Current Science*, 812, 314–17.

(2002). A new fossil mustelid from the Miocene of South Dakota, USA. *Naturwissenshaften*, 89, 270–4.

Lim, J.-D. and Miao, D.-S. (2000). New species of *Leptarctus* (Carnivora, Mustelidae) from the Miocene of Nebraska. *Vertebrate PalAsiatica*, 38, 52–7.

Lim, J.-D., Martin, L. D., and Wilson, R. (2001). A new species of *Leptarctus* (Carnivora, Mustelidae) from the late Miocene of Texas. *Journal of Paleontology*, 75, 1043–6.

Lindsay, E. H., Mou, Y., Downs, W., *et al.* (2002). Resolution of the Hemphillian/Blancan Boundary in Nevada. *Journal of Vertebrate Paleontology*, 22, 429–42.

Lucas, S. G. (1992). Redefinition of the Duchesnean land mammal "age," late Eocene of western North America. In *Eocene–Oligocene Climatic and Biotic Evolution*, ed. D. R. Prothero and W. A. Berggren, pp. 88–105. Princeton, NJ: Princeton University Press.

Lucas, S. G. and Emry, R. J. (2004). The entelodont *Brachyhyops* (Mammalia, Artiodactyla) from the upper Eocene of Flagstaff Rim, Wyoming. [In *Paleogene Mammals*, ed. S. G. Lucas, K. E. Zeigler, and P. E. Kondrashov.] *Bulletin of the New Mexico Museum of Natural History and Science*, 26, 97–100.

Lucas, S. G. and Holbrook, L. T. (2004). The skull of the Eocene perissodactyl *Lambdotherium* and its phylogenetic significance. [In *Paleogene Mammals*, ed. S. G. Lucas, K. E. Zeigler, and P. E. Kondrashov.] *Bulletin of the New Mexico Museum of Natural History and Science*, 26, 81–7.

Lucas, S. G. and Kondrashov, P. E. (2004a). A new species of *Deltatherium* (Mammalia, Tillodonta) from the Paleocene of New Mexico. [In *Paleogene Mammals*, ed. S. G. Lucas, K. E. Zeigler, and P. E. Kondrashov.] *Bulletin of the New Mexico Museum of Natural History and Science*, 26, 45–54.

(2004b). Taxonomic revision of the genus *Ectoconus* (Mammalia: Periptychidae) from the Paleocene of the western United States. [In *Paleogene Mammals*, ed. S. G. Lucas, K. E. Zeigler, and P. E. Kondrashov.] *Bulletin of the New Mexico Museum of Natural History and Science*, 26, 33–44.

Lucas, S. G. and Morgan, G. S. (1996). The Pliocene proboscidean *Rhyncotherium* (Mammalia: Gomphotheriidae) from south-central New Mexico. *Texas Journal of Science*, 48, 311–18.

Lucas, S. G. and Schoch, R. M. (1982). *Duchesneodus*, a new name for some titanotheres (Perissodactyla, Brontotheriidae) from the late Eocene of western North America. *Journal of Paleontology*, 56, 1018–23.

Lucas, S. G., Whistler, D. P., and Wagner, H. M. (1997). Giant entelodont (Mammalia, Artiodactyla) from the early Miocene of southern California. *Contributions in Science, Los Angeles County Museum of Natural History*, 466, 1–9.

Lucas, S. G., Foss, S. E., and Mihlbachler, M. C. (2004). *Achaenodon* (Mammalia, Artiodactyla) from the Eocene Clarno Formation, Oregon, and the age of the Hancock Quarry Local Fauna. *Bulletin of the New Mexico Museum of Natural History and Science*, 26, 89–95.

Ludtke, J. A. and Prothero, D. R. (2004). Taxonomic revision of the middle Eocene (Uintan–Duchesnean) protoceratid *Leptoreodon* (Mammalia: Artiodactyla). [In *Paleogene Mammals*, ed. S. G. Lucas, K. E. Zeigler, and P. E. Kondrasho.] *New Mexico Museum of Natural History and Science Bulletin*, 26, 101–11.

Lull, R. S. (1905). *Megacerops tyleri*, a new species of titanothere from the bad lands of South Dakota. *Journal of Geology*, 34, 443–56.

Macdonald, J. R. (1963). The Miocene faunas from the Wounded Knee area of western South Dakota. *Bulletin of the American Museum of Natural History*, 125, 141–238.

MacFadden, B. J. (2001). Three-toed browsing horse *Anchitherium clarencei* from the early Miocene (Hemingfordian) Thomas Farm, Florida. *Bulletin of the Florida Museum of Natural History*, 43, 79–109.

(2006). North American Miocene land mammals from Panama. *Journal of Vertebrate Paleontology*, 26, 720–34.

MacFadden, B. J. and Dobie, J. L. (1998). Late Miocene three-toed horse *Protohippus* (Mammalia: Equidae) from southern Alabama. *Journal of Paleontology*, 72, 149–51.

MacFadden, B. J. and Morgan, G. S. (2003). New oreodont (Mammalia; Artiodactyla) from the late Oligocene (early Arikareean) of Florida. *Bulletin of the American Museum of Natural History*, 279, 368–96.

MacPhee, R. D. E. and Flemming, C. (2000). New fossil mammals from the Cenozoic of Jamaica. *Journal of Vertebrate Paleontology*, 20(suppl. to no. 3), p. 55A.

Mader, B. J. (1989). The Brontotheriidae: a systematic revision and preliminary phylogeny of North American genera. In *The Evolution of Perissodactyls*, ed. D. R. Prothero and R. M. Schoch, pp. 458–84. Oxford: Oxford University Press.

(2000). *Pseudodiplacodon*, a new generic name for *Diplacodon progressum* Peterson (Mammalia, Perissodactyla, Brontotheriidae). *Journal of Vertebrate Paleontology*, 20, 164–6.

Mader, B. J. and Alexander, J. P. (1995). *Megacerops kuwagatarhinus* n. sp., an unusual brontothere (Mammalia, Perissodactyla) with distally forked horns. *Journal of Paleontology*, 69, 581–7.

Marsh, O. C. (1871). Notice of some new fossil mammals and birds from the Tertiary formation of the West: *American Journal of Science*, 2, 120–7.

(1872). Preliminary description of new Tertiary mammals. *American Journal of Science and Arts*, 4, 1–35.

(1873). Notice of new Tertiary mammals. *American Journal of Science and Arts*, 5, 1–9.

(1874). On the structure and affinities of the Brontotheriidae. *American Journal of Science and Arts*, 7, 1–8.

(1875). Notice of New Tertiary mammals. IV. *American Journal of Science and Arts*, 9, 239–50.

(1887). Notice of new fossil mammals. *American Journal of Science*, 34, 1–9.

(1890). Notice of new Tertiary mammals. *American Journal of Science*, 39, 523–5.

(1891). Notice of new vertebrate fossils. *American Journal of Science*, 42, 265–9.

McCullough, G., Skaff, A., Schirtzinger, E., and Thrasher, L. (2002). A faunal revision and overview of recent research of the Blanan 111 Ranch fossil beds in Graham County, Arizona. *Journal of Vertebrate Paleontology*, 22(suppl. to no. 3), p. 88A.

McKenna, M. C. and Lofgren, D. L. (2003). *Mimotricentes tedfordi*, a new arctocyonid from the late Paleocene of California. *Bulletin of the American Museum of Natural History*, 279, 632–43.

Meachen, J. A. (2005). A new species of *Hemiauchenia* (Artiodactyla, Camelidae) from the late Blancan of Florida. *Bulletin of the Florida Museum of Natural History*, 45, 435–48.

Mead, J. I. (1997). Blancan (Pliocene) caprine bovids (Artiodactyla) from Panaca Formation, Nevada, and 111 Ranch, Arizona. *Journal of Vertebrate Paleontology*, 17(suppl. to no. 3), p. 64A.

Meehan, T. J. and Wilson, R. W. (2002). New viverravids from the Torrejonian (middle Paleocene) of Kutz Canyon, New Mexico and the oldest skull of the order Carnivora. *Journal of Paleontology*, 76, 1091–101.

Mihlbachler, M. C. (2003). Demography of late Miocene rhinoceroses (*Teleoceras proterum* and *Aphelops malacorhinus*) from Florida: linking mortality and sociality in fossil assemblages. *Paleobiology*, 29, 412–28.

(2005a). Phylogenetic systematics of the Brontotheriidae (Mammalia, Perissodactyla). Ph.D. Thesis, Columbia University, New York.

(2005b). Linking sexual dimorphism and sociality in rhinoceroses: insights from *Teleoceras proterum* and *Aphelops malacorhinus* from the late Miocene of Florida. *Bulletin of the Florida Museum of Natural History*, 45, 495–520.

(2007). *Eubrontotherium clarnoensis*, a new genus and species of brontothere (Brontotheriidae, Perissodactyla) from the Hancock Quarry, Clarno Formation, Wheeler County, Oregon. *PaleoBios*, 27, 19–39.

(2008). Species taxonomy, phylogeny, and, biogeography of the Brontotheriidae (Perissodactyla, Mammalia). *Bulletin of the American Museum of Natural History* (in press).

Mihlbachler, M. C. and Solounias, N. (2002). Body size, dental microwear, and brontothere diets through the Eocene. *Journal of Vertebrate Paleontology*, 23(suppl. to no. 3), p. 88A.

Mihlbachler, M. C., Lucas, S. G., and Emry, R. J. (2004). The holotype specimen of *Menodus giganteus*, and the "insoluble" problem of Chadronian brontothere taxonomy. [In *Paleogene Mammals*, ed. S. G. Lucas, K. E. Zeigler, and P. E. Kondrashov.] *Bulletin of the New Mexico Museum of Natural History and Science*, 26, 129–35.

Morgan, G. S. and Lucas, S. G. (2003). Radiometrically calibrated oreodonts (Mammalia: Artiodactyla) from the late Oligocene of southwestern New Mexico. *Journal of Vertebrate Paleontology*, 23, 471–3.

Morlo, M. and Gunnell, G. F. (2005). New species of *Limnocyon* (Limnocyoninae, Hyaenodontinae) from the middle Bridgerian (middle Eocene) of southwestern Wyoming. *Journal of Vertebrate Paleontology*, 25, 251–5.

Morlo, M., Gunnell, G. F., and Alexander, J. A. (2000). Small Limnocyoninae (Hyaenodontidae, Creodonta) from the Bridgerian (latest early and middle Eocene) of Wyoming. *Journal of Vertebrate Paleontology*, 20(suppl. to no. 3), p. 59A.

Morlo, M., Peigné, S., and Nagel, D. (2004). A new species of *Prosansanosmilus*: implications for the systematic relationships of the family Barbourofelidae new rank (Carnivora, Mammalia). *Zoological Journal of the Linnean Society*, 140, 43–61.

Naples, V. L., Martin, L. D., and Babiarz, J. B. (1998). A new type of sabertoothed cat. *Journal of Vertebrate Paleontology*, 18(suppl. to no. 3), p. 66A.

Norris, C. A. (1999). The cranium of *Bunomeryx* (Artiodactyla: Homacodontidae) from the Upper Eocene Uinta deposits of Utah and its implication for tylopod systematics. *Journal of Vertebrate Paleontology*, 19, 742–51.

(2000). The cranium of *Leptotragulus* (Artiodactyla: Protoceratidae) from the middle Eocene of North America. *Journal of Vertebrate Paleontology*, 20, 341–8.

Osborn, H. F. (1889). The Mammalia of the Uinta Formation. Part III. The Perissodactyla. *Transactions of the American Philosophical Society*, 16, 505–30.

(1895). Fossil mammals of the Uinta Basin. Expedition of 1894. *Bulletin of the American Museum of Natural History*, 7, 71–105.

(1896). The cranial evolution of *Titanotherium*. *Bulletin of the American Museum of Natural History*, 8, 157–97.

(1902). The four phyla of Oligocene titanotheres. *Bulletin of the American Museum of Natural History*, 16, 91–109.

(1908). New or little known titanotheres from the Eocene and Oligocene. *Bulletin of the American Museum of Natural History*, 24, 599–617.

(1913). Lower Eocene titanotheres. Genera *Lambdotherium*, *Eotitanops*. *Bulletin of the American Museum of Natural History*, 21, 407–15.

(1916). Two new Oligocene titanotheres. *Bulletin of the American Museum of Natural History*, 35, 721–3.

(1919). New titanotheres of the Huerfano. *Bulletin of the American Museum of Natural History*, 41, 557–69.

(1929). Titanotheres of Ancient Wyoming, Dakota, and Nebraska. *United States Geological Survey Monographs*, 55, 1–894.

Osborn, H. F., W. B. Scott, and F. Spier. (1878). Palaeontological report of the Princeton scientific expedition of 1877. *Contributions from the Museum of Geology and Archaeology of Princeton College*, 1, 1–106.

O'Sullivan, J. A. (2000). Stable isotopes reveal progenesis in the evolution of the Miocene dwarf horse *Archaeohippus* (Mammalia, Equidae). *Journal of Vertebrate Paleontology*, 20(suppl. to no. 3), pp. 61A–2A.

(2003). A new species of *Archaeohippus* (Mammalia, Equidae) from the Arikareean of central Florida. *Journal of Vertebrate Paleontology* 23, 877–85.

Owen, D. D., Norwood, J. G., and Evans, J. (1850). Notice of fossil remains brought from Mr. J. Evans from the Mauvais Terres or badlands of White River, 150 miles west of the Missouri. *Proceedings of the Philadelphia Academy of Natural Sciences*, 5, 66.

Owen, P. R. (2002). Primitive American badgers (Mustelidae, Taxidiinae) from the late Clarendonian of Nebraska. *Journal of Vertebrate Paleontology*, 22(suppl. to no. 3), p. 94A.

(2006). Description of a new late Miocene badger (Taxidiinae), utilizing high-resolution X-ray computed tomography. *Palaeontology*, 49, 999–1011.

Pagnac, D. (2004). New camels (Artiodactyla: Cameliae) from the mid-Miocene Barstow Formation, Mojave desert, California. *Journal of Vertebrate Paleontology*, 24(suppl. to no. 3), p. 100A.

(2006). *Scaphohippus*, a new genus of horse (Mammalia: Equidae) from the Barstow Formation of California. *Journal of Mammalian Evolution*, 13, 37–61.

Peigné, S. (2003). Systematic review of European Nimravinae (Mammalia, Carnivora, Nimravidae) and the phylogenetic relationships of Palaeogene Nimravidae. *Zoologica Scripta*, 32, 199–229.

Penkrot, T., R., Zack, S., Rose, K. D., and Block, J. (2003). Postcrania of early Eocene *Apheliscus* and *Haplomylus* (Mammalia: "Condylarthra"). *Journal of Vertebrate Paleontology*, 23(suppl. to no. 3), p. 86A.

Peterson, O. A. (1914a). A small titanothere from the lower Uinta beds. *Annals of the Carnegie Museum*, 9, 53–7.

(1914b). A new titanothere from the Uinta Eocene. *Annals of the Carnegie Museum*, 9, 29–52.

(1931). New species of the genus *Teleodus* from the Upper Uinta of Northeastern Utah. *Annals of the Carnegie Museum*, 20, 307–12.

(1934). New titanotheres from the Uinta Eocene in Utah. *Annals of the Carnegie Museum*, 22, 351–61.

Pomel A. (1849). Description d'un os maxillaire fossile de *Palaeotherium* par Hiram Prout. *American Journal of Science* by Silliman's [sic] and J. Dana, 2e Série, Vol. 3, p. 248. *Bibliothéque Universelle de Geneve Archives Des Sciences Physiques et Naturelles*, 10, 73–5.

Price, S. A., Bininda-Edmonds, O. R. P., and Gittleman, J. L. (2005). A complete phylogeny of the whales, dolphins and even-toed hoofed mammals (Cetartiodactyla). *Biological Reviews*, 80, 445–73.

Prothero, D. R. (2005). *The Evolution of North American Rhinoceroses*. Cambridge, UK: Cambridge University Press.

(2007). Family Moschidae (musk deer). In *The Evolution of Artiodactyls*, ed. D. R. Prothero and S. E. Foss, pp. 221–40. Baltimore, MD: Johns Hopkins University Press.

(2008). Systematics of the musk deer (Artiodactyla: Moschidae: Blastomerycinae) from the Miocene of North America. *Bulletin of the New Mexico Museum of Natural History and Science* (in press).

Prothero, D. R. and Liter, M. (2008). Review of the Dromomerycidae (Mammalia: Artiodactyla). *Bulletin of the New Mexico Museum of Natural History and Science* (in press).

Prothero, D. R. and Ludtke, J. A. (2007). Family Protoceratidae. In *The Evolution of Artiodactyls*, ed. D. R. Prothero and S. E. Foss, pp. 169–76. Baltimore, MD: Johns Hopkins University Press.

Prout, H. A. (1860). On a tooth found in Virginia. *Transactions of the Saint Louis Academy of Science*, 1, 699–700.

Randall, K. A. and Jefferson, G. T. (2002). A re-examination of the Plio-Pleistocene camelids from Anza-Borego Desert State Park, California. *Journal of Vertebrate Paleontology*, 22(suppl. to no. 3), p. 98A.

Riggs, E. S. (1912). New or little known titanotheres from the lower Uintah formations with notes on the stratigraphy and distribution of fossils. *Field Museum of Natural History, Geological Series*, 4, 17–41.

Rothwell, T. (2003). Phylogenetic systematics of North America *Pseudaelurus* (Carnivora: Felidae). *American Museum Novitates*, 3403, 1–64.

Ruez, D. R., Jr. (2005). Earliest record of *Palaeolama* (Mammalia, Camelidae) with comments on "*Palaeolama*" *guanajuatensis*. *Journal of Vertebrate Paleontology*, 25, 741–4.

Russell, L. S. (1934). Revision of the lower Oligocene Vertebrate fauna of the Cypress Hills, Saskatchewan. *Transactions of the Royal Society of Canada*, 20, 49–67.

Schlaikjer, E. M. (1935). Contributions to the stratigraphy and paleontology of the Goshen Hole area, Wyoming IV: new vertebrates and the stratigraphy of the Oligocene and early Miocene. *Bulletin of the Museum of Comparative Zoology*, 76, 71–93.

Scott, C. (2005). New species of *Cyriacotherium* (Mammalia: Pantodonta) from the Paleocene of Alberta, Canada. *Journal of Vertebrate Paleontology*, 25(suppl. to no. 3), p. 112A.

Scott, C. S., Webb, M. W., and Fox, R. C. (2006). *Horolodectes sunae*, an enigmatic mammal from the late Paleocene of Alberta, Canada. *Journal of Paleontology*, 80, 1009–25.

Scott, W. B. and H. F. Osborn. (1887). Preliminary account of the fossil mammals from the White River Formation contained in the Museum of Comparative Zoology. *Bulletin of the Museum of Comparative Zoology*, 13, 151–71.

Semprebon, G. (2005). Miocene savannas and the Great Transformation: testing dietary shifts within the Equidae using low-magnification stereomicroscopy. *Journal of Vertebrate Paleontology*, 25(suppl. to no. 3), p. 114A.

Semprebon G. and Rivals, S. (2007). Was grass more prevalent in the pronghorn past? An assessment of the dietary adaptations of Miocene to Recent Antilocapridae. *Palaeogeography, Palaeoclimatology, Palaeoecology* 253, 332–47.

Semprebon, G., Janis, C., and Solounias, N. (2004). The diets of the Dromomerycidae (Mammalia: Artiodactyla) and their response to Miocene vegetational change. *Journal of Vertebrate Paleontology*, 24, 427–44.

Silcox, M. T. and Rose, K. D. (2001). Unusual vertebrate microfaunas from the Willwood Formation, early Eocene of the Bighorn Basin, Wyoming. In *Eocene Biodiversity: Unusual Occurrences and Rarely Sampled Habitats*, ed. G. F. Gunnell, pp. 131–64. New York: Kluwer Academic/Plenum.

Solounias, N. and Semprebon, G. (2002). Advances in the reconstruction of ungulate ecomorphology with application to early fossil equids. *American Museum Novitates*, 3366, 1–49.

Stevens, M. S. and Stevens, J. B. (1996). Merycoidodontidae and Miniochoerinae. In *The Terrestrial Eocene–Oligocene Transition in North America*, ed. D. R. Prothero and R. J. Emry, pp. 498–573. Cambridge, UK: Cambridge University Press.

(2003). Carnivora (Mammalia, Felidae, Canidae, and Mustelidae) from the earliest Hemphillian Screw Bean Local Fauna, Big Bend National Park, Brewster County, Texas. *Bulletin of the American Museum of Natural History*, 279, 177–211.

(2007). Family Merycoidodontidae. In *The Evolution of Artiodactyls*, ed. D. R. Prothero and S. E. Foss, pp. 157–68. Baltimore, MD: Johns Hopkins University Press.

Stock, C. (1933). Carnivora from the Sespe of Las Posas Hills, California. *Carnegie Institution of Washington Publication, Contributions to Paleontology*, 440, 29–42.

(1935). Titanothere remains from the Sespe of California. *Proceedings of the National Academy of Sciences*, 21, 456–62.

(1936). Titanotheres from the Titus Canyon Formation, California. *Proceedings of the National Academy of Sciences*, 22, 656–61.

(1937). An Eocene titanothere from San Diego County, California, with remarks on the age of the Poway Conglomerate. *Proceedings of the National Academy of Sciences*, 23, 48–53.

(1938). A titanothere from the type Sespe of California. *Proceedings of the National Academy of Sciences*, 24, 507–12.

Storer, J. E. (1996). Eocene–Oligocene faunas of the Cypress Hills Formation, Saskatchewan. In *The Terrestrial Eocene–Oligocene Transition in North America*, ed. D. R. Prothero and R. J. Emry, pp. 240–61. Cambridge, UK: Cambridge University Press.

Stovall, J. W. (1948). Chadron vertebrate fossils from below the Rim Rock of Presidio County, Texas. *American Journal of Science*, 246, 78–95.

Strait, S. G. (2003). New mammalian fossils from the earliest Eocene (Wa-0), Bighorn Basin, Wyoming. *Journal of Vertebrate Paleontology*, 23(suppl. to no. 3), p. 101A.

Strömberg, C. A. E. (2006). Evolution of hypsodonty in equids: testing a hypothesis of adaptation. *Paleobiology*, 32, 236–58.

Stucky, R. K., Prothero, D. R., Lohr, W. G., and Snyder, J. R. (1996). Magnetic stratigraphy, sedimentology, and mammalian faunas of the early Uintan Washakie Formation, Sand Wash Basin, northwestern Colorado. In *The Terrestrial Eocene–Oligocene Transition in North America*, ed. D. R. Prothero and R. J. Emry, pp. 40–51. Cambridge, UK: Cambridge University Press.

Tabrum, A. R. (1998). First record of a hypertragulid artiodactyl from the Chadronian of western Montana. *Journal of Vertebrate Paleontology*, 18(suppl. to no. 3), p. 81A.

Tabrum, A. R., Prothero, D. R., and Garcia, D. (1996). Magnetostratigraphy and biostratigraphy of the Eocene–Oligocene transition, southwestern Montana. In *The Terrestrial Eocene–Oligocene Transition in North America*, ed. D. R. Prothero and R. J. Emry, pp. 75–117. Cambridge, UK: Cambridge University Press.

Tedford, R. H. (2004). Miocene mammalian faunas, Ogallala Group, Pawnee Buttes Area, Weld County, Colorado. [In *Fanfare for an Uncommon Paleontologist: Essays in Honor of Malcolm C. McKenna*, ed. M. R. Dawson and J. A. Lillegraven.] *Bulletin of the Carnegie Museum of Natural History*, 36, 277–90.

Tedford, R. H. and Harington, C. R. (2003). An Arctic mammal fauna from the Early Pliocene of North America. *Nature*, 425, 388–90.

Theodor, J. M. (1999). *Protoreodon walshi*, a new species of agriochoerid (Oreodonta, Artiodactyla, Mammalia) from the late Uintan of San Diego County, California. *Journal of Paleontology*, 73, 1179–90.

Theodor, J. M. and Foss, S. E. (2005). Deciduous dentitions of Eocene cebochoerid artiodactyls and cetartiodactyl relationships. *Journal of Mammalian Evolution*, 12, 161–81.

Thornton, M. L. and Rasmussen, D. T. (2001). Taphonomic interpretation of Gnat-Out-of-Hell, an early Uintan small mammal locality in the Uinta Formation, Utah. In *Eocene Biodiversity: Unusual Occurrences and Rarely Sampled Habitats*, ed. G. F. Gunnell, pp. 299–316. New York: Kluwer Academic/Plenum.

Tucker, S. (2003). Carnivores and microtine-like rodents from a new late Miocene (Hemphillian) locality in north-central Nebraska. *Journal of Vertebrate Paleontology*, 23(suppl. to no. 3), p. 105A.

Van Valkenburgh, B., Sacco, T., and Wang, X. (2003). Pack hunting in Miocene borophagine dogs: evidence from craniodental morphology and body size. *Bulletin of the American Museum of Natural History*, 279, 147–62.

Van Valkenburgh, B., Wang, X., and Damuth, J. (2004). Cope's rule, hypercarnivory, and extinction in North American canids. *Science*, 306, 101–4.

Voorhies, M. R. and Perkins, M. E. (1998). Odocoileine deer (?*Bretzia*) from the Santee Ash Locality, late Hemphillian, Nebraska: the oldest New World cervid? *Journal of Vertebrate Paleontology*, 18(suppl. to no. 3), p. 84A.

Wagner, H. M. (1999). A new saber-toothed carnivore from the middle Eocene of San Diego County, California. *Journal of Vertebrate Paleontology*, 19(suppl. to no. 3), p. 82A.

Wallace, S. C. and Wang, X. (2004). Two new carnivores from an unusual late Tertiary forest biota in eastern North America. *Nature*, 431, 556–9.

Walsh, S. L. (1996). Middle Eocene mammal faunas of San Diego County, California. In *The Terrestrial Eocene–Oligocene Transition in North America*, ed. D. R. Prothero and R. J. Emry, pp. 75–117. Cambridge, UK: Cambridge University Press.

(1998). Notes on the anterior dentition and skull of *Proteroixoides* (Mammalia: Insectivora: Dormaaliidae), and a new dormaaliid genus from the early Uintan (middle Eocene) of southern California. *Proceedings of the San Diego Society of Natural History*, 34, 1–26.

(2001). New specimens of *Merycobunodon* (Artiodactyla: Oromerycidae) from the early Eocene of San Diego County, California. *Journal of Vertebrate Paleontology*, 21(suppl. to no. 3), p. 111A.

(2005). Revision of *Simimeryx* (Artiodactyla, Hypertragulidae). *Journal of Vertebrate Paleontology*, 25(suppl. to no. 3), p. 127A.

Wang, X. (2003). New material of *Osbornodon* from the early Hemingfordian of Nebraska and Florida. *Bulletin of the American Museum of Natural History*, 279, 163–76.

Wang, X. Tedford, R. H., and Taylor, B. E. (1999). Phylogenetic systematics of the Borophaginae (Carnivora: Canidae). *Bulletin of the American Museum of Natural History*, 243, 1–391.

Wang, X., Wideman, B. C., Nichols, R., and Hanneman, D. L. (2004). A new species of *Aelurodon* (Carnivora, Canidae) from the Barstovian of Montana. *Journal of Vertebrate Paleontology*, 24, 445–52.

Wang, X., Whistler, D., and Takeuchi, G. (2005). A new species of *Martinogale* (Carnivora, Mephitinae) from the Late Miocene Dove Spring Formation, California, and the origin of New World mephitines. *Journal of Vertebrate Paleontology*, 25, 936–49.

Webb, S. D. (2000). Evolutionary history of the New World Cervidae. In *Antelopes, Deer, and Relatives*, ed. E. S. Vrba and G. B. Schaller, pp. 38–64. New Haven, CT: Yale University Press.

Webb, S. D., MacFadden B. J., and Baskin, J. A. (1981). Geology and paleontology of the Love Bone Bed from the late Miocene of Florida. *American Journal of Science*, 281, 513–44.

Webb, S. D., Beatty, B. L., and Poinar, G., Jr. (2003). New evidence of Miocene Protoceratidae, including a new species from Chiapas, Mexico. *Bulletin of the American Museum of Natural History*, 279, 348–67.

Weinstock, J., Willerslev, E., Sher, A., *et al.* (2005). Evolution, systematics, and phylogeography of Pleistocene horses in the New World: a molecular perspective. *PLoS Biology*, 3, 241.

Wesley, G. D. and Flynn, J. J. (2003). A revision of *Tapocyon* (Carnivoramorpha), including analysis of the first cranial specimens and identification of a new species. *Journal of Paleontology*, 77, 769–83.

Wheatly, P. V. and Ruez, D. R., Jr. (2006). Pliocene *Odocoileus* from Hagerman Fossil Beds National Monument, Idaho, and comments on the taxonomic status of *Odocoileus brachyodontus*. *Journal of Vertebrate Paleontology*, 26, 463–5.

Whistler, D. P. and Webb, S. D. (2005). A new goat-like camelid from the late Pliocene of Tecopa Lake Beds, California. *Contributions in Science, Los Angeles Museum of Natural History*, 503, 1–40.

Williamson, T. (2004). *Microclaenodon* (Mammalia) revisited. *Journal of Vertebrate Paleontology*, 24(suppl. to no. 3), p. 130A.

Williamson, T. E., Lucas, S. G., and Stucky, R. K. (1996). *Megalesthonyx hopsoni* (Mammalia: Tillodonta) from the early Bridgerian

(Garnerbuttean) of the Wind River Formation, northeastern Wind River Basin, Wyoming. *Proceedings of the Denver Museum of Natural History*, 3, 1–3.

Wilson, J. A. (1977). Early Tertiary vertebrate faunas Big Bend Area Trans-Pecos Texas: Brontotheriidae. *The Pearce-Sellards Series*, 25, 1—17.

Woodburne, M. O. (1996). Reappraisal of the *Cormohipparion* from the Valentine Formation, Nebraska. American Museum Novitates, 3163, 1–56.

(2003). Craniodental analysis of *Merychippus insignis* and *Cormohipparion goorisi* (Mammalia, Equidae), Barstovian, North America. *Bulletin of the American Museum of Natural History*, 279, 397–468.

(ed.), (2004). *Late Cretaceous and Cenozoic Mammals of North America*. New York: Columbia University Press.

(2005). A new occurrence of *Cormohipparion*, with implications for Old World *Hippotherium* datum. *Journal of Verrtebrate Paleontology*, 25, 256–7.

Zack, S. P. (2004a). Skull and partial skeleton of the rare early Eocene mammal *Wyolestes* (Mammalia: Eutheria) from the Bighorn Basin, Wyoming. *Journal of Vertebrate Paleontology*, 24(suppl. to no. 3), p. 133A.

(2004b). An early Eocene arctostylopid (Mammalia: Arctostylopida) from the Green River Basin, Wyoming. *Journal of Vertebrate Paleontology*, 24, 498–501.

Zack, S. P., Penkrot, T. A., Krause, D. W., and Maas, M. C. (2005). A new apheliscine "condylarth" mammal from the late Paleocene of Montana and Alberta and the phylogeny of "hyopsodontids." *Acta Palaeontologica Polonica*, 50, 809–30.

Zonneveld, J.-P. and Gunnell, G. P. (2003). A new species of cf. *Dilphodon* (Mammalia, Perissodactyla) from the early Bridgerian of southwestern Wyoming. *Journal of Vertebrate Paleontology*, 23, 652–8.

Zonneveld, J.-P., Gunnell, G. F, and Bartels, W. S. (2000). Early Eocene fossil vertebrates from the southwestern Green River Basin, Lincoln and Uinta Counties, Wyoming. *Journal of Vertebrate Paleontology*, 20, 369–86.

Appendices

Appendix I: Unified locality listing

New localities, and emendments to previous localities, are given in bold face. See the Introduction for further explanation.

WEST COAST MARINE

ALASKA

WM1. Unalaska Formation: (Arikareean)
Unalaska Island.

WM1II. Poul Creek Formation: (Arikareean–Barstovian)
Lawrence Creek, northwest of Icy Bay.

WM1III. Gubik Formation: (Blancan)
Fish Creek Local Fauna (L. Bl).
Ref: Repenning, 1981.

WM1IV. Cape Deceit Formation: (Blancan)
Fort Selkirk Local Fauna (L. Bl–E. Ir).
Refs: Storer, 2003.

BRITISH COLUMBIA

WM2. Sooke Formation: (Arikareean)
Vancouver Island.

WM2II. Hesquiat Formation: (Arikareean)
Southwest coast of Vancouver Island.
Refs: Russell, 1968; Barnes *et al.*, 1994.

WASHINGTON

WM3. Pysht Formation: (Arikareean)
Clallam County.
Refs: Prothero, Streig, and Barnes, 2001a.

WM3II. Clallam Formation: (Arikareean)
Merrick Bay and Slip Point, Clallam Bay, Clallam County (Ar3–4).
Refs: Barnes and Goedert, 1996; Tedford *et al.*, 1994.

WM4. Astoria Formation: (Arikareean–Barstovian) (Ar 3/4–Ba 1)

OREGON

WM4II. Alsea Formation: (Orellan–Whitneyan)
Near Toledo on the Yaquina River (Or1–Wh2).
Refs: Prothero, Streig, and Barnes, 2001a; Fordyce, 2002.

WM5. Yaquina Formation: (Arikareean)
A. Ona Beach, Lincoln County (Ar1–2).
B. Beach between Beaver Creek and Seal Rock State Park (Ar1–2).
Refs: Emlong, 1966; Berta, 1991; Barnes, 1994.

WM6. Nye mudstone: (Arikareean–Hemingfordian)
Near Newport, Lincoln County (Ar3/4–He1).
Refs: Reinhart, 1982; Berta, 1994.

WM6II. Astoria Formation: (Arikareean)
Jump Off Joe, basal Astoria Formation (Ar3–4).
Refs: Prothero *et al.*, 2001b.

WM7. Astoria Formation: (Hemingfordian)
Inc. "Iron Mountain Beds" (He2).
Refs: Barnes, 1989a; Prothero *et al.*, 2001c; Deméré and Berta, 2001, 2002.

WM8. Empire Formation: (Hemphillian)
Refs: Orr and Miller, 1983.

WM8II. Port Orford Formation: (Blancan)
1.5 miles southeast of Cape Blanco Lighthouse.
Refs: Barnes and Mitchell, 1975.

CALIFORNIA

WM9. Skooner Gulch Formation: (Arikareean)
Schooner Gulch, near Iverson Point, south of Point Arena, Mendocino County (Ar3–4).
Refs: Phillips, Welton, and Welton, 1976; Clark, 1990.

WM10. Jewett Sand: (Arikareean)
Woody Local Fauna (Ar3–4).
Refs: Mitchell and Tedford, 1973; Barnes (1991, pers. comm.).

WM11. Vaqueros Formation: (Arikareean)
Unnamed local fauna (Ar4).
Refs: Barnes (1991, pers. comm.).

WM12. Unnamed formation: (Hemingfordian–Barstovian; He2–Ba) **(note emended dating, previously Barstovian)**
Refs: Barnes and Hirota, 1985.

WM13. Round Mountain Silt: (Barstovian)
A. Sharktooth Hill Local Fauna (Ba1) **(previously just WM13).**
Refs: Tedford *et al.*, 1987; Barnes, 1991 (pers. comm.).
B. Olcese Sand (Ba1).
C. Below Shark Tooth Hill Bonebed (Ba1).

WM14. Briones Formation: (Barstovian; Ba2)
Refs: Trask, 1922.

WM15. Monterey Shale: (Barstovian–Clarendonian; Ba2–Cl1)
Refs: Barnes, 1989b.

WM15II. Ladera Formation: (Ladera Sandstone) (Barstovian–Clarendonian)
Santa Mateo County (Bl2–Cl1).
Refs: Brabb, Graymer, and Jones, 1998.

WM16. Santa Margarita Formation: (Clarendonian)
A. Scotts Valley, Santa Cruz County (Cl1–2) **(previously just WM16).**
Refs: Barnes (1991, pers. comm.).
B. San Raphael Mountains, Santa Barbara County (Cl1–2).
C. Stanford Locality, Pine Mountain, Sespe Valley, Santa Barbara County (Cl1–2).

WM17. Monterey Formation (= Sisquoc Formation) (Clarendonian)
A. Alameda Creek, near San Francis County.
Refs: Rensberger, 1969.
B. Lompoc, Santa Barbara County.
Refs: Kellogg, 1925.
C. Near El Toro, Orange County.
D. Palos Verdes Peninsula (Valmonte Diatomite Member).
Refs: E. Wilson, 1973.
E. Near Santa Maria.
Refs: Lull, 1914.
F. East of Monterey, type section of Monterey Formation.
Refs: Hanna and McLellan, 1924.
G. Laguna Niguel District, San Joaquin Hills, Orange County.
Refs: Barnes, 1985.
H. Altamira Shale Member, Palos, Verde Peninsula.
Refs: Downs, 1956a.

WM17II. Monterey Formation (referred): (Clarendonian)
Crocker Springs Creek, Kern County (Cl1).
Refs: Wang, Tedford, and Taylor, 1999.

WM18. Modelo Formation: (Clarendonian)
Refs: Barnes, 1985.

WM18II. Santa Cruz mudstone: (Clarendonian–Hemphillian)
Bolinas Point, Point Reyes Peninsula, Marin County (Cl3–Hh1).
Refs: Zeigler, Chan, and Barnes, 1997.

WM19. Towsley Formation: (Hemphillian) (Hh1–2)
Refs: Berkoff and Barnes, 1998.

WM20. Purisima Formation (= Drakes Bay Formation): (Hemphillian)
Inc. Lava Mountain Fauna (Hh1–2).
Refs: Perry, 1977.

WM21. San Mateo Formation: (Hemphillian)
A. San Luis Rey River Local Fauna (Hh1–3).
Refs: Barnes, Hutchinson, and Welton, 1981.
B. Lawrence Canyon Local Fauna (Hh1–3).
Refs: Barnes, Hutchinson, and Welton, 1981.

WM22. Imperial Formation: (Hemphillian)
Refs: Deméré, 1993.

WM23. Paso Robles Formation: (Hemphillian)
Refs: Kellogg, 1921.

WM23II. Unnamed formation (Hemphillian)
Point Reyes, Marin County (Hh4).
Refs: Domning, 1978.

WM24. Capistrano Formation: (Hemphillian–Blancan)
A. Lower Part (Hemphillian).
Refs: Barnes and Raschke, 1991.
B. Upper Part (Blancan).

WM25. Pismo Formation: (E. Blancan)
Refs: Barnes, 1998a.

WM26. San Diego Formation: (Blancan)
Chula Vista District (inc. Otay Ranch) and Poggi Canyon (E. Bl).
Refs: Deméré, 1994; Wagner, Riney, and Prothero, 2000; Wagner *et al.*, 2001.

WM26II. Merced Formation: (Blancan–Irvingtonian)
Coast Range North of San Francis County.
Refs: Lucas, 1995.

WM26III. Pico Formation: (Blancan)
Near Humphreys, Los Angeles County.
Refs: Kellogg, 1929.

WM26IV. Unnamed formation: (Blancan)
Little River Beach State Park.
Refs: Barnes and Mitchell, 1975.

BAJA CALIFORNIA, MEXICO

WM27. San Gregorio Formation: (Arikareean)
North and northwest of La Paz (Ar1–2).
Refs: González-Barba, Olivares-Bañuelos, and Goedert, 2001.

WM27II. El Cien Formation, San Juan Member: (Arikareean)
A. San Juan de las Costa, San Juan de la Costa Local Fauna (Ar2).
Refs: Barnes, 1998b.

B. Ten Minute Locality, El Cien Local Fauna (Ar2).
Refs: Barnes, 1998b
C. San Hilario, San Hilario Local Fauna (Ar2).
Refs: Barnes, 1998b.
WM28. Ysidro Formation: (Hemingfordian)
La Purisima, San Ignacio, and Punta Pequena Local Faunas.
Refs: Barnes, 1998b.
WM28II. Tortugas Formation: (Barstovian–Clarendonian)
Arroyo Las Chivas Local Fauna.
Refs: Barnes, 1998b.
WM29. Unnamed formation: (Hemphillian)
El Rosario Arriba.
WM30. Almejas Formation: (Hemphillian)
A. Isla Cedros.
Refs: Barnes, 1998b.
B. Vizcaiño Peninsula.
Refs: Barnes, 1998b.
WM31. Tirabuzon Formation: (Hemphillian–Blancan)
Santa Rosalia Local Fauna (Hh4–Bl1).
Refs: Barnes, 1998b.
WM32. Salada Formation: (Blancan)
Near Santa Rita and Bahia Magadelena.
Refs: Barnes, 1998b.
WM33. San Diego Formation: (Blancan)
La Joya (L. Bl).
Refs: Berta and Deméré, 1986.

CENTRAL AMERICA

MEXICO GULF COAST

CA0. Balumtum Sandstone: (Arikareean)
Simojovel area, Chiapas (Ar3–4).
Refs: Webb, Beatty, and Poinar, 2003.
CA1. Suchilquintongo Formation: (Hemingfordian)
Valle de Oaxaca Local Fauna, Oaxaca Valley (He1) **(note emended dating, previously Ba1).**
Refs: Ferrusquía-Villafranca, 1984, 1990, 2003.
CA2. Unnamed unit: (Barstovian)
El Gramal Local Fauna, Oaxaca (Ba1).
Refs: Ferrusquía-Villafranca, 1984.
CA2II. El Camarón Formation: (Barstovian)
Nejapa Fauna, southeastern Oaxaca (Ba1).
Refs: Ferrusquía-Villafranca, 2003.
CA2III. Matatlán Formation: (Barstovian)
Matatlán Fauna, central Oaxaca (Ba2).
Refs: Ferrusquía-Villafranca, 2003.
CA3. Ixtapa Formation: (Barstovian)
Ixtapa Local Fauna, west central Chiapas (?Ba1) **(note emended dating, previously ?Cl3).**
Refs: Ferrusquía-Villafranca, 1984, 1990, 2003.
CA4. Unnamed unit: (Hemphillian–Blancan) **(now SB55II)**
CA5. Unnamed unit: (Hemphillian) **(now SB56II)**
CA6. Carrillo Puerto Formation: (Hemphillian)
A. Rancho Chapas, Yucatan (Hh3).
Refs: Domning, 1989, 1990.
B. Noc Ac, near Merida, Yucatan (Hh3).
Refs: Domning, 1989.

OTHER CENTRAL AMERICAN AREAS

CA7. Cucaracha Formation: (Hemingfordian)
Gaillard Cut Local Fauna, Canal Zone, Panama (He1) (or Ar4–Ba1, see MacFadden, 2006).
Refs: Slaughter 1981; Ferrusquía-Villafranca, 1984; MacFadden, 2006.
CA8. Unnamed unit: (Barstovian)
Carbonaras Creek, Guatamala (Ba1).
CA9. Gracias Formation: (Hemphillian)
Gracias Fauna, Honduras (Hh1).
Refs: Webb and Perrigo, 1984.
CA10. Unnamed unit: (Hemphillian)
Corinto Fauna, El Salvadore (Hh1).
Refs: Webb and Perrigo, 1984.

GULF COAST

TEXAS GULF COAST

GC1. Yegua Formation: (Duchesnean)
A. Unnamed site, Houston County.
B. Unnamed site, Burleson County.
Refs: Kellogg, 1936.
GC2. Unnamed formation: (Arikareean)
Dinohyus site (Ar4).
Refs: Tedford *et al.*, 1987.
GC3. Oakville Formation: (Arikareean–Hemingfordian)
Garvin Gulley Fauna:
A. Derrick Farm Local Fauna (= Cedar Run Local Fauna) and Cedar Creek (He1) **(note emended dating, previously Ar3).**
Refs: Wood and Wood, 1937; J. Wilson, 1960; Tedford *et al.*, 1987.
B. Garvin Farm Local Fauna, near Navasota, and Hildalgo Bluff (He1).
Refs: Hesse, 1943; Wilson, 1960; Patton, 1969a; Forstén, 1975; Slaughter 1981; Tedford *et al.*, 1987.
GC3II. Unknown formation:
(?Arikareean–Hemingfordian)
Independence, Washington County (?Ar4–He1).
GC4. Fleming Formation: (Hemingfordian–Barstovian)
A. Aiken Hill Local Fauna (He1) **(note emended dating, previously Ar3)**.
Refs: Tedford *et al.*, 1987.
B. Pointblank Local Fauna (inc. Bennett Farm and Gay Hill) (Ba1).
Refs: Hesse, 1943; Wilson, 1960; Patton, 1969a; Forstén, 1975; Hinderstein and Boyce, 1977; Prothero and Sereno,

1982; MacFadden, 1984; Prothero and Manning, 1987; Tedford *et al.*, 1987.
C. Trinity River Local Fauna (inc. Trinity River Pit 1, McMurray Pits 1 and 2), Stephen Creek, Kelly Farm, and Moscow Local Faunas (Ba1).
Refs: Woodburne, 1969; Patton and Taylor, 1971, 1973; Taylor and Webb, 1976; Slaughter 1981; Baskin, 1982; Prothero and Sereno, 1982; MacFadden, 1984; Prothero and Manning, 1987; Tedford *et al.*, 1987; Hulbert, 1988a; Bryant, 1991; Wang, Tedford, and Taylor, 1999; Baskin, 2003.
D. Burkeville Fauna (Ba1) **(note emended dating, previously Ba1–2)**.
Refs: Patton, 1969a; Tedford *et al.*, 1987.
DD. Castor Creek Member, LA (Ba2).
Inc. DISC area sites (Discovery, Gully, Stonehenge, and Persimmon Barrow Sites) and TVOR area sites.
Refs: Schiebout and Ting, 2000.
E. Cold Spring Fauna (Ba2).
Inc. Noble Farm Fauna, Grimes Prairie, Stephen Prairie, Sam Houston, and Goodrich Local Faunas, and McMurray Pits).
Refs: Hesse, 1943; Wilson, 1960; Patton, 1969a; Forstén, 1975; Prothero and Sereno, 1982; Prothero and Manning, 1987; Tedford *et al.*, 1987; Hulbert, 1988a; Baskin, 2003.

GC5. Lower Fleming Formation: (Arikareean–Hemingfordian)
A. Toledo Bend Local Fauna (Ar3).
Refs: Albright, 1998a,b, 1999a; Hayes, 2000.
B. Carnahan Bayou Member, Toledo Bend (He1) **(previously just GC5).**
Refs: Manning, 1990.

GC6. Goliad Formation: (Clarendonian–Hemphillian)
A. Lapara Member, Bridge Ranch (= Normanna) and Buckner Ranch Local Faunas (Cl2).
Refs: Patton, 1969a; Forstén, 1975; Wilson, 1978; Webb and Hulbert, 1986; Hulbert, 1987a, 1988a,b.
B. Lapara Member, Lapara Creek Fauna, inc. Farish Ranch Fauna (= Berclair Local Fauna) and Navasota (Cl2).
Refs: Patton, 1969a; Forstén, 1975; Wilson, 1978; Webb and Hulbert, 1986; Hulbert, 1987a, 1988a,b; Tedford *et al.*, 1987.
C. Labahia Member, Labahia Mission Local Fauna (Hh1–3).
D. Nueces River Valley (inc. Fordyce Quarry and Sorensen Ranch sand and gravel pit) (Hh4).
Refs: Baskin, 1991.

FLORIDA

GC7. Unnamed fissure fill: (Whitneyan or Arikareean)
I-75 Local Fauna (Wh2 or Ar1).
Refs: Patton, 1969b; Hayes, 2000; MacFadden and Morgan, 2003; Morgan and Czaplewski, 2003.

GC7II. St. Mark's or Torreya Formation: (Arikareean)
St. Mark's River, Leon or Wakulla County (exact location unknown) (?Ar3–4).
Refs: Coombs *et al.*, 2001.

GC8. Hawthorn Group, Penny Farms, Parachucla, Coosawatchie, Marks Head, and Statenville Formations, or unnamed fissure fills of equivalent age: (Arikareean–Clarendonian).
AA. Cow House Slough Local Fauna (**previously GC7**) and Brooksville 2 Local Fauna (**previously GC8B**) (Ar2).
Refs: Morgan, 1989, 1993; Albright, 1998b; Hayes, 2000; Czaplewski *et al.*, 2003; MacFadden and Morgan, 2003.
AB. White Springs Local Fauna (Ar2).
Inc. White Springs 1A, 3A, and 3B Sites.
Refs: Morgan, 1989, 1993; Albright, 1998b; Hayes, 2000; MacFadden and Morgan, 2003.
AC. Curlew Creek Locality (Ar2–3).
Refs: O'Sullivan, 2003.
A. SB-1A (= Live Oak) Local Fauna (Ar2–3).
Refs: Frailey, 1978; Albright, 1998b; Hayes, 2000; MacFadden and Morgan, 2003.
B. Buda Local Fauna (**previously GC8A**), Franklin Phosphate Pit No. 2 (**previously GC8C**) and Brooksville 1 Local Fauna (**previously GC8C)** (Ar3).
Refs: Simpson, 1930; Patton, 1967; Frailey, 1979; Albright, 1998b; Hayes, 2000; MacFadden and Morgan, 2003.
C. Martin–Anthony Oreodont Site and Martin–Anthony Local Fauna (**previously GC8B**) (Ar4).
Refs: MacFadden, 1980a; Morgan, 1993; Albright, 1998b; Hayes, 2000; MacFadden and Morgan, 2003.
D. Thomas Farm Local Fauna (He1–2).
Refs: Black, 1963; Maglio, 1966; Tedford and Frailey, 1976; MacFadden and Webb, 1982; Tedford *et al.*, 1987; Pratt, 1989, 1990; Pratt and Morgan, 1989; Czaplewski and Morgan, 2000; Czaplewski, Morgan, and Naeher, 2003; Morgan and Czaplewski, 2003.
DA. Colclough Hill Site and Devil's Millhopper (He1).
Refs: Olsen, 1964a; MacFadden and Webb, 1982.
DB. Brooks Sink Local Fauna (He2).
Refs: Morgan and Pratt, 1988; Morgan, 1993.
E. Ashville Local Fauna and Roaring Creek Local Fauna (Ba2).
Refs: Olsen, 1964b; Yon, 1965; MacFadden and Webb, 1982; Hulbert, 1988a; Morgan, 1989, 1993.
F. Occidental Fauna (Cl2) (**Leisey 1C locality now = GC13A**).
Refs: MacFadden and Webb, 1982; Morgan, 1989, 1993.

GC9. Hawthorn Group, Torreya Formation: (Hemingfordian–Barstovian)
A. Seabord Local Fauna, Griscom Plantation Local Fauna, Tallahassee Waterworks Site, and Jim Woodruff Dam Site (= Chattahoochee Bluff Local Fauna) (He1–2).
Refs: Sellards, 1916; Colbert, 1932; Bryant, 1991;

Bryant, MacFadden, and Mueller, 1992; Morgan, 1993.
B. Midway Local Fauna and Quincy Local Fauna (He2).
Refs: Simpson, 1930; MacFadden and Webb, 1982; Bryant, MacFadden, and Mueller, 1992; Morgan, 1993.
C. Willacoochee Creek Fauna (Ba1).
Inc. Eglehard La Camelia Mine 1 and 2, Milwhite Gunn Farm Mine, Floridin Corry Mine, and Floridin Smith Mine.
Refs: Bryant, 1991; Bryant, MacFadden, and Mueller, 1992; Morgan, 1993, 1994.

GC9II. Alum Bluff Group: (Hemingfordian–Blancan)
Alum Bluff (He1-Ba1).
Refs: Olsen, 1964a; Webb, Beatty, and Poinar, 2003.

GC9III. Unknown formation: (Hemingfordian)
Miller Local Fauna, Suwannee River, Dixie County (He1).
Refs: Baskin, 2003.

GC10. Hawthorn Group, Arcadia and Lower Bone Valley Formations: (Barstovian–Hemphillian)
AA. Bird Branch Local Fauna (= Nichols Mine Site) and bedrock at Leisey Shell Pit (Ba1).
Refs: Hulbert and MacFadden, 1991; Morgan, 1993; Tedford *et al.*, 2004.
A. Sweetwater Branch Local Fauna (Ba1).
Refs: Hulbert and MacFadden, 1991; Morgan, 1993; Tedford *et al.*, 2004.
B. Bradley Fauna (Ba2).
Inc. Red Zone of Phosphoria Mine.
Refs: Webb and Hulbert, 1986; Hulbert, 1988a,b; Morgan, 1993, 1994; Tedford *et al.*, 2004.
C. Agricola Fauna (Cl1–2).
Inc. Agricola Road Site and Gray Zone of Phosphoria Mine.
Refs: Webb and Hulbert, 1986; Hulbert, 1988a,b; Morgan, 1993, 1994; Tedford *et al.*, 2004.
D. Four Corners Fauna (Hh1).
Inc. Stream Matrix Locality in Nichols Mine, Brewster Phosphate Mine.
Refs: Hulbert, 1988a,b; Hulbert, Webb, and Morgan, 2003.

GC11. Hawthorn Group, "Alachua Formation" and unnamed fissure and channel fills of equivalent age (Clarendonian–Hemphillian)
A. Love Bone Bed Local Fauna (Cl3).
Refs: Webb, MacFadden, and Baskin, 1981; Baskin, 2005.
B. Archer Fauna and Gainesville Creeks Fauna (Hh1).
Inc. Mixson's Bone Bed and McGehee Farm Local Faunas, Emathela Site, Haile 5B, Haile 6A, Haile 19A, and Coffrin Creek.
Refs: Webb, 1966, 1969b, 1990; Hirschfeld and Webb, 1968; Webb and Hulbert, 1986; Hulbert, 1988a,b, 1993; Morgan, 1993.
C. Moss Acres Local Fauna (Hh2).
Refs: Hulbert, 1987b, 1988a,b, 1993; Lambert, 1997.
D. Tyner Farm Local Fauna (late Hh1 or early Hh2).
Refs: Hulbert, Poyner, and Webb, 2002; Hulbert *et al.*, 2002.

GC12. Formation uncertain: (Hemphillian)
Waccasassa River 3B (?Hh2).
Refs: Domning, Morgan, and Ray, 1982; Domning, 1990; UF collections.

GC12II. Hawthorn Group or fissure fills of equivalent age: (Hemphillian)
A. Withlacoochee River Sites 3C, 4A, 4X, 5D, and 5E (Hh2) **(previously just GC12II).**
Refs: Webb, 1969a, 1990; Becker, 1985; Hulbert, 1987b, 1993.
B. Dunnellon Phosphate Mines (Hh2).
Refs: Simpson, 1930; Hulbert, 1987b, 1988b.

GC13. Hawthorn Group, Upper Bone Valley Formation: (Hemphillian)
A. Manatee Fauna (Hh2).
Inc. Manatee County Dam Local Fauna, Leisey 1C, Lockwood Meadows Local Fauna (**previously GC13C**), Port Manatee Dredgings, Braden River Site, and SR 64 Site.
Refs: Webb and Tessman, 1968; MacFadden, 1986; Hulbert, 1988b, 1993; Morgan, 1993, 1994; Tedford *et al.*, 2004.
B. Palmetto Fauna (= Upper Bone Valley Fauna) (Hh4).
Inc. TRO Quarry, Whidden Creek Local Fauna, and Palmetto Mine Microsite.
Refs: Simpson, 1930; Webb, 1969a, 1981; MacFadden, 1986; Webb and Hulbert, 1986; Domning, 1990; Morgan, 1994; Webb, Hulbert, and Lambert, 1995; Hulbert, Webb, and Morgan, 2003.

GC14. Unnamed fluvial deposits: (Blancan)
A. Santa Fe River Blancan Sites (L. Bl) **(previously just GC14).**
B. Inc. Santa Fe 1A, 1B, 4A, 8A, and 15A (L. Bl).
Refs: Webb, 1974; MacFadden and Waldrop, 1980; Ray, Anderson, and Webb, 1981; Morgan and Ridgeway, 1987; Morgan and Hulbert, 1995.
BB. Devils Elbow 1 and 2 Sites in Saint Johns River (L. Bl).
Refs: UF collections.
C. Waccasassa River 9A (L. Bl).
Refs: UF collections.
D. Withlacooche River IA (L. Bl).

GC15. Unnamed fissure fill deposits: (Blancan–Irvingtonian)
A. Haile 15A (**previously just GC15**) (L. Bl).
Refs: Robertson, 1976; Morgan and Hulbert, 1995.
B. Haile 7C (L. Bl).
Refs: Morgan and Hulbert, 1995; Hulbert, 1997; Emslie, 1998.
C. Inglis 1A, 1B, 1C, 1D, 1E, 1F, and IG (L. Bl).
Refs: Martin, 1979; Webb and Wilkins, 1984; Morgan and Hulbert, 1995; Emslie, 1998; Ruez, 2001; UF collections.

D. Haile 16A. (L. Bl–E. Ir).
Refs: Morgan, 1991.
E. Haile 21A, Alachua County (E. Ir).
Refs: Morgan and Hulbert, 1995.
F. McLeod Limerock Mine (M. Ir).
Refs: Morgan and Hulbert, 1995.

GC16. Unnamed deposit above Tamiami Formation: (Blancan)
St. Petersburg Times Site (L. Bl).
Refs: Morgan and Ridgeway, 1987; Morgan and Hulbert, 1995.

GC17. Tamiami Formation, Pinecrest Beds: (Blancan)
A. Macasphalt Shell Pit Local Fauna **(previously just GC17)** and Richardson Road Shell Pit (= Quality Aggregates Shell Pit) (L. Bl).
Refs: Hulbert, 1987a; Morgan and Ridgeway, 1987; Jones *et al.*, 1991; Morgan, 1993; Morgan and Hulbert, 1995; Czaplewski, Morgan, and Naether, 2003.
B. Acline Shell Pit **(previously just GC17)**, Bass Point Waterway 1 **(previously GC18A)**, Mule Pen Quarry **(previously GC18A)**, Brighton Canal, Copely Road Shell Pit, McQueen Shell Pit, El Jobean Shell Pit **(previously GC18)**, and Kissimmee River (L. Bl).
Refs: Morgan and Ridgeway, 1987; Morgan and Hulbert, 1995.

GC18II. Caloosahatchee Formation: (Blancan)
A. Lehigh Acres Rock Pit (L. Bl).
Refs: Morgan and Hulbert, 1995; Missimer and Tobias, 2004.
B. De Soto, Forsberg, DM, Pelican Road, Davis, Arcadia, Charlotte County Shell Pits and CoCo Plum Waterway (L. Bl).
Refs: Morgan, 1993, 1994; Morgan and Hulbert, 1995; Emslie, 1998.

GC18III. Bermont Formation: (Irvingtonian)
A. Leisey Shell Pit Local Fauna and Apollo Beach (E. Ir).
Refs: Hulbert and Morgan, 1989; Morgan and Hulbert, 1995; Morgan and White, 1995.
B. Crystal River Power Plant Site (E. Ir).
Refs: Morgan and Hulbert, 1995.
C. Rigby Shell Pit, Punta Gorda, and Port Charlotte area (E. Ir).
Refs: Morgan and Hulbert, 1995.

GC181V. Nashua Formation: (Irvingtonian)
Payne Creek Mine, Pool Branch Site, Peace River Mine, and Tucker Burrow Pit. (E. Ir).
Refs: Morgan (2005, pers. comm.).

SOUTHERN EAST COAST

GC19. Unnamed formation, Louisiana: (Torrejonian)
Junior Oil Company Beard No. 1 well (To2).
Refs: Archibald *et al.*, 1987.

GC20. Williamsburg Formation, South Carolina; (Clarkforkian) Black Mingo Group, Santee Rediversion Canal, nr. St. Stephens.
Refs: Schoch, 1985.

GC20II. Hatchetigbee Formation, Alabama: (Wasatchian)
Hatchetigbee Bluff Local Fauna (Wa2–3).
Refs: Westgate, 2001.

GC21. Bashi Formation, Mississippi: (Wasatchian; Wa2–3)
Refs: Beard and Tabrum, 1991.

GC21II. Tuscahoma Formation, Mississippi: (Wasatchian)
Red Hot Local Fauna (Wa2–3).
Refs: Beard, Dawson, and Tabrum, 1995; Beard and Dawson, 2001.

GC21III. Gosport Sand, Alabamba: (Uintan)
Little Stave Creek Local Fauna.
Refs: Westgate, 2001.

GC21IV. Lisbon Formation, Alabama and Mississippi: (Uintan–Duchesnean)
A. Lutetian Marine series, Mississippi (Ui3–Du) **(previously EM2)**.
Refs: Krishtalka *et al.*, 1987.
B. Isney, Choctaw County, Alabama (Du).
Refs: Westgate, 2001, University of Alabama collections.
C. Coffeeville Landing, Tombigbee River, Clark County, Alabama (Du).
Refs: Westgate, 2001, USNM collections.

GC21V. Kosciusko Formatino, Doby's Bluff Tounge: (Uintan–Duchesnean)
Doby's Bluff, Clarke County, Mississippi (Ui3–Du).
Refs: Ivany, 1998.

GC22I. Clinchfield Formation: (Duchesnean–Chadronian)
A. Hardy Kaolin Mine Local Fauna (Du).
Refs: Westgate 2001.
B. Houston County (inc. Pennsylvania-Dixie Cement County Quarry, Clinchfield, Perry) (Du–Ch1).
Refs: Kellogg, 1936; Domning, Morgan, and Ray, 1982.
C. Twiggs Clay, Twiggs County (Du–Ch1) **(previously just GC22)**.
Refs: Cooke and Shearer, 1918; Westgate, 1990.

GC22II. McBean Formation: (Uintan)
Plant Vogtle, Georgia (U13).
Refs: Petkewich and Lancaster, 1984; Hulbert *et al.*, 1998.

GC23. Jackson Group, Arkansas and Louisiana: (Uintan–Chadronian)
A. Crow Creek Local Fauna. Arkansas (Ui3–Du) **(previously just GC23)**.
Refs: Westgate and Emry, 1985, Westgate, 2001.
B. Lover's Leap Local Fauna, Arkansas (Ui1–Du).
Refs: Westgate, 2001.
C. Unnamed site(s), Caldwell Parish, Louisiana (Du–Ch1).
Refs: Kellogg, 1936.
D. Red River, Grant Parish, Louisiana (Du–Ch1).
Refs: Kellogg, 1936.

GC23II. Cook Mountain Formation, Texas and Louisiana: (Duchesnean)
A. Unnamed site(s), Bossier Parish, Lousiana.
Refs: LSUMG colls.
B. Unnamed site(s), Natchitoches Parish, Louisiana.
Refs: Maher and Jones, 1949; Uhen, 1998a.
C. Unnamed site(s), Leon County, Texas.
Refs: Kellogg, 1936.

GC23III. Moodys Branch Formation, Mississippi: (Duchesnean)
Thompson Creek, Yazoo County.
Refs: Daly, 1999.

GC23IV. Yazoo Formation, Alabama, Mississippi, and Louisiana: (Chadronian)
A. Near Melvin, Choctaw County (inc. Cocoa, Isney, and Fail, Alabama).
Refs: Kellogg, 1936.
B. Clarksville, Clark County (inc. "The Rocks" near Dead Level, Alabama).
Refs: Kellogg, 1936.
C. Old Court House, Washington County, Alabama.
Refs: Kellogg, 1936.
D. West of Melvin, AL, Clarke County, Mississippi.
Refs: Kellogg, 1936; Daly, 1992.
E. Milsaps College, Hinds County, Mississippi.
Refs: Kellogg, 1936; Daly, 1992.
F. Goshen Springs, Rankin County, Mississippi.
Refs: Kellogg, 1936; USNM collections.
G. West of Bucatunna River, Wayne County, Mississippi.
Refs: Kellogg, 1936; Daly, 1992.
H. Hunter's Point, Madison, Madison County, Mississippi.
Refs: Dockery and Johnston, 1986.
J. Tullos Member, Montgomery Landing, Grant Parish, Louisiana.
Refs: Lancaster, 1986.

GC23V. Crystal River Formation, Alabama and Florida: (Duchesnean–Chadronian)
A. Unnamed site(s), Covington County, Alabama.
Refs: Kellogg, 1936.
B. Point A Dam, Escambia County, Alabama.
Refs: Uhen, 1998b.
C. Unnamed site(s), Lafayette County, Florida.
Refs: Kellogg, 1936; Uhen, 1998b.

GC23VI. White Bluff Formation, Arkansas: (Chadronian)
A. Crow Creek, St. Francis County.
Refs: Palmer, 1939; Westgate, 2001.
B. Red Bluff, Jefferson County.
Refs: Westgate and Salazar, 1996.

GC23VII. Ocala Formation: (Duchesnean–Chadronian)
A. Suggsville, Alabama.
Refs: Kellogg, 1936.
B. Cedar Reach Rock, Conecuh River, Escambia County, Alabama.
Refs: Kellogg, 1936.
C. Unnamed sites, Bibb County, Georgia.
Refs: Kellogg, 1936.
D. Cedar Creek, Crisp County, Georgia.
Refs: Kellogg, 1936; Uhen, 1998b.
E. Georgia Limerock Products Company Quarry, near Perry, Houston County, Georgia.
Refs: Kellogg, 1936.

GC23VIII. Ocalla Formation: (Duchesnean–Chadronian)
A. Dolime Quarry, Citrus County, Florida.
Refs: UF collections
B. Oakhurst, Lime Company Quarry, near Ocalla, Marion County, Florida.
Refs: Kellogg, 1938; Morgan, 1978; Uhen, 1998b.
C. Haile Quarry.
Refs: UF collections
D. Devil's Eye Spring.
Refs: UF collections.
E. Santa Fe River 17.
Refs: UF collections.
F. Suwannee River.
Refs: UF collections.
J. Stinhatchee River.
Refs: UF collections.
K. Chipola River 1.
Refs: UF collections.
L. Dowling Park Quarry.
Refs: UF collections.
M. Dell Limerock Mine.
Refs: UF collections.
N. D. H. Bell site.
Refs: UF collections.
O. Coon Cave.
Refs: UF collections.
P. Waccassasa River.
Refs: UF collections.

GC23IX. Jackson Group: (Chadronian?)
Unnamed sites, Lauderdale County, Tennessee.
Refs: Corgan, 1976.

GC23X. Inglis Formation: (Duchesnean?)
Citrus County, Florida.
Refs: Randazzo *et al.*, 1990; Uhen, 1998b.

GC24. Byram Formation, Mississippi: (Orellan; Or1–2) **(note emended dating, previously Chadronian)**
Refs: Manning, 1997.

GC24II. Unnamed formation, Alabama: (Arikareean)
Conecuh River (Ar2).
Refs: Westgate, 1992.

GC24III. Parachucla Formation: (Arikareean; Ar1–2)

GC25. Hawthorn Formation, South Carolina: (Hemingfordian) **(now EM6II)**

GC26. Fleming Formation, Louisiana: (Barstovian)
Fort Polk, Vernon Parish (Ba2).
Refs: Morgan, 1991.

GC27. Egor Rouge Sand (previously Citronelle Formation), Alabama: (Clarendonian–Hemphillian)

Mobile Local Fauna (= Mauvilla Local Fauna) (Cl3–Hh1).
Refs: Tedford and Hunter, 1984: MacFadden and Dobie, 1998; Hulbert and Whitmore, 2006.

GC28. Yorktown Formation, North Carolina: (Hemphillian) **(now EM8B)**

GC29. Citronelle Formation, Louisiana: (Hemphillian)
Tunica Hills (Hh4).
Refs: Manning and MacFadden, 1989.

GC30. ?Yorktown Formation, North Carolina: (Blancan) **(now EM8D).**

GC31. Unnamed formation, Georgia: (Hemphillian)
Near Reynolds (Hh4).
Refs: Voorhies, 1974

CALIFORNIA CENTRAL REGION AND COASTAL RANGES

CALIFORNIA

CC1. Unnamed formation: (Wasatchian)
Morena Boulevard Local Fauna (Wa5–6).
Refs: Walsh, 1991a.

CC2. Delmar Formation: (Bridgerian)
Swami's Point (Br2).
Refs: SDMNH collections.

CC3. Scripps Formation: (Bridgerian–Uintan)
Black's Beach Local Fauna (Br3–Ui1).
Refs: Walsh, 1991a.

CC4. Friars Formation: (Uintan)
Lower sandstone–mudstone tongue, middle conglomerate tongue, and upper sandtone–mudstone tongue (Ui1) **(previously CC5, Poway Fauna, = Friars-Misson Valley Fauna).** Including State Route 56 East Sites 1 and 2, Azuaga II Sites 1 and 5, Scripps Ranch North Site 20-B, State Route 52 West Site 2, Mission Terrace Site 1 and 5, San Diego Mission.
Refs: Lillegraven, 1976, 1980; Golz and Lillegraven, 1977; Golz, 1979; Lillegraven, McKenna, and Krishtalka, 1981; Krishtalka and Stucky, 1983b; Korth 1985; Walsh, 1991a, 1996; Gunnell, 1995; Chiment and Korth, 1996; Walsh *et al.*, 1996; Colbert, 2006.

CC5. "Cypress Canyon Unit": **(now included with CC4A)**

CC6. Stadium Conglomerate: (Uintan)
A. Lower Member, Murray Canyon Local Fauna (Ui2).
B. Upper Member, Stonecrest Local Fauna (Ui3).
Refs: Walsh, 1991a, 1996.

CC7. Santiago Formation: (Uintan–Duchesnan)
A. Mesa Drive Local Fauna (Ui2) **(note emended dating, previously Ui1)**.
Refs: Walsh, 1991a, 1996, 1997.
B. Jeff's Discovery, Kelley's Ranch, College Boulevard and Rancho del Oro Local Faunas (Ui2).
Refs: Walsh, 1991a, 1996, 1997, 1998, 2000.
C. Laguna Riviera (Ui3) **(note emended dating, previously Ui3–Du)**.
Refs: Golz and Lillegraven, 1977; Golz, 1979; Walsh, 1991a, 1996.
D. Chestnut Avenue Local Fauna (inc. Water Tower Locality, Blue Bone Locality, and Mission del Oro) (Ui3–Du).
Refs: Golz, 1979; Walsh, 1996; SDMNH collections.
E. Camp San Onofre Local Fauna (Du).
Refs: Lillegraven, 1976, 1977, 1980; Golz and Lillegraven, 1977; Walsh, 1991a, 1996, 1997; Chiment and Korth, 1996.

CC8. Mission Valley Formation: (Ui3)
Cloud 9 Fauna (inc. La Mesa, San Carlos, Fletcher Hills, and East San Diego).
Refs: Lillegraven, 1976, 1980; Golz and Lillegraven, 1977; Walsh, 1991a, 1996, 1997, 1998.

CC8II. Pomerado Conglomerate (Uintan–Chadronian)
A. Lower Conglomerate and Miramar Sandstone Members, and Eastview Local Fauna **(previously included in CC6B)** (Ui3).
Refs: Walsh, 1991a, 1996, 1997, 1998.
B. Upper Conglomerate Member (Du–Ch1).
SNSHN Localities 4041, 4042.
Refs: Walsh and Gutzler, 1999.

CC9. Sespe Formation, Simi Valley Area: (Uintan–Arikareean)
A0. Unnamed local fauna, Lower Member (Ui3).
Refs: Kelly *et al.*, 1991.
A. Tapo Canyon Local Fauna (Ui3).
Refs: Golz and Lillegraven, 1977; Krishtalka and Stucky, 1983a; Kelly, 1990, 1992; Kelly and Whistler, 1994.
AA. Brea Canyon Local Fauna and Kelly Quarry (Ui3).
Refs: Kelly, 1990, 1992; Kelly and Whistler, 1994.
B. Strathern Local Fauna (Ui3–Du).
Refs: Kelly, 1990; Kelly and Whistler, 1994.
BB. Pearson Ranch Local Fauna (inc. Alamos Canyon Sites) (Du).
Refs: Golz and Lillegraven, 1977; Krishtalka and Stucky, 1983a; Kelly, 1990; Lucas, 1992; Kelly and Whistler, 1994.
B2. Simi Valley Landfill Local Fauna (Du).
Refs: Kelly, 1990, 1992; Kelly and Whistler, 1994, 1998.
C. Kew Quarry Local Fauna, Ventura County (Ar1).
Refs: Lander, 1983; Emry, Russell, and Bjork, 1987.
D. Lower and Upper South Mountain, Grimes Canyon, Shiells Canyon, Alamos Canyon, and Sycamore Canyon Local Faunas, unnamed local fauna (Little Simi Valley), and East Fork of Maria Ygnacio Creek (Ar1).
Refs: Lander, 1983.
E. Canton Canyon, Piru Creek–Santa Clara River drainage (Ar2).

CC9II. Sespe Formation, Sespe Gorge Area: (Uintan–Duchesnean)

Hartman Ranch Local Fauna (Ui1–Du).
Refs: Lindsay, 1968; Golz, 1979; Kelly, 1990.

CC9III. Sespe Formation, Los Angeles and Ventura Counties: (Arikareean–Hemingfordian)
A. Blakeley Local Fauna (Ar1).
Refs: Whistler and Lander, 2003.
B. Lower and Upper Piuma Road Local Fauna (He1).
Refs: Whistler and Lander, 2003.

CC9IV. Undifferentiated Sespe/Vaqueros Formations, Orange County: (Uintan–Hemingfordian)
A. Basal Orange County Sespe/Vaqueros Fauna (Ui3).
Refs: Whistler and Lander, 2003.
B. Lower Bowerman Local Fauna (Ar1).
Refs: Whistler and Lander, 2003.
C. Lower Orange County Sespe/Vaqueros Fauna (inc. Parkridge, Lower ETC Windy Ridge, ETC Jamboree Road, ETC Santiago Road, Lower ETC Rattlesnake Canyon, Upper Bowerman Landfill, San Joaquin Hills, and the FTC/Oso Local Faunas) (Ar3–4).
Refs: Whistler and Lander, 2003.
D. Upper Orange County Sespe/Vaqueros Fauna (inc. Upper ETC Windy Ridge, Santiago Canyon Landfill, Upper ETC Rattlesnake Canyon, ETC Bee Canyon, Bolero Lookout, and the Upper Oso Dam Local Faunas) (He1).
Refs: Whistler and Lander, 2003.

CC10. "Sweetwater" Formation: (Duchesnean)
Bonita Local Fauna.
Refs: Walsh, 1991b, 1996.

CC10II. Lower Temblor Formation: (Orellan) **(previously CC13II)**
Temblor Range.
Refs: Prothero, 2005.

CC10III. Unknown formation: (?Oligocene)
Chili Gulch, Calaveras County.
Refs: Prothero, 2005.

CC11. Alegria Formation: (Arikareean)
Canada de la Gaviota Local Fauna, Santa Ynez Mountains (Ar1).

CC12. Otay Formation: (Arikareean)
Eastlake fossil Localities, San Diego Area (Ar1).
Refs: Deméré, 1988.

CC13. Tecuya Formation Sites: (Arikareean)
Tecuya Canyon Local Fauna (Ar2).
Refs: Stock, 1920; Lander, 1983; Emry, Russell, and Bjork, 1987; Deméré, 1988; Walsh and Deméré 1991.

CC14. Jewett Sand: (Arikareean)
Pyramid Hill Local Fauna (Ar3).
Refs: Lander, 1983.

CC15. Tick Canyon Formation: (Arikareean)
Tick Canyon Local Fauna (Ar4).
Refs: Jahns, 1940; Whistler, 1967; Whistler and Lander, 2003.

CC16. Vaqueros Formation: (Arikareean–Hemingfordian)
"A" sites (?He1) (**note emended dating, previously Ar3–4).**
Refs: Repenning and Vedder, 1961; Tedford *et al.*, 1987; Lucas, Whistler, and Wagner, 1997.

CC17. Caliente Formation: (Arikareean–Hemphillian)
A. Below Lower Triple Basalt, Unit 2 and "A" sites (He1).
Refs: Repenning and Vedder, 1961.
B. Hidden Treasure Spring Local Fauna (He2).
Refs: Kelly and Lander, 1988.
C. West Dry Canyon Local Fauna, Jerd Spring Fauna, Padrones Spring Local Fauna, Road Gap Canyon 2, and CIT Site 315 (Ba1) (**note emended dating, previously He2**).
Refs: James, 1963; Kelly and Lander, 1988; LACM collections.
D. Upper Dry Canyon Fauna, Lower Dome Springs Fauna, Eastern Caliente Range, CIT Site 322–323, and Kent Quarry (Ba1).
Refs: Kelly and Lander, 1988; Wang, Tedford, and Taylor, 1999.
E. Upper Dome Spring Fauna (Ba1).
Refs: James, 1963; Barnosky, 1986; Kelly and Lander, 1988.
F. Doe Spring Fauna (Ba2).
Refs: James, 1963; Kelly and Lander, 1988.
G. Mathews Ranch Fauna (inc. Cuyama Sites, Apache Canyon, and Hedgehog Quarry) (Cl2).
Refs: James, 1963; Barnosky, 1986.
H. Nettle Spring Fauna (inc. Remnant Hill and Big Cat Quarry) (Cl3).
Refs: James, 1963; Barnosky, 1986; UCB collections.
I. "Hh" sites and Sequence Canyon Fauna (inc. Klipstein Ranch) (Hh3–4) (**note emended dating, previously Hh2–3**).
Refs: James, 1963; Tedford *et al.*, 1987; UCB collections.

CC18. Topanga Formation: (Hemingfordian)
A. Orange County (He2).
Refs: Whistler and Lander, 2003.
B. Santa Ana Mountains (He1) (**note emended dating, previously Ba1**).
Refs: Raschke, 1984.

CC19. Branch Canyon Sandstone: (Hemingfordian)
Vedder Local Fauna (He2).
Refs: Lindsay, 1974; Munthe, 1979; Whistler and Lander, 2003; UCB collections.

CC20. Olcese Sand: (Hemingfordian)
Barkers Ranch Local Fauna (He2).
Refs: Tedford *et al.*, 1987.

CC21. Bopesta Formation (= Kinnick Formation): (Hemingfordian–Barstovian)
A. Phillips Ranch Local Fauna (*Merychippus carrizoensis* zone) (He2).
Refs: Buwalda, 1916; Quinn, 1987.
B. Cache Peak Fauna (*Merychippus stylodontus* zone) (Ba1).
Refs: Buwalda, 1916; Quinn, 1987; UCB collections.

C. Cache Peak Fauna (*Merychippus* cf. *M. intermontanus* zone) (Ba2).
Refs: Quinn, 1987.

CC22. Crowder Formation (= Punchbowl Formation or Cajon Formation): (Hemingfordian–Barstovian)
A. UCR RV-6989, Cajon Valley (Crowder Unit 1) (He2).
Refs: Woodburne and Golz, 1972; Reynolds, Reynolds, and Korth, 1991.
B. B4, Cajon Valley (Crowder Unit 2) (Ba2).
Refs: Reynolds, Reynolds, and Korth, 1991.

CC23. Upper Temblor Formation: (Arikareean–Barstovian)
A. North Coalinga Local Fauna and Domengine Creek (Ba1) (**previously just CC23**) (**note emended dating, previously Ba2**).
Refs: Merriam, 1915; Bode, 1935; Woodburne, 1969; Barnosky, 1986; LACM collections.
B. San Luis Obispo (?Ba2).

CC24. Santa Margarita Formation: (Clarendonian)
Comanche Point Local Fauna (Cl2).
Refs: Savage, 1955.

CC25. Mehrten Formation: (Barstovian–Hemphillian)
A. Two Mile Bar Local Fauna (Ba2).
Refs: Stirton and Goeriz, 1942.
AA. Siphon Canal (Hh1).
Refs: Hirschfeld, 1981.
B. Turlock Lake, Modesto Reservoir Local Fauna (inc. Rhino Island), and Black Rascal Creek (Hh4) (**note emended dating, previously Hh3**).
Refs: Stirton and Gooritz, 1942; Hirschfeld and Webb, 1968; Wagner, 1976; Wang, Tedford, and Taylor, 1999, Prothero, 2005; LACM and UCB collections.

CC26. Green Valley Formation: (Clarendonian–Blancan)
A. Sycamore Creek Local Fauna (inc. Green Valley Site) (Cl3) (**note emended dating, previously Cl2**).
B. Black Hawk Ranch Local Fauna (inc. Mount Diablo) (Cl3).
Refs: Macdonald, 1948; Ritchie, 1948; Savage, 1955; Baskin, 1981; Wang, Tedford, and Taylor, 1999; Prothero and Tedford, 2000; UCB collections.
C. Maxum Local Fauna, Contra Costa County (E. Bl).
Refs: May and Repenning, 1982.

CC27. Orinda Formation: (Clarendonian)
Orinda sites (Cl2).
Refs: Richey, 1940.

CC28. Moraga Formation: (Clarendonian) (Cl2)

CC30. Contra Costa Group: (Clarendonian)
A. Kendall–Mallory Local Fauna (Cl3).
B. Layfayette Ridge and Wills Pit (Cl3).

CC31. San Pablo Group: (Clarendonian)
Inc. Tesla, UCMP V5019, Las Trampas Ridge, and Ingram Creek (Cl2 or Cl3).
Refs: Wang, Tedford, and Taylor, 1999.

CC32. Chanac Formation: (Clarendonian–Hemphillian)
A. South Tejon Hills Local Fauna (Cl2).
Refs: Drescher, 1942; Savage, 1955; Tedford *et al.*, 1987.
B. North Tejon Hills Local Fauna (Cl3).
Refs: Drescher, 1942; Tedford *et al.*, 1987; LACM collections.

CC33. Mint Canyon Formation: (Clarendonian)
Mint Canyon Local Fauna (Cl3).
Refs: Maxson, 1930.

CC34. Lake Mathews Formation: (Clarendonian)
Cajalco Local Fauna (Cl2).

CC35. Siesta Formation: (Clarendonian–Hemphillian)
A. Siesta sites (Cl3).
Refs: UCB collections.
B. Norris Canyon (Hh1).
Refs: Webb, 1983.

CC36. Mulholland Formation: (Hemphillian)
Mulholland Fauna (Hh3).
Refs: Tedford *et al.*, 1987; UCB collections.

CC37. Pinole Formation: (Hemphillian)
Inc. Hemme Hills and Pinole Local Faunas (Hh4).
Refs: Stirton, 1939; Tedford *et al.*, 1987; UCB collections.

CC38. Kern River Formation: (Hemphillian–Blancan)
Kern River sites (Hh2).
Refs: LACM collections.

CC39. Etchegoin Formation sites: (Barstovian–Blancan)
A. Kettleman Hills (Ba3).
B. North Dome, Kettleman Hills (Hh2).
Refs: Merriam, 1915; Tedford *et al.*, 1987.

CC40. Mount Eden Formation: (Hemphillian)
Mount Eden Local Fauna (Hh4).
Refs: Frick, 1921; May and Repenning, 1982; Albright 1999b; UCB collections.

CC41. Hungry Valley Formation and Peace Valley Beds: (Hemphillian)
Inc. Kinsey Ranch Local Fauna and Coyote Canyon (Hh3–4).
Refs: LACM collections.

CC41II. Petaluma Formation: (Hemphillian)
Sonoma County (Hh3–4).

CC42. San Joaquin Formation: (Blancan)
Inc. Buttonwillow and Kettleman Hills (E. Bl).

CC43. Tehama Formation: (Blancan)
A. Tehama Local Fauna (= Black Ranch) (E. Bl) (**previously just CC43**).
Refs: Kurtén and Anderson, 1980.
B. Northern part of Sacramento Valley (L. Bl).
Refs: Wang, Tedford, and Taylor, 1999.

CC44. Sonoma Volcanics: (Blancan)
Ritchie Creek (E. Bl).
Refs: Woodburne, 1966.

CC45. Tulare Formation: (Blancan)
Asphalto (E. Bl).
Refs: Kurtén and Anderson, 1980.

CC45II. Unnamed formation: (Blancan)
Two miles southeast of Cornwall, Contra Costa County.
Refs: Wang, Tedford, and Taylor, 1999.

CC46. Unnamed unit: (Blancan)
Aguanga Horizon, Riverside County.

CC46II. Unnamed formation: (Blancan)
Radec, Temecula Arkose (L. Bl).
Refs: Golz *et al.*, 1977.

CC46III. Unnamed sandstone: (Blancan)
Nutmeg (L.? Bl).

CC47. San Timoteo Formation: Eden Hot Springs–Jack Rabbit Trail: (Hemphillan–Blancan)
Refs: Albright, 1999b.
A. Sites 157, 160A, 160B (E. Bl).
B. Site 176B, 177, 181A, 182 (E. Bl).
C. Site 187 (E. Bl).
D. Sites 195, 202A (L. Bl).

CC48. San Timoteo Formation: Riverside County Landfill–El Casco composite section: (Blancan–Irvingtonian)
Refs: Albright, 1999b.
A. Site 304 (E. Bl).
B. Site 311, 313 (E. Bl).
C. El Casco Local Fauna and Site 350 (E. Ir).

CC49. Old Woman Sandstone: (Blancan)
San Bernadino County Museum Locality, southeast of Lucerne Valley, San Bernadino County.
Refs: Czaplewski, 1993.

BAJA CALIFORNIA

CC50. Tepetate Formation: (Wasatchian–Bridgerian)
Lomas Las Tetas de Cabra Fauna (= Punta Prieta–Rancho Rosarito Locality, and Occidental Buttes (Wa1)).
Refs: Novacek *et al.*, 1991.

CC51. Isidro Formation: (Hemingfordian)
La Purísima, Baja California Sur (?He1).
Refs: Ferrusquía-Villafranca, 1984, 2003.

CC52. Rosarito Beach Formation: (Hemingfordian–Barstovian)
La Mision, Baja California Norte (He1–Ba1).
Refs: Ferrusquía-Villafranca, 1984, 2003.

CC52II. Unnamed formation: (Blancan)
San Jose del Cabo Basin, Baja California Sur (E. Bl).
Refs: Miller and Carranza-Casteñeda, 2002.

CC52III. Unnamed formation: (Blancan)
Miraflores, Baja California Sur.
Refs: Carranza-Casteñeda, Miller, and Kowallis, 1998; Carranza-Castañeda and Miller, 2004.

CC53. Unnamed unit: (Blancan)
Las Tunas Local Fauna, Baja California Sur (E. Bl) (**note emended dating, previously L. Bl**).
Refs: Miller and Carranza-Casteñeda, 1984; Carranza-Castañeda and Miller, 2004.

NORTHERN GREAT BASIN

CALIFORNIA GREAT BASIN

NB1. Goler Formation: (Torrejonian–Tiffanian)
A. Laudate Locality (inc. Edentulous Jaw and Land of Oz sites) (To3–Ti1).
Refs: McKenna, 1960a,b; Archibald *et al.*, 1987; Lofgren *et al.*, 2002.
B. RAMV 98012, Member 4d (Tiffanian).
Refs. Lofgren *et al.*, 2002, 2004.

NB2. Lower Red Beds: (Duchesnean)
Titus Canyon, Death Valley, Grapevine Mountains.
Refs: Stock, 1949.

NB2II. Unnamed formation: (Chadronian)
Lower Cedarville Flora, Modoc County.
Refs: Prothero, 2005.

NB3. Hector Formation: (Arikareean–Hemingfordian)
A. Black Butte Mine Local Fauna (Ar3).
Refs: Woodburne *et al.*, 1974.
B. Logan Mine Fauna (He1) (**note emended dating, previously Ar4**).
Refs: Woodburne *et al.*, 1974; Lander, 1985.
C. Lower Cady Mountain Local Fauna (Ar4–He1).
Refs: Lander, 1985.
D. Upper Cady Mountain Local Fauna (He2).
Refs: Lander, 1985; Tedford *et al.*, 1987.

NB4. Upper part of the Tropico Group: (Hemingfordian)
Boron Local Fauna (He1).
Refs: Whistler, 1984; Whistler and Lander, 2003.

NB5. Diligencia Formation: (Hemingfordian)
Unit 2 (He1).

NB6. Barstow Formation: (Hemingfordian–Barstovian)
A. Owl Conglomerate Member, Red Division Fauna (He2).
Refs: Lander, 1985; Woodburne, Tedford, and Swisher, 1990.
B. Middle Member, Rak Division Fauna (= Barstow, Third Division Fauna) (He2).
Inc. Ledge 3, Coon Canyon, Waterfall Canyon and Unknown Quarry.
Refs: Woodburne, Tedford, and Swisher, 1990; Lindsay, 1991.
C. Middle Member, Green Hills Fauna (Ba1).
Inc. Steepside, Sunset, Rak, Turbin, Oreodon, Camp, Deep, Raven, Sunrise, Sunder Ridge, and Yermo Quarries, Rainbow Canyon, and Daggett Ridge Local Fauna, Alvord Mountain, Slippery Slide, and Doc's Steepside.
Refs: Lander and Reynolds, 1985; Reynolds and Lander, 1985; Barnosky, 1986; Woodburne, Tedford, and Swisher, 1990; AMNH collections.
D. Middle Member, Second Division Fauna (Ba1).
Inc. Valley View, Mud Hills, Sandstone, and Hailstone Quarries, Saucer Butte, and Coon Canyon.

Refs: Kelly and Lander, 1988; Woodburne, Tedford, and Swisher, 1990; Lindsay, 1991; Wang, Tedford, and Taylor, 1999.

E. Upper Member, Barstow Fauna (= Barstow Fauna, First Division) (Ba2).

Inc. Starlight, Skyline, Mayday, Yule Tide, New Hope, Leader, Hidden Hollow, Sunnyside, Hell Gate Basin No. 4, New Year, *Hemicyon*, and Easter Quarries, Slug Bed, East of Owl Canyon, North End, Rodent Hill Basin, Rodent Hill, Carnivore Canyon 4, 5, and 6, Lake Bed, Eureka, Coon Canyon 5, Camel B.3, and Bonanza.

Refs: Merriam, 1919; Kelly and Lander, 1988; Woodburne, Tedford, and Swisher, 1990; Lindsay, 1991; Wang, Tedford, and Taylor, 1999.

F. Cronese Local Fauna (Ba2).

Refs: Lander, 1985.

NB7. Dove Spring Formation (= Ricardo Formation): (Clarendonian–Hemphillian)

A. Iron Canyon Fauna (*Ustatochoerus profectus/Copemys russelli* zone), Last Chance Gulch, and Nightmare Gulch Fauna (Cl2) (**note emended dating, previously Ba2–Cl1**).

Refs: Whistler and Burbank, 1992.

B. Iron Canyon Fauna (*Cupidinimus avawatzensis/Paracosoryx furlongi* zone) (Cl2) (**note emended dating, previously Cl1–Cl2**).

Refs: Whistler and Burbank, 1992.

C. Ricardo Fauna (*Epicyon aphobus/Hipparion forcei* zone) (inc. *S. burnhami* sites and Red Rock Canyon) (Cl2–Cl3).

Refs: Whistler and Burbank, 1992; Wang, Tedford, and Taylor, 1999.

D. Dove Spring Fauna (*Epicyon aphobus/Hipparion forcei* zone) (inc. Opal Canyon) (Hh1) (**note emended dating, previously Cl3**).

Refs: Whistler and Burbank, 1992; Wang, Tedford, and Taylor, 1999.

E. Dove Spring Fauna (*Paronychomys/Osteoborus diabloensis* zone) (Hh1).

Refs: Whistler and Burbank, 1992.

NB8. Avawatz Formation: San Bernadino County: (Clarendonian)

Avawatz Mountain Fauna (Cl2).

Refs: Henshaw, 1939; R. W. Wilson, 1940; Barnosky, 1986; LACM collections.

NB9. Bedrock Springs Formation: (Hemphillian)

Lava Mountains Fauna and Red Rock Canyon (Hh1).

Refs: Tedford *et al.*, 1987; Wang, Tedford, and Taylor, 1999.

NB10. Horned Toad Formation: (Hemphillian)

Warren Local Fauna (Hh4).

Refs: May and Repenning, 1982; Tedford *et al.*, 1987.

NB11. Alturas Formation: (Hemphillian–Blancan)

Alturas sites (Hh1–2).

Refs: Repenning, 2003; UCB collections.

NB12. Coso Formation: (Hemphillian–Blancan)

A0. South of Coso Mountains (Cl or Hh).

Refs: Wang, Tedford, and Taylor, 1999.

A. Coso Mountains (Hh3–4).

Refs: Kurtén and Anderson, 1980.

B. Coso Mountains (E. Bl).

C. Coso Mountains (L. Bl).

Refs: Wang, Tedford, and Taylor, 1999.

NB12II. Unknown formation: (Blancan)

Tecopa Lake Beds (L. Bl).

Refs: Whistler and Webb, 2005.

NB13. Palm Springs Formation: (Blancan–Irvingtonian)

A. Layer Cake Local Fauna (E. Bl).

Refs: Downs and White, 1968; Kurtén and Anderson, 1980; Cassiliano, 1999.

B. Arroyo Seco Local Fauna (E. Bl).

Refs: Downs and White, 1968; Kurtén and Anderson, 1980; Cassiliano, 1999.

C. Anza-Borrego Desert/Vallecito Creek Fauna (L. Bl–E. Ir).

Refs: Downs and White, 1968; MacFadden and Skinner, 1979; Kurtén and Anderson, 1980; Becker and White, 1981; White, 1984; Albright, 1999b; Cassiliano, 1999.

NEVADA

NB14. Sheeps Pass Formation: (Bridgerian)

Elderberry Canyon Local Fauna (Br2).

Refs: Emry and Korth, 1989; Emry, 1990.

NB15. Unnamed formation: (Arikareean)

Rizzi Ranch Local Fauna, Allegheny Creek, Wild Horse Range (Ar1).

NB15II. Unnamed formation: (Hemingfordian)

Hackberry Local Fauna, Lanfair Valley (He2).

Refs: Wang, Tedford, and Taylor, 1999.

NB16. Unnamed formations: (Hemingfordian)

Inc. Horse Camp Springs and Red Mountain Local Faunas, and Southwestern Fish Creek Mountain (He2).

NB17. Unnamed formation: (Hemingfordian) (**on Columbia Plateau**)

Massacre Lake Fauna (He2).

Refs: Morea, 1981; Smith, Cifelli, and Czaplewski, 2004.

NB18. Virgin Valley Formation: (Barstovian) (**on Columbia Plateau**)

A. Virgin Valley Local Fauna (Ba1) (**previously just NB18**).

Refs: Merriam, 1911; Merriam and Stock, 1928; Tedford *et al.*, 2004; LACM collections; UCB collections.

B. High Rock Lake Local Fauna (Ba2).

Refs: Tedford *et al.*, 2004.

NB19. Carlin Formation (Humboldt Group): (Hemingfordian–Hemphillian)

A. Humbolt Group (inc. Carlin area [Hadley Ranch]) (Ba1).

Refs: UCB collections.

B. UCMP Loc. V4709 (= Fossil Hill) (Ba2).
C. Chalk Springs Fauna (Cl2).
Inc. Trilolite, North Carlin, Reese River Crossing, South Reese River, North Carlin, Chalk Spring, South Pine Valley, Mount Lewis, and Elko Wells Localities.
Refs: Macdonald and Pellatier, 1958.
D. South Pine Valley (Hh1).

NB20. Washoe Formation: (Barstovian)
A. High Rock Canyon Local Fauna (Ba1).
Refs: Stirton, 1940; UCB collections.
B. Late Barstovian site (Ba2).

NB21. Raine Ranch Formation: (Barstovian)
Inc. Lone Mountain, Camp Creek, and Woodruff Creek Local Faunas (Ba1).
Refs: Macdonald, 1966.

NB22. Monarch Mill Formation: (Barstovian–Hemphillian)
A. Near Eastgate (Ba1).
B. Eastgate (?Hh).

NB23. Esmerelda Formation (= Siebert Formation): (Barstovian–Clarendonian)
A. Early Barstovian site (Ba1).
B. Stewart Spring Fauna, Tonopah and Savage Canyon Local Faunas (Ba1) (**note emended dating, previously Ba2**).
Refs: Henshaw, 1942; Barnosky, 1986; Tedford *et al.*, 1987; AMNH, LACM, and UCB collections.
C. Cedar Mountain Local Fauna (inc. Snowball Valley North), Fish Lake Valley Local Fauna, and near Warrior Mine (Cl2).
Refs: Macdonald and Pelletier, 1958; Clark, Dawson, and Wood, 1964; Barnosky, 1986; Korth, 1999a; Wang, Tedford, and Taylor, 1999; UCB collections.

NB24. Stewart Valley Formation: (Barstovian)
UCMP Loc. V6020, Mineral County (Ba2).

NB24II. Unknown formation: (Barstovian)
Home Station Pass, Rock Corral Wash, Pershing County (?Ba2).
Refs: Prothero, 2005.

NB25. Unnamed unit: (Barstovian–Clarendonian)
A. Jersey Valley Local Fauna (Ba2).
B. Jersey Valley Local Fauna (Dacies Pass) (Cl2).
Refs: UCB collections.

NB26. Unnamed unit. Antelope Valley Sites: (Barstovian–Clarendonian)
A. Antelope Creek Local Fauna (Ba2).
B. L. E. Clarendonian site (Cl2).

NB26II. Desert Peak Formation (= ?Kate Peak Formation): (Barstovian–Clarendonian)
Eagle–Pilcher Mine Local Site (Ba2–Cl1).
Refs: Kelly 1998.

NB27. Coal Valley Formation: (Clarendonian–Hemphillian)
A. Coal Valley Fauna (Cl2).
Refs: Axelrod, 1956; Macdonald and Pelletier, 1958.
A1. Churchill Valley Local Fauna (?Cl).
Refs: Kelly, 1998.
A2. Churchill Narrows Site (Cl3–Hh1).
Refs: Kelly, 1998.
B. Smiths Valley Fauna (inc. Upper Petrified Tree Canyon, Wilson Canyon, and Yerington Locality) (Hh1) (**note emended dating, previously Hh2**).
Refs: Macdonald and Pelletier, 1958; Macdonald, 1959; Tedford *et al.*, 1987; Becker and McDonald, 1998.

NB27II. Unnamed formation: (Clarendonian)
Seven Miles west of Blair Junction, Esmeralda County.

NB28. Unnamed formation: (Clarendonian)
Roglove Local Fauna (Cl2).
Refs: Nelson and Madsen, 1987.

NB29. Truckee Formation: (Clarendonian–Hemphillian)
A. Inc. Brady Pocket and Nightingale Road Local Faunas, Kawosh Mountain, Hazen, Kate Peak, and Verdi Localities (Cl3).
Refs: Macdonald, 1956; Macdonald and Pelletier, 1958; UCB collections.
B. Silver Springs Local Fauna (Hh3).
Refs: Kelly, 1998.

NB30. Unnamed unit: (Hemphillian)
Rabbit Hole, Loc. V73117 (Hh1).

NB30II. Unnamed formation: (Hemphillian)
Lime Quarry, Humbolt County (Hh1).
Refs: Prothero, 2005.

NB31. Thousand Creek Formation: (Hemphillian) (**on Columbia Plateau**)
Thousand Creek Fauna (Hh1) (**note emended dating, previously Hh2**).
Refs: Merriam, 1911; Macdonald and Pelletier, 1958; Becker and McDonald, 1998; Korth, 1999a; LACM collections.

NB32. Muddy Creek Formation: (Hemphillian–Blancan)
A. Mormon Mesa, Clark County (Hh1).
B. Inc. Double Butte, Limestone Corner, Rodent Quarry and near Reno (E. Bl).

NB33. Panaca Formation: (Blancan)
A. Panaca Local Fauna (= Meadow Valley Fauna) (E. Bl) (**previously just NB33; note emended dating, previously Hh3–4**).
Refs: Macdonald and Pelletier, 1958; Mou, 1997; Lindsay *et al.*, 2002.
B. Channel Sands Pocket (L. Bl).
Refs: Wang, Tedford, and Taylor, 1999.

NB33II. Unnamed formation: (Hemphillian)
Carlin Area, Elko County (inc. Hi Level Quarry, Lake Quarry, and Goose Creek) (Hh1–2).
Refs. Prothero, 2005.

NB34. Unnamed unit: (Hemphillian)
Golgotha Watermill Pothole Quarry (Hh3–4).

NB34II. Unnamed formation: (Hemphillian–Irvingtonian)
Hoye Canyon Local Fauna, Wellington Hills, Douglas County (Hh3–4).
Refs: Kelly, 2000.

NB35. Hay Ranch Formation: (Hemphillian)
Carlin High Quarry (Hh3–4).

NB35II. Horse Spring Formation: (Hemphillian–Blancan)
Moapa, Clark County, San Bernadino County Museum Locality 2.12.16, Bed 1 (Hh3–4).

NB35III. Sunrise Pass Formation, Carson Valley/Pine Nut Mountains, Douglas County: (Hemphillian–Blancan)
A. Washoe Local Fauna (Hh4).
Refs: Kelly, 1997.
B. Buckeye Creek Local Fauna (E. Bl).
Refs: Kelly, 1994, 1997.
C. Fish Springs Flat Local Fauna (L. Bl).
Refs: Kelly, 1994, 1997.

NB35IV. White Narrows Formation: (Blancan)
White Narrows Locality, Clark County (E. Bl).
Refs: Reynolds and Lindsay, 1999; Bell *et al.*, 2004.

NB36. Unnamed formation: (Blancan)
Wichman Faunule.
Refs: Macdonald and Pelletier, 1958; Kurtén and Anderson, 1980.

NB36II. Unnamed formation: (Blancan)
Peavine Creek, near Reno (E.? Blancan).

NB36III. Unnamed formation, Antelope Valley/Wellington Hills, Douglas County: (Blancan–Irvingtonian)
A. Wellington Hills Local Fauna (L. Bl).
Refs: Kelly, 1997.
B. Topaz Lake Local Fauna (E. Ir).
Refs: Kelly, 1997.

SOUTHERN GREAT BASIN

ARIZONA

SB1. "Chalk Canyon" Formation: (Orellan)
Cottonwood Creek, Cave Creek Area (Or1).
Refs: Lindsay and Tessman, 1974.

SB2. Unnamed formation: (Whitneyan–Arikareean)
Atravesada, Pima County (Wh2–Ar1).
Refs: Lindsay and Tessman, 1974.

SB3. Unnamed unit: (Arikareean)
Muggins Mountains, near Wellton, Yuma County (?Ar3).
Refs: Lindsay and Tessman, 1974.

SB4. Unnamed formation: (L. Arikareean–Hemingfordian)
Anderson Mine, Yavapai County (He1–2).
Refs: Lindsay and Tessman, 1974.

SB5. Unnamed formation: (Hemingfordian)
Black Canyon, Aqua Fria River Drainage (He1).

SB6. Milk Creek Formation: (Barstovian–Clarendonian)
Inc. Shields Ranch Quarry, Walnut Grove, Hazelwood Ranch, Iron Springs, and Prescott Southeast (Ba2–Cl1).
Refs: Lindsay and Tessman, 1974.

SB7. Unnamed formation: (Clarendonian)
Greywater Wash, Apache County (Cl2).
Refs: Lindsay and Tessman, 1974.

SB8. Unnamed formation: (Hemphillian)
Bartlette Reservoir, Maricopa County (Hh 1–2).
Refs: Lindsay and Tessman, 1974.

SB9. Big Sandy Formation: (Hemphillian)
Wikieup Local Fauna (Hh3–4).
Inc. Clay Bank, Birdbone, Horseshoe, Concretion and Split Cliff Quarries, and Ward's Prospect.
Refs: MacFadden, Johnson, and Opdyke, 1979; MacFadden and Skinner, 1979.

SB10. Quibiris Formation: (Hemphillian)
Inc. Old Cabin Quarry, Camel Canyon, Redington Local Fauna (Reddington Quarry = Bingham Ranch), San Pedro Valley, and Turtle Pocket (Hh3–4).
Refs: Lindsay, Opdyke, and Johnson, 1984; Czaplewski, 1993.

SB11. Bidahochi Formation: (Clarendonian–Hemphillian)
A. President Wilson Springs, Jeddito Valley (Cl).
Refs: Wang, Tedford, and Taylor, 1999.
B. Inc. White Cone Local Fauna, Jeddito Local Fauna, Taylor Gravel and Keams Canyon Local Fauna (Hh3–4) **(previously just SB11).**
Refs: Baskin, 1979; Lindsay, Opdyke, and Johnson, 1984; Lindsay and Tessman, 1984.

SB12. Verde Formation: (Blancan)
Verde Local Fauna (E. Bl).
Inc. Clarkdale, Buckboard Wash, House Mountain, and Sacred Mountain.
Refs: Lindsay and Tessman, 1984; Lindsay, Opdyke, and Johnson, 1984; Czaplewski, 1987, 1993.

SB13. Unnamed formation: (Blancan)
Bear Springs and Matthew Wash (E. Bl).
Refs: Lindsay and Tessman, 1974; Tomida, 1987.

SB14. St. David Formation Sequence: (Blancan–Irvingtonian)
A. Benson Local Fauna (E. Bl).
Inc. Gomphothere 6947, Rabbit, Rat Fink, Carnivore, Betonite, Gray Point, Post Ranch, East Side Is., Devil's Lawn, and Green Saddle Quarries.
Refs: Lindsay and Tessman, 1974; Opdyke *et al.*, 1977; Dalquest, 1978; Harrison, 1978; Kurtén and Anderson, 1980.
B. Mendeville Ranch Local Fauna, Cochise County and Noye's Bonanza (E. Bl).
Refs: Lindsay and Tessman, 1974; Opdyke *et al.*, 1977.
C. McCrae Wash, Honey's Hummock and Horsey Green Bed (E. Bl).
Refs: Lindsay and Tessman, 1974; Opdyke *et al.*, 1977.
D. Wolf Ranch and Cal Tech (L. Bl) (**note emended dating, previously E. Bl**).
Refs: Lindsay and Tessman, 1974; Opdyke *et al.*, 1977; Harrison, 1978.
E. California Wash and Johnson Pocket (L. Bl).
Refs: Lindsay and Tessman, 1974; Opdyke *et al.*, 1977; Kurtén and Anderson, 1980.
F. Curtis Ranch, *Glyptotherium*, and Gidley Localities (L. Bl).

Refs: Hager 1974; Lindsay and Tessman, 1974; Opdyke *et al.*, 1977; Harrison, 1978; Kurtén and Anderson, 1980.
G. Three and one-half mile section (near Benson), and Post Ranch Fauna, Carnivore Site (L. Bl).
Refs: Wang, Tedford, and Taylor, 1999.

SB15. Gila Conglomerate: (Blancan)
A. Duncan, Greenlee County (**previously just SB15**) (E. Bl).
Refs: Lindsay and Tessman, 1974; Tomida, 1987.
B. Pearson Mesa (extends into New Mexico as SB34IVA) (E. Bl).
Refs: Lindsay (2006, pers. comm.).
C. Country Club, Graham County (E. Bl).
Refs: Tomida, 1987; Bell *et al.*, 2004.

SB16. Unnamed formation: (Blancan)
Cosomi Wash, Santa Cruz County (E. Bl).
Refs: Lindsay and Tessman, 1974; Kurtén and Anderson, 1980.

SB17. Fissure Fillings in Kaibab Limestone: (L. Blancan–E. Irvingtonian)
Anita Local Fauna (inc. Cal Verde Mine).
Refs: Hay, 1921; Lindsay and Tessman, 1974; Kurtén and Anderson, 1980.

SB18. Gila Conglomerate, 111 Ranch Beds: (Blancan)
A0. Duncan section, Duncan Basin (E. Bl).
Refs: Tomida, 1981; Bell *et al.*, 2004.
A. Flat Tire Local Fauna (L. Bl).
Refs: Lindsay and Tessman, 1974.
B. Tusker Local Fauna (L. Bl).
Refs: Lindsay and Tessman, 1974; Kurtén and Anderson, 1980.
C. Henry Ranch and South Red Knoll Quarry (L. Bl).
Refs: Lindsay and Tessman, 1974.
D. Dry Mountain Locality and Whitlock Hills (L. Bl).
Refs: Lindsay, 1978; Wang, Tedford, and Taylor, 1999.
E. Artesia Road Fauna (L. Bl).
Refs: Bell *et al.*, 2004.

SB19. Unnamed formation: (Blancan)
Snowflake, Navajo County (L. Bl).
Refs: Lindsay and Tessman, 1974.

SOUTHERN COLORADO

SB20. Animas Formation, San Juan Basin: (Tiffanian–Clarkforkian)
A. Mason Pocket near Tiffany (Ti4).
Refs: Rose, 1975; Krishtalka and Stucky, 1983a; Archibald *et al.*, 1987.
B. Bayfield and others (Ti5).
Refs: Archibald *et al.*, 1987.
C. North of Mason Pocket (Cf2).

SB21. Cuchara Formation: (Wasatchian)
Huerfano Basin (Wa7).
Refs: Robinson, 1963.

SB22. Huerfano Formation: (Wasatchian–Bridgerian)
A. Lower Huerfano A, Localities VIII, IX, XII (Wa6).
Refs: Robinson, 1966; Korth, 1984; Krishtalka and Stucky, 1985.
B. Upper Huerfano A, Localities IV, VI, XI (Wa7).
Refs: Robinson, 1966; Bown and Kihm, 1981; Schoch and Lucas, 1981; Korth, 1984; Stucky, 1984a; Krishtalka and Stucky, 1985.
C. Huerfano B and Uppermost Huerfano A, Localities I, II, III,(?IV), V, VII (Br1).
Refs: Robinson, 1966; McKenna, 1976; Korth, 1984, 1985; Stucky, 1984a.

SB22II. Farasita Formation (Bridgerian) (Br1?)

NEW MEXICO

SB23. Nacimiento Formation, San Juan Basin: (Puercan–Torrejonian)
A. *Ectoconus* zone (= *Hemithaleus* facies) (Pu2).
Inc. De-na-zin Wash (Sinclair and Granger Locality 2, lower part, Williamson and Lucas Locality 7) and lower Kimbetoh and Betonnie–Tsosie Washes (inc. Sinclair and Granger Localities 5, 6, and 7, Black Toe, Black Stripe, Old Dolan Ranch, Tsosie, Eduardo Arroyo, Mammalon Hill, plus others, Williamson and Lucas Locality 11).
Refs: Tsentas, 1981; Archibald *et al.*, 1987; Williamson, 1993; Williamson and Lucas, 1993.
B. *Taeniolabis* zone (Pu3).
Inc. West Fork of Gallegos Canyon (inc. Sinclair and Granger Locality 4, Williamson and Lucas Locality 6), De-na-zin Wash (inc. Barrel Spring Arroyo, Coal Creek, Sinclair and Granger Locality 2, upper part, Williamson and Lucas Locality 8), and Split Lips Flat.
Refs: Lucas, 1984; Archibald *et al.*, 1987; Williamson, 1993; Williamson and Lucas, 1993; Williamson and Weil, 2002.
C. Lower Kutz Canyon sites (Williamson and Lucas Locality 1) (To1).
Inc. De-na-Zin Wash (Sinclair and Granger Locality 3), *Periptychus* site (Williamson and Lucas Locality 9), Kimbetoh and Betonnie–Tsosie Washes (inc. Upper Kimbetoh, Arroyo Head of Kimbetoh, Arroyo, and Dick's Ditch), and upper Betonnie–Tsosie Arroyo (Power Line), Williamson and Lucas Localities 12 and 13.
Refs: Tomida, 1981; Archibald *et al.*, 1987; Williamson, 1993; Williamson and Lucas, 1993.
D. Mesa de Cuba: type area of Cope's "Puerco marls" (= Simpson's Locality 222) (Williamson and Lucas Locality 25) (To1–2).
Refs: Simpson, 1959; Archibald *et al.*, 1987; Williamson, 1993; Williamson and Lucas, 1993.
DD. Mesa de Cuba (Simpson's Localities 222, 226, and 229) (Williamson and Lucas Locality 26) (To2).

Refs: Simpson, 1959; Archibald *et al.*, 1987; Williamson, 1993; Williamson and Lucas, 1993.
E. *Deltatherium* zone (To2).
Inc. Kutz Canyon (Williamson and Lucas Locality 2) and Head of Gallagos Canyon (Sinclair and Granger Locality 1, "Chico Springs," Williamson and Lucas Locality 10).
Refs: Archibald *et al.*, 1987; Williamson, 1993; Williamson and Lucas, 1993, 1997.
F. West Kutz Canyon Sites (To2).
Inc. Bob's Jaw, Taylor Mound, "Angel's Peak Pocket" (= Big Pocket, KU Locality 13), Bob's Basin and others) (Williamson and Lucas Locality 3).
Refs: Taylor, 1981; Archibald *et al.*, 1987; Williamson, 1993; Williamson and Lucas, 1993; Meehan and Wilson, 2002.
G. South Kutz Canyon Sites (To3).
Inc. O'Neill *Pantolambda*, Coprolite Point and others (Williamson and Lucas Locality 4), lower horizon of Sinclair and Granger Locality 14 ("cliffs at the head of Escaveda Wash," Williamson and Lucas Locality 16), lower part of West Flank Torreon Wash (Williamson and Lucas Localities 18–20), and lower part of East Flank Torreon Wash (Sinclair and Granger Locality 11 [lower horizon] Williamson and Lucas Localities 22 and 23).
Refs: Tsentas, 1981; Archibald *et al.*, 1987; Williamson, 1993; Williamson and Lucas, 1993.
GG. Animas River Valley (?To3: may span entire range).
Inc. Cedar Hill, upper Kimbeto and Betonnie–Tsosie Washes (inc. Sinclair and Granger Localities 8 and 9 [lower part], and KU Locality 9 [= Little Pocket]) Williamson and Lucas Localities 14 and 15).
Refs: Archibald *et al.*, 1987; Williamson, 1993; Williamson and Lucas, 1993.
H. *Pantolambda* zone, in part (To3) (**previously To3–Ti1**).
Inc. Upper Kutz Canyon sites (Williamson and Lucas Locality 5); Head of Escaveda Wash (Sinclair and Granger Locality 14) (Williamson and Lucas Locality 17), upper part of West Flank Torreon Wash (= Sinclair and Granger Locality 10 ["west branch of Arroyo Torreon"]) (Williamson and Lucas Locality 21), upper part of East Flank Torreon Wash (= Sinclair and Granger Locality 1 ["east branch of Arroyo Torreon"]) (Williamson and Lucas Locality 24); plus ?microsite of Tsentas.
Refs: Archibald *et al.*, 1987; Williamson, 1993; Williamson and Lucas, 1993.

SB24. San Jose Formation, San Juan Basin: (Wasatachian)
Regina and Tapicitos Members, Almagre and Largo Beds (inc. *Meniscotherium* Quarry) (Wa6).
Refs: Lucas *et al.*, 1981; Korth, 1984; Flanagan, 1986; Krishtalka and Stucky, 1986.

SB25. Galisteo Formation: (Wasatchian–Duchesnean)
A. Cerillos Local Fauna (Wa6).
Refs: Lucas and Kues, 1979; Lucas *et al.*, 1981.
B. Tonque Local Fauna and Windmill Hill (Du).
Refs: Lucas and Kues, 1979; Lucas *et al.*, 1981; Lucas, 1992.

SB26. Baca Formation: (Duchesnean)
A. Carthage Coal Field and east of Rio Grande (Du) (**note emended dating, previously Br**).
Refs: Lucas *et al.*, 1981; Lucas, 1983, 1992.
B. White Mesa and Mariano Mesa, west of Rio Grande (Du).
Refs: Lucas *et al.*, 1981, 1983.

SB26II. Hart Mine Formation: (Bridgerian)
Carthage Local Fauna (Br3).
Refs: Lucas and Williamson, 1993; Lucas, 1997.

SB27. Rubio Peak Formation: (Duchesnean-Chadronian)
A. Black Range (Du).
Refs: Lucas 1986a,b.
B. Black Range (Ch1).
Refs: Lucas 1986a,b.

SB27II. Palm Park Formation: (Chadronian) (Ch1)
Refs: Lucas and Williamson, 1993; Lucas, Estep, and Froehlich, 1997.

SB28. Abiquiu Formation: (Arikareean)
Abiquiu Sites (Ar3).
Refs: Tedford, 1981.

SB28II. Sandstone of Inman Ranch: (Arikareean)
Seventyfour Draw (Ar1–2).
Refs: Morgan and Lucas, 2003a.

SB29. Zia Formation: (Arikareean–Hemingfordian)
A. Piedra Parada Member (inc. Standing Rock Quarry) (Ar4).
Refs: Tedford, 1981.
B. Chamisa Mesa Member (inc. Sandoval County site and Blick and *Cynarctoides* Quarries) (He1).
Refs: Tedford, 1981.
C. Canyarda Pilares Member (inc. Straight Cliff Prospect and Chamisa Mesa Member, Jeep Quarry, and Mesa Prospect) (He1).
Refs: Tedford, 1981; Wang, Tedford, and Taylor, 1999.
D. Kiva Quarry and Arroyo Chamisa Prospect, Chamisa Mesa Member (He2).
Refs: Tedford, 1981.

SB30. "Upper Zia Sand": (Barstovian–Clarendonian)
A. *Rakomylus* Site, inc. Jemez Creek area (Ba1).
Refs: Tedford, 1981; Wang, Tedford, and Taylor, 1999.
B. Cerro Conejo Member: Rincon Quarry, Jemez Creek Area, Red Cliff Prospect, Canyada Piedra Parada, and Pilares Quarry (Ba2)
Refs: Wang, Tedford, and Taylor, 1999.
B2. Santa Ana Wash and Arroyo Chamisa Prospect (Ba2–Cl1).
Refs: Wang, Tedford, and Taylor, 1999; Prothero, 2005.
C. Middle Red Sites (Cl2).
Refs: Tedford, 1981.

SB31. Gila Group: (Arikareean–Blancan)
A. Black Range, head of Taylor Creek. (Ar3–4).
Refs: Tedford, 1981.
B. Magnus Trench and Dry Creek (Hh3–4).
Refs: Tedford, 1981.
C. Magnus Trench and Duncan Basin (L. Bl).
Refs: Tedford, 1981.
D. San Simon Local Fauna (L. Bl).
E. Pima (L. Bl).
Refs: Wang, Tedford, and Taylor, 1999.

SB32. Tesuque Formation: (Hemingfordian–Hemphillian)
A. Nambé Member (He2).
Refs: Tedford, 1981; Tedford and Barghoorn, 1993.
AA. Nambé Member (Ba1).
Refs: Prothero, 2005.
B. Skull Ridge Member, Skull Ridge Sites (inc. White Operation Quarry) (Ba1).
Refs: Galusha and Blick, 1971; Tedford, 1981; Tedford and Barghoorn, 1993.
C. Chama-El Rito Member, Rio del Oso-Abiquiu Sites (inc. *Aelurodon* Wash) (Ba2).
Refs: Tedford, 1981; Tedford and Barghoorn, 1993.
D. Pojoaque Member (inc. Camel Area, Pojoaque Bluffs Area, Santa Cruz Sites, and Jacona Microfossil Quarry) (Ba2).
Refs: Galusha and Blick, 1971; Tedford, 1981; Tedford and Barghoorn, 1993; Korth and Chaney, 1999.
E. Chama-El Rito Member (inc. Rinconada Sites) (Ba2).
Refs: Tedford and Barghoorn, 1993.
EE. Dixon Member, Dixon Sites (Ba2).
Refs: Tedford and Barghoorn, 1993.
F. Chama-El Rito Member, Chama-El Rito Sites (Ba2).
Refs: Tedford, 1981; Tedford and Barghoorn, 1993.
FF. Ojo-Calliente Member (inc. Conical Hill Quarry) (Ba2).
Refs: Tedford and Barghoorn, 1993; Wang, Tedford, and Taylor, 1999.
G. Pojoaque Member (Cl1–2).
Refs: Tedford and Barghoorn, 1993; Wang, Tedford, and Taylor, 1999.
H. Ojo-Caliente Member (Cl2).
Refs: Tedford and Barghoorn, 1993.
HH. Pojoaque Member, Santa Cruz Area and Santa Clara District (Cl2).
Refs. Prothero, 2005.
I. Ojo-Caliente Member, Black Mountain Site (Hh1).
Refs: Tedford and Barghoorn, 1993.

SB33. Santa Fe Formation: (Clarendonian–Blancan)
A. Early site (Cl1).
Refs: Lozinsky and Tedford, 1991.
B. Middle Red Member and Santa Fe Sites (Cl3).
Refs: Tedford, 1981; Lozinsky and Tedford, 1991.
C. Ceja Member and Los Lunas Sites (L. Bl).
Refs: Tedford, 1981.

SB33II. Popotosa Formation: (Hemphillian)
Gabaldon Badlands site (Hh1).
Refs: Lozinsky and Tedford, 1991.

SB34. Chamita Formation: (Clarendonian–Hemphillian)
A. Round Mountain Quarry and Battleship Mountain, San Ildefonso sites (inc. Black Mesa) (Cl2).
Refs: Tedford, 1981; Barnosky 1986; Tedford and Barghoorn, 1993; Wang, Tedford, and Taylor, 1999; USNM collections.
B. Lower Chamita sites (Hh1).
Refs: Tedford and Barghoorn, 1993.
C. San Juan and Rak Camel Quarries, *Osbornoceros* Quarry, and Leyden Quarry (Hh2).
Refs: MacFadden, 1977; MacFadden, Johnson, and Opdyke, 1979; Tedford and Barghoorn, 1993.
D. Santa Cruz (Hh3).
Refs: Prothero, 2005.

SB34II. Gila Group, Mangas Basin: (Hemphillian–Blancan)
A. Walnut Canyon Local Fauna (Hh4).
Refs: Morgan *et al.*, 1997; Morgan and Lucas, 2003b.
B. Buckhorn Local Fauna (E. Bl).
Refs: Morgan and Lucas, 2003b.

SB34III. Puyé Formation: (Hemphillian) **(previously SB34II)**
Puyé Formation Sites (Hh4) (**note emended dating, previously E. Bl**).
Refs: Tedford and Barghoorn, 1993; Morgan and Lucas, 2003b.

SB34IV. Gila Group, Duncan Basin: (Blancan)
A. Pearson Mesa Local Fauna (extends into Arizona as Locality SB15B) (L. Bl).
Refs: Morgan and Lucas, 2003b.
B. Virden Local Fauna (L. Bl).
Refs: Morgan and Lucas, 2003b.

SB35. Palomas Formation: (Blancan)
A. Truth or Consequences Local Fauna (E. Bl).
Refs: Jacobs and Lindsay, 1984; Repenning and May, 1987; Morgan and Lucas, 2003b.
B. Rio Cuchillo Negro Creek (**previously just SB35**) and Elephant Butte Lake Local Fauna (**previously included in SB38**) (E. Bl) (**note emended dating, previously L. Bl**).
Refs: Morgan and Lucas, 2003b.
C. Arroyo de la Parida Local Fauna (E. Bl) (**previously included in SB36; note emended dating, previously L. Bl**).
Refs: Morgan and Lucas, 2003b.
D. Las Palomas Creek sites (E. Bl) (**previously just SB35; note emended dating, previously L. Bl**).
Refs: Tedford, 1981; Morgan and Lucas, 2003b.

SB36. Sierra Ladrones Formation: (Blancan–Irvingtonian)
A. Lemitar 1 and 2, Person Power Plant, and Laguna (L. Bl) (**previously just SB36**).
Refs: Tedford, 1981.

B. Santa Domingo Local Fauna (L. Bl).
Refs: Morgan and Lucas, 2003b.
C. Western Mobile Gravel Pit, Tijeras Arroyo Local Fauna, and Fite Ranch (E. Ir).
Refs: Morgan and Lucas, 2003b.
SB37. Camp Rice Formation: (Blancan–Irvingtonian)
A. Tonuci Mountain Local Fauna (E. Bl).
Refs: Morgan and Lucas, 2003b.
B. Hatch Local Fauna and Rincon Arroyo (E. Bl).
Refs: Morgan and Lucas, 2003b.
C. Mesilla Basin Faunas A (L. Bl).
Refs: Morgan and Lucas, 2003b.
D. Mesilla Basin Fauna B plus Chamberino/Canutillo Sites (L. Bl–E. Ir) **(previously just SB37)**.
Inc. E. Anthony, Vicinity La Union, Vicinity Anapra, UTEP 33.
Refs: Tedford, 1981; Harris, 1993; Morgan, Lucas, and Estep, 1998; Morgan and Lucas, 2003b.
E. Mesilla Basin Faunas C (E. Ir).
Refs: Morgan and Lucas, 2003b.
F. Tortugas Mountain Local Fauna (E. Ir).
Refs: Morgan and Lucas, 2003b.
SB37II. Ancha Formation: (Blancan)
Ancha Formation Sites, Española Basin (L. Bl).
Refs: Morgan and Lucas, 2003b.
SB38. Axial River Facies: (Blancan)
Santo Domingo sites (L. Bl) (**Elephant Butte now included in SB35A**).
Refs: Tedford, 1981.
SB38II. Arroyo Ojito Formation: (Blancan)
A. Loma Colorado De Abajo Local Fauna (E. Bl).
Refs: Morgan and Lucas, 2003b.
B. Mountainview Local Fauna and Sevilleta Local Fauna (E. Bl).
Refs: Morgan and Lucas, 2003b.
C. Inc. Loma Colorado de Abajo, Belen, and Pajarito Local Faunas, Los Lunas, Isleta, and Veguita Sites (E. Bl).
Refs: Morgan and Lucas, 2000, 2003b.

TEXAS BIG BEND AREA

SB39. Black Peaks Formation: (Torrejonian–Wasatchian)
A. Schiebout–Reeves Quarry (Western Tornillo Flats) (Ti1).
Refs: Schiebout, 1974.
B. Eastern Tornillo Flats Washing Site, Ray's Bonebed, and Ray's Annex (Ti3).
Refs: Schiebout, 1974; Archibald *et al.*, 1987.
C. Joe's Bone Bed (Ti5).
Refs: Schiebout, 1974; Archibald *et al.*, 1987.
D. New Taeniodont site (Ti6-Cf1).
Refs: Schiebout, 1974.
E. Southwall (Cf2–Wa0).
Refs: Schiebout, 1974.
F. Big Bend National Park (Wa0).
Refs: Rose, 1981.
SB39II. Lower Tornillo Formation: (Puercan–Torrejonian)
A. Dogie Locality (Pu2–3) (**previously just SB39II**).
Refs: Standhardt, 1995.
B. Tom's Top (LSUMG Locality VL-111) (Pu3–To1).
Refs: Williamson, 1996,
SB40. Hannold Hill Formation: (Wasatchian)
Jack's Locality, Big Bend National Park (Wa7) (**note emended dating, previously early Wa**).
Refs: Rapp, MacFadden, and Schiebout, 1983.
SB41. Canoe Formation: (Uintan)
Big Yellow Sandstone Member, Big Bend National Park (Ui1) (**note emended dating, previously Br**).
Refs: J. A. Wilson, 1967; Wilson and Schiebout, 1984.
SB42. Laredo Formation: (Uintan)
Casa Blanca Local Fauna (inc. Spillway Quarry) (Ui3) (**previously SB42B; note emended dating, previously Ui1**).
Refs: Westgate, 1990, 2001; Walton, 1993a,b.
SB43. Devil's Graveyard Formation, Buck Hill Group: (Uintan– Chadronian)
AA. Junction (Ui1).
A. Whistler Squat Local Fauna (inc. "Pruett Formation," Hen Egg Mountain, Wax Camp, and Boneanza) (Ui1–2).
Refs: Gustafson, 1986a; J. A. Wilson, 1986; Walton, 1993a.
B. Serendipity Local Fauna (inc. Purple Bench Locality, Titanothere Channels, Stone Creek and Agua Fria-Green Valley Area) (Ui3).
Refs: J. A. Wilson, 1986; Walton, 1993a.
C. Skyline Channels and Cotter Channels, Bandera Mesa Member (Du).
Refs: Gustafson, 1986a; J. A. Wilson, 1986; Wilson and Stevens, 1986.
D. Montgomery Bonebed, Bandera Mesa Member (Du).
Refs: J. A. Wilson, 1986.
E. Bandera Mesa Member: Coffee Cup Local Fauna, Trans Pecos (Ch1).
Refs: J. A. Wilson 1984, 1986.
SB44. Vieja Group (= Chambers Tuff Formation): (Uintan–Chadronian)
A. Candelaria Local Fauna, Colemena Tuff (Ui3).
Refs: J. A. Wilson, 1978; Wilson and Schiebout, 1984; Gustafson, 1986a; Walton, 1993a.
B. Porvenir Local Fauna, Chambers Tuff (Du).
Refs: J. A. Wilson, 1978; Wilson and Schiebout, 1984; Gustafson, 1986a, Lucas, 1992; Heaton, 1993.
C. Little Egypt Local Fauna, Chambers Tuff, Trans-Pecos (inc. Reeves Bone Bed and Chalk Gap Draw) (Ch1).
Refs: J. A. Wilson, 1977, 1978; Korth 1985; Gustafson, 1986a .

D. Airstrip Local Fauna, Presido County (inc. Red Mound North) (= Capote Mountain Tuff Formation) (Ch2).
Refs: J. A. Wilson, 1971a,b, 1974, 1977, 1978; Emry, 1978; Gustafson, 1986a.
E. Ash Springs Local Fauna (Ch3–4).
Refs: J. A. Wilson, 1971a,b, 1974, 1977, 1978, 1984; Emry, 1978.

SB45. Unnamed formation: (Arikareean)
University of Texas Site 40225 (Ar1).

SB46. Delaho Formation: (Arikareean–Hemingfordian)
Castolon Local Fauna (Ar3).
Refs: Stevens, Stevens, and Dawson, 1969; Stevens and Stevens, 1989.

SB47. Closed Canyon Formation **(previously Rawls Formation)**: (Ar3) **(note emended dating, previously Ar4)**
Santana Mesa Local Fauna, Member 9, Rio Grande Valley.
Refs: Stevens, Stevens, and Dawson, 1969; Stevens and Stevens, 1989.

SB47II. Couch Formation, Yellowhorse Canyon: (Clarendonian)
A. Lower Member (Cl1).
Refs: Winkler, 1990.
B. Upper Member (Cl1).
Refs: Winkler, 1990.

SB48. Banta Shut-In Formation: (Hemphillian)
Screw Bean Local Fauna (Hh1) **(note emended dating, previously Cl3)**.
Refs: Stevens, Stevens, and Dawson, 1969; Stevens and Stevens, 1989, 2003.

SB48II. Unnamed formation: (Hemphillian)
Janes Quarry, Yellowhorse Canyon (Hh2).
Refs: Winkler, 1990.

SB48III. Bridwell Formation: (Hemphillian)
Yellowhorse Canyon (Hh3–4).
Refs: Winkler, 1990.

SB49. Camp Rice Formation: (Blancan)
Huspeth Local Fauna (L. Bl).
Refs: Akersten, 1972; Kurtén and Anderson, 1980.

SB50. Love Formation: (Blancan)
Red Light Local Fauna (L. Bl).
Refs: Akersten, 1972; Kurtén and Anderson, 1980.

NORTHERN AND CENTRAL MEXICO

SB51. Unnamed formation: (Bridgerian–Uintan)
Marfil Faunule, Guanajuato (Bridgerian) **(note emended dating, previously Du)**.
Refs: Ferrusquía-Villafranca, 1984, 2005.

SB52. Prietos Formation: (Chadronian)
Rancho Gaitan Local Fauna, Chihuahua (Ch1).
Refs: J. A. Wilson, 1978; Ferrusquía-Villafranca, 1984, 1993.

SB53. Unnamed formation: (? Hemingfordian)
Yécora, Sonora.
Refs: Ferrusquía-Villafranca, 1984, 2003.

SB54. Unnamed formation: (Hemingfordian)
Tubutama, Sonora (He2).
Refs: Ferrusquía-Villafranca, 1984, 2003.

SB55. El Zoyatal Tuft: (Hemingfordian)
Zoyatal Local Fauna, Arroyo Cedazo, southeastern Aguascalientes (He?1) **(note emended dating, previously He2)**.
Refs: Dalquest and Mooser, 1974; Ferrusquía -Villafranca, 1984, 2003.

SB55II. Unnamed unit: (Hemphillian–Blancan) **(previously CA4)**
A. Tehuichila, Hidalgo (Hh1).
Refs: Hulbert, 1988a; Carranza-Casteñeda and Miller, 1993.
AA. Potero Zietla Fauna assemblage (Hh2).
Refs: Castillo-Creon, 2000.
B. Tehuichila, Hildago (Bl).
BB. Chilcuahutla, Hidalgo (Bl–Ir).
Refs: Cabral-Perdomo, Bravo-Cuevas, and Castillo-Ceron, 2005.

SB55III. Unnamed unit: (Hemphillian)
La Nopalera, Querétaro (Hh1–2).
Refs: Carranza-Casteñeda, Miller, and Kowallis, 1998.

SB55IV. Unnamed unit: (Hemphillian)
Landa de Matamoros, Querétaro.
Refs: Carranza-Casteñeda, Miller, and Kowallis, 1998.

SB55V. Unnamed unit: (Hemphillian)
Cinqua, Michoacan.
Refs: Carranza-Casteñeda, Miller, and Kowallis, 1998.

SB55VI. Unnamed formation: (Hemphillian)
Colotlan, Jalis County.
Refs: Carranza-Casteñeda, Miller, and Kowallis, 1998.

SB56. Unnamed unit: (Hemphillian)
Tlaxcala (Hh3).
Refs: Miller and Carranza-Casteñeda, 1984.

SB56II. Unnamed unit: (Hemphillian) **(previously CA5)**
Las Golondrinas, Hidalgo (Hh3).
Refs: Carranza-Casteñeda, Miller, and Kowallis, 1998.

SB57. Unnamed unit: (Hemphillian)
Basuchil, Chihuahua (Hh3).
Refs: Lindsay, 1984.

SB57II. Unnamed unit: (Hemphillian–Blancan)
Tula, Hidalgo (Hh4–E. Bl).
Refs: Carranza-Casteñeda, Miller, and Kowallis, 1998.

SB57III. Unnamed formation: (Hemphillian)
San Miguel de Allende, Guanajuato state (Hh2–4).
Refs: Jiménez-Hidalgo and Carranza-Casteñeda, 2005a.

SB58. Unnamed unit: (Hemphillian–Blancan)
Ocote Local Fauna, Rancho El Ocote, Guanajuato.

A. L. Hemphillian unit (Hh3–4).
Refs: Dalquest and Mooser, 1980; Miller and Carranza-Casteñeda, 1984; Montellano, 1989.
B. Latest Hemphillian unit (Hh4).
Refs: Dalquest and Mooser, 1980; Miller and Carranza-Casteñeda, 1984; Montellano, 1989; Flynn *et al.*, 2005.
C. E. Blancan unit (inc. Arroyo el Tanque) (E. Bl).
Refs: Dalquest and Mooser, 1980; Miller and Carrenza-Casteñeda, 1984; Montellano, 1989; Flynn *et al.*, 2005.

SB58II. Unnamed formation, Tecolotlen Basin, Jalisco: (Hemphillian–Blancan/Irvingtonian)
Refs: Carranza-Casteñeda and Miller, 2002, 2004.
A. Santa Mara-Los Corrales Beds (Hh3–4).
B. San Jos-La Hacienda Beds (Hh4).
C. Las Gravas Beds (L. Bl–E. Ir).

SB59. Unnamed unit: (Hemphillian)
Coecillo Fauna, Rancho Viejo area (Hh4).
Refs: Miller and Carrenza-Casteñeda, 1984; Carranza-Casteñeda and Miller, 1993.

SB60. Unnamed unit: (Hemphillian)
Yepómera/Rincon Local Fauna, Chihuahua (Hh4).
Refs: Furlong, 1941; Lindsay, 1984; Lindsay and Jacobs, 1985; Carranza-Castañeda and Miller, 2004.

SB60II. Unnamed unit: (Hemphillian-Irvingtonian)
Nexpa, Tlalquitenango, Morelos (Hh3–L. Bl).
Refs: Corona-M and Alberdi, 2006.

SB61. Unnamed unit: (Blancan)
Matachic, Chihuahua (E. Bl).
Refs: Lindsay, 1984.

SB62. Unnamed unit: (Blancan)
Concha Local Fauna, Chihuahua (E. Bl).
Refs: Lindsay, 1984; Lindsay and Jacobs, 1985; Carranza-Castañeda and Miller, 2004.

SB63. Unnamed unit: (Blancan)
Minaca Mesa and Hearst Ranch, Chihuahua (E. Bl).
Refs: Kurtén and Anderson, 1980; Lindsay, 1984.

SB64. Unnamed unit: (Blancan–Irvingtonian)
La Goleta, Michoacan, lower beds (E. Bl).
Refs: Kurtén and Anderson, 1980; Miller and Carranza-Casteñeda, 1984.

SB64II. Unnamed unit: (Blancan)
Las Arcinas, Hidalgo.
Refs: Carranza-Casteñeda, Miller, and Kowallis, 1998.

SB65. Unnamed unit: (Blancan)
Rancho Viejo, Guanajuato (inc. Rinconada Locality) (E. Bl).
Refs: Miller and Carrenza-Casteñeda, 1984; Wang, Tedford, and Taylor, 1999; Jimenez-Hildago and Carranza-Casteñeda, 2002; Carranza-Castañeda and Miller, 2004; Flynn *et al.*, 2005.

SB65II. Unnamed unit: (Blancan)
La Pantera, Guanajuato (E. Bl).
Refs: Flynn *et al.*, 2005.

SB66. Unnamed formation: (Blancan)
Jal-Teco (E. Bl).
Refs: MacFadden and Carranza-Casteñeda, 2002.

SB67. Unnamed formation: (Blancan)
Vicinity of town of Tequixquiac, Valley of Mexico (L.? Bl).
Refs: Wang, Tedford, and Taylor, 1999.

SB68. Unnamed formation: (Blancan-Rancholabrean)
Cedazo Fauna (lower unit), Aguascalientes (L. Bl).
Refs: Montellano-Ballesteros 1992; Bell *et al.*, 2004.

SB69. Unnamed unit: (Blancan–Irvingtonian)
San Juan Union, Taxco, Guerrero.
Refs: Corona-M and Alberdi, 2006.

SOUTHERN GREAT PLAINS

TEXAS/OKLAHOMA PANHANDLE

SP1. Ogallala Formation: (Clarendonian–Blancan)
A0. Bivins Ranch, Amarillo Area (TX) (Cl2).
Refs: Wang, Tedford, and Taylor, 1999.
A. Laverne (Beaver) Local Fauna (Cl2) (**note emended dating, previously Cl3**).
Inc. Wisenhunt Quarry (OK), Excell Local Fauna (TX), Durham Local Fauna (OK), Coetas Creek Local Fauna (TX), TMM 1176, Red River Drainage (TX), Cole Highway Pit Fauna, and Staked Plains (TX).
Refs: Dalquest and Hughes, 1966; Schultz, 1977; Dalquest, Baskin, and Shultz, 1996.
B. Capps Pit Local Fauna (George New Pitts), Arnett Local Fauna (= Adair Ranch Quarry) (OK), Sanford Pit, Fritzler Lower Pit, Fritzler South Pit, and Port of Entry Pit (TX) (Hh1).
Refs: Schultz, 1977; Tedford *et al.*, 1987; Wang, Tedford, and Taylor, 1999; Prothero, 2005.
C. Higgins Local Fauna (= Sebits Ranch) (Hh1) (**note emended dating, previously Hh2**).
Inc. V. V. Parker Place, Schwab Place, Rentfro Pit 1, Sanford Pit, North Pit, Canadian River Sites, and Burson Ranch Pit (TX).
Refs: Schultz, 1977; MacFadden and Skinner, 1979; Tedford *et al.*, 1987; Dalquest and Patrick, 1989; Wang, Tedford, and Taylor, 1999; Prothero, 2005.
D. Optima (Guymon) Local Fauna, Ogallala Group (OK), Turkey Track Canyon, Parcell Ranch, and Canadian River Sites (TX) (Hh3).
Refs: Schultz, 1977.
E. Virgil Clark Pit, Campell Pit, Miller Pit and Nation Pit (OK) (Hh4).
Refs: Tedford *et al.*,1987.
F. Beck Ranch Local Fauna (TX) (E. Bl).
Refs: Dalquest, 1978; Madden and Dalquest, 1990.
G. Red Corral Local Fauna and Channing, Oldham County (TX) (E. Bl).

Inc. Proctor Pit A, C, D, Bevins Pit 1, Collins Pit 2.
Refs: Kurtén and Anderson, 1980; Wang, Tedford, and Taylor, 1999.
H. Cita Canyon Local Fauna (TX) (L. Bl).
Refs: Schultz, 1977; Kurtén and Anderson, 1980.

SP2. Clarendon Beds (TX): (Clarendonian)
A. Clarendon Local Fauna (Cl2).
Inc. MacAdams Quarry, White Fish Creek Quarry, Mill Iron Ranch, Turkey Creek, Blocker Ranch, Bluff Creek Quarry Quarries 1–4, Bromley Ranch, Spade Flats, Charles and Rowe-Lewis Ranch, Stanton Ranch, Dilli Place, Cliff Creek Pit, Shannon Ranch, Grant Quarry, Rizley Quarry, Adam Rizley Farm, and Skillet Creek (Cl2).
Refs: Schultz, 1977; Wang, Tedford, and Taylor, 1999; Prothero, 2005; AMNH collections.
B. Clarendon Local Fauna (Cl2) (**note emended dating, previously Cl3**).
Inc. Gidley Horse Quarry.

SP3. Hemphill Beds (TX): (Hemphillian)
A. Box T Local Fauna (inc. Bridge Creek, Box T Ranch) and McGehee Place Quarry (Hh2).
Refs: Schultz, 1977; Tedford *et al.*, 1987; Wang, Tedford, and Taylor, 1999; Prothero, 2005.
B. Coffee Ranch Local Fauna (= Miami Quarry) (Hh2) (**note emended dating, previously Hh3**).
Refs: Schultz, 1977; MacFadden, Johnson, and Opdyke, 1979; Dalquest, 1983; Tedford *et al.*, 1987; Dalquest and Patrick, 1989; AMNH collections.

SP4. Goodnight Beds (TX): (Hemphillian)
A. Goodnight Local Fauna (Hh3).
Inc. Horace Baker Pit, Hill Quarry, Center Hill Pit, Christian Pit 2, and Hubbard Place Quarry.
Refs: Hibbard, 1954; Schultz, 1977; Dalquest, 1983; Wang, Tedford, and Taylor, 1999; AMNH collections.
B. Axtel (= Palo Duro Canyon), Currie Ranch, Smart Ranch, and Christian Ranch Local Faunas (TX), Buis Ranch Local Fauna (OK) (Hh4).
Refs: Hibbard, 1954; Schultz, 1977.

SP5. Blanco Formation (TX): (Blancan–Irvingtonian)
Blanco Local Fauna (inc. Low, Marmot, Carter, Mead, and Red Quarries, Mt. Blanco, Blanco Canyon) (L. Bl).
Refs: Dalquest, 1975; Schultz, 1977; Kurtén and Anderson, 1980; Wang, Tedford, and Taylor, 1999.

CENTRAL GREAT PLAINS

UTAH

CP1. North Horn Formation, Wasatch Plateau: (Puercan–Torrejonian)
A. Gas Tank Hill Local Fauna (= Flagstaff Peak) (Pu2).
Refs: Robison, 1980.
B. Wagonroad (= lower part of Locality 4 of Gazin) (Pu3).
Refs: Robison, 1980; Tomida and Butler, 1980.
C. Dragon Canyon (= Locality 2 and upper part of Locality 4 of Gazin), and Sage Flats (To1).
Refs: Robison, 1980; Tomida and Butler, 1980; Archibald *et al.*, 1987; Gunnell, 1989.

CP2. Flagstaff Formation: (Tiffanian)
Spanish Fork Canyon.

CP3. Flagstaff Limestone: (?Wasatchian)
Wasatch Plateau.
Refs: Rich and Collinson, 1973.

CP4. Colton Formation: (Wasatchian)
Book Cliffs Area, Uinta Basin (Wa6–7).
Refs: Archibald *et al.*, 1987.

CP5. Green River Formation: (Wasatchian–Duchesnean)
A. Powder Wash, Uinta Basin (Br1).
Refs: Krishtalka and Stucky, 1984.
B. Central Utah, Sanpete County (Du).
Refs: Nelson, Madsen, and Stokes, 1980.

CP5II. "Brian Head Formation" Sevier Plateau: (Duchesnean)
Turtle Basin Local Fauna.
Refs: Eaton *et al.*, 1999; Korth and Eaton, 2004.

CP6. Uinta Formation: (Uintan)
AA. Coyote Basin: Gnat-Out-of-Hell, locality WU-18 (Ui2).
Refs: Thornton and Rasmussen, 2001.
A. Wagonhound Member, Uinta Basin (Ui2).
Inc. White River Pocket, Willow Creek, and Kennedy's Basin.
Refs: Kay, 1957; Korth 1985, 1988; Prothero, 1996.
B. Myton Member, Uinta Basin (Ui3).
Inc. Devil's Playground, Leland Bench Draw, Myton Pocket, Leota Ranch Quarry, Kennedy Hole, and *Protylopus* Quarry.
Refs: Kay, 1957; Korth 1985, 1988; Prothero, 1996.

CP7. Duchesne River Formation, Uinta Basin: (Uintan–Duchesnean)
A. Brennan Basin Member, Randlett Fauna (Ui3).
Refs: Anderson and Pickard, 1972.
B. Dry Gulch Member, Halfway Fauna (Ui3).
Refs: Kay, 1957; Anderson and Pickard, 1972.
C. LaPoint Member, Twelve Mile Creek (Du).
Refs: Anderson and Pickard, 1972; Emry, 1981; Lucas, 1992; Prothero, 1996.

CP8. Norwood Formation: (Duchesnean)
Morgan Valley.

CP9. Keetley Volcanics: (Chadronian)
Peoa Local Fauna (Ch1–2).
Refs: Nelson, 1976.

CP10. Salt Lake Formation: (Hemingfordidan)
Goose Creek Valley (He2).

CP10II. Unnamed formation: (Blancan)
Unnamed loess unit in Box Elder County (E. Bl).
Refs: Nelson and Miller, 1990.
CP11. Unnamed formation: (Blancan)
Beaver Local Fauna (L. Bl).

WYOMING

CP11II. Ferris Formation, Hanna Basin: (Lancian–Puercan)
A. Main Section, Localities V-91010, V-91014, V-91015, V-91016, V-91031, V-92014, and V-92016 (Pu1).
Refs: Eberle and Lillegraven, 1998a,b.
B. Windy Mudstone Section, Locality V-91004 (Pu1).
Refs: Eberle and Lillegraven, 1998a,b.
C. Main Section, Localities V-91031, V-92014, and V-92016 (Pu2).
Refs: Eberle and Lillegraven, 1998a,b.
D. Windy Mudstone Section, Locality V-91005 (Pu2).
Refs: Eberle and Lillegraven, 1998b.
E. Main Section, Localities V-91003, V-92031, V-91019, V-91020, V-91021, and V-92009 (Pu2).
Refs: Eberle and Lillegraven, 1998a,b.
F. Main Section, Localities V-91002, V-91018, V-91022, V-92037, and V-92021 (Pu3).
Refs: Eberle and Lillegraven, 1998a,b.
G. Main Section, Localities V-92022, V-91028, V-91027, V-92034, V-92035, V-91026, V-92024, V-92025, and V-92026 (Pu3).
Refs: Eberle and Lillegraven, 1998a,b.
H. Main Section, Locality V-91024 (Pu3).
Refs: Eberle and Lillegraven, 1998b.
CP12. Fort Union Formation, Bighorn Basin: (Puercan–Torrejonian)
A. Mantua Lentil Local Fauna (Polecat Bench) (Pu1).
Refs: Van Valen, 1978; Gingerich, Rose, and Krause, 1980; Archibald *et al.*, 1987.
B. Leidy Quarry (Pu1).
Refs: Van Valen, 1978.
CP13. Fort Union (Polecat Bench) Formation, Bighorn and Clark's Fork Basins: (Puercan–Clarkforkian)
A. UW Localities V-81084, 82007, 82010, 82046 (Puercan undifferentiated).
B. Rock Bench Quarry, plus Hunt Creek area and Eagle Quarry (MT) (To3).
Refs: Gingerich, Rose, and Krause, 1980; Rose, 1981; Butler, Krause, and Gingerich, 1987; Sloan, 1987; Gunnell, 1989.
C. UW Localities V81050-51, 81053, 81055-63, 82004-6 (To3).
Refs: Hartman, 1986.
D. UW Localities SC-263, V-81047, V-82026 (Ti2).
Refs: Hartman, 1986.
E. Seaboard Well, Cedar Point Quarry, Jepsen Quarry, and various UM Localities (Ti3).
Refs: Van Valen, 1978; Gingerich, Rose, and Krause, 1980; Rose, 1981; Gingerich, 1983.
EE. "False Lance" Locality (Ti3–5).
Refs: Secord, 1998.
F. Airport Locality (Long Draw Quarry), Lower Sand Draw, Witter (= Croc Tooth) Quarry, Divide Quarry, Sand Draw, Anthill Locality, UM SC-243, UM SC-261 (Ti4).
Refs: Rose, 1975: Gingerich, 1976; Archibald *et al.*, 1987.
G. Princeton, Schaff, Fritz, and Y2K Quarries, Fossil Hollow, Brice Canyon, Princeton Storm Locality, Jepsen Valley, Middle Sand Draw, Sunday Locality, and various UM Localities (Ti5).
Refs: Gingerich, Rose, and Krause, 1980; Rose, 1981; Schaff, 1985; Archibald *et al.*, 1987; Secord, 2002.
H. Reis Locality, Rough Gulch, Foster Gulch Cleopatra Reserve Locality, Foster Gulch Oil Well No. 1. (Cf2).
Refs: Rose, 1981.
CP13II. Hanna Formation (Carbon County): (Torrejonian–Tiffanian)
A. Grayson Ridge Fauna, lower part (inc. The Breaks) (To3–Ti1).
Refs: Secord, 1998.
AA. Grayson Ridge Fauna, middle part (Ti2).
Refs: Secord, 1998.
A1. Grayson Ridge Fauna, upper part (Ti3).
Refs: Secord, 1998.
B. Halfway Hill (To3–Ti1).
Refs: Secord, 1998.
C. Sand Creek Fauna (Ti3–Ti5).
Refs: Secord, 1998.
CP14. Fort Union Formation, Washakie Basin: (Torrejonian–Clarkforkian)
A. Swain Quarry (To3).
Refs: Rigby, 1980; Sloan, 1981; Archibald *et al.*, 1987; Krause, 1987; Gunnell, 1989.
B. Eastern Rock Springs Uplift (inc. UW Localities V-77009-10, 77012, 77014, 78055) (To3).
Refs: Winterfield, 1982.
C. Eastern Rock Springs Uplift (*Plesiadapis churchilli* zone) (inc. UW Localities V-7705-08, 7713, 77015-16, 77061) (Ti4).
Refs: Winterfield, 1982.
D. Eastern Rock Springs Uplift (*Plesiadapis simonsi* zone) (inc. UW Localities V-76008, 79059-60, 78052-54) (Ti5).
Refs: Winterfield, 1982.
E. Big Multi Local Fauna (Cf2).
Refs: Rose, 1981; Archibald *et al.*, 1987; Wilf *et al.*, 1998.
F. Ten Mile Draw; Twelve Mile Well Fauna (Cf2) **(previously CP17C)**.

Refs: Anemone, Johnson, and Rubick, 1999; Anemone *et al.*, 2000.

CP15. Fort Union Formation, Shotgun Member, Wind River Basin: (Tiffanian–Clarkforkian)

A. Shotgun Local Fauna (Keefer Hill, = Twin Buttes) (Ti1).

Refs: Sloan, 1987; Gunnell, 1989.

B. "Malcolm's Locality" (= the Badwater Locality) (Ti4).

Refs: Krishtalka, Black, and Riedal, 1975; Gingerich, 1979.

C. West Side of Shotgun Butte (Cf2).

Refs: Gazin, 1971.

CP16. Fort Union Group, Bison Basin: (Tiffanian)

A. Saddle Locality (Ti2).

Refs: Gazin, 1956, 1971; Beard, 2000.

B. Ledge Locality, Saddle Annex, and West End (Ti3).

Refs: Gazin, 1956, 1971; Beard, 2000.

C. *Titanoides* Locality (Ti5).

Refs: Gazin, 1956, 1971.

CP17. Fort Union and Willwood Formations, Bighorn-Clark's Fork Basins: (Tiffanian–Wasatchian)

A0. *Plesiadapis gingerichi* zone (inc. UW Localities SC-178, 191, 180, 193, and 181) (Ti6–Cf1).

Refs: Gingerich, 2001; Gingerich and Clyde, 2001.

A. Rodent Interval zone (**previously *Plesiadapis gingerichi* zone**) (Cf1).

Inc. Little Sand Coulee, various UM Localities (inc. Eastern Clark's Fork Basin and Badger Hill), and Bear Creek (MT) (inc. Silver Coulee Beds).

Refs: Rose, 1981; R. W. Wilson, 1986; Archibald *et al.*, 1987; Gingerich, 2001; Gunnell and Bartels, 2001a.

B. *Plesiadapis cookei* zone, plus *Phenacodus-Ectocion* Acme zone (Cf2–3).

Inc. *Franimys* Hill, Phil's Hill, Paint Creek, Krause Quarry, Holly's Microsite, Upper Sand Draw and various UM Localities (inc. Discovery site [SC-29], *Carpolestes* Skull Site, *Ectocion* Site, and Unionid Coquina site).

Refs: Rose, 1981; Korth 1984; Krause, 1986; Archibald *et al.*, 1987; Ivy, 1990; Gingerich, 2001; Gunnell and Bartels, 2001a.

CP18. Willwood Formation, Bighorn-Clark's Fork Basins: (Tiffanian–Clarkforkian)

A. Various UM Localities (Ti6).

Refs: Archibald *et al.*, 1987.

B. *Phenacodus–Ectocion* Acme zone (Cf3).

Inc. Granger Mountain, Rainbow Valley, and various UM Localities.

Refs: Rose, 1981; Archibald *et al.*, 1987; Ivy, 1990; Gingerich and Clyde, 2001.

CP19. Willwood Formation, Clark's Fork Basin Area: (Wasatchian–Bridgerian)

AA. Sand Coulee Sites (inc. Polecat Bench, McCullough Peaks area [*C. torresi* zone]) (Wa0).

Refs: Gingerich 1989, 2001.

A. Sandcouleean sites (Wa1–2) (**previously just CP19**). Inc. Bighorn River area near Worland, Foster Gulch, Hole-in-the-ground, McCullough Peaks, Sand Coulee area, and Twisty-Turn Hollow (*C. mckennai–C. ralstoni* zone).

Refs: Bown and Rose, 1987; Gingerich, 1987, 1989; Ivy, 1990.

B. Early Graybullian sites (*C. mckennai* zone) (Wa2–3).

Refs: Ivy, 1990.

C. Middle–late Graybullian sites (*C. trigonodus* zone) (Wa4–5).

Refs: Ivy, 1990; Gingerich, 2001.

D. Lysite sites, McCullough Peaks (*C. abditus* zone) (Wa6).

Refs: Clyde, 2001.

E. Lostcabin sites, McCullough Peaks (Wa7).

Refs: Clyde, 2001.

CP19II. Willwood Formation, Clark's Fork Basin: (Wasatchian)UM Locality SC-265, three miles south of Jepsen's Camp 1 (Wa4–5).

CP20. Willwood Formation, Bighorn Basin: (Wasatchian–Bridgerian)

AA. Castle Gardens, Honeycomb area (Wa0).

Refs: Strait, 2001, 2003.

A. "Sandcouleean" and "Gray Bull Beds," *Haplomylus–Ectocion* Range zone (Wa0–4).

Inc. Sand Creek Facies (= No Water Creek area) and Elk Creek Facies, Foster Gulch area.

Refs: Bown, 1979, 1980; Bown *et al.*, 1993.

B. "Gray Bull Beds," *Bunophorous* Interval zone (Wa5).

Refs: Schankler, 1980.

BB. Wasatchian Quarry localities (inc. Rose, McKinney, McNeil, and Dorsey Creek Quarries) (Wa4–5).

Refs: Silcox and Rose, 2001.

C. Lysite Beds, *Heptodon* Range zone (Wa6).

Refs: Schankler, 1980.

D. Lost Cabin Beds, Elk Creek Section, *Lambdotherium* Range zone (Wa7).

Refs: Schankler, 1980.

E. Wapiti I and II Faunas: North Fork of Shoshone River, Absaroka Range (Wa7).

Refs: Torres, 1985; Gunnell *et al.*, 1992; Gunnell and Bartels, 2001a.

F. Wapiti III Fauna: North Fork of Shoshone River, Absaroka Range (Br1).

Refs: Torres, 1985; Gunnell *et al.*, 1992; Gunnell and Bartels, 2001a.

CP21. Evanston Formation, Fossil Basin: (Tiffanian)

A. Little Muddy Creek faunule (Ti1).

Refs: Gazin, 1969.

B. Twin Creek (Ti3).

Refs: Gazin, 1969.

CP22. Hoback Formation, Hoback Basin: (Tiffanian–Clarkforkian)

A. Battle Mountain, UM-Sub-Wy Localities 21 and 22 (Ti3).
Refs: Dorr, 1978; Gingerich, 1983.
B. Dell Creek Quarry, UM-Sub-Wy Locality 1 (Ti5).
Refs: Dorr, 1978; Gingerich, 1980, 1983; Gingerich and Winkler, 1985.
C. UM-Sub-Wy Localities 7, 10, 20 (Cf2).
Refs: Dorr and Gingerich, 1980; Rose, 1981.

CP23. Wasatch and Pass Peak Formation, Hoback Basin: (Wasatchian)
A. UM-Sub-WY Localities 2, 4, 16, 28, 29 (Wa2–3).
Refs: Dorr, 1978.
B. UM-Sub-Wy Localities 23 and 27 (Wa4–5).
Refs: Dorr, 1978.

CP24. Wasatch Formation, Chappo Member: (Tiffanian–Wasatchian)
A. Chappo Type Locality (= Chappo-17) (Ti3).
Refs: Gingerich, 1979, 1983; Dorr and Gingerich, 1980.
B. Buckman Hollow Locality (= Chappo-1 and Chappo-12) (Cf2).
Refs: Dorr and Gingerich, 1980.
C. Chappo Oil Well (Wa1).
Refs: Dorr and Gingerich, 1980.

CP25. Wasatch Formation: (Wasatchian–Bridgerian)
A. Bitter Creek Area, Zonule 2, Washakie Basin, Bitter Creek Promontory section, plus Four Mile Area (also extends into Colorado) (Wa1–4).
Refs: Savage, Waters, and Hutchison, 1972; Korth, 1984.
B. Powder River Basin (inc. Dry Well and Bozeman Localities, Monument Blowout, and Reculusa Blowout) (Wa1–4).
Refs: Delson, 1971.
C. Bitter Creek and Red Desert Local Faunas, Knight Member (= Main body of Wasatch Formation), Washakie Basin (Wa1–4).
Refs: Gazin, 1962.
D. West of Elk Monument Local Fauna, Knight Member (= Main body of Wasatch Formation), Fossil Basin (Wa1–4).
Refs: Gazin, 1962.
E. Knight Station and Fossil Butte Local Faunas, Knight Member (= Main body of Wasatch Formation), Fossil Basin (Wa5–6).
Refs: Gazin, 1962.
F. Dad Local Fauna, Knight Member (= Main body of Wasatch Formation) (Wa7).
Inc. Washakie Basin, Niland Tounge, Patrick Draw, Zonule S and O, Turtle Graveyard, Table Rock, and *Arapahovius* Fauna.
Refs: Gazin, 1962 Savage, Waters, and Hutchison, 1972; Savage and Waters, 1978; Stucky, 1984b.
G. LaBarge Local Fauna, Knight Member (= Main Body of Wasatch Formation), and West of Rock Springs, Green River Basin (Wa7).
Refs: Gazin, 1962.
GG. Main Body of Wasatch Formation, South Pass (The Pinnacles): Freiegher Gap Fauna (Wa7).
Refs: Anemone *et al.*, 2000; Gunnell and Bartels, 2001b.
H. New Fork Tongue, La Barge area, Green River Basin (Wa7).
Refs: Gazin, 1962; West, 1973a; Stucky, 1984b.
I. Cathedral Bluffs Tongue, Washakie Basin (Br1) (**note emended dating, previously Wa7**).
Refs: Gazin, 1962; West and Dawson, 1973; Schoch and Lucas, 1981; Korth, 1984.
I1. Little Muddy Fauna I, inc. Desertion Point (Wa7).
Refs: Zonneveld, Gunnell, and Bartels, 2000; Zonneveld and Gunnell, 2003.
I2. Little Muddy Fauna II (Wa7).
Refs: Zonneveld, Gunnell, and Bartels, 2000.
J. Cathedral Bluffs Tongue, Green River Basin (inc. Streckfus Draw, Cottonwood Creek area, Green Locality, including little Muddy IV) (**not including Honeycomb Buttes or South Pass, now = J2**) (Br1).
Refs: West, 1973a; West and Dawson, 1973; Korth, 1984; Honey, 1988; Zonneveld, Gunnell, and Bartels, 2000.
J2. Cathedral Bluff Tongue, Green River Basin (inc. Little Muddy Fauna V, and South Pass, Honeycomb Buttes) (Br1).
Refs: Zonneveld, Gunnell, and Bartels, 2000; Gunnell and Bartels, 2001b.

CP26. Togwotee Pass Area, Wind River Basin: (Tiffanian–Wasatchian)
A. Love Quarry (Ti3).
Refs: McKenna, 1980a; Archibald *et al.*, 1987.
B. "Low Locality" and "Rohrer Locality" (Ti6–Cf1) (**note emended dating, previously just Ti6**).
Refs: McKenna, 1980a; Archibald *et al.*, 1987.
C. Lower variagated sequence: Purdy Basin, Sheridan Pass Quadrangle, and Red Creek (Cf2).
Refs: McKenna, 1980a; Archibald *et al.*, 1987.
D. Lower variagated sequence: Hardscrabble Creek, Lava Mountain, and Tripod Peak Quadrangle, north bank of Papoose Creek (Wa2–3).
Refs: McKenna, 1980a.

CP27. Wind River Formation: (Wasatchian–Bridgerian)
A. Cooper Creek Basin, Laramie Basin (Wa3–5).
Refs: Princhinello, 1971; Davidson, 1987.
B. Lysite Member (inc. Shirley Basin) (Wa6).
Refs: Guthrie, 1967; Harshman, 1972; Korth, 1984; Stucky, 1984b.
C. Lost Cabin Member, Lower part, Boysen Reservoir area (Wa7-Br1).
Refs: White, 1952.
D. Lost Cabin Member, Lower part = *Lambdotherium* Range zone, Wind River Basin (Wa7).
Inc. Red Creek facies, (lower) Deadman Butte section, Arminto Unit, Buck Spring, Thirty-One Cent, Pot Luck, Bunny Foot, and Halfway Draw Quarries, Pavillion Butte, and *Viverravus* Locality.

Refs: Guthrie, 1971; Korth and Evander, 1982; Stucky and Krishtalka, 1982; Stucky, 1984a,b; Stucky, Krishtalka, and Redline, 1990.

E. Lost Cabin Member, Upper part = *Palaeosyops* Assemblage-zone (Huerfano B Fauna) (Br0) (**note emended dating, previously Br1**).

Inc. (upper) Deadman Butte, Rainbow Butte and Davis Ranch (= Sullivan Ranch) sections.

Refs: Stucky 1984a,b; Dawson, Krishtalka, and Stucky, 1990.

CP28. Indian Meadows Formation: (Wasatchian)

A. Lysite Member, Northwest Wind River (Wa6).

Refs: Keefer, 1965.

B. Lost Cabin Member, Northwest Wind River (Wa7).

Refs: Keefer, 1965.

C. Shotgun Butte, Eastern Wind River Basin (Wa7).

Refs: Keefer, 1965; Archibald *et al.*, 1987.

CP29. Wagonbed Formation, Wind River Basin: (Bridgerian–Orellan or Whitneyan)

A. Localities 17 and 18, Badwater area (inc. Shirley Basin) (Br2) (**note emended dating, previously Br1**).

B. Locality 1, Badwater area (Ui2).

C. Hendry Ranch Member, Localities 5, 5A, 6, 7, and associated localities, Badwater area (Ui3).

Refs: Black and Dawson, 1966; Krishtalka and Setoguchi, 1977; Black, 1978; Krishtalka, 1979; Dawson, 1980; Krishtalka and Stucky, 1983a.

D. Hendry Ranch Member, Wood and Rodent Localities and Locality 20, Badwater area (Du) (**note emended dating, previously Ui3**).

Refs: Black and Dawson, 1966; Krishtalka and Setoguchi, 1977; Black, 1978; Krishtalka, 1979; Dawson, 1980; Krishtalka and Stucky, 1983a; Maas, 1985.

E. Unnamed formation. overlying Hendry Ranch Member (Or4–Wh1).

Refs: Korth, 1981, 1994.

CP30. Tatman Formation: (Wasatchian)

Western Bighorn Basin (= “Pitchfork Formation”) (Wa7).

CP31. Aycross Formation: (Wasatchian–Bridgerian)

A. L- 41, Love’s Locality, Togwotee Pass area (Wa7–Br1).

Refs: McKenna, 1980a.

B. Togwotee Summit Fauna (= Landslide Locality), Togwotee Pass (Wa7–Br1).

Refs: McKenna, 1980a.

C. Northwest Wind River Basin (Br1).

Refs: Stucky, 1984a.

D. Wapiti IV Fauna: North Fork of Shoshone River (Br1).

Refs: Torres, 1985; Gunnell *et al.*, 1992; Gunnell and Bartels, 2001a.

E. Owl Creek–Grass Creek area, Bighorn Basin, plus Wapiti V Fauna (Br2).

Refs: Bown, 1982; Eaton, 1982; Gunnell *et al.*, 1992; Gunnell and Bartels, 2001a.

CP32. Wapiti Formation: (Wasatchian–Bridgerian)

A. West of Cody (?Wa7).

B. Carter Mountain (Br2).

Refs: Eaton, 1982.

CP33. Green River Formation: (Wasatchian–Bridgerian)

A. Fossil Lake, Upper Fossil Butte Member (inc. Tynsky Quarry) (Wa7) (**previously just CP33**).

Refs: Froelich and Breithaupt, 1990, 1997.

AA. Red Desert and Tipton Bluffs (Wa7).

B. Upper LaClede Bed, Laney Member, including Little Muddy Creek Fauna VI (?Br1).

Refs: Roehler, 1991.

C. Gosuite Lake, and Parnell Creek, Wilkins Peak Member (inc. Little Muddy Creek Fauna III) (?Br1).

Refs: Jepsen, 1966; Grande, 1984; Zonneveld, Gunnell, and Bartels, 2000.

CP34. Bridger Formation, Green River Basin: (Bridgerian–Uintan)

A. Bridger A Localities (Br1).

Inc. Opal Fauna, Little Muddy Creek Fauna VII, South Pass (Continental Peak), and Big Island-Blue Rim.

Refs: McGrew and Sullivan, 1970; Krishtalka and Stucky, 1983a; Gunnell and Bartels, 1994, 2001b; Gunnell, 1998; Zonneveld, Gunnell, and Bartels, 2000.

B. Blacks Fork Member (inc. New Fork–Big Sandy Area) (Br2).

Refs: West, 1973a; Korth 1985.

C. Blacks Fork Member (Br2).

Inc. East Hill and Trap-Hutch Quarries, Trap 72, George’s Gorge, *Notharctus* Knob, Fort Bridger, Grizzly Buttes, Smith’s Fork, Millersville, Cottonwood Creek, Granger Station, Sugar White, and *Omomys* Quarry.

Refs: Gazin, 1976; Stucky, 1984a; Korth 1985; Alexander and Burger, 2001; Murphey *et al.*, 2001.

D. Upper Bridger Formation, Twin Buttes Member, Tabernacle Butte (Br3).

Inc. Henry’s Fork, *Hyopsodus* Hill, Morrow Creek Locality, Sage Mountain and Stuck Truck, plus Behunin Locality and Misery Quarry.

Refs: McGrew *et al.*, 1959; McKenna, Robinson, and Taylor, 1962; West and Atkins, 1970; Gazin, 1976; West and Hutchison, 1981; Krishtalka and Stucky, 1983b; Walton, 1993a.

E. Bridger E. Fauna, Turtle Bluffs Member (Ui1).

Refs: Evanoff *et al.*, 1994; Murphey, 2001, 2005 (pers. comm.).

CP35. Fowkes Formation: (Bridgerian)

Bulldog Hollow Member, State Line Quarry, Thomas Canyon Local Fauna (also extends into northeastern Utah) (Br3).

Refs: Nelson, 1973; Eaton, 1982.

CP36. Tepee Trail Formation, North Fork of Owl Creek (= Late Basic Breccia): (Bridgerian–Uintan)

A. Lower Tepee Trail Formation (inc. Lower Holy City Beds) (Br3).

Refs: Eaton, 1985.
B. Upper Tepee Trail Formation (= Bone Bed area, East Fork River, Wind River Basin) (inc. Upper Holy City Beds and Foggy Day Beds) (Ui1).
Refs: MacFadden, 1980b; McKenna, 1980a; Eaton, 1985.

CP37. Wiggins Formation, North Fork of Owl Creek, Absaroka Mountains: (Uintan)
A. Lower Wiggins Formation (Ui2).
Refs: Eaton, 1985.
B. Upper Wiggins Formation (Ui3).
Refs: Eaton, 1985.

CP38. Washakie Formation, Washakie Basin: (Bridgerian–Uintan)
A. Kinney Rim Member (Br2).
Refs: Granger, 1909; McCarroll, Flynn, and Turnbull, 1996.
B. Lower Adobe Town Member, Washakie A (Br3).
Refs: Turnbull, 1972; Roehler, 1973; Archibald *et al.*, 1987; McCarroll, Flynn, and Turnbull, 1996.
C. Middle Adobe Town Member, Washakie B (Ui1).
Refs: Turnbull, 1972; Roehler, 1973; McCarroll, Flynn, and Turnbull, 1996; Foss, Turnbull, and Barber, 2001.
D. Upper Adobe Town Member (Ui2).
Refs: Granger, 1909; McCarroll, Flynn, and Turnbull, 1996.

CP39. White River Formation, Bates Hole area, Fremont County and Natrona County: (Duchesnean–Whitneyan)
A. Flagstaff Rim, 95–175 m below Ash A (Du–Ch1).
Refs: Emry, 1992.
B. Flagstaff Rim, below Ash B (inc. Lone Tree Gulch/Dry Hole Quarry, Dilts Ranch [lower section]) (Ch2–3) (**note emended dating, previously Ch1 or 2**).
Refs: Emry, 1978, 1979, 1992; Korth, Wahlert, and Emry, 1991; Emry and Korth, 1993, 1996.
C. Flagstaff Rim, Ash B to F (Ch3) (**note emended dating, previously Ch1 or 2**).
Refs: Emry, 1973, 1975, 1992; Emry and Korth, 1996.
D. Shirley Basin (west flank of Laramie Range) (Ch2).
Refs: Emry, 1973; Emry and Korth, 1996.
E. Cameron Springs Local Fauna, Beaver Rim (Ch2–3).
Refs: Hough, 1955; J. A. Wilson, 1971a,b. Emry, 1973, 1992; Emry and Korth, 1996.
F. Flagstaff Rim, upper part (Ash F to J) (?inc. Dishpan Butte) (Ch3–4) (**note emended dating, previously Ch2 or 3**).
Refs: Emry, 1992; Emry and Korth, 1996.
G. Bates Hole/ Ledge Creek (inc. Harshman Quarry) (Ch3).
Refs: Skinner and Gooris, 1966; Emry, 1973; Asher *et al.*, 2002.
H. Bates Hole (Dilts Ranch, upper section) (Orellan).
I. Bates Hole, inc. Hall Ranch (= "Phinney Butte") (Whitneyan).

CP39II. White River Formation: (Chadronian)
A. Red Fauna and West Canyon Creek (Ch1).
Refs: Skinner and Gooris, 1966; Lillegraven and Tabrum, 1983.
B. Big Sandstone Draw Lentil, Beaver Divide Area (Ch2).
Refs: Emry, 1975; Heaton, 1993.
C. Beaver Divide Conglomerate (?inc. near Toltee, Laramie Mountains) (Ch4).

CP40. White River Formation, Douglas Area, Converse County: (Chadronian–Orellan)
A. North Platte River drainage (Ch4).
Inc. Southeast of Douglas and Irvine Bridge area.
B. North Platte River drainage (Orellan).
Inc. southeast of Douglas, south of Pine Bluffs, Irvine Quadrangle, Reno Ranch, Morton Ranch, W. R. Silver Ranch, and Herman Wulff Ranch.
Refs: Heaton, 1993.

CP41. White River Formation, Lusk Area, Niobrara County: (L. Chadronian–Arikareean)
A. Seaman Hills Area, Hat Creek Drainage (Ch4).
Inc. Nuttall Rhino Quarry, Shack Draw Indian Creek, Whitman Post Office, Emerson Hines Ranch, Keel Ranch, Kelinke Ranch, McGinnis Ranch, Kraft Ranch Buildings, Lance Creek Basin, Boner Brothers Ranch, Piper Ranch, Spring Draw, Hat Creek Store, *Titanotherium* Channels (inc. North of Whitman and Thompson Ranch), plus (some of these may also extend into Orellan) Roadstop, Anderson Ranch, Jim Christian Hill, Hermit Ranch, and Lost Springs.
B. Lusk Area/Hat Creek Drainage (Orellan).
Inc. northeast of Walker Ranch, Niobrara County, Kraft Ranch, S-Bar Creek, Bald Butte, Indian Creek, East Shack Draw, Spring Draw, Highway 85, Old Woman Creek, DeGerring Ranch, Ant Hill Locality, Road Stop Locality.
Refs: Hough and Alf, 1956; Clemens, 1964; Korth and Emry, 1991.
C. Sherrill Hills (= Seaman Hills), Niobrara County, plus Tea Kettle Rock, Three Tubs Locality, Phinney Ranch (Whitneyan).
D. Early Arikareean site (Ar1).
Refs: Korth, 1993b.

CP42. Goshen Hole Area, Goshen County (inc. Yoder, Brule and Arikaree Formations): (Chadronian–Arikareean)
A. Yoder Local Fauna (Ch2).
Refs: J. A. Wilson, 1971a,b, 1974; Kihm, 1987.
B. Manganese Pocket, Goshen Hole (Ch2–3).
Refs: Emry, 1978.
C. Goshen Hole (Orellan).
Inc. Brule Equivalent, Fort Laramie, and Harvard Fossil Reserve.
D. 66 Mountain (Wh2–Ar1).
Inc. Brule Equivalent, Fort Laramie, and Dog Skull Locality.
Refs: MacDonald, 1970.

CP43. Unnamed formation: (Chadronian)
Mink Creek Local Fauna (Ch3).
Refs: McKenna, 1972; Prothero and Shubin, 1989.

CP44. Unnamed formation: (Chadronian)
Pilgrim Creek Local Fauna (Ch3).
Refs: Sutton and Black, 1975; Krishtalka *et al.*, 1987.

CP45. Colter Formation, Jackson Hole: (Chadronian–Barstovian)
A. Emerald Lake Fauna, Washakie Range (could be in White River Formation) (Ch2).
Refs: Love, McKenna, and Dawson, 1976; Prothero and Shubin, 1989.
B. Emerald Lake Fauna, Washakie Range (Ar2).
Refs: Barnosky, 1986.
C. East Pilgrim 11 Local Fauna (Ar4) (**note emended dating, previously Ar3–4**).
Refs: Barnosky, 1986.
D. East Pilgrim 5 Local Fauna and Saunders Local Fauna (He1).
Refs: Barnosky, 1986.
E. Cunningham Hill Fauna, North Pilgrim 2, and Two Ocean Lake (Ba2).
Refs: Barnosky, 1986.

CP46. Tepee Trail or Wiggins Formation: (Orellan) (**note emended dating, previously Whitneyan**)
Cedar Ridge Local Fauna.
Refs: Setoguchi, 1978; Lillegraven and Tabrum, 1983.

CP46II. Brule Formation: (Whitneyan)
Indian Stronghold, White River Group.

CP47. Unnamed formation: (Arikareean)
Darton's Bluff Local Fauna (Ar1).
Refs: McKenna, 1980a.

CP48. Gering Formation (or Arikaree Formation): (Arikareean)
Inc. Little Muddy Creek, Willow Creek area, North Platte River drainage, Horse Creek drainage, and Spanish Diggings (Ar1).
Refs: McKenna, 1972.

CP49. Arikaree Formation: (Arikareean–Barstovian)
A. Split Rock area, *Hypsiops* Locality (Ar1).
B. Bear Creek Mountain, Niobrara County, and Fox Creek Gap (Ar2).
C. Joe's Quarry, Laramie County (He2–Ba1).
D. Gravel Quarry, Goshen Hole (Ba2).

CP50. Monroe Creek Formation: (Arikareean)
Big Muddy Creek (Ar2).

CP51. Upper Harrison Beds (= "Marsland Formation," or Anderson Ranch Formation): (Arikareean–Hemingfordian)
A. Inc. Marsland equivalent (Niobrara County), south of Lusk, 16 Mile District, 18 Mile District, 25 Mile District (Goshen County), Sand Gulch (Goshen County), Wheatland area (Platte County), Guernesy area (Platte County), Royal Valley District (Niobrara County), 5 miles southeast of Chugwater (Platte County), near Rawhide Buttes, Lay Ranch Beds, Spoon Buttes, and Roll Quarry (Ar4).
B. Arikaree Group, North Platte River drainage, and Van Tassel (He1).
Refs: LACM collections.

CP52. Harrison Formation: (Arikareean)
Inc. 77 Hill Quarry, Niobrara County (Harrison equivalent), Riggs Rawhide Creek, North Ridge Locality, Keeline area, Harrison equivalent (inc. Z Quarry), and Van Tassel Creek (Ar3).
Refs: LACM collections.

CP53. Upper Harrison equivalent: (Arikareean)
Inc. 7 miles south of Chugwater, and Jay-Em District (Ar4) (**note emended dating, previously He-1**).

CP53II. Unnamed formation: (Hemingfordian) northeast side of pass on top of west end of east Seaman Hills, Niobrara County (He1).
Refs: Prothero, 2005.

CP54. Split Rock Formation: (Hemingfordian–Clarendonian)
A. Split Rock Local Fauna, Sweetwater Drainage, Lower Porous Sandstone sequence and Silty Sandstone sequence (He1).
B. Split Rock Local Fauna, Sweetwater drainage, Granite Mountain and Devil's Gate Fauna, Arikaree group, Upper Porous Sandstone sequence (He2).
Refs: Munthe, 1988; Korth, 1999b.
C. *Aleurodon* Locality and Love's 23V (Cl2).
Refs: Munthe, 1988.

CP55. Unnamed unit: (Barstovian)
10 miles northeast of Saratoga, Carbon County.

CP56. Ash Hollow Formation: (Barstovian)
Inc. Trail Creek Quarry Local Fauna and Escarpment Quarry (Ba2).
Refs: Forstén, 1970; Cassiliano, 1980.

CP56II. Moonstone Formation: (Barstovian–Hemphillian)
A. Castle Basin Local Fauna (Ba2–Cl1).
Refs: Prothero *et al.*, 2008.
B. Beef Acre Local Fauna (inc. Big Blowout section) (Ba2-Cl1).
Refs: Prothero *et al.*, 2007

CP58. Unnamed formation: (Hemphillian)
Kelley Road Local Fauna (Hh1).

CP59. Teewinot Formation: (Hemphillian–Blancan)
A. Teewinot Local Fauna inc. Kelly Road (Hh2).
Refs: Hibbard, 1959.
B. Thayne Local Fauna, Lincoln County (L. Bl).

CP60. Shooting Iron Formation: (Blancan)
Boyle Ditch Local Fauna (L. Bl).
Refs: Barnosky, 1985.

NORTHERN COLORADO

CP61. Denver Formation, Denver Basin: (Puercan)
A. Littleton Local Fauna (inc. Alexander, South Table

Mountain Localities), *Oxyclaenodon* Site, and Nicole's Mammal Jaw Locality (Pu1).
Refs: Middleton, 1983; Archibald *et al.*, 1987; Eberle *et al.*, 2002.
B. Corral Bluffs, Jimmy Camp Creek, and West Bijou Creek 1 (Pu2–3).
Refs: Archibald, 1982; Archibald *et al.*, 1987; Eberle *et al.*, 2002.

CP62. Wasatch or Debeque Formation, Atwell Gulch Member: (Tiffanian–Clarkforkian)
A. Hummer Locality, Piceance Creek Basin (Ti3–4).
Refs: Rose, 1981.
A2. Big Rock Ranch and Hell's Half Acre (Ti6–Cf1).
B. Plateau Valley, and Dry Rock Local Fauna, Piceance Creek Basin (Cf1).
Refs: Kihm, 1984; Archibald *et al.*, 1987.
C. Late Clarkforkian site (Cf3).
Refs: Kihm, 1984.

CP63. Wasatch Formation, Sand Wash Basin: (Wasatchian)
Four Mile Area, Washakie Basin (also extends into Wyoming) (Wa1–3).
Inc. Sand, Timberlake, Despair, Anthill, and Kent Quarries, East Alheit and West Alheit Pockets.
Refs: McKenna, 1960b; Korth 1984; Archibald *et al.*, 1987.

CP64. DeBeque Formation, Piceance Creek Basin (= Shire Member of Wasatch Formation): (Wasatchian–Bridgerian)
A. Rifle area, Piceance Creek Basin (Wa2–3).
Refs: Kihm, 1984; Krishtalka and Stucky, 1986.
B. Piceance Creek Basin (Wa5–6).
Refs: Kihm, 1984; Korth, 1984
C. Piceance Creek Basin (Wa7).
Refs: Kihm, 1984.
D. Piceance Creek Basin (?Br1).
Refs: Kihm, 1984.

CP64II. Uinta Formation, Piceance Creek Basin: (Uintan)

CP64III. Green River Formation: (Bridgerian–Uintan)
A. Citadel Plateau (?Br1).
B. Parachute Creek Member (?Ui2).

CP65. Washakie Formation: (Uintan)
Sand Wash Basin (Ui1).
Refs: West and Dawson, 1975; Eaton, 1982; Stucky *et al.*, 1996.

CP66. Antero Formation and Tallahasse Creek Conglomerate: (Chadronian)
South Park, Rocky Mountains, Middle Member (Ch4).

CP67. Florissant Formation and Castle Rock Conglomerate: (Chadronian)
Upper Part, South Park, Rocky Mountains, and Denver Basin (Ch3).
Refs: Evanoff and DeToledo, 1999.

CP68. White River Formation: (Chadronian–Whitneyan)
A. Kings Canyon, Medicine Bow Mountains (Ch3).
B. Horsetail Creek Member (Ch4).
Inc. Horsetail Creek Member Kings Canyon, Pawnee Buttes (inc. *Trigonias* Quarry [= *Titanotherium* Beds, Horizon A]), Chalk Bluffs, Two Miles Creek, Peevy Ranch, George Creek.
Refs: Galbreath, 1953.
C. Cedar Creek Member (Orellan).
Inc. Pawnee Buttes (lower part) (Lower *Oreodon* Beds), Pawnee Creek (lower and upper parts), Chimney Canyon, ?Chalk Bluffs, Hirsh Small Jaw Locality, Mellinger Locality, Greenley/Pawnee Buttes area, Stone Ranch, Casement Ranch, Clyde Ward Ranch, Jack Casement Ranch, Micro Locality, Beaver Locality, Chimney Rock, Deadman's Draw, Oscar Olander Ranch, and Freemont Butte.
Refs: Galbreath, 1953, Korth, 1989a,b; Asher *et al.*, 2002.
D. Vista Member (Wh2).
Inc. Pawnee Buttes (*Leptauchenia* Beds, Horizon C), plus Castle Rock.
Refs: Galbreath, 1953.

CP69. ?Rosebud Formation: (Arikareean)
Gerry's Ranch (Ar3–4).

CP69II. Split Rock Formation: (Arikareean)
Yampa, near Routt County (Ar3).
Refs: Izett, 1975; Rich, 1981.

CP70. Arikaree Formation: (Arikareean)
Pawnee Buttes (Ar1).

CP71. Martin Canyon Beds: (Arikareean–Hemingfordian)
Martin Canyon Local Fauna (He1).
Inc. Quarry A Local Fauna (= Running Water equivalent), Clay Quarry, Grover area, Weld County, and unnamed local fauna.
Refs: Galbreath, 1953; Martin and Mengle, 1984; Barnosky, 1986; Tedford *et al.*, 1987.

CP72. North Park Formation: (Hemphillian)
A. Big Creek Park (He1).
B. Piney River Valley (USGS D532b), plus Rickstrew Ranch (USGS D523a), and Spruce Gulch (He2).
Refs: Wang, Tedford, and Taylor, 1999.

CP73. Browns Park Formation: (Arikareean–Clarendonian)
A. Near Sunbeam, lower part of Brown's Park Formation (Ar3).
B. Cross Mountain Local Fauna (He1–2).
Refs: Honey and Alzett, 1988.
C. Weller Ranch (Ba2).
D. Cedar Springs Draw Local Fauna (Ba2).
Refs: Honey and Alzett, 1988.
E. Near Craig, Swelter Draw (Cl3).
Refs: Wang, Tedford, and Taylor, 1999.

CP74. Troublesome Formation: (Arikareean–Clarendonian)
A0. Granby, Substation, and Junction Localities (Ar3).

Refs: Wang, Tedford, and Taylor, 1999.
A. Middle Part (He1).
B. Upper Part (inc. UCM Highway 40 Site) (Ba1).
C. Independence Mountain (Cl2).

CP75. Pawnee Creek Formation: (Hemingfordian–Barstovian)
A. KU Locality 48 (He1).
B. Lower Pawnee Creek Fauna (Ba1). Inc. Pawnee Quarry, Eubanks Fauna, Big Springs Quarry, R. Day Ranch, Court House Butte, Question Mark Pit, and Pawnee Buttes area.
Refs: Galbreath, 1953; Tedford *et al.*, 1987; Wang, Tedford, and Taylor, 1999; Tedford, 2004; UCB collections.
BB. Keota Fauna (Ba2).
Refs: Tedford, 2004.
C. Upper Pawnee Creek Fauna (Pawnee Buttes area) (Ba2).
Inc. Horse and Mastodon Quarries, and West Quarry.
Refs: Tedford *et al.*, 1987.

CP76. Ogallala Formation, lower part: (Barstovian)
A. Gerrys Ranch (?Ba1).
Refs: Wang, Tedford, and Taylor, 1999.
B. Upper Pawnee Creek Fauna (Ba2) (**previously just CP76**).
Inc. Sand Canyon Fauna, Kennesaw and Vim-Peetz Local Faunas, Tapir Hill, Uhl Pit, and Martin Canyon, Cedar Creek.
Refs: Galbreath, 1953; Tedford *et al.*, 1987; Wang, Tedford, and Taylor, 1999.

CP76II. Unnamed formation: (Clarendonian)
Dearby Peak Fauna, Garfield County (Cl1).
Refs: Prothero, 2005.

CP76III. Unnamed formation: (Barstovian)
Cap Rock Member: Quarry Hill, John Weiss Ranch, 8 miles northeast of Keota (Ba2).
Refs: Prothero, 2005.

CP77. Unnamed formation: (Hemphillian)
Sunny Side Ranch (Hh1–2).

CP78. Ogallala Group: (Hemphillian)
Wray Fauna and Beecher Island Fauna (Hh2).
Refs: Cook, 1922; Frye, Leonard, and Swineford, 1956; Tedford *et al.*, 1987; Wang, Tedford, and Taylor, 1999.

CP80. Unnamed formation: (Blancan)
Donnelly Ranch (E. Bl) (**note emended dating, previously L. Bl**).
Refs: Hager, 1974; Kurtén and Anderson, 1980.

CP81. Unknown formation: (Blancan–Irvingtonian)
Porcupine Cave, South Park (L. Bl).

SOUTH DAKOTA

CP82. Slim Buttes Formation: (Duchesnean)
Antelope Creek Local Fauna.
Refs: Bjork, 1967; Lucas, 1992.

CP83. Chadron Formation: (Chadronian)
A. Ahearn Member, Big Badlands, inc. Battle Creek Breaks (Ch1).
Refs: Clark, 1937; Clark, Beerbower, and Kietzke, 1967; J. A. Wilson, 1974.
B. Crazy Johnson Member, Big Badlands (Ch3).
Inc. Indian Creek, Phinney Breaks, and Quinn Draw, Lower *Titanotherium* Beds.
Refs: Clark, 1937; Clark, Beerbower, and Kietzke, 1967.
C. Peanut Peak Member, Big Badlands (Ch4).
Inc. Granath Quarry, Flat Top Buttes, Slim Buttes, South of Lead, Black Hills, Big Corral Draw, Little Corral Draw, Bohling Ranch, Quinn Draw, Stebbins Ranch, Hutenmacher Table, Section 12, T42N, R45W, and Upper *Titanotherium* beds.
Refs: Clark, 1937; Clark, Beerbower, and Kietzke, 1967; J. A. Wilson, 1974; Prothero and Shubin, 1989.

CP84. Brule Formation: (Orellan–Arikareean)
A. Scenic Member, Big Badlands (Orellan).
Inc. Short Pine Hills, Lower and Middle Oreodon Beds (= *Metamynodon* channels), northwest of Slim Buttes (Unit A), Slim Buttes Units A to E, Big Corral Draw, Little Corral Draw, "Mauvais Terraces," Cedar Pass, Harris Ranch Badlands (Unit A), Red Shirt Table, Spring Creek, Sheep Mountain area, South Arrow Wound Table, Cain Creek, Bear Creek Basin, Chamberlain Pass, Cottonwood Pass area, Quinn Draw Basin, Cuny Table, Harney Spring Range, Babby Butte, and "Lower Nodular zone and Tyree Basin Locality."
Refs: Clark, Beerbower, and Kietzke, 1967; Lillegraven, 1970; Lemley, 1971; Clark and Guensberg, 1972; Bjork and Macdonald, 1981; Damuth, 1982; Simpson, 1985; Prothero and Shubin, 1989; Korth and Emry, 1991.
B. Poleslide Member, Big Badlands (Whitneyan).
Inc. Upper *Oreodon* Beds, northwest of Slim Butte (Units B and C), Slim Buttes (Units F–H), *Protoceras* Channels, Indian Stronghold, Big Corral Draw, Cedar Pass, Red Shirt Table, Rockyford, Bloom Ranch, Harris Ranch Badlands (Units B–D), Palmer Creek area, Sheep Mountain Table, *Leptauchenia* clays, Big Corral Draw, Cottonwood Creek, Battle Creek Draw, Wolff Ranch Badlands, Wolff Lake, and Hay Creek.
Refs: Lillegraven, 1970; R. W. Wilson, 1984; Simpson, 1985; Prothero and Shubin, 1989.
C. Harris Ranch Local Fauna (= Harris Ranch Badlands Unit E) (Wh2–Ar1).
Refs: Simpson, 1985.

CP84II. Paha Sapa Sandstone: (Oligocene)
Refs: Macdonald, 1982.

CP85. Sharps Formation: (Whitneyan-Arikareean)
A. Sharps Fauna A and Blue Ash Local Fauna (Wh2).
Refs: Martin, 1974; Tedford *et al.*, 1985.
B. Sharps Fauna B (inc. Quiver Hill localities and ?Rockyford Ash) (Ar1).
Refs: Tedford *et al.*, 1985; AMNH collections.

C. Wounded Knee/Sharps Fauna = Sharps Fauna C (Ar1).
Refs: Macdonald, 1970; Rensburger, 1980; Martin and Green, 1984; Korth, 1987.
D. Godsell Ranch Channell, Blue Ash Channel and Cedar Pass (Ar1–2).
Refs: Tedford *et al.*, 1996.

CP86. Rosebud Formation (= Arikaree Formation): (Arikareean–Hemingfordian)
A. Rosebud Fauna, lower part (Ar1).
Refs: Skinner, Skinner, and Gooris, 1968; LACM collections.
B. Upper Part of Rosebud Formation and Wounded Knee/Monroe Creek Fauna (inc. Wounded Knee/ Porcupine Creek area) (Ar2).
Refs: Macdonald, 1970; Martin and Green, 1984.
C. Wounded Knee/Harrison Fauna (inc. Eagle Nest Butte and Squaw Butte Area) (Ar3).
Refs: Macdonald, 1970; Martin and Green, 1984.
D. Wounded Knee/Rosebud Fauna (= Upper Rosebud Fauna) (inc. Porcupine Buttes) (Ar4).
Refs: Macdonald, 1970; Martin and Green, 1984.
E. Wounded Knee (He1).
F. Black Bear Quarry II (He1–2).

CP87. Valentine Formation (= Turtle Butte Formation): (Arikareean–Barstovian)
A. Wewela Fauna, Turtle Butte (Ar2).
Refs: Skinner, Skinner, and Gooris, 1968.
B. Burge Member: Turtle Butte (Ba2).
Refs: Skinner, Skinner, and Gooris, 1968.

CP87II. Unknown formation: (Miocene)
Rice Ranch Locality.
Refs: Tabrum, 1981.

CP88. Batesland Formation: (Hemingfordian)
Flint Hill Local Fauna and Black Bear Quarry I (He1).
Refs: Harksen and Macdonald, 1967; Macdonald, 1970; Martin 1976; Martin and Green, 1984; Barnosky, 1986; Tedford *et al.*, 1987.

CP89. Fort Randall Formation: (Barstovian)
Bijou Hills Local Fauna (inc. Glenn Olson Quarry) (Ba2).
Refs: Green and Holman, 1977; Green, 1985.

CP89II. Valentine Formation: (Barstovian)
Burge Member, Bailey Ranch (Ba2).
Refs: Prothero, 2005.

CP90. Ash Hollow Formation (= Thin Elk Formation): (Clarendonian)
A. Inc. Little White River, Big Spring Canyon Local Fauna, Bennett County, Rosebud Agency Quarry, Hollow Horn Bear Quarry, and Joe Thin Elk Gravel Pit (= Mission Local Fauna) (Cl2).
Refs: Gregory, 1942; Harksen and Green, 1971; UCB collections.
B. Wolf Creek area (Cl2).
Refs: Skinner and Johnson, 1984.

CP94. Hieb Sand Pit: (Blancan)
Delmont Local Fauna (L. Bl).
Refs: Martin and Harksen, 1974; Pinsof, 1985.

CP95. Herrik Formation: (Blancan)
Burke and Irene Gravel Pit Localities (L. Bl).
Refs: Pinsof, 1985.

CP96. Medicine Root Gravels: (Blancan)
Medicine Root Gravel Locality (L. Bl).
Refs: Pinsof, 1985.

CP97. Western Alluvium: (Blancan)
Pickstown and Wood Sand Pit Localities (L. Bl).
Refs: Pinsof, 1985.

CP97II. Unnamed formation: (Blancan–Irvingtonian)
Java Local Fauna, Walworth County (L. Bl or E. Ir).
Refs: Martin, 1989.

NEBRASKA

CP98. Chadron Formation: (Chadronian)
A. Chadron A (Ch2).
Inc. Chadron area, Sioux and Dawes Counties, and Hat Creek Basin.
B. Chadron B (Ch3).
Inc. Norman Ranch, Raben Ranch, Bone Cove, Dirty Ridge Creek Flats, Dirty Ridge, Twin Buttes, and Ant Hill Local Faunas.
Refs: Hough and Alf, 1956; Ostrander, 1985, 1987.
BB. Middle–Late Chadronian Sites (Ch3–4).
Inc. Arner Ranch (Sioux County), Warbonnet Creek (Sioux County), and Section 2., T32N, R52W (Dawes County).
C. Chadron C (Ch4).
Inc. *Chadronia* Pocket Local Fauna, Brecht Ranch Local Fauna (inc. Bartlett West 1, Rook Ranch [lower levels], White's south end, Hawthorne A and B, Schomner Ranch [lower level]), Henry Morgan Quarry, Pine Ridge (Toadstool Park), and Geike Ranch.
Refs: Prothero, Denham, and Farmer, 1982; Ostrander, 1985; Gustafson, 1986b; Korth, 1987; Heaton, 1993.

CP99. Brule Formation, White River Group: (Orellan-E. E. Arikareean)
A. Orella Member (Orellan).
Inc. Everson Ranch, Warbonnet Creek, Meng's Ranch (= Harrison Ranch), *Oreodon* Beds, Roundtop area, Hat Creek Basin (inc. Badland Creek, Prairie Dog Creek, Cedar Creek, Dry Creek), Dout Ranch, Whitehead Creek, Brecht Ranch area, Warbonnet Creek, Geike Ranch, Munson Ranch, Dead Horse Local Fauna (inc. Schomner Ranch [upper levels], Bartlett Ranch, White Ranch, Rook Ranch [upper levels], Pasture 34, Hawthorn Island, Horse Island and Central Turtle level), Rabbit Graveyard Local Fauna (inc. Bartlett south central, Rabbit Graveyard, Scratching Post, Central Marker bed, Anderson's lumpy level, Bartlett's lumpy level, Midwest Valley), and Bartlett High Local Fauna (inc. Bartlett High, Bartlett

Dike Valley [highest level], and Anderson High), 2.5 miles northeast of Chadron, and north of Zerbst Ranch.
Refs: McGrew, 1937; Patterson and McGrew, 1937; Gustafson, 1986b; Korth, 1987, 1989a,b,1994; Korth and Emry, 1991.
B. Whitney Member (Whitneyan).
Inc. North Platte River drainage (Lower and Upper Ash), Faunas I and II (below Nonpariel Ash), Eagle Nest Butte, Ruby Ranch, Scottsbluff Monument, Roberts Draw, Roundhouse Rock, Crow Butte, Chimney Rock, and Weitzel Ranch.
Refs: Tedford *et al.*, 1985; Korth, 1989a,b; Korth and Emry, 1991.
C. Lower and Upper Fauna III (= lower Sharps Formation) (Ar1).
Inc. Hat Creek–White River drainages, Brown Siltstone Beds, Niobrara River drainage, and White River Valley.
Refs: Tedford *et al.*, 1985.
D. Horn Member (Ar1–2).
Refs: Wang, Tedford, and Taylor, 1999.

CP100. Whitney Formation: (Whitneyan)
A. West of White Clay (lower Whitney Formation).
B. Crow Butte Northeast.

CP100II. Arikaree Formation: (Whitneyan–Arikareean)
Birdcage Gap (Mo-106) (Wh2–Ar1).

CP100III. Near base of Arikaree group: (Arikareean)
University of Nebraska State Museum locality Dw-121 (Ar1).

CP101. Gering Formation: (Arikareean)
Gering Fauna (Ar1).
Inc. Horse Creek area, Helvas Canyon Member Sites (North Platte River drainage), Dunlap Ranch, Durnal Ranch, Lone Skull, Nipple Butte, and Nerud's Thrasher Quarries, No-Name, Mike's Main Channel, Pumice Channel Prospect, Redding Gap Ant Hill 1 and 2, Round Rock Gap, Black Hanks Canyon, Court House Rock, Pumpkin Seed Creek, Redington Gap, Schomp Ranch, Wildcat Hills, Ridgeview, and Middlebranch Local Fauna.
Refs: Voorhies, 1973; Swisher, 1982; Korth and Bailey, 1992; Albright, 1998a; Bailey, 2003.

CP102. Monroe Creek Formation: (Arikareean)
Inc. Squaw Butte area, Niobrara River drainage, Warbonnet Creek, Wildcat Hill, and Tunnell Hill Locality (Ar2).
Refs: Martin, 1987.

CP103. Harrison Formation: (Arikareean)
A. Harrison Fauna (Ar3) (**previously just CP103**).
Inc. Peterson's Quarry A, *Syndyoceros* Quarry, Eagle Crag Local Fauna, *Stenomylus* Quarry, south side of Dry Creek, north of Bear Creek, Van Tassel area, and Olson Ranch Locality.
Refs: McDonald, 1970; McKenna and Love, 1972; Rensberger, 1979; Hunt, 1981, 1985; Skinner and Johnson, 1984; Skolnick, 1985; Martin, 1987; Tedford *et al.*, 1987; Stevens, 1991; Korth, 1993b; Wang, Tedford, and Taylor, 1999.
B. McCann Canyon Local Fauna (Ar3).

CP103II. Unknown formation: (Arikareean)
Stage Hill Local Fauna (Ar3).
Refs: Bailey, 1999.

CP104. Marsland Formation (= Upper Harrison Beds, or Anderson Ranch Formation): (Arikareean–Hemingfordian)
A. Agate Springs Local Fauna (Ar4).
Inc. University, Carnegie, Harper, Agate Spring, American Museum-Cook, and Dicerathere Quarries, Quarries 1 and 2.
Refs: Hunt, 1985; Forstén, 1991; Wang, Tedford, and Taylor, 1999; AMNH collections.
B. Morava Ranch Quarry, *Stenomylus* Quarry (= Marsland Quarry), and Niobrara Canyon Local Fauna (Ar4).
Refs: Hunt, 1981, 1985; Coombs, 1984; Tedford *et al.*, 1987.
C. Marsland Local Fauna (He1).
Refs: Hunt, 1985.

CP105. Runningwater Formation: (Hemingfordian)
Runningwater Fauna (He1).
Inc. Runningwater, Marsland, Hemingford, Dunlap Camel, Apollo, Potter, Felton Dam, Woods Canyon, *Aletomeryx* (= Antelope Creek), Cottonwood Creek, "B," and Brown Quarries, Middle Cottonwood Creek, South of Antelope Valley, Warren Barnum Ranch, Hovarka Locality, *Cephalogale* Ash Locality, Gregory Ranch, Upper Loup Fork Beds, Agate area, Stamen Ranch, Paul Neeland Ranch, Clinton Highway Locality, Cross Cut Prospect, Dry Creek Prospect, Pebble Creek region, Sand Canyon region, Hay Springs Creek, and Whistle Creek–Niobrara River divide.
Refs: Cook, 1965; Hunt, 1981; Tedford *et al.*, 1987; Korth, 1996a; Wang, Tedford, and Taylor, 1999.

CP106. Runningwater Formation Equivalent: (Hemingfordian)
Inc. Bridgeport Quarries, Concretion Channel Prospect, Site 2 Prospect, Pepper Creek, and J. L. Ray Ranch (He1).
Refs: Hunt, 1981; Skinner and Johnson, 1984.

CP106II. Channel in Runningwater Formation: (Hemingfordian)
Perissodactyl Prospect (?He2).
Refs: Prothero, 2005.

CP107. Box Butte Formation: (Hemingfordian)
Red Valley Member, Box Butte Fauna (He2).
Inc. Foley Quarry, Dry Creek Prospect A, B, and D, Middle of the Road Quarry, and Sand Canyon region.
Refs: Galusha, 1975; Tedford *et al.*, 1987.

CP108. Sheep Creek Formation: (Hemingfordian)
A. Lower Sheep Creek Fauna (He2).

Inc. Greenside, Long, and Companion Quarries, and Antelope Draw.
Refs: Skinner, Skinner, and Gooris, 1977; Tedford *et al.*, 1987; Korth, 1999b; Bailey, 1999, 2003.
B. Upper Sheep Creek Fauna (He2).
Inc. Hilltop, Thistle, Target, Buck, Ashbrook, Vista, Rhino, Conference, Ravine, Thomson, Above Long, and Above Greenside Quarries, Watson Ranch, Stonehouse, *Aphelops*, and *Pliohippus* Draws.
Refs: Skinner, Skinner, and Gooris, 1977; Tedford *et al.*, 1987; Wang, Tedford, and Taylor, 1999.

CP109. Hay Springs area: (Hemingfordian–Barstovian)
A. Temporal equivalent of Sheep Creek Formation (He2).
Inc. Ginn Quarry, Antelope Valley, and "G" Quarry.
Refs: Galusha, 1975; Wang, Tedford, and Taylor, 1999.
B. Temporal equivalent of Valentine Formation (Ba2).

CP110. Olcott Formation: (Barstovian)
Lower Snake Creek Fauna (Ba1).
Inc. Camel, Snake, Trojan, North Wall, East Wall, Douglas, Sinclair, East Sand, West Sand, New Sand, Sand, Jenkins, East Jenkins, West Jenkins, AMNH 1908, *Mesoceras*, Surface, East Surface, West Surface, Far Surface, New Surface, Version, Echo (= Antelope Draw), Humbug, South Humbug, Mill, Boulder, Grass Root, Sheep Creek, and *Prosynthetoceras* Quarries, Princeton Locality 1000c, Pocket 34, Kilpatrick Pasture, Ashbrook Pasture, Quarry A, Quarry B, Quarry 2, East Sinclair and West Sinclair Draws (plus Olcott equivalent, Hay Springs area).
Refs: Skinner, Skinner, and Gooris, 1977; Wang, Tedford, and Taylor, 1999; AMNH collections.

CP111. Sand Canyon Beds (Olcott Formation equivalent): (Barstovian)
Inc. Observation, Survey, Surprise, Hay Springs, and Pioneer Quarries, and School House Prospect, Pepper Creek (Ba1).
Refs: Tedford *et al.*, 1987; Wang, Tedford, and Taylor, 1999; AMNH collections.

CP112. Unnamed formation: (Barstovian)
Lighthill Locality (Ba1).

CP112II. Unnamed formation: (Barstovian)
Home Station Pass (Ba2).
Refs: Wang, Tedford, and Taylor, 1999.

CP113. Ogallala Formation: (Barstovian–Hemphillian)
A0. Inc. Driftwood Creek, Republican River Beds (Hitchcock County), Spatz Quarry (Knox County), Devil's Nest Road (Knox County), and 2.9 miles east of White Clay (Nebraska/South Dakota boundary) (Ba2).
Refs: Wang, Tedford, and Taylor, 1999.
A. Inc. UNSM Locality Wt-15A (Webster County) and UNSM Locality Fr-20 (Franklin County) (Ba2).
A1. Inc. Hardin Bridge, Niobrara River, plus Hitchcock County Site (Cl1).
Refs: Wang, Tedford, and Taylor, 1999; Prothero, 2005.
A2. Inc. 0.25 miles west of Wounded Knee Creek (Cherry County), Driftwood Creek (Hitchcock County), and Patton Ranch (Cl2, Cap Rock equivalent).
Refs: Wang, Tedford, and Taylor, 1999.
B. UNSM Locality Bn-103 (Banner County) (Clarendonian).
B2. Inc. Head of Blue Creek (UNSM loc. 25890), Turtle-Carnivore Quarry, and Medicine Creek (UNSM loc. 1084) (Hh2).
Refs: Wang, Tedford, and Taylor, 1999.
B3. Turtle Locality (Cheyenne County) (Hh3).
Refs: Wang, Tedford, and Taylor, 1999.
C. UNSM Locality Sh-101 (Shannon County) (Hh4).

CP114. Valentine Formation: (Barstovian–Clarendonian).
A. Cornell Dam Member, Norden Bridge Quarry Local Fauna (Ba2).
Inc. Egelhoff, Achilles, Kuhre, Norden Bridge (= Tunnel Rock Locality), Penbrook, Rocky Ford, Lost Chance, Immense Journey, Zippy Lizard, Carrot Top, and Hottell Ranch Rhino Quarries, Quarry Without a Name, Site Cr 121, Lost Duckling Site, Rossetta Stone Locality, Welke Locality, High in the Saddle Locality, Small Falls Prospect, Miller Creek, Fairfield Creek No. 1, and Corner's Gap.
Refs: Klingener, 1968; Skinner and Johnson, 1984; Tedford *et al.*, 1987; Voorhies, 1987, 1990a,b; Voorhies, Holman, and Xiang-Xu, 1987; UNSM collections.
B. Crookston Bridge Member, Niobrara River Local Fauna (Ba2).
Inc. Crookston Bridge (= *Merycodus*), Railway A, Railway B, Devil's Jump Off, Nenzel, Ripple, Runlofson, Sawyer, Schoettger, Stewart, West Valentine, Yale, Jamber, Four Gate, and Hottell Ranch Horse Quarries, Hazard Homestead (= "Republican River Faunas"), Annies Geese Crossing, Deer Fly Locality, Myers Farm, Forked Hills of Hayden, Fairfield Creek IV, Dutch Creek, Jones Canyon Site (inc. "Loup Fork Beds, Merriman Bridge, Garner Bridge, Bone Creek, Fort Niobrara, and Tihen Locality).
Refs: Stirton and McGrew, 1935; Skinner and Johnson, 1984; Tedford *et al.*, 1987; Voorhies, 1990b; Wang, Tedford, and Taylor, 1999; UNSM collections.
C. Devil's Gulch Member (Ba2).
Inc. Elliot, Deep Creek, Devil's Gulch, Devil's Gulch Horse, Logan, Mizner Slide, Fairfield Falls, Fairfield Creek, Horse Thief Canyon, Rattlesnake Gulch, and Verdigree Quarries.
Refs: Skinner and Johnson, 1984; Barnosky, 1986; Voorhies, 1990b: Wang, Tedford, and Taylor, 1999.
D. Burge Member, Burge and Penny Creek Faunas (Cl1).
Inc. June, Lucht, Burge, Gordon Creek, Buzzard Feather, Ewert, West June, South Lucht, Moore Creek, Midway, Paleo Channel, 379, Quinn Mastodon, Swallow, *Tetrabelodon* Skull, White Point, White Face, and Yale

Mastodon Quarries, UCMP Locality V-3312, Fence Line Locality, Deep Creek, Meisner Slide, Bug Prospect (inc. "Ainsworth" near Burge and Forest Reservation, Williams Canyon, and Extension Quarry, Burge Equivalent).
Refs: Webb, 1969a; Skinner and Johnson, 1984; Barnosky, 1986; Voorhies 1990b; Korth 1997; Wang, Tedford, and Taylor, 1999; UNSM collections.

CP115. Snake Creek Formation:
(Clarendonian–Hemphillian)
A. Murphy Member (Cl2).
Unnamed local fauna.
Refs: Skinner, Skinner, and Gooris, 1977.
B. Laucomer Member, Snake Creek Fauna (Cl3).
Inc. Kilpatrick, West Jenkins, Olcott (= *Hipparion* Channel) Quarries, AMNH Quarry 7, Sinclair Draw, and *Hesperopithecus* Site (= Olcott Hill).
Refs: Skinner, Skinner, and Gooris, 1977; Wang, Tedford, and Taylor, 1999.
C. Johnson Member, *Aphelops* Draw Fauna (Hh2).
Inc. *Aphelops* Quarry and "The Pits."
Refs: Skinner, Skinner, and Gooris, 1977; Tedford *et al.*, 1987.
D. Johnson Member (Hh3).
Inc. ZX Bar Local Fauna and *Pliohippus* Draw.
Refs: Matthew and Cook, 1909; Skinner, Skinner, and Gooris, 1977.

CP116. Ash Hollow Formation: (Barstovian–Hemphillian)
A0. Andrew Hotell Ranch, Banner County (Ba2).
Refs: Voorhies, Holman, and Xiang-Xu, 1987.
A. Cap Rock Member, Minnechaduza Fauna (Cl2).
Inc. Little Beaver B, Rock Ledge Mastodon, Poison Ivy, Chokecherry, Clayton, East Clayton, Rhino Horizon 3, Quinn Rhino, Messinger, Garner, Big Beaver A, and Wilson Quarries, Medicine Creek, McGinley's Stadium, Fairfield Creek No. 3, Horse Thief Canyon 1 and 2, Johnson Rhino Site, Jonas Wilson Ranch, Horton Ranch, Martin Arent Ranch, Harvey Williams Ranch, Phalen Ranch, Plum Creek, Willow Creek, Deep Creek, Boiling Spring Flat, Thayer Ranch, Timm Ranch Site (= Prospect Locality 288), North Rim Locality, and Crazy Locality (inc. Box Butte Prospect, Davis Creek, and Deer Creek).
Refs: Webb, 1969a; Skinner and Johnson, 1984; Tedford *et al.*, 1987; Voorhies, 1990b; Korth 1997, 1998; Wang, Tedford, and Taylor, 1999; Prothero, 2005; UNSM collections.
AA. Ash Hollow Formation/Ogallala Formation (Clarendonian undifferentiated).
Inc. Lessig Canyon (Brown County), Dutch Canyon (Brown County), Turkey Creek (Keya Paha County), and Turtle Canyon (Sheridan County).
B. Merritt Dam Member (Cl2–3).
Inc. Eggers, Emry, Horn, Hurlburt, Jonas Wilson, *Platybelodon* (UNSM Cr22), Pratt, Pratt Slide, Wade, Blue Jay, Kepler, White Point, Fat Chance, Serendipity, Precarious, Bear Creek, Gallup Gulch, Miller, and Old Wilson Quarries, Jim Lessig Camel Site, Eli Ash Pit, North Shore Locality, Steer Creek High Channel, Dierex Brothers Ranch, Beed Place, Shoreline Local Fauna, Xmas-Kat Channels Fauna (inc. Balanced Rock, Xmas, Hans Johnson, *Machaerodus*, *Leptarctus*, Connection Kat, East Kat, Kat, West Line Kat, Quarter Line Kat, Trailside Kat, and West Kat Quarries), Broadwater and Harrisburg Localities, Lonergan Creek, and North Shore.
Refs: Skinner and Johnson, 1984; Leite, 1990; Voorhies, 1990b; Korth 1998; Wang, Tedford, and Taylor, 1999; UNSM collections.
C. Inc. Feltz Ranch, Lemoyne Local Fauna, and Ogallala Beach Local Fauna (Hh1).
Refs: Bown, 1980; Tedford *et al.*, 1987; Leite, 1990; Voorhies, 1990b; UNSM collections.
D. Inc. Kimball Fauna, Cambridge Local Fauna (= UNSM Locality Ft-40, Frontier County), Oshkosh Local Fauna, Potter Quarry, *Amebelodon fricki* Quarry, and Greenwood Canyon (Hh2).
Refs: Tedford *et al.*, 1987; Voorhies, 1990b; UNSM collections.
E. Inc. Bear Tooth, Mailbox, Honey Creek, Dalton, and Uptegrove Local Faunas (Hh3).
Refs: Tedford *et al.*, 1987; Voorhies, 1990b; UNSM collections.
F. Santee and Devil's Nest Airstrip Local Faunas (Hh4).
Refs: Voorhies, 1990b; UNSM collections.

CP116II. Unknown formation: (?Hemphillian)
Sand Canyon.

CP116III. Unknown formation: (Hemphillian)
Rich Irwin Site, Wyman Creek Local Fauna (Hh3).
Refs: Tucker, 2003.

CP117. Broadwater Formation: (Blancan)
A. Lisco Local Fauna (E. Bl).
Refs: Schultz and Stout, 1948; Kurtén and Anderson, 1980; Voorhies and Corner, 1986.
B. Broadwater Local Fauna (E. Bl) **(note emended dating, previously L. Bl)**.
Refs: Schultz and Stout, 1948; Hager, 1974; Kurtén and Anderson, 1980.

CP118. Kiem Formation: (Blancan)
Sand Draw Local Fauna (= *Stegomastodon* Quarry) (E. Bl) (**note emended dating, previously L. Bl**).
Refs: Hager, 1974; Kurtén and Anderson, 1980; Voorhies and Corner, 1986.

CP119. Unnamed formation: (Blancan)
Seneca Local Fauna (L. Bl).

CP120. Unnamed formation: (Blancan)
Mullen Local Fauna (L. Bl).

CP121. Long Pine Formation: (Blancan)
Big Springs Local Fauna and Hall Gravel Pit (L. Bl).
Refs: Voorhies, 1987; Wang, Tedford, and Taylor, 1999.

CP121II. Unknown formation: (Blancan)
Sappa Fauna, Harlan County (L. Bl).
Refs: Bell *et al.*, 2004.

KANSAS

CP123. Ogallala Formation: (Barstovian-Hemphillian)
A0. Silica Mine, Calvert (?Ba2).
Refs: Wang, Tedford, and Taylor, 1999.
A. North Fork of Solomon Creek (= Sappa Creek) (Cl2) (**Wakeeney Fauna now = CP123E**).
Refs: R. L. Wilson, 1968.
B. North Fork of Solomon Creek (inc. Republican River Beds) and Selby Ranch Quarry (Cl3).
Refs: Barnosky, 1986; Wang, Tedford, and Taylor, 1999.
C. Inc. Jack Swayze, John Dakin, Long Island, Arens, Rhino, Rhino Hill Quarries, Young Brothers Ranch, Sappho Creek, Hall Cope's Ranch, Beaver Creek area, and Sebastian Place (Hh1).
Refs: Hulbert, 1987a; Nelson and Madsen, 1987; Wang, Tedford, and Taylor, 1999.
D. Inc. Edson, Lost, and Found Quarries (Hh3).
Refs: Frye, Leonard, and Swineford, 1956; Bennett, 1979; Harrison, 1983.
E. Wakeeney Local Fauna (Hh3) (**previously CP123A; note emended dating, previously Hh2**).
Refs: R. L. Wilson, 1968.
F. Gretna (Ba2–Cl).
Refs: Zakrzewski, 1988.
G. Bemis (Hemphillian).
Refs: Zakrzewski, 1988.

CP124. Unnamed formation: (Clarendonian)
Ashland, Clark County Site (Cl2).

CP124II. Unnamed formation: (Clarendonian)
Densmore–Voss Quarry (Cl3).
Refs: Hulbert, 2005 (pers. comm.).

CP124III. Unnamed formation: (Clarendonian)
Keller Local Fauna (Cl3).
Refs: Hulbert, 2005 (pers. comm.).

CP125. Unnamed formation: (Hemphillian)
Smith County site and Trail Canyon, south fork of Driftwood Creek (Hh1–2).

CP126. Ash Hollow Formation: (Hemphillian)
Minium Quarry (Hh2).
Refs: Thomasson *et al.*, 1990.

CP26II. Unnamed formation: (Hemphillian)
Cogswell Quarry (Hh3–4).

CP127. Delmore Formation: (Hemphillian)
Kinkerman's sand pit and Moundridge gravel pit (Hh4).
Refs: Hibbard, 1952a.

CP128. Rexroad Formation: (Blancan)
AA. Saw Rock Canyon Local Fauna, Seward County (**previously CP123E**), plus Argonaut and Fallen Angel Local Faunas (? + XIT1 and 2) (E. Bl).
Refs: Hibbard, 1949, 1952a,b; Kurtén and Anderson, 1980; Martin, Honey, and Peláez-Campomanes, 2000.
A. Fox Canyon Local Fauna (E. Bl).
Refs: Hibbard, 1950; Dalquest, 1978; Kurtén and Anderson, 1980; Martin, Honey, and Peláez-Campomanes, 2000.
B. Keefe Canyon Local Fauna (E. Bl).
Inc. Wiens, Vasquez, Raptor 1A-C, Raptor 2, Bishop, Alien Canyon, East of Alien Canyon, Ripley, Newt, and Hornet.
Refs: Hibbard, 1950. Kurtén and Anderson, 1980; Martin, Honey, and Peláez-Campomanes, 2000.
C. "Rexroad Local Fauna" (E. Bl).
Inc. Wendell Fox Pasture (Meade County), Bender 1b and 1c (**previously CP128D**), Rexroad localities 2, 2a, 3, Deer Park Local Fauna (**previously CP130**), and "Seeger Local Fauna" (**previously CP132A**).
Refs: Hibbard, 1941a,b, 1950; Bjork, 1973; Dalquest, 1978; Kurtén and Anderson, 1980; Martin, Honey, and Peláez-Campomanes, 2000; Martin *et al.*, 2002.
D. Paloma Local Fauna (L. Bl).
Refs: R. A. Martin, (2006, pers. comm.).

CP129. Unnamed formation: (Blancan)
Hart Draw Local Fauna (E. Bl).

CP130. Ballard Formation: (Blancan)
Sanders Local Fauna (**previously CP130B, original CP130 now included in CP128C**) (E. Bl).
Refs: Hibbard, 1950, Hager 1970; Martin, Honey, and Peláez-Campomanes, 2000.

CP131. Belleville Formation: (Blancan)
White Rock Local Fauna, Republic County (L. Bl).
Inc. Hibbard, Sandpit and South Localities.
Refs: Eshelman, 1975; Kurtén and Anderson, 1980.

CP131II. Belleville equivalent: (Blancan)
Dixon Local Fauna (L. Bl) (**previously included in CP131**).
Refs: Hibbard, 1956.

CP132. Crooked Creek Formation: (Blancan–Irvingtonian)
A. Atwater Member: Margaret 1 and NBA (**Seger Local Fauna now included with CP128C**) (L. Bl).
Refs: Martin, Honey, and Peláez-Campomanes, 2000.
B. Atwater Member: Borchers Local Fauna (L. Bl).
Refs: Hibbard, 1941c; Hager, 1974; Bayne, 1976; Kurtén and Anderson, 1980; Zakrzewski, 1981; Martin, Honey, and Peláez-Campomanes, 2000; P. A. Martin (2006, pers. comm.).
C. Short Haul and Aries A (L. Bl–E. Ir).
Refs: Martin *et al.*, 2003.
D. Aries NE, Nash 72, Rick Forester, Aries B (L. Bl–E. Ir).
Refs: Martin *et al.*, 2003.

CP132II. Unnamed formation: (Blancan–Irvingtonian)
Unnamed fauna from Section 3, T1S, R10W, Jewell County (L. Bl–E. Ir).
Refs: Goodwin, 1995.

NORTHERN GREAT PLAINS

ALBERTA

NP1. Porcupine Hills Formation: (Puercan–Tiffanian)
A. Alberta Corehole 66-1 (= Balzac West) (?Pu2).
Refs: Fox, 1990.
B. Calgary 2E and Calgary 7E, Elbow River (To3).
Refs: Russell, 1929, 1958; Fox, 1990.
C. Cochrane I and II (Ti1).
Refs: Russell, 1929; Fox, 1990.
NP2. Coldspur Formation: (Torrejonian)
Diss Locality (?To1–2).
Refs: Fox, 1990.
NP3. Paskapoo Formation: (Torrejonian–Tiffanian)
A0. Who Nose? Locality (To3).
Refs: Scott, 1997.
A. Saunders Creek Locality, Nordegg Aaron's Locality, Hand Hills East, Birchwood Locality, and Hand Hills West (lower level) (Ti1).
Refs: Fox, 1990; Scott *et al.*, 2006.
B. Hand Hills West (upper level) (Ti3).
Refs: Fox, 1990; Webb, 1995.
C. Blindman River Localities: UADW-1, UADW-2, UADW-3, Mel's Place and Burbank (Ti3).
Refs: Archibald *et al.*, 1987; Fox, 1990.
D. Erikson's Landing, Red Deer, and Joffres Bridge Sites (Ti3).
Refs: Fox, 1990.
E. Canyon Ski Area, Crestomere School, and One-Jaw Gap (Ti4).
Refs: Fox, 1990.
F. Swan Hills Site 1 (Ti4).
Refs: Russell, 1967; Fox, 1990.
G. Poorly known general Tiffanian localities (inc. North Saskatchewan River, Dickson Dam, Gao Mine, and Wintering Hills).
Refs: Fox, 1990.
NP4. Ravenscrag Formation: (Tiffanian)
Police Point (Ti3–4).
Refs: Krishtalka, 1973; Krause, 1978; Fox, 1990.
NP5. Hand Hills Formation: (Clarendonian–Pleistocene)
A. Sinclair Site (Cl1).
Refs: Burns and Young, 1988.
B. Courtney Pit and Russell Pit (Hh3).
Refs: Burns and Young, 1988.

SASKATCHEWAN

NP6. Frenchman Formation: (Puercan)
Fr-1 site (Pu1).
Refs: Fox, 1990.
NP7. Ravenscrag Formation: (Puercan–Tiffanian)
A. Medicine Hat Brick and Tile Quarry (Long Fall Horizon) (Pu1).
Refs: Fox, 1990.
B. RAV W-1 (Pu1).
Refs: Russell, 1974; Archibald *et al.*, 1987; Fox, 1990.
B1. Pine Cree Park (Puercan 2) (**previously included in NP7B**).
C. Croc Pot (Pu2).
Refs: Fox, 1990.
D. Roche Percee (Ti4).
Refs: Fox, 1990.
NP8. Swift Current Creek Beds: (Uintan)
Swift Current Creek Local Fauna, Cypress Hills (Ui3).
Refs: Russell, 1965a; Storer, 1984a, 1996.
NP9. Cypress Hills Formation, Lac Pellatier assemblages: (Duchesnean–Chadronian)
A. LacPelletier Lower Fauna (Du).
Refs: Storer, 1988, 1990, 1996; Bryant, 1992: Walton, 1993b; Storer, 1995.
B. LacPelletier Upper Fauna (Du).
Refs: Storer, 1993, 1995, 1996.
C. Simmie Local Fauna (Du–Ch1).
Refs: Storer, 1993.
D. Blumenort Local Fauna (?Ch1).
Refs: Storer, 1993.
NP10. Cypress Hills Formation: (Duchesnean–Hemingfordian)
A. Southfork (Cypress Hills Local Fauna) (Du).
Refs: Storer, 1984b.
B. Cypress Hills Fauna (inc. Calf Creek Local Fauna and Hunter Quarry) (Ch1).
Refs: Russell, 1972, 1975, 1978, 1980, 1982, 1985; Hunt, 1974; Emry and Storer, 1981; Storer, 1981, 1984a, 1996; Kihm, 1987; Emry and Korth, 1996.
Bi. Carnagh Local Fauna (Ch3).
Refs: Storer, 1996.
Bii. Irish Spring Local Fauna (Ch?3).
Refs: Storer, 1996.
Biii. Anxiety Butte (Ch?3).
Refs: Storer, 1996.
B2. Kealey Springs West Local Fauna (Ch4).
Refs: Storer, 1994, 1996.
BB. Fossil Bush Local Fauna, Cypress Hills Plateau (Or2–3).
Refs: Storer, 1993, 1996.
B3. Anxiety Butte (Orellan).
Refs: Storer, 1996.
C. Anxiety Butte (Wh?1).
Refs: Storer and Bryant, 1993; Storer, 1996.
C2. Rodent Hill Local Fauna (Wh2).
Refs: Storer, 1996; Bell and Bryant, 2002.
CC. Kealey Springs Local Fauna, Eastend area (Ar2).
Refs: Storer, 1993, 1996, 2002; Williams and Storer, 1998.
C3. Anxiety Butte (Ar3).
Refs: Storer, 1996.
D. Anxiety Butte (?He1).
Refs: Storer and Bryant, 1993.

E. Topham Local Fauna (He2).
Refs: Skwara, 1988.
NP11. Wood Mountain Formation: (Barstovian)
Wood Mountain Local Fauna (Ba2).
Inc. Yost Farm, Kleinfelder Farm, Quantock and Four Corners Localities.
Refs: Storer, 1975; Madden and Storer, 1985.

BRITISH COLUMBIA

NP12. Unnamed formation: (Bridgerian)
Princeton (Br2–3).
Refs: Krishtalka *et al.*, 1987.
NP13. Kishenehn Formation: (Bridgerian–Chadronian)
Flathead River Valley (Br2–3).
Refs: Russell, 1965b.

MONTANA

NP15. Hell Creek/Tullock Formation: (Puercan)
A. Biochron bk1 (Pu1).
Inc Bug Creek Anthills.
Refs: Archibald, 1982.
B. Biochron bk2 (Pu1).
Inc. Bug Creek West and Scmenge Point.
Refs: Van Valen, 1978; Archibald, 1982.
C. Biochron bk3 (Pu1).
Inc. Harbicht Hill, Chris's Bone Bed, Ferguson Ranch, and Little Roundup Channel Local Fauna (inc. Eagle Nest Channel 1, Little Roundtop, Lower Level SW, Leafbranch Level).
Refs: Van Valen, 1978; Archibald, 1982; Lofgren, 1995.
NP16. Tullock Formation: (Puercan–Torrejonian)
A. McKeever Ranch Localities (Pu1).
Inc. Hell's Hollow Local Fauna (inc. Worm Coulee 5), McGuire Creek Localities, Jack's Channel Local Fauna (inc. Jack's Ridge N4 and Luck O Hatch), Z-line Channel Local Fauna (inc. Z-line Quarry and Z-line E Quarry), Black Spring Coulee Local Fauna (inc. Black Spring Coulee S, N1 and N3, Come Alive Condylarth, Condylarth Flats, and Late Celebration), Up-Up-the-Creek Local Fauna (inc. Up-the-Creek, Up-Up-The-Creek 1, 1A, 1B, 2, and 3), Brown-Grey Channel Local Fauna (inc. BC Bone-anza, Tedrow Quarries B, C, and D), Upper and Lower Tedrow Channel Local Faunas, Shiprock Local Fauna (inc. Grass Patch, Shiprock, North Edge, and Bad Mouth Turtle), and Second Level Channel Local Fauna (inc. Jaw Breaker, Eagle Nest Channel 2, Eagle High Extension, Eagle High, Eagle South, Three Buttes 1, Eagle Nest southeast, Rattlesnake Nest, Second Level, Second Level South, Eagle Nest Ridge West, Lone Juniper Tree North, and Juniper Tree West).
Refs: Archibald, 1982; Lofgren, 1995.
B. Garbani Quarry, Biscuit Butte, Biscuit Spring, and Yellow Sand Hill Localities (Pu3).
Refs: Archibald *et al.*, 1987.
C. Purgatory Hill (?Pu3).
Refs: Van Valen, 1978; Archibald *et al.*, 1987.
D. Horsethief Canyon and Mosquito Gulch Localities (To1).
Refs: Archibald *et al.*, 1987.
NP17. Bear Formation, Crazy Mountain Area: (Puercan)
Simpson Quarry (Pu2).
Refs: Hartman *et al.*, 1989.
NP17II. Fort Union Formation, Bighorn Basin: (Puercan–Torrejonian)
A. Ludlow Member, Hiatt Local Fauna (inc. Hiatt, Hiatt South, and Deer Crash Localities) (Pu2).
Refs: Hunter, Hartman and Krause, 1997.
B. Ludlow Member, School Well Local Fauna (To1–3?).
Refs: Hunter, Hartman and Krause, 1997.
NP18. Ludlow Formation: (Puercan)
Bechtold Site, Powder River Basin (Pu3).
Refs: Archibald *et al.*, 1987.
NP19. Lebo Formation, Crazy Mountain area: (Torrejonian–Tiffanian)
A. Simpson's Localities 9 and 78 (To1).
B. Simpson's Locality 25 and Lebo No. 2 (To2).
C. Silberling Quarry (Simpson's Locality 1), Gidley Quarry (Simpson's Locality 4), Simpson's Locality 9, Hartman and Azzara Quarries (To3).
Refs: Simpson, 1937b; Rose, 1981; Gingerich and Winkler, 1985; Archibald *et al.*, 1987; Sloan, 1987; McCullough *et al.*, 2004.
D. Bingo Quarry (Ti1).
Refs: McCullough *et al.*, 2004.
NP19II. Melville Formation: (Tiffanian)
A. Douglass Quarry (= Simpson's locality 63) (Ti3) **(note emended dating, previously To3)**.
Refs: Krause and Gingerich, 1983; Krause and Maas, 1990.
B. Bangtail Locality, Bangtail Plateau, and Willow Creek (Ti1).
Refs: Gingerich, Houde, and Krause, 1983; McCullough *et al.*, 2004.
C. Scarritt Quarry (= Simpson's Locality 56), Simpson's Locality 15 (Ti1).
Refs: Simpson, 1937a; Krishtalka, 1973; Krause and Maas, 1990.
D. Simpson's Localities 11 and 13 (= Melville Locality) (Ti3).
Refs: Rose, 1975; Archibald *et al.*, 1987.
E. Simpson's Locality 15 (Ti?).
Refs: Simpson, 1937a,b.
NP20. Tongue River Formation, Powder River Basin: (Torrejonian–Tiffanian)
A. Medicine Rocks 1, Mehling site (To3).
Refs: Rose, 1975; Gingerich, 1976; Archibald *et al.*, 1987.

B. Newell's Nook (USGS D-2003) (Ti1).
Refs: Robinson and Honey, 1987.
C. White Site, 7-UP Butte, and Highway Blowout (Ti2).
Refs: Gingerich, 1976; Archibald *et al.*, 1987.
D. Circle Locality (Ti3).
Refs: Wolberg, 1979.
E. Olive Locality (Ti4).
Refs: Wolberg, 1979.

NP20II. Fort Union Formation: (Clarkforkian)
Bear Creek.
Refs: Rose, 1981.

NP21. Sage Creek Formation, Sage Creek Basin: (Bridgerian)
A. Fields Draw Localities (Br2).
B. Type Sage Creek Formation (Br3).

NP22. Dell Beds, Sage Creek Basin: (Uintan)
Dunlap Draw, Douglass Draw, and Hough Draw Local Faunas, and "Sage Creek Beds" (Ui3).
Refs: Hough, 1955; Tabrum (1990, pers. comm.); Tabrum, Prothero, and Garcia, 1996.

NP23. Climbing Arrow Formation: (Duchenean–Chadronian)
A. Shoddy Springs Local Fauna, Mud Spring Gulch, Three Forks Basin (Du) **(note emended dating, previously Ui or Du)**.
Refs: Black, 1967; Krishtalka *et al.*, 1987.
B. Black Butte and Lone Mountain, Bozeman Group (Du).
C. Thompson Creek Fauna (Three Forks Basin) (in part) and Toston Fauna (Townsend Valley or Towstow area) (in part) (Ch1).
Refs: J. A. Wilson, 1971a,b.
D. Eureka Valley Road (Ch1).
Refs: Lillegraven and Tabrum, 1983.
E. Rahn Farm (Ch4).
Refs: Lillegraven and Tabrum, 1983.

NP24. Renova Formation, Climbing Arrow Member: (Uintan–Chadronian)
A. Lower Conley Ranch and Burnt Hills Local Faunas, Lower Ruby River Basin (?Ui3–Du).
B. Type locality of *Macrotarsius montanus*, North Boulder Valley (?Ui1-Du).
C. Pipestone Springs Local Fauna, Jefferson River Basin (inc. Fence Pocket) (Ch3).
Refs: Tabrum and Fields, 1980; Garcia, 1990 (pers. comm.); Emry and Korth, 1996; Tabrum, Prothero, and Garcia, 1996.
D. Little Pipestone Creek Local Faunas, Jefferson River Basin (Ch3–4).
Refs: Garcia (1990, pers. comm.); Tabrum, Prothero, and Garcia, 1996.
E. Miscellaneous Chadronian sites (precise age unknown).
Inc. Bull Mountain (Jefferson River Basin), Williams Creek North (Lower Ruby River Basin), Upper Conley Ranch (Upper Ruby River Basin), Colbert Creek, and Beaverhead Rock (Beaverhead Basin) Local Faunas, Lower Madison River Valley, and Sage Creek.
Refs: Garcia (1990, pers. comm.).

NP25. Renova Formation, undifferentiated, Beaverhead Basin: (Duchesnean–Chadronian)
A. Lower Mantle Ranch Local Fauna (Du).
Refs: Tabrum (1990, pers. comm.).
B. West McCarty's Mountain and Diamond-O Ranch Local Faunas (Du).
Refs: Tabrum (1990, pers. comm.); Tabrum, Prothero, and Garcia, 1996.
C. McCarty's Mountain Local Fauna (Ch2).
Refs: Tabrum, (1990, pers. comm.); Tabrum, Prothero, and Garcia, 1996.
D. Upper Mantle Ranch Local Fauna (Ch3).

NP26. Renova Formation, Bone Basin Member: (Chadronian)
Jefferson River Basin.

NP27. Renova Formation, Dunbar Creek Member: (Chadronian–Orellan)
A. Monforton Ranch Local Fauna, North Boulder Valley (Ch2).
B. Mud Spring Gulch, Hossfeldt Hills (Ch4).
C. Easter Lily Mine West Local Fauna, Jefferson River Basin (Ch4).
Refs: Garcia (1990, pers. comm.); Tabrum, Prothero, and Garcia, 1996; Tabrum, 1998.
C1. Easter Lily Local Fauna, Jefferson River Basin (Or1).
Refs: Tabrum, Prothero, and Garcia, 1996.
D. Jefferson River Basin Sites (inc. Palisade Cliff and Easter Lily Mine East Local Faunas) (Or1).
Refs: Garcia (1990, pers. comm.); Tabrum, Prothero, and Garcia, 1996.

NP28. Renova Formation, Passamari Member: (Whitneyan–Arikareean)
A. Williams Creek South Local Fauna, Lower Ruby River Basin (Wh1–2?).
B. Belmont Park Ranch Local Fauna and Chalk Buttes, Upper Ruby River Basin (Ar4).

NP29. Unmapped Beds, temporally equivalent to Renova Formation: (Chadronian–Orellan)
A. Prickly Pear Creek Local Fauna (?Ch2).
Refs: Ostrander, 1985.
B. Lower Thompson Creek Local Fauna (inc. North Boulder River Valley and Three Forks Basin) (?Ch3).
C. Dog Town Mine, Douglass Creek (Douglass Creek Basin), and Lower Winston–Hadcock Ranch Local Faunas and Canyon Ferry Reservoir (Ch3–4).
Refs: White, 1954; Konizeski, 1961; Ostrander, 1985; Prothero and Shubin, 1989.
D. Townsend Valley Sites (inc. Upper Winston–Hadcock Ranch Local Faunas [Canyon Ferry Sites]) (Ch4–Or1).
Refs: White, 1954.

E. Canyon Ferry area (Orellan).
F. Black Butte Local Fauna, Gravelly Range (Orellan).

NP29II. Unnamed formation: (Chadronian)
Capitol Rock Fauna, Capitol Rock, Custer National Forest, Carter County (?Ch1–2).
Refs: Mader and Alexander, 1995.

NP30. Dunbar Creek Formation: (Chadronian–Orellan)
A. Thompson Creek Fauna (in part), Three Forks Basin (Ch3).
B. Dry Hollow Local Fauna (= Upper Toston Local Fauna), Toston Area (inc. Cooper Gulch) (Orellan).
C. 10 N (Ch4).
Refs: Lillegraven and Tabrum, 1983.

NP31. Kishenehn Formation: (Chadronian)
Paola Sliding Locality (?Ch3).
Refs: McKenna, 1990.

NP32. Cook Ranch Formation, Sage Creek Basin: (L. Chadronian–Orellan)
A. Little Spring Gulch Local Fauna (Ch4).
B. Matador Ranch Local Fauna (Or3–4).
Refs: Tabrum (1990, pers. comm.); Tabrum, Prothero, and Garcia, 1996.
C. Cook Ranch Local Fauna (Or4) (**previously included in NP32B**).
Refs: Tabrum, Prothero, and Garcia, 1996.

NP33. Blacktail Deer Creek Formation, Sage Creek Basin: (Whitneyan–Arikareean)
A. White Hills Local Fauna (Wh1–2).
B. Blacktail Deer Creek Local Fauna, Blacktail Deer Creek Basin (Ar3).
Refs: Hibbard and Keenmon, 1950; Tedford *et al.*, 1987; Korth, 1996b.

NP34. Fort Logan Formation (= Deep River Beds): (Chadronian–Barstovian)
A. White Sulphur Springs (?Whitneyan).
B. White Sulphur Springs (inc. Spring Creek 1, Crab Tree Bluff, and Thompson Gulch North) (Ar1–2).
Refs: Douglass, 1901.
C. "Deep River Beds" (= Fort Logan Beds) (Ar3).
Refs: Koerner, 1940; Tedford *et al.*, 1987; UWA collections.
D. Deep River Local Faunas (= Smith Creek area, Spring Creek 2 (He2).
Refs: UWA collections.
E. Deep River Local Faunas (= Smith Creek area, or Smith River Valley), and *Cyclopidius* Beds) (Ba1).
Refs: Wang, Tedford, and Taylor, 1999.

NP35. Medicine Lodge Beds, Horse Prairie–Grasshopper Creek Basin: (Chadronian?–Arikareean)
A. Mill Point Local Fauna, Grasshopper Creek Basin (?Ar2).
B. Everson Creek Local Fauna, Horse Prairie Basin (Ar3).
C. Grasshopper Creek Local Fauna (Ar4).

NP36. Cabbage Patch Beds: (Arikareean)
A. Lower Cabbage Patch Fauna (Ar1).
Refs: Rasmussen, 1989.
B. Middle and Upper Cabbage Patch Faunas (inc. Bert Creek 2) (Ar2).
Refs: Rich and Rasmussen, 1973; Rasmussen, 1989.
C. Pikes Peak Creek Local Fauna (Ar2–3).
D. Tavenner Ranch Local Faunas (Ar2).
Refs: Rich and Rasmussen, 1973.

NP36II. Unknown formation: (Arikareean)
10 miles north of Bannack, Beaverhead County (Ar3).
Refs: Prothero, 2005.

NP37. Canyon Ferry Beds (? = Toston Formation): (L. E. Arikareean)
Canyon Ferry Local Fauna, Townsend Valley (Ar2).
Refs: White, 1954; Leiggi, 1991.

NP38. Six Mile Creek Formation: (Arikareean–Hemphillian)
A. Type Six Mile Creek Formation (Townsend Valley) and Roy Gulch–Sixteen Mile Creek drainage (Clarkstow Basin) (Ar3) (**note emended dating, previously Ar2**).
B. North Boulder Valley Fauna (= Nigger Hollow Local Fauna) (Ar3–4).
Refs: Tedford *et al.*, 1987.
C. Frank Ranch, Sant Ranch (Jefferson River Basin), McKanna Spring (North Boulder Valley), and Sweetwater Creek (Upper Ruby River Basin) Local Faunas (Ba1) (**note emended dating, previously He2-Ba1**).
Refs: Tedford *et al.*, 1987.
D. Antelope Hills Local Fauna (Jefferson River Basin) (Ba1).
E. Williams Ranch Local Fauna (Upper Ruby River Basin) (= Dorr and Wheeler Locality 5) (Ba1–2).
Refs: Dorr and Wheeler, 1964.
F. Old Windmill Local Fauna (?Cl).
G. Ruby River Fauna (Cl3–Hh1).
H. Mayflower Mine, Old Wagon Road (Jefferson River Basin), and North Silver Star Triangle Local Faunas (Hh1–2).
I. Tostow area sites, Townsend Valley (Hh1–2).

NP39. Divide Basin: (L. Arikareean–E. Barstovian)
Divide Local Faunas, Cabbage Patch Beds (Ar3–4).

NP40. Hepburn's Mesa Formation: (Hemingfordian–Barstovian)
A. Chalk Cliffs Local Fauna (Units 3–13), Paradise Valley (Ba1).
Refs: Barnosky and Labar, 1989.
B. Chalk Cliffs Local Fauna (Units 16–23), Paradise Valley (Ba2).
Refs: Barnosky and Labar, 1989.

NP40II. Unknown formation: (?Hemingfordian)
Red Rock Creek.

NP41. Madison Valley Formation, Three Forks Basin: (Arikareean–Clarendonian)

A. Middle Part, inc. near Belgrade (Ar3–4) (**note emended dating, previously He2**).
Refs: Korth, 1996b.
B. Anceney Local Fauna (= "Madison Valley Fauna") (Ba1) (**note emended dating, previously Ba2**).
Refs: Dorr, 1956; Sutton, 1977; Barnosky, 1986; Sutton and Korth, 1995.
C. Upper Part (Cl2).
D. Madison River (= "Loup Fork") (Cl3).
Refs: Wang, Tedford, and Taylor, 1999.

NP42. Flint Creek Beds, Flint Creek Basin: (Barstovian)
Flint Creek Local Fauna, inc. East of New Chicago (Ba1).
Refs: Barnosky, 1986; Tedford *et al.*, 1987; Pierce and Rasmussen, 1989; Wang, Tedford, and Taylor, 1999.

NP42II. Unknown formation: (Barstovian)
Aelurodon Cut, near Whitehall, Jefferson County (?Ba1).
Refs. Wang *et al.*, 2004.

NP43. Barnes Creek Beds, Flint Creek Basin: (Barstovian)
Barnes Creek Local Fauna (= East of New Chicago, plus Bert Creek Local Fauna) (Ba2).
Refs: Tedford *et al.*, 1987.

NP45. Unnamed formation, Deer Lodge Basin: (Hemphillian)
Deer Lodge Local Fauna, plus Dempsey Creek Gravel Pit and Drewsey Creek (Hh1).
Refs: Konizeski, 1957.

NP45II. Unnamed formation: (?Hemphillian)
Prison Gravel Pit, Powell County (UMMP Locality 7403).
Refs: Wang, Tedford, and Taylor, 1999.

NORTH DAKOTA

NP46. Bullion Creek Formation (= Fort Union Formation): (Torrejonian) Billings County

NP47. Tongue River Formation, Williston Basin: (Puercan–Tiffanian)
A0. Pita Flats (Pu2–3).
Refs: Hunter, 1999.
A. Donnybrook, Lloyd and Hares Site (= Heart Butte) (To3–Ti1).
Refs: Archibald *et al.*, 1987.
A1. X-X locality (Ti2).
Refs: Hunter, 1999.
B. Brisbane Local Fauna (Ti3).
Refs: Holtzman, 1978; Kihm and Hartman, 2004.
BB. Judson Local Fauna (Ti4) (**previously included with NP47B**).
Refs: Holtzman, 1978; Kihm and Hartman, 2004.
C. Wannagan Creek Quarry and Riverdale (Ti4).
Refs: Holtzman, 1978; Archibald *et al.*, 1987; Erickson, 1991.

NP48. Sentinel Butte Formation: (Tiffanian)
A. Type of *Titanoides primaevus* area near Fort Buford (?Ti3).
Refs: Russell, 1974.
B. Red Spring Locality (?Ti4).
Refs: Kihm, Hartman, and Krause, 1993.

NP49. Golden Valley Formation: (Wasatchian)
Camel Buttes Member, Williston Basin (inc. White Buttes and Turtle Valley Localities) (Wa3).
Refs: West, 1973b.

NP49II. Chadron Formation: (Chadronian)
Medicine Pole Hills Local Fauna (Ch3).
Refs: Pearson and Hoganson, 1995; Kihm *et al.*, 2001.

NP50. Brule Formation, White River Group: (Chadronian–Whitneyan)
A. Dickinson Member, Unit 3 (Ch4).
B. Dickinson Member, Units 4 and 5 (Or1–4).
Inc. White Butte, Little Badlands Locality, Kostelecky, Fitterer, and Meduna Ranches (Stark County), and 7 miles south of South Heart.
Refs: Korth and Bailey, 1992; Heaton, 1993; Korth, 1994.
C. Schefield Member, Unit 6, and Fitterer Ranch (Wh1–2).

NP50II. Lower Arikaree Group: (Arikareean)
White Buttes (Chalky Buttes), Slope County (Ar1–2).
Refs: Wang, Tedford, and Taylor, 1999.

NP51. Kildeer Formation: (Whitneyan–Hemingfordian)
A. Whitneyan sites, inc. Rainey Butte (Wh1–2).
AA. Arikareean sites (?Ar1–4).
B. Kildeer mountains (He2).

PACIFIC NORTHWEST

WASHINGTON

PN1. Keechelus Formation: (Arikareean)
A. Lower Wildcat Creek Local Fauna, Rimrock Lake area, Tieton River drainage (Ar1).
B. Upper Wildcat Creek Local Fauna, Rimrock Lake area, Tieton River drainage (Ar2).

PN1II. Unnamed formation: (Barstovian)
Roosevelt Local Fauna, Klickitat County (Ba2).
Refs: Prothero, 2005.

PN2. Ellensburg Formation: (Clarendonian–Hemphillian)
A. Granger Clay Pit, LACM 6431 (Cl3).
Refs: Martin and Tedrow, 1988.
B. Yakima Canyon (Hh3–4).

PN3. Ringold Formation: (Hemphillian–Blancan)
A. Lind Coulee Local Fauna (Hh4).
B. River Road Local Fauna (Hh4).
C. White Bluffs Local Fauna (inc. "Ringold Fauna" and Bluff Top Local Fauna) (E. Bl).
Refs: Gustafson, 1978, 1985; Kurtén and Anderson, 1980; Lindsay, Opdyke, and Johnson, 1984.
D. Taunton Substation (L. Bl).
Refs: Wang, Tedford, and Taylor, 1999.

PN4. Ringold Formation Equivalent: (Blancan)
Taunton and Haymaker's Orchard Local Faunas (E. Bl) (**note emended dating, previously L. Bl**).
Refs: Morgan and Morgan, 1995.

OREGON (ALL ON COLUMBIA PLATEAU)

PN5. Clarno Formation: (Bridgerian–Duchesnean)
A. Clarno Nut Bed (Br3).
Refs: Krishtalka *et al.*, 1987.
B. Hancock Quarry (Ui2) (**note emended dating, previously Duchesnean**).
Refs: Krishtalka *et al.*, 1987; Hanson, 1996.

PN6. John Day Formation: (Chadronian–Hemingfordian)
A. Big Basin Member, lower part (inc. Bridge Creek Floral Locality) (Chadronian).
Refs: Fremd, Bestland, and Retallack, 1997.
B. Big Basin Member, Middle Part (inc. Camp Creek, Silver Wells) (Orellan).
Refs: Merriam and Sinclair, 1907.
B1. Big Basin Member, upper part (inc. Carroll Rim Locality) (Wh2).
Refs: Fremd, Bestland, and Retallack, 1997.
C. Lower Turtle Cove Member (below Picture Gorge Ignimbrite) (Ar1).
Refs: Merriam and Sinclair, 1907; Eaton, 1922; Lull, 1922; Fisher and Rensberger, 1972; Fremd, Bestland, and Retallack, 1997.
C1. North Blue Basin (Units A, B) (Ar1).
Refs: Fremd, Bestland, and Retallack, 1997.
C2. Foree (Units D–F) (Ar1).
Refs: Fremd, Bestland, and Retallack, 1997.
D. Upper Turtle Cove and Kimberly Members, between Picture Gorge Ignimbrite and ATR Tuft (Ar1–2).
Refs: Merriam and Sinclair, 1907; Osborn, 1909; Lull, 1922; Macdonald, 1970; Fisher and Rensberger, 1972; Fremd, Bestland, and Retallack, 1997; LACM collections.
D1. Longview Ranch (Units G–J) (Ar1).
Refs: Fremd, Bestland, and Retallack, 1997.
D2. Roundup Flat (Units K1 and K2) (inc. South Haystack) (Ar2).
Refs: Fremd, Bestland, and Retallack, 1997.
D3. Bone Creek (Units L and M) (inc. Picture Gorge 36) (Ar2).
Refs: Fremd, Bestland, and Retallack, 1997.
E. Kimberly Member, *Meniscomys* zone (= "Middle John Day") (inc. sites UCMP 802, 858, 869, 904. V-4850, V-6632) (Ar2).
Refs: Merriam and Sinclair, 1907; Lull, 1922; Fisher and Rensberger, 1972; Prothero and Shubin, 1989.
F. Kimberly Member, *Entoptychus-Gregorymys* zone (= "Upper John Day") (Ar3–4).
Refs: Merriam and Sinclair, 1907; Osborn, 1909; Lull, 1922; Wood, 1935, 1936; Fisher and Rensberger, 1972.
G. Haystack Valley Member, *Mylagaulodon* zone (= "Uppermost John Day") (inc. Black Bone Hill Locality, locality JDNM-46 [V6668]) (Ar2) (**note emended dating, previously He1**).
Refs: Merriam and Sinclair, 1907; Wood, 1935, 1936; Fisher and Rensberger, 1972; Rensberger, 1973; Tedford *et al.*, 1987; Fremd, Bestland, and Retallack, 1997; LACM, UCB and UWA collections.
H. Warm Springs Local Fauna and Shitike Creek (inc. Mecca localities) (Ar3–He1).
Refs: Woodburne and Robinson, 1977; Tedford *et al.*, 1987.

PN7. Mascall Formation: (Barstovian)
Mascall Fauna (Ba1).
Inc. "John Day area," Picture Gorge, Crooked River, Cottonwood Creek, Carbonaras Creek, and Gateway assemblages, Rock Creek Locality, Tri-Creek Ranch (= McDonald Ranch), Birch Creek Locality, JDNM-70 (= V4827, Mascall 13), JDNM-3059 (V4943), V4824, V3059 (Mascall type locality), *Ticholeptus* Beds, Amyzon Beds, *Protolabis* Beds, Highway Locality (V3043), localities V4823, V4824, V4945.
Refs: Merriam and Sinclair, 1907: Downs, 1956b; Barnosky, 1986; Tedford *et al.*, 1987; Fremd, Bestland, and Retallack, 1997; McAfee, 2003; LACM and UCB collections.

PN7II. Unknown formation: (Barstovian)
Guano Ranch, Snyder Ranch, and Guano Lake Localities.
Refs: Repenning, 1967.

PN7III. Unknown formation: (Barstovian)
Bowden Hills.
Refs: Hutchison, 1968.

PN8. Butte Creek Volcanic Sandstone (= Battle Creek Formation): (Barstovian)
A. Beatty Buttes (Ba1).
Refs: Wallace, 1946; Barnosky, 1986.
B. Skull Spring Fauna and Red Basin Local Fauna (Ba1) (**note emended dating, previously Ba2**).
Refs: Shotwell, 1968; Barnosky, 1986; Korth, 1999a; LACM collections.

PN9. Sucker Creek Formation: (Barstovian)
A. Sucker Creek Fauna (Ba1–2) (**note emended dating, previously Ba2**).
Refs: Scharf, 1935; Korth, 1999a; LACM collections.
B. Owyhee River drainage and Quartz Basin Localities (= "Deer Butte Formation") (Ba2).
Refs: Shotwell, 1968; Barnosky, 1986.

PN10. Juntura Formation: (Clarendonian)
Black Butte Local Fauna (inc. Ironside sites) (Cl3).
Refs: Shotwell, 1970; Barnosky, 1986; UO collections.

PN10II. The Dalles Formation: (Clarendonian)
The Dalles Basin, Wasco County.
Refs: Wang, Tedford, and Taylor, 1999.

PN11. Drewsey Formation: (Hemphillian)
A. Rome, Drinkwater, Stinking Water Creek, Warm Springs, and Otis Basin Local Faunas, and Antelope Reservoir, (Hh1) (**previously just PN11**).
Refs: Shotwell, 1956, 1970; Becker and McDonald, 1998; Korth, 1999a.
B. Bartlett Mountain Local Fauna (Hh2) (**previously just PN11; note emended dating, previously Hh1**).
Refs: Shotwell, 1956, 1970; Becker and McDonald, 1998; Korth, 1999a.
PN12. Rattlesnake Formation: (Hemphillian)
Rattlesnake Fauna, Cottonwood Creek:
Refs: Fremd, Bestland, and Retallack, 1997; Martin, 1983; McAfee, 2003; UCB collections.
A. Lower Fanglomerate Member (Units 1–14) (Hh1).
B. Upper Fanglomerate Member (Unit 19) (inc. Rattlesnake Fauna, Cottonwood Creek) (Hh2).
PN13. Unnamed unit: (Hemphillian)
Juniper Creek Canyon (Hh2).
Refs: Shotwell, 1970.
PN14. Shutler Formation: (Hemphillian)
Inc. McKay Reservoir Local Fauna, Krebs Ranch, South of Arlington, and Ordinance (West End Blowout) (Hh3).
Refs: Shotwell, 1956, 1958; Martin, 1984; Barnosky, 1986; Korth, 1999a; LACM and UCB collections.
PN14II. Rome Formation: (Hemphillian)
Dry Creek Local Fauna (Hh3).
Refs: Prothero, 2005.
PN15. Chalk Buttes Formation: (Hemphillian)
Inc. Little Valley, Harper, and Christmas Valley Local Faunas (Hh3).
Refs: Shotwell, 1970; Barnosky, 1986; LACM collections.
PN15II. Unknown formation: (Hemphillian)
Enrico Ranch (Hh4).
Refs: Hutchison, 1968

IDAHO

PN16. Peterson Creek Beds, Salmon–Lemhi Basin: (Arikareean)
Peterson Creek Local Fauna (Ar1–2) (**note emended dating, previously Ar3**).
Refs: Nichols, 1976, 1998.
PN17. Starlight Formation: (Arikareean)
Cedar Ridge, Rockland Valley (Ar2).
PN17II. Imnaha Basalt: (Hemingfordian)
Eagle Creek interbed.
Refs: Tedrow and Martin, 1998.
PN17III. Geertson Formation: (Hemingfordian)
South of Salmon, Lemhi County (He1).
Refs: Wang, Tedford, and Taylor, 1999.
PN18. Mollie Gulch Beds, Salmon–Lehmi Basin: (Arikareean)
Mollie Gulch Local Fauna (inc. Boulder Creek) (Ar3–4) (**note emended dating, previously He2-Ba1**).
Refs: Nichols, 1998.
PN19. Railroad Canyon Beds, Medicine Lodge Basin: (Hemingfordian–Barstovian)
A. Railroad Canyon Local Fauna (= Maiden Creek Local Fauna) (Ba1) (**note emended dating, previously He2**).
Refs: Nichols, 1998.
B. Upper Railroad Canyon Local Fauna (Ba1).
See new stratigraphy (and faunal listings) in Barnosky *et al.* (2007).
PN20. Salt Lake Group: (Barstovian)
Marsh Creek, near Downata Hot Springs (Ba1).
Refs: MacFadden and Nelson, 1980.
PN21. Chalk Butte Formation: (Clarendonian–Hemphillian)
A. Chalk Basin East (Cl3).
B. Star Valley (Hh1–2).
Refs: Becker and McDonald, 1998.
PN22. Unnamed formation: (Hemphillian)
Stroud Claim (Hh1–2).
PN22II. Poison Creek Formation: (Hemphillian)
Reynolds Creek, Owyhee County (Hh1–2).
Refs: Wang, Tedford, and Taylor, 1999.
PN22III. Unnamed formation: (Hemphillian)
Notch Bluff Local Fauna, Owyhee County (Hh1–2).
Refs: Zakrzewski, 1998.
PN23. Glenns Ferry Formation: (Blancan)
A. Hagerman Local Fauna (inc. Horse Quarry) (E. Bl).
Refs: Zakrzewski, 1969; Bjork, 1970; Kurtén and Anderson, 1980; Lindsay, Opdyke, and Johnson, 1984; Thewissen and Smith, 1987; Morgan and Morgan, 1995.
B. Sand Point and Flat Iron Butte (E. Bl).
Refs: Morgan and Morgan, 1995.
C. Grand View Fauna (L. Bl).
Inc. Castle Butte, and Wild Horse Butte, Jackass Butte, Nine Foot Rapids, Shoofly Creek, Froman Ferry, and Birch Creek Local Faunas.
Refs: Shotwell, 1970; Kurtén and Anderson, 1980; Morgan and Morgan, 1995; Hearst, 1998.
D. Hammet, Owyhee County (L. Bl).
E. Fossil Creek Roadcut, Owyhee County (L. Bl).

CANADIAN HIGH ARCTIC

HA1. Eureka Sound Group: (Wasatchian–Duchesnean)
A0. Bay Fiord, CMN P7502, central Ellesmere Island (late Wasatchian).
Refs: Rose, Eberle, and McKenna, 2004; Eberle, 2005.
A. Iceberg Bay Formation, Ellesmere Island (Wa7–Br2) (**previously NP14**).
Refs: West and Dawson, 1977; McKenna, 1980b.
A2. Margaret Formation (inc. Localities 44 and D7), Ellesmere Island (Wasatchian).
Refs: Eberle, 2001.

B. Boulder Hills Formation, Axel Heiberg Island, and Mokka Fiord site (?Uintan).
C. Buchanan Lake Formation, Axel Heiberg Island, Geodetic Hills Area (Ui1–Du).
Refs: Eberle and Storer, 1999.

HA2. Haughton Formation, Devon Island, Nunavut: (Arikareean)
Haughton Fauna (Ar?3).
Refs: Dawson, 2003.

HA3. Unnamed formation, Ellesmere Island: (Blancan)
Beaver Pond Locality, Strathcona Fiord (E. Bl).
Refs: Harington, 2003; Tedford and Harington, 2003.

CENTRAL EAST AMERICA

TENNESSEE

CE1. Unnamed formation: (Hemphillian–Blancan)
Gray Fossil Site (Hh4–E. Bl).
Refs: Wallace and Wang, 2004.

INDIANA

CE2. Pipe Creek Sinkhole: (Blancan)
Pipe Creek Sinkhole Local Fauna (E. Bl).
Refs: Farlow *et al.*, 2001; Lindsay *et al.*, 2002; Martin, Goodwin, and Farlow, 2002.

NORTHERN EAST COAST

NC0. Aquia Formation, Maryland: (Tifffanian 3?)
Blue Banks.
Refs: Rose, 2000.

NC0II. Nanjemoy Formation, Virginia: (Wasatchian)
Fisher/Sullivan Site, east of Fredricksburg, Stafford County Virginia.
Refs: Rose, 1999.

NC1. Kirkwood Formation, Monmouth County, New Jersey: (Arikareean)
A. Farmingdale Fauna (Ar3) (**note emended dating, previously He1**).
Refs: Tedford and Hunter, 1984.
B. Shiloh Local Fauna (Ar3) **(note emended dating, previously Ba1–2)**.
Refs: Tedford and Hunter, 1984.

NC2. Lower Calvert Formation: (Hemingfordian)
A. Fairhaven Cliffs, Maryland (He1).
Refs: USNM collections.
B. Pollack Pit Local Fauna, Cheswold Sand, Delaware (He1) **(previously just NC2)**.
Refs: Emry and Eshelman, 1998.
C. Zone 6–8 north of Boy Scout camp, Calvert County, Maryland (He1).
D. Chesapeake Beach, Zones 3–10, Maryland (He1).
Refs: Myrick, 1979.
E. Near Pamunkey River, Hanover County, Virginia (He1).
Refs: Kellogg, 1968; Dooley, 2005.

NC2II. Unnamed formation, Martha's Vineyard, Massachusetts: (Barstovian)
Gay Head.

NC3. Calvert Formation, Plum Point Marl Member, Maryland and Virginia: (Hemingfordian-Barstovian)
A. Zones 10, 11, Maryland (Ba1).
Refs: Tedford and Hunter, 1984; Wright and Eshelman, 1987.
AA. Inc. Westmoreland State Park, Stratford Hall Plantation, Homini Cliffs, Stratford Harbor, Pope's Inn, Stratford Bluffs, Mill Pond, Gravitts Mill, Beach near Mill Wheel (Westmoreland County), South Bank of Pamunkey River (Hanover County), Zones 12–14, Virginia (Ba2).
B. Zones 12–16 (= Chesapeake Bay Fauna, in part), Maryland (Ba2).
Refs: Tedford and Hunter, 1984; Wright and Eshelman, 1987.
C. Tinker Creek, Matoaka Cottages, North of King's Creek, Zone 17 (Prince George's County, Maryland) (Ba2).
D. Martin Marietta Carmel Church Quarry (= Caroline Stone Quarry), Caroline County, Virgina (?Ba2).
Refs: Dooley, 1993; Dooley, Fraser, and Luo, 2004; Dooley and Frazer, 2005.
E. Richmond, Henrico County, Virginia (?Ba2).
Refs: Kellogg, 1957.

NC4. Choptank Formation: (Barstovian)
A. Maryland, Zone 17 (= BG and E Power Plant and Lisby) (= Chesapeake Bay Fauna, in part) (Ba2).
Refs: Tedford and Hunter, 1984; Wright and Eshelman, 1987.
B. Virginia, Zone 17.
Refs: Kellogg, 1969.

NC5. St. Mary's Formation, Maryland: (Clarendonian)
Little Cove Point (?Cl2).
Refs: Tedford and Hunter, 1984.

NC6. Yorktown Formation, Maryland: (L. Blancan)
Charles County, Maryland (L. Bl).

NC7. St. Mary's Formation, Virginia: (Clarendonian)
Chestnut Bluffs, Surry County, Virginia (Cl2–3).
Refs. Baum and Wheeler, 1977.

EAST COAST MARINE

NORTHERN EAST COAST

EM1. Shark River Marls, New Jersey: (Bridgerian)
Monmouth County.
Refs: Krishtalka *et al.*, 1987.

SOUTHERN EAST COAST

EM1II. Williamsberg Formation, South Carolina: (Clarkforkian) (**previously GC20**)
Black Mingo Group, Santee Rediversion Canal, near St. Stephen.
Refs: Schoch, 1985.

EM2. Lutetian Marine series, Mississippi: (Uintan–Duchesnean) (**now = GC22IVA**).

EM3. Castle Hayne Formation, North Carolina: (L. Uintan)
A. Various quarries, Jones County.
Refs: NCSM collections.
B. Martin Marietta Quarry, New Hanover County.
Refs: USNM collections.
C. Lanier's Pit, Maple Hill; Martin Marietta Quarry, Rocky Point, Pender County.
Refs: Uhen, 1999; NCSM collections.
D. Various quarries, Pitt County.
Refs: NCSM collections.

EM3II. New Bern Formation, North Carolina: (Uintan)
Martin Marietta Quarry, New Bern, Craven County (Ui3).
Refs: McLeod and Barnes, 1996

EM3III. Cross Formation, South Carolina: (Duchesnean)
A. Martin Marietta Cross Quarry, Berkeley County.
Refs: Geisler and Sanders, 1996.
B. Unnamed sites, Orangeburg County.

EM3IV. Harleyville Formation, South Carolina: (Chadronian)
Near Harleyville, Berkeley County.
Refs: Sanders, 1974.

EM4. Ashley Formation, South Carolina: (Arikareean, Ar1)
Refs: Weems and Lewis, 2002.

EM5. Chandler Bridge Formation, South Carolina: (Arikareean, Ar2)
A. Bed 1.
Refs: Weems and Lewis, 2002.
B. Bed 2.
Refs: Weems and Lewis, 2002.
C. Bed 3.
Refs: Weems and Lewis, 2002.

EM6. Charleston Phosphate Beds, South Carolina: (Oligocene–Pleistocene)
A. (L. Oligocene–E. Miocene).
B. (Arikareean–Hemingfordian).

EM6II. Hawthorn Formation, South Carolina: (Hemingfordian) (**previously GC25**)
Unnamed site, Dorchester County.

EM7. Pungo River Formation, North Carolina: (Hemingfordian–Barstovian)
Late Barstovian site (Ba2).

EM8. Yorktown Formation (North Carolina and Virginia): (Hemphillian–Blancan)
A. Unnamed site, North Carolina (Hh4).
B. Lee Creek Mine, North Carolina (Hh4) (**previously GC28**).
Refs: Tedford and Hunter, 1984; Eshelman (1996, pers. comm.).
C. Unnamed site, Virginia (Hh4–E. Bl) (**previously EM8B**).
D. Smith Mill Run, North Carolina (L. Bl) (**previously GC30**).
E. Quankey Creek, North Carolina (L. Bl).
Refs: Cope, 1870.
F. Rice's Pit, Viginia (Hh4).
Refs: Westgate and Whitmore, 2002.
G. Yorktown, Virginia (L. Bl).
Refs: Berry and Gregory, 1906.

EM8II. Unnamed formation, North Carolina: (Blancan)
Superior Stone Company Quarry, Neuse River, New Bern, Craven County.

EM9. Unnamed formation, South Carolina: (Blancan)
Dorchester County.

GULF COAST

CENTRAL AMERICA

EM12. Yellow Limestone Formation, Jamaica: (Bridgerian)
A. Freeman's Hall (**previously just EM12**).
B. Chapelton Formation, Guy's Hill Member (inc. Seven Rivers, Parish of St. James).
Refs: Domning, 2001.

EM13. San Sebastian Formation, Puerto Rico: (Arikareean, Ar3)
Refs: Scharlach, 1990.

EM14. Juana Diaz Formation, Puerto Rico: (Whitneyan–Arikareean).

EM15. Guines Formation, Cuba: (Hemingfordian or Barstovian).

Appendix II: References for localities in Appendix I

Akersten, W. A. (1972). Red Light Local Fauna (Blancan) of the Love Formation, Southeastern Huspeth County, Texas. *Bulletin of the Texas Memorial Museum*, 20, 1–52.

Albright, L. B., III (1998a). New genus of tapir (Mammalia: Tapiridae) from the Arikareean (earliest Miocene) of the Texas Coastal Plain. *Journal of Vertebrate Paleontology*, 18, 200–17.

(1998b). The Arikareean land mammal age in Texas and Florida: southern extension of Great Plains faunas and Gulf Coastal Plain endemism. [In *Depositional Environments, Lithostratigraphy, and Biostratigraphy of the White River and Arikaree groups (Late Eocene to Early Miocene, North America)*, ed. D. O. Terry, H. E. LaGarry, and R. M. Hunt.] *Geological Society of America Special Paper*, 325, 167–83.

(1999a). Ungulates of the Toledo Bend Local Fauna (late Arikareean, early Miocene) Texas Coastal Plain. *Bulletin of the Florida Museum of Natural History*, 42, 1–80.

(1999b). Biostratigraphy and vertebrate paleontology of the San Timoteo Badlands, Southern California. *University of California, Publications in Geological Sciences*, 144, 1–121.

Alexander, J. P. and Berger, B. J. (2001). Stratigraphy and taphonomy of Grizzly Buttes, Bridger Formation, and the middle Eocene of Wyoming. In *Eocene Biodiversity: Unusual Occurrences and Rarely Sampled Habitats*, ed. G. F. Gunnell, pp. 165–96. New York: Kluwer Academic/Plenum.

Anderson, D. W. and Picard, M. D. (1972). Stratigraphy of the Duchesne River Formation (Eocene–Oligocene), northern Uinta Basin, northeastern Utah. *Bulletin of Utah Geological and Mineral Survey*, 97, 1–29.

Anemone, R. L., Johnson, E. M., and Rubick, C. M. (1999). Primates and other mammals from the Great Divide Basin, SW Wyoming: systematics, geology, and chronology. *American Journal of Physical Anthropology*, 28(suppl.), p. 84.

Anemone, R. L., Johnson, E. M., Nachman, B. A., and Over, D. J. (2000). A new Clarkforkian primate fauna from the Great Divide Basin, SW Wyoming. *American Journal of Physical Anthropology*, 30(suppl.), p. 97.

Archibald, J. D. (1982). A study of Mammalia and geology across the Cretaceous–Tertiary boundary in Garfield County, Montana. *University of California, Publications in Geological Sciences*, 122, 1–286.

Archibald, J. D., Gingerich, P. D., Lindsay, E. H., *et al.* (1987). First North American Land Mammal Ages of the Cenozoic Era. In *Cenozoic Mammals of North America, Geochronology and Biostratigraphy*, ed. M. O. Woodburne, pp. 24–76. Berkeley, CA: University of California Press.

Asher, R. J., McKenna, M. C., Emry, R. J., Tabrum, A. R., and Kron, D. G. (2002). Morphology and relationships of *Apternodus* and other extinct, zalambdodont, placental mammals. *Bulletin of the American Museum of Natural History*, 273, 1–117.

Axelrod, D. I. (1956). Mio-Pliocene floras from West–Central Nevada. *University of California, Publications in Geological Sciences*, 33, 1–322.

Bailey, B. E. (1999). New Arikareean/Hemingfordian micromammal fauns from western Nebraska and their biostratigraphic significance. *Journal of Vertebrate Paleontology*, 19(suppl. to no. 3), pp. 30A–1A.

(2003). New fossil shrew remains from western Nevada, and a suggested subfamilial revision of the Soricidae (Mammalia: Insectivora). *Journal of Vertebrate Paleontology*, 23(suppl. to no. 3), p. 31A.

Barnes, L. G. (1985). The late Miocene dolphin *Pithanodelphis* Abel, 1905 (Cetacea: Kentriodontidae) from California. *Contributions in Science, Los Angeles County Museum of Natural History*, 367, 1–27.

(1989a). A new enaliarctine pinniped from the Astoria Formation, Oregon and a new classification of the Otariidae (Mammalia: Carnivora). *Contributions in Science, Los Angeles County Museum of Natural History*, 403, 1–26.

(1989b). A late Miocene dolphin, *Pithanodelphis nasalis*, from Orange County, California. *Memoirs of the Natural History Foundation of Orange County*, 2, 7–21.

(1998a). The sequence of fossil marine mammal assemblages in Mexico. In *Advances in Investigación, Paleontología de Vertebrados. Universidad Autonoma del Estado de Hidalgo*. Publicación Especial I, 26–79.

(1998b). Late Tertiary albireonid dolphins; (Cetacea, Odontoceti), from the North Pacific Ocean. *Journal of Vertebrate Paleontology*, 8(suppl. to no. 3), p. 8A.

Barnes, L. G. and Goedert, J. L. (1996). Marine vertebrate paleontology on the Olympic peninsula. *Washington Geology*, 24, 17–25.

Barnes, L. G. and Hirota, K. (1995). Miocene pinnipeds of the otariid subfamily Allodesminae in the North Pacific Ocean: systematics and relationships. *The Island Arc*, 3, 329–60.

Barnes, L. G. and Mitchell, E. D. (1975). Late Cenozoic northeast Pacific Phocidae. *Rapports et Procè-Verbauz des Réunions* (*Biology of the Seal*), 169, 34–42.

Barnes, L. G. and Raschke, R. E. (1991). *Gomphotaria pugnax*, a new genus and species of late Miocene Dusignathine otariid pinniped (Mammalia: Carnivora) from California. *Contributions in Science, Los Angeles County Museum of Natural History*, 426, 1–16.

Barnes, L. G., Howard, H., Hutchinson, J. H., and Welton, B. J. (1981). The vertebrate fossils of the marine Cenozoic San Mateo Formation at Oceanside, California. In *Geologic Investigation of the San Diego Coastal Plain*, ed. P. L. Abbott and S. O'Dunn, pp. 53–70. San Diego, CA: San Diego Association of Geologists.

Barnes, L. G., Kimura, M., Furusawa, H., and Sawamura, H. (1995). Classification and distribution of Oligocene Aetiocetidae (Mammalia; Cetacea; Mysticeti) from western North America and Japan. *The Island Arc*, 3, 392–431.

Barnosky, A. D. (1985). Late Blancan (Pliocene) microtine rodents from Jackson Hole, Wyoming. Biostratigraphy and biogeography. *Journal of Vertebrate Paleontology*, 5, 255–71.

(1986). Arikareean, Hemingfordian, and Barstovian mammals from the Miocene Colter Formation, Jackson Hole, Teton County, Wyoming. *Bulletin of the Carnegie Museum of Natural History*, 26, 1–69.

Barnosky, A. D. and Labar, W. J. (1989). Mid-Miocene (Barstovian) environmental and tectonic setting near Yellowstone Park, Montana and Wyoming. *Geological Society of America Bulletin*, 101, 1448–56.

Barnoksy, A. D., Bibi, F., Hopkins, S. S. B., and Nichols, R. (2007). Biostratigraphy and magnetostratigraphy of the mid-Miocene Railroad Canyon Sequence, Montana and Idaho, and the age of the mid-Tertiary unconformity west of the continental divide. *Journal of Vertebrate Paleontology*, 27, 204–24.

Baskin, J. A. (1979). Small mammals of the Hemphillian age White Cone local fauna, northeastern Arizona. *Journal of Paleontology*, 53, 695–708.

(1981). *Barbourofelis* (Nimravidae) and *Nimravides* (Felidae), with a description of two new species from the late Miocene of Florida. *Journal of Mammalogy*, 62, 122–39.

(1982). Tertiary Procyonidae (Mammalia: Carnivora) of North America. *Journal of Vertebrate Paleontology*, 2, 71–93.

(1991). Early Pliocene horses from late Pleistocene fluvial deposits, Gulf Coastal Plain, South Texas. *Journal of Paleontology*, 65, 995–1006.

(2003). New procyonines from the Hemingfordian and Barstovian of the Gulf Coast and Nevada, including the first fossil record of the Potosini. *Bulletin of the American Museum of Natural History*, 279, 125–46.

(2005). Carnivora from the late Miocene Love Bone Bed of Florida. *Bulletin of the Florida Museum of Natural History*, 45, 413–34.

Baum, G. R. and Wheeler, W. H. (1977). Cetaceans from the St. Marys and Yorktown Formations, Surry County, Virginia. *Journal of Paleontology*, 51, 492–504.

Bayne, C. K. (1976). *Guidebook for the 24th Annual Meeting of the Midwestern Friends of the Pleistocene: Stratigraphy and Faunal Sequence of the Meade County, Kansas*. Lawrence, KA: Kansas Geological Survey.

Beard, K. C. (2000). A new species of *Carpocristes* (Mammalia; Primatomorpha) from the middle Tiffanian of the Bison Basin, Wyoming, with notes on carpolestid phylogeny. *Annals of the Carnegie Museum*, 69, 195–208.

Beard, K. C. and Dawson, M. R. (2001). Early Wasatchian mammals from the Gulf Coastal Plain of Mississippi: biostratigraphic and paleobiogeographic implications. In *Eocene Biodiversity: Unusual Occurrences and Rarely Sampled Habitats*, ed. G. F. Gunnell, pp. 75–94. New York: Kluwer Academic/Plenum.

Beard, K. C. and Tabrum, A. R. (1991). The first early Eocene mammal from eastern North America: an omomyid primate from the Bashi Formation, Lauderdale County, Mississippi. *Mississippi Geology*, 11, 1–6.

Beard, K. C., Dawson, M. R., and Tabrum, A. R. (1995). First diverse land mammal fauna from the early Cenozoic of the southeastern United States: the early Wasatchian Red Hot Local Fauna, Lauderdale County, Mississippi. *Journal of Vertebrate Paleontology*, 19(suppl. to no. 3), p. 18A.

Becker, J. J. (1985) Fossil herons (Aves: Ardeidae) of the late Miocene and early Pliocene of Florida. *Journal of Vertebrate Paleontology*, 5, 24–31.

Becker, J. J. and McDonald, H. G. (1998). The Star Valley local fauna (early Hemphillian), southwestern Idaho. [In *And Whereas: Papers on the Vertebrate Paleontology of Idaho Honoring J. A. White*, Vol. 1, ed. W. A. Akersten, H. G. McDonald, D. J. Meldrum, and M. E. T. Flint.] *Idaho Museum of Natural History Occasional Papers*, 36, 25–49.

Becker, J. J. and White, J. A. (1981). Late Cenozoic geomyids (Mammalia: Rodentia) from the Anza-Borrego desert, southern California. *Journal of Vertebrate Paleontology*, 1, 211–18.

Bell, C. J., Lundelius, E. L., Jr., Barnosky, A. D., *et al.* (2004). The Blancan, Irvingtonian, and Rancholabrean Land Mammal Ages. In *Late Cretaceous and Cenozoic Mammals of North America*, ed. M. O. Woodburne, pp. 232–314. New York: Columbia University Press.

Bell, S. D. and Bryant, H. N. (2002). Early Oligocene rodents from the Rodent Hill Locality in the Cypress Hills Formation, southwest Saskatchewan. *Journal of Vertebrate Paleontology*, 22(suppl. to no. 3), p. 35A.

Bennett, D. K. (1979). The fossil fauna of Lost and Found Quarries (Hemphillian, latest Miocene), Wallace county, Kansas. *Occasional Papers of the Museum of Natural History, University of Kansas*, 79, 1–24.

Berkoff, M. and Barnes, L. G. (1998). The evolution of the dusignathines; pseudo-walruses of the late Miocene. *PaleoBios*, 18(suppl, to no. 3), p. 1–2.

Berry, E. W. and Gregory, W. K. (1906). *Prorosmarus alleni*, a new genus and species of walrus from the upper Miocene of Yorktown, Virginia. *American Journal of Science*, 21, 444–51.

Berta, A. (1991). New *Enaliarctos* (Pinnipedimorpha) from the Oligocene and Miocene of Oregon and the role of "enaliarctids" in pinniped phylogeny. *Smithsonian Contributions to Paleobiology*, 69, 1–33.

(1994). New specimens of the pinnipediform *Pteronarctos* from the Miocene of Oregon. *Smithsonian Contributions to Paleobiology*, 78, 1–30.

Berta, A. and Deméré, T. A. (1986). *Callorhinus gilmorei* n. sp., (Carnivora: Otariidae) from the San Diego Formation (Blancan) and its implications for otarid phylogeny. *Transactions of the San Diego Society of Natural History*, 21, 111–26.

Bjork, P. R. (1967). Latest Eocene vertebrates from Northwestern South Dakota. *Journal of Paleontology*, 41, 227–36.

(1970). The carnivores of the Hagerman Local Fauna (late Pliocene) of Southwestern Idaho. *Transactions of the American Philosophical Society*, 60, 1–54.

(1973). Additional carnivores from the Rexroad Formation (upper Pliocene) of Southwestern Kansas. *Transactions of the Kansas Academy of Science*, 76, 24–8.

Bjork, P. R. and Macdonald, J. R. (1981). Geology and paleontology of the Badland and Pine Ridge area, South Dakota. In Guidebook for the Annual Field Conference of the Rocky Mountain Section of the Geological Society of America: *Geology of the Black Hills, South Dakota and Wyoming*, ed. F. J. Rich, pp. 211–21. Rapid City, SD: Geological Society of America.

Black, C. C. (1963). Miocene rodents from the Thomas Farm local fauna, Florida. *Bulletin of the Museum of Comparative Zoology*, 128, 483–501.

(1967). Middle and late Eocene mammal communities: a major discrepancy. *Science* 156, 62–4.

(1978). Paleontology and geology of the Badwater Creek area, Central Wyoming. Part 14. The Artiodactyls. *Annals of the Carnegie Museum*, 47, 223–59.

Black, C. C. and Dawson, M. R. (1966). Paleontology and geology of the Badwater Creek area, central Wyoming. Part 1: History of field work and geological setting. *Annals of the Carnegie Museum*, 38, 297–307.

Bode, F. C. (1935). The fauna of the *Merychippus* Zone, North Coalinga district, California. *Carnegie Institute of Washington Publications*, 453, 66–96.

Bown, T. M. (1979). Geology and mammalian paleontology of the Sand Creek facies, lower Willwood Formation (Lower Eocene), Washakie county, Wyoming. *Geological Survey of Wyoming*, Memoir 2, 1–151.

(1980). The Willwood Formation (Lower Eocene) of the southern Bighorn Basin, Wyoming, and its mammalian fauna. [In *Early Cenozoic Paleontology and Stratigraphy of the Bighorn Basin, Wyoming*, ed. P. D. Gingerich.] *University of Michigan Papers in Paleontology*, 24, 127–36.

(1982). Geology, paleontology, and correlation of Eocene volcanistic rocks, southeast Absaroka Range, Hot Springs county, Wyoming. *United States Geological Survey, Professional Papers*, 1201-A, A1–A75.

Bown, T. M. and Kihm, A. J. (1981). *Xenicohippus*, an unusual new Hyracotherine (Mammalia, Perissodactyla) from lower Eocene rocks of Wyoming, Colorado, and New Mexico. *Journal of Paleontology*, 55, 257–70.

Bown, T. M. and Rose, K. D. (1987). Patterns of dental evolution in early Eocene anaptomorphine primates (Omomyidae) from the Bighorn Basin, Wyoming. *Journal of Paleontology*, 61(suppl. to no. 5, *Paleontological Society Memoir* 23), pp. 1–16.

Bown, T. M., Rose, K. D., Simons, E. L., and Wing, S. L. (1993). Distribution and stratigraphic correlation of upper Paleocene and lower Eocene fossil mammals and plant localities of the Fort Union, Willwood and Tatman Formations, southern Bighorn basin, Wyoming. *United States Geological Survey Professional Papers*, 1540, 1–269.

Brabb, E. E., Graymer, R. W., and Jones, D. L. (1998). Geology of the onshore part of San Mateo County, California: a digital database. *United States Geological Survey Open File Report*, OF 98-137, 1–9.

Bryant, H. D. (1992). The Carnivora of the Lac Pelletier lower fauna (Eocene: Duchesnean), Cypress Hills Formation, Saskatchewan. *Journal of Vertebrate Paleontology*, 66, 847–55.

Bryant, J. D. (1991). New early Barstovian (middle Miocene) vertebrates from the upper Torreya Formation, eastern Florida panhandle. *Journal of Vertebrate Paleontology*, 11, 472–89.

Bryant, J. D., MacFadden, B. J., and Mueller, P. A. (1992). Improved chronologic resolution of the Hawthorn and Alum Bluff groups in northern Florida: implications for Miocene chronostratigraphy. *Bulletin of the Geological Society of America*, 104, 208–18.

Burns, J. A. and Young, R. R. (1988). Stratigraphy and palaeontology of the Hand Hills Region. (Society of Vertebrate Paleontology 48th AGM Field Trip), *Occasional Papers of the Tyrell Museum of Paleontology*, 1–13.

Butler, R. F., Krause, D. W., and Gingerich, P. D. (1987). Magnetic polarity and biostratigraphy of middle-late Paleocene continental deposits of south-central Montana. *Journal of Geology*, 95, 647–57.

Buwalda, J. P. (1916). New mammalian faunas from Miocene sediments near Tehachapi Pass in the Southern Sierra Nevada. *University of California Publication, Bulletin of the Department of Geological Sciences*, 10, 75–85.

Cabral-Perdomo, M., Bravo-Cuevas, V., and Castillo-Ceron, J. (2005). A young gomphothere skull from the state of Hidalgo, Central Mexico. *Journal of Vertebrate Paleontology*, 25(suppl. to no. 3), 41A.

Carranza-Casteñeda, O. and Miller, W. E. (1993). Hemphillian and Blancan equids from Hildago, Mexico. *Journal of Vertebrate Paleontology*, 13(suppl. to no. 3), p. 29A.

(2002). Paleontology and stratigraphy of the Tecolotlen Basin, Jalisco, Mexico. *Journal of Vertebrate Paleontology* 23(suppl. to no. 3), pp. 41A–2A.

(2004). Late Tertiary terrestrial mammals from central Mexico and their relationship to South American immigrants. *Revista Brasileria de Paleontologia*, 7, 249–61.

Carranza-Castañada, O., Miller, W. E., and Kowallis, B. J. (1998). New vertebrate faunas from the Transmexican volcanic belt, Central Mexico. *Journal of Vertebrate Paleontology*, 18(suppl. to no. 3), p. 31A.

Cassiliano, M. L. (1980). Stratigraphy and vertebrate paleontology of the Horse Creek: Trail Creek area, Laramie county, Wyoming. *Contributions to Geology, University of Wyoming*, 19, 25–68.

(1999). Biostratigraphy of Blancan and Irvingtonian mammals in the Fish Creek: Vallecito Creek Section, southern California, and a review of the Blancan–Irvingtonian boundary. *Journal of Vertebrate Paleontology*, 19, 169–86.

Castillo-Ceron, J. M. (2000). Fossil vertebrates from the Miocene of Hidalgo, Mexico. *Journal of Vertebrate Paleontology*, 20(suppl. to no. 3), p. 34A.

Chiment, J. J. and Korth, W. W. (1996). A new genus of eomyid rodent (Mammalia) from the Eocene (Uintan–Duchesnean) of Southern California. *Journal of Vertebrate Paleontology*, 16, 116–24.

Clark, J. (1937). The stratigraphy and paleontology of the Chadron Formation in the Big Badlands of South Dakota. *Annals of the Carnegie Museum*, 25, 261–350.

Clark, J. B. and Guensburg, T. E. (1972). Arctoid genetic characters as related to the genus *Parictis*. *Fieldiana (Geology)*, 26, 1–73.

Clark, J. B., Dawson, M. R., and Wood, A. E. (1964). Fossil mammals from the lower Pliocene of the Fish Lake Valley, Nevada. *Bulletin of the Museum of Comparative Zoology*, 131, 27–63.

Clark, J. B., Beerbower, J. R., and Kietzke, K. K. (1967). Oligocene sedimentation, stratigraphy, paleoecology and paleoclimatology in the Big Badlands of South Dakota. *Fieldiana, Geology Memoirs*, 5, 1–158.

Clark, J. M. (1990). A new early Miocene species of *Paleoparadoxia* (Mammalia: Desmostylia) from California. *Journal of Vertebrate Paleontology*, 11, 490–508.

Clemens, W. A. (1964). Records of the fossil mammal *Sinclairella*, family Apatemyidae, from the Chadronian and Orellan. *Kansas University Museum of Natural History Publications*, 14, 483–91.

Clyde, W. C. (2001). Mammalian biostratigraphy of the McCullough Peaks area in the northern Bighorn Basin. [In *Paleocene–Eocene Stratigraphy and Biotic Change in the Bighorn and Clarks Fork Basins, Wyoming*, ed. P. D. Gingerich.] *University of Michigan Papers on Paleontology*, 33, 109–26.

Colbert, E. H. (1932). *Aphelops* from the Hawthorn Formation. *Bulletin of the Florida Geological Survey*, 10, 55–8.

Colbert, M. W. (2006). *Hesperaletes* (Mammalia: Perissodactyla). A new tapiroid from the middle Eocene of southern California. *Journal of Vertebrate Paleontology*, 26, 697–711.

Cook, H. J. (1922). A Pliocene fauna from Yuma County, Colorado, with notes on the closely related Snake Creek beds from Nebraska. *Proceedings of the Colorado Museum of Natural History*, 4, 1–29.

(1965). Runningwater Formation, middle Miocene of Nebraska. *American Museum Novitates*, 2227, 1–8.

(1984). Excavation of a late Arikareean vertebrate assemblage in northwest Nebraska. *National Geographic Research Reports*, 16, 145–52.

Cooke, C. W. and Shearer, H. K. (1918). Deposits of Claiborne and Jackson age in Georgia. *US Geological Survey Professional Paper*, 120(C), 41–81.

Coombs, M. C., Hunt, R. M., Jr., Stepleton, E., Albright, L B., III, and Fremd, T. J. (2001). Stratigraphy, chronology, biogeography and taxonomy of early Miocene small chalicotheres in North America. *Journal of Vertebrate Paleontology*, 2, 607–20.

Cope, E. D. (1870). Discovery of a huge whale in North Carolina. *The American Naturalist*, 4, 128.

Corgan, J. X. (1976). Vertebrate fossils of Tennessee. *Bulletin of the State of Tennessee Department of Conservation Division of Geology*, 77, 1–100.

Corona-M. E. and Alberdi, M. T. (2006). Two new records of Gompthotheriidae (Mammalia: Proboscidea) in southern México and some biogeographic implications. *Journal of Paleontology*, 80, 357–66.

Czaplewski, N. J. (1987) Middle Blancan vertebrate assemblage from the Verdi Formation, Arizona. *Contributions to Geology, University of Wyoming*, 25, 133–55.

(1993). Late Tertiary bats (Mammalia, Chiroptera) from the southwestern United States. *The Southwestern Naturalist*, 38, 111–18.

Czaplewski, N. J. and G. S. Morgan. 2000. A new vespertilionid bat (Mammalia: Chiroptera) from the early Miocene (Hemingfordian) of Florida, USA. *Journal of Vertebrate Paleontology*, 20, 736–42.

Czaplewski, N. J., Morgan, G. S., and Naeher, T. (2003). Molossid bats from the late Tertiary of Florida with a review of the Tertiary Molossidae of North America. *Acta Chiropterologica*, 5, 61–74.

Dalquest, W. W. (1975). Vertebrate fossils from the Blanco Local Fauna of Texas. *Occasional Papers of the Museum of Texas Technical University*, 30, 1–52.

(1978). Early Blancan mammals of the Beck Ranch Local Fauna of Texas. *Journal of Mammalogy*, 59, 269–98.

(1983). Mammals of the Coffee Ranch local fauna, Hemphillian of Texas. *Pearce-Sellards Series, Texas Memorial Museum*, 38, 1–41.

Dalquest, W. W. and Hughes, J. T. (1966). A new mammalian local fauna from the lower Pliocene of Texas. *Transactions of the Kansas Academy of Sciences*, 69, 79–87.

Dalquest, W. W. and Mooser, O. (1974). Miocene vertebrates from Aguascalientes, Mexico. *Pearce-Sellards Series, Texas Memorial Museum*, 21, 1–10.

(1980). The Late Hemphillian mammals of the Ocote Local Fauna, Guanajualo, Mexico. *Pearce-Sellards Series, Texas Memorial Museum*, 32, 1–25.

Dalquest, W. W. and Patrick, D. B. (1989). Small mammals from the early and medial Hemphillian of Texas, with descriptions of a new bat and gopher. *Journal of Vertebrate Paleontology*, 9, 78–88.

Dalquest, W. W., Baskin, J. A., and Schultz, G. E. (1996). Fossil mammals from a late Miocene (Clarendonian) site in Beaver County, Oklahoma. In *Contributions in Mammalogy: A Memorial Volume Honoring J. Knox Jones, Jr.*, ed. H. H. Genoways and R. J. Baker, pp. 107–37. Lubbock, TX: Museum of the Texas Technical University.

Daly, E. (1992). A list, bibliography and index of the fossil vertebrates of Mississippi. *Mississippi Department of Environmental Quality Office of Geology Bulletin*, 128, 1–47.

(1999). A middle Eocene *Zygorhiza* specimen from Mississippi (Cetacea, Archaeoceti). *Mississippi Geology*, 20, 21–31.

Damuth, J. (1982). Analysis of the preservation of community structure in assemblages of fossil mammals. *Paleobiology*, 8, 434–46.

Davidson, J. R. (1987). Geology and mammalian paleontology of the Wind River Formation, Laramie basin, southeastern Wyoming. *Contributions to Geology, University of Wyoming*, 25, 103–32.

Dawson, M. R. (1980). Geology and paleontology of the Badwater Creek area, central Wyoming. Part 20: The Late Eocene Creodonta and Carnivora. *Annals of the Carnegie Museum*, 49, 79–91.

(2003). Phylogenetic and geographic affinities of the early Miocene vertebrate fauna of Devon Island, Nunavut, Canada. *Journal of Vertebrate Paleontology*, 23, 44A.

Dawson, M. R., Krishtalka, L, and Stucky, R. K. (1990). Revision of the Wind River Faunas, early Eocene of Central Wyoming. Part 9: The oldest known hystricomorphous rodent (Mammalia: Rodentia). *Annals of the Carnegie Museum*, 59, 135–47.

Delson, E. (1971). Fossil mammals of the Early Wasatchian Powder Basin Local Fauna, Eocene of northeast Wyoming. *Bulletin of the American Museum of Natural History*, 146, 309–64.

Deméré, T. A. (1988). Early Arikareean (late Oligocene) vertebrate fossils and biostratigraphic correlations of the Otay Formation at Eastlake, San Diego County, California. [In *Paleogene Stratigraphy, West Coast of North America, Pacific Section*, ed. M. V. Filewicz and R. L. Squires.] *Society of Economic Paleontologists and Mineralogists, West Coast Paleogene Symposium*, 58, 35–43.

(1993). Fossil mammals from the Imperial Formation (upper Miocene–lower Pliocene), Coyote Mountains, Imperial County, California. *San Bernardino County Museum Association Special Publication*, 93, 182–5.

(1994). Two new species of fossil walruses from the Upper Pliocene San Diego Formation. *Proceedings of the San Diego Society of Natural History*, 29, 77–98.

Deméré, T. A. and Berta, A. (2001). A reevaluation of *Proneotherium repenningi* from the Miocene Astoria Formation of Oregon and its position as a basal odobenid (Pinnipedia: Mammalia). *Journal of Vertebrate Paleontology*, 21, 279–310.

(2002). The Miocene pinniped *Desmatophoca oregonensis* Condon, 1906 (Mammalia, Carnivora), from the Astoria Formation, Oregon. [In *Cenozoic Mammals of Land and Sea; Tributes to the Career of Clayton E. Ray*, ed. R. J. Emry.] *Smithsonian Contributions to Paleobiology*, 93, 113–47.

Dockery, D. T., III and Johnston, J. E. (1986). Excavation of an archaeocete whale, *Basilosaurus cetoides* (Owen), from Madison, Mississippi. *Mississippi Geology*, 6, 1–10.

Domning, D. P. (1978). Sirenian evolution in the North Pacific Ocean. *University of California, Publications in Geological Science*, 118, 1–176.

(1989). Fossil Sirenia of the West Atlantic and Caribbean region, III: *Xenosiren yucateca* gen. et sp. nov. *Journal of Vertebrate Paleontology*, 9, 429–37.

(1990). Fossil Sirenia of the West Atlantic and Caribbean region, IV: *Corystosiren varguezi*, gen. et sp. nov. *Journal of Vertebrate Paleontology*, 10, 361–71.

(2001). The earliest known fully quadrupedal sirenian. *Nature*, 413, 625–7.

Domning, D. P., Morgan, G. S., and Ray, C. E. (1982). North American Eocene Sea Cows (Mammalia: Sirenia). *Smithsonian Contributions to Paleobiology*, 52, 1–69.

Dooley, A. C., Jr. (1993). The vertebrate fauna of the Calvert Formation (Middle Miocene) at the Caroline Stone Quarry, Caroline County, VA. *Journal of Vertebrate Paleontology*, 13(suppl. to no. 3), p. 33A.

(2005). A new species of *Squaladon* (Mammalia, Cetacea) from the Middle Miocene of Virginia. *Virginia Museum of Natural History Memoirs*, 8, 1–14.

Dooley, A. C., Jr. and Frazer, N. C. (2005). A revised faunal list for the Carmel Church Quarry, Caroline County, Virginia. *Journal of Vertebrate Paleontology*, 25(suppl. to no. 3), p. 52A.

Dooley, A. C., Jr., Fraser, N. C., and Luo, Z-X. (2004). The earliest known member of the rorqual-gray whale clade (Mammalia, Cetacea). *Journal of Vertebrate Paleontology*, 24, 453–63.

Dorr, J. A., Jr. (1956). Anceny local mammal fauna, latest Miocene, Madison Valley Formation, Montana. *Journal of Paleontology*, 30, 62–74.

(1978). Revised and amended fossil vertebrate faunal lists, early Tertiary, Hoback basin, Wyoming. *Contributions to Geology, University of Wyoming*, 16, 79–84.

Dorr, J. A., Jr. and Gingerich, P. D. (1980). Early Cenozoic mammalian paleontology, geologic structure, and tectonic history in the overthrust belt near LaBarge, western Wyoming. *Contributions to Geology, University of Wyoming*, 18, 101–15.

Dorr, J. A., Jr. and Wheeler, W. H. (1964). Cenozoic paleontology, stratigraphy, and reconnaissance geology of the Upper Ruby River Basin, Southwestern Montana. *Contributions of the Museum of Paleontology, University of Michigan*, 13, 297–339.

Douglass, E. (1901). Fossil Mammalia of the White River Beds of Montana. *Transactions of the American Philosophical Society*, 20, 237–79.

Downs, T. (1956a). A new pinniped from the Miocene of Southern California, with remarks on the Otariidae. *Journal of Paleontology*, 30, 115–31.

(1956b). The Mascall fauna from the Miocene of Oregon. *University of California, Publications in Geological Sciences*, 31, 199–354.

Downs, T. and White, J. A. (1968). A vertebrate faunal succession in superposed sediments from late Pliocene to middle Pleistocene in California. *XXIII International Geological Congress*, 10, 41–7.

Drescher, A. B. (1942). Later Tertiary Equidae from the Tejon Hills, California. [In *Studies of Cenozoic Vertebrates of Western North America and of Fossil Primates*, ed. A. B. Drescher, E. L. Furlong, I. S. Demay, P. C., *et al.*] *Carnegie Institute of Washington Publications*, 530, 1–23.

Eaton, G. F. (1922). John Day Felidae in the Marsh Collection. *American Journal of Science*, 5, 425–52.

Eaton, J. G. (1982). Paleontology and correlation of Eocene volcanic rocks in the Carter Mountain area, Park county, southeastern Absaroka range, Wyoming. *Contributions to Geology, University of Wyoming*, 21, 153–94.

(1985). Paleontology and correlation of the Eocene Tepee Trail and Wiggins Formations in the North Fork of Owl Creek area, southeastern Absaroka range, Hot Springs county, Wyoming. *Journal of Vertebrate Paleontology*, 5, 345–70.

Eaton, J. G., Hutchison, J. H., Holroyd, P. A., Korth, W. W., and Goldstrand, P. M. (1999). Vertebrates of the Turtle Basin Local Fauna, middle Eocene, Sevier Plateau, south-central Utah. [In *Vertebrate Paleotology in Utah*, ed. D. D. Gillette.] *Utah Geological Survey Miscellaneous Publications*, 99-1, 463–8.

Eberle, J. J. (2001). Early Eocene Leptictida, Pantolesta, Creodonta, Carnivora, and Cete from Ellesmere Island: Arctic links to Europe and Asia. *Journal of Vertebrate Paleontology*, 21(suppl. to no. 3), p. 46A.

(2005). Early Eocene Brontotheriidae (Perissodactyla) from the Eureka Sound Group, Ellesmere Island, Canadian High Arctic: implications for brontothere origins and high-latitude dispersal. *Journal of Vertebrate Paleontology*, 25(suppl. to no. 3), p. 52A.

Eberle, J. J. and Lillegraven, J. A. (1998a). A new important record of earliest Cenozoic mammalian history: geologic setting, Multituberculata, and Peradectia. *Rocky Mountain Geology*, 33, 3–47.

(1998b). A new important record of earliest Cenozoic mammalian history: Eutheria and paleogeographic/biostratigraphic summaries. *Rocky Mountain Geology*, 33, 49–117.

Eberle, J. J. and Storer, J. E. (1999). Northernmost record of brontotheres, Axel Heiberg Island, Canada: implications for age of the Buchanan Lake Formation and brontothere paleobiology. *Journal of Paleontology*, 73, 979–83.

Eberle, J. J., Johnson, K., Raynolds, R., Hicks, J., and Nichols, D. (2002). New discoveries of Puercan mammals in the Denver Basin, Colorado: Revisions to local Puercan mammalian biostratigraphy that incorporate paleomagnetic and palynological zonations. *Journal of Vertebrate Paleontology*, 22(suppl. to no. 3), p. 50A.

Emlong, D. R. (1966). A new archaic cetacean from the Oligocene of northwest Oregon. *Bulletin of the Oregon University Museum of Natural History*, 3, 1–51.

Emry, R. J. (1973). Stratigraphy and preliminary biostratigraphy of the Flagstaff Rim Area, Natrona County, Wyoming. *Smithsonian Contributions to Paleobiology*, 18, 1–43.

(1975). Revised Tertiary stratigraphy and paleontology of the Western Beaver Divide, Fremont county, Wyoming. *Smithsonian Contributions to Paleontology*, 25, 1–20.

(1978). A new hypertragulid (Mammalia, Ruminantia) from the early Chadronian of Wyoming and Texas. *Journal of Paleontology*, 52, 1004–14.

(1979). Review of *Toxotherium* (Perissodactyla, Rhinoceratoidea) with new material from the early Oligocene of Wyoming. *Proceedings of the Biological Society of Washington*, 92, 28–41.

(1981). Additions to the mammalian faunas of the type Duchesnean, with comments on the status of the "Duchesnean" age. *Journal of Paleontology*, 55, 563–76.

(1990). Mammals of the Bridgerian (middle Eocene) Elderberry Canyon Local Fauna of eastern Nevada. In *Dawn of the Age of Mammals in the Northern Part of the Rocky Mountain Interior, North America*, ed. T. M. Bown and K. D. Rose. *Geological Society of America Special Papers*, 243, 187–209.

(1992). Mammalian range zones in the Chadronian White River Formation at Flagstaff Rim, Wyoming. In *Eocene–Oligocene Climatic and Biotic Evolution*, ed. D. R. Prothero and W. A. Berggren, pp. 106–15. Princeton, NJ: Princeton University Press.

Emry, R. J. and Eshelman, R. E. (1998). The Pollack Pit Local Fauna (early Hemingfordian, early Miocene): first Tertiary land mammals from Delaware. [In *Geology and Paleontology of the lower Miocene Pollack Farm fossil Site, Delaware*, ed. R. N. Benson.] *Special Publication of the Delaware Geological Survey*, 21, 153–73.

Emry, R. J. and Korth, W. W. (1989). Rodents of the Bridgerian (middle Eocene) Elderberry Canyon local fauna of eastern Nevada. *Smithsonian Contributions to Paleobiology*, 67, 1–14.

(1993). Evolution in Yoderimyinae (Eomyidae: Rodentia), with new material from the White River Formation (Chadronian) at Flagstaff Rim, Wyoming. *Journal of Paleontology*, 67, 1047–57.

(1996). The Chadronian squirrel "*Sciurus*" *jeffersoni* Douglass, 1901: a new generic name, new material, and its bearing on the early evolution of Sciuridae (Rodentia). *Journal of Vertebrate Paleontology*, 16, 775–80.

Emry, R. J. and Storer, J. E. (1981). The hornless protoceratid *Pseudoprotoceras* (Tylopoda: Artiodactyla) in the early Oligocene of Saskatchewan and Wyoming. *Journal of Vertebrate Paleontology*, 1, 101–10.

Emry, R. J., Russell, L. S., and Bjork, P. R. (1987). The Chadronian, Orellan, and Whitneyan North American Land Mammal Ages. In *Cenozoic Mammals of North America, Geochronology and Biostratigraphy*, ed. M. O. Woodburne, pp. 118–52. Berkeley, CA: University of California Press.

Emslie, S. D. (1998). Avian community, climate, and sea-level changes in the Plio-Pleistocene of the Florida Peninsula. *Ornithological Monographs*, 50, 1–113.

Erickson, B. R. (1991). Flora and fauna of the Wannagan Creek Quarry: late Paleocene of North America. *Scientific Publications of the Science Museum of Minnesota, New Series*, 7, 5–19.

Eshelman, R. E. (1975). Geology and paleontology of the early Pleistocene (late Blancan) White Rock Fauna from North Central Kansas. [In *Papers on Paleontology, Claude Hibbard Memorial*, Vol 4.] *Museum of Paleontology, University of Michigan*, 13, 1–60.

Evanoff, E. and Toledo, P. M., de (1999). Fossil mammals and biting insects from the upper Eocene Florissant Formation of Colorado. *Journal of Vertebrate Paleontology*, 19(suppl. to no. 3), p. 43A.

Evanoff, E., Robinson, P., Murphey, P., Kron, D., and Engard, D. (1994). An early Uintan fauna from the Bridger "E." *Journal of Vertebrate Paleontology*, 13(suppl. to no. 3), p. 24A.

Farlow. J. O., Sunderman, J. A., Havens, J. J., *et al.* (2001). The Pipe Creek Sinkhole biota, a diverse late Tertiary continental fossil assemblage from Grant County, Indiana. *The American Midland Naturalist*, 145, 367–78.

Ferrusquía-Villafranca, I. (1984). A review of the early and middle Tertiary faunas of Mexico. *Journal of Vertebrate Paleontology*, 4, 187–98.

(1990). Biostratigraphy of the Mexican continental Miocene. *Paleontologica Mexicana*, 56, 1–149.

(1993). Contributions to the knowledge of Mexico's Oligocene mammals: additions and revisions of the Chadronian Rancho Gaitan local fauna, Northeastern Chihuahua. *Journal of Vertebrate Paleontology*, 3(suppl. to no. 3), pp. 34A–5A.

(2003). Mexico's middle Miocene mammalian assemblages: an overview. *Bulletin of the American Museum of Natural History*, 279, 321–47.

(2005). The Marfil local fauna, Bridgerian of Guanajuato, central Mexico: review and significance. A progress report on the southernmost

Paleogene tetrapod assemblage of North America. *Journal of Vertebrate Paleontology*, 25(suppl. to no. 3), p. 58A.

Fisher, R. V. and Rensberger, J. M. (1972). Physical stratigraphy of the John Day Formation, Central Oregon. *University of California, Publications in Geological Sciences*, 101, 1–45.

Flanagan, K. M. (1986). Early Eocene rodents from the San Jose Formation, San Juan Basin, New Mexico. [In *Vertebrates, Phylogeny and Philosophy*, ed. K. M. Flanagan and J. A. Lillegraven.] *Contributions to Geology, University of Wyoming, Special Paper*, 3, 163–75.

Flynn, J. J., Kowallis, B. J., Nuñez, C., *et al.* (2005). Geochronology of Hemphillian–Blancan aged strata, Guanajuato, Mexico, and implications for the timing of the Great American Biotic Interchange. *Journal of Geology*, 113, 287–307.

Fordyce, R. E. (2002). *Simocetus rayi* (Odontoceti: Simocetidae, New Family): a bizarre new archaic Oligocene dolphin from the eastern North Pacific. *Smithsonian Contributions to Paleobiology*, 93, 185–222.

Forstén, A. (1970). The late Miocene Trail Creek mammalian fauna. *Contributions to Geology, University of Wyoming*, 9, 39–51.

(1975). The fossil horses of the Texas Gulf Coastal Plain: a revision. *Pearce-Sellards Series, Texas Memorial Museum*, 22, 1–86.

(1991). Size trends in Holarctic anchitherines (Mammalia, Equidae). *Journal of Paleontology*, 65, 147–59.

Foss, S. E., Turnbull, W. D., and Barber, L. (2001). Observations on a new specimen of *Achaenodon* (Mammalia, Artiodactyla) from the Eocene Washakie Formation of southern Wyoming. *Journal of Vertebrate Paleontology*, 21(suppl. to no. 3), p. 51A.

Fox, R. C. (1990). The succession of Paleocene mammals in Western Canada. *Geological Society of America Special Papers*, 243, 51–70.

Frailey, D. (1978). An early Miocene (Arikareean) fauna from North Central Florida (the SB-1A Local Fauna). *Occasional Papers of the Museum of Natural History, University of Kansas*, 75, 1–20.

(1979). The large mammals of the Buda Local Fauna (Arikareean, Alachua County, Florida). *Bulletin of the Florida State Museum, Biological Sciences*, 24, 123–73.

Fremd, T., Bestland. E. A., and Retallack, G. J. (1997). *John Day Basin Paleontology: Field Trip Guide and Road Log, for the 1994 Annual Meeting of the Society of Vertebrate Paleontology.* Seattle, WA: Northwest Interpretive Association in association with John Day Fossil Beds National Monument, Kimberly, OR.

Frick, C. (1921). Extinct vertebrate faunas of the Badlands of Bautista Creek and San Timoteo Canyon, southern California. *University of California Publications, Bulletin of the Department of Geological Sciences*, 12, 277–424.

Froelich, D. J. and Breithaupt, B. H, (1997). A *Lambdotherium* specimen from the Fossil Butte Member of the Green River Formation, with comments on its biostratigraphic and paleoenvironmental importance and the phylogenetic significance of its postcrania. *Journal of Vertebrate Paleontology*, 17(suppl. to no. 3), p. 47A.

(1990). Mammals from the Eocene epoch Fossil Butte Member of the Green River Formation, Fossil Basin, Wyoming. *Journal of Vertebrate Paleontology*, 18, 43–4.

Frye, J. C., Leonard, A. B., and Swineford, A. (1956). Stratigraphy of the Ogallala Formation (Neogene) of Northern Kansas. *Bulletin of the State Geological Survey of Kansas*, 118, 1–92.

Furlong, E. L. (1941). A new Pliocene antelope from Mexico with remarks on some known antilocaprids. [In *Studies of Cenozoic Vertebrates of Western North America and of Fossil Primates*, ed. A. B. Drescher, E. L. Furlong, I. S. Demay, *et al.*] *Carnegie Institute of Washington Publication*, 530, 25–33.

Galbreath, E. C. (1953). A contribution to the Tertiary geology and paleontology of Northeastern Colorado. *Paleontological Contributions*, 4, 1–120.

Galusha, T. (1975). Stratigraphy of the Box Butte Formation, Nebraska. *Bulletin of the American Museum of Natural History*, 56, 1–68.

Galusha, T. and Blick, J. C. (1971). Stratigraphy of the Santa Fe Group, New Mexico. *Bulletin of the American Museum of Natural History*, 144, 1–127.

Gazin, C. L. (1956). Paleocene mammalian faunas of the Bison Basin in South-Central Wyoming. *Smithsonian Miscellaneous Collections*, 131, 1–57.

(1962). A further study of the lower Eocene mammalian faunas of southwestern Wyoming. *Smithsonian Miscellaneous Collections*, 144, 1–98.

(1969). A new occurrence of Paleocene mammals in the Evanston Formation, southwestern Wyoming. *Smithsonian Contributions to Paleontology*, 2, 1–16.

(1971). Paleocene primates from the Shotgun Member of the Fort Union Formation in the Wind River Basin, Wyoming. *Proceedings of the Biological Society of Washington*, 84, 13–37.

(1976). Mammalian faunal zones of the Bridger middle Eocene. *Smithsonian Contributions to Paleontology*, 26, 1–25.

Geisler, J., Sanders, A. E., and Luo, Z. (1996). A new protocetid cetacean from the Eocene of South Carolina, USA: phylogenetic and biogeographic implications. *Paleontological Society Special Publications*, 8, 139.

Gingerich, P. D. (1976). Cranial anatomy and evolution of early Tertiary Plesiadapidae (Mammalia, Primates). *University of Michigan Papers on Paleontology*, 15, 1–141.

(1979). *Lambertocyon eximus*, a new arctocyonid (Mammalia, Condylarthra) from the late Paleocene of Western North America. *Journal of Paleontology*, 53, 524–9.

(1980). A new species of *Palaeosinopa* (Insectivora: Pantolestidae) from the Late Paleocene of Western North America. *Journal of Mammalogy*, 61, 449–54.

(1983). New Adapisoricidae, Pentacodontidae, and Hyposodontidae (Mammalia, Insectivora and Condylarthra), from the late Paleocene of Wyoming and Colorado. *Contributions from the Museum of Paleontology, University of Michigan*, 26, 197–225.

(1987). Early Eocene bats (Mammalia, Chiroptera) and other vertebrates in freshwater limestones of the Willwood Formation, Clark's Fork Basin, Wyoming. *Contributions from the Museum of Paleontology, University of Michigan*, 27, 275–320.

(1989). New earliest Wasatchian mammalian fauna from the Eocene of northwestern Wyoming; composition and diversity in a rarely sampled high-floodplain assemblage. *University of Michigan Papers on Paleontology*, 28, 1–97.

(2001). Biostratigraphy of the continental Paleocene–Eocene boundary interval on Polecat Bench in the northern Bighorn Basin. [In *Paleocene–Eocene Stratigraphy and Biotic Change in the Bighorn and Clarks Fork Basins, Wyoming*, ed. P. D. Gingerich.] *University of Michigan Papers on Paleontology*, 33, 37–71.

Gingerich, P. D. and Clyde, W. C. (2001). Overview of mammalian biostratigraphy in the Paleocene–Eocene Fort Union and Willwood Formations of the Bighorn and Clarks Fork Basins. [In *Paleocene–Eocene Stratigraphy and Biotic Change in the Bighorn and Clarks Fork Basins, Wyoming*, ed. P. D. Gingerich.] *University of Michigan Papers on Paleontology*, 33, 1–14.

Gingerich, P. D. and Winkler, D. A. (1985). Systematics of Paleocene Viverravidae (Mammalia, Carnivora) in the Bighorn Basin and Clark's Fork Basin, Wyoming. *Contributions from the Museum of Paleontology, University of Michigan*, 27, 87–128.

Gingerich, P. D., Rose, K. D., and Krause, D. W. (1980). Early Cenozoic mammalian fauna of the Clark's Fork Basin, Polecat Bench Area, Northwestern Wyoming. [In *Early Cenozoic Paleontology and Stratigraphy of the Bighorn Basin, Wyoming*, ed. P. D. Gingerich.] *University of Michigan Papers in Paleontology*, 24, 51–68.

Gingerich, P. D., Houde, P., and Krause, D. W. (1983). A new earliest Tiffanian (late Paleocene) mammalian fauna from Bangtail Plateau, western Crazy Mountain Basin, Montana. *Journal of Paleontology*, 57, 957–70.

Golz, D. J. (1979). Eocene Artiodactyla of Southern California. *Science Bulletin of the Natural History Museum of Los Angeles County*, 26, 1–85.

Golz, D. J. and Lillegraven, J. A. (1977). Survey of known occurrences of terrestrial vertebrates from Eocene strata of southern California. *Contributions to Geology, University of Wyoming*, 15, 43–64.

Golz, D. J., Jefferson, G. T., and Kennedy, M. P. (1977). Late Pliocene vertebrate fossils from the Elsinore fault zone, California. *Journal of Paleontology*, 51, 864–6.

González-Barba, G., Olivares-Bañuelos, N. C., and Goedert, J. L. (2001). Cetacean teeth from the Late Oligocene San Gegoria and El Cien Formations, Baja California Sur, Mexico. *Proceedings of the 97th Annual Meeting of the Geological Society of America Cordilleran Section and the Pacific Section of the American Association of Petroleum Geologists*, Abstract 3247.

Goodwin, H. T. (1995). Systematic revision of fossil prairie dogs with descriptions of two new species. *University of Kansas Museum of Natural History, Miscellaneous Publications*, 86, 1–38.

Grande, L. (1984). Paleontology of the Green River Formation, with a review of the fish fauna; second edition. *Bulletin of the Geological Survey of Wyoming*, 63, 1–333.

Granger, W. (1909). Faunal horizons of the Washakie Formation of southern Wyoming. *Bulletin of the American Museum of Natural History*, 26, 13–23.

Green, M. (1985). Micromammals from the Miocene Bijou Hills Local Fauna. *Dakoterra, South Dakota School of Mines*, 2, 141–54.

Green, M. and Holman, J. A. (1977). A late Tertiary stream-channel deposit from South Bijou Hills, South Dakota. *Journal of Paleontology*, 51, 543–7.

Gregory, J. T. (1942). Pliocene vertebrates from Big Spring Canyon, South Dakota. *University of California Publications, Bulletin of the Department of Geological Sciences*, 26, 307–446.

Gunnell, G. F. (1988). New species of *Unuchinia* (Mammalia: Insectivora) from the Middle Paleocene of North America. *Journal of Paleontology*, 62, 139–41.

(1989). Evolutionary history of Microsyopoidea (Mammalia,?Primates) and the relationship between Plesiadapiformes and Primates. *University of Michigan Papers on Paleontology*, 27, 1–157.

(1995). New notharctine (Primates, Adapiformes) skull from the Uintan (middle Eocene) of San Diego County, California. *American Journal of Physical Anthropology*, 98, 447–70.

(1998). Mammalian fauna from the lower Bridger Formation (Bridger A, early middle Eocene) of the southern Green River Basin, Wyoming. *Contributions from the Museum of Paleontology, University of Michigan*, 30, 83–130.

Gunnell, G. F. and Bartels, W. S. (1994). Early Bridgerian (middle Eocene) vertebrate paleontology and paleoecology of the southern Green River Basin, Wyoming. *Contributions to Geology, University of Wyoming*, 30, 57–70.

(2001a). Basin–margin vertebrate faunas on the Western flank of the Bighorn and Clarks Fork Basins. [In *Paleocene–Eocene Stratigraphy and Biotic Change in the Bighorn and Clarks Fork Basins, Wyoming*, ed. P. D. Gingerich.] *University of Michigan Papers on Paleontology*, 33, 145–55.

(2001b). Basin margins, biodiversity, evolutionary innovation, and the origin of new taxa. In *Eocene Biodiversity: Unusual Occurrences and Rarely Sampled Habitats*, ed. G. F. Gunnell, pp. 403–32. New York: Kluwer Academic/Plenum.

Gunnell, G. F., Bartels, W. S., Gingerich, P. D., and Torres, V. (1992). The Wapiti Valley faunas: early and middle Eocene fossil vertebrates from the North Fork of Shoshone River, Park County, Wyoming. *Contributions from the Museum of Paleontology, University of Michigan*, 28, 247–87.

Gustafson, E. P. (1978). The vertebrate fauna of the Pliocene Ringold Formation, South-Central Washington. *Bulletin of the Museum of Natural History, University of Oregon*, 23, 1–62.

(1985). Soricids (Mammalia, Insectivora) from the Blufftop Local Fauna, Blancan Ringold Formation of central Washington, and the correlation of Ringold Formation faunas. *Journal of Vertebrate Paleontology*, 5, 88–92.

(1986a). Carnivorous mammals of the late Eocene and early Oligocene of Trans-Pecos, Texas. *Bulletin of the Texas Memorial Museum*, 33, 1–66.

(1986b). Preliminary biostratigraphy of the White River Group (Oligocene, Chadron and Brule Formation) in the vicinity of Chadron, Nebraska. *Transactions of the Nebraska Academy of Sciences*, 14, 7–19.

Guthrie, D. A. (1967). The mammalian fauna of the Lysite Member, Wind River Formation (early Eocene) of Wyoming. *Memoirs of the Southern California Academy of Sciences*, 5, 1–53.

(1971). The mammalian fauna of the Lost Cabin Member, Wind River Formation (Lower Eocene) of Wyoming. *Annals of the Carnegie Museum*, 43, 47–113.

Hager, M. W. (1970). Fossils of Wyoming. *Bulletin of the Geological Survey of Wyoming*, 54, 1–50.

(1974). Late Pliocene and Pleistocene history of the Donnelly Ranch vertebrate site, Southeastern Colorado. *Contributions to Geology, University of Wyoming, Special Papers*, 2, 2–62.

Hanna, G. D. and McLellan, M. E. (1924). A new species of whale from the type locality of the Monterey Group. *Proceedings of the California Academy of Sciences*, 13, 237–41.

Hanson, C. B. (1996). Stratigraphy and vertebrate fauna of the Bridgerian–Duchesnean Clarno Formation, north-central Oregon. In *The Terrestrial Eocene–Oligocene Transition in North America*, ed. D. R. Prothero and R. J. Emry, pp. 206–39. Cambridge, UK: Cambridge University Press.

Harington, C. R. (2003). Life at an early Pliocene beaver pond in the Canadian High Arctic. *Journal of Vertebrate Paleontology*, 23(suppl. to no. 3), p. 59A.

Harksen, J. C. and Green, M. (1971). Thin Elk Formation, Lower Pliocene, South Dakota. *South Dakota Geological Survey Reports of Investigation*, 100, 1–7.

Harksen, J. C. and Macdonald, J. R. (1967). Miocene Batesland Formation named in Southwestern South Dakota. *South Dakota Geological Survey, Reports of Investigation*, 96, 1–10.

Harris, A. H. (1993). Quaternary vertebrates of New Mexico. [In *Vertebrate Paleontology in New Mexico*, ed. S. G. Lucas and J. Zikek.] *Bulletin of the New Mexico Museum of Natural History and Sciences*, 2, 179–97.

Harrison, J. A. (1978). Mammals of the Wolf Ranch Local Fauna, Pliocene of the San Pedro Valley, Arizona. *Occasional Papers of the Museum of Natural History, University of Kansas*, 73, 1–18.

(1983). The carnivores of the Edson Local Fauna (late Hemphillian), Kansas. *Smithsonian Contributions to Paleontology*, 54, 1–42.

Harshman, E. N. (1972). Geology and uranium deposits, Shirley Basin area, Wyoming. *United States Geological Survey Professional Papers*, 745, 1–82.

Hartman, J. E. (1986). Paleontology and biostratigraphy of lower part Polecat Bench Formation, southern Bighorn Basin, Wyoming. *Contributions to Geology, University of Wyoming*, 24, 11–63.

Hartman, J. H., Buckley, G. A., Krause, D. W., and Kroeger, T. J. (1989). Paleontology, stratigraphy, and sedimentology of Simpson Quarry (early Paleocene), Crazy Mountains Basin, South Central Montana. [In *1989 Field Conference Guidebook: Montana Centennial Edition*, ed. D. E. French and R. F. Grabb.] *Geologic Resources of Montana*, 1, 173–85.

Hay, O. P. (1921). Description of species of Pleistocene Vertebrata, types and specimens of most of which are preserved in the United States National Museum. *Proceedings of the United States National Museum*, 59, 599–642.

Hayes, F. G. (2000). The Brooksville 2 local fauna (Arikareean, latest Oligocene): Hernando County, Florida. *Bulletin of the Florida Museum of Natural History*, 43, 1–47.

Hearst, J. (1998). Depositional environments of the Birch Creek Local Fauna (Pliocene: Blancan), Owyhee County, Idaho. [In *And Whereas: Papers on the Vertebrate Paleontology of Idaho Honoring J. A. White*, Vol. 1, ed. W. A. Akersten, H. G. McDonald, D. J. Meldrum, and M. E. T. Flint.] *Idaho Museum of Natural History Occasional Papers*, 36, 56–93.

Heaton, T. H. (1993). The Oligocene rodent *Ischyromys* of the Great Plains: replacement mistaken for anagensis. *Journal of Paleontology*, 67, 297–308.

Henshaw, P. C. (1939). A Tertiary mammalian fauna from the Avawatz Mountains, San Bernadino County, California. [In *Studies of Cenozoic Vertebrates and Stratigraphy of Western North America*., ed. P. C. Henshaw, R. W. Wilson, H. Howard, *et al.*] *Carnegie Institute of Washington Publications*, 514, 1–30.

(1942). A Tertiary mammalian fauna from the San Antonio mountains near Tonopah, Nevada. [In *Studies of Cenozoic Vertebrates of Western North America and of Fossil Primates*, ed. A. B. Drescher, E. L. Furlong, I. S. Demay, *et al.*] *Carnegie Institute of Washington Publications*, 530, 77–168.

Hesse, C. J. (1943). A preliminary report on the Miocene vertebrate faunas of southeast Texas. *Transactions of the Texas Academy of Sciences, Proceedings*, 26, 157–79.

Hibbard, C. W. (1941a). Mammals of the Rexroad fauna from the upper Pliocene of Southwestern Kansas. *Transactions of the Kansas Academy of Science*, 44, 265–313.

(1941b). New mammals of the Rexroad fauna, upper Pliocene of Kansas. *The American Midland Naturalist*, 26, 357–68.

(1941c). The Borchers fauna, a new Pleistocene interglacial fauna from Meade County, Kansas. *State Geological Survey of Kansas Bulletin*, 38, 197–220.

(1949). Pliocene Saw Rock Canyon fauna in Kansas. *Contributions from the Museum of Paleontology, University of Michigan*, 7, 91–105.

(1950). Mammals of the Rexroad Formation from Fox Canyon, Kansas. *Contributions from the Museum of Paleontology, University of Michigan*, 8, 113–92.

(1952a). Vertebrate fossils from late Cenozoic deposits of Central Kansas. *University of Kansas Paleontological Contributions (Vertebrata)*, 2, 1–14.

(1952b). The Saw Rock Canyon fauna and its stratigraphic significance. *Papers of the Michigan Academy of Science, Arts and Letters*, 38, 387–411.

(1954). A new Pliocene vertebrate fauna from Oklahoma. *Papers of the Michigan Academy of Science, Arts and Letters*, 34, 339–59.

(1956). Vertebrate fossils from the Meade Formation of Southwestern Kansas. *Papers of the Michigan Academy of Science, Arts and Letters*, 41, 145–201.

(1959). Late Cenozoic microtine rodents from Wyoming and Idaho. *Papers of the Michigan Academy of Science, Arts and Letters*, 44, 3–40.

Hibbard, C. W. and Keenmon, K. A. (1950). New evidence of the Miocene age of the Blacktail Deer Creek Formation in Montana. *Contributions to the Museum of Paleontology, University of Michigan*, 8, 193–204.

Hinderstein, B. and Boyce, J. (1977). The Miocene salamander *Batracosauroides dissimulans* from east Texas. *Journal of Herpetology*, 11, 369–72.

Hirschfeld, S. E. (1981). *Pliometanastes protistus* (Edentata, Megalonychidae) from Knight's Ferry, California, with discussion of early Hemphillian megalonychids. *PaleoBios*, 36, 1–17.

Hirschfeld, S. E. and S. D. Webb. (1968). Plio-Pleistocene megalonychid sloths of North America. *Bulletin of the Florida State Museum*, 12, 213–96.

Holtzman, R. C. (1978). Late Paleocene mammals of the Tongue River Formation, western North Dakota. *North Dakota Geological Survey Report of Investigation*, 65, 1–68.

Honey, J. G. (1988). A mammalian fauna from the base of the Eocene Cathedral Bluffs Tongue of the Wasatch Formation, Cottonwood Creek area, Southeast Washakie basin, Wyoming. *United States Geological Survey Bulletin*, 1669-C, C1–C14.

Honey, J. G. and Alzett, G. (1988). Paleontology, taphonomy, and stratigraphy of the Browns Park Formation (Oligocene and Miocene) near Maybell, Moffat Co., Colorado. *United States Geological Survey Professional Papers*, 1358, 1–52.

Hough, J. (1955). An upper Eocene fauna from the Sage Creek area, Beaverhead county, Montana. *Journal of Paleontology*, 29, 22–36.

Hough, J. and Alf, R. (1956). A Chadron mammal fauna from Nebraska. *Journal of Paleontology*, 30, 132–40.

Hulbert, R. C., Jr. (1987a). A new *Cormohipparion* (Mammalia, Equidae) from the Pliocene (latest Hemphillian and Blancan) of Florida. *Journal of Vertebrate Paleontology*, 7, 451–68.

(1987b). Late Neogene *Neohipparion* (Mammalia, Equidae) from the Gulf Coastal Plain of Florida and Texas. *Journal of Paleontology*, 61, 809–30.

(1988a). *Calippus and Protohippus* (Mammalia, Perissodactyla, Equidae) from the Miocene (Barstovian–early Hemphillian) of the Gulf Coastal Plain. *Bulletin of the Florida State Museum*, 32, 221–340.

(1988b). *Cormohipparion* and *Neohipparion* (Mammalia, Perissodactyla, Equidae) from the late Neogene of Florida. *Bulletin of Florida State Museum, Biological Sciences*, 33, 229–338.

(1993). Late Miocene *Nannippus* (Mammalia, Perissodactyla) from Florida, with a description of the smallest hipparionine horse. *Journal of Vertebrate Paleontology*, 13, 350–66.

(1997). A new late Pliocene porcupine (Rodentia: Erethizontidae) from Florida. *Journal of Vertebrate Paleontology*, 17, 623–6.

Hulbert, R. C., Jr. and MacFadden, B. J. (1991). Morphologic transformation and cladogenesis at the base of the radiation of hypsodont horses. *American Museum Novitates*, 3000, 1–61.

Hulbert, R. C., Jr. and Morgan, G. S. (1989). Stratigraphy, paleoecology, and vertebrate faunas of the Leisey Shell Pit Local Fauna, early Pleistocene (Irvingtonian) of Southwestern Florida. *Papers in Florida Paleontology, Florida Museum of Natural History*, 2, 1–19.

Hulbert, R. C., Jr. and Whitmore, R. C., Jr. (2006). Late Miocene mammals from the Mauvilla Local Fauna, Alabama. *Bulletin of the Florida Museum of Natural History*, 46, 1–28.

Hulbert, R. C. Jr., Petkewich, R. M., Bishop, G. A., Bukry, D., and Aleshire, D. P. (1998). A new middle Eocene protocetid whale (Mammalia: Cetacea: Archaeoceti) and associated biota from Georgia. *Journal of Paleontology*, 72, 907–27.

Hulbert, R. C., Jr., Poyer, A. R., and Webb, S. D. (2002). Tyner Farm, a new early Hemphillian local fauna from north-central Florida. *Journal of Vertebrate Paleontology*, 22(suppl. to no. 3), p. 68A

Hulbert, R. C., Jr., Webb, S. D., and Morgan, G. S. (2003). Hemphillian terrestrial mammalian faunas from the south-central Florida Phosphate Mining District. *Journal of Vertebrate Paleontology*, 23(suppl. to no. 3), p. 63A.

Hunt, R. M., Jr. (1974). *Daphoenictis*, a cat-like carnivore (Mammalia, Amphicyonidae) from the Oligocene of North America. *Journal of Paleontology*, 48, 1030–47.

(1981). Geology and vertebrate paleontology of the Agate Fossil Beds National Monument and surrounding region, Sioux County, Nebraska. *National Geographic Society Reports*, 13, 263–85.

(1985). Faunal succession, lithofacies, and depositional environments in Arikaree rocks (lower Miocene) of the Hartville Table, Nebraska and Wyoming. *Dakoterra, South Dakota School of Mines*, 2, 335–52.

Hunter, J. P. (1999). The radiation of Paleocene mammals with the demise of the dinosaurs: Evidence from southwestern North Dakota. *Proceedings of the North Dakota Academy of Science*, 53, 141–4.

Hunter, J. P., Hartman, J. H., and Krause, D. W. (1997). Mammals and molluscs across the Cretaceous–Tertiary boundary from Makoshika

State Park and vicinity (Williston Basin), Montana. *Contributions to Geology, University of Wyoming*, 32, 61–114.

Hutchison, J. H. (1968). Fossil Talpidae (Insectivora, Mammalia) from the later Tertiary of Oregon. *Bulletin of the Museum of Natural History, University of Oregon*, 11, 1–117.

Ivany, L. C. (1998). Sequence stratigraphy of the Middle Eocene Claiborne Stage, US Gulf Coastal Plain. *Southeastern Geology*, 38, 1–20.

Ivy, L. D. (1990). Systematics of late Paleocene and early Eocene Rodentia (Mammalia) from the Clarks Fork Basin, Wyoming. *Contributions from the Museum of Paleontology, University of Michigan*, 28, 21–70.

Izett, G. A. (1975). Late Cenozoic sedimentation and deformation in northern Colorado and adjoining areas. [In *Cenozoic History of the Southern Rocky Mountains*, ed. B. F. Curtis.] *Memoirs of the Geological Society of America*, 44, 179–209.

Jacobs, L. L. and Lindsay, E. L. (1984). Holarctic radiation of Neogene muroid rodents and the origin of South American cricetids. *Journal of Vertebrate Paleontology*, 4, 265–72.

Jahns, R. H. (1940). Stratigraphy of the easternmost Ventura Basin, California, with a description of a new lower Miocene mammalian fauna from the Tick Canyon formation. [In *Studies of Cenozoic Vertebrates and Stratigraphy of Western North America*, ed. P. C. Henshaw, R. W. Wilson, H. Howard, *et al.*] *Carnegie Institute of Washington Publications*, 514, 145–94.

James, G. T. (1963). Paleontology and nonmarine stratigraphy of the Cuyama Valley Badlands, California Part 1: Geology, faunal interpretation, and systematic descriptions of Chiroptera, Insectivora and Rodentia. *University of California, Publications in Geological Sciences*, 45, 191–234.

Jepsen, G. L. (1966). Early Eocene Bat from Wyoming. *Science*, 154, 1333–9.

Jimenez-Hidalgo, E. and Carranza-Castañada, O. (2002). First Pliocene record of *Hemiauchenia blancoensis* (Mammalia, Camelidae) in Mexico. *Journal of Vertebrate Paleontology*, 22, 71–2.

(2005). Hemphillian camelids and protoceratids from San Miguel de Allende, Guanajuato state, Central Mexico. *Journal of Vertebrate Paleontology*, 25(suppl. to no. 3), p. 75A.

Jones, D. S., MacFadden, B, J., Webb, S. D., *et al.* (1991). Integrated geochronology of a classic Pliocene fossil site in Florida: linking marine and terrestrial biochronologies. *Journal of Geology*, 99, 637–48.

Kay, J. L. (1957). The Eocene vertebrates of the Uinta Basin. In *Eighth Annual Field Conference, Intermountain Association of Petroleum Geologists*, ed. O. G. Seal, pp. 110–14.

Keefer, W. R. (1965). Stratigraphy and geologic history of the uppermost Cretaceous, Paleocene, and lower Eocene rocks in the Wind River Basin, Wyoming. *United States Geological Survey, Professional Papers*, 495-A, A1–A77.

Kellogg, R. (1921). A new pinniped from the Upper Pliocene of California. *Journal of Mammalogy*, 2, 212–26.

(1925). On the occurrence of remains of fossil porpoises of the genus *Eurhinodelphis* in North America. *Proceedings of the United States National Museum*, 66, 1–40.

(1929). A new cetothere from southern California. *University of California, Publications in Geological Sciences*, 18, 449–57.

(1936). A review of the Archaeoceti. *Carnegie Institution of Washington Special Publication*, 482, 1–366.

(1938). On the Cetotheres figured by Vandelli. *Boletim do Laboratorio Mineralogico e Geologico da Universidade de Lisboa*, 13–23.

(1957). Two additional Miocene porpoises from the Calvert Cliffs Maryland. *Proceedings of the United States National Museum*, 107, 279–337.

(1968). Miocene Calvert mysticetes described by Cope. *Bulletin of the United States National Museum*, 247, 103–32.

(1969). Cetothere skeletons from the Miocene Choptank Formation of Maryland and Virginia. *Bulletin of the United States National Museum*, 294, 1–40.

Kelly, T. S. (1990). Biostratigraphy of Uintan and Duchesnean land mammal assemblages from the middle member of the Sespe Formation, Simi Valley, California. *Contributions in Science, Los Angeles County Museum of Natural History*, 419, 1–42.

(1994). Two Pliocene (Blancan) vertebrate faunas from Douglas County, Nevada. *PaleoBios*, 16, 1–23.

(1997). Additional late Cenozoic (latest Hemphillian to earliest Irvingtonian) mammals from Douglas County, Nevada. *PaleoBios*, 18, 1–31.

(1998). New Miocene mammalian faunas from west central Nevada. *Journal of Paleontology*, 72, 137–48.

(2000). A new Hemphillian (late Miocene) mammalian fauna from Hoye Canyon, west central Nevada. *Contributions in Science, Los Angeles County Museum of Natural History*, 481, 1–21.

Kelly, T. S. and Lander, E. B. (1988). Biostratigraphy and correlation of Hemingfordian and Barstovian Land Mammal assemblages, Caliente Formation, Cuyama Valley area, California. [In *Tertiary Tectonics and Sedimentation in the Cuyama Basin, San Luis Opisbo, Santa Barbara and Ventura Counties, California*.] *Pacific Section of the Society of Economic Paleontologists and Mineralogists*, 59, 1–19.

Kelly, T. S. and Whistler, D. P. (1994). Additional Uintan and Duchesnean (middle and late Eocene) mammals from the Sespe Formation, Simi Valley, California. *Contributions in Science, Los Angeles County Museum of Natural History*, 439, 1–29.

(1998). A new eomyid rodent from the Sespe Formation of southern California. *Journal of Vertebrate Paleontology*, 18, 440–3.

Kelly, T. S., Lander, E. B., Whistler, D. P., Roeder, M. A. and Reynolds, R. E. (1991). Preliminary report on a paleontologic investigation of the lower and middle members, Sespe Formation, Simi Valley Landfill, Ventura County, California. *PaleoBios*, 13, 1–13.

Kihm, A. J. (1984). Early Eocene mammalian fauna of the Piceance Creek Basin, Northwestern Colorado. Ph.D. Thesis, University of Colorado, Boulder.

(1987). Mammalian paleontology and geology of the Yoder Member, Chadron Formation, East-central Wyoming. *Dakoterra, South Dakota School of Mines*, 3, 28–45.

Kihm, A. J. and Hartman, J. H. (2004). A reevalution of the biochronology of the Brisbane and Judson Local Faunas (late Paleocene) of North Dakota. [In *Fanfare for an Uncommon Paleontologist: Essays in Honor of Malcolm C. McKenna*, ed. M. R. Dawson and J. A. Lillegraven.] *Bulletin of the Carnegie Museum of Natural History*, 36, 97–107.

Kihm, A. J., Hartman, J. H., and Krause, D. W. (1993). A new late Paleocene mammal Local Fauna from the Sentinel Butte Formation of North Dakota. *Journal of Vertebrate Paleontology*, 13(suppl. to no. 3), p. 44A.

Kihm, A. J., Schumaker, K. K., Warner-Evans, C., and Pearson, D. A. (2001). Marsupials from the Medicine Pole Hills Local Fauna (latest Eocene) of North Dakota. *Journal of Vertebrate Paleontology*, 21(suppl. to no. 3), p. 67A.

Klingener, D. (1968). Rodents of the Mio-Pliocene Norden Bridge local fauna, Nebraska. *The American Midland Naturalist*, 80, 65–74.

Koerner, H. E. (1940). The geology and vertebrate paleontology of the Fort Logan and Deep River Formations of Montana. Part 1: New vertebrates. *American Journal of Science*, 238, 837–62.

Konizeski, R. L. (1957). Paleoecology of the middle Pliocene Deer Lodge fauna, western Montana. *Bulletin of the Geological Society of America*, 68, 131–50.

(1961). Paleoecology of an early Oligocene biota from Douglass Creek, Montana. *Bulletin of the Geological Society of America*, 72, 1633–42.

Korth, W. W. (1981). New Oligocene rodents from western North America. *Annals of the Carnegie Museum*, 50, 289–318.

(1984). Earliest Tertiary evolution and radiation of rodents in North America. *Bulletin of the Carnegie Museum of Natural History*, 24, 1–71.

(1985). The rodents *Pseudotomus* and *Quadratomus* and the content of the tribe Mantishini (Paramyinae, Ischyromyidae). *Journal of Vertebrate Paleontology*, 5, 139–52.

(1987). Sciurid rodents (Mammalia) from the Chadronian and Orellan (Oligocene) of Nebraska. *Journal of Paleontology*, 61, 1247–55.

(1988). The rodent *Mytonomys* from the Unitan and Duchesnean (Eocene) of Utah, and the content of the Ailuravinae (Ischyromyidae, Rodentia). *Journal of Vertebrate Paleontology*, 8, 290–4.

(1989a). Aplodontid rodents (Mammalia) from the Oligocene (Orellan and Whitneyan) Brule Formation, Nebraska. *Journal of Vertebrate Paleontology*, 9, 400–14.

(1989b). Stratigraphic occurrence of rodents and lagomorphs in the Orella Member, Brule Formation (Oligocene), northwestern Nebraska. *Contributions to Geology, University of Wyoming*, 27, 15–20.

(1993a). The skull of *Hitonkala* (Florentiamyidae, Rodentia) and relationships within the Geomyidae. *Journal of Mammalogy*, 74, 168–74.

(1993b). Review of the Oligocene (Orellan and Arikareean) genus *Tenudomys* Rensberger (Geomyoidea: Rodentia). *Journal of Vertebrate Paleontology*, 13, 335–41.

(1994). A new species of the rodent *Prosciurus* (Aplodontidae, Prosciurinae) from the Orellan (Oligocene) of North Dakota and Nebraska. *Journal of Mammalogy*, 75, 478–82.

(1996a). A new species of *Pleurolicus* (Rodentia, Geomyidae) from the early Miocene (Arikareean) of Nebraska. *Journal of Vertebrate Paleontology*, 16, 781–4.

(1996b). A new genus of beaver (Mammalia: Castoridae: Rodentia) from the Arikareean (Oligocene) of Montana and its bearing on castorid phylogeny. *Annals of the Carnegie Museum*, 65, 167–79.

(1997). Additional rodents (Mammalia) from the Clarendonian (Miocene) of northcentral Nebraska, and a review of Clarendonian rodent biostratigraphy of that area. *Paludicola*, 1, 97–111.

(1998). Rodents and lagomorphs (Mammalia) from the late Clarendonian (Miocene) Ash Hollow Formation, Nebraska. *Annals of the Carnegie Museum*, 67, 299–348.

(1999a). *Hesperogaulus*, a new genus of mylagaulid rodent (Mammalia) from the Miocene (Barstovian to Hemphillian) of the Great Basin. *Journal of Paleontology*, 73, 945–51.

(1999b). A new genus of derived promylagauline rodent (Mylagaulidae) from the Miocene (late Hemingfordian–early Barstovian). *Journal of Vertebrate Paleontology*, 19, 752–6.

Korth, W. W. and Bailey, B. E. (1992). Additional species of *Leptodontomys douglassi* (Eomyidae, Rodentia) from the Arikareean (Late Oligocene) of Nebraska. *Journal of Mammalogy*, 73, 651–62.

Korth, W. W. and Cheney, D. S. (1999). A new subfamily of geomyoid rodents (Mammalia) and a possible origin of Geomyidae. *Journal of Paleontology*, 73, 1191–1200.

Korth, W. W. and Eaton, J. G. (2004). Rodents and a marsupial (Mammalia) from the Duchesnean (Eocene) Turtle Basin Local Fauna, Sevier Plateau, Utah. [In *Fanfare for an Uncommon Paleontologist: Essays in Honor of Malcolm C. McKenna*, ed. M. R. Dawson and J. A. Lillegraven.] *Bulletin of the Carnegie Museum of Natural History*, 36, 109–19.

Korth, W. W. and Emry, R. J. (1991). The skull of *Cedromus* and a review of the Cedromurinae (Rodentia: Sciuridae). *Journal of Paleontology*, 65, 984–94.

Korth, W. W. and Evander, R. L. (1982). A new species of *Orohippus* (Perissodactyla, Equidae) from the early Eocene of Wyoming. *Journal of Vertebrate Paleontology*, 2, 167–71.

Korth, W. W., Wahlert, J. H., and Emry, R. J. (1991). A new species of *Heliscomys* and recognition of the family Heliscomyidae (Geomyoidea: Rodentia). *Journal of Vertebrate Paleontology*, 11, 247–56.

Krause, D. W. (1978). Paleocene primates from western Canada. *Canadian Journal of Earth Sciences*, 15, 1250–71.

(1987). *Baiotomeus*, a new ptilodontid multituberculate (Mammalia) from the middle Paleocene of western North America. *Journal of Paleontology*, 61, 595–603.

Krause, D. W. and Gingerich, P. D. (1983). Mammalian fauna from Douglas Quarry, earliest Tiffanian (late Paleocene) of the eastern Crazy Mountain Basin, Montana. *Contributions from the Museum of Paleontology, University of Michigan*, 26, 157–96.

Krause, D. W. and Maas, M. C. (1990). The biogeographic origins of late Paleocene–early Eocene mammalian immigrants to the Western Interior of North America. *Geological Society of America Special Papers*, 243, 71–105.

Krishtalka, L. (1973). Late Paleocene mammals from the Cypress Hills, Alberta. *Special Publications of the Museum of Texas Technical University*, 2, 1–77.

(1979). Paleontology and geology of the Badwater Creek area, central Wyoming. Part 18: Revision of Late Eocene *Hyopsodus*. *Annals of the Carnegie Museum*, 48, 377–89.

Krishtalka, L. and Setoguchi, T. (1977). Paleontology and geology of the Badwater Creek area, central Wyoming Part 13: The Late Eocene Insectivora and Dermoptera. *Annals of the Carnegie Museum*, 46, 71–99.

Krishtalka, L. and Stucky, R. K. (1983a). Paleocene and Eocene marsupials of North America. *Annals of the Carnegie Museum*, 52, 229–63.

(1983b). Revision of the Wind River faunas, early Eocene of central Wyoming. Part 3: Marsupialia. *Annals of the Carnegie Museum*, 52, 205–27.

(1984). Middle Eocene marsupials (Mammalia) from Northeastern Utah and the mammalian fauna from Powder Wash. *Annals of the Carnegie Museum*, 53, 31–45.

(1985). Revision of the Wind River faunas, early Eocene of central Wyoming. Part 7: Revision of *Diacodexis* (Mammalia, Artiodactyla). *Annals of the Carnegie Museum*, 54, 413–86.

(1986). Early Eocene artiodactyls from the San Juan Basin, New Mexico. *Contributions to Geology, University of Wyoming*, 3, 183–96.

Krishtalka, L., Black, C. C., and Riedal, D. W. (1975). Paleontology and geology of the Badwater Creek area, Central Wyoming. Part 10: A late Paleocene mammal fauna from the Shotgun Member of the Fort Union Formation. *Annals of the Carnegie Museum*, 45, 179–212.

Krishtalka, L., West, R. M., Black, C. C., *et al.* (1987). Eocene (Wasatchian through Duchesnean) biochronology of North America. In *Cenozoic Mammals of North America, Geochronology and Biostratigraphy*, ed. M. O. Woodburne, pp. 77–117. Berkeley, CA: University of California Press.

Kurtén, B. and Anderson, E. (1980). *Pleistocene Mammals of North America*. New York: Columbia University Press.

Lambert, W. D. (1997). The osteology and paleoecology of the giant otter *Enhydritherium terraenovae*. *Journal of Vertebrate Paleontology*, 17, 738–49.

Lancaster, W. C. (1986). The taphonomy of an archaeocete skeleton and its associated fauna. *Proceedings of the 36th Annual Meeting of the Gulf Coast Association of Geological Societies, Montgomery Landing Site, Marine Eocene (Jackson) of Central LA*, pp. 119–31.

Lander, E. B. (1983). Continental vertebrate faunas from the upper member of the Sespe Formation, Simi Valley, California, and the terminal Eocene event. In *Fall Trip Volume and Guidebook: Cenozoic Geology of the Simi Valley Area, Southern California*, ed. R. R. Squires and M. V. Filewicz, pp. 124–44. Society of Economic Paleontologists and Mineralogists.

(1985). Early and middle Miocene continental vertebrate assemblages, Central Mohave Desert, San Bernadino County, California. In *Geological Investigations along Interstate 15, Cajon Pass to Manix Lake, California*, ed. R. E. Reynolds, pp. 127–44. Redlands, CA: San Bernadino County Museum.

Lander, E. B. and Reynolds, R. E. (1985). Fossil vertebrates from the Calico Mountain area, Central Mohave Desert, San Bernadino County,

California. In *Geological Investigations along Interstate 15, Cajon Pass to Manix Lake, California*, ed. R. E. Reynolds, pp. 153–6. Redlands, CA: San Bernadino County Museum.

Leiggi, P. (1991). Report. *Society of Vertebrate Paleontology News Bulletin*, 151, p. 65.

Leite, M. (1990). Stratigraphy and mammalian paleontology of the Ash Hollow Formation (Upper Miocene) on the north shore of Lake McConaughy, Keith County, Nebraska. *Contributions to Geology, University of Wyoming*, 28, 1–29.

Lemley, R. E. (1971). Notice of new finds in the Badlands. *Proceedings of the South Dakota Academy of Sciences*, 50, 70–4.

Lillegraven, J. A. (1970). Stratigraphy, structure and vertebrate fossils of the Oligocene Brule Formation, Slim Buttes, Northwestern South Dakota. *Bulletin of the Geological Society of America*, 81, 831–50.

(1976). Didelphids (Marsupialia) and *Uintasorex* (?Primates) from later Eocene sediments of San Diego County, California. *San Diego Society of Natural History Transactions*, 18, 1–20.

(1977). Small rodents (Mammalia) from Eocene deposits of San Diego County, California. *Bulletin of the American Museum of Natural History*, 158, 221–62.

(1980). Primates from later Eocene rocks of southern California. *Journal of Mammalogy*, 61, 181–204.

Lillegraven, J. A. and Tabrum, A. R. (1983). A new species of *Centetodon* (Mammalia, Insectivora, Geolabididae) from southwestern Montana and its biogeographical implications. *Contributions to Geology, University of Wyoming*, 22, 57–73.

Lillegraven, J. A., McKenna, M. C., and Krishtalka, L. (1981). Evolutionary relationships of middle Eocene and younger species of *Centetodon* (Mammalia, Insectivora, Geolabididae), with a description of the dentition of *Ankylodon* (Adapisoricidae). *University of Wyoming Publications*, 45, 1–115.

Lindsay, E. H. (1968). Rodents from the Hartman Ranch local fauna, California. *PaleoBios*, 6, 1–22.

(1974). The Hemingfordian mammal fauna of the Vedder locality, Branch Canyon Formation, Santa Barbara County, California, Part II: Rodentia (Eomyidae and Heteromyidae). *PaleoBios*, 16, 1–19.

(1978). *Eucricetodon asiaticus* (Matthew and Granger), an Oligocene rodent (Cricetidae) from Mongolia. *Journal of Paleontology*, 52, 590–5.

(1984). Late Cenozoic mammals from Northwestern Mexico. *Journal of Vertebrate Paleontology*, 4, 208–15.

(1991). Small mammals near the Hemingfordian/Barstovian boundary in the Barstow Syncline. *Quarterly Journal of the San Bernardino County Museum Association*, 38, 78–9.

Lindsay, E. H. and Jacobs, L. L. (1985). Pliocene small mammals from Chihuahua, Mexico. *Paleontologica Mexicana*, 51, 1–50.

Lindsay, E. H. and Tessman, N. T. (1974). Cenozoic vertebrate localities and faunas in Arizona. *Journal of the Arizona Academy of Science*, 9, 1–24.

Lindsay, E. H., Opdyke, V. D., and Johnson, N. M. (1984). Blancan–Hemphillian land mammal ages and late Cenozoic mammal dispersal events. *Annual Review of Earth and Planetary Science*, 12, 445–88.

Lindsay, E. H., Mou, Y., Downs, W., *et al.* (2002). Resolution of the Hemphillian/Blancan Boundary in Nevada. *Journal of Vertebrate Paleontology*, 22, 429–42.

Lofgren, D. L. (1995). The Bug Creek problem and the Cretaceous–Tertiary transition at McGuire Creek, Montana. *University of California Publications, Bulletin of the Department of Geological Sciences*, 140, 1–185.

Lofgren, D. L., McKenna, M. C., Walsh, S., *et al.* (2002). New records of Paleocene vertebrates from the Goler Formation of California. *Journal of Vertebrate Paleontology*, 22(suppl. to 3), p. 80A.

Lofrgen, D. L., McKenna, M., Nydam, R., and Hinkle, T. (2004). A phenacodont from Paleocene–Eocene marine beds of the uppermost Goler Formation, California. *Journal of Vertebrate Paleontology*, 24(suppl. to no. 3), p. 84A.

Love, J. D., McKenna, M. C., and Dawson, M. R. (1976). Eocene, Oligocene and Miocene rocks and vertebrate fossils at the Emerald Lake locality, three miles south of Yellowstone National Park, Wyoming. *United States Geological Survey Professional Papers*, 932-A, A1–A28.

Lozinsky, R. P. and Tedford, R. H. (1991). Geology and paleontology of the Santa Fe Group, southwestern Albuquerque Basin, Valencia County, New Mexico. *Bulletin of the New Mexico Bureau of Mines and Mineral Resources*, 132, 1–36.

Lucas, S. G. (1983). The Baca Formation and the Eocene–Oligocene boundary in New Mexico. In *Socorro Region II*, ed. C. E. Chapin and F. Callender, pp. 187–92. Albuquerque, NM: New Mexico Geological Society.

(1984). Early Paleocene vertebrates, stratigraphy, and biostratigraphy, West Fork of Gallegos Canyon, San Juan Basin, New Mexico. *New Mexico Geology*, 6, 56–60.

(1986a). The first Oligocene mammal from New Mexico. *Journal of Paleontology*, 60, 1274–6.

(1986b). Oligocene mammals from the Black Range, southwestern New Mexico. In *Truth or Consequences Region, Socorro*, ed. R. E. Clemons, W. E. King, and G. H. Mack, pp. 261–63. Albuquerque, NM: New Mexico Geological Society.

(1992). Redefinition of the Duchesnean land-mammal "age," late Eocene of western North America. In *Eocene–Oligocene Climatic and Biotic Evolution*, ed. D. R. Prothero and W. A. Berggren, pp. 88–105. Princeton, NJ: Princeton University Press.

(1995). The Thornton Beach mammoth and the antiquity of *Mammuthus* in North America. *Quaternary Research*, 43, 263–4.

(1997). Middle Eocene (Bridgerian) mammals from the Hart Mine Formation, south-central New Mexico. [In *New Mexico's Fossil Record*, ed. S. G. Lucas, J. W. Estrup, T. E. Williamson, and G. S. Morgan.] *Bulletin of the New Mexico Museum of Natural History and Science*, 11, 65–72.

Lucas, S. G. and Kues, B. S. (1979). Vertebrate biostratigraphy of the Eocene Galisteo Formation, North-Central New Mexico. *New Mexico Geological Society 30th Field Conference Guidebook for Santa Fe County*, pp. 225–9. Aberquerque, NM: New Mexico Geological Society.

Lucas, S. G. and Williamson, T. E. (1993). Eocene vertebrates and late Laramide stratigraphy of New Mexico. [In *Vertebrate Paleontology in New Mexico*. ed. S. G. Lucas and J. Zidek.] *Bulletin of the New Mexico Museum of Natural History and Science*, 2, 145–58.

Lucas, S. G., Schoch, R. M., Manning, E., and Tsentas, C. (1981). The Eocene biostratigraphy of New Mexico. *Bulletin of the Geological Society of America*, 92, 951–67.

Lucas, S. G., Estep, J. W., and Froehlich, J. W. (1997). *Mesohippus* (Mammalia: Perissodactyla) from the Chadronian (late Eocene) of south-central New Mexico. [In *New Mexico's Fossil Record 1*, ed. S. G. Lucas, J. W. Estrup, T. E. Williamson, and G. S. Morgan.] *Bulletin of the New Mexico Museum of Natural History and Science*, 11, 73–5.

Lull, R. S. (1914). Fossil dolphin from California. *American Journal of Science*, 37, 209–20.

(1922). Primitive Pecora in the Yale Museum. *American Journal of Science*, 5, 111–19.

Maas, M. C. (1985). Taphonomy of a late Eocene microvertebrate locality, Wind River Basin, Wyoming (USA). *Palaeogeography, Palaeoclimatology, Palaeoecology*, 52, 123–42.

Macdonald, J. R. (1948). The Pliocene carnivores of the Black Hawk Ranch fauna. *University of California Publications, Bulletin of the Department of Geological Sciences*, 28, 53–80.

(1956). A new Clarendonian mammalian fauna from the Truckee Formation of Western Nevada. *Journal of Paleontology*, 30, 186–202.

(1959). The middle Pliocene mammalian fauna from Smiths Valley, Nevada. *Journal of Paleontology*, 33, 872–87.

(1966). The Barstovian Camp Creek Fauna from Elko County, Nevada. *Contributions in Science, Los Angeles County Museum of Natural History*, 92, 1–18.

(1970). Review of the Miocene Wounded Knee faunas of Southwestern South Dakota. *Bulletin of the Los Angeles County Museum of Natural History, Science*, 8, 1–82.

(1982). Preliminary report on a late Oligocene fissure fill in the Paha Sapa Limestone near Rockerville, South Dakota. *Proceedings of the South Dakota Academy of Science*, 61, 172–3.

Macdonald, J. R. and Pelletier, W. J. (1958). The Pliocene mammalian faunas of Nevada. *US International Geological Congress, 20th. Session*, Section VII, pp. 365–88.

MacFadden, B. J. (1977). Magnetic polarity stratigraphy of the Chamita Formation stratotype (Mio-Pliocene) of North-Central New Mexico. *American Journal of Science*, 277, 769–800.

(1980a). An early Miocene land mammal (Oreodonta) from a marine limestone in northern Florida. *Journal of Paleontology*, 54, 93–101.

(1980b). Eocene perissodactyls from the type section of the Tepee Trail Formation of northeastern Wyoming. *Contributions to Geology, University of Wyoming*, 18, 135–43.

(1984). Systematics and phylogeny of *Hipparion, Neohipparion, Nannippus*, and *Cormohipparion* (Mammalia, Equidae) from the Miocene and Pliocene of the New World. *Bulletin of the American Museum of Natural History*, 179, 1–195.

(1986). Late Hemphillian monodactyl horses (Mammalia, Equidae) from the Bone Valley Formation of central Florida. *Journal of Paleontology*, 60, 466–75.

(2006). North American Miocene land mammals from Panama. *Journal of Vertebrate Paleontology*, 26, 720–34.

MacFadden, B. J. and Carranza-Casteñeda, O. (2002). Cranium of *Dinohippus mexicanus* (Mammalia: Equidae) from the early Pliocene (latest Hemphillian) of Central Mexico, and the origin of *Equus*. *Bulletin of the Florida Museum of Natural History*, 43, 163–85.

MacFadden, B. J. and Dobie, J. L. (1998). Late Miocene three-toed horse *Protohippus* (Mammalia, Equidae) from southern Alabama. *Journal of Paleontology*, 72, 149–52.

MacFadden, B. J. and Morgan, G. S. (2003). New oreodont (Mammalia, Artiodactyla) from the late Oligocene (early Arikareean) of Florida. *Bulletin of the American Museum of Natural History*, 279, 368–96.

MacFadden, B. J. and Nelson, M. E. (1980). Miocene three-toed horse from the Salt Lake group of southeastern Idaho. *Transactions of the Kansas Academy of Sciences*, 83, 20–5.

MacFadden, B. J. and Skinner, M. F. (1979). Diversification and biogeography of the one-toed horses *Onohippidium* and *Hippidium*. *Postilla, Yale Peabody Museum*, 175, 1–10.

MacFadden, B. J. and Waldrop, J. S. (1980). *Nannippus phlegon* (Mammalia, Equidae) from the Pliocene (Blancan) of Florida. *Bulletin of the Florida State Museum, Biological Sciences*, 25, 1–37.

MacFadden, B. J. and Webb, S. D. (1982). The succession of Miocene (Arikareean through Hemphillian) terrestrial mammal localities and faunas in Florida. In *Special Publication 25: Miocene of the Southwestern United States*, ed. T. M. Scott and S. B. Upchurch, pp. 186–99. Tallahassee, FL: Florida Department of Natural Resources, Bureau of Geology.

MacFadden, B. J., Johnson, N. M., and Opdyke, N. D. (1979). Magnetic polarity stratigraphy of the Mio-Pliocene mammal-bearing Big Sandy Formation of Western Arizona. *Earth and Planetary Science Letters*, 44, 349–64.

Madden, C. T. and Dalquest, W. W. (1990). The last rhinoceros in North America. *Journal of Vertebrate Paleontology*, 10, 266–7.

Madden, C. T. and Storer, J. E. (1985). The Proboscidea from the middle Miocene Wood Formation, Saskatchewan. *Canadian Journal of Earth Science*, 22, 1345–50.

Mader, B. J. and Alexander, J. P. (1995). *Megacerops kuwagatarhinus* n. sp., an unusual brontothere (Mammalia, Perissodactyla) with distally forked horns. *Journal of Paleontology*, 69, 581–7.

Maglio, V. J. (1966). A revision of the fossil selenodont artiodactyls from the middle Miocene of Thomas Farm, Gilchrist County, Florida. *Breviora, Museum of Comparative Zoology*, 225, 1–27.

Maher, J. C. and Jones, P. H. (1949). Ground-water exploration in the Natchitoches area Louisiana. *United States Geological Survey Water Supply Paper*, 968-D, 159–211.

Manning, E. M. (1990). The late early Miocene Sabine River. In *Transactions of the Gulf Coast Association of Geological Societies*, ed. E. Kinsland and T. Cagle, Vol. XL, pp. 531–49.

(1997). An early Oligocene Rhinoceros jaw from the marine Byram Formation of Mississippi. *Mississippi Geology*, 18, 1–31.

Manning, E. and MacFadden, B. J. (1989). Pliocene three-toed horses from Louisiana, with comments on the Citronelle Formation. *Tulane Studies in Geology and Paleontology*, 22, 35–46.

Martin, J. E. (1976). Small mammals from the Miocene Batesland Formation of South Dakota. *Contribution to Geology, University of Wyoming*, 14, 60–98.

(1984). A survey of Tertiary species of *Perognathus* (Perognathinae) and a description of a new species of Heteromyinae. *Carnegie Museum of Natural History Special Publication*, 9, 90–121.

Martin, J. E. and Green, M. (1984). Insectivora, Sciuridae, and Cricetidae from the early Miocene Rosebud Formation in South Dakota. [In *Papers in Vertebrate Paleontology Honoring Robert Warren Wilson*, ed. R. M. Mengel.] *Carnegie Museum of Natural History Special Publications*, 9, 28–40.

Martin, J. E. and James, E. (1983). Additions to the early Hemphillian (Miocene) Rattlesnake fauna from central Oregon. *Proceedings of the South Dakota Academy of Science*, 62, 23–33.

Martin, L. D. (1974). New rodents from the Lower Miocene Gering Formation of Western Nebraska. *Occasional Papers of the Museum of Natural History, University of Kansas*, 32, 1–12.

(1987). Beavers from the Harrison Formation (early Miocene) with a revision of *Euhapsis*. *Dakoterra, South Dakota School of Mines*, 3, 73–91.

Martin, L. D. and Mengel, R. (1984). A new cuckoo and chachalaca from the early Miocene of Colorado. [In *Papers in Vertebrate Paleontology Honoring Robert Warren Wilson*, ed. R. M. Mengel.] *Carnegie Museum of Natural History Special Publications*, 9, 171–7.

Martin, L. E. and Tedrow, A. R. (1988). Carnivora from the Ellensburg Formation (Miocene) of Central Washington. *Proceedings of the North Dakota Academy of Sciences*, 42, 15.

Martin, R. A. (1979). Fossil history of the rodent genus *Sigmodon*. *Evolutionary Monographs*, 2, 1–36.

(1989). Early Pleistocene zapodid rodents from the Java local fauna of north-central South Dakota. *Journal of Vertebrate Paleontology*, 9, 101–19.

Martin, R. A. and Harksen, J. C. (1974). The Delmont local fauna, Blancan of South Dakota. *Bulletin – New Jersey Academy of Science*, 19, 11–17.

Martin, R. A., Honey, J. G., and Peláez-Campomanes, P. (2000). The Meade Basin rodent project: a progress report. *Paludicola*, 3, 1–32.

Martin, R. A., Goodwin, H. T., and Farlow, J. O. (2002a). Late Neogene (late Hemphillian) rodents from the Pipe Creek Sinkhole, Grant County, Indiana. *Journal of Vertebrate Paleontology*, 22, 137–51.

Martin, R. A., Honey, J. G., Peláez-Campomanes, P., *et al.* (2002b). Blancan lagomorphs and rodents of the Deer Park assemblages, Meade County, Kansas. *Journal of Paleontology*, 76, 1072–90.

Martin, R. A., Hurt, R. T., Honey, J. G., and Peláez-Campomanes, P. (2003). Late Pliocene and early Pleistocene rodents from the northern Borchers Badlands (Meade County, Kansas), with comments on the Blancan–Irvingtonian boundary in the Meade Basin. *Journal of Paleontology*, 77, 985–1001.

Matthew, W. D. and Cook, H. J. (1909). A Pliocene fauna from Western Nebraska. *Bulletin of the American Museum of Natural History*, 26, 361–414.

Maxson, J. H. (1930). A Tertiary mammalian fauna from the Mint Canyon Formation of southern California. *Carnegie Institution of Washington Publications*, 41, 77–112.

May, S. R. and Repenning, C. A. (1982). New evidence for the age of the Mount Eden fauna, Southern California. *Journal of Vertebrate Paleontology*, 2, 109–13.

May, S. R., Walton, A. H., and Repenning, C. A. (1983). Uplift of the San Bernardino Mountains, California. *United States Geological Survey Professional Papers*, Report P1375, 79–180.

McAfee, R. (2003). Confirmation of the sloth genus *Megalonyx* (Xenarthra: Mammalia) from the John Day region and its implications. *Journal of Vertebrate Paleontology*, 23(suppl. to no. 3), p. 77A.

McCarroll, S. M., Flynn, J. J., and Turnbull, W. D. (1996). Biostratigraphy and magnetostratigraphy of the Bridgerian–Uintan Washakie Formation, Washakie Basin, Wyoming. In *The Terrestrial Eocene–Oligocene Transition in North America*, ed. D. R. Prothero and R. J. Emry, pp. 25–39. Cambridge, UK: Cambridge University Press.

McCullough, G., Silcox, M., Bloch, J., Boyer, D., and Krause, D. (2004). New palaechthonids (Mammalia, Primates) from the Paleocene of the Crazy Mountain Basin, Montana. *Journal of Vertebrate Paleontology* 24(suppl. to no. 3), p. 91A.

McGrew, P. (1937). New marsupials from the Tertiary of Nebraska. *Journal of Geology*, 45, 448–55.

McGrew, P. O. and Sullivan, R. (1970). The stratigraphy and paleontology of Bridger A. *Contributions to Geology, University of Wyoming*, 9, 66–85.

McGrew, P. O., Berman, J. E., Hecht, M. K., *et al.* (1959). The geology and paleontology of the Elk mountain and Tabernacle Butte area, Wyoming. *Bulletin of the American Museum of Natural History*, 117, 117–76.

McKenna, M. C. (1960a). A continental Paleocene vertebrate fauna from California. *American Museum Novitates*, 2024, 1–20.

(1960b). Fossil Mammalia from the early Wasatchian Four Mile fauna of northwest Colorado. *University of California, Publications in Geological Sciences*, 37, 1–130.

(1972). Vertebrate paleontology of the Togwotee Pass area, northwestern Wyoming. In *Guidebook for the Field Conference on Tertiary Biostratigraphy of Southern and Western Wyoming*, ed. R. M. West, pp. 80–101.

(1976). *Esthonyx* in the upper faunal assemblage, Huerfano Formation, Eocene of Colorado. *Journal of Paleontology*, 50, 354–61.

(1980a). Late Cretaceous and early Tertiary vertebrate paleontological reconnaissance, Towotee Pass area, northwestern Wyoming. In *Aspects of Vertebrate History*, ed. L. L. Jacobs, pp. 321–43. Flagstaff, AZ: Museum of Northern Arizona Press.

(1980b). Eocene paleolatitude, climate, and mammals of Ellesmere Island. *Palaeogeography, Palaeoclimatology, Palaeoecology*, 30, 349–62.

(1980c). Remaining evidence of Oligocene sedimentary rocks previously present across the Bighorn basin, Wyoming. [In *Early Cenozoic Paleontology and Stratigraphy of the Bighorn Basin, Wyoming*, ed. P. D. Gingerich.] *University of Michigan Papers in Paleontology*, 24, 143–6.

(1990). Plagiomenids (Mammalia, ?Dermoptera) from the Oligocene of Oregon, Montana, and South Dakota, and middle Eocene of northwestern Wyoming. *Geological Society of America Special Papers*, 243, 211–34.

McKenna, M. C. and Love, J. D. (1972). High-level strata containing early Miocene mammals on the Bighorn Mountains, Wyoming. *American Museum Novitates*, 2490, 1–31.

McKenna, M. C., Robinson, P., and Taylor, D. W. (1962). Notes on Eocene Mammalia and Mollusca from Tabernacle Butte, Wyoming. *American Museum Novitates*, 2102, 1–33.

McLeod, S. A. and Barnes, L. G. (1996). The systematic position of *Pappocetus lugardi* and a new taxon from North America (Archaeoceti: Protocetidae). [*Proceedings of the Sixth North American Paleontological Convention*.] *Paleontological Society Special Publication*, 8, 270.

Meehan, T. J., and Wilson, R. W. (2002). New viverravids from the Torrejonian (middle Paleocene) of Kutz Canyon, New Mexico and the oldest skull of the Order Carnivora. *Journal of Paleontology*, 76, 1091–101.

Merriam, J. C. (1911). Tertiary mammal beds of Virgin Valley and Thousand Creek in Northwestern Nevada. *University of California Publications, Bulletin of the Department of Geological Sciences*, 6, 199–304.

(1915). Tertiary vertebrate faunas of the North Coalinga region of California. *Transactions of the American Philosophical Society*, 22, 191–234.

(1919). Tertiary mammalian faunas of the Mojave Desert. *University of California Publications, Bulletin of the Department of Geological Sciences*, 11, 437–585.

Merriam, J. C. and Sinclair, W. J. (1907). Tertiary faunas of the John Day Region. *University of California Publications, Bulletin of the Department of Geological Sciences*, 5, 171–205.

Merriam, J. C. and Stock, C. (1928). A further contribution to the mammalian fauna of the Thousand Creek Pliocene, northwestern Nevada. *Carnegie Institute of Washington Publications*, 393, 5–21.

Middleton, M. D. (1983). Early Paleocene vertebrates of the Denver Basin, Colorado. Ph.D. Thesis, University of Colorado, Boulder.

Miller, W. E., and Carranza-Casteñeda, O. (1984). Late Cenozoic mammals from Central Mexico. *Journal of Vertebrate Paleontology*, 4, 216–36.

(2002). Late Tertiary vertebrates and sedimentation in the San Jose del Cabo Basin, Southern Baja, California. *Journal of Vertebrate Paleontology* 22(suppl. to no. 3), p. 88A.

Missimer, T. M. and Tobias, A. E. (2004). Geology and paleontology of a Caloosahatchee Formation deposit near Lehigh, Florida. *Florida Scientist*, 67, 48–62.

Mitchell, E. D. and Tedford, R. H. (1973). The Enaliarctinae: a new group of extinct aquatic carnivora and a consideration of the origin of the Otariidae. *Bulletin of the American Museum of Natural History*, 151, 203–84.

Montellano Ballesteros, M. (1992). Diversidad de los mamiferos en el registro fosil. Mammal diversity in the fossil record. *Revista de la Sociedad Mexicana de Historia Natural*, 43, 185–7.

Montellano, M. (1989). Pliocene Camelidae of Rancho El Ocote, central Mexico. *Journal of Mammalogy*, 70, 359–69.

Montellano, M. and Carranza-Castañada, O. (1981). Edentados pliocenicos de la region central de Mexico. Pliocene edentates of central Mexico. *Anais do Congresso Latino-Americano de Paleontologia*, 2, 683–95.

Morea, M. F. (1981). Massacre Lake Local Fauna (Miocene, Hemingfordian) from Northwestern Washoe County, Nevada. Ph.D. Thesis, University of California, Riverside.

Morgan, G. S. (1978). The fossil whales of Florida. *The Plaster Jacket*, 29, 1–20.

1989). Miocene vertebrate faunas from the Suwannee River basin of North Florida and South Georgia. [In *Miocene Paleontology and Stratigraphy of the Suwannee River basin of North Florida and South Georgia*, ed. G. S. Morgan.] *Southeastern Geological Society Field Conference Guidebook (Tallahassee)*, 30, 26–53.

(1991). Neotropical Chiroptera from the Pliocene and Pleistocene of Florida. *Bulletin of the American Museum of Natural History*, 206, 176–213.

(1993). Mammalian biochronology and marine-nonmarine correlations in the Neogene of Florida. [In *The Neogene of Florida and Adjacent Regions*, ed. V. A. Zullo, W. B. Harris, T. M. Scott, and R. W. Portell.] *Florida Geological Survey Special Publications*, 37, 55–66.

(1994). Miocene and Pliocene marine mammal faunas from the Bone Valley Formation of Central Florida. *Proceedings of the San Diego Society of Natural History*, 29, 239–68.

Morgan, G. S. and Czaplewski, N. J. (2003). A new bat (Chiroptera: Natalidae) from the early Miocene of Florida, with comments on natalid phylogeny. *Journal of Mammalogy*, 84, 729–52.

Morgan, G. S. and Hulbert, R. C., Jr. (1995). Overview of the geology and vertebrate biochronology of the Leisey Shell Pit Local Fauna, Hillsborough County, Florida. *Bulletin of the Florida Museum of Natural History*, 37, 1–92.

Morgan, G. S. and Lucas, S. G. (2000). Pliocene and Pleistocene vertebrate faunas from the Albuquerque Basin, New Mexico. [In *New Mexico's Fossil Record 2*, ed. S. G. Lucas.] *Bulletin of the New Mexico Museum of Natural History and Science*, 16, 217–40.

(2003a). Radiometrically calibrated oreodonts (Mammalia: Artiodactyla) from the late Oligocene of southwestern New Mexico. *Journal of Vertebrate Paleontology*, 23, 471–3.

(2003b). Mammalian biochronology of Blancan and Irvingtonian (Pliocene and early Pleistocene) faunas from New Mexico. *Bulletin of the American Museum of Natural History*, 279, 269–320.

Morgan, G. S. and Pratt, A. E. (1988). An early Miocene (late Hemingfordian) vertebrate fauna from Brooks Sink, Bradford County, Florida. *Southeastern Geological Society Field Trip Guide Book*, 29, 53–69.

Morgan, G. S. and Ridgeway, R. B. (1987). Late Pliocene (Late Blancan) vertebrates from the St. Petersburg Times site, Pinellas County, Florida, with a brief review of Florida Blancan faunas. *Papers in Florida Paleontology, Florida Museum of Natural History*, 1, 1–22.

Morgan, G. S. and White, J. A. (1995). Small mammals (Insectivora, Lagomorpha, and Rodentia) from the early Pleistocene (Irvingtonian) Leisey Shell Pit local fauna, Hillsborough County, Florida. *Bulletin of the Florida Museum of Natural History*, 37, 397–461.

Morgan, G. S., Sealey, P. L., Lucas, S. G., and Heckert, A. B. (1997). Pliocene (latest Hemphillian and Blancan) vertebrate fossils from the Mangas Basin, southwestern New Mexico. *Bulletin of the New Mexico Museum of Natural History and Science*, 11, 97–128.

Morgan, G. S., Lucas, S. G., and Estep, J. W. (1998). Pliocene (Blancan) vertebrate fossils from the Camp Rice Formation near Tonuco mountain, Dona Ana County, Southern New Mexico. In *Guidebook for the 49th Field Conference of the New Mexico Geological Society*, Las Cruces County, pp. 237–49. Alberquerque NM: New Mexico Geological Society.

Morgan, J. K. and Morgan, N. H. (1995). A new species of *Capromeryx* (Mammalia: Artiodactyla) from the Taunton Local Fauna of Washington, and the correlation with other Blancan faunas of Washington and Idaho. *Journal of Vertebrate Paleontology*, 15, 160–70.

Mou, Y. (1997). A new arvicoline species (Rodentia: Cricetidae) from the Pliocene Panaca Formation, southeast Nevada. *Journal of Vertebrate Paleontology*, 17, 376–83.

Munthe, J. (1979). The Hemingfordian mammalian fauna of Vedder Locality, Branch Canyon Sandstone, Santa Barbara County, California, Part III: Carnivores, perissodactyls, artiodactyls and summary. *PaleoBios*, 29, 1–22.

(1988). Miocene mammals of the Split Rock area, Granite Mountain Basin, central Wyoming. *University of California, Publications in Geological Sciences*, 126, 1–136.

Murphey, P. C. (2001). Stratigraphy, fossil distribution, and depositional environments of the upper Bridger Formation (middle Eocene) of southwestern Wyoming, and the taphonomy of an unusual Bridger microfossil assemblage. Ph.D. Thesis, University of Colorado, Boulder.

Murphey, P. C., Torick, L. L., Bray, E. S., Chandler, R., and Evanoff, E. (2001). Taphonomy, fauna, and depositional environment of the *Omomys* Quarry, and unusual accumulation from the Bridger Formation (middle Eocene) of southwestern Wyoming. In *Eocene Biodiversity: Unusual Occurrences and Rarely Sampled Habitats*, ed. G. F. Gunnell, pp. 361–402. New York: Kluwer Academic/Plenum.

Myrick, A. C., Jr. (1979). Variation, taphonomy, and adaptation of the Rhabdosteidae (= Eurhinodelphidae) (Odontoceti, Mammalia) from the Calvert Formation of Maryland and Virginia. Ph.D. Thesis, University of California, Los Angeles.

Nelson, M. E. (1973). Age and stratigraphic relationships of the Fowkes Formation, Eocene, of southwestern Wyoming and eastern Utah. *Contributions to Geology, University of Wyoming*, 12, 27–31.

(1976). A new Oligocene fauna from Northeastern Utah. *Transactions of the Kansas Academy of Sciences*, 79, 7–13.

Nelson, M. E. and Madsen, J. H. (1987). A new Clarendonian (late Miocene) fauna from Eastern Wyoming. *Contributions to Geology, University of Wyoming*, 25, 23–8.

Nelson, M. E. and Miller, D. M. (1990). A Pliocene record of the giant marmot, *Paenemarmota sawrockensis*, in northern Utah. *Contributions to Geology, University of Wyoming*, 8, 31–7.

Nelson, M. E., Madsen, J. H., Jr., and Stokes, W. L. (1980). A titanothere from the Green River Formation, central Utah: *Teleodus uintensis* (Perissodactyla, Brontotheriidae). *Contributions to Geology, University of Wyoming*, 12, 27–31.

Nichols, R. (1976). Early Miocene mammals from the Lemhi Valley of Idaho. *Tebiwa*, 18, 9–48.

(1998). The Lemhi Valley Oligo-Miocene: an overview. [In *And Whereas: Papers on the Vertebrate Paleontology of Idaho Honoring J. A. White*, Vol. 1, ed. W. A. Akersten, H. G. McDonald, D. J. Meldrum, and M. E. T. Flint.] *Idaho Museum of Natural History Occasional Papers*, 36, 10–12.

Novacek, M. J., Ferrusquía-Villafranca, I, Flynn, J. J., Wyss, A. R., and Norell, M. A. (1991). Wasatchian (early Eocene) mammals and other vertebrates from Baja California, Mexico: the Lomas Las Tetas de Cabra Fauna. *Bulletin of the American Museum of Natural History*, 208, 1–88.

O'Sullivan, J. A. (2003). A new species of *Archaeohippus* (Mammalia, Equidae) from the Arikareean of central Florida. *Journal of Vertebrate Paleontology*, 23, 877–85.

Olsen, S. J. (1964a). Vertebrate correlations and Miocene stratigraphy of north Florida fossil localities. *Journal of Paleontology*, 38, 600–4.

(1964b). An upper Miocene fossil locality in north Florida. *Quarterly Journal of the Florida Academy of Sciences*, 26, 307–14.

Opdyke, N. D., Lindsay, E. H., Johnson, N. M., and Downs, T. (1977). The paleomagnatism and magnetic polarity stratigraphy of the mammal-bearing section of the Anza-Borrego State Park, California. *Quaternary Research*, 7, 316–29.

Orr, W. N. and Miller, P. R. (1983) Fossil Cetacea (whales) in the Oregon western Cascades. *Oregon Geology*, 45, 95–8.

Osborn, H. F. (1909). Cenozoic mammal horizons of western North America. *Bulletin of the United States Geological Survey*, 361, 1–90.

Ostrander, G. E. (1985). Correlation of the early Oligocene (Chadronian) in northwestern Nebraska. *Dakoterra, South Dakota School of Mines*, 2, 205–31.

(1987). The early Oligocene (Chadronian) Raben Ranch local fauna, northwest Nebraska: Marsupialia, Insectivora, Dermoptera, Chiroptera, and Primates. *Dakoterra, South Dakota School of Mines*, 3, 92–104.

Patterson, B. and McGrew, P. O. (1937). A soricid and two erinaceids from the White River Oligocene. *Geological Series, Field Museum of Natural History*, 6, 245–72.

Patton, T. H. (1967). Oligocene and Miocene vertebrates from central Florida. [In *Miocene–Pliocene Problems of Peninsular Florida*, ed. H. K. Brooks and J. R. Underwood.] *Southeastern Geological Society Field Trip Guide Book*, 13, 3–10.

(1969a). Miocene and Pliocene artiodactyls of Texas Gulf Coastal Plain. *Bulletin of the Florida State Museum of Biological Sciences*, 14, 115–226.

(1969b). An Oligocene land vertebrate fauna from Florida. *Journal of Paleontology*, 43, 543–6.

Patton, T. H. and Taylor, B. E. (1971). The Synthetoceratinae (Mammalia, Tylopoda, Protoceratidae). *Bulletin of the American Museum of Natural History*, 145, 119–218.

(1973). The Protoceratinae (Mammalia, Tylopoda, Protoceratidae) and the systematics of the Protoceratidae. *Bulletin of the American Museum of Natural History*, 150, 351–413.

Pearson, D. A. and Hoganson, J. W. (1995). The Medicine Pole Hills local fauna: Chadron Formation (Eocene: Chadronian), Bowman County, North Dakota. *Proceedings of the North Dakota Academy of Sciences*, 49, 65.

Perry, F. A. (1977). *Fossils of Santa Cruz County*. Santa Cruz, CA: Santa Cruz City Museum.

Petkewich, R. M. and Lancaster, W. C. (1984). Middle Eocene archaeocete from the McBean Formation of Burke County, Georgia. *Georgia Journal of Science*, 42, 21.

Phillips, F. J., Welton, B., and Welton, J. (1976). Paleontologic studies of middle Tertiary Skooner Gulch and Gallaway formations at Point Arena, California. *Proceedings of the American Association of Petroleum Geologists Joint Meeting with the Society for Sedimentary Geologists and the Society for Exploratory Geologists, Pacific Sections;* Vol. 60, pp. 2187–8.

Pierce, H. G. and Rasmussen, D. L. (1989). New land snails (Archaeogastropoda, Helicinidae) from the Miocene (early Barstovian) Flint Creek beds of western Montana. *Journal of Paleontology*, 63, 646–51.

Pinsof, J. D. (1985). The Pleistocene vertebrate localities of South Dakota. *Dakoterra, South Dakota School of Mines*, 2, 233–64.

Pratt, A. E. (1989). Taphonomy of the microvertebrate fauna from the early Miocene Thomas Farm locality, Florida (USA). *Palaeogeography, Palaeoclimatology, Palaeoecology*, 76, 125–51.

(1990). Taphonomy of the large vertebrate fauna from the Thomas Farm locality (Miocene, Hemingfordian), Gilchrist County, Florida. *Bulletin of the Florida Museum of Natural History*, 35, 35–130.

Pratt, A. E. and Morgan, G. S. (1989). New Sciuridae (Mammalia: Rodentia) from the early Miocene Thomas Farm local fauna, Florida. *Journal of Vertebrate Paleontology*, 9, 89–100.

Princhinello, K. A. (1971). Earliest Eocene mammalian fossils from the Laramie Basin of southeastern Wyoming. *Contributions to Geology, University of Wyoming*, 10, 73–87.

Prothero, D. R. (1996). Magnetic stratigraphy and biostratigraphy of the middle Eocene Uinta Formation, Uinta Basin, Utah. In *The Terrestrial Eocene–Oligocene Transition in North America*, ed. D. R. Prothero and R. J. Emry, pp. 240–61. Cambridge, UK: Cambridge University Press.

(2005). *The Evolution of North American Rhinoceroses*. Cambridge, UK: Cambridge University Press.

Prothero, D. R. and Manning, E. M. (1987). Miocene rhinoceroses from the Texas Gulf Coastal Plain. *Journal of Paleontology*, 61, 388–423.

Prothero, D. R. and Sereno, P. C. (1982). Allometry and paleoecology of medial Miocene dwarf rhinoceroses from the Texas Gulf Coastal Plain. *Paleobiology*, 8, 16–30.

Prothero, D. R. and Shubin, N. (1989). The evolution of Oligocene horses. In *The Evolution of Perissodactyls*, ed. D. R. Prothero and R. M. Schoch, pp. 142–75. Oxford: Oxford University Press.

Prothero, D. R. and Tedford, R. H. (2000). Magnetic stratigraphy of the type Montediablan State (late Miocene), Black Hawk Ranch, Contra Costa County, California: Implications for regional correlations. *PaleoBios*, 20, 1–10.

Prothero, D. R., Denham, C. R., and Farmer, H. G. (1982). Oligocene calibration of the magnetic polarity time scale. *Geology*, 10, 650–3.

Prothero, D. R., Streig, A., and Burnes, C. (2001a). Magnetic stratigraphy and tectonic rotation of the upper Oligocene Pysht Formation, Clallam County, Washington. [In *Magnetic Stratigraphy of the Pacific Coast Cenozoic*, ed. D. R. Prothero.] *Pacific Section of the Society of Economic and Petroleum Mineralogists*, 91, 224–33.

Prothero, D. R., Bitboul, C. Z., Moore, G. W., and Moore, E. J. (2001b). Magnetic stratigraphy of the lower and middle Miocene Astoria Formation, Lincoln County, Oregon. [In *Magnetic Stratigraphy of the Pacific Coast Cenozoic*, ed. D. R. Prothero.] *Pacific Section of the Society of Economic and Petroleum Mineralogists*, 91, 272–83.

Prothero, D. R., Bitboul, C. Z., Moore, G. W., and Niem, A. R. (2001c). Magnetic stratigraphy and tectonic rotation of the Oligocene Alsea, Yaquina, and Nye Formations, Lincoln County, Oregon. [In *Magnetic Stratigraphy of the Pacific Coast Cenozoic*, ed. D. R. Prothero.] *Pacific Section of the Society of Economic and Petroleum Mineralogists*, 91, 184–94.

Prothero, D. R., Anderson, J. S., Chamberlain, K., and Ludtke, J. (2008). Magnetic stratigraphy and geochronology of the Barstovian–Clarendonian (middle to late Miocene) part of the Moonstone Formation, central Wyoming. *Bulletin of the New Mexico Museum of Natural History and Science*, in press.

Quinn, J. P. (1987). Stratigraphy of the middle Miocene Bopesta Formation, Southern Sierra Nevada, California. *Contributions in Science, Los Angeles County Museum of Natural History*, 393, 1–31.

Randazzo, A. F., Kosters, M., Jones, D. S., and Portell, R. W. (1990). Paleoecology of shallow-marine carbonate environments, middle Eocene of Peninsular Florida. *Sedimentary Geology*, 66, 1–11.

Rapp, S. D., MacFadden, B. J., and Schiebout, J. A. (1983). Magnetic polarity stratigraphy of the early Tertiary Black Peaks Formation, Big Bend National Park, Texas. *Journal of Geology*, 91, 555–72.

Raschke, R. E. (1984). Early and Middle Miocene vertebrates from the Santa Ana Mountains, California. [In *The Natural Science of Orange County: A Collection of Occasional Papers Concerning the Natural Science of Orange County, California, in Celebration of the 10 Anniversary of the Natural History Foundation of Orange County, California*, ed. B. Butler.] *Memoirs of the Natural History Foundation, Orange County*, 1, 61–7.

Rasmussen, D. L. (1989). Depositional environments, paleoecology, and biostratigraphy of Arikareean Bozeman Group Strata west of the Continental Divide in Montana. In *1989 Montana Geological Society Field Conference, Montana Centennial*, pp. 205–13. Billings, MT: Montana Geological Society.

Ray, C. E., Anderson, E., and Webb, S. D. (1981). The Blancan carnivore *Trigonictis* (Mammalia: Mustelidae) in the eastern United States. *Brimleyana*, 5, 1–36.

Reinhart, R. H. (1982). The extinct mammalian order Desmostylia. *Research Reports of the National Geographic Society*, 14, 549–55.

Rensberger, J. M. (1969). A new iniid cetacean from the Miocene of California. *University of California Publications in Geological Sciences*, 82, 1–36.

(1973). Pleurolicine rodents (Geomyoidea) of the John Day Formation, Oregon and their relationships to taxa from the early and middle Miocene, South Dakota. *University of California, Publications in Geological Sciences*, 102, 1–95.

(1979). *Promylagaulus*, progressive aplodontoid rodents of the early Miocene. *Contributions in Science, Los Angeles County Museum of Natural History*, 312, 1–16.

(1980). A primitive promylagauline rodent from the Sharp's Formation, South Dakota. *Journal of Paleontology*, 54, 1267–77.

Repenning, C. A. (1967). Subfamilies and genera of the Soricidae. *United States Geological Survey Professional Papers*, 565, 1–74.

(1981). Gubik Formation, Alaskan North Slope. *US Geological Survey Professional Papers*, P-1375, 1–180.

(2003). *Mimomys* in North America. *Bulletin of the American Museum of Natural History*, 279, 469–512.

Repenning, C. A. and May, S. R. (1987). New evidence for the age of the Palomas Formation Truth or Consequences, New Mexico. *Guidebook for the New Mexico Geological Society 37th Field Conference, Truth or Consequences*, pp. 257–60 Alberquerque, NM: New Mexico Geological Society.

Repenning, C. A. and Vedder, J. G. (1961). Continental vertebrates and their stratigraphic correlation with marine mollusks, Eastern Caliente Range, California. *United States Geological Survey Professional Papers*, 424-C, C235–9.

Reynolds, R. E. and Lander, E. B. (1985). Preliminary report on the Miocene Daggett Ridge Local Fauna, Central Mohave Desert, San Bernadino County, California. In *Geological Investigations Along Interstate 15, Cajon Pass to Manix Lake, California*, ed. R. E. Reynolds, pp. 105–10. Redlands, CA: San Bernadino County Museum.

Reynolds, R. E. and Lindsay, E. H. 1999. Late Tertiary basins and vertebrate faunas along the Nevada–Utah border. [In *Vertebrate Paleontology in Utah*, ed. D. D. Gillette.] *Miscellaneous Publications of the Utah Geological Survey*, 99-1, 469–78.

Reynolds, R. E., Reynolds, R. L., and Korth, W. W. (1991). Late Hemingfordian and early Barstovian Faunas from the Crowder Formation, Cajon Pass, San Bernadino County, California. *Journal of Vertebrate Paleontology*, 11(suppl. to no. 3), p. 52A.

Rich, T. H. V. (1981). Origin and history of the Erinaceinae and Brachyericinae (Mammalia, Insectivora) in North America. *Bulletin of the American Museum of Natural History*, 171, 1–116.

Rich, T. H. V. and Collinson, J. W. (1973). First mammalian fossil from the Flagstaff Limestone, Central Utah. *Vulpavus australis* (Carnivora, Miacidae). *Journal of Paleontology*, 47, 854–66.

Rich, T. H. V. and Rasmussen, D. L. (1973). New North American erinaceine hedgehogs (Mammalia, Insectivora) in North America. *Occasional Papers of the University of Kansas Museum of Natural History*, 21, 1–54.

Richey, K. A. (1943). A marine invertebrate fauna from the Orinda Formation California. *University of California, Publications in Geological Sciences*, 27, 25–36.

Rigby, K. J., Jr. (1980). Swain Quarry of the Fort Union Formation, middle Paleocene (Torrejonian), Carbon County, Wyoming. Geologic setting and mammalian fauna. *Evolutionary Monographs*, 3, 1–162.

Ritchie, K. A. (1948). Lower Pliocene horses from Black Hawk Ranch, Mount Diablo, California. *University of California Publications, Bulletin of the Department of Geological Sciences*, 28, 1–44.

Robertson, J. S. (1976). Latest Pliocene mammals from Haile XVA, Alachua County, Florida. *Bulletin of the Florida State Museum*, 20, 111–86.

Robinson, L. N. and Honey, J. G. (1987). Geology and setting of a new Paleocene mammal locality in the Northern Powder River Basin, Montana. *Palaios*, 2, 87–90.

Robinson, P. (1963). Fossil vertebrates and age of the Cuchara Formation of Colorado. *University of Colorado Studies, Series in Geology*, 1, 1–9.

(1966). Fossil Mammalia of the Huerfano Formation, Eocene, of Colorado. *Bulletin of the Museum of Natural History, Yale Peabody Museum*, 21, 1–95.

Robison, S. F. (1980). Paleocene (Puercan–Torrejonian) mammalian faunas of the North Horn Formation, central Utah. M.Sc. Thesis, Brigham Young University, Provo.

Roehler, H. W. (1973). Stratigraphy of the Washakie Formation in the Washakie Basin, Wyoming. *Geological Survey Bulletin*, 1369, 1–40.

(1991). Correlation and oil-shale assays of measured sections of the LaClede Bed of the Laney Member of the Green River Formation in outcrops along the western margins of Washakie Basin, Wyoming, and Sand Wash Basin, Colorado. [*US Geological Survey, Miscellaneous Investigations Series.] United States Geological Survey*, 1-2211.

Rose, K. D. (1975). The Carpolestidae, early Tertiary primates from North America. *Bulletin of the Museum of Comparative Zoology*, 147, 1–74.

(1981). The Clarkforkian land-mammal age and mammalian composition across the Paleocene–Eocene boundary. *University of Michigan Papers in Paleontology*, 26, 1–197.

(1999). Fossil mammals from the early Eocene Fisher/Sullivan site. In *Publication 152: Early Eocene Plants and Animals from the Fisher/Sullivan site (Nanjemoy Formation), Stafford County, Virginia*, ed. R. E. Weems and G. J. Grimsley, pp. 133–8. Charlottesville, VA: Virginia Division of Mineral Resources.

(2000). Land-mammals from the Late Paleocene Aquia Formation: the first early Cenozoic mammals from Maryland. *Proceedings of the Biological Society of Washington*, 113, 855–63.

Rose, K. D., Eberle, J. J., and McKenna, M. C. (2004). *Arcticanodon dawsonae*, a primitive new palaeanodont from the lower Eocene of Ellesmere Island, Canadian High Arctic. *Canadian Journal of Earth Sciences*, 41, 757–63.

Ruez, D. R. (2001). Early Irvingtonian (latest Pliocene) rodents from Inglis 1C, Citrus County, Florida. *Journal of Vertebrate Paleontology*, 21, 153–71.

Russell, L. S. (1929). Paleocene vertebrates from Alberta. *American Journal of Science*, 217, 162–76.

(1958). Paleocene mammal teeth from Alberta. *Bulletin of the National Museum of Canada*, 147, 96–103.

(1965a). Tertiary mammals of Saskatchewan. Part I: The Eocene fauna. *Contributions of the Royal Ontario Museum, Life Sciences*, 67, 1–33.

(1965b). The continental Tertiary of western Canada. [In *Vertebrate Paleontology in Alberta*, ed. R. E. Folinsbee and D. M. Ross.] *Bulletin of the University of Alberta Department of Geology*, 2, 41–52.

(1967). Paleontology of the Swan Hills area, North-central Alberta. *Life Sciences Contributions, Royal Ontario Museum*, 71, 1–30.

(1968). A new cetacean from the Oligocene Sooke Formation of Vancouver Island, British Columbia. *Canadian Journal of Earth Sciences*, 5, 929–33.

(1972). Tertiary mammals of Saskatchewan. Part II: The Oligocene fauna, non-ungulate orders. *Life Sciences Contributions, Royal Ontario Museum*, 84, 1–63.

(1974). Fauna and correlation of the Ravenscrag Formation (Paleocene) of southwestern Saskatchewan. *Life Sciences Contributions, Royal Ontario Museum*, 102, 1–52.

(1975). Revision of the fossil horses from the Cypress Hills Formation (lower Oligocene) of Saskatchewan. *Canadian Journal of Earth Sciences*, 12, 636–48.

(1978). Tertiary Mammals of Saskatchewan. Part IV: The Oligocene anthracotheres. *Life Sciences Contributions, Royal Ontario Museum*, 115, 1–16.

(1980). Tertiary Mammals of Saskatchewan. Part V: The Oligocene entelodonts. *Life Sciences Contributions, Royal Ontario Museum*, 122, 1–42.

(1982). Tertiary Mammals of Saskatchewan. Part VI: The Oligocene rhinoceroses. *Life Sciences Contributions, Royal Ontario Museum*, 113, 1–58.

Sanders, A. E. (1974). A paleontological survey of the Cooper Marl and Santee Limestone near Harleyville, South Carolina preliminary report. *Geologic Notes*, 18, 4–12.

Savage, D. E. (1955). Nonmarine lower Pliocene sediments in California. *University of California Publications, Bulletin of the Department of Geological Sciences*, 31, 1–26.

Savage, D. E. and Waters, B. T. (1978). A new omomyid primate from the Wasatch Formation of southern Wyoming. *Folia Primatologica*, 30, 1–29.

Savage, D. E., Waters, B. T., and Hutchison, J. H. (1972). Wasatchian succession at Bitter Creek Station, northwestern border of the Washakie basin, Wyoming. In *The Field Conference on Tertiary Biostratigraphy of Southern and Western Wyoming, Guidebook*, ed. R. M. West, pp. 32–9. Rapid City, SD: Geological Society of America.

Schaff, C. R. (1985). Paleocene mammals from the Beartooth region of Wyoming and Montana. *National Geographic Research Reports, 20*, 589–95.

Schankler, D. M. (1980). Faunal zonation of the Willwood Formation in the central Bighorn Basin, Wyoming. [In *Early Cenozoic Paleontology and Stratigraphy of the Bighorn Basin, Wyoming*, ed. P. D. Gingerich.] *University of Michigan Papers in Paleontology*, 24, 99–114.

Scharf, D. W. (1935). A Miocene mammalian fauna from Sucker Creek, southeastern Oregon. *Contributions to Paleontology, Carnegie Institute of Washington Publications*, 453, 97–118.

Scharlach, R. (1990). Depositional history of Oligocene–Miocene carbonate rocks of Northeastern Puerto Rico. *AAPG Bulletin*, 75, 757.

Schiebout, J. A. (1974). Vertebrate paleontology and paleoecology of Paleocene Black Peaks Formation, Big Bend National Park, Texas. *Bulletin of the Texas Memorial Museum*, 24, 1–88.

Schiebout, J. A. and Ting, S. (2000). *Paleofaunal Survey, Collecting, Processing and Documentation at Locations in the Castor Creek Member, Miocene Fleming Formation, Fort Polk, Louisiana*. [Contract No. DAC63-95-D-0051, Delivery Order No. 0010] Washington, DC: US Army Corps of Engineers, Fort Worth District.

Schoch, R. M. (1985). Preliminary description of a new late Paleocene land-mammal fauna from South Carolina, USA. *Postilla, Yale University Museum*, 196, 1–13.

Schoch, R. M. and Lucas, S. G. (1981). New conoryctines (Mammalia: Taeniodonta) from the middle Paleocene (Torrejonian) from western North America. *Journal of Mammalogy*, 62, 683–91.

Schultz, G. B. and Stout, T. M. (1948). Pleistocene mammals and terraces in the Great Plains. *Bulletin of the Geological Society of America*, 59, 553–88.

Schultz, G. E. (ed.) (1977). *Guidebook for the Field Conference on Late Cenozoic Biostratigraphy of the Texas Panhandle and Adjacent Oklahoma* [*Special Publication 1.*]. Canyon, TX: Killgore Research Center, Department of Geology and Anthropology, West Texas State University.

Scott, C. S. (1997). A new Paleocene mammal site from Calgary, Alberta. *Journal of Vertebrate Paleontology*, 17(suppl. to no. 3), p. 74A.

Scott, C. S., Webb, M. W., and Fox, R. C. (2006). *Horolodectes sunae*, an enigmatic mammal from the late Paleocene of Alberta, Canada. *Journal of Paleontology*, 80, 1009–25.

Secord, R. (1998). Paleocene mammalian biostratigraphy of the Carbon Basin, southeastern Wyoming, and age constraints on local phases of tectonism. *Rocky Mountain Geology*, 33, 119–54.

(2002). The Y2K Quarry, a new diverse latest Tiffanian (Late Paleocene) mammalian assemblage from the Fort Union Formation in the Northern Bighorn Basin, Wyoming. *Journal of Vertebrate Paleontology*, 22(suppl. to no. 3), p. 105A.

Sellards, E. H. (1916). Fossil vertebrates from Florida: a new Miocene fauna, new Pliocene species, the Pleistocene fauna. *Florida Geological Survey, Annual Reports*, 8, 77–119.

Setoguchi, T. (1978). Paleontology and geology of the Badwater Creek area, Central Wyoming. Part 16: The Cedar Ridge Local Fauna (late Oligocene). *Bulletin of the Carnegie Museum of Natural History*, 9, 1–61.

Shotwell, J. A. (1956). Hemphillian mammalian assemblage from Northeastern Oregon. *Bulletin of the Geological Society of America*, 67, 717–38.

(1958). Inter-community relationships in Hemphillian (Mio-Pliocene) mammals. *Ecology*, 29, 271–82.

(1968). Miocene mammals of southeast Oregon. *Bulletin of the Museum of Natural History, University of Oregon*, 14, 1–67.

(1970). Pliocene mammals of southeast Oregon and adjacent Idaho. *Bulletin of the Museum of Natural History, University of Oregon*, 17, 1–103.

Silcox, M. T. and Rose, K. D. (2001). Unusual vertebrate microfaunas from the Willwood Formation, early Eocene of the Bighorn Basin, Wyoming. In *Eocene Biodiversity: Unusual Occurrences and Rarely Sampled Habitats*, ed. G. F. Gunnell, pp. 131–64. New York: Kluwer Academic/Plenum.

Simpson, G. G. (1930). Tertiary land mammals of Florida. *Bulletin of the American Museum of Natural History*, 59, 149–211.

(1937a). Additions to the upper Paleocene fauna of Crazy Mountain Field. *American Museum Novitates*, 940, 1–15.

(1937b). The Fort Union of the Crazy Mountain Field, Montana and its mammalian faunas. *Bulletin of the United States National Museum*, 169, 1–287.

(1959). Fossil mammals from the type area of the Puerco and Nacimiento strata, Paleocene of New Mexico. *American Museum Novitates*, 1957, 1–22.

Simpson, W. D. (1985). Geology and paleontology of the Oligocene Harris Ranch Badlands, southwestern South Dakota. *Dakoterra, South Dakota School of Mines*, 2, 303–33.

Skinner, M. F. and Johnson, F. W. (1984). Tertiary stratigraphy and the Frick collection of fossil vertebrates from North-Central Nebraska. *Bulletin of the American Museum of Natural History*, 178, 215–368.

Skinner, M. F., Skinner, S. M., and Gooris, R. J. (1968). Cenozoic rocks and faunas of Turtle Butte, South Central South Dakota. *Bulletin of the American Museum of Natural History*, 138, 379–436.

(1977). Stratigraphy and biostratigraphy of late Cenozoic deposits in central Sioux County, Western Nebraska. *Bulletin of the American Museum of Natural History*, 158, 265–371.

Skinner, S. M., and Gooris, R. J. (1966). A note on *Toxotherium* (Mammalia, Rhinocerotoidea) from Natrona county, Wyoming. *American Museum Novitates*, 2261, 1–12.

Skolnick, R. (1985). Geology and paleontology of the Lay Ranch beds. *Proceedings of the National Academy of Sciences, USA*, 95, 54.

Skwara, T. (1988). Mammals of the Topham Local Fauna: early Miocene (Hemingfordian), Cypress Hills Formation, Saskatchewan. *Natural History Contributions, Museum of Natural History Regina*, 9, 1–169.

Slaughter, B. (1981). A new genus of geomyid rodent from the Pliocene of Texas and Panama. *Journal of Vertebrate Paleontology*, 1, 111–15.

Sloan, R. E. (1987). Paleocene and latest Cretaceous mammal ages, biozones, magnetozones, rates of sedimentation, and evolution. *Geological Society of America Special Papers*, 209, 165–95.

Smith, K., Cifelli, R., and Czaplewski, N. (2004). A new genus of eomyid (Mammalia: Rodentia) from the Miocene (late Hemingfordian and early Barstovian) of Nevada. *Journal of Vertebrate Paleontology*, 24(suppl. to no. 3), p. 115A.

Standhardt, B. R. (1995). Early Paleocene (Puercan) vertebrates of the Dogie locality, Big Bend National Park, Texas. In *National Park Service Technical Report* NPS/NPRO/NRTR-95/16, ed. V. L. Santucci and L. McClelland, pp. 46–48. Washington, DC: National Parks Service.

Stevens, M. S. (1991). Osteology, systematics, and relationships of earliest Miocene *Mesocyon venator* (Cook), Carnivora, Canidae. *Journal of Vertebrate Paleontology*, 11, 45–66.

Stevens, M. S. and Stevens, J. B. (1989). Neogene–Quaternary deposits and vertebrate faunas. In *Guidebook for the 4th Annual Meeting of the Society of Vertebrate Paleontology: Vertebrate Paleontology, Biostratigraphy and Depositional Environments, Latest Cretaceous and Tertiary, Big Bend Area, Texas*, ed. A. B. Busbey, III and T. M. Lehman, pp. 67–90. Austin, TX: Society of Vertebrate Paleontology, University of Texas,

(2003). Carnivora (Mammalia, Felidae, Canidae, and Mustelidae) from the earliest Hemphillian Screw Bean Local Fauna, Big Bend National Park, Brewster County, Texas. *Bulletin of the American Museum of Natural History*, 279, 177–211.

Stevens, M. S., Stevens, J. B., and Dawson, M. R. (1969). New early Miocene Formation and Vertebrate Local Fauna, Big Bend National Park, Brewster County, Texas. *Pearce-Sellards Series, Texas Memorial Museum*, 15, 1–53.

Stirton, R. A. (1939). Cenozoic mammal remains from the San Francisco Bay region. *University of California Publications, Bulletin of the Department of Geological Sciences*, 24, 339–410.

(1940). The Nevada Miocene and Pliocene mammalian faunas as faunal units. *Proceedings of the Sixth Pacific Science Congress*, 2, 627–40.

Stirton, R. A. and Goeriz, H. F. (1942). Fossil vertebrates from the superajacent deposits near Knights Ferry, California. *University of California*

Publications, Bulletin of the Department of Geological Sciences, 26, 447–72.

Stirton, R. A. and McGrew, P. O. (1935). A preliminary notice on the Miocene and Pliocene mammalian faunas near Valentine, Nebraska. *American Journal of Science*, 29,125–32.

Stock, C. (1920). An early Tertiary vertebrate fauna from the southern coast range of California. *University of California Publications, Bulletin of Geological Sciences*, 12, 267–76.

(1949). Mammalian fauna from the Titus Canyon Formation, California. *Carnegie Institution of Washington Publications*, 584, 229–44.

Storer, J. E. (1975). Tertiary mammals of Saskatchewan. Part III: The Miocene fauna. *Life Sciences Contributions, Royal Ontario Museum*, 103, 1–134.

(1981). Leptomerycid artiodactyls of the Calf Creek Local Fauna (Cypress Hills Formation, Oligocene, Chadronian) Saskatchewan. *Natural History Contributions, Museum of Natural History Regina*, 3, 1–32.

(1984a). Mammals of the Swift Current Creek Local Fauna (Eocene: Uintan, Saskatchewan). *Natural History Contributions, Museum of Natural History Regina*, 7, 1–158.

(1984b). Fossil mammals of the Southfork Local Fauna (early Chadronian) of Saskatchewan. *Canadian Journal of Earth Sciences*, 21, 1400–5.

(1988). The rodents of the Lac Pellatier lower fauna, late Eocene (Duchesnean) of Saskatchewan. *Journal of Vertebrate Paleontology*, 8, 84–101.

(1990). Primates of the Lac Pellatier lower fauna Eocene, Duchesnean, Saskatchewan. *Canadian Journal of Earth Sciences*, 27, 520–4.

(1993). Additions to the mammalian paleofauna of Saskatchewan. *Modern Geology*, 18, 475–87.

(1994). A latest Chadronian (Late Eocene) mammalian fauna from the Cypress Hills, Saskatchewan. *Canadian Journal of Earth Science*, 31, 1335–41.

(1995). Small mammals of the Lac Pelletier Lower Fauna, Duchesnean, of Saskatchewan, Canada: insectivorans and insectivore-like groups – a plagiomenid, a microsyopid and Chiroptera. In *Vertebrate Fossils and the Evolution of Scientific Concepts*, ed. W. A. S. Sarjeant, pp. 595–615. Melbourne, Australia: Gordon and Breach.

(1996). Eocene–Oligocene faunas of the Cypress Hills, Saskatchewan. In *The Terrestrial Eocene–Oligocene Transition in North America*, ed. D. R. Prothero and R. J. Emry, pp. 240–61. Cambridge, UK: Cambridge University Press.

(2002). Small mammals of the Kealey Springs Local Fauna (early Arikareean; late Oligocene) of Saskatchewan. *Paludicola*, 3, 105–33.

(2003). The Eastern Beringian vole *Microtus deceitensis* (Rodentia, Muridae, Arvicolinae). [In *Late Pliocene and Early Pleistocene Faunas of Alaska and Yukon*, ed. J. A. Westgate.] *Quaternary Research*, 60, 84–93.

Storer, J. E. and Bryant, H. N. (1993). Biostratigraphy of the Cypress Hills Formation (Eocene to Miocene), Saskatchewan: equid types (Mammalia: Perissodactyla) and associated faunal assemblages. *Journal of Paleontology*, 67, 660–9.

Strait, S. G. (2001). New Wa-0 mammalian fauna from Castle Gardens in the southeastern Bighorn Basin. *University of Michigan Papers on Paleontology*, 33, 127–43.

(2003). New mammalian fossils from the earliest Eocene (Wa-0), Bighorn Basin, Wyoming. *Journal of Vertebrate Paleontology*, 23,101A.

Stucky, R. K. (1984a). The Wasatchian–Bridgerian land mammal age boundary (early to middle Eocene) in western North America. *Annals of the Carnegie Museum*, 53, 347–82.

(1984b). Revision of the Wind River Faunas, Early Eocene of central Wyoming. Part 5: Geology and biostratigraphy of the upper part of the Wind River Formation, Northeastern Wind River Basin. *Annals of the Carnegie Museum*, 53, 231–94.

Stucky, R. K. and Krishtalka, K. L. (1982). Revision of the Wind River Faunas, Early Eocene of central Wyoming. Part 1: Introduction and multituberculates. *Annals of the Carnegie Museum*, 53, 231–94.

Stucky, R. K., Krishtalka, K. L., and Redline, A. D. (1990). Geology, vertebrate fauna, and paleoecology of the Buck Springs Quarries (early Eocene, Wind River Formation), Wyoming. *Geological Society of America Special Papers*, 243, 169–86.

Stucky, R. K., Prothero, D. R., Lohr, W. G., and Snyder, J. R. (1996). Magnetic stratigraphy, sedimentology, and mammalian faunas of the early Uintan Washakie Formation, Sand Wash Basin, Northwestern Colorado. In *The Terrestrial Eocene–Oligocene Transition in North America*, ed. D. R. Prothero and R. J. Emry, pp. 40–51. Cambridge, UK: Cambridge University Press.

Sutton, J. F. (1977). Mammals of the Anceney Local Fauna (late Miocene) of Montana. Ph.D. Thesis, Texas Technical University, Lubbock.

Sutton, J. F. and Black, C. C. (1975). Paleontology of the earliest Oligocene deposits in Jackson Hole, Wyoming. Part 1: Rodents exclusive of the Family Eomyidae. *Annals of the Carnegie Museum*, 45, 299–315.

Sutton, J. F. and Korth, W. W. (1995). Rodents (Mammalia) from the Barstovian (Miocene) Anceney local fauna, Montana. *Annals of the Carnegie Museum*, 64, 267–314.

Swisher, C. C. (1982). Stratigraphy and biostratigraphy of the Eastern portion of Wildcat Ridge, Western Nebraska. Ph.D. Thesis, University of Nebraska, Lincoln.

Tabrum, A. R. (1981). A contribution to the mammalian paleontology of the Ogallala Group of south-central South Dakota. M.Sc. Thesis, South Dakota School of Mines and Technology, Rapid City.

(1998). First record of a hypertragulid artiodactyl from the Chadronian of western Montana. *Journal of Vertebrate Paleontology*, 18(suppl. to no. 3), p. 81A.

Tabrum, A. R. and Fields, R. W. (1980). Revised mammalian faunal list for the Pipestone Springs Local Fauna (Chadronian, Early Oligocene), Jefferson County, Nebraska. *Northwest Geology*, 9, 45–51.

Tabrum, A. R., Prothero, D. R., and Garcia, D. (1996). Magnetostratigraphy and biostratigraphy of the Eocene–Oligocene transition, southwestern Montana. In *The Terrestrial Eocene–Oligocene Transition in North America*, eds. D. R. Prothero and R. J. Emry, pp. 75–117. Cambridge, UK: Cambridge University Press.

Taylor, B. E. and Webb, S. D. (1976). Miocene Leptomerycidae (Artiodactyla, Ruminantia) and their relationships. *American Museum Novitates*, 2596, 1–22.

Taylor, L. H. (1981). The Kutz Canyon Local Fauna, Torrejonian (middle Paleocene) of the San Juan Basin, New Mexico. In *Advances in San Juan Basin Paleontology*, ed. S. G. Lucas, J. K. Rigby, Jr., and B. S. Kues, pp. 242–63. Albuquerque, NM: University of New Mexico Press.

Tedford, R. H. (1981). Mammalian biochronology of the late Cenozoic basins of New Mexico. *Bulletin of the Geological Society of America*, 92, 1008–22.

(2004). Miocene mammalian faunas, Ogallala Group, Pawnee Buttes Area, Weld County, Colorado. [In *Fanfare for an Uncommon Paleontologist: Essays in Honor of Malcolm C. McKenna*, ed. M. R. Dawson and J. A. Lillegraven.] *Bulletin of the Carnegie Museum of Natural History*, 36, 277–90.

Tedford, R. H. and Barghoorn, S. (1993). Neogene stratigraphy and mammalian biochronology of the Espanola Basin, Northern New Mexico. [In *Vertebrate Paleontology in New Mexico*, ed. S. G. Lucas and J. Zidek.] *Bulletin of the New Mexico Museum of Natural History and Science*, 2, 159–68.

Tedford, R. H. and Frailey, D. (1976). Review of some Carnivora (Mammalia) from the Thomas Farm Local Fauna (Hemingfordian, Gilchrist County, Florida). *American Museum Novitates*, 2610, 1–9.

Tedford, R. H. and Harington, C. R. (2003). An Arctic mammal fauna from the Early Pliocene of North America. *Nature*, 425, 388–90.

Tedford, R. H. and Hunter, M. E. (1984). Miocene marine-non-marine correlations, Atlantic and Gulf Coastal Plains, North America. *Palaeogeography, Palaeoclimatology, Palaeoecology*, 47, 129–51.

Tedford, R. H., Swinehart, J. B., Hunt, R. M., Jr., and Voorhies, M. R. (1985). Uppermost White River and lowermost Arikaree rocks and faunas, White River Valley, Northwestern Nebraska, and their correlation with South Dakota. *Dakoterra, South Dakota School of Mines*, 2, 335–52.

Tedford, R. H., Skinner, M. S., Fields, R. S., *et al.* (1987). Faunal succession and biochronology of the Arikareean through Hemphillian (late Oligocene through earliest Pliocene epochs) in North America. In *Cenozoic Mammals of North America, Geochronology and Biostratigraphy*, ed. M. O. Woodburne, pp. 153–210. Berkeley, CA: University of California Press.

Tedford, R. H., Barnes, L. G., and Ray, C. E. (1994). The early Miocene littoral ursoid carnivoran *Kolponomos*; systematics and mode of life. [In *Contributions in Marine Mammal Paleontology Honoring Frank C. Whitmore, Jr.*, ed. A. Berta and T. A. Deméré.] *Proceedings of the San Diego Society of Natural History*, 29, 11–32.

Tedford, R. H., Swinehart, J. B., Swisher, C. C., III, *et al.* (1996). The Whitneyan–Arikareean transition in the high plains. In *The Terrestrial Eocene–Oligocene Transition in North America*, ed. D. R. Prothero and R. J. Emry, pp. 312–34. Cambridge, UK: Cambridge University Press.

Tedford, R. H., Albright, L. B., III, Barnosky, A. D., *et al.* (2004). Mammalian biochronology of the Arikareean through Hemphillian interval (late Oligocene through early Pliocene epochs). In *Late Cretaceous and Cenozoic Mammals of North America; Biostratigraphy and Geochronology*, ed. W. O. Woodburne, pp. 169–231. New York: Columbia University Press.

Tedrow, A. R. and Martin, J. E. (1998). *Plesiosorex* (Mammalia: Insectivora) from the Miocene Imnaha Basalts of western Idaho. *Idaho Museum of Natural History Occasional Papers*, 36, 21–4.

Thewissen, J. G. M. and Smith, G. R. (1987). Vespertilionid bats (Chiroptera, Mammalia) from the Pliocene of Idaho. *Contributions from the Museum of Paleontology, University of Michigan*, 27, 237–45.

Thomasson, J. R., Zakrzewski, J. R., Lagarry, H. E., and Mergen, D. C. (1990). Late Miocene (late early Hemphillian) biota from Northwestern Kansas. *National Geographic Research*, 6, 231–344.

Thornton, M. L. and Rasmussen, D. T. (2001). Taphonomic interpretation of Gnat-Out-of-Hell, an Early Uintan small mammal locality in the Uinta Formation, Utah. In *Eocene Biodiversity: Unusual Occurrences and Rarely Sampled Habitats*, ed. G. F. Gunnell, pp. 299–316. New York: Kluwer Academic/Plenum.

Tomida, Y. (1981). "Dragonian" fossils from the San Juan Basin and the status of the "Dragonian" land mammal "age." In *Advances in San Juan Basin Paleontology*, ed. S. G. Lucas, J. K. Rigby, Jr., and B. S. Kues, pp. 222–41. Albuquerque, NM: University of New Mexico Press.

(1987). Small mammal fossils and correlation of continental deposits, Safford and Duncan basins, Arizona, USA. *National Science Museum, Tokyo*, 1–141.

Tomida, Y. and Butler, R. F. (1980). Dragonian mammals and Paleocene magnetic polarity, North Horn Formation, Central Utah. *American Journal of Science*, 280, 787–811.

Torres, V. (1985). Stratigraphy of the Eocene Willwood, Aycross, and Wapiti Formations along the North Fork of the Shoshone River, north-central Wyoming. *Contributions to Geology, University of Wyoming*, 23, 83–97.

Trask, P. D. (1922). The Briones Formation of middle California. *University of California, Publications in Geological Sciences*, 13, 133–74.

Tsentas, C. (1981). Mammalian biostratigraphy of the middle Paleocene (Torrejonian) strata of the San Juan Basin: notes on Torreon Wash and the status of *Pantolambda* and *Deltatherium* faunal "zones." In *Advances in San Juan Basin Paleontology*, ed. S. G. Lucas, J. K. Rigby, Jr., and B. S. Kues, pp. 264–92. Albuquerque, NM: University of New Mexico Press.

Tucker, S. (2003). Carnivores and microtine-like rodents from a new late Miocene (Hemphillian) locality in north-central Nebraska. *Journal of Vertebrate Paleontology*, 23(suppl. to no. 3), p. 105A.

Turnbull, W. D. (1972). The Washakie Formation of Bridgerian–Uintan ages, and the related fauna. In *Guidebook for the Field Conference on the Tertiary Biostratigraphy of Southern and Western Wyoming*, ed. R. M. West, pp. 20–31. Rapid City, SD: Geological Society of America.

Uhen, M. D. (1998a). New protocetid (Mammalia, Cetacea) from the late middle Eocene Cook Mountain Formation of Louisiana. *Journal of Vertebrate Paleontology*, 18, 664–8.

(1998b). Middle to Late Eocene Basilosaurines and Dorudontines. In *The Emergence of Whales*, ed. J. G. M. Thewissen, pp. 29–61. New York: Plenum Press.

(1999). New species of protocetid archaeocete whale, *Eocetus wardii* (Mammalia: Cetacea) from the middle Eocene of North Carolina. *Journal of Paleontology*, 73, 512–28.

Van Valen, L. (1978). The beginning of the age of mammals. *Evolutionary Theory*, 4, 45–80.

Voorhies, M. R. (1973). Early Miocene mammals from northeast Nebraska. *Contributions to Geology, University of Wyoming*, 12, 1–10.

(1974). The Pliocene horse *Nannippus minor* in Georgia: geological implications. *Tulane Studies in Geology and Paleontology*, 11, 109–13.

(1987). Fossil armadillos in Nebraska: the northernmost record. *The Southwestern Naturalist*, 32, 237–43.

(1990a). *Vertebrate Paleontology of the Proposed Norden Reservoir area, Brown, Cherry, and Keya Paha Counties, Nebraska*. [*Technical Report 82-09*]. Denver, CO: Division of Archeological Research, United States Bureau of Land Reclamation.

(1990b). Vertebrate biostratigraphy of the Ogallala group in Nebraska. In *Geologic Framework and Regional Hydrology: Upper Cenozoic Blackwater Draw and Ogallala Formations, Great Plains*, ed. T. C. Gustavson, pp. 115–51. Austin, TX: Bureau of Economic Geology, University of Texas.

Voorhies, M. R. and Corner, R. G. (1986). *Megatylopus*(?) *cochrani* (Mammalia: Camelidae). A re-evaluation. *Journal of Vertebrate Paleontology*, 6, 65–75.

Voorhies, M. R., Holman, J. A., and Xiang-Xu, X. (1987). The Hotel Ranch rhino quarries (basal Ogallala, medial Barstovian), Banner County, Nebraska. Part 1: Geologic setting, faunal lists, lower vertebrates. *Contributions to Geology, University of Wyoming*, 25, 55–69.

Wagner, H. M. (1976). A new species of *Pliotaxidea* (Mustelidae: Carnivora) from California. *Journal of Paleontology*, 50, 107–27.

Wagner, H. M., Riney, B. O., and Prothero, D. R. (2000). A new terrestrial mammal assemblage of middle Blancan age from the San Diego Formation, California. *Journal of Vertebrate Paleontology*, 20(suppl. to no. 3), p. 76A.

Wagner, H. M., Riney, B. O., Deméré, T. A. and Prothero, D. R. (2001). Magnetic stratigraphy and land mammal biochronology of the nonmarine facies of the Pliocene San Diego Formation, San Diego County, California. [In *Magnetic Stratigraphy of the Pacific Coast Cenozoic*, ed. D. R. Prothero.] *Pacific Section of the Society of Economic Paleontologists and Mineralogists*, 91, 359–68.

Wallace, R. E. (1946). A Miocene mammalian fauna from Beatty Buttes, Oregon. *Contributions to Paleontology, Carnegie Institute of Washington*, 551, 113–34.

Wallace, S. C. and Wang, X. (2004). Two new carnivores from an unusual late Tertiary forest biota in eastern North America. *Nature*, 431, 556–9.

Walsh, S. L. (1991a). Eocene Mammal Faunas of San Diego County. [In *Eocene Geologic History of San Diego*, ed. P. L. Abbott and J. A. May.] *Pacific Section of the Society of Economic Paleontologists and Mineralogists*, 68, 161–78.

(1991b). Late Eocene mammals from the Sweetwater Formation, San Diego County, California. [In *Eocene Geologic History San Diego Region*, ed. P. L. Abbott and J. A. May.] *Pacific Section of the Society of Economic and Petroleum Mineralogists*, 68, 149–59.

(1996). Middle Eocene mammal faunas of San Diego County, California. In *The Terrestrial Eocene–Oligocene Transition in North America*, ed. D. R. Prothero and R. J. Emry, pp. 75–117. Cambridge, UK: Cambridge University Press.

(1997). New specimens of *Metanioamys, Pauromys*, and *Simimys* (Rodentia: Myomorpha) from the Uintan (middle Eocene) of San Diego County, California, and comments on the relationships of selected Paleogene Myomorpha. *Proceedings of the San Diego Society of Natural History*, 32, 1–20.

(1998). Notes on the anterior dentition and skull of *Proteroixoides* (Mammalia: Insectivora: Dormaaliidae), and a new dormaaliid genus from the early Uintan (middle Eocene) of southern California. *Proceedings of the San Diego Society of Natural History*, 34, 1–26.

(2000). Bunodont artiodactyls (Mammalia) from the Uintan (middle Eocene) of San Diego County, California. *Proceedings of the San Diego Society of Natural History*, 37, 1–27.

Walsh, S. L. and Deméré, T. A. (1991). Age and stratigraphy of the Sweetwater and Otay Formations, San Diego County, California. [In *Eocene Geologic History of San Diego*, ed. P. L. Abbott and J. A. May.] *Pacific Section of the Society of Economic and Petroleum Mineralogists*, 68, 131–48.

Walsh, S. L. and Gutzler, R. Q. (1999). Late Duchesnean–early Chadronian mammals from the upper member of the Pomerado Conglomerate, San Diego, California. *Journal of Vertebrate Paleontology*, 19(suppl. to no. 3), p. 82A.

Walsh S. L, Prothero, D. R., and Lundquist, D. J. (1996). Stratigraphy and paleomagnetism of the middle Eocene Friars Formation and Poway Group, southwestern San Diego County, California. In *The Terrestrial Eocene–Oligocene Transition in North America*, ed. D. R. Prothero and R. J. Emry, pp. 120–54. Cambridge, UK: Cambridge University Press.

Walton, A. H. (1993a). *Pauromys* and other small Sciuravidae (Mammalia, Rodentia) from the middle Eocene of Texas. *Journal of Vertebrate Paleontology*, 13, 243–61.

(1993b). A new genus of eutypomid (Mammalia: Rodentia) from the middle Eocene of the Texas Gulf Coast. *Journal of Vertebrate Paleontology*, 13, 262–6.

Wang, X, Tedford, R. H., and Taylor, B. E. (1999). Phylogenetic systematics of the Borophaginae (Carnivora, Canidae). *Bulletin of the American Museum of Natural History*, 243, 1–391.

Wang, X., Wideman, B. C., Nichols, R., and Hanneman, D. L. (2004). A new species of *Aelurodon* (Carnivora, Canidae) from the Barstovian of Montana. *Journal of Vertebrate Paleontology*, 24, 445–52.

Webb, S. D. (1966). A relict species of the burrowing rodent *Mylagaulus*, from the Pliocene of Florida. *Journal of Mammalogy*, 47, 401–12.

(1969a). The Pliocene Canidae of Florida. *Bulletin of the Florida State Museum*, 14, 273–308.

(1969b). The Burge and Minnechaduza Clarendonian mammalian faunas of North–Central Nebraska. *University of California, Publications in Geological Sciences*, 78, 1–191.

(1974). Chronology of Florida Pleistocene mammals. In *Pleistocene Mammals of Florida*, ed. S. D. Webb, pp. 5–31. Gainsville, FL: University of Florida Press.

(1981). *Kyptoceras amatorum*, new genus and species from the Pliocene of Florida, the last protoceratid artiodactyl. *Journal of Vertebrate Paleontology*, 1, 357–65.

(1983). A new species of *Pediomeryx* from the late Miocene of Florida, and its relationships within the subfamily Cranioceratinae (Ruminantia: Dromomerycidae). *Journal of Mammalogy*, 64, 261–76.

(1990). Osteology and relationships of *Thinobadistes segnis*, the first mylodont sloth in North America. In *Advances in Neotropical Mammalogy*, ed. J. F. Eisenberg and K. Redford, pp. 469–532. Gainesville, FL: Sandhill Crane Press.

(1995). A new Paleocene (Tiffanian) mammalian local fauna from near Drayton Valley, central Alberta, Canada. *Journal of Vertebrate Paleontology*, 15(suppl. to no. 3), p. 59A.

Webb, S. D. and Hulbert, R. C., Jr. (1986). Systematics and evolution of *Pseudhipparion* (Mammalia, Equidae) from the Late Neogene of the Gulf Coastal Plain and the Great Plains. [In *Vertebrates, Phylogeny, and Philosophy*, ed. K. M. Flanagan and J. A. Lillegraven.] *Contributions to Geology, University of Wyoming Special Papers*, 3, 237–72.

Webb, S. D. and Perrigo, S. C. (1984). Late Cenozoic vertebrates from Honduras and El Salvador. *Journal of Vertebrate Paleontology*, 4, 237–54.

Webb, S. D. and Tessman, N. (1968). A Pliocene vertebrate fauna from low elevation in Manatee County, Florida. *American Journal of Science*, 266, 777–811.

Webb, S. D. and Wilkins, K. T. (1984). Historical biogeography of Florida Pleistocene mammals. In *Contributions in Quaternary Vertebrate Paleontology: A Volume in Memorial to John E. Guilday*, ed. H. H. Genoways and M. R. Dawson.] *Carnegie Museum of Natural History, Special Publications*, 8, 370–83.

Webb, S. D., MacFadden, B. J., and Baskin, J. A. (1981). Geology and paleontology of the Love Bone Bed from the late Miocene of Florida. *American Journal of Science*, 281, 513–44.

Webb, S. D., Hulbert, R. C., and Lambert, W. D. (1995). Climatic implications of large-herbivore distributions in the Miocene of North America. In *Paleoclimate and Evolution, with Emphasis on Human Origins*, ed. E. S. Vrba, G. H. Denton, T. C. Partridge, and L. H. Burckle, pp. 91–108. New Haven, CT: Yale University Press.

Webb, S. D., B. L. Beatty, and G. Poinar, Jr. (2003). New evidence of Miocene Protoceratidae including a new species from Chiapas, Mexico. *Bulletin of the American Museum of Natural History*, 279, 348–67.

Weems, R. E. and Lewis, W. C. (2002). Structural and tectonic setting of the Charleston, South Carolina, region: evidence from the Tertiary stratigraphic record. *Bulletin of the Geological Society of America*, 114, 24–42.

West, R. M. (1973a). Geology and mammalian paleontology of the New Fork–Big Sandy area, Sublette County, Wyoming. *Fieldiana (Geology)*, 29, 1–193.

(1973b). New records of fossil mammals from the early Eocene Golden Valley Formation, North Dakota. *Journal of Mammalogy*, 54, 749–50.

West, R. M. and Atkins, E. G. (1970). Additional middle Eocene (Bridgerian) mammals from Tabernacle Butte, Sublette County, Wyoming. *American Museum Novitates*, 2404, 1–26.

West, R. M. and Dawson, M. R. (1973). Fossil mammals from the upper part of the Cathedral Bluffs Tongue of the Wasatch Formation (early Bridgerian), Northern Green River Basin, Wyoming. *Contributions to Geology, University of Wyoming*, 12, 33–41.

(1975). Eocene fossil Mammalia from the Sand Wash Basin, northwestern Moffet County, Colorado. *Annals of the Carnegie Museum*, 45, 231–53.

(1977). Mammals from the Palaeogene of the Eureka Sound Formation: Ellesmere Island, Arctic Canada. *Geobios Memoir Special*, 1, 107–24.

West, R. M. and Hutchison, J. H. (1981). Geology and paleontology of the Bridger Formation, southern Green River basin, southwestern Wyoming. Part 6: The fauna and correlations of Bridger E. *Milwaukee Museum Publications, Contributions in Biology and Geology*, 46, 1–8.

Westgate, J. W. (1990). Uintan land mammals (excluding rodents) from an esturine facies of the Laredo Formation (Middle Eocene,

Claibourne Group) of Webb County, Texas. *Journal of Paleontology*, 64, 454–64.
(1992). *Dinohyus* aff. *D. hollandi* (Mammalia, Entelodontidae) in Alabama. *Journal of Paleontology*, 66, 685–7.
(2001). Paleoecology and biostratigraphy of marginal marine Gulf Coast Eocene vertebrate localities. In *Eocene Biodiversity: Unusual Occurrences and Rarely Sampled Habitats*, ed. G. F. Gunnell, pp. 263–97. New York: Kluwer Academic/Plenum.
Westgate, J. W. and Emry, R. J. (1985). Land mammals of the Crow Creek Local Fauna, Late Eocene, Jackson Group, St. Francis County, Arkansas. *Journal of Paleontology*, 59, 242–8.
Westgate, J. W. and Salazar, A. (1996). Additions to the late Eocene (Jacksonian) cetacean and chondrichthyan faunas of Arkansas. In *Proceedings of the Geological Society of America Meeting, South-Central Section*, Vol. 28, p. 68. Manhatten, KS: Geological Society of America.
Westgate, J. W. and Whitmore, F. C., Jr. (2002). *Balaena ricei*, a new species of bowhead whale from the Yorktown Formation (Pliocene) of Hampton, Virginia. *Smithsonian Contributions to Paleobiology*, 93, 295–312.
Westgate, J. W., Gilette, C. N., and Rolater, E. (1994). Paleoecology of an Eocene coastal community from Georgia. *Journal of Vertebrate Paleontology*, 14(suppl. to no. 3), p. 52A.
Whistler, D. D. (1967). Oreodonts of the Tick Canyon Formation, southern California. *PaleoBios*, 1, 1–14.
(1984). An early Hemingfordian (early Miocene) fossil vertebrate fauna from Western Mohave Desert, California. *Contributions in Science, Los Angeles County Museum of Natural History*, 355, 1–36.
Whistler, D. D. and Burbank, D. W. (1992). Miocene biostratigraphy and biochronology of the Dove Spring Formation, Mojave Desert, California, and characterization of the Clarendonian mammal age (late Miocene) in California. *Bulletin of the Geological Society of America*, 104, 644–58.
Whistler, D. P. and Lander, E. B. (2003). New late Uintan to early Hemingfordian Land Mammal assemblages from the undifferentiated Sespe and Vaqueros Formations, Orange County, and from the Sespe and equivalent marine formations in Los Angeles, Santa Barbara, and Ventura counties, southern California. *Bulletin of the American Museum of Natural History*, 279, 231–68.
White, J. A. (1984). Late Cenozoic Leporidae (Mammalia, Lagomorpha) from the Anza-Borrego Desert, southern California. [In *Patterns in Vertebrate Paleontology, Honoring Robert Warren Wilson*, ed. R. M. Mengel.] *Carnegie Museum of Natural History Special Publication*, 9, 41–57.
(1991). North American Leporinae (Mammalia: Lagomorpha) from late Miocene (Clarendonian) to latest Pliocene (Blancan). *Journal of Vertebrate Paleontology*, 11, 67–89.
White, T. E. (1952). Preliminary analysis of the vertebrate fossil fauna of the Boysen Reservoir area. *Proceedings of the United States National Museum*, 102, 185–207.
(1954). Preliminary analysis of the fossil vertebrates of the Canyon Ferry Reservoir area. *Proceedings of the United States National Museum*, 103, 395–438.
Wilf, P., Beard, K. C., Davies-Vollum, K. S., and Norejko, J. W. (1998). Portrait of a late Paleocene (Clarkforkian) terrestrial ecosystem: Big Multi Quarry and associated strata, Washakie Basin, southwestern Wyoming. *Palaios*, 13, 514–32.
Williams, M. R. and Storer, J. E. (1998). Cricetid rodents of the Kealey Springs Local Fauna (Early Arikareean; Late Oligocene) of Saskatchewan. *Paludicola*, 1, 143–9.
Williamson, T. E. (1993). The beginning of the age of mammals in the San Juan Basin: biostratigraphy and evolution of Paleocene mammals of the Naciemiento Formation. Ph.D. Thesis, University of New Mexico, Albuquerque.
(1996). The beginning of the age of mammals in the San Juan Basin, New Mexico: biostratigraphy and evolution of Paleocene mammals of the Nacimiento Formation. *Bulletin of the New Mexico Museum of Natural History*, 8, 1–140.
Williamson, T. E. and Lucas, S. G. (1993). Paleocene vertebrate paleontology of the San Juan Basin, New Mexico. [In *Vertebrate Paleontology in New Mexico*, ed. S. G. Lucas and J. Zidek.] *Bulletin of the New Mexico Museum of Natural History and Science*, 2, 105–35.
(1997). The Chico Springs locality, Nacimiento Formation, San Juan Basin, New Mexico. *Guidebook for the 48th Field Conference for the New Mexico Geological Society: Mesozoic Geology and Paleontology of the Four Corners Region*, pp. 259–65. Albuquerque, NM: New Mexico Geological Society.
Williamson, T. E. and Weil, A. (2002). A late Puercan (Pu3) microfauna from the San Juan Basin, New Mexico. *Journal of Vertebrate Paleontology*, 22(suppl. to no. 3), pp. 119A–120A.
Wilson, J. A. (1960). Miocene carnivores, Texas coastal plain. *Journal of Paleontology*, 34, 983–1000.
(1967). Early Tertiary mammals. [In *Geology of Big Bend National Park*, ed. R. A. Maxwell, J. T. Lonsdale, R. T. Hazzard, and J. A. Wilson.] *University of Texas Publications*, 6711, 157–69.
(1971a). Early Tertiary vertebrate faunas, Vieja Group, Trans-Pecos, Texas: Agriochoeridae and Merycoidodontidae. *Bulletin of the Texas Memorial Museum*, 18, 1–83.
(1971b). Early Tertiary vertebrate faunas, Vieja Group, Trans-Pecos, Texas: Entelodontidae. *Pearce-Sellards Series, Texas Memorial Museum*, 17, 1–17.
(1974). Early Tertiary vertebrate faunas, Vieja Group and Buck Hill Group, Trans-Pecos, Texas: Protoceratidae, Camelidae, Hypertragulidae. *Bulletin of the Texas Memorial Museum*, 23, 1–34.
(1977). Early Tertiary vertebrate faunas, Big Bend area, Trans-Pecos, Texas: Brontotheriidae. *Pearce-Sellards Series, Texas Memorial Museum*, 25, 1–17.
(1978). Stratigraphic occurrence and correlation of early Tertiary vertebrate faunas, Trans-Pecos, Texas. Part I: Vieja area. *Bulletin of the Texas Memorial Museum*, 25, 1–42.
(1984). Vertebrate faunas 49 to 36 million years ago and additions to the species of *Leptoreodon* (Mammalia: Artiodactyla) found in Texas. *Journal of Vertebrate Paleontology*, 4, 199–207.
(1986). Stratigraphic occurrence and correlation of early Tertiary vertebrate faunas, Trans-Pecos Texas: Agua Fria-Green Valley areas. *Journal of Vertebrate Paleontology*, 6, 350–73.
Wilson, J. A. and Schiebout, J. A. (1984). Early Tertiary vertebrate faunas, Trans-Pecos, Texas: Ceratomorpha less Amynodontidae. *Pearce-Sellards Series, Texas Memorial Museum*, 39, 1–47.
Wilson, J. A. and Stevens, M. S. (1986). Fossil vertebrates from the latest Eocene, Skyline Channels, Trans-Pecos, Texas. *Contributions to Geology, University of Wyoming Special Papers*, 3, 221–35.
Wilson, L. E. (1973). A delphinid (Mammalia, Cetacea) from the Miocene of Palos Verdes Hills, California. *University of California, Publications in Geological Sciences*, 103, 1–33.
Wilson, R. L. (1968). Systematics and faunal analysis of a lower Pliocene vertebrate assemblage from Trego County, Kansas. *Contributions from the Museum of Paleontology, University of Michigan*, 22, 75–126.
Wilson, R. W. (1940). Two new Eocene rodents from California. *Carnegie Institution of Washington Publications*, 514, 85–95.
(1984). The National Geographic Society – South Dakota School of Mines and Technology expeditions into the Poleslide Member of the Big Badlands of South Dakota in 1969: a program of conservation collecting. *National Geographic Research Reports*, 10, 637–42.
(1986). The Paleogene record of the rodents: facts and interpretation. [In *Vertebrates, Phylogeny and Philosophy*, ed. K. M. Flanagan and J. A. Lillegraven.] *Contributions to Geology, University of Wyoming. Special Papers*, 3, 163–75.

Winkler, D. A. (1990). Sedimentary facies and biochronology of the Upper Tertiary Ogallala Group, Blanco and Yellow House Canyons, Texas Panhandle. In *Geologic Framework and Regional Hydrology; Upper Cenozoic Blackwater Draw and Ogallala Formations, Great Plains*, ed. T. C. Gustavson, pp. 39–55. Austin, TX: University of Texas at Austin, Bureau of Economic Geology.

Winterfield, G. F. (1982). Mammalian paleontology of the Fort Union Formation (Paleocene), eastern Rock Springs Uplift, Sweetwater County, Wyoming. *Contributions to Geology, University of Wyoming*, 21, 73–112.

Wolberg, D. L. (1979). Late Paleocene (Tiffanian) mammalian fauna of two localities in eastern Montana. *Northwest Geology*, 8, 83–93.

Wood, A. E. (1935). Two new rodents from the John Day Miocene. *American Journal of Science*, 30, 368–72.

(1936). Geomyid rodents from the Middle Tertiary. *American Museum Novitates*, 866, 1–31.

Wood, H. E. and Wood, A. E. (1937). Mid-Tertiary vertebrates from the Texas Coastal Plain: fact and fable. *The American Midland Naturalist*, 18, 129–46.

Woodburne, M. O. (1966). Equid remains from the Sonoma Volcanics, California. *Bulletin of the Southern California Academy of Sciences*, 65, 185–9.

(1969). Systematics, biogeography, and evolution of *Cynorca* and *Dyseohyus* (Tayassuidae). *Bulletin of the American Museum of Natural History*, 141, 271–356.

Woodburne, M. O. and Golz, D. J. (1972). Stratigraphy of the Punchbowl Formation, Cajon Valley, southern California. *University of California, Publications in Geological Sciences*, 92, 1–57.

Woodburne, M. O. and Robinson, P. T. (1977). A new late Hemingfordian mammal fauna from the John Day Formation, Oregon, and its stratigraphic implications. *Journal of Paleontology*, 51, 750–7.

Woodburne, M. O., Tedford, R. H., Stevens, M. S., and Taylor, B. E. (1974). Early Miocene mammalian faunas, Mohave desert, California. *Journal of Paleontology*, 48, 6–26.

Woodburne, M. O., Tedford, R. H., and Swisher, C. C., III (1990). Lithostratigraphy, biostratigraphy, and geochronology of the Barstow Formation, Mojave Desert, southern California. *Bulletin of the Geological Society of America*, 102, 459–77.

Wright, D. B. and Eshelman, R. E. (1987). Miocene Tayassuidae (Mammalia) from the Chesapeake Group of the Mid-Atlantic coast and their bearing on marine-nonmarine correlation. *Journal of Paleontology*, 61, 604–18.

Yon, J. W. (1965). The stratigraphic significance of an upper Miocene fossil discovery in Jefferson County, Florida. *Southeastern Geology*, 6, 167–76.

Zakrzewski, R. J. (1969). The rodents from the Hagerman local fauna, upper Pliocene of Idaho. *Contributions from the Museum of Paleontology, University of Michigan*, 23, 1–36.

(1981). Kangaroo rats from the Borchers local fauna, Blancan, Meade County, Kansas. *Transactions of the Kansas Academy of Science*, 84, 78–88.

(1988). Preliminary report on fossil mammals from the Ogallala (Miocene) of north-central Kansas. *Fort Hays Studies, 3rd Series (Science)*, 20, 117–27.

(1998). Additional records of the giant marmot *Paenemarmota* from Idaho. [In *And Whereas: Papers on the Vertebrate Paleontology of Idaho Honoring J. A. White*, Vol. 1, ed. W. A. Akersten, H. G. McDonald, D. J. Meldrum, and M. E. T. Flint.] *Idaho Museum of Natural History Occasional Papers*, 36, 50–5.

Zeigler, C. V., Chan, G. L., and Barnes, L. G. (1997). A new Late Miocene balaenopterid whale (Cetacea: Mysticeti), *Parabalaenoptera baulinensis*, (new genus and species), from the Santa Cruz mudstone, Point Reyes Peninsula, California. *Proceedings of the California Academy of Sciences*, 50, 115–38.

Zonneveld, J.-P. and Gunnell, G. F. (2003). A new species of cf. *Dilophodon* (Mammalia, Perissodactyla) from the early Bridgerian of southwestern Wyoming. *Journal of Vertebrate Paleontology*, 23, 652–8.

Zonneveld, J.-P., Gunnell, G. F., and Bartels, W. S. (2000). Early Eocene fossil vertebrates from the southwestern Green River Basin, Lincoln and Uinta Counties, Wyoming. *Journal of Vertebrate Paleontology*, 20, 369–86.

Appendix III: Museum acronyms

AM/ACM	Amherst College Museum, Amherst, MA
AMNH	Department of Vertebrate Paleontology, American Museum of Natural History, New York, NY
ANSP	Academy of Natural Sciences, Philadelphia, PA
BCPM	British Columbia Provincial Museum, Vancouver, Canada
BM(NH)	British Museum of Natural History, London, UK
CAS	California Academy of Sciences, San Francisco, CA
CGM	Cairo Geological Museum, Cairo, Egypt
ChM	Charleston Natural History Museum, Charleston, SC
CIT	California Institute of Technology, Pasadena, CA
CM	Carnegie Museum of Natural History, Pittsburgh, PA
CMM	Calvert Marine Museum, Solomons, MD
CMN	Canadian Museum of Nature, Ottawa, Canada
CRL	Collection Lemoine, Muséum National d'Histoire Naturelle, Paris, France
DMNH	Denver Museum of Natural History, Denver, CO
F:AM	Frick American Mammals, American Museum of Natural History, New York
FMNH	Field Museum of Natural History, Chicago, IL
GSM	Georgia Southern Museum, Statesboro, GA
HLMD	Hessisches Landesmuseum, Darmstadt, Germany
ICGU/IGM	Instituto de Geologia, Universidad Nacional Autonoma de Mexico, Mexico City, Mexico
IPHG	Institute für Paläontologie und Historische Geologie, Munich, Germany
IRSNB	Institut Royal des Sciences Naturelle de Belgique, Brussels, Belgium
JGU	Institut für Geowissenschaften–Paläontologie, Johannes Guttenberg-Universität, Mainz, Germany.
KUVP/ KUMNH	University of Kansas, Museum of Vertebrate Paleontology, Lawrence, KS
LACM	Natural History Museum of Los Angeles County, Los Angeles, CA
LACM/CIT	Los Angeles County Museum, California Institute of Technology collection, Los Angeles, CA
MACN	Museo Argentino de Ciencias Naturales, Buenos Aires, Argentina.
ME	Messel specimen, Forschungsinstitut Senckenberg, Frankfurt, Germany
MCZ	Museum of Comparative Zoology, Harvard University, Cambridge, MA
MDLP	Museuo de La Plata, Buenos Aires, Argentina
MHNH	Muséum National d'Histoire Naturelle, Paris, France
MMNS	Mississippi Museum of Natural Sciences, Jackson, MS
MNB	Museum für Naturkunde der Humboldt-Universität, Berlin, Germany
MPM	Milwaukee Public Museum, Milwaukee, WI
MRHN	Royal Museum of Natural History, Brussels, Belgium
MSUCFV	Midwestern State University Collection of Fossil Vertebrates, Wichita Falls, TX
NMMNH	New Mexico Museum of Natural History, Albuquerque, NM
NMC	National Museum of Canada, Ottawa, Canada
NR	Naturhistoriska Riksmuseet, Stockholm, Sweden
NSMT	National Science Museum, Tokyo, Japan
NUFV	Nunavut Fossil Vertebrate, temporarily housed in the Canadian Museum of Nature, Ottawa, Canada
OMSI	Oregon Museum of Science and Industry, Portland, OR
OUSM	Stovall Museum, University of Oklahoma, Norman, OK
PIN	Paleontological Institute, Russian Academy of Sciences, Moscow, Russia
PMUM	Paleontological Museum Uppsala, University of Uppsala, Sweden
PU/YPM-PU	Princeton University, now at Yale Peabody Museum, New Haven, CT
RCS	Royal College of Surgeons, London, UK
ROM	Royal Ontario Museum, Toronto, Canada.

SAM	South African Museum, Cape Town, South Africa
SBMNH	Santa Barbara Museum of Natural History, Santa Barbara, CA
SCSM	South Carolina State Museum, Colombia, SC
SDSM	South Dakota School of Mines, Rapid City, SD
SDSNH	San Diego Society of Natural History, San Diego, CA
SMNS	Staatliches Museum für Naturkunde, Stuttgart, Germany
SMU	Southern Methodist University, Dallas, TX
SMNH	Saskatchewan Museum of Natural History, Regina, SK
SPSM	St. Paul Science Museum, St. Paul, MN
TMM	Texas Memorial Museum, University of Texas at Austin, TX
UA	University of Alberta, Edmonton, Alberta, Canada
UALP	University of Arizona Laboratory of Paleontology, Tuscon, AZ
UCMP	University of California, Museum of Paleontology, Berkeley, CA
UCR	University of California, Riverside, CA
UF	University of Florida (Florida Museum of Natural History), Gainesville, FL
UF/FGS	Florida Geological Survey collection, Florida State Museum, University of Florida, Gainesville, FL
UM/UMMP	University of Michigan, Museum of Paleontology, Ann Arbor, MI
UMVP	University of Minnesota Vertebrate Paleontology, Minneapolis, MN
UNSM	University of Nebraska State Museum, Lincoln, NB
UO(MNH)	University of Oregon Museum of Natural History, Eugene, OR
USGS	United States Geological Survey, Denver, CO
USNM	United States National Museum, Washington DC
UTEP	University of Texas at El Paso, TX.
UTBEG	Bureau of Economic Geology, University of Texas, Austin, TX (now housed with the collections of the Texas Memorial Museum)
UW	University of Wyoming, Laramie, WY
VMNH	Virginia Museum of Natural History, Martinsville, VA
Wa	Walbeck Collection, Geologisch-Paläontologisches Institut, Halle, Germany
WFIS	Wagner Free Institute of Science, Philadelphia, PA
YPM	Peabody Museum of Natural History, Yale University, New Haven, CT
ZMUC	Zoological Museum of the University of Copenhagen, Copenhagen, Denmark

Index

Numbers in boldface refer to figures and tables.

Printed in the United States
By Bookmasters